ALGAE ABSTRACTS

A Guide to the Literature

Volume 2
1970 - 1972

Prepared from material supplied by
Water Resources Scientific Information Center
Office of Water Resources Research
Department of the Interior
Washington, D.C.

IFI/PLENUM · NEW YORK - WASHINGTON - LONDON

Library of Congress Catalog Card Number 73-84004
ISBN 0-306-67182-4

Published 1973 by IFI/Plenum Data Corporation
A Subsidiary of Plenum Publishing Corporation
227 West 17th Street, New York, N. Y. 10011

United Kingdom edition published by Plenum Press, London
A Division of Plenum Publishing Company, Ltd.
Davis House (4th Floor), 8 Scrubs Lane, Harlesden, NW10 6SE, London, England

Printed in the United States of America

INTRODUCTION

Algae Abstracts is the first in a series of bibliographies on water resources and pollution published by IFI/Plenum Data Corporation in cooperation with the Water Resources Scientific Information Center (WRSIC). It is produced wholly from the information base comprising material abstracted and indexed for *Selected Water Resources Abstracts.*

The bibliography is divided into volumes according to the publication dates of the source documents. Volume 1 contains 569 abstracts covering publication dates up to and including 1969; Volume 2 contains 730 abstracts covering the years 1970 to 1972. The material included in this bibliography represents computer selections based on the presence of a form of the word "alga" somewhere in the referenced citation. Substantively, the material typifies WRSIC's "centers of competence" approach to information support of the Office of Water Resources Research (OWRR) of the Department of the Interior. Most of the references in this bibliography are the work of the center of competence on eutrophication at the University of Wisconsin.

The indexes refer to the WRSIC accession number, which follows each abstract. The *Significant Descriptor Index* is made up of a fraction of the total descriptors and identifiers by which each paper has been indexed. It represents weighted terms that best describe the information content; this status is indicated by the asterisks which precede them. The *General Index* includes all the remaining descriptors and identifiers by which each paper in this bibliography has been indexed. Through permutation, each word in a multiple-word descriptor or identifier is made to file in its normal alphabetical order, giving access to each abstract by all conceivable routes. To use the indexes, scan the middle rank for a few keywords describing your subject matter interest, then note the WRSIC accession numbers at the right margin. These numbers locate the full record in the bibliography, which is arranged in ascending accession number order.

Each volume, including a complete *Author Index,* is a self-contained entity, offering ready access to the pertinent literature for the time period it spans.

CONTENTS

Abstracts 1

Significant Descriptor Index 411

General Index 453

Author Index 675

ABSTRACTS

Publication dates 1970 to 1972

The abstracts are presented in order of increasing accession number. This number appears as the last line of each entry.

CHEMISTRY OF NITROGEN AND PHOSPHORUS IN WATER,

P. L. MCCARTY.

JOURNAL AMERICAN WATER WORKS ASSOCIATION, VOL 62, NO 2, P 127-140, FEB 1970. 14 P, 6 FIG, 8 TAB, 64 REF.

DESCRIPTORS:
*NUTRIENTS, *EUTROPHICATION, *WATER QUALITY CONTROL, *NITROGEN COMPOUNDS, *PHOSPHORUS COMPOUNDS, ALGAE, NUTRIENT REQUIREMENTS, PRODUCTIVITY, ESSENTIAL NUTRIENTS, WATER QUALITY, WATER CHEMISTRY, CYCLING NUTRIENTS.

IDENTIFIERS:
NITROGEN REQUIREMENTS(ALGAE), PHOSPHORUS REQUIREMENTS(ALGAE), NUTRIENT CHEMISTRY(AQUATIC).

ABSTRACT:
OF THE MAJOR ELEMENTS ESSENTIAL TO ALGAL GROWTH, NITROGEN AND PHOSPHORUS ARE THE ONES MOST LIKELY TO BE OF CRITICALLY LIMITING AVAILABILITY IN NATURAL WATERS. BECAUSE THEY THEREFORE REPRESENT PROMISING WEAK LINKS IN ALGAL LIFE CYCLES, THEIR CHEMICAL STATES AND BEHAVIOR IN WATER ARE EXAMINED TO SEE HOW WATER TREATMENT MIGHT BENEFIT. LARGE SUPPLIES OF NITROGEN AND PHOSPHORUS ARE PRESENT IN MANY BODIES OF WATER EITHER IN THE SEDIMENTS, THE ATMOSPHERE ABOVE, OR IN THE FORM OF DISSOLVED GAS. THESE FORMS MAY BE AVAILABLE FOR THE GROWTH OF ALGAE AND OTHER AQUATIC PLANTS, BUT THE RATES AT WHICH THEY MAY BECOME AVAILABLE IS SLOW. THESE RATES ARE IMPORTANT, HOWEVER, AS THEY TEND TO CONTROL THE AMOUNT OF VEGETATIVE GROWTH WHICH CAN BE SUPPORTED. SOLUBLE NITROGEN AND PHOSPHORUS CONTAINED IN THE EFFLUENTS FROM WASTE TREATMENT PLANTS, ON THE OTHER HAND, ARE IN A READILY AVAILABLE FORM. IF DISCHARGED TO NATURAL BODIES OF WATER, THEY CAN STIMULATE GROWTH FAR IN EXCESS OF THAT WHICH WOULD OCCUR NATURALLY. (KNAPP-USGS)

FIELD 05C, 05A

ACCESSION NO. W70-04080

WISCONSIN WATER RESOURCE PROBLEMS,

WISCONSIN DEPT. OF NATURAL RESOURCES, MADISON.

L. P. VOIGT.

WISCONSIN CONSERVATION BULLETIN, P 3-5, JANUARY-FEBRUARY 1970. 2 PHOTOS.

DESCRIPTORS:
*WATER QUALITY, *WISCONSIN, PHOSPHORUS, PULP AND PAPER INDUSTRY, ALGAE, AQUATIC WEEDS, RECREATION, GROUNDWATER, FISH, FLOOD PLAINS, IRRIGATION, PESTICIDES.

IDENTIFIERS:
FOX RIVER BASIN(WIS), FWPCA.

ABSTRACT:
WISCONSIN'S WATER PROBLEMS RELATE TO WATER QUALITY AND NOT QUANTITY. THE SOUTHEASTERN AND EASTERN PARTS OF THE STATE ARE PARTICULARLY PROBLEMATIC DUE TO POPULATION AND INDUSTRIAL CONCENTRATIONS, AND SUBQUENTLY INCREASING DEMANDS FOR QUALITY WATER FOR RECREATION USE. THE PULP AND PAPER INDUSTRY REPRESENTS THE LARGEST WASTE SOURCE, WITH FOUR TIMES THE BIOCHEMICAL OXYGEN DEMAND OF MUNICIPAL WASTES. ANY SOLUTION WILL DEMAND THAT THE PRODUCT PRICES REFLECT TOTAL COSTS, INCLUDING ENVIRONMENTAL DAMAGES. FEDERAL ALLOCATION FUNDS FOR INDUSTRIAL AND MUNICIPAL WASTE TREATMENT PLANTS WERE ONLY 1/3 OF THE AUTHORIZED AMOUNT, THUS THE STATE'S EFFECTIVENESS IN POLLUTION CONTROL IS HAMPERED. IN THE PAST, THE PROBLEMS OF SILT, NUTRIENTS, AND PESTICIDES HAVE BEEN TACKED BY EDUCATION PROGRAMS, VOLUNTARY ACTION, AND COST SHARING BUT INTENSIFIED LAND USE MAY REQUIRE REGULATORY PROGRAMS. ONLY THE SYMPTOMS OF EUTROPHICATION HAVE BEEN TREATED TO DATE VIA WEED HARVESTING AND ALGAE POISONING, BUT THE STATE IS COMMITTED TO REDUCE PHOSPHORUS LOADINGS FROM MUNICIPAL AND INDUSTRIAL WASTES IN LAKE MICHIGAN BY 1972. FLOOD-PLAIN MANAGEMENT, FISH CONTROL, AND IRRIGATION ARE OTHER PROBLEM AREAS TO BE RESOLVED. (POWERS-WISCONSIN)

FIELD 05G

ACCESSION NO. W70-05103

ALGAL GROWTH AND DECOMPOSITION: EFFECTS ON WATER QUALITY, NUTRIENT UPTAKE AND CHEMICAL COMPOSITION OF ALGAE IN BATCH CULTURE,

KENTUCKY WATER RESOURCES INST., LEXINGTON.

EDWARD G. FOREE, AND JOHN S. TAPP, JR.

AVAILABLE FROM THE CLEARINGHOUSE AS PB-190 801, $3.00 IN PAPER COPY, $0.65 IN MICROFICHE. RESEARCH REPORT NO. 26, WATER RESOURCES INSTITUTE, UNIVERSITY OF KENTUCKY, LEXINGTON, KENTUCKY, MARCH, 1970. 76 P, 6 TAB, 19 FIG, 43 REF. OWRR PROJECT A-021-KY.

DESCRIPTORS:
*EUTROPHICATION, *ALGAE, *NUTRIENT REQUIREMENTS, CHLOROPHYTA, CYANOPHYTA, CYCLING NUTRIENTS, NITROGEN, PHOSPHORUS, CHEMICAL ANALYSIS, PROTEINS, CARBOHYDRATES, LIPIDS.

IDENTIFIERS:
*ALGAL GROWTH, NUTRIENT UPTAKE, CHEMICAL COMPOSITION, ALGAL DECOMPOSITION.

ABSTRACT:
THE CHEMICAL COMPOSITION OF ALGAE GROWN IN BATCH CULTURE DEPENDS MAINLY ON ENVIRONMENTAL CONDITIONS, NUTRIENT AVAILABILITY, PRESENCE OF PREDATORS, CELL AGE, AND SPECIES. THE EFFECTS OF NUTRIENT AVAILABILITY AND CELL AGE ON THE COMPOSITION OF THREE UNIALGAL CULTURES (ALGAE + BACTERIA) AND ONE HETEROGENEOUS CULTURE (ALGAE + BACTERIA + MICROSCOPIC ANIMALS) WERE EVALUATED. THE CULTURES WERE GROWN IN BATCH CULTURE UNDER BOTH NUTRIENT-ABUNDANT AND NUTRIENT-DEFICIENT CONDITIONS AND THE CHANGES IN COMPOSITIONS WERE OBSERVED. LUXURIOUS UPTAKE, WHERE NUTRIENTS ARE INCORPORATED INTO CELLULAR PROTOPLASM AT LEVELS GREATER THAN THOSE NECESSARY FOR GROWTH, AND SUPER-LUXURIOUS UPTAKE, WHERE SOME NUTRIENTS ARE STORED RATHER THAN CONVERTED INTO ALGAL PROTOPLASM, WERE OBSERVED. THE COMMONLY USED MODEL FOR CALCULATING THE WEIGHT PERCENTAGE OF PROTEIN WAS INACCURATE WHEN SUPER-LUXURIOUS UPTAKE OCCURRED. COMPOSITION OF THE CULTURES WAS GENERALLY CHARACTERIZED BY PROTEIN SYNTHESIS DURING THE NUTRIENT-ABUNDANT GROWTH PHASE, BY A FLUCTUATING COMPOSITION DURING TRANSITION FROM NUTRIENT-ABUNDANT TO NUTRIENT-DEFICIENT GROWTH, AND BY LIPID AND/OR CARBOHYDRATE SYNTHESIS AND THE ESTABLISHMENT OF A RELATIVELY CONSTANT COMPOSITION DURING THE NUTRIENT-DEFICIENT GROWTH PHASE.

FIELD 05C

ACCESSION NO. W70-05469

EFFECT OF LIGHT INTENSITY AND THICKNESS OF CULTURE SOLUTION ON OXYGEN PRODUCTION BY ALGAE,

NAVAL RESEARCH LABORATORY, WASHINGTON, D.C. CHEMISTRY DIV.

R. L. SHULER, AND W. A. AFFENS.

APPLIED MICROBIOLOGY, VOL 19, NO 1, P 76-86, 1970. 10 FIG, 3 TAB, 5 REF.

DESCRIPTORS:
*CHLORELLA, *LIGHT, *PHOTOSYNTHETIC OXYGEN, *ALGAE, CULTURES, VOLUME, MATHEMATICAL STUDIES, CORRELATION ANALYSIS.

IDENTIFIERS:
*CHLORELLA PYRENOIDOSA, OXYGEN PRODUCTION, LIGHT INTENSITY, CULTURE THICKNESS, CULTURE VOLUME.

ABSTRACT:
OBJECTIVE WAS TO OPTIMIZE OXYGEN PRODUCTION EFFICIENCY FOR USE IN CLOSED SYSTEMS, SUCH AS SUBMARINES. MATHEMATICAL RELATIONSHIP BETWEEN OXYGEN PRODUCTION RATE, BY CHLORELLA PYRENOIDOSA (SOROKIN STRAIN--OPTIMUM TEMPERATURE 39C), LIGHT INTENSITY, CULTURE THICKNESS, AND VOLUME OF SUSPENSION WERE WORKED OUT USING A SINGLE TYPE CULTURE VESSEL. THE VESSEL WAS AN UPRIGHT CYLINDER WITH THE INCANDESCENT LIGHT SOURCE (ALSO AN UPRIGHT CYLINDER) LOCATED CENTRALLY. THE CULTURE MEDIUM WAS PUMPED THROUGH CONTINUOUSLY. CULTURE CELL DENSITIES OF 0.5 TO 1.55% WET PACKED CELL VOLUMES WERE SUCH THAT EXPERIMENTAL DATA WERE COLLECTED ON LIGHT LIMITED CELLS UNDERGOING LOG GROWTH. LIGHT INTENSITIES OF 12,000-70,000-FOOT CANDLES WERE EMPLOYED. CULTURE THICKNESS WAS VARIED BY THE USE OF SEPARATE ANNULAR COMPARTMENTS AROUND THE SINGLE LIGHT SOURCE. COOLING WATER WAS PUMPED CONTINUOUSLY IN A LAYER BETWEEN LAMP AND CULTURE VESSEL SO AS TO MAINTAIN TEMPERATURE AT 38.5C. THE CONCLUSION WAS THAT OXYGEN EVOLUTION RATE WAS A LOG FUNCTION OF LIGHT INTENSITY AND THAT RATE OF OXYGEN EVOLUTION PER UNIT VOLUME OF CULTURE IS RELATED LINEARLY TO THE RECIPROCAL OF CULTURE THICKNESS. THE RELATIONSHIPS CITED FOR THE CYLINDRICAL VESSEL MAY HAVE WIDE APPLICABILITY. EFFICIENCIES WITH RESPECT TO ELECTRICAL (LIGHT) REQUIREMENTS WERE COMPUTED. (GERHOLD-WISCONSIN)

FIELD 02K, 05C

ACCESSION NO. W70-05547

RELATIVE CONTRIBUTIONS OF NUTRIENTS TO THE POTOMAC RIVER BASIN FROM VARIOUS SOURCES,

FEDERAL WATER POLLUTION CONTROL ADMINISTRATION, ANNAPOLIS, MD.; AND NEW YORK STATE DEPT. OF HEALTH, ALBANY. DIV. OF ENVIRONMENTAL HEALTH SERVICE.

NORBERT A. JAWORSKI, AND LEO J. HETLING.

CHESAPEAKE TECHNICAL SUPPORT LABORATORY TECHNICAL REPORT NO 31, FEDERAL WATER POLLUTION CONTROL ADMINISTRATION, JANUARY 1970. 36 P, 13 FIG, 6 TAB, 8 REF.

DESCRIPTORS:
*NUTRIENTS, *EUTROPHICATION, *WATER POLLUTION SOURCES, *ESTUARIES, RIVERS, PHOSPHATES, NITRATES, HUDSON RIVER, ALGAE, WATER QUALITY, SEWAGE DISPOSAL, WASTE WATER DISPOSAL, SEWAGE EFFLUENTS, FARM WASTES.

IDENTIFIERS:
POTOMAC RIVER BASIN, POTOMAC ESTUARY.

ABSTRACT:
THE UPPER POTOMAC ESTUARY IS HIGHLY EUTROPHIC. DURING THE SUMMER MONTHS, LARGE BLOOMS OF NUISANCE BLUE-GREEN ALGAE, MAINLY MICROCYSTIS, OCCUR IN THE FRESH-WATER PORTION OF THE UPPER ESTUARY. A RELATIONSHIP BETWEEN HIGH NUTRIENT CONTENT AND THE ACCELERATED EUTROPHICATION IN THE POTOMAC ESTUARY HAS BEEN ESTABLISHED. THE ANNUAL AVERAGE CONCENTRATION OF PHOSPHORUS VARIED FROM 0.09 MG/LITER IN THE SOUTH BRANCH TO 1.9 MG/LITER IN THE ANTIETAM WATERSHED. THE ANNUAL AVERAGE CONCENTRATION OF NITROGEN VARIED FROM 0.3 MG/LITER IN THE SOUTH BRANCH TO 2.2 MG/LITER IN OPEQUON CREEK. ABOUT 92,700 LBS/DAY OF TOTAL PHOSPHORUS ENTERED THE POTOMAC IN 1966, 87% FROM WASTEWATER. THE AVERAGE 1966 LOADING OF TOTAL NITROGEN WAS ABOUT 125,000 LBS/DAY, 51% FROM WASTEWATER. DURING LOW FLOW CONDITIONS A SIGNIFICANT PROPORTION OF THE PHOSPHORUS ENTERING THE SURFACE WATER FROM THE VARIOUS SOURCES IN THE UPPER BASIN IS RETAINED IN THE STREAM CHANNEL. AT HIGH STREAM FLOW, IT APPEARS THAT A LARGE PROPORTION OF THIS PHOSPHORUS IS 'FLUSHED' OUT OF THE STREAM CHANNEL AND TRANSPORTED DOWNSTREAM. A COMPARISON OF SOURCES OF NUTRIENTS IN THE HUDSON RIVER BASIN TO THOSE IN THE POTOMAC SUPPORTS THE CONTENTION THAT IN THE MIDDLE ATLANTIC REGION THE MAJOR SOURCE OF NUTRIENTS TO THE AQUATIC ECOSYSTEM IS FROM WASTEWATER DISCHARGES. (KNAPP-USGS)

FIELD 05B, 05C

ACCESSION NO. W70-06509

STABILIZATION OF DAIRY WASTES BY ALGAL-BACTERIAL SYMBIOSIS IN OXIDATION PONDS,

ALEXANDRIA UNIV. (EGYPT). HIGH INST. OF PUBLIC HEALTH.

F. M. EL-SHARKAWI, AND S. K. MOAWAD.

JOURNAL OF THE WATER POLLUTION CONTROL FEDERATION, VOL 42, NO 1, P 115-125, JANUARY 1970. 4 FIG, 5 TAB, 17 REF.

DESCRIPTORS:
*DAIRY INDUSTRY, *OXIDATION LAGOONS, *PILOT PLANTS, ALGAE, BIOCHEMICAL OXYGEN DEMAND, BIOLOGICAL TREATMENT, PHOTOSYNTHESIS, STABILIZATION, *WASTE WATER TREATMENT, *FARM WASTES.

IDENTIFIERS:
*ALEXANDRIA(EGYPT), PANDORINA, SOLUBLE ORGANIC SOLIDS.

ABSTRACT:
A PILOT-PLANT STUDY OF BOD REDUCTION OF MILK PROCESSING WASTES IS REPORTED. A SYNTHETIC DAIRY WASTE OF 750 MG/L BOD WAS FED CONTINUOUSLY TO RECTANGULAR CONCRETE BASINS WITH SLOPING SIDES TO MINIMIZE SLUDGING. THE DETENTION PERIOD WAS 10 DAYS. AN INFLUENT PH OF 9.8 WAS MAINTAINED TO KEEP THE PH AT A LEVEL CONDUCIVE TO ALGAL GROWTH. TANK DEPTH WAS IMPORTANT IN MAINTAINING BALANCE BETWEEN THE ALGAL AND BACTERIAL FRACTIONS OF THE SYSTEM. THE MICROFLORA SHOWED PLASTICITY IN ADAPTING TO ENVIRONMENTAL VARIATIONS. PANDORINA CONSTITUTED A MAJOR MEMBER OF THE FLORA HIGHLY ADAPTABLE TO INTERACTION WITH DAIRY WASTES. PANDORINA COULD TOLERATE WIDE TEMPERATURE VARIATIONS (11 DEG TO 32 DEG C) AT A CONSTANT DEPTH OF 75 CM. OTHER ORGANISMS WERE RESPONSIVE TO SPECIFIC CONDITIONS AND WHEN THE DOMINANT GROUPS SUFFERED A SERIOUS SETBACK, THE SUBDOMINANTS FLOURISHED. BIOCHEMICAL OXYGEN DEMAND (BOD) REDUCTIONS WERE 80 TO 90 PERCENT AT A BOD LOADING RATE OF 220 LBS/ACRE/DAY (246 KG/DIA/DAY). (AGUIRRE-TEXAS)

FIELD 05D

ACCESSION NO. W70-06619

DISPERSAL OF ALGAE, PROTOZOANS, AND FUNGI BY AQUATIC HEMIPTERA, TRICHOPTERA, AND OTHER AQUATIC INSECTS,

NORTH TEXAS STATE UNIV., DENTON. DEPT. OF BIOLOGY.

KENNETH W. STEWART, LARRY E. MILLIGER, AND BERNARD M. SCLON.

ANNALS OF THE ENTOMOLOGICAL SOCIETY OF AMERICA, VOL 63, NO 1, P 139-144, 1970. 1 FIG, 2 TAB, 13 REF.

DESCRIPTORS:
*ALGAE, *PROTOZOA, *FUNGI, *AQUATIC INSECTS, *DISPERSION, CADDISFLIES, STONEFLIES, DOBSONFLIES, MAYFLIES, CYANOPHYTA, CHLORCPHYTA, CHRYSOPHYTA, CHLORELLA, ROTIFERS, CHLAMYDOMONAS, PHYTOPLANKTON, ZOOPLANKTON, PERIPHYTON.

IDENTIFIERS:
*HEMIPTERA, *TRICHOPTERA, CORIXIDS, EUDORINA, NOTCNECTIDAE, BODO, HYDROPSYCHIDAE, PLECOPTERA, MAGALOPTERA, EPHEMERCPTERA, NANNOCHLORIS, CHLOROCCOCCUM, MICROCYSTIS, FUSARIUM, TRICHORIXA, RAMPHOCORIXA, SIGARA, BOENOA, SALDULA, CRYNELLUS, CHEUMATOPSYCHE, TRIAENCDES INUUSTA, OECETIS, LEPTOCELLA, ISONYCHIA, CAENIS, PERLESTA.

ABSTRACT:
OF 16 SPECIES OF AQUATIC INSECTS STUDIES (5 AQUATIC HEMIPTERA, 5 CADDISFLIES, 1 STONEFLY, 2 DOBSONFLIES, ANC 2 MAYFLIES) 15 WERE FOUND TO BE TRANSPORTING 27 GENERA OF VIABLE SMALL AQUATIC ORGANISMS. THREE CORIXIDS CARRIED 5 GENERA OF BLUE-GREEN ALGAE, 11 GREEN ALGAE, 1 YELLOW-GREEN ALGA, 2 PROTOZOANS, AND FUNGI. EUDORINA, REPRESENTING COLONIAL VOLVOCALEAN ALGAE, WAS FOUND FOR THE FIRST TIME ON FIELD-EXPOSED AQUATIC INSECTS. SUITABILITY OF THE CORIXIDAE AND NOTONECTIDAE AS PASSIVE DISPERSAL VEHICLES FOR SMALL ORGANISMS IS DISCUSSED. ONLY ONE OF SIX CADDISFLIES STUDIED CARRIED ALGAE AND PROTOZOANS, BUT THAT SPECIES, HYDROPSYCHE ORRIS ROSS, CARRIED 10 GENERA. ORGANISMS TAKEN IN CASUAL SAMPLINGS OF PLECOPTERA, MEGALOPTERA, AND EPHEMEROPTERA ARE INDICATED. EXCLUDING FUNGI, CHLORELLA WAS THE MOST COMMON VIABLE FORM TRANSPORTED, FOLLOWED BY BODO, NANNOCHLORIS, CHLOROCCOCCUM, AND MICROCYSTIS, IN ORDER OF FREQUENCY OF OCCURRENCE. UNIDENTIFIED FUNGI AND FUSARIUM WERE CARRIED BY MORE THAN 75% OF THE 16 INSECTS SAMPLED AND OCCURRED IN MORE THAN 50% OF THE 112 CULTURE SAMPLES. THESE DATA FURTHER SUBSTANTIATE ACCUMULATING EVIDENCE THAT AQUATIC INSECTS ARE AMONG THE MOST IMPORTANT OVERLAND PASSIVE TRANSPORT VEHICLES FOR SUCH ORGANISMS, CONTRIBUTING TO THEIR MAINTENANCE IN, AND DISTRIBUTION INTO, SUITABLE ISOLATED AQUATIC HABITATS. (JONES-WISCONSIN)

FIELD 05C, 02H

ACCESSION NO. W70-06655

TRANSFER OF RADIOISOTOPES BETWEEN DETRITUS AND BENTHIC MACROINVERTEBRATES IN LABORATORY MICROECOSYSTEMS,

OAK RIDGE NATIONAL LAB., TENN. RADIATION ECOLOGY SECTION.

JERRY L. WILHM.

HEALTH PHYSICS, VOL 18, NO 3, P 277-284, 1970. 3 TAB, 14 REF.

DESCRIPTORS:
*RADIOISOTOPES, *DETRITUS, *BENTHIC FAUNA, *INVERTEBRATES, *MICROENVIRONMENT, *ECOSYSTEMS, CESIUM, COBALT RADIOISOTOPES, ALGAE, RADIOACTIVITY, TROPHIC LEVEL, LIFE HISTORY STUDIES, TIME, SNAILS, RADIOACTIVE WASTES, TENNESSEE, CRAYFISH, MIDGES, GAMMA RAYS, BLOODWORMS, SLUDGE WORMS.

IDENTIFIERS:
*MACROINVERTEBRATES, RUTHENIUM-106, CESIUM RADIOISOTOPES, SPIROGYRA, LIMNODRILUS HOFFMEISTERI, STICTOCHIRONOMUS ANNULIORUS, PHYSA HETEROSTROPHA, PROCLADIUS, WHITE OAK LAKE(TENN), UPTAKE.

ABSTRACT:
RUTHENIUM-106, CESIUM-137 AND COBALT-60 TRANSFER BETWEEN DETRITUS OR ALGAE AND BENTHIC MACROINVERTEBRATES WAS STUDIED IN LABORATORY MICROCOSMS. DETRITUS OR ALGAE CONTAINING RADIOISOTOPES WAS COLLECTED FROM A LAKE SERVING AS SETTLING BASIN FOR PARTIALLY DECONTAMINATED WASTE WATER; FOUR SPECIES OF DETRITUS FEEDERS AND ONE CARNIVOROUS SPECIE WERE TAKEN FROM A CONSTANT TEMPERATURE SPRING. MEAN INITIAL ACTIVITY IN PICOCURIES PER ASH-FREE GRAMS IN WHOLE DETRITUS WAS 28,800 FOR RUTHENIUM-106, 6630 FOR CESIUM-137 AND 6870 FOR COBALT-60, WHILE VALUES IN SPIROGYRA WERE 135,000, 4930, AND 62,100, RESPECTIVELY. RADIOISOTOPES WERE PRESENT IN INTERMEDIATE CONCENTRATIONS IN PULVERIZED DETRITUS. WHEN WHOLE DETRITUS WAS USED, ACTIVITY OF ALL THREE RADIOISOTOPES WAS GREATER IN LIMNODRILUS HOFFMEISTERI THAN IN STICTOCHIRONOMUS ANNULIORUS; WHEN PULVERIZED DETRITUS OR ALGAE WAS USED AS NOURISHMENT SOURCE, THE REVERSE WAS TRUE. PHYSA HETEROSTROPHA WAS INTERMEDIATE BETWEEN LIMNODRILUS AND STICTOCHIRONOMUS IN ACTIVITY. PEAK CONCENTRATIONS OF RADIOISOTOPES IN DETRITUS FEEDERS WERE REACHED BY DAY 3, WHILE PROCLADIUS HAD DELAYED UPTAKE, INCREASING IN ACTIVITY UNTIL DAY 14. RUTHENIUM-106 WAS DETECTED IN ALL LIFE STAGES AND EXUVIAE OF STICTOCHIRONOMUS; CESIUM-137 AND COBALT-60 WERE DETECTED ONLY IN LARVAE. ACTIVITY OF RUTHENIUM-106 IN ADULTS WAS 1/3 LESS THAN IN LARVAE. (JONES-WISCONSIN)

FIELD 05C

ACCESSION NO. W70-06668

POLYMERS IN WATER TREATMENT,

UNION CARBIDE CORP. CHEMICAL DIV.

B. MANSFIELD.

PROCESS BIOCHEMISTRY, VOL 5, NO 2, P 28-30, 1970. 4 FIG, 3 TAB, 6 REF.

DESCRIPTORS:
*OXIDES, *FLUID FRICTION, SEWERS, SLURRIES, WASTE WATER TREATMENT, SLUDGE TREATMENT, TOXICITY, ALGAE, BACTERIA, FISH, SEWAGE SLUDGE, FLOW, FLOCCULATION, COAGULATION, FILTRATION, VISCOSITY.

IDENTIFIERS:
*HIGH-MOLECULAR-WEIGHT POLYMERS, TURBULANT FRICTICNAL DRAG, ALKYLENE OXIDE, POLYMERIZATION, ETHYLENE OXIDE, WATER CLARIFICATION, SLUDGE DRYING, POLYETHYLENE OXIDE POLYMERS, SEWAGE SETTLING, UNION CARBIDE CORP, POLYOX, CARBOWAX, FLOCCULANT DOSAGES, BAROCC CLAY SOILS, URANIUM ORE.

ABSTRACT:
THE DISCOVERY AND SUBSEQUENT PRODUCTION BY UNION CARBIDE OF HIGH-MOLECULAR-WEIGHT POLYMERS, SUCH AS POLYOX, PERMITS A REDUCTION OF SURCHARGES IN SEWER PIPELINES. INJECTION OF THESE POLYMERS AT CONCENTRATIONS OF 45 TO 200 PPM INCREASES MORE THAN 100% THE SEWAGE FLOW, THEREBY ELIMINATING THE NECESSITY OF PARALLEL PIPES. THE CHEMICALS DO NOT ADVERSELY INFLUENCE BACTERIA OR FISH, OR INCREASE ALGAL GROWTH. THEY HAVE NO MEASURABLE BIOCHEMICAL OXYGEN DEMAND, BUT IMPROVE SEWAGE SETTLING AND SLUDGE DRYING. ADDITION OF THE POLYMERS TO WATER-BASED SLURRY MARKEDLY DECREASES FLOW RESISTANCE. THE POLYMERS ARE HIGHLY EFFECTIVE AGENTS FOR INCREASING SETTLING AND FILTRATION RATES OF DISPERSED SOLIDS AND FACILITATE WATER CLARIFICATION. (WILDE-WISCONSIN)

FIELD 05D

ACCESSION NO. W70-06968

RELATIONSHIPS BETWEEN WATER QUALITY AND WATER RIGHTS,

R. B. ROBIE.

CONTEMPORARY DEVELOPMENTS IN WATER LAW, WATER RESOURCES SYMPOSIUM, VOL 4, P 72-82, 1970. 11 P, 28 REF.

DESCRIPTORS:
*CALIFORNIA, *WATER RIGHTS, *WATER UTILIZATION, *IMPAIRED WATER QUALITY, RIPARIAN RIGHTS, PRIOR APPROPRIATION, GROUNDWATER, WATER QUALITY, COMPETING USES, DIVERSION, RECIRCULATED WATER, WATER REUSE, LEGAL ASPECTS, JUDICIAL DECISIONS, LEGISLATION, ADMINISTRATION, NON-CONSUMPTIVE USE, ALGAE, EVAPORATION, REASONABLE USE, BRACKISH WATER, DOMESTIC WATER, GEOCHEMISTRY, WATER CEHMISTRY, SALT BALANCE.

ABSTRACT:
A STUDY OF THE INTERRELATIONSHIPS OF WATER QUALITY TO WATER RIGHTS INVOLVES MANY LEGAL ASPECTS--RIPARIAN, APPROPRIATIVE AND PRESCRIPTIVE RIGHTS, RIGHTS TO THE USE OF GROUNDWATER AND REUSE OF WATER. EVERY EXERCISE OF A WATER RIGHT HAS SOME EFFECT ON WATER QUALITY. THE RETURN FLOW OF DIVERTED WATERS MAY CONTAIN A HIGHER CONCENTRATION OF MINERALS AND SALT THAN PREVIOUSLY. THE REDUCED FLOW CAUSED BY DIVERSION MAY HAVE SEVERAL RESULTS INCLUDING: (1) A REDUCTION IN THE ASSIMILATIVE CAPACITY OF A STREAM; (2) LARGER EVAPORATION LOSSES; (3) INCREASED STRATIFICATIONS; (4) ACCELERATED ALGAE GROWTHS; AND (5) SALINE INTRUSION. COURTS HAVE LONG RECOGNIZED WATER QUALITY AS A WATER RIGHT. DOWNSTREAM RIPARIAN OWNERS, FOR EXAMPLE, HAVE BEEN ABLE TO ENJOIN THE IRRIGATION USE OF AN UPSTREAM RIPARIAN OWNER. JUNIOR APPROPRIATORS HAVE BEEN HELD LIABLE FOR POLLUTION THAT INTERFERES WITH THE RIGHTS OF A SENIOR APPROPRIATOR. IT IS POSSIBLE THAT THE RIGHT TO DEGRADE WATER COULD BE APPORTIONED AMONG USERS. A UNIFORM APPROACH IS NECESSARY HOWEVER. VARIOUS STATUTORY LAWS, CASE DECISIONS, AND ADMINISTRATIVE RULES AND REGULATIONS AFFECTING THE SUBJECT ARE DISCUSSED, ESPECIALLY AS RELATES TO CALIFORNIA. (MARSEE-FLORIDA)

FIELD 05G, 06E

ACCESSION NO. W70-07100

FACTORS INFLUENCING THE HERBICIDAL EFFICIENCY OF MCPA AND MCPB IN THREE SPECIES OF MICRO-ALGAE,

UNIVERSITY OF STRATHCLYDE, GLASGOW (SCOTLAND). DEPT. OF BIOLOGY.

R. C. KIRKWOOD, AND W. W. FLETCHER.

WEED RESEARCH, VOL 10, NO 1, P 3-10, 1970. 2 FIG, 3 TAB, 12 REF.

DESCRIPTORS:
*CHLAMYDOMONAS, *ALGAE, *HERBICIDES, PESTICIDES, CHLORELLA, RESPIRATION, PHOSPHORUS, TOXICITY, RADIOISOTOPES, INHIBITION.

IDENTIFIERS:
*CHLORINATED HERBICIDES, MCPB, MCPA, CHLAMYDOMONAS GLOBUSA, CHLORELLA PYRENOIDOSA, STICHOCOCCUS BACILLARIS, UNDISSOCIATED MOLECULES, ALGAL GROWTH.

ABSTRACT:
THE RELATIVE HERBICIDAL EFFECTIVENESS OF MCPA (4-CHLORO-2-METHYLPHENOXYACETIC ACID) AND MCPB (4-(4-CHLORO-2-METHYLPHENOXY) BUTYRIC ACID) WAS APPRAISED BY USING AS TEST PLANTS THREE SPP OF UNICELLULAR ALGAE. THE ALGAE WERE GROWN IN NUTRIENT SOLUTIONS OF PH 6.5 IN 10-1 ASPIRATORS. THE HERBICIDES WERE C-14-CARBOXYL-LABELLED. MCPB INHIBITED GROWTH, RESPIRATION, AND PHOSPHORUS UPTAKE MORE THAN MCPA AND WAS TOXIC AT THE MINIMUM CONCENTRATION OF 0.0025 MOLES. THE ENHANCED HERBICIDAL EFFECT OF MCPB COINCIDED WITH ITS INCREASED ABSORPTION. BOTH HERBICIDES INHIBITED PHOSPHORYLATION. OPTIMUM UPTAKE OF HERBICIDES WAS AT PH VALUES FAVORING MOVEMENT OF UNDISSOCIATED MOLECULES. (WILDE-WISCONSIN)

FIELD 05F, 05B

ACCESSION NO. W70-07255

ENHANCEMENT BY GLYCEROL OF PHOTOTROPHIC GROWTH OF MARINE PLANKTONIC ALGAE AND ITS SIGNIFICANCE TO THE ECOLOGY OF GLYCEROL POLLUTION,

FISHERIES RESEARCH BOARD OF CANADA, VANCOUVER, (BRITISH COLUMBIA). VANCOUVER LAB.

JOSEPH Y. CHENG, AND NAVAL J. ANTIA.

JOURNAL FISHERIES RESEARCH BOARD OF CANADA, VOL 27, NO 2, P 335-346, 1970. 2 FIG, 2 TAB, 16 REF.

DESCRIPTORS:
*POLLUTANTS, *SEA WATER, *ALGAE, *MARINE ALGAE, CULTURES, PHYTOPLANKTON, CYTOLOGICAL STUDIES, CYANOPHYTA, CHLOROPHYTA, CHRYSOPHYTA, RHODOPHYTA, TESTING, ECOLOGY.

IDENTIFIERS:
*GLYCEROL, PHOTOHETEROTROPHIC GROWTH, LUMINARIN-TYPE POLYSACCHARIDES, CRYPTOPHYCEA, DINOPHYCEA, BACCILARIOPHYCEA, PRYMNESIUM PARVUM, PHOTOAUTOTROPHS, CHROOMONAS SALINA.

ABSTRACT:
EIGHTEEN SPECIES OF MARINE PHYTOPLANKTERS WERE GROWN IN AXENIC CULTURE OF SEA WATER ENRICHED IN NITRATE, ORTHOPHOSPHATE, SILICATE, TRACE-METAL IONS, AND VITAMINES, AND BUFFERED AT PH 7.6-7.8. ASIDE FROM PRYMNESIUM PARVUM AND CHROOMONAS SALINA, NONE OF THE SPECIES SHOWED GROWTH ON GLYCEROL IN THE ABSENCE OF LIGHT. UNDER ILLUMINATION, GLYCEROL STIMULATED THE GROWTH OF 16 SPECIES PARTICULARLY THAT CHRYSOPHYCEA, CRYPTOPHYCEAE, PHAEDACTYLUM TRICORNUTUM, PORPHYRIDIUM CRUENTUM, AND NANNOCHLORIS OCULATA. THE GROWTH OF SOME SPECIES WAS INHIBITED BY GLYCEROL AT DIFFERENT CONCENTRATIONS. THE EFFECTS OF COMPOUND SUGGESTED ITS RELATION TO PHOTOHETEROTROPHIC GROWTH. IN SOME CASES GLYCEROL INDUCED OBVIOUS CYTOLOGICAL AND METABOLIC CHANGES. THE ECOLOGICAL IMPLICATIONS OF GLYCEROL POLLUTION OF SEA WATER ARE DISCUSSED. (WILDE-WISCONSIN)

FIELD 05B

ACCESSION NO. W70-07280

THE LANGE-KUENTZEL-KERR THESIS.

CANADIAN RESEARCH AND DEVELOPMENT, MARCH 1970. 8 P, 1 FIG, 1 TAB, 14 REF.

DESCRIPTORS:
*EUTROPHICATION, *PHOSPHORUS, *CARBON, BACTERIA, ALGAE, SYMBIOSIS, CARBON DIOXIDE, NITROGEN, LAKES, ESTUARIES, NUTRIENTS, DETERGENTS, WISCONSIN, LAKE ERIE, LAKE ONTARIO, ST LAWRENCE RIVER, INTERNATIONAL JOINT COMMISSION.

IDENTIFIERS:
*CARBONACEOUS MATERIAL, *CANADIAN PHOSPHATE DETERGENT BAN, WYANDOTTE CHEMICAL CORPORATION, FMC CORPORATION, SOAP AND DETERGENT ASSOCIATION, CARBOY TRIALS, CANADA, NITROGEN: PHOSPHORUS RATIO.

ABSTRACT:
REPORTS BY W LANGE (NATURE VOL. 215, NO. 5107: 1277-1278, SEP 17, 1967), L E KUENTZEL (JOURNAL WATER POLLUTION CONTROL FEDERATION: 1737-1747, OCT 1969), AND P C KERR (UNPUBLISHED) ARE CITED DEFENDING THE THESIS THAT CARBONACEOUS MATERIAL, NOT PHOSPHORUS, IS THE FACTOR CONTROLLING THE PROCESS OF EUTROPHICATION. AS ADDITIONAL PROOF OF THE MINOR ROLE OF PHOSPHORUS AND NITROGEN IN THE GROWTH OF ALGAE, THE REPORT INCLUDES THE RESULT OF AN EXPERIMENT CONDUCTED IN TWO SMALL OLIGOTROPHIC LAKES IN FLORIDA. ADDITION OF PHOSPHATE AND NITROGEN FERTILIZERS TO ONE OF THESE ORGANIC MATTER-FREE LAKES FAILED TO ALTER SIGNIFICANTLY THE TROPHIC STATE OF THE LAKE AND THE DENSITY OF PLANKTONIC ORGANISMS. THE ISSUE IN QUESTION IS OF A FAR-REACHING IMPORTANCE AS IT IS RELEVANT TO THE USA AND CANADIAN GOVERNMENT ACTION RESTRICTING PHOSPHATE ENRICHED EFFLUENTS OF SOAP AND DETERGENTS INTO ERIE AND ONTARIO LAKES AND THE INTERNATIONAL SECTION OF THE ST LAWRENCE RIVER. (WILDE-WISCONSIN)

FIELD 05B, 05C

ACCESSION NO. W70-07283

ALGAL PERIODICITY AND WASTE RECLAMATION IN A STABILIZATION POND ECOSYSTEM,

IOWA STATE UNIV., AMES.

RONALD L. RASCHKE.

JOURNAL OF THE WATER POLLUTION CONTROL FEDERATION, VOL 42, NO 4, P 518-530, APRIL 1970. 2 FIG, 6 TAB, 40 REF.

DESCRIPTORS:
*WASTE WATER TREATMENT, *PONDS, *ALGAE, LAGOONS, BIOCHEMICAL OXYGEN DEMAND, CHEMICAL OXYGEN DEMAND, PHOSPHATES, NITROGEN, STABILIZATION.

IDENTIFIERS:
*WATER RECLAMATION, VARIATIONS, COMPOSITION.

ABSTRACT:
IN AN ATTEMPT TO EVALUATE POSSIBLE PROCEDURES FOR MEETING NEW STANDARDS OF THE FEDERAL WATER QUALITY ACT, THE AMES WATER POLLUTION CONTROL DEPARTMENT BUILT A NUMBER OF EXPERIMENTAL PONDS. ONE OF THEM WAS MADE AVAILABLE FOR THE PRESENT INVESTIGATION IN WHICH THE INTENT HAS BEEN TO STUDY ALGAL COMPOSITION AND PERIODICITY, DISSOLVED OXYGEN, PH, TEMPERATURE, AND THE EXTENT OF REDUCTION IN 5-DAY BOD, COD, SOLIDS, PHOSPHATES, AMMONIA, AND NITRATES THROUGH THE PLANT POND COMPLEX. THE POND IS 0.12 ACRES IN AREA WITH A WATER HOLDING CAPACITY OF 105, 562 GALLONS AND AN AVERAGE DEPTH OF 2.53 FEET. THE POND DETENTION TIME RANGED BETWEEN 3.7 TO 4.2 DAYS. RESULTS SHOWED THAT ALGAL PERIODICITY AND BOD ANALYSIS MAY BE INFLUENCED BY ALGAL INHIBITORS. GREEN FAGELLATES DOMINATED IN WINTER AND SPRING, WHILE COCCOID GREENS DOMINATED IN SUMMER AND FALL. THE MAXIMUM PER CENT REMOVAL AT ANY ONE TIME WAS 78% AND 75% FOR BOD AND FILTERED BOD RESPECTIVELY. THE MAXIMUM COD REMOVAL WAS 63%. TOTAL SUSPENDED SOLIDS RANGED FROM 458 TO 687 WHICH IS LESS THAN A LIMIT OF 1000 MG/L SET BY THE U.S. PUBLIC HEALTH SERVICE STANDARDS. AMMONIA NITROGEN RANGED FROM 0.0 TO 21.0 MG/L AND NITRATE NITROGEN RANGED FROM 0.5 TO 1.8 MG/L. PH RANGED FROM 7.35 TO 9.09 AND TEMPERATURE FROM 0.8 DEG C TO 29.9 DEG C. THE EFFLUENT USUALLY MET THE IOWA WATER POLLUTION CONTROL COMMISSION'S STANDARDS FOR TEMPERATURE, PH, DISSOLVED OXYGEN CONCENTRATION, BUT PHOSPHORUS AND NITROGEN CONTENTS WERE HIGH. (HANCUFF-TEXAS)

FIELD 05D

ACCESSION NO. W70-07838

ENHANCEMENT BY GLYCEROL OF PHOTOTROPHIC GROWTH OF MARINE PLANKTONIC ALGAE AND ITS SIGNIFICANCE TO THE ECOLOGY OF GLYCEROL POLLUTION,

FISHERIES RESEARCH BOARD OF CANADA, VANCOUVER (BRITISH COLUMBIA). VANCOUVER LAB.

JOSEPH Y. CHENG, AND NAVAL J. ANTIA.

JOURNAL OF THE FISHERIES RESEARCH BOARD OF CANADA, VOL 27, P 335-346, 1970. 11 FIG, 2 TAB, 16 REF.

DESCRIPTORS:
ORGANIC WASTES, WATER POLLUTION EFFECTS, *ALGAE, PHYTOPLANKTON, DIATOMS, CHRYSOPHYTA, CHLOROPHYTA, *GROWTH RATES, INHIBITION, *PLANT PHYSIOLOGY, *PLANT GROWTH, PLANT GROWTH SUBSTANCES, PLANT MORPHOLOGY.

IDENTIFIERS:
*GLYCEROL, *PHOTOTROPHIC GROWTH, *MARINE PHYTOPLANKTON(PRYMNESIUM PARVUM)(CHROOMONAS SALINA), GLYCEROL POLLUTION.

ABSTRACT:
THE EFFECTS OF LOW (0.05 M) AND HIGH (0.5-1.0 M) CONCENTRATIONS OF GLYCEROL ON THE GROWTH OF 18 SPECIES OF MARINE PHYTOPLANKTERS WERE STUDIED. THE ALGAE WERE GROWN IN AXENIC CULTURE IN SEA WATER ENRICHED WITH NITRATE, ORTHOPHOSPHATE, SILICATE, TRACE-METAL IONS, VITAMIN B12, THIAMINE, BIOTIN, AND BUFFERED AT PH 7.6-7.8. APART FROM A CHRYSOMONAD (PRYMNESIUM PARVUM AND A CRYPTOMONAD (CHROOMONAS SALINA), NONE OF THE SPECIES SHOWED ANY SIGNIFICANT GROWTH ON GLYCEROL IN THE ABSENCE OF LIGHT. HOWEVER, IN THE PRESENCE OF LIGHT, GLYCEROL ENHANCED THE GROWTH OF 16 SPECIES, IN PARTICULAR MEMBERS OF THE CHRYSOPHYCYEAE AND CRYPTOPHYCEAE, ONE DIATOM (PHAEODACTYLUM TRICORNUTUM), ONE RHODOPHYTE (PORPHYRIDIUM CRUENTUM), AND ONE CHLOROPHYTE (NANNOCHLORIS OCULATA). THE ENHANCEMENT EFFECT WAS OBSERVED IN THE GROWTH RATE AND IN THE PEAK POPULATION DENSITY, WHICH IN MANY INSTANCES WAS SEVERAL TIMES THAT FROM NONGLYCERINATED CULTURES AND SUGGESTED PHOTOHETEROTROPHIC GROWTH. SOME SPECIES SHOWED OBVIOUS CYTOLOGICAL AND METABOLIC CHANGES FROM GROWTH ON GLYCEROL. (SJOLSETH-WASHINGTON)

FIELD 05C

ACCESSION NO. W70-07842

INSECTICIDE RESIDUES IN SOME COMPONENTS OF THE ST LAWRENCE RIVER ECOSYSTEM FOLLOWING APPLICATIONS OF DDD,

DEPARTMENT OF AGRICULTURE, SASKATOON (SASKATCHEWAN); AND SAINT DUNSTAN'S UNIV., CHARLOTTETOWN (PRINCE EDWARD ISLAND). DEPT. OF CHEMISTRY.

F. J. H. FREDEEN, AND J. REGIS DUFFY.

PESTICIDES MONITORING JOURNAL, VOL 3, NO 4, P 219-226, 1970. 4 TAB, 14 REF.

DESCRIPTORS:
*INSECTICIDES, *ST LAWRENCE RIVER, WATER POLLUTION SOURCES, WATER POLLUTION EFFECTS, MUD, MOLLUSKS, FISH, INSECTS, DDT, SILTS, SUCKERS, CATFISHES, BASS, PERCHES, PIKES, SAMPLING, SNAILS, ALGAE, DIATOMS, CLAMS, SHELLFISH, AMPHIPODA, WATER BIRDS, WISCONSIN, ANALYTICAL TECHNIQUES.

IDENTIFIERS:
*RESIDUES, *DDD, DDE, TDE, BLACK-FLY LARVAE, CLADOPHORA, CYPRINUS CARPIO, CATOSTOMUS COMMERSONI, AMEIURUS NEBULOSUS, PERCA FLAVESCENS, AMBLOPLITES RUPESTRIS, ESOX LUCIUS, CAMPELOMA, PSIDIUM, SHADFLY, CHIRONOMIDS, CLEAR LAKE(CALIF), RICHELIEU RIVER(QUEBEC), GREEN BAY(WIS), MACKEREL, MONTREAL(CANADA).

ABSTRACT:
DDD (TDE) RESIDUES WERE MEASURED IN WATER, MUD, MOLLUSKS, AND FISH OF THE ST LAWRENCE RIVER IN 1967 DURING AND AFTER APPLICATIONS OF DDD TO CONTROL NUISANCE INSECTS. CONCENTRATIONS DETECTED IN WATER (UP TO 0.0139 PPM) RANGED FROM 1 TO 17% OF THOSE APPLIED TO THE RIVER 10 MILES UPSTREAM. DDD CONCENTRATIONS IN MOLLUSKS 17 MILES UPSTREAM FROM THE POINT OF APPLICATION AND 10 AND 45 MILES DOWNSTREAM, AVERAGED 0.002, 0.101, AND 0.0 PPM, RESPECTIVELY. IN THE SAME SAMPLES, CONCENTRATIONS OF DDT AND DDE COMBINED (FROM UNKNOWN SOURCES) AVERAGED 0.030, 0.225, AND 0.027 PPM. IN EDIBLE FLESH FROM 216 FISH OF 5 SPECIES, DDD RESIDUES AVERAGED 0.156 PPM IN SAMPLES COLLECTED 17 MILES UPSTREAM AND 0.369 IN THE COMBINED SAMPLES FROM POINTS 10 AND 45 MILES DOWNSTREAM. RESIDUES OF DDT PLUS DDE IN THESE SAME SAMPLES AVERAGED 0.224 AND 0.227 PPM, RESPECTIVELY. THE HIGHEST CONCENTRATION OF DDD IN AN INDIVIDUAL FISH WAS 1.81 PPM. THE ULTIMATE FATE OF THE DDD, THAT IS ITS CONVERSION TO MATERIALS WITH REDUCED TOXICITIES, IS POORLY UNDERSTOOD. DDD IS ITSELF A PRODUCT OF THE REDUCTIVE DECHLORINATION OF DDT IN NATURE. (JONES-WISCONSIN)

FIELD 05C

ACCESSION NO. W70-08098

CURRENT WATER QUALITY CONDITIONS AND INVESTIGATIONS IN THE UPPER POTOMAC RIVER TIDAL SYSTEM,

FEDERAL WATER QUALITY ADMINISTRATION, ANNAPOLIS, MD. CHESAPEAKE TECHNICAL SUPPORT LAB.

JOHAN A. AALTO, NORBERT A. JAWORSKI, AND DONALD W. LEAR, JR.

FEDERAL WATER QUALITY ADMINISTRATION, CHESAPEAKE TECHNICAL SUPPORT LABORATORY TECHNICAL REPORT NO 41, MAY 1970. 38 P, 4 FIG, 6 TAB, 17 REF.

DESCRIPTORS:
*ESTUARIES, *WATER POLLUTION EFFECTS, *DISTRICT OF COLUMBIA, TIDAL EFFECTS, NUTRIENTS, EUTROPHICATION, WATER QUALITY, WATER POLLUTION CONTROL, SURVEYS, INVESTIGATIONS, PATH OF POLLUTANTS, SAMPLING, WATER ANALYSIS, ALGAE, COLIFORMS, DISSOLVED OXYGEN.

IDENTIFIERS:
*POTOMAC RIVER.

ABSTRACT:
BASED ON DATA OBTAINED BY FIELD INVESTIGATIONS AND FROM WASTE WATER TREATMENT AGENCIES IN THE WASHINGTON METROPOLITAN AREA, A STATEMENT ON CURRENT WATER CONDITIONS AND INVESTIGATIONS OF THE UPPER POTOMAC RIVER TIDAL SYSTEM WAS PREPARED AND SUMMARIZED. FECAL COLIFORM DENSITIES ARE LOWER THAN IN 1965 AS A RESULT OF THE INCREASED CHLORINATION OF TREATED WASTE DISCHARGES SINCE 1969. HIGH FECAL COLIFORM DENSITIES WERE PREVALENT AT TIMES OF HIGH STREAM FLOW ABOVE THE MAJOR WASTEWATER DISCHARGES. THESE HIGH DENSITIES CAN BE ATTRIBUTED TO A COMBINATION OF LAND RUNOFF FROM THE UPPER POTOMAC BASIN, URBAN RUNOFF, STORM SEWERS AND COMBINED SEWER OVERFLOWS. TRIBUTARIES ALSO CONTAINED VERY HIGH FECAL COLIFORM DENSITIES AT TIMES. EXTENSIVE PHYTOPLANKTON BLOOMS WERE DETECTED. SINCE THE LATE 1930'S THE AMOUNT OF PHOSPHORUS ENTERING THE POTOMAC HAS INCREASED ABOUT TENFOLD AND NITROGEN INCREASED ABOUT FIVEFOLD. THE MAJOR SHIFT TOWARD NUISANCE BLUE-GREEN ALGAL GROWTHS APPEARS TO BE RELATED TO INCREASES IN NITROGEN AND PHOSPHORUS, AND NOT BOD (CARBON). MOST OF THE PHOSPHORUS WHICH ENTERED THE TIDAL SYSTEM FROM THE UPPER BASIN, PLUS SOME FROM LOCAL WASTE WATER DISCHARGES, WAS ADSORBED AND DEPOSITED IN THE BOTTOM SEDIMENTS OF THE ESTUARY. DYE AND MATHEMATICAL MODEL INVESTIGATIONS INDICATE THAT WASTE WATER ASSIMILATION AND TRANSPORT RATES ARE VERY LOW. (KNAPP-USGS)

FIELD 05B, 02L

ACCESSION NO. W70-08248

PRELIMINARY ANALYSES OF THE WASTE WATER AND ASSIMILATION CAPACITIES OF THE ANACOSTIA TIDAL RIVER SYSTEM,

FEDERAL WATER QUALITY ADMINISTRATION, ANNAPOLIS, MD. CHESAPEAKE TECHNICAL SUPPORT LAB.

NORBERT A. JAWORSKI, LEO J. CLARK, AND KENNETH D. FEIGNER.

FEDERAL WATER QUALITY ADMINISTRATION, CHESAPEAKE TECHNICAL SUPPORT LAB REPORT NO 39, APRIL 1970. 57 P, 16 FIG, 4 TAB, 5 REF, 3 APPEND.

DESCRIPTORS:
*ESTUARIES, *WATER POLLUTION EFFECTS, *DISTRICT OF COLUMBIA, TIDAL EFFECTS, NUTRIENTS, EUTROPHICATION, WATER QUALITY, WATER POLLUTION CONTROL, SURVEYS, INVESTIGATIONS, PATH OF POLLUTANTS, SAMPLING, WATER ANALYSIS, ALGAE, COLIFORMS, DISSOLVED OXYGEN.

IDENTIFIERS:
*ANACOSTIA RIVER.

ABSTRACT:
A PRELIMINARY ANALYSIS OF WASTE WATER ASSIMILATION AND TRANSPORT CAPACITIES WAS MADE OF THE TIDAL PORTION OF THE ANACOSTIA RIVER, WASHINGTON, D.C. THE MEAN STREAMFLOW IN THE BASIN IS ABOUT 122 CFS AND THE 7-DAY LOW FLOW WITH A RECURRENCE INTERVAL OF ONCE-IN-TEN-YEARS IS 8 CFS. BASED ON 1969 SAMPLING DATA, THE WATER QUALITY CONDITIONS IN THE TIDAL RIVER UNDER SUMMER CONDITIONS ARE TYPIFIED BY: (A) LOW DISSOLVED OXYGEN CONCENTRATIONS OFTEN FALLING BELOW 2.0 MG/L, (B) HIGH FECAL COLIFORM DENSITIES OFTEN ABOVE 10,000 MPN/100 ML, (C) HIGH TURBIDITY LEVELS ESPECIALLY DURING PERIODS OF HIGH RUNOFF, AND (D) HIGH NUTRIENT CONCENTRATIONS. THE MOST PRONOUNCED DEGRADATION RESULTS FROM STORM SEWER AND COMBINED SEWER OVERFLOWS AND DEFECTIVE SANITARY SEWER SYSTEMS. MATHEMATICAL MODEL INVESTIGATIONS OF THE ANACOSTIA RIVER INDICATE THAT WASTE WATER ASSIMILATION AND TRANSPORT CAPABILITIES FOR LARGE ADVECTIVE FLOWS ARE MOST SENSITIVE TO THE DECAY RATE OF A POLLUTANT AND NOT TO THE DISPERSION EFFECT OF THE TIDAL SYSTEM. THE SENSITIVITY OF THE DECAY RATE IS A RESULT OF THE LONG DETENTION TIME, WHILE THE EFFECT OF DISPERSION IS DIMINISHED BY THE PRONOUNCED ADVECTIVE MOVEMENT. THE TIDAL SYSTEM CAPABILITY TO ASSIMILATE OXYGEN-DEMANDING WASTE WATER IS CURRENTLY BEING EXCEEDED. IF WASTE WATER WERE SUBJECTED TO HIGH CARBONACEOUS AND NITROGENOUS BOD REMOVAL AND IF THE EFFLUENT IS AERATED TO 6.0 MG/L, THE PRESENT WATER QUALITY OF THE ANACOSTIA RIVER WOULD BE ENHANCED. (KNAPP-USGS)

FIELD 05A, 05B, 02L

ACCESSION NO. W70-08249

BACTERICIDAL EFFECTS OF ALGAE ON ENTERIC ORGANISMS,

TEXAS UNIV., AUSTIN, CENTER FOR RESEARCH IN WATER RESOURCES.

ERNST DAVIS, AND EARNEST GLOYNA.

FWPCA GRANT 18050 DOL. TECHNICAL REPORT EHE-70-06, CRWR-55, 132 P, MAR 1970.
9 FIG, 144 TAB, 60 REF.

DESCRIPTORS:
*ALGAE, *CULTURES, *ENTERIC BACTERIA, *OXIDATION LAGOONS, *PATHOGENIC BACTERIA.

IDENTIFIERS:
*AUTOGOMISTIC EFFECTS, *AXEMIC CULTURES, *DIEOFF RATES, AFTERGROWTH, BLUE-GREEN ALGAE, FIELD STUDIES, GREEN ALGAE, LABORATORY STUDIES.

ABSTRACT:
A SERIES OF EXPERIMENTS INVOLVING THE EFFECTS OF BLUE-GREEN AND GREEN ALGAE ON THE DIEOFF RATES OF SELECTED BACTERIA WERE CONDUCTED. AXEMIC CULTURES OF ANABAENA CYLINDRICA, A. NIDULANS, OSCILLATORIA CHALYBIA, CHLORELLA PYRENOIDOSA AND SCENEDESMUS OBLUGUUS AMONG OTHERS. CULTURES OF ENTERIC BACTERIA SPECIES (ADCALIGEUES FAECALIS, ENTEROBACTER AEROGENES, E. COLI, PROTEUS VULGARIS, PSUDOMONAS AERARGINOSA, AND SERRATIA MARCERCENS) WERE ADDED TO THE OXENIC ALGAL CULTURES DURING DIFFERENT PERIODS OF THE ALGAL LIFE CYCLES. FILTRATE FROM ACTIVELY GROWING ALGAE WAS EXPOSED TO CULTURES OF ENTERICS TO DETERMINE WHETHER ANY ANTIBIOTIC COMPOUNDS WERE IMPARTED TO THE MEDIUM DURING LAG PHASE GROWTH OF ALGAE. TO DETERMINE AFTERGROWTH OF THE ENTERIC SPECIES, THE DURATION OF THE TESTS WAS EXTENDED TO ABOUT 90 DAYS. MIXED CULTURES OF GREEN AND BLUE-GREEN ALGAE WERE EXPOSED TO BOTH SINGLE SPECIES OF ENTERIC BACTERIA AND MIXED CULTURES. MIXED ALGAL CULTURES CAUSE A GREATER DIEOFF AMONG THE ENTERIC BACTERIA THAN DO INDIVIDUAL SPECIES OF ALGAE. THE DIEOFF CHARACTERISTICS OF PATHOGENIC SPECIES, NAMELY SALMONELLA TYPHOSA, S. PARATYPLIN, SHIGELLA DYSENTIRIAE, S. PARACYSENTESIAE, AND VIBRIO COMMA WERE ALSO DETERMINED. THE PATHOGENIC SPECIES DID NOT SURVIVE AS LONG AS THE ENTERIC TEST SPECIES UNDER SIMILAR TEST CONDITIONS. VIRTUALLY NO AFTERGROWTH WAS DETECTED ON THE PART OF THE PATHOGENES. (AGUIRRE-TEXAS)

FIELD 05D

ACCESSION NO. W70-08319

USES OF WASTE HEAT,

OAK RIDGE NATIONAL LAB., TENN.

SAM E. BEALL.

OAK RIDGE NATIONAL LABORATORY REVIEW, VOL 3, NO 4, P 9-14, SPRING 1970.

DESCRIPTORS:
*THERMAL POLLUTION, *GREENHOUSES, WATER TEMPERATURE, POWERPLANTS, DISTILLATION, ALGAE, AQUICULTURE, FISH, SHELLFISH, CRUSTACEANS, OYSTERS, SEASONAL.

IDENTIFIERS:
*WASTE HEAT, THERMAL EFFICIENCY, HEAT CONSUMPTION.

ABSTRACT:
WASTE HEAT FROM POWERPLANTS WHICH WOULD NORMALLY ADD TO THE PROBLEM OF THERMAL POLLUTION CAN BE REUSED FOR A PROFIT. ONE USE OF THIS WASTE HEAT WOULD BE AS A SUBSTITUTE FOR ELECTRICITY IN LOW-TEMPERATURE INDUSTRIAL USES. WASTE HEAT CAN BE USED FOR AIR COOLING IN AMMONIA OR LITHIUM BROMIDE AIR CONDITIONING SYSTEMS DURING THE SUMMER WHEN THERMAL RELEASES NEED TO BE REDUCED THE MOST. ANOTHER USE OF WASTE HEAT WOULD BE IN THE EVAPORATION OF WASTE WATER TO ALLOW REUSE OF THE VALUABLE WATER. SIMILARLY, DISTILLATION OF SALT WATER OR BRACKISH WATER COULD MAKE USE OF WASTE HEAT. THE BEST POSSIBLE USE WOULD BE IN THE HEATING AND COOLING OF GREENHOUSES AND POULTRY HOUSES LOCATED ON THE REACTOR EXCLUSION AREA. HOWEVER, AQUICULTURE POND HEATING PROMISES LARGE PROFITS IN THE HIGH YIELDS OF FISH, SHELLFISH, AND CRUSTACEANS. FURTHER STUDY ON THE SUBJECT OF WASTE HEAT REUSE IS BEING DONE IN THE PACIFIC NORTHWEST WHERE IRRIGATION USES ARE BEING CONSIDERED. (OSBORNE-VANDERBILT)

FIELD 05G, 08A

ACCESSION NO. W70-08832

ALGAL FLOCCULATION WITH ALUMINUM SULPHATE AND POLYELECTROLYTES,

NEW SOUTH WALES UNIV., KENSINGTON (AUSTRALIA).

MICHAEL G. MCGARRY.

JOURNAL WATER POLLUTION CONTROL FEDERATION, PART II, P R191-R207, MAY 1970, VOL 42, NO 5. 6 FIG, 4 TAB, 14 REF.

DESCRIPTORS:
*ALGAE, *FLOCCULATION, EQUIPMENT, COSTS, SEPARATION TECHNIQUES, COAGULATION, WASTE WATER TREATMENT.

IDENTIFIERS:
*ALUM, POLYELECTROLYTES, FACTORIAL DESIGN.

ABSTRACT:
THE HIGH RATE POND PROCESS IS BEING DEVELOPED TO TREAT WASTE WATER AND TO PROVIDE A NEW SOURCE OF EDIBLE PROTEIN IN THE FORM OF ALGAE. AS A FOOD PRODUCTION PROCESS, THE YIELD OF ALGAE EXCEEDS THE YIELDS OF ALL CURRENTLY KNOWN GRAIN CROPS. RECENT STUDIES IN THAILAND HAVE INDICATED THAT PROTEIN PRODUCTION RATES OF 25 TONS/YEAR/ACRE ARE QUITE EASILY ACHEIVED ON A PILOT PLANT SCALE. THIS YIELD COMPARES FAVORABLY WITH THAT OF WHEAT, 135 LB/YEAR/ACRE AND THAT OF SOY BEANS 576 LB/YEAR/ACRE. ALUMINUM SULFATE WAS TESTED AS A PRIMARY COAGULANT IN CONJUNCTION WITH A VARIETY OF POLYELECTROLYTES FOR CHEMICAL FLOCCULATION AS A MEANS OF HARVESTING THE DISPERSED ALGAE FROM HEAVILY LADEN POND WATER. THE INDEPENDENT VARIABLES STUDIED WERE ALUM AND POLYELECTROLYTE CONCENTRATIONS, TIME OF POLYELECTROLYTE ADDITION, FAST AND SLOW MIXING PERIOD AND DEGREE OF TURBULENCE. DEPENDENT VARIABLES CONSIDERED WERE: SUPERNATANT TRANSMISSION AND SETTLED VOLATILES AS A MEASURE OF HARVEST OF ALGAE, AFTER ONE HOUR OF SETTLEMENT. COST OF ALGAL HAVESTING BY ALUM ALONE AT CONCENTRATIONS OF 30 MG/L WOULD BE PROHIBITIVE. THE OVERALL MINIMUM COST PER UNIT OF ALGAL YIELD WAS ATTAINED WITH ALUM ALONG AT DOSAGE RANGE OF 75 TO 100 MG/L OF ALUM. (HANCUFF-TEXAS)

FIELD 05D

ACCESSION NO. W70-08904

BIOLOGICAL EFFECTS ON SEDIMENT-WATER NUTRIENT INTERCHANGE,

CALIFORNIA UNIV., BERKELEY. SANITARY ENGINEERING RESEARCH LAB; AND SUNN, LOW, TOM, AND HARA, INC., HONOLULU, HAWAII.

DONALD B. PORCELLA, JAMES S. KUMAGAI, AND E. JOE MIDDLEBROOKS.

ASCE PROCEEDINGS, JOURNAL OF THE SANITARY ENGINEERING DIVISION, VOL 96, NO SA4, PAPER 7460, P 911-926, AUGUST 1970. 5 FIG, 8 TAB, 10 REF. CONTRACT AT(11-1)-34 - PROJECT 100, AEC.

DESCRIPTORS:
*NUTRIENTS, *BOTTOM SEDIMENTS, *LIMNOLOGY, *ALGAE, *EUTROPHICATION, LAKES, WATER QUALITY, PHOSPHORUS, PHOSPHATES, PRODUCTIVITY, LEACHING.

IDENTIFIERS:
SEDIMENT-WATER NUTRIENT-EXCHANGE.

ABSTRACT:
DIFFERENT TYPES OF SEDIMENTS VARY IN THEIR ABILITY TO SUPPORT ALGAL GROWTH. THIS IS RELATED TO THE AMOUNT OF AVAILABLE PHOSPHORUS MEASURED IN THE SEDIMENTS. ALTHOUGH THE AMOUNT OF PHOSPHORUS RELEASED FROM THE SEDIMENTS VARIED WITH THE TYPE OF SEDIMENT, ALL OF THE AVAILABLE PHOSPHORUS EVENTUALLY SHOULD BE EXTRACTED IN THE 15-CM LAYER OF SEDIMENT STUDIED. THE DEVELOPMENT OF A THICK MAT OF OSCILLATORIA CAUSED AN INCREASE IN PRODUCTIVITY DUE TO THE INCREASED TRANSFER OF PHOSPHORUS FROM THE SEDIMENT; AND, IN GENERAL, THE MORE PRODUCTIVE SYSTEMS HAD SEDIMENTS CONTAINING GREATER AMOUNTS OF PHOSPHORUS. ALSO, THE EQUILIBRATION OF P-32 WITH STABLE PHOSPHATE OCCURRED MORE RAPIDLY IN THE MORE PRODUCTIVE SYSTEMS, AND THIS RATE OF EQUILIBRATION APPEARED TO BE ASSOCIATED WITH THE SEDIMENTS THEMSELVES. HENCE P-32 EXCHANGE WITH STABLE P WAS RELATED TO THE CONCENTRATION OF AVAILABLE PHOSPHORUS IN THE SYSTEM. (KNAPP-USGS)

FIELD 05C, 02H

ACCESSION NO. W70-08944

TRISODIUM NITRILOTRIACETATE AND ALGAE,

ONTARIO WATER RESOURCES COMMISSION, TORONTO.

A. E. CHRISTIE.

WATER AND SEWAGE WORKS, VOL 117, NO 2, P 58-59, FEBRUARY 1970.

DESCRIPTORS:
*ALGAE, *ACTIVATED SLUDGE, *TOXICITY, *LABORATORY TESTS, DETERGENTS, WASTE WATER TREATMENT.

IDENTIFIERS:
*TRISODIUM NITRILOTRIACETATE, *CHLORELLA PYRENCIDCSA, SOURCE OF NITROGEN, NTA.

ABSTRACT:
THE RELATIONSHIPS BETWEEN TRISODIUM NITRILOTRIACETATE, A POTENTIAL DETERGENT BUILDER, AND THE ALGAE CHLORELLA PYRENOIDOSA WERE INVESTIGATED, IN AN ACCLIMATED ACTIVATED SLUDGE SYSTEM, WITH RESPECT TO TOXICITY AND AS A SOURCE OF NITROGEN BOTH BEFORE AND AFTER TREATMENT OF THE MATERIAL. IT WAS REPORTED THAT TRISODIUM NITRILOTRIACETATE IS NON-TOXIC TO FISH AT CONCENTRATIONS UP TO 100 MG/L, TO MAMMALS FED UP TO SEVERAL THOUSAND MILLIGRAMS PER KILOGRAM BODY WEIGHT, AND NON-IRRITATING TO HUMAN SKIN. RESULTS INDICATED THAT NTANA3 AT CONCENTRATIONS UP TO 275 MG/L WAS NOT TOXIC TO THE TEST ORGANISM CHLORELLA PYRENOIDOSA AND COULD ACT AS A SOURCE OF NITROGEN FOR ALGAL GROWTH, ALTHOUGH, NOT TO THE SAME EXTENT AS EQUIVALENT QUANTITIES OF NITROGEN, FED AS NITRATE, UNDER THE SAME TEST CONDITIONS. TREATMENT OF 275 MG/L NTANA3 BY AN ACCLIMATIZED ACTIVATED SLUDGE SYSTEM DID NOT REDUCE THE ALGAL GROWTH POTENTIAL OF THE CHEMICAL. THE FEASIBILITY OF USING BIODEGRADABLE ORGANIC MATERIALS WHICH CONTAIN NEITHER NITROGEN OR PHOSPHORUS AS DETERGENT BUILDERS SHOULD BE THOROUGHLY CONSIDERED. (GALWARDI-TEXAS)

FIELD 05D

ACCESSION NO. W70-08976

THE EFFECT OF BIOLOGICAL LIFE ON THE DISSOLVED OXYGEN CONCENTRATION IN THE DELAWARE RIVER,

DREXEL UNIV., PHILADELPHIA, PA. DEPT. OF CIVIL ENGINEERING.

WILLIAM L. ZEMAITIS, AND GERALDINE V. COX.

PREPARED FOR THE CITY OF PHILADELPHIA WATER DEPARTMENT. DREXEL UNIVERSITY, DEPARTMENT OF CIVIL ENGINEERING, PHILADELPHIA, JULY 1970. 13 P, 10 FIG, 6 TAB, 131 REF. WORK NO. P-244, CONTRACT NO. 88-2239.

DESCRIPTORS:
*SLUDGE WORMS, *DISSOLVED OXYGEN, *EUTROPHICATION, *BIOCHEMICAL OXYGEN DEMAND, *CHEMICAL OXYGEN DEMAND, *WATER POLLUTION SOURCES, *WATER POLLUTION EFFECTS, *ORGANIC MATTER, *OXYGEN SAG, *IMPAIRED WATER QUALITY, *ESTUARIES, *DELAWARE RIVER, VEGETATION EFFECTS, SLUDGE, RIVER BEDS, ALGAE, MATHEMATICAL MODELS.

IDENTIFIERS:
*DELAWARE ESTUARY, *BIOLOGICAL LIFE, *DISSOLVED OXYGEN CONCENTRATION.

ABSTRACT:
DESCRIBES AN INVESTIGATION OF THE EFFECTS OF BIOLOGICAL LIFE ON THE DISSOLVED OXYGEN CONTENT IN THE DELAWARE ESTUARY. BIOLOGICAL, CHEMICAL, AND PHYSICAL TESTS WERE CONDUCTED FOR EIGHTEEN MONTHS. BENTHAL SURVEYS REVEALED THAT TURBIFICIDS (AQUATIC WORMS), ALGAE, ROOTED AQUATICS, AND TERRESTRIAL VEGETATION WERE PRESENT. THE TURBIFICIDS WERE PREDOMINANT INVERTEBRATE SPECIES IN THE SLUDGE DEPOSITS, AND THE POPULATION WAS GREAT WHERE THE SEDIMENT CONTAINED LARGE AMOUNTS OF ORGANIC COMPOUNDS. RESPIRATION OF MIXED COMMUNITIES OF INVERTEBRATES WAS DETERMINED TO BE AS HIGH AS 4.2 MG/L OF OXYGEN PER GRAM OF TURBIFICIDS. ONE DEAD GRAM OF DEAD TURBIFICIDS WAS FOUND TO CREATE A BOD5 OF 200 MG/L. TOXIC POLLUTION, WITH A RESULTING TOTAL KILL OF TURBIFICIDS IN A REGION, COULD CAUSE A 20,000 MG/L/M2 BOD5. THE BOD/COD RATIO INCREASES REPRESENT A VERY SIGNIFICANT SECONDARY POLLUTION OF THE ESTUARY. A SIGNIFICANT AMOUNT OF NATURAL ORGANICS IN THE FORM OF ALGAE AND OTHER VEGETATION WAS FOUND. SETTLEMENT OF THESE ORGANICS IN SLOWER MOVING WATERS CAUSES ENRICHMENT OF THE SEDIMENT AND PRODUCES A DISSOLVED OXYGEN SAG BELOW TRENTON DURING HIGH FLOW PERIODS WHEN BOTTOM SCOURING TAKES PLACE. MACROFAUNA METABOLISM AND ALGAE OXYGENATION AND DEOXYGENATION WERE NOT INCORPORATED INTO MATHEMATICAL MODELS (DEVELOPED BY OTHERS) FOR THE DELAWARE ESTUARY. THE AUTHORS STATE THAT THE WASTE ALLOCATIONS DETERMINED FROM THESE MODELS MIGHT HAVE DIFFERED GREATLY IF ALL PARAMETERS HAD BEEN CONSIDERED. RECOMMENDATIONS FOR FURTHER STUDIES ARE GIVEN. (POERTNER)

FIELD 05C

ACCESSION NO. W70-09189

USES OF WASTE HEAT,

OAK RIDGE NATIONAL LAB., TENN.

S. E. BEALL.

THIS ARTICLE DRAWS UPON SIX STUDIES CONDUCTED BY ORNL, SOME OF WHICH ARE SPONSORED BY THE U.S. DEPARTMENT OF HOUSING AND URBAN DEVELOPMENT. OAK RIDGE NATIONAL LABORATORY REVIEW, SPRING 1970, P 9-14. 4 FIG.

DESCRIPTORS:
*THERMAL POWERPLANTS, *POWERPLANTS, *ENERGY LOSSES, *WATER CONSERVATION, *AIR CONDITIONING, *HEATING, *COOLING, *THERMAL POWER, *MULTIPLE-PURPOSE PROJECTS, *WATER REUSE, *WATER SUPPLY, *POLLUTION ABATEMENT, *WATER POLLUTION CONTROL, *WATER COOLING, *STEAM TURBINES, *THERMAL POLLUTION, LONG TUBE VERTICAL DISTILLATION, EVAPORATORS, AQUATIC ALGAE, HORTICULTURAL CROPS, POULTRY, GREENHOUSES, FISH FARMING, IRRIGATION WATER.

IDENTIFIERS:
*WASTE HEAT, *USES OF WASTE HEAT, *CENTRAL HEATING AND COOLING, *SPACE HEATING, *FARM USES OF WASTE HEAT, STEAM POWERED BUSES.

ABSTRACT:
PRESENTS NEW CONCEPTS FOR BENEFICIAL USE OF A LARGE PART OF THE HEAT DISCHARGED FROM THE EXHAUSTS OF TURBINES OF FOSSIL FUEL OR NUCLEAR POWER PLANTS. BOTH URBAN AND RURAL SITES OFFER OPPORTUNITIES FOR APPLYING LOW TEMPERATURE HEAT TO USEFUL PURPOSES. WASTE STEAM OR HOT WATER FROM POWER PLANTS, BETWEEN 300 DEG F TO 380 DEG F, CAN BE COMPETITIVE WITH PRESENT URBAN SPACE HEATING SOURCES, EVEN IF THE POWER PLANT IS 10 MILES FROM THE HEAT DISTRIBUTION POINT. IN MOST CITIES IN THE UNITED STATES, MORE WASTE HEAT CAN BE USED FOR COOLING, (IN ABSORBTION SYSTEMS) DURING THE CRUCIAL SUMMER PERIOD THAN IS NEEDED IN THE WINTER FOR RESIDENTIAL AND COMMERCIAL HEATING. THE AVERAGE REDUCTION IN WASTE HEAT DISPOSAL THAT COULD BE ATTAINED IN THIS MANNER IN URBAN AREAS IS ESTIMATED TO BE 60 PER CENT. OTHER USES OF THE WASTE HEAT ARE DISCUSSED, INCLUDING HEATING AND COOLING OF GREENHOUSES AND POULTRY HOUSES, AND HEATING OF AQUACULTURE PONDS. EVAPORATION OF WASTE WATER AND DISTILLATION OF BRACKISH OR SALT WATER FOR POTABLE WATER SUPPLY OR IRRIGATION USES IS CONSIDERED FEASIBLE, BUT COSTLY AT PRESENT. STEAM STORAGE TANKS TO PROVIDE POWER FOR BUSES AND SWITCH ENGINES ARE CONSIDERED TECHNICALLY AND ECONOMICALLY FEASIBLE. THE USE OF 104 DEG F IRRIGATION WATER IS BEING STUDIED IN OREGON ON SEVERAL EXPERIMENTAL FARMS. SIMILAR STUDIES ARE UNDERWAY AT WASHINGTON STATE UNIVERSITY. A STUDY OF INSTALLATIONS OF UNDERGROUND PIPES HEATED WITH CONDENSER DISCHARGE WATER TO STIMULATE GROWTH OF CROPS IS UNDERWAY AT OREGON STATE UNIVERSITY. (POERTNER)

FIELD 03C, 08C, 05D

ACCESSION NO. W70-09193

A STATEMENT ON PHOSPHORUS,

MINNESOTA UNIV., MINNEAPOLIS.

JOSEPH SHAPIRO.

JOURNAL OF THE WATER POLLUTION CONTROL FEDERATION, VOL 42, NO 5, PART 1, PAGE 772-775, MAY 1970. 4 REF.

DESCRIPTORS:
*PHOSPHORUS, *DETERGENTS, *TERTIARY TREATMENT, *EUTROPHICATION, WASTE WATER TREATMENT, ALGAE, EFFLUENTS, NUTRIENTS, SEWAGE TREATMENT, LAKES.

ABSTRACT:
A PAPER BY L. E. KUENTZEL PUBLISHED IN THE OCTOBER 1969 ISSUE OF THE JWPCF IS DISCUSSED AND EVALUATED BY SHAPIRO. THE DISCUSSION IS BASICALLY A REFUTATION OF ARGUMENTS PRESENTED BY KUENTZEL IN HIS CHALLENGING THE EFFICACY OF PHOSPHORUS REMOVAL FOR EUTROPHICATION CONTROL. SEVERAL WORKS CITED, PREVIOUS TO KUENTZEL'S, SUPPORTS THE FACT THAT REDUCTION OF PHCRPHORUS IN EFFLUENTS THROUGH CURTAILED USE OF DETERGENTS OR ADVANCE WASTE WATER TREATMENT WILL REDUCE EUTROPHICATION OF LAKES SIGNIFICANTLY AND WILL ALLOW THOSE ALREADY SEVERELY AFFECTED TO BEGIN THEIR RECOVERY THROUGH NATURAL PROCESSES. (HANCUFF-TEXAS)

FIELD 05D, 05C

ACCESSION NO. W70-09325

DETERGENTS, PHOSPHATES, AND WATER POLLUTION,

DEPARTMENT OF ENERGY, MINES AND RESOURCES, OTTAWA (ONTARIO). INLAND WATERS BRANCH.

P. D. GOULDEN, W. J. TRAVERSY, AND G. KERR.

CANADA DEPARTMENT OF ENERGY, MINES AND RESOURCES INLAND WATERS BRANCH TECHNICAL BULLETIN NO 22, 1970. 8 P.

DESCRIPTORS:
*WATER POLLUTION EFFECTS, *PHOSPHATES, *NITRATES, *NUTRIENTS, *EUTROPHICATION, ALGAE, SURFACTANTS, WATER POLLUTION SOURCES, SURFACE WATERS, WATER POLLUTION, DOMESTIC WASTES, MUNICIPAL WASTES.

IDENTIFIERS:
CANADA.

ABSTRACT:
THIS REPORT WAS PREPARED TO PROVIDE THE GENERAL PUBLIC WITH TECHNICAL INFORMATION ON DETERGENTS AND THE EFFECTS OF PHOSPHATES ON THE WATER ENVIRONMENT. IT ANSWERS SOME OF THE MORE COMMONLY-ASKED QUESTIONS REGARDING PHOSPHATE-BASED DETERGENTS--WHAT THEY ARE, WHAT THEY CONTAIN, HOW THEY WORK AND HOW THEY AFFECT THE QUALITY OF OUR WATER RESOURCES. THE PROPERTY OF BIODEGRADABILITY AS IT APPLIES TO DETERGENTS IS EXPLAINED. WHILE PHOSPHATES AND NITRATES DO NOT IN THEMSELVES POSE ANY THREAT TO HEALTH AT THE CONCENTRATIONS INVOLVED AFTER DIFFUSION IN RECEIVING WATERS, THEY DO POSSESS NUTRITIVE PROPERTIES WHICH ENCOURAGE THE EXCESSIVE GROWTH OF ALGAE AND OTHER FORMS OF UNDESIRABLE AQUATIC VEGETATION. WHEN LARGE AMOUNTS OF THIS VEGETATION DECOMPOSE AT THE END OF THE GROWING PERIOD, DEPLETION OF THE VITAL LIFE-SUSTAINING DISSOLVED OXYGEN IN WATER OCCURS. THIS FORM OF POLLUTION, WHICH IS CAUSED BY EXCESS NUTRIENT ENRICHMENT, IS ALSO REFERRED TO AS EUTROPHICATION AND IN ITS EXTREME FORM RESULTS IN THE ACCELERATED AGING OR DYING OF LAKES. (KNAPP-USGS)

FIELD 05C, 05B

ACCESSION NO. W70-09388

TOXICITY STUDIES WITH AN OIL-SPILL EMULSIFIER AND THE GREEN ALGA PRASINOCLADUS MARINUS,

UNIVERSITY COLL. OF WALES, ABERYSTWYTH. DEPT. OF BOTANY.

A. D. BONEY.

JOURNAL OF THE MARINE BIOLOGICAL ASSOCIATION OF THE UNITED KINGDOM, (1970), P 461-473. 14 REF, 4 TAB, 4 FIG.

DESCRIPTORS:
BIOASSAY, TOXICITY, *EMULSIFIERS, DETERGENTS, OIL, *OIL WASTES, OILY WATER, OIL-WATER INTERFACES, WATER POLLUTION TREATMENT, WATER POLLUTION EFFECTS, ENVIRONMENTAL EFFECTS, ALGAL POISONING, ALGAE, SOLVENTS, SURFACTANTS, WATER TEMPERATURE, PLANT PIGMENTS, PLANT PHYSIOLOGY, *PLANT PATHOLOGY, PLANT GROWTH REGULATORS, SALINITY, PLANT POPULATIONS.

IDENTIFIERS:
*BP 1002, *GREEN ALGAE, PRASINOCLADUS MARINUS, TEMPERATURE EFFECTS.

ABSTRACT:
CYST PHASES OF THE GREEN ALGA PRASINOCLADUS MARINUS HAVE BEEN USED IN AN INVESTIGATION OF THE TOXIC PROPERTIES OF AN OIL-SPILL EMULSIFIER BP 1002, AND OF ITS SOLVENT AND SURFACTANT FRACTIONS. VARIOUS ASPECTS OF A REJUVENATION PROCESS HAVE ALL BEEN UTILIZED AS A MEANS OF ASSAY IN ADDITION TO OBSERVATIONS ON CELL VIABILITY. THE 'AGED' CYSTS WERE MORE TOLERANT OF ALL TYPES OF TOXIC AGENTS THAN WERE THE YOUNG NON-MOTILE CELLS. THE SURFACTANT FRACTIONS WERE MORE TOXIC WHEN USED ALONE, AND THE SOLVENT FRACTION ALONE MORE TOXIC THAN THE COMPOUNDED BP 1002. THE APPLICATION OF ANY OF THE TOXIC AGENTS AT LOW TEMPERATURE (4 DEG. C) RESULTED IN A MARKED REDUCTION IN THEIR EFFECTS AT HIGH CONCENTRATIONS (E.G. 500 PPM). THE TOXIC EFFECT WAS APPRECIABLY INCREASED WITH BOTH 'AGED' AND 'YOUNG' CELLS WHEN ACCOMPANIED BY A LOWERING IN SALINITY. AERATION OF THE TOXIC SOLUTIONS CAUSED A SIGNIFICANT LOWERING OF TOXICITY WITH BOTH BP 1002 AND THE SOLVENT FRACTION. CHLOROPLAST PIGMENT REGENERATION IN 'RECOVERING' CYSTS WAS A SENSITIVE MEANS OF ASSAYING TOXIC EFFECTS. (SJOLSETH-WASHINGTON)

FIELD 05C

ACCESSION NO. W70-09429

THERMAL STANDARDS IN THE UNITED STATES OF AMERICA,

FEDERAL WATER POLLUTION CONTROL ADMINISTRATION, DULUTH, MINN. NATIONAL WATER QUALITY LAB.

D. I. MOUNT.

SYMPOSIUM ON ENVIRONMENTAL ASPECTS OF NUCLEAR POWER STATIONS, NEW YORK, AUGUST 10-14, 1970. VOL SM-146, NO 15, PREPRINT, 5 P.

DESCRIPTORS:
*TEMPERATURE CONTROL, *STANDARDS, *WATER TEMPERATURE, SEASONAL, MONTHLY, AQUATIC LIFE, INTERSTATE RIVERS, HYPOLIMNION, VOLUME, MIXING, SAFETY, WATER COOLING, ANAEROBIC CONDITIONS, ALGAL CONTROL, SPAWNING, WARM-WATER FISH.

ABSTRACT:
TEMPERATURE STANDARDS SPECIFYING PERMISSIBLE WATER TEMPERATURES FOR INTERSTATE RIVERS AND LAKES OF THE UNITED STATES HAVE BEEN ESTABLISHED. STANDARDS FOR AQUATIC LIFE MUST INSURE A SEASONAL TEMPERATURE CYCLE, AND SO TWELVE MAXIMUM MONTHLY TEMPERATURES ARE THE MINIMUM TO DESCRIBE DESIRED CONDITIONS. IN STREAMS, A 5F RISE MAY BE PERMITTED AS LONG AS THE RESULTING TEMPERATURE DOES NOT EXCEED THE APPROPRIATE MONTHLY MAXIMUM. IN LAKES, A 3F RISE MAY BE USED WITH AN ADDITIONAL REQUIREMENT THAT THE HYPOLIMNION CANNOT BE WARMED OR CHANGED IN VOLUME. MIXING ZONES POSE A DIFFICULT REGULATORY PROBLEM. THEY ARE DIFFICULT TO RESTRICT AS TO SIZE AND SHAPE ESPECIALLY BECAUSE THE EFFECTS OF MIXING ZONES ON RECEIVING WATER DEPENDS, IN PART, ON THE NUMBER OF MIXING ZONES AND THE SIZE OF THE LAKE OR RIVER. WHERE COOLING DEVICES ARE REQUIRED, THE STANDARDS DO NOT SPECIFY THE TYPE TO BE USED. THE FREQUENCY THAT DEVICES WILL NOT MEET THE STANDARDS HAS NOT BEEN WELL DEFINED EXCEPT IN A GENERAL WAY. IF NO SAFETY MARGIN IS PERMITTED, STANDARDS MUST BE MORE COMPLEX IN ORDER TO INSURE ACCEPTABLE CONDITIONS. (OSBORNE-VANDERBILT)

FIELD 05G, 06E

ACCESSION NO. W70-09595

POSSIBILITIES FOR BENEFICIAL USES OF HEATED WATER DISCHARGES INTO COOLING RESERVOIRS,

MISSOURI UNIV., COLUMBIA. DEPT. OF ZOOLOGY.

ROBERT S. CAMPBELL, ARTHUR WITT, JR., AND JAMES R. WHITLEY.

FIFTH ANNUAL WATER RESOURCES RESEARCH CONFERENCE, FEB. 3-4, 1970, WASHINGTON, D.C., U.S. DEPT. OF INTERIOR, OFFICE OF WATER RESOURCES RESEARCH, JUNE 1970. P 57-64, 27 REF.

DESCRIPTORS:
*HEATED WATER, *BENEFICIAL USE, *COOLING WATER, *FISHERIES, WATER QUALITY, FISH ATTRACTANTS, FISH HARVEST, PLANKTON, BENTHIC FAUNA, ALGAE, ECOLOGY, MISSOURI.

IDENTIFIERS:
COOLING RESERVOIR, ARTIFICIAL RESERVOIR, MONTROSE LAKE, THOMAS LAKE.

ABSTRACT:
MONTROSE LAKE AND THOMAS LAKE IN MISSOURI WERE STUDIED. BOTH WERE ARTIFICIAL COOLING RESERVOIRS. STUDIES INCLUDED WATER QUALITY, PRIMARY PRODUCTION OF ALGAE, PLANKTON AND BOTTOM FAUNA AND FISH. IMPACT OF HEATED WATER UPON THE ECOLOGY OF A RESERVOIR WILL BE LESSENED AS THE VOLUME OF THE RESERVOIR IS INCREASED. EMPLOYMENT OF A LONG COOLING CANAL RESULTS IN LOSS OF HEAT TO THE ATMOSPHERE PRIOR TO DISCHARGE INTO THE RESERVOIR. UNDER THE CONDITIONS EXISTING IN THOMAS HILL RESERVOIR, NO LETHAL EFFECTS FROM HEATED WATER INFLOW WAS OBSERVED. BENEFICIAL EFFECTS WHICH MAY BE DERIVED FROM CONTROLLED DISCHARGE OF HEATED WATER ARE: (A) MIGRATION OF FISHES MAY RESULT IN A BUILDUP OF FISH DENSITIES. (B) THE INCREASED DENSITY OF FISHES MAY PROVIDE AN EXTENDED HARVEST INTO THE WINTER PERIOD. (C) CONTINUED GROWTH OF GAME FISHES MAY OCCUR DURING THE WINTER PERIOD. (D) AN INCREASED PRODUCTION OF COMMERCIALLY RAISED FISH MAY BE POSSIBLE. (E) THE LOCATION OF THE POWER PLANT MIGHT BE PLANNED TO PROVIDE OPTIMAL CONDITIONS FOR FISH HARVEST AND FOR USE OF HEATED WATER EFFLUENT IN HATCHERY AND REARING POOLS. (HSIEH-VANDERBILT)

FIELD 05C, 06G

ACCESSION NO. W70-09596

THERMAL EFFECTS AND NUCLEAR POWER STATIONS IN THE U.S.A.,

ATOMIC ENERGY COMMISSION, WASHINGTON, D.C.

R. E. NAKATANI, D. MILLER, AND J. V. TOKAR.

SYMPOSIUM ON ENVIRONMENTAL ASPECTS OF NUCLEAR POWER STATIONS, NEW YORK, AUGUST 12, 1970. PAPER IAEA-SM-146/30. 14 P, 1 TAB, 21 REF. PREPRINT.

DESCRIPTORS:
*THERMAL POLLUTION, *NUCLEAR POWER PLANTS, *COOLING WATER, *WATER TEMPERATURE, CONDENSERS, DESIGN, STANDARDS, FISH KILL, ALGAE, COOLING TOWERS, DENSITY, VISCOSITY, FISH BEHAVIOR, FERTILITY, SEASONAL, DIFFUSIVITY.

IDENTIFIERS:
*WASTE HEAT, SUBLETHEL EFFECTS, MIXING ZONES, MODELING.

ABSTRACT:
NUCLEAR POWER PLANTS IN THE U.S. WERE SURVEYED THAT ARE EITHER OPERABLE, BEING BUILT, OR PLANNED AS OF DECEMBER 31, 1969. OVER SIXTY PERCENT OF THE PLANTS USED THE ONCE-THROUGH SYSTEM WHERE THE CONDENSER COOLING WATER WAS TAKEN FROM NEARBY RIVERS, LAKES, ESTUARIES, OR THE OCEAN AND THEN USUALLY RETURNED TO THE SAME SOURCE. THE AVERAGE MAXIMUM TEMPERATURE RISE ACROSS THE NUCLEAR POWER PLANT CONDENSER WAS APPROXIMATELY 10 DEGREE C. GENERAL OBSERVATIONS AND RECOMMENDATIONS BASED ON STATEMENTS DESCRIBING THE PLANNING, DESIGN, AND PROCEDURES FOR WASTE HEAT RELEASE AND INFORMATION OBTAINED FROM PERSONAL PLANT VISITS INCLUDE: (1) THERMAL QUALITY STANDARDS ARE BECOMING INCREASINGLY STRINGENT; (2) FISH KILLS ARE RARE AND THEIR SEVERITY LIMITED; (3) STUDIES OF THE SHORT- AND LONG-RANGE MOVEMENT AND EFFECTS OF HEAT SHOULD BE UNDERTAKEN; (4) THE EFFORT SHOULD BE MADE TO ACQUIRE KNOWLEDGE OF THE SUBLETHEL EFFECTS OF RELEASING HEAT; (5) ALTERNATE METHODS FOR RELEASING WASTE HEAT SHOULD AND ARE BEING EMPLOYED; (6) THE NEED FOR STOCHASTIC AND ANALYTIC MODELS OF PLUME BEHAVIOR EXISTS. (OSBORNE-VANDERBILT)

FIELD 05B, 05C

ACCESSION NO. W70-09612

THERMAL EFFECTS STUDIES IN NEW YORK STATE,

STOLLER (S. M.) ASSOCIATES, NEW YORK; AND NIAGARA MOHAWK POWER CORP., BUFFALO, N.Y.

THOMAS W. PHILBIN, AND HOWARD D. PHILLIPP.

SYMPOSIUM ON ENVIRONMENTAL ASPECTS OF NUCLEAR POWER STATIONS, NEW YORK, AUGUST 12, 1970. PAPER IAEA-SM-146/32. 13 P, 3 TAB, 10 REF. PREPRINT.

DESCRIPTORS:
*POWER PLANTS, *THERMAL POLLUTION, *THERMAL STRATIFICATION, HYDRAULIC MODELS, FISH, ALGAE, FATHOMETERS, BENTHIC FAUNA, BENTHIC FLORA, LAKE ONTARIO, HUDSON RIVER, STANDARDS, SURFACE WATERS, EPILIMNION, INTAKE STRUCTURES, DISCHARGE(WATER), WATER TEMPERATURE.

IDENTIFIERS:
LAKE CAYUGA, CLADOPHORA, ECOLOGICAL STUDIES.

ABSTRACT:
SEVERAL THERMAL EFFECTS STUDIES HAVE BEEN AND ARE PRESENTLY UNDERWAY TO DETERMINE WHAT THE EFFECTS OF THERMAL DISCHARGES ARE ON THE WATER ENVIRONMENT OF NEW YORK STATE. AT PRESENT, THE RESULTS ARE INCOMPLETE; BUT THEY DO YIELD ENOUGH INFORMATION TO ALLOW SOME INITIAL CONCLUSIONS. ON LAKE ONTARIO, THE GINNA AND NINE MILE POINT STUDIES SHOWED THAT FISH TENDED TO GATHER AROUND BOTH INTAKE AND DISCHARGE STRUCTURES, SLIGHTLY ALTERING THE LOCAL FISH AND BENTHIC DISTRIBUTIONS. ON LAKE CAYUGA, THE BELL NUCLEAR STATION HAS BEEN POSTPONED WITH NO IMMEDIATE PLANS FOR RESUMPTION. PHYSICAL EFFECTS STUDIES ON THE LAKE, IF THE BELL STATION PLANT WERE OPERATING, HAVE YIELDED A PROBABLE 0.7 DEGREE F RISE IN THE AVERAGE SURFACE TEMPERATURE ALONG WITH AN EIGHT TO TEN DAY LONGER STRATIFICATION PERIOD. MOREOVER, IN OCTOBER, WATER COULD POSSIBLY BE DRAWN FROM THE EPILIMNION, INCREASING DISCHARGED WATER ANOTHER 5 DEGREE F. HOWEVER, IF NEW YORK STATE DISCHARGE STANDARDS ARE ADHERED TO, NO ACUTE EFFECTS ARE ANTICIPATED EVEN CLOSE TO THE OUTFALL. ON THE HUDSON RIVER, ON THE OTHER HAND, INCREASED WATER TEMPERATURES AS HIGH AS 91 DEGREE F DO NOT APPEAR TO INFLUENCE THE ABUNDANCE OF FISH. (OSBORNE-VANDERBILT)

FIELD 05C, 08C

ACCESSION NO. W70-09614

THE RELATIVE SIGNIFICANCE OF PHOSPHORUS AND NITROGEN AS ALGAL NUTRIENTS,

NORTH CAROLINA WATER RESOURCES RESEARCH INST., RALEIGH.

CHARLES M. WEISS.

AVAILABLE FROM THE NTIS AS PB-194 054, $3.00 IN PAPER COPY, $0.65 IN MICROFICHE. REPORT NO. 34, WATER RESOURCES RESEARCH INSTITUTE OF THE UNIVERSITY OF NORTH CAROLINA, JUNE 1970. 54 P, 27 FIG, 8 REF, 2 APPEND. OWRR PROJECT B-003-NC(2).

DESCRIPTORS:
EUTROPHICATION, *ALGAL BLOOMS, *NUTRIENTS, *PHOSPHORUS, *NITROGEN, *NORTH CAROLINA.

IDENTIFIERS:
NUTRIENT INTERCHANGE, HAW RIVER, NEW HOPE RIVER.

ABSTRACT:
BY THE EXAMINATION OF THE INTERACTION BETWEEN NITROGEN AND PHOSPHORUS SPECIES RELATIVE TO ALGAL GROWTH IN SEVERAL FRESH WATER ENVIRONMENTS OF DIFFERING TROPHIC STATE, IT HAS BEEN POSSIBLE TO ESTABLISH THE RELATIVE SIGNIFICANCE OF THESE ELEMENTS AS ALGAL NUTRIENTS. THE ALGAL ASSAY WAS CARRIED OUT USING MEMBRANE FILTERED SAMPLES DERIVED FROM A SERIES OF OXIDATION PONDS RECEIVING SECONDARY EFFLUENT FROM A TRICKLING FILTER PLANT AND FROM SAMPLES DERIVED FROM SAMPLING POINTS ON THE NEW HOPE AND HAW RIVERS. THE LATTER REPRESENTED A SERIES OF CHANGING RIVER QUALITIES WITH PARTICULAR RESPECT TO THE OXIDATION STATES OF NITROGEN AND PHOSPHORUS. THE ALGAL ASSAY USED PURE CULTURES OF FIVE SPECIES; EUGLENA ROSTIFERA, CHLAMYDOMONAS REINHARDTII, PANDORINA MORUM, SCENEDUSMUS QUADRICAUDA, AND CHLORELLA ELLIPSOIDEA. EACH OF THESE SPECIES HAS BEEN DESCRIBED AS NORMALLY ASSOCIATED WITH POLLUTED WATERS. THE RESULTS OF BOTH CHEMICAL AND BIOLOGICAL ASSAY EXAMINED THROUGH MULTIPLE REGRESSION ANALYSIS OF THE INDEPENDENT VARIABLE INVOLVED IN THE ALGAL ASSAY AS WELL AS A QUADRATIC ANALYSIS OF COVARIANCE OF NH3-N, NO3N AND PO4-P ESTABLISHED THE RELATIVE SIGNIFICANCE OF THE NITROGEN AND PHOSPHORUS SPECIES IN WATER CONTAINING HIGH CONCENTRATIONS OF THESE ELEMENTS, CONCENTRATIONS NORMALLY FOUND IN DISCHARGES FROM BIOLOGICAL WASTE TREATMENT PLANTS AND FOLLOWING DILUTION IN RECEIVING STREAMS. UNDER THESE CIRCUMSTANCES IT HAS BEEN SHOWN THAT THE QUANTITY OF NITROGEN RATHER THAN THAT OF PHOSPHORUS DETERMINES THE BIOMASS OF ALGAE THAT MIGHT BE EXPECTED TO GROW. THIS RESPONSE HOLDS TRUE FOR THE SEVERAL SPECIES OF ALGAL THAT WERE USED. IT WOULD THUS APPEAR THAT THE QUESTION OF ALGAL NUTRIENTS TAKES ON A SOMEWHAT MORE FORMIDABLE DIMENSION DUE TO THE CONSIDERABLY GREATER DIFFICULTY IN REMOVING IN SIGNIFICANT AMOUNTS THE NITROGEN THAT IS NORMALLY FOUND IN MUNICIPAL WASTE WATERS.

FIELD 05C

ACCESSION NO. W70-09669

THE FATE OF NITROGENOUS COMPOUNDS THROUGH SEWAGE TREATMENT PLANTS,

WASHINGTON UNIV., ST. LOUIS, MO. CENTER FOR THE BIOLOGY OF NATURAL SYSTEMS.

NAVA NARKIS.

WASHINGTON UNIV, ST LOUIS, MISSOURI, CENTER FOR THE BIOLOGY OF NATURAL SYSTEMS, ENVIRONMENTAL FIELD PROGRAM, FEBRUARY 1970. 52 P, 1 FIG, 18 TAB.

DESCRIPTORS:
*NITROGEN COMPOUNDS, *SEWAGE TREATMENT, *DOMESTIC WASTES, CHEMICAL ANALYSIS, EFFLUENTS, TRICKLING FILTERS, HYDROLYSIS, BIOLOGICAL TREATMENT, ACTIVATED SLUDGE, TERTIARY TREATMENT, ION EXCHANGE, ELECTROCHEMISTRY, REVERSE OSMOSIS, ELECTRODIALYSIS, DISTILLATION, NITRIFICATION, DENITRIFICATION, SEPTIC TANKS, OXIDATION LAGOONS, LAND APPLICATION, BACTERIA, ALGAE, SYMBIOSIS, PUBLIC HEALTH, CHLORINE, CORROSION.

IDENTIFIERS:
*LITERATURE SURVEY, PRIMARY SEWAGE TREATMENT, SECONDARY SEWAGE TREATMENT, PHYSICOCHEMICAL METHODS, AMMONIA STRIPPING, GUGGENHEIM PROCESS.

ABSTRACT:
LITERATURE CONCERNING NITROGENOUS COMPOUNDS IN SEWAGE TREATMENT PLANTS IS REVIEWED, ESPECIALLY EMPHASIZING THE MODIFICATION OR REPLACEMENT OF CLASSICAL TREATMENT OPERATIONS TO REMOVE DISSOLVED INORGANIC MATERIALS MORE EFFICIENTLY. IN PRIMARY TREATMENT, NITROGEN REMOVAL IS LOW AND SEDIMENTATION OF SUSPENDED SOLIDS IS THE ONLY PHYSICAL PROCESS CONTRIBUTING TO ITS TOTAL REDUCTION. RETURNING SUPERNATANT LIQUOR FROM SLUDGE DIGESTERS TO PRIMARY CLARIFIER DECREASES NITROGEN REMOVAL EFFICIENCIES. CLARIFIED WASTE WATER FROM PRIMARY TREATMENT MAY RECEIVE SECONDARY TREATMENT, NORMALLY A BIOLOGICAL PROCESS, DESIGNED PRINCIPALLY TO REDUCE AMOUNT OF SUSPENDED SOLIDS, OXYGEN DEMANDING MATERIAL AND BACTERIA. ACTIVATED SLUDGE AND TRICKLING FILTERS REMOVE 20% TO 50% OF THE INCOMING NITROGEN. EFFLUENTS FROM TRICKLING FILTERS CONTAIN HIGH ORGANIC AND AMMONIA NITROGEN AS WELL AS NITRATES AND NITRITES. IT IS POSSIBLE TO OPERATE ACTIVATED SLUDGE PROCESSES TO EFFECT HIGH NITRATE CONCENTRATIONS, WHILE AMMONIA NITRITE AND ORGANIC NITROGEN CONCENTRATIONS ARE MINIMAL. IN THE TERTIARY TREATMENT, AMONG THE PHYSICOCHEMICAL METHODS FOR REMOVING NITROGENOUS COMPOUNDS FROM EFFLUENTS, ION EXCHANGE AND AMMONIA STRIPPING ARE THE ONLY SUITABLE METHODS PRESENTLY EFFICIENT. NITROGEN REMOVAL BY MICROBIAL DENITRIFICATION SEEMS ECONOMICALLY FEASIBLE. SEPTIC TANK EFFLUENTS ARE ANALYZED. STABILIZATION PONDS DO NOT REMOVE NITROGEN COMPOUNDS ADEQUATELY. (JONES-WISCONSIN)

FIELD 05D, 10

ACCESSION NO. W70-09907

APPLIED RESEARCH CN MARINE SEWAGE DISPOSAL FROM THE GOTHENBURG REGION, SWEDEN,

STOCKHOLM UNIV. (SWEDEN). INTERNATIONAL METEOROLOGICAL INST.

L. ARNBORG, AND E. ERIKSSON.

FIFTH INTERNATIONAL WATER POLLUTION RESEARCH CONFERENCE, SAN FRANCISCO, JULY 26-AUGUST 1, 1970, PREPRINT. 7 P, 7 FIG.

DESCRIPTORS:
*RESEARCH AND DEVELOPMENT, *WATER POLLUTION, PHOSPHORUS, THERMAL STRATIFICATION, CURRENTS(WATER), ALGAE, LAKES, RIVERS, SEWAGE DISPOSAL, SEWAGE TREATMENT, WASTE WATER TREATMENT.

IDENTIFIERS:
*MARINE DISPOSAL, *GOTHENBRUG, SWEDEN.

ABSTRACT:
THE CITY OF GOTHENBURG AT THE WEST COAST OF SWEDEN PLANS TO CONSTRUCT A SYSTEM OF INTERCEPTING SEWERS AND ROCK TUNNELS BY WHICH ALL THE SEWAGE FROM THE CITY CAN BE CONVEYED TO A TREATMENT PLANT. AFTER BIOLOGICAL TREATMENT, THE SEWAGE CAN BE DISCHARGED EITHER INTO THE RIVER MOUTH OR INTO THE SEA OUTSIDE A NEARBY ARCHIPELAGO. IN ORDER TO COMPARE THE DIFFERENT OUTFALL SITES, AN INVESTIGATION OF THE AREA WAS UNDERTAKEN FROM THE MOUTH OF THE RIVER, GOTA ALV, THROUGH THE ARCHIPELAGO INTO THE OPEN SEA. THE MEAN VALUES AND STANDARD DEVIATIONS OF THE VERTICAL DISTRIBUTION OF TEMPERATURE, SALINITY AND DENSITY DURING SUMMER AND WINTER IN THE AREA OUTSIDE THE ARCHIPELAGO ARE PRESENTED GRAPHICALLY. LARGE VARIATIONS OF CURRENT PATTERNS TAKE PLACE DUE TO TIDES, WINDS, AND VARIATIONS IN THE ATMOSPHERIC PRESSURE. THE MAIN DIRECTIONS OF THE CURRENTS ARE NORTH AND SOUTH WITH THE RESULTANT TO THE NORTH. A FIGURE IS PRESENTED SHOWING THE FREQUENCIES SPECTRUM DENSITY. THE STUDY OF THE USE OF A DIFFUSER SYSTEM FOR THE DISPERSAL OF TREATED SEWAGE RESULTED IN THE FOLLOWING DESIGN AND CONCLUSIONS: THE AREA OF RELEASE IS 3 KM WEST OF THE ARCHIPELAGO. THE DIFFUSER PORTHOLES ARE 0.1 M IN DIAMETER AND LOCATED 30 TO 35 M BELOW THE SURFACE. THE INITIAL VELOCITY THROUGH THE PORTHOLES WILL BE 2.5 METERS/SECOND. WITH THIS ARRANGEMENT AND THE EXISTING VERTICAL DENSITY STRATIFICATION THE PLUM WILL RISE TO A LEVEL OF 14 TO 20 METERS BELOW THE SURFACE. (HANCUFF-TEXAS)

FIELD 05B, 05D

ACCESSION NO. W70-09947

FILTRATION OF ALGAL SUSPENSIONS,

BOHNA ENGINEERING AND RESEARCH INC., SAN FRANCISCO, CALIF.

I. J. WRIGHT, AND W. M. LUIZ.

BOHNA ENGINEERING AND RESEARCH, INCORPORATED, SAN FRANCISCO, CALIFORNIA, BOHNA INTERNAL REPORT F170, 1970. 31 P, 7 FIG, 10 TAB.

DESCRIPTORS:
*WASTE WATER TREATMENT, *FILTRATION, *ALGAE, *SUSPENSION, HARVESTING, PONDS, CALIFORNIA, SCENEDESMUS, SUSPENDED LOAD, NITRATES, CALCIUM CARBONATE, PROTEINS, FILTERS, ENGINEERS ESTIMATES, ECONOMICS, FEASIBILITY, COSTS, ANNUAL COSTS, PILOT PLANTS.

IDENTIFIERS:
FIREBAUGH(CALIF), VOLATILE SOLIDS, SANBORN FILTER, BACKWASHING.

ABSTRACT:
THE SANBORN FILTER, PROPRIETARY OF BOHNA ENGINEERING AND RESEARCH, INC, A FABRIC AND GRANULAR-MEDIA DEVICE, HAS BEEN EVALUATED FOR USE IN HARVESTING ALGAE FROM PONDS OF THE AGRICULTURAL WASTE WATER TREATMENT CENTER AT FIREBAUGH, CALIFORNIA. REMOVALS OF UP TO 98% HAVE BEEN ACHIEVED WITHOUT THE USE OF FLOCCULANTS OR OTHER CHEMICAL ADDITIVES. THE BACKWASH FROM THE FILTER HAS SUSPENDED SOLIDS CONCENTRATIONS UP TO 1.7% WHICH SHOWS EVIDENCE OF INCREASED COAGULATION, ALLOWING THE CONCENTRATION OF SOLIDS TO A LEVEL OF 8% TO 12% BY SIMPLE SEDIMENTATION. AN ECONOMIC ANALYSIS OF THE USE OF ALGAE FOR NITROGEN STRIPPING OF SAN JOAQUIN VALLEY DRAINAGE WATER WAS MADE. ASSUMING A FLOW SHEET WHICH INCLUDES LINED, MIXED GROWTH PONDS FOLLOWED BY FILTRATION WITH THE SANBORN FILTER, CONCENTRATION OF THE BACKWASH BY SEDIMENTATION OR CENTRIFUGATION AND AIR DRYING OF THE ALGAE, AN APPROXIMATE TOTAL COST OF $37.01 PER MILLION GALLONS TREATED IS PROJECTED, BASED ON AMORTIZATION AND INTEREST CHARGES OF 10% PER YEAR OF THE INITIAL INVESTMENT. CAPACITY OF 700 MILLION GALLONS PER DAY WAS ASSUMED. (JONES-WISCONSIN)

FIELD 05D

ACCESSION NO. W70-10174

THE FULL-SCALE RECLAMATION OF PURIFIED SEWAGE EFFLUENT FOR THE AUGMENTATION OF THE DOMESTIC SUPPLIES OF THE CITY OF WINDHOEK,

NATIONAL INST. FOR WATER RESEARCH, PRETORIA (SOUTH AFRICA).

L. R. J. VAN VUUREN, M. R. HENZEN, G. J. STANDER, AND A. J. CLAYTON.

FIFTH INTERNATIONAL WATER POLLUTION CONFERENCE, SAN FRANCISCO, JULY 26-AUGUST 1, 1970. PREPRINT, PAPER I-32. 9 P, 5 FIG, 4 TAB, 8 REF.

DESCRIPTORS:
*SEWAGE TREATMENT, *WASTE WATER TREATMENT, *TERTIARY TREATMENT, *WATER REUSE, *WATER QUALITY, WATER SUPPLY, ALGAE, ACTIVATED CARBON, ADSORPTION, FILTRATION, CHLORINATION, CHEMICAL PRECIPITATION, FLOCCULATION, FLOTATION.

IDENTIFIERS:
WINDHOEK, SOUTH AFRICA.

ABSTRACT:
IN EARLY SEPTEMBER 1968, A CRITICAL SITUATION IN WATER SUPPLY AND DEMAND DEVELOPED WHICH NECESSITATED THE EXPEDITOUS CONDITIONING OF A FULL-SCALE WATER RENOVATION PLANT. THE RECLAIMABLE SEWAGE EFFLUENT IS DERIVED FROM A CONVENTIONAL SEWAGE TREATMENT PLANT COMPRISED OF PRIMARY SEDIMENTATION BY FILTRATION, SECONDARY SEDIMENTATION FOLLOWED BY FURTHER BIOLOGICAL PURIFICATION IN 9 MATURATION PONDS IN SERIES. TOTAL HYDRAULIC RETENTION TIME IS 14 DAYS. THE RECLAMATION PLANT CONSTITUTES AN INTEGRAL PART OF THE EXISTING WATER TREATMENT WORKS AND IS COMPRISED OF RECARBONATION, ALGAE FLOTATION, FOAM FRACTIONATION, CHEMICAL FLOCCULATION, BREAK POINT CHLORINATION, CLARIFICATION, SAND FILTRATION AND ACTIVATED CARBON. THE RECLAIMED WATER AFTER CARBON ABSORPTION IS MIXED WITH PURIFIED RAW WATER AND THE CLEAR WATER SUPPLY. THE ADD MIXED STRAINS ARE THEN POST-CHLORINATED TO A FREE RESIDUAL OF 0.2 PPM AND THEN PUMPED TO THE CITY'S SERVICE RESERVOIRS. THE TOTAL COST FOR THE PRODUCTION OF RECLAIMED WATER INCLUDING FIXED CAPITOL, CHEMICAL AND RUNNING COSTS, AMOUNT TO 11.9 CU PM, WHILE THE AVERAGE COSTS FOR PURIFIED DAM WATER ARE 10.7 CU PM. THE COST FIGURE FOR CARBON TREATMENT IS BASED ON A SINGLE USE OF THE CARBON MEDIUM WITHOUT REGENERATION RESULTING IN A HIGH COST. IF REGENERATION FACILITIES ARE PROVIDED IT IS ESTIMATED THAT THE TOTAL COST FOR RECLAIMED WATER WILL BE 9.53 CU PM. (HANCUFF-TEXAS)

FIELD 05D, 03D

ACCESSION NO. W70-10381

ASSAYING ALGAL GROWTH WITH RESPECT TO NITRATE CONCENTRATION BY A CONTINUOUS FLOW TURBIDOSTAT,

CALIFORNIA UNIV., BERKELEY; AND MINISTRY OF HEALTH, JERUSALEM (ISRAEL).

G. SHELEF, W. J. OSWALD, AND C. C. GOLUEKE.

FIFTH INTERNATIONAL WATER POLLUTION RESEARCH CONFERENCE, SAN FRANCISCO, JULY 26-AUGUST 1, 1970. PREPRINT. 9 P.

DESCRIPTORS:
*ANALYTICAL TECHNIQUES, *ALGAE, *ASSAY, CHLORELLA, NITRATES, CHEMICAL ANALYSIS, CHLOROPHYTA, DIATOMS, TEMPERATURE, OPTICAL PROPERTIES, GROWTH RATES.

IDENTIFIERS:
*SAN JOAQUIN DELTA(CALIF), *CONTINUOUS FLOW TURBIDOSTAT, *GROWTH KINETICS, CHEMOSTATS, FLAGELLATE ALGAE, OPTICAL DENSITY SENSOR.

ABSTRACT:
THIS STUDY EMPLOYED A CONTINUOUS FLOW TURBIDOSTAT AND A CHEMOSTAT TO APPRAISE THE EFFECT OF NITRATE-NITROGEN CONCENTRATION ON THE GROWTH OF TWO STRAINS OF CHLORELLA PYRENOIDOSA (TX 71105 AND EMERSON). IN ONE TRIAL A MIXTURE OF ALGAE INDIGENOUS TO SAN JOAQUIN DELTA (CALIFORNIA)--GREEN, GREEN FLAGELLATES AND DIATOMS--WAS USED AS TEST ORGANISMS. THE SPECIFIC GROWTH RATE INCREASED WITH TEMPERATURE AND CONCENTRATION OF NITROGEN. THE CONCENTRATION OF NITRATE-NITROGEN DECREASED WITH TEMPERATURE ABOVE 35C. THE GROWTH OPTIMUM WAS NOT NECESSARILY OPTIMUM FOR NITRATE UTILIZATION. THE MIXED CULTURE OF ALGAE SHOWED SUBSTANTIAL SPECIFIC GROWTH RATE AT MUCH LOWER NITROGEN CONCENTRATION THAN C PYRENOIDOSA. RESULTS OF THE STUDY EMPHASIZE THE WIDE DIFFERENCE IN NITROGEN REQUIREMENTS OF VARIOUS ALGAL SPECIES. THE TURBIDOSTATIC METHOD PROVED MORE EXPENSIVE, BUT MORE ADVANTAGEOUS THAN THE CHEMOSTATIC METHOD. (WILDE-WISCONSIN)

FIELD 05A, 07B

ACCESSION NO. W70-10403

THE EXTENT OF NITROGEN AND PHOSPHORUS REGENERATION FROM DECOMPOSING ALGAE,

KENTUCKY UNIV., LEXINGTON. DEPT. OF CIVIL ENGINEERING; TEXAS UNIV., AUSTIN. DEPT. OF CIVIL ENGINEERING; AND STANFORD UNIV., CALIF. DEPT. OF ENVIRONMENTAL ENGINEERING.

E. G. FOREE, W. J. JEWELL, AND P. L. MCCARTY.

FIFTH INTERNATIONAL WATER POLLUTION RESEARCH CONFERENCE, SAN FRANCISCO, JULY 26-AUGUST 1, 1970. PREPRINT. 15 P.

DESCRIPTORS:
*EUTROPHICATION, *ALGAE, *DECOMPOSING ORGANIC MATTER, NUTRIENTS, SURFACE WATERS, NITROGEN, WATER QUALITY, PHOSPHORUS, AEROBIC CONDITIONS, ANAEROBIC CONDITIONS, CHLORELLA, CHLAMYDOMONAS, SCENEDESMUS, EFFLUENTS, OXIDATION LAGOONS.

IDENTIFIERS:
SAN PABLO ESTUARY(CALIF), SACRAMENTO RIVER(CALIF), SEARSVILLE LAKE(CALIF), SAN JOAQUIN RIVER(CALIF), NITZSCHIA.

ABSTRACT:
DECOMPOSITION OF MONOSPECIFIC AND HETEROGENOUS ALGAE AND SUBSEQUENT RELEASE OF NUTRIENTS WERE INVESTIGATED IN DARK-AEROBIC AND DARK-ANAEROBIC CULTURES. UNDER AEROBIC CONDITIONS, 50% ON THE AVERAGE OF THE INITIAL PARTICULATE NITROGEN AND PHOSPHORUS WERE REGENERATED, BUT THE RECOVERY VARIED FROM ZERO TO NEARLY 100%. UNDER ANAEROBIC CONDITIONS, THE AVERAGE REGENERATION OF NITROGEN AND PHOSPHORUS WAS 40% AND 60%, RESPECTIVELY, WITH VARIATION NEARLY AS GREAT AS THAT FOR AEROBIC DECOMPOSITION. A GENERAL AGREEMENT BETWEEN EQUATION-PREDICTED NITROGEN AND PHOSPHORUS REFRACTORY VALUES AND MEASURED VALUES WAS OBTAINED. THE RESULTS SUGGEST THAT CHANCES FOR A SUCCESSFUL CONTROL OF EUTROPHICATION ARE MUCH GREATER BEFORE REFRACTORY REMAINS OF ALGAE HAVE BEEN PERMITTED TO ACCUMULATE. (WILDE-WISCONSIN)

FIELD 05C, 05G

ACCESSION NO. W70-10405

WATER POLLUTION AND THE TEXTILE INDUSTRY,

W. J. COSGROVE.

CANADIAN TEXTILE JOURNAL, VOL 87, NO 1, 1970, P 63-66. 1 TAB.

DESCRIPTORS:
*ST. LAWRENCE RIVER, BIOCHEMICAL OXYGEN DEMAND, HYDROGEN ION CONCENTRATION, ALGAE, *ENTROPHICATION, NEUTRALIZATION, SEDIMENTATION, TRICKLING FILTERS, WASTE WATER TREATMENT.

IDENTIFIERS:
*TEXTILE MILL WASTES, MONTREAL, BALANCING TANKS.

ABSTRACT:
THERE WAS A DRAMATIC REDUCTION IN THE INCIDENCE OF TYPHOID FEVER DUE TO IMPROVEMENTS IN WATER SUPPLIES IN THE YEARS 1906 TO 1923, AT WHICH TIME THE IMPORTANCE OF TYPHOID CARRIERS WAS DISCOVERED. THE CITY OF MONTREAL NOW FACES PROBLEMS OF SEWAGE DISPOSAL AND THE FORMATION OF ALGAL BLOOMS IN NEAR BY RIVERS AND LAKES. THE TEXTILE INDUSTRY IS ONE OF SEVERAL SOURCES OF WASTES WHICH ADDS TO THE PROBLEM. THE BOD'S AND PH'S OF TYPICAL TEXTILE WASTES ARE TABULATED. ABOUT THREE QUARTERS OF THE TEXTILE PLANTS DISCHARGE THEIR WASTES TO SEWERS AND THE REMAINDER TO RIVERS. TREATMENTS USED FOR TEXTILE WASTES DEPEND UPON BALANCING TANKS, NEUTRALIZATION, SEDIMENTATION, TRICKLING FILTERS AND THE USE OF SPECIFIC OPERATIONS AND CHEMICAL TREATMENTS DECTATED BY THE PARTICULAR PROCESSING OPERATIONS AND CHEMICALS INVOLVED. (WORK-NORTH CAROLINA STATE)

FIELD 05D, 05B

ACCESSION NO. W70-10427

HYDROBIOLOGICAL PROPSPECTIVES OF THE MEURTHE: MINERAL POLLUTION AND ALGAL VEGETATIONS (IN FRENCH),

LABORATOIRE DE BIOLOGIE VEGETALE, NANCY (FRANCE).

JEAN-FRANCOIS PIERRE.

COMPTES RENDUS ACADEMIE DES SCIENCES, PARIS, VOL 270, NC 17, SERIE D, P 2101-2102, 1970. 3 REF.

DESCRIPTORS:
*WATER TREATMENT, *CHLORINATION, *ALGAL CONTROL, RIVERS, WATER POLLUTION EFFECTS, ALGAE, DIATOMS.

IDENTIFIERS:
RHINE-MAAS BASIN(FRANCE), MEURTHE RIVER(FRANCE), CLADOPHORA, PHORMIDIUM, OSCILLATCRIA, OEDOGONIUM.

ABSTRACT:
ONE OF THE FLOWING CULTURES CONTAINED WATER CF THE RIVER MEURTHE WITH CHLORINE ION CONCENTRATION IN THE PROXIMITY OF 37 MILLIGRAM/LITER. IN ANOTHER CULTURE THE CONCENTRATION OF CHLORINE WAS INCREASED TO 760 MILLIGRAM/LITER. INOCULA INCLUDEC CLACOPHORA GLOMERATA, OEDOGONIUM SPP, PHORMIDIUM TENUE, OSCILLATCRIA SPP, ANC NUMEROUS DIATOMS. THE INCREASED CONCENTRATION OF CHLORINE FAILED TO INFLUENCE SIGNIFICANTLY THE ALGAL POPULATION THUS DEMONSTRATING CONSIDERABLE ECOLOGICAL PLASTICITY OF THESE ORGANISMS. THE RESULTS SUGGESTED THAT THE CCNCENTRATION OF 800 MILLIGRAM/LITER OF CHLORINE IONS IN THE RIVER MEURTHE WOULD NOT PRODUCE A HARMFUL SUPPRESSION OF ALGAL GROWTH. (WILDE-WISCCNSIN)

FIELD 05F

ACCESSION NO. W71-00099

CONDENSED PHOSPHATES IN LAKE WATER AND WASTE WATER,

TORONTO UNIV. (ONTARIO). DEPT. OF CIVIL ENGINEERING; AND MCMASTER UNIV., HAMILTON (ONTARIO). DEPT. OF CHEMICAL ENGINEERING.

G. W. HEINKE, AND J. D. NORMAN.

FIFTH INTERNATIONAL WATER POLLUTION RESEARCH CONFERENCE, SAN FRANCISCO, JULY 26-AUGUST 2, 1970. PREPRINT. 6 P, 2 FIG, 2 TAB, 12 REF.

DESCRIPTORS:
*PHOSPHATES, *LAKES, *WASTE WATER(POLLUTION), *HYDROLYSIS, DETERGENTS, GREAT LAKES, TEMPERATURE, HYDROGEN ION CONCENTRATION, ACTIVATED SLUDGE, TREATMENT, EUTROPHICATION, SURFACTANTS, ENZYMES, LAKE ONTARIO, SEASONAL, ALGAE, CULTURES, MICROORGANISMS, EFFLUENTS, SETTLING BASINS, WASTE WATER TREATMENT, LABORATORY TESTS.

IDENTIFIERS:
*CONDENSED PHOSPHATES, SODIUM TRIPOLYPHOSPHATE, SCDIUM PYROPHOSPHATE.

ABSTRACT:
HYDROLYSIS RATE OF SODIUM TRIPOLYPHOSPHATE AND SODIUM PYROPHOSPHATE IN GREAT LAKES WATER AND WASTE WATER UNDER CONDITIONS OF TEMPERATURE, PH AND CONCENTRATION OCCURRING IN THE ENVIRONMENT ARE EVALUATED. CONDENSED PHOSPHATES FROM DETERGENTS ARE MAJOR PHOSPHORUS CONTRIBUTORS TO WASTE WATER--ABOUT DOUBLE THE HUMAN WASTE CONTRIBUTION IN THE UNITED STATES AND ABOUT EQUAL TO IT IN CANADA. CONDENSED PHOSPHATES HYDROLYZE EXTENSIVELY DURING ACTIVATED SLUDGE TREATMENT, WITH ABOUT 15% REMAINING IN THE EFFLUENT, HYDROLYZING IN WASTE WATER THREE TIMES FASTER IN SUMMER THAN WINTER. CNCE DISCHARGEC TO SURFACE WATERS, HYDROLYSIS PROCEEDS AT A MUCH SLOWER RATE, CEPENDING ON THE EXTENT OF MICROBIAL ACTIVITY IN THE SURFACE WATER. STUDIES ON THE EFFECT OF PH CHANGES SHOW FASTEST RATES OF HYDRCLYSIS OCCURRING AT THE NATURAL PH OF WASTE WATER OR LAKE WATER. EXPERIMENTAL DATA AT LOW LEVELS OF CONCENTRATION, BELOW 15 MILLIGRAMS/LITER PHOSPHATES, ARE BEST FITTED BY A ZERO-ORDER MODEL. A GRAPHICAL PRESENTATION ALLOWS AN ORDER OF MAGNITUDE PREDICTION OF HYDROLYSIS RATE TO BE EXPECTED IN A PARTICULAR SITUATION. RESULTS OF THIS WORK ARE OF VALUE IN RESEARCH ON PHOSPHORUS REMCVAL METHODS AND EUTROPHICATION STUDIES. (JONES-WISCONSIN)

FIELD 02K, 05C

ACCESSION NO. W71-00116

BIOLOGICAL AND CHEMICAL INVESTIGATIONS ON THE EFFECT OF SEWAGE ON THE EUTROPHICATION OF BAVARIAN LAKES,

H. LIEBMANN.

FIFTH INTERNATIONAL WATER POLLUTION RESEARCH CONFERENCE, SAN FRANCISCO, JULY 26-AUGUST 2, 1970. PREPRINT. 7 P, 5 FIG, 1 TAB, 27 REF.

DESCRIPTORS:
*SEWAGE, *EUTROPHICATION, *LAKES, AQUATIC ENVIRONMENT, CHEMICAL PROPERTIES, PHOSPHORUS, NITROGEN, POTASSIUM, DIVERSION, ALGAE, NITROGEN FIXATION, PLANKTON, PHYTOPLANKTON, BACTERIA, ZOOPLANKTON, SALMONIDS, SEASON, OXYGEN, HYDROGEN SULFIDE, DISTRIBUTION, NANNOPLANKTON, TURBIDITY, HYDROGEN ION CONCENTRATION, ALKALINITY, CARBONATES, HARDNESS(WATER), SULFATES, CONDUCTIVITY, SEDIMENTS, TERTIARY TREATMENT, NUTRIENTS.

IDENTIFIERS:
*BAVARIA, SCHLIERSEE, TEGERNSEE, ANABAENA, CCREGONS, LYNGBYA LIMNETICA, OSCILLATORIA REDECKEI, OSCILLATORIA RUBESCENS, SYNEDRA ACUS, RHODOMONAS, LAKE CONSTANCE, SEWAGE CANALS.

ABSTRACT:
LAKES SCHLIERSEE AND TEGERNSEE IN UPPER BAVARIA WERE ASSESSED FOR EUTROPHICATION BY COMPARING CHEMICAL AND BIOLOGICAL CHARACTERISTICS BEFORE AND AFTER SEWAGE DIVERSION. BY TERTIARY TREATMENT WITH CHEMICAL FLOCCULATION ABOUT 90% PHOSPHORUS REMOVAL IS POSSIBLE, BUT A PHOSPHORUS RESIDUE AND A RELATIVELY HIGH AMOUNT OF NITROGEN REMAIN. THUS, FULL CANALIZATION AROUND THE LAKES AND DISCHARGE OF THE RAPIDLY FLOWING EFFLUENT AFTER BIOLOGICAL TREATMENT APPEARED THE MOST EFFECTIVE METHOD TO ACHIEVE IMPROVEMENT. BECAUSE CLIMATIC CONDITIONS PRECLUDE PUMPING SEWAGE FOR GREAT DISTANCES IN MIDDLE EUROPE, A PROPOSAL WAS MADE THAT LOCAL PURIFIED SEWAGE EFFLUENTS BE PUMPED INTO PLASTIC PIPES, SUBMERGED IN LAKE CONSTANCE FOR THE SHORTEST ROUTING, WHERE A TERTIARY TREATMENT WITH PHOSPHORUS REMOVAL SHOULD BE ESTABLISHED FOR THE WHOLE SEWAGE. TERTIARY TREATMENT MAY BE ESPECIALLY ADVANTAGEOUS IF THERE ARE OTHER LAKES AND IMPOUNDMENTS DOWNSTREAM. DANGER THAT LEAKAGE FROM PLASTIC PIPELINES MAY INTRODUCE PURIFIED SEWAGE INTO THE LAKE SEEMS MINIMAL WHEN COMPARED TO THE FACT THAT OTHER POSSIBLE SOLUTIONS TO THE PROBLEM MAY DEFER THE REMEDY FOR SO LONG THAT THE EUTROPHICATION PROCESSES BECOME ALMOST IRREVERSIBLE. (JONES-WISCONSIN)

FIELD 02H, 05C

ACCESSION NO. W71-00117

SURFACE PROPERTIES AND ION EXCHANGE IN ALGAE REMOVAL,

CALIFORNIA UNIV., RICHMOND.

C. G. GOLUEKE, AND W. J. OSWALD.

JOURNAL OF THE WATER POLLUTION CONTROL FEDERATION, VOL 42, NO 8, PART II, P R304-R314, AUGUST 1970. 4 FIG, 12 REF.

DESCRIPTORS:
*ALGAE, *HARVESTING OF ALGAE, *ION EXCHANGE, DISSOLVED SOLIDS, FLOCCULATION, HEAD LOSS, HYDROGEN ION CONCENTRATION, CHLORELLA, SCENEDESMUS, WASTE WATER TREATMENT.

IDENTIFIERS:
*SURFACE CHARGE, REGENERATION, SUSPENDED SOLIDS.

ABSTRACT:
A FACET OF ION EXCHANGE YET TO BE REPORTED IN THE LITERATURE IS THE USE OF ION EXCHANGE COLUMNS FOR THE REMOVAL OF CELLULAR MATERIAL HAVING A SURFACE CHARGE. IN EXPERIMENTS ON THE REMOVAL OF ALGAE FROM POND EFFLUENT IT WAS FOUND (1) IN THE PRESENCE OF EITHER STRONG OR WEAK CATION ION EXCHANGE RESINS, ALGAL CELLS AGGREGATE INTO TIGHT CLUMPS AND SETTLE IN A CONE-SHAPED MASS ON THE TOP OF EACH RESIN BEAD, (2) THE AGGREGATED CELLS ARE EASILY REMOVED FROM THE COLUMN BY BACKWASHING AND RETAIN THEIR TENDENCY TO AGGREGATE AND SETTLE IN THE DISPLACED BACKWASH, AND (3) THE EXCHANGE EFFICIENCY OF THE RESIN WITH RESPECT TO ALGAL CELLS IS BUT A SMALL FRACTION OF THAT WITH RESPECT TO CATIONS IN THAT THE CAPACITY FOR REMOVING MG+2, CA+2, AND OTHER CATIONS CONTINUES LONG AFTER THE CAPACITY FOR REMOVING ALGAL CELLS IS EXHAUSTED. AGGREGATION WAS ITS STRONGEST WITHIN THE PH RANGE OF 2.8 TO 3.5 AND DROPPED OFF SHARPLY ON EITHER SIDE OF THIS RANGE. CONCENTRATION OF NA+, CA+2, AND MG+2 FROM 0.5 TO 4.0 MILLIMOLES DID NOT AFFECT THE PRECIPITATION OF THE CELLS. DESPITE THE SIMPLICITY OF THE PROCESS, THE PURITY OF THE ALGAL PRODUCT, AND THE HIGH QUALITY OF THE RECLAIMED WATER ATTAINABLE WITH IT, UNTIL A REGENERANT LESS EXPENSIVE THAN H2SO4 OR HCL IS FOUND, THE INTEREST IN THE PROCESS MUST REMAIN ACADEMIC RATHER THAN ECONOMIC. (AGUIRRE-TEXAS)

FIELD 05D

ACCESSION NO. W71-00138

RELATIONSHIP OF CHLOROPHYLL A TO ALGAL COUNT AND CLASSIFICATION IN OXIDATION PONDS,

LAMAR STATE COLL. OF TECHNOLOGY, BEAUMONT, TEX.

JOSEPH V. DUST, AND ADAM SHINDALA.

JOURNAL OF THE WATER POLLUTION CONTROL FEDERATION, VOL 42, NO 7, P 1362-1369, JULY 1970. 7 FIG, 7 REF.

DESCRIPTORS:
*ALGAE, *CHLOROPHYLL, *CLASSIFICATION, *OXIDATION LAGOONS, CHLORELLA, OXYGEN DEMAND, PHOTOSYNTHESIS, RESPIRATION, SCENEDESMUS, WASTE WATER TREATMENT.

IDENTIFIERS:
*ENUMERATION, ANACYSTIS.

ABSTRACT:
A STUDY WAS CONDUCTED AIMED AT OBTAINING A QUANTITATIVE COMPARISON OF CHLOROPHYLL A WITH THE TOTAL NUMBER OF ORGANISMS IN THE TRINITY RIVER AUTHORITY WASTE WATER TREATMENT PLANT AS WELL AS A COMPARISON WITH THE VARIOUS SPECIES OF ALGAE EXISTING FROM TIME TO TIME IN THESE OXIDATION PONDS. THE LITERATURE INDICATES THAT THE ANALYTICAL DETERMINATION OF CHLOROPHYLL PIGMENTS CAN BE ACCOMPLISHED SPECTROPHOTOMETRICALLY WITH THE FOLLOWING EQUATION: CHLOROPHYLL A IN MG/L = (14.3) D665, WHERE D655 IS THE LIGHT ABSORBANCE AT A WAVE LENGTH OF 655 MILLIMICRONS AND A SLIT WIDTH OF 0.3 MM. EIGHT MAJOR TYPES OF ALGAE WERE FOUND IN SIGNIFICANT NUMBERS: OOCYSTIS, ANACYSTIS, CHLORELLA, DESMIDS, SCENEDESMUS, COELASTRUM, ZYGNEMA, AND CLOSTERIUM. THE ANACYSTIS ORGANISM WAS FOUND MORE FREQUENTLY AND IN GREATER NUMBERS THAN ANY OTHER ORGANISM. THE AVERAGE WEEKLY COUNT OF ANACYSTIS VARIED FROM ABOUT 1 TO 4 MILLION PER ML DURING THE FIRST 15 WEEKS OF THE STUDY COMPRISING THE SUMMER AND EARLY FALL MONTHS OF THE YEAR. CHLORELLA AND SCENEDESMUS WERE THE PREDOMINANT ORGANISMS DURING THE LATE FALL, WINTER, AND EARLY SPRING MONTHS, WITH CELL COUNTS OF 10,000 TO 100,000 CELLS/ML. A RELATIONSHIP BETWEEN CHLOROPHYLL A AND TOTAL ALGAL COUNT WAS NOT FOUND. HOWEVER, A DEFINITE RELATIONSHIP EXISTED BETWEEN THE AMOUNT OF CHLOROPHYLL A AND THE NUMBER OF ANACYSTIS ORGANISMS PRESENT. A SIMILAR RELATIONSHIP WAS FOUND WITH ZYGNEMA. (AGUIRRE-TEXAS)

FIELD 05D

ACCESSION NO. W71-00139

AUTOMATED POTENTIOMETRIC TECHNIQUES FOR THE ON-SITE MONITORING OF ION CONCENTRATIONS IN WATER,

MISSOURI UNIV., COLUMBIA. DEPT. OF CHEMISTRY.

STANLEY E. MANAHAN.

AVAILABLE FROM NTIS AS PB-195 167, $3.00 IN PAPER COPY, $0.95 IN MICROFICHE. OWRR COMPLETION REPORT, 1970. OWRR PROJECT A-024-MO(11).

DESCRIPTORS:
*ANALYTICAL TECHNIQUES, *WATER ANALYSIS, *NITRATES, *ELECTRODES, IONS, FLUORIDES, ALGAE, VOLUMETRIC ANALYSIS, WATER CHEMISTRY.

IDENTIFIERS:
ION-SELECTIVE ELECTRODES, STANDARD ADDITION, POTENTIOMETRY, INTERFERING IONS.

ABSTRACT:
THE CHARACTERISTICS OF THE NITRATE ELECTRODE - FLUORIDE ELECTRODE CELL WITHOUT LIQUID JUNCTION WERE INVESTIGATED FOR APPLICABILITY TO NITRATE ION ANALYSIS. IN ADDITION TO ENABLING ACCURATE NITRATE ANALYSIS, THE CELL WAS FOUND TO BE USEFUL FOR EXAMINING THE EFFECTS OF INTERFERING IONS. AN ATTEMPT WAS MADE TO USE THE NITRATE ELECTRODE IN FOLLOWING NITRATE ION UPTAKE BY ALGAL CULTURES. THE ALGAE INTERFERE TO A CERTAIN EXTENT WITH THE ELECTRODE RESPONSE. USING KNOWLEDGE GAINED FROM RESEARCH UNDER THIS GRANT A POTENTIOMETRIC NITRATE ANALYSIS CAPABILITY HAS BEEN SET UP AT THE UNIVERSITY OF MISSOURI ENVIRONMENTAL TRACE SUBSTANCES LABORATORY.

FIELD 05A, 02K

ACCESSION NO. W71-00474

THE IRTELATION OF CARBON AND PHOSPHORUS IN REGULATING HETEROTROPHIC AND AUTOTROPHIC POPULATICNS IN AQUATIC ECOSYSTEMS,

FEDERAL WATER QUALITY ADMINISTRATION, ATHENS, GA. NATICNAL POLLUTANTS FATE RESEARCH PROGRAM.

PAT C. KERR, DORIS F. PARIS, AND D. L. BROCKWAY.

AVAILABLE FROM NTIS AS PB-195 195, $0.95 IN MICROFICHE. ALSO FOR SALE BY SUPERINTENDENT OF DOCUMENTS, U.S. GOVERNMENT PRINTING OFFICE, WASHINGTON, D.C. 20402, PRICE $0.60, I67.13/4:16050 FGS 07/70. WATER POLLUTICN CONTROL RESEARCH SERIES 16050 FGS 07/70, (JULY 1970). 53 P, 4 TAB, 17 FIG, 51 REF. FWQA PROJECT 16050 FGS.

DESCRIPTORS:
CYCLING NUTRIENTS, *CARBON, *PHOSPHORUS, *CYANOPHYTA, AQUATIC ENVIRONMENT, CARBON DIOXIDE, ECOSYSTEMS, SOUTH CAROLINA, ALGAE, TROPHIC LEVEL.

IDENTIFIERS:
*ANTOTROPHIC POPULATIONS, *HETERCTROPHIC POPULATICNS, LAKE HARTWELL(SOUTH CAROLINA), ANACYSTIS NIDULANS, AQUATIC ECOSYSTEMS, ALGAE GROWTH, ALGAL POPULATIONS.

ABSTRACT:
LABORATORY AND FIELD INVESTIGATIONS WERE CONDUCTED ON THE FATE AND CYCLING OF CARBON AND PHOSPHORUS IN SELECTED AQUATIC ECOSYSTEMS. INORGANIC CARBON, AS CO2, SUPPLIED BY BOTH BACTERIAL CULTURES AND CYLINDER GASES, STIMULATED THE GROWTH OF THE BLUE-GREEN ALGA ANACYSTIS NIDULANS. THE CARBON REQUIREMENT (10-5 UG CO2 PER CELL) FOR THIS ALGA WAS DETERMINED FOR A SINGLE SET OF EXPERIMENTAL CONDITIONS. THE ADDITION OF CO2 TO NATURAL WATER LOW IN PHOSPHORUS (5 UG P) AND NITROGEN (5 UG N) IN THE LABORATORY STIMULATED THE GROWTH OF INDIGENOUS ALGAL POPULATIONS. THE LIMITING AND LUXURY CELLULAR CONCENTRATIONS OF PHOSPHORUS FOR STARVED ANACYSTIS NIDULANS WERE FOUND TO BE 0.3 X 10-8 UG P AND 3.0 X 10-8 UG P PER CELL, RESPECTIVELY. DIEL STUDIES OF A STREAM WHICH RECEIVED BIOLOGICALLY-TREATED SEWAGE DEMONSTRATED THAT THE DISSOLVED CO2 AND HCO3 CONTINUALLY PRODUCED IN THE SYSTEM WERE ESSENTIALLY DEPLETED BY THE AUTOTROPHIC ORGANISMS DURING DAYLIGHT HOURS, WHILE THE CONCENTRATION OF PHOSPHORUS (1.3-2.2 MG/L P) REMAINED UNCHANGED. ADDITION OF ORGANIC CARBON AND INORGANIC NITROGEN AND PHOSPHORUS ALONG AND IN COMBINATION TO THE WATERS STUDIED DIRECTLY STIMULATED THE OXIDATIVE METABOLISM OF THE HETERCTROPHIC POPULATION, WHICH RESULTED IN INCREASED DISSOLVED CO2 AND HCO3-. THIS INCREASED AVAILABILITY OF INORGANIC CARBON, RATHER THAN THE DIRECT METABOLIC REMOVAL OF DISSOLVED PHOSPHORUS BY THE ALGAE, APPEARED TO BE DIRECTLY RESPONSIBLE FOR THE GROWTH OF THE ALGAL POPULATIONS IN THE WATERS STUDIED.

FIELD 05C, 02I, 02H

ACCESSION NO. W71-00475

PROCEEDINGS OF THE CONFERENCE WATER RESOURCES RESEARCH - 1970.

CONNECTICUT UNIV., STORRS. INST. OF WATER RESOURCES.

AVAILABLE FROM NTIS AS PB-195 667, $3.00 IN PAPER COPY, $0.95 IN MICROFICHE. WILLIAM C. KENNARD, EDITOR. CONNECTICUT UNIVERSITY INSTITUTE OF WATER RESOURCES REPORT NO 13, SEPTEMBER 1970. 20 P. OWRR PROJECT A-999-CONN(8).

DESCRIPTORS:
*WATER RESOURCES RESEARCH ACT, CONNECTICUT, *UNIVERSITIES, *ABSTRACTS, WATER POLLUTION TREATMENT, INSTRUMENTATION, WASTE DISPOSAL, AERATION, EROSION CONTROL, WATER POLLUTION SOURCES, WATER LAW, ALGAE, DISSOLVED OXYGEN.

IDENTIFIERS:
WATER RESOURCES RESEARCH.

ABSTRACT:
THESE PROCEEDINGS INCLUDE ABSTRACTS OF PAPERS PRESENTED AT THE CONFERENCE HELD BY THE INSTITUTE OF WATER RESOURCES AT THE UNIVERSITY OF CONNECTICUT ON MAY 18, 1970. THE TITLE OF EACH PRESENTATION IS THE SAME AS THAT OF THE RESEARCH PROJECT. ALSO GIVEN FOR EACH ARE THE PERSONNEL, ACADEMIC DEPARTMENTS, PLANNED DURATION AND THE OBJECTIVES OF THE INVESTIGATIONS. EACH ABSTRACT IS A BRIEF SUMMARY OF ACCOMPLISHMENTS TO DATE AND PLANS FOR THE FUTURE. THE PROGRAM OF THE INSTITUTE HAS BECOME INCREASINGLY DIVERSE IN THE APPROXIMATELY SIX YEARS THAT IT HAS BEEN IN EXISTENCE. PROJECTS IN THE AGRICULTURAL, BIOLOGICAL, ENGINEERING, EARTH, SOCIAL AND PHYSICAL SCIENCES HAVE BEEN OR NOW ARE ACTIVE. CONTINUED EXPANSION, BOTH IN SCOPE OF THE RESEARCH AND IN THE NUMBER OF PROJECTS, IS PLANNED. THE RESULTS OF THESE INVESTIGATIONS WILL, IN MANY CASES, HAVE DIRECT APPLICATION TO SOLVING PROBLEMS OF WATER RESOURCES USE AND DEVELOPMENT IN CONNECTICUT AND WILL RESULT IN IMPORTANT CONTRIBUTIONS TO THE FUND OF SCIENTIFIC INFORMATION ABOUT WATER IN A BROAD RANGE OF DISCIPLINES. (KNAPP-USGS)

FIELD 09A, 06B, 05G

ACCESSION NO. W71-01192

BIOLOGICAL UPTAKE OF PHOSPHORUS BY ACTIVATED SLUDGE,

ARIZONA UNIV., TUCSON. DEPT. OF MICROBIOLOGY AND MEDICAL TECHNOLOGY.

IRVING YALL, WILLIAM H. BOUGHTON, RICHARD C. KNUDSEN, AND NORVAL A. SINCLAIR.

APPLIED MICROBIOLOGY, VOL 20, NO 1, P 145-150, 1970. 5 FIG, 2 TAB, 10 REF. FWQA SUPPORTED.

DESCRIPTORS:
*PHOSPHORUS, *ACTIVATED SLUDGE, PHOSPHORUS RADIOISOTOPES, SEWAGE, RADIOACTIVITY, PHOSPHATES, CHEMICAL ANALYSIS, ALGAE, MICROORGANISMS, HYDROGEN ION CONCENTRATION, PHENOLS, ARIZONA.

IDENTIFIERS:
*BIOLOGICAL UPTAKE, CALCIUM RADIOISOTOPES, BLOOMS, 2,4 DINITRO PHENOL, TUCSON(ARIZ).

ABSTRACT:
EXPERIMENTS TO REPRODUCE CONDITIONS POSSIBLY EXISTING IN AN OPTIMALLY AERATED, PLUG FLOW ACTIVATED SLUDGE UNIT ARE REPORTED. ABILITY OF ACTIVATED SLUDGE TO REMOVE PHOSPHATES WAS STUDIED BY ADDING CARRIER-FREE PHOSPHORUS-32 TO RAW SEWAGE AND MEASURING INCORPORATION OF RADIOACTIVITY INTO CELLS. RADIOISOTOPE DETERMINATIONS INDICATED THAT 48% OF PHOSPHORUS-32 RADIOACTIVITY WAS REMOVED IN 12 HOURS. CHEMICAL METHODS INDICATED THAT ONLY 30% OF ORTHOPHOSPHATE APPARENTLY DISAPPEARED FROM THE SEWAGE DURING THIS PERIOD. EXPERIMENTS WITH SLUDGE RELABELED WITH PHOSPHORUS INDICATED CONSIDERABLE PHOSPHATE TURNOVER. CELLS RELEASED LARGE AMOUNTS OF RADIOACTIVITY AS THEY WERE INCORPORATING FRESH PHOSPHATES. STARVATION IN ISOTONIC SALINE FOR 18 HOURS CAUSED SLUDGE TO DUMP PHOSPHATE. WHEN INTRODUCED INTO FRESH SEWAGE CONTAINING PHOSPHORUS-32, THE STARVED SLUDGE REMOVED ABOUT 60% OF THE RADIO-ACTIVITY IN 6 HOURS WITH LITTLE PHOSPHATE TURNOVER. ABILITY OF SLUDGE TO REMOVE PHOSPHORUS-32 WAS INHIBITED APPROXIMATELY 83% BY 0.001 MOLAR 2,4-DINITRO PHENOL. THIS INHIBITION WAS AT THE EXPENSE OF THE CELL FRACTION THAT CONTAINED RIBO-NUCLEIC ACID AND DEOXYRIBONUCLEIC ACID. THE SLUDGE CELLS RELEASED ORTHOPHOSPHATE WHEN EXPOSED TO THE CHEMICAL AGENT. EXPERIMENTS USING CALCIUM-45 INDICATED THAT CALCIUM PHOSPHATE PRECIPATION PLAYS A MINOR ROLE IN PHOSPHATE REMOVEL UNDER EXPERIMENTAL CONDITIONS. (JONES-WISCONSIN)

FIELD 05D

ACCESSION NO. W71-01474

SOME ECOLOGICAL EFFECTS OF DISCHARGED WASTES ON MARINE LIFE,

CALIFORNIA UNIV., SAN DIEGO; AND SCRIPPS INSTITUTION OF OCEANOGRAPHY, SAN DIEGO, CALIF.

RICHARD W. GRIGG, AND ROBERT S. KIWALA.

CALIFORNIA FISH AND GAME, VOL 56, NO 3, P 145-155, 1970. 2 FIG, 4 TAB, 10 REF.

DESCRIPTORS:
*ECOLOGY, *WASTES, *MARINE ANIMALS, *MARINE PLANTS, DEPTH, CALIFORNIA, SANDS, SEDIMENTS, PRODUCTIVITY, TOXICITY, CRABS, ALGAE, INVERTEBRATES, FISH, POLLUTANTS, TURBIDITY, ORGANIC LOADING, CORAL, MOLLUSKS, WORMS.

IDENTIFIERS:
SAN PEDRO(CALIF), EPIBENTHIC SPECIES, ABALONE, TUNICATES, ARTHROPODS, COELENTERATES, BRYOZOA, ECHINODERMS.

ABSTRACT:
APPROXIMATELY ONE BILLION GALLONS OF SEWAGE ARE DISCHARGED DAILY INTO THE SHALLOW NEARSHORE MARINE ENVIRONMENT OFF SOUTHERN CALIFORNIA. A SURVEY WAS MADE OF THE EFFECT OF EFFLUENT ON THE BIOLOGY OF THE AREA IN 1954. THE RESULTS OF A 1969 SURVEY TO COLLECT COMPARABLE DATA, AND TO EXAMINE THESE TO DETECT POSSIBLE LONG TERM ECOLOGICAL CHANGES ARE DESCRIBED. THE NUMER OF MACROSCOPIC SPECIES PRESENT AT 5 DIVING STATIONS RANGING IN DEPTHS FROM 45 TO 65 FEET, OFF THE PALOS VERDES PENINSULA, NEAR SAN PEDRO, WAS NEGATIVELY CORRELATED TO THE AMOUNT OF FINE GRAIN ORGANIC-LADEN SAND PRESENT IN THE SEDIMENT. ORGANIC RICH SEDIMENTS WERE THICKEST AT STATIONS NEAR THE OUTFALL. ACCUMULATION OF THIS MATERIAL AT THESE DEPTHS APPEARS TO HAVE MODIFIED OR COVERED SUBSTRATES OTHERWISE SUITABLE FOR THE SETTLEMENT OF MANY EPIBENTHIC SPECIES. FISH, KELP, ABALONE AND SPINY LOBSTER ARE PARTICULARLY AFFECTED. SINCE BOTTOM TOPOGRAPHY AT WHITE POINT HAS NOT CHANGED, DECLINE OF FISHES, IF NOT CAUSED DIRECTLY BY TOXIC WASTE PRODUCTS, MAY INDICATE THAT RELIEF IS MORE IMPORTANT AS A SUBSTRATE FOR FOOD RATHER THAN A SOURCE OF SHELTER OR POINT OF ORIENTATION. (JONES-WISCONSIN)

FIELD 05C

ACCESSION NO. W71-01475

UTROPHICATION CONTROL BY PLANT HARVESTING,

FLORIDA STATE BOARD OF HEALTH, WINTER HAVEN. MIDGE RESEARCH LAB.

JAMES L. YOUNT, AND ROY A. CROSSMAN, JR.

JOURNAL WATER POLLUTION CONTROL FEDERATION, VOL 42, NO 5, PART 2, P R173-R183, 1970. 2 FIG, 1 TAB, 28 REF. FWQA GRANT WP-00216.

DESCRIPTORS:
*EUTROPHICATION, *WATER POLLUTION CONTROL, WATER POLLUTION EFFECTS, *AQUATIC PLANTS, HARVESTING, PONDS, WATER HYACINTH, NUTRIENTS, MIDGES, FISH, LAKES, CONDUCTIVITY, ALGAE, STANDING CROP, CHEMICAL ANALYSIS, MOSQUITOES, BACTERIA, FLOATING PLANTS, ON-SITE INVESTIGATIONS, TURTLES.

IDENTIFIERS:
*CHIRONOMIDS, SAWDUST, EICHHORNIA CRASSIPES, SALVINIA ROTUNDIFOLIA, MILORGANITE, GAMBUSIA AFFINIS, PRODUCTIVITY MEASUREMENTS, NUTRIENT REMOVAL.

ABSTRACT:
DATA REPORTED INDICATE THAT GROWING ORGANISMS IN HYPERTROPHIC PONDS AND HARVESTING THEM TO REMOVE NUTRIENTS, REDUCE THE PONDS' PRIMARY PRODUCTIVITY. THIS CONCLUSION IS CONFIRMED BY EXPERIMENTS SHOWING THAT PLANT PRODUCTIVITY WAS REDUCED IN TEST PONDS, WHILE THE CONTROL PONDS, WHERE NUTRIENTS WERE RETURNED, PRODUCTIVITY REMAINED RELATIVELY HIGH. LARGE-SCALE HARVESTING FROM NATURAL WATERS CAN BE EXPECTED TO REDUCE THE PRODUCTIVITY OF THOSE WATERS, AND PROBABLY REVERSE THE TREND TOWARD HYPERTROPHY, ESPECIALLY IN POLLUTED WATERS. THE METHOD OF CONTROLLING THE 'PEST' WATER HYACINTH AND OTHER PLANTS BY CHEMICAL SPRAYS IS RECIRCULATING THEIR NUTRIENTS TO LAKES AND EXACERBATING HYPERTROPHY. GIVEN A CONSTANT INFLOW OF POLLUTANTS, IF TOO MUCH VEGETATION IS REMOVED, THEN AVAILABILITY OF THESE POLLUTANTS TO OTHER ORGANISMS WOULD INCREASE. THE PROBLEM IS RESOLVED BY MANAGING A POPULATION ON A SUSTAINED-YIELD BASIS. HARVESTING HYACINTHS FROM EFFLUENT STREAMS MIGHT PROVE A MORE PRACTICAL METHOD FOR HANDLING EXCESSIVE NUTRIENTS THAN BY HARVESTING FROM LAKES. (JONES-WISCONSIN)

FIELD 05C

ACCESSION NO. W71-01488

EFFECTS OF FOREST CUTTING AND HERBICIDE TREATMENT ON NUTRIENT BUDGETS IN THE HUBBARD BROOK WATERSHED-ECCSYSTEM,

DARTMOUTH COLL., HANOVER, N.H. DEPT. OF BIOLOGICAL SCIENCES; YALE UNIV., NEW HAVEN, CONN. SCHOOL OF FORESTRY; GEOLOGICAL SURVEY, WASHINGTON, D.C.; AND FOREST SERVICE, (USDA), DURHAM, N.H. NCRTHEASTERN FOREST EXPERIMENT STATION.

GENE E. LIKENS, F. HERBERT BORMANN, NOYE M. JOHNSON, D. W. FISHER, AND ROBERT S. PIERCE.

ECCLOGICAL MONOGRAPHS, VOL 40, NO 1, P 23-47, 1970. 15 FIG, 8 TAB, 54 REF.

DESCRIPTORS:
*FORESTS, *CUTTING MANAGEMENT, *HERBICIDES, *NUTRIENTS, *ECOSYSTEMS, *WATERSHEDS(BASINS), STREAMFLOW, IONS, NITRATES, SULFATES, DRAINAGE WATER, DISSOLVED SOLIDS, NITROGEN CYCLE, HYDROGEN ION CONCENTRATION, TEMPERATURE, ELECTRICAL CONDUCTANCE, DISSOLVED OXYGEN, AIR POLLUTION, CHEMICAL ANALYSIS, EUTROPHICATION, TRANSPIRATION, PRECIPITATION(ATMOSPHERIC), RUNOFF, EVAPORATION, GEOLOGIC FORMATIONS, DUSTS, TURBIDITY, EROSION, CONDUCTIVITY, CHLORIDES, SILICA, AMMONIA, MICROORGANISMS, ALGAE, CALCIUM, MAGNESIUM, POTASSIUM, SODIUM, ALUMINUM, BICARBONATES, NITRIFICATION, SEASONAL.

IDENTIFIERS:
*HUBBARD BROOK EXPERIMENTAL FOREST, PARTICULATE MATTER, NITROSOMONAS, NITROBACTER, THIOBACILLUS THIOOXIDANS, DESULFOVIBRIO, ULOTHRIZ ZONATA.

ABSTRACT:
QUANTITY AND QUALITY OF DRAINAGE WATERS WERE SIGNIFICANTLY ALTERED FOLLOWING DEFORESTATION OF A NORTHERN HARDWOODS WATERSHED-ECOSYSTEM. ANNUAL WATER RUNOFF EXCEEDED EXPECTED VALUE (BASED ON UNDISTURBED WATERSHED) BY 39% DURING FIRST WATER-YEAR AFTER DEFORESTATION AND 28% DURING THE SECOND. DEFORESTATION RESULTED IN LARGE INCREASES IN CONCENTRATIONS OF ALL MAJOR IONS EXCEPT AMMONIUM, SULFATE, AND CARBONIC ACID, WITH GREATEST INCREASE FOR NITRATE IN STREAMWATER. SULFATE WAS THE ONLY MAJOR ION THAT DECREASED AFTER DEFORESTATION. IN UNDISTURBED WATERSHEDS, STREAMWATER HAS A PH OF ABOUT 5.1 FROM SULFURIC ACID, WHEREAS AFTER DEFORESTATION IT BECAME A NITRIC ACID SOLUTION OF PH 4.3 ENRICHED IN METALLIC IONS AND DISSOLVED SILICA. THE INCREASE IN NITRATE CONCENTRATION IN PRECIPITATION MAY SOMEWHAT INCREASE AIR POLLUTION. GREATLY INCREASED EXPORT OF DISSOLVED NUTRIENTS FROM THE DEFORESTED ECOSYSTEM WAS DUE TO AN ALTERATION OF THE ECOSYSTEM NITROGEN CYCLE. INCREASED AVAILABILITY OF NITRATE AND HYDROGEN IONS RESULTED FROM NITRIFICATION. TOTAL NEW EXPORT OF DISSOLVED INORGANIC SUBSTANCES WAS 14-15 TIMES GREATER THAN FROM NATURAL ECOSYSTEMS. THE DEFORESTATION EXPERIMENT RESULTED IN SIGNIFICANT POLLUTION OF THE DRAINAGE STREAM, WITH NITRATE CONCENTRATION EXCEEDING THE MAXIMUM RECOMMNDED FOR DRINKING WATER. A BLOOM OF ALGAE APPEARED EACH SUMMER.
(JONES-WISCONSIN)

FIELD 02A, 05C

ACCESSION NO. W71-01489

BIOLOGICAL AND CHEMICAL ASPECTS OF RHINE WATER IN THE BERENPLAAT RESERVOIR,

DRINKWATERLEIDING DER GEMEENTE ROTTERDAM (NETHERLANDS).

JOHN J. ROOK, AND GIJSBERT OSKAM.

JOURNAL AMERICAN WATER WORKS ASSOCIATION, VOL 62, NO 4, P 249-259, 1970. 10 FIG, 7 TAB, 30 REF.

DESCRIPTORS:
*WATER PURIFICATION, *DETENTION RESERVOIRS, RESERVOIRS, PHOTOSYNTHESIS, DISSOLVED OXYGEN, PHOSPHATES, WATER YIELD IMPROVEMENT, WATER POLLUTION TREATMENT, ALGAE, POTABLE WATER, TASTE, AMMONIA, SILICA.

IDENTIFIERS:
RHINE RIVER, ASTERIONELLA FORMOSA, CRYPTOMONAS, CHLAMIDOMONAS, BERENPLAAT RESERVOIR(ROTTERDAM), THE NETHERLANDS.

ABSTRACT:
A CONSIDERABLE PURIFICATION OF THE HIGHLY POLLUTED WATER OF THE RHINE RIVER IS ACHIEVED BY ITS RELATIVELY BRIEF STORAGE IN A RETENTION RESERVOIR. THE HIGH LEVEL OF RIVER'S FERTILITY INDUCES A HEAVY GROWTH OF ALGAE, PARTICULARLY ASTERIONELLA FORMOSA. THE STORAGE AND RE-AERATION OF THE ALGAL SUSPENSION PROMOTES PHOTOSYNTHESIS AND RESTORES THE CONTENT OF DISSOLVED OXYGEN. AN OCCASIONAL INCREASE IN DISSOLVED ORGANIC MATTER IN THE RESERVOIR HAS FAILED TO PRODUCE UNFAVORABLE EFFECTS ON POTASSIUM PERMANGANATE CONSUMPTION. THE GROWING ALGAE UTILIZE LARGE AMOUNTS OF PHOSPHATES AND SILICA; THE LATTER APPEARS TO ACT AS A LIMITING FACTOR. THE AMMONIA CONTENT IS REDUCED BY LOSS TO THE ATMOSPHERE AND BY NITRIFICATION, THE NITRATES PRESUMABLY CONTRIBUTING TO THE DISSIMILATION OF SETTLED ORGANIC MATTER. THE MOST IMPORTANT BENEFIT OF THE XTORAGE OF RHINE WATER IS THE STRIKING REDUCTION IN THRESHOLD TASTE VALUE. THE ALGAE ARE WITHDRAWN FROM THE RESERVOIR PRIOR TO THEIR COMPLETION OF THE LIFE CYCLE. AN OPINION IS EXPRESSED THAT THE NETHERLANDS WILL BUILD MANY NEW RHINE RESERVOIRS. (WILDE-WISCONSIN)

FIELD 05F

ACCESSION NO. W71-01491

WATER QUALITY CONTROL WITH SYNTHETIC POLYMERIC FLOCCULANTS,

CONNECTICUT UNIV., STORRS. INST. OF WATER RESOURCES.

J. K. DIXON.

AVAILABLE FROM NTIS AS PB-195 983, $3.00 IN PAPER COPY, $0.95 IN MICROFICHE. TECHNICAL COMPLETION REPORT, NOVEMBER 1970. 4 P. OWRR PROJECT A-016-CONN(1).

DESCRIPTORS:
*FLOCCULATION, FILTRATION, ELECTROPHORESIS, SETTLING, *ALGAE, *SILICA, *BACTERIA, WATER POLLUTION CONTROL, WASTE WATER TREATMENT.

IDENTIFIERS:
POLYACRYLAMIDE, POLYETHYLENIMINE.

ABSTRACT:
STUDIES HAVE BEEN MADE OF THE FLOCCULATION OF SILICE, ALGAE AND BACTERIA BY MEANS OF LOW CONCENTRATIONS OF SYNTHETIC POLYMERS. THE EFFECTS OF POLYMER MOLECULAR WEIGHT AND STRUCTURE HAVE BEEN EMPHASIZED, USING THE SILICA, AND PURE CULTURES OF GREEN ALGAE AND E. COLI UNDER A VARIETY OF CONDITIONS OF PH AND COLLOID CONCENTRATION. FILTRATION RATE, LIGHT TRANSMISSION AND ELECTROPHORETIC MOBILITY HAVE BEEN EMPLOYED TO MEASURE DEGREE OF FLOCCULATION. NON-IONIC AND ANIONIC POLYMERS DID NOT PRODUCE FLOCCULATION, BUT CATIONIC POLYETHYLENEIMINES WERE EFFECTIVE AT CONCENTRATIONS OF 0.1 TO 100 PPM, THE OPTIMUM BEING DEPENDENT ON PH, COLLOID CONCENTRATION AND POLYMER MOLECULAR WEIGHTS ABOVE ABOUT 2000. ALGAE REQUIRED SOME 30 TO 500 TIMES HIGHER SOLUTION CONCENTRATION OF POLYMER THAN DID THE BACTERIA AND SILICA, RESPECTIVELY. TWO TYPES OF POLYETHYLENEIMINE POLYMER TAGGED WITH RADIOACTIVE C14 WERE PREPARED WHICH FLOCCULATED THE THREE KINDS OF COLLOIDS IN THE SAME WAY AS UNTAGGED POLYMERS. IN THE REGION WHERE FLOCCULATION WAS OPTIMUM THE CONCENTRATIONS OF POLYMER ADSORBED ON THE COLLOID SURFACES STOOD IN ROUGHLY THE RATIO1:10:200 FOR SILICA, E. COLI AND ALGAE, RESPECTIVELY. POOR FLOCCULATION EFFICIENCY WAS NOT TO BE ASSOCIATED WITH LOW ADSORPTION, BUT TO SPECIFIC POLYMER-SURFACE INTERACTIONS WHICH REDUCED THE EFFICIENCY OF THE ADSORBED POLYMER. THE RESULTS HAVE BEEN SHOWN TO CONFORM WITH CURRENT THEORIES OF FLOCCULATION OF OTHER COLLOIDS. THEY SHOULD ASSIST IN ATTAINING BETTER WATER QUALITY CONTROL BY THE USE OF THE SYNTHETIC POLYMERS IN NUMEROUS PRACTICAL OPERATIONS. (DIXON-CONN)

FIELD 05D, 05G

ACCESSION NO. W71-01899

TEMPERATURE EFFECTS ON THE SORPTION OF RADIONUCLIDS BY FRESHWATER ALGAE,

DU PONT DE NEMOURS (E. I.) AND CO., AIKEN, S.C. SAVANNAH RIVER LAB.

R. S. HARVEY,

HEALTH PHYSICS, PERGAMON PRESS, VOL 19, AUG 1970, P 293-297. 4 FIG, 3 TAB, 7 REF.

DESCRIPTORS:
TEMPERATURE, SORPTION, RADIOISOTOPES, ALGAE, GROWTH RATES, THERMAL POLLUTION.

IDENTIFIERS:
SAVANNAH RIVER PLANT, CULTURE MEDIA, DRY WEIGHT, THERMAL EFFECTS.

ABSTRACT:
THE SPECIES STUDIED WERE COLLECTED FROM THE REACTOR EFFLUENT STREAMS AT SAVANNAH RIVER PLANT. UNIALGAL CULTURES WERE DEVELOPED IN INORGANIC MEDIA. ALL TESTS WERE CONDUCTED USING THE CONTINUOUS FLOW CULTURE SYSTEM DESCRIBED BY WATTS AND HARVEY. WATER TEMPERATURES OF 23, 26, 29 AND 32C HAD NO SIGNIFICANT EFFECT ON THE SORPTION OF 137CS, 85SR, 65ZN, 59FE, 57CO AND 54MN BY THE FILAMENTOUS GREEN ALGA STIGEOCLONIUM LUBRICUM. RADIONUCLIDE CONCENTRATIONS IN THE UNICELLULAR DIATOM NAVICULA SEMINULUM WERE 2-5 TIMES HIGHER AT 32C THAN THOSE OBTAINED AT LOWER TEMPERATURES. WATER TEMPERATURES OF 25, 30, 35, 40C HAD NO SIGNIFICANT EFFECT ON THE SORPTION OF 137CS, 85SR, 65ZN AND 59FE BY THE FILAMENTOUS BLUE-GREEN ALGA PLECTONEMA BORYANUM. HOWEVER, 57CO CONCENTRATIONS IN P. BORYANUM DECREASED WITH TEMPERATURE, AND 54MN CONCENTRATIONS INCREASED FROM 25 TO 35C. GROWTH RATES OF N. SEMINULUM AND P. BORYANUM WERE INHIBITED AT 32 AND 25C, RESPECTIVELY. GROWTH OF S. LUBRICUM WAS NOT INFLUENCED BY THE TEMPERATURES TESTED. THESE DATA SHOW THAT NONLETHAL CHANGES IN WATER TEMPERATURE HAD NO MAJOR INFLUENCE ON THE SORPTION OF ESSENTIALL ELEMENTS BY THE ALGAE STUDIED. (UPADHYAYA-VANDERBILT)

FIELD 05C

ACCESSION NO. W71-02075

PHOSPHORUS, NITROGEN, AND ALGAE IN LAKE WASHINGTON AFTER DIVERSION OF SEWAGE,

WASHINGTON UNIV., SEATTLE. DEPT. OF ZOOLOGY.

W. T. EDMONDSON,

SCIENCE, VOL 169, NO 3946, P 690-691, AUGUST 14, 1970. 2 FIG, 8 REF.

DESCRIPTORS:
SEWAGE EFFLUENTS, *ALGAE, *PHYTOPLANKTON, *AQUATIC PRODUCTIVITY, *PHOSPHATES, *NITRATES, *CHLOROPHYLL, DIVERSION, CARBON DIOXIDE, ALKALINITY, EPILIMNION, WASHINGTON, PHOSPHORUS, NITROGEN, NUTRIENTS.

IDENTIFIERS:
LAKE WASHINGTON(WASHINGTON).

ABSTRACT:
BECAUSE OF THE LARGE AMOUNTS OF PHOSPHORUS AND NITROGEN BEING ADDED TO LAKE WASHINGTON FROM SEWAGE EFFLUENT, A PROGRAM WAS SET UP TO DIVERT ALL SEWAGE FROM THE LAKE. THE FIRST DIVERSION OF 11 TREATMENT PLANTS OCCURRED IN 1963. FROM 1963 TO 1969, PHOSPHATE DECREASED TO 28% OF THE 1963 CONCENTRATION, BUT NITRATE REMAINED AT MORE THAN 80% OF THE 1963 VALUE. FREE CARBON DIOXIDE AND ALKALINITY REMAINED RELATIVELY HIGH. THE AMOUNT OF PHYTOPLANKTONIC CHLOROPHYLL IN THE SUMMER WAS VERY CLOSELY RELATED TO THE MEAN WINTER CONCENTRATION OF PHOSPHATE, BUT NOT TO THAT OF NITRATE OR CARBON DIOXIDE. PHYTOPLANKTON COUNTS HAVE NOT BEEN COMPLETED, BUT DATA ARE AVAILABLE ON THE CHLOROPHYLL CONTENT OF THE PHYTOPLANKTON IN THE EPILIMNION. THE RELATIONSHIP BETWEEN THE SUMMER MEAN AND THE CONCENTRATION OF PHOSPHATE IN THE SURFACE WATER DURING THE PREVIOUS WINTER STRONGLY SUGGESTS THAT PHOSPHORUS IS THE MOST IMPORTANT LIMITING ELEMENT IN LAKE WASHINGTON. TWO INCLUDED GRAPHS SHOW THE MEAN WINTER VALUES OF PHOSPHATE, NITRATE, CHLOROPHYLL, AND SURFACE PHYTOPLANKTON; AND CORRELATION BETWEEN SURFACE VALUES OF PHOSPHATE AND NITRATE DURING THE SPRING INCREASE OF PHYTOPLANKTON WHEN THE CONCENTRATIONS OF NUTRIENTS ARE DECREASING. (LITTLE-BATTELLE)

FIELD 05C, 05G, 02H

ACCESSION NO. W71-02681

CHEMICAL, PHYSICAL AND BIOLOGICAL DYNAMICS OF NORTHERN PRAIRIE LAKES,

SOUTH DAKOTA STATE UNIV., BROOKINGS.

JOHN G. NICKUM.

AVAILABLE FROM NTIS AS PB-196 357, $3.00 IN PAPER COPY, $0.95 IN MICROFICHE. COMPLETION REPORT, WATER RESOURCES INSTITUTE, SOUTH DAKOTA STATE UNIVERSITY, BROOKINGS, DECEMBER 1970. 38 P, 3 TAB, 10 FIG, 23 REF. OWRR PROJECT B-002-S DAK(4).

DESCRIPTORS:
*WATER QUALITY, *LAKES, *EUTROPHICATION, PHYTOPLANKTON, PLANKTON, PRODUCTIVITY, PRIMARY PRODUCTIVITY, ICED LAKES, IONS, SULFATES, CARBONATES, BICARBONATES, DIATONICS, CYANOPHYTA, ALGAE, SOUTH DAKOTA, WATER POLLUTION EFFECTS, WATER POLLUTION SOURCES, CULTIVATION.

IDENTIFIERS:
LAKE HERMAN(S. DAK), ENEMY SWIM LAKE(S. DAK).

ABSTRACT:
WATER SAMPLES AND SAMPLES OF PLANKTON WERE COLLECTED FROM LAKE HERMAN AND ENEMY SWIM LAKE, BOTH LOCATED IN EASTERN SOUTH DAKOTA, FROM JULY, 1967 THROUGH JUNE, 1970. ALL SAMPLES WERE ANALYZED FOR SELECTED CHEMICAL CHARACTERISTICS AND FOR PHYTOPLANKTON POPULATION. RESULTS INDICATED THAT BOTH LAKES ARE EUTROPHIC BUT LAKE HERMAN IS CONSIDERABLY MORE PRODUCTIVE THAN ENEMY SWIM. CONCENTRATIONS OF MOST IONS WERE FROM 50 PER CENT TO 100 PER CENT GREATER IN LAKE HERMAN THAN IN ENEMY SWIM AND PHYTOPLANKTON POPULATIONS WERE OFTEN 100 TIMES MORE DENSE IN LAKE HERMAN. RELATIVE ABUNDANCE OF MAJOR IONS WERE SIMILAR IN BOTH LAKES EXCEPT THAT SULFATE WAS THE DOMINANT ANION IN LAKE HERMAN AND BICARBONATE - CARBONATE IN ENEMY SWIM. BOTH LAKES SHOWED UNIFORM CHEMICAL CONDITIONS FROM ONE SAMPLING STATION TO ANOTHER AND SIMILAR ANNUAL DYNAMICS MARKED BY A SHARP INCREASE TO PEAK LEVELS OF ALL MAJOR IONS UNDER ICE COVER. PHYTOPLANKTON POPULATIONS WERE DOMINATED BY BLUE-GREEN ALGAE IN LAKE HERMAN, AND BY DIATOMS IN ENEMY SWIM. NUMBERS AND KINDS OF ALGAE VARIED FROM SEASON-TO-SEASON AND FROM YEAR-TO-YEAR INDICATING THAT INTENSIVE LONG TERM SAMPLING IS NECESSARY IN ANY STUDY OF PLANKTON DYNAMICS IN THESE LAKES. THE GREATER EUTROPHICATION OF LAKE HERMAN APPEARS RELATED TO THE INTENSIVE CULTIVATION OF ITS WATERSHED.

FIELD 05C, 02H, 05B

ACCESSION NO. W71-02880

LAKE EUTROPHICATION--WATER POLLUTION CAUSES, EFFECTS AND CONTROL.

MINNESOTA UNIV., MINNEAPOLIS. WATER RESOURCES RESEARCH CENTER.

AVAILABLE FROM NTIS AS PB-196 479, $3.00 IN PAPER COPY, $0.95 IN MICROFICHE. 'SAVE THE LAKES SYMPOSIUM' HELD AT DETROIT LAKES, MINNESOTA, AUGUST 18-19, 1969. WATER RESOURCES RESEARCH CENTER, UNIVERSITY OF MINNESOTA, MINNEAPOLIS, BULLETIN NO 22, 1970. 61 P. OWRR PROJECT A-999-MINN(13).

DESCRIPTORS:
*EUTROPHICATION, *LAKES, *WATER POLLUTION, MINNESOTA, WATER QUALITY, NUTRIENTS, SEDIMENTS, COLIFORMS, MUNICIPAL WASTES, AGRICULTURE, FALLOUT, DRAINAGE, MANAGEMENT, WISCONSIN, PRODUCTIVITY, ALGAE, ECONOMICS, POLITICAL ASPECTS, GEOLOGY, TOPOGRAPHY, GROUNDWATER, BIOCHEMICAL OXYGEN DEMAND, PHOSPHORUS, FISH, WEEDS, CYCLES, BACTERIA, CHEMISTRY, PHYSICS, STRATIFICATION, BIOLOGICAL TREATMENT, VIRUSES, CYANOPHYTA, LEGISLATION, CHEMCONTROL.

IDENTIFIERS:
DETROIT LAKES(MINN), COMMUNITY ACTION, LAKE SALLIE(MINN), LAKE MELISSA(MINN), NUTRIENT REMOVAL.

ABSTRACT:
PAPERS PRESENTED AT 'SAVE THE LAKES SYMPOSIUM' HELD AT DETROIT LAKES, MINNESOTA, AUGUST 18-19, 1969, FOCUSED ON LAKE POLLUTION AND EUTROPHICATION: WHAT IT IS, WHAT CAUSES IT, THE TECHNOLOGY NEEDED AND AVAILABLE TO CONTROL IT, AND THE TYPE OF COMMUNITY ACTION THAT CAN AND MUST BE TAKEN FOR EFFECTIVE CONTROL AND IMPROVEMENT. THE PURPOSE IS TO MAKE FACTS AVAILABLE TO THE PUBLIC AND TO CLARIFY THE COOPERATIVE ROLE OF CITIZENS AND GOVERNMENT. THE LIFE CYCLE OF A LAKE IS DESCRIBED. IMPROVED METHODS OF LAKE MANAGEMENT MAY BE UNCOVERED IN FUNDAMENTAL STUDIES DEALING WITH PRESENT SOURCES OF POLLUTION, MUNICIPAL SEWAGE AND DIFFUSE SOURCES, SUCH AS AGRICULTURAL OPERATIONS, FALLOUT AND WASHOUT FROM AIR, DRAINAGE, AND THE TECHNICAL PROBLEMS INVOLVED. THE HISTORY AND DIVERSION OF WASTES AT MADISON, WISCONSIN, LAKE TAHOE AND LAKE WASHINGTON IS DESCRIBED. BESIDES THE SCIENTIFIC AREAS, THE MAJOR PROBLEMS ARE LACK OF POLITICAL LEADERSHIP NECESSARY FOR PROPER EVALUATION OF WATER QUALITY AND PROVIDING FINANCING NECESSARY FOR SOLUTION. THE GEOLOGICAL HISTORY OF THE DETROIT LAKES IS GIVEN. THERE ARE MANY MEANS OF ALLEVIATING EUTROPHICATION: ECOLOGICAL CONTROL, BIOLOGICAL CONTROL, CHEMICAL CONTROL, AND MECHANICAL CONTROL. (JONES-WISCONSIN)

FIELD 05G

ACCESSION NO. W71-03012

THE EFFECTS OF COPPER SULFATE ON CERTAIN ALGAE AND ZOOPLANKTERS IN WINNISQUAM LAKE, NEW HAMPSHIRE,

NEW HAMPSHIRE UNIV., DURHAM. WATER RESOURCES RESEARCH CENTER.

PHILIP J. SAWYER.

AVAILABLE FROM NTIS AS PB-196 481, $3.00 IN PAPER COPY, $0.95 IN MICROFICHE. NEW HAMPSHIRE WATER RESOURCES RESEARCH CENTER, WRR-2, 1970. 28 P, 2 FIG, 5 TAB, 17 REF. OWRR PROJECT A-004-NH(4).

DESCRIPTORS:
*COPPER SULFATE, *ALGAE, *ZOOPLANKTERS, *CHEMCONTROL, NEW HAMPSHIRE, COPEPODS, DAPHNIA, SAMPLING, DEPTH, CYANOPHYTA, CHRYSOPHYTA, CHLOROPHYTA, PRIMARY PRODUCTIVITY, CARBON RADIOISOTOPES, PHOTOSYNTHESIS.

IDENTIFIERS:
*WINNISQUAM LAKE(N H), ANABAENA CIRCINALIS, BOSMINA LONGIROSTRIS, SYNURA UVELLA, DINOBRYON, OSCILLATORIA LIMNETICA, DICTYOSPHAERIUM, CELL COUNTS, FRAGILARIA, ASTERIONELLA, CRYPTOMONAS, SPHAEROCYSTIS, DAPHNIA GALEATA MENDOTAE, DAPHNIA MAGNA.

ABSTRACT:
IN A RELATIVELY UNBUFFERED LAKE SUCH AS WINNISQUAM, CONTROL OF CYANOPHYCEAN BLOOM CAN BE ACCOMPLISHED WITH RELATIVELY SMALL AMOUNTS OF COPPER SULFATE --4 LBS/ACRE. DATA ON 27 ALGAL SPECIES PRESENT BEFORE, DURING, AND AFTER ADDITION OF COPPER SULFATE INDICATE THAT A RAPID SHIFT IN SPECIES RELATIONSHIPS OCCURRED. CARBON FIXATION RATE WAS FOUND TO BE A MORE SENSITIVE MEASURE OF BLOOM POTENTIAL THAN CELL COUNTS. SUSCEPTIBLE SPECIES DECREASED IN NUMBERS; OTHER FORMS INCREASED IN POPULATION; SOME APPEARED UNAFFECTED BY TREATMENT. THE LATTER GROUP MAY HAVE BEEN PROTECTED BY NATURAL RESISTANCE OR BY THEIR POSITION IN THE WATER COLUMN. ZOOPLANKTON REACTED TO COPPER TREATMENT IN VARIOUS WAYS. THE EFFECTS ON MICROCRUSTACEANS SUCH AS COPEPODS AND DAPHNIDS SEEMED TO FOLLOW A PREDICTABLE PATTERN. COPEPODS AND BOSMINA LONGIROSTRIS WERE MORE RESISTANT THAN DAPHNIA. BOSMINA MULTIPLIED TO A CONCENTRATION OF 154/LITER IN THE TOP METER OF THE WATER COLUMN. THE RISE AND FALL OF SPECIE NUMBERS OF PLANTS AND ANIMALS AND THEIR RAPID RESPONSE TO CONDITIONS AFTER COPPER SULFATE TREATMENT ILLUSTRATE THEIR COMPLEX RELATIONSHIPS. THE CRUDE TREATMENT WITH COPPER SULFATE DID PRODUCE CONTROL OF AN OBNOXIOUS BLUE-GREEN ALGA, ANABAENA CIRCINALIS. (JONES-WISCONSIN)

FIELD 05F, 02H

ACCESSION NO. W71-03014

BIOGEOCHEMICAL MODELING OF EUTROPHIC LAKES FOR WATER QUALITY IMPROVEMENT,

NOTRE DAME UNIV., IND. DEPT. OF CIVIL ENGINEERING.

M. W. TENNEY, W. F. ECHELBERGER, JR., P. C. SINGER, F. H. VERHOFF, AND W. A. GARVEY.

FIFTH INTERNATIONAL CONFERENCE ON WATER POLLUTION RESEARCH, SAN FRANCISCO, CALIFORNIA, JULY 26-AUGUST 1, 1970. PREPRINT. 22 P, 4 FIG, 2 TAB, 17 REF.

DESCRIPTORS:
*SYSTEMS ANALYSIS, *WATER QUALITY, *EUTROPHICATION, *MATHEMATICAL MODELS, SIMULATION, ANALYTICAL TECHNIQUES, PHOSPHORUS, ALGAE, LAKES, MICROORGANISMS, TERTIARY TREATMENT, EFFLUENTS, MODEL STUDIES.

IDENTIFIERS:
HETEROTROPHS, AUTOTROPHS.

ABSTRACT:
A MATHEMATICAL SIMULATION OF BIOGEOCHEMICAL EXCHANGES IN A LAKE SYSTEM AND ITS USE IN SYNTHESISING THE EUTROPHICATION HISTORY ARE DESCRIBED. TAKEN AS A SINGLE-STAGE, CONTINUOUS-FLOWING REACTOR SYSTEM WITH VARIABLE INPUTS AND OUTPUTS, GENERAL TRENDS ARE PREDICTABLE. THE CONCENTRATION OF ORGANIC CARBON AND PHOSPHORUS DETERMINE THE EXTENT OF HETEROTROPHIC AND AUTOTROPHIC ACTIVITY IN THE MODEL, AND THE RESULTANT ACTIVITY IS MEASURED AS SUSPENDED SOLIDS. INTERNAL EXCHANGES BETWEEN STATE AND LOCATION OF POLLUTANTS IS CHARACTERIZED BY RATE FUNCTIONS, AND WHEN NECESSARY, A STOCHIOMETRY OF EACH OF THE BIO- AND PHYSICAL REACTIONS IS INCLUDED. A DETERMINISTIC SIMULATION MODEL WAS APPLIED TO STONE LAKE, CASSOPOLIS, MICHIGAN, AND REASONABLE PREDICTIONS WERE MADE WITH REGARD TO ORGANIC MATERIAL, INORGANIC PHOSPHORUS, AND SUSPENDED MICROORGANISMS. THE MODEL IS APPLICABLE TO PREDICTION OF FUTURE WATER QUALITY AND CERTAIN MANIPULATION PROCEDURES TO ENHANCE WATER QUALITY. ECOLOGICAL MODELING DEVELOPMENTS WILL ASSIST POLITICAL DECISIONS LEADING TO DEVELOPMENT OF SUCCESSFUL CONCEPTS IN WATER QUALITY MANAGEMENT. (AUEN-WISCONSIN)

FIELD 05G, 05C, 02H

ACCESSION NO. W71-03021

COPPER IONS AS POISON IN THE SEA AND IN FRESHWATER,

COPENHAGEN UNIV., HILLEROD (DENMARK). FRESHWATER BIOLOGICAL LAB.

E. STEEMANN NIELSEN, AND S. WIUM-ANDERSEN.

MARINE BIOLOGY, VOL 6, NO 2, P 93-97, 1970. 4 FIG, 20 REF.

DESCRIPTORS:
*COPPER, *IONS, *POISONS, *SEA WATER, *FRESH WATER, ALGAE, PHOTOSYNTHESIS, ORGANIC MATTER, PLANKTON, CHELATION, CARBON RADIOISOTOPES, CHLORELLA, GROWTH RATES, DIATOMS, PHYTOPLANKTON, TOXICITY, IRON, LIGHT, HYDROGEN ION CONCENTRATION.

IDENTIFIERS:
CHLORELLA PYRENOIDOSA, NITZSCHIA PALEA, SKELETONEMA COSTATUM.

ABSTRACT:
DURING THE LAST TWO YEARS INTENSIVE EXPERIMENTAL WORK HAS BEEN APPLIED IN INVESTIGATIONS OF THE INFLUENCE OF IONIC COPPER ON PHOTOSYNTHESIS AND GROWTH OF GREEN ALGA, CHLORELLA PYRENOIDOSA AND THE DIATOM, NITZSCHIA PALEA. COPPER IN IONIC FORM IS FOUND TO BE VERY TOXIC TO PHOTOSYNTHESIS AND GROWTH OF UNICELLULAR ALGAE AT CONCENTRATIONS USUALLY FOUND IN NATURAL WATERS, INDICATING THAT COPPER IS NOT ORDINARILY PRESENT IN IONIC FORM BUT IS COMPLEXED BY ORGANIC MATTER SUCH AS POLYPEPTIDES. THE AFFINITY OF COPPER TO DIETHYL-DITHIOCARBAMINATE IS VERY MUCH HIGHER THAN TO THE ORGANIC MATTER WHICH COMPLEXES COPPER IN NATURE, THUS IT IS NOT POSSIBLE TO DISTINGUISH THE TWO FORMS OF COPPER DURING ANALYSIS. COMPLEXED COPPER IS NOT POISONOUS TO ALGAE. IT HAS RECENTLY BEEN SHOWN THAT OCEAN WATER IN THE CENTERS OF UPWELLING BECOMES SUITABLE FOR PLANKTON GROWTH ONLY AFTER THE ADDITION OF A CHELATOR, SUGGESTING THAT A LARGE PART OF THE COPPER FOUND IN SUBSURFACE WATERS OF OCEANS IS PRESENT IN IONIC FORM. SOME MANUFACTURERS OF CARBON-14 AMPOULES HAVE USED ORDINARY DISTILLED WATER WHICH OFTEN HAS A CONTENT OF ABOUT 250 MICROGRAMS COPPER PER LITER; THUS, IT IS PROBABLE THAT SOME PRODUCTIVITY MEASUREMENTS HAVE BEEN INFLUENCED. (JONES-WISCONSIN)

FIELD 05B

ACCESSION NO. W71-03027

NUTRIENTS AND QUALITY IN IMPOUNDED WATER,

ILLINOIS STATE WATER SURVEY, PEORIA. WATER QUALITY SECTION.

WUM-CHENG WANG, AND RALPH L. EVANS.

JOURNAL AMERICAN WATER WORKS ASSOCIATION, VOL 62, NO 8, P 510-514, 1970. 8 FIG, 3 TAB, 15 REF.

DESCRIPTORS:
*NUTRIENTS, *WATER QUALITY, *IMPOUNDED WATERS, STREAMS, FLOW, SEASONAL, STRATIFICATION, NITRATES, SILICA, AMMONIA, PHOSPHATES, RUNOFF, ILLINOIS, HYDROGEN ION CONCENTRATION, ALKALINITY, HARDNESS(WATER), SAMPLING, DIATOMS, NITROGEN, BACTERIA, ALGAE, THERMAL STRATIFICATION, DISSOLVED OXYGEN, HYPOLIMNION.

IDENTIFIERS:
LAKE BLOOMINGTON(ILL), CHEMOSTRATIFICATION.

ABSTRACT:
AN EXTENSIVE STUDY WAS MADE OF IMPOUNDED LAKE BLOOMINGTON (ILLINOIS), AS SIGNIFICANT CHANGES OCCUR IN STREAM WATER QUALITY WHEN THEY ARE IMPOUNDED. WITH FEW EXCEPTIONS CONCENTRATION OF EACH CONSTITUENT DECREASES WITH PASSAGE THROUGH THE IMPOUNDMENT. MAJOR INFLUENCES WITHIN THE SYSTEM OBSERVED WERE STREAM FLOW, SEASONAL PATTERNS, AND STRATIFICATION. NITRATE NITROGEN AND SILICA CONCENTRATION INCREASED WITH INCREASING STREAM FLOW. AMMONIA-NITROGEN AND ORTHOPHOSPHATE CONCENTRATION WERE MINIMAL. SINCE THESE CONSTITUENTS WERE PRINCIPALLY DERIVED FROM AGRICULTURAL RUNOFF THE RELATIVE MOBILITY OF EACH IN SOIL, APPEARS CLEAR. NITRATE NITROGEN AND PHOSPHORUS CONTRIBUTIONS TO THE WATER SYSTEM REPRESENTED 8.3% AND 0.1%, RESPECTIVELY, OF THAT APPLIED TO THE DRAINAGE AREA. DURING WARM WEATHER, ORGANIC NITROGEN CONCENTRATIONS WERE HIGHEST WHILE NITRATE-NITROGEN WAS MINIMAL, SUGGESTING INFLUENCE OF BIOLOGICAL ACTIVITY. DIFFERENCE BETWEEN RATE OF NITRATE-NITROGEN AND SILICA CONCENTRATION REDUCTION, THOUGH OF THE SAME ORIGIN, ALSO SUGGESTED INFLUENCE OF BIOLOGICAL UPTAKE. WITHIN THE IMPOUNDMENT, STRATIFICATION APPEARED THE GOVERNING INFLUENCE, AS EVIDENCED BY CHANGES IN NITRATE NITROGEN AND SILICA CONCENTRATION WITH DEPTH. FURTHER INSIGHT INTO THE INTERRELATIONSHIPS WOULD APPEAR TO PERMIT INTELLIGENT FLEXIBILITY ON THE PART OF WATER WORKS MANAGEMENT. (JONES-WISCONSIN)

FIELD 05C, 02H

ACCESSION NO. W71-03028

SIMULATING THE EFFECT OF SINKING AND VERTICAL MIXING ON ALGAL POPULATION DYNAMICS,

OREGON STATE UNIV., CORVALLIS. DEPT. OF CIVIL ENGINEERING.

DAVID A. BELLA.

JOURNAL WATER POLLUTION CONTROL FEDERATION, VOL 42, NO 5, PART 2, P R140-R152, 1970. 1 FIG, 1 TAB, 17 REF.

DESCRIPTORS:
*MATHEMATICAL MODELS, *ALGAE, *POPULATION, DYNAMICS, LAKES, DISSOLVED OXYGEN, NUTRIENTS, CYANOPHYTA, MIXING, PHYTOPLANKTON.

IDENTIFIERS:
*ALGAL SINKING, *VERTICAL MIXING, EUPHOTIC ZONE, DIFFUSION COEFFICIENT, LAKE SAMMAMISH(WASH).

ABSTRACT:
THE DEVELOPED SIMULATION MODEL DESCRIBES THE COMBINED EFFECT OF GROWTH, SINKING, AND VERTICAL MIXING ON DYNAMICS OF ALGAL POPULATION OF LAKES. SMALL DIFFERENCES IN SINKING VELOCITIES MAY SIGNIFICANTLY INCREASE THE GROWTH OF ALGAE. A LOWERING OF SINKING OFTEN LEADS TO AN INVASION OF A LESS OBJECTIONABLE PHYTOPLANKTON. VERTICAL MIXING INCREASES THE DENSITY OF ALGAE AND MAY RESULT IN TRANSFORMATION OF THE ENTIRE COMMUNITY. ARTIFICAL MIXING MAY SERVE AS AN AID IN CONTROL OF ALGAE, BUT THE HOMOGENIZATION OF ENVIRONMENT IS APT TO REDUCE THE SPECIES DIVERSITY AT ALL TROPHIC LEVELS. (WILDE-WISCONSIN)

FIELD 05C, 07B

ACCESSION NO. W71-03034

POLLUTION OF THE NORTH SEA AND REARING EXPERIMENTS ON MARINE PHYTOFLAGELLATES AS AN INDICATION OF RESULTANT TOXICITY,

H. KAYSER.

FIFTH INTERNATIONAL WATER POLLUTION RESEARCH CONFERENCE, SAN FRANCISCO, JULY 26-AUGUST 1, 1970. PREPRINT, 7 P, 5 FIG, 5 REF.

DESCRIPTORS:
*WATER POLLUTION EFFECTS, WATER POLLUTION SOURCES, INDUSTRIAL WASTES, DOMESTIC WASTES, DINOFLAGELLATES, *LABORATORY TESTS, CULTURES, *GROWTH RATES, PLANT GROWTH REGULATORS, PLANT GROWTH SUBSTANCES, ALGAE, IRON, BIOASSAY, BIOINDICATORS, FOOD CHAINS, TITANIUM, PLANT POPULATIONS, TOXICITY.

IDENTIFIERS:
*NORTH SEA, PROROCENTRUM MICANS, CERATIUM FURCA, TURBIDOSTAT CULTURES.

ABSTRACT:
THE EXTREME POLLUTION OF THE NORTH SEA WAS DESCRIBED. SINCE THIS TREND IS LIKELY TO CONTINUE, THE AUTHOR DEVELOPED METHODS FOR DETERMINATION OF THE POSSIBLE BIOLOGICAL CONSEQUENCES. THIS METHOD INVOLVES THE CULTURE OF PRIMARY PRODUCERS, NAMELY DINOFLAGELLATES, IN WASTE WATERS. MASS CULTURES PROVIDED INFORMATION ABOUT THE GROWTH RATE AND MAXIMUM DENSITY OF POPULATIONS UNDER THE INFLUENCE OF WASTE WATER. THIS METHOD IS ABLE TO INDICATE THE SUBLETHAL INFLUENCES ON CHARACTERISTIC MEMBERS OF THE MARINE FOOD CHAIN. IT IS HOPED THAT FURTHER ADVANCEMENT OF THE TEST METHOD WILL ALLOW CALCULATION OF THE SUBLETHAL LEVEL OF WASTE WATER DISCHARGE INTO THE NORTH SEA. (SJOLSETH-WASHINGTON)

FIELD 05C

ACCESSION NO. W71-03095

DESIGN CONSTRUCTION AND MAINTENANCE OF WASTE STABILIZATION LAGOONS,

FARMERS HOME ADMINISTRATION, CHAMPAIGN, ILL.

DAVID H. STOLTENBERG.

PUBLIC WORKS, VOL 101, NO 9, P 103-106, SEPTEMBER 1970. 4 FIG, 3 TAB, 17 REF.

DESCRIPTORS:
*OXIDATION LAGOONS, *STABILIZATION, *ALGAE, AEROBIC TREATMENT, BACTERIA, MUNICIPAL WASTES, BIOCHEMICAL OXYGEN DEMAND, WASTE WATER TREATMENT.

IDENTIFIERS:
DEPTH, SLOPE, *BOD LOADING, *WASTE STABILIZATION.

ABSTRACT:
THERE ARE SEVERAL IMPORTANT FACTORS INVOLVED IN WASTE STABILIZATION USING LAGOONS. THE FACT THAT OXYGEN IS PRODUCED BY ALGAE DURING THE DAY AND CONSUMED BY ALGAE AT NIGHT IS PRESENTED AS A CRITICAL FACTOR IN LAGOON DESIGN AND MAINTENANCE. THE VARIANCE OF ALLOWABLE BIOCHEMICAL OXYGEN DEMAND LOADING WITH GEOGRAPHICAL LOCATION IS ALSO DISCUSSED. CONSTRUCTION COSTS ARE ALSO PRESENTED. ESTIMATES RANGE FROM $11 TO $18 PER CAPITA FOR INSTALLATIONS SERVING 1000 PEOPLE, AND FROM $30 TO $36 PER CAPITA FOR INSTALLATIONS SERVING 100 PEOPLE. OPERATION AND MAINTENANCE COSTS ARE ESTIMATED TO BE ONE-FOURTH, OR ONE-FIFTH OF THE COST FOR CONVENTIONAL WASTE TREATMENT. (LOWRY-TEXAS)

FIELD 05D

ACCESSION NO. W71-03896

RESPIRATORY RELATIONSHIPS OF A SYMBIOTIC ALGAL-BACTERIAL CULTURE FOR WASTE WATER NUTRIENT REMOVAL,

OHIO STATE UNIV., COLUMBUS. DEPT. OF CIVIL ENGINEERING.

F. J. HUMENIK, AND G. P. HANNA, JR.

BIOTECHNOLOGY AND BIOENGINEERING, VOL 12, P 541-560, 1970. 8 FIG, 3 TAB, 17 REF.

DESCRIPTORS:
*SYMBIOSIS, *ALGAE, *BACTERIA, *RESPIRATION, ACTIVATED SLUDGE, WASTE WATER TREATMENT, NUTIRIENTS, AEROBIC CONDITIONS, AERATION, MICROBIOLOGY, BIOMASS, CHLORELLA.

IDENTIFIERS:
*NUTRIENT REMOVAL, OXYGEN DECLINE, RESPIRATORY RATES, PHOTOSYNTHETIC OXYGENATION, ACTIVATED ALGAE.

ABSTRACT:
RESEARCH WAS CONDUCTED TO INVESTIGATE THE RESPIRATORY RELATIONSHIPS OF A MIXED ALGAL-BACTERIAL CULTURE AND THE EFFECT OF ITS NATURAL ECOLOGICAL ASSOCIATIONS ON THE STABILIZATION OF A DOMESTIC WASTE WATER. IN ORDER TO DO THIS THE AUTHORS DEVELOPED A CONTINUOUS SYMBIOTIC ALGAL-BACTERIAL SYSTEM OF A MIXED CHLORELLA-ACTIVATED SLUDGE CULTURE WHICH WOULD EFFICIENTLY REMOVE NUTRIENTS FROM WASTE WATER UNDER AEROBIC CONDITIONS AND WITHOUT ADDITIONAL AERATION. DAILY HARVESTING OF EXCESS BIOMASS WAS CONDUCTED TO PROVIDE STABLE RELATIVE BIOLOGICAL POPULATIONS AND A DISSOLVED OXYGEN CONCENTRATION OF APPROXIMATELY 2 MILLIGRAMS PER LITER WAS MAINTAINED THROUGHOUT STEADY STATE OPERATIONS. TO PREDICT RESPIRATORY RATES, PHOTOSYNTHETIC OXYGENATION AND STEADY-STATE OXYGEN CONCENTRATIONS OXYGEN DECLINE DATA WERE FITTED TO MATHEMATICAL MODELS. THE RESPIRATORY AND PHYSIOLOGICAL RELATIONSHIPS INDICATED THAT THE CARBON DIOXIDE-OXYGEN BALANCE MAY BE A PRIMARY CONTROL GOVERNING THE STEADY-STATE OPERATION OF A SYMBIOTIC ALGAL-BACTERIAL CULTURE. MICROSCOPIC EXAMINATIONS SHOWED THAT BACTERIA WERE ATTACHED TO ALGAL CELLS AND THE CHLORELLA WERE INTEGRALLY ENMESHED WITHIN THE BACTERIAL MIX WHICH ALLOWED THIS ALGAL-BACTERIAL FLOC TO SETTLE QUITE RAPIDLY AND YIELD A CLEAR SUPERNATANT. (ELLIS-TEXAS)

FIELD 05D

ACCESSION NO. W71-04079

ACCUMULATION OF PHOSPHATES IN WATER,

AGRICULTURAL RESEARCH SERVICE, MORRIS, MINN. NORTH CENTRAL SOIL CONSERVATION RESEARCH CENTER.

ROBERT F. HOLT, DONALD R. TIMMONS, AND JOSEPH J. LATTERELL.

JOURNAL OF AGRICULTURE AND FOOD CHEMISTRY, VOL 18, NO 5, P 781-784, 1970. 1 TAB, 37 REF.

DESCRIPTORS:
*PHOSPHATES, *PATH OF POLLUTANTS, *EUTROPHICATION, *LEACHING NUTRIENTS, FERTILIZERS, PRECIPITATION(ATMOSPHERIC), RUNOFF, ANIMAL WASTES, SURFACE WATERS, SOIL EROSION, NUTRIENTS, ALGAE, LAKE SUPERIOR, WATER POLLUTION SOURCES, MINNESOTA, WASHINGTON, CONNECTICUT.

IDENTIFIERS:
ILLINOIS RIVER, ST. LOUIS RIVER, BLACK RIVER(MINNESOTA), LAKE WASHINGTON(WASHINGTON), LINSLEY POND(CONNECTICUT), LAKE ZOAR(CONNECTICUT), LAKE MINNETONKA(MINNESOTA), BIG STONE LAKE(MINNESOTA), LAKE CRYSTAL(MINNESOTA).

ABSTRACT:
NATURAL AND AGRICULTURAL SOURCES OF PHOSPHORUS TO SURFACE WATERS INCLUDE PERCIPITATION, ANIMAL WASTES, FERTILIZERS, AND LAND RUNOFF. THE ACTUAL CONTRIBUTION FROM THESE SOURCES IS SHOWN TO BE QUITE LOW. HOWEVER, THE CONCENTRATION OF PHOSPHORUS REQUIRED TO SUPPORT PROFUSE ALGAL BLOOMS IS SO LOW THAT THE LIMITED AMOUNTS SUPPLIED ARE SUFFICIENT TO EXCEED THIS REQUIREMENT. ERODED SOIL DELIVERS APPRECIABLE AMOUNTS OF PHOSPHORUS TO SURFACE WATERS, BUT THE SOIL MATERIALS CAPACITY TO SORB PHOSPHORUS RESULTS IN LITTLE TENDENCY FOR RELEASE OF THIS SOURCE INTO THE WATER. BOTTOM SEDIMENTS APPEAR TO BE A SINK FOR DISSOLVED ORTHOPHOSPHATE THAT IS SUPPLIED TO SURFACE WATERS. LEACHING OF VEGETATION CAN SUPPLY RELATIVELY LARGE AMOUNTS OF PHOSPHORUS TO LAKES AND STREAMS. DEEP INCORPORATION OF PHOSPHATIC FERTILIZERS MATERIALLY REDUCES THE CONCENTRATION OF PHOSPHORUS IN RUNOFF WATERS AS COMPARED TO SHALLOW INCORPORATION. PHOSPHORUS CONCENTRATIONS IN SEVERAL LAKES AND STREAMS ARE SHOWN. (MCCANN-BATTELLE)

FIELD 05B

ACCESSION NO. W71-04216

LABORATORY STREAM STUDIES OF A BENTHIC COMMUNITY,

MISSOURI UNIV., COLUMBIA. DEPT. OF SANITARY ENGINEERING.

GALE ALLEN WRIGHT.

MS THESIS, 1970. 70 P, 32 FIG, 27 REF, APPEND. OWRR PROJECT A-011-MO(2).

DESCRIPTORS:
*SELF-PURIFICATION, *ALGAE, *STREAMS, LABORATORY TESTS, DISSOLVED OXYGEN, HYDROGEN ION CONCENTRATION, ALKALINITY, RESPIRATION, PHOTOSYNTHESIS, CARBON DIOXIDE, PRIMARY PRODUCTIVITY, BIODEGRADATION, ORGANIC MATTER, OXYGEN, CHEMICAL OXYGEN DEMAND, BENTHOS.

IDENTIFIERS:
*OXYGEN BALANCE.

ABSTRACT:
THE AIM OF THIS STUDY WAS TO DETERMINE THE ROLE OF ALGAE IN BIODEGRADATION OF ORGANIC WASTES AND SELFPURIFICATION OF STREAMS. THE MONITORING OF THE SYSTEM IN A LABORATORY STREAM MICROCOSM WAS ACCOMPLISHED BY DETERMINATION OF DISSOLVED OXYGEN, PH, AND BICARBONATE-CARBONATE ALKALINITY. THE RESULTS INDICATED A LESSER CONSTANCY OF PRODUCTION PROCESS IN COMPARISON WITH THAT OF RESPIRATION. THE CARBON:OXYGEN RATIO WAS INFLUENCED BY THE AGE OF ALGAL CELLS. THE OXIDATION OF ORGANIC SUBSTANCES OCCURRED DURING PERIODS OF DARKNESS. PHOTOSYNTHETIC PRODUCTION WAS LIMITED BY THE AVAILABLE CARBON DIOXIDE. THE CONTRIBUTION OF ALGAE TO SELFPURIFICATION OF WATER APPEARED TO BE VERY LOW. THE DETERMINATION OF PHOTOSYNTHETICALLY PRODUCED OXYGEN COULD NOT BE RELIABLY ACCOMPLISHED BEFORE THE SYNTHESIZED MATERIAL HAD BEEN OXIDIZED. (WILDE-WISCONSIN)

FIELD 05G

ACCESSION NO. W71-04518

CONTRIBUTION TO THE EPIPHYTIC ALGAL FLORA OF THE LAKE NEUSIEDLER, (IN GERMAN),

VIENNA UNIV. (AUSTRIA). BOTANISCHES INSTITUT UND BOTANISCHER GARTEN.

LOTHAR GEITLER.

OSTERREICHISCHE BOTANISCHE ZEITSCHRIFT, VOL 118, NC 1/2, P 17-29, 1970. 3 FIG, 24 REF.

DESCRIPTORS:
*EPIPHYTOLOGY, *LAKES, *ALGAE, ALKALINE WATER, MAGNESIUM, SODIUM SULFATES, CYANOPHYTA, RHODCPHYTA.

IDENTIFIERS:
*LAKE NEUSIEDLER(AUSTRIA), CHAMAESIPHON SUBAEQUALIS, APISTONEMA EXPANSUM, PORPHYRIDIUM GRISEUM, CHRYSOTRICHALAE.

ABSTRACT:
DETAILED DESCRIPTIONS ARE GIVEN OF THE FOLLOWING NEW SPECIES OF ALGAE REVEALED IN THE ALKALINE LAKE NEUSIEDLER: CHAMAESIPHON SUBAEQUALIS OF CYANOPHYCEAE, APISTONEMA EXPANSUM OF CHRYSOTRICHALAE, AND PORPHYRIDUM GRISEUM OF RHODOPHYCEA. THE PAPER INCLUDED A RECCRD OF SOME PREVIOUSLY IDENTIFIED ALGAE OF THE LAKE. (WILDE-WISCONSIN)

FIELD 02H, 05C

ACCESSION NO. W71-04526

A REVIEW OF HERBIVOROUS FISH FOR WEED CONTROL,

BUREAU OF SPORT FISHERIES AND WILDLIFE, WARM SPRINGS, GA. SOUTHEASTERN FISH CONTROL LAB.

JOE B. SILLS.

THE PROGRESSIVE FISH-CULTURIST, VOL 32, NO 3, P 158-161, 1970. 12 REF.

DESCRIPTORS:
*FISH, *HERBIVORES, *AQUATIC WEED CONTROL, TILAPIA, CARP, ALGAE, ROOTED AQUATIC PLANTS, PLANKTON, ARKANSAS, INSECTS, TURBIDITY, CHARA, PONDWEEDS.

IDENTIFIERS:
TILAPIA NILOTICA, TILAPIA MOSSAMBICA, TILAPIA MELANOPLEURA, CYPRINUS CARPIO, CTENOPHARYNGCDON IDELLUS.

ABSTRACT:
THERE IS GROWING INTEREST IN BIOLOGICAL PEST CCNTRCL AS A SUBSTITUTE FOR PESTICIDES. FISH RECEIVING MOST ATTENTION FOR AQUATIC WEED CONTROL POSSIBILITIES ARE TILAPIA NILOTICA, T MOSSAMBICA, AND T MELANOPLEURA; THE ISRAELI STRAIN OF CARP (CYPRINUS CARPIO); AND THE CHINESE OR GRASS CARP (CTENOPHARYNGODCN IDELLUS). RESULTS OBTAINED WITH EACH ARE REVIEWED. ALL TESTS IN WHICH TILAPIA WERE STOCKED IN PONDS RESULTED IN OVERCROWDED POPULATIONS AND UNSATISFACTORY WEED CONTROL. ALTHOUGH THE CARP IS EFFECTIVE IN REDUCING OR CONTROLLING AQUATIC PLANTS, DETRIMENTAL ASPECTS OF ITS PRESENCE MAKE ITS USAGE UNDESIRABLE IN RECREATION AND FISHING WATERS. TESTS SHOWED THAT THE ISRAELI STRAIN IS NOT PRIMARILY HERBIVOROUS, BUT ROOTS IN THE BOTTOM MUCH AS WILD CARP DO. OBSERVATIONS ARE CONTINUED TO LEARN WHY THE ISRAELI STRAIN DOES NOT REPRODUCE IN PONDS WITH MIXED POPULATIONS. ALL OBSERVATIONS TO DATE ON THE GRASS CARP ARE MOST FAVORABLE. IT DOES FEED ON AQUATIC VEGETATION. THERE IS NO INDICATION OF NATURAL REPRODUCTION, BUT THE FISH CAN BE SPAWNED. AS ITS NATIVE HABITAT IS SIMILAR TO SOME OF OUR LARGE RIVER SYSTEMS, THE GRASS CARP MAY ADAPT TO THEM. MORE INFORMATION IS NEEDED FOR ITS USE EXCEPT IN STRICTLY CONTROLLED ENVIRONMENTS. (JONES-WISCONSIN)

FIELD 05G

ACCESSION NO. W71-04528

STABLE CARBON ISOTOPES IN BLUE-GREEN ALGAL MATS,

TEXAS UNIV., PORT ARANSAS. INST. OF MARINE SCIENCE; AND TEXAS UNIV., PORT ARANSAS. DEPT. OF GEOLOGICAL SCIENCES.

E. W. BEHRENS, AND S. A. FRISHMAN.

JOURNAL OF GEOLOGY, VOL 79, NO 1, P 94-100, JANUARY 1971. 7 P, 3 FIG, 15 REF. NSF GRANT GA 911.

DESCRIPTORS:
*ALGAE, *STABLE ISOTOPES, *CARBON, *ANAEROBIC BACTERIA, *TEXAS, BIODEGRADATION, TRACERS, LAGOONS, CYANOPHYTA.

IDENTIFIERS:
BAFFIN BAY(TEX).

ABSTRACT:
BLUE-GREEN ALGAL MATS HAVE ACCUMULATED FOR OVER 1,000 YEARS IN KLEBERG POINT LAGOON ON THE MARGIN OF BAFFIN BAY, TEXAS. ORGANIC CARBON PRODUCED WITHIN THE MATS IS ISOTOPICALLY HEAVIER THAN THE CARBON ASSOCIATED WITH BAY SEDIMENTS. THE DIFFERENCE BETWEEN BAY AND MAT ORGANIC CARBON IS PRESERVED IN THE OLDER, BURIED SEDIMENTS, AND SEEMS TO BE INCREASED BY FRACTIONATION ASSOCIATED WITH ANAEROBIC BACTERIAL DECOMPOSITION OF ALGAL MAT ORGANIC MATTER. (KNAPP-USGS)

FIELD 02L, 02K, 05A

ACCESSION NO. W71-04873

POTENTIAL ALGICIDES FOR THE CONTROL OF ALGAE,

LOS ANGELES COUNTY SANITATION DISTRICT, CALIF.

JAMES C. GRATTEAU.

WATER AND SEWAGE WORKS, 1970 REFERENCE NUMBER, VOL 117, P R24-R61, NOVEMBER 28, 1970. 17 P, 6 TAB, 27 REF.

DESCRIPTORS:
*ALGICIDES, *WATER QUALITY CONTROL, *AQUATIC WEED CONTROL, *DOMESTIC WATER, *SURFACE WATERS, WATER SUPPLY, EUTROPHICATION, ALGAE, NUTRIENTS, WATER POLLUTION, BIOLOGICAL TREATMENT, CHEMICALS, COPPER SULFATE, CHLORINE, WATER RESOURCES, METHODOLOGY, EVALUATION, CHEMICAL ANALYSIS.

IDENTIFIERS:
ROSIN AMINE D ACETATE(RADA).

ABSTRACT:
PROBLEMS CONCERNING ALGAE IN WATER SUPPLIES AND METHODS OF CONTROLLING ALGAE GROWTH ARE DISCUSSED. IN GENERAL, THERE ARE THREE BASIC METHODS OF ALGAE CONTROL: PHYSICAL, ECOLOGICAL AND CHEMICAL. PHYSICAL METHODS OF CONTROL, RAKING, DRAGGING, PULLING, AND UNDERWATER MOWING HAVE BEEN USED MAINLY ON FLOATING MASSES OF ALGAE AND ROOTED AQUATICS. ECOLOGICAL CONTROL OF ALGAE INVOLVES LIMITATION OF ONE OR MORE OF THE FACTORS NECESSARY FOR GROWTH AND REPRODUCTION: LIGHT OR NUTRIENT MATERIAL. ONLY TWO CHEMICALS ARE WIDELY USED TO CONTROL ALGAE IN WATER SUPPLIES, COPPER SULFATE AND CHLORINE. IN EXCESSIVE CONCENTRATIONS COPPER SULFATE MAY POISON FISH AND OTHER AQUATIC LIFE. APPROXIMATELY 300 CHEMICALS WERE SCREENED FOR TOXICITY AND THEIR EFFECTIVENESS COMPARED WITH THAT OF COPPER SULFATE. DATA FOR SOME OF THE MOST PROMISING CHEMICAL COMPOUNDS (FOR EXAMPLE, ROSIN AMINE D ACETATE) AS DETERMINED FROM PRELIMINARY SCREENING TESTS AND TOXICITY STUDIES ARE TABULATED. A BIBLIOGRAPHY WITH 418 REFERENCES IS INCLUDED. (WOODARD-USGS)

FIELD 05G, 04A, 10

ACCESSION NO. W71-05083

BACTERICIDAL EFFECTS OF ALGAE ON ENTERIC ORGANISMS,

TEXAS UNIV., AUSTIN. CENTER FOR RESEARCH IN WATER RESOURCES.

ERNST M. DAVIS, AND EARNEST F. GLOYNA.

COPY AVAILABLE FROM GPO SUP DOC AS I67.13/4:18050 DOL 03/70, $1.25; MICROFICHE FROM NTIS AS PB-197 862, $0.95. WATER POLLUTICN CONTROL RESEARCH SERIES 18050 DOL 03/70, MARCH 1970. 132 P, 114 TAB, 9 FIG, 60 REF, 3 APPEND. FWQA PROGRAM 18050 DOL.

DESCRIPTORS:
*ALGAE, *ENTERIC BACTERIA, *BACTERICIDES, GROWTH STAGES, PATHOGENIC BACTERIA, CYANOPHYTA, CHLOROPHYTA, SCENEDESMUS, ALGAL TOXINS, ANTIBIOTICS(PESTICIDES), DATA PROCESSING, COLIFORMS, MICROORGANISMS, HYDROGEN ION CONCENTRATION, WASTE WATER TREATMENT, WATER POLLUTION EFFECTS.

IDENTIFIERS:
*MIXED CULTURES, DIEOFF RATES, AFTERGROWTH.

ABSTRACT:
VARIOUS ENTERIC ORGANISMS WERE EXPOSED TO AXENIC CULTURES OF GREEN AND BLUE-GREEN ALGAE TO DETERMINE WHAT EFFECT THE ALGAE MIGHT HAVE. TESTS WERE EXTENDED TO 90 DAYS TO DETERMINE WHETHER AFTERGROWTH WAS POSSIBLE. MIXED CULTURES OF BOTH GREEN AND BLUE-GREEN ALGAE WERE EXPOSED TO BOTH SINGLE SPECIES AND MIXED CULTURES OF ENTERIC BACTERIA, AT VARYING STAGES OF THE ALGAE GROWTH PERIOCS. RESULTS INDICATED THAT MIXED CULTURES DO HAVE A PRONOUNCED BACTERICIDAL EFFECT CN ENTERIC ORGANISMS, WHILE FOR SINGLE ALGAL SPECIES, THE EFFECT IS LESS PRONOUNCED. ALSO, THE BACTERICIDAL EFFECTS ARE MORE SPECIFIC FOR PATHOGENS, WITH VIRTUALLY NO AFTERGROWTH OF PATHOGENS CETECTED. (LOWRY-TEXAS)

FIELD 05C, 05D

ACCESSION NO. W71-05155

MODELING OF THE NITROGEN AND ALGAL CYCLES IN ESTUARIES,

MANHATTAN COLL., BRONX, N.Y. ENVIRCNMENTAL ENGINEERING AND SCIENCE PROGRAM.

R. V. THOMANN, D. J. O'CONNOR, AND D. M. DITORO.

PROCEEDINGS, 5TH INTERNATIONAL WATER POLLUTION RESEARCH CONFERENCE, PAPER III-9, JULY-AUGUST 1970. 14 P, 9 FIG, 1 TAB, 10 REF.

DESCRIPTORS:
*MATHEMATICAL MODELS, *NITROGEN CYCLE, *OXIDATION, AMMONIA, NITRITES, OXYGEN, PHYTOPLANKTON, BIOCHEMICAL OXYGEN DEMAND, DISSOLVED OXYGEN, ESTUARIES, WATER POLLUTION EFFECTS, WATER POLLUTICN SOURCES, NITRIFICATION, ALGAE.

IDENTIFIERS:
ALGAL CYCLE, DELAWARE ESTUARY(DEL-NJ), SAN JOAQUIN DELTA(CALIF), STEADY-STATE MODELS.

ABSTRACT:
TWO MATHEMATICAL MODELS WERE CONSTRUCTED TO ADDRESS THE PROBLEMS OF NITROGEN AND ALGAL CYCLES. THE STEADY-STATE, MULTI-DEMENSICNAL MODEL WAS USED FOR ANALYZING NITRIFICATION AND ALGAL UTILIZATION OF AVAILABLE NITROGEN. THE DYNAMIC MODEL INCORPORATED THE GROWTH AND DEATH OF PHYTOPLANKTON AND HERBIVOROUS ZOOPLANKTON AND THE UTILIZATION OF INORGANIC NITROGEN. APPLICATION OF THE FIRST MODEL TO THE DELAWARE ESTUARY INDICATED THE RATE OF AMMONIA OXIDATION OF ABOUT 0.1/DAY AT 20C WITH NITRIFICATION INHIBITION AT DO LESS THAN 1-2 MG/L. A LOWER DO RESULTED FROM NITROGEN OXIDATION. THE SAME MODEL APPLIED TO THE POTOMAC ESTUARY INDICATED ALGAL UTILIZATION OF NITRATES AT 0.1/DAY AT 20C. THE DYNAMIC NON-LINEAR MODEL, APPLIED TO SACRAMENTO-SAN JOAQUIN DELTA, (CALIF) ADEQUATELY DESCRIBED THE ALGAL GROWTH AND UTILIZATION OF NUTRIENTS. THE NET ALGAL GROWTH COEFFICIENT RANGED UP TO 0.3/DAY DURING THE SPRING. (WILD-WISCONSIN)

FIELD 02L, 05C

ACCESSION NO. W71-05390

WATER RESOURCES POTENTIAL OF AN URBAN ESTUARY (SAUGUS RIVER, PINES RIVER AND LYNN HARBOR COMPLEX),

NORTHEASTERN UNIV., BOSTON, MASS. DEPT. OF CIVIL ENGINEERING; AND NORTHEASTERN UNIV., BOSTON, MASS. DEPT. OF ENVIRONMENTAL SCIENCE.

JOHN J. COCHRANE, CONSTANTINE J. GREGORY, AND GERALD L. ARONSON.

AVAILABLE FROM NTIS AS PB-197 991, $3.00 IN PAPER COPY, $0.95 IN MICROFICHE. MASSACHUSETTS WATER RESOURCES RESEARCH REPORT, JUNE 1970. 83 P, 29 FIG, 7 TAB, 44 REF. OWRR PROJECT A-028-MASS(1).

DESCRIPTORS:
*ESTUARIES, *WATER POLLUTION EFFECTS, *WATER RESOURCES DEVELOPMENT, *MASSACHUSETTS, *EUTROPHICATION, NUTRIENTS, ALGAE, PHOSPHATES, NITRATES, SURVEYS, PESTICIDES, WASTE DISPOSAL, PATH OF POLLUTANTS, WATER QUALITY.

IDENTIFIERS:
*BOSTON(MASS), LYNN HARBOR(MASS).

ABSTRACT:
THE WATER RESOURCES POTENTIAL OF THE URBAN ESTUARINE COMPLEX COMPRISED OF THE SAUGUS AND PINES RIVERS AND LYNN HARBOR, MASSACHUSETTS, WAS EVALUATED. IMPAIRMENT OF RECREATIONAL USAGE IS CAUSED BY EUTROPHICATION RESULTING FROM RAW WASTE DISCHARGES AND NUTRIENT RESERVES IN SEDIMENTS. A LABORATORY STUDY OF THE GROWTH OF MARINE ALGAE, ULVA LATISSIMA, MADE IN FLOWING SEA WATER, INDICATES AN OPTIMUM GROWTH AT NITROGEN-PHOSPHOROUS RATIOS OF BETWEEN 40 AND 60 TO 1, AT A PHOSPHATE CONCENTRATION OF 120 MICROGRAMS PER LITER. AVERAGE VALUES FOR NUTRIENTS IN EUTROPHIC AREAS WERE 268 MICROGRAMS/LITER TOTAL ORTHOPHOSPHATE AND 0.513 MG/L NITRATES. SEDIMENTS FROM THESE AREAS AVERAGED 5.98% VOLATILE SOLIDS, 0.532 MG/L EXTRACTED ORTHOPHOSPHATES, AND 2.177 MG/L TOTAL KJELDAHL NITROGEN. IN CONTRAST, OLIGOTROPHIC AREAS, INCLUDING PARTS OF THE PINES RIVER, HAD AVERAGED VALUES OF 164 MICROGRAMS/L TOTAL ORTHOPHOSPHATE, 0.175 MG/L NITRATES, 0.79% VOLATILE SOLIDS, 0.349 MG/L EXTRACTED ORTHOPHOSPHATES AND 0.294 MG/L TOTAL KJELDAHL NITROGEN. (KNAPP-USGS)

FIELD 05C, 02L

ACCESSION NO. W71-05553

SEDIMENT-WATER NUTRIENT RELATIONSHIPS - PARTS 1 AND 2,

FEDERAL WATER QUALITY ADMINISTRATION, CINCINNATI. FIELD INVESTIGATIONS BRANCH.

GERALD D. MCKEE, LOYS P. PARRISH, CARL R. HIRTH, KENNETH M. MACKENTHUM, AND LOWELL E. LEUP.

WATER AND SEWAGE WORKS, VOL 117, NO 6, P 203-206, JUNE 1970. 23 REF; NO 7, P 246-250, JULY 1970. 4 FIG, 27 REF.

DESCRIPTORS:
*NUTRIENTS, *SEDIMENTS, *WATER ANALYSIS, LAKE MORPHOLOGY, SEDIMENTATION, PLANKTON, DIEL MIGRATION, PROTOZOA, HYDROGEN ION CONCENTRATION, WATER TEMPERATURE, BACTERIA, ALGAE, ALKALINITY, IRON, PHOSPHATE, PHOSPHORUS, CARBON, NITROGEN, CURRENTS(WATER), WINDS, NEVADA, CALIFORNIA, NEW YORK, OREGON, WISCONSIN, MAINE.

IDENTIFIERS:
FECAL MATERIAL, LAKE TAHOE, LAKE SEBASTICOOK(MAINE), FAYETTEVILLE GREEN LAKE(NEW YORK), UPPER KLAMATH LAKE(OREGON), LAKE MENDOTA(WISCONSIN).

ABSTRACT:
A THOROUGH REVIEW OF THE MECHANISMS OF NUTRIENT DEPOSITION AND EXCHANGE IN AQUATIC ENVIRONMENTS LED TO THE FOLLOWING CONCLUSIONS: (1) A PORTION OF THE NUTRIENTS, SOLUBLE AND INSOLUBLE, IN A WATER BODY ARE ULTIMATELY DESTINED TO BECOME PART OF THE SEDIMENTS. (2) ORGANISMS ARE PRIMARY CONCENTRATORS OF DISSOLVED NUTRIENTS. (3) THE INITIAL AREAS OF DEPOSITION MAY BE ONLY TEMPORARY, THE NUTRIENT-CONTAINING SEDIMENTS MAY BE RE-DISSOLVED OR MAY BE PHYSICALLY TRANSPORTED. (4) WIND-INDUCED CURRENTS ARE A MAJOR FACTOR THAT DETERMINES THE RATE AND AREA OF FINAL DEPOSITION AND THE CONTACT TIME BETWEEN SUSPENDED SEDIMENTS AND WATER. (5) MORPHOLOGY OF THE WATER BODY AFFECTS THE ULTIMATE AREA OF DEPOSITION. (6) AQUATIC ORGANISMS CONVERT INORGANIC NUTRIENTS TO ORGANIC NUTRIENTS AND VICE VERSA. (7) SUSPENSION OF SEDIMENTS INCREASES THEIR EFFECT ON THE OVERLYING WATER. (8) THE SEDIMENTS ACT AS RESERVOIRS OF NUTRIENTS FOR THE OVERLYING WATER. (9) THE RAPIDITY OF SEDIMENT BUILD-UP WILL AFFECT THE DEGREE OF INFLUENCE ON THE OVERLYING WATER. (LITTLE-BATTELLE)

FIELD 05B, 05C, 02H

ACCESSION NO. W71-05626

SOME ALGAE OF THE UPPER CUYAHOGA RIVER SYSTEM IN OHIO,

KENT STATE UNIV., OHIO. DEPT. OF BIOLOGICAL SCIENCES.

RUSSELL G. RHODES, AND ANTHONY J. TERZIS.

THE OHIO JOURNAL OF SCIENCE, VOL 70, NO 5, P 295-299, SEPTEMBER 1970. 1 FIG, 1 TAB, 16 REF.

DESCRIPTORS:
*ALGAE, OHIO, CHLOROPHYTA, CHRYSOPHYTA, RHODOPHYTA, CYANOPHYTA, EUGLENOPHYTA.

IDENTIFIERS:
*CUYAHOGA RIVER(OHIO), CLADOPHORA, APHANOCHAETE, RHIZOCLONIUM, VAUCHERIA, TRIBONEMA, OSCILLATORIA.

ABSTRACT:
SIXTY-FOUR SPECIES OF ALGAE WERE FOUND IN A QUALITATIVE SURVEY MADE DURING JUNE AND SEPTEMBER, 1967, IN THREE TRIBUTARIES OF THE CUYAHOGA RIVER IN GEAUGA COUNTY: WEST BRANCH, EAST BRANCH, AND TARE CREEK. EIGHT SPECIES WHICH WERE COLLECTED AT THE MAJORITY OF THE 14 STATIONS SAMPLED ARE CLADOPHORA GLOMERATA, APHANOCHAETE REPENS, RHIZOCLONIUM HIEROGLYPHICUM, EUGLENA GRACILIS, VAUCHERIA SESSILIS, TRIBONEMA BOMBYCINUM, OSCILLATORIA NIGRA, AND O. LIMOSA. (LITTLE-BATTELLE)

FIELD 05A, 05C

ACCESSION NO. W71-05629

BANGIA ATROPURPUREA (ROTH) A. IN WESTERN LAKE ERIE,

OHIO STATE UNIV., COLUMBUS. DEPT. OF BOTANY.

JACK KISHLER, AND CLARENCE E. TAFT.

THE OHIO JOURNAL OF SCIENCE, VOL 70, NO 1, P 56-57, JANUARY 1970. 1 FIG, 5 REF.

DESCRIPTORS:
*RHODOPHYTA, *LAKE ERIE, OHIO, ALGAE.

ABSTRACT:
BANGIA ATROPURPUREA WAS COLLECTED 2 MARCH 1969 AT THE STATE HIGHWAY PARK ON THE EAST SHORE OF MARBLEHEAD PENINSULA, OTTAWA COUNTY, OHIO. THIS IS THE FIRST RECORD OF BANGIA IN WESTERN LAKE ERIE. IT APPEARED AS LAX RED-PURPLE TUFTS, 3/4 INCH LONG, COVERING A FLAGSTONE ON THE SHORELINE WHERE THERE WAS AN OPENING IN THE ICE. (LITTLE-BATTELLE)

FIELD 05A, 05C, 02H

ACCESSION NO. W71-05630

TOXICITY OF ZINC, COPPER AND LEAD TO CHLOROPHYTA FROM FLOWING WATERS,

DURHAM UNIV. (ENGLAND). DEPT. OF BOTANY.

B. A. WHITTON.

ARCHIV FUR MIKROBIOLOGIE, VOL 72, P 353-360, 1970. 1 TAB, 7 REF.

DESCRIPTORS:
*INDICATORS, *ALGAE, *HEAVY METALS, *TOXICITY, CCPPER, CHLOROPHYTA, WATER POLLUTION, NUTRIENTS, RIVERS, LAKES, FARM PONDS.

IDENTIFIERS:
*ZINC, *LEAD, STIGEOCLONIUM TENUE, CLADOPHORA GLCMERATA, ULOTRICHALES, ZYGNEMALES, OEDOGONIUM, MOUGCTIA, SPOROTETRAS PYRIFORMIS, GONGROSIRA, MICROSPORA, SPIROGYRA, ULCTHRIX.

ABSTRACT:
THE BIOLOGICAL AND ECONOMIC INTEREST OF HEAVY METALS IN THE RELATION TO FRESHWATER ALGAE LED TO A SURVEY OF TOXICITY OF ZINC, COPPER, AND LEAD TO CHLOROPHYTA IN FLCWING WATERS. TWENTY POPULATICNS EACH OF STIGEOCLONIUM TENUE AND CLADCPHORA GLOMERATA WERE TESTED FOR VARIATIONS IN METAL RESISTANCE. THE ONLY INDICATION FOUND WAS A SLIGHT INCREASE IN RESISTANCE TO ZINC OF ONE STIGEOCLONIUM POPULATICN FROM A METAL-POLLUTED STREAM. THIRTY-FIVE OTHER ALGAL PCPULATIONS, REPRESENTING ABOUT 25 SPECIES, WERE ALSO TESTED. CCMPARISON OF THESE POPULATIONS SHOWED THAT IN THE TEST MEDIUM USED CLADOPHORA GLOMERATA WAS THE MOST OR ALMOST THE MOST SENSITIVE TO ALL THREE METALS. IT WAS ABSENT FROM STREAMS KNOWN TO BE POLLUTED BY LEAD OR LEAD AND ZINC. THE ULOTRICHALES AND MOST OF THE ZYGNEMALES WERE RELATIVELY RESISTANT TO ZINC, WHILE ALL THE CEDOGONIUM SPECIES TAKEN FROM THE FIELD WERE SENSITIVE TO ZINC. HOWEVER, AN OEDCGONIUM POPULATION HIGHLY ZINC RESISTANT WAS OBTAINED FROM A ZINC ENRICHED LABORATORY TANK. BECAUSE THE MEDIUM USED IN THE PRESENT EXPERIMENTS IS RELATIVELY RICH IN NUTRIENTS SUCH AS PHOSPHATES IT MAY WELL MASK EFFECTS THAT WOULD SHOW IN A LESS FAVORABLE ENVIRONMENT. (JONES-WISCONSIN)

FIELD 05C

ACCESSION NO. W71-05991

EVALUATION OF SOME STABILIZATION PONDS IN INDIA,

CENTRAL PUBLIC HEALTH ENGINEERING RESEARCH INST., NAGPUR (INDIA).

R. H. SIDDIQI, AND B. K. HANDA.

JOURNAL OF THE SANITARY ENGINEERING DIVISION, AMERICAN SOCIETY OF CIVIL ENGINEERS, VOL 97, NO SA1, P 91-100, FEBRUARY 1971. 7 FIG, 1 TAB, 11 REF.

DESCRIPTORS:
*OXIDATION LAGOCNS, *AEROBIC TREATMENT, *ANAERCBIC CONDITIONS, PONDS, CLIMATIC ZONES, TEMPERATURE, TURBULENCE, OXYGENATION, ALGAE, PHOTOSYNTHESIS, MUNICIPAL WASTES, BIOCHEMICAL OXYGEN DEMAND, WASTE WATER TREATMENT, MICROORGANISMS.

IDENTIFIERS:
FACULTATIVE MICROORGANISMS, INDIA.

ABSTRACT:
CLIMATIC CONDITIONS IN INDIA ARE FAVORABLE TO THE OPERATION OF ENGINEERED WASTE STABILIZATION PONCS. CONSEQUENTLY, THE TREATMENT OF WASTES IN SUCH PONDS IS ECONOMICALLY COMPETITIVE WITH CONVENTIONAL BIOLOGICAL TREATMENT. DATA COLLECTED FROM SEVERAL INSTALLATIONS IN INDIA WERE ANALYZED TO DETERMINE WHAT USEFUL PARAMETERS OF OPERATION COULD BE IDENTIFIED. POND LOADING WAS FOUND TO BE BEST EXPRESSED BY A LOAD FACTOR, L SUB F, WHICH IS THE RATIO OF BCD LCAD TO ALGAL PRODUCED OXYGEN. FOR L SUB F BETWEEN .44 AND 8.0, THE PERFORMANCE WAS DETERMINED FROM THE FOLLOWING EQUATION: E = 100 OVER (1 + 0.188 L SUB F OR EXPLANENTIAL TO 0.48). FROM THE PRECEDING INVESTIGATIONS, IT WAS DETERMINED THAT THE MAJORITY OF ORGANIC MATTER DESTROYED IS DESTROYED ANAEROBICALLY. THEREFORE, PONDS OF DEPTH GREATER THAN 5 FEET ARE MORE EFFICIENT IN THEIR OPERATION SINCE THERE IS LESS TURBULENCE AND LESS CHANGE OF THE ANAEROBES BEING EXPOSED TO OXYGEN. IT WAS ALSO DETERMINED THAT SINGLE CELL REACTORS, OR THE FIRST CELL OF A MULTI-CELL ARRANGEMENT MAINTAIN A HIGHER DESTRUCTION RATE CCNSTANT, WITH THE RATE OF BOD REDUCTION BEING DESCRIBED AS A FIRST ORDER EQUATION. THE RATE DROPS OFF WITH EACH SUCCESSIVE CELL IN A MULTI-CELL ARRANGEMENT. (LOWRY-TEXAS)

FIELD 05D

ACCESSION NO. W71-06033

ALGAECIDAL EVALUATION AND ENVIRONMENTAL STUDY OF MAT PRODUCING BLUEGREEN ALGAE,

BUREAU OF RECLAMATION, DENVER, COLO. OFFICE OF CHIEF ENGINEER.

NAMAN E. OTTO.

AVAILABLE FROM NTIS AS PB-194 808, $3.00 IN PAPER COPY, $0.95 IN MICROFICHE. BUR OF RECLAM REPORT REC-OCE-70-25, JUNE 1970. 27 P, 13 FIG, 7 TAB, 27 REF.

DESCRIPTORS:
*ALGAE, *AQUATIC WEEDS, *WEED CONTROL, GROWTH, *CULTURES, *ALGICIDES, BIBLIOGRAPHIES, *ECOLOGY, AQUATIC LIFE, IRRIGATION SYSTEMS, WATER ANALYSIS, TEMPERATURE, LIMNOLOGY, CYANOPHYTA, IRRIGATION CANALS, ALGAL CONTROL.

IDENTIFIERS:
*ALGAE CULTURE TECHNIQUE, AQUATIC WEED STUDY FACILITY, PRODUCT EVALUATION, BLUE-GREEN ALGAE.

ABSTRACT:
ALGAECIDAL TESTS OF 74 COMPOUNDS SHOWED THAT ONLY 9 WERE MORE ACTIVE THAN COPPER SULFATE, AND ONE PROVIDED TOTAL CONTROL OF THE BLUE-GREEN ALGAE MAT COLONIES. RESULTS OF PRELIMINARY ALGAECIDAL FIELD TESTS WITH COPPER SULFATE AND ADMIXTURES OF ENHANCE ACTIVITY ARE GIVEN. OBSERVATIONS OF ALGAECIDAL TREATMENTS ON AN IRRIGATION CANAL SUGGEST THAT COPPER SULFATE APPLICATIONS OF 0.48 LB/CFS APPLIED EVERY 2 WK SUPPRESSED BLUE-GREEN MAT GROWTH. WATER TEMPERATURE APPEARS CRITICAL TO TREATMENT SUCCESS. RESULTS OF STUDIES CONDUCTED IN 1967-1969 TO DETERMINE ENVIRONMENTAL PARAMETERS OF IRRIGATION CANALS ARE PRESENTED. DATA SHOW THAT MAT-TYPE, BLUE-GREEN ALGAE GROW IN A WIDE RANGE OF WATER QUALITY AND TEMPERATURE CONDITIONS AND DO NOT REQUIRE ENRICHED CONDITIONS FOR GROWTH. COLORATION SUBSTANCES IN THE WATER MAY INFLUENCE THE AVAILABILITY OF MICRONUTRIENT METALS AND MAY STIMULATE DIRECTLY BLUE-GREEN ALGAE GROWTH.

FIELD 04A, 05C

ACCESSION NO. W71-06102

PHYSICAL PROPERTIES AND PROCESSING CHARACTERISTICS OF MACROPHYTES AS RELATED TO MECHANICAL HARVESTING,

WISCONSIN UNIV., MADISON. WATER RESOURCES CENTER.

H. D. BRUHN, D. F. LIVERMORE, AND F. O. ABOABA.

AVAILABLE FROM NTIS AS PB-198 129, $3.00 IN PAPER COPY, $0.95 IN MICROFICHE. REPRINT, AMERICAN SOCIETY OF AGRICULTURAL ENGINEERS, ST JOSEPH, MICHIGAN 49085, PAPER 70-582, 1970. 17 P, 12 FIG, 30 REF. OWRR PROJECT B-018-WIS(2).

DESCRIPTORS:
*HARVESTING OF ALGAE, *AQUATIC WEED CONTROL, *MECHANICAL CONTROL, *DEWATERING.

IDENTIFIERS:
FARM MACHINERY, INDUSTRIAL MACHINERY, UTILIZATION, MODIFICATION.

ABSTRACT:
SELECTIVE MECHANICAL HARVESTING OF UNDESIRABLE AQUATIC VEGETATION APPEARS TO BE A FEASIBLE AND ECOLOGICALLY SOUND APPROACH TO CONTROL IN RECREATIONAL WATERS. THE TRANSPORTATION FROM THE HARVEST AREA TO THE DISPOSAL SITE IS A MAJOR EXPENSE IN THE OVERALL HARVEST OPERATION. SINCE AQUATIC VEGETATION IS APPROXIMATELY 90 PER CENT WATER PROCESSING AND DEWATERING THIS MATERIAL AS A PART OF THE INITIAL HARVESTING OPERATION GREATLY FACILITATES ITS TRANSPORTATION AND ULTIMATE DISPOSAL OR UTILIZATION. INTENSIVE MECHANICAL DEWATERING BASED ON PROCEDURES DEVELOPED IN THIS RESEARCH RESULTS IN REDUCING THE FIBROUS FRACTION OF AQUATIC VEGETATION TO 16 PER CENT OF ITS ORIGINAL VOLUME AND 32 PER CENT OF ITS ORIGINAL WEIGHT, WHILE REMOVING FROM THE HARVESTED AREA 90 PER CENT OF THE ORIGINAL DRY MATTER, 85 PER CENT OF THE PROTEIN, 60 PER CENT OF THE POTASSIUM, AND 80 PER CENT OF THE PHOSPHORUS PRESENT IN THE GROWING VEGETATION. THE REDUCTION IN FIBER LENGTH RESULTING FROM THE DEWATERING PROCESS INCREASES THE EASE WITH WHICH THE VEGETATIVE MATERIAL CAN BE TRANSFERRED THROUGH CONVEYING SYSTEMS THUS FURTHER FACILITATING HANDLING AND DISPOSAL. THE MAJOR PORTION OF THE EQUIPMENT REQUIRED FOR THE PREPROCESSING AND DEWATERING CAN BE DEVELOPED BY UTILIZATION AND MODIFICATION OF COMPONENTS OF ALL READILY AVAILABLE FARM AND INDUSTRIAL MACHINERY.

FIELD 04A, 05G

ACCESSION NO. W71-06188

ALGICIDES,

WISCONSIN UNIV., MADISON. WATER RESOURCES CENTER.

GEORGE P. FITZGERALD.

AVAILABLE FROM NTIS AS PB-198 130, $3.00 IN PAPER COPY, $0.95 IN MICROFICHE. WISCONSIN UNIV, MADISON, WATER RESOURCES CENTER, EUTROPHICATION INFORMATION PROGRAM, LITERATURE REVIEW NO 2, 1971. 50 P, 1 FIG, 9 TAB, 153 REF. OWRR PROJECT W-117NO 1614)(3).

DESCRIPTORS:
*ALGICIDES, *ALGAE, *ALGAL CONTROL, TESTING, LABORATORY TESTS, CHEMICALS, BACTERIOCIDES, CULTURES, TEST PROCEDURES, PESTICIDE TOXICITY, RESISTANCE, LAKES, RESERVOIRS, SWIMMING POOLS, CHELATION, APPLICATION METHODS, COOLING TOWERS, WATER POLLUTION EFFECTS.

IDENTIFIERS:
DETOXIFICATION, SYNERGISM, FISH PONDS, ALGISTATIC ACTIVITY, ALGICIDAL ACTIVITY.

ABSTRACT:
AN EVALUATION OF ALGICIDES--WHAT THEY ARE, THEIR USAGE, HOW THEY ARE TESTED, THEIR EFFICIENCY, AND METHODS OF APPLICATION--IS PRESENTED. DETAILED ARE APPROPRIATE TEST ORGANISMS, CULTURE AND TEST MEDIA, THE IMPORTANCE OF ALGICIDAL VERSUS ALGESTATIC EFFECTS, AND EFFECTIVE CHEMICAL CONCENTRATIONS REQUIRED TO ACHIEVE POTENTIAL TOXICITY. CHARACTERISTICS OF AN ALGAL PROBLEM, TOGETHER WITH ITS ENVIRONMENT, INFLUENCE THE METHODOLOGY OF APPLICATION PROCESSES AS WELL AS DETOXIFICATION, RESISTANCE, AND SYNERGISM. ALGAL PROBLEMS OF WATER SUPPLY RESERVOIRS AND RECREATIONAL LAKES, FISH PONDS, SWIMMING POOLS, AND COOLING TOWERS ARE SPECIFICALLY DISCUSSED WITH SUGGESTIONS FOR EFFECTIVE APPLICATION OF TOXICANTS FOR THEIR CONTROL. (AUEN-WISCONSIN)

FIELD 05C, 05F

ACCESSION NO. W71-06189

DEVELOPMENT OF PHOSPHATE-FREE HOME LAUNDRY DETERGENTS,

IIT RESEARCH INST., CHICAGO, ILL.

KARL A, ROSEMAN, AND WARNER M. LINFIELD.

COPY AVAILABLE FROM GPO SUP DOC AS EPA-WQO REPORT NO 16080 DVF, DECEMBER 1970, $1.00; MICROFICHE FROM NTIS AS PB-198 222, $0.95. 103 P, 4 TAB, 10 REF. EPA-WQO CONTRACT 14-12-575.

DESCRIPTORS:
*DETERGENTS, *ALGAL CONTROL, *FORMULATION, *SURFACTANTS, *EUTROPHICATION, LINEAR ALKYLATE SULFONATES, CHELATION, PHOSPHATES, ORGANIC COMPOUNDS, WATER POLLUTION CONTROL.

IDENTIFIERS:
PHOSPHATE-FREE DETERGENTS, TRISODIUM NITRILOTRIACETATE, SODIUM CITRATE.

ABSTRACT:
BASIC STUDIES WERE PERFORMED TOWARDS THE DEVELOPMENT OF PHOSPHATE-FREE HOME LAUNDRY DETERGENTS. FIVE SURFACTANTS WERE SYNTHESIZED WITH THE IDEA THAT THEY MIGHT POSSESS HARD ION CHELATING PROPERTIES. THE CLEANING ABILITIES OF THESE MATERIALS WERE COMPARED TO THE WIDELY USED LINEAR ALKYLBENZENE SULFONATE AS INCORPORATED INTO THE SAME FORMULATIONS. THE DETERGENT COMPOSITIONS CONTAINED 2% CARBOXYMETHYLCELLULOSE AND THE SILICATE CONTENT WAS VARIED. SODIUM ACETATE AND SODIUM CARBONATE WERE INVESTIGATED AS POSSIBLE RESERVOIRS OF ALKALINITY. SURFACTANT COMPATIBILITY WITH SODIUM CHLORIDE AND SODIUM SULFATE WAS EXAMINED. OTHER ADDITIVES INCLUDED TRISODIUM NITRILOTRIACETATE AND SODIUM CITRATE AT MODERATE LEVELS. FIFTEEN DETERGENT FORMULATIONS WERE SCREENED AND THE RESULTS LEAVE LITTLE DOUBT THAT ACCEPTABLE PHOSPHATE-FREE HOME LAUNDRY DETERGENTS CAN BE DEVELOPED.

FIELD 05G

ACCESSION NO. W71-06247

SOURCES OF NITROGEN IN WATER SUPPLIES,

GEOLOGICAL SURVEY, DENVER, COLO.

MARVIN C. GOLDBERG.

AGRICULTURAL PRACTICES AND WATER QUALITY, IOWA STATE UNIVERSITY PRESS, AMES, IOWA, 1970, CHAPTER 7, P 94-124. 4 FIG, 8 TAB, 72 REF.

DESCRIPTORS:
*NITROGEN, *NITRATES, GROUNDWATER, AMMONIA, PRECIPITATION, SEDIMENTS, DENITRIFICATION, RUNOFF, UREAS, FERTILIZERS, DRAINAGE WATER, IRRIGATION, RETURN FLOW, WATER SUPPLY, LIVESTOCK, SEWAGE, INFILTRATION, INDUSTRIAL WASTES, ALGAE, PONDS, FARM WASTES.

IDENTIFIERS:
*SURFACE WATERS, GEOLOGICAL SOURCES, MINERALIZATION, NITROGEN SOURCES, WELL WATER, FEEDLOTS.

ABSTRACT:
WATER SUPPLIES CAN BE CATEGORIZED AS SURFACE WATERS OR GROUNDWATERS. THIS PAPER EXAMINES REPRESENTATIVE STUDIES OF NITRATE ENTRANCE TO BOTH TYPES OF WATER SUPPLIES, WITH SUMMARIES OF SOME OF THE MANY LABORATORY AND FIELD STUDIES DESCRIBED IN THE CURRENT LITERATURE. SOME OF THE SOURCES OF NITROGEN ENTRANCE TO WATER SUPPLIES INCLUDE ATMOSPHERIC, GEOLOGIC, RURAL AND URBAN RUNOFF, SEWAGE, IRRIGATION, ANIMAL WASTES, AND INDUSTRIAL WASTES AMONG MANY OTHERS. SOURCES OF MAJOR IMPORTANCE TO BOTH SURFACE AND GROUNDWATER SUPPLIES ARE POINTED OUT AND FIELD OR LABORATORY STUDIES ARE REPORTED. (WHITE-IOWA STATE)

FIELD 05B

ACCESSION NO. W71-06435

EFFECTS OF AGRICULTURAL POLLUTION ON EUTROPHICATION,

WISCONSIN UNIV., MADISON. DEPT. OF SANITARY ENGINEERING, AND WISCONSIN UNIV., MADISON. DEPT. OF WATER CHEMISTRY.

D. E. ARMSTRONG, AND G. A. ROHLICH.

AGRICULTURAL PRACTICES AND WATER QUALITY, IOWA STATE UNIVERSITY PRESS, AMES, IOWA, 1970, CHAPTER 23, P 314-330. 14 TAB, 2 FIG, 26 REF.

DESCRIPTORS:
*EUTROPHICATION, *NITROGEN, *PHOSPHORUS, NUTRIENTS, ALGAE, NITRATES, SURFACE RUNOFF, BASE FLOW, PERCOLATION, LEACHING, DRAINAGE, FARM WASTES, SOIL MANAGEMENT, GROUNDWATER, WISCONSIN, WATER SUPPLY.

IDENTIFIERS:
*AGRICULTURAL DRAINAGE, LAKE METABOLISM, MOBILITY, PARTICULATE FORM, FEEDLOTS, NUTRIENT SOURCES.

ABSTRACT:
THE PAPER DISCUSSES NITROGEN AND PHOSPHORUS TRANSPORT IN AGRICULTURAL DRAINAGE SINCE THESE ARE THE MOST IMPORTANT NTURIENTS INVOLVED IN EUTROPHICATION. IT IS GENERALLY EXPECTED THAT INORGANIC NITROGEN IS TRANSPORTED MAINLY AS NITRATE BY PERCOLATING WATER, ALTHOUGH THE AMOUNTS OF AMMONIUM AND NITRATE CARRIED IN RUNOFF WATERS MAY BE HIGHLY SIGNIFICANT IN TERMS OF THE RECEIVING WATER. SIMILARLY, THE LARGEST AMOUNT OF PHOSPHORUS IS LIKELY TRANSPORTED IN PARTICULATE FORM IN RUNOFF WATERS, BUT THE AMOUNT OF DISSOLVED PHOSPHORUS IN RUNOFF WATER MAY BE OF EQUAL OR GREATER IMPORTANCE EVEN THOUGH LOWER IN QUANTITY. THE CONTRIBUTION OF AGRICULTURAL DRAINAGE TO THE NITROGEN AND PHOSPHORUS STATUS OF WATERS IS NEXT EXAMINED. THE DATA PRESENTED SUGGEST THAT AGRICULTURAL LAND IS AN IMPORTANT CONTRIBUTOR OF NITROGEN AND PHOSPHORUS TO WATER. ABOUT 60% OF THE NITROGEN AND 42% OF THE PHOSPHORUS WERE ESTIMATED TO COME FROM AGRICULTURAL LAND. NUTRIENT BUDGET ESTIMATIONS WERE BASED ON DATA OBTAINED ON A SMALL SCALE AND EXTRAPOLATED AND THUS HAVE A LOW RELIABILITY. NUTRIENT SOURCES ARE NUMEROUS AND GENERALIZATIONS AS TO WHICH SOURCE IS THE MOST IMPORTANT CANNOT BE MADE. THE CONTRIBUTION OF AGRICULTURE SHOULD BE REDUCED BY IMPROVED AND MORE EFFICIENT AGRICULTURAL MANAGEMENT PRACTICES. (WHITE-IOWA STATE)

FIELD 05C, 02H

ACCESSION NO. W71-06443

EFFECTS OF AGRICULTURAL POLLUTANTS ON RECREATIONAL USES OF SURFACE WATERS,

MISSOURI UNIV., COLUMBIA. DEPT. OF ZOOLOGY; AND MISSOURI DEPT. OF CONSERVATION, COLUMBIA.

ROBERT S. CAMPBELL, AND JAMES R. WHITLEY.

AGRICULTURAL PRACTICES AND WATER QUALITY, IOWA STATE UNIVERSITY PRESS, AMES, IOWA, 1970, CHAPTER 24, P 331-343. 3 TAB, 1 FIG, 43 REF.

DESCRIPTORS:
*POLLUTANTS, *RECREATION, LAKES, STREAMS, ALGAE, PESTICIDES, DDT, RETURN FLOW, SEDIMENTS, SOIL EROSION, FERTILIZERS, FARM WASTES, FISH, FISH EGGS, TURBIDITY, NUTRIENTS, DISSOLVED OXYGEN, NITROGEN, PHOSPHORUS, WATER QUALITY ACT, WATER QUALITY.

ABSTRACT:
UNQUESTIONABLY MANY AGRICULTURAL POLLUTANTS AFFECT RECREATION THROUGH ALTERATION OF WATER QUALITY AND DEGRADATION OF FISH AND AQUATIC LIFE. THE MORE SERIOUS POLLUTING AGENTS ARE ERODED SOIL, AGRICULTURAL FERTILIZERS, ANIMAL WASTES, AND PESTICIDES. WHILE THE PROBLEMS RELATING TO AGRICULTURAL POLLUTION ARE COMPLEX, AND THE SOLUTIONS WILL NOT EASILY BE ATTAINED, IT SEEMS REASONABLE THAT IN MANY INSTANCES ALTERNATIVE PROCEDURES CAN BE DEVELOPED. POLLUTION CONTROL MEASURES ARE AVAILABLE WHICH WILL ALLOW CONTINUATION OF AGRICULTURAL PRODUCTION AND ENHANCE AND PROTECT WATER QUALITY AND RECREATION. WHILE THESE PROCEDURES MAY BE COSTLY TO APPLY, THE EXPENDITURE SHOULD BE JUDGED IN LIGHT OF ITS CONTRIBUTION TOWARD THE PRESERVATION OF MAN'S ENVIRONMENT. ESPECIALLY IN THE INSTANCE OF PESTICIDE USE, PROTECTION OF WATER QUALITY MAY BE REQUISITE TO PROTECTION OF THE HEALTH OF MAN FROM UNKNOWN LONG-TERM EFFECTS OF PESTICIDES. REDUCTION AND CONTROL OF AGRICULTURAL POLLUTANTS ARE ESSENTIAL TO DEVELOP AND MAINTAIN A HIGH QUALITY ENVIRONMENT. QUALITY OF LIFE AND QUALITY OF ENVIRONMENT ARE SYNONYMOUS. (WHITE-IOWA STATE)

FIELD 05C

ACCESSION NO. W71-06444

EFFECTS OF SURFACE RUNOFF ON THE FEASIBILITY OF MUNICIPAL ADVANCED WASTE TREATMENT,

IOWA STATE UNIV., AMES. DEPT. OF CIVIL ENGINEERING.

ROBERT E. BAUMANN, AND SHELDON KELMAN.

AGRICULTURAL PRACTICES AND WATER QUALITY, IOWA STATE UNIVERSITY PRESS, AMES, IOWA, 1970, CHAPTER 25, P 344-362. 7 FIG, 1 TAB, 19 REF.

DESCRIPTORS:
*SURFACE RUNOFF, *SEWAGE TREATMENT, TERTIARY TREATMENT, BIOCHEMICAL OXYGEN DEMAND, INDUSTRIAL WASTES, FARM WASTES, POLLUTANTS, FERTILIZERS, NITROGEN, NITRATES, PHOSPHORUS, PHOSPHATES, ALGAE, EUTROPHICATION, WATER QUALITY, RIVERS, IOWA, CORN, DISCHARGE, CHLOROPHYLL.

IDENTIFIERS:
INDUSTRIAL WATER POLLUTION, DES MOINES RIVER, PACKING PLANTS.

ABSTRACT:
THE PROTECTION OF THE QUALITY OF WATER IN IOWA STREAMS REQUIRES THAT ATTENTION BE DIRECTED AT THE VARIOUS CONTRIBUTORS OF THE SIGNIFICANT POLLUTANTS. ATTENTION IS CURRENTLY BEING DIRECTED AT MUNICIPAL AND INDUSTRIAL WASTES DISCHARGES, SINCE THESE ENTER STREAMS THROUGH A POINT SOURCE AND ARE EASILY CONTROLLED. ALL SUCH WASTES MUST BE GIVEN SECONDARY TREATMENT PRIOR TO DISCHARGE TO IOWA'S STREAMS. AS MORE STRINGENT TREATMENT REQUIREMENTS ARE DEMANDED IN THE FUTURE, THERE IS SOME QUESTION AS TO WHETHER NUTRIENT REMOVALS FROM MUNICIPAL AND INDUSTRIAL WASTES WILL BE SUFFICIENT TO PROTECT THE STREAM. THIS STUDY INDICATED THAT DURING PERIODS OF DRY WEATHER WHEN LIGHT AND TURBIDITY CONDITIONS ARE FAVORABLE FOR PHYTOPLANKTON GROWTH, THE PRINCIPAL SOURCE OF THE N AND P REQUIRED TO SUPPORT SUCH GROWTH IS DERIVED FROM MUNICIPAL AND INDUSTRIAL WASTE WATER DISCHARGES. REMOVAL OF N AND P FROM SUCH WASTE WATER DISCHARGES WILL HELP REDUCE PHYTOPLANKTON GROWTH. IN PERIODS OF HIGH STREAM FLOW, WHEN TURBIDITY LEVELS ARE HIGH ENOUGH TO BE UNFAVORABLE TO PHYTOPLANKTON GROWTH, RUNOFF FROM URBAN AND RURAL LANDS AND CHANNEL EROSION ARE PROBABLY THE PRINCIPAL CONTRIBUTORS OF N AND P TO THE STREAM. REMOVAL OF N AND P FROM MUNICIPAL AND INDUSTRIAL WASTES DURING THESE PERIODS WILL NOT REDUCE NUTRIENT LEVELS SIGNIFICANTLY. UNDER THE LATTER CONDITIONS, TERTIARY TREATMENT OF MUNICIPAL AND INDUSTRIAL WASTES WILL BE OF LESS BENEFIT UNTIL RUNOFF CONTRIBUTIONS OF N AND P ARE ALSO CONTROLLED.
(WHITE-IOWA STATE)

FIELD 05D, 05B

ACCESSION NO. W71-06445

DYNAMIC BEHAVIOUR OF OXIDATICN PONDS,

CAPE TOWN UNIV. (SOUTH AFRICA).

G. V. R. MARAIS.

2ND INTERNATIONAL SYMPOSIUM FOR WASTE TREATMENT LAGOONS, JUNE 23-25, 1970, KANSAS CITY, MISSOURI, P 15-46. 28 FIG, 14 REF, APPEND.

DESCRIPTORS:
*OXIDATION LAGOONS, *MATHEMATICAL MODELS, *COMPUTER SIMULATION, MIXING, TURBULENCE, TEMPERATURE, STRATIFICATION, SOLAR RADIATION, ALGAE, PHOTOSYNTHESIS, ANAEROBIC CONDITIONS, AEROBIC CONCITIONS, SLUDGE, WASTE WATER TREATMENT, BIODEGRADATION, BIOCHEMICAL OXYGEN DEMAND.

ABSTRACT:
A COMPUTER MODEL OF OXIDATION PONDS STIMULATED INTEREST IN SEVERAL PARAMETERS WHICH WERE NOT PREVIOUSLY CONSIDERED TC BE IMPORTANT. DATA WAS COLLECTED FROM THE MATERS NORTH PCND IN LUSAKA, ZAMBIA, AND THEN ANALYZED AND USED TO DEVELOP THE COMPUTER MODEL. THIS MODEL CLEARLY DEMONSTRATED THAT (1) MIXING IN THE POND DEMANDS MUCH GREATER ATTENTION AS AN INFLUENTIAL PARAMETER WITH RESPECT TO BOTH THE THEORETICAL AND THE PRACTICAL ASPECTS, (2) ADVENT OF ANAEROBIC CCNDITIONS IN AEROBIC OXIDATION PONDS IS A MUCH MORE CCMPLEX PHENOMENON THAN WAS PREVIOUSLY SUPPOSED, BEING DEPENDENT UPON TEMPERATURE, MIXING, ALGAL GROWTH AND BOD IN THE POND, AND POSSIBLY OTHER PARAMETERS, (3) THE SLUDGE LAYER OCCUPIES AN IMPORTANT POSITION IN THE CEGRADATICN PRCCESS AND DIRECTS ATTENTION TO THE BENEFITS IN PARTIAL SEPARATION BY ANAEROBIC PRETREATMENT. (LOWRY-TEXAS)

FIELD 05D

ACCESSION NO. W71-07084

CHLORINATION OF WASTE POND EFFLUENTS,

SACRAMENTO STATE COLL., CALIF.

LECNARD W. HOM.

2ND INTERNATIONAL SYMPOSIUM FOR WASTE TREATMENT LAGOONS, JUNE 23-25, 1970, KANSAS CITY, MISSOURI, P 151-159. 7 FIG, 4 TAB, 15 REF. NATIONAL SCIENCE FOUNDATION GRANT GY3799, NATIONAL INSTITUTE OF HEALTH, USPHS, WP00026-RS.

DESCRIPTORS:
*OXIDATION LAGCONS, *DISINFECTION, DEGRADATION, CHLORINE, CONTACT TIME, SAMPLING, COLIFORMS, BACTERIA, OXIDATION, ORGANIC LOADING, ALGAE, WASTE WATER TREATMENT, BIOCHEMICAL OXYGEN DEMAND.

IDENTIFIERS:
*RESIDUAL, MOST PROBABLE NUMBER.

ABSTRACT:
A TWO YEAR SERIES OF TESTS WERE PERFORMED CN EFFLUENT FROM EXPERIMENTAL STABILIZATION LAGOONS OF THE CITY OF CONCORD, CALIFORNIA, IN AN ATTEMPT TO DETERMINE METHODS FOR CONTROLLING ALGAE KILL BY CCNTROLLING CHLORINE DOSAGE. THREE DAY COMPOSITE SAMPLES, OF 500 ML VCLUME, WERE TAKEN AT THREE-HOUR INTERVALS, AND THE CHLORINE DEMAND WAS DETERMINED BY THE OT TEST USING A 30 MIN. CONTACT TIME. IN ADDITION, THE SAMPLES WERE ANALYZED FOR BOD LEVELS AND FOR COLIFORM POPULATICN. SELECTIVE CHLORINATION OF STABILIZATION LAGOON EFFLUENTS CAN BE ACCOMPLISHED. CONTROL OF REACTION AND CHLORINE CONCENTRATION IS CRITICAL, SINCE EXCESSIVE CHLORINE CAN RELEASE NUTRIENTS FROM ALGAL CELLS, THEREBY INCREASING THE BOD. THE SOLUTION LIES IN EXPERIMENTAL OPTIMIZATION OF BOTH CHLORINE DOSAGE AND RESIDUAL CHLORINE CONCENTRATIONS. TIME-CONCENTRATION RELATIONSHIPS REPORTED HERE PROVIDE A RATIONAL SET OF PROCESS DESIGN PARAMETERS FOR CHLORINE CISINFECTICN IN ALGAL-BACTERIAL SYSTEMS. (LOWRY-TEXAS)

FIELD 05D, 05F

ACCESSION NO. W71-07096

EFFECT OF LAGOON EFFLUENT ON A RECEIVING STREAM,

MISSOURI UNIV., COLUMBIA.

DARRELL L. KING, ALLEN J. TOLMSOFF, AND MICHAEL J. ATHERTON.

2ND INTERNATIONAL SYMPOSIUM FOR WASTE TREATMENT LAGOONS, JUNE 23-25, 1970, KANSAS CITY, MISSOURI, P 159-167. 5 FIG, 4 TAB, 3 REF.

DESCRIPTORS:
*OXIDATION LAGOONS, *ALGAE, *TRICKLING FILTERS, EUTROPHICATION, PHOTOSYNTHESIS, DISSOLVED OXYGEN, SEDIMENTATION, SAMPLING, CHEMICAL OXYGEN DEMAND, BIOCHEMICAL OXYGEN DEMAND, STABILIZATION PONDS, MICROORGANISMS, BIODEGRADATION, WASTE WATER TREATMENT, WATER POLLUTION EFFECTS.

IDENTIFIERS:
*RIFFLES, *CALORIC CONTENT.

ABSTRACT:
BEAR CREEK, A SMALL STREAM WHOSE FLOW IS 21.5% LAGOON EFFLUENT AND SEEPAGE, WAS THE SITE OF A 90 DAY STUDY WHICH ATTEMPTED TO DETERMINE WHAT STREAM PARAMETERS OR CHARACTERISTICS WERE ALTERED BY LAGOON EFFLUENT. PARAMETERS MONITORED WERE, BOD, COD, AND ALSO VOLATILE, SUSPENDED, AND OTHER SOLIDS MEASUREMENTS. THE OXYGEN DEMANDS OF THE VARIOUS CONSTITUENTS WERE THEN CALCULATED AND ANALYZED. FROM THESE EXPERIMENTS, IT WAS DETERMINED THAT RECEIVING STREAMS FOR ALGAE-LADEN EFFLUENTS CAN BE SIGNIFICANTLY INFLUENCED FROM SEVERAL FEET TO SEVERAL MILES DOWNSTREAM. THE AMOUNT OF FLOW IN THE STREAM, AND THE RIFFLE-POOL RATIO WERE FOUND TO BE THE FACTORS WHICH DETERMINE THE DISTANCE DOWNSTREAM WHICH A STREAM WILL BE AFFECTED. SMALLER STREAMS MAY BE BROKEN DOWN INTO RIFFLES, WHICH ACT MUCH LIKE TRICKLING FILTERS, AND POOLS, WHICH ARE MERELY SEDIMENTATION AND DIGESTION UNITS. A MAJOR POINT OF THESE EXPERIMENTS WAS DEMONSTRATION OF THE FACT THAT THE RECEIVING STREAM, IN MANY CASES, IS AN INTEGRAL PART OF THE TREATMENT FACILITY AND MUST BE CONSIDERED AS SUCH BEFORE THE TOTAL SYSTEM CAN BE EVALUATED. (LOWRY-TEXAS)

FIELD 05D, 05C

ACCESSION NO. W71-07097

EFFECTS OF OXIDATION POND EFFLUENT ON RECEIVING WATER IN THE SAN JOAQUIN RIVER ESTUARY,

FEDERAL WATER QUALITY ADMINISTRATION, PACIFIC SOUTHWEST REGION, CALIFORNIA/NEVADA BASINS OFFICE.

RICHARD C. BAIN, JR., PERRY L. MCCARTY, JAMES A. ROBERTSON, AND WILLIAM H. PIERCE.

2ND INTERNATIONAL SYMPOSIUM FOR WASTE TREATMENT LAGOONS, JUNE 23-25, 1970, KANSAS CITY, MISSOURI, P 168-180. 8 FIG, 6 TAB, 8 REF.

DESCRIPTORS:
*OXIDATION LAGOONS, *OXYGENATION, *ALGAE, DEPTH, TEMPERATURE, PHOTOSYNTHESIS, RE-AERATION, LIGHT PENETRATION, NUTRIENTS, PHOSPHORUS, NITROGEN, TIDAL WATERS, CHEMICAL OXYGEN DEMAND, BIOCHEMICAL OXYGEN DEMAND, WASTE WATER TREATMENT, WATER POLLUTION EFFECTS.

IDENTIFIERS:
*SAN JOAQUIN RIVER.

ABSTRACT:
OXIDATION POND EFFLUENT ENTERING THE SAN JOAQUIN RIVER ESTUARY WAS ASSUMED TO BE A MAJOR CAUSE OF LOW DISSOLVED OXYGEN LEVELS, AND RESULTANT FISH KILLS. IN AN EFFORT TO PROVE OR DISPROVE THIS ASSUMPTION, FACTORS CONSIDERED RELATED TO DISSOLVED OXYGEN CONCENTRATION WERE STUDIED. THESE FACTORS INCLUDED WATER TEMPERATURE, ALGAL POPULATION, OXYGEN DEMANDS, NUTRIENTS, AND CHANNEL AND FLOW CHARACTERISTICS. SAMPLES WERE TAKEN FROM THE RIVER AT VARIOUS TIMES, FROM INFLUENT AND EFFLUENT FROM THE TREATMENT PLANT, AND FROM THE OXIDATION LAGOON. THEY WERE THEN ANALYZED FOR NH3-NITROGEN, NO3-NITROGEN, ORGANIC NITROGEN, ORGANIC CARBON, CARBONATE ALKALINITY, BICARBONATE ALKALINITY, ORTHOPHOSPHORUS, TOTAL PHOSPHORUS, 5 DAY BOD, 30 DAY BOD, AND COD. FROM THE PRECEDING INVESTIGATIONS, IT WAS DETERMINED THAT DEPRESSED OXYGEN LEVELS WERE THE RESULT OF BOTH PHYSICAL AND BIOLOGICAL FACTORS. PHYSICALLY, THE DEEPENING OF THE CHANNEL REDUCED TIDAL VELOCITY AND THEREBY REDUCED TURBULENCE AND RATE OF NATURAL RE-OXYGENATION, LEADING TO LOWER OXYGEN TRANSFER RATES. BIOLOGICALLY, ALGAE THRIVES IN BOTH THE OXIDATION PONDS AND THE SHALLOW RIVER, BUT IS TRAPPED IN THE DEEPER CHANNEL WHERE LIGHT PENTRATION IS INSUFFICIENT TO SUPPORT IT. THE ALGAE THEN DECOMPOSES AND REQUIRES OXYGEN. THE PROBLEM THEN, IS A COMPLEX COMBINATION OF FACTORS WHICH REQUIRES THE SYSTEM APPROACH, IF A FULLY COMPREHENSIVE SOLUTION IS TO BE OBTAINED. (LOWRY-TEXAS)

FIELD 05D, 05C

ACCESSION NO. W71-07098

MATHEMATICAL SIMULATION OF WASTE STABILIZATION PONDS,

FEDERAL WATER QUALITY ADMINISTRATION, CINCINNATI, OHIO. ADVANCED WASTE TREATMENT LABS.

JOSEPH F. ROESLER, AND HERBERT C. PREUL.

2ND INTERNATIONAL SYMPOSIUM FOR WASTE TREATMENT LAGOONS, JUNE 23-25, 1970, KANSAS CITY, MISSOURI, P 180-185. 20 REF.

DESCRIPTORS:
*OXIDATION LAGOONS, *MATHEMATICAL MODELS, MIXING, WATER RE-USE, LIGHT INTENSITY, EVAPORATION, RE-AERATION, DISSOLVED OXYGEN, ALGAE, ANAEROBIC CONDITIONS, AEROBIC CONDITIONS, BIODEGRADATION, SEEPAGE, SLUDGE, COST ANALYSIS, *BIOCHEMICAL OXYGEN DEMAND, WASTE WATER TREATMENT.

ABSTRACT:
THE MASS BALANCE APPROACH WAS USED TO OBTAIN EQUATIONS RELATING TO OXIDATION LAGOON PROCESSES. THE MODEL WAS BASED ON THE ASSUMPTION THAT THE INFLUENT WAS DOMESTIC SEWAGE, AND THAT AN ANAEROBIC SLUDGE LAYER WAS FORMED AT THE BOTTOM OF THE POND. EQUATIONS WERE DERIVED FOR BOD REMOVAL BY ALGAE, ANAEROBIC DECOMPOSITION, AND OXIDATION POND DEPTH. IN ORDER TO VERIFY EQUATIONS DERIVED FROM THE MASS BALANCE APPROACH, EXPERIMENTAL DATA ON EFFLUENT BOD, DETENTION TIME, SUNLIGHT INTENSITY, TEMPERATURE, EVAPORATION RATE, AND DISSOLVED OXYGEN DEFICIENCY WERE OBTAINED FROM THE LITERATURE. A COMPUTER PROGRAM WAS THEN DESIGNED TO EVALUATE THE CONSTANTS AND THE DISPOSITION OF ALL BOD'S. FROM THE PRECEDING EQUATION AND CALCULATIONS, IT WAS CONCLUDED THAT: (1) OXIDATION LAGOONS ARE ECONOMICAL TO BUILD AND MAINTAIN, (2) THE ROLE OF ALGAE HAS BEEN OVEREMPHASIZED, WITH RE-AERATION AND ANAEROBIC PROCESSES EQUALLY, IF NOT MORE, IMPORTANT, AND (3) SINCE ALGAE OCCASIONALLY INTERFERE WITH RECEIVING STREAM WATER QUALITY, INVESTIGATIONS SHOULD BE CONDUCTED IN HOW TO MAXIMIZE ANAEROBIC DECOMPOSITION AT THE BOTTOM AND MINIMIZE ALGAE PRODUCTION. (LOWRY-TEXAS)

FIELD 05D

ACCESSION NO. W71-07099

DESIGNING WASTE PONDS TO MEET WATER QUALITY CRITERIA,

CALIFORNIA UNIV., BERKELEY. DEPT. OF SANITARY ENGINEERING AND PUBLIC HEALTH.

WILLIAM J. OSWALD, AARON MERON, AND MARIO D. ZABAT.

2ND INTERNATIONAL SYMPOSIUM FOR WASTE TREATMENT LAGOONS, JUNE 23-25, 1970, KANSAS CITY, MISSOURI, P 186-194, 3 FIG. 4 TAB, 23 REF.

DESCRIPTORS:
*OXIDATION LAGOONS, *DESIGN CRITERIA, *WATER QUALITY, ALGAE, HARVESTING, DEPTH, TEMPERATURE, LIGHT INTENSITY, CHEMICAL PRECIPITATION, HYDROGEN ION CONCENTRATION, COLIFORMS, BACTERIA, PHOTOSYNTHESIS, ORGANIC COMPOUNDS, PHOSPHATES, NITRATES, NUTRIENTS, ANAEROBIC CONDITIONS, AEROBIC CONDITIONS, BIOCHEMICAL OXYGEN DEMAND, WASTE WATER TREATMENT.

ABSTRACT:
THE OVERALL PROCESS OF WASTE STABILIZATION IN OXIDATION LAGOONS WAS EXAMINED FROM A LOGICAL DESIGN STANDPOINT. SIGNIFICANT FACTORS WERE EXAMINED FROM THE LITERATURE, AND THESE INCLUDED INFECTIOUS AGENTS, PLANT NUTRIENTS, ORGANIC CHEMICALS, AND LAND EROSION AND SUBSEQUENT ADDITION OF MINERALS, CHEMICALS, AND SILTS, RADIOACTIVE SUBSTANCES, AND HEAT POLLUTION. IN EACH CASE, DESIGN PRACTICES BOTH FROM LITERATURE RESEARCH AND PRACTICAL EXPERIENCE WERE PRESENTED. A SUMMARY OF THE THREE YEARS OF OPERATION OF THE SAINT HELENA WASTE POND SYSTEM IS PRESENTED IN SUPPORT OF THE DESIGN PRACTICES PREVIOUSLY ADVOCATED. WATER QUALITY CONTROL OF A HIGH ORDER IS CONSISTENTLY OBTAINED AT THE SAINT HELENA PLANT, AND THIS CONTROL IS ATTRIBUTED MAINLY TO THESE IMPROVED DESIGN PRACTICES. FURTHER IMPROVEMENT MAY BE POSSIBLE WITH ADDITION OF ALGAE HARVESTING FACILITIES WHICH WILL ALSO REMOVE PHOSPHATES, BUT SINCE THERE IS LITTLE OR NO EFFLUENT FROM THE PONDS NOW THE PRESENT SYSTEM MORE THAN COMPLIES WITH CURRENT STANDARDS. (LOWRY-TEXAS)

FIELD 05D

ACCESSION NO. W71-07100

NEW EXPERIMENTAL POND DATA,

TEXAS UNIV., AUSTIN. ENVIRONMENTAL HEALTH ENGINEERING RESEARCH LAB.

E. F. GLOYNA, AND J. AGUIRRE.

2ND INTERNATIONAL SYMPOSIUM FOR WASTE TREATMENT LAGOONS, JUNE 23-25, 1970, KANSAS CITY, MISSOURI, P 200-210. 3 FIG, 10 TAB, 10 REF.

DESCRIPTORS:
*OXIDATION LAGOONS, *SLUDGE, ANAEROBIC CONDITIONS, AEROBIC CONDITIONS, PILOT PLANTS, ALGAE, HYDROGEN ICN CONCENTRATION, CHEMICAL OXYGEN DEMAND, *BIOCHEMICAL OXYGEN DEMAND, MIXING, SAMPLING, ORGANIC LOADING.

IDENTIFIERS:
LABORATORY SCALE, *TOTAL ORGANIC CARBON.

ABSTRACT:
THREE LABORATORY-SCALE UNITS AND THREE PILOT-SCALE UNITES WERE CONSTRUCTED AND OPERATED FOR A PERIOD OF EIGHT MONTHS AT THE CONVENTIONALLY ACCEPTED LOADING RATES OF 70 LBS BODU/ACRE/DAY AND 65 DAY DETENTION TIME FOR THE PILOT SCALE PLANTS, WITH ONLY 10 LBS BODU/ACRE/DAY AND 90 DAYS DETENTION TIME IN THE LAB SCALE UNITS. THE ONLY DIFFERENCE IN THE THREE UNITS WAS THE LOCATION OF THE ANAEROBIC SLUDGE DEPOSITION ZONE. ORGANIC REMOVAL EFFICIENCIES WERE DETERMINED BY TESTS ON BOD5, COD, AND TOC. SUSPENDED SOLIDS REMOVAL WAS DETERMINED BY MEMBRANE FILTER ANALYSIS, AND GENERAL POND CHARACTERISTICS WERE EVALUATED THROUGH MEASUREMENT OF PH, DO, TEMPERATURE, NITROGEN, PHOSPHORUS, AND ALGAE TYPING. FROM THIS EXPERIMENTAL EVALUATION, IT WAS DETERMINED THAT PILOT SCALE SYSTEMS ARE OF CONSIDERABLE VALUE IN ARRIVING AT DESIGN CRITERIA FOR A PARTICULAR WASTE IN A PARTICULAR LOCALITY. PILOT PLANTS LACK MANY OF THE DISADVANTAGE OF LAB SCALE PLANTS MAINLY DUE TO THE INCREASED SIZE, AND THEREFORE, DECREASED SENSITIVITY OF THE PILOT PLANTS. ALSO, SERIES ARRANGEMENT OF A MULTIPLE POND SYSTEM HAD A MUCH GREATER EFFECT THAN DID CHANGING THE LOCATION OF THE SLUDGE DEPOSITION ZONE. ONE OF THE POND SYSTEMS UTILIZED FOR THIS STUDY OPERATED ON THE SAME PRINCIPAL AS A SERIES POND ARRANGEMENT, AND THE ANAEROBIC PRETREATMENT PORTION OF THE SYSTEM PROVIDED 50 TO 76% REDUCTION OF ORGANIC MATERIAL IN FROM 3 TO 5 DAYS BETWEEN 12 AND 24C. (LOWRY-TEXAS)

FIELD 05D

ACCESSION NO. W71-07102

A RATIONAL APPROACH TO THE DESIGN OF AERATED LAGOONS,

CORPS OF ENGINEERS, ANCHORAGE, ALASKA. SANITARY AND CIVIL ENGINEERING SECTION.

EDWARD F. POHL.

2ND INTERNATIONAL SYMPOSIUM FOR WASTE TREATMENT LAGOONS, JUNE 23-25, 1970, KANSAS CITY, MISSOURI, P 231-243. 9 FIG, 3 TAB, 26 REF.

DESCRIPTORS:
*OXIDATION LAGOONS, *OXYGENATION, DISSOLVED OXYGEN, SATURATION, ALGAE, METHANE BACTERIA, EUTROPHICATION, NITRATES, PHOSPHATES, BIODEGRADATION, ANAEROBIC CONDITIONS, AEROBIC CONDITIONS, SLUDGE, LIGHT INTENSITY, COLIFORMS, BIOCHEMICAL OXYGEN DEMAND, WASTE WATER TREATMENT.

IDENTIFIERS:
*AERATED LAGOONS, SUSPENDED SOLIDS.

ABSTRACT:
THE VARIOUS PARAMETERS INVOLVED IN THE DESIGN OF AERATED LAGOONS WERE INVESTIGATED WITH PARTICULAR REFERENCE TO THE USE OF AIR DIFFUSED INTO THE LAGOON THROUGH LENGTHS OF TUBING. IT WAS DETERMINED THAT A DESIGN FOR AN AERATED LAGOON SHOULD EVOLVE IN THREE STAGES, EACH TO BE EVALUATED FOR SUMMER AND WINTER CONDITIONS. TO DESIGN AN AERATED LAGOON, THE FIRST STAGE INVOLVES EVALUATION OF THE RECEIVING BODY BY BOTH THE WASTE PRODUCERS, AND THE AGENCY RESPONSIBLE FOR THE PROTECTION OF THE QUALITY OF THE RECEIVING WATER. THE REQUIRED EFFLUENT QUALITY DETERMINATION IS BASED ON BOD, SUSPENDED SOLIDS, EFFLUENT COLIFORM ORGANISM COUNT. DETERMINATION OF THE DETENTION TIME REQUIRED TO ACHIEVE THE STAGE I OBJECTIVES IS THE SECOND STAGE, AND THE FINAL STAGE INVOLVES AN EVALUATION OF THE OXYGEN NEEDED TO SUPPLY BOTH THE AERATION TANK MIXED LIQUOR SOLIDS RESPIRATION, AND THE BENTHAL OXYGEN DEMAND. CONSERVATIVE OVERALL DESIGN VALUES ARE RECOMMENDED BECAUSE OF THE GAPS IN THE RESEARCH DATA. UNTIL MORE RESEARCH HAS BEEN DONE TO INVESTIGATE THE CONSTANTS AND THE MECHANISMS OF THE PROCESS, CONSERVATIVE ESTIMATES WILL CONTINUE TO BE RECOMMENDED IN ORDER TO PROVIDE A LARGER MARGIN OF SAFETY. (LOWRY-TEXAS)

FIELD 05D

ACCESSION NO. W71-07106

A COMPARATIVE STUDY OF AERATED LAGOON TREATMENT OF MUNICIPAL WASTE WATERS,

METROPOLITAN CORP. OF GREATER WINNIPEG (MANITOBA). WATER WORKS AND WASTE DISPOSAL DIV.

G. E. BURNS, R. M. GIRLING, A. R. PICK, AND D. W. VAN ES.

2ND INTERNATIONAL SYMPOSIUM FOR WASTE TREATMENT LAGOONS, JUNE 23-25, 1970, KANSAS CITY, MISSOURI, P 258-276. 14 FIG, 3 TAB, 6 REF.

DESCRIPTORS:
*OXIDATION LAGOONS, *CLIMATIC ZONES, *COST ANALYSIS, TEMPERATURE, NUTRIENTS, DISSOLVED OXYGEN, ALGAE, SLUDGE, OXYGENATION, ANAEROBIC CONDITIONS, AEROBIC CONDITIONS, BIOCHEMICAL OXYGEN DEMAND, WASTE WATER TREATMENT.

IDENTIFIERS:
*AERATED LAGOONS, SURFACE AERATORS.

ABSTRACT:
THREE PILOT SCALE AERATED LAGOONS WERE CONSTRUCTED AT WINNIPEG, CANADA TO TEST THE EFFECT OF THE CANADIAN PRAIRIE CLIMATIC CONDITIONS ON LAGOON OPERATION. THE PILOT CELLS WERE OF THE ANAEROBIC-AEROBIC TYPE. EACH LAGOON WAS EQUIPPED WITH A DIFFERENT TYPE OF AERATION SYSTEM, DIFFUSED AIR, SURFACE AERATORS, OR A COMBINATION OF BOTH. TEST WERE CONDUCTED OVER A 20 MONTH PERIOD ON AN INFLUENT AVERAGING 175 MG/L BOD AND 188 MG/L OF SUSPENDED SOLIDS. FROM THE RESULTS OF THIS INVESTIGATION, IT WAS CONCLUDED THAT AERATED LAGOONS CAN SATISFACTORILY PROVIDE SECONDARY TREATMENT UNDER PRAIRIE CLIMATIC CONDITIONS. BOD REMOVAL EFFICIENCY AND DISSOLVED OXYGEN CONCENTRATION BOTH UNDERGO A SEASONAL SUMMER DECLINE DUE TO THE BUILD-UP OF SLUDGE DURING THE SUMMER MONTHS. THE PREVAILING CONSIDERATION FOR ECONOMIC FEASIBILITY OF AN AERATED LAGOON SYSTEM IS THE EXTENT AND COST IMPLICATIONS OF SLUDGE REMOVAL AND DISPOSAL FACILITIES. ALSO THE USE OF SURFACE AERATORS WAS DEMONSTRATED TO BE IMPRACTICAL IN AREAS WHICH SUSTAIN LARGE AMOUNTS OF ICE COVER. RESEARCH IS CONTINUING ON THE ECONOMIC FACTORS INVOLVED IN SLUDGE REMOVAL AND DISPOSAL. (LOWRY-TEXAS)

FIELD 05D

ACCESSION NO. W71-07109

TERTIARY TREATMENT BY AERATED LAGOON,

V. N. WAHBEH, AND L. W. WELLER.

2ND INTERNATIONAL SYMPOSIUM FOR WASTE TREATMENT LAGOONS, JUNE 23-25, 1970, KANSAS CITY, MISSOURI, P 293-299. 11 FIG, 3 TAB, 5 REF.

DESCRIPTORS:
*OXIDATION LAGOONS, *TERTIARY TREATMENT, *DESIGN CRITERIA, TRICKLING FILTERS, ALGAE, ORGANIC LOADING, ODOR, COLIFORMS, TEMPERATURE, DISSOLVED OXYGEN, AERATION, OXYGENATION, NUTRIENTS, SEDIMENTATION, CHLORINATION, SLOPES, BIOCHEMICAL OXYGEN DEMAND, HYDROGEN ION CONCENTRATION, WASTE WATER TREATMENT, KANSAS.

IDENTIFIERS:
*AERATED LAGOONS, SUSPENDED SOLIDS.

ABSTRACT:
AN OXIDATION POND, ORIGINALLY DESIGNED AS THE TREATMENT FACILITY FOR A SMALL COMMUNITY AND LATER SUPERCEDED BY A HIGH-RATE TRICKLING FILTER PLANT, WAS REMODELED AND EQUIPPED WITH DIFFUSED AIR AERATION USING THREE 600 CFM BLOWERS. THIS UNIT PROVIDED TREATMENT MAINLY FOR STORM BYPASS, UNTIL IT WAS UTILIZED AS A TERTIARY STAGE TO FOLLOW THE TRICKLING FILTER. TESTS WERE THEN CONDUCTED, USING AN AUTOMATIC SAMPLING DEVICE, TO DETERMINE THE EFFICIENCY OF SUCH AN ARRANGEMENT DURING SEVERAL SHORT TESTING PERIODS. IT WAS DISCOVERED THAT CONVENTIONAL OXIDATION PONDS LACKING AERATION WERE INEFFECTIVE IN REDUCTION OF BOD, SUSPENDED SOLIDS, AND INORGANIC NUTRIENTS UNLESS SOME METHOD OF ALGAE HARVESTING WAS EMPLOYED. AERATED LAGOONS, HOWEVER, PROVIDED IN EXCESS OF 60% REDUCTION OF BOD AND SUSPENDED SOLIDS, ALTHOUGH NO DETECTABLE REDUCTION IN INORGANIC NUTRIENTS WAS EVIDENCED. SOME CRITICAL PARAMETERS FOR DESIGN OF AERATED LAGOONS ARE (1) GREATER THAN 4:1 AND PREFERABLY A 2:1 SIDE SLOPE TO PREVENT SOLIDS DEPOSITION ON THE SLOPE, AND (2) MINIMIZATION OF WATER SURFACE AREA PREVENTS UNWANTED ALGAL CELL GROWTH WHICH OTHERWISE WOULD DEFEAT THE PURPOSE OF THE USE OF AERATED LAGOONS INSTEAD OF OXIDATION LAGOONS. (LOWRY-TEXAS)

FIELD 05D

ACCESSION NO. W71-07113

CHALLENGE FOR WASTE WATER LAGOONS,

FEDERAL WATER QUALITY ADMINISTRATION, CINCINNATI, OHIO. ADVANCED WASTE TREATMENT RESEARCH LAB.

FRANCIS M. MIDDLETOWN, AND ROBERT L. BUNCH.

2ND INTERNATIONAL SYMPOSIUM FOR WASTE TREATMENT LAGOONS, JUNE 23-25, 1970, KANSAS CITY, MISSOURI, P 364-366.

DESCRIPTORS:
*OXIDATION LAGOONS, *WASTE WATER TREATMENT, *WATER RE-USE, BIOCHEMICAL OXYGEN DEMAND, ORGANIC LOADING, ODORS, PHOTOSYNTHESIS, ALGAE, BACTERIA, CHLORINATION, GROUNDWATER, COST ANALYSIS.

IDENTIFIERS:
SUSPENDED SOLIDS.

ABSTRACT:
INCREASING USE OF WATER HAS NECESSITATED GREATER AND GREATER USAGE OF RECONDITIONED WASTE WATER AS DRINKING WATER. WITH THIS GREATER USAGE HAS COME INCREASINGLY STRICTER LAWS REGARDING THE QUALITY OF THE EFFLUENT WHICH A TREATMENT FACILITY IS PERMITTED TO DISCHARGE TO A RECEIVING WATER. THEREFORE TREATMENT METHODS MUST BECOME MORE SOPHISTICATED AS TIME GOES ON IN ORDER THAT EFFLUENTS NOT ONLY PRESERVE THE QUALITY OF A RECEIVING STREAM, BUT MUST HELP TO RESTORE IT. IN LIGHT OF THESE FACTS LAGOON PERFORMANCE MUST BE EVALUATED WITH RESPECT TO FUTURE CONDITIONS. LAGOONS HAVE SEVERAL DRAW BACKS FROM A WATER POLLUTION STANDPOINT. THESE ARE: (1) ALTHOUGH COLIFORMS MAY BE REDUCED AS MUCH AS 98% IN A LAGOON, THE WATER QUALITY STANDARDS FOR NUMBER OF ORGANISMS PER MILLILITER MAY STILL BE VIOLATED; (2) IF ALGAE IS ALLOWED TO PASS TO THE RECEIVING STREAM, IT MAY BECOME AS MUCH OF A POLLUTIONAL FACTOR AS THE RAW WASTE SINCE IT CONTAINS THE SAME AMOUNT OF ORGANIC MATTER; (3) LAGOON EFFLUENT IS HIGH IN SUSPENDED SOLIDS WHICH ARE BOTH AESTHETICALLY UNPLEASING AND OXYGEN DEMANDING; (4) LAGOONS ARE A THREAT TO GROUNDWATER QUALITY BECAUSE OF THE DIFFICULTY IN SEALING THEM. UNTIL NOW, THE MOST ATTRACTIVE FEATURE OF LAGOONS HAS BEEN THEIR LOW COST. AS TREATMENT MUST BECOME MORE SOPHISTICATED, LAGOONS AS PRESENTLY CONSTRUCTED WILL NEED ADDITIONAL TREATMENT WHICH WILL MORE THAN OFFSET THEIR COST ADVANTAGE. THEREFORE, UNLESS SIGNIFICANT ADVANCES ARE MADE IN LAGOON TECHNOLOGY, IT IS DOUBTFUL IF LAGOONS WILL HAVE A PLACE IN THE FUTURE. (LOWRY-TEXAS)

FIELD 05D

ACCESSION NO. W71-07123

STATE OF THE ART-OXICATION PONDS,

NORTH DAKOTA UNIV., GRAND FORKS. DEPT. OF MICROBIOLOGY.

JOHN W. VENNES.

2ND INTERNATIONAL SYMPOSIUM FOR WASTE TREATMENT LAGOONS, JUNE 23-25, 1970, KANSAS CITY, MISSOURI, P 366-376. 3 TAB, 67 REF.

DESCRIPTORS:
*OXIDATION LAGOONS, *WASTE WATER TREATMENT, ORGANIC LOADING, BIOCHEMICAL OXYGEN DEMAND, SOLAR RADIATION, NUTRIENTS, TEMPERATURE, COLIFORMS, CHLORINATION, SEEPAGE, PERCOLATION, EVAPORATION, INSECTS, EUTROPHICATION, ALGAE.

ABSTRACT:
AN INTENSIVE LITERATURE SEARCH OF THE WASTE STABILIZATION POND METHOD OF TREATING WASTE WATER WAS CONDUCTED TO DETERMINE THE PRESENT STATUS OF SUCH UNITS WITH REGARD TO DESIGN CRITERIA, OPERATIONAL PARAMETERS, AND COMPLIANCE WITH WATER POLLUTION PREVENTION REGULATIONS. DESIGN CONSIDERATIONS ARRIVED AT WERE: (1) 1 LANGLEY/DAY OF RADIATION IS SUFFICIENT TO TREAT 1 LB BOD/ACRE/DAY; (2) MULTI-STAGE SERIES PONDS PROVIDE THE BEST ORGANIC AND BACTERIAL REMOVALS; (3) SEEPAGE AND EVAPORATION MUST BE CONSIDERED IN LAGOON DESIGN SINCE IN MANY CASES THESE ARE THE ONLY FORMS OF EFFLUENT; (4) CHLORINATION WHILE KILLING THE BACTERIA, WILL ALSO KILL THE ALGAE WHICH WILL THEN EFFECT AN IMMEDIATE ORGANIC LOAD ON THE SYSTEM OR THE RECEIVING STREAM; (5) ALGAE REMOVAL TECHNIQUES MUST BE DEVELOPED SINCE MOST OF THE ORGANIC MATERIAL IS NOT DESTROYED, BUT MERELY CONVERTED TO ALGAE, AS ARE THE INORGANIC NUTRIENTS AS WELL. WITHOUT ALGAE REMOVAL, OXIDATION POND EFFLUENT IS HIGH IN ORGANIC CONTENT WHICH MAY CAUSE DETERIORATION OF THE RECEIVING STREAM QUALITY. MORE STUDY IS NEEDED ON THE MECHANISMS WHICH ARE UTILIZED IN OXIDATION PONDS FOR THE REMOVAL OF ORGANICS, AND ALSO ON THE ACTUAL MEASUREMENT OF THE ORGANIC LOADINGS ON THE FACILITIES BEFORE ORGANIC POLLUTANTS AND MICRO-ORGANISMS PRESENT IN POND EFFLUENTS CAN BE RELIABLY PREDICTED. (LOWRY-TEXAS)

FIELD 05D

ACCESSION NO. W71-07124

A COMPUTER BASED FLORISTIC ANALYSIS OF PAMLICO RIVER PHYTCPLANKTON,

EAST CAROLINA UNIV., GREENVILLE, N.C. DEPT. OF BIOLOGY.

VINCENT J. BELLIS.

COPIES AVAILABLE FROM WATER RESOURCES RESEARCH INST, 124 RIDDICK BLDG, NORTH CAROLINA STATE UNIV, RALEIGH, NC, 27607, PRICE $2.50. NORTH CAROLINA UNIVERSITY WATER RESOURCES RESEARCH INSTITUTE REPORT NO 46, JAN 1971. 28 P, 7 FIG, 5 TAB, 13 REF. OWRR PROJECT A-044-NC(1) AGREEMENT NO 14-31-0001-3033, FY 1970.

DESCRIPTORS:
*ESTURARIES, *PHYTOPLANKTON, *DISTRIBUTION PATTERNS, *NORTH CAROLINA, *COMPUTER PROGRAMS, DATA PROCESSING, DATA COLLECTIONS, VARIABILITY, PROBABILITY, WATER POLLUTION EFFECTS, ALGAE, AQUATIC HABITATS, AQUATIC ENVIRONMENT.

IDENTIFIERS:
*PAMLICO RIVER(NC).

ABSTRACT:
A COMPUTER BASED TECHNIQUE FOR ANALYZING DISTRIBUTION PATTERNS AMONG ESTUARINE ORGANISMS WAS DEVELOPED AND TESTED ON PHYTOPLANKTON COLLECTIONS FROM PAMLICO RIVER, NORTH CAROLINA. JACCARD'S COEFFICIENT OF SIMILARITY WAS USED TO GENERATE A SIMILARITY MATRIX WITH FINAL PRINTOUT IN DENDROGRAM FORM SHOWING CLUSTERS OF COLLECTIONS HAVING INHERENT SIMILARITY. USE OF A SIMILARITY INDEX BASED SOLELY UPON PRESENCE OR ABSENCE OF SPECIES WITHIN ALL POSSIBLE PAIRS OF DATA COLLECTIONS PRODUCED A DENDROGRAM WHICH DESCRIBED THE SEASONAL PERIODICITY OF PHYTOPLANKTON IN THE CENTRAL PAMLICO RIVER. COMPUTER PROGRAMS DEVELOPED IN THIS PROJECT ARE WRITTEN IN FORTRAN IV AND HAVE BEEN DEPOSITED IN THE PROGRAM LIBRARY OF THE EAST CAROLINA UNIVERSITY COMPUTER CENTER AND CAN BE OBTAINED UPON REQUEST FROM ITS DIRECTOR. (KNAPP-USGS)

FIELD 02L, 07C, 05C

ACCESSION NO. W71-07337

NOTES ON FRESHWATER ALGAE, (IN GERMAN),

VIENNA UNIV, (AUSTRIA), BOTANISCHES INSTITUT UND BOTANISCHER GARTEN.

LOTHAR GEITLER.

OSTERREICHISCHE BOTANISCHE ZEITSCHRIFT, NO 118, P 306-310, 1970. 16 REF.

DESCRIPTORS:
*ALGAE, *FRESH WATER, DISTRIBUTION PATTERNS, BEHAVIOR.

IDENTIFIERS:
MOUGEOTIA GENUFLEXA, ANKISTRODESMUS FALCATUS, PODCHEDRA.

ABSTRACT:
THIS PAPER DESCRIBES DISTRIBUTION AND BEHAVIOR OF FRESHWATER ALGAE, INCLUDING MOUGEOTIA GENUFLEXA, ANKISTRODESMUS FALCATUS, VAR STIPITATUS, AND MEMBERS OF THE GENUS PODCHEDRA. (WILDE-WISCONSIN)

FIELD 05C

ACCESSION NO. W71-07360

LIGHT INTENSITY AND THE VERTICAL DISTRIBUTION OF ALGAE IN TERTIARY OXIDATION PONDS,

NORTH CAROLINA UNIV., CHAPEL HILL. DEPT. OF ENVIRONMENTAL SCIENCES AND ENGINEERING.

WILLIAM R. HARTLEY, AND CHARLES M. WEISS.

WATER RESEARCH, VOL 4, NO 11, NOV 1970, P 751-763. 7 FIG, 2 TAB, 18 REF.

DESCRIPTORS:
*ALGAE, *OXIDATION LAGOONS, *LIGHT INTENSITY, PONDS, ROTIFERS, PROTOZOA, PHOTOSYNTHESIS, CHLOROPHYLL, TEMPERATURE, HYDROGEN ION CONCENTRATION, DISSOLVED OXYGEN, STRATIFICATION, SAMPLING, MICROORGANISMS, SLUDGE, TERTIARY TREATMENT, BIOCHEMICAL OXYGEN DEMAND, WASTE WATER TREATMENT, *NORTH CAROLINA.

IDENTIFIERS:
DURHAM(NC).

ABSTRACT:
FIVE OXIDATION PONDS AT THE THIRD FORK TREATMENT PLANT IN DURHAM, NORTH CAROLINA, WERE INVESTIGATED TO DETERMINE THE OPTIMUM LIGHTING CONDITIONS, THE TYPES OF ORGANISMS PRESENT IN DIFFERENT SITUATIONS, AND THEIR VERTICAL DISTRIBUTION IN THE POND LEVELS. SAMPLES IN THE PONDS WERE TAKEN AT 3,6,9,12,15,18,21,24 INCHES, AND THE SLUDGE LAYER. THESE SAMPLES WERE TAKEN EVERY TWO HOURS OVER THE 24 HOUR DIURNAL CYCLE. THE SAMPLES WERE PRESERVED IN 3% FORMALIN, AND THEN TAKEN TO THE LABORATORY FOR ANALYSIS. LIGHT INTENSITY, PH, DISSOLVED OXYGEN, AND TEMPERATURE WERE ALSO RECORDED AT THE TIME OF SAMPLING. THE SAMPLES TAKEN WERE COUNTED FOR NUMBER OF ORGANISMS PRESENT, AND, ALSO CENTRIFUGED FOR IDENTIFICATION OF SPECIES PRESENT IN LESSER NUMBERS. THE SURFACE WEIRS INSTALLED IN THE PONDS AFFECT THE CHEMICAL AND PHYSICAL PARAMETERS OF POND PERFORMANCE WITH PARTICULAR REFERENCE TO THE DISSOLVED OXYGEN CONCENTRATION, IT BEING MUCH LOWER IN THE POND WHICH ALLOWED THE MAT TO FORM AT THE SURFACE. EUGENA ROSTIFERA WILL CHANGE THEIR VERTICAL POSITION IN THE POND TO REACH THE POINT OF SUBMERGENCE WHERE THE LIGHT INTENSITY IS 75 CANDLES/FT2. WHERE THEY WERE NOT ABLE TO MOVE TO RETREAT FROM HIGH LIGHT INTENSITY CYSTS WERE FORMED. (LOWRY-TEXAS)

FIELD 05D

ACCESSION NO. W71-07382

A POLYCHLORINATED BIPHENYL (AROCLOR 1254) IN THE WATER, SEDIMENT, AND BIOTA OF ESCAMBIA BAY, FLORIDA,

BUREAU OF COMMERCIAL FISHERIES, GULF BREEZE, FLA. CENTER FOR ESTUARINE AND MENHADEN RESEARCH.

T. W. DUKE, J. I. LOWE, AND A. J. WILSON, JR.

BULLETIN OF ENVIRONMENTAL CONTAMINATION AND TOXICOLOGY, VOL 5, NO 2, P 171-180, MAR-APR 1970. 2 FIG, 3 TAB, 7 REF.

DESCRIPTORS:
*WATER POLLUTION EFFECTS, *BIOASSAY, *TOXICITY, *CHLORINATED HYDROCARBONS, *BIOINDICATORS, INDUSTRIAL WASTES, FLORIDA, SESSILE ALGAE, ESTUARIES, CHEMICAL WASTES, GAS CHROMATOGRAPHY, SEDIMENTS, OYSTERS, SHRIMP, CRABS, WATER POLLUTION SOURCES, SEA WATER, TROUT, TOXINS.

IDENTIFIERS:
*POLYCHLORINATED BIPHENYLS, AROCLOR 1254, ESCAMBIA RIVER, PINFISH, FLOUNDER, CROAKER, MENHADEN.

ABSTRACT:
AROCLOR 1254, A POLYCHLORINATED BIPHENYL, HAS BEEN DETECTED IN THE BIOTA, SEDIMENT, AND WATER OF ESTUARINE AREAS NEAR PENSACOLA, FLORIDA. ONLY ONE SOURCE OF THE CHEMICAL, AN INDUSTRIAL PLANT ON THE ESCAMBIA RIVER, HAS BEEN FOUND. WATER, SEDIMENT, AND FISH, CRAB, OYSTER, AND SHRIMP SAMPLES WERE COLLECTED FROM APRIL THROUGH OCTOBER 1969 AND ANALYZED USING GAS CHROMATOGRAPHY PROCEDURES. IN ADDITION, BIOASSAYS ON FISH, SHRIMP, AND OYSTERS WERE CONDUCTED UNDER CONTROLLED LABORATORY CONDITIONS TO DETERMINE THE TOXIC EFFECTS OF AROCLOR 1254. THESE STUDIES SHOWED THAT JUVENILE SHRIMP WERE THE MOST SENSITIVE AND WERE KILLED WHEN EXPOSED TO 5.0 PPB OF AROCLOR 1254 IN FLOWING SEA WATER. THE AROCLOR CONTENT IN WATER FROM ESCAMBIA BAY, EVEN NEAR THE MOUTH OF THE RIVER, WAS LESS THAN 1 PPB. SHRIMP COLLECTED FROM THE BAY CONTAINED A MAXIMUM OF 2.5 P.P.M. THUS, SHRIMP IN THE BAY PROBABLY WERE NOT EXPOSED TO LETHAL LEVELS DURING THE SAMPLING PERIOD. HIGHEST CONCENTRATIONS IN THE WATER OCCURRED DURING AUGUST AND DECREASED WHEN LEAKAGE FROM THE PLANT WAS CORRECTED. (MORTLAND-BATTELLE)

FIELD 05C, 05B, 05A

ACCESSION NO. W71-07412

A COMPREHENSIVE APPRAISAL OF THE EFFECTS OF COOLING WATER DISCHARGE ON AQUATIC ECOSYSTEMS,

BATTELLE MEMORIAL INST., COLUMBUS, OHIO. COLUMBUS LABS.

ARTHUR A. LEVIN, THOMAS J. BIRCH, ROBERT E. HILLMAN, AND GILBERT E. RAINES.

PRESENTED AT THE PUBLIC AFFAIRS WORKSHOP 'WHY NUCLEAR POWER', HILTON HEAD ISLAND, SOUTH CAROLINA, SEPT 13-16, 1970. 45 P, 9 FIG, 2 TAB, 137 REF.

DESCRIPTORS:
*POWER PLANTS, *THERMAL POLLUTION, *REVIEWS, WATER TEMPERATURE, DISSOLVED OXYGEN, BIOINDICATORS, E. COLI, AQUATIC ALGAE, SUNFISHES, SNAILS, EELS, CHINOOK SALMON, BASS, TROUT, MINNOWS, CRAYFISH, MOSQUITOES, FOOD CHAIN, HERRING, OYSTERS, CLAMS, COLUMBIA RIVER, CALIFORNIA, WASHINGTON, CONNECTICUT, FLORIDA, DELAWARE RIVER, INDIANA.

IDENTIFIERS:
BARNACLES, FLOUNDER, HUMBOLDT BAY, CONNECTICUT RIVER, SAN JOAQUIN RIVER(CALIFORNIA), BISCAYNE BAY, WHITE RIVER(INDIANA), PATUXENT RIVER(CANADA), ONTARIO.

ABSTRACT:
A SELECTIVE REVIEW WAS MADE OF THE LITERATURE, AND BASED UPON THE INFORMATION ACQUIRED, A COMPREHENSIVE APPRAISAL OF THE EFFECTS OF COOLING WATER DISCHARGE ON FRESHWATER AND MARINE ECOSYSTEMS IS MADE. THE AQUATIC ORGANISMS CONSIDERED INCLUDE FISH, ALGAE, DIATOMS, ENTERIC BACTERIA, PATHOGENS, SNAILS, AND LARVAE. (LITTLE-BATTELLE)

FIELD 05B, 05C

ACCESSION NO. W71-07417

FEEDING VALUE OF ANIMAL WASTES,

AGRICULTURAL RESEARCH SERVICE, BELTSVILLE, MD. ANIMAL SCIENCE RESEARCH.

L. W. SMITH.

IN: ANIMAL WASTE REUSE--NUTRITIVE VALUE AND POTENTIAL PROBLEMS FROM FEED ADDITIVES--A REVIEW. ARS 44-224, P 5-13, FEB 1971. 1 TAB.

DESCRIPTORS:
*FEEDS, *RUMINANTS, ALGAE, CATTLE, POULTRY, HOGS, ANIMAL DISEASES, CATFISH, WASTE TREATMENT, FARM WASTES, DEHYDRATION, FEASIBILITY.

IDENTIFIERS:
*ANIMAL MANURE, *LITERATURE REVIEW, FEEDING VALUE.

ABSTRACT:
THIS PAPER REVIEWS THE LITERATURE CONCERNED WITH FEEDING ANIMAL WASTE TO LIVESTOCK. FIBER IN DIETS FOR RUMINANTS IS NOT DIGESTED TO THE MAXIMUM POSSIBLE EXTENT DURING THE INITIAL PASS THROUGH THE DIGESTIVE TRACT. OTHER NUTRIENTS ALSO ESCAPE DIGESTION. FEEDING FECES IS NOT A NEW CONCEPT. EARLY IN THE 1940'S COW MANURE WAS LOOKED UPON AS A SOURCE OF B-COMPLEX VITAMINS. POULTRY AND CATFISH HAVE BEEN SUCCESSFULLY FED RATIONS CONTAINING FEEDLOT MANURE. THERE HAVE BEEN MANY ARTICLES CONCERNING THE USE OF POULTRY LITTER IN RUMINANT FEEDING PROGRAMS. FEEDING POULTRY FECES TO POULTRY WAS REPORTED TO HAVE NO ADVERSE EFFECT ON BIRD MORTALITY OR EGG TASTE. ALGAE GROWN ON SEWAGE HAS BEEN FED TO RATS. THE AUTHORS INDICATE THAT ALGAE IS A POTENTIALLY VALUABLE LIVESTOCK FEED. (CHRISTENBURY-IOWA STATE)

FIELD 05G

ACCESSION NO. W71-07544

FARM WASTE DISPOSAL SYSTEMS,

PURDUE UNIV., LAFAYETTE, IND. DEPT. OF AGRICULTURAL ENGINEERING.

A. C. DALE.

COOPERATIVE EXTENSION SERVICE, PURDUE UNIVERSITY, AE-8C, FEB 1971. 10 P, 1 TAB, 5 FIG, 34 REF.

DESCRIPTORS:
*FARM WASTES, *WASTE DISPOSAL, *LAGOONS, *OXIDATION LAGOONS, AEROBIC CONDITIONS, ANAEROBIC CONDITIONS, ALGAE, ORGANIC MATTER, VOLUME, DRYING, ODOR, NITROGEN, AERATION, HOGS, CATTLE, POULTRY, RESEARCH AND DEVELOPMENT, SOIL, SOIL CONTAMINATION.

IDENTIFIERS:
*DISPOSAL SYSTEMS, *LAND DISPOSAL, OXIDATION DITCHES, AERATED LAGOONS, COMPOSTING, REFEEDING, ANHYDROUS AMMONIA, WASTE CHARACTERIZATION.

ABSTRACT:
IN THIS PUBLICATION THE PRESENT AVAILABLE ALTERNATIVES FOR ANIMAL WASTE DISPOSAL AND CRITERIA FOR SELECTION OF THESE METHODS ARE PRESENTED. A BRIEF LITERATURE REVIEW TELLS OF RESEARCH BEING DONE IN ALL AREAS AND ASPECTS OF ANIMAL WASTE DISPOSAL. LAND DISPOSAL STILL REMAINS THE MOST SUITABLE AND MOST WIDELY USED DISPOSAL METHOD. RESEARCH INDICATES THAT APPROXIMATELY 250 POUNDS OF NITROGEN CAN BE ADDED TO EACH ACRE OF SOIL WITHOUT UNDULY POLLUTING IT. OTHER METHODS OF DISPOSAL DISCUSSED ARE AEROBIC, ANAEROBIC, AND MECHANICALLY AERATED LAGOONS, AS WELL AS OXIDATION DITCHES, COMPOSTING, AND DRYING. RECOMMENDATIONS ARE MADE FOR THE CHEMICAL TREATMENT OF ANIMAL WASTES TO REDUCE ODORS WHILE SPREADING. (WHITE-IOWA STATE)

FIELD 05D

ACCESSION NO. W71-07551

DETOXICATION OF SIMAZINE BY MICROSCOPIC ALGAE,

VSESOYUZNYI NAUCHNO-ISSLEDOVATELSKII INSTITUT SELSKOKHOZYAISTVENNOI MIKROBIOLOGII, LENINGRAD.

YU V. KRUGLOV, AND L. N. PAROMENSKAYA.

TRANS FROM MIKROBIOLOGIYA, VOL 39, NO 1, JAN-FEB 1970. MICROBIOLOGY, VOL 39, NO 1, P 139-142, 1970. 1 FIG, 2 TAB, 10 REF.

DESCRIPTORS:
*HERBICIDES, *PESTICIDES, *ALGAE, *TOXICITY, SOILS, TRIAGINE PESTICIDES, RESISTANCE, CULTURES, METABOLISM.

IDENTIFIERS:
*SIMAZINE, *DETOXICATION, CHLOROSARCINA, ANKISTRODESMUS BRAUNII, CHLORELLA VULGARIS.

ABSTRACT:
INOCULATION OF SIMAZINE-TREATED SOILS WITH A CULTURE OF CHLOROSARCINA SP SUGGESTED A DECREASE IN TOXICITY OF THE HERBICIDE, THE PROCESS BEING PARTLY INFLUENCED BY DIRECT SOLAR RADIATION. THE CHEMICAL ABSORBED BY CHLOROSARCINA AND ANKISTRODESMUS BRAUNII WAS PARTLY METABOLIZED BY CELLS AND FORMED SOME PHYSIOCOCHEMICAL BOND WITH PROTEIN. AS SHOWN BY C-14 ANALYSES, THE STRENGTH AND NATURE OF THIS BOND VARY WITH THE SPECIES OF ALGAE. CELLS OF CHLOROSARCINA, RESISTANT TO SIMAZINE, RETAINED CONSIDERABLY GREATER AMOUNT OF THE CHEMICAL THAN DID THE CELLS OF ANKISTRODESMUS, SENSITIVE TO THE HERBICIDE. (WILDE-WISCONSIN)

FIELD 05G

ACCESSION NO. W71-07675

STUDIES OF PRIMARY PRODUCTIVITY IN COASTAL WATERS OF SOUTHERN LONG ISLAND, NEW YORK,

DOW CHEMICAL CO., FREEPORT, TEX; AND PUERTO RICO UNIV., MAYGUEZ. DEPT. OF MARINE SCIENCES; AND TOWN OF HEMPSTEAD, N.Y. DEPT. OF CONSERVATION AND WATERWAYS; AND VIRGIN ISLANDS COLL., ST. THOMAS.

E. F. MANDELLI, P. R. BURKHOLDER, T. E. DOHENY, AND

MARINE BIOLOGY, VOL 7, NO 2, P 153-160, 1970. 9 FIG, 2 TAB, 34 REF.

DESCRIPTORS:
*PRIMARY PRODUCTIVITY, *COASTS, *PHYTOPLANKTON, *ESTUARIES, *STANDING CROP, SEASONAL, DIATOMS, DINOFLAGELLATES, CHLOROPHYLL, PHOTOSYNTHESIS, TEMPERATURE, SAMPLING, SALINITY, DISSOLVED OXYGEN, NUTRIENTS, CARBON RADIOISOTOPES, PHOSPHATES, SILICATES, SOLAR RADIATION, DEPTH, THERMOCLINE, NITRATES, LIGHT INTENSITY, EUTROPHICATION, MARINE ALGAE, SEA WATER, NEW YORK.

IDENTIFIERS:
*COASTAL WATERS, *SOUTHERN LONG ISLAND, SKELETONEMA COSTATUM, THALASSIOSIRA, CHAETOCEROS, RHIZOSOLENIA ALATA, PERIDINIUM DEPRESSUM, PERIDINIUM PALLIDUM, CERATIUM MASSILENSE, CERATIUM FURCA, CERATIUM TRIPOS, CERATIUM MACROCEROS, NOCTILUCA MILIARIS, GYMNODINIUM SPLENDENS, EXUVIELLA MARINA, EUTREPTIA, NITZSCHIA SERIATA, THALASSIONEMA NITZSCHOIDES, ASTERIONELLA JAPONICA, CHLOROPHYLL-A.

ABSTRACT:
IN CONSERVATION OF COASTAL MARINE RESOURCES, MAINTENANCE OF NATURAL PRIMARY PRODUCTIVITY OF COASTAL WATERS IS SIGNIFICANT. DURING 1966, MONTHLY DETERMINATIONS OF PHYTOPLANKTON PRODUCTIVITY OF TIDAL ESTURARIES AND COASTAL WATERS OF SOUTHERN LONG ISLAND WERE MADE. SUSTAINED BLOOMS OF GREEN FLAGELLATES AND DINOFLAGELLATES WERE FOUND DURING SPRING AND SUMMER IN ESTUARINE WATERS; IN COASTAL AREAS ALTERNATING ABUNDANCES OF DIATOMS AND DINOFLAGELLATES DOMINATED THE STANDING CROP DURING LATE WINTER, EARLY FALL, AND SUMMER. THE CHLOROPHYLL-A DISTRIBUTION IN ESTUARIES EXHIBITED TWO PATTERNS, LASTING ABOUT SIX MONTHS EACH, WITH CONCENTRATION RANGING FROM 1.0 TO 27.6 MG/CU M. IN THE COASTAL AREAS ITS DISTRIBUTION SHOWED A REVERSED PATTERN WITH RANGE FROM 1.45 TO 10.15 MG/CU M. VERTICAL DISTRIBUTION OF CHLOROPHYLL-A WITHIN THE COASTAL REGION EUPHOTIC ZONE SHOWED SIMILAR PATTERNS BOTH NEARSHORE AND OFFSHORE. MEAN PHOTOSYNTHESIS RATES PER UNIT OF CHLOROPHYLL-A VARIED FROM 3.1 TO 3.5 MG CARBON/MG CHLOROPHYLL-A/HOUR; AT LIGHT SATURATION, RATIO VARIED WITH WATER TEMPERATURE AND SPECIES COMPOSITION. MEAN PRIMARY PRODUCTIVITY VALUES WERE 0.35, 0.22, 0.16 G CARBON/CU M PER DAY FOR ESTUARIES, NEARSHORE, AND OFFSHORE AREAS, RESPECTIVELY, DECREASING SEAWARD. (JONES-WISCONSIN)

FIELD 02L, 05C

ACCESSION NO. W71-07875

HIGH MOLECULAR WEIGHT ALGAL SUBSTANCES IN THE SEA,

NAVAL UNDERSEA RESEARCH AND DEVELOPMENT CENTER, PASADENA, CALIF.

J. W. HOYT.

MARINE BIOLOGY, VOL 7, NO 2, P 93-99, 1970. 3 FIG, 3 TAB, 19 REF.

DESCRIPTORS:
*MOLECULAR STRUCTURES, *ALGAE, *FLUID FRICTION, *LITTORAL, PLANT GROWTH SUBSTANCES, PHYTOPLANKTON, FLUID MECHANICS, EXUDATION, OCEANS, CHLOROPHYTA, PHAEOPHYTA, RHODOPHYTA, MARINE BACTERIA.

IDENTIFIERS:
*HIGH MOLECULAR WEIGHT COMPOUNDS, *MOLECULAR SIZE, *ALGAL EXUDATES, EXTRACELLULAR MATERIAL, TURBULENT FRICTION, SEAWEED, PORPHYRIDIUM CRUENTUM, PORPHYRA, GIGARTINA.

ABSTRACT:
TURBULENT-FLOW MEASUREMENTS IN A SPECIAL RHEOMETER REVEALED THAT COMMERCIAL SEAWEED EXTRACTS AND CERTAIN MEMBERS OF PHYTOPLANKTON HAVE EXTRACELLULAR MATERIALS OF LARGE MOLECULES WITH WEIGHTS EXCEEDING 50,000. EXTRACTS FROM A VARIETY OF LITTORAL ALGAE, PARTICULARLY OF PORPHYRA AND GIGARTINA GENERA, HAVE CHANGED THE TURBULENT FRICTION THEREBY INDICATING THE PRESENCE OF HIGH-POLYMER SUBSTANCES. PORPHYRIDIUM CRUENTUM PRODUCED FRICTION REDUCTION AS HIGH AS 60%. RESULTS SUGGEST THAT ALGAL EXUDATES ARE THE SOURCE OF HIGH MOLECULAR COMPOUNDS IN THE SEA. (WILDE-WISCONSIN)

FIELD 02L, 05C

ACCESSION NO. W71-07878

MICRCBIOLOGY OF STREAMS,

KENTUCKY UNIV., LEXINGTON. DEPT. OF MICROBIOLOGY.

R. H. WEAVER, AND H. D. NASH.

IN: INFLUENCES OF STRIP MINING ON THE HYDROLOGIC ENVIRONMENT OF PARTS OF BEAVER CREEK BASIN, KENTUCKY, 1955-66, GEOLOGICAL SURVEY PROFESSIONAL PAPER 427-C, P C53-C57, 1970. 5 P, 1 FIG, 4 TAB.

DESCRIPTORS:
*AQUATIC MICROORGANISMS, *ACID MINE WATER, *STRIP MINES, *WATER POLLUTION EFFECTS, *KENTUCKY, SULFUR BACTERIA, WATER QUALITY, SULFATES, HYDROGEN ION CONCENTRATION, ALGAE, BACTERIA, WATER POLLUTION SOURCES.

IDENTIFIERS:
*BEAVER CREEK BASIN(KY).

ABSTRACT:
DRAINAGE FROM STRIP-MINED AREAS APPEARS TO HAVE AFFECTED THE MICROFLORA OF CANE BRANCH, KENTUCKY. CHEMICAL OXIDATICN OF PYRITIC COMPOUNDS FOUND EXTENSIVELY IN SPOIL BANKS HAS RESULTEC IN THE FORMATICN OF FERROUS SULFATE AND SULFURIC ACID. THIS APPEARS TO HAVE LED TO THE ESTABLISHMENT IN THE MINED PART OF THE CANE BRANCH STUDY AREA OF FERROBACILLUS FERROOXIDANTS, WHICH CONTRIBUTES TO THE PRODUCTION OF ACID ENTERING THE STREAM. THE LOWERING OF PH HAS ENABLED THIS CRGANISM TO EXIST THROUGHOUT THE STREAM FROM THE VICINITY CF THE SPOIL BANKS DOWNSTREAM TO THE GAGING STATION. STANDARD PLATE COUNTS SHOW A MUCH SMALLER NUMBER OF SAPROPHYTIC BACTERIA IN CANE BRANCH THAN IN HELTON BRANCH. THIS, TOO, CAN BE ATTRIBUTED TO THE LOW PH OF CANE BRANCH. THE FILAMENTOUS FUNGI ARE MORE NUMERCUS AND DIVERSIFIED IN CANE BRANCH THAN IN HELTON BRANCH. IN ADDITION, THE YEAST, RHCDOTCRULA, WHICH IS ASSOCIATED WITH INCREASED ACID PRODUCTION BY THIOBACILLUS FERROOXIDANS, ANC THE ALGA BUMILLERIA WERE ISOLATED ONLY FROM CANE BRANCH. (KNAPP-USGS)

FIELD 05C

ACCESSION NO. W71-07942

DEMONSTRATING THE EFFECTS OF NUTRIENTS IN BIO-OXIDATICN PCND RECEIVING STREAMS,

OKLAHOMA UNIV., NORMAN. BUREAU OF WATER RESOURCES RESEARCH.

GEORGE W. REID, LEALE E. STREEBIN, AND OLIVER T. LOVE JR.

AVAILABLE FROM THE NATIONAL TECHNICAL INFORMATION SERVICE AS PB-199 269, $3.00 IN PAPER COPY, $0.95 IN MICROFICHE. MAR 1971. 72 P, 16 FIG, 10 TAB, 49 REF, APPEND. ENVIRONMENTAL PROTECTION AGENCY PROGRAM 16010--03/71, DEMONSTRATION GRANT WPD-98-01-66.

DESCRIPTORS:
*EUTROPHICATION, *OXIDATION LAGOONS, *SEWAGE LAGOONS, BICDEGRADATION, *NUTRIENTS, ALGAE, *DEGRADATION(STREAM), WATER PCLLUTION EFFECTS, WASTE WATER TREATMENT, CYANOPHYTE, ENGLENOPHYTA, ORGANIC LCADING.

ABSTRACT:
THIS STUDY CONSIDERED THE RECEIVING STREAM AS AN INTEGRAL PART OF THE BIO-OXIDATION POND METHOD OF TREATMENT WITH THE CBJECTIVE BEING TO PROVIDE A BETTER UNDERSTANDING OF THE BIO-OXIDATION POND - RECEIVING STREAM SYSTEM. AS REPRESENTATIVE OF THIS 'REAL WORLD' SITUATION WITH ALL OF ITS VARIABLES, FIVE EXISTING CENTRAL OKLAHCMA BIO-OXIDATION PONDS WHICH HAD DIVERSE LOADINGS AND DESIGNS WERE UTILIZED. BY OBSERVING THESE SYSTEMS UNDER VARYING CLIMATIC CCNDITICNS, THE EFFECTS OF THE BIO-OXIDATION POND NUTRIENTS ALCNG WITH OTHER PCLLUTIONAL PARAMETERS WHICH WERE DISCHARGED INTO INTERMITTENT RECEIVING STREAMS WERE EVALUATED. EXCEPT FOR SCOURING, BIO-OXIDATICN PONDS AND BIO-OXIDATION POND RECEIVING STREAMS WERE FOUND TO BEHAVE ESSENTIALLY THE SAME AS THE STREAMS BECAME A CONTINUATION CF THE POND. IN ADDITION TO MAKING BIOCHEMICAL ADJUSTMENTS, THE STREAMS LCST MUCH OF THEIR BIOLOGICAL IDENTITY AND ASSUMED CHARACTERISTICS MORE CLOSELY ASSOCIATED WITH THE BIOLOGICAL LOADINGS FROM THE POND EFFLUENT. THE MOST PERSISTENT ALGAE IN THE SYSTEMS WERE THE FLAGELLATES (EUGLENOPHYTA) AND THE BLUE-GREEN ALGAE (CYANOPHYTA) AS THESE PLANKTERS HAD LITTLE DIFFICULTY MAKING THE TRANSITION FROM THEIR ACCLIMATED LIFE IN THE POND TO THE STREAM.

FIELD 05C, 05D

ACCESSION NO. W71-07973

POPULATION DYNAMICS AND SALINITY TOLERANCE OF HYADESIA FUSCA (LOHMAN) (ACARINA, SARCOPTIFORMES) FROM BRACKISH WATER ROCKPOOLS, WITH NOTES ON THE MICROENVIRONMENT INSIDE ENTEROMORPHA TUBES,

STOCKHOLM UNIV. (SWEDEN). ASKO LAB.; AND STOCKHOLM UNIV. (SWEDEN). DEPT. OF ZOOLOGY.

BJORN GANNING.

OECOLOGIA (BERL), VOL 5, P 127-137, 1970. 5 FIG, 16 REF.

DESCRIPTORS:
*AQUATIC ANIMALS, *AQUATIC MICROBIOLOGY, *ACARICIDES, BRACKISH WATER, MICROENVIRONMENT, LITTORAL, OXYGEN, CHEMICAL REACTIONS, HYDROGEN ION CONCENTRATION, ALGAE, REPRODUCTION, SAMPLING, SEASONAL, BOTTOM SEDIMENTS, ICE, AMPHIPODA.

IDENTIFIERS:
*HYADESIA FUSCA, *SARCOPTIFORMES, *ROCKPOOLS, *ENTEROMORPHA, SWEDEN, TIGRIOPUS BREVICORNIS, GAMMARUS DUEBENI, FUCUS SERRATUS, FUCUS VESICULOSUS, CLADOPHORA GLOMERATA, HETEROCYPRIS SALINUS, GASTEROSTEUS ACULEATUS, BALTIC SEA, NITOCRA SPINIPES.

ABSTRACT:
THE SALT WATER MITE HYADESIA FUSCA HAS BEEN RECORDED IN SCANDINAVIA FOR THE FIRST TIME. IT IS ENTIRELY RESTRICTED TO THE GREEN ALGA ENTEROMORPHA SPP IN LITTORAL ZONES OR IN ROCKPOOLS, AND IS THE DOMINANT FAUNA OF MANY BRACKISH WATER ROCKPOOLS. THE ALGAE SERVE AS SUBSTRATUM, FOOD AND BREEDING ROOM AND GIVE RISE TO VIOLENT FLUCTUATIONS IN ABIOTIC ENVIRONMENTAL PARAMETERS. OXYGEN AVAILABILITY AND HYDROGEN-ION ACTIVITIES DO NOT DIFFER GREATLY INSIDE AND OUTSIDE THE ALGAL THALLI. DURING BREEDING IN JUNE AND JULY MORE THAN 900 ANIMALS, MOSTLY LARVAE, MAY BE FOUND ON 0.1 GRAM ENTEROMORPHA DRY WEIGHT. IN WINTER AND AVERAGE OF 13 HIBERNATING ANIMALS WERE FOUND PER 0.1 GRAM ALGAE. COPULATION FOLLOWS AFTER A LONG PRECOPULA; NEWBORN LARVAE ARE OFTEN FOUND NEAR THE RHIZOMES INSIDE ALGAL TUBES. THE MITE IS VERY TOLERANT OF SALINITY VARIATIONS. HYADESIA FUSCA IS CHARACTERISTIC OF ROCKPOOL ECOSYSTEMS, OFTEN VERY ABUNDANT, ONE OF THE FEW ENTEROMORPHA GRAZERS OF THE ECOSYSTEM AND SERVES AS FOOD FOR FISHES IN THE SYSTEM. TOGETHER WITH SOME BETTER KNOWN CRUSTACEANS THIS MITE, BECAUSE OF ITS TOLERANCE AND SURVIVAL CAPABILITIES, MUST BE CONSIDERED ONE OF THE HARDIENT MEMBERS OF THE UNSTABLE ROCKPOOL ECOSYSTEMS. (JONES-WISCONSIN)

FIELD 05C, 02L

ACCESSION NO. W71-08026

MICROFLORA OF CHLORELLA K DURING A PROLONGED CULTIVATION OF THE ALGAE IN A PERFUSION UNIT WITH A CONTINUAL RECYCLING OF THE CULTURE MEDIUM, (IN RUSSIAN),

MOSCOW STATE UNIV. (USSR). FACULTY OF BIOLOGY AND SOIL SCIENCE.

M. N. PIMENOVA, I. V. MAXIMOVA, G. I. MELESHKO, E. K. LEBEDEVA, AND T. B. GALKINA.

ENGLISH SUMMARY. MIKROBIOLOGIYA, VOL 39, NO 4, P 645-650, 1970. 1 FIG, 3 TAB, 17 REF.

DESCRIPTORS:
*LABORATORY TESTS, *CHLORELLA, *CULTURES, *AQUATIC MICROORGANISMS, SPORES, ALGAE, CYTOLOGICAL STUDIES, MYCOBACTERIUM, PSEUDOMONAS, BACTERIA.

IDENTIFIERS:
PERFUSION UNIT, FLAVOBACTERIUM, MICROCCCCUS.

ABSTRACT:
THE COMPOSITION AND DENSITY OF MICROFLORA ACCOMPANYING CHLORELLA K WERE DETERMINED DURING A PROLONGED CULTIVATION OF THE ALGAE IN A PERFUSION UNIT WITH A CONTINUAL RECYCLING OF THE NUTRIENT MEDIUM. THE PREDOMINANT ORGANISMS BELONGED TO FOUR GENERA: MYCOBACTERIUM, FLAVOBACTERIUM, MICROCOCCUS, AND PSEUDOMONAS. THE LATTER WERE PARTICULARLY NUMEROUS AND DIVERSIFIED. THE SPORE-FORMING CELLS DID NOT EXCEED 0.01% OF THE TOTAL NUMBER OF BACTERIA. THE FLUCTUATION IN THE DENSITY OF ASSOCIATED MICROORGANISMS AND SPORES WERE RELATED TO PERFORMANCE OF ALGAE AND THE RATIO BETWEEN THE PHOTOSYNTHETICALLY ACTIVE CELLS, PARENT CELLS, AND AUTOSPORES. (WILDE-WISCONSIN)

FIELD 05C, 02H

ACCESSION NO. W71-08027

CONCENTRATION OF STRONTIUM-85 AND CESIUM-137 FROM WATER SOLUTIONS BY CLADOPHORA AND PITHOPHORA,

ALABAMA UNIV., UNIVERSITY. DEPT. OF BIOLOGY.

LOUIS G. WILLIAMS.

JOURNAL OF PHYCOLOGY, VOL 6, NO 3, P 314-316, 1970. 1 TAB, 4 REF.

DESCRIPTORS:
*RADIOISOTOPES, *ALGAE, *ABSORPTION, MONITORING, ANALYTICAL TECHNIQUES, LABORATORY TESTS, CESIUM, STRONTIUM RADIOISOTOPES, IONS, RIVERS, CULTURES, ORGANIC MATTER, BACTERIA, WATER ANALYSIS.

IDENTIFIERS:
*CLADOPHORA GLOMERATA, *PITHOPHORA OEDOGONIA.

ABSTRACT:
WITH THE USE OF PACKAGED LIVE AND PRESERVED GREEN ALGAE AND OTHER ORGANIC MATERIALS, RADIONUCLIDES IN NATURAL WATERWAYS, NOT FOUND BY THE USUAL METHODS OF ANALYSIS OF RAW WATER CAN BE DETECTED. LIVE ALGAE CONCENTRATE CESIUM AND STRONTIUM IONS MUCH MORE THAN ION-EXCHANGE RESINS OR DEAD MATERIALS. THIS IS SIGNIFICANT IN THE ECOLOGY OF FOOD WEBS WHERE RADIONUCLIDES MAY BE PASSED ON TO HIGHER TROPHIC LEVELS. THE CONCENTRATION FACTORS OF NONLIVING ORGANIC MATERIALS WERE MUCH LOWER AND VARIED WITH THE MATERIAL, NOT WITH THE CALCIUM/POTASSIUM RATIO. THE EFFECT OF HIGH CALCIUM IONS AND HIGH POTASSIUM IONS WAS TO LOWER THE UPTAKE OF CESIUM MORE THAN STRONTIUM. CLADOPHORA AND PITHOPHORA WERE GROWN IN AQUARIA, PACKAGED LIVE IN PERFORATED POLYETHYLENE BAGS AND TIED SUBMERGED IN VARIOUS RIVERS FOR PERIODS OF 2 TO 7 DAYS, AFTER WHICH THE CONTENTS SHOWED HIGH CONCENTRATIONS OF MANY FISSION PRODUCTS, INCLUDING STRONTIUM AND CESIUM. WHERE AQUATIC ENVIRONMENTS ARE UNFAVORABLE FOR LIVE PITHOPHORA OR CLADOPHORA, THESE ALGAE, FIXED WITH MERTHIOLATE-IODINE PRESERVATIVE, HAVE BEEN SHOWN TO BE ABOUT TWICE AS EFFECTIVE AS MIXED ION-EXCHANGE RESINS. (JONES-WISCONSIN)

FIELD 05A

ACCESSION NO. W71-08032

DETERMINATION OF THE ACTIVITY OF HETEROTROPHIC MICROFLORA IN THE OCEAN USING C-14-CONTAINING ORGANIC MATTER,

AKADEMIYA NAUK SSSR, MOSCOW. INSTITUT BIOLOGII VNUTRENNYKH VOD.

YU. I. SOROKIN.

TRANS FROM MIKROBIOLOGIYA, VOL 3, NO 1, JAN-FEB 1970. MICROBIOLOGY (USSR) VOL 39, NO 1, P 133-138, 1970. 5 FIG, 4 TAB, 22 REF.

DESCRIPTORS:
*OCEANS, *ORGANIC MATTER, *CARBON RADIOISOTOPES, *ANALYTICAL TECHNIQUES, *MARINE MICROORGANISMS, MARINE PLANTS, BACTERIA, POPULATION, HAWAII, ABSORPTION, TROPICAL REGIONS, PACIFIC OCEAN, COLUMNS, BOTTOM SEDIMENTS, ALGAE, CHLORELLA, BENTHIC FLORA, MUD, MICROORGANISMS, PROTEINS, RADIOACTIVITY TECHNIQUES, FOOD CHAINS.

IDENTIFIERS:
*ALGAL HYDROLYSATE, TONGA ISLANDS, TUTUILA ISLAND, HONSHU ISLAND, OPEN OCEAN, SAPROPHYTES, HETEROTROPHIC BACTERIA, C-14.

ABSTRACT:
THE POSSIBILITY OF USING CARBON-LABELED DISSOLVED ORGANIC MATTER (ALGAL HYDROLYSATE) FOR ESTIMATING THE COMPARATIVE ACTIVITY OF NATURAL POPULATIONS OF MICROFLORA IN THE WATER COLUMN AND BOTTOM DEPOSITS WAS INVESTIGATED IN THE TROPICAL ZONE OF THE PACIFIC OCEAN. THE CRITERION FOR MICROFLORAL ACTIVITY WAS THE RATE OF BACTERIAL ASSIMILATION OF CARBON-14 IN EXPERIMENTS OF SHORT DURATION. THE RATE AND UTILIZATION EFFICIENCY OF LABELED ORGANIC MATTER BY SEA WATER MICROFLORA DEPEND ON ITS INITIAL CONCENTRATION. THE MAXIMUM UTILIZATION EFFICIENCY OF DISSOLVED ORGANIC MATTER BY AQUATIC MICROFLORA FOR BIOSYNTHESIS WAS 45%, STAYING QUITE HIGH EVEN WHEN THE INITIAL CONCENTRATION OF LABELED HYDROLYSATE ADDED WAS 1 TO 2 MG/L. SUPPORTED IS THE IDEA THAT BACTERIAL DESTRUCTION OF ORGANIC MATERIAL IN THE OCEAN IS LIMITED TO THE UPPER WATER LAYER TO A DEPTH OF 600 TO 800 METERS, WITH THE MAXIMUM IN THE TROPICAL ZONE AT DEPTHS OF 400-600. AT GREAT OCEAN DEPTHS THE HETEROTROPHIC MICROFLORA IS SCANT AND SLIGHTLY ACTIVE. INFORMATION ABOUT THE EFFICIENCY OF UTILIZATION OF DISSOLVED ORGANIC MATTER BY THE AQUATIC MICROFLORA IS OF PARAMOUNT IMPORTANCE FOR ESTIMATING ITS ROLE AS A LINK IN THE FOOD CHAIN. (JONES-WISCONSIN)

FIELD 05C, 07B

ACCESSION NO. W71-08042

HOW CAN PORK PRODUCERS COMPLY WITH ENVIRONMENTAL QUALITY STANDARDS,

IOWA STATE UNIV., AMES. DEPT. OF AGRICULTURAL ENGINEERING.

J. RONALD MINER.

AMERICAN PORK CONGRESS-PROCEEDINGS, ENVIRONMENTAL QUALITY WORKSHOP, DES MOINES, IOWA, MAR 3, 1971. P 98-102.

DESCRIPTORS:
*FARM WASTES, *HOGS, *ENVIRONMENT, *POLLUTION ABATEMENT, WATER QUALITY, STANDARDS, WATER POLLUTION, AIR POLLUTION, ODOR, CONFINEMENT PENS, ORGANIC MATTER, NUTRIENTS, NITROGEN, PHOSPHORUS, EUTROPHICATION, ALGAE, PATHOGENIC BACTERIA, EFFLUENT, IRRIGATION, STORAGE, WASTE DISPOSAL.

IDENTIFIERS:
*ENVIRONMENTAL QUALITY, AIR CONTAMINANT, STREAM QUALITY, WASTE MANAGEMENT, MANURE COLLECTION, MANURE TRANSPORT.

ABSTRACT:
TO PREVENT WATER AND AIR POLLUTION WHILE MAINTAINING ENVIRONMENTAL QUALITY IS A COMPLEX PROBLEM. IT BECOMES MORE COMPLICATED BY TRYING TO DESIGN WASTE MANAGEMENT SYSTEMS WHICH CONTRIBUTE MATERIALLY TO OUR EFFECTIVENESS AS PORK PRODUCERS WITH POLLUTION CONTROL AS A SIDE BENEFIT. A SWINE MANURE MANAGEMENT SYSTEM MIGHT INCLUDE A COLLECTION DEVICE, A MANURE TRANSPORT SYSTEM, SOME MEANS OF MANURE STORAGE AND/OR TREATMENT, AND FINALLY, A MANURE OR EFFLUENT DISPOSAL SYSTEM. IN SOME CASES MORE THAN ONE OF THESE COMPONENTS MAY BE INCLUDED IN A SINGLE COMPONENT. GIVING INITIAL CONSIDERATION TO THE DISPOSAL SCHEME WILL HELP DETERMINE DECISIONS TO BE MADE CONCERNING THE OTHER ASPECTS OF THE SYSTEM. THERE IS MUCH REMAINING TO BE LEARNED RELATIVE TO THE CONTROL AND MEASUREMENT OF ODORS. VARIOUS ODOR LEVELS CAN BE ACHIEVED BY THE JUDICIOUS SELECTION OF MANURE HANDLING TECHNIQUES. (WHITE-IOWA STATE)

FIELD 05D, 05G

ACCESSION NO. W71-08214

NITRATE REMOVAL FROM AGRICULTURAL WASTE WATER,

FEDERAL WATER POLLUTION CONTROL ADMINISTRATION, FRESNO, CALIF.; AND CALIFORNIA DEPT. OF WATER RESOURCES, FRESNO.

PERCY P. ST. AMANT, AND LOUIS A. BECK.

IN: WATER QUALITY MANAGEMENT PROBLEMS IN ARID REGIONS, WATER POLLUTION CONTROL RESEARCH SERIES, 13030 DYY, 6/69, OCT 1970, USDI, FEDERAL WATER QUALITY ADMINISTRATION, P 1-8. 1 TAB, 1 FIG.

DESCRIPTORS:
*RETURN FLOW, *NITRATES, WATER POLLUTION, DESALINATION, ALGAE, DENITRIFICATION, CALIFORNIA, FILTERS, PARTICLE SIZE, ANAEROBIC CONDITIONS, WASTE WATER TREATMENT.

IDENTIFIERS:
*NITRATE REMOVAL, ALGAE STRIPPING, POND DENITRIFICATION, FILTER DENITRIFICATION, METHANOL, BACTERIAL DENITRIFICATION, SAN JOAQUIN VALLEY.

ABSTRACT:
THE PROBLEM OF DISPOSING OF IRRIGATION WASTE WATER FROM THE SAN JOAQUIN VALLEY OF CALIFORNIA IS A VERY LARGE ONE. THE MOST SERIOUS POTENTIAL POLLUTANT IS NITROGEN IN THE NITRATE FORM. A WASTE WATER TREATMENT CENTER AT FIREBAUGH, CALIFORNIA HAS ORGANIZED AND IS CARRYING OUT RESEARCH IN THE AREAS OF DESALINATION, ALGAE STRIPPING, AND BACTERIAL DENITRIFICATION. ALGAE STRIPPING SIMPLY INVOLVES GROWING A CROP OF ALGAE TO REMOVE NITROGEN FROM THE WATER, AND THEN HARVESTING THE ALGAE. VARIOUS MARKETS HAVE BEEN PROPOSED FOR THE USE OF ALGAE. TWO METHODS OF BACTERIAL DENITRIFICATION BEING EXPLORED ARE POND DENITRIFICATION, AND FILTER DENITRIFICATION. THE THREE DENITRIFICATION METHODS ARE COMPARED AS TO LAND REQUIREMENTS AND PROJECT COSTS. EACH IS NEARLY THE SAME IN COST - AROUND $10 PER ACRE FOOT, HOWEVER THE ALGAE STRIPPING METHOD REQUIRES MUCH MORE LAND. (WHITE-IOWA STATE)

FIELD 05D, 05G

ACCESSION NO. W71-08223

TRANSFER OF TOXIC ALGAL SUBSTANCES IN MARINE GOOD CHAINS,

HAWAII UNIV., HONOLULU. DEPT. OF BOTANY.

MAXWELL S. DOTY, AND GERTRUDES AGUILAR-SANTOS.

PACIFIC SCIENCE, VOL 24, P 351-355, JULY 1970. 1 FIG, 2 TAB, 11 REF. USPHS GRANT NO FD-00101-03, NIH GRANT NO 5-R01-GM-151-98-03, AEC CONTRACT NO AT (04-3)-235.

DESCRIPTORS:
*FOOD WEBS, *FOOD CHAINS, *TOXICITY, *ALGAL TOXINS, *CHLOROPHYTA, *PHYTOTOXINS, ECOLOGY, FISH FOOD ORGANISMS, PATH OF POLLUTANTS, BIOASSAY, TOXINS, POISONS, POISONOUS PLANTS, CORAL, INVERTEBRATES, ALGAE, MARINE ALGAE, ALGAL POISONING, CHROMATOGRAPHY, CHEMICAL ANALYSIS, BIOCHEMISTRY, PLANT PHYSIOLOGY, BIOLOGICAL COMMUNITIES.

IDENTIFIERS:
*CIGUATERA, CAULERPA SP., CAULERPICIN, CAULERPIN, PALMITIC ACID, BETA-SITOSTEROL, THIN-LAYER CHROMATOGRAPHY.

ABSTRACT:
ALCOHOLIC AND ETHER EXTRACTS OF OBLIGATE HERBIVORES, OMNIVORES AND DETRITUS FEEDERS COMMON ON CAULERPA OR IN ITS COMMUNITIES WERE SOMETIMES FOUND BY THIN-LAYER CHROMATOGRAPHY TO CONTAIN VARYING AMOUNTS OF CAULERPICIN, CAULERPIN, PALMITIC ACID, AND BETA-SITOSTEROL. CERITHIUM AND SOFT CORALS, WHICH MAY BE EITHER OMNIVOROUS OR CARNIVOROUS, ON OCCASION CONTAIN CAULERPICIN. THE CRUSTACEAN DETRITUS FEEDERS DID NOT SEEM TO PRESERVE EITHER CAULERPICIN OR CAULERPIN. CAULERPICIN AND CAULERPIN, WHICH, AS PRODUCED BY CAULERPA ARE PHYSIOLOGICALLY ACTIVE AND TOXIC TO RATS AND TO MICE, ARE APPARENTLY TRANSFERRED ALONG THE FOOD CHAINS AND CONCENTRATED IN AT LEAST SOME HERBIVORES. (LEGORE-WASHINGTON)

FIELD 05C

ACCESSION NO. W71-08597

LAKE MINNETONKA: NUTRIENTS, NUTRIENT ABATEMENT, AND THE PHOTOSYNTHETIC SYSTEM OF THE PHTOPLANKTON,

MINNESOTA UNIV., MINNEAPOLIS. LIMNOLOGICAL RESEARCH CENTER.

ROBERT O. MEGARD.

AVAILABLE FROM THE NATIONAL TECHNICAL INFORMATION SERVICE AS PB-199 915, $3.00 IN PAPER COPY, $0.95 IN MICROFICHE. FINAL REPORT, DEC 1970. 210 P, 21 FIG, 10 TAB, 41 REF, 3 APPEND. OWRR PROJECT A-016-MINN(4).

DESCRIPTORS:
*LIMNOLOGY, *LAKES, *PHOTOSYNTHESIS, *PHYTOPLANKTON, *PRODUCTIVITY, ALGAE, NITROGEN, PHOSPHORUS, MATHEMATICAL MODELS, NUTRIENTS, MEASUREMENT, POPULATION, CHLOROPHYLL, WATER POLLUTION EFFECTS, TEMPERATURE, CARBON, DIATOMS, LIGHT, OXYGEN, MINNESOTA.

IDENTIFIERS:
*LAKE MINNETONKA(MINN).

ABSTRACT:
DENSITY OF PLANKTONIC ALGAE, RATES OF PHOTOSYNTHESIS, NUTRIENT CONCENTRATIONS, AND OTHER LIMNOLOGICAL CHARACTERISTICS WERE RECODED DURING 1968-1970 IN DIFFERENT PARTS OF LAKE MINNETONKA, MINN. IN THE COURSE OF THE PAST 30 YEARS, THE DENSITY OF ALGAE INCREASED 2 OR 3 TIMES, AND THE DOMINANCE OF DIATOMS WAS REPLACED BY THAT OF BLUE-GREEN ALGAE. A LINEAR RELATIONSHIP WAS DISCLOSED BETWEEN THE ALGAE DENSITY AND THE CONTENT OF TOTAL PHOSPHORUS WITHIN THE RANGE OF 50 TO 200 MG/CU M. A DECREASE OF PHOSPHORUS BELOW 45 MG/CU M PROMISES A RAPID IMPROVEMENT OF THE LAKE QUALITY. THE NUTRIENT ENRICHMENT OF THE LAKE IS LARGELY DUE TO THE LAND RUNOFF AND STORM DRAINAGE, RATHER THAN THE DISCHARGE OF SEWAGE. A MATHEMATICAL MODEL OF THE PHOTOSYNTHETIC SYSTEM WAS BASED ON THE RELATION BETWEEN THE INTENSITY OF ILLUMINATION AND THE BIOMASS OF ALGAE. THE REPORT INCLUDES A DETAILED RECORD TO TAXA AND DENSITIES OF ALGAE DURING DIFFERENT MONTHS OF 1968 AND 1969. (WILDE-WISCONSIN)

FIELD 05C

ACCESSION NO. W71-08670

METABOLICALLY ACTIVE SPHEROPLASTS OF BLUE-GREEN ALGAE (IN RUSSIAN),

MOSCOW STATE UNIV (USSR). DEPT. OF MICROBIOLOGY.

M. V. GUSEV, K. A. NIKITINA, AND T. G. KORZHENEVSKAYA.

ENGLISH SUMMARY. MIKROBIOLOGIYA, VOL 39, NO 5, P 862-868, 1970. 5 FIG, 2 TAB, 15 REF.

DESCRIPTORS:
*ALGAE, *METABOLISM, *CYTOLOGICAL STUDIES, PHOTOSYNTHESIS, RESPIRATION.

IDENTIFIERS:
*LYSOZYME, *SPHEROPLASTS, BLUE-GREEN ALGAE, ANABAENA VARIABILIS, ANACYSTIS NIDULANS.

ABSTRACT:
THE USE OF LYSOZYME PERMITTED A QUANTITATIVE SEPARATION OF SPHEROPLASTS FROM THE CELLS OF BLUE-GREEN ALGAE ANABAENA VARIABILIS AND ANACYSTIS NIDULANS. THE YIELD OF SPHEROPLASTS WAS CORRELATED WITH THE CONCENTRATION OF LYSOZYME, THE VOLUME OF REACTION MIXTURE, CONDITIONS OF MIXING, AND THE PERIOD OF INCUBATION. THE TOLERANCE OF CELLS TO LYSOZYME WAS DEPENDENT ON THE SPECIES AND THE GROWTH PHASE OF ALGAE. SPHEROPLASTS OF A VARIABILIS EXHIBITED ENDOGENOUS RESPIRATION AT THE RATE SIMILAR TO THAT OF THE INTACT CELLS. A DECREASE IN THE CONTENT OF PHYCOCYANIN IMMOBILIZED THE ABILITY OF SPHEROPLASTS TO EMIT OXYGEN IN LIGHT, THUS CONFIRMING THE CRITICAL ROLE OF PHYCOCYANIN IN PHOTOCHEMICAL REACTIONS. (WILDE-WISCONSIN)

FIELD 05C

ACCESSION NO. W71-08683

STIMULATORY PROPERTIES OF FILTRATE FROM THE GREEN ALGA HORMOTILA BLENNISTA. I DESCRIPTION,

PROVIDENCE COLL., R.I. DEPT. OF BIOLOGY; AND CONNECTICUT UNIV., STORRS. BIOLOGICAL SCIENCES.

THOMAS J. MONAHAN, AND FRANCIS R. TRAINOR.

JOURNAL OF PHYCOLOGY, VOL 6, NO 3, P 263-269, 1970. 3 FIG, 4 TAB, 37 REF. OWRR PROJECT A-014-CONN(3).

DESCRIPTORS:
*ALGAE, *GROWTH RATES, *PRODUCTIVITY, *PLANT GROWTH SUBSTANCES, PHOTOSYNTHESIS, METABOLISM, BACTERIA, HYDROGEN ION CONCENTRATION, LABORATORY TESTS.

IDENTIFIERS:
*GREEN ALGAE, *HORMOTILA BLENNISTA, FILTRATES, GROWTH AUTOSTIMULATION, GROWTH REGULATION, EXTRACELLULAR PRODUCTS, ALGAL EXCRETIONS.

ABSTRACT:
AUTOSTIMULATION OF GROWTH BY FILTRATES OF HORMOTILA BLENNISTA IS ATTRIBUTED TO SECRETION OF ORGANIC METABOLITES. THE MAXIMUM STIMULATION IN EXCESS OF 100% WAS EXERTED BY FILTRATES OBTAINED FROM 1 TO 4 WEEK OLD ACTIVELY GROWING CULTURES. THE FILTRATES SUPPORTED BACTERIAL GROWTH AND STIMULATED THE GROWTH OF SCENEDESMUS AT PH 6.3, BUT NOT AT PH 7.7. STIMULATORY PROPERTIES OF FILTRATES WERE TERMINATED BY AUTOCLAVE TREATMENT. A SUGGESTION IS MADE THAT EXTRACELLULAR SECRETIONS OF H BLENNISTA INFLUENCE THE SURVIVAL OF THE ALGA AND THE GROWTH OF OTHER ORGANISMS. (WILDE-WISCONSIN)

FIELD 05C

ACCESSION NO. W71-08687

WATER POLLUTION FROM PHOSPHATE,

ALICE Q. HOWARD.

ENFO, JANUARY 1971, 4 P, 1 TAB.

DESCRIPTORS:
*PHOSPHATES, *DETERGENTS, *WATER POLLUTION EFFECTS, *WATER POLLUTION CONTROL, *BIODEGRADATION, FISHKILL, EUTROPHICATION, ALGAL BLOOMS, WATER QUALITY.

IDENTIFIERS:
ENZYME PRESOAKS, DISHWATER DETERGENTS, NTA.

ABSTRACT:
AN EVALUATION OF WATER POLLUTION BY PHOSPHATES IN HOUSEHOLD DETERGENTS IS PRESENTED. THE DISTINCTION IS MADE BETWEEN BIODEGRADABILITY, OR THE BREAKDOWN ABILITY OF THE CLEANING AGENT OF DETERGENTS IN WASTE WATER, AND THE PHOSPHATE PROBLEM. PHOSPHATES ARE PLACED IN DETERGENTS TO ACT AS WATER SOFTENERS TO MAKE THE CLEANING AGENT MORE EFFECTIVE. THE CYCLE OF POLLUTION EFFECTS FROM PHOSPHATE-INDUCED PLANT GROWTH INCLUDE EXCESSIVE ALGAE BLOOMS, EXCESSIVE OXYGEN USE AS THE PLANTS LATER DIE, AND FINALLY SUFFOCATION OF ANIMAL LIFE SUCH AS FISHES. THE USDI ESTIMATES THAT 50-70% OF PHOSPHORUS IN CITY SEWAGE COMES FROM DETERGENTS. PHOSPHATE SUBSTITUTES SUCH AS NTA REQUIRE EXTENSIVE RESEARCH TO BE SURE THEY DO NOT CAUSE PROBLEMS OF THEIR OWN. ENZYME PRESOAKS AND AUTOMATIC DISHWATER DETERGENTS HAVE HIGH PHOSPHATE LEVELS. SUGGESTIONS FOR CONSUMER ACTION TO REDUCE PHOSPHATE USE ARE GIVEN, INCLUDING A LIST OF DETERGENTS AND THEIR PHOSPHATE PER LOAD CONTENT. (MCENTYRE-PAI)

FIELD 05C, 05G

ACCESSION NO. W71-08768

NUTRIENT MANAGEMENT IN THE POTOMAC ESTUARY,

FEDERAL WATER QUALITY ADMINISTRATION, ANNAPOLIS, MD. CHESAPEAKE SUPPORT LAB.

NORBERT A. JAWORSKI, DONALD W. LEAR, JR., AND ORTERIO VILLA, JR.

TECHNICAL REPORT 45, JANUARY 1971, 69 P, 21 FIG, 3 TAB.

DESCRIPTORS:
*ESTUARIES, *NUTRIENTS, WATER QUALITY CONTROL, *ECOLOGY, *ALGAE, *EUTROPHICATION, NUTRIENT REQUIREMENTS, MUNICIPAL WASTES.

IDENTIFIERS:
*POTOMAC RIVER ESTUARY.

ABSTRACT:
THE WATER QUALITY OF THE UPPER POTOMAC ESTUARY HAS BEEN DEGRADED AS A RESULT OF MUNICIPAL WASTEWATER FROM THE METROPOLITAN AREA OF WASHINGTON. PAST STUDIES INDICATED HIGH COLIFORM DENSITIES, LOW DISSOLVED OXYGEN CONTENT, AND LARGE POPULATIONS OF BLUE-GREEN ALGAE AS MAJOR WATER QUALITY MANAGEMENT PROBLEMS OF THE UPPER AND MIDDLE AREAS OF THE ESTUARY. STUDIES AND CONCEPTS USED TO FORMULATE A NUTRIENT MANAGEMENT PROGRAM FOR THE POTOMAC ESTUARY ARE PRESENTED. CURRENT WATER QUALITY CONDITIONS AND ECOLOGICAL TRENDS ARE DISCUSSED. THE SOURCES, CONTROLLABILITY, TRANSPORT AND CRITERIA ESTABLISHMENT OF NUTRIENTS ARE DISCUSSED AND SUGGESTIONS FOR WASTEWATER TREATMENT REQUIREMENTS AND A WATER QUALITY MANAGEMENT PROGRAM ARE MADE. (ENSIGN-PAI)

FIELD 05C, 02L

ACCESSION NO. W71-08775

THE SANTA BARBARA OIL SPILLS OF 1969: A POST-SPILL SURVEY OF THE ROCKY INTERTIDAL,

UNIVERSITY OF SOUTHERN CALIFORNIA, LOS ANGELES. ALLAN HANCOCK FOUNDATION.

NANCY L. NICHOLSON, AND ROBERT L. CIMBERG.

IN: BIOLOGICAL AND OCEANOGRAPHICAL SURVEY OF THE SANTA BARBARA CHANNEL OIL SPILL, 1969-1970, VOL 1, BIOLOGY AND BACTERIOLOGY, P 325-399, 1971. 18 TAB, 25 REF, 9 PLATES.

DESCRIPTORS:
*MARINE PLANTS, *OILY WATER, WATER POLLUTION EFFECTS, *INTERTIDAL AREAS, *INVERTEBRATES, *ALGAE, CALIFORNIA.

IDENTIFIERS:
*SANTA BARBARA CHANNEL.

ABSTRACT:
DIFFERENCES BETWEEN THE JANUARY 1969 SANTA BARBARA OIL SPILL AND EARLIER SPILLS, AS THE 1957 TAMPICO MARU ACCIDENT AND THE 1969 TORREY CANYON ACCIDENT ARE COMPARED. IN THE TWO SHIP INCIDENTS, MASSIVE KILLS OF INTERTIDAL ORGANISMS WERE DUE TO TOXIC COMPONENTS IN DIESEL OIL OR DISPERSANTS. SPECTULAR EFFECTS WERE NOT FOUND AT THE SANTA BARBARA INTERTIDAL, EVEN AFTER A YEAR STUDY. COMPARISONS ARE MADE OF MARINE PLANT POPULATIONS FROM MONTHLY SAMPLINGS TAKEN AT TEN ROCKY INTERTIDAL STATIONS IN AND OUTSIDE OF THE SPILL AREA. SELECTION OF SAMPLING SITES, SAMPLING METHODS USED, AND TREATMENT OF THE DATA ACCOMPANY DETAILED DISCUSSION OF STUDY FINDINGS. MANY SPECIES ARE LISTED IN EXTENSIVE TABLES THAT CHART THE STUDY FINDINGS. (MOE-PAI)

FIELD 05C

ACCESSION NO. W71-08792

WHAT HAS BEEN THE EFFECT OF THE SPILL ON THE ECOLOGY IN THE SANTA BARBARA CHANNEL,

UNIVERSITY OF SOUTHERN CALIFORNIA, LOS ANGELES. ALLAN HANCOCK FOUNDATION.

DALE STRAUGHAN.

IN: BIOLOGICAL AND OCEANOGRAPHICAL SURVEY OF THE SANTA BARBARA CHANNEL OIL SPILL, 1969-1970, VOL 1, BIOLOGY AND BACTERIOLOGY, P 401-426, 1971. 1 FIG, 5 TAB, 11 REF.

DESCRIPTORS:
*OIL, SEEPAGE, WATER POLLUTION EFFECTS, ENVIRONMENTAL EFFECTS, *BENTHIC FAUNA, *INVERTEBRATES, *FISH POPULATIONS, *MORTALITY, *ALGAE, CALIFORNIA.

IDENTIFIERS:
*SANTA BARBARA CHANNEL.

ABSTRACT:
BECAUSE MANY QUESTIONS LACK ANSWERS, THE EFFECT ON ECOLOGY OF THE 1969 SANTA BARBARA CHANNEL OIL SPILL MAY NEVER BE KNOWN. ENVIRONMENTAL FACTORS SUCH AS PRIOR NATURAL SEEPAGE, HEAVY RAINS, INCREASED SEDIMENTATION AND A POSSIBLE INCREASE IN PESTICIDES, CONTRIBUTE TO THE PROBLEMS OF ISOLATING OIL SPILL EFFECTS. EVIDENCE SUGGESTS INCREASED PRODUCTIVITY OF BENTHIC FAUNA IN INSHORE WATERS AFTER THE SPILL, DUE POSSIBLY TO INCREASED NUTRIENTS. STUDIES OF SANDY BEACH FAUNA SHOWED NO DIRECT EFFECTS. INVERTEBRATES WHOSE SHELLS WERE COVERED WITH OIL APPEARED HEALTHY. FISH SURVEYS INDICATE A STABLE FISH POPULATION AND THERE IS NO PROOF THAT OIL WAS DIRECTLY RESPONSIBLE FOR MARINE MAMMAL DEATHS. HIGH MORTALITY RATES WERE RECORDED IN PELAGIC BIRD POPULATIONS, MARINE GRASS, BARNACLE SPECIES AND IN MARINE ALGAE. HYPOTHESES ARE PRESENTED AS TO WHY THE SPILL CAUSED SO LITTLE DAMAGE TO THE AREA'S ECOLOGY, ESPECIALLY WHEN COMPARED TO OTHER SPILLS. (MOE-PAI)

FIELD 05C

ACCESSION NO. W71-08793

WATER RECLAMATION AND ALGAE HARVESTING,

ASIAN INST. OF TECH., BANGKOK (THAILAND).

M. G. MCGARRY, AND C. TONGKASAME.

JOURNAL WATER POLLUTION CONTROL FEDERATION, VOL 43, NO 5, MAY 1971, P 824-835. 9 FIG, 3 TAB, 9 REF.

DESCRIPTORS:
*PONDS, *WATER REUSE, RECLAMATION, ALGAE, PROTEIN, *OXIDATION LAGOONS, CLIMATIC ZONES, PRECIPITATION, COAGULATION, LIME, HYDROGEN ION CONCENTRATION, TEMPERATURE, CHLORINATION, FLOATATION, DEWATERING, DRYING, COST ANALYSES, WASTE WATER TREATMENT.

IDENTIFIERS:
*ALGAE HARVESTING.

ABSTRACT:
APPLICATION OF HIGH RATE OXIDATION PONDS EQUIPPED WITH ALGAE HARVESTING MAY HELP TO AUGMENT THE DWINDLING WATER SUPPLIES OF LARGE METROPOLITAN AREAS BY RECLAIMING WASTE WATER FOR VARIED USES, WHILE AT THE SAME TIME PRODUCING A USABLE ALGAL FEED SUPPLEMENT. CONDITIONS FOR OPTIMAL OPERATION INCLUDE: (1) 200 LB BOD/ACRE/DAY; (2) 17.7 IN. DEPTH; AND (3) 1 DAY DETENTION TIME. PONDS OPERATED IN THIS MANNER PROVIDE AN AVERAGE EFFLUENT BOD (AFTER ALGAE REMOVAL) OF LESS THAN 10 MG/L, AND ONE ACRE OF POND CAN PRODUCE 100,000 LB PER YER OF ALGAE CONTAINING 60% PROTEIN. CHEMICAL COAGULATION AND PRECIPITATION CHEMICALS STUDIED INCLUDED LIME, ALUM, AND 50 DIFFERENT POLYELECTROLYTES. POLYCATIONS WERE FOUND TO BE MOST ECONOMIC, BUT USAGE OF POLYELECTROLYTES AS AIDS CONTRIBUTED A GREATER CHEMICAL COST THAN USAGE OF ALUM ALONG AT PH 6.5. THE DOWNFLOW SOLIDS CONTACT SYSTEM WAS EXAMINED FOR REMOVAL OF THE ALGAE, EITHER BY THE SPLIT FLOW, DISSOLVED AIR, OR SUPERSATURATED OXYGEN PRINCIPLES. THE ALGAL PASTE WAS SUN DRIED ON UNDRAINED FLAT PLATES, TO LESS THAN 10% MOISTURE. AT SOLAR ENERGY LEVELS OF 480 G CAL/SQ CM/DAY, 2800 LBS/DAY OF DRIED ALGAE COULD BE PROCESSED ON ONE ACRE. THE RESULTS OF THE INVESTIGATIONS WERE INCORPORATED INTO AN URBAN MODEL WHICH INCLUDES RECYCLING OF CLARIFIED POND EFFLUENT FOR HOUSEHOLD CLEANING PURPOSES. A 67% REDUCTION IN RAW WATER NEEDS WAS PREDICTED FOR SUCH A SYSTEM, BUT MUCH FURTHER INVESTIGATION AND REFINEMENT IS NECESSARY. (LOWRY-TEXAS)

FIELD 05D

ACCESSION NO. W71-08951

NUTRIENTS AND NUTRIENT BUDGET IN THE BAY OF QUINTE, LAKE ONTARIO,

ONTARIO WATER RESOURCES COMMISSION, TORONTO.

M. G. JOHNSON, AND G. E. OWEN.

JOURNAL WATER POLLUTION CONTROL FEDERATION, VOL 43, NO 5, MAY 1971, P 836-853. 8 FIG, 7 TAB, 37 REF.

DESCRIPTORS:
*NUTRIENTS, *NITROGEN, *PHOSPHORUS, ALGAE, *EUTROPHICATION, AQUATIC ENVIRONMENTS, GREAT LAKES, BAYS, WATERSHEDS(BASINS), SURFACE DRAINAGE, MUNICIPAL WASTES, INDUSTRIAL WASTES, SAMPLING, WATER QUALITY CONTROL, COST ANALYSIS, WASTE WATER TREATMENT, NUTRIENT REQUIREMENTS.

IDENTIFIERS:
BAY OF QUINTE, *LAKE ONTARIO.

ABSTRACT:
ALGAE BLOOMS IN THE BAY OF QUINTE, LAKE ONTARIO, HAVE BEEN INCREASING IN SEVERITY AND LENGTH OVER THE PAST SEVERAL YEARS. EXAMINATION OF AQUATIC LIFE IN THE BAY REVEALED AN OVERWHELMING PREDOMINANCE OF ORGANISMS WHICH READILY FUNCTION IN ORGANICALLY RICH WATERS. INPUT TO THE BAY WAS MONITORED IN 1968. DURING THAT TIME, THE BAY RECEIVED 9.7 MILLION LBS OF NITROGEN AND 731,000 LBS OF PHOSPHATES FROM ALL SOURCES. APPROXIMATELY 90% OF THE ENTERING NITROGEN ENTERED VIA THE RIVERS WHICH EMPTY INTO THE BAY, AND THESE SAME RIVERS ACCOUNTED FOR 60% OF THE PHOSPHORUS INPUT. THE REMAINING 10% OF THE NITROGEN AND 40% OF THE PHOSPHORUS WAS CONTRIBUTED FROM MUNICIPAL-INDUSTRIAL SOURCES. HOWEVER, MUNICIPAL-INDUSTRIAL FLOWS ARE HIGH STRENGTH LOW VOLUME EFFLUENTS WHICH DISPLACE ONLY A SMALL VOLUME OF WATER FROM THE LAKE. ON THE OTHER HAND, RIVER FLOWS ARE EXTREMELY LOW STRENGTH HIGH VOLUME ADDITIONS TO THE BAY WHICH DISPLACE LARGE VOLUMES OF WATER. THEREFORE, IT WAS ESSENTIAL TO CONSIDER THE NET ADDITIONS OF P AND N RATHER THAN THE TOTAL ADDITIONS, THE NET ADDITION BEING THE AMOUNT OF NUTRIENT CONTAINED IN AN INPUT WHICH IS IN EXCESS OF THE AMOUNT OF NUTRIENTS CONTAINED IN THE VOLUME OF WATER DISPLACED AT THE OUTLET. ON THIS BASIS, 50% OF THE NET NITROGEN AND 85% OF THE NET PHOSPHORUS WERE CONTRIBUTED BY MUNICIPAL-INDUSTRIAL SOURCES. PHOSPHORUS REMOVAL FROM HIGH CONCENTRATION, LOW VOLUME INPUTS, AT A COST OF $200,000 PER YEAR WAS RECOMMENDED. (LOWRY-TEXAS)

FIELD 05C

ACCESSION NO. W71-08953

PILGRIM NUCLEAR POWER STATION. PRELIMINARY SAFETY ANALYSIS REPORT, AMENDMENT 16.

BOSTON EDISON CO., MASS.

AVAILABLE FROM THE NATIONAL TECHNICAL INFORMATION SERVICE AS DOCKET-50293-31, $3.00 IN PAPER COPY, $0.95 IN MICROFICE. REPORT DOCKET-50293-31, OCT 2, 1970. 189 P.

DESCRIPTORS:
*WATER POLLUTION CONTROL, *NUCLEAR POWER PLANTS, *POTABLE WATER, *BIOTA, SAMPLING, MARINE FISH, MARINE ANIMALS, MARINE ALGAE, TRITIUM, IODINE RADIOISOTOPES, STRONTIUM RADIOISOTOPES, ZINC RADIOISOTOPES, COBALT RADIOISOTOPES, ABSORPTION, WATER POLLUTION SOURCES.

IDENTIFIERS:
MANGANESE RADIOISOTOPES, CESIUM RADIOISOTOPES.

ABSTRACT:
THE SURVEILLANCE PROGRAM OUTSIDE THE STATION SECURITY FENCE INCLUDES THE AIR, DOMESTIC AND SEA WATER, MARINE LIFE, SILT, MILK, AND CROPS. WATER PUMPING STATIONS WHICH CONTROL WATER SUPPLY TO MOST AREA RESIDENTS ARE SAMPLED AND ANALYZED FOR GROSS BETA AND GAMMA BIWEEKLY, AND FOR TRITIUM AND STRONTIUM MONTHLY. SEA WATER INTAKE AND DISCHARGE CANALS ARE ANALYZED MONTHLY FOR GROSS BETA AND GROSS GAMMA (WITH GAMMA SPECTRUM ANALYSIS IF THE GROSS GAMMA INCREASES), AND TRITIUM. QUARTERLY COMPOSITES OF THE MONTHLY SAMPLES ARE ANALYZED FOR MANGANESE-54, ZINC-65, COBALT-58, AND COBALT-60. FISH, LOBSTERS, AND MOLLUSKS ARE SAMPLED QUARTERLY; IRISH SEA MOSS, DURING ITS HARVEST PERIOD. ALL MARINE LIFE SAMPLES ARE TAKEN FROM THE VICINITY OF THE DISCHARGE CANAL OUTFALL AND OFFSHORE CURRENT PATTERNS AND ARE ANALYZED FOR GROSS BETA, GROSS GAMMA, STRONTIUM-90, CESIUM-137, MANGANESE-54, COBALT-58, COBALT-60, ZINC-65, AND IODINE-131. EQUIPMENT FAILURES ARE ANALYZED WITH RESPECT TO RELEASE OF RADIONUCLIDES TO THE ENVIRONMENT. (BOPP-NSIC)

FIELD 05B

ACCESSION NO. W71-09018

SHORELINE ALGAE OF WESTERN LAKE ERIE,

OHIO STATE UNIV., COLUMBUS. GRADUATE STUDIES IN BOTANY.

RACHEL COX DOWNING.

THE OHIO JOURNAL OF SCIENCE, VOL 70, NO 5, P 257-276, 1970. 97 FIG, 37 REF.

DESCRIPTORS:
*LAKE SHORES, *ALGAE, *LAKE ERIE, AQUATIC HABITATS, LAKES, AQUATIC ENVIRONMENT.

IDENTIFIERS:
*ALGAL SPECIES, WESTERN LAKE ERIE, ARNOLDIELLA CONCHOPHILA MILLER.

ABSTRACT:
IN SPITE OF SOME 70 YEAR INVESTIGATIONS OF ALGAE INHABITING WESTERN LAKE ERIE, ALMOST NOTHING WAS KNOWN PRIOR TO THIS STUDY OF THE SHORELINE AS A SPECIFIC HABITAT OF THESE ORGANISMS. THIS SITE HARBORS 61 TAXA, 39 OF WHICH ARE NEW RECORDS FOR THIS PART OF THE LAKE, AND ONE, ARNOLDIELLA CONCHOPHILA MILLER, WAS PREVIOUSLY REPORTED ONLY FROM CENTRAL RUSSIA. (WILDE-WISCONSIN)

FIELD 05C

ACCESSION NO. W71-09156

FUTURE PROSPECTS OF ALGAE AND MAN,

VERMONT UNIV., BURLINGTON. DEPT. OF BOTANY.

RICHARD M. KLEIN.

ANNALS NEW YORK ACADEMY OF SCIENCES, VOL 175, P 778-781, 1970. 9 REF.

DESCRIPTORS:
*ALGAE, *HUMAN POPULATION, FERTILIZERS, DIATOMACEOUS EARTH, FOOD CHAINS, OXYGEN, ECOLOGY, RHODOPHYTA, PHAEOPHYTA, CHLORELLA, PLANT PATHOLOGY, INDUSTRIES, IONS, CYANOPHYTA, RADIOISOTOPES, TEMPERATURE, ECOSYSTEMS, CALIFORNIA, IRRIGATION, ECONOMIC FEASIBILITY, WATER POLLUTION CONTROL, CALIFORNIA, ARIZONA.

IDENTIFIERS:
FODDER, GLUE, AGAR, POLYSIPHONIA, ANTIBIOTICS, IODINE, ATOMIC ENERGY, REACTOR COOLING, IMPERIAL VALLEY(CALIF), LOS ANGELES(CALIF), TUCSON(ARIZ).

ABSTRACT:
ECONOMIC CRITERIA OF HOW AND WHERE ALGAE CAN BE USED TO SERVE MAN'S BIOLOGICAL, PHYSICAL, AND ESTHETIC NEEDS IS CONSIDERED. ALTHOUGH THE ACKNOWLEDGED BASE OF ALL FOOD CHAINS AND THE PRIMARY SOURCE FOR MOST OF THE EARTH'S OXYGEN, THE ECONOMIC POTENTIAL OF ALGAE HAS BARELY BEEN EXPLORED. THE LARGER RED AND BROWN ALGAE ARE INCREASINGLY NEEDED FOR FERTILIZERS AND FODDER; ANTIBIOTICS ARE DERIVED FROM CHLORELLA, POLYSIPHONIA, AND OTHER SPECIES; IODINE HAS BEEN OBTAINED BY BURNING MARINE ALGAE. THE POTENTIALS OF UTILIZING THE ABILITY OF ALGAE TO CONCENTRATE MINERAL IONS HUNDREDS OF TIMES THE CONCENTRATION IN WATER, THE CONTROL OF SPECIATION AND ITS EFFECT ON POLLUTION, POSSIBLE UTILIZATION OF BLUE-GREEN ALGAL PARASITES FOR INCREASING CONCENTRATION OF NITRATES AND PHOSPHATES SUFFICIENTLY THAT ALGAE MAY BE HARVESTED FOR COMMERCIAL USAGE; USE OF HALOPHYTIC ALGAE IN LOWERING EXCESSIVE CONCENTRATION OF SODIUM CHLORIDE IN DRINKING WATER. THE QUALITATIVE AND QUANTITATIVE ALTERATION IN SPECTRUM OF ALGAE SPECIES AS THE RESULT OF ATOMIC ENERGY PLANT OPERATIONS AND THEIR EFFECTS ON MAN ARE UNKNOWN, PARTICULARLY WHETHER ALGAE WILL ACCUMULATE RADIOACTIVE IONS EFFECTIVELY. ALGAL TAXONOMISTS, ALGAL PHYSIOLOGISTS, AND ALGAL HISTOCHEMISTS MUST CONSIDER THEMSELVES ECOLOGISTS AND POOL THEIR KNOWLEDGE TO CONTRIBUTE TO HUMANITY'S WELFARE. (AUEN-WISCONSIN)

FIELD 05C, 06B

ACCESSION NO. W71-09157

UTILIZATION OF HERBIVOROUS FISH IN FISH MANAGEMENT AND MELIORATION OF WATER BASINS, (IN RUSSIAN).

VESTNIK AKADEMII NAUK SSSR NO 11, P 26-30, 1970.

DESCRIPTORS:
*WATER QUALITY CONTROL, *ALGAE CONTROL, *FORAGE FISH, BIOCONTROL, REMEDIES, FOODS, FISH, HERBIVORES, FISH MANAGEMENT.

IDENTIFIERS:
USSR.

ABSTRACT:
AN ADDRESS BY G B NIKOLSKI IS REVIEWED WHICH WAS DELIVERED AT THE MEETING OF THE PRESIDIUM OF THE USSR ACADEMY OF SCIENCE. AT THIS TIME, THE CATCH OF OCEAN INHABITING FISH IS INADEQUATE TO MEET HUMAN POPULATION DEMANDS AND AN ARTIFICIAL CULTURE OF HERBIVOROUS FISH CAN PROVIDE AN INEXPENSIVE SOURCE OF PROTEINS. IN ADDITION TO INCREASING THE FOOD SUPPLY, SUCH CULTURE WILL HELP TO ERADICATE THE DETRIMENTAL GROWTHS OF ALGAE IN IRRIGATION AND DRAINAGE SYSTEMS AND RESERVOIRS OF HYDRO-ELECTRIC STATIONS. HERBIVOROUS FISH OF DESIRABLE PROPERTIES ARE ABSENT IN BOTH EUROPE AND CENTRAL ASIA, BUT WATERS OF FAR-EASTERN SIBERIA HARBOR THREE HERBIVOROUS (COMMON AND MOTTLED THICKHEADS AND WHITE AMUR), ATTAINING WEIGHTS OVER 20 AND EVEN 30 KG. THE TWO VARIETIES ARE NOT COMPETITORS AS THEY DERIVE THEIR SUSTENANCE FROM EITHER NEAR-BOTTOM OR SURFACE PHYTOPLANKTON. THE PRESIDIUM APPROVED A BROAD PROGRAM OF INVESTIGATION OF HERBIVOROUS FISH IN ASIA, AFRICA, AND SOUTH AMERICA, AND A SYSTEMATIC ACCLIMATIZATION OF SUITABLE STOCK IN DIFFERENT PARTS OF THE USSR. (WILDE-WISCONSIN)

FIELD 05G, 02H

ACCESSION NO. W71-09163

ON THE GROSS AND NET PRODUCTION OF PERIPHYTON AND PLANKTON ALGAE, (IN RUSSIAN),

IRKUTSKII GOSUDARSTVENYI UNIVERSITET (SSSR).

O. M. KOZHOVA.

DOKLADY AKADEMII NAUK SSSR, VOL 195, NO 4, P 965-968, 1970. 3 TAB, 5 REF.

DESCRIPTORS:
*PERIPHYTON, *PLANKTON, *PRODUCTIVITY, PHOTOSYNTHESIS, BIOMASS, OXYGEN, ALGAE, COMPARATIVE PRODUCTIVITY, EUTROPHICATION.

IDENTIFIERS:
BRATSK RESERVOIR(SIBERIA), ANGARA RIVER(SIBERIA).

ABSTRACT:
THIS STUDY OF THE BRATSK BASIN INCLUDED DETERMINATIONS OF PHOTOSYNTHESIS OF PERIPHYTON BY LIGHT AND DARK BOTTLES, AND PRODUCTION OF PHYTOPLANKTON BY THE VINBERG OXYGEN METHOD. THE REPORTED RESULTS PRESENT AVERAGES FOR VEGETATION SEASONS OF 1964 AND 1967 FOR PHYTOPLANKTON, AND OF 1965, 1967, AND 1968 FOR PERIPHYTON. THE PHYTOPLANKTON CONSISTED LARGELY OF MELOSIRA, STEPHANODISCUS, ASTERIONELLA, AND CRYPTOMONADACEA SP IN THE SPRING, AND OF APHANIZOMENON AND CERATIUM SP IN SUMMER. THE PREDOMINANT MEMBERS OF PERIPHYTON INCLUDED CLADOPHORA, ULOTHRIX, OEDOGONIUM, SPIROGYRA, AND MOUGEOTIA SP. THE DIURNAL GROSS AND NET PRODUCTION OF PERIPHYTONIC ALGAE WAS 44.2 AND 33.8 MG OXYGEN PER G OF FRESH WEIGHT, RESPECTIVELY. THE PRODUCTION TO BIOMASS RATIO (P/B COEFFICIENT) OF PHYTOPLANKTON, RECALCULATED ON CARBON, WAS APPRECIABLY HIGHER THAN THAT OF THE PERIPHYTON. (WILDE-WISCONSIN)

FIELD 05C

ACCESSION NO. W71-09165

INITIAL UPTAKE, DISTRIBUTION AND LOSS OF SOLUBLE RU-106 IN MARINE AND FRESHWATER ORGANISMS IN LABORATORY CONDITIONS,

CENTRE D'ETUDE DE L'ENERGIE NUCLEAIRE, BRUSSELS (BELGIUM). DEPT. OF RADIOBIOLOGY.

O. VAN DER BORGHT, AND S. VAN PUYMBROECK.

HEALTH PHYSICS, VOL 19 (DEC), P 801-811, 1970. 10 FIG, 4 TAB, 15 REF.

DESCRIPTORS:
*ABSORPTION, *DISTRIBUTION, *RADIOISOTOPES, *MARINE ANIMALS, *MARINE PLANTS, AQUATIC ANIMALS, LABORATORY TESTS, NITRATES, MUSSELS, SNAILS, FISH, ALGAE, GASTROPODS, ADSORPTION, TEMPERATURE, METABOLISM, KINETICS, WATER POLLUTION EFFECTS, NUCLEAR WASTES.

IDENTIFIERS:
RUTHENIUM, FIXATION, TISSUES, ALBURNUS LUCIDUS, BELGIUM.

ABSTRACT:
IN WASTES FROM NUCLEAR REPROCESSING PLANTS, RUTHENIUM-106 IS APPARENTLY THE MOST IMPORTANT RADIOISOTOPE CONTAMINATING ALGAE AND FISH. THIS INVESTIGATION WAS CONDUCTED TO UNDERSTAND THE FACTORS, SUCH AS ENVIRONMENTAL CONDITIONS, AND PHYSICOCHEMICAL STATE OF RUTHENIUM ON THE RADIOACTIVE CONTAMINATION OF ORGANISMS, AND THE EARLY KINETICS OF FIXATION, DISTRIBUTION AND RELEASE OF SOLUBLE RUTHENIUM ORGINATING FROM NITRATE NITROSYL COMPLEXES BY SOME MARINE AND FRESHWATER ORGANISMS. THE EARLY UPTAKE, DISTRIBUTION AND LOSS BY MUSSELS, SNAILS, FISH, ALGAE, AND TWO GASTROPODS OF RUTHENIUM-106 WERE STUDIED FOR SHORT PERIODS UNDER LABORATORY CONDITIONS. INITIAL UPTAKE AND DISTRIBUTION OF RUTHENIUM-106 IS SIMILAR IN THE MARINE AND FRESHWATER LAMELLIBRANCH (MYTILUS AND ANODONTA), BUT FIXATION IS MUCH HIGHER IN THE FRESHWATER SNAIL, VIVIPARUS, THAN IN THE FISH, ALBURNUS. IN THE RADIOCONTAMINATION OF ORGANISMS BY SOLUBLE RUTHENIUM, THIN STRUCTURES, SUCH AS BYSSUS, IN DIRECT CONTACT WITH WATER, SHOW HIGH RUTHENIUM CONTENT, INDICATING THE IMPORTANT ROLE OF SURFACE ADSORPTION. TEMPERATURE-DEPENDANCE SUGGESTS METABOLIC INTERFERENCE, ALTHOUGH TEMPERATURE-DEPENDANCE AND GENERAL KINETICS OF THE UPTAKE ARE SIMILAR BOTH IN ALGAE AND IN ANIMALS STUDIED. SLOWER UPTAKE OF ISOTOPES AT LOWER TEMPERATURES INDICATES THE ADVISABILITY TO RELEASE RADIOACTIVE WASTES DURING PERIODS OF LOW METABOLIC ACTIVITY. (JONES-WISCONSIN)

FIELD 05C

ACCESSION NO. W71-09168

CONTEMPORANEOUS DISEQUILIBRIUM, A NEW HYPOTHESIS TO EXPLAIN THE 'PARADOX OF THE PLANKTON',

CALIFORNIA UNIV., DAVIS. DEPT. OF ZOOLOGY; AND CALIFORNIA UNIV., DAVIS. INST. OF ECOLOGY.

PETER RICHERSON, RICHARD ARMSTRONG, AND CHARLES R. GOLDMAN.

PROCEEDINGS OF THE NATIONAL ACADEMY OF SCIENCES, VOL 67, NO 4, P 1710-1714, 1970. 2 FIG, 1 TAB, 10 REF.

DESCRIPTORS:
*PHYTOPLANKTON, *LAKES, EPILIMNION, ZOOPLANKTON, DISTRIBUTION, PRIMARY PRODUCTIVITY, ALGAE, BIOMASS, OLIGOTROPHY, DIATOMS, EUGLENA, DAPHNIA, HABITATS, NICHES, COMPETITION, CALIFORNIA.

IDENTIFIERS:
*CONTEMPORANEOUS DISEQUILIBRIUM, LAKE TAHOE(CALIF), CASTLE LAKE(CALIF), DIVERSITY, PATCHES.

ABSTRACT:
UNEXPECTEDLY HIGH DIVERSITY FOUND IN EVEN SMALL SAMPLES OF LAKE PHYTOPLANKTON HAS BEEN TERMED 'THE PARADOX OF THE PLANKTON.' A SMALL VOLUME OF WATER, FOR EXAMPLE 10 ML, USUALLY SHOWS SOME TENS OF SPECIES WHERE THE COMPETITIVE EXCLUSION PRINCIPLE MIGHT LEAD TO EXPECTATION OF ONLY ONE OR A FEW SPECIES. PATCHES OF WATER MAY EXIST IN WHICH ONE SPECIES COMPETES ADVANTAGEOUSLY RELATIVE TO THE OTHERS. THEY ARE STABLE ENOUGH FOR CONSIDERABLE PATCHINESS AMONG PHYTOPLANKTON, BUT ARE OBLITERATED FREQUENTLY ENOUGH TO PREVENT EXCLUSIVE OCCUPATION OF EACH NICHE BY A SINGLE SPECIES. WITH EPILIMNION MIXING, ONLY ONE OR AT MOST, A FEW NICHES FOR PRIMARY PRODUCERS MIGHT BE EXPECTED. IN SAMPLES FROM CASTLE LAKE, CALIFORNIA, A HIGH DEGREE OF PATCHINESS FOR MANY PHYTOPLANKTON SPECIES WAS FOUND, INDICATING MIXING RATE IS SUFFICIENTLY SLOW IN RELATION TO THE ALGAL PRODUCTIVITY RATE, FOR MANY DIFFERENT NICHES TO EXIST SIMULTANEOUSLY. IN LAKE TAHOE, PRODUCTIVITY PER UNIT BIOMASS RATIOS SHOW THAT TURNOVER TIMES FOR CARBON ARE OFTEN LESS THAN ONE DAY. HIGH DIVERSITY IS ASSOCIATED WITH HIGH PRODUCTIVITY PER UNIT BIOMASS AND HIGH ZOOPLANKTON POPULATIONS. A CONTEMPORANEOUS DISEQUILIBRIUM MODEL IS A PLAUSIBLE EXPLANATION OF THE DIVERSITY. (JONES-WISCONSIN)

FIELD 05C

ACCESSION NO. W71-09171

PHOTOSYNTHESIS IN THE ALGAE,

BRANDEIS UNIV., WALTHAM, MASS. DEPT. OF BIOLOGY.

MARTIN GIBBS, ERWIN LATZKO, MICHAEL J. HARVEY, ZVI PLAUT, AND

ANNALS NEW YORK ACADEMY OF SCIENCES, VOL 175, P 541-554, 1970. 5 FIG, 6 TAB, 40 REF.

DESCRIPTORS:
*BIOCHEMISTRY, *PHOTOSYNTHESIS, *ALGAE, PHYLOGENY, CARBON CYCLE, ENZYMES, METABOLISM, HYDROGEN, EUGLENA, CHLORELLA, SCENEDESMUS, CHLAMYDOMONAS, CYANOPHYTA, CHLOROPHYTA, OXYGEN, LIGHT, CHRYSOPHYTA, OCHROMONAS, RESPIRATION, CHLOROPHYLL, CARBON RADIOISOTOPES, CARBON DIOXIDE, RHODOPHYTA.

IDENTIFIERS:
WARBURG EFFECT, EUGLENA GRACLIS, CALVIN CYCLE, ANACYSTIS NIDULANS.

ABSTRACT:
THE PHOTOSYNTHETIC CARBON REDUCTION CYCLE IN ALGAE IS DISCUSSED FROM EVIDENCE DESCRIBED IN THE LITERATURE; THIS CYCLE IS NOW FIRMLY ESTABLISHED. TWELVE ENZYMICALLY CATALYZED STEPS ARE INVOLVED. AT LEAST RIBULOSE 1,5-DIP CARBOXYLASE AND FRUCTOSE 1,6-DIPHOSPHATASE ARE INVOLVED IN CARBON DIOXIDE FIXATION. THE BULK OF THE EVIDENCE SUGGESTS THAT THE CARBON REDUCTION CYCLE IS THE MAJOR PATHWAY FOR THE ENTRY OF PHOTOSYNTHETICALLY ASSIMILATED CARBON DIOXIDE IN THE MAJORITY OF CHLOROPHYLLOUS ALGAE. ON THE BASIS OF PRESENT KNOWLEDGE OF PHOTOSYNTHETIC CARBON ASSIMILATION IN ALGAE, THE AUTOTROPHIC PHASE CEASES WITH THE LIGHT-CATALYZED CONVERSION OF CARBON DIOXIDE INTO FRUCTOSE 6-P AND GLUCOSE 6P. THE HETEROTROPHIC PHASE COMMENCES WITH THEIR CONVERSION INTO OTHER CELLULAR COMPONENTS, SOMETIMES REFERRED TO AS SECONDARY PRODUCTS. IMPORTANT EVIDENCE FOR METABOLIC REGULATION OF THE PHOTOSYNTHETIC CARBON REDUCTION CYCLE IN ALGAE INDICATES THE CYCLE IS NOT ISOLATED FROM THE REST OF THE CELL METABOLISM. ASSESSMENT OF THE TRUE VALUE OF THESE RESULTS WAS HINDERED BY POSSIBLE INTERCHANGE OF INTERMEDIATES COMMON TO BOTH PHOTOSYNTHESIS AND RESPIRATION. SUCCESSFUL ISOLATION OF ALGAL CHLOROPLASTS FROM THE GIANT CELL OF ACETABULARIA MEDITERRANEA, ABLE TO CARRY OUT NORMAL PHOTOSYNTHESIS, WAS RECENTLY ACCOMPLISHED. (JONES-WISCONSIN)

FIELD 05C

ACCESSION NO. W71-09172

HYPNODINIUM-LIKE ALGAL BLOOMS IN GEORGIA LAKES,

GEORGIA STATE UNIV., ATLANTA. DEPT. OF BIOLOGY; AND GEORGIA WATER QUALITY CONTROL BOARD, ATLANTA.

D. G. AHEARN, EDWARD T. HALL, JR., AND DONALD J. REINHARDT.

BIOSCIENCE, (RESEARCH REPORTS), P 115, FEBRUARY 1, 1971. 1 FIG, 1 REF.

DESCRIPTORS:
*ALGAE, *LAKES, *GEORGIA, SEWAGE EFFLUENTS, WATER POLLUTION EFFECTS, WATER POLLUTION SOURCES, EUTROPHICATION, COLIFORMS, SPORES, LIFE CYCLES, ODOR, RECREATION, SYSTEMATICS.

IDENTIFIERS:
*WHITE ALGAL BLOOMS, HYPNODINIUM, LAKE SIDNEY LANIER(GEORGIA), LAKE JACKSON(GEORGIA), LAKE CARDINAL(GEORGIA), LAKE BUCKHORN(GEORGIA).

ABSTRACT:
LAKES ON DIFFERENT RIVER SYSTEMS IN NORTHWEST GEORGIA PRODUCED A WHITE ALGAL BLOOM OF VARYING INTENSITY IN JUNE, LASTING 7-10 DAYS. AT ITS PEAK, DENSE SURFACE FLOCS WERE FOUND WHERE SEWAGE EFFLUENTS WERE RECEIVED, IN ALL THE 1970 BLOOM SITES. PRIOR TO, DURING, AND AFTER THE BLOOM, HYDROGRAPHIC CONDITIONS IN THE FLAT CREEK AREA OF LAKE LANIER WERE MONITORED WITH OBSERVATIONS ON THERMOCLINE, TEMPERATURE, PH, DISSOLVED OXYGEN. FOUR DAYS PRIOR TO THE APPEARANCE OF THE BLOOM, THE FECAL COLIFORM DENSITY INCREASED FROM LESS THAN 4/100 ML TO MORE THAN 240/100 ML. THE MILKY WHITE SURFACE SCUM FROM ALL LAKES WAS COMPOSED OF BODIES SURROUNDED WITH A CLEAR ENVELOPE CONTAINING LARGE PROTOPLASMIC BODIES. A DARK PROTOPLASTIC BODY MATURED INTO A DARK MACROSPORE WHICH SUBDIVIDED PRODUCING 2 TO 15 OR MORE BI-PORED MICROSPORES. NO EXIT OF PROTOPLASMIC BODIES THROUGH THE PORES WAS OBSERVED, ALTHOUGH BI-FLAGELLATED ZOOSPORES OCCURRED IN LARGE NUMBERS IN THE MICROSCOPE PREPARATIONS. THE BI-PORED CELLS OF THIS ALGA MAKE IT DISTINCT FROM HYPNODINIUM SPHAERICUM KLEBS. IT HAS NOT BEEN DIRECTLY ASSOCIATED WITH FISH KILLS. A DISAGREEABLE ODOR ACCOMPANIES THE BLOOM AND THE THICK SURFACE FLOCS INTERFERE WITH RECREATION. (JONES-WISCONSIN)

FIELD 05C, 02H

ACCESSION NO. W71-09173

INITIAL UPTAKE, DISTRIBUTION AND LOSS OF SOLUBLE RUTHENIUM-106 IN MARINE AND FRESHWATER ORGANISMS IN LABORATORY CONDITIONS,

CENTRE D'ETUDE DE L'ENERGIE NUCLEAIRE, BRUSSELS (BELGIUM). DEPT. OF RADIOBIOLOGY.

O. VAN DER BORGHT, AND S. VAN PUYMBROECK.

HEALTH PHYSICS, VOL. 19, P 801-811, DEC. 1970.

DESCRIPTORS:
*RADIOISOTOPES, *MARINE ANIMALS, *ABSORPTION, AQUARIA, MOLLUSKS, SNAILS, MUSSELS, MARINE ALGAE, AQUATIC ANIMALS, FRESH WATER, FRESHWATER FISH, TEMPERATURE, KINETICS, NUCLEAR WASTES, SOLUBILITY, ADSORPTION, WATER POLLUTION EFFECTS, FOOD CHAINS.

IDENTIFIERS:
*RUTHENIUM RADIOISOTOPES.

ABSTRACT:
RUTHENIUM UPTAKE EXTENDING TO LONG-TERM EQUILIBRIA WERE STUDIED SINCE THIS ELEMENT IS OFTEN THE LIMITING FACTOR IN THE DISPOSAL OF WASTES FROM NUCLEAR REPROCESSING PLANTS. THE GENERAL KINETICS INCLUDING THE TEMPERATURE DEPENDENCE WERE SIMILAR FOR FOUR MARINE ORGANISMS (AN ALGA, A MUSSEL, AND TWO SNAILS); AND THE INITIAL UPTAKE WAS SIMILAR FOR MARINE AND FRESHWATER MUSSELS. A RELATIVELY HIGH CONCENTRATION IN THIN ORGANS IN DIRECT CONTACT WITH THE WATER SUGGESTED THAT SURFACE ABSORPTION PLAYS A ROLE. RELEASE OF WASTES DURING PERIODS OF LOW TEMPERATURE (WINTER) WILL MINIMIZE BIOLOGICAL CONCENTRATION BEFORE DILUTION IN THE BIOSPHERE. (BOPP-NSIC)

FIELD 05C

ACCESSION NO. W71-09182

INTERACTION OF TRACE ELEMENTS WITH THE ORGANIC CONSTITUENTS IN THE MARINE ENVIRONMENT,

BHABHA ATOMIC RESEARCH CENTRE, BOMBAY (INDIA).

M. V. M. DESAI, AND A. K. CANGULY.

AVAILABLE FROM THE NATIONAL TECHNICAL INFORMATION SERVICE AS BARC-488, $3.00 IN PAPER COPY, $0.95 IN MICROFICHE. REPORT BARC-488, 1970. 117 P, 15 FIG, 41 TAB, 47 REF.

DESCRIPTORS:
*RADIOISOTOPES, *ORGANIC COMPCUNDS, *MARINE ALGAE, PHYTOPLANKTON, HUMIC ACIDS, FULVIC ACIDS, TRACE ELEMENTS, CHEMICAL ANALYSIS, ION EXCHANGE, ABSORPTION, STRONTIUM RADIOISOTOPES, ZINC RADIOISOTOPES, COBALT RADIOISOTOPES, URANIUM RADIOISOTOPES, RADIUM RAICDISOTOPES, ALUMINUM, IRON, COPPER, MAGNESIUM, MANAGNESE, CALCIUM, CHROMIUM, POTASSIUM, CHELATION, METABOLISM, NUCLEAR WASTES, WATER POLLUTION EFFECTS.

IDENTIFIERS:
MANGANESE RADIOISOTOPES, CESIUM RADIOISOTOPES, RUTHENIUM RADIOISOTOPES, CERIUM RADIOISOTOPES, THORIUM RADICISOTOPES, IRON RADICISOTOPES, ZIRCONIUM RADIOISOTOPES, ZINC.

ABSTRACT:
LOW-LEVEL NUCLEAR WASTES DISHCHARGE INTO THE ENVIRCNMENT THE RADIOISOTOPES OF ZINC, MANGANESE, IRON, COBALT, RADIUM, THORIUM, URANIUM, ETC. IN THE PRESENT WORK IT IS SHOWN THAT THE RADIONUCLIDES ARE SOLUBILIZED BY GROWTH PRODUCTS OF MARINE ORGANISMS. COMPLEXATION OF RADIONUCLIDES WAS STUDIED WHEN PHYTOPLANKTCN WAS GROWN IN A MEDIUM SPIKED WITH RADIONUCLIDES IN DIFFERENT PHASES. HUMIC AND FULVIC ACIDS WERE EXTRACTED FROM SEA WATER AND ANALYZED FOR TRACE ELEMENTS. THE STUDIES SHOW MECHANISMS BY WHICH THE AVAILABILITY MAY BE INCREASED THROUGH AQUATIC FOOD CHAINS TO MAN. (BOPP-NSIC)

FIELD 05C

ACCESSION NO. W71-09188

PATHWAYS OF TRACE ELEMENTS IN ARCTIC LAKE ECOSYSTEMS. PROGRESS REPORT, APRIL 15, 1970-APRIL 14, 1971,

ALASKA UNIV., COLLEGE. INST. OF MARINE SCIENCES.

R. J. BARSDATE.

AVAILABLE FROM THE NATIONAL TECHNICAL INFORMATION SERVICE AS SAN-310-P-4-10, $3.00 IN PAPER COPY, $0.95 IN MICROFICHE. REPORT SAN-310-P-4-10, JAN. 1971, 151 P, AEC CONTRACT AT(04-3)-310.

DESCRIPTORS:
*ECOSYSTEMS, *ALASKA, *LAKES, ARCTIC, PRODUCTIVITY, EUTROPHICATION, RED TIDE, SILICA, ALGAE, PHYTOPLANKTON, FREEZING, TEMPERATURE, CHELATION, HYDROGEN ION CONCENTRATION, DYSTROPHY, ZINC RADICISOTOPES, COBALT RADIOISOTOPES, PHOSPHORUS RADIOISOTOPES, SEDIMENTS, TRACE ELEMENTS, VOLCANOES, BENTHIC FLORA, ALKALINITY, PHOTOSYNTHESIS, HYDROLOGIC ASPECTS, MAGMATIC WATER, WATER POLLUTION SOURCES.

IDENTIFIERS:
ZINC, MANGANESE RADICISOTOPES.

ABSTRACT:
THIS REPORT CONSIDERS FIVE SUBJECTS: THE FIELD PROGRAM AT POINT BARROW; CHARACTER AND RESIDENCE TIME FOR ORGANIC COMPLEXES OF TRACE METALS ASSOCIATED WITH A RED TIDE; A STUDY OF ZINC COMPLEXATION WITH ORGANIC MATERIAL FROM A NATURAL WATER SYSTEM; PHYSICAL LIMNOLOGY, CHEMISTRY AND PLANT PRODUCTIVITY OF A TAIGA LAKE; AND EFFECTS OF VOLCANIC ASHFALLS ON ALASKAN LAKES. (BOPP-NSIC)

FIELD 05C, 05B, 02H

ACCESSION NO. W71-09190

MECHANISMS OF THE ACCUMULATION OF PLUTONIUM-239 AND POLONIUM-210 BY THE BROWN ALGA ASCOPHYLLUM NODOSUM AND MARINE PHYTOPLANKTON,

POLYARNYI NAUCHNO-ISSLEDOVATELSKII I PROEKTNYI INSTITUT MORSKOGO RYBNOGO KHOZYAISTVA I OKEANOGRAFII, MURMANSK (USSR).

V. S. ZLOBIN, AND O. V. MOKANU.

AVAILABLE FROM THE NATIONAL TECHNICAL INFORMATION SERVICE AS AEC-TR-7205, $3.00 IN PAPER COPY, $0.95 IN MICROFICHE. (RADIOBIOLOGY), VOL. 10, NO. 4, P 584-589, JULY-AUGUST 1970, 3 FIG, 4 TAB, 18 REF) AEC-TR-7205 FROM RADIOBIOLOGIYA.

DESCRIPTORS:
*RADIOISOTOPES, *PHYTOPLANKTON, *METABOLISM, *ABSORPTION, MARINE PLANTS, MARINE ALGAE, PHAEPHYTA, AQUARIA, PARTICLE SIZE, COLLOIDS, WATER POLLUTION EFFECTS.

IDENTIFIERS:
*POLONIUM, *PLUTONIUM.

ABSTRACT:
RADIONUCLIDE UPTAKE BY A BROWN ALGA WAS STUDIED IN LABORATORY AQUARIA; BY PHYTOPLANKTON CULTURED FROM SURFACE SEA WATER, IN SHIPBOARD AQUARIA. THE EFFECT OF INHIBITORS OF THE CYTOCHROME SYSTEM (CADMIUM CHLORIDE, SODIUM CYANIDE, AND AMMONIUM NITRATE) INDICATED THAT UPTAKE WAS AFFECTED BY CELL RESPIRATION. THE RADIONUCLIDES WERE SHOWN TO BE IN THE FORM OF COLLOIDS. THE PARTICLE SIZE OF THE PLUTONIUM SALT WAS 0.1-0.3 MICRONS; OF THE POLONIUM SALT, 0.3-15 MICRONS. (BOPP-NSIC)

FIELD 05C

ACCESSION NO. W71-09193

ECOLOGY AND POLLUTION OF THE ENVIRONMENT (IN FRENCH).

BULLETIN D'INFORMATIONS SCIENTIFIQUES ET TECHNIQUES, NO 151, P 3-73 (SEPT 1970).

DESCRIPTORS:
*STRONTIUM RADIOISOTOPES, *WATER POLLUTION EFFECTS, NUCLEAR WASTES, RIVERS, ABSORPTION, FISH, MOLLUSKS, MARINE ANIMALS, COBALT RADIOISOTOPES, FOOD CHAINS, IRRIGATION, FLOW PROFILES, MARINE ALGAE, WINDS, SEA WATER, PHYSICOCHEMICAL PROPERTIES.

IDENTIFIERS:
*CESIUM RADIOISOTOPES, *RUTHENIUM RADIOISOTOPES, ZIRCONIUM RADIOISOTOPES.

ABSTRACT:
A COLLECTION OF ARTICLES WHICH INCLUDES THE FOLLOWING TOPICS: PLANTS AND PARTICULATE ATMOSPHERIC POLLUTION (WIND TUNNEL EXPERIMENTS), UPTAKE OF RADIONUCLIDES BY MARINE BIOTA, IN-SITU CONTAMINATION OF MARINE BIOTA BY RUTHENIUM-106 AND ZIRCONIUM-95, PHYSICOCHEMICAL BEHAVIOR OF RUTHENIUM IN SEA WATER, A CESIUM SOURCE FOR IRRADIATION STUDIES OF MARINE ORGANISMS, SIMULATION OF NATURAL FLOW IN RIVER POLLUTION STUDIES, MIGRATION OF FISSION PRODUCTS THROUGH SOIL, UPTAKE OF RADIONUCLIDES BY VEGETABLES FROM IRRIGATION WATER. (BOPP-NSIC)

FIELD 05C

ACCESSION NO. W71-09225

TEMPERATURE EFFECTS ON THE SORPTION OF RADIONUCLIDES BY FRESHWATER ALGAE,

SAVANNAH RIVER LAB., AIKAN, S.C.

R. S. HARVEY.

HEALTH PHYSICS, VOL 19, NO 2, P 293-297 (AUG 1970). 4 FIG, 3 TAB, 7 REF.

DESCRIPTORS:
*AQUATIC ALGAE, *RADIOISOTOPES, ABSORPTION, TEMPERATURE, STRONTIUM RADIOISOTOPES, ZINC RADIOISOTOPES, COBALT RADIOISOTOPES, WATER POLLUTION EFFECTS, FRESH WATER.

IDENTIFIERS:
*CESIUM RADIOISOTOPES, IRON RADIOISOTOPES, MANGANESE RADIOISOTOPES.

ABSTRACT:
WATER TEMPERATURES OF 23-32 DEG C DID NOT AFFECT THE SORPTION OF CESIUM-137, STRONTIUM-85, ZINC-65, IRON-59, COBALT-57, AND MANGANESE-54 BY THE FILAMENTCUS GREEN ALGA, STIGEOCLONIUM LUBRICUM. RADIONUCLIDE CONCENTRATIONS IN THE UNICELLULAR DIATOM, NAVUCILA SEMINULUM, WERE 2-5 TIMES HIGHER AT 32 DEG C THAN AT 23 DEG C. WATER TEMPERATURES OF 25-40 DEG C HAD NO SIGNIFICANT EFFECT ON THE SORPTION OF CESIUM-137, STRONTIUM-85, ZINC-65, AND IRON-59 BY THE FILMENTCUS BLUE-GREEN ALGA, PLECTONEMA BORYANUM. HOWEVER, COBALT-57 CONCENTRATIONS IN P. BORYANUM DECREASED WITH TEMPERATURE, AND MANGANESE-54 CONCENTRATIONS INCREASED FROM 25-35 DEG C. GROWTH OF S. LUBRICUM WAS NOT AFFECTED BY THE TEMPERATURE. (BOPP-NSIC)

FIFLD 05C

ACCESSION NO. W71-09250

THE EFFECTS OF ENVIRONMENTAL STRESS ON THE COMMUNITY STRUCTURE AND PRODUCTIVITY OF SALT MARSH EPIPHTIC COMMUNITIES,

CITY COLLEGE, NEW YORK.

J. J. LEE,

AVAILABLE FROM NTIS. ATOMIC ENERGY COMMISSION TECHNICAL REPORT NYO-3995-14, JAN 1970. 46 P.

DESCRIPTORS:
*ECOSYSTEMS, *SALT MARSHES, *PRODUCTIVITY, RADIOISOTOPES, STRONTIUM RADIOISOTOPES, ALGAE, PLANT GROWTH, TEMPERATURE, PROTOZOA, CHLOROPHYTA, ABSORPTION.

ABSTRACT:
THE FOLLOWING TOPICS ARE CONSIDERED: REMOVAL OF CALCIUM-45 AND STRONTIUM-89, -90 FROM COASTAL ENVIRONMENTS BY CALCAREOUS FORAMINIFERS (PROTOZOA); GROWTH AND REPRODUCTION OF FORAMIFERS AND EPIPHYTIC ALGAE; EFFECT OF TEMPERATURE AND LIGHT ON ENTEROMORPHA (ALGAE) AND ITS EPIPHYTES; STIMULATION OF GROWTH OF 9 OUT OF 11 REPRESENTATIVE EIPIHYTIC ALGAE BY EITHER ARGININE OR ALANINE; AND PROGRESS IN OTHER SUBPROJECTS. (BOPP-NSIC)

FIELD 05C

ACCESSION NO. W71-09254

SOME OBSERVATIONS ON ALGAE INVADING A CESIUM-137 CONTAMINATED POND,

ATCMIC ENERGY OF CANADA LTD., PINAWA (MANITOBA).

JANET R. DUGLE, ANC J. E. GUTHRIE.

AVAILABLE FROM NTIS. ATOMIC ENERGY OF CANADA TECHNICAL REPORT AECL-3463, (1970). 12 P.

DESCRIPTORS:
*AQUATIC ALGAE, *RADIOACTIVITY EFFECTS, *PONDS, WATER POLLUTION EFFECTS, RADIOISOTOPES, COMPETITION, SPECIATION.

IDENTIFIERS:
*CESIUM RADIOISOTOPES.

ABSTRACT:
A COMPARISON OF THE SPECIES OF ALGAE FOUND AT VARIOUS COLLECTION SITES WITHIN TWO PONDS, ONE CONTAMINATED WITH CESIUM-137, WAS MADE IN OCTOBER 1968. THESE OBSERVATIONS WILL FORM THE BASIS OF A STUDY OF THE SUCCESSION OF ALGAL SPECIES IN POND COMMUNITIES. (BOPP-NSIC)

FIELD 05C

ACCESSION NO. W71-09256

THE ADHESIVE PROPERTIES OF 'CHLORELLA VULGARIS',

PUERTO RICO UNIV., MAYAGUEZ. DEPT. OF MARINE SCIENCES.

THCMAS R. TOSTESON, AND LUIS R. ALMODOVAR.

AVAILABLE FROM THE NATIONAL TECHNICAL INFORMATION SERVICE AS AD-721 114, $3.00 IN PAPER COPY, $0.95 IN MICROFICHE. ONR TECHNICAL REPORT, NO. TR-1, APR. 1971, 35 P.

DESCRIPTORS:
*CHORELLA, *ALGAE, *FOULING, *NUISANCE ALGAE, FLCATING PLANTS, SEAWATER, PLANKTON, BIODEGRADATION, LIGHT, TEMPERATURE.

IDENTIFIERS:
ACRYLIC RESINS, GLASS, THYMIDINES, CELLS(BIOLOGY), *BIODETERIORATION, DEMOCOLCINE.

ABSTRACT:
AMBIENT SEA WATER CONTAINS MATERIAL THAT PROMOTES THE ADHESION OF THE PLANKTONIC ALGAE CHLORELLA VULGARIS TO PLASTIC SURFACES. THIS ADHESION TAKES PLACE WITHIN 3 TP 6 HOURS, AND IS INHIBITED BY THE ABSENCE OF LIGHT. THE EFFECT OF LIGHT ON THE RESPONSE OF CHLCRELLA TO THE OCEAN TREATED SURFACE IS DUE TO THE EFFECT OF LIGHT ON THE RATE OF GROWTH OF THE ALGAE POPULATION. THE ADHESION OF CHLORELLA TC GLASS SURFACES IS SIGNIFICANTLY INCREASED IN THE PRESENCE OF THYMIDINE AND COLCEMIDE. THYMIDINE ACCELERATES THE RATE OF GROWTH OF THE ALGAL CELLS AND COLCEMIDE BLOCKS THIS GROWTH DURING MITOSIS. THE EFFECT OF OCEAN DEPOSITED MATERIALS ON THE ADHESION OF CHLORELLA TO PLASTIC APPEARS SIMILAR TO THE EFFECT OF AGENTS THAT INCREASE THE RELATIVE NUMBER OF CELLS TO THE G2 PHASE OF THE CELL CYCLE ON THE ACHESION OF THIS CELL TO GLASS.

FIELD 05C

ACCESSION NO. W71-09261

PHOSPHATES IN DETERGENTS AND THE EUTROPHICATION OF AMERICA'S WATERS.

AVAILABLE FROM SUP OF DOC, US GOVERNMENT PRINTING OFFICE, WASH DC 20402 PRICE $0.40, HOUSE REPORT NO 91-1004, TWENTY-THIRD REPORT BY THE COMM ON GOV'T OPERATIONS, 91ST CONG, 2D SESS (1970). 88 P, 2 FIG, 9 TAB, 87 REF, 2 APP.

DESCRIPTORS:
*EUTROPHICATION, *PHOSPHATES, *DETERGENTS, *POLLUTION ABATEMENT, WATER POLLUTION EFFECTS, NUTRIENTS, PHOSPHOROUS, WATER SOFTENING, LEGAL ASPECTS, HARD WATER, PUBLIC HEALTH, COST COMPARISONS, AQUATIC ALGAE, GRANTS, MODEL STUDIES.

ABSTRACT:
PHOSPHOROUS POLLUTION CAUSES EUTROPHICATION, AND OVER HALF OF THE PHOSPHORUS POLLUTION IN THE UNITED STATES IS CAUSED BY DETERGENTS. IN THESE HEARINGS THE DETERGENT INDUSTRY CONTENDED THAT PHOSPHATE DETERGENTS DID NOT HARM AND WERE ESSENTIAL TO MAINTAIN CLEANLINESS AND SANITATION STANDARDS, SINCE THERE WAS NO SUITABLE REPLACEMENT. REBUTTALS TO THESE DEFENSES SUGGEST SEVERAL ALTERNATIVES TO PHOSPHATE DETERGENTS. THE REMOVAL OF NUTRIENTS FROM WASTE WATERS INSTEAD OF PHOSPHATES FROM DETERGENTS AS SUGGESTED BY THE INDUSTRY, WOULD BE COSTLY, SLOW AND CREATE ADDITIONAL POLLUTION PROBLEMS. THE COMMITTEE CONCLUDED THAT NO PROGRESS HAD BEEN MADE IN COMBATING EUTROPHICATION FROM DETERGENTS AND RECOMMENDED A SERIES OF GRADUAL REDUCTIONS IN THE USE OF PHOSPHATE DETERGENTS, WITH A COMPLETE ELIMINATION OF THEIR USE BY THE END OF 1972. ADDITIONALLY, THE COMMITTEE RECOMMENDED: THE MANDATORY LABELING OF PHOSPHATE CONTENT DURING THE INTERIM PERIOD, EXPANDED RESEARCH BY THE FEDERAL WATER QUALITY ADMINISTRATION TO DEVELOP LOW-PHOSPHORUS OR PHOSPHORUS-FREE DETERGENTS, AND THE IMMEDIATE ELIMINATION OF THE INDUSTRY-GOVERNMENT TASK FORCE, WHOSE ACTUAL PURPOSE WAS TO SECURE GOVERNMENT COOPERATION IN RETAINING PHOSPHATE DETERGENTS. (GALLAGHER-FLORIDA)

FIELD 05C, 06E, 05G

ACCESSION NO. W71-09429

NUTRIENT REMOVAL AND ADVANCED WASTE TREATMENT.

PITTSBURG UNIV., PA. GRADUATE SCHOOL OF PUBLIC HEALTH.

PROCEEDINGS, PITTSBURG SANITARY ENGINEERING CONFERENCE, 8, PITTSBURG, PENNSYLVANIA, FEB 1970. 265 P.

DESCRIPTORS:
*EUTROPHICATION, *TERTIARY TREATMENT, *WATER QUALITY CONTROL, AMMONIA, NITROGEN, PHOSPHORUS, ALGAE, FEASIBILITY STUDIES, TECHNICAL FEASIBILITY, ECONOMIC FEASIBILITY, PILOT PLANTS, ION EXCHANGE, ELECTROLYSIS, CHEMICAL PRECIPITATION, NITRIFICATION, DENITRIFICATION, *WASTE WATER TREATMENT.

ABSTRACT:
THE 8TH PITTSBURG SANITARY ENGINEERING CONFERENCE WAS CONDUCTED IN FEBRUARY OF 1970 TO ASSEMBLE CURRENT TECHNIQUES AND METHODS OF NITROGEN AND/OR PHOSPHORUS REMOVAL. A FEW OF THE METHODS PRESENTED ARE: (1) CATION EXCHANGE; (2) ELECTROLYSIS; (3) BREAKPOINT CHLORINATION; (4) AMMONIA STRIPPING; AND (5) MANY OTHERS. REMOVAL EFFICIENCIES FOR AMMONIA, IN PARTICULAR WERE REPORTED AS HIGH AS 93%. SYSTEMS WERE TESTED AND RETESTED USING VARIOUS FEED RATES, SYSTEM LOADINGS, PH RANGES, TEMPERATURE, ETC. MANY OF THE TECHNIQUES REPORTED NOT ONLY REMOVAL EFFICIENCIES BUT COSTS IN TERMS OF CENTS/1000 GALLONS OF THROUGHPUT. FOR EXAMPLE, COLUMN DENITRIFICATION OF WATER CONTAINING 20 MGL1 AS NO3-N, AND USING A 3/1 RATIO OF METHANOL TO NITROGEN, WAS REPORTED AT 2.8 CENTS/1000 GALLONS FOR CHEMICALS. BY GATHERING TOGETHER ALL OF THIS INFORMATION AND DISCUSSING AND REPRODUCING IT, IT WAS HOPED THAT FUTURE LAKE, RIVER, AND STREAM EUTROPHICATION COULD BE PREVENTED BY TAKING PREVENTIVE ACTION BEFORE THE WATERS ARE DESTROYED OR IRREPARABLY DAMAGED. (LOWRY-TEXAS)

FIELD 05D, 05C

ACCESSION NO. W71-09450

REUSE OF CHEMICAL FIBER PLANT WASTE WATER AND COOLING WATER BLOWDOWN,

FIBER INDUSTRIES INC., CHARLOTTE, N.C.

WILLIAM J. DAY.

COPY AVAILABLE FROM GPO SUP DOC AS EP2.10:12090EOX10/70, $0.70; MICROFICHE FROM NTIS AS PB-200 695, $0.95. WATER POLLUTION CONTROL RESEARCH SERIES, OCTOBER 1970. 66 P, 15 FIG, 5 TAB, 15 REF. EPA GRANT NC WPRD-100-01-68, PROGRAM NO 12090 EUX 10/70.

DESCRIPTORS:
*WATER REUSE, *DOMESTIC WASTES, *INDUSTRIAL WASTES, ACTIVATED SLUDGE, TRICKLING FILTERS, LAGOONS, TERTIARY TREATMENT, ACTIVATED CARBON, FLOCCULATION, ALGAE, SLUDGE, CHROMIUM, TOXICITY, NEUTRALIZATION, HYDROGEN ION CONCENTRATION, TEMPERATURE, BIOCHEMICAL OXYGEN DEMAND, CHEMICAL OXYGEN DEMAND, COOLING TOWERS, *WASTE WATER TREATMENT.

IDENTIFIERS:
*MICRO SCREENING.

ABSTRACT:
WASTE WATERS FROM A FORTREL POLYESTER MANUFACTURING PLANT CONSISTED OF ORGANIC CHEMICAL PROCESS WASTES, COOLING SYSTEM BLOWDOWN, AND SANITARY WASTES FROM THE PLANT. A WATER REUSE PROGRAM WAS INSTITUTED WHICH CONSISTED OF: (1) PRETREATMENT OF COOLING WATERS FOR REMOVAL OF HEAVY METALS; (2) IN-PLANT MODIFICATIONS AND ADDITIONS TO THE EXISTING SYSTEM TO INCREASE TREATMENT PLANT CAPACITY; AND (3) A POST TREATMENT SYSTEM FOR EFFLUENT POLISHING PRIOR TO SELECTED REUSE. THE FINAL SYSTEM CONSISTED OF: (1) A CHROMATE REDUCTION UNIT RATED AT 120 GPM FOR CONCENTRATIONS OF UP TO 300 MG/1 CRO4 AND DESIGNED FOR CONTINUOUS OPERATION; (2) EQUALIZATION BASINS HAVING A COMBINED CAPACITY OF 195,000 GALLONS AND CONTAINING SUFFICIENT MIXING CAPACITY TO PREVENT SHORT-CIRCUITING AND STRATIFICATION; (3) A PLASTIC MEDIA ROUGHING FILTER CONSISTING OF TWO TIERS OF POLY-VINYL CHLORIDE MEDIA 10 FEET THICK AND 25 FT IN DIAMETER; (4) AN AERATION BASIN EQUIPPED WITH 175 HP OF AERATION AND MIXING CAPACITY AND USING 100% RECYCLE OF CLARIFIER SLUDGE; (5) A PERIPHERAL FLOW TYPE CLARIFIER; (6) TWO SERIES CONNECTED POLISHING PONDS; (7) A MICRO SCREEN OR ALGAE SCREEN; (8) A FLOCCULANT AND/OR CARBON UNIT; (9) A SLUDGE POND; AND (10) A DIGESTER. CHROMIUM WAS REMOVED FROM THE COOLING TOWER BLOWDOWN FOR $.21 PER POUND OF CHROMATE REMOVED. THE PLASTIC MEDIA TRICKLING FILTER, USED AS A ROUGHING FILTER, PROVIDED 40% BOD REMOVAL OVER A WIDE RANGE OF LOADING RATES. THE 0.33 MGD INDUSTRIAL AND DOMESTIC WASTE WATER WAS TREATED AND REUSED AT A RATE OF 0.10 MGD FOR APPROXIMATELY 40 CENTS/1000 GALLONS. (LOWRY-TEXAS)

FIELD 05D

ACCESSION NO. W71-09524

EFFECTS OF PLANNEC FRESH WATER DIVERSIONS ON THE SAN FRANCISCO BAY AND SACRAMENT=SAN JOAQUIN ESTUARY,

CALIFORNIA UNIV., CAVIS. DEPT. OF CIVIL ENGINEERING.

RAY B. KRONE.

REPORT TO THE SAN FRANCISCO BAY CONSERVATION AND DEVELOPMENT COMMISSION, MAY 21, 1970. 7 P, 1 FIG.

DESCRIPTORS:
*DELTAS, *BAYS, *WATER QUALITY, *WATER POLLUTION CONTROL, *DIVERSION, SEDIMENTS, WATER CIRCULATICN, ABSORPTICN, NUTRIENTS, PESTICIDES, ALGAE, DREDGING, ESTUARINE ENVIRONMENT, CALIFCRNIA WATER CONVEYANCE, CANALS, HEAVY METALS.

IDENTIFIERS:
*SAN FRANCISCO BAY-DELTA.

ABSTRACT:
THE EFFECTS ARE DISCUSSED OF THE PLANNED DIVERSICN OF FRESH WATER FLOWS AND SUSPENDED SEDIMENTS FRCM THE SAN FRANCISCO BAY AND SACRAMENTO-SAN JOAQUIN ESTUARY. SEDIMENT IS CARRIED INTO THE BAY SYSTEM WITH FRESH WATER INFLOWS. BAY SEDIMENTS HAVE THREE OUTSTANDING EFFECTS: THEY IMPAIR THE PENETRATICN OF LIGHT, THEY ABSORB TREMENDCUS AMOUNTS OF TOXIC MATERIALS, AND CLOG NAVIGABLE WATERWAYS. THE TURBIDITY CAUSED BY SUSPENDED SEDIMENT MATERIALS LIMITS THE DEPTH TO WHICH THERE IS SUFFICIENT LIGHT FOR ALGAE TO MULTIPLY. SINCE NUTRIENTS FOR ALGAE GROWTH ARE PLENTIFUL IN THE BAY, THE LIMITED LIGHT PENETRATION DUE TO SEDIMENT IS THE LIMITING FACTOR TO THE RAPID MULTIPLICATION OF ALGAE IN THE BAY. THE SEDIMENTS ALSO ABSORB TOXIC COMPOUNDS, SUCH AS HEAVY METALS, PESTICIDES, AND RADIOACTIVE FALLOUT. THE CNLY FORESEEABLE BENEFIT OF THE REDUCED SEDIMENTS ARE RECUCED DREDGING COSTS. THE REPORT RECOMMENDS THAT COMPREHENSIVE STUDIES BE MADE OF THE REDUCED WATER AND SEDIMENT INFLOWS; THAT CRITERIA BE ESTABLISHED AS A BASIS FOR REGULATING FRESH WATER DIVERSION; THAT AN ADEQUATE MCNITORING SYSTEM BE MAINTAINED; THAT THE CONSTITUENT LEVELS AND LOCATICNS OF WASTE DISCHARGES TO THE SYSTEM BE REGULATED; THAT NEW METHCDS CF TREATMENT OF WASTE WATERS FOR REMOVAL OF NUTRIENTS AND TOXIC MATERIALS BE DEVELOPED AND SUPPORTED. (POERTNER)

FIELD 06G, 05G

ACCESSION NO. W71-09581

NUTRIENTS AND ALGAL REMOVAL FROM OXIDATICN PONDS EFFLUENTS,

MISSISSIPPI STATE UNIV., STATE COLLEGE. DEPT. OF SANITARY ENGINEERING.

ADNAN SHINDALA.

CONFERENCE HELD APRIL 13-14, 1971, VICKSBURG, WATER RESOURCES RESEARCH INSTITUTE, MISSISSIPPI STATE UNIVERSITY, P 1-7, 1971. 7 P, 1 FIG, 1 TAB, 3 REF.

DESCRIPTORS:
*ALGAE, *WASTE WATER TREATMENT, *TERTIARY TREATMENT, *COAGULATION, *OXIDATION LAGCONS, AEROBIC TREATMENT, OXIDATICN, CRGANIC MATTER.

IDENTIFIERS:
NUTRIENT REMOVAL.

ABSTRACT:
CHEMICAL COAGULATION IS AN EFFECTIVE POST TREATMENT PROCESS FOR ALGAL REMOVAL AND FOR IMPRCVING THE QUALITY OF EFFLUENTS FROM STABILIZATION PONDS. OF THE COAGULANTS TESTED, ALUM WAS THE BEST. THE OPTIMUM DOSAGE FOR BEST REMOVAL OF THE PARAMETERS STUDIED WAS IN THE RANGE OF 75 TO 100 MG/LITER. USING THIS DOSAGE, THE SUPERNATANT FROM THE CHEMICAL COAGULATION PROCESS WAS FOUND TO CONTAIN 2.5 MG/LITER BOD, 22.9 MG/LITER COD, 1.5 MG/LITER TOTAL PHOSPHATES, 3.5 MG/LITER TOTAL PHOSPHATES, 3.5 MG/LITER TCTAL NITROGEN, 500 TO 1000 ALGAL CELLS/ML AND APPROXIMATELY 5,000 COLIFORMS/100 ML. THE ALGAE IN THE POND EFFLUENTS CONTRIBUTE HEAVILY TO THE BOD, COD, AND NITROGEN IN THE EFFLUENT, WHILE THE CONTRIBUTION TO THE PHOSPHATES CONCENTRATION WAS LESS IMPORTANT. (KNAPP-USGS)

FIELD 05G, 05D

ACCESSION NO. W71-09629

ON THE SIGNIFICANCE OF METAL COMPLEXING AGENTS IN SECONDARY SEWAGE EFFLUENTS,

MICHIGAN UNIV., ANN ARBOR, DEPT. OF ENVIRONMENTAL HEALTH.

MICHAEL E. BENDER, WAYNE R. MATSON, AND ROBERT A. JORDAN.

ENVIRONMENTAL SCIENCE AND TECHNOLOGY, VOL 4, NO 6, P 520-521, JUNE 1970. 1 FIG, 1 TAB, 11 REF. FWPCA GRANT 1-FL-WP-26-294-01, NIH GRANT 5-501-FR-05447-07.

DESCRIPTORS:
*WATER POLLUTION EFFECTS, *ORGANIC WASTES, *EUTROPHICATION, *CHELATION, *SEWAGE, *ORGANIC COMPOUNDS, WATER POLLUTION TREATMENT, WATER POLLUTION SOURCES, ENVIRONMENTAL SANITATION, SECONDARY TREATMENT, AQUATIC ALGAE, PRIMARY PRODUCTIVITY, TRACE ELEMENTS, PHOSPHATES, NITRATES, LABORATORY TESTS.

IDENTIFIERS:
*CHELATORS, ANODIC STRIPPING, ORGANIC CHELATORS.

ABSTRACT:
SEVERAL COMPONENTS OF SECONDARY SEWAGE OTHER THAN PHOSPHOROUS AND NITROGEN COULD BE ABETTING LAKE EUTROPHICATION. THE CHELATION OF METALS BY ORGANIC COMPOUNDS HAS BEEN DEMONSTRATED TO CAUSE SIGNIFICANT INCREASES IN ALGAL PRODUCTION. THIS STUDY WAS AN ATTEMPT TO FIND SUCH CHELATORS IN SECONDARY SEWAGE EFFLUENTS. TWO DISTINCT METAL-COMPLEXING MOLECULAR WEIGHT FRACTIONS WERE DEMONSTRATED USING RECENT ADVANCES IN ANODIC STRIPPING TECHNIQUES. ONE FRACTION HAS A MOLECULAR WEIGHT SIMILAR TO THAT OF SYNTHETIC CHELATORS AND HAS PROVEN EFFECTIVE IN STIMULATING ALGAL GROWTH. SUBSEQUENT INVESTIGATIONS ARE ATTEMPTING TO RULE OUT THE POSSIBILITY THAT OTHER SUBSTANCES IN SEWAGE FRACTIONS COULD BE RESPONSIBLE FOR THE APPARENT STIMULATION. (LEGORE-WASHINGTON)

FIELD 05B, 05A

ACCESSION NO. W71-09674

A WATER RESOURCE-WATER SUPPLY STUDY OF THE POTOMAC ESTUARY,

ENVIRONMENTAL PROTECTION AGENCY, ANNAPOLIS, MD. CHESAPEAKE TECHNICAL SUPPORT LAB.

NORBERT A. JAWORSKI, LEO J. CLARK, AND KENNETH D. FEIGNER.

ENVIRONMENTAL PROTECTION AGENCY, WATER QUALITY OFFICE, TECHNICAL REPORT 35, APRIL 1971, 263 P, 39 TAB.

DESCRIPTORS:
*ESTUARIES, *WATER QUALITY, *WATER RESOURCES DEVELOPMENT, *WATER SUPPLY, WASTE WATER, RUNOFF, DISSOLVED OXYGEN, ALGAE, NUTRIENTS, BACTERIA, VIRUSES, HEAVY METALS, HUMAN POPULATION, WASTE WATER TREATMENT, WATER QUALITY CONTROL.

IDENTIFIERS:
POTOMAC ESTUARY.

ABSTRACT:
A DETAILED INVESTIGATION OF THE WATER QUALITY AND WATER RESOURCES OF THE POTOMAC ESTUARY WAS CONDUCTED BY THE CHESAPEAKE TECHNICAL SUPPORT LABORATORY. INCLUDED IN THE STUDY WERE AN EVALUATION OF POLLUTION SOURCES INCLUDING NUTRIENTS; THE DEVELOPMENT OF MATHEMATICAL MODELS TO PREDICT POLLUTANT EFFECTS ON WATER QUALITY; THE PROJECTION OF WATER SUPPLY NEEDS AND WASTEWATER LOADINGS; AN EVALUATION AS A POTENTIAL WATER SUPPLY SOURCE; THE DETERMINATION OF THE MAXIMUM POUND LOADINGS BY ZONE FOR CERTAIN POLLUTANTS UNDER DIFFERENT FLOW CONDITIONS; ALTERNATIVE WASTE TREATMENT PLANS AND COST ANALYSIS OF WASTEWATER TREATMENT. (ENSIGN-PAI)

FIELD 05B, 06D, 02L

ACCESSION NO. W71-09788

EVALUATING OIL SPILL CLEANUP AGENTS, DEVELOPMENT OF TESTING PROCEDURES AND CRITERIA,

CALIFORNIA STATE WATER RESOURCES CONTROL BOARD, SACREMENTO.

CHARLES R. HAZEL, FRED KOPPERDAHL, NORMAN MORGAN, AND WALTER THOMSEN.

CALIFORNIA WATER RESOURCES CONTROL BOARD PUBLICATION NO. 43, JULY 6, 1970, 150 P, 67 FIG, 17 TAB, 100 REF.

DESCRIPTORS:
*OIL WASTES, *CLEANING, *DISPERSION, *CHEMICALS, *SURFACTANTS, *TOXICITY, *BIOASSAY, FISH, SHELLFISH, ALGAE, DEGRADATION.

IDENTIFIERS:
*DISPERSANTS, *COLLECTING AGENTS, *SINKING AGENTS, EFFECTIVENESS, TESTING PROCEDURES, CRITERIA, CLASSIFICATIONS.

ABSTRACT:
CRITERIA FOR LICENSING AND REGULATING THE USE OF OIL SPILL CLEANUP AGENTS WAS STUDIED AND DEVELOPED. CRITERIA INCLUDED TOXICITY, PERFORMANCE EFFECTIVENESS AND PHYSICAL-CHEMICAL DESCRIPTIONS OF THE OSCA. BIOASSAYS WERE PERFORMED TO ESTABLISH 96-HOUR TLM VALUES FOR DISPERSANTS, OIL AND COMBINATIONS OF OIL AND DISPERSANTS. BIODEGRADATION OF DISPERSANTS WAS TESTED BY BOD, FOAM CALIBRATION AND TOXICITY DECAY BIOASSAY. MISCIBILITY WITH SEAWATER, PERCENT OF OIL EMULSIFIED, OIL SINKING AND DISPERSION AFTER SEVERAL HOURS WERE THE CRITERIA FOR PERFORMANCE EFFECTIVENESS TESTS. FEASIBILITY OF 'FINGERPRINTING' DISPERSANTS WERE TESTED QUANTITATIVELY. LICENSING CRITERIA AND TEST PROCEDURES FOR PRODUCT FLASH POINT, P.H, TRACE SUBSTANCES AND OCCUPATIONAL HEALTH HAZARDS WERE DETERMINED. (ENSIGN-PAI)

FIELD 05G, 07B

ACCESSION NO. W71-09789

LAMINARIA SACCHARINA AND MARINE POLLUTION IN NORTH-EAST ENGLAND,

LIVERPOOL UNIV. (ENGLAND). DEPT. OF BOTANY.

E. M. BURROWS, AND C. PYBUS.

MARINE POLLUTION BULLETIN, VOL. 2, NO. 4, P 53-56, APRIL 1971.

DESCRIPTORS:
*POLLUTANT IDENTIFICATION, *BIOINDICATORS, *ALGAE, GROWTH RATES, CULTURES, SUSPENDED LOAD, SEWAGE, INDUSTRIAL WASTES, SILTS, WATER POLLUTION EFFECTS.

IDENTIFIERS:
*ENGLAND, *LAMINARIA SACCHARINA.

ABSTRACT:
THE GROWTH RATE OF LAMINARIA SACCHARINA IN BOTH THE LABORATORY AND IN THE FIELD HAS BEEN STUDIED AS AN INDICATOR FOR THE EFFECTS OF POLLUTION ON THE NORTH-EAST COAST OF ENGLAND. THIS SPECIES PROVIDES A CONSTANT SUBSTRATE FOR OTHER MARINE ORGANISMS OVER A LARGE AREA OF THE COAST MAKING IT AN IMPORTANT COMPONENT OF THE COASTAL ECOSYSTEM. LAMINARIA SACCHARINA IS VERY SENSITIVE TO ITS ENVIRONMENT AND THEREFORE IS CAPABLE OF GIVING A GRADED RESPONSE TO DIFFERENT TYPES AND DEGREES OF POLLUTION. THE FACTOR THOUGHT TO INHIBIT GROWTH MOST WAS THE PRESENCE OF SILT; HOWEVER POLLUTION OF THE WATERS THEMSELVES HAVING DELETERIOUS EFFECTS ON GROWTH CANNOT BE RULED OUT. (ENSIGN-PAI)

FIELD 05C

ACCESSION NO. W71-09795

RESEARCH ON ECOLOGICAL STUDIES ON THE DYNAMICS OF PLANKTONIC BLUE-GREEN ALGAE WITH SPECIAL REFERENCE TO THEIR MICROSTRATIFICATION.

MINNESOTA UNIV., MINNEAPOLIS.

COO-1820-2 (1970), 6 P.

DESCRIPTORS:
*AQUATIC ALGAE, *LAKES, *PHYTOPLANKTON, *DENSITY STRATIFICATION, SPECIATION, VERTICAL MIGRATION, AQUARIA, GROWTH RATES, SEDIMENTATION, BUOYANCY, WIND VELOCITY, CONVECTION, WATTER POLLUTION EFFECTS.

ABSTRACT:
SPECIES ASSOCIATIONS AT VARIOUS DEPTHS ARE FORTUITOUS, TRANSIENT, AND DEPENDENT UPON PHYSICAL CONTROL MANIFEST BY THE ABIOTIC ENVIRONMENT, RATHER THAN UPON ANY DIRECT BIOLOGICAL CONTROL. INDIVIDUALS COMPRISING THE PHYTOPLANKTON COMMUNITY ARE NOT BOUND TO A SUBSTRATE, AND ARE ENTIRELY MOBILE. VERTICAL DISTRIBUTION IS INFLUENCED BY DIFFERENTIAL GROWTH RATES AT VARIOUS DEPTHS, SEDIMENTATION, BUOYANCY, WIND, AND THERMAL CONVECTION. CURRENTLY, PROPORTIONATELY MORE TIME IS BEING DEVOTED WITH SEVERAL LABORATORY MODELS OF THE STUDY LAKES. (BOPP-NSIC)

FIELD 05C

ACCESSION NO. W71-09852

SECONDARY EFFECTS OF NITROGEN IN WATER,

MICHIGAN UNIV., ANN ARBOR, DEPT. OF SANITARY ENGINEERING.

JACK A. BORCHARDT.

PROCEEDINGS 12TH SANITARY ENGINEERING CONFERENCE ON NITRATE AND WATER SUPPLY: SOURCE AND CONTROL, FEBRUARY 11-12, 1970, UNIVERSITY OF ILLINOIS, URBANA: ILLINOIS UNIVERSITY, COLLEGE OF ENGINEERING PUBLICATION, P 66-77, 1970. 12 P, 9 FIG, 1 TAB, 10 REF.

DESCRIPTORS:
*WATER POLLUTION EFFECTS, *NITRATES, *NUTRIENTS, *ALGAE, *BACTERIA, AESTHETICS, COLOR, TASTE, CORROSION, DISEASES, EUTROPHICATION, FOULING, TOXICITY, ODOR, OXYGEN SAG, TURBIDITY, WATER QUALITY, WATER TREATMENT, NITROGEN CYCLE.

ABSTRACT:
ONE OF THE PRIMARY EFFECTS OF INORGANIC NITROGEN IN NATURAL WATERS IS ITS INFLUENCE ON THE GROWTH OF FLORA. ALGAE IN WATER SUPPLIES CAN CAUSE OBNOXIOUS CONDITIONS IN STORAGE RESERVOIRS, CLOGGING OF INTAKE SCREENS, CAN SERIOUSLY AFFECT PRECHLORINATION, MAY MAKE ADDITIONAL CHEMICALS NECESSARY FOR PURIFICATION, MAY CAUSE SHORTER FILTER RUNS, AND MAY RESULT IN SERIOUS TASTE AND ODOR PROBLEMS. THESE PROBLEMS ARE DISCUSSED IN SOME DETAIL. (KNAPP-USGS)

FIELD 05B, 05C

ACCESSION NO. W71-09958

ENERGY FLOW IN A WOODLAND STREAM ECOSYSTEM: I. TISSUE SUPPORT TROPHIC STRUCTURE OF THE AUTUMNAL COMMUNITY, (WITH GERMAN SUMMARY),

PITTSBURG UNIV., PA. PYMATUNING LAB. OF ECOLOGY; AND MICHIGAN STATE UNIV., HICKORY CORNERS. W. K. KELLOGG BIOLOGICAL STATION.

WILLIAM P. COFFMAN, KENNETH W. CUMMINS, AND JOHN C. WUYCHECK.

ARCHIV FUR HYDROBIOLOGIE, VOL 68, NO 2, P 232-276, 1971. 8 FIG, 13 TAB, 36 REF. OWRR PROJECT NOS A-032-MICH(1) AND B-008-MICH(1).

DESCRIPTORS:
*STREAMS, *STANDING CROP, *TROPHIC LEVEL, *ENERGY BUDGET, PRODUCTIVITY, HERBIVORES, CARNIVORES, ECOSYSTEMS, AQUATIC LIFE, SAMPLING, ALGAE, LABORATORY TESTS, DETRITUS, BENTHOS, AQUATIC ANIMALS, DIATOMS, INSECTS, BACTERIA, FUNGI, TURNOVERS.

IDENTIFIERS:
*RIFFLE BENTHOS, WOODLAND STREAM, ECOSYSTEM ANALYSIS, MACROCONSUMERS, AUTUMNAL COMMUNITY, CALORIES, DETRIVORES, LINESVILLE CREEK(PA).

ABSTRACT:
TO DETERMINE AN ENERGY BUDGET FOR A LOTIC ECOSYSTEM, A 2-YEAR INTENSIVE INVESTIGATION OF A WOODLAND STREAM WAS CONDUCTED OF THE AUTUMNAL BENTHIC MACROCONSUMER COMMUNITY OF THE RIFFLE AREA. STANDING CROP AND TROPHIC STRUCTURE WERE DETERMINED AND PRODUCTION PARAMETERS ESTIMATED. THE PROCEDURE WAS DEVELOPED TO EXPRESS LEVELS ON THE BASIS OF TISSUE SUPPORT RATHER THAN SPECIES CATEGORIES. TROPHIC ANALYSES WERE MADE FROM GUT CONTENTS, AND THE DATA DETERMINED THE PROPORTION OF EACH SIZE CLASS OF EACH SPECIES TO BE ASSIGNED TO ONE OF THREE TROPHIC LEVELS. ON AN INSTANTANEOUS INGESTION BASIS, 17% TO 21% OF THE STANDING CROP TISSUE WAS SUPPORTED BY ALGAL CALORIES, 2% TO 5% BY DETRIAL CALORIES, AND 71% TO 73% BY ANIMAL CALORIES. THE ASSIMILATIVE EFFICIENCY FOR INGESTED FOOD IS DEPENDENT UPON SPECIES ENZYME SYSTEMS AND VARIES IN RELATION TO DIFFERENCES IN INGESTIVE MECHANISMS. THE HIGH SUPPORT BY ANIMAL CALORIES DEMANDS A RAPID BIOMASS TURNOVER (LESS THAN A MONTH) BY THE HERBIVORE-DETRITIVORE COMPONENT AND A SLOWER TURNOVER (SEVERAL MONTHS) BY THE CARNIVORE TISSUE. THE STANDING CROP AND TROPHIC STRUCTURE RECORDED YIELD A CONSISTENT PICTURE OVER TIME. (JONES-WISCONSIN)

FIELD 05C

ACCESSION NO. W71-10064

A CHECKLIST OF MARINE ALGAE OF EASTERN CANADA,

MEMORIAL UNIV. OF NEWFOUNDLAND, ST JOHN'S. DEPT. OF BIOLOGY; AND LAVAL UNIV. QUEBEC. DEPARTEMENT DE BIOLOGIE.

G. ROBIN SOUTH, AND ANDRE CARDINAL.

CANADIAN JOURNAL OF BOTANY, VOL 48, P 2077-2095, 1970. 1 FIG, 132 REF, INDEX.

DESCRIPTORS:
*MARINE ALGAE, *SYSTEMATICS, *COASTS, BENTHIC FLORA, RHODOPHYTA, PHAEOPHYTA, CHLOROPHYTA, MAINE, SPECIATION.

IDENTIFIERS:
*CHECKLIST, *EASTERN CANADA, CAPE CHIDLEY(LABRADOR), NEW BRUNSWICK, ANTICOSTI ISLAND(QUEBEC), MAGDALEN ISLAND(QUEBEC), SABLE ISLAND(NOVA SCOTIA), ST PIERRE(CANADA), MIQUELON(CANADA), MIKROSYPHAR PORPHYRAE, PROTECTOCARPUS SPECIOSUS.

ABSTRACT:
A COMPREHENSIVE CHECKLIST DOCUMENTING MARINE BENTHIC ALGAE WITHIN THE POLITICAL BOUNDARIES OF EASTERN CANADA IS PRESENTED. TAYLOR'S SECOND EDITION (1957) FORMS THE BASIC REFERENCE FOR THIS LIST WHICH INCLUDES ALL PREVIOUSLY PUBLISHED RECORDS TOGETHER WITH SOME NEW ONES, INCORPORATING NOMENCLATURAL CHANGES THAT HAVE OCCURRED SINCE 1957. IT CONSISTS OF 371 SPECIES, SUBSPECIES, AND VARIETIES FROM EASTERN CANADA, INCLUDING 157 RHODOPHYCEAE, 127 PHAEOPHYCEAE, AND 87 CHLOROPHYCEAE. RECORDS FOR THE ENTIRE COASTLINE FROM CAPE CHIDLEY, LABRADOR, IN THE NORTH TO THE NEW BRUNSWICK-MAINE BORDER IN THE SOUTH ARE INCLUDED, AS WELL AS FROM ANTICOSTI ISLAND, MAGDALEN ISLAND, SABLE ISLAND, AND ST PIERRE AND MIQUELON. MIKROSYPHAR PORPHYRAE KUCK, AND PROTECTOCARPUS SPECIOSUS (BERG) KUCK ARE NEW RECORDS FOR THE AREA. THE COMPLETE AUTHOR CITATION FOR EACH GENUS AND SPECIES IS GIVEN TOGETHER WITH EXTENSIVE EXPLANATORY NOTES GIVING SYNONYMS AND PREVIOUS AUTHOR NOTES, 132 REFERENCES, AND AN INDEX TO THE GENERA. (JONES-WISCONSIN)

FIELD 05C, 07B, 02L

ACCESSION NO. W71-10066

SOLUBILIZATION OF TRICALCIUM PHOSPHATE BY BLUE-GREEN ALGAE,

INDIAN AGRICULTURAL RESEARCH INST., NEW DELHI. DIV. OF MICROBIOLOGY.

PAROMITA BOSE, UJJAL SINGH NAGPAL, G. S. VENKATARAMAN, AND S. K. GOYAL.

CURRENT SCIENCE, VOL 40, NO 7, P 165-166, 1971. 1 TAB, 22 REF.

DESCRIPTORS:
*PHOSPHATES, *SOLUBILITY, *CYANOPHYTA, ALGAE, CULTURES.

IDENTIFIERS:
*TRICALCIUM PHOSPHATES, ANABAENA, NOSTOC, TOLYPOTHRIX TENUIS, AULOSIRA FERTILISSIMA, ANACYSTIS NIDULANS.

ABSTRACT:
TWENTY-SEVEN STRAINS OF NITROGEN-FIXING BLUE-GREEN ALGAE AND ONE NON-NITROGEN-FIXING ALGA, ANACYSTIS NIDULANS, WERE INCUBATED FOR 15 DAYS IN A NUTRIENT MEDIUM CONTAINING TRICALCIUM PHOSPHATE. RESULTING DETERMINATIONS REVEALED THAT 17 STRAINS, BELONGING TO THE GENERA ANABAENA, NOSTOC, TOLYPOTHRIX, AND AULOSIRA, WERE EFFECTIVE IN SOLUBILIZATION OF ROCK PHOSPHATE. THE SOLUBILIZATION EFFICIENCIES INDICATED BY TOLYPOTHRIX TENUIS AND AULOSIRA FERTILISSIMA ARE OF PARTICULAR INTEREST. (WILDE-WISCONSIN)

FIELD 05C

ACCESSION NO. W71-10070

DISSOLVED ORGANIC PHOSPHORUS EXCRETION BY MARINE PHYTOPLANKTON,

NORTH CAROLINA UNIV., CHAPEL HILL. DEPT. OF ENVIRONMENTAL SCIENCES AND ENGINEERING.

EDWARD J. KUENZLER.

JOURNAL OF PHYCOLOGY, VOL 6, NO 1, P 7-13, 1970. 4 FIG, 2 TAB, 33 REF. AEC CONTRACT AT-(40-1)-3549.

DESCRIPTORS:
*ORGANIC COMPOUNDS, *PHOSPHORUS, *MARINE PLANTS, *PHYTOPLANKTON, ALGAE, LIGHT INTENSITY, SALINITY, NUTRIENT REQUIREMENTS, CHLORELLA, GROWTH RATES, TEMPERATURE, SEA WATER, PHOSPHATES, PHOSPHORUS RADIOISOTOPES, DIATOMS, HYDROGEN ION CONCENTRATION, MARINE ANIMALS, BACTERIA, ABSORPTION, CULTURES.

IDENTIFIERS:
*DISSOLVED ORGANIC PHOSPHORUS, *EXCRETION, ALKALINE PHOSPHATASE, REASSIMILATION, GREEN FLAGELLATE, CYCLOTELLA CRYPTICA, THALASSIOSIRA FLUVIATILIS, DUNALIELLA TERTIOLECTA, SYNECHOCOCCUS, RHODOMONAS, COCCOLITHUS HUXLEYI, AUTOLYSIS, PHAEODACTYLUM.

ABSTRACT:
EXPERIMENTS PERFORMED ON LABORATORY CULTURES SHOW THAT ALGAE ELIMINATE DISSOLVED ORGANIC PHOSPHORUS (DOP) COMPOUNDS UNDER VARYING CONDITIONS FOR DIFFERENT SPECIES, BUT CERTAIN PATTERNS EMERGE. THE AMOUNTS OF EXTRACELLULAR DOP PRODUCED BY EIGHT SPECIES OF MARINE PLANKTONIC ALGAE (AXENIC CULTURES) UNDER VARIOUS CONDITIONS OF LIGHT, SALINITY, AND NUTRITION WERE COMPARED. AFTER INITIAL LABELING, P-32 WAS MEASURED IN THE FRACTIONS. DIFFERENT SPECIES TAKE DIFFERENT FRACTIONS OF THE TOTAL EXCRETED DOP. MORE THAN 20% OF TOTAL PHOSPHORUS IN THE SYSTEM WAS EXCRETED BY CYCLOTELLA CRYPTICA, THALASSIOSIRA FLUVIATILIS, DUNALIELLA TERTIOLECTA, AND SYNECHOCOCCUS UNDER ONE OR MORE OF THE EXPERIMENTAL CONDITIONS. EXCRETION OF DOP WAS PROPORTIONAL TO LIGHT INTENSITY IN DUNALIELLA, RHODOMONAS, CHLORELLA, AND COCCOLITHUS HUXLEYI. PHOSPHORUS LIMITATION REDUCED DOP PRODUCTION BY CYCLOTELLA AND THALASSIOSIRA, NITROGEN LIMITATION REDUCED DOP BY PHAEODACTYLUM, DUNALIELLA, AND RHODOMONAS, AND LACK OF IRON REDUCED DOP LEVELS IN CYCLOTELLA CULTURES. SALINITY AFFECTED GROWTH, BUT NO CLEAR RELATIONSHIP TO DOP EXCRETION WAS EVIDENT. THE DOP ELIMINATED DURING GROWTH WAS REASSIMILATED BY THE SPECIES THAT PRODUCED IT AND BY OTHER SPECIES, BUT LACK OF ALKALINE PHOSPHATASE REDUCED THE AMOUNT OF DOP AVAILABLE TO CERTAIN ALGAE. (JONES-WISCONSIN)

FIELD 05C

ACCESSION NO. W71-10083

PRIMARY PRODUCTION: DEPRESSICN OF OXYGEN EVOLUTION IN ALGAL CULTURES BY ORGANOPHOSPHORUS INSECTICIDES,

NORTHWESTERN UNIV., BOSTON, MASS. DEPT. OF BIOLOGY.

SYLVIA B. DERBY, AND ERNEST RUBER.

BULLETIN OF ENVIRONMENTAL CONTAMINATION AND TOXICOLOGY, VOL 5, NO 6, P 553-557, 1971. 1 TAB, 10 REF.

DESCRIPTORS:
*INSECTICIDES, *ALGAE, *PRIMARY PRODUCTIVITY, CHLORINATED HYDROCARBON PESTICIDES, DDT, ORGANOPHOSPHORUS PESTICIDES, PHYTOPLANKTON, TOXICITY, DISSOLVED OXYGEN, CULTRUES, ENDRIN, DIELDRIN, CXYGENATION, CARBAMATE PESTICIDES.

IDENTIFIERS:
ABATE, BAYTEX, BAYGONE, CYCLOTELLA, DUNALIELLA, SKELETONEMA.

ABSTRACT:
THE EFFECT OF DDT, ORGANOPHOSPHATES (BAYTEX, ABATE), AND CARBAMATE (BAYGONE) INSECTICIDES WAS APPRAISED ON THE BASIS OF OXYGEN PRODUCED BY ALGAE IN ACETONE CULTURE MEDIA. THE INSECTICIDES WERE USED IN CONCENTRATIONS OF 1.0, 0.1, AND 0.01 PPM. THE RESULTS VARIED WITH ALGAL SPECIES AND THE NATURE OF ERADICANTS. THE ORDER FROM THE MOST TO LEAST TOXIC COMPOUNDS WAS: BAYTEX, BAYGONE, DDT, AND ABATE. THE VULNERABILITY OF ALGAE TO INSECTICIDES INCREASED IN THE FOLLOWING ORDER: CYCLOTELLA NANA, PHAEODACTYLUM TRICORNUTUM, SKELETONEMA COSTATUM, AND DUNALIELLA EUCHLORA. (WILDE-WISCONSIN)

FIELD 05C

ACCESSION NO. W71-10096

ASSESSMENT OF POLLUTICN EFFECTS BY THE USE OF ALGAE,

LIVERPOOL UNIV. (ENGLAND). HARTLEY BOTANICAL LABS.

ELSIE M. BURROWS.

PROCEEDINGS OF THE ROYAL SOCIETY OF LONDON B, VOL 177, P 295-306, 1971. 5 FIG, 22 REF.

DESCRIPTORS:
*POLLUTANT IDENTIFICATION, *BIOASSAY, *INDICATORS, *WATER POLLUTION EFFECTS, *ALGAE, *MARINE ALGAE, *BIOINDICATORS, *PHOTOSYNTHESIS, *PLANT PHYSIOLOGY, POLLUTANTS, ANALYTICAL TECHNIQUES, WATER POLLUTION SOURCES, TOXINS, WASTES, CHLOROPHYTA, SESSILE ALGAE, WATER ANALYSIS.

IDENTIFIERS:
*LAMINARIA SP., *ULVA SP., MACRCCYSTIS SP.

ABSTRACT:
AN ATTEMPT WAS MADE TO USE THE GROWTH RATE IN CULTURE OF SOME OF THE LARGER ATTACHED ALGAE FOR INDICATION OF POLLUTION. ULVA LACTUA HAS A HIGH POTENTIAL FROM THIS POINT OF VIEW BECAUSE OF THE EASE WITH WHICH IT CAN BE CULTURED AND ALSO BECAUSE OF ITS REACTIONS TO POLLUTION BY SEWAGE. THE GROWTH RATE OF LAMINARIA SACCHARINA SHOWS A GRADED RESPONSE TO CHANGES IN TOTAL MEDIUM AND PHYSICAL CONDITIONS AND TO CONCENTRATIONS OF ADDED SINGLE SUBSTANCES. THIS SPECIES COULD BE A USEFUL POLLUTION INDICATOR, NOT ONLY BECAUSE OF ITS SENSITIVITY TO CHANGES, BUT ALSO BECAUSE OF THE PART IT PLAYS IN THE ECOSYSTEMS OF THE BRITISH COASTS. (LEGORE-WASHINGTON)

FIELD 05A, 05C

ACCESSION NO. W71-10553

PESTICIDE EFFECT OF GROWTH AND C-14 ASSIMILATION IN A FRESHWATER ALGA,

MISSOURI UNIV., COLUMBIA. DIV. OF BIOLOGY; AND BUREAU OF SPORT FISHERIES AND WILDLIFE, COLUMBIA, MO. FISH-PESTICIDE RESEARCH LAB.

LELYN STADNYK, ROBERT S. CAMPBELL, AND B. THOMAS JOHNSON.

BULLETIN OF ENVIRONMENTAL CONTAMINATION AND TOXICOLOGY, VOL 6, NO 1, P 1-8, 1971. 1 FIG, 1 TAB, 16 REF.

DESCRIPTORS:
*PESTICIDES, *CHLORINATED HYDROCARBON PESTICIDES, *CARBAMATE PESTICIDES, *ORGANOPHOSPHOROUS PESTICIDES, *2-4-D, *DDT, *DIELDRIN, *DIAZINON, *WATER POLLUTION EFFECTS, *PHOTOSYNTHESIS, PESTICIDE TOXICITY, PLANT GROWTH REGULATORS, INHIBITION, INHIBITORS, PHYTOPLANKTON, PHYTO-TOXICITY, HERBICIDES, INSECTICIDES, BIOMASS, PRODUCTIVITY, SCENEDESMUS, ALGAL.

IDENTIFIERS:
*DIURON, *CARBARYL, *TOXAPHENE.

ABSTRACT:
THE EFFECTS WERE EVALUATED OF PESTICIDES (DIURON, CARBARYL, 2,4-D, DDT, DIELDRIN, TOXAPHENE, AND DIAZINON) ON LOW DENSITY POPULATIONS OF THE FRESHWATER ALGA, SCENEDESMUS QUADRICAUDATA, IN TERMS OF CHANGES IN GROWTH AND METABOLISM RATHER THAN DEATH. OF THE COMPOUNDS TESTED, THE SOIL STERILANT AND HERBICIDE DIURON WAS THE MOST TOXIC TO THE ALGAL POPULATION, ALTHOUGH APPARENT BACTERIAL DETOXIFICATION OF THIS SUBSTANCE WAS INDICATED. WHILE DDT, TOXAPHENE AND DIELDRIN GENERALLY CAUSED A DECREASE IN CELL NUMBER, BIOMASS AND CARBON ASSIMILATION, CARBARYL AND DIAZINON EITHER HAD NO EFFECT OR STIMULATED CARBON FIXATION AND CELL DIVISION. LONG TERM CHRONIC EFFECTS OF PESTICIDES IN TERMS OF PLANT LIFE AS WELL AS ANIMAL SPECIES MUST BE CONSIDERED. (LEGORE-WASHINGTON)

FIELD 05C

ACCESSION NO. W71-10566

SALTON SEA, CALIFORNIA--WATER QUALITY AND ECOLOGICAL MANAGEMENT CONSIDERATIONS,

FEDERAL WATER QUALITY ADMINISTRATION, SAN FRANCISCO, CALIF. PACIFIC SOUTHWEST REGION.

R. C. BAIN, A. M. CALDWELL, R. H. CLAWSON, H. L. SCOTTEN, AND R. G. WILLS.

FEDERAL WATER QUALITY ADMINISTRATION PACIFIC SOUTHWEST REGION REPORT, JULY 1970. 54 P, 15 FIG, 10 TAB, 16 REF.

DESCRIPTORS:
*SURFACE WATERS, *SEA WATER, *WATER QUALITY, *ECOLOGY, *CALIFORNIA, HYDROLOGIC DATA, CHEMICAL ANALYSIS, DATA COLLECTIONS, NUTRIENTS, SALINITY, SEDIMENTS, EUTROPHICATION, FISH, ALGAE, RECREATION, ENVIRONMENTAL EFFECTS, INVERTEBRATES, WATER RESOURCES DEVELOPMENT.

IDENTIFIERS:
*SALTON SEA(CALIF), ECOLOGICAL MANAGEMENT.

ABSTRACT:
THIS REPORT PRESENTS DATA RELATED TO SALINITY AND NUTRIENT RELATED PROBLEMS AND CONTROL MEASURES TO ALLEVIATE ADVERSE WATER QUALITY CONDITIONS IN THE SALTON SEA. THE SALTON SEA LIES IN A LOW-LYING DESERT SINK AREA APPROXIMATELY 85 MILES EAST OF SAN DIEGO, CALIFORNIA. FORMED IN 1905-06, THE 230,000 ACRE SEA IS THREATENED WITH RAPIDLY RISING SALINITY LEVELS WHICH IF UNCONTROLLED ARE EXPECTED TO ELIMINATE THE CURRENTLY VALUABLE SPORT FISHERY WITHIN THE NEXT DECADE. FLUCTUATING WATER LEVELS AND EUTROPHICATION SYMPTOMS, SUCH AS DISSOLVED OXYGEN DEFICIENCIES IN DEEPER WATERS, DISCOLORATIONS AND TURBIDITY OF THE WATER AND OFFENSIVE ODORS CAUSED BY DENSE PHYTOPLANKTON POPULATIONS, ARE ALSO MAJOR SALTON SEA PROBLEMS. AN ANNUAL INFLOW OF APPROXIMATELY 1.2 MILLION ACRE FEET PRINCIPALLY FROM THE NEW AND ALAMO RIVERS BRINGS SALTS, NUTRIENTS, PESTICIDES AND FECAL BACTERIA TO THE SEA. MUCH OF THIS WATER IS AGRICULTURAL DRAINAGE FROM THE IMPERIAL VALLEY AND SEWAGE FROM VALLEY COMMUNITIES AND FROM MEXICO. EVAPORATION LOSSES WITHIN THE SALTON SEA APPROXIMATE ANNUAL INFLOW. THUS A HYDRODYNAMIC BALANCE EXISTS. RECENT DATA SHOW SALT LEVELS ARE APPROXIMATELY EQUAL TO OCEANIC SALINITY ALTHOUGH IONIC COMPOSITION IS SOMEWHAT DIFFERENT. (WOODARD-USGS)

FIELD 05G, 02H, 05A

ACCESSION NO. W71-10577

BIOLCGY OF CLADOPHORA IN FRESHWATERS,

DURHAM UNIV. (ENGLAND). DEPT. CF BOTANY.

B. A. WHITTON.

WATER RESEARCH, VOL 4, NO 7, P 457-476, JULY 1970. 20 P, 1 TAB, 80 REF.

DESCRIPTORS:
*EUTROPHICATION, *ALGAE, *SURFACE WATERS, *REVIEWS, *WATER POLLUTION SOURCES, NUTRIENTS, PLANT PHYSIOLOGY, LAKES, PONDS, AQUATIC ALGAE, ECOLOGY, BIOLOGICAL PROPERTIES, FRESH WATER, WATER QUALITY.

IDENTIFIERS:
*CLADOPHORA.

ABSTRACT:
THE LITERATURE ON CLADOPHORA GROWING IN FRESHWATERS IS SUMMARIZED, ESPECIALLY THOSE SITUATIONS WHERE NUTRIENT ENRICHMENT BY MAD HAS LED TO THE PRESENCE OF CONSPICUOUS AND SOMETIMES TROUBLESOME GROWTHS. IT IS REASONABLE TO ASSUME THAT CLADOPHORA PLAYED A RELATIVELY MINOR ROLE IN AQUATIC COMMUNITIES BEFORE THE ACTIVITIES OF MAN LED TO WIDESPREAD NUTRIENT ENRICHMENT. MASSIVE GROWTHS PROBABLY COULD NOT DEVELOP IN FLOWING WATERS WITHOUT MAN'S ACTIVITIES. THEREFORE, IT IS POSSIBLE THAT THE ALGAE ITSELF MAY BE UNDERGOING EVOLUTICNARY CHANGES IN RESPONSE TO THESE NEW ENVIRONMENTS. CLADOPHORA AS A GENUS IS FAVORED BY HIGH LIGHT INTENSITIES, HIGH WATER TURBULENCE, HIGH NUTRIENT LEVELS, HIGH PH VALUES, HARD WATERS. NOT ALL SPECIES SHARE ALL THESE CHARACTERS, BUT C. GLOMERATA, THE SPECIES WHICH HAS INCREASED THE MOST AS A RESPONSE OF MAN'S ACTIVITIES, IS THE ORGANISM WHICH COMBINES ALL THESE CHARACTERS WITHIN A SINGLE SPECIES. (WOODARC-USGS)

FIELD 05C, 05B

ACCESSION NO. W71-10580

NUTRITIONAL ECOLOGY ANC COMMUNITY STRUCTURE OF THE PHYTOPLANKTON OF GREEN BAY,

WISCONSIN UNIV., MADISON. WATER RESOURCES CENTER.

PAUL E. SAGER.

AVAILABLE FROM THE NATIONAL TECHNICAL INVORMATION SERVICE AS PB-201 696, $3.00 IN PAPER COPY, $0.95 IN MICROFICHE. TECHNICAL CCMPLETION REPORT, WATER RESOURCES CENTER, THE UNIVERSITY OF WISCONSIN-MADISON, WISCONSIN, 1971, 31 P, 7 FIG, 11 TAB, 21 REF, OWRR PROJECT A-017-WIS(1).

DESCRIPTORS:
*PHYTOPLANKTON, *AQUATIC ALGAE, *PHOSPHORUS, *CYANOPHYTA, *EUTROPHICATION, BIOMASS, WATER QUALITY, CHLOROPHYLL, NITROGEN, WISCONSIN, NUTRIENTS, WATER POLLUTION EFFECTS, INTERFACES, DISSOLVED OXYGEN, TEMPERATURE, ORGANCPHOSPHOROUS COMPOUNDS, ALGAE.

IDENTIFIERS:
GREEN BAY, FOX RIVER, SPECIFIC CONDUCTANCE, TRANSPARENCY, ORTHOPHOSPHATES, DRY WEIGHT, PHOSPHATE UPTAKE.

ABSTRACT:
THE INFLUENCE OF THE FOX RIVER ON CERTAIN CHEMICAL AND PHYSICAL PARAMETERS OF WATER QUALITY IN LOWER GREEN BAY HAS BEEN STUDIED WITH RESPECT TO RESPONSES IN THE STRUCTURE AND PHCSPHCRUS NUTRITION OF THE PHYTOPLANKTON COMMUNITY. SECCHI DISC TRANSPARENCY, DISSOLVED OXYGEN, SPECIFIC CONDUCTANCE, TEMPERATURE, ORTHOPHOSPHATE, NITRATE, CHLOROPHYLL A, PHYTOPLANKTON DRY WEIGHT, AND EXTRACTABLE PHOSPHATE (LUXURY UPTAKE BY ALGAE) WERE ANALYZED WEEKLY AT NINE STATIONS ALCNG A TRANSECT RUNNING 13.5 MILES UP THE BAY FROM THE MOUTH OF THE FOX RIVER. THE EXISTENCE OF TWO WATER MASSES CAN BE OBSERVED IN THE LOWER BAY, ONE CHARACTERIZED BY FOX RIVER PARAMETERS AND THE CTHER REPRESENTING THE BAY WATER. THE DIFFUSE INTERFACE BETWEEN THE TWO MASSES CAN BE LOCATED IN THE VICINITY OF LCNG TAIL POINT, APPROXIMATELY FIVE MILES FROM THE MOUTH OF THE RIVER. IN THE EXTREME LOWER BAY, THE PHYTOPLANKTON, DOMINATED BY RIVER ALGAL SPECIES, EXHIBIT HIGH BICMASS AND LOW LUXURY UPTAKE OF ORTHCPHOSPHATE IN THE PRESENCE OF LOW CCNCENTRATIONS OF AVAILABLE PHOSPHATE.

FIELD 05C, 02H

ACCESSION NO. W71-10645

EFFECTS OF HYDROSTATIC PRESSURE ON PHOTOSYNTHESIS AND GROWTH OF UNICELLULAR MARINE ALGAE AND DIATOMS,

HAWAII UNIV., HONOLULU. DEPT. OF MICROBIOLOGY.

LESLIE RALPH BERGER.

AVAILABLE FROM THE NATIONAL TECHNICAL INFORMATION SERVICE AS AD-720 401, $3.00 IN PAPER COPY, $0.95 IN MICROFICHE. ONR PROGRESS REPORT, APR 1971. 11 P. ONR PROJECT NO 306820.

DESCRIPTORS:
*ALGAE, *PHOTOSYNTHETIC OXYGEN, *PHOTOSYNTHESIS, INSTRUMENTATION, HYDROSTATIC PRESSURE, DIATOMS.

IDENTIFIERS:
GROWTH, OXYGEN, MONITORS.

ABSTRACT:
LIGHT-DEPENDENT OXYGEN PRODUCTION AND GROWTH OF ALGAL CULTURES HAVE BEEN MEASURED AT 25C AT VARIOUS LIGHT INTENSITIES AND HYDROSTATIC PRESSURES. A DEVICE WHICH MAINTAINS A DESIRED CONCENTRATION OF DISSOLVED OXYGEN DURING GROWTH AND OXYGEN EVOLUTION BY PHOTOSYNTHETIC ORGANISMS IS DESCRIBED. THE SYSTEM USES A MODIFIED RATE-MEASURING OXYGEN ELECTRODE SYSTEM IN CONJUNCTION WITH AN OXYGEN CONCENTRATION MONITORING UNIT.

FIELD 05C

ACCESSION NO. W71-10791

THE ROLE OF INORGANIC IONS IN THE EUTROPHICATION OF FARM PONDS,

AGRICULTURAL RESEARCH SERVICE, UNIVERSITY PARK, PA. NORTHEAST WATERSHED RESEARCH CENTER.

RICHARD W. TERKELTOUB.

TYPESCRIPT; TO BE PUBLISHED IN THE PROCEEDINGS OF THE SYMPOSIUM ON MAN-MADE LAKES, KNOXVILLE, TENNESSEE, MAY 1971. 4 TAB, 11 REF.

DESCRIPTORS:
*EUTROPHICATION, *IONS, *AQUATIC PRODUCTIVITY, FARM PONDS, NUTRIENTS, SURFACES, TEMPERATURE, PHOSPHORUS, NITRATES, CHLORIDES, CALCIUM, MAGNESIUM, SODIUM, POTASSIUM, IRON, MANGANESE, COPPER, ALGAE, SAMPLING, CHLOROPHYTA.

IDENTIFIERS:
*INORGANIC IONS, ZINC, VOLVOX, HYDRODICTYON, SPODYLOSIUM, SPIROGYRA.

ABSTRACT:
SEVEN EASTERN PENNSYLVANIA FARM PONDS WERE STUDIED TO ASCERTAIN WHETHER ANY PARTICULAR CONCENTRATIONS OR RATIOS OF THE PRINCIPAL AQUEOUS INORGANIC IONS ARE ASSOCIATED WITH EUTROPHICATION. THE SURFACE AND BOTTOM WATERS WERE SAMPLED FROM MAY 7 TO NOVEMBER 18, 1969. TEMPERATURE WAS MEASURED, AND CALCIUM, MAGNESIUM, SODIUM, POTASSIUM, CHLORIDE, NITRATE, ORTHOPHOSPHATE, IRON, MANGANESE, ZINC, AND COPPER CONCENTRATIONS WERE DETERMINED. THE ALGAE FOUND WERE SPODYLOSIUM, SPIROGYRA, AND VOLVOX. WATER TEMPERATURES WERE USUALLY A FEW DEGREES CENTIGRADE LOWER IN THE BOTTOM WATERS THAN IN THE SURFACE WATERS. THERE WERE NO INDICATIONS OF THERMAL STRATIFICATION IN ANY POND. THE CHEMICAL CONDITIONS THAT SPURRED ALGAL GROWTH IN THE THREE PONDS, 1, 2, AND 7, WERE NOT DISTINGUISHABLY DIFFERENT FROM THOSE PRESENT IN THE ALGAE-FREE PONDS. SIMILARLY, THE SUBSEQUENT DECAY AND DISAPPEARANCE OF THE ALGAE WERE NOT RELATABLE TO ANY OF THE CHEMICAL PARAMETERS MEASURED. NO DISCERNIBLE SET OF CHEMICAL CHARACTERISTICS DIFFERENTIATED THE THREE PONDS THAT SUPPORTED ALGAE DURING THE SUMMER FROM THE FOUR PONDS THAT DID NOT. (JONES-WISCONSIN)

FIELD 05C

ACCESSION NO. W71-11001

POLLUTION POTENTIAL OF SALMONID FISH HATCHERIES,

KRAMER, CHIN AND MAYO, SEATTLE, WASH.

PAUL B. LIAO.

WATER AND SEWAGE WORKS, VOL 117, NO 12, P 291-297, 1970. 6 FIG, 3 TAB, 15 REF.

DESCRIPTORS:
*WATER POLLUTION SOURCES, *FISH HATCHERIES, *SALMONIDS, WATER REQUIREMENTS, TEMPERATURE, NUTRIENTS, ALGAE, WEEDS, TASTE, ODOR, PATHOGENIC BACTERIA, ORGANIC WASTES, SOLID WASTES, CHEMICALS, MICHIGAN, CALIFORNIA, TUBIFICIDS, COLORADO, WATER POLLUTION CONTROL, WASHINGTON, CHEMICAL OXYGEN DEMAND, BIOCHEMICAL OXYGEN DEMAND, DISSOLVED OXYGEN, HYDROGEN ION CONCENTRATION, AMMONIA, NITRATES, PHOSPHATES, SUSPENDED LOAD, DISSOLVED SOLIDS, EFFLUENTS.

IDENTIFIERS:
SETTLEABLE SOLIDS, HATCHERY EFFLUENTS, PARASITES, JORDAN RIVER(MICH), SAN JOAQUIN RIVER(CALIF), RIFLE FALLS(COLO), FISH FECAL WASTES, RESIDUAL FOOD, GREEN RIVER(WASH), COWLITZ TROUT HATCHERY(WASH).

ABSTRACT:
WATER POLLUTION PROBLEMS ASSOCIATED WITH SALMONID HATCHERY OPERATIONS INCLUDE NUTRITIONAL ENRICHMENT, ALGAE AND WEED GROWTH, TASTE, ODOR, SETTLEABLE SOLIDS, PATHOGENIC BACTERIA, PARASITES, ORGANIC MATTER, CHEMICALS AND DRUGS. FISH FECAL WASTES AND RESIDUAL FOOD ARE MOST SERIOUS BECAUSE THEY ARE ENCOUNTERED CONTINUOUSLY UNDER NORMAL OPERATING PROCEDURES; AFTER FIELD TESTING, THESE WASTES ARE CLASSIFIED INTO ORGANIC, NUTRIENT AND SOLID POLLUTANTS. THE AVERAGE BIOCHEMICAL OXYGEN DEMAND (BOD) CONCENTRATION OF HATCHERY EFFLUENTS DURING POND CLEANING IS SEVERAL TIMES GREATER THAN DURING NORMAL OPERATION; CLOSELY RELATED TO BOD, IS THE DISSOLVED OXYGEN LEVEL. THE NUTRIENT POLLUTANTS, NITRATE AND PHOSPHATE, ARE END-PRODUCTS OF DECOMPOSITION OF FISH FOOD. THE HATCHERY EFFLUENT TESTED MAY STIMULATE ALGAL GROWTH AND CAUSE ALGAL BLOOMS UNDER CERTAIN CONDITIONS. THE HIGH PERCENTAGE OF SUSPENDED AND SETTLEABLE SOLIDS INDICATES THAT MOST SOLIDS IN THE CLEANING WATER WILL BE DEPOSITED ON THE STREAM BOTTOM BELOW THE HATCHERY. PROPER FEEDING WOULD GREATLY REDUCE RATE OF POLLUTANT PRODUCTION. THE POLLUTION POTENTIAL OF HATCHERY CLEANING WATER IS COMPARABLY TO DOMESTIC SEWAGE WHEN DILUTED WITH INFILTRATION WATER. HATCHERY OPERATING IMPROVEMENTS SHOULD INCLUDE PROPER FISH LOADING TECHNIQUES, PROPER FEEDING PROCEDURES, AND WATER SUPPLY ADJUSTMENTS. (JONES-WISCONSIN)

FIELD 05C, 05B

ACCESSION NO. W71-11006

TOXICITY OF AMMONIA TO MARINE DIATOMS,

ALASKA UNIV., COLLEGE. INST. OF MARINE SCIENCE.

K. V. NATARAJAN.

JOURNAL OF WATER POLLUTION CONTROL FEDERATION, VOL 42, NO 5, PART 2, P R184-R190, 1970. 2 FIG, 2 TAB, 15 REF.

DESCRIPTORS:
*TOXICITY, *AMMONIA, *MARINE ALGAE, *DIATOMS, CHEMICAL ANALYSIS, HYDROGEN ION CONCENTRATION, SEA WATER, PHOTOSYNTHESIS, PHYTOPLANKTON, AMMONIUM COMPOUNDS, NUTRIENTS, PRIMARY PRODUCTIVITY, FERTILIZERS, EFFLUENTS, WATER POLLUTION SOURCES, WATER POLLUTION EFFECTS, SAMPLING, TEMPERATURE, COPEPODS, NITROGEN, BACTERIA, OXIDATION-REDUCTION POTENTIAL, LABORATORY TESTS.

IDENTIFIERS:
COOK INLET(ALASKA), CYCLOTELLA NANA, LEPTOCYLINDRUS, PACIFIC OCEAN, ALEUTIAN ISLANDS, PRYMNESIUM PARVUM.

ABSTRACT:
ALTHOUGH THE WORLD'S FERTILIZER SUPPLY CONSISTS LARGELY OF AMMONIUM COMPOUNDS, THESE COMPOUNDS ARE ALSO KNOWN TO BE TOXIC. AMMONIA CAN BECOME TOXIC AT CERTAIN CONCENTRATIONS IN THE AQUATIC ENVIRONMENT. EFFECT ON MARINE PHYTOPLANKTON OF A FERTILIZER PLANT EFFLUENT CONSISTING ESSENTIALLY OF AMMONIA COMPOUNDS WAS INVESTIGATED WITH BOTH FIELD AND LABORATORY STUDIES. THE RESULTS OF THE FIELD EXPERIMENTS SHOWED A CLOSE PARALLEL WITH THE LABORATORY EXPERIMENTS WITH AXENIC CULTURES OF MARINE DIATOMS. LEVELS OF THE EFFLUENT TOXIC TO THE ENDEMIC PHYTOPLANKTON WERE BETWEEN 0.1 AND 1% (1.1 AND 11 MG/1 AMMONIA). IT IS EVIDENT THAT THE TOXICITY OF AMMONIA AND AMMONIUM SALTS DEPENDS ON THE CONCENTRATION OF MOLECULAR NONIONIZED AMMONIA OR AMMONIA HYDROXIDE, SINCE THE AMMONIUM ION IS COMPARATIVELY NONTOXIC; (DIRECT EXPERIMENTAL OBSERVATIONS HAVE NOT BEEN OBTAINED). TOXIC EFFECTS OF THE AMMONIA EFFLUENT TO PHYTOPLANKTON OF THIS INVESTIGATION CAN BE EXPLAINED BY THIS PHENOMENON. (JONES-WISCONSIN)

FIELD 05C

ACCESSION NO. W71-11008

NUTRIENTS AND NUTRIENT BUDGETS IN THE BAY OF QUINTE, LAKE CNTARIO,

ONTARIO WATER RESOURCES COMMISSION, TORONTO.

M. G. JOHNSON, AND G. E. OWEN.

JOURNAL OF WATER POLLUTION CONTROL FEDERATION, VOL 43, NC 5, P 836-853, 1971.
8 FIG, 7 TAB, 37 REF.

DESCRIPTORS:
*NUTRIENTS, *LAKE CNTARIC, *EUTROPHICATION, LAKE ERIE, ALGAE, NITROGEN, PHOSPHORUS, DRAINAGE, INDUSTRIES, MUNICIPAL WASTES, TURNOVERS, SEDIMENTS, HUMAN POPULATION, TOURISM, WATERSHEDS(BASINS), TROPHIC LEVELS, TURBIDITY, OXYGEN, FISH, RECREATION, INVERTEBRATES, RIVERS, CYANOPHYTA, GEOLOGIC FORMATICNS, BENTHOS, MAYFLIES, TUBIFICIDS, DIATOMS, WATER POLLUTION EFFECTS, WATER POLLUTICN CONTROL, DISSOLVED OXYGEN, CARBON, SAMPLING, EQUATIONS, WATER POLLUTICN SOURCES, RAINFALL, MIDGES, INPUT-OUTPUT ANALYSIS, MUD, BACTERIA.

IDENTIFIERS:
*NUTRIENT BUDGETS, *BAY OF QUINTE(ONTARIO), NUTRIENT INPUTS, CHIRONOMUS PLUMOSUS, CHIRONOMUS ATTENTUATUS, CHIRONOMUS ANTHRACINUS, LIMNODRILUS HOFFMEISTERI, TUBIFEX TUBIFEX, APHANIZOMENON, CLADCPHCRA.

ABSTRACT:
ALGAL BLOOMS, TURBIDITY, DEPLETICN OF DEEP-WATER CXYGEN, AND CHANGES IN COMPOSITION OF THE BIOTA ARE INCREASINGLY OBVIOUS IN BAY OF QUINTE, LAKE ONTARIO. CLARIFICATION OF RESPECTIVE SIGNIFICANCE OF NUTRIENT CONTRIBUTIONS FROM TRIBUTARY RIVERS AND FRCM MUNICIPAL-INDUSTRIAL SOURCES ARE DESCRIBED. THE BAY RECEIVED ABOUT 9,700,000 POUNDS OF NITROGEN AND 700,000 PCUNDS OF PHCSPHORUS IN 1968. 89% OF THE NITROGEN AND 60% OF THE PHOSPHORUS WERE ATTRIBUTABLE TO LAND DRAINAGE AND THE REMAINDER TO MUNICIPAL-INDUSTRIAL SOURCES. COMPARISONS BASED ON 'NET INPUTS', THE AMCUNT CF NUTRIENT CONTAINED IN AN INPUT IN EXCESS OF THE AMOUNT OF NUTRIENT IN THE EQUIVALENT VOLUME OF WATER DISPLACED AT THE OUTLET, ARE PROPOSED. ABOUT 50% OF THE 'NET INPUT' OF NITRCGEN AND 85% OF PHOSPHORUS WERE CCNTRIBUTED BY MUNICIPAL-INDUSTRIAL SOURCES IN 1968. IT IS RECCMMENDED THAT PHOSPHORUS BE REMOVED FROM THESE SOURCES. THE WATER TURNOVER RATE IN THE BAY, FIVE TIMES ANNUALLY, TRANSLOCATES RESUSPENDED NUTRIENTS IN SEDIMENTS OF THE INNER BAY SHALLOW WATERS TO SEDIMENTS IN DEEPER WATERS OF THE CUTER BAY AND LAKE CNTARIO, THUS IMPROVING WATER QUALITY. ESTIMATED PHOSPHORUS REMCVAL COST, $200,000 DOLLARS/YEAR, IS JUSTIFIED ON THE BASIS CF ECONOMICS INCOME FROM RECREATION AND TOURISM. (JONES-WISCONSIN)

FIELD 05C

ACCESSION NO. W71-11009

THE PERIPHYTON OF THE SUBMERGED MACROPHYTES OF MIKOLAJSKIE LAKE,

WARSAW UNIV. (POLAND). CHAIR OF HYDROBIOLOGY.

LUCYNA JOANNA BOWNIK.

EKOLOGIA POLSKA, VOL 18, NO 24, P 503-520, 1970. 6 FIG, 2 TAB, 16 REF.

DESCRIPTORS:
*PERIPHYTON, *SUBMERGED PLANTS, BIOMASS, NEMATODES, ALGAE, OLIGOCHAETES, LIFE CYCLES, SEASONAL, DENSITY, DIATOMS, LAKES, PROTOZOA, ROTIFERS, CRUSTACEANS, DIPTERA, MAYFLIES, DRAGONFLIES, CADDISFLIES, MOLLUSKS, CYANOPHYTA, CHRYSOPHYTA, CHLOROPHYTA, PYRROPHYTA, INVERTEBRATES, AQUATIC ANIMALS, LITTORAL, WAVES(WATER).

IDENTIFIERS:
*MACROPHYTES, *MIKOLAJSKIE LAKE(POLAND), COLONIZATION, POTAMOGETON LUCENS, POTAMOGETON PERFOLIATUS, POTAMOGETON PECTINATUS, MYRIOPHYLLUM SPECATUM, ELODEA CANADENSIS, PHRAGMITES, FONTINALIS, CHARALES, CERATOPHYLLUM, LEMNA.

ABSTRACT:
AN ANALYSIS WAS MADE OF THE PERIPHYTON COLONIZING POTAMOGETON LUCENS, P PERFOLIATUS, MYRIOPHYLLUM SPICATUM AND ELODEA CANADENSIS IN MIKOLAJSKIE LAKE, POLAND. COMPOSITION AND THE NUMBERS DYNAMICS OF THE DOMINANT PERIPHYTON GROUPS WERE RECORDED INDICATING THAT NEMATODES ARE DOMINANT WITH CHIRONOMIDS AND OLIGOCHAETA NUMEROUS ON SUBMERGED VEGETATION. A GREAT DIFFERENCE WAS NOTICED BETWEEN THE NUMBERS DYNAMICS OF PERIPHYTON FAUNA COLONIZING PLANTS LIVING IN THE LAKE DURING WINTER. THE LOWEST NUMBERS OF PERIPHYTON FAUNA WERE FOUND ON PLANTS WITH A SHORT LIFE CYCLE, IN SPRING, SHOWING A GRADUAL COLONIZATION OF A GROWING PLANT. ON PLANTS WITH A LONGER LIFE CYCLE, PERIPHYTON FAUNA OCCUR IN LARGE NUMBERS ALREADY IN SPRING BUT MAXIMUM IS REACHED IN AUTUMN. THE LARGEST DENSITY OF DOMINANT GROUPS OF PERIPHYTON ORGANISMS WAS FOUND ON POTAMOGETON LUCENS, DECREASING ON MYRIOPHYLLUM SPICATUM AND POTAMOGETON PERFOLIATUS; THE SMALLEST ON ELODEA CANADENSIS. IN A SHORE OVERGROWN BY PLANTS WITHERING DURING WINTER, CHANGES OF PLANT BIOMASS HAVE A DECISIVE INFLUENCE ON PERIPHYTON ABUNDANCE. ON PLANTS WITH LONGER LIFE CYCLES, CHANGES OF THE FAUNA DENSITY DETERMINED ABUNDANCE OF PERIPHYTON. ALGAE ARE MORE NUMEROUS THAN ANIMAL COMPONENTS OF THE PERIPHYTON AND DIATOMS ARE THE MOST NUMEROUS ALGAE. (JONES-WISCONSIN)

FIELD 05C

ACCESSION NO. W71-11012

WASTEWATER LOADING GUIDELINES FOR THE GRAND RIVER BASIN.

ONTARIO WATER RESOURCES COMMISSION, TORONTO.

ONTARIO WATER RESOURCES COMMISSION, CANADA, INTERIM REPORT, JANUARY 1971. 1 FIG, 1 TAB, APPEND.

DESCRIPTORS:
*WASTE WATER(POLLUTION), *WASTE ASSIMILATION CAPACITY, *ORGANIC LOADS, NUTRIENTS, DISSOLVED OXYGEN, AQUATIC LIFE, MUNICIPAL WASTES, PHOTOSYNTHESIS, RESPIRATION, WATER QUALITY, STREAMFLOW, AGRICULTURE, INDUSTRIES, WASTE TREATMENT, WATERSHEDS(BASINS), PHOSPHORUS, ALGAE, ORGANIC MATTER, LAKE ERIE, EFFLUENTS, SEWAGE, DETERGENTS, PLANTS, LAND USE, BIOCHEMICAL OXYGEN DEMAND, FISHERIES, RESERVOIRS, PIPELINES, ESTIMATING, ANALYSIS.

IDENTIFIERS:
*LOADING GUIDELINES, *GRAND RIVER BASIN(ONTARIO), CANADA, BUFFER CAPACITY.

ABSTRACT:
IN CONSIDERING WATER QUALITY OF THE GRAND RIVER BASIN, ONTARIO, ACCEPTABLE LOADINGS, BASED ON THE NEW DISSOLVED OXYGEN CRITERIA ADAPTED BY ONTARIO WATER RESOURCES COMMISSION IN 1970, UPGRADE THE MINIMUM DISSOLVED OXYGEN LEVEL FROM 4.0 TO 5.0 MG/L IN ALL STREAMS EXCEPT THOSE SUPPORTING COLDWATER FISHERIES WHERE MINIMUM DISSOLVED OXYGEN CRITERIA IS 6.0 MG/LITER. AS A RESULT, PREVIOUSLY ACCEPTABLE WASTE DISCHARGES NOW EXCEED PRESENT LOADING GUIDELINES. THE PHOSPHORUS INPUT, CONSIDERED THE CONTROLLING NUTRIENT IN ALGAL PRODUCTION, FROM THE MUNICIPAL SEWAGE TREATMENT PLANTS IS ESTIMATED AS 70% TO 80% OF THE TOTAL ANNUAL INPUT OF THIS NUTRIENT INTO THE BASIN. REDUCTION OF NUTRIENT AND ORGANIC LOADINGS ARE REQUIRED FOR WATER QUALITY IMPROVEMENT AND PROTECTION OF LAKE ERIE. ALTERNATIVES, INCLUDING EFFLUENT POLISHING AND STREAMFLOW AUGMENTATION, WHICH CAN BE UTILIZED TO INCREASE THE POTENTIAL VARIOUS RIVER USES, WHILE REDUCING POLLUTION PRESSURES ARE CONSIDERED. IN MAKING ESTIMATES OF ACCEPTABLE ORGANIC LOADINGS, THE WATERSHED WAS DIVIDED INTO EIGHT SUB-BASINS. THE LOADINGS FOR EACH SUB-BASIN, EXPRESSED IN TERMS OF FIVE-DAY BIOCHEMICAL OXYGEN DEMAND FROM TREATMENT SOURCES, WERE DETERMINED ON BASIC ASSUMPTIONS, DISSOLVED OXYGEN CRITERIA, DESIGN STREAMFLOWS, AND EXISTING WASTEWATER LOADINGS. RESULTS ARE TABULATED; DETAILS APPENDED. (JONES-WISCONSIN)

FIELD 05C, 06B

ACCESSION NO. W71-11017

PERIPHYTON OF THE EXPERIMENTAL LAKES AREA, NORTHWESTERN ONTARIO,

FISHERIES RESEARCH BOARD OF CANADA, WINNIPEG (MANITOBA). FRESHWATER INST.

JOHN G. STOCKNER, AND F. A. J. ARMSTRONG.

JOURNAL FISHERIES RESEARCH BOARD OF CANADA, VOL 28, NO 2, P 215-229, 1971. 9 FIG, 2 TAB, 37 REF.

DESCRIPTORS:
*PERIPHYTON, *ALGAE, *LITTORAL, LIGHT PENETRATION, BENTHIC FLORA, LAKES, DEPTH, DIATOMS, CHLOROPHYTA, CYANOPHYTA, BIOMASS, CHEMICAL ANALYSIS, STATISTICAL METHODS, NITROGEN, PHOSPHORUS, CHLOROPHYLL, DISTRIBUTION, CARBON, SEASONAL, CADDISFLIES, GASTROPODS, SANDS, CHRYSOPHYTA.

IDENTIFIERS:
*EXPERIMENTAL LAKES AREA, *NORTHWESTERN ONTARIO, EPILITHIC ALGAE, ACHNANTHES MINUTISSIMA, ACHNANTHES FLEXELLA, EUNCTIA PECTINALIS, DESMIDS.

ABSTRACT:
SINCE THE LITTORAL ZONE OF THE FOUR LAKES STUDIED IN THE EXPERIMENTAL LAKES AREA (ELA) OF NORTHWESTERN ONTARIO WAS COMPOSED CHIEFLY OF LARGE BOULDERS AND ROCK SHELFS, THE EPILITHIC ALGAL ASSEMBLAGE OF PERIPHYTON WAS DOMINANT. DIATOMS WERE THE DOMINANT GROUP WITHIN THE EPILITHIC ASSEMBLAGE. BENTHIC ALGAL GROWTH IN MOST LAKES WAS NEGLIGIBLE AT DEPTHS OVER 10 METERS. FILAMENTOUS GREEN AND BLUE-GREEN ALGAE INCREASE SIGNIFICANTLY IN JULY AND AUGUST BUT NEVER CONSTITUTED MORE THAN 40% OF TOTAL ALGAL BIOMASS. A WELL=DEFINED DIATOM SUCCESSION OCCURRED. ACHNANTHES MINUTISSIMA WAS THE MOST ABUNDANT DIATOM IN LITTORAL ZONES OF ALL LAKES. VERTICAL DISTRIBUTION OF LITTORAL DIATOMS WAS EXAMINED IN LAKE 240 AND SPECIES DIFFERENCES ARE DISCUSSED IN LIGHT OF POSSIBLE REGULATING MECHANISMS. ACHNANTHES FLEXELLA AND EUNCTIA PECTINALIS WERE FOUND ONLY IN THE PSAMMONAL HABITAT OF LAKE 240. STATISTICAL TREATMENT OF CHEMICAL ANALYSES OF NITROGEN, PHOSPHORUS, AND CHLOROPHYLL-A CONTENT OF PERIPHYTON SHOWED NO SIGNIFICANT DIFFERENCE IN AMOUNTS OF TOTAL NITROGEN, BUT DISTRIBUTION OF BOTH TOTAL PHOSPHORUS AND CHLOROPHYLL-A WAS SIGNIFICANTLY DIFFERENT. COMPARISONS AMONG LAKES ARE MADE, AND ROLE OF NITROGEN AND PHOSPHORUS AS REGULATORS OF PERIPHYTON GROWTH IN ELA LAKES IS DISCUSSED. THIS PAPER CONTAINS 37 REFERENCES. (JONES-WISCONSIN)

FIELD 05C

ACCESSION NO. W71-11029

ALGAL GROWTH ASSESSMENT IN NATURAL WATERS,

ILLINOIS STATE WATER SURVEY, PEORIA. WATER QUALITY SECTION.

WUN-CHENG WANG, AND WILLIAM T. SULLIVAN.

TYPESCRIPT; PRESENTED AT 14TH CONFERENCE ON GREAT LAKES RESEARCH, APRIL 19-21, 1971.

DESCRIPTORS:
*FORECASTING, *ALGAE, *PLANT GROWTH, *EUTROPHICATION, GROWTH RATES, SAMPLING, LAKE MICHIGAN, MEASUREMENT, SUSPENDED LOAD, ALKALINITY, HARDNESS(WATER), ANALYTICAL TECHNIQUES, AMMONIA, NITRATES, PHOSPHATES, WATER QUALITY, NITROGEN, NUTRIENTS, TROPHIC LEVEL.

IDENTIFIERS:
ALGAL GROWTH POTENTIAL, ILLINOIS RIVER(ILL), DIAGNOSTIC CRITERION.

ABSTRACT:
METHODS ARE DESCRIBED FOR DIAGNOSTIC AND PREDICTIVE EVALUATION OF NATURAL WATER QUALITY BASED ON ALGAL GROWTH POTENTIAL. TECHNIQUES FOR MEASUREMENT OF ALGAL GROWTH WERE INVESTIGATED AND ALGAL NUTRITION EXAMINED. AFTER REMOVAL OF PLANKTON BY MEMBRANE FILTRATION, SAMPLES WERE INOCULATED WITH A NATURAL, MIXED ALGAL CULTURE; GROWTH WAS SATISFACTORY MEASURED BY INCREASES IN LIGHT ABSORBANCE AND FILTERABLE ORGANIC AND INORGANIC MASS AND BY DECREASES IN ALKALINITY AND HARDNESS. ALGAL PIGMENT FLUORESCENCE WAS NOT COMPARABLE TO THE OTHER PARAMETERS ABOVE OLIGOTROPHIC LEVELS, POSSIBLY DUE TO INADEQUATE EXTRACTION OF FLUORESCENT COMPOUNDS. MAXIMUM DAILY ALGAL GROWTH WAS ATTAINED IN THREE TO FIVE DAYS FOLLOWING INOCULATION; AMMONIA WAS PREFERRED TO NITRATES AS A NITROGEN SOURCE. RATIO OF FILTERABLE INORGANIC MASS TO ORGANIC MASS INCREASED WITH HIGHER INITIAL SAMPLE CONCENTRATIONS OF ALKALINITY AND HARDNESS. 'ALGAL GROWTH POTENTIAL', REPRESENTING THE TROPHIC LEVEL OF A NATURAL WATER SOURCE, WAS BEST REPRESENTED BY WEIGHT OF FILTERABLE ORGANIC MASS PRODUCED AFTER SEVEN DAYS OF INCUBATION. LAKE MICHIGAN AT THE CHICAGO CENTRAL WATER FILTRATION PLANT WAS JUDGED ESSENTIALLY OLIGOTROPHIC, WHILE THE ILLINOIS RIVER NEAR PEORIA IS EUTROPHIC. THIS PAPER CONTAINS 17 REFERENCES. (JONES-WISCONSIN)

FIELD 05C

ACCESSION NO. W71-11033

DETERMINATION OF MERCURY IN BIOLOGICAL AND ENVIRONMENTAL SAMPLES BY NEUTRON ACTIVATION ANALYSIS,

WESTERN NEW YORK NUCLEAR RESEARCH CENTER, INC., BUFFALO.

K. K. SIVASANKARA PILLAY, CHARLES C. THOMAS, JR., JAMES A. SONDEL, AND CAROLYN M. HYCHE.

TPYESCRIPT; PRESENTED AT 161ST NATIONAL MEETING OF THE AMERICAN CHEMICAL SOCIETY AT LOS ANGELES, CALIFORNIA, MARCH 31, 1971.

DESCRIPTORS:
*SAMPLING, *ENVIRONMENT, *NEUTRON ACTIVATION ANALYSIS, RADIOACTIVITY, ADSORPTION, FISH, LAKE ERIE, SEDIMENTS, SILTS, PLANKTON, ALGAE, TRACE ELEMENTS, VOLATILITY, FREEZE DRYING, X-RAYS, ELECTROLYSIS, COMPUTER PROGRAMS, GAMMA RAYS, POLLUTANT IDENTIFICATION, COLORIMETRY, TRACERS.

IDENTIFIERS:
*MERCURY, *BIOLOGICAL TISSUES, MERCURY ISOTOPES, HUMAN BRAIN, LOW-TEMPERATURE OVEN DRYING, ASHING, EXTRACTIVE DIGESTION, NEUROANATOMY, OXYGEN PLASMA ASHING, ATOMIC ABSORPTION, TITRATION.

ABSTRACT:
THE MINUTE QUANTITIES AND VOLATILE NATURE OF MERCURY CREATES PROBLEMS IN SAMPLING AND ANALYSIS. OF VARIOUS MERCURY DETERMINATION PROCEDURES, ONLY A LIMITED NUMBER CAN BE READILY ACAPTED TC BIOLOGICAL AND ENVIRONMENTAL SAMPLING FOR MONITORING. THE FOLLOWING PROCEDURES WERE DEVELOPED FOR THE INVESTIGATION OF THE MERCURY POLLUTION OF LAKE ERIE AND ITS AQUATIC LIFE. PRE-IRRADIATION PROCESSING OF SAMPLES HAS BEEN SYSTEMATICALLY INVESTIGATED. AFTER REACTOR IRRADIATION, THE SAMPLES ARE WET ASHED WITH MERCURY CARRIER UNDER REFLUX CONDITIONS. A PRELIMINARY SULFIDE PRECIPITATION IS FOLLOWED BY FURTHER PURIFICATION AND EVENTUAL ISOLATION OF MERCURY BY ELECTRODEPOSITION OR BY PRECIPITATION AS MERCURIC OXIDE. THE RADIOACTIVITIES FROM MERCURY-197 AND MERCURY-197-M ISOTOPES ARE MEASURED BY SCINTILLATION GAMMA RAY SPECTROMETRY USING A THIN SODIUM IODIDE DETECTOR TO DETERMINE THE MERCURY LEVELS IN VARIOUS SAMPLES. THESE ANALYTICAL PROCEDURE RESULTS ARE COMPARED WITH OTHER TECHNIQUES. TRACER STUDIES INDICATED THAT THE ERRORS OF THIS PROCEDURE WERE LESS THAN 15% AT 0.01 PPM LEVEL AND LESS THAN 5% AT 2 PPM LEVEL OF MERCURY IN BIOLOGICAL TISSUES. ANALYSIS OF FISH SAMPLES AND SEDIMENTS SAMPLES CONTAINING NATURAL FORMS OF MERCURY SHOWED A STANDARD DEVIATION OF LESS THAN 5% AT 5 PPM LEVELS, LESS THAN 7% AT 1.5 PPM LEVELS AND LESS THAN 17% AT 0.01 PPM LEVELS. (JONES-WISCONSIN)

FIELD 05B, 07B

ACCESSION NO. W71-11036

DETERMINATION OF MICROBIAL BIOMASS IN DEEP OCEAN WATER,

CALIFORNIA UNIV., SAN DIEGO, LA JOLLA. INST. OF MARINE RESOURCES.

OSMUND HOLM-HANSEN.

IN: SYMPOSIUM ON ORGANIC MATTER IN NATURAL WATERS, UNIVERSITY OF ALASKA, COLLEGE, SEPT 2-4, 1968: INSTITUTE OF MARINE SCIENCE OCCASIONAL PUBLICATION NO 1, P 287-300, JUNE 1970. 14 P, 5 FIG, 17 REF. USAEC CONTRACT AT(11-1)GEN 10, PA 20.

DESCRIPTORS:
*BIOMASS, *ORGANIC MATTER, *SEA WATER, WATER ANALYSIS, ALGAE, CHEMICAL ANALYSIS, SAMPLING, ANALYTICAL TECHNIQUES, NUTRIENTS, FOOD CHAINS, BIOLUMINESCENCE, POLLUTANT IDENTIFICATION.

IDENTIFIERS:
*ATP ANALYSIS, *ADENOSINE TRIPHOSPHATE.

ABSTRACT:
THE CONTENT OF ATP IN THE PARTICULATE FRACTION WAS DETERMINED IN FOUR PROFILES IN THE PACIFIC OCEAN DOWN TO 3500 M. THE ATP VALUES WERE CONVERTED TO CELLULAR ORGANIC CARBON VALUES AND COMPARED TO THE TOTAL ORGANIC CARBON IN THE PARTICULATE FRACTION. THE MICROBIAL BIOMASS AS ESTIMATED BY THE ATP DATA IS VERY HIGH IN THE EUPHOTIC ZONE AND DECREASES RAPIDLY TO 1 TO 2 MICROGRAMS OF CELLULAR C/LITER AT ABOUT 200 M. AT THE LOWER DEPTHS SAMPLED, THE CALCULATED BIOMASS CONTAINED ABOUT 0.1 MICROGRAM OF ORGANIC C/LITER. MICROSCOPIC EXAMINATION OF DEEP SAMPLES SHOWED THE PRESENCE OF LARGE NUMBERS OF SMALL FLAGELLATED ALGAL-LIKE CELLS. (KNAPP-USGS)

FIELD 05A, 05B

ACCESSION NO. W71-11234

IMPROVEMENT AND APPLICATION OF BENTHIC ALGAL ISOTOPE PRODUCTIVITY MEASURING METHODS,

HAWAII UNIV., HONOLULU. DEPT. OF BOTANY.

MAXWELL S. DOTY.

AVAILABLE FROM THE NATIONAL TECHNICAL INFORMATION SERVICE AS UH-235P4-4, $3.00 IN PAPER COPY, $0.95 IN MICROFICHE, ATOMIC ENERGY COMMISSION CONTRACT REPORT UH-235P4-4, JUNE 1970. 54 P, 9 FIG, 5 TAB, 47 REF.

DESCRIPTORS:
*RADIOACTIVITY TECHNIQUES, *BENTHIC FLORA, *GROWTH RATE, *AQUATIC PRODUCTIVITY, RADIOISOTOPES, TRACERS, MARINE ALGAE, HAWAII, WATER TEMPERATURE, WAVES, LIGHT INTENSITY.

IDENTIFIERS:
PHILIPPINES, ENIWETOK, MICRONESIA, FANNING ISLANDS.

ABSTRACT:
RADIOISOTOPES HAVE BEEN USED AS QUANTITATIVE TRACERS IN MEASURING RATES OF BENTHIC ALGAL PRODUCTION. BASIC ECOLOGICAL STUDIES ON BENTHIC REEF ALGAE HAVE BEEN COMPLETED OR ARE UNDERWAY. THESE STUDIES HAVE BEEN CONDUCTED IN HONOLULU, ENIWETOK, MICRONESIA, THE PHILIPPINES, AND FANNING ISLAND. IT WAS SHOWN THAT THE CROP OF SOME SPECIES VARY SEASONALLY WITH LIGHT, SOME WITH MEAN WAVE SIZE, AND SOME WITH TEMPERATURE. STUDIES AT WAIKIKI INDICATED THAT AMONG BENTHIC ALGAE, STANDING CROPS OF THE CRUSTOSE FORMS VARY THE LEAST. (MORTLAND-BATTELLE)

FIELD 05A, 05C

ACCESSION NO. W71-11486

AIRBORNE AMMONIA EUTROPHIES LAKES.

AGRICULTURAL RESEARCH (USDA), VOL. 19, NO. 2, P 8-9, AUGUST 1970.

DESCRIPTORS:
*AMMONIA, *EUTROPHICATION, *NITROGEN, ALGAE, WATER POLLUTION SOURCES, COLORADO, CATTLE, URINE, PATH OF POLLUTANTS, FARM WASTES.

IDENTIFIERS:
FEEDLOTS.

ABSTRACT:
AMMONIA TRAPS AND RAIN GAGES WERE INSTALLED AT FIVE SITES AND IN TWO CONTROL AREAS IN COLORADO TO DETERMINE THE RATE AT WHICH AMMONIA IS ADSORBED DIRECTLY FROM THE AIR BY WATER SURFACES UNDER DIFFERENT CONDITIONS OF TEMPERATURE AND CLIMATE AT VARIOUS DISTANCES AND DIRECTIONS FROM CATTLE FEEDLOTS. IN ONE NORTHEAST COLORADO LAKE A LITTLE OVER A MILE FROM A LARGE FEEDLOT, THE SURFACE ADSORBED ABOUT 30 POUNDS OF NITROGEN AS AMMONIA PER ACRE PER YEAR. THIS AMOUNT IS SUFFICIENT TO EUTROPHY A LAKE AVERAGING 20 FEET IN DEPTH TO TWO OR THREE TIMES THE CONCENTRATION NEEDED FOR ALGAL BLOOMS. INDICATIONS ARE THAT EVEN SMALL FEEDLOTS MAY RELEASE ENOUGH AMMONIA TO HAVE AN EFFECT ON NEARBY WATER SURFACE AND THAT AIRBORNE AMMONIA FROM FEEDLOTS MAY CONTRIBUTE MORE NITROGEN THAN RUNOFF AND DEEP PERCOLATION FROM THE SAME SOURCES. (MORTLAND-BATTELLE)

FIELD 05B, 05C

ACCESSION NO. W71-11496

PESTICIDES IN WATER: COPPER SULFATE IN FLOODED CRANBERRY BOGS,

MASSACHUSETTS UNIV., EAST WAREHAM. CRANBERRY EXPERIMENT STATION.

KARL H. DEUBERT, AND IRVING E. DEMORANVILLE.

PESTICIDES MONITORING JOURNAL, VOL. 4, NO. 1, P 11-13, JUNE 1970. 1 TAB, 7 REF.

DESCRIPTORS:
*COPPER SULFATE, *ALGAE, *FLOODWATER, *PATH OF POLLUTANTS, DIELDRIN, DDT, ORGANIC MATTER, ANALYTICAL TECHNIQUES, SAMPLING, MASSACHUSETTS, POLLUTANT IDENTIFICATION.

IDENTIFIERS:
*CRANBERRY BOGS.

ABSTRACT:
CRANBERRY BOGS ARE TREATED WITH COPPER SULFATE TO CONTROL ALGAL GROWTH. IN ORDER TO ASSESS POSSIBLE WATER POLLUTION AFTER RELEASE OF TREATED FLOODWATER INTO STREAMS AND PONDS, THE RATE AT WHICH COPPER DISAPPEARED FROM THE WATER AFTER TREATMENT WAS MONITORED IN TWO SEPARATE BOGS. IN BOTH BOGS THE CONCENTRATION OF COPPER 25 HOURS AFTER APPLICATION WAS HIGHER THAN EXPECTED DUE TO SMALLER VOLUMES OF FLOODWATER. DURING THE FIRST 6 DAYS AFTER TREATMENT, COPPER CONCENTRATIONS DECREASED RAPIDLY, AND AFTER 10 DAYS ABOUT 95 PERCENT OF THE COPPER HAD DISAPPEARED. WHEN FLOODWATER WAS RELEASED ABOUT 4 WEEKS AFTER TREATMENT, THE CONCENTRATION OF COPPER WAS AT THE SAME LEVEL FOUND IN THE WATER PRIOR TO TREATMENT. (MCCANN-BATTELLE)

FIELD 05A, 05B, 05C

ACCESSION NO. W71-11498

A VISUAL DEMONSTRATION OF THE BENEFICIAL EFFECTS OF SEWAGE TREATMENT FOR PHOSPHATE REMOVAL ON PARTICULATE MATTER PRODUCTION IN WATERS OF LAKES ERIE AND ONTARIO,

FISHERIES RESEARCH BOARD OF CANADA, WINNIPEG (MANITOBA). FRESHWATER INST.; AND ONTARIO WATER RESOURCES COMMISSION, TORONTO.

J. R. VALLENTYNE, W. E. JOHNSON, AND A. J. HARRIS.

JOURNAL OF THE FISHERIES RESEARCH BOARD OF CANADA, VOL 27, NO 8, P 1493-1496, AUGUST 1970. 1 FIG, 1 TAB, 5 REF.

DESCRIPTORS:
*SEWAGE TREATMENT, *PHOSPHATES, *EUTROPHICATION, *WATER POLLUTION CONTROL, SEWAGE EFFLUENTS, AQUATIC ALGAE, LAKE ERIE, LAKE ONTARIO, NUTRIENTS, NITROGEN, FILTRATION, MICROSCOPY, INDUSTRIAL WASTES, DETERGENTS, AMMONIA, NITRITES, NITRATES.

IDENTIFIERS:
ORTHOPHOSPHATES.

ABSTRACT:
A PROGRAM TO CONTROL EUTHROPHICATION IN LAKES ERIE AND ONTARIO BY DECREASING THE SUPPLY OF PHOSPHORUS COMPOUNDS HAS BEEN DEVELOPED. THIS STUDY WAS CONDUCTED TO DETERMINE THE EFFECT OF REMOVING PHOSPHATE FROM SEWAGE ON ALGAL GROWTH. FILTERED SAMPLES OF RAW SEWAGE, BIOLOGICALLY TREATED SEWAGE, AND SEWAGE TREATED CHEMICALLY FOR PHOSPHATE REMOVAL WERE ADDED TO UNFILTERED WATERS FROM LAKES ERIE AND ONTARIO, AND PARTICULATE RESIDUES (PR) ON MILLIPORE FILTERS PHOTOGRAPHED AFTER INCUBATION IN LIGHT FOR 10 AND 30 DAYS. PR LEVELS IN THE SEWAGE-ENRICHED FLASKS WERE LEAST IN THE CASE OF SEWAGE TREATED FOR REMOVAL OF PHOSPHATES. ADDITION OF PHOSPHATE TO THE PHOSPHATE-DEPLETED EFFLUENT INCREASED ITS PR GENERATING ABILITY TO THAT OF RAW AND BIOLOGICALLY TREATED SEWAGE. THE REMOVAL OF PHOSPHATES FROM SEWAGE WASTES THUS APPEARS TO ELIMINATE THEIR FERTILIZING EFFECT. (MORTLAND-BATTELLE)

FIELD 05C, 05D, 02H

ACCESSION NO. W71-11507

ANALYSIS OF TRACE ELEMENTS IN SEAWEED,

CENTRAL INST. FOR INDUSTRIAL RESEARCH, OSLO (NORWAY).

G. LUNDE.

JOURNAL OF THE SCIENCE OF FOOD AND AGRICULTURE, VOL 21, NO 8, P 416-418, AUGUST 1970. 4 TAB, 18 REF.

DESCRIPTORS:
*AQUATIC ALGAE, *NEUTRON ACTIVATION ANALYSIS, *TRACE ELEMENTS, COPPER, MOLYBDENUM, MANGANESE, COBALT, IRON, INDUSTRIAL WASTES, SEWAGE EFFLUENTS, POLLUTANT IDENTIFICATION.

IDENTIFIERS:
*ATOMIC ABSORPTION, ARSENIC, SELENIUM, ZINC, ANTIMONY, NORWAY, ASHING, BIOLOGICAL MAGNIFICATION, PELVETIA CANALICULATA, FUCUS SERRATUS, FUCUS SPIRALIS, FUCUS VESICULOSUS, LAMINARIA DIGITATA, LAMINARIA HYPERBOREA, ASCOPHYLLUM NODOSUM, GIGARTINA MAMILLOSA, RHODYMENIA PALMATA.

ABSTRACT:
THE FOLLOWING TRACE ELEMENTS HAVE BEEN ANALYZED BY NEUTRON ACTIVATION AND ATOMIC ABSORPTION IN VARIOUS ALGAE SAMPLES: ARSENIC, COPPER, MOLYBDENUM, MANGANESE, ZINC, COBALT, ANTIMONY, SELENIUM AND IRON. THE ASH CONTENT HAS ALSO BEEN DETERMINED. THE SAMPLES WERE COLLECTED FROM TWO DIFFERENT LOCALITIES IN NORWAY, ONE (TRONDHEIMSFJORD) CHARACTERIZED BY A RELATIVELY STRONG INFLUENCE OF RIVER WATER, INDUSTRIAL WASTE ETC.; THE OTHER (REINE IN LOFOTEN) IS FREE FROM THIS TYPE OF CONTAMINATION. IN THE CASE OF TWO SPECIES, ASCOPHYLLUM NODOSUM AND LAMINARIA HYPERBOREA, THE VARIATION IN THE CONTENT OF TRACE ELEMENTS HAS BEEN STUDIED DURING A ONE-YEAR PERIOD, SAMPLES BEING TAKEN EVERY SECOND MONTH. CONSIDERABLE DIFFERENCES IN THE CONTENT OF TRACE ELEMENTS WAS FOUND BETWEEN THE LAMINARIACEAE AND FUCACEAE. THE RESULTS INDICATE THAT THE MARINE ALGAE GENERALLY CONTAIN LARGER CONCENTRATIONS OF THE TRACE ELEMENTS ANALYZED, ESPECIALLY ARSENIC, THAN DO TERRESTRIAL PLANTS. ARSENIC IS ENRICHED IN MARINE ALGAE BY A FACTOR OF 200 TO 500. MORE ATTENTION SHOULD BE FOCUSED ON HOW THIS ELEMENT OCCURS AND WHETHER OR NOT IT HAS ANY PHYSIOLOGICAL ROLE IN MARINE ALGAE. (MORTLAND-BATTELLE)

FIELD 05A, 05B

ACCESSION NO. W71-11515

DETERGENT POLLUTION CONTROL ACT OF 1971 (A BILL TO AMEND THE FEDERAL WATER POLLUTION CONTROL ACT TO ESTABLISH STANDARDS AND PROGRAMS TO ABATE AND CONTROL WATER POLLUTION BY SYNTHETIC DETERGENTS).

HOUSE BILL 7725, 92D CONG, 1ST SESS (1971). 14 P.

DESCRIPTORS:
*WATER POLLUTION CONTROL, *EUTROPHICATION, *DETERGENTS, *PHOSPHATES, LEGISLATION, WATER LAW, FEDERAL GOVERNMENT, WATER POLLUTION, STANDARDS, WATER POLLUTION EFFECTS, ALKYLBENZENE SULFONATES, WATER POLLUTION SOURCES, AQUATIC ALGAE, PUBLIC HEALTH, BIODEGRADATION, TOXICITY, CONTRACTS, GRANTS, RESEARCH AND DEVELOPMENT, INSPECTION, FEDERAL JURISDICTION.

ABSTRACT:
SURFACE AND GROUNDWATERS IN THE UNITED STATES ARE BEING SERIOUSLY POLLUTED AND DEGRADED BY THE DISCHARGE INTO SUCH WATERS OF SYNTHETIC DETERGENTS. IN ORDER TO ABATE SUCH POLLUTION THE FEDERAL WATER POLLUTION CONTROL ACT WOULD BE AMENDED BY THE DETERGENT POLLUTION CONTROL ACT OF 1971. THE FEDERAL GOVERNMENT WOULD PROVIDE ASSISTANCE IN THE DEVELOPMENT OF POLLUTION-FREE DETERGENTS. THE ENVIRONMENTAL PROTECTION AGENCY WOULD ESTABLISH STANDARDS OF WATER EUTROPHICATION ABILITY, BIODEGRADABILITY, TOXICITY, AND OF EFFECTS ON THE PUBLIC HEALTH AND WELFARE WHICH WOULD HAVE TO BE MET BY ALL DETERGENTS. PROCEDURES FOR ESTABLISHING SUCH STANDARDS ARE HEREIN SET FORTH. SUCH STANDARDS WOULD HAVE TO BE MET BY JUNE 30, 1973. DETERGENTS NOT IN COMPLIANCE WOULD BE PROCEEDED AGAINST AND CONDEMNED. FURTHER ENFORCEMENT PROVISIONS AND JUDICIAL PROCEDURES ARE SET FORTH. PROVISION IS ALSO MADE FOR THE INSPECTION OF MANUFACTURING FACILITIES, AND THE SAMPLING OF BOTH DOMESTIC AND IMPORTED DETERGENTS. (ROBINSON-FLORIDA)

FIELD 06E, 05C

ACCESSION NO. W71-11516

BIOLOGICAL ASPECTS OF THERMAL POLLUTION. I: ENTRAINMENT AND DISCHARGE CANAL EFFECTS,

OAK RIDGE NATIONAL LAB., TENN. ECOLOGICAL SCIENCES DIV.; AND MINNESOTA UNIV., MINNEAPOLIS. DEPT. OF ECOLOGY AND BEHAVIORAL BIOLOGY.

CHARLES C. COUTANT, AND ALAN J. BROOK.

CRITICAL REVIEWS IN ENVIRONMENTAL CONTROL, VOL 1, NO 3, P 341-381, NOVEMBER 1970. 16 FIG, 2 TAB, 140 REF.

DESCRIPTORS:
*THERMAL POLLUTION, *BIOLOGICAL COMMUNITIES, *THERMAL POWER PLANTS, *DISCHARGE(WATER), AQUATIC HABITATS, ENVIRONMENTAL EFFECTS, HEATED WATER, NUCLEAR POWER PLANTS, STEAM, ECOLOGY, PHYTOPLANKTON, ZOOPLANKTON, LARVAE, EGGS, CHINOOK SALMON, STRIPED BASS, MINNOWS, FISHKILL, HERRING, COPEPODS, CRUSTACEANS, DIATOMS, CYANOPHYTA, PHOTOSYNTHESIS, OYSTERS, CLAMS, TROUT, MUSSELS, MATHEMATICAL MODELS, DISSOLVED OXYGEN, COOLING TOWERS, LAKE MICHIGAN, CONDENSERS, DIPTERA, PREDATION, ALGAE, NITROGEN FIXATION, FOOD CHAINS, EUTROPHICATION, CATFISHES, BULLHEADS, CARP, PERCHES, METABOLISM, OXYGEN REQUIREMENTS, RESPIRATION, SUCKERS, SEXUAL MATURITY, SPAWNING, RESISTANCE, AQUATIC PRODUCTIVITY, CANALS.

IDENTIFIERS:
*ENTRAINMENT, WHITE RIVER(INDIANA), CONNECTICUT RIVER(CONNECTICUT), GREEN RIVER(KENTUCKY), CAYUGA LAKE(NEW YORK), PATUXENT ESTUARY, BISCAYNE BAY(FLORIDA), HUMBOLT BAY(CALIFORNIA), POTOMAC RIVER, COLUMBIA RIVER(WASHINGTON), COCKLES, GOLDFISH, MERRIMACK RIVER(NEW HAMPSHIRE), DELAWARE RIVER, ILLINOIS RIVER, HOLSTON RIVER(TENNESSEE).

ABSTRACT:
THIS REVIEW ATTEMPTS TO CRITICALLY EVALUATE SOME THERMAL EFFECTS SEEN AT OPERATING THERMAL POWER PLANTS, TO GROUP THESE INTO SEVERAL 'PROBLEMS' ASSOCIATED WITH (1) ENTRAINMENT AND (2) DISCHARGE CANALS, AND TO INDICATE PERTINENT FIELD AND LABORATORY EXPERIMENTS THAT CAN ASSIST IN DEVELOPING INFORMATION OF PREDICTIVE UTILITY. MOST POWER PLANT SURVEYS LACK DETAIL OF OBSERVATION AND DEFINITION OF GOALS SUFFICIENT TO PROVIDE MORE THAN CIRCUMSTANTIAL EVIDENCE FOR ECOLOGICAL PROCESSES. ON THE OTHER HAND, LABORATORY EXPERIMENTS ARE OFTEN UNREALISTIC SIMULATIONS OF COMPLEX PHENOMENA. TRUE PREDICTABILITY WILL REQUIRE JUDICIOUS APPLICATION OF DATA FROM BOTH SOURCES. UNTIL COMPLETE INFORMATION IS AVAILABLE, CERTAIN LABORATORY TESTS PROVIDE CONSERVATIVE APPROXIMATIONS THAT CAN GUIDE POWER PLANT SITING AND DESIGN SO THAT SAFE ENVIRONMENTS CAN BE MAINTAINED FOR AQUATIC LIFE. (MCCANN-BATTELLE)

FIELD 05C, 05B

ACCESSION NO. W71-11517

THE ROLE OF CARBON IN EUTROPHICATION,

MISSOURI UNIV., COLUMBIA. DEPT. OF CIVIL ENGINEERING.

DARRELL L. KING.

JOURNAL WATER POLLUTION CONTROL FEDERATION, VOL 42, NO 12, P 2035-2051, DECEMBER 1970. 8 FIG, 65 REF.

DESCRIPTORS:
*EUTROPHICATION, *CARBON, *PHOTOSYNTHESIS, *ALGAE, CARBON DIOXIDE, ALKALINITY, NUTRIENTS, AMMONIA, HYDROGEN ION CONCENTRATION, CYANOPHYTA, CHLORELLA, CHLOROPHYTA, SCENEDESMUS, EUGLENA, DISSOLVED OXYGEN, DIURNAL DISTRIBUTION, AQUATIC PRODUCTIVITY, NITROGEN, PHOSPHORUS, LAKES.

IDENTIFIERS:
PHORMIDIUM, CHLORELLA PYRENOIDOSA, CHLAMYDOMONAS, FONTINALIS, ASTINASTRUM, ANACYSTIS NIDULANS, MICROCYSTIS.

ABSTRACT:
THIS LITERATURE REVIEW SURVEYS THE ROLE OF CARBON IN LAKE EUTROPHICATION. THE IMPORTANCE OF CARBONATE ALKALINITY AND THE ROLE OF CARBON DIOXIDE IN PHOTOSYNTHESIS ARE DISCUSSED IN DETAIL. IT IS CONCLUDED THAT ALTHOUGH NITROGEN, PHOSPHORUS, AND A VARIETY OF OTHER NUTRIENTS ARE REQUIRED BY ALGAE, EUTROPHICATION SEEMS TO BE ULTIMATELY A CARBON-ACCUMULATION PHENOMENON. INITIAL ALGAL RESPONSE TO ADDITIONS OF NITROGEN AND PHOSPHORUS TO AN OLIGOTROPHIC LAKE INDICATES THAT ONE OF THESE BASIC NUTRIENTS IS THE FACTOR LIMITING PHOTOSYNTHESIS IN THAT LAKE. HOWEVER, IN A LAKE WITH LOW ALKALINITY, BIOLOGICAL INDICATORS OF EUTROPHICATION, SUCH AS MIDSUMMER BLUE-GREEN ALGAL BLOOMS, OFTEN ARE NOTED FOLLOWING ONLY SLIGHT ADDITIONS OF NITROGEN AND PHOSPHORUS. THE AMOUNT OF NITROGEN, PHOSPHORUS, AND OTHER PLANT NUTRIENTS REQUIRED TO PRODUCE MIDSUMMER BLUE-GREEN ALGAL DOMINANCE IN ANY LAKE SEEMS TO BE RELATED DIRECTLY TO THE BICARBONATE-CARBONATE ALKALINITY OF THAT LAKE. ANY ATTEMPT TO MAXIMIZE ALGAL PRODUCTIVITY WHILE MAINTAINING ALGAL SPECIES SUITABLE FOR HUMAN OR ANIMAL FOOD WILL REQUIRE CONTROL OF CARBON DIOXIDE AVAILABILITY IN ADDITION TO THE SUPPLY OF SUFFICIENT QUANTITIES OF OTHER PLANT NUTRIENTS TO ALLOW UNHINDERED ALGAL PHOTOSYNTHESIS. (MORTLAND-BATTELLE)

FIELD 05C

ACCESSION NO. W71-11519

TAXONOMY AND DISTRIBUTION PATTERNS OF PLANKTONIC DIATOMS IN THE EQUATORIAL PACIFIC,

AKADEMIYA NAUK SSSR, MOSCOW. INSTITUT OKEANOLOGII.

T. V. BELYAYEVA.

OCEANOLOGY (USSR), VOL 10, NO 1, P 101-107, 1970. 6 FIG, 1 TAB, 15 REF.

DESCRIPTORS:
*DIATOMS, *PHYTOPLANKTON, *PACIFIC OCEAN, *ALGAE.

IDENTIFIERS:
JUDAY NET, ASTEROLAMPRA MARYLANDICA, ASTEROLAMPRA HEPTACTIS, BACTERIASTRUM COMOSUM, CHAETOCEROS ATLANTICUS VAR. NEAPOLITANA, CHAETOCEROS COARCTATUS, CHAETOCEROS LORENZIANUS, CHAETOCEROS MESSANENSIS, CHAETOCEROS PERUVIANUS, CLADOGRAMMA KOLBEI, CLADOGRAMMA CRENULATUS, CLADOGRAMMA LINEATUS-EXCENTRICUS, CLADOGRAMMA NODULIFER, DACTYLIOSOLEN MEDITERRANEUS, ETHMODISCUS GAZELLAE, ETHMODISCUS REX, HEMIAULUS HAUCKII, HEMIDISCUS CUNEIFORMIS, PLANKTONIELLA SOL, PODOSIRA SP., RHIZOSOLENIA ALATA, RHIZOSOLENIA BERGONII, RHIZOSOLENIA CALCAR-AVIS, RHIZOSOLENIA CASTRACANEI, RHIZOSOLENIA STYLIFORMIS, ROPERIA TESSELLATA, ROPERIA TESSELLATA VAR. OVATA, THALASSIOSIRA DECIPIENS, THALASSIOSIRA SUBTILIS, TRICERATIUM CINNAMOMEUN VAR. MINOR, AMPHORA SP., NITZSCHIA AEQUATORIALIS, NITZSCHIA BICAPITATA, NITZSCHIA BRAARUDII, NITZSCHIA DELICATISSIMA, NITZSCHIA MARINA, NITZSCHIA SICULA, PSEUDOEUNOTIA DOLIOLUS, STIGMAPHORA ROSTRATA, THALASSIONEMA NITZSCHIOIDES VAR. OBTUSA, THALASSIONEMA NITZSCHIOIDES VAR. PARVA, THALASSIOTHTIX GIBBERULA, THALASSIOTHTIX VANHOFFENII, TROPIDONEIS SP.

ABSTRACT:
THE CONTENT AND DISTRIBUTION OF DIATOMS IN THE PLANKTON OF THE EQUATORIAL PART OF THE PACIFIC OCEAN WERE STUDIED ON FOUR MERIDIONAL SECTIONS ALONG 160 E LONG. AND 176, 154 AND 140 W DURING SEPTEMBER-DECEMBER 1961. THE PHYTOPLANKTON COLLECTIONS WERE MADE WITH A JUDAY NET TO A DEPTH OF 100 M. ALL THE DIATOMS FOUND IN THIS REGION ARE LISTED TOGETHER WITH AN INDICATION OF THE FREQUENCY OF OCCURRENCE OF EACH SPECIES ON THE SECTIONS. NINETY-SEVEN PLANKTONIC DIATOMACEOUS ALGAE WERE FOUND IN THE CENTRAL AND EASTERN PARTS OF THE PACIFIC OCEAN: 86 OF THEM WERE IDENTIFIED AS TO SPECIES, 9 ONLY AS TO GENUS, AND TWO OF THE DIATOMS HAVE NOT BEEN IDENTIFIED AS TO GENUS. MOST OF THE DIATOMS IN OUR REGION ARE TROPICAL SPECIES. EIGHTEEN OF THE SPECIES OCCUR NOT ONLY IN THE TROPICAL REGIONS BUT ALSO IN THE BOREAL REGION (12 SPECIES) OR IN THE ANTARCTIC REGION (6 SPECIES). THE WIDELY DIFFUSED (COSMOPOLITAN) SPECIES FOUND AMOUNTED TO 13. (MCCANN-BATTELLE)

FIELD 05A

ACCESSION NO. W71-11527

CALCIUM AND STRONTIUM DISCRIMINATION BY AQUATIC PLANTS,

ATOMIC ENERGY OF CANADA LTD., CHALK RIVER (ONTARIO). CHALK RIVER NUCLEAR LABS.

I. L. OPHEL, AND C. D. FRASER.

ECOLOGY, VOL 51, NO 2, P 324-327, EARLY SPRING 1970. 1 FIG, 1 TAB, 15 REF.

DESCRIPTORS:
*CALCIUM, *FOOD CHAINS, *SPECTROPHOTOMETRY, *POLLUTANT IDENTIFICATION, ONTARIO, ALGAE, MOSSES, AQUATIC PLANTS.

IDENTIFIERS:
*STRONTIUM, *SAMPLE PRESERVATION, *FLAME EMISSION SPECTROPHOTOMETRY, PERCH LAKE(ONTARIO), ASHING, BIOLOGICAL MAGNIFICATION.

ABSTRACT:
AQUATIC PLANTS WERE COLLECTED OVER A 3-YEAR PERIOD FROM PERCH LAKE, ONTARIO. THE PLANTS WERE TAKEN TO THE LABORATORY IN PLASTIC BAGS, WASHED, DRIED TO A CONSTANT WEIGHT AT 110 C, AND ASHED IN PORCELAIN CRUCIBLES. CALCIUM AND STRONTIUM WERE ANALYZED USING A FLAME EMISSION SPECTROPHOTOMETER. SR/CA RATIOS ARE GIVEN FOR 22 SPECIES OF ALGAE, MOSS, AND VASCULAR PLANTS. WHEN COMPARED WITH THE LAKE WATER, 16 SPECIES WERE FOUND TO DISCRIMINATE AGAINST CALCIUM RELATIVE TO STRONTIUM. THESE SPECIES INTRODUCE A STRONTIUM ENRICHMENT STEP INTO ANY FOOD CHAIN IN WHICH THEY PARTICIPATE. ONE SPECIES (NYMPHAEA ODORATA) SHOWED A REMARKABLE DISCRIMINATION AGAINST STRONTIUM. (MORTLAND-BATTELLE)

FIELD 05C, 05A

ACCESSION NO. W71-11529

CERTAIN ASPECTS OF PRODUCTION AND STANDING STOCK OF PARTICULATE MATTER IN THE SURFACE WATERS OF THE NORTHWEST ATLANTIC OCEAN,

BEDFORD INST., DARTMOUTH (NOVA SCOTIA). MARINE ECOLOGY LAB.

W. H. SUTCLIFFE, JR., R. W. SHELDON, AND A. PRAKASH.

JOURNAL FISHERIES RESEARCH BOARD OF CANADA, VOL 27, NO 11, P 1917-1926, NOVEMBER 1970. 7 FIG, 1 TAB, 25 REF.

DESCRIPTORS:
*CARBON, *CHLOROPHYLL, *SEAWATER, *AQUATIC PRODUCTION, *BIOMASS, ORGANIC MATTER, ATLANTIC OCEAN, GROWTH RATES, ALGAE, INCUBATION, PHYTOPLANKTON.

IDENTIFIERS:
*ADENOSINE TRIPHOSPHATE, *DEOXYRIBONUCLEIC ACID, SAMPLE PRESERVATION, COULTER COUNTER.

ABSTRACT:
SAMPLES COLLECTED DURING JANUARY AND OCTOBER 1968 FROM 44 DEGREES TO 19 DEGREES N IN THE WESTERN NORTH ATLANTIC WERE ANALYZED FOR PARTICULATE ORGANIC CARBON, ATP, DNA, CHLOROPHYLL A AND TOTAL PARTICULATE VOLUME BETWEEN 1 AND 40 MICRON DIAMETER. ALTHOUGH THE RATIO OF CHLOROPHYLL TO ATP DECREASED FROM NORTH TO SOUTH, THE RATIO OF PARTICULATE CARBON TO ATP DID NOT SHOW ANY SYSTEMATIC TREND. A GOOD CORRELATION WAS FOUND BETWEEN ATP AND DNA EVEN THOUGH THE DNA CONCENTRATION WAS TOO GREAT TO BE ENTIRELY REPRESENTATIVE OF LIVING MATERIAL. GROWTH OR PRODUCTION OF PARTICULATE MATERIAL AS MEASURED BY THE COULTER COUNTER METHOD WAS HIGHER THAN THAT MEASURED BY C-14 METHOD. A TREND WAS NOTED TOWARDS HIGHER SPECIFIC GROWTH RATE WITH SMALLER STANDING CROP IN SOUTHERN WATERS. (LITTLE-BATTELLE)

FIELD 05C, 05A

ACCESSION NO. W71-11562

OCCURRENCE AND DISTRIBUTION OF DIATOMS AND OTHER ALGAE IN THE UPPER POTOMAC RIVER,

VIRGINIA POLYTECHNIC INST. AND STATE UNIV., BLACKSBURG; KANSAS UNIV., LAWRENCE; AND ACADEMY OF NATURAL SCIENCES OF PHILADELPHIA, PA.

JOHN CAIRNS, JR., ROGER L. KAESLER, AND RUTH PATRICK.

NOTULAE NATURAE OF THE ACADEMY OF NATURAL SCIENCES OF PHILADELPHIA, NO. 436, DEC 22, 1970. 12 P, 3 FIG, 2 TAB, 12 REF.

DESCRIPTORS:
*WATER POLLUTION EFFECTS, *DIATOMS, WATER POLLUTION SOURCES, INDUSTRIAL WASTES, AQUATIC ALGAE, BIOINDICATORS, BIOASSAY, ALGAE, POWERPLANTS, EFFLUENTS, DISCHARGE(WATER), STATISTICAL MODELS.

IDENTIFIERS:
*SPECIES DIVERSITY, *POTOMAC RIVER.

ABSTRACT:
TEN LIMNOLOGICAL SURVEYS WERE MADE OF A PORTION OF THE UPPER POTOMAC RIVER IN CONJUNCTION WITH INSTALLATION AND EXPANSION OF THE DICKERSON POWER STATION OF THE POTOMAC ELECTRIC POWER COMPANY. CLUSTER ANALYSES OF MATRICES OF JACCARD COEFFICIENTS COMPUTED FROM PRESENCE-ABSENCE DATA ON (1) 344 SPECIES OF DIATOMS IN 46 AGGREGATIONS, AND (2) 161 SPECIES OF OTHER ALGAE IN 50 AGGREGATIONS YIELDED RESULTS IN AGREEMENT WITH THOSE OBTAINED IN SIMILAR STUDIES OF PROTOZOANS AND INSECTS AND ALSO WITH RESULTS OF SPECIES DIVERSITY STUDIES. THESE STUDIES INDICATE THAT THE POPULATIONS OF DIATOMS IN THESE AREAS WERE VERY SIMILAR BEFORE AND AFTER PLANT OPERATIONS STARTED, AND THERE WAS NO EVIDENCE FROM THE DIVERSITY OR KINDS OF SPECIES OF ADVERSE EFFECTS DUE TO POWER PLANT DISCHARGE. (LEGORE - WASHINGTON)

FIELD 05C

ACCESSION NO. W71-11583

HOW AERIAL PHOTOGRAPHS CAN IDENTIFY AND MEASURE POLLUTANTS,

CORNELL AERONAUTICAL LAB., BUFFALO, N.Y.

KENNETH R. PIECH.

PUBLIC WORKS, VOL. 101, NO. 3, P 68-71, MARCH 1970. 4 FIG.

DESCRIPTORS:
*POLLUTANT IDENTIFICATION, *AERIAL PHOTOGRAPHY, *MATHEMATICAL METHODS, MUNICIPAL WASTES, INDUSTRIAL WASTES, NEW YORK, ALGAE, WATER POLLUTION.

IDENTIFIERS:
NIAGARA RIVER.

ABSTRACT:
AERIAL PHOTOGRAPHY CAN BE USED TO BOTH MEASURE AND QUANTIFY POLLUTANTS IN WATERWAYS. IT HAS THE POTENTIAL OF PROVIDING POLLUTION DATA FOR AN ENTIRE WATERWAY RATHER THAN ISOLATED POINTS AND IS OFTEN ABLE TO PINPOINT THE SOURCE OF A PARTICULAR POLLUTANT. THE METHOD IS BASED ON THE FACT THAT EVERY SUBSTANCE HAS A CHARACTERISTIC SPECTRAL CURVE BY WHICH IT CAN BE IDENTIFIED. BY USING SPECIALLY CONSTRUCTED SPECTRAL FILTERS, PHOTOGRAPHS CAN BE TAKEN IN WHICH LIGHT OUTSIDE A GIVEN RANGE OF WAVELENGTHS(THE SPECTRAL CURVE OF THE POLLUTANT TO BE IDENTIFIED FALLS WITHIN THE RANGE) IS FILTERED OUT. ON THE DEVELOPED PHOTOGRAPH, THE POLLUTANT THEN SHOWS UP LIGHTER AND ITS QUANTITY CAN BE DETERMINED BY MATHEMATICAL ANALYSIS, IN PARTS PER MILLION, FOR EXAMPLE. IN REALITY, IT WOULD BE IMPRACTICAL TO CONSTRUCT FILTERS FOR EACH POLLUTANT, SINCE MANY SUBSTANCES HAVE VERY SIMILAR SPECTRAL CURVES. THUS, THE PRACTICALITY OF SINGLING OUT A SPECIFIC POLLUTANT IS LARGELY GOVERNED BY ITS TYPE COMPARED WITH THE COMBINATIONS OF OTHER POLLUTANTS THAT MAY ALSO BE PRESENT. BECAUSE THE TYPE OF POLLUTANT IS OFTEN KNOWN, THE ABILITY TO TRACK IT BY PHOTOGRAPHY IS PROBABLY OF MORE IMPORTANCE. THE METHOD CAN ALSO BE USED TO MEASURE ALGAL CONCENTRATIONS. (MORTLAND-BATTELLE)

FIELD 05A, 05B, 07B

ACCESSION NO. W71-11685

METHODS OF REMOVING NITRATES FROM WATER,

INTERAGENCY AGRICULTURAL WASTE WATER TREATMENT CENTER, FIREBAUGH, CALIF.

PERCY P. ST. AMANT, AND LOUIS A. BECK.

JOURNAL OF AGRICULTURAL AND FOOD CHEMISTRY, VOL. 18, NO. 5, P 785-788, SEPTEMBER-OCTOBER 1970. 1 TAB, 7 REF.

DESCRIPTORS:
NITRATES, ALGAE, DENTRIFICATION, FILTRATION, DESALINATION, DRAINAGE WATER, AGRICULTURAL CHEMICALS, FARM WASTES, WASTE WATER TREATMENT, CALIFORNIA.

ABSTRACT:
AN INTERAGENCY NITROGEN REMOVAL GROUP COMPOSED OF THREE AGENCIES, THE FEDERAL WATER POLLUTION CONTROL ADMINISTRATION, THE U.S. BUREAU OF RECLAMATION, AND THE CALIFORNIA DEPARTMENT OF WATER RESOURCES, WAS FORMED TO ATTEMPT TO FIND A METHOD OF REMOVING NITRATES FROM AGRICULTURAL WASTE WATERS IN CALIFORNIA. THREE METHODS OF NITROGEN REMOVAL ARE BEING EVALUATED: POND AND FILTER DENITRIFICATION, WHICH DEPEND UPON BACTERIAL ACTION, AND ALGAE STRIPPING. DESALINATION OF TILE DRAINAGE IS ALSO BEING EVALUATED. DETAILS AND PRELIMINARY RESULTS OF THE STUDIES ARE GIVEN. (LITTLE-BATTELLE)

FIELD 05C, 05D, 05F, 05G

ACCESSION NO. W71-11698

TRACE MATERIALS IN WASTES DISPOSED TO COASTAL WATERS-FATES, MECHANISMS AND ECOLOGICAL GUIDANCE AND CONTROL,

FEDERAL WATER QUALITY ADMINISTRATION, CORVALLIS, OREG. NATIONAL COASTAL RESEARCH PROGRAM.

MILTON H. FELDMAN.

AVAILABLE FROM THE NATIONAL TECHNICAL INFORMATION SERVICE AS PB-202 346, $3.00 IN PAPER COPY, $0.95 IN MICROFICHE. WORKING PAPER, NO. 78, JULY 1970. 102 P, 20 FIG, 17 TAB, 137 REF. EPA PROGRAM 16070---07/70.

DESCRIPTORS:
*OCEANS, *COASTS, *WATER QUALITY CONTROL, BIOMASS, TOXICITY, EUTROPHICATION, TRACE ELEMENTS, NITROGEN, PHOSPHORUS, METABOLISM, ALGAE, SLUDGE, ANALYTICAL TECHNIQUES, INDUSTRIAL WASTES, MUNICIPAL WASTES, METALS, HEAVY METALS, WASTE DISPOSAL, WATER POLLUTION EFFECTS.

IDENTIFIERS:
*TRACE ORGANIC CONTAMINANTS, *TRACE MATERIALS, BIOINHIBITION, BIOSTIMULATION.

ABSTRACT:
WASTES CURRENTLY BEING DISCHARGED TO THE COASTAL WATERS OF THE UNITED STATES INCLUDE TRACE ORGANIC CONTAMINANTS (TC), TRACE ELEMENTS (TE), AND OTHER TRACE MATERIALS (TM). THOSE TM DEALT WITH INCLUDED: (1) KNOWN VIOLENTLY NOXIOUS MATERIALS; (2) MATERIALS WHICH ARE BIOSTIMULATORY TO SOME SPECIES (COBALAMIN, IRON CHELATES, THIAMIN, BIOTIN, MN, MG; AND (3) MATERIALS WHICH ARE BIOINHIBITORY FOR AT LEAST SOME SPECIES IN VARIOUS MECHANISMS (DDT, SE, MN, MG). A DIFFERENT VIEW OF THE PROBLEM WAS TAKEN IN WHICH WAYS OF SELECTING THE OPTIMUM ORGANISM TO UTILIZE A PARTICULAR WASTE WERE PREFERRED TO METHODS OF REMOVING IT CONVENTIONALLY. A THOROUGH LITERATURE SEARCH REVEALED A SIGNIFICANT LACK OF LITERATURE IN WASTE BREAKDOWN FOR BOTH QUALITATIVE AND QUANTITATIVE EVALUATION. FURTHER STUDIES SHOULD BE PERFORMED FOR EACH KNOWN TRACE MATERIAL TO DETERMINE: (1) THE MECHANISM WHEREBY IT IS SECLUDED; (2) THE METABOLIC THRESHOLD; (3) THE ACTIVE LEVEL; (4) THE HARMFUL LEVEL; (5) THE SPECIATION REQUIREMENTS; (6) THE ABSOLUTE RATES IN AND OUT OF THE COMPARTMENTS OF WHICH COASTAL WATERS, SEDIMENTS, CHEMICAL SYSTEM PHASES, AND BIOTA MAY BE CONSIDERED AS COMPOSED. EVALUATIONS SUCH AS THESE WERE PERFORMED FOR DDT, AND THEY MUST BE PERFORMED FOR THE KNOWN SET OF WASTE CONSTITUENTS BEFORE RATIONAL ACTION TO PREVENT DAMAGE TO THE OCEANS IS POSSIBLE. (LOWRY-TEXAS)

FIELD 05B, 05C, 02L, 05G

ACCESSION NO. W71-11793

THE ISOLATION AND OCCURRENCE OF HYALCCHLORELLA MARINA,

CALIFORNIA UNIV. BERKELEY, MICROBIOLOGY GROUP; AND CALIFORNIA UNIV; BERKELEY. DEPT. OF BOTANY.

R. O. POYTON.

JOURNAL OF GENERAL MICROBIOLOGY, VOL. 62, PART 2, P 189-194, AUGUST 1970. 2 TAB, 13 REF.

DESCRIPTORS:
*MARINE FUNGI, *ISOLATION, ALGAE, SEPARATION TECHNIQUES, ATLANTIC OCEAN, PACIFIC OCEAN, GULF OF MEXICO.

IDENTIFIERS:
*HYALOCHLORELLA MARINA, RHODOMELA SP., ENDOCLADIA SP., PHYLLOSPADEX SP., MICROCLADIA SP., CLADOPHORA SP., POLYSIPHONIA SP., GELIDIUM SP., CHLORELLA, AGARS.

ABSTRACT:
FIVE METHODS WERE EVALUATED FOR ISOLATION OF HYALOCHLORELLA MARINA FROM A MARINE ENVIRONMENT. THE METHODS WERE: (1) FILAMENTOUS ALGAE COLLECTED FROM INTERTICAL AND EULITTORAL ZONES WAS PLACED IN ISOLATION MEDIUM CONTAINING 370 UNITS OF STREPTOMYCIN SULFATE/ML AND 825 UNITS PENICILLIN 'G' SODIUM/ML. (2) SEA WATER WAS CENTRIFUGED AND THE CONCENTRATE PLATED ONTO THE ISOLATION MEDIUM. (3) SEA WATER AND MARINE MUD SUSPENSIONS WERE PREFILTERED TO REMOVE LARGE MUD PARTICLES AND DEBRIS, FILTERED THROUGH A STERILE NUCLEOPORE MEMBRANE FILTER, AND THE FILTERS INCUBATED ON THE ISOLATION MEDIUM. (4) SEA WATER WAS SEEDED WITH POLLEN GRAINS AND OBSERVED AFTER 4, 7, AND 10 DAYS FOR HYALOCHLORELLA FILAMENTS. (5) MICROSCOPE SLIDES WERE IMMERSED AT VARIOUS DEPTHS IN THE MARINE ENVIRONMENT AND EXAMINED FOR HYALOCHLORELLA GROWTH AT VARIOUS TIME INTERVALS. THE LATTER TWO METHODS PROVIDED NO SATISFACTORY RESULTS IN THE EXPERIMENTS. THE FIRST THREE METHODS PROVED EFFECTIVE FOR ISOLATION OF HYALOCHLORELLA MARINA. THESE METHODS SHOWED THAT THIS ORGANISM OCCURS IN NATURE BOTH ATTACHED TO ALGAL FILAMENTS AND FREE FLOATING. HOWEVER, THE CONCENTRATION OF ORGANISMS ON ALGAL SURFACES WAS 300 TO 13,000 TIMES THAT OF FREE-FLOATING ORGANISMS IN AN EQUIVALENT AMOUNT OF SEA WATER. ADDITIONAL INVESTIGATION OF THE EFFECT OF ANTIBIOTICS ON THE VIABILITY OF HYACHLORELLA MARINA SHOWED THAT VIABLE COUNTS WERE THE SAME WITH OR WITHOUT PENICILLIN AND/OR STREPTOMYCIN. (LITTLE-BATTELLE)

FIELD 05A, 05C

ACCESSION NO. W71-11851

COMPARISON OF THE POLLUTED HOOGHLY ESTUARY WITH THE UNPOLLUTED MATLAH ESTUARY, INDIA,

CENTRAL INLAND FISHERIES RESEARCH INST., BARRACKPORE (INDIA).

A. K. BASU, B. B. GHOSH, AND R. N. PAL.

JOURNAL WATER POLLUTION CONTROL FEDERATION, VOL 42, NO 10, P 1771-1781, OCTOBER 1970. 7 FIG, 1 TAB, 17 REF.

DESCRIPTORS:
*INDUSTRIAL WASTES, *WATER QUALITY, *DISSOLVED OXYGEN, *PHYTOPLANKTON, ESTUARIES, WATER TEMPERATURE, TURBIDITY, PHOTOSYNTHESIS, HYDROGEN ION CONCENTRATION, NITRITE, AMMONIA, SALINITY, AQUATIC ALGAE, ZOOPLANKTON, BIOCHEMICAL OXYGEN DEMAND, COPEPODS.

IDENTIFIERS:
COSCINODISCUS, NITZSCHIA, OSCILLATORIA, PLEUROSIGMA, NAUPLII, LAMELLIBRANCH, SPIROGYRA, EUDORINA, CYCLOPS, KERATELLA, ASPLANCHNA, LITHODESMIUM, HARPESTICOIDS, NOCTILUCA, CENTROPYXIS, BIDDULPHIA, CHAETOCEROS, THALASSISTHRIX, INDIA.

ABSTRACT:
A SHORT PRELIMINARY STUDY TO ASSESS THE GENERAL PHYSICOCHEMICAL AND BIOLOGICAL CONDITIONS OF THE HOOGHLY ESTUARY, INTO WHICH A LARGE QUANTITY OF INDUSTRIAL POLLUTANTS IS BEING DISCHARGED FROM DIFFERENT FACTORIES, WAS ATTEMPTED, AND A COMPARISON HAS BEEN DRAWN WITH THE CONDITIONS OF THE UNPOLLUTED MATLAH ESTUARY. THE RESULTS REVEALED THE FOLLOWING POINTS: (1) THE GENERAL CONDITION OF THE HOOGHLY WITH RESPECT TO DISSOLVED OXYGEN, OXYGEN CONSUMPTION VALUE, AND TEMPERATURE WAS INFERIOR TO THAT OF THE MATLAH. (2) TURBIDITY OF THE HOOGHLY WAS GENERALLY HIGHER THAN THAT OF THE MATLAH. (3) PH, NITRITE N, AND FREE AMMONIA VALUES WERE MORE OR LESS THE SAME IN THE MIDSTREAM OF BOTH THE ESTUARIES, ALTHOUGH A LOW PH VALUE OF 4.3 WAS OBSERVED NEAR A DISCHARGE POINT IN THE HOOGHLY. (4) THE QUANTUM OF PLANKTON, WHICH SERVES AS A FISH FOOD, WAS LESS IN THE HOOGHLY THAN THE MATLAH. (5) SOME OF THE PHYSICOCHEMICAL CONDITIONS, MAINLY TEMPERATURE, TURBIDITY, PH, NITRITE NITROGEN, FREE AMMONIA, DO, AND CC, NEAR A FEW DISCHARGE OUTLETS IN THE HOOGHLY ESTUARY INDICATED CONDITIONS UNFAVORABLE FOR FISH AND FISHERIES OR AT LEAST TENDING TO MAKE THE ADJOINING AREA UNINHABITABLE. (MORTLAND-BATTELLE)

FIELD 05C, 02L

ACCESSION NO. W71-11876

HYPNODINIUM-LIKE ALGAL BLOOMS IN GEORGIA LAKES,

GEORGIA WATER QUALITY CONTROL BOARD, ATLANTA; AND GEORGIA STATE UNIV., ATLANTA. DEPT. OF BIOLOGY.

EDWARD T. HALL, JR., DONALD G. AHEARN, AND DONALD J. REINHARDT.

BIOSCIENCE, VOL 21, NO 3, P 115, FEBRUARY 1, 1971. 1 FIG, 1 REF.

DESCRIPTORS:
*EUTROPHICATION, *ALGAE, *SEWAGE EFFLUENTS, GEORGIA, THERMOCLINE, HYDROGEN ION CONCENTRATION, WATER TEMPERATURE, DISSOLVED OXYGEN, COLIFORMS, RECREATION, ODOR.

IDENTIFIERS:
*LAKE SIDNEY LANIER(GEORGIA), LAKE JACKSON(GEORGIA), LAKE CARDINAL(GEORGIA), LAKE BUCKHORN(GEORGIA), HYPNODINIUM SPHAERICUM.

ABSTRACT:
AN ALGAL BLOOM IS DESCRIBED SIMILAR TO HYPNODINIUM SPHAERICUM THAT HAS APPEARED EACH JUNE SINCE 1968 IN LAKE SIDNEY LANIER, GEORGIA. THIS BLOOM OF VARYING INTENSITY LASTS 7-10 DAYS. IN 1969, THE SURFACE OF THE LAKE TURNED MILKY WHITE. IN 1970 THE BLOOM OCCURED SHORTLY AFTER THE WATER STRATIFIED WITH A THERMOCLINE NEAR 9 M. THE SURFACE WATER TEMPERATURE WAS 28 C, THE PH 9.1, AND THE DISSOLVED OXYGEN 8.2 MG/LITER. AT THE THERMOCLINE, THE WATER TEMPERATURE WAS ABOUT 17 C, THE PH 8.3, AND THE DISSOLVED OXYGEN 3.4 MG/LITER. FOUR DAYS PRIOR TO THE RECORDING OF THE BLOOM, THE FECAL COLIFORM DENSITY INCREASED FROM LESS THAN 4/100 ML TO GREATER THAN 240/100 ML. FIFTEEN DAYS LATER THE COLIFORM DENSITY WAS LESS THAN 3/100 ML. THE ALGA HAS NOT BEEN DIRECTLY ASSOCIATED WITH ANY FISH KILLS. A DISAGREEABLE ODOR ACCOMPANIES THE BLOOM AND THE THICK SURFACE FLOCS INTERFERE WITH RECREATIONAL ACTIVITIES. THE SIGNIFICANCE AND LONG-TERM EFFECTS OF THIS ALGA ARE UNKNOWN. (MCCANN-BATTELLE)

FIELD 05A, 05B, 05C

ACCESSION NO. W71-11914

DETERMINATION OF ATP IN CHLORELLA WITH THE LUCIFERIN-LUCIFERASE ENZYME SYSTEM,

UNITED STATES DEPARTMENT OF AGRICULTURE, BELTSVILLE, MD. CROPS RESEARCH DIV.

J. B. ST. JOHN.

ANALYTICAL BIOCHEMISTRY, VOL 37, P 409-416, OCTOBER 1970. 1 FIG, 4 TAB, 18 REF.

DESCRIPTORS:
*CHLORELLA, *BIOASSAYS, ENZYMES, LIGHT INTENSITY, ALGAE.

IDENTIFIERS:
*ADENOSINE TRIPHOSPHATE, *LUCIFERIN-LUCIFERASE ENZYME SYSTEM, PLANT EXTRACTS, CHLORELLA SOROKINIANA.

ABSTRACT:
A METHOD FOR MEASURING ATP BASED ON THE FIREFLY LUCIFERIN-LUCIFERASE REACTION IS DESCRIBED. THE METHOD INVOLVES A 30 SECOND INTEGRATION OF THE LIGHT PRODUCED FOLLOWING MIXING OF THE SAMPLE AND ENZYME. THE RELATIONSHIP BETWEEN LIGHT MEASURED AND QUANTITY OF ATP PRESENT IS LINEAR OVER AT LEAST A THOUSANDFOLD CONCENTRATION RANGE, WITH LESS THAN A 2 PERCENT RELATIVE STANDARD DEVIATION. THE METHOD OF LIGHT MEASUREMENT INCLUDES THE PEAK INTENSITY OF INITIAL FLASH AND A PORTION OF THE SIGNAL WHICH FOLLOWS. THEREFORE, THE PEAK INTENSITY OF THE INITIAL FLASH ALONE IS NOT NECESSARILY THE MOST RELIABLE BASIS FOR MEASUREMENT OF ATP. THE METHOD USES A COMMERCIALLY AVAILABLE LUCIFERIN-LUCIFERASE PREPARATION WITHOUT FURTHER PURIFICATION. THE LEVEL OF SENSITIVITY IS EQUAL TO THAT REPORTED FOR PURIFICATION. THE LEVEL OF SENSITIVITY IS EQUAL TO THAT REPORTED FOR PURIFIED LUCIFERIN-LUCIFERASE PREPARATIONS. THE METHOD OF CONSTANT ADDITION FOR QUANTITATIVE DETERMINATION OF ATP IN PLANT EXTRACTS IS DESCRIBED. THE LACK OF SPECIFICITY ASSOCIATED WITH THE LUCIFERIN-LUCIFERASE REACTION IS CONTROLLED BY THIS METHOD. (MORTLAND-BATTELLE)

FIELD 05A

ACCESSION NO. W71-11916

EUTROPHICATION OF NORTHEASTERN OHIO LAKES: I INTRODUCTION, MORPHOMETRY, AND CERTAIN PHYSICO-CHEMICAL DATA OF DOLLAR LAKE,

KENT STATE UNIV., OHIO. INST. OF LIMNOLOGY.

G. D. COOKE, AND ROBERT L. KENNEDY.

THE OHIO JOURNAL OF SCIENCES, VOL. 70, NO. 3, P 150-161, MAY 1970. 7 FIG, 1 TAB, 14 REF.

DESCRIPTORS:
*EUTROPHICATION, *LAKES, *WATER POLLUTION EFFECTS, OHIO, NUTRIENTS, WATER POLLUTION SOURCES, DISSOLVED OXYGEN, LAKE MORPHOLOGY, AQUATIC ALGAE, MOSSES, WATER TEMPERATURE, WATER CHEMISTRY, CHLOROPHYLL, HYPSOMETRIC ANALYSIS, HYPOLIMNION, EPILIMNION, CYANOPHYTA, DINOFLOGELLATES, SEASONAL, ZOOPLANKTON, PHYTOPLANKTON.

IDENTIFIERS:
DOLLAR LAKE(OHIO), SPHAGNUM, APHANIZOMENON FLOS-AQUAE, CERATIUM, OSCILLATORIA, TRANSPARENCY.

ABSTRACT:
THREE NORTHEASTERN OHIO LAKES (EAST TWIN, WEST TWIN, AND DOLLAR) ARE BEING STUDIED TO COMPARE LAKES IN DIFFERENT STAGES OF EUTROPHICATION WITHIN THE SAME AREA AND TO EVOLVE LONG-TERM BEFORE-AND-AFTER COMPARISONS. THIS IS THE FIRST OF A SERIES OF ARTICLES TO BE PREPARED FROM THESE STUDIES. BASIC EUTROPHICATION DATA ON DOLLAR LAKE ARE PRESENTED. BECAUSE IT IS MOST STRONGLY INFLUENCED BY HUMAN HABITATION AND IS THE SMALLEST OF LAKES, IT WAS SELECTED AS THE MODEL LAKE FROM WHICH COMPARATIVE OBSERVATIONS CAN BE DRAWN. DOLLAR LAKE HAS A VOLUME OF 86,400 CUBIC METERS AND AN AREA OF 22,212 SQUARE METERS. THE AVERAGE DEPTH IS 3.89 METERS, THE MAXIMUM DEPTH IS 7.5 METERS. MAXIMUM WIDTH OF THE LAKE IS 140 METERS, THE MAXIMUM LENGTH 215 METERS. EXCEPT FOR BRIEF PERIODS IN SPRING AND FALL, DEEP WATERS ARE DEPLETED OF DISSOLVED OXYGEN. SECCHI DISC TRANSPARENCY IS FREQUENTLY BELOW ONE METER. MASSIVE BLOOMS OF ALGAE ARE OFTEN OBSERVED. THE LAKE STRATIFIES THERMALLY IN APRIL, CIRCULATES IN OCTOBER, AND RESTRATIFIES IN DECEMBER AFTER ICE FORMATION. SPRING CIRCULATION OCCURS IN MARCH. DOLLAR LAKE IS A DIMICTIC, SECOND-CLASS EUTROPHIC LAKE. (MORTLAND-BATTELLE)

FIELD 05C, 05B

ACCESSION NO. W71-11949

TRISODIUM NITRILOTRIACETATE AND ALGAE,

A. E. CHRISTIE.

WATER AND SEWAGE WORKS, VOL 117, NO 2, P 58-59, FEBRUARY 1970. 4 TAB, 6 REF.

DESCRIPTORS:
*ALGAE, *DETERGENTS, *TOXICITY, *ACTIVATED SLUDGE, *NUTRIENTS, NITROGEN, WATER POLLUTION EFFECTS, WATER POLLUTION SOURCES.

IDENTIFIERS:
*CHLORELLA PYRENOIDOSA, *TRISODIUM NITRILOTRIACETATE.

ABSTRACT:
RELATIONSHIPS BETWEEN TRISODIUM NITRILOTRIACETATE, A POTENTIAL DETERGENT BUILDER, AND THE ALGA CHLORELLA PYRENOIDOSA WITH RESPECT TO TOXICITY AND AS A NITROGEN SOURCE WERE EXAMINED. PRELIMINARY EXPERIMENTS ON THE ROLE OF NTANA3 AS A SOURCE OF NITROGEN FOR ALGAE WERE CARRIED OUT BY MEASURING GROWTH IN A MINERAL MEDIA. EQUIVALENT QUANTITIES OF NITROGEN WERE ADDED EITHER AS NTANA3 OR SODIUM NITRATE, BOTH WITH AND WITHOUT THE ADDITION OF A SMALL QUANTITY OF RAW SEWAGE. PRELIMINARY TESTING OF NTANA3 TOXICITY TO ALGAE WAS ACCOMPLISHED BY ADDING THE COMPOUND TO WEEK-OLD CULTURES GROWN IN A COMPLETE MINERAL MEDIUM. EXAMINATION OF NTANA3 AS A NITROGEN SOURCE FOR ALGAE WAS ALSO INVESTIGATED BY CULTURING CHLORELLA IN THE FILTERED SUPERNATANTS OF ALIQUOTS OF ACTIVATED SLUDGE WHICH HAD RECEIVED VARIOUS CONCENTRATIONS OF NTANA3. THE RESULTS INDICATE THAT NTANA3 AT CONCENTRATIONS UP TO 275 MG/L IS NOT TOXIC TO THE TEST ORGANISM AND CAN ACT AS A SOURCE OF NITROGEN FOR ALGAL GROWTH. NO SIGNIFICANT DIFFERENCES WERE FOUND IN ALGAL RESPONSES IN THE SUPERNATANTS OF THE ACTIVATED SLUDGE EXCEPT IN THOSE RELATED TO THE HIGHEST CONCENTRATIONS OF THE NTANA3. THE HIGHEST CONCENTRATION DID NOT REDUCE THE ALGAL GROWTH POTENTIAL OF THE CHEMICAL. SINCE CARBON IS GENERALLY THE NUTRIENT LIMITING BACTERIAL ACTION IN SEWAGE TREATMENT PLANTS, THE FEASIBILITY OF BIODEGRADABLE ORGANIC MATERIALS AS DETERGENT BUILDERS WHICH CONTAIN NEITHER NITROGEN NOR PHOSPHORUS SHOULD BE THOROUGHLY CONSIDERED. (LITTLE-BATTELLE)

FIELD 05C, 05B

ACCESSION NO. W71-11972

COLOR INFRARED PHOTOGRAPHY FROM 60,000 FEET ALTITUDE AS A TOOL FOR DELINEATING AQUATIC VEGETATION,

NATIONAL AERONAUTICS AND SPACE ADMINISTRATION, HOUSTON, TEX. MANNED SPACECRAFT CENTER.

CURTIS C. MASON.

IN: HYDROBIOLOGY, 'BIORESOURCES OF SHALLOW WATER ENVIRONMENTS', PROCEEDINGS OF SYMPOSIUM, MIAMI BEACH, FLORIDA, JUNE 24-27, 1970; AMERICAN WATER RESOURCES ASSOCIATION, URBANA, ILLINOIS, PROCEEDINGS SERIES NO 8, P 317-328, 1970. 12 P, 6 FIG.

DESCRIPTORS:
*REMOTE SENSING, *AERIAL PHOTOGRAPHY, *INFRARED RADIATION, *AQUATIC ALGAE, *SURFACE WATERS, ANALYTICAL TECHNIQUES, PHOTOMETRY, WATER POLLUTION.

IDENTIFIERS:
*INFRARED PHOTOGRAPHY, *AQUATIC VEGETATION.

ABSTRACT:
COLOR INFRARED (IR) PHOTOGRAPHY WITH ITS ABILITY TO EXTEND THE RANGE OF PHOTOGRAPHIC DETECTABILITY INTO THE NEAR IR WAVELENGTH IS AN EXCELLENT TOOL FOR MAPPING AQUATIC VEGETATION. THIS VEGETATION APPEARS IN A BROADER RANGE OF COLOR THAN ON CONVENTIONAL COLOR PHOTOGRAPHY. WATER APPEARS AS A VERY DISTINCT BLUE OR BLACK SHARPLY OUTLINING THE AQUATIC VEGETATION. USING HIGH ALTITUDE AIRCRAFT PERMITS COVERAGE OF LARGE AREAS ON A SINGLE MISSION. (WOODARD-USGS)

FIELD 05B, 07B

ACCESSION NO. W71-12034

THE EFFECT OF CARBON ON ALGAL GROWTH--ITS RELATIONSHIP TO EUTROPHICATION,

UTAH STATE UNIV., LOGAN. UTAH WATER RESEARCH LAB.

JOEL C. GOLDMAN, DONALD B. PORCELLA, E. JOE MIDDLEBROOKS, AND DANIE F. TOERIEN.

COLLEGE OF ENGINEERING, OCCASIONAL PAPER 6, APRIL 1971. 56 P, 7 FIG, 7 TAB, 242 REF.

DESCRIPTORS:
*CARBON, *ALGAE, *EUTROPHICATION, *LIMITING FACTORS, PHOTOSYNTHESIS, CARBON CYCLE, ABSORPTION, DECOMPOSING ORGANIC MATTER, ALKALINITY, HYDROGEN ION CONCENTRATION, NUTRIENTS, AQUATIC ENVIRONMENT, CARBON DIOXIDE.

IDENTIFIERS:
CHEMISTRY OF INORGANIC CARBON, CARBON TRANSFORMATION, CARBON SOURCES, ORGANIC CARBON, CARBON UTILIZATION.

ABSTRACT:
AVAILABILITY OF INORGANIC AND ORGANIC CARBON FOR PHOTOSYNTHESIS OF ALGAL CELLS DEPENDS UPON IONIC COMPOSITION OF THE MEDIUM AND THE POPULATION OF THE AQUATIC SYSTEM. IN SEWAGE LAGOONS, EUTROPHIC LAKES, OR LABORATORY CULTURES CONTAINING AN EXCESS OF NUTRIENTS, CARBON CAN BE THE FACTOR LIMITING THE GROWTH OF ALGAL MASS. HOWEVER, IN MOST LAKES INORGANIC SOURCES, ATMOSPHERE, AND BACTERIAL DEGRADATION PROVIDE AN ADEQUATE SUPPLY OF CARBON FOR THE AMOUNT OF AVAILABLE LIGHT, NITROGEN, AND PHOSPHORUS. IN TERMS OF PRACTICAL TECHNOLOGY, THE CONTROL OF ALGAL GROWTH IS USUALLY RELATED TO THE CONTENT OF PHOSPHORUS. (WILDE-WISCONSIN)

FIELD 05C, 10, 02H

ACCESSION NO. W71-12068

THE RELATION OF ACETYLENE REDUCTION TO HETEROCYST FREQUENCY IN BLUE-GREEN ALGAE,

WESTFIELD COLL., LONDON (ENGLAND). DEPT. OF BOTANY.

WILLIAM J. JEWELL, AND S. A. KULASOORIYA.

JOURNAL OF EXPERIMENTAL BOTANY, VOL 21, NO 69, P 874-880, 1970. 1 FIG, 3 TAB, 13 REF.

DESCRIPTORS:
*CYANOPHYTA, *ALGAE, *NITROGEN FIXATION, ENZYMES, BACTERIA, SPORES, NITROGEN, METABOLISM.

IDENTIFIERS:
*ACETYLENE REDUCTION, *HETEROCYSTS, ANABAENA CYLINDRICA, ANABAENOPSIS, ACETYLENE TEST, CYLINDROSPERMUM LICHENEFORME, MASTIGOCLADUS LAMINOSUS.

ABSTRACT:
IF NITROGEN-FIXING ENZYME SYSTEM IS LOCATED EXCLUSIVELY IN THE HETEROCYSTS, AND, IF ACETYLENE-REDUCTION TECHNIQUE GIVES A QUANTITATIVE MEASURE OF THIS ACTIVITY, IT MAY BE POSSIBLE TO ESTIMATE NITROGEN FIXATION FROM THE FREQUENCY OF HETEROCYSTS AND THE ACETYLENE REDUCTION CAPACITY OF HETEROCYSTS. THIS STUDY ATTEMPTED TO DETERMINE VARIATION, WITH CULTURE AGE, OF ACETYLENE-REDUCTION ACTIVITY AND HETEROCYST FREQUENCY IN FIVE SPECIES OF BLUE-GREEN ALGAE USING BACTERIA-FREE CULTURES OF ANABAENA CYLINDRICA (NONSPORING), ANABAENA CYLINDRICA (SPORING), ANABAENOPSIS, CYLINDROSPERMUM LICHENIFORME, AND MASTIGOCLADUS LAMINOSUS. VARIATION AND AVERAGE HETEROCYST FREQUENCY ARE SUMMARIZED EXCEPT FOR CYLINDROSPERMUM LICHENIFORME WHICH GREW WITH TIGHTLY INTERTWINED FILAMENTS SO THAT IT WAS IMPOSSIBLE TO COUNT A LARGE NUMBER OF SAMPLES. THE HETEROCYST FREQUENCY VARIED BETWEEN 2.5% TO 12.3% OF TOTAL CELLS; AND THE ACETYLENE REDUCTION ACTIVITY VARIED 2.6-FOLD (AVERAGE). CONSISTENCY OF THE VALUES OF THE COEFFICIENT OF ACETYLENE REDUCTION INDIRECTLY SUPPORTS THE INITIAL ASSUMPTION THAT SITE OF NITROGEN FIXATION IS THE HETEROCYST. ESTIMATES OF NITROGEN FIXATION RATES BY BLUE-GREEN ALGAE MAY BE MADE BY IN SITU HETEROCYST COUNT AND ACETYLENE REDUCTION MEASUREMENTS. (JONES-WISCONSIN)

FIELD 05C

ACCESSION NO. W71-12070

TOLERANCE OF BLUE-GREEN ALGAE TO PESTICIDES,

INDIAN AGRICULTURAL RESEARCH INST., NEW DELHI. DIV. OF MICROBIOLOGY.

G. S. VENKATARAMAN, AND B. RAJYALAKSHMI.

CURRENT SCIENCE, VOL 40, NO 6, P 143-144, 1971. 1 TAB, 8 REF.

DESCRIPTORS:
*CYANOPHYTA, *PESTICIDES, *FERTILIZATION, NUTRIENTS, ALGAE, RICE, NITROGEN, NITROGEN FIXATION, 2-4-D, DALAPON, HERBICIDES, CROP PRODUCTION.

IDENTIFIERS:
*TOLERANCE LEVELS, TOLYPOTHRIX TENUIS, AULOSIRA FERTILISSIMA, ANACYSTIS NIDULANS, CERASON, DIATHENE, PROPAZINE, COTORON, DIURON, LINURON.

ABSTRACT:
INFLUENCE OF ALGALIZATION ON HIGH-YIELDING RICE VARIETIES UNDER HIGH LEVELS OF FERTILIZATION HAS PROMPTED A STUDY ON TOLERANCE OF NITROGEN-FIXING BLUE-GREEN ALGAE TO SOME PESTICIDES. TWENTY-EIGHT STRAINS OF BLUE-GREEN ALGAE BELONGING TO FIVE DIFFERENT GENERA WERE TESTED AGAINST VARYING CONCENTRATIONS OF EIGHT PESTICIDES. THE DIFFERENCE IN GROWTH OF TREATED AND UNTREATED CULTURES WAS USED TO ASSESS THE EFFECT OF PESTICIDES. A LARGE PERCENTAGE OF THE STRAINS TESTED SHOWED A HIGH TOLERANCE LIMIT TO ALL PESTICDES APPLIED MANY TIMES MORE THAN THE RECOMMENDED DOSES OF APPLICATION. TOLYPOTHRIX TENUIS, A POWERFUL NITROGEN-FIXER, WIDELY USED IN INOCULATION TRIALS, GREW IN HIGH CONCENTRATIONS OF ALL PESTICIDES, EXCEPT DIURON. THE RELATED AULOSIRA FERTILISSIMA, WHILE TOLERATING HIGH LEVELS OF CERASAN, DIATHENE, 2,4-D AND DALAPON, WAS SENSITIVE TO THE SUBSTITUTED UREA COMPOUNDS LIKE COTORON, DIURON, AND LINURON AS WELL AS TO THE TRIAZINE HERBICIDE, PROPAZINE. THE NON-NITROGEN-FIXING ANACYSTIS NIDULANS TOLERATED HIGH CONCENTRATIONS OF ALL THE PESTICIDES EXAMINED. THE PRESENT INVESTIGATION INDICATES THAT THESE ALGAE CAN BE USED IN PRESENCE OF PESTICIDES AND ARE ABLE TO FUNCTION IN PRESENCE OF HIGH LEVELS OF NITROGENOUS FERTILIZERS. (JONES-WISCONSIN)

FIELD 05C

ACCESSION NO. W71-12071

ROLE OF PHOSPHORUS IN EUTROPHICATION,

WISCONSIN UNIV., MADISON. WATER CHEMISTRY PROGRAM.

G. FRED LEE.

PRESENTED AT THE SYMPOSIUM OF AMERICAN CHEMICAL SOCIETY, DIVISION OF WATER, AIR AND WASTE CHEMISTRY, LOS ANGELES, CALIFORNIA, APRIL 1970.

DESCRIPTORS:
*PHOSPHORUS, *EUTROPHICATION, FERTILIZATION, AQUATIC PLANTS, MATHEMATICAL MODELS, SELF-PURIFICATION, ALGAE, GREAT LAKES, NUTRIENTS, SEDIMENTS, LAKE MICHIGAN, WATER POLLUTION SOURCES, WATER POLLUTION CONTROL, TROPHIC LEVELS, PHOSPHATES, DETERGENTS, BIOASSAY, ANALYTICAL TECHNIQUES, FORECASTING, SILICA.

IDENTIFIERS:
FLUSHING, GREEN BAY(WIS), NITRILOTRIACETIC ACID, ALGAL ASSAY PROCEDURES, ALGAL GROWTH, IN-LAKE NUTRIENT CONTROL.

ABSTRACT:
SIGNIFICANCE OF PHOSPHORUS AS THE KEY ELEMENT IN EXCESSIVE FERTILIZATION OF NATURAL WATERS IS PRESENTED AND ITS ROLE ON PLANT GROWTH IN LAKES. TOOLS TO ASSESS PHOSPHORUS AND OTHER ELEMENTS FERTILIZING NATURAL WATERS ARE MATHEMATICAL MODELS, ENZYMATIC AND TISSUE ASSAY PROCEDURES. APPRAISAL OF NUTRIENT STATUS OF LAKES CAN BE MADE DURING FEBRUARY AND MARCH OF EACH YEAR IN TEMPERATE LAKES TO SHOW THE POTENTIAL PROBLEMS AND STEPS INITIATED TO CONTROL EXCESSIVE DISCHARGE OF NUTRIENTS. SEDIMENTS SERVE AS A SINK FOR PHOSPHORUS WITH THE NET FLUX OF PHOSPHORUS FROM LAKE WATER TO SEDIMENTS. CONTEMPLATED NUTRIENT REMOVAL PROJECTS SHOULD BE ASSOCIATED WITH SOME LABORATORY LEACHING TESTS ON LAKE SEDIMENTS AND RESULTS COMPARED WITH LAKE RECOVERY RATE UPON NUTRIENT REDUCTION. DEVELOPMENT OF MODELS FOR AQUEOUS ENVIRONMENTAL CHEMISTRY OF AQUATIC NUTRIENTS IN NATURAL WATERS IS ESSENTIAL TO IMPROVE THE PREDICTABILITY OF RELATIONSHIPS BETWEEN THE FLUX OF AQUATIC PLANT NUTRIENTS AND GROWTH OF ALGAE AND OTHER AQUATIC PLANTS. IN REPLACING PHOSPHATES IN DETERGENTS, THE PRIMARY PROBLEM IS THE PROCEDURE USED TO EVALUATE THE REPLACEMENT. (JONES-WISCONSIN)

FIELD 05C

ACCESSION NO. W71-12072

EUTROPHICATION AND THE CURRENT CONTROVERSY OVER ITS CAUSES AND CURES,

FISHERIES RESEARCH BOARD OF CANADA, WINNIPEG (MANITOBA). FRESHWATER INST.

A. L. HAMILTON.

PRESENTED AT 22ND ANNUAL CONVENTION, WESTERN CANADA WATER AND SEWAGE CONFERENCE, SEPTEMBER 23-25, 1970, HELD AT WINNIPEG, P 67-71, 1970. 1 FIG, 11 REF.

DESCRIPTORS:
*WATER POLLUTION CONTROL, *EUTROPHICATION, *WATER POLLUTION SOURCES, POPULATION, INDUSTRIES, WASTE DISPOSAL, MONITORING, NUTRIENTS, ALGAE, INSECTICIDES, HERBICIDES, HEAVY METALS, TASTE, ODOR, RECREATION, PHOSPHORUS, CARBON, LAKE ERIE, DETERGENTS, LAKE ONTARIO, TERTIARY TREATMENT, INTERNATIONAL JOINT COMMISSION.

IDENTIFIERS:
MERCURY, PHOSPHORUS REMOVAL, LAKE WASHINGTON(WASH).

ABSTRACT:
ACUTENESS OF THE WASTE DISPOSAL PROBLEM IS DUE TO BOTH POPULATION AND INDUSTRIAL GROWTH. WASTES EMPTIED INTO AQUATIC ENVIRONMENTS BELONG TO TWO BROAD GROUPS: NUTRIENTS WHICH STIMULATE LIFE AND TOXIC SUBSTANCES DEPRESSING LIFE. THE FIRST POLLUTANTS, GROWTH STIMULATORS, CONTRIBUTE TO EUTROPHICATION, THE MAJOR SYMPTOMS OF WHICH ARE ACCUMULATION OF ALGAE, REDUCED TRANSPARENCY, UNPLEASANT TASTE AND ODOR, OXYGEN DEFICITS, AND REDUCTION IN RECREATION POTENTIAL. MAJORITY OF LIMNOLOGISTS SUPPORT THE ARGUMENT THAT PHOSPHORUS IS THE KEY LIMITING NUTRIENT AND ITS INPUT TO LAKES CAN BE CONTROLLED MORE EFFECTIVELY THAN THAT OF OTHER NUTRIENTS. EFFICIENT AND RELATIVELY INEXPENSIVE METHODS CAN REMOVE PHOSPHORUS DURING SEWAGE TREATMENT, AND ELIMINATION OF PHOSPHATES FROM DETERGENTS WOULD LOWER COSTS OF REDUCING PHOSPHORUS LEVELS IN EFFLUENTS. THE SINGLE MOST CONVINCING PROGRAM OF PRACTICAL BENEFITS IN REDUCING PHOSPHORUS IS EXEMPLIFIED BY THE DIVERSION OF SEWAGE AROUND LAKE WASHINGTON, PROVIDING STRONG EVIDENCE THAT PHOSPHATES WERE THE LIMITING NUTRIENT AND A CONCLUSIVE DEMONSTRATION THAT EUTROPHICATION WAS, AT LEAST IN THIS INSTANCE, A REVERSIBLE PROCESS. VARIOUS ARGUMENTS FAVORING REDUCTION OF PHOSPHORUS INPUTS MAY NOT BE ENTIRELY CONCLUSIVE BUT COMBINED THEY CONSTITUTE A VERY STRONG CASE FOR TAKING MEANINGFUL REMEDIAL ACTIONS - NOW. (JONES-WISCONSIN)

FIELD 05C

ACCESSION NO. W71-12091

EFFECTS OF THERMAL DISCHARGES ON THE CHEMICAL PARAMETERS OF WATER QUALITY AND EUTROPHICATION,

WISCONSIN UNIV., MADISON. WATER CHEMISTRY PROGRAM.

FRED G. LEE, AND GILMAN D. VEITH.

PRE-PRINT: PROCEEDINGS INTERNATIONAL SYMPOSIUM ON THE IDENTIFICATION AND MEASUREMENT OF POLLUTANTS IN THE ENVIRONMENT, HELD AT OTTAWA, ONTARIO, CANADA, JUNE 1971. 27 P, 23 REF.

DESCRIPTORS:
*THERMAL POLLUTION, *CHEMICAL REACTIONS, *WATER QUALITY, *CHEMICAL PROPERTIES, EUTROPHICATION, WATER TEMPERATURE, WATER POLLUTION EFFECTS, WATER POLLUTION CONTROL, ENERGY, DIFFUSION, THERMODYNAMICS, SOLUBILITY, SORPTION, PESTICIDES, TOXICITY, DISSOLVED OXYGEN, LAKE MICHIGAN, ALGAE.

IDENTIFIERS:
*THERMAL DISCHARGES.

ABSTRACT:
DEPENDING ON THE HEAT ASSIMILATIVE CAPACITY OF THE RECEIVING WATER, THERMAL DISCHARGES ARE DIVIDED INTO TWO TYPES: THOSE ELEVATING WATER TEMPERATURE BY 5 TO 10C FOR ONLY A FEW HOURS, AND THOSE INCREASING THE TEMPERATURE FOR SEVERAL DAYS. THE LATTER DISCHARGES MAY IN CERTAIN INSTANCES LEAD TO DETERIORATION OF WATER QUALITY--INTENSIFY OBJECTIONABLE TASTE AND ODOR, INCREASE ALGAL BLOOMS, AND REDUCE DO CONCENTRATION. THE ADVERSE EFFECTS OF THERMAL DISCHARGES CAN PARTICULARLY BE EXPECTED WHEN THE TEMPERATURE IS INCREASED TO ABOUT 15 TO 20C IN NUTRIENT-ENRICHED WATERS. (WILDE-WISCONSIN)

FIELD 05C

ACCESSION NO. W71-12092

FACTORS INFLUENCING PHOSPHOROUS REMOVAL BY BIOLOGICAL TREATMENT,

FEDERAL WATER QUALITY ADMINISTRATION, CINCINNATI, OHIO.

ROBERT L. BUNCH.

CHEMICAL ENGINEERING PROGRESS, SYMPOSIUM SERIES, VOL 67, NO 107, P 90-94, 1971. 3 TAB, 33 REF.

DESCRIPTORS:
*BIOLOGICAL TREATMENT, *PHOSPHOROUS, *WASTE WATER TREATMENT, NUTRIENTS, SYNTHESIS, SLUDGE, EFFLUENT, ALGAE, SECONDARY TREATMENT, BIOCHEMICAL OXYGEN DEMAND, CHEMICAL OXYGEN DEMAND, CARBON, FERTILIZATION, DISSOLVED OXYGEN, SEWAGE, HYDROGEN ION CONCENTRATION.

IDENTIFIERS:
SUSPENDED SOLIDS.

ABSTRACT:
IN EWECENT YEARS THE PHOSPHATE CONTENT OF WASTE WATER TREATMENT PLANT EFFLUENTS HAS BEEN RISING, PRESUMABLY TRACEABLE TC THE USE OF PHOSPHATES IN DETERGENTS. CONVENTIONAL BIOLOGICAL TREATMENT DOES NOT EFFECTIVELY REMOVE NUTRIENTS FROM WASTE WATER. THE INFLUENCE OF AN INCREASED USE OF PHOSPHATES, DIGESTER RECYCLE, AND IMBALANCE OF NUTRIENTS IN WASTE WATER WAS RECENTLY STUDIED. PHOSPHOROUS REMOVAL BY SYNTHESIS CAN BE HIGHER THAN NORMALLY ENCOUNTERED IF DIGESTER SUPERNATANT IS NOT RECYCLED AND PRIMARY CLARIFICATION IS ELIMINATED. THE ELIMINATION OF THESE 2 PRACTICES WILL INCREASE THE CARBON TO PHOSPHOROUS RATIO, CORRECTING THE NUTRIENT IMBALANCE. IF PHOSPHOROUS IS TO BE REMOVED FROM WASTE WATER ON A SUSTAINED BASIS, THEN CHEMICAL OR CHEMICAL-BIOLOGICAL METHODS MUST BE USED AND EFFORTS MUST BE CONTINUALLY MADE TO IMPROVE THE DEPENDABILITY AND APPLICABILITY OF THE BIOLOGICAL PROCESS. (GUTIERREZ-TEXAS)

FIELD 05D

ACCESSION NO. W71-12188

PHOSPHATE REMOVAL BY FOAM FRACTIONATION,

GARRETT RESEARCH AND DEVELOPMENT CO., INC., LA VERNE, CALIF.

DONALD E. GARRETT.

CHEMICAL ENGINEERING PROGRESS, SYMPOSIUM SERIES, VOL 67, NO 107, P 296-303, 1971. 10 FIG, 4 TAB, 7 REF.

DESCRIPTORS:
*PHOSPHATES, *WASTE WATER TREATMENT, *FOAM FRACTIONATION, LIME, HYDROGEN ION CONCENTRATION, ECONOMICS, ALGAE, SOLUBILITY, SEWAGE, BIOCHEMICAL OXYGEN DEMAND, ION EXCHANGE, CHEMICAL OXYGEN DEMAND, FERTILIZER, TURBIDITY.

IDENTIFIERS:
*FRACTIONATION-FLOTATION PROCESS.

ABSTRACT:
A PROCESS HAS BEEN DEVELOPED BY WHICH AIR AND WATER ARE CONTACTED AT HIGH PRESSURES, ADDITIVES ARE MIXED WITH THE SOLUTION, AND VERY FINELY DIVIDED BUBBLES THAT ARE FORMED ON THE RELEASE OF PRESSURE ARE SEPARATED. THE PROCESS ALSO INVOLVES A COMBINATION OF A NEW VARIATION OF THE FOAM FRACTIONATION PROCESS COUPLED WITH A PRECIPITATION STEP TO ALLOW EFFECTIVE REMOVAL OF PHOSPHATES IN AN INEXPENSIVE AND SIMPLE MANNER. THE ADDITION OF IRON AND LIME RESULT IN A COMPLETE, INEXPENSIVE SEPARATION OF PHOSPHATE FROM THE WATER TO BE TREATED. THE FOAM FRACTIONATION-FLOTATION PROCESS IS MORE THAN COMPETITIVE WITH CONVENTIONAL TREATMENT SCHEMES. THE FOAMING PROCESS OFFERS THE ADVANTAGE OF REQUIRING LESS LAND THAN CONVENTIONAL TREATMENT, WHICH IS A SIGNIFICANT FACTOR IN LAND-SHORT AREAS IN HELPING ALLEVIATE DIFFICULT PHOSPHATE REMOVAL PROBLEMS. ITS GREATEST ADVANTAGES ARE ITS SIMPLICITY AND HIGH REMOVAL EFFICIENCY FOR ALL THE CONTAMINANTS, COD, TURBIDITY, AND PHOSPHATES. FLOWSHEETS AND ECONOMIC EVALUATIONS OF THE PROCESS WERE ALSO DISCUSSED IN THIS PAPER. (GUTIERREZ-TEXAS)

FIELD 05D

ACCESSION NO. W71-12200

TREATMENT OF WASTEWATER RESULTING FROM THE PRODUCTION OF POLYHYDRIC ORGANIC COMPOUNDS,

DOW CHEMICAL CO., FREEPORT, TEX.

M. A. ZEITOUN, W. F. MCLLHENNY, AND W. A. TABER.

CHEMICAL ENGINEERING PROGRESS, SYMPOSIUM SERIES, VOL 67, NO 107, P 495-503, 1971. 18 FIG, 2 TAB, 4 REF.

DESCRIPTORS:
*WASTE WATER TREATMENT, *BIOLOGICAL TREATMENT, *ACTIVATED SLUDGE, EFFLUENT, *ORGANIC WASTES, ECONOMICS, CHEMICAL OXYGEN DEMAND, BIOCHEMICAL OXYGEN DEMAND, AERATION, *OXIDATION, HYDROGEN ION CONCENTRATION, MICROORGANISMS, FERMENTATION, ALKALINE HYDROLYSIS, ALGAE BLOOMS, UTAH, GREAT SALT LAKE, ORGANIC COMPOUNDS, BRINES.

IDENTIFIERS:
FUSARIUM NO 83, BACTERIUM NO 52, ENDOGENOUS RESPIRATION, TOTAL OXYGEN DEMAND, GLYCOLS.

ABSTRACT:
UNDER A GRANT FROM THE FEDERAL WATER QUALITY ADMINISTRATION A DEVELOPMENT PROGRAM WAS UNDERTAKEN TO INVESTIGATE THE TREATMENT OF WASTE WATERS RESULTING FROM THE PRODUCTION OF POLYHYDRIC ORGANIC COMPOUNDS. THE DOW CHEMICAL COMPANY AS THE RECIPIENT OF THE GRANT HAS SUBCONTRACTED WITH TEXAS A AND M UNIVERSITY FOR SUPPORT OF INVESTIGATIONS ON BIOLOGICAL OXIDATION OF ORGANIC COMPOUNDS. FOUR OVERALL OBJECTIVES WERE SET UP: (1) TO DEVELOP A METHOD OF TREATMENT OF WASTES FROM A POLYHYDRIC MANUFACTURING PROCESSES; (2) EXAMINE SEVERAL ALTERNATIVE TREATMENT METHODS; (3) DETERMINE TECHNICAL, ENGINEERING AND ECONOMIC FEASIBILITY, AND (4) DEVELOP SUFFICIENT INFORMATION FOR THE CONCEPTUAL DESIGN OF A WASTE TREATMENT FACILITY. A BIOLOGICAL OXIDATION PROCEDURE USING AN ACTIVATED SLUDGE TECHNIQUE HAS BEEN ABLE TO REDUCE ORGANIC CONTENTS OF SODIUM CHLORIDE BRINES TO ACCEPTABLE LEVELS. A PILOT FACILITY OPERATING ON FULL STRENGTH WASTES HAS BEEN DESIGNED AND IS BEING USED TO OBTAIN DESIGN INFORMATION FOR A LARGE SCALE WASTE CONTROL UNIT. THE PROGRAM INVOLVED IS PRESENTLY STILL UNDERWAY. (GUTIERREZ-TEXAS)

FIELD 05D

ACCESSION NO. W71-12223

PROPERTIES OF TILE DRAINAGE WATER,

IOWA STATE UNIV., AMES. DEPT. OF AGRICULTURAL ENGINEERING.

T. L. WILLRICH.

PAPER NO 70-752 PRESENTED AT 1970 WINTER MEETING OF THE AMERICAN SOCIETY OF AGRICULTURAL ENGINEERS, DECEMBER 8-11, 1970, CHICAGO, ILLINOIS. 10 P, 9 TAB. OWRR PROJECT A-013-IA(4).

DESCRIPTORS:
*TILE DRAINAGE, *WATER PROPERTIES, *FERTILIZATION, *CULTIVATED LANDS, *IOWA, WATER QUALITY, FERTILIZERS, NUTRIENTS, NITROGEN COMPOUNDS, PHOSPHORUS COMPOUNDS, POTASH, ALGAE, AQUATIC PLANTS, CHEMICAL ANALYSIS, WATER POLLUTION SOURCES.

ABSTRACT:
A STUDY OF PLANT NUTRIENT CONTENT AND POLLUTIONAL CHARACTERISTICS OF IOWA TILE DRAINAGE WATER WAS CONDUCTED FROM JULY 1965 THROUGH JUNE 1969. PLANT NUTRIENTS AND SOIL AMENDMENTS ARE LEACHED FROM THE SOIL BY PERCOLATING WATER AND ARE CONVEYED TO RECEIVING STREAMS THROUGH TILE DRAINS WHEN DRAINAGE FLOW OCCURS. SOME OF THESE NUTRIENTS HAVE THE POTENTIAL TO DEGRADE WATER QUALITY TO THE EXTENT THAT BENEFICIAL USES OF THE WATER MAY BE IMPAIRED. THE CONCENTRATIONS OF NITROGEN (MEDIAN VALUES RANGED FROM 12.4 TO 27.4 MG/LITER) AND PHOSPHORUS (MEDIAN VALUES OF 0.2 TO 0.3 MG/LITER) FOUND IN THE TILE DRAINAGE WATER SAMPLES WERE SUFFICIENTLY HIGH TO BE CONDUCTIVE TO THE GROWTH OF ALGAE AND OTHER AQUATIC PLANTS. (WOODARD-USGS)

FIELD 05B, 02G, 02K

ACCESSION NO. W71-12252

DDT UPTAKE AND GROWTH OF EUGLENA GRACILIS,

NATIONAL RESEARCH COUNCIL OF CANADA, OTTAWA (ONTARIO). DIV. OF BIOLOGY.

H. W. DE KONING, AND D. C. MORTIMER.

BULLETIN OF ENVIRONMENTAL CONTAMINATION AND TOXICOLOGY, VOL 6, NO 3, P 244-248, 1971. 2 FIG, 2 TAB, 3 REF.

DESCRIPTORS:
*CHLORINATED HYDROCARBON PESTICIDES, *DDT, *PESTICIDE TOXICITY, *WATER POLLUTION EFFECTS, PESTICIDES, AGRICULTURAL CHEMICALS, PESTICIDE RESIDUES, PESTICIDE KINETICS, ANIMAL PHYSIOLOGY, PLANT PHYSIOLOGY, PHYSIOLOGICAL ECOLOGY, BIOASSAY, MICROORGANISMS, *EUGLENA, ALGAE.

ABSTRACT:
THE EFFECT OF DDT ON THE GROWTH RATE OF EUGLENA GRACILIS AND THE RATE OF DDT UPTAKE BY ACTIVELY GROWING E. GRACILIS CULTURES WERE EXAMINED. DDT SUPPRESSED GROWTH WHEN ADDED IN 1.0 ML OF ETHANOL, BUT HAD NO EFFECT WHEN ADDED ALONE OR IN 0.1 ML OF ETHANOL. THE INHIBITED CULTURES RECOVERED AFTER 3 DAYS, PROBABLY AFTER UTILIZATION OF THE ETOH. DDT ACCUMULATION HAS NO OBSERVABLE EFFECT ON CELL DIVISION, AND NO INTRACELLULAR DEGRADATION PRODUCTS OF DDT WERE DETECTED. (LEGORE-WASHINGTON)

FIELD 05C

ACCESSION NO. W71-12303

FUNDAMENTAL VARIATIONS IN THE WATER QUALITY WITH PERCOLATION IN INFILTRATION BASINS,

INSTITUTE FOR WATER RESEARCH LTD., DORTMUND (WEST GERMANY).

W. H. FRANK.

PAPER NO 7 OF ARTIFICIAL GROUNDWATER RECHARGES CONFERENCE, UNIVERSITY OF READING, ENGLAND, SEPTEMBER 21-24, 1970: THE WATER RESEARCH ASSOCIATION, MARLOW, ENGLAND. 22 P, 25 FIG.

DESCRIPTORS:
*ARTIFICIAL RECHARGE, *INDUCED INFILTRATION, *PIT RECHARGE, *WATER TREATMENT, *FILTATION, ALLUVIA CHANNELS, SURFACE-GROUNDWATER RELATIONSHIPS, PERMEABILITY, ALGAE, DISSOLVED OXYGEN.

IDENTIFIERS:
*RUHR VALLEY.

ABSTRACT:
THE DORTMUNDER STADTWERKE AG SUPPLIES ABOUT 100 MILLION CU M OF MUNICIAL AND INDUSTRIAL WATER PER YEAR FROM THE GROUNDWATER COLLECTED IN THE ALLUVIUM OF THE RUHR VALLEY. THE AQUIFER IS 4 TO 5 M DEEP AND IS COVERED BY A LAYER OF MEADOW LOAM 0.5 TO 2 M THICK. THE PERMEABILITY OF THE GRAVEL IS IN GENERAL VERY GOOD, ABOUT .001 TO 0.01 M/SEC. MORE THAN 90% OF THE GROUNDWATER CONSISTS OF VARYING PROPORTIONS OF BANK INFILTRATION AND ARTIFICIALLY RECHARGED GROUNDWATER. IN DRY YEARS WITH LITTLE WATER FLOW, MORE THAN 75% OF THE REQUIRED WATER MUST BE ARTIFICIALLY RECHARGED. THE INFILTRATION TAKES PLACE IN BASINS MEASURING 25 X 200 M. A LAYER OF SAND 50 TO 70 CM THICK IS LAID DOWN AS A FILTER BED. THIS SAND HAS AN EFFECTIVE PARTICLE DIAMETER (D 10%) OF 0.12 MM, AND AT (D 60%) OF 0.3 MM A COEFFICIENT OF NON-UNIFORMITY OF U = 2.5. ITS PERMEABILITY IN MOST AREAS IS LESS THAN THAT OF THE SUBSOIL. BOTH THE BANK INFILTRATION AND THE ARTIFICIALLY RECHARGED GROUNDWATER ARE RE-COLLECTED AT A DEPTH OF 7 M AFTER A PASSAGE OF AT LEAST 50 M THROUGH THE SOIL AND PASSED DIRECTLY INTO THE MAINS WITHOUT FURTHER TREATMENT AFTER A LIGHT PRECAUTIONARY CHLORINATION. SUSPENDED ORGANIC AND INORGANIC MATTER IS LARGELY RETAINED ON THE FILTER SURFACE. TOGETHER WITH THE ALGAE WHICH GROW THERE, THEY CAUSE AN OBSTRUCTION IN THE FILTER. THE DISSOLVED ORGANIC MATERIAL IN THE WATER IS MINERALIZED WITH THE HELP OF THE BACTERIA WHICH INHABIT THE FILTER, OXYGEN BEING CONSUMED AND CARBON DIOXIDE LIBERATED. IN THE WATER ABOVE AN EXPOSED SAND FILTER, ALGAE CAUSE A DISTINCT DROP IN THE PHOSPHATE ION CONCENTRATION. IN THE UPPER MOST LAYERS OF SAND MORE PHOSPHATE IONS ARE LIBERATED BY MICROBIOL DECOMPOSITION THAN THE BACTERIA NEED FOR THEIR METABOLISM. (KNAPP-USGS)

FIELD 04B, 03A, 05F

ACCESSION NO. W71-12410

PHYSICAL ASPECTS OF ELECTROFILTRATION,

JADAVPUR UNIV., CALCUTTA (INDIA). DEPT. OF PHYSICAL CHEMISTRY.

S. P. MOULIK,

ENVIRONMENTAL SCIENCE AND TECHNOLOGY, VOL 5, NO 9, P 771-776, SEPTEMBER 1971.
7 FIG, 2 TAB, 19 REF.

DESCRIPTORS:
*WATER PURIFICATION, *FILTRATION, *ELECTROCHEMISTRY, *WATER POLLUTION TREATMENT, ANALYTICAL TECHNIQUES, SEPARATION TECHNIQUES, SEDIMENTS, OILY WATER, ALGAE.

IDENTIFIERS:
*ELECTROFILTRATION.

ABSTRACT:
THE PRIMARY PHYSICAL PHENOMENA INVOLVED IN ELECTROFILTRATION AND A THEORY FOR THE PROCESS ARE DISCUSSED. ELECTROPHORETIC FILTRATION WAS EFFECTIVE IN SUSPENSION CLARIFICATION, ESPECIALLY FOR WATER AND WASTE PURIFICATION. SYSTEMS SUCH AS BENTONITE CLAY SUSPENSIONS, SUSPENSIONS OF PURE CULTURE OF ALGAE, OIL-IN-WATER EMULSION, RIVER WATER, ETC. WERE CHOSEN AS REPRESENTATIVE MODELS FOR COMMONLY OCCURRING POLLUTED WATER SAMPLES. THE ELECTROFILTRATION PROCESS INCLUDES ELECTROPHORESIS AND FILTRATION. THE FILTER-CAKE BUILDUP ENCOUNTERED IN NORMAL FILTRATION CAN BE COMPLETELY AVOIDED IN THIS METHOD. CONSIDERATIONS OF FILTER MEDIUM ELECTRO-OSMOSIS, FILTER-CAKE ELECTRO-OSMOSIS, AND PARTICLE ELECTROPHORESIS HAVE MODIFIED THE FILTRATION EQUATION OF SPERRY TO PREDICT ELECTROFILTRATION RESULTS. FOR EFFICIENT ELECTROFILTRATION A CRITICAL VOLTAGE MUST BE ATTAINED. THIS VOLTAGE COMPLETELY PREVENTS THE FILTRATION HINDRANCE OF A DEPOSITED FILTER-CAKE. (WOODARD-USGS)

FIELD 05D, 03A

ACCESSION NO. W71-12467

SHORELINE ALGAE OF WESTERN LAKE ERIE,

RACHEL COX DOWNING.

OHIO J SCI. 70(5): 257-276. ILLUS. MAPS. 1970 (RECD. 1971).

DESCRIPTORS:
*ALGAE, *CHLOROPHYTA, *CYANOPHYTA, *LAKES, RHODOPHYTA, SHORES.

IDENTIFIERS:
ARNOLDIELLA-CONCHOPHILA, LAKE ERIE.

ABSTRACT:
THE ALGAE OF WESTERN LAKE ERIE HAVE BEEN EXTENSIVELY STUDIED FOR MORE THAN 70 YR, BUT, UNTIL THIS STUDY, CONDUCTED BETWEEN APRIL AND OCT., 1967; ALMOST NOTHING WAS KNOWN OF THE SHORELINE AS A SPECIFIC ALGAL HABITAT. A TOTAL OF 61 TAXA WERE IDENTIFIED FROM THE SHORELINES. THE (23 CHLOROPHYTA, 37 CYANOPHYTA, 1 RHODOPHYTA) FOUND 39 ARE NEW RECORDS FOR WESTERN LAKE ERIE, AND ONE, ARNOLDIELLA CONCHOPHILA MILLER, APPEARS TO BE A NEW USA RECORD, HAVING BEEN PREVIOUSLY REPORTED ONLY FROM CENTRAL RUSSIA.--COPYRIGHT 1971, BIOLOGICAL ABSTRACTS, INC.

FIELD 02H, 05C

ACCESSION NO. W71-12489

PRIMARY PRODUCTION AND STANDING CROPS OF EPIPSAMMIC AND EPIPELIC ALGAE,

BRISTOL UNIV. (ENGLAND). DEPT. OF BOTANY.

M. HICKMAN, AND F. E. ROUND.

BRITISH PHYCOLOGICAL JOURNAL, VOL 5 NO 2, P 247-255, 1970. 4 FIG, 6 TAB, 14 REF.

DESCRIPTORS:
*PRIMARY PRODUCTIVITY, *STANDING CROP, *ALGAE, DIATOMS, MEASUREMENT, DETRITUS, METHODOLOGY, SEDIMENTS, LAKES, CHLOROPHYLL, HYDROGEN ION CONCENTRATION, CARBON DIOXIDE.

IDENTIFIERS:
*EPIPSAMMIC ALGAE, *EPIPELIC ALGAE, SHEAR WATER(ENGLAND), AMPHORA OVALIS, OPEPHORA MARTYI, CELL COUNTS, NAVICULA, NITZSCHIA, STEPHANODISCUS, MICROCYSTIS, APHANIZOMENON, ANABENA SPIROIDES.

ABSTRACT:
A METHOD DEVISED FOR DETERMINING PRIMARY PRODUCTION INVOLVED AN ADDITIONAL STEP OF REMOVAL OF EPIPSAMMIC ALGAE FROM THE SAND GRAINS. THE EPIPELIC ALGAE WERE ALSO SEPARATED FROM THE SEDIMENT. SINCE THE EXPERIMENTS WERE CARRIED OUT IN THE LABORATORY AND NOT IN SITU, PRIMARY PRODUCTION IS DESIGNATED 'POTENTIAL' PRODUCTION. THE 'POTENTIAL' PRIMARY PRODUCTION OF THE EPIPELON AND EPIPSAMMON OF SHEAR WATER (ENGLAND) HAS BEEN ESTIMATED AND RELATED TO STANDING CROP PARAMETERS OVER A THIRTY MONTH PERIOD. ALL RESULTS ARE EXPRESSED PER UNIT AREA OF HABITAT TO BE COMPARABLE WITH RESULTS OF OTHER WORKERS. THE EPIPSAMMIC ALGAE COMPRISE MAINLY SMALL DIATOMS. THE EPIPSAMMON CAN BE SEPARATED FROM THE EPIPELON AND DETRITUS BY REPEATEDLY SWIRLING THE SEDIMENT WITH LAKE WATER FILTERED THROUGH WHATMAN GLASS FIBRE FILTER PAPER REMOVING PHYTOPLANKTONIC PRIMARY PRODUCERS AND DETRITAL MATERIAL. DATA HAVE ALSO BEEN OBTAINED FOR THE 'POTENTIAL PRODUCTION' OF THE FLORA IN SHORT CORES FROM THE SAME SITE. STANDING CROP AND PRODUCTION OF THE EPIPSAMMON WAS ALWAYS GREATER THAN THAT OF THE EPIPELON. CORRELATION BETWEEN STANDING CROP AND PRODUCTION IS CLOSER FOR THE EPIPELON THAN FOR THE EPIPSAMMON OWING TO RETENTION OF DEAD CELLS IN THE LATTER COMMUNITY. (JONES-WISCONSIN)

FIELD 05C

ACCESSION NO. W71-12855

LAKE EUTROPHICATION: A LABORATORY INVESTIGATION,

ONTARIO WATER RESOURCES COMMISSION, TORONTO. DIV. OF RESEARCH.

S. A. BLACK.

ONTARIO WATER RESOURCES COMMISSION, DIVISION OF RESEARCH PAPER NO 2026, 1970. 26 P, 3 FIG, 5 TAB.

DESCRIPTORS:
*LABORATORY TESTS, *EUTROPHICATION, *NUTRIENTS, *ALGAE, SEWAGE, PRODUCTIVITY, WATER POLLUTION EFFECTS, PHOSPHORUS, NITROGEN, AMMONIA, CARBON, CHLORELLA.

IDENTIFIERS:
*LIMITING NUTRIENTS, CANADA.

ABSTRACT:
WATER FROM AN OLIGOTROPHIC LAKE WAS SEEDED TO NATIVE ALGAE AND PLACED INTO FOUR 35-LITER PLEXIGLASS CONTAINERS. THE CULTURES WERE FED WITH A MIXTURE OF LAKE WATER, DISTILLED WATER, AND VARIOUSLY TREATED SEWAGE, AND SUBSEQUENTLY ENRICHED IN DISODIUM PHOSPHATE AND SODIUM CARBONATE HYDROGEN. ANALYSES OF THE MEDIA AND ALGAL COUNTS INDICATED THAT 0.015 MG/L OF P AND 0.5 MG/L OF N ARE REQUIRED TO PRODUCE GROWTH RESPONSE OF ALGAE IN THE LAKE INVESTIGATED. CARBON BELOW 8 MG/L INHIBITED THE ALGAL GROWTH. AMMONIA WAS USED BY CHLORELLA IN PREFERENCE TO NITRATES. REMOVAL OF PHOSPHORUS FROM SEWAGE WILL GREATLY REDUCE THE GROWTH EFFECT PRODUCED BY CONVENTIONALLY TREATED ACTIVATED SLUDGE. (WILDE-WISCONSIN)

FIELD 05C

ACCESSION NO. W71-12857

GROWTH OF CHLORELLA IN RELATION TO BORON SUPPLY,

ARGONNE NATIONAL LAB., ILL. DIV. OF BIOLOGICAL AND MEDICAL RESEARCH.

LANDY MCBRIDE, WILLIAM CHORNEY, AND JOHN SKOK.

BOTANICAL GAZETTE, VOL 132, NO 1, P 10-13, 1971. 4 TAB, 18 REF.

DESCRIPTORS:
*ALGAE, *CHLORELLA, *BORON, LABORATORY TESTS, CYTOLOGICAL STUDIES, PLANT GROWTH, SCENEDESMUS.

IDENTIFIERS:
NOSTOC, CYLINDROTHECA.

ABSTRACT:
THE GROWTH OF CHLORELLA VULGARIS, COLUMBIA STRAIN, EITHER IN LOW-BORON OR IN PURIFIED MEDIA WAS NOT INFLUENCED BY BORON CONCENTRATION WITHIN THE RANGE OF 0.001 TO 10.0 MG/L. NO CORRELATION WAS DETECTED BETWEEN THE TRACES OF BORON IN THE ALGAL CELLS AND THE CONCENTRATION OF THE ELEMENT IN THE NUTRIENT SOLUTION. (WILDE-WISCONSIN)

FIELD 05C

ACCESSION NO. W71-12858

ASSESSMENT OF POLLUTION EFFECTS BY THE USE OF ALGAE,

LIVERPOOL UNIV. (ENGLAND). HARTLEY BOTANICAL LABS.

ELSIE M. BURROWS.

PROCEEDINGS OF THE ROYAL SOCIETY OF LONDON, B, VOL 177, P 295-306, 1971. 5 FIG, 22 REF.

DESCRIPTORS:
*ALGAE, *WATER POLLUTION EFFECTS, *INDICATORS, *SEA WATER, MARINE PLANTS, ASSESSMENTS, MONITORING, SEWAGE, ESTUARIES, GROWTH RATES, BIOLOGICAL COMMUNITIES.

IDENTIFIERS:
*ULVA, *LAMINARIA, LIVERPOOL BAY(ENGLAND), IRISH SEA.

ABSTRACT:
INCREASES IN THE DIAMETER OF THALLUS DISCS OF ULVA AND IN THE SURFACE AREA OF LAMINARIA ALGAE SERVED AS THE CRITERION FOR APPRAISAL OF SEWAGE POLLUTION. RELATIVELY SIMPLE CULTURE TRIALS WITH SAMPLES OF PURE AND POLLUTED WATER FROM THE IRISH SEA DISCLOSED THAT ULVA LACTUA AND LAMINARIA SACCHARINA HAVE A HIGH POTENTIAL AS INDICATORS OF POLLUTION EFFECTS BECAUSE OF THEIR EASE OF CULTIVATION AND SENSITIVITY TO CHANGES IN WATER COMPOSITION, PARTICULARLY THOSE CAUSED BY SEWAGE DISCHARGES. THE BEHAVIOR OF DIFFERENT ALGAE SPECIES WITH THEIR ALTERNATION OF GENERATION OFFERS WIDE POSSIBILITIES FOR ASSESSING BOTH THE EUTROPHYING AND TOXIC EFFECTS OF URBAN AND INDUSTRIAL POLLUTION. (WILDE-WISCONSIN)

FIELD 05C

ACCESSION NO. W71-12867

STATISTICAL MODEL FOR THE SELECTION OF UNICELLULAR ALGAL STRAINS BY THE RATE OF INCREASE IN NUMBERS (STATISTICHESKAYA MODEL CTBORA SHTAMMOV CDNOKLETOCHNYKH VODOROSLEI PO SKOROSTI UVELICHENIYA CHISLENNOSTI),

AGROFIZICHESKII NAUCHNO-ISSLEDOVATELSKII INSTITUT, LENINGRAD (USSR).

I. A. SHVYTOV.

ENGLISH SUMMARY, ZHURNAL OBSHCHEI BIOLOGII, VOL 32, NO 3, P 260-265, 1971. 2 TAB, 5 REF.

DESCRIPTORS:
*MODEL STUDIES, *STATISTICAL METHODS, *ALGAE, CYTCLOGICAL STUDIES, SELECTIVITY, PRODUCTIVITY.

IDENTIFIERS:
ALGAL GROWTH.

ABSTRACT:
THE GENETIC SELECTION AIMS TO ISOLATE CELLS FROM A LARGE PHENOTYPIC POPULATION THAT BELONG TO A DEFINITE GENOTYPE OF A HIGH PRODUCTIVE POTENTIAL. THE INTRODUCED STATISTICAL MODEL FOR THE COMPETITIVE SELECTION OF TWO-COMPONENT MICROALGAL POPULATICNS PREDICTS THAT DURING A LONG PERIOD ONE OF THE COMPONENTS WILL UNDERGO DEGENERATION. THE GIVEN FORMULAS DELINEATE THE PROBABILITIES OF ELIMINATION OF STRAINS IN ACCORDANCE WITH THEIR SPECIFIC RATE OF GROWTH AND THE INITIAL DENSITIES. A PREDOMINANCE OF CELLS OF INFERIOR STRAINS MAY CAUSE DEGENERATION OF THE RAPIDLY GROWING CELLS. (WILDE-WISCONSIN)

FIELD 07B, 05C

ACCESSION NO. W71-12868

THE PHYTOPLANKTON OF MINNESOTA LAKES - A PRELIMINARY SURVEY,

MINNESOTA UNIV., MINNEAPOLIS. WATER RESOURCES RESEARCH INST.

A. J. BROOK.

MINNESOTA WATER RESOURCES RESEARCH CENTER, BULLETIN 36, JUNE 1971. 12 P, 1 TAB, 16 REF. OWRR PROJECT A-007-MINN(2).

DESCRIPTORS:
*PHYTOPLANKTON, *EUTROPHICATION, *OLIGOTROPHY, MINNESOTA, LAKES, GLACIAL DRIFT, NUTRIENTS, SALANITY, ALGAE, DIATOMS.

IDENTIFIERS:
*DESMIDS, *EUPLANKTON, *LIMNOPLANKTON.

ABSTRACT:
BETWEEN THE YEARS 1965 AND 1967, PHYTOPLANKTON CCLLECTIONS WERE TAKEN IN THE SUMMER FROM NEARLY 200 LAKES IN A DIVERSITY OF AREAS THROUGHOUT THE STATE. ABOUT 220 TAXA OF EUPLANKTONIC ALGAE WERE IDENTIFIED. THE ANALYSIS OF THE MINNESOTA LAKE PHYTOPLANKTON INDICATES THERE IS A DIMINUTION IN SPECIES DIVERSITY IN THE COURSE OF THE EVOLUTIONARY PROGRESSION AS LAKES CHANGE IN CHARACTER FROM OLIGCTROPHY TO EUTROPHY. MANY OF THE MARKEDLY EUTROPHIC LAKES HAVE SUFFERED SEVERE DISTURBANCE OF THE NATURAL SYSTEM DUE TO ARTIFICIAL ENRICHMENT. EUTROPHIC LAKES IN MINNESOTA ARE TYPICALLY DOMINATED IN SUMMER AND EARLY FALL BY WATER BLOOMS OF BLUE GREEN ALGAE AS IS USUAL IN MOST PRODUCTIVE LAKES OF TEMPERATE REGIONS. MICROCYSTIS AERUGINOSA, M. WESENBERGII, COELOSPHAERIUM NAEGELIANIUM, ALPHANIZCMENON FLOS AQUAE, NUMEROUS SPECIES OF ANABAENA, LYNGBYA BIRGEI, AND GLOEOTRICHIA ECHINULATA ARE MOST COMMON. (WALTON-MINNESOTA)

FIELD 05C, 05A, 02H

ACCESSION NO. W71-13149

REVIEW OF NATIONAL RESEARCH POLICY ON EUTROPHICATION PROBLEMS,

WATER POLLUTION RESEARCH LAB., STEVENAGE (ENGLAND).

A. L. DOWNING.

JOURNAL OF THE SOCIETY FOR WATER TREATMENT AND EXAMINATION, VOL 19, PART 3, P 223-238, 1970. DISCUSSION.

DESCRIPTORS:
*EUTROPHICATION, *ALGAL CONTROL, FINANCING, PLANT GROWTH, TOXICITY, NUTRIENTS, WATER QUALITY, NITRATES, PLANNING, NITROGEN, PHOSPHORUS, RESERVOIRS, FISHERIES, WATER SUPPLY, COSTS, WATER DEMAND, ECONOMIC JUSTIFICATION, AQUATIC WEED CONTROL, RIVERS, AGRICULTURE, ROOTED AQUATIC PLANTS, DRAINAGE, RUNOFF, FERTILIZERS, LIVESTOCK, ECONOMICS, FISH KILLS, INHIBITORS, WATER POLLUTION SOURCES, HUMAN DISEASES.

IDENTIFIERS:
*RESEARCH POLICY, *UNITED KINGDOM, *FUTURE TRENDS, CLADOPHORA, RESEARCH STRATEGY, THAMES RIVER(ENGLAND), LEE RIVER(ENGLAND).

ABSTRACT:
FOR NEW WATER SUPPLIES AND FOR GREATER RECREATIONAL EXPLOITATION OF NATURAL WATERS IN THE UNITED KINGDOM, APPROPRIATE LEVELS OF ACTIVITY AND LINES OF INQUIRY ARE NEEDED. COST INCURRED AS A RESULT OF EUTROPHICATION WILL PROBABLY NOT INCREASE TO MORE THAN ABOUT DOUBLE THE PRESENT EXPENDITURES BY THIS CENTURY'S END. IF ALGAL AND WEED GROWTH WERE FULLY UNDERSTOOD, PERHAPS SOME COMPARATIVELY SIMPLE PREVENTIVE MEASURES COULD BE APPLIED FOR ELIMINATING THESE NUISANCES. A BALANCED PROGRAM IS REQUIRED FOR BASIC RESEARCH AND FOR EMPIRICAL INVESTIGATIONS, ESPECIALLY ON STATIC WATERS. SINCE FACTORS INFLUENCING ALGAL GROWTH VARY GEOGRAPHICALLY, EXAMINING THE INFLUENCES OF LOCATION OF WATER MAY SHOW THAT ONE METHOD OF CONTROL MAY BE MORE APPROPRIATE THAN ANOTHER. THOUGH IT IS UNLIKELY ELIMINATION OF PHOSPHATES FROM DETERGENTS WOULD MATERIALLY REDUCE ALGAL PROBLEMS, SEARCH FOR TROUBLE-FREE SUBSTITUTES SHOULD CONTINUE. PROCESSES USED IN REMOVING NITROGEN AND PHOSPHORUS FROM EFFLUENTS SHOULD BE EXAMINED FOR SUITABILITY OF REMOVING OTHER SUBSTANCES (CARBON ADSORPTION, OZONIATION), AND ABILITY OF EFFLUENTS TO SUPPORT ALGAL GROWTH BY UNSUSPECTED COMPONENTS IS PROFOUNDLY IMPORTANT. GROWTH OF WEEDS IN RIVERS AND EFFECT OF SEWAGE EFFLUENT ON CLADOPHORA NEEDS INVESTIGATION. INSURING THAT PRESENT EFFORTS ARE WELL COORDINATED TAKES PRECEDENCE OVER EMBARKING ON MANY NEW INITIATIVES. (JONES-WISCONSIN)

FIELD 05C, 06B

ACCESSION NO. W71-13172

PRIMARY PRODUCTION,

FRESHWATER BIOLOGICAL ASSOCIATION, AMBLESIDE (ENGLAND).

J. W. G. LUND.

JOURNAL OF THE SOCIETY FOR WATER TREATMENT AND EXAMINATION, VOL 19, PART 4, P 332-358, 1970. 6 FIG, 1 TAB, 36 REF, DISCUSSION.

DESCRIPTORS:
*WATER QUALITY, *EUTROPHICATION, *PRIMARY PRODUCTIVITY, *ALGAE, NITROGEN, PHOSPHORUS, THERMAL STRATIFICATION, WATER SUPPLY, SUCCESSION, ECONOMICS, NUTRIENTS, NITROGEN FIXATION, FORECASTING, PHOTOSYNTHESIS, BIOMASS, NITRATES, RESERVOIRS, LIGHT PENETRATION, CYANOPHYTA, CHRYSOPHYTA.

IDENTIFIERS:
*UNITED KINGDOM, ESTHWAITE WATER(ENGLAND), BLELHAM TARN(ENGLAND), LUXURY UPTAKE, LOCH LEVEN(SCOTLAND), LAKE WINDERMERE(ENGLAND).

ABSTRACT:
EUTROPHICATION IS NOT YET A MAJOR ECONOMIC PROBLEM IN THE BRITISH WATER INDUSTRY BUT COULD BECOME ONE AND FURTHER SUPPORT FOR ECOLOGICAL INVESTIGATIONS IS NEEDED. THE MAJOR NUTRIENTS, NITROGEN AND PHOSPHORUS, ARE STUDIED WITH REFERENCE TO ECOLOGICAL FACTORS CONTROLLING ALGAL POPULATION SIZE AND PRODUCTION RATES. TYPICALLY, THE DEGREE TO WHICH THE NUTRIENT CONTENT OF A RESERVOIR APPROACHES THAT OF WINTER INFLOWS IS AFFECTED BY RETENTION TIME AND NUMBER OF ALGAE PRESENT. PRODUCTION RATE IS A GUIDE TO HOW TROUBLES MAY INCREASE AND TO DANGERS AHEAD WHEN ALGAE ARE RELATIVELY UNCOMMON. THE STRONG CHEMICAL STRATIFICATION ACCOMPANYING THERMAL STRATIFICATION IN EUTROPHIC LAKES MAKES IT UNCERTAIN HOW GREAT THE NUTRIENT SUPPLY CAN BE TO ALGAE IN THE EPILIMNION OR EUPHOTIC ZONE EVEN IN ABSENCE OF ACTIVE VERTICAL MIGRATION. THERMAL STRATIFICATION MAY BE ADVANTAGEOUS OR DISADVANTAGEOUS TO WATER SUPPLY. BIOASSAY IS A USEFUL METHOD FOR COMPARING POTENTIAL FERTILITIES OF WATERS AND FOR ELUCIDATING SPECIFIC, LIMITED PROBLEMS. OTHER NUTRIENTS ARE IMPORTANT BESIDES NITROGEN AND PHOSPHORUS. FORECASTING DEVELOPMENT OF ALGAL POPULATIONS AND SEASONAL SPECIES' SUCCESSIONS IS POSSIBLE IN GENERAL TERMS BUT DETAILED FORECASTING OVER RELATIVELY LONG PERIODS IS NOT YET POSSIBLE. (JONES-WISCONSIN)

FIELD 05C

ACCESSION NO. W71-13183

THE NUISANCE ALGAE: CURIOSITIES IN THE BIOLOGY OF PLANKTONIC BLUE-GREEN ALGAE,

WESTFIELD COLL., LONDON (ENGLAND). DEPT. OF BOTANY.

A. E. WALSBY.

JOURNAL OF THE SOCIETY FOR WATER TREATMENT AND EXAMINATION, VOL 19, PART 4, P 359-373, 1970. 2 FIG, 1 TAB, 43 REF, DISCUSSION.

DESCRIPTORS:
*NUISANCE ALGAE, *CYANOPHYTA, *ALGAL CONTROL, LABORATORY TESTS, BUOYANCY, EUTROPHICATION, NIRTROGEN FIXATION, BIOCONTROL, PLANT PHYSIOLOGY, MOVEMENT, VIRUSES, PHOTOSYNTHESIS, TURBULENCE, LIGHT INTENSITY.

IDENTIFIERS:
*GAS VACUOLES, MOTILE FORMS, CYANOPHAGES.

ABSTRACT:
BLUE-GREEN ALGAE, QUITE DIFFERENT FROM ALL OTHER ALGAL GROUPS, HAVE CERTAIN CHARACTERISTICS WHICH CONTRIBUTE TO FORMATION OF BLOOMS; ONE IS THE ABILITY TO FIX ATMOSPHERIC NITROGEN. THEY POSSESS GAS FILLED VACUOLES WHICH PROVIDE A BUOYANCY-REGULATING MECHANISM ENABLING THEM TO OCCUPY WATER LAYERS BELOW THE SURFACE WHERE CONDITIONS ARE OPTIMAL FOR GROWTH, BUT ALSO CAUSE THICK ALGAL SCUMS ON THE SURFACE OF LAKES AND RESERVOIRS. ON THE CREDIT SIDE THEY PRODUCE OXYGEN WHICH KEEPS THE WATER IN THE AEROBIC CONDITION ESSENTIAL FOR PREVENTING GROWTH OF ANAEROBIC BACTERIA AND FOR SUPPORTING FISH AND OTHER ANIMAL POPULATIONS. THEY TAKE UP INORGANIC NUTRIENTS IN GROWTH AND CLEANSE WATER OF NITRATE AND OTHER SUBSTANCES, POTENTIALLY TOXIC. ON THE DEBIT SIDE, THEY CONTRIBUTE TO THE SUSPENDED SOLIDS WHICH MUST BE REMOVED BY FILTERING; THEIR BREAKDOWN PRODUCTS, WHICH PASS FILTERING PROCESSES, MAY PRODUCE TASTES AND ODORS IN TREATED WATER. BEFORE BREAKDOWN OCCURS, SOLUBLE EXTRACELLULAR PRODUCTS, SUCH AS MUCOPOLYSACCHARIDES, OCCASIONALLY CAUSE NUISANCE. SOME PRODUCE TOXIC SUBSTANCES. METHODS OF DESTROYING GAS VACUOLES OR THEIR EFFECTS WOULD PROVIDE MEANS OF CONTROLLING ALGAL BLOOMS; ANOTHER METHOD MIGHT BE TO USE CYANOPHAGES, VIRUSES WHICH ATTACK THESE ALGAE SELECTIVELY. (JONES-WISCONSIN)

FIELD 05C

ACCESSION NO. W71-13184

EUTROPHICATION IN RELATICN TO WATER SUPPLIES,

BCLTON INST. OF TECH. (ENGLAND)

E. G. BELLINGER.

JOURNAL OF THE SOCIETY FOR WATER TREATMENT AND EXAMINATION, VOL 19, PART 4, P 400-409, 1970. 1 FIG, 3 TAB, 21 REF, DISCUSSION.

DESCRIPTORS:
*EUTROPHICATION, *WATER SUPPLY, WATER POLLUTION EFFECTS, NUTRIENTS, RESERVOIR OPERATION, WATER REUSE, POTABLE WATER, ALGAE, NUTRIENT REQUIREMENTS, CYANOPHYTA, CHLAMYDOMONAS.

IDENTIFIERS:
*UNITED KINGDOM, ESTHWAITE WATER(ENGLAND), LAKE WINDERMERE(ENGLAND), GREAT OUSE(ENGLAND), ALGAL GROWTH.

ABSTRACT:
THE GREASTEOT INCREASES IN EUTROPHICATION HAVE TAKEN PLACE IN BRITIAN SINCE 1910 AND ARE CORRELATED WITH POPULATION GRCWTH. DISSOLVED SOLIDS AND PHOSPHATE CONCENTRATIONS HAVE ALSO INCREASED SIGNIFICANTLY COUPLED WITH BIOLOGICAL PRODUCTIVITY--WITH ALGAE THE MAIN OFFENDERS. REMOVAL OF NUTRIENTS BY VARIOUS SEWAGE TREATMENT PROCESSES WCULD NOT BE LCW ENOUGH TO PREVENT ALGAL GROWTH, BUT SUFFICIENTLY LOW TO SLOW DOWN ARTIFICIAL EUTROPHICATION. THE EFFECTS OF EUTROPHICATION ANC CONSEQUENT DIATOM GROWTHS AND BLUE-GREEN ALGAE CAUSE PROBLEMS IN WATERWORKS AND UNPLEASANT OLFACTORY RESPONSES TC POTABLE WATER AS WELL AS ANAEROBIC CONDITIONS DELETERIOUS TO WATER SUPPLIES. RESERVCIR MANAGEMENT SHOULD PROVIDE AT THE OUTLET TOWER FOR A NUMBER OF ABSTRACTION POINTS AT DIFFERENT DEPTHS TO PERMIT USE OF THE BEST QUALITY WATER AVAILABLE. CONTROL OF THE THERMCCLINE LEVEL BY PUMPING, THUS INCREASING THE DEPTH OF THE EPILIMNION, WCULD PROVIDE A GREATER DEPTH CF BETTER QUALITY WATER. INASMUCH AS ALGAE IS LIGHT DEPENDENT, PLASTIC SHEETING TO EXCLUDE LIGHT AND INCREASING TURBIDITY BY PUMPING AND JETTING WOULD REDUCE LIGHT PENETRATION. IT SEEMS CERTAIN THAT MCRE AND MCRE WASTEWATERS WILL HAVE TO BE REUSED AND THAT WASTEWATER SHOULD BE CONSIDERED AS AN INTEGRAL PART OF THE NATICN'S RESOURCES. (JONES-WISCONSIN)

FIELD 05F

ACCESSION NO. W71-13185

CHLORINATION OF ODORANTS FROM ALGAL BLOOMS,

TECHNION - ISRAEL INST. OF TECH., HAIFA. SANITARY ENGINEERING LAB.

M. REBHUN, M. A. FOX, AND J. B. SLESS.

JOURNAL AMERICAN WATER WORKS ASSCCIATICN, VOL 63, NO 4, P 219-224, 1971. 8 FIG, 7 TAB, 17 REF.

DESCRIPTORS:
*WATER QUALITY, *CHLORINATION, *WATER QUALITY CONTROL, *ODOR, ALGAE, EUTROPHICATION, TASTE, RESERVOIRS, SEASONAL, ACTINOMYCETES, CYANOPHYTA, PLANKTON, ON-SITE TESTS, LABCRATORY TESTS.

IDENTIFIERS:
*ISRAEL, *LAKE KINNERET(ISRAEL), PERIDINIUM WESTI, OSCILLATORIA PROLIFICA, THRESHOLD ODOR NUMBER.

ABSTRACT:
THE WATER SUPPLIES DISTRIBUTED BY ISRAEL'S NATIONAL WATER CARRIER FREQUENTLY DEVELOP TASTE AND ODOR PROBLEMS. A STUDY WAS CONDUCTED TO DETERMINE THE RELATICNSHIPS BETWEEN THE PLANKTCNIC ALGA PERIDINIUM WESTI AND THE BENTHONIC ALGA OSCILLATORIA TO ISOLATE THE CHARACTERIZE THE ODORANTS AND TO DETERMINE WHETHER A QUANTITATIVE RELATIONSHIP EXISTS BETWEEN ODOR AND CHLORINATION LEVELS. FIELD TESTS EVALUATED INTERRELATIONSHIPS BETWEEN ALGAL NUMBERS AND SPECIES COMPOSITION, AND THRESHOLD ODOR NUMBER AND CHLORINE DEMAND. ODORANTS PRODUCED BY PERIDINIUM AND OSCILLATORIA WERE CONCENTRATED ANC COLLECTED FOR CHARACTERIZATION AND QUANTITATIVE ANALYSIS OF EFFECT OF CHLORINATION ON THRESHOLD ODOR NUMBER IN THE LABORATORY. FLUCTUATICNS IN LEVELS OF ALGAL NUMBERS, CHLORINE DEMAND, ANC THRESHCLD CDCR NUMBER OF RESERVOIR WATER WERE SEASCNAL AND INTERRELATED. CHLCRINE DEMAND WAS EXERTED BY SUBSTANCES IN SOLUTICN AND NOT BY ALGAL CELLS AT NORMAL CONCENTRATIONS. ODOR CONTROL CAUSED BY PERIDINUM WAS CEPENCENT CN PROPER CHLORINATION OF FREE RESIDUAL. EARTHLY ODORS CAUSEC BY OSCILLATCRIA CCULD NOT BE COMPLETELY REMOVED BY CHLORINATION; THE LEVEL CF CDOR REDUCTION WAS LINEARLY CEPENDENT ON INITIAL OCOR LEVEL. A METHCD OF COLLECTING ODORANTS FROM LABORATORY CULTURES CF AQUATIC ORGANISMS WAS DEVELOPED. (JONES-WISCONSIN)

FIELD 05F, 05G

ACCESSION NC. W71-13187

OBSERVATIONS CONCERNING THE EXPERIMENTAL CONTAMINATIONS AND THE 'IN SITU' CONTAMINATIONS OF MARINE SPECIES BY RUTHENIUM 106, (OBSERVATIONS CONCERNANT LES CONTAMINATIONS EXPERIMENTALES ET LES CONTAMINATIONS 'IN SITU' D'ESPECES MARINES PAR LE'RUTHENIUM 106),

COMMISSARIAT A L'ENERGIE ATOMIQUE, CHERBOURG (FRANCE). CENTRE DE LA HAGUE.

J. ANCELLIN, AND P. BOVARD.

REVUE INTERNATIONALE D'OCEANOGRAPHIE MEDICALE, NICE, FRANCE, VOL 21, P 85-92, 1971. 8 REF.

DESCRIPTORS:
*WATER POLLUTION EFFECTS, *PLANTS, *ANIMALS, *ALGAE, *INVERTEBRATES, *RADIOACTIVITY EFFECTS, *RADIOISOTOPES.

IDENTIFIERS:
*RUTHENIUM 106, SOLUBLE FORMS, INSOLUBLE FORMS, LABORATORY EXPERIMENTS.

ABSTRACT:
CONTAMINATION OF MARINE FLORA AND FAUNA BY RUTHENIUM 106 WAS STUDIED UNDER BOTH LABORATORY AND IN SITU CONDITIONS. ALGAE AND INVERTEBRATES CONTAMINATED BY SOLUBLE AND INSOLUBLE FORMS OF RUTHENIUM 106 INDICATED THAT THE SOLUBLE RU 106 WAS 5-10 CONCENTRATION FACTORS LOWER THAN THE INSOLUBLE RU 106 FORMS. THE HIGH AMOUNTS OF CONCENTRATION FACTORS OBSERVED IN SITU WERE USUALLY COMPARABLE TO THOSE EXPERIMENTALLY DETERMINED FOR THE SOLUBLE FORMS IN BOTH ALGAE AND INVERTEBRATES. (ENSIGN-PAI)

FIELD 05C

ACCESSION NO. W71-13233

THE ROLE OF ALGAE IN CYCLING OF RADIONUCLIDES,

COMITATO NAZIONALE PER L'ENERGIA NUCLEARE, LA SPEZIA (ITALY). LABORATORIO PER LO STUDIO DELLA CONTAMINAZIONE RADIOATTIVE DEL MARE.

A. LATTERA, AND M. BERNHARD.

REVUE INTERNATIONALE D'OCEANOGRAPHIE MEDICALE, NICE, FRANCE, VOL 20, 1970, P 29-52. 67 REF.

DESCRIPTORS:
*RADIOISOTOPES, *ALGAE, *SEA WATER, *FOOD CHAINS, *ABSORPTION, BIOCHEMISTRY, PHYSIOLOGICAL ECOLOGY, CESIUM, STRONTIUM RADIOISOTOPES, LINC RADIOISOTOPES, MANGANESE, COBALT RADIOISOTOPES, CADMIUM RADIOISOTOPES, METABOLISM, TROPHIC LEVELS.

IDENTIFIERS:
CONCENTRATION, ABUNDANCE, IRON, CHROMIUM, COPPER, NICKEL, RUTHENIUM, CERIUM, LINCONIUM.

ABSTRACT:
UPTAKE, LOSS AND STABLE CONTENT FOR ELEMENTS WHICH POSSESS POSSIBLE DANGEROUS RADIOISOTOPES, AND THE FACTORS INFLUENCING THESE ARE DISCUSSED. THE ACCUMULATION OF TRACE METALS BY ALGAE WAS PROPORTIONAL TO THEIR EXTERNAL CONCENTRATIONS. UPTAKE WAS RELATED TO PHYSICO-CHEMICAL FORMS OF RADIO-NUCLIDES. CONCENTRATION FACTORS FOR THE DIFFERENT ELEMENTS WERE FOUND TO VARY, SOMETIMES TO MORE THAN A THOUSAND TIMES OVER THEIR CONCENTRATION IN THE ENVIRONMENT. (ENSIGN-PAI)

FIELD 05C

ACCESSION NO. W71-13243

ECONOMIC ASPECTS OF ALGAE,

LOUISVILLE UNIV., KY. DEPT. OF BIOLOGY.

V. E. WIEDEMAN.

IN: SYMPOSIUM ON HYDROBIOLOGY, 'BIORESOURCES OF SHALLOW WATER ENVIRONMENTS', JUNE 24-27, 1970, MIAMI BEACH, FLORIDA, P 25-33. 2 TAB, 60 REF.

DESCRIPTORS:
*MARINE ALGAE, *ECONOMICS, *PRODUCTIVITY, ALGAE, RESOURCES, FOOD CHAINS, FISH FOOD ORGANISMS, OXYGEN, ECOSYSTEMS.

ABSTRACT:
ALGAE PROVIDE THE MAIN FOOD SUPPLY IN THE SEA AND HAVE GREAT ECONOMIC IMPORTANCE. THE BLUE-GREEN ALGAE (CYANOPHYCOPHYTE) ARE PRIMITIVE ORGANISMS RESEMBLING BACTERIA AND INCORPORATING A NITROGEN FIXING CAPABILITY WHICH CAUSES HIGH BIOMASS PRODUCTION. THEY SHOULD BE FULLY INVESTIGATED FOR ECONOMICALLY USEFUL METABOLITES. THE GREEN ALGAE (CHLOROPHYCOPHYTA) ARE OF GREAT DIVERSITY BUT CERTAIN GENERA SHOW POTENTIAL AS A PROTEIN-RICH FOOD SOURCE, AS SOIL FERTILIZER, AND AS A LEAD TO FURTHER UNDERSTANDING OF THE PHOTOSYNTHETIC PROCESS AND CYTOLOGY. EUGLENOID FLAGELLATES (EUGLENOPHYCOPHYTA) CAN BE USED AS BIOASSAY ORGANISMS FOR VITAMIN B12, AND SOME OCCUR AS A MAJOR COMPONENT IN WASTE-STABILIZATION PONDS. CERTAIN OF THE GOLDEN ALGAE (CHRYSOPHYCOPHYTA) ARE UTILIZED IN VITAMIN ASSAY BUT HAVE AN UNPLEASANT ODOR AND TASTE IN WATER. THE DIATOMS (BACILLARIOPHYCOPHYTA) CAN BE USED FOR INSULATING, SCOURING AND FILTERING MATERIALS, AND SOME SPECIES HAVE ANTIBIOTIC PROPERTIES. THE BROWN ALGAE (PHAEOPHYCOPHYTA), WHICH INCLUDE KELP, ARE A SOURCE OF IODINE, DRIED FOOD, FODDER, FERTILIZER AND ALGIN, WHICH HAS MANY USES IN MODERN PRODUCTS. THE RED ALGAE (RHODOPHYCOPHYTA) HAVE SIMILAR USEFUL PROPERTIES, WHEREAS THE DINOFLAGELLATES (PYRROPHYCOPHYTA) NEED TO BE CONTROLLED BECAUSE OF TOXIC PROPERTIES. ALGAE THUS CONSTITUTE THE BASE OF MOST FOOD CHAINS AND CONTRIBUTE MOST OF THE WORLD'S OXYGEN SUPPLY. (SMITH-PAI)

FIELD 05C, 06B

ACCESSION NO. W71-13246

THE MEASUREMENT OF OUR NATURAL RESOURCES,

NEW ENGLAND AQUARIUM, BOSTON, MASS.

G. C. MCLEOD.

IN: SYMPOSIUM ON HYDROBIOLOGY, 'BIORESOURCES OF SHALLOW WATER ENVIRONMENTS', JUNE 24-27, 1970, MIAMI BEACH, FLORIDA, P 88-98. 2 FIG, 30 REF. N00014-69-C0380.

DESCRIPTORS:
*MARINE ALGAE, *LIGHT, *GROWTH RATES, PHYTOPLANKTON, NATURAL RESOURCES, DEPTH, DISTRIBUTION PATTERNS, MEASUREMENT, PHOTOSYNTHESIS, PLANT PHYSIOLOGY, PIGMENTS.

IDENTIFIERS:
SUBMARINE LIGHT DATA SPHERE.

ABSTRACT:
OVER SHORT PERIODS OF TIME ALGAE PHOTOSYNTHESIZE FASTER THAN THEY GROW. THE LIGHT-INTENSITY CURVE FOR GROWTH OF ALGAE IS AN INTRINSIC CHARACTERISTIC OF THE ALGAE USED AND THE TEMPERATURE, WHEREAS THAT OF PHOTOSYNTHESIS IS NOT. MEASUREMENTS OF PRODUCTIVITY OF ALGAE MUST CONSIDER THE VARIABILITY IN SYNTHESIS OF STORAGE MATERIALS AND IMPORTANT PIGMENTS. THE QUESTION OF LIGHT RESPONSE MAY INFLUENCE THE DOMINANCE OF CERTAIN PHYTOPLANKTON IN SELECTED ENVIRONMENTS. SOME STUDIES OF THE EFFECT OF LIGHT INTENSITY ON THE GROWTH, PIGMENT COMPOSITION AND METABOLISM OF FRESH WATER AND MARINE PHYTOPLANKTON ARE DESCRIBED. THE INTENSITY AND SPECTRAL COMPOSITION OF LIGHT IN THE MARINE ENVIRONMENT IS VARIABLE. THE EFFECTS OF SPECTRUM-LIMITED LIGHT ON GROWTH, PIGMENTATION AND PHOTOSYNTHESIS HAVE ONLY RECENTLY BEEN STUDIED IN CHROMATIC LIGHT AND DARKNESS. THE EFFECTS OF BLUE LIGHT CAN BE REVERSED BY INCREMENTS OF RED LIGHT. THE KINETICS OF THE ADAPTATION TO BLUE LIGHT AND A POSSIBLE MECHANISM OF ACTION ARE CONSIDERED. SPECIES MUST ADAPT TO THE CHANGES IN SPECTRAL COMPOSITION OF SUBMARINE LIGHT, THEREFORE THE DISTRIBUTION OF CERTAIN SPECIES AT CERTAIN TIMES OF YEAR WILL VARY. A SUBMARINE LIGHT DATA SPHERE IS SPECIFIED, WITH SYSTEMS FOR MEASURING TOTAL INTEGRATED INTENSITY OF LIGHT OVER TIME AS A FUNCTION OF DEPTH. (SMITH-PAI)

FIELD 05C, 05A

ACCESSION NO. W71-13252

EUTROPHICATION: A THREAT TO WATER RESOURCES,

ENVIRONMENTAL PROTECTION AGENCY, CORVALLIS, OREG. PACIFIC NORTHWEST WATER LAB.

A. F. BARTSCH.

IN: SYMPOSIUM ON HYDROBIOLOGY, 'BIORESOURCES OF SHALLOW WATER ENVIRONMENTS', JUNE 24-27, 1970, MIAMI BEACH, FLORIDA, P 127-135. 15 REF.

DESCRIPTORS:
*WATER QUALITY CONTROL, *LAKES, *EUTROPHICATION, AQUATIC ALGAE, NUTRIENTS, WATER RESOURCES, WASTE ASSIMILATIVE CAPACITY, RESEARCH AND DEVELOPMENT, WASTE WATER TREATMENT.

IDENTIFIERS:
NUTRIENT CONTROL METHODS.

ABSTRACT:
QUALITY DETERIORATION OF WATER RESOURCES, ESPECIALLY LAKES, THROUGH EUTROPHICATION IS A WORLDWIDE PROBLEM AND REQUIRES IMMEDIATE ACTION. THE NATURAL PROCESS HAS BEEN AGGRAVATED AND SPEEDED UP BY MAN'S ACTIVITIES IN AUGMENTING THE FERTILITY OF WATER. ARTIFICIALLY CREATED LAKES QUICKLY DEVELOP EUTROPHICATION PROBLEMS. INDUSTRIAL WASTES, SEWAGE WASTES AND AGRICULTURAL RUNOFF ALL CONTRIBUTE NUTRIENTS TO WATERCOURSES. THE QUESTION OF LAKE RESTORATION IS BEING STUDIED BY SCIENTIFIC COMMUNITIES AND THE FWQA. ONE RESTORATIVE AND PREVENTIVE APPROACH IS TO CURB NUTRIENT INPUT, WHICH IS ALREADY BEING DONE IN SOME PLACES BY DIVERTING SEWAGE FROM LAKES. A SECOND CONTROL POSSIBILITY IS TO TREAT SEWAGE AND WASTES TO STRIP THEM OF THEIR NUTRIENT CONTENT, ESPECIALLY PHOSPHORUS. HOWEVER, ALL LAKES ARE DIFFERENT AND THERE IS NEED TO BE ABLE TO PREDICT HOW QUICKLY AND HOW MUCH A LAKE WILL CHANGE AFTER RESTORATIVE EFFORTS. DETERGENTS ARE ANOTHER SOURCE OF PHOSPHORUS IN SEWAGE WHICH MUST BE CONTROLLED. IMPROVED HANDLING OF AGRICULTURAL WASTES CAN ALSO MINIMIZE NUTRIENT INPUT TO WATER. STUDIES ARE BEING MADE OF THE USE OF CERTAIN CHEMICALS FOR NUTRIENT INACTIVATION. OTHER IDEAS UNDER CONSIDERATION ARE THE FEEDING AND HARVESTING OF ALGAE, CONTROLLED DREDGING, STIMULATING DISEASES AND PARASITES TO DESTROY PLANT POPULATIONS, AND, AS A LAST RESORT, THE USE OF ALGICIDES AND HERBICIDES. (SMITH-PAI)

FIELD 05C

ACCESSION NO. W71-13256

BACTERIA-ALGAE SYMBIOSIS - A CAUSE OF ALGAL BLOOMS,

WYANDOTTE CHEMICALS CORP., MICH.

L. E. KUENTZEL.

IN: SYMPOSIUM ON HYDROBIOLOGY, 'BIORESOURCES OF SHALLOW WATER ENVIRONMENTS', JUNE 24-27, 1970, MIAMI BEACH, FLORIDA, P 136-145. 23 REF.

DESCRIPTORS:
*ALGAE, BACTERIA, *LAKES, SYMBIOSIS, ALGAL CONTROL, EUTROPHICATION, WATER POLLUTION, BIOCHEMICAL OXYGEN DEMAND, ORGANIC MATTER, PHOSPHORUS, CARBON DIOXIDE.

ABSTRACT:
THE EXISTENCE OF THE MUTUALLY BENEFICIAL SYMBIOTIC RELATIONSHIP BETWEEN BLUE-GREEN ALGAE AND CERTAIN BACTERIA IS WELL ESTABLISHED. THE PRESENCE OF BIODEGRADABLE ORGANIC MATTER WILL ALWAYS CONTRIBUTE TO MASSIVE ALGAL BLOOM GROWTH IF ALL OTHER FACTORS PERMIT, AND THERE HAS BEEN A GENERAL INCLINATION TO TURN TO PHOSPHORUS AS A CONTROLLING FACTOR. THE LOGISTICS OF SUPPLY FOR CO2 AND P ARE EXAMINED IN THE LIGHT OF THE ESTABLISHED REQUIREMENTS FOR THE GROWTH OF MASSIVE ALGAL BLOOMS IN INCREASINGLY POLLUTED LAKES, AND FINDINGS FROM PREVIOUS AND NEW STUDIES ARE CITED. NUTRIENT CONTROL HAS LITTLE EFFECT ON LAKES WHICH HAVE BEEN SUPPORTING ALGAL GROWTH FOR SEVERAL YEARS BECAUSE OF THE ACCUMULATION OF ORGANIC DEBRIS. IT APPEARS THAT ORGANIC MATTER AND BACTERIAL ACTION PLAY A MAJOR ROLE IN SUPPORTING ALGAL BLOOM DEVELOPMENT. ATTEMPTS TO CONTROL ALGAL GROWTH VIA CONTROL OF CARBON (BOD) OFFER DISTINCT ADVANTAGES, WHEREAS PHOSPHATE REMOVAL DOES NOT SEEM TO HAVE MUCH EFFECT. WASTEWATERS NEED TO RECEIVE OPTIMUM BOD AND AMMONIA N REMOVAL BY TREATMENT PLANTS TO HELP PREVENT EUTROPHICATION OF RIVERS AND LAKES. (SMITH-PAI)

FIELD 05C, 02H

ACCESSION NO. W71-13257

A LABORATORY EVALUATION OF THE PERFORMANCE OF THREE CELL OXIDATION PONDS IN SERIES,

MISSISSIPPI STATE UNIV., STATE COLLEGE.

ROBERT JACK FREEMAN, JR.

MASTER'S THESIS, MAY 1970. 91 P, 36 FIG, 6 TAB, 26 REF.

DESCRIPTORS:
*OXIDATION PONDS, *ALGAE, LABORATORY TESTS, AEROBIC CONDITIONS, ANAEROBIC CONDITIONS, ORGANIC LOADING, *BIOCHEMICAL OXYGEN DEMAND, NITROGEN, PHOSPHOROUS, PHOTOSYNTHESIS, LIGHT INTENSITY, OXYGENATION, DENITRIFICATION, OXIDATION, ADSORPTION, COLIFORMS, BACTERIA, CENTRIFUGATION, *WASTE WATER TREATMENT.

IDENTIFIERS:
*DETENTION TIME.

ABSTRACT:
THE RELATIVE EASE WITH WHICH THE CRITICAL PARAMETERS COULD BE VARIED PROMPTED THE USE OF LABORATORY SCALE EQUIPMENT FOR EVALUATIONS OF 3 CELL OXIDATION PONDS. 2 SETS OF 3 SERIES PONDS WERE USED, ONCE AT A 300 LB BOD5/ACRE/DAY LOADING RATE AND A 2 DAY PER POND DETENTION TIME. SUBSEQUENT RUNS INVOLVED CONSTANT LOADING RATES OF 50 AND 100 LBS BOD5/ACRE/DAYS AND DETENTION TIMES OF 2,3, AND 4 DAYS PER POND. A SYNTHETIC SUBSTRATE, CONTINUOUS FLOW LOADING, AND CONTROLLED LIGHTING WERE USED. BOTH UNCENTRIFUGED AND CENTRIFUGED SAMPLES WERE ANALYZED FOR BOD, TOTAL NITROGEN, ALGAL CONCENTRATION, SOLUBLE ORTHOPHOSPHATE, PH, AND DISSOLVED OXYGEN. FOR UNCENTRIFUGED SAMPLES, BOD, TOTAL NITROGEN, AND SOLUBLE ORTHOPHOSPHATE REMOVALS RANGED FROM 90 TO 95%, 20 TO 50% AND 16 TO 32% RESPECTIVELY. CENTRIFUGATION OF THE EFFLUENT FOR REMOVAL OF THE ALGAE CELLS EFFECTED A 40 TO 70% REDUCTION OF EFFLUENT BOD, A 25% TO 40% REDUCTION OF TOTAL NITROGEN IN THE EFFLUENT, WHILE LITTLE OR NO REDUCTION OF SOLUBLE ORTHOPHOSPHATE WAS OBSERVED. ALSO, AT CONSTANT LOADING RATE, VARYING DETENTION TIMES PRODUCED NO EFFECTS ON REMOVALS. A SIGNIFICANT PERCENTAGE REMOVAL OF COLIFORMS WAS OBTAINED, BUT EFFLUENT COLIFORM CONCENTRATIONS WERE STILL GREATLY IN EXCESS OF ACCEPTABLE STANDARDS DUE TO THE EXTREMELY HIGH INFLUENT CONCENTRATIONS. (LOWRY-TEXAS)

FIELD 05D

ACCESSION NO. W71-13326

PHOSPHOROUS EXTRACTION FROM SOLUTION BY HETEROGENOUS BIOLOGICAL SYSTEMS,

MAINE UNIV., ORONO, DEPT. OF CIVIL ENGINEERING.

RICHARD D. ROX.

MASTER'S THESIS, UNIVERSITY OF MAINE, JANUARY 1970. 67 P, 20 FIG, 43 REF.

DESCRIPTORS:
*EUTROPHICATION, *PHOSPHOROUS, ALGAE, NITROGEN, ACTIVATED SLUDGE, CHEMICAL OXYGEN DEMAND, OXYGEN, AERATION, DISSOLVED OXYGEN, HYDROGEN ION CONCENTRATION, SETTLING, MICROORGANISMS, ACCLIMATION, CALCIUM, SEDIMENTATION, WASTE WATER TREATMENT.

IDENTIFIERS:
*CHEMICAL-BIOLOGICAL TREATMENT, MIXED LIQUOR, SUSPENDED SOLIDS, UPTAKE.

ABSTRACT:
PHOSPHOROUS HAS BECOME THE MAJOR CONTROLLABLE NUTRIENT IN THE FIGHT AGAINST EUTROPHICATION. AN ATTEMPT HAS BEEN MADE TO FIND EFFECTIVE MEANS OF PHOSPHOROUS REMOVAL THROUGH A CHEMICAL-BIOLOGICAL TREATMENT SCHEME. BENCH SCALE ACTIVATED SLUDGE UNITS OF THE PHOSPHOROUS UPTAKE AND RELEASE BY ACTIVATED SLUDGE SYSTEMS. SOLUBLE COD UPTAKE, MLSS GROWTH, OXYGEN UTILIZATION, INITIAL F:M RATIOS AND DIVALENT CATION CONCENTRATIONS WERE STUDIED TO DETERMINE THEIR EFFECT ON REMOVAL. PHOSPHOROUS RELEASE WAS STUDIED UNDER EXTENDED PERIODS OF AERATION, LOW CONCENTRATIONS OF DISSOLVED OXYGEN, QUIESCENT SETTLING CONDITIONS AND LOW LEVELS OF PH. ACTIVATED SLUDGE MICROORGANISMS WERE SHOWN TO PRODUCE REPRODUCEABLE PHOSPHOROUS UPTAKE DATA WHILE BEING ACCLIMATED TO A SYNTHETIC WASTE. A PHOSPHOROUS UPTAKE OF 0.02 POUNDS P/LB.COD/DAY WAS NOTED FOR BACTERIAL CULTURES OPERATING AT SEVERAL F:M RATIOS. A CLOSE CORRELATION OF THE MICROBIAL OXYGEN UTILIZATION AND THE TOTAL SOLUBLE PHOSPHOROUS UPTAKE WAS OBSERVED. PHOSPHORUS RELEASE WAS SHOWN TO BE INDEPENDENT OF DISSOLVED OXYGEN CONCENTRATIONS AND THE PERIOD OF AERATION. LOW PH CONDITIONS, HOWEVER, SHOWED SIGNIFICANT LEVELS OF PHOSPHOROUS SECRETION. CALCIUM ADDED TO THE ACTIVATED SLUDGE MIXED LIQUOR SHOWED VARYING DEGREES OF ENHANCED PHOSPHOROUS REMOVAL WITH NO SIGNIFICANT DETERIORATION IN BIOLOGICAL ACTIVITY. (ATKINS-TEXAS)

FIELD 05D, 05C

ACCESSION NO. W71-13334

SEVERAL METHODS OF ALGAE REMOVAL IN MUNICIPAL OXIDATION PONDS,

KANSAS UNIV., LAWRENCE.

DONALD M. MARTIN.

MASTER'S THESIS, UNIVERSITY OF KANSAS, DECEMBER 1970. 82 P, 22 FIG, 6 TAB, 19 REF.

DESCRIPTORS:
*OXIDATION LAGOONS, *ALGAE, MUNICIPAL WASTES, *FILTRATION, *FILTERS, BIOCHEMICAL OXYGEN DEMAND, CHEMICAL OXYGEN DEMAND, PHOSPHORUS, NITRATES, AMMONIA, COAGULATION, FLOCCULATION, SEDIMENTATION, FILTRATION, LIME, ROCKS, METABOLISM, DESIGN CRITERIA, *WASTE WATER TREATMENT, EFFLUENTS.

IDENTIFIERS:
*SUSPENDED SOLIDS, SODIUM HYDROXIDE, *ROCK FILTERS.

ABSTRACT:
EFFLUENT FROM AN OXIDATION POND WAS COLLECTED AND USED BOTH FOR A JAR TEST AND FOR RUNS THROUGH AN UPFLOW ROCK FILTER. A WIDE VARIETY OF COAGULANTS AND THE NEW POLYELECTROLYTES WERE USED IN THE HOPE OF DISCOVERING SOME ECONOMICAL MEANS OF REMOVING ALGAE FROM OXIDATION POND EFFLUENTS. LIME, ALUM AND SODIUM HYDROXIDE WERE TESTED FOR USE AS FLOCCULENTS, WHILE CALGON'S CAT-FLOC WAS USED AT A CONCENTRATION OF 8 MG/L AS A FLOCCULENT AID. NO COST FIGURES WERE REPORTED, BUT THE ADDITION OF AN EXTENSIVE DOSING, MIXING, AND SEDIMENTATION SYSTEM, PLUS THE FACT THAT AS MUCH AS 120 MG/L OF FLOCCULENT WOULD BE REQUIRED, INDICATED THE PROCESS WOULD DEFEAT THE MAIN PURPOSE OF OXIDATION PONDS, NAMELY LOW COST AND LITTLE OR NO MAINTENANCE. THE ROCK FILTERS, HOWEVER, SHOWED MUCH MORE PROMISE. OPERATED ONLY BY THE HYDRAULIC HEAD OF THE POND LEVEL, THE FILTERS REQUIRED NO MECHANICAL, ELECTRICAL, OR CHEMICAL EQUIPMENT AND LITTLE OR NO MAINTENANCE. SINCE COD WAS SHOWN TO BE PRESENT BOTH IN DISSOLVED AND SUSPENDED FORMS, THEN THE DIFFERENCE BETWEEN SOLUBLE COD AND TOTAL COD WOULD INDICATE THE EFFICIENCY OF REMOVAL OF SUSPENDED MATTER IN THE ROCK FILTER. ON THIS BASIS, ROCK FILTERS WERE SHOWN TO BE CAPABLE OF REMOVING 80 TO 90% OF THE SUSPENDED MATERIAL COD. FURTHER FIELD SCALE TESTS WERE RECOMMENDED FOR THE ESTABLISHMENT OF OPERATIONAL PARAMETERS AND DESIGN CRITERIA. (LOWRY-TEXAS)

FIELD 05D

ACCESSION NO. W71-13341

A LABORATORY EVALUATION OF CHEMICAL COAGULATION AS A METHOD OF TREATING STABILIZATION POND EFFLUENT,

MISSISSIPPI STATE UNIV., STATE COLLEGE. DEPT. OF CIVIL ENGINEERING.

JERRY STEWART.

MASTER'S THESIS, AUGUST 1970. 73 P, 7 FIG, 5 TAB, 14 REF.

DESCRIPTORS:
*COAGULATION, *WASTE WATER TREATMENT, ANALYTICAL TECHNIQUES, COALESCENCE, FLOCCULATION, LIQUID WASTES, SEPARATION TECHNIQUES, WATER PURIFICATION, *OXIDATION LAGOONS, STABILIZATION, INDUSTRIAL WASTES, MUNICIPAL WASTE, BIOCHEMICAL OXYGEN DEMAND, CHEMICAL OXYGEN DEMAND, *PHOSPHATES, NITROGEN, COLIFORMS, MATHEMATICAL MODELS, ALGAE, SAMPLING, STATISTICAL METHODS.

IDENTIFIERS:
*ALUMINUM SULFATE, *FERRIC CHLORIDE, *FERRIC SULFATE, *STABILIZATION PONDS, ALUM, POST-TREATMENT.

ABSTRACT:
A LABORATORY STUDY WAS PERFORMED TO EVALUATE THE SUITABILITY OF CHEMICAL COAGULATION AS A POST TREATMENT METHOD FOR EFFLUENTS FROM WASTE STABILIZATION PONDS. THE DEGREE OF EFFECTIVENESS WAS MEASURED IN TERMS OF BIOCHEMICAL OXYGEN DEMAND, CHEMICAL OXYGEN DEMAND, TOTAL PHOSPHATES, TOTAL NITROGEN, COLIFORM COUNTS, AND ALGAL CELL COUNTS. THREE COAGULANTS: ALUMINUM SULFATE, FERRIC CHLORIDE, AND FERRIC SULFATE WERE EVALUATED. A MATHEMATICAL MODEL WAS CONSTRUCTED WHICH ALLOWED FOR THE COMPUTATION OF THE REQUIRED DOSAGE OF COAGULANT FOR VARIOUS INFLUENT AND EFFLUENT CONCENTRATION OF PHOSPHATES. THE RESULTS INDICATED THAT CHEMICAL COAGULATION WITH ALUM CAN BE EFFECTIVELY USED AS A POST TREATMENT METHOD OF STABILIZATION POND EFFLUENTS. NEARLY COMPLETE ALGAE REMOVAL, OVER 90 PERCENT BOD AND PHOSPHATE REMOVALS, OVER 70 PERCENT COD REMOVALS, AND AN AVERAGE OF 5,000 COLIFORMS/1000 ML., WERE OBTAINED USING AN ALUM DOSAGE OF 1000 MG/L. SETTLING CHARACTERISTICS OF THE ALUM-ALGAE SLUDGE WERE ALSO INVESTIGATED. (ATKINS-TEXAS)

FIELD 05D

ACCESSION NO. W71-13356

OCCURRENCE OF PHOSPHORUS AND NITROGEN IN SALT CREEK AT LINCOLN, NEBRASKA,

GEOLOGICAL SURVEY, LINCOLN, NEBR.

R. A. ENGBERG, AND T. O. RENSCHLER.

GPO. WASHINGTON, D.C. 20402-PRICE $2.75. GEOLOGICAL SURVEY RESEARCH 1971, CHAPTER C, PROFESSIONAL PAPER 750-C, P C223-C227, 1971. 1 FIG, 2 TAB, 11 REF.

DESCRIPTORS:
*WATER POLLUTION SOURCES, *PHOSPHATES, *NITRATES, *NEBRASKA, URBANIZATION, CITIES, STORM RUNOFF, SEWAGE, WASTE WATER DISPOSAL, PATH OF POLLUTANTS, STORM DRAINS, DETERGENTS, EUTROPHICATION, ALGAE, NUTRIENTS.

IDENTIFIERS:
*URBAN RUNOFF, URBAN HYDROLOGY, LINCOLN(NEBR).

ABSTRACT:
CONCENTRATIONS OF PHOSPHORUS AND NITROGEN IN SALT CREEK INCREASE MARKEDLY IN THE 6-MILE REACH OF THE CREEK WITHIN THE CITY OF LINCOLN, NEBR. MOST OF THE INCREASE IS DUE TO INFLOW FROM THE LINCOLN SEWAGE-TREATMENT PLANT AND FROM STORM-SEWER AND OTHER URBAN RUNOFF ENTERING THE REACH. THE CITY CONTRIBUTES AVERAGE AMOUNTS OF 0.94 TONS PHOSPHORUS AND 1.8 TONS NITROGEN PER DAY TO THE STREAM, INDICATING ANNUAL PER CAPITA CONTRIBUTIONS OF PHOSPHORUS AND NITROGEN OF 4.5 AND 8.7 POUNDS PER YEAR, RESPECTIVELY. (KNAPP-USGS)

FIELD 05B, 05C

ACCESSION NO. W71-13466

KINETIC ASSESSMENT OF ALGAL GROWTH,

CALIFORNIA UNIV., BERKELEY.

E. A. PEARSON, E. J. MIDDLEBROOKS, M. TUNZI, A. ADINARAYANA, AND P. H. MCGAUHEY.

CHEMICAL ENGINEERING PROGRESS, SYMPOSIUM SERIES, VOL 67, NO 107, P 5-14, 1971. 8 FIG, 5 TAB, 9 REF.

DESCRIPTORS:
KINETICS, *BIOLOGICAL TREATMENT, PLANKTON, *EUTROPHICATION, *NUTRIENTS, MICROORGANISMS, EFFLUENT, BIOCHEMICAL OXYGEN DEMAND, SATURATION, OXYGEN, METABOLISM, NITROGEN, PHOSPHOROUS, *ALGAL CONTROL, CALIFORNIA.

IDENTIFIERS:
*ALGAL GROWTH, *LAKE TAHCE(CALIF).

ABSTRACT:
IT IS WELL KNOWN THAT NUTRIENTS SUCH AS NITROGEN AND PHOSPHORUS ARE REQUIRED FOR ALGAL METABOLISM. HOWEVER, THE QUESTION REMAINS, IS NITROGEN OR PHOSPHORUS USUALLY OR FREQUENTLY LIMITING THE GROWTH RATE OR STANDING STOCK IN MOST RECEIVING WATERS. TO EFFECTIVELY CONTROL THE RATE OF EUTROPHICATION OF OUR NATURAL WATERS, THIS QUESTION MUST BE ANSWERED. IT IS NOT ENOUGH TO REMOVE NITROGEN AND PHOSPHORUS FROM WASTEWATERS IN HOPES OF REDUCING THE GROWTH RATE OR STANDING STOCK OF PLANKTON IN WATERS CONCERNED. THERE IS AN URGENT NEED TO DEVELOP METHODS OR TECHNIQUES FOR ASSESSING THE BIOSTIMULATORY CHARACTER OF WASTE EFFLUENTS, NATURAL RUNOFF, ETC., ESTIMATING THE LEVEL OF CONSEQUENCES ONE MIGHT EXPECT FOR A GIVEN LEVEL OF RATE-LIMITING NUTRIENT OR SUBSTANCE. IN NO SINGLE CASE IS THIS NEED MORE URGENT THAN IN THE CASE OF LAKE TAHOE. ALTHOUGH ACTION TO REMOVE SEWAGE EFFLUENTS FROM THE BASIN IS WELL UNDERWAY, HUMAN ACTIVITY IN THE AREA IS RAPIDLY INCREASING AND IT IS NOT KNOWN WHETHER OTHER INPUTS TO THE LAKE SHOULD BE CONTROLLED TO LIMIT THE RATE OF EUTROPHICATION OR ENRICHMENT OF THE LAKE. A LABORATORY METHOD OF EVALUATING THE CAPACITY OF AQUATIC ENVIRONMENTS TO GROW ALGAE IS DEFINITELY NEEDED. (GUTIERREZ-TEXAS)

FIELD 05C, 02H

ACCESSION NO. W71-13553

REPORT OF THE FAO TECHNICAL CONFERENCE ON MARINE POLLUTION AND ITS EFFECTS ON LIVING RESOURCES AND FISHING.

FOOD AND AGRICULTURE ORGANIZATION (UN), NEW YORK. FISHERY RESOURCES DIV.

FAO FISHERIES REPORTS, NO. 99, FIRM/R99 (EN), ROME, ITALY, 9-18 DECEMBER 1970. 188 P. 132 REF.

DESCRIPTORS:
*POLLUTION ABATEMENT, *WATER POLLUTION EFFECTS, *WATER POLLUTION SOURCES, *FISHERIES, OIL WASTES, FISH REPRODUCTION, WASTE DISPOSAL, MARINE MICROORGANISMS, SEDIMENTATION, ALGAE, INVERTEBRATES, RADIATION, ECOSYSTEMS, ESTUARIES, CORAL, REEFS, WATER RESOURCES, SYSTEMS ANALYSIS, WASTE TREATMENT, AIR POLLUTION, LEGISLATION, BENTHIC FLORA, FORECASTING, LARVAE, LAGOONS, DDT, ENZYMES, MOLLUSKS, NUTRIENTS, RADIOISOTOPES, VIRUSES, SEWAGE, EUTROPHICATION, TOXICITY, LEGAL ASPECTS, PHYTOPLANKTON.

ABSTRACT:
THE FAO TECHNICAL CONFERENCE INCLUDED IN PRINCIPLE THE CONSIDERATION OF ALL TYPES OF POLLUTANTS, FROM WHATEVER SOURCE, AND IN ALL SEA AREAS WITH REFERENCE TO THE CONFERENCE'S DEFINITION OF 'MARINE POLLUTION.' GENERAL OBJECTIVES WERE TO CONCENTRATE IN ONE PLACE AS MUCH AS POSSIBLE OF WORLD-WIDE KNOWLEDGE OF MARINE POLLUTION, TO CONSIDER WAYS OF PREVENTING POLLUTION IN MARINE WATERS, TO FOCUS ATTENTION ON SCIENTIFIC PROBLEMS WHERE INTERNATIONAL COOPERATION WAS IMPERATIVE, AND TO PROVIDE SCIENTIFIC AND TECHNICAL GUIDELINES FOR THE INTERNATIONAL CONTROL OF MARINE POLLUTION. (CURRER-PAI)

FIELD 05C, 05B, 05G

ACCESSION NO. W71-13723

THE KENT COAST IN 1970,

BRITISH MUSEUM LONDON (ENGLAND). DEPT. OF BOTANY.

I. TITTLEY.

MARINE POLLUTION BULLETIN, VOL 2, NO 8, AUGUST 1971, P 120-122. 3 FIG, 22 REF.

DESCRIPTORS:
*WATER POLLUTION EFFECTS, *DATA COLLECTIONS, ALGAE, ECOTYPES, INDUSTRIAL PLANTS, INDUSTRIAL WASTES, OIL WASTES, ESTUARIES, MONITORING.

IDENTIFIERS:
*KENT COAST, INDUSTRIAL EXPANSION, ISLE OF THANET, MEDWELL ESTUARY.

ABSTRACT:
A POLLUTION RECORD IS COMPILED FOR THE KENT COAST. THE EFFECTS OF COAST 'PROTECTION' ON THE CLIFF FLORA AND THE PROBLEMS OF INDUSTRIAL EXPANSION ON THE MEDWAY ARE DESCRIBED. (CURRER-PAI)

FIELD 05C

ACCESSION NO. W71-13746

GROWTH INHIBITORS PROCUCED BY THE GREEN ALGAE (VOLVOCACEAE),

KENTUCKY UNIV., LEXINGTON. DEPT. OF BOTANY.

D. O. HARRIS.

ARCHIV FUR MIKROBIOLOGIE, VOL 76, P 47-50, 1971. 2 TAB, 9 REF. OWRR A-018-KY(1).

DESCRIPTORS:
*CHLOROPHYTA, *INHIBITORS, ALGAL CONTROL, PLANT GROWTH REGULATORS, PLANKTON, CULTURES, WATER POLLUTION EFFECTS.

IDENTIFIERS:
HETEROINHIBITORS, AUTOINHIBITORS, VOLVCCACEAE, PANDORINA MORUM, PLATYDORINA CAUDATA, EUDORINA CALIFORNICA, EUDORINA ILLINOISENSIS.

ABSTRACT:
CELLS OF 12 STRAINS CF VOLVOCACEAE GREEN ALGAE WERE REMOVED BY FILTRATION FROM THE NUTRIENT MEDIUM AFTER 21 DAY GROWTH. THE CELL-FREE MEDIA WERE INOCULATED WITH TEST GENERA AND INCUBATED FOR 14 DAYS. THE GROWTH INHIBITORY EFFECT WAS DETERMINED ON THE BASIS OF OPTICAL DENSITY OF MILLIPORE-FILTERED AND AUTOCLAVED CULTURES. CULTURE FILTRATES OF PANDORINA MORUM, VOLVULINA PRINGSHEIMII AND EUDORINA CALIFCRNICA WERE INHIBITORY TO MOST MEMBERS OF THE FAMILY. A HOPE IS EXPRESSED THAT THESE NATURAL INHIBITORY SUBSTANCES WILL EVENTUALLY SERVE TO CONTROL THE UNDESIRABLE GROWTH OF ALGAE IN WATER BASINS. (WILDE-WISCONSIN)

FIELD 05C

ACCESSION NO. W71-13793

ADENOSINE TRIPHOSPHATE CONTENT OF SELENASTRUM CAPRICCRNUTUM,

WISCONSIN UNIV., MADISON. DEPT. OF SOIL SCIENCE; AND WISCONSIN UNIV., MADISON. WATER CHEMISTRY LAB.

C. C. LEE, R. F. HARRIS, J. K. SYERS, AND D. E. ARMSTRCNG.

APPLIED MICROBIOLOGY, VOL 21, NO 5, P 957-958, 1971. 1 TAB, 6 REF.

DESCRIPTORS:
*TROPHIC LEVELS, *BIOASSAY, PHOSPHORUS, NUTRIENTS, MEASUREMENT, BACTERIA, ALGAE, EUTROPHICATION, DATA COLLECTIONS, TEST PROCEDURES, CYCLING NUTRIENTS, ANALYTICAL TECHNIQUES, WATER PCLLUTION EFFECTS, POLLUTANT IDENTIFICATION.

IDENTIFIERS:
*ADENOSINE TRIPHOSPHATE, *SELENASTRUM CAPRICORNUTUM.

ABSTRACT:
TO QUANTIFY THE LIVING BIOMASS OF HETEROGENEOUS MICROBIAL POPULATIONS IN SEEKING ACCURATE EVALUATION AND EVENTUAL UNDERSTANDING AND CONTROL OF NUTRIENT CYCLES AND OTHER ENVIRCNMENTAL PHENOMENA MEDIATED BY MICRO-ORGANISMS, A STUDY WAS MADE OF ADENOSINE TRIPHOSPHATE AS A PARAMETER OF LIVING MICROBIAL BIOMASS. THE POTENTIAL OF ATP AS A PARAMETER DEPENDS ON THE ASSUMPTION THAT ATP IS PRESENT IN A RELATIVELY CONSTANT OR PREDICTABLE AMOUNT IN DIVERSE MICROORGANISMS. SELENASTRUM CAPRICORNUTUM WAS CHOSEN AS TEST ORGANISM FOR BICASSAY; ITS UNICELLULAR CHARACTERISTICS FACILITATED ACCURATE MEASUREMENT OF VIABLE AND TOTAL ALGAL CELLS FOR CRITICAL EVALUATION OF ATP-LIVING BIOMASS RELATIONSHIPS. BASED ON THE USE OF THE 4-DAY BIOMASS VALUES AS VALID REFLECTIONS OF LIVING BIOMASS, S CAPRICORNUTUM CONTAINED 3.4, 3.1 AND 1.4 MICROGRAMS OF ATP/MG (DRY WEIGHT) OF LIVING BIOMASS UNDER PHOSPHORUS-RICH, BALANCED, AND PHOSPHORUS-DEFICIENT CONDITIONS, RESPECTIVELY. THE EFFECT OF GROWTH STAGE ON ATP CELL CONTENT WAS APPARENTLY MUCH LESS PRONOUNCED FOR S CAPRICCRNUTUM THAN FOR BACTERIA. THESE DATA TEND TO SUPPORT THE PCTENTIAL OF ATP MEASUREMENTS FOR PROVIDING VALID APPROXIMATIONS OF FLUCTUATIONS IN LIVING MICROBIAL BIOMASS IN SURFACE WATERS WHERE ALGAE ARE MAJOR CONTRIBUTORS TO THE TOTAL MICROBIAL POPULATION. (JONES-WISCONSIN)

FIELD 05C, 05A

ACCESSION NO. W71-13794

BIOLOGICAL PROCESSES FOR NITROGEN REMOVAL--THEORY AND APPLICATION,

STANFORD UNIV., CALIF.

PERRY L. MCCARTY.

AVAILABLE FROM ENGINEERING PUBLICATIONS OFFICE, 112 ENGINEERING HALL, UNIVERSITY OF ILLINOIS 61801 - PRICE $4.50. IN: PROCEEDINGS 12TH SANITARY ENGINEERING CONFERENCE ON NITRATE AND WATER SUPPLY: SOURCE AND CONTROL, FEBRUARY 11-12, 1970, UNIVERSITY OF ILLINOIS, URBANA: ILLINOIS UNIVERSITY, COLLEGE OF ENGINEERING PUBLICATION, P 136-152, 1970. 17 P, 3 FIG, 29 REF.

DESCRIPTORS:
*WASTE WATER TREATMENT, *NITRIFICATION, *DENITRIFICATION, *NITRATES, NITROGEN COMPOUNDS, AMMONIA, ALGAE, SEWAGE TREATMENT, TERTIARY TREATMENT, WATER POLLUTION CONTROL, BIODEGRADATION.

IDENTIFIERS:
*NITROGEN REMOVAL.

ABSTRACT:
NITROGEN CAN BE REMOVED BIOLOGICALLY FROM WASTE WATER BY THREE DIFFERENT PROCESSES: BACTERIAL ASSIMILATION, ALGAE HARVESTING, AND NITRIFICATION-DENITRIFICATION. BACTERIAL NITRIFICATION-DENITRIFICATION IS PERHAPS THE MOST PROMISING OF THE PROCESSES. AMMONIA NITROGEN REMOVAL IS THE RESULT OF TWO STAGES. IN NITRIFICATION, AMMONIA-NITROGEN IS CONVERTED TO NITRATE-NITROGEN BY TWO DIFFERENT GROUPS OF AUTOTROPHIC NITRIFYING BACTERIA. DENITRIFICATION IS THE REDUCTION OF NITRATES AND NITRITES TO NITROGEN GAS BY A WIDE VARIETY OF FACULTATIVE BACTERIA UNDER ANAEROBIC CONDITIONS. AN ADDITIONAL ORGANIC SOURCE IS NECESSARY FOR EFFICIENT CONVERSION TO NITROGEN GAS, AND METHANOL APPEARS TO BE THE LEAST EXPENSIVE. IN GENERAL FOR AGRICULTURAL AND DOMESTIC WASTE WATER, COSTS FOR NITROGEN REMOVAL MAY RANGE FROM 2 TO 10 CENTS PER 1,000 GALLONS. OF THE THREE PROCESSES CONSIDERED, NITRIFICATION-DENITRIFICATION IS PERHAPS THE MOST GENERALLY APPLICABLE BECAUSE OF GOOD RELIABILITY, SUITABILITY TO A VARIETY OF CONDITIONS, LOW AREA REQUIREMENTS AND MODERATE COST. (KNAPP-USGS)

FIELD 05D

ACCESSION NO. W71-13939

COMBINED WASTEWATER COLLECTION AND TREATMENT FACILITY, MOUNT CLEMENS, MICHIGAN,

SPALDING, DEDECKER AND SSOCIATES, INC, MADISON HEIGHTS, MICH.

VIJAYSINH U. MAHIDA.

PREPRINT, 44TH ANNUAL CONFERENCE, WATER POLLUTION CONTROL FEDERATION, SESSION 4, NO 4, OCTOBER 3-8, 1971, SAN FRANCISCO, CAL 11 P.

DESCRIPTORS:
*STORM RUN-OFF, *SUSPENDED SOLIDS, *RECREATION, *TREATMENT FACILITIES, SAMPLING, MONITORING, AUTOMATIC CONTROLS, AERATION, SEDIMENTATION, FILTRATION, ANAEROBIC DIGESTION, CHLORINATION, DISINFECTION, ULTRAVIOLET LIGHT, ALGAE, OXYGENATION, BIOCHEMICAL OXYGEN DEMAND, WATER SPORTS, PARKS, COST ANALYSIS, MICHIGAN, WASTE WATER TREATMENT.

IDENTIFIERS:
*COMBINED SEWERS, *SUSPENDED SOLIDS, BACKWASHING, *MOUNT CLEMENS(MICH).

ABSTRACT:
THE CITY OF MOUNT CLEMENS, MICHIGAN, HAS 50 MILES OF COMBINED SEWERS SERVING 3.06 SQUARE MILES, AND 10 MILES OF SANITARY SEWERS PLUS 6 MILES OF STORM SEWERS SERVING AN ADDITIONAL 0.8 SQUARE MILES. INTERCEPTORS WERE DESIGNED TO COLLECT STORMWATER OVERFLOWS FROM TWO LOCATIONS ON THE COMBINED SEWERS, SERVING 212 ACRES AND PUMP IT TO A SMALL LAKE. THIS FIRST LAKE PROVIDED SETTLING, NATURAL AND MECHANICAL SURFACE AERATION, AND BOTH AEROBIC AND ANAEROBIC DIGESTION. STORM OVERFLOWS COULD THEN BE DRAWN OFF AT A CONTROLLED 1.0 MGD RATE TO A MICROSTRAINER. ULTRAVIOLET RADIATION EQUIPMENT FOR ALGAE CONTROL WAS ALSO INSTALLED WITH THE MICROSTRAINER, SO THE TWO COMBINED PROVIDED MECHANICAL FILTRATION, SUSPENDED SOLIDS AND INCIDENTAL BOD REMOVAL, AND ALGAE REMOVAL. BEFORE ENTERING THE SECOND LAKELET, THE EFFLUENT ALSO IS SUBJECTED TO CHLORINE-CHLORINE DIOXIDE DISINFECTION. LAKELET 2 WAS DESIGNED TO PROVIDE CHLORINE CONTACT TIME, NATURAL SURFACE AERATION, AND PHOTOSYNTHETIC OXYGENATION, WHILE LAKELET 3 PROVIDED MECHANICAL AS WELL AS SURFACE AERATION. A PRESSURE SAND FILTER WAS THEN DESIGNED AS A POLISHING STEP. THE ENTIRE TREATMENT SYSTEM WAS INCLUDED AS PART OF AN INNER-CITY PARK. COST OF THIS SYSTEM WAS $7000 PER ACRE BENEFITED, WHEREAS COST OF SEWER SEPARATION ONLY WAS ESTIMATED AT $15,000 PER ACRE. (LOWRY-TEXAS)

FIELD 05D

ACCESSION NO. W72-00042

EFFECTS OF ZOOPLANKTON ON ALGAE IN WESTHAMPTON LAKE,

RICHMOND UNIV., VA. DEPT. OF BIOLOGY.

JOHN W. BISHOP.

AVAILABLE FROM THE NATIONAL TECHNICAL INFORMATION SERVICE AS PB-203 961, $3.00 IN PAPER COPY, $0.95 IN MICROFICHE. VIRGINIA POLYTECHNIC INSTITUTE, WATER RESOURCES RESEARCH CENTER BULLETIN 43, 1971. 33 P, 17 FIG, 8 TAB, 22 REF. OWRR A-022-VA (2).

DESCRIPTORS:
*ZOOPLANKTON, *ALGAE, *PHOTOSYNTHESIS, RAINFALL, DIURNAL, FLUCTUATIONS, CHLOROPHYTA, EUGLENOPHYTA, CYANOPHYTA, CHRYSOPHYTA, COPEPODS, CRUSTACEANS, LABORATORY TESTS, VIRGINIA.

IDENTIFIERS:
WESTHAMPTON LAKE (VA).

ABSTRACT:
ASSEMBLAGES OF ALGAE WITH DIFFERENT CONCENTRATIONS OF ZOOPLANKTERS WERE KEPT IN CYLINDERS OF POLYETHYLENE AND FLEXIGLASS, SUSPENDED IN THE LAKE. ANALYSES OF WATER AND BOTH GROUPS OF ORGANISMS WERE CONDUCTED INSIDE AND OUTSIDE THESE AQUARIUMS. ROTIFERS EXHIBITING NO CONSISTENT VERTICAL MIGRATION DECREASED THE DAILY PHOTOSYNTHESIS OF ALGAE BY 22%. VERTICALLY MIGRATING COPEPODS AND CLADOCERANS EXERTED NO SIGNIFICANT EFFECT ON THE PHOTOSYNTHESIS. NONE OF THE ANIMAL GROUPS INFLUENCED THE PHOTOSYNTHETIC EFFICIENCY OF ALGAE, THEIR CARBOHYDRATE PRODUCTION PER UNIT OF CHLOROPHYLL. PHOTOSYNTHESIS AND PHOTOSYNTHETIC EFFICIENCY WERE INCREASED SEVERAL DAYS AFTER RAINFALL. (WILDE-WISCONSIN)

FIELD 05C, 02H

ACCESSION NO. W72-00150

TOXICITY OF SELENIUM TO THE BLUE-GREEN ALGAE, ANACYSTIS NIDULANS AND ANABAENA VARIABILIS,

UDAIPUR UNIV. (INDIA). DEPT. OF BOTANY.

H. D. KUMAR, AND G. PRAKASH.

ANNALS OF BOTANY, VOL 35, P 697-705, 1971. 5 FIG, 12 REF.

DESCRIPTORS:
*TOXICITY, *CYANOPHYTA, SULFUR, ALGAE, GRANULES, INHIBITION, METABOLISM, CULTURES, ENVIRONMENTAL EFFECTS, PLANT GROWTH, SULFATE, LABORATORY TESTS, CYTOLOGICAL STUDIES.

IDENTIFIERS:
*SELENIUM, *ANABAENA VARIABILIS, *ANACYSTIS NIDULANS, GROWTH INHIBITION, SELENATE, SELENITE, SELENOMETHIONINE, SELENOPURINE, ANTIMETABOLITES.

ABSTRACT:
COMPARISON WAS MADE OF TOXIC EFFECTS OF SELENATE, SELENITE, SELENOMETHIONINE, AND SELENOPURINE ON DIFFERENT GROWTH PHASES OF ANACYSTIS NIDULANS AND ANABAENA VARIABILIS IN CONSIDERING SELENIUM TOXICITY. A PROTECTIVE ROLE OF SULFUR AGAINST SELENIUM TOXICITY WAS SUGGESTED AS THESE COMPOUNDS WERE LESS TOXIC IN CULTURE MEDIUM CONTAINING SULFATE THAN IN SULFUR-FREE MEDIUM. GROWTH INHIBITORY EFFECTS OF SELENATE AND SELENITE WERE STUDIED ON ALGAE GROWN ON AGAR MEDIUM AND IN LIQUID MINERAL MEDIUM. SELENOMETHIONINE AND SELENOPURINE WERE STUDIED ONLY IN LIQUID CULTURE. GROWTH WAS ESTIMATED BY MEASURING OPTICAL DENSITY OF CULTURE TUBES AND PERCENT SURVIVAL SCORED BY COLONY COUNTING. DOSE-RESPONSE CURVES OF BOTH ALGAE SUGGEST COMPARATIVELY GREATER KILLING BY SELENITE THAN BY SELENATE. ANACYSTIS NIDULANS IS THREE TIMES MORE TOLERANT TO SELENITE AND SELENATE KILLING THAN ANABAENA VARIABILIS. SELENITE IS MORE TOXIC IN AGAR PLATE CULTURES AND LESS TOXIC IN LIQUID CULTURES THAN SELENATE. CELLS GROWN IN THE HIGHEST GROWTH-PERMITTING CONCENTRATION OF SELENITE IN LIQUID MEDIUM FORM A VARIABLE NUMBER OF RED GRANULES. NO SUCH GRANULATION OCCURS IN CELLS TREATED WITH OTHER SELENO-COMPOUNDS, INDICATING THAT THE MODE OF SELENITE ACTION IN BLUE-GREEN ALGAE DIFFERS FROM THAT OF OTHER SELENIUM COMPOUNDS. (JONES-WISCONSIN)

FIELD 05C

ACCESSION NO. W72-00153

SOME ECOLOGICAL EFFECTS ON THE USE OF PARAQUAT FOR THE CONTROL OF WEEDS IN SMALL LAKES,

NATURE CONSERVANCY, ABBOTS RIPTON (ENGLAND). MONKS WOOD EXPERIMENTAL STATION; AND IMPERIAL CHEMICAL INDUSTRIES LTD., JEALOTT'S HILL (ENGLAND). JEALOTT'S HILL RESEARCH STATION; AND NOTTINGHAM UNIV. (ENGLAND). DEPT. OF BOTANY.

J. M. WAY, J. F. NEWMAN, N. W. MOORE, AND F. W. KNAGGS.

THE JOURNAL OF APPLIED ECOLOGY, VOL 8, NO 2, P 509-532, 1971.

DESCRIPTORS:
*HERBICIDES, *AQUATIC WEED CONTROL, *PARAQUAT, *ENVIRONMENTAL EFFECTS, DIQUAT, ALGAE, ALGAL CONTROL, DIATOMS, PERSISTENCE, BACTERIA, BIRDS, FISH, ON-SITE INVESTIGATION, METHODOLOGY, INVERTEBRATES, AMPHIBIANS, TOXICITY, CHARA, SUBMERGED PLANTS, PROTOZOA, EUGLENA, CHLAMYDOMONAS, CHLOROPHYTA, SCENEDESMUS, CYANOPHYTA, MOLLUSKS, MITES, ANNELIDS, CRUSTACEANS.

IDENTIFIERS:
BIOLOGICAL EFFECTS, NOTTINGHAM(ENGLAND), POLYGONUM, TYPHA, ELODEA.

ABSTRACT:
THE RESULTS OF PARAQUAT APPLICATIONS TO A SERIES OF SMALL LAKES WERE STUDIED AND THE BIOLOGICAL EFFECTS AND PARAQUAT RESIDUE DETERMINATIONS DESCRIBED. IN TWO EXPERIMENTS IN LAKES AT OXTON, NOTTINGHAMSHIRE (ENGLAND), 0.5 MG/L PARAQUAT ERADICATED ALL SUBMERGED AND FLOATING PLANTS, EXCEPT POLYGONUM AMPHIBIUM AND CHARA, WITHIN 32 DAYS OF APPLICATION. ONE LAKE REMAINED SUBSTANTIALLY FREE OF VEGETATION FOR TWO YEARS AFTER TREATMENT. NO MATS OF FILAMENTOUS ALGAE DEVELOPED IN THE TREATED LAKES ALTHOUGH THEY WERE WIDESPREAD IN ADJACENT UNTREATED LAKES. CHANGES IN SPECIES AND POPULATIONS OF PLANKTONIC ALGAE AND DIATOMS AND INCREASES IN POPULATION OF BACTERIA SOON AFTER TREATMENT WERE OBSERVED. SURVIVAL OF CAPTIVE POPULATIONS OF ASELLUS, LYMNAEA AND SIALIS LARVAE UP TO FOUR DAYS AFTER TREATMENT AND OBSERVATIONS ON OTHER FREE-LIVING INVERTEBRATES WERE RECORDED. THE CHEMICAL DID NOT APPARENTLY CAUSE ANY IMPORTANT MORTALITY OF INVERTEBRATES. PARAQUAT DISAPPEARED RAPIDLY FROM THE WATER. SIGNIFICANT QUANTITIES OF THE CHEMICAL WERE FOUND IN THE WEEDS WITHIN 24 HOURS OF APPLICATION. THERE WAS A GRADUAL BUILDUP OF PARAQUAT RESIDUES IN BOTTOM DEPOSITS UP TO 32 AND SUBSEQUENT 197 DAYS AFTER APPLICATION WITH A MARKED FALL-OFF AT 364 DAYS. (JONES-WISCONSIN)

FIELD 04A, 05C

ACCESSION NO. W72-00155

THE USE OF RADIOACTIVE CARBON METHOD IN STUDIES OF TROPHIC RELATIONSHIPS OF PLANKTON (PRIMENENIE RADIOUGLERODNOGO METODA DLYA ISUCHENIYA TROFICHESKIKH VZAIMOOTNOSHENII V PLANKTONE),

AKADEMIYA NAUK SSSR, LENINGRAD. ZOOLOGICHESKII INSTITUT.

S. M. VARDAPETYAN, B. L. GUTELMACHER, AND N. G. OZERETSKOVSKAYA.

DOKLADY AKADEMII NAUK SSSR, VOL 197, NO 3, P 705-707, 1971. 1 FIG, 2 TAB, 5 REF.

DESCRIPTORS:
*ANALYTICAL TECHNIQUES, *RADIOACTIVITY TECHNIQUES, *PLANKTON, PRIMARY PRODUCTIVITY, FOOD HABITS, ALGAE, ZOOPLANKTON, CARNIVORES, DAPHNIA.

IDENTIFIERS:
BOSMINA, CERIODAPHNIA, EUDIAPTOMUS.

ABSTRACT:
SAMPLES OF PLANKTON FROM TWO LAKES WERE PLACED IN 20 LITER CONTAINERS WITH LAKE WATER AND C-14 AS BICARBONATE OF SODIUM IN CONCENTRATION OF 10 MICRO/LITER. IN INTERVALS FROM 1 TO 8 DAYS, THE RADIOACTIVITY OF PHYTOPLANKTON WAS DETERMINED ON 300 ML ALIQUOTS. THE REMAINDER WAS FILTERED THROUGH NO 62 SIEVE. THE RETAINED ZOOPLANKTON WAS FIXED IN FORMALIN AND DISTRIBUTED ACCORDING TO SPECIES. AFTER 24 HOURS DRYING IN A DESICCATOR, THE RADIOACTIVITY OF DIFFERENT SPECIES WAS DETERMINED. THE RESULTS PROVIDED A PICTURE OF UTILIZATION OF THE PRIMARY PRODUCTION BY BOSMINA, DAPHNIA, CERIODAPHNIA, AND EUDIAPTOMUS SPECIES. RESULTS OBTAINED WITH NUTRITION OF PREDATORS WERE LESS WELL PRONOUNCED. (WILDE-WISCONSIN)

FIELD 05C, 07A, 05A

ACCESSION NO. W72-00161

ENVIRONMENTAL CONTROL OF GROWTH AND DIFFERENTIATION IN MULTICELLULAR PHCTOSYNTHETICS,

WASHINGTON UNIV., ST. LOUIS, MC.

H. WAYNE NICHOLS.

AVAILABLE FROM THE NATIONAL TECHNICAL INFORMATION SERVICE AS AD-725 497, $3.00 IN PAPER COPY, $0.95 IN MICROFICHE. FINAL REPORT NOV 17, 1970. 6 P. CNR GRANT NR 104-801.

DESCRIPTORS:
*ALGAE, *ECOLOGY, ENVIRONMENT, PHOTOPERIODISM, GROWTH RATES, TEMPERATURE, BIOCHEMISTRY, PHYSIOLOGY, LIGHT.

IDENTIFIERS:
PHOTOSENSITIVITY.

ABSTRACT:
A SUMMARY IS GIVEN OF A RESEARCH PROJECT WHICH CONCERNED ITSELF WITH BASIC DEVELOPMENT AND BEHAVIORAL ASPECTS OF UNICELLULAR AND MULTICELLULAR RED ALGAE. VARIOUS PHYSIOLOGICAL AND BIOCHEMICAL CHARACTERISTICS OF THE ALGAE HAVE BEEN STUDIED, AND TECHNIQUES FOR CONTROL OF CERTAIN DEVELOPMENTAL ASPECTS OF THE CRGANISMS IN THE LABORATORY HAVE BEEN DEVELCPED. TO DATE THE RESEARCH HAS BASICALLY INVOLVED PHYSICAL ASPECTS OF THE ENVIRONMENT. HOWEVER, IT IS NOW CLEARLY EVIDENT THAT CHEMICAL EVENTS OCCURRING WITHIN THE DIFFERNTIATING CELLULAR SYSTEMS MUST BE STUDIED AND RELATED TO SUCH PHYSICAL CONTROL MECHANISMS AS PHOTOPERIODICITY, TEMPERATURE AND LIGHT QUALITY.

FIELD 05B

ACCESSION NO. W72-00377

BIOLCGICAL--GAMMA-RADIATION SYSTEM FOR SEWAGE PROCESSING,

ENERGY SYSTEMS, INC., MELBCURNE, FLA.

L. A. MANN.

ISCTOPES AND RADIATION TECHNOLOGY, VOL 8, NO 4, P 439-444, SUMMER, 1971. 4 FIG, 1 TAB, 4 REF.

DESCRIPTORS:
*SEWAGE EFFLUENTS, *SEWAGE TREATMENT, *ECONOMIC JUSTIFICATION, *COBALT, *RADIATION, COSTS, SEWAGE BACTERIA, ALGAE, CCLIFORMS, INSECTICIDES.

IDENTIFIERS:
*BIOLOGICAL--GAMMA-RADIATION SYSTEM, MODULAR SYSTEM, COBALT-60.

ABSTRACT:
A BIOLOGICAL--GAMMA-RADIATION METHOD FOR PURIFYING SEWAGE WAS DEVELOPED, AND BOTH A PILOT AND COMMERCIAL PLANT WERE DESIGNED, CONSTRUCTED, AND TESTED. THE PLANT DESIGN INCORPCRATES SEVERAL MODULES INCLUDING A WET WALL, A BIOLCGICAL TREATMENT UNIT, AN IRRADIATOR, A PRIMARY FILTER SYSTEM, AND AN ACTIVATED-CHARCOAL (POLISHING) FILTER. AT DIFFERENT FLOW RATES ON NORMAL SEWAGE, TESTS INDICATE THAT ALMOST ALL COLIFORM BACTERIA WERE KILLED, AND BOTH BIODEGRADABLE AND NONBIODEGRADABLE DETERGENTS WERE MORE THAN 90% DESTROYED. LIMITED TESTS ON PARATHION IN WATER RESULTED IN 25-30% DESTRUCTION OF THIS INSECTICIDE. THERE IS SOME INDICATION THAT ALGAE WILL NOT GROW IN THE EFFLUENT. FURTHERMORE, SETTLEABLE-SOLIDS CONCENTRATION AND TURBIDITY WERE DECREASED, AND A SATISFACTORY BOD LEVEL WAS MAINTAINED IN THE TWO PLANTS. THE COST OF BUILDING AND OPERATING A PLANT USING A BIOLOGICAL--GAMMA-RADIATION SYSTEM DEPENDS ON THE QUALITY OF WATER DESIRED AND THE MODULES SELECTED FOR THE PLANT. ESTIMATES INDICATE THAT THE COST OF TREATMENT PLANTS CONSISTING OF A CONVENTIONAL SECONDARY TREATMENT PLUS IRRADIATOR AND PRIMARY FILTRATION MODULES IS LESS THAN THE COST OF CONVENTIONAL ADVANCED-TREATMENT PLANTS. (SETTLE-WISCONSIN)

FIELD 05D

ACCESSION NO. W72-00383

DISINFECTION OF ALGAL LADEN WATERS,

NOTRE DAME UNIV., IND.

WAYNE F. ECHELBERGER, JR., JOSEPH L. PAVONI, PHILIP C. SINGER, AND MARK W. TENNEY.

JOURNAL OF THE SANITARY ENGINEERING DIVISION, PROCEEDING OF ASCE, VOL 97, NO SA 5, OCTOBER, 1971. P 721-730, 9 FIG, 10 REF.

DESCRIPTORS:
*CHLORINATION, *ALGAE, *PHOTOSYNTHESIS, SOLUBILITY, CHEMICAL OXYGEN DEMAND, BIOCHEMICAL OXYGEN DEMAND, FILTRATION, DISINFECTION, FLOCCULATION, *WASTE WATER TREATMENT.

IDENTIFIERS:
*LYSIS, *CONTACT TIME, CELLULAR METABOLITES.

ABSTRACT:
BATCH LABORATORY SCALE EXPERIMENTS WERE USED TO DEMONSTRATE THE EFFECT OF ALGAE ON THE CHLORINATION PROCESS. TOTAL CHLORINE RESIDUALS WERE MEASURED BY THE STANDARD ORTHO TOLIDINE PROCEDURE. STANDARD JAR TEST FLOCCULATION STUDIES USING CHLORINE AND A CATIONIC POLYAMINE AS COAGULANTS WERE ALSO CONDUCTED. RESULTS INDICATED THAT ALGAL CELLS DO EXERT A CHLORINE DEMAND, AND AS THE ALGAL CONCENTRATION INCREASES, THE CHLORINE DOSAGE REQUIRED TO MAINTAIN A CERTAIN FREE RESIDUAL LEVEL INCREASES. CELLULAR CHLOROPHYLL FRACTIONS OF THE ALGAE WERE REDUCED BY CHLORINE DISINFECTION, RESULTING IN IMPAIRMENT OF THE PHOTOSYNTHETIC CAPABILITY OF THE SLUDGE. APPARENT ALGAL CELL LYSING A CELLULAR METABOLITE LEAKAGE FOLLOWING CHLORINATION TO A DESIRABLE RESIDUAL LEVEL SIGNIFICANTLY INCREASED THE SOLUBLE ORGANIC CONCENTRATION OF THE SUSPENDING MEDIUM. FROM THE ALGAE REMOVAL STANDPOINT, CHLORINATION DID ENHANCE FLOCCULATION, PROBABLY BECAUSE SOME OF THE CELLULAR METABOLITES RELEASED WERE NATURAL ALGAL FLOCCULANTS. WIDESPREAD USAGE OF CHLORINATION FOR DISINFECTION MAY SOON REQUIRE ALGAE REMOVAL PRIOR TO DISINFECTION TO INSURE ADEQUATE DISINFECTIONS. (LOWRY-TEXAS)

FIELD 05D

ACCESSION NO. W72-00411

CHLORINATION DYNAMICS IN WASTEWATER EFFLUENTS,

TECHION - ISRAEL INST. OF TECH., HAIFA (ISRAEL). SANITARY ENGINEERING LAB.

YEHUDA KOTT.

JOURNAL OF THE SANITARY ENGINEERING DIVISION, PROCEEDINGS OF ASCE, VOL 97, NO SA 5, OCTOBER 1971, P 647-659. 2 FIG, 10 TAB, 17 REF.

DESCRIPTORS:
*CHLORINATION, *DISINFECTION, *OXIDATION LAGOONS, COLIFORMS, ALGAE, TEMPERATURE, ANALYTICAL TECHNIQUES, WATER QUALITY CONTROL, *WASTE WATER TREATMENT.

IDENTIFIERS:
*CYSTS, ISRAEL.

ABSTRACT:
THE RESPONSE OF COLIFORMS, ENDAMOEBA HISTOLYTICA CYSTS, AND ALGAE TO CHLORINATION WAS INVESTIGATED ON A LABORATORY SCALE. EFFLUENT USED WAS OBTAINED FROM A TRICKLING FILTER PLANT, AND FROM AN ARTIFICIAL OXIDATION POND. APPLICATION OF 8 MG/L OF CHLORINE TO THE TRICKLING FILTER EFFLUENT REDUCED THE COLIFORM COUNT FROM 10 TO THE 7TH POWER ORGANISMS PER 100 ML TO LESS THAN 100 ORGANISMS PER 100 ML. APPLICATION OF 14 MG/L ACTUALLY PRODUCED A GREATER NUMBER OF COLIFORM ORGANISMS IN THE EFFLUENT THAN WAS NOTICED FOR THE 8 MG/L DOSAGE. THE PROBABLE CAUSE WAS ASSUMED TO BE THE BREAK-DOWN OF ORGANIC MATERIAL AND RELEASE OF ENTRAPPED COLIFORM ORGANISMS. APPLICATION OF THE 8 MG/L DOSAGE TO SOLUTIONS CONTAINING THE ENDAMOEBA HISTOLYTICA CYSTS ALSO PRODUCED A RAPID AND THOROUGH KILL, ALTHOUGH THE DEATH RATE WAS QUITE TEMPERATURE DEPENDENT. ALGAE WAS UNAFFECTED FOR A TWO HOUR INTERVAL, EXCEPT FOR THE CESSATION OF GROWTH. SINCE THE CHLORINE DID NOT ATTACK THE ALGAE FOR 2 HOURS, MUCH LONGER THAN THE NORMAL CONTACT TIME, THEN, CHLORINATION SHOULD BE WELL SUITED FOR DISINFECTION OF OXIDATION POND EFFLUENT. (LOWRY-TEXAS)

FIELD 05D

ACCESSION NO. W72-00422

PLANKTON ASSOCIATIONS AND RELATED FACTORS IN A HYPEREUTROPHIC LAKE,

WASHINGTON UNIV., SEATTLE. WATER AND AIR RESOURCES DIV.

RONALD M. BUSH, AND EUGENE B. WELCH.

AVAILABLE FROM NATIONAL TECHNICAL INFORMATION SERVICE AS PB-204 230, $3.00 IN PAPER COPY, $0.95 IN MICROFICHE. PARTIAL PROJECT COMPLETION REPORT, AUGUST 15, 1971. 34 P, 3 FIG, 2 TAB, 37 REF. OWRR A-034-WASH(1).

DESCRIPTORS:
WATER POLLUTION EFFECTS, *EUTROPHICATION, LAKES, *NUTRIENTS, ALGAL CONTROL, *CYANPHYTA, *NUISANCE ALGAE, PLANKTON, WASHINGTON, DIATOMS, CHLOROPHYTA.

IDENTIFIERS:
HYPEREUTROPHIC LAKES, *MOSES LAKE(WASH).

ABSTRACT:
CLUSTER ANALYSIS WAS USED TO GROUP SAMPLES COLLECTED FROM TEN STATIONS IN MOSES LAKE, WASHINGTON, ACCORDING TO THE SIMILARITY OF THEIR CONTAINED ALGAL SPECIES. DURING THE PERIOD 1968 TO 1970, NINE RECURRING DISTINCT SAMPLE GROUPS, OR ALGAL POPULATIONS, WERE IDENTIFIED. OF THE NINE, THREE WERE MOST DISTINCT; THEY CONSISTENTLY RECURRED AT THE SAME STATIONS, AND WERE DOMINATED BY DIATOM, GREEN, AND BLUE-GREEN ALGAE, RESPECTIVELY. OF THE SIX SPECIES OF BLUE-GREENS THAT CHARACTERIZED THAT POPULATION, THE RECREATIONALLY NUISANCE FORMS, APHANISOMENON FLOS-AQUAE AND MICROCYSTIS ACRUGINOSA, WERE DOMINANT. THE BLUE-GREEN POPULATION WAS THE MOST WIDE-SPREAD IN THE LAKE AND OCCURRED IN WATERS THAT WERE WARMEST AND CONTAINED THE LOWEST CONCENTRATIONS OF INORGANIC NUTRIENTS: NITROGEN, PHOSPHORUS, AND CARBON. GREEN ALGAE DOMINATED IN WATERS THAT RECEIVED TREATED SEWAGE EFFLUENT AND CONTAINED RELATIVELY HIGH CONCENTRATIONS OF NUTRIENTS. AS THE NUTRIENT CONTENT DECLINED PROCEEDING AWAY FROM THAT AREA, BLUE-GREEN ALGAE BECAME DOMINANT. TEMPORAL VARIATION IN BIOMASS (CHLOROPHYLL CONTENT) OF THE BLUE-GREEN POPULATION WAS INVERSELY RELATED TO PHOSPHATE CONTENT, BUT NOT TO THE OTHER NUTRIENTS. THESE RESULTS SUPPORT THE HYPOTHESIS THAT NUISANCE BLUE-GREEN ALGAE DOMINATE IN SHALLOW EUTROPHIC LAKES DURING WARM SUMMER MONTHS WHEN AMBIENT NUTRIENT CONTENT IS LOW BECAUSE, UNDER THESE CONDITIONS, THEY APPARENTLY OUT-COMPETE OTHER FORMS FOR NUTRIENTS. (SEE ALSO W72-00799)

FIELD 05C, 05G

ACCESSION NO. W72-00798

DILUTION AS A CONTROL FOR NUISANCE ALGAL BLOOMS,

WASHINGTON UNIV., SEATTLE. WATER AND AIR RESOURCES DIV.

EUGENE B. WELCH, JAMES A. BUCKLEY, AND RONALD M. BUSH.

PARTIAL PROJECT COMPLETION REPORT, AUGUST 31, 1971. 39 P, 14 FIG, 8 TAB, 18 REF. OWRR A-034-WASH(2).

DESCRIPTORS:
*WASTE DILUTION, WATER POLLUTION CONTROL, *NUISANCE ALGAE, *EUTROPHICATION, LAKES, WASHINGTON, *ALGAL CONTROL, NUTRIENTS, BIOMASS, *CYANOPHYTA, PLANT GROWTH, PLANKTON.

IDENTIFIERS:
*MOSES LAKE(WASH), HYPEREUTROPHIC LAKES.

ABSTRACT:
EXPERIMENTS WERE CONDUCTED IN SITU IN HYPEREUTROPHIC MOSES LAKE IN EASTERN WASHINGTON TO DETERMINE THE EFFECT OF LOW-NUTRIENT DILUTION WATER ON GROWTH AND DOMINANCE OF NUISANCE BLUE-GREEN ALGAE, THE MECHANISM OF EFFECT AND SUGGEST POSSIBLE CONTROL MEASURES BY COMPARISON WITH RESULTS FROM A SUCCESSFUL CONTROL PROGRAM IN GREEN LAKE IN SEATTLE. THE ADDITION OF LOW-NUTRIENT COLUMBIA RIVER WATER TO MOSES LAKE WATER REDUCES THE SUBSEQUENT MAXIMUM BIOMASS OF NUISANCE BLUE-GREEN ALGAE ATTAINED IN THE DILUTION WATER-LAKE WATER MIXTURE IN DIRECT PROPORTION TO THE AMOUNT OF DILUTION WATER ADDED. IN CONTRAST TO BIOMASS, THE RELATIONSHIP BETWEEN DILUTION WATER ADDITION AND MAXIMUM GROWTH RATE OF BLUE-GREEN ALGAE IS NON-LINEAR. GROWTH RATE REMAINED HIGH (TWO DOUBLINGS PER DAY) AND RELATIVELY CONSTANT AT LAKE WATER CONCENTRATIONS DOWN TO ABOUT 50% AND THEN DROPPED RAPIDLY TO NEGATIVE RATES AT 0% LAKE WATER. COMPARISON OF RESULTS OF DILUTION WATER ADDITION TO GREEN LAKE IN SEATTLE WITH EXPERIMENTAL AND CALCULATED POTENTIAL EFFECTS OF DILUTION WATER IN PARKER HORN OF MOSES LAKE ALLOWED ESTIMATES OF DILUTION RATES THAT MIGHT OFFER SOME CONTROL OF BLUE-GREEN ALGAE BLOOMS. ADDITION OF LOW-NUTRIENT RIVER WATER AT ABOUT 1% PER DAY WOULD BE EXPECTED TO RESULT IN A LAKE WATER--RIVER WATER MIXTURE THAT WOULD REDUCE MAXIMUM GROWTH RATE BY ONE HALF IN ABOUT 150 DAYS AND REACH A STEADY STATE BIOMASS IN ABOUT 200 DAYS. (SEE ALSO W72-00798)

FIELD 05G, 05C

ACCESSION NO. W72-00799

NITROGEN REMOVAL AND IDENTIFICATION FOR WATER QUALITY CONTROL,

WASHINGTON UNIV., SEATTLE. DEPT. OF CIVIL ENGINEERING.

DALE A. CARLSON.

AVAILABLE FROM NATIONAL TECHINAL INFORMATION SERVICE AS PB-204 231, $3.00 IN PAPER COPY, $0.95 IN MICROFICHE. AUGUST 15, 1971. 52 P, 12 FIG, 8 TAB, 85 REF. OWRR A-040-WASH(1).

DESCRIPTORS:
*NITRIFICATION, *DENITRIFICATION, WATER QUALITY CONTROL, EUTROPHICATION, ACTIVATED SLUDGE, DESIGN, NITROGEN, DIFFUSION, *ALGAL CONTROL, AERATION, *WASTE WATER TREATMENT.

IDENTIFIERS:
*ION-SELECTIVE CATHODES.

ABSTRACT:
REMOVAL OF NITROGEN FROM WASTE WATER SYSTEMS BY BIOLOGICAL NITRIFICATION AND DENITRIFICATION IN AN ACTIVATED SLUDGE SYSTEM IS CONSIDERED. ONE OF THE MAJOR METHODS PROPOSED FOR CONTROLLING EXCESS ALGAL PRODUCTION IN RECEIVING WATERS, THE PROCESS CONSISTS OF AERATING ACTIVATED SLUDGE MIXED LIQUOR SOLIDS FOR DETENTION TIMES SUFFICIENT FOR NITRIFYING BACTERIA TO OXIDIZE AMMONIA TO NITRATE. THE MIXED LIQUOR THEN GOES TO AN ANAEROBIC DENITRIFICATION ZONE WHERE THE NITRATE IS BIOLOGICALLY CONVERTED TO NITROGEN GAS. PILOT PLANT STUDIES PRECEDING THE CURRENT STUDIES ARE DISCUSSED PRIOR TO PRESENTATION OF THE WORK ON PURE CULTURE KINETIC STUDIES AND DEVELOPMENT OF ION SELECTIVE CATHODES FOR IDENTIFICATION OF NITROGEN FRACTIONS. FOR THE CONDITIONS USED, THE RATE OF DENITRIFICATION BY PSEUDOMONAS DENITRIFICANS FOLLOWED THE MICHAELIS PATTERN, NOT BEING INHIBITED BY HIGH CONCENTRATIONS OF CH3COONA AS THE SUBSTRATE. THE THEORETICAL MAXIMUM RATE OF DENITRIFICATION (V) WAS ABOUT 10 MICRON 1N2/HR/MG BACTERIAL CELL AND THE THEORETICAL VALUE OF KS WAS 5.8 MM WITH A CELL CONCENTRATION OF 3400 MG/L. THE OBTAINED KS VALUE PROMISES A REASONABLE RATE OF DENITRIFICATION UNDER THE GIVEN CONDITIONS. ALTHOUGH THE RATES DENITRIFICATION AT THE VARIOUS KNO3 CONCENTRATIONS DID NOT FOLLOW THE MICHAELIS PATTERN, INHIBITING EFFECTS OF NITRATE WERE NOT OBSERVED AT KNO3 CONCENTRATIONS AS HIGH AS 8.0 MM/L. THE USE OF ION SELECTIVE CATHODES FOR DETECTING IONS SUCH AS THE NITROGENS WAS INVESTIGATED. IT APPEARS THAT THE SELECTIVITY OF ION-SELECTIVE CATHODES CAN BE INCREASED GREATLY BY THE ADDITION OF TIME DEPENDENT RETARDING VOLTAGES.

FIELD 05D

ACCESSION NO. W72-00800

BIOASSAYS TO DETERMINE ALGAL GROWTH POTENTIAL OF MICRONUTRIENTS,

WASHINGTON STATE UNIV., PULLMAN. WATER RESEARCH CENTER.

WILLIAM H. FUNK, RICHARD J. CONDIT, AND WAYNE T. CRANEY.

AVAILABLE FROM NATIONAL TECHNICAL INFORMATION SERVICE AS PB-204 232, $3.00 IN PAPER COPY, $0.95 IN MICROFICHE. JULY 1971. 66 P, 22 FIG, 12 TAB, 45 REF, 3 APPEND. OWRR A-033-WASH(1).

DESCRIPTORS:
WATER POLLUTION EFFECTS, *MICROORGANISMS, ALGAL CONTROL, *POLLUTANT IDENTIFICATION, ANALYTICAL TECHNIQUES, *ALGAE, *NUTRIENTS, LAKES, PLANT GROWTH, *MOLYBDENUM, BIOASSAY, NEUTRON ACTIVATION ANALYSIS.

IDENTIFIERS:
*MICRO-NUTRIENTS, *ALGAL GROWTH.

ABSTRACT:
ADAPTING CONTINUOUS CULTURE TECHNIQUES FOR THE MEASUREMENT OF MICRO-NUTRIENT EFFECTS UPON ALGAE GROWTH HAS BEEN SHOWN TO BE A SATISFACTORY METHOD. SEVERAL REFINEMENTS AND MODIFICATIONS OF PREVIOUSLY UTILIZED EQUIPMENT FOR THE STUDY OF ALGAE MACRONUTRIENT UPTAKE MADE IT POSSIBLE TO TEST THE EFFECTS OF MOLYBDENUM UPON SCENEDESMUS QUADRICAUDA. HOWEVER, WITH THE PROGRESSIVELY REDUCED SUBSTRATE LEVELS OF MOLYBDENUM TWO PROBLEMS BECAME INCREASINGLY APPARENT: (1) MOLYBDENUM CONTAMINATION OF THE MEDIA BY MAGNESIUM SULFATE APPROACHED OR SURPASSED THE EXPERIMENTAL LEVELS IN THE DILUTION CULTURES AND PREVENTED THE ATTAINMENT OF A MOLYBDENUM DEFICIENT CONTROL; AND (2) THE EXTREMELY LOW SUBSTRATE LEVELS OF MOLYBDENUM COULD NOT BE DETECTED BY THE NEUTRON ACTIVATION ANALYSIS DUE TO INTERFERENCE BY THE HIGH PHOSPHATE CONCENTRATIONS IN THE MEDIA. HOWEVER, THE ANALYSES WERE USEFUL WHERE EITHER THE PHOSPHATE CONCENTRATIONS WERE LOW (LAKE WATER AND REAGENT CHEMICALS) OR MOLYBDENUM CONCENTRATIONS WERE RELATIVELY HIGH (TRIAL IV SUBSTRATES). THE CONTAMINATION PROBLEM MAY BE RESOLVED THROUGH THE PURIFICATION OF THE REAGENT CHEMICALS--HOWEVER, GREAT CARE MUST BE EXERCISED, SINCE FURTHER CONTAMINATION CAN RESULT THROUGH CARELESS TECHNIQUES. IMPROVED MOLYBDENUM ANALYSIS BY NEUTRON ACTIVATION WILL ALLOW THE FUTURE USE OF CONTINUOUS FLOW KINETICS ANALYSIS--A VALUABLE TOOL IN ASSESSING THE EFFECTS OF TRACE ELEMENTS IN ALGAL GROWTH.

FIELD 05C, 05A

ACCESSION NO. W72-00801

INFLUENCE OF MEDIUM PH AND CARBON DIOXIDE ON ASSIMILATION OF SOME ORGANIC ACIDS BY SCENEDESMUS QUADRICAUDA,

MOSCOW STATE UNIV. (USSR), DEPT. OF MICROBIOLOGY; AND MOSCOW STATE UNIV. (USSR). DEPT. OF SOIL BIOLOGY.

N. F. PISKUNKOVA, AND M. N. PIMENOVA.

MIKROBIOLOGIA (USSR), VOL 39, NO 6, P 854-857, 1970. 2 TAB, 18 REF.

DESCRIPTORS:
*CYANOPHYTA, *GROWTH RATES, *ORGANIC ACIDS, *HYDROGEN ION CONCENTRATION, METABOLISM, CARBON DIOXIDE, ALGAE, CULTURES, ACIDITY, ORGANIC COMPOUNDS.

IDENTIFIERS:
SCENEDESMUS QUADRICAUDA, ACETATE, PYRUVATE, LACTATE.

ABSTRACT:
THE NUTRIENT MEDIUM FOR A PURE CULTURE OF GREEN ALGA, SCENEDESMUS QUADRICAUDA, WAS SUPPLEMENTED BY ACETIC, PYRUVIC AND LACTIC ACIDS, ADDED AS SODIUM SALTS IN CONCENTRATION OF 0.05%. THE EFFECT OF ACIDS WAS RECORDED IN TERMS OF DRY BIOMASS ACCRETIONS. IN THE PRESENCE OF LIGHT ALL THREE ACIDS STIMULATED THE ALGAL GROWTH, BUT THEIR UTILIZATION OF THE ACIDS WAS INFLUENCED BY THE PH OF THE MEDIUM. IN THE DARK, GOOD GROWTH OF SCENEDESMUS OCCURRED ONLY IN THE PRESENCE OF ACETATE. PASSAGE OF AIR WITH 5% CARBON DIOXIDE STIMULATED THE GROWTH OF TEST PLANTS IN ACID MEDIA (PH 5.0) ENRICHED IN ACETATE AND PYRUVATE, BUT REDUCED THE CONSUMPTION OF ORGANIC ACIDS. THE STUDY SUGGESTS THAT ALGAE IN ACID WATER BASINS DEFICIENT IN CARBON DIOXIDE ADAPT THEMSELVES TO UTILIZING ORGANIC COMPOUNDS. (WILDE-WISCONSIN)

FIELD 05C

ACCESSION NO. W72-00838

MANAGING OUR ENVIRCNMENT.

AGRICULTURAL RESEARCH SERVICE, WASHINGTON, D. C.

DEPT OF AGRICULTURE, WASHINGTON D C, AGRICULTURE INFORMATION BULLETIN NO 351, APRIL 1971. 48 P.

DESCRIPTORS:
*MANAGEMENT, *ENVIRONMENT, *AGRICULTURE, *WATER POLLUTION CONTROL, SEDIMENTS, FARM WASTES, NUTRIENTS, PHOSPHORUS, ALGAE, NITRATES, WATER REUSE, SALINITY, PESTICIDES, LIVESTOCK, WASTE DISPOSAL, OXIDATION LAGOONS, DEHYDRATION, RUNOFF, RADIOACTIVITY, FALLCUT, BIOCONTROL, INSECT CONTROL, IRRIGATION, PREDATION, PARASITISM, INSECT RESISTANCE, INSECT ATTRACTANTS, PRECIPITATION(ATMOSPHERIC), GENETICS, EROSION CONTROL, AIR POLLUTION, TREES.

IDENTIFIERS:
FEEDLOTS, COMPOSTING, PLANT RESIDUES, RECYCLING FCCD, PROCESSING WASTES, PATHOGENS, BIOENVIRONMENTAL CONTROLS.

ABSTRACT:
SOME OF THE MAJOR PRCBLEMS IN AGRICULTURAL RESEARCH DEALING WITH NEW AND OLDER METHODS OF ENVIRONMENTAL MANAGEMENT ARE DESCRIBED IN AN EFFORT TOWARD INTERESTING THE PUBLIC IN PRESERVATION OF THE QUALITY OF OUR ENVIRONMENT. GENERAL MATERIAL IS PRESENTED UNDER THE SUBJECTS 'PROTECTING LAND, WATER AND WATERWAYS,' 'MANAGEMENT OF FARM WASTES,' 'RECYCLING FOOD PROCESSING WASTES,' 'NEW WAYS TO FIGHT PESTS--ALTERNATIVES TO PESTICIDES,' AND 'A GREEN WORLD--A CLEAN WORLD.' AMONG THE PROBLEMS DISCUSSED ARE PREVENTION OF ANIMAL WASTES REACHING WATERS, PHOSPHORUS FROM HUMAN WASTES AND DETERGENTS, MULTIPLE WATER REUSE, AND SALINITY IN IRRIGATED LANDS OF THE SOUTHWEST. SCIENTISTS ARE TRYING TO PREVENT PESTICIDE RESIDUES IN SOIL AND WATER AND AVOID PESTICIDE OVERUSE. FALLCUT FROM NUCLEAR WEAPON TESTING CALLS FOR VARIOUS DECONTAMINATION TREATMENTS; FOOD PROCESSING WASTE DISPOSAL AND RECYCLING IS DESCRIBED, AND RECOVERY OF POTABLE WATER FROM SEAWATER BY REVERSE OSMOSIS. ALTERNATIVES TO PESTICIDES ARE DESTRUCTION OF INSECTS AND WEEDS BY INTRODUCTION OF PREDATORS, PARASITES, AND PATHOGENS WHICH FEED ON OR INFECT PESTS; RESISTANT VARIETIES, ATTRACTANTS, GENETIC CONTROL, BIOENVIRONMENTAL CONTROLS, AND HORMONE AND DAYLIGHT MANIPULATION. (JONES-WISCONSIN)

FIELD 05G, 05C

ACCESSION NO. W72-00846

BOTTCM DEPOSITS,

G. W. FOESS, AND T. H. FENG.

JOURNAL WATER POLLUTICN CONTROL FEDERATION, VOL 43, NO 6, P 1257-1266, 1971. 80 REF.

DESCRIPTORS:
*REVIEWS, *BOTTOM SEDIMENTS, *AQUATIC ENVIRONMENT, SURFACE WATERS, NUTRIENTS, ALGAE, PHOSPHORUS, NITROGEN, EUTROPHICATION, SEDIMENT TRANSPORT, CYCLING NUTRIENTS, DEGRADATION(DECOMPOSITION), SAMPLING, BENTHIC FAUNA, SEDIMENTATION, SEDIMENT DISTRIBUTICN, PHYSICOCHEMICAL PROPERTIES, PESTICIDES, TROPHIC LEVEL, GASES, MUDS, ADSORPTION, COPPER, CLAYS, HUMIC ACIDS, SALINITY, OXIDATION-REDUCTION POTENTIAL, RADIOISOTOPES, DREDGING, ESTUARIES, BIOCHEMICAL OXYGEN DEMAND, DISSOLVED OXYGEN, WATER POLLUTION SOURCES.

ABSTRACT:
EIGHTY PAPERS ON BOTTOM DEPOSITS ARE REVIEWED. STUDIES, COVERING THE FIELD EXTENSIVELY ARE REPORTED FROM A WIDE RANGE, INCLUDING THE GREAT LAKES, FINLAND, LONG ISLAND SOUND, ITALY, ALASKA, COLUMBIA RIVER, OHIO, AND LAKE GENEVA. BOTTOM DEPOSITS ARE INCREASINGLY RECOGNIZED FOR THEIR EFFECT ON THE QUALITY OF OVERLYING WATERS. DIFFERENT SEDIMENT TYPES, PROBABLY RELATED TO AMOUNT OF AVAILABLE PHOSPHORUS PRESENT, VARY IN THEIR ALGAL SUPPORT. THE SORPTION OF PHOSPHORUS BY LAKE MUDS UNDER AEROBIC CONDITICNS MAY BE USED TO REMOVE PHOSPHORUS FROM LAKE WATER. SEDIMENT SAMPLING EQUIPMENT SHOULD NOT ONLY PROVICE AN UNDISTURBED SAMPLE BUT ALSO PERMIT RELIABLE SUBSAMPLING OF THE RETRIEVED MATERIAL. THE MOVEMENT OF SEDIMENTS HAS IMPORTANT PRACTICAL RAMIFICATIONS, AND DIVERSE METHODS FOR MEASURING THIS WERE DEVELOPED. THE SURFACE CHEMISTRY OF BOTTOM DEPOSITS RECEIVED ATTENTION FRCM SEVERAL INVESTIGATORS. INTACT SEDIMENT CCRES WERE USED TC MEASURE EPIBENTHIC ALGAL PRODUCTION AND BENTHIC COMMUNITY RESPIRATICN BY FOLLOWING CHANGES IN DO IN THE WATER OVER THE CORES. THE BIOLOGICAL INTERACTIONS OF BOTTOM DEPOSITS WITH PESTICIDES, HEAVY METALS, AND OTHER TRACE CONSTITUENTS IS INCREASINGLY IMPORTANT. TUBIFICID POPULATION HAS BEEN USED AS AN INDICATOR OF AMOUNT CF ORGANIC POLLUTICN. (JONES-WISCONSIN)

FIELD 05C, 10

ACCESSION NO. W72-00847

ECOLOGICAL STUDIES OF THE CONNECTICUT RIVER, VERNON, VERMONT.

WEBSTER-MARTIN, INC., SOUTH BURLINGTON, VT.

AVAILABLE FROM NATIONAL TECHNICAL INFORMATION SERVICE AS NP-18953, $3.00 IN PAPER COPY, $0.95 IN MICROFICHE. AEC CONTRACT REPORT NP-18953, JULY 1971. 24 P, 17 FIG, 2 TAB.

DESCRIPTORS:
*NUCLEAR WASTES, *RADIOACTIVITY EFFECTS, *THERMAL POLLUTION, CONNECTICUT, RIVERS, POST-IMPOUNDMENT, NUCLEAR POWERPLANTS, ECOLOGY, RADIOECOLOGY, MONITORING, POLLUTION ABATEMENT, ENVIRONMENTAL EFFECTS, AQUATIC LIFE, FISH POPULATIONS, PHYTOPLANKTON, BENTHIC FAUNA, PERIPHYTON, AQUATIC ALGAE, PLANTS, ENVIRONMENTAL ENGINEERING, COOLING TOWERS, VERMONT.

IDENTIFIERS:
CONNECTICUT RIVER.

ABSTRACT:
ASPECTS OF VERMONT-YANKEE (A NUCLEAR POWERPLANT) OPERATION OF SPECIAL CONCERN TO THE GENERAL PUBLIC ARE POSSIBLE EFFECTS OF WASTE-HEAT DISPOSAL ON THE ECOSYSTEM OF THE RIVER, AND DISCHARGE OF RADIOACTIVE WASTES. THE PLANT DESIGN PERMITS OPERATION OF THE COOLING SYSTEM IN AN OPEN OR A CLOSED-CYCLE MODE OR A HYBRID OF THESE MODES. ALLOWABLE INCREASES IN THE WATER TEMPERATURE OF THE IMPOUNDMENT CREATED BY VERNON DAM HAVE BEEN ESTABLISHED BY THE VERMONT WATER RESOURCES BOARD. A PRE-OPERATIONAL ENVIRONMENTAL SURVEY INCLUDED WATER PROPERTIES (TEMPERATURE, PH, DISSOLVED OXYGEN, AND CONDUCTIVITY), FISH, PHYTOPLANKTON, ALGAL PERIPHYTON, BENTHIC FAUNA, AND VASCULAR PLANTS. SAMPLES OF WATER, MUD, FISH AND VEGETATION ARE COLLECTED ROUTINELY AND ANALYZED FOR RADIOACTIVITY. FISH SAMPLED FROM BOTH ABOVE AND BELOW VERNON DAM INCLUDED 25 SPECIES. (BOPP-NSIC)

FIELD 05C

ACCESSION NO. W72-00939

RADIONUCLIDES AND SELECTED TRACE ELEMENTS IN MARINE PROTEIN CONCENTRATES,

WASHINGTON UNIV., SEATTLE, LAB. OF RADIATION ECOLOGY.

T. M. BEASLEY, T. A. JOKELA, AND R. J. EAGLE.

AVAILABLE FROM NATIONAL TECHNICAL INFORMATION SERVICE AS RLO-2225-T-14-2, $3.00 IN PAPER COPY, $0.95 IN MICROFICHE. AEC CONTRACT REPORT RLO-2225-T-14-2, 1970. 15 P, 4 TAB, 18 REF.

DESCRIPTORS:
*FOOD CHAINS, *TRACE ELEMENTS, *LEAD RADIOISOTOPES, HEAVY METALS, AIR POLLUTION EFFECTS, SURFACE WATERS, PATH OF POLLUTANTS, MARINE ALGAE, MARINE FISH, ABSORPTION, WATER POLLUTION SOURCES, RADIOACTIVITY EFFECTS, RADIOACTIVITY TECHNIQUES.

ABSTRACT:
CONSUMPTION OF 10-30 G/DAY OF CONCENTRATES PREPARED FROM SURFACE FEEDING FISHES WOULD SUBSTANTIALLY INCREASE INTAKE OF PB-210; PO-210; AND STABLE PB, CO, AND AG. THE STABLE PB IS ATTRIBUTED LARGELY TO AUTOMOTIVE EXHAUST POLLUTION OF SURFACE WATERS. IT IS UNCERTAIN WHETHER AG, CO, CD, AND CU ARE DERIVED SOLELY FROM ECOLOGICAL CONCENTRATION PROCESSES SINCE CONTAMINATION MAY OCCUR DURING PROCESSING. SR-90 IN THE CONCENTRATES WAS LESS THAN 0.05 DISINTEGRATIONS PER MINUTE PER GRAM OF DRY WEIGHT. (BOPP-NSIC)

FIELD 05C

ACCESSION NO. W72-00940

EFFECTS OF ENVIRONMENTAL STRESS ON THE COMMUNITY STRUCTURE, PRODUCTIVITY, ENERGY FLOW, AND MINERAL CYCLING IN SALT MARSH EPIPHYTIC COMMUNITIES,

CITY COLL., NEW YORK. DEPT OF BIOLOGY.

JOHN J. LEE, JOHN H. TIETJEN, AND ROBERT J. STONE.

AVAILABLE FROM NATIONAL TECHNICAL INFORMATION SERVICE $3.00 IN PAPER COPY, $0.95 IN MICROFICHE. AEC REPORT NYO-3995-18 (1970). 31 P, 7 FIG, 1 TAB, 21 REF. (CONF 710501-9). AEC CONTRACT AT(30-1)-3995.

DESCRIPTORS:
*WATER POLLUTION EFFECTS, *PROTOZOA, *MARINE ALGAE, *SALT MARSHES, NEW YORK, NEMATODES, EPIPHYTOLOGY, PRIMARY PRODUCTIVITY, PHYSIOLOGICAL ECOLOGY, RADIOECOLOGY, RADIOACTIVITY EFFECTS, ABSORPTION, FOOD WEBS, STRONTIUM RADIOISOTOPES, MARINE BACTERIA, COASTAL MARSHES, MICROORGANISMS.

ABSTRACT:
THE PURPOSE WAS TO GAIN INSIGHT INTO CHANGES IN PRODUCTIVITY AND COMMUNITY STRUCTURE RESULTING FROM RADIONUCLIDE, THERMAL, ORGANIC, PESTICIDE OR HERBICIDE POLLUTION. FORAMINIFERA (PROTOZOA) WERE ISOLATED IN CULTURE AND THEIR PHYSIOLOGICAL ECOLOGY, ENERGY BUDGETS, MINERAL CYCLING, AND LIFE CYCLES STUDIED. ENTEROMORPHA (ALGAE) HAD HIGH PRODUCTIVITY (ABOUT 10% OF ITS DRY WEIGHT/HR AT THE PEAK PRIMARY PRODUCTIVITY DURING THE SUMMER). REMOVAL RATES OF CA-45 AND SR-89, 90 FROM COASTAL MARINE ENVIRONMENTS BY INCORPORATION IN FORAMINIFERAN SHELLS WERE ESTIMATED. RADIONUCLIDES ARE EXCELLENT TOOLS FOR MEASURING COMPETITION, FEEDING, GROWTH, AND MINERAL CYCLING RATES OF FORAMINIFERA AND NEMATODES. (BOPP-NSIC)

FIELD 05C

ACCESSION NO. W72-00948

THE RAIN FOREST PROJECT. ANNUAL REPORT, JUNE 1970,

PUERTO RICO NUCLEAR CENTER, MAYAGUEZ.

RICHARD G. CLEMENTS, GEORGE E. DREWRY, AND ROBERT J. LAVIGNE.

AVAILABLE FROM NATIONAL TECHNICAL INFORMATION CENTER AS PRNC-147, $3.00 IN PAPER COPY, $0.95 IN MICROFICHE. AEC REPORT PRNC-147, ISSUED JUNE 1971. 142 P, 21 FIG, 30 TAB, 98 REF. AEC CONTRACT AT(40-1)-1833.

DESCRIPTORS:
*RADIOACTIVITY EFFECTS, *ECOSYSTEMS, *RAIN FORESTS, CYCLING NUTRIENTS, SOIL-WATER-PLANT RELATIONSHIPS, RADIOISOTOPES, ABSORPTION, RADIOACTIVITY TECHNIQUES, TRACERS, PATH OF POLLUTANTS, ON-SITE INVESTIGATIONS, FOOD WEBS, FOOD CHAINS, FROGS, LICHENS, ALGAE, FUNGI, SMALL ANIMALS(MAMMALS), CYTOLOGICAL STUDIES, SOIL CHEMISTRY, RADIOSENSITIVITY, RADIOECOLOGY, FALLOUT, INSECT, SOIL WATER MOVEMENT.

IDENTIFIERS:
CHEMISTRY OF THE ATMOSPHERE.

ABSTRACT:
CURRENT RESEARCH IN THE TERRESTRIAL ECOLOGY PROGRAM (INTERIM REPORTS AND COMPLETED SHORT TERM STUDIES, 24 IN NUMBER) INCLUDES STUDIES ON INSECT ECOLOGY; MOVEMENT OF ISOTOPES THROUGH THE ANIMAL FOOD WEB; ELEMENT INPUT THROUGH RAINFALL AND SUBSEQUENT DISTRIBUTION THROUGH THE FOREST; RECOVERY IN IRRADIATED AREAS; MOVEMENT AND DISTRIBUTION OF PREVIOUSLY APPLIED RADIOISOTOPES IN SOIL, PLANTS AND ANIMALS; AND RESEARCH IN THE VISITING SCIENTIST PROGRAM. INCREASED EMPHASIS WILL BE PLACED ON THE PHYSICAL AND CHEMICAL PROPERTIES OF FOREST SOILS AND THE MOVEMENT OF MACRO- AND TRACE ELEMENTS VIA SOIL WATER TO THE STREAM. (BOPP-NSIC)

FIELD 05C

ACCESSION NO. W72-00949

PRELIMINARY ECOLOGICAL STUDIES AT JERVIS BAY,

AUSTRALIAN ATOMIC ENERGY COMMISSION RESEARCH ESTABLISHMENT, LUCAS HEIGHTS.

M. S. GILES.

AVAILABLE FROM NATIONAL TECHNICAL INFORMATION SERVICE AS CONF-710511-3, $3.00 IN PAPER COPY, $0.95 IN MICROFICHE. AEC REPORT CONF-710511-3 (FROM FORTY-THIRD CONGRESS OF THE AUSTRALIAN AND NEW ZEALAND ASSOCIATION FOR THE ADVANCEMENT OF SCIENCE; BRISBANE, AUSTRALIA, 24 MAY 1971). 22 P, 6 FIG, 5 TAB, 8 REF.

DESCRIPTORS:
*FOOD CHAINS, *RADIOACTIVITY EFFECTS, *NUCLEAR POWERPLANTS, MONITORING, PATH OF POLLUTANTS, AIR POLLUTION EFFECTS, WATER POLLUTION EFFECTS, BAYS, WATERSHEDS(BASINS), SMALL WATERSHEDS, CROPS, EGGS, FISHERIES, ON-SITE INVESTIGATIONS, ENVIRONMENTAL EFFECTS, ESTUARINE ENVIRONMENT, REGULATION, FALLOUT, OYSTERS, MUSSELS, PLANKTON, MARINE ALGAE.

IDENTIFIERS:
ENVIRONMENTAL STATEMENT, CRITICAL NUCLIDE PATHWAY, ABALONE.

ABSTRACT:
ATMOSPHERIC CRITICAL NUCLIDE PATHWAYS INCLUDE HONEY PRODUCTION, WATER CATCHMENT, FREE-RANGE EGG PRODUCTION, BACKYARD GARDENING, AND COLLECTION OF RAINWATER FOR DRINKING. LIQUID-EFFLUENT PATHWAYS ARE SEAFOODS (ABALONE, ROCK OYSTERS, AND DREDGE OYSTERS), SEAWEED MULCHES, AND AGAR PRODUCTS. THE SHALLOWER WATERS WILL HAVE THE HIGHEST RADIOACTIVITY SINCE THE WARM EFFLUENT HAS RELATIVELY LOW DENSITY. THIS MAKES ABALONE, WHICH SPENDS SEVERAL YEARS IN ONE (KELP) FEEDING AREA, A MAJOR PATHWAY. DREDGE OYSTERS ARE ANOTHER PATHWAY, BUT LARGE BEDS OF ROCK OYSTERS ARE NOT PRESENT. (BOPP-NSIC)

FIELD 05C

ACCESSION NO. W72-00959

AMCHITKA BIOENVIRONMENTAL PROGRAM. AMCHITKA BIOLOGICAL INFORMATION SUMMARY,

BATTELLE MEMORIAL INST., COLUMBUS, OHIO.

R. GLEN FULLER.

AVAILABLE FROM NATIONAL TECHNICAL INFORMATION SERVICE, AS BMI-171-132, $3.00 IN PAPER COPY, $0.95 IN MICROFICHE. AEC CONTRACT REPORT BMI-171-132, MAY 1971. 16 P, 1 FIG, 18 REF. AEC CONTRACT AT(26-1)-171.

DESCRIPTORS:
*RADIOACTIVITY EFFECTS, *ENVIRONMENTAL EFFECTS, *NUCLEAR EXPLOSIONS, BIRDS, SALMON, STICKLEBACKS, CRABS, OTTERS, ZOOPLANKTON, PHYTOPLANKTON, MAMMALS, MARINE ALGAE, FOOD CHAINS, MARINE FISHERIES, LAKES, POLLUTION ABATEMENT.

IDENTIFIERS:
RADIONUCLIDE UPTAKE.

ABSTRACT:
EARLY IN 1967, THE UNITED STATES ATOMIC ENERGY COMMISSION NEVADA OPERATIONS OFFICE BEGAN A DETAILED EVALUATION OF AMCHITKA ISLAND IN THE WESTERN ALEUTIANS AS A POTENTIAL SITE FOR UNDERGROUND NUCLEAR TESTING. IN OCTOBER 1969, AN UNDERGROUND TEST, PROJECT MILROW, INVOLVED THE DETONATION OF A DEVICE OF ABOUT 1 MEGATON YIELD, ABOUT 4,000 FT BELOW THE SURFACE. STUDIES WERE DESIGNED TO: (1) PREDICT, DOCUMENT, AND EVALUATE THE EFFECTS OF AEC TESTING ACTIVITIES ON THE ENVIRONMENT; (2) RECOMMEND MEASURES TO MINIMIZE EFFECTS; AND (3) PREDICT RADIONUCLIDE UPTAKE BY MARINE FOOD CHAINS, SHOULD THERE BE AN INADVERTENT RELEASE. RESULTS ARE REPORTED OF STUDIES OF SOILS, VEGETATION, FRESHWATER LIFE, BIRDS, AND MARINE LIFE. (BOPP-NSIC)

FIELD 05C

ACCESSION NO. W72-00960

THE USE OF A RUBBLE CHIMNEY FOR DENITRIFICATION OF IRRIGATION RETURN WATERS,

STANFORD UNIV., PALO ALTO, CALIF.

ROY B. EVANS, AND PAUL KRUGER.

AVAILABLE FROM THE NATIONAL TECHNICAL INFORMATION SERVICE AS CONF 700101, VOL 2, $6.00/SET, $0.95 MICROFICHE. AEC CONFERENCE PAPER, CONF 700101, VOL 2, P 1222-1245, JAN 14-18, 1970.

DESCRIPTORS:
*NUCLEAR EXPLOSION, *FILTERS, *DIATOMACEOUS EARTH, *POROUS MEDIA, *WASTE WATER TREATMENT, *DENITRIFICATION, *NITRATES, EUTROPHICATION, IRRIGATION WATER, AQUATIC ALGAE, BIOLOGICAL PROPERTIES, WATER POLLUTION EFFECT, WATER POLLUTION SOURCES, COST COMPARISONS, BENEFICIAL USE, WATER QUALITY, CALIFORNIA, BIOCONTROL, COST-BENEFIT THEORY.

IDENTIFIERS:
RUBBLE CHIMNEY, SAN JOAQUIN VALLEY.

ABSTRACT:
THE USE OF WATER FOR IRRIGATION SERIOUSLY REDUCES THE QUALITY OF THE WATER. IRRIGATION RETURN WATER MAY HAVE HIGH CONCENTRATIONS OF SALTS, PESTICIDES, AND PLANT NUTRIENTS, AND DISCHARGE OF THE RETURN FLOWS INTO RIVERS OR OTHER RECEIVING WATERS CAN SERIOUSLY AFFECT THE QUALITY OF THE RECEIVING WATERS. THE INCREASED AVAILABILITY OF CHEAP FERTILIZERS HAS HAD THE UNDESIRABLE SIDE-EFFECT OF INCREASING CONCENTRATIONS OF NITROGEN AND PHOSPHORUS IN IRRIGATION RETURN FLOWS. THE PRESENCE OF THESE ELEMENTS CAN BE TROUBLESOME. BOTH ARE NUTRIENTS NECESSARY FOR THE GROWTH OF ALGAE, AND IN SOME SITUATIONS THEIR PRESENCE MAY RESULT IN SEVERE ALGAL BLOOMS IN THE RECEIVING WATERS. THEIR REMOVAL FROM WASTE WATER IS CONSIDERED DIFFICULT AND EXPENSIVE. BIOLOGICAL DENITRIFICATION HAS BEEN PROPOSED AS A MEANS OF REMOVING NITRATES FROM WASTE WATERS TO CONTROL EUTROPHICATION IN RECEIVING WATERS. A POTENTIAL USE FOR THIS METHOD IS THE TREATMENT OF IRRIGATION RETURN WATERS CONTAINING HIGH CONCENTRATIONS OF NITRATE-NITROGEN, SINCE DIRECT DISCHARGE OF SUCH WASTES MAY CAUSE OBJECTIONABLE ALGAL GROWTH IN THE RECEIVING WATERS. FOR EXAMPLE, THE PROCESS MAY BE USED TO TREAT AGRICULTURAL WASTE WATERS IN THE SAN JOAQUIN VALLEY IN CALIFORNIA, WHERE AN ESTIMATED 580,000 ACRE-FEET/YEAR OF RETURN WATERS, CONTAINING 20 MG/L OF NITRATE-NITROGEN, WILL REQUIRE DISPOSAL BY A.D. 2020. THIS PAPER PRESENTS THE RESULTS OF A PRELIMINARY INVESTIGATION OF THE FEASIBILITY OF USING A RUBBLE CHIMNEY AS A BIOLOGICAL FILTER FOR DENITRIFICATION. (HOUSER-NSIC)

FIELD 08H, 05D

ACCESSION NO. W72-00974

DETERGENT POLLUTION CONTROL ACT OF 1971 (A BILL TO AMEND THE FEDERAL WATER POLLUTION CONTROL ACT TO BAN POLYPHOSPHATES IN DETERGENTS AND TO ESTABLISH STANDARDS AND PROGRAMS TO ABATE AND CONTROL WATER POLLUTION BY SY NTHETIC DETERGENTS).

HOUSE BILL 6180, 92D CONG. 1ST SESS. (1971). 14 P.

DESCRIPTORS:
*UNITED STATES, *DETERGETNTS, *PHOSPHATES, *WATER POLLUTION CONTROL, *WATER POLLUTION SOURCES, EUTROPHICATION, LEGISLATION, LEGAL ASPECTS, REGULATION, POLLUTION ABATEMENT, WASTE DISPOSAL, STANDARDS, PUBLIC HEALTH, INSPECTION, WATER SOFTENING, SOAPS, AQUATIC ALGAE, WATER POLLUTION EFFECTS, WATER QUALITY, CHEMICAL DEGRADATION.

IDENTIFIERS:
WATER POLLUTION CONTROL ACT.

ABSTRACT:
THIS BILL IS INTENDED TO ABATE AND CONTROL THE POLLUTION AND DEGRADATION OF WATER CAUSED BY THE CONTINUING DISCHARGE OF SYNTHETIC DETERGENTS. RESEARCH HAS DEMONSTRATED THAT POLYPHOSPHATES IN DETERGENTS ARE A MAJOR CAUSE OF WATER POLLUTION, ACCELERATING BEYOND CONTROL THE GROWTH OF ALGAE WHICH INTERFERES WITH THE USE OF THE WATER AND DEGRADES WATER QUALITY. THE BILL RECOGNIZES THAT CERTAIN NONPHOSPHORUS BASED INGREDIENTS PERFORM AT LEAST AS WELL AS PRESENT POLYPHOSPHATE INGREDIENTS IN SYNTHETIC DETERGENTS AND SOAPS. THE BILL WOULD MAKE IT UNLAWFUL TO MANUFACTURE OR IMPORT ANY DETERGENT CONTAINING PHOSPHORUS AFTER JUNE 30, 1972. PROVISIONS ARE MADE FOR THE SEIZURE AND DISPOSAL OF SUCH DETERGENTS AND THE ADJUDICATION OF ALLEGED VIOLATIONS. THE SECRETARY OF THE INTERIOR WOULD BE AUTHORIZED TO MAKE APPROPRIATE INSPECTIONS. STANDARDS OF WATER EUTROPHICATION ABILITY, BIODEGRADIBILITY, TOXICITY, AND OF EFFECTS ON THE PUBLIC HEALTH AND WELFARE WOULD BE ESTABLISHED BY THE SECRETARY. THE SECRETARY IS ALSO AUTHORIZED TO INVENTORY EXISTING TECHNOLOGY AND ASSIST IN THE DEVELOPMENT OF POLYPHOSPHATE SUBSTITUTES. (JOHNSON-FLORIDA)

FIELD 06E, 05G

ACCESSION NO. W72-01085

CHICAGO LEADS FIGHT AGAINST PHOSPHATE POLLUTICN--NOW CONGRESS MUST ACT,

HOUSE OF REPRESENTATIVES, WASHINGTON, D.C.

R. C. PUCINSKI.

CONGRESSIONAL RECORD, VOL 117, NO 8, P H326-29 (DAILY ED. FEBRUARY 1, 1971). 4 P.

DESCRIPTORS:
*DETERGENTS, *WATER POLLUTION CONTROL, *PHOSPHATES, *FEDERAL GOVERNMENT, UNITED STATES, ALGAL CONTROL, STANDARDS, LEGISLATION, LEGAL ASPECTS, WATER POLLUTION, ADJUDICATION PROCEDURE, ADMINISTRATIVE AGENCIES, ADMINISTRATION, ALGAE, BIODEGRADATION, LOCAL GOVERNMENTS, WASTE DISPOSAL, WATER POLLUTION SOURCES, SOAPS, TOXICITY, PUBLIC HEALTH, WATER POLLUTION EFFECTS, FEDERAL JURISDICTION.

IDENTIFIERS:
*WATER POLLUTION CONTROL ACT.

ABSTRACT:
CONGRESSMAN PUCINSKI INTRODUCED FEDERAL LEGISLATION, AMENDING THE FEDERAL WATER POLLUTION CONTROL ACT, TO BAN POLYPHOSPHATES IN DETERGENTS AND ESTABLISH STANDARDS AND PROGRAMS TO ABATE AND CONTROL WATER POLLUTION FROM SYNTHETIC DETERGENTS. PHOSPHATES USED IN MOST DETERGENTS PROMOTE THE GROWTH OF ALGAE; THE UNRESTRICTED GROWTH OF ALGAE, BY CONSUMING AVAILABLE OXYGEN, INEVITABLY SUFFOCATES FISH AND OTHER MARINE LIFE UNTIL WATER LITERALLY BEGINS TO DIE. NOTING A NEW CHICAGO ORDINANCE WHICH BANS DETERGENTS HAVING A SPECIFIC PHOSPHOROUS CONTENT. PUCINSKI EMPHASIZED THAT FEDERAL LEGISLATION IS ESSENTIAL TO EFFECTIVELY COMBAT WATER POLLUTION. HIS BILL, IN ADDITION TO MAKING THE MANUFACTURING AND IMPORTING OF DETERGENTS CONTAINING PHOSPHORUS UNLAWFUL, IMPOSES IN REM LIABILITY AND ALLOWS CONDEMNATION IN FEDERAL COURT. ADJUDICATORY PROCEDURES FOR THE DISPOSITION OF SUCH DETERGENTS ARE DETAILED IN THE BILL. THE BILL AUTHORIZES THE SECRETARY OF THE TREASURY TO CONDUCT EXAMINATIONS AND INSPECTIONS. ADDITIONALLY, THE SECRETARY MUST ESTABLISH STANDARDS OF WATER AND EUTROPHICATION ABILITY, BIODEGRADABILITY, TOXICITY, AND OF EFFECTS ON PUBLIC HEALTH. THESE STANDARDS MUST BE MET BY ALL SYNTHETIC DETERGENTS. TO ACCELERATE THE DEVELOPMENT OF PHOSPHATE-FREE DETERGENTS THE SECRETARY MUST REPORT EXISTING TECHNOLOGY OR SUBSTITUTES AND IS AUTHORIZED TO MAKE GRANTS AND CONTRACTS FOR RESEARCH AND DEVELOPMENT.

FIELD 06E, 05G

ACCESSION NO. W72-01093

PROCEEDINGS THIRTEENTH CONFERENCE ON GREAT LAKES RESEARCH, PART I AND II.

GREAT LAKES RESEARCH CENTER, DETROIT, MICH.

AVAILABLE FROM TREASURER, P. O. BOX 640, ANN ARBOR, MICH. 48107. PRICE $18.00 A SET. 1970. 1063 P.

DESCRIPTORS:
*GREAT LAKES, *LAKES, EUTROPHICATION, ALGAE, *LAKE SUPERIOR, *LAKE MICHIGAN, *LAKE HURON, *LAKE ONTARIO, *LAKE ERIE, WATER POLLUTION EFFECTS, WATER POLLUTION SOURCES, LIMNOLOGY.

ABSTRACT:
THE THIRTEENTH CONFERENCE ON GREAT LAKES RESEARCH WAS HELD 1-3 APRIL, 1970 AT BUFFALO, NEW YORK AND CO-HOSTED BY CORNELL AERONAUTICAL LABORATORY INC. AND THE GREAT LAKES LABORATORY OF THE STATE UNIVERSITY COLLEGE AT BUFFALO. (SEE ALSO W72-01095 THRU W72-01108)

FIELD 02H, 05C

ACCESSION NO. W72-01094

LASER INDUCED FLUORESCENCE IN RHODAMINE B AND ALGAE,

SYRACUSE UNIV. RESEARCH CORP., N.Y.

G. DANIEL HICKMAN, AND RICHARD B . MOORE.

INTERNATIONAL ASSOCIATION FOR GREAT LAKES RESEARCH, PROCEEDINGS 13TH CONFERENCE ON GREAT LAKES RESEARCH, PART 1, P 1-14, 1970. 11 FIG, 2 TAB, 22 REF.

DESCRIPTORS:
*ALGAE, *REMOTE SENSING, *TRACKING TECHNIQUES, MEASUREMENT, CURRENTS(WATER), CHLOROPHYLL, FLUORESCENCE, BATHYMETRY.

IDENTIFIERS:
LASER, RHODAMINE B.

ABSTRACT:
MAPPING OF THREE-DIMENSIONAL CURRENTS AND OF DISTRIBUTION OF NEAR-SURFACE ALGAE WAS INVESTIGATED BY RECORDING FLUORESCENCE STIMULATED BY THE RECENTLY DEVELOPED OCEANIC PULSED LASER. A 3 NSEC PULSED NEON LASER (540 NM) WAS EMPLOYED TO EXCITE THE EMISSION OF RHODAMINE B (580 NM). THE SAME LASER WAS FILLED WITH NITROGEN (337 NM) TO STIMULATE THE EMISSION OF ALGAE. RHODAMINE FLUORESCENCE WAS OBSERVED FOR CONCENTRATIONS AS LOW AS 0.1 PPB, WHEREAS THAT FROM ANACYSTIS AND CHLORELLA SPECIES, FROM 6 TO 12 MG/CU M. AIRBORNE LASER-RECEIVER UNIT SHOULD DETECT THE FLUORESCENCE FROM AN ALTITUDE OF 100 METERS. (SEE ALSO W72-01094) (WILDE-WISCONSIN)

FIELD 05C, 05A, 02H

ACCESSION NO. W72-01095

BLUE-GREEN ALGAE AND HUMIC SUBSTANCES,

CINCINNATI UNIV., OHIO. TANNER'S COUNCIL LAB.

WILLY LANGE.

INTERNATIONAL ASSOCIATION FOR GREAT LAKES RESEARCH, PROCEEDINGS 13TH CONFERENCE ON GREAT LAKES RESEARCH, PART 1, P 58-70, 1970. 4 FIG, 6 TAB, 17 REF.

DESCRIPTORS:
*CYANOPHYTA, *HUMIC ACIDS, ORGANIC MATTER, ALGAE, EUTROPHICATION, CHELATION, DECOMPOSING ORGANIC MATTER, PHYTOPLANKTON, PRODUCTIVITY, IRON.

IDENTIFIERS:
*FULVIC ACID, GROWTH STIMULATION, ANABAENA, GLOEOTRICHIA, MICROCYSTIS, NOSTOC, SLAKED LIME.

ABSTRACT:
A REPLACEMENT OF EDTA + CITRATE IN THE ZEHNDER-GORHAM MEDIA WITH SYNTHETIC FULVIC ACID INCREASED GREATLY THE CELL NUMBERS OF ANABAENA CIRCINALIS, GLOEOTRICHIA ECHINULATA, MICROCYSTIS AERUGINOSA, AND NOSTOC MUSCORUM. THE GROWTH STIMULATION IS ATTRIBUTED TO CHELATING EFFECTS OF FULVIC ACID AND THE RESULTING INCREASED AVAILABILITY OF IRON AND OTHER TRACE ELEMENTS. RESULTS INDICATE THAT SEEPAGE OF FULVIC ACID FROM RAW HUMUS AND PEAT SOILS, OR ITS RELEASE FROM SEWAGE EFFLUENTS PLAY AN IMPORTANT PART IN THE NUISANCE GROWTH OF CYANOPHYTA. SLAKED LIME IS SUGGESTED AS A PARTIALLY EFFICIENT REMEDY FOR CIRCUM-NEUTRAL WATERS. (SEE ALSO W72-01094) (WILDE-WISCONSIN)

FIELD 05C, 02H

ACCESSION NO. W72-01097

NEUTRON ACTIVATION ANALYSIS OF SEDIMENTS IN WESTERN LAKE ERIE,

OHIO STATE UNIV., COLUMBUS. COLL. OF BIOLOGICAL SCIENCES.

PAUL L. ZUBKOFF, AND WALTER E. CAREY.

INTERNATIONAL ASSOCIATION FOR GREAT LAKES RESEARCH, PROCEEDINGS 13TH CONFERENCE ON GREAT LAKES RESEARCH, PART 1, P 319-325, 1970. 2 FIG, 3 TAB, 11 REF.

DESCRIPTORS:
*SEDIMENTS, *ANALYTICAL TECHNIQUES, *NEUTRON ACTIVATION ANALYSIS, *CHEMICAL ANALYSIS, GREAT LAKES, EUTROPHICATION, NUTRIENTS, ALGAE, BACTERIA, LAKE ERIE, ALUMINUM, MANGANESE, SODIUM, CHROMIUM, IRON.

IDENTIFIERS:
VANADIUM, LANTHANUM, SCANDIUM.

ABSTRACT:
THE CENTERS OF 1 CM LATERAL SECTIONS OF 15 CM SEDIMENT CORES WERE WASHED FREE OF INTERSTITIAL WATER AND SUBJECTED TO A 2.0 X 10 CM 11 POWER NEUTRON 0.01 CM 0.1 SEC FLUX IN RESEARCH REACTOR. ANALYSIS OF GAMMA-RAY SPECTRA, OBTAINED WITH A NAI(TL) CRYSTAL, INDICATED A UNIFORM CONCENTRATION OF AL, V, MN, NA, LA, CR, AND SC. THE CONTENTS OF VANADIUM AND CHROMIUM ARE AT LEAST THREE TIMES AS GREAT AS FOUND IN SOILS. (SEE ALSO W72-01094) (WILDE-WISCONSIN)

FIELD 05C, 02H

ACCESSION NO. W72-01101

AQUEOUS PHOSPHATE AND LAKE SEDIMENT INTERACTION,

GREAT LAKES RESEARCH CENTER, DETROIT, MICH.

R. C. GUMERMAN.

INTERNATIONAL ASSOCIATION FOR GREAT LAKES RESEARCH, PROCEEDINGS 13TH CONFERENCE ON GREAT LAKES RESEARCH, PART 2, P 673-682, 1970. 8 FIG, 1 TAB, 8 REF.

DESCRIPTORS:
*PHOSPHATES, *SEDIMENT-WATER INTERFACES, ADSORPTION, LAKES, NUTRIENTS, LAKE ERIE, LAKE SUPERIOR, LABORATORY TESTS, TEMPERATURE, ALGAE, PHOSPHORUS, ION EXCHANGE, OXIDATION-REDUCTION POTENTIAL, HYDROGEN ION CONCENTRATION.

IDENTIFIERS:
SEDIMENTARY PHOSPHORUS RELEASE.

ABSTRACT:
THIS LABORATORY STUDY OF STERILE SEDIMENTS FROM LAKE ERIE AND LAKE SUPERIOR DISCLOSED THAT THE PHOSPHORUS-SEDIMENT COMPLEX IS FORMED UNDER THE INFLUENCE OF BOTH PHYSICAL AND CHEMICAL ADSORPTION. MAXIMUM REMOVAL OF AQUEOUS P OCCURS WITHIN THE PH RANGE OF 4.5 TO 5.5. LOWERING REDOX POTENTIAL TO ZERO MAY OR MAY NOT EFFECT A RELEASE OF P FROM THE SEDIMENT. THE MAXIMUM ADSORBING CAPACITY OF SEDIMENTS IS IN SURFACE LAYERS LESS THAN 3.5 MM DEEP, AND IS ZERO AT A DEPTH EXCEEDING 14 MM BELOW SEDIMENT-WATER INTERFACE. AS LONG AS THE SEDIMENT CONTAINS SOME ADSORBED PHOSPHORUS, ITS RELEASE WILL MAINTAIN A MINIMUM CONCENTRATION OF 0.1 MG/L OF AQUEOUS PHOSPHATE RADICAL. IN TURN, UNDER SUCH CONDITIONS CESSATION OF PHOSPHATE INPUT MAY NOT EFFECT A REDUCTION OF NUISANCE ALGAL GROWTH FOR A LONG TIME. (SEE ALSO W72-01094) (WILDE-WISCONSIN)

FIELD 05C, 02H

ACCESSION NO. W72-01108

PHYTCPLANKTON SPECIES AND POPULATICNS IN THE PAMLICO RIVER ESTUARY OF NORTH CAROLINA,

NORTH CAROLINA STATE UNIV., RALEIGH. DEPT. OF ZOCLOGY.

J. E. HOBBIE.

AVAILABLE FROM THE NATIONAL TECHNICAL INFORMATION SERVICE AS PB-204 489, $3.00 IN PAPER COPY, $0.95 IN MICROFICHE. NORTH CAROLINA WATER RESOURCES RESEARCH INSTITUTE, RALEIGH. REPCRT NO 56, SEPTEMBER 1971. 147 P, 37 FIG, 2 TAB, 27 REF, 2 APPEND. OWRR B-004-NC(11).

DESCRIPTORS:
*ESTUARIES, *EUTROPHICATION, *ALGAL BLOOMS, *PHYTOPLANKTON, *NUTRIENTS, WATER POLLUTION EFFECTS, AQUATIC ALGAE, NORTH CARCLINA, *ALGAE, DIATOMS, *DINOFLAGELLATES, *RED TIDE, WATER POLLUTION SOURCES.

IDENTIFIERS:
*PAMLICO RIVER ESTUARY.

ABSTRACT:
THE PAMLICO RIVER ESTUARY EXTENDS SOME 35 MILES FRCM WASHINGTON, N. C. TO PAMLICO SOUND. THE PHYTOPLANKTON CYCLE OF THIS ESTUARY IS COMPLETELY DOMINATED BY DINOFLAGELLATES. DIATOMS, HOWEVER, BECOME MORE AND MORE IMPORTANT IN THE LOWER REACHES OF THE RIVER CLOSE TO THE POINT WHERE IT EMPTIES INTO PAMLICO SOUND. THE DOMINANT ORGANISM IS PERIDINIUM TRIQUETRUM, THAT CREATES A RED TICE DURING JANUARY, FEBRUARY, AND MARCH. THE PERIDINIUM IS ACCOMPANIED BY OTHER DINCGLAGELLATES. THIS BLOOM LASTS UNTIL LATE MARCH AND THEN POPULATICNS REMAIN LOW UNTIL A LATE SUMMER PEAK OF ALGAE DOMINATED BY G. AUREOLUM, G. ESTUARIALE, K. ROTUNDATUM, POLYKRIKOS SP., AND CALYCOMONAS OVALIS. THIS INCREASE TAKES PLACE IN LATE AUGUST AND EARLY SEPTEMBER AND IS FOLLOWED BY A FALL LOW THAT LASTS UNTIL THE EARLY SPRING BLOOM BEGINS IN DECEMBER OR EARLY JANUARY. THIS YEARLY CYCLE IS SIMILAR TO THAT FOUND IN SEVERAL OF THE RIVER ESTUARIES THAT ENTER CHESAPEAK BAY. EVEN THE RED TIDE FORMING ORGANISM, P. TRIQUETRUM, IS THE SAME. PERIDINIUM TRIQUETRUM IS AN INDICATOR OF EXTREMELY RICH OR POLLUTEC CONDITIONS. IN THE PAMLICO RIVER ESTUARY, THE ENTICHMENT COMES FROM SEWAGE FROM SEVERAL SMALL CITIES AND FROM FARM AND SWAMP RUNOFF. IT IS LIKELY THAT THE RED TIDE IS THE RESULT OF LARGE AMOUNTS OF NITRATE THAT REACH THE MIDDLE PARTS OF THE ESTUARY IN MID-WINTER. PHOSPHORUS IS IN AMPLE SUPPLY THE YEAR AROUND.

FIELD 05C, 02L

ACCESSION NO. W72-01329

BLUE-GREEN ALGAE BLOOMS--A CURRENT HYDROBIOLOGICAL PROBLEM (ZAKWITY SINIC--AKTUALNY PROBLEM HYDROBIOLOGII),

POLISH ACADEMY OF SCIENCES, WARSAW. INST. OF ECOLOGY; AND POLISH ACADEMY OF SCIENCES, WARSAW. DEPT. OF HYDROBIOLOGY.

IRENA SPODNIEWSKA.

WIADOMOSCI EKOLOGICZNE, VOL 17, NO 2, P 157-163, 1971. 3 REF. ENGLISH SUMMARY.

DESCRIPTORS:
*CYANOPHYTA, *NUISANCE ALGAE, *EUTROPHICATION, REVIEWS.

IDENTIFIERS:
USSR.

ABSTRACT:
BLUE-GREEN ALGAE APPEARS MOST FREQUENTLY IN STAGNANT WATERS AND WATERS WITH A SLUGGISH CURRENT, RICH IN ORGANIC SUBSTANCES. THEIR PERSISTENT BLOOMS, WHICH ACCOMPANY PROGRESSIVE EUTROPHICATION, ARE WIDESPREAD PHENOMENON. THREE RUSSIAN LANGUAGE VOLUMES--'ECOLOGY AND PHYSIOLOGY OF BLUE-GREEN ALGAE' EDITED BY L P BRAGINSKII, MOSCOW AND LENINGRAD 1965, 272 PAGES, 'WATER BLOOMS' PART I, EDITED BY A V TOPACHEVSKII, KIEV, 1968, 386 PAGES AND PART II, 1969, 267 PAGES, DESCRIBING RESEARCH CONDUCTED ON BLUE-GREEN ALGAE DURING THE LAST TEN YEARS IN THE SOVIET UNION AND WHICH CONTAIN A DISCUSSION OF WORLD LITERATURE ARE REVIEWED. EMPHASIS WAS ON THE SPECIFIC PROPERTIES OF BLUE-GREEN ALGAE WHICH ENABLE THEM TO OCCUR UNDER CONDITIONS UNFAVORABLE TO DEVELOPMENT OF OTHER PLANT ORGANISMS. (AUEN-WISCONSIN)

FIELD 05C

ACCESSION NO. W72-01358

AN ECOLOGICAL STUDY OF THREE FRESHWATER PONDS OF HYDERABAC-INDIA. I. THE ENVIRONMENT,

OSMANIA UNIV., HYDERABAD (INDIA). DEPT. OF BOTANY; AND CSMANIA UNIV., HYDERABAD (INCIA). HYDROBIOLOGY LAB.

V. S. RAO.

HYDROBIOLOGIA, VOL 38, NO 2, P 213-223, 1971. 2 FIG, 3 TAB, 30 REF.

DESCRIPTORS:
*ECOLOGY, *PONDS, *CHEMICAL ANALYSIS, TEMPERATURE, SEASONAL, IONS, ALGAE, SALINITY, PHYTOPLANKTON, THERMAL PROPERTIES, HYDROGEN ION CONCENTRATION, DISSOLVED OXYGEN, CARBON DIOXIDE, AMMONIA, ORGANIC MATTER.

IDENTIFIERS:
*HYDERABAD(INDIA), EICHHCRNIA CRASSIPES, TYPHA ANGUSTATA, IPOMOEA AQUATICA, TOTAL SOLIDS.

ABSTRACT:
THE DIVERSITY AND COMPLEXITY OF THE ECOLOGICAL PHENOMENA OF INDIAN FRESH WATERS CALL FOR MORE DATA, PARTICULARLY ON IONIC COMPOSITION AND THE BEHAVIOR OF BASIC ELEMENTS. THREE PONDS SUBJECT TO DIVERSE POLLUTION NEAR HYDERABAD WERE STUDIED. CHEMICAL ANALYSES WERE MADE OF WATER FROM EACH POND FOR TWO YEARS. DATA WERE COLLECTED ON WATER TEMPERATURE IN RELATION TO THE AIR TEMPERATURE IN SUMMER, RAINY SEASON, AND IN WINTER. THE TEMPERATURE DIFFERENCES WERE, TO A DEGREE, ATTRIBUTED TO WATER QUANTITY BUT THE INTERMEDIATE SIZED POND DID NOT BEAR OUT THIS CONCLUSION; THE HIGH SALT CONTENT AND DENSE ALGAL POPULATION OF THAT POND WAS POSSIBLY RESPONSIBLE FOR KEEPING IT COOLER. THE IONIC COMPOSITION OF THESE PONDS WAS DETERMINED; THEIR PROPORTIONS DIFFERED CONSIDERABLY FROM RODHE'S (1949) STANDARD COMPOSITION. THE IONIC COMPOSITION OF THE SAME BODY OF WATER ALSO DIFFERED FROM SEASON TO SEASON. THE SEASONAL FLUCTUATIONS, WHICH HAVE SECONDARY IMPORTANCE IN LAKES, BECOME MORE SIGNIFICANT IN SMALL FRESHWATER BODIES. IT IS STRESSED THAT THE IONIC COMPOSITION ASSESSED FROM A LIMITED NUMBER OF OBSERVATIONS MAY LEAD TO WRONG CONCLUSIONS. (JONES-WISCONSIN)

FIELD 05C

ACCESSION NO. W72-01363

ON THE LIFE HISTORIES OF SOME BROWN ALGAE FROM EASTERN CANADA,

NATIONAL RESEARCH COUNCIL OF CANADA, HALIFAX (NOVA SCOTIA). ATLANTIC REGIONAL LAB.

T. EDELSTEIN, L C-M CHEN, AND J. MCLACHLAN.

CANADIAN JOURNAL OF BOTANY, VOL 49, NO 7, P 1247-1251, 1971. 16 FIG, 22 REF.

DESCRIPTORS:
*LIFE HISTORY STUDIES, *PHAEOPHYTA, EPIPHYTOLOGY, ALGAE, HABITATS, CULTURES.

IDENTIFIERS:
*EASTERN CANADA, NOVA SCOTIA, ISTHMOPLEA SPHAEROPHORA, MELANOSIPHON INTESTINALIS, HECATONEMA MACULANS, ELACHISTA LUBRICA, RALFSIA VERRUCOSA.

ABSTRACT:
THE COMPLETE LIFE CYCLES OF FIVE PHAEOPHYCEAN SPECIES, THREE FROM THE ECTOCARPALES AND TWO FROM THE DICTYOSIPHONALES, FROM NOVA SCOTIA, CANADA, WERE COMPLETED IN CULTURE. ISTHMOPLEA SPHAEROPHORA, MELANOSIPHON INTESTINALIS, HECATONEMA MACULANS, ELACHISTA LUBRICA, AND RALFSIA VERRUCOSA REPLICATED DIRECTLY THE PARENT PLANTS. DETAILED DESCRIPTIONS OF THE ALGAE ARE GIVEN, BUT CYTOLOGICAL STUDIES WERE NOT MADE. APART FROM HECATONEMA MACULANS, WHERE PLURILOCULAR SPORANGIA ONLY SERVED AS THE INOCULUM, CULTURES WERE ESTABLISHED FROM SINGLE UNILOCULAR SPORANGIA. CULTURES OF MELANOSIPHON WERE DERIVED FROM BOTH UNI- AND PLURILOCULAR SPORANGIA. NO FUSION WAS NOTED AND THE LIFE CYCLE OBTAINED IN CULTURE WAS PROBABLY ASEXUAL. THE POSSIBILITY IS RECOGNIZED THAT UNDER CONDITIONS OF CULTURE A SPECIES MAY UNDERGO AN ABBREVIATED CYCLE BY ASEXUAL REPRODUCTION ONLY. THE ORIGINAL HABITATS ARE RECORDED. THREE OF THE SPECIES ARE RECORDED AS EPIPHYTIC ON OTHER ALGAE. IN CULTURING, THE MOST SATISFACTORY CONDITIONS OF INCUBATION WERE 13C WITH A PHOTOPERIOD OF 10 HRS LIGHT AND AN INTENSITY OF ABOUT 4000 LUX. THE CULTURE MEDIUM WAS SWM-3 OF CHEN ET AL (1969), CHANGED WEEKLY. (JONES-WISCONSIN)

FIELD 05C

ACCESSION NO. W72-01370

BIOLCGICAL ASPECTS OF WATER POLLUTION,

NATIONAL INST. OF WATER RESEARCH, PRETORIA (SOUTH AFRICA).

R. G. NOBLE, W. A. PRETORIUS, AND F. M. CHUTTER.

SOUTH AFRICAN JOURNAL OF SCIENCE, VOL. 67, NO. 3, P. 132-136, MARCH 1971.

DESCRIPTORS:
*WATER POLLUTION EFFECTS, *ORGANIC WASTES, *NUTRIENTS, *POLLUTANT IDENTIFICATION, ALGAE, AQUATIC BACTERIA, SEWAGE BACTERIA, ENVIRONMENTAL EFFECTS, BIOLOGICAL COMMUNITIES, HEAVY METALS, HERBICIDES, TOXINS, WASTE DILUTION, NITROGEN, PHOSPHORUS, EUTROPHICATION, SURFACE WATERS.

ABSTRACT:
ALTHOUGH ORGANIC WASTES ARE MAINLY NON-TOXIC, THE PROCESSES INVOLVED IN THEIR DECOMPOSITION IN WATER MAY CREATE CONDITIONS SUCH AS ANAEROBIOSIS AND AMMONIA AND SULPHIDE ACCUMULATIONS WHICH ARE HIGHLY TOXIC. SEWAGE, INDUSTRIAL WASTES AND AGRICULTURAL RUNOFF ALSO CONTAIN HIGH LEVELS OF NITRATES AND PHOSPHATES WHICH PROMOTE THE GROWTH OF ALGAE AND HYDROPHYTES, LEADING TO HIGHLY DESTRUCTIVE EFFECTS IN NATURAL WATERS. IN RECENT YEARS TOXIC SUBSTANCES SUCH AS HERBICIDES AND HEAVY METAL IONS HAVE APPEARED IN HIGH QUANTITIES. ALTHOUGH THEY ACT IN DIFFERENT WAYS, POISONS AND NUTRIENTS HAVE STRIKINGLY SIMILAR EFFECTS ON BIOLOGICAL COMMUNITIES. IN BOTH CASES, LOCAL ENVIRONMENTS ARE ALTERED RADICALLY ENOUGH SO THAT ONLY RELATIVELY FEW SPECIES SURVIVE. IN AQUATIC EXOSYSTEMS SUCH LOW SPECIES DIVERSITIES RESULT IN LOW ECOSYSTEM STABILITY SO THAT MANY OF THE PROCESSES IN WATER SELF-PURIFICATION ARE NO LONGER OPERATIVE. BACTERIOLOGICAL, ALGOLGICAL AND FAUNAL MEASURES OF POLLUTION ARE DISCUSSED. SINCE THE COMBATTING OF POLLUTANTS IS EXTREMELY DIFFICULT ONCE THEY HAVE BEEN RELEASED INTO WATERWAYS; THEIR EARLY DETECTION IS IMPORTANT. (CASEY-ARIZONA)

FIELD 05C, 05B

ACCESSION NO. W72-01472

BIOLOGICAL LIFE IN WATER,

NATIONAL INST. FOR WATER RESEARCH, PRETORIA (SOUTH AFRICA).

B. J. CHOLNOKY.

SOUTH AFRICAN JOURNAL OF SCIENCE, VOL 67, NO. 3, P. 128-131, MARCH 1971.

DESCRIPTORS:
*POLLUTANT IDENTIFICATION, *AQUATIC PLANTS, *AQUEOUS SOLUTIONS, *BIOLOGICAL PROPERTIES, MOLECULAR STRUCTURE, PERMEABILITY, PLANT POPULATIONS, ALGAE, CHEMICAL PROPERTIES, WATER POLLUTION, PROTEIN, PRIMARY PRODUCTIVITY.

IDENTIFIERS:
*CHEMICAL EVOLUTION, *MACROMOLECULES.

ABSTRACT:
THE BROAD TRENDS OF CHEMICAL EVOLUTION ARE REVIEWED. THE PROCESSES INVOLVED IN MACROMOLECULE SYNTHESIS DEVELOPED AND OPERATE ONLY IN AQUEOUS SOLUTIONS. AS LARGE ELECTRICALLY CHARGED MOLECULES ACCUMULATED, WATER DIPOLES BECAME ORIENTED, CREATING GROUPS OF PARTICLES OR MICELLES, AND MOLECULARLY DISPERSE SOLUTIONS WERE PREVENTED. MECHANISMS OF 'DUPLICATION' BY THE INTEGRATION OF DISSOLVED MOLECULES OF DIFFERENT TYPES INTO THE STRUCTURES OF NEW MACROMOLECULES WERE DEVELOPED THROUGH THE HARNESSING OF RADIANT ENERGY LEADING TO THE PRIMARY PRODUCERS, THE GREEN PLANTS. EVENTUALLY, OTHER PLANT FORMS EVOLVED WHICH WERE AT LEAST PARTIALLY DEPENDENT ON DISSOLVED ORGANIC MOLECULES FROM PRIMARY PRODUCERS, AND FINALLY, ANIMALS EVOLVED, WHO ARE TOTALLY DEPENDENT ON PRIMARY PRODUCERS. A MAJOR DIFFERENCE BETWEEN AQUATIC PRIMARY PRODUCERS, SUCH AS ALGAE, AND ALL OTHER LIFE FORMS, IS THE NECESSITY FOR EXTREMELY SELECTIVE PERMEABILITY MECHANISMS IN THE PRIMARY PRODUCERS TO FACILITATE SELECTIVE UPTAKE OF THE IMPORTANT INORGANIC MOLECULES THAT SERVE AS BUILDING BLOCKS OF MACROMOLECULES. IT IS THEREFORE ARGUED THAT THIS PROPERTY CONFERS A GREATER RELIABILITY ON GREEN PLANTS AS WATER POLLUTION INDICATORS. WE ARE STILL INCLINED TO OVERESTIMATE THE IMPORTANCE OF BACTERIA AND TO UNDERESTIMATE THE IMPORTANCE OF BACTERIA AND TO UNDERESTIMATE THE IMPORTANCE OF ALGAE. (CASEY-ARIZONA)

FIELD 05C, 02I

ACCESSION NO. W72-01473

A BILL TO AMEND THE FEDERAL WATER POLLUTION CONTROL ACT AS AMENDED, AND FOR OTHER PURPOSES. REFERRED TO THE COMMITTEE ON PUBLIC WORKS,

SENATE, WASHINGTON, D.C.

W. F. MONDALE.

CONGRESSIONAL RECORD, VOL 117, NO 24, P S 2079-81 (DAILY ED. FEBRUARY 26, 1971). 3 P.

DESCRIPTORS:
*LAKES, *WATER POLLUTION CONTROL, *WATER QUALITY CONTROL, *FEDERAL GOVERNMENT, *GRANTS, EUTROPHICATION, GOVERNMENT FINANCE, ENVIRONMENTAL SANITATION, ALGAE, NUTRIENTS, INDUSTRIAL WASTES, MUNICIPAL WASTES, FARM WASTES, WASTE DISPOSAL, PESTICIDE REMOVAL, OXYGEN SAG, DREDGING, STATE GOVERNMENTS, TREATMENT FACILITIES, ADMINISTRATIVE AGENCIES, STANDARDS, REMEDIES.

ABSTRACT:
MANY OF THE NATION'S FRESH WATER LAKES ARE DETERIORATING. A BILL ENTITLED THE 'CLEAN LAKES ACT OF 1971' AMENDS THE FEDERAL WATER POLLUTION CONTROL ACT TO PROVIDE FOR A COORDINATED REHABILITATION PROGRAM FOR LAKES. THE PROGRAM INVOLVES INCREASED WASTE TREATMENT AND LAKE CLEANSING, UTILIZING THE LATEST TECHNOLOGY. IT HAS FOUR MAJOR POINTS: (1) FEDERAL GRANTS FOR TREATMENT WORKS LOCATED NEAR OR ADJACENT TO A LAKE AND WHICH DISCHARGE TREATED WASTES INTO THE LAKE OR TRIBUTARY WATERS WOULD BE INCREASED TO A MAXIMUM OF 65%, IF THE STATE PAYS AT LEAST 20%; (2) TECHNICAL AND FINANCIAL ASSISTANCE TO STATES AND CITIES WOULD BE PROVIDED, INCLUDING THE USE OF HARMLESS CHEMICALS TO DESTROY ALGAE, THE DREDGING OF LAKE BOTTOMS TO REMOVE DECAYING SLUDGE AND OTHER NOXIOUS POLLUTANTS, AND THE RECOVERY OF OVERGROWTH OF ALGAE FROM THE SURFACE; (3) THE USE OF FEDERAL WATER RESOURCE AGENCIES TO EXECUTE THE PROGRAM UNDER AGREEMENTS WITH THE STATES; AND (4) MEASURES TO ENFORCE WATER QUALITY STANDARDS, INCLUDING PENALITIES AND INJUNCTIVE RELIEF. ANNUAL APPROPRIATIONS FOR TREATMENT WORKS WOULD BE $150 MILLION FOR FISCAL YEARS 1972-1975. (REES-FLORIDA)

FIELD 06E, 05G

ACCESSION NO. W72-01659

STATEMENT IN SUPPORT OF THE CLEAN LAKES ACT OF 1971,

SENATE, WASHINGTON, D.C.

H. W. CANNON.

CONGRESSIONAL RECORD, VOL 117, NO 24, P S 2110 (DAILY ED. FEBRUARY 26, 1971). 1 P.

DESCRIPTORS:
*LAKES, *WATER POLLUTION CONTROL, *ENVIRONMENTAL SANITATION, *LEGISLATION, *FEDERAL GOVERNMENT, EUTROPHICATION, ENVIRONMENTAL ENGINEERING, WATER TREATMENT, WATER QUALITY CONTROL, STANDARDS, WATER POLLUTION SOURCES, WATER POLLUTION EFFECTS, RECREATION DEMAND, POPULATION, COMMUNITY DEVELOPMENT, ENVIRONMENTAL EFFECTS, GOVERNMENT FINANCE, TECHNOLOGY, ZONING, ALGAL CONTROL, SOIL EROSION, CITIES, ADMINISTRATIVE AGENCIES.

IDENTIFIERS:
*LAKE TAHOE.

ABSTRACT:
THE PROPOSED CLEAN LAKES ACT OF 1971 WOULD BE OF GREAT VALUE IN IMPROVING AND MAINTAINING THE UNIQUE PURITY AND CLARITY OF LAKE TAHOE. THIS GLACIER LAKE IS 22 MILES LONG AND 12 1/2 MILES WIDE, BUT ITS DEPTH OF 1645 FEET IS BEING DIMINISHED BY ALGAE AND MUD EXPANDING IN ITS WATERS AS A RESULT OF MAN'S DISTURBANCE OF WATERSHED SOIL. DUE TO THE INCREASING POPULATION AND RECREATION DEMANDS UPON LAKE TAHOE, IT IS IN DANGER OF PREMATURELY AGING--A PROCESS CALLED EUTROPHICATION. IN THE LAST 10 YEARS THE FERTILITY OF THE LAKE HAS INCREASED BY 72%. THIS FERTILIZATION PROCESS HAS BEEN SPEEDED UP BY COMMUNITY DEVELOPMENT PROJECTS, LUMBERING, ROADBUILDING, AND LAND CLEARING. A REGIONAL PLANNING AGENCY IS ALREADY ESTABLISHING STANDARDS AND CONTROLS OVER AIR AND WATER POLLUTION, ZONING, BUILDING AND THE GENERAL DEVELOPMENT OF THE LAKE AND THE SURROUNDING LAND AREA. HOWEVER, THE LAKE NEEDS IMMEDIATE ATTENTION FROM PAST NEGLECT. THE PROPOSED ACT WOULD DIRECT THE ENVIRONMENTAL PROTECTION AGENCY TO PROVIDE TECHNICAL AND FINANCIAL ASSISTANCE TO STATES AND CITIES AND TO CONDUCT A COMPREHENSIVE PROGRAM OF POLLUTION CONTROL AND REDEVELOPMENT OF THE LAKE AND THE SURROUNDING LAND. (REES-FLORIDA)

FIELD 06E, 05C

ACCESSION NO. W72-01660

THE DETERGENT POLLUTION CONTROL ACT OF 1970,

SENATE, WASHINGTON, D. C.

G. NELSON.

CONGRESSIONAL RECORD, VOL 116, NO 27, P S 2444-47 (DAILY ED FEBRUARY 26, 1970), 4 P.

DESCRIPTORS:
*WATER POLLUTION CONTROL, *PHOSPHATES, *DETERGENTS, *LEGISLATION, CHEMICALS, CHEMICAL WASTES, CLEANING, DOMESTIC WASTES, EUTROPHICATION, WATER POLLUTION, WATER POLLUTION SOURCES, WATER POLLUTION TREATMENT, NUTRIENTS, OXYGEN REQUIREMENTS, AGING(BIOLOGICAL), AQUATIC ALGAE, WASTE ASSIMILATIVE CAPACITY, WATER POLLUTION EFFECTS, POLLUTANTS, POLLUTION ABATEMENT, FEDERAL GOVERNMENT.

IDENTIFIERS:
*WATER POLLUTION CONTROL ACT.

ABSTRACT:
LEGISLATION TO BAN POLYHOSPHATES FROM DETERGENTS IN THE UNITED STATES IS NECESSARY TO HALT WATER POLLUTION. THE DETERGENT INDUSTRY HAS SHOWN NO INTENTION OF CORRECTING PHOSPHATE POLLUTION WITHOUT LEGISLATION. THE DETERGENT POLLUTION CONTROL ACT OF 1970, WHICH WOULD AMEND THE FEDERAL WATER POLLUTION CONTROL ACT, REQUIRES THE ELIMINATION OF POLYHOSPHATES IN DETERGENTS AND THE ESTABLISHMENT OF NATIONAL ENVIRONMENTAL STANDARDS ON DETERGENT INGREDIENTS. DETERGENTS ARE NOT CONTROLLED BY SEWAGE TREATMENT BECAUSE PHOSPHATES PASS THROUGH MOST TREATMENT SYSTEMS. ONCE ENTERING LAKES AND RIVERS, PHOSPHATES EXCESSIVELY ENTICH THEM WITH NUTRIENTS. THIS CAUSES EXCESSIVE ALGAE GROWTH AND EUTROPHICATION. SEVERAL NON-POLLUTION PHOSPHATE SUBSTITUTES WHICH ARE FEASIBLE FOR USE IN INDUSTRIAL AND HOUSEHOLD DETERGENTS HAVE BEEN DEVELOPED. ENZYMES ARE ANOTHER POLLUTANT USED IN MOST HOUSEHOLD DETERGENTS. THE PROPOSED ACT MAKES IT UNLAWFUL TO MANUFACTURE DETERGENTS WITH PHOSPHATE AFTER JUNE 30, 1972. IT ALSO DIRECTS THE ESTABLISHMENT OF STANDARDS OF TOXICITY, ABILITY, AND BIODEGRADABILITY FOR ALL DETERGENTS. A FEDERAL ASSISTANCE PROGRAM WOULD BE USED TO ACCELERATE THE DEVELOPMENT OF EFFECTIVE PHOSPHATE SUBSTITUTES. (HART-FLORIDA)

FIELD 06E, 05G

ACCESSION NO. W72-01685

POTENTIOMETRIC TECHNIQUES FOR MONITORING IONS INVOLVED IN WATER POLLUTION,

MISSOURI UNIV., COLUMBIA. DEPT. OF CHEMISTRY.

S. E. MANAHAN, M. J. SMITH, D. ALEXANDER, AND P. ROBINSON.

AVAILABLE FROM THE NATIONAL TECHNICAL INFORMATION SERVICE AS PB-204 890, $3.00 IN PAPER COPY, $0.95 IN MICROFICHE. MISSOURI WATER RESOURCES RESEARCH CENTER, COLUMBIA, COMPLETION REPORT, AUG 1, 1971. 26 P, 4 FIG, 2 TAB, 4 REF. OWRR-B-040-MO(1).

DESCRIPTORS:
*POLLUTANT IDENTIFICATION, *NITROGEN, ELECTROCHEMISTRY, TRACE ELEMENTS, HEAVY METALS, ELECTRODES, NITRATES, ALGAE, COPPER, IONS, ION TRANSPORT.

IDENTIFIERS:
POTENTIOMETRY, CADMIUM IONS, COPPER IONS, OOCYSTIS.

ABSTRACT:
THE USE OF ION-SELECTIVE ELECTRODES, PARTICULARLY THE NITRATE ELECTRODE, WAS EXPLORED FOR THE ANALYSIS OF IONIC SPECIES IN NATURAL AQUATIC SYSTEMS. ATTEMPTS TO COMPENSATE QUANTITATIVELY FOR THE EFFECTS OF INTERFERING IONS WERE UNSUCCESSFUL. ION-SELECTIVE ELECTRODES SHOULD NOT BE USED, THEREFORE, WHEN SUBSTANTIAL INTERFERENCES ARE PRESENT. STANDARD ADDITION IS GENERALLY THE PREFERRED TECHNIQUE WITH ION-SELECTIVE ELECTRODES. THE CADMIUM ELECTRODE WAS USED TO DETERMINE THE FORMATION CONSTANT OF THE CITRATE COMPLEX OF CADMIUM ION. THE LOG OF THE FORMATION CONSTANT WAS FOUND TO BE 3.76 PLUS OR MINUS 0.04, 95% CONFIDENCE LEVEL. COPPER ION DEFICIENCY IN ALGAL CULTURES WAS STUDIED AND CORRELATED WITH COPPER ION ACTIVITY AS MEASURED BY THE COPPER ELECTRODE. IT WAS FOUND THAT A MINIMUM LEVEL OF APPROXIMATELY 40 PARTS PER BILLION OF COPPER WAS REQUIRED FOR OPTIMUM GROWTH OF A CULTURE OF OOCYSTIS. THE GROWTH OF THE ALGAE AT MINIMUM COPPER LEVELS COULD BE SUPPRESSED BY THE ADDITION OF CHELATING AGENT.

FIELD 05A

ACCESSION NO. W72-01693

PHYTOPLANKTONIC NITROGEN AS AN INDEX OF CULTURAL EUTROPHICATION,

MICHIGAN STATE UNIV., HICKORY CORNERS. W. K. KELLOGG BIOLOGICAL STATION.

R. G. WETZEL, AND BRUCE A. MANNY.

AVAILABLE FROM THE NATIONAL TECHNICAL INFORMATION SERVICE AS PB-204 707, $3.00 IN PAPER COPY, $0.95 IN MICROFICHE. COMPLETION REPORT, (NOVEMBER, 1971). 15 P, 1 FIG, 36 REF. OWRR-B-009-MICH(1)

DESCRIPTORS:
*NITROGEN, PLANKTON, *EUTROPHICATION, POLLUTANT IDENTIFICATION, ALGAE, ANALYTICAL TECHNIQUES, *ULTRAVIOLET RADIATION, DISSOLVED OXYGEN, LAKES, PHYTOPLANKTON, NUTRIENTS.

IDENTIFIERS:
*POLLUTION INDEX.

ABSTRACT:
ALGAE LESS THAN 10 MICRONS IN DIAMETER ARE NITROGEN RICH ON A CELL VOLUME BASIS COMPARED TO ALGAE LARGER THAN 10 MICRONS IN DIAMETER. A PROCEDURE FOR DISSOLVED ORGANIC NITROGEN (DON) DETERMINATION IN NATURAL WATERS WAS DEVELOPED. THE PROCEDURE UTILIZED HIGH INTENSITY ULTRAVIOLET LIGHT TO DESTROY THE DON AND IS 100-FOLD MORE SENSITIVE THAN THE MICRO-KJELDAHL PROCEDURE. THE PROCEDURE CAN DIFFERENTIATE UV-LABILE AND UV-REFRACTORY DON WITHIN THE DON POOL PRESENT IN A WIDE VARIETY OF NATURAL WATERBODIES. ALLOCHTHONOUS DISSOLVED ORGANIC CARBON (DOC) AND NITROGEN (DON) COMPRISE ABOUT HALF THE DOC AND DON LEAVING THE LAKE OUTLET IMPLYING ABOUT HALF THE DOC AND DON LEAVING THE LAKE ORIGINATES WITHIN THE LAKE AS A RESULT OF PLANT PHOTOSYNTHESIS AND DECOMPOSITION PROCESSES. DURING TRANSPORT IN A HARDWATER STREAM, UV-LABILE DON WAS REMOVED FROM THE WATER AND UV-REFRACTORY DON ACCUMULATED IN THE WATER UNTIL A STABLE EQUILIBRIUM RATIO OF ABOUT 7:3 REFRACTORY TO LABILE DON WAS ATTAINED. SECRETION OF DOC AND DON BY TWO AQUATIC MACROPHYTES IN AXENIC CULTURE WAS DIRECTLY PROPORTIONAL TO INCREASING CARBON FIXATION RATES, LIGHT INTENSITY, PH, CATIONIC CONCENTRATION AND ORGANIC CARBON CONCENTRATION. ORGANIC ENRICHMENT MAY ACCELERATE NUTRIENT CYCLES AND EUTROPHICATION RATES IN HARDWATER LAKES BY STIMULATING INCREASED SECRETION OF DOC AND DON BY THE LITTORAL FLORA.

FIELD 05C, 05A

ACCESSION NO. W72-01780

THE DIVERSITY OF PIGMENTS IN LAKE SEDIMENTS AND ITS ECOLOGICAL SIGNIFICANCE,

MINNESOTA UNIV., MINNEAPOLIS. DEPT. OF BOTANY.

JON E. SANGER, AND EVILLE GORHAM.

LIMNOLOGY AND OCEANOGRAPHY, VOL 15, NO 1, P 59-69, 1971. 2 FIG, 3 TAB, 21 REF. NSF G-23309.

DESCRIPTORS:
*PIGMENTS, *LAKES, *SEDIMENTS, *INDICATORS, CHROMATOGRAPHY, CHLOROPHYLL, TROPHIC LEVEL, EUTROPHICATION, ALGAE, DECOMPOSING ORGANIC MATTER, MINNESOTA, CYANOPHYTA, TREES, GRASSES, LEAVES, FOREST SOILS, SAMPLING, MUD, PHYTOPLANKTON, FLUORESCENCE, PHOTOSYNTHETIC BACTERIA.

IDENTIFIERS:
CAROTENOIDS, XANTHOPHYLLS, MINNESOTA LAKES, LUTEIN.

ABSTRACT:
DIVERSITY OF SEDIMENTARY PIGMENTS IS PARTICULARLY A CONSEQUENCE OF DECOMPOSITION AND SOURCE MATERIAL HAS A PRONOUNCED INFLUENCE. CONCENTRATIONS OF CHLOROPHYLL DERIVATIVES IN THE ORGANIC MATTER OF SURFACE SEDIMENTS IN THE ENGLISH LAKE DISTRICT PROVED SENSITIVE INDICES OF LAKE FERTILITY. THIS STUDY INVESTIGATES FURTHER SOURCES OF ORGANIC MATTER IN LAKE SEDIMENTS, BY EXAMINING PLANT PIGMENT DIVERSITY OF WIDELY DIFFERING TROPHIC, MORPHOLOGICAL, AND CHEMICAL CHARACTERISTICS, AND IN TERRESTRIAL AND AQUATIC PLANT MATERIAL IN VARYING STAGES OF DECOMPOSITION. LAKES CHOSEN (18 IN MINNESOTA AND 6 IN ENGLISH LAKE DISTRICT) INCLUDED THOSE WHOSE PRIMARY PRODUCTIVITY AND WATER AND SEDIMENT CHEMISTRY ARE NOW BEING STUDIED. THIN-LAYER CHROMATOGRAPHY SHOWS A LARGE NUMBER OF PIGMENTS (CHLOROPHYLL DERIVATIVES AND CAROTENOIDS) IN PROFUNDAL LAKE SEDIMENTS, DIVERSITY BEING SOMEWHAT GREATER IN EUTROPHIC THAN IN OLIGOTROPHIC LAKES. WHILE QUANTITY OF PIGMENTS PER GRAM ORGANIC MATTER IS MUCH LOWER IN OLIGOTROPHIC THAN EUTROPHIC LAKES, PIGMENT DIVERSITY IS ONLY SLIGHTLY LOWER. SEDIMENTARY PIGMENTS ARE MUCH MORE NUMEROUS (24-27)THAN THOSE OF UPLAND VEGETATION (7-8), AQUATIC MACROPHYTES (12-15), AND PLANKTONIC ALGAE (10-21). ALGAL DECOMPOSITION, WHICH IS ACCOMPANIED BY A MARKED INCREASE IN PIGMENT NUMBER, SEEMS THE MOST LIKELY CAUSE FOR THE EXTREME DIVERSITY OF SEDIMENTARY PIGMENTS. (JONES-WISCONSIN)

FIELD 05C, 02H

ACCESSION NO. W72-01784

AEROBIC DECOMPOSITION OF ALGAE,

STANFORD UNIV., CALIF.

WILLIAM J. JEWELL, AND PERRY L. MCCARTY.

ENVIRONMENTAL SCIENCE AND TECHNOLOGY, VOL 5, NO. 10, OCTOBER 1971, P 1023-1031, 7 FIG, 4 TAB, 40 REF.

DESCRIPTORS:
*OXYGEN REQUIREMENTS, *DECOMPOSING ORGANIC MATTER, AERATION, *ALGAE, *BIODEGRADATION, LABORATORY TESTS, FILTRATION, AMMONIA, HYDROGEN ION CONCENTRATION, TEMPERATURE, LIGHT INTENSITY, SEDIMANTATION, OXIDATION, STORAGE, BIOCHEMICAL OXYGEN DEMAND, CHEMICAL OXYGEN DEMAND, PHOSPHORUS, NITROGEN, EUTROPHICATION, *AEROBIC CONDITIONS, *AEROBIC TREATMENT.

IDENTIFIERS:
REFRACTORY MATERIALS.

ABSTRACT:
5 GAL. WATER SAMPLES FROM VARIOUS LAKE, STREAMS, AND RESERVOIRS, AS WELL AS EFFLUENTS FROM SEVERAL SEWAGE TREATMENT FACILITIES WERE OBTAINED. EACH SAMPLE WAS FILTERED AND AERATED, TO ALLOW OXIDATION OF ALL REDUCED MATERIALS, EXPOSED TO DIURNAL FLOURESCENT LIGHT, AND AERATED WITH A MIXTURE OF 1% CO2 IN AIR. LIGHTING, PH AND TEMPERATURE WERE HELD CONSTANT DURING ANY GIVEN EXPERIMENT. AT PREDETERMINED TIME INTERVALS, SAMPLES WERE TAKEN FROM THE ILLUMINATED VESSELS AND PLACED IN DARK VESSELS AND AERATED WITH THE SAME AERATION MIXTURE. RESULTS OF THESE TESTS INDICATED THAT ALGAE AND ALGAL-DERIVED ORGANIC MATTER CONSISTED OF THREE FRACTIONS. THE FIRST FRACTION CONSISTED OF DEGRADABLE STORAGE PRODUCTS THAT DISAPPEAR WITHIN A FEW HOURS AFTER THE ORGANISMS ARE PLACED IN THE DARK. OXYGEN DEMAND OF THE STORAGE PRODUCTS WAS DEEMED INSIGNIFICANT IN LONG-TERM CONSIDERATIONS, BUT ITS DECOMPOSITION WAS DETERMINED TO BE A POSSIBLE SIGNIFICANT FACTOR IN INFLUENCING DIURNAL OXYGEN VARIATIONS IN NATURAL WATERS. BIODEGRADABLE ORGANIC MATERIAL, THE SECOND FRACTION, COMPRISED SOME 30% OF THE TOTAL MASS OF ORGANIC ALGAL MATERIALS. A FAIRLY CONSTANT, BUT INTERMEDIATE, FIRST ORDER RATE CONSTANT (.01 TO .06 DAY-1) WAS DETERMINED, INDICATING THAT ALL OF THE BIODEGRADABLE ORGANICS SHOULD BE DECOMPOSED WITHIN ONE YEAR. THE REMAINING FRACTION, REFRACTORY MATERIAL WAS FOUND TO DECOMPOSE BY ONLY A FEW PERCENT PER YEAR, HAVING A SIGNIFICANT OXYGEN DEMAND ONLY BY ACCUMULATION. (LOWRY-TEXAS)

FIELD 05D, 05C

ACCESSION NO. W72-01881

IODINE AND ALGAE IN SEDIMENTARY ROCKS ASSOCIATED WITH IODINE-RICH BRINES,

BUREAU OF MINES, BARTLESVILLE, OKLA. PETROLEUM RESEARCH CENTER.

A. G. COLLINS, J. H. BENNETT, AND O. K. MANUEL.

GEOLOGICAL SOCIETY OF AMERICA BULLETIN, VOL 82, NO 9, P 2607-2610, SEPTEMBER 1971. 2 FIG, 2 TAB, 7 REF. NSF GRANT GA-12099.

DESCRIPTORS:
*SEDIMENTARY ROCKS, *BRINES, *ALGAE, *OKLAHOMA, *IODINE RADIOISOTOPES, URANIUM RADIOISOTOPES, ANALYTICAL TECHNIQUES, PALEOZOIC ERA, WATER CHEMISTRY.

IDENTIFIERS:
NEUTRON ACTIVATION ANALYSIS.

ABSTRACT:
NEUTRON ACTIVATION ANALYSES OF IODINE AND URANIUM IN PALEOZOIC SEDIMENTARY ROCKS FROM THE NORTHERN OKLAHOMA PLATFORM OF THE ANADARKO BASIN SHOWED 0.9 TO 12.3 PPM I AND 0.07 TO 8.7 PPM U. THE SAMPLES WERE TAKEN FROM STRATA IN KINGFISHER COUNTY, OKLAHOMA, WHERE ANOMALOUSLY HIGH CONCENTRATIONS OF IODINE WERE FOUND IN ASSOCIATED SUBSURFACE BRINES. MICROPALEONTOLOGICAL EXAMINATIONS REVEALED ALGAL STRANDS IN THE MORE IODINE-RICH ROCKS. (WOODARD-USGS)

FIELD 02K, 02F, 05A

ACCESSION NO. W72-02073

BLUE-GREEN ALGAL EFFECTS ON SOME HYDROLOGIC PROCESSES AT THE SOIL SURFACE,

ARIZONA UNIV., TUCSON. WATER RESOURCES RESEARCH CENTER.

W. F. FAUST.

IN: HYDROLOGY AND WATER RESOURCES IN ARIZONA AND THE SOUTHWEST, PROCEEDINGS, ARIZONA SECTION-AMERICAN WATER RESOURCES ASSOCIATION AND THE HYDROLOGY SECTION-ARIZONA ACADEMY OF SCIENCE, APRIL 22-23, 1971, TEMPE, VOL 1, P 99-105, 1971. 1 FIG, 2 TAB, 4 REF.

DESCRIPTORS:
*SOIL ALGAE, *SOIL SURFACES, *SEDIMENT YIELD, *SIMULATED RAINFALL, *SOIL TYPES, SOIL TEXTURE, CLAYS, LABORATORY TESTS, STATISTICAL METHODS, SOIL MICROORGANISMS, RUNOFF, HYDROLIC DATA.

IDENTIFIERS:
*SUSPENDED SEDIMENT PRODUCTION, *BLUE-GREEN ALGAE.

ABSTRACT:
PREVIOUS STUDIES HAVE INDICATED THAT BLUE-GREEN ALGAE MAY AFFECT RUNOFF, INFILTRATION AND EROSION AT SOIL SURFACES. USING SOIL PLOTS UPON WHICH BLUE-GREEN ALGAE WERE GROWN UNDER AN ARTIFICIAL WETTING REGIME, STUDIES WERE MADE USING SIMULATED RAINFALL. A 30% CLAY CONTENT PIMA SOIL AND A CONTRASTING 8% CLAY CONTENT RIVER-BOTTOM ANTHONY SOIL WERE USED. SCYTONEMA HOFFMANII AND MICROCOLEUS VAGINATUS GREW ON THE PIMA SOIL WHILE SCHIZOTHRIX CALCICOLA DEVELOPED ON THE ANTHONY SOIL. THE RESULTS SHOWED THAT BLUE-GREEN ALGAL GROWTHS SIGNIFICANTLY REDUCED THE AMOUNT OF SUSPENDED SOIL MATERIAL IN RUNOFF WATER AS COMPARED WITH BARE SOILS. DIFFERENCES IN RUNOFF SUSPENDED SEDIMENTS WERE ALSO RELATED TO DIFFERENCES IN SOIL TYPE AND SIMULATED RAINFALL INTENSITY. AN ANALYSIS OF VARIANCE OF THE EFFECTS OF THESE 3 FACTORS AND THEIR INTERACTIONS SHOWED THAT THE SMALLER DIFFERENCES IN SUSPENDED SEDIMENT PRODUCTION ON THE ANTHONY SOIL DUE TO THE MICROVEGETATION TREATMENT WAS VERIFIED BY A HIGHLY SIGNIFICANT SOILS-MICROVEGETATION INTERACTION, PROBABLY BECAUSE THE FINER PIMA SOILS WASH AWAY MORE EASILY WITHOUT STABILIZING MICROVEGETATION. ALSO, LESS VEGETATION SEEMS TO GROW ON THE ANTHONY SOIL. DIFFERENCES IN RUNOFF AND INFILTRATION VOLUMES AND IN SETTLEABLE SEDIMENT AMOUNTS WERE NOT DETECTED.
(CASEY-ARIZONA)

FIELD 02G, 05C, 04A

ACCESSION NO. W72-02218

EVALUATION OF EFFECT OF IMPOUNDMENT ON WATER QUALITY IN CHENEY RESERVOIR,

COLORADO STATE UNIV., FORT COLLINS.

J. C. WARD, AND S. KARAKI.

BUREAU OF RECLAMATION RESEARCH REPORT NO 25, 1971. 69 P, 38 FIG, 18 TAB, 23 REF, APPEND. BUR. RECLAM CONTRACT 14-06-D-6578.

DESCRIPTORS:
*WATER QUALITY, *IMPOUNDED WATERS, *RESERVOIRS, *HYDROLOGIC DATA, *KANSAS, WATER CHEMISTRY, CHEMICAL ANALYSIS, EVAPORATION, SEEPAGE, TURBIDITY, HYDROLOGIC BUDGET, SOLAR RADIATION, WATER TEMPERATURE, SALT BALANCE, DISSOLVED OXYGEN, ALGAE, BACTERIA, ODOR, INFLOW, DISSOLVED SOLIDS, DATA COLLECTIONS.

IDENTIFIERS:
*CHENEY RESERVOIR(KANSAS).

ABSTRACT:
THE EFFECT OF IMPOUNDMENT ON THE QUALITY OF WATER IN CHENEY RESERVOIR NEAR WICHITA, KANS. IS PRESENTED. THE RESERVOIR DID NOT STRATIFY DURING THE PERIOD OF DATA COLLECTION. THE INCREASE IN THE DISSOLVED SOLIDS CONCENTRATION WAS DIRECTLY RELATED TO EVAPORATION. ON AN ANNUAL BASIS, 42% OF THE TOTAL INFLOW WAS EVAPORATED FROM THE RESERVOIR. BYPASSING THE POOREST QUALITY WATERS OF THE STREAM SERVING THE RESERVOIR IS SUGGESTED TO REDUCE THE DISSOLVED SOLIDS CONCENTRATION IN THE RESERVOIR AND IN THE STREAM BELOW THE RESERVOIR. THE BIOLOGICAL ACTIVITY WITHIN THIS RESERVOIR DID NOT SEEM TO AFFECT THE WATER QUALITY MATERIALLY. ODOR APPEARS TO HAVE STABILIZED AT A THRESHOLD ODOR NUMBER OF ABOUT 5. THE EFFECT OF THE INTERACTION BETWEEN THE MICROORGANISMS AND NUTRIENTS WAS CHARACTERIZED IN THE ANALYSIS OF PHOSPHATES, NITRATES, AND SILICA CONCENTRATIONS IN THE RESERVOIR. THE DISSOLVED OXYGEN PERCENT SATURATION DECREASED FROM 100% AT THE WATER SURFACE TO 82% AT A 25-FOOT DEPTH. (WOODARD-USGS)

FIELD 05C, 02H, 04A

ACCESSION NO. W72-02274

A THERMODYNAMIC ANALYSIS OF A PRIMARY WASTE STABILIZATION POND,

UTAH STATE UNIV., LOGAN. UTAH WATER RESEARCH LAB.

D. W. HENDRICKS, W. D. POTE, AND J. G. ANDREW.

AVAILABLE FROM THE NATIONAL TECHNICAL INFORMATION SERVICE AS PB-205 282, $3.00 IN PAPER COPY, $0.95 IN MICROFICHE. REPORT NO PRCWRR 16-1, SEPTEMBER 1970, 63 P, 14 FIG, 7 TAB, 40 REF. OWRR A-006-UTAH(1).

DESCRIPTORS:
*OXIDATION LAGOONS, *ALGAE, *OXIDATION, PHOTOSYNTHESIS, OXYGEN, *THERMODYNAMICS, ANALYTICAL TECHNIQUES, DESIGN CRITERIA, KINETICS, *WASTE WATER TREATMENT.

IDENTIFIERS:
*SOLAR INSOLATION, *STOICHIOMETRY.

ABSTRACT:
A 97.5 ACRE OXIDATION POND WITH AN AVERAGE DEPTH OF 1 1/2 TO 1 2/3 METERS, IN OPERATION SINCE 1967, WAS SAMPLED IN SEPTEMBER 1969 AND IN JUNE 1970. THE PRIMARY OBJECTIVE WAS TO QUANTITATE THE ACTUAL ENERGY TRADE-OFF, IN TERMS OF ALGAE PRODUCED VS. AMOUNT OF WASTE DEGRADED, FOR OXIDATION PONDS. SUCH QUANTITATION WAS ACCOMPLISHED BY: (1) DEFINING THE CHEMICAL REACTIONS INVOLVED-BOTH STOICHIOMETRICALLY AND THERMODYNAMICALLY (THE LATTER IN TERMS OF EQUILIBRIUM CONDITIONS); (2) MEASURING TERMS IN A DAILY MASS BALANCE MODEL OF AN OPERATING PRIMARY POND; AND (3) EVALUATING THE 'ALGAE PRODUCTION POTENTIAL' FOR THE POND STUDIED, BASED UPON AVAILABLE SOLAR INSOLATION. THESE RESULTS DEFINED RESPECTIVELY: (1) THE CALCULATED ABSOLUTE LOWER LIMIT OF DAILY ALGAL SYNTHESIS NECESSARY FOR PRODUCTION OF THE STOICHIOMETRIC OXYGEN TO SATISFY THE DAILY INFLUENT BOD REQUIREMENT; (2) A MEASURED DAILY SYNTHESIS RATE OF ALGAE TO COMPARE WITH THE DAILY INFLUENT TOC, UNDER CONDITIONS OF MAXIMUM SUNSHINE IN THE ANNUAL CYCLE; AND (3) THE CALCULATED ABSOLUTE UPPER LIMIT OF DAILY ALGAL SYNTHESIS, THROUGH THE ANNUAL CYCLE, IF ALL USABLE SOLAR ENERGY WERE UTILIZED. RESULTS ESTABLISHED: (1) ALGAE PRODUCTION IS SIGNIFICANT IN PROPORTION TO WASTE DEGRADED, EVEN IN THE LOWER LIMIT; (2) ACTUAL PRODUCTION WAS OVER 100 TIMES THE STOICHIOMETRIC AMOUNT; AND (3) THE UPPER PRODUCTION LIMIT WAS OVER 3 TIMES THE ACTUAL PRODUCTION. ALL RESULTS INDICATED A VAST ENERGY OVERTURN WITH LITTLE OR NO NET EFFECT. (LOWRY-TEXAS)

FIELD 05D

ACCESSION NO. W72-02363

BIOLOGICAL RESPONSE TO TERTIARY TREATED EFFLUENT IN INDIAN CREEK RESERVOIR,

UTAH STATE UNIV., LOGAN. DEPT. OF ENVIRONMENTAL BIOLOGY.

D. B. PORCELLA, P. H. MCGAUHEY, AND G. L. DUGAN.

PREPRINT, PRESENTED AT 44TH ANNUAL CONFERENCE OF WATER POLLUTION CONTROL FEDERATION, SESSION 25, OCTOBER 3-8, 1971. 31 P, 5 FIG, 9 TAB, 13 REF.

DESCRIPTORS:
*WATER DILUTION, *NITROGEN, *PHOSPHORUS, *DETENTION RESERVOIRS, TERTIARY TREATMENT, ALGAE, PLANTS, TROUT, AQUATIC ENVIRONMENT, EUTROPHICATION, NITRIFICATION, RESERVOIR EVAPORATION, RESERVOIRS, NUTRIENTS, WATER REUSE.

IDENTIFIERS:
SOUTH TAHOE PUBLIC UTILITIES DISTRICT, *INDIAN CREEK RESERVOIR, NUTRIENT BALANCE.

ABSTRACT:
INDIAN CREEK RESERVOIR RECEIVES TERTIARY EFFLUENT FROM THE SOUTH TAHOE PUBLIC UTILITY DISTRICT. AN ON-GOING STUDY IS OBSERVING THE EFFECTS OF NUTRIENT REMOVAL (MAINLY PHOSPHORUS) AND THE POSSIBILITY OF IMPOUNDMENTS AS A PROCESS FOR RECLAIMING WATER FROM SEWAGE. THE RESERVOIR IS CHANGING FROM A LOW DIVERSITY AND HIGH PRODUCTION TO A HIGHER DIVERSITY AND LOWER PRODUCTION AQUATIC ECOSYSTEM, A MORE BALANCED SYSTEM LESS LIKELY TO BE DISTURBED. HIGH LEVELS OF NITROGEN AND PHOSPHORUS EXIST, ALTHOUGH CONSIDERABLE REMOVAL OF BOTH NITROGEN AND PHOSPHORUS OCCURS IN THE RESERVOIR. A BALANCE ACCOUNTED FOR PHOSPHORUS, BUT A DEFICIT WAS FOUND FOR NITROGEN THAT WAS ATTRIBUTED TO NITRIFICATION-DENITRIFICATION. ALGAL GROWTH WAS OBSERVED THAT WAS ASSOCIATED WITH BENTHIC ALGAE. THE DIVERSITY OF BENTHIC ORGANISMS WAS LOWER THAN EXPECTED IN A MATURE RESERVOIR, BUT IS APPARENTLY INCREASING. FISH SURVIVAL DURING THE 1971 SEASON INDICATES THAT THE RESERVOIR CONTINUES TO BE A SUITABLE ENVIRONMENT FOR PLANTED TROUT SPECIES. AQUATIC VASCULAR PLANTS ARE PRESENT IN SOME ABUNDANCE, PARTICULARLY IN THE SHALLOW WATER AREAS. THOUGH SUFFICIENT PHOSPHORUS IS AVAILABLE NO OBJECTIONABLE ALGAL BLOOMS HAVE OCCURRED. WATER QUALITY, BOTH CHEMICALLY AND BIOLOGICALLY, DIFFERS DRAMATICALLY FROM THE TERTIARY EFFLUENT, FROM WHICH IT PREDOMINANTLY DERIVES, SUGGESTS IMPOUNDMENT AS AN ECONOMICAL PROCESS TO REDUCE NITROGEN FOLLOWING PHOSPHORUS REMOVAL. (MORGAN-TEXAS)

FIELD 05C

ACCESSION NO. W72-02412

PHOTOSYNTHETIC RECLAMATION OF AGRICULTURAL SOLID AND LIQUID WASTE,

CALIFORNIA UNIV., BERKELEY. SANITARY ENGINEERING RESEARCH LAB.

G. L. DUGAN, C. G. GOLUEKE, W. J. OSWALD, AND C. E. RIXFORD.

SERL REPORT NO 70-1, MAY 1970. 165 P, 55 TAB, 24 FIG, 51 REF. USPHS 5 R01 U1 00566-03.

DESCRIPTORS:
*FARM WASTE, *POULTRY, *OXIDATION LAGOONS, NUTRIENTS, AQUATIC ALGAE, NITROGEN, ANAEROBIC DIGESTION, AEROBIC, ALKALINITY, ACIDITY, LABORATORY, DIGESTION TANK, PONDAGE, PILOT PLANT, SETTLING BASIN.

IDENTIFIERS:
*DETENTION PERIOD, FLUSHING OPERATION, ACID INJECTION PUMP, ALUM INJECTION PUMP, MANURE GRINDER,

ABSTRACT:
THE RESEARCH PLAN ON WHICH THE GRANT WAS BASED AND REPORTED CALLED FOR LABORATORY AND PILOT PLANT STUDIES TO DEVELOP A PARTIALLY-CLOSED SYSTEM OF ANIMAL WASTE MANAGEMENT BASED ON THE INTEGRATION OF AN ANAEROBIC AND AN AEROBIC PHASE, THE RECYCLING OF WATER, AND THE RECLAMATION OF A USABLE PRODUCT. CONTAINED HEREIN ARE: A REVIEW OF THE LABORATORY AND 'PRE-PILOT PLANT' STUDIES DESCRIBED IN A PROGRESS REPORT (1) ISSUED DURING THE SECOND YEAR OF THE STUDY, AND A FULL ACCOUNT OF PILOT PLANT STUDIES COMPLETED AT THE TIME OF THIS WRITING. THE PILOT PLANT INCLUDES A POULTRY ENCLOSURE, A HYDRAULIC SYSTEM FOR HANDLING THE WASTES, A HEATED ANAEROBIC DIGESTER WITH AUXILIARY EQUIPMENT, AND AN ALGAE-PRODUCTION POND. (BUNDY-IOWA STATE)

FIELD 05D

ACCESSION NO. W72-02850

REMOVAL OF NITRATE BY AN ALGAL SYSTEM,

CALIFORNIA STATE DEPT. OF WATER RESOURCES, FRESNO. SAN JOAQUIN DISTRICT.

RANDALL L. BROWN.

COPY AVAILABLE FROM GPC SUP DOC FOR $1.25; MICROFICHE FROM NTIS AS PB-205 425, $0.95. ENVIRONMENTAL PROTECTION AGENCY - WATER QUALITY OFFICE, WATER POLLUTION CONTROL RESEARCH SERIES, APRIL, 1971, 132 P, 58 FIG, 27 TAB, 59 REF. EPA PROGRAM 13030 ELY.

DESCRIPTORS:
AGRICULTURAL WASTES, WATER POLLUTION CONTROL, *BIOLOGICAL TREATMENT, *NITRATES, TREATMENT FACILITIES, ALGAE, *WASTE WATER TREATMENT, *ALGAL CONTROL, *AQUATIC WEED CONTROL, CALIFORNIA.

IDENTIFIERS:
*ALGAE STRIPPING, SCENEDESMUS, ALGAL GROWTH, ALGAL HARVESTING, *SAN JOAQUIN VALLEY(CALIF).

ABSTRACT:
AN ALGAL SYSTEM CONSISTING OF ALGAE GROWTH, HARVESTING AND DISPOSAL WAS EVALUATED AS A POSSIBLE MEANS OF REMOVING NITRATE-NITROGEN FROM SUBSURFACE AGRICULTURAL DRAINAGE IN THE SAN JOAQUIN VALLEY OF CALIFORNIA. THE STUDY OF THIS ASSIMILATORY NITROGEN REMOVAL PROCESS WAS INITIATED TO DETERMINE OPTIMUM CONDITIONS FOR GROWTH OF THE ALGAL BIOMASS, SEASONAL VARIATIONS IN ASSIMILATION RATES, AND METHODS OF HARVESTING AND DISPOSAL OF THE ALGAL PRODUCT. A SECONDARY OBJECTIVE OF THE STUDY WAS TO OBTAIN PRELIMINARY COST ESTIMATES AND PROCESS DESIGN. THE GROWTH STUDIES SHOWED THAT ABOUT 75 TO 90 PERCENT OF THE 20 MG/L INFLUENT NITROGEN WAS ASSIMILATED BY SHALLOW (12-INCH CULTURE DEPTH) ALGAL CULTURES RECEIVING 2 TO 3 MG/L ADDITIONAL IRON AND PHOSPHORUS AND A MIXTURE OF 5 PERCENT CO2. THEORETICAL HYDRAULIC DETENTION TIMES REQUIRED FOR THESE ASSIMILATION RATES VARIED FROM 5 TO 16 DAYS, DEPENDING ON THE TIME OF THE YEAR. THE TOTAL NITROGEN REMOVAL BY THE ALGAL SYSTEM, ASSUMING 95 PERCENT REMOVAL OF THE ALGAL CELLS, RANGED FROM 70 TO 85 PERCENT OF THE INFLUENT NITROGEN. THE MOST ECONOMICAL AND EFFECTIVE ALGAL HARVESTING SYSTEM TESTED WAS FLOCCULATION AND SEDIMENTATION FOLLOWED BY FILTRATION OF THE SEDIMENT. THE ALGAL CAKE FROM THE VACUUM FILTER, CONTAINING ABOUT 20 PERCENT SOLIDS, WAS THEN AIR- OR FLASH-DRIED TO ABOUT 90 PERCENT SOLIDS. THE MARKET VALUE FOR THIS PRODUCT AS A PROTEIN SUPPLEMENT WAS ESTIMATED TO BE ABOUT $80 TO $100 PER TON. MINER-IOWA STATE)

FIELD 05D

ACCESSION NO. W72-02975

LYSOGENY OF A BLUE-GREEN ALGA, PLECTONEMA BORYANUM,

DELAWARE UNIV., NEWARK. DEPT. OF BIOLOGICAL SCIENCES.

R. E. CANNON, M. S. SHANE, AND VALERIE N. BUSH.

VIROLOGY, VOL. 45, NO. 1, P 149-153, JULY 1971. 1 FIG, 1 TAB, 9 REF. OWRR A-016-DEL(1).

DESCRIPTORS:
*CYANOPHYTA, ALGAE, *VIRUSES.

IDENTIFIERS:
*LYSOGENY, *PLECTONEMA BORYANUM, LPP-1, LPP-1D, PHYCOVIRUS, MITOMYCIN C.

ABSTRACT:
EVIDENCE FOR LYSOGENY OF A FILAMENTOUS BLUE-GREEN ALGA, PLECTONEMA BORYANUM, IS PRESENTED. A SUSPECTED LYSOGENIC STRAIN WHICH CARRIES PHYCOVIRUS DESIGNATED LPP-1D (DELAWARE STRAIN) HAS BEEN SUBCULTURED FOR FOUR YEARS. VIRAL NEUTRALIZATION TESTS SHOW THAT LPP-1D IS ANTIGENICALLY SIMILAR TO LPP-1. INDUCTION EXPERIMENTS OF LYSOGENIC PLECTONEMA WITH AN ANTIBIOTIC, MITOMYCIN C, RESULT IN 100-FOLD INCREASE IN PHYCOVIRUS TITER 4-5 HR AFTER TREATMENT. GROWTH OF LYSOGENIC CULTURES WITH ANTIPHYCOVIRUS SERUM ELIMINATES ALL FREE PHYCOVIRUS IN THE CULTURE. TEN DAYS LATER, PHYCOVIRUS IS AGAIN PRESENT IN THE MEDIUM. LPP-1 APPEARS TO BE A VIRULENT STRAIN WHILE LPP-1D MAY BE A TEMPERATE STRAIN OF THE SAME TYPE OF PHYCOVIRUS WHICH HAS LYSOGENIZED PLECTONEMA BORYANUM.

FIELD 05C

ACCESSION NO. W72-03188

SAMPLING AND MEASUREMENT IN THE AQUATIC ENVIRONMENT,

WASHINGTON STATE UNIV., PULLMAN. DEPT. OF SANITARY ENGINEERING.

SURINDER K. BHAGAT, DONALD E. PROCTOR, AND WILLIAM H. FUNK.

PRESENTED AT THE 25TH PURDUE INDUSTRIAL WASTE CONFERENCE, LAFAYETTE, INDIANA, MAY 5-7, 1970, MIMEO (UNDATED), 22 P. 13 FIG, 1 TAB, 14 REF. 15-12-68 16080 ERO.

DESCRIPTORS:
*EUTROPHICATION, *MEASUREMENT, *SAMPLING, WATER RESOURCES, AREA REDEVELOPMENT, BIOLOGICAL CHARACTERISTICS, ALGAE, ZOOPLANKTON, BACTERIA, NUTRIENTS, SEDIMENTS, ENVIRONMENTAL EFFECTS, TRACE ELEMENTS, ANALYTICAL TECHNIQUES, COLUMBIA RIVER, WASHINGTON, OREGON.

IDENTIFIERS:
*VANCOUVER LAKE(WASH), VANCOUVER(WASH), PORTLAND(ORE), WATER QUALITY ANALYZER.

ABSTRACT:
VANCOUVER LAKE (WASHINGTON) PRESENTLY POLLUTED, WHICH HAS THE POTENTIAL OF BECOMING A USEFUL MULTIPURPOSE RESOURCE, WAS STUDIED TO DETERMINE THE PRESENT WATER QUALITY CONDITIONS, EVALUATE POLLUTION SOURCES, AND EXPLORE WAYS TO IMPROVE ITS USEFULNESS. SUCH SAMPLING AND MEASUREMENTS WHICH WERE CONSIDERED NECESSARY IN ATTAINING VANCOUVER LAKE PROJECT OBJECTIVES ARE DISCUSSED IN DETAIL. MEASUREMENTS WERE MADE OF BIOLOGICAL AND BACTERIOLOGICAL ACTIVITIES, NUTRIENT LEVELS, AND THE EXISTING ENVIRONMENTAL CONDITIONS. THE MOST PROMINENT ALGA IS APHANIZONMENON FLOS-AQUAE, ONE OF THE MORE UNSIGHTLY AND ODORIFEROUS. A BIOLOGICAL INVENTORY OF THE BOTTOM ORGANISMS SHOWED AQUATIC EARTHWORMS CHARACTERISTIC OF SHALLOW AND TURBID WATERS. BACTERIOLOGICAL EXAMINATION INDICATED THAT BACTERIAL CONTAMINATION IS EXCESSIVELY HIGH PRECLUDING RECREATIONAL USE. SPHAEROTILUS, RESPONSIBLE FOR SLIME GROWTHS IN STREAMS AND FOR DESTROYING HABITATS FOR VARIOUS AQUATIC ANIMALS WAS MEASURED IN THE ADJACENT COLUMBIA RIVER. THE NUTRIENT LEVELS IN VANCOUVER LAKE ARE QUITE HIGH AND RESPONSIBLE FOR EXCESSIVE ALGAL POPULATIONS. EQUIPMENT FOR MEASUREMENTS OF ENVIRONMENTAL FACTORS IS DELINEATED. TRACE ELEMENTS WERE MEASURED BY NEUTRON ACTIVATION AND HIGH RESOLUTION SPECTROMETRY. (JONES-WISCONSIN)

FIELD 05C

ACCESSION NO. W72-03215

PRIMARY PRODUCTIVITY, CHEMO-ORGANOTROPHY, AND NUTRITIONAL INTERACTIONS OF EPIPHYTIC ALGAE AND BACTERIA ON MACROPHYTES IN THE LITTORAL OF A LAKE,

MICHIGAN STATE UNIV., HICKORY CORNERS, W. K. KELLOGG BIOLOGICAL STATION.

HAROLD L. ALLEN.

ECOLOGICAL MONOGRAPHS, VOL 41, NO 2, P 97-127, 1971. 33 FIG, 13 TAB, 18 REF. AEC AT (11-1)-1599 NSF GB-6538.

DESCRIPTORS:
*PRIMARY PRODUCTIVITY, *EUTROPHICATION, *NUTRIENTS, *BIOLOGICAL COMMUNITIES, *AQUATIC PLANTS, ALGAE, BACTERIA, LITTORAL, LAKES, PERIPHYTON, ORGANIC MATTER, BIOMASS, METABOLISM, PHYTOPLANKTON, PLANT PHYSIOLOGY, PHYSIOLOGICAL ECOLOGY, ORGANIC COMPOUNDS, ON-SITE TESTS, METHODOLOGY, CHARA.

IDENTIFIERS:
*EPIPHYTIC ALGAE, *EPIPHYTIC BACTERIA, *MACROPHYTE-EPIPHTYE METABOLISM, LAWRENCE LAKE(MICH), SCIRPUS ACUTUS, NAJAS FLEXILIS.

ABSTRACT:
COMMUNITY METABOLISM OF MACROPHYTE-EPIPHYTE SYSTEMS AND NUTRITIONAL RELATIONSHIPS OF EPIPHYTIC ALGAE AND BACTERIA WERE INVESTIGATED BY C-14 TECHNIQUES IN THE LITTORAL ZONE OF A SMALL LAKE. THE STUDY INDICATED THAT EPIPHYTIC ALGAE CONTRIBUTED 31.3% TO THE TOTAL LITTORAL PRODUCTION AND THAT THE ALGAL EPIPHYTES MAY BE AMONG THE DOMINANT PRODUCERS IN SHALLOW-WATER ECOSYSTEMS WITH SUBMERGED MACROPHYTES. THE CHEMO-ORGANOTROPHY OF EPIPHYTIC BACTERIA WAS EVALUATED ON GLUCOSE AND ACETATE SUBSTRATES BY ENZYME KINETIC ANALYSIS. THE NATURE OF EXTRACELLULAR RELEASE SUGGESTS NUTRITIONAL INTERACTIONS IN MACROPHYTE-EPIPHYTE SYSTEMS. THE METABOLISM OF THE TWO GROUPS OF ORGANISMS MAY BE A SOURCE OF DISSOLVED ORGANIC MATTER THAT CONTRIBUTES TO THE PRIMARY PRODUCTION. (WILDE-WISCONSIN)

FIELD 05C

ACCESSION NO. W72-03216

STUDIES ON THE PHYSIOLOGY OF HETEROCYST PRODUCTION IN THE NITROGEN-FIXING BLUE-GREEN ALGA ANABAENA SP L-31 IN CONTINUOUS CULTURE,

BHABHA ATOMIC RESEARCH CENTRE, BOMBAY (INDIA). BIOLOGY DIV.

J. THOMAS, AND K. A. V. DAVID.

JOURNAL OF GENERAL MICROBIOLOGY, VOL 66, NO 1, P 127-131, 1971. 3 FIG, 1 TAB, 17 REF.

DESCRIPTORS:
*NITROGEN FIXATION, *CYANOPHYTA, PLANT PHYSIOLOGY, CYTOLOGICAL STUDIES, CULTURES, AMMONIA, POTASSIUM COMPOUNDS, NITRATES, DENSITY, GROWTH RATES, LABORATORY TESTS, CYANOPHYTO, ALGAE.

IDENTIFIERS:
*HETEROCYSTS, *ANABAENA SP L-31, CONTINUOUS CULTURE, BATCH CULTURES.

ABSTRACT:
COMPOUNDS WHICH INHIBIT HETEROCYST PRODUCTION IN FILAMENTOUS BLUE-GREEN ALGAE INCLUDE NITRATE, NITRITE, AMMONIA, AND OTHER NITROGENEOUS SUBSTANCES, AMMONIA BEING THE MOST EFFECTIVE. MECHANISM OF THIS INHIBITION HAS NOT BEEN CLEARLY UNDERSTOOD. STUDIES IN BATCH AND CONTINUOUS CULTURES OF ANABAENA SP L-31 ARE REPORTED TO ELUCIDATE FURTHER THE PHYSIOLOGY OF INHIBITION OF HETEROCYSTS. DAILY MEASUREMENTS WERE MADE OF TURBIDITY, CELL NUMBER, CELL SIZE, HETEROCYST FREQUENCY, FILAMENT LENGTH, AND EXTRACELLULAR AMMONIA. INDUCTION OF HETEROCYSTS IN THIS ALGA IS TOTALLY INHIBITED BY POTASSIUM NITRATE IN BATCH CULTURES, WHEREAS IN CONTINUOUS CULTURES NO INHIBITION IS OBSERVED AT HIGH DILUTION RATES. WHEN NITRATE IS UTILIZED, AMMONIA ACCUMULATES IN THE GROWTH MEDIUM, THE QUANTITY OF EXTRACELLULAR AMMONIA DECLINING WITH INCREASING DILUTION RATE. THE RATE OF RELEASE OF AMMONIA PER CELL INCREASES WITH DECREASING DENSITY OF ORGANISMS, AND INDUCTION OF HETEROCYSTS IS CONSISTENTLY OBSERVED WHEN AMMONIA RELEASE PER CELL EXCEEDS 2 X 10 TO THE 10TH POWER MICROGRAMS. IT IS INFERRED THAT SUCH EXCESSIVE RELEASE DEPLETES THE LEVEL OF INTRACELLULAR AMMONIA CAUSING THE INDUCTION OF HETEROCYSTS. THE PRESENT RESULTS SUPPORT THE VIEW THAT HETEROCYSTS ARE THE POSSIBLE SITES OF NITROGEN FIXATION IN BLUE-GREEN ALGAE. (JONES-WISCONSIN)

FIELD 05C

ACCESSION NO. W72-03217

ESTIMATING EUTROPHIC POTENTIAL OF POLLUTANTS,

MONSANTA CO., ST LOUIS, MO; AND WASHINGTON UNIV., ST LOUIS, MO. DEPT. OF ENVIRONMENTAL AND SANITARY ENGINEERING.

DEE MITCHELL, AND JAMES C. BUZZELL, JR.

JOURNAL OF SANITARY ENGINEERING DIVISION, PROCEEDINGS OF THE AMERICAN SOCIETY OF CIVIL ENGINEERS, VOL 97, NO SA 4, P 453-465, 1971. 5 FIG, 3 TAB, 22 REF.

DESCRIPTORS:
*EUTROPHICATION, *LABORATORY TESTS, *WATER POLLUTION EFFECTS, *ALGAE, *PLANKTON, WATER POLLUTION CONTROL, METHODOLOGY, MICRCENVIRONMENT, MEASUREMENT.

IDENTIFIERS:
*MICROCOSM ALGAL ASSAY PROCEDURE, SPECIES DIVERSITY INDEX.

ABSTRACT:
THIS STUDY IS CONCERNED WITH EFFECTS OF VARIOUS CHEMICALS AND WASTEWATER ON THE COMPOSITION AND GROWTH OF ALGAL MICROCOSMS OF LAKE WATER AND BOTTOM MUD CULTURES. THE BIOASSAY WAS CONDUCTED IN NINE LITER PYREX BOTTLES FILLED WITH 1:7 MUD-LAKE WATER SUSPENSION. THE TREATMENTS INCLUDED 10% SOLUTION OF DOMESTIC WASTEWATER, 10% SECONDARY TREATMENT EFFLUENT, DIFFERENT CONCENTRATIONS OF 23-19-17 FERTILIZER, AND THE ALGISTAT HERBICIDE. EFFECTS OF THESE TREATMENTS WERE RECORDED IN TERMS OF THE NUMBER OF ALGAL GENERA AND THE TOTAL ALGAL CELLS COUNT. NEARLY ALL ENRICHMENTS OF THE MEDIA CAUSED A MARKED REDUCTION OF THE DIVERSITY INDEX OF MICROORGANISMS. THE RESULTS SUGGESTED THAT THE LABORATORY PROCEDURE MAY SERVE FOR AN APPRAISAL OF THE EFFECT OF DIFFERENT POLLUTANTS. (WILDE-WISCONSIN)

FIELD 05C, 07B, 05A

ACCESSION NO. W72-03218

FRICTION REDUCTION BY ALGAL AND BACTERIAL POLYMERS,

NAVAL UNDERSEA RESEARCH AND DEVELOPMENT CENTER, SAN DIEGO, CALIF.

PAUL R. KENIS, AND J. W. HOYT.

AVAILABLE FROM THE NATIONAL TECHNICAL INFORMATION SERVICE AS AD-726 181, $3.00 IN PAPER COPY, $0.95 IN MICROFICHE. REPORT NUC TP 240, JUNE 1971, 34 P, 18 FIG, 6 TAB, 20 REF. ONR NR 137-699.

DESCRIPTORS:
*HYDRODYNAMICS, *FRICTION, *TURBULENT FLOW, ALGAE, BACTERIA, MARINE PLANTS, ENGINEERING, WATER PROPERTIES, PHYTOPLANKTON, CHLOROPHYTA, PHAEOPHYTA, RHODOPHYTA, DIATOMS, PYRROPHYTA, CHLORELLA, PSEUDOMONAS, TESTING, MEASUREMENT, PHEOLOGY.

IDENTIFIERS:
*BIOLOGICAL POLYMERS, *FRICTION REDUCTION, POLYSACCHARIDES, DRAG REDUCTION.

ABSTRACT:
FRICTION REDUCTION PHENOMENA, ATTRIBUTED TO DISSOLVED LONG-CHAIN POLYSACCHARIDES EXUDED BY ALGAE AND BACTERIA, STIMULATED THE INVESTIGATION OF THE ABILITY OF BIOLOGICAL POLYMERS PRODUCED BY SEAWEEDS, MICROSCOPIC ALGAE, AND BACTERIA TO REDUCE FRICTION IN WATER FLOW, INCLUDING THE EFFECT OF POLYMERS OF BIOLOGICAL ORIGIN ON THE REPORTED UNEXPLAINABLE VARIATIONS IN HYDRODYNAMIC TEST FACILITIES, THEIR USABILITY FOR FRICTION-REDUCTION APPLICATION AND FOR MEASUREMENTS TO QUANTITATE AND CHARACTERIZE BIOLOGICAL POLYMERS. FRICTION-REDUCING MATERIALS HAVE BEEN FOUND IN CULTURE MEDIA. ALL WATER SAMPLES TESTED FROM INLAND AND MARINE SOURCES GAINED FRICTION-REDUCTION ABILITY WHEN ENRICHED WITH SUGAR, AS A CONSEQUENCE OF POLYSACCHARIDE SYNTHESIS BY BACTERIA, WHICH LEADS TO THE ASSUMPTION THAT BIOLOGICAL POLYMERS ARE THE PROBABLE CAUSE OF THE UNEXPLAINABLE VARIATIONS IN HYDRODYNAMIC TEST FACILITIES. BACTERIAL POLYSACCHARIDES WERE MORE EFFECTIVE THAN SEAWEED EXTRACT AT LOW CONCENTRATIONS FOR FRICTION REDUCTION, BUT BOTH WERE MUCH LESS EFFECTIVE THAN SYNTHETIC POLYMERS. TURBULENT-FLOW FRICTIONAL MEASUREMENTS WERE FOUND TO BE SENSITIVE FOR DETECTION, MEASUREMENT, AND PARTIAL CHARACTERIZATION OF LONG-CHAIN POLYMERS. THE FRICTION-REDUCTION TECHNIQUE IS A RAPID AND EFFECTIVE PROCEDURE FOR THE DETECTION AND QUANTIFICATION OF LONG-CHAIN POLYMERS. (JONES-WISCONSIN)

FIELD 05A

ACCESSION NO. W72-03220

REGULATION OF NITRATE REDUCTASE IN CHLORELLA VULGARIS,

CORNELL UNIV., ITHACA, N.Y.; AND AGRICULTURAL RESEARCH SERVICE, ITHACA, N.Y. PLANT, SOIL AND NUTRITION LAB.

F. W. SMITH, AND JOHN F. THOMPSON.

PLANT PHYSIOLOGY, VOL 48, P 224-227, 1971. 4 FIG, 2 TAB, 15 REF.

DESCRIPTORS:
*BIOCHEMISTRY, *NITRATES, *CHLORELLA, ENZYMES, PROTEINS, LABORATORY TESTS, AMMONIA, ASSAY, INHIBITION, UREAS, SYNTHESIS, ALGAE.

IDENTIFIERS:
CHLORELLA VULGARIS, NITRATE REDUCTASE, RIBONUCLEIC ACID.

ABSTRACT:
SEVERAL AMINO ACIDS WERE TESTED AS POSSIBLE INHIBITORS OF NITRATE REDUCTASE. THE RESULTS OF ASSAYS AND INDUCTION STUDIES DISCLOSED THAT NITRATE REDUCTASE IS INCREASED IN CHLORELLA VULGARIS BY AN ADDITION OF NITRATE WITH MAXIMUM INDUCTION ABOVE MM. ACTINOMYCIN D, CYCLOHEXIMIDE, AND PUROMYCIN ANNULED THE EFFECT OF NITRATE THUS INDICATING THAT DE NOVO SYNTHESIS OF MESSENGER RNA AND PROTEIN IS REQUIRED. TESTS OF NITROGENOUS COMPOUNDS REVEALED THAT AMMONIUM CHLORIDE, UREA, AND SEVERAL AMINO ACIDS ARRESTED AN INCREASE OF THE REDUCTASE IN VIVO, BUT NOT IN VITRO. THE STUDY SUGGESTED THAT NITRATE REDUCTION IN CHLORELLA IS CONTROLLED BY REPRESSION OF ENZYME SYNTHESIS. (WILDE-WISCONSIN)

FIELD 05C

ACCESSION NO. W72-03222

STUDIES ON FRESHWATER BACTERIA: FACTORS WHICH INFLUENCE THE POPULATION AND ITS ACTIVITY,

FRESHWATER BIOLOGICAL ASSOCIATION, AMBLESIDE (ENGLAND).

J. G. JONES.

THE JOURNAL OF ECOLOGY, VOL 59, NO 2, P 593-613, 1971. 12 FIG, 7 TAB, 42 REF.

DESCRIPTORS:
*AQUATIC BACTERIA, *POPULATION, *ALGAE, *STRATIFICATION, *ENZYMES, THERMOCLINE, EPILIMNION, MUD, PHYSICOCHEMICAL PROPERTIES, HYDROGEN ION CONCENTRATION, TEMPERATURE, DISSOLVED OXYGEN, EUTROPHICATION, OLIGOTROPHY, PHYTOPLANKTON, ANALYTICAL TECHNIQUES, TURBIDITY, NITRATES, PHOSPHATES, WIND VELOCITY, RAINFALL.

IDENTIFIERS:
ESTHWAITE WATER(ENGLAND), LAKE WINDERMERE(ENGLAND), EXOENZYME PRODUCERS, CHLOROPHYLL A.

ABSTRACT:
THE FACTORS WERE STUDIED CONTROLLING BACTERIAL POPULATIONS DURING THE PERIOD OF WATER STRATIFICATION. THE ANALYZED PARAMETERS INCLUDED TEMPERATURE, PH, DISSOLVED OXYGEN, TURBIDITY, PARTICULATE MATTER, PHOSPHATE, NITRATE, RAINFALL, WIND VELOCITY, AND CHLOROPHYLL A. THE DENSITY OF VIABLE BACTERIA, EXOENZYME PRODUCING BACTERIA, AND THE LEVELS OF CERTAIN ENZYMES WERE ALSO ESTIMATED. IN THE EUTROPHIC ESTHWAITE WATER THE MAJOR FACTORS CONTROLLING BACTERIA APPEARED TO BE TEMPERATURE, CO CONCENTRATION, AND PH. IN THE LAKE WINDERMERE A POSITIVE CORRELATION WAS OBSERVED BETWEEN BACTERIAL NUMBERS AND PH, TEMPERATURE, PARTICULATE MATTER, AND RAINFALL. BACTERIAL POPULATION AND ENZYMATIC ACTIVITY OCCASIONALLY INCREASED WITH PHYTOPLANKTON MAXIMA. (WILDE-WISCONSIN)

FIELD 05C

ACCESSION NO. W72-03224

PHYTOPLANKTON PRIMARY PRODUCTION IN SOME FINNISH COASTAL AREAS IN RELATION TO POLLUTION,

INSTITUTE OF MARINE RESEARCH, HELSINKI (FINLAND); AND HELSINKI UNIV. (FINLAND). DEPT. OF LIMNOLOGY.

PAULI BAGGE, AND PASO O. LEHMUSLUOTO.

MERENTUTKIMUSLAIT JULK/HAVSFORSKNINGSINT SKR NO 235, P 3-18, 1971. 7 FIG, 4 TAB, 24 REF.

DESCRIPTORS:
*PRIMARY PRODUCTIVITY, *PHYTOPLANKTON, *SEA WATER, COASTS, BIORTHYMS, ON-SITE TESTS, WATER POLLUTION SOURCES, WATER POLLUTION EFFECTS, FLUCTUATION, COMPARATIVE PRODUCTIVITY, MUNICIPAL WASTES, INDUSTRIAL WASTES, ANNUAL TURNOVER, ENERGY, CYANOPHYTA, ALGAE, EUTROPHICATION, SUMMER, ENERGY, SEASONAL.

IDENTIFIERS:
*FINLAND, *BOTHNIAN BAY, *GULF OF FINLAND, DARK FIXATION, COASTAL WATERS, TROPHOGENIC LAYER.

ABSTRACT:
CARBON-14 ANALYSES DISCLOSED SIGNIFICANT ANNUAL AND SEASONAL FLUCTUATIONS OF PHYTOPLANKTON PRIMARY PRODUCTION IN THE POLLUTED AND NONPOLLUTED COASTAL AREAS OF FINLAND. IN SOME AREAS, RECEIVING WASTEWATER, THE ANNUAL PRODUCTION EXCEEDS 150 G C/SQ M, IN COMPARISON WITH 15 TO 60 G C/SQ M OF UNPOLLUTED WATERS. BECAUSE OF THE SHADING PLANKTON, THE THICKNESS OF THE TROPHOGENIC LAYER DURING THE SUMMER IN SEWAGE POLLUTED WATERS IS USUALLY LESS THAN 2 M; IN WATERS POLLUTED BY PAPER MILL WASTES, THE THICKNESS IS UNDER 1 M. THE EUTROPHIC WATERS ARE DOMINATED BY BLUE-GREEN ALGAE, WHEREAS OLIGOTROPHIC WATERS BY DIATOMS AND DINOPHYCEAE. DARK FIXATION VALUES OF EUTROPHIC WATERS MAY BE 10 TIMES AS HIGH AS THOSE OF UNPOLLUTED WATERS. (WILDE-WISCONSIN)

FIELD 05C, 02L

ACCESSION NO. W72-03228

NEED FOR BETTER INFORMATION OF EFFECTS OF COASTAL REACTORS,

SCRIPPS INSTITUTION OF OCEANOGRAPHY, LA JOLLA, CALIF.

T. R. FOLSOM.

AVAILABLE FROM THE NATIONAL TECHNICAL INFORMATION SERVICE AS TID-25774, $3.00 IN PAPER COPY, $0.95 IN MICROFICHE. REPORT TID-25774 REV JULY 15, 1971. 16 P.

DESCRIPTORS:
*NUCLEAR WASTES, *MONITORING, *PACIFIC COAST REGION, TRITIUM, OCEAN CURRENTS, COBALT RADIOISOTOPES, NUCLEAR POWERPLANTS, SITES, MOLLUSKS, MARINE ALGAE, ON-SITE INVESTIGATIONS, CALIFORNIA, FOOD CHAINS, PUBLIC HEALTH, WATER POLLUTION SOURCES.

ABSTRACT:
UPTAKE BY ORGANISMS IS THE MOST SENSITIVE KNOWN INDICATOR OF RADIONUCLIDE POLLUTION. CO-58 AND PROBABLY CO-60 AND AG-110M WERE DETECTABLE IN A SPECIES OF RED ALGAE AND IN A MOLLUSK THAT FEEDS UPON THE ALGAE, WHICH WERE COLLECTED FROM AN AREA NEAR THE SAN ONOFRE NUCLEAR POWERPLANT. THE TRITIUM DISCHARGE REACHES 11,000 TU AT OUTFALL, AN AMOUNT SUFFICIENT TO SERVE AS A TRACER OF THE EFFLUENT IN THE LA JOLLA-SAN DIEGO COASTAL REGION. (BOPP-ORNL)

FIELD 05C

ACCESSION NO. W72-03324

THE ALGO-AGRO-INDUSTRIAL COMPLEX. AN AGRO-INDUSTRIAL COMPLEX AT NUCLEAR ENERGY CENTERS WITH ASSOCIATED PRODUCTION OF AUTOTROPHIC MICROORGANISMS,

CESKOSLOVENSKA AKADEMIE VED, PRAGUE. INST. OF MICROBIOLOGY; AND CESKOSLOVENSKA AKADEMIE VED, TREBON. LAB. OF ALGOLOGY.

I. MALEK, J. BARTOS, J. SIMMER, AND B. PROKES.

AVAILABLE FROM THE NATIONAL TECHNICAL INFORMATION SERVICE AS A/CONF.49/P/552. $3.00 PER COPY, $0.95 MICROFICHE. REPORT A/CONF.49/P/552, MAY 1971. 20 P, 7 FIG, 2 TAB, 10 REF.

DESCRIPTORS:
*NUCLEAR POWERPLANTS, *DESALINATION PLANTS, *ELECTRIC POWER PRODUCTION, *IRRIGATION PROGRAMS, *AGRICULTURE, *ECONOMICS, CYANOPHYTA, ALGAE, PLANTS, PROTEINS, NUTRIENTS, BIOMASS, HARVESTING OF ALGAE, PRODUCTIVITY.

IDENTIFIERS:
AGRO-INDUSTRIAL COMPLEX, ELECTRIC ENERGY CONSUMPTION, MODEL TESTING.

ABSTRACT:
THIS PAPER CONSIDERS THE PRODUCTION OF THE BIOMASS OF AUTOTROPHIC MICROORGANISMS (AM), FIRST OF ALL THE GREEN AND BLUE-GREEN ALGAE. THE COMPLEX OF PRODUCTION TYPES CONSIDERED HERE WAS HENCE CALLED THE ALGO-AGRO-INDUSTRIAL COMPLEX (AAIC). THE HETEROGENEITY OF THE STRUCTURE OF THE AAIC PRESENTED OBSTACLES THAT COULD NOT BE ALWAYS OVERCOME SUCCESSFULLY; IT IS HARDLY POSSIBLE IN THIS BRIEF REVIEW TO PRESENT MORE THAN THE BASIC INFORMATION ON THE INDIVIDUAL PARTS OF THE AAIC AND ON THEIR MUTUAL RELATIONS. A FULL-LENGTH PUBLICATION ON THE SUBJECT IS BEING PREPARED. THE MOSAIC OF DATA COMPILED HERE IS MEANT AS A BASIS FOR FURTHER DEVELOPMENT. (HOUSER-ORNL)

FIELD 03A, 03F, 05C

ACCESSION NO. W72-03327

SOILS AS COMPONENTS OF ECOSYSTEMS,

OAK RIDGE NATIONAL LAB. TENN. ECOLOGICAL SCIENCES DIV.

M. WITKAMP.

IN: ANNUAL REVIEW OF ECOLOGY AND SYSTEMATICS; ANNUAL REVIEWS INC.; PALO ALTO, CALIFORNIA, P 85-110, 1971.

DESCRIPTORS:
*SOIL-WATER-PLANT RELATIONSHIPS, *FALLOUT, *CYCLING NUTRIENTS, ECOSYSTEMS, RAIN FORESTS, MATHEMATICAL MODELS, LINEAR PROGRAMMING, SYSTEMS ANALYSIS, DECIDUOUS FORESTS, DECOMPOSING ORGANIC MATERIAL, FERTILITY, RHIZOSPHERE, ROOT SYSTEMS, SOIL FUNGI, SOIL ENVIRONMENT, SOIL TEMPERATURE, SOIL CHEMICAL PROPERTIES, SOIL BACTERIA, SOIL ALGAE.

ABSTRACT:
MINERAL CYCLING LITERATURE OF THE LAST 2-3 YEARS IS REVIEWED TO PROVIDE INSIGHT INTO DRIVING FORCES AND RATES OF TRANSFER, AS WELL AS THE IMPORTANCE OF THE VARIOUS PATHWAYS. BIOTA ARE THE PRIME AGENTS IN MAINTAINING PRODUCTIVITY WHICH IS OFTEN LIMITED BY MINERAL AVAILABILITY AND IS DEPENDENT ON RECYCLING THROUGH PLANT AND ANIMAL DEBRIS. MUCH OF THE INTERNAL TURNOVER IN SUBSYSTEMS CAN BE TREATED AS BLACK BOXES BY CONSIDERING ONLY TURNOVER RATES AND POOL SIZE. TURNOVER TIMES IN A TROPICAL RAIN FOREST CALCULATED AS TOTAL POOL/TURNOVER RATE VARY FROM 0.08 YR FOR K TO 133 YR FOR N, AND DECREASE IN THE ORDER MG, CA, H, C, MN, NA AND P. A LARGE P INFLUX STEMS FROM RAIN. INPUTS OF SODIUM AND CHLORINE BY RAIN DEPEND TO A LARGE EXTENT ON OCEANIC STORMS THAT TRANSPORT SPRAY FAR INLAND. SIMILARLY, ABOUT HALF OF THE POTASSIUM INPUT IS IN RAIN. THE POTASSIUM INPUT IS ONLY ABOUT 1% OF THAT RECYCLED. EFFICIENT RECYCLING ACCOUNTS FOR THE LOWER NUTRIENT REQUIREMENTS WITH FORESTS AS COMPARED TO AGRICULTURE. MICROBIAL ACTIVITY, AND CONSEQUENTLY REMINERALIZATION IF TEMPERATURE DEPENDENT. INTERRUPTION OF MINERALIZATION DURING WINTER MAY ALLOW LOSS OF NUTRIENTS BY LEACHING. LOW RATES OF EVAPORATION AND RESULTING ANAEROBIC CONDITIONS ALSO TEND TO INHIBIT MINERALIZATION. (BCPP-ORNL)

FIELD 02I, 05B, 02G

ACCESSION NO. W72-03342

AMCHITKA RADIOBIOLOGICAL PROGRAM,

WASHINGTON UNIV., SEATTLE. COLL. OF FISHERIES.

E. E. HELD.

AVAILABLE FROM NATIONAL TECHNICAL INFORMATION SERVICE, $3.00 IN PAPER COPY, $0.95 IN MICROFICHE. PROGRESS REPORT, JULY 1970-APRIL 1971, NVO-269-11, JULY 1971. 38 P, 5 FIG, 9 TAB, 13 REF. AEC CONTRACT NO. AT(26-1)-269.

DESCRIPTORS:
*NUCLEAR EXPLOSIONS, *WATER POLLUTION EFFECTS, *RADIOACTIVITY EFFECTS, UNDERGROUND, ALASKA, PONDS, WELLS, FISH, BAYS, AIR POLLUTION EFFECTS, AIR POLLUTION, WATER POLLUTION, LICHENS, SOIL-WATER-PLANT RELATIONSHIPS, MARINE ALGAE, MOSSES, CRABS, ALGAE, SNAILS, MUSSELS, ISOPODS, AMPHIPODA, RADIOISOTOPES, ABSORPTION, PACIFIC OCEAN, FOOD CHAINS, FALLOUT, TRITIUM, ECOLOGY, ON-SITE INVESTIGATIONS, MONITORING.

IDENTIFIERS:
RADIONUCLIDE UPTAKE.

ABSTRACT:
ORGANISMS, WATER, AND AIR SAMPLES WERE ANALYZED FOR RADIONUCLIDES TO DETERMINE WHETHER THERE WAS ANY RELEASE AFTER UNDERGROUND NUCLEAR TESTS. EXCEPTING TRITIUM (3600 T.U. IN PONDS AND TEST WELLS) AND SHORT-LIVED SCANDIUM-46 WHICH COULD BE FROM A NUCLEAR-POWERED VESSEL, THE RADIONUCLIDES DETECTED WERE FROM WORLDWIDE FALLOUT. IN FISH, IRON-55 WAS AS GREAT AS 152 PCI/G DRY WEIGHT (OR 254 PCI/MG FE SPECIFIC ACTIVITY) AND OTHER RADIONUCLIDES (EXCEPTING K-40) WERE LESS THAN 5 PCI/G DRY WEIGHT. IN LICHENS, CESIUM-137 WAS 20 PCI/G DRY WEIGHT. IN FRESHWATER PLANTS, ZIRCONIUM-95, NIOBIUM-95 WAS 12 PCI/G DRY WEIGHT. GENERALLY, THE CONCENTRATIONS OF THE RADIONUCLIDES WERE SMALL AND WITHIN THE RANGE OF VALUES REPORTED FOR SIMILAR SAMPLES FROM OTHER PARTS OF THE NORTHERN HEMISPHERE. (BCPP-ORNL)

FIELD 05C

ACCESSION NO. W72-03347

TRITIUM: DISCRIMINATION AND CONCENTRATION IN FRESH WATER MICROCOSMS,

ARGONNE NATIONAL LAB., ILL.

M. L. STEWART, G. M. ROSENTHAL, AND J. R. KLINE.

AVAILABLE FROM NATIONAL TECHNICAL INFORMATION SERVICE AS CONF-710501-26, $3.00 IN PAPER COPY, $0.95 IN MICROFICHE. REPORT CONF-710501-26, MAY 1971. 29 P, 9 FIG, 1 TAB, 10 REF.

DESCRIPTORS:
*SNAILS, *FROGS, *TRITIUM, ABSORPTION, WATER POLLUTION EFFECTS, RADIOACTIVITY EFFECTS, RADIOACTIVITY TECHNIQUES, AQUARIA, ALGAE, MATHEMATICAL MODELS, PATH OF POLLUTANTS, TRACERS, ANIMAL PHYSIOLOGY, NUCLEAR WASTES.

ABSTRACT:
TRITIUM MOBILITY WAS STUDIED IN AN AQUATIC ENVIRONMENT AND IN AQUATIC ANIMALS (TWO SPECIES OF SNAIL AND ONE SPECIES OF TADPOLE). THE TRITIUM MOBILITY IN SNAILS WAS DESCRIBED BY AN EMPIRICAL, COMPARTMENTAL MODEL. COMPARTMENTS AND RESIDENCE TIMES OF TRITIUM ARE: EXTRACELLULAR, UNBOUND WATER (80% OF THE UNBOUND WATER WITHIN THE SNAIL SHELL), 4 MINUTES; CELLULAR, UNBOUND WATER (20% OF THE UNBOUND WATER WITHIN THE SNAIL SHELL), 330 MINUTES; CELLULAR, BOUND TRITIUM (ACCOUNTING FOR 28% OF THE EXCHANGEABLE HYDROGEN), 2,650 MINUTES; EXTRACELLULAR, BOUND TRITIUM (ACCOUNTING FOR 28% OF THE EXCHANGEABLE HYDROGEN), 8 MINUTES; SURFACE, BOUND TRITIUM (WHICH IS DIRECTLY EXPOSED TO ENVIRONMENTAL TRITIUM AND ACCOUNTS FOR 4% OF THE EXCHANGEABLE HYDROGEN), 2 MINUTES. COMPARTMENTAL SIZES AND RATE TRANSFER COEFFICIENTS ARE GIVEN. (BOPP-ORNL)

FIELD 05C

ACCESSION NO. W72-03348

QUALITATIVE COMPOSITION OF PHYTOPLANKTON IN THE VICINITY OF NEWFOUNDLAND,

ALL-UNION RESEARCH INST. OF MARINE FISHERIES AND OCEANOGRAPHY, MOSCOW (USSR).

O. A. MOVCHAN.

OKEANOLOGIYA, VOL. 10, NO. 3, P 381-387, MAY-JUNE 1970. 5 FIG, 16 REF.

DESCRIPTORS:
*PHYTOPLANKTON, *ALGAE, *DIATOMS, *CURRENTS, *DISTRIBUTION PATTERNS, ATLANTIC OCEAN, PACIFIC OCEAN, TROPICAL REGIONS, POLAR REGIONS, CYANOPHYTA, CHRYSOPHYTA, PYRROPHYTA, SAMPLING, ECOLOGY, WATER TEMPERATURE.

IDENTIFIERS:
NEWFOUNDLAND, PERIDINIANS, QUALITATIVE COMPOSITION, COSCINOSIRA POLYCHORDA, THALASSIOSIRA DECIPIENS, THALASSIOSIRA GRAVIDA, THALASSIOSIRA NORDENSKIOLDII, THALASSIOSIRA ROTULA, SCHRODERELLA DELICATULA, DACTYLIOSOLEN MEDITERRANEUS, LEPTOCYLINDRUS DANICUS, RHIZOSOLENIA FRAGILISSIMA, RHIZOSOLENIA HEBETATA F. SEMISPINA, CHAETOCEROS AFFINIA, CHAETOCEROS ATLANTICUS, CHAETOCEROS BOREALIS.

ABSTRACT:
AMONG THE PHYTOPLANKTON COLLECTED IN THE 590 SAMPLES TAKEN BY THE SHIP R/V MIKHAIL LOMOSOV FROM ITS STATIONS OFF NEWFOUNDLAND WERE 131 SPECIES AND VARIETIES OF ALGAE, 25 GENERA AND 67 SPECIES OF DIATOMS, 15 GENERA AND 50 SPECIES OF PERIDINIANS, 6 GENERA OF COCCOLITHINES, 4 SPECIES OF PROTOCOCCUCI, 2 SPECIES OF SILICO FLAZELLATES, 1 SPECIES OF HETEROCONTAE AND 1 SPECIES OF BLUE-GREEN ALGAE. FIFTY PERCENT OF THE TYPES OF DIATOMS ARE MASSIVE (OVER 1 MILLION CELLS/CU M IN THE 0- TO 100-M LAYER). STUDY OF THE ECOLOGY AND THE GEOGRAPHIC DISTRIBUTION OF THE PHYTOPLANKTON SHOWED ABUNDANCE OF TROPICAL SPECIES BUT FEW ARCTIC-BOREAL SPECIES SOUTH OF 42 DEGREES N. NORTH OF THIS WAS A ZONE OF MIXING OF TROPICAL AND ARCTIC-BOREAL SPECIES. THE COSMOPOLITE SPECIES WAS WIDELY SPREAD IN THE NEWFOUNDLAND REGION. SIMILARITIES WERE FOUND IN THE COMPOSITION OF THE PHYTOPLANKTON IN THE REGION OF NEWFOUNDLAND AND IN ADJACENT WATERS IN THE NORTHWESTERN ATLANTIC AND NORTHWESTERN PACIFIC. INVESTIGATION OF THE DISTRIBUTION OF THE NERITIC AND OCEANIC FORMS OF DIATOMS AND A FEW SPECIES OF PERIDINIANS SHOWED THE INFLUENCE OF CURRENTS AND THE MIXING OF WATERS FROM POLAR AND TROPICAL ZONES ON ABUNDANCE. MAPS SHOW THE CONCENTRATION OF SPECIES, THEIR GEOGRAPHICAL LOCATION, AND SEASONAL VARIATION. (LITTLE-BATTELLE)

FIELD 05A, 05C

ACCESSION NO. W72-03543

GROWTH RESPONSE OF BLUE-GREEN ALGAE TO ALDRIN, DIELDRIN ENDRIN AND THEIR METABOLITES,

WISCONSIN UNIV., MADISON. DEPT. OF ENTOMOLOGY.

J. C. BATTERTON, G. M. BOUSH, AND F. MATSUMURA.

BULLETIN OF ENVIRONMENTAL CONTAMINATION AND TOXICOLOGY, VOL. 6, NO. 6, P 589-594, NOVEMBER/DECEMBER 1971. 2 TAB, 17 REF.

DESCRIPTORS:
*CHLORINATED HYDROCARBON PESTICIDES, *ALDRIN, *DIELDRIN, *ENDRIN, *GROWTH RATES, *CYANOPHYTA, *TOXICITY, PESTICIDES, PESTICIDE RESIDUES, ALGAE, DDT, PHYTOPLANKTON, MARINE ALGAE, AQUATIC ALGAE, COLORIMETRY, TRACES.

IDENTIFIERS:
PHOTODIELDRIN, ANACYSTIS NIDULANS, AGMENELLUM QUADRUPLICATUM STRAIN PR-6, PHOTOALDRIN, KETOENDRIN, METABOLITES, SKELETONEMA, ANABAENA CYLINDRICA, DUNALIELLA.

ABSTRACT:
THE EFFECTS OF DIELDRIN, ENDRIN, ALDRIN, AND FIVE CORRESPONDING METABOLITES ON THE GROWTH RATES OF TWO BACTERIA-FREE BLUE-GREEN ALGAL SPECIES WERE TESTED. THE TEST SPECIES WERE ANACYSTIS MIDULANS (TX20), A FRESHWATER ISOLATE, AND AGMENELLUM QUADRUPLICATUM STRAIN PR-6, A MARINE ISOLATE, BOTH OF WHICH WERE GROWN IN DIFFERENT MEDIA AT THE SAME PH, TEMPERATURE, AND LIGHT INTENSITY. A COMPARISON OF THE RESPONSES OF THE TEST ALGAE SHOWED THAT THE MARINE ISOLATE (PR-6) WAS GENERALLY MORE TOLERANT TO THE INSECTICIDE COMPOUNDS THAN THE FRESHWATER ISOLATE (TX20). THERE WAS AN OVERALL DEPRESSION OF GROWTH RATES WITH HIGH CONCENTRATIONS (475 PPB, 950 PPB) OF DIELDRIN AND ITS METABOLITES. OF THE PARENT COMPOUNDS TESTED, ALDRIN HAD THE LEAST EFFECT ON EITHER TEST SPECIES. THE SELECTED LIGHT INTENSITY AND TEMPERATURE FOR OPTIMUM GROWTH RATES REMAINED UNCHANGED IN THE PRESENCE OF THE INSECTICIDE COMPOUNDS. THAT PR-6 WAS GENERALLY THE MORE TOLERANT OF THE TWO TEST SPECIES RAISES THE QUESTION OF THE EFFECT OF MEDIUM COMPOSITION UPON INSECTICIDE TOXICITY. TABULAR DATA SHOW GROWTH RATES FOR EXPERIMENTS USUALLY LASTING 30-36 HOURS. (HOLOMAN-BATTELLE)

FIELD 05C

ACCESSION NO. W72-03545

PRODUCTION AND BIOENERGETIC ROLE OF THE MIDGE GLYPTOTENDIPES BARBIPES (STAEGER) IN A WASTE STABILIZATION LAGOON,

OREGON STATE UNIV., CORVALLIS. DEPT. OF ENTOMOLOGY.

R. A. KIMERLE, AND N. H. ANDERSON.

LIMNOLOGY AND OCEANOGRAPHY, VOL. 16, NO. 4, P 646-659, JULY 1971. 5 FIG, 5 TAB, 27 REF.

DESCRIPTORS:
*ENERGY BUDGET, *SEWAGE, *AQUATIC PRODUCTIVITY, *MIDGES, LAGOONS, BIOMASS, OREGON, SLUDGE, ALGAE, DIPTERA, DISSOLVED OXYGEN, DREDGING, BOTTOM SEDIMENTS, STABILIZATION, PRIMARY PRODUCTIVITY, GROWTH RATES, RESPIRATION.

IDENTIFIERS:
GLYTOTENDIPES BARBIPES, EKMAN DREDGE.

ABSTRACT:
THE BIOENERGETIC ROLE OF A POPULATION OF GLYPTOTENDIPES BARBIPES IN THE PROCESS OF WASTE STABILIZATION IN TWO SEWAGE LAGOONS WAS STUDIED. WEEKLY PRODUCTION RATES OF THE MULTIVOLTINE MIDGE WERE COMPUTED. ANNUAL PRODUCTION OF G. BARBIPES WAS 808 KCAL/SQ METER IN A NARROW BAND NEARSHORE OF THE SECONDARY LAGOON CONTAINING 90 PERCENT OF THE BIOMASS. BIOMASS DATA FROM BOTH LAGOONS IN 1966 AND 1967 WERE USED TO ESTIMATE PRODUCTION USING A TURNOVER RATIO (TR) OF 8.49 (RATIO OF PRODUCTION:MEAN BIOMASS) FROM DEFINITIVE DATA COLLECTED IN 1967. PRODUCTION IN THE SECONDARY LAGOON WAS 459 KCAL/SQ METER IN 1966 AND 37 IN 1967; IN THE PRIMARY LAGOON IT WAS 165 AND 18 RESPECTIVELY. THE FACTORS CAUSING THESE DIFFERENCES IN PRODUCTION WERE PROBABLY THE DISSOLVED OXYGEN CONCENTRATIONS DURING THE GROWING SEASON, PERCENT OF THE TOTAL LAGOON BOTTOM INHABITABLE BY MIDGE LARVAE, AND THE CONDITION OF THE SLUDGE SUBSTRATE. THE TOTAL ENERGY REMOVED BY EMERGENCE AND RESPIRATION OF G. BARBIPES WAS COMPARED WITH THE ENERGY IN BOTH PATHWAYS IN THE LAGOON: IMPORT OF SEWAGE, PRIMARY PRODUCTION, COMMUNITY RESPIRATION, STORAGE, AND EXPORT. IN 1966, G. BARBIPES REMOVED ABOUT 6.6 PERCENT OF THE NET PRIMARY PRODUCTION IN THE SECONDARY LAGOON AND 0.5 PERCENT IN 1967. (MORTLAND-BATTELLE)

FIELD 05C

ACCESSION NO. W72-03556

EFFECT OF WIDE TEMPERATURE FLUCTUATION ON THE BLUE-GREEN ALGAE OF BEAD GEYSER, YELLOWSTONE NATIONAL PARK,

INDIANA UNIV., INDIANAPOLIS. DEPT. OF MICROBIOLOGY.

J. L. MOSSER, AND T. D. BROCK.

LIMNOLOGY AND OCEANOGRAPHY, VOL. 16, NO. 4, P 640-645, JULY 1971. 3 FIG, 2 TAB, 9 REF.

DESCRIPTORS:
*THERMAL POLLUTION, *WATER TEMPERATURE, *CYANOPHYTA, GEYSERS, PHOTOSYNTHESIS, CULTURES, CHLOROPHYLL, PROTEINS, ALGAE, GROWTH RATES, TRACERS.

IDENTIFIERS:
YELLOWSTONE PARK, SYNECHOCOCCUS, FLEXIBACTERIA, PHORMIDIUM, MASTIGOCLADUS, PHEOPHYTIN, GRASSLAND SPRING.

ABSTRACT:
MEASUREMENT OF PHOTOSYNTHESIS AND GROWTH OF CYANOPHYTA IN THE DRAINWAY OF A SMALL GEYSER IN YELLOWSTONE INDICATES THAT FLUCTUATING TEMPERATURES LIMIT THE EXTENT OF ALGAL GROWTH. AN ALGAL MAT OF 'GRASSLAND SPRING' WAS USED IN THESE EXPERIMENTS AS AN EXAMPLE OF A MAT OCCURRING IN A THERMALLY CONSTANT ENVIRONMENT. PHOTOSYNTHESIS WAS MEASURED BY THE CARBON 14-BICARBONATE METHOD. MAT SAMPLES WERE ALSO ANALYZED FOR CHLOROPHYLL, PROTEIN, AND PHEOPHYTIN. ALGAE FROM THE MAT WERE CULTURED IN MEDIUM D. THE MAT WAS EXPOSED TO HIGH TEMPERATURES ONLY DURING THE BRIEF ERUPTIONS OF THE GEYSER. SINCE OPTIMAL TEMPERATURES FOR THE ALGAE WERE HIGHER THAN THE MEAN ENVIRONMENTAL TEMPERATURE, TEMPERATURES OPTIMAL FOR PHOTOSYNTHESIS AND GROWTH OCCURRED ONLY DURING A SMALL FRACTION OF THE TIME. IT WAS INDICATED THAT THESE ALGAE WERE RESISTANT TO THE HIGH TEMPERATURE THAT OCCURRED DURING ERUPTION, WHEREAS ORGANISMS HAVING THE SAME OPTIMAL TEMPERATURE BUT INHABITING A THERMALLY CONSTANT ENVIRONMENT WERE MORE HEAT SENSITIVE. THE DATA SUGGEST THAT ORGANISMS MAY ADAPT TO SUDDEN TEMPERATURE SHOCKS, SUCH AS THOSE EXPERIENCED NEAR STEAM POWER PLANTS, BUT THAT THEY CANNOT EVOLVE THE ABILITY TO GROW WELL THROUGHOUT THE TEMPERATURE CYCLE. SUCH DATA MAY DIFFER, HOWEVER, IN A MORE RANDOM SITUATION, SINCE THIS WAS A GEYSER THAT ERUPTED PREDICTABLY. FURTHER STUDY OF RANDOMLY ERUPTING GEYSERS IS SUGGESTED. (MORTLAND-BATTELLE)

FIELD 05C

ACCESSION NO. W72-03557

ABUNDANCE OF ETHMODISCUS IN PACIFIC PLANKTON,

AKADEMIYA NAUK SSSR, MOSCOW. INSTITUT OKEANOLOGII.

T. V. BELYAYEVA.

OKEANOLOGIYA, VOL 10, NO 5, P 672-675, SEPTEMBER-OCTOBER 1970. 1 FIG, 1 TAB, 11 REF.

DESCRIPTORS:
*PLANKTON, *DIATOMS, *SAMPLING, *DISTRIBUTION PATTERNS, *PHOSPHATES, *AQUATIC POPULATIONS, MAPS, MAPPING, PHYTOPLANKTON, PACIFIC OCEAN, BOTTOM SEDIMENTS, STATIONS, DATA COLLECTIONS, ALGAE.

IDENTIFIERS:
*ETHMODISCUS REX, ETHMODISCUS GAZELLAE, BAJA CALIFORNIA, HEMIAULUS HAUCKII, STIGMAPHORA ROSTRATA, ETHMODISCUS OOZES.

ABSTRACT:
A MAP WAS COMPILED BRINGING TOGETHER ALL THE SCATTERED INFORMATION ON THE DISTRIBUTION OF THE PLANKTON DIATOM ETHMODISCUS REX IN THE PACIFIC OCEAN. THE DATA WERE ACQUIRED MAINLY FROM ANALYSIS OF SAMPLES COLLECTED DURING SEVERAL VOYAGES OF THE R/V VITYAZ' IN EASTERN, CENTRAL, AND WESTERN PARTS OF THE PACIFIC FROM 1955 TO 1967. MORE THAN 400 SAMPLING LOCATIONS WERE USED FOR THE MAP SHOWING THE QUANTITATIVE DISTRIBUTION OF ETHMODISCUS. THE DISTRIBUTION PATTERN IS EXTREMELY MOSAIC WITH THE MAXIMUM POPULATION OCCURRING BETWEEN THE EQUATOR AND 5 DEGREES N. A RELATIONSHIP WAS SHOWN BETWEEN THE DISTRIBUTION OF ETHMODISCUS AND PHOSPHATE CONCENTRATION, THE MAXIMUM POPULATIONS NOT COINCIDING WITH REGIONS OF HIGHEST PHOSPHATE CONCENTRATION AS IS USUALLY THE CASE WITH MOST DIATOMS. THIS UNUSUAL DISTRIBUTION OF ETHMODISCUS IS UNEXPLAINED, BUT ALSO OCCURS WITH THE DIATOMS, HEMIAULUS HAUCKII AND STIGMAPHORA ROSTRATA. THE ETHMODISCUS POPULATION IN PACIFIC PLANKTON IS INSIGNIFICANT, AVERAGING FROM 0.01 TO 0.5 KL/CU M IN THE 0- TO 100-M LAYER. THEREFORE, THE PRESENT DATA IS INCONCLUSIVE ON THE ORIGIN OF THE ETHMODISCUS OOZES, THEIR CONCENTRATION IN THE WESTERN PACIFIC, OR THE MOSAIC PATTERN OF THEIR DISTRIBUTION. PHYTOPLANKTON SAMPLES WERE COLLECTED MOSTLY WITH A DZHOM NET WITH MOUTH OPENING 80 CM IN DIAMETER AND APERTURE SIZE IN THE FILTERING CONE OF ABOUT 180 MICRONS. (JEFFERIS-BATTELLE)

FIELD 05A, 05C

ACCESSION NO. W72-03567

BIOLCGICAL REMOVAL OF PHOSPHATES FROM AQUATIC MEDIA,

ARIZONA UNIV., TUCSON. DEPT. OF BIOLOGICAL SCIENCES.

DOUGLAS EDWIN GREER.

MASTER'S THESIS, MAY 1971. 29 P, 4 FIG, 4 TAB, 23 REF. OWRR A-019-ARIZ(2).

DESCRIPTORS:
*PHOSPHATES, *CLAMS, *ALGAE, CHEMICAL PRECIPITATION, HYDROGEN ION CONCENTRATION, FILTRATION, COLLOIDS, CARBON DIOXIDE, CALCIUM, MAGNESIUM, NUTRIENTS, *WASTE WATER TREATMENT, TERTIARY TREATMENT, EUTROPHICATION.

IDENTIFIERS:
*ANOXIA.

ABSTRACT:
THE PRECIPITATION OF EXCESS PHOSPHATES IN THE FORM OF HYDROXYL-APATITE HAS BEEN ACCOMPLISHED WITHOUT THE ADDITION OF CHEMICAL REAGENTS. THE PROCESS REQUIRES ELEVATION OF THE OH(-) ION CONCENTRATION THROUGH THE REMOVAL OF CO2 FROM THE WATER BY ALGAE, RESULTING IN A HYDROXYL-APATITE ALGAL SUSPENSION. THIS SUSPENSION IS REMOVED FROM THE WATER BY BEING FILTERED THROUGH BEDS OF CLAMS, CORBICULA FLUMINEA. THESE CLAMS ARE ABLE TO SURVIVE IN EUTROPHIC WATER PROVIDED THE WATER IS CONTINUOUSLY RECIRCULATED, AND THE TEMPERATURE IS KEPT BELOW 30C. ESTUARINE PHOSPHATES MAY BE EVEN MORE SUSCEPTIBLE TO 'AUTOMATIC' REMOVAL SINCE SEA WATER CONTAINS AN AVERAGE OF 410 MG/L CA(+2), WHICH SHOULD RESULT IN MORE COMPLETE PRECIPITATION. OVERALL QUALITY OF WATER RECEIVING CLAM-ALGAE PHOSPHATE-REMOVAL TREATMENT IS ENHANCED, SINCE PH ELEVATION WITHOUT LIME ADDITION RESULTS IN CA(+2) AND MG(+2) PRECIPITATION. TERTIARY TREATED WASTEWATER, SUBJECTED TO NUTRIENT REMOVAL AND CONTAINING LESS THAN 0.30 MG/L TOTAL PHOSPHATES, WILL ALLOW HIGH LEVELS OF PRIMARY PRODUCTIVITY WITHOUT THE DANGER OF ANOXIA FROM EXCESSIVE BUILDUPS OF THE ALGAL STANDING CROP. (LOWRY-TEXAS)

FIELD 05D

ACCESSION NO. W72-03613

PLANT-AVAILABLE PHOSPHORUS STATUS OF LAKES,

WISCONSIN UNIV., MADISON. DEPT. OF WATER CHEMISTRY; AND WISCONSIN UNIV., MADISON. DEPT. OF SOIL SCIENCE.

D. E. ARMSTRONG, R. F. HARRIS, AND J. K. SYERS.

AVAILABLE FROM THE NATIONAL TECHNICAL INFORMATION SERVICE AS PB-206 156, $3.00 IN PAPER COPY, $0.95 IN MICROFICHE. TECHNICAL COMPLETION REPORT, WISCONSIN WATER RESOURCES CENTER, MADISON, 1971. 29 P, 2 TAB, 20 REF, 10 APPEND. OWRR B-022-WIS(6).

DESCRIPTORS:
*PHOSPHORUS, *SEDIMENTS, *ALGAE, *EUTROPHICATION, LAKES, AQUATIC PLANTS, IRON, MICROORGANISMS, COLORIMETRY, WISCONSIN.

IDENTIFIERS:
AVAILABLE PHOSPHORUS, EXCHANGEABLE PHOSPHORUS, A*DENOSINE TRIPHOSPHATE(ATP).

ABSTRACT:
THE FACTORS CONTROLLING THE AVAILABLE P STATUS OF LAKES WERE INVESTIGATED. THE INFLUENCE OF SEDIMENT P ON THE P STATUS OF THE OVERLYING WATER WAS GIVEN PRIMARY ATTENTION. SEDIMENTS FROM SEVERAL NONCALCAREOUS AND CALCAREOUS WISCONSIN LAKES WERE INVESTIGATED. SEDIMENTS CONTAINED LARGE AMOUNTS OF INORGANIC AND ORGANIC P; GENERALLY MORE THAN ONE-HALF OF THE INORGANIC P ASSOCIATED WITH A GEL COMPLEX DOMINATED BY SHORT-RANGE ORDER HYDRATED IRON OXIDES AND HYDROUS OXIDES. MEASUREMENTS OF EXCHANGEABILITY USING 32P SHOWED THAT A LARGE PORTION OF THE SEDIMENT INORGANIC P EXCHANGED RAPIDLY WITH P IN THE SURROUNDING WATER. INVESTIGATION OF THE NUTRIENT STATUS OF SEDIMENT MICROORGANISMS (ATP PER CELL RELATIONSHIPS) AND THE ABILITY OF ALGAE INOCULATED INTO SEDIMENT-WATER SYSTEMS TO UTILIZE SEDIMENT P AS THEIR SOLE SOURCE OF P INDICATED THAT A SUBSTANTIAL PORTION OF SEDIMENT P IS AVAILABLE TO ORGANISMS IN CLOSE CONTACT WITH SEDIMENTS. EVIDENCE BASED IN PART ON COMPARISON OF 32P AND INORGANIC PHOSPHATE MEASUREMENTS INDICATED THAT THE MURPHY AND RILEY (1962) COLORIMETRIC METHOD PROVIDED VALID INFORMATION ON THE LEVELS OF DISSOLVED INORGANIC P IN LAKE WATERS.

FIELD 05C, 02H

ACCESSION NO. W72-03742

REVIEW OF STARCH PROBLEMS AS RELATED TO STREAM POLLUTION,

WESTERN MICHIGAN UNIV., KALAMAZOO. DEPT. OF PAPER TECHNOLOGY.

W. J. GILLESPIE, C. A. MAZZOLA, AND D. W. MARSHALL.

PAPER TRADE JOURNAL, VOL. 154, NO. 9, MARCH 2, 1970, P 29-32, 6 FIG, 16 REF.

DESCRIPTORS:
*WASTE WATER TREATMENT, *TURBIDITY, BIOCHEMICAL OXYGEN DEMAND, DISSOLVED OXYGEN, ALGAE, THERMAL STRATIFICATION, COAGULATION, BENTONITE, LIME, *PULP WASTES, TEOTIARY TREATMENT, EFFLUENTS.

IDENTIFIERS:
*STARCH REMOVAL, STARCH SUBSTITUTION, POYELECTROLYTES.

ABSTRACT:
ALTHOUGH THERE ARE OPERATING LOSSES OF STARCH DURING PRODUCTION, THE PRIMARY ADVERSE EFFECT RESULTS FROM THE REPULPING OF STARCH BEARING BROKE OR WASTE PAPER. THIS EXCESS STARCH, PARTICULARLY OXIDIZED STARCH, IN THE SYSTEM ADVERSELY AFFECTS THE PAPER MAKING PROCESS AND INCREASES TURBIDITY OF THE EFFLUENTS THAT MUST BE TREATED PRIOR TO DISCHARGE. STARCH DISPERSIONS ARE DIFFICULT TO COAGULATE WITH CONVENTIONAL FLOCCULANTS, HOWEVER, BENTONITE CLAYS HAVE BEEN USED SUCCESSFULLY AT DOSAGES OF 200 MG/L. STARCH DISPERSED IN THE MILL DISCHARGES EXERTS TWO DIBILITATING EFFECTS ON THE RECEIVING STREAM, A HIGH BOD AND AN INCREASED TURBIDITY WHICH LIMITS LIGHT PENETRATION. THOUGH STARCH SUBSTITUTION AND PROCESS MODIFICATIONS OFFER CONSIDERABLE PROMISE FOR ELIMINATION OF STARCH INDUCED DISPERSIONS, IT IS PROBABLE THAT UNIVERSAL APPLICATION OF SUCH ALTERNATIVES CANNOT BE ACHIEVED. SECONDARY TREATMENT DOES PROVIDE A MECHANISM FOR STARCH TURBIDITY REMOVAL. HOWEVER, ITS UTILIZATION FOR ONLY THE AESTHETIC IMPROVEMENT OF WASTE DISCHARGE REPRESENTS A COSTLY APPROACH IN AREAS WHERE EFFLUENT STANDARDS DO NOT OTHERWISE REQUIRE IT. THE TASK THAT REMAINS IS FOR TECHNOLOGY TO DEVISE A MEANS OF STARCH PROBLEM RESOLUTION IN THE PRIMARY CLARIFIER. (GOESSLING-TEXAS)

FIELD 05D

ACCESSION NO. W72-03753

TECHNICAL PHOSPHORUS POSITION PAPER, A RESPONSE TO THE IPCB PROPOSED WATER QUALITY STANDARDS REVISIONS, NO. 71-14.

METROPOLITAN SANITARY DISTRICT OF GREATER CHICAGO, ILL. RESEARCH AND DEVELOPMENT DEPT.

SEPTEMBER 1971. 14 P, 37 REF.

DESCRIPTORS:
*PHOSPHORUS, *WASTE WATER TREATMENT, *ALGAL CONTROL, *AQUATIC ALGAE, *EUTROPHICATION, MUNICIPAL WASTES, CITIES, PHOSPHATES, ILLINOIS, ECOLOGY, TERTIARY TREATMENT, SEWAGE EFFLUENTS, WATER REUSE.

IDENTIFIERS:
*PHOSPHORUS REMOVAL, CHICAGO(ILLINOIS).

ABSTRACT:
THE EXACT RELATIONSHIP BETWEEN PHOSPHORUS CONCENTRATION AND AQUATIC GROWTH IS NOT KNOWN. INVESTIGATORS REPORT QUITE DIVERGENT OPINIONS CONCERNING THE AMOUNT OF PHOSPHORUS NECESSARY FOR AQUATIC GROWTH. SOME HAVE SUGGESTED THAT VERY LOW CONCENTRATIONS (LESS THAN .01 MG/L) ARE LIMITING, AND SOME HAVE SUGGESTED THAT EVEN UNDETECTABLE AMOUNTS WILL CAUSE GROWTH. SOME INVESTIGATORS HAVE SUGGESTED THAT OTHER FACTORS MAY BE LIMITING. PHOSPHATE REMOVAL PROCESSES ARE QUESTIONABLE AND VERY DIFFICULT TO CONTROL. ASSUMING THAT IT MAY BE POSSIBLE TO MEET THE LIMITING PERMISSABLE PHOSPHORUS OF 0.1 MG/L PROPOSED BY THE ILLINOIS POLLUTION CONTROL BOARD (USING COAGULATION-SEDIMENTATION OF SECONDARY EFFLUENT PLUS MICRO-FILTRATION) THE CHEMICAL COSTS ALONE HAVE BEEN ESTIMATED TO BE IN EXCESS OF $18 MILLION PER YEAR FOR THE MSDGC. CAPITAL COSTS FOR REQUIRED DOSING EQUIPMENT AND SETTLING TANKS WOULD EXCEED $50 MILLION. THE INCREASED SLUDGE DISPOSAL COST TO HANDLE THE ADDITIONAL SOLIDS GENERATED BY REMOVING PHOSPHORUS WOULD INVOLVE AN ANNUAL COST OF AT LEAST $9,000,000. CONSEQUENTLY, AN ADDITIONAL ESTIMATED ANNUAL OPERATING COST OF $27,000,000 MUST BE ASSIGNED TO PHOSPHORUS REMOVAL COMPARED TO THE PRESENT TOTAL ANNUAL COST OF TREATMENT OF 10-14 MILLION DOLLARS. THERE IS NO SUBSTANTIAL EVIDENCE TO INDICATE THAT THE LIMITING NUTRIENT IS PHOSPHORUS. PHOSPHORUS REMOVAL AT WASTEWATER TREATMENT PLANTS WHICH DO NOT DISCHARGE TO LAKE MICHIGAN DOES NOT APPEAR AT THIS TIME TO BE JUSTIFIED. (POERTNER)

FIELD 05C

ACCESSION NO. W72-03970

FINAL REPORT, PROVISIONAL ALGAL ASSAY PROCEDURES.

CALIFORNIA UNIV., BERKELEY. SANITARY ENGINEERING RESEARCH LAB.

D. F, TOERIEN, C. H. HUANG, J. RADIMSKY, E. A. PEARSON, AND J. SCHERFIG.

AVAILABLE FROM THE NATIONAL TECHNICAL INFORMATION SERVICE AS PB-206 140, $3.00 IN PAPER COPY, $0.95 IN MICROFICHE. ENVIRONMENTAL PROTECTION AGENCY, WATER POLLUTION CONTROL RESEARCH SERIES, SERL REPORT 71-6, OCTOBER 1971. 211 P, 51 FIG, 37 TAB, 111 REF, 4 APPEND. EPA PROJECT 16010 DQB 10/71.

DESCRIPTORS:
*BIOASSAY, CULTURES, *BICMASS, NUTRIENTS, *NUTRIENT REQUIREMENTS, NUISANCE ALGAS, *ENTROPHICATION, CYTOLCGICAL STUCIES, ANALYTICAL TECHNIQUES, POLLUTANT IDENTIFICATION.

IDENTIFIERS:
*CHEMOSTAL ASSAYS, *ALGAL ASSAY PROCEDURES, SELENASTRUM CAPRICORNUTUM.

ABSTRACT:
BATCH AND CONTINUOUS FLOW (CHEMOSTAT) ASSAYS WERE INVESTIGATED AS PART OF A JOINT INDUSTRY-GOVERNMENT SPONSORED, MULTILABCRATORY EFFORT TO DEVELOP A STANDARDIZED ALGAL ASSAY PROCEDURE FOR NUTRIENT LEVEL ASSESSMENT. ASSAYS WERE CONDUCTED WITH SELENASTRUM CAPRICORNUTUM AS A STANDARD ASSAY CRGANISM. BATCH CULTURE ASSAYS WERE FOUND TO HAVE A LOWER LEVEL CF PRECISION THAN CHEMOSTAT ASSAYS IN THE ASSESSMENT OF GROWTH RESPONSE AS A FUNCTION OF NUTRIENT CONCENTRATION. THE BIOMASS PARAMETER, MAXIMUM CELL CCNCENTRATION, X, OF THE BATCH ASSAY GENERALLY RESPONDED TO THE NUTRIENT CONCENTRATION OF THE SAMPLE; HOWEVER, THE CHEMOSTAT BICMASS PARAMETER, STEADY STATE CELL CCNCENTRATION, X SUB 1, ALWAYS WAS FOUND TO BE PROPORTIONAL TO THE NUTRIENT CONCENTRATION OF THE SAMPLES. THE RESULTS OF SPIKING TESTS WITH BATCH ASSAYS GENERALLY WERE INCONCLUSIVE, WITH RESPECT TO IDENTIFICATION CF THE GROWTH RATE LIMITING NUTRIENT, WHEREAS THE RESULTS OF SPIKING TESTS WITH CHEMOSTATS INDICATED CLEARLY THE GROWTH RATE LIMITING NUTRIENT. IT IS RECOMMENDED THAT BATCH TYPE ALGAL ASSAYS BE USED ONLY FOR CRUDE SCREENING OR ROUTINE MONITORING PURPOSES AND THAT THE CHEMOSTAT SHCULD BE USED FOR THE QUANTITATIVE ASSESSMENT CF THE ALGAL GROWTH SUPPORTING PROPERTIES OF WATERS AS WELL AS FOR THE DEVELOPMENT OF KINETIC DESCRIPTIONS FOR NUISANCE ALGAE AND THE RATE LIMITING NUTRIENTS OF CONCERN. A KINETIC DESCRIPTION OF SELENASTRUM CAPRICORNUTUM INDICATED A LOW HALF SATURATION CONSTANT, K SUB S, (THE CONCENTRATION CF NUTRIENT SUPPORTING ONE-HALF THE MAXIMUM GRCWTH RATE) OF ABOUT 5 MICRCGRAM P/L FOR PHOSPHATE PHOSPHORUS AND A YIELC COEFFICIENT, Y, THAT VARIED AS A FUNCTION CF GROWTH RATE. A THEORETICAL MODEL WAS PROPOSED AND EVALUATED WHICH DESCRIBES THE VARYING YIELD COEFFICIENT (THE RESULT OF 'ECESS' UPTAKE) AS A FUNCTION OF THE GROWTH RATE (MEAN CELL RESIDENCE TIME). THE FUNCTION WAS VERIFIED EXPERIMENTALLY AT A VERY HIGH STATISTICAL CONFIDENCE LEVEL. THE SIGNIFICANCE OF THESE FINDINGS AND THEIR APPLICATION TO THE PRACTICAL PROBLEM OF EUTRCPHICATION ASSESSMENT IS PRESENTED. (EPA ABSTRACT)

FIELD 05C, 05A

ACCESSION NC. W72-03985

NUTRIENT LIMITATION,

MINISTRY OF AGRICULTURE, CRUMLIN (NORTHERN IRELAND). FRESHWATER BIOLOGICAL INVESTIGATION UNIT.

C. E. GIBSON.

JOURNAL WATER POLLUTICN CONTROL FEDERATICN, VOL 43, NO 12, DECEMBER 1971, P 2436-2440, 2 TAB, 12 REF.

DESCRIPTORS:
*NUTRIENTS, *ALGAE, *EUTROPHICATION, *PHOSPHORUS, *NITROGEN, EVAPORATION, STORM RUNOFF, BIOASSAYS, BIO-INDICATCRS, STORAGE, FLOW RATES, INHIBITION, ANALYTICAL TECHNIQUES, ENZYMES, COLOR, WATER QUALITY CONTROL, RESEARCH AND DEVELCPMENT.

ABSTRACT:
DETERMINATION OF LIMITING LEVELS OF NITROGEN AND PHOSPHORUS HAS BEEN EXCEEDINGLY DIFFICULT FOR SEVERAL REASONS. INPUTS OF PHOSPHATES AND NITRATES ARE EXTREMELY VARIABLE WITH SEASON OR WITH SHIFTS IN POPULATION. SECONDLY, THE RATIO CF CARBON TO NITRCGEN TO PHOSPHORUS IS JUST AS IMPORTANT AS THE ACTUAL AMOUNTS OF EACH PRESENT. THIRDLY, CERTAIN ALGAE HAVE A CAPACITY FOR STORING PHOSPHATES, AND THRIVING ALGAL GROWTHS MAY APPEAR IN AREAS WHERE PHOSPHATE IS SERIOUSLY DEPLETED. PERHAPS THE GREATEST DIFFICULTY ENCOUNTERED IS THE INABILITY OF RESEARCHERS TO DEVELOP A SYNTHETIC MEDIUM UPON WHICH ALGAE WILL PRODUCE GROWTH PATTERNS SIMILAR TO THOSE OF ALGAE GRCWN IN STREAMS OR LAKES. PHCSPHORUS AND NITROGEN REQUIREC BY ALGAE GROWN ON SYNTHETIC MEDIA FAR EXCEED THE SUPPLY AVAILABLE IN ANY WATERS LESS THAN GROSSLY POLLUTED. HOWEVER, ADVANCEMENTS IN BIOASSAY TECHNIQUES ARE CONTINUING, AND THERE IS NOW A SIMPLE SPCT TEST FOR PHOSPHORUS LIMITATION. PRESENCE OF ALKALINE PHOSPHATES HAS BEEN SHOWN TO BE AN INDICATOR OF PHOSPHATE LIMITATION, AND A RAPID COLOR TEST HAS BEEN DEVELCPED. TEAMS OF RESEARCHERS ARE CONTINUING TO DEVOTE THEIR ATTENTION TO THIS AND OTHER NEW METHODS FOR OBTAINING INFORMATION ABOUT NUTRIENT LIMITATION. (LOWRY-TEXAS)

FIELD 05C

ACCESSION NO. W72-04070

NITRCGEN REMOVAL FROM WASTEWATERS--STATEMENT OF THE PROBLEM,

FEDERAL WATER QUALITY ADMINISTRATION, CINCINNATI, OHIO. ADVANCED WASTE TREATMENT RESEARCH LAB.

EDWIN F. BARTH, AND ROBERT B. DEAN.

IN: NITROGEN REMOVAL FROM WASTEWATERS, PAPER NO 1, MAY 1970. 6 P. FWQA PROJECT 17010---05/70.

DESCRIPTORS:
*NITROGEN, *AMMONIA, NITRATES, NITRITES, TOXICITY, CORROSION, PUBLIC HEALTH, ODORS, OXIDATION-REDUCTION POTENTIAL, BIOCHEMICAL OXYGEN DEMAND, NITRIFICATION, NUTRIENTS, SLUDGE, ALGAE, BACTERIA, *WASTE WATER TREATMENT, EUTROPHICATION.

IDENTIFIERS:
*METHEMOGLOBINEMIA.

ABSTRACT:
RELEASE OF AMMONIA TO A RECEIVING WATER BODY MAY CAUSE A VARIETY OF UNWANTED RESULTS. THE TOXICITY OF AMMONIA TO ADULT HUMANS IS SO SLIGHT AS TO REQUIRE NO LIMITS IN DRINKING WATER, ALTHOUGH THE PRESENCE OF AMMONIA IS REGARDED AS EVIDENCE OF POLLUTION. HOWEVER, INFANT HUMANS, BECAUSE OF THE LACK OF SUFFICIENT INTESTINAL FLORA, ARE NOT ABLE TO REDUCE NITRITES TO NITROGEN GAS. THE NITRITE THEN COMBINES WITH THE HEMOGLOBIN WHICH RESULTS IN METHEMOGLOBINEMIA. HIGH AMMONIA LEVELS IN DRINKING WATER WILL ALSO DISSOLVE COPPER PLUMBING. FOR THESE REASONS, LIMITS HAVE BEEN SET AT NC MCRE THAN 10 MG OF NITRATE NITROGEN IN WATER TO BE FED TO BABIES, ALTHOUGH NO STANDARDS HAVE BEEN SET FOR AMMONIA. THE OXYGEN DEMAND OF NITROGENOUS MATERIAL HAS SIGNIFICANTLY ACCELERATED EUTROPHICATION IN MANY AREAS, INCLUDING THE POTOMAC ESTUARY, THE GRAND AND CLINTON RIVERS IN MICHIGAN AND KANAWHA RIVER IN WEST VIRGINIA. A NITRIFIED EFFLUENT, HOWEVER, WILL PROVIDE OXYGEN TC SLUDGE BEDS AND PREVENT SEPTIC ODORS, BE MCRE EFFICIENTLY AND EFFECTIVELY DISINFECTED BY CHLORINE, AND WILL CONTAIN LESS SOLUBLE ORGANIC MATTER THAN THE SAME EFFLUENT BEFORE CHLORINATION. (LOWRY-TEXAS)

FIELD 05D

ACCESSION NO. W72-04086

ACCUMULATION AND DISTRIBUTION OF CHLORINATED HYDROCARBONS IN CULTURES OF (CHLOROPHYCEAE),

LUND UNIV. (SWEDEN). DEPT. OF ANIMAL ECOLOGY.

A. SODERGREN.

OIKOS, VOL 22, P 215-220, 1971. 7 FIG, 1 TAB, 20 REF.

DESCRIPTORS:
*CHLORINATED HYDROCARBON PESTICIDES, *PESTICIDE RESIDUES, *ALGAE, CHLOROPHYTA, CHLORELLA, TOXICITY, WATER ANALYSIS, ABSORPTION, CULTURES, DDT, ANALYTICAL TECHNIQUES, BIOMASS, GAS CHROMATOGRAPHY, CENTRIFUGATION, PESTICIDES, BIOASSAY, DISTRIBUTION PATTERNS, CHROMATOGRAPHY, GROWTH RATES, POLLUTANT IDENTIFICATION.

IDENTIFIERS:
*CHLORELLA PYRENOIDOSA, LINDANE, P P'-DDE, CLOPHEN A 50, ELECTRON CAPTURE GAS CHROMATOGRAPHY, POLYCHLORINATED BIPHENYLS, BIOLOGICAL MAGNIFICATION, CHLAMYDOMONAS, METABOLITES, HAEMATOCRIT TUBE.

ABSTRACT:
THE ACCUMULATION AND DISTRIBUTION OF LINDANE, P,P'-DDE AND CLOPHEN A 50 (PCB) HAVE BEEN STUDIED IN CONTINUOUS FLOW CULTURES OF CHLORELLA PYRENOIDOSA CHICK. IN ESTIMATING THE CELL VOLUMES OF ALGAE IN THE CULTURE A TECHNIQUE HAS BEEN DEVELOPED USING CENTRIFUGATION IN A MODIFIED HAEMATOCRIT TUBE. THE NUMBER OF CELLS IN THE CULTURE WAS CALCULATED BY CALIBRATING THE CELL VOLUMES VERSUS THE CELL COUNTS. SAMPLES OF ALGAE, AIR, AND WATER FROM WITHIN THE CULTURE SYSTEM WERE ANALYZED FOR RESIDUES OF THE CHLORINATED HYDROCARBON PESTICIDES ON A VARIAN AEROGRAPH 204 GAS CHROMATOGRAPH WITH AN ELECTRON CAPTURE DETECTOR. IT WAS FOUND THAT: (1) LINDANE WAS ACCUMULATED BY THE ALGAE BUT WAS ALSO DISTRIBUTED IN THE WATER AND AIR OF THE CULTURE. (2) P,P'-DDE AND CLOPHEN A 50 WERE TAKEN UP BY THE ALGAE TO A GREAT EXTENT AND ONLY A SMALL PART REMAINED IN THE WATER. AT THE END OF THE EXPERIMENTS 82 PERCENT AND 88 PERCENT RESPECTIVELY OF P,P'-DDE AND CLOPHEN A 50 WERE ACCUMULATED BY THE ALGAE. (3) IN THE EXPERIEMENTS THERE WERE NO DISTURBANCES OF THE GROWTH RATE OF THE CULTURED ALGA DUE TO THE PRESENCE OF THE TEST SUBSTANCE. THE UPTAKE AND DISTRIBUTION OF THE PESTICIDES WITHIN THE CULTURE SYSTEM AND THE RELATIONSHIP BETWEEN CELL VOLUMES AND CELL COUNTS ARE SHOWN GRAPHICALLY. (HOLOMAN-BATTELLE)

FIELD 05A

ACCESSION NO. W72-04134

PHARMACOLOGICAL TESTING OF BLUE-GREEN ALGAE FOR CONSTITUENTS HAVING THERAPEUTIC VALUE,

WORLD LIFE RESEARCH INST., COLTON, CALIF.

BRUCE W. HALSTEAD.

AVAILABLE FROM SOD EP2.10:16010DCU06/70, $0.95 IN MICROFICHE. ENVIRONMENTAL PROTECTION AGENCY, WATER QUALITY OFFICE, WATER POLLUTION CONTROL RESEARCH SERIES, JUNE 1970. 17 P, 1 TAB, 9 REF. EPA PROGRAM 16010DOU06/70, 14-12-535.

DESCRIPTORS:
*CYANOPHYTA, *LABORATORY TESTS, ALGAL TOXINS, HUMAN PATHOLOGY.

IDENTIFIERS:
*THERAPEUTIC VALUE, *APHANIZOMENON FLOS-AQUAE, *PHARMACOLOGICAL TESTS, ANTIMICROBIOTICS, ESCHERICHIA COLI, BETA STREPTOCOCCUS, PSEUDOMONAS AERUGINOSA, MYCOBACTERIUM FURTUITUM, STAPHYLOCOCCUS AUREUS, CANDIDA ALBICANS.

ABSTRACT:
THE BLUE-GREEN ALGA APHANIZOMENON FLOS-AQUAE (LINNAEUS) RALFS, WAS COLLECTED DURING THE PEAK OF ITS PRIMARY BLOOM AND PREPARED WITH ACETONE AND ETHANOLIC EXTRACTS FOR PHARMALOGICAL AND MICROBIOLOGICAL TESTING. THE FRACTIONS EXHIBITED LITTLE ACUTE OVERTSYMPTOMATOLOGY AND THUS PROBABLY DO NOT CONTAIN PRODUCTS THAT EXERT PHARMACOLOGICAL EFFECTS SIMILAR TO THE TYPE SEEN BY DRUGS WHICH ORDINARILY DIRECTLY PRODUCE PSYCHOMATIC, NEUROLOGICAL, OR CARDIOVASCULAR EFFECTS. GENERALLY THE ACTIVITY OF THE FRACTIONS RESULTED IN DEPRESSION OF FUNCTION, ALERTNESS, AND APPETITE, AND DECREASED BODY WEIGHT IN RATS, CHARACTERISTIC OF AN ANTIMETABOLITE ACTION COMMON TO MANY NATURALLY OCCURRING SUBSTANCES. THERE WAS LOW POTENCY, NO FRACTIONAL LOCALIZATION OF ACTIVITY, NO SIGNIFICANT NEUROLOGICAL, CARDIOVASCULAR, OR ANY OUTSTANDING BIOLOGICAL ACTIVITY. ANTIMICROBIAL TESTS PERFORMED ON ESCHERICHIA COLI, BETA STREPTOCOCCUS, PSEUDOMONAS AERUGINOSA, MYCOBACTERIUM FURTUITUM, STAPHYLOCOCCUS AEREUS, AND CANDIDA ALBICANS. NO ANTIMICROBIAL ACTIVITY WAS OBSERVED, HOWEVER THE ALGAE COULD BE CLASSIFIED AS 'TOXIC' IF SUFFICIENT QUANTITIES ARE INGESTED. A SAMPLE WAS SCREENED, TESTED, AND EVALUATED FOR ANTITUMOR ACTIVITY BUT WAS NOT FOUND TO HAVE ANY SIGNIFICANT ANTITUMOR ACTIVITY. THE TESTS FAILED TO SHOW EVIDENCE OF PHARMACOLOGICAL PROPERTIES HAVING COMMERCIAL PHARMACEUTICAL POTENTIAL. FURTHER INVESTIGATION AS TO THE COMMERCIAL POTENTIAL OF THE PHYTOCHEMICAL CONSTITUENTS OF APHANIZOMENON FLOS-AQUAE DOES NOT APPEAR WARRANTED. (AUEN-WISCONSIN)

FIELD 05C, 05G

ACCESSION NO. W72-04259

WATER QUALITY CONTROL THROUGH FLOW AUGMENTATION,

HEIDELBERG COLL., TIFFIN, OHIO. DEPT. OF BIOLOGY.

DAVID B. BAKER, AND JACK W. KRAMER.

AVAILABLE FROM SOD EP2.10:16080DF001/71, $0.95 IN MICROFICHE. ENVIRONMENTAL PROTECTION AGENCY, WATER QUALITY OFFICE, WATER POLLUTION CONTROL RESEARCH SERIES, JANUARY 1971. 156 P, 27 FIG, 8 TAB, 14 REF, 12 APPEND. EPA PROGRAM 16080DF001/71.

DESCRIPTORS:
*WATER QUALITY CONTROL, *FLOW AUGMENTATION, WATER QUALITY, RESERVOIRS, WATER POLLUTION SOURCES, ALGAE, RIVER FLOW, PHOSPHORUS, NITRATES, POTASSIUM, DISSOLVED OXYGEN, FORECASTING, FLUORIDES, OHIO.

IDENTIFIERS:
SANDUSKY RIVER(OHIO).

ABSTRACT:
A 60-MILE SECTION OF THE SANDUSKY RIVER, OHIO, WAS INVESTIGATED TO EVALUATE THE RELATIONSHIP BETWEEN THE VOLUME AND THE QUALITY OF FLOWING WATER. THE CONTENT OF FLUORIDE, CALCIUM, MAGNESIUM, AND SODIUM WERE DIRECTLY AND THOSE OF TOTAL AND SOLUBLE PHOSPHORUS INDIRECTLY RELATED TO THE FLOW VOLUME. NO CORRELATION WAS OBSERVED BETWEEN THE FLOW AND CONCENTRATION OF EITHER POTASSIUM OR NITRATES. OXYGEN CONTENT WAS HIGH AT AN ABUNDANT FLOW, BUT SHOWED CONSIDERABLE VARIATION, BEING INFLUENCED BY RESPIRATION OF ALGAE. THE STUDY SUGGESTED THAT AN INCREASED CURRENT VELOCITY REDUCES THE DENSITY OF ALGAL POPULATIONS. (WILDE-WISCONSIN)

FIELD 05G

ACCESSION NO. W72-04260

PHOSPHATE NUTRIENT OCCURRENCE AND DISTRIBUTION IN GREAT LAKES SEDIMENTS,

ILLINOIS UNIV., CHICAGO. SOIL MECHANICS LAB.

MARSHALL L. SILVER, AND CHARLES A. MOORE.

SM PUBLICATION NO 12, 1971. 27 P, 7 FIG, 5 TAB, 55 REF. OWRR A-053-ILL(1).

DESCRIPTORS:
*PHOSPHATES, *GREAT LAKES, *SEDIMENTS, *NUTRIENTS, EUTROPHICATION, NITROGEN, LITTORAL, HARBORS, SPATIAL DISTRIBUTION, ALGAE, TRIBUTARIES, CYCLING NUTRIENTS.

IDENTIFIERS:
*NUTRIENT LEVELS, CUYAHOGA RIVER(OHIO).

ABSTRACT:
PUBLISHED AND UNPUBLISHED SOURCES ARE COMBINED TO PROVIDE INSIGHT INTO THE SOURCE AND FATE OF POLLUTANTS IN THE GREAT LAKES. DATA ARE PRESENTED SUMMARIZING THE QUANTITY OF PHOSPHORUS PRESENT IN CENTRAL LAKE WATERS AND CENTRAL LAKE SEDIMENTS AS WELL AS PHOSPHORUS CONTENTS IN SEDIMENTS AND WATERS OF RIVERS AND HARBORS TRIBUTARY TO THE LAKES, AND NUTRIENT CONTENTS IN THE GREAT LAKES MARGINS. AVAILABLE LITERATURE IS SUMMARIZED DESCRIBING THE EFFECT OF PHOSPHORUS ON THE GROWTH RATE OF AQUATIC PLANTS WHOSE ACCELERATED GROWTH IS RESPONSIBLE FOR EUTROPHICATION. AN ATTEMPT IS MADE TO ORGANIZE MECHANISMS CONTRIBUTING TO THE CYCLING OF PHOSPHORUS AND OTHER NUTRIENTS IMPORTANT TO GREAT LAKES STUDIES. GENERALLY, THE MINIMUM PHOSPHORUS CONTENT NEEDED FOR ALGAL GROWTH IS BELOW 0.01 MG/L AND MAY BE LESS THAN 0.001 MG/L. PHOSPHORUS LEVELS IN EXCESS OF 0.1 MG/L OFTEN CAUSE ACCELERATED GROWTH OF VARIOUS PHYTOPLANKTON AND ALGAE. NITROGEN CONTENTS GREATER THAN 0.1 MG/L ARE NECESSARY FOR GROWTH WHILE CONCENTRATIONS GREATER THAN 1 MG/L LEAD TO ACCELERATED PLANT GROWTH. (AUEN-WISCONSIN)

FIELD 05C

ACCESSION NO. W72-04263

DETERGENTS A STATUS REPORT.

NATIONAL INDUSTRIAL POLLUTION CONTROL COUNCIL, WASHINGTON, D.C.

AVAILABLE FROM THE NATIONAL TECHNICAL INFORMATION SERVICE AS COM-71-50084, $3.00 IN PAPER COPY, $0.95 IN MICROFICHE. SUB-COUNCIL REPORT MARCH 1971. 16 P.

DESCRIPTORS:
*DETERGENTS, *EUTROPHICATION, *ECONOMIC EFFICIENCY, *PUBLIC HEALTH, WATER POLLUTION, HAZARDS, NUTRIENTS, ALGAE.

IDENTIFIERS:
NITRILOTRIACETIC ACID, HOUSEHOLD DETERGENTS, SOAP AND DETERGENT INDUSTRY, NTA.

ABSTRACT:
THE USE OF THE MOST PROMISING SUBSTITUTE FOR PHOSPHATES IN DETERGENTS, NITRILOTRIACETIC ACID (NTA) HAS BEEN SUSPENDED PENDING FURTHER STUDY AT THE REQUEST OF THE U S SURGEON GENERAL AND THE ENVIRONMENTAL PROTECTION AGENCY. THE SEARCH FOR PHOSPHATE SUBSTITUTES CONTINUES. EUTROPHICATION IS DEFINED AS SIMPLY THE OVERABUNDANCE OF A NATURAL AND NECESSARY PROCESS--TROUBLESOME INDEED WHEN IT OCCURS, BUT, BY NO STRETCH OF THE IMAGINATION, A FORM OF WATER POLLUTION. THERE IS NO EVIDENCE WHATEVER THAT THE REMOVAL OF PHOSPHATES FROM DETERGENTS WILL STOP ACCELERATED EUTROPHICATION. ADEQUATE TECHNOLOGY EXISTS TO REMOVE PHOSPHORUS FROM SEWAGE BY MEANS OF CHEMICAL TREATMENT. FOR RELATIVELY MODEST COSTS, CHEMICAL SEWAGE TREATMENT PERMITS THE REMOVAL OF PHOSPHORUS FROM ALL SOURCES WITHOUT THE CAPITAL COST OF SECONDARY OR TERTIARY SEWAGE PLANTS AND IT IS HIGHLY FLEXIBLE, IN THAT IT CAN BE EMPLOYED IN ONLY THOSE LOCALITIES WHERE CULTURAL EUTROPHICATION IS A PROBLEM. (JONES-WISCONSIN)

FIELD 05C

ACCESSION NO. W72-04266

LIMNOLOGICAL CHARACTERISTICS OF NORTH AND CENTRAL FLORIDA LAKES,

FLORIDA UNIV., GAINESVILLE. DEPT. OF ENVIRONMENTAL ENGINEERING.

EARL E. SHANNON, AND PATRICK L. BREZONIK.

RESEARCH REPORT, 1971. 36 P, 9 FIG, 3 TAB, 31 REF. OWRR B-004-FLA(5). FWQA DON 16010.

DESCRIPTORS:
*LIMNOLOGY, *FLORIDA, *LAKES, PONDS, SAMPLING, THERMAL STRATIFICATION, HARDNESS(WATER), COLOR, OLIGOTROPHY, EUTROPHICATION, ALGAE, CYANOPHYTA, PHYTOPLANKTON, LAKE MORPHOMETRY, GEOLOGY, LIMESTONE, TEMPERATURE, DEPTH, CHEMICAL PROPERTIES, ACIDITY, ALKALINITY, CONDUCTIVITY, CALCIUM, SCENEDESMUS, WATER HYACINTH, LIGHT PENETRATION, OXYGEN, SILICA, AQUATIC PLANTS.

IDENTIFIERS:
MESOTROPHIC, MICROCYSTIS, ANABAENA, LYNGBYA.

ABSTRACT:
AN EXTENSIVE SURVEY OF THE PHYSICAL, CHEMICAL, AND BIOLOGICAL CHARACTERISTICS OF 55 NORTH-CENTRAL FLORIDA LAKES AND PONDS WAS MADE. RESULTS OF THE ONE-YEAR INVESTIGATION, WITH PARTICULAR REFERENCE TO THEIR TYPOLOGIC AND TROPHIC FEATURES ARE DESCRIBED. THEY ARE TYPICALLY SHALLOW, WITH MEAN DEPTHS RANGING FROM 0.7 TO 8.1 M AND DERIVED FROM SOLUTION OF LIMESTONE FORMATIONS WHICH UNDERLY THE FLORIDA PENINSULA. STABLE THERMAL STRATIFICATION OCCURS IN 13 LAKES. THEY ARE TYPICALLY SOFT WATER, LOW IONIC STRENGTH WATERS. HIGH ORGANIC COLOR, DERIVED PRIMARILY FROM PINE LITTER, IS COMMON. TROPHIC CONDITIONS OBTAIN FROM ULTRAOLIGOTROPHIC TO HYPEREUTROPHIC. MULTI-VARIATE CLUSTER ANALYSIS CONSIDERING SEVEN QUANTITATIVE TROPHIC INDICATORS GROUPED THE CLEAR LAKES INTO THREE READILY INTERPRETABLE GROUPS (OLIGOTROPHIC, MESOTROPHIC, AND EUTROPHIC) AND THE COLORED LAKES INTO FIVE GROUPS. HIGHEST EUTROPHIC CONDITIONS ARE ASSOCIATED WITH ALKALINE (HARDWATER) CLEAR LAKES, WHEREAS THE SOFT-WATER CLEAR LAKES WERE ALL OLIGOTROPHIC; COLOR APPEARS TO DAMPEN THE RANGE OF TROPHIC CONDITIONS. ALGAL BLOOMS CAN OCCUR ALMOST CONTINUOUSLY THROUGHOUT THE YEAR IN EUTROPHIC LAKES BECAUSE OF FAVORABLE GROWTH CONDITIONS; THE BLUE-GREEN BLOOM-FORMING ALGAE, MICROCYSTIS, ANABAENA, AND LYNGBYA ARE THE MOST DOMINANT PHYTOPLANKTON OF EUTROPHIC FLORIDA LAKES. (JONES-WISCONSIN)

FIELD 05C, 02H

ACCESSION NO. W72-04269

PRIMARY PRODUCTIVITY IN FIVE SALT VALLEY RESERVOIRS,

NEBRASKA UNIV., LINCOLN. DEPT. OF ZOOLOGY.

JOHN L. ANDERSEN.

MS THESIS, AUGUST 1971. 54 P, 12 FIG, 5 TAB, 22 REF. OWRR A-014-NEB(1).

DESCRIPTORS:
*PRIMARY PRODUCTIVITY, *RESERVOIRS, *NEBRASKA, *EUTROPHICATION, PLANKTON, CHLOROPHYLL, RECREATION, RUNOFF, MEASUREMENT, TEST PROCEDURES, PHOTOSYNTHESIS, SEASONAL, LIGHT PENETRATION, LIGHT INTENSITY, DISSOLVED SOLIDS, PHYTOPLANKTON, ALGAE, NUTRIENTS.

IDENTIFIERS:
*SALT VALLEY RESERVOIRS(NEB), ULTRA PLANKTON, C-14.

ABSTRACT:
FIVE SALT VALLEY RESERVOIRS, LOCATED WITHIN A 20-MILE RADIUS OF LINCOLN, NEBRASKA, WERE CONSTRUCTED TO PROVIDE FLOOD PROTECTION BUT RECREATIONAL FACILITIES HAVE ALSO BEEN EXTENSIVELY DEVELOPED. THESE RESERVOIRS DEPEND UPON RUNOFF FROM AGRICULTURAL LANDS AS THEIR SOURCE OF WATER AND ARE RICH IN NUTRIENTS. MORPHOMETRY AND PHYSICAL-CHEMICAL PARAMETERS ARE RECORDED. THE IN SITU CARBON-14 TECHNIQUES WERE USED FOR STUDYING PRODUCTIVITY. ABSOLUTE ACTIVITY OF THE CARBON-14 MUST BE DETERMINED SINCE THERE OFTEN EXISTS A WIDE DIFFERENCE BETWEEN THE LABELED AND ACTUAL VALUE OF COMMERCIALLY PREPARED RADIOISOTOPE SOLUTIONS, THE LAKES ARE DISCUSSED INDIVIDUALLY AND RANKED ACCORDING TO PRODUCTIVITY INCLUDING PRODUCTIVITY VALUES, CHLOROPHYLL VALUES, AND THE ULTRA PLANKTON CONTRIBUTION, WHICH RANGED FROM 0 TO 13.3% OF THE TOTAL PRODUCTIVITY. PRIMARY PRODUCTIVITY OCCURRING UNDER THE ICE CONSTITUTES A LARGE PORTION OF THE ANNUAL PRODUCTIVITY AND CANNOT BE IGNORED. IN COMPARING THE PRODUCTIVITY OF THE SALT VALLEY LAKES WITH OTHER LAKES AROUND THE WORLD, IT CAN BE SEEN THAT THE SALT VALLEY LAKES ARE QUITE EUTROPHIC. IF THE RATE OF YEARLY PRODUCTIVITY INCREASES IN THESE RESERVOIRS (500-600 MG/YEAR) CONTINUES, IT MAY BE A SHORT TIME UNTIL THESE LAKES ARE IN THE PRODUCTIVITY RANGE OF SEWAGE LAGOONS. (JONES-WISCONSIN)

FIELD 05C

ACCESSION NO. W72-04270

PHYTOCHEMICAL, PHARMACOLOGICAL AND ANTIMICROBIAL SCREENING OF MINNESOTAN AQUATIC PLANTS,

MINNESOTA UNIV., MINNEAPOLIS.

KWEI LEE SU.

PHD THESIS, JUNE 1971. 158 P, 13 FIG, 23 TAB, 342 REF. OWRR A-025-MINN(1).

DESCRIPTORS:
*AQUATIC PLANTS, *ALGAE, *PUBLIC HEALTH, ANALYTICAL TECHNIQUES, TOXICITY, CHEMICAL PROPERTIES, MINNESOTA, LIPIDS, HUMAN PATHOLOGY, DISEASES.

IDENTIFIERS:
ANTIMICROBIOTICS, PHARMACOLOGICAL PROPERTIES, PHYTOCHEMICAL SCREENINGS, NYMPHAEA TUBEROSA, NUPHAR VARIEGATUM, NUPHAR JAPONICUM, NUPHAR LUTEN, ANTICANCER POTENTIAL, ANTICOAGULANT POTENTIAL, SPARGANIUM FLUCTUANS, ANTINEOPLASTIC ACTIVITY, ALKALOIDS, FLAVONOIDS, STEROIDS, TANNINS, SAPONINS.

ABSTRACT:
SEVENTEEN MONOCOTS, 4 DICOTS, AND 1 ALGA, COLLECTED FROM MINNESOTA LAKES WERE EXTRACTED WITH SKELLYSOLVE F, CHLOROFORM, 80% ETHANOL, AND ACIDIC AND BASIC WATER. AS REVEALED BY THIN-LAYER CHROMATOGRAPHY AND OTHER ANALYSES, MANY PLANTS CONTAIN ALKALOIDS, FLAVONOLES, FLAVONONES, BETA-SITOSTEROL, LIPIDS, TANNINS, AND SAPONINS. THE TOXICITY, INCLUDING LETHAL DOSE, MEDIAN TEST, AND ANTINEOPLASTIC ACTIVITY IN VIRO WERE DETERMINED USING SWISS WEBSTER MICE AND SYRIAN HAMSTERS. SCREENING WAS PERFORMED ON HUMAN EPIDERMOID CARCINOMA AND LYMPHOID LEUKEMIA. ANTIMICROBIAL EFFECTS OF EXTRACTS WERE STUDIED ON BOTH ANIMAL AND PLANT PATHOGENES BY THE DISC DIFFUSION METHOD. SOME EXTRACTS WERE EFFECTIVE AGAINST STAPHYLOCOCCUS AUREUS, MYCOBACTERIUM SMEGMATIS, AND CANDIDA ALBICANS. RECOMMENDATIONS FOR ADDITIONAL STUDIES INCLUDE IDENTIFICATION OF CERTAIN ALKALOIDS, ANTICANCER POTENTIAL AGAINST AMELANOMA MALIGNANCY, ANTICOAGULANT EFFECTS OF SEVERAL AQUATIC PLANTS, AND INHIBITION OF CANDIDA ALBICANS. (WILDE-WISCONSIN)

FIELD 05C

ACCESSION NO. W72-04275

WORKSHOP-CONFERENCE ON RECLAMATION OF MAINE'S DYING LAKES.

MAINE UNIV., BANGOR.

AVAILABLE FROM THE NATIONAL TECHNICAL INFORMATION SERVICE AS PB-206 432, $3.00 IN PAPER COPY, $0.95 IN MICROFICHE. UNIVERSITY OF MAINE AT ORONO, MAINE, JUNE 1971. 113 P, 4 FIG, 1 MAP, 6 PHOTO, 39 REF, 2 APPEND. OWRR A-999-ME(8).

DESCRIPTORS:
*MAINE, *EUTROPHICATION, *WATER QUALITY CONTROL, *NON-STRUCTURAL ALTERNATIVES, AGING(BIOLOGICAL), AQUATIC ALGAE, AQUATIC PRODUCTIVITY, FISH, FISHKILL, LIMNOLOGY, NUTRIENTS, OXYGEN, OXYGEN REQUIREMENTS, OXYGEN SAG, WASTE ASSIMILATIVE CAPACITY, WATER PROPERTIES, WATER POLLUTION, WATER POLLUTION SOURCES, WATER POLLUTION EFFECTS, WATER POLLUTION CONTROL, WATER POLLUTION TREATMENT, WATER QUALITY, WATER QUALITY CONTROL, LAND USE, LAND TENURE, LAND DEVELOPMENT, LAND MANAGEMENT.

ABSTRACT:
METHODS OF CORRECTING AND PREVENTING EUTROPHIC LAKES IN MAINE ARE COMPREHENSIVELY DISCUSSED IN THIS REPORT OF A CONFERENCE CONDUCTED AT THE UNIVERSITY OF MAINE. EUTROPHICATION WAS THE FIRST CONFERENCE TOPIC CONSIDERED; THE EUTROPHICATION PROCESS, EFFECTS, AND REMEDIES WERE DISCUSSED. PARTICULAR EMPHASIS WAS PLACED UPON LAKE FERTILIZATION, CULTURAL EUTROPHICATION AND ITS CAUSES, LAND DRAINAGE, AND WASTEWATER DISCHARGE. GENERALLY, THE TOPICS WERE EXAMINED BY MEANS OF PANEL DISCUSSIONS WHICH SOLICITED QUESTIONS FROM THE AUDIENCE. VARIOUS TECHNIQUES FOR CORRECTING EUTROPHICATION TO ACHIEVE INLAND LAKE RENEWAL WERE DISCUSSED; THE ADDRESS DESCRIBED THE PROBLEM, RECLAMATION PROJECTS, AND DESTRATIFICATION METHODS. LAND USE AND PLANNING CONSIDERATIONS IN CONTROLLING EUTROPHICATION WERE DESCRIBED, WITH EMPHASIS UPON AN INTEGRATED ZONING, PLANNED UNIT DEVELOPMENT, AND SANITATION PROGRAM TO PREVENT EUTROPHICATION. A HIGHLY TECHNICAL ADDRESS CONCERNING LAKE FLUSHING RATES FOR WATER QUALITY CONTROL WAS ALSO PRESENTED. METHODS OF SAMPLING LAKE WATER QUALITY WERE DISCUSSED, ALONG WITH THE FUNCTION OF THE MAINE INLAND FISH AND GAME DEPARTMENT IN WATER QUALITY CONTROL. THE CONCLUSION ADDRESSES THE ROLE OF LAW IN MAINTAINING LAKE WATER QUALITY. (SEE ALSO W72-04280) (HART-FLORIDA)

FIELD 05C, 06E, 05B, 06F, 05G

ACCESSION NO. W72-04279

EUTROPHICATION - THE PROBLEM - DETECTION - EXTENT,

MAINE UNIV., ORONO. WATER RESOURCES RESEARCH CENTER.

MILLARD W. HALL.

PROCEEDINGS, WORKSHOP--CONFERENCE ON RECLAMATION OF MAINE'S DYING LAKES, HELD AT UNIVERSITY OF MAINE, BANGOR, MARCH 24-25, 1971. P 5-13.

DESCRIPTORS:
*EUTROPHICATION, *LAKES, *MAINE, RECREATION, WATER SUPPLY, WATER POLLUTION EFFECTS, WATER POLLUTION SOURCES, ALGAE, NUTRIENTS.

IDENTIFIERS:
*LAKE ANNABESSACOOK(ME), *KENNEBEC RIVER(ME), CULTURAL EUTROPHICATION.

ABSTRACT:
NATURAL AND CULTURAL EUTROPHICATION ARE DEFINED AND STRESS IS PLACED ON THE LIMITED POSSIBILITIES AVAILABLE TO MAN IN HIS EFFORTS TO REDUCE THE ACCUMULATION OF NUTRIENTS IN WATER BASINS AND TO LIQUIDATE THE RESULTING BIOLOGICAL 'IMBALANCE.' PARTICULAR EMPHASIS IS PLACED ON THE FUTILITY OF ALGAL CONTROL BY COPPER SULFATE AND OTHER CHEMICALS. DIVERSION OF WASTEWATERS MAY PREVENT THE CULTURAL EUTROPHICATION OF CERTAIN LAKES, E.G., LAKE ANNABESSACOOK, BUT ONLY AT THE COST OF POLLUTING A RIVER. THE USE OF INOCULA OF PARASITIC VIRUSES MAY ACHIEVE CONTROL OF BLUE-GREEN ALGAE. (SEE ALSO W72-04279) (WILDE-WISCONSIN)

FIELD 05C

ACCESSION NO. W72-04280

ADENOSINE TRIPHOSPHATE IN LAKE SEDIMENTS,

WISCONSIN UNIV., MADISON. DEPT. OF SOIL SCIENCE.

C. C.LEE.

WISCONSIN WATER RESOURCES CENTER, MADISON, PH.D. THESIS, 1971. 98 P, 5 FIG, 14 TAB, 111 REF. OWRR B-022-WIS(7).

DESCRIPTORS:
*SOIL BACTERIA, *LAKE SOILS, *PHOSPHORUS, *EUTROPHICATION, LAKES, AQUATIC ALGAE, BACTERIA, BIOCHEMISTRY, WISCONSIN.

IDENTIFIERS:
*ADENOSINE TRIPHOSPHATE, *MICROBIAL BIOMASS, *GROWTH STAGE, *DETERMINATION, *ORIGIN, *SIGNIFICANCE, *EXTRACTION, LUCIFERIN-LUCIFERASE METHOD, SEDIMENT-WATER INTERACTIONS, BIOLUMINESCENCE, SELENASTRUM CAPRICONUTUM, AEROBACTER AEROGENES.

ABSTRACT:
A MODIFICATION OF THE LUCIFERIN-LUCIFERASE BIOLUMINESCENCE TECHNIQUE WAS DEVELOPED FOR DETERMINING ADENOSINE TRIPHOSPHATE (ATP) IN SEDIMENTS. ATP RECOVERY RANGED FROM 20 TO 85%, WAS A CHARACTERISTIC, REPRODUCIBLE PROPERTY OF A GIVEN SEDIMENT, BUT WAS NOT RELATED CONSISTENTLY TO ANY OTHER SEDIMENT PROPERTY. GENERALLY, THE DETECTION LIMITS OF THE METHOD WERE ABOUT 0.05 MICROGRAM ATP/G OVEN DRY SEDIMENT. THE ATP CONTENTS OF NINE SEDIMENT SAMPLES OBTAINED FROM DIFFERENT LAKES IN WISCONSIN RANGED FROM 0.34 TO 9.5 MICROGRAM ATP/G SEDIMENT. THE ATP CONTENT OF SELENASTRUM CAPRICONUTUM WAS EVALUATED IN RELATION TO GROWTH STAGE, SUBSTRATE P STATUS, AND THE APPLICABILITY OF ATP AS AN INDEX OF LIVING ALGAL CELL MATERIAL IN BIOASSAY SYSTEMS. GENERALLY, THE ATP CONTENT PER VIABLE CELL IN PAAP MEDIUM WAS MAINTAINED RELATIVELY CONSTANT WITHIN A 4 TO 6.5 X 10 TO THE MINUS 8TH POWER MICROGRAM/CELL RANGE OVER A 32-DAY INCUBATION PERIOD. VIABLE CELLS COMPRISED ONLY 40 TO 50% OF THE TOTAL CELLS FROM 8 TO 32 DAYS. ATP PER LIVING BIOMASS WAS 3 MICROGRAM/MG IN PAAP MEDIUM. IN P-DEFICIENT PAAP MEDIUM THE ATP CONTENT OF S. CAPRICONUTUM DECLINED TO ABOUT 2 X 10 TO THE MINUS 8TH POWER MICROGRAM/VIABLE CELL AND 1.5 MICROGRAM/MG LIVING BIOMASS. NEVERTHELESS, PRECISE CONVERSION OF ATP TO LIVING ALGAL CELL MATERIAL NECESSITATES CONSIDERATION OF THE EFFECT OF NUTRIENT, PARTICULARLY P, DEFICIENCIES ON THE ALGAL ATP POOL.

FIELD 05C, 02H

ACCESSION NO. W72-04292

ENZYMATIC PATTERNS IN TWO DENITRIFYING MICROBIAL POPULATIONS,

NATIONAL INST. FOR WATER RESEARCH, PRETORIA (SOUTH AFRICA).

P. J. DUTOIT, D. F. TOERIEN, AND T. R. DAVIES.

WATER RESEARCH, VOL 4, 1970, P 149-156. 1 FIG, 4 TAB, 23 REF.

DESCRIPTORS:
*WASTE WATER TREATMENT, *DENITRIFICATION, WATER REUSE, *NITROGEN, PHOSPHORUS, CARBON, *SEWAGE BACTERIA, SEWAGE, ALGAE, HYDROGEN ION CONCENTRATION, CARBON DIOXIDE, NITRATES, NITRITES, *ENZYMES.

IDENTIFIERS:
*GLUCOSE, *MALATE, CITRIC ACID CYCLE, CYTOCHROMES, PHOSPHORYLATION, CLYCOLYSIS.

ABSTRACT:
IN VIEW OF THE INCREASED ATTENTION GIVEN TO DENITRIFICATION AS A PROCESS IN WASTEWATER RENOVATION, IT WAS NECESSARY ALSO TO STUDY THE MICROBIOLOGY AND BIOCHEMISTRY OF THE MICROBIAL POPULATIONS ACTIVE IN THIS PROCESS. THE DENITRIFYING BACTERIA WERE OBTAINED FROM A LABORATORY-SCALE, DENITRIFYING UNIT RECEIVING THE SUPERNATANT FLUID OF SETTLED DOMESTIC SEWAGE. TO ENSURE THAT AN ACTIVELY DENITRIFYING POPULATION WOULD BE USED, THE ORIGINAL POPULATION WAS ENRICHED IN A SUITABLE MEDIUM. GLUCOSE AND MALATE WERE DEGRADED BY THE ENRICHED DENITRIFYING CULTURES WITH AN ACCOMPANYING DECREASE IN NITRATE CONTENT. NITRATE ACCUMULATED TOWARDS THE END OF THE EXPERIMENT BUT CONCENTRATIONS WERE LOW COMPARED TO THE AMOUNT OF NITRATE REMOVED IN BOTH CULTURES. BOTH CULTURES WERE STARTED FROM A SMALL INOCULUM OF ONLY 1 ML IN 6.1. WITHIN 36 HOURS THE CULTURES WERE TURBID AND GROWTH WAS PRONOUNCED AFTER 80 HOURS AT 30 DEG C. CONCLUSIONS WERE: (1) DENITRIFICATION TOOK PLACE WITH EITHER GLUCOSE OR MALATE AS THE SOURCE OF CARBON, (2) THE INTERMEDIARY METABOLIC PATTERNS OF THE TWO CULTURES WERE DIFFERENT, (3) GLUCOSE DEGRADATION OCCURRED VIA GLYCOLYSIS, THE PENTOSE PHOSPHATE SHUNT AND THE CITRIC ACID CYCLE, (4) DEGRADATION OF MALATE OCCURRED MAINLY THROUGH THE CITRIC ACID CYCLE, (5) CYTOCHROMES WERE MORE ABUNDANT IN THE MALATE CULTURES THAN IN THE GLUCOSE CULTURE. SUBSTRATE PHOSPHORYLATIONS AS WELL AS OXIDATIVE PHOSPHORYLATION APPEARED TO OCCUR IN THE GLUCOSE CULTURE, BUT ONLY OXIDATIVE PHOSPHORYLATION WAS DETECTED IN THE MALATE CULTURES. (BIGGS-TEXAS)

FIELD 05D

ACCESSION NO. W72-04309

CAN BOD CONTRIBUTE TO ALGAL MASS,

WASHINGTON STATE UNIV., PULLMAN.

NING-HSI TANG, AND SURINDER K. BHAGAT.

WATER AND SEWAGE WORKS, VOL 118, NO 12, P 396-401, DECEMBER 1971. 9 FIG, 3 TAB, 4 REF.

DESCRIPTORS:
*ALGAE, *BIOCHEMICAL OXYGEN DEMAND, *EUTROPHICATION, PHOSPHOROUS, NITROGEN, CHEMICAL OXYGEN DEMAND, LABORATORY TESTS, ALKALINITY, CARBONATES, BICARBONATES, CARBON DIOXIDE, HYDROGEN ION CONCENTRATION, WATER QUALITY CONTROL, WASTE WATER TREATMENT, CHLOROPHYTA, GROWTH RATES.

IDENTIFIERS:
*ORGANIC CARBON, *ALGAL MASS, ALGAL GROWTH.

ABSTRACT:
THE EFFECT OF ORGANIC CARBON ON THE GROWTH OF ALGAE WAS DEMONSTRATED USING GLUCOSE AS A CARBON SOURCE AND COD AS A MEASURE OF CARBON PRESENT. ALGAE DERIVE CARBON FROM THREE SOURCES, CO2 FROM THE ATMOSPHERE, FROM CARBONATES AND BICARBONATES, AND FROM ORGANIC MATERIAL IN THE WATER. ON A SEMICONTINUOUS FLOW BASIS, ONCE THE ALGAL-BACTERIAL SYMBIOSIS WAS ESTABLISHED, MORE ALGAE GREW IN THE FLASKS WITH ORGANIC CARBON THAN GREW IN THE FLASKS WITHOUT ORGANIC CARBON, EVEN THOUGH ALL FLASKS WERE FED WITH EQUAL, ADEQUATE AMOUNTS OF N AND P. RESULTS INDICATE THAT THE AMOUNT OF EXCESSIVE ALGAL GROWTH IN AN AQUATIC SYSTEM WILL BE ABOUT THE SAME AS THE AMOUNT OF INFLUENT BOD FOR A GIVEN WASTEWATER INFLUENT BOD OF LESS THAN 50 MG/L. FOR AN INFLUENT WASTEWATER BOD IN EXCESS OF 50 MG/L THE AMOUNT OF EXCESSIVE ALGAL GROWTH WILL BE SOMEWHAT LESS THAN THE INCOMING BOD. A GRAPH WAS CONSTRUCTED TO SERVE AS A GUIDE FOR PREDICTING THE AMOUNT OF EXCESSIVE ALGAL GROWTH FOR A GIVEN EFFLUENT BOD. (LOWRY-TEXAS)

FIELD 05C

ACCESSION NO. W72-04431

EVALUATION OF ALGAL ASSAY PROCEDURES--PAAP BATCH TEST,

CALIFORNIA UNIV., IRVINE. SCHOOL OF BIOLOGICAL SCIENCES; AND CALIFORNIA UNIV., IRVINE. SCHOOL OF ENGINEERING.

STEVEN MURRAY, JAN SCHERFIG, AND PETER S. DIXON.

JOURNAL WATER POLLUTION CONTROL FEDERATION, VOL 43, NO 10, P 1991-2003, OCTOBER 1971. 4 FIG, 13 TAB, 3 REF.

DESCRIPTORS:
*BIOASSAY, *LABORATORY TESTS, RESEARCH AND DEVELOPMENT, TRACE ELEMENTS, ALGAE, NUTRIENTS, LIGHT INTENSITY, HYDROGEN ION CONCENTRATION, CARBON DIOXIDE, AERATION, MIXING, *EUTROPHICATION, *ANALYTICAL TECHNIQUES, *ALGAL CONTROL.

IDENTIFIERS:
SELENASTRUM CAPRICORNUTUM.

ABSTRACT:
PRELIMINARY INVESTIGATIONS HAVE DEMONSTRATED FOUR IMPORTANT DIFFICULTIES INVOLVED IN THE PROVISIONAL ALGAL ASSAY PROCEDURE (PAAP) BATCH TEST. THE FOUR DIFFICULTIES WERE: (1) THE EFFECTS OF MEDIUM PREPARATION ON THE FINAL COMPOSITION OF THE MEDIUM; (2) PH CHANGES DURING THE TEST PERIOD IN RELATION TO CARBON DIOXIDE AVAILABILITY; (3) THE METHOD OF GAS ADDITION IN THE BATCH TEST UNITS; AND (4) THE EFFECTS OF LIGHT INTENSITY ON ALGAL GROWTH. BECAUSE OF THESE LIMITATIONS, CERTAIN MODIFICATIONS TO THE STANDARDIZED PROCEDURES WERE DEVELOPED. A NEW CULTURE MEDIA PREPARATION WHICH ELIMINATES THE PRECIPITATION OF ESSENTIAL TRACE ELEMENTS WAS DEVELOPED AND PROVIDES A RELIABLE REFERENCE MEDIUM FOR ALGAL ASSAY WORK. THE ADDITION OF CO2 -ENRICHED AIR REDUCED THE GROWTH-LIMITING EFFECTS OF CO2 AND STABILIZES THE PH AT 7.5 TO 8.0. AERATION WAS SHOWN TO BE MORE EFFECTIVE THAN VENTILATION FOR CO2 ADDITION AND GASEOUS AS A RESULT OF BICARBONATE UTILIZATION. LIGHT INTENSITIES OF 500 FT-C REDUCED GROWTH RATES OF SELENASTRUM CAPRICORNUTUM IN BOTTLE TEST EXPERIMENTATION COMPARED WITH 350 FT-C. (LOWRY-TEXAS)

FIELD 05C, 05G

ACCESSION NO. W72-04433

ECOLOGICAL ASPECTS OF PLUTONIUM DISSEMINATION IN AQUATIC ENVIRONMENTS; WHAT HAS PU DATA TO TELL US ABOUT OTHER TRANSURANICS,

WOODS HOLE OCEANOGRAPHIC INSTITUTION, MASS.

VICTOR E. NOSHKIN.

AVAILABLE FROM THE NATIONAL TECHNICAL INFORMATION SERVICE, SPRINGFIELD, VA. 22151 AS NYO-2174-132, $3.00 IN PAPER COPY, $0.95 IN MICROFICHE. REPORT NYO-2174-132, 1971. 31 P, 2 FIG, 8 TAB, 40 REF. AEC CONTRACT AT(30-1)-2174.

DESCRIPTORS:
*RADIOACTIVITY EFFECTS, *FALLOUT, *FOOD CHAINS, *AQUATIC LIFE, RADIOISOTOPES, ABSORPTION, NUCLEAR WASTES, WATER POLLUTION EFFECTS, MOLLUSKS, MARINE PLANTS, MARINE ALGAE, PUBLIC HEALTH, ESTUARINE ENVIRONMENT, ON-SITE INVESTIGATIONS, REVIEWS.

IDENTIFIERS:
PLUTONIUM.

ABSTRACT:
THIS REVIEW INDICATES THAT FALLOUT PLUTONIUM CONTRIBUTES MORE THAN STRONTIUM OR CESIUM TO THE ARTIFICIAL RADIATION EXPOSURE OF MANY MARINE SPECIES. ORGANISMS AT HIGHER TROPHIC LEVELS AND THOSE WHICH FEED ON SEDIMENTS HAVE MORE PLUTONIUM. BONE AND LIVER HAVE MUCH MORE THAN MUSCLE TISSUE. AS USE OF NUCLEAR POWER INCREASES, SUCH FOOD ITEMS AS ANCHOVIES, SARDINES, CANNED SALMON, ETC. WHICH ARE CONSUMED WHOLE, AND FISH PROTEIN CONCENTRATES WILL REPRESENT IMPORTANT TRANSFER VECTORS FOR PLUTONIUM. THE RELATIVELY UNIFORM PLUTONIUM UPTAKE BY THE MUSSEL (CONCENTRATION FACTOR OF 200-500) SUGGESTS USE AS A BIOLOGICAL INDICATOR. THE MUSSEL INDICATES THE PLUTONIUM MORE RECENTLY INTRODUCED INTO THE WATER IN SUSPENSION RATHER THAN THAT PRESENT IN SEDIMENTS. (BOPP-ORNL)

FIELD 05C

ACCESSION NO. W72-04461

INFLUENCE OF THE PHYSICO-CHEMICAL FORMS OF RADIONUCLIDES AND STABLE TRACE ELEMENTS IN SEAWATER IN RELATION TO UPTAKE BY THE MARINE BIOSPHERE,

BATTELLE-NORTHWEST, RICHLAND, WASH. PACIFIC NORTHWEST LAB.

D. E, ROBERTSON.

AVAILABLE FROM THE NATIONAL TECHNICAL INFORMATION SERVICE, SPRINGFIELD, VA. 22151, AS BNWL-SA-4048, $3.00 IN PAPER COPY, $0.95 IN MICROFICHE. BNWL-SA-4048, SEPTEMBER 8, 1971. 60 P, 105 REF. AEC CONTRACT NO. AT(45-1)-1830.

DESCRIPTORS:
*RADIOACTIVITY EFFECTS, *NUCLEAR WASTES, *FALLOUT, *OCEANS, WATER POLLUTION EFFECTS, FOOD CHAINES, PUBLIC HEALTH, ESTUARINE ENVIRONMENT, ZINC RADIOISOTOPES, PHYSICOCHEMICAL PROPERTIES, FISH, MARINE PLANTS, MARINE ALGAE, HYDROLYSIS, CHELATION, SEDIMENTS, SETTLING VELOCITY, RADIOISOTOPES, ABSORPTION, REVIEWS.

IDENTIFIERS:
PLUTONIUM RADIOISOTOPES, RUTHENIUM RADIOISOTOPES, IRON RADIOISOTOPES.

ABSTRACT:
THIS REVIEW PRESENTS EXAMPLES OF DISSIMILAR BIOLOGICAL UPTAKE BETWEEN IMPORTANT ARTIFICIAL RADIONUCLIDES (FE-55, ZN-65, AND RU-106) AND THEIR STABLE ISOTOPES IN THE MARINE ENVIRONMENT. ALTHOUGH DEFINITIVE WORK IS NEEDED, THERE IS EVIDENCE THAT ZN, FE, AND CU ARE PRESENT AS SOLUBLE METAL CHELATES. RU, FE, AND MN HYDROLYZE TO FORM POLYMERIC SPECIES. MOST PU IS IN PARTICULATE FORM. THE RELATIVELY HIGH UPTAKE OF PU BY SEAWEEDS AND CERTAIN CRUSTACEANS APPEARS TO BE A SURFACE ABSORPTION PROCESS. THE MAXIMUM PERMISSIBLE CONCENTRATION OF PU IN SEAWATER IS ABOUT 6 X 0.01 MICROCI/ML, BASED UPON A SINGLE INTAKE OF 200 GMS OF FISH PER INDIVIDUAL. (BOPP-ORNL)

FIELD 05C

ACCESSION NO. W72-04463

REMOVAL OF TRACE METALS FROM MARINE CULTURE MEDIA,

NATIONAL MARINE WATER QUALITY LAB., WEST KINGSTON, R.I.

EARL W. DAVEY, JOHN H. GENTILE, STANTON J. ERICKSON, AND PETER BETZER.

LIMNOLOGY AND OCEANOGRAPHY, VOL 15, NO 3, P 486-488, MAY 1970. 1 FIG, 5 REF.

DESCRIPTORS:
*CHELATION, CHROMATOGRAPHY, HEAVY METALS, *TRACE ELEMENTS, RADIOCHEMICAL ANALYSIS, RESINS, *SEA WATER, BIOASSAY, COPEPODS, OYSTERS, DINOFLAGELLATES, DIATOMS, ALGAE CULTURES, COPPER, TOXICITY.

IDENTIFIERS:
CHELEX 100, ARTEMIA SALINA, BIO-RAD, CYCLOTELLA NANA.

ABSTRACT:
THE SODIUM FORM OF PURIFIED CHELEX-100 CAN BE USED TO REMOVE TRACE METALS FROM ARTIFICIAL AND NATURAL SEAWATER WITHOUT ALTERING THE MAJOR CATION-ANION COMPOSITION OR CONTRIBUTING ORGANIC TOXICANTS OR CHELATORS. SPECIFIC CONTAMINANT TRACE METALS, OFTEN PRESENT IN ARTIFICIAL SEAWATER AT TOXIC LEVELS, ARE REDUCED TO LEVELS THAT PERMIT THE GROWTH OF SENSITIVE MARINE PHYTOPLANKTON AND THE DEVELOPMENT OF CERTAIN INVERTEBRATES FROM THE EGG THROUGH EARLY LARVAL STAGES. THE CONCENTRATIONS OF ESSENTIAL TRACE METALS ARE REDUCED BELOW THE NUTRITIONAL REQUIREMENTS OF SELECTED SPECIES OF MARINE PHYTOPLANKTON. (EPA ABSTRACT)

FIELD 05A

ACCESSION NO. W72-04708

STRATEGIES FOR CONTROL OF MAN-MADE EUTROPHICATION,

COMMITTEE ON PUBLIC WORKS (U.S. SENATE).

R. D. GRUNDY.

ENVIRONMENTAL SCIENCE AND TECHNOLOGY, VOL 5, NO 12, P 1184-1190, DECEMBER 1971. 6 FIG, 5 REF.

DESCRIPTORS:
*EUTROPHICATION, *PHOSPHATES, *WATER POLLUTION CONTROL, *WASTE WATER TREATMENT, *COST-BENEFIT ANALYSIS, DETERGENTS, PHOSPHORUS, NITROGEN, NUTRIENTS, SEWAGE, WATER TEMPERATURE, LIGHT PENETRATION, CARBON, LAKE ERIE, RUNOFF, EROSION, CHLORELLA, SODIUM COMPOUNDS, ALKALINITY, MINING, BIOCHEMICAL OXYGEN DEMAND, ECONOMICS, ALGAE, AGRICULTURAL CHEMICALS, LEGISLATION.

IDENTIFIERS:
POTOMAC RIVER, ALAFIA RIVER, CHLORELLA PYRENOIDOSA, SODIUM SILICATE, SODIUM METASILICATE, LAKE WASHINGTON, NITRILOTRIACETATE, HUMAN FECES.

ABSTRACT:
SOURCES OF PHOSPHATES, AND OTHER NUTRIENTS IN THE AQUATIC ECOSYSTEM INCLUDE NOT ONLY DETERGENTS, BUT ALSO SEWAGE, EROSION, AND AGRICULTURAL RUNOFF. ON A NATIONAL BASIS, DETERGENTS PROVIDE 30 TO 40 PERCENT OF ALL THE PHOSPHORUS ENTERING THE AQUATIC ENVIRONMENT. HOWEVER, THIS FIGURE IS SHOWN TO VARY ON A REGIONAL BASIS. BECAUSE THERE ARE MANY OTHER SOURCES OF PHOSPHATES, THEIR CONTROL IN DETERGENTS IS NOT IN ITSELF A SUFFICIENT STRATEGY TO CONTROL EUTROPHICATION. ALSO, ANY REGULATION ON PHOSPHATES IN DETERGENTS SHOULD NOT BE UNDERTAKEN WITHOUT CAREFUL CONSIDERATION OF THE PUBLIC HEALTH AND ENVIRONMENTAL IMPLICATIONS OF ALTERNATIVE FORMULATIONS. VALID CONTROL STRATEGIES SHOULD INCLUDE ADVANCED WASTE WATER TREATMENTS, DIVERSION, DILUTION, AND LAND DISPOSAL. WASTE WATER TREATMENT USING CHEMICAL PRECIPITATION PROCESSES, NOT ONLY REMOVES PHOSPHORUS BUT ALSO FACILITATES THE REMOVAL OF BOD, TOXICANTS, AND OTHER NUTRIENTS. ECONOMIC CONSIDERATIONS SHOW THAT 90 PERCENT OF MUNICIPAL WASTE WATER COULD BE TREATED FOR PHOSPHORUS REMOVAL AT LESS COST TO THE CONSUMER THAN THE INCREASED PRODUCT COSTS FOR DETERGENT PHOSPHATE SUBSTITUTES. IT IS ALSO SIGNIFICANT THAT TREATMENT REDUCES TOTAL PHOSPHATE LEVELS WHILE PRODUCT CONTROLS AFFECT DETERGENT PHOSPHATE ALONE. (MORTLAND-BATTELLE)

FIELD 05D, 05G, 05C

ACCESSION NO. W72-04734

DISTRIBUTION OF BLUE-GREEN ALGAL VIRUSES IN VARIOUS TYPES OF NATURAL WATERS,

DELAWARE UNIV., NEWARK. DEPT. OF BIOLOGICAL SCIENCES.

M. S. SHANE.

WATER RESEARCH, VOL 5, NO 9, P 711-716, SEPTEMBER 1971. 1 FIG, 3 TAB, 10 REF.

DESCRIPTORS:
*VIRUSES, *CYANOPHYTA, *DISTRIBUTION PATTERNS, *WATER ANALYSIS, *BIOINDICATORS, ALGAE, RIVERS, STREAMS, DELAWARE RIVER, OXIDATION LAGOONS, PONDS, FARM PONDS, MARYLAND, QUARRIES, OHIO RIVER, HUDSON RIVER, INDICATORS, DELAWARE, PATH CF PCLLUTANTS, HCSTS, PENNSYLVANIA, NEW YORK, NEW HAMPSHIRE, OHIO, MICHIGAN.

IDENTIFIERS:
*LPP VIRUS, *BLUE-GREEN ALGAL VIRUSES, *PLECTONEMA BCRYANIUM, SUSQUEHANNA RIVER, BRANDYWINE RIVER, RED CLAY RIVER, WHITE CLAY RIVER, CHRISTINA RIVER, BIG ELK RIVER, LITTLE ELK RIVER, CCTORARA RIVER, PLECTONEMA, LYNGBYA, PHORMIDIUM, CYANOPHYCEAE, OSCILLATORIACEAE, SCHIZOTHRIX CALCICOLA, ELKTON POND, RISING SUN PCND, INDUSTRIAL STORAGE TANKS.

ABSTRACT:
A SURVEY WAS CONDUCTED IN THE DELAWARE-MARYLAND AREA TO ASCERTAIN THE DISTRIBUTION OF LPP BLUE-GREEN ALGAL VIRUSES IN ALL TYPES OF NATURAL WATERS. THEIR PRESENCE WAS DETERMINED ON THE BASIS OF THE ABILITY OF SAMPLES TO PRODUCE LYSIS IN CULTURES OF THE ALGAL TEST ORGANISM, PLECTONEMA BORYANUM (UI581), AND TC FORM PLAQUES CN THE HOST ALGAL LAWN USING THE SOFT-AGAR TECHNIQUE OF SAFFERMAN AND MCRRIS. THE VIRAL CONCENTRATION WAS DETERMINED BY THE DIRECT COUNT CF THE NUMBER OF PLAQUES FORMED PER ML OF WATER SAMPLE. THE VIRAL STRAIN, LPP-1, WAS DETERMINED FOR SELECTED VIRAL CULTURES ISOLATED FRCM THE CHRISTINA, RED CLAY, AND WHITE CLAY RIVERS AND THE ELKTON AND RISING SUN CXIDATION PONDS. LPP-VIRUSES EXISTED IN EVERY TYPE OF WATER. THE OXIDATION PONDS GAVE THE HIGHEST PERCENTAGE OF SAMPLES CONTAINING AT LEAST ONE VIRUS AND THE HIGHEST CONCENTRATION OF VIRUSES. ALL RESERVOIRS AND INDUSTRIAL STORAGE TANKS WHICH CONTAINED VIRUSES HAC INFLUENTS WHICH ALSO CONTAINED VIRUSES. TEST DATA ON THE RIVERS SHOWEC THESE VIRUSES PRESENT IN MOST STREAMS. THE VIRUSES WERE CONSISTENTLY ABSENT FROM THE HEADWATERS OF THE THREE RIVERS LISTED ABOVE BUT INCREASED AS THESE RIVERS FLOWED THROUGH MORE POPULATED AREAS. THE HEADWATERS FOR ALL THREE ARE IN LOW POPULATICN DENSITY AND RURAL ENVIRONMENTS. IT IS SUGGESTED THAT THE ALGAL HCSTS OF THESE VIRUSES THRIVE IN AREAS CF HIGH ORGANIC MATERIAL SUCH AS MAY BE ASSOCIATED WITH PCLLUTION. TABLES SHOW THE DISTRIBUTION AND INCIDENCE OF THE LPP-VIRUSES IN THE WATERS. (HOLOMAN-BATTELLE)¬28(U)

FIELD 05B, 05C

ACCESSION NO. W72-04736

EARTH RESOURCE TECHNOLOGY USED IN POLLUTION DETECTION,

B. M. ELSON.

AVIATION WEEK AND SPACE TECHNOLOGY, VOL 95, NO 24, P 46-49, DECEMBER 13, 1971.

DESCRIPTORS:
*REMOTE SENSING, *WATER POLLUTION CONTROL, *SEA WATER, *OILS, *NUTRIENTS, AIRCRAFT, SATELLITES(ARTIFICIAL), PHOSPHORUS, NITROGEN, HEAVY METALS, ORGANOPHOSPHORUS PESTICIDES, INDUSTRIAL WASTES, VIRUSES, ENERIC BACTERIA, TRITIUM, RADIOACTIVE WASTES, MONITORING, PHOTOGRAPHY, ULTRAVIOLET RADIATION, FLUORESCENCE, INFRARED RADIATION, RADAR, CHLOROPHYLL, WATER PROPERTIES, ALGAE.

IDENTIFIERS:
SPACE PLATFORMS, LEAD, MERCURY, LASERS, MICROWAVE RADIOMETRY, VIDEO, SCANNERS.

ABSTRACT:
VARIOUS REMOTE SENSING DEVICES, CARRIED IN AIRCRAFT OR SATELLITES, ARE BEING DEVELOPED TO MONITOR OCEAN POLLUTANTS. PRIMARY EMPHASIS IN THIS PAPER IS ON THE DETECTION, IDENTIFICATION,MAPPING, AND VOLUMETRIC MEASUREMENT OF OIL SPILLS. ONE MEANS OF DETECTING SOME KINDS OF OILS ON WATER IS PHOTOGRAPHY IN THE BLUE AND ULTRAVIOLET REGIONS. FLUORESCENCE UNDER ULTRAVIOLET LIGHT IS ANOTHER MEANS. OIL SPILL BOUNDARIES HAVE BEEN ACCURATELY DEFINED USING INFRARED SCANNERS, BUT CERTAIN ANOMALIES ARE ALSO OCCURRING WHICH LIMIT THE USEFULNESS OF THIS TECHNIQUE. MICROWAVE RADIOMETRY WAS USED TO MEASURE OIL SPILLS OFF THE SOUTHERN CALIFORNIA COAST. THE NAVY HAS USED SYNTHETIC APERTURE RADAR TO PERFORM WIDE-AREA SURVEILLANCE OVER OIL SPILLS. MANY POLLUTANTS OTHER THAN OIL ALSO CAN BE REMOTELY DETECTED. A MULTISPECTRAL RADIOMETRIC SCANNER CAN BE USED TO DETERMINE OPTICAL AND THERMAL PROPERTIES OF WATER AND TO MONITOR ALGAL GROWTHS AND SHORELINE VEGETATION. AN AIRBORNE SYSTEM FOR DETERMINING CHLOROPHYLL CONTENT OF WATER HAS BEEN DEVELOPED. THIS TECHNIQUE UTILIZES A DIFFERENTIAL RADIOMETER THAT MEASURES CHANGES IN SPECTRAL RADIANCE BETWEEN TWO OR MORE WAVELENGTH REGIONS OF THE SPECTRUM. (MORTLAND-BATTELLE)

FIELD 05A, 07B

ACCESSION NO. W72-04742

ISOLATION AND COUNTING OF ATHIORHODACEAE WITH MEMBRANE FILTERS,

PENNSYLVANIA STATE UNIV., UNIVERSITY PARK. DEPT. OF MICROBIOLOGY.

W. C. SWOAGER, AND E. S. LINSTROM.

APPLIED MICROBIOLOGY, VOL 22, NO 4, P 683-687, OCTOBER 1971. 3 TAB, 13 REF.

DESCRIPTORS:
*CULTURES, *ANAEROBIC BACTERIA, *ISOLATION, *MEMBRANE PROCESSES, BACTERIA, SULFUR BACTERIA, PHOTOSYNTHETIC BACTERIA, AQUATIC BACTERIA, LAKES, FILTERS, COLIFORMS, METABOLISM, CARBON, SULFUR, SAMPLING, WATER ANALYSIS, ALGAE, SPECTROPHOTOMETRY, PENNSYLVANIA, INHIBITORS, ECOLOGY, ECOTYPES, FRESH WATER.

IDENTIFIERS:
*ATHIORHODACEAE, WHIPPLE DAM, STONE VALLEY DAM(PA.), ATRAZINE, BIOTYPES, POUR PLATE METHOD, MOST PROBABLE-NUMBER TEST, MEMBRANE FILTRATION, RHODOPSEUDOMONAS GELATINOSA, THIORHODACEAE, CHLOROBACTERIACEAE.

ABSTRACT:
A MEMBRANE FILTER TECHNIQUE WAS DEVELOPED TO GROW, ISOLATE, AND ENUMERATE ATHIROHODACEAE FROM AQUATIC ECOSYSTEMS. WATER SAMPLES WERE TAKEN FROM THE SPILLWAYS OF TWO CENTRAL PENNSYLVANIA ARTIFICIAL LAKES DURING SPRING, SUMMER, AND FALL OF 1969. UPPER LAYER AND DEPTH SAMPLES WERE TAKEN AT 1-M INTERVALS FROM STONE VALLEY DAM IN MAY, JUNE, AND JULY 1970. THE SAMPLES WERE FILTERED THROUGH A 47-MM HA MEMBRANE FILTER AND THE FILTRATE CULTURED ON PERFORATED PETRI DISHES IN A CARBON-BASED MEDIUM TO PROMOTE GROWTH OF ATHIORHODACEAE. ATRAZINE WAS ADDED TO THE BASAL CULTURE MEDIUM TO INHIBIT THE GROWTH OF ALGAE. ABSORPTION SPECTRA WERE PREPARED USING A BECKMAN DK-2A SPECTROPHOTOMETER. PURE CULTURES WERE ISOLATED BY REPEATED STREAKING ON PLATES INCUBATED ANAEROBICALLY IN 200-400 FT-C OF INCANDESCENT LIGHT. IDENTIFICATION OF THE ISOLATES AS ATHIORHODACEAE WAS CORROBORATED BY THE FACT THAT THEY USED NO SULFIDE OR THIOSULFATE IN CULTURE MEDIA AND ALL (40) HAD SPECTRAL CHARACTERISTICS OF ATHIORHODACEAE. QUANTITATIVE DATA AND A WIDE RANGE OF BIOTYPES CAN BE OBTAINED BY THE CULTURING METHOD DESCRIBED. ACCURACY WAS VALIDATED BY PARALLEL COUNTS USING THE POUR PLATE AND MOST PROBABLE NUMBERS METHODS. BY USING A SULFUR-BASED MEDIUM AND MORE ANAEROBIC CONDITIONS, THE METHOD COULD PROBABLY BE USED TO COUNT AND ISOLATE SULFUR PHOTOSYNTHETIC BACTERIA SUCH AS CHLOROBACTERIACEAE AND THIROHADACEAE. (JEFFERIS-BATTELLE)

FIELD 05A

ACCESSION NO. W72-04743

CULTURAL ESTIMATION OF YEASTS ON SEAWEEDS,

RHODE ISLAND UNIV., KINGSTON. NARRAGANSETT MARINE LAB.

R. SESHADRI, AND J. M. SIEBURTH.

APPLIED MICROBIOLOGY, VOL 22, NO 4, P 507-512, OCTOBER 1971. 6 TAB, 31 FIG.

DESCRIPTORS:
*YEASTS, *CULTURES, *MARINE ALGAE, RHODOPHYTA, PHAEOPHYTA, CHLOROPHYTA, INHIBITORS, ANTIBIOTICS, SEA WATER, RHODE ISLAND.

IDENTIFIERS:
*AGARS, EPIPHYTES, CANDIDA PARAPSILOSIS, CHONDRUS CRISPUS, RHODOTORULA GLUTINIS, SARGASSUM NATANS, ASCOPHYLLUM NODOSUM, FUCUS VESICULOSUS, LAMINARIA DIGITATA, LAMINARIA AGARDHII, ULVA LACTUCA, RHODYMENIS PALMATA, POLYSIPHONIA LANOSA, POLYSIPHONIA HARVEYI, RHODOTORULA RUBRA, RHODOTORULA LACTOSA, CANDIDA ZEYLANOIDES, CAMP VARNUM BEACH.

ABSTRACT:
FIVE PERCENT SUSPENSIONS OF FRESHLY HARVESTED SEAWEEDS WERE USED AS AN INOCULUM TO DEVELOP A SELECTIVE MEDIUM FOR EPIPHYTIC YEASTS. CONDITIONS FOR SATISFACTORY YEAST GROWTH AND VISUALIZATION AS RED COLONIES ON MEMBRANE FILTERS WERE OBTAINED BY SUPPLEMENTING A BASAL GLUCOSE-TRYPTICASE-YEAST EXTRACT-AGAR AT PH 7.0 WITH 100 MG EACH OF CHLORAMPHENICOL AND 2,3,5-TRIPHENYL TETRAZOLIUM CHLORIDE PER LITER. MAXIMAL COUNTS WERE OBTAINED BY TRITURATING THE ALGAE IN PRECHILLED (4 C) SEAWATER WITH A BLENDER FOR 2 TO 5 MIN. INHIBITORY PHENOLIC MATERIALS RELEASED FROM PHAEOPHYTES DURING THIS PROCESS WERE REMOVED WITH A MODIFIED CHOLODNY FILTRATION. A PRELIMINARY SURVEY INDICATED THAT YEASTS WERE EPIPHYTIC ON ALL NINE SPECIES OF SEAWEEDS AND THAT MAXIMAL POPULATIONS OCCURRED ON THE CHLOROPHYTES AND RHODOPHYTES ESPECIALLY DURING THE PERIODS OF WARMER WATER. (HOLOMAN-BATTELLE)

FIELD 05A

ACCESSION NO. W72-04748

EUTROPHICATION OF SMALL RESERVOIRS IN THE GREAT PLAINS,

NEBRASKA UNIV., LINCOLN. DEPT. OF CIVIL ENGINEERING; AND NEBRASKA UNIV., LINCOLN. DEPT. OF ZOOLOGY.

MARK J. HAMMER, AND GARY L. HERGENRADER.

NEBRASKA ENGINEER, JUNE 1971. 5 P, 3 FIG, 1 TAB. OWRR A-014-NEB(4).

DESCRIPTORS:
*EUTROPHICATION, *RESERVOIRS, *GREAT PLAINS, NUTRIENTS, RECREATION, LIGHT PENETRATION, NEBRASKA, WASTEWATER TREATMENT, TURBIDITY, RUNOFF, PHOSPHORUS, ALGAE, AQUATIC PLANTS, CLAYS, ODOR, WATER QUALITY, FOOD CHAINS, DISSOLVED OXYGEN, COPPER SULFATE, HERBICIDES, WATER POLLUTION EFFECTS, WATER POLLUTION SOURCES, WATER POLLUTION CONTROL.

IDENTIFIERS:
SALT VALLEY RESERVOIRS(NEB), MADISON(WIS), LAKE WASHINGTON(WASH), DYEING WATER.

ABSTRACT:
RESERVOIRS BUILT WITHIN A 20-MILE RADIUS OF LINCOLN, NEBRASKA, FOR FLOOD CONTROL AND SOIL CONSERVATION WITH RECREATION AS A SECONDARY BENEFIT HAVE DETERIORATED DUE TO EUTROPHICATION. THE RUNOFF WATERS ENTERING ARE PRINCIPALLY FROM CULTIVATED FARMLAND AND CONTAIN SUFFICIENT NUTRIENT SALTS TO SUPPORT ABUNDANT GROWTHS OF AQUATIC PLANTS. THERE IS NO READY SOLUTION FOR REMOVAL OF NUTRIENTS FROM LAND RUNOFF. IT IS DOUBTFUL THAT SOIL AND WATER CONSERVATION PRACTICES CAN REDUCE THE NUTRIENT LEVELS IN RUNOFF SUFFICIENTLY TO PREVENT EUTROPHICATION. THE RAINFALL-RUNOFF PATTERNS OF TYPICAL NEBRASKA WEATHER MAKE NUTRIENT CONTROL IN DRAINAGE WATER IMPRACTICAL. USE OF COPPER SULFATE FOR ALGAL BLOOMS HAS BEEN ABANDONED DUE TO ITS TOXICITY IN BOTTOM MUDS TO AQUATIC LIFE; HERBICIDES MUST BE USED WITH CAUTION TO PREVENT UNWANTED BIOLOGICAL DAMAGE. PLANT GROWTH HARVESTING IN MOST SITUATIONS CAN REMOVE ONLY SMALL QUANTITIES OF NUTRIENTS. A NEW AND NOVEL APPROACH TO CONTROLLING WEED AND ALGAE GROWTH IS BEING EVALUATED--INHIBITING SUNLIGHT PENETRATION THROUGH APPLICATION OF CHEMICAL DYES INTO THE WATER OR ONTO THE SURFACE. COMMERCIALLY AVAILABLE DYES ARE BEING TESTED WITH CULTURES OF BLUE-GREEN ALGAE, COMMONLY BLOOMING IN THE SALT VALLEY RESERVOIRS. (JONES-WISCONSIN)

FIELD 05C, 02H

ACCESSION NO. W72-04759

THE DISTRIBUTION AND NET PRODUCTIVITY OF SUBLITTORAL POPULATICNS OF ATTACHED MACROPHYTIC ALGAE IN AN ESTUARY ON THE ATLANTIC COAST OF SPAIN,

DURHAM UNIV. (ENGLAND). DEPT. OF BOTANY.

D. M. JOHN.

MARINE BIOLOGY, VOL 11, NO 1, P 90-97, 1971. 3 FIG, 3 TAB, 39 REF.

DESCRIPTORS:
*AQUATIC PLANTS, *DISTRIBUTION, *MARINE ALGAE, *ESTUARIES, PRODUCTIVITY, POPULATION, ATLANTIC OCEAN, DEPTH, STANDING CROP, CURRENTS(WATER), WAVES(WATER), SAMPLING, BENTHOS, GRAZING, GASTROPODS, LITTORAL, AGE, BIOMASS, LIGHT QUALITY, LIGHT PENETRATION, EFFICIENCES, COMPETITION, GROWTH RATES, PLANT MORPHOLOGY, WATER CIRCULATION.

IDENTIFIERS:
*SUBLITTORAL BENTHIC MACROPHYTES, RIA DE ALDAN(SPAIN), LAMINARIA OCHROLEUCA, SACCORHIZA POLYSCHIDES.

ABSTRACT:
DISTRIBUTION OF SUBLITTORAL POPULATIONS OF LAMINARIA OCHROLEUCA AND SACCORHIZA POLYSCHIDES WAS STUDIED AT SEVEN LOCALITIES ON THE NORTHWEST COAST OF SPAIN. POSSIBLE FACTORS CONTROLLING THE EQUILIBRIUM BETWEEN THE TWO POPULATIONS WERE ASSESSED AND THE NET PRODUCTIVITY OF THESE POPULATIONS MEASURED BY IN SITU CROPPING USING SCUBA DIVING TECHNIQUES. A NARROW TRANSITION ZONE BETWEEN THE TWO POPULATIONS WAS FOUND AND THE DEPTH OF THIS ZONE WAS GOVERNED BY AMOUNT OF WATER MOVEMENT. TOTAL STANDING CROP AND PRODUCTIVITY PER UNIT AREA DECREASED WITH BOTH AN INCREASE IN DEPTH AND WAVE ACTION WHILE THE HIGHEST VALUES OF ALL WERE IN TWO LOCALITIES WHERE THERE WAS CONSIDERABLE CURRENT SURGE. THE MAXIMUM PRODUCTIVITY OF L OCHROLEUCA WAS FOUND IN THE MOST SHELTERED LOCALITY, WHILE THE HIGHEST VALUE FOR S POLYSCHIDES WAS FOUND WHERE CURRENT SURGE WAS GREATEST BUT WAVE ACTION ONLY MODERATE. THE NET PERCENTAGE EFFICIENCY OF ENERGY FIXATION SHOWS A NEARLY LINEAR RELATIONSHIP WITH DEPTH WHEN BASED ON SURFACE RADIATION. WHEN EFFICIENCY IS BASED ON THE RADIATION REACHING EACH DEPTH, THERE IS A FALL-OFF ABOVE AND BELOW 8.4 METERS; AN INVERSE RELATIONSHIP EXISTS BETWEEN PRODUCTIVITY AND EFFICIENCY DOWN TO THIS DEPTH BUT NOT BELOW. (JONES-WISCONSIN)

FIELD 05C, 02L

ACCESSION NO. W72-04766

ABCS OF CULTURAL EUTROPHICATION AND ITS CONTROL. PART I--CULTURAL CHANGES,

METCALF AND EDDY, INC., BOSTON, MASS.

CLAIR N. SAWYER.

WATER AND SEWAGE WORKS, P 278-281, SEPTEMBER 1971. 5 FIG, 2 TAB, 2 REF.

DESCRIPTORS:
*EUTROPHICATION, *CARBON, *WATER POLLUTION CONTROL, WATER POLLUTION EFFECTS, NUTRIENTS, ALGAE, CYANOPHYTA, PHOSPHORUS, WATER POLLUTION SOURCES, HYDROGEN ION CONCENTRATION, CARBON DIOXIDE, NITROGEN, CHLOROPHYTA, NITROGEN FIXATION, LAND MANAGEMENT, FERTILIZATION, PRIMARY PRODUCTIVITY.

ABSTRACT:
DRASTIC CHANGES IN NUTRIENT INPUT MUST BE ACCOMPLISHED IN ORDER TO CONTROL CULTURAL EUTROPHICATION. ALTHOUGH INDICATIONS ARE THAT GREEN ALGAE AND MANY BLUE-GREEN ALGAE ARE DEPENDENT UPON FIXED FORMS OF NITROGEN, AT LEAST FOUR GENERA OF BLUE-GREENS (ANABAENA, GLOEOTRICHIA, APHANIZOMENON AND NOSTIC) ARE CAPABLE OF FIXING ATMOSPHERIC NITROGEN. SUCCESS OF NITROGEN CONTROL IN SOME AREAS COULD DEPEND TO A GREAT EXTENT ON THE COOPERATION OF FARMERS IN CURTAILING BAD LAND MANAGEMENT AND FERTILIZATION PRACTICES. PHOSPHORUS GAINS ACCESS TO NATURAL WATERS MAINLY THROUGH THE DISCHARGE OF WASTEWATERS AND TO SOME EXTENT THROUGH SURFACE WASH FROM FARMLANDS. WASTEWATER TREATMENT METHODS FOR PHOSPHORUS REMOVAL FROM EFFLUENTS HAVE BEEN KNOWN FOR SEVERAL YEARS, AND A LARGE PART OF PHOSPHORUS IN DOMESTIC WASTEWATERS AND ESSENTIALLY ALL PHOSPHORUS IN SOME INDUSTRIAL WASTES, IS CONTRIBUTED BY SYNTHETIC DETERGENTS. PHOSPHORUS LIMITATION IN LAKES AND STREAMS SEEMS TO BE THE ONLY KNOWN MEANS TO CONTROL NITROGEN-FIXING BLUE-GREEN ALGAE. THUS PHOSPHORUS REMOVAL MUST BE PART OF ANY PLAN TO CONTROL EUTROPHICATION IN ADDITION TO NITROGEN REMOVAL, WHICH IS THE SECOND MOST IMPORTANT NUTRIENT IN PRIMARY PRODUCTION. (JONES-WISCONSIN)

FIELD 05C, 05G

ACCESSION NO. W72-04769

SOME OBSERVATIONS ON THE LIMNOLOGY OF A POND RECEIVING ANIMAL WASTES,

OKLAHOMA STATE UNIV., STILLWATER. DEPT. OF ZOOLOGY.

DALE W. TOETZ.

PROCEEDINGS OKLAHOMA ACADEMY OF SCIENCE, VOL 51, P 30-35, 1971. 4 FIG, 2 TAB, 11 REF.

DESCRIPTORS:
*OXIDATION LAGOONS, *WATER POLLUTION EFFECTS, *FARM WASTES, OKLAHOMA, CYANOPHYTA, CHLORELLA, IONS, PHYTOPLANKTON, NITRATES, NITROGEN FIXATION, ALGAE, CONDUCTIVITY, CHLOROPHYLL, PIGMENTS, DISPERSION, RUNOFF.

IDENTIFIERS:
*FEEDLOTS, LEMNA,

ABSTRACT:
A SMALL POND RECEIVING RUNOFF FROM A HOG YARD WAS DOMINATED BY A LARGE POPULATION OF BLUE-GREEN ALGAE AND PHYTOFLAGELLATES DURING SUMMER AND BY CHLORELLA SP DURING WINTER. HEAVY RAINFALLS DECREASED THE IONIC CONCENTRATION OF WATER AND ALTERED THE PHYTOPLANKTON COMPOSITION. OCCASIONAL CONCENTRATION OF OXYGEN BELOW 5 MG/L AND HIGH AMMONIA CONTENT SUGGESTED THAT THE POND IS UNSUITABLE FOR WARM-WATER FISH CULTURE. (WILDE-WISCONSIN)

FIELD 05C

ACCESSION NO. W72-04773

KINETICS OF ALGAL SYSTEMS IN WASTE TREATMENT--AMMONIA-NITROGEN AS A GROWTHLIMITING FACTOR AND OTHER PERTINENT TOPICS,

CALIFORNIA UNIV., BERKELEY. SANITARY ENGINEERING RESEARCH LAB.

M. SOBSEY, J. E. HARRISON, H. GEE, G. SHELET, AND J. C. GOLDMAN.

AVAILABLE FROM THE NATIONAL TECHNICAL INFORMATION SERVICE AS PB-206 811, $3.00 IN PAPER COPY, $0.95 IN MICROFICHE. FINAL REPORT, PART II, SEPTEMBER 1970, 156 P. 45 FIG, 11 TAB, 102 REF. FWQA PROGRAM 17010 DZQ 09/70.

DESCRIPTORS:
*ALGAE, *KINETICS, *NITROGEN, *NUTRIENT REQUIREMENTS, AMMONIA, *GROWTH RATES, *WASTE ASSIMILATIVE CAPACITY, OXIDATION, DISSOLVED OXYGEN, HYDROGEN ION CONCENTRATION, DESIGN CRITERIA, MATHEMATICAL MODELS, *BIOASSAY, *WASTE DILUTION, LAGOONS, *WASTE WATER TREATMENT, ALGAL CONTROL, EUTROPHICATION, MODEL STUDIES, MATHEMATICAL MODELS.

IDENTIFIERS:
*OSTRACODS, PREDATOR-PREY RELATIONSHIPS, *AMMONIA-NITROGEN, *GROWTH REACTORS, ALGAL PONDS, ALGAL GROWTH RATES, *POTAMOCYPRIS.

ABSTRACT:
4 SMALL OUTDOOR REACTORS, A PREDATOR INFESTED OUTDOOR POND, AND NUMEROUS INDOOR SYSTEMS WERE DESIGNED, CONSTRUCTED AND OPERATED TO PROVIDE DATA DESCRIBING THE KINETICS ASPECTS OF ALGAL SYSTEMS USED AS WASTEWATER TREATMENT SCHEMES. THE STUDIES INCLUDED DETERMINATIONS OF: (1) THE KINETICS OF AMMONIA-NITROGEN AS A GROWTH LIMITING FACTOR; (2) THE EFFECT OF NUTRIENTS IN WASTEWATER EFFLUENTS ON ALGAL GROWTH; (3) SUGGESTED DESIGNS FOR BOTH INDOOR AND OUTDOOR GROWTH REACTORS; AND (4) PREDATOR-PREY RELATIONSHIPS IN OUTDOOR ALGAL SYSTEMS. SPECIFIC GROWTH RATE DATA FOR NITROGEN WERE QUITE CONSISTENT, ALLOWING THE USE OF THE MAXIMUM SPECIFIC NET GROWTH RATE AS AN EXTREMELY RELIABLE PARAMETER OF SYSTEM OPERATION. PARAMETERS KA, KS, AND SN ALL DIRECTLY RELATED TO THE MAXIMUM SPECIFIC NET GROWTH RATE, WERE SHOWN TO BE USEFUL BOTH IN MATHEMATICAL INTERPRETATION OF SYSTEM KINETICS AND IN DETERMINING TO WHAT PERCENT OF CAPACITY AN ALGAL SYSTEM IS OPERATING. OTHER STUDIES DEMONSTRATED THAT PERMISSIBLE DILUTION RATIOS FOR WASTE WATER DISCHARGES INTO RECEIVING WATERS CAN BE FORMULATED EMPIRICALLY. DESIGN AND OPERATIONAL INFORMATION OF VARIOUS SYSTEMS IS PRESENTED, AS WELL AS PRELIMINARY BIOASSAY RESULTS ON THE MICROBIAL POPULATIONS OF ALGAL PONDS. (LOWRY-TEXAS)

FIELD 05D, 05C

ACCESSION NO. W72-04787

KINETICS OF ALGAL SYSTEMS IN WASTE TREATMENT--PHOSPHORUS AS A GROWTH LIMITING FACTOR,

CALIFORNIA UNIV., BERKELEY. SANITARY ENGINEERING RESEARCH LAB.

MARIO D. ZABAT, WILLIAM J. OSWALD, CLARENCE G. GOLUEKE, AND HENRY GEE.

AVAILABLE FROM THE NATIONAL TECHNICAL INFORMATION SERVICE AS PB-206 810, $3.00 IN PAPER COPY, $0.95 IN MICROFICHE. FINAL REPORT PART I, SEPTEMBER 1970, 210 P. 57 FIG, 15 TAB, 210 REF. FWQA PROGRAM 17010 DZQ 09/70.

DESCRIPTORS:
*ALGAE, *GROWTH RATES, *KINETICS, *PHOSPHORUS, *NUTRIENT REQUIREMENTS, TEMPERATURE, HYDROGEN ION CONCENTRATION, EFFICIENCIES, WATER QUALITY CONTROL, EUTROPHICATION, ALGAL CONTROL, *CHLORELLA, *WASTE WATER TREATMENT.

IDENTIFIERS:
*CHEMOSTATIC CONTINUOUS CULTURES, MAXIMUM SPECIFIC GROWTH RATE, SATURATION CONSTANT, DILUTION, CHLORELLA PYRENOIDOSA, MONOD GROWTH MODEL, ALGAL GROWTH RATES.

ABSTRACT:
THE KINETICS AND CHARACTERISTICS OF PHOSPHATE-LIMITED ALGAL CULTURES WERE STUDIED ON A LABORATORY SCALE IN ORDER TO: (1) EVALUATE KINETIC PARAMETERS OF ALGAL GROWTH IN RELATION TO PHOSPHORUS CONCENTRATION; (2) DERIVE AN EXPRESSION DESCRIBING THE GROWTH RATE-LIMITING NUTRIENT RELATIONSHIP; AND (3) DETERMINE DESIRABLE OR OPTIMUM DESIGN AND OPERATING PARAMETERS FOR EFFECTIVE PHOSPHORUS REMOVAL BY ALGAL SYSTEMS. PHOSPHORUS LIMITING ALGAL GROWTH WAS DESCRIBED BY THE MONOD GROWTH MODEL FOR SPECIFIC GROWTH RATE. MAXIMUM SPECIFIC GROWTH RATE FOR THE ALGAE TESTED WAS 4.19/DAY, AND ALL GROWTH CHARACTERISTICS WERE SIGNIFICANTLY AFFECTED BY VARIATIONS IN PH AND TEMPERATURE OF THE CULTURE. REMOVALS OF MODERATE CONCENTRATIONS OF PHOSPHORUS (LESS THAN OR EQUAL TO 7 MG/L AS P) IN EXCESS OF 85% CAN BE ACCOMPLISHED BY CONTINUOUS ALGAL CULTURES. (LOWRY-TEXAS)

FIELD 05D, 05C

ACCESSION NO. W72-04788

KINETICS OF ALGAL SYSTEMS IN WASTE TREATMENT FIELD STUDIES,

CALIFORNIA UNIV., BERKELEY. SANITARY ENGINEERING RESEARCH LAB.

AARON MERON, WILLIAM J. OSWALD, AND HENRY GEE.

AVAILABLE FROM THE NATIONAL TECHNICAL INFORMATION SERVICE AS PB-206 812, $6.00 IN PAPER COPY, $0.95 IN MICROFICHE. FINAL REPORT PART III, MAY 1971, 334 P. 46 FIG, 44 TAB, 122 REF. FWQA PROGRAM 17010 DZQ 05/71.

DESCRIPTORS:
*ALGAE, *KINETICS, *WASTE WATER TREATMENT, *NUTRIENT REQUIREMENTS, TEMPERATURE, *NITROGEN, *PHOSPHATES, ANAEROBIC CONDITIONS, AEROBIC CONDITIONS, METABOLISM, SLUDGE, BIOCHEMICAL OXYGEN DEMAND, CHEMICAL OXYGEN DEMAND, CARBON DIOXIDE, LININGS, DISSOLVED OXYGEN, LIGHT INTENSITY, DESIGN CRITERIA, *OXIDATION LAGOONS, SEWAGE TREATMENT, ALGAL CONTROL, BIOLOGICAL TREATMENT, *GROWTH RATES.

IDENTIFIERS:
ALGAL GROWTH RATES.

ABSTRACT:
THREE DOMESTIC SEWAGE STABILIZATION POND SYSTEMS WERE STUDIED TO EVALUTE THE RELATIONSHIP BETWEEN DESIGN CRITERIA AND PERFORMANCE CRITERIA FOR POND SYSTEMS IN ACCOMPLISHING SPECIFIC WASTE TREATMENT STEPS. SYSTEM 1 CONSISTED OF A LINED HIGH-RATE POND FOLLOWING SEDIMENTATION, SYSTEM 2 WAS COMPRISED OF CONVENTIONAL SECONDARY WASTE TREATMENT FOLLOWED BY A POND SYSTEM, AND SYSTEM 3 CONSISTED SOLELY OF SEVERAL DEEP PONDS IN SERIES. STUDY RESULTS INDICATED THAT SYSTEM 3, CONSISTING OF AN ANEROBIC POND, AND UNLINED HIGH-RATE POND, AND DEEP HIGH-RATE POLISHING POND, WAS EFFECTIVE IN BOTH BOD AND NUTRIENT REMOVALS, WITH MEAN ANNUAL REMOVALS OF 97.3%, 93.2%, 91.6%, AND 64.5% FOR BOD, COD, TOTAL NITROGEN, AND PHOSPHATE. RESULTS CONFIRMED THAT POND FUNCTIONING IS DETERMINED BY POSITION WITH RESPECT TO OTHER PONDS AND UNIT PROCESSES, AS WELL AS THEIR INDIVIDUAL DESIGN CHARACTERISTICS. OTHER STUDY RESULTS DEMONSTRATED THAT: (1) EFFICIENT N AND P REMOVAL DEPENDS UPON CARBON AVAILABILITY; (2) HIGH-RATE PONDS DEVELOP A SLUDGE-SLIME BOTTOM LAYER WHICH PREVENTS EROSION; (3) PROPER NUTRIENT BALANCE, AS MEASURED BY C:N AND C:P IS ESSENTIAL FOR EFFECTIVE ALGAL GROWTH AND NUTRIENT REMOVAL; (4) DEEP ANAEROBIC PONDS SHOULD BE OPERATED AS AN AEROBIC LAYER ABOVE AN ANAEROBIC LAYER; AND (5) NUTRIENT BALANCE AS WELL AS PH CAN BE ADJUSTED IN HIGH RATE PONDS BY CO2 ADDITION. (LOWRY-TEXAS)

FIELD 05D

ACCESSION NO. W72-04789

WATER QUALITY OFFICE.

ENVIRONMENTAL PROTECTION AGENCY, SAN FRANCISCO, CALIF. WATER QUALITY OFFICE.

ENVIRONMENTAL PROTECTION AGENCY, PACIFIC SOUTHWEST REGION WATER QUALITY OFFICE REPORT, DECEMBER 1970. 59 P, 3 FIG, 10 TAB, 11 REF, 3 APPEND.

DESCRIPTORS:
*WATER QUALITY, *WATER POLLUTION EFFECTS, *LAKES, *COLORADO, *RECREATION FACILITIES, WATER TEMPERATURE, CHEMICAL ANALYSIS, ALGAE, NUTRIENTS, WATER ANALYSIS, DOMESTIC WASTES.

IDENTIFIERS:
GRAND LAKE(COLO), SHADOW MOUNTAIN LAKE(COLO), LAKE GRANBY(COLO).

ABSTRACT:
LABORATORY ANALYSES OF SAMPLES COLLECTED FROM GRAND LAKE, SHADOW MOUNTAIN LAKE, AND LAKE GRANBY (ONE OF THE MOST POPULAR RECREATIONAL AREAS IN COLORADO) SHOW THAT THE INORGANIC CONCENTRATIONS WERE UNIFORM THROUGHOUT THE LAKES AND AVERAGE 0.107 MG/L. THIS CONCENTRATION IS APPROXIMATELY ONE-THIRD OF THAT CONSIDERED NECESSARY TO INITIATE AN ALGAE BLOOM. THE ORTHOPHOSPHATE CONCENTRATION IN THE THREE LAKES AVERAGED 0.0255 MG/L. THIS AVERAGE CONCENTRATION EXCEEDS THE MINIMUM LEVEL OF ORTHOPHOSPHATE CONSIDERED AS NECESSARY TO SUSTAIN AN ALGAE 'BLOOM' (0.010 MG/L). LAKE GRANBY AND GRAND LAKE HAD THERMALLY STRATIFIED LAYERS ALTHOUGH THERMOCLINES WERE NOT IN EVIDENCE. SHADOW MOUNTAIN LAKE WAS GENERALLY ISOTHERMAL BECAUSE OF ITS SHALLOW DEPTH. THE DISSOLVED OXYGEN CONCENTRATION IN THE NEAR SURFACE WATERS OF ALL THREE LAKES EXCEEDED THE 6.0 MG/L STANDARD FOR COLD WATER FISHERIES AT ALL STATIONS EXCEPT FOR ONE STATION ON SHADOW MOUNTAIN LAKE. THE WATER QUALITY OF THE THREE LAKES WAS EQUAL TO OR BETTER THAN THE STANDARDS ESTABLISHED FOR THE THREE LAKES EXCEPT THAT THE DISSOLVED OXYGEN CONCENTRATIONS IN THE HYPOLIMNION OF LAKE GRANBY WERE LESS THAN THE ESTABLISHED STANDARDS OF 6.0 MG/L. (WOODARD-USGS)

FIELD 05C, 02H

ACCESSION NO. W72-04791

NATURAL CONDITIONS AND AQUATIC LIFE OF THE BRATSK RESERVOIR (FORMIROVANIYE PRIRODNYKH USLOVIY I ZHIZNI BRATSKOGO VODOKHRANILISHCHA).

IZDATEL ' STVO 'NAUKA', MOSCOW, G. I. GALAZIY, EDITOR, 1970. 280 P.

DESCRIPTORS:
*AQUATIC LIFE, *AQUATIC BACTERIA, *AQUATIC ANIMALS, *AQUATIC PLANTS, *RESERVOIR STAGES, EARLY IMPOUNDMENT, BIOMASS, AQUATIC ALGAE, PHYTOPLANKTON, ZOOPLANKTON, FISH POPULATIONS, WATER TEMPERATURE, THERMAL STRATIFICATION, OXIDATION, ORGANIC MATTER, WATER QUALITY, WATER CHEMISTRY, WATER POLLUTION SOURCES, METEOROLOGICAL DATA, SEASONAL.

IDENTIFIERS:
*USSR, *IRKUTSK OBLAST, *BRATSK RESERVOIR, ANGARA RIVER, OKA RIVER, HYDROBIOLOGY, BACTERIOLOGY, SAPROPHYTES, STURGEON, METALIMNION.

ABSTRACT:
THIS COLLECTION OF 8 PAPERS DEALS PRIMARILY WITH BIOLOGICAL ASPECTS OF AQUATIC LIFE IN THE BRATSK RESERVOIR, EASTERN SIBERIA, (IRKUTSK OBLAST)--THE LARGEST MANMADE LAKE IN THE WORLD. THE PAPERS, MAY BE USED AS A BASIS FOR FORECASTING WATER QUALITY OF THE RESERVOIR FOR DRINKING-WATER SUPPLY AND INDUSTRIAL USE, AND FOR RECLAIMING THE RESERVOIR AS A FISHERY. SUBJECTS EXAMINED INCLUDE: (1) SEASONAL METEOROLOGICAL CONDITIONS IN THE REGION OF THE RESERVOIR; (2) FORMATION OF PHYTOPLANKTON IN THE RESERVOIR; (3) AQUATIC BACTERIA IN THE RESERVOIR (1965); (4) ESTABLISHMENT OF ZOOPLANKTON IN THE RESERVOIR DURING THE FIRST TWO YEARS OF IMPOUNDMENT (1962-63); (5) FISH POPULATION AND DISTRIBUTION IN THE OKA RIVER SECTION OF THE RESERVOIR; (6) STERLET IN THE RESERVOIR AND ANGARA RIVER; (7) LONG-TERM WATER-QUALITY CHARACTERISTICS OF THE RESERVOIR AND FORECAST OF ORGANIC MATTER ACCUMULATION; AND (8) WATER QUALITY OF THE RESERVOIR DURING THE FIRST YEARS OF IMPOUNDMENT (1962-64). THE COLLECTION IS OF PARTICULAR VALUE TO HYDROBIOLOGISTS, ICHTHYOLOGISTS, HYDROLOGISTS, CLIMATOLOGISTS, ZOOLOGISTS AND BOTANISTS. (SEE ALSO W72-04855)
(JOSEFSON-USGS)

FIELD 02H, 05C

ACCESSION NO. W72-04853

FORMATION OF PHYTOPLANKTON IN THE BRATSK RESERVOIR (FORMIROVANIYE FITOPLANKTONA BRATSKOGO VODOKHRANILISHCHA),

LIMNOLOGICHESKII INSTITUT, IRKUTSK (USSR).

O. M. KOZHOVA.

IN: FORMIROVANIYE PRIRODNYKH USLOVIY I ZHIZNI BRATSKOGO VODOKHRANILISHCHA; IZDATEL 'STVO 'NAUKA', MOSCOW, P 26-160, 1970. 24 FIG, 37 TAB, 68 REF.

DESCRIPTORS:
*AQUATIC PLANTS, *AQUATIC ALGAE, *PHYTOPLANKTON, *RESERVOIRS, *RESERVOIR STAGES, EARLY IMPOUNDMENT, BIOMASS, DIATOMS, CHLOROPHYTA, CYANOPHYTA, PYRROPHYTA, CHRYSOPHYTA, EUGLENOPHYTA, WATER TEMPERATURE, THERMAL STRATIFICATION, ICE, EUTROPHICATION, OXYGEN, SEASONAL.

IDENTIFIERS:
*USSR, *IRKUTSK OBLAST, *BRATSK RESERVOIR, *HYDROBIOLOGY, METALIMNION, BACILLARIOPHYTA, VOLVOCINEAE, PROTOCOCCINEAE, ULOTRICHINEAE, DESMIDIALES.

ABSTRACT:
SEASONAL CHANGES IN THE COMPOSITION, PRODUCTION, AND VERTICAL AND HORIZONTAL DISTRIBUTION OF PHYTOPLANKTON IN THE BRATSK RESERVOIR, EASTERN SIBERIA, (AREA--5,500 SQ KM; VOLUME--179 CU KM) DURING ITS EARLY IMPOUNDMENT (1963-65) ARE EXAMINED. THE PHYTOPLANKTON IS REPRESENTED BY 118 ALGAL SPECIES, CONSISTING OF 47 CHLOROPHYTA, 20 CYANOPHYTA, 17 BACILLARIOPHYTA, 14 PYRROPHYTA, 11 CHRYSOPHTYA, AND 9 EUGLENOPHYTA. A TOTAL OF 17 DOMINANT SPECIES WERE IDENTIFIED IN 1964 AND 13 IN 1965. THE MOST MASSIVE ALGAL POPULATIONS ARE APHANIZOMENON FLOS-AQUAE, STEPHANODISCUS HANTZSCHII, MELOSIRA ISLANDICA HELVETICA, AND ASTERIONELLA FORMOSA. NO WATER BLOOMS ARE FOUND UNDER ICE (ALTHOUGH THE BIOMASS IS SEVERAL MG/CU M). BACILLARIOPHYTA ARE PREDOMINANT IN SPRING (SEVERAL G/CU M); AN ANNUAL BIOMASS MAXIMUM OF 320 G OF DOMINANT APHANIZOMENON FLOS-AQUAE/CU M IS OBSERVED IN SUMMER AND FALL. MAXIMUM BIOMASSES INCREASED FROM 12 TO 320 G/CU M BETWEEN 1963 AND 1965; AVERAGE AREAL INDICES FOR THE RESERVOIR DURING THE SUMMER MAXIMUM WERE 55 G/SQ M IN 1964 AND 195 G/SQ M IN 1965. MAXIMUM BIOMASS INDEX WAS 816 G/SQ M AND APPROACHED EUTROPHIC CONDITIONS. TOTAL PHYTOPLANKTON PRODUCTION OF OXYGEN ALSO CHANGED. IN 1964 IT WAS 100 G/SQ M OR 370 KCAL/SQ M DURING THE GROWING PERIOD AND IN 1965-66 IT WAS 389 G/SQ M. RIVER INFLOW CONTAINS ABOUT 13,000 METRIC TONS OF PHYTOPLANKTON. A TOTAL OF 30,000 METRIC TONS IS DISCHARGED FROM THE RESERVOIR. PHYTOPLANKTON COMPOSITION, PRODUCTION, VERTICAL DISTRIBUTION, AND BIOMASS ARE DETERMINED BY THE MORPHOMETRIC HETEROGENEITY OF THE RESERVOIR, ITS ENORMOUS SIZE, AND BY DIFFERENCES IN RESERVOIR DEPTHS, WHICH RANGE FROM 3 TO 100 M. (SEE ALSO W72-04853) (JOSEFSON-USGS)

FIELD 05C, 02H

ACCESSION NO. W72-04855

ALGAL-BACTERIAL SYMBIOSIS FOR REMOVAL AND CONSERVATION OF WASTEWATER NUTRIENTS,

NORTH CAROLINA STATE UNIV., RALEIGH. DEPT. OF BIOLOGICAL AND AGRICULTURAL ENGINEERING; AND NORTH CAROLINA STATE UNIV., RALEIGH. DEPT. OF CIVIL ENGINEERING.

FRANK J. HUMENIK, AND GEORGE P. HANNA, JR.

JOURNAL WATER POLLUTION CONTROL FEDERATION, VOL 43, NO 4, APRIL 1971, P 580-594. 9 FIG, 1 TAB, 27 REF.

DESCRIPTORS:
*ALGAE, *BACTERIA, *SYMBIOSIS, PHOTOSYNTHESIS, OXIDATION, RESPIRATION, OXYGENATION, TEMPERATURE, LIGHT INTENSITY, MIXING, DISSOLVED OXYGEN, METABOLISM, CARBON DIOXIDE, PHOSPHATES, NITRATES, *NUTRIENTS, ALKALINITY, CHEMICAL OXYGEN DEMAND, ANALYTICAL TECHNIQUES, TERTIARY TREATMENT, *WASTE WATER TREATMENT.

ABSTRACT:
A LABORATORY-SCALE CONTINUOUS SYMBIOTIC ALGAL-BACTERIAL SYSTEM WAS DEVELOPED THAT REMOVED NUTRIENTS FROM WASTEWATER AND REMAINED WITHOUT SUPPLEMENTAL AERATION. THE BIOMASS GROWTH UNIT SUPPORTED A MIXED CHLORELLA-ACTIVATED SLUDGE CULTURE. MAXIMUM AND MOST CONSISTENT NUTRIENT REMOVAL AND CONSERVATION OCCURRED DURING UNAERATED OPERATION WITH SOLIDS CONTROL BY DAILY HARVESTING. PHOSPHORUS REMOVALS WERE NOT SIGNIFICANT. THE BIOMASS CULTURE WAS AN INTIMATELY ASSOCIATED MIXTURE OF ALGAE AND ACTIVATED SLUDGE, AND THE RESULTING SYMBIOTIC BENEFITS PRESENTED MORE FAVORABLE CONDITIONS FOR THE REMOVAL AND CONSERVATION OF WASTEWATER NUTRIENTS. THE REESTABLISHMENT OF THE SYMBIOTIC ALGAL-BACTERIA SYSTEM AFTER AN EQUIPMENT FAILURE PROVIDES EVIDENCE THAT SUCH A PROCESS COULD BE RECOVERED AFTER SOME OPERATIONAL MISHAP HAD MADE IT NECESSARY TO REVERT TO SUPPLEMENTAL AERATION. (LOWRY-TEXAS)

FIELD 05D

ACCESSION NO. W72-04983

PREDICTING QUALITY EFFECT OF PUMPED STORAGE,

WATER RESOURCES ENGINEERS, INC., WALNUT CREEK, CALIF.

C. W. CHEN, AND G. T. ORLOB.

PAPER, INTERNATIONAL CONFERENCE ON PUMPED STORAGE DEVELOPMENT AND ENVIRONMENTAL EFFECTS, MILWAUKEE, WIS, SEPT 1971. 21P, 7FIG, 2 TAB, 13 REF.

DESCRIPTORS:
*ECOSYSTEMS, *ECOLOGY, *PUMPED STORAGE, *WATER QUALITY, *MODEL STUDIES, ZOOPLANKTON, BIOTA, BACTERIA, BENTHIC FAUNA, FISH, ENVIRONMENTAL EFFECTS, RESERVOIR STORAGE, NUTRIENTS, TEMPERATURE, ALGAE, HYDROLOGIC DATA, DISSOLVED OXYGEN, WEATHER, RESERVOIR OPERATION, WATER TEMPERATURE, PHYTOPLANKTON, EUTHROPHICATION.

IDENTIFIERS:
SMITH MOUNTAIN PROJECT(VA), WEATHER EFFECT.

ABSTRACT:
WHILE MUCH ATTENTION HAS BEEN PAID TO THE ECONOMIC ASPECTS OF PUMPED STORAGE, THE ENVIRONMENTAL EFFECTS, PARTICULARLY THE WATER QUALITY WITHIN THE RESERVOIR, HAVE SELDOM BEEN FULLY ASSESSED. THE PUMPED STORAGE IMPOUNDMENT IS INVARIABLY A COMPLEX ECOSYSTEM HOSTING A WIDE SPECTRUM OF BIOTA--BACTERIA, PHYTOPLANKTON, ZOOPLANKTON, FISH AND BENTHIC ANIMALS. NUTRIENTS FROM RUNOFF CAN BE CONCENTRATED AND RETAINED IN THE ECOSYSTEM THROUGH ALGAL ASSIMILATION AND SEDIMENTATION. ACCELERATED EUTHROPHICATION OF THE ENVIRONMENT MAY RESULT. A PREDICTIVE TECHNIQUE HAS BEEN DEVELOPED FOR CALCULATING THE ENVIRONMENTAL CHANGES AND EVALUATING THE ALTERNATIVES OF PUMPED STORAGE OPERATION THAT MAY ENHANCE A NORMAL ECOLOGIC SUCCESSION AND FORESTALL ADVERSE ENVIRONMENTAL PROBLEMS. THE MODEL, APPLIED TO SMITH MOUNTAIN RESERVOIR, VA, REVEALS THAT IMPROVED WATER QUALITY CAN BE ACHIEVED BY MODIFYING THE PUMPING-RETURN SCHEDULE. (USBR)

FIELD 05F, 06B

ACCESSION NO. W72-05282

LIQUID WASTE TREATMENT PROCESS,

J. M. VALDESPINO.

U. S. PATENT NO. 3,625,883, 5 P, 3 FIG, 11 REF. OFFICIAL GAZETTE, VOL. 893, NO. 1, P. 254, DECEMBER 7, 1971.

DESCRIPTORS:
*PATENTS, *WASTE WATER TREATMENT, *LIQUID WASTES, SEWAGE TREATMENT, SEPARATION TECHNIQUES, *CHEMICAL WASTES, INDUSTRIAL WASTES, CHLORINATION, PHOSPHATES, NITRATES, FILTRATION, POLLUTION ABATEMENT, ALGAE, NUTRIENT, *BIODEGRADATION, WATER PURIFICATION, *AEROBIC TREATMENT, *OXIDATION LAGOONS.

IDENTIFIERS:
AEROBIC BIOLOGICAL DEGRADATION.

ABSTRACT:
A METHOD IS DESCRIBED FOR THE TREATMENT OF LIQUID WASTE FROM DOMESTIC AND INDUSTRIAL SOURCES. THE LIQUID WASTE IS FIRST FED TO A COMMINUTOR FOR CHOPPING AND MIXING SOLID WASTE WITH THE LIQUID. THE PRODUCT IS THEN FED TO A SURGING TANK FOR PROVIDING AN INTERMITTENT FLOW INTO A CENTRIFUGING STEP. THE WASTE IS CENTRIFUGED THROUGH A FILTER MEDIUM SUCH AS SAND OR DIATOMACECUS EARTH. THE SEPARATED EFFLUENT IS FED TO AN OXIDATION LAGOON FOR AEROBIC BIOLOGICAL DEGRADATION. THE GENERATION OF ALGAE MAY BE PROVIDED TO REMOVE PHOSPHATES, NITRATES AND OTHER NUTRIENTS. THE EFFLUENT THEN IS FED TO A SECOND CENTRIFUGE FILTER TO SEPARATE REMAINING SOLIDS WHICH ARE BURNED. CHLORINATION PROVISIONS ARE INCLUDED. (SINHA-OEIS)

FIELD 05D

ACCESSION NO. W72-05315

AQUATIC AND MARINE MICROORGANISMS, INTERRELATIONSHIPS IN ENRICHMENT CULTURES,

WASHINGTON UNIV., SEATTLE. DEPT. OF MICROBIOLOGY.

E. J. ORDAL.

AVAILABLE FROM THE NATIONAL TECHNICAL INFORMATION SERVICE AS AD-731 409, $3.00 IN PAPER COPY, $0.95 IN MICROFICHE. FINAL TECHNICAL REPORT 71-1, OCTOBER 14, 1971. 18 P, 2 TAB, 10 REF. PROJ. NC. NR306-487, N00014-67-A-0103-0006.

DESCRIPTORS:
*BACTERIA, *CULTURES, *SEA WATER, *NITRIFICATION, AMMONIA, NITRITE, SPHAEROTILUS, NUTRIENTS, PACIFIC OCEAN, NITRATE, CHEMICAL ANALYSIS, SODIUM CHLORIDE, COLUMBIA RIVER, ORGANIC COMPOUNDS, VITAMINS, YEASTS, SULFUR, FOULING, ALGAE, SULFATES, PERCOLATION, HERRING, SALMON, TROUT, TOXICITY, WASHINGTON, MARINE ALGAE, MARINE BACTERIA.

IDENTIFIERS:
HYPHOMICROBIUM, VIBRIO ANGUILLARUM, ENRICHMENT CULTURES, AGARS, CHEMOSTATS, NITROSOMONAS, NITROBACTER, NITROSOCYSTIS OCEANUS, SPHAEROTILUS NATANS, GLUCOSE, GALACTOSE, SUCROSE, MALTOSE, MANNITAL, SORBITAL, SUCCINATE, FUMARATE, BUTYRATE, BUTANOL, GLYCEROL, LACTATE, PYRUVATE, ACETATE, ETHANOL, BENZOATE, PROPIANATE, PROPANOL, LAURATO, PALMITATE, N-HEPTYLATE, XYLOSE, CITRATE, YEAST EXTRACT, MINERAL SALTS, CAULOBACTER, OLIGOCARBOPHILIC BACTERIA, ULVA, DESULFOVIBRIO, MACROMONAS, THIOSPIRA, VIBRIO PARAHAEMOLYTICUS, NITRIFYING BACTERIA.

ABSTRACT:
THE METHOD OF STEADY-STATE ENRICHMENT CULTURES WAS APPLIED TO THE STUDY OF MARINE NITRIFYING BACTERIA. WHEN LOW CONCENTRATIONS OF AMMONIA OR NITRITE WERE EMPLOYED, THE NITRIFYING BACTERIA WERE FOUND ATTACHED TO THE WALLS OF THE CULTURE VESSELS. WHEN CONCENTRATIONS OF AMMONIA OR NITRITE WERE INCREASED, THE NITRIFYING BACTERIA APPEARED ALSO IN THE CULTURE FIELDS ENABLING THE ISOLATION OF PURE CULTURES IN LIQUID MEDIA. AGAR WAS TOXIC AND THE NITRIFYING BACTERIA WERE FOUND TO REQUIRE AN UNKNOWN COMPONENT IN SEA WATER IN ADDITION TO SODIUM CHLORIDE. CONTROL FLOW-THROUGH CULTURES OF OCEAN WATER ALONE SUPPORTED LARGE POPULATIONS OF THE BUDDING, BRANCHING BACTERIUM HYPHOMICROBIUM, WHICH ALSO APPEARED IN CULTURES WITH LOW CONCENTRATIONS OF AMMONIA OR NITRITE. HYPHOMICROBIUM FED BY SMALL CONCENTRATIONS OF ORGANIC MATTER IN SEA WATER QUICKLY APPEARED ON GLASS, CELLOPHANE AND METAL SURFACES AND MAY BE IMPORTANT AS AN INITIAL STEP IN FOULING OF VESSELS. STEADY-STATE ENRICHMENT CULTURES WERE EMPLOYED TO DETERMINE THE RANGE OF ORGANIC COMPOUNDS USED BY SPHAEROTILUS. CULTURES GROWN ON GLUCOSE PLUS MINERAL SALTS GREW WELL IN ENRICHMENT CULTURES BUT FAILED TO GROW IN PURE CULTURE UNLESS VTAMINS WERE ADDED. IT WAS CONCLUDED THAT ASSOCIATED BACTERIA WERE PROVIDING THE NECESSARY VITAMINS FOR GROWTH OF SPHAEROTILUS. A GROUP OF MARINE VIBRIOS ISOLATED FROM DISEASED SALMON AND HERRING WERE INVESTIGATED BY SEVERAL METHODS. STRAINS OF VIBRIO ANGUILLARUM FROM THE PACIFIC NORTHWEST WERE CLOSELY RELATED TO STRAINS FROM SCOTLAND AND DENMARK. HOWEVER, A NUMBER OF OTHER VIBRIOS WERE CLEARLY DISTINCT, INDICATING THAT THERE ARE SEVERAL KINDS OF MARINE VIBRIOS. (MORTLAND-BATTELLE)

FIELD 05C, 05A

ACCESSION NO. W72-05421

TWO SIMPLE DURABLE EPIFAUNAL COLLECTORS,

LAVAL UNIV., QUEBEC. DEPARTEMENT DE BIOLOGIE.

E. BOURGET, AND G. LACROIX.

JOURNAL OF THE FISHERIES RESEARCH BOARD OF CANADA, VOL. 28, NO. 8, P 1205-1207, AUGUST 1971. 3 FIG, 1 TAB, 5 REF.

DESCRIPTORS:
*BENTHOS, *SAMPLING, *SESSILE ALGAE, *INVERTEBRATES, ST. LAWRENCE RIVER, ALGAE, NEMATODES, GASTROPODS, MOLLUSKS, AMPHIPODA, MECHANICAL EQUIPMENT.

IDENTIFIERS:
*VERTICAL COLLECTORS, *HORIZONTAL COLLECTORS, BRYOZOANS, FORAMINIFERA, HYDROIDS, POLYCHAETA, OSTRACODS, HARPACTICOIDS, CIRRIPEDIA, ACARINA.

ABSTRACT:
TWO TYPES OF EPIFAUNAL SAMPLERS (VERTICAL AND HORIZONTAL) FOR USE IN THE INFRALITTORAL ZONE WERE DEVELOPED AND EVALUATED AT FOUR STATIONS IN THE ST. LAWRENCE ESTUARY. THE VERTICAL SAMPLER CONSISTED OF AN IRON PEDESTAL WITH A PROJECTING ARM DESIGNED TO HOLD 12 PLYWOOD COLLECTION PANELS. THE ARM WAS CONSTRUCTED TO ALLOW IT TO ROTATE TO OFFER MINIMAL RESISTANCE TO THE WATER FLOW AND AVOID TOPPLING THE STRUCTURE. THE HORIZONTAL SAMPLER CONSISTED OF A BOX-LIKE FRAME OF IRON WITH A HINGED TOP TO WHICH 16 COLLECTION PANELS WERE ATTACHED. BOTH SAMPLERS WERE DESIGNED FOR EASY REMOVAL OF THE SAMPLE PANELS BY A SCUBA DIVER. NO BUOYS WERE ATTACHED DIRECTLY TO THE SAMPLERS SO THAT ALGAL COLLECTIONS ON MOORING LINES WOULD NOT INTERFERE WITH THE SAMPLES. EVALUATION OF THE TWO TYPES OF SAMPLERS SHOWED THAT THE VERTICAL SAMPLER IS MORE SUITABLE FOR STUDY OF EPIBENTHIC COMMUNITIES SINCE IT DRAWS A WIDER RANGE OF SPECIES. THE HORIZONTAL SAMPLER, ON THE OTHER HAND, COLLECTED A LARGER NUMBER OF SPECIMENS AND IS CONCLUDED TO BE PREFERABLE IF ONLY SESSILE AND SEDIMENTARY ORGANISMS ARE TO BE STUDIED. FOR COMPLETE SAMPLING OF THE EPIFAUNA, USE OF BOTH SAMPLERS IS RECOMMENDED. (JEFFERIS-BATTELLE)

FIELD 05A, 07B, 02L

ACCESSION NO. W72-05432

BACTERICAL PATHOGENS OF FRESHWATER BLUE-GREEN ALGAE,

DUNDEE UNIV. (SCOTLAND). DEPT. OF BIOLOGICAL SCIENCES.

M. J. DAFT, AND W. D. P. STEWART.

NEW PHYTOLOGIST, VOL 70, NO 5, P 819-829, 1971. 6 FIG, 3 TAB, 2 PLATES, 12 REF.

DESCRIPTORS:
*BIOCONTROL, *PATHOGENIC BACTERIA, *ALGAL CONTROL, FRESH WATER, CYANOPHYTA, MYXOBACTERIA, CULTURES, METABOLISM, SEWAGE, CARBOHYDRATES, HYDROGEN ION CONCENTRATION, TEMPERATURE, ENZYMES, CELLULOSE, VIRUSES, HUMAN PATHOLOGY, SYSTEMATICS.

IDENTIFIERS:
LYTIC BACTERIA, ENGLAND, METABOLIC INHIBITION, HOSTS, MYXOPHYCEAE, SCOTLAND.

ABSTRACT:
EVIDENCE IS PRESENTED THAT HETEROTROPHIC BACTERIA WHICH ARE PATHOGENIC TO CERTAIN BLUE-GREEN ALGAE OCCUR IN BRITISH FRESHWATER HABITATS. FOUR BACTERIAL ISOLATES WERE OBTAINED WHICH CAUSE LYSIS IN LABORATORY CULTURES AND NATURAL POPULATIONS OF BLUE-GREEN ALGAE. THE ORGANISMS ARE TENTATIVELY IDENTIFIED AS MEMBERS OF THE MYXOBACTERALES. THE HOST RANGE OF THESE STRAINS IS WIDE AND SPECIES BELONGING TO ALL ORDERS OF MYXOPHYCEAE ARE SUSCEPTIBLE. THE ISOLATES ALSO LYSE A VARIETY OF GRAM-NEGATIVE AND GRAM POSITIVE BACTERIA AND ARE CAPABLE OF GROWTH FREE FROM HOSTS, AND THEIR INHIBITORY EFFECT IS RAPID (IN SOME INSTANCES WITHIN 2-5 HOURS). SOME BACTERIA SUSCEPTIBLE ARE HUMAN PATHOGENS THUS IT MAY BE THAT THREE OF THE LYTIC BACTERIA OBTAINED FROM SEWAGE PONDS DECREASE BACTERIAL NUMBERS WHICH ARE PATHOGENIC TO MAN. THE FINDINGS, IF EXTRAPOLATED TO NATURAL POPULATIONS, SUGGEST THAT GIVEN SUFFICIENT LYTIC BACTERIA, ALGAL BLOOMS WILL NOT OCCUR, UNLESS RESISTANT STRAINS ARE PRESENT. POSSIBLY ALGAL FORMATION COULD BE REGULATED TO SOME DEGREE BY THE EXTENT TO WHICH ALGAL GROWTH OUTSTRIPS, OR IS OUTSTRIPPED BY, THE GROWTH OF ALGAL PATHOGENS. THE ISOLATES CAN BE PROPAGATED READILY IN THE ABSENCE OF HOSTS. (AUEN-WISCONSIN)

FIELD 05C

ACCESSION NO. W72-05457

PHYTOPLANKTON ENERGETICS IN A SEWAGE-TREATMENT LAGOON,

SOUTHERN ILLINOIS UNIV., CARBONDALE. DEPT. OF BOTANY.

JACOB VERDUIN.

ECOLOGY, VOL 52, NO 4, P 626-631, 1971. 5 FIG, 3 TAB, 20 REF. NSF GE-2649.

DESCRIPTORS:
*PHYTOPLANKTON, *METABOLISM, *SEWAGE TREATMENT, *SEWAGE LAGOONS, OHIO, OXYGEN, DISSOLVED OXYGEN, LIGHT, DIFFUSIVITY, CARBON DIOXIDE, LIGHT INTENSITY, PHOTOSYNTHESIS, EQUATIONS, LIGHT PENETRATION, ALGAE, HYDROGEN ION CONCENTRATION.

IDENTIFIERS:
EXTINCTION COEFFICIENT, LAMBERT-BEER EQUATION, EDDY DIFFUSIVITY, DESHLER(OHIO).

ABSTRACT:
INTENSIVE INVESTIGATIONS WERE CARRIED OUT FROM MARCH 21 THROUGH JUNE 10, 1964 OF A SEWAGE-TREATMENT LAGOON SYSTEM NEAR DESHLER, OHIO--A PERIOD CRITICAL FOR SEWAGE PONDS IN THE TEMPERATE ZONE. THE PONDS WERE WELL MIXED VERTICALLY, EXHIBITED DAYTIME AVERAGE EDDY DIFFUSIVITIES OF ABOUT 1 SQ CM/SEC AND HAD A HIGH TITRATABLE BASE. LOW OXYGEN LEVELS WERE ENCOUNTERED FOR ONLY A SHORT PERIOD IN THE SECOND STAGE OF THE TWO-POND SYSTEM. LIGHT ABSORPTION AVERAGED ABOUT 90%/M. THE PHYTOPLANKTON COMMUNITIES, DOMINATED BY SCENEDESMUS, ROSE TO 50 MICROLITERS/L. RATES OF CARBON DIOXIDE UPTAKE IN THE ORDER OF 1 M/CM PER DAY WERE OBSERVED, BUT OXYGEN PRODUCTION AVERAGED ONLY 1/5 AS HIGH. A GRAPH OF CARBON DIOXIDE UPTAKE VERSUS LIGHT INTENSITY REVEALED A CURVE CLOSELY SIMILAR IN SHAPE TO THE TYPICAL PHOTOSYNTHESIS VERSUS LIGHT CURVE, BUT A SPEARATE SET OF DATA, BASED ON C-14 UPTAKE, SHOWED ANOMALOUSLY HIGH VALUES AT SURFACE LIGHT INTENSITIES. CARBON DIOXIDE UPTAKE BY A PHOTOSYNTHETIC METABOLIC PROCESS THAT DOES NOT PRODUCE OXYGEN IS POSTULATED TO EXPLAIN THE IMBALANCE BETWEEN CARBON DIOXIDE AND OXYGEN BUDGETS. (JONES-WISCONSIN)

FIELD 05D

ACCESSION NO. W72-05459

LAKE KINNERET: THE NUTRIENT CHEMISTRY OF THE SEDIMENTS,

KINNERET LIMNOLOGICAL LAB., TIBERIAS (ISRAEL).

C. SERRUYA.

LIMNOLOGY AND OCEANOGRAPHY, VOL 16, NO 3, P 510-521, 1971. 8 FIG, 7 TAB, 15 REF.

DESCRIPTORS:
*TROPHIC LEVEL, *LAKES, *NUTRIENTS, *CHEMICAL PROPERTIES, *SEDIMENTS, CLAYS, CORES, SEASONAL, ALGAE, MUD-WATER INTERFACES, IRON, MANGANESE, PHOSPHORUS, NITROGEN, ORGANIC MATTER, CALCIUM CARBONATE, DETRITUS, PLANKTON, ADSORPTION, EUTROPHICATION, TEMPERATURE, LIGHT PENETRATION, AMMONIA, PHOSPHATES, SALINE WATER.

IDENTIFIERS:
*LAKE KINNERET(ISRAEL), BOTTOM WATERS, RIVER JORDAN(ISRAEL).

ABSTRACT:
TO UNDERSTAND EUTROPHICATION PROCESSES IN LAKE KINNERET, ISRAEL, IMPORTANT AS A WATER SUPPLY, BOTTOM DEPOSITS WERE STUDIED. THE BOTTOM WATERS WERE SAMPLED ONCE A WEEK FOR TWO YEARS, MUD SAMPLES WERE OBTAINED WITH A DREDGE, AND CORES TAKEN. SINCE THE SEDIMENTS MAY ACTIVELY CONTRIBUTE TO THE NUTRIENT SUPPLY CAUSING WINTER ALGAL BLOOM, THE MUD-WATER EXCHANGES WERE INVESTIGATED. TEMPERATURE WAS RECORDED WITH RELATION TO THE SEASONS, COLDWATER BROUGHT IN BY THE RIVER JORDAN, DISCHARGE OF THE SUBMARINE HOT SPRINGS, AND INTERNAL WAVES. LIGHT TRANSMISSION WAS MEASURED. THE CONCENTRATIONS OF IRON, MANGANESE, PHOSPHORUS, NITROGEN, ORGANIC CARBON, AND CALCIUM CARBONATE IN RECENT SEDIMENTS WERE DETERMINED FOR THE WHOLE LAKE AND THE RESULTS DISCUSSED IN THE LIGHT OF THE CHEMICAL CONDITIONS OF BOTTOM WATERS, THE DETRITAL INPUT, AND THE PLANKTON COMPOSITION. ONLY 4% OF THE TOTAL AUTOTROPHIC CARBON IS WITHDRAWN FROM THE LAKE CYCLE BY THE SEDIMENTS. THE LAKE MUDS CONCENTRATE PHOSPHORUS IN SPITE OF THE PREVAILING REDUCING CONDITIONS, WHICH SUGGESTS AN ACTIVE PHYSICAL ADSORPTION. THE SPECIFIC CLAY SURFACES COULD BE USED AS NUTRIENT FIXERS TO SLOW EUTROPHICATION. (JONES-WISCONSIN)

FIELD 05C, 02H

ACCESSION NO. W72-05469

ABCS OF CULTURAL EUTROPHICATION AND ITS CONTROL: PART 2--WASTEWATERS,

METCALF AND EDDY, INC., BOSTON, MASS.

CLAIR N. SAWYER.

WATER AND SEWAGE WORKS, P 322-327, OCTOBER 1971. 9 FIG, 1 TAB, 4 REF.

DESCRIPTORS:
*EUTROPHICATION, *WATER POLLUTION CONTROL, *WASTE WATER(POLLUTION), VIRGINIA, NUTRIENTS, NITROGEN, PHOSPHORUS, SEWAGE EFFLUENTS, SEWAGE TREATMENT, ALGAE, WATER POLLUTION SOURCES, CHLOROPHYTA, CYANOPHYTA, WASHINGTON, DIVERSION, LAKE ERIE, NITROGEN FIXATION, DETERGENTS, CARBON DIOXIDE, ALKALINITY, BURNING, DOMESTIC WASTES, INDUSTRIAL WASTES, AGRICULTURE, SURFACE RUNOFF, WISCONSIN.

IDENTIFIERS:
*NITROGEN:PHOSPHORUS RATIO, LAKE WAUBESA(WIS), OCCOQUAN RESERVOIR(VA), MADISON(WIS), LAKE MENDOTA(WIS), LAKE MONONA(WIS), LAKE KEGONSA(WIS), LAKE WASHINGTON(WASH), GREEN LAKE(WASH).

ABSTRACT:
WASTEWATERS AS A NUTRIENT SOURCE BECOMES APPARENT IN THE CULTURAL EUTROPHICATION PROBLEM. DATA PRIOR TO THE ADVENT OF SYNTHETIC DETERGENTS CONTAINING PHOSPHATES WERE COLLECTED FROM LAKE WAUBESA, WISCONSIN. OCCOQUAN RESERVOIR, VIRGINIA, INDICATED A LARGER RELATIVE INCREASE IN PHOSPHORUS, AS COMPARED TO NITROGEN THAN THAT ENTERING LAKE WAUBESA, PROBABLY DUE TO THE HIGHER PHOSPHORUS CONTENT OF MODERN SEWAGES DUE TO PHOSPHATE-BEARING SYNTHETIC DETERGENTS. CULTURAL EUTROPHICATION CONTROL DEPENDS UPON LIMITING THE INPUT OF BOTH PHOSPHORUS AND NITROGEN. OF THE MAJOR INORGANIC NUTRIENT SOURCES, DOMESTIC AND INDUSTRIAL WASTEWATERS ARE EASIEST TO CONTAIN AND TREAT WHILE AGRICULTURAL SOURCES ARE PROBABLY THE MOST DIFFICULT TO CONTROL; PHOSPHORUS CONTROL ALONE MAY SUFFICE IN SOME LOCATIONS. LAKES WAUBESA AND KEGONSA (WISCONSIN) AND LAKE WASHINGTON HAVE RECOVERED AFTER DIVERSION OF SEWAGE. THE BASIC PHILOSOPHY IS SCIENTIFICALLY CORRECT THAT, IF THE DEGREE OF CULTURAL EUTROPHICATION IS RELATED TO THE DEGREE OF FERTILIZATION, ANY REDUCTION IN WASTEWATER QUANTITY OR IN EQUIVALENT NUTRIENTS SHOULD REDUCE PRIMARY PRODUCTIVITY. ELIMINATION OF PHOSPHATES IN DETERGENTS WOULD BE BENEFICIAL TO THE GREAT LAKES; REMOVAL OF 80% OF PHOSPHORUS FROM WASTEWATERS MAY OR MAY NOT BE ADEQUATE, DEPENDING UPON LAKE SIZE AND DETENTION TIME. (JONES-WISCONSIN)

FIELD 05C

ACCESSION NO. W72-05473

RELATIONSHIP BETWEEN PHOSPHATES AND ALKALINE PHOSPHATASE OF ANABAENA FLOS-AQUAE IN CONTINUOUS CULTURE,

QUEEN'S UNIV. KINGSTON (ONTARIO). DEPT. OF CHEMICAL ENGINEERING.

DEREK H. BONE.

ARCHIV FUR MIKROBIOLOGIE, VOL 80, P 147-153, 1971. 3 TAB, 16 REF.

DESCRIPTORS:
*ALGAE, *PHOSPHATES, *ENZYMES, *CULTURES, NITRATES, PLANT PHYSIOLOGY, PLANT GROWTH, BIOMASS, GROWTH RATES, LIGHT INTENSITY, CARBON DIOXIDE, NITROGEN, LABORATORY TESTS, CYANOPHYTA, CYTOLOGICAL STUDIES.

IDENTIFIERS:
*ALKALINE PHOSPHATASE, *ANABAENA FLOS-AQUAE, *CONTINUOUS CULTURE, CHEMOSTATS, PHOSPHATASE ACTIVITY.

ABSTRACT:
TO DEFINE INTERRELATIONSHIPS BETWEEN GROWTH AND IMPORTANT PHYSIOLOGICAL CHARACTERISTICS, THE CONTINUOUS CULTIVATION TECHNIQUE WAS APPLIED. ANABAENA FLOS-AQUAE WAS GROWN IN CHEMOSTATS WITH PHOSPHATE-LIMITING GROWTH AND VARIOUS DILUTION RATES TO ASCERTAIN THE INFLUENCE OF DILUTION RATES, PHOSPHATE AND NITRATE CONCENTRATIONS ON THE YIELDS AND ALKALINE PHOSPHATASE ACTIVITY OF THE ALGA. STEADY STATE CONDITIONS WERE ASSUMED TO EXIST WHEN THE DRY WEIGHT ENZYME ACTIVITIES AND CHEMICAL ANALYSIS WERE CONSTANT FOR FOUR CONSECUTIVE DAYS. THE CELL YIELDS WERE DEPENDENT ON DILUTION RATE AND A TWO-FOLD INCREASE OBTAINED BY GROWTH IN THE PRESENCE OF 15 MILLIMOLE POTASSIUM NITRATE. ALKALINE PHOSPHATASE ACTIVITY VARIED 20-FOLD, LOWEST ACTIVITY WITH EXCESS PHOSPHATE LIGHT-LIMITED CELLS AND HIGHEST ACTIVITY WITH CELLS GROWN IN THE PRESENCE OF 15 MILLIMOLE POTASSIUM NITRATE. THERE WAS NO CORRELATION BETWEEN HOT WATER SOLUBLE PHOSPHATE OF CELLS AND ALKALINE PHOSPHATASE ACTIVITY. INCREASING THE INTENSITY OF LIGHT 1.5 FOLD DID NOT AFFECT THE YIELD OF THE ALGAL CULTURES. THE ORTHOPHOSPHATE CONTENT OF CELLS FROM THE CHEMOSTATS, THROUGHOUT THESE EXPERIMENTS WAS IN THE RANGE OF 10-15% OF THE TOTAL PHOSPHATE INPUT, AND ORTHOPHOSPHATE WAS NEVER DETECTED IN THE GROWTH MEDIUM. (JONES-WISCONSIN)

FIELD 05C

ACCESSION NO. W72-05476

INVESTIGATION OF FOAM AS A MEANS OF DETECTING SMALL CONCENTRATIONS OF MINERAL AND ORGANIC SUBSTANCES IN NATURAL WATERS (ISSLEDOVANIYE PENY KAK METOD OBNARUZHENIYA MINERAL'NYKH I ORGANICHESKIKH VESHCHESTV, SODERZHASHCHIKHSYA V PRIRODNYKH VODAKH V MALYKH KOLICHESTVAKH),

AKADEMIYA NAUK SSSR, MOSCOW. INSTITUT BIOLOGII VNUTRENNYKH VOD.

S. M. DRACHEV, AND A. A. BYLINKINA.

IN: KHIMICHESKIYE RESURSY MOREY I OKEANOV; 'NAUKA', MOSCOW, P 202-208, 1970. 6 TAB, 7 REF.

DESCRIPTORS:
*WATER CHEMISTRY, *FOAM SEPARATION, *FOAM FRACTIONATION, *ORGANIC MATTER, *INORGANIC COMPOUNDS, IRON, MANGANESE, PHOSPHORUS, CARBON, NITROGEN, OXIDATION, BIOCHEMICAL OXYGEN DEMAND, RADIOACTIVITY, RADIOACTIVE WASTES, WASTE WATER(POLLUTION), WATER POLLUTION SOURCES, ALGAE, PLANKTON, SEASONAL.

IDENTIFIERS:
*USSR, MOSCOW RIVER, OKA RIVER, FOAM, NATURAL WATERS.

ABSTRACT:
FORMATION OF FOAM ON NATURAL AND POLLUTED WATERS OF THE MOSCOW AND OKA RIVERS IN THE FALL OF 1964 AND SUMMER OF 1965 IS DESCRIBED. THE CONTENT OF ORGANIC MATTER IN DRY FOAM RESIDUE FORMED DURING FOAM BREAKAGE IS USUALLY HIGH AND OFTEN EXCEEDS 50%. ORGANIC CARBON-NITROGEN RATIOS AND CONCENTRATIONS IN A POLLUTED REACH OF THE MOSCOW RIVER DECREASED WITH INCREASING DISTANCE FROM THE SOURCE OF POLLUTION. A HIGH FE, MN, AND P CONTENT WAS OBSERVED IN THE FOAM. THE LEVEL OF RADIOACTIVITY IN FOAM SAMPLES COLLECTED FROM NATURAL WATERS WAS HIGHER THAN THAT IN WATER AND INDICATED CONSIDERABLE CONCENTRATION OF URANIUM FISSION PRODUCTS. HIGHER RADIOACTIVITY IN FOAM IS PRESUMED TO BE RELATED TO ACCUMULATION OF IRON HYDROXIDES AND SURFACE-ACTIVE SUBSTANCES INTRODUCED WITH DOMESTIC AND INDUSTRIAL WASTES. (FOSEFSON-USGS)

FIELD 02K, 05A

ACCESSION NO. W72-05506

LOPEZ WATER SUPPLY PROJECT,

KOEBIG AND KOEBIG, INC., LOS ANGELES, CALIF.

C. H. LAWRANCE, K. G. TRANBARGER, AND R. A. CRAHN.

JOURNAL OF THE AMERICAN WATER WORKS ASSOCIATION, VOL. 63, NO. 11, P 711-727, NOVEMBER 1971. 16 FIG, 7 TAB, 15 REF.

DESCRIPTORS:
*WATER SUPPLY, *WATER RESOURCES DEVELOPMENT, *DISTRIBUTION SYSTEMS, *WATER QUALITY, WATER CONSERVATION, WILDLIFE CONSERVATION, FLOOD CONTROL, RECREATION, HYDROLOGIC DATA, WATER TREATMENT, STREAM GAGES, GROUNDWATER, SAFE YIELD, ALGAE, MINERALOGY, TURBIDITY, COLOR, ALKALINITY, VOLUMETRIC ANALYSIS, FISH, INTAKE STRUCTURES, CARBON DIOXIDE, CALCIUM CARBONATE, CALCIUM, MAGNESIUM, SODIUM, IRON, MANGANESE, SULFATES, CHLORIDES, CHLORINE, IONS, BICARBONATES, CARBONATES, COPPER, NITRATES, FLUORIDES, FLUORINE, CALIFORNIA.

IDENTIFIERS:
*LOPEZ WATER SUPPLY PROJECT, LOPEZ CREEK, ARROYO GRANDE CREEK.

ABSTRACT:
THE LOPEZ WATER SUPPLY PROJECT, A MULTIPURPOSE WATER-SUPPLY SYSTEM, HAS BEEN RECENTLY COMPLETED IN SAN LUIS OBISPO COUNTY, CALIFORNIA AT A COST OF $17,714,000. THIS PROJECT INCORPORATES WATER CONSERVATION, FLOOD CONTROL, DOMESTIC WATER SUPPLY, FISH AND WILDLIFE PRESERVATION AND RECREATION. THE LOPEZ AND ARROYO GRANDE CREEKS ARE THE SOURCES OF WATER UTILIZED BY THIS PROJECT. THEY WERE SAMPLED AND QUALITY CHECKED FOR MINERAL CONTENT, ALKALINITY, HARDNESS, COLOR AND TURBIDITY PRIOR TO AND AFTER CONSTRUCTION OF THE LOPEZ RESERVOIR. THE LABORATORY ANALYTICAL DATA FOR THE CREEKS AND THE RESERVOIR ARE SUMMARIZED IN TABULAR FORM. THE FEATURES OF THE ENTIRE PROJECT ARE SUMMARIZED INCLUDING INFORMATION ON THE INTAKE AND OUTLET FACILITIES (PIPING AND INSTRUMENTATION) OF THE RESERVOIR. ALGAL PROBLEMS AND CONTROL AND THE EFFECTS OF THE PROJECT ON FISH AND WILDLIFE ARE DISCUSSED. (HOLOMAN-BATTELLE)

FIELD 04A, 05A, 02K

ACCESSION NO. W72-05594

ALGAL DISTRIBUTION IN SIX THERMAL SPRING EFFLUENTS,

SOUTHEAST MISSOURI STATE COLL., CAPE GIRARDEAU. DEPT. OF BIOLOGY.

R. G. KULLBERG.

THE AMERICAN MICROSCOPICAL SOCIETY TRANSACTIONS, VOL 90, NO 4, P 412-434, OCTOBER 1971. 9 FIG, 6 TAB, 20 REF.

DESCRIPTORS:
*ALGAE, *DISTRIBUTION PATTERNS, *THERMAL SPRINGS, CYANOPHYTA, ALKALINITY, NUTRIENTS, WATER QUALITY, DIATOMS, CHLOROPHYTA, THERMAL WATER, *MONTANA, PHOSPHATES, NITRATES, NITRITES, IRON, SILICON, SULFATES, SODIUM, POTASSIUM, CALCIUM, MAGNESIUM, CHLORIDES, HYDROGEN ION CONCENTRATION, PERIPHYTON, WATER TEMPERATURE.

IDENTIFIERS:
ALHAMBRA HOT SPRINGS, BOULDER HOT SPRINGS, JACKSON HOT SPRINGS, LOLO HOT SPRINGS, PIPESTONE HOT SPRINGS, SLEEPING CHILD HOT SPRINGS.

ABSTRACT:
THE DISTRIBUTION OF ALGAE WAS STUDIED IN SIX THERMAL SPRINGS OF WESTERN MONTANA. ALHAMBRA (NORTH), 54.4 C; ALHAMBRA (SOUTH) 48.0 C; BOULDER, 61.3 C; JACKSON, 61.5 C; LOLO, 46.0 C; PIPESTONE, 59.5 CP AND SLEEPING CHILD, 52.0 C. THE BLUE-GREEN ALGAE (MYXOPHYCEAE) WERE THE ONLY ALGAE NEAR THE SOURCES OF THE STREAMS. THE MEAN MAXIMUM TEMPERATURE ENDURED BY THE DIATOMS (BACILLARIOPHYCEAE) WAS 43.2 C; BY THE GREEN ALGAE (CHLOROPHYCEAE), 40.9 C. PRESENCE LISTS OF THE ALGAE ALONG A TEMPERATURE GRADIENT INDICATE THE ORDER IN WHICH THE ALGAE APPEARED IN THE STREAM AS THE WATER COOLED. THE PER CENT VOLUMES FOR THE MAJOR AND INTERMEDIATE SPECIES WERE PLOTTED ALONG THE TEMPERATURE GRADIENT TO SHOW THE INTERACTIONS AMONG THE POPULATIONS OF THE CONTINUUM. THE STREAMS CONTAINING THE GREATEST VARIATIONS OF HABITATS RESULTED IN MORE ERRATIC CURVES OF SEVERAL MODES AMONG THE MAJOR POPULATIONS AND SCATTERED OCCURRENCES AMONG THE INTERMEDIATE SPECIES. VARIATIONS IN TYPES OF HABITATS DOWNSTREAM PROMOTED THE RELOCATION OF TRANSLOCATED ALGAL MASSES UNTIL OVERWHELMED BY THE PREVIOUSLY ESTABLISHED POPULATIONS. FIVE NEW TAXA ARE DESCRIBED: CHAMAESIPHON PRESCOTTI N. SP.; PSEUDANABAENA OBLONGA N. SP.; SYNECHOCOCCUS LIVIDUS COPELAND VAR. NANUM N. VAR; OSCILLATORIA GEMINATA COPELAND VAR. FRAGILIS FORMA BREVE N. FORMA; OSCILLATORIA GEMINATA COPELAND VAR. TENELLA FORMA MINOR N. FORMA. (MORTLAND-BATTELLE)

FIELD 05B, 05C

ACCESSION NO. W72-05610

POLYCHLORINATED BIPHENYLS: TOXICITY TO CERTAIN PHYTOPLANKTERS,

STATE UNIV. OF NEW YORK, STONY BROOK. MARINE SCIENCES RESEARCH CENTER.

J. L. MOSSER, N. S. FISHER, T. C. TENG, AND C. F. WURSTER.

SCIENCE, VOL 175, NO 4018, P 191-192, JANUARY 14, 1972. 1 FIG, 18 REF.

DESCRIPTORS:
*POLYCHLORINATED BIPHENYLS, *ALGAE, *WATER POLLUTION EFFECTS, *PESTICIDES, DIATOMS, PHYTOPLANKTON, TOXICITY, CULTURES, DDT, GROWTH RATES.

IDENTIFIERS:
THALASSIOSIRA PSEUDONANA, SKELETONEMA COSTATUM, DUNALIELLA TERTIOLECTA, EUGLENA GRACILIS, CHLAMYDOMANAS REINHARDTII, LETHAL DOSAGE.

ABSTRACT:
POLYCHLORINATED BIPHENYLS (PCBS) AND DDT WERE ADDED TO CULTURES OF FIVE SPECIES OF UNICELLULAR ALGAE TO INVESTIGATE THEIR EFFECTS ON GROWTH RATE. THE GROWTH RATES OF TWO CENTRIC DIATOMS (THALASSIOSIRA PSEUDONANA AND SKELETONEMA COSTATUM) WERE REDUCED BY PCB CONCENTRATIONS AS LOW AS 25 PPB FOR THE FORMER AND 10 PPB FOR THE LATTER. FOR BOTH THESE SPECIES, PCBS WERE MORE TOXIC THAN THE EQUIVALENT AMOUNT OF DDT. BY CONTRAST, A MARINE GREEN ALGAE (DUNALIELLA TERTIOLECTA) AND TWO FRESHWATER ALGAE (EUGLENA GRACILIS AND CHLAMYDOMONAS REINHARDTII) WERE NOT INHIBITED AT THESE, OR HIGHER, CONCENTRATIONS. THIS SENSITIVITY PARALLELED THAT TO DDT. (MORTLAND-BATTELLE)

FIELD 05C

ACCESSION NO. W72-05614

GREENE COUNTY SURFACE WATER RESOURCES,

ILLINOIS DEPT. OF CONSERVATION, SPRINGFIELD. DIV. OF FISHERIES.

R. LOCKART.

ILLINOIS DIVISION OF FISHERIES REPORT, AUGUST 1971. 44 P, 5 FIG, 12 TAB, 12 REF.

DESCRIPTORS:
*SURFACE WATERS, *LAKES, *STREAMS, *REVIEWS, *ILLINOIS, RECREATION, FISHING, SWIMMING, BOATING, DOCUMENTATION, WATER QUALITY, SILTING, AGRICULTURAL CHEMICALS, WATER POLLUTION SOURCES, ALGAE, GEOLOGY, DATA COLLECTIONS.

IDENTIFIERS:
*GREENE COUNTY(ILL).

ABSTRACT:
GREENE COUNTY, ILLINOIS, HAS A TOTAL SURFACE WATER INVENTORY OF 1,136.0 ACRES OF LAKES AND PONDS AND 494.70 ACRES OF NAMED STREAMS. THE LARGEST SINGLE WATER AREA (58.8 ACRES) IN GREENE COUNTY IS GREENFIELD CITY LAKE. THE MAIN DRAINAGE SYSTEMS ARE APPLE CREEK IN THE NORTHERN PART OF THE COUNTY AND MACOUPIN CREEK IN THE SOUTHERN PART OF THE COUNTY. MOST OF THE STREAMS IN GREENE COUNTY ARE SLOW MOVING. THE TWO EXCEPTIONS TO THIS ARE HURRICANE CREEK AND WOLF RUN. FLOODING ON ALL STREAMS IS A YEARLY OCCURRENCE DURING THE SPRING RAINS. ALONG THE ILLINOIS RIVER AND APPLE CREEK FLOODPLAIN DRAINAGE IS A MAJOR PROBLEM. MANY CANALS AND DITCHES HAVE BEEN CONSTRUCTED IN THESE TWO AREAS. SILTATION OCCURS IN ALL OF THE SLOWER MOVING STREAMS AND IS THE MAIN SOURCE OF WATER POLLUTION IN GREENE COUNTY. THE IMPOUNDMENTS AND STREAMS ARE MEDIUM HARD IN ALKALINITY. THE HARDNESS RANGES FROM 125 TO 350 PPM. PH VALUES RANGE FROM 7.4 TO 8.0. MOST OF THE LAKES AND PONDS CONTAIN SOME SPECIES OF SPORT FISH SUCH AS LARGEMOUTH BASS, BLUEGILL, CHANNEL CATFISH, AND BULLHEADS. (WOODARD-USGS)

FIELD 02H, 02E, 05B

ACCESSION NO. W72-05869

MERCURY: ITS OCCURRENCE AND EFFECTS IN THE ECOSYSTEM,

CORNELL UNIV., ITHACA, N.Y. ECOLOGY AND SYSTEMATICS SECTION.

D. B. PEAKALL, AND R. J. LOVETT.

BIOSCIENCE, VOL. 22, NO. 1, P 20-25, JANUARY 1972. 1 FIG, 3 TAB, 69 REF.

DESCRIPTORS:
*WATER POLLUTION, *FOOD CHAINS, *ECOSYSTEMS, ABSORPTION, FISH, INDUSTRIAL WASTES, ALGAE, AQUATIC PLANTS, SEA WATER, TRACE ELEMENTS, TOXICITY, CHROMOSOMES, FUNGICIDES, PESTICIDES, ECOLOGY, BIRD TYPES, BIRDS, AQUATIC ANIMALS, MOSSES, INVERTEBRATES, DIATOMS, PHYTOPLANKTON, PHOTOSYNTHESIS, WATER POLLUTION SOURCES, WATER POLLUTION EFFECTS, DDT, OCEANS, AIR POLLUTION, VOLCANOES, BIOASSAY, DIETS, SOIL MICROORGANISMS, ANIMALS, URINE, GENETICS, PATH OF POLLUTANTS.

IDENTIFIERS:
*MERCURY, *METHYLMERCURY, BRAIN, ORGANOMERCURIAL COMPOUNDS, ESOX LUCIUS, ALKYL MERCURY, EAGLES, SALMO GAIRDNERII, IRIS PSEUDACORUS, LYSIMACHIA NUMMULARIA, BLUE HERON, STERNA HIRUNDO, RATS, PARTRIDGE, DDE, KIDNEYS, LIVER, BLOOD, GASTEROSTEUS ACULEATUS, DAPHNIA MAGNA, DAPHNIA, GASTEROSTEUS.

ABSTRACT:
THE WORLD PRODUCTION AND INDUSTRIAL USES OF MERCURY, ITS NATURAL AND INDUSTRIAL ROUTES OF CONTAMINATION OF THE WATER AND OF THE ATMOSPHERE, AND ITS TOXICITY TO ANIMALS AND MAN ARE REVIEWED. MERCURY CAN FORM STABLE ORGANOMERCURY COMPOUNDS. OF THESE METHYL MERCURY IS OF SPECIAL IMPORTANCE BECAUSE OF ITS TOXICITY, ITS SLOW EXCRETION FROM THE BODY, AND ITS RELATIVELY LONG BIOLOGICAL HALF-LIFE. THE HUMAN HAZARD OF ENVIRONMENTAL MERCURY SEEMS TO BE FOR PERSONS WHOSE MAIN DIET IS FISH WHICH CONTAIN METHYL MERCURY OBTAINED THROUGH THE AQUATIC FOOD CHAIN. MERCURY IN THE ECOSYSTEM APPEARS TO MAINTAIN A BALANCED LEVEL. MERCURY LEVELS IN THE OCEAN HAVE NOT BEEN CHANGED BY MAN'S ACTIVITY. HOWEVER, MAN'S USE OF MERCURY CAN HAVE A SIGNIFICANT EFFECT ON FRESHWATER AND THE LAND ENVIRONMENT. (JEFFERIS-BATTELLE)

FIELD 05B, 05A, 05C

ACCESSION NO. W72-05952

RADIOCHEMICAL DETERMINATION OF PLUTONIUM IN SEA WATER, SEDIMENTS AND MARINE ORGANISMS,

WOODS HOLE OCEANOGRAPHIC INSTITUTION, MASS.

K. M. WONG.

ANALYTICA CHIMICA ACTA, VOL 56, NO 3, P 355-364, OCTOBER 1971. 1 FIG, 5 TAB, 18 REF.

DESCRIPTORS:
*RADIOCHEMICAL ANALYSIS, *SEA WATER, *SEDIMENTS, *MARINE ANIMALS, *MARINE ALGAE, RADIOISOTOPES, FALLOUT, CHEMICAL PRECIPITATION, IRON, BIOASSAY, LEACHING, VOLUMETRIC ANALYSIS, SOIL ANALYSIS, SPECTROMETERS, ZOOPLANKTON, MUSSELS, CLAMS, FISH, SHELLFISH, AQUEOUS SOLUTIONS, PHAEOPHYTA, ATLANTIC OCEAN, SHARKS, RADIOACTIVITY TECHNIQUES.

IDENTIFIERS:
*PLUTONIUM, DETECTION LIMITS, BIOLOGICAL SAMPLES, SARGASSO WEED, STARFISH, ASTERIAS FORBESI, PLUTONIUM 236, PLUTONIUM 238, PLUTONIUM 239, POLYMERIZATION, LIVER, BONE, DETECTION LIMITS, CHEMICAL RECOVERY, SAMPLE PREPARATION.

ABSTRACT:
A RADIOCHEMICAL PROCEDURE FOR PLUTONIUM IS DESCRIBED WHICH HAS A SENSITIVITY OF 0.004 D.P.M. PER 100 LITER OF SEA WATER (FOR A 50-LITER SAMPLE), 0.02 D.P.M. PER KG OF SEDIMENTS (100-G SAMPLE) AND 0.002 D.P.M. PER KG OF MARINE ORGANISMS (1-KG SAMPLE). AN IRON(II) HYDROXIDE COPRECIPITATION METHOD IS USED FOR THE CONCENTRATION OF PLUTONIUM IN SEA WATER. A NITRIC-HYDROCHLORIC ACID LEACHING METHOD IS ADAPTED FOR THE TREATMENT OF SEDIMENTS AND ASHED ORGANISMS. OF 30 SAMPLES OF 5-60 LITERS OF SEA WATER ANALYZED RADIOCHEMICALLY, THE AVERAGE PLUTONIUM RECOVERY WAS 34-70 PERCENT AS COMPARED WITH 11-39 PERCENT RECOVERY ON 38 SAMPLES BY THE IRON(III) HYDROXIDE METHOD. A POSSIBLE REASON FOR THE LOW RECOVERY WITH THE IRON(III) HYDROXIDE PROCEDURE IS THAT PLUTONIUM IS LOST THROUGH THE FORMATION OF PLUTONIUM(IV) POLYMERS. THE AVERAGE RECOVERY OF 75 SAMPLES OF SEDIMENTS AND ASHED ORGANISMS WAS 43-83 PERCENT BY THE LEACHING METHOD. THE RESULTS OF THE VARIOUS REPLICATE ANALYSES ARE TABULATED, AND THE FACTORS INFLUENCING THE RECOVERY, CONTAMINATION, AND BLANK ACTIVITY ARE DISCUSSED. (HOLOMAN-BATTELLE)

FIELD 05A, 02K, 02L

ACCESSION NO. W72-05965

ECOLOGICAL INVESTIGATIONS ON THE PLANKTON OF THE RIVERIS RESERVOIR (OKOLOGISCHE UNTERSUCHUNGEN AM PLANKTON DER RIVERISTALSPERRE),

BONN UNIV. (WEST GERMANY).

H. SCHNITZLER.

ARCHIV FUR HYDROBIOLOGIE, VOL 69, NO 1, P 60-94, AUGUST 1971. 18 FIG, 17 REF.

DESCRIPTORS:
*PLANKTON, *RESERVOIRS, *AQUATIC ALGAE, *AQUATIC ANIMALS, *BIOMASS, *PRIMARY PRODUCTIVITY, *SECONDARY PRODUCTIVITY, PHYTOPLANKTON, ZOOPLANKTON, EUTROPHICATION, OLIGOTROPHY, PROTOZOA, CHLORELLA, CHLAMYDOMONAS, CYANOPHYTA, CHLOROPHYTA, CHRYSOPHYTA, DIATOMS, DINOFLAGELLATES, SCENEDESMUS, ROTIFERS, CRUSTACEANS, COPEPODS, NUTRIENTS, NITROGEN, PHOSPHORUS, ECOLOGY, BIOINDICATORS, WATER QUALITY, NITRATES, PHOSPHATES, SEASONAL, DISTRIBUTION PATTERNS.

ABSTRACT:
FROM NOVEMBER 1966 TO OCTOBER 1968 THE QUANTITY AND QUALITY OF THE ZOOPLANKTON AND PHYTOPLANKTON OF THE RIVERIS RESERVOIR NEAR TRIER WERE INVESTIGATED IN RELATION TO THE ABIOTIC FACTORS IN THE ENVIRONMENT. THE QUANTITY OF INORGANIC NUTRIENTS COMING INTO THE RESERVOIR WAS CALCULATED FROM THE RESULTS OF CHEMICAL ANALYSIS OF THE TRIBUTARIES. THE RESULTS OBTAINED BY MEASURING NITROGENOUS AND PHOSPHOROUS COMPOUNDS AND OXYGEN QUALIFY THE RIVERIS RESERVOIR AS AN OLIGOTROPHIC WATER. THE PHYTOPLANKTON AND ZOOPLANKTON SPECIES FOUND ARE TABULATED AND THE SEASONAL DISTRIBUTION OF THE MORE COMMON SPECIES IS DIAGRAMMED. MOST OF THE PHYTOPLANKTON SPECIES ARE CHARACTERISTIC OF OLIGOTROPHIC WATERS. THE BIOMASS OF THE PHYTOPLANKTON WAS CALCULATED AND OWING TO THE OCCURRENCE OF THE LARGE GYMNODINIUM UBERRIMUM THERE WAS FROM TIME TO TIME IN SUMMER 1967 A COMPARATIVELY HIGH TOTAL BIOMASS WHICH, HOWEVER, SHOULD NOT NECESSARILY BE INTERPRETED AS A SIGN OF INCIPIENT EUTROPHY. (HOLOMAN-BATTELLE)

FIELD 05C, 05A, 02H

ACCESSION NO. W72-05968

MERCURY INHIBITION ON LIPID BIOSYNTHESIS IN FRESHWATER ALGAE,

WESTERN WASHINGTON STATE COLL., BELLINGHAM, WASHINGTON. DEPT. OF CHEMISTRY.

ROBERT S. MATSON, GEORGE E. MUSTOE, AND S. B. CHANG.

ENVIRONMENTAL SCIENCE AND TECHNOLOGY, VOL 6, NO 2, FEBRUARY 1972, P 158-160. 2 FIG, 2 TAB, 12 REF.

DESCRIPTORS:
*ALGAE, *INHIBITION, *LIPIDS, TOXICITY, HEAVY METALS, LABORATORY TESTS, PHOTOSYNTHESIS, CHLOROPHYLL, ENZYMES, PERMEABILITY, WATER QUALITY CONTROL, WATER POLLUTION EFFECTS.

IDENTIFIERS:
*MERCURY, *SYNTHESIS, MERCURIC CHLORIDE, METHYL MERCURIC CHLORIDE, ANKISTRODESMUS BRAUNII, EUGLENA GRACILIS.

ABSTRACT:
WHOLE CELLS OF SPECIMENS OF UNICELLULAR ALGAE, ANKISTRODESMUS BRAUNII AND EUGLENA GRACILIS, WERE EXPOSED TO VARIOUS CONCENTRATIONS OF INORGANIC MERCURIC AND METHYL MERCURIC CHLORIDE TO DETERMINE IF INHIBITION OF LIPID BIOSYNTHESIS WAS ONE OF THE TOXIC EFFECTS OF MERCURY COMPOUNDS. AT A MERCURIC CHLORIDE CONCENTRATION OF 3.5 PPM, CHLOROPHYLL SYNTHESIS WAS 98% INHIBITED AND GALACTOLIPID SYNTHESIS WAS 50% INHIBITED IN A. BRAUNII. HOWEVER, SIGNIFICANT INHIBITION OF BOTH SYNTHESIS PROCESSES WAS DETECTED AT MERCURIC CHLORIDE LEVELS LESS THAN 1.0 PPM. FOR METHYL MERCURIC CHLORIDE, A 2.0 PPM LEVEL INHIBITED 98% OF CHLOROPHYLL SYNTHESIS AND 85% OF GALACTOLIPID SYNTHESIS. THESE MERCURY COMPOUNDS WERE ALSO SHOWN TO SPECIFICALLY INHIBIT THE GALACTOSYL TRANSFERASE ACTIVITY IN EUGLENA CHLOROPLASTS. THESE RESULTS COULD LOGICALLY BE EXTENDED TO OTHER PHOTOSYNTHETIC ORGANISMS, INCLUDING PHYTOPLANKTON AND OTHER SPECIES OF UNICELLULAR ALGAE. (LOWRY-TEXAS)

FIELD 05C

ACCESSION NO. W72-06037

THE ROLE OF BENTHIC PLANTS IN A FERTILIZED ESTUARY,

HARVARD UNIV., CAMBRIDGE, MASS. LAB. OF APPLIED MICROBIOLOGY.

T. WAITE, AND R. MITCHEL.

PREPRINT, PRESENTED AT 44TH ANNUAL CONFERENCE OF WATER POLLUTION CONTROL FEDERATION, SESSION 13, NO. 1, SAN FRANCISCO, CALIFORNIA, OCTOBER 5, 1971, 17 P, 1 FIG, 6 TAB, 17 REF.

DESCRIPTORS:
*NUTRIENTS, *BENTHIC FAUNA, *BENTHIC FLORA, *PHYTOPLANKTON, ALGAE, GROWTH RATES, BIOASSAY, DISSOLVED OXYGEN, CARBON DIOXIDE, AMMONIA, PHOSPHATE, ANALYTICAL TECHNIQUES, *PHOTOSYNTHESIS, CARBON FIXATION, EUTROPHICATION, WATER QUALITY CONTROL, MASSACHUSETTS, *NUTRIENTS, OUTLETS.

IDENTIFIERS:
*WOODS HOLE(MASS).

ABSTRACT:
FIELD PRODUCTIVITY TESTS WERE CONDUCTED USING CARBON-14 TECHNIQUES TO COMPARE PHYTOPLANKTON AND BENTHIC PLANTS. TEST STATIONS WERE CHOSEN AT VARIOUS DISTANCES FROM THE WOODS HOLE, MASSACHUSETTS, SEWAGE OUTFALL. THE BOTTOM FLORA OF THE AREA GENERALLY CONSISTS OF A LITTORAL ZONE, AND THE PRODUCTIVITY DETERMINATIONS WERE RUN ONLY ON THE BENTHIC ALGAL ZONE OR NON-ROOTED PLANTS. ON AN AREA BASIS THE BENTHIC PLANTS IN THE LITTORAL ZONE FIX CARBON AT ABOUT 40 TIMES THE RATE OF PHYTOPLANKTON. ALSO, THE WATER MOVEMENT ASSOCIATED WITH ESTUARINE OR COASTAL AREAS MIXED THE PHYTOPLANKTON DILUTING THE EFFECT OF ENRICHMENT, BUT DID NOT MIX BENTHIC MACROPHYTES WHICH WERE CONTINUALLY AFFECTED BY ENRICHMENT. THE BENTHIC PLANTS ARE BETTER INDICATORS OF NUTRIENT STIMULATION. IT IS PROPOSED THAT THE BAULEP MITSCHERLICH EQUATION BE USED TO PREDICT THE CONTRIBUTION TO PHOTOSYNTHETIC YIELD OF THE BOTTOME FLORA AS A FUNCTION OF NUTRIENT ENRICHMENT, DISCOUNTING THE CONCEPT OF A 'SINGLE' LIMITING NUTRIENT. (MORGAN-TEXAS)

FIELD 05C, 02L, 05B

ACCESSION NO. W72-06046

METHODS IN MICROBIOLOGY-VOLUME 5B.

ACADEMIC PRESS INC., NEW YORK, N.Y. 1971. J. R. NORRIS AND D. W. RIBBONS, EDITORS. 695 P, 21 REF.

DESCRIPTORS:
*SEPARATION TECHNIQUES, *ALGAE, *BACTERIA, *FUNGI, *BACTERIOPHAGE, *VIRUSES, ABSORPTION, ELECTROPHORESIS, DETERGENTS, AMINO ACIDS, SPECTROPHOTOMETRY, MAGNESIUM, SULFATES, SPORES, AZOTOBACTER, CYANOPHYTA, YEASTS, CARBON DIOXIDE, CHLORELLA, CLOSTRIDIUM, E. COLI, EUGLENA, EUGLENOPHYTA, CHLOROPHYTA, FERROBACILLUS, GAS CHROMATOGRAPHY, IONS, CALCIUM, CARBON, SNAILS, LACTOBACILLUS, MAGNESIUM, MANGANESE, MYCOBACTERIUM, HYDROGEN ION CONCENTRATION, PHOSPHATES, POTASSIUM, SALMONELLA, SHIGELLA, PROTOZOA, SODIUM, CENTRIFUGATION, ULTRASONICS, CULTURES, HEAVY METALS, IRON, COLIFORMS, ISOLATION.

ABSTRACT:
THE DISINTEGRATION OF CELLS, THEIR CHEMICAL ANALYSIS, AND THE TECHNIQUES USED TO SEPARATE AND CHARACTERIZE THEIR COMPONENTS ARE DISCUSSED. TOPICS INCLUDED ARE: FREE-FLOW ELECTROPHORESIS, DISC ELECTROPHORESIS, PREPARATIVE ZONAL ELECTROPHORESIS, REFLECTANCE SPECTROPHOTOMETRY, ISOELECTRIC FOCUSING AND SEPARATION OF PROTEINS, CHEMICAL EXTRACTION METHODS OF MICROBIAL CELLS, CHEMICAL ANALYSIS OF MICROBIAL CELLS, CENTRIFUGAL TECHNIQUES FOR ISOLATION AND CHARACTERIZATION OF SUB-CELLULAR COMPONENTS FROM BACTERIA. (HOLOMAN-BATTELLE)

FIELD 05A, 02K

ACCESSION NO. W72-06124

MICROSCOPIC WATER QUALITY AND FILTRATION EFFICIENCY,

NEW YORK STATE DEPT. OF HEALTH, ALBANY. BUREAU OF WATER AND WASTEWATER UTILITIES MANAGEMENT.

S. SYROTYNSKI.

JOURNAL AMERICAN WATER WORKS ASSOCIATION, VOL. 63, NO. 4, P 237-245, APRIL 1971. 8 FIG, 6 TAB, 3 REF.

DESCRIPTORS:
*WATER TREATMENT, *FILTRATION, *MICROORGANISMS, *SURFACE WATERS, *PLANKTON, *ALGAE, TURBIDITY, *NEW YORK, *WATER QUALITY.

IDENTIFIERS:
*MICROSCOPIC COUNTS, *AMORPHOUS MATTER.

ABSTRACT:
HISTORICAL RECORDS OF THE NEW YORK STATE DEPARTMENT OF HEALTH FOR THE PERIOD 1946-1967 WERE STUDIED. IN SURFACE WATERS THE MEAN VALUES FOR TOTAL MICROSCOPIC COUNT, AMORPHOUS MATTER, TOTOAL PLANKTON AND TOTAL ALGAE WERE 2,000, 1,500, 350, AND 140 ASU/ML, RESPECTIVELY. THE RANGE OF TOTAL MICROSCOPIC COUNTS WAS FROM 14 TO 195,900 ASU/ML; VALUES ABOVE 20,000 GENERALLY WERE ASSOCIATED WITH HIGH TURBIDY (TO 30 UNITS). OF 172 TURBIDITY VALUES WITHIN THE RANGE 1,500-2,500 ASU/ML, 91/8 PERCENT WERE LESS THAN 10, AND 65.7 PERCENT LESS THAN 5 TURBIDITY UNITS. A MEAN VALUE OF 140 ASU/ML WAS DETERMINED FOR TOTAL ALGAE, THOUGH MAXIMUM VALUES WERE AS HIGH AS 10,700. THE HISTORICAL DATA FOR FILTERED-WATER TOTAL COUNTS SHOW 80 PERCENT EQUAL TO OR LESS THAN 200 ASU/ML. A VALUE OF 200 ASU/ML TOTAL MICROSCOPIC COUNT IS RECOMMENDED AS A LIMIT IN FILTER EFFLUENT QUALITY. OF UNTREATED SURFACE WATERS 3.65 PERCENT CONTAINED LESS THAN 200 ASU/ML AND 7 PERCENT LESS THAN 300. (BEAN-AWWARF)

FIELD 05F, 05A

ACCESSION NO. W72-06191

PHOSPHATE AND TURBIDITY CONTROL BY FLOCCULATION AND FILTRATION,

WAHNBACHTALSPERRENVERBAND, SIEGBURG (WEST GERMANY).

H. BERNHARDT, J. CLASEN, AND H. SCHELL.

JOURNAL OF AMERICAN WATER WORKS ASSOCIATION, VOL. 63, NO. 6, P 355-368, JUNE 1971. 16 FIG, 12 TAB, 40 REF.

DESCRIPTORS:
*WATER TREATMENT, *PHOSPHATES, *TURBIDITY, *FLOCCULATION, *FILTRATION, *ALGAE, ELECTROLYTES, ACTIVATED CARBON, HYDROGEN ION CONCENTRATION, NUTRIENTS.

IDENTIFIERS:
*FERRIC CHLORIDE, MULTI-LAYER FILTERS, ANTHRACITE, *WAHNBACH RESERVOIR(GERMANY), OSCILLATORA RUBESCENS.

ABSTRACT:
HIGH NUTRIENT LOAD ENTERING THE WAHNBACH RESERVOIR HAS LED TO INCREASES IN BIO-PRODUCTION AND A CHANGE IN PREDOMINANT SPECIES FROM DIATOMACEAE TO BLUE-GREEN ALGAE, THE MAIN ALGAE BEING OSCILLATORIA RUBESCENS, WHICH OCCUR IN MASSES. BECAUSE 70 PERCENT OF THE NUTRIENTS ORIGINATE FROM AGRICULTURAL ACTIVITY, EXCESS FERTILIZATION CANNOT BE COMBATTED BY TREATING ONLY THE DOMESTIC SEWAGE OF THE 8,000 INHABITANTS. ALSO IRON COMPOUNDS ARE PRESENT IN THE RUNOFF. THE RIVER WATER WILL BE TREATED AS IT FLOWS INTO THE RESERVOIR BY FLOCCULATION WITH FERRIC IRON AND FILTRATION, SO THAT TOTAL PHOSPHORUS WILL BE REDUCED TO A MAXIMUM OF 10 PPB. ALSO TURBIDITY WILL BE REMOVED. REMOVAL OF PHOSPHATE IS OPTIMUM AT A PH OF UP TO 7.2 USING 40 TIMES THE STOICHIOMETRIC FERRIC IRON REQUIREMENT. USE OF AN ANIONIC FLOCCULANT AID EXTENDED FILTER RUNS TO APPROXIMATELY DOUBLE THE ORIGINAL LENGTH. USE OF A THREE-LAYER FILTER PRODUCED BETTER FILTER RUNS THAN A ONE-LAYER OR TWO-LAYER FILTER AND BETTER USE OF THE FILTER BED. BY COMBINING ACTIVATED CARBON (THE LIGHTEST LAYER), HYDROANTHRACITE (CENTRAL LAYER), AND QUARTZ SAND (BOTTOM) A STABLE, THREE LAYER FILTER WAS CONSTRUCTED WITHOUT USE OF EXPENSIVE HEAVY MATERIALS. (BEAN-AWWARF)

FIELD 05F, 05C

ACCESSION NO. W72-06201

THE BEHAVIOR OF CHLORELLA PYRENOIDOSA IN STEADY STATE CONTINUOUS CULTURE,

CALIFORNIA UNIV., BERKELEY. LAWRENCE RADIATION LAB.

J. N. DABES, C. R. WILKE, AND K. H. SAUER.

AVAILABLE FROM THE NATIONAL TECHNICAL INFORMATION SERVICE AS UCRL-19958, $3.00 IN PAPER COPY, $0.95 IN MICROFICHE. AUGUST 1970. 184 P, 43 FIG, 3 TAB, 138 REF.

DESCRIPTORS:
*ALGAE, *CULTURES, PHOTOSYNTHESIS, BIOMASS, LIGHT INTENSITY, GROWTH RATES, MATHEMATICAL MODELS, WATER TEMPERATURE, WATER QUALITY, CARBON DIOXIDE, HYDROGEN ION CONCENTRATION, CARBON, OXYGEN, NITROGEN, HYDROGEN, PRIMARY PRODUCTIVITY, RESPIRATION, PLANT PHYSIOLOGY, NUTRIENTS, NITRATES, SULFATES, PHOSPHATES, CHLORIDES, BORATES, ABSORBANCE.

IDENTIFIERS:
DNA, CHLOROPHYLL A, CHLORELLA PYRENOIDOSA, SPIROGIRA, CHLORELLA ELLIPSOIDEA, RNA, CHLOROPHYLL B, ASHING, LABORATORY TECHNIQUES.

ABSTRACT:
THE GROWTH OF CHLORELLA PYRENOIDOSA IN STEADY STATE CONTINUOUS CULTURE WAS NEVER LIMITED BY CO2, MINERALS, PH, OR TEMPERATURE. THE EFFECTS OF THE TWO REMAINING INDEPENDENT VARIABLES, SPECIFIC GROWTH RATE AND INCIDENT LIGHT INTENSITY, ON ALGAL BIOMASS PRODUCTIVITY AND ALGAL PHYSIOLOGY WERE EXAMINED. IT WAS FOUND THAT OPTIMUM ALGAL BIOMASS PRODUCTIVITY WAS OBTAINED AT A SPECIFIC GROWTH RATE OF APPROXIMATELY 1.6 DAY, WHEN THE INCIDENT LIGHT INTENSITY WAS 8.05 MW/SQ CM. THIS OPTIMUM SPECIFIC GROWTH RATE IS NOT EXPECTED TO CHANGE SIGNIFICANTLY AS A FUNCTION OF INCIDENT LIGHT INTENSITY. TOTAL CHLOROPHYLL CONTENT, CHLOROPHYLL A/CHLOROPHYLL B RATIO, LIGHT SATURATED RATE OF PHOTOSYNTHESIS, DARK RESPIRATION RATES, AND RNA CONTENT WERE STRONG FUNCTIONS OF SPECIFIC GROWTH RATE. ON THE OTHER HAND, MAXIMUM QUANTUM EFFICIENCY, LIGHT SATURATED RATE AND MAXIMUM QUANTUM EFFICIENCY OF THE QUINONE HILL REACTION, AND DNA CONTENT CHANGED LITTLE, IF AT ALL, AS A FUNCTION OF SPECIFIC GROWTH RATE. A MATHEMATICAL EXPRESSION FOR THE LIGHT RESPONSE CURVE OF PHOTOSYNTHESIS WAS FORMULATED AND A MATHEMATICAL MODEL FOR THE PERFORMANCE OF OPTICALLY DENSE ALGAL SYSTEMS, WHICH ARE OF INTEREST FOR THE MASS CULTURE OF ALGAE, IS PRESENTED. THIS MODEL DIFFERS FROM PREVIOUS MODELS, SINCE IT USES THE ABOVE-MENTIONED LIGHT RESPONSE CURVE TO DESCRIBE THE LOCAL RATE OF PHOTOSYNTHESIS AND ALSO COUNTS FOR CHANGES IN THE PHYSIOLOGY OF THE ALGAE. (MORTLAND-BATTELLE)

FIELD 05C, 05A

ACCESSION NO. W72-06274

IMPROVEMENT AND APPLICATION OF BENTHIC ALGAL ISOTOPE PRODUCTIVITY MEASURING METHODS,

HAWAII UNIV., HONOLULU. DEPT. OF BOTANY.

M. S. DOTY.

AVAILABLE FROM THE NATIONAL TECHNICAL INFORMATION SERVICE AS UH-235-P-4-5, $3.00 IN PAPER COPY, $0.95 IN MICROFICHE. JUNE 1971. 94 P, 12 FIG, 14 TAB, 51 REF. CONT. NO. AT(04-3)-235.

DESCRIPTORS:
*MARINE ALGAE, *BENTHIC FLORA, *PRIMARY PRODUCTIVITY, *ENERGY CONVERSION, STANDING CROP, CHLOROPHYTA, CYANOPHYTA, RHODOPHYTA, PHAEOPHYTA, CARBON RADIOISOTOPES, MARINE FISH, INSTRUMENTATION, ABSORPTION, SEDIMENT DISCHARGE, REGRESSION ANALYSIS, BIOMASS, METHODOLOGY, REEFS, SAMPLING, OXYGEN, TEMPERATURE, LIGHT, WAVES(WATER), PHYTOPLANKTON, LIMESTONES, GASTROPODS, MOLLUSKS, INVERTEBRATES, CRABS, PHOTOSYNTHESIS, MARINE ANIMALS, BENTHIC FAUNA, PACIFIC OCEAN, RESPIRATION, RADIOACTIVITY TECHNIQUES, REPRODUCTION, ABSORPTION, CARBON, TRACERS.

IDENTIFIERS:
SEAWEEDS, APHIDS, SPONGES, CARBON-14.

ABSTRACT:
THERE IS NEED FOR A QUANTIFICATION OF THE DIFFERENT FACTORS THAT CAUSE VARIATION IN BENTHIC ALGAL POPULATIONS AND THE PASSAGE OF ENERGY AND MATERIALS THROUGH THEM. A REVIEW OF THREE YEARS' WORK ON THE ROLES OF BENTHIC ALGAE (PRIMARY PRODUCTIVITY, SEDIMENT DEPOSITION, LIMESTONE PRODUCTION) IN THE TROPICAL PACIFIC IS GIVEN. THE WORK CARRIED OUT IN THE 3-YEAR PERIOD INCLUDED (1) TECHNIQUES FOR ELUCIDATING THE ROLES OF BENTHIC ALGAE; (2) QUANTITATIVE DESCRIPTIONS OF THE ALGAL CROPS, THEIR COMPONENTS, AND COMPOSITION, AS WELL AS PRODUCTIVITY; AND (3) DETERMINATIONS OF STANDING CROP OR BIOMASS(C), PRIMARY PRODUCTION OR PHOTOSYNTHESIS(P), RESPIRATION AND REPRODUCTION(R), AND LOSSES TO THE ALGAL CROP(D). NEW TECHNIQUES HAVE BEEN DEVELOPED AND APPLIED FOR RADIOISOTOPIC ANALYSES (FOR EXAMPLE, RADIOACTIVE TRACER AND CHEMICAL OXYGEN PRODUCTIVITY MEASUREMENT), AND INSTRUMENTATION FOR RADIATION AND OTHER MEASUREMENTS HAS BEEN ACCUMULATED. TABULATED AND GRAPHIC DATA INCLUDE EFFECTS OF TEMPERATURE, LIGHT, AND WAVE ACTION ON STANDING ALGAL CROP, STATISTICS DEALING WITH CROP VARIATIONS, PRIMARY PRODUCTIVITY, REDUCED OXYGEN IN RELATION TO PRIMARY PRODUCTIVITY, UPTAKE AND UTILIZATION OF RADIOCARBON, AND ALGAL-ANIMAL ASSOCIATIONS. SOME OF THE INCLUDED RESEARCH REPORTS ARE 'PHYSICAL FACTORS IN THE PRODUCTION OF TROPICAL BENTHIC MARINE ALGAE', 'ABRASION BY WATER-BORNE SEDIMENTS AS AN ERROR PRODUCING FACTOR IN THE MEASUREMENT OF DIFFUSION GRADIENTS THROUGH DISSOLUTION OF CALCIUM SULFATE', 'THE EFFECT OF OXGEN CONCENTRATION ON PRODUCTIVITY MEASUREMENT', 'BENTHIC ALGAL PRODUCTIVITY MEASUREMENTS CORRELATED WITH STANDING CROP HARVESTS'.
(HOLOMAN-BATTELLE)

FIELD 05C, 05A

ACCESSION NO. W72-06283

THE SUSPENSION AND SINKING OF PHYTOPLANKTON IN THE SEA,

RHODE ISLAND UNIV., KINGSTON. GRADUATE SCHOOL OF OCEANOGRAPHY.

THEODORE J. SMAYDA.

OCEANOGR. MAR. BIOL. ANN. REV., VOL. 8, (HAROLD BARNES, EDITOR), 1970, P 353-414. 7 FIG, 9 TAB, 226 REF.

DESCRIPTORS:
*PHYTOPLANKTON, *MARINE ALGAE, *SEDIMENTATION RATES, *REVIEWS, BIOCHEMISTRY, GEOCHEMISTRY, RADIOISOTOPES, PATH OF POLLUTANTS, PHYSIOLOGICAL ECOLOGY, ENVIRONMENTAL EFFECTS, PRODUCTIVITY, RADIOECOLOGY, OCEANOGRAPHY.

ABSTRACT:
FAECEL-PELLET TRANSPORT MAY EXPLAIN RADIONUCLIDE SINKING RATES OBSERVED BY OSTERBERG AND OTHERS (NATURE, 1963, 200, 1276-71). THIS REVIEW CONSIDERS: SUSPENSION MECHANISMS (MORPHOLOGICAL, PHYSIOLOGICAL, PHYSICAL); SINKING-RATES (DETERMINATION METHODS, RESULTS, AND EFFECTS ON POPULATION DISTRIBUTIONS); EFFECTS ON ECOLOGY; BIOGEOCHEMICAL CONSEQUENCES (NON-CONSERVATIVE-SUBSTANCE DISTRIBUTIONS, MATTER TRANSPORT TO DEPTH, PHYTOPLANKTON REMAINS IN SEDIMENTS); ACCELERATED SINKING (DENSITY INVERSION CURRENTS, AGGREGATES, DOWNWELLING, FAECAL PELLETS). (BOPP-ORNL)

FIELD 05B, 05C

ACCESSION NO. W72-06313

LABORATORY STREAM RESEARCH: OBJECTIVES, POSSIBILITIES, AND CONSTRAINTS,

OREGON STATE UNIV., CORVALLIS. DEPT. OF FISHERIES AND WILDLIFE.

CHARLES E. WARREN, AND GERALD E. DAVIS.

IN: ANNUAL REVIEW OF ECOLOGY AND SYSTEMATICS, VOLUME 2; ANNUAL REVIEWS INC.; PALO ALTO, CALIFORNIA. R. F. JOHNSTON, EDITOR, 1971, P 111-144. 8 FIG, 88 REF.

DESCRIPTORS:
*WATER POLLUTION EFFECTS, *MODEL STUDIES, *STREAMS, *ECOSYSTEMS, AQUATIC HABITATS, BIOMASS, AQUATIC ALGAE, AQUATIC ANIMALS, BALANCE OF NATURE, ECOLOGY, TROPHIC LEVEL, HERBIVORES, NATURAL STREAMS, PULP WASTES, PRODUCTIVITY, DIELDRIN, PATH OF POLLUTANTS, REVIEWS, FOOD CHAINS, FORECASTING, CONSTRAINTS, RESEARCH FACILITIES.

ABSTRACT:
RELATIVELY SMALL, CONSTRUCTED CHANNELS HAVING CONTROLLED WATER FLOW, WHICH HAVE BEEN DEVELOPED MAINLY OVER THE PAST 15 YEARS, HAVE POTENTIAL FOR DETAILED EXAMINATION OF THE FUNCTION OF ECOSYSTEM PARTS AS THEY LEAD TO THE BEHAVIOR OF THE WHOLE. FOOD ORGANISM BIOMASS MAY SERVE AS A 'SUFFICIENT' (R. LEVINS, AM. SCI. 54, 421-431(1966)) PARAMETER FOR SOME TROPHIC MODELS. A VERY WIDE RANGE OF CIRCUMSTANCES MAY BE MODELLED WITH REGARD TO POLLUTION, HOWEVER CONSTRAINTS, HAVING TO DO WITH THE SUCCESSION, DIVERSITY, AND STABILITY OF COMMUNITIES MAY BE DIFFERENT FROM NATURAL ECOSYSTEMS, INTENTIONALLY OR NOT. (BOPP-ORNL)

FIELD 05C

ACCESSION NO. W72-06340

TRITIATION OF AQUATIC ANIMALS IN AN EXPERIMENTAL FRESHWATER POOL,

CALIFORNIA UNIV., LIVERMORE. LAWRENCE RADIATION LAB.

FLORENCE L. HARRISON, AND JOHN J. KORANDA.

AVAILABLE FROM THE NATIONAL TECHNICAL INFORMATION SERVICE AS UCRL-72930, $3.00 IN PAPER COPY, $0.95 IN MICROFICHE. AUGUST 3, 1971. 35 P.

DESCRIPTORS:
*FOOD CHAINS, *RADIOACTIVITY EFFECTS, *TRITIUM, *NUCLEAR WASTES, WATER POLLUTION EFFECTS, CATTAILS, AQUATIC ANIMALS, CLAMS, CARP, CRAYFISH, AQUATIC ALGAE, ABSORPTION, AQUARIA, ANIMAL PHYSIOLOGY, WATER BALANCE, TRACERS, PLANT PHYSIOLOGY, AQUATIC PLANTS, RADIOECOLOGY.

ABSTRACT:
FRESHWATER ANIMALS (CLAMS, CRAYFISH, AND GOLDFISH) AND ALGAE SHOWED HALF TIMES OF LESS THAN ONE DAY FOR EQUILIBRATION OF MORE THAN 95% OF TISSUE WATER TO THE SPECIFIC ACTIVITY OF POOL WATER. THE SPECIFIC ACTIVITY OF THE TISSUE WATER OF A PLANT (CATTAILS) REACHED ONLY 60-70% OF THAT OF THE POOL WATER SINCE THERE WAS INTERCHANGE WITH ATMOSPHERIC WATER THROUGH THE LEAVES. RATIOS OF THE SPECIFIC ACTIVITY OF THE ORGANICALLY BOUND TRITIUM TO THE SPECIFIC ACTIVITY OF TISSUE WATER WERE 1.0 FOR ALGAE, 0.62-0.99 FOR CATTAILS, 0.60 FOR THE VISCERAL ORGANS OF CLAMS AND FISHES, AND 0.30 FOR THE MUSCLE TISSUE. CONSUMPTION OF 1 KILOGRAM OF THE ANIMAL TISSUE WOULD GIVE A WHOLE BODY DOSE OF ABOUT 4 MREM, WHEN THE RADIOACTIVITY OF THE POOL WATER WAS ABOUT 0.05 MICROCURIES/ML. (BOPP-ORNL)

FIELD 05C

ACCESSION NO. W72-06342

A PRELIMINARY ECOLOGICAL STUDY OF AREAS TO BE IMPOUNDED IN THE SALT RIVER BASIN OF KENTUCKY,

KENTUCKY WATER RESOURCES INST., LEXINGTON.

LOUIS A. KRUMHOLZ.

AVAILABLE FROM THE NATIONAL TECHNICAL INFORMATION SERVICE AS PB-207 868, $3.00 IN PAPER COPY, $0.95 IN MICROFICHE. RESEARCH REPORT NO. 43, SEPTEMBER 1971, 34 P, 2 FIG, 1 TAB, 8 REF. OWRR B-005-KY(1).

DESCRIPTORS:
*ECOLOGY, *PREIMPOUNDMENTS, *AQUATIC HABITATS, *SAMPLING, *ECOSYSTEMS, *ALGAE, *CHLORAPHYTA, *AQUATIC INSECTS, *CHLORAPHYTA, *AQUATIC INSECTS, *CRUSTACEANS, *MOLLUSKS, *AQUATIC ANIMALS, *BIOTA, *AQUATIC PLANTS, WATER QUALITY, ENVIRONMENTAL EFFECTS, LIMNOLOGY, PLANNING, EUTROPHICATION, EVALUATION, WATER CHEMISTRY, WATER TEMPERATURE, DISSOLVED OXYGEN, TURBIDITY, ALKALINITY, NITRATES, IRON, MANGANESE, ANIONS, CATIONS, DIPTERA, CLAMS, GASTROPODS, BOTTOM FAUNA, KENTUCKY.

IDENTIFIERS:
*SALT RIVER(KENT), VASCULAR PLANTS, NITRATE NITROGEN, ORTHOPHOSPHATES, *OLIGASAPROBIC STREAMS, FINGERNAIL CLAMS, UNIONIDS, ORCONECTES RUSTICUS, LIRCEUS LINEATUS, GAMMARUS.

ABSTRACT:
A SERIES OF 25 SAMPLING STATIONS WAS ESTABLISHED IN THE MAINSTREAM AND TRIBUTARIES OF THE SALT RIVER. SAMPLING FOR WATER CHEMISTRY AND BIOTA WAS CARRIED OUT SEMIMONTHLY. PHYSICAL AND CHEMICAL DATA, ALONG WITH THE FLORA AND FAUNA PRESENT THE CHARACTERISTICS OF A RELATIVELY HEALTHY ECOSYSTEM. WATER TEMPERATURES REFLECT AIR TEMPERATURES CLOSELY AND DISSOLVED OXYGEN VALUES ARE NEAR SATURATION. TURBIDITY INCREASED WITH RUNOFF, THE STREAM FLOW INCREASING RAPIDLY DURING RAINY PERIODS AND FALLING TO A MINIMUM DURING DRY PERIODS. TOTAL ALKALINITIES RANGED FROM 135 TO 210 MG/1 AS CACO3 WITH RANGES IN PH FROM 6.3 TO 8.2. NITRATE NITROGEN RANGED FROM 2.0 TO 11.3 MG/1 AND ORTHOPHOSPHATE FROM 0.25 TO 2.78 MG/1. IRON AND MANGANESE RANGED FROM 0.07 TO 0.46 AND 0.09 TO 0.39 MG/1, RESPECTIVELY. A TOTAL OF 74 SPECIES OF ALGAE REFERABLE TO 35 FAMILIES WERE IDENTIFIED. GREEN ALGAE (CHLOROPHYTA) WERE REPRESENTED BY 38 SPECIES. MORE THAN 200 SPECIES OF VASCULAR PLANTS REFERABLE TO 50 FAMILIES HAVE BEEN COLLECTED FROM THE RIPARIAN VEGETATION. BOTTOM FAUNA INCLUDES 98 SPECIES OF INSECTS REPRESENTING 8 ORDERS AND 42 FAMILIES. PROMINENT AMONG THESE ARE THE 23 SPECIES OF CHIRONOMIDS THAT HAVE BEEN IDENTIFIED TO DATE. THE MOST COMMON CRUSTACEANS ARE ORCONECTES RUSTICUS AND LIRCEUS LINEATUS ALONG WITH SEVERAL SPECIES OF GAMMARUS. MOLLUSCS INCLUDE GASTROPODS, FINGERNAIL CLAMS, AND UNIONIDS. MORE THAN 50 SPECIES OF FISHES HAVE BEN COLLECTED.

FIELD 05A, 05C, 05B

ACCESSION NO. W72-06526

CHEMISTRY OF NITROGEN AND PHOSPHORUS IN WATER.

AMERICAN WATER WORKS ASSOCIATION, NEW YORK. WATER QUALITY DIV.; AND AMERICAN WATER WORKS ASSOCIATION, NEW YORK. COMMITTEE ON NUTRIENTS IN WATER.

JOURNAL OF AMERICAN WATER WORKS ASSOCIATION, VOL. 62, NO. 2, P 127-140, FEB. 1970. 6 FIG, 8 TAB, 64 REF.

DESCRIPTORS:
*WATER QUALITY, *AQUATIC PRODUCTIVITY, *AQUATIC MICRORGANISMS, *ALGAE, *NITROGEN, *PHOSPHORUS, SEDIMENTS, ORGANIC WASTES.

ABSTRACT:
OF THE MAJOR ELEMENTS ESSENTIAL TO ALGAL GROWTH, NITROGEN AND PHOSPHORUS ARE MOST LIKELY TO BE FOUND IN LIMITED AMOUNTS IN NATURAL WATERS. SINCE THEY THEREFORE REPRESENT PROMISING WEAK LINKS IN THE ALGAL LIFE CYCLES, THEIR CHEMICAL STATES AND BEHAVIOR IN WATER ARE EXAMINED TO SEE HOW WATER TREATMENT MIGHT BENEFIT. SUPPLIES OF NITROGEN AND PHOSPHORUS ARE PRESENT IN BODIES OF WATER EITHER IN THE SEDIMENTS, THE ATMOSPHERE ABOVE, OR IN THE FORM OF DISSOLVED GASES. THESE MAY BE AVAILABLE FOR THE GROWTH OF ALGAE AND OTHER AQUATIC PLANTS, BUT THE RATES AT WHICH THEY MAY BECOME AVAILABLE IS SLOW. THESE RATES ARE IMPORTANT AS THEY TEND TO CONTROL THE AMOUNT OF GROWTH WHICH CAN BE SUPPORTED. SOLUBLE NITROGEN AND PHOSPHORUS CONTAINED IN EFFLUENTS FROM WASTE TREATMENT PLANTS ARE IN READILY SOLUBLE FORM AND THEY CAN STIMULATE GROWTH FAR IN EXCESS OF THAT WHICH WOULD OCCUR NATURALLY. IN ASSESSING THE EXTENT OF NUTRIENT-RELATED PROBLEMS AND THEIR CONTROL, THE WATER MANAGER MUST EVALUATE THE SIGNIFICANCE BOTH OF THE READILY AVAILABLE FORMS AND THE FORMS WHICH MAY BE SLOWLY RELEASED FROM SUSPENDED PARTICLES AND FROM SEDIMENTS. (BEAN-AWWARF)

FIELD 05C, 05B, 05G

ACCESSION NO. W72-06532

BIOLCGICAL PROBLEMS ENCOUNTERED IN WATER SUPPLIES,

FEDERAL WATER QUALITY ADMINISTRATION, WASHINGTON, D.C. DIV. OF TECHNICAL SUPPORT.

K. M. MACKENTHUN, AND L. E. KEUP.

JOURNAL OF AMERICAN WATER WORKS ASSOCIATION, VOL. 62, NO. 8, P 520-526, AUGUST 1970. 2 FIG, 5 TAB, 36 REF.

DESCRIPTORS:
*WATER TREATMENT, *MICROORGANISMS, *ALGAE, *ALGICIDES, MICROSTRAINING, AERATION, PHOSPHATES, CLAY, TASTE, ODOR, *SURFACE WATERS, *GROUNDWATERS, SURVEYS, *WATER SUPPLY.

IDENTIFIERS:
*FILTER CLOGGING, IRON BACTERIA, POND WEEDS, PERMANGANATE.

ABSTRACT:
THE TYPES OF ORGANISMS GENERALLY RECOGNIZED AS CAUSING DIFFICULTIES IN WATER SUPPLIES AND SYSTEMS, AND THE TYPE OF PROBLEMS ENCOUNTERED ARE DISCUSSED, INCLUDING TASTES AND ODORS, FILTER CLOGGING, ALGAE ON RESERVOIR WALLS, IRON BACTERIA, AND ANIMALS. ALSO, VARIOUS CONTROLS ARE DESCRIBED. THE RESULTS OF A 1969 SURVEY ARE REPORTED. REPLIES WERE RECEIVED FROM 869 MANAGERS REPRESENTING OVER 1,372 MUNICIPAL AND 24 INDUSTRIAL SYSTEMS; THE MUNICIPAL SYSTEMS SERVE 80 MILLION PEOPLE. ORGANISMS HAVE CREATED PROBLEMS FOR 25 PERCENT OF THE MANAGERS WITHIN THE PAST FIVE YEARS; THE FREQUENCY IN THE LARGER SYSTEMS WAS TWICE THE AVERAGE. THE MOST FREQUENTLY REPORTED PROBLEMS WERE ALGAE AND POND WEEDS IN SURFACE SOURCES AND IRON BACTERIA IN WELLS AND DISTRIBUTION SYSTEMS. WITH SURFACE WATERS, ALGAL PROBLEMS WERE REPORTED 14 TIMES AS OFTEN AS IRON BACTERIA, WHEREAS WITH GROUNDWATERS THE IRON BACTERIA WERE REPORTED FOUR TIMES AS OFTEN, BUT 75 PERCENT OF THESE WERE IN WELLS. NON-CHEMICAL METHODS OF PROBLEM CONTROL ARE SCREENING, MICROSTRAINING, MECHANICAL CLEANING, FLUSHING, REDESIGN OF SYSTEM, AERATION OF RESERVOIR, CUTTING POND WEEDS, AND ADJUSTMENT OF FILTER RATES. CHEMICALS USED IN CONTROLLING ORGANISMS ARE CHLORINE, COPPER SULFATE, CARBON, POTASSIUM PERMANGANATE, ACID, SODIUM METAPHOSPHATE, AMMONIA, 2,4,-D, AROMATIC SOLVENTS, CLAY, 2,4,5-T, DOW-PON, AND PITT CHLOR. THE FIRST FOUR WERE MOST FREQUENTLY USED. CARBON IS ALMOST UNIVERSALLY USED TO REMOVE OBJECTIONABLE BY-PRODUCTS THROUGH ABSORPTION. (BEAN-AWWARF)

FIELD 05F

ACCESSION NO. W72-06536

FACTORS REGULATING THE GROWTH OF ALGAE IN CONTINUOUS CULTURE IN DILUTED SECONDARY SEWAGE TREATMENT PLANT EFFLUENT AND SUBSEQUENT BIODEGRADABILITY,

KENTUCKY WATER RESOURCES INST., LEXINGTON.

EDWARD G. FOREE, AND CAROLINE P. WADE.

AVAILABLE FROM THE NATIONAL TECHNICAL INFORMATION SERVICE AS PB-208 030, $3.00 IN PAPER COPY, $0.95 IN MICROFICHE. RESEARCH REPORT NUMBER 45, JANUARY 1972, 56 P, 10 FIG, 5 TAB, 29 REF. OWRR A-023-KY(1).

DESCRIPTORS:
*ALGAE, *NITROGEN, HYDROGEN ION CONCENTRATION, BIOCHEMICAL OXYGEN DEMAND, *SEWAGE EFFLUENTS, *SEWAGE TREATMENT, CULTURE, *WASTE-WATER TREATMENT, *CARBON DIOXIDE, *GROWTH RATES, *AEROBIC CONDITIONS, *GROWTH STAGES, *BIODEGRADATION, *BIOCHEMICAL OXYGEN DEMAND, CHEMICAL OXYGEN DEMAND.

IDENTIFIERS:
*ALGAE GROWTH, *CHEMOSTATS.

ABSTRACT:
HETEROGENEOUS ALGAL CULTURES WERE GROWN IN LABORATORY CONTINUOUS CULTURE IN CONTINUOUS FLOW, COMPLETELY MIXED CHEMOSTATS IN SECONDARY SEWAGE TREATMENT PLANT EFFLUENT DILUTED TO GIVE AN AMMONIA NITROGEN CONCENTRATION OF 10 MG/L. VARIABLES WERE LIGHTING, PH, CARBON DIOXIDE AVAILABILITY, AND HYDRAULIC RESIDENCE TIME. OPTIMUM GROWTH OCCURRED UNDER PH.7.0, EXCESS CO2, AND CONTINUOUS LIGHTING CONDITIONS. THE AVAILABILITY OF ARTIFICIALLY SUPPLIED EXCESS CO2 GREATLY INCREASED THE MASS (STANDING CROP) AT STEADY-STATE OVER THAT PRODUCED UNDER OTHERWISE IDENTICAL CONDITIONS FOR ALL RESIDENCE TIMES STUDIED. FOR THE CASE OF EXCESS CO2 AVAILABILITY, THE NITROGEN CONCENTRATION IN THE ALGAL CELLS REGULATED GROWTH RATHER THAN THE CONCENTRATION OF NUTRIENTS IN SOLUTION. A MATHEMATICAL EXPRESSION WAS HYPOTHESIZED TO DESCRIBE THIS PHENOMENON AND WAS CONFIRMED BY THE EXPERIMENTAL RESULTS. UNDER DARK-AEROBIC CONDITIONS, THE ALGAL CULTURES EXERTED A TWO-STAGE BOD, THE SECOND STAGE APPARENTLY BEGINNING AFTER THE DEATH OF THE ALGAL CELLS. LONGER CHEMOSTAT RESIDENCE TIMES DURING GROWTH PRODUCED CULTURES WITH LOWER PERCENTAGE BIODEGRADABILITY. CARBON DIOXIDE ENRICHED GROWTH CONDITIONS PRODUCED CULTURES WITH LOWER PERCENTACE BIODEGRADABILITY THAN CULTURES GROWN IN A CARBON DIOXIDE DEFICIENT MEDIUM.

FIELD 05C, 05D

ACCESSION NO. W72-06612

PHOSPHATES IN DETERGENTS: THE CHICAGO-TYPE ORDINANCE AND OTHER REMEDIES,

B. NIEHOFF.

CINCINNATI LAW REVIEW, VOL. 40, NO. 3, P 548-568, 1971. 115 REF.

DESCRIPTORS:
*EUTROPHICATION, *DETERGENTS, *PHOSPHATES, *LOCAL GOVERNMENTS, *WATER POLLUTION CONTROL, WATER QUALITY CONTROL, PHOSPHORUS, AQUATIC ENVIRONMENT, WATER POLLUTION EFFECTS, ALGAE, NUTRIENTS, WATER POLLUTION SOURCES, SEWAGE TREATMENT, PUBLIC HEALTH, WASTE WATER TREATMENT, RESEARCH AND DEVELOPMENT, MUNICIPAL WASTES, LEGISLATION, REGULATION, PHOSPHORUS COMPOUNDS, FEDERAL GOVERNMENT.

ABSTRACT:
PHOSPHATE DETERGENTS ARE A MAJOR CAUSE OF ACCELERATED EUTROPHICATION. THIS PROCESS FOULS LAKES AND RIVERS WITH MATS OF ALGAE AND WEEDS. A NUMBER OF LOCAL GOVERNMENTS HAVE ENACTED CHICAGO-TYPE ORDINANCES TO CONTROL THE USE OF PHOSPHATE DETERGENTS. UNDER THE CHICAGO ORDINANCE IT IS ILLEGAL TO SELL DETERGENTS CONTAINING PHOSPHATES. THE CHICAGO ORDINANCE ALLOWS SOME EXEMPTIONS FOR MACHINES BUILT TO USE ONLY PHOSPHATE DETERGENTS, AND FOR HEALTH AND SANITATION REASONS. MANY ORDINANCES HAVE LABELING REQUIREMENTS. A FEDERAL TRADE COMMISSION PROPOSAL WOULD GO FURTHER AND REQUIRE A WARNING ON THE CONTAINER. THIS PROPOSAL MAY, HOWEVER, BE CONFUSING AND RAISE UNNECESSARY CONSUMER DOUBTS. THE DETERGENT INDUSTRY INSISTS LOCAL ORDINANCES ARE AN UNCONSTITUTIONAL RESTRAINT OF INTERSTATE COMMERCE. HOWEVER, THEY ARE PROBABLY A CONSTITUTIONAL EXERCISE OF LOCAL POLICE POWERS. AT LEAST FOUR PHOSPHATE-DETERGENT CONTROL LAWS ARE NOW BEFORE CONGRESS. THE SOUNDEST OF THESE PROHIBITS THE MANUFACTURING OR IMPORTATION OF ANY DETERGENT CONTAINING PHOSPHATE AFTER JUNE 30, 1973. (BRACKINS-FLORIDA)

FIELD 05G, 06E

ACCESSION NO. W72-06638

LEGISLATIVE RESPONSE TO 'SOFT SOAP' ON DETERGENTS,

CONGRESS, WASHINGTON, D.C.; AND SENATE, WASHINGTON, D.C.

T. J. MCINTYRE.

CONGRESSIONAL RECORD, VOL. 117, NO. 170, S 18015-18023 (DAILY ED.) NOVEMBER 10, 1971.

DESCRIPTORS:
*DETERGENTS, *PHOSPHATES, *WATER POLLUTION SOURCES, *POLLUTION ABATEMENT, EUTROPHICATION, LEGISLATION, PUBLIC HEALTH, WATER QUALITY CONTROL, ALGAL CONTROL, FEDERAL GOVERNMENT, NUTRIENTS, SEWAGE TREATMENT, WATER POLLUTION, WATER POLLUTION EFFECTS, RESEARCH AND DEVELOPMENT, WASTE WATER TREATMENT, BIODEGRADATION, LOCAL GOVERNMENTS, MUNICIPAL WASTES, ADMINISTRATIVE AGENCIES, GREAT LAKES.

ABSTRACT:
PHOSPHATE DETERGENTS ARE A MAJOR CAUSE OF THE DEATH OF MANY LAKES AND RIVERS. ALTHOUGH THE EFFECTS OF PHOSPHATE DETERGENTS ARE WELL KNOWN, THE FEDERAL GOVERNMENT HAS BEEN RECALCITRANT IN TAKING ACTION TO CONTROL THEIR USE. THE MOST OSTENSIBLE REASON IS THE LACK OF A SAFE AND EFFECTIVE SUBSTITUTE. ANOTHER REASON IS THE NORMAL DELAY WHEN THERE ARE POWERFUL INTERESTS OR OPPOSITION TO SUCH ACTION. A PROPOSAL BEFORE THE SENATE COMMERCE COMMITTEE'S ENVIRONMENTAL SUBCOMMITTEE WOULD ACCOMPLISH THE FOLLOWING: (1) LIMIT THE PHOSPHATE CONTENT IN DETERGENTS FOR THE PRESENT, WITH A CLEAR MANDATE FOR THEIR EVENTUAL TOTAL REMOVAL; (2) ESTABLISH TEST PROTOCOLS, STANDARDS, AND REGULATIONS FOR ALL DETERGENT INGREDIENTS; AND (3) ESTABLISH A FEDERAL PROGRAM FOR THE DEVELOPMENT OF SAFE SUBSTITUTES WHICH WILL NOT HARM THE ENVIRONMENT. SOME, INCLUDING THE DETERGENT INDUSTRY, CLAIM ELIMINATION OF PHOSPHATES AT THE SEWAGE TREATMENT STAGE IS THE MOST EFFECTIVE METHOD OF DEALING WITH THE PROBLEM. STRICTER LABELING REQUIREMENTS, DISCLOSING THE AMOUNT OF PHOSPHATE IN A DETERGENT, ARE ALSO OFFERED AS AN IMMEDIATE STEP WHICH COULD BE TAKEN TO PARTIALLY ALLEVIATE THE PROBLEM. (BRACKINS-FLORIDA)

FIELD 06E, 05G

ACCESSION NO. W72-06655

MICROBIOLOGICAL STUDIES OF OXYGEN DEPLETION IN THE LAKE ERIE CENTRAL BASIN,

DEPARTMENT OF NATIONAL HEALTH AND WELFARE, KINGSTON (ONTARIO). DIV. OF PUBLIC HEALTH ENGINEERING.

A. S. MENON, AND A. A. JURKOVIC.

MANUSCRIPT REPORT KR 70-4, 1970. 51 P, 15 FIG, 13 TAB, 18 REFS.

DESCRIPTORS:
*AQUATIC BACTERIA, *SULFUR BACTERIA, *OXYGEN, *SEDIMENT-WATER INTERFACES, *LAKE ERIE, HYPOLIMNION, AQUATIC ALGAE.

ABSTRACT:
THE SIGNIFICANCE OF BACTERIAL ACTIVITY IN THE OVER-ALL PROCESSES OF OXYGEN DEPLETION AND NUTRIENT REGENERATION IN THE CENTRAL BASIN OF LAKE ERIE WAS ASSESSED. MOST INTENSIVE BACTERIAL ACTIVITY OCCURRED AT THE SEDIMENT-WATER INTERFACE. BACTERIAL DECOMPOSITION OF ORGANIC MATTER ACCUMULATING AT THE INTERFACE RESULTED IN THE FORMATION OF REDUCED PRODUCTS OF LOW MOLECULAR WEIGHT AND DEPLETION OF OXYGEN IN THE HYPOLIMNION. THESE COMPOUNDS WERE SUBSEQUENTLY OXIDIZED BY CHEMOAUTOTROPHIC BACTERIA WITH FURTHER LOSS OF O2. REDUCING CONDITIONS ON THE BOTTOM ADVERSELY AFFECTED NITRIFYING BACTERIAL DENSITIES. HOWEVER, ACTIVELY PHOTOSYNTHESIZING ALGAE FRESHLY DEPOSITED ON THE BOTTOM STIMULATED MULTIPLICATION OF NITRIFYING BACTERIA AND NITRIFICATION. LARGE BACTERIAL POPULATIONS WERE ABSENT IN THE THERMOCLINE, SUGGESTING THAT THIS ZONE WAS NOT A SITE FOR INTENSIVE BACTERIAL ACTIVITY. QUANTITATIVE ANALYSIS INDICATED THAT THE HIGH BACTERIAL DENSITIES IN THE HYPOLIMNION, ESPECIALLY AT THE SEDIMENT-WATER INTERFACE, RESPIRING AT THE RATE OF 2.4 X 10-11 MG O2 PER CELL PER HOUR COULD ACCOUNT FOR OXYGEN DEPLETION IN THE LAKE. (CCIW)

FIELD 05C, 02H

ACCESSION NO. W72-06690

EFFECTS OF DETERGENTS ON WATER SUPPLIES,

ONTARIO WATER RESOURCES COMMISSION, TORONTO.

A. J. HARRIS, K. J. ROBERTS, AND A. E. CHRISTIE.

JOURNAL AMERICAN WATER WORKS ASSOCIATION, VOL. 63, NO. 12, P 795-799, DECEMBER 1971. 2 TAB, 56 REF.

DESCRIPTORS:
*DETERGENTS, *PHOSPHATES, *POTABLE WATER, EUTROPHICATICN, ALKYLBENZENE SULFONATES, SURFACTANTS, LINEAR ALKYLATE SULFONATES, SEWAGE, NUTRIENTS, COAGULATION, SEDIMENTATION, TOXICITY, TASTE, ODOR, CORROSION, ENZYMES, CHLEATION, ALGAE, WATER QUALITY, BIODEGRADATION, SEWAGE TREATMENT PLANTS, DIATOMS, PHYTOPLANKTCN, HYDORGEN ICN TEMPERATURE, WATER POLLUTION EFFECTS.

IDENTIFIERS:
ARSENIC, NITRILOTRIACETATE, POLYSILICATES, POLYCARBOXYLATES.

ABSTRACT:
THE DEVELOPMENT OF DETERGENTS AND THEIR DIRECT AND INDIRECT EFFECTS ON WATER SUPPLIES ARE REVIEWED. THE MAJORITY OF PRESENT-DAY PROBLEMS ARE ASSOCIATED WITH PHOSPHATES IN THE DETERGENTS. THESE PROBLEMS INCLUDE EUTROPHICATION OF LAKES, WATER QUALITY AND TREATMENT PROBLEMS, AND UNDESIRABLE TASTES AND ODORS ASSOCIATED WITH ALGAL GROWTH. SEVERAL MATERIALS SUCH AS POLYSILICATES, POLYCARBOXYLATES, NTA, AND STARCH DERIVATIVES HAVE BEEN STUDIED AS POSSIBLE REPLACEMENTS FOR PHOSPHATE. LACK OF BIODEGRADABILITY, UNDESIRABLE CHELATING PROPERTIES, AND CORROSION PROBLEMS ARE SOME OF THE FAULTS THAT HAVE BEEN DETECTED IN THESE MATERIALS. NTA HAS BEEN MOST PROMISING BUT HAS BEEN WITHDRAWN BECAUSE OF A POSSIBLE HEALTH HAZARD. BECAUSE IT IS NOW POSSIBLE TO REMOVE PHOSPHATE FROM WASTE WATER BY CHEMICAL PRECIPITATION AND BECAUSE DETERGENT PHOSPHATE IS ONLY ONE PART OF THE TOTAL PHOSPHATE IN SEWAGE EFFLUENTS, REMOVAL FROM THE WASTE WATER RATHER THAN THE DETERGENT PRODUCT SHOULD BE CONSIDERED. (MORTLANC-BATTELLE)

FIELD 05C, 05G

ACCESSION NO. W72-06837

SUPPLEMENTARY AERATION OF LAGOONS IN RIGOROUS CLIMATE AREAS,

WYCMING UNIV., LARAMIE. DEPT. OF CIVIL ENGINEERING.

R. L. CHAMPLIN.

COPY AVAILABLE FROM GPO SUP DOC AS EP2.10:17050 DVO-10/71, $0.75; MICROFICHE FROM NTIS AS PB-2C8 204, $0.95.ENVIRONMENTAL PROTECTION AGENCY WATER POLLUTION CONTROL RESEARCH SERIES, OCTOBER 1971, 73 P. 24 FIG, 15 TAB, 7 REF. EPA PROGRAM 17050DVC--10/71.

DESCRIPTORS:
*ON-SITE INVESTIGATION, *CCLC REGIONS, *OXIDATION LAGOCNS, ORGANIC LOADING, AERATION, MIXING, ALGAE, TEMPERATURE, ALTITUDE, CLIMATES, METABOLISM, BIOCHEMICAL OXYGEN DEMAND, PILOT PLANTS, *WASTE WATER TREATMENT, WYOMING, *AERATED LAGOONS.

ABSTRACT:
A PILOT SCALE FIELD INVESTIGATION OF THE EFFECTS OF SUPPLEMENTAL AERATION ON WASTE STABILIZATION LAGOONS WAS CONDUCTED AT LARAMIE, WYOMING, A LOW TEMPERATURE, HIGH ALTITUDE AREA. BCTH BATCH AND COMPLETE MIXED EXPERIMENTS WERE CONDUCTED USING CONSTANT AIR FLCWS. LOADING RATES, BOTH HYDRAULIC AND PROCESS , WERE VARIED FRCM 160 LBS BOD5/ACRE/DAY (0.725 LBS/1000 FT3/DAY) TO 900 LBS BOD5/ACRE/DAY (4.08 LBS/1000 FT3/DAY). THE SUPPLEMENTAL AERATION PROVIDED BOTH AERATION AND MIXING, THEREBY INCREASING METABOLIC RATES. BOD REDUCTIONS VARIED FROM 72 TO 85% UNDER THREE DIFFERENT LOADINGS, AT TEMPERATURES OF LESS THAN 12 DEG C. LOADING BELOW 320 LBS/ACRE/DAY AND SECONDARY CELL OPERATION PRODUCED SIGNIFICANT ALGAL GROWTH EVEN AT TEMPERATURES AROUND 6 DEG C. NO SETTLEABLE SOLIDS WERE FOUND IN THE EFFLUENT FROM THE AERATED SYSTEM. SERIES OPERATION WAS DEMONSTRATED TO HAVE THE ADVANTAGES OF DAMPING VARIATIONS IN QUALITY PARAMETERS, PROVIDING FOR SHOCK LOADING, AND REDUCING COLIFORM COUNTS TO MINIMUM LEVELS. SHORTER DETENTION PERIODS ALSO TAKE GREATER ADVANTAGE OF THE WARMER INFLUENT TEMPERATURES IN ORDER TO SATISFY EASILY OXIDIZED ORGANIC MATERIAL. (LOWRY-TEXAS)

FIELD 05D

ACCESSION NO. W72-06838

ECOSYSTEM ALTERNATION BY MOSQUITOFISH (GAMBUSIA AFFINIS) PREDATION,

SAN DIEGO STATE COLL., CALIF. DEPT. OF BIOLOGY.

S. H. HURLBERT, J. ZEDLER, AND D. FAIRBANKS.

SCIENCE, VOL. 175, NO. 4022, P 639-641, FEBRUARY 11, 1972.

DESCRIPTORS:
*ECOSYSTEMS, FOOD CHAINS, *ZOOPLANKTON, *PREDATION, *BIOASSAY, BIOLOGICAL COMMUNITIES, DOMINANT ORGANISMS, AQUATIC ENVIRONMENT, AQUATIC ALGAE, AQUATIC ANIMALS, AQUATIC INSECTS, DAPHNIA, AQUATIC POPULATIONS, FISH DIETS, FRESHWATER FISH, PONDS, LAKES, FOOD HABITS, EUTROPHICATION, LARVAE, ROTIFERS, AQUATIC MICROORGANISMS, DIPTERA, PHOSPHORUS, WATER TEMPERATURE, WATER CHEMISTRY, WATER ANALYSIS, COPEPODS, CYANOPHYTA, MAYFLIES, PHYTOPLANKTON, SESSILE ALGAE, CYCLING NUTRIENTS, BIOMASS, CHARA, LIGHT PENETRATION, BENTHIC FLORA, MODEL STUDIES, AQUATIC LIFE, CRUSTACEANS, FISH, ANALYTICAL TECHNIQUES.

IDENTIFIERS:
*GAMBUSIA, CYCLOPS NAUPLII, COCCOCHLORIS PENIOCYSTIS, SPIROGYRA, EPHEMEROPTERA BAETIDAE, EPHYDRA, CHAETOGASTER, AEOLOSOMA, CHAOBORUS, *GAMBUSIA AFFINS, EPHEMEROPTERA, MENIDIA, BRACHYDEUTERA, CHYDORUS SPHAERICUS, CHYDORUS, MONOSTYLA, LEPADELLA, TRICHOCERCA, HAEMATOCOCCUS LACUSTRIS.

ABSTRACT:
A MODEL ECOSYSTEM CONSISTING OF EIGHT PLASTIC POOLS, 2 M IN DIAMETER AND 30 CM DEEP, WAS SET UP OUTDOORS TO DETERMINE THE EFFECT OF FISH ON BIOLOGICAL POPULATIONS AND PHOSPHORUS CYCLE. A 3 CM LAYER OF SAND, TAP WATER TO A DEPTH OF 20 CM, A LITER OF DRY ALFALFA PELLETS AS A SOURCE OF NUTRIENTS, AND INOCULA OF PLANKTON FROM A LAKE AND FROM A LABORATORY COLONY OF DAPHNIA PULEX WERE ADDED TO EACH POOL. MOSQUITOFISH (GAMBUSIA AFFINS) WERE ADDED TO THREE POOLS WITH THE REMAINING POOLS SERVING AS CONTROLS. THE STATUS OF THE SYSTEM WAS DETERMINED BY REGULAR SAMPLING OF PHYTOPLANKTON, ZOOPLANKTON, INSECT POPULATION, BENTHOS, ALGAE, AND BY DETERMINATION OF PHOSPHORUS CONTENT, WATER TEMPERATURE, AND TRANSPARENCY. GAMBUSIA AFFINS GREATLY REDUCED ROTIFER, CRUSTACEAN, AND INSECT POPULATIONS AND THUS PERMITTED EXTRAORDINARY DEVELOPMENT OF PHYTOPLANKTON POPULATIONS (2 X 108 CELLS PER MILLILITER). OTHER EFFECTS INCLUDED DECREASED OPTICAL TRANSMISSIVITY AND INCREASED TEMPERATURE OF THE WATER, DECREASED AMOUNTS OF DISSOLVED INORGANIC PHOSPHORUS, AND INCREASED AMOUNTS OF DISSOLVED ORGANIC PHOSPHORUS, INHIBITION OF SPIROGYRA, AND REPLACEMENT OF ONE ANNELID, CHAETOGASTER, BY ANOTHER, AEOLOSOMA. THE RESULTS OF THE EXPERIMENT ALSO INDICATED FISH CAN BE USED AS A POSSIBLE CONTROLLING AGENT FOR EUTROPHICATION. FURTHER, THE BIOLOGIC CHANGES IN AQUATIC SYSTEMS ARE PIMARILY A RESULT OF CHANGES IN THE AQUATIC FOOD CHAIN. (SNYDER-BATTELLE)

FIELD 05C

ACCESSION NO. W72-07132

COMPARATIVE STUDY OF SPECIES NAUTOCOCCUS MAMMILATUS AND NAUTOCOCCUS PYRIFORMIS (TETRASPORALES),

CESKOSLOVENSKA AKADEMIE VED, TREBON. LAB. OF ALGOLOGY.

J. LUKAVSKY.

ARCHIV FUR HYDROBIOLOGIE, VOL. 39, NO. 4, P 245-258, OCTOBER 1971. 70 FIG, 18 REF.

DESCRIPTORS:
*ALGAE, *SYSTEMATICS, *PLANT MORPHOLOGY, REPRODUCTION, LIFE CYCLES, AQUATIC ALGAE, PERIPHYTON.

IDENTIFIERS:
EPINEUSTONT, NAUTOCOCCUS MAMMILATUS, NAUTOCOCCUS PYRIFORMIS NAUTOCOCCUS, APIOCOCCUS, OSCILLATORIA, NAUTOCOCCUS CAUDATUS, NAUTOCOCCUS CONSTRICTUS, NAUTOCOCCUS EMERSUS.

ABSTRACT:
THE MORPHOLOGICAL VARIABILITY AND THE KIND OF REPRODUCTION OF TETRASPORAL ALGA NAUTOCOCCUS MAMMILATUS KORSH. IN NATURE AND IN THE CULTURE WERE STUDIED AND COMPARED WITH THE CULTURE OF TWO STRAINS OF NAUTOCOCCUS PYRIFORMIS KORSH. THE STABILITY AND TAXONOMICAL VALUE OF SOME FEATURES, NAMELY THE MORPHOLOGY AND THE CELL STRUCTURE, THE KIND OF FLOATING, THE MORPHOLOGY OF THE FLOATING CAP AND THE MORPHOLOGY OF ZOOSPORES WERE STUDIED. FOR THE FIRST TIME THE ISOGAMICAL SEXUAL PROCESS OF ONE STRAIN, NAUTOCOCCUS PYRIFORMIS, WAS OBSERVED. THE SURFACE FILM OF WATER IS CONSIDERED NOT TO BE THE REASON OF FLOATING OF THE CELLS OF NAUTOCOCCUS; THE CELLS ARE LIFTED BY GAS BUBBLES. THE DIAGNOSIS OF THE GENUS NAUTOCOCCUS KORSH. WAS EXTENDED, THE GENUS APIOCOCCUS KORSH. WAS INCLUDED AS A SYNONYM. (HOLCMAN-BATTELLE)

FIELD 05C, 05A, 02I

ACCESSION NO. W72-07141

ION TRANSPORT STUDIES ON PHYTOPLANKTON OF A FISH POND AT ILE-IFE,

IFE UNIV. (NIGERIA). DEPT. OF BIOLOGICAL SCIENCES.

A. M. A. IMEVBORE, AND Z. BOSZORMENYI.

ARCHIV FUR HYDROBIOLOGIE, VOL. 69, NO. 2, P 200-209, OCTOBER 1971. 4 FIG, 3 TAB, 41 REF.

DESCRIPTORS:
*PHYTOPLANKTON, *ION TRANSPORT, *ABSORPTION, *PHOSPHORUS, *PHOSPHORUS RADIOISOTOPES, ALGAE, FRESH WATER, FILTRATION, LEACHING, FUNGI, LIGHT, PHOSPHATES, NUTRIENTS, WATER POLLUTION EFFECTS, RADIOACTIVITY TECHNIQUES.

IDENTIFIERS:
CYANIDES, PHOSPHORUS-32.

ABSTRACT:
THE ION TRANSPORT OF PHOSPHORUS BY PHYTOPLANKTON WAS INVESTIGATED BY EXPOSING MICROORGANISMS TO WATER SAMPLES FROM A FISHPOND AT ILE-IFE (NIGERIA) WHICH WERE LABELED WITH PHOSPHORUS-32. THE ANALYSES WERE RESTRICTED TO THE PROCESSES ACTING BETWEEN THE PHYTOPLANKTON AND THE EXTERNAL INORGANIC PHOSPHATE POOL. THE STUDIES PRODUCED DATA ON THE RATE OF UPTAKE, EFFECT OF PHOSPHORUS CONCENTRATION ON UPTAKE, EFFECT OF WASHING WITH INACTIVE PHOSPHORUS, AND THE EFFECTS OF LIGHT, CYANIDE, AND SHAKING ON UPTAKE. THE QUALITATIVE AND QUANTITATIVE DISSIMILARITY BETWEEN THE ABSORPTION AT THE HIGH (1MM) AND THE LOW (0.01MM) PHOSPHORUS-32 CONCENTRATIONS SEEM TO EXCLUDE THE POSSIBILITY THAT THE ABSORPTION PROCESSES ACTING IN BOTH RANGES ARE THE SAME. THEREFORE THERE ARE MORE LIKELY TWO DIFFERENT MECHANISMS IN ION TRANSPORT. AT THE LOW PHOSPHORUS-32 CONCENTRATIONS, CYANIDE WAS ABLE TO BLOCK THE PHOSPHATE UPTAKE ALMOST COMPLETELY; AT THE HIGH PHOSPHORUS-32 CONCENTRATION THERE WAS LITTLE OR NO INHIBITION. THE OBSERVED UPTAKE AT THE LOW CONCENTRATION WAS IN ACCORDANCE WITH THE GENERAL VIEW THAT PHOSPHATE ABSORPTION WAS DEPENDENT UPON METABOLISM AND THAT IT WAS A BIOLOGICAL TRANSPORT PROCESS. ON THE OTHER HAND, IT WAS PROBABLE THAT THE ACTION OF CYANIDE AT HIGH PHOSPHORUS-32 CONCENTRATION DEPENDED UPON THE DAMAGE IT CAUSED TO THE STRUCTURE OF BIOLOGICAL MEMBRANES. THEN THE INCREASE IN UPTAKE MUST BE DUE ENTIRELY TO A PHYSICAL PERMEATION PROCESS. CONTRARY TO EXPECTATION, SHAKING TENDED TO REDUCE THE ABSORPTION RATE. (HOLOMAN-BATTELLE)

FIELD 05B, 05C

ACCESSION NO. W72-07143

THE MICROCHAMBER CULTIVATION OF ALGAE,

CESKOSLOVENSKA AKADEMIE VED, BRNO. LAB. OF SCIENTIFIC FILM.

J. HRIB, AND V. BREZINA.

ARCHIV FUR HYDROBIOLOGIE, VOL. 39, NO. 4, P 349-354, OCTOBER 1971. 2 FIG, 13 REF.

DESCRIPTORS:
*AQUATIC ALGAE, *CHLOROPHYTA, *GROWTH CHAMBERS, *LIFE CYCLES, CULTURES, WATER QUALITY, SCENEDESMUS, LABORATORY EQUIPMENT, PHOTOGRAPHY.

IDENTIFIERS:
*SYNCHRONOUS CULTURES, SCENEDESMUS SOLI, SCENEDESMUS QUADRICAUDA, MOUGEOTIA, ULOTHRIX AEQUALIS, STEPHANOPYXIS TURRIS.

ABSTRACT:
A CULTIVATION METHOD FOR ALGAE BY THE SYSTEM OF MICROCHAMBER CULTIVATION IS PRESENTED. THE METHOD ENABLES CONTINUAL CULTIVATION OF ALGAL CELLS WITH THE POSSIBILITY OF CINEMATOGRAPHIC REGISTRATION, ESPECIALLY OF TIME LAPSE MICROCINEMATOGRAPHY. THE SYSTEM OF THE CULTIVATION MICROCHAMBER IS DIVIDED INTO FOUR SECTIONS: A CULTIVATION SECTION, A MICROMANIPULATION SECTION, A LIGHT SECTION AND A CINEMATOGRAPHIC SECTION. THE MAIN PART OF THE SYSTEM IS A CULTIVATION CHAMBER WHICH IS OF FLOWING CHARACTER WITH THE POSSIBILITY OF SIMPLE MICROMANIPULATION. THIS SYSTEM PROVED TO BE USEFUL IN STUDYING THE ONTOGENETIC CYCLE OF THE ALGA SCENEDESMUS QUADRICAUDA, STRAIN GREIFSWALD/15. (HOLOMAN-BATTELLE)

FIELD 05C, 05A

ACCESSION NO. W72-07144

OPTICAL DENSITY PROFILES AS AN AID TO THE STUDY OF MICROSTRATIFIED PHYTOPLANKTON POPULATIONS IN LAKES,

MINNESOTA UNIV., MINNEAPOLIS. LIMNOLOGICAL RESEARCH CENTER.

A. L. BAKER, AND A. J. BROOK.

ARCHIV FUR HYDROBIOLOGIE, VOL. 69, NO. 2, P 214-233, OCTOBER 1971. 12 FIG, 1 TAB, 14 REF.

DESCRIPTORS:
*PHYTOPLANKTON, *LAKES, *TURBIDITY, *PROFILES, PHOTOMETER, HYPOLIMNION, EPILIMNION, LAKE MORPHOMETRY, DISTRIBUTION PATTERNS, AQUATIC ALGAE, LIMNOLOGY, CHRYSOPHYTA, EUGLENOPHYTA, PYRROPHYTA, BIOMASS, HYDROGEN ION CONCENTRATION, MINNESOTA, WATER TEMPERATURE, DISSOLVED OXYGEN.

IDENTIFIERS:
*MICROSTRATIFICATION, *OPTICAL DENSITY, MICROAMMETER, OSCILLATORIA AGARDHII VAR ISOTHRIX, LAKE ITASCA, LAKE ELK, LAKE MARY, LAKE JOSEPHINE, LAKE ARCO, LAKE DEMING, LOWER LA SALLE LAKE, LAKE LONG, LAKE SQUAW, LAKE BUDD, OSCILLATORIA REDEKEI, LYNGBYA LIMNETICA, ANABAENA SPP., CHROMATIUM, MERISMOPEDIA TROLLERI, CRYPTOMONADS, STICHOGLOEA DEODERLEINII, APHANIZOMENON FLOS-AQUAE, MALLOMONAS ACAROIDES, OSCILLATORIA ORNATA, TURBIDIMETER, METALIMNION.

ABSTRACT:
TURBIDITY PROFILES OF TEN NORTH-CENTRAL MINNESOTA LAKES HAVE BEEN DETERMINED WITH A SCHENK-TURBIDIMETER. THE LAKES WERE STUDIED AT RELATIVELY CLOSE TIME INTERVALS AND EACH TIME A TURBIDITY PROFILE WAS TAKEN, TEMPERATURE AND DISSOLVED OXYGEN PROFILES AND IN MANY CASES PH AND CONDUCTIVITY PROFILES WERE ALSO DETERMINED. SOME OF THESE DATA ARE GRAPHED WITH THE OPTICAL DENSITY PROFILES FOR EACH LAKE. OBSERVED DIFFERENCES IN TURBIDITY WERE CORRELATED WITH THE MICROSTRATIFICATION OF SPECIFIC POPULATIONS OF PHYTOPLANKTON. THE TURBIDITY PROFILES DEMONSTRATED THE FOLLOWING: (1) MOVEMENT AND DISRUPTION OF MICROSTRATA DURING VERNAL AND AUTUMNAL OVERTURN, THEIR REAPPEARANCE UNDER ICE IN WINTER AND ESPECIALLY IN THE METALIMNION IN EARLY SUMMER; (2) PRONOUNCED MICROSTRATIFICATION ESPECIALLY IN SMALL LAKES WITH RELATIVE DEPTH VALUES GREATER THAN 5.0 METERS; AND (3) INTEGRITY AND STABILITY OF THE MICROSTRATA. THERE IS A DISCUSSION OF THE TURBIDITY PROFILES AS COMPARED WITH THOSE RECORDED BY OTHER INVESTIGATORS AND THE CAUSES OF OPTICAL DENSITY MAXIMA OTHER THAN BY PHYTOPLANKTON. (HOLOMAN-BATTELLE)

FIELD 05C, 05A, 02H

ACCESSION NO. W72-07145

PHOSPHORUS UPTAKE BY PLANKTONIC ALGAE IN THE DARK AND UNDER FAINT LIGHT,

INSTITUTE OF BIOLOGY OF THE SOUTHERN SEAS, SEVASTOPOL (USSR).

D. K. KRUPATKINA.

OKEANOLOGIYA, VOL 11, NO 2, P 221-226, 1971. 3 TAB, 18 REF.

DESCRIPTORS:
*ALGAE, *PHYTOPLANKTON, *ABSORPTION, *AQUATIC ALGAE, *LIGHT INTENSITY, *PHOSPHORUS, PLANKTON, CHLOROPHYTA, PYRROPHYTA, DINOFLAGELLATES, PRIMARY PRODUCTIVITY, MINERALOGY, LIGHT, CULTURES, BIOASSAY.

IDENTIFIERS:
PHAEODACTYLUM TRICORNUTUM, GONYAULAX POLYEDRA, PLATYMONAS VIRIDIS, CHAETOCEROS CURVISETUS, SKELETONEMA COSTATUM, GLENODINIUM FOLIACUM, GYRODINIUM FISUM.

ABSTRACT:
THE UPTAKE OF MINERAL PHOSPHORUS AND ORGANIC PHOSPHORUS IN THE DARK AND UNDER FAINT LIGHT BY PLANKTONIC ALGAE NOT DEFICIENT IN PHOSPHORUS HAS BEEN INVESTIGATED. SIX SPECIES OF PLANKTONIC ALGAE (GREEN, DIATOM, DINOFLAGELLATE) WERE STUDIED. MINERAL PHOSPHORUS WAS TAKEN UP IN THE DARK BY CELLS NOT DEFICIENT IN PHOSPHORUS IN MOST SPECIES (EXCEPT PHAEODACTYLUM TRICORNUTUM AND GONYAULAX POLYEDRA). THE PHOSPHORUS CONTENT INCREASED 1.5- TO 2.5-FOLD WHEN THE CELLS (PERCENT OF DRY WEIGHT) WERE KEPT IN THE DARK FOR 3 DAYS. ORGANIC PHOSPHORUS WAS TAKEN UP BOTH IN THE DARK AND IN THE PRESENCE OF LIGHT. SOME SPECIES (PLATYMONAS VIRIDIS AND CHAETOCEROS CURVISETUS) TAKE UP MINERAL PHOSPHORUS AND ORGANIC PHOSPHORUS SIMULTANEOUSLY BOTH IN THE DARK AND IN THE PRESENCE OF FAINT LIGHT, WHILE OTHERS (PHAEODACTYLUM TRICORNUTUM AND GONYAULAX POLYEDRA) TAKE UP ONLY ORGANIC PHOSPHORUS. THE SOURCES OF ORGANIC PHOSPHORUS IN THE MEDIUM ARE THE INTRAVITAL EXCRETIONS OF CELLS AND DECOMPOSITION PRODUCTS OF DEAD ALGAE. ORGANIC PHOSPHORUS IN THE MEDIUM CONSTITUTES 2 PERCENT OF THE DISSOLVED AND TOTAL P. ALL SUPPORTING DATA ARE TABULATED. (HOLOMAN-BATTELLE)

FIELD 05C, 05A

ACCESSION NO. W72-07166

THE EFFECTS OF HEATED WASTE WATERS ON SOME MICROORGANISMS,

VIRGINIA POLYTECHNIC INST. AND STATE UNIV., BLACKSBURG. WATER RESOURCES RESEARCH CENTER.

J. CAIRNS, AND G. R. LANZA.

AVAILABLE FROM THE NATIONAL TECHNICAL INFORMATION SERVICE AS PB-208 414, $3.00 IN PAPER COPY, $0.95 IN MICROFICHE. WATER RESOURCES RESEARCH CENTER BULLETIN 48, FEBRUARY 1972. 101 P, 18 FIG, 19 TAB, 48 REF. OWRR B-017-VA(6).

DESCRIPTORS:
*THERMAL POLLUTION, *WATER POLLUTION EFFECTS, *AQUATIC MICROORGANISMS, *INVESTIGATIONS, *ANALYTICAL TECHNIQUES, DATA COLLECTIONS, ALGAE, PROTOZA, FUNGI, EQUIPMENT, INSTRUMENTATION, FLUORESCENCE, WATER TEMPERATURES, THERMAL STRESS.

IDENTIFIERS:
RESEARCH, CHYTRIDIACEOUS FUNGI, HEAT SHOCK.

ABSTRACT:
THE EFFECTS OF HEATED WASTE WATER ON AQUATIC MICROORGANISMS WERE STUDIED. NUMEROUS CATEGORIES OF TEMPERATURE SHOCK WERE INVESTIGATED WITH EMPHASIS ON (1) GENERAL SIMULATION OF THE PASSAGE OF WATER THROUGH A CONDENSING SYSTEM OF A STEAM ELECTRIC POWER PLANT TO DETERMINE THE EFFECTS OF THIS EXPOSURE AND (2) THE EFFECTS OF SUCH ENTRAINMENT UPON THE MICROBIAL SYSTEM BELOW THE DISCHARGE POINT. THREE GENERAL GROUPS OF ORGANISMS WERE EXAMINED: PROTOZOANS, ALGAE, AND CHYTRIDIACEOUS FUNGI. NEW EQUIPMENT DEVELOPED SPECIFICALLY TO AID IN THESE STUDIES (A SIMPLE APPARATUS FOR DELIVERING HEAT SHOCK TO MICROORGANISMS) AND A NEW FLUORESCENT SURVEY TECHNIQUE TO CHARACTERIZE STRESS INDUCED CELLULAR ALTERATIONS ARE DESCRIBED IN DETAIL. RESULTS OF THE STUDY ARE SUMMARIZED IN TABLES AND GRAPHS. (WOODARD-USGS)

FIELD 05C

ACCESSION NO. W72-07225

AQUATIC PLANTS FROM MINNESOTA, PART 2 - TOXICITY, ANTI-NEOPLASTIC, AND COAGULANT EFFECTS,

MINNESOTA UNIV., MINNEAPOLIS. WATER RESOURCES RESEARCH CENTER.

K. LEE SU, AND E. JOHN STABA.

AVAILABLE FROM THE NATIONAL TECHNICAL INFORMATION SERVICE AS PB-208 609, $3.00 IN PAPER COPY, $0.95 IN MICROFICHE. MINNESOTA WATER RESOURCES RESEARCH CENTER BULL. 47, FEBRUARY 1972. 24 P, 4 FIG, 5 TAB, 73 REF. OWRR A-025-MINN(3).

DESCRIPTORS:
*AQUATIC PLANTS, *MINNESOTA, *TOXICITY, COAGULATION, WATER POLLUTION EFFECTS, *ALGAL TOXINS.

IDENTIFIERS:
*PROTHROMBIN TIME, *PHARMACOLOGICAL PROPERTIES, ANTICANCER, ANTINEOPLASTICS.

ABSTRACT:
TOXICITY, ANTINEOPLASTIC, COAGULANT AND ANTICOAGULANT EFFECTS OF THE FOLLOWING 22 MINNESOTAN AQUATIC PLANTS WERE EVALUATED IN TERMS OF PHARMACOLOGICAL PROPERTIES: ANACHARIS CANADENSIS, CALLA POLUSTRIS, CAREX LACUSTRIS, CERATOPHYLLUM DEMERSUM, CHARA VULGARIS, ELEOCHARIS SMALLII, LEMNA MINOR, MYRIOPHYLLUM EXALBESCENS, NUPHAR VARIEGATUM, NYMPHAEA TUBEROSA, POTAMOGETON AMPLIFOLIUS, P. NATANS, P. PECTINATUM, P. RICHARDSONU, P. ZOSTERIFORMIS, SAGITTARIA CUNEATA, S. LATIFOLIA, SPARGANIUM EURYCARPUM, S. FLUCTUANS, TYPHA ANGUSTIFOLIA, VALLISNERIA AMERICANA, AND ZIZANIA AQUETICA. TOXICITY OF SKELLYSOLVE F, CHLOROFORM, 80% ETHANOL AND WATER EXTRACTS OF THESE AQUATIC PLANTS WERE EVALUATED IN A NUMBER OF ANIMAL EXPERIMENTS IN SWISS WEBSTER MICE. ANTINEOPLASTIC EXPERIMENTS INVOLVED AMELANOMA TUMOR CELLS. IN VIVO PROTHROMBIN TIME (PT) AND PARTIAL THROMBOPLASTIN TIME (PTT) WERE ASSAYED IN ANTI-COAGULATION EXPERIMENTS. THE TOXICITY OF THE AQUATIC PLANTS IN GENERAL WAS FOUND TO BE RELATIVELY LOW. THE LD50 FOR THE MOST TOXIC ONE, I.E., N. TUBEROSA (STEM), IN MICE WAS 3 GM OF DRY PLANT MATERIAL/KG (CA. 25.4 GM OF WET PLANT MATERIAL/KG). ONLY NUPHAR VARIEGATUM INDICATED AN ANTICANCER POSSIBILITY, THE REMAINING AQUATIC PLANTS HAD NO SIGNIFICANT INHIBITION ACTIVITY AT THE DOSES SELECTED. NORMAL PARTIAL THROMBOPLASTIN TIME FOR MICE WAS 51 SECONDS AND ONLY THE PROLONGATION OF PTT (LONGER THAN 61 SECONDS) WAS OBSERVED IN 50% OF THE AQUATIC PLANTS TESTED. AMONG THESE PLANTS, THE MOST SIGNIFICANT INCREASE OF PTT (MORE THAN 20 MINUTES) WAS OBSERVED IN CAREX LACUSTRIS, MYRIOPHYLUM EXALBESCENS, NUPHAR VARIEGATUM AND NYMPHAEA TUBEROSA.

FIELD 05C

ACCESSION NO. W72-07360

STUDIES ON THE REGULATION OF ALGAL GROWTH BY GIBBERELLIN,

WESTERN AUSTRALIA UNIV., NEDLANDS.

R. C. JENNINGS.

AUSTRALIAN JOURNAL OF BIOLOGICAL SCIENCES, VOL 24, P 1115-1124, 1971. 3 FIG, 4 TAB, 18 REF.

DESCRIPTORS:
*ALGAE, *PLANT GROWTH, RHODOPHYTA, INHIBITORS, PLANT TISSUES, BIOASSAY, PHAEPHYTA.

IDENTIFIERS:
*BIOSYNTHESIS, *GIBBERELLIN, HYPNEA MUSCIFORMIS, GRACILARIA VERUCOSA, ECKLONIA RADIATA, GROWTH REGULATION.

ABSTRACT:
THAT ENDOGENOUS GIBBERELLINS MAY BE INVOLVED IN GROWTH REGULATION OF BROWN ALGA, ECKLONIS RADIATA, AND THE GREEN ALGA, ENTEROMORPHA PROLIFERA, PROMPTED STUDIES OF RED ALGAE, HYPNEA MUSCIFORMIS AND GRACILARIA VERUCOSA. CCC AND AMO-1618, AT RELATIVELY HIGH CONCENTRATIONS ONLY, INHIBITED GROWTH OF EXCISED BRANCH APICES OF H MUSCIFORMIS. NEITHER GA-3 NOR GA-7 STIMULATED GROWTH OF THE ALGA IN PRESENCE OR ABSENCE OF THESE COMPOUNDS, AND GIBBERELLIN-LIKE MATERIAL EXTRACTED FROM H MUSCIFORMIS FAILED TO STIMULATE GROWTH. BOTH GIBBERELLIS STIMULATED GROWTH OF SLOW-GROWING, BUT NOT FAST-GROWING, BRANCH APICES OF GRACILARIA VERUCOSA. IT IS CONCLUDED THAT ENDOGENOUS GIBBERELLINS MAY NOT REGULATE GROWTH OF H MUSCIFORMIS, BUT THIS MAY BE A SPECIES PECULIARITY AND NOT A GENERAL PHENOMENON IN RED ALGAE. CCC INHIBITED GAMETOPHYTE GROWTH OF ECKLONIA RADIATA, AND GA-3 SIGNIFICANTLY OVERCAME THIS INHIBITION IN A MANNER COMPATIBLE WITH THE CONCEPT OF CCC INHIBITING GIBBERELLIN BIOSYNTHESIS. THE COMPLEMENT OF ACIDIC, ETHYL ACETATE-SOLUBLE GIBBERELLINS, EXTRACTED FROM THOSE REGIONS OF E RADIATA SPOROPHYTES ACTIVE IN CELL DIVISION, WAS CHROMATOGRAPHICALLY SIMILAR BUT DIFFERED FROM THOSE EXTRACTED FROM A RELATIVELY QUIESCENT REGION OF THE ALGA. THESE DATA SUPPORT CONCLUSIONS THAT ENDOGENOUS GIBBERELLINS ARE INVOLVED IN GROWTH REGULATION OF E RADIATA. (JONES-WISCONSIN)

FIELD 05C

ACCESSION NO. W72-07496

EXTRUSION OF CARBON ACCOMPANYING UPTAKE OF AMINO ACIDS BY MARINE PHYTOPLANKTERS,

CALIFRONIA UNIV., IRVINE. DEPT. OF DEVELOPMENTAL AND CELL BIOLOGY.

C. G. STEPHENS, AND B. B. NORTH.

LIMNOLOGY AND OCEANOGRAPHY, VOL 16, NO 5, P 752-757, 1971. 4 FIG, 1 TAB, 22 REF.

DESCRIPTORS:
*ANALYTICAL TECHNIQUES, *CARBON, *AMINO ACIDS, *MARINE ALGAE, PHYTOPLANKTON, NITROGEN, ABSORPTION, SEA WATER.

IDENTIFIERS:
PLATYMONAS, NITZSCHIA OVALIS, CARBON EXTRUSION, CARBON-14, NITROGEN SOURCE.

ABSTRACT:
TO GAIN INFORMATION ABOUT CONCENTRATIONS OF FREE AMINO ACIDS IN THE OCEAN, SHORT-TERM UPTAKE MEASUREMENTS WITH C-14-LABELED AMINO ACIDS OF CELL NITROGEN, DISSOLVED C-14, AND DISSOLVED AMINO NITROGEN WERE MADE. IT IS INDICATED THAT AMBIENT AMINO ACIDS ENTER PHYTOPLANKTON CELLS, THE AMINO NITROGEN IS RETAINED, AND SOME CARBON IS RETURNED TO THE MEDIUM. PLATYMONAS AND NITZSCHIA OVALIS WERE USED THROUGHOUT. THE DISTRIBUTION OF C-14 WAS FOLLOWED BY DETERMINING RADIO-ACTIVITY IN MEDIUM SAMPLES AND IN CELL SAMPLES COLLECTED ON MILLIPORE FILTERS. UPTAKE OF THE NITROGENOUS PORTION OF THE AMINO ACID MOLECULE COULD BE FOLLOWED DIRECTLY IN TWO WAYS. CHROMATOGRAPHY AND AUTORADIOGRAPHY OF A PORTION OF THE ELUENT PRODUCED A SINGLE SPOT AT THE PROPER POSITION FOR THE AMINO ACID CONCERNED. RESULTS IMPLY THAT RADIOCHEMICAL MEASUREMENTS OF UPTAKE RATES USING C-14-LABELED AMINO ACIDS MAY PRODUCE SERIOUS UNDERESTIMATES OF ACTUAL UPTAKE RATES, PARTICULARLY WITH LONG INCUBATION TIMES. ALSO, IT IS CONSISTENT WITH THE HYPOTHESIS THAT AMINO ACIDS MAY BE A SIGNIFICANT NITROGEN SOURCE FOR PHYTOPLANKTERS IN THE OCEAN. THESE OBSERVATIONS MAKE IT NECESSARY TO RE-EVALUATE PROCEDURES FOR STUDY OF AMINO ACID UPTAKE BY ALGAE. (JONES-WISCONSIN)

FIELD 05C

ACCESSION NO. W72-07497

PRODUCTION OF BENTHIC MICROALGAE IN THE LITTORAL ZONE OF A EUTROPHIC LAKE,

COPENHAGEN UNIV., HILLEROD (DENMARK). FRESHWATER-BIOLOGICAL LAB.

C. HUNDING.

OIKOS, VOL 22, NO 3, P 389-397, 1971. 8 FIG, 5 REF.

DESCRIPTORS:
*PRIMARY PRODUCTIVITY, *BENTHIC FLORA, MICROORGANISMS, ALGAE, LAKES, EUTROPHICATION, LITTORAL, SANDS, SHALLOW WATER, PERIPHYTON, SEDIMENTS, DIATOMS, LIGHT INTENSITY, PHOTOSYNTHESIS, SESSILE ALGAE, TEMPERATURE, DEPTH, LIGHT PENETRATION.

IDENTIFIERS:
LAKE FURESO (DENMARK), COCCONEIS, RHOICOSPHENIA, ACHANTHES, FRAGILARIA CONSTRUENTS, NAVICULA, NITZSCHIA, EPIPYTIC ALGAE.

ABSTRACT:
EFFECT OF VARYING LIGHT INTENSITY AND TEMPERATURE ON MICROBENTHIC COMMUNITY FLORA WERE STUDIED THROUGHOUT THE YEAR SIMULTANEOUSLY WITH PRIMARY PRODUCTION, WHICH WAS MEASURED ON BENTHIC MICROALGAE OF THE SANDY LITTORAL ZONE IN DANISH EUTROPHIC LAKE FURESO BY THE C-14 METHOD. ANNUAL POTENTIAL GROSS PRODUCTION WAS CALCULATED SHOWING HIGHEST VALUES IN JULY AND ANNUAL TRUE GROSS PRODUCTION ESTIMATED. DARK FIXATION WAS RELATIVELY HIGH. ALTHOUGH THE FOUR STATIONS DIFFERED SOMEWHAT IN WAVE ACTION, NO SIGNIFICANT DIFFERENCES BETWEEN PRODUCTION WERE NOTED. DAILY PRODUCTION FOR ALL STATIONS WAS AVERAGED. SESSILE ALGAE ACCOUNTED FOR THE BULK OF PRIMARY PRODUCTION (70 TO 90%). ALGAE WERE MAINLY PENNATE DIATOMS. DIATOMS WITH INTACT CHROMATOPHORES WERE FOUND DOWN TO A 10 CM DEPTH IN SEDIMENT. WHEN CELLS FROM THE ANAEROBIC ZONE WERE BROUGHT INTO LIGHT, PHOTOSYNTHESIS STARTED IMMEDIATELY. DURING SUMMER, LIGHT INTENSITY VALUE WAS 9 TO 10 WHILE THE AUTUMN VALUE WAS 3 TO 5 KLUX. PHOTOINHIBITION WAS DEMONSTRATED FOR THE AUTUMN POPULATION AT LIGHT INTENSITIES EXCEEDING 25 TO 30 KLUX. IT SEEMS TO BE CHARACTERISTIC FOR BENTHIC ALGAE THAT LIGHT SATURATED PHOTOSYNTHESIS TAKES PLACE OVER A WIDE RANGE OF LIGHT INTENSITIES. BENTHIC SESSILE DIATOM POPULATIONS APPEAR RATHER STABLE. (JONES-WISCONSIN)

FIELD 05C

ACCESSION NO. W72-07501

A PRELIMINARY INVESTIGATION OF A POTENTIAL NEW ALGICIDE,

HATFIELD POLYTECHNIC (ENGLAND). DEPT. OF BIOLOGICAL SCIENCES.

K. H. GOULDING.

PROCEEDINGS 6TH BRITISH INSECTICIDE AND FUNGICIDE CONFERENCE, P 621-629, 1971. 4 REF.

DESCRIPTORS:
*ALGICIDES, *ALGAL CONTROL, WATER POLLUTION, EUTROPHICATION, NUTRIENTS, CHLORELLA, CHLAMYDOMONAS, DIATOMS, SILICATES, CYANOPHYTA, TOXICITY, COSTS, CARBON DIOXIDE.

IDENTIFIERS:
*DACONIL, *CHLOROPHTHALONIL, ULOTHRIX, ANABAENA, OSCILLATORIA, MICROCYSTIS, ALGISTAT.

ABSTRACT:
SINCE THE PROBLEM OF INCREASED ALGAL GROWTH INVOLVES SUCH A WIDE RANGE OF SPECIES ANY SUCCESSFUL ALGICIDE SHOULD HAVE THE WIDEST RANGE OF EFFECTIVENESS; IT MUST BE PARTICULARLY ACTIVE AGAINST BLUE-GREEN ALGAE. A LARGE NUMBER OF CRITERIA SHOULD BE SATISFIED BEFORE AN ALGICIDE IS USED ON A WIDE SCALE AND SHOULD REALLY ACT MORE AS AN ALGISTATIC AGENT SLOWING DOWN GROWTH AND SPREADING OUT AN ALGAL 'BLOOM' OVER A LONGER TIME PERIOD. IF THIS COULD BE ACHIEVED WATER TREATMENT PLANTS COULD COPE WITH THE ALGAL PROBLEMS. SOME OF THE WORK SO FAR CARRIED OUT IN EXAMINING THE ALGICIDAL PROPERTIES OF THE COMPOUND CHLOROPHTHALONIL INDICATES THE POTENTIAL OF THIS COMPOUND. IT IS EFFECTIVE AGAINST A RANGE OF ALGAE INCLUDING CHLORELLA, CHLAMYDOMONAS, ULOTHRIX, ANABAENA, OSCILLATORIA, AND MICROCYSTIS AT LOW CONCENTRATIONS--OFTEN LESS THAN 0.001 PPM. THE EFFECT IS GENERALLY LESS AFTER 300 HOURS THAN AFTER 150 HOURS AND IS DEPENDENT UPON THE SIZE OF THE INITIAL CELL INOCULUM. THE WORK CARRIED OUT EVALUATES CHLOROPHYTHALONIL WITH A LIST OF CRITERIA FOR AN IDEAL ALGICIDE. (JONES-WISCONSIN)

FIELD 05C

ACCESSION NO. W72-07508

ALGAL GROWTH EXCITERS,

KAPPE ASSOCIATES, INC., ROCKVILLE, MD.

D. S. KAPPE, AND S. E. KAPPE.

WATER AND SEWAGE WORKS, VOL 118, NO 8, P 245-248, 1971. 14 REF.

DESCRIPTORS:
*ALGAL CONTROL, *ESSENTIAL NUTRIENTS, *DEFICIENT NUTRIENTS, *AQUATIC ALGAE, *CARBON DIOXIDE, NUISANCE ALGAE, EUTROPHICATION, PHOSPHATES, PHOSPHORUS, CARBON, ORGANIC MATTER, MANGANESE, IRON, SULFUR, POTASSIUM, MOLYBDENUM, DETERGENTS, SEWAGE TREATMENT, CARBONATES, BICARBONATES, HARDNESS(WATER), ALKALINITY, ENZYMES, TEMPERATURE, PHOTOSYNTHESIS, DECOMPOSING ORGANIC MATTER, LIGHT INTENSITY, CALCIUM, CYANOPHYTA, SCENEDESMUS, SODIUM, MANGANESE, CHLOROPHYLL, POTASSIUM, COPPER, ALGICIDES, COBALT.

IDENTIFIERS:
LAKE TAHOE(CALIF), ZINC, ASCORBIC ACID, PYRIDOXINE, INDAL COMPOUNDS.

ABSTRACT:
NO APPARENT CONSISTENT RELATIONSHIP APPEARS TO EXIST BETWEEN LAKE EUTROPHICATION AND PHOSPHORUS RETENTION CAPACITY AND EUTROPHICATION CANNOT BE EXPLAINED SOLELY ON THE BASIS OF SEDIMENTS' CAPACITY TO RETAIN PHOSPHORUS. CURRENT FINDINGS INDICATE THAT EUTROPHICATION IS MORE DIRECTLY RELATED TO BICARBONATE-CARBONATE ALKALINITIES OF WATER THAN TO PHOSPHATE CONCENTRATIONS AND DISCLOSE THAT WATER TEMPERATURE PLAYS AN IMPORTANT PART IN THE ALGAL CYCLE; THAT THE BICARBONATE-CARBONATE ALKALINITY OF WATER IS ESSENTIAL TO ALGAL GROWTH; THE CONCENTRATIONS OF NUTRIENTS ARE CRITICAL FOR PHOTOSYNTHESIS; THE INTENSITY OF SUNLIGHT, THE CLARITY OF WATER, AND THE ABILITY OF ALGAE TO ADSORB AND ABSORB ENERGY FROM SUNLIGHT HAVE A GREAT EFFECT ON ALGAL GROWTH; AND THAT ALGAL GROWTH IS CONTROLLED AND STIMULATED BY MANY METAL-ACTIVATED ENZYMES. THE CURRENT CONCEPT IS THAT ALTHOUGH NITROGEN AND PHOSPHORUS ARE MAJOR FACTORS IN THE EUTROPHIC CONDITION THAT LEADS TO ALGAL BLOOMS, THEIR PRESENCE IS NOT ALWAYS THE CRITICAL FACTOR. NITROGEN IS THE CRITICAL LIMITING FACTOR TO ALGAL GROWTH AND EUTROPHICATION IN COASTAL MARINE WATERS AND REMOVAL OF PHOSPHATES FROM DETERGENTS IS NOT LIKELY TO SLOW EUTROPHICATION. PHOTOSYNTHETIC BACTERIA, PHOTOCHEMICAL BACTERIA, AND OTHER AUTOTROPHIC AND HETEROTROPHIC BACTERIA COULD BE THE PRIMARY REACTORS THAT PRODUCE NUTRIENTS AND ESSENTIAL ELEMENTS DIRECTLY OR INDIRECTLY IN A FORM RESPONSIBLE FOR ALGAL GROWTH. PHOSPHORUS CANNOT BE THE GROWTH LIMITING NUTRIENT IN ALL WATERS. (AUEN-WISCONSIN)

FIELD 05C

ACCESSION NO. W72-07514

THE EFFECTS OF HEATED WASTE WATERS UPON MICROBIAL COMMUNITIES,

VIRGINIA POLYTECHNIC INST. AND STATE UNIV., BLACKSBURG. WATER RESOURCES RESEARCH CENTER.

J. CAIRNS, JR., AND R. A. PATERSON.

AVAILABLE FROM THE NATIONAL TECHNICAL INFORMATION SERVICE AS PB-208 697, $3.00 IN PAPER COPY, $0.95 IN MICROFICHE. VIRGINIA WATER RESOURCES RESEARCH CENTER, BLACKSBURG, PROJECT COMPLETION REPORT SEPTEMBER 1971 24 P, 2 FIG, 9 REF. OWRR B-017-VA(7).

DESCRIPTORS:
*THERMAL STRESS, *THERMAL POLLUTION, *MICROORGANISMS, *FLUORESCENCE, FOOD CHAINS, EQUIPMENT, ANALYTICAL TECHNIQUES, HEATED WATER, CONDENSERS, PROTOZOA, ALGAE, FUNGI, WATER POLLUTION EFFECTS.

ABSTRACT:
THE EFFECTS OF THERMAL STRESS ON SELECTED AQUATIC MICROORGANISMS WERE INVESTIGATED. NUMEROUS CATEGORIES OF TEMPERATURE SHOCK WERE STUDIED WITH EMPHASIS ON (1) SIMULATION OF THE PASSAGE OF WATER THROUGH A CONDENSING SYSTEM TO DETERMINE THE EFFECTS OF THIS EXPOSURE AND (2) THE EFFECTS OF SUCH ENTRAINMENT UPON THE MICROBIAL SYSTEM BELOW THE DISCHARGE POINT. THREE GENERAL GROUPS OF ORGANISMS WERE EXAMINED: PROTOZOANS, ALGAE, AND CHYTRIDIACEOUS FUNGI. NEW EQUIPMENT DEVELOPED SPECIFICALLY TO AID IN THESE STUDIES (A SIMPLE APPARATUS FOR DELIVERING HEAT SHOCK TO MICROORGANISMS) AND A NEW FLUORESCENT SURVEY TECHNIQUE TO CHARACTERIZE STRESS INDUCED CELLULAR ALTERATIONS ARE DESCRIBED IN DETAIL.

FIELD 05C, 05B

ACCESSION NO. W72-07526

CARBON, NITROGEN, AND PHOSPHORUS AND THE EUTROPHICATION OF FRESHWATER LAKES,

FISHERIES RESEARCH BOARD OF CANADA, WINNIPEG (MANITOBA). FRESHWATER INST.

D. W. SCHINDLER.

JOURNAL OF PHYCOLOGY, VOL. 7, NO. 4, P 321-329, DECEMBER 1971. 1 FIG, 2 TAB, 83 REF.

DESCRIPTORS:
*EUTROPHICATION, *LAKES, *NUTRIENTS, *FRESHWATER, *BIOASSAY, *STANDING CROP, CARBON, NITROGEN, PHOSPHORUS, PHYTOPLANKTON, CARBON DIOXIDE, SESTON, FERTILIZATION, WATER QUALITY, RADON RADIOISOTOPES, PRIMARY PRODUCTIVITY, CULTURES, BIOMASS, CARBON RADIOISOTOPES, LIMITING FACTORS, PHOTOSYNTHESIS, GREAT LAKES, LAKE ERIE, LAKE ONTARIO, HYDROGEN ION CONCENTRATION, TURBIDITY, OLIGOTROPHY, CHLOROPHYTA, ALGAE, CHRYSOPHYTA, CYANOPHYTA, EPILIMNION, OCHROMONAS, LAKES.

IDENTIFIERS:
SECCHI DISC, DISSOLVED NITROGEN, DISSOLVED PHOSPHORUS, OSCILLATORIA, RADON RADIOISOTOPES, CHROMULINA, CRYPTOMONAS, MALLOMONAS, STAURASTRUM, PHACOMYXA, SPONDYLOSIUM, LYNGBYA, PSEUDOANABAENA, DATA INTERPRETATION.

ABSTRACT:
THE QUESTION OF NUTRIENTS RESPONSIBLE FOR EUTROPHICATION OF FRESHWATER LAKES IS REVIEWED, AND RECENT ADDITIONS TO THE LITERATURE ON NUTRIENT LIMITATION ARE DISCUSSED. THE PAPER BY LANGE IS CRITICIZED ON SEVERAL GROUNDS, INCLUDING THE FACTS THAT UTILIZATION OF BICARBONATE BY PHYTOPLANKTON AND THE INVASION OF LAKE WATERS BY ATMOSPHERIC CO2 ARE IGNORED AS SOURCES OF PHOTOSYNTHETIC CARBON. THE PHOSPHORUS AND NITROGEN CONCENTRATIONS USED IN LANGE'S EXPERIMENTS ARE FAR HIGHER THAN VALUES PUBLISHED BY OTHERS FOR LAKES ERIE AND ONTARIO. PRELIMINARY RESULTS OF FERTILIZING A SMALL OLIGOTROPHIC LAKE WITH NITROGEN AND PHOSPHORUS ARE DESCRIBED. THE STANDING CROP OF PHYTOPLANKTON INCREASED BY 30-50 TIMES, WHILE THE P:N:C RATIO IN SESTON DID NOT CHANGE FROM RATIOS FOUND IN UNFERTILIZED LAKES. OTHER EXPERIMENTS DONE IN WATER COLUMNS ISOLATED WITH POLYETHYLENE FILM SHOWED THAT ADDITION OF CARBON DID NOT INCREASE THE PHYTOPLANKTON STANDING CROP. SINCE THE FERTILIZED LAKE WAS INITIALLY LOWER IN TOTAL CO2 THAN ANY OTHER RECORDED IN THE LITERATURE, IT IS CONCLUDED THAT CARBON IS UNLIKELY TO LIMIT THE STANDING CROP OF PHYTOPLANKTON IN ALMOST ANY SITUATION. MEASUREMENTS OF INVASION OF ATMOSPHERIC GASES TO THE FERTILIZED LAKE BY THE RADON-222 TECHNIQUE WERE COMPARED WITH PHYTOPLANKTON PRODUCTION MEASUREMENTS, REVEALING THAT ATMOSPHERIC INVASION OF CO2 IS SUFFICIENT TO SUPPORT THE HIGH PHYTOPLANKTON STANDING CROP IN THE EPILIMNION OF THE LAKE. POSSIBLE ERRORS IN INTERPRETATION OF CULTURE AND BOTTLE-BIOASSAY EXPERIMENTS WITH RESPECT TO EUROPHICATION ARE DISCUSSED. (MORTLAND-BATTELLE)

FIELD 05C, 02H

ACCESSION NO. W72-07648

EFFECT OF MERCURY ON ALGAL GROWTH RATES,

NAVAL RESEARCH LAB., WASHINGTON, D.C.

P. J. HANNAN, AND C. PATOUILLET.

BIOTECHNOLOGY AND BIOENGINEERING, VOL 14, P 93-101, 1972. 3 FIG, 1 TAB, 12 REF.

DESCRIPTORS:
*HEAVY METALS, *ALGAE, *GROWTH RATES, *TOXICITY, *FLUORESCENCE, TRACE ELEMENTS, COPPER, CHLORELLA, CULTURES, NUTRIENTS, RADICISOTOPES, CHLOROPHYLL, PIGMENT, MAGNESIUM, PHOSPHATES, NITROGEN, SEA WATER, FLUOROMETRY, EQUIPMENT, INHIBITION, IONS, ABSORPTION, BIOASSAY, *MERCURY.

IDENTIFIERS:
SILVER, CADMIUM, LEAD, CHLCRELLA PHYRENOIDOSA, PHAEODACTYLUM TRICORNUTUM, CYCLOTELLA NANA, CHAETOCEROS GALVESTONENSIS, MERCURIC CHLORIDE, DIMETHYL MERCURY, FLUOROMICROPHOTOMETER.

ABSTRACT:
COMPARISONS CF THE EFFECTS OF 0.1 PPM OF MERCURY AND OTHER METALLIC IONS (SILVER, CADMIUM, LEAD, COPPER) CN THE GROWTH RATES OF ONE FRESHWATER (CHLORELLA PYRENOIDOSA) AND THREE MARINE ALGAE (PHAEODACTYLUM TRICORNUTUM, CYCLOTELLA NANA, CHAETOCEROS GALVESTONENSIS) HAVE BEEN MADE UNDER CONTROLLED LABORATORY CONDITIONS. GROWTH, DETERMINED BY AN INCREASE IN FLUORESCENCE, WAS MONITORED USING A FLURO MICROPHOTOMETER. FLUORESCENCE WAS MEASURED INITIALLY AND ONCE EACH DAY FOR THREE DAYS. GRAPHIC DATA SHOW THAT MERCURY WAS MORE TOXIC THAN THE OTHER METALS TESTED. MERCURY AS MERCURIC CHLORIDE WAS MORE TOXIC THAN AS DIMETHYL MERCURY. THE TOXICITY OF MERCURY WAS FOUND TO BE COMPARATIVELY IRREVERSIBLE AND TC VARY INVERSELY WITH THE CONCENTRATION OF NUTRIENTS PRESENT IN THE GROWTH MEDIA. JUST HOW MUCH MERCURY ALGAE CAN ABSORB AND STILL GROW IS YET TO BE DETERMINED. (JEFFERIS-BATTELLE)

FIELD 05C, 05A

ACCESSION NO. W72-07660

EFFECTS OF DETERGENT PROTEASE ENZYMES ON SEWAGE OXIDATICN POND PHYTOPLANKTON,

VIRGINIA POLYTECHNIC INST., BLACKSBURG. DEPT. OF BIOLOGY.

B. C. PARKER, G. L. SAMSEL, AND E. K. OBENG-ASAMOA.

BIOSCIENCE, VOL 21, NO 20, P 1035-1042, OCTOBER 15, 1971. 4 FIG, 4 TAB, 15 REF.

DESCRIPTORS:
*ENZYMES, *DETERGENTS, *ALGAE, *PHYTOPLANKTON, *SEWAGE LAGOONS, BIODEGRADATION, SEWAGE TREATMENT, WASTE WATER TREATMENT, PONDS, SCENDESOMUS, CYANOPHYTA, BIOASSAY, SAMPLING, OXIDATICN LAGOONS, EUGLENA, CHLAMYDOMONAS, EUGLENOPHYTA, PROTOZOA, EQUIPMENT, WATER POLLUTION EFFECTS.

IDENTIFIERS:
*PROTEASE ENZYMES, CHLOROGONIUM, OOCYSTIS, OSCILLATORIA, SPIRULINA, PANDORINA, PHOCUS, STAURASTRUM, TRACHELOMONAS.

ABSTRACT:
FIELD INVESTIGATIONS ON THE EFFECTS OF DETERGENT PROTEASE ENZYMES ON ALGAL COMMUNITIES OF SEWAGE CXIDATION PONDS SUGGEST THAT ADDITIONS UP TO 1.0 MG/L CAUSE ONLY RELATIVELY SMALL CHANGES IN PHYTOPLANKTON POPULATIONS AND ALGAL COMMUNITY STRUCTURE. AS CONCENTRATIONS APPROACH 10 MG/L, THE COMMUNITY STRUCTURE MIGHT BE DISTURBED SIGNIFICANTLY IN SOME PONDS, DEPENDING ON THE PARTICULAR ENZYME PREPARATION. ENZYME PREPARATIONS NUMBERS 1 AND 2 (EP-1 AND EP-2) FROM THE SOAP AND DETERGENT ASSOCIATION WERE USED IN CONCENTRATICNS OF 10, 1.0, AND 0.1 MG/L ON SEVERAL BIOASSAY SYSTEMS. EITHER ONE LITER, OPEN POLYETHYLENE BAGS, FLOATING AT THE SURFACE IN POLYSTYRENE FRAMES, OR TRANSPARENT ACRYLIC PLASTIC CYLINDERS, WITH THEIR BASES STUCK IN THE MUD AND TOPS PROTRUDING ABOVE THE WATER LINE, WERE USED IN SAMPLING. EP-2 FAVORED GROWTH OF BLUE-GREEN ALGAE, INDICATING THAT WIDESPREAD USE OF SUCH PREPARATIONS SHOULD BE AVCIDED. (MORTLANC-BATTELLE)

FIELD 05C

ACCESSION NO. W72-07661

COMPOSITION OF PHYTOPLANKTON OFF THE SOUTHEASTERN COAST OF THE UNITED STATES,

OLD DOMINION UNIV., NORFOLK, VA.

H. G. MARSHALL.

BULLETIN OF MARINE SCIENCE, VOL 21, NO 4, P 806-825, DECEMBER 1971. 1 FIG, 12 TAB, 12 REF.

DESCRIPTORS:
*PHYTOPLANKTON, *SYSTEMATICS, *DIATOMS, SEA WATER, SEASONAL, FLUCTUATIONS, CONTINENTAL SHELF, VERTICAL DISTRIBUTION, NORTH CAROLINA, SOUTH CAROLINA, SAMPLING, MICROSCOPY, ELECTRON MICROSCOPY, PYRROPHYTA, DISTRIBUTION PATTERNS, WATER TEMPERATURE, WATER QUALITY, PLANKTON, ANALYTICAL TECHNIQUES, ALGAE.

IDENTIFIERS:
COCCOLITHOPHORES, SILICOFLAGELLATES, SARGASSO SEA, SHELF WATERS, GULF STREAM, CAPE HATTERAS, NANSEN BOTTLE, SPECIES DIVERSITY, SAMPLE PRESERVATION, COUNTING.

ABSTRACT:
PHYTOPLANKTON COMPOSITION OFF THE SOUTHEASTERN COAST OF THE UNITED STATES WAS STUDIED DURING A 42-MONTH PERIOD FROM 1964 TO 1968. COLLECTIONS WERE MADE USING NANSEN BOTTLES AT NINE DEPTHS FROM 0 TO 300 FEET IN CONTINENTAL SHELF WATERS, THE GULF STREAM, AND THE SARGASSO SEA. A 500-ML WATER SAMPLE WAS PRESERVED FOR ANALYSIS IMMEDIATELY AFTER SAMPLING USING NEUTRALIZED FORMALIN. THE SUPERNATANT LIQUID WAS SIPHONED OFF 4-6 WEEKS LATER TO OBTAIN A 10 - 40-ML CONCENTRATE. MICROSCOPIC TECHNIQUES WERE THEN USED TO COUNT PHYTOPLANKTON. CONSIDERABLE SPECIES DIVERSITY AND SEASONAL FLUCTUATIONS WERE NOTED. DIATOMS WERE PREDOMINANT IN SHELF WATERS AND TO THE WESTERN BOUNDARY OF THE GULF STREAM, BUT THEIR NUMBERS DECLINED RAPIDLY INTO THE SARGASSO SEA. SKELETONEMA COSTATUM AND RHIZOSOLENEA ALATA WERE THE MAJOR SPECIES FOUND. COCCOLITHOPHORES, PYRRHOPHYCEANS, AND SILICOFLAGELLATES WERE FOUND IN GREATEST NUMBERS IN THE GULF STREAM. IN THE SARGASSO SEA, COCCOLITHOPHORES AND PYRROHPHYCEANS PREDOMINATED. HOWEVER, THE TOTAL CONCENTRATION OF PHYTOPLANKTON AND THE NUMBER OF SPECIES WERE LESS IN THE SARGASSO SEA THAN IN THE SHELF WATERS OR GULF STREAM. (MORTLAND-BATTELLE)

FIELD 05B, 05A, 02L

ACCESSION NO. W72-07663

BACTERIAL DIEOFF IN PONDS,

TEXAS UNIV., HOUSTON. SCHOOL OF PUBLIC HEALTH.

E. M. DAVIS, AND E. F. FLCYNA.

JOURNAL OF THE SANITARY ENGINEERING DIVISION, AMERICAN SOCIETY OF CIVIL ENGINEERS, VOL 98, NO SA 1, P 59-69, FEBRUARY 1972. 2 FIG, 5 TAB, 14 REF.

DESCRIPTORS:
*ALGAE, *ENTERIC BACTERIA, *PATHOGENIC BACTERIA, CULTURES, E. COLI, BIOINDICATORS, PONDS, ANAEROBIC CONDITIONS, WASTE WATER TREATMENT, CHLOROPHYTA, CYANOPHYTA, AEROBIC CONDITIONS, BIOASSAY, GROWTH RATES, MORTALITY.

IDENTIFIERS:
STABILIZATION PONDS, ANABAENA CYLINDRICA, ANACYSTIS NIDULANS, GLOEOCAPSA ALPICOLA, OSCILLATORIA CHALYBIA, OSCILLATORIA FORMOSA, PHORMIDIUM FAVEOLARUM, ANKISTRODESMUS BRAUNII, CHLORELLA PYRENOIDOSA, CHLORELLA VULGARIS, SCENEDESMUS OBLIQUUS, ALCALIGENES FAECALIS, ENTEROBACTER AEROGENES, PROTEUS VULGARIS, PSEUDOMONAS AERUGINOSA, SERRATIA MARCESCENS, SALMONELLA PARATYPHI, SALMONELLA TYPHOSA, SHIGELLA PARADYSENTERIAE, SHIGELLA DYSENTERIAE, VIBRIO COMMA.

ABSTRACT:
LABORATORY AND FIELD TESTS WERE CONDUCTED TO EVALUATE SPECIFIC INTERACTIONS BETWEEN ALGAE AND TYPICAL ENTERIC AND SELECTED PATHOGENIC BACTERIA IN WASTE STABILIZATION PONDS. BACTERIAL DIEOFF FROM EXPOSURE TO SINGLE ALGAL SPECIES AND MIXED ALGAL ENVIRONMENTS WAS EVALUATED. RESULTS FROM STUDIES OF BOTH LABORATORY AND PILOT-SCALE STABILIZATION PONDS SHOWED THAT SINGLE ALGAL SPECIES HAD LITTLE EFFECT ON DIEOFF RATES, BUT AS THE COMPLEXITY OF THE ALGAL ENVIRONMENT INCREASED SO DID THE DIEOFF RATE. THIS WAS TRUE FOR BOTH ENTERIC AND PATHOGENIC BACTERIA. ESCHERISCHIA, PSEUDOMONAS, AND SERRATIA EXHIBITED AFTERGROWTH POTENTIAL BUT PROTEUS, ALCALIGENES, ENTEROBACTER, SALMONELLA, SHIGELLA, AND VIBRIO DID NOT. ANAEROBIC PRETREATMENT IN COMBINATION WITH FACULTATIVE AND MATURATION PONDS RESULTED IN HIGHER DIEOFFS OF ENTERIC BACTERIA WHEN COMPARED TO AEROBIC CONDITIONS. (MORTLAND-BATTELLE)

FIELD 05C, 05B

ACCESSION NO. W72-07664

A LIST OF NEW GENERA AND TYPE SPECIES OF FLAGELLATES AND ALGAE PUBLISHED IN 1966-1968. PART IV,

B. V. SKVORTZOV.

HYDROBIOLOGIA, VOL. 39, ISSUE 1, P 1-7, JANUARY 31, 1972. 24 FIG, 9 REF.

DESCRIPTORS:
*ALGAE, *CHLOROPHYTA, *EUGLENOPHYTA, *PYRROPHYTA, *SYSTEMATICS, DINOFLAGELLATES, SPECIATION.

IDENTIFIERS:
KOLBEANA OVOIDEA, PALMERIAMONAS PLANCTONICA, MANCHUDINIUM SINICUM, SINAMONAS STAGNALIS, KOFOIDIELLA UNIFLAGELLATA, TSUMURAIA NUMEROSA, HIROSEIA QUINQUELOBATA, TROITSKIELLA TRIANGULATA, AKIYAMAMONAS TERRESTRIS, BRASILOBIA PISCIFORMIS, PROWSEMONAS TROPICA, STIGMOBODO BRASILIANA, PAVLOVIAMONAS FRUTICOSA, NODAMASTIX SPIROGYRAE, COLEMANIA VERRUCOSA, GORDYMONAS VITALIS, PROCTORMONAS CHARACOLA, IOLYA PLANCTONICA, PROTOACEROMONAS SPINOSA, ANGULOMONAS TRIQUETRA, SWIRENKOIAMONAS UNIFLAGELLATA, PROTOCHROCMONAS GRANULATA, ROTUNDOMASTIX PLUVIALIS.

ABSTRACT:
TWENTY-THREE NEW GENERA AND TYPE SPECIES ARE LISTED. OTHER DATA INCLUDE ILLUSTRATIONS OF THE TYPE SPECIES, SIZE OF THE CELLS AND THEIR HABITATS. (HOLOMAN-BATTELLE)

FIELD 05A, 05C

ACCESSION NO. W72-07683

COMPARATIVE METABOLISM OF DDT, METHYLCHLOR, AND ETHOXYCHLOR IN MOUSE, INSECTS, AND IN A MODEL ECOSYSTEM,

ILLINOIS UNIV., URBANA. DEPT. OF ENTOMOLOGY; AND ILLINOIS UNIV., URBANA. DEPT. OF ZOOLOGY.

I. P. KAPOOR, R. L. METCALF, A. S. HIRWE, P.-Y. LU, AND J. R. COATS.

JOURNAL OF AGRICULTURAL AND FOOD CHEMISTRY, VOL 20, NO 1, P 1-6, JANUARY/FEBRUARY 1971. 5 FIG, 7 TAB, 14 REF.

DESCRIPTORS:
*BIODEGRADATION, *METABOLISM, *DDT, *RADIOACTIVITY TECHNIQUES, PESTICIDES, ECOSYSTEMS, POLLUTION SOURCES, PESTICIDE TOXICITY, FOOD CHAINS, FISH, BIOASSAY, ALGAE, SNAILS, MOSQUITOES, MODEL STUDIES.

IDENTIFIERS:
MICE, INSECTS, *ETHOXYCHLOR, *METHYLCHLOR, METABOLITES, BIOLOGICAL MAGNIFICATION, BIOLOGICAL SAMPLES, LIVER, URINE, OEDOGONIUM, PHYSA, CULEX, GAMBUSIA, FATE OF POLLUTANTS.

ABSTRACT:
METABOLIC PATHWAYS FOR ETHOXYCHLOR AND METHYLCHLOR IN INSECTS, MICE, AND IN A MODEL ECOSYSTEM DEMONSTRATE THAT THESE COMPOUNDS ARE PERSISTENT, INSECT TOXIC, BIODEGRADABLE ANALOGS OF DDT. THE EXPERIMENTAL PROCEDURE INVOLVED RADIO 'TAGGING' THE PESTICIDES AND FOLLOWING THEM AND THEIR REACTION PRODUCTS THROUGH THE METABOLIC PATHWAYS. ANALYSIS SHOWED THAT ETHOXYCHLOR IS OXYGEN DEALKYLATED TO MONO AND BISPHENOLIC PRODUCTS WITH DEHYDROCHLORINATION BEING A MAJOR METABOLIC PATHWAY. METHYLCHLOR IS OXIDIZED TO BENZYL ALCOHOL AND BENZOIC ACID ANALOGS AND CONJUGATES. EXCRETION-RECOVERY RATES INDICATE INITIAL RAPID ELIMINATION OF BOTH ANALOGS OVER DDT, BUT OVER A LONGER PERIOD OF TIME THEIR RETENTION SUPERCEDES THAT OF DDT. METABOLISM BY THE MOUSE ALSO INDICATES DEGRADATION TO ALCOHOLIC AND CARBOXYLIC ACID COMPONENTS. IN THE MODEL ECOSYSTEM, PESTICIDES WERE FOUND IN ALGAE, SNAILS, MOSQUITOES, AND FISH. ETHOXYCHLOR WAS FOUND IN FISH AT 1500 TIMES THE AMOUNT IN WATER, METHYLCHLOR WAS FOUND AT 1400 TIMES THE AMOUNT IN WATER, AND DDT WAS FOUND CONCENTRATED AT 85,000 TIMES THE AMOUNT IN WATER, ILLUSTRATING THE BIODEGRADATIVE CHARACTERISTICS OF THE ANALOGS. (MACKAN-BATTELLE)

FIELD 05C, 05B

ACCESSION NO. W72-07703

GROWTH AND THE PRODUCTION OF EXTRACELLULAR SUBSTANCES BY TWO STRAINS OF PHAEOCYSTIS POUGHETI,

WOODS HOLE OCEANOGRAPHIC INSTITUTION, MASS.

R. R. L. GUILLARD, AND J. A. HELLEBUST.

JOURNAL OF PHYCOLOGY, VOL 7, NO 4, P 330-338, DECEMBER 1971. 2 FIG, 7 TAB, 27 REF.

DESCRIPTORS:
*ALGAE, *CARBOHYDRATES, *ORGANICS ACIDS, *PLANT PHYSIOLOGY, *GROWTH RATES, PLANT MORPHOLOGY, CULTURES, PHOTOSYNTHESIS, CARBON RADIOISOTOPES, CHROMATOGRAPHY, AMINO ACIDS, GAS CHROMATOGRAPHY, PHAEOPHYTA, EUTROPHICATION, PLANKTON, ATLANTIC OCEAN, SEA WATER, SALINITY, PHENOLS, WATER TEMPERATURE, CHEMICAL ANALYSIS, SAMPLING, ECOLOGY, PHYTOPLANKTON, ALCOHOLS, SIEVE ANALYSIS, INHIBITORS, SURFACE WATERS, BIOASSAY, TRACERS.

IDENTIFIERS:
*EXCRETION, DISACCHARIDES, MONOSACCHARIDES, OLIGOSACCHARIDES, CLONES, PHAEOCYSTIS POUCHETI, GLUCOSE, MANNOSE, RHAMNOSE, ACRYLIC ACID, SODIUM, PYRIDINE, DEXTRAM, RAFFINOSE, SUCROSE, SENESCENT CULTURES, XYLOSE, RIBOSE, ARABINOSE, GALACTOSE, HEXURONIC ACID, VINEYARD SOUND, ORGANIC CARBON, ACID-VOLATILE COMPOUNDS, WOODS HOLE, HAPTAPHYCEAE, UNIALGAL CULTURES, GEL FILTRATION, AXENIC CULTURES, AUTORADIOGRAPHY, C-14, SURVIVAL.

ABSTRACT:
THE GROWTH AND RELEASE OF EXTRACELLULAR SUBSTANCES BY COLD-WATER STRAINS OF PHAEOCYSTIS, ISOLATED FROM THE WINTER SURFACE WATERS AT WOODS HOLE, WERE COMPARED WITH A WARM-WATER STRAIN FROM THE TROPICAL ATLANTIC NEAR THE COAST OF SURINAM. THE COLD-WATER STRAINS WERE CULTURED AT 4-8C UNDER ABOUT 4.5 KLUX OF LIGHT FOR 14 HOURS/DAY, AND THE TROPICAL CLONE, AT ABOUT 25 DEGREES C IN 14-HR. LIGHT - 10-HR. DARK CYCLE UNDER 3-4.5 KLUX OF LIGHT. NORTHERN STRAINS SURVIVED ONLY UP TO 14C, WHILE THE TROPICAL STRAIN SURVIVED ONLY AS LOW AS 17C. COLONY SHAPES OF THE NORTHERN AND TROPICAL CLONES DIFFERED SOMEWHAT, BUT THE MOTILE AND NONMOTILE SINGLE CELLS OF BOTH STRAINS SEEMED IDENTICAL IN THE LIGHT MICROSCOPE. BY CURRENT TAXONOMIC CRITERIA BOTH STRAINS BELONG TO THE SPECIES P. POUCHETI (HARIOT) LAGERHEIM. WHEN GROWING IN THE FORM OF COLONIES, BOTH STRAINS EXCRETED 16-64 PERCENT OF THEIR PHOTOASSIMILATED CARBON INTO THE MEDIUM, MAINLY AS CARBOHYDRATES OF VARYING MOLECULAR WEIGHTS. HOWEVER, CULTURES PREDOMINANTLY IN THE FORM OF SINGLE CELLS RELEASED ONLY ABOUT 3 PERCENT OF THEIR PHOTOASSIMILATED CARBON. THE QUALITATIVE COMPOSITION OF THE CARBOHYDRATES RELEASED IS SIMILAR FOR THE 2 STRAINS, CONSISTING OF SOME 8 SUGARS OR SUGAR DERIVATIVES WITH GLUCOSE, MANNOSE, AND RHAMNOSE AS THE DOMINANT COMPONENTS. THE PRODUCTION OF ACRYIC ACID WAS CONFIRMED. IT HAS BEEN ESTIMATED THAT AS MUCH AS 7 MICROGRAMS/LITER OF ACRYLIC ACID, AND AT LEAST 0.3 MG/LITER OF POLYSACCHARIDES CAN BE LIBERATED IN A PHAEOCYSTIS BLOOM. (SNYDER-BATTELLE)

FIELD 05C, 05A

ACCESSION NO. W72-07710

ORGANIC MATERIAL ANALYZER MONITORS SHORELINE SEWAGE DISPOSAL,

CHEMURGIC COUNCIL, NEW YORK.

J. W. TICKNOR.

WATER AND SEWAGE WORKS, VOL 119, NO 3, P 50-51, MARCH 1972. 3 FIG.

DESCRIPTORS:
*CHEMICAL ANALYSIS, *MONITORING, *CYCLING NUTRIENTS, *FOOD CHAINS, *ORGANIC COMPOUNDS, SHELLFISH, INSTRUMENTATION, PHYTOPLANKTON, AUTOMATION, EQUIPMENT, ANALYTICAL TECHNIQUES, SEWAGE, RESEARCH EQUIPMENT, SEWAGE EFFLUENTS, SEWAGE TREATMENT, SEWAGE DISPOSAL, MARINE ALGAE, CLAMS, MUSSELS, OYSTERS, EUTROPHICATION, NITROGEN, PHOSPHOROUS, CARBON, HYDROGEN, MARINE ANIMALS, NUTRIENTS, MOLLUSKS, CULTURES.

IDENTIFIERS:
ELEMENTAL ANALYZERS, MACROINVERTEBRATES.

ABSTRACT:
SCIENTISTS AT THE WOODS HOLE OCEANOGRAPHIC INSTITUTION, MASSACHUSETTS, ARE WORKING ON A PROJECT CONCERNED WITH THE HEALTHFUL DISPOSAL OF TREATED SEWAGE AND PROVIDING NUTRIENTS FOR INCREASED SHELLFISH CULTURE ALONG THE EASTERN SEABOARD. EXPERIMENTATION INVOLVES FEEDING TREATED SEWAGE TO ALGAE AND FEEDING THE ALGAE TO SHELLFISH (OYSTERS, MUSSELS, SCALLOPS) UNDER CONTROLLED CONDITIONS. IN ORDER TO DETERMINE WHAT NUTRITIVE VALUES THE SHELLFISH DERIVE FROM THE SEWAGE-PHYTOPLANKTON-SUNLIGHT SYSTEM, THE FOOD CHAIN IS CONSTANTLY MONITORED AT SELECTED POINTS USING A PERKIN-ELMER MODEL 240 ELEMENTAL ANALYZER. THIS INSTRUMENT AUTOMATICALLY ANALYZES ORGANIC COMPOUNDS FOR CARBON, HYDROGEN, AND NITROGEN CONTENT SIMULTANEOUSLY, PERFORMING A COMPLETE ANALYSIS WITHIN 13 MINUTES. IT IS USED, IN THIS CASE, TO DETERMINE LEVELS AND RATIOS OF N AND P PRESENT IN ALGAE, MOLLUSCS, AND MOLLUSCS' DETRITUS. THE ANALYTICAL METHOD CONSISTS OF: (1) BURNING THE ORGANIC SAMPLE IN AN O2 ATMOSPHERE AIDED BY CHEMICALS SUCH AS SILVER TUNGSTATE AND MAGNESIUM OXIDE, (2) FLUSHING THE GASEOUS COMBUSTION PRODUCTS WITH A HELIUM STREAM THROUGH A REDUCTION TUBE, AND (3) REMOVING H, C, AND N AS THE GASEOUS PRODUCTS MOVE THROUGH A SERIES OF TRAPS. THE ANALYZER CAN BE READILY CONVERTED FOR OXYGEN ANALYSIS. (MACKAN-BATTELLE)

FIELD 05A, 05C, 05E

ACCESSION NO. W72-07715

CLUSTER ANALYSIS OF FISH IN A PORTION OF THE UPPER POTOMAC RIVER,

VIRGINIA POLYTECHNIC INST. AND STATE UNIV., BLACKSBURG. DEPT. OF BIOLOGY.

J. CAIRNS, JR., AND R. L. KAESLER.

TRANSACTIONS OF THE AMERICAN FISHERIES SOCIETY, VOL. 100, NO. 4, P 750-756, OCTOBER 1971. 2 FIG, 3 TAB, 10 REF.

DESCRIPTORS:
*SURVEYS, *STATISTICAL METHODS, *POTOMAC RIVER, DATA COLLECTIONS, EVALUATION, ON-SITE DATA COLLECTIONS, RIVERS, *ELECTRIC POWERPLANTS, WATER POLLUTION EFFECTS, WATER POLLUTION SOURCES, *FISH, DIATOMS, PROTOZOA, AQUATIC INSECTS, AQUATIC ANIMALS, AQUATIC ALGAE.

IDENTIFIERS:
*CLUSTER ANALYSIS(FISH), *DICKERSON PLANT.

ABSTRACT:
A SERIES OF SURVEYS WERE CARRIED OUT FROM 1956 UNTIL 1965 IN THE VICINITY OF THE DICKERSON PLANT OF THE POTOMAC ELECTRIC POWER CO, LOCATED BELOW THE CONFLUENCE OF THE POTOMAC AND MONOCACY RIVERS. THESE SURVEYS WERE MADE AT VARIOUS STAGES OF GROWTH OF THE POWER STATION AND AT BOTH HIGH AND LOW WATER LEVELS OF THE RIVER. CLUSTER ANALYSES WERE MADE OF THE FISH DATA IN ORDER TO PROVIDE A RAPID AND EXPLICIT MEANS OF ANALYZING THE DATA. A SECONDARY PURPOSE WAS TO MEASURE THE REDUNDANCY OF INFORMATION OBTAINED BY ANALYSIS OF OTHER GROUPS, NAMELY ALGAE, PROTOZOANS, AQUATIC INSECTS, AND OTHER INVERTEBRATES. ON THE BASIS OF THE CLUSTER ANALYSIS OF DATA ON OCCURRENCE AND DISTRIBUTION OF FISH, NO CHANGES IN THE ENVIRONMENT OF THE UPPER POTOMAC RIVER CAN BE ASCRIBED TO THE OPERATION OF THE DICKERSON PLANT. THIS CONCLUSION AGREES WITH OTHER RESULTS OBTAINED BY ANALYSIS OF PROTOZOANS, ALGAE, AND INSECTS. JACCARD COEFFICIENTS AMONG AGGREGATES OF SPECIES OF FISH ARE APPRECIABLY HIGHER THAN FOR OTHER GROUPS OF ORGANISMS, INDICATING THE RELATIVE MOBILITY OF FISH. THERE IS MUCH INFORMATION REDUNDANCY WHEN ALL GROUPS OF ORGANISMS ARE STUDIED SIMULTANEOUSLY. THIS REDUNDANCY IS CONSIDERED DESIRABLE FOR CONFIRMATION OF RESULTS. (LEGORE-WASHINGTON)

FIELD 05C

ACCESSION NO. W72-07791

CONCENTRATIONS OF PU, CO AND AG RADIONUCLIDES IN SELECTED PACIFIC SEAWEEDS,

SCRIPPS INSTITUTION OF OCEANOGRAPHY, LA JOLLA, CALIF.

K. M. WONG, V. F. HODGE, AND T. R. FOLSOM.

AVAILABLE FROM THE NATIONAL TECHNICAL INFORMATION SERVICE AS CONF 710817-1 $3.00 IN PAPER COPY, $0.95 IN MICROFICHE. (1971) 11P.

DESCRIPTORS:
*RADIOACTIVITY EFFECTS, *MONITORING, *MARINE ALGAE, RHODOPHYTA, PHAEOPHYTA, *COBALT RADIOISOTOPES, NUCLEAR WASTES, PATH OF POLLUTANTS, ESTUARINE ENVIRONMENT, RADIOACTIVITY TECHNIQUES, ABSORPTION.

IDENTIFIERS:
*PLUTONIUM, *SILVER RADIOISOTOPES.

ABSTRACT:
RED ALGAE NEAR A NUCLEAR PLANT SHOWED EVIDENCE OF ENHANCED CO-58, CO-60, AND AG-110 (TYPICAL RADIOACTIVITIES WERE 2200, 180, AND 260 DPM/KG WET SAMPLE, RESPECTIVELY.), BUT NO ENHANCEMENT OF PU-239. ALL SPECIES OF SEAWEED WERE CONTAMINATED WITH PU-239 FROM (WORLD-WIDE) FALLOUT, BUT FURTHER WORK IS NEEDED TO SHOW THE CORRELATION BETWEEN THE AMOUNT IN THE SEA WATER AND THAT IN THE SEAWEED. (BUPP-ORNL)

FIELD 05C, 05B

ACCESSION NO. W72-07826

SECOND ANNUAL REPORT, ENVIRONMENTAL STUDIES, MAINE YANKEE ATOMIC POWER COMPANY.

MAINE YANKEE ATOMIC POWER CO., WESTBORO, MASS.

AVAILABLE FROM THE NATIONAL TECHNICAL INFORMATION SERVICE AS DOCKET 50309-45, $6.00 IN PAPER COPY, $0.95 IN MICROFICHE. DOCKET 50309-45, 1971. 364 P.

DESCRIPTORS:
*MONITORING, *SURVEYS, *MEASUREMENT, *DATA COLLECTIONS, *ENVIRONMENT, *RADIOACTIVITY, *RADIOACTIVITY EFFECTS, RIVERS, EFFLUENT, RADIOECOLOGY, AQUATIC POPULATIONS, HYDROLOGIC ASPECTS, NUTRIENTS, CHEMICALS, PLANKTON, BENTHOS, ALGAE, FISH, WATER POLLUTION, WATER POLLUTION SOURCES.

IDENTIFIERS:
CONCENTRATION, RADIATION SAFETY, RADIATION CONTROL.

ABSTRACT:
THE ENVIRONMENTAL SURVEY OF THE MONTSWEAG BAY-BACK RIVER AREA WAS INITIATED AS A FULL-SCALE INTEGRATED PROGRAM IN OCTOBER, 1969. PORTIONS OF THE PROGRAM HAD BEEN STARTED AS EARLY AS THE SUMMER OF 1968, BUT SAMPLING WAS NOT CONDUCTED ON A REGULAR SCHEDULE UNTIL THE FALL AND SPRING OF 1969-1970. INITIAL EFFORTS INVOLVED ASSEMBLING AND TESTING EQUIPMENT, EVALUATING SAMPLING PROCEDURES, AND SELECTING PERMANENT SAMPLING STATIONS. ALMOST ALL PORTIONS OF THE PROGRAM HAVE NOW BEEN OPERATIVE FOR A YEAR OR MORE AND SOME CONCLUSIONS CAN BE DRAWN FROM THE FINDINGS. HYDROLOGY AND NUTRIENT CHEMISTRY, PLANTON, BENTHOS, SOME COMMERCIAL INVERTEBRATES, FINFISH, AND MARINE ALGAE ARE DISCUSSED. (HOUSER-ORNL)

FIELD 05A, 05B

ACCESSION NO. W72-07841

ALGAE CONTROL BY MIXING, STAFF REPORT ON KEZAR LAKE IN SUTTON, NEW HAMPSHIRE.

NEW HAMPSHIRE WATER SUPPLY AND POLLUTION CONTROL COMMISSION, CONCORD.

AVAILABLE FROM THE NATIONAL TECHNICAL INFORMATION SERVICE AS COM-71-01087, $3.00 IN PAPER COPY, $0.95 IN MICROFICHE. DECEMBER 1970. 103 P, 28 FIG, 30 TAB, 76 REF.

DESCRIPTORS:
WATER QUALITY CONTROL, *CYANOPHYTA, *WATER ANALYSIS, LAKES, SAMPLING, DIATOMS, TURBIDITY, WATER POLLUTION SOURCES, NUTRIENTS, SEWAGE, WATER QUALITY, AERATION, WATER PROPERTIES, *NEW HAMPSHIRE, ALGAE, COLOR, COPPER SULFATE, NETS, HYDROGEN ION CONCENTRATION, PHOSPHATES, NITROGEN, HARDNESS(WATER), CONDUCTIVITY, ZINC, WATER TEMPERATURE, ALKALINITY, CHLORIDES, COPPER, MANGANESE, DISSOLVED OXYGEN, IRON, LIGHT PENETRATION, CHLOROPHYTA, PIPES, MECHANICAL EQUIPMENT, NITRATES, EUTROPHICATION, CARBONATES, CARBON DIOXIDE, WATER POLLUTION EFFECTS, MIXING.

IDENTIFIERS:
*KEZAR LAKE(N.H.), APHANIZOMENON, ANABAENA SPP, MICROCYSTIS AERUGINOSA, APHANIZOMENON HOLSATICUM, ICTALURUS NEBULOSUS, NOTEMIGONUS CRYSOLEUCAS, CATOSTOMUS COMMERSONII COMMERSONII, ASTERIONELLA SPP, ORTHOPHOSPHATES, SECCHI DISC, NEWFOUND LAKE, SQUAM LAKE, APHANIZOMENON FLOW-AQUAE, DESTRATIFICATION, ANABAENA FLOW-AQUAE, HOMOGENIZING, TRANSPARENCY.

ABSTRACT:
KEZAR LAKE IS A NEW HAMPSHIRE RECREATIONAL LAKE, WHICH IN 1964 WAS OBVIOUSLY SUFFERING FROM OBJECTIONABLE ALGAE BLOOM. AFTER SEVERAL COPPER SULFATE TREATMENTS FAILED TO SOLVE THE PROBLEM, A DESTRATIFICATION PROCESS WAS ATTEMPTED. THIS WAS ACCOMPLISHED BY FORCING COMPRESSED AIR, FROM SHORE-LOCATED COMPRESSORS, THROUGH P.V.C., 2-INCH DIAMETER PIPING TO THE DEEPEST PORTION OF THE LAKE WHERE IT WAS RELEASED THROUGH CERAMIC DIFFUSERS, AND ALLOWED TO BUBBLE UP TO THE LAKE SURFACE THROUGH THE WATER COLUMN. CLARITY OF THE WATER WAS VISIBLY AND MEASURABLY IMPROVED; THE POPULATIONS OF NOXIOUS ALGAE, SO OBJECTIONABLE TO RECREATION INTERESTS, WERE DECREASED IN NUMBER; AND NO HARMFUL EFFECTS WERE DETECTED DURING 1968 AND 1969 SUMMER MIXING OF KEZAR LAKE. MANY IMPROVEMENTS IN VARIOUS WATER QUALITY PARAMETERS WERE ALSO NOTED. OPERATING PRESSURES FOR THE COMPRESSORS ARE LOW. THE EQUIPMENT IS CONVENIENT FOR NECESSARY MAINTENANCE, AND THE OPERATING AND STUDY BUDGET IS MODEST. (MORTLAND-BATTELLE)

FIELD 05G, 05C, 04A

ACCESSION NO. W72-07890

THE ECOLOGY OF THE PLANKTON OFF LA JOLLA, CALIFORNIA, IN THE PERIOD APRIL THROUGH SEPTEMBER, 1967.

CALIFORNIA UNIV., BERKELEY.

AVAILABLE FROM THE NATIONAL TECHNICAL INFORMATION SERVICE AS UCSD-10-P20-54, \$3.00 IN PAPER COPY, \$0.95 IN MICROFICHE. BULLETIN OF THE SCRIPPS INSTITUTION OF OCEANOGRAPHY OF THE UNIVERSITY OF CALIFORNIA, LA JOLLA, J.D.H. STRICKLAND, EDITOR, VOL 17, NOV 16, 1970. 103 P, 42 FIG, 17 TAB, 98 REF.

DESCRIPTORS:
*PLANKTON, *BIOMASS, *SAMPLING, *NUTRIENTS, *HYDROGRAPHY, *PACIFIC OCEAN, ECOLOGY, WATER TEMPERATURE, PHYTOPLANKTON, ZOOPLANKTON, CHLOROPHYLL, SOLAR RADIATION, VITAMIN B, PIGMENTS, GROWTH RATES, ALGAE, FOOD WEBS, COPEPODS, ORGANOPHOSPHORUS COMPOUNDS, PRIMARY PRODUCTIVITY, SALINITY, PROFILES, PHOSPHATES, NITRATES, SILICATES, HYDROGRAPHS, ISOTHERMS, CARBON, NITROGEN, PHOSPHORUS, RED TIDE, PHOTOSYNTHESIS, DEPTH, SYSTEMATICS, DINOFLAGELLATES, DIATOMS, TROPHIC LEVELS, PROTOZOA, CRUSTACEANS, FLUORESCENCE, ANALYTICAL TECHNIQUES, DETRITUS, STANDING CROP, DISTRIBUTION PATTERNS, STATISTICAL METHODS, MECHANICAL EQUIPMENT, SAMPLING, PUMPS, MORTALITY.

IDENTIFIERS:
ORGANIC CARBON, CHLOROPHYLL A, ORGANIC NITROGEN, THIAMINE, BIOTIN, VITAMIN B 12, GONYAULAX POLYEDRA, PERIDINIUM DEPRESSUM, METAZOA, CALANUS HELGOLANDICUS(PACIFICUS), THERMISTORS, DATA INTERPRETATION, COUNTING.

ABSTRACT:
WEEKLY OBSERVATIONS OF NEARSHORE PLANKTON AND RELATED HYDROGRAPHIC VARIABLES WERE MADE FROM MID-APRIL TO MID-SEPTEMBER, 1967, AT THREE STATIONS, 1.4, 4.6, AND 12.1 KM OFFSHORE, JUST NORTH OF LA JOLLA. DAILY WATER-TEMPERATURE MEASUREMENTS WERE OBTAINED FROM THE NAVY ELECTRONIC LABORATORY OCEANOGRAPHIC RESEARCH (NEL) TOWER AND THE SCRIPPS INSTITUTION OF OCEANOGRAPHY PIER. THE AMOUNT OF INCOMING SOLAR RADIATION WAS MEASURED AT THE SCRIPPS INSTITUTION OF OCEANOGRAPHY. AT EACH WEEKLY STATION, MEASUREMENTS WERE MADE OF TEMPERATURE, SALINITY, SUBMARINE LIGHT ATTENUATION, PHYTOPLANKTON, MICROZOOPLANKTON, CHLOROPHYLL A, PHOSPHATE, NITRATE, AND SILICATE, THE LAST FOUR USING AUTOMATED METHODS OF ANALYSIS. IN ADDITION, AN ESTIMATE WAS OBTAINED OF THE TOTAL AMOUNT OF PARTICULATE AND DISSOLVED ORGANIC CARBON, NITROGEN, AND PHOSPHORUS OVER THE 'PLANT PIGMENT DEPTH'. HYDROGRAPHY AND CHEMISTRY; VITAMIN B12, THIAMINE, AND BIOTIN; ESTIMATES OF PHYTOPLANKTON CROP SIZE, GROWTH RATE, AND PRIMARY PRODUCTION; RELATIONSHIPS OF PHYTOPLANKTON SPECIES DISTRIBUTION TO THE DEPTH DISTRIBUTION OF NITRATE; PHYTOPLANKTON TAXONOMY AND STNDING CROP; NUMERICAL ABUNDANCE AND ESTIMATED BIOMASS OF MICROZOOPLANKTON, PRODUCTION OF THE PLANKTONIC COPEPOD, CALANUS HELGOLANDICUS ARE DISCUSSED. BECAUSE THERE WAS A CLOSE CORRELATION AMONG EACH OF THE THREE PLANT NUTRIENTS AND TEMPERAUTURE, 'UPWELLING', CAUSED LAYERS OF HIGH NUTRIENT CONCENTRATION TO MOVE NEARER THE AREA SURFACE. THE SHOALING OF THE 'TROPHOCLINE' WAS THE FEATURE PROBABLY MOST RESPONSIBLE FOR QUALITATIVE AND QUANTITATIVE CHANGES OF PRODUCTIVITY. ALTHOUGH THE AMOUNT OF DETRITUS IN THE WATER APPEARED TO DEPEND ON THE LEVEL OF PRIMARY PRODUCTION, THE PRODUCTION HAD LITTLE EFFECT ON THE AMOUNT OF DISSOLVED ORGANIC MATERIAL EXCEPT PERHAPS AT THE STATION CLOSEST TO THE COAST, WHERE THE PLANT-CELL CONCENTRATION WAS DENSEST.
(JEFFERIS-BATTELLE)

FIELD 05C, 05A, 02L

ACCESSION NO. W72-07892

A HYDROBIOLOGICAL STUDY OF THE POLLUTED RIVER LIEVE (GHENT, BELGIUM),

RIJKSUNIVERSITAIR CENTRUM ANTWERPEN (BELGIUM). LAB. OF ECOLOGY.

W. H. O. DE SMET, AND F. M. J. C. EVENS.

HYDROBIOLOGIA, VOL 39, ISSUE 1, P 91-154, JANUARY 31, 1972. 3 FIG, 38 TAB, 75 REF.

DESCRIPTORS:
*RIVERS, *BACTERIA, IPLANKTON, PHYSICOCHEMICAL PROPERTIES, *ALGAE, *HYDROBIOLOGY, ENTERIC BACTERIA, SAMPLING, ZOOPLANKTON, PHYTOPLANKTON, WATER ANALYSIS, POLLUTANTS, AQUATIC ALGAE, AQUATIC MICROORGANISMS, AQUATIC ANIMALS, NUTRIENTS, WATER POLLUTION EFFECTS, WATER QUALITY, SEASONAL, SUSPENDED LOAD, BIOINDICATORS, WATER POLLUTION, MICROBIOLOGY, WATER TEMPERATURE, HYDROGEN ION CONCENTRATION, DISSOLVED OXYGEN, BIOCHEMICAL OXYGEN DEMAND, NITROGEN, NITRATES, NITRITES, CHLORIDES, PHOSPHATES, COLIFORMS, CYANOPHYTA, PYRROPHYTA, EUGLENOPHYTA, CHRYSOPHYTA, CHLOROPHYTA, PROTOZOA, COPEPODS, CRUSTACEANS, ROTIFERS, EUGLENA, DIATOMS, SCENEDESMUS, ODOR, AGARS, ODOR-PRODUCING ALGAE, VOLUMETRIC ANALYSIS, COLORIMETRY, PLANKTON NETS, CULTURES, COLOR, CHLAMYDOMONAS, DAPHNIA, PRIMARY PRODUCTIVITY, BIOMASS, SECONDARY PRODUCTIVITY, AERATION, TURBIDITY, SESTON, DOMESTIC WASTES, CHEMICAL WASTES, INDUSTRIAL WASTES, SUMMER, WINTER, AUTUMN, METHODOLOGY.

IDENTIFIERS:
AMOEBOBACTER ROSEUS CHROMATIUM OKENII, CHROMATIUM, MACROMONAS MOBILIS, THIOVULUM MAJUS, THIOSPIRA WINOGRADSKY, ZOOGLEA RAMIGERA, BEGGIATOA ALBA, BEGGIATOA, THIOTHRIX NIVEA, THIOTHRIX TENUIS, MERISMOPEDIA GLAUCA, DACTYLOCOCCOPSIS SMITHII, OSCILLATORIA PRINCEPS, OSCILLATORIA AMOENA, OSCILLATORIA TENUIS, OSCILLATORIA CHLORINA, OSCILLATORIA, LYNGBYA, ANABAENA CONSTRICTA, LIEVE RIVER, POTASSIUM PERMANGANATE-CONSUMPTION, SAPROBIC VALENCY, SARCODINA, RHIZOPODA, QUANTITATIVE ANALYSIS, SUCTORIANS, HARPACTOIDEA, COUNTING CHAMBERS, FECES, *BELGIUM, BACILLARIOPHYTA, BACTERIAPHYTA, CRYPTOPHYTA, XANTHOPHYTA, CILIATA, SUCTOREA, ROTATORIA, *LIEVE RIVER(BELG.).

ABSTRACT:
PHYSICO-CHEMICAL, BACTERIOLOGICAL, AND PLANKTON LEVELS WERE DETERMINED ABOUT EVERY TWO WEEKS (FROM 14 JULY 1964 TO 15 MARCH 1965) AT SIX SITES ON THE LIEVE RIVER. ALL SAMPLES WERE TAKEN FROM THE UPPER 20 CM OF WATER AT THE MIDDLE OF THE RIVER. THREE OF THE SAMPLING STATIONS WERE LOCATED OUTSIDE AND THREE INSIDE THE AREA OF THE RIVER WHICH WAS AERATED. FROM THE CHEMICAL AND BACTERIOLOGICAL POINT OF VIEW, THE LIEVE WAS CHARACTERIZED AS HEAVILY POLLUTED. A POLLUTION GRADIENT CAN BE DEMONSTRATED AS WELL B PHYSICO-CHEMICAL INVESTIGATIONS AS BY THE BIOLOGICAL RESULTS. THE PRESENCE AND THE PRODUCTION OF THE PLANKTON IN THE DIFFERENT PLACES ARE DISCUSSED AND SOME GENERAL CONSIDERATIONS LEADING TO A NEW WORK HYPOTHESIS ARE MADE. (HOLOMAN-BATTELLE)

FIELD 05C, 05A

ACCESSION NO. W72-07896

STUDIES AT OYSTER BAY IN JAMAICA, WEST INDIES. V. QUALITATIVE OBSERVATIONS ON THE PLANKTONIC ALGAE AND PROTOZOA,

JOHNS HOPKINS UNIV., BALTIMORE, MD. MCCOLLUM-PRATT INST.; AND JOHNS HOPKINS UNIV., BALTIMORE, MD. DEPT. OF BIOLOGY.

R. J. BUCHANAN.

BULLETIN OF MARINE SCIENCE, VOL. 21, NO. 4, P 914-937, DECEMBER 1971. 1 FIG, 3 TAB, 45 REF.

DESCRIPTORS:
*ALGAE, *PROTOZOA, *PRIMARY PRODUCTIVITY, INVERTEBRATES, *PLANKTON, *WATER TEMPERATURE, SYSTEMATICS, SAMPLING, NETS, CENTRIFUGATION, MICROSCOPY, DIATOMS, NUTRIENTS, BACTERIA, CHRYSOPHYTA, PHYTOPLANKTON, ZOOPLANKTON, PYRROPHYTA, DINOFLAGELLATES, CYANOPHYTA, CHLOROPHYTA, EUGLENOPHYTA, GRAZING, WATER QUALITY.

IDENTIFIERS:
*OYSTER BAY, *JAMAICA, FALMOUTH HARBOR, *TINTINNIDS, *TYCHOPELAGIC, *TEMPERATURE TOLERANCES, SAMPLE PRESERVATION, COUNTING CHAMBERS, NUTRIENT CYCLING.

ABSTRACT:
PLANKTON SAMPLES WERE COLLECTED FROM OYSTER BAY, JAMAICA OVER A 15-MONTH PERIOD TO MAKE A QUALITATIVE ANALYSIS OF PLANKTONIC ALGAE AND PROTOZOA. ONE HUNDRED FIFTY-TWO ORGANISMS WERE IDENTIFIED FROM THE ENTIRE COLLECTION SERIES. OF THESE, 28 TAXA WERE FOUND TO BE MOST COMMON AND ABUNDANT. MANY TAXA (ALL OF THE SARCODINA AND OVER HALF THE BACILLARIOPHYCEAE) WERE TYCHOPELAGIC AND IT IS SUGGESTED THAT THESE SPECIES ARE IMPORTANT IN THE PRIMARY PRODUCTION OF OYSTER BAY. ALL THE CILIATED PROTOZOANS IDENTIFIED WERE TINTINNIDS, WHICH, BEING PLANKTONIC, ARE PROBABLY SIGNIFICANT GRAZERS ON BACTERIA AND THE SMALLEST ALGAE. THEY MIGHT ALSO PLAY AN IMPORTANT ROLE IN THE REGENERATION OF NUTRIENTS IN OYSTER BAY. EXAMINATION OF THE LITERATURE ON THE KNOWN TEMPERATURE TOLERANCES OF THE 28 MOST COMMON AND ABUNDANT TAXA SHOWED THAT ONLY PYRODINIUM BAHAMENSE AND CERATIUM HIRCUS WERE RESTRICTED TO THE TEMPERATURE RANGE FOUND IN OYSTER BAY. THIS, AND OTHER EVIDENCE, INDICATES THAT OYSTER BAY SELECTS AGAINST HIGHLY SPECIALIZED ORGANISMS AND IN FAVOR OF HIGHLY ADAPTABLE ONES. (MORTLAND-BATTELLE)

FIELD 05B, 05C

ACCESSION NO. W72-07899

BENTHIC ALGAL COMMUNITIES OF THE METOLIUS RIVER,

OREGON STATE UNIV., CORVALLIS.

B. J. SHERMAN, AND H. K. PHINNEY.

JOURNAL OF PHYCOLOGY, VOL. 7, NO. 4, P 269-273, DECEMBER 1971. 1 FIG, 1 TAB, 10 REF.

DESCRIPTORS:
*ALGAE, *BENTHIC FLORA, *OREGON, RIVERS, BIOLOGICAL COMMUNITIES, SEASONAL, DIATOMS, *LIGHT, SOLAR RADIATION, *PHOTOPERIODISM, PERIPHYTON, *BENTHOS.

IDENTIFIERS:
*CLADOPHORA, *ACHNANTHES, *SPIROGYRA, CLODOPHORA GLOMERATA, *METOLIUS RIVER(ORE).

ABSTRACT:
BENTHIC ALGAE IN THE METOLIUS RIVER (OREGON) WERE SAMPLED AT REGULAR INTERVALS OVER A NEARLY 1-YEAR PERIOD TO DETERMINE CAUSAL FACOTRS IN THE OCCURRENCE AND DISTRIBUTION OF ALGAE SPECIES. THIS RIVER IS PRACTICALLY FREE OF SEASONAL CHANGES IN CURRENT, TURBIDITY, DISSOLVED SUBSTANCE, AND TEMPERATURE. THUS PHOTOPERIOD AND TOTAL LIGHT ENERGY ARE THE ONLY VARIABLE ENVIRONMENTAL FACTORS TO BE CONSIDERED. THREE GENERA, CLODOPHORA, ACHNANTHES, AND SPIROGYRA, PREDOMINATED THROUGHOUT THE YEAR. CLADAPHORA GLOMERATA WAS THE MOST ABUNDANT FILAMENTOUS SPECIES. IT DECREASED AT ONE STATION DURING WINTER MONTHS WHEN LESS LIGHT WAS AVAILABLE, INDICATING ITS DEPENDENCE ON ABUNDANT LIGHT. OTHER SPECIES WERE PRESENT ONLY AT CERTAIN TIMES OF THE YEAR, ATTRIBUTED ALSO TO CHANGE IN AVAILABLE SOLAR RADIATION. HOWEVER, OF APPROXIMATELY 60 SPECIES OF ALGAE IDENTIFIED, ONLY 9 SHOWED A DEFINITE SEASONAL DISTRIBUTION. THE PRESENCE OF RELATIVELY FEW SPECIES WAS ATTRIBUTED TO THE PREDOMINANCE OF CLADOPHORA GLOMERATA. (MORTLAND-BATTELLE)

FIELD 05A

ACCESSION NO. W72-07901

THERMAL DISCHARGES: ECOLOGICAL EFFECTS,

BATTELLE MEMORIAL INST., COLUMBUS, OHIO.

A. A. LEVIN, T. J. BIRCH, R. E. HILLMAN, AND G. E. RAINES.

ENVIRONMENTAL SCIENCE AND TECHNOLOGY, VOL. 6, NO. 3, P 224-230, MARCH 1972. 1 FIG, 1 TAB.

DESCRIPTORS:
*THERMAL POLLUTION, *CRUSTACEANS, *FISH, HEATED WATER, NUCLEAR POWER PLANTS, ELECTRICAL POWER PLANTS, HEAT TRANSFER, WATER POLLUTION EFFECTS, WATER TEMPERATURE, ALGAE, BULLHEADS, SALMON, SHELLFISH, CLAMS, MARINE ANIMALS, AQUATIC LIFE, ECOSYSTEMS, SALMONIDS.

IDENTIFIERS:
CHALK POINT, COLUMBIA RIVER, PATUXENT RIVER, CONTRA COSTA POWER PLANT, SAN JOAQUIN RIVER, MORRO BAY POWER PLANT, HUMBOLDT BAY, CONNETICUT YANKEE NUCLEAR PLANT, CONNECTICUT RIVER, TURKEY POINT, BISCAYNE BAY, FLORIDA, BALANUS, BARNACLES, EPIFAUNA, SAGARTIA, ANEMONE, MOLGULA, TUNICATE, PISMO CLAM, TIVELA STULTORUM, OSTREA LURIDA, CARDIUM CORBIS, COCKLES, PROTOTHACA STAMINAE, LITTLENECK CLAMS, SAXIDOMUS GIGANTEUS, BUTTER CLAMS, TRESUS NUTTALLI, GAPER CLAMS, SPECIES DIVERSITY, MACROINVERTEBRATES, SURVIVAL.

ABSTRACT:
BY USING PROJECTIONS OF BOTH FOSSIL AND UNCLEAR FUELED ELECTRICAL GENERATION CAPACITY, DATA ON THERMAL EFFICIENCY AND WATER WITHDRAWAL AS WELL AS THE QUANTITY OF WASTE HEAT THAT WILL BE DISSIPATED INTO CONDENSER COOLING WATERS BY THE ELECTRICAL UTILITY INDUSTRY CAN BE ESTIMATED. BASED ON STUDIES CONDUCTED AT GENERATOR STATION SITES, DEGRADATION OF AQUATIC ECOSYSTEMS APPEARS TO VARY WITH THE GENERATOR SYSTEM AND ITS OUTPUT AND THE FLORA AND FAUNA OF THE SITE. SOME FISH AND CRUSTACEANS TOLERATE TEMPERATURE CHANGES EVEN TO AN INSTANT INCREASE OF 25 DEGREES F WITH NO MORTALITY. CERTAIN BIVALVES AND SHELLFISH FIND DISCHARGE CANALS SUPPORTIVE OR FAVORABLE. HEATED EFFLUENTS HAVE BEEN SHOWN TO REDUCE THE DIVERSITY AND ABUNDANCE OF PHYTOPLANKTON, ALGAE, AND ANIMALS IN SOME AREAS SUGGESTING THAT AN INCREASED OUTPUT AND EXPANSION OF THE INDUSTRY MAY INCREASE THE DELETERIOUS EFFECTS AND MAKE WASTE MANAGEMENT MORE DIFFICULT. NO INFORMATION IS AVAILABLE YET ON SUBLETHAL EFFECTS OF THERMAL DISCHARGE. ALTHOUGH NO MAJOR DAMAGE HAS BEEN OBSERVED, ECOLOGICAL CHANGES HAVE TAKEN PLACE. RECOMMENDATIONS FOR STANDARDS FOR LIMITING THERMAL LOAD ON AQUATIC ECOSYSTEMS ARE SUGGESTED. (MACKAN-BATTELLE)

FIELD 05C, 05B

ACCESSION NO. W72-07907

OIL POLLUTION DAMAGE OBSERVED IN TROPICAL COMMUNITIES ALONG THE ATLANTIC SEABOARD OF PANAMA,

SMITHSONIAN INSTITUTION, WASHINGTON, D.C. DEPT. OF INVERTEBRATE ZOOLOGY.

K. RUTZLER, AND W. STERRER.

POLLUTION - FOUNDATIONS FOR TODAY, VOL. 2, P 70-73, 8 FIG, 1 TAB, 3 REF, 1971.

DESCRIPTORS:
*WATER POLLUTION EFFECTS, *OIL WASTES, *CRUSTACEANS, OILY WATER, ALGAE, NEMATODES, OYSTERS, MUSSELS, PROTOZOA, MANGROVE SWAMPS, LITTORAL, SHORES, TIDES, BEACHES, SANDS, INTERTICAL AREAS, ANNELIDS, COPEPODS, CRABS, AQUATIC LIFE, BACTERIA.

IDENTIFIERS:
*OIL SPILLS, S.S. WITWATER, GALETA ISLAND, CANAL ZONE, MACROINVERTEBRATES, DIESEL FUEL, BUNKER C OIL, AVICENNIA, RHIZOPHORA, BOSTRYCHIETUM, FIDDLER CRABS, MESOFAUNA, TURBELLARIA, SPONGES, TUNICATES, BRYOZOANS, SEA TURTLES.

ABSTRACT:
THE RUPTURE OF THE TANKER S.S. WITWATER IN DECEMBER, 1968, SPILLED NEARLY 20,000 BARRELS OF DIESEL AND BUNKER C OIL NEAR THE SHORELINE OF GALETA ISLAND, CANAL ZONE. MUCH OF THE OIL WAS REMOVED BY BURNING AND PUMPING. HOWEVER, SUBSEQUENT INVESTIGATIONS CONDUCTED TO STUDY THE EFFECTS OF THE REMAINING OIL SHOWED A NUMBER OF DETRIMENTAL EFFECTS. IN ROCKY SHORE AREAS, POLLUTED WATER SPRAY KILLED TREES AND SHRUBS. SUPRALITTORAL SPRAY POOLS AND UPPER MESOLITTORAL TIDAL POOLS WHICH WERE COVERED WITH AN OIL LAYER WERE DEVOID OF LIFE. DAMAGE TO GASTROPOD AND BARNACLE POPULATIONS WAS ASSUMED. SUBTIDAL CORAL REEFS AND ASSOCIATED ORGANISMS SHOWED NO EFFECTS SINCE THEY WERE EXPOSED TO THE OIL. BEACH MEIOFAUNA SUCH AS TURBELLARIA, NEMATODES, ANNELIDA, COPEPODS AND OTHERS WERE RADICALLY DESTROYED BECAUSE OF THE COMPLETE PERMEATION OF THE OIL. CRUSTACEANS WERE FIRST TO DISAPPEAR. SMALL CILIATES THRIVED ON THE INCREASING NUMBERS OF OIL DEGRADING BACTERIA. IN THE MANGROVES, TREES WERE SEVERELY DAMAGED OR KILLED, THE FIDDLER CRAB POPULATION WAS REDUCED AND THE INTERTIDAL ALGAE COMMUNITY 'BOSTRYCHIETUM' AND ITS MICROFAUNA WERE PRACTICALLY ELIMINATED, AS WERE OYSTERS, MUSSELS, SPONGES, TUNICATES, BRYOZOANS, SEA TURTLES, AND SEA BIRDS. (MACKAN-BATTELLE)

FIELD 05C

ACCESSION NO. W72-07911

PRELIMINARY NOTES ON CHANGES IN ALGAL PRIMARY PRODUCTIVITY FOLLOWING EXPOSURE TO CRUDE OIL IN THE CANADIAN ARCTIC,

OTTAWA UNIV. (ONTARIO). DEPT. OF BIOLOGY.

MIKE DICKMAN.

CAN FIELD NAT. 85(3): 249-251. 1971.

DESCRIPTORS:
*PRIMARY PRODUCTIVITY, *OIL WASTES, *ARCTIC, *ALGAE.

IDENTIFIERS:
*CANADIAN ARCTIC, *CHLAMYDOMONAS-SPP, *CRYPTOMONAS-SPP.

ABSTRACT:
MACKENZIE VALLEY CRUDE OIL WHICH HAD BEEN EXPOSED FOR 2 MO. TO NATURAL ARCTIC SUMMER CONDITIONS WAS ADDED TO BOTTLES CONTAINING ALGAE TAKEN FROM A MARSH NEAR INUVIK, N. W. T. (NORTHWEST TERRITORIES). C-14 PRIMARY PRODUCTIVITY WAS 10 TIMES LOWER IN THE OIL TREATED SAMPLES (0.59 + 0.30 MGC/M3/HR) THAN IN THE UNTREATED CONTROL SAMPLES (5.12 + 1.2 MGC/M3/HR) AFTER A 4 HR INCUBATION PERIOD. SMALL FLAGELLATES SUCH AS CRYPTOMONAS SPP. AND CHLAMYDOMONAS SPP. COMPRISED NEARLY 80% OF THE PRIMARY PRODUCERS IN THE INUVIK MARSH SAMPLES. SOME IMPLICATIONS OF THE SIGNIFICANCE OF THESE PRELIMINARY FINDINGS ARE DISCUSSED IN VIEW OF THE PROPOSED 800 MI. MACKENZIE VALLEY PIPELINE ROUTE.--COPYRIGHT 1972, BIOLOGICAL ABSTRACTS, INC.

FIELD 05C, 02C

ACCESSION NO. W72-07922

MICROBIOLOGICAL ASPECTS OF THE POLLUTION OF FRESH WATER WITH INORGANIC NUTRIENTS,

WISCONSIN UNIV., MADISON. DEPT. OF SOIL SCIENCE; AND EDINBURGH UNIV. (SCOTLAND). DEPT. OF MICROBIOLOGY.

D. R. KEENEY, R. A. HERBERT, AND A. J. HOLDING.

IN: MICROBIAL ASPECTS OF POLLUTION, G. SYKES AND F. H. SKINNER, EDITORS, SOCIETY FOR APPLIED BACTERIOLOGY SYMPOSIUM SERIES NO 1, P 181-200, 1971. ACADEMIC PRESS, LONDON. 2 FIG, 3 TAB, 99 REF. EPA PROGRAM 16010 EHR.

DESCRIPTORS:
*WATER POLLUTION, MICROORGANISMS, *INORGANIC COMPOUNDS, *NUTRIENTS, FRESH WATER, ALGAE, BACTERIA, NITROGEN, PHOSPHORUS, OXIDATION-REDUCTION POTENTIAL, LAKES, SOILS, EUTROPHICATION, AMMONIFICATION, NITRIFICATION, DENITRIFICATION, NITROGEN FIXATION, SEDIMENTS, NITRATES, WISCONSIN.

IDENTIFIERS:
*LIMITING NUTRIENTS, LAKE MENDOTA(WIS).

ABSTRACT:
THIS LITERATURE REVIEW CONSIDERS BOTH PHOSPHORUS AND NITROGEN BUT EMPHASIZES MICROBIAL PROCESSES GOVERNING NITROGEN AVAILABILITY, PRIMARILY IN LAKE SYSTEMS. HOWEVER, MANY PRINCIPLES ARE APPLICABLE TO RIVER SYSTEMS WHEN COMPARABLE ENVIRONMENTAL CONDITIONS PREVAIL. DUE TO THE MANY COMPLEX COMPETING BIOLOGICAL REACTIONS OCCURRING WITHIN AN ECOSYSTEM IT IS EXTREMELY DIFFICULT TO DETERMINE RELATIVE IMPORTANCE OF EACH INDIVIDUAL MICROBIAL PROCESS. THE NITRATE-NITROGEN BALANCE AT ANY GIVEN TIME IS GOVERNED BY THE RELATIVE RATE OF NITRATE-NITROGEN LOSS (RESULTING FROM DENITRIFICATION AND IMMOBILIZATION) TO THE RATE OF NITRATE-NITROGEN REGENERATION (FROM GROUNDWATER SEEPAGE, STREAM DRAINAGE, AND AMMONIFICATION/NITRIFICATION). ALGAL GROWTH IS LARGELY GOVERNED BY NITROGEN AND PHOSPHORUS AVAILABILITY. NITROGEN FIXATION BY BLUE-GREEN ALGAE PROBABLY CONSTITUTES ONLY 1-2% OF THE INPUT INTO FRESH-WATERS BUT IT MAY BE IMPORTANT TO BIOLOGICAL PRODUCTIVITY WHEN INORGANIC NITROGEN LEVELS BECOME DEPLETED DURING THE SUMMER. THE SMALL BACTERIAL POPULATIONS FOUND IN OLIGOTROPHIC WATERS SUGGEST THEIR ROLE IN NUTRIENT CYCLING IS PERHAPS MINIMAL. IN EUTROPHIC WATERS, OXYGEN DEPLETION BY MICROORGANISMS CAN LEAD TO A SERIES OF EVENTS LARGELY GOVERNED BY THE REDOX POTENTIAL. VALID MODEL SYSTEMS, WHEREBY INDIVIDUAL PARAMETERS CAN BE CONTROLLED ARE INVALUABLE IN DIFFERENTIATING SPECIFIC MICROBIAL PROCESSES TO AID IN PREDICTING FRESH WATER BEHAVIOR IN RESPONSE TO VARYING NUTRIENT LOADS. (JONES-WISCONSIN)

FIELD 05C, 10F

ACCESSION NO. W72-07933

A REVIEW OF THE FACTORS LIMITING THE GROWTH OF NUISANCE ALGAE,

MICHIGAN WATER RESOURCES COMMISSION, LANSING.

ALBERT MASSEY, AND JOHN ROBINSON.

WATER AND SEWAGE WORKS, VOL 118, NO 11, P 352-355, 1971. 38 REF.

DESCRIPTORS:
*ALGAL CONTROL, *NUISANCE ALGAE, EUTROPHICATION, REVIEWS, TROPHIC LEVEL, PHOSPHORUS, PHOSPHATES, CARBON, VITAMIN B, NITROGEN, CYANOPHYTA, LAKE MICHIGAN, NUTRIENTS.

IDENTIFIERS:
*LIMITING FACTORS, SILICON.

ABSTRACT:
ISOLATING THE KEY FACTOR WHICH LIMITS ALGAL GROWTH HAS PROVEN CONFUSING BECAUSE THE AQUATIC ECOSYSTEM IS A MULTIFACTOR SYSTEM IN DYNAMIC EQUILIBRIUM ESTABLISHED BY THE PARTICULAR GEOCHEMICAL CHARACTER AND BIOTA OF THE LAKE. THAT PHOSPHORUS IS THE LIMITING ELEMENT HAS BEEN THEORIZED SINCE LIMNOLOGY'S INFANCY. THROUGH INDEPENDENT RESEARCH NUMEROUS INVESTIGATORS HAVE ESTABLISHED THAT PHOSPHORUS IS THE ELEMENT WHICH USUALLY LIMITS ALGAL GROWTH. SINCE SHORTAGE OF ANY OF 15 ELEMENTS MAY LIMIT ALGAE IF PHOSPHORUS OR NITROGEN IS ADDED, PRIMARY PRODUCTION INCREASES UNTIL SOME OTHER ELEMENT BECOMES LIMITING. CONTINUED DEPLETION OF SILICON WILL FAVOR PROLIFERATION OF PHYTOPLANKTON OTHER THAN DIATOMS AND COULD RESULT IN BLOOMS OF UNDESIRABLE ALGAE. BECAUSE SOME ALGAE FIX ATMOSPHERIC NITROGEN, CONTROL OF CULTURAL EUTROPHICATION BY LIMITING NITROGEN SUPPLY IS HIGHLY QUESTIONABLE. THREE RECENT PAPERS PROPOSING THAT CARBON RATHER THAN PHOSPHORUS IS THE GROWTH-LIMITING FACTOR, CONTAIN NO ORIGINAL RESEARCH. THE CARBON SUPPLY FROM INORGANIC SOURCES, FROM THE ATMOSPHERE, FROM BACTERIAL DEGRADATION, WOULD BE MORE THAN ADEQUATE TO FORCE SOME OTHER FACTOR TO BE LIMITING. IN TERMS OF PRACTICAL TECHNOLOGY ONE WOULD ALMOST HAVE TO ANSWER THAT IT WOULD BE NECESSARY TO CONTROL PHOSPHORUS. WITH THE REDUCTION OF PHOSPHATE INPUTS, CULTURAL EUTROPHICATION OF LAKES MAY BE SLOWED, STOPPED OR EVEN IN SOME CASES REVERSED. (JONES-WISCONSIN)

FIELD 05C, 10F

ACCESSION NO. W72-07937

FOOD QUALITY AND ZOOPLANKTON NUTRITION,

OXFORD UNIV. (ENGLAND). DEPT. OF ZOOLOGY.

JAMES E. SCHINDLER.

JOURNAL OF ANIMAL ECOLOGY, VOL 40, NO 3, P 589-595, 1971. 2 FIG, 3 TAB, 30 REF.

DESCRIPTORS:
*ZOOPLANKTON, *FOODS, DAPHNIA, COPEPODS, ALGAE, NUTRIENT REQUIREMENTS, CYANOPHYTA, CHLOROPHYTA, PHYTOPLANKTON, GRAZING.

IDENTIFIERS:
DAPHNIA LONGISPINA, DIAPTOMUS GRACILIS, CYCLOPS STRENUUS, INGESTION RATES, ASSIMILATION EFFICIENCES, CRYPTOMONAS, ANKISTRODESMUS, ELAKOTOTHRIX, OSCILLATORIA, ANABAENA, MICROCYSTIS, TRIBONEMA, APHANIZOMENON, OOCYSTIS, GLOEOCYSTIS, COELASTRUM, ASTERIONELLA, CRYPTOMONAS.

ABSTRACT:
FOOD SUPPLY HAS BEEN CONSIDERED AMONG THE POSSIBLE FACTORS LEADING TO SEASONAL CHANGES IN ZOOPLANKTON POPULATIONS AND A QUANTITATIVE DETERMINATION OF THE DEGREE OF NUTRITION THAT CAN BE SUPPLIED BY DIFFERENT FOODS PERMITS A VERY RAPID ASSESSMENT OF INGESTION AND ASSIMILATION RATES. IN AN ATTEMPT TO DETERMINE THE AMOUNTS AS WELL AS KIND OF FOOD INGESTED AND ASSIMILATED BY THREE SPECIES OF ZOOPLANKTON, DAPHNIA LONGISPINA, DIAPTOMUS GRACILIS, AND CYCLOPS STRENUUS, C-14 WAS USED TO LABEL 11 DIFFERENT TYPES OF ALGAE. FRESHLY COLLECTED ZOOPLANKTON WAS PRECONDITIONED IN THE DARK AT A CONSTANT TEMPERATURE OF 15C AND WITH THE FOOD TYPE TO BE USED IN THE EXPERIMENTS. ONLY THE LARGER ADULT FEMALES WITHOUT EGGS WERE SELECTED. IN GENERAL THIS WORK TESTS THE RELATIONSHIP BETWEEN SOME CRUDE MEASURE OF PHYTOPLANKTON CONCENTRATION AND SOME MEASURE OF ZOOPLANKTON DENSITY. THE RESULTS SHOW HOW FOOD QUALITY AFFECTS THE ASSIMILATION AND INGESTION RATES OF ZOOPLANKTON. A RELATIONSHIP WAS FOUND BETWEEN THE INGESTION RATES AND ASSIMILATION EFFICIENCIES OF DIAPTOMUS. THE ECOLOGICAL IMPLICATIONS OF THE RELATIONSHIP BETWEEN FOOD QUALITY AND ZOOPLANKTON NUTRITION ARE DISCUSSED. (JONES-WISCONSIN)

FIELD 05C

ACCESSION NO. W72-07940

KINETICS OF SYNTHESIS OF NITROGENASE IN BATCH AND CONTINUOUS CULTURE OF ANABAENA FLOS-AQUAE,

QUEEN'S UNIV. KINGSTON (ONTARIO), DEPT. OF CHEMICAL ENGINEERING.

H. BONE.

ARCHIV FUR MICROBIOLOGIE, VOL 80, NO 3, P 242-251, 1971. 2 FIG, 3 TAB, 28 REF.

DESCRIPTORS:
*NITROGEN FIXATION, *SYNTHESIS, *CULTURES, *NITROGEN, OXYGEN, BIOCHEMISTRY, INHIBITORS, ALGAE, PROTEINS.

IDENTIFIERS:
*ANABAENA FLOS-AQUA, *NITROGENASE ACTIVITY, OXYGEN INACTIVATION.

ABSTRACT:
BACTERIAL NITROGENASES ARE OXYGEN SENSITIVE AND CAN BE SEPARATED INTO TWO FRACTIONS, AN IRON-MOLYBDENUM-PROTEIN AND AN IRON-PROTEIN. FOR A FEW AEROBIC FACULTATIVE NITROGEN-FIXING BACTERIA, THE IRON-PROTEIN IS OXYGEN SENSITIVE. NITROGENASE OF ANABAENA FLOS-AQUAE WAS INACTIVATED BY OXYGEN AND RECOVERY OF ACTIVITY WAS MEASURED IN BATCH AND IRON, PHOSPHATE AND UREA-LIMITED CONTINUOUS CULTURES. A FLOS-AQUAE WAS SHOWN TO ACTIVELY SYNTHESISE NEW NITROGENASE COMPONENTS AFTER OXYGEN INACTIVATION. IN BATCH CULTURE, CANAVANINE, CHLORAMPHENICAOL, METHYLAMINE, PROFLAVINE, PUROMYCIN AND UREA INHIBITED THE RECOVERY PROCESS. A FLOS-AQUAE GROWING AT A DILUTION RATE OF 0.03 PER HOUR HAS 180% THE SPECIFIC ACTIVITY OF NITROGENASE COMPARED TO PHOSPHATE-LIMITED CELLS WHICH SUGGEST THAT PHYTOFLAVIN MIGHT BE ACTIVE IN A FLOS-AQUAE. THE KINETICS OF OXYGEN INACTIVATION OF NITROGENASE OF IRON AND PHOSPHATE-LIMITED A FLOS-AQUAE ARE THE SAME AND ALL THESE PIECES OF EVIDENCE INDICATE THAT THE OXYGEN SENSITIVE SITE IS A NITROGENASE COMPONENT. THE RATE OF RECOVERY OF NITROGENASE ACTIVITY IN CONTINUOUS CULTURES WAS DEPENDENT ON LIGHT INTENSITY, CONCENTRATION OF UREA, AMMONIUM SALTS AND NITRATE, AND INDEPENDENT OF GROWTH RATE. STEADY-STATE NITROGENASE ACTIVITIES SEEM TO BE MAINTANINED BY BALANCING NITROGENASE SYNTHESIS AND INACTIVATION PROCESSES. (JONES-WISCONSIN)

FIELD 05C

ACCESSION NO. W72-07941

ALIGICIDAL NONFRUITING MYXOBACTERIA WITH HIGH G + C RATIOS,

TEXAS UNIV. MEDICAL SCHOOL, SAN ANTONIO. DEPT. OF BIOCHEMISTRY; AND NORTH CAROLINA UNIV., CHAPEL HILL. DEPT. OF BOTANY.

J. R. STEWART, AND R. M. BROWN, JR.

ARCHIV FUR MICROBIOLOGIE, VOL 80, NO 2, P 176-190, 1971. 6 FIG, 2 TAB, 52 REF. FWQA PROGRAM 18050 DBR.

DESCRIPTORS:
*ALGAL TOXINS, *MYXOBACTERIA, CHLOROPHYTA, CYANOPHYTA, SYSTEMATICS, VITAMINS, ANAEROBIC BACTERIA, ELECTRON MICROSCOPY, PIGMENTS, CARBOHYDRATES, PROTEINS, ACTINOMYCETES, CHLORELLA.

IDENTIFIERS:
*G + C RATIOS, *NONFRUITING MYXOBACTERIA, DNA, MOTILITY, CYST FORMATION, ANTIBIOTIC SENSITIVITY, CYTOPHAGA JOHNSONII, SPOROCYTOPHAGA MYXOCOCCOIDES.

ABSTRACT:
BACTERIA WHICH KILL ALGAE HAVE BEEN REPORTED ON GLIDING ORGANISMS IN THE ORDER MYXOBACTERALES OF FRUITING AND NONFRUITING GROUPS BUT ONLY THE LATTER APPEARS TO CONTAIN ORGANISMS KNOWN TO KILL BOTH GREEN AND BLUE-GREEN ALGAE. DEOXYRIBONUCLEIC ACID BASE RATIOS ARE GENERALLY AGREED TO BE IN THE 70'S FOR FRUITING MYXOBACTERIA AND IN THE 30'S FOR NONFRUITING. A FEW NONFRUITING MYXOBACTERIA WITH HIGH G + C RATIOS HAVE BEEN REPORTED. FIVE NONFRUITING MYXOBACTERIA WITH G + C RATIOS RANGING FROM 69-71 MOLE PERCENT, THREE BEING NEW ISOLATES AND TWO OBTAINED ELSEWHERE ARE REPORTED. THESE BASE RATIOS OF THE DEOXYRIBONUCLEIC ACIDS OF THESE ALGICIDAL MYXOBACTERIA WERE DETERMINED BY THERMAL DENATURATION TEMPERATURES; NO UNUSUAL NUCLEIC ACID BASES WERE DETECTED. THESE ORGANISMS ARE DESCRIBED AS AMICRO-CYSTOGENOUS, GLIDING, GRAM-NEGATIVE BACILLI CAPABLE OF DEGRADING GELATIN, CASEIN, STARCH, CELLULOSE, CHITIN, AND ALGINATE. ALL HAVE BEEN SHOWN TO BE ALGICIDAL. POLY-BETA-HYDROXYBUTYRATE IN EACH WAS INDICATED BY CONVERSION TO CROTONIC ACID. ANTIBIOTIC SENSITIVITY WAS SIMILAR TO THAT OF KNOWN NONFRUITING MYXOBACTERIA. FINE STRUCTURE OF ONE, MYXOBACTER 44, REVEALED A TRIPLE-LAYERED CELLULAR ENVELOPE WHOSE MIDDLE LAYER IS LYSOZYME SENSITIVE. RUTHENIUM RED-POSITIVE SLIME MATERIAL ADHERED TO THE OUTER SURFACE. (JONES-WISCONSIN)

FIELD 05C

ACCESSION NO. W72-07951

INTERACTIONS OF DISSOLVED AND PARTICULATE NITROGEN IN LAKE METABOLISM,

MICHIGAN STATE UNIV., HICKORY CORNERS. W. K. KELLOGG BIOLOGICAL STATION; AND MICHIGAN STATE UNIV., HICKORY CORNERS. DEPT. OF BOTANY AND PLANT PATHOLOGY.

BRUCE A. MANNY.

PHD THESIS, 1971. 189 P, 32 FIG, 7 TAB, 268 REF. B-009-MICH(2).

DESCRIPTORS:
*NITROGEN, *LAKES, ANALYTICAL TECHNIQUES, MICHIGAN, SESTON, ALGAE, PHYTOPLANKTON, NANNOPLANKTON, AMMONIA, NITRATES, NITRITES, HARDNESS(WATER), NITROGEN CYCLE, ORGANIC MATTER, PHOTOSYNTHESIS, EUTROPHICATION, PHOSPHORUS, CHLOROPHYTA, CYANOPHYTA, CHRYSOPHYTA, EUGLENOPHYTA, DISSOLVED OXYGEN, CARBON, LAKE MORPHOMETRY, CHEMICAL PROPERTIES, OXYGEN, CHLAMYDOMONAS, AQUATIC PLANTS, ZOOPLANKTON, DAPHNIA, SEDIMENTS, CONDUCTIVITY, BACTERIA, PSEUDOMONAS, NITROGEN COMPOUNDS.

IDENTIFIERS:
*LAKE METABOLISM, *DISSOLVED ORGANIC NITROGEN, *PARTICULATE ORGANIC NITROGEN, CALCAREOUS LAKES, ULTRA-VIOLET LABILE NITROGEN, ULTRA-VIOLET REFRACTORY NITROGEN, NETPLANKTON, LAWRENCE LAKE(MICH), WINTERGREEN LAKE(MICH), CARBONATE PARTICLES, PHOSPHORUS CYCLE.

ABSTRACT:
TO COMPARE RATES OF ALLOCHTHONOUS NITROGEN ENTRY, RATES OF NITROGEN REGENERATION AND ORGANIC NITROGEN PRODUCTION, TWO LAKES, REPRESENTING EXTREMES OF HARDWATER, WERE SAMPLED WEEKLY. MEASUREMENTS OF DISSOLVED AMMONIA, NITRITE, AND NITRATE FOR 22 MONTHS AND ORGANIC NITROGEN FOR 12 MONTHS WERE MADE AND SEASONALLY IN FOUR OTHER LAKES REPRESENTING CHEMICAL AND TROPHIC SPECTRUM IN GLACIATED MICHIGAN. SESTON AT 1 M DEPTH WAS FRACTIONATED WEEKLY FOR 14 MONTHS; NET AND NANNO FRACTIONS ANALYZED. SEASONAL CHANGES IN ALL NITROGENOUS PARAMETERS WERE RELATED TO CHANGES IN 15 OTHER CHEMICAL AND BIOLOGICAL PARAMETERS ASSAYED SIMULTANEOUSLY. NANNOPHYTOPLANKTON CONTAIN MORE NITROGEN PER UNIT CELL VOLUME THAN NETPHYTOPLANKTON. DISSOLVED ORGANIC NITROGEN COMPOUNDS AT NATURAL CONCENTRATIONS WERE MEASURED AND UV-LABILE FRACTION DIFFERENTIATED FROM UV-REFRACTORY FRACTION WITHIN THE NITROGEN POOL IN SIX LAKES. MEASUREMENTS OF ALLOCHTHONOUS DISSOLVED ORGANIC NITROGEN ENTERING THE MOST CALCAREOUS LAKE IN SPRING REVEALED THE TWO FRACTIONS ENTERED THE LAKE. THE MOST CALCAREOUS LAKE REVEALED THE UV-REFRACTORY FRACTION ORIGINATED LARGELY IN THE SURROUNDING WATERSHED, WHEREAS THE UV-LABILE FRACTION ORIGINATED LARGELY WITHIN THE LAKE. PELAGIC DISSOLVED ORGANIC NITROGEN INTERACTIONS SEEM TO EXPLAIN HOW AQUATIC PHOTOSYNTHESIS IN HARDWATER LAKES IS REGULATED. EUTROPHICATION RATES STEMMING FROM VARIOUS COMBINATIONS OF THESE INTERACIONS ARE DISCUSSED. (JONES-WISCONSIN)

FIELD 05C, 02H

ACCESSION NO. W72-08048

EFFECTS OF LOW NUTRIENT DILUTION WATER AND MIXING ON THE GROWTH OF NUISANCE ALGAE,

WASHINGTON UNIV., SEATTLE.

JAMES A. BUCKLEY.

MS THESIS, 1971. 116 P. 54 FIG, 20 TAB, 33 REF, 3 APPEND. OWRR A-034-WASH(3).

DESCRIPTORS:
*ALGAL CONTROL, *DIFFUSION, *NUTRIENTS, *NUISANCE ALGAE, WASHINGTON, PHYTOPLANKTON, CYANOPHYTA, NITRATES, PHOSPHATES, LAKES, STRATIFICATION, MIXING, BOTTOM SEDIMENTS, SUCCESSION, CHLOROPHYTA, DIATOMS, CHLOROPHYLL, ALKALINITY, CARBON DIOXIDE, TEMPERATURE, DISSOLVED OXYGEN, HYDROGEN ION CONCENTRATION, SCENEDESMUS, BIOMASS.

IDENTIFIERS:
*MOSES LAKE(WASH), NUTRIENT DILUTION, LIMITING NUTRIENTS, ALGAL CELL WASHOUT.

ABSTRACT:
THE PURPOSE OF THE DILUTION EXPERIMENTS WAS TO STUDY THE EFFECT OF LOW NUTRIENT DILUTION WATER ON ALGAL GROWTH IN SITU. OF PRIMARY INTEREST WAS THE POSSIBILITY OF CHANGING THE ALGAL COMMUNITY STRUCTURE FROM PREDOMINATELY BLUE-GREEN ALGAE TO A COMMUNITY DOMINATED BY LESS OBNOXIOUS FORMS BY ADDING LOW NUTRIENT WATER TO THE LAKE. ADDITIONAL DILUTION EXPERIMENTS, EMPLOYING NUTRIENT ADDITIONS, WERE DESIGNED TO RELATE OBSERVED ALGAL GROWTH CHANGES TO DILUTION OF NITROGEN OR PHOSPHORUS AT GROWTH LIMITING LEVELS. THE EXPERIMENTS, CONDUCTED IN MOSES LAKE, WASHINGTON, FROM JUNE THROUGH SEPTEMBER 1970, SHOWED THAT DILUTION RESULTED IN REDUCTION OF THE YIELD AND GROWTH RATE OF BLUE-GREEN ALGAE WITHOUT INCREASING OR DECREASING THE ABUNDANCE OF OTHER ALGAL FORMS. NITRATE WAS FOUND TO STIMULATE ONLY BLUE-GREEN ALGAL GROWTH WHEN ADDED TO THE DILUTION WATER WHILE ADDITIONS OF ORTHOPHOSPHATE GENERALLY HAD NO EFFECT. IN SITU EXPERIMENTS WERE ALSO CONDUCTED TO DETERMINE THE EFFECT OF LAKE STRATIFICATION AND MIXING ON ALGAL GROWTH. RESULTS INDICATE WHEN THE LAKE IS STRATIFIED AND CONDITIONS ARE AEROBIC THE WATER NEAR THE BOTTOM HAS THE GREATEST ALGAL GROWTH POTENTIAL. DURING LAKE MIXING RESULTS MAY BE UNPREDICTABLE. (JONES-WISCONSIN)

FIELD 05C

ACCESSION NO. W72-08049

ARE SOME BACTERIA TOXIC FOR MARINE ALGAE,

CENTRE D'OCEANOGRAPHIE, MARSEILLE (FRANCE). STATION MARINE D'ENDOUME.

B. R. BERLAND, D. J. BONIN, AND S. Y. MAESTRINI.

MARINE BIOLOGY, VOL 12, P 189-193, 1972. 4 FIG, 1 TAB, 18 REF.

DESCRIPTORS:
*BACTERIA, *TOXICITY, *MARINE ALGAE, SEA WATER, CHLOROPHYTA, CHLAMYDOMONAS, LABORATORY TESTS, CHRYSOPHYTA, OPTICAL PROPERTIES.

IDENTIFIERS:
*ANTIBIOSIS, OPTICAL DENSITY, PRASINOPHYCEAE, BACCILARIOPHYCEAE, XANTHOPHYCEAE, VIBRIO, ACHROMOBACTER, FLAVOBACTERIA.

ABSTRACT:
ALGAL AND BACTERIAL ENVIRONMENT IN THE SEA IS NOT ONLY RULED BY A SIMPLE TROPHIC RELATIONSHIP, BUT ANTIBIOSIS MAY ALSO PLAY AN IMPORTANT ROLE IN MARINE ECOLOGY. IN VITRO EXPERIMENTS WERE CARRIED ON WITH 13 ALGAL SPECIES, 20 BACTERIAL STRAINS, AND THREE BACTERIAL UNDETERMINED STRAINS ISOLATED FROM ALGAL CULTURES. EXPERIMENTS REVEALED AN OBVIOUS INHIBITION OF SEVERAL MARINE ALGAE BY SOME BACTERIA. PSEUDOMONAS AERUGINOSA AND RELATED STRAINS ARE NOT MARINE ORGANISMS, BUT THEY CAN BE TRANSPORTED TO THE SEA. INSOFAR AS THERE IS NO QUANTITATIVE SIMILARITY BETWEEN IN SITU AND CULTURE DENSITIES OF BACTERIAL POPULATIONS, BACTERIAL POISONS MAY ONLY BE IMPORTANT WHEN CONCENTRATED; FOR EXAMPLE, IN NARROW WATERS RICH IN SUSPENDED OR DISSOLVED ORGANIC MATTER. THE IN VITRO EXPERIMENTS INDICATE THAT THE PIGMENTED POISON OF THE BACTERIUM PSEUDOMONAS AERUGINOSA IS A STRONG GROWTH INHIBITOR OF THE ALGA TETRASELMIS STRIATA. SEVERAL BACTERIA STRAINS FROM DIFFERENT ORIGINS ARE RECOGNIZED TO HAVE THE SAME TOXICITY AGAINST VARIOUS MARINE ALGAE. TAKING INTO ACCOUNT THE VERY GREAT DIFFERENCES BETWEEN IN VITRO EXPERIMENTS AND IN SITU ENVIRONMENTAL CONDITIONS, IT IS NOT AT PRESENT POSSIBLE TO STATE THAT BACTERIAL POISONS REALLY PLAY A ROLE IN THE ALGAE-BACTERIA RELATIONSHIPS IN THE SEA. (JONES-WISCONSIN)

FIELD 05C, 02I

ACCESSION NO. W72-08051

NITROGEN FIXATION IN LAKE ERKEN,

UPPSALA UNIV (SWEDEN). INST. OF LIMNOLOGY.

U. GRANHALL, AND A. LUNDGREN.

LIMNOLOGY AND OCEANOGRAPHY, VOL 16, NO 5, P 711-719, 1971. 8 FIG, 3 TAB, 26 REF.

DESCRIPTORS:
*NITROGEN FIXATION, *ALGAE, MEASUREMENT, CYANOPHYTA, PHYTOPLANKTON, PHOTOSYNTHESIS, PRIMARY PRODUCTIVITY, EUTROPHICATION, SOLAR RADIATION, DISTRIBUTION, DEPTH, DIURNAL, SEASONAL.

IDENTIFIERS:
*LAKE ERKEN(SWEDEN), *DARK NITROGEN FIXATION.

ABSTRACT:
TO DETERMINE SEASONAL NITROGEN FIXATION BY PLANKTONIC ALGAE IN THE WHOLE PELAGIAL OF A LAKE AND ESTIMATE ITS IMPORTANCE IN RELATION TO OTHER SOURCES OF COMBINED NITROGEN, LAKE ERKEN, EAST OF UPPSALA, UNPOLLUTED AND MODERATELY EUTROPHIC, WAS STUDIED. IN SITU FIXATION OF MOLECULAR NITROGEN WAS MEASURED IN 1970 BY THE ACETYLENE REDUCTION TECHNIQUE EVERY TWO WEEKS DURING FIVE MONTHS. SAMPLES WERE TAKEN FROM 20 RANDOMLY DISTRIBUTED STATIONS. SUBSAMPLES WERE WITHDRAWN FOR NITROGEN FIXATION EXPERIMENTS, PRIMARY PRODUCTION MEASUREMENTS, ALGAL COUNTS OR CHEMICAL ANALYSES. THE ALGAL DIURNAL CYCLES AND VERTICAL DISTRIBUTION WERE INVESTIGATED. FIXATION WAS CORRELATED WITH THE PRESENCE OF HETEROCYSTOUS BLUE-GREEN ALGAE, ESPECIALLY APHANIZOMENON, IN THE PHYTOPLANKTON, AND WAS LIGHT DEPENDENT, THOUGH APPRECIABLE DARK FIXATION ALSO OCCURRED, OWING TO ENDOGENOUS UTILIZATION OF PHOTOSYNTHETIC PRODUCTS FORMED DURING PREVIOUS LIGHT PERIODS. ANNUAL CONTRIBUTION OF NITROGEN FIXATION IN THE PELAGIAL WAS AROUND 0.5 G N/SQ M. HIGHEST VALUES WERE OBTAINED BEFORE MASS DEVELOPMENT OF APHANIZOMENON AND MAXIMAL PRIMARY PRODUCTION. HIGHEST NITROGENASE ACTIVITY WAS SHOWN IN THE FIRST ALGAL DEVELOPMENT PERIOD. CONTRIBUTION OF EASILY AVAILABLE COMBINED NITROGEN BY PELAGIC NITROGEN FIXATION INCREASES ANNUAL COMBINED NITROGEN LOADING BY 40% AND IS IMPORTANT. (JONES-WISCONSIN)

FIELD 05C

ACCESSION NO. W72-08054

THE DISTRIBUTION OF UREA IN COASTAL AND OCEANIC WATERS,

WOODS HOLE OCEANOGRAPHIC INSTITUTION, MASS.

C. C. REMSEN.

LIMNOLOGY AND OCEANOGRAPHY, VOL 16, NO 5, P 732-740, 1971. 5 FIG, 3 TAB, 19 REF.

DESCRIPTORS:
*DISTRIBUTION, *UREAS, *COASTS, *OCEANS, SURFACE WATERS, CONTINENTAL SHELF, DEPTH, NITROGEN, PHYTOPLANKTON, HYDROGRAPHY, NITRITES, NITRATES, AMMONIA, ALGAE, CHLORELLA.

IDENTIFIERS:
*COASTAL WATERS, PANAMA, CALLAO(PERU), CAPE COD(MASS), CAPE MAY(N J), SARGASSO SEA.

ABSTRACT:
IT HAS BEEN SUGGESTED THAT UREA SHOULD BE CONSIDERED A PART OF THE NITROGEN RESERVE IN COASTAL WATERS AND PERHAPS IN OCEANIC WATERS AS WELL. THE DISTRIBUTION WAS DETERMINED FOR CERTAIN COASTAL AND OCEANIC WATERS, AS FOLLOWS: UREA-NITROGEN IN SURFACE WATERS OFF THE CONTINENTAL SHELF BETWEEN PANAMA AND CALLAO, PERU, WAS EXTREMELY PATCHY AND VARIED IN CONCENTRATION FROM 0.54 TO 5.00 MICROGRAM-ATOM UREA-N/LITER. HIGHER VALUES WERE GENERALLY FROM SAMPLES COLLECTED WITHIN A FOAM SLICK OR WINDROW. SURFACE WATERS IN NONUPWELLING WATERS NORTH OF CALLAO AVERAGED 1.83 MICROGRAM-ATOM UREA-N/LITER WHILE SURFACE WATERS IN UPWELLING WATERS SOUTH OF CALLAO AVERAGED 3.46. ALONG THE CONTINENTAL SHELF OF THE NORTHEAST UNITED STATES BETWEEN CAPE COD AND CAPE MAY, THE CONCENTRATION OF UREA RANGED FROM 0.25 MICROGRAM-ATOM UREA-N/LITER ON THE 1000 FATHOM (1830 METER) LINE TO A HIGH OF 11.20 WITHIN NEW YORK HARBOR. THE VERTICAL DISTRIBUTION OF UREA IN PERUVIAN WATERS, ALONG THE NORTHEAST UNITED STATES, AND SARGASSO SEA FLUCTUATED CONSIDERABLY WITH DEPTH BUT THERE WERE INDICATIONS OF PEAKS. THE SUGGESTION THAT UREA MAY SERVE AS AN AVAILABLE SOURCE OF NITROGEN FOR PHYTOPLANKTON GROWTH IS SUPPORTED. (JONES-WISCONSIN)

FIELD 05C

ACCESSION NO. W72-08056

DISTRIBUTIONAL PATTERNS IN ASSEMBLAGES OF ATTACHED DIATOMS FROM YAQUINA ESTUARY, OREGON,

OREGON STATE UNIV., CORVALLIS. DEPT. OF BOTANY; AND OREGON STATE UNIV., CORVALLIS. DEPT. OF STATISTICS.

C. D. MC INTIRE, AND W. S. OVERTON.

ECOLOGY, VOL. 52, NO. 5, P 758-777, LATE SUMMER 1971. 9 FIG, 11 TAB, 25 REF.

DESCRIPTORS:
*DIATOMS, *RESEARCH EQUIPMENT, ENVIRONMENTAL EFFECTS, ESTUARIES, *DISTRIBUTION PATTERNS, *AQUATIC ALGAE, STATISTICAL METHODS, MARINE ALGAE, *OREGON, *SEASONAL, CHRYSOPHYTA, SYSTEMATICS, *ESTUARINE ENVIRONMENT, SEA WATER, SALINITY, WATER TEMPERATURE, CHEMICAL ANALYSIS, WATER ANALYSIS, TIDAL EFFECTS, STABILITY, SOLAR RADIATION, PHOTOPERIODISM, FRESH WATER, SAMPLING, ANALYTICAL TECHNIQUES, SPECIES DIVERSITY, *DIATOMS.

IDENTIFIERS:
*YAQUINA BAY(ORE), YAQUINA ESTUARY, DATA INTERPRETATION, MACROALGAE, SPECIES DIVERSITY INDEX, FRAGILARIA CONSTRUENS, COCCONEIS PLACENTULA, NAVICULA SRYTOCEPHALA, ACHNANTHES BREVIPES, ACHNANTHES JAVANICA, LICMOPHORA JURGENSII, AMPHORA OVALIS, SYNEDRA FASCIULATA, EUNOTIA PECTINALIS, LITHOPHYTES, ENTEROMORPHA, INSOLATION, TEXTURE, SUBSTRATES, NAVICULA MUTICA, SYNEDRA FASCICULATA, FRAGILARIA STRIATULA VAR CALIFORNIA, MELOSIRA MONILIFORMIS, NAVICULA DISERTA, NITZSCHIA FRUSTULUM VAR PERPUSILLA, MELOSIRA NUMMULOIDES, NAVICULA SPP, ACHNANTHES SPP, ACHNANTHES LANCEOLATA.

ABSTRACT:
SUMMER AND WINTER DISTRIBUTIONAL PATTERNS OF ATTACHED DIATOMS WERE INVESTIGATED IN YAQUINA BAY AND ESTUARY, OREGON. DIFFERENCES IN SPECIES COMPOSITION AND DIVERSITY OF DIATOM ASSEMBLAGES AT SELECTED STATIONS FROM FRESH WATER JUST BELOW ELK CITY, OREGON, TO THE MARINE WATERS OF LOWER YAQUINA BAY WERE RELATED TO ENVIRONMENTAL GRADIENTS. A TOTAL OF 16,475 DIATOMS FROM 30 SAMPLES WAS SEPARATED INTO 256 SPECIES AND VARIETIES, OF WHICH 97 WERE FOUND IN ONLY ONE SAMPLE, AND 72 WERE REPRESENTED BY A SINGLE INDIVIDUAL. THE MOST ABUNDANT DIATOMS IN THE AUGUST SAMPLES WERE FRAGILARIA STRIATULA VAR. CALIFORNIA, MELOSIRA MONILIFORMIS, MELOSIRA NUMMULOIDES, NAVICULA MUTICA, AND SYNEDRA FASCICULATA, WHILE IN THE FEBRUARY SAMPLES ACHNANTHES NO. 2 AND NO. 4, NAVIVULA DISERTA, NAVICULA MUTICA, AND NITZSCHIA FRUSTULUM VAR. PERPUSILLA WERE DOMINANT. OF THE MOST ABUNDANT TAXA, NAVIVULA NO. 2, NAVICULA DISERTA, NAVICULA GREGARIA, NITZSCHIA FRUSTULUM VAR PERPUSILLA, SYNEDRA FASCICULATA, AND THALASSIONEMA NITZSCHIODES WERE THE MOST EVENLY DISTRIBUTED AMONG THE STATIONS. THE MEAN SPECIES DIVERSITY FOR DIATOM ASSEMLAGES SAMPLE IN FEBRUARY WAS SLIGHTLY HIGHER THAN THAT FOR ASSEMBLAGES COLLECTED IN AUGUST. IN FEBRUARY THE MEAN SPECIFIC DIVERSITY WITHIN A GENUS WAS HIGHER AND THE MEAN GENERIC DIVERSITY SLIGHTLY LOWER THAN IN AUGUST. IN GENERAL, DIFFERENCES IN ASSEMBLAGES WERE CLOSELY RELATED HORIZONTALLY TO THE SALINITY GRADIENT AND VERTICALLY TO THE DESICCATION AND INSOLATION GRADIENTS. HOWEVER, BIOLOGICAL FACTORS WERE NORE IMPORTANT IN ACCOUNTING FOR DIFFERENCES AMONG ASSEMBLAGES IN THE SUMMER THAN IN THE WINTER, AND THESE FACTORS WERE PRIMARILY SPECIES INTERACTIONS BETWEEN DIATOMS AND MACRO-ALGAE. (HOLOMAN-BATTELLE)

FIELD 05B, 05C, 02L

ACCESSION NO. W72-08141

INFESTATION OF BENTHIC CRUSTACES, FISH EGGS, AND TROPICAL ALGAE,

RHODE ISLAND UNIV., KINGSTON. DEPT. OF BACTERIOLOGY AND BIOPHYSICS.

P. W. JOHNSON, J. MCN. SIEBURTH, A. SASTRY, C. R. ARNOLD, AND M. S. DOTY.

LIMNOLOGY AND OCEANOGRAPHY, VOL. 16, NO. 6, P 962-969, NOVEMBER 1971. 1 FIG, 3 TAB, 17 REF.

DESCRIPTORS:
*CRUSTACEANS, *MARINE BACTERIA, *WATER QUALITY, BENTHOS, CRABS, COPEPODS, MARINE ALGAE, SHRIMP, FISH EGGS, FISH, BACTERIA, FISH PARASITES, ANTIBIOTICS(PESTICIDES), BACTERICIDES, ATOLLS, CULTURES, SPHAEROTILUS, CLADOPHORA.

IDENTIFIERS:
PLEOPODS, URCPODS, ENRICHMENT, LEUCOTHRIX MUCCR, MAJURO ATOLL, MARCHALL ISLAND, POLYSIPHONIA LANOSA, CHONDRUS CRISPUS, COD, FLOUNDER, CRAB EGGS, PELAGIC EGGS, GADUS MORHUA, GRASS SHRIMP, GREEN CRAB, PENICILLIUM, STREPTOMYCIN, EPIFLORA, WALLEYE LARVAE, SHAD, MACROINVERTEBRATES, CYANOPHYCEAE.

ABSTRACT:
CHARACTERISTIC FILAMENTS OF THE BACTERIUM LEUCOTHRIX MUCOR ARE OFTEN FOUND ON APPENDAGES AND EGGS OF BENTHIC MARINE CRUSTACEANS AND ON A WIDE VARIETY OF ALGAE. PLANKTONIC CRUSTACEA AND FISH EGGS CAN BECOME INFESTED IN AQUARIA IN THE ABSENCE OF ANTIBIOTICS. DEATH OF COD, FLOUNDER, BENTHIC INVERTEBRATES, AND CRAB EGGS HAS BEEN ATTRIBUTED TO LARGE POPULATIONS OF L. MUCOR. COPEPODS, SHRIMP, AND A VARIETY OF CRABS CAN BE INFECTED. ALTHOUGH L. MUCOR IS NOT PATHOGENIC, IT MAY CAUSE DEATH BY CAUSING PELAGIC EGGS TO SINK BELOW THE SURFACE AND BY INTERFERING WITH THE FILTERING APPARATUS OF CRUSTACEAN LARVAL FORMS, E.G., PLEOPODS AND UROPODS. ANTIBIOTICS SUCH AS PENICILLIN AND STREPTOMYCIN PREVENT L. MUCOR DEVELOPMENT AND DECREASE ASSOCIATED MORTALITY. EXAMINATION OF 48 MARINE ALGAE SAMPLES FROM THE LAGOON AT MAJURO ATOLL IN THE MARSHALL ISLANDS SHOWED THAT 81 PERCENT WAS INFECTED BY THE BACTERIA. EIGHTEEN RANDOM SAMPLES SHOWED 100 PERCENT INFESTATION UPON ENRICHMENT. THESE OBSERVATIONS ARE AT VARIANCE WITH PREVIOUS REPORTS OF L. MUCOR'S ABSENCE OR RARITY IN WARM WATERS. SPHAEROTILUS NATANS IS THE FRESHWATER COEFFICIENT OF THE ORGANISM, DISPLAYING THE SAME EFFECTS ON FLORA AND FAUNA OF THE AREAS INFESTED. (MACKAN-BATTELLE)

FIELD 05C, 05A

ACCESSION NO. W72-08142

GRANULAR CARBON FILTERS FOR TASTE AND ODOR REMOVAL,

WATER PURIFICATION PLANT AND PUMPING STATION, MT. CLEMENS, MICH.

R. E. HANSEN.

JOURNAL OF THE AMERICAN WATER WORKS ASSOCIATION, VOL 64, NO. 3, P 176-181, MARCH 1972, 11 FIG, 2 REF.

DESCRIPTORS:
*TASTE, *ODOR, *POTABLE WATER, ACTIVATED CARBON, ADSORPTION, FILTRATION, ALGAE, PHENOLS, DETERGENTS, PESTICIDES, OPERATION AND MAINTENANCE, *WATER TREATMENT, *MICHIGAN.

IDENTIFIERS:
*CARBON REGENERATION, *BACKWASHING, *MT CLEMENS(MICH).

ABSTRACT:
THE MT. CLEMENS, MICHIGAN WATER PURIFICATION PLANT EXPERIENCED TASTE AND ODOR PROBLEMS FOR 33 YEARS, STARTING WITH SEPTIC AND PHENOL TASTES AND ODORS AND GRADUALLY MOVING TOWARDS ACTINOMYCETES AND ALGAL PRODUCED TASTES AND ODORS. ALTHOUGH POWDERED ACTIVATED CARBON SLURRIES WERE SOMEWHAT EFFECTIVE IN KEEPING DOWN THE TASTE AND ODOR PROBLEMS, THERE WAS NOT ENOUGH SLURRY HANDLING CAPACITY TO ELIMINATE THE PROBLEM. EFFORTS TOWARD IMPROVING AND EXPANDING CARBON SLURRY OPERATIONS WERE TERMINATED WHEN IT WAS FOUND THAT GRANULAR ACTIVATED CARBON WITH AN EFFECTIVE SIZE OF 0.55 TO 0.65 MM AND A UNIFORMITY COEFFICIENT OF 1.4-1.7 COULD BE USED IN THE EXISTING SANDFILTERS. SINCE PLACING THE ACTIVATED CARBON FILTERS ON LINE, NO FURTHER TASTE AND ODOR PROBLEMS HAVE BEEN ENCOUNTERED. BASED ON A 3 YEAR LIFE, A TOTAL INSTALLED COST OF $4.97/MILLION GALLONS OF WATER TREATED OR $0.05/MONTH/HOUSEHOLD WAS COMPUTED. IN ADDITION TO REMOVING TASTE AND ODORS, AS WELL AS DOING AN EXCELLENT FILTERING JOB, THE CARBON FILTERS ALSO REMOVE ALL ORGANICS, DETERGENTS, AND PESTICIDES. (LOWRY-TEXAS)

FIELD 05F

ACCESSION NO. W72-08357

KINETICS OF ALGAL GROWTH IN AUSTERE MEDIA,

ALBANY COUNTY SEWER DISTRICT, N.Y.

G. C. MCDONALD, R. D. SPEAR, R. J. LAVIN, AND N. L. CLESCERI.

IN: PROPERTIES AND PRODUCTS OF ALGAE, PLENUM PRESS 1970, P 97-105, 5 FIG, 2 TAB, 5 REF.

DESCRIPTORS:
*ALGAE, *GROWTH RATES, *NUTRIENTS, CULTURES, INHIBITION, OLIGOTROPHY, PHOSPHORUS, NITROGEN, *EUTROPHICATION, LABORATORY TESTS, WATER QUALITY CONTROL, *KINETICS, *CHLOROPHYTA, ALGAL CONTROL, *NEW YORK.

IDENTIFIERS:
*SELENASTRUM CAPRICORNUTUM, *ALGAL GROWTH, *LAKE GEORGE(NY).

ABSTRACT:
SELENASTRUM CAPRICORNUTUM IS A UNI-CELLULAR GREEN ALGA WHICH IS EASY TO CULTURE IN THE LABORATORY AND HAS BEEN DOCUMENTED AS HAVING PRODUCED NUISANCE BLOOMS IN EUROPEAN LAKES. S. CAPRICORNUTUM WAS CULTURED IN LABORATORY USING THE BASIC ASM MEDIUM OF THE PROVISIONAL ALGAL ASSAY PROCEDURE (PAAP) AND A 10 TO 1 DILUTION OF GORHAM'S MEDIUM WITH AN INCREASE IN SODIUM CARBONATE CONCENTRATION TO 50 MG/L AS A PH CONTROL AID. DILUTION WATER FOR BOTH MEDIA PREPARATIONS WAS GLASS DISTILLED WATER AND/OR 0.45 MICRON MEMBRANE FILTERED LAKE GEORGE WATER (AN OLIGOTROPHIC SOFT WATER). AFTER GROWTH RATES HAD BEEN ESTABLISHED IN THE VARIOUS MEDIA, DISTILLED WATER, AND LAKE GEORGE WATER COMBINATIONS, THE CONCENTRATIONS OF NITROGEN AND PHOSPHORUS WERE REDUCED TO ONE HALF AND ONE QUARTER OF THE FULL AMOUNT AND NEW GROWTH RATE LEVELS AT EACH CONCENTRATION WERE ASCERTAINED. RESULTS DEMONSTRATED AN INHIBITORY EFFECT, AS EVIDENCED BY REDUCED GROWTH RATES, WHEN LAKE GEORGE WATER WAS USED AS DILUTION WATER IN EITHER MEDIUM. THE EXTENT AND CAUSATIVE FACTORS FOR SUCH INHIBITION BECAME THE FOCAL POINT FOR FURTHER INVESTIGATION. RESULTS ALSO SHOWED THAT THE NITROGEN CONCENTRATION IN MODIFIED TENTH GORHAM'S MEDIUM MAY BE REDUCED TO ONE-HALF THE POSITED LEVEL WITHOUT ANY SIGNIFICANT GROWTH RATE CHANGES FOR SELENASTRUM CAPRICORNUTUM. (LOWRY-TEXAS)

FIELD 05C, 02H

ACCESSION NO. W72-08376

RECYCLING SYSTEM FOR POULTRY WASTES,

LAKE TAHOE AREA COUNCIL, TAHOE CITY, CALIF.

G. L. DUGAN, C. G. GOLUEKE, AND W. J. OSWALD.

JOURNAL WATER POLLUTION CONTROL FEDERATION, VOL. 44, NO. 3 P 432-440, MARCH 1972, 3 FIG, 2 TAB, 9 REF. EPA GRANT 5R01 U100566-03.

DESCRIPTORS:
*FARM WASTES, POULTRY, NITROGEN, LAGOONS, PUMPING, AEROBIC TREATMENT, BIOCHEMICAL OXYGEN DEMAND, CHEMICAL OXYGEN DEMAND, NUTRIENTS, *WASTE TREATMENT, *ALGAE, COSTS.

IDENTIFIERS:
*OXIDATION DITCH, HYDRAULIC MANURE HANDLING.

ABSTRACT:
AN INTEGRATED WASTE MANAGEMENT SYSTEM WAS DEVELOPED IN WHICH ANIMAL ENCLOSURE SANITATION WAS INTEGRATED WITH WASTE TREATMENT. IT WAS A LARGELY CLOSED HYDRAULIC SYSTEM INVOLVING AN ANAEROBIC PHASE AND AN AEROBIC PHASE IN WHICH OXYGENATION COULD BE ACCOMPLISHED EITHER BY THE PHOTOSYNTHETIC ACTIVITY OF ALGAE OR BY MECHANICAL AERATION. WHEN PHOTOSYNTHETIC OXYGENATION WAS USED, ALGAE WERE HARVESTED. THE RANGE OF APPLICATION OF THE PROCESS IS FROM SMALL-SCALE TO LARGE-SCALE OPERATIONS. ALGAE RECLAMATION WOULD BE PRACTICED IN LARGE-SCALE OPERATIONS AND INDUCED AERATION IN SMALLER ONES. AN IMPORTANT OPERATIONAL FEATURE OF THE SYSTEM IS TO KEEP THE SOLIDS CONTENT OF THE MANURE SLURRY TO LESS THAN 3 PERCENT, WET WEIGHT. AT CONCENTRATIONS OF 3 PERCENT OR LESS, 70 PERCENT OR MORE OF SUSPENDED SOLIDS IN MANURE SLURRIES SETTLE OUT OF SUSPENSION IN LESS THAN 30 MIN. POND DEPTH SHOULD NOT EXCEED 12 IN. (30.5 CM). THE INDICATED POND AREA PER BIRD WAS 2 SQ. FT. (0.19 SQ. M.). AN ECONOMIC EVALUATION BASED ON AN INTEGRATED SYSTEM OF 100,000 EGG LAYERS AND THE APPLICATION OF THE LOW-LOADING, HIGH-COST, AND OVERDESIGNED COMPONENTS USED IN THE RESEARCH INDICATES THAT THE WASTE-HANDLING COSTS OF THE SYSTEM WOULD BE AT THE MOST, $0.02/DOZEN EGGS. IF THE VALUE OF THE ALGAL CROP WERE CREDITED TO THE OPERATION, THE NET WASTE-HANDLING COST WOULD BE ABOUT $0.01/DOZEN EGGS. (BUNDY-IOWA STATE)

FIELD 05D

ACCESSION NO. W72-08396

CHARACTERISTICS AND EFFECTS OF CATTLE FEEDLOT RUNOFF,

ROBERT S. KERR WATER RESEARCH CENTER, ADA, OKLA.

M. R. SCALF, W. R. DUFFER, AND R. D. KREIS.

IN: PROCEEDINGS, INDUSTRIAL WASTE CONFERENCE, 25TH, MAY 5, 6, AND 7, 1970. PURDUE UNIVERSITY, ENGINEERING EXTENSION SERIES NO. 137, PART 2, P 855-864, 10 FIG, 3 TAB, 6 REF.

DESCRIPTORS:
*FARM WASTES, *RUNOFF, *FISHKILL, CATTLE, DISSOLVED OXYGEN, DIVERSION STRUCTURES, SEDIMENTATION, BIOCHEMICAL OXYGEN DEMAND, ALGAE, CONFINEMENT PENS, IMPOUNDMENTS, *WATER POLLUTION SOURCES, *AGRICULTURAL RUNOFF, *WATER POLLUTION EFFECTS, *CATTLE, *FEED LOTS.

IDENTIFIERS:
ALGAL BLOOMS.

ABSTRACT:
CATTLE FEEDLOT CAPACITY IN THE UNITED STATES HAS BEEN INCREASING AT ABOUT 10 PERCENT ANNUALLY IN RECENT YEARS. ESSENTIALLY, ALL THIS GROWTH HAS BEEN IN THE FORM OF LARGE SCALE FEEDLOTS OF 5000 TO 100,000 HEAD CAPACITY. AS WITH THE CONCENTRATIONS OF PEOPLE, THE CONCENTRATION OF THOUSANDS OF ANIMALS IN A SMALL AREA PRODUCES MASSIVE ENVIRONMENTAL PROBLEMS. RAINFALL RUNOFF MAY CONTAIN POLLUTANT CONCENTRATIONS 10 TO 100 TIMES THOSE OF RAW MUNICIPAL SEWAGE, AND UNCONTROLLED ACCESS TO STREAMS CAN RESULT IN OXYGEN DEPLETION, FISH KILLS, AND OTHER LONG TERM, UNDESIRABLE ECOLOGICAL CONDITIONS FOR MILES DOWNSTREAM. THIS STUDY WAS DESIGNED TO MEASURE THE QUANTITY OF RAINFALL RUNOFF AND ITS POLLUTIONAL CHARACTERISTICS FROM A COMMERCIAL FEEDLOT AND EVALUATE THE EFFECT OF THIS WASTEWATER ON SMALL IMPOUNDMENTS. LESS THAN TWO WEEKS OF SEDIMENTATION IN RUNOFF COLLECTION PONDS PRODUCED ON EFFLUENT WITH POLLUTANT CONCENTRATIONS OF 10 TO 30 PER CENT OF THE MEAN DIRECT RUNOFF CONCENTRATIONS. THE NECESSITY OF FURTHER TREATMENT WAS DEMONSTRATED WHEN THE FEEDLOT OPERATOR PUMPED COLLECTION POND EFFLUENT THROUGH AN INADEQUATE TREATMENT SYSTEM INTO A 45-ACRE FLOOD CONTROL RESERVOIR. ESSENTIALLY, ALL GAME FISH IN THE RESERVOIR WERE KILLED DUE TO DISSOLVED OXYGEN STRESS AND HIGH AMMONIA CONCENTRATIONS. (DORLAND-IOWA STATE)

FIELD 05B, 05C

ACCESSION NO. W72-08401

POLYCHLORINATED BIPHENYLS: TOXICITY TO CERTAIN PHYTOPLANKTERS,

STATE UNIV. OF NEW YORK, STONY BROOK. MARINE SCIENCE RESEARCH CENTER.

N. S. FISCHER, J. L. MOSSER, T. C. TENG, AND C. F. WURSTER.

SCIENCE, VOL. 175, P 191-192, 14 JANUARY 1972. 1 FIG, 18 REF.

DESCRIPTORS:
*POLYCHLORINATED BIPHENYLS, *CHLORNATED HYDROCARBON PESTICIDES, *PESTICIDE TOXICITY, *DDT, PESTICIDES, PHYTOPLANKTON, DIATOMS, ALGAE, MARINE ALGAE, AQUATIC ALGAE, EUGLENA, CHLAMYDOMONAS.

IDENTIFIERS:
FRESHWATER ALGAE, THALASIOSIRA SPP., SKELETONEMA SPP., DUNALIELLA SPP.

ABSTRACT:
THE GROWTH RATES OF TWO SPECIES OF MARINE DIATOMS WERE REDUCED BY POLYCHLORINATED BIPHENYLS (PCB'S), WIDESPREAD POLLUTANTS OF THE MARINE ENVIRONMENT, AT CONCENTRATIONS AS LOW AS 10-25 MICROGRAMS/1. IN CONTRAST, A MARINE GREEN ALGA AND TWO SPECIES OF FRESHWATER ALGA WERE NOT INHIBITED AT THESE OR HIGHER CONCENTRATIONS. THE SENSITIVITY OF THESE SPECIES TO PCB'S PARALLELED THEIR SENSITIVITY TO DDT (1,1,1-TRICHLORO-2,2-BIS(P-CHLOROPHENYL)ETHANE). (SVENSSON-WASHINGTON)

FIELD 05C

ACCESSION NO. W72-08436

TOXICITY OF 2,4-D AND PICLORAM TO FRESH WATER ALGAE,

PURDUE UNIV., LAFAYETTE, IND.

J. H. ELDER, C. A. LEMBI, AND D. J. MORRE.

PURDUE UNIV. AND INDIANA HIGHWAY COMM. JOINT PUBLICATION NO. 23, 1970. 10 P, 4 TAB, 7 REF.

DESCRIPTORS:
*PESTICIDE TOXICITY, *WATER POLLUTION EFFECTS, BIOASSAY, PESTICIDES, HERBICIDES, 2-4-D, PLANT GROWTH REGULATORS, ALGAE, AQUATIC ALGAE, CHLORELLA.

IDENTIFIERS:
TORDON, FRESHWATER ALGAE, PICLORAM, PEDIASTRUM, 4-AMINO-3,5,6-TRICHLOROPICOLINIC ACID, 2,4-DICHLOROPHENOXYACETIC ACID.

ABSTRACT:
THE SOLUBILITY OF 2,4-D ACID IN WATER IS APPROXIMATELY 0.0025 M, THAT OF PICLORAM 0.0018 M. THEREFORE, THE HIGHEST CONCENTRATION TESTED WAS 0.001 M. AT THIS CONCENTRATION, NO EFFECT WAS OBSERVED ON MOST OF THE ORGANISMS WITH 2,4-D AND NONE OF THE ORGANISMS WITH PICLORAM. WITH TECHNICAL PICLORAM, MOTILE SPECIES WERE FOUND TO LOSE MOTILITY AT 0.001 M AND 0.005 M, BUT NOT AT 0.0001 M. THE TOXIC PRINCIPLE IS AN IMPURITY IN THE TECHNICAL PICLORAM TENTATIVELY IDENTIFIED AS 2(3,4,5,6-TETRACHLORO-2-PYRIDYL) GUANIDINE. THE RESULTS SHOW THAT THE POTENTIAL HAZARD OF 2,4-D OR PICLORAM TO BOTH FRESHWATER AND MARINE ALGAE FROM TERRESTRIAL RUNOFF WATER OR FROM DIRECT OR INDIRECT CONTAMINATION IS NIL. CERTAIN 2,4-D DERIVATIVES (PARTICULARLY ESTERS) MAY BE SUBSTANTIALLY MORE TOXIC THAN THE PARENT ACID. IN THIS REGARD, IT WILL BE NECESSARY TO EXAMINE A NUMBER OF 2,4-D DERIVATIVES IN DIFFERENT FORMULATIONS TO SEEK THOSE WHICH WILL RESULT IN MINIMUM DAMAGE TO ALGAE, FISH AND OTHER AQUATIC ORGANISMS. THERE IS NO EVIDENCE FOR BIOLOGICAL MAGNIFICATION OF EITHER 2,4-D OR PICLORAM IN ALGAE. (SVENSSON-WASHINGTON)

FIELD 05C

ACCESSION NO. W72-08440

THE FATE OF NITROGEN IN AQUATIC ECOSYSTEMS,

WISCONSIN UNIV., MADISON. WATER RESOURCES CENTER.

DENNIS R. KEENEY.

AVAILABLE FROM THE NATIONAL TECHNICAL INFORMATION SERVICE AS PB-209 217, $3.00 PAPER COPY, $0.95 IN MICROFICHE. WISCONSIN WATER RESOURCES CENTER, MADISON. EUTROPHICATION INFORMATION PROGRAM, LITERATURE REVIEW NO 3, 1972. 59 P. 4 FIG, 13 TAB, 198 REF. OWRR W-117(NO 1614)(4).

DESCRIPTORS:
*NITROGEN, *AQUATIC ENVIRONMENT, PRODUCTIVITY, SEDIMENTS, NITRIFICATION, OXIDATION-REDUCTION POTENTIAL, DENITRIFICATION, NITROGEN FIXATION, NITROGEN CYCLE, AMMONIA, NITRATES, MODEL STUDIES, DISSOLVED OXYGEN, CARBON, DISTRIBUTION, PHOSPHORUS, EUTROPHICATION, ALGAE, AQUATIC WEEDS, PHYTOPLANKTON, OLIGOTROPHY, HARDNESS(WATER), SOILS, AMINO ACIDS, DREDGING, SULFUR, ORGANIC MATTER, HYDROGEN ION CONCENTRATION, BACTERIA, FUNGI, RUNOFF, WATER POLLUTION SOURCES, PRECIPITATION(ATMOSPHERIC), GROUNDWATER, STREAMS, SEEPAGE, DRAINAGE, MARSHES, DIVERSION, WASTE WATER TREATMENT, ION EXCHANGE.

IDENTIFIERS:
LAKE MENDOTA(WIS), BANTAM LAKE(CONN), TROUT LAKE(WIS), SEDIMENT COMPACTION.

ABSTRACT:
THIS REVIEW OF 198 PAPERS CONSIDERS THE MANY ASPECTS OF NITROGEN IN AQUATIC ECOSYSTEMS. EVIDENCE INDICATES THAT ALGAE CAN USE ORGANIC AND INORGANIC FORMS OF NITROGEN AND THAT AMMONIUM IS USED PREFERENTIALLY BY SOME SPECIES. BECAUSE OF THE MANY TRANSFORMATIONS INVOLVED, THE CONCENTRATIONS OF VARIOUS NITROGEN SPECIES IN LAKE WATERS VARY WIDELY AND ARE THE NET RESULT OF NUMEROUS ENVIRONMENTAL FACTORS INFLUENCING RATE OF NITROGEN IMMOBILIZATION, MINERALIZATION, NITRIFICATION, AND DENITRIFICATION. THE LARGE POTENTIAL SOURCE OF NITROGEN TO LAKES FROM SEDIMENTS IS ILLUSTRATED. NITROGEN REQUIREMENTS FOR ALGAL GROWTH MUST BE MET FROM SOURCES OTHER THAN JUST THE NITROGEN IN WATER. THE OXIDATION-REDUCTION SYSTEM IN LAKES IS DISCUSSED EXTENSIVELY. THE RELATIVE ROLES OF VARIOUS BIOLOGICAL ENTITIES (BACTERIA, FAUNA, AND FLORA) RESPONSIBLE FOR NITROGEN ASSIMILATION AND RELEASE WILL VARY WITH THE SYSTEM AND WITH CHANGES IN MORPHOLOGICAL, PHYSICAL AND CHEMICAL VARIABLES WITHIN A GIVEN SYSTEM. NITRIFICATION, DENITRIFICATION, AND NITROGEN FIXATION WITH THE ORGANISMS RELATED TO THEM ARE DETAILED. (JONES-WISCONSIN)

FIELD 05C, 10F

ACCESSION NO. W72-08459

NITROGEN FIXATION BY BLUE-GREEN ALGAE IN YELLOWSTONE THERMAL AREAS,

DUNDEE UNIV. (SCOTLAND). DEPT. OF BIOLOGICAL SCIENCES.

W. D. P. STEWART.

PHYCOLOGIA, VOL 9, NO 3-4, P 261-268, DECEMBER 1970. 3 FIG, 5 TAB, 22 REF. NSF GRANT GB-5258.

DESCRIPTORS:
*NITROGEN FIXATION, *ALGAE, *CYANOPHYTA, *THERMAL WATER, *WYOMING, NATIONAL PARKS, HOT SPRINGS, WATER TEMPERATURES, NOSTOC STREAMS, DATA COLLECTIONS, ANALYTICAL TECHNIQUES, NITROGEN FIXING BACTERIA, THERMAL POLLUTION.

IDENTIFIERS:
*YELLOWSTONE NATIONAL PARK(WYO), BLUE-GREEN ALGAE, MASTIGOCLADUS, CALOTHRIX.

ABSTRACT:
POTENTIAL NITROGEN-FIXING BLUE-GREEN ALGAE WERE COMMON IN THREE HOT SPRING STREAMS IN YELLOWSTONE NATIONAL PARK. IN TWO STREAMS, WHERE THE DOMINANT NITROGEN-FIXING ALGAE WERE SPECIES OF CALOTHRIX, NITROGEN FIXATION, AS MEASURED BY UPTAKE OF N-15 WAS DETECTED IN SITU IN THE TEMPERATURE RANGE 28-46 C. AT HIGHER TEMPERATURES NITROGEN FIXATION WAS NOT DETECTED, ALTHOUGH THE ALGAE MAY HAVE RECEIVED FIXED NITROGEN FROM A GROWTH OF CALOTHRIX, NOSTOC, AND UNICELLULAR ALGAE WHICH OCCURRED AT LOWER TEMPERATURES ON THE SIDES OF THE STREAMS. IN THE THIRD STREAM, WHERE MASTIGOCLADUS WAS ABUNDANT, NITROGEN FIXATION WAS DETECTED AT TEMPERATURES UP TO 54 C, ALTHOUGH THE OPTIMUM FOR FIXATION WAS NEAR 42.5 C. THE OVERALL DATA IMPLY THAT IN SITU NITROGEN FIXATION CONTRIBUTES TO THE PRODUCTIVITY OF YELLOWSTONE HOT SPRINGS REGIONS AND THAT MASTIGOCLADUS AND CALOTHRIX ARE THE MOST IMPORTANT NITROGEN-FIXING BLUE-GREEN ALGAE. (WOODARD-USGS)

FIELD 05C, 05B

ACCESSION NO. W72-08508

A KINETIC MODEL OF PHYTOPLANKTON GROWTH, AND ITS USE IN ALGAL CONTROL BY RESERVOIR MIXING,

MUNICIPAL WATERWORKS OF ROTTERDAM (NETHERLANDS).

F. OSKAM.

PRESENTED AT INTERNATIONAL SYMPOSIUM ON MAN-MADE LAKES, THEIR PROBLEMS AND ENVIRONMENTAL EFFECTS, KNOXVILLE, TENNESSEE, MAY 3-7, 1971. 21 P, 5 FIG, 36 REF.

DESCRIPTORS:
*PHYTOPLANKTON, *MATHEMATICAL MODELS, *ALGAL CONTROL, *RESERVOIRS, DESTRATIFICATION, MIXING, PHOSPHORUS, NITROGEN, TURBULENCE, PHOTOSYNTHESIS, DEPTH, LIGHT PENTRATION, RESPRIATION, NUTRIENTS, TURBIDITY, PRODUCTIVITY, PLANKTON, STANDING CROP, LIMNOLOGY, EUTROPHICATION.

IDENTIFIERS:
RHINE RIVER(THE NETHERLANDS), MEUSE RIVER(THE NETHERLANDS), BIESBOSCH RESERVOIR(THE NETHERLANDS), KINETIC EQUATION.

ABSTRACT:
AN IN-RESERVOIR METHOD OF ALGAL CONTROL BY TURBULENT MIXING IS CONSIDERED. BASED ON THE CONCLUSION THAT LIGHT IS THE BASIC CONTROLLER OF PLANKTON DEVELOPMENT AND BY CONSIDERING NET GROWTH AS THE BALANCE BETWEEN GROSS PHOTOSYNTHESIS AND RESPIRATION, A SIMPLIFIED MATHEMATICAL EXPRESSION WAS DEVELOPED FOR THE RELATION BETWEEN WATER TRANSPARENCY, LIGHT PENTRATION, AND MIXING DEPTH. A MIXED LAYER OF ABOUT 20 M MIGHT SEVERLY RESTRICT ALGAL GROWTH AT HIGH LEVELS OF NATURAL TURBIDITY, WHEREAS PRODUCTIVITY IN A CLEAR WATER RESERVOIR WOULD HARDLY BE AFFECTED. IN THE LATTER CASE NUTRIENT LIMITATION CAN BE EXPECTED TO BE OPERATIVE LONG BEFORE LIGHT WOULD BECOME LIMITING. THIS EXPLAINS THE FACT THAT UNTIL RECENTLY THE CONNECTION BETWEEN TURBULENCE AND ALGAL GROWTH HAS BEEN NEGLECTED IN LIMNOLOGICAL STUDIES. DEEPENING THE MIXED LAYER MAY ALSO BE AT LEAST PARTLY RESPONSIBLE FOR THE DECREASE IN STANDING CROP, OBSERVED IN SEVERAL DESTRATIFICATION EXPERIMENTS. AN EVEN DISTRIBUTION OF ALGAE IS AN ESSENTIAL REQUIREMENT FOR THE MODEL TO BE VALID. MAINTAINING A HOMOGENEOUS STATE ASKS FOR A CONTINUOUS LEVEL OF TURBULENCE, WHICH MUST BE INTRODUCED ARTIFICIALLY. THE CONDITIONS FOR PRARCTICAL REALIZATION OF EFFECTIVE TURBULENT MIXING HAVE TO BE ESTABLISHED BEFORE THE RELIABILITY OF ALGAL CONTROL BY THIS METHOD CAN BE ASCERTAINED. (AUEN-WISCONSIN)

FIELD 05C, 02H, 04A

ACCESSION NO. W72-08559

ALGAL POPULATIONS IN MOSES LAKE, WASHINGTON: TEMPORAL AND SPATIAL DISTRIBUTION AND RELATIONSHIP WITH ENVIRONMENTAL PARAMETERS,

WASHINGTON UNIV., SEATTLE.

R. M. BUSH.

MS THESIS, 1971. 113 P. 9 FIG, 2 PLATES, 8 TAB, 35 REF, 2 APPEND. OWRR A-034-WASH(2).

DESCRIPTORS:
*ALGAE, *POPULATION, *TEMPORAL DISTRIBUTION, *SPATIAL DISTRIBUTION, ENVIRONMENT, SAMPLING, CHEMICAL ANALYSIS, PLANKTON, COMPUTER PROGRAMS, BIOMASS, TEMPERATURE, CONDUCTIVITY, PHOSPHATES, NITRATES, ALKALINITY, DIATOMS, CYANOPHYTA, CHLOROPHYTA, SEASONAL, NUTRIENTS, THERMAL STRATIFICATION, CARBON, SEDIMENTS, CARBON DIOXIDE, CHLORPHYLL, DOMESTIC WASTES, EUTROPHICATION, TURBIDITY, SESTON, HYDROGEN ION CONCENTRATION, SCENEDESMUS, DINOFLAGELLATES, CHRYSOPHYTA, PROTOZOA, *WASHINGTON.

IDENTIFIERS:
*MOSES LAKE(WASH).

ABSTRACT:
NATURALLY OCCURRING ALGAL POPULATIONS IN MOSES LAKE, WASHINGTON WERE STUDIED RELATIVE TO ENVIRONMENTAL PARAMETERS INFLUENCING GROWTH AND DISTRIBUTION, ESPECIALLY THOSE LIMITING OR CAUSING NUISANCE GROWTHS. USING CLUSTER ANALYSIS, THE DIFFERENT ALGAL POPULATIONS WERE DELINEATED FROM PLANKTON SAMPLES COLLECTED OVER TWO YEARS. SAMPLES WERE GROUPED ON THE BASIS OF HIGH JOINT OCCURRENCES OF ALGAL SPECIES. THE THREE MOST PREVALENT POPULATIONS WERE (1) COMPOSED PREDOMINANTLY OF DIATOM GENERA AND OFTEN DOMINATED DIATOMS, (2) DOMINATED BY BLUE-GREEN ALGEE, AND (3) COMPOSED PREDOMINATLY OF GREEN ALGAE AND DOMINATED BY GREEN ALGAE. ALGAL POPULATIONS 1 AND 3 WERE FOUND CONSISTENTLY IN SPECIFIC LAKE AREAS AND DEMONSTRATED LITTLE TEMPORAL VARIATION. THE REMAINING LAKE AREAS FOLLOWED A SUCCESSIONAL PATTERN IN ALGAL POPULATIONS RESULTING IN ONE DOMINATED BY BLUE-GREEN ALGAE IN THE SUMMER. MULTIPLE REGRESSION ANALYSIS INDICATED THAT ORTHPHOSPHATE AND CONDUCTIVITY BEST EXPLAINED THE VARIATION IN BLUE-GREEN POPULATION BIOMASS. THIS OCCURRED DESPITE THE FACT THAT NITROGEN SEEMED TO BE THE LIMITING NUTRIENT. AS WITH THE BLUE-GREEN ALGAL POPULATION, THE GREEN ALGAL BIOMASS WAS CORRELATED NEGATIVELY WITH CONDUCTIVITY AND PHOSPHATE-PHOSPHORUS INDICATING AND UPTAKE OF PHOSPHATE-PHOSPHORUS AND DISSOLVED SALTS (REPRESENTED BY CONDUCTIVITY) AS THE BIOMASS INCREASED. (JONES-WISCONSIN)

FIELD 05C, 02H

ACCESSION NO. W72-08560

SEWAGE FUNGUS IN RIVERS IN THE UNITED KINGDOM: THE SLIME COMMUNITY AND ITS CONSTITUENT ORGANISMS,

WATER POLLUTION RESEARCH LAB., STEVENAGE (ENGLAND).

E. J. C. CURTIS, AND C. R. CURDS.

WATER RESEARCH, VOL 5, P 1147-1159, 1971. 3 FIG, 4 TAB, 27 REF.

DESCRIPTORS:
*SEWAGE, *FUNGI, *SLIME, *BIOLOGICAL COMMUNITIES, RIVERS, BACTERIA, WATER POLLUTION, ACTIVATED SLUDGE, ALGAE, PROTOZOA, DIATOMS, EUGLENA, MATHEMATICAL STUDIES.

IDENTIFIERS:
*UNITED KINGDOM, *SEWAGE FUNGUS, SPHAEROTOTILUS NATANS, ZOOGLOEAL BACTERIA.

ABSTRACT:
BIOLOGICAL COMPOSITION OF SLIMES EXAMINED DURING A SURVEY OF SLIME OUTBREAKS IN THE UNITED KINGDOM IS DESCRIBED AND COMPARED WITH OTHER POLLUTED WATER BIOTA. SLIMES WERE FREQUENTLY DOMINATED BY SPHAEROTILUS NATANS OR ZOOGLOEAL BACTERIA, AND THE INTERRELATIONSHIPS OF THESE AND OTHER SPECIES WERE STUDIED USING CLUSTER ANALYSIS AND DIVERSITY INDICES. COMPARED WITH S NATANS AND ZOOGLOEAL FORMS, ALL OTHER ORGANISMS CONTRIBUTING TO THE ACTUAL SLIME MATRIX OCCURRED ONLY INFREQUENTLY. THE MOST COMMON TO OCCUR AS DOMINANT ORGANISMS WERE THE FUNGI, LEPTOMITUS LACTEUS AND GEOTRICHUM CANDIDUM OCCURRING IN LARGE AMOUNTS IN 3.4 AND 4.5% OF SITES RESPECTIVELY, AND THE GLIDING BACTERIUM BEGGIATOA ALBA (5.6%). OTHER FILAMENTOUS ORGANISMS WERE MORE WIDELY FOUND AS MINOR COMPONENTS OF SLIMES FORMED PRIMARILY BY S NATANS OR ZOOGLOEAL BACTERIA. FILAMENTS OF FLEXIBACTERIUM AND FLAVOBACTERIUM WERE SOMETIMES ATTACHED TO FILAMENTS OF S NATANS OR OTHER MATERIAL WITHOUT SPECIFIC HOLDFAST. FILAMENTOUS ALGAE ONLY RARELY OCCURRED IN SEWAGE FUNGUS SLIMES. MOST SAMPLES CONTAINED PROTOZOA AND ALL CILIATED PROTOZOA. SEVENTY-SEVEN CILIATE SPECIES WERE RECORDED. IN ADDITION TO CLUSTER ANALYSIS, ANOTHER TECHNIQUE DESCRIBING ASSOCIATION BETWEEN ORGANISMS AND GROUPS OF ORGANISMS IS CONSTRUCTION OF A MINIMUM-SPANNING TREE. (JONES-WISCONSIN)

FIELD 05C

ACCESSION NO. W72-08573

POLLUTION CONTROLLED CHANGES IN ALGAL AND PROTOZOAN COMMUNITIES,

VIRGINIA POLYTECHNIC INST., BLACKSBURG. CENTER FOR ENVIRONMENTAL STUDIES; AND VIRGINIA POLYTECHNIC INST., BLACKSBURG. DEPT. OF BIOLOGY.

J. CAIRNS, JR., AND G. R. LANZA.

IN: WATER POLLUTION MICROBIOLOGY, P 245-272. JOHN WILEY AND SONS, INC, NEW YORK, 1972. OWRR B-017-VA(5).

DESCRIPTORS:
*AQUATIC MICROORGANISMS, *WATER POLLUTION EFFECTS, *ALGAE, *PROTOZOA, *BIOLOGICAL COMMUNITIES, ECOLOGICAL DISTRIBUTION, BIOINDICATORS, WATER POLLUTION SOURCES.

IDENTIFIERS:
RIDLEY CREEK(PA), LITITZ CREEK(PA), SPECIES DIVERSITY, DIRECT EFFECTS, INDIRECT EFFECTS.

ABSTRACT:
STRESSES CAUSED BY WASTES AFFECT ALGAL AND PROTOZOAN COMMUNITIES BY REDUCTION IN SPECIES NUMBER, INCREASE IN RANGE OF NUMBERS OF INDIVIDUALS PER SPECIES, REDUCTION IN COLONIZATION RATES, CHANGES IN SELECTIVE PREDATOR OR PARASITE PRESSURE, AND SHIFT IN DOMINANCE WITHIN THE COMMUNITY. A RATHER SIMPLE RESPONSE PATTERN TO STRESS IS CHARACTERISTIC OF BOTH POPULATIONS AND COMMUNITIES AND MAY BE USED TO ASSESS THE STRESS EFFECT. ALGAL AND PROTOZOAN SPECIES FREQUENTLY HAVE COMPLEX REQUIREMENTS THAT MAY RIVAL THOSE OF HIGHER ORGANISMS. IT IS BECOMING INCREASINGLY EVIDENT THAT THE SAME ECOLOGICAL PRINCIPLES APPLICABLE TO HIGHER PLANTS AND ANIMALS ARE ALSO VALID FOR ALGAE AND PROTOZOA. MICROBIAL COMMUNITIES HAVE A STRUCTURE WHICH IS MAINTAINED DESPITE SUCCESSION; OTHER HOMEOSTATIC MECHANISMS ARE ALSO OPERATIVE. GENERAL EFFECTS OF STRESS CAUSED BY INDUSTRIAL AND MUNICIPAL WASTES AND AGRICULTURAL RUNOFF ARE DISCUSSED. THE BASIC CATEGORIES OF POLLUTION ARE: NONTOXIC ORGANIC AND INORGANIC SUBSTANCES, THERMAL CHANGES, TOXIC SUBSTANCES, SUSPENDED SOLIDS, AND RADIOACTIVE MATERIALS. WITH A FINITE ECOLOGICAL BASE AND INCREASED PRESSURES ON THIS BASE, STANDARDS MUST BE DEVELOPED TO PROTECT THE SYSTEM AND KNOWLEDGE OF ITS OPERATIONAL PREREQUISITES MUST BE GAINED TO MANAGE IT WELL. (JONES-WISCONSIN)

FIELD 05C

ACCESSION NO. W72-08579

PHOTCRESPIRATION AND NITROGENASE ACTIVITY IN THE BLUE-GREEN ALGA, ANABAENA CYLINDRICA,

DUNDEE UNIV. (SCOTLAND). DEPT. OF BIOLOGICAL SCIENCES.

M. LEX, W. B. SILVESTER, AND W. D. STEWART.

PROCEEDINGS OF THE ROYAL SOCIETY OF LONDON, B, VCL 180, P 87-102, 1972. 7 FIG, 7 TAB, 41 REF.

DESCRIPTORS:
*LIGHT, *PHOTOSYNTHESIS, *CYANOPHYTA, NITROGEN FIXATICN, OXYGEN, CARBON DIOXIDE, BIOCHEMISTRY, ALGAE, INHIBITICN, LABORATORY TESTS.

IDENTIFIERS:
*PHOTORESPIRATION, *ANABAENA CYLINDRICA, ACETYLENE REDUCTICN, WARBURG EFFECT.

ABSTRACT:
GROWN UNDER ASEPTIC CONDITIONS IN NITROGEN-FREE MEDIUM UNDER CONTINUOUS LIGHT, THE PHOTORESPIRATION OF ANABAENA CYLINDRICA AND ITS EFFECT ON PHOTOSYNTHESIS AND NITROGENASE ACTIVITY WERE STUDIED. IN LIGHT, OXYGEN UPTAKE MAY BE UP TO 20 TIMES DARK RESPIRATION RATE; UPTAKE RATE IN LIGHT INCREASES LINEARLY WITH INCREASING OXYGEN PRESSURE, WHILE DARK RESPIRATION IS SATURATED AT CXYGEN PRESSURE NEAR 0.05 ATMOSPHERES. DCMU, NOT KCN, INHIBITS PHOTORESPIRATION. EXOGENCUSLY SUPPLIED HYDROXYETHANE SULPHONATE, A GLYCCLLATE OXIDASE ACTIVITY INHIBITOR, AND GLYCOLLATE DO NOT AFFECT RESPIRATION, ALTHOUGH C-14-LABELED GLYCOLLATE IS ASSIMILATED IN BOTH LIGHT AND DARK. PHOTORESPIRATICN IS HIGHLY SENSITIVE TO CARBON DIOXIDE PRESSURE AND TO BICARBONATE CONCENTRATION AND APPROACHES TRUE PHOTOSYNTHETIC OXYGEN PRODUCTICN AT THE CARBON DIOXIDE COMPENSATION POINT OF 10 PARTS/1,000,000. CARBON DIOXIDE CONCENTRATION (0.02 ATMOSPHERES) COMPLETELY INHIBITS PHOTORESPIRATION, WHEREAS TRUE PHOTOSYNTHESIS IS SCARCELY AFFECTED. CONDITIONS WHICH STIMULATE PHOTORESPIRATION (LOW PRESSURE CARBON DIOXIDE AND HIGH PRESSURE OXYGEN) PROGRESSIVELY INHIBIT ACETYLENE REDUCTION. IN SHORT-TERM STUDIES DCMU INHIBITS ACETYLENE REDUCTION UNDER CONDITIONS STIMULATING PHOTORESPIRATION BUT HAS LITTLE EFFECT UNDER CONDITIONS INHIBITING PHOTORESPIRATICN. PHCTORESPIRATION AND NITROGENASE ACTIVITY APPARENTLY COMPETE INDIRECTLY FOR REDUCING POWER AND AT LEAST ONE MECHANISM OF OXYGEN INHIBITICN OF NITROGENASE ACTIVITY IS VIA A STIMULATION OF PHOTORESPIRATION. (JONES-WISCONSIN)

FIELD 05C

ACCESSION NO. W72-08584

AQUATIC PLANTS FROM MINNESOTA, PART 3 - ANTIMICROBIAL EFFECTS,

MINNESOTA UNIV., MINNEAPOLIS. WATER RESOURCES RESEARCH CENTER.

K. LEE SU, E. J. STABA AND Y. ABUL-HAJJ.

AVAILABLE FROM THE NATIONAL TECHNICAL INFORMATION SERVICE AS PB-209 530. MINNESOTA WATER RESOURCES RESEARCH CENTER, BULLETIN 48, FEBURARY 1972. 36 P, 6 FIG, 9 TAB, 145 REF. OWRR A-025-MINN(4).

DESCRIPTORS:
*AQUATIC PLANTS, *MINNESOTA, BACTERIA, FUNGI, PROTOZCA, PATHOGENS, MICROORGANISMS, ALGAL TOXINS.

IDENTIFIERS:
*ANTIMICROBIAL ACTIVITY, *SKELLYSOLVE, *FILTER PAPER DISC DIFFUSION METHOD, *ZONES OF INHIBITION, ANTIFUNGAL ACTIVITY.

ABSTRACT:
THE ANTIMICROBIAL ACTIVITY OF THE FOLLOWING 22 MINNESOTAN AQUATIC PLANTS WAS INVESTIGATED: ANACHARIS CANADENSIS, CALLA PCLUSTRIS, CAREX LACUSTRIS, CERATOPHYLLUM DEMERSUM, CHARA VULGARIS, ELECCHARIS SMALLII, LEMNA MINOR, MYRIOPHYLLUM EXALBESCENS, NUPHAR VARIEGATUM, NYMPHAEA TUBEROSA, POTAMOGETON AMPLIFCLUS, P. NATANS, P. PECTINATUS, P. RICHARDSONU, P. ZCSTERIFORMIS, SAGITTARIA CUNEATA, S. LATIFOLIA, SPARGANIUM EURYCARPUM, S. FLUCTUANS, TYPHA ANGUSTIFOLIA, VALLISNERIA AMERICANA, AND ZIZANIA AQUETICA. THE CHEMICAL CONSTITUENTS RESPONSIBLE FOR THE SIGNIFICANT ANTIMICRCBIAL EFFECT WERE ISOLATED AND IDENTIFIED. THE SKELLYSOLVE F, CHLOROFORM, 80% ETHANOL AND FRESH WATER EXTRACTS OF PLANT SPECIES WERE TESTED FOR ANTIMICROBIAL ACTIVITY EMPLOYING THE QUALITATIVE FILTER PAPER DISC DIFFUSION METHOD AND REFERENCE ANTIBIOTIC DISCS. ETHANOL (80%) EXTRACTS OF MYRIOPHYLLUM EXALBESCENS (ACTIVITY RATIO OF 0.34 AS COMPARED TO THE 30 MCG CHLORAMPHENICOL DISCS), NYMPHAEA TUBEROSA (LEAF: 0.40, STEM: 0.38) AND NUPHAR VARIEGATUM COLLECTED IN LAKE MINNETONKA (LEAF: 0.43, STEM: 0.45) WERE MODERATELY ACTIVE AGAINST S. AUREUS. ETHANOL (80%) EXTRACTS CF CAREX LUCUSTRIS (ACTIVITY RATIO OF 0.34 AS CCMPARED TO THE 10 MCG STREPTCMYCIN DISCS), NUMPHAEA TUBEROSA (LEAF: 1.01, STEM: 1.10) AND NUPHAR VARIEGATUM COLLECTED IN LAKE MINNETCNKA (LEAF: 0.73, STEM: 0.58) WERE ACTIVE AGAINST M. SMEGMATIS. ALL EXTRACTS WERE RELATIVELY INACTIVE AGAINST E. COLI EXCEPT THE WATER EXTRACT OF POTAMOGETON NATANS WHERE A LOW ACTIVITY RATIO OF 0.10 AS COMPARED TO THE 30 MCG CHLORAMPHENICOL DISCS WAS INDICATED. (SEE ALSC W72-05877) (WALTON-MINNESOTA)

FIELD 05C, 05A

ACCESSION NO. W72-08586

BIOLCGICAL EFFECTS OF COOLING TOWER BLOWDOWN,

NATIONAL ENVIRONMENTAL RESEARCH CENTER, CORVALLIS, OREG.

R. R. GARTON.

PREPRINT, OF PAPER PRESENTED AT AMERICAN INSTITUTE OF CHEMICAL ENGINEERS NATIONAL MEETING, 71ST, DALLAS, TEXAS, FEBRUARY 20-23, 1972, 25 P, 5 FIG, 9 TAB, 6 REF.

DESCRIPTORS:
*COOLING TOWERS, *TOXICITY, *BIOASSAYS, LABORATORY TESTS, FISH, ALGAE, CHEMICAL WASTES, ALGICIDES, BACTERICIDES, CHROMIUM, ZINC, TEMPERATURE, HYDROGEN ION CONCENTRATION, CORRCSION, WATER POLLUTICN EFFECTS.

IDENTIFIERS:
*SLIMICIDES.

ABSTRACT:
A SIMULATED COOLING TOWER BLCWDOWN WASTE WAS ASSEMBLED, USING THOSE CHEMICALS AND CONCENTRATICNS PRESENTED IN A WASTE DISCHARGE PERMIT APPLICATION TO THE ENVIRONMENTAL PROTECTION AGENCY AS MAXIMUMS TO BE RELEASED BY A 1130 MWE NUCLEAR POWER PLANT UNDER CONSTRUCTION. THIS SIMULATED BLOWDOWN MIXTURE AND DILUTIONS THEREOF WERE THEN USED IN BIOASSAY EXPERIMENTS TO DETERMINE TOXICITY TO REPRESENTATIVES FROM TWO TROPHIC LEVELS, A GREEN ALGA AND A SALMONID FISH. BICASSAY RESULTS POINTED OUT ONLY TWO OF THE TOXIC COMPCUNDS, ZINC AND CHROMIUM. ZINC WAS TOXIC TO ALGAE AT 0.064 MG/L AND THE 96 HOUR LC50 TO JUVENILE STEEL HEAD WAS 0.09 MG/L. FOR CHROMATE, NO LETHAL EFFECT ON JUVENILE STEEL HEAD WAS NOTED IN 96 HRS. AT 31 MG/L CRO4, BUT ALGAL PRODUCTIVITY WAS DECREASED AT LEVELS OF 0.14 MG/L CRO4. THESE RESULTS INDICATE: (1) TOXICITY MUST BE DETERMINED FOR EACH INDIVIDUAL SPECIES EXPOSED TO BLOWDOWN MIXTURES; (2) SAFE EXPOSURE LEVELS MAY BE AS LITTLE AS 1% OF THE 96-HOUR LC50; (3) CONCENTRATIONS OF CHEMICALS CF KNCWN TOXICITY MUST BE MAINTAINED WITHIN SAFE LEVELS. (LOWRY-TEXAS)

FIELD 05C

ACCESSION NO. W72-08683

METHCDS, MODELS AND INSTRUMENTS FOR STUDIES OF AQUATIC PCLLUTICN. AN ANNOTATED BIBLIOGRAPHY,

OCEAN ENGINEERING INFORMATION SERVICE, LA JOLLA, CALIF.

E. SINHA.

AVAILABLE FROM OCEAN ENGINEERING INFORMATION SERVICE, LA JOLLA, CAL. 92037, PRICE: $6.00. OCEAN ENGINEERING INFORMATION SERIES VOL 5, 1971. 29 P.

DESCRIPTORS:
*BIBLIOGRAPHIES, *ANALYTICAL TECHNIQUES, *WATER POLLUTION CONTROL, *INSTRUMENTATION, MATHEMATICAL MODELS, WATER ANALYSIS, POLLUTANT IDENTIFICATION, DETERGENTS, NUTRIENTS, PESTICIDES, OIL, METALS, THERMAL POLLUTION, SAMPLING, BACTERIA, TOXICITY, BEHAVIOR, BITUMINOUS MATERIALS, ANIMAL WASTES, FISH, FLUORESCENCE, INFRARED RADIATION, IRRADIATION, SPECTROPHOTOMETRY, SPECTROSCOPY, HEAVY METALS, LIGNINS, FARM WASTES, POLAROGRAPHIC ANALYSIS, SEDIMENTOLOGY, SEDIMENTS, SURVEYS, PHOSPHORUS, DISSOLVED OXYGEN, BIOCHEMICAL OXYGEN DEMAND, CLAMS, CRUSTACEANS, PHOSPHATES, ANION EXCHANGE, LAKES, RIVERS, NITROGEN, GREAT LAKES, VOLUMETRIC ANALYSIS, ENTROPHICATION, E. COLI, BIOASSAY, BOTTOM SEDIMENTS, MEMBRANE PROCESSES, DIGITAL COMPUTERS, ALGAE, RADIOACTIVITY TECHNIQUES, SEPARATION TECHNIQUES.

IDENTIFIERS:
VOLTAMMETRY, RADIOGRAPHY, ION SELECTIVE ELECTRODES, LASERS, ORTHOPHOSPHATES.

ABSTRACT:
THIS BIBLIOGRAPHY CONTAINS 204 ABSTRACTS OF LITERATURE PROVIDING SUBSTANTIAL SCIENTIFIC AND TECHNICAL INFORMATION ON METHODS, MODELS AND INSTRUMENTS USED IN STUDIES OF ACUATIC POLLUTION AND MEANS OF ABATEMENT. THESE DEAL WITH THE DETECTION, IDENTIFICATION AND MEASUREMENT OF THE PARAMETERS OF POLLUTION, BIOTIC CONSTITUENTS, DETERGENTS AND NUTRIENTS, PESTICIDES, OIL, METALS, AND NON-METALLIC TOXICANTS. VARIOUS ASPECTS OF WATER QUALITY MANAGEMENT ARE ENCOMPASSED. PERTINENT PATENTS, A BIBLIOGRAPHY OF BIBLIOGRAPHIES, A SUBJECT OUTLINE, A KEYTERM INDEX, AND AN INDEX CITING ALL AUTHORS AND CO-AUTHORS ARE INCLUDED. (MORTLAND-BATTELLE)

FIELD 05A, 07B, 10B

ACCESSION NO. W72-08790

BENTHIC ALGAE IN POLLUTED ESTUARIES,

DURHAM UNIV. (ENGLAND). DEPT. OF BOTANY.

P. EDWARDS.

MARINE POLLUTION BULLETIN, VOL 3, NO 4, P 55-60, APRIL 1972. 4 FIG, 4 TAB, 13 REF.

DESCRIPTORS:
*ALGAE, *ESTUARIES, *WASTE DISPOSAL, *SEWAGE, *INDUSTRIAL WASTES, WATER POLLUTION SOURCES, ON-SITE INVESTIGATIONS, METHODOLOGY, FOREIGN RESEARCH, MARINE ALGAE, PRODUCTIVITY, POLLUTANTS, CHEMICAL WASTES, SEWAGE EFFLUENTS, WATER POLLUTION EFFECTS.

IDENTIFIERS:
*ENGLAND, *POLLUTED ESTUARIES, DURHAM COUNTY RIVERS.

ABSTRACT:
ON-SITE STUDIES WERE MADE OF THREE ESTUARIES IN NORTHEASTERN ENGLAND IN WHICH DIFFERENT CONDITIONS OF POLLUTION EXIST. THE RIVERS, LOCATED IN COUNTY DURHAM, ARE THE WEAR, A RELATIVELY UNPOLLUTED STREAM, THE TYNE, WHICH RECEIVES A LARGE VOLUME OF UNTREATED SEWAGE, AND THE TEES, WHICH IS MIXED WITH INDUSTRIAL WASTES. THESE SYSTEMS PROVIDE A HUGE NATURAL EXPERIMENT SINCE THE DEGREE OR TYPE OF POLLUTION IS PROBABLY THE ONLY ENVIRONMENTAL FACTOR THAT VARIES SIGNIFICANTLY BETWEEN THE THREE ESTUARIES. THE ALGAL FLORA OF THE ESTUARIES IS COMPARED TO REVEAL THE EFFECTS OF DIFFERENT KINDS OF CONTAMINATION. A TOTAL OF 69 STATIONS AT ABOUT 1 KM INTERVALS REACHING FROM THE MOUTH TO THE TIDAL LIMIT OF EACH ESTUARY WERE USED TO DETERMINE THE VARIOUS SPECIES OF ALGAE. VEGETATION IDENTIFIED CONSISTS OF 54 SPECIES FROM THE THREE ESTUARIES; THESE ARE LISTED. A TABLE ALSO GIVES THE SPECIES OF BENTHIC MARINE ALGAE IN THE TEES ESTUARY FOR 1935. A DECREASE IN ALGAL VEGETATION SINCE THE 1930'S IN THE TEES IS PROBABLY DUE TO GROWTH OF THE CHEMICAL INDUSTRY AND THE ASSOCIATED DISCHARGE OF TOXIC CHEMICAL WASTES. (LANG-USGS)

FIELD 05C, 05B, 02L

ACCESSION NO. W72-08804

EUTROPHICATION FACTORS IN NORTH CENTRAL FLORIDA LAKES,

FLORIDA UNIV., GAINSVILLE. DEPT. OF ENVIRONMENTAL ENGINEERING.

H. D. PUTNAM, W. H. MORGAN, P. L. BREZONIK, E. E. SHANNON, AND P. E. MASLIN.

COPY AVAILABLE FROM GPO SUP DOC, $1.25; MICROFICHE FROM NTIS AS PB-209 863.
ENVIRONMENTAL PROTECTION AGENCY WATER POLLUTION CONTROL RESEARCH SERIES,
FEBRUARY 1972, 141 P, 36 TAB, 40 FIG, 87 REF. EPA PROGRAM 16010 DON 02/72.

DESCRIPTORS:
*EUTROPHICATION, *MATHEMATICAL MODELS, *ESSENTIAL NUTRIENTS, *PRIMARY PRODUCTIVITY, WATER QUALITY, TROPHIC LEVEL, AQUATIC ALGAE, FISH POPULATIONS, FLORIDA, WATER POLLUTION EFFECTS, NITROGEN, PHOSPHORUS.

IDENTIFIERS:
ANDERSON-CUE LAKE, MELROSE(FLORIDA).

ABSTRACT:
A SMALL FLORIDA LAKE HAS BEEN RECEIVING A REGIMEN OF NUTRIENT ADDITION EQUIVALENT TO 500 MG/CU M-YEAR N AND 43 MG/CU M-YEAR P SINCE 1967. DATA WERE ACCUMULATED THROUGH 1969. THE EFFECT ON THE LACUSTRINE ECOSYSTEM OF VARIOUS BIOGENES INCLUDES PRODUCTION BY PRIMARY PRODUCERS, SPECIES DIVERSITY OF PLANKTON AND CERTAIN PRODUCTION ESTIMATES AT THE SECONDARY TROPHIC LEVEL USING NATURAL POPULATIONS OF PLANKTIVOROUS FISH. PLANKTON PRODUCTION USING ISOTOPIC CARBON IS CA. 58 G/SQ M-YEAR. SPECIES DIVERSITY IS SLOWLY CHANGING TO A MIXED CHLOROPHYCEAN AND YELLOW-GREEN. BIOMASS OF BENTHIC GREEN FILAMENTOUS TYPES HAS INCREASED SLIGHTLY. NUTRIENT ADDITION HAS HAD LITTLE INFLUENCE ON ZOOPLANKTON PRODUCTION. RELATED STUDIES ON 53 OTHER REGIONAL LAKES HAVE BEEN DONE USING A MULTI-DIMENSIONAL HYBRID CONCEPT AS DEFINED BY SEVERAL TROPHIC STATE INDICATORS. THIS TROPHIC STATE INDEX HAS PROVIDED A MEANS FOR RANKING THE LAKES ON AN ARBITRARY SCALE. CLUSTER ANALYSIS UTILIZING PERTINENT CHARACTERISTICS RESULTED IN CLASSIFICATION OF OTHER LAKES. LAND USE PATTERNS AND POPULATION CHARACTERISTICS WERE DETERMINED PHOTOGRAPHICALLY AND N AND P BUDGETS ESTIMATED. USING MULTIPLE REGRESSION AND CANONICAL ANALYSIS, SEVERAL SIGNIFICANT RELATIONSHIPS WERE FOUND BETWEEN LAKE TROPHIC STATE, LAKE BASIN, LAND USE, AND POPULATION CHARACTERISTICS. IN GENERAL, TROPHIC STATE OF LAKES CAN BE EXPRESSED AS A SIMPLE RELATIONSHIP INCORPORATING N AND P INFLUX RATES. (EPA ABSTRACT)

FIELD 05B, 05C

ACCESSION NO. W72-08986

VITAMIN B12 PRODUCTION AND DEPLETION IN A NATURALLY OCCURRING EUTROPHIC LAKE,

OREGON STATE UNIV., CORVALLIS. DEPT. OF MICROBIOLOGY; AND OREGON STATE UNIV., CORVALLIS. DEPT. OF OCEANOGRAPHY.

A. GILLESPIE, AND Y. MORITA.

APPLIED MICROBIOLOGY VOL. 23, NO. 2, P 341-348, 1972, 11 FIG, 3 TAB, 9 REF.
EPA PROGRAM NO. 16010 EBB 02/72.

DESCRIPTORS:
*VITAMIN B, *EUTROPHICATION, *AQUATIC MICROBIOLOGY, OREGON, *SEDIMENT-WATER INTERFACES, SEDIMENTS, AQUATIC PRODUCTIVITY, PHYTOPLANKTON, ALGAE.

IDENTIFIERS:
UPPER KLAMATH LAKE(OREGON).

ABSTRACT:
THE DISTRIBUTION OF VITAMIN B12 WITHIN UPPER KLAMATH LAKE WAS SURVEYED AT APPROXIMATELY MONTHLY INTERVALS DURING A PERIOD FROM SEPTEMBER 1968 TO NOVEMBER 1969. HIGH CONCENTRATIONS (UP TO 1.8MICRO G/G OF DRY SEDIMENT) CHARACTERISTICALLY OCCURRED AT THE WATER-SEDIMENT INTERFACE, WITH A SHARP DECLINE BELOW THIS AREA. A HEAVY BLOOM OF APHANIZOMENON FLOS-AQUAE OCCURRED FROM THE LATTER PART OF MAY THROUGH OCTOBER 1969. B12 CONCENTRATIONS OF THE UPPERMOST SEDIMENTS, FROM ALL BUT ONE SAMPLING SITE, INCREASED GRADUALLY THROUGH THE BLOOM FOLLOWED BY A DRASTIC INCREASE DURING THE DIE-OFF PERIOD. B12 IS PROBABLY NOT A LIMITING FACTOR FOR PRIMARY PRODUCTIVITY, SINCE SUFFICIENT LEVELS OF THIS VITAMIN WERE FOUND TO OCCUR THROUGHOUT THE YEAR. OF 42 CULTURES ISOLATED FROM UPPER KLAMATH LAKE WATER AND SEDIMENTS, 20 WERE FOUND CAPABLE OF PRODUCING 50PG OR MORE OF B12/ML OF MEDIUM. PHYTOPLANKTON SAMPLES WERE FOUND TO CONTAIN UP TO 5 MICRO G OF B12/G OF DRY MATERIAL. DEGRADATION OF B12 OCCURRED IN STERILIZED AS WELL AS FRESH SEDIMENT SAMPLES. (EPA ABSTRACT)

FIELD 05C, 02H

ACCESSION NO. W72-08989

SOME ECOLOGICAL EFFECTS OF ARTIFICIAL CIRCULATION ON A SMALL EUTROPHIC NEW HAMPSHIRE LAKE,

NEW HAMPSHIRE UNIV., DURHAM. DEPT. OF ZOOLOGY.

R. C. HAYNES.

PHD THESIS, 1971. 166 P, 13 FIG, 21 TAB, 93 REF. OWRR A-004-NH(6).

DESCRIPTORS:
*AERATION, *EUTROPHICATION, *ECOLOGICAL DISTRIBUTION, *AQUATIC ALGAE, *CYANOPHYTA, *BEGGIATOA, PHYTOPLANKTON, *NEW HAMPSHIRE, MIXING, ENVIRONMENTAL EFFECTS.

IDENTIFIERS:
*ARTIFICIAL CIRCULATION, *KEZAR LAKE(NH).

ABSTRACT:
FOR SEVERAL YEARS ANNUAL NOXIOUS BLUE-GREEN ALGAL BLOOMS PLAGUED KEZAR LAKE, N. H. ARTIFICIAL CIRCULATION WAS TRIED IN AN ATTEMPT TO IMPROVE LAKE CONDITIONS. THIS STUDY WAS MADE TO HELP UNDERSTAND SOME OF THE ECOLOGICAL EFFECTS OF MIXING THE LAKE. SAMPLES COLLECTED WEEKLY WERE ANALYZED BY STANDARD METHODS. ARTIFICIAL CIRCULATION COMPLETELY DESTRATIFIED KEZAR LAKE AND ISOTHERMAL CONDITIONS WERE MAINTAINED THROUGHOUT THE TEST PERIODS OF 1968 AND 1969. THE SUPPLY AND DISTRIBUTION OF OXYGEN, CARBON DIOXIDE, ALKALINITY AND MANY ALGAL NUTRIENTS WERE MEASURED BEFORE AND AFTER LAKE MIXING. CONCENTRATIONS OF THE BLOOM-FORMING ALGAE OCCURRED AFTER LAKE MIXING IN BOTH 1968 AND 1969, BUT SECCHI DISK READINGS IMPROVED. CHLOROPHYLL CONCENTRATION IN RELATION TO BLOOMING OF A. FLOS=AQUAE WAS DETERMINED AND RATE OF CARBON FIXATION BY PHYTOPLANKTON WAS EXPLORED. A DISCUSSION OF THE RESULTS IS INCLUDED AND A COMPARISON MADE WITH THE FINDINGS OF OTHER AUTHORS.

FIELD 05C, 05G, 02H

ACCESSION NO. W72-09061

KAISER REFRACTORIES ENVIRONMENTAL STUDIES,

MOSS LANDING MARINE LABS. CALIF.

J. P. HARVILLE.

AVAILABLE FROM THE NATIONAL TECHNICAL INFORMATION SERVICE AS CCM-71-01107, $3.00 IN PAPER COPY, $0.95 IN MICROFICHE. REPORT NO. TP-71-3, NOAA-71112214, JUNE 1, 1971. 201 P, 61 FIG, 33 TAB, 31 REF. GRANT NO. GH-94.

DESCRIPTORS:
*INDUSTRIAL WASTES, *WATER POLLUTION EFFECTS, EFFLUENTS, CALIFORNIA, SEDIMENTS, SAMPLING, CHLOROPHYLL, SEDIMENT TRANSPORT, BENTHIC FAUNA, PLANKTON, BIOINDICATORS, CURRENTS(WATER), TIDES, WIND, LIGHT, WAVES, CLAMS, CRUSTACEANS, WORMS, MOLLUSKS, WATER TEMPERATURE, SALINITY, SEA WATER, WATER QUALITY, ON-SITE TESTS, HYDROGEN ION CONCENTRATION, CALCIUM, MAGNESIUM, ALGAE, CYANOPHYTA, DIATOMS, MUSSELS, PHYTOPLANKTON, LABORATORY STUDIES, MORTALITY, CRABS, BIOASSAY, LITTORAL DRIFT, INVERTEBRATES, CALCITE, SILICON, SODIUM, POTASSIUM, PERIPHYTON, ALUMINUM, IRON, QUARTZ, SEWAGE, ANNELIDS, TOPOGRAPHY, COPEPODS, ISOPODS, NEMATODES, GASTROPODS, ORGANIC MATTER, ZOOPLANKTON, FISH EGGS, AMPHIPODS, PERCHES, CHRYSOPHYTA, PYRROPHYTA, DINOFLAGELLATES, PROTOZOA, ROTIFERS, LARVAE, X-RAY DIFFRACTION.

IDENTIFIERS:
*MOSS LANDING, BARNACLES, LETHAL DOSAGE, CALCIUM ION CONCENTRATION, VERTEBRATES, ABSORBANCE, MAGNESIUM ION CONCENTRATION, BRUCITE, BOTTOM TOPOGRAPHY, PRIONOSPIO, LUMBRINEREIS, GLYCERA, NEPHTYS, CAPITELLA, MAGELONA, ARMANDIA, TELLINA BUTTONI, VENUS, TIVELA STULTORUM, EMERITA ANALOGA, DIASTYLOPSIS, OXYUROSTYLIS, HEMILAMPROPS, LAMPROPS, COLUROSTYLIS, CUMELLA, BATHYCUMA, LEPTOSTYLIS, LEUCON, BRACHYURA, ANOMURA, OLIVELLA, METZGERIA, NASSARIUS, VELUTINA, POLINICES, COLUS, CLINOCARDIUM, ZIRFAEA, COSCINODISCUS.

ABSTRACT:
A LONG-RANGE STUDY WAS CONDUCTED OF POTENTIAL EFFECTS OF THE KAISER REFRACTORIES INDUSTRIAL EFFLUENT ON THE MARINE ENVIRONMENT OF THE MOSS LANDING AREA. THE INVESTIGATIONS FOLLOWED THREE PRIMARY THRUSTS: (1) DETERMINATION OF PHYSICAL CHARACTERISTICS AND DYNAMICS OF THE WATER MASS IN THE PROPOSED OUTFALL AREA, AND ASSESSMENT OF BOTTOM STRUCTURE, SEDIMENTS, AND SEDIMENT TRANSPORT CHARACTERISTICS OF THE AREA. (2) BIOLOGICAL INVESTIGATIONS OF BOTTOM FAUNA OF THE PROPOSED OUTFALL AREA, WITH ANCILLARY STUDIES OF PLANKTON, FISHES, AND INTERTIDAL FAUNA. (3) FIELD AND LABORATORY STUDIES OF THE IMPACT OF VARIOUS DILUTIONS OF KAISER EFFLUENT UPON SELECTED BIOINDICATOR PLANT AND ANIMAL SPECIES. SELECTED CONCLUSIONS ARE: (A) INVESTIGATIONS OF BENTIC COMMUNITIES INDICATE GREAT VARIABILITY IN THE DISTRIBUTION AND POPULATION DENSITIES OF ORGANISMS. THE FOLLOWING GENERALIZATIONS APPEAR SIGNIFICANT: (1) THE PISMO CLAM (TIVELA STULTORUM), THE BUTTON CLAM (TELLINA BUTTONI), AND CERTAIN POLYCHAETE WORMS ARE DOMINANT. (2) CRUSTACEANS APPEAR TO CONSTITUTE AN IMPORTANT COMPONENT OF THE SURFACE FAUNA, AND POLYCHAETE WORMS USUALLY DOMINATE THE SEDIMENTS BELOW THE SURFACE. (3) DUMPING OF DREDGE SPOIL HAS AN IMMEDIATE DEVASTATING EFFECT UPON THE BENTHIC COMMUNITY, BUT THE COMMUNITY APPEARS TO RECOVER IN THE YEAR FOLLOWING. (B) THE KAISER EFFLUENT IS CHARACTERIZED CHEMICALLY BY HIGH PH, HIGH CALCIUM ION CONCENTRATIONS, LOW MAGNESIUM ION CONCENTRATIONS, AND IS CHARGED WITH A MILKY-WHITE PRECIPITATE WHICH IS PRINCIPALLY CALCITE (CACO3) AND BRUCITE (MG(OH)2). (C) THE KAISER EFFLUENT SHARPLY RESTRICTED THE AMOUNT AND VARIABILITY OF BIOTA. (D) THE PHOTOSYNTHETIC RATE OF MIXED PHYTOPLANKTON CULTURES INDICATED A LINEAR DEPRESSION OF THAT RATE BY THE KAISER EFFLUENT. (E) UNDILUTED EFFLUENT (APPROXIMATELY 1700 PPM CA IONS) WAS LETHAL TO COPEPODS TO VARIOUS DILUTIONS OF THE EFFLUENT FOR 24 HOUR PERIODS, BUT GRADUAL ADAPTATION TO UP TO 50 PERCENT EFFLUENT WAS POSSIBLE. (F) THE SIZE AND ABUNDANCE OF MUSSELS ARE DIRECTLY RELATED TO DISTANCES FROM THE KAISER OUTFALL. (G) THE COMMON SAND CRAB, EMERITA ANALOGA, IS SENSITIVE PHYSIOLOGICALLY TO ABNORMALLY HIGH CALCIUM-MAGNESIUM RATIOS. (H) THE PISMO CLAM, TIVELA STULTORUM, SHOWED BEHAVIORAL AND PHYSIOLOGICAL DAMAGE WHEN SUBJECTED TO HIGH CONCENTRATIONS OF KAISER EFFLUENT UNDER LABORATORY CONDITIONS. (MORTLAND-BATTELLE)

FIELD 05C, 05A

ACCESSION NO. W72-09092

RELATIONSHIP BETWEEN LIGHT CARBON DIOXIDE FIXATION AND DARK CARBON DIOXIDE FIXATION BY MARINE ALGAE,

NOVA UNIV., DANIA, FLA. PHYSICAL OCEANOGRAPHIC LAB.

I. MORRIS, C. M. YENTSCH, AND C. S. YENTSCH.

LIMNOLOGY AND OCEANOGRAPHY, VOL. 16, NO. 6, P 854-858, NOVEMBER 1971. 2 FIG, 2 TAB, 6 REF.

DESCRIPTORS:
*LIGHT INTENSITY, *PRIMARY PRODUCTIVITY, *RADIOACTIVITY TECHNIQUES, *PHOTOSYNTHESIS, CARBON DIOXIDE, PHYTOPLANKTON, ALGAE, CULTURES, *FLORIDA, SEA WATER, CHLOROPHYLL, NITRATES, NITRITES, PHOSPHATES, NUTRIENTS.

IDENTIFIERS:
*FLORIDA STRAIT, CARBON-14, DUNALIELLA TERTIOLECTA, PHAEODACTYLUM TRICORNUTUM.

ABSTRACT:
EXPERIMENTS WERE CONDUCTED ON NATURAL POPULATIONS (IN THE FLORIDA STRAIT) AND CULTURED MARINE PHYTOPLANKTON (DUNALIELLA TERTIOLECTA AND PHAEODACTYLUM TRICORNUTUM) TO DETERMINE RATES OF DARK FIXATION OF CARBON DIOXIDE. SUCH INFORMATION IS IMPORTANT TO A BETTER UNDERSTANDING OF THE CARBON-14 TECHNIQUE FOR ASSESSING PRIMARY PRODUCTIVITY. THE RATIO OF CARBON DIOXIDE FIXATION IN LIGHT TO THAT IN DARK INCREASES WITH INCREASING CONCENTRATIONS OF PHYTOPLANKTON. THIS IS TRUE IN BOTH NATURAL POPULATIONS AND CULTURES. POSSIBLE EXPLANATIONS OF THIS ARE PRESENTED AND DISCUSSED AND THE USE OF DARK FIXATION VALUES IN THE CARBON-14 TECHNIQUE IS CONSIDERED. THE C-14 TECHNIQUE IS ONLY AS GOOD AS THE KNOWLEDGE OF HOW THE CARBON IS FIXED. (SEE ALSO W72-09103) (MORTLAND-BATTELLE)

FIELD 05C, 05B

ACCESSION NO. W72-09102

THE PHYSIOLOGICAL STATE WITH RESPECT TO NITROGEN OF PHYTOPLANKTON FROM LOW-NUTRIENT SUBTROPICAL WATER AS MEASURED BY THE EFFECT OF AMMONIUM ION ON DARK CARBON DIOXIDE FIXATION,

NOVA UNIV., DANIA, FLA. PHYSICAL OCEANOGRAPHIC LAB.

I. MORRIS, C. M. YENTSCH, AND C. S. YENTSCH.

LIMNOLOGY AND OCEANOGRAPHY, VOL. 16, NO. 6, P 859-868, NOVEMBER 1971. 4 FIG, 3 TAB, 11 REF.

DESCRIPTORS:
*PHYTOPLANKTON, *NITROGEN, *NUTRIENTS, *PHOTOSYNTHESIS, LIGHT INTENSITY, AMMONIA, CARBON DIOXIDE, FLORIDA, CULTURES, *ALGAE, INCUBATION, CENTRIFUGATION, RADIOACTIVITY TECHNIQUES, CHLOROPHYLL, SEA WATER, IONS, NUTRIENT REQUIREMENTS, AQUATIC POPULATIONS, PRIMARY PRODUCTIVITY.

IDENTIFIERS:
CARBON-14, *FLORIDA STRAIT, DUNALIELLA TERIOLECTA, PHAEODACTYLUM TRICORNUTUM, SKELETONEMA COSTATUM, CARBOY, OSCILLATORIA ERYTHRAEA.

ABSTRACT:
STUDIES CONDUCTED WITH THREE SPECIES OF MARINE ALGAE (DUNALIELLA TERTIOLECTA, PHAEODACTYLUM TRICORNUTUM, AND SKELETONEMA COSTATUM) INDICATE THAT THERE IS NO INCREASE IN DARK FIXATION IN NON-NITROGEN DEFICIENT CULTURES AFTER THE ADDITION OF AMMONIUM ION. THE PRESENCE OF AMMONIUM IONS HAD NO EFFECT ON THE CAPACITY FOR LIGHT FIXATION OF CARBON-14. THE EXPERIMENTS WERE CONDUCTED IN LOW-NITROGEN, SUBTROPICAL WATERS OF THE FLORIDA STRAIT, AND THE RESULTS INDICATED THAT THE PHYTOPLANKTON CELLS WERE NOT PHYSIOLOGICALLY NITROGEN DEFICIENT, ALTHOUGH NITROGEN MAY LIMIT POPULATION SIZE. THE AMMONIUM ION ENHANCEMENT WHICH WAS USED OFFERS THE ADVANTAGE OF ASSESSING THE POPULATION AT TIME ZERO. WHEN ENHANCEMENT WAS NOT NOTED AT TIME ZERO, INCUBATION OF THE WATER IN A CARBOY RESULTED IN DEPLETION OF THE AVAILABLE NITROGEN AND A PHYSIOLOGICAL STATE OF NITROGEN DEFICIENCY SUBSEQUENTLY BECAME MEASURABLE BY THIS METHOD. THE EVENTUAL POSITIVE RESULTS FROM THE TEST THEREFORE CONFIRM THE NEGATIVE RESULTS AT TIME ZERO. (SEE ALSO W72-09102) (MORTLAND-BATTELLE)

FIELD 05C, 05A

ACCESSION NO. W72-09103

THE SEASONAL VARIATION OF DISSOLVED ORGANIC CARBON IN THE INSHORE WATERS OF THE MENAI STRAIT IN RELATION TO PRIMARY PRODUCTION,

UNIVERSITY COLL. OF NORTH WALES, MENAI BRIDGE. MARINE SCIENCE LABS.

A. W. MORRIS, AND P. FOSTER.

LIMNOLOGY AND OCEANOGRAPHY, VOL. 16, NO. 6, P 987-989, NOVEMBER 1971. 1 FIG, 6 REF.

DESCRIPTORS:
*ANALYTICAL TECHNIQUES, *ORGANIC COMPOUNDS, *CARBON, GRAVIMETRIC ANALYSIS, PRODUCTIVITY, SEASONAL, CHLOROPHYLL, ALGAE, PHYTOPLANKTON, WATER ANALYSIS, CARBON CYCLE, ORGANIC MATTER, NUTRIENTS.

IDENTIFIERS:
*DISSOLVED ORGANIC CARBON, *MENAI STRAIT(WALES).

ABSTRACT:
A REGULAR ANNUAL CYCLE FOR DISSOLVED ORGANIC CARBON WAS FOUND IN THE MENAI STRAIT FROM MEASUREMENTS MADE OVER A TWO-YEAR PERIOD. ULTRAVIOLET PHOTOOXIDATION OF 1-LITER SAMPLES OF WATER FOLLOWED BY GRAVIMETRIC ESTIMATION OF THE LIBERATED CARBON DIOXIDE WAS USED FOR DETERMINATIONS. WINTER LEVELS OF ABOUT ONE MG C/LITER ARE FOLLOWED BY A STEADY INCREASE THROUGH SPRING AND SUMMER, REACHING MAXIMUM VALUES OF 3-4 MG C/LITER IN AUTUMN FOLLOWED BY A SHARP RETURN TO WINTER VALUES. STUDIES INDICATED A HIGH RATE OF PRODUCTION IS MAINTAINED THROUGHOUT THE SUMMER PERIOD ALTHOUGH THE STANDING CROP MAY VARY CONSIDERABLY. (SNYDER-BATTELLE)

FIELD 05A, 05B

ACCESSION NO. W72-09108

DISTRIBUTION OF PHYTOPLANKTON IN A POLLUTED SALINE LAKE, ONONDAGA LAKE, NEW YORK,

CORNELL UNIV., ITHACA, N.Y. DIV. OF BIOLOGICAL SCIENCES.

P. SZE, AND J. M. KINGSBURY.

JOURNAL OF PHYCOLOGY, VOL. 8, NO. 1, P 25-37, MARCH 1972. 4 FIG, 2 TAB, 42 REF.

DESCRIPTORS:
*PHYTOPLANKTON, *DISTRIBUTION PATTERNS, *SYSTEMATICS, *SALINE LAKES, *EUTROPHICATION, WATER SAMPLING, WATER POLLUTION EFFECTS, BIOMASS, NUTRIENTS, ALGAE, EUGLENOPHYTA, CHRYSOPHYTA, CHLOROPHYTA, PYRROPHYTA, CYANOPHYTA, MANGANESE, ZINC, METALS, SECCHI DISKS, HEAVY METALS, BIOINDICATORS, SILICA, CHROMIUM, COPPER, *NEW YORK, DIATOMS, PHOSPHORUS, NITROGEN, CHLORELLA, CHLAMYDOMONAS, SCENEDESMUS, DINOFLAGELLATES, EUGLENA, E. COLI, PHOSPHATES, DISSOLVED OXYGEN, BIOCHEMICAL OXYGEN DEMAND, HYDROGEN ION CONCENTRATION, CALCIUM, SODIUM, POTASSIUM, MAGNESIUM, IRON.

IDENTIFIERS:
*ONONDAGA LAKE, OEDOGONIUM, RHIZOCLONIUM, COSMARIUM, PHACUS, CHAETOCEROS, MOUGEOTIA, NAVICULA, PINNULARIA, SURIRELLA, PERIDINIUM, SENECA RIVER, NINEMILE CREEK, VAN DORN SAMPLER, KEMMERER SAMPLER, CHLAMYDOMONAS EPIPHYTICA, CHLAMYDOMONAS SPP, CARTERIA FRITSCHII, PANDORINA MORUM, SPHAEROCYSTIS SCHROETERI, ULOTHRIX SPP, MICROTHAMNION KUETZINGIANUM, MICRACTINIUM PUSILLUM, ERRERELLA BORNHEMIENSIS.

ABSTRACT:
DURING 1969, ONONDAGA LAKE (NEW YORK) WAS REGULARLY SAMPLED AT FIVE SITES IN ORDER TO STUDY THE PHYTOPLANKTON. SAMPLES OBTAINED FOR QUANTITATIVE EXAMINATIONS WERE TAKEN FROM THE SURFACE WATERS AT ALL STATIONS AND AT DEPTHS OF 3, 6, AND 12 M AT TWO STATIONS USING DIPPERS AND VAN DORN OR KEMMERER SAMPLERS. SAMPLES FOR GROSS QUALITATIVE ANALYSES WERE OBTAINED BY MAKING VERTICAL AND HORIZONTAL HAULS WITH A NO. 20 NET AT EACH STATION. AN IBM 1130 COMPUTER WAS USED FOR ANALYZING THE RAW DATA. THE LAKE IS RELATIVELY SALINE AND HAS BEEN FOUND TO SUPPORT AN ALGAL FLORA CHARACTERISTIC OF A EUTROPHIC LAKE WITH AN ADMIXTURE OF SALINE SPECIES. SEASONS IN THE LAKE CAN BE IDENTIFIED BY FLORAL SUCCESSION WITH CERTAIN SPECIES OF ALGAE APPEARING FIRST AT THE OUTFLOW AND THEN SPREADING THROUGH THE LAKE. PHOSPHORUS AND NITROGEN WERE NEVER LIMITING, BUT SILICA DIMINUTION WAS LIMITING FOR FURTHER DIATOM POPULATION GROWTH AND WAS RELATED TO DIATOM BLOOMS. SINCE THE EUTROPHIC LAYER WAS MUCH SHALLOWER THAN THE THERMOCLINE, TURBULENCE AND MIXING WERE THOUGHT TO PLAY AN IMPORTANT ROLE IN CONTROLLING CERTAIN POPULATIONS. NO OBVIOUS CONTROLLING RELATIONSHIP EXISTED BETWEEN HERBIVORES AND PHYTOPLANKTON POPULATIONS. CR AN CU WERE HIGH AS A RESULT OF INDUSTRIAL DISCHARGES AND MAY BE RESPONSIBLE FOR INHIBITING BLOOMS. BIOMASS WAS CALCULATED FOR THE MAJOR PHYTOPLANKTERS. (HOLOMAN-BATTELLE)

FIELD 05C, 02H

ACCESSION NO. W72-09111

A LIST OF NEW GENERA AND TYPE SPECIES OF FLAGELLATES AND ALGAE PUBLISHED IN 1969, PART V,

B. V. SKVORTZOV.

HYDROBIOLOGIA, VOL 39, NO. 2, P 241-245, FEBRUARY 29, 1972. 15 FIG, 4 REF.

DESCRIPTORS:
*PROTOZOA, *SYSTEMATICS, *ALGAE, CHLOROPHYTA, HABITATS, AQUATIC HABITATS, EUGLENOPHYTA, PYRROPHYTA, FRESHWATER ALGAE.

IDENTIFIERS:
PROTOCRYPTOCHRYSIS OBOVATA, GUTTULA BACILLARIOPHAGA, REFRACTOCYSTIS PLUVIALIS, GOMESIAMONAS STAGNALIS, REFRACTODES BRASILIANA, OYEMONAS PULSULAE, REFRACTOMONAS BRASILIANA, HORTOBAGGIAMONAS PLICATA, TRICHOCYANELLA SPIRALIS, STROMIA SUBSPHAERICA, TETRACULAMONAS NATANS, MARINIAMONAS SAUPAULENSIS, ENEIDAMONAS APPLANATA, BICUDOMONAS CYANOPHORA, LIANGIANA TERRESTRIS, CHLOROMONADS, CRYPTOMONADS, BICUDOMONAS, *FLAGELLATES.

ABSTRACT:
A LIST OF 15 NEW GENERA OF FLAGELLATES AND ALGAE IS GIVEN. INCLUDED ARE ILLUSTRATIONS OF THE ORGANISMS, SIZES OF THE CELLS, TYPE SPECIES, AND THE HABITATS IN WHICH THEY MAY BE FOUND. (MACKAN-BATTELLE)

FIELD 05A, 05C

ACCESSION NO. W72-09119

SURVEY OF MACROPHYTE RESOURCES IN THE COASTAL WATER OF ALASKA,

ALASKA UNIV., COLLEGE. INST. OF MARINE SCIENCE.

C. P. MCROY, M. MUELLER, S. STOKER, J. J. GOERING, AND M. T. GOTTSCHALK.

AVAILABLE FROM THE NATIONAL TECHNICAL INFORMATION SERVICE AS COM-71-01141, $3.00 IN PAPER COPY, $0.95 IN MICROFICHE. REPORT NO. 3, R71-6, (NOAA-71101806), MAY 1971. 43 P, 14 FIG, 3 TAB, 40 REF.

DESCRIPTORS:
*MARINE PLANTS, *ALASKA, SAMPLING, *SYSTEMATICS, BENTHOS, SCUBA DIVING, DEPTH, DISTRIBUTION PATTERNS, AERIAL PHOTOGRAPHY, REMOTE SENSING, SATELLITES(ARTIFICIAL), ALGAE, CHLOROPHYTA, PHAEOPHYTA, RHODOPHYTA.

IDENTIFIERS:
*SEA WEEDS, *MACROPHYTES, *SEA GRASSES, TELEVISION, DEPTH SOUNDER, VERTICAL DISTRIBUTION, COLD BAY, BERMERS BAY, ALARIA, NEREOCYSTIS, LAMINARIA, THALASSIOPHYLLUM, FUCUS, SCHIZYMENIA, RHODOMELA, RHODOMENIA, MARMION ISLAND, MORRIA REEF, BEAR BAY, NEVA ISLAND, BAIRD ISLAND, SHOLIN ISLAND, WINIFRED ISLAND, IZEMBEK LAGOON, CHAETOMORPHA CANNABINA, CHAETOMORPHA MELAGONIUM, CLADOPHORA GLAUCESCENS, CLADOPHORA STIMPSONII.

ABSTRACT:
A PROJECT WAS INITIATED TO QUANTITATIVELY ASSESS NATURAL STOCKS OF SEAWEEDS AND SEAGRASSES IN ALASKA COASTAL WATERS. THE FIRST YEAR'S PROGRESS IS REPORTED. A TECHNIQUE UTILIZING SCUBA DIVERS AND SUBMARINE TELEVISION HAS BEEN DEVISED IN WHICH A DIVER MAKES PARALLEL TRANSECTS AT RIGHT ANGLES TO SHORE ON A SLED TOWED BY A SURFACE VESSEL OR WINCH. ON THE SLED IS ALSO MOUNTED THE TV CAMERA THAT MAKES A PERMANENT RECORD OF THE TRANSECT. A DEPTH PROFILE IS ALSO MADE WITH A RECORDING DEPTH SOUNDER MOUNTED ON A BOSTON WHALER. IN ADDITION THIS CAN BE USED TO IDENTIFY CERTAIN TYPES OF KELP BEDS. THIS TYPE OF TRANSECT IS RAPID AND GIVES A GOOD RECORD OF THE VERTICAL DISTRIBUTION OF THE MAJOR SPECIES. FOLLOWING THE VISUAL SURVEY, DIVERS TAKE REPLICATE SAMPLES OF ALL PLANTS WITHIN A QUADRAT. THE SURVEY TO DATE INDICATES THAT SEAWEED SPECIES IN THE FOLLOWING GENERA ATTAIN HIGH ENOUGH STANDING STOCKS IN ALASKA TO SUSTAIN EXPLOITATION: ALARIA, NEREOCYSTIS, LAMINARIA, THALASSIOPHYLLUM, FUCUS, SCHIZYMENIA, RHODOMELA, AND RHODOMENIA. REPRINTS OF JOURNAL ARTICLES COVERING THIS RESEARCH AND A SURVEY OF EELGRASS DISTRIBUTION ON THE ALASKA COAST ARE INCLUDED. (MORTLAND-BATTELLE)

FIELD 02L, 05A

ACCESSION NO. W72-09120

CARBON DIOXIDE AND PRIMARY PRODUCTIVITY IN THE GLACIAL FIORD SYSTEM OF SOUTHEAST ALASKA,

ALASKA UNIV., COLLEGE. INST. OF MARINE SCIENCE.

L. L. LONGERICH, M. BILLINGTON, V. ALEXANDER, J. J. KELLEY, AND D. W. HOOD.

AVAILABLE FROM THE NATIONAL TECHNICAL INFORMATION SERVICE AS AD-734 672, $3.00 IN PAPER COPY, $0.95 IN MICROFICHE. REPORT NO. R71-19, SEPTEMBER 1971. 24 P, 1 FIG, 1 TAB, 5 REF. ONR CONTRACT N00014-67-AC-317-0001AB.

DESCRIPTORS:
*CARBON DIOXIDE, *PRIMARY PRODUCTIVITY, ALASKA, FIORDS, SURFACE WATERS, NITROGEN, SEA WATER, WINDS, SAMPLING, INCUBATION, RUNOFF, SALINITY, DEPTH, ABSORPTION, ALGAE, SEASONAL, WEATHER, WATER ANALYSIS, TRACERS, RADIOACTIVITY TECHNIQUES.

IDENTIFIERS:
AUKE BAY, GLACIAL WATER, BECKMAN ANALYZER, VAN DORN SAMPLERS, C-14, BIOCARBONATES.

ABSTRACT:
EARLY SPRING STUDIES OF CARBON DIOXIDE CONCENTRATIONS AND PRIMARY PRODUCTIVITY WERE MADE IN THE GLACIAL FIORD SYSTEM OF SOUTHEASTERN ALASKA. THE SURFACE WATERS OF AUKE BAY AT THE BEGINNING OF THE SPRING BLOOM HAD A LOWER CARBON DIOXIDE CONCENTRATION AND HIGHER PARTICULATE NITROGEN AND CARBON-14-HCO3 UPTAKE THAN THE MORE OPEN WATERS SURROUNDING IT. SURFACE WATER CARBON DIOXIDE IN A GLACIAL OUTFLOW AREA WAS UNEXPECTEDLY LOW COMPARED TO VALUES FOR SURFACE SEAWATER IN A NORMAL FRESH WATER STREAM OUTFLOW. AN EXAMPLE OF WIND INDUCED UPWELLING WAS NOTED WHICH PRODUCED INCREASED SURFACE CARBON DIOXIDE CONCENTRATIONS OVER THE DURATION OF A STORM. IT WAS EVIDENT FROM PREVIOUS CARBON DIOXIDE STUDIES AND FROM THIS SURVEY THAT SURFACE CARBON DIOXIDE CONCENTRATIONS ARE DEPENDENT ON LOCATION, PRIMARY PRODUCTIVITY, PROXIMITY TO FRESH WATER RUNOFF, THE SEASON, AND THE WEATHER. PAST WORK SHOWS THAT LONG-TERM, ON-SITE MULTIDISCIPLINARY OBSERVATIONS ARE ESSENTIAL TO AN UNDERSTANDING OF HOW CARBON DIOXIDE INTERACTS IN THE MARINE ENVIRONMENT. (MORTLAND-BATTELLE)

FIELD 05C, 05A, 02C

ACCESSION NO. W72-09122

NUTRIENTS AND EUTROPHICATION: THE LIMITING NUTRIENT CONTROVERSY.

AMERICAN SOCIETY OF LIMNOLOGY AND OCEANOGRAPHY SPECIAL SYMPOSIA, VOL I, ALLEN PRESS, LAWRENCE, KANSAS, 1972. 328 P. OWRR X-109(NO. 3415)(1).

DESCRIPTORS:
*EUTROPHICATION, *CARBON, *PHOSPHORUS, NITROGEN, ALGAE.

IDENTIFIERS:
*LIMITING NUTRIENTS.

ABSTRACT:
THE INTENT OF THE AMERICAN SOCIETY OF LIMNOLOGY AND OCEANOGRAPHY TO PROVIDE AN OPEN FORUM FOR COMMUNICATION BETWEEN ACADEME, STATE AND FEDERAL AGENCIES, AND INDUSTRY, CULMINATED IN THIS SYMPOSIUM HELD AT W. K. KELLOGG BIOLOGICAL STATION OF MICHIGAN STATE UNIVERSITY ON FEBRUARY 11-12, 1971. IT WAS THE HOPE TO PROVIDE A CLEAR STATEMENT OF CURRENT IDEAS RELATIVE TO THE IMPORTANCE OF VARIOUS REGULATING OR LIMITING NUTRIENTS IN THE EUTROPHICATION OF AQUATIC ECOSYSTEMS. ONLY ONE FACTOR--NUTRIENT AVAILABILITY--WAS SELECTED AS THE SUBJECT. THE FOCUS WAS ON THE RELATIVE IMPORTANCE OF CARBON AND PHOSPHORUS IN REGULATING EUTROPHICATION. BECAUSE OF THE RECENT CARBON VS. PHOSPHORUS CONTROVERSY (CENTERED ON THE PROPOSAL THAT CARBON RATHER THAN PHOSPHORUS OR NITROGEN LIMITS ALGAL PRODUCTIVITY) THIS SUBJECT CARRIES POLITICAL AND ECONOMIC OVERTONES. SINCE PHOSPHORUS IN DETERGENTS IS LINKED TO CULTURAL EUTROPHICATION, THE CONTROVERSY IS NOW EMOTIONALLY CHARGED FOLLOWING LEGISLATIVE PROPOSALS TO REMOVE PHOSPHORUS FROM DETERGENT FORMULATIONS. THE PROCEEDINGS OF THIS SYMPOSIUM WHERE IDEAS AND DATA WERE OPENLY AND AUTHORITATIVELY DISCUSSED, QUESTIONED, AND DEBATED, ARE INTENDED TO PROVIDE THE PUBLIC AND POLITICIANS WITH SOME USEFUL GUIDELINES TO THIS PROBLEM. IT IS UP TO EACH READER TO EVALUATE THE CONTROVERSY AND TO DETERMINE WHETHER THERE WAS ONE. (SEE W72-09159 THRU W72-09171) (AUEN-WISCONSIN)

FIELD 05C

ACCESSION NO. W72-09155

THE INTERRELATION OF CARBON AND PHOSPHORUS IN REGULATING HETEROTROPHIC AND AUTOTROPHIC POPULATIONS IN AN AQUATIC ECOSYSTEM, SHRINER'S POND,

FEDERAL WATER POLLUTION CONTROL ADMINISTRATION, ATHENS, GA. SOUTHEAST WATER LAB.

P. C. KERR, D. L. BROCKWAY, D. F. PARIS, AND J. T. BARNETT, JR.

IN: NUTRIENTS AND EUTROPHICATION: THE LIMITING NUTRIENT CONTROVERSY, P 41-62, 9 FIG, 58 REF. AMERICAN SOCIETY OF LIMNOLOGY AND OCEANOGRAPHY SPECIAL SYMPOSIA VOL I, ALLEN PRESS, LAWRENCE, KANSAS, 1972.

DESCRIPTORS:
*CARBON, *PHOSPHORUS, *POPULATION, AQUATIC ENVIRONMENT, ECOSYSTEMS, GEORGIA, BIORTHYTHMS, LABORATORY TESTS, ALGAE, CARBON DIOXIDE, CYCLING NUTRIENTS, BICARBONATES, PLANT GROWTH, DURNAL.

IDENTIFIERS:
*LIMITING NUTRIENTS, POTASSIUM CHLORIDE, SHRINER'S POND(GA).

ABSTRACT:
ANY OF THE TYPES OF BIOLOGICAL REGULATION--PHYSICAL, CHEMICAL, GENETIC, AND NUTRITIONAL--CAN REGULATE BIOLOGICAL ACTIVITY OF ORGANISMS AT DIFFERENT TIMES AND PLACES. SHRINER'S POND, GEORGIA WAS STUDIED TO ASCERTAIN CHEMICAL AND BIOLOGICAL CHANGES ASSOCIATED WITH ADDITION OF NITROGEN AND PHOSPHORUS OR CARBON DIOXIDE. WITHIN 12 HOURS AFTER REAGENT-GRADE NITROGEN, PHOSPHORUS, AND POTASSIUM CHLORIDE WERE ADDED, THE BACTERIAL POPULATION INCREASED. ALGAL POPULATION INCREASED 36-48 HOURS AFTER FERTILIZATION. DIEL CYCLING OF CARBON DIOXIDE AND BICARBONATE WAS MEASURED. ALTHOUGH THESE DATA DO NOT CLEARLY INDICATE THAT INCREASED AVAILABILITY OF INORGANIC CARBON WAS RESPONSIBLE FOR INITIATION OF ALGAL GROWTH, CONTINUED BIOLOGICAL PRODUCTION OF CARBON DIOXIDE APPEARED TO PROLONG THE BLOOM DURATION. DATA INDICATE MORE NITROGEN AND PHOSPHORUS WERE REMOVED DURING NIGHT THAN DURING DAY; THE HIGHER REMOVAL WAS ASSOCIATED WITH OXYGEN REMOVAL, DECREASING PH, AND ACCUMULATION OF CARBON DIOXIDE AND BICARBONATE, INDICATING HETEROTROPHIC ACTIVITY. CARBON DIOXIDE WAS REMOVED AND BICARBONATE WAS DEPLETED FROM THE WATER DURING LIGHT HOURS. ALGAL GROWTH WAS STIMULATED BY BUBBLING 5% AND 0.03% CARBON DIOXIDE IN AIR THROUGH WATER. THESE EXPERIMENTS INDICATE IMPORTANCE OF CARBON IN REGULATING ALGAL GROWTH. (SEE ALSO W72-09155) (JONES-WISCONSIN)

FIELD 05C

ACCESSION NO. W72-09159

CARBON LIMITATION IN SEWAGE LAGOONS,

MISSOURI UNIV., COLUMBIA. DEPT. OF CIVIL ENGINEERING.

D. L. KING.

IN: NUTRIENTS AND EUTROPHICATION: THE LIMITING NUTRIENT CONTROVERSY, P 98-110, 8 FIG, 16 REF. AMERICAN SOCIETY OF LIMNOLOGY AND OCEANOGRAPHY SPECIAL SYMPOSIA VOL I, ALLEN PRESS, LAWRENCE, KANSAS, 1972.

DESCRIPTORS:
*CARBON, *SEWAGE LAGOONS, *LIMITING FACTORS, ALGAE, PRODUCTIVITY, PHOTOSYNTHESIS, NUTRIENTS, EUTROPHICATION, CYANOPHYTA, ALKALINITY, DIURNAL, NITROGEN.

IDENTIFIERS:
*LIMITING NUTRIENTS.

ABSTRACT:
OBSERVATIONS OF SEWAGE LAGOONS OFFER SIGNIFICANT INSIGHT INTO THE PROCESSES INVOLVED IN LAKE EUTROPHICATION. IT IS CONCLUDED THAT BOTH THE QUALITATIVE AND QUANTITATIVE ASPECTS OF EUTROPHICATION MUST BE CONSIDERED. ADDITIONS OF REQUIRED ALGAL NUTRIENTS TO A LAKE ALLOW INCREASES IN QUANTITY OF ALGAE AND CAN STRAIN THE CARBON AVAILABILITY TO THE POINT WHERE THERE IS ALSO A CHANGE IN ALGAL QUALITY. ESTABLISHMENT OF SUMMER BLUE-GREEN ALGAL BLOOMS USUALLY IS OF GREATER CONCERN THAN THE PRECURSORY INCREASE IN QUANTITY OF THE MORE DESIRABLE ALGAL FORMS. CALCULATIONS SUGGEST THAT AMOUNTS OF ALGAL NUTRIENTS, OTHER THAN CARBON, REQUIRED TO PROMOTE SUMMER BLOOMS OF BLUE-GREEN ALGAE ARE DETERMINED BY ALKALINITY OF THE WATER IN QUESTION. ATTEMPTS TO CONTROL EUTROPHICATION BY LIMITING JUST CARBON AVAILABILITY PROBABLY WOULD RESULT IN ESTABLISHMENT OF BLUE-GREEN ALGAE DOMINANCE BUT PERHAPS IN LOWER QUANTITIES. THE BLUE-GREEN ALGAE PROBABLY WOULD ACCELERATE AND RATE OF CARBON DIOXIDE GAIN FROM THE ATMOSPHERE. ATTEMPTS TO LIMIT AVAILABLE NITROGEN MAY RESULT IN ESTABLISHMENT OF BLOOMS BLUE-GREENS WHICH CAN FIX ATMOSPHERIC NITROGEN. LIMITATION OF PHOSPHORUS APPEARS TO OFFER THE BEST CHANCE OF CONTROLLING BOTH QUALITATIVE AND QUANTITATIVE ASPECTS OF EUTROPHICATION. (SEE ALSO W72-09155) (JONES-WISCONSIN)

FIELD 05C

ACCESSION NO. W72-09162

CHARACTERIZATION OF PHOSPHORUS-LIMITED PLANKTON ALGAE,

NEW YORK STATE DEPT. OF HEALTH, ALBANY. ENVIRONMENTAL HEALTH CENTER.

G. W. FUHS, S. C. DEMMERLE, E. CANELLI, AND M. CHEN.

IN: NUTRIENTS AND EUTROPHICATION: THE LIMITING NUTRIENT CONTROVERSY, P 113-133, 16 FIG, 3 TAB, 59 REF. AMERICAN SOCIETY OF LIMNOLOGY AND OCEANOGRAPHY SPECIAL SYMPOSIA VOL I, ALLEN PRESS, LAWRENCE, KANSAS, 1972.

DESCRIPTORS:
*PHOSPHORUS, *PLANKTON, *ALGAE, *LIMITING FACTORS, GROWTH RATES, DIATOMS, BACTERIA, LAKES, NITROGEN, ANALYTICAL TECHNIQUES, CARBON, EUTROPHICATION, STANDING CROPS, LABORATORY TESTS, CHEMICAL ANALYSIS, BIOASSAY, CYTOLOGICAL STUDIES, NUTRIENTS, ENZYMES.

IDENTIFIERS:
*LIMITING NUTRIENTS, CANADARAGO LAKE(NY), LAKE GEORGE(NY).

ABSTRACT:
GROWTH RATE OF MICROORGANISMS AS A FUNCTION OF CONCENTRATION OF PHOSPHORUS SOURCE IS CALCULATED FROM CONTINUOUS CULTURE AND PHOSPHORUS UPTAKE EXPERIMENTS. FOR TWO DIATOMS AND THREE BACTERIA, VALUES OF THE MINIMUM PHOSPHORUS CONTENT, GROWTH RATE WITH PHOSPHORUS NONLIMITING, THE MICHAELIS CONSTANT, AND RATE FOR UPTAKE OF THE PHOSPHORUS SOURCE ARE GIVEN. PHOSPHORUS CONTENT OR ORGANISMS DEPENDS ON PROTOPLASMA VOLUME ALTHOUGH MAXIMUM GROWTH RATES VARY WIDELY. MAXIMUM UPTAKE RATES PER UNIT AREA OF CELL SURFACE ARE SIMILAR. BACTERIA WITH MORE FAVORABLE SURFACE-TO-VOLUME RATIO MAY OUTGROW ALGAE EVEN THOUGH THEY SHOW LOWER AFFINITY TOWARD ORTHOPHOSPHATE. PHOSPHORUS-LIMITED DIATOMS SHOW INCREASES IN MEAN CELL VOLUME, CELL CARBON, REFRACTILITY OF CELLS, AND ALKALINE PHOSPHATASE CONTENT, WHEREAS CELL NITROGEN AND CELL PROTEIN DECREASE. THE CARBON:PHOSPHORUS ATOMIC RATIO REFLECTS AVAILABILITY OF PHOSPHORUS AND, ALTHOUGH LESS DRAMATICALLY, SO DOES THE NITROGEN:PHOSPHORUS RATIO. THESE RATIOS ARE IMPORTANT IN DETECTING NUTRIENT LIMITATION. FOR PRESERVATION OF CANADARAGO LAKE AND LAKE GEORGE, NEW YORK FEASIBILITY OF REMOVING LIMITING OR NEAR-LIMITING ELEMENTS FROM TRIBUTARY WATERS SHOULD BE CONSIDERED. KINETIC MODELS AND MEASUREMENT OF SPECIES CONSTANTS UNDER LABORATORY AND FIELD CONDITIONS ARE REQUIRED FOR PREDICTION OF SHORT-TERM CHANGES AND TURNOVER OF LAKE BIOMASS. (SEE ALSO W72-09155) (JONES-WISCONSIN)

FIELD 05C

ACCESSION NO. W72-09163

ALGAL RESPONSES TO NUTRIENT ADDITIONS IN NATURAL WATERS. I. LABORATORY ASSAYS,

ENVIRONMENTAL PROTECTION AGENCY, CORVALLIS, OREG.

T. E. MALONEY, W. E. MILLER, AND T. SHIROYAMA.

IN: NUTRIENTS AND EUTROPHICATION: THE LIMITING NUTRIENT CONTROVERSY, P 134-140, 12 FIG, 2 TAB, 8 REF. AMERICAN SOCIETY OF LIMNOLOGY AND OCEANOGRAPHY SPECIAL SYMPOSIA VOL I, ALLEN PRESS, LAWRENCE, KANSAS, 1972.

DESCRIPTORS:
*ALGAE, *NUTRIENTS, *LABORATORY TESTS, *BIOASSAY, OREGON, LAKES, NITROGEN, PHOSPHORUS, CARBON, GROWTH RATES, OLIGOTROPHY, EUTROPHICATION.

IDENTIFIERS:
*LIMITING NUTRIENTS.

ABSTRACT:
IN THE FALL OF 1970 WATERS OF NINE OREGON LAKES OF VARYING WATER QUALITY WERE STUDIED IN A SERIES OF LABORATORY ALGAL ASSAYS ON A QUARTERLY BASIS TO DETERMINE EFFECTS OF SEASONAL CHANGES AND OF WATER CHEMISTRY UPON ALGAL GROWTH AND TO EVALUATE POTENTIAL EFFECTS ON ALGAL GROWTH OF VARIOUS NUTRIENT ADDITIONS. SELANASTRUM CAPRICORNUTUM WAS USED AS THE TEST SPECIES. ADDITIONS OF NITROGEN, PHOSPHORUS, AND CARBON, SINGLY AND IN COMBINATION, WERE MADE TO THE WATERS AND ALGAL GROWTH RATES WERE DETERMINED. IN FOUR OF THE WATERS ADDITION OF PHOSPHORUS ALONE GREATLY STIMULATED ALGAL GROWTH AND IN TWO OF THE LAKE WATERS ADDITION OF NITROGEN ALONE SLIGHTLY STIMULATED ALGAL GROWTH RATES. THREE OF THE TEST WATERS WERE CAPABLE OF SUPPORTING RELATIVELY HIGH ALGAL GROWTH RATES WITHOUT NUTRIENT ADDITIONS, AND IN ONE HIGHLY OLIGOTROPHIC LAKE WATER THE ADDITION OF NITROGEN, PHOSPHORUS, AND CARBON HAD NO EFFECT ON ALGAL GROWTH RATES. IN ALL CASES ALGAL GROWTH RATES WERE DIRECTLY PROPORTIONAL TO THE AMOUNTS OF DISSOLVED PHOSPHORUS IN THE WATERS, BUT THERE WAS NO OBVIOUS CORRELATION BETWEEN ALGAL GROWTH RATES AND CONCENTRATIONS OF NITROGEN AND CARBON. (SEE ALSO W72-09155) (JONES-WISCONSIN)

FIELD 05C

ACCESSION NO. W72-09164

ALGAL RESPONSES TO NUTRIENT ADDITIONS IN NATURAL WATERS. II. FIELD EXPERIMENTS,

PACIFIC NORTHWEST WATER LAB., CORVALLIS, OREG.; AND SHAGAWA LAKE EUTROPHICATION CONTROL PROJECT, ELY, MINN.

C. F. POWERS, D. W. SCHULTS, K. W. MALUEG, R. M. BRICE, AND M. D. SCHULDT.

IN: NUTRIENTS AND EUTROPHICATION: THE LIMITING NUTRIENT CONTROVERSY, P 141-156, 9 FIG, 5 TAB, 6 REF. AMERICAN SOCIETY OF LIMNOLCGY AND OCEANOGRAPHY SPECIAL SYMPOSIA, VOL I, ALLEN PRESS, LAWRENCE, KANSAS, 1972.

DESCRIPTORS:
*EUTROPHICATION, *ON-SITE TESTS, *NUTRIENTS, ALGAE, LAKES, PHOSPHORUS, CARBON, NITROGEN.

IDENTIFIERS:
*LIMITING NUTRIENTS, SHAGAWA LAKE(MINN), WALDO LAKE(ORE).

ABSTRACT:
RESULTS OF THE IN SITU ASSAY EXPERIMENTS HAVE SOME DEFINITE IMPLICATIONS APPLICABLE TO EUTROPHICATION AND LAKE RESTORATION. WHILE EUTROPHIC SHAGAWA LAKE, MINNESOTA RECEIVES ABOUT 80% OF LOW NUTRIENT CONTENT AND EXCELLENT QUALITY WATER FROM BURNTSIDE RIVER ABOUT 70% OF THE TOTAL PHOSPHORUS AND 25% OF THE TOTAL NITROGEN ORIGINATES FROM DISCHARGES OF THE ELY MUNICIPAL WASTE TREATMENT PLANT. NEITHER PHOSPHORUS NOR NITROGEN ALONE STIMULATED ALGAL GROWTH IN BURNTSIDE WATER; CARBON LIKEWISE HAD NO POSITIVE EFFECT. ALL DISSOLVED MATERIALS WERE PRESENT IN EXTREMELY LOW CONCENTRATIONS IN PRISTINE WALDO LAKE, OREGON; A POSITIVE RESPONSE TO PHOSPHORUS ENRICHMENT ON THREE OF THE FOUR TESTING DATES INDICATED THAT INCREASING RATES OF PRODUCITIVITY MIGHT RAPIDLY RESULT FROM RELATIVELY SMALL ADDITIONS OF THIS OBVIOUSLY CRITICAL ELEMENT. AT NO TIME DID THIS LAKE RESPOND TO ADDITIONS OF EITHER NITROGEN OR CARBON. TESTS AT MODERATELY PRODUCTIVE TRIANGLE LAKE, OREGON AND EUTROPHIC CLINE'S POND SHOWED VARIABLE REACTIONS TO NUTRIENT ENRICHMENT SERVING TO POINT UP THE IMPORTANCE OF NUMEROUS OBSERVATIONS OVER DIFFERENT SEASONS. FREQUENTLY MORE THAN ONE NUTRIENT ELEMENT APPEARS TO BE LIMITING PHYTOPLANKTON AT THE SAME TIME. EVEN THOUGH POSITIVE GROWTH RESPONSE OCCASIONALLY FOLLOWED ADDITIONS OF OTHER NUTRIENTS, PHOSPHORUS WAS THE MOST FREQUENTLY LIMITING. (SEE ALSO W72-09155) (AUEN-WISCONSIN)

FIELD 05C

ACCESSION NO. W72-09165

NUTRIENTS AND PHYTOPLANKTON IN LAKE WASHINGTON,

WASHINGTON UNIV., SEATTLE. DEPT. OF ZOOLOGY.

W. T. EDMONDSON.

IN: NUTRIENTS AND EUTROPHICATION: THE LIMITING NUTRIENT CONTROVERSY, P 172-193, 8 FIG, 3 TAB, 34 REF. AMERICAN SOCIETY OF LIMNOLOGY AND OCEANOGRAPHY SPECIAL SYMPOSIA VOL 1, ALLEN PRESS, LAWRENCE, KANSAS, 1972.

DESCRIPTORS:
*NUTRIENTS, *PHYTOPLANKTON, WATER POLLUTION EFFECTS, EUTROPHICATION, PRODUCTIVITY, ALGAE, PHOSPHATES, NITROGEN, SEWAGE, CARBON DIOXIDE, SEASONAL, PHOSPHORUS, WATER POLLUTION SOURCES, BIOCHEMICAL OXYGEN DEMAND, WATER POLLUTION CONTROL, NITRATES, EFFLUENTS, DETERGENTS, SODIUM CHLORIDE, CHEMICAL PROPERTIES, SECCHI DISKS, METHODOLOGY.

IDENTIFIERS:
*LIMITING NUTRIENTS, *LAKE WASHINGTON(WASH).

ABSTRACT:
LAKE WASHINGTON RECEIVED INCREASING AMOUNTS OF RAW SEWAGE UNTIL A DIVERSION SYSTEM TO PUGET SOUND WAS COMPLETED ABOUT 1963. INFORMATION ABOUT THE LAKE'S CHANGES PERTINENT TO THE RELATIVE IMPORTANCE OF DIFFERENT ELEMENTS IN CONTROL OF PRODUCTIVITY IN NATURAL LAKES IS SUMMARIZED. RESULTING FROM ENRICHMENT (1941 TO 1963) DUE TO EFFLUENT FROM SECONDARY SEWAGE TREATMENT PLANTS, PRODUCTION AND ABUNDANCE OF ALGAE INCREASED SEVERAL FOLD. THE WINTER PHOSPHATE CONCENTRATION INCREASED PROPORTIONALLY MUCH MORE THAN DID NITRATE OR CARBON DIOXIDE. AFTER DIVERSION, BY 1969, WINTER PHOSPHATE DECREASED TO 28% OF ITS 1963 VALUE, SUMMER CHLOROPHYLL CONCENTRATIONS DECREASED ABOUT AS MUCH, BUT NITRATE AND CARBON DIOXIDE FLUCTUATED YEAR TO YEAR AT RELATIVELY HIGH VALUES. THUS PHOSPHORUS AND PHYTOPLANKTON SHOWED THE MAJOR CHANGE, NOT NITRATE OR CARBON DIOXIDE. SIMILARLY AT SUMMER'S END, PHOSPHATE CONCENTRATION IN HYPOLIMNION CHANGED MORE THAN DID NITRATE AND CARBON DIOXIDE. THIS IS CONSISTENT WITH SECONDARY EFFLUENT, WHICH CONTAINS MORE PHOSPHORUS THAN NITROGEN OR CARBON. THE BIOLOGICAL OXYGEN DEMAND OF EFFLUENTS REACHING THE HYPOLIMNION IS SMALL RELATIVE TO OBSERVED CHANGES IN HYPOLIMNETIC OXYGEN. PHYTOPLANKTON WAS NOT APPRECIABLY CHANGED BY A SMALL SALT WATER INVASION. EFFECTIVENESS OF A GIVEN PHOSPHATE SUPPLY MAY BE AFFECTED BY NITRATE. (SEE ALSO W72-09155) (JONES-WISCONSIN)

FIELD 05C

ACCESSION NO. W72-09167

NATURAL CARBON SOURCES, RATES OF REPLENISHMENT, AND ALGAL GROWTH,

WARF INST., INC., MADISON, WIS.

S. D. MORTON, R. SERNAU, AND P. H. DERSE.

IN: NUTRIENTS AND EUTROPHICATION: THE LIMITING NUTRIENT CONTROVERSY, P 197-204, 2 FIG, 3 TAB, 15 REF. AMERICAN SOCIETY OF LIMNOLOGY AND OCEANOGRAPHY SPECIAL SYMPOSIA VOL I, ALLEN PRESS, LAWRENCE, KANSAS, 1972.

DESCRIPTORS:
*CARBON, *ALGAE, *EUTROPHICATION, *GROWTH RATES, CHLORELLA, CARBON DIOXIDE, BICARBONATES, ALGAL CONTROL, AERATION.

IDENTIFIERS:
*LIMITING NUTRIENTS, ANABAENA, CARBON SOURCES.

ABSTRACT:
IN RESEARCH ORIGINALLY UNDERTAKEN TO DETERMINE WHETHER EUTROPHICATION IN SMALL AREAS COULD BE CONTROLLED BY SWEEPING CARBON DIOXIDE OUT OF WATER BY AERATION WITH AIR CONTAINING LOW CARBON DIOXIDE AMOUNTS, THREE AREAS WERE STUDIED: THE STEADY STATE (ALGAL GROWTH RATES AT VARIOUS CONSTANT, MAINTAINED, DISSOLVED CARBON DIOXIDE CONCENTRATIONS), THE NONEQUILIBRIUM CASE (NATURAL ATMOSPHERIC REPLENISHMENT THE SOLE CARBON SOURCE), AND ALGAL GROWTH WITH INORGANIC BICARBONATE AS SOLE CARBON SOURCE. CARBON AVAILABILITY WAS STUDIED BY OBSERVATIONS OF GROWTH RATES OF CHLORELLA, MYCROCYSTIS, AND ANABAENA. ALGAE CAN EFFICIENTLY UTILIZE CARBON DIOXDE AT CONCENTRATIONS MUCH LOWER THAN THOSE PRESENT FROM ATMOSPHERIC EQUILIBRIA. IT IS VERY DIFFICULT TO CONTROL GROWTH BY CARBON DIOXIDE CONTROL IN SYSTEMS OPEN TO THE ATMOSPHERE. BICARBONATE IS A GOOD SOURCE OF CARBON AND IS AT LEAST 50% UTILIZED AT GROWTH RATES OF AT LEAST 5 MG/L PER DAY. MAY LAKES CAN HAVE MASSIVE ALGAL BLOOMS, USING NATURALLY PRESENT BICARBONATE AS SOLE CARBON SOURCE. THE ATMOSPHERE, WITHOUT ANY VIGOROUS OR CONTINUOUS WIND MIXING, IS AN ADEQUATE SOURCE OF CARBON DIOXIDE FOR DEPTHS TO AT LEAST 1.7 M, PERMITTING ALGAL GROWTH RATES UP TO 2 MG/L PER DAY. (SEE ALSO W72-09155) (JONES-WISCONSIN)

FIELD 05C

ACCESSION NO. W72-09168

DETERGENTS: NUTRIENT CONSIDERATIONS AND TOTAL ASSESSMENT,

PROCTER AND GAMBLE CO., CINCINNATI, OHIO. ENVIRONMENTAL WATER QUALITY RESEARCH DEPT.

J. R. DUTHIE.

IN: NUTRIENTS AND EUTROPHICATION: THE LIMITING NUTRIENT CONTROVERSY, P 205-216, 2 FIG, 24 REF. AMERICAN SOCIETY OF LIMNOLOGY AND OCEANOGRAPHY SPECIAL SYMPOSIA VOL I, ALLEN PRESS, LAWRENCE, KANSAS, 1972.

DESCRIPTORS:
*DETERGENTS, *PHOSPHATES, *EUTROPHICATION, NUTRIENTS, ASSESSMENTS, ALGAE, FORMULATION, WASTE TREATMENT, INDUSTRIES, WATER QUALITY, TECHNOLOGY, SURFACTANTS, TOXICITY, HAZARDS, BIODEGRADATION.

IDENTIFIERS:
NITRILOTRIACETATE.

ABSTRACT:
PROCTOR AND GAMBLE INTENDS TO TAKE PHOSPHATES OUT OF LAUNDRY DETERGENTS AS RAPIDLY AS POSSIBLE AND IN A THOROUGHLY REASONABLE MANNER. SEARCH FOR AND QUALIFICATION OF A PHOSPHATE REPLACEMENT HAS BEEN AND CONTINUES TO BE THE PRIME TECHNICAL OBJECTIVE OF THEIR INDUSTRY. THE DETERGENT INDUSTRY DOES NOT EXPECT THOSE IN THE LIMNOLOGY FIELD TO SHARE EQUALLY THE BURDEN OF FINDING ALTERNATIVES FOR PHOSPHATES, BUT THERE ARE CERTAIN THINGS THAT SCIENTISTS CAN DO INDIVIDUALLY AND AS A GROUP TO CREATE THE CLIMATE IN WHICH RESPONSIBLE AND PRODUCTIVE CHANGES CAN BE MADE. WHEN TALKING TO LAYMEN, DETERGENT PHOSPHATES CAN BE PUT INTO PROPER PERSPECTIVE MAKING IT CLEAR THAT PHOSPHATE IS AN IMPORTANT DETERGENT INGREDIENT AND, ALTHOUGH AN ALGAL NUTRIENT, IT IS NOT A POLLUTANT OR POISON. ENVIRONMENTAL ASSESSMENTS OF NTA, NOW UNDER WAY, SHOULD BE COMPLETED. NTA STILL REPRESENTS A VERY REAL HOPE AS A PHOSPHATE REPLACEMENT; IT IS THE MOST TESTED REPLACEMENT MATERIAL. IF, AS EXPECTED, FURTHER TESTING CLEARS THE TERATOLOGY QUESTIONS RAISED FROM THE PRELIMINARY TESTING OF EXAGGERATED LEVELS OF HEAVY METALS AND NTA, ITS USE IS SURE TO BE CONSIDERED. REALISTIC PLANNING OF WASTE TREATMENT PROCESSES SHOULD BE CONTINUED AS NUTRIENT REMOVAL IS CERTAIN TO BE A NECESSITY OF THE FUTURE. (SEE ALSO W72-09155) (JONES-WISCONSIN)

FIELD 05C

ACCESSION NO. W72-09169

NUTRIENT MANAGEMENT IN THE POTOMAC ESTUARY,

ENVIRONMENTAL PROTECTION AGENCY, ANNAPOLIS, MD. CHESAPEAKE TECHNICAL SUPPORT LAB.

N. A, JAWORSKI, D. W. LEAR, JR., AND O. WILLA, JR.

IN: NUTRIENTS AND EUTROPHICATION: THE LIMITING-NUTRIENT CONTROVERSY, P 246-273, 11 FIG, 17 TAB, 35 REF. AMERICAN SOCIETY OF LIMNOLOGY AND OCEANOGRAPHY SPECIAL SYMPOSIA, VOL I, ALLEN PRESS, LAWRENCE, KANSAS, 1972.

DESCRIPTORS:
*ESTUARIES, *EUTROPHICATION, *WATER POLLUTION CONTROL, MANAGEMENT, ANALYSIS, WATER QUALITY, CYANOPHYTA, NUISANCE ALGAE, COMPREHENSIVE PLANNING, AQUATIC WEEDS, WATER POLLUTION SOURCES, MATHEMATICAL MODELS, NITROGEN, ABATEMENT, COSTS, ANALYTICAL TECHNIQUES, MUNICIPAL WATER, PHOSPHORUS.

IDENTIFIERS:
*POTOMAC ESTUARY.

ABSTRACT:
WATER QUALITY STUDIES WERE UNDERTAKEN TO DEFINE WASTEWATER TREATMENT REQUIREMENTS OF UPPER POTOMAC ESTUARY SINCE 1965. STUDIES AND CONCEPTS USED TO FORMULATE A NUTRIENT MANAGEMENT PROGRAM ARE PRESENTED. CAUSES AND CONTROL NEEDS WERE STUDIED RELATIVE TO THE CHANGES IN NUTRIENT ENRICHMENT, INCLUDING APPEARANCE OF NUISANCE BLUE-GREEN ALGAE. DATA FROM ALGAL COMPOSITION ANALYSIS, ANNUAL NUTRIENT CYCLES AND LONGITUDINAL PROFILES, BIOASSAY STUDIES, ALGAL MODELING, COMPARISON WITH A NONEUTROPHIC ESTUARY, AND REVIEW OF HISTORICAL MATERIAL WERE USED TO ESTABLISH NUTRIENT CRITERIA. BASED ON A SUBJECTIVE ANALYSIS, DESIRED UPPER LIMITS OF CHLOROPHYLL A CONCENTRATIONS WERE DETERMINED FOR ESTABLISHING DEGREE OF EUTROPHICATION CONTROL REQUIRED TO MINIMIZE DETRIMENTAL EFFECTS ON WATER QUALITY AND WATER USES. ALTHOUGH AT THE PRESENT TIME NO SPECIFIC CRITERIA RELATIVE TO REQUIREMENTS FOR WASTEWATER TREATMENT HAVE BEEN ESTABLISHED FOR THE MESOHALINE PORTION OF THE ESTUARY, SPECIFIC NUTRIENT CRITERIA HAVE BEEN DEVELOPED FOR THE FRESHWATER PORTION. WITH A PROPERLY DESIGNED FACILITY THE DISSOLVED OXYGEN CONCENTRATION MAY BE ENHANCED AND ALGAL GROWTH REDUCED. THE WATER QUALITY MANAGEMENT PROGRAM BEING DEVELOPED WILL NOT ONLY IMPROVE THE WATER QUALITY TO MEET MINIMUM DESIGNATED STANDARDS, BUT WILL RENDER IT A FEASIBLE SOURCE OF MUNICIPAL WATER SUPPLY. (SEE ALSO W72-09155) (JONES-WISCONSIN)

FIELD 05C

ACCESSION NO. W72-09171

PHOSPHORUS IN WASTEWATER EFFLUENTS AND ALGAL GROWTH,

INDIAN INST. OF SCIENCE, BANGALORE. DEPT. OF BIOCHEMISTRY.

E. G. SRINATH, AND S. C. PILLAI.

JOURNAL OF WATER POLLUTION CONTROL FEDERATION, VOL 44, NO 2, P 303-308, 1972. 1 FIG, 4 TAB, 7 REF.

DESCRIPTORS:
*PHOSPHORUS, *SEWAGE EFFLUENTS, *ALGAE, WASTE WATER TREATMENT, ACTIVATED SLUDGE, LIME, AEROBIC TREATMENT, DOMESTIC WASTES.

IDENTIFIERS:
BANGALORE(INDIA).

ABSTRACT:
SAMPLES FROM THE WASTEWATER TREATMENT PLANT AT THE INDIAN INSTITUTE OF SCIENCE AND FROM THREE OUTFALLS AT BANGALORE, INDIA WERE EXAMINED OVER A PERIOD OF 12 YEARS AND EVIDENCE COLLECTED ON THE CONCENTRATION OF PHOSPHORUS IN DOMESTIC WASTEWATER, PHOSPHORUS CONTENTS IN EFFLUENTS FROM DIFFERENT WASTEWATER TREATMENT PROCESSES, EXTENT OF ALGAL GROWTH IN VARIOUS WASTEWATER EFFLUENTS, INFLUENCE OF DIFFERENT CONCENTRATIONS OF PHOSPHORUS ON THE GROWTH OF ALGAE IN WATER, AND METHODS OF REMOVAL OF PHOSPHORUS FROM WASTEWATER EFFLUENTS. PHOSPHORUS IN WASTEWATER AND EFFLUENTS WAS DETERMINED BY THE METHOD RECOMMENDED BY FISKE AND SUBBAROW AND MODIFIED BY KING. THE OTHER ITEMS OF ANALYSIS, EXCEPT TURBIDITY AND ALGAL GROWTH, WERE CARRIED OUT BY STANDARD METHODS. AMONG THE AEROBIC BIOLOGICAL METHODS OF WASTEWATER TREATMENT, THE ACTIVATED SLUDGE PROCESS REMOVES THE MAXIMUM AMOUNT OF PHOSPHORUS; THE REMAINING PHOSPHORUS IS RELATIVELY SMALL, BUT IT MUST BE REMOVED IF ALGAL GROWTH IN THE RECEIVING WATER IS TO BE CONTROLLED. IT CAN PREFERABLY BE ELIMINATED BY THE ADDITION OF CALCULATED AMOUNTS OF A MIXTURE OF ALUM AND LIME. EXPERIMENTS INDICATED THAT THE CONCENTRATION OF PHOSPHORUS IN WATER SHOULD BE LESS THAN 0.05 MGP/L FOR CONTROL OF ALGAL GROWTH. (JONES-WISCONSIN)

FIELD 05C

ACCESSION NO. W72-09176

SPECIES AND INDIVIDUAL PRODUCTIVITY IN PHYTOPLANKTON COMMUNITIES,

TEXAS UNIV., AUSTIN. DEPT. OF ZOOLOGY.

B. MAGUIRE, JR., AND W. E. NEILL.

ECOLOGY, VOL. 52, NO. 5, P 903-907, LATE SUMMER 1971. 1 FIG, 2 TAB.

DESCRIPTORS:
*ALGAE, *PHYTOPLANKTON, *PRIMARY PRODUCTIVITY, MICROSCOPY, CHLOROPHYTA, SCENEDESMUS, *CYANOPHYTA, CHLORELLA, BIOLOGICAL COMMUNITIES, DOMINANT ORGANISMS, AQUATIC PRODUCTIVITY, BIOMASS, CARBON RADIOISOTOPES, SUCCESSION, RADIOACTIVITY TECHNIQUES, CULTURES.

IDENTIFIERS:
CARBON-14, CARTERIA EUGAMETOS, STICHOCOCCUS, CHLOROCOCCUM, HYALOTHECA, LAKE LIVINGSTON, AUTORADIOGRAPHY, SAMPLE PREPARATION, SCINTILLATION COUNTING.

ABSTRACT:
THE RELATIVE PRODUCTIVITY OF INDIVIDUAL CELLS IN A MIXED PHYTOPLANKTON COMMUNITY CAN BE DETERMINED BY C-14 AUTORADIOGRAPHY. IN GENERAL, THE TECHNIQUE INVOLVES PLACING C-14-LABELLED ALGAL CELLS ON MICROSCOPE SLIDES AND COVERING THEM WITH A RADIOSENSITIVE EMULSION. AFTER SUITABLE EXPOSURE AND PHOTOGRAPHIC DEVELOPMENT, THE CELLS AND THE SILVER GRAINS PRODUCED BY THE RADIOACTIVE DECAY OF INCORPORATED C-14 ARE CLEARLY VISIBLE. THE PROPORTION OF GRAINS PRODUCED BY EACH SPECIES IS THEN USED TO PARTITION COMMUNITY-WIDE MEASUREMENTS OF PRIMARY PRODUCTION. ALGAL SAMPLES FROM ARTIFICIAL LAKE WATER IN THE LABORATORY AND LAKE LIVINGSTON WERE STUDIED. LABORATORY DATA SHOWED CARTERIA, BECAUSE OF ITS SIZE (685 CUBIC MICRONS), TO BE MORE PRODUCTIVE PER CELL THAN CHLORELLA (41 CUBIC MICRONS) OR STICHOCOCCUS (27 CUBIC MICRONS) DURING BOTH THE 1-HR AND 24 HR PERIODS TESTED. HOWEVER, IN TERMS OF PRODUCTIVITY PER UNIT VOLUME PER 24 HRS, CHLORELLA WILL BECOME MOST DOMINANT, STICHOCOCCUS SUBDOMINANT, AND CARTERIA PROGRESSIVELY LESS IMPORTANT IN NUMBER AND ECOLOGICAL EFFECT. LAKE LIVINGSTON DATA SHOWED CHLORELLA AND BLUE-GREEN ALGAE PREDOMINATING NUMERICALLY BUT WITH CHLORELLA HAVING THE GREATEST BIOVOLUME. ALTHOUGH CHLORELLA PRODUCED FAR MORE PHOTOSYNTHATE THAN ALL OTHER SPECIES COMBINED, ITS PRODUCTIVITY PER UNIT VOLUME WAS CONSIDERABLY LOWER THAN THAT OF SCENEDESMUS, HYALOTHECA, AND CHLOROCOCCUM. CHLORELLA COULD THUS BE REPLACED FROM THE DOMINANT POSITION AND UNICELLULAR BLUE-GREEN ALGAE ELIMINATED AS A SIGNIFICANT PART OF THE COMMUNITY. (MACKAN-BATTELLE)

FIELD 05C

ACCESSION NO. W72-09239

SOME QUANTITATIVE FEEDING PATTERNS EXHIBITED BY THE COPEPOD CALANUS,

AKADEMIYA NAUK SSSR, KALININGRAD. INSTITUT OKEANOLOGII.

YE. G. PERUYEVA.

OKEANOLOGIYA, VOL. 11, NO. 2, P 232-239, 1971. 5 FIG, 1 TAB, 29 REF.

DESCRIPTORS:
*COPEPODS, CRUSTACEANS, ZOOPLANKTON, ALGAE, FEEDING RATES, *LARVAL GROWTH STAGE, PHYTOPLANKTON, NUTRIENT REQUIREMENTS, INVERTEBRATES, CHLAMYDOMONAS, MATHEMATICAL STUDIES, BIOLOGICAL COMMUNITIES, FOOD CHAIN, DIATOMS, LARVAE, DIETS.

IDENTIFIERS:
*FEEDING PATTERNS, *CALANUS HELGOLANDICUS, COSCINODISCUS, COPEPODIDS, *CALANUS GLACIALIS, DITYLUM, PROROCENTRUM, NITZSCHIA, LAUDERIA, CHAETOCEROS.

ABSTRACT:
A STUDY WAS MADE OF THE FEEDING OF TWO WIDESPREAD SPECIES-CALANUS GLACIALIS AND C HELGOLANDICUS - ON THE LARGE DIATOM COSCINODISCUS SP. THE DIFFERENCES NOTED IN THE SHAPING OF THE DIET OF THE TWO SPECIES - ASYMPTOTIC GROWTH IN STAGE V C. GLACIALIS AND LINEAR GROWTH IN FEMALE C. HELGOLANDICUS IN RELATION TO THE TIME OF FEEDING - ARE CAUSED BY THE DIFFERENT FOOD REQUIREMENTS IN THE VARIOUS STAGES OF CALANUS DEVELOPMENT. SATIATION CONDITIONS (STAGE V C. GLACIALIS) ARE CHARACTERIZED BY THE CONSTANT RATE OF MOVEMENT OF THE CRUSTACEANS IN THE RANGE OF CONCENTRATIONS USED AND BY INCREASED SELECTIVITY WITH INCREASING FOOD CONCENTRATION. WHEN C. HELGOLANDICUS FEMALES FEED IN PLACES OF INSUFFICIENT CONCENTRATIONS, THEIR SPEED OF MOVEMENT CHANGES AT DIFFERENT FOOD CONCENTRATIONS AND SELECTIVITY BECOMES INDEPENDENT OF THE CONCENTRATION. (MACKAN-BATTELLE)

FIELD 05C

ACCESSION NO. W72-09248

ALGAE CONTROL BY MIXING, STAFF REPORT ON KEZAR LAKE IN SUTTON, N.H.

NEW HAMPSHIRE WATER SUPPLY AND POLLUTION CONTROL COMMISSION, CONCORD.

REPORT TO THE NEW ENGLAND REGIONAL COMMISSION, APRIL 1971. 103 P, 28 FIG, 30 TAB, 76 REF.

DESCRIPTORS:
*ALGAL CONTROL, *EUTROPHICATION, *DESTRATIFICATION, *WATER QUALITY CONTROL, *LAKE MORPHOLOGY, AERATION, MIXING, ALGAL BLOOMS, ALGAL TOXINS, WATER CIRCULATION, WATER POLLUTION CONTROL, FISHKILL, NEW HAMPSHIRE, WATER POLLUTION EFFECTS, COPPER SULFATE.

IDENTIFIERS:
*KEZAR LAKE, AIR INJECTION, COMPRESSED AIR.

ABSTRACT:
THE CONTROL OF ALGAE IN KEZAR LAKE BY DESTRATIFICATION WAS DEMONSTRATED IN FIELD EXPERIMENTAL WORK CONDUCTED IN 1968 AND 1969. KEZAR LAKE HAS SUFFERED FROM INCREASINGLY VIRULENT BLOOMS OF ALGAE WHICH APPARENTLY HAVE BEEN INDUCED BY NUTRIENTS CONTAINED IN WASTEWATER REACHING THE LAKE. A GROWING NUMBER OF NEW ENGLAND LAKES SHARES THIS PROBLEM, WHICH THREATENS ONE OF THE REGION'S MAJOR RECREATIONAL RESOURCES. BEYOND THE EUTROPHICATION OF THE LAKES THEMSELVES, THE RESULTS CAN BE MEASURED IN DECLINING PROPERTY VALUES AND DECREASES IN RECREATION BUSINESS WHICH ARE DIRECTLY ATTRIBUTABLE TO THE UNAESTHETIC APPEARANCE OF AN ALGAE-CHOKED LAKE. THE NEW HAMPSHIRE WATER SUPPLY AND POLLUTION CONTROL COMMISSION APPEARS TO HAVE DEVELOPED AN EFFECTIVE AND INEXPENSIVE METHOD OF BRINGING SUCH ALGAL BLOOMS UNDER CONTROL BY MIXING THE WATER IN THE LAKE AND ALLOWING NATURAL PROCESSES TO TAKE OVER. HOMOGENIZING, MIXING OR DESTRATIFYING KEZAR LAKE WAS ACCOMPLISHED BY FORCING COMPRESSED AIR, FROM SHORE-LOCATED COMPRESSORS, THROUGH P.V.C., 2-INCH-DIAMETER PIPING TO THE DEEPEST PORTION OF THE LAKE WHERE IT WAS RELEASED THROUGH CERAMIC DIFFUSERS, AND ALLOWED TO BUBBLE UP TO THE LAKE SURFACE THROUGH THE WATER COLUMN. CLARITY OF THE WATER WAS VISIBLY AND MEASUREABLY IMPROVED; THE POPULATIONS OF NOXIOUS ALGAE, SO OBJECTIONABLE TO RECREATION INTERESTS, WERE DECREASED IN NUMBER; AND NO HARMFUL EFFECTS WERE DETECTED DURING 1968 AND 1969 SUMMER MIXING OF KEZAR LAKE. OPERATING PRESSURES FOR THE COMPRESSORS ARE LOW. THE EQUIPMENT IS CONVENIENT FOR NECESSARY MAINTENANCE, AND THE OPERATING AND STUDY BUDGET IS MODEST. (POERTNER)

FIELD 05G

ACCESSION NO. W72-09304

GAS CHROMATOGRAPHIC ANALYSIS OF THE HIGHER FATTY ACIDS OF THE ALGA CHLORELLA VULGARIS (PYRENOIDOSA),

INSTITUTE FOR RESEARCH, PRODUCTION AND USES OF RADIOISOTOPES, PRAGUE (CZECHOSLOVAKIA).

M. MATUCHA, L. ZILKA, AND K. SVIHEL.

JOURNAL OF CHROMATOGRAPHY, VOL. 65, NO. 2, P 371-376, FEBRUARY 23, 1972. 5 FIG, 1 TAB, 20 REF.

DESCRIPTORS:
CHEMICAL ANALYSIS, *ALGAE, COLUMNS, ORGANIC ACIDS, SEPARATION TECHNIQUES, *CHROMATOGRAPHY.

IDENTIFIERS:
*FATTY ACIDS, *GAS LIQUID CHROMATOGRAPHY, METHYL ESTERS, ESTERS, CHLORELLA VULGARIS, SAMPLE PREPARATION.

ABSTRACT:
LABELED (C-14) FATTY ACIDS WERE PREPARED FROM RADIOACTIVE CHLORELLA VULGARIS FOR GAS-LIQUID CHROMATOGRAPHY BY THE FOLLOWING PROCEDURE. LIPIDS WERE EXTRACTED FROM EXTRACTED AND WASHED CELLS WITH 96 PERCENT ETHANOL AND AN ETHANOL-DIETHYL ETHER MIXTURE; REMAINING SUGARS WERE REMOVED FROM THE LIPID EXTRACT BY WATER EXTRACTION. THE DRIED LIPIDS WERE THEN TRANSESTERIFIED WITH METHANOLIC HYDROCHLORIC ACID. AFTER EXTRACTION OF THE METHYLESTERS BY PETROLEUM ETHER, THEY WERE PURIFIED BY VACUUM SUBLIMATION. TO IDENTIFY CHLORELLA FATTY ACIDS WITH A RELATIVE ABUNDANCE HIGHER THAN 0.5 PERCENT, A POLAR POLYESTER DIETHYLENE GLYCOL SUCCINATE (DEGS) AND A NON-POLAR GREASE, APIEZON-L, WERE CHOSEN AS STATIONARY PHASES. G-L CHROMATOGRAPHY WAS CARRIED OUT ON A COLUMN 2 M LONG AND 4MM I.D. FILLED WITH 15 PERCENT DEGS WITH AN ARGON-IONIZATION DETECTOR AND ON A COLUMN 1.5 M LONG AND 3 MM I.D. FILLED WITH 15 PERCENT APIEZON WITH A FLAME-IONIZATION DETECTOR. A 2 ML GAS PROPORTIONAL COUNTER GAVE A SIMULTANEOUS DETERMINATION OF RADIOACTIVITY. IN ADDITION TO THE MAJOR ACIDS (PALMITIC, OLEIC, LINOLEIC, AND LINOLENIC), EVEN-NUMBERED, STRAIGHT-CHAIN FATTY ACIDS AND ODD-NUMBERED, SATURATED FATTY ACIDS WERE FOUND. (MACKAN-BATTELLE)

FIELD 05A, 05C

ACCESSION NO. W72-09365

CONTROLLED EUTROPHICATION-INCREASING FOOD PRODUCTION FROM THE SEA BY RECYCLING HUMAN WASTES,

WOODS HOLE OCEANOGRAPHIC INSTITUTION, MASS.

J. H. RYTHER, W. M. DUNSTAN, K. R. TENORE, AND J. E. HUGUENIN.

BIOSCIENCE, VOL. 22, NO. 3, P 144-152, MARCH 1972. 4 FIG, 11 REF.

DESCRIPTORS:
*EUTROPHICATION, *RECYCLING, WATER QUALITY CONTROL, SEWAGE TREATMENT, SEWAGE EFFLUENTS, SANITARY ENGINEERING, SECONDARY TREATMENT, MARINE ALGAE, PHYTOPLANKTON, BIOMASS, GROWTH RATES, PRODUCTIVITY, ANIMAL GROWTH, PLANT GROWTH, OYSTERS, SHELLFISH, MUSSELS, SHELLFISH FARMING, AQUACULTURE, COMMERCIAL SHELLFISH, BIOLOGICAL COMMUNITIES, FOODCHAINS, HABITATS, NICHES, TURNOVERS, CULTURES, WASTE WATER TREATMENT, *CYCLING NUTRIENTS, NITROGEN, CARBON DIOXIDE, CHLORELLA, NUTRIENTS, TERTIARY TREATMENT, FLOW RATES, WATER POLLUTION EFFECTS, PRIMARY PRODUCTIVITY.

IDENTIFIERS:
GROWTH KINETICS, BIOENERGETICS, ENRICHMENT, CULTURE MEDIA, MACROINVERTEBRATES.

ABSTRACT:
THE ESSENTIAL FEATURE OF 'CONTROLLED EUTROPHICATION' IS THE PHYSICAL SEPARATION AND COMPARTMENTALIZATION OF THE PRODUCER AND CONSUMER LEVELS OF A BIOLOGICAL COMMUNITY. FOLLOWING THESE GUIDELINES, LABORATORY EXPERIMENTS WERE BEGUN IN THE SUMMER OF 1970 ON THE GROWTH KINETICS OF MARINE PLANKTON ALGAE GROWN IN SEAWATER ENRICHED WITH EFFLUENT FROM A SECONDARY SEWAGE TREATMENT PLANT. IN GENERAL, DILUTED SEWAGE WAS FOUND TO BE AN EXCELLENT CULTURE MEDIUM FOR THE MARINE PHYTOPLANKTON. AT CONCENTRATIONS OF 10 PERCENT SEWAGE, THE YIELD OF ALGAE INCREASED WITH FLOW RATE THROUGH THE SYSTEM UP TO A 'TURNOVER RATE' OF 50 PERCENT OF THE CULTURE PER DAY. THESE ALGAE WERE SUBSEQUENTLY FED TO MONITORED OYSTER CULTURES, THUS COMPLETING THE PRODUCER-CONSUMER FOOD CHAIN. IN ONE SUCH EXPERIMENT, A NATURAL POPULATION OF DIATOMS GROWN ON 10 PERCENT SEWAGE WAS PASSED THROUGH A 3M X 1.5M X .5 IN. TANK CONTAINING SUSPENDED STRINGS OF OYSTER SPAT ATTACHED TO SCALLOP SHELLS. OVER 30 DAYS, AT 7-10C, THE OYSTERS REMOVED 77 PERCENT OF THE ALGAE AND CONVERTED 22 PERCENT OF THE CELLS INTO NEW OYSTER FLESH. THESE AND SIMILAR EXPERIMENTS HAVE PROVIDED BASIC DATA ON THE KINETICS AND BIOENERGETICS OF A SMALL SCALE 'CONTROLLED EUTROPHICATION PROGRAM' AND REVEALED THE VALUE FOR DEVELOPING APPLICATIONS IN ADVANCED SEWAGE TREATMENT AND COMMERCIAL AQUACULTURE. (MACKAN-BATTELLE)

FIELD 05G, 05C

ACCESSION NO. W72-09378

WASTE TREATMENT LAGOONS-STATE OF THE ART.

MISSOURI BASIN ENGINEERING HEALTH COUNCIL, CHEYENNE, WYO.

COPY AVAILABLE FROM GPO SUP DOC, $1.25; MICROFICHE FROM NTIS AS PB-209 937, $0.95, ENVIRONMENTAL PROTECTION AGENCY WATER POLLUTION CONTROL RESEARCH SERIES, JULY 1971. 152 P, 9 TAB, 42 REF. EPA PROGRAM 17090 EHX 07/71.

DESCRIPTORS:
*SEWAGE LAGOONS, *AERATED LAGOONS, *OXIDATION LAGOONS, *DESIGN CRITERIA, BIODEGRADATION, ALGAE, PHOTOSYNTHESIS, SUSPENDED SOLIDS, SEPARATION TECHNIQUES, FILTRATION, MUNICIPAL WASTES, INDUSTRIAL WASTES, OPERATION AND MAINTENANCE, *WASTE WATER TREATMENT, WATER QUALITY CONTROL, *REVIEWS.

IDENTIFIERS:
*FACULTATIVE LAGOONS, *ANAEROBIC LAGOONS.

ABSTRACT:
A REVIEW OF PUBLISHED LITERATURE AND FIELD EVALUATIONS REVEALED THE PRESENCE OF OVER 3500 WASTE TREATMENT LAGOONS CURRENTLY IN OPERATION IN THE UNITED STATES. THE THREE TYPES OF LAGOONS IN USE INCLUDE: (1) OXIDATION LAGOONS; (2) AERATED LAGOONS, AND (3) ANAEROBIC LAGOONS. OXIDATION LAGOONS DEPEND UPON ALGAE TO SUPPLY OXYGEN BY PHOTOSYNTHESIS AND DEGRADE THE WASTE PRODUCTS. EFFLUENT QUALITY IS DETERMINED BY THE QUANTITY OF ALGAE IN THE EFFLUENT AND SEVERAL METHODS OF ALGAE REMOVAL ARE CURRENTLY UNDER INVESTIGATION. AERATED LAGOONS MAY BE MERELY OXIDATION PONDS WITH SUPPLEMENTAL AERATION, PARTIALLY MIXED ACTIVATED SLUDGE (FACULTATIVE AERATED) OR COMPLETE MIX ACTIVATED SLUDGE (CMAS) SYSTEMS. HIGH QUALITY EFFLUENTS FROM AERATED LAGOONS CAN BE ACHIEVED ONLY BE REMOVING EFFLUENT MICROBIAL SOLIDS. ANAEROBIC LAGOONS CAN PROVIDE UP TO 80% BOD REMOVALS, BUT MUST BE FOLLOWED BY SOME TYPE OF AEROBIC TREATMENT TO PRODUCE A HIGH QUALITY EFFLUENT. THIS REVIEW HAS DEMONSTRATED THAT LAGOONS DO HAVE APPLICABILITY TO THE TOTAL WASTE TREATMENT PROBLEM, BUT THE FUTURE OF LAGOONS DEPENDS UPON PROPER DESIGN AND OPERATION IN RELATIONSHIP WITH THE FUNDAMENTAL BIOCHEMISTRY OF THE MICROBES IN THE VARIOUS SYSTEMS. (LOWRY-TEXAS)

FIELD 05D, 10F

ACCESSION NO. W72-09386

FILTRATION OF WATER AND WASTEWATER,

UNIVERSITY COLL., LONDON (ENGLAND).

K. J. IVES.

CRC CRITICAL REVIEWS IN ENVIRONMENTAL CONTROL, VOL 2, NO. 2, AUGUST 1971. P 293-335, 8 FIG, 9 TAB, 99 REF.

DESCRIPTORS:
*FILTRATION, *OPERATION AND MAINTENANCE, *COST ANALYSIS, *FILTERS, HEADLOSS, PRESSURE, TURBIDITY, ALGAE, FLOCCULATION, CLEANING, *WATER TREATMENT, TERTIARY TREATMENT, *WASTE WATER TREATMENT.

IDENTIFIERS:
*MICROSTRAINING, *PRECOAT FILTRATION, BACKWASHING, SPECIFIC RESISTANCE.

ABSTRACT:
A REVIEW OF THE RECENT ADVANCEMENTS IN FILTRATION TECHNOLOGY WAS CONDUCTED, DEALING PRIMARILY WITH SLOW SAND FILTERS, RAPID SAND FILTERS, PRECOAT FILTERS, AND MICROSTRAINERS. APPLICATIONS, MODES OF OPERATION, MAINTENANCE REQUIRED, AND COST FIGURES WERE ASSEMBLED FOR EACH TYPE OF FILTER. SOME GENERAL RULES FOR FILTRATION WHICH WERE SELECTED FROM THE TECHNICAL LITERATURE INCLUDE: (1) HEAVY TURBIDITY LOADS CANNOT BE SUSTAINED BY EITHER SLOW SAND OR PRECOAT FILTERS; (2) ALUM FLOC WILL CLOG SLOW SAND FILTERS WHEREAS MICROSTRAINERS WILL NOT RETAIN IT; (3) SLOW SAND AND PRECOAT FILTERS RETAIN FINE TURBIDITY, RAPID SAND FILTERS RETAIN IT IF IT IS PREVIOUSLY FLOCCULATED, AND MICROSTRAINERS CANNOT RETAIN IT; (4) ALL 4 FILTERS RETAIN PLANKTON, ALTHOUGH SMALLER ALGAE ARE NOT RETAINED BY MICROSTRAINERS AND RAPID FILTERS; (5) ONLY MICROSTRAINERS AND RADIAL FLOW RAPID FILTERS OPERATE CONTINUOUSLY; (6) MICROSTRAINER HEADLOSS IS 0.15 M COMPARED TO 1 TO 20 M FOR THE OTHER FILTER TYPES; AND (7) RAPID FILTERS HAVE THE WIDEST APPLICABILITY TO BOTH WATER AND WASTEWATER TREATMENT. COSTS PRESENTED WERE NOT FOR OPTIMUM FILTERS, SINCE TREATMENT SYSTEMS, NOT TREATMENT COMPONENTS, ARE OPTIMIZED. (LOWRY-TEXAS)

FIELD 05D, 05F, 10F

ACCESSION NO. W72-09393

ALGAL INFLUENCES ON DIEOFF RATES OF INDICATOR BACTERIA,

TEXAS UNIV., AUSTIN. COLL. OF ENGINEERING.

E. M. DAVIS, AND E. F. GLOYNA.

PROCEEDINGS, INDUSTRIAL WASTE CONFERENCE, 25TH, MAY 5-7, 1970, PURDUE UNIVERSITY ENGINEERING EXTENSION SERIES NO. 137, P 266-273, 1 FIG, 4 TAB, 17 REF.

DESCRIPTORS:
*ENTERIC BACTERIA, *WATER QUALITY, *DOMESTIC WASTES, ALGAE, INDUSTRIAL WASTES, METABOLISM, GROWTH RATES, OXIDATION LAGOONS, COLIFORMS, NUTRIENTS, ANALYTICAL TECHNIQUES, MICROBIOLOGY, *WASTE WATER TREATMENT.

IDENTIFIERS:
*AFTERGROWTH.

ABSTRACT:
DESTRUCTION OF COLIFORM BACTERIA IN WASTE TREATMENT FACILITIES HAS BEEN ATTRIBUTED TO A NUMBER OF PHYSICAL, BIOLOGICAL, AND CHEMICAL FACTORS. HOWEVER, SEVERAL SPECIES OF ALGAE APPEAR TO BE CAPABLE OF STIMULATING SPECIFIC COLIFORM ORGANISMS, THROUGH ALGAL METABOLIC EXUDATES WHICH ARE RELEASED DURING LOG GROWTH CONDITIONS AND FROM NUTRIENTS WHICH MAY BE RELEASED AFTER DEATH OF THE ALGAE. INVESTIGATIONS OF BEHAVIORAL PATTERNS OF SELECTED SPECIES OF ENTERIC BACTERIA WHEN IN THE PRESENCE OF SINGLE AXENIC CULTURES AND MIXED CULTURES OF SELECTED BLUE-GREEN AND GREEN ALGAE, AND INVESTIGATIONS OF THE BEHAVIOR OF ENTERIC BACTERIA IN THE PRESENCE OF ALGAE AND KNOWN QUANTITIES OF INDUSTRIAL WASTE WERE CONDUCTED. RESULTS INDICATED THAT ALCALIGENES FAECALIS, ENTEROBACTER AEROGENES, AND PROTEUS VULGARIS EXHIBITED ONLY LIMITED TENDENCIES FOR AFTERGROWTH, WHILE ESCHERICHIA COLI, PSEUDOMONAS, AND SERRATIA MARCESCENS SHOWED DISTINCT AFTERGROWTH CAPABILITIES IN ALL TEST ENVIRONMENTS USED. THESE INVESTIGATIONS DEMONSTRATED THAT THE ORGANISMS WHICH ARE USED AS POLLUTION INDICATORS MAY APPEAR IN NUMBERS SEEMING TO INDICATE DOMESTIC WASTEWATER POLLUTION WHEN NONE IS PRESENT. (LOWRY-TEXAS)

FIELD 05D

ACCESSION NO. W72-09590

PROBLEMS IN WATER TOXICOLOGY (VOPROSY VODNOY TOKSIKOLOGII).

'NAUKA', MMOSCOW, 1970. 224 P.

DESCRIPTORS:
*HYDROBIOLOGY, *PATHOLOGY, *TOXICITY, *WATER QUALITY CONTROL, *WATER POLLUTION EFFECTS, WATER POLLUTION SOURCES, AQUATIC LIFE, AQUATIC ANIMALS, AQUATIC MICROORGANISMS, AQUATIC PLANTS, AQUATIC ALGAE, PESTICIDES, INSECTICIDES, HERBICIDES, PHOSPHORUS, PHENOLS, METALS, POLLUTANTS, INDUSTRIAL WASTES, CONFERENCES.

IDENTIFIERS:
*USSR, *TOXICOLOGY, *TOXICANTS, *HYDROBIONTS, ICHTHYOLOGY, ICHTHYOFAUNA, MACROPHYTES, SAPROPHYTES, CERIUM.

ABSTRACT:
THIS COLLECTION OF 37 PAPERS CONTAINS REPORTS PRESENTED AT THE ALL-UNION CONFERENCE ON PROBLEMS IN WATER TOXICOLOGY, HELD IN MOSCOW JANUARY 30 TO FEBRUARY 2, 1968. THE BOOK IS DIVIDED INTO 5 SECTIONS: (1) GENERAL PROBLEMS IN WATER TOXICOLOGY (10 PAPERS); (2) EFFECTS OF PESTICIDES ON HYDROBIONTS (8 PAPERS); (3) EFFECTS OF PHOSPHORUS AND PHOSPHORUS COMPOUNDS ON HYDROBIONTS (6 PAPERS); (4) EFFECTS OF PHENOLS ON HYDROBIONTS (7 PAPERS); AND (5) EFFECTS OF METALS AND METALLIC COMPOUNDS ON HYDROBIONTS (6 PAPERS). SAMPLES COLLECTED FOR ANALYSIS INCLUDED BACTERIA, ALGAE, AND A VARIETY OF INVERTEBRATES AND FISHES. CHANGES IN BIOLOGICAL PROCESSES IN SEA WATER AND FRESHWATER UNDER INFLUENCE OF DIFFERENT POLLUTANTS AND INDUSTRIAL WASTES WERE EXAMINED TOGETHER WITH METHODS FOR FUTURE INVESTIGATION OF THE TOXICITY OF THE AQUATIC ENVIRONMENT. (SEE W72-09649 AND W72-09650) (JOSEFSON-USGS)

FIELD 05C, 02H, 05G

ACCESSION NO. W72-09646

DETERMINATION OF THE TOXICITY OF CONTAMINATED FRESHWATER WITH RESPECT TO CERTAIN HYDROBIONTS (OPREDELENIYE TOKSICHNOSTI ZAGRYAZNENNYKH PRESNYKH VOD V OTNOSHENII NEKOTORYKH GIDROBIONTOV),

MOSCOW STATE UNIV. (USSR).

G. D. LEBEDEVA.

IN: VOPROSY VODNOY TOKSIKOLOGII; 'NAUKA', MOSCOW, P 57-61, 1970. 2 FIG, 10 REF.

DESCRIPTORS:
*HYDROBIOLOGY, *TOXICITY, *WATER POLLUTION, *FRESHWATER, *AQUATIC LIFE, AQUATIC ANIMALS, CARP, DAPHNIA, AQUATIC PLANTS, AQUATIC ALGAE, SCENEDESMUS, RADIOISOTOPES, STRONTIUM RADIOISOTOPES, PHOSPHORUS RADIOISOTOPES, CESIUM, RADIOACTIVITY EFFECTS, LABORATORY TESTS.

IDENTIFIERS:
*USSR, *HYDROBIONTS, *CERIUM, RADIOMETRY.

ABSTRACT:
THE EFFECTS OF RADIOACTIVE ELEMENTS ON BIOLOGICAL FUNCTIONS OF HYDROBIONTS WERE BASED ON LABORATORY ANALYSES OF FISH (CYPRINUS CARPIO), CRUSTACEANS (DAPHNIA MAGNA), HIGHER AQUATIC PLANTS (LEMNA MINOR AND ELODEA CANADENSIS), AND GREEN ALGAE (SCENEDESMUS QUADRICAUDA). THE RADIOACTIVE ISOTOPES USED WERE SR-90, SR-89, CS-137, CE-144, AND P-32. PLANT ORGANISMS, WHICH CONCENTRATE LARGE AMOUNTS OF RADIOACTIVE STRONTIUM, ARE LESS SENSITIVE TO WATER CONTAMINATION BY THIS ELEMENT. OF THE ANIMAL ORGANISMS EXAMINED, DAPHNIA ARE THE MOST SENSITIVE TO WATER CONTAMINATION BY RADIOSTRONTIUM. THE TOXICITY CRITERIA FOR FISH WERE THE GROWTH RATE AND AMOUNT OF EDIBLE FOOD; FOR DAPHNIA--THE SURVIVAL RATE, REPRODUCTION AND FERTILITY; AND FOR AQUATIC PLANTS--THE POPULATION AND INCREASE IN THE BIOMASS. (SEE ALSO W72-09646) (JOSEFSON-USGS)

FIELD 05C, 02H

ACCESSION NO. W72-09649

PROBLEM OF PESTICIDES IN WATER TOXICOLOGY (PROBLEMA PESTITSIDOV V VODNOY TOKSIKOLOGII),

AKADEMIYA NAUK URSR, KIEV. INSTYTUT HIDROBIOLOGII.

L. P. BRAGINSKIY.

IN: VOPROSY VODNOY TOKSIKOLOGII; 'NAUKA', MOSCOW, P 81-88, 1970. 2 FIG, 3 TAB, 4 REF.

DESCRIPTORS:
*HYDROBIOLOGY, *PESTICIDES, *PESTICIDE RESIDUES, *PESTICIDE TOXICITY, *AQUATIC PLANTS, AQUATIC ALGAE, PHYTOPLANKTON, AQUATIC MICROORGANISMS, AQUATIC ANIMALS, DAPHNIA, APPLICATION METHODS, GRANULES, PLANT GROWTH REGULATORS, INHIBITORS, HERBICIDES, MONURON, PHOTOSYNTHESIS.

IDENTIFIERS:
*USSR, *TOXICOLOGY, *HYDROBIONTS, MACROPHYTES, SAPROPHYTES.

ABSTRACT:
PESTICIDES IN AN AQUATIC ENVIRONMENT AFFECT HYDROBIOLOGICAL PROCESSES BY (1) ALTERING THE HABITATS OF HYDROBIONTS BY INHIBITING PLANT PHOTOSYNTHESIS (OXYGEN DEPLETION) AND BY STIMULATING SAPROPHYTE GROWTH; (2) ACUTE POISONING OF HYDROBIONTS BY TOXIC CONCENTRATIONS OF PESTICIDES; AND (3) PRODUCING A VARIETY OF ACUTE AND CHRONIC EFFECTS ON HYDROBIONTS BY THE PRESENCE OF PESTICIDE RESIDUES ACCUMULATED THROUGH FOOD CHAINS. PROSPECTS FOR THE USE OF PESTICIDES TO REGULATE ALGAL AND MACROPHYTE GROWTH ARE BASED PRIMARILY ON THE APPLICATION OF GRANULES TO PLANT PROTECTION. METHODS ARE GIVEN FOR STUDYING THE EFFECTS OF PESTICIDES ON WATER BODIES AND ON THE BIOLOGICAL PROCESSES OCCURRING IN THEM. (SEE ALSO W72-09646) (JOSEFSON-USGS)

FIELD 05C, 05G

ACCESSION NO. W72-09650

POSSIBLE EFFECTS OF ORGANOCHLORINE PESTICIDES ON PRIMARY PRODUCTIVITY AND SKELETOGENESIS OF NEW ENGLAND ESTUARINE AND COASTAL MARINE ALGAE,

MASSACHUSETTS UNIV., AMHERST. WATER RESOURCES RESEARCH CENTER.

J. P. SEARS, AND C. YENTSCH.

AVAILABLE FROM THE NATIONAL TECHNICAL INFORMATION SERVICE AS PB-210 146, $3.00 IN PAPER COPY, $0.95 IN MICROFICHE. PROJECT COMPLETION REPORT, FEBRUARY 1972. 13 P, 6 FIG, 2 REF. OWRR A-035 MASS.(1).

DESCRIPTORS:
*ANALYTICAL TECHNIQUES, RATES, *PHOTOSYNTHESIS, *MARINE ALGAE, DDT, *PESTICIDES, *PRIMARY PRODUCTIVITY, WATER POLLUTION EFFECTS, ESTUARINE ENVIRONMENT.

IDENTIFIERS:
RHODYMENIA, FUCUS, ULVA.

ABSTRACT:
TECHNIQUES HAVE BEEN DEVELOPED TO ESTIMATE RATES OF PHOTOSYNTHESIS IN MACROSCOPIC MARINE ALGAE BY MONITORING OXYGEN PRODUCTION. ATTEMPTS TO USE 14C TECHNIQUES WERE UNSECCESSFUL BECAUSE OF THE DIFFICULTY OF RECOVERING ASSIMILATED 14C BY MACERATION, ACID HYDROLOSIS AND STRONG BASE SOLUBILIZATION OF THALLI. THE PROBLEM OF 14C RECOVERY WOULD BE SOLVED IF THE PLANT MATERIAL COULD BE COMPLETELY COMBUSTED AND TOTAL CARBON RECOVERED IN ETHANOLAMINE. BASE LINE DATA ON THE INFLUENCE OF TEMPERATURE AND LIGHT INTENSITY ON RATES OF PHOTOSYNTHESIS WERE OBTAINED FOR RHODYMENIA, FUCUS AND ULVA. PHOTOSYNTHESIS BY MACROSCOPIC ALGAE WAS STRONGLY INFLUENCED BY WATER MOVEMENT. UNDER CONDITIONS OF STANDING WATER, OXYGEN PRODUCTION DECREASED TO A FRACTION OF OXYGEN PRODUCTION IN A STIRRED SYSTEM. THE REDUCED RATE OF PHOTOSYNTHESIS WAS PROBABLY DUE TO DEPLETION OF CO2 AT THE THALLUS SURFACE, AND A SUBSEQUENT RISE IN PH. FUTURE ESTIMATES OF PHOTOSYNTHESIS IN MACROALGAE SHOULD INCLUDE SOME MECHANISM FOR WATER AGITATION, ESPECIALLY DURING ESTIMATES OF PRODUCTIVITY. THE EFFECT OF DDT ON PHOTOSYNTHESIS OF RHODYMENIA, FUCUS AND ULVA COULD NOT BE CONSISTENTLY DEMONSTRATED. IF ANY EFFECT DID OCCUR, ITS MAGNITUDE WAS BELOW THE LEVEL OF DETECTION BECAUSE OF THE PHOTOSYNTHETIC VARIABILITY AMONG PLANTS. THESE FINDINGS ARE IN CONTRAST TO THE PRONOUNCED EFFECTS OF DDT ON PHYTOPLANKTON PHOTOSYNTHESIS REPORTED BY MENZEL, 1970; AND WURSTER, 1969.

FIELD 05C, 02L

ACCESSION NO. W72-09653

THE RELATIONSHIPS BETWEEN 32P ACCUMULATICN IN ALGAE, BACTERIA AND TUBIFICIDS,

WESTERN MICHIGAN UNIV., KALAMAZOO, DEPT. OF BIOLOGY.

W. L. STROMBERG, AND C. J. GOODNIGHT.

AVAILABLE FROM THE NATIONAL TECHNICAL INFORMATION SERVICE AS COO-1803-6, $3.00 IN PAPER COPY, $0.95 IN MICROFICHE. FINAL REPORT 1970. 27 P, 8 FIG, 2 TAB, 18 REF. CONTRACT NO. AT(11-1)-1803.

DESCRIPTORS:
*PHOSPHORUS RADIOISOTOPES, *TUBIFICIDS, *FOOD CHAINS, *PATH OF POLLUTANTS, *ABSORPTION, ALGAE, CHORELLA, TRACERS, PHOSPHATES, ECOSYSTEMS, AQUATIC LIFE, ANTIBICTICS(PESTICIDES), ECOLOGY, BENTHOS, INVERTEBRATES, PHYTOPLANKTON, PLANKTON, WATER TEMPERATURE, E. COLI, BENTHIC FAUNA, CELLULOSE, ENZYMES, PULP WASTES, RADIOECOLOGY, WATER POLLUTION, PHOSPHORUS, WATER TEMPERATURE, HYDROGEN ION CONCENTRATION, NUTRIENTS, RADIOISOTOPES, ALGAE, WATER QUALITY, ANNELIDS, ADSORPTION.

IDENTIFIERS:
*CHLORELLA PYRENOIDOSA, *CHLORAMPHENICAL, LIMNODRILUS SPP., TUBIFEX, UNIALGAL CULTURES, OUABAIN, BIOLCGICAL MAGNIFICATION, PHOSPHORUS-32, LIMNODRILUS.

ABSTRACT:
KINETICS OF ACCUMULATION AND RETENTION OF P-32 WERE DETERMINED FOR UNIALGAL CULTURES OF CHLORELLA PYRENOIDOSA IN CARRIER (52 MG NA2HPO4/1.) AND CARRIER-FREE KNOP'S SOLUTION (MODIFIED) USING 10 MICROCURIES P-32/1 AS TRACER AND A GEIGER-MUELLER COUNTING SYSTEM. THE NUMBER OF COUNTS WAS SUFFICIENT TO ASSURE ACCURACY WITHIN 5 PERCENT. ACCUMULATION IN A CARRIER-FREE SOLUTION WAS EXPONENTIAL AND APPROACHED A 90 PERCENT EFFICIENCY OF REMOVAL WITHIN 30 MINUTES. IN THE CARRIER SOLUTION, ACCUMULATION WAS APPROXIMATELY LINEAR AND DID NOT REACH EQUILIBRIUM WITHIN A 27-HOUR PERIOD. USING OUABAIN AS AN INHIBITOR, IT WAS SHOWN THAT THE OBSERVED SORPTION OF RADIOPHOSPHATE IN CARRIER-FREE SOLUTION BY THIS ALGA WAS INDEPENDENT OF METABOLIC ACTIVITY. MOST LOSS OF THE PHOSPHATE TAKEN UP IN A 27-HOUR INTERVAL OCCURRED WITHIN ONE HOUR IN AGITATED CARRIER AND CARRIER-FREE KNOP'S SOLUTION. ACCUMULATION OF P-32 BY TUBIFICID WORMS WAS ALSO STUDIED USING SIMILAR COUNTING METHODS. UPTAKE BY THESE WORMS WAS APPROXIMATELY EXPONENTIAL IN MEDIA WITH A BACTERIAL SOURCE OF RADIOSPHSPHATE, AND WAS REDUCED BY ADDITION OF CHLORAMPHENICOL AND REMOVAL OF P-32 FROM SOLUTION BY ALGAE. PRELIMINARY RESULTS INDICATED THAT C. PYRENOIDOSA COMPETED EFFECTIVELY WITH THE WORMS FOR INORGANIC PHOSPHATE, AND THAT ACCUMULATION OF P-32 BY INGESTION OF ALGAE AMONG TUBIFICIDS MAY BE SLIGHT. THE ECOLOGICAL IMPLICATIONS OF THESE RESULTS ARE DISCUSSED. (JEFFERIS-BATTELLE)

FIELD 05C, 05B

ACCESSION NO. W72-09668

SUMMER CONFERENCE OF SOCIETY FOR APPLIED BACTERIOLOGY, LIVERPOOL, 13-15 JULY 1971,

OFFICE OF NAVAL RESEARCH, LONDON (ENGLAND).

G. A. HOTTLE.

AVAILABLE FROM THE NATIONAL TECHNICAL INFORMATION SERVICE AS AD731-726, $3.00 IN PAPER COPY, $0.95 IN MICROFICHE. REPORT NO. ONRL-C-19-71, 31 AUGUST 1971. 12 P, 11 REF.

DESCRIPTORS:
*MICROORGANISMS, *WATER POLLUTION CONTROL, *WASTES, *PATHOGENIC BACTERIA, CONFERENCES, *PUBLIC HEALTH, RIVERS, *SEWAGE, BIOCHEMICAL OXYGEN DEMAND, SOLID WASTES, AMMONIA, NITROGEN COMPOUNDS, NITRATES, PHOSPHATES, SLUDGE TREATMENT, DISSOLVED OXYGEN, BACTERIA, FUNGI, PROTOZOA, FERMENTATION, BACTERIOPHAGE, NUTRIENTS, FILTERS EQUIPMENT, OXIDATION, LAKES, WISCONSIN, ANAEROBIC DIGESTIION, ALGAE, EUTROPHICATION, BIODEGRADATION, PLASTICS, PHOSPHORUS COMPOUNDS, SOIL FUNGI, PSEUDOMONAS, DAIRY INDUSTRY, MUNICIPAL WASTES, HERBICIDES, SOIL BACTERIA, PESTICIDES, INDUSTRIAL WASTES, DDT, TRACE ELEMENTS, WATER QUALITY, DETERGENTS, FARM WASTES, URINE, AEROBIC BACTERIA, CARBOHYDRATES, BACTERICIDES, CLOSTRIDIUM, STREPTOCOCCUS, SALMONELLA, WATER PURIFICATION, SHEEP, YEASTS, FOODS, ANAEROBIC BACTERIA, SOIL CONTAMINATION, WASTE TREATMENT, ORGANIC MATTER, DIGESTION, ACTIVATED SLUDGE, ACTINOMYCETES, HYDROCARBON PESTICIDES, VIRUSES, WATER POLLUTION SOURCES, WATER POLLUTION EFFECTS, PATH OF POLLUTANTS.

IDENTIFIERS:
BDELLOVIBRID, PELODICTYON, POLIOVIRUSES, PLASTICIZERS, PHTHALATE, THERMOPHILIC FUNGI, POLYETHYLENE, POLYPROPYLENE, BACILLUS, CORYNEBACTERIA, STREPTOTHRIX HYORHINA, CADMIUM, DIMETHYLNITROSAMINE, AMINES, CARCINOGENS, TORULOPSIS SPP., ASPERGILLUS, PENICILLIUM, FUSARIA, VIBRIS.

ABSTRACT:
THE ANNUAL CONFERENCE OF THE SOCIETY FOR APPLIED BACTERIOLOGY, HELD AT THE UNIVERSITY OF LIVERPOOL, 13-15 JULY 1971, INCLUDED A TWO-DAY SYMPOSIUM ON 'MICROBIOL ASPECTS OF POLLUTION' AND A ONE-DAY SESSION DEVOTED TO PAPERS ON INDIVIDUAL RESEARCH. THE FOLLOWING TOPICS WERE DISCUSSED: (1) WATER POLLUTION BY DOMESTIC, AGRICULTURAL AND INDUSTRIAL WASTES, (2) SEWAGE TREATMENT USING COMBINED AEROBIC-ANAEROBIC SYSTEMS, (3) MICROBIAL ECOLOGY OF THE ACTIVATED SLUDGE PROCESS, (4) MICROBIAL ASPECTS OF POLLUTION IN THE FOOD AND DAIRY INDUSTRY, (5) POLLUTION OF FRESHWATERS WITH INORGANIC NUTRIENTS, (6) MICROBIAL DEGRADATION OF PLASTICS, HERBICIDES, AND PESTICIDES, (7) AEROBIC METHODS FOR THE TREATMENT OF FARM WASTES, (8) FACTORS AFFECTING ALGAL BLOOMS, (9) THE ROLE OF OBLIGATE ANAROLS IN THE DIGESTION OF ORGANIC MATERIAL, (10) HEALTH HAZARD OF POLLUTION, AND (11) SEWAGE POLLUTION OF NATURAL WATERS. INDIVIDUAL RESEARCH TOPICS INCLUDED (1) 'AN EVALUATION OF PROCEDURES FOR ENUMERATING BACTERIA IN ACTIVATED SLUDGE', (2) 'THE MICROBIAL POLLUTION OF WATER COURSES AS A RESULT OF THE SEWAGE AND ANIMAL WASTES AND THE APPLICATION OF ANIMAL SLURRY TO LAND', AND (3) 'METHODS FOR ANALYZING THE MICROBIAL DECAY OF SOLID WASTES'. (JEFFERIS-BATTELLE)

FIELD 05C, 05B, 05D

ACCESSION NO. W72-09675

A MICROBIOLOGICAL SURVEY IN LAKE ERIE NEAR CLEVELAND, OHIO,

CONNECTICUT UNIV., STORRS. BIOLOGICAL SCIENCES GROUP.

R. P. COLLINS.

COPY AVAILABLE FROM GPO SUP DOC, $0.50; MICROFICHE FROM NTIS AS PB-210 324, $0.95. ENVIRONMENTAL PROTECTION AGENCY, WATER POLLUTION CONTROL RESEARCH SERIES, OCTOBER 1971. 31 P, 14 TAB, 12 REF. EPA PROGRAM 16020 GDQ 10/71.

DESCRIPTORS:
*LAKE ERIE, *POTABLE WATER, *TASTE, *ODOR, SURVEYS, MICROBIOLOGY, WATER TREATMENT, PHYTOPLANKTON, FUNGI, BACTERIA, ALGAE.

IDENTIFIERS:
*CLEVELAND(OHIO), CERATIUM, COELOSPHAERIUM, DINOBRYON, FRAGILARIA, PEDIASTRUM, STAURASTRUM, TABELLARIA, MOUGEOTIA.

ABSTRACT:
PERIODIC TASTE AND ODOR AT THE CLEVELAND, OHIO CROWN WATER TREATMENT PLANT PROMPTED INVESTIGATION OF THE ROLE MICROORGANISMS PLAY IN THIS PROBLEM. FUNGI, BACTERIA, AND ALGAE COLLECTED NEAR THE PLANT INTAKE WERE STUDIED DURING JUNE THROUGH AUGUST 1971. DURING THE THREE MONTHS OF SAMPLING, NO VERTICAL DISTRIBUTION PATTERN WAS NOTED IN QUANTITATIVE ANALYSIS OF THE PHYTOPLANKTON. IN JUNE THE MAXIMUM CONCENTRATION OF PHYTOPLANKTON WAS CONSIDERABLY LOWER THAN THAT OBSERVED IN JULY AND AUGUST. MAXIMUM CONCENTRATION OCCURRED AT THE SURFACE WITH 13,674 ORGANISMS PER LITER; SIMILAR CONCENTRATIONS WERE AT THE 3 AND 9 METER DEPTHS; LOWEST CONCENTRATIONS WERE AT THE 6 AND 12 METER DEPTHS. STUDIES SHOWED THAT FUNGI AND BACTERIA PLAYED LITTLE, IF ANY, ROLE IN THE PROBLEM. A NUMBER OF ALGAE, REPORTED TO INDUCE TASTE AND ODOR IN WATER, WERE IDENTIFIED. WHATEVER THE SOURCE OF THESE ODORS, THEY WERE NOT DUE TO BENTHIC OR PERIPHYTON ALGAE, BUT COULD HAVE BEEN ASSOCIATED WITH THE PHYTOPLANKTON COMMUNITY AS THE REPORTED 'LAKE ERIE ODOR' COINCIDED WITH PHYTOPLANKTON INCREASE. THE TASTE AND ODOR PRODUCING ALGAE INCLUDED: COELOSPHAERIUM, DINOBRYON, FRAGILARIA, PEDIASTRUM, STAURASTRUM, TABELLARIA, AND MOUGEOTIA. THERE WAS NO EVIDENCE THAT BENTHIC ORGANISMS PLAYED ANY SIGNIFICANT ROLE. (JONES-WISCONSIN)

FIELD 05C

ACCESSION NO. W72-10076

STUDY OF THE POSSIBLE ROLE OF POLLUTION IN THE PREVALENCE OF SEA NETTLES IN THE CHESAPEAKE BAY AND THE DEVELOPMENT OF A CENSUS TAKING METHOD.

BIOSPHERICS INC., ROCKVILLE, MD.

AVAILABLE FROM THE NATIONAL TECHNICAL INFORMATION SERVICE AS N72-102 73, $3.00 IN PAPER COPY, $0.95 IN MICROFICHE. FINAL REPORT, OCTOBER 15, 1971. 135 P, 29 FIG, 12 TAB, 26 REF. NASW-2115.

DESCRIPTORS:
*SEA NETTLES, *POLLUTANTS, *MAINTENANCE, *PLANT MORPHOLOGY, *PHOSPHATES, *NITRATES, *AMMONIUM COMPOUNDS, SEWAGE EFFLUENTS, TEMPERATURE CONTROL, ASSAY, REMOTE SENSING, ALGAE, ANTIBIOTICS, MERCURY, ARSENIC COMPOUNDS, CHESAPEAKE BAY.

IDENTIFIERS:
*CHRYSAORA QUINQUECIRRHA, *MEDUSAE MAINTENANCE, *POLYP MAINTENANCE, *POLYP METABOLISM, TETRACYCLINE, SULFANILAMIDE, PENICILLIN G, POLYMYXIN B, PODOCYSTS.

ABSTRACT:
THE EFFECT OF POLLUTANTS ON THE POLYP STAGE OF THE SEA NETTLE, CHRYSAORA QUINQUECIRRHA, AND A MEANS TO DETECT THE MEDUSAE FORM BY REMOTE SENSING WERE INVESTIGATED. PHOSPHATE, NITRATE, AMMONIUM, COMBINATIONS OF THESE, PH FROM 6-8, AND SYNTHETIC SEWAGE EFFLUENTS WERE EXAMINED FOR MAINTENANCE AND MORPHOLOGY OF THE POLYPS. PHOSPHATE, NITRATE, AND THEIR COMBINATIONS WERE FOUND TO CONTRIBUTE TO THE PROLIFERATION OF POLYPS, PH HAD NO EFFECT WHILE AMMONIUM, AMMONIUM COMBINATIONS AND SEWAGE WERE FOUND TO BE DETRIMENTAL TO POLYPS. PHOSPHATE AND NITRATE SEEMED TO ACT AS A PROTECTION AGAINST THE LETHAL EFFECTS OF AMMONIUM. AN ASSAY BASED ON THE LABELED RELEASE TECHNIQUE WAS USED TO MEASURE POLYP METABOLISM. THE RESULTS SUPPORT THOSE OBTAINED IN THE MAINTENANCE STUDY SUGGESTING A POTENTIAL TO PREDICT AND SUPPORT THE MORPHOLOGICAL EFFECTS OF POLLUTANTS ON POLYPS. (ENSIGN-PAI)

FIELD 05C, 02L

ACCESSION NO. W72-10162

BIOCHEMICAL CHANGES IN OXIDATION PONDS,

MAHARAJA SAYAJIRAO UNIV. OF BARODA (INDIA). DEPT. OF BIOCHEMISTRY.

P. M. AMIN, AND S. V. GANAPATI.

JOURNAL WATER POLLUTION CONTROL FEDERATION, VOL 44, NO 2, FEBRUARY 1972, P 183-200, 11 FIG, 5 TAB, 68 REF.

DESCRIPTORS:
*OXIDATION LAGOONS, *ON-SITE INVESTIGATIONS, *BACTERIA, *ALGAE, OXIDATION, METABOLISM, RESPIRATION, LIPIDS, AMMONIA, CARBOHYDRATES, CARBON DIOXIDE, POLYMERS, HYDROGEN ION CONCENTRATION, PROTOZOA, DISSOLVED OXYGEN, SLUDGE, WASTE WATER TREATMENT.

ABSTRACT:
A LABORATORY STUDY OF THE FIRST OPERATIONAL STAGE OF A WASTEWATER LAGOON, NAMELY, THE 28 DAY PERIOD OF NON-CHANGING ENVIRONMENT WHEN ALGAE ARE ALLOWED TO DEVELOP NATURALLY, SHOWED THAT THIS STAGE IS SEPARATED INTO TWO DISTINCT PHASES, THE BACTERIAL PHASE I AND THE ALGAL PHASE II. THE MOST NOTABLE FINDINGS WERE: (1) ABSENCE OF ACIDITY IN BOTH PHASES; (2) ABSENCE OF DISSOLVED OXYGEN IN PHASE I AND ITS ABUNDANCE IN PHASE II; (3) LARGE REDUCTIONS IN COLIFORM DENSITY; (4) LARGE INCREASES AND DECREASES IN BIOCHEMICAL CONSTITUENTS, SUCH AS SUGARS, IN PHASE I AND PHASE II, RESPECTIVELY; (5) INCREASES IN FATTY SUBSTANCES AND CHLOROPHYLLS; (6) LARGE PROTOZOA POPULATION IN PHASE I COMPARED TO LARGE ALGAL POPULATIONS IN PHASE II; AND (7) LACK OF APPRECIABLE SLUDGE FORMATION. CARBOHYDRATE USAGE MOSTLY FOR SYNTHESIS OF FATTY SUBSTANCES AND MUCH LESS FOR POLYMER ACCUMULATIONS WITHIN CELLS, RESULTED IN MORE FAT IN THE CLEAR FINAL EFFLUENT, THUS EXPLAINING THE ABSENCE OF SLUDGE IN THE ECOSYSTEM. (LOWRY-TEXAS)

FIELD 05D, 05G

ACCESSION NO. W72-10233

REMOVAL OF ALGAE FROM WASTE STABILIZATION POND EFFLUENTS--A STATE OF THE ART,

ILLINOIS STATE WATER SURVEY, URBANA.

V. KOTHANDARAMAN, AND R. L. EVANS.

STATE OF ILLINOIS DEPARTMENT OF REGISTRATION AND EDUCATION CIRCULAR NO 108, 1972. 9 P, 3 FIG, 1 TAB, 16 REF.

DESCRIPTORS:
*OXIDATION LAGOONS, *HARVESTING OF ALGAE, *SEPARATION TECHNIQUES, CHEMICAL PRECIPITATION, COAGULATION, SEDIMENTATION, POLYELECTROLYTES, LIME, NEUTRALIZATION, FILTRATION, CENTRIFUGATION, VACUUM DRYING, FLOTATION, WASTEWATER TREATMENT.

IDENTIFIERS:
*AGGLOMERATION.

ABSTRACT:
UNI-ALGAL CELLS HAVE BEEN FOUND TO CARRY A NEGATIVE CHARGE AT PH BETWEEN 2 AND 11. AT PH 2 AND PH 11, THE ALGAL CELLS POSSESS A VERY HIGH CHARGE DENSITY, WHILE AT PH 7 A VERY LOW NEGATIVE CHARGE DENSITY IS EXHIBITED. CHEMICAL PRECIPITATION OF ALGAE HAS BEEN POSTULATED AS BEING DUE TO CHARGE NEUTRALIZATION, AGGLOMERATION, AND SEDIMENTATION. ALGAE HANDLING IN THE REMOVAL PROCESS CONSISTS OF CONCENTRATION OF ALGAL CELLS FROM 200-4000 MG/L TO 1 TO 4% BY WEIGHT, DEWATERING THE 1 TO 4% MIXTURE TO BETWEEN 8 AND 20%, AND DRYING THE 8 TO 20% MIXTURE TO AN 85 TO 90% MIXTURE OF ALGAL CELLS. CHEMICAL COAGULANTS, PARTICULARLY LIME, ALUM, FERRIC SALTS AND CATIONIC POLYMERS, ARE EFFECTIVE IN CAUSING COAGULATION AND SEDIMENTATION OF ALGAL CELLS. DEWATERING AND DRYING OF ALGAL SLURRIES FROM THE CONCENTRATION STEP CAN BE MOST ECONOMICALLY ACHIEVED BY SAND BED APPLICATION. VACUUM FILTRATION HAS BEEN ONLY PARTIALLY SUCCESSFUL, WHILE CENTRIFUGATION HAS BEEN TOO EXPENSIVE. A MARKET FOR SEWAGE-GROWN ALGAE AS LIVESTOCK FEED HAS YET TO DEVELOP, BUT USE AS A SOIL CONDITIONER HAS BEEN SUGGESTED AND OTHER CONVENTIONAL DISPOSAL METHODS ARE BEING INVESTIGATED. (LOWRY-TEXAS)

FIELD 05D

ACCESSION NO. W72-10237

NUTRIENT REMOVAL BY NATURAL GAS FERMENTATION,

BRITISH COLUMBIA RESEARCH COUNCIL, VANCOUVER. DIV. OF APPLIED BIOLOGY.

J. C. MUELLER.

JOURNAL WATER POLLUTION CONTROL FEDERATION, VOL 44, NO 1, JANUARY 1972, P 25-33, 4 FIG, 10 TAB, 13 REF.

DESCRIPTORS:
*NUTRIENTS, *MUNICIPAL WASTES, *SEPARATION TECHNIQUES, FERMENTATION, NITROGEN, PHOSPHORUS, AMMONIA, METHANE, TEMPERATURE, HYDROGEN ION CONCENTRATION, PROTEIN, LABORATORY TESTS, COST ANALYSIS, ALGAE, WASTE WATER TREATMENT.

ABSTRACT:
NITROGEN AND PHOSPHORUS WERE REMOVED FROM SECONDARY WASTE TREATMENT PLANT EFFLUENTS BY FERMENTATION OF GASEOUS HYDROCARBONS TO ASSIMILATE BOTH NITROGEN AND ORTHOPHOSPHATE INTO A BACTERIA CELL MASS. IN LABORATORY TESTS, 93 TO 100% OF THE NITROGEN AND 98 TO 99% OF THE PHOSPHORUS WERE RECOVERED IN THE FORM OF A PROTEIN RICH BIOMASS. THIS BIOMASS WAS SHOWN TO CONTAIN 60 TO 70% PROTEIN, INDICATING THAT ONE POSSIBLE OUTLET FOR THE PRODUCT COULD BE ANIMAL FODDER OR FEED SUPPLEMENTS. AT PRESENT EFFICIENCIES, AS INDICATED IN THE LABORATORY TESTS, PRODUCTION COSTS OF THE BIOMASS PROTEIN WERE ESTIMATED AT 25% MORE THAN THE COST OF COMMERCIALLY AVAILABLE ANIMAL FEED SUPPLEMENTS. (LOWRY-TEXAS)

FIELD 05D

ACCESSION NO. W72-10390

SEQUENTIAL PROCESSING IN WASTEWATER LAGOONS,

ARIZONA STATE UNIV., TEMPE.

J. W. KLOCK.

JOURNAL WATER POLLUTION CONTROL FEDERATION, VOL 44, NO 2, FEBRUARY 1972, P 241-254. 12 FIG, 3 TAB, 5 REF.

DESCRIPTORS:
*OXIDATION LAGOONS, *CHANNELS, *HEAT TRANSFER, *PLASTICS, DOMESTIC WASTES, BACTERIA, ALGAE, NUTRIENTS, OXIDATION-REDUCTION POTENTIAL, ODORS, OPERATION AND MAINTENANCE, WASTE WATER TREATMENT.

ABSTRACT:
A RELIABLE AND SIMPLY OPERATED LAGOON, INCORPORATING THE PRINCIPLES OF HEAT CONSERVATION, SEQUENTIAL-PHASE PROCESSING, AND UTILIZATION OF THIN PLASTIC FILMS TO FORM CHANNEL BARRIERS AND HEAT TRANSFER SURFACES, WAS DEVELOPED. THE CONSTRUCTION OF THE SYSTEM INCLUDED THE USE OF CHANNELS CONSTRUCTED OF BLACK POLYETHYLENE SHEETS. THE INFLUENT TO THE SYSTEM FLOWED INTO THE LAGOON IN THE LOWER PHASE I, AND THE PHASE II CHANNELS OF THE BLACK POLYETHYLENE FLOATED ON THE SURFACE OF THE PHASE I LIQUID. THIS ARRANGEMENT EFFECTIVELY PREVENTED HEAT LOSS DIRECTLY TO THE ATMOSPHERE ABOVE THE PHASE I LIQUID, AND THE THIN PLASTIC MATERIAL ALLOWED HEAT EXCHANGE FROM THE LOWER PHASE TO THE UPPER PHASE. PHASE I WAS DESIGNED AS STRICTLY A BACTERIAL PHASE TO PREVENT PREMATURE ALGAE GROWTH AND SUBSEQUENT NUTRIENT FIXATION, WITH PHASE II INTENDED AS AN ALGAL PHASE FOLLOWED BY LIMITED GROWTH OF CRUSTACEANS AND AQUATIC INSECTS. AVERAGE BOD REDUCTION WAS 76.8%, WITH MINIMAL SURFACE SOLIDS AND NO ODORS. ASSOCIATED OPERATIONAL PROBLEMS AND SOLUTIONS ARE PRESENTED. (LOWRY-TEXAS)

FIELD 05D

ACCESSION NO. W72-10401

LIMNOLOGICAL STUDIES OF LAKE JACOMO, JACKSON COUNTY, MISSOURI. I. WATER QUALITY AND SURFACE PLANKTON, 1970 - 1971,

MISSOURI WATER RESOURCES RESEARCH CENTER, COLUMBIA.

D. H. STERN, AND M. S. STERN.

AVAILABLE FROM THE NATIONAL TECHNICAL INFORMATION SERVICE AS PB-210 587, $3.00 IN PAPER COPY, $0.95 IN MICROFICHE. MISSOURI WATER RESOURCES RESEARCH CENTER COMPLETION REPORT MAY 1972. 61 P, 14 FIG, 4 TAB, 18 REF. PROJECT NO. OWRR A-041-MO(1). 14-31-0001-3225.

DESCRIPTORS:
*ALGAE, *EUTROPHICATION, *NUTRIENTS, *PLANKTON, *RECREATION, *WATER QUALITY, CYANOPHYTA, LAKES, MISSOURI, NITROGEN, PHOSPHORUS.

IDENTIFIERS:
LAKE JACOMO(MO).

ABSTRACT:
WATER QUALITY AND SURFACE PLANKTON IN LAKE JACOMO, THE PRIMARY RECREATIONAL WATER FOR KANSAS CITY, WERE STUDIED SYNOPTICALLY FROM JUNE 1970 THROUGH MAY 1971. THE LAKE, WHICH HAS 2 MILLION VISITORS PER YEAR, HAS BEEN CULTURALLY EUTROPHICATED BY NUTRIENT-RICH RUNOFF FROM WATERSHED LAND CONTAINING FARMS, FEEDLOTS, AND URBAN AREAS. NUISANCE BLOOMS ARE PROMOTED BY ITS DISSECTED SHORELINE AND RELATIVELY SMALL SIZE. AMMONIA NITROGEN CONCENTRATIONS RANGED FROM 0.25 PPM TO 2.64 PPM, NITRATE NITROGEN FROM 0.02 PPM TO 1.00 PPM, AND ORTHOPHOSPHATE FROM 0.02 PPM TO 1.79 PPM. ONE-HUNDRED AND EIGHTY PLANKTON TAXA WERE IDENTIFIED. CYANOPHYTA, IN PARTICULAR APHANIZOMENON FLOS-AQUAE, WERE DOMINANT, EXCEPT IN JUNE AND FROM MID-NOVEMBER THROUGH JANUARY WHEN CHRYSOPHYTA DOMINATED THE PLANKTON. RECOMMENDATIONS FROM THIS STUDY INCLUDE THE FOLLOWING: (1) PURCHASE OR ZONING CONTROL OF THE LAKE'S WATERSHED; (2) LAKE BANK STABILIZATION BY THE PARK DEPARTMENT; (3) ENCOURAGEMENT OF APPROPRIATE POWERBOAT OPERATION TO MINIMIZE WAKES ALONG UNSTABLE BANKS; (4) REDUCTION OF NUTRIENT-RICH RUNOFF FROM AGRIBUSINESSES; (5) EROSION CONTROL BY WISE CONSTRUCTION AND FARMING PRACTICES; (6) REDUCTION OF BLUE-GREEN ALGAE BY REPEATED APPLICATIONS OF COPPER SULFATE OR COPPER-CONTAINING ALGICIDES; AND (7) LIMITED DREDGING OF PRODUCTIVE SHALLOW COVES. AERATION AND DESTRATIFICATION ARE NOT RECOMMENDED AT THIS TIME. (MISSOURI ABSTRACT)

FIELD 05C

ACCESSION NO. W72-10431

THE CARBON DIOXIDE SYSTEM AND EUTROPHICATION,

WARF INST., INC., MADISON, WIS.

S. D. MORTON, P. H. DERSE, AND R. C. SERNAU.

COPY AVAILABLE FROM GPO SUP DOC, $0.75; MICROFICHE FROM NTIS AS PB-210 706, $0.95. ENVIRONMENTAL PROTECTION AGENCY, WATER POLLUTION CONTROL RESEARCH SERIES NOVEMBER 1971. 72 P, 2 FIG, 8 TAB, 15 REF, APPEND. EPA PROGRAM 16010 DXV 11/71.

DESCRIPTORS:
*NUISANCE ALGAE, *CARBON DIOXIDE, *ALGAL CONTROL, CYANOPHYTA, CARBON, CHLORELLA, BICARBONATES, AERATION, LIMITING FACTORS, EUTROPHICATION.

IDENTIFIERS:
*ATMOSPHERIC CARBON, MICROCYSTIS, ANABAENA.

ABSTRACT:
TO DETERMINE THE FEASIBILITY OF EUTROPHICATION CONTROL BY CONTROLLING CARBON, THREE MAJOR AREAS WERE STUDIED: THE STEADY STATE, IN WHICH THE GROWTH RATES OF ALGAE AT VARIOUS CONSTANT, MAINTAINED DISSOLVED CARBON DIOXIDE CONCENTRATIONS WERE DETERMINED; THE NON-EQUILIBRIUM, WHERE NATURAL ATMOSPHERIC REPLENISHMENT WAS THE SOLE CARBON SOURCE; AND ALGAL GROWTH WITH INORGANIC BICARBONATE AS THE SOLE CARBON SOURCE. IN STUDYING GROWTH RATES OF CHLORELLA, MICROCYSTIS, AND ANABAENA WITH RESPECT TO CARBON AVAILABILITY, IT WAS FOUND THAT ALGAE CAN UTILIZE DISSOLVED CONCENTRATIONS OF CARBON DIOXIDE MUCH LOWER THAN THOSE FROM ATMOSPHERIC EQUILIBRIA. ALGAL GROWTH CONTROL BY SWEEPING CARBON DIOXIDE OUT BY AERATION WITH AIR CONTAINING VERY LOW CARBON DIOXIDE CONCENTRATIONS IS DIFFICULT BECAUSE OF ATMOSPHERIC REPLENISHMENT OF CARBON DIOXIDE. BICARBONATE IS AT LEAST 50% UTILIZED AT GROWTH RATES AS HIGH AS 7 MG/L PER DAY. ATMOSPHERIC REPLENISHMENT OF CARBON DIOXIDE, WITHOUT ANY WIND MIXING, CAN SUSTAIN GROWTH RATES OF 1.5-2 MG/L PER DAY FOR DEPTHS OF AT LEAST 1.7 METERS. (JONES-WISCONSIN)

FIELD 05C, 05G

ACCESSION NO. W72-10607

NUTRIENT SOURCES FOR ALGAE AND THEIR CONTROL,

WISCONSIN UNIV., MADISON. WATER RESOURCES CENTER.

G. P. FITZGERALD.

COPY AVAILABLE FROM GPO SUP DOC, $1.00; MICROFICHE FROM NTIS AS PB-210 707, $0.95. ENVIRONMENTAL PROTECTION AGENCY, WATER POLLUTION CONTROL RESEARCH SERIES AUGUST 1971. 77 P, 4 FIG, 26 TAB, 53 REF. EPA PROGRAM 16010 EHR 08/71.

DESCRIPTORS:
*ANALYTICAL TECHNIQUES, *NUTRIENTS, *ALGAE, *BIOASSAY, PHOSPHORUS COMPOUNDS, NITROGEN COMPOUNDS, CYCLING NUTRIENTS, ESSENTIAL NUTRIENTS, LAKES, MUD, WISCONSIN, AQUATIC WEEDS, EUTROPHICATION.

IDENTIFIERS:
*NUTRIENT SOURCES, ACETYLENE REDUCTION, LAKE MENDOTA(WIS), LAKE MONONA(WIS).

ABSTRACT:
BIOASSAYS FOR NUTRIENT AVAILABILITY WERE EVALUATED TO DEFINE CONDITIONS AND LIMITS UNDER WHICH EACH METHOD CAN GIVE MEANINGFUL RESULTS. THE BIOLOGICAL AVAILABILITY OF ALGAL NUTRIENTS IN A WATER SAMPLE AND THE ALGAL RESPONSE TO CHANGES IN THE GROWTH-LIMITING NUTRIENT WERE MEASURED. FACTORS OTHER THAN INSOLUBILITY PREVENT THE NITROGEN OR PHOSPHORUS OF CERTAIN SAMPLES OF AEROBIC LAKE MUDS FROM BEING READILY AVAILABLE FOR ALGAL GROWTH; P-LIMITED SPIROGYRA HAS BEEN FOUND GROWING THROUGH MUD LAYERS CONTAINING 0.1% TOTAL PHOSPHORUS. THE FACTS, THAT LIVE ALGAE AND AQUATIC WEEDS DO NOT SHARE THEIR ADEQUATE OR SURPLUS NUTRIENTS WITH NUTRIENT-LIMITED ALGAE AND THAT LAKE MUDS DO NOT PROVIDE READILY AVAILABLE NITROGEN OR PHOSPHORUS, INDICATE THAT ONCE LAKE WATERS ARE STRIPPED OF AVAILABLE NUTRIENTS BY PLANT PRODUCTION, FURTHER PLANT PRODUCTION WILL DEPEND UPON NUTRIENTS FROM CONTINUOUS SOURCES, SUCH AS WASTEWATER EFFLUENTS. PHOSPHORUS-STARVED CELLS OF ANABAENA RAPIDLY INCREASE THEIR CAPACITY TO REDUCE ACETYLENE TO ETHYLENE WHEN THEY RECEIVE PHOSPHORUS. THIS RESPONSE MAY BE USED AS A BIOASSAY FOR DETECTING AVAILABLE PHOSPHORUS IN AQUATIC ECOSYSTEMS. SENSITIVITY OF THE METHOD COMPARES FAVORABLY WITH CONVENTIONAL METHODS FOR MEASURING DISSOLVED ORTHOPHOSPHATE, AND HAS THE ADVANTAGE THAT IT MEASURES AVAILABLE PHOSPHORUS. (JONES-WISCONSIN)

FIELD 05C, 02H

ACCESSION NO. W72-10608

PLANKTONIC BLUE-GREEN ALGAE: GROWTH AND ODOR-PRODUCTION STUDIES,

NORTH TEXAS STATE UNIV., DENTON. DEPT. OF BIOLOGICAL SCIENCES; AND TELEDYNE BROWN ENGINEERS, HUNTSVILLE, ALA. AND EAST TENNESSEE STATE UNIV., JOHNSON CITY. DEPT. OF BIOLOGY.

J. K. SILVEY, D. E. HENLEY, AND J. T. WYATT.

JOURNAL OF THE AMERICAN WATER WORKS ASSOCIATION, P 35-39, JANUARY 1972. 2 FIG, 47 REF.

DESCRIPTORS:
*DOMESTIC WATER, *CYANOPHYTA, RESERVOIRS, ODOR-PRODUCING ALGAE, SOUTHWEST U.S., TEMPERATURE, OXYGEN, LIGHT INTENSITY, NUTRIENTS, ACTINOMYCETES, TASTE, EUTROPHICATION.

IDENTIFIERS:
ANABAENA CIRCINALIS, ANABAENA CYLINDRICA, APHANIZOMENON FLOS-AQUAE.

ABSTRACT:
PRODUCTION AND CONTROL OF BLUE-GREEN ALGAE, ESPECIALLY THOSE RESPONSIBLE FOR TASTE AND ODOR IN WATER SUPPLIES, ARE SURVEYED. BLUE-GREEN ALGAL BLOOMS IN SOUTHWESTERN RESERVOIRS ALMOST ALWAYS CAUSE TASTE AND ODOR PROBLEMS. MAJOR OFFENDERS ARE THE HETEROCYSTOUS BLUE-GREEN FORMS, PARTICULARLY ANABAENA CIRCINALIS AND APHANIZOMENON FLOS-AQUAE. THE BOTTOM WATERS OF A RESERVOIR SHOULD BE EXAMINED TO DETERMINE THE PHYSICOCHEMICAL AND BIOLOGICAL PARAMETERS INFLUENCING THESE BLOOMS. CONTROL HAS BEEN PRIMARILY LIMITED TO CHEMICAL TREATMENT OF THE WATER IN TREATMENT PLANTS. ACTIVATED CARBON REMOVES SOME ODOR; CHLORINE REMOVES SOME TASTES AND ODORS, BUT IT IS KNOWN TO INTENSIFY OTHERS. THE MOST LOGICAL METHOD OF CONTROLLING TASTES AND ODORS IN WATER SUPPLIES IS TO RESEARCH THEIR SOURCE. NEW APPROACHES OF ECONOMIC IMPORTANCE INCLUDE USE OF WEAK ELECTROLYTES TO CAUSE LYSIS OF ALGAL CELLS, USE OF VIRAL AGENTS AND BACTERIA THAT LYSE BLUE-GREEN ALGAE, AND DEVELOPMENT OF AGENTS THAT WOULD GRADUALLY RELEASE ALGICIDES WHILE FLOATING OR AFTER SINKING TO THE BOTTOM. MANIPULATION OF ECOLOGICAL CONDITIONS MAY PROVE TO BE THE MOST DESIRABLE CONTROL METHOD. EXPERIMENTATION IN THIS AREA HAS INCLUDED FORCED AERATION AND CIRCULATION AND CONTINUOUS LOW-LEVEL SUPPLIES OF NITROGEN COMPOUNDS. (JONES-WISCONSIN)

FIELD 05C, 02H, 05G

ACCESSION NO. W72-10613

A TECHNIQUE FOR BIOASSAY OF FRESHWATER, WITH SPECIAL REFERENCE TO ALGAL ECOLOGY,

FRESHWATER BIOLOGICAL ASSOCIATION, WINDERMERE (ENGLAND).

J. W. G. LUND, G. H. M. JAWORSKI, AND H. BUCKA.

ACTA HYDROBIOLOGICA, VOL. 13, NO. 3, P 235-249, 1971. 3 FIG, 2 REF.

DESCRIPTORS:
*ANALYTICAL TECHNIQUES, *TESTING PROCEDURES, *BIOASSAY, *ALGAE, FRESHWATER, MEASUREMENT, DIATOMS, EUTROPHICATION.

IDENTIFIERS:
BRITISH ISLES.

ABSTRACT:
THE USE OF BIOASSAY ENLARGES KNOWLEDGE OF ALGAL ECOLOGY AS OBTAINED ON THE BASIS OF THEIR GROWTH IN CULTURES. AMONG OTHER USES IT CAN BE APPLIED FOR DETERMINATION OF THE POTENTIAL FERTILITY OF WATERS, AS WELL AS UTILIZED FOR THEIR DIFFERENTIATION. A TECHNIQUE FOR BIOASSAY OF FRESHWATER AND POSSIBILITIES OF ITS PRACTICAL APPLICATION ARE DESCRIBED. THE GROWTH OF THE ALGA IN THE SAMPLES IS A MEASURE OF THE RELATIVE POTENTIAL, QUALITATIVE AND QUANTITATIVE, FERTILITIES OF THESE WATERS. THE METHOD DESCRIBED HAS ALREADY BEEN OF VALUE IN UNPUBLISHED TESTS ON WATER FROM BRITISH LAKES. IT CONSISTS BASICALLY OF THE COLLECTION OF A WATER SAMPLE, FILTRATION OF THE WATER THROUGH GLASS FIBER INTO FLASKS, ADDITION OF A TEST ALGA TO ALL THESE FLASKS AND OF CERTAIN SUBSTANCES, SUCH AS NUTRIENTS, TO SOME OF THE FLASKS. THE WATER IN THE FLASKS IS TRANSFERRED TO EXPERIMENTAL TRANSPARENT OR TRANSLUCENT CONTAINERS WHICH ARE THEN EXPOSED FOR GIVEN TIMES TO CONSTANT CONDITIONS OF LIGHT AND TEMPERATURE. TO REDUCE CHEMICAL CONTAMINATION OR ALTERATION OFTHE WATER SAMPLE TO A MINIMUM, ALL GLASSWARE AND OTHER EQUIPMENT IS CLEANED AND WASHED VERY CAREFULLY. (JONES-WISCONSIN)

FIELD 05C, 05A

ACCESSION NO. W72-10619

EUTROPHICATION,

FRESHWATER BIOLOGICAL ASSOCIATION, AMBLESIDE (ENGLAND).

J. W. G. LUND.

PROCEEDINGS OF THE ROYAL SOCIETY OF LONDON, B, VOL. 180, P 371-382, 1972. 3 FIG, 25 REF.

DESCRIPTORS:
*EUTROPHICATION, *NUTRIENTS, *ALGAE, SEWAGE TREATMENT, WATER POLLUTION SOURCES, WATER POLLUTION EFFECTS, AGRICULTURAL RUNOFF, PHOSPHORUS, NITROGEN, PHOSPHATES, DETERGENTS, POTABLE WATER.

IDENTIFIERS:
ASTERIONELLA FORMOSA, BLELHAM TARN(ENGLAND), WINDERMERE(ENGLAND).

ABSTRACT:
ALTHOUGH THE MAIN CAUSES OF EUTROPHICATION ARE WELL KNOWN, THERE IS LITTLE DETAILED UNDERSTANDING OF THEIR INTERACTION WITH OTHER ENVIRONMENTAL FACTORS; IT IS NOT YET POSSIBLE TO FORECAST THE EXACT CHANGES TO BE EXPECTED IN AQUATIC ECOSYSTEMS. OVER THE LAST 27 YEARS, OBSERVATIONS WERE MADE ON TWO ENGLISH LAKE DISTRICT WATERS--WINDERMERE, RECEIVING THE MAIN SOURCE OF URBAN SEWAGE, AND BLELHAM TARN; PHYTOPLANKTON WAS EXAMINED, USUALLY AT WEEKLY INTERVALS; CONCENTRATIONS OF PHOSPHATES, NITRATES, AND SILICATES WERE ALSO DETERMINED WEEKLY NEARLY ALL THROUGH THIS PERIOD; OTHER CHEMICAL ANALYSES HAVE BEEN CARRIED OUT OVER SHORTER PERIODS. THE MAXIMUM CONCENTRATION OF PHOSPHATES HAS RISEN IN BOTH WATERS. A COMPARISON OF THE CHANGES IN THE ABUNDANCE AND RATE OF GROWTH OF SOME MAJOR ALGAE WAS MADE. RESULTS SHOWED THAT CHANGES IN THE TWO WATERS HAVE NOT FOLLOWED THE SAME PATH. MOREOVER, CERTAIN ASPECTS OF THE CHEMICAL CHANGES IN THE WATER ARE DIFFICULT TO UNDERSTAND. THE INCREASE IN PHOSPHATE OVER THE YEARS HAS FOLLOWED A SIMILAR COURSE IN EACH BODY OF WATER BUT THE AVERAGE NUMBER OF CELLS OF ASTERIONELLA HAS DECREASED IN WINDERMERE AND INCREASED IN BLELHAM TARN. (JONES-WISCONSIN)

FIELD 05C

ACCESSION NO. W72-10625

SURVEY OF TOXAPHENE LEVELS IN GEORGIA ESTUARIES,

GEORGIA UNIV., SAPELO ISLAND. MARINE INST.

R. J. REIMOLD, AND C. J. DURANT.

AVAILABLE FROM THE MARINE RESOURCE CENTER, SAVANNAH, GA. TECHNICAL REPORT SERIES NO. 72-2, FEB. 1972, 51 P, 2 FIG, 25 TAB, 10 REF.

DESCRIPTORS:
*SURVEYS, *DATA COLLECTIONS, *HYDROLOGIC DATA, *MEASUREMENT, *ASSAY, ANALYTICAL TECHNIQUES, EVALUATION, WATER POLLUTION, POLLUTANT IDENTIFICATION, SAMPLING, BIOASSAY, ENVIRONMENT, TOXICITY, POISONS, ALGAL TOXINS, FISH TOXINS, SEDIMENTS, SALT MARSHES, ECOLOGY, PESTICIDES, ESTUARIES, INTERTIDAL AREAS, ESTUARINE FISHERIES.

ABSTRACT:
DATA COLLECTED AND PROCESSED FROM 1 AUGUST 1970 TO 31 JULY 1971 ARE INCLUDED. AN EFFORT IS MADE TO COVER EACH PORTION OF THE RESEARCH WITH EXPLANATIONS OF METHODOLOGY AND RESULTS. THIS PROCEDURE DEVIATES FROM THE NORMAL INTRODUCTION, METHODS, RESULTS AND DISCUSSION TO PROVIDE THE READER WITH A COHERENT SUMMARY OF THE RESEARCH FINDINGS. THE SECTIONS ARE SUBDIVIDED IN THE FOLLOWING BROAD CATEGORIES: (1) ENVIRONMENTAL TOXAPHENE RESIDUE RESULTS (EXCEPT SEDIMENT); (2) SEDIMENT ANALYSIS; (3) BIOASSAYS; (4) TRAWL DATA; AND (5) SUMMARY OF FINDINGS RELATED TO THE PROBLEMS OF TOXAPHENE CONTAMINATION IN THE SALT MARSH. (HOUSER-ORNL)

FIELD 05B, 05C, 02L

ACCESSION NO. W72-10678

RADIUM IN AQUATIC FOOD CHAINS: RADIUM UPTAKE BY FRESH WATER ALGAE,

ATOMIC ENERGY OF CANADA LTD., CHALK RIVER (ONTARIO).

B. HAVLIK.

RADIATION RESEARCH VOL 46, NO. 3, JUNE 1971, P 490-505 10 FIG, 3 TAB, 31 REF.

DESCRIPTORS:
*RADIOISOTOPES, *RADIUM, *ABSORPTION, *PATH OF POLLUTANTS, *WATER POLLUTION, WATER POLLUTION SOURCES, AQUATIC ALGAE, FOOD CHAINS, MEASUREMENT, ADSORPTION.

IDENTIFIERS:
CONCENTRATION, UPTAKE.

ABSTRACT:
ACCUMULATION OF 226RA WAS STUDIED IN FOUR SPECIES OF GREEN ALGAE AND TWO SPECIES OF BLUE-GREEN ALGAE. ALGAE WERE CULTIVATED IN INORGANIC CULTURE. RADIUM WAS ADDED IN AMOUNTS OF 1 MICRO CI/1., 0.1 MICRO CI/1. AND 0.01 MICRO CI/1. AFTER 14 DAYS THE AMOUNT OF RADIUM IN THE MEDIUM IN DEAD AND LIVING ALGAL CELLS, AND IN WASHES FROM THE ALGAE, WAS DETERMINED AT INTERVALS USING A LIQUID SCINTILLATION COUNTER. FROM 50 TO 80% OF THE RADIUM WAS ABSORBED WITHIN THE CELLS AND THE AMOUNT ABSORBED WAS PROPORTIONAL TO THE LENGTH OF EXPOSURE. RADIUM WAS MOSTLY ADSORBED ON THE CELL SURFACE (25-50%) AND ONLY 1-8% WAS PRESENT WITHIN THE CELLS. THE HIGHEST ACCUMULATIVE FACTOR WAS REACHED AFTER 24 HOURS EXPOSURE. THE ACCUMULATIVE FACTOR OF THE RADIUM ABSORBED BY THE ALGAE WAS INVERSELY PROPORTIONAL TO CONCENTRATION OF RADIUM IN THE MEDIUM AND WAS DEPENDENT ON: THE SPECIES OF ALGAE; THE CONCENTRATION RADIUM IN THE MEDIUM; THE GROWTH RATE OF THE ALGAE AND THEIR PHYSIOLOGICAL CONDITION; THE PERIOD OF EXPOSURE. THE FACTORS RESPONSIBLE FOR RADIUM ACCUMULATION WERE ADSORPTION, ABSORPTION AND INCORPORATION, IN THAT ORDER OF IMPORTANCE. (HOUSER-ORNL)

FIELD 05C, 05B

ACCESSION NO. W72-10686

EFFECT OF A MIXTURE OF URANIUM FISSION PRODUCTS ON THE SANITARY CONDITIONS AND HYDROBIONTS OF WEAKLY-MINERALIZED FRESH-WATER BASINS, (IN RUSSIAN),

V. N. GUSKOVA, A. N. BRAGINA, A. A. ZASEDATELEV, B. N. ILYIN, AND V. M. KUPRIYANOVA.

GIDROBIOL. ZH., VOL 6, NO. 4, P 5-11, 1970, 3 TAB, 3 FIG.

DESCRIPTORS:
*RADIOISOTOPES, *RADIOACTIVITY EFFECTS, *RADIOECOLOGY, *RESERVOIRS, *FRESHWATER, HYDROBIOLOGY, ALGAE, FISH, ABSORPTION, BIOASSAY, METABOLISM.

IDENTIFIERS:
CONCENTRATION, MUSCLE, BONE PHYSIOLOGY/METABOLISM.

ABSTRACT:
EXPERIMENTAL INVESTIGATIONS DID NOT ESTABLISH A CONSIDERABLE NEGATIVE EFFECT OF TWO MIXTURES OF FISSION RADIONUCLIDES IN CONCENTRATIONS OF 2.0 X 10-1 AND 1.0 X 10-5 CI/1 ON SANITARY CONDITIONS AND HYDROBIANTS OF A RESERVOIR. A CONCENTRATION OF THE MIXTURES OF 1.0 X 10-3 CI/1 CAUSED INHIBITION OF BIOCHEMICAL CONSUMPTION OF OXYGEN AND DEVELOPMENT OF FISH SPAWN. A DECREASE IN CONCENTRATION OF MIXTURES (RADIONUCLIDES) IN WATER OCCURRED MOST INTENSIVELY (BY 50 TO 60% AS COMPARED WITH THE INITIAL ONE) DUE TO PROTOCOCCUS ALGAE AND DUCKWEED. THE LOWEST VALUES FOR THE ACCUMULATION COEFFICIENTS OF THE SEPARATE ISOTOPE MIXTURES, THE CONCENTRATION BEING 1.0 X 10-8 CI/1 IN WATER, WERE FOUND IN THE MUSCLES OF FISH. AXIS SKELOION ACCUMULATED MAINLY 89SR AND 140BA. (HOUSER-ORNL)

FIELD 05C

ACCESSION NO. W72-10694

THE INFLUENCE OF LIVING AND DEAD CELLS OF CHLORELLA VULGARIS AND SCENEDESMUS OBLIQUUS ON AQUATIC MICROORGANISMS,

WYZSZA SZKOLA ROLNICZA, OLSZTYN-KORTOWA (POLAND).

S. NIEWOLAK.

POLSKIE ARCHIWUM HYDROBIOLOGII, VOL 18, NO 1, P 43-54, 1971. 8 FIG, 38 REF.

DESCRIPTORS:
*CHLORELLA, *SCENEDESMUS, ALGAE, MICROORGANISMS, ENVIRONMENTAL EFFECTS, GROWTH RATES, E. COLI, AZOTOBACTER, PSEUDOMONAS, WATER POLLUTION EFFECTS.

IDENTIFIERS:
*MICROBIAL GROWTH, MICROCOCCUS, BACILLUS, VIBRIO, RHODOTORULA, AEROMONAS.

ABSTRACT:
THE STUDY CONCERNED THE INFLUENCE OF LIVING AND DEAD CELLS OF CHLORELLA VULGARIS AND SCENEDESMUS OBLIQUUS UPON SOME SPECIES OF SAPROPHYTIC BACTERIA SAMPLED RANDOMLY FROM AMONG A FEW HUNDRED SPECIES ISOLATED FROM ILAWA LAKES WATER. FOUR DIFFERENT PATTERNS OF BEHAVIOR OF MICROORGANISMS HAVE BEEN FOUND IN THE PRESENCE OF LIVING AND DEAD CELLS OF THESE TWO ALGAL SPECIES: (1) DYING OF MICROORGANISMS IN LIVING AND DEAD CULTURES OF ALGAE (MICROCOCCUS UREAE); (2) DYING OF MICROORGANISMS IN LIVING CULTURES OF ALGAE AND DEVELOPMENT IN DEAD CULTURES OF ALGAE (BACILLUS MYCOIDES, ESCHERICHIA COLI, VIBRIO SP.); (3) DYING OF MICROORGANISMS IN LIVING AND DEAD CULTURES OF ALGAE IN THE INITIAL PERIOD OF JOINT CULTIVATION AND DEVELOPMENT IN THE LATER PERIOD (AZOTOBACTER SP., PSEUDOMONAS FLUORESCENS, RHODOTORULA SP.); (4) INTENSIVE DEVELOPMENT OF MICROORGANISMS IN KILLED CULTURES OF ALGAE AND DYING IN LIVING CULTURES IN THE INITIAL PERIOD OF JOINT CULTIVATION, FOLLOWED BY DEVELOPMENT IN THE LATER PERIOD (AEROMONAS SP.). (LEGORE-WASHINGTON)

FIELD 05C

ACCESSION NO. W72-10861

GREAT LAKES ALGAE MONITORING PROGRAM, 1969.

MICHIGAN WATER RESOURCES COMMISSION, LANSING. DEPT. OF NATURAL RESOURCES.

REPORT, FEBRUARY 1970. 16 P, 7 FIG, 5 TAB, 2 REF.

DESCRIPTORS:
*ALGAE, *GREAT LAKES, *MONITORING, *BIOINDICATORS, THERMAL POLLUTION, MICROSCOPY, DIATOMS, NUTRIENTS, PHYTOPLANKTON, WATER QUALITY, MICHIGAN, SURFACE WATERS, COASTS, SAMPLING, BIOLOGICAL COMMUNITIES, LAKE MICHIGAN, LAKE ERIE, LAKE HURON, LAKE SUPERIOR, WATER TEMPERATURE, HYDROGEN ION CONCENTRATION, HARDNESS, ALKALINITY, CONDUCTIVITY, SUSPENDED SOLIDS, NITRATES, NITRITES, NITROGEN, PHOSPHORUS, CHLORIDE, CYANOPHYTA, CURRENTS(WATER), CHLOROPHYTA, SCENEDESMUS, *EUTROPHICATION, WATER ANALYSIS.

IDENTIFIERS:
SAMPLE PRESERVATION, FORMALIN, COUNTING, CYCLOTELLA, OSCILLATORIA, APHANOTHECE, ANABAENA, APHANIZOMENON, NAVICULA, GLENODINIUM, MICROSPORA, SYNEDRA, DINOBRYON, OEDOGONIUM, FRAGILARIA, DIATOMA, MELOSIRA, TABELLARIA, ACTINASTRUM, PHYTOCONIS, CYMBELLA, STEPHANODISCUS, ANKISTRODESMUS, TRACHELOMONAS, SAGINAW BAY, ORGANIC NITROGEN.

ABSTRACT:
WATER SAMPLES CONTAINING ALGAE WERE COLLECTED DURING 1969 FROM 49 STATIONS IN THE GREAT LAKES TO ATTEMPT TO CORRELATE THE ALGAL SPECIES WITH TROPHIC CONDITIONS IN EACH LAKE. EACH SAMPLE WAS PRESERVED WITH FORMALIN AND SENT TO THE LANSING LABORATORY OF THE WATER RESOURCES COMMISSION FOR MICROSCOPIC SORTING AND COUNTING. WATER TEMPERATURE, SUSPENDED SOLIDS, NITRATE-N, AMMONIA-N, ORGANIC N, PHOSPHATES, CHLORINE, SULFATES, PH, HARDNESS, CARBONATES, ALKALINITY, AND CONDUCTIVITY WERE ALSO DETERMINED AT THE SAMPLING SITES. IN GENERAL, THE ALGAL COMPOSITION OF LAKE SUPERIOR, LAKE MICHIGAN, AND LAKE HURON (WITH THE EXCEPTION OF SAGINAW BAY) SHOW LOW AVERAGE ALGAE COUNTS WITH GENERA CHARACTERISTIC OF OLIGOTROPHIC CLEAN WATER CONDITIONS. LAKE SUPERIOR AND LAKE HURON BOTH SUPPORT A CENTRIC DIATOM POPULATION WITH THE MOST COMMON GENUS BEING CYCLOTELLA. IN LAKE MICHIGAN ALGAL SAMPLES, PENNATE DIATOMS PREDOMINATED WITH SYNEDRA BEING THE MOST COMMON GENUS. BLUE-GREEN ALGAE WERE OBSERVED IN THE SOUTHERN LAKE MICHIGAN SAMPLES WHICH MAY INDICATE HIGH NUTRIENT LEVELS. LAKE ERIE AND SAGINAW BAY, ON THE OTHER HAND, HAVE HIGH ALGAL POPULATIONS WITH GENERA ASSOCIATED WITH EUTROPHIC CONDITIONS. IT IS CONCLUDED THAT ALGAE OF CERTAIN GENERA, ALONG WITH OTHER BIOLOGICAL ORGANISMS, ARE INDICATORS OF WATER QUALITY AND ARE USEFUL IN DETERMINING THE QUALITY OF MICHIGAN'S INSHORE WATERS. (MORTLAND-BATTELLE)28(U)

FIELD 05A, 05C

ACCESSION NO. W72-10875

RECENT DEVELOPMENTS IN THE MEASUREMENT OF THE RESPONSE OF PLANKTON AND PERIPHYTON TO CHANGES IN THEIR ENVIRONMENT,

NATIONAL ENVIRONMENTAL RESEARCH CENTER, CINCINNATI, OHIO. ANALYTICAL QUALITY CONTROL LAB.

C. I. WEBER.

PAPER PRESENTED AT THE SYMPOSIUM ON BIOASSAY TECHNIQUES IN ENVIRONMENTAL CHEMISTRY 162ND NATIONAL MEETING, WASHINGTON, D.C., SEPTEMBER 15, 1972. 29 P, 3 FIG, 7 TAB, 74 REF.

DESCRIPTORS:
*ANALYTICAL TECHNIQUES, CHEMICAL ANALYSIS, WATER ANALYSIS, EVALUATION, WATER QUALITY CONTROL, *BIOASSAY, *BIOMASS, CHLOROPHYLL, NITROGEN FIXATION, *PERIPHYTON, *PHYTOPLANKTON, BIOLOGICAL COMMUNITIES, DOMINANT ORGANISMS, REDUCTION(CHEMICAL), TROPHIC LEVELS, PROTOZOA, ALGAE, BACTERIA, DIATOMS, OCEANS, ACTIVATED SLUDGE, SOIL BACTERIA, PHOTOSYNTHESIS, FLUOROMETRY, ABSORPTION, CYANOPHYTA, CHLOROPHYTA, YEASTS.

IDENTIFIERS:
*ADENOSINE TRIPHOSPHATE, *AUTOTROPHIC INDEX, MACROINVERTEBRATES, ACETYLENE, ETHYLENE, NITROGEN RADIOISOTOPES, N-15, LUCIFERIN, LUCIFERASE, CHLOROPHYLL A, OPTICAL DENSITY, ACETYLENE REDUCTION, SCHIZOTHRIX CALCIOCOLA, CLOSTERIUM MONILIFERUM, MELOSIRA VARIANS, NAVICULA TRIPUNCTATA, GOMPHONEMA PARVULUM, BACILLARI PARADOXA, SPHAEROTILUS NATANS, EPILIMNION, AMPHIDINIUM CARTERI, CHLORELLA VULGARIS, CYCLOTELLA NANA, DITYLUM BRIGHTWELLII.

ABSTRACT:
WATER QUALITY IS REFLECTED IN BIOASSAY ANALYSIS BASED UPON BIOMASS, POPULATION DENSITY, AND SPECIES COMPOSITION AND DIVERSITY OF AQUATIC ORGANISMS, E.G. PLANKTON, PERIPHYTON, MACROINVERTEBRATES, FISH. OF THE MORE RECENTLY DEVELOPED METHODS OF DETERMINING PLANKTON AND PERIPHYTON BIOMASS AND CONDITION, SPECIAL INTEREST HAS BEEN FOCUSED ON CHLOROPHYLL A, ATP, AND NITROGEN FIXATION AS INDEXES OF PRODUCTIVITY, RESPIRATION, AND SUCCESS IN THE ENVIRONMENT. THE PRIMARY PHOTOSYNTHETIC PIGMENT, CHLOROPHYLL A, HAS BEEN USED IN DEVELOPMENT OF AN 'AUTOTROPHIC INDEX' RELATIONSHIP WHICH HAS PROVEN VALUABLE IN DETERMINING THE DOMINANCE AND TROPHIC LEVELS WITHIN THE PLANKTON AND PERIPHYTON COMMUNITIES OF WATERWAYS. ADENOSINE TRIPHOSPHATE (ATP) OFFERS CONSIDERABLE PROMISE AS AN INDEX OF TOTAL VIABLE PLANKTON BIOMASS AND ALSO AS AN INDEX OF TOXIC SUBSTANCES AND THEIR EFFECTS UPON THE AQUATIC SYSTEM. SINCE NITROGEN IS A MAJOR CELL COMPONENT, ANALYSIS OF NITROGEN FIXATION, AS INDICATED BY ACETYLENE TO ETHYLENE REDUCTION, HAS LENT ITSELF TO RAPID, ACCURATE EVALUATIONS OF NITROGEN BUDGETS OF AQUATIC ORGANISMS AND PLANKTON POPULATION DYNAMICS. ORGANISMS AND SPECIFIC METHODOLOGY ARE INCLUDED IN THE DISCUSSION OF THESE PARAMETERS OF ANAYSIS AS WELL AS DATA COLLECTED TO SUBSTANTIATE THEIR VALUE AS ANALYTICAL TECHNIQUES. (MACKAN-BATTELLE)

FIELD 05A, 05C

ACCESSION NO. W72-10883

RADIOECOLOGY AND CHEMOECOLOGY IN THE SERVICE OF THE PROTECTION OF NATURE,

INSTITUTE OF BIOLOGY OF THE SOUTHERN SEAS, SEVASTOPOL (USSR).

G. G. POLIKARPOV, L. G. KULEBAKINA, V. G. TSITSUGINA, AND V. V. ANDRUOTSCHENKO.

PRESENTED AT THE COMMISSION OF THE EUROPEAN COMMUNITIES INTERNATIONAL SYMPOSIUM, RADIOECOLOGY APPLIED TO THE PROTECTION OF MAN AND HIS ENVIRONMENT, ROME, SEPT. 1971. 6 P, 1 TAB, 14 REF.

DESCRIPTORS:
*RADIOISOTOPES, *ABSORPTION, *ENVIRONMENTAL EFFECTS, NUCLEAR WASTES, PATH OF POLLUTANTS, MARINE ALGAE, CHROMOSOMES, EMBRYONIC GROWTH STAGE, CRUSTACEANS, MONITORING, WATER POLLUTION EFFECTS, MOLLUSKS.

IDENTIFIERS:
BLACK SEA.

ABSTRACT:
CONCENTRATION FACTORS FOR SR-90 UPTAKE IN THE BLACK SEA ARE 60-250 FOR BROWN ALGAE AND 70-300 FOR MOLLUSCS AND CRUSTACEA. INCUBATION OF MARINE ORGANISMS IN RADIOACTIVE SOLUTIONS INCREASES CHROMOSOMAL ABERRATIONS. LETHAL DDT CONCENTRATIONS IN SEAWATER ARE 0.1 MG/LITER FOR AN ALGA AND 0.01 MG/LITER FOR A CRUSTACEAN. (BOPP-ORNL)

FIELD 05B, 05C

ACCESSION NO. W72-10955

ZINC AND COBALT UPTAKE BY THE BROWN SEAWEED FUCUS SPIRALIS (L.),

REACTOR CENTRUM NEDERLAND, PETTEN.

A. W. VAN WEERS,

PRESENTED AT THE COMMISSION OF THE EUROPEAN COMMUNITIES INTERNATIONAL SYMPOSIUM, RADIOECOLOGY APPLIED TO THE PROTECTION OF MAN AND HIS ENVIRONMENT, ROME, SEPT. 1971. 11 P, 5 FIG, 5 REF.

DESCRIPTORS:
*ZINC RADIOISOTOPES, *COBALT RADIOISOTOPES, *MARINE ALGAE, HEAVY METALS, ABSORPTION, BIOINDICATORS, PATH OF POLLUTANTS, FOOD CHAINS, NUCLEAR WASTES, ENVIRONMENTAL EFFECTS, EFFLUENTS, MONITORING, CHELATION.

ABSTRACT:
HAZARDS FROM UPTAKE OF BOTH RADIOACTIVE AND NON-RADIOACTIVE, HEAVY-METAL POLLUTANTS WERE STUDIED; AND A BIOLOGICAL INDICATOR OF POLLUTION WAS EVALUATED. ADDITION OF 500 PPB OF STABLE CO REDUCED THE CONCENTRATION FACTOR FOR CO-60 UPTAKE (MEASURED AFTER 6 DAYS) FROM ABOUT 350 TO 60; 500 PPB OF STABLE ZN REDUCED THE CONCENTRATION FACTOR FOR ZN-65 FROM ABOUT 1000 TO 300. A CHELATING AGENT (EDTA) REDUCED UPTAKE OF ZN-65 TO 10% OF THE CONTROL AT AN EDTA CONCENTRATION OF 1 MICROMOLE/LITER; A SIMILAR REDUCTION OF CO-60 UPTAKE OCCURRED AT AN EDTA CONCENTRATION OF 30 MICROMOLES/LITER. A SEASONAL VARIATION IN THE ZN CONTENT OF SEAWEED OBSERVED OVER A PERIOD OF SEVERAL YEARS IS PROBABLY RELATED TO ITS GROWTH CYCLE. (BOPP-ORNL)

FIELD 05B, 05C

ACCESSION NO. W72-10957

INDICATOR BASED SURVEILLANCE PROGRAM (MARINE) AT A NUCLEAR SITE,

BHABHA ATOMIC RESEARCH CENTRE, BOMBAY (INDIA). HEALTH PHYSICS DIV.

I. S. BHAT, AND P. R. KAMATH.

PRESENTED AT THE COMMISSION OF THE EUROPEAN COMMUNITIES INTERNATIONAL SYMPOSIUM, RADIOECOLOGY APPLIED TO THE PROTECTION OF MAN AND HIS ENVIRONMENT, ROME, SEPT. 1971. 13 P, 4 TAB, 6 REF.

DESCRIPTORS:
*IODINE RADIOISOTOPES, *CESIUM RADIOISOTOPES, *COBALT RADIOISOTOPES, *BIOINDICATORS, ANALYTICAL TECHNIQUES, POLLUTANT IDENTIFICATION, NUCLEAR POWERPLANTS, EFFLUENTS, NUCLEAR WASTES, PATH OF POLLUTANTS, ASIA, MARINE ALGAE, SEDIMENTS, ESTUARINE ENVIRONMENT, MONITORING, CRABS, SHRIMP.

IDENTIFIERS:
INDIA, CESIUM RADIOISOTOPES.

ABSTRACT:
EFFLUENTS FROM THE TARAPUR (BWR REACTOR) POWERPLANT CONTAIN IODINE, CESIUM, AND COBALT RADIOISOTOPES. ENVIRONMENTAL MONITORING IS PARTICULARLY USEFUL FOR DETECTING AN OCCASIONAL LARGE DISCHARGE MADE UNWITTINGLY. RESEARCH WAS CONDUCTED TO OPTIMIZE TECHNIQUES FOR SAMPLING AND ANALYSIS. TRACE ELEMENT CONTENT WAS MEASURED FOR SEVERAL ORGANISMS (LOBSTER, OYSTER, CLAM, CRAB, BOMBAY DUCK, SINGHALA, AND SEA WEED). PRAWNS AND CRABS ARE SUITABLE INDICATORS FOR CS-137, AND SEA WEED (FOR IODINE AND COBALT RADIOISOTOPES) AND SEA SILT ARE VERSATILE INDICATORS. THE HIGH ISOTOPIC DILUTION OF SR-90 IN SEA WATER RESULTS IN LOW SENSITIVITY OF INDICATORS IN THIS CASE.
(BOPP-ORNL)

FIELD 05B, 05A

ACCESSION NO. W72-10958

QUANTITATIVE CHARACTERISTICS OF THE MEANS OF PENETRATION OF SR-90 INTO THE BODIES OF GASTROPOD MOLLUSKS,

V. M. B. NYANISHKENE, AND G. G. POLIKARPOV.

AVAILABLE FROM NTIS AS AEC-TR-7225, $3.00 IN PAPER COPY, $0.95 MICROFICHE. RADIOBIOLOGY, VOL. 10, NO. 6, P. 198-202, 1970. 3 TAB, 5 REF. TRANSLATION FROM RADIOBIOLOGIYA, VOL. 10, P. 928-930, 1970.

DESCRIPTORS:
*STRONTIUM RADIOISOTOPES, *ABSORPTION, *SNAILS, GASTROPODS, PATH OF POLLUTANTS, NUCLEAR WASTES, WATER POLLUTION EFFECTS, FOOD CHAINS, PUBLIC HEALTH, AQUATIC LIFE, AQUATIC ALGAE, CALCIUM, RADIOACTIVITY EFFECTS.

ABSTRACT:
FOOD-INTAKE AND ABSORPTION-FROM-WATER PATHWAYS OF SR-90 UPTAKE WERE COMPARED IN ABOUT ONE-MONTH EXPERIMENTS. BOTH THE SNAILS AND THEIR FOOD ORGANISMS (ELODEA) WERE KEPT IN WATER CONTAINING 10 MICROCI/LITER OF SR-90 AND 50 MG/LITER OF CA. EXPERIMENTS WERE CONDUCTED IN THE SPRING, SUMMER, AND AUTUMN PERIODS UNDER LABORATORY CONDITIONS IN TWO REPETITIONS AND FOUR VARIATIONS. IN VARIATION I THE SNAILS WERE KEPT IN WATER WITHOUT SR-90 AND WERE FED ELODEA THAT HAD ACCUMULATED SR-90; IN VARIATION II THEY WERE KEPT IN WATER WITH SR-90 AND WERE FED ELODEA THAT HAD ACCUMULATED SR-90; IN VARIATION III THEY WERE KEPT IN WATER WITH SR-90 AND WERE FED ELODEA THAT DID NOT CONTAIN SR-90; IN VARIATION IV THEY WERE KEPT IN WATER WITH SR-90 WITHOUT FEEDING. UPTAKE FROM THE WATER WAS ONE (FOR THE FLESH) AND TWO (FOR THE SHELL) ORDERS OF MAGNITUDE GREATER THAN FROM FOOD. (BOPP-ORNL)

FIELD 05C

ACCESSION NO. W72-10978

A SYMBIOTIC SYSTEM FOR ADVANCED TREATMENT OF WASTEWATER,

OHIO STATE UNIV., COLUMBUS. DEPT. OF CIVIL ENGINEERING.

F. J. HUMENIK, AND G. P. HANNA, JR.

WATER AND SEWAGE WORKS, VOL 117, NO 6, JUNE 1970, P 198-202. 6 FIG, 11 REF.

DESCRIPTORS:
*WASTE WATER TREATMENT, *NUTRIENTS, *ALGAE, *BACTERIA, *SYMBIOSIS, ACTIVATED SLUDGE, NITROGEN, PHOSPHORUS, CHEMICAL OXYGEN DEMAND, CHLORELLA, PERFORMANCE, EFFICIENCIES, LABORATORY TESTS, LABORATORY EQUIPMENT, AERATION, DISSOLVED OXYGEN, AEROBIC CONDITIONS, BIOMASS, RESPIRATION, PHOTOSYNTHETIC OXYGEN, *TERTIARY TREATMENT.

IDENTIFIERS:
ORGANIC NITROGEN, SETTLING CHARACTERISTICS, SOLIDS RECYCLE, SOLIDS CONTROL.

ABSTRACT:
UNDER CONTROLLED ENVIRONMENTAL CONDITIONS A SYMBIOTIC ALGAL-BACTERIAL CULTURE WAS DEVELOPED AND MAINTAINED WHICH WOULD EFFICIENTLY REMOVE NUTRIENTS FROM WASTEWATER. THE CULTURE OF THE BIOMASS GROWTH SYSTEM WAS A NATURAL MIXTURE OF ALGAE, PREDOMINANTLY CHLORELLA, AND ACTIVATED SLUDGE IN WHICH THE ALGAE BECAME ENMESHED WITHIN THE BACTERIAL MATRIX. THE ALGAL-BACTERIAL FLOC EXHIBITED GOOD SETTLING CHARACTERISTICS UNDER QUIESCENT CONDITIONS. MAXIMUM AND MOST CONSISTENT REMOVAL OF INFLUENT COD AND ORGANIC NITROGEN WAS OBTAINED DURING UNAERATED OPERATION WITH SOLIDS CONTROL AND THROUGH DAILY HARVEST OF EXCESS BIOMASS. COD REMOVAL FOR THE RECLARIFIED SUPERNATANT AVERAGED 82.5 PERCENT WITH 88.2 PERCENT AS THE MAXIMUM. ORGANIC NITROGEN REMOVAL AVERAGED 85.3 PERCENT. HOWEVER, NO APPRECIABLE REMOVAL OF PHOSPHORUS WAS RECORDED. THE SYMBIOTIC ALGAL-BACTERIAL SYSTEM REMAINED AEROBIC DURING UNAERATED PERIODS AS A RESULT OF A BALANCE BETWEEN RESPIRATION AND PHOTOSYNTHETIC OXYGENATION. AN AVERAGE VALUE OF 2 MG/L DISSOLVED OXYGEN WAS REPORTED. SUPPLEMENTAL AERATION DID NOT RESULT IN INCREASED NUTRIENT REMOVAL, AND WAS THEREFORE CONSIDERED AS A WASTED ENERGY INPUT. (GALWARDI-TEXAS)

FIELD 05D

ACCESSION NO. W72-11100

PHOSPHATES AND PHOSPHATE SUBSTITUTES IN DETERGENTS (PART 2).

COMMITTEE ON GOVERNMENT OPERATIONS (U.S. HOUSE).

HEARINGS--COMM. ON GOVERNMENT OPERATIONS, U.S. HOUSE OF REPRESENTATIVES, 92D CONG, 1ST SESS, OCTOBER 29, 1971. 394 P, 3 FIG, 6 ILLUS, 1 PHOTO, 38 TAB, 5 APPEND.

DESCRIPTORS:
*PHOSPHATES, *WATER POLLUTION SOURCES, *EUTROPHICATION, *DETERGENTS, WATER QUALITY CONTROL, DOMESTIC WASTES, RUNOFF, WATER SOFTENING, ENVIRONMENTAL EFFECTS, AQUATIC ALGAE, ALGAL CONTROL, ALGAL POISONING, AQUATIC ENVIRONMENT, AQUATIC FUNGI, ACQUIFER CHARACTERISTICS, GENETICS, DISSOLVED OXYGEN.

ABSTRACT:
TESTIMONY IS PRESENTED TO THE SUBCOMMITTEE ON NATURAL RESOURCES OF THE HOUSE GOVERNMENT OPERATIONS COMMITTEE CONCERNING THE EFFECTS OF USING PHOSPHATES IN DETERGENTS. TESTIMONY CONCERNS CONSUMER MARKETING ATTEMPTS TO ELIMINATE PHOSPHATE DETERGENTS BOTH VOLUNTARILY AND BY PROHIBITING THEIR USE THROUGH LAW. TESTIMONY IS ALSO PRESENTED CONCERNING THE MANNER BY WHICH PHOSPHATE DETERGENT DISCHARGES SPEED THE NORMAL EUTROPHICATION OF A WATER BODY. THE COMPLETE BAN ON PHOSPHATE DETERGENT SALE IN DADE COUNTY, FLORIDA, IS DISCUSSED ALONG WITH THE UNIQUE ECOLOGICAL CONDITONS PRESENT IN THAT AREA. EFFORTS ON THE PART OF THE ENVIRONMENTAL PROTECTION AGENCY TO SPEED UP THE REMOVAL OF PHOSPHATE DISCHARGES IN WATER BODIES ARE DESCRIBED. THE DANGEROUS HEALTH ASPECTS OF PHOSPHATE SUBSTITUTES ARE ALSO DISCUSSED. OPTICAL BRIGHTENERS USED IN DETERGENTS ARE EXAMINED AND A SCIENTIFIC PAPER IS APPENDED. ONE PROBLEM WITH BRIGHTENERS IS THAT NOT ENOUGH IS KNOWN ABOUT THEIR EFFECT ON THE ENVIRONMENT. ONE SCIENTIST BELIEVES THEY MAY CAUSE GENETIC MUTATIONS IN SOME ORGANISMS. APPENDICES CONTAIN INFORMATION ON EUTROPHICATION, DETERGENT CONTENT AND ANALYSIS OF THEIR EFFECTS ON THE ENVIRONMENT, AND SURVEY STUDIES ON POLLUTION AND EUTROPHICATION IN THE NATION'S LAKES. (GRANT-FLORIDA)

FIELD 05C, 05G, 06E

ACCESSION NO. W72-11186

BENTHIC MARINE ALGAE FROM WATERS ADJACENT TO THE CRYSTAL RIVER ELECTRIC POWER PLANT (1969 AND 1970),

FLORIDA DEPT. OF NATURAL RESOURCES, ST. PETERSBURG.

K. A. STEIDINGER, AND J. F. VAN BREEDVELD.

PROFESSIONAL PAPERS SERIES, NO 16, JUNE 1971. 46 P, 1 FIG, 22 TAB, 14 REF.

DESCRIPTORS:
*BENTHIC FLORA, *MARINE ALGAE, *GULF OF MEXICO, *ELECTRIC POWERPLANTS, *THERMAL POLLUTION, *AQUATIC ENVIRONMENT, FLORIDA, ELECTRIC GENERATORS, WATER TEMPERATURE, RHODOPHYTA, PHAEOPHYTA, CHLOROPHYTA, COOLING WATER, HYDROLOGIC DATA.

IDENTIFIERS:
*SPECIES DIVERSITY, CRYSTAL RIVER.

ABSTRACT:
ONE HUNDRED SIX TAXA OF MARINE ALGAE WERE IDENTIFIED FROM GULF OF MEXICO WATERS ADJACENT TO THE FLORIDA POWER CORPORATION ELECTRICAL GENERATING PLANT AT CRYSTAL RIVER, FLORIDA IN 1969 AND 1970. OF THE 106 TAXA, 19 BELONG TO CHLOROPHYTA, 24 TO PHAEOPHYTA, AND 63 TO RHODOPHYTA. REDUCTIONS IN SPECIES DIVERSITY, AS WELL AS INCIDENCE OF OCCURRENCE, WERE NOTED IN 1970. CAUSES OF THESE REDUCTIONS ARE NOT KNOWN. HOWEVER, INCREASED TEMPERATURE DOES NOT APPEAR TO BE A FACTOR SINCE DEEPER WATER STATIONS, WHICH WERE ALSO AFFECTED, ARE NOT EXPOSED TO THE THERMAL PLUME. THE STUDY AREA IS SEMI-TROPICAL OR WARM-TEMPERATE, WITH A PRIMARILY SEASONAL FLORA. WINTER APPEARED TO BE THE SEASON OF LOWEST SPECIES DIVERSITY. SEVERAL NEW DISTRIBUTION RECORDS ARE REPORTED. (SVENSSON-WASHINGTON)

FIELD 05C

ACCESSION NO. W72-11252

MICROBIOLOGICAL STUDIES ON NITROGEN FIXATION IN AQUATIC ENVIRONMENTS--VII. SOME ECOLOGICAL ASPECTS OF NITROGEN FIXING BACTERIA,

KYOTO UNIV. (JAPAN). RESEARCH INST. OF FOOD SCIENCE; AND MIE PREFECTURAL UNIV. TSU (JAPAN).

A. KAWAI, AND I. SUGAHARA.

BULLETIN OF THE JAPANESE SOCIETY OF SCIENTIFIC FISHERIES, VOL 38, NO 3, P 291-297, 1972. 3 FIG, 1 TAB, 7 REF.

DESCRIPTORS:
*MICROBIOLOGY, *NITROGEN FIXING BACTERIA, *NITROGEN CYCLE, ENVIRONMENTAL EFFECTS, ANAEROBIC CONDITIONS, NITRATES, DENITRIFICATION, BOTTOM SEDIMENTS, SEA WATER, FRESHWATER, AMMONIA, EUTROPHICATION, OXIDATION, SYMBIOSIS, AZOTOBACTER, MARINE ALGAE, ORGANIC MATTER.

IDENTIFIERS:
NITROSOMONAS, NITROBACTER, AZOTOBACTER.

ABSTRACT:
THE OCCURRENCE AND ABUNDANCE OF BACTERIA RESPONSIBLE FOR NITROGEN METABOLISM FORM A PECULIAR PATTERN FOR EACH WATER REGION. THE NUMBER OF EACH BACTERIA GROUP HAVING THE ABILITY TO PERFORM THE FOLLOWING PROCESSES IN THE NITROGEN CYCLE WAS COUNTED IN THE WATER AND BOTTOM SEDIMENTS OF VARIOUS REGIONS: AMMONIFICATION, NITRATE REDUCTION, DENITRIFICATION, NITRIFICATION. THE RELATIONSHIP IS CLOSE BETWEEN THE NUMBER OF NITROGEN FIXING BACTERIA OCCURRING IN THE WATER AND BOTTOM SEDIMENTS OF VARIOUS FRESH WATER REGIONS AND THAT OF NITRATE REDUCERS AND DENITRIFIERS. THE GROWTH OF THE THREE BACTERIA GROUPS MAY BE CONTROLLED BY THE SAME ENVIRONMENTAL FACTORS, PRESUMABLY CONCENTRATION OF AVAILABLE ORGANIC MATTER. THIS RELATIONSHIP WAS NOT OBTAINED IN SEA WATER REGIONS WHERE AMOUNT OF ORGANIC MATTER IS CONSIDERABLY POORER. BACTERIAL ADHESION ON THE SUSPENDED MATTER PRESENT IN WATER REGIONS WAS STUDIED; WATER SAMPLES WERE TREATED WITH SONIC OSCILLATION AND THE BACTERIA COUNTED. IT MAY BE THAT MANY NITROGEN BACTERIA OCCUR, TOGETHER WITH THE OTHER HETEROTROPHS, ON OR IN THE SUSPENDED PARTICLES AND UTILIZE THE PARTICLE COMPONENTS AS NUTRITIONAL SOURCES. IN GENERAL, A DENSE BACTERIAL BIOMASS IS FOUND IN FRESH WATER ENVIRONMENTS, THE POPULATION DECREASING GRADUALLY FROM COASTAL TO OFFSHORE REGIONS. (JONES-WISCONSIN)

FIELD 05C

ACCESSION NO. W72-11563

A SHOCK-WAVE TECHNIQUE TO COLLAPSE THE VACUOLES OF BLUE-GREEN ALGAE,

UNIVERSITY OF WALES INST. OF SCIENCE AND TECH., CARDIFF.

D. C. MENDAY, AND A. A. BUCK.

WATER RESEARCH, VOL 6, NO 3, P 279-284, 1972. 5 FIG, 18 REF.

DESCRIPTORS:
*ALGAL CONTROL, *CYANOPHYTA, EXPLOSIVES, COSTS, FISH, FISHKILL, RESERVOIRS, FISHERIES, PRESSURE.

IDENTIFIERS:
*SHOCK-WAVES, *GAS-VACUOLE DEFLATION, SOUTH WALES, MICROCYSTIS AERUGINOSA, ANABAENA.

ABSTRACT:
A FIELD SYSTEM DESIGNED TO CONTROL BLUE-GREEN ALGAE BY PRODUCING PRESSURE WAVES TO BURST THEIR GAS-VACUOLES, CAUSING THEM TO SINK, IS DESCRIBED. THE PRESSURES REQUIRED TO BURST GAS-VACUOLES OF MICROCYSTIS AERUGINOSA, THE DESIGN OF AN EXPLOSIVE SYSTEM, TOGETHER WITH METHODS TO RECORD PRESSURE DISTRIBUTION, AND THE ADVERSE EFFECTS OF SUCH PRESSURES ON FISH ARE CONSIDERED. THE PRIMARY CONCERN IS THE POSSIBLE APPLICATION IN AN INDUSTRIAL WATER-SUPPLY RESERVOIR IN SOUTH WALWES WHERE THE PRINCIPAL ALGAE IS MICROCYSTIS AERUGINOSA AND WHERE BROWN TROUT (SALMO TRUTTA) HAVE BEEN INTRODUCED FOR ANGLING. IT WAS FOUND THAT A PRESSURE OF ABOUT 4.5 KG/SQ CM WAS REQUIRED TO BURST VACUOLES IN ALL CELLS OF M. AERUGINOSA. THE TECHNIQUE DEVELOPED PROVIDED A QUICK AND CHEAP CONTROL METHOD WHEN A BLUE-GREEN ALGAL BLOOM IS SERIOUSLY AFFECTING A WATER SUPPLY. IT IS NOT SUGGESTED THAT EXPLOSIVES SHOULD BE USED AS A REGULAR MEANS OF ALGAL CONTROL. IT SITUATIONS WHERE ALGAE ACCUMULATE NEAR ABSTRACTION POINTS, THIS TECHNIQUE HAS PRACTICAL ADVANTAGES WITH LOW OPERATIONAL COSTS AND AVOIDANCE OF CAPITAL CHARGES, PARTICULARLY APPLICABLE WHERE BLOOMS ARE INFREQUENT OR OF SHORT DURATION. (JONES-WISCONSIN)

FIELD 05G, 05C

ACCESSION NO. W72-11564

THE EFFECT OF CUPRIC AND FLUORIDE IONS IN THE RESPIRATION OF CHLORELLA,

UNIVERSITY OF WESTERN ONTARIO, LONDON (ENGLAND). DEPT. OF BIOPHYSICS.

D. F. SARGENT, AND C. P. S. TAYLOR.

CANADIAN JOURNAL OF BOTANY, VOL. 50, P 905-907, 1972. 2 TAB, 7 REF.

DESCRIPTORS:
*ALGAL CONTROL, *RESPIRATION, *CHLORELLA, *INHIBITION, COPPER, FLUORIDES, IONS, CYTOLOGICAL STUDIES, WATER POLLUTION EFFECTS.

IDENTIFIERS:
CHLORELLA PYRENOIDOSA.

ABSTRACT:
SIMULATANEOUS APPLICATION OF CUPRIC AND FLUORIDE IONS VIRTUALLY STOPS RESPIRATION OF CHLORELLA VULGARIS, ALTHOUGH EITHER CUPRIC OR FLUORIDE IONS ALONE CAUSE NO INHIBITION. TESTS WITH VARIOUS COMBINATIONS OF CUPRIC AND FLUORIDE IONS WERE MADE FOR ENDOGENOUSLY RESPIRING CHLORELLA PYRENOIDOSA. THE SUM OF THE INHIBITIONS CAUSED BY CUPRIC AND FLUORIDE IONS INDIVIDUALLY IS MUCH LESS THAN THE INHIBITION FOUND WHEN BOTH AGENTS ARE PRESENT. THIS WORK INDICATES THAT CUPRIC IONS ALONE DO NOT BLOCK THE ALTERNATE DISSIMILATION PATHWAY, BUT THAT BOTH CUPRIC AND FLUORIDE IONS ARE REQUIRED TO INHIBIT COMPLETELY. THE RESPONSE DEPENDS ON THE ORDER OF ADDITION AND THE TIME BETWEEN ADDITIONS OF THE TWO POISONS. THE RESPONSE TO ADDED CUPRIC DECREASES AS THE TIME SINCE THE ADDITION OF FLUORIDE IONS INCREASES. IN CHLORELLA POISONED BY DICHLOROPHENYLDIMETHYLUREA THERE IS A LIGHT-INDUCED RESPIRATORY INHIBITION WHICH DEPENDS UPON THE FUNCTIONING OF GLYCOLYSIS AND WHICH IS SUPPRESSED BY SUCH GLYCOLYTIC INHIBITORS AS FLUORIDE IONS AND IODOACETATE. WHEN FLUORIDE IONS ARE USED, THE LIGHT-INDUCED INHIBITION GRADUALLY REAPPEARS, UNTIL, AFTER AN 80-MINUTE INCUBATION WITH FLUORIDE, THE EFFECT IS AS GREAT AS BEFORE FLUORIDE WAS ADDED. CLEARLY GLYCOLYSIS IS NO LONGER INHIBITED. (JONES-WISCONSIN)

FIELD 05C

ACCESSION NO. W72-11572

CHARACTERIZATION OF TASTE AND ODORS IN WATER SUPPLIES,

CONNECTICUT UNIV., STORRS. REGULATORY BIOLOGY SECTION.

R. P. COLLINS.

COPY AVAILABLE FROM GPO SUP DOC EP2.10:16040 DGH 08/71, $0.35; MICROFICHE FROM NTIS AS PB-211 025, $0.95. ENVIRONMENTAL PROTECTION AGENCY WATER POLLUTION CONTROL RESEARCH SERIES. AUGUST, 1971, 21 P, 1 FIG, 8 REF. EPA PROGRAM 16040 DGH 08/71.

DESCRIPTORS:
*TASTE, *ODOR, *ORGANIC COMPOUNDS, ANALYTICAL TECHNIQUES, SEPARATION TECHNIQUES, GAS CHROMATOGRAPHY, SEPCTROSCOPY, ALGAE, ACTINOMYCETES, FUNGI, LABORATORY TESTS, WATER QUALITY CONTROL.

IDENTIFIERS:
*ODOR-PRODUCING ALGAE.

ABSTRACT:
CULTURED SAMPLES OF STREPTOMYCES ODORIFER, SYNURA PETERSENII, AND TRICHODERMA VIRIDE, AN ACTINOMYCETE, AN ALGAE, AND A FUNGUS, RESPECTIVELY, WERE STEAM DISTILLED AND THE AQUEOUS DISTILLATES WERE THEN EXTRACTED WITH AN ORGANIC SOLVENT AND REDUCED TO A SMALL VOLUME FOR ANALYSIS TO DETERMINE THEIR ODOROUS CONSTITUENTS. ANALYSIS OF THE EXTRACTS WAS BY GAS CHROMATOGRAPHY, INFRARED, MASS, AND NUCLEAR MAGNETIC SPECTROSCOPY. THE MAJOR ODOROUS CONSTITUENTS OF STREPTOMYCES ODORIFER WERE IDENTIFIED AS TRANS-1, 10-DIMETHYL-TRANS-9-DECALOL (GEOSIM), 2-EXO-HYDROXY-2-METHYLBORNANE, AND CADIN-4-ENE-1-OL. A LARGE NUMBER OF ODORUS COMPOUNDS WERE IDENTIFIED FROM SYNURA PETERSENII, WITH HEPTANOL CONTAINING THE ODOR NORMALLY ASSOCIATED WITH SYNURA PETERSENII. THE MAJOR ODOROUS COMPOUND PRODUCED BY TRICHODERMA VIRIDE WAS 6-PENTYL-2-PYRONE. (LOWRY-TEXAS)

FIELD 05A, 05D

ACCESSION NO. W72-11604

AEROBIC LAKE MUDS FOR THE REMOVAL OF PHOSPHORUS FROM LAKE WATERS,

WISCONSIN UNIV., MADISON. WATER CHEMISTRY LAB.

G. P. FITZGERALD.

LIMNOLOGY AND OCEANOGRAPHY, VOL 15, NO 4, 1970, P 550-555. 5 TAB, 14 REF.

DESCRIPTORS:
*PHOSPHORUS, *LAKE BEDS, *ALGAE, ALGAL CONTROL, NUTRIENT REQUIREMENTS, SORPTION, EUTROPHICATION, DEFICIENT ELEMENTS, ESSENTIAL NUTRIENTS, *AEROBIC CONDITIONS, WATER POLLUTION CONTROL.

IDENTIFIERS:
LAKE MUD, PHOSPHORUS SORPTION, PHOSPHORUS REMOVAL, RATE OF SORPTION.

ABSTRACT:
BIOASSAYS OF ALGAE INDICATED THAT THEY BECAME PHOSPHORUS OR NITROGEN-LIMITED DESPITE THE FACT THAT THE ALGAE WERE CLOSE TO MUDS CONTAINING BOTH PHOSPHORUS AND NITROGEN COMPOUNDS. LABORATORY TESTS WERE CONDUCTED TO DETERMINE IF THE PHOSPHORUS OF LAKE MUDS IS READILY AVAILABLE TO ALGAE UNDER AEROBIC CONDITIONS. PHOSPHORUS-LIMITED SELENASTRUM AND CLADOPHORA WILL RESPOND BY GROWTH OR CHANGES IN EXTRACTABLE PO4-P TO AS LITTLE AS 0.02 MG PO4-P IN SOLUTION; THESE SPECIES DID NOT RESPOND WHEN EXPOSED FOR A PERIOD OF 1 OR 2 WEEKS TO AS MUCH AS 2 MG OF PHOSPHORUS AS LAKE MUDS UNDER AEROBIC CONDITIONS. STUDIES OF THE RATE OF SORPTION OF PHOSPHORUS BY LAKE MUDS UNDER AEROBIC CONDITIONS FROM TWO LAKES, AND FROM THREE DEPTHS IN ONE LAKE, INDICATED THAT AS LITTLE AS 0.4 G (DRY WEIGHT) OF MUD COULD SORB ABOUT 0.05 MG PO4-P IN LESS THAN 30 MIN. THESE FINDINGS SUGGEST THAT THE SORPTION OF PHOSPHORUS BY LAKE MUDS UNDER AEROBIC CONDITIONS CAN BE USED TO REMOVE PHOSPHORUS FROM LAKE WATER. (MORGAN-TEXAS)

FIELD 05G

ACCESSION NO. W72-11617

THE DAN REGION LARGE SCALE OXIDATION PONDS,

WATER PLANNING FOR ISRAEL LTD., TEL-AVIV. DAN REGION SEWAGE RECLAMATION PROJECT.

Y. FOLKMAN, M. KREMER, AND P. G. J. MEIRING.

PREPRINT, PRESENTED AT SIXTH INTERNATIONAL WATER POLLUTION RESEARCH CONFERENCE, SESSION 7, HALL C, PAPER NO 15, JUNE 21, 1972. 12 P, 4 FIG, 2 TAB, 14 REF.

DESCRIPTORS:
*OXIDATION LAGOONS, *MUNICIPAL WASTES, *WATER REUSE, ALGAE, PHOTOSYNTHESIS, DISSOLVED OXYGEN, AEROBIC CONDITIONS, ANAEROBIC CONDITIONS, ORGANIC LOADING, ODOR, SLUDGE, COST ANALYSIS, NITROGEN, CLAYS, *WASTE WATER TREATMENT.

IDENTIFIERS:
*RECIRCULATION RATIO, *ISRAEL(DAN REGION).

ABSTRACT:
ALTHOUGH MOST DESIGNERS RELEGATE OXIDATION LAGOONS TO ONLY THE SMALLEST INSTALLATIONS, LARGE SCALE OXIDATION LAGOONS DESIGNED TO HANDLE 22,000 CU. M. PER DAY OF RAW SEWAGE HAVE BEEN GIVING SATISFACTORY PERFORMANCE WITH EXCELLENT REDUCTIONS OF ALL INDICATORS OF POLLUTION. THE PONDS ARE PRESENTLY OPERATING AT THEIR OVERALL DESIGN LOAD OF 230 KG BOD/DAY/HECTARE OF SURFACE AREA, BUT LOADING IS TO BE INCREASED QUITE GRADUALLY. THE PERFORMANCE OF THE PRIMARY PONDS DEPENDS ALMOST ENTIRELY ON THE 1.5 TO 1 RECIRCULATION RATE OF EFFLUENT FROM THE LAST POND IN THE RECIRCULATION POND SYSTEM BACK TO THE PRIMARY POND. NO ODOR PROBLEMS OR SLUDGE BUILD-UP PROBLEMS HAVE BEEN ENCOUNTERED. COST OF SEWAGE TREATMENT IN THE SECONDARY PONDS WAS INCREASED TO 2.6 U. S. CENTS PER CUBIC METER TREATED BECAUSE OF THE QUANTITIES OF SAND DUNES TO BE MOVED, AS WELL AS THE NECESSITY OF LINING THE BOTTOMS. HOWEVER, A 65 TO 70% REDUCTION OF AN AVERAGE 90 MG/L TOTAL NITROGEN INFLUENT IS ACCOMPLISHED AT THIS PRICE, FAR LESS THAN THE COST OF REMOVING NITROGEN BY ANY OTHER BIOLOGICAL MEANS. (LOWRY-TEXAS)

FIELD 05D

ACCESSION NO. W72-11642

MANMADE POLLUTION ANC AMERICA'S 100,000 LAKES,

CALIFORNIA UNIV., BERKELEY. SANITARY ENGINEERING RESEARCH LAB.

P. H. MCGAUHEY, E. J. MIDDLEBROOKS, ANC C. B. PORCELLA.

PUBLIC WORKS, VOL. 103, NO. 3, P 87-88, MARCH 1972.

DESCRIPTORS:
*SEWAGE TREATMENT, *EFFICIENCIES, *EUTROPHICATION, *LAKES, *WATER POLLUTION EFFECTS, RECLAIMED WATER, SEWAGE EFFLUENTS, RESEARCH AND DEVELOPMENT, NUTRIENTS, NITRCGEN, PHOSPHORUS, ALGAE, PLANT GROWTH, WATER POLLUTION TREATMENT, EVALUATION, WATER QUALITY.

IDENTIFIERS:
FATE OF POLLUTANTS.

ABSTRACT:
TO COMBAT THE PROBLEM OF POLLUTICN OF OUR LAKES, IT HAS BEEN ADVISED THAT RESEARCH BE DONE TO DETERMINE SEVERAL FACTORS IN RELATION TO SEWAGE EFFLUENTS AND THEIR EFFECTS ON THE EUTROPHICATION PROCESS: (1) THE CONCENTRATION AT WHICH NITROGEN AND PHOSPHORUS WILL TRIGGER OR SUPPORT SUBSTANTIAL ALGAL GRCWTH, (2) THE GROWTH STIMULATING EFFECT OF SEWAGE THAT HAS HAD ITS P AND N PARTIALLY REMOVED, AND (3) THE ABILITY OF NUTRIENT REMOVAL PROCESSES TO REDUCE N AND P TC LEVELS BELOW THAT CRITICAL TO ALGAL GROWTH. IN ADDITION, IT IS FELT THAT SINCE INCREASING POPULATIONS TEND TO ACCELERATE POLLUTION IN LAKE BASINS, THE NEED FOR POLLUTION RESEARCH TAKES ON OTHER ASPECTS: (1) THE RESIDUAL ABILITY OF TERTIARY SEWAGE EFFLUENTS TO STIMULATE ALGAL GROWTH; (2) THE ULTIMATE FATE OF NUTRIENTS REMOVED FROM SEWAGE IN A BASIN; AND (3) THE EFFECT OF MAN'S NEAR-SHORE AND SHORELINE MODIFICATIONS ON THE BEAUTY AND BIOLOGY OF THE SYSTEM. LAKE TAHOE IS CITED AS AN EXAMPLE CF EFFICIENT WASTEWATER RECLAMATION AND TREATMENT. (MACKAN-BATTELLE)

FIELD 05C, 02H

ACCESSION NO. W72-11702

OSCILLATORY VARIATION OF A PHYTOPLANKTON POPULATION IN A TROPICAL OCEAN,

MCGILL UNIV., MONTREAL (QUEBEC). MARINE SCIENCES CENTRE.

D. M. STEVEN, AND R. GLOMBITZA.

NATURE, VOL 237, NO 5350, P 105-107, MAY 12, 1972. 4 FIG, 1 TAB, 9 REF.

DESCRIPTORS:
*VARIABILITY, *PHYTOPLANKTON, *DEPTH, ON-SITE INVESTIGATIONS, *POPULATION, TEMPORAL DISTRIBUTION, CHLOROPHYLL, SAMPLING, MATHEMATICAL STUDIES, ATLANTIC OCEAN, *TROPICAL REGIONS, MARINE ALGAE, SPATIAL DISTRIBUTION, FOURIER ANALYSIS, BICRHYTHMS, CORRELATION ANALYSIS, PRIMARY PRODUCTIVITY.

IDENTIFIERS:
*PERIODICITY, TRICHODESMIUM THIEBAUDII, BARBADCS.

ABSTRACT:
TRICHODESMIUM THIEBAUDII, A TROPICAL PHYTOPLANKTER, EXHIBITED OSCILLATORY VARIATIONS IN PRCDUCTIVITY WHILE UNDER OBSERVATION IN THE TROPICAL WESTERN ATLANTIC. SIXTY-FIVE COLLECTIONS WERE MADE AT REGULAR INTERVALS (AUGUST 29, 1967-JUNE 4, 1970) AT AN OCEANOGRAPHIC STATION ABOUT 9 KM WEST OF SPEIGHTSTOWN, BARBADOS, AT A DEPTH OF 460 M. CHLOROPHYLL CONCENTRATIONS WERE MEASURED AT 5, 25, 50, 75 AND 100 M IN 2 LITER SAMPLES THE NUMBERS CF PHYTOPLANKTON WERE ESTIMATED AT 5, 50, AND 100 M IN 0.5 LITER SAMPLES OF WATER FIXEC WITH ONE PERCENT NEUTRALIZED FORMALIN. THE NUMBER OF FILAMENTS OF TRICHODESMIUM AND OTHER COMMON PHYTOPLANKTON WERE COUNTED. RAW DATA INDICATE THAT CHLOROPHYLL CONCENTRATIONS AND THE NUMBER OF FILAMENTS OF TRICHODESMIUM AT 5 M FLUCTUATED REGULARLY WITH A PERIODICITY OF BETWEEN 3 AND 4 MONTHS. OTHER PHYTOPLANKTON WAS USUALLY LESS NUMEROUS AND DID NOT SHOW SUCH VARIATION. THE ASSOCIATION BETWEEN CHLOROPHYLL AND THE TRICHODESMIUM WAS STUDIED BY HARMONIC ANALYSIS AND BY AUTO- AND CROSS-CORRELATIONS. IT WAS SHOWN THAT CHLOROPHYLL AND TRICHODESMIUM OSCILLATE IN PHASE WITH A PERIODICITY OF ABOUT 105-120 DAYS. A SIMILAR ANALYSIS APPLIED TO THE CCMBINED DATA FROM ALL DEPTHS TO 100 M REVEALED CHLOROPHYLL CONCENTRATION INCREASES WITH DEPTH. THIS INCREASE IS DEEMED TO BE ASSOCIATED WITH OTHER ORGANISMS, NOT TRICHODESMIUM. (SNYDER-BATTELLE)

FIELD 05C

ACCESSION NO. W72-11719

DIURNAL PH PATTERNS AS PREDICTORS OF CARBON LIMITATION IN ALGAL GROWTH,

CLEMSON UNIV., S.C. DEPT. OF ENVIRONMENTAL SYSTEMS ENGINEERING.

L. G. RICH, J. F. ANDREWS, AND T. M. KEINATH.

WATER AND SEWAGE WORKS, VOL 119, NO 5, P 126-130, MAY 1972. 5 FIG, 1 TAB, 12 REF.

DESCRIPTORS:
*HYDROGEN ION CONCENTRATION, *ALGAE, *GROWTH RATES, *LIMITING FACTORS, *DIURNAL DISTRIBUTION, CARBON DIOXIDE, NITROGEN, PHOSPHORUS, LIGHT, NUTRIENTS, MATHEMATICAL STUDIES, EQUATIONS, AQUATIC ENVIRONMENT, WATER PROPERTIES, ALGAL CONTROL, WATER ANALYSIS, CHEMCONTROL, WATER POLLUTION EFFECTS, WATER QUALITY, CARBON.

ABSTRACT:
A MATHEMATICAL STUDY OF DIURNAL PH PATTERNS IS PRESENTED AS THEY RELATE TO ALGAL GROWTH LIMITING FACTORS, WHETHER THEY ARE CARBON DIOXIDE, NITROGEN, PHOSPHORUS, OR LIGHT. SUCH INFORMATION MIGHT BE OF USE IN ANALYZING DIURNAL PH PATTERNS IN NATURE, THEREBY ELUCIDATING THAT PARTICULAR FACTOR CONTROLLING LOCAL ALGAL GROWTH. DATA FROM COMPUTATIONS CONCLUDE THAT FAIRLY SIGNIFICANT CHANGES IN PH MAGNITUDE IN A DIURNAL PERIOD (OF THE ORDER OF SEVERAL TENTHS OF A PH UNIT) CAN OCCUR ONLY WHEN CO2 IS THE LIMITING FACTOR FOR AT LEAST A PORTION OF THE DIURNAL CYCLE. HOWEVER, IT MUST BE REMEMBERED THAT TIDAL ACTION AND FLOW PHENOMENA MAY ALSO DISTORT THE PH PATTERNS. (MACKAN-BATTELLE)

FIELD 05C

ACCESSION NO. W72-11724

TOXICITY OF MERCURY TO PHYTOPLANKTON,

NEW YORK OCEAN SCIENCE LAB., MONTAUK.

R. NUZZI.

NATURE, VOL 237, NO 5349, P 38-39, MAY 5, 1972. 3 FIG.

DESCRIPTORS:
*PHYTOPLANKTON, *MERCURY, *TOXICITY, CULTURES, *CHLORELLA, AQUATIC ALGAE, CHLOROPHYTA, HEAVY METALS, *CHLAMYDOMONAS, INHIBITORS, *LETHAL LIMIT, WATER POLLUTION EFFECTS, WATER POLLUTION SOURCES, INHIBITION.

IDENTIFIERS:
PHAEODACTYLUM TRICORNUTUM, PHENYLMERCURIC ACETATE, MERCURY CHLORIDE, PHENYLACETATE, MERCURY COMPOUNDS.

ABSTRACT:
EXPERIMENTS WERE PERFORMED TO DETERMINE THE TOXICITY OF MERCURY TO THREE PHYTOPLANKTON SPECIES AT CONCENTRATIONS AS LOW AS 0.06 MICROGRAMS/LITER. PHAEODACTYLUM TRICORNUTUM, CHLORELLA SP., AND CHLAMYDOMONAS SP. WERE GROWN AND TESTED AXENICALLY IN CHEMICALLY DEFINED MEDIA. EACH WAS INOCULATED INTO TWO SERIES OF TUBES CONTAINING HGCL2 AT CONCENTRATIONS RANGING FROM 0.74-66.6 MICROGRAMS HG/LITER AND PHENYLMERCURIC ACETATE (PMA) CONCENTRATIONS RANGING FROM 0.66-15.0 MICROGRAMS HG/LITER. P. TRICORNUTUM WAS ALSO TESTED AGAINST PHENYLACETATE CONCENTRATIONS EQUIVALENT TO THE PHENYLACETATE CONCENTRATION IN THE PMA SERIES OF TUBES. CELLS WERE COUNTED AFTER 16 DAYS INCUBATION AT 18 C AND IT WAS FOUND THAT ALL THE ORGANISMS HAD BEEN INHIBITED BY THE MERCURY COMPOUNDS. PMA WAS MORE TOXIC THAN MERCURY CHLORIDE, CAUSING INHIBITION AT CONCENTRATIONS AS LOW AS 0.06 MICROGRAMS HG PER LITER. THE INHIBITION OF P. TRICORNUTUM BY PMA AT 3 MICROGRAMS HG/L WAS ALMOST COMPLETELY REVERSED BY THE ADDITION OF 5 PERCENT GLUTATHIONE. CELLS GROWN AT SUBLETHAL CONCENTRATIONS SHOWED INCREASED INCIDENCE OF MORPHOLOGICAL ABNORMALITIES WITH AN INCREASE IN MERCURY CONTENT. (SNYDER-BATTELLE)

FIELD 05C

ACCESSION NO. W72-11727

A NEW MOUNTING MEDIUM FOR DIATOMS,

BEMIDJI STATE COLL., MINN. DEPT. OF BIOLOGY.

D. B. CZARNECKI, AND H. D. WILLIAMS.

TRANSACTIONS AMERICAN MICROSCOPICAL SOCIETY, VOL. 91, NO. 1, P 73, JANUARY 1972. 2 REF.

DESCRIPTORS:
*DIATOMS, *MICROSCOPY, CHRYSOPHYTA, AQUATIC ALGAE.

IDENTIFIERS:
*MOUNTING MEDIA, METHYLENE IODIDE, TOLUENE, POLYSTYRENE, SLIDE PREPARATION, SAMPLE PREPARATION.

ABSTRACT:
AN IMPROVED INEXPENSIVE MOUNTING MEDIUM, CONSISTING OF A MIXTURE OF POLYSTYRENE AND METHYLENE IODIDE IN TOLUENE, HAS BEEN USED IN PREPARING DIATOM SLIDES. THE MEDIUM HAS A REFRACTIVE INDEX OF 1.75 AT 24 C AND PROVIDES FOR DETAIL RESOLUTION EXCEEDING THAT OF HYRAX MOUNTS. FOR MIXING THE MEDIUM: ADD 15.75 G OF POLYSTYRENE IN SOME TOLUENE, ALLOW TO STAND UNTIL CLEAR, THEN MIX IN 200 G OF METHYLENE IODIDE. (SYNDER-BATTELLE)

FIELD 05A, 07B

ACCESSION NO. W72-11738

MARINE ALGAE OF THE SMITHSONIAN-BREDIN EXPEDITION TO YUCATAN-1960,

MICHIGEN UNIV., ANN ARBOR. DEPT. OF BOTANY.

W. R. TAYLOR.

BULLETIN OF MARINE SCIENCE, VOL. 28, NO. 1, P 34-44, MARCH 1972. 19 REF.

DESCRIPTORS:
*SYSTEMATICS, *MARINE ALGAE, *CYANOPHYTA, *CHLOROPHYTA, PHAEOPHYTA, RHODOPHYTA, CLADOPHORA.

IDENTIFIERS:
*YUCATAN, LYNGBYA CONFERVOIDES, LYNGBYA MAJUSCULA, GOMONTIA POLYRHIZA, ULVA RIGIDA, CLADOPHORA FULIGINOSA, BATOPHORA OERSTEDI, ACETABULARIA CRENULATA, VALONIA VENTRICOSA, VALONIA OCELLATA, ERNODESMIS VERTICILLATA, DICTYOSPHAERIA VANBOSSEAE, DICTYOSPHAERIA CAVERNOSA, CLADOPHOROPSIS MEMBRANACEA, BOODLEA COMPOSITA, ANADYOMENE STELLATA, BRYOPSIS PENNATA, CAULERPA SPECIES, AVRAINVILLA SPECIES, CLADOCEPHALUS LUTEOFUSCUS, UDOTEA FLABELLUM, PENICILLUS SPECIES, HALIMEDA SPECIES, DILOPHUS SPECIES, DICTYOTA SPECIES, POCOCKIELLA VARIEGATA, STYPODIUM ZONALE, PADINA SANCTAE-CRUCIS, SARGASSUM FILIPENDULA, TURBINARIA TURBINATA, GALAXAURA LAPIDESCENS, FOSLIELLA FARINOSA, GONIOLITHON STRICTUM, AMPHIROA FRAGILISSIMA, AMPHIROA BRASILIANA, CORALLINA CUBENSIS, CORALLINA SUBULATA, JANIA CAPILLACEA, JANIA ADHERENS, HYPNEA CERVICORNIS, HYPNEA MUSCIFORMIS, CRYPTARACHNE PLANIFRONS, WRANGELIA PENICILLATA, GRIFFITHSIA SCHOUSBOEI, CERAMIUM BYSSOIDEUM, CENTROCERAS CLAVULATUM, SPYRIDIA FILAMENTOSA, SPYRIDIA ACULEATA, HETEROSIPHONIA WURDEMANNI, HETEROSIPHONIA GIBBESII, POLYSIPHONIA SUBTILISSIMA, BRYOTHAMNION TRIQUETRUM, DIGENIA SIMPLEX, LOPHOCLADIA TRICHOCLADOS, WRIGHTIELLA TUMANOWICZII, BOSTRYCHIA MONTAGNEI, HERPOSIPHONIA TENELLA, ACANTHOPHORA MUSCOIDES, ACANTHOPHORA SPICIFERA, LAURENCIA SPECIES.

ABSTRACT:
OVER 80 SPECIES OF MARINE ALGAE THAT WERE COLLECTED BY THE FOURTH SMITHSONIAN-BREDIN EXPEDITION OF 1960 CAME FROM THE TERRITORY OF QUINTANNA ROO ON THE EAST SIDE OF THE PENINSULA OF YUCATAN, MEXICO, AND SUBSTANIALLY ADD TO THE RECORDS OF MEXICAN MARINE ALGAE. THE PRINCIPAL SET OF SPECIMENS HAS BEEN DEPOSITED IN THE U. S. NATIONAL HERBARIUM. (HOLOMAN-BATTELLE)

FIELD 05A, 05C

ACCESSION NO. W72-11800

THE FLOCCULATION OF ALGAE WITH SYNTHETIC POLYMERIC FLOCCULANTS,

CONNECTICUT UNIV., HARTFORD. SCHOOL OF MEDICINE.

R. C. TILTON, J. MURPHY, AND J. K. DIXON.

WATER RESEARCH, VOL 6, NO 2, P 155-164, FEBRUARY 1972. 3 FIG, 26 REF.

DESCRIPTORS:
*ALGAE, *FLOCCULATION, WATER POLLUTION TREATMENT, COLLOIDS, *POLYMERS, ANIONS, CATIONS, ORGANIC COMPOUNDS, SEPARATION TECHNIQUES, CALCIUM, MAGNESIUM, E. COLI, SILICA, *FILTRATION, LIGHT PENTRATION, PHYSICAL PROPERTIES, ELECTROPHORESIS, CHEMICAL PROPERTIES, HYDROGEN ION CONCENTRATION, *WASTE WATER TREATMENT.

IDENTIFIERS:
POLYACRYLAMIDES, POLYSTYRENE SULFONATE, POLYETHYLENEIMINE, *CHLORELLA ELLIPSOIDIA.

ABSTRACT:
SAMPLES OF A PURE ALGAL CULTURE OF CHLORELLA ELLIPSOIDIA, AT CONCENTRATIONS OF 50-3000 MG/L, WERE SUBJECTED TO POLYMER CONCENTRATIONS OF 0.01-1000 MG/L AT PH 4-7. A NUMBER OF ANALYTICAL METHODS WERE USED IN ORDER TO MEASURE THE FLOCCULATION OF THE ALGAL CULTURE BY THE SYNTHETIC POLYMERIC FLOCCULANTS. BY EMPLOYING A 9.6 SQ CM MILLIPORE MEMBRANE FILTER, THE UPPER LIMIT OF POLYMER CONCENTRATION ABOVE WHICH FILTER BLOCKAGE OCCURRED WAS DETERMINED. IN ANOTHER TEST THE DEGREE OF FLOCCULATION WAS MEASURED BY MEASURING THE LIGHT TRANSMITTED BY ALGAL DISPERSION AFTER POLYMER ADDITION AND SETTLING. THE FLOCCULATION EFFICIENCY WAS FURTHER ANALYZED BY MEASUREMENT OF THE ELECTROPHORETIC MOBILITIES OF ALGAL DISPERSIONS IN A MODIFIED BRIGGS-TYPE CELL. A BRIEF DISCUSSION WAS ALSO PRESENTED ON A COMPARISON OF RESULTS ON ALGAE WITH THOSE OF OTHER COLLOIDS, IT BEING ESPECIALLY SIGNIFICANT THAT AT THE SAME CONCENTRATION OF BIOCOLLOID THE ALGAE NEED ABOUT 100 TIMES HIGHER CONCENTRATION OF POLYMER THAN DOES E. COLI. (MACKAN-BATTELLE)

FIELD 05D, 05A

ACCESSION NO. W72-11833

RESPONSE OF THE ALGA CHLORELLA SOROKINIANA TO CO-60 GAMMA RADIATION,

GRUMMAN AEROSPACE CORP., BETHPAGE, N.Y. RESEARCH DEPT.

W. F. KUNZ.

NATURE, VOL 236, NO 5343, P 178-179, MARCH 24, 1972. 2 FIG, 11 REF.

DESCRIPTORS:
*COBALT RADIOISOTOPES, *RADIOSENSITIVITY, *RADIOACTIVITY EFFECTS, *ALGAE, CHLORELLA, IRRADIATION, RADIOACTIVITY TECHNIQUES, RESISTANCE, CULTURES, *CHLOROPHYTA.

IDENTIFIERS:
*CHLORELLA SOROKINIANA, CO-60, GAMMA RADIATION, SURVIVAL, CHLORELLA PYRENOIDOSA.

ABSTRACT:
THE RESPONSE AND SURVIVAL CHARACTERISTICS OF CHLORELLA SOROKINIANA TO IONIZING RADIATION HAD NOT BEEN DETERMINED, ALTHOUGH THE THERMOPHILIC, BLUE-GREEN ALGAE HAD PREVIOUSLY BEEN USED IN BIOREGENERATION STUDIES IN RADIATION ENVIRONMENTS. THE SURVIVAL OF THE CELLS TO CO-60 GAMMA RADIATION WAS THEREFORE MEASURED IN TERMS OF COLONY-FORMING ABILITY AFTER EXPOSING ALIQUOTS OF THE STOCK CULTURE TO CO-60 RADIATION AT AN ABSORBED DOSE RATE OF 15 KRAD/MIN IN FULLY AEROBIC CONDITIONS. THE EXPOSED CELLS WERE PLATED AND INCUBATED ON GLUCOSE-SUPPLEMENTED KNOP'S AGAR, AND SCORED 6 DAYS LATER FOR VISIBLE CULTURES. IN ADDITION, THE SURVIVAL-ABSORPTION RELATIONSHIP WAS DETERMINED FOR 1-HR. OLD CELLS BY EXPOSURE TO DIFFERENT SINGLE DOSES AND MEASURING THE SURVIVING FRACTIONS. SUBLETHAL DAMAGE REPAIR, REPAIR RATE, AND THE INFLUENCE OF SUBLETHAL DAMAGE REPAIR ON AGE AND RESPONSE FUNCTIONS WERE ALL TABULATED. FROM THE RESULTS IT IS MOST OBVIOUS THAT SURVIVAL WITH REPAIR IS ALWAYS HIGHER THAN THAT WITHOUT REPAIR, REGARDLESS OF THE CELL'S AGE. SURVIVAL CURVES ARE INCLUDED FOR ALL THE TESTS. (MACKAN-BATTELLE)

FIELD 05C

ACCESSION NO. W72-11836

ECOLOGICAL STUDIES ON MACROINVERTEBRATE POPULATIONS ASSOCIATED WITH POLLUTED KELP FORESTS IN THE NORTH SEA,

DURHAM UNIV. (ENGLAND). DEPT. OF BOTANY.

D. J. JONES.

HELGOLAENDER WISSENSCHAFTLICHE MEERESUNTERSUCHUNGEN, VOL 22, P 417-441, 1971. 10 FIG, 7 TAB, 44 REF.

DESCRIPTORS:
WATER POLLUTION EFFECTS, ECOLOGY, ECOSYSTEMS, *BIOLOGICAL COMMUNITIES, *SUCCESSION, AQUATIC ENVIRONMENT, *ECOLOGICAL DISTRIBUTION, *ESTUARINE ENVIRONMENT, HABITATS, AQUATIC LIFE, HEAVY METALS, INDUSTRIAL WASTES, INVERTEBRATES, ALGAE, *KELP.

IDENTIFIERS:
*NORTH SEA, *LAMINAREA SPP., SEWAGE POLLUTION, EPIFAUNA.

ABSTRACT:
TWO GRADIENTS OF POLLUTION, ONE ESTUARINE AND ONE OFF THE OPEN COAST, ARE DESCRIBED. THE INTERVENING SEACOAST HAS LITTLE OR NO POLLUTION. A COMPARATIVE METHOD OF POLLUTION SURVEYING IS PRESENTED. ECOLOGICAL COMPARISON IS MADE OF THE COMMUNITY DEVELOPMENT DESCRIBED FOR CLEAN AND POLLUTED STATIONS. TWO ECOLOGICAL BARRIERS TO NORMAL COMMUNITY DEVELOPMENT IN THE POLLUTED ENVIRONMENT ARE POSTULATED. (SVENSSON-WASHINGTON)

FIELD 05C, 02L

ACCESSION NO. W72-11854

ENVIRONMENTAL CHANGES ASSOCIATED WITH A FLORIDA POWER PLANT,

ROSENSTIEL SCHOOL OF MARINE AND ATMOSPHERIC SCIENCES, MIAMI, FLA.

M. A. ROESSLER.

MARINE POLLUTION BULLETIN, VOL 2, NO 6, P 87-90, JUNE 1971. 3 FIG, 1 TAB, 9 REF.

DESCRIPTORS:
BAYS, *MARINE ALGAE, *MARINE ANIMALS, *THERMAL POLLUTION, *THERMAL POWER, ATLANTIC OCEAN, MARINE BIOLOGY, BENTHIC FLORA, BIOASSAY, CYANOPHYTA, *FLORIDA, CORAL, SALINITY, CURRENTS(WATER), MOLLUSKS, ENVIRONMENTAL EFFECTS.

IDENTIFIERS:
*BISCAYNE BAY(FLORIDA), HEATED EFFLUENT, TURKEY POINT(FLORIDA), TURTLE GRASS, THALASSIA SPP., UDOTEA SPP., PENICILLUS SPP., LAURENCIA SPP., SPONGES.

ABSTRACT:
DAMAGE TO THE BIOTA OF BISCAYNE BAY BY THE HEATED EFFLUENT OF A POWER PLANT IS DEMONSTRATED QUANTITATIVELY AND QUALITATIVELY. ALGAE AND GRASSES ARE REPLACED BY BLUE-GREEN FILAMENTOUS ALGAL MATS; SEASONAL RECOVERY IS SLOW AND THE AFFECTED AREAS CONTAIN FEWER KINDS AND SMALLER NUMBERS OF ANIMALS. INCREASED TEMPERATURE IS THE CHIEF CAUSE OF THE CHANGES. (KATZ-WASHINGTON)

FIELD 05C, 02L

ACCESSION NO. W72-11876

A QUANTITATIVE STUDY OF FACTORS AFFECTING ALGAL DIVISION SYNCHRONY MEASUREMENTS,

SAINT LOUIS UNIV., MO. DEPT. OF BIOLOGY.

D. W. ROONEY.

MATHEMATICAL BIOSCIENCES, VOL. 13, NO. 3/4, P 205-211, APRIL 1972. 2 FIG, 1 TAB, 3 REF.

DESCRIPTORS:
*ALGAE, *MODEL STUDIES, MATHEMATICAL STUDIES, DIGITAL COMPUTERS, DATA ANALYSIS, EQUATIONS, REPRODUCTION, BIORHYTHMS, MEASUREMENT, POPULATIONS, CYTOLOGICAL STUDIES, STATISTICAL METHODS, CULTURES.

IDENTIFIERS:
*ERRORS, *SYNCHRONOUS CULTURES, AUTOSPORES, VARIABILITY, BIAS, SYNCHRONY INDEX, COUNTING.

ABSTRACT:
A DIGITAL COMPUTER WAS USED TO SIMULATE ERRORS IN ALGAL CELL NUMBER MEASUREMENTS USED IN SYNCHRONY INDEX COMPUTATIONS. THE INDEX S SUB R IS BASED ON CELL COUNTS OVER A TIME PERIOD IN WHICH THE NUMBER OF CELLS PER UNIT VOLUME INCREASES. REPEATED SIMULATIONS DEMONSTRATED THAT THE RESULTING DEGREES OF BIAS AND VARIABILITY IN THE MEASURED SYNCHRONY INDEX S SUB R DEPEND NOT ONLY ON THE MAGNITUDE OF ERROR IN CELL NUMBER MEASUREMENTS, BUT ALSO ON THE NUMBER OF SUCH MEASUREMENTS AND THE AUTOSPORE YIELD AT DIVISION. THE SIMULATIONS DEMONSTRATED THAT BIAS AND VARIABILITY IN MEASURED S SUB R DECREASE WITH INCREASES IN AUTOSPORE YIELD. THE VARIABILITY IN MEASURED S SUB R INCREASES WITH INCREASING FREQUENCY OF PERIODIC CELL COUNTS. MINIMAL BIAS OF S SUB R OCCURS AT INTERMEDIATE CELL COUNT FREQUENCIES. (SNYDER-BATTELLE)

FIELD 05C, 05A

ACCESSION NO. W72-11920

EYES BENEATH THE WAVES,

DURHAM UNIV. (ENGLAND). DEPT. OF BOTANY.

D. BELLAMY.

NEW SCIENTIST, VOL. 54, NO. 791, P 76-68, APRIL 13, 1972. 3 FIG.

DESCRIPTORS:
*ON-SITE INVESTIGATIONS, *SCUBA DIVING, *SEA WATER, *DATA COLLECTIONS, COPPER, LEAD, HEAVY METALS, MUSSELS, *KELPS, STARFISH, CRUSTACEANS, FOOD CHAINS, FOOD WEBS, WATER POLLUTION EFFECTS, AQUATIC ENVIRONMENT, TURBIDITY, WATER POLLUTION, PHAEOPHYTA, MOLLUSKS, MARINE ALGAE, MARINE ANIMALS.

IDENTIFIERS:
ENGLAND, BIOACCUMULATION, BIOLOGICAL MAGNIFICATION, FATE OF POLLUTANTS.

ABSTRACT:
DATA COLLECTED BY AMATEUR DIVERS ALONG THE EAST COAST OF BRITAIN SHOWS THAT IN POLLUTED SITES KELP FAILS TO GROW AT DEPTHS BELOW 4 M, WHEREAS IN UNPOLLUTED WATERS IT NORMALLY GROWS TO AT LEAST TWICE THAT DEPTH. SINCE KELP DEPTH RANGES CAN BE CONTROLLED BY AT LEAST 2 MAIN FACTORS, LIGHT AND BARE ROCK AREA, IT IS PRESUMED THAT INCREASED TURBIDITY IS THE CONTROLLING FACTOR IN POLLUTED WATER. STUDIES IN NATURALLY TURBID WATER SUPPORT THIS CONTENTION. IN ANOTHER STUDY, THE BACKGROUND COUNT OF HEAVY METALS OF AN 'INSHORE FOOD WEB' WAS ASCERTAINED BY THE COLLECTION OF ALGAE, MUSSELS AND STARFISH FOR ANALYSIS OF LEAD AND COPPER IN THEIR TISSUES. AT THE TWO UNPOLLUTED SITES TESTED, BOTH METALS APPEARED TO BE BIOLOGICALLY CONCENTRATED ALONG THE FOOD CHAIN, WHILE AT THE 2 POLLUTED SITES NO OVERALL BIOLOGICAL CONCENTRATION WAS INDICATED. THE ANIMALS FROM THE POLLUTED AREAS DISPLAY CONCENTRATION OF THE METALS AT 1-2 TIMES THE CONCENTRATION IN UNPOLLUTED SAMPLES. (MACKAN-BATTELLE)

FIELD 05C, 05A

ACCESSION NO. W72-11925

CHEMICAL CHARACTERISTICS OF HUMIC COMPOUNDS ISOLATED FROM SOME DECOMPOSED MARINE ALGAE,

BEDFORD INST., DARTMOUTH (NOVA SCOTIA). ATLANTIC OCEANOGRAPHIC LAB.

M. A. RASHID, AND A. PRAKASH.

JOURNAL OF FISHERIES RESEARCH BOARD OF CANADA, VOL. 29, NO. 1, P 55-60, JANUARY 1972. 3 TAB, 30 REF.

DESCRIPTORS:
*CHEMICAL ANALYSIS, *DECOMPOSING ORGANIC MATTER, *MARINE ALGAE, HUMUS, ORGANIC COMPOUNDS, POLLUTANT IDENTIFICATION, AQUEOUS SOLUTIONS, ION EXCHANGE, METHODOLOGY, ORGANIC ACIDS, FULVIC ACIDS, KELPS, WATER ANALYSIS, EXUDATION, WATER POLLUTION, PHYTOPLANKTON, WATER POLLUTION SOURCES, PHAEOPHYTA, OXIDATION.

IDENTIFIERS:
*HUMIC COMPOUNDS, COLUMN CHROMATOGRAPHY, FUCUS VESICULOSUS, LAMINARIA DIGITATA, ST. MARGARET'S BAY, SCOUDOUC RIVER, HUMIC ACIDS, OXIMATION, SAMPLE PREPARATION.

ABSTRACT:
THE CHEMICAL CHARACTERISTICS OF HUMIC COMPOUNDS ISOLATED FROM DECOMPOSED MARINE ALGAE WERE DETERMINED AND COMPARED WITH THOSE EXTRACTED FROM RIVER WATER. THE CHEMICAL NATURE OF HUMIC COMPOUNDS ISOLATED FROM THE EXUDATES OF LIVE, INTACT PLANTS OF FUCUS VESICULOSUS AND LAMINARIA DIGITATA WAS DETERMINED BY ANALYSES OF THEIR FUNCTIONAL GROUPS, MOLECULAR WEIGHT, ELEMENTAL COMPOSITION, AND CERTAIN SPECTRAL PROPERTIES. WATER SAMPLES WITH A HIGH, DISSOLVED HUMIC CONTENT WERE OBTAINED FROM ST. MARGARET'S BAY, NOVA SCOTIA AND SCOUDOUC RIVER, NEW BRUNSWICK. AFTER THE IMPURITIES WERE REMOVED, THE MOLECULAR WEIGHT WAS DETERMINED BY A COLUMN CHROMATOGRAPHIC TECHNIQUE USING SEPHADEX GELS AND TOTAL ACIDITY MEASURED BY BARIUM HYDROXIDE SOLUTION IN N2. THE CARBOXYLS WERE DETERMINED BY ION-EXCHANGE WITH CALCIUM ACETATE SOLUTION, AND THE CARBONYLS BY OXIMATION. THE DATA FOR OXYGEN-CONTAINING FUNCTIONAL GROUPS INDICATED THAT AS COMPARED TO HUMIC ACIDS, THE FULVIC ACIDS SHOWED RELATIVELY LARGE VALUES OF TOTAL ACIDITY AND 1.5 - 2.5 TIMES HIGHER CARBOXYL CONTENT. THE PHENOLIC HYDROXYL GROUP CONTENT WAS LOW (0.0-1.0 MILLIEQUIVALENTS PER GRAM) IN ALL SAMPLES AND APPEARED TO BE CHARACTERISTIC OF HUMIC COMPOUNDS ORIGINATING ORIGINATING IN THE MARINE ENVIRONMENT. EXCEPT FOR LAMINARIA HUMIC ACIDS, CARBONYL GROUPS WERE HIGH IN ALL SAMPLES, PARTICULARLY THE FULVIC ACIDS ISOLATED FROM RIVER WATER AND LAMINARIA EXUDATE. IN GENERAL, THE MOLECULAR WEIGHT PROPERTIES OF HUMIC AND FULVIC ACIDS WERE SIMILAR TO THE CORRESPONDING FRACTIONS PREVIOUSLY OBTAINED FROM MARINE SEDIMENTS. OPTICAL DENSITY TESTS INDICATED A HIGHER DEGREE OF CONDENSATION FOR MARINE AS COMPARED WITH FRESHWATER HUMIC COMPOUNDS. (BYRD-BATTELLE)

FIELD 05B, 05A

ACCESSION NO. W72-12166

USEFULNESS OF CULTURES IN THE TAXONOMY OF BLUE-GREEN ALGAE, (BENUTZUNG VON KULTUREN IN DER BLAUALGENTAXONOMIE),

CESKOSLOVENSKA AKADEMIE VED, TREBON. BOTANICKY USTAV.

J. KOMAREK.

SCHWEIZERISCHE ZEITSCHRIFT FUR HYDROLOGIE, VOL. 33, NO. 2, P 553-565, DECEMBER 1971. 3 FIG, 2 TAB, 17 REF.

DESCRIPTORS:
*CYANOPHYTA, *CULTURES, *SYSTEMATICS, CLASSIFICATION, *AQUATIC ALGAE.

IDENTIFIERS:
CULTUTURING TECHNIQUES, CULTURE MEDIA, ALGOLOGY, TAXON.

ABSTRACT:
CULTIVATION, BEHAVIOR PATTERN AND OTHER DIFFICULTIES ARISING FROM THE STUDY OF BLUE-GREEN ALGAE IN CULTURES ARE STUDIED. SOME OF THE PROPERTIES OF THE CYANOPHYTA WHICH CAUSE COMPLICATIONS IN TAXONOMY INCLUDE THE FOLLOWING: (1) SPECIAL NUTRIENT SOLUTIONS ARE NEEDED FOR MANY STRAINS; (2) THE GREAT VARIETY OF ECOLOGICAL NEEDS PLACES PARTICULAR DEMANDS ON CULTIVATION AND PRESUPPOSES TECHNICAL EQUIPMENT WHICH CANNOT BE STANDARDIZED; (3) ATYPICAL FORMS THAT DO NOT CORRESPOND TO ECOTYPES IN NATURE RESULT WHEN MANY OF THE STRAINS ARE CULTIVATED; (4) IT IS DIFFICULT TO DRAW GENERAL CONCLUSIONS FROM THE RELATIVELY SMALL AMOUNTS OF COMPARABLE DATA OBTAINED FROM PARALLEL EXPERIMENTS. HOWEVER, CULTIVATION IS RECOMMENDED FOR THE FOLLOWING CASES: (A) STUDY OF POTENTIAL VARIABILITY IN THE CLONAL POPULATION; (B) COMPARISON OF INDUCED VARIABILITY OF ONE FEATURE IN SEVERAL STRAINS; (C) DEFINITION OF QUANTITATIVE OR QUALITATIVE DIFFERENCES AMONG SOME STRAINS OR VERIFICATION OF THEIR TAXONOMICAL IDENTITY; (D) DETERMINATION OF ULTRASTRUCTURAL, PHYSIOLOGICAL AND BIOCHEMICAL PROPERTIES OF A TAXON. (SNYDER-BATTELLE)

FIELD 05C, 05A

ACCESSION NO. W72-12172

OCEANOGRAPHY OF THE NEARSHORE COASTAL WATERS OF THE PACIFIC NORTHWEST RELATING TO POSSIBLE POLLUTION,

OREGON STATE UNIV., CORVALLIS. DEPT. OF OCEANOGRAPHY.

COPY AVAILABLE FROM GPO SUP DOC EP2.10:16070 EOK 07/71 SN5501-0140 VOL I, $5.25, VOL II, $6.00; MICROFICHE FROM NTIS AS PB-211 275, $0.95/VOL. 1. ENVIRONMENTAL PROTECTION AGENCY, WATER POLLUTION CONTROL RESEARCH SERIES, JULY 1971, VOL I. 615 P; VOL II. 744 P. EPA PROGRAM 16070 EOK 07/71.

DESCRIPTORS:
*OCEANOGRAPHY, *PACIFIC NORTHWEST, *COAST REVIEW, *BIBLIOGRAPHIES, WATER POLLUTION SOURCES, THERMAL POLLUTION, REVIEWS, *DATA COLLECTIONS, TOXICITY, PATH OF POLLUTANTS, *METALS, TRACE ELEMENTS, PESTICIDES, SALINITY, NUTRIENTS, ALGAE, AQUATIC ANIMALS, RADIONUCLIDES.

IDENTIFIERS:
*LITERATURE REVIEWS.

ABSTRACT:
THIS STUDY IS LIMITED TO THE COASTAL ZONE OF THE PACIFIC NORTHWEST FROM HIGH TIDE TO TEN KILOMETERS FROM SHORE, AND DOES NOT INCLUDE ESTUARIES AND BAYS. THE LITERATURE HAS BEEN REVIEWED IN 21 CHAPTERS INCLUDING CHAPTERS ON GEOLOGY, HYDROLOGY, WINDS, TEMPERATURE AND SALINITY, HEAT BUDGET, WAVES, COASTAL CURRENTS, CARBON DIOXIDE AND PH, OXYGEN, NUTRIENTS, AND BIOLOGY. SPECIAL CHAPTERS DEAL WITH FIELD STUDIES ON THERMAL DISCHARGES, HEAT DISPERSION MODELS, PULP AND PAPER INDUSTRIAL WASTES, TRACE METALS, RADIOCHEMISTRY, PESTICIDES AND CHLORINE, THERMAL ECOLOGY, AND BIOLOGY OF 20 SELECTED SPECIES. A SUMMARY CHAPTER IS ENTITLED 'THE NEARSHORE COASTAL ECOSYSTEM: AN OVERVIEW.' THE BIBLIOGRAPHY CONTAINS MORE THAN 3100 ENTRIES, MOST FROM THE OPEN LITERATURE, BUT SOME FROM UNPUBLISHED REPORTS. A SEPARATE VOLUME (VOL II), INCLUDES THE FOLLOWING APPENDICES: (1) WIND DATA; (2) TEMPERATURE AND SALINITY DATA; (3) WAVE DATA; (4) TRACE METALS (INCLUDING TRACE METAL TOXICITIES); (5) PESTICIDE TOXICITIES; (6) OXYGEN, NUTRIENT, AND PH DATA; (7) RADIONUCLIDES; AND (8) AN ANNOTATED CHECKLIST OF PLANTS AND ANIMALS (INCLUDING MORE THAN 4400 SPECIES).

FIELD 05B, 05A, 10F

ACCESSION NO. W72-12190

LAKE SUPERIOR PERIPHYTON IN RELATION TO WATER QUALITY,

MINNESOTA UNIV., MINNEAPOLIS.

T. A. OLSON, AND T. O. ODLAUG.

COPY AVAILABLE FROM GPO SUP DOC AS EPA 18050 DBM 02/72 FOR $2.00; MICROFICHE FROM NTIS AS PB-211 185, $0.95. ENVIRONMENTAL PROTECTION AGENCY, WATER POLLUTION CONTROL RESEARCH SERIES, FEBRUARY 1972. 253 P, 171 FIG, 41 TAB, 7 REF, 4 APPEND. EPA PROGRAM 18050 DBM 02/72 (FORMERLY WP-00828).

DESCRIPTORS:
*CHLOROPHYLL, *PLANT PIGMENTS, AQUATIC PRODUCTIVITY, *PRIMARY PRODUCTIVITY, *ALGAE, *DIATOMS, *PERIPHYTON, *PHYTOPLANKTON, *SESSILE ALGAE, *LAKE SUPERIOR, LIMNOLOGY, FERTILITY, PHOTOSYNTHESIS, RESPIRATION, EUTROPHICATION, PLANT POPULATION, CLASSIFICATION, SAMPLING, CHLOROPHYTA, *CHRYSOPHYTA, CYANOPHYTA.

ABSTRACT:
LABORATORY AND FIELD STUDIES WERE CONDUCTED TO EVALUATE THE IMPORTANCE OF PERIPHYTON IN WESTERN LAKE SUPERIOR WITH SPECIAL REFERENCE TO THE MAKE-UP AND DISTRIBUTION OF THE PERIPHYTON GROWTHS AND TO THE OVERALL IMPORTANCE OF PRODUCTIVE CAPACITY OF THIS ASSEMBLAGE OF ORGANISMS. THE TAXONOMIC PORTION OF THE INVESTIGATION INDICATED THAT OVER 90% OF THE TOTAL NUMBER OF ORGANISMS WERE DIATOMS AND THAT THE PHYLA TO WHICH THESE DIATOMS BELONGED WERE THE CHRYSOPHYTA, THE CHLOROPHYTA, AND THE CYANOPHYTA. PREDOMINANT GENERA WERE SYNEDRA, ACHNANTHES, NAVICULA, CYMBELLA, AND GOMPHONEMA. IN MANY RESPECTS, THE PERIPHYTON OF LAKE SUPERIOR WAS SIMILAR TO THAT FOUND IN STREAMS AND THERE WAS EVIDENCE THAT THE INTERRELATED FACTORS THAT AFFECTED PERIPHYTON GROWTHS WERE TEMPERATURE, LIGHT INTENSITY, DEPTH OF WATER, WATER MOVEMENTS, NUTRIENT LEVELS, AND THE TYPE OF SUBSTRATE. ARTIFICIALLY DENUDED ROCKS DEMONSTRATED DEFINITE RE-GROWTH BUT AFTER 46 DAYS THIS GROWTH LEVEL WAS ONLY 18% OF THAT OCCURRING NATURALLY. THE MEAN TOTAL COUNTS OF ORGANISMS IN THE PRIMARY SAMPLING AREA RANGED FROM 497,000 TO 1,470,000 PER SQUARE CENTIMETER OF ROCK SURFACE. STUDIES OF THE PIGMENT CONCENTRATIONS SHOWED THAT THE BIOMASS OF PERIPHYTON ALONG THE NORTH SHORE OF LAKE SUPERIOR RESEMBLE THOSE OF OTHER OLIGOTROPHIC BODIES OF WATER AND RANGE FROM 0.338 TO 3.59 MG OF TOTAL PIGMENT PER 100 SQUARE CENTIMETERS OF ROCK SURFACE. THE AVERAGE WAS 1.36 MG PER 100 SQUARE CENTIMETERS OF ROCK SURFACE. PIGMENT RATIOS INDICATED THAT THE LAKE SUPERIOR PERIPHYTON WAS DOMINATED BY THE CHRYSOPHYTA.

FIELD 05C, 02H

ACCESSION NO. W72-12192

FIXATION OF SR, ZN AND CE RADIONUCLIDES BY SODIUM ALGINATE AND ALGINIC ACID FROM SEA WATER,

INSTITUTE OF BIOLOGY OF THE SOUTHERN SEAS, SEVASTOPOL (USSR).

G. E. LAZORENKO.

PAPER PRESENTED AT THE COMMISSION OF THE EUROPEAN COMMUNITIES INTERNATIONAL SYMPOSIUM, RADIOECOLOGY APPLIED TO THE PROTECTION OF MAN AND HIS ENVIRONMENT, ROME, SEPT. 1971. 8 P, 5 FIG, 5 REF.

DESCRIPTORS:
*MARINE ALGAE, *STRONTIUM RADIOISOTOPES, *ZINC RADIOISOTOPES, ABSORPTION, SEDIMENTATION, ANALYTICAL TECHNIQUES, MOLECULAR STRUCTURE, CATION EXCHANGE, RADIOECOLOGY, *PATH OF POLLUTANTS.

IDENTIFIERS:
*CERIUM RADIOISOTOPES.

ABSTRACT:
SEDIMENTATION AND LIGHT SCATTERING SHOWED THAT SODIUM ALGINATE OBTAINED FROM BLACK SEA CYSTOSEIRA HAS TWO MOLECULAR WEIGHT FRACTIONS, 800,000 AND 37,000. ION EXCHANGE STUDIES SHOWED THAT ZN IS BOUND BY ONLY THE LOWER, CE IS BOUND TO A GREATER EXTENT BY THE HIGHER, AND SR IS BOUND BY BOTH. (BOPP-ORNL)

FIELD 05B

ACCESSION NO. W72-12203

EXPERIMENTAL STUDY OF TH ISOTOPES ACCUMULATION BY MARINE ORGANISMS,

INSTITUTE OF BIOLOGY OF THE SOUTHERN SEAS, SEVASTOPOL (USSR).

A. B. NAZAROV, AND A. YA. ZESENKO.

PAPER PRESENTED AT THE COMMISSION OF THE EUROPEAN COMMUNITIES INTERNATIONAL SYMPOSIUM, RADIOECOLOGY APPLIED TO THE PROTECTION OF MAN AND HIS ENVIRONMENT, ROME, SEPT. 1971. 9 P, 4 FIG, 11 REF.

DESCRIPTORS:
*RADIOISOTOPES, *MARINE ALGAE, *ABSORPTION, *MINE WASTES, RADIOECOLOGY, *PATH OF POLLUTANTS, PHYSICOCHEMICAL PROPERTIES.

IDENTIFIERS:
THORIUM RADIOISOTOPES.

ABSTRACT:
TO DETERMINE THE ROLE OF MARINE ORGANISMS IN MIGRATION OF TH IN MINE WASTES, UPTAKE WAS STUDIED BY FOUR SPECIES OF ALGAE FROM SEAWATER CONTAINING OXALIC AND HYDROCHLORIC TH COMPOUNDS. POSSIBLE REASONS FOR DIFFERENCES BETWEEN THE CONCENTRATION FACTORS FOUND AND LITERATURE VALUES ARE: (1) UNDER NATURAL CONDITIONS THE TIME OF ALGAE CONTACT WITH SEA WATER IS CONSIDERABLY MORE THAN IN THE PRESENT EXPERIMENT, (2) THE POSSIBLE EXISTENCE OF REGIONS OF THE OCEAN WITH HIGH TH CONCENTRATION, AND (3) DIFFERENCE IN THE PHYSICOCHEMICAL STATE. (BOPP-ORNL)

FIELD 05B

ACCESSION NO. W72-12204

NON-BIOLOGICAL UPTAKE OF ZINC-65 FROM A MARINE ALGAL NUTRIENT MEDIUM,

OREGON STATE UNIV., CORVALLIS.

R. D. TOMLINSON.

AVAILABLE FROM NTIS, SPRINGFIELD, VA. AS RLO-1750-83, $3.00 PAPER COPY; $0.95 MICROFICHE. RLO-1750-83, SEPT. 1970, 73 P.

DESCRIPTORS:
*ASSAY, *ZINC, *ADSORPTION, *SURFACES, LABORATORY EQUIPMENT, *LABORATORY TESTS, *MEASUREMENT, *MARINE ALGAE TRACERS, CHEMICAL PRECIPITATION, HYDROGEN ION CONCENTRATION.

IDENTIFIERS:
NON-BIOLOGICAL UPTAKE.

ABSTRACT:
BOTH QUALITATIVE AND QUANTITATIVE EVALUATION WERE MADE OF THE NON-BIOLOGICAL ADSORPTION OF ZINC IN A LABORATORY SYSTEM DESIGNED TO MEASURE ZINC UPTAKE BY A MARINE ALGA. CARRIER-FREE 65ZN WAS USED AS A ZINC RADIOTRACER. FOUR GENERAL AREAS OF INVESTIGATION RECEIVED SPECIAL EMPHASIS. THESE WERE: (1) THE PHYSIOCHEMICAL NATURE OF THE NUTRIENT MEDIUM PRECIPITATE, (2) ADSORPTION OF 65ZN BY THE NUTRIENT MEDIUM PRECIPITATE, (3) ADSORPTION OF 65ZN BY LABORATORY GLASSWARE, AND, AS A FACTOR POTENTIALLY INFLUENCING ADSORPTION, (4) PH LEVELS IN ALGAL CULTURE CONDITIONS. (HOUSER-ORNL)

FIELD 05A, 05B

ACCESSION NO. W72-12210

EXPERIMENTAL-ECOLOGICAL INVESTIGATIONS ON PHAEOCYSTIS POUCHETI (HAPTOPHYCEAE): CULTIVATION AND WASTE WATER TEST,

BIOLOGISCHE ANSTALT HELGOLAND (WEST GERMANY).

H. KAYSER.

HELGOLAENDER WISSENSCHAFTLICHE MEERESUNTERSUCHUNG, VOL 20, P 195-212, 1970. 13 FIG, 6 REF.

DESCRIPTORS:
*MARINE ALGAE, *LABORATORY TESTS, *CHEMICAL WASTE, *GROWTH RATES, *SEAWATER, *TOXICITY, *NUTRIENTS, LETHAL LIMITS, METHODOLOGY, INDUSTRIAL WASTES, DOMESTIC WASTES, BIOASSAY, SULFATES, IRON COMPOUNDS, SEWAGE, NITRATES, PHOSPHATES.

IDENTIFIERS:
SILT-WATER-PLANT RELATIONSHIPS, NORTH SEA, PHAEOCYSTIS SPP., SOIL CULTURES, TITANIUM DIOXIDE WASTES, HELGOLAND.

ABSTRACT:
THE INFLUENCE OF LIGHT, TEMPERATURE, NUTRIENTS, INDUSTRIAL AND DOMESTIC WASTES ON THE ALGA, PHAEOCYSTIS POUCHETI, UNDER LABORATORY CONDITIONS WAS OBSERVED. THE MULTIPLICATION RATES OF THE VARIOUS STAGES IN SOIL, NITRATE AND PHOSPHATE SOLUTIONS, AND SEAWATER SOLUTIONS UNDER VARIOUS CULTURAL METHODS WAS DETERMINED. INDUSTRIAL WASTE WATER (CONSISTING PRIMARILY OF H2SO4 AND FESO4) FROM A TITANIUM DIOXIDE FACTORY FAVORS THE GROWTH OF THE COLONY STAGE OF P. POUCHETI IN A DILUTION OF 1 PART WASTE WATER TO 100,000 PARTS OF NUTRIENT MEDIUM. A DILUTION OF 1:4000 SIGNIFICANTLY REDUCES THE MULTIPLICATION RATES OF COLONIES. A DILUTION OF 1:2250 IS LETHAL. UNFILTERED DOMESTIC SEWAGE IN CONCENTRATIONS OF 1-5 PARTS OF SEWAGE TO 1000 PARTS OF SEA WATER RESULTS IN A VIGOROUS DEVELOPMENT OF THE COLONY STAGE, WHICH IS FOLLOWED BY DAMAGE TO THE COLONIES. THE SINGLE CELL STAGE SHOWS SLIGHTLY INCREASED MULTIPLICATION RATES ALL OF THE TIME. TEN PARTS OF SEWAGE TO 1000 PARTS SEAWATER RESULTS IN TOXICITY TO BOTH STAGES. (KATZ-WASHINGTON)

FIELD 05C

ACCESSION NO. W72-12239

THE INFLUENCE OF ALCOHOL EXTRACTS OF SOME ALGAE (CHLORELLA AND SCENEDESMUS) ON AQUATIC MICROORGANISMS,

WYZSZA SZKOLA ROLNICZA, OLSZTYN-KORTOWA (POLAND).

S. NIEWOLAK.

POLSKIE ARCHIWUM HYDROBIOLOGII, VOL 18, NO 2, P 31-42, 1971. 1 FIG, 3 TAB, 25 REF.

DESCRIPTORS:
*ALGAL TOXINS, *INHIBITION, *CHLORELLA, *SCENEDESMUS, AQUATIC MICROORGANISMS, AQUATIC BACTERIA, AQUATIC ALGAE, ALGAE, CHLOROPHYTA, PSEUDOMONAS, ENTERIC BACTERIA.

IDENTIFIERS:
BACILLUS, MICROCOCCUS, GRAM NEGATIVE BACTERIA.

ABSTRACT:
ALCOHOL EXTRACTS OF ALGAE OF THE GENERAL CHLORELLA AND SCENEDESMUS INHIBIT THE DEVELOPMENT OF ABOUT 20% OF WATER MICROORGANISMS. PARTICULARLY VULNERABLE ARE GRAM-POSITIVE BACTERIA OF THE GENERA MICROCOCCUS AND BACILLUS. AMONG THE GRAM-NEGATIVE BACTERIA, THE GREATEST NUMBER OF VULNERABLE STRAINS CAN BE FOUND IN THE PSEUDOMONAS GROUP AND IN THE ENTEROBACTERIACEAE FAMILY. CHLORELLA EXTRACTS HAVE MORE EXTENSIVE SCOPE OF ANTAGONISTIC ACTION THAN HAVE SCENEDESMUS EXTRACTS. THE LATTER, HOWEVER, ARE MORE ACTIVE AND THEY INHIBIT THE GROWTH OF BACTERIA IN LARGER AREAS. (LEGORE-WASHINGTON)

FIELD 05C

ACCESSION NO. W72-12255

TOXICITY OF MERCURY COMPOUNDS TO AQUATIC ORGANISMS AND ACCUMULATION OF THE COMPOUNDS BY THE ORGANISMS,

FRESHWATER FISHERIES RESEARCH LAB., TOKYO (JAPAN).

Y. MATIDA, H. KUMADA, S. KIMURA, Y. SAIGA, AND T. NOSE.

BULLETIN OF THE FRESHWATER FISHERIES RESEARCH LABORATORY, VOL 21, NO 2, P 197-227, 1971. 4 FIG, 11 TAB, 33 REF.

DESCRIPTORS:
*MERCURY, *PUBLIC HEALTH, *SHELLFISH, WATER POLLUTION EFFECTS, TOXICITY, WATER POLLUTION, FOOD CHAINS, PATH OF POLLUTANTS, BIOASSAY, FISH, RAINBOW TROUT, AQUATIC LIFE, AQUATIC ENVIRONMENT, FISH PHYSIOLOGY, ALGAE, DAPHNIA.

IDENTIFIERS:
*MERCURY COMPOUNDS, *MINAMATA DISEASE, BIOLOGICAL MAGNIFICATION, MINAMATA BAY(JAPAN), ORGANIC MERCURY COMPOUNDS, BIOSYNTHESIS, GUPPIES.

ABSTRACT:
THIS STUDY CONFIRMED THAT FISH FEEDING ON MERCURY-CONTAMINATED ORGANISMS FROM MINAMATA BAY, JAPAN, SUFFER FROM MERCURY POISONING. THE RESULTS OF STUDIES OF CHRONIC TOXICITY TO FISH OF TOXIC SHELLFISH CONTAINING SOME METHYL MERCURY COMPOUNDS AND OF VARIOUS MERCURY COMPOUNDS ARE DISCUSSED. IN ADDITION, THE RESULTS OF SOME EXPERIMENTS ON BIOLOGICAL MAGNIFICATION OF MERCURY COMPOUNDS THROUGH THE FOOD CHAIN FROM PHYTOPLANKTON TO FISH, AND ON BIOSYNTHESIS OF ORGANIC MERCURY FROM INORGANIC MERCURY ARE CONSIDERED. (SVENSSON-WASHINGTON)

FIELD 05C

ACCESSION NO. W72-12257

PREDICTION OF PHOTOSYNTHETIC BIOMASS PRODUCTION IN ACCELERATED ALGAL-BACTERIAL WASTEWATER TREATMENT SYSTEMS,

HEBREW UNIV., JERUSALEM (ISRAEL). DEPT. OF MEDICAL ECOLOGY.

G. SHELEF, M. SCHWARZ, AND H. SCHECHTER.

PREPRINT, PRESENTED AT 6TH INTERNATIONAL WATER POLLUTION RESEARCH CONFERENCE, SESSION 5, HALL A, PAPER NO 9, JUNE 20, 1972. P A/5/9/1-A/5/9/10, 6 FIG, 2 TAB, 23 REF.

DESCRIPTORS:
*OXIDATION LAGOONS, *ALGAE, *WASTE WATER TREATMENT, *MATHEMATICAL MODELS, GROWTH RATES, SOLAR RADIATION, FLOTATION, FLOCCULATION, PERFORMANCE, EFFICIENCIES, *BIOMASS, ORGANIC MATTER, NUTRIENTS, PHOTOSYNTHESIS.

IDENTIFIERS:
ALUM, FERRIC CHLORIDE, *JERUSALEM(ISRAEL).

ABSTRACT:
THE USE OF SPECIALLY DESIGNED PONDS AS A CONTROLLED WASTE WATER TREATMENT WITH MAXIMIZED PHOTOSYNTHETIC ACTIVITY OF ALGAE WAS DEMONSTRATED. THE KINETICS OF ALGAE PRODUCTION WITH RESPECT TO SOLAR INCIDENT IRRADIANCE SERVED FOR THE CONSTRUCTION OF A MATHEMATICAL MODEL FOR PREDICTING ALGAE BIOMASS PRODUCTION AND THE CONCENTRATION OF ALGAE AS A FUNCTION OF HYDRAULIC DILUTION RATE UNDER GIVEN LEVELS OF SOLAR IRRADIANCE. THE PREDICTED LEVELS OF ALGAE PRODUCTION AND CONCENTRATIONS WERE CLOSE TO ACTUAL LEVELS, ALTHOUGH IN DETERMINING THE OPTIMAL DILUTION RATE, SOME CORRECTIONS HAD TO BE MADE. THE REMOVAL EFFICIENCY OF ORGANIC MATTER AND NUTRIENTS IN THE POND SYSTEM WAS RELATIVELY HIGH, PROVIDED THE ALGAE BIOMASS WAS SEPARATED FROM THE EFFLUENT. A 1,750 GAL/HR RECTANGULAR FLOTATOR WAS USED FOR SEPARATING THE ALGAL BIOMASS FROM THE POND EFFLUENT AFTER TREATMENT WITH ALUM OR FERRIC CHLORIDE. PERFORMANCE DATA OF THE ACCELERATED PHOTOSYNTHETIC POND FOR MONTHS JUNE AND DECEMBER WERE ARRANGED IN TABULAR FORM. (GALWARDI-TEXAS)

FIELD 05D

ACCESSION NO. W72-12289

EFFECT OF SPECTRAL COMPOSITION ON PHOTOSYNTHESIS IN TURBID RESERVOIRS--PHOTOSYNTHETIC PRODUCTION IN A TURBID RESERVOIR II. DETAILS OF AN INCUBATION MODEL AND COMMENTS ON THE EFFECT OF LIGHT QUALITY ON PHOTOSYNTHESIS,

KANSAS WATER RESOURCES RESEARCH INST., MANHATTAN.

J. A. OSBORNE, AND G. R. MARZOLF.

AVAILABLE FROM THE NATIONAL TECHNICAL INFORMATION SERVICE AS PB-211 368, $3.00 IN PAPER COPY, $0.95 IN MICROFICHE. CONTRIBUTION NO 106, JUNE 1972. 102 P, 8 FIG, 11 TAB, 69 REF, 2 APPEND. OWRR A-047-KAN(1).

DESCRIPTORS:
*RESERVOIRS, *PRIMARY PRODUCTIVITY, *TURBIDITY, *PHOTOSYNTHESIS, LIMNOLOGY, *EUTROPHICATION, *KANSAS, MODEL STUDIES, ECOSYSTEMS, ALGAE, *LIGHT QUALITY, EUTROPHIC ZONE, CHLOROPHYLL, PHOSPHATES, NITRATES, SEASONAL, REGRESSION ANALYSIS, CORRELATION ANALYSIS.

IDENTIFIERS:
*PHOTOSYNTHETIC MODEL, *TUTTLE CREEK RESERVOIR(KAN), GREAT PLAINS, AQUATIC ECOSYSTEM.

ABSTRACT:
A NUMERICAL MODEL, BASED ON SOME SIMPLIFYING ASSUMPTIONS, WAS USED TO EVALUATE SPATIAL AND SEASONAL PATTERNS IN PRIMARY PRODUCTION IN TUTTLE CREEK RESERVOIR, KANSAS, FROM JUNE 1970 THROUGH MAY 1971. THE ASSUMPTIONS OF THE MODEL BEST FIT TURBID, HOMOTHERMAL LAKES, SUCH AS THOSE IN KANSAS AND THE GREAT PLAINS PROVINCE. THE MODEL DESCRIBES THE PHOTOSYNTHESIS-DEPTH PROFILE ON THE BASIS OF LIGHT ATTENUATION WITH DEPTH AND ALLOWS INTEGRAL PHOTOSYNTHESIS TO BE OBTAINED BY A SIMPLE ANALYTICAL CALCULATION. THE USE OF THE MODEL HAD ITS GREATEST ADVANTAGE IN ALLOWING PELAGIC ALGAL PHOTOSYNTHESIS TO BE MONITORED IN SHIPBOARD INCUBATORS. VARIOUS CHEMICAL, PHYSICAL, AND BIOLOGICAL FEATURES WERE MEASURED SIMULTANEOUSLY WITH PRIMARY PRODUCTION TO DESCRIBE THE DIVERSITY OF THE RESERVOIR, TO EVALUATE HORIZONTAL PRODUCTION PATTERNS, AND TO VALIDATE THE ASSUMPTIONS OF THE MODEL.

FIELD 05C, 02H, 02J

ACCESSION NO. W72-12393

COLOR INFRARED (CIR) PHOTOGRAPHY: A TOOL FOR ENVIRONMENTAL ANALYSIS,

DARTMOUTH COLL., HANOVER, N.H. DEPT. OF GEOGRAPHY.

D. T. LINDGREN.

AVAILABLE FROM THE NATIONAL TECHNICAL INFORMATION SERVICE AS PB-204 472, $3.00 IN PAPER COPY, $0.95 IN MICROFICHE. FINAL REPORT, AUGUST 1971. 42 P, 8 REF. CONTRACT NO. DI-14-08-0001-12958.

DESCRIPTORS:
*REMOTE SENSING, *POLLUTANT IDENTIFICATION, WATER POLLUTION, *MONITORING, AERIAL PHOTOGRAPHY, *INFRARED RADIATION, WATER QUALITY, ALGAE, WATER QUALITY CONTROL, EUTROPHICATION, WASTE WATER TREATMENT, SEDIMENTATION RATES, OIL SPILLS, WATER POLLUTION SOURCES.

IDENTIFIERS:
*COLOR INFRARED PHOTOGRAPHY.

ABSTRACT:
THE NATURE OF COLOR INFRARED FILM (CIR), ITS CAPABILITIES AND LIMITATIONS, ARE DESCRIBED AND ITS POTENTIAL AS A TOOL IN ENVIRONMENTAL STUDIES IS DISCUSSED. INCLUDED IS A SECTION ON THE USE OF CIR IN WATER QUALITY ANALYSIS. THE GROWTH OF ALGAE IS EASILY DETECTED BY THIS TECHNIQUE, WHICH CAN BE USED AS AN INDICATOR OF UNDESIRABLE CONDITIONS. IN TREATMENT FACILITIES, CIR HAS BEEN USED TO IDENTIFY TRICKLING FILTERS THAT ARE OVERLOADED SINCE THEY REFLECT GREATER AMOUNTS OF INFRARED THAN THOSE THAT ARE OPERATING EFFICIENTLY. CIR CAN ALSO BE USED TO DETERMINE THE DEGREE OF SEDIMENTATION OF BODIES OF WATER. FOR DETECTION OF OIL SPILLS, CIR HAS NOT PROVEN AS RELIABLE AS OTHER TYPES OF FILM BECAUSE OF ITS SENSITIVITY TO SUN ANGLES. FINALLY, CIR CAN BE USED FOR DEAD-FISH COUNTS WHERE SOME TOXIC SUBSTANCE HAS BEEN RELEASED IN THE WATER. (MORTLAND-BATTELLE)

FIELD 05A, 07B

ACCESSION NO. W72-12487

CAPACITY OF DESERT ALGAL CRUSTS TO FIX ATMOSPHERIC NITROGEN,

ARIZONA UNIV., TUCSON. DEPT. OF AGRICULTURAL CHEMISTRY AND SOILS.

A. N. MACGREGOR, AND D. E. JOHNSON.

SOIL SCIENCE SOCIETY OF AMERICA PROCEEDINGS, VOL 35, NO 5, P 843-844, SEPTEMBER-OCTOBER 1971. 7 REF.

DESCRIPTORS:
*ALGAE, *NITROGEN FIXATION, *NITROGEN FIXING BACTERIA, *RAINFALL, *ARID LANDS, *ARIZONA, GRASSLANDS, ON-SITE DATA COLLECTIONS, PLANT GROWTH.

ABSTRACT:
IN ADDITION TO THEIR ROLE IN INCREASING NITROGEN, ALGAL CRUSTS, WHICH CONTAIN A VARIETY OF MICROORGANISMS, ARE REPORTED TO ENHANCE SOIL WATER INFILTRATION AND TO PROVIDE A LAYER OF ORGANIC MATTER SUITABLE FOR SEED GERMINATION. A STUDY WAS UNDERTAKEN TO DETERMINE HOW SOON ALGAL CRUSTS FROM A SEMIARID SOIL ARE ABLE TO FIX N AFTER BEING MOISTENED AND THE RATE OF N FIXATION PER UNIT AREA OF MOISTENED CRUSTS. THE STUDY WAS CONDUCTED IN A GRASSLAND AREA IN THE TUCSON BASIN OF SOUTHERN ARIZONA WHERE APPROXIMATELY 4% OF THE SOIL SURFACE POSSESSED ALGAL CRUST FORMATIONS. THREE HOURS AFTER BEING MOISTENED, ALGAL CRUSTS PRODUCED DETECTABLE LEVELS OF N FIXATION AS MEASURED BY THE ACETYLENE-ETHYLENE METHOD. IT WAS ESTIMATED THAT 1 HA OF DESERT GRASSLAND RECEIVES A N INPUT OF 3-4 G/HR FOLLOWING A RAINFALL. THIS IS FORTUNATE IN VIEW OF THE BURST OF HERBACEOUS PLANT GROWTH AND SUBSEQUENT PLANT N REQUIREMENTS DURING THE RAINY SEASON IN THE SONORAN DESERT. (CASEY-ARIZONA)

FIELD 05B, 02G

ACCESSION NO. W72-12505

INTERRELATIONS AMONG PLANKTON, ATTACHED ALGAE, AND THE PHOSPHORUS CYCLE IN ARTIFICIAL OPEN SYSTEMS,

ITHACA COLL., N.Y. DEPT. OF BIOLOGY.

J. L. CONFER.

ECOLOGICAL MONOGRAPHS, VOL 42, NO 1, P 1-22, 1972. 10 FIG, 9 TAB, 32 REF.

DESCRIPTORS:
*PLANKTON, *ALGAE, *PHOSPHORUS, *MODEL STUDIES, LAKES, CYCLING NUTRIENTS, TRACERS, LITTORAL, KINETICS, PHOSPHATES.

IDENTIFIERS:
*PHOSPHORUS CIRCULATION, SMALL LAKES.

ABSTRACT:
PHOSPHORUS CIRCULATION DURING SUMMER STRATIFICATION WAS STUDIED IN 200-LITER AQUARIA, CONTINUOUSLY SUPPLIED WITH TAP WATER, BY MEANS OF ANALYTICAL AND TRACER METHODS. THIS MODEL OF PHOSPHORUS CIRCULATION IS BELIEVED TO APPLY TO SMALL LAKES WITH EXTENSIVE LITTORAL VEGETATION, THE PHOSPHORUS INFLUX TO VARIOUS BIOLOGICAL COMPARTMENTS BEING EQUALED BY A CORRESPONDING OUTFLOW. A MAJOR MEANS BY WHICH PHOSPHORUS WAS REMOVED FROM THE OPEN WATER WAS BY TRAPPING OF PARTICLES BY THE COMMUNITY ASSOCIATED WITH THE AQUARIA SIDES. AFTER THE FIRST RUN THE PONDS WERE INOCULATED WITH WATER FROM THE PREVIOUS PONDS. A BIOTIC SUCCESSION DEVELOPED FROM A PURELY PLANKTONIC COMMUNITY TO A MORE COMPLEX, TWO-COMMUNITY SYSTEM OF ATTACHED ORGANISMS ON THE SIDES AND THE PLANKTON. THE P REMOVAL RATE VARIED WIDELY, DEPENDING ON EXTENT OF 'LITTORAL' GROWTH AND NATURE OF THE PARTICLES. CONSIDERING SUCCESSION TO BE THE TOTAL CHANGE IN PHYSICAL AND BIOLOGICAL CONDITIONS OVER PROLONGED TIME, THIS OPEN-SYSTEM DESIGN DEVELOPED FROM A ONE-COMMUNITY, FEW-SPECIES SYSTEM INTO A TWO-COMMUNITY, SEVERAL-SPECIES SYSTEM. NUTRIENT CIRCULATION RATE GREATLY INCREASED WITH TIME AND DEVELOPED INTO THIS STEADY STATE, NOT AN EQUILIBRIUM SYSTEM WHICH DEPENDED ON A CONTINUAL PHOSPHORUS INFLUX TO MAINTAIN CONCENTRATIONS AND CIRCULATION RATES. (JONES-WISCONSIN)

FIELD 05C, 02H

ACCESSION NO. W72-12543

NUTRIENTS LIMITING ALGAL GROWTH,

DSM'S CENTRAL LAB., GELEEN (NETHERLANDS). BIOLOGICAL DEPT.

J. W. WOLDENDORP.

STRIKSTOF, DUTCH NITROGENOUS FERTILIZER REVIEW, NO 15, P 16-27, 1972. 3 FIG, 12 PHOTOS, 26 REF.

DESCRIPTORS:
*EUTROPHICATION, *NUTRIENTS, *ALGAE, *LIMITING FACTORS, PHOSPHATES, CHLOROPHYTA, CYANOPHYTA, NITROGEN, CARBON, DIATOMS, BACTERIA, CARBON DIOXIDE, CHLAMYDOMONAS, SCENEDESMUS, CHLORELLA, SILICA, LIGHT INTENSITY, TEMPERATURE, MUD, GROWTH RATES, ANABAENA, NITROGEN FIXATION, DENITRIFICATION, WATER POLLUTION EFFECTS, WATER POLLUTION CONTROL.

IDENTIFIERS:
NETHERLANDS, LIMITING NUTRIENTS.

ABSTRACT:
LITERATURE ON EXCESSIVE ALGAL GROWTH IS REVIEWED. ALTHOUGH LIMNOLOGISTS ARGUE THAT ALGAL GROWTH IS GENERALLY LIMITED BY PHOSPHORUS, NITROGEN CAN ALSO BE LIMITING IN SOME CASES; IT IS ALSO ARGUED THAT CARBON, NOT PHOSPHORUS, IS THE LIMITING NUTRIENT. IF A CORRELATION IS FOUND BETWEEN PHOSPHORUS CONCENTRATION AND ALGAL YIELD, IT IS NOT PROPER TO CONCLUDE THAT PHOSPHORUS IS THE LIMITING NUTRIENT. IT WILL TAKE A CONSIDERABLE TIME BEFORE ANY EFFECT OF PHOSPHATE REMOVAL IS OBSERVED, CONSIDERING THE VAST QUANTITIES OF PHOSPHATES STORED IN THE MUD. MICROBIAL PROCESSES IN TOTO PROBABLY HAVE A REGULATING EFFECT ON NITROGEN CONCENTRATION OF SURFACE WATERS. IT IS UNLIKELY THAT CARBON WOULD BE THE LIMITING ELEMENT IN VIEW OF HIGH BICARBONATE CONTENT OF MOST WATERS. WHETHER BICARBONATES OR CARBON DIOXIDE ARE UTILIZED BY ALGAL POPULATIONS IS DISCUSSED. LITERATURE CONCERNING THE INFLUENCE OF N, P AND C ON ALGAL GROWTH IN SURFACE WATERS DOES NOT INDICATE CLEARLY WHICH ELEMENT HAS LIMITING EFFECT UNDER THE PREVAILING NETHERLAND CONDITIONS. ALL FORMS OF POLLUTION BY HEAT, CHEMICALS, PATHOGENIC MICROORGANISMS, ETC., MUST BE DIMINISHED IF THE WATER IS TO CONTINUE TO SUPPORT LIFE. (JONES-WISCONSIN)

FIELD 05C

ACCESSION NO. W72-12544

SYNOPTIC STUDY OF ACCELERATED EUTROPHICATION IN LAKE TAHCE--AN ALPINE LAKE,

CALIFORNIA UNIV., DAVIS. INST. OF ECOLOGY.

C. R. GOLDMAN, G. MOSHIRI, AND E. DE AMEZAGA.

IN: INTERNATIONAL SYMPOSIUM ON WATER POLLUTION CONTROL IN COLD CLIMATES, JULY 22-24, 1970, UNIVERSITY OF ALASKA, COLLEGE, P 1-21. 9 FIG, 2 TAB, 28 REF.

DESCRIPTORS:
*EUTROPHICATION, *LAKES, *ALPINE, *SYNOPTIC ANALYSIS, COLD REGIONS, NUTRIENTS, BIOLOGICAL COMMUNITIES, ALGAE, PERIPHYTON, PHYTOPLANKTON, PHOTOSYNTHESIS, INVERTEBRATES, BIOMASS, BENTHIC FAUNA, BENTHIC FLORA, HUMAN POPULATION, PRIMARY PRODUCTIVITY, WATER CHEMISTRY, *CALIFORNIA, *NEVADA.

IDENTIFIERS:
*LAKE TAHOE(CALIF-NEV).

ABSTRACT:
LAKE TAHOE WAS STUDIED BY THE SYNOPTIC APPROACH WHICH PROVIDES A NEARLY INSTANTANEOUS EVALUATION OF CONDITIONS EXISTING ON A GIVEN DAY, ALLOWING NUTRIENT SOURCES TO BE LOCATED ACCURATELY. INCREASED FERTILITY WAS EVIDENT AT THE SOUTH SHORE UNDER THE INFLUENCE OF THE TRUCKEE RIVER AND HIGH RESIDENT POPULATION, IN CRYSTAL BAY WHICH CONTAINED HIGHLY DISTURBED LAND DRAINAGE, AND NEAR THE LAKE OUTFLOW WHERE THERE WERE HIGH RESIDENT POPULATION AND FAIRLY EXTENSIVE SHALLOW WATER AREAS. ALTHOUGH OCCASIONAL HIGH PERIPHYTON VALUES WERE ENCOUNTERED NEAR TRIBUTARIES, THERE WAS LESS CORRELATION WITH TRIBUTARIES THAN WAS FOUND FOR PHYTOPLANKTON PRODUCTIVITY AND BIOMASS; DISTRIBUTION OF PERIPHYTON WAS FAIRLY UNIFORM AROUND THE LAKE. THE ABUNDANCE AND DIVERSITY OF BENTHIC ORGANISMS MAY NOT BE FUNCTIONS OF THE SAME ENVIRONMENTAL PROPERTY AS THE ABUNDANCE, PRODUCTIVITY, AND DIVERSITY OF THE PHYTOPLANKTON. THE PRIMARY PRODUCERS MUST BE VIEWED AS A MORE SENSITIVE INDICATION OF INCREASED FERTILITY THAN CHEMICAL PARAMETERS, SINCE ANY ADDITIONAL NUTRIENTS APPEAR TO MOVE RAPIDLY INTO THE PHYTOPLANKTON. LITTLE OR NO MEASURABLE CHANGE IN WATER CHEMISTRY WAS FOUND WHILE PHYTOPLANKTON PHOTOSYNTHESIS SHOWED A VERY SIGNIFICANT CHANGE. (JONES-WISCONSIN)

FIELD 05C, 02H

ACCESSION NO. W72-12549

MICROBIOLOGIC INDICATORS OF THE EFFICIENCY OF AN AERATED, CONTINUOUS-DISCHARGE, SEWAGE LAGOON IN NORTHERN CLIMATES,

NORTH DAKOTA UNIV., GRAND FORKS. SCHOOL OF MEDICINE.

J. W. VENNES, AND O. O. OLSON.

IN: INTERNATIONAL SYMPOSIUM ON WATER POLLUTION CONTROL IN COLD CLIMATES, JULY 22-24, 1970, UNIVERSITY OF ALASKA, COLLEGE, P 286-311. 4 TAB, 5 REF, APPEND.

DESCRIPTORS:
*SEWAGE BACTERIA, *WASTE WATER TREATMENT, *SEWAGE LAGOONS, *COLD REGIONS, AERATED LAGOONS, *NORTH DAKOTA, BIOCHEMICAL OXYGEN DEMAND, BACTERIA, TEMPERATURE, COLIFORMS, ENTERIC BACTERIA, NITROGEN, VITAMINS, SUSPENDED SOLIDS, LACTOBACILLUS, CHLORELLA, SULFUR BACTERIA, ALGAE, SALMONELLA, E. COLI.

IDENTIFIERS:
*CONTINUOUS DISCHARGE LAGOON.

ABSTRACT:
THE AERATED LAGOON, A DEVELOPMENT IN BIOLOGICAL WASTE TREATMENT WAS STUDIED IN HARVEY, NORTH DAKOTA. COLIFORM, FECAL COLIFORM, AND ENTEROCOCCI WERE DETERMINED AS WELL AS BOD, NITROGEN, PH, TOTAL AND SUSPENDED SOLIDS; AND TOTAL BACTERIAL POPULATIONS ENUMERATED. LAGOON EFFICIENCY DEPENDS ON TEMPERATURE AND OXYGEN; THEIR EFFECT ON BIOLOGIC STABILIZATION IS DETERMINED AND THE FINDINGS ARE REFLECTED IN RELATIVE ABUNDANCE OF SEVERAL MICROBIAL SPECIES AND IN BOD. COLIFORM, FECAL COLIFORM, AND ENTEROCOCCAL NUMBERS IN THE SECONDARY LAGOON DURING ZERO CENTIGRADE TEMPERATURES WERE DIRECTLY RELATED TO BOD AND TOTAL NITROGEN. DURING SUMMER TEMPERATURES, LITTLE CORRELATION BETWEEN THESE ORGANISMS AND BOD AND TOTAL NITROGEN WAS NOTED. THERE WAS A CORRELATION BETWEEN THE TOTAL MICROBIAL POPULATION AND BOD AT SUMMER TEMPERATURES. SINCE ONLY 1% OR LESS OF THE TOTAL MICROBIAL POPULATION IS REPRESENTED BY THE ENTERIC ORGANISMS STUDIED, IT IS APPARENT THAT OTHER ORGANISMS MUST BE STUDIED TO DEFINE BETTER THE ROLE OF MICROBIOLOGIC INDICATORS IN THE EFFICIENCY OF THIS SEWAGE TREATMENT SYSTEM. A SYSTEM CONCERNED WITH PRODUCTION AND UTILIZATION OF THE VITAMIN B, BIOTIN, IS BEING STUDIED, SINCE IT RELATES TO SEVERAL ORGANISMS THAT THRIVE IN SEWAGE OXIDATION LAGOONS. (JONES-WISCONSIN)

FIELD 05D, 05C, 02C

ACCESSION NO. W72-12565

STRUCTURE AND FUNCTIONING OF ESTUARINE ECOSYSTEMS EXPOSED TO TREATED SEWAGE WASTES.

NORTH CAROLINA UNIV., CHAPEL HILL. INST. OF MARINE SCIENCES.

AVAILABLE FROM THE NATIONAL TECHNICAL INFORMATION SERVICE AS COM-71-00688, $6.00 IN PAPER COPY, $0.95 IN MICROFICHE. ANNUAL REPORT FOR 1970-1971, FEBRUARY 1971. 345 P, 62 FIG, 50 TAB, 110 REF. NOAA GH 103.

DESCRIPTORS:
*ECOSYSTEMS, *PRODUCTIVITY, *ESTUARINE ENVIRONMENT, *DOMESTIC WASTES, *SALT MARSHES, *CARBON CYCLE, *PHOSPHORUS COMPOUNDS, *NITROGEN CYCLE, *CRABS, *FISHES, ECOLOGY, FOOD CHAINS, ESTUARINE FISHERIES, MARINE FISHERIES, SEWAGE DISPOSAL, SEWAGE LAGOONS, BRACKISH WATER FISH, BIOLOGICAL TREATMENT, OXIDATION LAGOONS, PHYTOPLANKTON, AQUATIC INSECTS, WATERFOWL, MARINE ALGAE, AQUICULTURE.

IDENTIFIERS:
MEIOFAUNA, FORAMINIFERA.

ABSTRACT:
THIS IS THE THIRD ANNUAL REPORT FROM AN INVESTIGATION OF THE ECOLOGICAL SYSTEMS WHICH DEVELOP WHEN ESTUARINE WATERS ARE ENRICHED WITH SEWAGE WASTES. VARIOUS PHASES OF COMMUNITY STRUCTURE AND METABOLISM OF SIX EXPERIMENTAL BRACKISH-WATER PONDS, THREE OF WHICH RECEIVED TREATED SEWAGE WASTES, AND OF A SMALL TIDAL CREEK AND ITS SALT MARSHES WERE STUDIED. INCLUDED ARE CHAPTERS ON PRODUCTIVITY, CARBON METABOLISM, THE PHOSPHOROUS BUDGET, NITROGEN, AND BACTERIAL HETEROTROPHY; ON THE STANDING CROPS OF PHYTOPLANKTON, DECAPOD CRUSTACEANS, FISHES, MEIOFAUNA, FORAMINIFERA, INSECTS, MOLLUSKS, AND BIRDS; ON CALCIUM ANALYSIS; AND ON GROWTH AND REPRODUCTION OF ALGAE. THE WASTES PONDS HAVE DEVELOPED INTO PRODUCTIVE, WELL INTEGRATED, BUT SLIGHTLY UNSTABLE SYSTEMS. THEY PERFORM SOME OF THE FUNCTIONS OF TERTIARY TREATMENT AND HOLD PROMISE FOR PRODUCTION OF HARVESTABLE SEAFOOD PROTEIN. (KATZ-WASHINGTON)

FIELD 05C, 02L

ACCESSION NO. W72-12567

THE EFFECT OF MARINE POLLUTANTS ON LAMINARIA HYPERBOREA,

MARINE BIOLOGICAL STATION, PORT ERIN, ISLE OF MAN (ENGLAND).

R. HOPKINS, AND J. M. KAIN.

MARINE POLLUTION BULLETIN, VOL. 2, NO. 5, P 75-77, MAY 1971. 1 TAB, 3 REF.

DESCRIPTORS:
*ALGAE, *BIOASSAY, *PLANT PHYSIOLOGY, POLLUTANTS, *TOXICITY, WATER POLLUTION EFFECTS, RESPIRATION, *HEAVY METALS, CHEMICAL WASTES, MERCURY, COPPER, ZINC, PESTICIDES, PESTICIDE TOXICITY, *HERBICIDES, DALAPON, 2-4-D, INSECTICIDES, PHENOLS, *DETERGENTS, *LETHAL LIMIT, ALKYLBENZENE SULFONATES.

IDENTIFIERS:
*LAMINARIA, SUBLITTORAL REGION, CULTURE EXPERIMENTS, ATRAZINE, MCPA, ENDOSULFAN, FAIRY LIQUID, BLUSYL, SODIUM LAURYL, ETHER SULFATE, SODIUM DODECYL, BENZENE SULFONATE, COCONUT FATTY ACID, PLURONICS, DIETHANOLAMIDE.

ABSTRACT:
A STUDY HAS BEEN MADE OF THE TOXIC CONCENTRATIONS OF 15 CHEMICAL POLLUTANTS, INCLUDING EXAMPLES OF HEAVY METALS, DETERGENTS, AND HERBICIDES, USING LAMINARIA HYPERBOREA AS THE TEST ORGANISM. THIS PLANT IS ECOLOGICALLY THE MOST IMPORTANT IN THE SUBLITTORAL REGION AROUND MUCH OF THE COAST OF BRITAIN. TWO DIFFERENT BIOASSAY TECHNIQUES WERE USED, ONE OF A CULTURE EXPERIMENT TYPE, AND THE OTHER UTILIZING RESPIRATION MEASUREMENTS. (SVENSSON-WASHINGTON)

FIELD 05C

ACCESSION NO. W72-12576

FIVE NEW SPECIES OF CHLAMYDOMONAS,

WISCONSIN STATE UNIV., LA CROSSE. DEPT. OF BIOLOGY.

J. M. KING.

JOURNAL OF PHYCOLOGY, VOL. 8, NO. 1, P 120-126, MARCH 1972. 23 FIG, 24 REF.

DESCRIPTORS:
*SYSTEMATICS, *CHLAMYDOMONAS, CULTURES, CHLOROPHYTA, PROTOZOA, *ALGAE.

IDENTIFIERS:
CHLAMYDOMONAS ISABELIENSIS, CHLAMYDOMONAS PALLIDOSTIGMATICA, CHLAMYDOMONAS FOTTII, CHLAMYDOMONAS TEXENSIS, CHLAMYDOMONAS PSEUDOMICROSPHAERA, CULTURE MEDIA, CULTURING TECHNIQUES, MORPHOLOGY, FLAGELLATES.

ABSTRACT:
FIVE NEW SPECIES OF CHLAMYDOMONAS, C. ISABELIENSIS, C. PALLIDOSTIGMATICA, C. FOTTII, C. PSEUDOMICROSPHAERA AND C. TEXENSIS WERE ISOLATED INTO AXENIC CULTURE DURING AN INVESTIGATION OF GREEN MICROALGAE. THE AXENIC CULTURES WERE MAINTAINED ON BOLDO BASAL MEDIUM (BBM) SOLIDIFIED WITH 1.5 PERCENT AGAR AND THE MORPHOLOGY OF THE ORGANISMS STUDIED WITH WET MOUNTS AND HANGING DROP SLIDES FROM ISOLATES ON BBM AGAR OR BBM LIQUID MEDIUM. EXPERIMENTS WERE PERFORMED IN A CONTROLLED ENVIRONMENT CULTURE ROOM WITH TEMPERATURE AT 22 C AND INCIDENT LIGHT PROVIDED BY FLUORESCENT BULBS. SUPPLEMENTARY ATTRIBUTES OF THE ALGAE MAY SERVE AS POSSIBLE TAXONOMIC AIDS. MORPHOLOGICAL DESCRIPTIONS AND CULTURING CHARACTERISTICS ARE GIVEN FOR EACH OF THE NEW SPECIES. (SNYDER-BATTELLE)

FIELD 05A

ACCESSION NO. W72-12632

STUDIES ON THE BIOLOGY OF BROWN ALGAE ON THE ATLANTIC COAST OF VIRGINIA. I. PORTERINEMA FLUVIATILE (PORTER) WAERN,

KENT STATE UNIV., OHIO. DEPT. OF BIOLOGICAL SCIENCES.

R. G. RHODES.

JOURNAL OF PHYCOLOGY, VOL. 8, NO. 1, P 117-119, MARCH 1972. 11 FIG, 8 REF.

DESCRIPTORS:
*PHAEOPHYTA, *TIDAL MARSHES, *MARINE ALGAE, CULTURES, SALINITY, SALT MARSHES, SYSTEMATICS, *VIRGINIA, *ATLANTIC OCEAN, NORTH AMERICA.

IDENTIFIERS:
*PORTERINEMA FLUVIATILE, CULTURE MEDIA, HUMMOCK CHANNEL, CHLOROPLASTS, MORPHOLOGY, EPIPHYTES.

ABSTRACT:
PORTERINEMA FLUVIATILE FOUND AS A MICROSCOPIC FILAMENT ON A CULM OF SPARTINA ON THE ATLANTIC COAST OF VIRGINIA, REPRESENTS THE FIRST REPORT OF THIS MARINE ALGA IN NORTH AMERICA. THE COLLECTION WAS MADE FROM A TIDAL MARSH WITH WATER TEMPERATURE OF 19.5 C AND SALINITY OF ADJACENT WATER BEING 2.98 PERCENT. SINGLE FILAMENTS WERE ISOLATED AND PLACED INTO CULTURES CONTAINING MODIFIED SCHREIBER'S SOLUTION. CULTURES WERE MAINTAINED AT 20 PLUS OR MINUS 1 DEGREE C WITH CONTINUOUS ILLUMINATION FROM FLUORESCENT LAMPS. PHOTOGRAPHS WERE TAKEN WITH BRIGHT FIELD ILLUMINATION AND CHLOROPLAST MORPHOLOGY AND NUMBER WERE EXAMINED IN LIGHT PASSING THROUGH A WRATTEN FILTER NO. 48 (EASTMAN KODAK). (SNYDER-BATTELLE)

FIELD 05A

ACCESSION NO. W72-12633

RADIOISOTOPIC STUDY OF CALCIFICATION IN THE ARTICULATED CORALLINE ALGA BOSSIELLA ORBIGNIANA,

CALIFORNIA INST. OF TECH., PASADENA. DIV. OF GEOLOGICAL AND PLANETARY SCIENCES.

V. B. PEARSE.

JOURNAL OF PHYCOLOGY, VOL. 8, NO. 1, P 88-97, MARCH 1972. 10 FIG, 39 REF.

DESCRIPTORS:
*PHOTOSYNTHESIS, CALCIUM CHLORIDE, RADIOACTIVITY TECHNIQUES, METABOLISM, TRACERS, RADIOACTIVITY, CARBON DIOXIDE, SALTS, MARINE ALGAE.

IDENTIFIERS:
*CALCIFICATION, *CORALLINE ALGAE, CALCIUM RADIOISOTOPES, *BOSSIELLA ORBIGNIANA, PAPER CHROMATOGRAPHY, SCINTILLATION COUNTING, CA-45.

ABSTRACT:
IN AN ATTEMPT TO LINK CALCIFICATION TO PHOTOSYNTHETIC PROCESSES IN PLANTS, 2-3 CM BRANCHES OF BOSSIELLA ORBIGNIANA WERE SEPARATED FROM THE PLANTS AND INCUBATED IN LIGHT OR DARK IN MILLIPORE FILTERED SEAWATER CONTAINING 2 MICROCURIES/ML CA-45 AS CALCIUM CHLORIDE. AFTER 32 HR INCUBATION, THE BRANCHES WERE WASHED, POST-INCUBATED IN MEDIA FREE OF CA-45, AND THE MEDIA PERIODICALLY ASSAYED FOR CA-45 ACTIVITY BY METHODS OF DECALCIFICATION, SCINTILLATION COUNTING, AND PAPER CHROMATOGRAPHY. DATA INDICATE THAT CALCIFICATION IN BOSSIELLA ORBIGNIANA (CORALLINACEAE) IS CHARACTERIZED BY A SERIES OF FEATURES WHICH VARY ACCORDING TO THE AGE OF THE SEGMENT (INTERGENICULUM). FROM THE YOUNGEST SEGMENT AT THE BRANCH TIP TO OLDER, MORE BASAL SEGMENTS: (1) WEIGHT AND DEGREE OF MINERALIZATION INCREASE, WHILE RATE OF WEIGHT GAIN DECREASES; (2) RATE AND STABILITY OF CA-45 LABELING DECREASE; (3) EFFECT OF LIGHT ON CA-45 LABELING RATE DECREASES, WHILE EFFECT OF LIGHT ON STABILITY OF LABEL INCREASES; HOWEVER, FOR ALL SEGMENTS, CA-45 LABELING IS MORE RAPID AND MORE STABLE IN THE LIGHT THAN IN THE DARK; (4) RATE OF CA-45 LABELING IN KILLED CONTROLS DECREASES. COMPARATIVE STUDIES WITH THE TROPICAL CORALLINE ALGA AMPHIROA YIELDED SIMILAR RESULTS. CHARACTERISTICS OF THE GRADIENTS IN CALCIFICATION AND EFFECT OF LIGHT IN THESE CORALLINE ALGAE ARE VERY SIMILAR TO THOSE FOUND IN A REEF-BUILDING CORAL CONTAINING SYMBIOTIC ALGAE, AND THE DATA SUGGEST THAT ORGANIC PRODUCTS OF PHOTOSYNTHESIS MAY BE OF GENERAL IMPORTANCE TO CALCIFICATION. (MACKAN-BATTELLE)

FIELD 05C, 05A

ACCESSION NO. W72-12634

CHARACTERIZATION OF NATURALLY OCCURRING DISSOLVEC ORGANOPHOSPHORUS COMPOUNDS,

WASHINGTON UNIV., SEATTLE. DEPT. OF CIVIL ENGINEERING.

R. A. MINEAR.

ENVIRONMENTAL SCIENCE AND TECHNOLOGY, VOL. 6, NO. 5, P 431-437, MAY 1972. 11 FIG, 6 TAB, 30 REF.

DESCRIPTORS:
*ALGAE, *PHOSPHORUS, *LAKES, CULTURES, PHOSPHATES, NUTRIENTS, CHLOROPHYTA, *ORGANOPHOSPHORUS COMPOUNDS, WASHINGTCN, SEPARATION TECHNIQUES, DISTILLATION, SPECTRCPHOTOMETRY, CHLOROPHYLL, HYDROLYSIS, PHYTOPLANKTON, FLUORESCENCE, PHOSPHORUS COMPOUNDS, CHLAMYDOMONAS, WATER ANALYSIS, FLUORCMETRY, PIGMENTS.

IDENTIFIERS:
*ORTHOPHOSPHATES, CHLAMYDOMONAS REINHARDTII, MOSES LAKE, PINE LAKE, LAKE WASHINGTON, CHLCROPHYLL A, *DNA, *DISSOLVED CRGANCPHOSPHORUS COMPOUNDS, NATURAL ORGANICS.

ABSTRACT:
TWO SOURCES OF DISSOLVED ORGANIC PHOSPHORUS COMPCUNDS (DOP) WERE USED IN A STUDY CONCERNED WITH THE FORMATION AND MECHANISMS OF RELEASE OF DOP COMPOUNDS IN NATURAL WATERS: (1) PURE CULTURES OF THE GREEN ALGA, CHLAMYDOMAS REINHARDTII, GROWN IN A CARBON-DIOXIDE ENRICHED ATMOSPHERE; AND (2) NATURAL WATER SAMPLES TAKEN FROM THREE SEPARATE LAKES. SPECTROPHOTOMETRY WAS USED FCR ORTHOPHOSPHATE CETERMINATION AND DNA ANALYSES. CHLOROPHYLL A AND TOTAL PIGMENT WERE DETERMINED FLUORIMETRICALLY. IN THE PURE ALGAL CULTURES, HIGH LEVELS CF SOLUBLE ORGANIC PHOSPHORUS WERE OBTAINED WITH ACCOMPANYING LOW RESIDUAL ORTHOPHOSPHATE. MOLECULAR SIEVE SAMPLES OF THE NATURAL LAKE WATER SAMPLES SHOWED THAT UP TO 20 PERCENT OF THE RECOVERABLE CRGANIC PHOSPHORUS IS HIGH-MOLECULAR-WEIGHT MATERIAL. A SIZABLE PERCENTAGE (UP TO 50 PERCENT) OF THIS MATERIAL APPEARED, IN MOST CASES, TO BE DNA OR ITS FRAGMENTS. THE NATURAL ORIGIN OF THE DOP, ESPECIALLY, DNA, AND OTHER HIGH-MOLECULAR-WEIGHT COMPONENTS WAS SUBSTANTIATED IN SPECIAL CULTURE STUDIES WHICH UTILIZED DUAL CHAMBER VESSELS DIVIDED BY 0.22 MICRON MEMBRANES. THE PRESENCE OF DNA IN THE PURE ALGAL CULTURES AND THE CORRELATION WITH CHLOROPHYLL AND TOTAL PIGMENT FOR NATURAL WATER SAMPLES SUGGEST THAT THIS MATERIAL ORIGINATES FROM THE PHYTOPLANKTON. (MORTLAND-BATTELLE)

FIELD 05C, 05B

ACCESSION NO. W72-12637

NATURAL ABUNDANCE OF THE STABLE ISOTOPES OF CARBON IN BICLCGICAL SYSTEMS,

TEXAS UNIV., AUSTIN. DEPT. OF BOTANY.

B. N. SMITH.

BICSCIENCE, VOL 22, NO 4, P 226-231, APRIL 1972. 4 FIG, 1 TAB, 64 REF.

DESCRIPTORS:
*ISOTOPE STUDIES, *STABLE ISOTOPES, *CARBON RADICISOTOPES, ENVIRONMENTAL EFFECTS, METHODOLOGY, ALGAE, AQUATIC PLANTS, ISOTOPE FRACTIONATION, CARBON DIOXIDE, RESPIRATION, MARINE PLANTS, BICARBONATES, MASS SPECTRCMETRY, DISTRIBUTION, SEPARATION TECHNIQUES.

IDENTIFIERS:
*C-13, *C-12, BIOLOGICAL SYSTEMS, BELEMNITELLA AMERICANA, TERRESTRIAL PLANTS, COMBUSTION, ISOTOPE RATIOS, SPECTRCMETRY, LICHENS, MOSSES, GYMNOSPERMS, ANGIOSPERMS.

ABSTRACT:
THE GASEOUS CO2 USED IN MEASURING THE STABLE CARBCN ISOTOPES, C-12 AND C-13, IS COLLECTED WITH SPECIFIC APPARATUS FROM THE COMBUSTION OF ORGANIC MATERIAL AT 800-900 DEGREES C IN EXCESS OXYGEN OR FROM RESPIRATION OF LIVING TISSUE. AFTER COLLECTION, THE GASECUS CO2 IS ANALYZED ON AN ISOTOPE RATIO MASS SPECTROMETER. THE C-13/C-12 RATIO IN ANY GIVEN SAMPLE IS THEN CCMPARED WITH A STANDARD - CO2 FROM THE FOSSIL CARBONATE SKELETON OF BELEMNITELLA AMERICANA (PDB SUB L) ACCORDING TO THE FUNCTION: DELTA C-13 PER MIL IS EQUAL TO 1000 TIMES THE C-13/C-12 OF THE SAMPLE MINUS THE C-13/C-12 CF THE STANDARD DIVIDED BY C-13/C-12 OF THE STANDARD. THE PRECISION OF MEASURING DELTA C-13 WITH THE MASS SPECTROMETER IS PLUS OR MINUS 0.1 PER MIL. RESULTS OF THESE METHODS HAVE SHOWN MARINE AND FRESHWATER PLANTS TO HAVE RELATIVELY MORE C-13 THAN MOST TERRESTRIAL PLANTS. THE PRESUMED DIFFERENCE ALLOWED FOR DISTINGUISHING BETWEEN MARINE AND FRESHWATER SEDIMENTS, PLANTS AND PETROLEUM, AND HAS SUGGESTED THAT ENVIRONMENTAL EFFECTS (E.G., TEMPERATURE) MAY ACCCUNT FCR SOME OF THE ISOTOPIC FRACTIONATIONS IN ORGANISMS AND PLANTS. (MACKAN-BATTELLE)

FIELD 05B, 05A

ACCESSION NO. W72-12709

SKEWED ALGAL DIVISION PATTERNS: EFFECTS OF AUTOSPORE YIELD ON COMPUTED SYNCHRONY INDICES,

SAINT LOUIS UNIV., MO. DEPT. OF BIOLOGY.

D. W. ROONEY.

MATHEMATICAL BIOSCIENCIES, VOL 12, NO 3/4, P 367-373, DECEMBER 1971. 2 FIG, 3 REF.

DESCRIPTORS:
*MATHEMATICAL STUDIES, *ALGAE, EQUATIONS, BIOLOGICAL PROPERTIES, REPRODUCTION, ANALYTICAL TECHNIQUES, NUMERICAL ANALYSIS, MODEL STUDIES, COMPUTERS, CULTURES.

IDENTIFIERS:
*DIVISION PATTERNS, *AUTOSPORES, *SYNCHRONOUS CULTURES, SYNCHRONY INDEX.

ABSTRACT:
REPORTED ARE ALGEBRAIC AND NUMERICAL ANALYSES OF THE ALGAL DIVISION SYNCHRONY INDEX S SUB T WHICH REFLECTS THE DEGREE OF NONLINEARITY OF PLOTS OF LOG CELL NUMBER VERSUS TIME. IT HAS BEEN FOUND THAT FOR SKEWED DISTRIBUTIONS OF DIVISION TIMES, S SUB T IS AFFECTED BY THE NUMBER OF AUTOSPORES PRODUCED PER DIVISION. A NEW ALGAL DIVISICN SYNCHRONY INDEX S SUB R IS PRESENTED. FOR A CULTURE OF ANY GIVEN AUTCSPORE YIELD, S SUB R REFLECTS THE NONLINEARITY OF A SEMILOGARITHMIC PLOT VERSUS TIME OF CELL NUMBER IN A CULTURE OF IDENTICAL DIVISION TIME PATTERN YIELDING TWO AUTOSPORES PER DIVISION. THE INDEX S SUB R, INDEPENDENT OF AUTOSPORE YIELD FOR ANY DIVISION TIME DISTRIBUTICN, ASSUMES HIGHER VALUES FOR DISTRIBUTIONS SKEWED TOWARD LONGER TIMES THAN FOR OPPOSITELY SKEWED DISTRIBUTIONS. (MACKAN-BATTELLE)

FIELD 05A, 05C

ACCESSION NC. W72-12724

THE GRADUAL DESTRUCTION OF SWEDEN'S LAKES,

NATIONAL SWEDISH ENVIRCNMENT PROTECTION BOARD, STOCKHOLM.

T. WILLEN.

AMBIO, VOL 1, NO 1, P 6-14, FEBRUARY 1972. 9 FIG, 3 REF.

DESCRIPTORS:
*LAKES, *WATER POLLUTION EFFECTS, *LIMNOLOGY, WATER POLLUTION, WATER POLLUTION SOURCES, SEWAGE, INDUSTRIAL WASTES, BIOINDICATCRS, CHEMICAL PROPERTIES, PHYSICAL PROPERTIES, PHYTOPLANKTON, ZOOPLANKTON, CHLOROPHYTA, DIATOMS, CYANCPHYTA, BENTHIC FAUNA, AQUATIC ANIMALS, INVERTEBRATES, AQUATIC ALGAE, AERIAL PHOTOGRAPHY, PHOSPHORUS, NITRATES, NITROGEN, *EUTROPHICATION, HYDROGEN ION CONCENTRATION, NUTRIENTS, CHLOROPHYLL, PIKES, PERCHES, PHYSICOCHEMICAL PROPERTIES, TROUT, SALMON, FRESHWATER FISH, ANABAENA, *TURBIDITY, CHRYSOPHYTA, AQUATIC PLANTS, MIDGES, OLIGOCHAETES, ANNELIDS, OLIGOTROPHY, BACTERIA.

IDENTIFIERS:
*CHLOROPHYLL A, MACRCPHYTES, *SWEDEN, MICROCYSIS, CSCILLATORIA, SYNURA, AROGLENA, CRYPTOPHYCEAE, PCNTOPOREIA AFFINIS, ESCX LUCIUS, PERCA FLUVIATILIS, SALVELINUS ALPINUS, SALMO TRUTTA, VENDACE, SALMCN TROUT, LOTA LOTA, LAKE MALOGEN, SWEDEN, CHAR, LUCIOPERCA LUCISPERCA, BURBOT, COREGONUS LAVARETUS, COREGONUS ALBULA, WHITEFISH, BRANCHIURA SOWERBYI, PIKE PERCH, LAKE HJALMAREN, LAKE VATTERN, LAKE VANERN.

ABSTRACT:
DETRIMENTAL CHANGES ARE TAKING PLACE IN A GRCUP CF 4 SWEDISH LAKES CAUSED BY INCREASING SEWAGE DISPOSAL AND INDUSTRIAL POLLUTION OF THEIR WATERS. WATER SAMPLES COLLECTED 6 TIMES A YEAR AND BIOLOGICAL SAMPLES (BACTERIA, PLANKTON, CHLOROPHYLL) COLLECTED MONTHLY WERE TRESTED BY PHYSICO-CHEMICAL METHODS FOR PH, ORGANIC MATTER, OXYGEN CONTENT, SILICA, TURBIDITY, AND CONDUCTIVITY. THE FLORA ANC FAUNA WERE ALSO STUDIED. LONG TERM STUDIES REVEAL INCREASING DETERIORATION AROUND HIGHLY POPULATED OR INDUSTRIAL AREAS. LAKE MALAREN, TAKEN AS A REPRESENTATIVE LAKE, SHOWS INCREASING TURBIDITY, HIGH CONCENTRATIONS OF PLANT NUTRIENTS RESULTING IN RELATIVELY HIGH PHYTOPLANKTON LEVELS, AND HIGH VOLUMES OF ZOOPLANKTON FEEDING OR BOTH. ALTHCUGH ORGANISM QUANTITY APPEARS TO BE INCREASING, QUALITY AND COMPOSITION ARE NOT. CHLOROPHYLL A STUDIES SHOW CONSIDERABLE YEARLY VARIATICN BUT APPEAR TO INVERSELY CORRELATE WITH TURBIDITY LEVELS. (MACKAN-BATTELLE)

FIELD 05C, 05B, 02H

ACCESSION NO. W72-12729

AUTOANTAGONISM, HETEROANTAGONISM AND OTHER CONSEQUENCES OF THE EXCRETIONS OF ALGAE FROM FRESH OR THERMAL WATER, (AUTO., HETEROANTAGONISME ET AUTRES CONSEQUENCES DES EXCRETIONS D'ALGUES D'EAU DOUCE OU THERMALE),

CENTRE NATIONAL DE LA RECHERCHE SCIENTIFIQUE, PARIS (FRANCE). LABORATOIRE DE CYTOLOGIE ET DE CYTOPHYSIOLOGIE DE LA PHOTOSYNTHESE.

M. TASSIGNY, AND M. LEFEVRE.

INTERNATIONALE VEREINIGUNG FUR THEORETISCHE UND ANGEWANDTE LIMNOLOGIE, MITTEILUNGEN NO 19, P 26-38, NOVEMBER 1971. 6 FIG, 4 TAB, 39 REF.

DESCRIPTORS:
*AQUATIC ALGAE, *COMPETITION, *NUTRIENTS, *GROWTH RATES, *INHIBITION, FUNGI, THERMAL SPRINGS, CYANOPHYTA, BACTERIA, ECOLOGY, WATER POLLUTION EFFECTS.

IDENTIFIERS:
*EXCRETORY PRODUCTS, *AUTOANTAGONISM, *HETEROANTAGONISM, AXENIC CULTURES, EXCRETION, APHANIZOMENON GRACILE.

ABSTRACT:
DIFFERENT SPECIES OF ALGAE COMPETE DIRECTLY FOR NUTRIENTS, OR INDIRECTLY BY THE PRODUCTION OF SUBSTANCES THAT INHIBIT THE GROWTH OF COMPETITORS (HETEROANTAGONISM). LEFEVRE ALSO RECOGNIZES AUTOANTAGONISM, THE INHIBITION OF THE GROWTH OF A SPECIES BY ITS OWN EXCRETORY PRODUCTS. CERTAIN ALGAE IN CULTURE REACH A STAGE WHERE GROWTH CEASES ALTHOUGH THERE APPEARS TO BE NO LACK OF NUTRIENTS AND GROWTH CANNOT BE RESTARTED BY THE ADDITION OF NUTRIENTS. IT IS OBSERVED IN NATURE THAT, AS ONE SPECIES BECOMES PARTICULARLY ABUNDANT, OTHER SPECIES BECOME SCARCE. WHEN THE ABUNDANT SPECIES DECLINES, THE OTHERS RESUME ACTIVE MULTIPLICATION. EXPERIMENTALLY IT WAS SHOWN THAT SPECIES CULTURED IN THE WATER FEEDING A CERTAIN CANAL FLOURISHED, WHEREAS THOSE CULTURED IN FILTERED WATER FROM THE CANAL, WHERE THERE WAS A LARGE POPULATION OF APHANIZOMENON GRACILE, DID NOT. BACTERIA AND FUNGI WERE ELIMINATED IN A DENSE POPULATION OF AN ALGA, AND HETEROANTAGONISM WAS DEMONSTRATED IN THE LABORATORY WITH STRAINS CULTURED FREE OF BACTERIA AND FUNGI. THE SUBSTANCES PRODUCED BY SOME SPECIES OCCASIONALLY STIMULATE THE GROWTH OF OTHERS. THE EFFECT OF ONE SPECIES UPON ANOTHER VARIES ACCORDING TO THE CONDITIONS OF THE MEDIUM. (HOLOMAN-BATTELLE)

FIELD 05C

ACCESSION NO. W72-12734

ACTICN OF CALCIUM ON THE GROWTH OF AXENIC DESMIDS, (ACTION DU CALCIUM SUR LA CROISSANCE DE DESMIDIEES AXENIQUES),

CENTRE NATIONAL DE LA RECHERCHE SCIENTIFIQUE, PARIS (FRANCE). LABORATOIRE DE CYTOLOGIE ET DE CYTOPHYSIOLOGIE DE LA PHOTOSYNTHESE.

M. TASSIGNY.

INTERNATIONALE VEREINIGUNG FUR THEORETISCHE UND ANGEWANDTE LIMNOLOGIE, MITTEILUNGEN NO 19, P 292-313, NOVEMBER 1971. 9 FIG, 3 TAB, 31 REF.

DESCRIPTORS:
*CALCIUM, *GROWTH RATES, *CULTURES, *AQUATIC ALGAE, CHLOROPHYTA, INHIBITORS, WATER POLLUTION EFFECTS, ENVIRONMENTAL EFFECTS, COMPETITION, AQUATIC POPULATIONS, NUTRIENTS, NUTRIENT REQUIREMENTS.

IDENTIFIERS:
*DESMIDS, *AXENIC CULTURES, CULTURING TECHNIQUES, STAURASTRUM PARADOXUM, STAURASTRUM SEBALDII VAR. ORNATUM, MICRASTERIAS CRUX-MELITENSIS, CLOSTERIUM STRIGOSUM, CULTURE MEDIA.

ABSTRACT:
IN ORDER TO STUDY THE EFFECT OF CALCIUM ON THE GROWTH OF PURE CULTURES, EXPERIMENTS WERE PERFORMED IN A MEDIUM BUFFERED WITH SODIUM PHOSPHATE TO PREVENT EXCESSIVE CHANGES IN PH AND 300 MG/L POTASSIUM NITRATE WAS ADDED TO THE CULTURE MEDIUM TO MAINTAIN GROWTH. STAURASTRUM PARADOXUM, WHICH WAS TAKEN ORIGINALLY FROM A CALCIUM-RICH POLLUTED POND, GREW EQUALLY WELL IN ALL FOUR CONCENTRATIONS OF CALCIUM USED. POPULATIONS OF THREE OTHER SPECIES TESTED INCREASED MORE SLOWLY THE HIGHER THE CONCENTRATION OF CALCIUM IN THE CULTURE MEDIUM, THOUGH THE EFFECT ON EACH WAS NOT IDENTICAL. S. SEBALDII VAR. ORNATUM GREW SLOWLY IN ALL MEDIA; THE RATE AT WHICH THE POPULATION INCREASED WAS DEPRESSED BY CALCIUM, THOUGH THE TOTAL NUMBER ATTAINED VARIED LESS THAN DID THAT OF SOME OTHER SPECIES. CALCIUM NOT ONLY DEPRESSES THE TIME BETWEEN DIVISIONS OF MICRASTERIAS CRUX-MELITENSIS BUT REDUCES MARKEDLY THE FINAL NUMBERS ATTAINED. CLOSTERIUM STRIGOSUM DIFFERS FROM MICRASTERIAS MAINLY IN THAT 13.5 MG/L IS THE CONCENTRATION BELOW WHICH EFFECT ON THE FINAL SIZE OF THE POPULATION IS SLIGHT. WHEN M. CRUS-MELITENSIS AND S. SEBALDII VAR. ORNATUM ARE CULTURED TOGETHER IN 17.1 MG/L CA, THE PROPORTION OF STAURASTRUM FALLS AT FIRST BECAUSE IT TAKES LONGER TO ADAPT TO THE NEW CONDITIONS OF THE CULTURE MEDIUM. AT CA CONCENTRATIONS OF LESS THAN 17.1 MG/L THE RISE IN THE PROPORTION OF STAURASTRUM TAKES PLACE MORE SLOWLY BECAUSE COMPETITION WITH M. CRUX-MELITENSIS IS MAXIMUM. THERE IS NO HETEROANTAGONISTIC REACTION OF EITHER TO THE OTHER, THOUGH THIS APPEARS TO BE THE FACTOR INHIBITING THE GROWTH OF M. CRUX-MELITENSIS IN CULTURE WITH CLOSTERIUM STRIGOSUM.
(HOLOMAN-BATTELLE)

FIELD 05C

ACCESSION NO. W72-12736

PHYTOPLANKTON OF TWO DANISH LAKES, WITH SPECIAL REFERENCE TO SEASONAL CYCLES OF THE NANNOPLANKTON,

COPENHAGEN UNIV. (DENMARK). INST. OF PLANT ANATOMY AND CYTOLOGY.

J. KRISTIANSEN.

INTERNATIONALE VEREINIGUNG FUER THEORETISCHE UND ANGEWANDTE LIMNOLOGIE, MITTEILUNGEN NO 19, P 253-265, NOVEMBER 1971. 12 FIG, 15 REF.

DESCRIPTORS:
*PHYTOPLANKTON, *NANNOPLANKTON, LAKES, *SEASONAL, *PRIMARY PRODUCTIVITY, AQUATIC ALGAE, ZOOPLANKTON, CHRYSOPHYTA, CHLOROPHYTA, BIOMASS, PYRROPHYTA, SPRING, SUMMER, AUTUMN, SAMPLING, PLANKTON NETS, DIATOMS, FOOD CHAINS, ANABAENA, SCENEDESMUS, SEASONAL, DINOFLAGELLATES.

IDENTIFIERS:
LAKE ESROM, TYSTRUP LAKE, BAVELSE LAKE, CRYPTOPHYCEAE, *DENMARK, CYCLOTELLA COMTA, STEPHANODISCUS HANTZSCHII, APHANIZOMENON FLOS-AQUAE, CHROOMONAS ACUTA, MELOSIRA GRANULATA, CERATIUM, MICROCYSTIS, RHODOMONAS MINUTA, ASTERIONELLA FORMOSA, STEPHANODISCUS ASTREA, ANKISTRODESMUS, CHRYSOCOCCUS MINUTUS, DICTYOSPHAERIUM, LAGERHEIMIA.

ABSTRACT:
THE PHYTOPLANKTON OF TYSTRUP-BAVELSE LAKES AND LAKE ESROM (DENMARK) WAS EXAMINED IN RELATIONSHIP TO COMPOSITION, QUANTITY, AND SEASONAL CYCLES OF NANNOPLANKTON. FOR MOST OF THE YEAR, ABOUT 80 PERCENT OF THE TOTAL CELL NUMBER WAS NANNOPLANKTON. IN THE TYSTRUP-BAVELSE LAKES DIATOMS WERE DOMINANT IN SPRING, CHLOROPHYCEAE IN SUMMER, AND CRYPTOPHYCEAE IN AUTUMN AND WINTER. IN LAKE ESROM, CHRYSOPHYCEAE AND CHLOROPHYCEAE WERE DOMINANT IN WINTER AND SPRING, CRYPTOPHYCEAE IN SUMMER AND AUTUMN. NANNOPLANKTON CONSTITUTED 25 PERCENT OF THE PHYTOPLANKTON BIOMASS IN BOTH LAKES WITH A VERNAL MAXIMUM (50-80 PERCENT). THE SEASONAL CYCLES OF NANNOPLANKTON BIOMASS WERE DUE PRIMARILY TO DIATOMS. BIOMASS VALUES WERE CALCULATED FROM VOLUME ESTIMATIONS OF EACH SPECIES CONCERNED. RESULTS WERE EXPRESSED IN ONE MILLION CUBIC MICRONS/ML, WITH VALUES BEING TRANSFORMED INTO WEIGHT VALUES, MG/L. (SNYDER-BATTELLE)

FIELD 05C, 02H

ACCESSION NO. W72-12738

MINERAL NUTRITION OF PLANKTONIC ALGAE: SOME CONSIDERATIONS, SOME EXPERIMENTS,

KOHLENSTOFFBIOLOGISCHE FORSCHUNGSSTATION, DORTMUND (WEST GERMANY).

C. J. SOEDER, H. MULLER, H. D. PAYER, AND H. SCHULLE.

INTERNATIONALE VEREINIGUNG FUR THEORETISCHE UND ANGEWANDTE LIMNOLOGIE, MITTEILUNGEN NO 19, P 39-58, NOVEMBER 1971. 6 FIG, 5 TAB, 53 REF.

DESCRIPTORS:
*MINERALOGY, *NUTRIENT REQUIREMENTS, *PHYTOPLANKTON, *AQUATIC ALGAE, CULTURES, PHOSPHATES, NITRATES, NUTRIENTS, SALTS, TRACE ELEMENTS, LIGHT INTENSITY, SCENEDESMUS, CHLORELLA, GROWTH RATES, RESISTANCE, PHOSPHORUS, IRON, MANGANESE, ZINC, NICKEL, COBALT, COPPER SULFATE, HEAVY METALS, METALS, SULFATES.

IDENTIFIERS:
*CHLORELLA FUSCA, *SCENEDEMUS ACUTUS VAR ALTERNANS, *NITZSCHIA ACTINASTROIDES, *STAURASTRUM PINGUE, SCENEDESMUS QUADRICAUDA, CHLORELLA PYRENOIDOSA, LITHIUM CHLORIDE, LITHIUM, ASTERIONELLA FORMOSA, ASTERIONELLA JAPONICA, STAURASTRUM PARADOXIUM, CULTURE MEDIA, ZINC SULFATE, MANGANESE CHLORIDE, NICKEL SULFATE, COBALT SULFATE, POTASSIUM BROMIDE.

ABSTRACT:
SEVERAL ASPECTS OF MINERAL NUTRITION IN PLANKTONIC ALGAE ARE COMPARED BASED ON BOTH A LITERATURE REVIEW AND SEVERAL EXPERIMENTS. UPPER LIMITS OF NUTRIENT CONCENTRATIONS REPRESENTING THE NUTRIENT CONCENTRATION TOLERANCE WERE ESTABLISHED FOR EACH OF FOUR ORGANISMS: CHLORELLA FUSCA, SCENEDESMUS ACUTUS VAR. ALTERNANS, NITZSCHIA ACTINASTROIDES, AND STAURASTRUM PINGUE. FROM THESE EXPERIMENTS IT WAS TENTATIVELY CONCLUDED THAT THE RELATIVE SALT TOLERANCE OF FRESHWATER ALGAE IS INVERSELY PROPORTIONAL TO THE GROWTH RATE WHICH IS ATTAINED UNLESS THE SALT CONCENTRATION BECOMES SUPRAOPTIMAL. THE RELATIVITY OF THE OPTIMUM OF NUTRIENT CONCENTRATIONS WAS ALSO EVIDENT IN STUDIES ON TRACE ELEMENT DOSAGE. SUSPENSION DENSITY WAS FOUND TO INFLUENCE OPTIMAL TRACE ELEMENT CONCENTRATIONS, THE AVERAGE REQUIREMENTS OF THE DENSE CULTURES BEING HIGHER THAN THESE OF THE MORE DILUTE SYNCHRONOUS CULTURES. A LENGTHY DISCUSSION ON THE ROLE OF PHOSPHATE IS INCLUDED. (MORTLAND-BATTELLE)

FIELD 05C

ACCESSION NO. W72-12739

DICHCTOMOSIPHON IN FLORIDA SPRINGS,

FLCRIDA UNIV., GAINESVILLE. DEPT. CF BOTANY.

J. S. DAVIS, AND W. F. GWOREK.

JOURNAL OF PHYCOLOGY, VOL 8, NO 1, P 130-131, MARCH 1972. 6 REF.

DESCRIPTORS:
*CHLOROPHYTA, *SPRINGS, DIATCMS, BIOTA, PROTOZOA, *AMPHIPODA, *FLORIDA, STREAMS, ISOPODS, SCENEDESMUS, AQUATIC HABITATS, CLASSIFICATION, AQUATIC ALGAE.

IDENTIFIERS:
*DICHOTOMOSIPHON TUBEROSUS, *EPIPHYTES, GROWTH MEDIA, COCCONEIS PLACENTULA, CLOSTERIUM, GCMPHONEMA LCNGICEPS, CYMBELLA, VAUCHERIA.

ABSTRACT:
DICHOTOMOSIPHON AND ITS ASSOCIATED BIOTA WERE OBTAINED FROM SEVERAL FLORIDA SPRINGS. COLLECTIONS WERE MADE SEVERAL TIMES BY DIVING WITH SCUBA EQUIPMENT AND WADING. TO AID IN IDENTIFICATION, PIECES OF THE DICHOTOMOSIPHON MAT WERE PLACED IN 3 DIFFERENT GRCWTH MEDIA: SOIL AND WATER, SPRING WATER, AND A DEFINED MINERAL MEDIUM. THEY WERE INCUBATED AT 70F UNDER FLORESCENT LIGHT (350 FT-C INTENSITY WITH A 16 HR-LIGHT, 8 HR-DARK CYCLE). AFTER 12-15 DAYS, THE TUBES IN SPRING WATER PRCDUCED HIGHLY CHARACTERISTIC SEX ORGANS AND AKINETES. IN ITS NATURAL HABITAT THIS TUBULAR ALGA PRODUCES EXTENSIVE BRIGHT GREEN MATS WHICH HARBOR EPIPHYTES (COCCONEIS PLACENTULA, GCMPHONEMA LONGICEPS), UNICELLULAR ALGAE (CLCSTERIUM, SCENEDESMUS, AND CYMBELLA), CILIATES, AND AMPHIPODS. (SNYDER-BATTELLE)

FIELD 05A

ACCESSION NO. W72-12744

ACCUMULATION AND PERSISTENCE OF DDT IN A LCTIC ECOSYSTEM,

MAINE UNIV., ORONO. DEPT. OF ENTCMOLOGY; AND MAINE UNIV., CRONO. DEPT. OF BIOCHEMISTRY.

J. B. DIMOND, A. S. GETCHELL, AND J. A. BLEASE.

JOURNAL OF FISHERIES RESEARCH BOARD OF CANADA, VOL 28, NO 12, P 1877-1882, DECEMBER 1971. 3 TAB, 20 REF.

DESCRIPTORS:
*PESTICIDE RESIDUES, *LOTIC ENVIRONMENT, WATERSHECS(BASINS), *DDT, *SMALL WATERSHEDS, ECOSYSTEMS, PESTICIDE KINETICS, *MAINE, MUDS, INVERTEBRATES, FISH, BIRDS, STREAMS, AQUATIC PLANTS, AQUATIC INSECTS, CRUSTACEANS, MUSSELS, COMMON MERGANSER DUCK, BIOASSAY, WATER POLLUTION EFFECTS, GAS CHROMATOGRAPHY, BROCK TROUT, SAMPLING, SEPARATION TECHNIQUES, ALGAE, CRAYFISH, FOOD CHAINS, PERSISTENCE, RUNOFF, SEDIMENTS.

IDENTIFIERS:
CHUBS, KINGFISHES, SALVELINUS FONTINALIS, SEMOTILUS ATROMACULATUS, CLEANUP, SAMPLE PRESERVATION, CAMBARUS BARTONI, BIOLCGICAL SAMPLES, MERGUS MERGANSER, MEGACERYLE ALCYCN, BIOACCUMULATION.

ABSTRACT:
SMALL WATERSHEDS IN MAINE THAT HAD RECEIVED VARIOUS NUMBERS OD DDT TREATMENTS STARTING IN 1958 WERE STUDIED FCR PERSISTENCE OF DDT RESIDUES WITH ADJACENT UNTREATED WATERSHEDS BEING USED AS CONTROLS. MUDS, PLANTS, INVERTEBRATES, AND FISH WERE COLLECTED IN 1967 AND 1968 AND BIRDS IN 1969. SAMPLES WERE FROZEN WITHIN A FEW HOURS AFTER COLLECTION AND HELD AT MINUS 15 C UNTIL EXTRACTICN OF PESTICIDES. VARIOUS COLLECTIONS AND EXTRACTICN TECHNIQUES WERE USED, DEPENDING UPON THE TYPE OF SAMPLE. DDT WAS FOUND TO PERSIST IN THE STREAMS FOR AT LEAST 10 YEARS FOLLOWING LIGHT APPLICATION TO THE FOREST; HOWEVER, RESIDUES DID DECLINE SHARPLY AFTER 2 - 3 YEARS. THE PROLONGED PERSISTENCE LED TO CUMULATIVE LEVELS IN STREAMS SPRAYED MORE THAN ONCE AND RESIDUE CONCENTRATION THROUGHOUT THE FOOD CHAIN WAS EVIDENT. (MORTLAND-BATTELLE)

FIELD 05B

ACCESSION NO. W72-12930

PHYSIOLOGICAL INVESTIGATIONS ON THE TOLERANCE OF FUCUS VIRSOIDES (DON) J. AG., (PHYSIOLOGISCHE UNTERSUCHUNGEN UBER DIE TOLERANZ VON FUCUS VIRSOIDES (DON) J. AG),

ZAVOD ZA BIOLOGIJU MORA, ROVINJ (YUGOSLAVIA).

F. GESSNER, AND L. HAMMER.

INTERNATIONALE REVIE DER GESAMTEN HYDROBIOLOGIE, VOL 56, NO 4, P 581-597, 1971. 9 FIG, 7 TAB, 16 REF.

DESCRIPTORS:
*MARINE ALGAE, *KELPS, *PHYSIOLOGICAL ECOLOGY, PHAEOPHYTA, STANDING CROPS, RESISTANCE, PRIMARY PRODUCTIVITY, BIOMASS, ANAEROBIC CONDITIONS, WATER TEMPERATURE, CHLORIDES, PHOTOSYNTHESIS, LITTORAL, RESPIRATION, PLANT PHYSIOLOGY, BICARBONATES, WATER POLLUTION EFFECTS.

IDENTIFIERS:
*FUCUS VIRSOIDES, FUCUS SPIRALIS, MEDITERRANEAN SEA.

ABSTRACT:
THE WORK OF LINARDIC IS REPORTED, WHO PROVED THAT FUCUS VIRSOIDES, THROUGH ISOLATION HAS BECOME A SEPARATE SPECIES, ENDEMIC TO THE MEDITERRANEAN; IT IS POSSIBLY A DESCENDANT OF FUCUS SPIRALIS. THE ECOLOGICAL HABITAT CONDITIONS ARE ANALYZED AND IT IS NOTED THAT FUCUS VIRSOIDES, AVOIDS VERY EXPOSED COASTS. ACCORDING TO LINARDIC, THE MAIN GROWTH PERIODS ARE WINTER AND SPRING; WHEN COMPARED TO NORTHERN LITTORAL ALGAL POPULATIONS, 'STANDING CROP' AND PRIMARY PRODUCTION SEEM TO BE VERY LIMITED. THE TEMPERATURE LIMIT FOR PHOTOSYNTHESIS AT 12 HOURS EXPOSURE IS ABOUT 34C. AFTER EXPOSURE TO TEMPERATURES A FEW DEGREES ABOVE ZERO, PHOTOSYNTHESIS REMAINS NORMAL, BUT FUCUS DOES NOT TOLERATE FREEZING. IN RESPECT TO TEMPERATURE, RESPIRATION PROVES TO BE MORE TOLERANT THAN PHOTOSYNTHESIS; HOWEVER, EXPOSURE TO SUN THROUGHOUT THE DAY DOES NOT LEAD TO A REDUCTION IN PHOTOSYNTHESIS BUT A RISE IN THE BICARBONATE CONTENT CAUSES A MAINFOLD INCREASE IN RATE. AIR-DRY CONDITIONS CAN BE TOLERATED FOR 2-3 DAYS WITHOUT DAMAGE; HOWEVER, SUBSEQUENT PROCESSES IN FUCUS EXPOSED TO DRYNESS LEAD TO THE DEATH OF THE ALGAE. THIS WAS PROVED BY THE INCREASE OF CL-LOSS AND THE WINKLER-TITRATION. FUCUS VIRSOIDES CAN TOLERATE 1 WEEK IN DISTILLED WATER ALMOST WITHOUT DAMAGE AND IS AMAZINGLY TOLERANT TO OXYGEN-FREE CONDITIONS (ANAEROBIOSIS). PHOTOSYNTHESIS DOES NOT DECREASE UNTIL SEVERAL DAYS IN ANAEROBIC CONDITIONS; IN CONTRAST, RESPIRATION REMAINS UNAFFECTED. FUCUS VIRSOIDES IS NOT AN INDICATOR FOR MINIMAL EFFECTS OF POLLUTED WATER; HOWEVER, IT DISAPPEARS RAPIDLY AFTER STRONG POLLUTION OF COASTAL AREAS. (IN GERMAN) (HOLOMAN-BATTELLE)

FIELD 05C

ACCESSION NO. W72-12940

INTERDISCIPLINARY MONITORING OF THE NEW YORK BIGHT,

GRUMMAN AEROSPACE CORP., BETHPAGE, N.Y. RESEARCH DEPT.

W. G. EGAN, J. M. CASSIN, AND M. E. HAIR.

AVAILABLE FROM THE NATIONAL TECHNICAL INFORMATION SERVICE AS AD-737 506, $3.00 IN PAPER COPY, $0.95 IN MICROFICHE. GRUMMAN RESEARCH DEPARTMENT MEMORANDUM NO RM-534J, JANUARY 1972. 15 P, 5 FIG, 7 REF.

DESCRIPTORS:
*INSTRUMENTATION, *MONITORING, *WATER PROPERTIES, *ON-SITE DATA COLLECTIONS, NEW YORK, BIOLUMINESCENCE, BIOLOGICAL PROPERTIES, CHLOROPHYLL, TURBIDITY, SAMPLING, MATHEMATICAL MODELS, ESTUARINE ENVIRONMENT, SALINITY, CHEMICAL ANALYSIS, BIOMASS, DISSOLVED OXYGEN, THERMOCLINE, HYDROGEN ION CONCENTRATION, FLOW, CHEMICAL PROPERTIES, PHYSICAL PROPERTIES, HUDSON RIVER, DEPTH, WATER TEMPERATURE, NUTRIENTS, ALGAE, NITROGEN, PHOSPHORUS, DINOFLAGELLATES, LABORATORY TESTS, WATER QUALITY CONTROL, ON-SITE INVESTIGATIONS, PHYSIOCHEMICAL PROPERTIES, WATER QUALITY.

IDENTIFIERS:
*NEW YORK BIGHT, COLLABORATIVE STUDIES.

ABSTRACT:
INTERDISCIPLINARY PHYSICAL, CHEMICAL, AND BIOLOGICAL MEASUREMENTS WERE MADE IN THE NEW YORK BIGHT DURING 1969-70 USING NEW IN SITU ELECTRONIC INSTRUMENTATION, WITH ASSOCIATED MONITORING BY CONVENTIONAL PROCEDURES TO CHECK THE PERFORMANCE. THIS NEW EQUIPMENT WAS PARTLY AN ORIGINAL DESIGN AND PARTLY MODIFIED COMMERCIAL INSTRUMENTATION AND INCLUDED PHOTOMETRIC SENSORS AND A MODIFIED BECKMAN CONDUCTIVITY CELL, PHOTOVOLT FLOW TYPE ELECTRODE SENSOR, AND DELTA SCIENTIFIC OXYGEN PROBE. THIS EQUIPMENT WAS FOUND TO BE FEASIBLE FOR THE MEASUREMENT OF CHLOROPHYLL (AND THE RELATED BIOMASS), BIOLUMINESCENCE, GELBSTOFF, PH, DISSOLVED O2, SALINITY, AND THE LOCATION OF THE THERMOCLINE. IT APPEARS THAT THE IN SITU INSTRUMENTATION MAY BE ADAPTED TO CONTINUOUS SYNOPTIC MONITORING OF THE ESTUARINE AND OCEANOGRAPHIC PARAMETERS NECESSARY FOR MATHEMATICAL MODELING. (SNYDER-BATTELLE)

FIELD 05A, 02L

ACCESSION NO. W72-12941

EUTROPHICATION OF SURFACE WATERS--LAKE TAHOE.

LAKE TAHOE AREA COUNCIL, SOUTH LAKE TAHOE, CALIF.

COPY AVAILABLE FROM GPO SUP DOC EP2.10:16010 DJW 05/71, $1.25; MICROFICHE FROM NTIS AS PB-211 460, $0.95. ENVIRONMENTAL PROTECTION AGENCY, WATER POLLUTION CONTROL RESEARCH SERIES, MAY 1971, 154 P. 17 FIG., 45 TAB., 25 REF., 5 APPEND. EPA PROGRAM 16010 DJW 05/71.

DESCRIPTORS:
*WATER POLLUTION SOURCES, *EUTROPHICATION, *LAKES, SEWAGE EFFLUENTS, SEEPAGE, SEPTIC TANKS, BIOASSAY, CHEMICAL ANALYSES, NITROGEN, NUTRIENTS, SURFACE WATERS, ALGAE, CHEMICAL PROPERTIES, LANDFILLS, PERCOLATION, DRAINAGE, RUNOFF, PRECIPITATION(ATMOSPHERIC).

IDENTIFIERS:
*LAKE TAHOE(CALIF.).

ABSTRACT:
A SURVEY WAS MADE OF NUTRIENT AND OTHER CHEMICAL CONSTITUENTS OF SURFACE WATERS FROM DEVELOPED AND UNDEVELOPED LAND AREAS, SEWAGE EFFLUENTS, SEEPAGE FROM SEPTIC TANK PERCOLATION SYSTEM AND REFUSE FILLS, DRAINAGE FROM SWAMPS, PRECIPITATION, AND LAKE TAHOE WATER. ALGAL GROWTH STIMULATING POTENTIAL OF THE SAMPLES WERE BIOASSAYED WITH SELENASTRUM GRACILE AS A TEST ORGANISM. ALGAL RESPONSE TO NUTRIENTS WAS MEASURED BY MAXIMUM GROWTH RATE AND MAXIMUM CELL COUNT IN A 5-DAY GROWTH PERIOD. PONDS SIMULATING THE SHALLOW PORTIONS OF THE LAKE WERE USED FOR CONTINUOUS FLOW ASSAY OF THE BIOMASS OF INDIGENOUS LAKE ORGANISMS PRODUCED BY SEWAGE EFFLUENT. FLASK ASSAYS AND CHEMICAL ANALYSES WERE MADE OVER TWO YEARS ON THREE MAJOR CREEKS. TWENTY-EIGHT OTHER CREEKS AND PRECIPITATIONS WERE MONITORED BY CHEMICAL ANALYSIS. EVALUATING THE EUTROPHICATION POTENTIAL, LAKE TAHOE IS NITROGEN SENSITIVE AND RESPONDS TO IT IN PROPORTION TO ITS CONCENTRATION. CREEKS DRAINING DEVELOPED LAND CARRIED TWICE THE NITROGEN AS THOSE DRAINING RELATIVELY UNDISTURBED WATERSHEDS. HUMAN ACTIVITY DOUBLES NITROGEN INFLOW TO THE LAKE. EXPORTING ALL SEWAGE WOULD REMOVE 70% OF THE NITROGEN. (JONES-WISCONSIN)

FIELD 05C, 02H

ACCESSION NO. W72-12955

USE OF ALGAL ASSAYS IN STUDYING EUTROPHICATION PROBLEMS,

NATIONAL ENVIRONMENTAL RESEARCH CENTER, CORVALLIS, OREG. NATIONAL EUTROPHICATION RESEARCH PROGRAM.

T. E. MALONEY.

PREPRINT, PRESENTED AT 6TH INTERNATIONAL WATER POLLUTION RESEARCH CONFERENCE, SESSION 3, NO. 6, HALL C, JUNE 20, 1972, 10 P, 6 TAB, 3 REF.

DESCRIPTORS:
*EUTROPHICATION, *ASSAY, *ALGAE, ANALYTICAL TECHNIQUES, WASTE WATER TREATMENT, NUTRIENTS, NUTRIENT REQUIREMENTS, GROWTH RATES, PHOSPHORUS, NITROGEN, *REGRESSION ANALYSIS, VARIABILITY.

IDENTIFIERS:
DISSOLVED PHOSPHORUS, *ALGAL ASSAYS, ASSAY PROCEDURE.

ABSTRACT:
THE USE OF ALGAL ASSAYS TO ASSIST IN THE SOLUTION OF PRACTICAL EUTROPHICATION PROBLEMS WAS ILLUSTRATED. THE BOTTLE TEST ASSAY, AS DEVELOPED BY THE NATIONAL EUTROPHICATION RESEARCH PROGRAM WAS UTILIZED DURING THESE STUDIES. PRACTICAL APPLICATION OF THE ASSAY PROCEDURE INCLUDED ASSESSMENT OF AFFECTS OF CHANGES IN WASTE TREATMENT PROCESSES ON RECEIVING WATERS, IDENTIFICATION OF ALGAL GROWTH-LIMITING NUTRIENTS AND ASSESSMENT OF RECEIVING WATERS TO DETERMINE THEIR NUTRIENT STATUS AND SENSITIVITY TO CHANGE. A MULTIPLE REGRESSION ANALYSIS, USING THE VARIOUS CHEMICAL PARAMETERS AS THE INDEPENDENT VARIABLES AND THE MAXIMUM ALGAL YIELD AS THE DEPENDENT VARIABLE WAS RUN TO DETERMINE THE NUTRIENT HAVING THE MOST INFLUENCE ON ALGAL GROWTH. DISSOLVED PHOSPHORUS ALONE EXPLAINED 79.84 PERCENT OF THE VARIABILITY IN ALGAL GROWTH IN THE SAMPLES, AND THE SIX NUTRIENTS ONLY INCREASED THIS BY 6.29%. (GALWARDI-TEXAS)

FIELD 05C, 05D

ACCESSION NO. W72-12966

PROJECT HYPO,

DEPARTMENT OF ENERGY, MINES AND RESOURCES, BURLINGTON (ONTARIO). CANADA CENTRE FOR INLAND WATERS; AND ENVIRONMENTAL PROTECTION AGENCY, FAIRVIEW PARK, OHIO. OHIO DISTRICT BASIN OFFICE.

NOEL M. BURNS, AND C. ROSS.

CANADA CENTRE FOR INLAND WATERS PAPER NO 6, AND ENVIRONMENTAL PROTECTION AGENCY TECHNICAL REPORT TS-05-71-208-24, FEBRUARY 1972. 182 P, 6 APPEND.

DESCRIPTORS:
*LAKE ERIE, *EUTROPHICATION, *HYPOLIMNION, *DISSOLVED OXYGEN, *ANAEROBIC CONDITIONS, ALGAE, SEDIMENTS, OXYGEN DEMAND, PHOSPHORUS, NITROGEN, NUTRIENTS, ON-SITE DATA COLLECTIONS, LAKES.

ABSTRACT:
INTERDISCIPLINARY FINDINGS AND ESTIMATES RESULTING FROM PROJECT HYPO, AN INVESTIGATION OF THE CENTRAL BASIN OF LAKE ERIE, LEAD TO ONE DEFINITE CONCLUSION: PHOSPHORUS INPUT TO LAKE ERIE MUST BE REDUCED IMMEDIATELY; IF THIS IS DONE A RAPID IMPROVEMENT CAN BE EXPECTED; IF IT IS NOT DONE, THE RATE OF DETERIORATION WILL BE MUCH GREATER THAN IT HAS BEEN IN RECENT YEARS. CONTRIBUTIONS DISCUSS IN DETAIL THE EFFECTS OF ALGAE, CAUSES AND SITE OF OXYGEN DEPLETION, SEDIMENT OXYGEN DEMAND, BUDGET CALCULATIONS, HYPOLIMNION VOLUME INCREASE, UPWELLING OF WATER MASSES, PROXIMITY OF A PROCESS OF CONTINUAL SELF-FERTILIZATION, AND PHOSPHORUS AND NITROGEN ELIMINATION FROM THE LAKE SYSTEM. EVIDENCE IS PRESENTED WHICH SUGGESTS THAT 76% OF THE PHOSPHORUS AND 57% OF THE NITROGEN IN ALGAL MATERIAL WHICH SEDIMENTS TO THE LAKE FLOOR IS RETAINED THERE IF OXIC CONDITIONS ARE MAINTAINED IN THE OVERLYING WATER. APPROXIMATELY 80% OF THE PHOSPHORUS AND 56% OF THE NITROGEN LOADED INTO LAKE ERIE FROM EXTERNAL SOURCES WILL BE REMOVED FROM THE WATER IF OXYGENATED CONDITIONS ARE MAINTAINED. APPENDIX I DESCRIBES AN AUTOMATIC UNDERWATER CAMERA SYSTEM DESIGNED FOR USE DURING THIS PROJECT AND APPENDIX II DESCRIBES A SUBMERSIBLE AUTOMATIC DISSOLVED OXYGEN-TEMPERATURE MONITORING SYSTEM. (SEE W72-12992 THRU W72-12997) (AUEN-WISCONSIN)

FIELD 05C, 02H

ACCESSION NO. W72-12990

PHYSICAL PROCESSES AFFECTING THE HYPOLIMNION OF THE CENTRAL BASIN OF LAKE ERIE,

DEPARTMENT OF ENERGY, MINES AND RESOURCES, BURLINGTON (ONTARIO). CANADA CENTRE FOR INLAND WATERS; AND ENVIRONMENTAL PROTECTION AGENCY, FAIRVIEW PARK, OHIO.

J. O. BLANTON, AND A. R. WINKLHOFER.

IN: PROJECT HYPO: CANADA CENTRE FOR INLAND WATERS PAPER NO 6, AND ENVIRONMENTAL PROTECTION AGENCY TECHNICAL REPORT TS-05-71-208-24, FEBRUARY 1972, P 9-37. 22 FIG, 4 TAB, 9 REF.

DESCRIPTORS:
*HYPOLIMNION, *LAKE ERIE, *DISSOLVED OXYGEN, TEMPERATURE, CURRENTS(WATER), WINDS, THERMOCLINE, ALGAE, SEDIMENTS.

IDENTIFIERS:
*LAKE ERIE CENTRAL BASIN, POINTE PELEE(ONTARIO), OXYGEN DEPLETION.

ABSTRACT:
IN SITU AUTOMATIC MONITORING DEVICES, EXISTING LAND MONITORING STATIONS, AND CRUISE SAMPLING FURNISH DATA TO DETERMINE RELATIONSHIP BETWEEN DOMINANT WINDS, DOMINANT MOTIONS IN THE HYPOLIMNION, AND THERMOCLINE RESPONSE TO THESE MOTIONS. HYPOLIMNION VOLUME INCREASES WERE EXPLAINED BY LOW WIND ENERGY PERIODS FOLLOWED BY BRIEF PERIODS OF HIGH WIND ENERGY. THIS WIND ENERGY CYCLING DOES NOT NECESSARILY OCCUR ANNUALLY. HYPOLIMNION OXYGEN ENTRAINMENT FROM ABOVE WOULD BE MINIMAL RESULTING IN THE HYPOLIMNION BECOMING ANOXIC AT AN UNPRECEDENTED RATE UNDER CONDITIONS OF LONG DURATION HIGH ENERGY WINDS WITH FEW INTERVENING CALM PERIODS. THE NET HYPOLIMNION WATER MOVEMENT IS THE RESULT OF THE DOMINANT SOUTHWEST WINDS CAUSING THE CANADIAN NEARSHORE AREAS, EAST OF POINTE PELEE, TO BE A POTENTIAL STAGING AREA FOR PROFUSE ALGAL BLOOMS DUE TO ACCUMULATION OF NUTRIENT-RICH ANOXIC WATER IN UPWELLED AREAS. DISSOLVED OXYGEN DEPLETION RATES AS DETERMINED, CONFIRM THE LONGTERM TREND OF INCREASING DEPLETION RATE. MEASURED HYPOLIMNION DISSOLVED OXYGEN FLUCTUATIONS SOLELY ATTRIBUTABLE TO ALGAL PRESENCE WERE NOT DETECTED. SEDIMENT OXYGEN DEMAND MIGHT ACCOUNT FOR THE MEASURED HYPOLIMNION RATE; HOWEVER, DUE TO THE VARIETY AND COMPLEXITY OF THE LAKE SYSTEM, IT CANNOT BE ASSUMED THAT THIS SINGLE PARAMETER CAN DEFINE A SITUATION. (SEE ALSO W72-12990) (JONES-WISCONSIN)

FIELD 05C

ACCESSION NO. W72-12992

SEDIMENT OXYGEN DEMAND IN LAKE ERIE'S CENTRAL BASIN, 1970.

ENVIRONMENTAL PROTECTION AGENCY, CINCINNATI, OHIO. DIV. OF FIELD INVESTIGATIONS.

A. M. LUCAS, AND N. A. THOMAS.

IN: PROJECT HYPO: CANADA CENTRE FOR INLAND WATERS PAPER NO 6, AND ENVIRONMENTAL PROTECTION AGENCY TECHNICAL REPORT TS-05-71-208-24, FEBRUARY 1972, P 45-50. 3 FIG, 1 TAB, 2 REF.

DESCRIPTORS:
*SEDIMENTS, *LAKE ERIE, *DISSOLVED OXYGEN, OXYGEN DEMAND, HYPOLIMNION, AQUATIC PLANTS, ALGAE.

IDENTIFIERS:
LAKE ERIE CENTRAL BASIN.

ABSTRACT:
THE BIOCHEMICAL AND CHEMICAL OXYGEN DEMAND IN LAKE ERIE'S CENTRAL BASIN HYPOLIMNETIC WATERS WAS TOO SMALL TO EXPLAIN THE OXYGEN DEPLETION RATES; TESTS INDICATED THAT DISSOLVED OXYGEN DEPLETION WAS CAUSED PRIMARILY BY THE OXYGEN DEMAND OF BOTTOM SEDIMENTS. SEDIMENT OXYGEN DEMAND RATES WERE MEASURED AT FIVE LOCATIONS IN JUNE, AUGUST, AND SEPTEMBER 1970. THE RATES WERE DETERMINED FROM CHANGES IN THE DISSOLVED OXYGEN CONCENTRATION OF WATER SEALED AND CIRCULATED WITHIN BLACK AND CLEAR PLEXIGLASS CHAMBERS IMBEDDED IN THE LAKE BOTTOM. SEDIMENT OXYGEN DEMAND RATES RECORDED IN JUNE VARIED FROM 1.2 TO 2.2 GM OXYGEN/SQ M PER DAY AND WERE INDICATIVE OF EUTROPHIC CONDITIONS. RATES MEASURED IN AUGUST DURING DAYLIGHT HOURS WITH A CLEAR CHAMBER WERE LESS THAN THOSE MEASURED AT NIGHT WITH THE CLEAR CHAMBER OR DURING THE DAY WITH THE BLACK CHAMBER. OXYGEN PRODUCED BY ALGAL PHOTOSYNTHETIC ACTIVITY ON THE LAKE BOTTOM OFFSET THE SEDIMENT OXYGEN DEMAND DURING PART OF THE DAY RESULTING IN DAILY SOD RATES OF 0.4 TO 0.7 GM OXYGEN/SQ M PER DAY. RATES MEASURED IN SEPTEMBER WITH OXYGENATED SURFACE WATER TRAPPED AND CARRIED TO THE BOTTOM IN THE CHAMBERS RANGED FROM 1.0 TO 2.4 GM OXYGEN/SQ M PER DAY. (SEE ALSO W72-12990) (AUEN-WISCONSIN)

FIELD 05C

ACCESSION NO. W72-12994

BIOLOGICAL STUDIES RELATED TO OXYGEN DEPLETION AND NUTRIENT REGENERATION PROCESSES IN THE LAKE ERIE CENTRAL BASIN,

ENVIRONMENTAL PROTECTION AGENCY, FAIRVIEW PARK, OHIO.

T. BRAIDECH, P. GEHRING, AND C. KLEVENO.

IN: PROJECT HYPO: CANADA CENTRE FOR INLAND WATERS PAPER NO 6, AND ENVIRONMENTAL PROTECTION AGENCY, TECHNICAL REPORT TS-05-71-208-24, FEBRUARY 1972, P 51-70. 17 FIG, 3 PLATES, 4 REF.

DESCRIPTORS:
*OXYGEN DEMAND, *ANAEROBIC CONDITIONS, *LAKE ERIE, *ALGAE, *BENTHIC FAUNA, *BOTTOM SEDIMENTS, *HYPOLIMNION, PHYTOPLANKTON, LIGHT PENETRATION, DISSOLVED OXYGEN, BACTERIA, THERMOCLINE, SEDIMENT-WATER INTERFACES, CHRYSOPHYTA, CHLOROPHYTA, CYANOPHYTA, BIOCHEMICAL OXYGEN DEMAND.

IDENTIFIERS:
*LAKE ERIE CENTRAL BASIN, TRIBONEMA, OEDOGONIUM, ANACYSTIS, CERATIUM, COSMARIUM.

ABSTRACT:
A COMPREHENSIVE ATTEMPT WAS MADE TO DEFINE PHYTOPLANKTON CONDITIONS THROUGHOUT THE WATER COLUMN AND ON THE SEDIMENTS OF LAKE ERIE CENTRAL BASIN. SPECIAL TECHNIQUES WERE EMPLOYED TO EXPLAIN PRESENCE, ORIGIN, AND VIABILITY OF HERETOFORE UNOBSERVED, APPARENTLY METABOLIZING ALGAE. TRIBONEMA AND OEDOGONIUM, DEPOSITED ON THE BOTTOM, WERE THE MAJOR SOURCE OF ORGANIC CARBON UTILIZED IN CONSUMPTION OF HYPOLIMNETIC OXYGEN AS A RESULT OF BACTERIAL ACTIVITY AT THE WATER-SEDIMENT INTERFACE. THE ALGAE ON THE SEDIMENTS MAINTAIN GROWTH AFTER LIGHT BECOMES LIMITING FOR OTHER SEDIMENTAL FORMS. THE MIXING OF NUTRIENT-RICH HYPOLIMNION WATER INTO THE THERMOCLINE AND LOWER EPILIMNION STIMULATED ALGAL GROWTH, PRIMARILY ANACYSTIS, AT THESE LEVELS, PARTICULARLY AT STATIONS WHERE DISSOLVED OXYGEN DEPLETION IN THE HYPOLIMNION HAD BEEN RECORDED. FROM APPROXIMATELY JULY 21 TO AUGUST 10 THERE WAS MAXIMUM PHOTOSYNTHETIC ACTIVITY ON THE BOTTOM RESULTING IN A REDUCED SEDIMENT OXYGEN DEMAND. INCREASING PHYTOPLANKTON VOLUMES IN THE OVERLYING WATERS AND THE DECREASING PHOTOPERIOD REDUCED LIGHT ON THE BOTTOM TO BIOLOGICALLY LIMITING LEVELS IN MID-AUGUST. ALGAE DO CONTRIBUTE OXYGEN TO THE HYPOLIMNION FOR A PERIOD OF TIME; HOWEVER, IMPACT AND MAGNITUDE OF THIS CONTRIBUTION APPEARS TO BE MASKED BY A STRONGER OXYGEN DEMANDING PHYSICAL AND CHEMICAL PHENOMENA. (SEE ALSO W72-12990) (JONES-WISCONSIN)

FIELD 05C

ACCESSION NO. W72-12995

MICROBIOLOGICAL STUDIES RELATED TO OXYGEN DEPLETION AND NUTRIENT REGENERATION PROCESSES IN THE LAKE ERIE CENTRAL BASIN,

DEPARTMENT OF ENERGY, MINES AND RESOURCES, BURLINGTON (ONTARIO). CANADA CENTRE FOR INLAND WATERS; AND ENVIRONMENTAL PROTECTION AGENCY, FAIRVIEW PARK, OHIO. MICROBIOLOGY LAB.

A. S. MENON, C. V. MARION, AND A. N. MILLER.

IN: PROJECT HYPO: CANADA CENTRE FOR INLAND WATERS PAPER NO 6, AND ENVIRONMENTAL PROTECTION AGENCY TECHNICAL REPORT TS-05-71-208-24, FEBRUARY 1972, P 71-84. 10 FIG, 4 TAB, 17 REF.

DESCRIPTORS:
*DISSOLVED OXYGEN, *LAKE ERIE, *BACTERIA, *SEDIMENT-WATER INTERFACES, *HYPOLIMNION, NITRIFICATION, ALGAE, THERMOCLINE, ANAEROBIC BACTERIA, AEROBIC BACTERIA, SULFUR BACTERIA, IRON COMPOUNDS.

IDENTIFIERS:
*DEOXYGENATION, *NUTRIENT REGENERATION, LAKE ERIE CENTRAL BASIN.

ABSTRACT:
THIS BACTERIOLOGICAL STUDY AIMED TO EVALUATE DISTRIBUTION OF BACTERIAL DENSITIES AND BIOTYPES AT THE SEDIMENT-WATER INTERFACE AND OVERLYING WATERS IN RELATION TO TIME, CHEMICAL, PHYSICAL, AND BIOLOGICAL DATA AND TO ASSESS THEIR ROLE IN OVERALL OXYGEN DEPLETION AND NUTRIENT REGENERATION PROCESSES IN THE HYPOLIMNION OF LAKE ERIE'S CENTRAL BASIN. MAJOR SITE OF INTENSIVE BACTERIAL ACTIVITY WAS THE SEDIMENT-WATER INTERFACE. ORGANIC DEPOSITS FROM ALGAL RAINS AND OTHER SOURCES, ACCUMULATED AT THE BOTTOM, UNDERWENT BACTERIAL DECOMPOSITION RESULTING IN OXYGEN DEPLETION AND FORMATION OF REDUCED PRODUCTS OF LOW MOLECULAR WEIGHT. THE REDUCED PRODUCTS WERE SUBSEQUENTLY OXIDIZED BY CHEMOAUTOTROPHIC BACTERIA AT THE SEDIMENT-WATER INTERFACE, OR IN THE OVERLYING WATERS, RESULTING IN ADDITIONAL OXYGEN DEPLETION. THIS PROCESS REPEATED ITSELF AFTER EACH ALGAL RAIN CAUSING FURTHER OXYGENLOSS. PRECIPITATION OF PARTICULATE MATTER THROUGH THE HYPOLIMNION FROM INTERMITTENT ALGAL RAINS WAS PRIMARILY RESPONSIBLE FOR HIGH BACTERIAL DENSITIES IN THE HYPOLIMNION BECAUSE PHYTOPLANKTON CONSTITUTED A LOCUS FOR BACTERIAL ATTACHMENT AND PRODUCED SOLUBLE ORGANIC SUBSTRATES FOR BACTERIAL GROWTH. A SIGNIFICANT CORRELATION WAS OBTAINED BETWEEN DESULFOVIBRIO DENSITIES AND DEGREE OF OXYGEN DEPLETION. THESE FACTORS STRONGLY SUGGEST THAT HETEROTROPHIC AND CHEMOAUTOTROPHIC BACTERIA ARE THE PRINCIPAL FACTORS IN DEPLETING OXYGEN. (SEE ALSO W72-12990) (JONES-WISCONSIN)

FIELD 05C

ACCESSION NO. W72-12996

OXYGEN-NUTRIENT RELATIONSHIPS WITHIN THE CENTRAL BASIN OF LAKE ERIE,

DEPARTMENT OF ENERGY, MINES AND RESOURCES, BURLINGTON (ONTARIO). CANADA CENTRE FOR INLAND WATERS; AND ENVIRONMENTAL PROTECTION AGENCY, FAIRVIEW PARK, OHIO.

N. M. BURNS, AND C. ROSS.

IN: PROJECT HYPO: CANADA CENTRE FOR INLAND WATERS PAPER NO 6, AND ENVIRONMENTAL PROTECTION AGENCY TECHNICAL REPORT TS-05-71-208-24, FEBRUARY 1972, P 85-119. 19 FIG, 13 TAB, 3 PLATES, 10 REF.

DESCRIPTORS:
*DISSOLVED OXYGEN, *NUTRIENTS, *LAKE ERIE, *HYPOLIMNION, MATHEMATICAL MODELS, IRON, MANGANESE, PHOSPHORUS, ANAEROBIC CONDITIONS, CHLOROPHYLL, EPILIMNION, SEDIMENTS, DECOMPOSING ORGANIC MATTER, BACTERIA, CHEMICAL PROPERTIES, NITROGEN, HEAT TRANSFER, PHOSPHATES, ALKALINITY, OXYGEN DEMAND, HEAT BUDGET, ALGAE, EUTROPHICATION.

IDENTIFIERS:
*LAKE ERIE CENTRAL BASIN, NUTRIENT REGENERATION, DEOXYGENATION, ALGAL SEDIMENTATION.

ABSTRACT:
THE CHEMICAL BUDGET COMPONENTS IN LAKE ERIE'S CENTRAL BASIN WERE CALCULATED. A LARGE EROSION OF THE HYPOLIMNION DURING JULY 1970 SEEMED THE INITIAL FACTOR IN STARTING THE ALGAL GROWTH PERIOD LASTING TO OCTOBER; THE FIRST HEAVY BLOOM RESULTED IN ALGAL FALL-OUT ONTO THE LAKE FLOOR AT THE END OF JULY. FOR A WHILE THESE SEDIMENTED ALGAE WERE PHOTOSYNTHETIC, AMELIORATING OXYGEN DEPLETION BUT NOT PREVENTING NET OXYGEN DEPLETION CAUSED BY ORGANIC DECAY. AEROBIC HETEROTROPHIC AND SULFATE-REDUCING BACTERIA INCREASED STEADILY WHILE THE ALGAE ON THE BOTTOM DIED AND MATTED DOWN. LOSS OF OXYGEN FROM PHOTOSYNTHESIS PLUS ACTIVITY OF THE LARGE BACTERIAL POPULATIONS CAUSED HIGH OXYGEN DEPLETION RATE WITH ANOXIC CONDITIONS APPEARING ON AUGUST 12TH. ANOTHER PERIOD OF ALGAL RAINS, ABOUT AUGUST 17TH, AGAIN DIMINISHED THE OXYGEN DEPLETION RATE BUT WAS INSUFFICIENT TO PREVENT SPREAD OF ANOXIA WHICH, BY AUGUST 25TH, EXTENDED ACROSS THE HYPOLIMNION AREA. THE ANOXIC CONDITION CAUSED LARGE SCALE NUTRIENT REGENERATION BY DISSOLUTION OF INORGANIC FORMS. A MASSIVE BLOOM RESULTED WHEN THESE NUTRIENTS WERE MIXED WITH SURFACE WATER DURING SEPTEMBER. OXYGENATED CONDITIONS SHOULD BE MAINTAINED IN THE WATER AS A SIMPLE MECHANISM FOR ENSURING THAT LITTLE OF THE PHOSPHORUS IN THE SEDIMENTS RETURNS TO THE OVERLYING WATER. (SEE ALSO W72-12990) (JONES-WISCONSIN)

FIELD 05C

ACCESSION NO. W72-12997

UPTAKE AND METABOLISM OF 2,2-BIS-(P-CHLOROPHENYL)-1,1,1-TRICHLOROETHANE(DDT) BY MARINE PHYTOPLANKTON AND ITS EFFECT ON GROWTH AND CHLOROPLAST ELECTRON TRANSPORT,

SCRIPPS INSTITUTION OF OCEANOGRAPHY, LA JOLLA, CALIF.

G. W. BOWES.

PLANT PHYSIOLOGY, VOL 49, P 172-176, 1972. 3 FIG, 1 TAB, 31 REF.

DESCRIPTORS:
*OCEANS, *MATABOLISM, *MARINE ALGAE, *CHLORINATED HYDROCARBON PESTICIDES, *CYTOLOGICAL STUDIES, ABSORPTION, DDT, PHYTOPLANKTON, PLANT GROWTH, CHLOROPHYLL, DDE, INHIBITION.

IDENTIFIERS:
*ELECTRON TRANSPORT, DUNALIELLA TERTIOLECTA, CYCLOTELLA NANA, THALASSIOSIRA FLUVIATILIS, AMPHIDINIUM CARTERI, COCCOLITHUS HUXLEYI, PORPHYRIDIUM, SKELETONEMA COSTATUM.

ABSTRACT:
MARINE PHYTOPLANKTERS WERE STUDIED RELATIVE TO THEIR INTERACTION WITH DDT. AT 80 PPB DDT, GROWTH OF DUNALIELLA TERTIOLECTA WAS UNAFFECTED, AND THERE WAS SLIGHT, IF ANY, INFLUENCE ON DEVELOPMENT OF CYCLOTELLA NANA, THALASSIOSIRA FLUVIATILIS, AMPHIDINIUM CARTERI, COCCOLITHUS HUXLEYI, AND PORPHYRIDIUM. SKELETONEMA COSTATUM EXHIBITED A 9-DAY LAG BEFORE CELL DIVISION, GROWTH SUBSEQUENTLY BEING THE SAME AS THE CONTROL. ABILITY OF MARINE PHYTOPLANKTON TO METABOLIZE DDT VARIED. DDE WAS THE ONLY SIGNIFICANT HEXANE-SOLUBLE METABOLITE DETECTED. IT OCCURRED IN CELLS OF S. COSTATUM, C. NANA, T. FLUVIATILIS AND D. TERTIOLECTA. MAXIMUM DEGREE OF CONVERSION WAS 7.5% AND WAS BASED ON DDT FOUND IN CELL-WATER SYSTEM OF 9-DAY D. TERTIOLECTA CULTURES. THE TOTAL RECOVERED FROM CULTURES IN 2- TO 3-WEEK EXPERIMENTS RANGED FROM 63.5% FOR T. FLUVIATILIS TO 90.7% FOR S. COSTATUM. AMOUNT OF DDT FOUND ASSOCIATED WITH THE CELLS, COLLECTED BY CENTRIFUGATION, IN THE CELL-WATER SYSTEM RANGED FROM 70.8 TO 99.5%. NONCYCLIC ELECTRON FLOW, MEASURED BY FERRICYANIDE REDUCTION WAS INHIBITED BY DDT AND DDE, AND COULD EXPLAIN GROWTH INHIBITION. PHYTOPLANKTON SENSITIVITY TO TOXIC HYDROPHOBIC CHLORINATED HYDROCARBONS MAY BE DEPENDENT UPON PENETRATION OF THE MOLECULES TO ACTIVE SITES WITHIN MEMBRANES. (JONES-WISCONSIN)

FIELD 05C, 05B

ACCESSION NO. W72-12998

ATTACHED ALGAE ON ARTIFICIAL AND NATURAL SUBSTRATES IN LAKE WINNIPEG, MANITOBA,

FISHERIES RESEARCH BOARD OF CANADA, WINNIPEG (MANITOBA). FRESHWATER INST.

D. EVANS, AND J. G. STOCKNER.

JOURNAL OF FISHERIES RESEARCH BOARD OF CANADA, VOL 29, NO 1, P 31-44, JANUARY 1972. 7 FIG, 4 TAB, 21 REF.

DESCRIPTORS:
*BIOMASS, *AQUATIC ALGAE, *AQUATIC PRODUCTIVITY, LIGHT, TURBIDITY, NUTRIENTS, HETEROGENEITY, CHLOROPHYTA, CHRYSOPHYTA, BIOINDICATORS, PHYSICOCHEMICAL PROPERTIES, MICROSCOPY, LIMNOLOGY, WATER QUALITY, ECOLOGY, AQUATIC HABITATS, CLADOPHORA, LIMITING FACTORS, ECOLOGICAL DISTRIBUTION.

IDENTIFIERS:
*LAKE WINNIPEG, *SUBSTRATES, ACHNANTHES MINUTISSIMA, ACHNANTHES CF. BIRGIANI, AMPHIPLEURA PELLUCIDA, AMPHORA OVALIS, AMPHORA OVALIS V. PEDICULUS, COCCONEIS PEDICULUS, COCCONEIS PLACENTULA, CYCLOTELLA CF. SOCIALIS, CYCLOTELLA STELLIGERA, CYMBELLA SPP, DIATOMA SSP.

ABSTRACT:
THE COMPOSITION AND DISTRIBUTION OF ATTACHED ALGAE ON ARTIFICIAL (NAVIGATIONAL BUOYS) AND NATURAL SUBSTRATES IN LAKE WINNIPEG IN 1969-70, WAS STUDIED, IN THE FALL AFTER ABOUT 145 DAYS' GROWTH, WITH SPECIAL EMPHASIS ON THE DIATOMS, THE MAJOR COMPONENT OF THE ATTACHED ALGAL FLORA. ATTACHED ALGAE FROM KNOWN AREAS OF THE SURFACE OF REPRESENTATIVE SUBSTRATES WERE WEIGHED (DRY-WEIGHT) AND COUNTED. TOTAL DIATOM VOLUMES (CU MM/SQ CM) WERE CALCULATED USING AN EQUATION FOR THE GEOMETRICAL SHAPE THAT CLOSELY RESEMBLED INDIVIDUAL SPECIES, AND BY APPLYING SUITABLE CORRECTIONS TO EACH CALCULATION. THE PERCENT BY VOLUME OF OTHER ALGAL GROUPS WAS ESTIMATED BY MICROSCOPY. ON BOTH NATURAL AND ARTIFICIAL SUBSTRATES A DISTINCT ZONATION PATTERN RELATED TO WATER TRANSPARENCY OR AVAILABLE LIGHT WAS PREVALENT. PRELIMINARY RESULTS INDICATED THAT LIGHT WAS THE MOST IMPORTANT PHYSICAL FACTOR AFFECTING GROWTH AND SPECIFIC COMPOSITION OF ATTACHED ALGAE ON SUBSTRATES. BIOMASS VALUES (DRY WEIGHT) RANGED FROM 1.7 TO 29.1 MG/SQ CM, WITH THE GREATEST VALUES CONSISTENTLY OCCURRING BETWEEN A 10 AND 25 CM DEPTH ON BUOYS IN THE SOUTH BASIN. DIATOM DENSITY AND DIATOM VOLUME WERE ESTIMATED FROM BUOY SAMPLES AND VARIED RESPECTIVELY, WITHIN THE RANGE 200,000-6,650,000 DIATOMS/SQ CM AND 0.32-10.00 CU MM/SQ CM. IT APPEARS THAT NUTRIENTS MAY LIMIT GROWTH OF ATTACHED ALGAE IN THE NORTH BASIN, WHEREAS IN THE SOUTH BASIN, LIGHT IS THE MAJOR LIMITING FACTOR. THE SPECIFIC COMPOSITION OF THE ALGAL ASSEMBLAGES ON PARTICULAR BUOYS WAS MORE RELATED TO THE PHYSICAL-CHEMICAL FEATURES OF A MAJOR RIVER PLUME IN CLOSE PROXIMITY TO THE BUOY THAN TO A HOMOGENEOUS LAKE WINNIPEGWATER MASS. THIS FACTOR, TOGETHER WITH GREATER TURBIDITY IN THE SHALLOWER SOUTH BASIN, IS LARGELY RESPONSIBLE FOR THE OBSERVED HETEROGENEITY AMONG THE ATTACHED ALGAL ASSEMBLAGES IN LAKE WINNIPEG. (BYRD-BATTELLE)

FIELD 05A

ACCESSION NO. W72-13142

A PILOT PROGRAM TO DETERMINE THE EFFECT OF SELECTED NUTRIENTS (DISSOLVED ORGANICS, PHOSPHORUS, AND NITROGEN) ON NUISANCE ALGAL GROWTH IN AMERICAN FALLS RESERVOIR,

IDAHO UNIV., MOSCOW. WATER RESOURCES RESEARCH INST.

F. L. ROSE, AND G. W. MINSHALL.

AVAILABLE FROM THE NATIONAL TECHNICAL INFORMATION SERVICE AS PB-211 612, $3.00 IN PAPER COPY, $0.95 IN MICROFICHE. MOSCOW, COMPLETION REPORT, JUNE 1972, 37 P, 15 FIG, 5 TAB, 9 REF, APPEND. OWRR A-039IDA(1).

DESCRIPTORS:
*ALGAL CONTROL, *PHOSPHORUS, *NITROGEN, *IDAHO, *CHLOROPHYLL, *ANABAENA, NUTRIENTS, DISSOLVED SOLIDS, PHYTOPLANKCN, WATER POLLUTION EFFECTS.

IDENTIFIERS:
*GLUCOSE, *SNAKE RIVER(IDAHO), *DISSOLVED ORGANIC CARBON, *AMERICAN FALLS RESERVOIR, *APHANIZCMENON, ALGAL GROWTH.

ABSTRACT:
A FIELD INVESTIGATION WAS CONDUCTED TO DETERMINE THE FEASIBILITY OF USING POLYETHYLENE TUBES AS IN SITU CULTURE CHAMBERS IN ASSESSING THE INFLUENCE OF SELECTED NUTRIENT FACTORS ON ALGAL GROWTH. THE STUDY WAS CARRIED OUT IN AMERICAN FALLS RESERVOIR, A LARGE MULTIPURPOSE IMPOUNDMENT ON THE SNAKE RIVER IN S.E. IDAHO, DURING THE SUMMER OF 1971. THE TUBES CONTAINED ADDED AMOUNTS OF PHOSPHORUS OR PHOSPHORUS AND ORGANIC CARBON (GLUCOSE) AND WERE SUBSEQUENTLY 'SEEDED' WITH NATURALLY OCCURRING ALGAL SPECIES. PERIODIC WATER SAMPLES TAKEN FROM THE TUBES WERE ANALYZED FOR TOTAL AND SOLUBLE PHOSPHORUS, TOTAL AND SOLUBLE ORTHOPHOSPHORUS, AMMONIUM-N, NITRITE-N, NITRATE-N, PARTICULATE AND DISSOLVED ORGANIC CARBON, TOTAL DISSOLVED SOLIDS, CHLOROPHYLL A, AND ABUNDANCE OF MAJOR PHYTOPLANKTON SPECIES. RESULTS WERE COMPARED WITH THOSE FROM AMBIENT WATER COLLECTED BOTH AT THE LOCATION OF THE EXPERIMENTAL WORK AND AT OTHER SITES IN THE RESERVOIR. GROWTH OF PLANKTONIC ORGANISMS, ESPECIALLY APHANIZOMENON AND ANABAENA, REACHED LEVELS 10 TIMES MORE NUMEROUS IN THE EXPERIMENTAL TUBES THAN IN THE AMBIENT WATER. ALTHOUGH MODIFICATION OF SOME PARTS OF THE TECHNIQUE ARE SUGGESTED THE USE OF POLYETHYLENE TUBES APPEARS TC BE OF CONSIDERABLE PROMISE IN FUTURE INVESTIGATIONS.

FIELD 05C, 04A, 05A

ACCESSION NO. W72-13300

CONTAMINATION OF MARINE LITTCRAL AND SEA FOODS BY PESTICIDES, (CONTAMINATION DU LITTORAL MARITIME ET DES FRUITS DE MER PAR LES PESTICIDES),

POITIERS UNIV. (FRANCE). FACULTE DE MEDICINE ET DE PHARMACIE.

J. BRISOU.

MEDEDELINGEN FACULTEIT LANDBOUW-WETENSCHAPPEN GENT, VOL 35, NO 2, P 739-743, 1970. 3 TAB, 3 REF. ENGLISH SUMMARY.

DESCRIPTORS:
*PESTICIDE RESIDUES, *AQUATIC LIFE, *AQUATIC ENVIRONMENT, *WATER POLLUTION, *PATH OF POLLUTANTS, *PUBLIC HEALTH, ENZYMES, ANALYTICAL TECHNIQUES, PESTICIDES, FUNGICIDES, ORGANOPHOSPHOROUS PESTICIDES, CHLORINATED HYDROCARBON PESTICIDES, CARBAMATE PESTICIDES, ORGANIC PESTICIDES, HERBICIDES, WATER POLLUTION SOURCES, PLANKTON, OYSTERS, SHELLFISH, ALGAE, MUD, SEDIMENTS, GAS CHROMATOGRAPHY, COLORIMETRY.

IDENTIFIERS:
MERCURIAL PESTICIDES, THIN LAYER CHROMATOGRAPHY, GROWTH INHIBITION.

ABSTRACT:
A SYSTEMATIC INVESTIGATION OF THE FRENCH COASTS OF THE CHANNEL, THE ATLANTIC AND THE MEDITERRANEAN SEA SHOWS A SIGNIFICANT POLLUTION OF MUDS, SEDIMENTS, ALGAE, PLANKTON, AND SHELLFISH BY PESTICIDES. THE PER CENT OF CONTAMINATION BY ONE OR SEVERAL PRODUCTS IS BY APPROXIMATION: OYSTERS - 72%, OTHER SHELLFISH - 67%, ALGAE - 66%, PLANKTON- 100%, AND WATER - 0%. THE PLANKTON IS ONE OF THE MOST CONTAMINATED MATERIALS BECAUSE MERCURIAL FUNGICIDES ARE MORE PREVALENT THAN OTHER POLLUTANT PRODUCTS, AND THEY ARE CONCENTRATED BY THE PLANKTON. DATA HAVE BEEN OBTAINED USING SEVERAL TECHNIQUES: GROWTH INHIBITIONS, ANTICHOLINESTERASE ACTIVITY OF EXTRACTS, THIN LAYER CHROMATOGRAPHY, SPECTROCOLORIMETRY, AND IN A FEW INSTANCES GAS CHROMATOGRAPHY. (SVENSSON-WASHINGTON)

FIELD 05C, 05A

ACCESSION NO. W72-13347

PHOTOSYNTHETIC YIELDS AND BYPRODUCT RECOVERY FROM SEWAGE OXIDATION PONDS,

ASIAN INST. OF TECH., BANGKOK (THAILAND). DEPT. OF ENVIRONMENTAL ENGINEERING.

M. G. MCGARRY, C. D. LIN, AND J. L. MERTO.

PREPRINT, PRESENTED AT 6TH INTERNATIONAL WATER POLLUTION RESEARCH CONFERENCE, SESSION 4, HALL C, PAPER NO. 7, JUNE 20, 1972, P C/4/7/1-C/4/7/11. 6 FIG, 2 TAB, 17 REF.

DESCRIPTORS:
*OXIDATION PONDS, *ALGAE, *WASTE WATER TREATMENT, *BYPRODUCTS, MIXING, SOLAR RADIATION, ORGANIC LOADING, DIURNAL DISTRIBUTION, HYDROGEN ION CONCENTRATION, DISSOLVED OXYGEN, BIOCHEMICAL OXYGEN DEMAND, PERFORMANCE, FLOCCULATION, FLOTATION, ECONOMIC FEASIBILITY, COSTS, ESTIMATED COSTS.

IDENTIFIERS:
ALUM, RECYCLING, TREATMENT COSTS, ASIA.

ABSTRACT:
ALGAL GROWTH AND SEWAGE TREATMENT UNDER TROPICAL CONDITIONS AS AFFECTED BY POND MIXING, SOLAR RADIATION AND DIURNAL VARIATIONS IN POND LOADING WERE DESCRIBED. FOR MASS ALGAL CULTURE, PONDS SHOULD BE THOROUGHLY MIXED DAILY AND THE MIXING PERIOD LIMITED IN ORDER TO MINIMIZE TURBIDITY. DIFFERENT SHADING MATERIAL OVER POND SURFACES ALLOWED A PREDETERMINED PERCENTAGE OF TOTAL INCIDENT RADIATION TO REACH EACH SURFACE. ALGAL CONCENTRATIONS INCREASED IN A NEAR LINEAR FASHION WITH RADIATION AVAILABLE. EFFECTS OF DIURNAL VARIATIONS IN POND LOADING AND EFFLUENT REMOVAL REQUIRED ON POND CHARACTERISTICS WERE PRESENTED IN TABULAR FORM. NORMAL PH, D.O., AND ALKALINITY VARIATIONS WERE EXHIBITED BY ALL PONDS AND FILTERED BOD'S DID NOT EXCEED 20 MG/L. ALUM FLOCCULATION AND DOWNFLOW FLOTATION WERE USED TO CLARIFY THE POND EFFLUENT AND CONCENTRATE THE ALGAE. ALUMINUM RECOVERY AND RECYCLING WERE INVESTIGATED. THE POTENTIAL USE OF ALGAE AS A LIVESTOCK FEED WAS INVESTIGATED. ESTIMATES INDICATED THAT IF SEWAGE TREATMENT COSTS WERE SET AGAINST ALGAE PRODUCTION, ALGAE COULD BE PRODUCED FOR APPROXIMATELY 26 CENTS/KG. (GALWARDI-TEXAS)

FIELD 05D

ACCESSION NO. W72-13508

ESTURINE ECOSYSTEMS AND HIGH TEMPERATURES,

NORTH CAROLINA WATER RESOURCES RESEARCH INST., RALEIGH.

B. J. COPELAND, AND H. LEE DAVIS.

AVAILABLE FROM THE NATIONAL TECHNICAL INFORMATION SERVICE AS PB-211 808, $3.00 IN PAPER COPY, $0.95 IN MICROFICHE. NORTH CAROLINA WATER RESOURCES RESEARCH INSTITUTE, RALEIGH, REPORT NO 68, JUNE 1972 (UNC-WRRI-72-68). 90 P, 39 FIG, 9 TAB, 47 REF. OWRR A-041-NC(1).

DESCRIPTORS:
*WATER POLLUTION EFFECTS, *HEATED WATER, *WATER TEMPERATURE, *WATER QUALITY STANDARDS, *ESTURARIES, *THERMAL POLLUTION, *NORTH CAROLINA, NUTRIENTS, SEWAGE EFFLUENTS, EUTNOPHICATION, ALGAE, CYANOPHYTA, CHLOROPHYTA, METABOLISM, NEKTON, BENTHOS, BIOMASS.

IDENTIFIERS:
*PAMLICO RIVER ESTUARY, PHOTOSYNTHESIS/RESPIRATION RATIOS.

ABSTRACT:
RESPONSES OF ESTUARINE COMMUNITY STRUCTURE, RESPIRATION AND PRODUCTION TO ADDED HEAT, SEWAGE AND THEIR COMBINATION WERE INVESTIGATED. PLASTIC POOLS CONTAINING TRANSPLANTED ECOSYSTEMS FROM SOUTH CREEK ESTUARY, N. C. WERE USED. TEMPERATURE REPLICATION ACHIEVED. TEMPERATURE IN THE HEATED POOLS WAS REGULATED AT 5 C(9F) ABOVE THAT OF THE AMBIENT POOLS. THERMAL TREATMENT INCREASED NUTRIENT REGENERATION RATES, YIELDING SLIGHTLY HIGHER ALGAL BIOMASS; ALTHOUGH, SEASONAL DIFFERENCES WERE MORE SIGNIFICANT. GROSS COMMUNITY PRODUCTIVITY WAS REGULATED BY AMMONIA, LIGHT, AND TEMPERATURE LEVELS AND TOTAL RESPIRATION BY TEMPERATURE AND PRIMARY PRODUCTIVITY. SEWAGE ADDITION SUBSTANTIALLY INCREASED AMMONIA LEVELS, PARTICULARLY DURING THE WINTER. COMMUNITY METABOLISM RESPONDED POSITIVELY TO THERMAL TREATMENT, BUT NOT TO SEWAGE TREATMENT. THERMAL TREATMENT AND THE COMBINATION OF SEWAGE AND THERMAL TREATMENTS INCREASED THE PHOTOSYNTHESIS/RESPIRATION RATIOS (P/R) DURING SPRING AND SUMMER, BUT DECREASED THE P/R WHEN TEMPERATURE WAS LIMITING DURING WINTER. TEMPERATURE HAD VERY LITTLE EFFECT ON PHYTOPLANKTON COMPOSITION DURING THE SPRING. BLUE-GREEN ALGAE AND COCCOID GREEN ALGAE DOMINATED IN THE HEATED AND SEWAGE-TREATED POOLS DURING SUMMER. NEKTON AND BENTHIC INCREASED TO HIGHER BIOMASS IN HEATED POOLS DURING SPRING AND ACHIEVED A LOWER BIOMASS IN THE HEATED POOLS DURING SUMMER THAN IN THE AMBIENT POOLS. NO SIGNIFICANT DIFFERENCES WERE OBSERVED DURING WINTER AMONG HEATED AND AMBIENT POOLS. SEWAGE ADDITION DID NOT SUBSTANTIALLY ALTER PATTERNS BETWEEN HEATED AND AMBIENT SYSTEMS. OYSTERS, BAY CLAMS AND WIDGEON GRASS REACHED HIGHER BIOMASS IN HEATED POOLS DURING WINTER THAN IN AMBIENT POOLS. A FLOW-THROUGH EXPERIMENT WAS CONDUCTED DURING THE 1971 SUMMER TO TST MORE REALISTIC ESTUARINE CONDITIONS. RESULTS WERE SUBSTANTIALLY THE SAME AS OBTAINED UNDER QUIESCENT CONDITIONS.

FIELD 05C, 02L

ACCESSION NO. W72-13636

AVAILABILITY OF PHOSPHORUS FOR CLADOPHORA GROWTH IN LAKE MICHIGAN,

WISCONSIN UNIV., MILWAUKEE. DEPT. OF BOTANY; AND WISCONSIN UNIV., MILWAUKEE. CENTER FOR GREAT LAKES STUDIES.

CHANG-KWEI LIN.

IN: PROCEEDINGS 14TH CONFERENCE ON GREAT LAKES RESEARCH, UNIVERSITY OF TORONTO, ONTARIO, APRIL 19-21, 1971, P 39-43. 4 FIG, 15 REF.

DESCRIPTORS:
*PHOSPHORUS, *CLADOPHORA, *PLANT GROWTH, *LAKE MICHIGAN, ALGAE, PHOSPHATES, HYDROLYSIS, STORM WATER, RAINFALL, WASTE WATER(POLLUTION), EUTROPHICATION, WATER POLLUTION SOURCES, NUTRIENTS.

IDENTIFIERS:
CLADOPHORA GLOMERATA, MILWAUKEE HARBOR(WISCONSIN), POLYPHOSPHATES, ORTHOPHOSPHATES.

ABSTRACT:
THE PROFUSE GROWTH OF THE ALGA, CLADOPHORA, IN THE GREAT LAKES HAS SERIOUSLY CURTAILED SHORELINE RECREATION. SAMPLED FROM MILWAUKEE HARBOR AND FROM SITES ALONG LAKE MICHIGAN NORTH OF THE HARBOR, CLADOPHORA GLOMERATA WAS ANALYZED FOR EXTRACTABLE INORGANIC ORTHOPHOSPHATE AND FOR ITS ABILITY TO HYDROLYZE POLYPHOSPHATES. THOSE ALGAE GROWING IN THE HARBOR AREA SHOWED A CONSISTENTLY HIGH VALUE OF SURPLUS STORED PHOSPHORUS AND THE VALUES DECREASED AS THE DISTANCE FROM THE HARBOR INCREASED. LONGER, DENSER, AND GREENER CLADOPHORA WAS FOUND NEAR THE HARBOR THAN FROM MORE DISTANT SITES. SAMPLES FROM ALL LOCATIONS TAKEN AFTER HEAVY RAINFALLS CONTAINED SUBSTANTIAL INCREASES IN STORED PHOSPHORUS. STORM SEWER OVERFLOW AND INCREASED RIVER DISCHARGE DUE TO PRECIPITATION COULD SPORADICALLY PROVIDE EXCESSIVE SUPPLIES OF PHOSPHORUS. GREAT VARIATION IN ABILITY TO HYDROLYZE POLYPHOSPHATES WAS FOUND IN ALGAE SAMPLED FROM DIFFERENT LOCATIONS. THE YIELD OF ORTHOPHOSPHATES FROM POLYPHOSPHATES WAS INVERSELY PROPORTIONAL TO THE AMOUNT OF SURPLUS STORED PHOSPHORUS IN THE ALGAE. THE ALGAE GROWING NEAR THE HARBOR ACCUMULATED PHOSPHATE PHOSPHORUS WHILE THE ALGAE GROWING AT GREATER DISTANCE WERE CONCURRENTLY PHOSPHORUS STARVED, PERHAPS THEY ADAPTED TO OBTAIN THEIR PHOSPHORUS PRIMARILY FROM HYDROLYSIS OF POLYPHOSPHATE. (JONES-WISCONSIN)

FIELD 05C

ACCESSION NO. W72-13644

DEVELOPMENT OF A THEORETICAL SEASONAL GROWTH RESPONSE CURVE OF CLADOPHORA GLOMERATA TO TEMPERATURE AND PHOTOPERIOD,

STATE UNIVERSITY OF NEW YORK, BUFFALO, DEPT. OF BIOLOGY; AND STATE UNIV. COLL., BUFFALO. GREAT LAKES LAB.

J. R. STORR, AND R. A. SWEENEY.

IN: PROCEEDINGS 14TH CONFERENCE ON GREAT LAKES RESEARCH, UNIVERSITY OF TORONTO, ONTARIO, APRIL 19-21, 1971, P 119-127. 4 FIG, 4 TAB, 14 REF.

DESCRIPTORS:
*ALGAE, *SEASONAL, *CLADOPHORA, *TEMPERATURE, *PHOTOPERIODISM, GREAT LAKES, MODEL STUDIES, NUTRIENTS, PLANT GROWTH, ANALYTICAL TECHNIQUES, WATER TEMPERATURE.

IDENTIFIERS:
*CLADOPHORA GLOMERATA, GROWTH CURVES.

ABSTRACT:
LABORATORY AND FIELD INVESTIGATIONS ASCERTAINED EFFECTS OF HEAT ON CLADOPHORA GROWTH WHEN THIS PLANT IS CULTURED UNDER SPRING AND SUMMER PHOTOPERIODS. RESULTS INDICATED THAT AT THE LEVEL OF NUTRIENTS PRESENT, OPTIMUM GROWTH WAS AROUND 18C. LOW GROWTH LEVELS OCCURRED AT THE LOW TEMPERATURE RANGE (14.5C) BUT RESULTS WERE INCONCLUSIVE. IN THE UPPER TEMPERATURE RANGE, GROWTH CESSATION APPEARED AT ABOUT 25C. GROWTH RESPONSE TO PHOTOPERIOD LEVELS IN THE EXPERIMENTS APPEARED TO BE EXPONENTIAL IN CHARACTER WHILE THE GROWTH RESPONSE CURVES GENERATED BY THE TEMPERATURE EXPERIMENTS WERE VERY SIMILAR IN FORM AT THE DIFFERENT LEVELS OF PHOTOPERIOD. A FAMILY OF TEMPERATURE RELATED GROWTH CURVES WAS CONSTRUCTED FOR DIFFERENT PHOTOPERIODS THROUGHOUT THE MAY-OCTOBER PERIOD. RECORDED LAKE WATER TEMPERATURE DATA WAS PLOTTED ON THESE CURVES FOR THE SAME PERIOD AND A SEASONAL GROWTH RESPONSE CURVE WAS GENERATED RELATING SEASONAL PHOTOPERIOD AND TEMPERATURE. THIS RESULTANT PLOT APPEARED TO BE IN CLOSE AGREEMENT WITH OBSERVED CLADOPHORA GROWTH THROUGHOUT THE SEASON. A THEORETICAL GROWTH CURVE WAS ALSO PLOTTED USING THIS TECHNIQUE AND THE TEMPERATURE DATA FOR THE SAME AREA AND TIME PERIOD. THE THEORETICAL MODEL OF GROWTH RESPONSE RATE MAY BE USEFUL IN PREDICTING PROBABLE SEASONAL GROWTH AND APPROXIMATE STANDING CROP, ESPECIALLY IN LOCALIZED AREAS AROUND THERMAL DISCHARGES. (JONES-WISCONSIN)

FIELD 05C

ACCESSION NO. W72-13651

CHLOROPHYLL A - TOTAL PHOSPHORUS RELATIONSHIPS IN LAKE ERIE,

ONTARIO WATER RESOURCES COMMISSION, REXDALE (ONTARIO).

THOMAS G. BRYDGES.

IN: PROCEEDINGS 14TH CONFERENCE ON GREAT LAKES RESEARCH, UNIVERSITY OF TORONTO, ONTARIO, APRIL 19-21, 1971, P 185-190. 5 FIG, 4 REF.

DESCRIPTORS:
*CHLOROPHYLL, *PHOSPHORUS, *LAKE ERIE, NITROGEN, ALGAE, INORGANIC COMPOUNDS, STANDING CROPS, LIMITING FACTORS, EUTROPHICATION.

ABSTRACT:
SINCE 1966 AN EXTENSIVE WATER QUALITY MONITORING PROGRAM HAS BEEN CONDUCTED ON THE GREAT LAKES. SOME LAKE WATER SAMPLES HAVE BEEN ANALYZED FOR CHLOROPHYLL A AS A ROUTINE SINCE 1967 TO ESTIMATE ALGAL STANDING CROPS. EXAMINATION OF CHOROPHYLL RESULTS FOR 1967, 1968, AND 1969 IS PRESENTED WITH PARTICULAR REFERENCE TO THEIR RELATIONSHIPS WITH INORGANIC NITROGEN AND PHOSPHORUS CONCENTRATIONS. IN 1967 AVERAGE CHLOROPHYLL A AND TOTAL PHOSPHORUS CONCENTRATIONS AT 87 OPEN LAKE STATIONS (THREE TO FIVE MEASUREMENTS AT EACH) WERE DIRECTLY PROPORTIONAL. HIGH TOTAL PHOSPHORUS AND RELATIVELY LOWER CHLOROPHYLL A CONCENTRATIONS CHARACTERIZED THE DETROIT, RAISIN, AND MAUMEE RIVER STATIONS. LAKE STATIONS INCLUDED THE WESTERN BASIN AND A FIVE MILE WIDE BAND ALONG NORTH SHORE FROM POINT PELEE TO BUFFALO. THERE WERE NO APPARENT TRENDS BETWEEN CHLOROPHYLL A AND INORGANIC NITROGEN CONCENTRATIONS. THE CONCENTRATIONS OF CHLOROPHYLL A AND TOTAL PHOSPHORUS MEASURED IN 1968 AND 1969 WERE DIRECTLY PROPORTIONAL. EIGHT STATIONS IN WESTERN BASIN WERE SAMPLED FOR NINE SUCCESSIVE DAYS IN JULY 1968. IN 1969, 97 STATIONS (422 MEASUREMENTS) WERE SAMPLED OVER THE FULL LENGTH OF THE LAKE. THAT PHOSPHORUS IS AN ALGAL GROWTH LIMITING FACTOR IN LAKE ERIE IS CONCLUDED FROM THESE EXTENSIVE EMPIRICAL OBSERVATIONS. (JONES-WISCONSIN)

FIELD 05C

ACCESSION NO. W72-13653

THE EFFECT OF DDT AND DIELDRIN UPON C-14 UPTAKE BY IN SITU PHYTOPLANKTON IN LAKES ERIE AND ONTARIO,

DEPARTMENT OF ENERGY MINES AND RESOURCES, BURLINGTON (ONTARIO). CANADA CENTRE FOR INLAND WATERS.

WALTER A. GLOOSCHENKO.

IN: PROCEEDINGS 14TH CONFERENCE ON GREAT LAKES RESEARCH, UNIVERSITY OF TORONTO, ONTARIO, APRIL 19-21, 1971, P 219-223. 6 FIG, 12 REF.

DESCRIPTORS:
*PESTICIDES, *DDT, *DIELDRIN, *PHYTOPLANKTON, CARBON RADIOISOTOPES, LAKE ERIE, LAKE ONTARIO, ADSORPTION, AGRICULTURAL RUNOFF, ALGAE, FOOD CHAINS, CHLORINATED HYDROCARBON PESTICIDES, FISHERIES.

ABSTRACT:
SINCE PREVIOUS WORK WITH CHLORINATED HYDROCARBONS HAS BEEN MAINLY RESTRICTED TO MARINE SPECIES OF PHYTOPLANKTON, THIS STUDY INVESTIGATED THE EFFECTS OF DDT AND DIELDRIN UPON C-14 UPTAKE BY NATURAL PHYTOPLANKTON COMMUNITIES IN LAKES ONTARIO AND ERIE. IN SITU STUDIES WERE PERFORMED, BUT THE CONCENTRATIONS OF DDT AND DIELDRIN USED WERE HIGHER THAN NORMALLY FOUND IN THE GREAT LAKES. THE LOWEST CONCENTRATION USED, 1 PPB, IS AT LEAST 200 TIMES THAT FOUND IN SITU. IF DEPRESSION OF CARBON FIXATION BY PHYTOPLANKTON IN THE GREAT LAKES IS OCCURRING, IT WILL BE NEGLIGIBLE. POSSIBLY AGRICULTURAL RUNOFF IN LOCAL AREAS MAY BE MUCH HIGHER IN PESTICIDE CONCENTRATION WHICH MAY LEAD TO DDT OR DIELDRIN LIMITING CARBON FIXATION IN LOCALIZED INSHORE AREAS. THE ADAPTATION OF ALGAE TO PESTICIDES NEEDS CONSIDERATION. THE MOST SERIOUS HAZARD RESULTING FROM PESTICIDES IN WATER IS BIOLOGICAL CONCENTRATION; WHILE ALGAE THEMSELVES MAY NOT BE SERIOUSLY AFFECTED, THEY MAY CONCENTRATE HIGH AMOUNTS OF PESTICIDE BY ACTIVE UPTAKE ON SURFACE ADSORPTION AND TRANSFER IT TO HIGHER TROPHIC LEVELS. CONCENTRATION AND TRANSFER OF THESE TWO COMPOUNDS TO HIGHER TROPHIC LEVELS IS OF MAJOR CONCERN TO GREAT LAKES FISHERIES.
(JONES-WISCONSIN)

FIELD 05C

ACCESSION NO. W72-13657

HYPOTHESIS FOR DISSOLVED OXYGEN DEPLETION IN THE CENTRAL BASIN HYPOLIMNION OF LAKE ERIE,

LAKE ERIE BASIN OFFICE, FAIRVIEW PARK, OHIO.

CONRAD KLEVENO, O. BRAIDECH, E. THOMAS, AND PHILIP E. GEHRING.

IN: PROCEEDINGS 14TH CONFERENCE ON GREAT LAKES RESEARCH, UNIVERSITY OF TORONTO, ONTARIO, APRIL 19-21, 1971, P 252-255. 1 FIG, 1 TAB.

DESCRIPTORS:
*DISSOLVED OXYGEN, *HYPOLIMNION, *LAKE ERIE, BENTHIC FLORA, ALGAE, LIGHT PENETRATION, PLANKTON, NUTRIENTS, BIOCHEMICAL OXYGEN DEMAND, INORGANIC COMPOUNDS, THERMAL STRATIFICATION, DIATOMS, EPILIMNION, EUTROPHICATION.

IDENTIFIERS:
*LAKE ERIE CENTRAL BASIN, *OXYGEN DEPLETION.

ABSTRACT:
IN 1968 A RAPID DEPLETION OF DISSOLVED OXYGEN IN THE HYPOLIMNION OF THE CENTRAL BASIN OF LAKE ERIE WAS OBSERVED FROM 8.2 MGL ON AUGUST 6 TO 1.9 ON AUGUST 8. IN THE SUMMER OF 1969 A STUDY WAS INITIATED TO DETERMINE THE RATE AND DESCRIBE THE MECHANICS OF DISSOLVED OXYGEN DEPLETION IN THE HYPOLIMNION IN THE BASIN. THE DECLINE OF THE DISSOLVED OXYGEN WAS GRADUAL DURING EARLY SUMMER, BUT IN AUGUST THE RATE OF DEPLETION INCREASED RAPIDLY. THE DEATH AND DECOMPOSITION OF BENTHIC ALGAE, TRIBONEMA UTRICULOSUM AND OEDOGONIUM MAY EXPLAIN THIS RELATIVELY SUDDEN DECLINE. THESE ALGAE WERE FOUND GROWING IN PROFUSION AT THE BOTTOM DURING THE SUMMER OF 1969. IT IS POSTULATED THAT THEY ARE KILLED BY A REDUCTION IN LIGHT CAUSED BY AN INCREASE OF PLANKTON IN OVERLYING WATERS. THE INCREASED PLANKTON IS CAUSED BY AN INCREASED VERTICAL CIRCULATION OF NUTRIENTS WHEN THE LAKE BEGINS TO COOL. THE DEATH OF THE BENTHIC ALGAE RESULTS IN A TREMENDOUS INCREASE IN BIOCHEMICAL OXYGEN DEMAND IN HYPOLIMNION WATERS, THUS THE RAPID DEPLETION OF DISSOLVED OXYGEN. (JONES-WISCONSIN)

FIELD 05C

ACCESSION NO. W72-13659

A CASE OF NUTRIENT ENRICHMENT IN AN INSHORE AREA OF GEORGIAN BAY,

ONTARIO WATER RESOURCES COMMISSION, REXDALE (ONTARIO).

DENIS M. VEAL, AND M. F. P. MICHALSKI.

IN: PROCEEDINGS 14TH CONFERENCE ON GREAT LAKES RESEARCH, UNIVERSITY OF TORONTO, ONTARIO, APRIL 19-21, 1971, P 277-292. 4 FIG, 5 TAB, 25 REF.

DESCRIPTORS:
*EUTROPHICATION, *NUTRIENTS, LAKE HURON, EFFLUENTS, WATER QUALITY, PHOSPHORUS, PHYTOPLANKTON, SEWAGE TREATMENT, NITROGEN, PHYSOCHEMICAL PROPERTIES, DIATOMS, ALGAE, CLADOPHORA, PRODUCTIVITY, WATER POLLUTION, STANDING CROPS.

IDENTIFIERS:
*GEORGIAN BAY(ONTARIO).

ABSTRACT:
SURVEYS AND PUBLIC COMPLAINTS INDICATED THAT PHYTOPLANKTON POPULATIONS IN PENETANG HARBOR, GEORGIAN BAY, WERE REACHING EXCESSIVE LEVELS. A GRADATION OF WATER QUALITY CONDITIONS NOT UNLIKE THOSE OF LAKE ERIE WERE REVEALED IN A STUDY OF THE PENETANGUISHENE TO WAUBAUSHENE AREA IN 1969. MAJOR EMPHASIS WAS PLACED ON AN EVALUATION OF NITROGEN AND PHOSPHORUS CONCENTRATIONS, STANDING STOCKS OF PHYTOPLANKTON AND ALGAL PRODUCTIVITY AND AN ASSESSMENT OF NUTRIENT LOADING. MEAN TOTAL PHOSPHORUS CONCENTRATIONS RANGED FROM 0.02 MG/L (AS P) OVER MOST OF THE STUDY AREA TO 0.03 MG/L IN MIDLAND BAY AND 0.06 MG/L IN THE SOUTHERN END OF PENETANG HARBOR. PHYTOPLANKTON VALUES INCREASED WITH HIGHER TOTAL PHOSPHORUS CONCENTRATIONS. THE MIDLAND AND PENETANG SEWAGE TREATMENT PLANTS WERE RESPONSIBLE FOR APPROXIMATELY 53% OF THE NET PHOSPHORUS LEADING. THE FLUSHING RATE AS WELL AS EXHANGE OF HIGH QUALITY WATER FROM THE OPEN LAKE WILL PROMOTE RAPID RECOVERY WHEN SIGNIFICANT MUNICIPAL NUTRIENT SOURCES ARE ELIMINATED. IT IS EVIDENT THAT MUNICIPAL WASTE DISCHARGES TO SMALL BAYS ALONG THE UPPER GREAT LAKES WILL REQUIRE AN IMPROVED DEGREE OF TREATMENT, INCLUDING NUTRIENT REMOVAL, IN ORDER TO PRESERVE WATER QUALITY AND TO MAINTAIN HIGH RECREATIONAL POTENTIAL OF THE SHORELINES.
(JONES-WISCONSIN)

FIELD 05C

ACCESSION NO. W72-13662

A QUANTITATIVE COMPARISON OF PIGMENT EXTRACTION BY MEMBRANE AND GLASS-FIBER FILTERS,

KENT STATE UNIV., OHIO. DEPT. OF BIOLOGICAL SCIENCES.

E. B. LONG, AND G. D. COOKE.

LINMOLOGY AND OCEANOGRAPHY, VOL. 16, NO. 6, P 990-992, NOVEMBER 1971. 1 FIG, 1 TAB, 3 REF.

DESCRIPTORS:
*PHYTOPLANKTON, *CHLOROPHYLL, *EVALUATION, *SEPARATION TECHNIQUES, SPECTROPHOTOMETRY, LAKES, PLANT PIGMENTS, DETRITUS, CHLOROPHYTA, CHRYSOPHYTA, SOLVENT EXTRACTIONS, DIATOMS, AQUATIC ALGAE, FILTRATION, CYANOPHYTA, *EUTROPHICATION, *OHIO.

IDENTIFIERS:
*GLASS FIBER FILTERS, *MEMBRANE FILTERS, GRANGE LAKE, EAST TWIN LAKE, LAKE ROCKWELL, LAKE HODGSON, CLEARWATER QUARRY, DOLLAR LAKE, CHLOROPHYLL A, ABSORBANCE.

ABSTRACT:
WATER SAMPLES COLLECTED FROM SIX EUTROPHIC NORTHEASTERN OHIO LAKES WERE USED IN A STUDY CONCERNED WITH A QUANITATIVE COMPARISON OF PIGMENT EXTRACTION BY MEMBRANE AND GLASS-FIBER FILTERS. THE PHYTOPLANKTON-CONTAINING WATER SAMPLES WERE MEASURED INTO 500-ML SUBSAMPLES AND FILTERED AT 0.25-0.33 ATM. THE FILTER TYPES TESTED WERE 47-MM-DIAMETER TYPE HA MILLIPORE MEMBRANE FILTERS (0.45 MICRON PORE SIZE; CELLULOSE ACETATE) AND 42.5-MM-DIAMETER TYPES GF/C AND GF/A WHATMAN GLASS-FIBER FILTERS. THE PHYTOPLANKTON WAS EXTRACTED INTO ACETONE AND ABSORBANCES MEASURED SPECTROPHOTOMETRICALLY AT 665 MILLIMICRONS. DATA WERE ANALYZED BY 'STUDENT'S' T STATISTIC. ABSORBANCES OF THE ACETONE EXTRACTS FROM LAKE PHYTOPLANKTON COLLECTED ON GLASS-FIBER FILTERS, GROUND AND EXTRACTED 2 HR, WERE 6-18 PERCENT GREATER TAHN FROM PHYTOPLANKTON COLLECTED ON MEMBRANE FILTERS AND PROCESSED IDENTICALLY. USE OF THE GLASS-FIBER FILTERS CUT FILTERING TIME BY A FACTOR OF 10 AND MATERIAL COST BY A FACTOR OF 4 OR MORE.
(SNYDER-BATTELLE)

FIELD 05A, 05C

ACCESSION NO. W72-13673

NUTRIENT STUDIES IN TEXAS IMPOUNDMENTS,

TEXAS UNIV., AUSTIN. DEPT. OF ENVIRONMENTAL HEALTH AND ENGINEERING.

V. H. HUANG, J. R. MASE, AND E. G. FRUH.

DECEMBER 1970. 39 P., 10 FIG, 11 TAB, 6 REF. OWRR B-040-TEX(4).

DESCRIPTORS:
*PHYTOPLANKTON, *ALGAE, *NUTRIENTS, *NITROGEN FIXATION, LAKES, NITROGEN, PHOSPHORUS, IRON, *TEXAS, LABORATORY TESTS, CARBON RADIOISOTOPES, *CYANOPHYTA, EUTROPHICATION, WATER POLLUTION EFFECTS.

IDENTIFIERS:
*LAKE TRAVIS, *LAKE LIVINGSTON, FIELD MEASUREMENTS, CARBON-14 MEASUREMENTS, ACETYLENE REDUCTION MEASUREMENTS, NUTRIENT ENRICHMENT TESTS.

ABSTRACT:
BLUE-GREEN ALGAE BECAME DOMINANT IN LATE SUMMER 1970 IN LAKE LIVINGSTON ON THE TRINITY RIVER. PRELIMINARY BATCH LABORATORY NUTRIENT ENRICHMENT TESTS INDICATED NITROGEN TO BE THE LIMITING NUTRIENT. LABORATORY AND IN-SITU CARBON- 14 MEASUREMENTS SUBSTANTIATED THE CONCLUSIONS THAT NITROGEN WAS THE LIMITING NUTRIENT. IN-SITU ACETYLENE REDUCTION MEASUREMENTS INDICATED THAT THE NITROGEN FIXATION RATE WAS UNMEASURABLE IN AUGUST BUT WAS 0.74 MILLIMICRO MOLES C2H4/MG PROTEIN/ MIN IN SEPTEMBER. THE RATE OF NITROGEN FIXATION SIGNIFICANTLY INCREASED WITH THE ADDITION OF PHOSPHORUS AFTER 2 WEEKS OF INCUBATION. BLUE-GREEN ALGAE WERE ALSO ONE OF THE DOMINANT GROUPS OF PHYTOPLANKTON IN LAKE TRAVIS ON THE COLORADO RIVER. PRELIMINARY BATCH LABORATORY NUTRIENT ENRICHMENT TESTS INDICATED THAT NITROGEN, PHOSPHORUS, AND IRON ALL COULD BE LIMITING PHYTOPLANKTON GROWTH. LABORATORY CARBON- 14 ANALYSES INDICATED IRON TO BE MOST CRITICAL BUT FIELD TESTS WERE INCONCLUSIVE. IN-SITU ACETYLENE REDUCTION MEASUREMENTS INDICATED THAT NO NITROGEN FIXATION WAS OCCURRING. FIXATION IN THE LABORATORY DID OCCUR AFTER 33 DAYS OF INCUBATION OF SAMPLES ENRICHED WITH PHOSPHORUS AND PHOSPHORUS AND IRON WITH THE LATTER SAMPLE SHOWING A HIGHER RATE. HOWEVER, THIS RATE WAS SIGNIFICANTLY LOWER THAN THAT FOUND IN LAKE LIVINGSTON SAMPLES. (GALWARDI-TEXAS)

FIELD 05C, 02H

ACCESSION NO. W72-13692

SALMONELLA.

DEFENSE DOCUMENTATION CENTER, ALEXANDRIA, VA.

AVAILABLE FROM THE NATIONAL TECHNICAL INFORMATION SERVICE AS AD 737 900. $3.00 IN PAPER COPY, $0.95 IN MICROFICHE. REPORT NO. DDC-TAS-71-62, FEBRUARY 1972. 274 P, 171 REF.

DESCRIPTORS:
*SALMONELLA, *BIBLIOGRAPHIES, BIOLOGY, EPIDEMIOLOGY, CULTURES, METHODOLOGY, DOCUMENTATION, ABSTRACTS, DATA COLLECTIONS, MICROBIOLOGY, SYSTEMATICS, POLLUTANTS, PATHOGENIC BACTERIA, COLIFORMS, TOXINS, ANTIBIOTICS(PESTICIDES), ENTERIC BACTERIA, DISEASES, E. COLI, SHIGELLA, VIRUSES, CHROMATOGRAPHY, GAS CHROMATOGRAPHY, ANALYTICAL TECHNIQUES, SEPARATION TECHNIQUES, TEMPERATURE, OXYGEN, ENVIRONMENT, ENZYMES, GEOGRAPHICAL REGIONS, ALGAE, MICROORGANISMS, IRON, RADIATION, METABOLISM, INSTRUMENTATION, EQUIPMENT, LABORATORY EQUIPMENT, MICROBIOLOGY, BACTERIOPHAGE, AEROSOLS, GENETICS, WATER POLLUTION, RADIOACTIVITY EFFECTS.

IDENTIFIERS:
DETECTION, FLUORESCENT ANTIBODY TECHNIQUES, SALMONELLA TYPHIMURIUM, AMPICILLIN, VIBRIOS, SALMONELLA TYPHOSA, TYPHOID, COLIPHAGE, IMMUNOLOGY, PASTEURELLA PSEUDOTUBERCULOSIS, PENICILLIN, DRUGS, CORTISONE, HORMONES, CHLORAMPHENICOL, SALMONELLA ENTEROCOLITIS, SHIGELLA DYSENTERIAE, SHIGELLA FLEXNERI, BIOCHEMICAL STUDIES, ARTHOPODS, STREPTOMYCIN, SALMONELLA PARATYPHI, STAPHYLOCOCCUS, NUCLEAR WARFARE, CHOLERA, PASTEURELLA PESTIS, CHARACTERIZATION, BIOSYNTHESIS.

ABSTRACT:
THIS BIBLIOGRAPHY IS A COLLECTION OF UNCLASSIFIED REFERENCES ON SALMONELLA. THE DISEASE, THE MODES OF TRANSMISSION AND THE VECTORS ARE PRESENTED. INDEXES FOR CORPORATE AUTHOR-MONITORING AGENCY, SUBJECT, TITLE, AND PERSONAL AUTHOR ARE INCLUDED. (LONG-BATTELLE)

FIELD 05C, 05A, 05B, 10C

ACCESSION NO. W72-13800

DDT: INHIBITION OF SODIUM CHLORIDE TOLERANCE BY THE BLUE-GREEN ALGA ANACYSTIS NIDULANS,

WISCONSIN UNIV., MADISON. DEPT. OF ENTOMOLOGY.

J. C. BATTERTON, G. M. BOUSH, AND F. MATSUMURA.

SCIENCE, VOL 176, P 1141-1143, JUNE 9, 1972. 2 TAB, 14 REF.

DESCRIPTORS:
*DDT, *SODIUM CHLORIDE, *INHIBITION, *SALT TOLERANCE, WATER POLLUTION EFFECTS, CHLORINATED HYDROCARBON PESTICIDES, *CYANOPHYTA, SODIUM, POTASSIUM, CULTURES, RESISTANCE, AQUATIC ALGAE, GROWTH RATES, BIOASSAY, LABORATORY TESTS, CALCIUM.

IDENTIFIERS:
*ANACYSTIS NIDULANS, *ADENOSINE TRIPHOSPHATASE, BIOCHEMICAL TESTS, OUABAIN, POTASSIUM CHLORIDE.

ABSTRACT:
THE EFFECTS OF DDT ON THE SODIUM CHLORIDE TOLERANCE OF THE BLUE-GREEN ALGA, ANACYSTIS NIDULANS WERE INVESTIGATED BY GROWTH-RATE STUDIES. ALGAL CULTURES WERE GROWN AT 30 DEGREES C ON PREVIOUSLY DEFINED MEDIA WITH ADDITIONS OF SODIUM CHLORIDE (1 PERCENT BY WEIGHT) AND/OR DDT (800 PPB) AND GROWTH RATE CONSTANTS WERE CALCULATED FOR EACH CONDITION. A. NIDULANS WAS ABLE TO TOLERATE SODIUM CHLORIDE AND DDT ALONE BUT NOT IN COMBINATION. WHEN NACL WAS REPLACED BY KCL, NO GROWTH WAS OBSERVED WITH KCL IN COMBINATION WITH DDT. GROWTH DID OCCUR IN CULTURES WITH LOWER CONCENTRATIONS OF EITHER KCL OR DDT. THAT DDT LOWERED ALGAL TOLERANCE TO SALTS BY THE INHIBITION OF ADENOSINE TRIPHOSPHATASE ACTIVITY AND THE SODIUM-PUMP WAS SUBSTANTIATED BY (1) ADDITIONS OF CALCIUM, A TRANSPORT INHIBITOR, TO THE GROWTH CULTURES (THE CALCIUM REDUCED THE PERMEABILITY OF THE CELLS TO NACL, THUS REDUCING NACL STRESS) AND (2) ENZYME ASSAYS. TWO TYPES OF ATPASES WERE USED IN A COMPARISON OF INHIBITION BY DDT AND OUABAIN. THE DEGREE OF INHIBITION BY DDT WAS FOUND TO EXCEED THAT DUE TO OUABAIN. GROWTH EXPERIMENTS CARRIED OUT WITH GLYCEROL INDICATED THAT THE NACL PLUS DDT INHIBITION IS PROBABLY DUE TO AN IONIC EFFECT AND NOT TO OSMOTIC STRESS. (LONG-BATTELLE)

FIELD 05C, 05A

ACCESSION NO. W72-13809

VARIATION IN THE TOXICITY OF ARSENIC COMPOUNDS TO MICROORGANISMS AND THE SUPPRESSION OF THE INHIBITORY EFFECTS BY PHOSPHATE,

COMMONWEALTH SCIENTIFIC AND INDUSTRIAL RESEARCH ORGANIZATION, MELBOURNE (AUSTRALIA). DIV. OF FOREST PRODUCTS.

E. W. B. DA COSTA.

APPLIED MICROBIOLOGY, VOL 23, NO 1, P 46-53, JANUARY 1972. 2 FIG, 5 TAB, 11 REF.

DESCRIPTORS:
*TOXICITY, *ARSENIC COMPCUNDS, *PHOSPHATES, *INHIBITIONS, *MICROORGANISMS, BACTERIA, FUNGI, METHODOLOGY, CULTURES, BIOASSAY, ENVIRONMENTAL EFFECTS, WATER POLLUTION EFFECTS, ALGAE, HEAVY METALS, GROWTH, PSEUDOMONAS, POTASSIUM COMPOUNDS, POLLUTANT IDENTIFICATION, CHLOROPHYTA, CHLORELLA.

IDENTIFIERS:
ACREMONIELLA, DIMETHYL SODIUM ARSQNATE, ARSENATES, SODIUM CACODYLATE, CULTURE MEDIA, PORIA MONTICOLA, PORIA COCOS, PORIA VAILLANTII, CONIOPHORA OLIVACEA, LENZITES TRABEA, SCOPULARIOPSIS BREVICAULIS, CLADOSPORIUM HERBARUM, TRAMETES VERSICOLOR, FOMES ANNOSUS, FUSCOPORIA CONTIGUA, TRAMETES LILACINO-GILVA, XYLOBOLUS FRUSTULATUS, ASPERGILLUS NIGER, CHAETOMIUM GLOBOSUM, FUSARIUM SOLANI, MUCOR MICROSPORUS, PENICILLIUM SPINULOSUM, STREPTOMYCES GRISEUS, BACILLUS SUBTILIS, PSEUDOMONAS AERUGINOSA, CHLORELLA PYRENOIDOSA.

ABSTRACT:
TWO TESTS WERE USED TO MEASURE THE TOXICITY OF ARSENIC COMPOUNDS TO MICROORGANISMS AND THE SUPPRESSION OF THE INHIBITCRY EFFECTS BY PHOSPHATE. THE TOXICITY PHIAL TEST MEASURED THE GROWTH OF FUNGI ON AGAR CONTAINING VARIOUS CONCENTRATIONS OF ARSENATES AND PHOSPHATES. THE SECOND, A SEEDED PLATE INHIBITION ZONE TEST, MEASURED THE EFFECTS OF PHOSPHATE ON THE ARSENATE GROWTH INHIBITION ZONES. THE TOXICITY OF POTASSIUM ARSENATE, AS MEASURED BY RETARDATION OR INHIBITION OF GROWTH ON SOLID NUTRIENT MEDIA, SHOWED WIDE VARIATION AMONG DIFFERENT FUNGI BUT WAS CONSISTENTLY REDUCED BY THE ADDITICN OF LARGE AMOUNTS OF POTASSIUM PHOSPHATE, WITH BOTH ARSENIC-SENSITIVE AND ARSENIC-TOLERANT FUNGI. PORIA MONTICOLA WAS COMPLETELY INHIBITED BY 0.0025 M ARSENATE BUT WAS PROGRESSIVELY LESS INHIBITED AS THE PHOSPHATE CONTENT OF THE MEDIUM INCREASED AND GREW SLOWLY AT 0.04 M ARSENATE WHEN 0.16 M KH2PO4 WAS ADDED. CLADOSPORIUM HERBARUM SHOWED 36 PERCENT REDUCTION IN GROWTH AT 0.08 M ARSENATE IN A LOW-PHOSPHATE MEDIUM, BUT WHEN 0.01 M KH2PO4 WAS ADDED, ARSENATE CONCENTRATIONS UP TO 0.64 M (AT WHICH THE MEDIUM CONTAINS 4.8 PERCENT AS) CAUSED NO REDUCTION IN GRCWTH RATE. ADDITION OF PHOSPHATE ALSO REDUCED THE TOXICITY OF POTASSIUM ARSENITE BUT NOT THAT OF DIMETHYL SODIUM ARSONATE (SODIUM CACODYLATE). THE COUNTERACTING EFFECT OFPHOSPHATE ON ARSENATE TOXICITY WAS FOUND TO OCCUR WITH EVERY ONE OF A WIDE VARIETY OF MICROORGANISMS TESTED. THE PRACTICAL IMPLICATIONS OF THE COUNTER-INHIBITION PHENOMENON IN LABORATORY INVESTIGATIONS AND STANDARD TESTS OF ARSENICAL FUNGICIDES, IN BIOCHEMICAL RESEARCH, AND IN THE CCMMERCIAL USE OF ARSENICAL BIOCIDES ARE SET OUT. (LONG-BATTELLE)

FIELD 05C

ACCESSION NO. W72-13813

CHANGES IN PLANKTON SPECIES COMPOSITION AND DIVERSITY IN A CONTROLLED NUTRIENT ENRICHMENT STUDY,

KANSAS UNIV., LAWRENCE. DEPT. OF SYSTEMATICS AND ECOLOGY.

W. J. O'BRIEN.

TRANSACTIONS OF THE AMERICAN MICROSCOPICAL SOCIETY, VOL 91, NO 1, P 77-91, JANUARY 1972. 2 FIG, 3 REF.

DESCRIPTORS:
*PHYTOPLANKTON, *NUTRIENT REQUIREMENTS, NUTRIENTS, BIOINDICATORS, WATER POLLUTION EFFECTS, FERTILIZERS, NITROGEN, BIOASSAY, PHOSPHORUS, CYANOPHYTA, OLIGOTROPHY, CHLAMYDOMONAS, DEFICIENT ELEMENTS, MESOTROPHY, EUTROPHICATION, TROPHIC LEVEL, PROTOZOANS, AQUATIC ALGAE, ENVIRONMENTAL EFFECTS, CHLOROPHYTA, PRIMARY PRODUCTIVITY, CHRYSOPHYTA.

IDENTIFIERS:
ENRICHMENT, CHROMULINA, CRYPTOMONAS EROSA, ERKINIA SUBEQUICILIATA, UROGLENOPSIS AMERICANA, CRYPTOMONAS PULSILLA, MICROCYSTIS AERUGINOSA, CHROOMONAS CAUDATA, CHLAMYDOMONAS PERTUSA, PANDORINA MORUM.

ABSTRACT:
AN INVESTIGATION WAS MADE OF CHANGES IN PLANKTON SPECIES COMPOSITION AND DIVERSITY AS BIOLOGICAL INDICATORS BY CONTROLLING THE NUTRIENT CONTENT/DENSITY OF THEIR ENVIRONMENT. CONTROLLED EXPERIMENTATION WAS CARRIED OUT FOR TWO YEARS AT EIGHT PONDS WHICH WERE ORGANIZED INTO FOUR TREATMENT LEVELS (CONTROL, LOW, MEDIUM, HIGH) BY ADDING SPECIFIED AMOUNTS OF HIGH QUALITY INORGANIC FERTILIZER. THE CONTROL PONDS WERE OLIGOTROPHIC, THE LOW AND MEDIUM PONDS MESOTROPHIC, AND THE HIGH PONDS EUTROPHIC. IN THE FIRST YEAR THE PHYTOPLANKTON DENSITY WAS QUITE VARIABLE IN THE TREATED PONDS WITH THE HIGH TREATMENT LEVEL PONDS AT TIMES HAVING THE LOWEST PHYTOPLANKTON DENSITY AND AT OTHER TIMES THE HIGHEST. BY MID-SUMMER OF THE SECOND YEAR, HOWEVER, THE PHYTOPLANKTON DENSITY INCREASED PROPORTIONATELY WITH INCREASING TREATMENT LEVEL. IN 1969, THE PRIMARY PRODUCTIVITY WAS MEASURED AND ITS RESPONSE TO TREATMENT LEVEL WAS SIMILAR TO THAT OF THE PHYTOPLANKTON DENSITY. IN A LABORATORY STUDY, WATER FROM EACH TREATMENT LEVEL WAS COLLECTED AND PLACED IN FLASKS TO WHICH WAS ADDED DISSOLVED INORGANIC NITROGEN OR PHOSPHORUS OR NITROGEN AND PHOSPHORUS TOGETHER. THE TEST ORGANISM, PANDORINA MORUM, WAS INCUBATED IN THESE SEPARATE MEDIA FOR 2 WEEKS, AFTER WHICH THE YIELD WAS MEASURED. THE FINAL YIELD OF P. MORUM INCREASED WITH INCREASING TREATMENT LEVEL. THE ADDITION OF NITROGEN STIMULATED FINAL YIELD AT ALL TREATMENT LEVELS BUT WAS PROPORTIONATELY LESS STIMULATORY WITH INCREASING TREATMENT LEVEL. THE ADDITION OF PHOSPHORUS DID NOT STIMULATE FINAL YIELD AT ALL. (BYRD-BATTELLE)

FIELD 05C

ACCESSION NO. W72-13816

LINMCLOGY OF ONEIDA LAKE WITH EMPHASIS ON FACTORS CONTRIBUTING TC ALGAL BLOOMS,

GECLOGICAL SURVEY, ALBANY, N.Y.

P. E. GREESON.

NEW YORK STATE CEPT OF ENVIRONMENT CONSERVATION, ALBANY, (GEOLOGICAL SURVEY OPEN-GILE REPORT), 1971. 185 P, 51 FIG, 32 TAB, 145 REF, APPEND.

DESCRIPTORS:
*LIMNOLOGY, *LAKES, *EUSTROPHICATICN, *HYDROLOGY, *NEW YORK, WATER QUALITY, NUTRIENTS, ALGAE, INFLCW, DISCHARGE(WATER), WATER BALANCE, SEDIMENTS, CHEMICAL PROPERTIES, DISSOLVED SOLIDS, HYDROLOGIC DATA, DATA COLLECTIONS, LAKE MORPHOLOGY, ECOLOGY, BIOLOGICAL PROPERTIES.

IDENTIFIERS:
*ONEIDA LAKE(NY).

ABSTRACT:
ONEIDA LAKE, IN THE STATE OF NEW YORK, IS A NATURALLY EUTROPHIC LAKE THAT HAS EXISTED FOR ABOUT 10,500 YEARS. IT HAS BEEN IN A EUTROPHIC STATE FOR AT LEAST 350 YEARS, AND THE GEOCHEMICALLY DERIVED DISSOLVED MATERIALS ENTERING THE LAKE FROM THE DRAINAGE BASIN ARE OF SUFFICIENT QUANTITY (449,700 TONS PER YEAR) TO SUPPORT ANNUAL ALGAL BLOOMS. THE LAKE RETAINS 50,000 TONS OF DISSOLVED SOLIDS EACH YEAR. THESE MATERIALS BECOME INCORPORATED IN THE BOOTTOM SEDIMENTS. WATER IN ONEIDA LAKE REPRESENTS A HYDROLOGIC EQUILIBRIUM. WATER ENTERING THE LAKE IS FROM DIRECT PRECIPITATION (149,600 ACRE-FEET PER YEAR) AND FROM SURFACE WATER INFLOW (1,729,000 ACRE-FEET PER YEAR(. WATER LEAVES THE LAKE BY OUTFLOW THROUGH THE CNEIDA RIVER (1,730,400 ACRE-FEET PER YEAR) AND BY EVAPORATION (148,200 ACRE-FEET PER YEAR). SIXTY-SEVEN PERCENT OF ALL WATER ENTERING ONEIDA LAKE ORIGINATES IN THE NORTHERN PART OF THE DRAINAGE BASIN. THE FOUR MOST IMPORTANT FACTORS AFFECTING THE ECOLOGICAL PROCESSES IN THE LAKE ARE: (1) HIGH FERTILITY OF THE DRAINAGE BASIN, (2) PHYSICAL POSITION AND SHALLOWNESS OF THE LAKE, (3) MIXING AS CAUSED BY WIND, AND (4) FERTILITY OF THE BOTTOM SEDIMENTS. (WOODARD-USGS)

FIELD 05C, 02H

ACCESSION NO. W72-13851

DYNAMICS OF PHYTOPLANKTCN PRIMARY PRODUCTION AND BIOMASS IN LOVIISA ARCHIPELAGO (GULF OF FINLAND),

INSTITUTE OF MARINE RESEARCH, HELSINKI (FINLAND).

P. BAGGE, AND A. NIEMI.

MERENTUTKIMUSLAIT. JULK./HAVSFORSKNINGSINST. SKR., NO 233, P 19-41, 1971. 7 FIG, 2 TAB, 35 REF, 1 APP.

DESCRIPTORS:
*PHYTOPLANKTON, *PRIMARY PRODUCTIVITY, *EUTROPHICATION, *BIOMASS, *HYDROGRAPHY, ALGAE, AQUATIC ALGAE, OLIGOTROPHY, DISSOLVED OXYGEN, WATER TEMPERATURE, HYDROGEN ION CONCENTRATION, STRATIFICATICN.

IDENTIFIERS:
ORTHOPHOSPHATE CONCENTRATICNS, GULF OF FINLAND.

ABSTRACT:
PHYTOPLANKTON PRIMARY PRODUCTION AND BIOMASS AT TWO SAMPLING LOCALITIES SITUATED IN THE LOVIISA ARCHIPELAGO WERE STUDIED DURING 1967-1969. PRIMARY PRODUCTION WAS MEASURED MAINLY BY THE IN SITU CARBON-14 TECHNIQUE AND QUANTITATIVE ANALYSES OF PHYTOPLANKTON WERE USUALLY MADE ONCE A MONTH DURING THE GROWING SEASON. NO LINEAR CORRELATION WAS OBSERVED BETWEEN PRODUCTION AND BIOMASS VALUES. THE DATA CF PRIMARY PRODUCTION AND PHYTOPLANKTON OBTAINED IN THE LOVIISA AREA ARE COMPARED WITH THOSE AVAILABLE FROM CTHER AREAS SITUATED ON THE NORTHERN COAST OF THE GULF OF FINLAND. THE MAGNITUDE OF PRODUCTION (CA. 30-40 G C(ASS.)/SQUARE METER/YEAR) AND THE SUCCESSION AND BIOMASS OF PHYTOPLANKTON IN THE LOVIISA AREA WERE FOUND TO BE TYPICAL OF THE OLIGOTROPHIC WATERS OF THE SOUTH COAST OF FINLAND, HOWEVER THE OCCURRENCE OF SOME ALGAL SPECIES POINTS TO A WEAK EUTROPHICATION OF THE WATERS STUDIED. (SVENSSON-WASHINGTON)

FIELD 05C

ACCESSION NO. W72-13871

A STUDY OF PHOSPHATE INDUCED ALGAL GROWTH IN ORDER TO SUPPRESS OR ELIMINATE THIS PHENOMENON,

NEW MEXICO STATE UNIV., UNIVERSITY PARK. WATER RESOURCES RESEARCH INST.

N. E. VANDERBORGH, AND A. G. BUYERS.

AVAILABLE FROM THE NATIONAL TECHNICAL INFORMATION SERVICE AS PB-212 026, $3.00 IN PAPER COPY, $0.95 IN MICROFICHE. NEW MEXICO WATER RESOURCES RESEARCH INSTITUTE, LAS CRUCES, COMPLETION REPORT 003, JUNE 1972, 35 P, 9 FIG, 7 TAB, 22 REF. OWRR A-035-NMEX (1).

DESCRIPTORS:
*ALGAL CONTROL, *PHOSPHATES, *ALGAL NUTRIENTS, *CHLORELLA, ALGAE, *EUTROPHICATION, NEW MEXICO, BIODEGRADATION.

IDENTIFIERS:
*ALGAL GROWTH, *RADIOISOTOPE P32, SODIUM PYROPHOSPHATE, RIO GRANDE VALLEY.

ABSTRACT:
THE KINETICS OF THE RADIOACTIVE ISOTOPE P32 UPTAKE BY THE FRESH WATER ALGAE, CHLORELLA, WERE INVESTIGATED IN SOIL-WATER CULTURES, NATURAL SYSTEMS, AND SYNTHETIC CULTURE MEDIA. LABORATORY INVESTIGATIONS WERE CONDUCTED AT PH VALUES OF 4, 7 AND 10, AT A CONSTANT TEMPERATURE 20 PLUS OR MINUS 2C. PORTIONS OF REAGENT GRADE SODIUM PYROPHOSPHATE WERE IRRADIATED WITH A FAST NEUTRON FLUX CONVERTING A PORTION OF THE SAMPLE INTO THE RADIOISOTOPE P32. (P32 EMITS A 1.7 MEV BETA WITH A HALF-LIFE OF 14.31 DAYS.) TAGGING WAS DONE AT A RATE SUFFICIENT TO PRODUCE AN ACTIVITY OF 0.03 MICROCURIES/GRAM. SOLUTIONS WERE THEN PREPARED CONTAINING THE TAGGED PYROPHOSPHATE AND USED TO SPIKE CHLORELLA CULTURES. THE UPTAKE OF THE ISOTOPE WAS THEN MONITORED AS A FUNCTION OF TIME, AFTER VARIOUS TIMES THE CULTURES WERE SAMPLED, FILTERED AND COUNTED FOR P32 ACTIVITY. RESULTS INDICATED THAT THE KINETIC INTERPRETATION OF P32 UPTAKE MUST ALLOW FOR FIRST ORDER KINETICS WITH RESPECT TO P32 AT A PH OF 10, AND A SECOND ORDER KINETIC SCHEME AT PH 7 AND 4. THIS EVIDENCE INDICATED THAT ALGAL SYSTEMS ONLY UTILIZE PHOSPHORUS IN THE ORTHOPHOSPHATE FORM. PHOSPHORUS EXCHANGE RATES WITH SOIL AND INFORMATION ABOUT DEGRADATION PATHWAYS ARE INCLUDED. TWO EXPERIMENTS, ONE WITH AN OXYGEN RICH ENVIRONMENT AND THE SECOND WITH A CARBON DIOXIDE RICH ENVIRONMENT, WERE CARRIED OUT WITH SODIUM PYROPHOSPHATE-CHLORELLA, SOIL-WATER AND CHEMICAL CULTURES. IN THE OXYGEN EXPERIMENT P32 REMOVAL WAS ELIMINATED IN THE CHEMICAL CULTURES AND REMOVAL WAS SLOWED FOR THE SOIL-WATER CULTURES. CO2 ENRICHMENT GREATLY ACCELERATED THE REMOVAL. (CREEL-NMEX STATE)

FIELD 05A, 05G

ACCESSION NO. W72-13992

COMPARATIVE EFFECTS OF AQUATIC BIOTOXINS ON CARCIAC SYSTEMS,

NEW HAMPSHIRE UNIV., DURHAM. DEPT. OF ZOOLOGY.

F. P. THURBERG.

PHD THESIS, JANUARY 1972. 74 P, 9 FIG, 3 TAB, 100 REF. OWRR A-013-NH(6) AND A-021-NH(3).

DESCRIPTORS:
*ALGAL TOXINS, *POISONOUS PLANTS, *CYANOPHYTA, *NUISANCE ALGAE, *AQUATIC ALGAE, *CRUSTACEANS, *FISH TOXINS, *ALGAL POISONING, DINOFLAGELLATES, BEGGIATOA, PHYTOPLANKTON BLOOMS, PHYSIOLOGICAL ECOLOGY, MORTALITY, RED TIDE, TOXICITY.

IDENTIFIERS:
AQUATIC BIOTOXINS, *BIOTOXINS.

ABSTRACT:
TOXINS FROM 3 MARINE DINOFLAGELLATES, GYMNODIUM BREVE, AMPHIDINIUM CARTERI AND GONYAULAX CATENALLA, AND A FRESHWATER BLUE-GREEN ALGA, APHANIZOMENON FLOS-AQUAE WERE OBTAINED BY LABORATORY CULTURE, FIELD COLLECTION AND CORRESPONDENCE WITH OTHER INVESTIGATORS. IN VIVO AND ISOLATED HEARTS OF DECAPOD CRUSTACEANS, CANCER IRRORATUS AND CARCINUS MAENUS, THE BIVALVE MOLLUSCS, MYA ARENARIA AND MERCENARIA MERCENARIA AND THE GRASS FROG RANA PIPIENS WERE EXPOSED TO THESE TOXINS, AND MECHANICAL AND ELECTRICAL ACTIVITY WERE MEASURED. GYMNODINIUM BREVE TOXIN EXCITED CRUSTACEAN HEARTS, DEPRESSED FROG HEARTS AND HAD NO EFFECT ON MOLLUSCAN HEARTS. THESE OBSERVATIONS AND EXPERIMENTS WITH HUMAN BLOOD CHOLINESTERASE AND MAMMALIAN INTESTINE SUGGEST ANTICHOLINESTERASE-LIKE ACTIVITY AS ONE ACTION OF G. BREVE TOXIN. AMPHIDINIUM CARTERI TOXIN EXCITED CRUSTACEAN HEARTS AND DEPRESSED MOLLUSCAN AND FROG HEARTS. THIS CHOLINE-LIKE ACTION WAS FURTHER DEMONSTRATED WITH MAMMALIAN INTESTINE PREPARATIONS AND THE USE OF THE CHOLINE BLOCKING COMPOUND, MYTOLON CHLORIDE. APHANIZOMENON FLOS-AQUAE AND GONYAULAX CATENELLA TOXINS DEPRESSED FROG AND CRUSTACEAN HEARTS BUT HAD NO EFFECT ON MOLLUSCAN HEARTS. THIS EVIDENCE SUPPORTS REPORTED PHYSIOLOGICAL AND CHEMICAL SIMILARITIES OF THESE TWO TOXINS.

FIELD 05C, 05A

ACCESSION NO. W72-14056

LIMNOLOGICAL EFFECTS OF SIMULATED PUMPED-STORAGE OPERATICN AT YARDS CREEK,

DELAWARE RIVER BASIN COMMISSION, TRENTON, N.J.

C. F. BAREN.

DECEMBER 1971. 158 P, 36 FIG, 31 TAB, 103 REF.

DESCRIPTORS:
*WATER LEVEL FLUCTUATIONS, *PUMPED STORAGE, PHYSICOCHEMICAL PROPERTIES, *PLANKTON, *AQUATIC PLANTS, *AQUATIC ANIMALS, *MODEL STUDIES, LIMNOLOGY, BIOLOGICAL PROPERTIES, MINERALOGY, FISH, FISH BEHAVIOR, CHEMICAL ANALYSIS, SAMPLING, VOLUMETRIC ANALYSIS, ALGAE, RESERVOIR STORAGE, IMPOUNDMENTS, DISSOLVED OXYGEN, CHLORIDES, INVERTEBRATES, HARDNESS(WATER), TURBIDITY, ALKALINITY, CARBCN DIOXIDE, PHOSPHORUS, NITROGEN, SILICA, COPPER, SULFATES, IRON, MANGANESE, SPAWNING, FISH REPRODUCTION, LIFE HISTORY STUDIES, PHYTOPLANKTON, STANDING CROP, BIOMASS, SPORT FISH, ZOOPLANKTON, WATER TEMPERATURE, AIR TEMPERATURE, JUVENILE FISHES, FRY, PLANKTON NETS, NETS, SUNFISHES, BASS, ROCK BASS, DIATOMS, CHRYSCPHYTA, BULLHEADS, SHINERS, YELLOW PERCH, WALLEYE, MINNOWS, PROTOZOA, ROTIFERS, BROOK TROUT, RAINBOW TROUT, PIKES, STRIPED BASS, CHLOROPHYTA, EUGLENOPHYTA, PHRROPHYTA, EUGLENA, CYANOPHYTA, CRUSTACEANS, COPEPODS, NESTING, BROWN TROUT, SCENEDESMUS, DINOFLAGELLATES, LARVAE, DAPHNIA, MIDGES, CADDISFLIES, GASTROPODS, PONDWEEDS, NEW JERSEY, DELAWARE RIVER, NUTRIENTS, PRIMARY PRODUCTIVITY, SECONDARY PRODUCTIVITY.

IDENTIFIERS:
MACROINVERTEBRATES, *TOCKS ISLAND RESERVOIR(NJ), AMERICAN SHAD, EKMAN DREDGE, PUMPKINSEED, BLACK CRAPPIE, CHAIN PICKEREL, TENDIPES TENTANS, TRICHOPTERA, LIMNEPHILUS, GLCSSOSOMA, PHYSA GYRINA, KEMMERER SAMPLER, SECCHI *YARDS CREEK(NJ), RELATIVE CONDITION FACTOR.

ABSTRACT:
A PUMPED-STORAGE POWER GENERATING SYSTEMS WHICH WOULD ALTER WATER LEVELS HAS BEEN PROPOSED FOR INCLUSION IN THE TOCKS ISLAND RESERVOIR PROJECT AT BLAIRSTOWN, NEW JERSEY. RESULTS ARE REPORTED OF A TWO-YEAR STUDY TO DETERMINE WHAT EFFECTS CHANGING WATER LEVELS MIGHT HAVE ON AQUATIC BIOTA. FOUR T DISC,EST PONDS WERE CONSTRUCTED IN THE VICINITY OF THE PLANNED PROJECT TO INVESTIGATE: (1) THE EFFECTS OF DIURNALLY FLUCTUATING WATER LEVELS ON THE LIFE HISTORY OF SELECTED, NATIVE NEST-BUILDING AND NON-NEST-BUILDING FISH; (2) THE PHYSICAL AND CHEMICAL CHARACTERISTICS OF VARIOUSLY FLUCTUATED TEST PONDS; (3) THE RESULTING EFFECTS OF FLUCTUATING WATER LEVELS ON MACROINVERTEBRATES, PLANKTON, AND AQUATIC PLANTS; AND (4) THE CONSEQUENT COMBINED EFFECTS OF PHYSICOCHEMICAL CHARACTERISTICS AND CHANGING WATER LEVELS ON THE BEHAVIOR OF TH SELECTED FISHES. IT WAS FOUND THAT (1) FISH ADAPTED TO THE UNIFORM REGIME OF WATER LEVEL FLUCTUATIONS AND WERE SUCCESSFUL IN SPAWNING AND HATCHING EGGS AND (2) THE MEASURABLE EFFECTS OBSERVED IN PLANKTON AND MACROINVERTEBRATE PRODUCTIVITY, PLANT COLCNIZATION, RELATIVE CONDITION FACTORS OF FISH, AND THE PHYSICAL AND CHEMICAL PROPERTIES OF THE TEST POND HAD NO MARKED EFFECT ON THE FISH POPULATIONS. THE RESULTS OF THE PHYSIOCHEMICAL ANALYSES AND PROFILES, THE PLANTS AND ANIMALS STUDIED, AND SAMPLING PROCEDURES UTILIZED IN THE STUDY ARE INCLUDED IN FIGURES, TABLES AND APPENDICES.
(HOLOMAN-BATTELLE)

FIELD 05C

ACCESSION NO. W72-14280

CONFERENCE IN THE MATTER OF POLLUTION OF LAKE ERIE AND ITS TRIBUTARIES-INDIANA-MICHIGAN-NEW YORK-OHIO-PENNSYLVANIA: VOLUMES 1 AND 2.

FEDERAL WATER QUALITY ADMINISTRATION, WASHINGTON, D.C.

HELD IN DETROIT, MICHIGAN, JUNE 3 AND 4 1970, 740 P. 21 FIG, 25 TAB, 18 APPEND.

DESCRIPTORS:
*WATER POLLUTION, *LAKE ERIE, *MERCURY, *MUNICIPAL WASTES, *INDUSTRIAL WASTES, WATER POLLUTION CONTROL, WATER POLLUTION SOURCES, TRIBUTARIES, FRESHWATER FISH, YELLOW PERCH, SEWAGE EFFLUENTS, OIL SPILLS, THERMAL POLLUTION, NUTRIENTS, BIOASSAY, *WATER QUALITY, ALGAE, PESTICIDES, HEAVY METALS, WASTEWATER TREATMENT, SMELTS, MICHIGAN, OHIO, NEW YORK, PENNSYLVANIA, TOXICITY, DISCHARGE(WATER), PHENOLS, ELECTRIC POWERPLANTS, NUCLEAR WASTES, PHOSPHORUS, DISSOLVED OXYGEN, CLADOPHORA, ACIDS, ORGANIC WASTES, BIOCHEMICAL OXYGEN DEMAND, SALMON, CARP, WHITE BASS, CHANNEL CATFISH, SUCKERS, BULLHEADS, BUFFALO FISH, PIKE, SULFATE, CHLORIDE, SODIUM, POTASSIUM, CALCIUM, DRUMS, SAUGER, WALLEYE, WATER QUALITY STANDARDS, WATER ANALYSIS.

IDENTIFIERS:
MAUMEE RIVER, ST. CLAIR RIVER, LAKE ST. CLAIR, DETROIT RIVER, BLACK RIVER, PINE RIVER, CLINTON RIVER, HURON RIVER, BELL RIVER, SALT RIVER, MICROCYSTIS, APHANIZOMENON, CYANIDES, COHO SALMON, SHEEPSHEAD, CHUBS, NORTHERN PIKE, GIZZARD SHAD.

ABSTRACT:
A VERBATIM TRANSCRIPT OF THE PROCEEDINGS OF A CONFERENCE CONCERNED WITH THE WATER POLLUTION PROBLEMS OF LAKE ERIE AND ITS TRIBUTARIES AND SUMMARIES OF THE VARIOUS REPORTS PRESENTED HAVE BEEN COMPILED INTO A TWO-VOLUME REPORT. STATEMENTS OF COMPLIANCE WITH THE LAKE ERIE ENFORCEMENT CONFERENCE ABATEMENT SCHEDULES FOR MUNICIPALITIES AND INDUSTRIES AND DETAILED DISCUSSIONS ON THE SOURCES OF POLLUTION ON THE LAKES, INCLUDING INDUSTRIAL WASTES FROM MAJOR CITIES SUCH AS DETROIT AND CLEVELAND, AND MERCURY STANDARDS, SOURCES, AND LEVELS ARE ALSO INCLUDED. PARTICIPANTS IN THE CONFERENCE WERE FROM WATER POLLUTION CONTROL AGENCIES OF THE STATES OF MICHIGAN, INDIANA, OHIO, PENNSYLVANIA AND NEW YORK, AND OF THE U. S. DEPARTMENT OF THE INTERIOR. (MORTLAND-BATTELLE)

FIELD 05B, 05C, 05G

ACCESSION NO. W72-14282

RESEARCH ON AND CULTURE OF CALCAREOUS GREEN ALGAE,

OHIO STATE UNIV. RESEARCH FOUNDATION, COLUMBUS.

L. HILLIS-COLINVAUX.

FINAL REPORT, FEBRUARY 11, 1972. 30 P, 1 FIG, 3 TAB, 27 REF.

DESCRIPTORS:
*CULTURES, *PLANT GROWTH, *CHLOROPHYTA, *MARINE ALGAE, *BIOASSAY, REPRODUCTION, SEA WATER, MARINE PLANTS, PRIMARY PRODUCTIVITY, CORAL, REEFS, ANTIBIOTICS(PESTICIDES), CHLORINATED HYDROCARBON PESTICIDES, PESTICIDE TOXICITY, PLANT POPULATIONS, AQUARIA, LABORATORY EQUIPMENT, PLANT PHYSIOLOGY.

IDENTIFIERS:
EPIPHYTES, CULTURING VESSELS, PENICILLIN, LINDANE, DIADEMA ANTILLARUM, AANDOMENE, SIPHONALES, *HALIMEDA SPP., PENICILLUS CAPITATUS, PENICILLUS DUMENTOSUS, RHIPOCEPHALUS PHOENIX, UDOTEA FLABELLUM, BRYOPSIS, CAULERPA, ACROPORA PALMATA, PENICILLUS LAMOUROUXII.

ABSTRACT:
FOUR GENERA OF CORAL REEF SIPHONALES, HALIMEDA, PENICILLUS, RHIPOCEPHALUS AND UDOTEA WERE GROWN IN LABORATORY AQUARIA UNDER LIGHT INTENSITIES OF 650, 200-375 AND 125-200 FT-CANDLES. THE COMMERCIAL PREPARATION 'INSTANT OCEAN' WAS TESTED FOR USE WITH PLANTS. PENICILLIN AND LINDANE WERE TESTED FOR THEIR EFFECT ON EPIPHYTES AND OTHER NUISANCES. A 15 PPM SOLUTION OF THE INVERTEBRATE KILLER LINDANE DID NOT KILL HALIMEDAE WHICH WERE IMMERSED IN IT FOR 15 MINUTES. PENICILLIN AT A CONCENTRATION OF 3000 UNITS PER ML WAS INITIALLY EFFECTIVE IN CONTROLLING THE WEEDS, BUT WITHIN 2 MONTHS THEY WERE ABUNDANT AGAIN AND PENICILLIN WAS THEREAFTER INEFFECTIVE. THE GROWTH AND DEVELOPMENT OF ALL 4 GENERA FROM TINY PROTUBERANCES ABOVE THE SAND TO WHITE, DYING AND DISINTEGRATING INDIVIDUALS WAS FOLLOWED. ALL PRODUCED NEW INDIVIDUALS FROM RHIZOIDAL-LIKE FILAMENTS THAT EXTENDED OUTWARDS THROUGH THE SAND FROM THE HOLDFAST OF AN OLDER PLANT. IT WAS DEMONSTRATED THAT VEGETATIVE REPRODUCTION MIGHT OCCUR WITH PARTIAL BURYING OF INDIVIDUALS BY SHIFTING SANDS IN A REEF, OR FROM PORTIONS OF PLANTS BROKEN OFF BY GRAZING OR OTHER NATURAL ACTIVITIES. PLANTS PRODUCED VEGETATIVELY ARE NOT INITIALLY EPIPHYTIZED AND SO MAY BE A SOURCE OF CLEAN PLANTS FOR LABORATORY EXPERIMENT. THE FIRST SWARMER-PRODUCING PENICILLUS PLANTS SO FAR KNOWN WERE PRODUCED DURING THIS STUDY. THE RATHER UNDISTINGUISHED STRUCTURES ASSOCIATED WITH THEM, THE RELEASE OF SWARMERS FROM HALIMEDA, AND ASSOCIATED OBSERVATIONS ARE DESCRIBED. INSIGHT INTO THE ROLE OF CALCAREOUS PLANTS IN A REEF WAS OBTAINED BY A PRODUCTIVITY STUDY WHICH COMBINED A CENSUS OF SUCH PLANTS IN A REEF AND A LABORATORY MEASURE OF OXYGEN CHANGES IN AN ENTIRE AQUARIUM. SOME HALIMEDA REEF POPULATIONS APPEARED TO BE GRAZED BYTHE URCHIN DIADEMA. A CONSERVATIVE FIGURE FOR THE PRODUCTIVITY OF LABORATORY HALIMEDA PLANTS IS 2.5 MG C/PLANT/DAY (NET) OR 4.5 MG C/PLANT/DAY (GROSS). (LONG-BATTELLE)

FIELD 05C, 05A

ACCESSION NO. W72-14294

TECHNIQUE FOR MEASURING C-1402 UPTAKE BY SOIL MICROORGANISMS IN SITU,

WISCONSIN UNIV., MADISON. DEPT. OF BACTERIOLOGY.

D. W. SMITH, C. B. FLIERMANS, AND T. D. BROCK.

APPLIED MICROBIOLOGY, VOL, 23, NO. 3, P 595-600, MARCH 1972. 3 FIG, 1 TAB, 12 REF.

DESCRIPTORS:
*SOIL MICROORGANISMS, *ABSORPTION, *CARBON DIOXIDE, *RADIOACTIVITY TECHNIQUES, SOIL ALGAE, *METHODOLOGY, MEASUREMENT, OXIDATION, CARBON RADIOISOTOPES, *ON-SITE TESTS, SULFUR BACTERIA, PRODUCTIVITY, RADIOCHEMICAL ANALYSIS, THERMOPHILIC BACTERIA, ORGANIC MATTER, CULTURES, SOIL BACTERIA.

IDENTIFIERS:
CHEMICAL INTERFERENCE, SCINTILLATION COUNTING, CYANIDIUM CALDARIUM, THIOBACILLUS, SULFOLOBUS, GLUCOSE, URACIL.

ABSTRACT:
UPTAKE OF C-1402 IN SOILS DUE TO ALGAE OR SULFUR-OXIDIZING BACTERIA WAS EXAMINED BY INCUBATION OF SOIL SAMPLES WITH GASEOUS C-1402 AND SUBSEQUENT CHEMICAL OXIDATION OF THE BIOLOGICALLY FIXED RADIOACTIVE ISOTOPE TO C-1402 FOR DETECTION WITH A LIQUID SCINTILLATION COUNTING SYSTEM. THE C-1402 WAS ADDED TO THE SOIL IN THE GAS PHASE SO THAT NO ALTERATION OF THE MOISTURE OR IONIC STRENGTH OF THE SOIL OCCURRED. WET OXIDATION OF RADIOACTIVE ORGANIC MATTER WAS CARRIED OUT IN SEALED AMPOULES, AND THE C-1402 PRODUCED WAS TRANSFERRED TO A PHENETHYLAMINE-LIQUID SCINTILLATION COUNTING SYSTEM WITH A SIMPLY CONSTRUCTED APPARATUS. THE TECHNIQUE IS INEXPENSIVE AND EFFICIENT AND DOES NOT REQUIRE ELABORATE TRAPS SINCE SEVERAL POSSIBLE INTERFERING FACTORS WERE FOUND TO HAVE NO HARMFUL EFFECTS. THE EFFICIENCY OF THE TECHNIQUE IN OXIDIZING VARIOUS COMPOUNDS WAS EXAMINED USING VARYING CONCENTRATIONS OF UNIFORMLY LABELED GLUCOSE AND URACIL, AND LABELED CYANIDIUM CALDARIUM CELLS. IN ALL CASES, THE PROCEDURE WAS PERFORMED ON SAMPLES CONTAINING 0.5 G OF SOIL PER AMPOULE TO REPRODUCE THE TREATMENT OF NATURAL SAMPLES AS CLOSELY AS POSSIBLE. SIXTY-ONE TO EIGHTY-SEVEN PERCENT OF THE ADDED RADIOACTIVITY WAS RECOVERED DURING THE OXIDATION OF GLUCOSE AND URACIL, WHEREAS 98 PERCENT OF THE ADDED RADIOACTIVITY WAS RECOVERED DURING THE OXIDATION OF C. CALDARIUM. THE RECOVERY EFFICIENCY WAS INDEPENDENT OF THE AMOUNT OF RADIOACTIVITY PRESENT. EXPERIMENTS IN COAL MINE REGIONS AND IN GEOTHERMAL HABITATS HAVE DEMONSTRATED THE ECOLOGICAL APPLICABILITY OF THIS TECHNIQUE FOR MEASUREMENT OF CO2 FIXATION BY SULFUR OXIDIZING BACTERIA AND SOIL ALGAE. (LONG-BATTELLE)

FIELD 05A, 02G

ACCESSION NO. W72-14301

HEAVY METALS POLLUTE NATURE, MAY REDUCE PRODUCTIVITY,

LUND UNIV. (SWEDEN). DEPT. OF ECOLOGICAL BOTANY.

G. TYLER.

AMBIO, VOL. 1, NO. 2, P 52-59, APRIL 1972. 8 FIG, 1 TAB, 11 REF.

DESCRIPTORS:
*TERRESTRIAL HABITATS, *HEAVY METALS, *PRODUCTIVITY, *ECOSYSTEMS, POTASSIUM, MAGNESIUM, SODIUM, CALCIUM, HYDROGEN, CHEMICAL WASTES, LEAD, NICKEL, TOXICITY, BIOINDICATORS, BIODEGRADATION, BIOMASS, LICHENS, MOSSES, FUNGI, ALGAE, SOIL CONTAMINATION, TRACE ELEMENTS, ION EXCHANGE, POLLUTANT IDENTIFICATION, COPPER, CADMIUM, CHROMIUM, NICKEL, MANGANESE, IRON, ZINC, ORGANIC COMPOUNDS, ADSORPTION.

IDENTIFIERS:
VANADIUM, BRYOPHYTES, BIOACCUMULATION, METAL COMPLEXES, PICEA ABIES, VACCINIUM VITIS IDAEA, VACCINIUM MYRTILLUS, DESCHAMPSIA FLEXUOSA, PARMELIA PHYSODES, HYPNUM CUPRESSIFORME.

ABSTRACT:
THE DEPOSITION, ACCUMULATION, AND POSSIBLE EFFECTS OF PB, CD, CU, ZN, NI, FE, AND MN ON TERRESTRIAL SITES IN SCANDINAVIA ARE DISCUSSED. BRYOPHYTES HAVE BEEN CITED AS THE MOST SENSITIVE BIOLOGICAL INSTRUMENTS FOR MEASURING THE DEPOSITION OF HEAVY METALS. ADDITIONAL WORK HAS SHOWN DEAD ORGANIC MATTER, LITTER, AND HUMUS THAT WAS DERIVED FROM MOSSES AND LICHENS TO HAVE A LARGE CAPACITY FOR CAPTURING HEAVY METALS THROUGH PASSIVE ION EXCHANGE. THE GREAT CAPACITY OF BRYOPHYTES AND TO A LESSER EXTENT, LICHENS, TO CAPTURE HEAVY METALS THROUGH ION EXCHANGE, IS ATTRIBUTABLE TO THE GREAT STABILITY OF THE CHEMICAL COMPLEXES FORMED BETWEEN HEAVY METAL IONS AND NEGATIVELY CHARGED ORGANIC GROUPS. ACCUMULATION IN MOSSES, LICHENS, AND IN LITTER AND HUMUS LAYERS OF NATURAL AND SEMI-NATURAL ECOSYSTEMS WAS FOUND TO BE GOVERNED BY ION EXCHANGE EQUILIBRIA DEPENDENT ON THE ABSORPTION COMPLEX BETWEEN IONS FROM PRECIPITATION, FOLIAGE LEACHING, AND THE DRY DEPOSITION PRESENT IN THE WATER PASSING THROUGH THESE COMPONENTS. INCREASING DEPOSITION OF HEAVY METALS DUE TO HUMAN ACTIVITY ADVERSELY AFFECTS BIODECOMPOSITION RATES, AFFECTING A DECREASE IN PRODUCTIVITY. (BYRD-BATTELLE)

FIELD 05C, 05B

ACCESSION NO. W72-14303

RELATIONSHIPS OF CHLOROPHYLL MAXIMA TO DENSITY STRUCTURE IN THE ATLANTIC OCEAN AND GULF OF MEXICO,

VICTORIA UNIV. (BRITISH COLUMBIA).

L. A. HOBSON, AND C. J. LORENZEN.

DEEP-SEA RESEARCH, VOL. 19, NO. 4, P 297-306, APRIL 1972. 7 FIG, 2 TAB, 19 REF.

DESCRIPTORS:
*CHLOROPHYLL, *SPATIAL DISTRIBUTION, *DENSITY, *DEPTH, MARINE ANIMALS, MARINE ALGAE, CYANOPHYTA, PYRROPHYTA, LARVAE, SAMPLING, INVERTEBRATES, *ATLANTIC OCEAN, *GULF OF MEXICO, PHYTOPLANKTON, CYTOLOGICAL STUDIES, SYSTEMATICS, BIOMASS, FLUOROMETRY, MICROSCOPY, ZOOPLANKTON, NUTRIENTS, PHOTOSYNTHESIS, STANDING CROP, ECOLOGICAL DISTRIBUTION, DINOFLAGELLATES, EQUIPMENT, LIGHT PENETRATION, NITROGEN, SALINITY, AMMONIA, NITRATES, NITRITES, COPEPODS, SEDIMENTATION, CARBON, PROTOZOA, RADIATION, ENZYMES, AQUATIC LIFE.

IDENTIFIERS:
*PYCNOCLINE, CHLOROPHYLL A, PHEOPHYTIN, TUNICATES, FORAMINIFERA, RADIOLARIA, TRICHODESMIUM, FLAGELLATES, CILIATES, NAUPLII, COPEPODIDS.

ABSTRACT:
SPATIAL DISTRIBUTION OF CHLOROPHYLL MAXIMA IN RELATION TO THE DEPTHS OF PYCNOCLINES WERE STUDIED IN THE ATLANTIC OCEAN AND IN THE GULF OF MEXICO. THE DISTRIBUTION OF DISSOLVED INORGANIC NITROGEN AND BIOMASS OF MICROORGANISMS AND THEIR TAXONOMIC COMPOSITION IN CHLOROPHYLL MAXIMA AND SURROUNDING WATER OF THE GULF OF MEXICO ALSO WERE EXAMINED. WATER SAMPLES WERE TAKEN FROM STATIONS ALONG THE SOUTHWEST COAST OF AFRICA, THE NORTHEASTERN U. S. COAST, AND THE GULF OF MEXICO AT DEPTHS TO 200 M USING NISKIN 5-LITER NON-METALLIC SAMPLING BOTTLES EQUIPPED WITH REVERSING THERMOMETERS. SALINITY MEASUREMENTS WERE TAKEN WITH A HYTECH MODEL 6210 LABORATORY SALINOMETER. THE START OF THE PYCNOCLINE WAS DETERMINED BY THE DEPTH AT WHICH THE STABILITY OF THE WATER COLUMN EXCEEDED .0001/DZ. STANDARD METHODS WERE USED TO DETERMINE AMMONIA, NITRATE, AND NITRITE AND A FLUORCMETRIC TECHNIQUE WAS USED TO MEASURE CHLOROPHYLL A. MICROORGANISMS WERE COUNTED BY MICROSCOPIC TECHNIQUES AND CELL VOLUMES WERE CALCULATED USING FORMULAS OF GEOMETRICAL SOLIDS THAT APPROXIMATED CELL SHAPES. CONVERSION FACTORS REPORTED EARLIER WERE USED TO CONVERT THESE VOLUMES TO CARBON. THE SPATIAL DISTRIBUTIONS OF CHLOROPHYLL MAXIMA WERE PATCHY AND THE MAXIMUM DEPTHS TO WHICH THEY FOLLOW PYCNOCLINES WERE VARIABLE. THIS VARIABILITY APPARENTLY RELATES TO LIGHT ADAPTATION BY PHYTOPLANKTON CELLS WHICH MAY BE A FUNCTION OF THE TAXONOMIC COMPOSITION OF THE PHYTOPLANKTON CROP. (MORTLAND-BATTELLE)

FIELD 05C, 05B

ACCESSION NO. W72-14313

ON PRODUCTIVITY ESTIMATIONS OF RIVER WATER SAMPLES BY MEANS OF ALGAE TEST PROCEDURES,

JENA UNIV. (EAST GERMANY). BIOLOGY SECTION.

W. BRAUNE.

INTERNATIONALE REVUE DER GESAMTEN HYDROBIOLOGIE, VOL. 56, NO. 5, P 795-810, 1971. 4 FIG, 2 TAB, 30 REF.

DESCRIPTORS:
*PRIMARY PRODUCTIVITY, BACTERIA, SCENEDESMUS, GROWTH RATES, CULTURES, AQUATIC PRODUCTIVITY, METABOLISM, ELECTRODES, ELECTROCHEMISTRY, PHOTOSYNTHESIS, RIVERS, PLANT PHYSIOLOGY, TESTING PROCEDURES, AQUATIC ALGAE, CHLOROPHYTA, LABORATORY TESTS, OXYGEN, WATER POLLUTION EFFECTS.

IDENTIFIERS:
SAALE RIVER, GERMANY, SCENEDESMUS OBLIQUUS.

ABSTRACT:
THE POSSIBLE INFLUENCE OF BACTERIA ON THE RESULTS OF LONG-TERM ALGAE TEST PROCEDURES (CULTIVATING OF ALGAE IN ISOLATED UNSTERILE WATER SAMPLES) WAS INVESTIGATED AND DIFFERENCES IN THE GROWTH RATES BETWEEN NONAXENIC AND AXENIC SCENEDESMUS CULTURES WERE OBSERVED, SHOWING DIFFERENCES DEPENDENT ON EXPERIMENTAL CONDITIONS. THE NECESSITY OF A LONG EXPERIMENTAL TIME IN SUCH TEST METHODS PROVED TO BE DISADVANTAGEOUS BECAUSE THERE WAS ALWAYS THE POSSIBILITY THAT IN THE UNSTERILE WATER SAMPLE SECONDARY ALTERATIONS (CAUSED BY BACTERIA) MAY OCCUR, THE ACTION OF WHICH IS HARD TO ESTIMATE UNDER TEST CONDITIONS. TO DIMINISH THESE DISADVANTAGES A PHYSIOLOGICAL TEST PROCEDURE WAS TRIED OUT. COMPARED TO METHODS USED TILL NOW THE NEW METHOD ALLOWED SHORT-TERM ESTIMATIONS (IN A FEW HOURS) OF THE PRODUCTIVITY OF ALGAE (UNIALGAL AND MIXED MATERIAL) IN WATER SAMPLES BY CONTINUOUSLY RECORDING THE O2-METABOLISM (ELECTROCHEMICAL O2-DETERMINATION BY MEANS OF MEMBRANE COVERED PT-ELECTRODES). IN WATER SAMPLES TAKEN FROM A POINT BELOW THE TOWN OF JENA IN THE RIVER SAALE THERE WERE REGULARLY HIGHER RATES OF NET-PHOTOSYNTHESIS THAN IN SAMPLES TAKEN FROM ABOVE THE TOWN (TEST ORGANISM: SCENEDESMUS OBLIQUUS). THE RESULTS AGREED WITH FINDINGS OBTAINED BY OTHER IN VITRO-METHODS AND ARE IN CONTRADICTION TO TESTING RESULTS IN SITU. (IN GERMAN) (MORTLAND-BATTELLE)

FIELD 05C, 05A

ACCESSION NO. W72-14314

PRODUCTIVITY OF THE BENTHIC MICROFLORA OF SHOAL ESTUARINE ENVIRONMENTS IN SOUTHERN NEW ENGLAND,

RHODE ISLAND UNIV., KINGSTON. GRADUATE SCHOOL OF OCEANOGRAPHY.

N. MARSHALL, C. A. OVIATT, AND D. M. SKAUEN.

INTERNATIONALE REVUE DER GESAMTEN HYDROBIOLOGIE, VOL. 56, NO. 6, P 947-956, 1971. 4 FIG, 1 TAB, 13 REF.

DESCRIPTORS:
*ESTUARINE ENVIRONMENT, *PRIMARY PRODUCTIVITY, *SHOALS, *BENTHIC FLORA, SEA WATER, SEDIMENTS, LIGHT INTENSITY, DIATOMS, CARBON RADIOISOTOPES, NEW ENGLAND, AQUATIC LIFE, AQUATIC MICROORGANISMS, MARINE ALGAE, SALINITY, MIGRATION, TIDES, LOW WATER MARK, CHRYSOPHYTA.

IDENTIFIERS:
*C-14 PRODUCTIVITY, LIQUID SCINTILLATION, SAMPLE PREPARATION, ZOSTREA, ULVA.

ABSTRACT:
CARBON-14 PRODUCTIVITY BY BENTHIC MICROFLORA FROM SHOAL ESTUARINE ENVIRONMENTS IN SOUTHERN NEW ENGLAND WAS MEASURED BY A C-14 UPTAKE METHOD. REPRESENTATIVE AND INTACT SEDIMENT SAMPLES WERE COLLECTED AND THE C-14 COUNTED BY LIQUID SCINTILLATION AFTER BEING PASSED THROUGH A PULVERIZED FILTER AND SUSPENDED IN CAB-O-SIL. PRODUCTIVITY WAS MEASURED BY CALCULATING THE DIFFERENCE IN C-14 UPTAKE MEASURED BETWEEN LIGHT AND DARK FLASKS. SEDIMENT VARIABILITY CAUSED EXTREMELY HIGH DEVIATIONS IN PRODUCTIVITY VALUES, PROBABLY AS A RESULT OF IRREGULAR MICROFLORA DISTRIBUTION. RESULTS OBTAINED WITH THE CAB-O-SIL SUSPENSION PROCESSING TECHNIQUE WHEN COMPARED WITH THE WET OXIDATION METHOD PROVED TO BE ACHIEVED MORE EASILY, DIRECTLY, AND CONSISTENTLY; HOWEVER, SIMILAR RESULTS WERE OBTAINED FROM BOTH METHODS. THE AVERAGE YEARLY C-14 PRODUCTIVITY OF ALL SAMPLED ESTUARINE SHOALS EQUALLED 20.1 MG/SQ M/HR WITH HIGH PEAKS OF PRODUCTIVITY OCCURRING IN THE WARMER MONTHS. THE LIMITED TESTS FOR LIGHT EFFECTS SEEMED TO INDICATE THAT THE STRONGEST MID-SUMMER LIGHT PEAKS ARE IN EXCESS OF THE OPTIMUM AND THAT AN UPWARD MIGRATION OF DIATOMS AT LOW TIDE MAY RESULT IN INCREASED PRODUCTIVITY. (BYRD-BATTELLE)

FIELD 05C, 05A, 02L

ACCESSION NO. W72-14317

APPARATUS FOR CONTINUOUS MEASUREMENT OF ACTIVE UPTAKE OF RADIOACTIVE SUBSTANCES,

REGENSBURG UNIV. (WEST GERMANY).

E. LOOS,

ANALYTICAL BIOCHEMISTRY, VOL 47, NO 1, P 90-101, MAY 1972. 6 FIG, 1 TAB, 8 REF.

DESCRIPTORS:
*LABORATORY EQUIPMENT, *RADIOISOTOPES, *MEASUREMENT, *KINETICS, *ABSORPTION, INSTRUMENTATION, RADIOACTIVITY TECHNIQUES, CARBON RADIOISOTOPES, PRODUCTIVITY, METABOLISM, FUNGI, ORGANIC COMPOUNDS, CHLORELLA, TRACERS, AQUATIC ALGAE.

IDENTIFIERS:
C-14, *CHLORELLA VULGARIS, NEUROSPORA CRASSA, SCINTILLATION COUNTING, GLUCOSE, MEMBRANE FILTERS, MONOSACCHARIDES, MEMBRANE FILTRATION.

ABSTRACT:
AN APPARATUS CONSISTING OF A MODIFIED SCINTILLATION CHAMBER, SAMPLE CELL, AND AN END-ON PHOTOMULTIPLIER HAS BEEN USED TO STUDY THE UPTAKE KINETICS OF RADIOACTIVE COMPOUNDS (C-14-LABELLED 3-O-METHYL-D-GLUCOSE) BY CHLORELLA VULGARIS AND CONIDIA OF NEUROSPORA CRASSA. THE APPARATUS IS CONSTRUCTED SO THAT THE CELLS ARE HELD IN A THIN LAYER CLOSE TO THE SCINTILLATION CRYSTAL WITH THE NUTRIENT MEDIUM DIFFUSING FROM ABOVE THROUGH A MEMBRANE FILTER. THE MEDIUM IS CONSTANTLY STIRRED TO MINIMIZE CONCENTRATION GRADIENTS DEVELOPED DURING UPTAKE. UPTAKE RATES ARE SEEN AS AN INCREASE OF MEASURED RADIOACTIVITY IN THE CELLS. FOR CHLORELLA A K SUB M OF 0.6 MM WAS FOUND FOR THE UPTAKE. PARALLEL EXPERIMENTS WITH THE MEMBRANE FILTER TECHNIQUE SHOW GOOD AGREEMENT AND CAN BE USED TO CALIBRATE THE APPARATUS. A DIAGRAM OF THE APPARATUS ACCOMPANIES THE DESCRIPTION. (MACKAN-BATTELLE)

FIELD 05A, 05C

ACCESSION NO. W72-14326

ISOLATION AND CHARACTERIZATION OF ULTRAVIOLET LIGHT-SENSITIVE MUTANTS OF THE BLUE-GREEN ALGA ANACYSTIS NIDULANS,

NATIONAL AERONAUTICS AND SPACE ADMINISTRATION, MOFFETT FIELD, CALIF. AMES RESEARCH CENTER.

Y. ASATO.

JOURNAL OF BACTERIOLOGY, VOL 110, NO 3, P 1058-1064, JUNE 1972. 8 FIG, 20 REF.

DESCRIPTORS:
*ISOLATION, *ULTRAVIOLET RADIATION, SEPARATION TECHNIQUES, CYANOPHYTA, CULTURES, ANTIBIOTICS(PESTICIDES), PHOTOACTIVATION, AQUATIC ALGAE, ENVIRONMENTAL EFFECTS.

IDENTIFIERS:
*WASTE CHARACTERIZATION, *ANACYSTIS NIDULANS, *MUTANTS, *SENSITIVITY, CHLORAMPHENICOL, RECOVERY, SURVIVAL.

ABSTRACT:
THREE UV-SENSITIVE MUTANTS OF THE BLUE-GREEN ALGA, ANACYSTIS NIDULANS WERE ISOLATED AND CHARACTERIZED AS TO THEIR ABILITY TO REPAIR UV DAMAGE (PHOTOREACTIVATION). UV LIGHT-SENSITIVE MUTANTS, PREPARED BY SUBJECTING CELLS TO UV RADIATION AND REPLATING THE SURVIVORS, WERE ISOLATED FOR STUDY BY TREATING EXPONENTIAL CULTURES WITH N-METHYL-N'-NITRO-N-NITROSOGUANIDINE FOLLOWED BY CENTRIFUGATION. THE PHOTOREACTIVE ABILITY OF VARIOUS A. NIDULANS MUTANTS WAS DETERMINED UNDER A VARIETY OF CONDITIONS WHICH INCLUDED EXPOSING CELLS TO CHLORAMPHENICOL, CAFFEINE, BLACK AND WHITE FLUORESCENT LIGHTS, AND RED LIGHTS. STRAIN UVS-1 WAS MOST SENSITIVE TO UV IN THE ABSENCE OF PHOTOREACTIVATION. PRETREATMENT WITH CAFFEINE SUPPRESSED THE DARK-SURVIVAL CURVE OF STRAIN UVS-1, INDICATING THE PRESENCE OF EXCISION ENZYMES INVOLVED IN DARK REPAIR. UNDER 'BLACK' AND 'WHITE' ILLUMINATION, STRAIN UVS-1 DISPLAYED PHOTOREACTIVATION PROPERTIES NEARLY COMPARABLE TO WILD-TYPE CULTURE. MUTANTS UVS-35 AND UVS-88 APPEARED TO HAVE PARTIAL PHOTORECOVERY CAPACITIES. UPON PRETREATMENT WITH CHLORAMPHENICOL, PHOTOREACTIVATION PROPERTIES OF STRAINS UVS-1 AND UVS-88 WERE NOT EVIDENT ALTHOUGH THE PARTIAL PHOTOREACTIVATION CHARACTERISTICS OF STRAIN UVS-35 REMAINED THE SAME. DATA INDICATE THAT STRAINS UVS-1, UVS-35, AND UVS-88 ARE PROBABLY GENETICALLY DISTINCT UV-SENSITIVE MUTANTS. (LONG-BATTELLE)

FIELD 05A, 05C

ACCESSION NO. W72-14327

WATER SUPPLY AND WASTE DISPOSAL SERIES: VOLUME I. TERMINCLOGY STANDARDIZATION AND MICROBIOLOGY.

FEDERAL HIGHWAY ADMINISTRATION, WASHINGTON, DC, OFFICE OF RESEARCH; AND FEDERAL HIGHWAY ADMINISTRATION, WASHINGTON, D.C. OFFICE OF DEVELOPMENT.

DEPT OF TRANSPORTATION-FEDERAL HIGHWAY ADMINISTRATION STAFF REPORT 72R-106S-1, NOVEMBER 1971. 51 P, 16 FIG, 12 TAB, 141 REF. APPEND.

DESCRIPTORS:
*WATER SUPPLY, *WASTE DISPOSAL, *RECREATION WASTES, *MICROORGANISMS, *WATER POLLUTION SOURCES, REVIEWS, MICROBIAL DEGRADATION, CLASSIFICATION, SEWAGE, SOIL PROPERTIES, GROUNDWATER MOVEMENT, OUTFALL SERVERS, STREAMS, SEPTIC TANKS, FUNGI, ALGAE, WATER POLLUTION CONTROL.

IDENTIFIERS:
*ROADSIDE REST AREAS.

ABSTRACT:
THIS IS THE FIRST VOLUME OF A SERIES OF REPORTS ON WATER SUPPLY AND WASTE DISPOSAL PUBLISHED BY THE FEDERAL HIGHWAY ADMINISTRATION. THIS SERIES EMPHASIZES WATER SUPPLIES, WATER-CARRIAGE WASTE TREATMENT, AND SOLID WASTE DISPOSAL, PARTICULARLY AS THEY RELATE TO ROADSIDE REST AREAS. BECAUSE DEFINITIONS, CLASSIFICATIONS, AND TERMINOLOGY CONCERNING MICROORGANISMS AND BIOLOGICAL TERMS SOMETIMES LACK UNIFORMITY AMONG THE DISCIPLINES, THIS VOLUME DEFINES AND STANDARDIZES THE TERMINOLOGY THAT WILL BE USED IN ALL VOLUMES OF THE SERIES. INFORMATION ON DISEASES THAT MIGHT BE TRANSMITTED AT REST OR RECREATION AREAS IS PRESENTED. THE CAPABILITY OF INDIVIDUAL SPECIES OF MICROORGANISMS TO SURVIVE IN THE SOIL, ON THE SOIL, IN WATER, AND IN OTHER SITUATIONS IS INDICATED. A BIBLIOGRAPHY CONTAINING 14 REFERENCES IS INCLUDED. (WOODARD-USGS)

FIELD 05B, 05D

ACCESSION NO. W72-14363

PREDICTING QUALITY EFFECTS OF PUMPED STORAGE,

WATER RESOURCES ENGINEERS, INC., WALNUT CREEK, CALIF.

C. W. CHEN, AND G. T. ORLOB.

JOURNAL OF THE POWER DIVISION, PROCEEDINGS OF ASCE, VOL. 98, NO. PO1, P 65-75, JUNE 1972. 7 FIG, 2 TAB, 13 REF.

DESCRIPTORS:
*WATER QUALITY, *HYDROELECTRIC PLANTS, *MATHEMATICAL MODELS, THERMAL STRATIFICATION, TEMPERATURE, DISSOLVED OXYGEN, COMPUTERS, ALGAE, AQUATIC LIFE, ENVIRONMENTAL EFFECTS, ENVIRONMENTAL ENGINEERING, *VIRGINIA, PROJECTIONS, MODEL STUDIES.

IDENTIFIERS:
*PUMPED STORAGE OPERATION, SMITH MOUNTAIN RESERVOIR(VA).

ABSTRACT:
PUMPED STORAGE OPERATION (PSO) AFFECTS THE WATER QUALITY AND IN TURN THE RESIDENT BIOTA. A PREDICTIVE TECHNIQUE WAS ADAPTED FROM A RECENTLY DEVELOPED ECOLOGIC MODEL TO CALCULATE THE ENVIRONMENTAL CHANGES AND TO EVALUATE THE ALTERNATIVES OF PSO THAT MIGHT ENHANCE A NORMAL ECOLOGICAL SUCCESSION, FORESTALLING ADVERSE ENVIRONMENTAL PROBLEMS. THE MODEL, IN AN APPLICATION TO SMITH MOUNTAIN RESERVOIR, VIRGINIA, PREDICTED ACCURATELY FOR AN ENTIRE YEAR THE DAILY TEMPERATURE, DISSOLVED OXYGEN AND OTHER QUALITY CONSTITUENTS OF THE WATER AT VARYING DEPTHS. PSO, UNDER SIMULATED CONDITIONS, WAS FOUND TO IMPROVE WATER QUALITY BY DEEPENING THE EPILIMNION. A PROLONGED PUMPING SCHEDULE AT A REDUCED RATE, HOWEVER, WAS A BETTER ALTERNATIVE, BECAUSE IT WOULD PROVIDE A MORE STABLE ECOSYSTEM WITHOUT SUBSTANTIALLY ALTERING THE THERMAL STRATIFICATION. THERMAL STRATIFICATION IS NECESSARY TO REGULATE ENERGY LOSS, AND THUS CONTROL EVAPORATION, AND TO ENSURE A VARIED ENVIRONMENT SUITABLE FOR HOSTING A DIVERSE AND STABLE BIOTA. (EAGLE-VANDERBILT)

FIELD 05C, 05G, 04A

ACCESSION NO. W72-14405

APPARATUS AND PROCESS FOR TREATING SEWAGE,

P. J. GRESHAM.

U. S. PATENT NO. 3,565,797, 7 P, 6 FIG, 6 REF; OFFICIAL GAZETTE OF THE UNITED STATES PATENT OFFICE, VOL. 883, NO. 4, P 1617, FEBRUARY 23, 1971.

DESCRIPTORS:
*PATENTS, *WASTE WATER TREATMENT, *FILTRATION, *SEWAGE TREATMENT, ALGAE, *AERATION, SEDIMENTATION, *CHLORINATION, OXIDATION, POLLUTION ABATEMENT, WATER POLLUTION TREATMENT, WATER QUALITY, STREAMS, LAKES, BIOCHEMICAL OXYGEN DEMAND, *OXYGENATION, *ANAEROBIC BACTERIA, BIOLOGICAL TREATMENT.

ABSTRACT:
SEWAGE IS FED INTO CONTACT WITH A BED OF LIVING FILAMENTOUS MACROSCOPIC ALGAE. THE SEWAGE IS OXYGENATED AND THE GROWTH OF ANAEROBIC BACTERIA IS INHIBITED. THE SEWAGE IS FILTERED THROUGH THE ALGAE BED OR SCREEN. SOLIDS ARE COLLECTED ON THE SCREENS AS SLUDGE FOR SEPARATE DISPOSAL AND THE SEWAGE LIQUID UNDERGOES CHLORINATION BEFORE DISPOSAL. THE SLUDGE IS RECYCLED FOR REMOVAL OF REMAINING LIQUIDS. (SINHA-OEIS)

FIELD 05D

ACCESSION NO. W72-14469

CHEMICAL STUDIES ON TOXINS FROM GYMNODINIUM BREVE AND APHANIZOMENON FLOS-AQUAE,

NEW HAMPSHIRE UNIV., DURHAM. DEPT. OF BIOCHEMISTRY.

M. ALAM.

PHD THESIS, NOVEMBER 1972. 95 P, 25 FIG, 12 TAB, 30 REF. OWRR A-013-NH(5) AND A-021-NH(2).

DESCRIPTORS:
*ALGAL TOXINS, *RED TIDE, *BEGGIATOA, *PHYTOTOXICITY, *CYANOPHYTA, *NEW HAMPSHIRE, MARINE ALGAE, PHYTOPLANKTON, AQUATIC PLANTS, CHEMICAL PROPERTIES, WATER QUALITY.

IDENTIFIERS:
*KEZAR LAKE(NH).

ABSTRACT:
PART I: THIS INVESTIGATION CONCERNS THE ISOLATION AND CHEMICAL CHARACTERIZATION OF THE TOXIN (S) FROM GYMNODINIUM BREVE WHICH IS LETHAL TO MARINE ORGANISMS. UNIALGAL CULTURES OF G. BREVE WERE GROWN AND THE TOXIC MATERIAL ISOLATED. THE PARTIALLY PURIFIED TOXIN ON COLUMN CHROMATOGRAPHY GAVE A TOXIC PALE YELLOW GLASSY RESIDUE. THE INFRA-RED SPECTRUM INDICATED PRESENCE OF A CARBONYL GROUP AND ABSENCE OF HYDROXYL AND AROMATIC GROUPS. ULTRA-VIOLET SPECTRUM INDICATED TOXIN POSSIBLY CONTAMINATED WITH VERY SMALL AMT OF PIGMENT. MOLECULAR WEIGHT DETERMINED BY MASS SPECTROMETRY AND OSMOMETRY. ON BASIS OF PRESENT INFORMATION IT IS DIFFICULT TO ASSIGN A DEFINITE FORMULA TO THE TOXIN FROM G. BREVE. PART II: THE PURPOSE OF THIS INVESTIGATION WAS TO ISOLATE AND CHARACTERIZE THE TOXIC COMPOUND(S) FROM NATURAL BLOOMS OF A.FLOS-AQUAE. THE TOXIN WAS PURIFIED AND OBTAINED AS A PALE YELLOW POWDER IN A CHROMATOGRAPHICALLY PURE FORM. THE A.FLOS-AQUAE TOXIN WAS A VERY BASIC SUBSTANCE. IT DIFFERED FROM SAXITOXIN IN ITS RF VALUES AND ITS BEHAVIOR TO SEVERAL SPRAY REAGENTS. PARTIALLY HYDRATED FORM HAD DIFFERENT INFRA-RED SPECTRUM THAN DOES SAXITIXIN. ON BASIS OF THIS AND OTHER INFORMATION IT IS CONCLUDED THAT A.FLOS-AQUAE IS DIFFERENT FROM SAXITOXIN.

FIELD 05C, 05A

ACCESSION NO. W72-14542

STUDIES ON THE EFFECTS OF A STEAM ELECTRIC GENERATING PLANT ON THE MARINE ENVIRONMENT AT NORTHPORT, NEW YORK,

STATE UNIV. OF NEW YORK, STONY BROOK. MARINE SCIENCES RESEARCH CENTER.

G. C. WILLIAMS, J. B. MITTON, T. H. SUCHANEK, JR., N. GEBELEIN, AND C. GROSSMAN.

TECHNICAL REPORT SERIES NO 9, NOVEMBER 1971. 123 P, 44 FIG, 9 TAB, 59 REF.

DESCRIPTORS:
*NEW YORK, *POWER PLANTS, *STREAM, *THERMAL POLLUTION, *ENVIRONMENTAL EFFECTS, *ECOLOGY, *PLANKTON, *ALGAE, BIOLOGICAL COMMUNITIES, PRODUCTIVITY, LIFE CYCLES, GROWTH STAGES, DISTRIBUTION, BEHAVIOR, SEASONAL, WATER POLLUTION EFFECTS.

IDENTIFIERS:
*NORTHPORT PLANT(LONG ISLAND).

ABSTRACT:
THE RESULTS OF A CONTINUING GROUP OF INVESTIGATIONS ON EFFECTS OF THE POWER PLANT AT NORTHPORT ARE SUMMARIZED. THE EFFECTS ON THE PLANKTON COMMUNITY ORIGINALLY IN THE WATER USED FOR COOLING OF STEAM CONDENSERS AND DISCHARGED BACK INTO LONG ISLAND SOUND WERE INVESTIGATED BY MEASURING PRODUCTIVITY CHANGES AND BY A DIRECT SEASONAL OBSERVATIONAL STUDY OF THE LIFE CYCLE STAGES, DISTRIBUTION AND BEHAVIOR OF PLANKTON ALGAE BEFORE AND AFTER USE. RESULTS LEFT LITTLE DOUBT THAT WASTE HEAT IS THE PRIMARY ECOLOGICAL DISTURBANCE IN THE DISCHARGE AREA. (ENSIGN-PAI)

FIELD 05C, 05B

ACCESSION NO. W72-14659

DISCOLORATION OF ALGAL BLOOMS AND SEAWEEDS (VEGETATIONSFARBUNGEN UND WASSERBLUTEN),

OESTERREICHISCHE GESELLSCHAFT FUER METEOROLOGIE, VIENNA (AUSTRIA).

F. WAWRIK.

WETTER UND LEBEN, VOL 23, NO 9-10, P 203-210, 1971. 11 REF.

DESCRIPTORS:
WATER POLLUTION EFFECTS, *ORGANIC MATTER, *DECOMPOSING ORGANIC MATTER, *EUTROPHICATION, *PHYTOPLANKTON, *ALGAE, *AQUATIC WEEDS, *WEEDS, *RED TIDE, COASTS, LAKES, PONDS, WATER POLLUTION CONTROL, AIRCRAFT, COPPER SULFATE.

ABSTRACT:
THE COMPLEX PHENOMENA OF ORGANIC WATER POLLUTION, OBSERVED IN POOLS, ALPINE LAKES AND COASTAL AREAS, ARE DISCUSSED. CAUSES AND CONSEQUENCES OF ORGANIC POLLUTION AND CONTROL OF HARMFUL SEAWEEDS ARE INCLUDED. THE FIRST FEW BLOOMS OF THE RED TIDE ARE CONTROLLED BY THE USE OF CUSO4-CRYSTALS DISPERSED FROM AIRPLANES. (ENSIGN-PAI)

FIELD 05C, 05B, 05G

ACCESSION NO. W72-14673

ENVIRONMENTAL LIMITS OF PLANTS IN FLOWING WATERS,

DURHAM UNIV. (ENGLAND). DEPT. OF BOTANY.

B. A. WHITTON.

SYMPOSIUM OF THE ZOOLOGICAL SOCIETY OF LONDON, NO 29, P 3-19, 1972. 60 REF.

DESCRIPTORS:
*RUNNING WATERS, *AQUATIC PLANTS, *AQUATIC ENVIRONMENT, WATER POLLUTION EFFECTS, STREAMS, RIVERS, ROOTED AQUATIC PLANTS, AQUATIC ALGAE, WATER TEMPERATURE, ACID STREAMS, HYDROGEN ION CONCENTRATION, HEAVY METALS, DISSOLVED OXYGEN, *PLANT GROWTH, LIGHT INTENSITY, TOXICITY, PRIMARY PRODUCTIVITY, STANDING CROPS, *PHOTOSYNTHESIS.

IDENTIFIERS:
DIVERSITY INDICES.

ABSTRACT:
A REVIEW IS PRESENTED SUMMARIZING THE LITERATURE AND SOME UNPUBLISHED DATA ON ATTACHED PHOTOSYNTHETIC PLANTS IN CERTAIN EXTREME FLOWING WATER ENVIRONMENTS. WITH UPPER TEMPERATURE LIMITS, IT IS POSSIBLE TO COMPARE SITUATIONS OCCURRING NATURALLY AND THOSE ARISING AS A RESULT OF POLLUTION. SUCH COMPARISONS SHOW UP OBVIOUS ANOMALIES. THE LIMITS OF PLANTS UNDER LOW PH CONDITIONS AND HIGH LEVELS OF HEAVY METALS ARE ALSO REVIEWED, ALTHOUGH WITH THESE EXTREMES THE DATA ARE INSUFFICIENT FOR COMPARISON OF NATURAL AND POLLUTED CONDITIONS. SECTIONS OF A MORE SPECULATIVE NATURE ARE ADDED ON LIGHT, CLIMATE, AND DISSOLVED OXYGEN AS FACTORS WHICH AT HIGH AND LOW LEVELS MAY LIMIT THE GROWTH OF PLANTS. THE QUANTITATIVE METHODS WHICH HAVE BEEN USED FOR COMPARING THE EFFECTS OF EXTREME ENVIRONMENTS ON PLANTS FALL INTO FOUR MAIN CATEGORIES: USE OF TOXICITY TESTS; USE OF DIVERSITY INDICES; MEASUREMENT OF PRIMARY PRODUCTION; MEASUREMENT OF STANDING CROPS. THESE METHODS DO NOT NECESSARILY GIVE SIMILAR RESULTS. AN IMPROVEMENT IN METHODS OF QUANTIFYING THE EXTENT TO WHICH AN ENVIRONMENT IS EXTREME IS IMPORTANT IF THE SIGNIFICANCE OF ENVIRONMENTAL LIMITS OF PLANTS IN NATURAL AND POLLUTED WATERS IS TO BE EVALUATED. (SVENSSON-WASHINGTON)

FIELD 05C, 10F

ACCESSION NO. W72-14690

FILAMENTOUS ALGAE AS WEEDS,

DURHAM UNIV. (ENGLAND). DEPT. OF BOTANY.

B. A. WHITTON.

IN: 3RD SYMPOSIUM ON AQUATIC WEEDS, EUROPEAN WEED RESEARCH COUNCIL, OXFORD, 1971, P 249-263. 40 REF.

DESCRIPTORS:
*EUTROPHICATION, *AQUATIC ALGAE, *ALGAL CONTROL, *AQUATIC WEED CONTROL, BIOCHEMICAL OXYGEN DEMAND, LAKE STAGES, LIMNOLOGY, NUISANCE ALGAE, NUTRIENT REMOVAL, NUTRIENTS, WATER POLLUTION EFFECTS, WATER QUALITY, HERBICIDES, ODOR-PRODUCING ALGAE.

IDENTIFIERS:
*FILAMENTOUS ALGAE, *AQUATIC ANGIOSPERM WEEDS, CLADOPHORA.

ABSTRACT:
IT STILL REMAINS DIFFICULT TO EVALUATE THE NUISANCE VALUE OF FILAMENTOUS ALGAE ON A WORLD SCALE, AND IT WOULD AT PRESENT BE IMPOSSIBLE TO ESTIMATE EVEN ROUGHLY THEIR ECONOMIC SIGNIFICANCE, EITHER AS AN ABSOLUTE VALUE OR IN COMPARISON WITH THAT OF AQUATIC ANGIOSPERM WEEDS. IT DOES, HOWEVER, SEEM CLEAR THESE ALGAE ARE A SPECIAL NUISANCE IN AT LEAST TWO RATHER DIFFERENT TYPES OF SITUATION. THEY ARE A PROBLEM AT THE EDGES OF SEVERAL LARGE LAKES, LIKE ONTARIO AND ZURICH, WHICH HAVE A HIGH AMENITY VALUE, AND WHICH HAVE BEEN SUBJECT TO CONSIDERABLE EUTROPHICATION IN RECENT YEARS. THEY ARE ALSO A PARTICULAR PROBLEM IN WARM DRY CLIMATES IN CHANNELS USED FOR THE TRANSPORT OF WATER FOR IRRIGATION AND OTHER PURPOSES. THEY MAY ALSO BE A PROBLEM IN MANY OTHER TYPES OF SITUATION, AS FOR INSTANCE THE DRAINAGE DITCHES OF FLAT, LOW-LYING AREAS LIKE PARTS OF THE NETHERLANDS, BUT THEY THEN OFTEN OCCUR TOGETHER WITH AQUATIC ANGIOSPERM WEEDS, AND BOTH THESE LATTER AND THE FILAMENTOUS ALGAE CAN NOW BE TREATED SIMULTANEOUSLY BY CHEMICAL MEANS. (SVENSSON-WASHINGTON)

FIELD 05C, 04A

ACCESSION NO. W72-14692

TOXICITY OF HEAVY METALS TO FRESHWATER ALGAE: A REVIEW,

DURHAM UNIV. (ENGLAND). DEPT. OF BOTANY.

B. A. WHITTON.

PHYKOS, VOL 9, NO 2, P 116-125, 1970. 38 REF.

DESCRIPTORS:
*REVIEWS, *HEAVY METALS, *ALGAE, *AQUATIC ENVIRONMENT, *TOXICITY, *WATER POLLUTION EFFECTS, ZINC, LEAD, COPPER, CHROMIUM, SILVER, CHLOROPHYTA, CYANOPHYTA, ALGAL CONTROL, ALGICIDES, COPPER SULFATE, SALINITY, HARDNESS(WATER), CHELATION, MINE DRAINAGE, WATER PROPERTIES.

IDENTIFIERS:
*FRESHWATER ALGAE, EDTA.

ABSTRACT:
A REVIEW IS PRESENTED OF THE LITERATURE ON TOXICITY OF HEAVY METALS TO FRESHWATER ALGAE. METALS CONSIDERED INCLUDE ZINC, LEAD, COPPER, CHROMIUM, AND SILVER. THE ACCOUNT IS CONCERNED MAINLY WITH FIELD OBSERVATIONS AND THOSE LABORATORY STUDIES WHICH ARE PARTICULARLY HELPFUL IN INTERPRETING FIELD DATA. EFFECTS OF SALINITY, WATER HARDNESS, AND ORGANIC COMPLEXATION ON TOXICITY ARE DISCUSSED. (SVENSSON-WASHINGTON)

FIELD 05C, 10F

ACCESSION NO. W72-14694

THE EFFECT OF CCC AND DIETHYLAMINE HYDROCHLORIDE ON CERTAIN SPECIES OF ALGAE BELONGING TO CYANOPHYCEAE, CHLOROPHYCEAE, AND DIATOMEAE,

BIALYSTOK MEDICAL ACADEMY (POLAND). DEPT. OF BIOLOGY.

R. CZERPACK.

ACTA HYDROBIOLOGIA, VOL 12, NO 2-3, P 143-151, 1970. 5 FIG, 33 REF.

DESCRIPTORS:
*INHIBITION, *PLANT GROWTH REGULATORS, *CHLORIDES, *CYANOPHYTA, *CHLOROPHYTA, *CHRYSOPHYTA, TOXICITY, INHIBITORS, TOXINS, INORGANIC COMPOUNDS, ALGAE, DIATOMS, CHLAMYDOMONAS, SCENEDESMUS, BIO-ASSAY, RESISTANCE, LETHAL LIMIT, POISONS, WATER POLLUTION EFFECTS.

IDENTIFIERS:
*CHLOROCHOLINE CHLORIDE, *CCC, *DIETHYLAMINE HYDROCHLORIDE, MERISOMOPEDIA SPP., ANABAENA SPP., GOMPHONEMA SPP., NITZCHIA SPP., SENSITIVITY.

ABSTRACT:
THE EFFECTS OF CHLOROCHOLINE CHLORIDE (CCC) AND DITETHYLAMINE HYDROCHLORIDE ON THE CONCENTRATION OF CELLS, THEIR SIZE, TOTAL CHLOROPHYLL CONTENT AND THE DYNAMICS OF THE GROWTH OF NINE SPECIES OF ALGAE BELONGING TO CYANOPHYCEAE (MERISMOPEDIA GLAUCA, ANABAENA CYLINDRICA), CHLOROPHYCEAE (CHLAMYDOMONAS NIVALIS, SCENEDESMUS QUADRICAUDA, S. ACUMINATUS, S. BASILIENSIS, S. BIJUGATUS), AND DIATOMEAE (GOMPHONEMA PARVULUM, NITZCHIA PALEA) WERE INVESTIGATED. DIETHYLAMINE, THE CHEMICAL ANALOG OF CCC, HAD SIMILAR EFFECT--EVEN A SOMEWHAT STRONGER ONE--TO THAT OF CCC AS A TYPICAL GROWTH RETARDANT AT SIMILAR CONCENTRATIONS. BOTH GROWTH REGULATORS HAD AN INHIBITORY EFFECT ON THE CELL CONCENTRATION, THEIR GROWTH DYNAMICS, AND THE TOTAL CHLOROPHYLL CONTENT OF THE ALGAE. THE DIATOMEAE AND FILAMENTOUS CYANOPHYCEAE WERE THE MOST SENSITIVE TO THE REGULATORS INVESTIGATED. ON THE OTHER HAND, A SLIGHT STIMULATORY EFFECT WAS OBSERVED WITH CHLOROPHYCEAE WHEN THE CHEMICALS WERE APPLIED IN CONCENTRATIONS OF 10 TO THE (-5) POWER TO 10 TO THE (-7) POWER M. (LEGORE-WASHINGTON)

FIELD 05C

ACCESSION NO. W72-14706

THE USE OF PHOTOGRAPHY IN WATER QUALITY RESEARCH,

WASHINGTON STATE UNIV., PULLMAN.

W. H. FUNK, AND D. C. FLAHERTY.

PHOTOGRAPHIC APPLICATIONS IN SCIENCE, TECHNOLOGY AND MEDICINE, VOL 7, NO 5, P 20-22, SEPTEMBER 1972. 7 FIG, 1 REF.

DESCRIPTORS:
SURVEYS, *AERIAL PHOTOGRAPHY, WATER POLLUTION EFFECTS, EUTROPHICATION, *LAKES, ALGAE, INFRARED RADIATION, *REMOTE SENSING, *POLLUTANT IDENTIFICATION, AQUATIC WEEDS, *WASHINGTON, *CHLOROPHYLL, WATER TEMPERATURE, ALKALINITY, DISSOLVED OXYGEN, CONDUCTIVITY, HYDROGEN ION CONCENTRATION.

IDENTIFIERS:
*SNAKE RIVER(WASH).

ABSTRACT:
THE FIRST STEP IN AERIAL PHOTOGRAPHIC STUDIES OF LAKES HAS BEEN TO MAKE AN AERIAL PHOTOGRAPHIC SURVEY OF THE PARTICULAR LAKE OR RESERVOIR WHICH HAS BEEN SCHEDULED FOR INVESTIGATION. THIS HELPS TO GIVE THE INVESTIGATORS A COMPREHENSIVE PICTURE OF THE LAKE ENVIRONMENT IN TERMS OF ITS DRAINAGE BASIN, SURFACE OUTLETS, AQUATIC GROWTH, POSSIBLE NUTRIENT SOURCES AND SIZE IN RELATION TO OTHER HYDROGRAPHIC FEATURES OF THE AREA. THIS INITIAL AERIAL SURVEY IS FOLLOWED BY CR CONDUCTED SIMULTANEOUSLY WITH INFRARED PHOTOGRAPHY OF THE LAKE AND ITS ENVIRONS. ONE OF THE CHARACTERISTICS OF THIS MATERIAL IS THAT THE CHLOROPHYL OF LIVING ALGAE AND AQUATIC WEEDS IS CAPTURED ON THE FILM IN VIVID PINKISH RED HUES, WHEREAS DEAD ALGAE OR VEGETATION EXHIBIT A WHITISH OR YELLOWISH ORANGE APPEARANCE. AERIAL RECONNAISSANCE HAS MADE IT POSSIBLE TO SURVEY EIGHT TO TEN LAKES IN A TWO-STATE AREA IN A SINGLE DAY. ON SAMPLING TRIPS, LIGHT PENETRATION, WATER TEMPERATURE, DISSOLVED OXYGEN, PH, ALKALINITY, AND CONDUCTIVITY TESTS ARE MADE AT VARIOUS DEPTHS. TWO-LITER WATER SAMPLES ARE ALSO COLLECTED FOR PLANKTON AND CHEMICAL ANALYSIS. A CURRENT PROJECT INVOLVING WATER QUALITY SURVEILLANCE IS A STUDY OF 150 MILES OF THE LOWER SNAKE RIVER DRAINAGE AREA LOCATED IN SOUTHEASTERN WASHINGTON. AERIAL PHOTOGRAPHY SHOWS THAT THIS STRETCH OF THE RIVER IS SUBJECTED TO EXCESSIVE RUNOFF OF TOPSOIL SHORTLY AFTER THE SNOW MELTS IN THE SPRING, AND ALSO DURING PERIODS OF HIGH RAINFALL. (KNAPP-USGS)

FIELD 05A, 07B, 05C, 02H

ACCESSION NO. W72-14728

THE PATHOLOGY OF LAKE ERIE,

Q. DADISMAN.

THE NATION, VOL. 214, P. 492-496, APRIL 17, 1972. 1 PLATE.

DESCRIPTORS:
*LAKE ERIE, *EUTROPHICATION, *CHEMICAL WASTES, *POLITICAL ASPECTS, LAKES, GREAT LAKES, ALGAL BLOOMS, AGING(BIOLOGICAL), LAKE STAGES, MESOTROPHY, OLIGOTROPHY, WATER POLLUTION SOURCES, WATER POLLUTION EFFECTS, MERCURY, SEWAGE DISPOSAL, PHOSPHATES, PHOSPHORUS, CHEMICAL WASTES, LIMNOLOGY, INTERNATIONAL JOINT COMMISSION, CANADA, FEDERAL GOVERNMENT, STATE GOVERNMENTS.

IDENTIFIERS:
*MERCURY POLLUTION, *ENVIRONMENTAL PROTECTION AGENCY.

ABSTRACT:
ALTHOUGH MOST REPORTS INDICATE THAT LAKE ERIE IS 'DEAD', A RECENT STUDY SHOWS THAT THE LAKE IN NOT DEAD AND COULD BE REVIVED IN ONE GENERATION. ONE SERIOUS PROBLEM IS THE INPUT OF PHOSPHORUS INTO THE LAKE. SCIENTISTS BELIEVE THAT IF THIS INPUT WERE REDUCED A QUICK IMPROVEMENT IN THE CONDITION OF THE LAKE COULD BE EXPECTED. OTHER PROBLEMS INCLUDE THE LARGE AMOUNTS OF MERCURY DISCHARGED BY CHLORALKALI PLANTS NEAR DETROIT AND THE DISCHARGES OF SEWAGE ALL ALONG THE COAST. ONE OF SEVERAL FACTORS PREVENTING IMPROVEMENT OF THE LAKE'S CONDITION IS THE LACK OF AN INSTITUTIONAL MECHANISM FOR COOPERATION BETWEEN THE UNITED STATES AND CANADA. THIS CAN BE TRACED TO THREE CONDITIONS IN AMERICA: INFERIOR AMERICAN TECHNOLOGY IN SEWAGE AND POLLUTION TREATMENT, THE U.S. HESITANCE TO ELIMINATE THE PHOSPHATE CONTENT OF HOUSEHOLD DETERGENTS, AND THE POLITICAL INABILITY OF THE U.S. TO ACT QUICKLY. AS A RESULT, PHOSPHATES AND OTHER POLLUTANTS CONTINUE TO ENTER LAKE ERIE WITH LITTLE HOPE REMAINING FOR THE LAKE. THE U.S. BUREAUCRACY, IN PARTICULAR THE ENVIRONMENTAL PROTECTION AGENCY, FAILS TO RECOGNIZE THAT STRICTER POLLUTION CONTROL STANDARDS ARE NEEDED IN THE GREAT LAKES THAN FOR THE OCEAN. (NIELSEN-FLORIDA)

FIELD 05G, 02H, 06E

ACCESSION NO. W72-14782

CARBON AND NITROGEN AS REGULATORS OF ALGAL GROWTH IN TREATED SEWAGE,

KENTUCKY WATER RESOURCES INST., LEXINGTON.

G. FOREE, AND R. SCROGGIN.

AVAILABLE FROM THE NATIONAL TECHNICAL INFORMATION SERVICE AS PB-212 375, $3.00 IN PAPER COPY, $0.95 IN MICROFICHE. RESEARCH REPORT NO. 49, MARCH 1972. 61 P, 15 FIG, 7 TAB, 32 REF. OWRR A-023-KY (2).

DESCRIPTORS:
*CARBON, *NITROGEN, *ALGAE, PLANT GROWTH, SEWAGE EFFLUENTS, REGULATION, CYANOPHYTA, CHLOROPHYTA, HYDROGEN ION CONCENTRATICN, CHEMICAL PORPERTIES, NITROGEN FIXATION, CULTURES, CARBON DIOXIDE, NUTRIENTS, LIMITING FACTORS, ALLOGENIC SUCCESSION, EUTROPHICATION.

IDENTIFIERS:
*ALGAL GROWTH, CONTINUOUS CULTURES, BATCH CULTURES.

ABSTRACT:
TO PROVIDE INFORMATICN ON ALGAL GROWTH REGULATORS THE FIRST PHASE OF THIS INVESTIGATION WAS A CONTINUCUS FLOW CULTURE STUDY IN WHICH ALGAE WERE GROWN UNDER VARIOUS CONDITIONS. WHEN THE CULTURES REACHED STEADY STATE, THEY WERE ASSAYED TO DETERMINE THE EXTENT AND RATE OF GROWTH, ALGAL GENERA PRESENT, NUTRIENT UTILIZATION, AND ALGAL COMPOSITION. THE SECOND PHASE WAS A BATCH CULTURE STUDY WHICH PROVIDED MEANS FOR COMPARING THE TWO TYPES OF CULTURES (I. E., BATCH AND CONTINUOUS) AND VERIFICATION OF SOME CONCLUSIONS OF THE CONTINUOUS CULTURE STUDY. EVEN IN SEWAGE EFFLUENTS CONTAINING SIGNIFICANT CONCENTRATIONS OF BOTH ORGANIC AND INORGANIC CARBON, ALGAL GROWTH WAS LIMITED BY AVAILABILITY OF CARBON DIOXIDE. THE RESULTS INDICATED THAT IN NATURAL SITUATIONS WHERE EXCESS CARBON DIOXIDE MIGHT BE AVAILABLE, BUT NITROGEN IS DEFICIENT IN SOLUTION, ALGAL SUCCESSION MAY OCCUR WITH SHIFTS TO CERTAIN BLUE-GREEN FORMS CONTAINING ATMOSPHERIC NITROGEN FIXING CAPABILITIES. A PREVIOUSLY DEVELOPED KINETIC THECRY WHICH DESCRIBES ALGAL GROWTH AS BEING PROPORTIONAL TO THE CELLULAR CONCENTRATION OF THE GROWTH REGULATING NUTRIENT WAS CONFIRMED WHEN APPLIED TO THE CARBON DIOXIDE ENRICHED CONDITIONS WITH 10 MG/L AMMONIA NITROGEN IN THE FEED SOLUTION. (JONES-WISCONSIN)

FIELD 05C

ACCESSION NO. W72-14790

GRAZING BY THE CILIATED PROTOZOON LOXODES MAGNUS ON THE ALGA SCENEDEMUS IN A EUTROPHIC POND,

FRESHWATER BIOLOGICAL ASSOCIATION, AMBLESIDE, (ENGLAND).

R. GOULDER.

OIKOS, VOL. 23, NO. 1, P. 109-115, 1972. 4 FIG., 3 TAB., 12 REF.

DESCRIPTORS:
*GRAZING, *PROTOZOA, *ALGAE, SCENEDESMUS, EUTROPHICATICN, PONDS, FEEDING RATES, BENTHOS, STANDING CROPS.

IDENTIFIERS:
*LOXODES MAGNUS.

ABSTRACT:
AN ASSESSMENT WAS MADE OF THE IMPORTANCE OF GRAZING BY LOXODES MAGNUS, ONE OF A GROUP OF BENTHIC CILIATED PROTOZOA FEEDING ON SCENEDESMUS DENTICULATUS, S. QUADRICAUDA, AND S. ACUMINATUS. THE RATE AT WHICH L. MAGNUS DIGESTS SCENEDESMUS WAS ESTIMATED BY MEANS OF A LABORATORY STARVATION PROCEDURE AND IT WAS ASSUMED THAT, UNDER NATURAL CONDITIONS, THE DIGESTION RATE EQUALS THE FEEDING RATE. NO CORRELATION WAS FOUND BETWEEN NUMBERS OF L. MAGNUS (AND ALSO THE CILIATE LOXODES STRIATUS) AND SCENEDESMUS CROPS. SCENEDESMUS CELLS PER SQUARE METER IN THE SEDIMENT DECREASED FROM S. QUADRICAUDA TO S. DENTICULATUS TO S. ACUMINATUS BUT INSIDE L. MAGNUS, NUMBERS OF S. DENTICULATUS WERE HIGHEST, DECREASING TO S. QUADRICAUDA AND FINALLY S. ACUMINATUS; THEREFORE, L. MAGNUS MAY BE ABLE TO DISTINGUISH BETWEEN SPECIES OF SCENEDESMUS IN ITS FEEDING. THE POPULATION FEEDING RATE OF L. MAGNUS REPRESENTED BETWEEN 0.003% AND 0.68% OF THE SCENEDESMUS CROP EATEN PER DAY; THEREFORE GRAZING BY L. MAGNUS PROBABLY HAD NO SIGNIFICANT EFFECT. OTHER CILIATES WERE EATING SCENEDESMUS, AND ALSO PERHAPS BENTHIC ROTIFERS AND CRUSTACEANS, CONSEQUENTLY GRAZING BY ALL INVERTEBRATES COULD BE SIGNIFICANT. L. MAGNUS UTILIZES OTHER FOOD THAN SCENEDESMUS--MEMBERS OF THIS GENUS ARE KNOWN TO FEED ON BOTH ALGAE AND BACTERIA. (JONES-WISCONSIN)

FIELD 05C

ACCESSION NO. W72-14796

QUANTITATIVE STUDIES ON THE PHYTOPLANKTON OF THE RIVERS THAMES AND KENNET AT READING,

READING UNIV. (ENGLAND). DEPT. OF ZOOLOGY.

T. J. LACK.

FRESHWATER BIOLOGY, VOL. 1, P. 213-224, 1971. 2 FIG., 2 TAB., 22 REF.

DESCRIPTORS:
*ALGAE, *PHYTOPLANKTON, *RIVERS, *DIATOMS, CHLORCPHYTA, BENTHIC FLORA, EUTROPHICATION, DOMINANT ORGANISMS, BIOLOGICAL COMMUNITIES, NUTRIENTS, SCENEDESMUS, CHRYSOPHYTA, SEASONAL, SILICA, DISCHARGE(WATER).

IDENTIFIERS:
*CENTRIC DIATOMS, *RIVER THAMES(ENGLAND), RIVER KENNET(ENGLAND), STEPHANODISCUS HANTZSCHII, CRYPTOMONAS, RHODOMONAS.

ABSTRACT:
SEASONAL CHANGES IN DENSITY AND SPECIES COMPOSITION OF RIVER THAMES PHYTOPLANKTON AND ITS TRIBUTARY, RIVER KENNET, AT READING (ENGLAND) WERE STUDIED. OBSERVATIONS EXTENDED FROM MAY 1966 TO MAY 1968. SAMPLES CONSISTING OF 250 ML OF SUB-SURFACE WATER WERE TAKEN WEEKLY FROM THE CENTER OF EACH RIVER APPROXIMATELY 200 M UPSTREAM OF THE CONFLUENCE. THESE WERE CONCENTRATED BY SEDIMENTATION AND COUNTED IN A HAEMOCYTOMETER. DATA ON DISCHARGE, TEMPERATURE, AND SILICA CONCENTRATION ARE FOLLOWED BY DESCRIPTIONS OF THE VARIATIONS IN NUMBER AND PHYTOPLANKTON COMPOSITION. IN BOTH RIVERS THERE WERE SPRING AND AUTUMN PEAKS OF THE CENTRIC DIATOM, STEPHANODISCUS HANTZSCHII. CHLOROPHYCEAE WERE MOST ABUNDANT DURING SUMMER. TWO CRYPTOPHYCEANS, CRYPTOMONAS AND RHODOMONAS, WERE SOMETIMES NUMEROUS. IN THE THAMES, POPULATION SIZE WAS CLOSELY CORRELATED WITH THE DISCHARGE, HIGHEST NUMBERS ALWAYS OCCURRING DURING LOW DISCHARGE PERIODS. IN THE KENNET, INCREASES IN DISCHARGE OFTEN BROUGHT ABOUT INCREASES IN CELL NUMBER DUE TO INFLUX OF BENTHIC FORMS. IT SEEMS HIGHLY PROBABLE THAT THE RIVERS MAINTAIN HIGH LEVELS OF NITRATES AND PHOSPHATES SINCE BOTH CARRY CONSIDERABLE QUANTITIES OF TREATED SEWAGE EFFLUENT. COMPARISONS WITH EARLIER STUDIES ON THE THAMES SHOWED DIFFERENCE IN PHYTOPLANKTON COMPOSITION WHICH WERE ALMOST CERTAINLY DUE TO EUTROPHICATION. (JONES-WISCONSIN)

FIELD 05C

ACCESSION NO. W72-14797

LIGHT-INDUCED FLUORESCENCE CHANGES IN CHLORELLA, AND THE PRIMARY PHOTOREACTIONS FOR THE PRODUCTION OF OXYGEN,

ROCKEFELLER UNIV., NEW YORK.

D. MAUZERALL.

PROCEEDINGS OF THE NATIONAL ACADEMY OF SCIENCES, VOL 69, NO 6, P 1358-1362, 1972. 3 FIG, 27 REF.

DESCRIPTORS:
*ALGAE, *FLUORESCENCE, *CHLORELLA, *PHOTOACTIVATION, *OXYGEN, LIGHT, PHOTOSYNTHESIS, OPTICAL PROPERTIES, ABSORPTION, MATHEMATICAL STUDIES, CHLOROPHYLL, OXIDATION REDUCTION POTENTIAL.

IDENTIFIERS:
CHLORELLA VULGARIS.

ABSTRACT:
THE INCREASE OF THE QUANTUM FLUORESCENCE YIELD IN GREEN PLANTS AND ALGAE WITH INCREASING ILLUMINATION HAS LONG BEEN AN INDICATOR OF THE STATE OF THE PHOTOREACTIVE CENTERS IN PHOTOSYNTHESIS. THERE ARE DIFFERENT PARTS OF THIS NONLINEAR FLUORESCENCE; SOME PARTS ARE FAIRLY SLOW (SECONDS TO MINUTES); OTHERS ARE FAST (MILLISECONDS). THE LIGHT-INDUCED INCREASES OF THE EFFECTIVE FLUORESCENCE YIELD IN CHLORELLA ARE TOO SLOW TO BE A PRIMARY PROCESS IN PHOTOSYNTHESIS. THE FAST TRANSIENT STATE IS ATTRIBUTED TO A PRIMING REACTION FOR THE PHOTOSYSTEM THAT MAKES OXYGEN. THE SLOWER CYCLICAL PROCESS IS ATTRIBUTED TO THE DARK REACTIONS THAT MAKE OXYGEN AFTER PHOTOEXCITATION OF THIS SYSTEM. THE SLOWER CYCLICAL PROCESS IS ALSO DISTINGUISHED BY A NARROWER EMISSION SPECTRUM THAT PEAKS AT A SHORTER WAVELENGTH THAN THE DARK ADAPTED OR FAST TRANSIENT STATE. A MINIMUM OF SIX DIFFERENT FLUORESCENT STATES ARE REQUIRED TO EXPLAIN THE DATA. IN ADDITION TO THE USUAL ASSUMPTION ABOUT CHANGING QUANTUM YIELD OF FLUORESCENCE IN THESE PROCESSES, THE DATA SUGGEST THAT CHANGES IN CROSS SECTION OF OPTICAL ABSORPTION MUST ALSO BE CONSIDERED. THE SLOWEST RELAXATION TIMES OBSERVED ARE WELL CORRELATED WITH THE SLOW STEPS DETECTED IN EVOLUTION OF OXYGEN. (JONES-WISCONSIN)

FIELD 05C

ACCESSION NO. W72-14800

SOME DATA ON THE ROLE OF FOOD IN THE BIOLOGY OF EUDIAPTOMUS ZACHARIASI POPPE,

INSTYTUT RYBACTWA SRODLADOWEGO, WARSAW (POLAND).

E. GRYGIEREK.

EKOLOGIA POLSKA, VOL 19, NO 21, P 277-292, 1971. 8 FIG, 38 REF.

DESCRIPTORS:
*CRUSTACEANS, *FOODS, *LIFE HISTORY STUDIES, COPEPODS, DIATOMS, CHLOROPHYTA, TEMPERATURE, SIZE, FECUNDITY, ALGAE.

IDENTIFIERS:
*EUDIAPTOMUS ZACHARIASI.

ABSTRACT:
THE ROLE OF FOOD IN THE DEVELOPMENT OF THE CRUSTACEAN, EUDIAPTOMUS ZACHARIASI, IN A NATURAL HABITAT WAS STUDIED. BY DAILY SAMPLING FROM SIX SIMILAR FRY PONDS, EXAMINATION WAS MADE OF THE RELATION BETWEEN DURATION OF DEVELOPMENT, SIZE, AND FECUNDITY AND THE AMOUNT OF ITS FOOD. ANALYSIS WAS MADE OF TOTAL ABUNDANCE, OF VARIATION IN AGE AND SEX STRUCTURE, IN LENGTH OF INDIVIDUALS AT DIFFERENT STAGES OF DEVELOPMENT, ABUNDANCE OF EGGS AND NUMBERS OF FEMALES WITH EGGS. EUDIAPTOMUS ZACHARIASI FED ON ROTIFERS, BUT WHEN THERE WERE FEW ROTIFERS, ITS FOOD WAS ALGAE, MEASURING UP TO 50 MICRONS, MORE DIFFICULT TO DIGEST THAN ROTIFERS. THE DEVELOPMENT CYCLE OF THE GENERATION OF E. ZACHARIASI HATCHED FROM RESTING EGGS LASTED FROM 13 TO 23 DAYS, AS DID THAT OF THE NEXT GENERATION. DIFFERENCES WERE NOT GREATER THAN A FEW DAYS AND MIGHT HAVE BEEN CAUSED BY AMOUNT OF FOOD DURING DEVELOPMENT PERIOD OF THE GENERATION. IN EXTREME CASES THE FOOD LEVEL LIMITS THE NUMBER OF GENERATIONS. THE SIZE OF INDIVIDUALS, SEX STRUCTURE AND PARTICULARLY INDIVIDUAL FECUNDITY OF E. ZACHARIASI DEPENDED ON AMOUNT OF FOOD. INDIVIDUAL FECUNDITY DID NOT ALWAYS DETERMINE THE FECUNDITY OF THE WHOLE POPULATION. (JONES-WISCONSIN)

FIELD 05C, 02H

ACCESSION NO. W72-14805

RELATIVE TOLERANCE OF NITROGEN-FIXING BLUE-GREEN ALGAE TC PESTICIDES,

INDIAN AGRICULTURAL RESEARCH INST., NEW DELHI.

G. S. VENKATARAMAN, AND B. RAJYALAKSHMI.

INDIAN JOURNAL OF AGRICULTURAL SCIENCES, VOL 42, NO 2, P 119-121, 1972. 1 TAB, 11 REF.

DESCRIPTORS:
*ALGAE, *CYANOPHYTA, *PESTICIDES, FUNGICIDES, HERBICIDES, IRRIGATION, AGRICULTURE, RICE, CROP RESPONSE.

IDENTIFIERS:
*TOLERANCE, ANABAENA, NOSTOC, AULOSIRA, TOLYPOTHRIX, ANACYSTIS NIDULANS.

ABSTRACT:
IN-VITRO TOLERANCE OF BLUE-GREEN ALGAE TO PESTICIDES WAS STUDIED. TWENTY-SEVEN STRAINS OF NITROGEN-FIXING CYANOPHYTA FROM FOUR GENERA WERE TESTED. COMMERCIAL PREPARATIONS OF CERESAN M, DITHANE, 2,4-D, DELAPON, PROPAZINE, COTORON, DIURON, AND LINURON WERE USED AT CONCENTRATIONS VARYING FROM 0.01 TO 2000 PPM. MOST NITROGEN-FIXING BLUE-GREEN ALGAE COULD TOLERATE HIGH PESTICIDE LEVELS ALTHOUGH THEY SHOWED WIDE VARIATION IN RELATIVE TOLERANCE. LEVELS OF DIFFERENT PESTICIDES ARE DESCRIBED IN RELATION TO SPECIFIC STRAINS; FOR EXAMPLE, THOUGH MOST OF THE STRAINS OF ANABAENA COULD TOLERATE 100 PPM OF CERESAN, STRAIN 310 WAS SENSITIVE TO A CONCENTRATION OF 0.01 PPM; THOUGH DITHANE WAS LETHAL TO SOME STRAINS OF ANABAENA AND NOSTOC AT THE LOWEST CONCENTRATION USED, SOME STRAINS COULD GROW WELL AT 50 PPM. HIGH CONCENTRATIONS OF CERESAN, DITHANE, 2,4-D AND DELAPON WERE TOLERATED BY TOLYPOTHRIX TENUIS AND AULOSIRA FERTILISSIMA. PROPAZINE, COTORON, AND LINURON WERE LETHAL, AND DIURON LETHAL TO BOTH ALGAE. SPECIFIC STRAINS SHOULD BE USED WITH DIFFERENT PESTICIDES AND A MIXTURE OF DIFFERENT STRAINS--NOT A SINGLE STRAIN--SHOULD BE USED AS SEEDING MATERIAL WHEN MORE THAN ONE PESTICIDE IS APPLIED TO THE CROP. (JONES-WISCONSIN)

FIELD 05C

ACCESSION NO. W72-14807

OBSERVATIONS ON THE CYTOLOGY AND ULTRASTRUCTURE OF THE NEW ALGAL CLASS, EUSTIGMATOPHYCEAE,

LEEDS UNIV. (ENGLAND), DEPT. OF BOTANY.

D. J. HIBBERD, AND G. F. LEEDALE.

ANNALS OF BOTANY, VOL 36, P 49-71, 1972. 2 FIG, 7 PLATES, 65 REF.

DESCRIPTORS:
*CYTOLOGICAL STUDIES, *SYSTEMATICS, *ALGAE, CHRYSOPHYTA, PHAEOPHYTA.

IDENTIFIERS:
*EUSTIGMATOPHYCEAE, XANTHOPHYCEAE, POLYEDRIELLA HELVETICA, PLEUROCHLORIS COMMUTATA, PLEUROCHLORIS MAGNA, VISCHERIA PUNCTATA, VISCHERIA STELLATA, ELLIPSOIDION ACUMINATUM.

ABSTRACT:
THE FIRST DETAILED ACCOUNT OF A NEW CLASS OF ALGAE, EUSTIGMATOPHYCEAE IS GIVEN. ITS UNIQUE FEATURES ARE EMPHASIZED IN COMPARISON WITH THOSE COCCOID FORMS WHICH MUST BE CONSIDERED AS MEMBERS OF THE XANTHOPHYCEAE SENSU STRICTO. THE ZOOSPORES, AND TO A LESSER EXTENT, THE VEGETATIVE CELLS OF SIX SPECIES OF EUSTIGMATOPHYCEAE SPECIES WERE STUDIED WITH LIGHT AND ELECTRON MICROSCOPY. THE FEATURES OF ORGANIZATION ARE COMPARED IN DETAIL WITH THE SAME COMPONENTS IN THE XANTHOPHYCEAE AND OTHER CLASSES OF ALGAE, WITH PARTICULAR REFERENCE TO THE MOTILE CELLS. IT IS CONCLUDED THAT EUSTIGMATOPHYCEAE IS LESS LIKE THE XANTHOPHYCEAE SENSU STRICTO THAN THE LATTER IS LIKE THE CHRYSOPHYCEAE AND PHAEOPHYCEAE, AND IN MOST CHARACTERISTICS SHOWS FUNDAMENTAL DIFFERENCES FROM ALL OTHER CLASSES OF ALGAE. THE TAXONOMIC DIFFICULTY RAISED BY EUSTIGMATOPHYCEAN POSSESSION OF AN ANTERIOR HAIRY FLAGELLUM OF TYPICALLY 'HETEROKONT' CONSTRUCTION IS DISCUSSED. THE TAXONOMIC AND NOMENCLATURAL PROBLEMS RESULTING FROM THE REMOVAL OF SEVERAL SPECIES FROM THE XANTHOPHYCEAE INTO THE EUSTIGMATOPHYCEAE ARE BRIEFLY CONSIDERED AND PREVIOUS OBSERVATIONS BY OTHER AUTHORS ON SPECIES NOW PLACED IN THE EUSTIGMATOPHYCEAE ARE REVIEWED. (JONES-WISCONSIN)

FIELD 05C

ACCESSION NO. W72-14808

THE INFLUENCE OF EXPERIMENTAL INCREASE OF BIOMASS OF THE BLUE-GREEN ALGAE GLOEOTRICHIA ECHINULATA (SMITH) RICHTER ON PHYTOPLANKTON PRODUCTION,

POLISH ACADEMY OF SCIENCES, WARSAW. INST. OF ECOLOGY.

I. SPODNIEWSKA.

EKOLOGIA POLSKA, VOL 19, NO 31, P 476-483, 1971. 2 FIG, 5 TAB, 4 REF.

DESCRIPTORS:
*ALGAE, *BIOMASS, *CYANOPHYTA, PHYTOPLANKTON, PRODUCTIVITY, PHOTOSYNTHESIS, EPILIMNION, LIGHT INTENSITY, RESPIRATION, LIMITING FACTORS.

IDENTIFIERS:
*GLOEOTRICHIA ECHINULATA, MIKOLAJSKIE LAKE(POLAND), CERATIUM HIRUNDINELLA.

ABSTRACT:
INVESTIGATIONS IN MIKOLAJSKIE LAKE, POLAND WERE MADE OF THE POSSIBILITIES OF INCREASING PHYTOPLANKTON PRODUCTION IN PARTICULAR TROPHIC CONDITIONS BY INCREASING BIOMASS OF THE PLANKTONIC BLUE-GREEN ALGA GLOEOTRICHIA ECHINULATA AND OF THE ESTIMATION OF NET PRODUCTION OF THIS SPECIES IN NATURAL CONDITIONS. THIS SPECIES WAS CHOSEN BECAUSE OF THE GOOD POSSIBILITY OF ISOLATING IT FROM OTHER PHYTOPLANKTON. SURFACE WATER PLANKTON WAS CONDENSED AND COLONIES SEPARATED. PRIMARY PRODUCTION WAS ESTIMATED. OF THREE EXPERIMENTAL SERIES, THE FIRST AND SECOND WERE JUST UNDER THE WATER SURFACE, AND THE THIRD AT VARIOUS DEPTHS OF THE TROPHOGENIC LAYER (0-6 M). PHYTOPLANKTON PRODUCTION OF AN UNCHANGED BIOMASS WAS COMPARED WITH PHYTOPLANKTON PRODUCTION TO WHICH A KNOWN NUMBER OF COLONIES OF G. ECHINULATA WAS ADDED. PRODUCTION WAS THEN CALCULATED FROM THE DIFFERENCE. IN EACH ONE OF THE EXPERIMENTAL SERIES, INDEPENDENT OF THE PERIOD THEY WERE CARRIED OUT, AND THUS INDEPENDENT OF THE POSSIBLE EXISTENCE OF DIFFERENCES IN PHYSIOLOGICAL ALGAL PROPERTIES AND/OR IN ENVIRONMENTAL CONDITIONS AT THE TIME, IT WAS FOUND THAT PRODUCTION INCREASES AS A RESULT OF AN INCREASE OF PHYTOPLANKTON DENSITY DUE TO ADDITION OF G. ECHINULATA COLONIES. (JONES-WISCONSIN)

FIELD 05C

ACCESSION NO. W72-14812

WATER QUALITY CONTROL WITH SYNTHETIC POLYMERIC FLOCCULANTS: EFFECT OF METAL IONS ON FLOCCULATION OF BIOCOLLOIDS,

CONNECTICUT UNIV., STORRS. INST. OF WATER RESOURCES.

J. K. DIXON, AND R. C. TILTON.

AVAILABLE FROM THE NATIONAL TECHNICAL INFORMATION SERVICE AS PB-212 364, $3.00 IN PAPER COPY, $0.95 IN MICROFICHE. COMPLETION REPORT 1972. 9 P. OWRR A-031-CONN(2).

DESCRIPTORS:
*FLOCCULATION, *ALGAE, *SILICA, *E. COLI, *IONS, *WASTEWATER TREATMENT, *POLYMERS, CATIONS, ANIONS, CHLORELLA, FILTRATION, ELECTROPHORESIS, TRACERS.

IDENTIFIERS:
*MULTIVALENT METALLIC CATIONS, *POLYACRYLAMIDE, *POLYETHYLENEMINE, *POLYSTYRENE SULFONATE, CHLORELLA ELLIPSOIDIA.

ABSTRACT:
EFFECT OF THE ADDITION OF MULTIVALENT METALLIC CATIONS UP TO 0.02M ON THE RATE OF FLOCCULATION OF PURE ALGAL CULTURES OF CHLORELLA ELLIPSOIDIA BY SYNTHETIC POLYMERIC CATIONIC POLYETHYLENEIMINES (PEI) WAS FOUND TO BE NEGLIGIBLE. A HUNDREDFOLD VARIATION IN THE MG (++) CONCENTRATION PRESENT IN NUTRIENT USED DURING THE GROWTH OF THESE ALGAE ALSO HAD NO EFFECT ON EFFICIENCY OF FLOCCULATION OF THE ALGAE BY THE PEI POLYMERS. ADDITION OF NON-IONIC POLYACRYLAMIDE TO THE PEI POLYMERS TO THE BACTERIA, E. COLI, HAD NO EFFECT ON THEIR GOOD PERFORMANCE. ADDITION OF ANIONIC POLYSTYRENE SULFONATE POLYMERS, WHICH WERE INEFFECTIVE ALONE, TO PEI POLYMERS DECREASED THE FLOCCULATION PERFORMANCE OF THE LATTER FOR E. COLI. EFFECTS OF THE MULTIVALENT METALLIC CATIONS AND/OR THE USE OF ANIONIC-CATIONIC POLYMER MIXTURES FOR FLOCCULATING THE COMPLEX DISPERSIONS PRESENT IN COMMERCIAL OPERATIONS ARE OFTEN SUCCESSFUL. THE RESULTS ON THE SIMPLE ALGAL AND BACTERIA SYSTEMS APPARENTLY CANNOT BE TRANSLATED INTO SOME GUIDING PRINCIPLES USEFUL DURING THE FLOCCULATION OF THE MORE COMPLEX COMMERCIAL SYSTEMS. ON THE OTHER HAND, THE RESULTS OF PREVIOUS STUDIES ON FLOCCULATION OF E. COLI, ALGAE AND SILICA WITH RADIOACTIVE TAGGED PEI POLYMERS SHOULD BE QUITE USEFUL FOR APPLICATION TO COMMERCIAL OPERATIONS.

FIELD 05G, 05D

ACCESSION NO. W72-14840

IRRIGATION AND FERTILIZATION WITH WASTEWATER,

IDAHO UNIV., MOSCOW. DEPT. OF CIVIL ENGINEERING; AND IDAHO UNIV., MOSCOW. DEPT. OF HYDROGEOLOGY.

D. D. EIER, A. T. WALLACE, AND R. E. WILLIAMS.

COMPOST SCIENCE JOURNAL OF WASTE RECYCLING, VOL 12, NO 3, MAY-JUNE 1971, P 26-29. 2 FIG, 1 REF.

DESCRIPTORS:
*IRRIGATION, *FERTILIZATION, *WASTE WATER(POLLUTION), *WASTE WATER DISPOSAL, *WATER POLLUTION CONTROL, ECONOMIC FEASIBILITY, STREAMS, SEWAGE EFFLUENTS, ALGAE, TOURISM, RECREATION, FISHERIES, HYDROELECTRIC POWER, NAVIGATION, CONSUMPTIVE USE, WATER POLLUTION, OXYGEN, SOIL PROPERTIES, IDAHO.

ABSTRACT:
AN ECONOMICALLY PALATABLE METHOD OF WATER POLLUTION CONTROL IS PRESENTED. THE PLAN PRESENTS THE ENTRY INTO STREAMS OF THE $18 WORTH OF NUTRIENTS PER ACRE FOOT OF SECONDARY TREATMENT PLANT EFFLUENT. DRIFTING ISLANDS OF ALGAE IN RIVERS, AND FERTILIZER DEMANDS CAN BE REDUCED. TOURISM, RECREATION, FISHERY PRODUCTION, HYDROELECTRIC POWER, NAVIGATION, IRRIGATION AGRICULTURE, AND DOMESTIC AND INDUSTRIAL USE OF THE SNAKE RIVER IN SOUTHERN IDAHO ARE AFFECTED BY WATER POLLUTION. PARTICULAR PROBLEMS INCLUDE TASTE, ODOR, BACTERIAL CONTAMINATION, AQUATIC GROWTH AND THERMAL EFFECTS. THE PRINCIPAL CAUSE OF THIS POLLUTION IS THE DISPOSAL OF UNTREATED OR INADEQUATELY TREATED WASTES. THE WASTES REDUCE AVAILABLE OXYGEN AND ENCOURAGE GROWTH OF UNDESIRABLE ALGAE. CONTROL OF NUTRIENT INPUT WOULD GO A LONG WAY IN REDUCING THIS POLLUTION. CROP IRRIGATION OR SURFACE APPLICATION OF DOMESTIC AND INDUSTRIAL WASTE WATER FOR SOIL RENOVATION IS SUGGESTED. INVESTIGATIONS AT THE UNIVERSITY OF IDAHO INDICATE IRRIGATION AND FERTILIZATION WITH WASTE WATER ARE FEASIBLE AND ECONOMICAL. (POPKIN-ARIZONA)

FIELD 03F, 05D

ACCESSION NO. W72-14878

MICROBIAL CRITERIA OF ENVIRONMENT QUALITIES,

COPENHAGEN UNIV. (DENMARK). INST. OF HYGIENE.

E. FJERDINGSTAD.

ANNUAL REVIEW OF MICROBIOLOGY, VOL. 25, P 563-582, 1971. 3 TAB, 58 REF.

DESCRIPTORS:
*EUTROPHICATION, *BIOLOGICAL PROPERTIES, *WATER QUALITY, *BIOINDICATORS, *BIOLOGICAL COMMUNITIES, AQUATIC ALGAE, AQUATIC PRODUCTIVITY, BIOCHEMICAL OXYGEN DEMAND, LAKE STAGES, NUTRIENTS, MESTROPHY, OLIGOTROPHY, OXYGEN SAG, AQUATIC LIFE, BIOASSAY, CLASSIFICATION.

IDENTIFIERS:
BIOTESTS.

ABSTRACT:
AFTER A BRIEF DISCUSSION OF EUTROPHICATION AND A SURVEY OF BIOTESTS, BIOLOGICAL ASSESSMENT OF EUTROPHICATION AND POLLUTION IS DISCUSSED. A NUMBER OF CLASSIFICATION SYSTEMS FOR DENOTING THE DEGREE TO WHICH WATER IS POLLUTED, SOME DEFINED IN TERMS OF BIOLOGICAL POPULATIONS, ARE GIVEN. THE SYSTEM MOST DISCUSSED RELATES SAPROBICITY LEVEL AND STRUCTURES OF THE COMMUNITIES OF ORGANISMS. (SVENSSON-WASHINGTON)

FIELD 05C, 05B

ACCESSION NO. W73-00002

USE OF TESTS FOR LIMITING OR SURPLUS NUTRIENTS TO EVALUATE SOURCES OF NITROGEN AND PHOSPHORUS FOR ALGAE AND AQUATIC WEEDS,

WISCONSIN UNIV., MADISON. WATER CHEMISTRY LAB.

G. P. FITZGERALD, AND G. F. LEE.

REPORT JULY 1, 1971. 35 P, 4 FIG, 3 TAB, 23 REF.

DESCRIPTORS:
*ALGAE, *AQUATIC WEEDS, *NUTRIENTS, BIOASSAY, LIMITING FACTORS, NITROGEN, PHOSPHORUS, RAIN, LAKES, CLADOPHORA, ANALYSIS, PHYTOPLANKTON, CHEMICAL ANALYSIS, DIATOMS, CYANOPHYTA, TURNOVERS, EUTROPHICATION.

IDENTIFIERS:
LAKE MENDOTA(WIS.), LAKE MONONA(WIS.), LAKE WINGRA(WIS.), LIMITING NUTRIENTS, SURPLUS NUTRIENTS, NUTRIENT SOURCES.

ABSTRACT:
SOURCES OF NITROGEN AND PHOSPHORUS AVAILABLE TO ALGAE AND AQUATIC WEEDS AND THEIR NUTRITIONAL STATUS WERE DETERMINED BY A SIMPLE BIOASSAY PROCEDURE. THE ALGAE AND AQUATIC WEEDS WERE COLLECTED FROM LAKE MENDOTA'S SOUTH SHORE, THE WEST BAY OF LAKE MONONA, OR FROM LAKE WINGRA, ALL IN MADISON, WISCONSIN WITHIN ABOUT ONE MILE OF EACH OTHER, ALTHOUGH VARYING WIDELY. RAIN COULD BE THE MAJOR SOURCE OF AVAILABLE NITROGEN TO CLADOPHORA IN LAKE MENDOTA DURING SUMMER; INCREASES IN PHOSPHORUS ASSOCIATED WITH RAINFALLS WERE DETECTED BUT WERE NOT AS DRAMATIC AS NITROGEN INCREASES. MIXED BLOOMS OF PLANKTONIC ALGAE DO NOT SHARE THEIR NUTRIENTS WITH OTHER ALGAE EVEN WHEN ONE SPECIES MAY HAVE A SURPLUS AND ANOTHER IS NUTRIENT-LIMITED. IN COMPARING SURFACE AND SUBSURFACE PHYTOPLANKTON, AT CERTAIN TIMES SURFACE PLANKTON CAN BE NUTRIENT LIMITED WHILE THE SAME SPECIES FROM SUBSURFACE SOURCES HAS ADEQUATE OR SURPLUS NUTRIENTS. FALL OVERTURN AS A NUTRIENT SOURCE DEMONSTRATES HOW SIMILAR CHANGES CAN TAKE PLACE IN THE NUTRITION OF DIFFERENT ALGAL TYPES. FOR ACCURATE EVALUATION OF THE NUTRIENT LEVEL OF LAKES, BIOASSAYS OF SURFACE WATERS SHOULD BE CONTRASTED WITH ANALYSES OF WATERS FROM THE THERMOCLINE OR HYPOLIMNION AND WITH WATERS OBTAINED IN THE SPRING BEFORE MAXIMUM PLANT PRODUCTION. (AUEN-WISCONSIN)

FIELD 05C, 02H

ACCESSION NO. W73-00232

THE IMPORTANCE OF ALGAL CULTURES FOR THE ASSESSMENT OF THE EUTROPHICATION OF THE OSLOFJORD,

NORWEGIAN INST. FOR WATER RESEARCH, OSLO.

O. M. SKULBERG.

HELGOLANDER WISSENSCHAFTLICHE MEERESUNTERSUCHUNGEN, VOL 20, P 111-125, 1970. 10 FIG, 3 TAB, 30 REF.

DESCRIPTORS:
*ALGAE, *CULTURES, *INDICATORS, *EUTROPHICATION, FJORDS, BIOASSAY, CHLORELLA, NUTRIENTS, WATER SAMPLING, ORGANIC MATTER, WATER POLLUTION EFFECTS, SEWAGE, NITROGEN, PHOSPHORUS, IRON, PHYTOPLANKTON, DIATOMS, DINOFLAGELLATES.

IDENTIFIERS:
*OSLOFJORD(NORWAY).

ABSTRACT:
THE RELATIVE SIGNIFICANCE IN EUTROPHICATION OF THE DIFFERENT CONTRIBUTING WATERS OF THE OSLOFJORD SEA WATER, RUNOFF WATER FROM CATCHMENT AREAS, BOTTOM WATERS OF THE FJORD, AND SEWAGE, AND OTHER POLLUTING MATTER OF URBAN ORIGIN, ARE EVALUATED BY APPLYING ALGAL CULTURE ASSAYS. IN 1962 TO 1965 UNIALGAL CULTURES OF TEST ALGAE, INCLUDING SELENASTRUM CAPRICORNUTUM, CHLORELLA OVALIS, SKELETONEMA COSTATUM, AND PHAEODACTYLUM TRICORNUTUM WERE USED FOR BIOASSAYS. THE AMOUNTS AND AVAILABILITY OF PLANT NUTRIENTS IN THE WATER ARE THUS MEASURED AND THE QUALITY OF THE WATER FOR ALGAL GROWTH DETERMINED. THE RESULTS INDICATED THAT THE WATER OF THE INNER PART OF THE FJORD WAS IN A CONDITION WHERE A SIGNIFICANT INCREASE IN SEWAGE LOAD GAVE A RELATIVELY SMALL INCREASE IN ALGAL GROWTH. CONVERSELY, IN WATER SAMPLES FROM THE OUTER OSLOFJORD, A SMALL ADDITION OF SEWAGE WOULD GIVE RELATIVELY HIGH INCREASE IN GROWTH. IT WAS SHOWN, IN ENRICHMENT EXPERIMENTS WITH ESSENTIAL NUTRIENTS, THAT THE SUPPLY OF COMPOUNDS OF NITROGEN, PHOSPHORUS, AND IRON WAS IMPORTANT FOR THE RESULTING YIELD OF ALGAE. THE MAIN FERTILIZING EFFECT COULD NOT BE ASCRIBED TO ANY SINGLE COMPONENT. (JONES-WISCONSIN)

FIELD 05C

ACCESSION NO. W73-00234

ECOLOGICAL STUDIES ON DISSOLVED OXYGEN AND BLOOM OF MICROCYSTIS IN LAKE SUWA--I. HORIZONTAL DISTRIBUTION OF DISSOLVED OXYGEN IN RELATION TO DRIFTING OF MICROCYSTIS BY WIND,

SHINSHU UNIV., SUWA (JAPAN). SUWA HYDROBIOLOGICAL STATION.

H. YAMAGISHI, AND K. AOYAMA.

BULLETIN OF THE JAPANESE SOCIETY OF SCIENTIFIC FISHERIES, VOL 38, NO 1, P 9-16, 1972. 6 FIG, 1 TAB, 7 REF.

DESCRIPTORS:
*CYANOPHYTA, *DISSOLVED OXYGEN, *EUTROPHICATION, *DISTRIBUTION, FLOATING, WINDS, ALGAE, SURFACE WATERS, HYDROGEN ION CONCENTRATION, LIGHT PENETRATION, PHOTOSYNTHESIS, PHYTOPLANKTON, STRATIFICATION, DIATOMS, CURRENTS(WATER), FISH FARMING.

IDENTIFIERS:
*MICROCYSTIS, *LAKE SUWA(JAPAN).

ABSTRACT:
THE SEVERE WATER-BLOOM OF BLUE-GREEN ALGAE DOMINATED BY MICROCYSTIS APPEARS ALMOST EVERY SUMMER AND SUMMER KILL OFTEN ATTACKS THE CARP CULTURED IN FLOATING NETS. OBSERVATIONS WERE MADE ON HORIZONTAL DISTRIBUTION OF DISSOLVED OXYGEN IN THE SURFACE WATER ACROSS LAKE SUWA, JAPAN IN RELATION TO WIND INDUCED DRIFT OF MICROCYSTIS. TEMPERATURE, PH, AND CELL NUMBER OF MICROCYSTIS IN THE SURFACE WATER TOGETHER WITH TRANSPARENCY AND WIND DIRECTION WERE RECORDED. THE STUDIES DEMONSTRATED THAT DISSOLVED OXYGEN IN THE LAKE INCREASED FROM THE WINDWARD TO THE LEEWARD WITH CONCOMITANT INCREASES IN THE DENSITY OF DRIFTING MICROCYSTIS AND PH VALUES. OXYGEN WAS MARKEDLY UNDERSATURATED ON THE WINDWARD (LESS THAN 50% SATURATION), BUT EXTREMELY SUPERSATURATED ON THE LEEWARD (MORE THAN 250% SATURATION). CELL NUMBERS OF MICROCYSTIS REACHED 28 MILLION/ML AT AREAS OF HIGHEST DENSITY. LABORATORY TESTS AND IN SITU OBSERVATIONS INDICATE THAT A VERY THICK LAYER OF MICROCYSTIS IN THE TOPMOST WATER OF THE LAKE INHIBIT PHOTOSYNTHESIS OF UNDERLYING ALGAE BY SHADING BUT THE REASON FOR THE CONSTANT PH VALUE BEFORE AND AFTER ILLUMINATION REMAINED UNKNOWN. (JONES-WISCONSIN)

FIELD 05C, 02H

ACCESSION NO. W73-00236

ALGAE AS INDICATORS OF PESTICIDE,

STATE UNIV., COLL., BUFFALO, N.Y. GREAT LAKES LAB.

R. A, SWEENEY.

SPECIAL REPORT NO. 4, FEBRUARY 1970. 10 P. 1 FIG, 28 REF. (PRESENTED AT 'ALGAE AS INDICATIONS' SYMPOSIUM OF THE PHYCOLOGICAL SOCIETY OF AMERICA, SEPTEMBER 4, 1968, OHIO STATE UNIVERSITY, COLUMBUS).

DESCRIPTORS:
*ALGAE, *INDICATORS, *PESTICIDES, CHLORINATED HYDROCARBON PESTICIDES, INSECTICIDES, RODENTICIDES, HERBICIDES, FUNGICIDES, ALGICIDES, CARBAMATE PESTICIDES, ORGANOPHOSPHORUS PESTICIDES, ACARICIDES, FOOD CHAINS, WATER POLLUTION SOURCES.

ABSTRACT:
THE NATURE AND EXTENT OF PESTICIDE CONTAMINATION OF AQUATIC HABITATS IN THE UNITED STATES ARE REVIEWED. A ROUTINE SAMPLING PROGRAM FOR PESTICIDES HAS BEEN INITIATED AT APPROXIMATELY 100 SURVEILLANCE STATIONS ON THE MAJOR RIVER BASINS WITHIN THE UNITED STATES. THIS IS CONDUCTED THROUGH COORDINATED AND COOPERATIVE EFFORTS OF FEDERAL, STATE, LOCAL, AND PRIVATE AGENCIES. SPECIAL ATTENTION HAS BEEN GIVEN TO ALDRIN, BENZENE HEXACHLORIDE, DIELDRIN, ENDRIN, HEPTACHLOR, HEPTACHLOR EPOXIDE, TOXAPHENE, AND DDT AND ITS DERIVATIVES--DDE AND DDD, ALL CHARACTERIZED BY LONG RESIDUAL ACTION. SINCE ALGAE RAPIDLY ACCUMULATE NUMEROUS PESTICIDES AND COMPRISE THE BASE OF MOST AQUATIC TROPHIC PYRAMIDS, THE ROUTINE COLLECTION AND ANALYSIS OF ALGAE SHOULD BE PART OF OUR NATIONWIDE MONITORING SYSTEM AND SHOULD ENLIST THE ASSISTANCE OF ALGAL TAXONOMISTS. THIS WOULD GIVE A MORE MEANINGFUL INDICATION OF THE THREAT OF PESTICIDE CONTAMINATION OF THE HIGHER AQUATIC ORGANISMS AND MAN. BIOLOGICAL DETERMINATIONS OF PESTICIDE UPTAKE ARE NECESSARY AS A CHECK BECAUSE CHEMICAL ANALYSES, USING THIN LAYER AND GAS CHROMATOGRAPHY, ARE NOT ALWAYS RELIABLE. ALGAE WHICH ACCUMULATE PESTICIDES CAN BE EMPLOYED IN THE COLLECTION, IDENTIFICATION, AND QUANTIFICATION OF PESTICIDES. (JONES-WISCONSIN)

FIELD 05C, 05B

ACCESSION NO. W73-00238

THE EFFECTS OF INCREASING LIGHT AND TEMPERATURE ON THE STRUCTURE OF DIATOM COMMUNITIES,

ACADEMY OF NATURAL SCIENCES OF PHILADELPHIA, PA. DEPT. OF LIMNOLOGY.

R. PATRICK.

LIMNOLOGY AND OCEANOGRAPHY, VOL 16, NO 2, P 405-421, MARCH 1971. 1 FIG, 4 TAB, 17 REF.

DESCRIPTORS:
*LIGHT, *TEMPERATURE, *DIATOMS, BIOLOGICAL COMMUNITIES, SYSTEMATICS, CYANOPHYTA, CHEMICAL ANALYSIS, SEASONAL, BIOMASS, ON-SITE INVESTIGATIONS, ALGAE.

IDENTIFIERS:
GOMPHONEMA OLIVACEOIDES, WHITE CLAY CREEK(PA.).

ABSTRACT:
THE DIFFERENCES IN TEMPERATURE OR LIGHT REQUIREMENTS, OR BOTH, APPARENTLY INFLUENCE SEASONAL SUCCESSION OF SPECIES. THE PROBLEM ARISES WHETHER RAISING THE TEMPERATURE A FEW DEGREES NEAR THE ENDS OF THEIR RANGE OF TOLERANCE HAS A GREATER EFFECT ON DIATOM COMMUNITIES THAN RAISING THE TEMPERATURE ABOVE AMBIENT IN THE MIDDLE PORTION OF THE TOLERANCE RANGE. EFFECTS ON COMMUNITY STRUCTURE OF VARIOUS DAY LENGTHS AND OF ARTIFICIAL TEMPERATURE INCREASE COMPARED WITH NATURAL TEMPERATURE INCREASE WERE STUDIED. SUMMARY OF CHEMICAL ANALYSIS OF THE STREAM, THE CHARACTERISTICS OF THE DIATOM COMMUNITIES AND THE PERCENTAGE OCCURRENCE OF THE MORE COMMON SPECIES ARE TABULATED. NATURALLY INCREASING DAY LENGTH WAS MORE FAVORABLE FOR COMMUNITY DEVELOPMENT THAN INCREASING DAY LENGTH BY ARTIFICIAL LIGHT. INCREASING TEMPERATURE IS MOST BENEFICIAL WHEN TEMPERATURES ARE NEAR OC. MOVING AWAY FROM THE TOLERANCE LIMITS AT EITHER END OF THE RANGE PRODUCED THE GREATEST CHANGES IN DIATOM COMMUNITY STRUCTURE. INTERMEDIATE CHANGES NEAR THE OPTIMUM RANGE PRODUCED LESS PREDICTABLE RESULTS. ONE NEW TAXON NAMED GOMPHONEMA OLIVACEOIDES VAR. HUTCHINSONIANA VAR. NOV., IS DESCRIBED. (JONES-WISCONSIN)

FIELD 05C

ACCESSION NO. W73-00242

COMPARATIVE STUDIES ON ALGAL TOXINS,

NEW HAMPSHIRE UNIV., DURHAM. DEPT. OF ZOOLOGY; AND NEW HAMPSHIRE UNIV., DURHAM. JACKSON ESTUARINE LAB.

J. J. SASNER, JR.

(1971), 98 P. 13 FIG., 3 TAB., 121 REF. OWRR A-021-NH (1) AND A-013-NH (4).

DESCRIPTORS:
*ANALYTICAL TECHNIQUES, *ALGAL TOXINS, AQUATIC PLANTS, TOXINS, RED TIDE, PLANT PHYSIOLOGY, BIOASSAY, CYTOLOGICAL STUDIES, DINOFLAGELLATES, CYANOPHYTA.

IDENTIFIERS:
*TOXICOLOGY, BIOTOXINS, PHARMACOLOGY.

ABSTRACT:
CHANGES WERE DEVELOPED IN THE THEORIES CONCERNING THE IONIC BASIS OF CELL ELECTROGENESIS AND JUNCTIONAL TRANSMISSION BETWEEN CELLS WHICH TOXICOLOGISTS HAVE USED TO STUDY SITES AND MODES OF ACTION OF AQUATIC POISONS. THE COMPARATIVE PHYSIOLOGIST MAY FIND BIOTOXINS USEFUL IN DETERMINING SIMILARITIES AND DIFFERENCES IN ELECTROGENIC AND TRANSMITTER MECHANISMS IN A WIDE VARIETY OF ORGANIC SYSTEMS. TETRODOTOXIN HAS ALREADY RECEIVED WIDE USAGE IN BLOCKING NA+ CONDUCTANCE ACROSS NERVE AND MUSCLE MEMBRANES; THE EFFECT OF THIS TOXIN IS MORE SPECIFIC THAN PROCAINE AND OTHER ANAESTHETICS WHICH ALSO AFFECT K+ CONDUCTANCE. TO DETERMINE WHETHER DINOFLAGELLATE TOXINS THAT CAUSE GENERAL MEMBRANE DEPOLIRIZATION PRODUCE THEIR EFFECTS BY ALTERING MEMBRANE PERMEABILITY OR BY INACTIVATING THE IONIC PUMP ACTIVE TRANSPORT MECHANISM AND WHETHER DEPOLARIZING TOXINS CAN ACT AS METABOLIC POISONS, OR CAN PROVEN DEPOLARIZING EFFECTS BE ANTAGONIZED BY TETRODOTOXIN SAXITOXIN OR THEIR ANALOGUES WHICH BLOC IONIC CHANNELS WILL BE OF INTEREST. THE APPARENT IMMUNITY TO POTENT BIOTOXINS OF TRANSVECTORS WHO APPARENTLY ACT AS 'BIOLOGICAL STORAGE DEPOTS' FOR ORGANISMS HIGHER IN THE FOOD CHAIN IS OF EQUAL INTEREST. IT IS CLEAR THE FLAGELLATE AND BLUE-GREEN ALGAL TOXINS AFFECT IONIC MECHANISMS OF ELECTROGENESIS ASSOCIATED WITH THE MEMBRANE AND THE TRANSMITTER SYSTEM EXERTING CHEMICAL CONTROL BETWEEN CONTIGUOUS CELLS. (AUEN-WISCONSIN)

FIELD 05C

ACCESSION NO. W73-00243

APHANIZOMENON FLOS-AQUAE: INFECTION BY CYANOPHAGES,

LANTBRUKHOGSKOLAN, UPPSALA (SWEDEN). DEPT. OF MICROBIOLOGY.

U. GRANHALL.

PHYSIOLOGIA PLANTARUM, VOL 26, P 332-337, 1972. 9 FIG, 18 REF.

DESCRIPTORS:
*ALGAL CONTROL, *INFECTION, *VIRUSES, CYANOPHYTA.

IDENTIFIERS:
*CYANOPHAGES, *APHANIZOMENON FLOS-AQUAE, LAKE ERKEN(SWEDEN).

ABSTRACT:
NITROGEN-FIXING BLUE-GREEN ALGAE WERE STUDIED IN LAKE ERKEN, SWEDEN DURING THE SUMMER OF 1970. A LYTIC AGENT WAS PRESENT IN LAKE ERKEN AND ELECTRON MICROGRAPHS OF ALGAL SAMPLES SHOWED THAT CYANOPHAGE-LIKE PARTICLES OCCURRED IN LARGE NUMBERS, SPECIFICALLY ASSOCIATED WITH VEGETATIVE CELLS OF APHANIZOMENON FLOS-AQUAE. ELECTRON MICROSCOPY, STERILE FILTRATIONS (INCLUDING STEPS TO EXCLUDE BACTERIA) AND PLAQUE TECHNIQUES WERE USED TO CORRELATE THE VIRUS FLUCTUATIONS WITH THE WATER-BLOOM AND DEGRADATION OF APHANIZOMENON. ONLY VEGETATIVE CELLS OF APHANIZOMENON WERE LYSED, NOT HETEROCYSTS AND AKINETES. INFECTED CELLS OF APHANIZOMENON AND EXTRACELLULAR VIRIONS COULD NOT BE DETECTED BY ELECTRON MICROSCOPY IN LAKE SAMPLES UNTIL THE LATE STAGE OF THE WATER-BLOOM; THE HOST SHOWED NORMAL ULTRASTRUCTURE UP TO THE PERIOD OF MOST EXTENSIVE WATER-BLOOM. ALGAL SAMPLES, TAKEN SOME WEEKS LATER SHOWED QUITE DIFFERENT ULTRASTRUCTURES DUE TO CELL INFECTION BY VIRUS PARTICLES. IT IS SUGGESTED THAT A CYANOPHAGE NAMED AP-1 REGULATES TERMINATION OF THE A. FLOS-AQUAE BLOOM. IF THE ISOLATED VIRUSES COULD BE USED IN ENVIRONMENTAL CONTROL AFTER PROPAGATION IN THE LABORATORY IT WOULD CONSTITUTE A NEW APPROACH IN CONTROLLING APHANIZOMENON BLOOMS. (JONES-WISCONSIN)

FIELD 05C

ACCESSION NO. W73-00246

ENDOGENOUS RHYTHM OF THE PRODUCTIVITY IN CHLORELLA AND THE INFLUENCE OF LIGHT (ENDOGENE RHYTHMIK DER PRODUKTION-SFAHIGKEIT BEI CHORELLA UND IHRE BEEINFLUSSUNG DURCH LICHT),

GOETTINGEN UNIV. (WEST GERMANY). INST. FOR PLANT PHYSIOLOGY.

M. HESSE.

ZEITSCHRIFT FUR PFLANZENPHYSIOLOGIE, VOL. 67, P 58-77, 1972. 12 FIG, 2 TAB, 32 REF. ENGLISH SUMMARY.

DESCRIPTORS:
*BIORTHYMS, *METABOLISM, *CHLORELLA, *LIGHT, ALGAE, GROWTH RATES, CHLOROPHYLL.

IDENTIFIERS:
*CHLORELLA PYRENOIDOSA.

ABSTRACT:
SINCE CHLORELLA IS UNIVERSALLY USED FOR ALGAL RESEARCH, CHARACTERISTIC PHYSIOLOGICAL CHANGES WERE STUDIED. UNDER 54 HOURS OF CONSTANT DARK CONDITIONS THE AUTOSPORES OF CHLORELLA PYRENOIDOSA STRAIN 211-8B WERE FOLLOWED BY MEASURING DRY-WEIGHT, CARBOHYDRATE, PROTEIN, CHLOROPHYLL, AND CELL NUMBER. THE CHANGES IN ACTIVITY ARE ENDOGENOUS AND OF A CIRCADIAN NATURE AND ARE INDEPENDENT OF THE DEVELOPMENTAL STAGE OF THE CELLS. AFTER REMAINING IN CONTINUOUS DIM WHITE LIGHT DURING 'WAITING' THE RHYTHM COULD NOT BE DEMONSTRATED AND THE PHYSIOLOGICAL ACTIVITY WAS MINIMUM; TRANSFER FROM DIM LIGHT TO CONTINUOUS DARKNESS RESTORED THE ENDOGENOUS RHYTHM. THIS CHANGE INITIATES A RHYTHM IN THE INDIVIDUAL CELL AND IS NOT A SYNCHRONIZING FACTOR. WHEN THE DARK PERIOD OF THE LIGHT-DARK-CHANGE IS REPLACED BY DIM LIGHT, MINIMUM PRODUCTION FOLLOWS. THIS IS ONE REASON WHY THE PRODUCTIVITY PER LIGHT-HOUR IN CONTINUOUS LIGHT IS NOT AS HIGH AS UNDER LIGHT-DARK-CHANGE. APPEARANCE OF MAXIMUM AND MINIMUM PRODUCTION AFTER DIFFERENT INTERVALS IS DUE TO THE DIFFERENCE IN GROWTH RATE WHICH OCCURS ABOUT 8 HOURS AFTER THE START OF LIGHT. SIMILARLY THERE ARE LARGE DIFFERENCES IN THE RATE OF INCREASE OF PROTEIN AND CARBOHYDRATE PRODUCTION AT THE 2ND AND 4TH HOURS AFTER START OF LIGHT DEPENDING ON THE DURATION OF 'WAITING' RECEIVED BY THE AUTOSPORES. (JONES-WISCONSIN)

FIELD 05C

ACCESSION NO. W73-00251

THE INTERLABORATORY PRECISION TEST. AN EIGHT LABORATORY EVALUATION OF THE PROVISIONAL ALGAL ASSAY PROCEDURE BOTTLE TEST,

NORTH CAROLINA UNIV., CHAPEL HILL. DEPT. OF ENVIRONMENTAL SCIENCES AND ENGINEERING.

C. M. WEISS, AND R. W. HELMS.

ENVIRONMENTAL PROTECTION AGENCY, NATIONAL EUTROPHICATION RESEARCH PROGRAM, OCTOBER 1971. 70 P, 10 FIG, 16 TAB, 3 APPEND. EPA PROGRAM 16010 DQT.

DESCRIPTORS:
*ANALYTICAL TECHNIQUES, *ALGAE, EVALUATION, TESTING, ASSAY, NUTRIENTS, CULTURES, MEASUREMENT, LABORATORY TESTS, BIOASSAY.

IDENTIFIERS:
*PAAP TEST, *ALGAL GROWTH, BOTTLE TEST.

ABSTRACT:
A COMPARATIVE EXPERIMENT TO EVALUATE THE 'BOTTLE TEST' OF THE PROVISIONAL ALGAL ASSAY PROCEDURE BY EIGHT LABORATORIES REQUIRED ALL PARTICIPANTS TO FOLLOW A SET OF PRESCRIBED CONDITIONS AND PROCEDURES USING COMMON MATERIALS AND CULTURES WITH THEIR RESULTS BEING SUBJECTED TO STATISTICAL ANALYSIS FOR THE PURPOSES OF ANSWERING THE QUESTION OF WHAT IS THE INHERENT VARIABILITY THAT MIGHT BE EXPECTED IN AN ALGAL ASSAY BOTTLE TEST. ALL LABORATORIES PREPARED THEIR TEST MEDIA FROM A COMMON BATCH OF REAGENTS SUPPLIED BY ONE LABORATORY AND ALL USED FRESH SUBCULTURES OF SELENASTRUM CAPRICORNUTUM OBTAINED FROM THE NATIONAL EUTROPHICATION PROGRAM. A REGRESSION ANALYSIS FOR EACH OF THE LABORATORIES, INDIVIDUALLY AND FOR ALL EIGHT COMBINED, WAS CARRIED OUT. IN THE FINAL ASSESSMENT THE RESULTS OF ALL EIGHT LABORATORIES CAN BE CONSIDERED AS A TEST UNIT AND THE DEGREE OF VARIATION OR COEFFICIENT OF VARIATION COMPUTED TO ESTABLISH THE EXPECTED ORDER OF PRECISION FOR THIS TYPE OF ASSAY. THE DATA ON THE AVERAGE VALUES AND COEFFICIENT OF VARIATION OF THE MEAN INDICATED A CERTAIN DEGREE OF CONSISTENCY AMONG VARIOUS MEASURING PARAMETERS AS WELL AS WHICH WOULD BE PREFERABLE. (JONES-WISCONSIN)

FIELD 05C

ACCESSION NO. W73-00255

ECOLOGY OF PHYTOPLANKTON OF THE VOLTA LAKE,

GHANA UNIV., LEGON. VOLTA BASIN RESEARCH PROJECT.

S. BISWAS.

HYDROBIOLOGIA, VOL 39, NO 2, P 277-288, 1972. 8 FIG, 6 TAB, 19 REF.

DESCRIPTORS:
*PHYTOPLANKTON, *ALGAE, *DISSOLVED OXYGEN, PHOTOSYNTHESIS, TURBIDITY, LIGHT PENETRATION, IMPOUNDMENTS, SEASONAL, COLOR, IRON, PHYSICOCHEMICAL PROPERTIES, DIATOMS, DOMINANT ORGANISMS.

IDENTIFIERS:
*VOLTA LAKE(GHANA), ACTINASTRUM GRACILIMUM, CRYPTOMONAS EROSA, PERIDINIUM AFRICANUM, NITZCHIA ACICULARIS, SYNEDRA ACUS.

ABSTRACT:
THE CHANGES WHICH TOOK PLACE IN THE PHYSICOCHEMICAL CONDITIONS IN A NEW IMPOUNDMENT OF THE VOLTA RIVER, GHANA ARE EXAMINED IN RELATION TO THE PHYTOPLANKTON. A CORRELATION BETWEEN THE DISSOLVED OXYGEN AND PHYTOPLANKTON, NOT OBSERVED BEFORE, WAS FOUND. INITIALLY FILLED WITH OPALESCENT BROWN FLOOD WATER RICH IN IRON, A DECREASE IN TRANSPARENCY LOWERED THE PHOTOSYNTHETIC ACTIVITY OF THE PHYTOPLANKTON. DATA FROM THE LOWER REACHES OF THE VOLTA LAKE ON TRANSPARENCY, COLOR, TOTAL IRON, AND DISSOLVED OXYGEN REVEALED SIGNIFICANT CHANGES FROM 1965 TO 1966, BUT NOT FROM 1966 TO 1967 THOUGH A COMMON SEASONAL PATTERN WAS EVIDENT. THESE FACTORS SHOWED CORRELATIONS OF SUCH SIGNIFICANCE THAT MANY OF THEM COULD BE ESTIMATED FROM A SIMPLE MEASUREMENT OF THE TRANSPARENCY ALONE. THE PHYTOPLANKTON DENSITY DID NOT CHANGE SIGNIFICANTLY EITHER FROM 1965 TO 1966 OR FROM 1966 TO 1967. THE CHIEF CONSTITUENTS, HOWEVER, PASSED THROUGH A SERIES OF CHANGES FROM A GREEN ALGA (ACTINASTRUM) IN EARLY 1965 TO THE FLAGELLATES (CRYPTOMONAS AND PERIDINIUM) DURING 1965 TO 1966, TO THE DIATOMS (NITZSCHIA AND SYNEDRA) DURING 1966 TO 1967. THE PHYTOPLANKTON DENSITY CORRELATED SIGNIFICANTLY ONLY WITH THE DISSOLVED OXYGEN FROM WHICH IT COULD BE ROUGHLY ESTIMATED. (JONES-WISCONSIN)

FIELD 05C, 02H

ACCESSION NO. W73-00256

A SLIDING-CHAMBER PHYTOPLANKTON SETTLING TECHNIQUE FOR MAKING PERMANENT QUANTITATIVE SLIDES WITH APPLICATIONS IN FLUORESCENT MICROSCOPY AND AUTORADIOGRAPHY,

ALASKA UNIV., COLLEGE. INST. OF MARINE SCIENCE.

C. COULON, AND V. ALEXANDER.

LIMNOLOGY AND OCEANOGRAPHY, VOL 17, NO 1, P 149-152, JANUARY 1972. 3 FIG, 1 TAB, 4 REF.

DESCRIPTORS:
*PHYTOPLANKTON, LABORATORY EQUIPMENT, SEDIMENTATION, METHODOLOGY, POLLUTANT IDENTIFICATION, PROTOZOA, CHRYSOPHYTA, CHLOROPHYTA, AQUATIC ALGAE, MARINE ALGAE, SESSILE ALGAE.

IDENTIFIERS:
*FLUORESCENT MICROSCOPY, *AUTORADIOGRAPHY, GLUTARALDEHYDE, SAMPLE PREPARATION, COUNTING, SAMPLE PRESERVATION, FLAGELLATES, RHODOMONAS, IKROAVICH LAKE, UTERMOHL METHOD.

ABSTRACT:
A SYSTEM HAS BEEN DEVISED FOR PREPARING PERMANENT QUANTITATIVE PHYTOPLANKTON SLIDES USING SEDIMENTATION TECHNIQUES. THESE SLIDES CAN BE USED WITH A CONVENTIONAL MICROSCOPE. A THREE-PART SETTLING CHAMBER ALLOWS SEPARATION OF THE SUPERNATANT FROM THE SETTLED MATERIAL. WATER-SOLUBLE MOUNTING MEDIUM ADDED TO THE SAMPLE BEFORE SETTLING PREVENTS DESICCATION AND DISTORTION OF THE ORGANISMS. THE PIGMENT-PRESERVING AND MOUNTING-MEDIUM QUALITIES OF GLUTARALDEHYDE COMBINED WITH THIS METHOD HAVE APPLICATIONS IN PLANKTON COUNTING AND AUTORADIOGRAPHY. (LONG-BATTELLE)

FIELD 05A

ACCESSION NO. W73-00273

RESISTANCE TO DDT OF A FRESHWATER ALGA,

OBERLIN COLL., OHIO. DEPT. OF BIOLOGY.

D. A. EGLOFF, AND R. PARTRIDGE.

THE OHIO JOURNAL OF SCIENCE, VOL 72, NO 1, P 6-10, JANUARY 6, 1972. 1 FIG, 1 TAB, 18 REF.

DESCRIPTORS:
DESCRIPTORS:*DDT, *RESISTANCE, *PLANT PHYSIOLOGY, CULTURES, PHOTOSYNTHESIS, AQUATIC ALGAE, CHLOROPHYLL, OXYGEN, RESPIRATION, CHLORINATED HYDROCARBON PESTICIDES, GROWTH RATES, ENVIRONMENTAL EFFECTS, PHYTOPLANKTON, CHLAMYDOMONAS, CHLOROPHYTA, WATER POLLUTION EFFECTS, ABSORPTION, PESTICIDE TOXICITY, LABORATORY TESTS.

IDENTIFIERS:
*CHLAMYDOMONAS REINHARDTII, CULTURE MEDIA, CHLORINATED HYDROCARBONS.

ABSTRACT:
LABORATORY CULTURES OF A FRESHWATER ALGA (CHLAMYDOMONAS REINHARDTII) WERE EXPOSED TO DDT(100 - 1,000 PPB) FOR 16 - 96 HOURS AT 18 - 22 C IN AN INORGANIC BASAL MEDIUM WITH AND WITHOUT ACETATE TO DETERMINE ITS EFFECTS ON GROWTH PHOTOSYNTHESIS, RESPIRATION, AND CHLOROPHYLL CONCENTRATION. EXPERIMENTAL DATA SHOWED THAT THE AMOUNT OF CHLOROPHYLL, OXYGEN EVOLUTION IN THE LIGHT, AND OXYGEN UPTAKE IN THE DARK WERE UNAFFECTED BY EXPOSURE TO DDT. GROWTH RATE AND FINAL CELL DENSITY WERE IDENTICAL IN CONTROL AND EXPERIMENTAL CULTURES EXPOSED TO 1000 PPB DDT IN THE GROWTH MEDIUM FOR NINE DAYS AT 22 C. (SNYDER-BATTELLE)

FIELD 05C

ACCESSION NO. W73-00281

NORTH CAROLINA MARINE ALGAE. I. THREE NEW SPECIES FROM THE CONTINENTAL SHELF,

DUKE UNIV., DURHAM, N.C. DEPT. OF BOTANY.

R. B. SEARLES.

PHYCOLOGIA, VOL 11, NO 1, P 19-24, MARCH 1972. 4 FIG, 10 REF.

DESCRIPTORS:
*DISTRIBUTION PATTERNS, *CHLOROPHYTA, *MARINE ALGAE, *RHODOPHYTA, *SYSTEMATICS, PHOTOGRAPHY, SEA WATER, *NORTH CAROLINA, MICROSCOPY, ATLANTIC OCEAN.

IDENTIFIERS:
TREMATOCARPUS PAPENFUSSII, GLOIODERMA ATLANTICA, CODIUM CAROLINIANUM, *ONSLOW BAY(NC).

ABSTRACT:
THREE NEW SPECIES OF MARINE ALGAE WERE IDENTIFIED FROM COLLECTIONS MADE FROM THE R/V EASTWARD USING A CERAME VIVAS ROCK DREDGE OR A CAPE TOWN DREDGE IN ONSLOW BAY, NORTH CAROLINA. WHOLE PLANTS WHICH HAD BEEN PRESERVED IN FORMALIN SEAWATER WERE PHOTOGRAPHED WHILE IMMERSED IN WATER. RED ALGAE WERE STAINED WITH AQUEOUS ANILINE BLUE DYE FOR MICROSCOPIC STUDY; DRAWINGS AND PHOTOGRAPHS OF THEM EMPHASIZE CYTOPLASMIC BOUNDARIES RATHER THAN CELL WALLS. DRAWINGS WERE MADE WITH THE AID OF A CAMERA LUCIDA. THE THREE SPECIES WERE: ONE GREEN ALGA, CODIUM CAROLINIANUM SP. NOV. OF THE CODIALES, AND TWO RED ALGAE, TREMATOCARPUS PAPENFUSSII SP. NOV. OF THE GIGARTINALES AND GLOIODERMA ATLANTICA SP. NOV. OF THE RHODYMENIALES. THE GENUS GLOIODERMA HAS NOT PREVIOUSLY BEEN REPORTED FROM THE ATLANTIC OCEAN. ALL THREE PLANTS GROW OFFSHORE ON THE NORTH CAROLINA CONTINENTAL SHELF. (LONG-BATTELLE)

FIELD 05A

ACCESSION NO. W73-00284

BORON IN SWEDISH AND NORWEGIAN FRESH WATERS,

UPPSALA UNIV. (SWEDEN). INST. OF LIMNOLOGY.

TH. AHL, AND E. JONSSON.

AMBIO, VOL 1, NO 2, P 66-70, APRIL 1972. 4 FIG, 6 TAB, 10 REF.

DESCRIPTORS:
FRESHWATER, *BORON, *PATH OF POLLUTANTS, METHODOLOGY, RIVERS, WATER POLLUTION SOURCES, ION EXCHANGE, *CHEMICAL ANALYSIS, WATER POLLUTION, WATER ANALYSIS, BIOASSAY, CYANOPHYTA, RESINS, AQUATIC PLANTS, AQUATIC ALGAE, NITRATES, NITRITES, EUTROPHICATION.

IDENTIFIERS:
SAMPLE PREPARATION, CHEMICAL INTERFERENCE, INTERLABORATORY STUDIES, PRECISION, *SWEDEN, LAKE MALAREN, MICROCYSTIS FLOS-AQUAE, PHRAGMITES COMMUNIS, *NORWAY, MACROPHYTES, FYRISAN RIVER, ARBOGAAN RIVER, BOTORPSSTROMMEN RIVER, ATRAN RIVER, MOTALA STROM RIVER, SPARGANIUM ERECTUM, SCHOENOPLECTUS LACUSTRIS, ACORUS CALAMUS, TYPHA ANGUSTIFOLIA.

ABSTRACT:
MONTHLY WATER SAMPLES WERE COLLECTED FROM SWEDISH AND NORWEGIAN RIVERS AND ANALYZED BY A MODIFIED CURCUMIN METHOD. SAMPLES WERE PASSED THROUGH A SODIUM-LOADED DOWEX 50-W-X4 COLUMN AND THEN CENTRIFUGED 5 MIN BEFORE ABSORPTION MEASUREMENT TO AVOID INTERFERENCE FROM PRECIPITATES. A LINEAR RELATIONSHIP WAS OBTAINED BETWEEN ABSORPTION AND CONCENTRATION IN THE RANGE 0-100 MICROGRAMS/L. TESTS SHOWED INTERFERENCE STILL OCCURRING FROM NITRATE AND NITRITE SO SAMPLES HIGH IN THESE COMPOUNDS WERE NO LONGER TESTED. THE BORON CONTENT IN THE RIVERS RANGED FROM 1 TO 1,046 MICROGRAMS B/L, WITH A MEDIAN VALUE FOR ALL RIVERS TESTED AT 13 MICROGRAMS/L. ANY REGIONAL DIFFERENCES WERE ATTRIBUTED TO DIFFERENCES IN GEOLOGY, LAND USE, AND POPULATION DENSITY. AQUATIC PLANTS WERE SUBJECTED TO ANALYSIS, AND THE RESULTS SHOWED THAT THE ENRICHMENT OF BORON IN AQUATIC PLANTS IS VERY SMALL COMPARED WITH THE ENRICHMENT OF PHOSPHORUS AND NITROGEN. SEWAGE WATER WAS ANALYZED AND SHOWED A FAR LOWER BORON CONTENT THAN ENGLISH SEWAGE WATER. THE DATA SHOW THAT BORON CONTENT IN THESE FRESHWATERS IS LOW, WITH A MEAN CONCENTRATION CLOSE TO THAT OF THE RIVERS OF THE WORLD. (MACKAN-BATTELLE)

FIELD 05B, 05C

ACCESSION NO. W73-00286

DISTRIBUTION AND ECOLOGY OF CERATIUM EGYPTIACUM HALIM AND ITS VALIDITY AS INDICATOR OF THE CURRENT REGIME IN THE SUEZ CANAL,

ALEXANDRIA UNIV. (EGYPT). DEPT. OF OCEANOGRAPHY.

N. M. DOWIDAR.

INTERNATIONALE REVUE DER GESAMTEN HYDROBIOLOGIE, VOL 56, NO 6, P 957-966, 1971. 3 TAB, 23 REF.

DESCRIPTORS:
*BIOINDICATORS, *ECOLOGICAL DISTRIBUTION, *CURRENTS(WATER), *DINOFLAGELLATES, ECOLOGY, DISTRIBUTION PATTERNS, MARINE ANIMALS, SALINITY, SAMPLING, SEA WATER, BODIES OF WATER, PYRROPHYTA, PROTOZOA, MARINE ALGAE, ZOOPLANKTON, PHYTOPLANKTON.

IDENTIFIERS:
*CERATIUM EGYPTIACUM, *SUEZ CANAL, FLAGELLATES, MEDITERRANEAN SEA, RED SEA, DRIFT ORGANISMS.

ABSTRACT:
CERATIUM EGYPTIACUM, AN ERYTHRAEAN DINOFLAGELLATE SPECIES, WAS RECORDED FOR THE FIRST TIME FROM THE MEDITERRANEAN WATERS IN 1966. THE SPECIES IS INDIGENOUS AND PERENNIAL TO THE RED SEA AND THE SOUTHERN PART OF THE SUEZ CANAL. ECOLOGICALLY THE SPECIES PROVED TO BE A STRICTLY NERITIC SURFACE WATER FORM AVOIDING THE OCEANIC AND DEEP WATERS. IT IS ALSO HIGHLY TOLERANT; ITS SALINITY AND TEMPERATURE RANGES RECORDED IN THIS INVESTIGATION ARE RESPECTIVELY 3.3-4.7 PERCENT AND 14.2 C - 33 C. THE OCCURRENCE OF THE SPECIES IN THE MEDITERRANEAN WATERS SUGGESTS A RECENT IMMIGRATION TO THAT SEA THROUGH THE SUEZ CANAL AND PROVIDES A GOOD EXAMPLE FOR STUDYING THE CURRENT REGIME IN THE CANAL, PARTICULARLY AFTER THE CONSTRUCTION OF THE ASWAN HIGH DAM. THE DISTRIBUTION OF THE SPECIES IN THE MEDITERRANEAN IS DETAILED. THE CURRENT IN THE CANAL IS NORTHWARD THROUGHOUT THE WHOLE YEAR AND IS NOT REVERSED IN AUGUST AND SEPTEMBER AS WAS PREVIOUSLY NORMAL. THE IMPORTANCE OF THE SUEZ CANAL AS A BIOLOGICAL LINK BETWEEN THE MEDITERRANEAN AND THE RED SEA, THE MIGRATION OF RED SEA PLANKTON ORGANISMS THROUGH THE CANAL AND THE FUTURE PROSPECTS OF THE NEW CONDITIONS ARE DISCUSSED. (LONG-BATTELLE)

FIELD 05B, 05C

ACCESSION NO. W73-00296

OCCURRENCE OF ELECTRA CRUSTULENTA (BRYOZOA) IN RELATION TC LIGHT,

STOCKHOLM UNIV. (SWEDEN). ASKO LAB.; AND STOCKHOLM UNIV. (SWEDEN). DEPT. OF ZOOLOGY.

L. SILEN, AND B-O. JANSSON.

OIKOS, VOL 23, NO 1, P 59-62, 1972. 2 FIG, 10 REF.

DESCRIPTORS:
*LIGHT, *GROWTH RATES, *ENVIRONMENTAL EFFECTS, LARVAE, ON-SITE TESTS, INVERTEBRATES, WATER POLLUTION EFFECTS, ECOLOGY, SEA WATER, LABORATORY EQUIPMENT, INHIBITION, ECOLOGICAL DISTRIBUTION, MARINE ALGAE, MARINE ANIMALS.

IDENTIFIERS:
*ELECTRA CRUSTULENTA, *BRYOZCA.

ABSTRACT:
EXPERIMENTS WERE PERFORMED TO RELATE THE OCCURRENCE OF NEW BRYOZOAN (ELECTRA CRUSTULENTA) COLONIES AND THE REACTIONS CF LARVAE TO LIGHT. TO TEST THE EFFECTS OF LIGHT CN BRYOZCAN COLONIES, VARICUS SETTLING EXPERIMENTS WERE CARRIED OUT AT DIFFERENT TIMES OF THE YEAR BY ALLOWING THE INVERTEBRATES TO SETTLE ON A SUBSTRATE HAVING BOTH AN ILLUMINATED AND SHADOWED SURFACE. A BLACK POLYETHENE TUBE WAS USED AS A SUBSTRATE FOR THESE SESSILE ORGANISMS, SINCE ILLUMINATED ORGANISMS LIVING ON THE TUBE'S SURFACE COULD BE DIRECTLY COMPARED TO ORGANISMS ON A PERMANENTLY SHADOWED INTERIOR. PHOTIC RESPONSES OF LARVAE, IN DIFFERENT STAGES OF DEVELOPMENT, WERE EXAMINED BY PLACING THE BRYOZCANS IN SEA WATER (15C) AND EXPOSING THEM TO LIGHT. RESULTS SHOWED THAT BRYOZOANS WERE ABLE TO GROW ON EITHER LIGHT OR DARK SURFACES; HOWEVER, LARVAE SHOWED A WEAK PREFERENCE FOR DARK SURFACES. DIRECT ILLUMINATION CF LARVAE APPEARED TO HAVE NO EFFECT. CONVERSELY, LIGHT COULD INDIRECTLY AFFECT THE SETTLING OF BRYOZOAN LARVAE OR THE GROWTH OF COLONIES BY RESULTING IN ALGAL GROWTH WHICH IS INHIBITORY TO THE BRYOZOANS. (LONG-BATTELLE)

FIELD 05C

ACCESSION NO. W73-00348

METABOLIC TRANSFORMATION OF DDT, DIELDRIN, ALDRIN, AND ENDRIN BY MARINE MICROORGANISMS,

WISCONSIN UNIV., MADISON. DEPT. CF ENTOMOLOGY.

K. C. PATIL, F. MATSUMURA, AND G. M. BOUSH.

ENVIRONMENTAL SCIENCE AND TECHNOLOGY, VOL 6, NO 7, P 629-632, JULY 1972. 2 FIG, 3 TAB, 16 REF.

DESCRIPTORS:
*CHLORINATED HYDROCARBON PESTICIDES, SEA WATER, *METABOLISM, BIODEGRADATION, *MARINE MICROORGANISMS, *MICROBIAL DEGRADATION, *DDT, *DIELDRIN, *ALDRIN, *ENDRIN, MARINE ALGAE, BOTTOM SEDIMENTS, PLANKTON, RADIOACTIVITY TECHNIQUES, MONITORING, SURFACE WATERS, PERSISTENCE, PATH OF POLLUTANTS, OCEANS, ESTUARIES, CULTURES, SAMPLING, HAWAII, SEPARATION TECHNIQUES, FILMS.

IDENTIFIERS:
*BIOTRANSFORMATION, BIOLOGICAL SAMPLES, FATE OF POLLUTANTS, METABOLITES, DUNALIELLA, AGMENELLUM QUADRAPLICATUM, C-14, THIN LAYER CHROMATOGRAPHY, CLEANUP, RADIOAUTOGRAPHY, SUBSTRATE UTILIZATION.

ABSTRACT:
A STUDY WAS MADE OF THE METABOLISM OF CHLORINATED HYDROCARBON PESTICIDES IN SEAWATER TO DETERMINE THEIR BIODEGRADATION IN A MARINE ENVIRONMENT. SAMPLES OF SEAWATER, BOTTOM SEDIMENTS, SURFACE FILMS, ALGAE, AND MARINE PLANKTON COLLECTED IN OAHU, HAWAII WERE TREATED WITH RADIOLABELED INSECTICIDES AT THE COLLECTION SITE AND CULTURED FOR 30 DAYS AT 23 C IN THE LABORATORY. MICROORGANISMS ISOLATED FROM THE SAMPLES WERE MONITORED FOR METABOLIC ACTIVITY ALONG WITH LABORATORY CULTURES OF UNICELLULAR ALGAE. THESE MICROBIAL CULTURES WERE ALSO LABELLED AND KEPT FOR 30 DAYS IN AN INCUBATOR AT 30 C. THIN-LAYER CHROMATOGRAPHIC TECHNIQUES WERE USED TO IDENTIFY METABOLITES AFTER CHLOROFORM EXTRACTION OF WATER SAMPLES. THE MOST SIGNIFICANT RESULT OF THE STUDY WAS THAT THE INSECTICIDES WERE NOT METABOLIZED IN PLAIN SEAWATER; EVEN IN RELATIVELY POLLUTED WATERS THEY WERE NOT DEGRADED. SEA BOTTOM SEDIMENTS SHOWED ONLY SLIGHT DEGRADATION ACTIVITY. MOST DEGRADATION ACTIVITY WAS ASSOCIATED WITH BIOLOGICAL SAMPLES SUCH AS ALGAE, PLANKTON, AND SURFACE FILMS. A NUMBER OF MICROORGANISMS IN PURE CULTURE ALSO SHOWED METABOLIC ACTIVITIES WITH PATTERNS SIMILAR TO THOSE OBSERVED IN FIELD COLLECTED SAMPLES. (MORTLAND-BATTELLE)

FIELD 05B, 05C

ACCESSION NO. W73-00361

ECOLOGICAL ENERGETICS OF THE SEA-WEED ZONE IN A MARINE BAY ON THE ATLANTIC COAST OF CANADA. II. PRODUCTIVITY OF THE SEAWEEDS,

BEDFORD INST., DARTMOUTH (NOVA SCOTIA).

K. H, MANN,

MARINE BIOLOGY, VOL 14, NO 3, P 199-209, JUNE 1972. 6 FIG, 5 TAB, 22 REF.

DESCRIPTORS:
*GROWTH RATES, *PRIMARY PRODUCTIVITY, ENERGY BUDGET, *MARINE ALGAE, MARINE PLANTS, BIOMASS, CANADA, COASTS, ATLANTIC OCEAN, SEASONAL, BAYS, SEA WATER, ORGANIC MATTER, MORTALITY, WATER TEMPERATURE, PHAEOPHYTA, SAMPLING, SCUBA DIVING, ECOLOGY, STANDING CROPS, PATH OF POLLUTANTS.

IDENTIFIERS:
*SEAWEEDS, LAMINARIA LONGICRURIS, LAMINARIA DIGITATA, AGARUM CRIBROSUM, *NOVA SCOTIA, *ST. MARGARET'S BAY.

ABSTRACT:
SCUBA DIVING TECHNIQUES WERE USED TO FOLLOW THE GROWTH OF SEAWEEDS IN ST. MARGARET'S BAY, NOVA SCOTIA OVER A 2-YEAR PERIOD. IT WAS FOUND THAT THE BLADES OF LAMINARIA LONGICRURIS, L. DIGITATA, AND AGARUM CRIBROSUM TURN OVER THEIR BIOMASS MANY TIMES A YEAR AND HAVE AMONG THE HIGHEST PRODUCTIVITIES OF ANY NATURAL COMMUNITY. THE SCUBA DIVERS PUNCHED HOLES OF ABOUT 5 MM DIAMETER AT A DISTANCE OF 10 CM FROM THE JUNCTION OF STRIPE AND BLADE. THE MOVEMENT OF THESE HOLES WAS FOLLOWED THROUGHOUT THE SEASONS AND NEW HOLES WERE PUNCHED WHEN THE OLD ONES APPROACHED THE TIPS OF THE BLADES. THE BLADES WERE FOUND TO BEHAVE LIKE MOVING BELTS OF TISSUE, GROWING AT THE BASE AND ERODING AT THE TIPS. ALL SIZES OF PLANTS GREW ABOUT 200 CM ANNUALLY WITH THE YEAR'S GROWTH AMOUNTING TO 1-5 TIMES THE INITIAL LENGTH. BIOMASS INCREASE WAS APPROXIMATELY PROPORTIONAL TO THE SQUARE OF THE LENGTH INCREMENT. MAXIMUM GROWTH RATES WERE ACHIEVED BETWEEN JANUARY AND APRIL WHEN WATER TEMPERATURE WAS CLOSE TO 0 C. MINIMUM RATES OCCURRED IN THE JULY-OCTOBER PERIOD. THESE PLANTS MAKE A CONSIDERABLE CONTRIBUTION TO THE PRODUCTIVITY OF COASTAL WATERS. SINCE THEIR MAXIMUM GROWTH IS AT LOWER TEMPERATURES, THIS CONTRIBUTION MAY BE MORE SIGNIFICANT IN NORTHERN WATERS. (MORTLAND-BATTELLE)

FIELD 05C, 05B

ACCESSION NO. W73-00366

ENVIRONMENTAL FACTORS AFFECTING THE STANDING CROP OF FORAMINIFERA IN SUBLITTORAL AND PSAMMOLITTORAL COMMUNITIES OF A LONG ISLAND SALT MARSH,

CITY COLL., NEW YORK. DEPT. OF BIOLOGY.

N. J. MATERA, AND J. J. LEE.

MARINE BIOLOGY, VOL 14, NO 2, P 89-103, MAY 1972. 9 FIG, 6 TAB, 38 REF.

DESCRIPTORS:
*SALT MARSHES, *PROTOZOA, *BIOLOGICAL COMMUNITIES, *ENVIRONMENTAL EFFECTS, *STANDING CROPS, TIDAL EFFECTS, MARINE ALGAE, SEA WATER, SALINITY, SAMPLING, CORES, DISTRIBUTION PATTERNS, CURRENTS(WATER), PARTICLE SIZE, SEDIMENTS, SEPARATION TECHNIQUES, SPATIAL DISTRIBUTION, ECOLOGICAL DISTRIBUTION, BENTHIC FAUNA, SIEVES, WATER TEMPERATURE, FLUORESCENT DYE, CHLOROPHYTA, PHAEOPHYTA.

IDENTIFIERS:
*FORAMINIFERA, SUBSTRATES, EPIPHYTES, SAMPLE PRESERVATION, SAMPLE PREPARATION, SUBLITTORAL, LONG ISLAND, ENTEROMORPHA INTESTINALIS, POLYSIPHONIA, ULVA LACTUCA, ZANICHELLIA PALUSTRIS, CODIUM, FUSCUS, MACROPHYTES, AMMOBACULITES DILATATUS, AMMONIA BECCARII, AMMOTIUM SALSUM, ELPHIDIUM ADVENUM, ELPHIDIUM CLAVATUM, ELPHIDIUM GALVESTONENSE, ELPHIDIUM GUNTERI, ELPHICIUM INCERTUM, ELPHIDIUM TRANSLUCENS, PROTELPHIDIUM TISBURYENSIS, QUINQUELOCUTINA LATA, QUINQUELOCUTINA SEMINULUM, TROCHAMMINA INFLATA, TROCHAMMINA MACRESCENS.

ABSTRACT:
FORAMINIFERA SAMPLES WERE COLLECTED WEEKLY IN THE TOWD POINT SALT MARSH ON LONG ISLAND 13 JUNE - 3 SEPTEMBER 1968. ALGAL SAMPLES WERE TAKEN AT LOW TIDE AND DEPOSITED IN TEST TUBES CONTAINING MILLIPORE FILTERED SEA WATER. ONE HOUR AFTER THE ADDITION OF ROSE BENGAL TO EACH TUBE, EACH SAMPLE WAS PRESERVED WITH A NEUTRAL FORMALIN SOLUTION. IN THE LABORATORY, A VORTEX MIXER WAS USED TO SEPARATE THE EPIPHYTIC COMMUNITY FROM THE MACROALGAL SUBSTRATE. THE EPIPHYTIC COMMUNITY WAS TRANSFERRED INTO ETHANOL AND FORAMINIFERA WERE ISOLATED WITH A PASTEUR PIPETTE AND PLACED ON SLIDES FOR STORAGE AND STUDY. CORE SEDIMENT SAMPLES WERE TAKEN AT BOTH HIGH AND LOW TIDES AND PREPAREC AND PRESERVED IN THE SAME MANNER. IN THE LABORATORY, THESE SAMPLES WERE SIEVED AND FRACTIONS GREATER THAN 0.074 MM WERE TRANSFERRED TO ETHANOL. CURRENT MOVEMENT WITHIN THE SAMPLING AREA WAS STUDIED WITH A FLUORESCIN DYE. SOME MACROPHYTES PROVIDE BETTER SUBSTRATES FOR FORAMINIFERA THAN OTHERS. DOMINANT SPECIES AMONG FORAMINIFERA WERE ELPHIDIUM INCESTUM AND PROTELPHIDIUM TISBURYENSIS. OVERALL DISTRIBUTION PATTERNS WERE DETERMINED BY SMALL RIVULETS FLOWING THROUGH THE MARSH WHICH ALTERED WATER TEMPERATURE, SALINITY, AND SEDIMENT GRAIN SIZE. THERE WAS NO EVIDENCE TOSUGGEST THAT FORAMINIFERA MIGRATE IN THE SEDIMENTS AS A FUNCTION OF TIDAL CHANGES. (MORTLAND-BATTELLE)

FIELD 05C, 05B, 05A

ACCESSION NO. W73-00367

INFLUENCE OF ENVIRONMENTAL PARAMETERS ON INTRASPECIFIC VARIATION IN FUCUS VESICULOSUS,

MAINE UNIV., ORONO. DEPT. OF OCEANOGRAPHY; AND MAINE UNIV., ORONO. DEPT. OF ZOOLOGY.

A. J. JORDAN, AND R. L. VADAS.

MARINE BIOLOGY, VOL 14, NO 3, P 248-252, JUNE 1972. 1 FIG, 2 TAB, 12 REF.

DESCRIPTORS:
*PHAEOPHYTA, *SALINITY, *ENVIRONMENTAL EFFECTS, *PLANT MORPHOLOGY, *GENETICS, *VARIABILITY, WAVES(WATER), MAINE, STATISTICAL METHODS, SEA WATER, COASTS, MARINE ALGAE.

IDENTIFIERS:
*FUCUS VESICULOSUS, VESICULATION, SAMPLE PREPARATION, BRANCHING.

ABSTRACT:
THE VESICULATION AND BRANCHING OF THE MARINE BROWN ALGAE, FUCUS VESICULOSUS, WERE EXAMINED FOR SAMPLES TAKEN FROM THREE LOCATIONS ON THE MAINE COAST. THESE SITES WERE A NORMAL SALINITY EXPOSED AREA, A LOW SALINITY PROTECTED AREA, AND A NORMAL SALINITY PROTECTED AREA. AFTER COLLECTION, THE TOP AND BOTTOM OF EACH PLANT WERE CUT AND EACH SEGMENT WAS NUMBERED AND WEIGHED AND THE VESICLES AND DICHOTOMIES WERE COUNTED. THE VESICLES WERE MEASURED AND THEIR VOLUMES CALCULATED. A TOTAL OF 10 MEASUREMENTS OF MORPHOMETRIC CHARACTERISTICS AND RATIOS OF THESE MEASUREMENTS WERE COMPARED FOR EACH OF 300 PLANTS BY STATISTICAL ANALYSES ON A COMPUTER. THE RESULTS INDICATED THAT DECREASED SALINITY CORRELATED WITH INCREASED VESICULATION AND BRANCHING. INCREASED WAVE ACTION CORRELATED WITH DECREASED VESICULATION. (MORTLAND-BATTELLE)

FIELD 05C, 05A

ACCESSION NO. W73-00369

DENSITY-GRADIENT CENTRIFUGATION AS AN AID TO SORTING PLANKTONIC ORGANISMS. I. GRADIENT MATERIALS,

MARINE BIOLOGICAL LAB., WOODS HOLE, MASS.

R. A. BOWEN, J. M. ST. ONGE, J. B. COLTON, JR., AND C. A. PRICE.

MARINE BIOLOGY, VOL 14, NO 3, P 242-247, JUNE 1972. 5 FIG, 11 REF.

DESCRIPTORS:
*ZOOPLANKTON, *LARVAE, *FISH EGGS, *SEPARATION TECHNIQUES, *MARINE FISH, SILICA, MARINE ALGAE, INVERTEBRATES, ISOLATION, REFRACTIVITY, COPEPODS, MARINE ANIMALS, MOLLUSKS, GASTROPODS, PHAEOPHYTA.

IDENTIFIERS:
*DENSITY-GRADIENT CENTRIFUGATION, ISOPYCNIC CENTRIFUGATION, SAMPLE PRESERVATION, SALPA, LUDOX AM, SUCROSE, DENSITY GRADIENTS, TUNICATES, PHAEOCYSTIS, SORTING, SODIUM BROMIDE, DEXTRAN, OIKOPLEURA, SAGITTA, PTEROPODS, EUPHAUSIIDS, GADUS MORHUA, UROPHYCIS.

ABSTRACT:
DENSITY-GRADIENT CENTRIFUGATION WAS STUDIED AS AN AID TO THE SORTING OF ICHTHYOPLANKTON AND IT WAS SHOWN THAT FISH EGGS AND LARVAE CAN BE SEPARATED BY THIS METHOD USING GRADIENTS OF SUCROSE OR SILICA. PRESERVED SAMPLES WERE LAYERED OVER LINEAR GRADIENTS IN 100 CC SWINGING BUCKETS AND CENTRIFUGED FOR 1 HR AT 1000 RPM. IN THE SUCROSE GRADIENTS, ZOOPLANKTON WERE CONFINED TO TWO ENDS OF THE GRADIENT AND THE FISH EGGS TO AN INTERMEDIATE ZONE. USING SILICA GRADIENTS, IN THE FORM OF LUDOX AM, THE FISH EGGS BANDED IN ONE NARROW ZONE, THE FISH LARVAE IN ANOTHER, AND NEARLY ALL OF THE INVERTEBRATE PLANKTON WERE BETWEEN. BOTH SODIUM BROMIDE AND DEXTRAN GRADIENTS WERE ALSO TRIED BUT PROVED UNSUITABLE. SIX CLASSES OF ZOOPLANKTON WERE STUDIED AND ONLY SALPA OVERLAPPED APPRECIABLY WITH FISH EGGS. THERE WAS NO OVERLAP WITH LARVAE. THE LUDOX AM GRADIENT WAS FOUND TO OFFER THE MOST ADVANTAGES. (MORTLAND-BATTELLE)

FIELD 05A

ACCESSION NO. W73-00371

STUDIES OF MARINE FOULING AND BORING OFF KODIAK ISLAND, ALASKA,

NAVAL OCEANOGRAPHIC OFFICE, WASHINGTON, D. C.

E. R. LONG,

MARINE BIOLOGY, VOL 14, NO 1, P 52-57, MAY 1972. 3 FIG, 3 TAB, 15 REF.

DESCRIPTORS:
*FOULING, *ARTIFICIAL SUBSTRATES, DEPTH, *SAMPLING, *PERIPHYTON, EQUIPMENT, TUBIFICIDS, MOLLUSKS, WATER TEMPERATURE, MARINE ANIMALS, ALASKA, BIOMASS, SALINITY, BENTHIC FAUNA, ASBESTOS, INORGANIC COMPOUNDS, WORMS, INVERTEBRATES, ANNELIDS, SEASONAL, CRUSTACEANS, ISOPODS, MARINE ALGAE, *ALASKA.

IDENTIFIERS:
*WOOD BORERS, BARNACLES, BRYOZOAS, HYDROIDS, SPONGES, COELENTERATES, PORIFERA, POLYCHAETES, SERPULIDS.

ABSTRACT:
MARINE FOULING AND BORING OFF KODIAK ISLAND, ALASKA, WAS STUDIED AT DEPTHS OF 5, 15, AND 30 M BY EXPOSING WOOD/ASBESTOS TEST PANELS AND RETRIEVING THEM AT MONTHLY AND CUMULATIVELY LONGER INTERVALS. FOULING WAS MODERATE BETWEEN JUNE AND OCTOBER AND NEGLIGIBLE BETWEEN NOVEMBER AND MAY. THE FOULING COMMUNITIES AT 15 AND 30 M WERE QUITE DISSIMILAR, PROBABLY DUE TO A LARGE SURFACE-TO-BOTTOM SALINITY GRADIENT. THE COMMUNITY AT THE 30 M LEVEL WAS DOMINATED BY BALANUS CRENATUS AND PSEUDOCHITINOPOMA OCCIDENTALIS, THAT AT 15 M BY ALCYONIDIUM POLYOUM AND OBELIA BOREALIS. AT THE 5 M DEEP, PIERSIDE SITE, THE MOST COMMON ORGANISMS WERE B. CRENATUS, O. BOREALIS AND DESMACYSTIS SANDALIA. BORER ATTACK WAS A TRACE AT 5 AND 15 M AND SLIGHTLY MORE SEVERE AT 30 M. (SNYDER-BATTELLE)

FIELD 05A

ACCESSION NO. W73-00374

PRODUCTION OF DISSOLVED ORGANIC MATTER FROM DEAD GREEN ALGAL CELLS. I. AEROBIC MICROBIAL DECOMPOSITION,

TOKYO METROPOLITAN UNIV. (JAPAN). DEPT. OF CHEMISTRY.

A. OTSUKI, AND T. HANYA.

LIMNOLOGY AND OCEANOGRAPHY, VOL 17, NO 2, P 248-257, MARCH 1972. 12 FIG, 3 TAB, 28 REF.

DESCRIPTORS:
*AEROBIC BACTERIA, *AEROBIC CONDITIONS, *KINETICS, *SCENEDESMUS, *DISSOLVED SOLIDS, *MICROBIAL DEGRADATION, CARBON, ORGANIC MATTER, WATER POLLUTION SOURCES, CHLOROPHYTA, AQUATIC PRODUCTIVITY, AQUATIC ALGAE, NITROGEN, POLLUTANT IDENTIFICATION, ENVIRONMENTAL EFFECTS, LABORATORY EQUIPMENT, FREEZE DRYING, SEPARATION TECHNIQUES, SOLVENT EXTRACTION, MATHEMATICAL STUDIES, SYSTEMATICS, METABOLISM.

IDENTIFIERS:
FATE OF POLLUTANTS, SUBSTRATE UTILIZATION, ORGANIC NITROGEN, ORGANIC CARBON, CULTURE MEDIA, PAPER CHROMATOGRAPHY, INFRARED SPECTROSCOPY, ASSIMILATION, KJEDAHL PROCEDURE, SAMPLE PREPARATION.

ABSTRACT:
THE PRODUCTION OF DISSOLVED ORGANIC MATTER (DOM) FROM DEAD GREEN ALGAL CELLS (SCENEDESMUS) BY AEROBIC BACTERIA WAS INVESTIGATED WITH PARTICULAR EMPHASIS ON THE PATTERNS AND KINETICS OF C AND N. DOM WAS PREPARED FOR EXPERIMENTAL USE BY FREEZE-DRYING AND ASHING APPROXIMATELY 99 PERCENT PURE CULTURES OF SCENEDESMUS. AEROBIC DECOMPOSITION OF THE DOM WAS INITIATED BY INOCULATING MICROFLORA INTO A SPECIALLY PREPARED DECOMPOSITION APPARATUS CONTAINING CULTURE MEDIA AND DOM AT PH 7. AFTER INCUBATION, DISSOLVED ORGANIC CARBON (DOC) AND NITROGEN (DON) WERE MEASURED BY AN ELEMENTAL ANALYZER AND BY THE MICRO-KJELDAHL METHOD. ORGANIC MATERIALS, SUCH AS PROTEIN AND AMINO ACIDS, WERE EXTRACTED BY THIN-LAYER CHROMATOGRAPHY FOR IR-SPECTROSCOPIC ANALYSIS. THE DOM PRODUCTION, BY THE 30TH DAY OF INCUBATION, WAS ABOUT 7 PERCENT C AND 6 PERCENT N. AEROBIC MICROBIAL DECOMPOSITION APPROXIMATED A FIRST ORDER REACTION DURING THIS PERIOD. KINETIC CONSIDERATIONS OF THE DECOMPOSITION PATTERN OF THE CELL NITROGEN AND THE PRODUCTION OF DON SUGGEST THAT DEAD ALGAL CELL SUBSTANCE MAY BE DIVIDED INTO LABILE AND REFRACTORY CONSTITUENTS BY THEIR RELATIVE RESISTANCE TO THE ACTION OF BACTERIA. THE DISSOLVED ORGANIC NITROGENOUS MATERIAL PRODUCED IS COMPOSED OF TWO MAJOR FRACTIONS: ONE IS PRODUCED WITH THE DECOMPOSITION OF ALGAL CELLS, AND THE OTHER IS PROBABLY EXCRETED BY BACTERIA THROUGH REASSIMILATION OF MINERALIZED NITROGEN. (LONG-BATTELLE)

FIELD 05C, 05B

ACCESSION NO. W73-00379

UTILIZATION OF UREA BY SOME MARINE PHYTOPLANKTERS,

WOODS HOLE OCEANOGRAPHIC INSTITUTION, MASS.

E. J. CARPENTER, C. C. REMSEN, AND S. W. WATSON.

LIMNOLOGY AND OCEANOGRAPHY, VOL 17, NO 2, P 265-269, MARCH 1972. 1 FIG, 3 TAB, 19 REF.

DESCRIPTORS:
*PHYTOPLANKTON, *MARINE ALGAE, CHEMICAL ANALYSIS, *NITRATES, *AMMONIA, DIATOMS, GROWTH RATES, NITROGEN, WATER POLLUTION EFFECTS, RADIOACTIVITY, METABOLISM, PLANT PHYSIOLOGY, CHRYSOPHYTA, CULTURES, SEA WATER, NUTRIENT REQUIREMENTS, WATER POLLUTION.

IDENTIFIERS:
*SUBSTRATE UTILIZATION, *SARGASSO SEA, COCCOLITHOPHORIDS, FLAGELLATES, FATE OF POLLUTANTS, STEPHANOPYXIS COSTATA, ASSIMILATION, EMILIANA(COCCOLITHUS) HUXLEYI, AMPHIPRORA ALATA, CHRYSOCHROMULINA, SKELETONEMA, CHAETOCEROS SIMPLEX.

ABSTRACT:
THE POSSIBLE EXISTENCE OF ADDITIONAL UREA-UTILIZING PHYTOPLANKTERS IN OFFSHORE AND INSHORE AREAS WAS INVESTIGATED AND THEIR GROWTH RATES ON UREA, NITRATE, AND AMMONIA WERE COMPARED. THE CONCEPT OF WHETHER THESE SPECIES CAN ASSIMULATE UREA AT SIGNIFICANT RATES AT THE CONCENTRATIONS PRESENT IN THEIR NORMAL HABITATS WAS ALSO CONSIDERED. THREE DIATOMS FROM THE SARGASSO SEA AND TWO DIATOMS AND A HAPTOPHYTE FLAGELLATE FROM AN INSHORE AREA NEAR WOODS HOLE WERE FOUND TO EXHIBIT SIMILAR GROWTH RATES ON UREA, NITRATE, AND AMMONIA. A COCCOLITHOPHORE EMILIANA (COCCOLITHUS) HUXLEYI FROM THE SARGASSO SEA DID NOT GROW ON UREA. THE HALF-SATURATION CONSTANT FOR UREA DETERMINED FOR ONE INSHORE DIATOM, STEPHANOPYXIS COSTATA (SKELETONEMA COSTATUM), WAS 8.5 MICROGRAMS-ATOM UREA-N/LITER. AT THE UREA CONCENTRATIONS OF ITS HABITATS, THE CALCULATED DIVISION RATE FOR THIS SPECIES, GROWING ON UREA AS THE SOLE NITROGEN SOURCE, WAS 2.2 DAYS, SIMITAR TO THAT OF DIATOMS GROWING IN INSHORE HABITATS. IT APPEARS, FROM DATA ON NATURAL UREA CONCENTRATIONS AND UREA UPTAKE RATES BY S. COSTATA, THAT UREA CAN BE A SIGNIFICANT N SOURCE FOR AT LEAST ONE COMMON INSHORE PHYTOPLANKTER. ACCORDING TO GROWTH RATE STUDIES, AT LEAST THREE INSHORE AND THREE OFFSHORE MARINE PHYTOPLANKTON CAN GROW AS RAPIDLY ON UREA AS ON NITRATE OR AMMONIA. (BYRD-BATTELLE)

FIELD 05C, 05B

ACCESSION NO. W73-00380

SIGNIFICANCE TO EUTROPHICATION OF SPATIAL DIFFERENCES IN NUTRIENTS AND DIATOMS IN LAKE MICHIGAN,

WISCONSIN UNIV., MILWAUKEE. CENTER FOR GREAT LAKES STUDIES.

R. E. HOLLAND, AND A. M. BEETON.

LIMNOLOGY AND OCEANOGRAPHY, VOL 17, NO 1, P 88-96, JANUARY 1972. 4 FIG, 1 TAB, 22 REF.

DESCRIPTORS:
*SAMPLING, *NUTRIENTS, *DIATOMS, *LAKE MICHIGAN, *EUTROPHICATION, *SPATIAL DISTRIBUTION, PHOSPHORUS, NITRATES, SILICA, PATH OF POLLUTANTS, ORGANIC COMPOUNDS, PHYTOPLANKTON, CARBON, WATER TEMPERATURE, GREAT LAKES REGION, WISCONSIN, MICHIGAN, CHRYSOPHYTA, PLANT PIGMENTS, AQUATIC ALGAE, NITROGEN.

IDENTIFIERS:
CHLOROPHYLL A, CHLOROPHYLL C, CAROTENOIDS, FRAGILARIA CROTONENSIS, TABELLARIA FLOCCULOSA, STEPHANODISCUS HANTZSCHII, CYCLOT MICHIGANIANA, RHIZOSOLENIA ERIENSIS, MELOSIRA ISLANDICA, MELOSIRA ITALICA, MELOSIRA AMBIGUA, SYNEDRA FILIFORMIS, FRAGILARIA CAPUCINA, CYCLOTELLA STELLIGERA, STEPHANODISCUS.

ABSTRACT:
WATER SAMPLES WERE TAKEN FROM A WATER-COOLING INTAKE AT A DEPTH OF ABOUT 4 M FROM A RAILROAD FERRY BETWEEN MILWAUKEE, WISCONSIN, AND LUDINGTON, MICHIGAN, 27 MAY 1970 - 6 JANUARY 1971, TO DETERMINE INSHORE-OFFSHORE DIFFERENCES IN SPECIES AND ABUNDANCE OF DIATOMS, AND CONCENTRATIONS OF PHOSPHORUS, SILICA, NITRATE, AND PIGMENTS. THE VARIOUS NUTRIENTS AND PIGMENTS WERE DETERMINED, AND THE DIATOMS IDENTIFIED BY PREVIOUSLY DESCRIBED METHODS. INSHORE WATERS (WITHIN 16 KM OF SHORE) HAD GREATER DIATOM POPULATIONS, DIFFERENT SPECIES COMPOSITION WITHIN THE DIATOM COMMUNITIES, AND DIFFERENT CONCENTRATIONS OF MAJOR NUTRIENTS AND PIGMENTS THAN OFFSHORE WATERS (GREATER THAN OR EQUAL TO 16 KM FROM SHORE). RESULTS INDICATE THAT DATA FROM WATER INTAKES WHICH HAVE BEEN USED IN THE PAST TO DOCUMENT CHANGES IN THE LAKE MAY HAVE REPRESENTED ONLY LOCAL CONDITIONS. (SNYDER-BATTELLE)

FIELD 05C, 05B

ACCESSION NO. W73-00384

DISTRIBUTION AND TAXONOMY OF LAMINARIA SINCLAIRII AND L. LONGIPES (PHAEOPHYCEAE, LAMINARIALES),

BRITISH COLUMBIA UNIV., VANCOUVER. DEPT. OF BOTANY.

J. W. MARKHAM.

PHYCOLOGIA, VOL 11, NO 2, P 147-157, JUNE 1972. 6 FIG, 3 TAB, 79 REF.

DESCRIPTORS:
*DISTRIBUTION PATTERNS, *SYSTEMATICS, *ECOLOGICAL DISTRIBUTION, *SPATIAL DISTRIBUTION, GEOGRAPHICAL REGIONS, PHAEOPHYTA, MARINE ALGAE, ENVIRONMENTAL EFFECTS, MARINE PLANTS, CANADA, WASHINGTON, OREGON, CALIFORNIA, ALASKA, SEA WATER.

IDENTIFIERS:
*LAMINARIA SINCLAIRII, *LAMINARIA LONGIPES, BRITISH COLUMBIA, RUSSIA, LAMINARIA RODRIGUEZII.

ABSTRACT:
LAMINARIA SINCLAIRII (HARV.) FARL., AND., AND EAT. OCCURS FROM CENTRAL BRITISH COLUMBIA TO SOUTHERN CALIFORNIA. L. LONGIPES BORY OCCURS FROM THE KURILE ISLANDS THROUGHOUT THE ALEUTIAN ISLANDS TO SOUTHEAST ALASKA. DETAILED DISTRIBUTION RECORDS ARE PRESENTED. BOTH SPECIES ARE DISTINGUISHED FROM OTHER SPECIES OF LAMINARIA IN HAVING MULTIPLE STIPES FROM AN EXPANDED, RHIZOME-LIKE HOLDFAST. THE TWO SPECIES ARE VERY SIMILAR IN APPEARANCE. MOST AUTHORS HAVE DISTINGUISHED THEM ON THE BASIS OF INTERNAL STIPE ANATOMY; L. SINCLAIRII HAS MUCILAGE DUCTS IN THE STIPE, WHEREAS L. LONGIPES DOES NOT. ECOLOGICAL STUDIES WERE CARRIED OUT ON BOTH SPECIES IN THE FIELD AND IN THE LABORATORY. THE RESULTS SHOW THE SPECIES ALSO DIFFER IN SEVERAL OTHER POINTS, INCLUDING LENGTH OF STIPES, WIDTH OF BLADES, WINTER LOSS OF BLADES, MORPHOLOGY OF GAMETOPHYTES, AND HABITAT. THE EVIDENCE CONFIRMS THAT THEY SHOULD BE RETAINED AS TWO SEPARATE SPECIES. (LONG-BATTELLE)

FIELD 05A, 05C

ACCESSION NO. W73-00428

PROCEDURES ADOPTED FOR THE LABORATORY CULTIVATION OF TRICHODESMIUM ERYTHRAEUM,

CENTRE FOR ADVANCED STUDY IN MARINE BIOLOGY, PORTO NOVA (INDIA).

V. D. RAMAMURTHY.

MARINE BIOLOGY, VOL 14, NO 3, P 232-234, JUNE 1972. 11 REF.

DESCRIPTORS:
*CULTURES, *CYANOPHYTA, *MARINE ALGAE, *ISOLATION, MARINE PLANTS, EUTROPHICATION, OCEANS, SAMPLING, SEPARATION TECHNIQUES, CULTIVATION, VITAMIN B, PLANT GROWTH REGULATORS, NITRATES, PHOSPHATES, NUTRIENT REQUIREMENTS, ESSENTIAL NUTRIENTS.

IDENTIFIERS:
*TRICHODESMIUM ERYTHRAEUM, *CULTURING TECHNIQUES, CULTURE MEDIA, *BAY OF BENGAL, INDIAN OCEAN, SOIL EXTRACTS, ERDSCHREIBER MEDIUM, ESTUARINE SOILS, AXENIC CULTURES, GIBBERELIC ACID.

ABSTRACT:
VARIOUS CULTURING TECHNIQUES WERE INVESTIGATED FOR THE AXENIC OR PURE CULTIVATION OF THE MARINE BLUE-GREEN ALGA TRICHODESMIUM ERYTHRAEUM ISOLATED A DEPTH OF 9-12 M IN THE BAY OF BENGAL DURING THE PRE-BLOOM PERIOD IN MARCH, 1965. MODIFIED, ENRICHED, ERDSCHREIBER MEDIUM, CONTAINING A DIFFERENTIAL CONCENTRATION OF STREPTOMYCIN, TETRACYCLINE, AND SULFADIAZINE FOR INHIBITING MICROBIAL GROWTH, WAS USED FOR CULTIVATING THE ALGAE. SUB-CULTURES WERE FORMED EVERY 7-9 DAYS BY INOCULATING FRESH MEDIUM WITH 10-15 CELLS PER 10 ML. THE INGREDIENTS IN THE CULTURE MEDIUM INCLUDE: NITRATE, PHOSPHATE, A 50 ML ESTUARINE SOIL EXTRACT, SEAWATER (3.3 PERCENT SALINITY), VITAMIN B12, GIBBERELLIC ACID, AND ANTIBIOTICS. INSTRUCTIONS FOR THE ISOLATION OF THE ALGAE AND THE PREPARATION OF THE CULTURE MEDIA ARE PRESENTED. (LONG-BATTELLE)

FIELD 05C, 05A

ACCESSION NO. W73-00430

PRIMARY PRODUCTION AT THE THERMOCLINE LEVEL IN THE NERITIC ZONE OF THE NORTH-OCCIDENTAL MEDITERRANEAN, (PRODUCTION PRIMAIRE AU NIVEAU DE LA THERMOCLINE EN ZONE NERITIQUE DE MEDITERRANEE NORD-OCCIDENTALE),

ARAGO LAB., BANYULS-SUR-MER (FRANCE).

G. CAHET, M. FIALA, G. JACQUES, AND M. PANOUSE.

MARINE BIOLOGY, VOL 14, NO 1, P 32-40, MAY 1972. 7 FIG, 3 TAB, 26 REF.

DESCRIPTORS:
*PHYTOPLANKTON, *THERMOCLINE, *PRIMARY PRODUCTIVITY, *NERITIC, NITRATES, RADIOACTIVITY TECHNIQUES, EUTROPHICATION, MARINE ALGAE, WATER POLLUTION SOURCES, SPATIAL DISTRIBUTION, BIOMASS, PHOTOSYNTHESIS, NUTRIENTS, LIMITING FACTORS, SALINITY, CARBON RADIOISOTOPES, SESTON, ORGANIC MATTER, DISTRIBUTION PATTERNS, THERMAL PROPERTIES, THERMAL STRATIFICATION, CHLOROPHYLL, MAGNESIUM COMPOUNDS, PIGMENT, SURFACE WATERS, SEDIMENTATION, PHOSPHATES.

IDENTIFIERS:
NITZSCHIA DELICATISSIMA, NITZSCHIA PUNGENS, C-14, CHLOROPHYLL A, MEDITERRANEAN SEA, FRANCE, PYCNOCLINE.

ABSTRACT:
VERTICAL PROFILES OF PHYSICAL, CHEMICAL AND PHYTOPLANKTONIC PARAMETERS ARE DESCRIBED, AT THE LEVEL OF THE THERMOCLINE, IN THE AREA OF BANYULS-SUR MER, FRANCE. THE RESULTS SHOW THAT THE THERMOCLINE DIVIDES TWO MASSES OF WATER: (1) MEDITERRANEAN SURFACE WATER WITH LOW NUTRIENT CONCENTRATIONS AND A SALINITY BELOW 3.8 PERCENT; (2) DEEP, NUTRIENT-RICH UPWELLED WATER (N-NO3 GREATER THAN 3 MICROAT-G/L, P-PO4 GREATER THAN 0.3 MICROAT-G/L, GREATER THAN 3.83 PERCENT S), WHICH COMES FROM THE UPPER LIMIT OF THE MEDITERRANEAN INTERMEDIATE WATER, USUALLY LOCATED AT THE 200 M LEVEL. CONSEQUENTLY, CONDITIONS ARE SUITABLE FOR HIGH PRODUCTION RATES AT THE BOTTOM OF THE THERMOCLINE, WHERE CHL A IS ABOVE 0.5 MG/CU M; DOMINANT SPECIES ARE NITZSCHIA DELICATISSIM AND N. PUNGENS. A DIAGRAM IS PRESENTED EXPLAINING THE DIFFERENT EFFECTS OF THE PYCNOCLINES ON PRIMARY PRODUCTION: EUTROPHICATION AT THE PYCNOCLINE LEVELS IS THE RESULT OF PASSIVE ACCUMULATION OF PHYTOPLANKTON AND ORGANIC MATTER DURING SEDIMENTATION, AND/OR OF REDUCED DIFFUSION OF NUTRIENTS FROM DEEP WATERS TOWARDS THE SURFACE. (LONG-BATTELLE)

FIELD 05C, 05B

ACCESSION NO. W73-00431

STEPHANODISCUS BINDERANUS (KUTZ.) KRIEGER OR MELOSIRA BINDERANA KUTZ. (BACILLARIOPHYTA, CENTRALES),

BRISTOL UNIV. (ENGLAND). DEPT. OF BOTANY.

F. E. ROUND.

PHYCOLOGIA, VOL 11, NO 2, P 109-116, JUNE 1972. 27 FIG, 22 REF.

DESCRIPTORS:
*DIATOMS, *BIOINDICATORS, *EUTROPHICATION, *POLLUTANT IDENTIFICATION, *LAKE MICHIGAN, CHRYSOPHYTA, AQUATIC ALGAE, ELECTRON MICROSCOPY, PHOTOGRAPHY, WATER POLLUTION EFFECTS, WATER PURIFICATION, *SYSTEMATICS.

IDENTIFIERS:
*STEPHANODISCUS BINDERANUS, *MELOSIRA BINDERANA, SCANNING ELECTRON MICROSCOPY, TRANSMISSION ELECTRON MICROSCOPY, LIGHT MICROSCOPY.

ABSTRACT:
THE CENTRIC DIATOM WHICH HAS BEEN IDENTIFIED EITHER AS A SPECIES OF MELOSIRA C. A. AG. OR OF STEPHANODISCUS C. G. EHRENB IS AN INDICATOR OF INCREASING EUTROPHICATION IN LAKES, AND IS ALSO RESPONSIBLE FOR SERIOUS PROBLEMS IN WATER PURIFICATION PLANTS. WATER SAMPLES WERE OBTAINED FROM LAKE MICHIGAN TO DETERMINE THE GENUS TO WHICH THIS SPECIES BELONGS. THE ALGAE WERE INVESTIGATED BY LIGHT, SCANNING, AND TRANSMISSION MICROSCOPY. THIS REPRESENTS THE FIRST DETAILED STUDY OF THIS SPECIES AND CONFIRMED ITS POSITION IN STEPHANODISCUS. (LONG-BATTELLE)

FIELD 05C, 05A, 02H

ACCESSION NO. W73-00432

INTERREGULATION OF MARINE PLANKTONIC DIATOMS IN MONO-AND MIXED CULTURES, (IN RUSSIAN),

MOSCOW STATE UNIV. (USSR).

V. D. FEDOROV, AND N. G. KUSTENKO.

OKEANOLOGIYA, VOL 12, NO 1, P 111-122, JANUARY/FEBRUARY 1972. 10 FIG, 2 TAB, 14 REF.

DESCRIPTORS:
*DIATOMS, *CULTURES, *NUTRIENT REQUIREMENTS, SEA WATER, NITRATES, PHOSPHORUS, NUTRIENTS, CHRYSOPHYTA, COMPETITION, WATER POLLUTION SOURCES, POLLUTANT IDENTIFICATION, PHYTOPLANKTON, MARINE ALGAE, ESSENTIAL NUTRIENTS.

IDENTIFIERS:
*INTERREGULATION, SKELETONEMA COSTATUM, THALASSIONEMA NITZSCHIOIDES, MIXED CULTURES, METABOLITES, SUBSTRATE UTILIZATION, GROWTH.

ABSTRACT:
A METHOD OF NUTRIENT SALT ADDITIONS IS USED IN COMBINATION WITH THE DESIGN OF THE FULL FACTOR EXPERIMENTS TO SHOW THE PREVAILING DEPENDENCE OF THE GROWTH OF THE ALGOLOGICALLY PURE CULTURE OF SKELETONEMA COSTATUM ON NITRATE NITROGEN CONCENTRATIONS AND OF THALASSIONEMA NITZSCHIOIDES ON PHOSPHORUS CONCENTRATIONS. BY VARYING THE CONCENTRATIONS OF THESE TWO ELEMENTS IN THE NUTRIENT MEDIUM (THE SEA WATER WITH NITROGEN AND PHOSPHORUS ADDITIONS), AS WELL AS THE SOWING TITER OF EACH DIATOM SPECIES, THE EXPERIMENTERS SUCCEEDED IN CHANGING THE DIRECTION OF THE EXCLUSION OF SOME SPECIES BY THE COMPETITIVE ONE WHEN THE SPECIES ARE CULTIVATED TOGETHER. THE MECHANISM OF THE QUANTITATIVE INTERREGULATION OF BOTH CULTURES IS SHOWN TO BE BASED ON THE ACTIVITY OF EXTRACELLULAR DIATOM METABOLITES THE ACCUMULATION AND ACTION OF WHICH IS VERY SPECIFIC (I.E., CONNECTED WITH A CERTAIN STAGE OF DEVELOPMENT) FOR EACH DIATOM. (LONG-BATTELLE)

FIELD 05C, 05B

ACCESSION NO. W73-00441

SEASONAL CHARACTERISTICS OF PHYTOPLANKTON IN THE AMURSKY BAY OF THE SEA OF JAPAN, (IN RUSSIAN),

AKADEMIYA NAUK SSSR, VLADIVOSTOK. INSTITUT BIOLOGII.

G. V. KONOVALOVA.

OKEANOLOGIYA, VOL 12, NO 1, P 123-128, JANUARY/FEBRUARY 1972. 2 FIG, 14 REF.

DESCRIPTORS:
*PHYTOPLANKTON, *ENVIRONMENTAL EFFECTS, *SYSTEMATICS, *SEASONAL, SEA WATER, SAMPLING, BIOMASS, DIATOMS, POLLUTANT IDENTIFICATION, MARINE ALGAE, CHRYSOPHYTA, AUTUMN, WINTER, SUMMER, PRIMARY PRODUCTIVITY, DINOFLAGELLATES, PYRROPHYTA.

IDENTIFIERS:
SEA OF JAPAN, *AMURSKY BAY, PERIDINIANS, EUCAMPIA ZOODIACUS, CHAETOCEROS AFFINIS, SKELETONEMA COSTATUM, CHAETOCEROS SOCIALIS.

ABSTRACT:
SEASONAL NUMBER AND BIOMASS DYNAMICS OF THE PHYTOPLANKTON, AS WELL AS OF ITS CONSTITUENT SYSTEMATIC GROUPS AND MASS SPECIES, WERE DETERMINED ON THE BASIS OF THE ANNUAL PHYTOPLANKTON COLLECTIONS IN THE AMURSKY BAY OF THE SEA OF JAPAN. THE SPECIFIC COMPOSITION OF ALGAE WAS DETERMINED IN THE WATER-BOTTLE COLLECTIONS. A MAJOR PART OF THE BIOMASS CALCULATIONS WAS BASED ON THE TABLE OF VOLUMES COMPILED FOR THE MASS PHYTOPLANKTON SPECIES. THREE MAXIMA OF ALGAL BIOMASS ARE ESTABLISHED: IN WINTER, SUMMER AND AUTUMN. DIATOMS PLAY A LEADING ROLE IN THE FORMATION PHYTOPLANKTON BIOMASS AND ITS NUMBERS. PERIDINIANS STAND SECOND IN BIOMASS. THE GREATEST BIOMASS WAS OBSERVED IN EUCAMPIA ZOODIACUS AND CHAETOCEROS AFFINIS, THE GREATEST NUMBERS IN SKELETONEMA COSTATUM, CHAETOCEROS AFFINIS AND CHAETOCEROS SOCIALIS. AT ALL THE STATIONS THE BIOMASS MAXIMUM WAS RECORDED AT THE END OF SEPTEMBER AND IN OCTOBER, WHEREAS THE MINIMUM WAS IN APRIL. THERE WERE DISTINCT CHANGES OF FORMS IN THE PLANKTONIC ALGAE. (LONG-BATTELLE)

FIELD 05C, 02L

ACCESSION NO. W73-00442

TOXIC SUBSTANCES CONTROL ACT OF 1971 AND AMENDMENT.

HEARINGS ON S.1478--SUBCOMM. ON THE ENVIRONMENT--COMM. ON COMMERCE, U.S. SENATE, 92ND CONGRESS, AUGUST 3, 4, 5; OCTOBER 4 AND NOVEMBER 5, 1971. 3 VOLUMES, 1251 P.

DESCRIPTORS:
*LEGISLATION, *TOXINS, *ENVIRONMENTAL EFFECTS, FEDERAL GOVERNMENT, AIR POLLUTION, WATER POLLUTION, TOXICITY, CHEMICALS, REGULATION, ALGAL TOXINS, MERCURY, DETERGENTS, WATER POLLUTION SOURCES, ASBESTOS, EUTROPHICATION, POLYCHLORINATED BIPHENYLS, PUBLIC HEALTH, ENVIRONMENT CONTROL, LEGAL ASPECTS, REGULATION, PERMITS.

IDENTIFIERS:
*LEGISLATION(PROPOSED), ENVIRONMENTAL PROTECTION AGENCY.

ABSTRACT:
THE PURPOSE OF THE PROPOSED LEGISLATION IS TO REGULATE TOXIC SUBSTANCES, RECOGNIZING THAT GAPS EXIST IN THE REGULATORY FRAMEWORK THAT PERMIT PUBLIC HEALTH AND ENVIRONMENTAL HAZARDS TO OCCUR WITHOUT REGULATION. FOUR CHEMICAL SUBSTANCES WHICH WILL BE PRIME CANDIDATES FOR REGULATION ARE DISCUSSED IN GREAT DETAIL. THESE ARE MERCURY, POLYCHLORINATED BIPHENYLS (PCB'S), ASBESTOS, AND DETERGENTS. UNDER THE ACT THE ADMINISTRATOR OF THE ENVIRONMENTAL PROTECTION AGENCY WILL BE EMPOWERED TO RESTRICT THE USE OR DISTRIBUTION, INCLUDING A TOTAL PROHIBITION, OF A CHEMICAL SUBSTANCE, IF SUCH RESTRICTIONS ARE NECESSARY TO PROTECT THE ENVIRONMENT. THE BILL PROVIDES FOR SETTING TEST STANDARDS ON VARIOUS CLASSES AND USES OF NEW CHEMICALS. A NEW COMPOUND WHICH DOES NOT MEET THESE STANDARDS WILL BE KEPT OFF THE MARKET. THE AMENDMENT'S PRECLEARANCE CERTIFICATION PROCEDURE IS CRITIZED AS BEING NOT ONLY TOO BURDENSOME UPON THE GOVERNMENT, BUT IMPOSING A SUBSTANTIAL IMPEDIMENT TO TECHNOLOGICAL INNOVATION. THE BILL'S IMPACT ON EXPORTS AND IMPORTS IS CONSIDERED. SETTING STANDARDS IN THIS COUNTRY WILL CONSTITUTE HIDDEN TRADE BARRIERS. IF ANY EXPORT HAS AN EFFECT ON THE GLOBAL ENVIRONMENT, IT WILL COME WITHIN THE PROPOSED LEGISLATION. INCLUDED ARE NUMEROUS DETAILED CHEMICAL RESEARCH REPORTS ON THE TOXIC SUBSTANCES UNDER DISCUSSION. (BEARDSLEY-FLORIDA)

FIELD 05G, 06E, 05C

ACCESSION NO. W73-00703

AMCHITKA BIOENVIRONMENTAL PROGRAM. RESEARCH PROGRAM ON MARINE ECOLOGY, AMCHITKA ISLAND, ALASKA. JULY 1, 1970-JUNE 30, 1971,

WASHINGTON UNIV., SEATTLE. FISHERIES RESEARCH INST.

R. L. BURGNER, AND R. E. NAKATANI.

AVAILABLE FROM NTIS, SPRINGFIELD, VA., AS BMI-171-144; $3.00 IN PAPER COPY; $0.95 IN MICROFICHE. REPORT BMI-171-144, 1972. 79 P, 10 FIG, 18 TAB, 40 REF.

DESCRIPTORS:
*NUCLEAR EXPLOSIONS, *ALASKA, *MARINE ALGAE, *FISHERIES, FISH POPULATIONS, FISH BEHAVIOR, INVERTEBRATES, OTTERS, ROCKSLIDES, INTERTIDAL AREAS, ON-SITE INVESTIGATIONS, FOOD CHAINS, SPECIATION, REPRODUCTION.

IDENTIFIERS:
AMCHITKA ISLAND, MILROW TEST, CANNIKIN TEST.

ABSTRACT:
EFFECTS WERE ESTIMATED ON THE MARINE ENVIRONMENT OFF AMCHITKA ISLAND FROM THE MILROW TEST (1969), AND BASELINE DATA WERE OBTAINED PRIOR TO THE CANNIKIN TEST. SEA OTTER STOMACH CONTENTS, FISH, INVERTEBRATES, AND ALGAE WERE SAMPLED. NO MILROW EFFECTS ON NEARSHORE FISHES WERE OBSERVED; BUT OCEAN PERCH WERE NOT FOUND IN AREAS OF PREVIOUS ABUNDANCE, AND PREVIOUSLY UNSEEN JUVENILE OFFSHORE FISHES WERE PREDOMINANT. MILROW-PRODUCED ROCK FALLS WERE COLONIZED WITHIN 1 YEAR. ADDITIONAL CONTINUING STUDIES INCLUDE ALGAL AND INVERTEBRATE SUCCESSION ON ROCKS DISPLACED BY MILROW, AND TAXONOMY AND REPRODUCTIVE CYCLES OF ALGAE. (BOPP-ORNL)

FIELD 05C

ACCESSION NO. W73-00764

PLUTONIUM CONCENTRATION ALONG FRESH-WATER FOOD CHAINS OF THE GREAT LAKES. PROGRESS, 1971-72.

WOODS HOLE OCEANOGRAPHIC INSTITUTION, MASS.

AVAILABLE FROM NTIS, SPRINGFIELD, VA., AS COO-3568-1; $3.00 IN PAPER COPY, $0.95 IN MICROFICHE. REPORT COO-3568-1, 1972. 26 P, 5 FIG, 6 TAB, 7 REF.

DESCRIPTORS:
*GREAT LAKES, *STRONTIUM RADIOISOTOPES, *FALLOUT, *FOOD CHAINS, ABSORPTION, SEDIMENTS, CORES, ANALYTICAL TECHNIQUES, RADIOACTIVITY TECHNIQUES, RADIOECOLOGY, TRACERS, MONITORING, NUCLEAR POWERPLANTS, NUCLEAR WASTES, PATH OF POLLUTANTS, WATER POLLUTION SOURCES, PUBLIC HEALTH, ON-SITE INVESTIGATIONS, WASTE ASSIMILATIVE CAPACITY, WASTE DILUTION, BOTTOM SEDIMENTS, DECOMPOSING ORGANIC MATTER, RADIOCHEMICAL ANALYSIS, ALGAE.

IDENTIFIERS:
*PLUTONIUM, CESIUM RADIOISOTOPES.

ABSTRACT:
LESS THAN 10% OF THE PU239 DEPOSITED BY FALLOUT IS PRESENTLY DETECTABLE IN LAKE ONTARIO WATER; AND ALTHOUGH ENVIRONMENTAL FACTORS DIFFER, THE PU CONTENT OF THE WATER IS SIMILAR IN LAKE MICHIGAN AND FALMOUTH RESERVOIR. THE HYPOTHESIS THAT BIOLOGICAL SEDIMENTATION IS CONTROLLING ACCOUNTS FOR INCREASED PU OBSERVED AT DEPTHS. AN EAST-WEST CONCENTRATION GRADIENT SUGGESTS THAT NIAGRA OUTFALL PLAYS A KEY ROLE. THERE WAS NO INDICATION OF INCREASED PU IN THE WATER OR ALGAE NEAR A NUCLEAR POWERPLANT, ALTHOUGH CS134 WAS DETECTABLE IN ALGAE NEAR THE POWERPLANT AND IN ALEWIFE FISH. THE CONCENTRATION FACTOR FOR PU239 UPTAKE BY ALGAE WAS 1-2 THOUSAND WITH A MEAN WATER CONCENTRATION OF 1.7 DPM/1000 KG. THE RATIO PU238/PU239 WAS HIGHER FOR THE WATER THAN FOR SEDIMENT (CORRESPONDING TO THE SNAP-9A BURNUP IN 1964). PRELIMINARY MEASUREMENTS INDICATED 50% MORE SR90 IN OPEN LAKE ONTARIO WATER THAN IN LAKE MICHIGAN. SR90 FOUND IN SEDIMENT WAS ONLY 1% OF THAT ESTIMATED FOR THE WATER COLUMN. (BOPP-ORNL)

FIELD 05B, 02H

ACCESSION NO. W73-00790

EFFECTS OF ENVIRONMENTAL STRESS ON THE COMMUNITY STRUCTURE AND PRODUCTIVITY OF SALT MARSH EPIPHYTIC COMMUNITIES. PROGRESS REPORT, SEPTEMBER 1, 1971-AUGUST 31, 1972,

CITY COLL., NEW YORK.

J. J. LEE.

AVAILABLE AS COO-3054-3(PT. 1) FROM NTIS, SPRINGFIELD, VA., $3.00 IN PAPER COPY, $0.95 IN MICROFICHE. REPORT COO-3054-3 (PT. 1), 1972. 76 P, 45 FIG, 9 TAB, 4 REF.

DESCRIPTORS:
*MARINE ALGAE, *SALT MARSHES, *PROTOZOA, *BACTERIA, FLORIDA, RADIATION SENSITIVITY, RADIOECOLOGY, ECOLOGY, CYCLING NUTRIENTS, THERMAL POLLUTION, TROPHIC LEVEL, TROPICAL REGIONS, ULTRAVIOLET RADIATION, FOOD CHAINS, GAMMA RAYS, RESISTANCE, BIOLOGICAL COMMUNITIES, TIDAL MARSHES, FOOD WEBS, INTERTIDAL AREAS, CONNATE WATER.

IDENTIFIERS:
FORAMINIFERA.

ABSTRACT:
BY MEANS OF STUDIES ON LABELED UPTAKE AND THE GROWTH OF ALGAE AND BACTERIA, AN AUXOTROPIC PROFILE OVER A GROWING SEASON WAS OBTAINED. RECYCLED ORGANIC NUTRIENTS PLAY A CENTRAL ROLE IN THE NUTRITION, REGULATION, AND SUCCESSION WITHIN THE ALGAL COMPARTMENT. SENSITIVITY TO IONIZING RADIATION AND THERMAL STRESS VARIES WIDELY. AT 40 C, SOME OF THE THERMALLY MORE RESISTANT SPECIES HAVE A MINIMAL LETHAL DOSE (MLD) GREATER THAN 135 MIN., AND ARE ALSO MORE RESISTANT TO ULTRAVIOLET (MLD GREATER THAN 45,000 ERGS) AND GAMMA (MLD ABOUT 10 MILLION ERGS) IRRADIATION. THIS COMBINATION OF RESISTANCE TO STRESSES SEEMS TO BE AN ADAPTATION TO GROWTH OR REPEATED EXPOSURE AT EBB TIDE IN HOT SHALLOW TIDE POOLS OR IN INTERSTITIAL WATERS. LIGHT QUALITY IS ALSO IMPORTANT IN SELECTIVELY REGULATING GROWTH AND SUCCESSION AMONG THE ALGAE. THE TROPHIC DYNAMICS, MINERAL CYCLING, AND NICHES OF SUBLITTORAL FORAMINIFERA WERE ALSO STUDIED. (BOPP-ORNL)

FIELD 05C, 02L

ACCESSION NO. W73-00796

DISTRIBUTION OF RADIOACTIVE AND STABLE ZINC IN AN EXPERIMENTAL MARINE ECOSYSTEM,

NATIONAL MARINE FISHERIES SERVICE, BEAUFORT, N.C. ATLANTIC COASTAL FISHERIES CENTER.

F. A. CROSS, J. N. WILLIS, AND J. P. BAPTIST.

JOURNAL FISHERIES RESEARCH BOARD OF CANADA, VOL 28, NO 11, P 1783-1788, 1971. 4 TAB, 7 REF.

DESCRIPTORS:
*ZINC RADIOISOTOPES, *ABSORPTION, *MARINE ALGAE, *PHYTOPLANKTON, BACTERIA, AQUARIA, DECOMPOSING ORGANIC MATTER, WASTE ASSIMILATIVE CAPACITY, EUTROPHICATION, NUTRIENT REMOVAL, *ZINC, CHELATION, FOOD CHAINS, PATH OF POLLUTANTS, PUBLIC HEALTH, WATER POLLUTION EFFECTS, RADIOECOLOGY, RADIOACTIVITY TECHNIQUES, ANALYTICAL TECHNIQUES.

ABSTRACT:
THE BEHAVIOR OF ZN65 TRACER WAS STUDIED IN A PLANKTONIC BLOOM IN 2000 LITERS OF SEAWATER IN AN AQUARIUM. THE ZN65 SPECIFIC ACTIVITY WAS SIGNIFICANTLY LESS FOR THE PARTICULATE FRACTION WHICH CONSISTED MAINLY OF CHLORELLA SP., NITZSCHIA CL., AND BACTERIA. POSSIBLY EXCHANGE OF ZN WAS RELATIVELY SLOW, OR ORGANICALLY-COMPLEXED ZN WAS PRESENT WHICH DID NOT EXCHANGE READILY WITH DISSOLVED INORGANIC ZN BUT WAS AVAILABLE TO PHYTOPLANKTON. CONCENTRATION FACTORS WERE 980 FOR ZN65 AND 1400 FOR TOTAL ZN (FOR THE MIXED PHYTOPLANKTON COMMUNITY ALONG WITH BACTERIA AND DETRITUS). (BOPP-ORNL)

FIELD 05B, 05C

ACCESSION NO. W73-00802

RADIOECOLOGY OF CERTAIN MOLLUSCS IN INDIAN COASTAL WATERS,

BHABHA ATOMIC RESEARCH CENTER, BOMBAY (INDIA).

B. PATEL, P. G. VALANJU, C. D. MULAY, M. C. BALANI, AND

AVAILABLE FROM UNIPUB INC., P.O. BOX 433, NEW YORK, N.Y. 10016. IN: SYMPOSIUM ON THE INTERACTION OF RADIOACTIVE CONTAMINANTS WITH THE CONSTITUENTS OF THE MARINE ENVIRONMENT, JULY 10-14, 1972, SEATTLE, WASHINGTON. PAPER IZEA/SM-158/13. 34 P, 5 FIG, 11 TAB, 61 REF.

DESCRIPTORS:
*REVIEWS, *NUCLEAR WASTES, *BIOINDICATORS, *MARINE ANIMALS, ASIA, MOLLUSKS, COBALT RADIOISOTOPES, IODINE RADIOISOTOPES, MARINE ALGAE, TRACE ELEMENTS, PATH OF POLLUTANTS, ABSORPTION, RADIOISOTOPES, STABLE ISOTOPES, ESTUARINE ENVIRONMENT, COASTS, ANIMAL PHYSIOLOGY, SEDIMENTS, ANIMAL METABOLISM.

IDENTIFIERS:
CESIUM RADIOISOTOPES.

ABSTRACT:
UPTAKE BY MARINE ORGANISMS, WHICH MAY SERVE AS BIOINDICATORS OF CERTAIN RADIONUCLIDES AND STABLE ELEMENTS, IS REVIEWED. MEASUREMENTS ARE REPORTED FOR THE MOLLUSC, APLYSIA BENEDICTI, WHICH MOVES SHOREWARDS IN COASTAL WATERS DURING COLDER MONTHS. HEPATOPANCREAS AND GUT CONTAINED MOST CO; THE HORNY INTERNAL SHELL AND THE SPAWN CHORDS, MOST I. UPTAKE OF CS137 WAS FOUND WHEN ALGAL FOOD WAS PRESENT. CONCENTRATION FACTORS WERE (IN UNITS OF TEN THOUSANDS): CO, 5-11, NI, 1-2; FE, 5-12; MN, 2-3; CU, 2-8; ZN, 2-4; SR, 0.001-0.002; CO60, 5; CO58, 2. THE DIFFERENCES BETWEEN THE CO ISOTOPES RESULT FROM DIFFERENCES IN HALF-LIVES. (BOPP-ORNL)

FIELD 05B, 02L

ACCESSION NO. W73-00811

BIOLOGICAL PHYSICOCHEMICAL ASPECTS OF RADIOACTIVE CONTAMINATION OF MARINE ORGANISMS AND OF MARINE SEDIMENTS (ASPECTS BIOLOGIQUES ET PHYSICO-CHIMIQUES DE LA CONTAMINATION RADIOACTIVE D'ESPECES ET DE SEDIMENTS MARINE),

COMMISSARIAT A L'ENERGIE ATOMIQUE, CHERBOURG (FRANCE). CENTRE DE LA HAGUE.

J. ANCELLIN.

AVAILABLE FROM UNIPUB INC., P.O. BOX 433, NEW YORK, N.Y. 10016. IN: SYMPOSIUM ON THE INTERACTION OF RADIOACTIVE CONTAMINANTS WITH THE CONSTITUENTS OF THE MARINE ENVIRONMENT, JULY 10-14, 1972, SEATTLE, WASHINGTON. PAPER IAEA/SM-158/19. 32 P, 4 FIG, 4 TAB, 60 REF.

DESCRIPTORS:
*REVIEWS, *RADIOISOTOPES, *MARINE ANIMALS, *MARINE ALGAE, SEDIMENTS, NUCLEAR WASTES, PATH OF POLLUTANTS, SEA WATER, WATER ANALYSIS, RADIOACTIVITY TECHNIQUES, RADIOECOLOGY, ENVIRONMENTAL EFFECTS, ABSORPTION, SORPTION, WASTE DILUTION, WASTE ASSIMILATIVE CAPACITY, EUROPE, ANIMAL PHYSIOLOGY, TRACE ELEMENTS, HEAVY METALS, PHYSICOCHEMICAL PROPERTIES, FOOD CHAINS, PUBLIC HEALTH.

ABSTRACT:
UPTAKE BY MARINE LIFE OF RADIONUCLIDES FROM SEA WATER AND SEDIMENTS IS REVIEWED, INCLUDING SOME ASPECTS OF FOOD CHAINS. THE CONCENTRATION FACTORS MAY DECREASE WITH INCREASING CONCENTRATION IN THE WATER, IN THE CASE OF TRACE ELEMENTS WHICH HAVE PHYSIOLOGICAL FUNCTIONS (ZN, CU, FE, MG, MN, CO, NI), BECAUSE OF METABOLIC REGULATION. HOWEVER, THE EFFECT MAY DIFFER BETWEEN SPECIES, AND BETWEEN ORGANS OF A GIVEN ORGANISM. MANY WORKS ARE CITED WHICH SHOW THAT ENVIRONMENTAL RADIONUCLIDES WERE MORE READILY ASSIMILATED THAN THE STABLE ISOTOPES OF THE SAME ELEMENT, SINCE THE LATTER ARE PRESENT IN A DIFFERENT PHYSICOCHEMICAL FORM, HAVING UNDERGONE MORE AGING. THE PERMANENCE OF RADIONUCLIDE FIXATION BY SEDIMENTS REQUIRES STUDY, ESPECIALLY WITH REGARD TO SEDIMENT-FEEDING ORGANISMS. (BOPP-ORNL)

FIELD 05B, 02J

ACCESSION NO. W73-00817

FALLOUT MN54 ACCUMULATED BY BAY SCALLOPS ARGOPECTEN IRRADIANS (LAMARCK) NEAR BEAUFORT, NORTH CAROLINA,

MICHIGAN UNIV., ANN ARBOR. GREAT LAKES RESEARCH DIV.

C. L. SCHELSKE.

AVAILABLE FROM UNIPUB INC., P.O. BOX 433, NEW YORK, N.Y. 10016. IN: SYMPOSIUM ON THE INTERACTION OF RADIOACTIVE CONTAMINANTS WITH THE CONSTITUENTS OF THE MARINE ENVIRONMENT, JULY 10-14, 1972, SEATTLE, WASHINGTON. PAPER IAEA/SM-158/20. 19 P, 3 FIG, 5 TAB, 16 REF.

DESCRIPTORS:
*BIOINDICATORS, *FALLOUT, *MOLLUSKS, *ATLANTIC COASTAL PLAIN, NORTH CAROLINA, ABSORPTION, SOUNDS, ESTUARINE ENVIRONMENT, PATH OF POLLUTANTS, FOOD CHAINS, PHYTOPLANKTON, MARINE ALGAE, ENVIRONMENTAL EFFECTS, TRACE ELEMENTS, COLLOIDS, NUCLEAR WASTES, RADIOECOLOGY.

IDENTIFIERS:
MANGANESE RADIOISOTOPES, BOGUE SOUND, CORE SOUND.

ABSTRACT:
FALLOUT MN54 IN SCALLOP KIDNEYS PEAKED IN 1964, AND DECREASED FROM 1964-1966 WITH AN EFFECTIVE HALF-LIFE (240 DAYS) GREATER THAN THE BIOLOGICAL HALF-LIFE (30-50 DAYS) MEASURED IN THE LABORATORY. THE DIFFERENCE BETWEEN THE TWO IS RELATED TO THE SUPPLY FROM ENVIRONMENTAL RESERVOIRS. EXPERIMENTS SHOWED THAT MN54 TAKEN UP BY INGESTION OF LABELLED PHYTOPLANKTON WAS RETAINED LONGER (21% REDUCTION IN 21 DAYS) THAN THAT TAKEN UP DIRECTLY FROM SEA WATER (50% REDUCTION IN THE SAME TIME). SCALLOPS SAMPLED FROM SEVERAL LOCATIONS IN BOGUE SOUND AND AT BELLS ISLAND IN CORE SOUND CONTAINED MORE MN54 THAN THOSE FROM OTHER LOCATIONS IN CORE SOUND. (BOPP-ORNL)

FIELD 05B, 02L

ACCESSION NO. W73-00818

RADICSTRONTIUM UPTAKE BY MARINE ORGANISMS (INCORPORACION DE RADIOESTRONCIO POR ORGANISMOS MARINOS),

COMISION NACIONAL DE ENERGIA ATOMICA, BUENOS AIRES (ARGENTINA).

D. CANCIO, J. A. LLAURO, N. R. CIALLELLA, AND C. J. BENINSCN.

AVAILABLE FROM UNIPUB INC., P.O. BOX 433, NEW YORK, N.Y. 10016. IN: SYMPOSIUM ON THE INTERACTION OF RADIOACTIVE CONTAMINANTS WITH THE CONSTITUENTS OF THE MARINE ENVIRONMENT, JULY 10-14, 1972, SEATTLE, WASHINGTON. PAPER IAEA/SM-158/21. 14 P, 8 TAB, 17 REF.

DESCRIPTORS:
*STRONTIUM RADIOISOTOPES, *MARINE ANIMALS, *MARINE ALGAE, *FOOD CHAINS, ABSORPTION, CALCIUM, PUBLIC HEALTH, SOUTH AMERICA, ATLANTIC COASTAL PLAIN, COASTS, CRUSTACEANS, MARINE FISH, MOLLUSKS, POLLUTANT IDENTIFICATION.

IDENTIFIERS:
ARGENTINA.

ABSTRACT:
UPTAKE OF SR AND CA BY ORGANISMS (FISH, CRUSTACEANS, MOLLUSKS, AND ALGAE) OF THE ARGENTINA ATLANTIC COAST WAS STUDIED IN THE LABORATORY, BOTH BY DOSING WITH CA AND SR AND BY A DOUBLE-TRACER TECHNIQUE. DISCRIMINATION FACTORS (ACTIVITY OF SR/G CA, IN THE ORGANISM)/(ACTIVITY OF SR/G CA, IN SEA WATER) WERE NEARLY THE SAME FOR ALL PARTS OF A SPECIES OF FISH (0.23-0.34) AND FOR THE EXOSKELETON AND SOFT PARTS OF CRUSTACEANS (0.71-0.95), BUT DIFFERED BETWEEN THE SHELL (0.07-0.12) AND THE SOFT PARTS (0.36-0.54) OF MOLLUSKS. FIELD RESULTS GAVE REASONABLE AGREEMENT. CONCENTRATION FACTORS FOR UPTAKE OF SR WERE: FISH, 2-10, CRUSTACEANS AND MOLLUSKS, 13-70; ALGAE, 0.2-70. (BOPP-ORNL)

FIELD 05B, 05A

ACCESSION NO. W73-00819

THE STATE OF COBALT IN SEA WATER AND ITS UPTAKE BY MARINE ORGANISMS AND SEDIMENTS,

PUERTO RICO NUCLEAR CENTER, RIO PIEDRAS. RADIOECOLOGY DIV.

F. G. LOWMAN, AND R. Y. TING.

AVAILABLE FROM UNIPUB INC., P.O. BOX 433, NEW YORK, NY 10016. IN: SYMPOSIUM ON THE INTERACTION OF RADIOACTIVE CONTAMINANTS WITH THE CONSTITUENTS OF THE MARINE ENVIRONMENT, JULY 10-14, 1972, SEATTLE, WASHINGTON. PAPER IAEA/SM-158/23. 25 P, 10 FIG, 2 TAB, 6 REF.

DESCRIPTORS:
*COBALT RADIOISOTOPES, *COBALT, *ABSORPTION, *MARINE ANIMALS, MARINE ALGAE, SEDIMENTS, PUERTO RICO, ESTUARINE ENVIRONMENT, PATH OF POLLUTANTS, FALLOUT, NUCLEAR WASTES, RADIOECOLOGY, RADIOACTIVITY TECHNIQUES, ANALYTICAL TECHNIQUES, CYCLING NUTRIENTS, TRACE ELEMENTS, CHELATION, ORGANIC COMPOUNDS, FOOD CHAINS, PUBLIC HEALTH, PHYTOPLANKTON, ZOOPLANKTON, WATER ANALYSIS, CATION EXCHANGE, RADIOCHEMICAL ANALYSIS.

ABSTRACT:
ANALYTICAL TECHNIQUES WERE TESTED TO MEASURE IONIC AND COMPLEXED CO IN SEA WATER USING IONIC CO58 AND VITAMIN B12 TAGGED WITH CO57 AS TRACERS. SEA WATER COLLECTED NEAR PUERTO RICO, WHICH CONTAINED 0.043 MICROG/LITER OF IONIC AND 0.03 MICROG/LITER OF COMPLEXED CO AND THE ABOVE TRACERS ADDED IN AMOUNTS TO GIVE ROUGHLY THE SAME RADIOACTIVITIES FOR CO58 AND CO57, WAS USED IN RADIONUCLIDE UPTAKE EXPERIMENTS WITH MARINE ORGANISMS. AFTER ABOUT 17 DAYS, RATIOS OF UPTAKE OF ORGANIC TRACER TO IONIC TRACER WERE: PHYTOPLANKTON, 15; ZOOPLANKTON, 15; BRINE SHRIMP, 15; CRAB LARVAE, 5; CLAM SOFT PARTS, 4; PERIPHYTON, 4; PELAGIC MACRURAN CRUSTACEA, 0.9; ESTUARINE SEDIMENTS, 0.5; SPINY LOBSTER MOLT, 0.5; SPINY LOBSTER (WHOLE), 0.4; FRESH WATER PHYTOPLANKTON, 0.2; CRAB SHELL, 0.03. (BOPP-ORNL)

FIELD 05A, 05B

ACCESSION NO. W73-00821

THE KINETICS OF AND A PRELIMINARY MODEL FOR THE UPTAKE OF RADIOZINC BY PHAEODACTYLUM TRICORNUTUM IN CULTURE,

INTERNATIONAL LAB. OF MARINE RADIOACTIVITY, MONTE CARLC (MCNACO).

A. G, DAVIES.

AVAILABLE FROM UNIPUB INC., P.O. BCX 433, NEW YORK, N.Y. 10016. IN: SYMPOSIUM ON THE INTERACTION OF RADIOACTIVE CONTAMINANTS WITH THE CONSTITUENTS OF THE MARINE ENVIRONMENT, JULY 10-14, 1972, SEATTLE, WASHINGTON. PAPER IAEA/SM-158/25. 20 P, 7 FIG, 4 TAB, 14 REF.

DESCRIPTORS:
*ZINC RADIOISOTOPES, *ABSORPTION, *PHYTOPLANKTON, *MARINE ALGAE, FOOD CHAINS, PUBLIC HEALTH, NUCLEAR WASTES, PATH CF POLLUTANTS, DIFFUSION, CYTOLOGICAL STUDIES, ADSORPTION, PENETRATION, CATION ADSORPTION, CHELATION, MATHEMATICAL MODELS.

ABSTRACT:
INTEGRATION OF A MATHEMATICAL MODEL BY COMPUTER GAVE SATISFACTORY AGREEMENT WITH EXPERIMENT. IT WAS ASSUMED THAT THE RATE OF UPTAKE BY INTRACELLULAR PROTEIN IS DIFFUSION-CONTROLLED, AND THAT THE EQUILIBRIUM CONTENT OF INTRACELLULAR PROTEIN DECREASES WITH DEPLETION OF NUTRIENT SALTS OF P AND EXPECIALLY N. LANGMUIR ADSORPTICN ISOTHERMS WERE USED TO REPRESENT EQUILIBRIA. THE EVIDENCE INDICATES THAT THE UPTAKE OF ZINC BY PHAEDACTYLUM TRICORNUTUM IS LARGELY PASSIVE AND DIFFUSION CONTROLLED AND THAT THE ZINC CONTENT OF THE CELLS IS PROBABLY RELATED TO INTRA-CELLULAR PROTEIN LEVEL. (BOPP-ORNL)

FIELD 05B

ACCESSION NO. W73-00823

ACCUMULATION AND LCSS OF COBALT AND CESIUM BY THE MARINE CLAM, MYA ARENARIA, UNDER LABORATORY AND FIELD CONDITIONS,

CALIFORNIA UNIV., LIVERMORE. LAWRENCE LIVERMORE LAB.

F. L. HARRISON,

AVAILABLE FROM UNIPUB INC., P.O. BCX 433, NEW YORK, N.Y. 10016. IN: SYMPOSIUM ON THE INTERACTION OF RADIOACTIVE CONTAMINANTS WITH THE CONSTITUENTS OF THE MARINE ENVIRONMENT, JULY 10-14, 1972, SEATTLE, WASHINGTON. PAPER IAEA/SM-158/28. 35 P, 10 FIG, 6 TAB, 24 REF.

DESCRIPTORS:
*MARINE ANIMALS, *CLAMS, *FOOD CHAINS, *NUCLEAR WASTES, *COBALT RADIOISOTOPES, PATH OF POLLUTANTS, WASTE ASSIMILATIVE CAPACITY, *CALIFORNIA, RADICECOLOGY, MARINE ALGAE, ESTUARINE ENVIRCNMENT, ABSORPTION, ANIMAL POPULATIONS, ANIMAL PHYSIOLOGY, RADIOACTIVITY TECHNIQUES, CYCLING NUTRIENTS, TRACE ELEMENTS, PUBLIC HEALTH, RADIOECOLOGY.

IDENTIFIERS:
*HUMBOLDT BAY, *CESIUM RADIOISOTOPES.

ABSTRACT:
UPTAKE OF RADICNUCLIDES FROM DIET AND WATER ARE BEING COMPARED BETWEEN AQUARIA AND THE HUMBOLDT BAY, CALIFORNIA, REACTOR DISCHARGE CANAL. STEADY-STATE CONCENTRATION FACTORS IN EDIBLE TISSUES (ABOUT 5 FOR CS AND 200 FOR CO), BIOLOGICAL HALF-TIMES FOR UPTAKE (2-15 DAYS FOR CS AND 50-120 DAYS FOR CO), BIOLOGICAL LOSS-RATE CONSTANTS, AND TURNOVER RATES IN THE VARIOUS BODY PARTS ARE TABULATEC. LITTLE METABOLIC REGULATION OF CS UPTAKE WAS OBSERVED IN THE RANGE 0.5-12.5 MICROGRAMS/LITER, BUT SOME REGULATION OF CO UPTAKE OCCURRED. FOOD IN THE FORM OF SUSPENSICNS OF RADIOACTIVE DIATOMS INCREASED UPTAKE OVER THAT FROM WATER ALONE WITH CO60, BUT NOT WITH CS137. (BOPP-ORNL)

FIELD 05C, 02L

ACCESSION NO. W73-00826

LEVELS OF RADIOACTIVITY IN THE MARINE ENVIRONMENT AND THE DOSE COMMITMENT TO MARINE ORGANISMS,

MINISTRY OF AGRICULTURE, FISHERIES AND FOOD, LOWESTOFT (ENGLAND). FISHERIES RADIOBIOLOGICAL LAB.

D. S. WOODHEAD.

AVAILABLE FROM UNIPUB INC., P.O. BOX 433, NEW YORK, N.Y. 10016. IN: SYMPOSIUM ON THE INTERACTION OF RADIOACTIVE CONTAMINANTS WITH THE CONSTITUENTS OF THE MARINE ENVIRONMENT, JULY 10-14, 1972, SEATTLE, WASHINGTON. PAPER IAEA/SM-158/31. 34 P, 2 FIG, 15 TAB, 63 REF.

DESCRIPTORS:
*BACKGROUND RADIATION, *RADIOACTIVITY EFFECTS, *FALLOUT, *MARINE ANIMALS, MARINE PLANTS, MARINE ALGAE, NUCLEAR WASTES, PATH OF POLLUTANTS, RADIOECOLOGY, MATHEMATICAL MODELS, WASTE ASSIMILATIVE CAPACITY, ESTUARINE ENVIRONMENT, FOOD CHAINS, PUBLIC HEALTH, EUROPE, RADIOISOTOPES, EFFLUENTS.

IDENTIFIERS:
WINDSCALE PROCESSING PLANT.

ABSTRACT:
DATA ARE REVIEWED ON SEA-WATER AND SEA-BED RADIOACTIVITY; AND, BASED ON IDEALIZED MODELS, DOSE RATES (MICRORADS/HOUR) ARE CALCULATED FOR FIVE GROUPS OF MARINE BIOTA (1) PHYTOPLANKTON, (2) ZOOPLANKTON, (3) MOLLUSCA, (4) CRUSTACEA, AND (5) FISH. VALUES ARE OBTAINED FOR: (A) NATURAL SOURCES OF RADIOACTIVITY - (1) 2.8-8.2, (2) 3.3-16.4, (3) 9.5-31.5, (4) 10.0-38.0, (5) 3.3-20.8; (B) FALLOUT - (1) 0.26-24.5, (2) 1.4-147, (3) 0.10-8.0, (4) 0.36-0.46, (5) 0.14-42.7; (C) WASTE DISPOSAL OPERATIONS IN THE WINDSCALE PROCESSING PLANT VICINITY - (1) 200-2100, (2) 530-6900, (3) 51.8-3400, (4) 43.3-3410, (5) 0.6-3340. ALTHOUGH THE PRACTICE OF MARINE DISPOSAL LEADS TO LOCALLY HIGH DOSE RATES, IN POPULATION OR GLOBAL TERMS THE DOSE RATE CONTRIBUTION IS AND WILL PROBABLY REMAIN NEGLIGIBLE. IN THE CASE OF SR90, CS137, H3, C14, AND PU239, FOR WHICH THERE ARE EXTENSIVE DATA, APPROXIMATE TISSUE LEVELS CAN BE DERIVED USING THE APPROPRIATE CONCENTRATION FACTORS. (BOPP-ORNL)

FIELD 05B, 05C

ACCESSION NO. W73-00828

RADIONUCLIDE TRANSPORT STUDIES IN THE HUMBOLDT BAY MARINE ENVIRONMENT,

CALIFORNIA UNIV., LIVERMORE. LAWRENCE LIVERMORE LAB.

R. E. HEFT, W. A. PHILLIPS, H. R. RALSTON, AND W. A. STEEL.

AVAILABLE FROM UNIPUB INC., PO BOX 433, NEW YORK, NY 10016. IN: SYMPOSIUM ON THE INTERACTION OF RADIOACTIVE CONTAMINANTS WITH THE CONSTITUENTS OF THE MARINE ENVIRONMENT, JULY 10-14, 1972, SEATTLE, WASHINGTON. PAPER IAEA/SM-158/37. 25 P, 3 FIG, 10 TAB, 6 REF.

DESCRIPTORS:
*PATH OF POLLUTANTS, *NUCLEAR POWERPLANTS, *RADIOACTIVITY EFFECTS, *MATHEMATICAL MODELS, FORECASTING, COBALT RADIOISOTOPES, MARINE ALGAE, ABSORPTION, SEDIMENTS, WASTE DILUTION, WASTE ASSIMILATIVE CAPACITY, CALIFORNIA, TRITIUM, MARINE PLANTS, MONITORING, SAMPLING, WATER ANALYSIS, SEA WATER, STABLE ISOTOPES, RADIOISOTOPES, NUCLEAR WASTES, ESTUARINE ENVIRONMENT, ON-SITE INVESTIGATIONS, LABORATORY TESTS, TURNOVERS, WATER CIRCULATION.

IDENTIFIERS:
CESIUM RADIOISOTOPES.

ABSTRACT:
FROM JUNE 1971 TO MARCH 1972 AT 5-6 WEEK INTERVALS, SAMPLES OF BOTTOM SEDIMENT, WATER PLUS SUSPENDED SEDIMENT, EELGRASS, AND ENTEROMORPHA (ALGAE) WERE ANALYSED FOR GAMMA-EMITTING RADIONUCLIDES AND STABLE ELEMENTS. ABOUT 1/4 THE TOTAL ESTIMATED ANNUAL RELEASE FROM THE WASTE TREATMENT SYSTEM OF A 65-MW BOILING-WATER REACTOR WAS FOUND IN THE BAY FOR CO60 AND OTHER RADIONUCLIDES (EXCEPTING TRITIUM WHICH WAS NEAR OCEANIC BACKGROUND AND CS137 WHICH WAS LARGELY FROM FALLOUT). EXPERIMENTS ARE PLANNED IN WHICH THE RADIOACTIVITY IN THE WATER IN THE DISCHARGE CANAL WILL BE MONITORED, AND UPTAKE WILL BE DETERMINED BY VARIOUS BOTTOM SEDIMENTS AND ALGAL SPECIES. AUXILIARY EXPERIMENTS WILL BE CONDUCTED IN THE LABORATORY. THE LONG-RANGE OBJECTIVE IS TO INCORPORATE PARAMETERS INTO AN ADVECTION/DIFFUSION MODEL FOR THE BAY IN ORDER TO PREDICT COMPARTMENTAL RADIONCULIDE CONCENTRATION. (BOPP-ORNL)

FIELD 05B, 02L, 05C

ACCESSION NO. W73-00831

ACCUMULATION OF CERTAIN TRACE ELEMENTS IN MARINE ORGANISMS FROM THE SEA AROUND THE CAPE OF GOOD HOPE,

ATOMIC ENERGY BOARD, PELINDABA, PRETORIA (SOUTH AFRICA).

D. VAN AS, H. O. FOURIE, AND C. M. VLEGGAAR.

AVAILABLE FROM UNIPUB INC., P.O. BOX 433, NEW YORK, N.Y. 10016. IN: SYMPOSIUM ON THE INTERACTION OF RADIOACTIVE CONTAMINANTS WITH THE CONSTITUENTS OF THE MARINE ENVIRONMENT, JULY 10-14, 1972, SEATTLE, WASHINGTON. PAPER IAEA/SM-158-39. 12 P, 1 FIG, 3 TAB, 12 REF.

DESCRIPTORS:
*NUCLEAR WASTES, *MARINE ALGAE, *MARINE ANIMALS, *MARINE FISH, MUSSLES, LOBSTERS, ABSORPTION, COBALT RADIOISOTOPES, ZINC RADIOISOTOPES, MANGANESE, CHROMIUM, IRON, ANALYTICAL TECHNIQUES, FOOD CHAINS, AFRICA, EFFLUENTS, PATH OF POLLUTANTS, NEUTRON ACTIVATION ANALYSIS, SPECTROSCOPY, TRACE ELEMENTS, SAMPLING, WATER ANALYSIS, SEA WATER.

IDENTIFIERS:
ANTIMONY, CESIUM, *CAPE OF GOOD HOPE.

ABSTRACT:
IN CONNECTION WITH PREDICTING EFFECTS FROM THE RELEASE OF RADIOACTIVE EFFLUENTS TO THE SEA, EDIBLE MARINE LIFE WAS ANALYZED AND CONCENTRATION FACTORS WERE CALCULATED. PARTICULAR ATTENTION WAS PAID TO THE INFLUENCE OF CONTAINER MATERIAL ON LIQUID SAMPLES AND TO THE EFFECT OF VARIOUS METHODS OF SAMPLE TREATMENT. WITH ANALYSIS OF FE BY ATOMIC ABSORPTION, BETTER RESULTS WERE OBTAINED FOR AN ORGANIC EXTRACT THAN BY PERFORMING THE ANALYSIS DIRECTLY ON A SULFURIC ACID MEDIUM CONTAINING THE DIGESTED SAMPLE. THE CONCENTRATION FACTORS FOUND FELL IN THE RANGES: CRUSTACEANS (CR, 3-10 THOUSAND; FE AND ZN, 2-27 THOUSAND; CO, 6-50; MN, 300-2400; SB, 200-500; CS, 2-22), SEAWEEDS (CR, 600-12,000; FE AND ZN, 1-133 THOUSAND; CO, 8-94; MN, 400-9200; SB, 160-480); FISH (CR, 40-60,000; FE AND ZN, 2000-17000; CO, 2-15; MN, 100-1500; SB, 40-420, CS, 12-90). (BOPP-ORNL)

FIELD 05C

ACCESSION NO. W73-00832

CONTRIBUTIONS FROM THE ALPHA EMITTER, POLONIUM-210, TO THE NATURAL RADIATION ENVIRONMENT OF THE MARINE ORGANISMS,

SCRIPPS INSTITUTION OF OCEANOGRAPHY, LA JOLLA, CALIF.

T. R. FOLSOM, AND T. M. BEASLEY.

AVAILABLE FROM UNIPUB INC., P.O. BOX 433, NEW YORK, N.Y. 10016. IN: SYMPOSIUM ON THE INTERACTION OF RADIOACTIVE CONTAMINANTS WITH THE CONSTITUENTS OF THE MARINE ENVIRONMENT, JULY 10-14, 1972, SEATTLE, WASHINGTON. PAPER IAEA/SM-158/41. 10 P, 1 FIG, 5 TAB, 12 REF.

DESCRIPTORS:
*BACKGROUND RADIATION, *NUCLEAR WASTES, *ABSORPTION, *FOOD CHAINS, PUBLIC HEALTH, PATH OF POLLUTANTS, RADIOECOLOGY, RADIOACTIVITY EFFECTS, NUCLEAR POWERPLANTS, MARINE ALGAE, MARINE ANIMALS, MARINE PLANTS, HUMAN POPULATION, RADIOACTIVITY TECHNIQUES, RADIUM RADIOISOTOPES.

IDENTIFIERS:
*POLONIUM, PLUTONIUM.

ABSTRACT:
PO210 ACCUMULATES TO RELATIVELY HIGH LEVELS IN SOME MARINE ECOSYSTEMS WHICH ALSO ACCUMULATE PU EFFECTIVELY; THUS STUDY OF THE NATURAL RADIONUCLIDE MAY SERVE TO INDICATE THE PROBABLE FATE OF PU. THE DOSE FROM PO210 IN TISSUES IS ABOUT 12 MRADS/YEAR TO MAN (5-20% OF THE DOSE FROM NATURAL RADIATION), AND FROM 0.004 TO 800 MRADS/YEAR FOR THE TISSUES OF VARIOUS MARINE ORGANISMS. PREDICTION OF THE DOSE TO CERTAIN CRITICAL TISSUES IS DIFFICULT OWING TO THE EXTREMELY SHORT RANGE OF ALPHA PARTICLES. (BOPP-ORNL)

FIELD 05B

ACCESSION NO. W73-00833

HEAT TOLERANCE OF REEF ALGAE AT LA PARGUERA, PUERTO RICO,

GUSTAVUS ADOLPHUS CCLL., ST. PETER, MINN.

S. L. SCHWARTZ, AND L. R. ALMODOVAR.

NOVA HEDWIGIA, VOL. 21, NO. 1, P 231-240, 1971. 4 FIG, 1 TAB, 12 REF.

DESCRIPTORS:
*HEAT RESISTANCE, *MARINE ALGAE, LIGHT INTENSITY, *ECOLOGICAL DISTRIBUTION, SURFACE WATERS, LAGCCNS, CHLCROPHYTA, RHODOPHYTA, PHYSICAL PROPERTIES, WATER TEMPERATURE, CATA COLLECTIONS, MARINE PLANTS, ENVIRCNMENTAL EFFECTS, REEFS, LABORATORY TESTS, SAMPLING, PHAEOPHYTA.

IDENTIFIERS:
SURVIVAL, PENICILLUS, CHONDRIA, CLADOPHOROPSIS MEMBRANACEA, PENICILLUS CAPITATUS, CERAMIUM NITENS, WRANGELIA ARGUS, VALCNIA VENTRICOSA, BRYOPSIS PENNATA, SPYRIDIA FILAMENTOSA, DICTYOTA BARTAYRESII, LAURENCIA OBTUSA, LAURENCIA PAPILLOSA, COELOTHRIX IRREGULARIS, CENTROCERAS CLAVULATUM, HYPENA MUSCIFORMIS, CERAMILM BUSSOIDEUM, CHONDRIA LITTORALIS, HYPNEA SPINELLA.

ABSTRACT:
HEAT AND LIGHT TOLERANCES ARE DESCRIBED FOR REEF ALGAE COLLECTED FROM ENRIQUE REEF AT MAGUEYES ISLAND, LA PARGUERA. SIXTEEN SPECIES OF ALGAE WERE TESTED INCLUDING ELEVEN REDS, FOUR GREENS, AND ONE BROWN. ALGAE WERE COLLECTED IN GROUPS OF FOUR FROM 3 SPECIFIC AREAS: GROUP I - SHALLOW WATER AROUND BREAKING WAVES; GROUPS II AND III - SHALLCW, CALM WATER AT HIGH TIDE; AND GROUP IV - LAGOONS AT 6-8 FEET DEPTH. ALGAL SAMPLES WERE PLACED ON A SEAWATER-MOISTENED TOWEL CONTAINED IN A WHITE ENAMEL PAN; THE TOWEL WAS FREQUENTLY SPRAYED WITH SEAWATER TO AVOID DRYING. THE ALGAE WERE EXPOSED TO DIRECT SUNLIGHT AND THE LENGTH OF EXPOSURE, TEMPERATURE, AND RANGE FROM FIRST DEATH SIGNS TO DEATH WERE MEASURED FOR EACH. THE RESULTS SHOWED THAT MEMBERS OF RHODOPHYCOPHYTA, THE SHALLOW WATER ORGANISMS, SURVIVED EXTREME TEMPERATURES BETTER THAN THOSE MEMBERS OF CHLOROPHYCOPHYTA, DEEP LIVING ORGANISMS. FURTHERMORE, THE SPECIMENS ALLOWED TO DRY OUT IN THE SHADE SURVIVED LCNGER THAT THOSE EXPOSED TO THE SUN. (LCNG-BATTELLE)

FIELD 05C

ACCESSION NO. W73-00838

EUTROPHICATION,

NEW YORK STATE DEPT. OF ENVIRONMENTAL CONSERVATION, RONKCNKOMA.

J. FOEHRENBACH.

JOURNAL WATER POLLUTICN CONTROL FEDERATION, VOL 44, NO 6, P 1150-1159, JUNE 1972. 65 REF.

DESCRIPTORS:
*REVIEWS, *EUTROPHICATION, *NUTRIENTS, *WATER POLLUTICN CONTROL, ORGANIC WASTES, INDUSTRIAL WASTES, WASTE WATER(PCLLUTION), BIOASSAY, ECOSYSTEMS, BICDEGRADATION, ALGAE, WATER PCLLUTICN EFFECTS, HEAVY METALS, CYCLING NUTRIENTS, CARBON, NITROGEN, PHOSPHORUS, SULFUR, CALCIUM, MANGANESE, POTASSIUM, SODIUM, CARBON DICXIDE, CYANOPHYTA, FISH, OXYGEN, NITRATES, CHEMICAL PRECIPITATION, IRCN, PHOSPHATES, ANAEROBIC CONDITIONS, AMMONIA, MUD-WATER INTERFACES, POLLUTANTS, COPPER, TRACE ELEMENTS, WATER POLLUTION, OLIGOTRCPHY, COAGULATION, SEA WATER, FRESHWATER, BIOCHEMICAL OXYGEN DEMAND, FILTRATION, HYDROGEN ION CONCENTRATION.

IDENTIFIERS:
BAY OF QUINTE, SALT CREEK, CHARA, ELODEA, CALLITRICHE, ANABAENA FLOSAQUAE, SPIRULINA, GYMNODINIUM BREVE, PERIDINIUM, OSCILLATORIA RUBECENS, ANABAENA, CRUSSCSTREA, OSCILLATORIA.

ABSTRACT:
THE LITERATURE IS REVIEWED ON SEVERAL ASPECTS OF THE EUTROPHICATION PROCESS FOR SOME NATURAL BODIES OF WATER. SOME OF THE MORE PERTINENT ASPECTS OF THE EUTROPHICATION PROCESS INCLUDED ARE THE FOLLOWING: A REVIEW OF THE RCLE OF VARIOUS NUTRIENTS; ANALYSIS OF THE EFFECT OF DIFFERENT WASTE PRODUCTS ON SEVERAL BOCIES OF WATER; METHODS TO CONTROL THE RATE OF EUTROPHICATION OF A BODY OF WATER BY PROPER MANAGEMENT OF ITS WATERSHED; AND APPROACHES FOR THE REMOVAL CF NUTRIENTS FROM BODIES OF WATERS. (BYRD-BATTELLE)

FIELD 05C, 10F

ACCESSION NO. W73-00840

LABORATORY STUDIES OF ASSEBLAGES OF ATTACHED ESTUARINE DIATOMS,

OREGON STATE UNIV., CORVALLIS. DEPT. OF BOTANY.

B. L. WULFF, AND C. D. MCINTIRE.

LIMNOLOGY AND OCEANOGRAPHY, VOL 17, NO 2, P 200-214, MARCH 1972. 7 FIG, 8 TAB, 18 REF.

DESCRIPTORS:
*ESTUARIES, *DIATOMS, *SESSILE ALGAE, LIGHT INTENSITY, TIDAL EFFECTS, SALINITY, HEATED WATER, BIOMASS, THERMAL STRESS, GROWTH RATES, SPATIAL DISTRIBUTION, PRIMARY PRODUCTIVITY, CHRYSOPHYTA, BIOLOGICAL COMMUNITIES, ESTUARINE ENVIRONMENT, WATER POLLUTION SOURCES, BIODEGRADATION, ECOSYSTEMS, WATER POLLUTION.

IDENTIFIERS:
DESSICATION, SPECIES DIVERSITY, MELOSIRA NUMMULOIDES, CHLOROPHYLL A, *YAQUINA BAY, AMPHORA SPP, MELOSIRA SULCATA, NAVICULA SPP, SYNEDRA FASCICULATA, PLAGIOGRAMMA BROCKMANNI, AMPHIPLEURA RUTILANS, THALASSIOSIRA, THALASSIONEMA NITZSCHIOIDES, PLAGIOGRAMMA VANHEURCKII, PLEUROSIGMA, GYROSIGMA FASCIOLA, NITZSCHIA SPP, BACILLARIA PAXILLIFER, LICMOPHORA PARADOXA, FRAGILIARIA STRIATULA VAR CALIFORNICA, COCCONEIS SCUTELLUM VAR PARVA, ACHNANTHESE TEMPEREI.

ABSTRACT:
EFFECTS OF LIGHT INTENSITY, EXPOSURE TO DESICCATION, SALINITY, AND HEATED WATER ON THE VERTICAL DISTRIBUTION AND GROWTH OF POPULATIONS OF ATTACHED ESTUARINE DIATOMS WERE STUDIED IN A LABORATORY MODEL ECOSYSTEM. OF THE 35 MOST ABUNDANT DIATOM TAXA FOUND IN 36 SAMPLES FROM THE LABORATORY SYSTEM, ALL BUT 8 WERE ALSO ABUNDANT IN SAMPLES OBTAINED FROM YAQUINA BAY AND ESTUARY. VERTICAL DISTRIBUTION OF DIATOMS WAS MORE CLOSELY RELATED TO LIGHT INTENSITY AND PERIOD OF EXPOSURE TO DESICCATION IN THE SUMMER EXPERIMENTS THAN IN THE WINTER EXPERIMENTS. A SUDDEN, UNREASONABLE DECREASE IN SALINITY OR INCREASE IN WATER TEMPERATURE HAD A MUCH GREATER EFFECT ON DIATOM ASSEMBLAGES EXPOSED TO 12,270 LUX THAN ON THOSE THAT DEVELOPED AT EITHER 1,030 OR 4,710 LUX; THE MOST NOTICEABLE CHANGES IN COMMUNITY STRUCTURE INCLUDED A DECREASE IN SPECIES DIVERSITY AND A RAPID GROWTH OF A POPULATION OF MELOSIRA NUMMULOIDES, A FILAMENTOUS SPECIES. PRIMARY PRODUCTIVITY IN DIATOM ASSEMBLAGES EXPOSED TO PERIODS OF DESICCATION WAS LESS UNDER WINTER CONDITIONS THAN UNDER CORRESPONDING CONDITIONS IN SUMMER. PRIMARY PRODUCTIVITY IN ASSEMBLAGES NOT EXPOSED TO DESICCATION WAS STRONGLY AFFECTED BY LIGHT INTENSITY DURING BOTH SUMMER AND WINTER EXPERIMENTS, AND THE RATIO PRIMARY PRODUCTIVITY: CHLOROPHYLL A WAS GREATER DURING WINTER THAN SUMMER. (BYRD-BATTELLE)

FIELD 05C, 02L

ACCESSION NO. W73-00853

FARTHEST SOUTH ALGAE AND ASSOCIATED BACTERIA,

CALIFORNIA INST. OF TECH., PASADENA. BIOSCIENCE AND PLANETOLOGY SECTION.

R. E. CAMERON.

PHYCOLOGIA, VOL 11, NO 2, P 133-139, JUNE 1972. 3 FIG, 1 TAB, 18 REF.

DESCRIPTORS:
*ANTARCTIC, *DISTRIBUTION PATTERNS, *SYSTEMATICS, *AQUATIC ALGAE, *SOIL BACTERIA, SAMPLING, CULTURES, ECOLOGY, SOIL MICROBIOLOGY, CYANOPHYTA, FROZEN SOILS, IONS, NOSTOC, ENVIRONMENTAL EFFECTS, POLLUTANT IDENTIFICATION, PHYSICAL PROPERTIES, BULK DENSITY, POROSITY, COLOR, HYDROGEN ION CONCENTRATION, NITROGEN, CARBON, SODIUM, CALCIUM, POTASSIUM, MAGNESIUM, IRON, ALUMINUM, NITRATES, NITRITES, PHOSPHATES, BICARBONATES, SULFATES, CHLORIDES.

IDENTIFIERS:
NEOCHLORIS AQUATICA, SCHIZOTHRIX CALCICOLA, BORATES, ARTHROBACTER, NOSTOC COMMUNE, PORPHYROSIPHON NOTARISII, PSYCHROPHILIC, BIOTIC ASSOCIATIONS, LA GORCE MOUNTAINS, AMMONIUM IONS.

ABSTRACT:
A NEW RECORD IS REPORTED FOR ALGAE COLLECTED FROM THE HIGHEST LATITUDE, A FROZEN POND IN THE LA GORCE MOUNTAINS, ANTARCTICA (86 DEGREES 45 MINUTE SOUTH, 146 DEGREES ZERO MINUTES WEST). CULTURABLE ALGAE INCLUDED NEOCHLORIS AQUATICA STARR AND SCHIZOTHRIX CALCICOLA (AG.) GOM. PORPHYROSIPHON NOTARISII (MENEGH.) KUTZ. WAS NOT RECOVERABLE IN CULTURE. ASSOCIATED BACTERIA WERE SOIL DIPHTHEROIDS OF THE GENUS ARTHROBACTER. SOIL CONDITIONS WHICH SUPPORTED THE GROWTH OF ARTHROBACTER ARE PRESENTED. THE OCCURRENCE OF HIGH LATITUDE PHOTOSYNTHETIC ORGANISMS IS IMPORTANT IN THE SEARCH FOR POSSIBLE EXTRA-TERRESTRIAL LIFE BECAUSE ENVIRONMENTAL CONDITIONS, IN SOME ASPECTS, APPROACH THOSE OF MARS. (LONG-BATTELLE)

FIELD 05B, 05A

ACCESSION NO. W73-00854

HYDROBIOLOGICAL NOTES ON THE HIGH-SALINITY WATERS OF THE SINAI PENINSULA,

HEBREW UNIV., JERUSALEM (ISRAEL). DEPT. OF ZOOLOGY.

F. D. POR,

MARINE BIOLOGY, VOL 14, NO 2, P 111-119, MAY 1972. 2 FIG, 4 TAB, 42 REF.

DESCRIPTORS:
*SALINITY, *LAGOONS, *HYDROBIOLOGY, *HALOPHILIC ANIMALS, *HALOPHYTES, SEA WATER, MARINE ANIMALS, BIOLOGICAL COMMUNITIES, CHLORINE, MOLLUSKS, COPEPODS, DIPTERA, GULFS, GEOLOGIA HISTORY, PLEISTORENE EPOCH, ADAPTATION, ENVIRONMENTAL EFFECTS, CHLOROPHYTA, MARINE ALGAE , SALT TOLERANCE, CRUSTACEANS, MARINE FISH, MARSH PLANTS.

IDENTIFIERS:
*SINAI PENINSULA, CHLORINITY, METAHALINE, HYPERHALINE, EURYHALINE, RED SEA, GULF OF SUEZ, DECAPODS, COLEOPTERA, MEDITERRANEAN SEA, COELENTERATES, ECHINODERMS, POLYCHAETES, CHIRONOMIDS, BARNACLES, ARTHROPODS, SARGASSUM CRISPUM, DIGENIA SIMPLEX, HALOPHILA STIPULACEA, DIPLANTHERA UNINERVIS, CASSIOPEA ANDROMEDA, AKABARIA, STROMBUS TRICORNIS, FUSUS MARMORATUS, BERTHELLA CITRINA, MACTRA OLORINA.

ABSTRACT:
A 4-YEAR SURVEY WAS MADE OF THE HIGH-SALINITY SEAS SURROUNDING THE SINAI PENINSULA AND THE SALINE LAGOONS BORDERING ON IT. THESE HIGH-SALINITY BODIES OF WATER HAVE IONIC CONTENT SIMILAR TO THAT OF THE SEA. THE RED SEA HAS FORMED A METHALINE MARINE FAUNA BY SUCCESSFULLY ADAPTING TO SALINITIES OF ABOUT 45 PERCENT. THE SALINE LAGOONS HAVING SALINITIES OF 6-8 PERCENT ARE ALSO INHIBITED BY METABOLINE FAUNA. LAGOONS OF GREATER SALINITY ARE INHABITED BY A SMALL NUMBER OF EURYHALINE MARINE ORGANISMS. HYPERSALINE LAGOONS ARE THOSE ISOLATED FROM THE SEA OR WITH SALINITY VALUES GREATER THAN 100 PERCENT. HYPERSALINE CONTINENTAL FRESH-WATER ELEMENTS INHABIT THESE POOLS. THE SPECIFIC OBSERVATIONS MADE ARE ALSO DISCUSSED IN LIGHT OF THEIR POSSIBLE GENERAL APPLICATION TO OTHER NEARSHORE AND LAGOON AREAS. (MORTLAND-BATTELLE)

FIELD 05A, 05C

ACCESSION NO. W73-00855

CHLOROPHYLL AND THE MARGALEF PIGMENT RATIO IN A MOUNTAIN LAKE,

BRADLEY UNIV., PEORIA, ILL. DEPT. OF BIOLOGY.

B. J. MATHIS.

THE AMERICAN MIDLAND NATURALIST, VOL 88, NO 1, P 232-235, JULY 1972. 2 TAB, 13 REF.

DESCRIPTORS:
*LAKES, *PIGMENTS, *SUCCESSION, *EUTROPHICATION, CHLOROPHYLL, PLANKTON, BIOLOGICAL COMMUNITIES, ECOSYSTEMS, SPECTROPHOTOMETRY, VIRGINIA, SAMPLING, SEPARATION TECHNIQUES, CENTRIFUGATION, TURBIDITY, WATER TEMPERATURE, ALKALINITY, HYDROGEN ION CONCENTRATION, WATER QUALITY, NUTRIENTS, COLORIMETRY, DEFICIENT ELEMENTS, AQUATIC ALGAE, BIOINDICATORS, VIRGINIA, TROPHIC LEVEL.

IDENTIFIERS:
MARGALEF PIGMENT RATIO, PIGMENT DIVERSITY INDEX, CHLOROPHYLL A, CARTENOIDS, SAMPLE PREPARATION, MOUNTAIN LAKE.

ABSTRACT:
MARGALEF'S PIGMENT DIVERSITY INDEX WAS USED TO ESTIMATE THE SUCCESSIONAL STAGE OF A PLANKTON COMMUNITY IN A MOUNTAIN LAKE. IN THIS METHOD, OPTICAL DENSITY OF PLANT PIGMENTS IN 90 PERCENT ACETONE IS MEASURED WITH A SPECTROPHOTOMETER TO ESTABLISH A YELLOW/GREEN RATIO. CHLOROPHYLL IS LOST AS CELLS BECOME NITROGEN DEFICIENT AND THUS, ACCORDING TO MARGALEF, THE YELLOW/GREEN RATIO INCREASES, IN GENERAL, AS SUCCESSION ADVANCES. IN THIS STUDY, 1000-ML WATER SAMPLES WERE TAKEN AT A DEPTH OF 3.5 M FROM MOUNTAIN LAKE, VIRGINIA. PIGMENTS WERE EXTRACTED, THE SAMPLES WERE CENTRIFUGED, THEIR OPTICAL DENSITY WAS DETERMINED COLORIMETRICALLY, AND CHLOROPHYLL A DETERMINATIONS WERE MADE. WATER TEMPERATURE, ALKALINITY, PH, AND SPECIFIC CONDUCTANCE WERE DETERMINED IN THE FIELD AND TURBIDITY WAS MEASURED IN THE LABORATORY. THESE LATTER DATA INDICATED AN OLIGOTROPHIC LAKE WITH AN IMMATURE PLANKTON COMMUNITY. HOWEVER, THE MARGALEF RATIO SUGGESTED AN OLDER MORE STABLE COMMUNITY. IT IS SUGGESTED THAT MORE COMPARATIVE STUDIES ARE NEEDED BEFORE THE INDEX CAN BE VALUABLE IN ESTIMATING SUCCESSIONAL STAGES IN AQUATIC COMMUNITIES. (MORTLAND-BATTELLE)

FIELD 05C, 02H

ACCESSION NO. W73-00916

ECOLOGICAL STUDIES OF EUGLENINEAE IN CERTAIN POLLUTED AND UNPOLLUTED ENVIRONMENTS,

OSMANIA UNIV., HYDERABAD (INDIA).

M. MUNAWAR.

HYDROBIOLOGIA, VOL 39, NO 3, P 307-320, MARCH 1972. 3 FIG, 3 TAB, 28 REF.

DESCRIPTORS:
*EUGLENOPHYTA, *BIOINDICATORS, *PHYTOPLANKTON, *POLLUTANT IDENTIFICATION, PROTOZOA, PONDS, LIMNOLOGY, ECOLOGY, CHEMICAL ANALYSIS, AQUATIC ALGAE, NUTRIENTS, WATER QUALITY, WATER ANALYSIS, FREQUENCY, FLUCTUATIONS, BIOLOGICAL COMMUNITIES, PHYSIOCHEMICAL PROPERTIES, DISTRIBUTION PATTERNS, ECOLOGICAL DISTRIBUTION, WATER POLLUTION EFFECTS, FRESHWATER, NITRATES, AMMONIA, CARBON DIOXIDE, PHOSPHATES, SULFATES, SODIUM, POTASSIUM, MAGNESIUM, CHLORIDES, NITROGEN, ALKALINITY, SILICA, CALCIUM, DISSOLVED OXYGEN, ORGANIC MATTER, SULFIDES.

IDENTIFIERS:
*INDIA, ORGANIC NITROGEN, EUGLENA ACUS, EUGLENA CAUDATA, EUGLENA LIMNOPHYLA, EUGLENA OBLONGA, EUGLENA OXYURIS, EUGLENA POLYMORPHA, EUGLENA SPIROGYRA, LEPOCINCLIS OVUM, PHACUS CIRCUMFLEXUS, PHACUS ONYX, PHACUS PLEURONECTES, PHACUS PYRUM, TRACHELOMONAS ARMATA, TRACHELOMONAS HISPIDA VAR CRENULATOCOLLIS.

ABSTRACT:
VARIOUS PHYSICOCHEMICAL PROPERTIES OF THREE FRESHWATER PONDS IN INDIA WERE STUDIED IN RELATION TO THE DISTRIBUTION AND PERIODICITY OF EUGLENINEAE. TWO PONDS, SEWAGE AND GARDEN, WERE CONSIDERABLY POLLUTED AND THE THIRD, TYPHA, WAS COMPARATIVELY PURE. THE HIGHEST PERCENTAGE OF EUGLENINEAE WERE FOUND IN SEWAGE POND, WHICH CONTAINED 13 DIFFERENT SPECIES. GARDEN POND HARBORED JUST 2 SPECIES, BUT IN LARGE NUMBERS WHILE THE TRACHELOMONAS HISPIDA AND EUGLENA SP. FOUND IN TYPHA POND WERE TOTALLY ABSENT IN THE OTHER TWO. SEWAGE POND HAD HIGH AVERAGE CONCENTRATIONS OF FREE CARBON DIOXIDE, WHICH SEEMED TO FAVOR EUGLENOID GROWTH. NITRATE FLUCTUATIONS IN ALL THREE PONDS COINCIDED WITH FLUCTUATIONS IN EUGLENOID GROWTH SUGGESTING THAT INORGANIC SOURCES OF NITROGEN MIGHT BE IMPORTANT IN THE ECOLOGY OF THESE ALGAE. THE FLAGELLATES WERE ABUNDANT DURING PERIODS OF LOW TOTAL SULPHIDES AND CONCENTRATIONS GREATER THAN 2.0 PPM ADVERSELY AFFECTED THEIR DEVELOPMENT. A TEMPERATURE RANGE OF 27-39 C FAVORED THEIR GROWTH AND THEY USUALLY DEVELOPED AFTER RAINS OR DURING INTERMITTENT SHOWERS. IN SEWAGE POND, THERE WAS A DIRECT RELATIONSHIP BETWEEN EUGLENOID POPULATION AND HIGHER CONCENTRATIONS OF OXIDIZABLE ORGANIC MATTER. THIS SUGGESTS THE USE OF CERTAIN EUGLENOID SPECIES AS BIOINDICATORS. (MORTLAND-BATTELLE)

FIELD 05C, 05A

ACCESSION NO. W73-00918

SUBLITTORAL ECOLOGY OF THE KELP BEDS OFF DEL MONTE BEACH, MONTEREY, CALIFORNIA,

NAVAL POSTGRADUATE SCHOOL, MONTEREY, CALIF.

C. S. MINTER, III.

AVAILABLE FROM THE NATIONAL TECHNICAL INFORMATION SERVICE AS AD-738 875, $3.00 IN PAPER COPY, $0.95 IN MICROFICHE. MASTER'S THESIS, SEPTEMBER 1971. 181 P, 102 FIG, 5 APPEND, 31 REF.

DESCRIPTORS:
*KELPS, *ECOLOGICAL DISTRIBUTION, *MARINE ANIMALS, *SYSTEMATICS, *ECOLOGY, *MARINE ALGAE, ANNELIDS, MOLLUSKS, RHODOPHYTA, PHAEOPHYTA, PHOTOGRAPHY, SCUBA DIVING, SEA WATER, SAMPLING, ENVIRONMENTAL EFFECTS, MARINE FISH, PERIPHYTON, ON-SITE DATA COLLECTIONS, CRUSTACEANS, GASTROPODS, SNAILS, CRABS, BENTHIC FLORA, BENTHIC FAUNA, DISTRIBUTION PATTERNS, PACIFIC OCEAN, AQUATIC HABITATS, ISOPODS, CLAMS, CALIFORNIA, SHELLFISH.

IDENTIFIERS:
COELENTERATES, ARTHROPODS, NEMERTEANS, ECHINODERMS, SIPUNCULIDS, BRYOZOA, INTOPROCTA, TUNICATES, SPONGES, VERTEBRATES, MACROINVERTEBRATES, SPECIES DIVERSITY, PORIFERA, SEA CUCUMBERS, STARFISH, CHITONS, RHABDODERMELLA, LOPHOPANOPEUS, EUPENTACTA, CNEMIDOCARPA, ENHYDRA LUTRIS NEREIS, SEA OTTER, MACROCYSTIS PYRIFERA, CHORDATES, POLYCHAETES, BARNACLES, DECAPODS, SEA URCHINS, ACARNUS ERITHACUS, CRANIELLA, HYMENAMPHIASTRA CYANOCRYPTA.

ABSTRACT:
IN ORDER TO ASSESS THE ECOLOGICAL EFFECTS OF WAVE BARRIERS IN MONTEREY HARBOR, AN OVERALL STUDY OF THE AREA WAS CONDUCTED. THIS PART OF THE STUDY DEALT WITH THE IDENTIFICATION OF MACROSCOPIC ORGANISMS EXISTING IN THE KELP BEDS OFF DEL MONTE BEACH AND THE MAPPING AND COUNTING OF BENTHIC PLANTS AND ANIMALS LIVING WITHIN TWO CAREFULLY SELECTED AND PERMANENTLY MARKED STATIONS ON SHALE SUBSTRATE. THESE SPECIMENS WERE COLLECTED BY SCUBA DIVERS, AND MORE THAN 160 SPECIES WERE FOUND TO EXIST. COLLECTION METHODS AND TECHNIQUES UTILIZED BY DIVERS WERE DOCUMENTED AND NUMEROUS UNDERWATER PHOTOGRAPHS WERE TAKEN. THE PERMANENTLY MARKED AREAS WERE FOUND TO BE OF GENERALLY SIMILAR BIOLOGICAL POPULATION BUT OF MARKEDLY DIFFERENT SPECIES DISTRIBUTION AND RELATIVE ABUNDANCE. (LONG-BATTELLE)

FIELD 05C

ACCESSION NO. W73-00932

ECOLOGICAL ASPECTS OF PLANKTON PRODUCTION,

CENTRE OF ADVANCED STUDY IN MARINE BIOLOGY, PORTO NOVO (INDIA).

R. C. SUBBARAJU, AND K. KRISHNAMURTHY.

MARINE BIOLOGY, VOL 14, NO 1, P 25-31, MAY 1972. 4 FIG, 4 TAB, 42 REF.

DESCRIPTORS:
*ECOLOGY, *AQUATIC PRODUCTIVITY, *ZOOPLANKTON, *PHYTOPLANKTON, NERITIC, ESTUARINE ENVIRONMENT, SEA WATER, DIATOMS, COPEPODS, DINOFLAGELLATES, CRUSTACEANS, ANNELIDS, WATER POLLUTION EFFECTS, STANDING CROPS, MOLLUSKS, LARVAE, CHRYSOPHYTA, CYANOPHYTA, PYRROPHYTA, MARINE ALGAE, GASTROPODS.

IDENTIFIERS:
SECONDARY PRODUCTIVITY, CEPHALOCHORDATES, HEMICHORDATES, ARTHROPODS, *BAY OF BENGAL, COELENTERATES, CTENOPHORA, CHAETOGNATHS, CYPHONAUTES, ECHINODERMS, TUNICATES, NOCTILUCA, SIPHONOPHORES, MEDUSAE, PILIDIUM, CLADOCERA, LINGULA, ACTINOTROCHA, VELIGER, PTEROPODS, HETEROPODS, OSCILLATORIA, STOMATOPODA, POLYCHAETES, AMPHIOXUS, SALPS, BALANOGLOSSUS, DOLIOLIDS, EUCALANUS ELONGATUS, EUCHAETA MARINA, PONTELLA PRINCEPS, PONTELLOPSIS HERDMANNI, NANNOCALANUS MINOR, CANTHOCALANUS PAUPER, RHINCALANUS SPP, UNDINULA VULGARIS, OITHONA SPP, SAPPHRINA NIGROMACULATA, ACROCALANUS GRACILIS, PARACALANUS PARVUS, EUTERPINA ACUTIFRONS, PSEUDODIAPTOMUS SPP, ACARTIA SPP.

ABSTRACT:
A STUDY WAS CARRIED OUT IN THE NERITIC AND ESTUARINE WATERS OF PORTO NOVO, COROMANDEL COAST, BAY OF BENGAL, INDIA BETWEEN 1960 AND 1967 ON THE ECOLOGICAL ASPECTS OF PLANKTON PRODUCTION. REPRESENTATIVE SAMPLES WERE COLLECTED FROM BOTH NEARSHORE AND ESTUARINE WATERS OF PORTO NOVO, AND THE PLANKTON VOLUMES RECORDED BY THE DISPLACEMENT METHOD TO THE NEAREST 0.10 CC AND EXPRESSED IN CC/CU M. THE AVERAGE DISPLACEMENT VOLUME OF PLANKTON USUALLY VARIED BETWEEN 2 AND 4 CC/CU M. DURING SUMMER, WITH A SEASON OF HIGH PLANKTON PRODUCTIVITY, THE AVERAGE PLANKTON DISPLACEMENT VOLUME ROSE TO 8 CC/CU M. GENERALLY SPEAKING, THE AVERAGE ZOOPLANKTON DENSITY (STANDING CROP) WAS USUALLY BETWEEN 80,000 AND 100,000 ORGANISMS/CU M, OF WHICH COPEPODS ALONE COMPRISED USUALLY BETWEEN 70,000 AND 90,000 ORGANISMS/CU M. THE AVERAGE COPEPOD DENSITY PER SAMPLE VARIED FROM 30,000 TO 50,000 ORGANISMS/CU M. HOWEVER, IN THE SUMMER MONTHS, THE COPEPOD DENSITY WAS USUALLY NOT LESS THAN 100,000 ORGANISMS/CU M; IN SOME YEARS THIS WAS EVEN HIGHER (FROM 125,000 TO 170,000 ORGANISMS/CU M). COPEPODS COMPRISED BETWEEN 80 AND 95 PERCENT OF THE ZOOPLANKTON POPULATION. THE MAXIMUM NON-COPEPOD POPULATION IN THE ZOOPLANKTON SELDOM REACHED 30 PERCENT, WAS OFTEN BELOW 25 PERCENT, AND USUALLY LESS THAN 20 PERCET. DURING THE PERIOD MARCH TO OCTOBER (IN SOME YEARS AS EARLY AS FEBRUARY, AND IN SOME YEARS UP TO NOVEMBER), EITHER AN INCREASING OR A STEADY TREND OF PLANKTON PRODUCTION WAS EVIDENT. IT WOULD APPEAR THAT SALINITY AND RAINFALL DETERMINE THE OCCURRENCE AND DISTRIBUTION OF PLANKTON IN PORTO NOVO. (BYRD-BATTELLE)

FIELD 05C

ACCESSION NO. W73-00935

EFFECT OF DREDGING ON THE NUTRIENT LEVELS AND BIOLOGICAL POPULATIONS OF A LAKE,

SOUTH DAKOTA STATE UNIV., BROOKINGS. WATER RESOURCES INST.

C. L. CHURCHILL, AND C. K. BRASHIER.

AVAILABLE FROM THE NATIONAL TECHNICAL INFORMATION SERVICE AS PB-212 718, $3.00 IN PAPER COPY, $0.95 IN MICROFICHE. COMPLETION REPORT, AUGUST 1972. 155 P, 15 FIG, 9 TAB, 16 REF, APPEND. OWRR B-013-SDAK(1).

DESCRIPTORS:
*NUTRIENTS, *DREDGING, DESILTING, SEDIMENTATION, *EUTROPHICATION, BIOLOGICAL COMMUNITIES, FOOD CHAINS, LAKE MORPHOLOGY, RESERVOIR SILTING, LAKES, *SOUTH DAKOTA, PLANKTON, *ALGAE, *PHOSPHORUS COMPOUNDS, CYANOPYTA, CHLOROPHYTA.

IDENTIFIERS:
PRAIRIE LAKES, *LAKE HERMAN(SO. DAKOTA), *ORTHOPHOSPHATES.

ABSTRACT:
LAKE HERMAN, A SHALLOW WARM WATER PRAIRIE LAKE, HAS BEEN MONITORED FOR CHEMICAL NUTRIENTS AND PLANKTONIC ALGAE BEFORE AND DURING A DREDGING PROJECT. THE MAJOR CHANGE IN THE LAKE WITH THE COMMENCEMENT OF DREDGING ACTIVITY WAS A 300% INCREASE IN BOTH ORTHO PHOSPHATE AND TOTAL PHOSPHORUS. CALCIUM AND TOTAL HARDNESS DECREASED SLIGHTLY AND SILICA AND TURBIDITY WERE JUDGED TO HAVE INCREASED SLIGHTLY. THERE WERE APPARENTLY NO ACCOMPANYING CHANGES IN PLANKTONIC ALGAE, EITHER IN POPULATION DENSITIES OR GENERA PRESENT. THE SILT AND WATER REMOVED WITH DREDGING WAS DEPOSITED IN A LOW LYING AREA NEAR THE LAKE, WHERE THE SILT SETTLED OUT AND THE WATER WAS RETURNED TO THE LAKE. THE WATER IN THIS SILT DEPOSIT AREA WAS ALSO MONITORED. IT WAS AS MUCH AS 2 PH UNITS CLOSER TO NEUTRALITY THAN THE LAKE WATER AND HAD A HIGHER CONDUCTIVITY (UP TO 10% HIGHER) THAN THE LAKE WATER. EARLY IN THE DREDGING SEASON, THIS WATER IN THE SILT DEPOSIT AREA HAD HIGH LEVELS OF ORTHO PHOSPHATE (0.90 MG PO4/1), BUT OVER A FOUR MONTH PERIOD, AS DREDGING PROCEEDED, THE ORTHO PHOSPHATE IN THE SILT DEPOSIT AREA WATER DECREASED UNTIL IT WAS ONLY ABOUT HALF AS GREAT (0.19 MG PO4/1) AS THE LEVEL IN THE LAKE BEFORE DREDGING BEGAN. (WIERSMA-SOUTH DAKOTA)

FIELD 05C, 02H

ACCESSION NO. W73-00938

I. CHEMORECEPTION IN MARINE BACTERIA. II. CHEMICAL DETECTION OF MICROBIAL PREY BY BACTERIAL PREDATORS,

HARVARD UNIV., CAMBRIDGE, MASS. DIV. OF ENGINEERING AND APPLIED PHYSICS.

S. FOGEL, I. CHET, AND R. MITCHELL.

AVAILABLE FROM THE NATIONAL TECHNICAL INFORMATION SERVICE AS AD-736 982, $3.00 IN PAPER COPY, $0.95 IN MICROFICHE. TECHNICAL REPORT NO. 2, NR-306-025, SEPTEMBER 1971. 28 P, 10 TAB, 20 REF. CONTRACT NO. N00014-67A-0298-0026.

DESCRIPTORS:
*MARINE BACTERIA, *ATTRACTANTS, *PREDATION, ORGANIC COMPOUNDS, AMINO ACIDS, VITAMINS, CARBOHYDRATES, ALCOHOLS, POLLUTANT IDENTIFICATION, SYSTEMATICS, NUTRIENTS, PSEUDOMONAS, CULTURES, VITAMIN B, BIOASSAY, CELLULOSE, DIALYSIS, SEA WATER, ORGANIC ACIDS, CHRYSOPHYTA, INHIBITION, MARINE ALGAE, DIATOMS, BEHAVIOR.

IDENTIFIERS:
*CHEMORECEPTION, *CHEMOTAXIS, *PREY, PYTHIUM DEBARYANUM, SKELETONEMA COSTATUM, NUCLEOTIDES, ETHANOL, MONOSACCHARIDES, CULTURE MEDIA, DISACCHARIDES, THREONINE, PROLINE, GLUTAMATE, LEUCINE, ARGININE, SERINE, TYROSINE, TRYPTOPHAN, HISTIDINE, ASPARTATE, METHIONINE, PHENYLALANINE, ALANINE, ISOLEUCINE, VALINE, LYSINE, ASPARAGINE, CYSTINE, CYSTEINE, GLYCINE, GLYCEROL, RIBOSE, MALTOSE, FRUCTOSE, GLUCOSE 6-P, GALACTOSE, GLUCOSAMINE, ARABINOSE, GLUCOSE, RAFFINOSE, SUCROSE, INOSITOL, LACTOSE, ADENINE, URACIL, ACETATE, CITRATES, PYRUVATE, GLUCURONIC ACID, THIAMINE, NIACIN, BIOTIN, NUTRIENT BROTH.

ABSTRACT:
A WIDE VARIETY OF MOTILE MARINE BACTERIA WERE TESTED FOR THEIR ABILITY TO DETECT AND BE ATTRACTED TO ORGANIC CHEMICALS. ALL BACTERIA TESTED DISPLAYED CHEMOTACTIC RESPONSES TO MANY DIFFERENT MATERIALS. THESE RESPONSES WERE HIGHLY SPECIFIC FOR EACH BACTERIUM. THE ATTRACTANT WAS NOT NECESSARILY METABOLIZED. THE ECOLOGICAL SIGNIFICANCE OF THIS WIDESPREAD PHENOMENON OF CHEMORECEPTION AMONG MARINE BACTERIA IS DISCUSSED. A MOTILE, PREDACIOUS BACTERIUM WHICH DEGRADED PYTHIUM DEBARYANUM WAS STRONGLY ATTRACTED TO SUBSTANCES RELEASED INTO THE MEDIUM BY THE FUNGUS. A NONPREDACIOUS BACTERIUM WAS NOT ATTRACTED TO THESE SUBSTANCES. THE PREDATOR BACTERIUM WAS SPECIFICALLY ATTRACTED TO THESE SUBSTANCES. THE PREDATOR BACTERIUM WAS SPECIFICALLY ATTRACTED TO CELLULOSE AND ITS OLIGOMERS WHICH ARE KNOWN TO BE COMPONENTS OF THE CELL WALL OF PYTHIUM. ETHANOL INHIBITED CHEMOTAXIS OF THE BACTERIUM WITHOUT AFFECTING EITHER ITS MOTILITY OR ITS ABILITY TO DEGRADE CELLULOSE. A SECOND PREDACIOUS BACTERIUM WAS ISOLATED FOR THE ALGA, SKELETONEMA COSTATUM. THE ROLE OF CHEMORECEPTION IN THE DETECTION OF MICROBIAL PREY BY BACTERIAL PREDATORS IN NATURAL HABITATS IS DISCUSSED. (LONG-BATTELLE)

FIELD 05A, 05C

ACCESSION NO. W73-00942

PRODUCTION OF DISSOLVED ORGANIC MATTER FROM DEAD GREEN ALGAL CELLS. II. ANAEROBIC MICROBIAL DECOMPOSITION,

TOKYO METROPOLITAN UNIV., (JAPAN). DEPT. OF CHEMISTRY.

A. OTSUKI, AND T. HANYA.

LIMNOLOGY AND OCEANOGRAPHY, VOL 17, NO 2, P 258-264, MARCH 1972. 9 FIG, 2 TAB, 19 REF.

DESCRIPTORS:
*ANAEROBIC BACTERIA, *ANAEROBIC CONDITIONS, *KINETICS, *DISSOLVED SOLIDS, ORGANIC MATTER, NITROGEN, WATER POLLUTION SOURCES, ENVIRONMENTAL EFFECTS, AEROBIC CONDITIONS, CHLOROPHYTA, SCENEDESMUS, AQUATIC ALGAE, AQUATIC PRODUCTIVITY, POLLUTANT IDENTIFICATION, LABORATORY EQUIPMENT, AEROBIC BACTERIA, SOLVENT EXTRACTIONS, FREEZE DRYING, SYSTEMATICS, ORGANIC COMPOUNDS, WATER ANALYSIS.

IDENTIFIERS:
ORGANIC CARBON, ORGANIC NITROGEN, SUBSTRATE UTILIZATION, PARTICULATE MATTER, FATE OF POLLUTANTS, PAPER CHROMATOGRAPHY, INFRARED SPECTROSCOPY, CULTURE MEDIA, FATTY ACIDS, SAMPLE PREPARATION, NATURAL ORGANICS.

ABSTRACT:
THE PRODUCTION OF DISSOLVED ORGANIC MATTER (DOM) FROM DEAD GREEN ALGAL CELLS (SCENEDESMUS) BY ANAEROBIC BACTERIA WAS INVESTIGATED AND COMPARED WITH THOSE PRODUCED UNDER AEROBIC CONDITIONS. DOM WAS PREPARED FOR EXPERIMENTAL USE BY FREEZE-DRYING AND ASHING APPROXIMATELY 99 PERCENT PURE CULTURES OF SCENEDESMUS. ANAEROBIC DECOMPOSITION OF THE DOM WAS INITIATED BY INOCULATING MICROFLORA INTO A SPECIALLY PREPARED DECOMPOSITION APPARATUS CONTAINING CULTURE MEDIA AND DOM AT PH 7. AFTER INCUBATION, DISSOLVED ORGANIC CARBON (DOC) AND NITROGEN (DON) WERE MEASURED BY AN ELEMENTAL ANALYZER AND BY THE MICRO-KJELDAHL METHOD. ORGANIC ACIDS, PARTICULARLY FATTY ACIDS, WERE EXTRACTED BY THIN-LAYER CHROMATOGRAPHY FOR IR SPECTROPHOTOMETRIC DETERMINATION. THIRTY PERCENT OF ADDED ALGAL CELL CARBON WAS TRANSFORMED INTO DOC AND 20 PERCENT MINERALIZED; 50 PERCENT REMAINED AS PARTICULATE MATTER. ON THE OTHER HAND, 8 PERCENT OF THE ADDED ALGAL CELL NITROGEN WAS TRANSFORMED INTO DON, 48 PERCENT WAS MINERALIZED, AND 44 PERCENT REMAINED IN PARTICULATE FORM. THE DISSOLVED ORGANIC COMPOUNDS CONSISTED MAINLY OF LOWER FATTY ACIDS AND YELLOWISH ACIDIC SUBSTANCES. SOME PROTEINACEOUS MATERIAL WAS FOUND. ANAEROBIC DECOMPOSITION PATTERNS AS COMPARED WITH THOSE UNDER AEROBIC CONDITIONS SUGGEST THE PRESENCE OF RELATIVELY HIGH CONCENTRATIONS OF DOM IN ANAEROBIC NATURAL ENVIRONMENTS. (SEE ALSO W73-00379) (LONG-BATTELLE)

FIELD 05C, 05B

ACCESSION NO. W73-01066

INVESTIGATIONS ON THE INFLUENCE OF OIL POLLUTIONS ON MARINE ALGAE. I. THE EFFECT OF CRUDE-OIL FILMS ON THE CO2 GAS EXCHANGE OUTSIDE THE WATER (UNTERSUCHUNGEN UBER DEN EINFLUSS VON OLVERSCHMUTZUNGEN AUF MEERESALG EN. I. DIE WIRKUNG VON ROHOFILMEN AUF DEN CO2 - GASWECHSEL AUSSERHALB DES WASSERS),

KIEL UNIV. (WEST GERMANY). INST. FOR HYDROGRAPHY.

W. SCHRAMM.

MARINE BIOLOGY, VOL 14, NO 3, P 189-198, JUNE 1972. 9 FIG, 1 TAB, 9 REF.

DESCRIPTORS:
*MARINE ALGAE, *WATER POLLUTION EFFECTS, *PHOTOSYNTHESIS, *OIL POLLUTION, *TOXICITY, CHEMICAL REACTIONS, CARBON DIOXIDE, WATER POLLUTION SOURCES, OIL SPILLS, SEA WATER, MARINE PLANTS, PHAEOPHYTA, OILY WATER, EVAPORATION, DIFFUSION, RHODOPHYTA.

IDENTIFIERS:
FUCUS VESICULOSUS, LAMINARIA DIGITATA, PORPHYRA UMBILICALIS, SAMPLE PREPARATION, CRUDE OIL.

ABSTRACT:
OIL POLLUTION IN THE SEA IS GENERALLY RESTRICTED TO THIN OIL FILMS FLOATING ON THE WATER SURFACE. SUCH OIL FILMS TEND TO COAT LITTORAL PLANTS OR ANIMALS DURING LOW TIDE. THE EFFECTS OF COATING WITH CRUDE OIL ON THE CO2-UPTAKE OF VARIOUS MARINE ALGAE HAVE BEEN INVESTIGATED UNDER CONDITIONS OF EMERSION. IN EMERSED ALGAE, CO2-UPTAKE IS MORE OR LESS DEPRESSED, DEPENDING ON THE THICKNESS OF THE OIL FILM (0.1 TO 0.0001 MM) AND THE TYPE OF CRUDE OIL (IRAN, LIBYA, VENEZUELA). ON THE OTHER HAND, WATER LOSS DURING EXPOSURE IS REDUCED, SO THAT THE OIL-COVERED ALGAE ARE ABLE TO PHOTOSYNTHESIZE OVER A LONGER PERIOD THAN ALGAE WITHOUT OIL COVER. AFTER RETRANSFER TO OIL-FREE SEA WATER, IN MOST CASES PHOTOSYNTHESIS RATES REMAINED DEPRESSED THROUGHOUT THE PERIOD OF OBSERVATION. THERE ARE TWO EFFECTS WHICH PROBABLY INTERFERE WITH GAS EXCHANGE: (1) LOWERING OF DIFFUSION RATES OF PHOTOSYNTHETIC GASES AND OF WATER EVAPORATION BY THE OIL FILMS; (2) TOXIC EFFECTS OF CRUDE-OIL COMPONENTS. (LONG-BATTELLE)

FIELD 05C

ACCESSION NO. W73-01074

STUDIES OF A SIMPLE LABORATORY MICROSYSTEM: EFFECTS OF STRESS,

GEORGIA UNIV., ATHENS. INST. OF ECOLOGY.

M. C. FERENS, AND R. J. BEYERS.

REPORT AVAILABLE FROM THE NATIONAL TECHNICAL INFORMATION SERVICE AS SRO-310-1 (PT 1), $3.00 IN PAPER COPY, $0.95 IN MICROFICHE. IN: SAVANNAH RIVER ECOLOGY LABORATORY ANNUAL REPORT NO SRO-310-1 (PT 1), AUGUST 1971, P 8-24. 6 FIG, 1 TAB, 11 REF. AEC AT (38-1) 310.

DESCRIPTORS:
*STRESS, *PRIMARY PRODUCTIVITY, *AQUATIC ALGAE, *COMPARATIVE PRODUCTIVITY, *ECOSYSTEMS, *BIOLOGICAL COMMUNITIES, BIOMASS, PRODUCTIVITY, METABOLISM, CHLORELLA, RESPIRATION, CHLOROPHYLL, LABORATORY TESTS, IRRADIATION, CHLOROPHYTA, INHIBITION, GAMMA RAYS.

IDENTIFIERS:
CHLOROPHYLL A, SCHIZOTHRIX CALCICOLA.

ABSTRACT:
INOCULATION FROM AN ESTABLISHED MICROCOSM WAS SUBJECTED TO ACUTE GAMMA RADIATION LEVELS OF 0.798 KILO- AND 79.8 MEGARADS. TWO SETS OF 240 FLASKS WERE PREPARED WITH ONE HALF RECEIVING IRRADIATED INOCULA AND THE ONE-HALF CONTROLLED INOCULA. THESE WERE SAMPLED OVER A 40-DAY PERIOD AND MEASURED FOR CHLOROPHYLL A CONTENT, DIURNAL COMMUNITY METABOLISM, AND PARTICULATE BIOMASS. BIOMASS AND CHLOROPHYLL SAMPLES WERE OBTAINED BY FILTRATION WITH WHATMAN GF/C GLASS FIBER FILTERS. SIGNIFICANT REDUCTION IN CHLOROPHYLL A WAS FOUND WHEN EACH TREATMENT LEVEL WAS COMPARED TO ITS CONTROL. BIOMASS WAS SIGNIFICANTLY REDUCED THROUGHOUT SUCCESSION FOR BOTH TREATMENT LEVELS EXCEPT FOR THE LAST SAMPLE OF 79.8 MEGARAD MICROCOSMS. NET DAYTIME PRODUCTIVITY AND NIGHTTIME RESPIRATION VARIED AND WERE REDUCED DURING EARLY SUCCESSION AT BOTH TREATMENT LEVELS. 0.798 KILORAD LEVELS SHOWED MORE STRESS THAN 7.98 MEGARAD LEVELS WHEN COMPARING RADIATION TREATMENT EFFECTS AT THE COMMUNITY LEVEL AS RELATED TO OTHER FORMS OF STRESS WITHIN THE COMMUNITY. (SNYDER-BATTELLE)

FIELD 05C

ACCESSION NO. W73-01080

MARSH PRODUCTION: A SUMMARY OF THE LITERATURE,

MARYLAND UNIV., SOLOMONS. NATURAL RESOURCES INST.

C. W. KEEFE.

CONTRIBUTIONS IN MARINE SCIENCE, VOL 16, P 163-181, MARCH 1972. 1 FIG, 3 TAB, 86 REF.

DESCRIPTORS:
*PRIMARY PRODUCTIVITY, *FRESHWATER, *MARSHES, SALINE WATER, BIOMASS, NUTRIENTS, FOOD WEBS, DETRITUS, ECOSYSTEMS, ENERGY BUDGETS, NUTRIENTS, REVIEWS, BIOLOGICAL COMMUNITIES, AQUATIC PLANTS, LIMITING FACTORS, ROOTED AQUATIC PLANTS, CYCLING NUTRIENTS, RESPIRATION, MARSH PLANTS, CHEMICAL ANALYSIS, LEAVES, SAMPLING, PHYTOPLANKTON, IRON, MANGANESE, IONS, SALINITY, PHOSPHORUS, NITROGEN, ALGAE, ORGANIC MATTER, CARBON, CATTAILS.

IDENTIFIERS:
ENGLAND, SWEDEN, GERMANY, SPARTINA SPP, DISTICHLIS SPICATA, FIMBRISTYLIS, BORRICHIA, PHRAGMITES COMMUNIS, JUNCUS SPP, ATRIPLEX HASTATA, SCIRPUS SPP, TYPHA SPP, ZIZANIA AQUATICA, CAREX SPP, BUTOMUS UMBELLATUS, SPARGANIUM RAMOSUM, LEERSIA ORYZOIDES, NUPHAR ADVENA, EICHHORNIA CRASSIPES, JUSTICIA AMERICANA, ALTERNANTHERA PHILOXEROIDES, GLYCERIA MAXIMA, SAGITTARIA SPP, GLYCERIA FLUITANS, SPARGANIUM SPP, ELEOCHARIS PALUSTRIS, PHALARIS ARUNDINACEA, MOLINIA CAERULEA, MANYANTHES TRIFOLIATA, CALAMOGROSTIS LANCEOLATA, CLADIUM MARISCUS, JUSTICIA AMERICANA.

ABSTRACT:
STUDIES OF PRIMARY PRODUCTION IN BOTH SALT AND FRESHWATER MARSHES ARE REVIEWED. METHODS ARE DISCUSSED FOR MEASURING PRODUCTION WHICH RELY PRIMARILY ON THE AERIAL PORTIONS OF THE PLANT SINCE ROOT MATERIAL IS DIFFICULT TO SAMPLE. ONE METHOD IS TO CLIP AERIALS IN A UNIT AREA SELECTED AT RANDOM FROM A LARGER AREA AND TO SEPARATE LIVING AND DEAD COMPONENTS. THE TOTAL OF THE WEIGHT INCREASES IN MATERIAL PROVIDES A MEASURE OF PRODUCTION. IN A SECOND METHOD, ONLY LIVING MATERIAL IS CLIPPED AND THE WEIGHT OF AN AVERAGE MATURE LEAF IS DETERMINED. PRODUCTION IS CALCULATED FROM THE WEIGHT OF MATERIAL PRESENT PLUS MATERIAL ESTIMATED TO HAVE BEEN REMOVED BEFORE HARVEST. SEVERAL REASONS FOR THE HIGH PRODUCTIVITY OF MARSH COMMUNITIES AS COMPARED TO TERRESTRIAL COMMUNITIES ARE DISCUSSED. IT IS CONCLUDED THAT THE RELATIVE CONTRIBUTIONS OF PRIMARY PRODUCERS TO THE FOOD SUPPLY MAY AFFECT THE POPULATIONS OF CONSUMERS AND THE ENERGY FLOW PATTERNS OF THE ENTIRE ECOSYSTEM. (MORTLAND-BATTELLE)

FIELD 05C, 05A, 02L, 10F

ACCESSION NO. W73-01089

TAXONOMY OF AUSTRALIAN FRESHWATER ALGAE. 2. SOME PLANKTIC STAURASTRA FROM TASMANIA,

AQUATIC BIOLOGY INST., UPPSALA (SWEDEN)

K. THOMASSON, AND P. A. TYLER.

NOVA HEDWIGIA, VOL 21, NO 1, P 287-319, 1971. 12 FIG, 1 TAB, 57 REF.

DESCRIPTORS:
*PHYTOPLANKTON, *AQUATIC ALGAE, *SYSTEMATICS, *DISTRIBUTION PATTERNS, LAKES, AQUATIC PLANTS, AUSTRALIA, SAMPLING, MICROSCOPY, ECOLOGICAL DISTRIBUTION, PLANKTON NETS, CHLOROPHYTA, PLANKTON.

IDENTIFIERS:
*STAURASTRUM, DESMIDS, STAURASTRUM LONGIPES VAR CONTRACTUM, ARTHURS LAKE, LAKE SORELL, LAKE CRESCENT, TOOMS LAKE, WOODS LAKE, LAGOON OF ISLANDS, STAURASTRUM SPP.

ABSTRACT:
THE TAXONOMY AND DISTRIBUTION OF AUSTRALIAN FRESHWATER ALGAE OF THE GENUS, STAURASTRUM ARE DESCRIBED. PLANKTON SAMPLES WERE TAKEN WITH A 10-MICRON OR 60-MICRON PLANKTON NET FROM LAKES IN TASMANIA, AUSTRALIA, MICROSCOPICALLY IDENTIFIED AND DIAGRAMMED FOR REFERENCE. FIFTY-NINE TAXA OR VARIATIONS THEREOF WERE FOUND. (LONG-BATTELLE)

FIELD 05A, 02I

ACCESSION NO. W73-01094

AMINC ACID FLUX IN A NATURALLY EUTHROPHIC LAKE,

OREGON STATE UNIV., CORVALLIS.

B. K. BURNISON.

AVAILABLE FROM UNIV. MICROFILMS, INC., ANN ARBOR, MICH., 48106, ORDER NO. 71-27, 850. PH D DISSERTATION, 1971. 89 P.

DESCRIPTORS:
*AMINO ACIDS, *EUTROPHICATION, *PLANKTCN, OREGCN, CRGANIC COMPOUNDS, ABSORPTION, KINETICS, WATER TEMPERATURE, RESPIRATION, ALGAE, CYANOPHYTA, ANALYTICAL TECHNIQUES, WATER POLLUTICN EFFECTS, NUTRIENT REQUIREMENTS.

IDENTIFIERS:
UPPER KLAMATH LAKE, SUBSTRATE UTILIZATION.

ABSTRACT:
THE YEAR'S STUDY MADE ON THE IN SITU ACTIVITIES OF THE HETEROTROPHIC PLANKTON IN UPPER KLAMATH LAKE, OREGON USING THE WRIGHT-HOBBIE KINETIC APPROACH OF MEASURING THE UPTAKE OF ORGANIC COMPCUNDS WAS INITIATED TO TEST THE APPLICABILITY OF THE KINETIC APPROACH TO A HIGHLY EUTROPHIC SYSTEM. SIXTEEN AMINO ACIDS WERE USED AS THE ORGANIC SUBSTRATES AND IT WAS SHOWN THAT THE MAXIMUM VELOCITY, V SUB MAX, FOR THE UPTAKE OF ALL THE AMINO ACIDS IS PROPORTIONAL TO TEMPERATURE. THE TURNOVER TIME, T SUB T, AND THE SUM OF A TRANSPORT CONSTANT AND THE NATURAL SUBSTRATE CONCENTRATION, (K SUB T PLUS S SUB N), ARE DIFFICULT TO INTERPRET BECAUSE OF POSSIBLE COMPETITIVE INHIBITION EFFECTS AMONG THE AMINO ACIDS. THE V SUB MAX IS UNAFFECTED BY COMPETITIVE INHIBITION, BUT THE TURNOVER TIME AND SUBSTRATE CONCENTRATION VALUES ARE INCREASED. THESE LATTER VALUES THEN REFLECT THE TOTAL NATURAL CONCENTRATION AND AFFINITIES OF THE AMINO ACIDS TRANSPORTED BY A PARTICULAR TRANSPORT SYSTEM. THOSE AMINO ACIDS EXHIBITING THE HIGHEST V SUB MAX VALUES HAD THE HIGHEST PERCENT RESPIRED AS CO2. THE PERCENT RESPIRED APPEARS TO BE INVERSELY PROPORTIONAL TO TEMPERATURE. THE AMINO ACID CONCENTRATIONS DETERMINED FOR UPPER KLAMATH LAKE SURFACE WATERS PROBABLY ARE THE SUM OF THE FREE AND ADSORBED AMINO ACIDS. BIOCHEMICAL ANALYSES WERE MADE OF THE PREDOMINANT BLUE-GREEN ALGAE, FOUND IN THE NUISANCE ALGAL BLOOMS, TO BETTER UNDERSTAND THEIR CONTRIBUTION TO THE NUTRITION OF THE HETROTROPHIC PLANKTON. (MORTLAND-BATTELLE)

FIELD 05C, 05B

ACCESSION NO. W73-01269

AN INVESTIGATION INTO THE DEVELOPMENT OF ELECTROPHORETIC AND ELECTROCHEMICAL WATER PURIFICATION SYSTEMS,

TEXAS A AND M UNIV., COLLEGE STATION.

W. M. LYLE.

AVAILABLE FROM UNIVERSITY MICROFILMS, ANN ARBOR, MICHIGAN, ORDER NO. 71-17, 816. PH. D. DISSERTATION, 1970. 127 P.

DESCRIPTORS:
*SUSPENSION, *ELECTROPHORESIS, *ELECTROCHEMISTRY, *WATER PURIFICATION, CLAY, ALGAE, BACTERIA, WASTE WATER TREATMENT, WATER TREATMENT, MODEL STUDIES.

IDENTIFIERS:
*PARALLEL PLATE MODEL.

ABSTRACT:
ELECTROPHORETIC AND ELECTROCHEMICAL SYSTEMS APPEARED TO BE USEFUL IN THE REMOVAL OF ELECTRICALLY CHARGED SUSPENDED POLLUTANTS SUCH AS CLAY, ALGAE, AND BACTERIA. DESIGN CONCEPTS OF A PARALLEL PLATE MODEL (ENTIRELY FOR ELECTROPHORETIC REMOVAL) AND POROUS FILLER AND ELECTRODE GRID MODELS (INCORPORATING BOTH ELECTROPHORETIC AND ELECTROCHEMICAL REMOVAL) WERE TESTED AS LABORATORY MODELS. THE PARALLEL PLATE MODEL ACHIEVED SUCCESSFUL WATER CLARIFICATION ONLY WHEN THE INFLUENT WATER HAD A VERY LOW ELECTRICAL CONDUCTIVITY. ELECTROCHEMICAL WATER PURIFICATION NOT ONLY OVERCAME THIS PROBLEM BUT WAS ALSO SUCCESSFUL BOTH ECONOMICALLY AND OPERATIONALLY. AN EXAMPLE DESIGN OF A SMALL SEMIAUTOMATED ELECTRCCHEMICAL WATER SYSTEM WHICH INCCRPORATED ELECTROCHEMICAL FLOCCULATION, SETTLING AND DISINFECTICN OPERATIONS IS PRESENTED. (ALBERT-TEXAS)

FIELD 05D, 05F

ACCESSION NO. W73-01362

PARAMETERS INFLUENCING PHOSPHORUS ELIMINATION BY ALGAE,

STUTTGART UNIV. (WEST GERMANY). INST. FOR SANITARY ENGINEERING.

K-H. HUNKEN, AND I. D. SEKCULOV.

WATER RESEARCH, VOL 6, P 1087-1096, 1972. 11 FIG, 7 REF.

DESCRIPTORS:
*PHOSPHORUS, *ALGAE, *WASTE WATER TREATMENT, HYDROGEN ION CONCENTRATION, CYANOPHYTA.

IDENTIFIERS:
*PHOSPHORUS ELIMINATION.

ABSTRACT:
THE PROCESS OF BIOPRECIPITATION OF PHOSPHORUS USES ALGAL PHOTOSYNTHESIS TO REMOVE CARBONATE CARBON TO ALTER THE PH. TO ENHANCE THE PHOTOSYNTHETIC ACTIVITY, ARTIFICIAL LIGHTS WERE USED, MAINTAINING THE PH ABOVE 9. USING THE ACCOMPANYING DIAGRAM, IT IS POSSIBLE TO CALCULATE DETENTION TIMES NEEDED TO REACH SUITABLE PH LEVELS, THUS DETERMINING TECHNICAL AND ECONOMIC FEASIBILITY. IN THE BATCH TESTS STUDIED, THE FINAL PH DETERMINED THE FINAL P CONCENTRATION. MIXING BY AERATION PREVENTED OVERSATURATION WITH CO2 AND STARTED PHOTOSYNTHESIS AT A PH OF 8.3. THE SUBMERGED ILLUMINATION MADE THE REACTION CHAMBER INDEPENDENT OF SURFACE AREA, ALLOWING MORE CONSTRUCTION FREEDOM. FOR AN ALGAE-FREE EFFLUENT, A FILTER AND A SETTLING BASIN WERE SATISFACTORY. (ANDERSON-TEXAS)

FIELD 05D

ACCESSION NO. W73-01367

RELATIONSHIP OF TRACE ELEMENTS TO ALGAE GROWTH,

WASHINGTON STATE UNIV., PULLMAN.

J. J. KOELLING.

AVAILABLE FROM UNIV. MICROFILMS, INC., ANN ARBOR, MICH., 48106, ORDER NO. 71-28,778. PH. D. THESIS, 1971, 142 P.

DESCRIPTORS:
*ALGAE, *GROWTH RATES, *PRIMARY PRODUCTIVITY, *HEAVY METALS, *TRACE ELEMENTS, NUTRIENTS, MERCURY, CHROMIUM, COBALT, FRESH WATER, LAKES, PHYSICAL PROPERTIES, CHEMICAL PROPERTIES, GOLD, IRON, CESIUM, ALKALI METALS, SEASONAL, ZINC, ABSORPTION, LIGHT PENETRATION, ALKALINITY, HALOGENS, HYDROGEN ION CONCENTRATION, PHOSPHATES, NITRATES, DISSOLVED OXYGEN, TEMPERATURE, SPECTROMETERS, TOXICITY, DIATOMS, CHRYSOPHYTA, CYANOPHYTA, METALS, IRRADIATION, AUTOMATION, EUTROPHICATION, NUCLEAR REACTORS, COMPUTERS, DATA PROCESSING, *WASHINGTON, SAMPLING.

IDENTIFIERS:
*WILLIAMS LAKE(WASH), URANIUM, THORIUM, ANTIMONY, MACRONUTRIENTS, MICRONUTRIENTS, SCANDIUM, RUBIDIUM, BROMINE, ACTINIDES, RADIOACTIVE DECAY, GAMMA-RAY SPECTROMETRY, COUNTING.

ABSTRACT:
RESEARCH DIRECTED TOWARD FINDING CORRELATIONS BETWEEN ALGAL POPULATIONS AND TRACE ELEMENTS IN A NATURAL AQUATIC ENVIRONMENT, WILLIAMS LAKE (WASHINGTON), BASICALLY INVOLVED THE IDENTIFICATION AND QUANTIFICATION OF ALGAL SPECIES AS WELL AS THE MEASUREMENT OF SOME 14 TRACE ELEMENTS PRESENT IN THE LAKE WATER. EIGHTY-TWO SAMPLES OF WATER WITH ALGAE WERE COLLECTED OVER A 13-MONTH PERIOD TO OBSERVE SEASONAL FLUCTUATIONS IN BOTH ALGAL SPECIES AND TRACE ELEMENTS. COLLECTION WAS MADE IN POLYETHYLENE BOTTLES FROM THREE DIFFERENT DEPTHS AT ONE LOCATION WHERE THE LAKE WAS THE DEEPEST. THESE SAMPLES ALONG WITH APPROPRIATE STANDARDS WERE IRRADIATED IN THE WASHINGTON STATE UNIVERSITY TRIGA III RESEARCH REACTOR FOR 12-15 HOURS. AFTER A DECAY PERIOD OF APPROXIMATELY 2 WEEKS, EACH SAMPLE AND STANDARD WAS COUNTED ON A NUCLEAR DATA 2200 GAMMA-RAY SPECTROMETER SYSTEM USING A HIGH RESOLUTION GE(LI) DETECTOR FOR APPROXIMATELY 17 HOURS TO DETERMINE TRACE ELEMENT CONCENTRATIONS. ALGAE CELL IDENTIFICATION AND QUANTITATIVE ESTIMATES WERE DETERMINED BY A MICROSCOPIC COUNT. DATA ANALYSIS AND A LINEAR CORRELATION STUDY BETWEEN ALGAL SPECIES AND TRACE ELEMENT CONCENTRATION WAS PERFORMED UTILIZING AN IBM 360/67 COMPUTER. SEVERAL PHYSICAL AND CHEMICAL MEASUREMENTS WERE ALSO MADE INCLUDING LIGHT TRANSMISSION, ALKALINITY, ACIDITY, PH, DISSOLVED OXYGEN, TEMPERATURE, PHOSPHATES, AND NITRATES. RESULTS INDICATE THAT: (1) URANIUM, THORIUM, AND CESIUM MAY BE PRESENT IN EXCESS OF GROWTH REQUIREMENTS OR MAY POSSIBLY HAVE NO AFECT ON THE GROWTH RATE OF ANY OF THE ALGAL FORMS STUDIED. (2) MERCURY, ALTHOUGH TOXIC IN LARGE QUANTITIES, MAY BE UTILIZED BY DIATOMS IN SMALL QUANTITIES. (3) CHROMIUM, GOLD, COBALT, AND ANTIMONY APPEAR TO BE TAKEN UP BY DIATOMS. (4) ZINC AND IRON APPEAR TO BE UTILIZED BY BLUE-GREEN ALGAE. (5) UPTAKE OF SCANDIUM, RUBIDIUM, AND BROMINE BY ALGAE IS NOT APPARENT, BUT MAY BE POSSIBLE. (HOLOMAN-BATTELLE)

FIELD 05C

ACCESSION NO. W73-01434

LIMNOLOGICAL STUDIES ON BIGHORN LAKE (YELLOWTAIL DAM) AND ITS TRIBUTARIES,

MONTANA STATE UNIV., BOZEMAN.

R. A. SOLTERO.

AVAILABLE FROM UNIV. MICROFILMS, INC., ANN ARBOR, MICH., 48106, ORDER NO. 71-28,873. PH. D. THESIS, 1971, 290 P.

DESCRIPTORS:
*LIMNOLOGY, *PHYTOPLANKTON, *WATER ANALYSIS, *STANDING CROP, *PRIMARY PRODUCTIVITY, LAKES, SAMPLING, RESERVOIRS, TRIBUTARIES, DAMS, CHEMICAL ANALYSIS, BIOMASS, ALGAE, PHYSICOCHEMICAL PROPERTIES, DIATOMS, SUCCESSION, NUTRIENTS, RIVERS, INFLUENT STREAMS, EFFLUENT STREAMS, LIGHT INTENSITY, NITROGEN, CHRYSOPHYTA, NITRATES, CHLOROPHYLL, PHOSPHATES, TURBIDITY, CONDUCTIVITY, OXYGEN, HYDROGEN ION CONCENTRATION, TEMPERATURE, FRESH WATER, WATER QUALITY, NUTRIENTS, *MONTANA.

IDENTIFIERS:
*BIGHORN LAKE(MONT), BIGHORN RIVER, SHOSHONE RIVER, YELLOWTAIL DAM, FRAGILARIA CROTONENSIS, CRYPTOMONAS OVATA, STEPHANODISCUS NIAGARAE, ASTERIONELLA FORMOSA, RHODOMONAS LACUSTRIS, COUNTING, ORTHOPHOSPHATES, CHLOROPHYLL A.

ABSTRACT:
THE PHYTOPLANKTON COMMUNITY OF BIGHORN LAKE IN RELATION TO ITS PHYSICAL AND CHEMICAL ENVIRONMENT WAS STUDIED DURING 1968-70. SAMPLES AND IN SITU MEASUREMENTS WERE TAKEN AT SIX PERMANENT SAMPLING STATIONS ON THE RESERVOIR DURING 56 CRUISES. CHARACTERIZATION OF THE INFLUENT AND EFFLUENT WATERS OF THE RESERVOIR REVEALED THAT BIGHORN LAKE WAS FERTILIZED BY THE BIGHORN AND SHOSHONE RIVERS. THE RELATIONSHIPS OF CONDUCTIVITY, TURBIDITY, NITROGEN, AND PHOSPHATE TO TRIBUTARY DISCHARGE WERE EXAMINED. UNDER AVERAGE CONDITIONS TOTAL VISIBLE LIGHT WAS REDUCED TO 1 PERCENT OF SURFACE INTENSITY AT A DEPTH OF APPROXIMATELY 10 METERS NEAR THE DAM, BUT ONLY 1 METER AT STATION 5 (80.5 KM UPSTREAM FROM THE DAM). RESULTS OF CHEMICAL ANALYSES SHOWED BIGHORN LAKE TO BE PREDOMINANTLY A CALCIUM-SODIUM-SULFATE-BICARBONATE WATER. NITRATE-NITROGEN AND ORTHOPHOSPHATE AS WELL AS THE OTHER CHEMICAL CONSTITUENTS DETERMINED WERE RELATIVELY HIGH AND APPEARED NOT TO BE LIMITING. STANDING CROPS OF THE PHYTOPLANKTON TAXA WERE DETERMINED BY DIRECT COUNT. FRAGILARIA CROTONENSIS, CRYPTOMONAS OVATA, STEPHANODISCUS NIAGARAE AND ASTERIONELLA FORMOSA REACHED THE LARGEST STANDING CROPS ACCORDING TO ABSOLUTE MEAN CELL VOLUMES, WHEREAS RHODOMONAS LACUSTRIS, CRYPTOMONAS OVATA, ASTERIONELLA FORMOSA AND FRAGILARIA CROTONENSIS WERE MORE IMPORTANT ON A PRESENCE BASIS. THE MEAN TOTAL ALGAL STANDING CROPS AND MEAN CHLOROPHYLL A FOR THE RESERVOIR WERE 2.3-23.2 CU MM/1 AND 0.6-10.4 MG/CU M, RESPECTIVELY. THE CHLOROPHYLL TO ALGAL CELL OLUME RATIO WAS 3.4 MICROGRAMS CHLOROPHYLL A TO 1.0 CU MM OF ALGAL CELLS. A DEFINITE ALGAL SUCCESSION WAS OBSERVED FOR ALL THREE YEARS. NET PRIMARY PRODUCTIVITY WAS CALCULATED AND A MEAN FOR ALL STATIONS SHOWED A RANGE OF 0.51 G C/SQ M/DAY - 1.42 G C/SQ M/DAY DURING 1968 AND 1970, RESPECTIVELY. (HOLOMAN-BATTELLE)

FIELD 05C, 02H

ACCESSION NO. W73-01435

THE ORGANIC GEOCHEMISTRY OF HYDROCARBONS IN COASTAL ENVIRONMENTS,

TEXAS UNIV., AUSTIN.

J. R. SEVER.

AVAILABLE FROM UNIV. MICROFILMS, INC., ANN ARBOR, MICH., 48106 ORDER NO. 72-2418. PH. D. THESIS, 1970. P 155.

DESCRIPTORS:
*COASTS, *ORGANIC COMPOUNDS, *GEOCHEMISTRY, CYANOPHYTA, BACTERIA, ALGAE, PLANKTON, PLANTS, SPECTROSCOPY, CHROMATOGRAPHY, GAS CHROMATOGRAPHY.

IDENTIFIERS:
NOSTOC MUSCORUM, LYNGBYA LAGERHAIMII, 7-METHYLHEPTADECANE, 8-METHYLHEPTADECANE, HYDROCARBONS, ALKANES, ALIPHATIC HYDROCARBONS, MASS SPECTROMETRY, INFRARED SPECTROSCOPY.

ABSTRACT:
SINCE HYDROCARBONS IN THE MARINE ENVIRONMENT ARE RELATIVELY BIOLOGICALLY AND CHEMICALLY INERT, A STUDY WAS CONDUCTED TO DETERMINE IF A PRECURSOR RELATIONSHIP COULD BE OBSERVED BETWEEN HYDROCARBONS IN THE SEDIMENTS AND HYDROCARBONS IN THE DOMINANT ORGANISMS WHICH CONTRIBUTE ORGANIC MATTER TO THE SEDIMENTS. BY EXAMINING THE HYDROCARBON DISTRIBUTIONS OF THE ORGANISMS AND ENVIRONMENTS OF THREE COASTAL AREAS - AN OPEN SHALLOW BAY, AN INTERTIDAL BLUE-GREEN ALGAL LAGOON, AND A SUPRATIDAL LAGOON, - IT WAS POSSIBLE TO OBSERVE DIRECT CORRELATIONS BETWEEN THE BIOLOGICAL AND GEOLOGICAL ALKANES. IDENTIFICATION OF THE HYDROCARBONS WAS DONE BY INFRARED SPECTROSCOPY, MASS SPECTROMETRY, AND GAS-LIQUID CHROMATOGRAPHY. FIVE CULTURED BLUE-GREEN ALGAE SHOWED A NARROW DISTRIBUTION OF NORMAL ALKANES WITH CHAIN LENGTHS BETWEEN C15 AND C19 WITH C17 PREDOMINATING. BACTERIAL ALKANE DISTRIBUTIONS WERE OF TWO TYPES: C14 - C18 RANGE AND BRANCHED HYDROCARBONS IN THE RANGE OF C15 - C20. ANALYSIS OF SEVEN DOMINANT HIGHER MARINE PLANTS REVEALED A DISTINCT ODD-CARBON NUMBER DISTRIBUTION IN THE C25 - C31 RANGE, WHILE A PLANKTON TOW YIELDED AN UNBIASED SMOOTH DISTRIBUTION OVER THE C14 - C28 RANGE. ANALYSIS OF EIGHT ANCIENT SHALES SHOWED THE UNIQUE 50:50 ALKANE MIXTURE OF 7-METHYL AND 8-METHYL HEPTADECANE WITH THE MAJORITY OF SAMPLES DISPLAYING THE C17 ALKANE AS ONE OF THE LARGEST HYDROCARBON COMPONENTS. THIS SUGGESTS LARGE DEPOSITS OF BLUE GREEN ALGAE AT THE TIME OF DEPOSITION. (MACKAN-BATTELLE)

FIELD 05B, 02L, 05C

ACCESSION NO. W73-01439

GROWTH AND PHOSPHATE REQUIREMENTS OF NITZSCHIA ACTINASTROIDES (LEMM.) V. GOOR IN BATCH AND CHEMOSTAT CULTURE UNDER PHOSPHATE LIMITATION, (WACHSTUM UND PHOSPHATBEDARF VON NITZSCHIA ACTINASTROICES (LEMM.) V GCCR IN STATISCHER UND HOMOKONTINUIERLICHER KULTUR UNTER PHOSPHATLIMITIERUNG),

FREIBURG UNIV. (WEST GERMANY). LIMNOLOGISCHES INSTITUT.

V. H. MULLER.

ARCHIV FUR HYDROBIOLOGIE, VOL 38, NO 4, P 399-484, MARCH 1972. 43 FIG, 14 TAB, 158 REF.

DESCRIPTORS:
*NUTRIENT REQUIREMENTS, *LIMITING FACTORS, *PHOSPHATES, DIATOMS, POLLUTANT IDENTIFICATION, GROWTH RATES, DEFICIENT ELEMENTS, WATER POLLUTION EFFECTS, AQUATIC ALGAE, CHRYSOPHYTA, LABORATORY EQUIPMENT, FRESHWATER, WATER POLLUTION SOURCES, INSTRUMENTATION, METHODOLOGY, NITROGEN, SILICON, ABSORPTION, PHYTOPLANKTON.

IDENTIFIERS:
*NITZSCHIA ACTINASTRCIDES, *BATCH CULTURES, *CHEMOSTATS, ORTHOPHOSPHATES, SUBSTRATE UTILIZATION, MONOD EQUATION, PENICILLIN G, TERRAMYCIN, CHLORAMPHENICOL, ACTIDON, CULTURE MEDIA, CHLOROPHYLL A, TERATOLOGY, NITZSCHIA.

ABSTRACT:
THE CONSTRUCTION OF A CHEMOSTAT WITH A CAPACITY OF 2 OR 4 L IS DESCRIBED, IT WAS USED TO DETERMINE GROWTH RESPONSES OF THE PLANKTONIC FRESHWATER DIATOM NITZSCHIA ACTINASTROIDES (LEMM.) V. GOOR TO LIMITING CONCENTRATIONS OF ORTHOPHOSPHATE. THE MAXIMAL SPECIFIC GROWTH-RATE WAS CALCULATED FROM THE DATA OBTAINED IN THE CHEMOSTAT AT 23 C WITH CONTINUOUS ILLUMINATION. IT IS 0.087 /HR AND AGREES WELL WITH THE FIGURE OF 0.083 /HR OBTAINED IN BATCH CULTURES WITHOUT SUBSTRATE LIMITATION. MONOD'S EQUATION DESCRIBES THE CORRELATICN BETWEEN GROWTH RATE AND SUBSTRATE CONCENTRATION OF PHOSPHORUS P SUB L AT MEDIUM AND HIGH GROWTH-RATES. THE SATURATION CONSTANT LIES BETWEEN 0.40 AND 0.44 MICROGRAM/L P SUB L. AT MEDIUM GROWTH-RATES, THE GROWTH-LIMITING FACTOR IS THE RATE OF P UPTAKE, AS IS INDICATED BY COMPARING UPTAKE RATES WITH THE KINETICS OF P UPTAKE IN P-STARVED CELLS. THE DISCREPANCIES OBSERVED AT LOW GROWTH RATES ARE CHARACTERIZED BY INCREASING P SUB L, HIGH DEATH RATES, AND HIGH FRACTIONS OF TERATOLOGICAL CELLS. THIS INDICATES THE APPEARANCE OF FACTORS OTHER THAN PHOSPHORUS LIMITING GROWTH. THRESHOLD VALUES OF THE DILUTION RATE EXIST, BELOW WHICH THE CULTURE IS WASHED OUT, THEY VARY WITH CONDITIONS IN THE CHEMOSTAT. THE CONTENTS IN P AND CHLOROPHYLL A OF CELLS WITH P-LIMITED GROWTH ARE DIRECTLY PROPORTIONAL TO P SUB L. P-STORAGE IN THE CELLS OCCURS IF GROWTH IS LIMITED BY FACTORS OTHER THAN P (E. G. CO2). THEREFORE THE YIELD COEFFICIENT IS CONTANT ONLY IF REFERRED TO THE CHLOROPHYLL A PRODUCED. REFERRED TO OTHER REFERENCE FIGURES, IT DECREASES WITH INCREASING GROWTH-RATE AND P-STORAGE, THE N-CONTENT OF THE CELLS INCREASES WITH INCREASING GROWTH RATE AND INCREASING STORAGE OF P. THE SI-CONTENT OF THE CELLS IS NOT INFLUENCED BY THESE FACTORS. THE RELIABILITY OF THE RESULTS AND THEIR TRANSFERABILITY TO NATURAL CONDITIONS ARE DISCUSSED. (LONG-BATTELLE)

FIELD 05C, 05A

ACCESSION NO. W73-01445

THE SPECIES COMPOSITION, SEASONAL SUCCESSION, REPRODUCTION AND DISTRIBUTION OF MARINE ALGAE FROM SCITUATE TO WOODS HOLE, MASSACHUSETTS,

NEW HAMPSHIRE UNIV., DURHAM.

D. C. COLEMAN.

AVAILABLE FROM UNIV. MICROFILMS, INC., ANN ARBOR, MICH., 48106 ORDER NO. 72-3735. PH. D. THESIS, 1971. P 143.

DESCRIPTORS:
*SEASONAL, *DISTRIBUTION PATTERNS, *MASSACHUSETTS, SYSTEMATICS, *REPRODUCTION, RHODOPHYTA, PHAEOPHYTA, WATER QUALITY, *MARINE ALGAE.

IDENTIFIERS:
*CAPE COD CANAL(MASS), BUZZARDS BAY, CAPE COD BAY, DATA INTERPRETATION, SPECIES DIVERSITY.

ABSTRACT:
BIMONTHLY COLLECTIONS WERE MADE FROM JANUARY TO DECEMBER, 1969 AND TOTAL OF 106 SPECIES WAS RECORDED. CONSPICUOUS DIFFERENCES IN SPECIES NUMBERS WERE EVIDENT AT EACH STATION. SCITUATE AND WOODS HOLE HAD THE HIGHEST NUMBERS, WHILE INTERMEDIATE VALUES WERE RECORDED IN THE CAPE COD CANAL. MOST OF THE SPECIES AT SCITUATE AND IN THE CANAL WERE PERENNIALS; ANNUALS WERE MOST ABUNDANT AT WINGS NECK AND WOODS HOLE. THE RHODOPHYCEAE ACCOUNTED FOR MOST OF THE SPRING AND SUMMER ANNUALS, WHILE THE BROWNS WERE THE MAJOR CONTRIBUTORS OF WINTER ANNUALS. SPRING AND EARLY SUMMER ANNUALS APPEARED SEVERAL WEEKS SOONER AT WINGS NECK AND WOODS HOLE THAN AT SCITUATE. SOME SPECIES APPEARED AS SPRING ANNUALS SOUTH OF THE CAPE, BUT AS SUMMER ANNUALS AT SCITUATE. SOME SPECIES REPRODUCED THROUGHOUT THE YEAR, WHILE OTHERS WERE RESTRICTED TO EITHER WARM OR COLD SEASONS. MOST SPECIES AT SCITUATE WERE FOUND IN THE INTERTIDAL AND/OR SUBTIDAL ZONE(S), WHILE THOSE AT ALL OTHER STATIONS WERE COLLECTED PRIMARILY IN THE SUBTIDAL ZONE. NUMEROUS FACTORS ARE RESPONSIBLE FOR THE VERTICAL DISPLACEMENT OF SPECIES RECORDED. AN INTERPRETATION OF THE DISTRIBUTIONAL PATTERNS IS GIVEN. (MORTLAND-BATTELLE)

FIELD 05C, 05B

ACCESSION NO. W73-01449

FIELD ASSESSMENT OF N2-FIXATION BY LEGUMES AND BLUE-GREEN ALGAE WITH THE ACETYLENE REDUCTION TECHNIQUE,

WISCONSIN UNIV., MADISON.

T. H. MAGUE.

AVAILABLE FROM UNIV. MICROFILMS, INC., ANN ARBOR, MICH., 48106 ORDER NO. 71-28,353. PH.D. THESIS, 1971, 200 P.

DESCRIPTORS:
*ALGAE, *LEGUMES, *NITROGEN FIXATION, *CYANOPHYTA, PLANTS, NITROGEN, REDUCTION(CHEMICAL), GAS CHROMATOGRAPHY, SEPARATION TECHNIQUES, *WISCONSIN, OLIGOTROPHY, EUTROPHICATION, LAKE SUPERIOR, LAKE MICHIGAN, LAKE HURON, LAKE ERIE, GREAT LAKES, SOYBEANS, LIGHT INTENSITY, AIR TEMPERATURE, SAMPLING, AQUATIC ALGAE, PHYTOPLANKTON, TROPHIC LEVELS.

IDENTIFIERS:
*ACETYLENE REDUCTION, RHIZOBIUM JAPONICUM, LAKE MENDOTA, CRYSTAL LAKE, TROUT LAKE, BIG ARBOR VITAE LAKE, LITTLE ARBOR VITAE LAKE, GREEN BAY, FOX RIVER, DETROIT RIVER, HETEROTROPHIC BACTERIA, ACETYLENE, ETHYLENE.

ABSTRACT:
FIELD STUDIES OF NITROGEN-FIXATION, BOTH IN AQUATIC AND TERRESTRIAL ENVIRONMENTS, UTILIZED A PROCEDURE WHICH CONSISTS OF (1) EXPOSING THE TEST MATERIAL (EXCISED AND INTACT SOYBEAN NODULES AND BLUE-GREEN ALGAE) TO APPROXIMATELY 0.1 ATM. OF ACETYLENE IN A CLOSED SYSTEM FOR 1/2 HOUR AND (2) QUANTITATING THE RESULTING ETHYLENE BY FLAME IONIZATION AFTER ITS GAS CHROMATOGRAPHIC SEPARATION FROM ACETYLENE. THE EXPERIMENTS WITH EXCISED SOYBEAN NODULES SHOWED THAT ACETYLENE REDUCTION WAS LINEAR FOR AT LEAST 80 MINUTES BUT WAS UNDETECTABLE IN THE ABSENCE OF OXYGEN. TWENTY SAMPLINGS FROM LAKE MENDOTA (MAY-NOVEMBER, 1970) REVEALED TWO MAJOR PEAKS IN BLUE-GREEN ALGAE N2-FIXING ACTIVITY, BUT NO CORRELATION BETWEEN DISSOLVED INORGANIC P OR N CONCENTRATIONS AND ACETYLENE REDUCTION RATES FOR THE SAME SAMPLES. OF THE FOUR WISCONSIN LAKES SAMPLED, (SUMMER, 1970) CRYSTAL AND TROUT LAKES (OLIGOTROPHIC) SUPPORTED LITTLE ACETYLENE REDUCTION/UNIT VOLUME; BIG AND LITTLE ARBOR VITAE LAKES (EUTROPHIC) SHOWED HIGH RATES OF ACETYLENE REDUCTION AND SUPPORTED HEAVY BLUE-GREEN ALGAL BLOOMS. SAMPLES (SEPTEMBER, 1970) FROM LAKES SUPERIOR, HURON, AND MICHIGAN SHOWED PRACTICALLY NO ACETYLENE REDUCTION. ACETYLENE REDUCTION BY SAMPLES TAKEN FROM GREEN BAY OF LAKE MICHIGAN VARIED AS A RESULT OF A NUTRIENT CONCENTRATION WHICH AT FIRST SUPPORTED GROWTH OF PHYTOPLANKTON REQUIRING COMBINED N, THEN WAS DILUTED SO THAT IT COULD ONLY SUPPORT PRIMARILY THOSE SPECIES ABLE TO FIX N2, AND FINALLY WAS DEPLETED TO THE POINT WHERE ORGANISMS TYPICAL OF OLIGOTROPHIC WATERS PREDOMINATED. THESE EXPERIMENT DEMONSTRATED THE SUITABILITY OF THE ACETYLENE REDUCTION ASSAY FOR FOLLOWING CHANGES IN RATES OF N2-FIXATION BY NODULATED LEGUMES IN THE FIELD AND BY BLUE-GREEN ALGAE IN LAKES. THE ASSAY ALSO SERVED TO DEFINE VARIOUS TROPHIC LEVELS IN LAKES AND TO LOCATE ZONES OF EUTROPHICATION IN THE GREAT LAKES. (HOLOMAN-BATTELLE)

FIELD 05C, 05A, 02H

ACCESSION NO. W73-01456

SELECTED REFERENCES CONCERNING THE ALGAE OF LAKE ERIE. II,

STATE UNIV., COLL., BUFFALO, N.Y. GREAT LAKES LAB.

R. A. SWEENEY.

SPECIAL REPORT NO 6, MARCH, 1970. 11 P.

DESCRIPTORS:
*BIBLIOGRAPHIES, *ALGAE, *LAKE ERIE, PLANKTON, BENTHIC FLORA, DIATOMS, PHYTOPLANKTON, PHOTOSYNTHESIS, PRIMARY PRODUCTIVITY, GREAT LAKES, NEW YORK, WATER SUPPLY, SYSTEMATICS, LIMNOLOGY, ECOLOGY, FISH, WATER POLLUTION, PROTOZOA, OHIO, EUTROPHICATION, LAKE ONTARIO, LIGHT PENETRATION, PHYSICOCHEMICAL PROPERTIES, AQUATIC PLANTS, INDUSTRIAL WASTES, FISH FOOD ORGANISMS, STREAMS.

IDENTIFIERS:
DESMIDS.

ABSTRACT:
THIS BIBLIOGRAPHY OF 127 REFERENCES IS COMPILED TO ASSIST THOSE CONDUCTING AND/OR CONTEMPLATING RESEARCH AND INSTRUCTION DEALING WITH ALGAE IN LAKE ERIE. IT IS A COMPILATION OF REFERENCES ON PLANKTONIC AND BENTHIC ALGAE PREVIOUSLY PUBLISHED IN VARIOUS JOURNALS, REPORTS, ETC., TO REDUCE LITERATURE SEARCH TIME. THE CITATIONS, LISTED ALPHABETICALLY, INCLUDE 13 PAPERS PUBLISHED BEFORE 1900, WITH THE EARLIEST DATED 1872. FORTY TITLES AND REFERENCES, DATED FROM 1900 TO 1949, RELATE TO TAXONOMY AND OF THAT GROUP FOUR INCLUDE POLLUTION PROBLEMS IN THE TITLES. THE PROBLEM OF PHOSPHORUS APPEARS IN PAPERS DATED 1951; THE EVIDENCE FOR EUTROPHICATION OF LAKE ERIE FROM PHYTOPLANKTON RECORDS IN 1964; REFERENCE TO INDUSTRIAL POLLUTION IN 1953; SEVERAL PAPERS ARE ON CLADOPHORA; NINE REFERENCES CONCERN PHOTOSYNTHESIS; AND SIX CONCENTRATE ON DIATOMS. (JONES-WISCONSIN)

FIELD 02H, 05C

ACCESSION NO. W73-01615

MORPHOGENESIS IN THE RED ALGA, GRIFFITHSIA PACIFICA: REGENERATION FROM SINGLE CELLS,

WASHINGTON UNIV., SEATTLE. DEPT. OF BOTANY.

C. S. DUFFIELD, S. D. WAALAND, AND R. CLELAND.

PLANTA (BERL.), VOL 105, P 185-195, 1972. 7 FIG, 27 REF. AEC AT(45-1)2225.

DESCRIPTORS:
*ALGAE, *RHODOPHYTA, *PLANT MORPHOLOGY, *PLANT GROWTH, CYTOLOGICAL STUDIES.

IDENTIFIERS:
*GRIFFITHSIA PACIFICA, *REGENERATION.

ABSTRACT:
A SYSTEM FOR STUDYING FORM DEVELOPMENT IN A RED ALGA IS DESCRIBED. PLANTS OF THE MARINE GIANT-CELLED GRIFFITHSIA PACIFICA REGENERATE FROM A SINGLE, ISOLATED SHOOT CELLS FOLLOWING A REGULAR AND PREDICTABLE PATTERN. THE CULTURE FOR THIS STUDY WAS COLLECTED IN MEXICO IN 1966 AND HAD BEEN MAINTAINED IN UNIALGAL CONDITION. REGENERATION OF PLANTS FROM ISOLATED SHOOT CELLS MAKES A SUITABLE SYSTEM FOR STUDY OF MORPHOGENESIS CONTROL IN RED ALGAE. THE CELLS' LARGE SIZE FACILITATES BOTH EXPERIMENTAL MANIPULATIONS AND OBSERVATIONS. BECAUSE OF MORPHOLOGY AND CELL SIZE, GRIFFITHSIA SPECIES HAVE BEEN WIDELY USED IN BIOCHEMICAL, CYTOLOGICAL, AND ELECTROPHYSIOLOGICAL STUDIES. IT HAS THE ADVANTAGE OF MULTICELLULARITY WITHOUT COMPLEXITIES SUCH AS CORTICATION FOUND IN MANY RED ALGAE. REGENERATION CAN BE INITIATED FROM ANY CELL OF THE PLANT. THE DEVELOPMENT IS SUFFICIENTLY RAPID THAT A PLANT OF 30-40 SHOOT-CELLS IS PRODUCED FROM A SINGLE CELL WITHIN A WEEK. THIS INVESTIGATION IS APPARENTLY THE FIRST IN WHICH THE PATTERN OF DEVELOPMENT OF INDIVIDUAL PLANTS HAS BEEN FOLLOWED WITH TIME AND IN WHICH KINETICS OF CELL DIVISION AND CELL ELONGATION HAVE BEEN ELUCIDATED. (JONES-WISCONSIN)

FIELD 05C

ACCESSION NO. W73-01622

RESPIRATORY ACTIVITIES OF CHLORELLA ELLIPSOIDEA IN VARIOUS NITRIENT MEDIA,

TAIWAN PROVINCIAL CHUNG HSING UNIV., TAICHUNG. INST. OF BOTANY.

J. TSCHEN, AND S. T. LIU.

BOTANICAL BULLETIN OF ACADEMIA SINICA, VOL 12, P 50-56, 1971. 2 FIG, 4 TAB, 27 REF.

DESCRIPTORS:
*LABORATORY TESTS, *RESPIRATION, *CHLORELLA, *NUTRIENTS, CULTURES, NITROGEN, POTASSIUM, PHOSPHATES, METABOLISM, ALGAE, SYNTHESIS, VITAMINS, SULFATES, IONS.

IDENTIFIERS:
*CHLORELLA ELLIPSOIDEA, CULTURE MEDIA.

ABSTRACT:
POPULATION DENSITY OF FRESH-WATER ALGAE IN A BATCH SYSTEM IN SMALL VESSELS DEPENDS ON CULTURAL CONDITIONS. IN SELECTING OPTIMAL CULTURAL MEDIA FOR FRESH-WATER ALGAE, CHLORELLA ELLIPSOIDEA WAS CULTURED IN VARIOUS NUTRIENT MEDIA, AND THE RESPIRATORY ACTIVITY DETERMINED BY A WARBURG RESPIROMETER. THE FOLLOWING MEDIA WERE STUDIED: BOLD'S BASAL MEDIUM AND MODIFICATION, 5N-BOLD'S BASAL MEDIUM, KANTZ' MODIFIED BBM AND OTT'S. OXYGEN UPTAKE WAS FOUND TO BE HIGHEST IN BBM WITH DECREASING ORDER IN 5N-BBM, KBBM, AND OTT'S MEDIUM. HIGHER RESPIRATORY ACTIVITY IN BBM MAY BE ATTRIBUTED TO NITRATE COMPONENTS. DECREASED RESPIRATION IN KBBM IS RELATED TO LOWER POTASSIUM AND PHOSPHATE CONTENTS. NO GROWTH TOOK PLACE IN OTT'S MEDIUM. NITROGEN IS ESSENTIAL IN RESPIRATION, PHOTOSYNTHESIS, AND IN THE SYNTHESIS OF PURINES AND PYRIMIDINES OF RNA AND DNA. NITROGEN SOURCE IN BBM IS SODIUM NITRATE, WHILE IN 5N-BBM, UREA IS USED INSTEAD. UREASE WAS NOT DETECTED IN C. ELLIPSOIDEA. THIS EXPLAINS WHY BBM IS SUPERIOR TO 5N-BBM. POTASSIUM IS ESSENTIAL AS AN ACTIVATOR FOR ENZYMES IN CERTAIN PEPTIDE BOND SYNTHESIS AND SO MAY AFFECT RESPIRATION AND PHOTOSYNTHESIS. PHOSPHORUS, FOUND IN MANY COENZYMES, BECOMES IMPORTANT IN PHOTOSYNTHESIS, GLYCOLYSIS, TCA CYCLE AND FATTY ACID SYNTHESIS. (JONES-WISCONSIN)

FIELD 05C

ACCESSION NO. W73-01626

COMPOSITION AND STRUCTURE OF ALGAL COMMUNITIES IN A TRIBUTARY STREAM OF LAKE ONTARIO,

GUELPH UNIV. (ONTARIO). DEPT. OF ZOOLOGY.

J. W. MOORE.

CANADIAN JOURNAL OF BOTANY, VOL 50, P 1663-1674, 1972. 6 FIG, 2 TAB, 33 REF.

DESCRIPTORS:
*ALGAE, *BIOLOGICAL COMMUNITIES, *STREAMS, LAKE ONTARIO, PLANKTON, DIATOMS, SEDIMENTS, HARDNESS(WATER), BENTHOS, GRAZING, SYSTEMATICS, CHEMICAL ANALYSIS, PHYSICAL PROPERTIES, CHLOROPHYTA, EUGLENOPHYTA, CYANOPHYTA, CHRYSOPHYTA, SCENEDESMUS, CHLAMYDOMONAS.

IDENTIFIERS:
SHELTER VALLEY CREEK(ONTARIO).

ABSTRACT:
SEASONAL SUCCESSION, COMMUNITY STRUCTURE, AND THE EFFECT OF HERBIVOROUS GRAZING ON PLANKTONIC AND EPIPELIC ALGAE IN SHELTER VALLEY CREEK, ONTARIO ARE DESCRIBED. THE COMPOSITION AND STRUCTURE OF THE ALGAL COMMUNITIES WERE STUDIED FOR ONE YEAR. THE OVERALL ASSEMBLAGE CONSISTED OF 388 TAXA OF WHICH 321 WERE BACILLARIOPHYCEAE. ON THE SEDIMENTS, SEASONAL SUCCESSION AND COMMUNITY STRUCTURE WAS, WITH SOME EXCEPTIONS, TYPICAL OF HARD-WATER STREAMS IN NORTHERN TEMPERATE ZONES. THE PLANKTONIC COMMUNITY WAS, FOR THE MOST PART, DERIVED FROM THE BENTHOS. HERBIVOROUS GRAZING BY LARVAL SEA LAMPREY, PETROMYZON MARINUS DID NOT NOTICEABLY AFFECT ALGAL NUMBERS IN EITHER THE PLANKTON OR SEDIMENTS. THE ALGAE COLLECTED ARE LISTED. THE 388 TAXA RECORDED CONTAINED 321 BACILLARIOPHYCEAE, 32 CHLOROPHYTA, 20 EUGLENOPHYTA, 14 CYANOPHYTA, AND 1 CHRYSOPHYCEAE. SIXTY-SIX TAXA ARE REPORTED AS NEW OCCURRENCES IN SOUTHERN ONTARIO. DIATOMS ACCOUNTED FOR 93.0 TO 99.5% BY NUMBERS OF THE EPIPELIC COMMUNITY. TAXA THAT OCCURRED FREQUENTLY THROUGHOUT THE YEAR WERE ACHANTHES MINUTISSIMA, COCCONEIS PLACENTULA VAR. EUGLYPTA, GOMPHONEMA OLIVACEUM VAR. BALTICUM, AND NAVICULA TRIPUNCTATA. (JONES-WISCONSIN)

FIELD 05C

ACCESSION NO. W73-01627

GENERAL CABLE THEORY FOR CELLS OF ALGAE CHARACEAE,

AGROFIZICHESKII NAUCHNO-ISSLEDOVATELSKII INSTITUT, LENINGRAD (USSR).

G. A. VOLKOV.

BIOCHIMICA ET BIOPHYSICA ACTA, VOL 255, P 709-719, 1972. 3 FIG, 3 REF.

DESCRIPTORS:
*BIOLOGICAL MEMBRANES, *ALGAE, *CYTYLOGICAL STUDIES, PLANT MORPHOLOGY, EQUATIONS, RESISTANCE, ELECTRICAL PROPERTIES.

IDENTIFIERS:
*CABLE THEORY, ELECTRONIC POTENTIAL, CHARACEAE.

ABSTRACT:
TO INVESTIGATE THE BIOLOGICAL MEMBRANE, THE ELECTRICAL CHARACTERISTICS OF MEMBRANES AND PARTICULARLY OF THE OUTER CYTOPLASMIC MEMBRANE OF THE INTERNODAL CELL OF ALGAE OF THE CHARACEAE ARE STUDIED. THE SO-CALLED CABLE PROPERTIES ARE CONSIDERED REFLECTION OF A CERTAIN MORPHOLOGICAL STRUCTURE OF THAT CELL. SOLUTION OF THE CABLE EQUATION IS OBTAINED TAKING INTO ACCOUNT THE DEFINITE RESISTANCE OF CELL NODES FOR THE GENERAL CASE OF AN APPLIED CURRENT, THE STRENGTH OF WHICH ALTERS WITH TIME. TWO IMPORTANT CASES FOR THE FUNCTION ARE CONSIDERED IN DETAIL. PRACTICAL FORMULAE FOR DETERMINING THE FUNDAMENTAL CHARACTERISTICS OF THE CELL ARE OBTAINED: MEMBRANE RESISTANCE AND CAPACITY (OF PLASMALEMMA), NODAL RESISTANCE, AND THE CHARACTERISTIC LENGTH OF THE CELL. FOR DETERMINATION OF THE MEMBRANE TIME CONSTANT AND HENCE THE MEMBRANE CAPACITY, A SIMPLIFIED METHOD FOR MEASURING MEMBRANE RESISTANCE, SUGGESTED BY OTHER AUTHORS, CAN BE USED. TAKING INTO ACCOUNT THE NODAL RESISTANCE OF THE CELL GIVES AN IMPROVED VALUE FOR THE ELECTRODE POSITION IN TERMS OF THE CRITICAL COORDINATE. THE CURRENT ELECTRODE HAS TO BE INSERTED IN THE MIDDLE OF THE CELL. (JONES-WISCONSIN)

FIELD 05C

ACCESSION NO. W73-01630

THERMAL POLLUTION AND ALGAE, (VARMVATTENUTSLAPP OCH ALGER),

UPPSALA UNIV. (SWEDEN). INST. FOR PHYSIOLOGIC BOTANY.

G. ERIKSSON, AND C. FORSBERG.

VATTEN, VOL 27, NO 4, P 441-448, 1971. 3 FIG, 41 REF.

DESCRIPTORS:
*INDUSTRIAL WASTES, *THERMAL POLLUTION, *COASTS, WATER POLLUTION EFFECTS, AQUATIC LIFE, HABITATS, ECOLOGY, *ECOSYSTEM, *ALGAE, *PHYTOPLANKTON, OXYGEN, PHOTOSYNTHESIS, RESPIRATION, SUCCESSION, PRODUCTIVITY, GROWTH RATES.

IDENTIFIERS:
*SWEDEN, SPECIES COMPOSITION.

ABSTRACT:
NUCLEAR POWER PLANTS BEING BUILT ALONG THE SWEDISH COASTS HAVE INCREASED INTEREST IN THE EFFECTS OF THERMAL POLLUTION. INCREASED TEMPERATURE FROM INSTALLATION DISCHARGES WILL CHANGE THE HABITATS OF MANY ORGANISMS. AS ALGAE, ESPECIALLY PHYTOPLANKTON, HAVE AN IMPORTANT ROLE IN THE ECOSYSTEM BECAUSE OF THEIR ABILITY TO PRODUCE O2 IN PHOTOSYNTHESIS, IT IS ESPECIALLY IMPORTANT TO KNOW THEIR REACTIONS TO TEMPERATURE CHANGES. THE EFFECTS OF TEMPERATURE CHANGE ON PHYTOPLANKTON PHOTOSYNTHESIS, RESPIRATION AND NET OXYGEN PRODUCTION ARE SUMMARIZED. THERMOPHILIC ALGAE, SPECIES COMPOSITION CHANGES, SUCCESSION, GROWTH RATE AND PRODUCTION ARE DISCUSSED ALONG WITH THE SOLUBILITY OF O2 IN WATER OF VARYING TEMPERATURES. (ENSIGN-PAI)

FIELD 05C

ACCESSION NO. W73-01632

CONCURRENT GROWTH OF BACTERIA AND ALGAE IN A CLOSED VESSEL,

NEBRASKA UNIV., LINCOLN. DEPT. OF CHEMICAL ENGINEERING.

P. J. REILLY.

AVAILABLE FROM THE NATIONAL TECHNICAL INFORMATION SERVICE AS PB-212 996, $3.00 IN PAPER COPY, $0.95 IN MICROFICHE. NEBRASKA WATER RESOURCES RESEARCH INSTITUTE TECHNICAL RESEARCH PROJECT COMPLETION REPORT, JULY 1972. 34 P, 13 TAB, 10 REF. OWRR A-020-NEB(1). 14-31-0001-3227.

DESCRIPTORS:
*LABORATORY TESTS, *BACTERIA, *ALGAE, *CULTURES, CYANOPHYTA, HYDROGEN ION CONCENTRATION, LIGHT INTENSITY, NUTRIENTS, PSEUDOMONAS, SYMBIOSIS.

IDENTIFIERS:
MICROCYSTIS AERUGINOSA, PSEUDOMONAS AERUGINOSA, LIMITING NUTRIENTS.

ABSTRACT:
GROWTH RESPONSES OF PURE CULTURES OF MICROCYSTIS AERUGINOSA TO CARBON DIOXIDE, PHOSPHATE, AND NITRATE ARE STUDIED TO IDENTIFY THE LIMITING NUTRIENT. EXPERIMENTS WERE CONDUCTED ON GROWTH OF THE OBLIGATE AEROBIC BACTERIUM, PSEUDOMONAS AERUGINOSA FOR USE AS A MODEL BACTERIUM IN INVESTIGATION OF SYMBIOTIC TWO-SPECIES BACTERIAL/ALGAL CULTURES WHICH MAY BE SIGNIFICANT IN THE EUTROPHICATION PROCESS. IN ADDITION, EVIDENCE WAS FOUND THAT BACTERIA REMOVES ALGAL TOXIC AGENTS WHICH MAY EXPLAIN WHY ALGAE BLOOMS SOMETIMES APPEAR SO SUDDENLY AND UNEXPECTEDLY. ONE BACTERIAL AND ONE ALGAL SPECIE WERE GROWN IN THE SAME VESSEL UNDER CONTROLLED CONDITIONS AFTER THE ALGAE WERE PURIFIED OF CONTAMINATING BACTERIA. ITS RESPONSES TO GLUCOSE, PH, LIGHT INTENSITY, AND AERATION, AND AGITATION RATE WERE DETERMINED. DURING THE PURIFICATION OF THE ALGAE OTHER OTHER INTERACTIONS OCCURRED IN ADDITION TO THE TRANSFER OF OXYGEN FROM ALGAE TO BACTERIA AND THE REVERSE TRANSFER OF CARBON DIOXIDE FROM BACTERIA TO ALGAE: TWO THAT WERE VERY OBVIOUS WERE THE TRANSFER OF ORGANIC EXCRETIONS TO THE BACTERIA AND THE BREAKDOWN OF AGENTS TOXIC TO THE ALGAE. (JONES-WISCONSIN)

FIELD 05C

ACCESSION NO. W73-01657

SPECIAL LAKE WATER TREATMENT PROBLEMS,

CHICAGO DEPT. OF WATER AND SEWERS, ILL. BUREAU OF WATER.

J. C. VAUGHN.

JOURNAL OF THE AMERICAN WATER WORKS ASSOCIATION, VOL 64, NO 9, P 585-589, SEPTEMBER 1972. 10 FIG, 2 REF.

DESCRIPTORS:
*WATER TREATMENT, WATER POLLUTION EFFECTS, *TREATMENT FACILITIES, *INFLUENT STREAMS, *GREAT LAKES, AQUATIC ALGAE, ICE, FISH, MICROORGANISMS, ODOR, TASTE, PLANKTON, CHEMICALS, PHENOLS, CHLORINE, OXIDATION, ADSORPTION, ACTIVATED CARBON, SLUDGE TREATMENT, DIATOMS, COAGULATION, CALCIUM CARBONATE, HYDROGEN ION CONCENTRATION, FILTRATION, PLANKTON NETS, NUISANCE ALGAE, CLADOPHORA.

IDENTIFIERS:
*INTERFERENCE, ALEWIVES, DICHOTOMOSIPHON, FRAGILARIA, TABELLARIA, ASTERIONELLA, SYNEDRA, DINOBRYON, MELOSIRA.

ABSTRACT:
WATER TREATMENT PLANTS DEPENDING ON SURFACE WATER FOR FILTRATION TREATMENT ARE LIABLE TO MANY INTAKE DELIVERY PROBLEMS. ICING UP OF INTAKES CAN BE AVOIDED BY DESIGN TECHNIQUES, BACKFLUSHING, TEMPERATURE CONTROL, AND CLEANING BY BLASTING. BLOCKAGE OF INTAKES BY FISH (ALEWIVES) HAS BEEN REMEDIED BY PROTECTIVE NETS AND SCREENS. ALGAL COLLECTIONS ON SCREENS OR NETS REQUIRE POSITIVE PRESSURE FOR REMOVAL (DICHOTOMOSIPHON, CLADOPHORA) OR INSTALLATION OF REVOLVING SCREENS AND PULVERIZING EQUIPMENT. PLANKTON (FRAGILLARIA, TABELLARIA, ASTERIONELLA, AND SYNEDRA) CAN BEST BE GOTTON RID OF BY A GOOD SURFACE-WASH SYSTEM SO THAT BROKEN-UP MATS CAN BE REMOVED FROM FILTERS BY BACKWASHING. PROBLEMS WITH TASTES AND ODORS CAN RESULT FROM MICROORGANISMS (DINOBRYON) OR CHEMICALS. TREATMENT USUALLY INVOLVES OXIDATION BY EXCESS CHLORINE TREATMENT AND/OR ADSORPTION ONTO ACTIVATED CARBON. MICROSTRAINERS AND SLUDGE BLANKET DEVICES REQUIRED NORMAL MAINTENANCE AND SOME SUPPLEMENTATION. THE MOST DIFFICULT AND EXPENSIVE PROBLEM INVOLVES WINTERTIME DIATOMS (MELOSIRA). A SIDE ISSUE OF THEIR PROLIFERATION IS DEVELOPMENT OF COLLOIDAL TURBIDITY; DOUBLING COAGULANT DOSAGE APPEARS SOMEWHAT REMEDIAL. CALCIUM CARBONATE AND PH CHANGES APPEAR RELATED TO MELOSIRA BLOOMS. (MACKAN-BATTELLE)

FIELD 05F, 02H

ACCESSION NO. W73-01669

PRELIMINARY RADIATION SURVEILLANCE OF AN AQUATIC SYSTEM NEAR THE NEVADA SITE, JUNE - JULY, 1967,

ENVIRONMENTAL PROTECTION AGENCY, LAS VEGAS, NEV. WESTERN ENVIRONMENTAL RESEARCH LAB.

W. L. KLEIN, AND R. A. BRECHBILL.

AVAILABLE FROM THE NATIONAL TECHNICAL INFORMATION SERVICE AS SWRHL-65-R, $3.00 IN PAPER COPY, $0.95 IN MICROFICHE. REPORT NO SWRHL-65R, FEBRUARY 1972. 23 P, 2 FIG, 3 TAB, 10 REF, 2 APPEND. MEMORANDUM OF UNDERSTANDING NO SF 54 373.

DESCRIPTORS:
*SURVEYS, *AQUATIC ENVIRONMENT, *RADIOECOLOGY, *RADIOCHEMICAL ANALYSIS, MEASUREMENT, *NEVADA, AQUATIC PLANTS, WATER POLLUTION SOURCES, SEDIMENTS, AQUATIC SOILS, PLANT TISSUES, FRESHWATER FISH, AQUATIC ALGAE, EVALUATION, SAMPLING, CARP, POTASSIUM RADIOISOTOPES, STRONTIUM RADIOISOTOPES, SOIL ANALYSIS, SOIL CONTAMINATION, CLADOPHORA, CHARA, FISH EGGS.

IDENTIFIERS:
*BIOLOGICAL SAMPLES, SR-90, K-40, CS-137, *PAHRANAGAT LAKE, GAMMA RAY SPECTROMETRY, ANIMAL TISSUES, MACROPHYTES, CYPRINUS CARPIO, MOSQUITO FISH, GAMBUSIA AFFINIS, CESIUM RADIOISOTOPES, CLADOPHORA SPP, PASPALUM DISTICHUM, POTAMOGETON PECTINATUS, ELEOCHARIS MONTEVIDENSES, CHARA ASPPRA, SCIRPUS SPP, GILLS, MUSCLE, VISCERA, EYES, SCALES(FISH), BONE.

ABSTRACT:
A THREE-MONTH PRELIMINARY RADIATION SURVEILLANCE STUDY WAS MADE OF AN AQUATIC SYSTEM IN UPPER PAHRANAGAT LAKE NEAR THE NEVADA TEST SITE. THE OBJECTIVES WERE TO DETERMINE THE CONCENTRATIONS OF FISSION PRODUCTS IN SELECTED SAMPLES AND TO ESTABLISH THE NECESSARY METHODOLOGY FOR RADIATION SURVEILLANCE IN AN AQUATIC ECOSYSTEM. BIOLOGICAL SAMPLES FROM A FRESHWATER LAKE NEAR THE NEVADA TEST SITE (NTS) WERE ANALYZED FOR THE PRESENCE OF SELECTED RADIONUCLIDES IN ORDER TO ESTABLISH A BASE LINE FOR THIS PARTICULAR SYSTEM AND TO DEVELOP METHODOLOGY NECESSARY FOR ANY FURTHER DEFINITIVE STUDIES OF THIS TYPE. RADIONUCLIDE CONCENTRATIONS WERE INSIGNIFICANT IN WATER, AQUATIC PLANT, AND FISH SAMPLES. SEDIMENT SAMPLES HAD DETECTABLE LEVELS OF CS-137, K-40, SR-90, AND U. STRONTIUM-90 LEVELS IN FISHBONE WERE LOW (2.38 PCI/G BONE ASH) COMPARED TO THOSE FOUND IN BOVINE FEMUR SAMPLES (6.9 PCI/G BONE ASH) COLLECTED DURING THE SAME PERIOD. (MACKAN-BATTELLE)

FIELD 05A, 02H

ACCESSION NO. W73-01673

SURVEY OF TECHNIQUES USED TO PRESERVE BIOLOGICAL MATERIALS,

STANFORD RESEARCH INSTITUTE, MENLO PARK, CALIF.

E. J. FEINLER, AND R. W. HUBBARD.

AVAILABLE FROM THE NATIONAL TECHNICAL INFORMATION SERVICE AS N72-18080, $6.00 IN PAPER COPY, $0.95 IN MICROFICHE. NASA CONTRACT REPORT NO. 114422, JANUARY 1972. 390 P, 78 TAB, 1350 REF. PROJECT NO. SRI LSU-8930. CONTRACT NO. NAS2-6201.

DESCRIPTORS:
*SURVEYS, *METHODOLOGY, FREEZE DRYING, REFRIGERATION, INCUBATION, DIALYSIS, CHEMICALS, DRYING, FREEZING, RADIATION, CYTOLOGICAL STUDIES, PLANT TISSUES, ALGAE, MICROORGANISMS, BACTERIA, INVERTEBRATES, FUNGI, ENZYMES, ANIMAL PARASITES, VIRUSES, MOLLUSKS, PROTOZOA, YEASTS, INSECTS, MILDEWS, E. COLI, FISH, FOODS, GRASSES, WORMS.

IDENTIFIERS:
*BIOLOGICAL MATERIALS, *BIOLOGICAL SAMPLES, *SAMPLE PRESERVATION, CHEMICAL PRESERVATION, HISTOCHEMISTRY, LYOPHILIZATION, ANIMAL TISSUES, HISTOLOGICAL STUDIES, BODY FLUIDS, VERTEBRATES, HEAT STERILIZATION, FIXATION, EMBEDDING, ASHING, BONE, ORGANS, BRAIN, BLOOD, KIDNEYS, LIVER, PLASMODIUM BERGHEI, BORRELIA KANSAS, BORRELIA ANSERINA.

ABSTRACT:
EXISTING TECHNIQUES USED TO PRESERVE BIOLOGICAL MATERIALS ARE DESCRIBED. THIS INFORMATION IS PRESENTED IN A HANDBOOK FORMAT THAT CATEGORIZES THE MOST IMPORTANT PRESERVATION TECHNIQUES AVAILABLE, AND INCLUDES A REPRESENTATIVE SAMPLING OF THOUSANDS OF APPLICATIONS OF THOSE TECHNIQUES TO BIOLOGICAL MATERIALS AND ORGANISMS. THE HANDBOOK IS DIVIDED INTO FOUR MAIN SECTIONS: (1) A REVIEW OF REVIEWS, (2) TABLES OF TECHNIQUES OF PRESERVATION, (3) INDEXES, AND (4) A COMPREHENSIVE BIBLIOGRAPHY. (HOLOMAN-BATTELLE)

FIELD 05A

ACCESSION NO. W73-01676

THE RESPONSES OF THE BIOTA OF LAKE WABAMUN, ALBERTA, TO THERMAL EFFLUENT,

ALBERTA UNIV., EDMONTON. DEPT. OF ZOOLOGY.

J. R. NURSALL, AND D. N. GALLUP.

IN: PROCEEDINGS OF THE INTERNATIONAL SYMPOSIUM ON THE IDENTIFICATION AND MEASUREMENT OF ENVIRONMENTAL POLLUTANTS, OTTAWA, P 295-304, JUNE 1971. 6 FIG, 2 TAB, 23 REF.

DESCRIPTORS:
*BIOTA, *TEMPERATURE, *CANADA, *LAKES, *THERMAL POLLUTION, WATER POLLUTION, COOLING, THERMAL POWERPLANTS, HEATED WATER, CIRCULATION, INFRARED RADIATION, HYDROGRAPHY, PLANKTON, ALGAE, ENVIRONMENTAL EFFECTS, BIOLOGICAL COMMUNITY, CANADA.

IDENTIFIERS:
*LAKE WABAMUN, BIOLOGICAL STUDIES, MACROPHYTES.

ABSTRACT:
TWO POWER-GENERATING PLANTS PUT WATER FROM CONDENSER-COOLING SYSTEMS INTO LAKE WABAMUN AT TEMPERATURES 8 - 14 C ABOVE AMBIENT. THE INPUT REPRESENTS ABOUT 0.5 PERCENT OF THE LAKE VOLUME PER DAY WHEN AVERAGED OVER THE YEAR. CIRCULATION RESTRICTS THE HEATED WATER TO THE EAST END OF THE LAKE, THEREBY LOCALIZING ITS EFFECTS. INFRA-RED IMAGERY HAS BEEN USED TO SUPPORT DETAILED DIRECT THERMOMETRY. DISCUSSED ARE THE EFFECTS OF HEATED EFFLUENT ON PLANKTONIC ALGAE, SEVERAL SPECIES OF MACROPHYTES, FIVE SPECIES OF ROTIFERS AND FOUR SPECIES OF PLANKTONIC CRUSTACEANS. NOTED IN PARTICULAR ARE THE WIDE DISTRIBUTION OF ELODEA CANADENSIS IN THE HEATED ZONE, CONCENTRATION OF POTAMOGETON PECTINATUS IN THE EFFLUENT CANAL, REDUCTION IN NUMBERS OF DIAPTOMUS OREGONENSIS IN THE HEATED ZONE, AND VARIATION IN NUMBERS AND EGG RATIOS FOR PLANKTONIC ROTIFERS BETWEEN THE HEATED AND UNHEATED REGIONS OF THE LAKE. (OLESZKIEWICZ-VANDERBILT)

FIELD 05C, 02H

ACCESSION NO. W73-01704

AN ECOSYSTEMATIC STUDY OF THE SOUTH RIVER, VIRGINIA,

VIRGINIA POLYTECHNIC INST. AND STATE UNIV., BLACKSBURG. WATER RESOURCES RESEARCHCENTER.

J. CAIRNS, JR., AND K. L. DICKSON.

AVAILABLE FROM THE NATIONAL TECHNICAL INFORMATION SERVICE AS PB-213 159, $3.00 IN PAPER COPY, $0.95 IN MICROFICHE. VIRGINIA WATER RESOURCES RESEARCH CENTER, BLACKSBURG, BULLETIN 54, JULY 1972. 104 P, 5 FIG, 11 TAB, 22 REF, APPEND. OWRR A-999-VA (15).

DESCRIPTORS:
*BIOLOGICAL PROPERTIES, *ECOSYSTEMS, *RIVERS, *VIRGINIA, *WATER POLLUTION EFFECTS, SURVEYS, SAMPLING, DATA COLLECTIONS, INDUSTRIAL WASTES, AGRICULTURAL RUNOFF, DOMESTIC WASTES, ECOLOGY, FISH, ALGAE, AQUATIC PLANTS, PROTOZOA, BACTERIA, STREAMFLOW, FLOW RATES, WATER TEMPERATURE, WATER CHEMISTRY, CHEMICAL ANALYSIS, NUTRIENTS.

IDENTIFIERS:
*SOUTH RIVER(VA).

ABSTRACT:
A BIOLOGICAL SURVEY OF THE FISH, MACROINVERTEBRATES, ALGAE, AQUATIC PLANTS, PROTOZOANS, AND BACTERIA WAS CONDUCTED ON THE SOUTH RIVER, VIRGINIA, IN SEPTEMBER 1970. THE PURPOSE OF THIS SURVEY WAS TO MEASURE THE ECOLOGICAL CONDITION OF A SYSTEM RECEIVING AGRICULTURAL, DOMESTIC, AND INDUSTRIAL WASTES. SAMPLES WERE COLLECTED AT 8 SITES TO DETERMINE THE EFFECTS OF VARIOUS WASTES ON THE BIOTA OF THE SOUTH RIVER. THE FAUNA AND FLORA AT EACH STATION WERE EXAMINED TO DETERMINE THE DIVERSITY, DENSITY, AND DISTRIBUTION OF AQUATIC LIFE IN RELATION TO PHYSICAL AND CHEMICAL WATER QUALITY. THE DISCHARGE OF DOMESTIC AND INDUSTRIAL WASTES IN WAYNESBORO, VIRGINIA (1) EXCEEDED THE WASTE ASSIMILATIVE CAPACITY OF THE RIVER AND CAUSED THE DISSOLVED OXYGEN TO BE ENTIRELY DEPLETED IN CERTAIN REACHES OF THE RIVER AT TIMES OF LOW FLOW AND HIGH TEMPERATURE; (2) ENRICHED THE SYSTEM BY ADDING NUTRIENTS SUCH AS CARBON, PHOSPHORUS, AND NITROGEN, CAUSING A DEFINITE SHIFT IN THE COMPOSITION OF THE FLORA; AND (3) DECREASED THE DIVERSITY OF FISH AND MACROINVERTEBRATES AND CAUSED QUALITATIVE SHIFTS IN ALGAE, HIGHER PLANTS, BACTERIA, AND PROTOZOANS WHEN COMPARED TO AREAS OF THE SOUTH RIVER UPSTREAM OF WAYNESBORO. THE BIOLOGICAL RECOVERY WAS NOT COMPLETE FOURTEEN MILES DOWNSTREAM OF WAYNESBORO AT HARRISTON, VIRGINIA. (WOODARD-USGS)

FIELD 05C, 05B

ACCESSION NO. W73-01972

WATER QUALITY CRITERIA DATA BOOK - VOLUME 3: EFFECTS OF CHEMICALS ON AQUATIC LIFE, SELECTED DATA FROM THE LITERATURE THROUGH 1968.

BATTELLE MEMORIAL INST., COLUMBUS, OHIO.

COPY AVAILABLE FROM GPO SUP DOC AS EP1.16:18050GWV 05/71/V3, $3.75; MICROFICHE FROM NTIS AS PB-213 210, $0.95. ENVIRONMENTAL PROTECTION AGENCY, WATER POLLUTION CONTROL RESEARCH SERIES, MAY 1971. 528 P, 1 FIG, 10 TAB, 961 REF, 4 APP. EPA PROGRAM 18050 GW VO 05/71 68-01-0007.

DESCRIPTORS:
*TOXICITY, REVIEWS, *BIOASSAY, *INDUSTRIAL WASTES, *PESTICIDES, *AQUATIC ORGANISMS, *PEST CONTROL, *CHEMCONTROL, *BIBLIOGRAPHIES, *DATA COLLECTIONS, *WATER POLLUTION EFFECTS, DOCUMENTATION, PUBLICATIONS, PESTICIDE TOXICITY, BIOINDICATORS, AGRICULTURAL CHEMICALS, FISH, CHEMICALS, CHEMICAL WASTES, BIOCHEMICAL OXYGEN DEMAND, BACTERIA, ALGAE, AQUATIC FUNGI, INVERTEBRATES, AQUATIC INSECTS, OYSTERS, SHRIMP.

IDENTIFIERS:
SUMMARIES.

ABSTRACT:
ORIGINAL DATA FROM MORE THAN 500 TECHNICAL PUBLICATIONS CONCERNING THE SPECIFIC EFFECTS OF CHEMICALS ON INDIVIDUALS SPECIES OF AQUATIC BIOTA WERE COLLECTED AND SUMMARIZED IN UNIFORM FORMAT. ALPHABETICAL ASSEMBLY OF THE DATA BY CHEMICAL ALLOWS RAPID ACCESS TO CONSIDERABLE DETAILED INFORMATION. A SPECIES INDEX FACILITATES SEARCH FOR INFORMATION ON THE TOXICITY OF CHEMICALS TO INDIVIDUAL AQUATIC SPECIES. THE DETAILS OF MAJOR PROCEDURES IN LABORATORY BIOASSAY AND FIELD ASSESSMENT OF CHEMICAL TOXICITY IN WATER ARE DISCUSSED. FRESHWATER AND MARINE PROCEDURES ARE INCLUDED. A TOTAL OF 961 REFERENCES WERE UTILIZED. RECOMMENDATIONS INCLUDE: (1) ESTABLISHMENT OF AN INFORMATION-ANALYSIS CENTER ON CHEMICAL WATER POLLUTION BASED TO SOME EXTENT ON THE REPORT PREPARED, (2) PREPARATION OF A LISTING OF CHEMICAL CONSTITUENTS OF EFFLUENTS AND CONTINUED UP-DATING OF THIS LIST, (3) DEVELOPMENT OF A PATTERN OF BIOASSAYS FOR EVALUATING THE EFFECTS OF A CHEMICAL ON AQUATIC LIFE . (DATA FROM THESE EVALUATIONS WOULD BE USED IN DEVELOPING MATHEMATICAL MODELS FOR PREDICTING CHEMICAL TOXICITY IN A WIDE RANGE OF ENVIRONMENTAL CIRCUMSTANCES). (4) DEVELOPMENT OF IN SITU BIOASSAY PROCEDURES FOR MORE REALISTIC ASSESSMENT OF CHEMICAL TOXICITY TO AQUATIC LIFE.(LEGORE-WASHINGTON)

FIELD 05C, 05A, 07C, 05B

ACCESSION NO. W73-01976

ECOLOGICAL EFFECTS OF OFFSHORE CONSTRUCTION,

MARINE SCIENCE INST., BAYOU LA BATRE, ALA.

G. A. ROUNSEFELL.

AVAILABLE FROM THE NATIONAL TECHNICAL INFORMATION SERVICE AS AD-739 704, $3.00 IN PAPER COPY, $0.95 IN MICROFICHE. JOURNAL OF MARINE SCIENCE, VOL 2, NO 1, 1972. 208 P, 1 TAB, 252 REF, 3 APPEND. CONTRACT NO DACW72-71-C-0002.

DESCRIPTORS:
*ECOLOGY, *ENVIRONMENTAL EFFECTS, *MARINE PLANTS, *MARINE ANIMALS, OFFSHORE PLATFORMS, NUCLEAR POWERPLANTS, WATER PROPERTIES, SALINITY, WATER TEMPERATURE, OXYGEN, TURBIDITY, BENTHIC FAUNA, WATER POLLUTION EFFECTS, STRIPED BASS, DISSOLVED OXYGEN, OIL, BRIDGES, BUOYS, HARBORS, TUNNELS, MARINE ALGAE, ANIMAL POPULATIONS, LOBSTERS, CRUSTACEANS, BIOMASS, SHRIMP, PINK SHRIMP, CLAMS, WATER POLLUTION, CHLOROPHYTA, ESTUARINE ENVIRONMENT, ISLANDS, GRASSES, INVERTEBRATES, MOLLUSKS, *REVIEWS.

IDENTIFIERS:
*OFFSHORE CONSTRUCTION, DELAWARE BAY, SCOLOPLOS ARMIGER, ALIMITOS BAY, LAKE PONTCHARTRAIN, PEARL RIVER, LAKE BORGNE, SUBSTRATES, MACROINVERTEBRATES, NEW YORK BIGHT, EELGRASS, ZOSTERA, PINAEUS DUORARUM, PINAEUS AZTECUS, PENAEUS SETIFERUS, MERCENARIA MERCENARIA, CLYMENELLA TORQUATA, CAPITELLA CAPITATA, MACOMA CALCAREA, ASTARTE BOREALIS, ENTEROMORPHA MINIMA, ULVA DACTYLIFERA, PANULIRUS INTERRUPTUS, MICROPOGON UNDULATUS, CROAKERS, CYNOSCION REGALIS, ROCCUS SAXATILIS, FUNDULUS HETEROCLITUS.

ABSTRACT:
AN EVALUATION OF CURRENT KNOWLEDGE OF THE PROBABLE ECOLOGICAL EFFECTS OF VARIOUS TYPES OF OFFSHORE CONSTRUCTION REVEALS SLIGHT DANGER FROM THE MAJORITY OF CONSTRUCTION PROGRAMS. THE GREATEST DANGERS LIE IN THE PLACEMENT OF ARTIFICIAL ISLANDS WITHIN OR TOO CLOSELY ADJACENT TO ESTUARIES WHERE THEY CAN SIGNIFICANTLY AFFECT WATER EXCHANGE, AND IN THE PROLIFERATION OF WATER COOLED NUCLEAR POWER PLANTS. PERHAPS THE MOST PRESSING NEED FOR ULTIMATE HUMAN SURVIVAL IS THE FURTHER DEVELOPMENT OF POWER FROM NATURAL FORCES TO REPLACE POWER FROM NUCLEAR AND FOSSIL FUEL SOURCES. (BYRD-BATTELLE)

FIELD 05C, 02L, 10F

ACCESSION NO. W73-02029

THE STRUCTURE AND FUNCTION OF FRESH-WATER MICROBIAL COMMUNITIES.

VIRGINIA POLYTECHNIC INST. AND STATE UNIV. BLACKSBURG.

RESEARCH DIVISION MONOGRAPH 3, 1971. J. CAIRNS, JR., EDITOR, 301 P.

DESCRIPTORS:
*AQUATIC PRODUCTIVITY, *ECOLOGY, *AQUATIC ENVIRONMENT, *BIOLOGICAL COMMUNITIES, *AQUATIC MICROORGANISMS, CYCLING NUTRIENTS, ECOLOGICAL DISTRIBUTION, ECOSYSTEMS, AQUATIC HABITATS, PRODUCTIVITY, AQUATIC BACTERIA, AQUATIC ALGAE, CARBON CYCLE, ON-SITE TESTS, AQUATIC ANIMALS, AQUATIC POPULATIONS, BIOASSAY, EUTROPHICATION.

IDENTIFIERS:
MINERAL CYCLE, CHARACTERIZATION, CHLAMYDOMONAS REINHARDTII, TETRAHYMENA VORAX, PSEUDOMONAS FLUORESCENS, AUTOTROPHIC BACTERIA, HETEROTROPHIC BACTERIA, GLUCOSE, ACETATES, BIOTIN, NIACIN, COBALAMINS, EUGLENA GRACILIS, COLONIZATION, CAULOBACTER, NAJAS FLEXILIS, SCIRPUS ACUTUS, C-14, DISSOLVED ORGANIC MATTER, MACROPHYTES, BIOLOGICAL SAMPLES.

ABSTRACT:
THE 1969 SYMPOSIUM ON 'THE STRUCTURE AND FUNCTION OF FRESH-WATER MICROBIAL COMMUNITIES' SPONSORED BY THE AMERICAN MICROSCOPICAL SOCIETY WAS DIRECTED TOWARD AN UNDERSTANDING OF THE RELATIONSHIPS AMONG LIVING THINGS IN THE AQUATIC ENVIRONMENT. THE TOPICS INCLUDED WERE: 'ADAPTATIONS FOR PHOTOREGENERATIVE CYCLING', 'CARBON FLOW IN THE AQUATIC SYSTEM'; 'AQUATIC LABORATORY MICROSYSTEMS AND COMMUNITIES'; 'THE ROLE OF LABORATORY EXPERIMENTATION IN ECOLOGICAL RESEARCH'; 'A CONTINUOUS GNOTOBIOTIC (SPECIES DEFINED) ECOSYSTEM'; 'COMMUNITY STRUCTURE OF PROTOZOANS AND ALGAE WITH PARTICULAR EMPHASIS ON RECENTLY COLONIZED BODIES OF WATER'; 'DIATOM COMMUNITIES'; 'MICROBIAL RELATIONSHIPS IN BIOLOGICAL WASTEWATER TREATMENT SYSTEMS'; 'HETEROTROPHIC BACTERIA IN AQUATIC ECOSYSTEMS; SOME RESULTS OF STUDIES WITH ORGANIC RADIOISOTOPES'; 'SEASONAL DISTRIBUTION OF COBALAMINS, BIOTIN AND NIACIN IN RAINWATER'; 'LACUSTRINE FUNGAL COMMUNITIES'; 'FACTORS AFFECTING THE NUMBER OF SPECIES IN FRESH-WATER PROTOZOAN COMMUNITIES'; 'THE INTERRELATIONSHIP BETWEEN FRESHWATER BACTERIA, ALGAE, AND ACTINOMYCETES IN SOUTHWESTERN RESERVORIS'; 'CHEMO-ORGANOTROPHY IN EPIPHYTIC BACTERIA WITH REFERENCE TO MACROPHYTIC RELEASE OF DISSOLVED ORGANIC MATTER'; AND 'BACTERIOLOGICAL PROFILES AND SOME CHEMICAL CHARACTERISTICS OF TWO PERMANENTLY FROZEN ANTARCTIC LAKES'. (LONG-BATTELLE)

FIELD 05C, 05B

ACCESSION NO. W73-02095

STUDIES ON ALGAL GROWTH, DEVELOPMENT, AND REPRODUCTION,

CALIFORNIA UNIV., IRVINE.

S. N. MURRAY.

AVAILABLE FROM UNIV. MICROFILMS, INC., ANN ARBOR, MICH. 48106. ORDER NO 72-14,661. PH D DISSERTATION, 1971. 119 P.

DESCRIPTORS:
*ALGAE, *GROWTH RATES, *SYSTEMATICS, *LIFE HISTORY STUDIES, *GROWTH STAGES, REPRODUCTION, CARBON DIOXIDE, CULTURES, LIMITING FACTORS, RHODOPHYTA, AQUATIC ALGAE, MICROORGANISMS, CHLOROPHYTA, BIOLOGY, POLLUTANT IDENTIFICATION, LIGHT INTENSITY, BIOASSAY.

IDENTIFIERS:
PORPHYROPSIS COCCINEA VAR. DAWSONII, CALLOPHYLLIS FIRMA, PLENOSPORIUM DASYOIDES, SELENASTRUM CAPRICORNUTUM, PAAP BATCH TEST, CULTURE MEDIA.

ABSTRACT:
THE LIFE HISTORY OF THE BANGIOPHYCEAN RED ALGAE PROPHYROPSIS COCCINEA VAR. DOWSONII HAS BEEN COMPLETED IN LABORATORY CULTURE. AN UNDESCRIBED FILAMENTOUS PHASE, DISTINCT IN CERTAIN RESPECTS FROM THE CONCHOCELIS OF BANGIA AND PORPHYRA, IS REPORTED. CALLOPHYLLIS FIRMA HAS BEEN SHOWN TO HAVE A POLYSIPHONIA-TYPE OF LIFE HISTORY IN LABORATORY CULTURE. THIS IS THE FIRST LABORATORY CONFIRMATION OF A TYPICAL POLYSIPHONIA-TYPE OF LIFE HISTORY IN THE CRYPTONEMIALES. THE RATES OF APICAL CELL DIVISION FOR THE MARINE RED ALGA PLEONOSPORIUM DASYOIDES WERE DETERMINED AT VARIOUS LIGHT INTENSITIES. RATES OF APICAL CELL DIVISION WERE FOUND TO BE CORRELATED WITH THE TOTAL AMOUNT OF ILLUMINATION RECEIVED PER 24 HOUR PERIOD. AN ANALYSIS AND DISCUSSION OF THE MECHANISMS OF APICAL CELL DIVISION IN FILAMENTS OF UNLIMITED AND LIMITED GROWTH OF P. DASYOIDES IS PRESENTED. THE PAAP BATCH TEST HAS BEEN EVALUATED IN TERMS IF ITS EFFEDTIVENESS. IT HAS BEEN DEMONSTRATED THAT THE PRESCRIBED METHOD OF MEDIUM PREPARATION RESULTS IN THE REMOVAL OF FE AND MN FROM THE MEDIUM DURING FILTER STERILIZATION AND CONSEQUENTLY REDUCES ALGAL YIELDS. CO2 LIMITS THE GROWTH OF THE GREEN ALGA SELENASTRUM CAPRICORNUTUM, THE PRESCRIBED TEST ORGANISM, AND THE CONCOMITANT INCREASE IN MEDIUM PH DURING THE PERIOD OF EXPERIMENTATION APPEARS TO HAVE AN EFFECT ON NUTRIENT AVAILABILITY IN THE TEST CULTURES. THE EFFECTS OF LIGHT INTENSITY AND THE METHOD OF AIR ADDITION ARE ALSO DISCUSSED. (LONG-BATTELLE)

FIELD 05C, 05A

ACCESSION NO. W73-02099

RELEASE OF DISSOLVED ORGANIC MATTER BY MARINE MACROPHYTES,

GEORGIA UNIV., ATHENS.

M. BRYLINSKY.

AVAILABLE FROM UNIV. MICROFILMS, INC., ANN ARBOR, MICH. 48106. ORDER NO. 72-10,923. PH D DISSERTATION, 1971. 125 P.

DESCRIPTORS:
*SEA WATER, *MARINE ALGAE, *CARBOHYDRATES, MARINE PLANTS, MARINE MICROORGANISMS, WATER ANALYSIS, BENTHIC FLORA, ORGANIC MATTER, DISSOLVED SOLIDS, BENTHOS, CARBON, PRODUCTIVITY, ORGANIC COMPOUNDS, CARBOHYDRATES, CYCLING NUTRIENTS, ALGAE.

IDENTIFIERS:
DISSOLVED ORGANIC CARBON, MACROPHYTES, SARGASSO SEA, SAMPLE PREPARATION, SPERMATOPHYTES, PHOTOASSIMILATION, GLYOXYLIC ACID, ASSIMILATION.

ABSTRACT:
THE AMOUNT OF PHOTOASSIMILATED CARBON RELAEASED AS DISSOLVED ORGANIC CARBON WAS INVESTIGATED FOR SIX SPECIES OF BENTHIC MARINE MACROPHYTES AND ONE SPECIES OF PELAGIC MARINE MACROPHYTE. RELEASE RATES RANGED BETWEEN 0.223 AND 1.805 MG C/GM HR. PERCENT RELEASE VALUES RANGED FROM 1.09 TO 3.82 PERCENT. SPERMATOPHYTES HAD SLIGHTLY HIGHER PERCENT RELEASE VALUES THAN ALGAE. THE RESULTS OF QUALITATIVE ANALYSIS PERFORMED ON THE SOLUBLE CARBOHYDRATES RELEASED SHOWED NEUTRAL CARBOHYDRATES TO BE LIBERATED IN THE LARGEST QUANTITY FOLLOWED BY LESSER AMOUNTS OF ACIDIC AND BASIC CARBOHYDRATES. GLYCOLIC ACID WAS NOT OBSERVED TO BE RELEASED IN SIGNIFICANT QUANTITIES. INVESTIGATIONS PERFORMED ON THE ABILITY OF RELEASE PRODUCTS TO BE UTILIZED BY HETEROTROPHIC ORGANISMS SHOWED THAT 20-30 PERCENT OF THE RELEASED CARBON WAS ASSIMILATED WITHIN ONE HOUR. PRELIMINARY CALCULATIONS ON THE CONTRIBUTION OF ORGANIC RELEASE PRODUCTS TO THE DISSOLVED ORGANIC CARBON POOL OF SEA WATER SHOW THIS TO BE RELATIVELY LOW IN TERMS OF ABSOLUTE AMOUNTS OF ORGANIC MATTER. A TECHNIQUE IS PRESENTED FOR THE CONCENTRATION AND MEASUREMENT OF DISSOLVED CARBOHYDRATE MATERIALS IN SEAWATER. ANALYSES OF SEAWATER SAMPLES COLLECTED IN VARIOUS INSHORE PLANT COMMUNITIES AND THE SARGASSO SEA INDICATE THAT THE BASIC PROCEDURE IS USEFUL IN OBTAINING DETAILED INFORMATION ON THE DISSOLVED CARBOHYDRATE MATERIALS PRESENT IN SEAWATER. (LONG-BATTELLE)

FIELD 05C, 05B

ACCESSION NO. W73-02100

DDT RESIDUES IN COASTAL MARINE PHYTOPLANKTON AND THEIR TRANSFER IN PELAGIC FOOD CHAINS,

STANFORD UNIV., CALIF.

J. L. COX,

AVAILABLE FROM UNIV. MICROFILMS, INC. ANN ARBOR, MICH. 48106. ORDER NO 72-5903. PH D DISSERTATION, 1971. 149 P.

DESCRIPTORS:
*DDT, *MARINE FISH, *PHYTOPLANKTON, *SHRIMP, *BIOASSAY, *ABSORPTION, PESTICIDE RESIDUES, WATER POLLUTION EFFECTS, *CALIFORNIA, SEA WATER, WATER ANALYSIS, MARINE ALGAE, CHEMICAL ANALYSIS, ADSORPTION, SEPARATION TECHNIQUES, FOOD CHAINS, CULTURES, FISH DIET, GAS CHROMATOGRAPHY, PARTICULATE MATTER, DETRITUS, ORGANIC COMPOUNDS, PATH OF POLLUTANTS.

IDENTIFIERS:
*ELECTRON CAPTURE GAS CHROMATOGRAPHY *BIOLOGICAL MAGNIFICATION, GAS LIQUID CHROMATOGRAPHY, ELECTRON CAPTURE DETECTOR, PRECONCENTRATION, BIOLOGICAL SAMPLES, LIQUID-LIQUID EXTRACTION, TRIPHOTURUS MEXICANUS, EUPHAUSIA PACIFICA, ENGRAULIS MORDAX.

ABSTRACT:
STUDIES WERE CONDUCTED ON THE ENTRY AND TRANSFER OF DDT RESIDUES IN PELAGIC MARINE FOOD CHAINS. ANALYSES WERE DONE ON PHYTOPLANKTON AND DETRITAL MATERIAL COLLECTED BY A NET OR BY CONTINUOUS-FLOW CENTRIFUGATION. SEAWATER SAMPLES WERE EXTRACTED BY CONTINUOUS FLOW, LIQUID-LIQUID EXTRACTION. GAS-LIQUID CHROMATOGRAPHY WITH ELECTRON CAPTURE DETECTION (GLC-EC) WAS EMPLOYED FOR THE ANALYSES. GLC-EC ANALYSES WERE ALSO PERFORMED ON SAMPLES OF SURFACE AND MIDWATER FISHES AND ZOOPLANKTON. EXPERIMENTAL WORK WITH C-14-DDT WAS DONE WITH PURE CULTURES OF PHYTOPLANKTON AND WITH A COMMON EUPHAUSIID SHRIMP. PHYTOPLANKTON SAMPLES COLLECTED IN MONTEREY BAY, CALIFORNIA, FROM 1955 TO 1969 CONTAINED COMPOUNDS IDENTIFIED AS P,P'-DDT, P,P'-DDD, AND P,P'-DDE. TOTAL CONCENTRATIONS OF THESE COMPOUNDS WERE APPROXIMATELY THREE TIMES GREATER IN THE LATER SAMPLES. UPTAKE STUDIES WITH PURE CULTURES OF MARINE PHYTOPLANKTON SHOWED THAT THE ALGAL CELLS, WHEN EXPOSED TO LOW PARTS PER TRILLION NOMINAL CONCENTRATIONS OF C-14-DDT IN THE MEDIUM, COULD CONCENTRATE THE LABELLED DDT BY FACTORS RANGING FROM 30,000 - 80,000. DDT RESIDUE CONCENTRATIONS IN WHOLE SEAWATER RANGED FROM 2.3 PG/ML OFF OREGON AND WASHINGTON, TO 5.6 PG/ML OFF SOUTHERN CALIFORNIA. DDT RESIDUE CONCENTRATIONS IN PARTICULATE MATERIAL RANGED FROM 1.2 TO 5.7 MICROGRAMS/G C (WITH ONE EXCEPTION). EXPERIMENTAL RESULTS ARE DESCRIBED WHICH IMPLICATE ADSORPTION AS THE UPTAKE MECHANISM FOR ALGAL CELLS; THESE EXPERIMENTS ALSO SUPPORT THE IDEA THAT SMALL PARTICLES CARRY MOST OF THE DDT RESIDUES IN WHOLE SEAWATER. EUPHAUSIA PACIFICA HENSEN CAN ACQUIRE SUFFICIENT DDT RESIDUES FROM ITS FOOD TO ACOUNT FOR AMOUNTS FOUND IN ITS TISSUES. DIRECT UPTAKE OF C-14-DDT FROM WATER IS PARTIALLY REVERSIBLE BY RETURNING ANIMALS TO UNLABELLED FLOWING SEAWATER. GLC-EC ANALYSES OF TRIPHOTURUS MEXICANUS, SHOWED THAT OLDER FISH HAD HIGHER DDT RESIDUE CONCENTRATIONS, SUGGESTING THAT FISH ACCUMULATE DDT RESIDUES FROM THE ENVIRONMENT DURING THEIR LIFE SPAN. THE DDT RESIDUE CONTENT OF DIFFERENT SIZE CLASSES OF ENGRAULIS MORDAX GIRARD RANGES FROM 0.2 TO 2.8 PPM, WET WEIGHT. THESE FINDINGS ARE DISCUSSED IN THE CONTEXT OF A SIMPLE MODEL OF DDT RESIDUE ASSIMILATION FOOD AND DDT RESIDUE LOSS VIA TRANSPORT IN THE REPRODUCTIVE MATERIALS. (SNYDER-BATTELLE)

FIELD 05C, 05B

ACCESSION NO. W73-02105

HEAVY METAL ION INTERACTION AND TRANSPORT WITH SYNTHETIC COMPLEXING AGENTS AND DETERGENT PHOSPHATE SUBSTITUTES IN AQUATIC SYSTEMS,

MISSOURI WATER RESOURCES RESEARCH CENTER, ROLLA.

S. E. MANAHAN, AND M. J. SMITH.

AVAILABLE FROM THE NATIONAL TECHNICAL INFORMATION SERVICE AS PB-213 252, $3.00 IN PAPER COPY, $0.95 IN MICROFICHE. MISSOURI WATER RESOURCES RESEARCH CENTER, COLUMBIA, COMPLETION REPORT, AUGUST 30, 1972. 181 P, 25 FIG, 14 TAB, 55 REF. OWRR A-049-MO.(1) 14-01-0001-3525.

DESCRIPTORS:
*COPPER, CULTURES, *CHELATION, *ION TRANSPORT, *HEAVY METALS, EUTROPHICATION, POLLUTANT IDENTIFICATION, TOXICITY, *IRON, *CHLORELLA, NUTRIENT REQUIREMENT, *ALGAE, DETERGENTS, PHOSPHATES.

IDENTIFIERS:
*ION-SELECTIVE ELECTRODES, EDTA, *POTENTIOMETRY, ELECTROANALYSIS, *ALGAL CULTURES, CHELATING AGENTS, OOCYSTIS MARSSCNII.

ABSTRACT:
THE CHEMICAL ASPECTS OF THE COPPER MICRONUTRIENT REQUIREMENT FOR ALGAE HAVE BEEN INVESTIGATED. A REPRODUCIBLE COPPER REQUIREMENT FOR CHLORELLA VULGARIS AND OOCYSTIS MARSSONII WAS DEMONSTRATED. OPTIMAL GROWTH WAS OBSERVED ABOVE 40 MICROGRAMS/L FOR OOCYSTIS AND 30 MICROGRAMS/L FOR CHLORELLA. A STUDY OF THE EFFECTS OF EDTA ON THE TOXICITYOOF COPPER TO CHLORELLA SHOWED THAT COPPER IN CHELATED FORM WAS NOT TOXIC TO THESE ALGAE AT CONCENTRATIONS UP TO 46 MG/L COPPER. WHEN ONLY SUFFICIENT CHELATING AGENT WAS PRESENT TO KEEP THE IRON (III) IN SOLUTION, HOWEVER, THE TOXIC EFFECTS OF COPPER WERE EVIDENT AT 7.00 MG/L OF COPPER. A SECOND ASPECT OF THE PROJECT INVOLVED THE DEVELOPMENT OF A SIMPLE, DIRECT MULTIPLE STANDARD ADDITION METHOD FOR THE POTENTIOMETRIC ANALYSIS OF COPPER IN WATER WITH A SOLID-STATE COPPER ION-SELECTIVE ELECTRODE. THE TECHNIQUE IS MORE SENSITIVE THAN CONVENTIONAL ATOMIC ABSORPTION ANALYSIS, THOUGH NOT SO RAPID. MEASUREMENTS ARE MADE IN A COMPLEXING ANTIOXIDANT BUFFER MEDIUM CONTAINING ACETATE (TO COMPLEX COPPER), FLUORIDE (TO COMPLEX IRON), AND FORMALDEHYDE (TO PROVIDE A REDUCING MEDIUM).

FIELD 05A, 05C, 02K

ACCESSION NO. W73-02112

ARTIFICIAL DESTRATIFICATION IN RESERVOIRS.

AMERICAN WATER WORKS ASSOCIATION, NEW YORK. QUALITY CONTROL IN RESERVOIRS COMMITTEE.

JOURNAL OF THE AMERICAN WATER WORKS ASSOCIATION, VOL 63, NO 9, P 597-604, SEPTEMBER 1971. 4 FIG, 3 TAB, 19 REF.

DESCRIPTORS:
*THERMAL STRATIFICATION, *TASTE, *ODOR, DISSOLVED SOLIDS, ALGAE, RESERVOIR STORAGE, *HYPOLIMNION, WATER QUALITY CONTROL.

IDENTIFIERS:
*MIXERS.

ABSTRACT:
THE QUALITY CONTROL IN RESERVOIRS COMMITTEE OF THE AMERICAN WATER WORKS ASSOCIATION CONDUCTED A STUDY OF WATER PURVEYORS USING ARTIFICIAL DESTRATIFICATION TECHNIQUES TO OVERCOME TASTES AND ODORS; HIGH CONCENTRATIONS OF IRON, MANGANESE, AND HYDROGEN SULFIDE; AND ALGAE DIFFICULTIES ASSOCIATED WITH ANAEROBIC CONDITIONS IN THE HYPOLIMNION LAYER IN RESERVOIRS. A HIGH RATE OF SUCCESS FROM BOTH HOMEMADE AND COMMERCIAL MIXING APPARATUS WAS RECORDED WITH AN ESTIMATED INITIAL COST PER UNIT VOLUME OF $3/MIL GAL AND AN OPERATING COST OF $0.45/MIL GAL/YR FOR A 10,000 ACRE-FT. RESERVOIR. MORE STUDY IS NEEDED TO EVALUATE THE VARIOUS TYPES OF APPARATUS, THE ACTUAL CHANGES IN WATER QUALITY, AND THE INFLUENCE OF ARTIFICIAL DESTRATIFICATION ON INDICATOR ORGANISMS AND ENTERIC PATHOGENS. (WEIR-AWWARF)

FIELD 05G, 02H, 05F

ACCESSION NO. W73-02138

COMMUNITY WATER POLLUTION R AND D NEEDS.

AMERICAN WATER WORKS ASSOCIATION, NEW YORK. COMMITTEE ON POLLUTION PARAMETERS.

JOURNAL OF THE AMERICAN WATER WORKS ASSOCIATION, VOL 64, NO 4, P 211-215, APRIL 1972. 9 REF.

DESCRIPTORS:
*WATER PURIFICATION, *WATER POLLUTION TREATMENT, *PUBLIC HEALTH, ANALYTICAL TECHNIQUES, WASTE WATER TREATMENT, TASTE, ODOR, POTABLE WATER, ACTIVATED CARBON, COLIFORMS, VIRUSES, CRUSTACEANS, NEMATODES, EPIDEMIOLOGY, PHOSPHATES, HEAVY METALS, PESTICIDES, TASTE-PRODUCING ALGAE, ODOR-PRODUCING ALGAE, ACTINOMYCETES, PHOSPHATES.

IDENTIFIERS:
*WATERBORNE DISEASES, *TOXIC CHEMICALS, *PUBLIC HEALTH STANDARDS, TOXICOLOGY, NTA,

ABSTRACT:
THE POLLUTION PARAMETERS COMMITTEE OF THE AWWA HAS PREPARED A REPORT ON THE CURRENT STATE OF THE ART IN THE FIELD OF ANTIPOLLUTION MEASURES AND HAS SET FORTH ITS RESEARCH RECOMMENDATIONS. AMONG THE WATER QUALITY PARAMETERS DISCUSSED ARE COLIFORM BACTERIA, VIRUSES, TASTES AND ODORS, PHOSPHATES, NTA, AND OTHER ORGANIC AND INORGANIC CHEMICALS. THE RESEARCH NEEDS RECOMMENDED BY THE COMMITTEE INCLUDE (1) IDENTIFICATION OF ALL POLLUTANTS PRESENTING A POTENTIAL HEALTH HAZARD AS WELL AS THOSE RESPONSIBLE FOR OBNOXIOUS TASTES AND ODORS; (2) INCREASING THE SENSITIVITY, PRECISION, AND RAPIDITY OF BOTH MICROBIOLOGICAL AND CHEMICAL TECHNIQUES FOR EXAMINING WASTEWATER; (3) TOXICOLOGICAL RESEARCH AND EPIDEMIOLOGICAL STUDIES THAT CORRELATE MORBIDITY DATA WITH SPECIFIC CHARACTERISTICS OF THE WATER; (4) LABORATORY RESEARCH AND PILOT-PLANT STUDIES TO DEVELOP AND EVALUATE PROCESSES THAT WILL CONTINUOUSLY TREAT DIRECTLY REUSED MUNICIPAL WASTEWATERS TO PRODUCE 'SAFE AND SATISFACTORY' DRINKING WATER. (NICHOLS-AWWARF)

FIELD 05G, 05D, 05F

ACCESSION NO. W73-02144

WHAT'S IT ALL ABOUT. ALGAE,

BUCK, SEIFEIT AND JOST, MORRISVILLE, PA.

J. M. FOULDS.

WATER AND WASTES ENGINEERING, VOL 9, NO 8, P 45-46, AUGUST 1972, 3 TAB.

DESCRIPTORS:
*ALGAE, *NITRIFICATION, WASTE TREATMENT, *PHOSPHORUS, *NITROGEN, COST ANALYSIS, OZONE, DENITRIFICATION, POLLUTION ABATEMENT, WATER POLLUTION CONTROL, COAGULATION.

IDENTIFIERS:
*ORGANIC NUTRIENTS, ALUM COAGULATION.

ABSTRACT:
BY COMMERCIAL GROWTH OF ALGAE FROM NITRIFICATION-DENITRIFICATION PLANTS OR FROM RAW SEWAGE, FERTILIZER MIGHT BE PRODUCED WITH ORGANIC NITROGEN AND PHOSPHATE IN SLOW RELEASE, HARD TO LEACH FORMS. HIGH RATE ALGAE FORMS SHOW 70% NITROGEN REMOVAL AND 50% PHOSPHORUS REMOVAL, INCREASING TO NEAR 100% PHOSPHORUS REMOVAL WITH ALUM COAGULATION HARVESTING. THE ESTIMATED COST IS $20,000 PER ACRE, INCLUDING LAND. LABORATORY BATCH AND CONTINUOUS REACTOR STUDIES IN AN OZONATED SYSTEM WERE MADE TO PREPARE SUITABLE EFFLUENTS FOR A SOURCE OF ORGANIC NUTRIENTS. OZONE OXIDIZES METALS TO HIGHER STATES AND STERILIZES THE EFFLUENT. THE OPTIMUM OZONE-OXYGEN RATES WERE DETERMINED AND THE FROTH COMPOSITION AND VOLUME AT DIFFERENT GAS RATES WERE DETERMINED. IT APPEARS THAT THE PROCESS IS FEASIBLE WITH SINGLE OR COMBINATIONS OF TRANSITION METALS-NECESSARY TRACE METALS FOR LAWN FERTILIZER. (ANDERSON-TEXAS)

FIELD 05G, 05C

ACCESSION NO. W73-02187

METABOLISM OF DDT BY FRESH WATER DIATOMS,

MANITOBA UNIV., WINNIPEG. DEPT. OF ENTOMOLOGY.

S. MIYAZAKI, AND A. J. THORSTEINSON.

BULLETIN OF ENVIRONMENTAL CONTAMINATION AND TOXICOLOGY, VOL 8, NO 2, P 81-83, 1972. 1 TAB, 13 REF.

DESCRIPTORS:
*DDT, *DDE, *PESTICIDE KINETICS, *PERSISTENCE, *WATER POLLUTION EFFECTS, *PLANT PHYSIOLOGY, *DIATOMS, PESTICIDES, DDD, PESTICIDE TOXICITY, PATH OF POLLUTANTS, PESTICIDE RESIDUES, CHLORINATED HYDROCARBON PESTICIDES, PESTICIDE DRIFT, METABOLISM, ALGAE, AQUATIC ALGAE, CHRYSOPHYTA, PLANKTON, PHYTOPLANKTON.

ABSTRACT:
REDUCTIVE DECHLORINATION OF DDT TO TDE (DDD) UNDER ANAEROBIC CONDITIONS HAS BEEN DEMONSTRATED IN BAKER'S YEAST, A FUNGUS, SEVERAL ACTINOMYCETES, AND BACTERIA. DDE, THE DEHYDROCHLORINATION PRODUCT OF DDT, ALSO OCCURS IN SEVERAL BACTERIA AND IN A MARINE DIATOM. A POSSIBLE ROLE OF DIATOMS IN DETOXIFYING DDT IN AQUATIC ENVIRONMENTS HAS BEEN SUGGESTED. RADIOLABELLING AND THIN-LAYER CHROMATOGRAPHY WERE USED IN THIS STUDY OF DDT METABOLISM BY A NITZSCHIA SP. AND AN UNIDENTIFIED DIATOM SPECIES. RESULTS SUGGEST THAT SOME SPECIES OF FRESHWATER DIATOMS MAY BE SIGNIFICANT IN THE DEGRADATION OF DDT TO THE NON-INSECTICIDAL METABOLITE, DDE, IN NATURE. (LEGORE-WASHINGTON)

FIELD 05C

ACCESSION NO. W73-02280

THE IMPACT OF REDUCED LIGHT PENETRATION ON A EUTROPHIC FARM POND,

NEBRASKA UNIV., LINCOLN. DEPT. OF ZOOLOGY.

E. G. BUGLEWICZ.

M.S. THESIS, AUGUST 1972. 99 P, 20 FIG, 11 TAB, 24 REF. OWRR A-014-NEB(7) 14-31-0001-3527.

DESCRIPTORS:
*EUTROPHICATION, AQUATIC ECOSYSTEMS, WATER QUALITY, *ALGAE, *LIGHT PENETRATION, *MACROPHYTES, *AQUATIC WEED CONTROL, *LIGHT INTENSITY, *NEBRASKA, *CYANOPHYTA, ALGAL CONTROL, *PONDWEEDS, PONDS, *THERMAL STRATIFICATION, DIATOMS, *CHARA.

IDENTIFIERS:
*ANALINE DYE.

ABSTRACT:
AN EXPERIMENTAL APPROACH TO THE CONTROL OF EUTROPHICATION PROBLEMS USING ANILINE DYES WAS ATTEMPTED ON AN ENRICHED FARM POND IN EASTERN NEBRASKA. PRIMARY PRODUCTIVITY WAS SHOWN TO DECREASE NOT ONLY IN THE DYED BOXES, BUT ALSO IN POND SAMPLES SUSPENDED IN THE BOXES. POTAMOGETON SP. WAS ELIMINATED FROM BLUE AND BROWN-DYED BOXES, WHILE CHARA SP. WAS ONLY ELIMINATED FROM THE BLUE-DYED BOXES. A STRONG TEMPERATURE STRATIFICATION OCCURRED IN ALL DYED BOXES DUE TO ABSORPTION OF INFRARED ENERGY IN THE FIRST FEW CM OF THE WATER COLUMNS. ALGAL POPULATIONS AND PULSES WITHIN THE BOXES DIFFERED FROM BOX TO BOX AND DID NOT REFLECT THE SAME CHANGES OBSERVED IN THE POND, BUT DID RESEMBLE FALL AND SPRING DIATOM PULSES. BLUE-GREEN BLOOMS PRESENT IN EXPERIMENTAL BOXES 2, 4, 5 AND 6 BEFORE DYE ADDITION WERE REPLACED BY POPULATIONS OF DIATOMS WITH SOME GREEN AND A FEW BLUE-GREEN ALGAE. BOX 3, WHICH WAS TURBID BEFORE DYE ADDITION, NEVER EXPERIENCED A BLUE-GREEN ALGAL PULSE, BUT MAINTAINED A HIGH POPULATION OF DIATOMS. LACK OF MIXING OF THE WATER COLUMNS MAY HAVE HAD A PROFOUND EFFECT ON THE RESULTS OF THE EXPERIMENT BY CAUSING TEMPERATURE STRATIFICATION, EXCESSIVE PRIMARY PRODUCTIVITY, PROLONGED ANAEROBIC CONDITIONS BELOW THE SURFACE OF THE WATER, AND INCREASED NUTRIENT RELEASE FROM THE BOTTOM MUDS.

FIELD 05C, 02H

ACCESSION NO. W73-02349

MULTIVARIATE APPROACHES TO ALGAL STRATEGIES AND TACTICS IN THE SYSTEMS ANALYSIS OF PHYTOPLANKTON,

WISCONSIN UNIV., MADISON. DEPT. OF BOTANY.

T. F. H. ALLEN, AND J. F. KOONCE.

MEMO REPORT NO 72-24 (PREPRINT), EASTERN DECIDUOUS FOREST BIOME, MAY, 1972. 52P, 12 FIG, 2 TAB, 32 REF. 16010 EHR.

DESCRIPTORS:
*ANALYTICAL TECHNIQUES, *COMPUTER PROGRAMS, *ALGAE, *PHYTOPLANKTON, BIOLOGICAL COMMUNITIES, STANDING CROPS, WISCONSIN, PRIMARY PRODUCTIVITY, SYSTEMATICS, GROWTH RATES, BIOMASS, DATA COLLECTIONS.

IDENTIFIERS:
*LAKE WINGRA(WIS.), MULTIVARIATE ANALYSIS, ORDINATION ANALYSIS.

ABSTRACT:
TRADITIONALLY, PHYTOPLANKTON DATA ARE ORGANIZED ON CHRONOLOGICAL AXES; THE POWER OF MULTIVARIATE ANALYSES OF THE TYPE EMPLOYED HERE LIES IN THEIR ABILITY TO ORGANIZE SPECIES DATA INTO NATURAL GROUPS NOT NECESSARILY CONFORMING TO CHRONOLOGICAL GROUPS OR, INDEED, ANY OTHER IMPOSED STRUCTURE. A LOGARITHMIC TRANSFORMATION IS OFTEN APPROPRIATE TO PHYTOPLANKTON DATA BECAUSE THESE PLANTS CAN GROW EXPONENTIALLY. A PRESENCE/ABSENCE TRANSFORMATION IGNORES STANDING CROPS AND GIVES INFORMATION AS TO SPECIES TOLERANCES RATHER THAN SPECIES OPTIMA. CERTAIN TRANSFORMATIONS, IN WHICH DATA ARE RELATIVIZED, GIVE EQUAL WEIGHT TO RARE AND COMMON SPECIES WHILE PRESERVING MANY QUANTITATIVE ASPECTS. BY USING SEVERAL TRANSFORMATIONS AND ANALYSES INSIGHT HAS BEEN GAINED INTO DIFFERENT ECOLOGICAL ASPECTS OF PHYTOPLANKTON, AND THE BIOLOGICAL IMPLICATIONS OF CERTAIN DATA TRANSFORMATIONS. AT WEEKLY INTERVALS, MARCH 1970 UNTIL FEBRUARY 1971, WATER SAMPLES WERE COLLECTED FROM THREE DEPTHS AT ONE STATION IN LAKE WINGRA, WISCONSIN FOR ESTIMATION OF PRIMARY PRODUCTION, IDENTIFICATION AND ENUMERATION OF PHYTOPLANKTON SPECIES, AND DETERMINATION OF IMPORTANT WATER CHEMISTRY PARAMETERS. A DISTINCTION IS DRAWN BETWEEN ALGAL TACTICS, WHICH GIVE THE PLANT THE ABILITY TO EXIST IN A PARTICULAR ENVIRONMENTAL SITUATION, AND ALGAL STRATAGEMS, WHICH DEFINE THE ORGANISMS' PLACE IN THE COMMUNITY. (JONES-WISCONSIN)

FIELD 05C, 07B

ACCESSION NO. W73-02469

ALGAL NITROGEN FIXATION IN TEMPERATE REGIONS,

UPPSALA UNIV. (SWEDEN). INST. OF PHYSIOLOGICAL BOTANY.

E. HENRIKSSON.

PLANT AND SOIL, SPECIAL VOLUME, P 415-419, 1971. 1 TAB, 16 REF.

DESCRIPTORS:
*ALGAE, *NITROGEN FIXATION, *SOILS, *TEMPERATE, CLIMATIC ZONES, CYANOPHYTA, SYMBIOSIS, LICHENS, NITROGEN, LIGHT INTENSITY, TEMPERATURE, ANABAENA, BACTERIA, PHOTOSYNTHESIS.

IDENTIFIERS:
SWEDEN, COLLEMA, PELTIGERA, NOSTOC.

ABSTRACT:
A FEW EARLIER INVESTIGATIONS OF NITROGEN-FIXATION BY BLUE-GREEN ALGAE IN TEMPERATE SOILS ARE REVIEWED INCLUDING RECENT STUDIES ON THE OCCURRENCE OF POTENTIAL NITROGEN-FIXING ALGAE IN SEVERAL DIFFERENT SOIL TYPES ON VARIOUS CONTINENTS. IN THE SUMMER OF 1969 AN INVESTIGATION WAS MADE OF NITROGEN FIXATION BY ALGAE IN SWEDISH SOILS. MEASUREMENTS WERE MADE FROM APRIL TO OCTOBER AT VARIOUS TIMES OF THE DAY. GOOD NITROGEN FIXATION BY BLUE-GREEN ALGAE WAS OBSERVED DURING THE WHOLE EXPERIMENTAL TIME. FIXATION OCCURRED AT ABOUT THE SAME RATE DURING BOTH THE LIGHT AND DARK PERIODS OF EACH DAY ON WHICH IT WAS MEASURED. NEITHER LIGHT NOR TEMPERATURE CONDITIONS GREATLY INFLUENCED RATE OF NITROGEN FIXATION. THE LOW RATES OF FIXATION FREQUENTLY OBSERVED IN SOILS CONTAINING FREE-LIVING ALGAE MAY BE EXPLAINED BY THE FACT THAT THE CELLS OF THE SPECIES ANABAENA AND NOSTOC IN SOILS CAN VERY EASILY BE CONVERTED TO SPORES (AKINETES) WHICH DO NOT FIX NITROGEN. DATA SHOW THAT BLUE-GREEN ALGAE CAN CONTRIBUTE TO THE TOTAL NITROGEN ECONOMY OF SOILS IN THE NORTHERN TEMPERATE ZONE BOTH AS FREE-LIVING CELLS AND IN SYMBIOTIC ASSOCIATIONS IN LICHENS. (JONES-WISCONSIN)

FIELD 05C, 02G

ACCESSION NO. W73-02471

NITROGEN FIXATION IN LAKES,

LONDON UNIV. (ENGLAND); AND WESTFIELD COLL., LONDON (ENGLAND). DEPT. OF BOTANY.

G. E. FOGG.

PLANT AND SOIL, SPECIAL VOLUME, P 393-401, 1971. 2 TAB, 24 REF.

DESCRIPTORS:
*NITROGEN FIXATION, *LAKES, CYANOPHYTA, LIGHT INTENSITY, EUTROPHICATION, PHYTOPLANKTON, PRODUCTIVITY, ANABAENA, AMMONIA, NITRATES, LIMITING FACTORS, NITROGEN, PLANKTON, DEPTH, ALGAE.

IDENTIFIERS:
*WINDERMERE(ENGLAND), HETEROCYSTS, BENTHIC BLUE-GREEN ALGAE.

ABSTRACT:
FOR VALID ESTIMATE OF TOTAL NITROGEN FIXED IN A LAKE IN A YEAR, DETERMINATIONS SHOULD BE MADE WITH ENOUGH REPRESENTATIVE SAMPLES TO TAKE ACCOUNT OF HORIZONTAL AND VERTICAL VARIATIONS IN DENSITY OF BLUE-GREEN PLANKTONIC ALGAE AT FREQUENT INTERVALS WHEN THE ALGAE ARE ABUNDANT. DETERMINATIONS IN OPEN WATERS USING N-15 TRACERS SHOW THAT NITROGEN FIXATION IS GENERALLY ASSOCIATED WITH PRESENCE OF HETEROCYSTOUS BLUE-GREEN ALGAE AND IS LIGHT DEPENDENT. FIXATION ITSELF IS NOT NECESSARILY INHIBITED BY PRESENCE OF NITRATE OR AMMONIA IN THE WATER, ALTHOUGH NITROGEN-FIXING BLUE-GREEN ALGAE TEND TO BE ABUNDANT WHEN THESE SOURCES OF COMBINED NITROGEN ARE LOW. ACTIVITY OF THESE ALGAE SHOWS A DIRECT RELATIONSHIP TO DISSOLVED ORGANIC NITROGEN CONCENTRATION. NITROGEN FIXATION PER UNIT AREA OF LAKE SURFACE PER YEAR TENDS TO BE GREATEST AT AN EARLY EUTROPHICATION STAGE. CONTRIBUTION OF BIOLOGICAL NITROGEN FIXATION IN RELATION TO THE TOTAL NITROGEN BUDGET OF A LAKE IS PROBABLY ALWAYS SMALL, BUT AT CERTAIN TIMES AND IN PARTICULAR WATER STRATA IT MAY CONTRIBUTE A MAJOR PART OF THE NITROGEN ASSIMILATED BY PHYTOPLANKTON. NITROGEN FIXATION MAY FREQUENTLY BE IMPORTANT IN EUTROPHIC LAKES IN ENABLING HIGHER PRODUCTION RATES THAN WOULD OTHERWISE BE POSSIBLE. (JONES-WISCONSIN)

FIELD 05C, 02H

ACCESSION NO. W73-02472

ALGAL NITROGEN FIXATION IN THE TROPICS,

SEIJO UNIV., TOKYO (JAPAN). BIOLOGICAL LAB.

A. WATANABE, AND Y. YAMAMOTA.

PLANT AND SOIL, SPECIAL VOLUME, P 403-413, 1971. 2 FIG, 4 TAB, 25 REF.

DESCRIPTORS:
*ALGAE, *NITROGEN FIXATION, *TROPICAL REGIONS, RICE, CYANOPHYTA, FERTILIZATION, TEMPERATURE, CROP PRODUCTION, FUNGI, SOILS, SYMBIOSIS, AFRICA, ASIA, ANABAENA.

IDENTIFIERS:
*GREEN MANURE, INDIA, JAPAN, RICE FIELDS.

ABSTRACT:
SOME RECENT DATA OBTAINED ON NITROGEN FIXATION IN RICE FIELDS, IN TROPICAL WATERS, AND IN SYMBIOTIC BLUE-GREEN ALGAE ARE SUMMARIZED. NITROGEN-FIXING BLUE-GREEN ALGAE SEEM TO GROW MOST ABUNDANTLY IN TROPICAL AND SUBTROPICAL REGIONS AND TO A LESSER EXTENT IN TEMPERATE AND SUB-TEMPERATE REGIONS. OF BLUE-GREEN ALGAE TESTED FOR NITROGEN-FIXING CAPACITY, TOLYPOTHRIX TENUIS WAS THE MOST EFFICIENT. THE USE OF TOLYPOTHRIX TENUIS AS A SOURCE OF GREEN MANURE IN AGRICULTURAL PRACTICE WAS TESTED DURING 1951-1956 ON ABOUT 40 RICE FIELDS IN VARIOUS PARTS OF JAPAN. OVER-ALL PRODUCTION SHOWS AN AVERAGE INCREASE IN RICE YIELDS EVERY YEAR. CERTAIN SPECIES OF BLUE-GREEN ALGAE CAN WITHSTAND EXTREME ENVIRONMENTAL CONDITIONS SUCH AS HIGH TEMPERATURE AND SALINITY. THOSE FIXING NITROGEN CONTRIBUTE TO THE NITROGEN FERTILITY OF THE SEAS AND LAKES. IN JAPANESE HOT SPRINGS, 320 SPECIES OF BLUE-GREEN WERE FOUND 10 OF WHICH FIXED NITROGEN. CERTAIN BLUE-GREEN ALGAE SPECIES FORM ASSOCIATIONS WITH OTHER ORGANISMS SUCH AS FUNGI, LIVERWORTS, FERNS, AND SEED PLANTS. THE RELATIONSHIP BETWEEN THESE TWO ORGANISMS IS SOMETIMES COMMENSAL AND OTHER TIMES SYMBIOTIC. CERTAIN SYMBIOTIC BLUE-GREEN ALGAE ARE PROVIDED WITH THE ABILITY TO FIX THE ATMOSPHERIC NITROGEN. (JONES-WISCONSIN)

FIELD 05C

ACCESSION NO. W73-02473

ROLE OF PHOSPHORUS IN EUTROPHICATION AND DIFFUSE SOURCE CONTROL,

WISCONSIN UNIV., MADISON. WATER CHEMISTRY PROGRAM.

G. F. LEE.

(PREPRINT) IN: CONFERENCE ON PHOSPHORUS IN FRESH WATER AND MARINE ENVIRONMENT, APRIL 1972, INTERNATIONAL ASSOCIATION ON WATER POLLUTION RESEARCH, LONDON (ENGLAND). 28 P, 1 TAB, 22 REF.

DESCRIPTORS:
*WATER POLLUTION CONTROL, *PHOSPHORUS, *EUTROPHICATION, CHEMICAL ANALYSIS, WATER POLLUTION SOURCES, BIOASSAY, ALGAE, DOMESTIC WASTES, NUTRIENTS, SEDIMENTS, ANALYTICAL TECHNIQUES, FERTILIZATION, AQUATIC PLANTS, NITROGEN, NUTRIENT REMOVAL, WATER TREATMENT, ASSAY, BACTERIA.

IDENTIFIERS:
ALGAL ASSAY, ORTHOPHOSPHATE, PHOSPHORUS SOURCES.

ABSTRACT:
WHAT IS NEEDED TO DEVELOP THE MOST MEANINGFUL CONTROL PROGRAM FOR EXCESSIVE FERTILIZATION OF NATURAL WATERS IS A METHOD OF ASSESSING THE ROLE OF PHOSPHORUS IN EUTROPHICATION OF A PARTICULAR LAKE. THE CURRENT TECHNOLOGY WITH RESPECT TO THE TESTS USED IS REVIEWED: BIOASSAY OF NUTRIENT STATUS, DETERMINATION OF CRITICAL NUTRIENT CONCENTRATIONS; FACTORS INFLUENCING EXCHANGE OF NUTRIENTS BETWEEN SEDIMENTS AND WATER; METHODS OF PHOSPHORUS CONTROL. AT THIS TIME THERE IS INSUFFICIENT EVIDENCE TO PREDICT QUANTITATIVELY THE ROLE OF PHOSPHORUS IN EXCESSIVE FERTILIZATION OF A GIVEN LAKE, THUS RESEARCH IS NEEDED TO ASSESS THIS WITH VARIOUS CHEMICAL AND BIOLOGICAL TECHNIQUES. EMPHASIS SHOULD BE PLACED ON DETERMINING PHOSPHORUS SOURCES, FORMS, RATES OF TRANSFORMATION, AND AVAILABILITY FOR AQUATIC PLANT GROWTH IN NATURAL WATERS. RESULTS SHOULD BE FORMULATED IN MATHEMATICAL MODELS ENABLING PREDICTIONS OF RELATIONSHIPS BETWEEN PHOSPHORUS INPUT AND EXCESSIVE GROWTHS OF AQUATIC PLANTS. EFFORTS SHOULD BE INCREASED TO CONTROL PHOSPHORUS FROM DIFFUSE SOURCES SUCH AS URBAN STORM WATER AND AGRICULTURAL RUNOFF. DETERGENTS WITHOUT PHOSPHORUS, PROPERLY EVALUATED WITH RESPECT TO POTENTIAL ENVIRONMENTAL IMPACT AND PERSONAL SAFETY HAZARDS, SHOULD BE FORMULATED. DIRECT LAKE WATER TREATMENT WITH PHOSPHORUS-PRECIPITATING CHEMICALS SHOULD BE STUDIED ON A LARGE SCALE. (JONES-WISCONSIN)

FIELD 05C, 05B

ACCESSION NO. W73-02478

WAYS IN WHICH A RESIDENT OF THE MADISON LAKES' WATERSHED MAY HELP TC IMPROVE WATER QUALITY IN THE LAKES,

WISCONSIN UNIV., MADISON. DEPT. OF CIVIL AND ENVIRCNMENTAL ENGINEERING.

G. F. LEE.

JULY 1972. 10 P.

DESCRIPTORS:
*EUTROPHICATION, *SCCIAL PARTICIPATION, *WATER PCLLUTICN CONTROL, WISCONSIN, LAKES, URBAN RUNOFF, AGRICULTURAL RUNCFF, FARM WASTES, ALGAE, AQUATIC WEEDS, FERTILIZATION, LEAVES, LEGISLATION, LAND USE, SHORES, MONITORING, MARSHES, SOIL EROSION, WASTE WATER TREATMENT.

IDENTIFIERS:
*MADISON(WIS), LAKE MENDOTA(WIS), LAWN MAINTENANCE.

ABSTRACT:
METHODS ARE RECCMMENDED BY WHICH MADISON RESIDENTS MAY PREVENT NUTRIENTS FROM ENTERING THE CITY'S LAKES, THEREBY REDUCING FREQUENCY AND SEVERITY OF EXCESSIVE ALGAL AND WEED GROWTH. A CENTRALIZED REGULATORY GOVERNMENTAL UNIT SHOULC BE DEVELOPED WITH RESPONSIBILITY FOR WATER QUALITY CONTROL IN THE COUNTY. BRIEFLY RECCMMENDATIONS ARE: FERTILIZER APPLICATICN BY HOMEOWNERS SHOULD NOT EXCEED THE MINIMUM NEEDED FOR A HEALTHY LAWN; GRASS CLIPPINGS, LEAVES, ETC., SHOULD BE WORKED INTO SOIL OR PLACED IN IMPERVIOUS CONTAINERS FOR COLLECTION; DOMESTIC AND RURAL IRRIGATION RUNOFF SHOULD BE PREVENTED; FUNDS SHOULD BE APPROPRIATED FOR MORE FREQUENT STREET SWEEPING; DAIRY FARMERS, LIVESTOCK OWNERS, AND FEEDLOT OPERATORS SHOULD NOT SPREAD MANURE ON FROZEN GROUND; CLOSE SCRUTINY SHOULD ASCERTAIN IF ANY FERTILIZING PROCESSES TEND TO INCREASE PHOSPHORUS RUNOFF; ORDINANCES SHOULD BE ADOPTED FOR PENALTY ASSESSMENT; ALL NEW URBAN DEVELOPMENTS SHOULD BE REQUIRED TO USE THE LATEST TECHNIQUES TO MINIMIZE PHOSPHORUS TRANSPORT; FUNDS SHOULD BE APPROPRIATED FOR ENFORCEMENT OF THE ORDINANCE PROHIBITING DISCHARGE TO STORM SEWERS OF COOLING TOWER BLOWDOWN WATER; SHORELINE DEBRIS SHOULD BE COLLECTED; FUNDS FOR LONG-TERM MONITORING PROGRAMS SHOULD BE APPROPRIATED; MARSH DRAINAGE SHOULD BE PROHIBITED; SOIL EROSION PREVENTED; EXCESSIVE GROWTHS OF AQUATIC WEEDS HARVESTED. (JONES-WISCONSIN)

FIELD 05C

ACCESSION NO. W73-02479

SALINITY-RELATED POLYMORPHISM IN THE BRACKISH-WATER DIATOM CYCLOTELLA CRYPTICA,

CONNECTICUT UNIVERSITY, STORRS. BIOLOGICAL SCIENCES GROUP.

M. E. SCHULTZ.

CANADIAN JOURNAL OF BOTANY, VOL 49, NO 8, P 1285-1289, 1971. 1 FIG, 1 TAB, 9 REF. OWRR A-014-CONN(5), 14-01-0001-901.

DESCRIPTORS:
*DIATOMS, *SALINITY, *BICINDICATORS, BRACKISH WATER, FRESHWATER, *ALGAE, WATER POLLUTICN EFFECTS.

IDENTIFIERS:
*CYCLOTELLA CRYPTICA, *CYCLOTELLA MENEGHINIANA, *PCLYMORPHISM, *VALVE MORPHOLOGY, *VALVE PATTERN CHARACTERISTICS, CLCNES.

ABSTRACT:
THE BRACKISH-WATER DIATOM CYCLOTELLA CRYPTICA IS A POLYMORPHIC SPECIES. NINE CLONES ARE CAPABLE OF PRODUCING THE VALVE PATTERN CHARACTERISTIC OF THE SPECIES C. MENEGHINIANA, AS WELL AS THE C. CRYPTICA PATTERN. A STUDY OF THE EFFECTS OF SALINITY AND FRESHWATER CCNDITIONS ON THE MORPHOLOGY OF THE VALVE SHOWS THAT THE CRYPTICA PATTERN IS PRODUCED IN SALINITIES OF ABOUT 4.3% TO FULL-STRENGTH SEAWATER, 28.7%. THE 'MENEGHINIANA' PATTERN IS THE FRESHWATER OR LOW SALINITY (1.4%) FORM. CHARACTERISTICS OF THE VALVE MORPHOLOGY AND LIFE HISTORY STAGES WHICH DISTINGUISH C. CRYPTICA FROM C. MENEGHINIANA AND CYCLOTELLA SP., CLONE 03A, ARE PRESENTED AND DISCUSSED.

FIELD 05C, 05A

ACCESSION NO. W73-02548

SIGNIFICANT DESCRIPTOR INDEX

The starred descriptors indexed here have been selected as best describing the content of each paper. If additional references to these or other terms are desired, consult the *General Index.* The numbers at the right are the accession numbers appearing at the end of the abstracts.

ERIA, DNA, MOTILITY, CYST FO/ *G	+ C RATIOS, *NONFRUITING MYXOBACT	W72-07951
*PHYTOPLANKTON, *ION TRANSPORT,	*ABSORPTION, *PHOSPHORUS, *PHOSPH	W72-07143
HT INTE/ *ALGAE, *PHYTOPLANKTON,	*ABSORPTION, *AQUATIC ALGAE, *LIG	W72-07166
INE ALGAE,/ *ZINC RADIOISOTOPES,	*ABSORPTION, *PHYTOPLANKTON, *MAR	W73-00823
OPLANKTON,/ *ZINC RADIOISOTOPES,	*ABSORPTION, *MARINE ALGAE, *PHYT	W73-00802
OUND RADIATION, *NUCLEAR WASTES,	*ABSORPTION, *FOOD CHAINS, PUBLIC	W73-00833
*COBALT RADIOISOTOPES, *COBALT,	*ABSORPTION, *MARINE ANIMALS, MAR	W73-00821
OTOPES, *MEASUREMENT, *KINETICS,	*ABSORPTION, INSTRUMENTATION, RAD	W72-14326
DIOACTIVI/ *SOIL MICROORGANISMS,	*ABSORPTION, *CARBON DIOXIDE, *RA	W72-14301
YTOPLANKTON, *SHRIMP, *BIOASSAY,	*ABSORPTION, PESTICIDE RESIDUES,	W73-02105
*RADIOISOTOPES, *MARINE ALGAE,	*ABSORPTION, *MINE WASTES, RADIOE	W72-12204
PATH/ *STRONTIUM RADIOISOTOPES,	*ABSORPTION, *SNAILS, GASTROPODS,	W72-10978
*WATE/ *RADIOISOTOPES, *RADIUM,	*ABSORPTION, *PATH OF POLLUTANTS,	W72-10686
TS, NUCLEAR WAS/ *RADIOISOTOPES,	*ABSORPTION, *ENVIRONMENTAL EFFEC	W72-10955
OOD CHAINS, *PATH OF POLLUTANTS,	*ABSORPTION, ALGAE, CHORELLA, TRA	W72-09668
ALGAE, *SEA WATER, *FOOD CHAINS,	*ABSORPTION, BIOCHEMISTRY, PHYSIO	W71-13243
OISOTOPES, *MARINE ANIMALS, *MA/	*ABSORPTION, *DISTRIBUTION, *RADI	W71-09168
ES, *PHYTOPLANKTON, *METABOLISM,	*ABSORPTION, MARINE PLANTS, MARIN	W71-09193
*RADIOISOTOPES, *MARINE ANIMALS,	*ABSORPTION, AQUARIA, MOLLUSKS, S	W71-09182
AL TECH/ *RADIOISOTOPES, *ALGAE,	*ABSORPTION, MONITORING, ANALYTIC	W71-08032
ACT, CONNECTICUT, *UNIVERSITIES,	*ABSTRACTS, WATER POLLUTION TREAT	W71-01192
ANIMALS, *AQUATIC MICROBIOLOGY,	*ACARICIDES, BRACKISH WATER, MICR	W71-08026
S, ANABAENA CYLINDRICA, ANABAEN/	*ACETYLENE REDUCTION, *HETEROCYST	W71-12070
APONICUM, LAKE MENDOTA, CRYSTAL/	*ACETYLENE REDUCTION, RHIZOBIUM J	W73-01456
RA GLOMERATA, *MET/ *CLADOPHORA,	*ACHNANTHES, *SPIROGYRA, CLODOPHO	W72-07901
WATER / *AQUATIC MICROORGANISMS,	*ACID MINE WATER, *STRIP MINES, *	W71-07942
IOISOTOPES, SEWAGE/ *PHOSPHORUS,	*ACTIVATED SLUDGE, PHOSPHORUS RAD	W71-01474
BORATORY TESTS, DETERGE/ *ALGAE,	*ACTIVATED SLUDGE, *TOXICITY, *LA	W70-08976
REATMENT, *BIOLOGICAL TREATMENT,	*ACTIVATED SLUDGE, EFFLUENT, *ORG	W71-12223
*ALGAE, *DETERGENTS, *TOXICITY,	*ACTIVATED SLUDGE, *NUTRIENTS, NI	W71-11972
IN-LUCIFERASE ENZYME SYSTEM, PL/	*ADENOSINE TRIPHOSPHATE, *LUCIFER	W71-11916
BONUCLEIC ACID, SAMPLE PRESERVA/	*ADENOSINE TRIPHOSPHATE, *DEOXYRI	W71-11562
*ATP ANALYSIS,	*ADENOSINE TRIPHOSPHATE.:	W71-11234
TRUM CAPRICORNUTUM.:	*ADENOSINE TRIPHOSPHATE, *SELENAS	W71-13794
PHIC INDEX, MACROINVERTEBRATES,/	*ADENOSINE TRIPHOSPHATE, *AUTOTRO	W72-10883
AL BIOMASS, *GROWTH STAGE, *DET/	*ADENOSINE TRIPHOSPHATE, *MICROBI	W72-04292
MICAL TEST/ *ANACYSTIS NIDULANS,	*ADENOSINE TRIPHOSPHATASE, BIOCHE	W72-13809
Y EQUIPMENT, *LA/ *ASSAY, *ZINC,	*ADSORPTION, *SURFACES, LABORATOR	W72-12210
S.:	*AERATED LAGOONS, SUSPENDED SOLID	W71-07113
S.:	*AERATED LAGOONS, SURFACE AERATOR	W71-07109
S.:	*AERATED LAGOONS, SUSPENDED SOLID	W71-07106
ONS, *DESIGN C/ *SEWAGE LAGOONS,	*AERATED LAGOONS, *OXIDATION LAGO	W72-09386
*WASTE WATER TREATMENT, WYOMING,	*AERATED LAGOONS.: /ILOT PLANTS,	W72-06838
ATION, *SEWAGE TREATMENT, ALGAE,	*AERATION, SEDIMENTATION, *CHLORI	W72-14469
OGICAL DISTRIBUTION, *AQUATIC A/	*AERATION, *EUTROPHICATION, *ECOL	W72-09061
L ME/ *POLLUTANT IDENTIFICATION,	*AERIAL PHOTOGRAPHY, *MATHEMATICA	W71-11685
DIATION, *AQUA/ *REMOTE SENSING,	*AERIAL PHOTOGRAPHY, *INFRARED RA	W71-12034
ION EFFECTS, EUTROPHIC/ SURVEYS,	*AERIAL PHOTOGRAPHY, WATER POLLUT	W72-14728
TIONS, *KINETICS, *SCENEDESMUS,/	*AEROBIC BACTERIA, *AEROBIC CONDI	W73-00379
SCENEDESMUS,/ *AEROBIC BACTERIA,	*AEROBIC CONDITIONS, *KINETICS, *	W73-00379
*CARBON DIOXIDE, *GROWTH RATES,	*AEROBIC CONDITIONS, *GROWTH STAG	W72-06612
T ELEMENTS, ESSENTIAL NUTRIENTS,	*AEROBIC CONDITIONS, WATER POLLUT	W72-11617
HORUS, NITROGEN, EUTROPHICATION,	*AEROBIC CONDITIONS, *AEROBIC: /P	W72-01881
NDITIONS, P/ *OXIDATION LAGOONS,	*AEROBIC TREATMENT, *ANAEROBIC CO	W71-06033
DEGRADATION, WATER PURIFICATION,	*AEROBIC TREATMENT, *OXIDATION LA	W72-05315
OPHICATION, *AEROBIC CONDITIONS,	*AEROBIC: /PHORUS, NITROGEN, EUTR	W72-01881
	*AFTERGROWTH.:	W72-09590
ILOSIS, CHONDRUS CRISPUS, RHODO/	*AGARS, EPIPHYTES, CANDIDA PARAPS	W72-04748
	*AGGLOMERATION.:	W72-10237

MENTS, *WATER POLLUTION SOURCES, *AGRICULTURAL RUNOFF, *WATER POLL W72-08401
RODUCTION, *IRRIGATION PROGRAMS, *AGRICULTURE, *ECONOMICS, CYANOPH W72-03327
NTRO/ *MANAGEMENT, *ENVIRONMENT, *AGRICULTURE, *WATER POLLUTION CO W72-00846
BOLISM, MOBILITY, PARTICULATE F/ *AGRICULTURAL DRAINAGE, LAKE META W71-06443
RGY LOSSES, *WATER CONSERVATION, *AIR CONDITIONING, *HEATING, *COO W70-09193
STACEANS, ISOPODS, MARINE ALGAE, *ALASKA.: /NNELIDS, SEASONAL, CRU W73-00374
S, FISH PO/ *NUCLEAR EXPLOSIONS, *ALASKA, *MARINE ALGAE, *FISHERIE W73-00764
VITY, EUTROPHICATI/ *ECOSYSTEMS, *ALASKA, *LAKES, ARCTIC, PRODUCTI W71-09190
BENTHOS, SCUBA / *MARINE PLANTS, *ALASKA, SAMPLING, *SYSTEMATICS, W72-09120
ORINATED HYDROCARBON PESTICIDES, *ALDRIN, *DIELDRIN, *ENDRIN, *GRO W72-03545
AL DEGRADATION, *DDT, *DIELDRIN, *ALDRIN, *ENDRIN, MARINE ALGAE, B W73-00361
LUBLE ORGANIC SOLIDS.: *ALEXANDRIA(EGYPT), PANDORINA, SO W70-06619
CONTROL/ *WATER QUALITY CONTROL, *ALGAE CONTROL, *FORAGE FISH, BIO W71-09163
WEED STUDY FACILITY, PRODUCT E/ *ALGAE CULTURE TECHNIQUE, AQUATIC W71-06102
*ALGAE GROWTH, *CHEMOSTATS.: W72-06612
*ALGAE HARVESTING.: W71-08951
GAL GROWTH, ALGAL HARVESTING, */ *ALGAE STRIPPING, SCENEDESMUS, AL W72-02975
*PHYTOPLANKTON, *PACIFIC OCEAN, *ALGAE.: *DIATOMS, W71-11527
DUCTIVITY, *OIL WASTES, *ARCTIC, *ALGAE.: *PRIMARY PRO W72-07922
CULTURES, CHLOROPHYTA, PROTOZOA, *ALGAE.: /ATICS, *CHLAMYDOMONAS, W72-12632
ATER TREATMENT, *MICROORGANISMS, *ALGAE, *ALGICIDES, MICROSTRAININ W72-06536
S, BIOASSAY, LIMITING FACTORS, / *ALGAE, *AQUATIC WEEDS, *NUTRIENT W73-00232
XICITY/ *REVIEWS, *HEAVY METALS, *ALGAE, *AQUATIC ENVIRONMENT, *TO W72-14694
*EUTROPHICATION, *PHYTOPLANKTON, *ALGAE, *AQUATIC WEEDS, *WEEDS, * W72-14673
LABORATORY TESTS, C/ *ALGICIDES, *ALGAE, *ALGAL CONTROL, TESTING, W71-06189
TROL, GROWTH, *CULTURES, *ALGIC/ *ALGAE, *AQUATIC WEEDS, *WEED CON W71-06102
ANALYTICAL TECH/ *RADIOISOTOPES, *ALGAE, *ABSORPTION, MONITORING, W71-08032
ITY, *LABORATORY TESTS, DETERGE/ *ALGAE, *ACTIVATED SLUDGE, *TOXIC W70-08976
ES, CHE/ *ANALYTICAL TECHNIQUES, *ALGAE, *ASSAY, CHLORELLA, NITRAT W70-10403
ACTIVATED SLUDGE, W/ *SYMBIOSIS, *ALGAE, *BACTERIA, *RESPIRATION, W71-04079
D, *EUTROPHICATION, PHOSPHOROUS/ *ALGAE, *BIOCHEMICAL OXYGEN DEMAN W72-04431
OTOSYNTHESIS, OXIDATION, RESPIR/ *ALGAE, *BACTERIA, *SYMBIOSIS, PH W72-04983
RIOPHAG/ *SEPARATION TECHNIQUES, *ALGAE, *BACTERIA, *FUNGI, *BACTE W72-06124
YTOPLANKTON, PRODUCTIVITY, PHOT/ *ALGAE, *BIOMASS, *CYANOPHYTA, PH W72-14812
RIVERS, BIOLOGICAL COMMUNITIES,/ *ALGAE, *BENTHIC FLORA, *OREGON, W72-07901
*STREAMS, LAKE ONTARIO, PLANKTO/ *ALGAE, *BIOLOGICAL COMMUNITIES, W73-01627
OGY, POLLUTANTS, *TOXICITY, WAT/ *ALGAE, *BIOASSAY, *PLANT PHYSIOL W72-12576
NAEROBIC CONDITIONS, *LAKE ERIE, *ALGAE, *BENTHIC FAUNA, *BOTTOM S W72-12995
EFFECTS, *NITRATES, *NUTRIENTS, *ALGAE, *BACTERIA, AESTHETICS, CO W71-09958
POSING ORGANIC MATTER, AERATION, *ALGAE, *BIODEGRADATION, LABORATO W72-01881
ALYTICAL TECHNIQUES, *NUTRIENTS, *ALGAE, *BIOASSAY, PHOSPHORUS COM W72-10608
STE WATER TREATMENT, *NUTRIENTS, *ALGAE, *BACTERIA, *SYMBIOSIS, AC W72-11100
ER, *WYOMIN/ *NITROGEN FIXATION, *ALGAE, *CYANOPHYTA, *THERMAL WAT W72-08508
, *LAKES, RHODOPHYTA, SHORES.: *ALGAE, *CHLOROPHYTA, *CYANOPHYTA W71-12489
ATORY TESTS, CYTOLOGICAL STUDIE/ *ALGAE, *CHLORELLA, *BORON, LABOR W71-12858
NT MORPH/ *BIOLOGICAL MEMBRANES, *ALGAE, *CYTYLOGICAL STUDIES, PLA W73-01630
D/ *LABORATORY TESTS, *BACTERIA, *ALGAE, *CULTURES, CYANOPHYTA, HY W73-01657
ACIDS, *PLANT PHYSIOLOGY, *GRO/ *ALGAE, *CARBOHYDRATES, *ORGANICS W72-07710
ABITATS, *SAMPLING, *ECOSYSTEMS, *ALGAE, *CHLORAPHYTA, *AQUATIC IN W72-06526
, BIOMASS, LIGHT INTENSITY, GRO/ *ALGAE, *CULTURES, PHOTOSYNTHESIS W72-06274
TA, *PYRROPHYTA, *SYSTEMATICS, / *ALGAE, *CHLOROPHYTA, *EUGLENOPHY W72-07683
EUTROPHICATION, FJORDS, BIOASSA/ *ALGAE, *CULTURES, *INDICATORS, * W73-00234
FUNGICIDES, HERBICIDES, IRRIGA/ *ALGAE, *CYANOPHYTA, *PESTICIDES, W72-14807
TION, *OXIDATION LAGOONS, CHLOR/ *ALGAE, *CHLOROPHYLL, *CLASSIFICA W71-00139
RIA, *OXIDATION LAGOONS, *PATHO/ *ALGAE, *CULTURES, *ENTERIC BACTE W70-08319
ER, NUTRIENTS,/ *EUTROPHICATION, *ALGAE, *DECOMPOSING ORGANIC MATT W70-10405
THERMAL SPRINGS, CYANOPHYTA, AL/ *ALGAE, *DISTRIBUTION PATTERNS, * W72-05610
TRIBUTION PATTE/ *PHYTOPLANKTON, *ALGAE, *DIATOMS, *CURRENTS, *DIS W72-03543
YNTHESIS, TURBI/ *PHYTOPLANKTON, *ALGAE, *DISSOLVED OXYGEN, PHOTOS W73-00256
ACTIVATED SLUDGE, *NUTRIENTS, N/ *ALGAE, *DETERGENTS, *TOXICITY, * W71-11972
UCTIVITY, *PRIMARY PRODUCTIVITY, *ALGAE, *DIATOMS, *PERIPHYTON, *P W72-12192
G FACTORS, PHOTOSYNTHE/ *CARBON, *ALGAE, *EUTROPHICATION, *LIMITIN W71-12068
AL, *SEWAGE, *INDUSTRIAL WASTES/ *ALGAE, *ESTUARIES, *WASTE DISPOS W72-08804
TS, *PLANKTON, *RECREATION, *WA/ *ALGAE, *EUTROPHICATION, *NUTRIEN W72-10431
RATES, CHLORELLA, CARB/ *CARBON, *ALGAE, *EUTROPHICATION, *GROWTH W72-09168
OTOPERIODISM, GROWTH RATES, TEM/ *ALGAE, *ECOLOGY, ENVIRONMENT, PH W72-00377
GENIC BACTERIA, CULTURES, E. CO/ *ALGAE, *ENTERIC BACTERIA, *PATHO W72-07664
RUS, *NITROGEN, EVA/ *NUTRIENTS, *ALGAE, *EUTROPHICATION, *PHOSPHO W72-04070
QUATIC/ *PHOSPHORUS, *SEDIMENTS, *ALGAE, *EUTROPHICATION, LAKES, A W72-03742
, *BOTTOM SEDIMENTS, *LIMNOLOGY, *ALGAE, *EUTROPHICATION, LAKES, W W70-08944
RICIDES, GROWTH STAGES, PATHOGE/ *ALGAE, *ENTERIC BACTERIA, *BACTE W71-05155
WATER QUALITY CONTROL, *ECOLOGY, *ALGAE, *EUTROPHICATION, NUTRIENT W71-08775
L, PLANT/ *MOLECULAR STRUCTURES, *ALGAE, *FLUID FRICTION, *LITTORA W71-07878
N PATTERNS, BEHAVIOR.: *ALGAE, *FRESH WATER, DISTRIBUTIO W71-07360
COSTS, SEPARATION TECHNIQUES, / *ALGAE, *FLOCCULATION, EQUIPMENT, W70-08904
, *PHOTOACTIVATION, *OXYGEN, LI/ *ALGAE, *FLUORESCENCE, *CHLORELLA W72-14800
LUTANTS, DIELD/ *COPPER SULFATE, *ALGAE, *FLOODWATER, *PATH OF POL W71-11498
, FLOATING PLANTS, S/ *CHORELLA, *ALGAE, *FOULING, *NUISANCE ALGAE W71-09261
UTION TREATMENT, COLLOIDS, *POL/ *ALGAE, *FLOCCULATION, WATER POLL W72-11833
FA/ *HYDROGEN ION CONCENTRATION, *ALGAE, *GROWTH RATES, *LIMITING W72-11724
CS, *LIFE HISTORY STUDIES, *GRO/ *ALGAE, *GROWTH RATES, *SYSTEMATI W73-02099
, *BIOINDICATORS, THERMAL POLLU/ *ALGAE, GREAT LAKES, *MONITORING W72-10875
*FLUORESCENCE, / *HEAVY METALS, *ALGAE, *GROWTH RATES, *TOXICITY, W72-07660
, CULTURES, INHIBITION, OLIGOTR/ *ALGAE, *GROWTH RATES, *NUTRIENTS W72-08376
RODUCTIVITY, *HEAVY METALS, *TR/ *ALGAE, *GROWTH RATES, *PRIMARY P W73-01434
ITY, *PLANT GROWTH SUBSTANCES, / *ALGAE, *GROWTH RATES, *PRODUCTIV W71-08687
*PHOSPHORUS, *NUTRIENT REQUIRE/ *ALGAE, *GROWTH RATES, *KINETICS, W72-04788
IZERS, DIATOMACEOUS EARTH, FOOD/ *ALGAE, *HUMAN POPULATION, FERTIL W71-09157
COPPER, CHLOROPHY/ *INDICATORS, *ALGAE, *HEAVY METALS, *TOXICITY, W71-05991
CHLORELLA, RESP/ *CHLAMYDOMONAS, *ALGAE, *HERBICIDES, PESTICIDES, W70-07255

N EXCHANGE, DISSOLVED SOLIDS, F/ *ALGAE, *HARVESTING OF ALGAE, *IO W71-00138
TON, PHYSICOCHEMICAL PROPERTIES, *ALGAE, *HYDROBIOLOGY, ENTERIC BA W72-07896
CHLORINATED HYDROCARBON PESTIC/ *ALGAE, *INDICATORS, *PESTICIDES, W73-00238
ICITY, HEAVY METALS, LABORATORY/ *ALGAE, *INHIBITION, *LIPIDS, TOX W72-06037
TION EFFECTS, *PLANTS, *ANIMALS, *ALGAE, *INVERTEBRATES, *RADIOACT W71-13233
REATMENT, *NUTRIENT REQUIREMENT/ *ALGAE, *KINETICS, *WASTE WATER T W72-04789
TRIENT REQUIREMENTS, AMMONIA, */ *ALGAE, *KINETICS, *NITROGEN, *NU W72-04787
ATS, LAKES, AQUAT/ *LAKE SHORES, *ALGAE, *LAKE ERIE, AQUATIC HABIT W71-09156
EFFLUENTS, WATER POLLUTION EFFE/ *ALGAE, *LAKES, *GEORGIA, SEWAGE W71-09173
RATES,/ *PHOSPHORUS, *PLANKTON, *ALGAE, *LIMITING FACTORS, GROWTH W72-09163
ION, BENTHIC FLORA/ *PERIPHYTCN, *ALGAE, *LITTORAL, LIGHT PENETRAT W71-11029
ION, *CYANOPHYTA, PLANTS, NITRO/ *ALGAE, *LEGUMES, *NITROGEN FIXAT W73-01456
THIC FLORA, DI/ *BIBLIOGRAPHIES, *ALGAE, *LAKE ERIE, PLANKTON, BEN W73-01615
UATIC ECOSYSTEMS, WATER QUALITY, *ALGAE, *LIGHT PENETRATICN, *MACR W73-02349
AT/ *EUTROPHICATION, *NUTRIENTS, *ALGAE, *LIMITING FACTORS, PHOSPH W72-12544
CAL STUDIES, DIGITAL COMPUTERS,/ *ALGAE, *MODEL STUDIES, MATHEMATI W72-11920
ATORS, *WATER POLLUTION EFFECTS, *ALGAE, *MARINE ALGAE, *BIOINDICA W71-10553
STUDIES, PHOTOSYNTHESIS, RESPI/ *ALGAE, *METABOLISM, *CYTOLOGICAL W71-08683
PHYTOP/ *POLLUTANTS, *SEA WATER, *ALGAE, *MARINE ALGAE, CULTURES, W70-07280
HLOROPHYTA, CY/ *EUTROPHICATION, *ALGAE, *NUTRIENT REQUIREMENTS, C W70-05469
ES, BACTERIA, SPOR/ *CYANOPHYTA, *ALGAE, *NITROGEN FIXATICN, ENZYM W71-12070
ESTS, *BIOASSAY, OREGON, LAKES,/ *ALGAE, *NUTRIENTS, *LABORATORY T W72-09164
FICATION, ANALYTICAL TECHNIQUES, *ALGAE, *NUTRIENTS, LAKES, PLANT W72-00801
OGEN FIXING BACTERIA, *RAINFALL/ *ALGAE, *NITROGEN FIXATION, *NITR W72-12505
ATION, LAKES, N/ *PHYTOPLANKTON, *ALGAE, *NUTRIENTS, *NITROGEN FIX W72-13692
ICAL REGIONS, RICE, CYANOPHYTA,/ *ALGAE, *NITROGEN FIXATICN, *TROP W73-02473
ATMENT, *PHOSPHORUS, *NITROGEN,/ *ALGAE, *NITRIFICATION, WASTE TRE W73-02187
S, *TEMPERATE, CLIMATIC ZONES, / *ALGAE, *NITROGEN FIXATICN, *SOIL W73-02471
ONCENTRATION, BIOCHEMICAL OXYGE/ *ALGAE, *NITROGEN, HYDROGEN ION C W72-06612
TIVITY, *AQUATIC MICROORGANISMS, *ALGAE, *NITROGEN, *PHOSPHORUS, S W72-06532
SO/ *SEDIMENTARY ROCKS, *BRINES, *ALGAE, *OKLAHOMA, *IODINE RADIOI W72-02073
S, OXYGEN, / *OXIDATION LAGOONS, *ALGAE, *OXIDATION, PHOTOSYNTHESI W72-02363
T INTENSITY, PONDS, ROTIFERS, P/ *ALGAE, *OXIDATION LAGOCNS, *LIGH W71-07382
IC INSECTS, *DISPERSION, CADDIS/ *ALGAE, *PROTOZOA, *FUNGI, *AQUAT W70-06655
PRODUCTIVITY,/ SEWAGE EFFLUENTS, *ALGAE, *PHYTOPLANKTON, *AQUATIC W71-02681
KES, DISS/ *MATHEMATICAL MODELS, *ALGAE, *POPULATION, DYNAMICS, LA W71-03034
ULTURES, NITRATES, PLANT PHYSIO/ *ALGAE, *PHOSPHATES, *ENZYMES, *C W72-05476
L TECHNIQUES, / *AQUATIC PLANTS, *ALGAE, *PUBLIC HEALTH, ANALYTICA W72-04275
TESTS, *WATER POLLUTION EFFECTS, *ALGAE, *PLANKTON, WATER POLLUTIO W72-03218
TY, CHEMICAL OXY/ *CHLORINATION, *ALGAE, *PHOTCSYNTHESIS, SOLUBILI W72-00411
, DIURNAL, FLUCTU/ *ZOOPLANKTON, *ALGAE, *PHOTOSYNTHESIS, RAINFALL W72-00150
PRODUCTIVITY, MICROSCOPY, CHLOR/ *ALGAE, *PHYTOPLANKTON, *PRIMARY W72-09239
STRIBUTION, *SPATIAL DISTRIBUTI/ *ALGAE, *POPULATICN, *TEMPORAL DI W72-08560
NISMS, *WATER POLLUTION EFFECTS, *ALGAE, *PROTOZOA, *BIOLOGICAL CO W72-08579
LORINATED HYDROC/ *INSECTICIDES, *ALGAE, *PRIMARY PRODUCTIVITY, CH W71-10096
PHOTOSYNTHESIS, INSTRUMENTATION/ *ALGAE, *PHOTOSYNTHETIC OXYGEN, * W71-10791
ATION, GROWTH RAT/ *FORECASTING, *ALGAE, *PLANT GROWTH, *EUTROPHIC W71-11033
CTIVITY, INVERTEBRATES, *PLANKT/ *ALGAE, *PROTOZOA, *PRIMARY PRODU W72-07899
, INHIBITORS, PLANT TISSUES, BI/ *ALGAE, *PLANT GROWTH, RHODOPHYTA W72-07496
AGOONS, / *ENZYMES, *DETERGENTS, *ALGAE, *PHYTOPLANKTON, *SEWAGE L W72-07661
ON, *AQUATIC ALGAE, *LIGHT INTE/ *ALGAE, *PHYTOPLANKTON, *ABSORPTI W72-07166
, HABITATS, ECOLOGY, *ECOSYSTEM, *ALGAE, *PHYTOPLANKTON, OXYGEN, P W73-01632
LAKES, *SOUTH DAKOTA, PLANKTON, *ALGAE, *PHOSPHORUS COMPOUNDS, CY W73-00938
*DIATOMS, CHLOROPHYTA, BENTHIC / *ALGAE, *PHYTOPLANKTON, *RIVERS, W72-14797
TECHNIQUES, *COMPUTER PROGRAMS, *ALGAE, *PHYTOPLANKTON, BIOLOGICA W73-02469
ES, LAKES, CYCLING N/ *PLANKTON, *ALGAE, *PHOSPHORUS, *MODEL STUDI W72-12543
URES, PHOSPHATES, NUTRIENTS, CH/ *ALGAE, *PHOSPHORUS, *LAKES, CULT W72-12637
OLOGY, *PLANT GROWTH, CYTOLOGIC/ *ALGAE, *RHODOPHYTA, *PLANT MORPH W73-01622
G TECHNIQUES, MEASUREMENT, CURR/ *ALGAE, *REMOTE SENSING, *TRACKIN W72-01095
*AQUATIC BACTERIA, *POPULATION, *ALGAE, *STRATIFICATION, *ENZYMES W72-03224
*ABSORPTION, B/ *RADIOISOTOPES, *ALGAE, *SEA WATER, *FOOD CHAINS, W71-13243
, *WATER POLLU/ *EUTROPHICATION, *ALGAE, *SURFACE WATERS, *REVIEWS W71-10580
A, THERMOCLINE/ *EUTROPHICATION, *ALGAE, *SEWAGE EFFLUENTS, GEORGI W71-11914
*WASTEWATER TRE/ *FLOCCULATION, *ALGAE, *SILICA, *E. COLI, *IONS, W72-14840
HOLOGY, REPRODUCTION, LIFE CYCL/ *ALGAE, *SYSTEMATICS, *PLANT MORP W72-07141
TEMPERATURE, *PHOTOPERIODISM, G/ *ALGAE, *SEASONAL, *CLADOPHORA, * W72-13651
TION, ELECTROPHORESIS, SETTLING, *ALGAE, *SILICA, *BACTERIA, WATER W71-01899
TE WATER TREATMENT, *FILTRATION, *ALGAE, *SUSPENSION, HARVESTING, W70-10174
, *ANAEROBIC BACTERIA, *TEXAS, / *ALGAE, *STABLE ISOTOPES, *CARBON W71-04873
S, DISSOLVE/ *SELF-PURIFICATION, *ALGAE, *STREAMS, LABORATORY TEST W71-04518
PHICATION, / *OXIDATION LAGOONS, *ALGAE, *TRICKLING FILTERS, EUTRO W71-07097
E PES/ *HERBICIDES, *PESTICIDES, *ALGAE, *TOXICITY, SOILS, TRIAGIN W71-07675
*P/ *POLYCHLORINATED BIPHENYLS, *ALGAE, *WATER POLLUTION EFFECTS, W72-05614
BYPRODUCTS, M/ *OXIDATION PONDS, *ALGAE, *WASTE WATER TREATMENT, * W72-13508
MATHEMATICA/ *OXIDATION LAGOONS, *ALGAE, *WASTE WATER TREATMENT, * W72-12289
YDROGEN ION CONCEN/ *PHOSPHORUS, *ALGAE, *WASTE WATER TREATMENT, H W73-01367
TERTIARY TREATMENT, *COAGULATIO/ *ALGAE, *WASTE WATER TREATMENT, * W71-09629
*INDICATORS, *SEA WATER, MARIN/ *ALGAE, *WATER POLLUTION EFFECTS, W71-12867
OL, NEW HAMPSH/ *COPPER SULFATE, *ALGAE, *ZOOPLANKTERS, *CHEMCONTR W71-03014
IDATION LAGOONS, *STABILIZATION, *ALGAE, AEROBIC TREATMENT, BACTER W71-03896
, SODIUM/ *EPIPHYTOLOGY, *LAKES, *ALGAE, ALKALINE WATER, MAGNESIUM W71-04526
EQUIRE/ *PHOSPHORUS, *LAKE BEDS, *ALGAE, ALGAL CONTROL, NUTRIENT R W72-11617
STE WA/ *EUTROPHICATION, *ASSAY, *ALGAE, ANALYTICAL TECHNIQUES, WA W72-12966
IS, ALGAL CONTROL, EUTROPHICATI/ *ALGAE, BACTERIA, *LAKES, SYMBIOS W71-13257
AL EFFECTS, *ECOLOGY, *PLANKTON, *ALGAE, BIOLOGICAL COMMUNITIES, P W72-14659
ATION, *CARBON, *PHOTOSYNTHESIS, *ALGAE, CARBON DIOXIDE, ALKALINIT W71-11519
, *FISH POPULATIONS, *MORTALITY, *ALGAE, CALIFORNIA.: /VERTEBRATES W71-08793
NTERTIDAL AREAS, *INVERTEBRATES, *ALGAE, CALIFORNIA.: /EFFECTS, *I W71-08792
PESTICIDES, *PESTICIDE RESIDUES, *ALGAE, CHLOROPHYTA, CHLORELLA, T W72-04134
YDROGEN IO/ *PHOSPHATES, *CLAMS, *ALGAE, CHEMICAL PRECIPITATION, H W72-03613

UATIC / *PROTOZOA, *SYSTEMATICS,	*ALGAE, CHLOROPHYTA, HABITATS, AQ	W72-09119
TOLOGICAL STUDIES, *SYSTEMATICS,	*ALGAE, CHRYSOPHYTA, PHAEOPHYTA.:	W72-14808
ITIVITY, *RADIOACTIVITY EFFECTS,	*ALGAE, CHLORELLA, IRRADIATION, R	W72-11836
ND, NUTRIENTS, *WASTE TREATMENT,	*ALGAE, COSTS.: /ICAL OXYGEN DEMA	W72-08396
EPARATION TE/ CHEMICAL ANALYSIS,	*ALGAE, COLUMNS, ORGANIC ACIDS, S	W72-09365
*LIGHT, *PHOTOSYNTHETIC OXYGEN,	*ALGAE, CULTURES, VOLUME, MATHEMA	W70-05547
L STUDIES, *STATISTICAL METHODS,	*ALGAE, CYTOLOGICAL STUDIES, SELE	W71-12868
OXIDATION LAGOONS, *OXYGENATION,	*ALGAE, DEPTH, TEMPERATURE, PHOTO	W71-07098
CHLORELLA, NUTRIENT REQUIREMENT,	*ALGAE, DETERGENTS, PHOSPHATES.: /	W73-02112
RY PRODUCTIVITY, *STANDING CROP,	*ALGAE, DIATOMS, MEASUREMENT, DET	W71-12855
, AQUATIC ALGAE, NORTH CAROLINA,	*ALGAE, DIATOMS, *DINOFLAGELLATES	W72-01329
ITY, *FLOCCULATION, *FILTRATION,	*ALGAE, ELECTROLYTES, ACTIVATED C	W72-06201
PERTIES,/ *MATHEMATICAL STUDIES,	*ALGAE, EQUATIONS, BIOLOGICAL PRO	W72-12724
Y, NUTR/ *ANALYTICAL TECHNIQUES,	*ALGAE, EVALUATION, TESTING, ASSA	W73-00255
*TESTING PROCEDURES, *BIOASSAY,	*ALGAE, FRESHWATER, MEASUREMENT,	W72-10619
IDENTIFICATION, *BIOINDICATORS,	*ALGAE, GROWTH RATES, CULTURES, S	W71-09795
RBON DIOXIDE, FLORIDA, CULTURES,	*ALGAE, INCUBATION, CENTRIFUGATIO	W72-09103
CONDITIONS, / *OXIDATION PONDS,	*ALGAE, LABORATORY TESTS, AEROBIC	W71-13326
*WASTE WATER TREATMENT, *PONDS,	*ALGAE, LAGOONS, BIOCHEMICAL OXYG	W70-07838
PHYTOPLANKT/ *NITROGEN FIXATION,	*ALGAE, MEASUREMENT, CYANOPHYTA,	W72-08054
TION, *FILT/ *OXIDATION LAGOONS,	*ALGAE, MUNICIPAL WASTES, *FILTRA	W71-13341
HICATION, *PRIMARY PRODUCTIVITY,	*ALGAE, NITROGEN, PHOSPHORUS, THE	W71-13183
PHYTA, RHODOPHYTA, CYANOPHYTA, /	*ALGAE, OHIO, CHLOROPHYTA, CHRYSO	W71-05629
-SITE INVESTIGATIONS, *BACTERIA,	*ALGAE, OXIDATION, METABOLISM, RE	W72-10233
*BIOCHEMISTRY, *PHOTOSYNTHESIS,	*ALGAE, PHYLOGENY, CARBON CYCLE,	W71-09172
WASTES, WATER POLLUTION EFFECTS,	*ALGAE, PHYTOPLANKTON, DIATOMS, C	W70-07842
UENTS, REGU/ *CARBON, *NITROGEN,	*ALGAE, PLANT GROWTH, SEWAGE EFFL	W72-14790
ON, PONDS,/ *GRAZING, *PROTOZOA,	*ALGAE, SCENEDESMUS, EUTROPHICATI	W72-14796
OL/ *EUTROPHICATION, *NUTRIENTS,	*ALGAE, SEWAGE TREATMENT, WATER P	W72-10625
TS, *EUTROPHICATION, *NUTRIENTS,	*ALGAE, SEWAGE, PRODUCTIVITY, WAT	W71-12857
SMS, *SURFACE WATERS, *PLANKTON,	*ALGAE, TURBIDITY, *NEW YORK, *WA	W72-06191
*PHOSPHORUS, *SEWAGE EFFLUENTS,	*ALGAE, WASTE WATER TREATMENT, AC	W72-09176
ORS, BRACKISH WATER, FRESHWATER,	*ALGAE, WATER POLLUTION EFFECTS.:	W73-02548
DISSOLVED PHOSPHORUS,	*ALGAL ASSAYS, ASSAY PROCEDURE.:	W72-12966
RUM CAPRICOR/ *CHEMOSTAL ASSAYS,	*ALGAL ASSAY PROCEDURES, SELENAST	W72-03985
HORUS, *NITROGE/ EUTROPHICATION,	*ALGAL BLOOMS, *NUTRIENTS, *PHOSP	W70-09669
UT/ *ESTUARIES, *EUTROPHICATION,	*ALGAL BLOOMS, *PHYTOPLANKTON, *N	W72-01329
/ *NUISANCE ALGAE, *CYANOPHYTA,	*ALGAL CONTROL, LABORATORY TESTS,	W71-13184
TABOLISM, NITROGEN, PHOSPHOROUS,	*ALGAL CONTROL, CALIFORNIA.: / ME	W71-13553
, ALGAE, *WASTE WATER TREATMENT,	*ALGAL CONTROL, *AQUATIC WEED CON	W72-02975
GE, DESIGN, NITROGEN, DIFFUSION,	*ALGAL CONTROL, AERATION, *WASTE	W72-00800
TROPHICATION, LAKES, WASHINGTON,	*ALGAL CONTROL, NUTRIENTS, BIOMAS	W72-00799
*DESTRATIFICATION, *WATER QUALI/	*ALGAL CONTROL, *EUTROPHICATION,	W72-09304
OSIVES, COSTS, FISH, FISHKILL, /	*ALGAL CONTROL, *CYANOPHYTA, EXPL	W72-11564
LORELLA, *INHIBITION, COPPER, F/	*ALGAL CONTROL, *RESPIRATION, *CH	W72-11572
NUISANCE ALGAE, *CARBON DIOXIDE,	*ALGAL CONTROL, CYANOPHYTA, CARBO	W72-10607
OPLANKTON, *MATHEMATICAL MODELS,	*ALGAL CONTROL, *RESERVOIRS, DEST	W72-08559
GROWTH, TOXICI/ *EUTROPHICATION,	*ALGAL CONTROL, FINANCING, PLANT	W71-13172
*WATER TREATMENT, *CHLORINATION,	*ALGAL CONTROL, RIVERS, WATER POL	W71-00099
SPHORUS, *WASTE WATER TREATMENT,	*ALGAL CONTROL, *AQUATIC ALGAE, *	W72-03970
IOCONTROL, *PATHOGENIC BACTERIA,	*ALGAL CONTROL, FRESH WATER, CYAN	W72-05457
ICATION, *ANALYTICAL TECHNIQUES,	*ALGAL CONTROL.: /IXING, *EUTROPH	W72-04433
RY TESTS, C/ *ALGICIDES, *ALGAE,	*ALGAL CONTROL, TESTING, LABORATO	W71-06189
RFACTANTS, *EUTROP/ *DETERGENTS,	*ALGAL CONTROL, *FORMULATION, *SU	W71-06247
ROGEN, *IDAHO, *CHLOROPHYLL, *A/	*ALGAL CONTROL, *PHOSPHORUS, *NIT	W72-13300
EUTROPHICATION, REVIEWS, TROPHI/	*ALGAL CONTROL, *NUISANCE ALGAE,	W72-07937
IENTS, *NUISANCE ALGAE, WASHING/	*ALGAL CONTROL, *DIFFUSION, *NUTR	W72-08049
EUTROPHICATION, NUT/ *ALGICIDES,	*ALGAL CONTROL, WATER POLLUTION,	W72-07508
NTS, *DEFICIENT NUTRIENTS, *AQU/	*ALGAL CONTROL, *ESSENTIAL NUTRIE	W72-07514
*EUTROPHICATION, *AQUATIC ALGAE,	*ALGAL CONTROL, *AQUATIC WEED CON	W72-14692
SES, CYANOPHYTA.:	*ALGAL CONTROL, *INFECTION, *VIRU	W73-00246
AL NUTRIENTS, *CHLORELLA, ALGAE/	*ALGAL CONTROL, *PHOSPHATES, *ALG	W72-13992
*POTENTIOMETRY, ELECTROANALYSIS,	*ALGAL CULTURES, CHELATING AGENTS	W73-02112
IGHT COMPOUNDS, *MOLECULAR SIZE,	*ALGAL EXUDATES, EXTRACELLULAR MA	W71-07878
HEMICAL COMPOSITION, ALGAL DECO/	*ALGAL GROWTH, NUTRIENT UPTAKE, C	W70-05469
SODIUM PYROPHOSPHATE, RIO GRAN/	*ALGAL GROWTH, *RADIOISOTOPE P32,	W72-13992
*PAAP TEST,	*ALGAL GROWTH, BOTTLE TEST.:	W73-00255
S, BATCH CULTURES.:	*ALGAL GROWTH, CONTINUOUS CULTURE	W72-14790
*SELENASTRUM CAPRICORNUTUM,	*ALGAL GROWTH, *LAKE GEORGE(NY).:	W72-08376
*MICRO-NUTRIENTS,	*ALGAL GROWTH.:	W72-00801
.:	*ALGAL GROWTH, *LAKE TAHOE(CALIF)	W71-13553
, TUTUILA ISLAND, HONSHU ISLAND/	*ALGAL HYDROLYSATE, TONGA ISLANDS	W71-08042
*ORGANIC CARBON,	*ALGAL MASS, ALGAL GROWTH.:	W72-04431
AE/ *ALGAL CONTROL, *PHOSPHATES,	*ALGAL NUTRIENTS, *CHLORELLA, ALG	W72-13992
GAE, *CRUSTACEANS, *FISH TOXINS,	*ALGAL POISONING, DINOFLAGELLATES	W72-14056
EUPHOTIC ZONE, DIFFUSION COEFF/	*ALGAL SINKING, *VERTICAL MIXING,	W71-03034
, ARNOLDIELLA CONCHOPHILA MILLE/	*ALGAL SPECIES, WESTERN LAKE ERIE	W71-09156
D WEBS, *FOOD CHAINS, *TOXICITY,	*ALGAL TOXINS, *CHLOROPHYTA, *PHY	W71-08597
*CYANOPHYTA, *NUISANCE ALGAE, /	*ALGAL TOXINS, *POISONOUS PLANTS,	W72-14056
TOA, *PHYTOTOXICITY, *CYANOPHYT/	*ALGAL TOXINS, *RED TIDE, *BEGGIA	W72-14542
XINS, R/ *ANALYTICAL TECHNIQUES,	*ALGAL TOXINS, AQUATIC PLANTS, TO	W73-00243
OROPHYTA, CYANOPHYTA, SYSTEMATI/	*ALGAL TOXINS, *MYXOBACTERIA, CHL	W72-07951
LATION, WATER POLLUTION EFFECTS,	*ALGAL TOXINS.: /*TOXICITY, COAGU	W72-07360
RELLA, *SCENEDESMUS, AQUATIC MI/	*ALGAL TOXINS, *INHIBITION, *CHLO	W72-12255
POLLUTION, EUTROPHICATION, NUT/	*ALGICIDES, *ALGAL CONTROL, WATER	W72-07508
ATMENT, *MICROORGANISMS, *ALGAE,	*ALGICIDES, MICROSTRAINING, AERAT	W72-06536
L, *AQUATIC WEED CONTROL, *DOME/	*ALGICIDES, *WATER QUALITY CONTRO	W71-05083
WEED CONTROL, GROWTH, *CULTURES,	*ALGICIDES, BIBLIOGRAPHIES, *ECOL	W71-06102
L, TESTING, LABORATORY TESTS, C/	*ALGICIDES, *ALGAE, *ALGAL CONTRO	W71-06189
FLOS-AQUAE, *CONTINUOUS CULTURE/	*ALKALINE PHOSPHATASE, *ANABAENA	W72-05476

REGIO/ *EUTROPHICATION, *LAKES,	*ALPINE, *SYNOPTIC ANALYSIS, COLD	W72-12549
L DESIGN.:	*ALUM, POLYELECTROLYTES, FACTORIA	W70-08904
DE, *FERRIC SULFATE, *STABILIZA/	*ALUMINUM SULFATE, *FERRIC CHLORI	W71-13356
AHO), *DISSOLVED ORGANIC CARBON,	*AMERICAN FALLS RESERVOIR, *APHAN	W72-13300
*ANALYTICAL TECHNIQUES, *CARBON,	*AMINO ACIDS, *MARINE ALGAE, PHYT	W72-07497
LANKTON, OREGON, ORGANIC COMPOU/	*AMINO ACIDS, *EUTROPHICATION, *P	W73-01269
DS, PREDATOR-PREY RELATIONSHIPS,	*AMMONIA-NITROGEN, *GROWTH REACTO	W72-04787
ICITY, CORROSION, PU/ *NITROGEN,	*AMMONIA, NITRATES, NITRITES, TOX	W72-04086
E, CHEMICAL ANALYSIS, *NITRATES,	*AMMONIA, DIATOMS, GROWTH RATES,	W73-00380
, CHEMICAL ANALYSIS,/ *TOXICITY,	*AMMONIA, *MARINE ALGAE, *DIATOMS	W71-11008
GEN, ALGAE, WATER POLLUTION SOU/	*AMMONIA, *EUTROPHICATION, *NITRO	W71-11496
PHOLOGY, *PHOSPHATES, *NITRATES,	*AMMONIUM COMPOUNDS, SEWAGE EFFLU	W72-10162
*MICROSCOPIC COUNTS,	*AMORPHOUS MATTER.:	W72-06191
RINGS, DIATOMS, BIOTA, PROTOZOA,	*AMPHIPODA, *FLORIDA, STREAMS, IS	W72-12744
IA ZOODIACUS, CHA/ SEA OF JAPAN,	*AMURSKY BAY, PERIDINIANS, EUCAMP	W73-00442
ACTIVITY, OXYGEN INACTIVATION./	*ANABAENA FLOS-AQUA, *NITROGENASE	W72-07941
EDUCTION, WA/ *PHOTORESPIRATION,	*ANABAENA CYLINDRICA, ACETYLENE R	W72-08584
NIDULANS, GROWTH INH/ *SELENIUM,	*ANABAENA VARIABILIS, *ANACYSTIS	W72-00153
TURE, BATCH CULTU/ *HETEROCYSTS,	*ANABAENA SP L-31, CONTINUOUS CUL	W72-03217
CULTURE/ *ALKALINE PHOSPHATASE,	*ANABAENA FLOS-AQUAE, *CONTINUOUS	W72-05476
*NITROGEN, *IDAHO, *CHLOROPHYLL,	*ANABAENA, NUTRIENTS, DISSOLVED S	W72-13300
	*ANACOSTIA RIVER.:	W70-08249
RIPHOSPHATASE, BIOCHEMICAL TEST/	*ANACYSTIS NIDULANS, *ADENOSINE T	W72-13809
*SELENIUM, *ANABAENA VARIABILIS,	*ANACYSTIS NIDULANS, GROWTH INHIB	W72-00153
ENSITI/ *WASTE CHARACTERIZATION,	*ANACYSTIS NIDULANS, *MUTANTS, *S	W72-14327
CAL OXYGEN DEMAND, *OXYGENATION,	*ANAEROBIC BACTERIA, BIOLOGICAL T	W72-14469
*DISSOLVE/ *ANAEROBIC BACTERIA,	*ANAEROBIC CONDITIONS, *KINETICS,	W73-01066
ONDITIONS, *KINETICS, *DISSOLVE/	*ANAEROBIC BACTERIA, *ANAEROBIC C	W73-01066
*FACULTATIVE LAGOONS,	*ANAEROBIC LAGOONS.:	W72-09386
, *ALGAE, *BENT/ *OXYGEN DEMAND,	*ANAEROBIC CONDITIONS, *LAKE ERIE	W72-12995
*HYPOLIMNION, *DISSOLVED OXYGEN,	*ANAEROBIC CONDITIONS, ALGAE, SED	W72-12990
*MEMBRANE PROCESSES,/ *CULTURES,	*ANAEROBIC BACTERIA, *ISOLATION,	W72-04743
ION LAGOONS, *AEROBIC TREATMENT,	*ANAEROBIC CONDITIONS, PONDS, CLI	W71-06033
LGAE, *STABLE ISOTOPES, *CARBON,	*ANAEROBIC BACTERIA, *TEXAS, BIOD	W71-04873
	*ANALINE DYE.:	W73-02349
PROGRAMS, *ALGAE, *PHYTOPLANKT/	*ANALYTICAL TECHNIQUES, *COMPUTER	W73-02469
C MATTER, *CARBON RADIOISOTOPES,	*ANALYTICAL TECHNIQUES, *MARINE M	W71-08042
RATION, MIXING, *EUTROPHICATION,	*ANALYTICAL TECHNIQUES, *ALGAL CO	W72-04433
ASSAY, CHLORELLA, NITRATES, CHE/	*ANALYTICAL TECHNIQUES, *ALGAE, *	W70-10403
ALYSIS, *NITRATES, *ELECTRODES,/	*ANALYTICAL TECHNIQUES, *WATER AN	W71-00474
HOTOSYNTHESIS, *MARINE ALGAE, D/	*ANALYTICAL TECHNIQUES, RATES, *P	W72-09653
S, *ALGAE, *BIOASSAY, PHOSPHORU/	*ANALYTICAL TECHNIQUES, *NUTRIENT	W72-10608
LLUTION CONTRO/ *BIBLIOGRAPHIES,	*ANALYTICAL TECHNIQUES, *WATER PO	W72-08790
ANALYSIS, WATER ANALYSIS, EVALU/	*ANALYTICAL TECHNIQUES, CHEMICAL	W72-10883
PROCEDURES, *BIOASSAY, *ALGAE, /	*ANALYTICAL TECHNIQUES, *TESTING	W72-10619
COMPOUNDS, *CARBON, GRAVIMETRIC/	*ANALYTICAL TECHNIQUES, *ORGANIC	W72-09108
IVITY TECHNIQUES, *PLANKTON, PR/	*ANALYTICAL TECHNIQUES, *RADIOACT	W72-00161
ACTIVATION ANALYSIS/ *SEDIMENTS,	*ANALYTICAL TECHNIQUES, *NEUTRON	W72-01101
XINS, AQUATIC PLANTS, TOXINS, R/	*ANALYTICAL TECHNIQUES, *ALGAL TO	W73-00243
VALUATION, TESTING, ASSAY, NUTR/	*ANALYTICAL TECHNIQUES, *ALGAE, E	W73-00255
*AMINO ACIDS, *MARINE ALGAE, PH/	*ANALYTICAL TECHNIQUES, *CARBON,	W72-07497
MICROORGANISMS, *INVESTIGATIONS,	*ANALYTICAL TECHNIQUES, DATA COLL	W72-07225
W, FEEDING VALUE.:	*ANIMAL MANURE, *LITERATURE REVIE	W71-07544
ATER POLLUTION EFFECTS, *PLANTS,	*ANIMALS, *ALGAE, *INVERTEBRATES,	W71-13233
	*ANOXIA.:	W72-03613
S, *SYSTEMATICS, *AQUATIC ALGAE/	*ANTARCTIC, *DISTRIBUTION PATTERN	W73-00854
SINOPHYCEAE, BACCILARIOPHYCEAE,/	*ANTIBIOSIS, OPTICAL DENSITY, PRA	W72-08051
OLVE, *FILTER PAPER DISC DIFFUS/	*ANTIMICROBIAL ACTIVITY, *SKELLYS	W72-08586
TROPHIC POPULATIONS, LAKE HARTW/	*ANTOTROPHIC POPULATIONS, *HETERO	W71-00475
ACOLOGICAL / *THERAPEUTIC VALUE,	*APHANIZOMENON FLOS-AQUAE, *PHARM	W72-04259
RKEN(SWEDEN).: *CYANOPHAGES,	*APHANIZOMENON FLOS-AQUAE, LAKE E	W73-00246
RBON, *AMERICAN FALLS RESERVOIR,	*APHANIZOMENON, ALGAL GROWTH.: /A	W72-13300
VITY, LIGHT, TURBIDIT/ *BIOMASS,	*AQUATIC ALGAE, *AQUATIC PRODUCTI	W72-13142
LCIUM, *GROWTH RATES, *CULTURES,	*AQUATIC ALGAE, CHLOROPHYTA, INHI	W72-12736
NT REQUIREMENTS, *PHYTOPLANKTON,	*AQUATIC ALGAE, CULTURES, PHOSPHA	W72-12739
TRIENTS, *GROWTH RATES, *INHIBI/	*AQUATIC ALGAE, *COMPETITION, *NU	W72-12734
S, *SYSTEMATICS, CLASSIFICATION,	*AQUATIC ALGAE.: /PHYTA, *CULTURE	W72-12172
HORA.: *FILAMENTOUS ALGAE,	*AQUATIC ANGIOSPERM WEEDS, CLADOP	W72-14692
AQUATIC WEED C/ *EUTROPHICATION,	*AQUATIC ALGAE, *ALGAL CONTROL, *	W72-14692
IES, *PLANKTON, *AQUATIC PLANTS,	*AQUATIC ANIMALS, *MODEL STUDIES,	W72-14280
S, *CYANOPHYTA, *NUISANCE ALGAE,	*AQUATIC ALGAE, *CRUSTACEANS, *FI	W72-14056
TUARIES, *DISTRIBUTION PATTERNS,	*AQUATIC ALGAE, STATISTICAL METHO	W72-08141
NUTRIENTS, *DEFICIENT NUTRIENTS,	*AQUATIC ALGAE, *CARBON DIOXIDE,	W72-07514
OWTH CHAMBERS, *LIFE CYCLES, CU/	*AQUATIC ALGAE, *CHLOROPHYTA, *GR	W72-07144
AE, *PHYTOPLANKTON, *ABSORPTION,	*AQUATIC ALGAE, *LIGHT INTENSITY,	W72-07166
NSECTS, *CRUSTACEANS, *MOLLUSKS,	*AQUATIC ANIMALS, *BIOTA, *AQUATI	W72-06526
RIBUTION PATTERNS, *SYSTEMATICS,	*AQUATIC ALGAE, *SOIL BACTERIA, S	W73-00854
STRIBUTION PATT/ *PHYTOPLANKTON,	*AQUATIC ALGAE, *SYSTEMATICS, *DI	W73-01094
*STRESS, *PRIMARY PRODUCTIVITY,	*AQUATIC ALGAE, *COMPARATIVE PROD	W73-01080
WATER TREATMENT, *ALGAL CONTROL,	*AQUATIC ALGAE, *EUTROPHICATION,	W72-03970
ON, *RESERVOIRS, *AQUATIC ALGAE,	*AQUATIC ANIMALS, *BIOMASS, *PRIM	W72-05968
*BIOMA/ *PLANKTON, *RESERVOIRS,	*AQUATIC ALGAE, *AQUATIC ANIMALS,	W72-05968
AQUATIC LIFE, *AQUATIC BACTERIA,	*AQUATIC ANIMALS, *AQUATIC PLANTS	W72-04853
RESERVOIRS, *R/ *AQUATIC PLANTS,	*AQUATIC ALGAE, *PHYTOPLANKTON, *	W72-04855
IOLOGY, *ACARICIDES, BRACKISH W/	*AQUATIC ANIMALS, *AQUATIC MICROB	W71-08026
BSORPTION, TEMPERATURE, STRONTI/	*AQUATIC ALGAE, *RADIOISOTOPES, A	W71-09250
ATION, *ECOLOGICAL DISTRIBUTION,	*AQUATIC ALGAE, *CYANOPHYTA, *BEG	W72-09061
NOPHYTA, *EUTRO/ *PHYTOPLANKTON,	*AQUATIC ALGAE, *PHOSPHORUS, *CYA	W71-10645
NKTON, *DENSITY STRATIFICATION,/	*AQUATIC ALGAE, *LAKES, *PHYTOPLA	W71-09852
FECTS, *PONDS, WATER POLLUTION /	*AQUATIC ALGAE, *RADIOACTIVITY EF	W71-09256

HOTOGRAPHY, *INFRARED RADIATION,	*AQUATIC ALGAE, *SURFACE WATERS,	W71-12034
ON ANALYSIS, *TRACE ELEMENTS, C/	*AQUATIC ALGAE, *NEUTRON ACTIVATI	W71-11515
ALGAE, *STRATIFICATION, *ENZYME/	*AQUATIC BACTERIA, *POPULATION, *	W72-03224
LS, *AQUATIC PLA/ *AQUATIC LIFE,	*AQUATIC BACTERIA, *AQUATIC ANIMA	W72-04853
IA, *OXYGEN, *SEDIMENT-WATER IN/	*AQUATIC BACTERIA, *SULFUR BACTER	W72-06690
*AQUATIC PRODUCTIVITY, *ECOLOGY,	*AQUATIC ENVIRONMENT, *BIOLOGICAL	W73-02095
GY, *RADIOCHEMICAL AN/ *SURVEYS,	*AQUATIC ENVIRONMENT, *RADIOECOLO	W73-01673
*REVIEWS, *HEAVY METALS, *ALGAE,	*AQUATIC ENVIRONMENT, *TOXICITY,	W72-14694
RUNNING WATERS, *AQUATIC PLANTS,	*AQUATIC ENVIRONMENT, WATER POLLU	W72-14690
ER/ *REVIEWS, *BOTTOM SEDIMENTS,	*AQUATIC ENVIRONMENT, SURFACE WAT	W72-00847
Y, SEDIMENTS, NITRIF/ *NITROGEN,	*AQUATIC ENVIRONMENT, PRODUCTIVIT	W72-08459
POWERPLANTS, *THERMAL POLLUTION,	*AQUATIC ENVIRONMENT, FLORIDA, EL	W72-11252
STICIDE RESIDUES, *AQUATIC LIFE,	*AQUATIC ENVIRONMENT, *WATER POLL	W72-13347
OSY/ *ECOLOGY, *PREIMPOUNDMENTS,	*AQUATIC HABITATS, *SAMPLING, *EC	W72-06526
COSYSTEMS, *ALGAE, *CHLORAPHYTA,	*AQUATIC INSECTS, *CHLORAPHYTA, *	W72-06526
*AQUATIC INSECTS, *CHLORAPHYTA,	*AQUATIC INSECTS, *CRUSTACEANS, *	W72-06526
DDIS/ *ALGAE, *PROTOZOA, *FUNGI,	*AQUATIC INSECTS, *DISPERSION, CA	W70-06655
*AQUATIC ANIMALS, *AQUATIC PLA/	*AQUATIC LIFE, *AQUATIC BACTERIA,	W72-04853
EFFECTS, *FALLOUT, *FOOD CHAINS,	*AQUATIC LIFE, RADIOISOTOPES, ABS	W72-04461
NT, *WATER/ *PESTICIDE RESIDUES,	*AQUATIC LIFE, *AQUATIC ENVIRONME	W72-13347
, *WATER POLLUTION, *FRESHWATER,	*AQUATIC LIFE, AQUATIC ANIMALS, C	W72-09649
OLLUTION EFFECTS, *ALGAE, *PROT/	*AQUATIC MICROORGANISMS, *WATER P	W72-08579
ED/ *VITAMIN B, *EUTROPHICATION,	*AQUATIC MICROBIOLOGY, OREGON, *S	W72-08989
NE WATER, *STRIP MINES, *WATER /	*AQUATIC MICROORGANISMS, *ACID MI	W71-07942
RY TESTS, *CHLORELLA, *CULTURES,	*AQUATIC MICROORGANISMS, SPORES,	W71-08027
S, BRACKISH W/ *AQUATIC ANIMALS,	*AQUATIC MICROBIOLOGY, *ACARICIDE	W71-08026
QUALITY, *AQUATIC PRODUCTIVITY,	*AQUATIC MICROORGANISMS, *ALGAE,	W72-06532
UTION, *WATER POLLUTION EFFECTS,	*AQUATIC MICROORGANISMS, *INVESTI	W72-07225
ONMENT, *BIOLOGICAL COMMUNITIES,	*AQUATIC MICROORGANISMS, CYCLING	W73-02095
*INDUSTRIAL WASTES, *PESTICIDES,	*AQUATIC ORGANISMS, *PEST CONTROL	W73-01976
*AQUATIC ENVIRONMENT, *BIOLOGIC/	*AQUATIC PRODUCTIVITY, *ECOLOGY,	W73-02095
ON, *PHYTOPLANKTON, N/ *ECOLOGY,	*AQUATIC PRODUCTIVITY, *ZOOPLANKT	W73-00935
ICITY, COAGULATION, WATER POLLU/	*AQUATIC PLANTS, *MINNESOTA, *TOX	W72-07360
ICROORGANISMS, / *WATER QUALITY,	*AQUATIC PRODUCTIVITY, *AQUATIC M	W72-06532
LUSKS, *AQUATIC ANIMALS, *BIOTA,	*AQUATIC PLANTS, WATER QUALITY, E	W72-06526
MENT, WATER PO/ *RUNNING WATERS,	*AQUATIC PLANTS, *AQUATIC ENVIRON	W72-14690
OCHEMICAL PROPERTIES, *PLANKTON,	*AQUATIC PLANTS, *AQUATIC ANIMALS	W72-14280
*PHYTOPLANKTON, *RESERVOIRS, *R/	*AQUATIC PLANTS, *AQUATIC ALGAE,	W72-04855
ATIC BACTERIA, *AQUATIC ANIMALS,	*AQUATIC PLANTS, *RESERVOIR STAGE	W72-04853
MARINE ALGAE, *ESTUARIES, PRODU/	*AQUATIC PLANTS, *DISTRIBUTION, *	W72-04766
HEALTH, ANALYTICAL TECHNIQUES, /	*AQUATIC PLANTS, *ALGAE, *PUBLIC	W72-04275
TRIBUTION PATTERNS, *PHOSPHATES,	*AQUATIC POPULATIONS, MAPS, MAPPI	W72-03567
AGOONS/ *ENERGY BUDGET, *SEWAGE,	*AQUATIC PRODUCTIVITY, *MIDGES, L	W72-03556
FLUENTS, *ALGAE, *PHYTOPLANKTON,	*AQUATIC PRODUCTIVITY, *PHOSPHATE	W71-02681
ONTROL, WATER POLLUTION EFFECTS,	*AQUATIC PLANTS, HARVESTING, POND	W71-01488
ERIA, FUNGI, PROTOZOA, PATHOGEN/	*AQUATIC PLANTS, *MINNESOTA, BACT	W72-08586
E RESIDUES, *PESTICIDE TOXICITY,	*AQUATIC PLANTS, AQUATIC ALGAE, P	W72-09650
RIENTS, *BIOLOGICAL COMMUNITIES,	*AQUATIC PLANTS, ALGAE, BACTERIA,	W72-03216
NS, / *POLLUTANT IDENTIFICATION,	*AQUATIC PLANTS, *AQUEOUS SOLUTIO	W72-01473
CARBON, *CHLOROPHYLL, *SEAWATER,	*AQUATIC PRODUCTION, *BIOMASS, OR	W71-11562
S, *BENTHIC FLORA, *GROWTH RATE,	*AQUATIC PRODUCTIVITY, RADIOISOTO	W71-11486
, NUTRI/ *EUTROPHICATION, *IONS,	*AQUATIC PRODUCTIVITY, FARM PONDS	W71-11001
BIDIT/ *BIOMASS, *AQUATIC ALGAE,	*AQUATIC PRODUCTIVITY, LIGHT, TUR	W72-13142
*INFRARED PHOTOGRAPHY,	*AQUATIC VEGETATION.:	W71-12034
WATER TREATMENT, *ALGAL CONTROL,	*AQUATIC WEED CONTROL, CALIFORNIA	W72-02975
*ENVIRONMENTAL EF/ *HERBICIDES,	*AQUATIC WEED CONTROL, *PARAQUAT,	W72-00155
LIGHT PENETRATION, *MACROPHYTES,	*AQUATIC WEED CONTROL, *LIGHT INT	W73-02349
GICIDES, *WATER QUALITY CONTROL,	*AQUATIC WEED CONTROL, *DOMESTIC	W71-05083
L CONTROL/ *HARVESTING OF ALGAE,	*AQUATIC WEED CONTROL, *MECHANICA	W71-06188
OWTH, *CULTURES, *ALGIC/ *ALGAE,	*AQUATIC WEEDS, *WEED CONTROL, GR	W71-06102
ARP, ALGAE,/ *FISH, *HERBIVORES,	*AQUATIC WEED CONTROL, TILAPIA, C	W71-04528
*AQUATIC ALGAE, *ALGAL CONTROL,	*AQUATIC WEED CONTROL, BIOCHEMICA	W72-14692
ICATION, *PHYTOPLANKTON, *ALGAE,	*AQUATIC WEEDS, *WEEDS, *RED TIDE	W72-14673
SAY, LIMITING FACTORS, / *ALGAE,	*AQUATIC WEEDS, *NUTRIENTS, BIOAS	W73-00232
IDENTIFICATION, *AQUATIC PLANTS,	*AQUEOUS SOLUTIONS, *BIOLOGICAL P	W72-01473
IMARY PRODUCTIVITY, *OIL WASTES,	*ARCTIC, *ALGAE.: *PR	W72-07922
OGEN FIXING BACTERIA, *RAINFALL,	*ARID LANDS, *ARIZONA, GRASSLANDS	W72-12505
ACTERIA, *RAINFALL, *ARID LANDS,	*ARIZONA, GRASSLANDS, ON-SITE DAT	W72-12505
*INHIBITIONS, *MICRO/ *TOXICITY,	*ARSENIC COMPOUNDS, *PHOSPHATES,	W72-13813
AMPLING, *PERIPHYTON,/ *FOULING,	*ARTIFICIAL SUBSTRATES, DEPTH, *S	W73-00374
FILTRATION, *PIT RECHARGE, *WAT/	*ARTIFICIAL RECHARGE, *INDUCED IN	W71-12410
AKE(NH).:	*ARTIFICIAL CIRCULATION, *KEZAR L	W72-09061
QUES, WASTE WA/ *EUTROPHICATION,	*ASSAY, *ALGAE, ANALYTICAL TECHNI	W72-12966
ACES, LABORATORY EQUIPMENT, *LA/	*ASSAY, *ZINC, *ADSORPTION, *SURF	W72-12210
*HYDROLOGIC DATA, *MEASUREMENT,	*ASSAY, ANALYTICAL TECHNIQUES, EV	W72-10678
*ANALYTICAL TECHNIQUES, *ALGAE,	*ASSAY, CHLORELLA, NITRATES, CHEM	W70-10403
NE VALLEY DAM(PA.), ATRAZINE, B/	*ATHIORHODACEAE, WHIPPLE DAM, STO	W72-04743
MARSHES, SYSTEMATICS, *VIRGINIA,	*ATLANTIC OCEAN, NORTH AMERICA.: /	W72-12633
LARVAE, SAMPLING, INVERTEBRATES,	*ATLANTIC OCEAN, *GULF OF MEXICO,	W72-14313
INDICATORS, *FALLOUT, *MOLLUSKS,	*ATLANTIC COASTAL PLAIN, NORTH CA	W73-00818
ANABAENA.:	*ATMOSPHERIC CARBON, MICROCYSTIS,	W72-10607
NIUM, ZINC, ANTIMONY, NORWAY, A/	*ATOMIC ABSORPTION, ARSENIC, SELE	W71-11515
PHATE.:	*ATP ANALYSIS, *ADENOSINE TRIPHOS	W71-11234
COMPOUNDS, A/ *MARINE BACTERIA,	*ATTRACTANTS, *PREDATION, ORGANIC	W73-00942
M, AXENIC / *EXCRETORY PRODUCTS,	*AUTOANTAGONISM, *HETEROANTAGONIS	W72-12734
LTURES, *DIEOFF RATES, AFTERGRO/	*AUTOGOMISTIC EFFECTS, *AXEMIC CU	W70-08319
SAMPL/ *FLUORESCENT MICROSCOPY,	*AUTORADIOGRAPHY, GLUTARALDEHYDE,	W73-00273
S, SYNCHRON/ *DIVISION PATTERNS,	*AUTOSPORES, *SYNCHRONOUS CULTURE	W72-12724
RATES,/ *ADENOSINE TRIPHOSPHATE,	*AUTOTROPHIC INDEX, MACROINVERTEB	W72-10883
AFTERGRO/ *AUTOGOMISTIC EFFECTS,	*AXEMIC CULTURES, *DIEOFF RATES,	W70-08319

IQUES, STAURASTRUM PA/ *DESMIDS,	*AXENIC CULTURES, CULTURING TECHN	W72-12736
VITY EFFECTS, *FALLOUT, *MARINE/	*BACKGROUND RADIATION, *RADIOACTI	W73-00828
ASTES, *ABSORPTION, *FOOD CHAIN/	*BACKGROUND RADIATION, *NUCLEAR W	W73-00833
*CARBON REGENERATION,	*BACKWASHING, *MT CLEMENS(MICH).:	W72-08357
AE, SEA WATER, CHLOROPHYTA, CHL/	*BACTERIA, *TOXICITY, *MARINE ALG	W72-08051
ICAL PROPERTIES, *ALGA/ *RIVERS,	*BACTERIA, IPLANKTON, PHYSICOCHEM	W72-07896
ALGAE, *SALT MARSHES, *PROTOZOA,	*BACTERIA, FLORIDA, RADIATION SEN	W73-00796
NOPHYTA, HYD/ *LABORATORY TESTS,	*BACTERIA, *ALGAE, *CULTURES, CYA	W73-01657
*DISSOLVED OXYGEN, *LAKE ERIE,	*BACTERIA, *SEDIMENT-WATER INTERF	W72-12996
ESIS, SETTLING, *ALGAE, *SILICA,	*BACTERIA, WATER POLLUTION CONTRO	W71-01899
ESIS, OXIDATION, RESPIR/ *ALGAE,	*BACTERIA, *SYMBIOSIS, PHOTOSYNTH	W72-04983
*NITRIFICATION, AMMONIA, NITRI/	*BACTERIA, *CULTURES, *SEA WATER,	W72-05421
*SEPARATION TECHNIQUES, *ALGAE,	*BACTERIA, *FUNGI, *BACTERIOPHAGE	W72-06124
D SLUDGE, W/ *SYMBIOSIS, *ALGAE,	*BACTERIA, *RESPIRATION, ACTIVATE	W71-04079
R TREATMENT, *NUTRIENTS, *ALGAE,	*BACTERIA, *SYMBIOSIS, ACTIVATED	W72-11100
AGOONS, *ON-SITE INVESTIGATIONS,	*BACTERIA, *ALGAE, OXIDATION, MET	W72-10233
, *NITRATES, *NUTRIENTS, *ALGAE,	*BACTERIA, AESTHETICS, COLOR, TAS	W71-09958
HOGE/ *ALGAE, *ENTERIC BACTERIA,	*BACTERICIDES, GROWTH STAGES, PAT	W71-05155
QUES, *ALGAE, *BACTERIA, *FUNGI,	*BACTERIOPHAGE, *VIRUSES, ABSORPT	W72-06124
HOPH/ *NITZSCHIA ACTINASTROIDES,	*BATCH CULTURES, *CHEMOSTATS, ORT	W73-01445
ANABAENA, COREGONS, LYNGBYA LIM/	*BAVARIA, SCHLIERSEE, TEGERNSEE,	W71-00117
ATES, HEMICHORDATES, ARTHROPODS,	*BAY OF BENGAL, COELENTERATES, CT	W73-00935
URING TECHNIQUES, CULTURE MEDIA,	*BAY OF BENGAL, INDIAN OCEAN, SOI	W73-00430
INPUTS, CHI/ *NUTRIENT BUDGETS,	*BAY OF QUINTE(ONTARIO), NUTRIENT	W71-11009
LUTION CONTROL, *DIVER/ *DELTAS,	*BAYS, *WATER QUALITY, *WATER POL	W71-09581
	*BEAVER CREEK BASIN(KY).:	W71-07942
ON, *AQUATIC ALGAE, *CYANOPHYTA,	*BEGGIATOA, PHYTOPLANKTON, *NEW H	W72-09061
OPHYT/ *ALGAL TOXINS, *RED TIDE,	*BEGGIATOA, *PHYTOTOXICITY, *CYAN	W72-14542
*FISHERIES, WATE/ *HEATED WATER,	*BENEFICIAL USE, *COOLING WATER,	W70-09596
MICR/ *RADIOISOTOPES, *DETRITUS,	*BENTHIC FAUNA, *INVERTEBRATES, *	W70-06668
EFFECTS, ENVIRONMENTAL EFFECTS,	*BENTHIC FAUNA, *INVERTEBRATES, *	W71-08793
PHYTOPLANKTON, ALGA/ *NUTRIENTS,	*BENTHIC FAUNA, *BENTHIC FLORA, *	W72-06046
LGA/ *NUTRIENTS, *BENTHIC FAUNA,	*BENTHIC FLORA, *PHYTOPLANKTON, A	W72-06046
*PRIMARY PRODUCTIVITY, *SHOALS,	*BENTHIC FLORA, SEA WATER, SEDIME	W72-14317
BIOLOGICAL COMMUNITIES,/ *ALGAE,	*BENTHIC FLORA, *OREGON, RIVERS,	W72-07901
VITY, *ENERGY CO/ *MARINE ALGAE,	*BENTHIC FLORA, *PRIMARY PRODUCTI	W72-06283
LGAE, LA/ *PRIMARY PRODUCTIVITY,	*BENTHIC FLORA, MICROORGANISMS, A	W72-07501
ULF OF MEXICO, *ELECTRIC POWERP/	*BENTHIC FLORA, *MARINE ALGAE, *G	W72-11252
UATI/ *RADIOACTIVITY TECHNIQUES,	*BENTHIC FLORA, *GROWTH RATE, *AQ	W71-11486
CONDITIONS, *LAKE ERIE, *ALGAE,	*BENTHIC FAUNA, *BOTTOM SEDIMENTS	W72-12995
ON, *PHOTOPERIODISM, PERIPHYTON,	*BENTHOS.: /*LIGHT, SOLAR RADIATI	W72-07901
AE, *INVERTEBRATES, ST. LAWRENC/	*BENTHOS, *SAMPLING, *SESSILE ALG	W72-05432
IE, PLANKTON, BENTHIC FLORA, DI/	*BIBLIOGRAPHIES, *ALGAE, *LAKE ER	W73-01615
MS, *PEST CONTROL, *CHEMCONTROL,	*BIBLIOGRAPHIES, *DATA COLLECTION	W73-01976
ACIFIC NORTHWEST, *COAST REVIEW,	*BIBLIOGRAPHIES, WATER POLLUTION	W72-12190
OLOGY, CULTURES, M/ *SALMONELLA,	*BIBLIOGRAPHIES, BIOLOGY, EPIDEMI	W72-13800
NIQUES, *WATER POLLUTION CONTRO/	*BIBLIOGRAPHIES, *ANALYTICAL TECH	W72-08790
R, SHOSHONE RIVER, YELLOWTAIL D/	*BIGHORN LAKE(MONT), BIGHORN RIVE	W73-01435
ESTICIDES, / *TOXICITY, REVIEWS,	*BIOASSAY, *INDUSTRIAL WASTES, *P	W73-01976
*LAKES, *NUTRIENTS, *FRESHWATER,	*BIOASSAY, *STANDING CROP, CARBON	W72-07648
HAINS, *ZOOPLANKTON, *PREDATION,	*BIOASSAY, BIOLOGICAL COMMUNITIES	W72-07132
TH, *CHLOROPHYTA, *MARINE ALGAE,	*BIOASSAY, REPRODUCTION, SEA WATE	W72-14294
ALUATION, WATER QUALITY CONTROL,	*BIOASSAY, *BIOMASS, CHLOROPHYLL,	W72-10883
TECHNIQUES, *TESTING PROCEDURES,	*BIOASSAY, *ALGAE, FRESHWATER, ME	W72-10619
TECHNIQUES, *NUTRIENTS, *ALGAE,	*BIOASSAY, PHOSPHORUS COMPOUNDS,	W72-10608
, *NUTRIENTS, *LABORATORY TESTS,	*BIOASSAY, OREGON, LAKES, NITROGE	W72-09164
MICALS, *SURFACTANTS, *TOXICITY,	*BIOASSAY, FISH, SHELLFISH, ALGAE	W71-09789
LLUT/ *POLLUTANT IDENTIFICATION,	*BIOASSAY, *INDICATORS, *WATER PO	W71-10553
MEASUREMENT, / *TROPHIC LEVELS,	*BIOASSAY, PHOSPHORUS, NUTRIENTS,	W71-13794
LUTANTS, *TOXICITY, WAT/ *ALGAE,	*BIOASSAY, *PLANT PHYSIOLOGY, POL	W72-12576
E FISH, *PHYTOPLANKTON, *SHRIMP,	*BIOASSAY, *ABSORPTION, PESTICIDE	W73-02105
N CRITERIA, MATHEMATICAL MODELS,	*BIOASSAY, *WASTE DILUTION, LAGOO	W72-04787
EARCH AND DEVELOPMENT, TRACE EL/	*BIOASSAY, *LABORATORY TESTS, RES	W72-04433
TRIENTS, *NUTRIENT REQUIREMENTS/	*BIOASSAY, CULTURES, *BIOMASS, NU	W72-03985
D HYD/ *WATER POLLUTION EFFECTS,	*BIOASSAY, *TOXICITY, *CHLORINATE	W71-07412
ITY, ALGAE,: *CHLORELLA,	*BIOASSAYS, ENZYMES, LIGHT INTENS	W71-11916
H, / *COOLING TOWERS, *TOXICITY,	*BIOASSAYS, LABORATORY TESTS, FIS	W72-08683
BIC CONDITIONS, ORGANIC LOADING,	*BIOCHEMICAL OXYGEN DEMAND, NITRO	W71-13326
TRATION, CHEMICAL OXYGEN DEMAND,	*BIOCHEMICAL OXYGEN DEMAND, MIXIN	W71-07102
SEEPAGE, SLUDGE, COST ANALYSIS,	*BIOCHEMICAL OXYGEN DEMAND, WASTE	W71-07099
OPHICATION, PHOSPHOROUS/ *ALGAE,	*BIOCHEMICAL OXYGEN DEMAND, *EUTR	W72-04431
SSOLVED OXYGEN, *EUTROPHICATION,	*BIOCHEMICAL OXYGEN DEMAND, *CHEM	W70-09189
*GROWTH STAGES, *BIODEGRADATION,	*BIOCHEMICAL OXYGEN DEMAND, CHEMI	W72-06612
ALGAE, PHYLOGENY, CARBON CYCLE,/	*BIOCHEMISTRY, *PHOTOSYNTHESIS, *	W71-09172
LLA, ENZYMES, PROTEINS, LABORAT/	*BIOCHEMISTRY, *NITRATES, *CHLORE	W72-03222
, *ALGAL CONTROL, FRESH WATER, /	*BIOCONTROL, *PATHOGENIC BACTERIA	W72-05457
TION ABATEMENT, ALGAE, NUTRIENT,	*BIODEGRADATION, WATER PURIFICATI	W72-05315
FECTS, *WATER POLLUTION CONTROL,	*BIODEGRADATION, FISHKILL, EUTROP	W71-08768
RGANIC MATTER, AERATION, *ALGAE,	*BIODEGRADATION, LABORATORY TESTS	W72-01881
OBIC CONDITIONS, *GROWTH STAGES,	*BIODEGRADATION, *BIOCHEMICAL OXY	W72-06612
T, *RADIOACTIVITY TECHNIQUES, P/	*BIODEGRADATION, *METABOLISM, *DD	W72-07703
ASS, THYMIDINES, CELLS(BIOLOGY),	*BIODETERIORATION, DEMOCOLCINE.: /	W71-09261
EFFECTS, *ALGAE, *MARINE ALGAE,	*BIOINDICATORS, *PHOTOSYNTHESIS,	W71-10553
TES,/ *POLLUTANT IDENTIFICATION,	*BIOINDICATORS, *ALGAE, GROWTH RA	W71-09795
LGAE, *GREAT LAKES, *MONITORING,	*BIOINDICATORS, THERMAL POLLUTION	W72-10875
ISOTOPES, *COBALT RADIOISOTOPES,	*BIOINDICATORS, ANALYTICAL TECHNI	W72-10958
*POLLUTANT IDENTIFICA/ *DIATOMS,	*BIOINDICATORS, *EUTROPHICATION,	W73-00432
ICAL PROPERTIES, *WATER QUALITY,	*BIOINDICATORS, *BIOLOGICAL COMMU	W73-00002
IBUTION, *CURRENTS(WATER), *DIN/	*BIOINDICATORS, *ECOLOGICAL DISTR	W73-00296
POLLUTANT IDENTI/ *EUGLENOPHYTA,	*BIOINDICATORS, *PHYTOPLANKTON, *	W73-00918

ASIA/ *REVIEWS, *NUCLEAR WASTES,	*BIOINDICATORS, *MARINE ANIMALS,	W73-00811
KS, *ATLANTIC COASTAL PLAIN, NO/	*BIOINDICATORS, *FALLOUT, *MOLLUS	W73-00818
CITY, *CHLORINATED HYDROCARBONS,	*BIOINDICATORS, INDUSTRIAL WASTES	W71-07412
UTION PATTERNS, *WATER ANALYSIS,	*BIOINDICATORS, ALGAE, RIVERS, ST	W72-04736
RESHWATER,/ *DIATOMS, *SALINITY,	*BIOINDICATORS, BRACKISH WATER, F	W73-02548
CTRON CAPTURE GAS CHROMATOGRAPHY	*BIOLOGICAL MAGNIFICATION, GAS LI	W73-02105
ON EFFECTS, ECOLOGY, ECOSYSTEMS,	*BIOLOGICAL COMMUNITIES, *SUCCESS	W72-11854
EN CONCENTRA/ *DELAWARE ESTUARY,	*BIOLOGICAL LIFE, *DISSOLVED OXYG	W70-09189
ISOTOPES, BLOOMS, 2,4 DINITRO P/	*BIOLOGICAL UPTAKE, CALCIUM RADIO	W71-01474
ERVATION/ *BIOLOGICAL MATERIALS,	*BIOLOGICAL SAMPLES, *SAMPLE PRES	W73-01676
CS-137, *PAHRANAGAT LAKE, GAMM/	*BIOLOGICAL SAMPLES, SR-90, K-40,	W73-01673
*ECOLOGY, *AQUATIC ENVIRONMENT,	*BIOLOGICAL COMMUNITIES, *AQUATIC	W73-02095
L SAMPLES, *SAMPLE PRESERVATION/	*BIOLOGICAL MATERIALS, *BIOLOGICA	W73-01676
MS, *RIVERS, *VIRGINIA, *WATER /	*BIOLOGICAL PROPERTIES, *ECOSYSTE	W73-01972
YTYLOGICAL STUDIES, PLANT MORPH/	*BIOLOGICAL MEMBRANES, *ALGAE, *C	W73-01630
, LAKE ONTARIO, PLANKTO/ *ALGAE,	*BIOLOGICAL COMMUNITIES, *STREAMS	W73-01627
ATIVE PRODUCTIVITY, *ECOSYSTEMS,	*BIOLOGICAL COMMUNITIES, BIOMASS,	W73-01080
*WATER QUALITY, *BIOINDICATORS,	*BIOLOGICAL COMMUNITIES, AQUATIC	W73-00002
ALITY, *BIOIND/ *EUTROPHICATION,	*BIOLOGICAL PROPERTIES, *WATER QU	W73-00002
MENTA/ *SALT MARSHES, *PROTOZOA,	*BIOLOGICAL COMMUNITIES, *ENVIRON	W73-00367
BACTER/ *SEWAGE, *FUNGI, *SLIME,	*BIOLOGICAL COMMUNITIES, RIVERS,	W72-08573
TION EFFECTS, *ALGAE, *PROTOZOA,	*BIOLOGICAL COMMUNITIES, ECOLOGIC	W72-08579
POWER PLAN/ *THERMAL POLLUTION,	*BIOLOGICAL COMMUNITIES, *THERMAL	W71-11517
OPES, HUMAN BRAIN, LO/ *MERCURY,	*BIOLOGICAL TISSUES, MERCURY ISOT	W71-11036
US, *WASTE WATER TREATMENT, NUT/	*BIOLOGICAL TREATMENT, *PHOSPHORO	W71-12188
SLUDGE/ *WASTE WATER TREATMENT,	*BIOLOGICAL TREATMENT, *ACTIVATED	W71-12223
ATIC PLANTS, *AQUEOUS SOLUTIONS,	*BIOLOGICAL PROPERTIES, MOLECULAR	W72-01473
TY, *EUTROPHICATION, *NUTRIENTS,	*BIOLOGICAL COMMUNITIES, *AQUATIC	W72-03216
EDUCTION, POLYSACCHARIDES, DRAG/	*BIOLOGICAL POLYMERS, *FRICTION R	W72-03220
WASTES, WATER POLLUTION CONTROL,	*BIOLOGICAL TREATMENT, *NITRATES,	W72-02975
*EUTROPHICATION, *NUT/ KINETICS,	*BIOLOGICAL TREATMENT, PLANKTON,	W71-13553
EM, MODULAR SYSTEM, COBALT-60.:	*BIOLOGICAL--GAMMA-RADIATION SYST	W72-00383
ATER, WATER ANALYSIS, ALGAE, CH/	*BIOMASS, *ORGANIC MATTER, *SEA W	W71-11234
*SEAWATER, *AQUATIC PRODUCTION,	*BIOMASS, ORGANIC MATTER, ATLANTI	W71-11562
ATER QUALITY CONTROL, *BIOASSAY,	*BIOMASS, CHLOROPHYLL, NITROGEN F	W72-10883
TON, PRODUCTIVITY, PHOT/ *ALGAE,	*BIOMASS, *CYANOPHYTA, PHYTOPLANK	W72-14812
Y PRODUCTIVITY, *EUTROPHICATION,	*BIOMASS, *HYDROGRAPHY, ALGAE, AQ	W72-13871
*HYDROGRAPHY, *PACIF/ *PLANKTON,	*BIOMASS, *SAMPLING, *NUTRIENTS,	W72-07892
QUIREMENTS/ *BIOASSAY, CULTURES,	*BIOMASS, NUTRIENTS, *NUTRIENT RE	W72-03985
AQUATIC ALGAE, *AQUATIC ANIMALS,	*BIOMASS, *PRIMARY PRODUCTIVITY,	W72-05968
TION, PERFORMANCE, EFFICIENCIES,	*BIOMASS, ORGANIC MATTER, NUTRIEN	W72-12289
C PRODUCTIVITY, LIGHT, TURBIDIT/	*BIOMASS, *AQUATIC ALGAE, *AQUATI	W72-13142
LA, *LIGHT, ALGAE, GROWTH RATES/	*BIORTHYMS, *METABOLISM, *CHLOREL	W73-00251
EA MUSCIFORMIS, GRACILARIA VERU/	*BIOSYNTHESIS, *GIBBERELLIN, HYPN	W72-07496
NS, *MOLLUSKS, *AQUATIC ANIMALS,	*BIOTA, *AQUATIC PLANTS, WATER QU	W72-06526
AKES, *THERMAL POLLUTION, WATER/	*BIOTA, *TEMPERATURE, *CANADA, *L	W73-01704
AR POWER PLANTS, *POTABLE WATER,	*BIOTA, SAMPLING, MARINE FISH, MA	W71-09018
AQUATIC BIOTOXINS,	*BIOTOXINS.:	W72-14056
MPLES, FATE OF POLLUTANTS, META/	*BIOTRANSFORMATION, BIOLOGICAL SA	W73-00361
FLUENT, TURKEY POINT(FLORIDA), /	*BISCAYNE BAY(FLORIDA), HEATED EF	W72-11876
ONEMA BORYANIUM, SU/ *LPP VIRUS,	*BLUE-GREEN ALGAL VIRUSES, *PLECT	W72-04736
*SUSPENDED SEDIMENT PRODUCTION,	*BLUE-GREEN ALGAE.:	W72-02218
N.: DEPTH, SLOPE,	*BOD LOADING, *WASTE STABILIZATIO	W71-03896
ODOLOGY, RIVERS, WA/ FRESHWATER,	*BORON, *PATH OF POLLUTANTS, METH	W73-00286
ICAL STUDIE/ *ALGAE, *CHLORELLA,	*BORON, LABORATORY TESTS, CYTOLOG	W71-12858
NE ALGAE, CALCIUM RADIOISOTOPES,	*BOSSIELLA ORBIGNIANA, PAPER CHRO	W72-12634
:	*BOSTON(MASS), LYNN HARBOR(MASS).	W71-05553
DARK FIXATION, COASTA/ *FINLAND,	*BOTHNIAN BAY, *GULF OF FINLAND,	W72-03228
ONMENT, SURFACE WATER/ *REVIEWS,	*BOTTOM SEDIMENTS, *AQUATIC ENVIR	W72-00847
LGAE, *EUTROPHICATI/ *NUTRIENTS,	*BOTTOM SEDIMENTS, *LIMNOLOGY, *A	W70-08944
KE ERIE, *ALGAE, *BENTHIC FAUNA,	*BOTTOM SEDIMENTS, *HYPOLIMNION,	W72-12995
ADUS MARINUS, TEMPERATURE EFFEC/	*BP 1002, *GREEN ALGAE, PRASINOCL	W70-09429
METALI/ *USSR, *IRKUTSK OBLAST,	*BRATSK RESERVOIR, *HYDROBIOLOGY,	W72-04855
OKA RIV/ *USSR, *IRKUTSK OBLAST,	*BRATSK RESERVOIR, ANGARA RIVER,	W72-04853
NE RADIOISO/ *SEDIMENTARY ROCKS,	*BRINES, *ALGAE, *OKLAHOMA, *IODI	W72-02073
*ELECTRA CRUSTULENTA,	*BRYOZOA.:	W73-00348
*ALGAE, *WASTE WATER TREATMENT,	*BYPRODUCTS, MIXING, SOLAR RADIAT	W72-13508
ITELLA AMERICANA, TERRES/ *C-13,	*C-12, BIOLOGICAL SYSTEMS, BELEMN	W72-12709
BELEMNITELLA AMERICANA, TERRES/	*C-13, *C-12, BIOLOGICAL SYSTEMS,	W72-12709
LLATION, SAMPLE PREPARATION, ZO/	*C-14 PRODUCTIVITY, LIQUID SCINTI	W72-14317
AL, CHARACEAE.:	*CABLE THEORY, ELECTRONIC POTENTI	W73-01630
ICUS, COSCINODISCUS, COPEPODIDS,	*CALANUS GLACIALIS, DITYLUM, PROR	W72-09248
SCUS, COPEPO/ *FEEDING PATTERNS,	*CALANUS HELGOLANDICUS, COSCINODI	W72-09248
CALCIUM RADIOISOTOPES, *BOSSIE/	*CALCIFICATION, *CORALLINE ALGAE,	W72-12634
S, *AQUATIC ALGAE, CHLOROPHYTA,/	*CALCIUM, *GROWTH RATES, *CULTURE	W72-12736
HOTOMETRY, *POLLUTANT IDENTIFIC/	*CALCIUM, *FOOD CHAINS, *SPECTROP	W71-11529
WATER, *WATER QUALITY, *ECOLOGY,	*CALIFORNIA, HYDROLOGIC DATA, CHE	W71-10577
Y PRODUCTIVITY, WATER CHEMISTRY,	*CALIFORNIA, *NEVADA.: /N, PRIMAR	W72-12549
SIDUES, WATER POLLUTION EFFECTS,	*CALIFORNIA, SEA WATER, WATER ANA	W73-02105
TS, WASTE ASSIMILATIVE CAPACITY,	*CALIFORNIA, RADIOECOLOGY, MARINE	W73-00826
R UTILIZATION, *IMPAIRED WATER /	*CALIFORNIA, *WATER RIGHTS, *WATE	W70-07100
*RIFFLES,	*CALORIC CONTENT.:	W71-07097
ON, WATER/ *BIOTA, *TEMPERATURE,	*CANADA, *LAKES, *THERMAL POLLUTI	W73-01704
SPP, *CRYPTOMONAS-SPP.:	*CANADIAN ARCTIC, *CHLAMYDOMONAS-	W72-07922
, WYAND/ *CARBONACEOUS MATERIAL,	*CANADIAN PHOSPHATE DETERGENT BAN	W70-07283
AY, CAPE COD BAY, DATA INTERPRE/	*CAPE COD CANAL(MASS), BUZZARDS B	W73-01449
ANTIMONY, CESIUM,	*CAPE OF GOOD HOPE.:	W73-00832
ORINATED HYDROCARBON PESTICIDES,	*CARBAMATE PESTICIDES, *ORGANOPHO	W71-10566
*DIURON,	*CARBARYL, *TOXAPHENE.:	W71-10566
*PLANT PHYSIOLOGY, *GRO/ *ALGAE,	*CARBOHYDRATES, *ORGANICS ACIDS,	W72-07710

RINE/ *SEA WATER, *MARINE ALGAE,	*CARBOHYDRATES, MARINE PLANTS, MA	W73-02100
*DOMESTIC WASTES, *SALT MARSHES,	*CARBON CYCLE, *PHOSPHORUS COMPOU	W72-12567
CULTURE, *WASTE-WATER TREATMENT,	*CARBON CIOXIDE, *GROWTH RATES, *	W72-06612
CIENT NUTRIENTS, *AQUATIC ALGAE,	*CARBON DIOXIDE, NUISANCE ALGAE,	W72-07514
OIL MICROORGANISMS, *ABSORPTICN,	*CARBON DIOXIDE, *RADIOACTIVITY T	W72-14301
IVITY, ALASKA, FIORDS, SURFACE /	*CARBON DIOXIDE, *PRIMARY PRODUCT	W72-09122
CYANOPHYTA, CA/ *NUISANCE ALGAE,	*CARBON DIOXIDE, *ALGAL CONTROL,	W72-10607
OTOPE STUDIES, *STABLE ISOTOPES,	*CARBON RADIOISOTCPES, ENVIRONMEN	W72-12709
L TEC/ *OCEANS, *ORGANIC MATTER,	*CARBON RADIOISOTOPES, *ANALYTICA	W71-08042
G, *MT CLEMENS(MICH).:	*CARBON REGENERATION, *BACKWASHIN	W72-08357
GAE, PH/ *ANALYTICAL TECHNIQUES,	*CARBON, *AMINO ACIDS, *MARINE AL	W72-07497
GROWTH, SEWAGE EFFLUENTS, REGU/	*CARBON, *NITROGEN, *ALGAE, PLANT	W72-14790
XAS, / *ALGAE, *STABLE ISOTOPES,	*CARBON, *ANAEROBIC BACTERIA, *TE	W71-04873
, AQUATIC EN/ CYCLING NUTRIENTS,	*CARBON, *PHOSPHORUS, *CYANOPHYTA	W71-00475
, WATER POLLUT/ *EUTROPHICATICN,	*CARBON, *WATER POLLUTION CONTROL	W72-04769
NG FACTORS, ALGAE, PRODUCTIVITY/	*CARBON, *SEWAGE LAGOONS, *LIMITI	W72-09162
, AQUATIC ENVIRONMENT, ECOSYSTE/	*CARBON, *PHOSPHORUS, *POPULATION	W72-09159
LGAE.: *EUTROPHICATICN,	*CARBON, *PHOSPHORUS, NITROGEN, A	W72-09155
*GROWTH RATES, CHLORELLA, CARB/	*CARBON, *ALGAE, *EUTROPHICATION,	W72-09168
*AQUATIC PRODUCTION, *BIOMASS,/	*CARBON, *CHLOROPHYLL, *SEAWATER,	W71-11562
CARBON DIOXID/ *EUTROPHICATION,	*CARBCN, *PHOTOSYNTHESIS, *ALGAE,	W71-11519
*LIMITING FACTORS, PHOTOSYNTHE/	*CARBON, *ALGAE, *EUTROPHICATION,	W71-12068
I/ *EUTROPHICATION, *PHOSPHORUS,	*CARBON, BACTERIA, ALGAE, SYMBIOS	W70-07283
TECHNIQUES, *ORGANIC COMPOUNDS,	*CARBON, GRAVIMETRIC ANALYSIS, PR	W72-09108
PHOSPHATE DETERGENT BAN, WYAND/	*CARBONACEOUS MATERIAL, *CANADIAN	W70-07283
UNOFF, *WATER POLLUTION EFFECTS,	*CATTLE, *FEED LOTS.: /CULTURAL R	W72-08401
, MERI/ *CHLOROCHOLINE CHLORIDE,	*CCC, *DIETHYLAMINE HYDROCHLORIDE	W72-14706
WASTE HEAT, *USES OF WASTE HEAT,	*CENTRAL HEATING AND COOLING, *SP	W70-09193
NGLAND), RIVER KENNET(ENGLAND),/	*CENTRIC DIATOMS, *RIVER THAMES(E	W72-14797
, FLAGELLATES, MEDITERRANEAN SE/	*CERATIUM EGYPTIACUM, *SUEZ CANAL	W73-00296
	*CERIUM RADIOISOTOPES.:	W72-12203
*USSR, *HYDROBIONTS,	*CERIUM, RADIOMETRY.:	W72-09649
DIOISOTO/ *IODINE RADIOISOTOPES,	*CESIUM RADIOISOTOPES, *COBALT RA	W72-10958
	*CESIUM RADIOISOTOPES.:	W71-09256
*HUMBOLDT BAY,	*CESIUM RADIOISOTCPES.:	W73-00826
RADIOISOTOPES, ZIRCONIUM RADIO/	*CESIUM RADIOISOTOPES, *RUTHENIUM	W71-09225
ISOTOPES, MANGANESE RADIOISOTOP/	*CESIUM RADIOISOTOPES, IRON RADIO	W71-09250
ICS, DOMEST/ *OXIDATION LAGOONS,	*CHANNELS, *HEAT TRANSFER, *PLAST	W72-10401
THERMAL STRATIFICATION, DIATOMS,	*CHARA.: /L, *PONDWEEDS, PONDS, *	W73-02349
CHIDLEY(LABRADOR), NEW BRUNSWI/	*CHECKLIST, *EASTERN CANADA, CAPE	W71-10066
ORGANIC WASTES, *EUTROPHICATION,	*CHELATION, *SEWAGE, *ORGANIC COM	W71-09674
Y METALS, EU/ *COPPER, CULTURES,	*CHELATION, *ION TRANSPORT, *HEAV	W73-02112
METALS, *TRACE ELEMENTS, RADIO/	*CHELATION, CHROMATOGRAPHY, HEAVY	W72-04708
ANIC CHELATORS.:	*CHELATORS, ANODIC STRIPPING, ORG	W71-09674
SULFATE, *ALGAE, *ZOOPLANKTERS,	*CHEMCONTROL, NEW HAMPSHIRE, COPE	W71-03014
QUATIC ORGANISMS, *PEST CONTROL,	*CHEMCONTROL, *BIBLIOGRAPHIES, *D	W73-01976
POLLUTION SOURCES, ION EXCHANGE,	*CHEMICAL ANALYSIS, WATER POLLUTI	W73-00286
CT/ *LAKE ERIE, *EUTROPHICATION,	*CHEMICAL WASTES, *POLITICAL ASPE	W72-14782
*CYCLING NUTRIENTS, *FOOD CHAIN/	*CHEMICAL ANALYSIS, *MONITORING,	W72-07715
ION, *BIOCHEMICAL OXYGEN DEMAND,	*CHEMICAL OXYGEN DEMAND, *WATER P	W70-09189
REATMENT, SEPARATION TECHNIQUES,	*CHEMICAL WASTES, INDUSTRIAL WAST	W72-05315
OPHIC LEVEL, *LAKES, *NUTRIENTS,	*CHEMICAL PROPERTIES, *SEDIMENTS,	W72-05469
TY, *CHEMIC/ *THERMAL POLLUTION,	*CHEMICAL REACTIONS, *WATER QUALI	W71-12092
MICAL REACTIONS, *WATER QUALITY,	*CHEMICAL PROPERTIES, EUTROPHICAT	W71-12092
LES.:	*CHEMICAL EVOLUTION, *MACROMOLECU	W72-01473
S, *NEUTRON ACTIVATION ANALYSIS,	*CHEMICAL ANALYSIS, GREAT LAKES,	W72-01101
SEASONAL, ION/ *ECOLOGY, *PONDS,	*CHEMICAL ANALYSIS, TEMPERATURE,	W72-01363
MARINE ALGAE, *LABORATORY TESTS,	*CHEMICAL WASTE, *GROWTH RATES, *	W72-12239
ORGANIC MATTER, *MARINE ALGAE, /	*CHEMICAL ANALYSIS, *DECOMPOSING	W72-12166
IXED LIQUOR, SUSPENDED SOLIDS, /	*CHEMICAL-BIOLOGICAL TREATMENT, M	W71-13334
WASTES, *CLEANING, *DISPERSION,	*CHEMICALS, *SURFACTANTS, *TOXICI	W71-09789
EY, PYTHIUM DEBARYANUM, SKELETO/	*CHEMORECEPTION, *CHEMOTAXIS, *PR	W73-00942
ROCEDURES, SELENASTRUM CAPRICOR/	*CHEMOSTAL ASSAYS, *ALGAL ASSAY P	W72-03985
MAXIMUM SPECIFIC GROWTH RATE, /	*CHEMOSTATIC CONTINUOUS CULTURES,	W72-04788
ACTINASTROIDES, *BATCH CULTURES,	*CHEMOSTATS, ORTHOPHOSPHATES, SUB	W73-01445
*ALGAE GROWTH,	*CHEMOSTATS.:	W72-06612
YANUM, SKELETO/ *CHEMORECEPTION,	*CHEMOTAXIS, *PREY, PYTHIUM DEBAR	W73-00942
	*CHENEY RESERVOIR(KANSAS).:	W72-02274
CRASSIPES, SALVINIA ROTUNDIFOL/	*CHIRONOMIDS, SAWDUST, EICHHORNIA	W71-01488
ES, PESTICIDES, CHLORELLA, RESP/	*CHLAMYDOMONAS, *ALGAE, *HERBICID	W70-07255
SPP.: *CANADIAN ARCTIC,	*CHLAMYDOMONAS-SPP, *CRYPTOMONAS-	W72-07922
RE MEDIA, CHLORINATED HYDROCARB/	*CHLAMYDOMONAS REINHARDTII, CULTU	W73-00281
LGAE, CHLOROPHYTA, HEAVY METALS,	*CHLAMYDOMONAS, INHIBITORS, *LETH	W72-11727
HYTA, PROTOZOA, */ *SYSTEMATICS,	*CHLAMYDOMONAS, CULTURES, CHLOROP	W72-12632
., TUBI/ *CHLORELLA PYRENOIDOSA,	*CHLORAMPHENICAL, LIMNODRILUS SPP	W72-09668
*SAMPLING, *ECOSYSTEMS, *ALGAE,	*CHLORAPHYTA, *AQUATIC INSECTS, *	W72-06526
*CHLORAPHYTA, *AQUATIC INSECTS,	*CHLORAPHYTA, *AQUATIC INSECTS, *	W72-06526
RATES/ *BIORTHYMS, *METABOLISM,	*CHLORELLA, *LIGHT, ALGAE, GROWTH	W73-00251
	*CHLORELLA PYRENOIDOSA.:	W73-00251
YGEN, LI/ *ALGAE, *FLUORESCENCE,	*CHLORELLA, *PHOTOACTIVATION, *OX	W72-14800
, *PHOSPHATES, *ALGAL NUTRIENTS,	*CHLORELLA, ALGAE, *EUTROPHICATIO	W72-13992
RASSA, SCINTILLATION COUN/ C-14,	*CHLORELLA VULGARIS, NEUROSPORA C	W72-14326
*LABORATORY TESTS, *RESPIRATION,	*CHLORELLA, *NUTRIENTS, CULTURES,	W73-01626
EDIA.:	*CHLORELLA ELLIPSOIDEA, CULTURE M	W73-01626
HENICAL, LIMNODRILUS SPP., TUBI/	*CHLORELLA PYRENOIDOSA, *CHLORAMP	W72-09668
F/ *ALGAL CONTROL, *RESPIRATION,	*CHLORELLA, *INHIBITION, COPPER,	W72-11572
MICROORGANISMS, ENVIRONMENTAL E/	*CHLORELLA, *SCENEDESMUS, ALGAE,	W72-10861
BORAT/ *BIOCHEMISTRY, *NITRATES,	*CHLORELLA, ENZYMES, PROTEINS, LA	W72-03222
STS, CYTOLOGICAL STUDIE/ *ALGAE,	*CHLORELLA, *BORON, LABORATORY TE	W71-12858
LIGHT INTENSITY, ALGAE.:	*CHLORELLA, *BIOASSAYS, ENZYMES,	W71-11916

M NITRILOTRIACETATE.: *CHLORELLA PYRENOIDOSA, *TRISODIU W71-11972
MI/ *ALGAL TOXINS, *INHIBITION, *CHLORELLA, *SCENEDESMUS, AQUATIC W72-12255
MMA RADIATION, SURVIVAL, CHLORE/ *CHLORELLA SOROKINIANA, CO-60, GA W72-11836
, *MERCURY, *TOXICITY, CULTURES, *CHLORELLA, AQUATIC ALGAE, CHLORO W72-11727
NE SULFONATE, POLYETHYLENEIMINE, *CHLORELLA ELLIPSOIDIA.: /LYSTYRE W72-11833
TUS VAR ALTERNANS, *NITZSCHIA A/ *CHLORELLA FUSCA, *SCENEDEMUS ACU W72-12739
IDENTIFICATION, TOXICITY, *IRON, *CHLORELLA, NUTRIENT REQUIREMENT, W73-02112
/ *TRISODIUM NITRILOTRIACETATE, *CHLORELLA PYRENOIDOSA, SOURCE OF W70-08976
IC OXYGEN, *ALGAE, CULTURES, VO/ *CHLORELLA, *LIGHT, *PHOTOSYNTHET W70-05547
ODUCTION, LIGHT INTENSITY, CULT/ *CHLORELLA PYRENOIDOSA, OXYGEN PR W70-05547
, EUTROPHICATION, ALGAL CONTROL, *CHLORELLA, *WASTE WATER TREATMEN W72-04788
P P'-DDE, CLOPHEN A 50, ELECTRO/ *CHLORELLA PYRENOIDOSA, LINDANE, W72-04134
ICROORGANISM/ *LABORATORY TESTS, *CHLORELLA, *CULTURES, *AQUATIC M W71-08027
ITION, *PLANT GROWTH REGULATORS, *CHLORIDES, *CYANOPHYTA, *CHLOROP W72-14706
ALGAE, *AERATION, SEDIMENTATION, *CHLORINATION, OXIDATION, POLLUTI W72-14469
ES, SEA WATER, *METABOLISM, BIO/ *CHLORINATED HYDROCARBON PESTICID W73-00361
N EFFECTS, *BIOASSAY, *TOXICITY, *CHLORINATED HYDROCARBONS, *BIOIN W71-07412
ES, *ALDRIN, *DIELDRIN, *ENDRIN/ *CHLORINATED HYDROCARBON PESTICID W72-03545
ES, *PESTICIDE RESIDUES, *ALGAE/ *CHLORINATED HYDROCARBON PESTICID W72-04134
PA, CHLAMYDOMONAS GLOBUSA, CHLO/ *CHLORINATED HERBICIDES, MCPB, MC W70-07255
VERS, WATER P/ *WATER TREATMENT, *CHLORINATION, *ALGAL CONTROL, RI W71-00099
ANS, *MATABOLISM, *MARINE ALGAE, *CHLORINATED HYDROCARBON PESTICID W72-12998
ES, *DDT, *PESTICIDE TOXICITY, / *CHLORINATED HYDROCARBON PESTICID W71-12303
ES, *CARBAMATE PES/ *PESTICIDES, *CHLORINATED HYDROCARBON PESTICID W71-10566
TROL, *ODOR, AL/ *WATER QUALITY, *CHLORINATION, *WATER QUALITY CON W71-13187
IDATION LAGOONS, COLIFORMS, ALG/ *CHLORINATION, *DISINFECTION, *OX W72-00422
HESIS, SOLUBILITY, CHEMICAL OXY/ *CHLORINATION, *ALGAE, *PHOTOSYNT W72-00411
S, / *POLYCHLORINATED BIPHENYLS, *CHLORNATED HYDROCARBON PESTICIDE W72-08436
IETHYLAMINE HYDROCHLORIDE, MERI/ *CHLOROCHOLINE CHLORIDE, *CCC, *D W72-14706
AENA, OSCILLATORIA, M/ *DACONIL, *CHLOROPHTHALONIL, ULOTHRIX, ANAB W72-07508
ROPHYTA, *SYSTEMATICS, / *ALGAE, *CHLOROPHYTA, *EUGLENOPHYTA, *PYR W72-07683
LIFE CYCLES, CU/ *AQUATIC ALGAE, *CHLOROPHYTA, *GROWTH CHAMBERS, * W72-07144
ATER QUALITY CONTROL, *KINETICS, *CHLOROPHYTA, ALGAL CONTROL, *NEW W72-08376
LATORS, *CHLORIDES, *CYANOPHYTA, *CHLOROPHYTA, *CHRYSOPHYTA, TOXIC W72-14706
ION, AQUATIC WEEDS, *WASHINGTON, *CHLOROPHYLL, WATER TEMPERATURE, W72-14728
DOPHYTA/ *DISTRIBUTION PATTERNS, *CHLOROPHYTA, *MARINE ALGAE, *RHO W73-00284
ASSAY/ *CULTURES, *PLANT GROWTH, *CHLOROPHYTA, *MARINE ALGAE, *BIO W72-14294
ON, *DENSITY, *DEPTH, MARINE AN/ *CHLOROPHYLL, *SPATIAL DISTRIBUTI W72-14313
CONTROL, PLANT GROWTH REGULATOR/ *CHLOROPHYTA, *INHIBITORS, ALGAL W71-13793
, RHODOPHYTA, SHORES.: *ALGAE, *CHLOROPHYTA, *CYANOPHYTA, *LAKES W71-12489
PRODUCTION, *BIOMASS,/ *CARBON, *CHLOROPHYLL, *SEAWATER, *AQUATIC W71-11562
*PHOSPHORUS, *NITROGEN, *IDAHO, *CHLOROPHYLL, *ANABAENA, NUTRIENT W72-13300
ATION TECHNIQUE/ *PHYTOPLANKTON, *CHLOROPHYLL, *EVALUATION, *SEPAR W72-13673
ERIE, NITROGEN, ALGAE, INORGANI/ *CHLOROPHYLL, *PHOSPHORUS, *LAKE W72-13653
DEN, MICROCYSIS, OSCILLATORIA, / *CHLOROPHYLL A, MACROPHYTES, *SWE W72-12729
BIOTA, PROTOZOA, *AMPHIPODA, *F/ *CHLOROPHYTA, *SPRINGS, DIATOMS, W72-12744
UATIC PRODUCTIVITY, *PRIMARY PR/ *CHLOROPHYLL, *PLANT PIGMENTS, AQ W72-12192
ECHNIQUES, RESISTANCE, CULTURES, *CHLOROPHYTA.: /, RADIOACTIVITY T W72-11836
ICS, *MARINE ALGAE, *CYANOPHYTA, *CHLOROPHYTA, PHAEOPHYTA, RHODOPH W72-11800
XIDATION LAGOONS, CHLOR/ *ALGAE, *CHLOROPHYLL, *CLASSIFICATION, *O W71-00139
CTIVITY, *PHOSPHATES, *NITRATES, *CHLOROPHYLL, DIVERSION, CARBON D W71-02681
HAINS, *TOXICITY, *ALGAL TOXINS, *CHLOROPHYTA, *PHYTOTOXINS, ECOLO W71-08597
SANCE ALGAE, FLOATING PLANTS, S/ *CHORELLA, *ALGAE, *FOULING, *NUI W71-09261
IC ACIDS, SEPARATION TECHNIQUES, *CHROMATOGRAPHY.: /COLUMNS, ORGAN W72-09365
E MAINTENANCE, *POLYP MAINTENAN/ *CHRYSAORA QUINQUECIRRHA, *MEDUSA W72-10162
FICATION, SAMPLING, CHLOROPHYTA, *CHRYSOPHYTA, CYANOPHYTA.: /LASSI W72-12192
IDES, *CYANOPHYTA, *CHLOROPHYTA, *CHRYSOPHYTA, TOXICITY, INHIBITOR W72-14706
ICIN, CAULERPIN, PALMITIC ACID,/ *CIGUATERA, CAULERPA SP., CAULERP W71-08597
A OEDOGONIA.: *CLADOPHORA GLOMERATA, *PITHOPHOR W71-08032
YRA, CLODOPHORA GLOMERATA, *MET/ *CLADOPHORA, *ACHNANTHES, *SPIROG W72-07901
VES.: *CLADOPHORA GLOMERATA, GROWTH CUR W72-13651
MICHIGAN, ALGAE, / *PHOSPHORUS, *CLADOPHORA, *PLANT GROWTH, *LAKE W72-13644
PERIODISM, G/ *ALGAE, *SEASONAL, *CLADOPHORA, *TEMPERATURE, *PHOTO W72-13651
*CLADOPHORA.: W71-10580
ATION, HYDROGEN IO/ *PHOSPHATES, *CLAMS, *ALGAE, CHEMICAL PRECIPIT W72-03613
STES, *COBALT / *MARINE ANIMALS, *CLAMS, *FOOD CHAINS, *NUCLEAR WA W73-00826
NS, CHLOR/ *ALGAE, *CHLOROPHYLL, *CLASSIFICATION, *OXIDATION LAGOO W71-00139
S, *SURFACTANTS, */ *OIL WASTES, *CLEANING, *DISPERSION, *CHEMICAL W71-09789
SPHAERIUM, DINOBRYON, FRAGILARI/ *CLEVELAND(OHIO), CERATIUM, COELO W72-10076
TEMPERATURE/ *OXIDATION LAGOONS, *CLIMATIC ZONES, *COST ANALYSIS, W71-07109
ON PLANT.: *CLUSTER ANALYSIS(FISH), *DICKERS W72-07791
TREATMENT, *TERTIARY TREATMENT, *COAGULATION, *OXIDATION LAGOONS, W71-09629
ENT, ANALYTICAL TECHNIQUES, COA/ *COAGULATION, *WASTE WATER TREATM W71-13356
CEANOGRAPHY, *PACIFIC NORTHWEST, *COAST REVIEW, *BIBLIOGRAPHIES, W W72-12190
ERU), CAPE COD(MASS), CAPE MAY(/ *COASTAL WATERS, PANAMA, CALLAO(P W72-08056
SLAND, SKELETONEMA COSTATUM, TH/ *COASTAL WATERS, *SOUTHERN LONG I W71-07875
ES, *STA/ *PRIMARY PRODUCTIVITY, *COASTS, *PHYTOPLANKTON, *ESTUARI W71-07875
CONTINE/ *DISTRIBUTION, *UREAS, *COASTS, *OCEANS, SURFACE WATERS, W72-08056
CHEMISTRY, CYANOPHYTA, BACTERIA/ *COASTS, *ORGANIC COMPOUNDS, *GEO W73-01439
BIOMASS, TOXICITY, EUT/ *OCEANS, *COASTS, *WATER QUALITY CONTROL, W71-11793
A,/ *MARINE ALGAE, *SYSTEMATICS, *COASTS, BENTHIC FLORA, RHODOPHYT W71-10066
RIAL WASTES, *THERMAL POLLUTION, *COASTS, WATER POLLUTION EFFECTS, W73-01632
ABSORPTION, *MARINE ANIMALS, MA/ *COBALT RADIOISOTOPES, *COBALT, * W73-00821
, *FOOD CHAINS, *NUCLEAR WASTES, *COBALT RADIOISOTOPES, PATH OF PO W73-00826
E ALGAE, RHODOPHYTA, PHAEOPHYTA, *COBALT RADIOISOTOPES, NUCLEAR WA W72-07826
GAE, HEAVY/ *ZINC RADIOISOTOPES, *COBALT RADIOISOTOPES, *MARINE AL W72-10957
ISOTOPES, *CESIUM RADIOISOTOPES, *COBALT RADIOISOTOPES, *BICINDICA W72-10958
ITIVITY, *RADIOACTIVITY EFFECTS/ *COBALT RADIOISOTOPES, *RADIOSENS W72-11836
ATMENT, *ECONOMIC JUSTIFICATION, *COBALT, *RADIATION, COSTS, SEWAG W72-00383
MALS, MA/ *COBALT RADIOISOTOPES, *COBALT, *ABSORPTION, *MARINE ANI W73-00821

, ORGAN/ *ON-SITE INVESTIGATION, *COLD REGIONS, *OXIDATION LAGOONS W72-06838
ATER TREATMENT, *SEWAGE LAGOONS, *COLD REGIONS, AERATED LAGOONS, * W72-12565
TS, EFFECTIVENESS/ *DISPERSANTS, *COLLECTING AGENTS, *SINKING AGEN W71-09789
*COLOR INFRARED PHOTOGRAPHY.: W72-12487
WATER POLLUTION EFFECTS, *LAKES, *COLORADO, *RECREATION FACILITIES W72-04791
DS, BACKWASHING, *MOUNT CLEMENS/ *COMBINED SEWERS, *SUSPENDED SOLI W72-00042
RY PRODUCTIVITY, *AQUATIC ALGAE, *COMPARATIVE PRODUCTIVITY, *ECOSY W73-01080
RATES, *INHIBI/ *AQUATIC ALGAE, *COMPETITION, *NUTRIENTS, *GROWTH W72-12734
OPLANKT/ *ANALYTICAL TECHNIQUES, *COMPUTER PROGRAMS, *ALGAE, *PHYT W73-02469
UTION PATTERNS, *NORTH CAROLINA, *COMPUTER PROGRAMS, DATA PROCESSI W71-07337
N LAGOONS, *MATHEMATICAL MODELS, *COMPUTER SIMULATION, MIXING, TUR W71-07084
POLYPHOSPHATE, SODIUM PYROPHOSP/ *CONDENSED PHOSPHATES, SODIUM TRI W71-00116
ES.: *LYSIS, *CONTACT TIME, CELLULAR METABOLIT W72-00411
LAKE TAHOE(CALIF), CASTLE LAKE(/ *CONTEMPORANEOUS DISEQUILIBRIUM, W71-09171
OWTH/ *SAN JOAQUIN DELTA(CALIF), *CONTINUOUS FLOW TURBIDOSTAT, *GR W70-10403
OSPHATASE, *ANABAENA FLOS-AQUAE, *CONTINUOUS CULTURE, CHEMOSTATS, W72-05476
*CONTINUOUS DISCHARGE LAGOON.: W72-12565
SSAYS, LABORATORY TESTS, FISH, / *COOLING TOWERS, *TOXICITY, *BIOA W72-08683
*HEATED WATER, *BENEFICIAL USE, *COOLING WATER, *FISHERIES, WATER W70-09596
OLLUTION, *NUCLEAR POWER PLANTS, *COOLING WATER, *WATER TEMPERATUR W70-09612
ON, *AIR CONDITIONING, *HEATING, *COOLING, *THERMAL POWER, *MULTIP W70-09193
ON, ALGAE, FEEDING RATES, *LARV/ *COPEPODS, CRUSTACEANS, ZOOPLANKT W72-09248
TER, *PATH OF POLLUTANTS, DIELD/ *COPPER SULFATE, *ALGAE, *FLOODWA W71-11498
KTERS, *CHEMCONTROL, NEW HAMPSH/ *COPPER SULFATE, *ALGAE, *ZOOPLAN W71-03014
TER, *FRESH WATER, ALGAE, PHOTO/ *COPPER, *IONS, *POISONS, *SEA WA W71-03027
ON TRANSPORT, *HEAVY METALS, EU/ *COPPER, CULTURES, *CHELATION, *I W73-02112
OTOPES, *BOSSIE/ *CALCIFICATION, *CORALLINE ALGAE, CALCIUM RADIOIS W72-12634
DATION LAGOONS, *CLIMATIC ZONES, *COST ANALYSIS, TEMPERATURE, NUTR W71-07109
ION, *OPERATION AND MAINTENANCE, *COST ANALYSIS, *FILTERS, HEADLOS W72-09393
CONTROL, *WASTE WATER TREATMENT, *COST-BENEFIT ANALYSIS, DETERGENT W72-04734
ORUS COMPOUNDS, *NITROGEN CYCLE, *CRABS, *FISHES, ECOLOGY, FOOD CH W72-12567
*CRANBERRY BOGS.: W71-11498
*CHLORAPHYTA, *AQUATIC INSECTS, *CRUSTACEANS, *MOLLUSKS, *AQUATIC W72-06526
POLLUTION EFFECTS, *OIL WASTES, *CRUSTACEANS, OILY WATER, ALGAE, W72-07911
, NUCLEAR P/ *THERMAL POLLUTION, *CRUSTACEANS, *FISH, HEATED WATER W72-07907
WATER QUALITY, BENTHOS, CRABS, / *CRUSTACEANS, *MARINE BACTERIA, * W72-08142
RY STUDIES, COPEPODS, DIATOMS, / *CRUSTACEANS, *FOODS, *LIFE HISTO W72-14805
*NUISANCE ALGAE, *AQUATIC ALGAE, *CRUSTACEANS, *FISH TOXINS, *ALGA W72-14056
DIAN ARCTIC, *CHLAMYDOMONAS-SPP, *CRYPTOMONAS-SPP.: *CANA W72-07922
ATER PROPERTIES, *FERTILIZATION, *CULTIVATED LANDS, *IOWA, WATER Q W71-12252
*NITROGEN FIXATION, *SYNTHESIS, *CULTURES, *NITROGEN, OXYGEN, BIO W72-07941
S, LIGHT INTENSITY, GRO/ *ALGAE, *CULTURES, PHOTOSYNTHESIS, BIOMAS W72-06274
PHYTA, *MARINE ALGAE, *BIOASSAY/ *CULTURES, *PLANT GROWTH, *CHLORO W72-14294
CATION, FJORDS, BIOASSA/ *ALGAE, *CULTURES, *INDICATORS, *EUTROPHI W73-00234
LGAE, *ISOLATION, MARINE PLANTS/ *CULTURES, *CYANOPHYTA, *MARINE A W73-00430
, SEA WATER, NITRATES/ *DIATOMS, *CULTURES, *NUTRIENT REQUIREMENTS W73-00441
RATORY TESTS, *BACTERIA, *ALGAE, *CULTURES, CYANOPHYTA, HYDROGEN I W73-01657
PHYTA,/ *CALCIUM, *GROWTH RATES, *CULTURES, *AQUATIC ALGAE, CHLORO W72-12736
CATION, *AQUATIC A/ *CYANOPHYTA, *CULTURES, *SYSTEMATICS, CLASSIFI W72-12172
YTA, PHAEOPHYTA, CHLOR/ *YEASTS, *CULTURES, *MARINE ALGAE, RHODOPH W72-04748
ISOLATION, *MEMBRANE PROCESSES,/ *CULTURES, *ANAEROBIC BACTERIA, * W72-04743
*ALGAE, *PHOSPHATES, *ENZYMES, *CULTURES, NITRATES, PLANT PHYSIO W72-05476
TION, AMMONIA, NITRI/ *BACTERIA, *CULTURES, *SEA WATER, *NITRIFICA W72-05421
*LABORATORY TESTS, *CHLORELLA, *CULTURES, *AQUATIC MICROORGANISM W71-08027
IC WEEDS, *WEED CONTROL, GROWTH, *CULTURES, *ALGICIDES, BIBLIOGRAP W71-06102
IDATION LAGOONS, *PATHO/ *ALGAE, *CULTURES, *ENTERIC BACTERIA, *OX W70-08319
DIA,/ *TRICHODESMIUM ERYTHRAEUM, *CULTURING TECHNIQUES, CULTURE ME W73-00430
ATORS, *ECOLOGICAL DISTRIBUTION, *CURRENTS(WATER), *DINOFLAGELLATE W73-00296
PHYTOPLANKTON, *ALGAE, *DIATOMS, *CURRENTS, *DISTRIBUTION PATTERNS W72-03543
*NUTRIENTS, *ECOSYST/ *FORESTS, *CUTTING MANAGEMENT, *HERBICIDES, W71-01489
, APHANOCHAETE, RHIZOCLONIUM, V/ *CUYAHOGA RIVER(OHIO), CLADOPHORA W71-05629
-AQUAE, LAKE ERKEN(SWEDEN).: *CYANOPHAGES, *APHANIZOMENON FLOS W73-00246
EUTROPHICATION, *DISTRIBUTION, / *CYANOPHYTA, *DISSOLVED OXYGEN, * W73-00236
ATION, MARINE PLANTS/ *CULTURES, *CYANOPHYTA, *MARINE ALGAE, *ISOL W73-00430
ALGAL TOXINS, *POISONOUS PLANTS, *CYANOPHYTA, *NUISANCE ALGAE, *AQ W72-14056
CTIVITY, PHOT/ *ALGAE, *BIOMASS, *CYANOPHYTA, PHYTOPLANKTON, PRODU W72-14812
IDE, *BEGGIATOA, *PHYTOTOXICITY, *CYANOPHYTA, *NEW HAMPSHIRE, MARI W72-14542
DES, HERBICIDES, IRRIGA/ *ALGAE, *CYANOPHYTA, *PESTICIDES, FUNGICI W72-14807
T GROWTH REGULATORS, *CHLORIDES, *CYANOPHYTA, *CHLOROPHYTA, *CHRYS W72-14706
E, *LEGUMES, *NITROGEN FIXATION, *CYANOPHYTA, PLANTS, NITROGEN, RE W73-01456
ES, SAMP/ WATER QUALITY CONTROL, *CYANOPHYTA, *WATER ANALYSIS, LAK W72-07890
NUTRIENTS, *CARBON, *PHOSPHORUS, *CYANOPHYTA, AQUATIC ENVIRONMENT, W71-00475
IELDRIN, *ENDRIN, *GROWTH RATES, *CYANOPHYTA, *TOXICITY, PESTICIDE W72-03545
LGAL TOXINS, HUMAN PATHOLOGY.: *CYANOPHYTA, *LABORATORY TESTS, A W72-04259
L POLLUTION, *WATER TEMPERATURE, *CYANOPHYTA, GEYSERS, PHOTOSYNTHE W72-03557
NS, *WATER ANALYSIS, / *VIRUSES, *CYANOPHYTA, *DISTRIBUTION PATTER W72-04736
ICS, CLASSIFICATION, *AQUATIC A/ *CYANOPHYTA, *CULTURES, *SYSTEMAT W72-12172
HY/ *SYSTEMATICS, *MARINE ALGAE, *CYANOPHYTA, *CHLOROPHYTA, PHAEOP W72-11800
ORINATED HYDROCARBON PESTICIDES, *CYANOPHYTA, SODIUM, POTASSIUM, C W72-13809
ORY TESTS, CARBON RADIOISOTOPES, *CYANOPHYTA, EUTROPHICATION, WATE W72-13692
OL, *LIGHT INTENSITY, *NEBRASKA, *CYANOPHYTA, ALGAL CONTROL, *POND W73-02349
SHORES.: *ALGAE, *CHLOROPHYTA, *CYANOPHYTA, *LAKES, RHODOPHYTA, W71-12489
XATION, ENZYMES, BACTERIA, SPOR/ *CYANOPHYTA, *ALGAE, *NITROGEN FI W71-12070
IZATION, NUTRIENTS, ALGAE, RICE/ *CYANOPHYTA, *PESTICIDES, *FERTIL W71-12071
ON, *AQUATIC ALGAE, *PHOSPHORUS, *CYANOPHYTA, *EUTROPHICATION, BIO W71-10645
*PHOSPHATES, *SOLUBILITY, *CYANOPHYTA, ALGAE, CULTURES.: W71-10070
DUCING ALGAE, / *DOMESTIC WATER, *CYANOPHYTA, RESERVOIRS, CDCR-PRO W72-10613
SCOPY, CHLOROPHYTA, SCENEDESMUS, *CYANOPHYTA, CHLORELLA, BIOLOGICA W72-09239
MIN/ *NITROGEN FIXATION, *ALGAE, *CYANOPHYTA, *THERMAL WATER, *WYO W72-08508
XYGEN,/ *LIGHT, *PHOTOSYNTHESIS, *CYANOPHYTA, NITROGEN FIXATION, O W72-08584

AL DISTRIBUTION, *AQUATIC ALGAE,	*CYANOPHYTA, *BEGGIATOA, PHYTOPLA	W72-09061
ISH, FISHKILL, / *ALGAL CONTROL,	*CYANOPHYTA, EXPLOSIVES, COSTS, F	W72-11564
NIC ACIDS, *HYDROGEN ION CONCEN/	*CYANOPHYTA, *GROWTH RATES, *ORGA	W72-00838
GAL CONTROL, NUTRIENTS, BIOMASS,	*CYANOPHYTA, PLANT GROWTH, PLANKT	W72-00799
LES, INHIBITION, MET/ *TOXICITY,	*CYANOPHYTA, SULFUR, ALGAE, GRANU	W72-00153
RATORY TESTS, / *NUISANCE ALGAE,	*CYANOPHYTA, *ALGAL CONTROL, LABO	W71-13184
TOLOGICAL S/ *NITROGEN FIXATION,	*CYANOPHYTA, PLANT PHYSIOLOGY, CY	W72-03217
	*CYANOPHYTA, ALGAE, *VIRUSES.:	W72-03188
TROPHICATION, REVIEWS.:	*CYANOPHYTA, *NUISANCE ALGAE, *EU	W72-01358
C MATTER, ALGAE, EUTROPHICATION/	*CYANOPHYTA, *HUMIC ACIDS, ORGANI	W72-01097
AKES, *NUTRIENTS, ALGAL CONTROL,	*CYANPHYTA, *NUISANCE ALGAE, PLAN	W72-00798
R-PLANT RELATIONSHIPS, *FALLOUT,	*CYCLING NUTRIENTS, ECOSYSTEMS, R	W72-03342
*CHEMICAL ANALYSIS, *MONITORING,	*CYCLING NUTRIENTS, *FOOD CHAINS,	W72-07715
RPHISM, */ *CYCLOTELLA CRYPTICA,	*CYCLOTELLA MENEGHINIANA, *POLYMO	W73-02548
MENEGHINIANA, *POLYMORPHISM, */	*CYCLOTELLA CRYPTICA, *CYCLOTELLA	W73-02548
	*CYSTS, ISRAEL.:	W72-00422
ORINATED HYDROCARBON PESTICIDES,	*CYTOLOGICAL STUDIES, ABSORPTION,	W72-12998
S, *ALGAE, CHRYSOPHYTA, PHAEOPH/	*CYTOLOGICAL STUDIES, *SYSTEMATIC	W72-14808
SIS, RESPI/ *ALGAE, *METABOLISM,	*CYTOLOGICAL STUDIES, PHOTOSYNTHE	W71-08683
*BIOLOGICAL MEMBRANES, *ALGAE,	*CYTYLOGICAL STUDIES, PLANT MORPH	W73-01630
HRIX, ANABAENA, OSCILLATORIA, M/	*DACONIL, *CHLOROPHTHALONIL, ULOT	W72-07508
NS, *PILOT PLANTS, ALGAE, BIOCH/	*DAIRY INDUSTRY, *OXIDATION LAGOO	W70-06619
*LAKE ERKEN(SWEDEN),	*DARK NITROGEN FIXATION.:	W72-08054
ITORING, *SURVEYS, *MEASUREMENT,	*DATA COLLECTIONS, *ENVIRONMENT,	W72-07841
, *CHEMCONTROL, *BIBLIOGRAPHIES,	*DATA COLLECTIONS, *WATER POLLUTI	W73-01976
IONS, *SCUBA DIVING, *SEA WATER,	*DATA COLLECTIONS, COPPER, LEAD,	W72-11925
CES, THERMAL POLLUTION, REVIEWS,	*DATA COLLECTIONS, TOXICITY, PATH	W72-12190
S, IN/ *WATER POLLUTION EFFECTS,	*DATA COLLECTIONS, ALGAE, ECOTYPE	W71-13746
TA, *MEASUREMENT, *AS/ *SURVEYS,	*DATA COLLECTIONS, *HYDROLOGIC DA	W72-10678
CLADOPHORA, CYPRINU/ *RESIDUES,	*DDD, DDE, TDE, BLACK-FLY LARVAE,	W70-08098
STENCE, *WATER POLLUTION / *DDT,	*DDE, *PESTICIDE KINETICS, *PERSI	W73-02280
*PERSISTENCE, *WATER POLLUTION /	*DDT, *DDE, *PESTICIDE KINETICS,	W73-02280
CARBON RADIOISOTOP/ *PESTICIDES,	*DDT, *DIELDRIN, *PHYTOPLANKTON,	W72-13657
OPHOSPHOROUS PESTICIDES, *2-4-D,	*DDT, *DIELDRIN, *DIAZINON, *WATE	W71-10566
GANISMS, *MICROBIAL DEGRADATION,	*DDT, *DIELDRIN, *ALDRIN, *ENDRIN	W73-00361
N, *SHRIMP, *BIOASSAY, *ABSORPT/	*DDT, *MARINE FISH, *PHYTOPLANKTO	W73-02105
ORINATED HYDROCARBON PESTICIDES,	*DDT, *PESTICIDE TOXICITY, *WATER	W71-12303
P/ *BIODEGRADATION, *METABOLISM,	*DDT, *RADIOACTIVITY TECHNIQUES,	W72-07703
OGY, CULTURES, PHOT/ DESCRIPTORS:	*DDT, *RESISTANCE, *PLANT PHYSIOL	W73-00281
ENVIRONMENT, WATERSHEDS(BASINS),	*DDT, *SMALL WATERSHEDS, ECOSYSTE	W72-12930
ON, *SALT TOLERANCE, WATER POLL/	*DDT, *SODIUM CHLORIDE, *INHIBITI	W72-13809
PESTICIDES, *PESTICIDE TOXICITY,	*DDT, PESTICIDES, PHYTOPLANKTON,	W72-08436
TION, *AL/ *OXYGEN REQUIREMENTS,	*DECOMPOSING ORGANIC MATTER, AERA	W72-01881
INE ALGAE, / *CHEMICAL ANALYSIS,	*DECOMPOSING ORGANIC MATTER, *MAR	W72-12166
LUTION EFFECTS, *ORGANIC MATTER,	*DECOMPOSING ORGANIC MATTER, *EUT	W72-14673
IENTS,/ *EUTROPHICATION, *ALGAE,	*DECOMPOSING ORGANIC MATTER, NUTR	W70-10405
L CONTROL, *ESSENTIAL NUTRIENTS,	*DEFICIENT NUTRIENTS, *AQUATIC AL	W72-07514
ODEGRADATION, *NUTRIENTS, ALGAE,	*DEGRADATION(STREAM), WATER POLLU	W71-07973
AIRED WATER QUALITY, *ESTUARIES,	*DELAWARE RIVER, VEGETATION EFFEC	W70-09189
FE, *DISSOLVED OXYGEN CONCENTRA/	*DELAWARE ESTUARY, *BIOLOGICAL LI	W70-09189
WATER POLLUTION CONTROL, *DIVER/	*DELTAS, *BAYS, *WATER QUALITY, *	W71-09581
WATER TREATMENT, *NITRIFICATION,	*DENITRIFICATION, *NITRATES, NITR	W71-13939
ONTROL, EUTROPH/ *NITRIFICATION,	*DENITRIFICATION, WATER QUALITY C	W72-00800
S MEDIA, *WASTE WATER TREATMENT,	*DENITRIFICATION, *NITRATES, EUTR	W72-00974
ITROGEN/ *WASTE WATER TREATMENT,	*DENITRIFICATION, WATER REUSE, *N	W72-04309
KE, BAVELSE LAKE, CRYPTOPHYCEAE,	*DENMARK, CYCLOTELLA COMTA, STEPH	W72-12738
C ALGAE, *LAKES, *PHYTOPLANKTON,	*DENSITY STRATIFICATION, SPECIATI	W71-09852
ISOPYCNIC CENTRIFUGATION, SAMP/	*DENSITY-GRADIENT CENTRIFUGATION,	W73-00371
OROPHYLL, *SPATIAL DISTRIBUTION,	*DENSITY, *DEPTH, MARINE ANIMALS,	W72-14313
ATION, LAKE ERIE CENTRAL BASIN./	*DEOXYGENATION, *NUTRIENT REGENER	W72-12996
ESERVA/ *ADENOSINE TRIPHOSPHATE,	*DEOXYRIBONUCLEIC ACID, SAMPLE PR	W71-11562
*SPATIAL DISTRIBUTION, *DENSITY,	*DEPTH, MARINE ANIMALS, MARINE AL	W72-14313
P/ *VARIABILITY, *PHYTOPLANKTON,	*DEPTH, ON-SITE INVESTIGATIONS, *	W72-11719
OWER PROD/ *NUCLEAR POWERPLANTS,	*DESALINATION PLANTS, *ELECTRIC P	W72-03327
TED LAGOONS, *OXIDATION LAGOONS,	*DESIGN CRITERIA, BIODEGRADATION,	W72-09386
ALGAE, HAR/ *OXIDATION LAGOONS,	*DESIGN CRITERIA, *WATER QUALITY,	W71-07100
ON LAGOONS, *TERTIARY TREATMENT,	*DESIGN CRITERIA, TRICKLING FILTE	W71-07113
KTON.:	*DESMIDS, *EUPLANKTON, *LIMNOPLAN	W71-13149
RING TECHNIQUES, STAURASTRUM PA/	*DESMIDS, *AXENIC CULTURES, CULTU	W72-12736
*ALGAL CONTROL, *EUTROPHICATION,	*DESTRATIFICATION, *WATER QUALITY	W72-09304
ILUTION, *NITROGEN, *PHOSPHORUS,	*DETENTION RESERVOIRS, TERTIARY T	W72-02412
TION, ACID INJECTION PUMP, ALUM/	*DETENTION PERIOD, FLUSHING OPERA	W72-02850
	*DETENTION TIME.:	W71-13326
, PHOTOSYN/ *WATER PURIFICATION,	*DETENTION RESERVOIRS, RESERVOIRS	W71-01491
*EUTROPHICATION, / *PHOSPHORUS,	*DETERGENTS, *TERTIARY TREATMENT,	W70-09325
MULATION, *SURFACTANTS, *EUTROP/	*DETERGENTS, *ALGAL CONTROL, *FOR	W71-06247
ECTS, *WATER POLLU/ *PHOSPHATES,	*DETERGENTS, *WATER POLLUTION EFF	W71-08768
ONOMIC EFFICIENCY, *PUBLIC HEAL/	*DETERGENTS, *EUTROPHICATION, *EC	W72-04266
TROL, *PHOSPHATES, *FEDERAL GOV/	*DETERGENTS, *WATER POLLUTION CON	W72-01093
POLLUTION CONTROL, *PHOSPHATES,	*DETERGENTS, *LEGISLATION, CHEMIC	W72-01685
HICATION, NUTRIENTS, ASSESSMENT/	*DETERGENTS, *PHOSPHATES, *EUTROP	W72-09169
LUTION SOURCES, *EUTROPHICATION,	*DETERGENTS, WATER QUALITY CONTRO	W72-11186
D SLUDGE, *NUTRIENTS, N/ *ALGAE,	*DETERGENTS, *TOXICITY, *ACTIVATE	W71-11972
LUTION CONTROL, *EUTROPHICATION,	*DETERGENTS, *PHOSPHATES, LEGISLA	W71-11516
,/ *EUTROPHICATION, *PHOSPHATES,	*DETERGENTS, *POLLUTION ABATEMENT	W71-09429
N, 2-4-D, INSECTICIDES, PHENOLS,	*DETERGENTS, *LETHAL LIMIT, ALKYL	W72-12576
ON, *SEWAGE LAGOONS, / *ENZYMES,	*DETERGENTS, *ALGAE, *PHYTOPLANKT	W72-07661
GOVERNMENTS, */ *EUTROPHICATION,	*DETERGENTS, *PHOSPHATES, *LOCAL	W72-06638
POLLUTION SOURCES, *POLLUTION A/	*DETERGENTS, *PHOSPHATES, *WATER	W72-06655
E WATER, EUTROPHICATION, ALKYLB/	*DETERGENTS, *PHOSPHATES, *POTABL	W72-06837

POLLUTION CONT/ *UNITED STATES, *DETERGETNTS, *PHOSPHATES, *WATER W72-01085
ICROBIAL BIOMASS, *GROWTH STAGE, *DETERMINATION, *ORIGIN, *SIGNIFI W72-04292
ISTRODESMUS BRAUNII,/ *SIMAZINE, *DETOXICATION, CHLOROSARCINA, ANK W71-07675
TEBRATES, *MICR/ *RADIOISOTOPES, *DETRITUS, *BENTHIC FAUNA, *INVER W70-06668
ED CONTROL, *MECHANICAL CONTROL, *DEWATERING.: /ALGAE, *AQUATIC WE W71-06188
A/ *NUCLEAR EXPLOSION, *FILTERS, *DIATOMACEOUS EARTH, *POROUS MEDI W72-00974
N PATTE/ *PHYTOPLANKTON, *ALGAE, *DIATOMS, *CURRENTS, *DISTRIBUTIO W72-03543
, IND/ *WATER POLLUTION EFFECTS, *DIATOMS, WATER POLLUTION SOURCES W71-11583
C OCEAN, *ALGAE.: *DIATOMS, *PHYTOPLANKTON, *PACIFI W71-11527
XICITY, *AMMONIA, *MARINE ALGAE, *DIATOMS, CHEMICAL ANALYSIS, HYDR W71-11008
N PATTERNS, *PHOSPHA/ *PLANKTON, *DIATOMS, *SAMPLING, *DISTRIBUTIO W72-03567
U/ *PHYTOPLANKTON, *SYSTEMATICS, *DIATOMS, SEA WATER, SEASONAL, FL W72-07663
VIRONMENTAL EFFECTS, ESTUARIES,/ *DIATOMS, *RESEARCH EQUIPMENT, EN W72-08141
QUIREMENTS, SEA WATER, NITRATES/ *DIATOMS, *CULTURES, *NUTRIENT RE W73-00441
HICATION, *POLLUTANT IDENTIFICA/ *DIATOMS, *BIOINDICATORS, *EUTROP W73-00432
HICATION/ *SAMPLING, *NUTRIENTS, *DIATOMS, *LAKE MICHIGAN, *EUTROP W73-00384
*ALGAE, *PHYTOPLANKTON, *RIVERS, *DIATOMS, CHLOROPHYTA, BENTHIC FL W72-14797
SYSTEMAT/ *LIGHT, *TEMPERATURE, *DIATOMS, BIOLOGICAL COMMUNITIES, W73-00242
NTENSITY, TIDAL EFF/ *ESTUARIES, *DIATOMS, *SESSILE ALGAE, LIGHT I W73-00853
A, AQUATIC ALGAE.: *DIATOMS, *MICROSCOPY, CHRYSOPHYT W72-11738
, *PRIMARY PRODUCTIVITY, *ALGAE, *DIATOMS, *PERIPHYTON, *PHYTOPLAN W72-12192
RS, BRACKISH WATER, FRESHWATER,/ *DIATOMS, *SALINITY, *BIOINDICATO W73-02548
TION EFFECTS, *PLANT PHYSIOLOGY, *DIATOMS, PESTICIDES, DDD, PESTIC W73-02280
ICIDES, *2-4-D, *DDT, *DIELDRIN, *DIAZINON, *WATER POLLUTION EFFEC W71-10566
HYTES, GROWTH MEDIA, COCCONEIS / *DICHOTOMOSIPHON TUBEROSUS, *EPIP W72-12744
*CLUSTER ANALYSIS(FISH), *DICKERSON PLANT.: W72-07791
S, *MICROBIAL DEGRADATION, *DDT, *DIELDRIN, *ALDRIN, *ENDRIN, MARI W73-00361
RADIOISOTOP/ *PESTICIDES, *DDT, *DIELDRIN, *PHYTOPLANKTON, CARBON W72-13657
HOROUS PESTICIDES, *2-4-D, *DDT, *DIELDRIN, *DIAZINON, *WATER POLL W71-10566
HYDROCARBON PESTICIDES, *ALDRIN, *DIELDRIN, *ENDRIN, *GROWTH RATES W72-03545
ISTIC EFFECTS, *AXEMIC CULTURES, *DIEOFF RATES, AFTERGROWTH, BLUE- W70-08319
*CHLOROCHOLINE CHLORIDE, *CCC, *DIETHYLAMINE HYDROCHLORIDE, MERI W72-14706
ALGAE, WASHING/ *ALGAL CONTROL, *DIFFUSION, *NUTRIENTS, *NUISANCE W72-08049
DISTRIBUTION, *CURRENTS(WATER), *DINOFLAGELLATES, ECOLOGY, DISTRI W73-00296
NORTH CAROLINA, *ALGAE, DIATOMS, *DINOFLAGELLATES, *RED TIDE, WATE W72-01329
MUNITIES, *THERMAL POWER PLANTS, *DISCHARGE(WATER), AQUATIC HABITA W71-11517
, COLIFORMS, ALG/ *CHLORINATION, *DISINFECTION, *OXIDATION LAGOONS W72-00422
INE, CONTAC/ *OXIDATION LAGOONS, *DISINFECTION, DEGRADATION, CHLOR W71-07096
*SINKING AGENTS, EFFECTIVENESS/ *DISPERSANTS, *COLLECTING AGENTS, W71-09789
ANTS, */ *OIL WASTES, *CLEANING, *DISPERSION, *CHEMICALS, *SURFACT W71-09789
TOZOA, *FUNGI, *AQUATIC INSECTS, *DISPERSION, CADDISFLIES, STONEFL W70-06655
, OXIDATION DITCHES, AERATED LA/ *DISPOSAL SYSTEMS, *LAND DISPOSAL W71-07551
AWARE ESTUARY, *BIOLOGICAL LIFE, *DISSOLVED OXYGEN CONCENTRATION.: W70-09189
N, *BIOCHEMICAL / *SLUDGE WORMS, *DISSOLVED OXYGEN, *EUTROPHICATIO W70-09189
XCRETION, ALKALINE PHOSPHATASE,/ *DISSOLVED ORGANIC PHOSPHORUS, *E W71-10083
DUSTRIAL WASTES, *WATER QUALITY, *DISSOLVED OXYGEN, *PHYTOPLANKTON W71-11876
STRAIT(WALES).: *DISSOLVED ORGANIC CARBON, *MENAI W72-09108
N, *DISTRIBUTION, / *CYANOPHYTA, *DISSOLVED OXYGEN, *EUTROPHICATIO W73-00236
, TURBI/ *PHYTOPLANKTON, *ALGAE, *DISSOLVED OXYGEN, PHOTOSYNTHESIS W73-00256
ITIONS, *KINETICS, *SCENEDESMUS, *DISSOLVED SOLIDS, *MICROBIAL DEG W73-00379
TICULATE ORGA/ *LAKE METABOLISM, *DISSOLVED ORGANIC NITROGEN, *PAR W72-08048
ANAEROBIC CONDITIONS, *KINETICS, *DISSOLVED SOLIDS, ORGANIC MATTER W73-01066
*GLUCOSE, *SNAKE RIVER(IDAHO), *DISSOLVED ORGANIC CARBON, *AMERI W72-13300
*LAKE ERIE, BENTHIC FLORA, ALGA/ *DISSOLVED OXYGEN, *HYPOLIMNION, W72-13659
ACTERIA, *SEDIMENT-WATER INTERF/ *DISSOLVED OXYGEN, *LAKE ERIE, *B W72-12996
HYPOLI/ *SEDIMENTS, *LAKE ERIE, *DISSOLVED OXYGEN, OXYGEN DEMAND, W72-12994
AKE ERIE, *HYPOLIMNION, MATHEMA/ *DISSOLVED OXYGEN, *NUTRIENTS, *L W72-12997
URREN/ *HYPOLIMNION, *LAKE ERIE, *DISSOLVED OXYGEN, TEMPERATURE, C W72-12992
, *EUTROPHICATION, *HYPOLIMNION, *DISSOLVED OXYGEN, *ANAEROBIC CON W72-12990
WASHINGTON, CHLOROPHYLL A, *DNA, *DISSOLVED ORGANOPHOSPHORUS COMPO W72-12637
ICS, *AQUATIC ALGAE/ *ANTARCTIC, *DISTRIBUTION PATTERNS, *SYSTEMAT W73-00854
N, *AQUATIC ALGAE, *SYSTEMATICS, *DISTRIBUTION PATTERNS, LAKES, AQ W73-01094
SETTS, SYSTEMATICS, / *SEASONAL, *DISTRIBUTION PATTERNS, *MASSACHU W73-01449
NVIRONMENTAL EFFECTS, ESTUARIES, *DISTRIBUTION PATTERNS, *AQUATIC W72-08141
OCEANS, SURFACE WATERS, CONTINE/ *DISTRIBUTION, *UREAS, *COASTS, * W72-08056
ICS, *ECOLOGICAL DISTRIBUTION, / *DISTRIBUTION PATTERNS, *SYSTEMAT W73-00428
YTA, *MARINE ALGAE, *RHODOPHYTA/ *DISTRIBUTION PATTERNS, *CHLOROPH W73-00284
SSOLVED OXYGEN, *EUTROPHICATION, *DISTRIBUTION, FLOATING, WINDS, A W73-00236
ICS, *SALINE LA/ *PHYTOPLANKTON, *DISTRIBUTION PATTERNS, *SYSTEMAT W72-09111
ON, *ALGAE, *DIATOMS, *CURRENTS, *DISTRIBUTION PATTERNS, ATLANTIC W72-03543
RO/ *ESTURARIES, *PHYTOPLANKTON, *DISTRIBUTION PATTERNS, *NORTH CA W71-07337
ARINE ANIMALS, *MA/ *ABSORPTION, *DISTRIBUTION, *RADIOISOTOPES, *M W71-09168
*PLANKTON, *DIATOMS, *SAMPLING, *DISTRIBUTION PATTERNS, *PHOSPHAT W72-03567
ALYSIS, / *VIRUSES, *CYANOPHYTA, *DISTRIBUTION PATTERNS, *WATER AN W72-04736
TUARIES, PRODU/ *AQUATIC PLANTS, *DISTRIBUTION, *MARINE ALGAE, *ES W72-04766
Y, *WATER RESOURCES DEVELOPMENT, *DISTRIBUTION SYSTEMS, *WATER QUA W72-05594
SPRINGS, CYANOPHYTA, AL/ *ALGAE, *DISTRIBUTION PATTERNS, *THERMAL W72-05610
ARIES, *WATER POLLUTION EFFECTS, *DISTRICT OF COLUMBIA, TIDAL EFFE W70-08249
ARIES, *WATER POLLUTION EFFECTS, *DISTRICT OF COLUMBIA, TIDAL EFFE W70-08248
GROWTH RATES, *LIMITING FACTORS, *DIURNAL DISTRIBUTION, CARBON DIO W72-11724
*DIURON, *CARBARYL, *TOXAPHENE.: W71-10566
ALITY, *WATER POLLUTION CONTROL, *DIVERSION, SEDIMENTS, WATER CIRC W71-09581
*SYNCHRONOUS CULTURES, SYNCHRON/ *DIVISION PATTERNS, *AUTOSPORES, W72-12724
LAKE WASHINGTON, CHLOROPHYLL A, *DNA, *DISSOLVED ORGANOPHOSPHORUS W72-12637
CTIVITY, *ESTUARINE ENVIRONMENT, *DOMESTIC WASTES, *SALT MARSHES, W72-12567
TES, ACTIVATED SL/ *WATER REUSE, *DOMESTIC WASTES, *INDUSTRIAL WAS W71-09524
NTERIC BACTERIA, *WATER QUALITY, *DOMESTIC WASTES, ALGAE, INDUSTRI W72-09590
ERVOIRS, ODOR-PRODUCING ALGAE, / *DOMESTIC WATER, *CYANOPHYTA, RES W72-10613
EN COMPOUNDS, *SEWAGE TREATMENT, *DOMESTIC WASTES, CHEMICAL ANALYS W70-09907
CONTROL, *AQUATIC WEED CONTROL, *DOMESTIC WATER, *SURFACE WATERS, W71-05083

ON, *EUTROPHICATION/ *NUTRIENTS, *DREDGING, DESILTING, SEDIMENTATI W73-00938
*FLOCCULATION, *ALGAE, *SILICA, *E. COLI, *IONS, *WASTEWATER TREA W72-14840
RADOR), NEW BRUNSWI/ *CHECKLIST, *EASTERN CANADA, CAPE CHIDLEY(LAB W71-10066
HMOPLEA SPHAEROPHORA, MELANOSIP/ *EASTERN CANADA, NOVA SCOTIA, IST W72-01370
C A/ *AERATION, *EUTROPHICATION, *ECOLOGICAL DISTRIBUTION, *AQUATI W72-09061
TS(WATER), *DIN/ *BIOINDICATORS, *ECOLOGICAL DISTRIBUTION, *CURREN W73-00296
RIBUTION PATTERNS, *SYSTEMATICS, *ECOLOGICAL DISTRIBUTION, *SPATIA W73-00428
ANIMALS, *SYSTEMATICS,/ *KELPS, *ECOLOGICAL DISTRIBUTION, *MARINE W73-00932
*MARINE ALGAE, LIGHT INTENSITY, *ECOLOGICAL DISTRIBUTION, SURFACE W73-00838
SUCCESSION, AQUATIC ENVIRONMENT, *ECOLOGICAL DISTRIBUTION, *ESTUAR W72-11854
, *MARINE ANIMALS, *SYSTEMATICS, *ECOLOGY, *MARINE ALGAE, ANNELIDS W73-00932
*ZOOPLANKTON, *PHYTOPLANKTON, N/ *ECOLOGY, *AQUATIC PRODUCTIVITY, W73-00935
BIOLOGIC/ *AQUATIC PRODUCTIVITY, *ECOLOGY, *AQUATIC ENVIRONMENT, * W73-02095
*MARINE PLANTS, *MARINE ANIMAL/ *ECOLOGY, *ENVIRONMENTAL EFFECTS, W73-02029
LLUTION, *ENVIRONMENTAL EFFECTS, *ECOLOGY, *PLANKTON, *ALGAE, BIOL W72-14659
TIC HABITATS, *SAMPLING, *ECOSY/ *ECOLOGY, *PREIMPOUNDMENTS, *AQUA W72-06526
SIS, TEMPERATURE, SEASONAL, ION/ *ECOLOGY, *PONDS, *CHEMICAL ANALY W72-01363
DISM, GROWTH RATES, TEM/ *ALGAE, *ECOLOGY, ENVIRONMENT, PHOTOPERIO W72-00377
ERS, *SEA WATER, *WATER QUALITY, *ECOLOGY, *CALIFORNIA, HYDROLOGIC W71-10577
RES, *ALGICIDES, BIBLIOGRAPHIES, *ECOLOGY, AQUATIC LIFE, IRRIGATIO W71-06102
UTRIENTS, WATER QUALITY CONTROL, *ECOLOGY, *ALGAE, *EUTROPHICATION W71-08775
S, *MARINE PLANTS, DEPTH, CALIF/ *ECOLOGY, *WASTES, *MARINE ANIMAL W71-01475
QUALITY, *MODEL S/ *ECOSYSTEMS, *ECOLOGY, *PUMPED STORAGE, *WATER W72-05282
L/ *DETERGENTS, *EUTROPHICATION, *ECONOMIC EFFICIENCY, *PUBLIC HEA W72-04266
GE EFFLUENTS, *SEWAGE TREATMENT, *ECONOMIC JUSTIFICATION, *COBALT, W72-00383
RIGATION PROGRAMS, *AGRICULTURE, *ECONOMICS, CYANOPHYTA, ALGAE, PL W72-03327
RESOURCES, FOOD/ *MARINE ALGAE, *ECONOMICS, *PRODUCTIVITY, ALGAE, W71-13246
NG NUTR/ *RADIOACTIVITY EFFECTS, *ECOSYSTEMS, *RAIN FORESTS, CYCLI W72-00949
ORAGE, *WATER QUALITY, *MODEL S/ *ECOSYSTEMS, *ECOLOGY, *PUMPED ST W72-05282
*WATER POLLUTION, *FOOD CHAINS, *ECOSYSTEMS, ABSORPTION, FISH, IN W72-05952
GEMENT, *HERBICIDES, *NUTRIENTS, *ECOSYSTEMS, *WATERSHEDS(BASINS), W71-01489
NVERTEBRATES, *MICROENVIRONMENT, *ECOSYSTEMS, CESIUM, COBALT RADIO W70-06668
TIC, PRODUCTIVITY, EUTROPHICATI/ *ECOSYSTEMS, *ALASKA, *LAKES, ARC W71-09190
UCTIVITY, RADIOISOTOPES, STRONT/ *ECOSYSTEMS, *SALT MARSHES, *PROD W71-09254
FECTS, *MODEL STUDIES, *STREAMS, *ECOSYSTEMS, AQUATIC HABITATS, BI W72-06340
NKTON, *PREDATION, *BIOASSAY, B/ *ECOSYSTEMS, FOOD CHAINS, *ZOOPLA W72-07132
S, *AQUATIC HABITATS, *SAMPLING, *ECOSYSTEMS, *ALGAE, *CHLORAPHYTA W72-06526
S, *HEAVY METALS, *PRODUCTIVITY, *ECOSYSTEMS, POTASSIUM, MAGNESIUM W72-14303
AQUATIC LIFE, HABITATS, ECOLOGY, *ECOSYSTEM, *ALGAE, *PHYTOPLANKTO W73-01632
*WATER / *BIOLOGICAL PROPERTIES, *ECOSYSTEMS, *RIVERS, *VIRGINIA, W73-01972
LGAE, *COMPARATIVE PRODUCTIVITY, *ECOSYSTEMS, *BIOLOGICAL COMMUNIT W73-01080
ARINE ENVIRONMENT, *DOMESTIC WA/ *ECOSYSTEMS, *PRODUCTIVITY, *ESTU W72-12567
LAKES, *WATE/ *SEWAGE TREATMENT, *EFFICIENCIES, *EUTROPHICATION, * W72-11702
*ELECTRA CRUSTULENTA, *BRYOZOA.: W73-00348
N-SITE DATA COLLECTIONS, RIVERS, *ELECTRIC POWERPLANTS, WATER POLL W72-07791
WERPLANTS, *DESALINATION PLANTS, *ELECTRIC POWER PRODUCTION, *IRRI W72-03327
*MARINE ALGAE, *GULF OF MEXICO, *ELECTRIC POWERPLANTS, *THERMAL P W72-11252
WATER PURIFICATION, *FILTRATION, *ELECTROCHEMISTRY, *WATER POLLUTI W71-12467
*SUSPENSION, *ELECTROPHORESIS, *ELECTROCHEMISTRY, *WATER PURIFIC W73-01362
UES, *WATER ANALYSIS, *NITRATES, *ELECTRODES, IONS, FLUORIDES, ALG W71-00474
*ELECTROFILTRATION.: W71-12467
ERTIOLECTA, CYCLOTELLA NANA, TH/ *ELECTRON TRANSPORT, DUNALIELLA T W72-12998
PHY *BIOLOGICAL MAGNIFICATION, / *ELECTRON CAPTURE GAS CHROMATOGRA W73-02105
RY, *WATER PURIFIC/ *SUSPENSION, *ELECTROPHORESIS, *ELECTROCHEMIST W73-01362
IL WASTES, / BIOASSAY, TOXICITY, *EMULSIFIERS, DETERGENTS, OIL, *O W70-09429
PESTICIDES, *ALDRIN, *DIELDRIN, *ENDRIN, *GROWTH RATES, *CYANOPHY W72-03545
ATION, *DDT, *DIELDRIN, *ALDRIN, *ENDRIN, MARINE ALGAE, BOTTOM SED W73-00361
PRODUCTIVITY, *MIDGES, LAGOONS/ *ENERGY BUDGET, *SEWAGE, *AQUATIC W72-03556
*STANDING CROP, *TROPHIC LEVEL, *ENERGY BUDGET, PRODUCTIVITY, HER W71-10064
IC FLORA, *PRIMARY PRODUCTIVITY, *ENERGY CONVERSION, STANDING CROP W72-06283
ERMAL POWERPLANTS, *POWERPLANTS, *ENERGY LOSSES, *WATER CONSERVATI W70-09193
*ENGLAND, *LAMINARIA SACCHARINA.: W71-09795
RHAM COUNTY RIVERS.: *ENGLAND, *POLLUTED ESTUARIES, DU W72-08804
, *DOMESTIC WASTES, ALGAE, INDU/ *ENTERIC BACTERIA, *WATER QUALITY W72-09590
OONS, *PATHO/ *ALGAE, *CULTURES, *ENTERIC BACTERIA, *OXIDATION LAG W70-08319
GROWTH STAGES, PATHOGE/ *ALGAE, *ENTERIC BACTERIA, *BACTERICIDES, W71-05155
CTERIA, CULTURES, E. CO/ *ALGAE, *ENTERIC BACTERIA, *PATHOGENIC BA W72-07664
CA, *SARCOPTIFORMES, *ROCKPOOLS, *ENTEROMORPHA, SWEDEN, TIGRIOPUS W71-08026
), CONNECTICUT RIVER(CONNECTICU/ *ENTRAINMENT, WHITE RIVER(INDIANA W71-11517
DROGEN ION CONCENTRATION, ALGAE, *ENTROPHICATION, NEUTRALIZATION, W70-10427
NT REQUIREMENTS, NUISANCE ALGAS, *ENTROPHICATION, CYTOLOGICAL STUD W72-03985
*ENUMERATION, ANACYSTIS.: W71-00139
T, WATER Q/ *FARM WASTES, *HOGS, *ENVIRONMENT, *POLLUTION ABATEMEN W71-08214
MINANT, STREAM QUALITY, WASTE M/ *ENVIRONMENTAL QUALITY, AIR CONTA W71-08214
ANALYSIS, RADIOACTI/ *SAMPLING, *ENVIRONMENT, *NEUTRON ACTIVATION W71-11036
AS/ *RADIOISOTOPES, *ABSORPTION, *ENVIRONMENTAL EFFECTS, NUCLEAR W W72-10955
AQUATIC WEED CONTROL, *PARAQUAT, *ENVIRONMENTAL EFFECTS, DIQUAT, A W72-00155
R POLLUTION CONTRO/ *MANAGEMENT, *ENVIRONMENT, *AGRICULTURE, *WATE W72-00846
EXPLOSI/ *RADIOACTIVITY EFFECTS, *ENVIRONMENTAL EFFECTS, *NUCLEAR W72-00960
LAKES, *WATER POLLUTION CONTROL, *ENVIRONMENTAL SANITATION, *LEGIS W72-01660
*MEASUREMENT, *DATA COLLECTIONS, *ENVIRONMENT, *RADIOACTIVITY, *RA W72-07841
N-SITE T/ *LIGHT, *GROWTH RATES, *ENVIRONMENTAL EFFECTS, LARVAE, O W73-00348
: *MERCURY POLLUTION, *ENVIRONMENTAL PROTECTION AGENCY. W72-14782
TS, *STREAM, *THERMAL POLLUTION, *ENVIRONMENTAL EFFECTS, *ECOLOGY, W72-14659
OVERNMEN/ *LEGISLATION, *TOXINS, *ENVIRONMENTAL EFFECTS, FEDERAL G W73-00703
OTOZOA, *BIOLOGICAL COMMUNITIES, *ENVIRONMENTAL EFFECTS, *STANDING W73-00367
ICS, *SEASONAL,/ *PHYTOPLANKTON, *ENVIRONMENTAL EFFECTS, *SYSTEMAT W73-00442
RPHOLOG/ *PHAEOPHYTA, *SALINITY, *ENVIRONMENTAL EFFECTS, *PLANT MO W73-00369
LANTS, *MARINE ANIMAL/ *ECOLOGY, *ENVIRONMENTAL EFFECTS, *MARINE P W73-02029
BON DIOXIDE, NITRATES, NITRITES, *ENZYMES.: /ON CONCENTRATION, CAR W72-04309

ANT PHYSIO/ *ALGAE, *PHOSPHATES,	*ENZYMES, *CULTURES, NITRATES, PL	W72-05476
HYTOPLANKTON, *SEWAGE LAGOONS, /	*ENZYMES, *DETERGENTS, *ALGAE, *P	W72-07661
LATION, *ALGAE, *STRATIFICATION,	*ENZYMES, THERMOCLINE, EPILIMNION	W72-03224
AND), AMPHOR/ *EPIPSAMMIC ALGAE,	*EPIPELIC ALGAE, SHEAR WATER(ENGL	W71-12855
IS / *DICHOTOMOSIPHON TUBEROSUS,	*EPIPHYTES, GROWTH MEDIA, COCCONE	W72-12744
ERIA, *MACROPHYTE-EPIPHTYE META/	*EPIPHYTIC ALGAE, *EPIPHYTIC BACT	W72-03216
EPIPHTYE META/ *EPIPHYTIC ALGAE,	*EPIPHYTIC BACTERIA, *MACROPHYTE-	W72-03216
KALINE WATER, MAGNESIUM, SODIUM/	*EPIPHYTOLOGY, *LAKES, *ALGAE, AL	W71-04526
E, SHEAR WATER(ENGLAND), AMPHOR/	*EPIPSAMMIC ALGAE, *EPIPELIC ALGA	W71-12855
UTOSPORES, VARIABILITY, BIAS, S/	*ERRORS, *SYNCHRONOUS CULTURES, A	W72-11920
PHICATION, *MATHEMATICAL MODELS,	*ESSENTIAL NUTRIENTS, *PRIMARY PR	W72-08986
NUTRIENTS, *AQU/ *ALGAL CONTROL,	*ESSENTIAL NUTRIENTS, *DEFICIENT	W72-07514
GAE, LIGHT INTENSITY, TIDAL EFF/	*ESTUARIES, *DIATOMS, *SESSILE AL	W73-00853
AGE, *INDUSTRIAL WASTES/ *ALGAE,	*ESTUARIES, *WASTE DISPOSAL, *SEW	W72-08804
ER POLLUTION CONTROL, MANAGEMEN/	*ESTUARIES, *EUTROPHICATION, *WAT	W72-09171
R RESOURCES DEVELOPMENT, *WATER/	*ESTUARIES, *WATER QUALITY, *WATE	W71-09788
AL BLOOMS, *PHYTOPLANKTON, *NUT/	*ESTUARIES, *EUTROPHICATION, *ALG	W72-01329
CTS, *WATER RESOURCES DEVELOPME/	*ESTUARIES, *WATER POLLUTION EFFE	W71-05553
TIVITY, *COASTS, *PHYTOPLANKTON,	*ESTUARIES, *STANDING CROP, SEASO	W71-07875
LITY CONTROL, *ECOLOGY, *ALGAE,/	*ESTUARIES, *NUTRIENTS, WATER QUA	W71-08775
S, *DISTRIBUTION, *MARINE ALGAE,	*ESTUARIES, PRODUCTIVITY, POPULAT	W72-04766
CTS, *DISTRICT OF COLUMBIA, TID/	*ESTUARIES, *WATER POLLUTION EFFE	W70-08248
EN SAG, *IMPAIRED WATER QUALITY,	*ESTUARIES, *DELAWARE RIVER, VEGE	W70-09189
CTS, *DISTRICT OF COLUMBIA, TID/	*ESTUARIES, *WATER POLLUTION EFFE	W70-08249
ATION, *WATER POLLUTION SOURCES,	*ESTUARIES, RIVERS, PHOSPHATES, N	W70-06509
SONAL, CHRYSOPHYTA, SYSTEMATICS,	*ESTUARINE ENVIRONMENT, SEA WATER	W72-08141
PRODUCTIVITY, *SHOALS, *BENTHIC/	*ESTUARINE ENVIRONMENT, *PRIMARY	W72-14317
NMENT, *ECOLOGICAL DISTRIBUTION,	*ESTUARINE ENVIRONMENT, HABITATS,	W72-11854
WA/ *ECOSYSTEMS, *PRODUCTIVITY,	*ESTUARINE ENVIRONMENT, *DOMESTIC	W72-12567
ATURE, *WATER QUALITY STANDARDS,	*ESTURARIES, *THERMAL POLLUTION,	W72-13636
TRIBUTION PATTERNS, *NORTH CARO/	*ESTUARIES, *PHYTOPLANKTON, *DIS	W71-07337
ELLAE, BAJA CALIFORNIA, HEMIAUL/	*ETHMODISCUS REX, ETHMODISCUS GAZ	W72-03567
OLITES, BIOLOGIC/ MICE, INSECTS,	*ETHOXYCHLOR, *METHYLCHLOR, METAB	W72-07703
	*EUDIAPTOMUS ZACHARIASI.:	W72-14805
OLOGY, BIOASSAY, MICROORGANISMS,	*EUGLENA, ALGAE.: /YSIOLOGICAL EC	W71-12303
EMATICS, / *ALGAE, *CHLOROPHYTA,	*EUGLENOPHYTA, *PYRROPHYTA, *SYST	W72-07683
HYTOPLANKTON, *POLLUTANT IDENTI/	*EUGLENOPHYTA, *BIOINDICATORS, *P	W73-00918
*DESMIDS,	*EUPLANKTON, *LIMNOPLANKTON.:	W71-13149
, POLYEDRIELLA HELVETICA, PLEUR/	*EUSTIGMATOPHYCEAE, XANTHOPHYCEAE	W72-14808
W YORK, WAT/ *LIMNOLOGY, *LAKES,	*EUSTROPHICATION, *HYDROLOGY, *NE	W72-13851
PLANKTON, *PRIMARY PRODUCTIVITY,	*EUTROPHICATION, *BIOMASS, *HYDRO	W72-13871
AL NUTRIENTS, *CHLORELLA, ALGAE,	*EUTROPHICATION, NEW MEXICO, BIOD	W72-13992
, *POLITICAL ASPECT/ *LAKE ERIE,	*EUTROPHICATION, *CHEMICAL WASTES	W72-14782
*ALGAL CONTROL, *AQUATIC WEED C/	*EUTROPHICATION, *AQUATIC ALGAE,	W72-14692
ER, *DECOMPOSING ORGANIC MATTER,	*EUTROPHICATION, *PHYTOPLANKTON,	W72-14673
IENTS, *DIATOMS, *LAKE MICHIGAN,	*EUTROPHICATION, *SPATIAL DISTRIB	W73-00384
IFICA/ *DIATOMS, *BIOINDICATORS,	*EUTROPHICATION, *POLLUTANT IDENT	W73-00432
ERTIES, *WATER QUALITY, *BIOIND/	*EUTROPHICATION, *BIOLOGICAL PROP	W73-00002
*CYANOPHYTA, *DISSOLVED OXYGEN,	*EUTROPHICATION, *DISTRIBUTION, F	W73-00236
*ALGAE, *CULTURES, *INDICATORS,	*EUTROPHICATION, FJORDS, BIOASSAY	W73-00234
DGING, DESILTING, SEDIMENTATION,	*EUTROPHICATION, BIOLOGICAL COMMU	W73-00938
ER POLLUTION CONTROL,/ *REVIEWS,	*EUTROPHICATION, *NUTRIENTS, *WAT	W73-00840
*LAKES, *PIGMENTS, *SUCCESSION,	*EUTROPHICATION, CHLOROPHYLL, PLA	W73-00916
N, ORGANIC COMPOU/ *AMINO ACIDS,	*EUTROPHICATION, *PLANKTON, OREGO	W73-01269
TS, *FRESHWATER, *BIOASSAY, *ST/	*EUTROPHICATION, *LAKES, *NUTRIEN	W72-07648
IGOTROPHY, PHOSPHORUS, NITROGEN,	*EUTROPHICATION, LABORATORY TESTS	W72-08376
OSPHATES, *LOCAL GOVERNMENTS, */	*EUTROPHICATION, *DETERGENTS, *PH	W72-06638
FINANCING, PLANT GROWTH, TOXICI/	*EUTROPHICATION, *ALGAL CONTROL,	W71-13172
IVITY, *ALGAE, / *WATER QUALITY,	*EUTROPHICATION, *PRIMARY PRODUCT	W71-13183
NNESOTA, LAKES,/ *PHYTOPLANKTON,	*EUTROPHICATION, *OLIGOTROPHY, MI	W71-13149
QUATIC PLANTS, MAT/ *PHOSPHORUS,	*EUTROPHICATION, FERTILIZATION, A	W71-12072
SOUR/ *WATER POLLUTION CONTROL,	*EUTROPHICATION, *WATER POLLUTION	W71-12091
AE, SEWAGE, / *LABORATORY TESTS,	*EUTROPHICATION, *NUTRIENTS, *ALG	W71-12857
LUTION EFFECTS, *ORGANIC WASTES,	*EUTROPHICATION, *CHELATION, *SEW	W71-09674
ENT, *WATER QUALITY CONTROL, AM/	*EUTROPHICATION, *TERTIARY TREATM	W71-09450
TERGENTS, *POLLUTION ABATEMENT,/	*EUTROPHICATION, *PHOSPHATES, *DE	W71-09429
, WATER POLLUTION SOU/ *AMMONIA,	*EUTROPHICATION, *NITROGEN, ALGAE	W71-11496
PRODUCTIVITY, FARM PONDS, NUTRI/	*EUTROPHICATION, *IONS, *AQUATIC	W71-11001
ALGAE, *PHOSPHORUS, *CYANOPHYTA,	*EUTROPHICATION, BIOMASS, WATER Q	W71-10645
ECASTING, *ALGAE, *PLANT GROWTH,	*EUTROPHICATION, GROWTH RATES, SA	W71-11033
WATERS, *REVIEWS, *WATER POLLU/	*EUTROPHICATION, *ALGAE, *SURFACE	W71-10580
, NI/ *NUTRIENTS, *LAKE ONTARIO,	*EUTROPHICATION, LAKE ERIE, ALGAE	W71-11009
*SEWAGE TREATMENT, *PHOSPHATES,	*EUTROPHICATION, *WATER POLLUTION	W71-11507
EFFLUENTS, GEORGIA, THERMOCLINE/	*EUTROPHICATION, *ALGAE, *SEWAGE	W71-11914
YNTHESIS, *ALGAE, CARBON DIOXID/	*EUTROPHICATION, *CARBON, *PHOTOS	W71-11519
S, PHOTOSYNTHE/ *CARBON, *ALGAE,	*EUTROPHICATION, *LIMITING FACTOR	W71-12068
OLLUTION EFFECTS, OHIO, NUTRIEN/	*EUTROPHICATION, *LAKES, *WATER P	W71-11949
OSPHA/ *WATER POLLUTION CONTROL,	*EUTROPHICATION, *DETERGENTS, *PH	W71-11516
PHYTOPLANKTON, *NUT/ *ESTUARIES,	*EUTROPHICATION, *ALGAL BLOOMS, *	W72-01329
FICATION, / *NITROGEN, PLANKTON,	*EUTROPHICATION, POLLUTANT IDENTI	W72-01780
*CYANOPHYTA, *NUISANCE ALGAE,	*EUTROPHICATION, REVIEWS.:	W72-01358
S, *WATER POLLUTION EFFECTS, *A/	*EUTROPHICATION, *LABORATORY TEST	W72-03218
LOGICAL / *PRIMARY PRODUCTIVITY,	*EUTROPHICATION, *NUTRIENTS, *BIO	W72-03216
AMPLING, WATER RESOURCES, AREA /	*EUTROPHICATION, *MEASUREMENT, *S	W72-03215
S, ALG/ WATER POLLUTION EFFECTS,	*EUTROPHICATION, LAKES, *NUTRIENT	W72-00798
LUTION CONTROL, *NUISANCE ALGAE,	*EUTROPHICATION, LAKES, WASHINGTO	W72-00799
GAE, NITROGEN, ACTIVATED SLUDGE/	*EUTROPHICATION, *PHOSPHOROUS, AL	W71-13334
ATER POLLUTION EFFECTS, NUTRIEN/	*EUTROPHICATION, *WATER SUPPLY, W	W71-13185
*BIOLOGICAL TREATMENT, PLANKTON,	*EUTROPHICATION, *NUTRIENTS, MICR	W71-13553
*WATER QUALITY CONTROL, *LAKES,	*EUTROPHICATION, AQUATIC ALGAE, N	W71-13256

*NUTRIENTS, ALGAE, LAKES, PHOSP/ *EUTROPHICATION, *ON-SITE TESTS, W72-09165
NS, *SYSTEMATICS, *SALINE LAKES, *EUTROPHICATION, WATER SAMPLING, W72-09111
CRUS, NITROGEN, ALGAE.: *EUTROPHICATION, *CARBON, *PHOSPH W72-09155
R QUALITY CONTROL, SEWAGE TREAT/ *EUTROPHICATION, *RECYCLING, WATE W72-09378
CONTROL, MANAGEMEN/ *ESTUARIES, *EUTROPHICATION, *WATER POLLUTION W72-09171
N, *WATER QUALI/ *ALGAL CONTROL, *EUTROPHICATION, *DESTRATIFICATIO W72-09304
HLORELLA, CARB/ *CARBON, *ALGAE, *EUTROPHICATION, *GROWTH RATES, C W72-09168
SMENT/ *DETERGENTS, *PHOSPHATES, *EUTROPHICATION, NUTRIENTS, ASSES W72-09169
DELS, *ESSENTIAL NUTRIENTS, *PR/ *EUTROPHICATION, *MATHEMATICAL MO W72-08986
RIBUTION, *AQUATIC A/ *AERATION, *EUTROPHICATION, *ECOLOGICAL DIST W72-09061
OLOGY, OREGON, *SED/ *VITAMIN B, *EUTROPHICATION, *AQUATIC MICROBI W72-08989
HATES, *WATER POLLUTION SOURCES, *EUTROPHICATION, *DETERGENTS, WAT W72-11186
AE, SEWAGE TREATMENT, WATER POL/ *EUTROPHICATION, *NUTRIENTS, *ALG W72-10625
SEWAGE TREATMENT, *EFFICIENCIES, *EUTROPHICATION, *LAKES, *WATER P W72-11702
NKTON, *RECREATION, *WA/ *ALGAE, *EUTROPHICATION, *NUTRIENTS, *PLA W72-10431
CTIVITY, *RESERVOIRS, *NEBRASKA, *EUTROPHICATION, PLANKTON, CHLORO W72-04270
*ALGAL CONTROL, *AQUATIC ALGAE, *EUTROPHICATION, MUNICIPAL WASTES W72-03970
ONTROL, *NON-STRUCTURAL/ *MAINE, *EUTROPHICATION, *WATER QUALITY C W72-04279
ENCY, *PUBLIC HEAL/ *DETERGENTS, *EUTROPHICATION, *ECONOMIC EFFICI W72-04266
*PHOSPHORUS, *SEDIMENTS, *ALGAE, *EUTROPHICATION, LAKES, AQUATIC P W72-03742
TROGEN, EVA/ *NUTRIENTS, *ALGAE, *EUTROPHICATION, *PHOSPHORUS, *NI W72-04070
POLLUTION CONTROL, WATER POLLUT/ *EUTROPHICATION, *CARBON, *WATER W72-04769
CONTROL, *WASTE WATER(POLLUTIO/ *EUTROPHICATION, *WATER POLLUTION W72-05473
EAT PLAINS, NUTRIENTS, RECREATI/ *EUTROPHICATION, *RESERVOIRS, *GR W72-04759
ARBON DIOXIDE, AERATION, MIXING, *EUTROPHICATION, *ANALYTICAL TECH W72-04433
GAE, *BIOCHEMICAL OXYGEN DEMAND, *EUTROPHICATION, PHOSPHOROUS, NIT W72-04431
RECREATION, WATER SUPPLY, WATER/ *EUTROPHICATION, *LAKES, *MAINE, W72-04280
TERIA, *LAKE SOILS, *PHOSPHORUS, *EUTROPHICATION, LAKES, AQUATIC A W72-04292
TER POLLUTION CONTROL, *WASTE W/ *EUTROPHICATION, *PHOSPHATES, *WA W72-04734
ALITY CONTROL, *ECOLOGY, *ALGAE, *EUTROPHICATION, NUTRIENT REQUIRE W71-08775
, *NITROGEN, *PHOSPHORUS, ALGAE, *EUTROPHICATION, AQUATIC ENVIRONM W71-08953
NS, *SEWAGE LAGOONS, BIODEGRADA/ *EUTROPHICATION, *OXIDATION LAGOO W71-07973
PHOSPHATES, *PATH OF POLLUTANTS, *EUTROPHICATION, *LEACHING NUTRIE W71-04216
ROL, *FORMULATION, *SURFACTANTS, *EUTROPHICATION, LINEAR ALKYLATE W71-06247
CES DEVELOPMENT, *MASSACHUSETTS, *EUTROPHICATION, NUTRIENTS, ALGAE W71-05553
PHORUS, NUTRIENTS, ALGAE, NITRA/ *EUTROPHICATION, *NITROGEN, *PHOS W71-06443
SOURCES, *ESTUARIE/ *NUTRIENTS, *EUTROPHICATION, *WATER POLLUTION W70-06509
T REQUIREMENTS, CHLOROPHYTA, CY/ *EUTROPHICATION, *ALGAE, *NUTRIEN W70-05469
ONTROL, *NITROGEN C/ *NUTRIENTS, *EUTROPHICATION, *WATER QUALITY C W70-04080
DETERGENTS, *TERTIARY TREATMENT, *EUTROPHICATION, WASTE WATER TREA W70-09325
M SEDIMENTS, *LIMNOLOGY, *ALGAE, *EUTROPHICATION, LAKES, WATER QUA W70-08944
OSPHATES, *NITRATES, *NUTRIENTS, *EUTROPHICATION, ALGAE, SURFACTAN W70-09388
SLUDGE WORMS, *DISSOLVED OXYGEN, *EUTROPHICATION, *BIOCHEMICAL OXY W70-09189
RBON, BACTERIA, ALGAE, SYMBIOSI/ *EUTROPHICATION, *PHOSPHORUS, *CA W70-07283
ENVIRONMENT, CHEMICAL / *SEWAGE, *EUTROPHICATION, *LAKES, AQUATIC W71-00117
SING ORGANIC MATTER, NUTRIENTS,/ *EUTROPHICATION, *ALGAE, *DECOMPO W70-10405
YSTEMS ANALYSIS, *WATER QUALITY, *EUTROPHICATION, *MATHEMATICAL MO W71-03021
LANKTON/ *WATER QUALITY, *LAKES, *EUTROPHICATION, PHYTOPLANKTON, P W71-02880
OLLUTION, MINNESOTA, WATER QUAL/ *EUTROPHICATION, *LAKES, *WATER P W71-03012
CONTROL, WATER POLLUTION EFFEC/ *EUTROPHICATION, *WATER POLLUTION W71-01486
HURON, EFFLUENTS, WATER QUALIT/ *EUTROPHICATION, *NUTRIENTS, LAKE W72-13662
C ALGAE, FILTRATION, CYANOPHYTA, *EUTROPHICATION, *OHIO.: / AQUATI W72-13673
AE, *LIMITING FACTORS, PHOSPHAT/ *EUTROPHICATION, *NUTRIENTS, *ALG W72-12544
ITY, *PHOTOSYNTHESIS, LIMNOLOGY, *EUTROPHICATION, *KANSAS, MODEL S W72-12393
*SYNOPTIC ANALYSIS, COLD REGIO/ *EUTROPHICATION, *LAKES, *ALPINE, W72-12549
ISSOLVED OXYGEN, *A/ *LAKE ERIE, *EUTROPHICATION, *HYPOLIMNION, *D W72-12990
FFLUE/ *WATER POLLUTION SOURCES, *EUTROPHICATION, *LAKES, SEWAGE E W72-12955
ANALYTICAL TECHNIQUES, WASTE WA/ *EUTROPHICATION, *ASSAY, *ALGAE, W72-12966
POLLUTION CONTROL, *PHOSPHORUS, *EUTROPHICATION, CHEMICAL ANALYSI W73-02478
MS, WATER QUALITY, *ALGAE, *LIG/ *EUTROPHICATION, AQUATIC ECOSYSTE W73-02349
ATION, *WATER POLLUTION CONTROL/ *EUTROPHICATION, *SOCIAL PARTICIP W73-02479
E/ *PHYTOPLANKTON, *CHLOROPHYLL, *EVALUATION, *SEPARATION TECHNIQU W72-13673
*DISSOLVED ORGANIC PHOSPHORUS, *EXCRETION, ALKALINE PHOSPHATASE, W71-10083
CCHARIDES, OLIGOSACCHARIDES, CL/ *EXCRETION, DISACCHARIDES, MONOSA W72-07710
ISM, *HETEROANTAGONISM, AXENIC / *EXCRETORY PRODUCTS, *AUTOANTAGON W72-12734
ESTERN ONTARIO, EPILITHIC ALGAE/ *EXPERIMENTAL LAKES AREA, *NORTHW W71-11029
INATION, *ORIGIN, *SIGNIFICANCE, *EXTRACTION, LUCIFERIN-LUCIFERASE W72-04292
LAGOONS.: *FACULTATIVE LAGOONS, *ANAEROBIC W72-09386
*SOIL-WATER-PLANT RELATIONSHIPS, *FALLOUT, *CYCLING NUTRIENTS, ECO W72-03342
TIVITY EFFECTS, *NUCLEAR WASTES, *FALLOUT, *OCEANS, WATER POLLUTIO W72-04463
LIFE, R/ *RADIOACTIVITY EFFECTS, *FALLOUT, *FOOD CHAINS, *AQUATIC W72-04461
LAKES, *STRONTIUM RADIOISOTOPES, *FALLOUT, *FOOD CHAINS, ABSORPTIO W73-00790
DIATION, *RADIOACTIVITY EFFECTS, *FALLOUT, *MARINE ANIMALS, MARINE W73-00828
ASTAL PLAIN, NO/ *BIOINDICATORS, *FALLOUT, *MOLLUSKS, *ATLANTIC CO W73-00818
ING AND COOLING, *SPACE HEATING, *FARM USES OF WASTE HEAT, STEAM P W70-09193
IZATION, *WASTE WATER TREATMENT, *FARM WASTES.: /SYNTHESIS, STABIL W70-06619
GOONS, *WATER POLLUTION EFFECTS, *FARM WASTES, OKLAHOMA, CYANOPHYT W72-04773
, *POLLUTION ABATEMENT, WATER Q/ *FARM WASTES, *HOGS, *ENVIRONMENT W71-08214
AGOONS, *OXIDATION LAGOONS, AER/ *FARM WASTES, *WASTE DISPOSAL, *L W71-07551
CATTLE, DISSOLVED OXYGEN, DIVE/ *FARM WASTES, *RUNOFF, *FISHKILL, W72-08401
LAGOONS, PUMPING, AEROBIC TREAT/ *FARM WASTES, POULTRY, NITROGEN, W72-08396
LAGOONS, NUTRIENTS, AQUATIC AL/ *FARM WASTE, *POULTRY, *OXIDATION W72-02850
OGRAPHY, METHYL ESTERS, ESTERS,/ *FATTY ACIDS, *GAS LIQUID CHROMAT W72-09365
CONTROL, *WATER QUALITY CONTROL, *FEDERAL GOVERNMENT, *GRANTS, EUT W72-01659
MENTAL SANITATION, *LEGISLATION, *FEDERAL GOVERNMENT, EUTROPHICATI W72-01660
POLLUTION CONTROL, *PHOSPHATES, *FEDERAL GOVERNMENT, UNITED STATE W72-01093
ATER POLLUTION EFFECTS, *CATTLE, *FEED LOTS.: /CULTURAL RUNOFF, *W W72-08401
LANDICUS, COSCINODISCUS, COPEPO/ *FEEDING PATTERNS, *CALANUS HELGO W72-09248
*FEEDLOTS, LEMNA.: W72-04773
, POULTRY, HOGS, ANIMAL DISEASE/ *FEEDS, *RUMINANTS, ALGAE, CATTLE W71-07544

, *STABILIZA/ *ALUMINUM SULFATE, *FERRIC CHLORIDE, *FERRIC SULFATE W71-13356
TERS, ANTHRACITE, *WAHNBACH RES/ *FERRIC CHLORIDE, MULTI-LAYER FIL W72-06201
MINUM SULFATE, *FERRIC CHLORIDE, *FERRIC SULFATE, *STABILIZATION P W71-13356
ILE DRAINAGE, *WATER PROPERTIES, *FERTILIZATION, *CULTIVATED LANDS W71-12252
RICE/ *CYANOPHYTA, *PESTICIDES, *FERTILIZATION, NUTRIENTS, ALGAE, W71-12071
UTION), *WASTE WAT/ *IRRIGATICN, *FERTILIZATION, *WASTE WATER(POLL W72-14878
OSPERM WEEDS, CLADOPHORA.: *FILAMENTOUS ALGAE, *AQUATIC ANGI W72-14692
*PIT RECHARGE, *WATER TREATMENT, *FILTATION, ALLUVIA CHANNELS, SUR W71-12410
POND WEEDS, PERMANGANATE.: *FILTER CLOGGING, IRON BACTERIA, W72-06536
ICROBIAL ACTIVITY, *SKELLYSOLVE, *FILTER PAPER DISC DIFFUSION METH W72-08586
AND MAINTENANCE, *COST ANALYSIS, *FILTERS, HEADLOSS, PRESSURE, TUR W72-09393
, MUNICIPAL WASTES, *FILTRATION, *FILTERS, BIOCHEMICAL OXYGEN DEMA W71-13341
OROUS MEDIA/ *NUCLEAR EXPLOSICN, *FILTERS, *DIATOMACEOUS EARTH, *P W72-00974
GOONS, *ALGAE, MUNICIPAL WASTES, *FILTRATION, *FILTERS, BIOCHEMICA W71-13341
ENANCE, *COST ANALYSIS, *FILTER/ *FILTRATION, *OPERATION AND MAINT W72-09393
WATER POLL/ *WATER PURIFICATION, *FILTRATION, *ELECTROCHEMISTRY, * W71-12467
ATES, *TURBIDITY, *FLOCCULATION, *FILTRATION, *ALGAE, ELECTROLYTES W72-06201
PATENTS, *WASTE WATER TREATMENT, *FILTRATION, *SEWAGE TREATMENT, A W72-14469
RFACE WATERS,/ *WATER TREATMENT, *FILTRATION, *MICROORGANISMS, *SU W72-06191
HARVES/ *WASTE WATER TREATMENT, *FILTRATION, *ALGAE, *SUSPENSION, W70-10174
IUM, MAGNESIUM, E. COLI, SILICA, *FILTRATION, LIGHT PENTRATICN, PH W72-11833
FINLAND, DARK FIXATION, COASTA/ *FINLAND, *BOTHNIAN BAY, *GULF OF W72-03228
S, *MARI/ *ZOOPLANKTON, *LARVAE, *FISH EGGS, *SEPARATION TECHNIQUE W73-00371
ER RE/ *WATER POLLUTION SOURCES, *FISH HATCHERIES, *SALMONIDS, WAT W71-11006
*BENTHIC FAUNA, *INVERTEBRATES, *FISH POPULATIONS, *MORTALITY, *A W71-08793
E, *AQUATIC ALGAE, *CRUSTACEANS, *FISH TOXINS, *ALGAL POISONING, D W72-14056
CONTROL, TILAPIA, CARP, ALGAE,/ *FISH, *HERBIVORES, *AQUATIC WEED W71-04528
FFECTS, WATER POLLUTION SOURCES, *FISH, DIATOMS, PROTOZOA, AQUATIC W72-07791
THERMAL POLLUTION, *CRUSTACEANS, *FISH, HEATED WATER, NUCLEAR POWE W72-07907
LOSIONS, *ALASKA, *MARINE ALGAE, *FISHERIES, FISH POPULATIONS, FIS W73-00764
*BENEFICIAL USE, *COOLING WATER, *FISHERIES, WATER QUALITY, FISH A W70-09596
FECTS, *WATER POLLUTION SOURCES, *FISHERIES, OIL WASTES, FISH REPR W71-13723
POUNDS, *NITROGEN CYCLE, *CRABS, *FISHES, ECOLOGY, FOOD CHAINS, ES W72-12567
EN, DIVE/ *FARM WASTES, *RUNOFF, *FISHKILL, CATTLE, DISSOLVED OXYG W72-08401
STRONTIUM, *SAMPLE PRESERVATION, *FLAME EMISSION SPECTROPHOTOMETRY W71-11529
ATMENT, *PHOSPHATES, *TURBIDITY, *FLOCCULATION, *FILTRATION, *ALGA W72-06201
E. COLI, *IONS, *WASTEWATER TRE/ *FLOCCULATION, *ALGAE, *SILICA, * W72-14840
EATMENT, COLLOIDS, *POL/ *ALGAE, *FLOCCULATION, WATER POLLUTION TR W72-11833
SEPARATION TECHNIQUES, / *ALGAE, *FLOCCULATION, EQUIPMENT, COSTS, W70-08904
OPHORESIS, SETTLING, *ALGAE, *S/ *FLOCCULATION, FILTRATION, ELECTR W71-01899
DIELD/ *COPPER SULFATE, *ALGAE, *FLOODWATER, *PATH OF POLLUTANTS, W71-11498
LECTA, PHAEODACTYLUM/ CARBON-14, *FLORIDA STRAIT, DUNALIELLA TERIO W72-09103
IELLA TERTIOLECTA, PHAEODACTYLU/ *FLORIDA STRAIT, CARBON-14, DUNAL W72-09102
PHYTOPLANKTON, ALGAE, CULTURES, *FLORIDA, SEA WATER, CHLOROPHYLL, W72-09102
, THERMAL STRATIFIC/ *LIMNOLOGY, *FLORIDA, *LAKES, PONDS, SAMPLING W72-04269
HIC FLORA, BIOASSAY, CYANOPHYTA, *FLORIDA, CORAL, SALINITY, CURREN W72-11876
MS, BIOTA, PROTOZOA, *AMPHIPODA, *FLORIDA, STREAMS, ISOPODS, SCENE W72-12744
, RESER/ *WATER QUALITY CONTROL, *FLOW AUGMENTATION, WATER QUALITY W72-04260
, WASTE WATER TREATMEN/ *OXIDES, *FLUID FRICTION, SEWERS, SLURRIES W70-06968
*MOLECULAR STRUCTURES, *ALGAE, *FLUID FRICTION, *LITTORAL, PLANT W71-07878
ACTIVATION, *OXYGEN, LI/ *ALGAE, *FLUORESCENCE, *CHLORELLA, *PHOTO W72-14800
IOGRAPHY, GLUTARALDEHYDE, SAMPL/ *FLUORESCENT MICROSCOPY, *AUTORAD W73-00273
ALGAE, *GROWTH RATES, *TOXICITY, *FLUORESCENCE, TRACE ELEMENTS, CO W72-07660
RMAL POLLUTION, *MICROORGANISMS, *FLUORESCENCE, FOOD CHAINS, EQUIP W72-07526
TER CHEMISTRY, *FOAM SEPARATION, *FOAM FRACTIONATION, *ORGANIC MAT W72-05506
SPHATES, *WASTE WATER TREATMENT, *FOAM FRACTIONATION, LIME, HYDROG W71-12200
TION, *ORGANI/ *WATER CHEMISTRY, *FOAM SEPARATION, *FOAM FRACTIONA W72-05506
TION, FISH, I/ *WATER POLLUTION, *FOOD CHAINS, *ECOSYSTEMS, ABSORP W72-05952
RADIOACTIVITY EFFECTS, *FALLOUT, *FOOD CHAINS, *AQUATIC LIFE, RADI W72-04461
OXINS, *CHLOROPHYTA/ *FOOD WEBS, *FOOD CHAINS, *TOXICITY, *ALGAL T W71-08597
*POLLUTANT IDENTIFIC/ *CALCIUM, *FOOD CHAINS, *SPECTROPHOTOMETRY, W71-11529
ORUS RADIOISOTOPES, *TUBIFICIDS, *FOOD CHAINS, *PATH OF POLLUTANTS W72-09668
DIOISOTOPES, *ALGAE, *SEA WATER, *FOOD CHAINS, *ABSORPTION, BIOCHE W71-13243
CTS, *NUCLEAR POWERPLANTS, MONI/ *FOOD CHAINS, *RADIOACTIVITY EFFE W72-00959
EAD RADIOISOTOPES, HEAVY METALS/ *FOOD CHAINS, *TRACE ELEMENTS, *L W72-00940
CTS, *TRITIUM, *NUCLEAR WASTES,/ *FOOD CHAINS, *RADIOACTIVITY EFFE W72-06342
*MONITORING, *CYCLING NUTRIENTS, *FOOD CHAINS, *ORGANIC COMPOUNDS, W72-07715
OBALT / *MARINE ANIMALS, *CLAMS, *FOOD CHAINS, *NUCLEAR WASTES, *C W73-00826
*MARINE ANIMALS, *MARINE ALGAE, *FOOD CHAINS, ABSORPTION, CALCIUM W73-00819
RONTIUM RADIOISOTOPES, *FALLOUT, *FOOD CHAINS, ABSORPTION, SEDIMEN W73-00790
N, *NUCLEAR WASTES, *ABSORPTION, *FOOD CHAINS, PUBLIC HEALTH, PATH W73-00833
TY, *ALGAL TOXINS, *CHLOROPHYTA/ *FOOD WEBS, *FOOD CHAINS, *TOXICI W71-08597
PEPODS, DIATOMS, / *CRUSTACEANS, *FOODS, *LIFE HISTORY STUDIES, CO W72-14805
NUTRIENT REQUIRE/ *ZOOPLANKTON, *FOODS, DAPHNIA, COPEPODS, ALGAE, W72-07940
QUALITY CONTROL, *ALGAE CONTROL, *FORAGE FISH, BIOCONTROL, REMEDIE W71-09163
TES, SAMPLE PRESERVATION, SAMPL/ *FORAMINIFERA, SUBSTRATES, EPIPHY W73-00367
TH, *EUTROPHICATION, GROWTH RAT/ *FORECASTING, *ALGAE, *PLANT GROW W71-11033
ERBICIDES, *NUTRIENTS, *ECOSYST/ *FORESTS, *CUTTING MANAGEMENT, *H W71-01489
OP/ *DETERGENTS, *ALGAL CONTROL, *FORMULATION, *SURFACTANTS, *EUTR W71-06247
NG PLANTS, S/ *CHORELLA, *ALGAE, *FOULING, *NUISANCE ALGAE, FLOATI W71-09261
DEPTH, *SAMPLING, *PERIPHYTON,/ *FOULING, *ARTIFICIAL SUBSTRATES, W73-00374
: *FRACTIONATION-FLOTATION PROCESS. W71-12200
NS, BEHAVIOR.: *ALGAE, *FRESH WATER, DISTRIBUTION PATTER W71-07360
ER, *IONS, *POISONS, *SEA WATER, *FRESH WATER, ALGAE, PHOTOSYNTHES W71-03027
GY, *TOXICITY, *WATER POLLUTION, *FRESHWATER, *AQUATIC LIFE, AQUAT W72-09649
CTS, *RADIOECOLOGY, *RESERVOIRS, *FRESHWATER, HYDROBIOLOGY, ALGAE, W72-10694
*FRESHWATER ALGAE, EDTA.: W72-14694
ROPHICATION, *LAKES, *NUTRIENTS, *FRESHWATER, *BIOASSAY, *STANDING W72-07648
ER, BIOM/ *PRIMARY PRODUCTIVITY, *FRESHWATER, *MARSHES, SALINE WAT W73-01089
DES, DRAG/ *BIOLOGICAL POLYMERS, *FRICTION REDUCTION, POLYSACCHARI W72-03220

, BACTERIA, MAR/ *HYDRODYNAMICS, *FRICTION, *TURBULENT FLOW, ALGAE W72-03220
ER POLLUTION EFFECTS, / *SNAILS, *FROGS, *TRITIUM, ABSORPTION, WAT W72-03348
SAMPLE PREPARATION, BRANCHING./ *FUCUS VESICULOSUS, VESICULATION, W73-00369
MEDITERRANEAN SEA.: *FUCUS VIRSOIDES, FUCUS SPIRALIS, W72-12940
ANABAENA, GLOEOTRICHIA, MICROC/ *FULVIC ACID, GROWTH STIMULATION, W72-01097
SION, CADDIS/ *ALGAE, *PROTOZOA, *FUNGI, *AQUATIC INSECTS, *DISPER W70-06655
N TECHNIQUES, *ALGAE, *BACTERIA, *FUNGI, *BACTERIOPHAGE, *VIRUSES, W72-06124
NITIES, RIVERS, BACTER/ *SEWAGE, *FUNGI, *SLIME, *BIOLOGICAL COMMU W72-08573
ESEARCH POLICY, *UNITED KINGDOM, *FUTURE TRENDS, CLADOPHORA, RESEA W71-13172
ACTERIA, DNA, MOTILITY, CYST FO/ *G + C RATIOS, *NONFRUITING MYXOB W72-07951
ETOGASTER, AEOLOSOMA, CHAOBORUS, *GAMBUSIA AFFINS, EPHEMEROPTERA, W72-07132
CHLORIS PENIOCYSTIS, SPIROGYRA,/ *GAMBUSIA, CYCLOPS NAUPLII, COCCO W72-07132
L ESTERS, ESTERS,/ *FATTY ACIDS, *GAS LIQUID CHROMATOGRAPHY, METHY W72-09365
OPHAGES.: *GAS VACUOLES, MOTILE FORMS, CYAN W71-13184
ES, MICROCYSTIS A/ *SHOCK-WAVES, *GAS-VACUOLE DEFLATION, SOUTH WAL W72-11564
NTAL EFFECTS, *PLANT MORPHOLOGY, *GENETICS, *VARIABILITY, WAVES(WA W73-00369
IA/ *COASTS, *ORGANIC COMPOUNDS, *GEOCHEMISTRY, CYANOPHYTA, BACTER W73-01439
POLLUTION EFFE/ *ALGAE, *LAKES, *GEORGIA, SEWAGE EFFLUENTS, WATER W71-09173
*GEORGIAN BAY(ONTARIO).: W72-13662
GRACILARIA VERU/ *BIOSYNTHESIS, *GIBBERELLIN, HYPNEA MUSCIFORMIS, W72-07496
ILTERS, GRANGE LAKE, EAST TWIN / *GLASS FIBER FILTERS, *MEMBRANE F W72-13673
SKIE LAKE(POLAND), CERATIUM HIR/ *GLOEOTRICHIA ECHINULATA, MIKOLAJ W72-14812
ISSOLVED ORGANIC CARBON, *AMERI/ *GLUCOSE, *SNAKE RIVER(IDAHO), *D W72-13300
CLE, CYTOCHROMES, PHOSPHORYLATI/ *GLUCOSE, *MALATE, CITRIC ACID CY W72-04309
WTH, LUMINARIN-TYPE POLYSACCHAR/ *GLYCEROL, PHOTOHETEROTROPHIC GRO W70-07280
*MARINE PHYTOPLANKTON(PRYMNESIU/ *GLYCEROL, *PHOTOTROPHIC GROWTH, W70-07842
*MARINE DISPOSAL, *GOTHENBRUG, SWEDEN.: W70-09947
DA, BUFFER/ *LOADING GUIDELINES, *GRAND RIVER BASIN(ONTARIO), CANA W71-11017
TY CONTROL, *FEDERAL GOVERNMENT, *GRANTS, EUTROPHICATION, GOVERNME W72-01659
EDESMUS, EUTROPHICATION, PONDS,/ *GRAZING, *PROTOZOA, *ALGAE, SCEN W72-14796
TOPES, *FALLOUT, *FOOD CHAINS, / *GREAT LAKES, *STRONTIUM RADIOISO W73-00790
T FACILITIES, *INFLUENT STREAMS, *GREAT LAKES, AQUATIC ALGAE, ICE, W73-01669
ION, ALGAE, *LAKE SUPERIOR, *LA/ *GREAT LAKES, *LAKES, EUTROPHICAT W72-01094
DICATORS, THERMAL POLLU/ *ALGAE, *GREAT LAKES, *MONITORING, *BIOIN W72-10875
NTS, EUTROPHICATIO/ *PHOSPHATES, *GREAT LAKES, *SEDIMENTS, *NUTRIE W72-04263
I/ *EUTROPHICATION, *RESERVOIRS, *GREAT PLAINS, NUTRIENTS, RECREAT W72-04759
US, TEMPERATURE EFFEC/ *BP 1002, *GREEN ALGAE, PRASINOCLADUS MARIN W70-09429
A, FILTRATES, GROWTH AUTOSTIMUL/ *GREEN ALGAE, *HORMOTILA BLENNIST W71-08687
FIELDS.: *GREEN MANURE, INCIA, JAPAN, RICE W73-02473
*GREENE COUNTY(ILL).: W72-05869
POWERPLANTS/ *THERMAL POLLUTION, *GREENHOUSES, WATER TEMPERATURE, W70-08832
ION.: *GRIFFITHSIA PACIFICA, *REGENERAT W73-01622
Y, TASTE, ODOR, *SURFACE WATERS, *GROUNDWATERS, SURVEYS, *WATER SU W72-06536
U/ *AQUATIC ALGAE, *CHLOROPHYTA, *GROWTH CHAMBERS, *LIFE CYCLES, C W72-07144
), *CONTINUOUS FLOW TURBIDOSTAT, *GROWTH KINETICS, CHEMOSTATS, FLA W70-10403
ATOMS, CHRYSOPHYTA, CHLOROPHYTA, *GROWTH RATES, INHIBITION, *PLANT W70-07842
ES, *LABORATORY TESTS, CULTURES, *GROWTH RATES, PLANT GROWTH REGUL W71-03095
ES, *ALDRIN, *DIELDRIN, *ENDRIN, *GROWTH RATES, *CYANOPHYTA, *TOXI W72-03545
*NUTRIENT REQUIREMENTS, AMMONIA, *GROWTH RATES, *WASTE ASSIMILATIV W72-04787
ORUS, *NUTRIENT REQUIRE/ *ALGAE, *GROWTH RATES, *KINETICS, *PHOSPH W72-04788
ANT GROWTH SUBSTANCES, / *ALGAE, *GROWTH RATES, *PRODUCTIVITY, *PL W71-08687
ATER TREATMENT, *CARBON DIOXIDE, *GROWTH RATES, *AEROBIC CONDITION W72-06612
SCENCE, / *HEAVY METALS, *ALGAE, *GROWTH RATES, *TOXICITY, *FLUORE W72-07660
ES, INHIBITION, OLIGOTR/ *ALGAE, *GROWTH RATES, *NUTRIENTS, CULTUR W72-08376
GANICS ACIDS, *PLANT PHYSIOLOGY, *GROWTH RATES, PLANT MORPHOLOGY, W72-07710
E HISTORY STUDIES, *GRO/ *ALGAE, *GROWTH RATES, *SYSTEMATICS, *LIF W73-02099
ITY, *HEAVY METALS, *TR/ *ALGAE, *GROWTH RATES, *PRIMARY PRODUCTIV W73-01434
ECTS, LARVAE, ON-SITE T/ *LIGHT, *GROWTH RATES, *ENVIRONMENTAL EFF W73-00348
ITY, ENERGY BUDGET, *MARINE ALG/ *GROWTH RATES, *PRIMARY PRODUCTIV W73-00366
C ALGAE, CHLOROPHYTA,/ *CALCIUM, *GROWTH RATES, *CULTURES, *AQUATI W72-12736
ALGAE, *COMPETITION, *NUTRIENTS, *GROWTH RATES, *INHIBITION, FUNGI W72-12734
BORATORY TESTS, *CHEMICAL WASTE, *GROWTH RATES, *SEAWATER, *TOXICI W72-12239
ROGEN ION CONCENTRATION, *ALGAE, *GROWTH RATES, *LIMITING FACTORS, W72-11724
CARBON, *ALGAE, *EUTROPHICATION, *GROWTH RATES, CHLORELLA, CARBON W72-09168
URAL RES/ *MARINE ALGAE, *LIGHT, *GROWTH RATES, PHYTOPLANKTON, NAT W71-13252
YDROGEN ION CONCEN/ *CYANOPHYTA, *GROWTH RATES, *ORGANIC ACIDS, *H W72-00838
VITY TECHNIQUES, *BENTHIC FLORA, *GROWTH RATE, *AQUATIC PRODUCTIVI W71-11486
ELATIONSHIPS, *AMMONIA-NITROGEN, *GROWTH REACTORS, ALGAL PONDS, AL W72-04787
RIPHOSPHATE, *MICROBIAL BIOMASS, *GROWTH STAGE, *DETERMINATION, *O W72-04292
TEMATICS, *LIFE HISTORY STUDIES, *GROWTH STAGES, REPRODUCTION, CAR W73-02099
OWTH RATES, *AEROBIC CONDITIONS, *GROWTH STAGES, *BIODEGRADATION, W72-06612
COASTA/ *FINLAND, *BOTHNIAN BAY, *GULF OF FINLAND, DARK FIXATION, W72-03228
*BENTHIC FLORA, *MARINE ALGAE, *GULF OF MEXICO, *ELECTRIC POWERP W72-11252
INVERTEBRATES, *ATLANTIC OCEAN, *GULF OF MEXICO, PHYTOPLANKTON, C W72-14313
TILLARUM, AANDOMENE, SIPHONALES, *HALIMEDA SPP., PENICILLUS CAPITA W72-14294
LINITY, *LAGOONS, *HYDROBIOLOGY, *HALOPHILIC ANIMALS, *HALOPHYTES, W73-00855
DROBIOLOGY, *HALOPHILIC ANIMALS, *HALOPHYTES, SEA WATER, MARINE AN W73-00855
TECHNIQUES/ *OXIDATION LAGOONS, *HARVESTING OF ALGAE, *SEPARATION W72-10237
ED CONTROL, *MECHANICAL CONTROL/ *HARVESTING OF ALGAE, *AQUATIC WE W71-06188
GE, DISSOLVED SOLIDS, F/ *ALGAE, *HARVESTING OF ALGAE, *ION EXCHAN W71-00138
LIGHT INTENSITY, *ECOLOGICAL DI/ *HEAT RESISTANCE, *MARINE ALGAE, W73-00838
*OXIDATION LAGOONS, *CHANNELS, *HEAT TRANSFER, *PLASTICS, DOMEST W72-10401
COOLING WATER, *FISHERIES, WATE/ *HEATED WATER, *BENEFICIAL USE, * W70-09596
, *WA/ *WATER POLLUTION EFFECTS, *HEATED WATER, *WATER TEMPERATURE W72-13636
CONSERVATION, *AIR CONDITIONING, *HEATING, *COOLING, *THERMAL POWE W70-09193
CHLOROPHY/ *INDICATORS, *ALGAE, *HEAVY METALS, *TOXICITY, COPPER, W71-05991
POLLUTION EFFECTS, RESPIRATION, *HEAVY METALS, CHEMICAL WASTES, M W72-12576
RES, *CHELATION, *ION TRANSPORT, *HEAVY METALS, EUTROPHICATION, PO W73-02112
TH RATES, *PRIMARY PRODUCTIVITY, *HEAVY METALS, *TRACE ELEMENTS, N W73-01434
OSYSTEMS/ *TERRESTRIAL HABITATS, *HEAVY METALS, *PRODUCTIVITY, *EC W72-14303

NVIRONMENT, *TOXICITY/ *REVIEWS, *HEAVY METALS, *ALGAE, *AQUATIC E W72-14694
TES, *TOXICITY, *FLUORESCENCE, / *HEAVY METALS, *ALGAE, *GROWTH RA W72-07660
S, EUDORINA, NOTONECTIDAE, BODO/ *HEMIPTERA, *TRICHOPTERA, CORIXID W70-06655
A, RESP/ *CHLAMYDOMONAS, *ALGAE, *HERBICIDES, PESTICIDES, CHLORELL W70-07255
*FORESTS, *CUTTING MANAGEMENT, *HERBICIDES, *NUTRIENTS, *ECOSYST W71-01489
*TOXICITY, SOILS, TRIAGINE PES/ *HERBICIDES, *PESTICIDES, *ALGAE, W71-07675
PESTICIDES, PESTICIDE TOXICITY, *HERBICIDES, DALAPON, 2-4-D, INSE W72-12576
L, *PARAQUAT, *ENVIRONMENTAL EF/ *HERBICIDES, *AQUATIC WEED CONTRO W72-00155
L, TILAPIA, CARP, ALGAE,/ *FISH, *HERBIVORES, *AQUATIC WEED CONTRO W71-04528
ETORY PRODUCTS, *AUTOANTAGONISM, *HETEROANTAGONISM, AXENIC CULTURE W72-12734
CONTINUOUS CULTURE, BATCH CULTU/ *HETEROCYSTS, *ANABAENA SP L-31, W72-03217
, ANABAEN/ *ACETYLENE REDUCTION, *HETEROCYSTS, ANABAENA CYLINDRICA W71-12070
HARTW/ *ANTOTROPHIC POPULATIONS, *HETEROTROPHIC POPULATIONS, LAKE W71-00475
*MOLECULAR SIZE, *ALGAL EXUDAT/ *HIGH MOLECULAR WEIGHT COMPOUNDS, W71-07878
TURBULANT FRICTIONAL DRAG, ALKY/ *HIGH-MOLECULAR-WEIGHT POLYMERS, W70-06968
BATEMENT, WATER Q/ *FARM WASTES, *HOGS, *ENVIRONMENT, *POLLUTION A W71-08214
, FORAMIN/ *VERTICAL COLLECTORS, *HORIZONTAL COLLECTORS, BRYOZOANS W72-05432
GROWTH AUTOSTIMUL/ *GREEN ALGAE, *HORMOTILA BLENNISTA, FILTRATES, W71-08687
T, PARTICULATE MATTER, NITROSOM/ *HUBBARD BROOK EXPERIMENTAL FORES W71-01489
IATOMACEOUS EARTH, FOOD/ *ALGAE, *HUMAN POPULATION, FERTILIZERS, D W71-09157
PES.: *HUMBOLDT BAY, *CESIUM RADIOISOTO W73-00826
AE, EUTROPHICATION/ *CYANOPHYTA, *HUMIC ACIDS, ORGANIC MATTER, ALG W72-01097
GRAPHY, FUCUS VESICULOSUS, LAMI/ *HUMIC COMPOUNDS, COLUMN CHROMATO W72-12166
*ROCKPOOLS, *ENTEROMORPHA, SWE/ *HYADESIA FUSCA, *SARCOPTIFORMES, W71-08026
SP., ENDOCLADIA SP., PHYLLOSPA/ *HYALOCHLORELLA MARINA, RHODOMELA W71-11851
SSIPES, TYPHA ANGUSTATA, IPOMOE/ *HYDERABAD(INDIA), EICHHORNIA CRA W72-01363
ITY, *WATER QUALITY CONTROL, *W/ *HYDROBIOLOGY, *PATHOLOGY, *TOXIC W72-09646
POLLUTION, *FRESHWATER, *AQUATI/ *HYDROBIOLOGY, *TOXICITY, *WATER W72-09649
ICIDE RESIDUES, *PESTICIDE TOXI/ *HYDROBIOLOGY, *PESTICIDES, *PEST W72-09650
KUTSK OBLAST, *BRATSK RESERVOIR, *HYDROBIOLOGY, METALIMNION, BACIL W72-04855
S, *HALOPH/ *SALINITY, *LAGOONS, *HYDROBIOLOGY, *HALOPHILIC ANIMAL W73-00855
SICOCHEMICAL PROPERTIES, *ALGAE, *HYDROBIOLOGY, ENTERIC BACTERIA, W72-07896
HYTES.: *USSR, *TOXICOLOGY, *HYDROBIONTS, MACROPHYTES, SAPROP W72-09650
*USSR, *TOXICOLOGY, *TOXICANTS, *HYDROBIONTS, ICHTHYOLOGY, ICHTHY W72-09646
.: *USSR, *HYDROBIONTS, *CERIUM, RADIOMETRY W72-09649
LENT FLOW, ALGAE, BACTERIA, MAR/ *HYDRODYNAMICS, *FRICTION, *TURBU W72-03220
CAL MODELS, THE/ *WATER QUALITY, *HYDROELECTRIC PLANTS, *MATHEMATI W72-14405
, *GROWTH RATES, *ORGANIC ACIDS, *HYDROGEN ION CONCENTRATION, META W72-00838
AE, *GROWTH RATES, *LIMITING FA/ *HYDROGEN ION CONCENTRATION, *ALG W72-11724
VITY, *EUTROPHICATION, *BIOMASS, *HYDROGRAPHY, ALGAE, AQUATIC ALGA W72-13871
*BIOMASS, *SAMPLING, *NUTRIENTS, *HYDROGRAPHY, *PACIFIC OCEAN, ECO W72-07892
*IMPOUNDED WATERS, *RESERVOIRS, *HYDROLOGIC DATA, *KANSAS, WATER W72-02274
AS/ *SURVEYS, *DATA COLLECTIONS, *HYDROLOGIC DATA, *MEASUREMENT, * W72-10678
OLOGY, *LAKES, *EUSTROPHICATION, *HYDROLOGY, *NEW YORK, WATER QUAL W72-13851
*LAKES, *WASTE WATER(POLLUTION), *HYDROLYSIS, DETERGENTS, GREAT LA W71-00116
*A/ *LAKE ERIE, *EUTROPHICATION, *HYPOLIMNION, *DISSOLVED OXYGEN, W72-12990
VED OXYGEN, TEMPERATURE, CURREN/ *HYPOLIMNION, *LAKE ERIE, *DISSOL W72-12992
OXYGEN, *NUTRIENTS, *LAKE ERIE, *HYPOLIMNION, MATHEMATICAL MODELS W72-12997
RIA, *SEDIMENT-WATER INTERFACES, *HYPOLIMNION, NITRIFICATION, ALGA W72-12996
ENTHIC FAUNA, *BOTTOM SEDIMENTS, *HYPOLIMNION, PHYTOPLANKTON, LIGH W72-12995
FLORA, ALGA/ *DISSOLVED OXYGEN, *HYPOLIMNION, *LAKE ERIE, BENTHIC W72-13659
OLIDS, ALGAE, RESERVOIR STORAGE, *HYPOLIMNION, WATER QUALITY CONTR W73-02138
CONTROL, *PHOSPHORUS, *NITROGEN, *IDAHO, *CHLOROPHYLL, *ANABAENA, W72-13300
ERS, *LAKES, *STREAMS, *REVIEWS, *ILLINOIS, RECREATION, FISHING, S W72-05869
ATER RIGHTS, *WATER UTILIZATION, *IMPAIRED WATER QUALITY, RIPARIAN W70-07100
S, *ORGANIC MATTER, *OXYGEN SAG, *IMPAIRED WATER QUALITY, *ESTUARI W70-09189
SE/ *NUTRIENTS, *WATER QUALITY, *IMPOUNDED WATERS, STREAMS, FLOW, W71-03028
HYDROLOGIC DATA/ *WATER QUALITY, *IMPOUNDED WATERS, *RESERVOIRS, * W72-02274
ACUS, EUGLENA CAUDATA, EUGLENA/ *INDIA, ORGANIC NITROGEN, EUGLENA W73-00918
TAHOE PUBLIC UTILITIES DISTRICT, *INDIAN CREEK RESERVOIR, NUTRIENT W72-02412
*PIGMENTS, *LAKES, *SEDIMENTS, *INDICATORS, CHROMATOGRAPHY, CHLO W72-01784
ALGAE, *WATER POLLUTION EFFECTS, *INDICATORS, *SEA WATER, MARINE P W71-12867
UTANT IDENTIFICATION, *BIOASSAY, *INDICATORS, *WATER POLLUTION EFF W71-10553
ATED HYDROCARBON PESTIC/ *ALGAE, *INDICATORS, *PESTICIDES, CHLORIN W73-00238
RDS, BIOASSA/ *ALGAE, *CULTURES, *INDICATORS, *EUTROPHICATION, FJO W73-00234
S, *TOXICITY, COPPER, CHLOROPHY/ *INDICATORS, *ALGAE, *HEAVY METAL W71-05991
RGE, *WAT/ *ARTIFICIAL RECHARGE, *INDUCED INFILTRATION, *PIT RECHA W71-12410
*WATER REUSE, *DOMESTIC WASTES, *INDUSTRIAL WASTES, ACTIVATED SLU W71-09524
Y, *DISSOLVED OXYGEN, *PHYTOPLA/ *INDUSTRIAL WASTES, *WATER QUALIT W71-11876
ARIES, *WASTE DISPOSAL, *SEWAGE, *INDUSTRIAL WASTES, WATER POLLUTI W72-08804
ION EFFECTS, EFFLUENTS, CALIFOR/ *INDUSTRIAL WASTES, *WATER POLLUT W72-09092
IE, *MERCURY, *MUNICIPAL WASTES, *INDUSTRIAL WASTES, WATER POLLUTI W72-14282
*TOXICITY, REVIEWS, *BIOASSAY, *INDUSTRIAL WASTES, *PESTICIDES, W73-01976
UTION, *COASTS, WATER POLLUTION/ *INDUSTRIAL WASTES, *THERMAL POLL W73-01632
: *ALGAL CONTROL, *INFECTION, *VIRUSES, CYANOPHYTA. W73-00246
EFFECTS, *TREATMENT FACILITIES, *INFLUENT STREAMS, *GREAT LAKES, W73-01669
EGETATION.: *INFRARED PHOTOGRAPHY, *AQUATIC V W71-12034
TE SENSING, *AERIAL PHOTOGRAPHY, *INFRARED RADIATION, *AQUATIC ALG W71-12034
*MONITORING, AERIAL PHOTOGRAPHY, *INFRARED RADIATION, WATER QUALIT W72-12487
SMUS, AQUATIC MI/ *ALGAL TOXINS, *INHIBITION, *CHLORELLA, *SCENEDE W72-12255
ER POLL/ *DDT, *SODIUM CHLORIDE, *INHIBITION, *SALT TOLERANCE, WAT W72-13809
*ARSENIC COMPOUNDS, *PHOSPHATES, *INHIBITIONS, *MICROORGANISMS, BA W72-13813
TION, *NUTRIENTS, *GROWTH RATES, *INHIBITION, FUNGI, THERMAL SPRIN W72-12734
NTROL, *RESPIRATION, *CHLORELLA, *INHIBITION, COPPER, FLUORIDES, I W72-11572
TORS, *CHLORIDES, *CYANOPHYTA, / *INHIBITION, *PLANT GROWTH REGULA W72-14706
EAVY METALS, LABORATORY/ *ALGAE, *INHIBITION, *LIPIDS, TOXICITY, H W72-06037
GROWTH REGULATOR/ *CHLOROPHYTA, *INHIBITORS, ALGAL CONTROL, PLANT W71-13793
DRODICTYON, SPODYLOSIUM, SPIROG/ *INORGANIC IONS, ZINC, VOLVOX, HY W71-11001
FRACTIONATION, *ORGANIC MATTER, *INORGANIC COMPOUNDS, IRON, MANGA W72-05506
WATER POLLUTION, MICROORGANISMS, *INORGANIC COMPOUNDS, *NUTRIENTS, W72-07933

, WATER POLLUTION SOURCES, WATE/	*INSECTICIDES, *ST LAWRENCE RIVER	W70-08098
RODUCTIVITY, CHLORINATED HYDROC/	*INSECTICIDES, *ALGAE, *PRIMARY P	W71-10096
IQUES, *WATER POLLUTION CONTROL,	*INSTRUMENTATION, MATHEMATICAL MO	W72-08790
ATER PROPERTIES, *ON-SITE DATA /	*INSTRUMENTATION, *MONITORING, *W	W72-12941
OSIPHON, FRAGILARIA, TABELLARIA/	*INTERFERENCE, ALEWIVES, DICHOTOM	W73-01669
TATUM, THALASSIONEMA NITZSCHIOI/	*INTERREGULATION, SKELETONEMA COS	W73-00441
WATER, WATER POLLUTION EFFECTS,	*INTERTIDAL AREAS, *INVERTEBRATES	W71-08792
TION EFFECTS, *INTERTIDAL AREAS,	*INVERTEBRATES, *ALGAE, CALIFORNI	W71-08792
NMENTAL EFFECTS, *BENTHIC FAUNA,	*INVERTEBRATES, *FISH POPULATIONS	W71-08793
OPES, *DETRITUS, *BENTHIC FAUNA,	*INVERTEBRATES, *MICROENVIRONMENT	W70-06668
THOS, *SAMPLING, *SESSILE ALGAE,	*INVERTEBRATES, ST. LAWRENCE RIVE	W72-05432
ECTS, *PLANTS, *ANIMALS, *ALGAE,	*INVERTEBRATES, *RADIOACTIVITY EF	W71-13233
EFFECTS, *AQUATIC MICROORGANISMS,	*INVESTIGATIONS, *ANALYTICAL TECH	W72-07225
CKS, *BRINES, *ALGAE, *OKLAHOMA,	*IODINE RADIOISOTOPES, URANIUM RA	W72-02073
DIOISOTOPES, *COBALT RADIOISOTO/	*IODINE RADIOISOTOPES, *CESIUM RA	W72-10958
E/ *ALGAE, *HARVESTING OF ALGAE,	*ION EXCHANGE, DISSOLVED SOLIDS,	W71-00138
SPHORUS, *PHOSP/ *PHYTOPLANKTON,	*ION TRANSPORT, *ABSORPTION, *PHO	W72-07143
*COPPER, CULTURES, *CHELATION,	*ION TRANSPORT, *HEAVY METALS, EU	W73-02112
*POTENTIOMETRY, ELECTROANALYSIS/	*ION-SELECTIVE ELECTRODES, EDTA,	W73-02112
	*ION-SELECTIVE CATHODES.:	W72-00800
M PONDS, NUTRI/ *EUTROPHICATION,	*IONS, *AQUATIC PRODUCTIVITY, FAR	W71-11001
SH WATER, ALGAE, PHOTO/ *COPPER,	*IONS, *POISONS, *SEA WATER, *FRE	W71-03027
TION, *ALGAE, *SILICA, *E. COLI,	*IONS, *WASTEWATER TREATMENT, *PO	W72-14840
ERTILIZATION, *CULTIVATED LANDS,	*IOWA, WATER QUALITY, FERTILIZERS	W71-12252
R, ANGARA RIVER, OKA RIV/ *USSR,	*IRKUTSK OBLAST, *BRATSK RESERVOI	W72-04853
R, *HYDROBIOLOGY, METALI/ *USSR,	*IRKUTSK OBLAST, *BRATSK RESERVOI	W72-04855
LUTANT IDENTIFICATION, TOXICITY,	*IRON, *CHLORELLA, NUTRIENT REQUI	W73-02112
NTS, *ELECTRIC POWER PRODUCTION,	*IRRIGATION PROGRAMS, *AGRICULTUR	W72-03327
TE WATER(POLLUTION), *WASTE WAT/	*IRRIGATION, *FERTILIZATION, *WAS	W72-14878
RES, *CYANOPHYTA, *MARINE ALGAE,	*ISOLATION, MARINE PLANTS, EUTROP	W73-00430
N, SEPARATION TECHNIQUES, CYANO/	*ISOLATION, *ULTRAVIOLET RADIATIO	W72-14327
HNIQUES, ATLANTI/ *MARINE FUNGI,	*ISOLATION, ALGAE, SEPARATION TEC	W71-11851
*CULTURES, *ANAEROBIC BACTERIA,	*ISOLATION, *MEMBRANE PROCESSES,	W72-04743
S, *CARBON RADIOISOTOPES, ENVIR/	*ISOTOPE STUDIES, *STABLE ISOTOPE	W72-12709
*RECIRCULATION RATIO,	*ISRAEL(DAN REGION).:	W72-11642
PERIDINIUM WESTI, OSCILLATORIA /	*ISRAEL, *LAKE KINNERET(ISRAEL),	W71-13187
NNIDS, *TYCHOPELAG/ *OYSTER BAY,	*JAMAICA, FALMOUTH HARBOR, *TINTI	W72-07899
ALUM, FERRIC CHLORIDE,	*JERUSALEM(ISRAEL).:	W72-12289
SIS, LIMNOLOGY, *EUTROPHICATION,	*KANSAS, MODEL STUDIES, ECOSYSTEM	W72-12393
, *RESERVOIRS, *HYDROLOGIC DATA,	*KANSAS, WATER CHEMISTRY, CHEMICA	W72-02274
AL WASTES, INVERTEBRATES, ALGAE,	*KELP.: /, HEAVY METALS, INDUSTRI	W72-11854
*MARINE ANIMALS, *SYSTEMATICS,/	*KELPS, *ECOLOGICAL DISTRIBUTION,	W73-00932
HAEOPHYTA, STAND/ *MARINE ALGAE,	*KELPS, *PHYSIOLOGICAL ECOLOGY, P	W73-12940
ER, LEAD, HEAVY METALS, MUSSELS,	*KELPS, STARFISH, CRUSTACEANS, FO	W72-11925
ROPHIC/ *LAKE ANNABESSACOOK(ME),	*KENNEBEC RIVER(ME), CULTURAL EUT	W72-04280
, ISLE OF THANET, MEDWELL ESTUA/	*KENT COAST, INDUSTRIAL EXPANSION	W71-13746
MINES, *WATER POLLUTION EFFECTS,	*KENTUCKY, SULFUR BACTERIA, WATER	W71-07942
*ARTIFICIAL CIRCULATION,	*KEZAR LAKE(NH).:	W72-09061
ESSED AIR.:	*KEZAR LAKE, AIR INJECTION, COMPR	W72-09304
ANABAENA SPP, MICROCYSTIS AERU/	*KEZAR LAKE(N.H.), APHANIZOMENON,	W72-07890
	*KEZAR LAKE(NH).:	W72-14542
T, *RADIOISOTOPES, *MEASUREMENT,	*KINETICS, *ABSORPTION, INSTRUMEN	W72-14326
C BACTERIA, *AEROBIC CONDITIONS,	*KINETICS, *SCENEDESMUS, *DISSOLV	W73-00379
RY TESTS, WATER QUALITY CONTROL,	*KINETICS, *CHLOROPHYTA, ALGAL CO	W72-08376
BACTERIA, *ANAEROBIC CONDITIONS,	*KINETICS, *DISSOLVED SOLIDS, ORG	W73-01066
, *NUTRIENT REQUIREMENT/ *ALGAE,	*KINETICS, *WASTE WATER TREATMENT	W72-04789
EQUIREMENTS, AMMONIA, */ *ALGAE,	*KINETICS, *NITROGEN, *NUTRIENT R	W72-04787
REQUIRE/ *ALGAE, *GROWTH RATES,	*KINETICS, *PHOSPHORUS, *NUTRIENT	W72-04788
EVELOPMENT, TRACE EL/ *BIOASSAY,	*LABORATORY TESTS, RESEARCH AND D	W72-04433
HUMAN PATHOLOGY.: *CYANOPHYTA,	*LABORATORY TESTS, ALGAL TOXINS,	W72-04259
ULTURES, *AQUATIC MICROORGANISM/	*LABORATORY TESTS, *CHLORELLA, *C	W71-08027
OMESTIC WASTES, DINOFLAGELLATES,	*LABORATORY TESTS, CULTURES, *GRO	W71-03095
E, *ACTIVATED SLUDGE, *TOXICITY,	*LABORATORY TESTS, DETERGENTS, WA	W70-08976
*CHLORELLA, *NUTRIENTS, CULTURE/	*LABORATORY TESTS, *RESPIRATION,	W73-01626
GAE, *CULTURES, CYANOPHYTA, HYD/	*LABORATORY TESTS, *BACTERIA, *AL	W73-01657
OPES, *MEASUREMENT, *KINETICS, /	*LABORATORY EQUIPMENT, *RADIOISOT	W72-14326
GON, LAKES,/ *ALGAE, *NUTRIENTS,	*LABORATORY TESTS, *BIOASSAY, ORE	W72-09164
ON EFFECTS, *A/ *EUTROPHICATION,	*LABORATORY TESTS, *WATER POLLUTI	W72-03218
N, *NUTRIENTS, *ALGAE, SEWAGE, /	*LABORATORY TESTS, *EUTROPHICATIO	W71-12857
*SURFACES, LABORATORY EQUIPMENT,	*LABORATORY TESTS, *MEASUREMENT,	W72-12210
E, *GROWTH RATES/ *MARINE ALGAE,	*LABORATORY TESTS, *CHEMICAL WAST	W72-12239
LIC ANIMALS, *HALOPH/ *SALINITY,	*LAGOONS, *HYDROBIOLOGY, *HALOPHI	W73-00855
*FARM WASTES, *WASTE DISPOSAL,	*LAGOONS, *OXIDATION LAGOONS, AER	W71-07551
C RIVER(ME), CULTURAL EUTROPHIC/	*LAKE ANNABESSACOOK(ME), *KENNEBE	W72-04280
, NUTRIENT REQUIRE/ *PHOSPHORUS,	*LAKE BEDS, *ALGAE, ALGAL CONTROL	W72-11617
E, *ODOR, SURVEYS, MICROBIOLOGY/	*LAKE ERIE, *POTABLE WATER, *TAST	W72-10076
GAN, *LAKE HURON, *LAKE ONTARIO,	*LAKE ERIE, WATER POLLUTION EFFEC	W72-01094
ES, AQUAT/ *LAKE SHORES, *ALGAE,	*LAKE ERIE, AQUATIC HABITATS, LAK	W71-09156
*RHODOPHYTA,	*LAKE ERIE, OHIO, ALGAE.:	W71-05630
RA, DI/ *BIBLIOGRAPHIES, *ALGAE,	*LAKE ERIE, PLANKTON, BENTHIC FLO	W73-01615
WASTES, *INDU/ *WATER POLLUTION,	*LAKE ERIE, *MERCURY, *MUNICIPAL	W72-14282
MICAL WASTES, *POLITICAL ASPECT/	*LAKE ERIE, *EUTROPHICATION, *CHE	W72-14782
GEN, *SEDIMENT-WATER INTERFACES,	*LAKE ERIE, HYPOLIMNION, AQUATIC	W72-06690
N DEMAND, *ANAEROBIC CONDITIONS,	*LAKE ERIE, *ALGAE, *BENTHIC FAUN	W72-12995
YGEN DEMAND, HYPOLI/ *SEDIMENTS,	*LAKE ERIE, *DISSOLVED OXYGEN, OX	W72-12994
MPERATURE, CURREN/ *HYPOLIMNION,	*LAKE ERIE, *DISSOLVED OXYGEN, TE	W72-12992
WATER INTERF/ *DISSOLVED OXYGEN,	*LAKE ERIE, *BACTERIA, *SEDIMENT-	W72-12996
*DISSOLVED OXYGEN, *NUTRIENTS,	*LAKE ERIE, *HYPOLIMNION, MATHEMA	W72-12997
OLIMNION, *DISSOLVED OXYGEN, *A/	*LAKE ERIE, *EUTROPHICATION, *HYP	W72-12990
PELEE(ONTARIO), OXYGEN DEPLETIO/	*LAKE ERIE CENTRAL BASIN, POINTE	W72-12992

MA, OEDOGONIUM, ANACYSTIS, CERA/	*LAKE ERIE CENTRAL BASIN, TRIBONE	W72-12995
*DISSOLVED OXYGEN, *HYPOLIMNION,	*LAKE ERIE, BENTHIC FLORA, ALGAE,	W72-13659
DEPLETION.:	*LAKE ERIE CENTRAL BASIN, *OXYGEN	W72-13659
T REGENERATION, DEOXYGENATION, /	*LAKE ERIE CENTRAL BASIN, NUTRIEN	W72-12997
GANI/ *CHLOROPHYLL, *PHOSPHORUS,	*LAKE ERIE, NITROGEN, ALGAE, INOR	W72-13653
EN FIXATION.:	*LAKE ERKEN(SWEDEN), *DARK NITROG	W72-08054
UM CAPRICORNUTUM, *ALGAL GROWTH,	*LAKE GEORGE(NY).: *SELENASTR	W72-08376
HOSPHATES.: PRAIRIE LAKES,	*LAKE HERMAN(SO. DAKOTA), *ORTHOP	W73-00938
*LAKE SUPERIOR, *LAKE MICHIGAN,	*LAKE HURON, *LAKE ONTARIO, *LAKE	W72-01094
M WESTI, OSCILLATORIA / *ISRAEL,	*LAKE KINNERET(ISRAEL), PERIDINIU	W71-13187
TERS, RIVER JORDAN(ISRAEL).:	*LAKE KINNERET(ISRAEL), BOTTOM WA	W72-05469
NTS, CARBON-14 ME/ *LAKE TRAVIS,	*LAKE LIVINGSTON, FIELD MEASUREME	W72-13692
NIC NITROGEN, *PARTICULATE ORGA/	*LAKE METABOLISM, *DISSOLVED ORGA	W72-08048
TION, *POLLUTANT IDENTIFICATION,	*LAKE MICHIGAN, CHRYSOPHYTA, AQUA	W73-00432
*SAMPLING, *NUTRIENTS, *DIATOMS,	*LAKE MICHIGAN, *EUTROPHICATION,	W73-00384
RUS, *CLADOPHORA, *PLANT GROWTH,	*LAKE MICHIGAN, ALGAE, PHOSPHATES	W72-13644
HICATION, ALGAE, *LAKE SUPERIOR,	*LAKE MICHIGAN, *LAKE HURON, *LAK	W72-01094
	*LAKE MINNETONKA(MINN).:	W71-08670
ICATION, *WATER QUALITY CONTROL,	*LAKE MORPHOLOGY, AERATION, MIXIN	W72-09304
SIPHON SUBAEQUALIS, APISTONEMA /	*LAKE NEUSIEDLER(AUSTRIA), CHAMAE	W71-04526
BAY OF QUINTE,	*LAKE ONTARIO.:	W71-08953
OR, *LAKE MICHIGAN, *LAKE HURON,	*LAKE ONTARIO, *LAKE ERIE, WATER	W72-01094
AKE ERIE, ALGAE, NI/ *NUTRIENTS,	*LAKE ONTARIO, *EUTROPHICATION, L	W71-11009
AQUATIC HABITATS, LAKES, AQUAT/	*LAKE SHORES, *ALGAE, *LAKE ERIE,	W71-09156
E JACKSON(GEORGIA), LAKE CARDIN/	*LAKE SIDNEY LANIER(GEORGIA), LAK	W71-11914
HICATION, LAKES/ *SOIL BACTERIA,	*LAKE SOILS, *PHOSPHORUS, *EUTROP	W72-04292
, *LAKES, EUTROPHICATION, ALGAE,	*LAKE SUPERIOR, *LAKE MICHIGAN, *	W72-01094
*PHYTOPLANKTON, *SESSILE ALGAE,	*LAKE SUPERIOR, LIMNOLOGY, FERTIL	W72-12192
*MICROCYSTIS,	*LAKE SUWA(JAPAN).:	W73-00236
	*LAKE TAHOE(CALIF.).:	W72-12955
	*LAKE TAHOE(CALIF-NEV).:	W72-12549
	*LAKE TAHOE.:	W72-01660
*ALGAL GROWTH,	*LAKE TAHOE(CALIF).:	W71-13553
IELD MEASUREMENTS, CARBON-14 ME/	*LAKE TRAVIS, *LAKE LIVINGSTON, F	W72-13692
, MACROPHYTES.:	*LAKE WABAMUN, BIOLOGICAL STUDIES	W73-01704
*LIMITING NUTRIENTS,	*LAKE WASHINGTON(WASH).:	W72-09167
ANALYSIS, ORDINATION ANALYSIS.:	*LAKE WINGRA(WIS.), MULTIVARIATE	W73-02469
ANTHES MINUTISSIMA, ACHNANTHES /	*LAKE WINNIPEG, *SUBSTRATES, ACHN	W72-13142
IS, COLD REGIO/ *EUTROPHICATION,	*LAKES, *ALPINE, *SYNOPTIC ANALYS	W72-12549
AGNESIUM, SODIUM/ *EPIPHYTOLOGY,	*LAKES, *ALGAE, ALKALINE WATER, M	W71-04526
ALITY, *WATER POLLUTION EFFECTS,	*LAKES, *COLORADO, *RECREATION FA	W72-04791
NKTON, PLANKTON/ *WATER QUALITY,	*LAKES, *EUTROPHICATION, PHYTOPLA	W71-02880
ALGAE, / *WATER QUALITY CONTROL,	*LAKES, *EUTROPHICATION, AQUATIC	W71-13256
OGY, *NEW YORK, WAT/ *LIMNOLOGY,	*LAKES, *EUSTROPHICATION, *HYDROL	W72-13851
S, WATER POLLUTION EFFE/ *ALGAE,	*LAKES, *GEORGIA, SEWAGE EFFLUENT	W71-09173
SUPPLY, WATER/ *EUTROPHICATION,	*LAKES, *MAINE, RECREATION, WATER	W72-04280
PERTIES, *SEDIM/ *TROPHIC LEVEL,	*LAKES, *NUTRIENTS, *CHEMICAL PRO	W72-05469
*BIOASSAY, *ST/ *EUTROPHICATION,	*LAKES, *NUTRIENTS, *FRESHWATER,	W72-07648
EUTROPHICATION, CHLOROPHYLL, PL/	*LAKES, *PIGMENTS, *SUCCESSION, *	W73-00916
ANKTON, *PRODUCTIVI/ *LIMNOLOGY,	*LAKES, *PHOTOSYNTHESIS, *PHYTOPL	W71-08670
STRATIFICATION,/ *AQUATIC ALGAE,	*LAKES, *PHYTOPLANKTON, *DENSITY	W71-09852
CHROMATOGRAPHY, CHLO/ *PIGMENTS,	*LAKES, *SEDIMENTS, *INDICATORS,	W72-01784
NOIS, RECREATI/ *SURFACE WATERS,	*LAKES, *STREAMS, *REVIEWS, *ILLI	W72-05869
*BIOTA, *TEMPERATURE, *CANADA,	*LAKES, *THERMAL POLLUTION, WATER	W73-01704
OTOMETER, HYPOL/ *PHYTOPLANKTON,	*LAKES, *TURBIDITY, *PROFILES, PH	W72-07145
TA, WATER QUAL/ *EUTROPHICATION,	*LAKES, *WATER POLLUTION, MINNESO	W71-03012
*HYDROLYSIS, DETER/ *PHOSPHATES,	*LAKES, *WASTE WATER(POLLUTION),	W71-00116
*WATER QUALITY CONTROL, *FEDER/	*LAKES, *WATER POLLUTION CONTROL,	W72-01659
*ENVIRONMENTAL SANITATION, *LE/	*LAKES, *WATER POLLUTION CONTROL,	W72-01660
OHIO, NUTRIEN/ *EUTROPHICATION,	*LAKES, *WATER POLLUTION EFFECTS,	W71-11949
*EFFICIENCIES, *EUTROPHICATION,	*LAKES, *WATER POLLUTION EFFECTS,	W72-11702
*LIMNOLOGY, WATER POLLUTION, W/	*LAKES, *WATER POLLUTION EFFECTS,	W72-12729
LLUTION EFFECTS, EUTROPHICATION,	*LAKES, ALGAE, INFRARED RADIATION	W72-14728
CHIGAN, SESTON, ALGA/ *NITROGEN,	*LAKES, ANALYTICAL TECHNIQUES, MI	W72-08048
ICAL / *SEWAGE, *EUTROPHICATION,	*LAKES, AQUATIC ENVIRONMENT, CHEM	W71-00117
ROPHICATI/ *ECOSYSTEMS, *ALASKA,	*LAKES, ARCTIC, PRODUCTIVITY, EUT	W71-09190
RIENTS, CH/ *ALGAE, *PHOSPHORUS,	*LAKES, CULTURES, PHOSPHATES, NUT	W72-12637
TY, EUTROPH/ *NITROGEN FIXATION,	*LAKES, CYANOPHYTA, LIGHT INTENSI	W73-02472
DISTRIBUTION, P/ *PHYTOPLANKTON,	*LAKES, EPILIMNION, ZOOPLANKTON,	W71-09171
AKE SUPERIOR, *LA/ *GREAT LAKES,	*LAKES, EUTROPHICATION, ALGAE, *L	W72-01094
STRATIFIC/ *LIMNOLOGY, *FLORIDA,	*LAKES, PONDS, SAMPLING, THERMAL	W72-04269
LGAE, *CHLOROPHYTA, *CYANOPHYTA,	*LAKES, RHODOPHYTA, SHORES.: *A	W71-12489
LUTION SOURCES, *EUTROPHICATION,	*LAKES, SEWAGE EFFLUENTS, SEEPAGE	W72-12955
EUTROPHICATI/ *ALGAE, BACTERIA,	*LAKES, SYMBIOSIS, ALGAL CONTROL,	W71-13257
, EPIFAUNA.: *NORTH SEA,	*LAMINAREA SPP., SEWAGE POLLUTION	W72-11854
ULTURE EXPERIMENTS, ATRAZINE, M/	*LAMINARIA, SUBLITTORAL REGION, C	W72-12576
), IRISH SEA.: *ULVA,	*LAMINARIA, LIVERPOOL BAY(ENGLAND	W71-12867
*ENGLAND,	*LAMINARIA SACCHARINA.:	W71-09795
YSTIS SP.:	*LAMINARIA SP., *ULVA SP., MACROC	W71-10553
MBIA, RU/ *LAMINARIA SINCLAIRII,	*LAMINARIA LONGIPES, BRITISH COLU	W73-00428
LONGIPES, BRITISH COLUMBIA, RU/	*LAMINARIA SINCLAIRII, *LAMINARIA	W73-00428
, AERATED LA/ *DISPOSAL SYSTEMS,	*LAND DISPOSAL, OXIDATION DITCHES	W71-07551
TECHNIQUES, *MARI/ *ZOOPLANKTON,	*LARVAE, *FISH EGGS, *SEPARATION	W73-00371
OPLANKTON, ALGAE, FEEDING RATES,	*LARVAL GROWTH STAGE, PHYTOPLANKT	W72-09248
OF POLLUTANTS, *EUTROPHICATION,	*LEACHING NUTRIENTS, FERTILIZERS,	W71-04216
*FOOD CHAINS, *TRACE ELEMENTS,	*LEAD RADIOISOTOPES, HEAVY METALS	W72-00940
PHORA GLOMERATA, ULOTRIC/ *ZINC,	*LEAD, STIGEOCLONIUM TENUE, CLADO	W71-05991
TROL, *ENVIRONMENTAL SANITATION,	*LEGISLATION, *FEDERAL GOVERNMENT	W72-01660
NTROL, *PHOSPHATES, *DETERGENTS,	*LEGISLATION, CHEMICALS, CHEMICAL	W72-01685
NTAL EFFECTS, FEDERAL GOVERNMEN/	*LEGISLATION, *TOXINS, *ENVIRONME	W73-00703

NTAL PROTECTION AGENCY.: *LEGISLATION(PROPOSED), ENVIRONME W73-00703
ANOPHYTA, PLANTS, NITRO/ *ALGAE, *LEGUMES, *NITROGEN FIXATION, *CY W73-01456
ECTICIDES, PHENOLS, *DETERGENTS, *LETHAL LIMIT, ALKYLBENZENE SULFO W72-12576
ALS, *CHLAMYDOMONAS, INHIBITORS, *LETHAL LIMIT, WATER POLLUTION EF W72-11727
*CHLOROPHYTA, *GROWTH CHAMBERS, *LIFE CYCLES, CULTURES, WATER QUA W72-07144
AE, *GROWTH RATES, *SYSTEMATICS, *LIFE HISTORY STUDIES, *GROWTH ST W73-02099
DIATOMS, / *CRUSTACEANS, *FOODS, *LIFE HISTORY STUDIES, COPEPODS, W72-14805
A, EPIPHYTOLOGY, ALGAE, HABITAT/ *LIFE HISTORY STUDIES, *PHAEOPHYT W72-01370
TIVITY, *RADIOACTIVITY TECHNIQU/ *LIGHT INTENSITY, *PRIMARY PRODUC W72-09102
ON, *ABSORPTION, *AQUATIC ALGAE, *LIGHT INTENSITY, *PHOSPHORUS, PL W72-07166
ROPHYTES, *AQUATIC WEED CONTROL, *LIGHT INTENSITY, *NEBRASKA, *CYA W73-02349
, P/ *ALGAE, *OXIDATION LAGOONS, *LIGHT INTENSITY, PONDS, ROTIFERS W71-07382
OSYSTEMS, WATER QUALITY, *ALGAE, *LIGHT PENETRATION, *MACROPHYTES, W73-02349
ODEL STUDIES, ECOSYSTEMS, ALGAE, *LIGHT QUALITY, EUTROPHIC ZONE, C W72-12393
NTAL EFFECTS, LARVAE, ON-SITE T/ *LIGHT, *GROWTH RATES, *ENVIRONME W73-00348
TON, NATURAL RES/ *MARINE ALGAE, *LIGHT, *GROWTH RATES, PHYTOPLANK W71-13252
YTA, NITROGEN FIXATION, OXYGEN,/ *LIGHT, *PHOTOSYNTHESIS, *CYANOPH W72-08584
ALGAE, CULTURES, VO/ *CHLORELLA, *LIGHT, *PHOTOSYNTHETIC OXYGEN, * W70-05547
IOLOGICAL COMMUNITIES, SYSTEMAT/ *LIGHT, *TEMPERATURE, *DIATOMS, B W73-00242
RTHYMS, *METABOLISM, *CHLORELLA, *LIGHT, ALGAE, GROWTH RATES, CHLO W73-00251
COMMUNITIES, SEASONAL, DIATOMS, *LIGHT, SOLAR RADIATION, *PHOTOPE W72-07901
*LIMITING FACTORS, SILICON.: W72-07937
(WIS).: *LIMITING NUTRIENTS, LAKE MENDOTA W72-07933
IATOMS,/ *NUTRIENT REQUIREMENTS, *LIMITING FACTORS, *PHOSPHATES, D W73-01445
GTON(WASH).: *LIMITING NUTRIENTS, *LAKE WASHIN W72-09167
*LIMITING NUTRIENTS.: W72-09155
(MINN), WALDO LAKE(ORE).: *LIMITING NUTRIENTS, SHAGAWA LAKE W72-09165
*LIMITING NUTRIENTS.: W72-09162
*LIMITING NUTRIENTS.: W72-09164
LORIDE, SHRINER'S POND(GA).: *LIMITING NUTRIENTS, POTASSIUM CH W72-09159
*PHOSPHORUS, *PLANKTON, *ALGAE, *LIMITING FACTORS, GROWTH RATES, W72-09163
RBON SOURCES.: *LIMITING NUTRIENTS, ANABAENA, CA W72-09168
AKE(NY), LAKE GEORGE(NY).: *LIMITING NUTRIENTS, CANADARAGO L W72-09163
IVITY/ *CARBON, *SEWAGE LAGOONS, *LIMITING FACTORS, ALGAE, PRODUCT W72-09162
*LIMITING NUTRIENTS, CANADA.: W71-12857
CARBON, *ALGAE, *EUTROPHICATION, *LIMITING FACTORS, PHOTOSYNTHESIS W71-12068
ROPHICATION, *NUTRIENTS, *ALGAE, *LIMITING FACTORS, PHOSPHATES, CH W72-12544
NTRATION, *ALGAE, *GROWTH RATES, *LIMITING FACTORS, *DIURNAL DISTR W72-11724
LAKES, *WATER POLLUTION EFFECTS, *LIMNOLOGY, WATER POLLUTION, WATE W72-12729
R ANALYSIS, *STANDING CROP, *PR/ *LIMNOLOGY, *PHYTOPLANKTON, *WATE W73-01435
ION, *HYDROLOGY, *NEW YORK, WAT/ *LIMNOLOGY, *LAKES, *EUSTROPHICAT W72-13851
*NUTRIENTS, *BOTTOM SEDIMENTS, *LIMNOLOGY, *ALGAE, *EUTROPHICATI W70-08944
IS, *PHYTOPLANKTON, *PRODUCTIVI/ *LIMNOLOGY, *LAKES, *PHOTOSYNTHES W71-08670
DS, SAMPLING, THERMAL STRATIFIC/ *LIMNOLOGY, *FLORIDA, *LAKES, PON W72-04269
*DESMIDS, *EUPLANKTON, *LIMNOPLANKTON.: W71-13149
LABORATORY/ *ALGAE, *INHIBITION, *LIPIDS, TOXICITY, HEAVY METALS, W72-06037
PATENTS, *WASTE WATER TREATMENT, *LIQUID WASTES, SEWAGE TREATMENT, W72-05315
.: *ANIMAL MANURE, *LITERATURE REVIEW, FEEDING VALUE W71-07544
E TREATMENT, SECONDARY SEWAGE T/ *LITERATURE SURVEY, PRIMARY SEWAG W70-09907
*LITERATURE REVIEWS.: W72-12190
CTURES, *ALGAE, *FLUID FRICTION, *LITTORAL, PLANT GROWTH SUBSTANCE W71-07878
THIC FLORA/ *PERIPHYTON, *ALGAE, *LITTORAL, LIGHT PENETRATION, BEN W71-11029
BASIN(ONTARIO), CANADA, BUFFER/ *LOADING GUIDELINES, *GRAND RIVER W71-11017
ATION, *DETERGENTS, *PHOSPHATES, *LOCAL GOVERNMENTS, *WATER POLLUT W72-06638
Z CREEK, ARROYO GRANDE CREEK.: *LOPEZ WATER SUPPLY PROJECT, LOPE W72-05594
SINS), *DD/ *PESTICIDE RESIDUES, *LOTIC ENVIRONMENT, WATERSHEDS(BA W72-12930
*LOXODES MAGNUS.: W72-14796
USES, *PLECTONEMA BORYANIUM, SU/ *LPP VIRUS, *BLUE-GREEN ALGAL VIR W72-04736
EM, PL/ *ADENOSINE TRIPHOSPHATE, *LUCIFERIN-LUCIFERASE ENZYME SYST W71-11916
ETABOLITES.: *LYSIS, *CONTACT TIME, CELLULAR M W72-00411
LPP-1, LPP-1D, PHYCOVIRUS, MITO/ *LYSOGENY, *PLECTONEMA BORYANUM, W72-03188
EEN ALGAE, ANABAENA VARIABILIS,/ *LYSOZYME, *SPHEROPLASTS, BLUE-GR W71-08683
6, CESIUM RADIOISOTOPES, SPIROG/ *MACROINVERTEBRATES, RUTHENIUM-10 W70-06668
*CHEMICAL EVOLUTION, *MACROMOLECULES.: W72-01473
YTIC ALGAE, *EPIPHYTIC BACTERIA, *MACROPHYTE-EPIPHTYE METABOLISM, W72-03216
OLAND), COLONIZATION, POTAMOGET/ *MACROPHYTES, *MIKOLAJSKIE LAKE(P W71-11012
ISION, DEPTH SOUNDE/ *SEA WEEDS, *MACROPHYTES, *SEA GRASSES, TELEV W72-09120
ITY, *ALGAE, *LIGHT PENETRATION, *MACROPHYTES, *AQUATIC WEED CONTR W73-02349
LAWN MAINTENANCE.: *MADISON(WIS), LAKE MENDOTA(WIS), W73-02479
UALITY CONTROL, *NON-STRUCTURAL/ *MAINE, *EUTROPHICATION, *WATER Q W72-04279
ECOSYSTEMS, PESTICIDE KINETICS, *MAINE, MUDS, INVERTEBRATES, FISH W72-12930
WATER/ *EUTROPHICATION, *LAKES, *MAINE, RECREATION, WATER SUPPLY, W72-04280
*PHO/ *SEA NETTLES, *POLLUTANTS, *MAINTENANCE, *PLANT MORPHOLOGY, W72-10162
HROMES, PHOSPHORYLATI/ *GLUCOSE, *MALATE, CITRIC ACID CYCLE, CYTOC W72-04309
ULTURE, *WATER POLLUTION CONTRO/ *MANAGEMENT, *ENVIRONMENT, *AGRIC W72-00846
ER POLLUTION EFFECTS, *PROTOZOA, *MARINE ALGAE, *SALT MARSHES, NEW W72-00948
TES, PHYTOPLANKTON, NATURAL RES/ *MARINE ALGAE, *LIGHT, *GROWTH RA W71-13252
CTIVITY, ALGAE, RESOURCES, FOOD/ *MARINE ALGAE, *ECONOMICS, *PRODU W71-13246
HNIQUES, RATES, *PHOTOSYNTHESIS, *MARINE ALGAE, DDT, *PESTICIDES, W72-09653
ELECTRIC POWERP/ *BENTHIC FLORA, *MARINE ALGAE, *GULF OF MEXICO, * W72-11252
ISOTOPES, *COBALT RADIOISOTOPES, *MARINE ALGAE, HEAVY METALS, ABSO W72-10957
ANALYSIS,/ *TOXICITY, *AMMONIA, *MARINE ALGAE, *DIATOMS, CHEMICAL W71-11008
STS, BENTHIC FLORA, RHODOPHYTA,/ *MARINE ALGAE, *SYSTEMATICS, *COA W71-10066
WATER POLLUTION EFFECTS, *ALGAE, *MARINE ALGAE, *BIOINDICATORS, *P W71-10553
HYTA, CHLOR/ *YEASTS, *CULTURES, *MARINE ALGAE, RHODOPHYTA, PHAEOP W72-04748
*AQUATIC PLANTS, *DISTRIBUTION, *MARINE ALGAE, *ESTUARIES, PRODUC W72-04766
ER, *SEDIMENTS, *MARINE ANIMALS, *MARINE ALGAE, RADIOISOTOPES, FAL W72-05965
*POLLUTANTS, *SEA WATER, *ALGAE, *MARINE ALGAE, CULTURES, PHYTOPLA W70-07280
DIOISOTOPES, *ORGANIC COMPOUNDS, *MARINE ALGAE, PHYTOPLANKTON, HUM W71-09188
ICAL ECOLOGY, PHAEOPHYTA, STAND/ *MARINE ALGAE, *KELPS, *PHYSIOLOG W72-12940

THERMAL POLLUTION, *THERM/ BAYS, *MARINE ALGAE, *MARINE ANIMALS, * W72-11876
ROPHYTA, PHAEOPHY/ *SYSTEMATICS, *MARINE ALGAE, *CYANOPHYTA, *CHLO W72-11800
*CHEMICAL WASTE, *GROWTH RATES/ *MARINE ALGAE, *LABORATORY TESTS, W72-12239
OTOPES, *ZINC RADIOISOTOPES, AB/ *MARINE ALGAE, *STRONTIUM RADIOIS W72-12203
IS, *DECOMPOSING ORGANIC MATTER, *MARINE ALGAE, HUMUS, ORGANIC COM W72-12166
*LABORATORY TESTS, *MEASUREMENT, *MARINE ALGAE TRACERS, CHEMICAL P W72-12210
WASTES, RADIOE/ *RADIOISOTOPES, *MARINE ALGAE, *ABSCRPTION, *MINE W72-12204
, / *PHAEOPHYTA, *TIDAL MARSHES, *MARINE ALGAE, CULTURES, SALINITY W72-12633
CARBON PE/ *OCEANS, *MATABOLISM, *MARINE ALGAE, *CHLORINATED HYDRO W72-12998
RINE PLANTS, MARINE/ *SEA WATER, *MARINE ALGAE, *CARBOHYDRATES, MA W73-02100
ES, *PLANT GROWTH, *CHLOROPHYTA, *MARINE ALGAE, *BIOASSAY, REPRODU W72-14294
ARY PRODUCTIVITY, ENERGY BUDGET, *MARINE ALGAE, MARINE PLANTS, BIO W73-00366
RIBUTION PATTERNS, *CHLOROPHYTA, *MARINE ALGAE, *RHODOPHYTA, *SYST W73-00284
*NITRATES, *AM/ *PHYTOPLANKTON, *MARINE ALGAE, CHEMICAL ANALYSIS, W73-00380
PLANTS/ *CULTURES, *CYANOPHYTA, *MARINE ALGAE, *ISOLATION, MARINE W73-00430
RIMARY PRODUCTIVITY, *ENERGY CO/ *MARINE ALGAE, *BENTHIC FLORA, *P W72-06283
ES, *REVIEWS, B/ *PHYTOPLANKTON, *MARINE ALGAE, *SEDIMENTATION RAT W72-06313
HYTA, CHL/ *BACTERIA, *TOXICITY, *MARINE ALGAE, SEA WATER, CHLOROP W72-08051
IOACTIVITY EFFECTS, *MONITORING, *MARINE ALGAE, RHODOPHYTA, PHAEOP W72-07826
CHNIQUES, *CARBON, *AMINO ACIDS, *MARINE ALGAE, PHYTOPLANKTON, NIT W72-07497
HYTA, PHAEOPHYTA, WATER QUALITY, *MARINE ALGAE.: /ODUCTION, RHODOP W73-01449
FFECTS, *PHOTOSYNTHESIS, *OIL P/ *MARINE ALGAE, *WATER POLLUTION E W73-01074
ECOLOGICAL DI/ *HEAT RESISTANCE, *MARINE ALGAE, LIGHT INTENSITY, * W73-00838
OTOZOA, *BACTERIA, FLORIDA, RAD/ *MARINE ALGAE, *SALT MARSHES, *PR W73-00796
O/ *NUCLEAR EXPLOSIONS, *ALASKA, *MARINE ALGAE, *FISHERIES, FISH P W73-00764
ES, *ABSORPTION, *PHYTOPLANKTON, *MARINE ALGAE, FOOD CHAINS, PUBLI W73-00823
RADIOISOTOPES, *MARINE ANIMALS, *MARINE ALGAE, *FOOD CHAINS, ABSO W73-00819
MARINE FISH, M/ *NUCLEAR WASTES, *MARINE ALGAE, *MARINE ANIMALS, * W73-00832
ZINC RADIOISOTOPES, *ABSORPTION, *MARINE ALGAE, *PHYTOPLANKTON, BA W73-00802
*RADIOISOTOPES, *MARINE ANIMALS, *MARINE ALGAE, SEDIMENTS, NUCLEAR W73-00817
ANIMALS, *SYSTEMATICS, *ECOLOGY, *MARINE ALGAE, ANNELIDS, MOLLUSKS W73-00932
KELPS, *ECOLOGICAL DISTRIBUTION, *MARINE ANIMALS, *SYSTEMATICS, *E W73-00932
AINS, *NUCLEAR WASTES, *COBALT / *MARINE ANIMALS, *CLAMS, *FOOD CH W73-00826
FOOD / *STRONTIUM RADIOISOTOPES, *MARINE ANIMALS, *MARINE ALGAE, * W73-00819
RADIOACTIVITY EFFECTS, *FALLOUT, *MARINE ANIMALS, MARINE PLANTS, M W73-00828
EDIME/ *REVIEWS, *RADIOISOTOPES, *MARINE ANIMALS, *MARINE ALGAE, S W73-00817
*NUCLEAR WASTES, *MARINE ALGAE, *MARINE ANIMALS, *MARINE FISH, MU W73-00832
OISOTOPES, *COBALT, *ABSORPTION, *MARINE ANIMALS, MARINE ALGAE, SE W73-00821
*NUCLEAR WASTES, *BIOINDICATORS, *MARINE ANIMALS, ASIA, MOLLUSKS, W73-00811
NMENTAL EFFECTS, *MARINE PLANTS, *MARINE ANIMALS, OFFSHORE PLATFOR W73-02029
ON, *THERM/ BAYS, *MARINE ALGAE, *MARINE ANIMALS, *THERMAL POLLUTI W72-11876
ARIA, MOLLUSKS,/ *RADIOISOTOPES, *MARINE ANIMALS, *ABSORPTION, AQU W71-09182
, *DISTRIBUTION, *RADIOISOTOPES, *MARINE ANIMALS, *MARINE PLANTS, W71-09168
DEPTH, CALIF/ *ECOLOGY, *WASTES, *MARINE ANIMALS, *MARINE PLANTS, W71-01475
NALYSIS, *SEA WATER, *SEDIMENTS, *MARINE ANIMALS, *MARINE ALGAE, R W72-05965
PREDATION, ORGANIC COMPOUNDS, A/ *MARINE BACTERIA, *ATTRACTANTS, * W73-00942
BENTHOS, CRABS, / *CRUSTACEANS, *MARINE BACTERIA, *WATER QUALITY, W72-08142
EDEN.: *MARINE DISPOSAL, *GOTHENBRUG, SW W70-09947
*MARINE ALGAE, *MARINE ANIMALS, *MARINE FISH, MUSSLES, LOBSTERS, W73-00832
SH EGGS, *SEPARATION TECHNIQUES, *MARINE FISH, SILICA, MARINE ALGA W73-00371
RIMP, *BIOASSAY, *ABSORPT/ *DDT, *MARINE FISH, *PHYTOPLANKTON, *SH W73-02105
SEPARATION TECHNIQUES, ATLANTI/ *MARINE FUNGI, *ISOLATION, ALGAE, W71-11851
ER, *METABOLISM, BIODEGRADATION, *MARINE MICROORGANISMS, *MICROBIA W73-00361
SOTOPES, *ANALYTICAL TECHNIQUES, *MARINE MICROORGANISMS, MARINE PL W71-08042
*GLYCEROL, *PHOTOTROPHIC GROWTH, *MARINE PHYTOPLANKTON(PRYMNESIUM W70-07842
OLOGY, *WASTES, *MARINE ANIMALS, *MARINE PLANTS, DEPTH, CALIFORNIA W71-01475
*RADIOISOTOPES, *MARINE ANIMALS, *MARINE PLANTS, AQUATIC ANIMALS, W71-09168
R POLLUTION EFFECTS, *INTERTIDA/ *MARINE PLANTS, *OILY WATER, WATE W71-08792
ECOLOGY, *ENVIRONMENTAL EFFECTS, *MARINE PLANTS, *MARINE ANIMALS, W73-02029
*ORGANIC COMPOUNDS, *PHOSPHORUS, *MARINE PLANTS, *PHYTOPLANKTON, A W71-10083
, *SYSTEMATICS, BENTHOS, SCUBA / *MARINE PLANTS, *ALASKA, SAMPLING W72-09120
IMARY PRODUCTIVITY, *FRESHWATER, *MARSHES, SALINE WATER, BIOMASS, W73-01089
EASONAL, *DISTRIBUTION PATTERNS, *MASSACHUSETTS, SYSTEMATICS, *REP W73-01449
S, *WATER RESOURCES DEVELOPMENT, *MASSACHUSETTS, *EUTROPHICATION, W71-05553
RINATED HYDROCARBON PE/ *OCEANS, *MATABOLISM, *MARINE ALGAE, *CHLO W72-12998
UATIONS, BIOLOGICAL PROPERTIES,/ *MATHEMATICAL STUDIES, *ALGAE, EQ W72-12724
*ALGAE, *WASTE WATER TREATMENT, *MATHEMATICAL MODELS, GROWTH RATE W72-12289
YCLE, *OXIDATION, AMMONIA, NITR/ *MATHEMATICAL MODELS, *NITROGEN C W71-05390
IMULATION, / *OXIDATION LAGOONS, *MATHEMATICAL MODELS, *COMPUTER S W71-07084
ER RE-USE, / *OXIDATION LAGOONS, *MATHEMATICAL MODELS, MIXING, WAT W71-07099
PULATION, DYNAMICS, LAKES, DISS/ *MATHEMATICAL MODELS, *ALGAE, *PO W71-03034
*WATER QUALITY, *EUTROPHICATION, *MATHEMATICAL MODELS, SIMULATION, W71-03021
RPLANTS, *RADIOACTIVITY EFFECTS, *MATHEMATICAL MODELS, FORECASTING W73-00831
QUALITY, *HYDROELECTRIC PLANTS, *MATHEMATICAL MODELS, THERMAL STR W72-14405
ROL, *RESERVOIR/ *PHYTOPLANKTON, *MATHEMATICAL MODELS, *ALGAL CONT W72-08559
NUTRIENTS, *PR/ *EUTROPHICATION, *MATHEMATICAL MODELS, *ESSENTIAL W72-08986
TIFICATION, *AERIAL PHOTOGRAPHY, *MATHEMATICAL METHODS, MUNICIPAL W71-11685
A COLLECTIONS, *HYDROLOGIC DATA, *MEASUREMENT, *ASSAY, ANALYTICAL W72-10678
SOURCES, AREA / *EUTROPHICATION, *MEASUREMENT, *SAMPLING, WATER RE W72-03215
ATORY EQUIPMENT, *RADIOISOTOPES, *MEASUREMENT, *KINETICS, *ABSORPT W72-14326
*ENVIRON/ *MONITORING, *SURVEYS, *MEASUREMENT, *DATA COLLECTIONS, W72-07841
RY EQUIPMENT, *LABORATORY TESTS, *MEASUREMENT, *MARINE ALGAE TRACE W72-12210
OF ALGAE, *AQUATIC WEED CONTROL, *MECHANICAL CONTROL, *DEWATERING. W71-06188
TENAN/ *CHRYSAORA QUINQUECIRRHA, *MEDUSAE MAINTENANCE, *POLYP MAIN W72-10162
CTR/ *STEPHANODISCUS BINDERANUS, *MELOSIRA BINDERANA, SCANNING ELE W73-00432
*ANAEROBIC BACTERIA, *ISOLATION, *MEMBRANE PROCESSES, BACTERIA, SU W72-04743
AST TWIN / *GLASS FIBER FILTERS, *MEMBRANE FILTERS, GRANGE LAKE, E W72-13673
*DISSOLVED ORGANIC CARBON, *MENAI STRAIT(WALES).: W72-09108
EASE, BIOLOGICAL MAGNIFICATION,/ *MERCURY COMPOUNDS, *MINAMATA DIS W72-12257
L PROTECTION AGENCY.: *MERCURY POLLUTION, *ENVIRONMENTA W72-14782

ION, IONS, ABSORPTION, BIOASSAY, *MERCURY.: /Y, EQUIPMENT, INHIBIT W72-07660
U/ *WATER POLLUTION, *LAKE ERIE, *MERCURY, *MUNICIPAL WASTES, *IND W72-14282
ISH, WATER POLLUTION EFFECTS, T/ *MERCURY, *PUBLIC HEALTH, *SHELLF W72-12257
HLORELLA, AQUAT/ *PHYTOPLANKTON, *MERCURY, *TOXICITY, CULTURES, *C W72-11727
RCURY ISOTOPES, HUMAN BRAIN, LO/ *MERCURY, *BIOLOGICAL TISSUES, ME W71-11036
ORGANOMERCURIAL COMPOUNDS, ESOX/ *MERCURY, *METHYLMERCURY, BRAIN, W72-05952
LORIDE, METHYL MERCURIC CHLORID/ *MERCURY, *SYNTHESIS, MERCURIC CH W72-06037
SEWAGE LAGOONS,/ *PHYTOPLANKTON, *METABOLISM, *SEWAGE TREATMENT, * W72-05459
*RADIOISOTOPES, *PHYTOPLANKTON, *METABOLISM, *ABSORPTION, MARINE W71-09193
, PHOTOSYNTHESIS, RESPI/ *ALGAE, *METABOLISM, *CYTOLOGICAL STUDIES W71-08683
ALGAE, GROWTH RATES/ *BIORTHYMS, *METABOLISM, *CHLORELLA, *LIGHT, W73-00251
DROCARBON PESTICIDES, SEA WATER, *METABOLISM, BIODEGRADATION, *MAR W73-00361
TECHNIQUES, P/ *BIODEGRADATION, *METABOLISM, *DDT, *RADIOACTIVITY W72-07703
S, TOXICITY, PATH OF POLLUTANTS, *METALS, TRACE ELEMENTS, PESTICID W72-12190
*METHEMOGLOBINEMIA.: W72-04086
ACTIVITY TECHNIQUES, SOIL ALGAE, *METHODOLOGY, MEASUREMENT, OXIDAT W72-14301
IGERATION, INCUBATION/ *SURVEYS, *METHODOLOGY, FREEZE DRYING, REFR W73-01676
IC/ MICE, INSECTS, *ETHOXYCHLOR, *METHYLCHLOR, METABOLITES, BIOLOG W72-07703
URIAL COMPOUNDS, ESOX/ *MERCURY, *METHYLMERCURY, BRAIN, ORGANOMERC W72-05952
SPIROGYRA, CLODOPHORA GLOMERATA, *METOLIUS RIVER(ORE).: /ANTHES, * W72-07901
D MAINTENANCE, *WATER TREATMENT, *MICHIGAN.: /ICIDES, OPERATION AN W72-08357
*MICRO SCREENING.: W71-09524
*MICRO-NUTRIENTS, *ALGAL GROWTH.: W72-00801
ACILLUS, VIBRIO, RHODOTORULA, A/ *MICROBIAL GROWTH, MICROCOCCUS, B W72-10861
ADATION, *MARINE MICROORGANISMS, *MICROBIAL DEGRADATION, *DDT, *DI W73-00361
*SCENEDESMUS, *DISSOLVED SOLIDS, *MICROBIAL DEGRADATION, CARBON, O W73-00379
, *DET/ *ADENOSINE TRIPHOSPHATE, *MICROBIAL BIOMASS, *GROWTH STAGE W72-04292
ACTERIA, *NITROGEN CYCLE, ENVIR/ *MICROBIOLOGY, *NITROGEN FIXING B W72-11563
SPECIES DIVERSITY INDEX.: *MICROCOSM ALGAL ASSAY PROCEDURE, W72-03218
*MICROCYSTIS, *LAKE SUWA(JAPAN).: W73-00236
*BENTHIC FAUNA, *INVERTEBRATES, *MICROENVIRONMENT, *ECOSYSTEMS, C W70-06668
*WATER TREATMENT, *FILTRATION, *MICROORGANISMS, *SURFACE WATERS, W72-06191
TE DISPOSAL, *RECREATION WASTES, *MICROORGANISMS, *WATER POLLUTION W72-14363
RMAL STRESS, *THERMAL POLLUTION, *MICROORGANISMS, *FLUORESCENCE, F W72-07526
ES, MICROSTRA/ *WATER TREATMENT, *MICROORGANISMS, *ALGAE, *ALGICID W72-06536
POLLUT/ WATER POLLUTION EFFECTS, *MICROORGANISMS, ALGAL CONTROL, * W72-00801
CONTROL, *WASTES, *PATHOGENIC / *MICROORGANISMS, *WATER POLLUTION W72-09675
UNDS, *PHOSPHATES, *INHIBITIONS, *MICROORGANISMS, BACTERIA, FUNGI, W72-13813
ALGAE.: *DIATOMS, *MICROSCOPY, CHRYSOPHYTA, AQUATIC W72-11738
ATTER.: *MICROSCOPIC COUNTS, *AMORPHOUS M W72-06191
ION, BACKWASHING, SPECIFIC RESI/ *MICROSTRAINING, *PRECOAT FILTRAT W72-09393
NSITY, MICROAMMETER, OSCILLATOR/ *MICROSTRATIFICATION, *OPTICAL DE W72-07145
*SEWAGE, *AQUATIC PRODUCTIVITY, *MIDGES, LAGOONS, BIOMASS, OREGON W72-03556
ZATION, POTAMOGET/ *MACROPHYTES, *MIKOLAJSKIE LAKE(POLAND), COLONI W71-11012
NIFICATION,/ *MERCURY COMPOUNDS, *MINAMATA DISEASE, BIOLOGICAL MAG W72-12257
PES, *MARINE ALGAE, *ABSORPTION, *MINE WASTES, RADIOECOLOGY, *PATH W72-12204
TS, *PHYTOPLANKTON, *AQUATIC AL/ *MINERALOGY, *NUTRIENT REQUIREMEN W72-12739
OZOA, PATHOGEN/ *AQUATIC PLANTS, *MINNESOTA, BACTERIA, FUNGI, PROT W72-08586
N, WATER POLLU/ *AQUATIC PLANTS, *MINNESOTA, *TOXICITY, COAGULATIO W72-07360
TERGROWTH.: *MIXED CULTURES, DIEOFF RATES, AF W71-05155
*MIXERS.: W73-02138
IES, DIGITAL COMPUTERS,/ *ALGAE, *MODEL STUDIES, MATHEMATICAL STUD W72-11920
*PLANKTON, *ALGAE, *PHOSPHORUS, *MODEL STUDIES, LAKES, CYCLING NU W72-12543
*PUMPED STORAGE, *WATER QUALITY, *MODEL STUDIES, ZOOPLANKTON, BIOT W72-05282
TEMS,/ *WATER POLLUTION EFFECTS, *MODEL STUDIES, *STREAMS, *ECOSYS W72-06340
QUATIC PLANTS, *AQUATIC ANIMALS, *MODEL STUDIES, LIMNOLOGY, BIOLOG W72-14280
ODS, *ALGAE, CYTOLOGICAL STUDIE/ *MODEL STUDIES, *STATISTICAL METH W71-12868
HIGH MOLECULAR WEIGHT COMPOUNDS, *MOLECULAR SIZE, *ALGAL EXUDATES, W71-07878
LUID FRICTION, *LITTORAL, PLANT/ *MOLECULAR STRUCTURES, *ALGAE, *F W71-07878
*AQUATIC INSECTS, *CRUSTACEANS, *MOLLUSKS, *AQUATIC ANIMALS, *BIO W72-06526
N, NO/ *BIOINDICATORS, *FALLOUT, *MOLLUSKS, *ATLANTIC COASTAL PLAI W73-00818
*NUTRIENTS, LAKES, PLANT GROWTH, *MOLYBDENUM, BIOASSAY, NEUTRON AC W72-00801
N, TRITIUM, OC/ *NUCLEAR WASTES, *MONITORING, *PACIFIC COAST REGIO W72-03324
MAL POLLU/ *ALGAE, *GREAT LAKES, *MONITORING, *BIOINDICATORS, THER W72-10875
NT, *DATA COLLECTIONS, *ENVIRON/ *MONITORING, *SURVEYS, *MEASUREME W72-07841
PHYTA, / *RADIOACTIVITY EFFECTS, *MONITORING, *MARINE ALGAE, RHODO W72-07826
*FOOD CHAIN/ *CHEMICAL ANALYSIS, *MONITORING, *CYCLING NUTRIENTS, W72-07715
IDENTIFICATION, WATER POLLUTION, *MONITORING, AERIAL PHOTOGRAPHY, W72-12487
ON-SITE DATA / *INSTRUMENTATION, *MONITORING, *WATER PROPERTIES, * W72-12941
OMS, CHLOROPHYTA, THERMAL WATER, *MONTANA, PHOSPHATES, NITRATES, N W72-05610
NVERTEBRATES, *FISH POPULATIONS, *MORTALITY, *ALGAE, CALIFORNIA.: / W71-08793
ION, LIMITING NUTRIENTS, ALGAL / *MOSES LAKE(WASH), NUTRIENT DILUT W72-08049
*MOSES LAKE(WASH).: W72-08560
LAKES.: *MOSES LAKE(WASH), HYPEREUTROPHIC W72-00799
HYPEREUTROPHIC LAKES, *MOSES LAKE(WASH).: W72-00798
DOSAGE, CALCIUM ION CONCENTRATI/ *MOSS LANDING, BARNACLES, LETHAL W72-09092
*SUSPENDED SOLIDS, BACKWASHING, *MOUNT CLEMENS(MICH).: /D SEWERS, W72-00042
, TOLUENE, POLYSTYRENE, SLIDE P/ *MOUNTING MEDIA, METHYLENE IODIDE W72-11738
RBON REGENERATION, *BACKWASHING, *MT CLEMENS(MICH).: *CA W72-08357
ATING, *COOLING, *THERMAL POWER, *MULTIPLE-PURPOSE PROJECTS, *WATE W70-09193
OLYACRYLAMIDE, *POLYETHYLENEMIN/ *MULTIVALENT METALLIC CATIONS, *P W72-14840
POLLUTION, *LAKE ERIE, *MERCURY, *MUNICIPAL WASTES, *INDUSTRIAL WA W72-14282
ALGAE, PHOT/ *OXIDATION LAGOONS, *MUNICIPAL WASTES, *WATER REUSE, W72-11642
CHNIQUES, FERMENTAT/ *NUTRIENTS, *MUNICIPAL WASTES, *SEPARATION TE W72-10390
TERIZATION, *ANACYSTIS NIDULANS, *MUTANTS, *SENSITIVITY, CHLORAMPH W72-14327
PHYTA, SYSTEMATI/ *ALGAL TOXINS, *MYXOBACTERIA, CHLOROPHYTA, CYANO W72-07951
*PRIMARY PRODU/ *PHYTOPLANKTON, *NANNOPLANKTON, LAKES, *SEASONAL, W72-12738
WEED CONTROL, *LIGHT INTENSITY, *NEBRASKA, *CYANOPHYTA, ALGAL CON W73-02349
SOURCES, *PHOSPHATES, *NITRATES, *NEBRASKA, URBANIZATION, CITIES, W71-13466
IMARY PRODUCTIVITY, *RESERVOIRS, *NEBRASKA, *EUTROPHICATION, PLANK W72-04270

RMOCLINE, *PRIMARY PRODUCTIVITY,	*NERITIC, NITRATES, RADICACTIVITY	W73-00431
DIMENTS, *ANALYTICAL TECHNIQUES,	*NEUTRON ACTIVATION ANALYSIS, *CH	W72-01101
IOACTI/ *SAMPLING, *ENVIRONMENT,	*NEUTRON ACTIVATICN ANALYSIS, RAD	W71-11036
ACE ELEMENTS, C/ *AQUATIC ALGAE,	*NEUTRON ACTIVATION ANALYSIS, *TR	W71-11515
Y, WATER CHEMISTRY, *CALIFORNIA,	*NEVADA.: /N, PRIMARY PRCDUCTIVIT	W72-12549
OCHEMICAL ANALYSIS, MEASUREMENT,	*NEVADA, AQUATIC PLANTS, WATER PO	W73-01673
OA, *PHYTOTOXICITY, *CYANOPHYTA,	*NEW HAMPSHIRE, MARINE ALGAE, PHY	W72-14542
ITY, AERATION, WATER PROPERTIES,	*NEW HAMPSHIRE, ALGAE, COLOR, COP	W72-07890
HYTA, *BEGGIATOA, PHYTOPLANKTON,	*NEW HAMPSHIRE, MIXING, ENVIRONME	W72-09061
UDIES.:	*NEW YORK BIGHT, COLLABCRATIVE ST	W72-12941
CS, *CHLOROPHYTA, ALGAL CONTROL,	*NEW YORK.: /ITY CONTROL, *KINETI	W72-08376
, *THERMAL POLLUTION, *ENVIRONM/	*NEW YORK, *POWER PLANTS, *STREAM	W72-14659
S, *EUSTROPHICATION, *HYDROLOGY,	*NEW YORK, WATER QUALITY, NUTRIEN	W72-13851
ATORS, SILICA, CHROMIUM, COPPER,	*NEW YORK, DIATOMS, PHOSP: /INDIC	W72-09111
S, *PLANKTON, *ALGAE, TURBIDITY,	*NEW YORK, *WATER QUALITY.: /ATER	W72-06191
, POND DENITRIFICATION, FILTER /	*NITRATE REMOVAL, ALGAE STRIPPING	W71-08223
INATICN, ALGAE, D/ *RETURN FLOW,	*NITRATES, WATER POLLUTICN, DESAL	W71-08223
PRECIPITATION, SEDIM/ *NITROGEN,	*NITRATES, GROUNDWATER, AMMONIA,	W71-06435
POLLUTION EFFECTS, *PHOSPHATES,	*NITRATES, *NUTRIENTS, *EUTROPHIC	W70-09388
UATIC PRODUCTIVITY, *PHOSPHATES,	*NITRATES, *CHLCROPHYLL, DIVERSIO	W71-02681
CAL TECHNIQUES, *WATER ANALYSIS,	*NITRATES, *ELECTRODES, ICNS, FLU	W71-00474
*PLANT MORPHOLOGY, *PHOSPHATES,	*NITRATES, *AMMONIUM COMPCUNDS, S	W72-10162
ACTER/ *WATER POLLUTION EFFECTS,	*NITRATES, *NUTRIENTS, *ALGAE, *B	W71-09958
NITRIFICATION, *DENITRIFICATICN,	*NITRATES, NITROGEN COMPOUNDS, AM	W71-13939
POLLUTION SOURCES, *PHOSPHATES,	*NITRATES, *NEBRASKA, URBANIZATIO	W71-13466
TER TREATMENT, *DENITRIFICATION,	*NITRATES, EUTROPHICATICN, IRRIGA	W72-00974
CONTROL, *BIOLOGICAL TREATMENT,	*NITRATES, TREATMENT FACILITIES,	W72-02975
ROTEINS, LABORAT/ *BIOCHEMISTRY,	*NITRATES, *CHLORELLA, ENZYMES, P	W72-03222
MARINE ALGAE, CHEMICAL ANALYSIS,	*NITRATES, *AMMONIA, DIATOMS, GRO	W73-00380
WATER QUALITY CONTROL, EUTROPH/	*NITRIFICATICN, *DENITRIFICATION,	W72-00800
*NITRA/ *WASTE WATER TREATMENT,	*NITRIFICATION, *DENITRIFICATION,	W71-13939
BACTERIA, *CULTURES, *SEA WATER,	*NITRIFICATION, AMMONIA, NITRITE,	W72-05421
*PHOSPHORUS, *NITROGEN,/ *ALGAE,	*NITRIFICATION, WASTE TREATMENT,	W73-02187
ERATE, CLIMATIC ZONES, / *ALGAE,	*NITROGEN FIXATION, *SOILS, *TEMP	W73-02471
IONS, RICE, CYANOPHYTA,/ *ALGAE,	*NITROGEN FIXATION, *TRCPICAL REG	W73-02473
PHYTA, LIGHT INTENSITY, EUTROPH/	*NITROGEN FIXATION, *LAKES, CYANO	W73-02472
ON CYCLE, *PHOSPHORUS COMPOUNDS,	*NITROGEN CYCLE, *CRABS, *FISHES,	W72-12567
ING BACTERIA, *RAINFALL/ *ALGAE,	*NITROGEN FIXATION, *NITROGEN FIX	W72-12505
ALL/ *ALGAE, *NITROGEN FIXATICN,	*NITROGEN FIXING BACTERIA, *RAINF	W72-12505
YTOPLANKTON, *ALGAE, *NUTRIENTS,	*NITROGEN FIXATION, LAKES, NITROG	W72-13692
TMENT, *DOMESTIC WASTES, CHEMIC/	*NITROGEN COMPOUNDS, *SEWAGE TREA	W70-09907
ICATICN, *WATER QUALITY CONTRCL,	*NITROGEN COMPOUNCS, *PHCSPHORUS	W70-04080
NIA, NITR/ *MATHEMATICAL MODELS,	*NITROGEN CYCLE, *OXIDATION, AMMO	W71-05390
	*NITROGEN REMOVAL.:	W71-13939
PLANT PHYSIOLOGY, CYTOLOGICAL S/	*NITROGEN FIXATION, *CYANOPHYTA,	W72-03217
ERIA, SPOR/ *CYANOPHYTA, *ALGAE,	*NITROGEN FIXATION, ENZYMES, BACT	W71-12070
OPHYTA, *THERMAL WATER, *WYOMIN/	*NITROGEN FIXATION, *ALGAE, *CYAN	W72-08508
GEN CYCLE, ENVIR/ *MICROBIOLOGY,	*NITROGEN FIXING BACTERIA, *NITRO	W72-11563
LOGY, *NITROGEN FIXING BACTERIA,	*NITROGEN CYCLE, ENVIRONMENTAL EF	W72-11563
CULTURES, *NITROGEN, OXYGEN, BI/	*NITROGEN FIXATION, *SYNTHESIS, *	W72-07941
REMENT, CYANOPHYTA, PHYTOPLANKT/	*NITROGEN FIXATION, *ALGAE, MEASU	W72-08054
PLANTS, NITRO/ *ALGAE, *LEGUMES,	*NITROGEN FIXATION, *CYANCPHYTA,	W73-01456
FIXATION, *SYNTHESIS, *CULTURES,	*NITROGEN, OXYGEN, BIOCHEMISTRY,	W72-07941
HNIQUES, MICHIGAN, SESTON, ALGA/	*NITROGEN, *LAKES, ANALYTICAL TEC	W72-08048
TION, BIOCHEMICAL OXYGE/ *ALGAE,	*NITROGEN, HYCROGEN ION CONCENTRA	W72-06612
*AQUATIC MICROORGANISMS, *ALGAE,	*NITROGEN, *PHOSPHORUS, SEDIMENTS	W72-06532
SEWAGE EFFLUENTS, REGU/ *CARBCN,	*NITROGEN, *ALGAE, PLANT GROWTH,	W72-14790
HESIS, LIGHT IN/ *PHYTOPLANKTON,	*NITROGEN, *NUTRIENTS, *PHOTOSYNT	W72-09103
PRODUCTIVITY, SEDIMENTS, NITRIF/	*NITROGEN, *AQUATIC ENVIRONMENT,	W72-08459
SOU/ *AMMONIA, *EUTROPHICATICN,	*NITROGEN, ALGAE, WATER PCLLUTION	W71-11496
N RESERVOIRS, / *WATER DILUTION,	*NITROGEN, *PHOSPHORUS, *DETENTIO	W72-02412
ION, POLLUTANT IDENTIFICATION, /	*NITROGEN, PLANKTCN, *EUTROPHICAT	W72-01780
E EL/ *POLLUTANT IDENTIFICATION,	*NITROGEN, ELECTROCHEMISTRY, TRAC	W72-01693
, ALGAE, NITRA/ *EUTROPHICATICN,	*NITROGEN, *PHOSPHORUS, NUTRIENTS	W71-06443
, AMMCNIA, PRECIPITATION, SEDIM/	*NITROGEN, *NITRATES, GRCUNDWATER	W71-06435
UTROPHICATION, AQUA/ *NUTRIENTS,	*NITROGEN, *PHOSPHORUS, ALGAE, *E	W71-08953
BLOOMS, *NUTRIENTS, *PHOSPHORUS,	*NITROGEN, *NORTH CARCLINA.: /AL	W70-09669
RIENT REQUIREMENTS, TEMPERATURE,	*NITROGEN, *PHOSPHATES, ANAEROBIC	W72-04789
, AMMONIA, */ *ALGAE, *KINETICS,	*NITROGEN, *NUTRIENT REQUIREMENTS	W72-04787
TRITES, TOXICITY, CORROSION, PU/	*NITROGEN, *AMMONIA, NITRATES, NI	W72-04086
E, *EUTROPHICATION, *PHOSPHORUS,	*NITROGEN, EVAPORATION, STORM RUN	W72-04070
, *DENITRIFICATION, WATER REUSE,	*NITROGEN, PHOSPHCRUS, CARBON, *S	W72-04309
*A/ *ALGAL CONTROL, *PHOSPHORUS,	*NITROGEN, *IDAHO, *CHLOROPHYLL,	W72-13300
N, WASTE TREATMENT, *PHOSPHORUS,	*NITROGEN, COST ANALYSIS, OZONE,	W73-02187
WAUBESA(WIS), OCCOQUAN RESERVOI/	*NITROGEN:PHOSPHORUS RATIO, LAKE	W72-05473
CTIVATION./ *ANABAENA FLOS-AQUA,	*NITROGENASE ACTIVITY, OXYGEN INA	W72-07941
CULTURES, *CHEMOSTATS, ORTHOPH/	*NITZSCHIA ACTINASTROIDES, *BATCH	W73-01445
SCENEDEMUS ACUTUS VAR ALTERNANS,	*NITZSCHIA ACTINASTROIDES, *STAUR	W72-12739
ICATION, *WATER QUALITY CONTROL,	*NON-STRUCTURAL ALTERNATIVES, AGI	W72-04279
OTILITY, CYST FO/ *G + C RATIOS,	*NONFRUITING MYXOBACTERIA, DNA, M	W72-07951
EMATICS, PHOTOGRAPHY, SEA WATER,	*NORTH CAROLINA, MICROSCOPY, ATLA	W73-00284
TRIENTS, *PHOSPHORUS, *NITRCGEN,	*NORTH CAROLINA.: /AL BLCOMS, *NU	W70-09669
LANKTCN, *DISTRIBUTION PATTERNS,	*NORTH CAROLINA, *COMPUTER PRCGRA	W71-07337
N DEMAND, WASTE WATER TREATMENT,	*NORTH CAROLINA.: /CHEMICAL OXYGE	W71-07382
*ESTURARIES, *THERMAL POLLUTICN,	*NORTH CAROLINA, NUTRIENTS, SEWAG	W72-13636
*COLD REGIONS, AERATED LAGOONS,	*NORTH DAKOTA, BIOCHEMICAL OXYGEN	W72-12565
GE POLLUTION, EPIFAUNA.:	*NORTH SEA, *LAMINAREA SPP., SEWA	W72-11854
CERATIUM FURCA, TURBIDOSTAT CUL/	*NORTH SEA, PROROCENTRUM MICANS,	W71-03095
	*NORTHPORT PLANT(LONG ISLAND).:	W72-14659
ALGAE/ *EXPERIMENTAL LAKES AREA,	*NORTHWESTERN ONTARIO, EPILITHIC	W71-11029

FLOS-AQUAE, PHRAGMITES COMMUNIS, *NORWAY, MACROPHYTES, FYRISAN RIV W73-00286
ARIA DIGITATA, AGARUM CRIBROSUM, *NOVA SCOTIA, *ST. MARGARET'S BAY W73-00366
RINE ALGAE, *FISHERIES, FISH PO/ *NUCLEAR EXPLOSIONS, *ALASKA, *MA W73-00764
TION EFFECTS, *RADIOACTIVITY EF/ *NUCLEAR EXPLOSIONS, *WATER POLLU W72-03347
EFFECTS, *ENVIRONMENTAL EFFECTS, *NUCLEAR EXPLOSIONS, BIRDS, SALMO W72-00960
ATOMACEOUS EARTH, *POROUS MEDIA/ *NUCLEAR EXPLOSION, *FILTERS, *DI W72-00974
CHAINS, *RADIOACTIVITY EFFECTS, *NUCLEAR POWERPLANTS, MONITORING, W72-00959
ON PLANTS, *ELECTRIC POWER PROD/ *NUCLEAR POWERPLANTS, *DESALINATI W72-03327
ITY EFFECT/ *PATH OF POLLUTANTS, *NUCLEAR POWERPLANTS, *RADIOACTIV W73-00831
ATER, *WATE/ *THERMAL POLLUTION, *NUCLEAR POWER PLANTS, *COOLING W W70-09612
ATER,/ *WATER POLLUTION CONTROL, *NUCLEAR POWER PLANTS, *POTABLE W W71-09018
S, WATE/ *RADIOACTIVITY EFFECTS, *NUCLEAR WASTES, *FALLOUT, *OCEAN W72-04463
MARINE ANIMALS, *MARINE FISH, M/ *NUCLEAR WASTES, *MARINE ALGAE, * W73-00832
*MARINE ANIMALS, ASIA/ *REVIEWS, *NUCLEAR WASTES, *BIOINDICATORS, W73-00811
OD CHAIN/ *BACKGROUND RADIATION, *NUCLEAR WASTES, *ABSORPTION, *FO W73-00833
E ANIMALS, *CLAMS, *FOOD CHAINS, *NUCLEAR WASTES, *COBALT RADIOISO W73-00826
RADIOACTIVITY EFFECTS, *TRITIUM, *NUCLEAR WASTES, WATER POLLUTION W72-06342
CIFIC COAST REGION, TRITIUM, OC/ *NUCLEAR WASTES, *MONITORING, *PA W72-03324
FFECTS, *THERMAL POLLUTION, CON/ *NUCLEAR WASTES, *RADIOACTIVITY E W72-00939
ENTS, ALGAL CONTROL, *CYANPHYTA, *NUISANCE ALGAE, PLANKTON, WASHIN W72-00798
LUTION, WATER POLLUTION CONTROL, *NUISANCE ALGAE, *EUTROPHICATION, W72-00799
REVIEWS.: *CYANOPHYTA, *NUISANCE ALGAE, *EUTROPHICATION, W72-01358
GAL CONTROL, LABORATORY TESTS, / *NUISANCE ALGAE, *CYANOPHYTA, *AL W71-13184
*ALGAL CONTROL, CYANOPHYTA, CA/ *NUISANCE ALGAE, *CARBON DIOXIDE, W72-10607
CONTROL, *DIFFUSION, *NUTRIENTS, *NUISANCE ALGAE, WASHINGTON, PHYT W72-08049
REVIEWS, TROPHI/ *ALGAL CONTROL, *NUISANCE ALGAE, EUTROPHICATION, W72-07937
*POISONOUS PLANTS, *CYANOPHYTA, *NUISANCE ALGAE, *AQUATIC ALGAE, W72-14056
S/ *CHORELLA, *ALGAE, *FOULING, *NUISANCE ALGAE, FLOATING PLANTS, W71-09261
OHIO).: *NUTRIENT LEVELS, CUYAHOGA RIVER(W72-04263
, CULTURES, *BIOMASS, NUTRIENTS, *NUTRIENT REQUIREMENTS, NUISANCE W72-03985
H RATES, *KINETICS, *PHOSPHORUS, *NUTRIENT REQUIREMENTS, TEMPERATU W72-04788
INETICS, *WASTE WATER TREATMENT, *NUTRIENT REQUIREMENTS, TEMPERATU W72-04789
*/ *ALGAE, *KINETICS, *NITROGEN, *NUTRIENT REQUIREMENTS, AMMONIA, W72-04787
, RESPIRATORY RATES, PHOTOSYNTH/ *NUTRIENT REMOVAL, OXYGEN DECLINE W71-04079
TA, CY/ *EUTROPHICATION, *ALGAE, *NUTRIENT REQUIREMENTS, CHLOROPHY W70-05469
, NITRATES/ *DIATOMS, *CULTURES, *NUTRIENT REQUIREMENTS, SEA WATER W73-00441
FACTORS, *PHOSPHATES, DIATOMS,/ *NUTRIENT REQUIREMENTS, *LIMITING W73-01445
CTION, LAKE MENDOTA(WIS), LAKE / *NUTRIENT SOURCES, ACETYLENE REDU W72-10608
(ONTARIO), NUTRIENT INPUTS, CHI/ *NUTRIENT BUDGETS, *BAY OF QUINTE W71-11009
, BIOINDICATORS/ *PHYTOPLANKTON, *NUTRIENT REQUIREMENTS, NUTRIENTS W72-13816
NKTON, *AQUATIC AL/ *MINERALOGY, *NUTRIENT REQUIREMENTS, *PHYTOPLA W72-12739
CENTRAL BASIN./ *DEOXYGENATION, *NUTRIENT REGENERATION, LAKE ERIE W72-12996
I/ *AQUATIC ALGAE, *COMPETITION, *NUTRIENTS, *GROWTH RATES, *INHIB W72-12734
ION, MATHEMA/ *DISSOLVED OXYGEN, *NUTRIENTS, *LAKE ERIE, *HYPOLIMN W72-12997
, WATER QUALIT/ *EUTROPHICATION, *NUTRIENTS, LAKE HURON, EFFLUENTS W72-13662
AKES, N/ *PHYTOPLANKTON, *ALGAE, *NUTRIENTS, *NITROGEN FIXATION, L W72-13692
WTH RATES, *SEAWATER, *TOXICITY, *NUTRIENTS, LETHAL LIMITS, METHOD W72-12239
TORS, PHOSPHAT/ *EUTROPHICATION, *NUTRIENTS, *ALGAE, *LIMITING FAC W72-12544
PHICATION, LAKE ERIE, ALGAE, NI/ *NUTRIENTS, *LAKE ONTARIO, *EUTRO W71-11009
BORATORY TESTS, *EUTROPHICATION, *NUTRIENTS, *ALGAE, SEWAGE, PRODU W71-12857
ER POLLUTION EFFECTS, *NITRATES, *NUTRIENTS, *ALGAE, *BACTERIA, AE W71-09958
S, *TOXICITY, *ACTIVATED SLUDGE, *NUTRIENTS, NITROGEN, WATER POLLU W71-11972
N, *WA/ *ALGAE, *EUTROPHICATION, *NUTRIENTS, *PLANKTON, *RECREATIO W72-10431
EPARATION TECHNIQUES, FERMENTAT/ *NUTRIENTS, *MUNICIPAL WASTES, *S W72-10390
OSPHORU/ *ANALYTICAL TECHNIQUES, *NUTRIENTS, *ALGAE, *BIOASSAY, PH W72-10608
T IN/ *PHYTOPLANKTON, *NITROGEN, *NUTRIENTS, *PHOTOSYNTHESIS, LIGH W72-09103
ENT, WATER POL/ *EUTROPHICATION, *NUTRIENTS, *ALGAE, SEWAGE TREATM W72-10625
YMBIOSI/ *WASTE WATER TREATMENT, *NUTRIENTS, *ALGAE, *BACTERIA, *S W72-11100
POLLUTION EFFECTS, EUTROPHICAT/ *NUTRIENTS, *PHYTOPLANKTON, WATER W72-09167
IOASSAY, OREGON, LAKES,/ *ALGAE, *NUTRIENTS, *LABORATORY TESTS, *B W72-09164
*EUTROPHICATION, *ON-SITE TESTS, *NUTRIENTS, ALGAE, LAKES, PHOSPHO W72-09165
LUTION EFFECTS, *ORGANIC WASTES, *NUTRIENTS, *POLLUTANT IDENTIFICA W72-01472
, *ALGAL BLOOMS, *PHYTOPLANKTON, *NUTRIENTS, WATER POLLUTION EFFEC W72-01329
, ANALYTICAL TECHNIQUES, *ALGAE, *NUTRIENTS, LAKES, PLANT GROWTH, W72-00801
EFFECTS, *EUTROPHICATION, LAKES, *NUTRIENTS, ALGAL CONTROL, *CYANP W72-00798
Y PRODUCTIVITY, *EUTROPHICATION, *NUTRIENTS, *BIOLOGICAL COMMUNITI W72-03216
MENT, PLANKTON, *EUTROPHICATION, *NUTRIENTS, MICROORGANISMS, EFFLU W71-13553
TESTS, *RESPIRATION, *CHLORELLA, *NUTRIENTS, CULTURES, NITROGEN, P W73-01626
ROL,/ *REVIEWS, *EUTROPHICATION, *NUTRIENTS, *WATER POLLUTION CONT W73-00840
SEDIMENTATION, *EUTROPHICATION/ *NUTRIENTS, *DREDGING, DESILTING, W73-00938
GAN, *EUTROPHICATION/ *SAMPLING, *NUTRIENTS, *DIATOMS, *LAKE MICHI W73-00384
CTORS, / *ALGAE, *AQUATIC WEEDS, *NUTRIENTS, BIOASSAY, LIMITING FA W73-00232
ING/ *ALGAL CONTROL, *DIFFUSION, *NUTRIENTS, *NUISANCE ALGAE, WASH W72-08049
OLIGOTR/ *ALGAE, *GROWTH RATES, *NUTRIENTS, CULTURES, INHIBITION, W72-08376
ORGANISMS, *INORGANIC COMPOUNDS, *NUTRIENTS, FRESH WATER, ALGAE, B W72-07933
*PLANKTON, *BIOMASS, *SAMPLING, *NUTRIENTS, *HYDROGRAPHY, *PACIFI W72-07892
Y, *ST/ *EUTROPHICATION, *LAKES, *NUTRIENTS, *FRESHWATER, *BIOASSA W72-07648
ER POLLUTION SOURCES, *ESTUARIE/ *NUTRIENTS, *EUTROPHICATION, *WAT W70-06509
ER QUALITY CONTROL, *NITROGEN C/ *NUTRIENTS, *EUTROPHICATION, *WAT W70-04080
UNDED WATERS, STREAMS, FLOW, SE/ *NUTRIENTS, *WATER QUALITY, *IMPO W71-03028
CUTTING MANAGEMENT, *HERBICIDES, *NUTRIENTS, *ECOSYSTEMS, *WATERSH W71-01489
EUTROPHICATION, *ALGAL BLOOMS, *NUTRIENTS, *PHOSPHORUS, *NITROGE W70-09669
EFFECTS, *PHOSPHATES, *NITRATES, *NUTRIENTS, *EUTROPHICATION, ALGA W70-09388
IMNOLOGY, *ALGAE, *EUTROPHICATI/ *NUTRIENTS, *BOTTOM SEDIMENTS, *L W70-08944
N DIOXIDE, PHOSPHATES, NITRATES, *NUTRIENTS, ALKALINITY, CHEMICAL W72-04983
*SEDIM/ *TROPHIC LEVEL, *LAKES, *NUTRIENTS, *CHEMICAL PROPERTIES, W72-05469
ON, *PHOSPHORUS, *NITROGEN, EVA/ *NUTRIENTS, *ALGAE, *EUTROPHICATI W72-04070
HATES, *GREAT LAKES, *SEDIMENTS, *NUTRIENTS, EUTROPHICATION, NITRO W72-04263
TION CONTROL, *SEA WATER, *OILS, *NUTRIENTS, AIRCRAFT, SATELLITES(W72-04742
HIC FLORA, *PHYTOPLANKTON, ALGA/ *NUTRIENTS, *BENTHIC FAUNA, *BENT W72-06046

QUALITY CONTROL, MASSACHUSETTS,	*NUTRIENTS, OUTLETS.: /ICN, WATER	W72-06046
S, ALGAE, *EUTROPHICATION, AQUA/	*NUTRIENTS, *NITROGEN, *PHOSPHORU	W71-08953
ALYSIS, LAKE MORPHOLOGY, SEDIME/	*NUTRIENTS, *SEDIMENTS, *WATER AN	W71-05626
, *ECOLOGY, *ALGAE,/ *ESTUARIES,	*NUTRIENTS, WATER QUALITY CONTROL	W71-08775
*SEWAGE LAGOONS, BIODEGRADATION,	*NUTRIENTS, ALGAE, *DEGRADATION(S	W71-07973
, *COAST REVIEW, *BIBLIOGRAPHIE/	*OCEANOGRAPHY, *PACIFIC NORTHWEST	W72-12190
AE, *CHLORINATED HYDROCARBON PE/	*OCEANS, *MATABOLISM, *MARINE ALG	W72-12998
RADICISOTOPES, *ANALYTICAL TEC/	*OCEANS, *ORGANIC MATTER, *CARBON	W71-08042
CONTROL, BIOMASS, TOXICITY, EUT/	*OCEANS, *COASTS, *WATER QUALITY	W71-11793
*DISTRIBUTION, *UREAS, *COASTS,	*OCEANS, SURFACE WATERS, CONTINEN	W72-08056
ECTS, *NUCLEAR WASTES, *FALLOUT,	*OCEANS, WATER POLLUTION EFFECTS,	W72-04463
	*ODOR-PRODUCING ALGAE.:	W72-11604
ICAL TECHNIQUES, SEPARA/ *TASTE,	*ODOR, *ORGANIC COMPOUNDS, ANALYT	W72-11604
CARBON, ADSORPTION, FIL/ *TASTE,	*ODOR, *POTABLE WATER, ACTIVATED	W72-08357
INATION, *WATER QUALITY CONTROL,	*ODOR, ALGAE, EUTROPHICATION, TAS	W71-13187
*THERMAL STRATIFICATION, *TASTE,	*ODOR, DISSOLVED SOLIDS, ALGAE, R	W73-02138
KE ERIE, *POTABLE WATER, *TASTE,	*ODOR, SURVEYS, MICROBIOLOGY, WAT	W72-10076
BAY, SCOLOPLOS ARMIGER, ALIMITO/	*OFFSHORE CONSTRUCTION, DELAWARE	W73-02029
ON, CYANOPHYTA, *EUTROPHICATION,	*OHIO.: / AQUATIC ALGAE, FILTRATI	W72-13673
LUTION EFFECTS, *PHOTOSYNTHESIS,	*OIL POLLUTION, *TOXICITY, CHEMIC	W73-01074
A ISLAND, CANAL ZONE, MACROINVE/	*OIL SPILLS, S.S. WITWATER, GALET	W72-07911
*PRIMARY PRODUCTIVITY,	*OIL WASTES, *ARCTIC, *ALGAE.:	W72-07922
ATER,/ *WATER POLLUTION EFFECTS,	*OIL WASTES, *CRUSTACEANS, OILY W	W72-07911
ON, *CHEMICALS, *SURFACTANTS, */	*OIL WASTES, *CLEANING, *DISPERSI	W71-09789
, *EMULSIFIERS, DETERGENTS, OIL,	*OIL WASTES, OILY WATER, OIL-WATE	W70-09429
FECTS, ENVIRONMENTAL EFFECTS, */	*OIL, SEEPAGE, WATER POLLUTION EF	W71-08793
R POLLUTION CONTROL, *SEA WATER,	*OILS, *NUTRIENTS, AIRCRAFT, SATE	W72-04742
CTS, *INTERTIDA/ *MARINE PLANTS,	*OILY WATER, WATER POLLUTION EFFE	W71-08792
IMENTARY ROCKS, *BRINES, *ALGAE,	*OKLAHOMA, *IODINE RADIOISOTOPES,	W72-02073
TRATE NITROGEN, ORTHOPHOSPHATES,	*OLIGASAPROBIC STREAMS, FINGERNAI	W72-06526
*PHYTOPLANKTON, *EUTROPHICATION,	*OLIGOTROPHY, MINNESOTA, LAKES, G	W71-13149
*MONITORING, *WATER PROPERTIES,	*ON-SITE DATA COLLECTIONS, NEW YO	W72-12941
IVING, *SEA WATER, *DATA COLLEC/	*ON-SITE INVESTIGATIONS, *SCUBA D	W72-11925
A, *ALGAE, / *OXIDATION LAGOONS,	*ON-SITE INVESTIGATIONS, *BACTERI	W72-10233
IONS, *OXIDATION LAGOONS, ORGAN/	*ON-SITE INVESTIGATION, *COLD REG	W72-06838
OXIDATION, CARBON RADIOISOTOPES,	*ON-SITE TESTS, SULFUR BACTERIA,	W72-14301
, LAKES, PHOSP/ *EUTROPHICATION,	*ON-SITE TESTS, *NUTRIENTS, ALGAE	W72-09165
	*ONEIDA LAKE(NY).:	W72-13851
CLONIUM, COSMARIUM, PHACUS, CHA/	*ONONDAGA LAKE, OEDOGONIUM, RHIZO	W72-09111
ATLANTICA, CODIUM CAROLINIANUM,	*ONSLOW BAY(NC).: /II, GLOIODERMA	W73-00284
ANALYSIS, *FILTER/ *FILTRATION,	*OPERATION AND MAINTENANCE, *COST	W72-09393
SCILLATOR/ *MICROSTRATIFICATION,	*OPTICAL DENSITY, MICROAMMETER, O	W72-07145
ATISTICAL METHODS, MARINE ALGAE,	*OREGON, *SEASONAL, CHRYSOPHYTA,	W72-08141
NITIES,/ *ALGAE, *BENTHIC FLORA,	*OREGON, RIVERS, BIOLOGICAL COMMU	W72-07901
CEN/ *CYANOPHYTA, *GROWTH RATES,	*ORGANIC ACIDS, *HYDROGEN ION CON	W72-00838
CHNIQUES, SEPARA/ *TASTE, *ODOR,	*ORGANIC COMPOUNDS, ANALYTICAL TE	W72-11604
IMETRIC/ *ANALYTICAL TECHNIQUES,	*ORGANIC COMPOUNDS, *CARBON, GRAV	W72-09108
*MARINE PLANTS, *PHYTOPLANKTON,/	*ORGANIC COMPOUNDS, *PHOSPHORUS,	W71-10083
OPHICATION, *CHELATION, *SEWAGE,	*ORGANIC COMPOUNDS, WATER POLLUTI	W71-09674
CYCLING NUTRIENTS, *FOOD CHAINS,	*ORGANIC COMPOUNDS, SHELLFISH, IN	W72-07715
, CYANOPHYTA, BACTERIA/ *COASTS,	*ORGANIC COMPOUNDS, *GEOCHEMISTRY	W73-01439
, PHYTOPLANKTON/ *RADIOISOTOPES,	*ORGANIC COMPOUNDS, *MARINE ALGAE	W71-09188
AL GROWTH.:	*ORGANIC CARBON, *ALGAL MASS, ALG	W72-04431
), *WASTE ASSIMILATION CAPACITY,	*ORGANIC LOADS, NUTRIENTS, DISSOL	W71-11017
R ANALYSIS, ALGAE, CH/ *BIOMASS,	*ORGANIC MATTER, *SEA WATER, WATE	W71-11234
SEPARATION, *FOAM FRACTIONATION,	*ORGANIC MATTER, *INORGANIC COMPO	W72-05506
TOPES, *ANALYTICAL TEC/ *OCEANS,	*ORGANIC MATTER, *CARBON RADIOISO	W71-08042
URCES, *WATER POLLUTION EFFECTS,	*ORGANIC MATTER, *OXYGEN SAG, *IM	W70-09189
ANIC M/ WATER POLLUTION EFFECTS,	*ORGANIC MATTER, *DECOMPOSING ORG	W72-14673
ION.:	*ORGANIC NUTRIENTS, ALUM COAGULAT	W73-02187
*CHE/ *WATER POLLUTION EFFECTS,	*ORGANIC WASTES, *EUTROPHICATION,	W71-09674
NT, *ACTIVATED SLUDGE, EFFLUENT,	*ORGANIC WASTES, ECONOMICS, CHEMI	W71-12223
LUTAN/ *WATER POLLUTION EFFECTS,	*ORGANIC WASTES, *NUTRIENTS, *POL	W72-01472
Y, *GRO/ *ALGAE, *CARBOHYDRATES,	*ORGANICS ACIDS, *PLANT PHYSIOLOG	W72-07710
STICIDES, *CARBAMATE PESTICIDES,	*ORGANOPHOSPHOROUS PESTICIDES, *2	W71-10566
SPHATES, NUTRIENTS, CHLOROPHYTA,	*ORGANOPHOSPHORUS COMPOUNDS, WASH	W72-12637
, *GROWTH STAGE, *DETERMINATION,	*ORIGIN, *SIGNIFICANCE, *EXTRACTI	W72-04292
EINHARDTII, MOSES LAKE, PINE LA/	*ORTHOPHOSPHATES, CHLAMYDOMONAS R	W72-12637
LAKES, *LAKE HERMAN(SO. DAKOTA),	*ORTHOPHOSPHATES.: PRAIRIE	W73-00938
	*OSLOFJORD(NORWAY).:	W73-00234
NSHIPS, *AMMONIA-NITROGEN, *GRO/	*OSTRACODS, PREDATOR-PREY RELATIO	W72-04787
URIFICATION, *AEROBIC TREATMENT,	*OXIDATION LAGOONS.: /ON, WATER P	W72-05315
IGHT INTENSITY, DESIGN CRITERIA,	*OXIDATION LAGOONS, SEWAGE TREATM	W72-04789
ION EFFECTS, *FARM WASTES, OKLA/	*OXIDATION LAGOONS, *WATER POLLUT	W72-04773
E, *CULTURES, *ENTERIC BACTERIA,	*OXIDATION LAGOONS, *PATHOGENIC B	W70-08319
, *CHLOROPHYLL, *CLASSIFICATION,	*OXIDATION LAGOONS, CHLORELLA, OX	W71-00139
N, *ALGAE, AEROBIC TREATMENT, B/	*OXIDATION LAGOONS, *STABILIZATIO	W71-03896
, ALGAE, BIOCH/ *DAIRY INDUSTRY,	*OXIDATION LAGOONS, *PILOT PLANTS	W70-06619
NS, BIODEGRADA/ *EUTROPHICATION,	*OXIDATION LAGOONS, *SEWAGE LAGOO	W71-07973
STES, *WASTE DISPOSAL, *LAGOONS,	*OXIDATION LAGOONS, AEROBIC CONDI	W71-07551
SE, RECLAMATION, ALGAE, PROTEIN,	*OXIDATION LAGOONS, CLIMATIC ZONE	W71-08951
MATICAL MODELS, *NITROGEN CYCLE,	*OXIDATION, AMMONIA, NITRITES, OX	W71-05390
TMENT, *ANAEROBIC CONDITIONS, P/	*OXIDATION LAGOONS, *AEROBIC TREA	W71-06033
DISSOLVED OXYGEN, SATURATION, /	*OXIDATION LAGOONS, *OXYGENATION,	W71-07106
ATMENT, *DESIGN CRITERIA, TRICK/	*OXIDATION LAGOONS, *TERTIARY TRE	W71-07113
MODELS, MIXING, WATER RE-USE, /	*OXIDATION LAGOONS, *MATHEMATICAL	W71-07099
ROBIC CONDITIONS, AEROBIC CONDI/	*OXIDATION LAGOONS, *SLUDGE, ANAE	W71-07102
ES, *COST ANALYSIS, TEMPERATURE/	*OXIDATION LAGOONS, *CLIMATIC ZON	W71-07109
TREATMENT, ORGANIC LOADING, BIO/	*OXIDATION LAGOONS, *WASTE WATER	W71-07124
KLING FILTERS, EUTROPHICATION, /	*OXIDATION LAGOONS, *ALGAE, *TRIC	W71-07097

ITY, PONDS, ROTIFERS, P/ *ALGAE,	*OXIDATION LAGOONS, *LIGHT INTENS	W71-07382
RIA, *WATER QUALITY, ALGAE, HAR/	*OXIDATION LAGOONS, *DESIGN CRITE	W71-07100
TREATMENT, *WATER RE-USE, BIOCH/	*OXIDATION LAGOONS, *WASTE WATER	W71-07123
, DEGRADATION, CHLORINE, CONTAC/	*OXIDATION LAGOONS, *DISINFECTION	W71-07096
MODELS, *COMPUTER SIMULATION, /	*OXIDATION LAGOONS, *MATHEMATICAL	W71-07084
*ALGAE, DEPTH, TEMPERATURE, PH/	*OXIDATION LAGOONS, *OXYGENATION,	W71-07098
E HANDLING.:	*OXIDATION DITCH, HYDRAULIC MANUR	W72-08396
TE INVESTIGATION, *COLD REGIONS,	*OXIDATION LAGOONS, ORGANIC LOADI	W72-06838
E WATER TREATMENT, *MATHEMATICA/	*OXIDATION LAGOONS, *ALGAE, *WAST	W72-12289
WATER TREATMENT, *BYPRODUCTS, M/	*OXIDATION PONDS, *ALGAE, *WASTE	W72-13508
ERTIARY TREATMENT, *COAGULATION,	*OXIDATION LAGOONS, AEROBIC TREAT	W71-09629
HEMICAL OXYGEN DEMAND, AERATION,	*OXIDATION, HYDROGEN ION CONCENTR	W71-12223
ATION, PHOTOSYNTHESIS, OXYGEN, /	*OXIDATION LAGOONS, *ALGAE, *OXID	W72-02363
G/ *CHLORINATION, *DISINFECTION,	*OXIDATION LAGOONS, COLIFORMS, AL	W72-00422
TECHNIQUES, WATER PURIFICATION,	*OXIDATION LAGOONS, STABILIZATION	W71-13356
ORY TESTS, AEROBIC CONDITIONS, /	*OXIDATION PONDS, *ALGAE, LABORAT	W71-13326
IPAL WASTES, *FILTRATION, *FILT/	*OXIDATION LAGOONS, *ALGAE, MUNIC	W71-13341
N, / *OXIDATION LAGOONS, *ALGAE,	*OXIDATION, PHOTOSYNTHESIS, OXYGE	W72-02363
UATIC AL/ *FARM WASTE, *POULTRY,	*OXIDATION LAGOONS, NUTRIENTS, AQ	W72-02850
STES, *WATER REUSE, ALGAE, PHOT/	*OXIDATION LAGOONS, *MUNICIPAL WA	W72-11642
EWAGE LAGOONS, *AERATED LAGOONS,	*OXIDATION LAGOONS, *DESIGN CRITE	W72-09386
EAT TRANSFER, *PLASTICS, DOMEST/	*OXIDATION LAGOONS, *CHANNELS, *H	W72-10401
STIGATIONS, *BACTERIA, *ALGAE, /	*OXIDATION LAGOONS, *ON-SITE INVE	W72-10233
F ALGAE, *SEPARATION TECHNIQUES/	*OXIDATION LAGOONS, *HARVESTING O	W72-10237
SLURRIES, WASTE WATER TREATMEN/	*OXIDES, *FLUID FRICTION, SEWERS,	W70-06968
	*OXYGEN BALANCE.:	W71-04518
*LAKE ERIE CENTRAL BASIN,	*OXYGEN DEPLETION.:	W72-13659
IONS, *LAKE ERIE, *ALGAE, *BENT/	*OXYGEN DEMAND, *ANAEROBIC CONDIT	W72-12995
G ORGANIC MATTER, AERATION, *AL/	*OXYGEN REQUIREMENTS, *DECOMPOSIN	W72-01881
LUTION EFFECTS, *ORGANIC MATTER,	*OXYGEN SAG, *IMPAIRED WATER QUAL	W70-09189
ATIC BACTERIA, *SULFUR BACTERIA,	*OXYGEN, *SEDIMENT-WATER INTERFAC	W72-06690
E, *CHLORELLA, *PHOTOACTIVATION,	*OXYGEN, LIGHT, PHOTOSYNTHESIS, O	W72-14800
AKES, BIOCHEMICAL OXYGEN DEMAND,	*OXYGENATION, *ANAEROBIC BACTERIA	W72-14469
ERATURE, PH/ *OXIDATION LAGOONS,	*OXYGENATION, *ALGAE, DEPTH, TEMP	W71-07098
ATURATION, / *OXIDATION LAGOONS,	*OXYGENATION, DISSOLVED OXYGEN, S	W71-07106
ARBOR, *TINTINNIDS, *TYCHOPELAG/	*OYSTER BAY, *JAMAICA, FALMOUTH H	W72-07899
TEST.:	*PAAP TEST, *ALGAL GROWTH, BOTTLE	W73-00255
C/ *NUCLEAR WASTES, *MONITORING,	*PACIFIC COAST REGION, TRITIUM, O	W72-03324
, *BIBLIOGRAPHIE/ *OCEANOGRAPHY,	*PACIFIC NORTHWEST, *COAST REVIEW	W72-12190
*DIATOMS, *PHYTOPLANKTON,	*PACIFIC OCEAN, *ALGAE.:	W71-11527
PLING, *NUTRIENTS, *HYDROGRAPHY,	*PACIFIC OCEAN, ECOLOGY, WATER TE	W72-07892
AL SAMPLES, SR-90, K-40, CS-137,	*PAHRANAGAT LAKE, GAMMA RAY SPECT	W73-01673
	*PAMLICO RIVER ESTUARY.:	W72-01329
HESIS/RESPIRATION RATIOS.:	*PAMLICO RIVER ESTUARY, PHOTOSYNT	W72-13636
	*PAMLICO RIVER(NC).:	W71-07337
	*PARALLEL PLATE MODEL.:	W73-01362
RBICIDES, *AQUATIC WEED CONTROL,	*PARAQUAT, *ENVIRONMENTAL EFFECTS	W72-00155
SM, *DISSOLVED ORGANIC NITROGEN,	*PARTICULATE ORGANIC NITROGEN, CA	W72-08048
*FILTRATION, *SEWAGE TREATMENT/	*PATENTS, *WASTE WATER TREATMENT,	W72-14469
*LIQUID WASTES, SEWAGE TREATME/	*PATENTS, *WASTE WATER TREATMENT,	W72-05315
ION, *LEACHING NUT/ *PHOSPHATES,	*PATH OF POLLUTANTS, *EUTROPHICAT	W71-04216
RIVERS, WA/ FRESHWATER, *BORON,	*PATH OF POLLUTANTS, METHODOLOGY,	W73-00286
ERPLANTS, *RADIOACTIVITY EFFECT/	*PATH OF POLLUTANTS, *NUCLEAR POW	W73-00831
ER SULFATE, *ALGAE, *FLOODWATER,	*PATH OF POLLUTANTS, DIELDRIN, DD	W71-11498
OPES, *TUBIFICIDS, *FOOD CHAINS,	*PATH OF POLLUTANTS, *ABSORPTION,	W72-09668
OISOTOPES, *RADIUM, *ABSORPTION,	*PATH OF POLLUTANTS, *WATER POLLU	W72-10686
C ENVIRONMENT, *WATER POLLUTION,	*PATH OF POLLUTANTS, *PUBLIC HEAL	W72-13347
, CATION EXCHANGE, RADIOECOLOGY,	*PATH OF POLLUTANTS.: / STRUCTURE	W72-12203
ION, *MINE WASTES, RADIOECOLOGY,	*PATH OF POLLUTANTS, PHYSICOCHEMI	W72-12204
ATER POLLUTION CONTROL, *WASTES,	*PATHOGENIC BACTERIA, CONFERENCES	W72-09675
. CO/ *ALGAE, *ENTERIC BACTERIA,	*PATHOGENIC BACTERIA, CULTURES, E	W72-07664
IC BACTERIA, *OXIDATION LAGOONS,	*PATHOGENIC BACTERIA.: /S, *ENTER	W70-08319
ROL, FRESH WATER, / *BIOCONTROL,	*PATHOGENIC BACTERIA, *ALGAL CONT	W72-05457
LITY CONTROL, *W/ *HYDROBIOLOGY,	*PATHOLOGY, *TOXICITY, *WATER QUA	W72-09646
AUDII, BARBADOS.:	*PERIODICITY, TRICHODESMIUM THIEB	W72-11719
PRODUCTIVITY, *ALGAE, *DIATOMS,	*PERIPHYTON, *PHYTOPLANKTON, *SES	W72-12192
CHLOROPHYLL, NITROGEN FIXATION,	*PERIPHYTON, *PHYTOPLANKTON, BIOL	W72-10883
IOMASS, NEMATODES, ALGAE, OLIGO/	*PERIPHYTON, *SUBMERGED PLANTS, B	W71-11012
IGHT PENETRATION, BENTHIC FLORA/	*PERIPHYTON, *ALGAE, *LITTORAL, L	W71-11029
VITY, PHOTOSYNTHESIS, BIOMASS, /	*PERIPHYTON, *PLANKTON, *PRODUCTI	W71-09165
AL SUBSTRATES, DEPTH, *SAMPLING,	*PERIPHYTON, EQUIPMENT, TUBIFICID	W73-00374
*DDT, *DDE, *PESTICIDE KINETICS,	*PERSISTENCE, *WATER POLLUTION EF	W73-02280
*PESTICIDES, *AQUATIC ORGANISMS,	*PEST CONTROL, *CHEMCONTROL, *BIB	W73-01976
, *BIOASSAY, *INDUSTRIAL WASTES,	*PESTICIDES, *AQUATIC ORGANISMS,	W73-01976
BON PESTIC/ *ALGAE, *INDICATORS,	*PESTICIDES, CHLORINATED HYDROCAR	W73-00238
ES, IRRIGA/ *ALGAE, *CYANOPHYTA,	*PESTICIDES, FUNGICIDES, HERBICID	W72-14807
TION EFFECTS, BIOASSAY, PESTICI/	*PESTICIDE TOXICITY, *WATER POLLU	W72-08440
LORNATED HYDROCARBON PESTICIDES,	*PESTICIDE TOXICITY, *DDT, PESTIC	W72-08436
, *WATER POLLUTION / *DDT, *DDE,	*PESTICIDE KINETICS, *PERSISTENCE	W73-02280
YTOPLANKTON, CARBON RADIOISOTOP/	*PESTICIDES, *DDT, *DIELDRIN, *PH	W72-13657
E, *AQUATIC ENVIRONMENT, *WATER/	*PESTICIDE RESIDUES, *AQUATIC LIF	W72-13347
ONMENT, WATERSHEDS(BASINS), *DD/	*PESTICIDE RESIDUES, *LOTIC ENVIR	W72-12930
OILS, TRIAGINE PES/ *HERBICIDES,	*PESTICIDES, *ALGAE, *TOXICITY, S	W71-07675
ALGAE, *WATER POLLUTION EFFECTS,	*PESTICIDES, DIATOMS, PHYTOPLANKT	W72-05614
ORINATED HYDROCARBON PESTICIDES,	*PESTICIDE RESIDUES, *ALGAE, CHLO	W72-04134
ED HYDROCARBON PESTICIDES, *DDT,	*PESTICIDE TOXICITY, *WATER POLLU	W71-12303
IENTS, ALGAE, RICE/ *CYANOPHYTA,	*PESTICIDES, *FERTILIZATION, NUTR	W71-12071
RBON PESTICIDES, *CARBAMATE PES/	*PESTICIDES, *CHLORINATED HYDROCA	W71-10566
PESTICIDES, *PESTICIDE RESIDUES,	*PESTICIDE TOXICITY, *AQUATIC PLA	W72-09650
*PESTICIDE TOXI/ *HYDROBIOLOGY,	*PESTICIDES, *PESTICIDE RESIDUES,	W72-09650

TOSYNTHESIS, *MARINE ALGAE, DDT, *PESTICIDES, *PRIMARY PRODUCTIVIT W72-09653
OXI/ *HYDROBIOLOGY, *PESTICIDES, *PESTICIDE RESIDUES, *PESTICIDE T W72-09650
HABITAT/ *LIFE HISTORY STUDIES, *PHAEOPHYTA, EPIPHYTOLOGY, ALGAE, W72-01370
INE ALGAE, CULTURES, SALINITY, / *PHAEOPHYTA, *TIDAL MARSHES, *MAR W72-12633
ENTAL EFFECTS, *PLANT MORPHOLOG/ *PHAEOPHYTA, *SALINITY, *ENVIRONM W73-00369
CANCER, ANTI/ *PROTHROMBIN TIME, *PHARMACOLOGICAL PROPERTIES, ANTI W72-07360
ALUE, *APHANIZOMENON FLOS-AQUAE, *PHARMACOLOGICAL TESTS, ANTIMICRO W72-04259
ENTS, *NUTRIENTS, EUTROPHICATIO/ *PHOSPHATES, *GREAT LAKES, *SEDIM W72-04263
ICAL PRECIPITATION, HYDROGEN IO/ *PHOSPHATES, *CLAMS, *ALGAE, CHEM W72-03613
AMPLING, *DISTRIBUTION PATTERNS, *PHOSPHATES, *AQUATIC POPULATIONS W72-03567
NITRATES, PLANT PHYSIO/ *ALGAE, *PHOSPHATES, *ENZYMES, *CULTURES, W72-05476
ATION, *FILTR/ *WATER TREATMENT, *PHOSPHATES, *TURBIDITY, *FLOCCUL W72-06201
REMENTS, TEMPERATURE, *NITROGEN, *PHOSPHATES, ANAEROBIC CONDITIONS W72-04789
TROL, *WASTE W/ *EUTROPHICATION, *PHOSPHATES, *WATER POLLUTION CON W72-04734
POLLUTION EFFECTS, *WATER POLLU/ *PHOSPHATES, *DETERGENTS, *WATER W71-08768
S, *E/ *WATER POLLUTION EFFECTS, *PHOSPHATES, *NITRATES, *NUTRIENT W70-09388
PLANKTON, *AQUATIC PRODUCTIVITY, *PHOSPHATES, *NITRATES, *CHLOROPH W71-02681
*EUTROPHICATION, *LEACHING NUT/ *PHOSPHATES, *PATH OF POLLUTANTS, W71-04216
(POLLUTION), *HYDROLYSIS, DETER/ *PHOSPHATES, *LAKES, *WASTE WATER W71-00116
*/ *EUTROPHICATION, *DETERGENTS, *PHOSPHATES, *LOCAL GOVERNMENTS, W72-06638
OPHICATION, ALKYLB/ *DETERGENTS, *PHOSPHATES, *POTABLE WATER, EUTR W72-06837
RCES, *POLLUTION A/ *DETERGENTS, *PHOSPHATES, *WATER POLLUTION SOU W72-06655
HLORELLA, ALGAE/ *ALGAL CONTROL, *PHOSPHATES, *ALGAL NUTRIENTS, *C W72-13992
REQUIREMENTS, *LIMITING FACTORS, *PHOSPHATES, DIATOMS, POLLUTANT I W73-01445
*TOXICITY, *ARSENIC COMPOUNDS, *PHOSPHATES, *INHIBITIONS, *MICRO W72-13813
GENTS, *WATER POLLUTION CONTROL, *PHOSPHATES, *FEDERAL GOVERNMENT, W72-01093
RFACES, ADSORPTION, LAKES, NUTR/ *PHOSPHATES, *SEDIMENT-WATER INTE W72-01108
ATION/ *WATER POLLUTION CONTROL, *PHOSPHATES, *DETERGENTS, *LEGISL W72-01685
T/ *UNITED STATES, *DETERGETNTS, *PHOSPHATES, *WATER POLLUTION CON W72-01085
, URB/ *WATER POLLUTION SOURCES, *PHOSPHATES, *NITRATES, *NEBRASKA W71-13466
DEMAND, CHEMICAL OXYGEN DEMAND, *PHOSPHATES, NITROGEN, COLIFORMS, W71-13356
*MAINTENANCE, *PLANT MORPHOLOGY, *PHOSPHATES, *NITRATES, *AMMONIUM W72-10162
RCES, *EUTROPHICATION, *DETERGE/ *PHOSPHATES, *WATER POLLUTION SOU W72-11186
RIENTS, ASSESSMENT/ *DETERGENTS, *PHOSPHATES, *EUTROPHICATION, NUT W72-09169
ION ABATEMENT,/ *EUTROPHICATION, *PHOSPHATES, *DETERGENTS, *POLLUT W71-09429
HYTA, ALGAE, CULTURES.: *PHOSPHATES, *SOLUBILITY, *CYANOP W71-10070
NT, *FOAM FRACTIONATION, LIME, / *PHOSPHATES, *WASTE WATER TREATME W71-12200
TER POLLUTIO/ *SEWAGE TREATMENT, *PHOSPHATES, *EUTROPHICATION, *WA W71-11507
L, *EUTROPHICATION, *DETERGENTS, *PHOSPHATES, LEGISLATION, WATER L W71-11516
ENT, NUT/ *BIOLOGICAL TREATMENT, *PHOSPHOROUS, *WASTE WATER TREATM W71-12188
TIVATED SLUDGE/ *EUTROPHICATION, *PHOSPHOROUS, ALGAE, NITROGEN, AC W71-13334
S, / *WATER DILUTION, *NITROGEN, *PHOSPHORUS, *DETENTION RESERVOIR W72-02412
TILIZATION, AQUATIC PLANTS, MAT/ *PHOSPHORUS, *EUTROPHICATION, FER W71-12072
*PHYTOPLANKTON, *AQUATIC ALGAE, *PHOSPHORUS, *CYANOPHYTA, *EUTROP W71-10645
TOPLANKTON,/ *ORGANIC COMPOUNDS, *PHOSPHORUS, *MARINE PLANTS, *PHY W71-10083
ENVIRONMENT, ECOSYSTE/ *CARBON, *PHOSPHORUS, *POPULATION, AQUATIC W72-09159
*EUTROPHICATION, *CARBON, *PHOSPHORUS, NITROGEN, ALGAE.: W72-09155
LIMITING FACTORS, GROWTH RATES,/ *PHOSPHORUS, *PLANKTON, *ALGAE, * W72-09163
ALGAE, WASTE WATER TREATMENT, A/ *PHOSPHORUS, *SEWAGE EFFLUENTS, * W72-09176
ALGAL CONTROL, NUTRIENT REQUIRE/ *PHOSPHORUS, *LAKE BEDS, *ALGAE, W72-11617
ICIDS, *FOOD CHAINS, *PATH OF P/ *PHOSPHORUS RADIOISOTOPES, *TUBIF W72-09668
, ALGAE, INORGANI/ *CHLOROPHYLL, *PHOSPHORUS, *LAKE ERIE, NITROGEN W72-13653
CHLOROPHYLL, *A/ *ALGAL CONTROL, *PHOSPHORUS, *NITROGEN, *IDAHO, * W72-13300
GROWTH, *LAKE MICHIGAN, ALGAE, / *PHOSPHORUS, *CLADOPHORA, *PLANT W72-13644
OSPHATES, NUTRIENTS, CH/ *ALGAE, *PHOSPHORUS, *LAKES, CULTURES, PH W72-12637
KES.: *PHOSPHORUS CIRCULATION, SMALL LA W72-12543
S, *SALT MARSHES, *CARBON CYCLE, *PHOSPHORUS COMPOUNDS, *NITROGEN W72-12567
S, CYCLING N/ *PLANKTON, *ALGAE, *PHOSPHORUS, *MODEL STUDIES, LAKE W72-12543
*NITRIFICATION, WASTE TREATMENT, *PHOSPHORUS, *NITROGEN, COST ANAL W73-02187
MICAL/ *WATER POLLUTION CONTROL, *PHOSPHORUS, *EUTROPHICATION, CHE W73-02478
TREATMENT, HYDROGEN ION CONCEN/ *PHOSPHORUS, *ALGAE, *WASTE WATER W73-01367
*PHOSPHORUS ELIMINATION.: W73-01367
*SOUTH DAKOTA, PLANKTON, *ALGAE, *PHOSPHORUS COMPOUNDS, CYANOPYTA, W73-00938
CROORGANISMS, *ALGAE, *NITROGEN, *PHOSPHORUS, SEDIMENTS, ORGANIC W W72-06532
ON, *ION TRANSPORT, *ABSORPTION, *PHOSPHORUS, *PHOSPHORUS RADIOISO W72-07143
SPORT, *ABSORPTION, *PHOSPHORUS, *PHOSPHORUS RADIOISOTOPES, ALGAE, W72-07143
AQUATIC ALGAE, *LIGHT INTENSITY, *PHOSPHORUS, PLANKTON, CHLOROPHYT W72-07166
TION, *ALGAL BLOOMS, *NUTRIENTS, *PHOSPHORUS, *NITROGEN, *NORTH CA W70-09669
EN/ CYCLING NUTRIENTS, *CARBON, *PHOSPHORUS, *CYANOPHYTA, AQUATIC W71-00475
HOSPHORUS RADIOISOTOPES, SEWAGE/ *PHOSPHORUS, *ACTIVATED SLUDGE, P W71-01474
RY TREATMENT, *EUTROPHICATION, / *PHOSPHORUS, *DETERGENTS, *TERTIA W70-09325
LGAE, SYMBIOSI/ *EUTROPHICATION, *PHOSPHORUS, *CARBON, BACTERIA, A W70-07283
TY CONTROL, *NITROGEN COMPOUNDS, *PHOSPHORUS COMPOUNDS, ALGAE, NUT W70-04080
ON, AQUA/ *NUTRIENTS, *NITROGEN, *PHOSPHORUS, ALGAE, *EUTROPHICATI W71-08953
TRA/ *EUTROPHICATION, *NITROGEN, *PHOSPHORUS, NUTRIENTS, ALGAE, NI W71-06443
ES/ *SOIL BACTERIA, *LAKE SOILS, *PHOSPHORUS, *EUTROPHICATION, LAK W72-04292
ALGAE, *GROWTH RATES, *KINETICS, *PHOSPHORUS, *NUTRIENT REQUIREMEN W72-04788
*EUTROPHICATION, LAKES, AQUATIC/ *PHOSPHORUS, *SEDIMENTS, *ALGAE, W72-03742
NT, *ALGAL CONTROL, *AQUATIC AL/ *PHOSPHORUS, *WASTE WATER TREATME W72-03970
RIENTS, *ALGAE, *EUTROPHICATION, *PHOSPHORUS, *NITROGEN, EVAPORATI W72-04070
NOIS).: *PHOSPHORUS REMOVAL, CHICAGO(ILLI W72-03970
LGAE, *FLUORESCENCE, *CHLORELLA, *PHOTOACTIVATION, *OXYGEN, LIGHT, W72-14800
IATOMS, *LIGHT, SOLAR RADIATION, *PHOTOPERIODISM, PERIPHYTON, *BEN W72-07901
ONAL, *CLADOPHORA, *TEMPERATURE, *PHOTOPERIODISM, GREAT LAKES, MOD W72-13651
NDRICA, ACETYLENE REDUCTION, WA/ *PHOTORESPIRATION, *ANABAENA CYLI W72-08584
PLANKTON, *NITROGEN, *NUTRIENTS, *PHOTOSYNTHESIS, LIGHT INTENSITY, W72-09103
VITY, *RADIOACTIVITY TECHNIQUES, *PHOTOSYNTHESIS, CARBON DIOXIDE, W72-09102
ROGEN FIXATION, OXYGEN,/ *LIGHT, *PHOTOSYNTHESIS, *CYANOPHYTA, NIT W72-08584
*ANALYTICAL TECHNIQUES, RATES, *PHOTOSYNTHESIS, *MARINE ALGAE, D W72-09653
*MARINE ALGAE, *BIOINDICATORS, *PHOTOSYNTHESIS, *PLANT PHYSIOLOG W71-10553

ZINON, *WATER POLLUTION EFFECTS, *PHOTOSYNTHESIS, PESTICIDE TOXICI W71-10566
THESIS, INSTRUMENTATION/ *ALGAE, *PHOTOSYNTHETIC OXYGEN, *PHOTOSYN W71-10791
*ALGAE, *PHOTOSYNTHETIC OXYGEN, *PHOTOSYNTHESIS, INSTRUMENTATION, W71-10791
IOXID/ *EUTROPHICATION, *CARBON, *PHOTOSYNTHESIS, *ALGAE, CARBON D W71-11519
L, FLUCTU/ *ZOOPLANKTON, *ALGAE, *PHOTOSYNTHESIS, RAINFALL, DIURNA W72-00150
ICAL OXY/ *CHLORINATION, *ALGAE, *PHOTOSYNTHESIS, SOLUBILITY, CHEM W72-00411
RIMARY PRODUCTIVITY, *TURBIDITY, *PHOTOSYNTHESIS, LIMNOLOGY, *EUTR W72-12393
EEK RESERVOIR(KAN), GREAT PLAIN/ *PHOTOSYNTHETIC MODEL, *TUTTLE CR W72-12393
, RADIOACTIVITY TECHNIQUES, MET/ *PHOTOSYNTHESIS, CALCIUM CHLORIDE W72-12634
RY PRODUCTIVITY, STANDING CROPS, *PHOTOSYNTHESIS.: /OXICITY, PRIMA W72-14690
ALGAE, *WATER POLLUTION EFFECTS, *PHOTOSYNTHESIS, *OIL POLLUTION, W73-01074
HOSPHATE, ANALYTICAL TECHNIQUES, *PHOTOSYNTHESIS, CARBON FIXATION, W72-06046
Y, CARBON CYCLE,/ *BIOCHEMISTRY, *PHOTOSYNTHESIS, *ALGAE, PHYLOGEN W71-09172
*PRODUCTIVI/ *LIMNOLOGY, *LAKES, *PHOTOSYNTHESIS, *PHYTOPLANKTON, W71-08670
ULTURES, VO/ *CHLORELLA, *LIGHT, *PHOTOSYNTHETIC OXYGEN, *ALGAE, C W70-05547
TOPLANKTON(PRYMNESIU/ *GLYCEROL, *PHOTOTROPHIC GROWTH, *MARINE PHY W70-07842
A, STAND/ *MARINE ALGAE, *KELPS, *PHYSIOLOGICAL ECOLOGY, PHAEOPHYT W72-12940
AKES, *SEASONAL, *PRIMARY PRODU/ *PHYTOPLANKTON, *NANNOPLANKTON, L W72-12738
ERALOGY, *NUTRIENT REQUIREMENTS, *PHYTOPLANKTON, *AQUATIC ALGAE, C W72-12739
MENTS, NUTRIENTS, BIOINDICATORS/ *PHYTOPLANKTON, *NUTRIENT REQUIRE W72-13816
P/ *PESTICIDES, *DDT, *DIELDRIN, *PHYTOPLANKTON, CARBON RADIOISOTO W72-13657
S, *NITROGEN FIXATION, LAKES, N/ *PHYTOPLANKTON, *ALGAE, *NUTRIENT W72-13692
ALUATION, *SEPARATION TECHNIQUE/ *PHYTOPLANKTON, *CHLOROPHYLL, *EV W72-13673
TY, CULTURES, *CHLORELLA, AQUAT/ *PHYTOPLANKTON, *MERCURY, *TOXICI W72-11727
, *ALGAE, *DIATOMS, *PERIPHYTON, *PHYTOPLANKTON, *SESSILE ALGAE, * W72-12192
Y, *ABSORPT/ *DDT, *MARINE FISH, *PHYTOPLANKTON, *SHRIMP, *BIOASSA W73-02105
UES, *COMPUTER PROGRAMS, *ALGAE, *PHYTOPLANKTON, BIOLOGICAL COMMUN W73-02469
VITY,/ SEWAGE EFFLUENTS, *ALGAE, *PHYTOPLANKTON, *AQUATIC PRODUCTI W71-02681
*PRIMARY PRODUCTIVITY, *COASTS, *PHYTOPLANKTON, *ESTUARIES, *STAN W71-07875
NOLOGY, *LAKES, *PHOTOSYNTHESIS, *PHYTOPLANKTON, *PRODUCTIVITY, AL W71-08670
N, ZOOPLANKTON, DISTRIBUTION, P/ *PHYTOPLANKTON, *LAKES, EPILIMNIO W71-09171
ORPTION, MARINE/ *RADIOISOTOPES, *PHYTOPLANKTON, *METABOLISM, *ABS W71-09193
TERNS, *NORTH CARO/ *ESTUARIES, *PHYTOPLANKTON, *DISTRIBUTION PAT W71-07337
*BENTHIC FAUNA, *BENTHIC FLORA, *PHYTOPLANKTON, ALGAE, GROWTH RAT W72-06046
*AQUATIC PLANTS, *AQUATIC ALGAE, *PHYTOPLANKTON, *RESERVOIRS, *RES W72-04855
AGE TREATMENT, *SEWAGE LAGOONS,/ *PHYTOPLANKTON, *METABOLISM, *SEW W72-05459
*STANDING CROP, *PR/ *LIMNOLOGY, *PHYTOPLANKTON, *WATER ANALYSIS, W73-01435
SYSTEMATICS, *DISTRIBUTION PATT/ *PHYTOPLANKTON, *AQUATIC ALGAE, * W73-01094
ATIC PRODUCTIVITY, *ZOOPLANKTON, *PHYTOPLANKTON, NERITIC, ESTUARIN W73-00935
*EUGLENOPHYTA, *BIOINDICATORS, *PHYTOPLANKTON, *POLLUTANT IDENTI W73-00918
TS, ECOLOGY, *ECOSYSTEM, *ALGAE, *PHYTOPLANKTON, OXYGEN, PHOTOSYNT W73-01632
ZINC RADIOISOTOPES, *ABSORPTION, *PHYTOPLANKTON, *MARINE ALGAE, FO W73-00823
PES, *ABSORPTION, *MARINE ALGAE, *PHYTOPLANKTON, BACTERIA, AQUARIA W73-00802
, CHLOROPHYTA, BENTHIC / *ALGAE, *PHYTOPLANKTON, *RIVERS, *DIATOMS W72-14797
ORGANIC MATTER, *EUTROPHICATION, *PHYTOPLANKTON, *ALGAE, *AQUATIC W72-14673
VITY, *EUTROPHICATION, *BIOMASS/ *PHYTOPLANKTON, *PRIMARY PRODUCTI W72-13871
IMARY PRODUCTIVITY, *NERITIC, N/ *PHYTOPLANKTON, *THERMOCLINE, *PR W73-00431
FECTS, *SYSTEMATICS, *SEASONAL,/ *PHYTOPLANKTON, *ENVIRONMENTAL EF W73-00442
EMICAL ANALYSIS, *NITRATES, *AM/ *PHYTOPLANKTON, *MARINE ALGAE, CH W73-00380
D OXYGEN, PHOTOSYNTHESIS, TURBI/ *PHYTOPLANKTON, *ALGAE, *DISSOLVE W73-00256
ENT, SEDIMENTATION, METHODOLOGY/ *PHYTOPLANKTON, LABORATORY EQUIPM W73-00273
ATOMS, SEA WATER, SEASONAL, FLU/ *PHYTOPLANKTON, *SYSTEMATICS, *DI W72-07663
ATIC ALGAE, *LIGHT INTE/ *ALGAE, *PHYTOPLANKTON, *ABSORPTION, *AQU W72-07166
*ENZYMES, *DETERGENTS, *ALGAE, *PHYTOPLANKTON, *SEWAGE LAGOONS, W72-07661
EDIMENTATION RATES, *REVIEWS, B/ *PHYTOPLANKTON, *MARINE ALGAE, *S W72-06313
ABSORPTION, *PHOSPHORUS, *PHOSP/ *PHYTOPLANKTON, *ION TRANSPORT, * W72-07143
Y, *PROFILES, PHOTOMETER, HYPOL/ *PHYTOPLANKTON, *LAKES, *TURBIDIT W72-07145
S, BIORT/ *PRIMARY PRODUCTIVITY, *PHYTOPLANKTON, *SEA WATER, COAST W72-03228
*CURRENTS, *DISTRIBUTION PATTE/ *PHYTOPLANKTON, *ALGAE, *DIATOMS, W72-03543
*EUTROPHICATION, *ALGAL BLOOMS, *PHYTOPLANKTON, *NUTRIENTS, WATER W72-01329
ATER QUALITY, *DISSOLVED OXYGEN, *PHYTOPLANKTON, ESTUARIES, WATER W71-11876
ALGAE.: *DIATOMS, *PHYTOPLANKTON, *PACIFIC OCEAN, * W71-11527
PHOSPHORUS, *CYANOPHYTA, *EUTRO/ *PHYTOPLANKTON, *AQUATIC ALGAE, * W71-10645
DS, *PHOSPHORUS, *MARINE PLANTS, *PHYTOPLANKTON, ALGAE, LIGHT INTE W71-10083
CATION,/ *AQUATIC ALGAE, *LAKES, *PHYTOPLANKTON, *DENSITY STRATIFI W71-09852
*OLIGOTROPHY, MINNESOTA, LAKES,/ *PHYTOPLANKTON, *EUTROPHICATION, W71-13149
TERNS, *SYSTEMATICS, *SALINE LA/ *PHYTOPLANKTON, *DISTRIBUTION PAT W72-09111
ELS, *ALGAL CONTROL, *RESERVOIR/ *PHYTOPLANKTON, *MATHEMATICAL MOD W72-08559
ENTS, *PHOTOSYNTHESIS, LIGHT IN/ *PHYTOPLANKTON, *NITROGEN, *NUTRI W72-09103
NVESTIGATIONS, *P/ *VARIABILITY, *PHYTOPLANKTON, *DEPTH, ON-SITE I W72-11719
NITROGEN FIXATION, *PERIPHYTON, *PHYTOPLANKTON, BIOLOGICAL COMMUN W72-10883
VITY, MICROSCOPY, CHLOR/ *ALGAE, *PHYTOPLANKTON, *PRIMARY PRODUCTI W72-09239
FFECTS, EUTROPHICAT/ *NUTRIENTS, *PHYTOPLANKTON, WATER POLLUTION E W72-09167
L TOXINS, *RED TIDE, *BEGGIATOA, *PHYTOTOXICITY, *CYANOPHYTA, *NEW W72-14542
TY, *ALGAL TOXINS, *CHLOROPHYTA, *PHYTOTOXINS, ECOLOGY, FISH FOOD W71-08597
CATION, CHLOROPHYLL, PL/ *LAKES, *PIGMENTS, *SUCCESSION, *EUTROPHI W73-00916
NDICATORS, CHROMATOGRAPHY, CHLO/ *PIGMENTS, *LAKES, *SEDIMENTS, *I W72-01784
RY INDUSTRY, *OXIDATION LAGOONS, *PILOT PLANTS, ALGAE, BIOCHEMICAL W70-06619
RECHARGE, *INDUCED INFILTRATION, *PIT RECHARGE, *WATER TREATMENT, W71-12410
*CLADOPHORA GLOMERATA, *PITHOPHORA OEDOGONIA.: W71-08032
NTHESIS, BIOMASS, / *PERIPHYTON, *PLANKTON, *PRODUCTIVITY, PHOTOSY W71-09165
MICROORGANISMS, *SURFACE WATERS, *PLANKTON, *ALGAE, TURBIDITY, *NE W72-06191
ALGAE, *AQUATIC ANIMALS, *BIOMA/ *PLANKTON, *RESERVOIRS, *AQUATIC W72-05968
DISTRIBUTION PATTERNS, *PHOSPHA/ *PLANKTON, *DIATOMS, *SAMPLING, * W72-03567
WATER POLLUTION EFFECTS, *ALGAE, *PLANKTON, WATER POLLUTION CONTRO W72-03218
QUES, *RADIOACTIVITY TECHNIQUES, *PLANKTON, PRIMARY PRODUCTIVITY, W72-00161
ORS, GROWTH RATES,/ *PHOSPHORUS, *PLANKTON, *ALGAE, *LIMITING FACT W72-09163
AE, *EUTROPHICATION, *NUTRIENTS, *PLANKTON, *RECREATION, *WATER QU W72-10431
*AMINO ACIDS, *EUTROPHICATION, *PLANKTON, OREGON, ORGANIC COMPOU W73-01269
ENVIRONMENTAL EFFECTS, *ECOLOGY, *PLANKTON, *ALGAE, BIOLOGICAL COM W72-14659

AGE, PHYSICOCHEMICAL PROPERTIES,	*PLANKTON, *AQUATIC PLANTS, *AQUA	W72-14280
NUTRIENTS, *HYDROGRAPHY, *PACIF/	*PLANKTON, *BIOMASS, *SAMPLING, *	W72-07892
ARY PRODUCTIVITY, INVERTEBRATES,	*PLANKTON, *WATER TEMPERATURE, SY	W72-07899
MODEL STUDIES, LAKES, CYCLING N/	*PLANKTON, *ALGAE, *PHOSPHORUS, *	W72-12543
GAE, / *PHOSPHORUS, *CLADOPHORA,	*PLANT GROWTH, *LAKE MICHIGAN, AL	W72-13644
TORS, PLANT TISSUES, BI/ *ALGAE,	*PLANT GROWTH, RHODOPHYTA, INHIBI	W72-07496
INE ALGAE, *BIOASSAY/ *CULTURES,	*PLANT GROWTH, *CHLOROPHYTA, *MAR	W72-14294
DES, *CYANOPHYTA, / *INHIBITION,	*PLANT GROWTH REGULATORS, *CHLORI	W72-14706
HEAVY METALS, DISSOLVED OXYGEN,	*PLANT GROWTH, LIGHT INTENSITY, T	W72-14690
*RHODOPHYTA, *PLANT MORPHOLOGY,	*PLANT GROWTH, CYTOLOGICAL STUDIE	W73-01622
ROWTH RAT/ *FORECASTING, *ALGAE,	*PLANT GROWTH, *EUTROPHICATION, G	W71-11033
E, *GROWTH RATES, *PRODUCTIVITY,	*PLANT GROWTH SUBSTANCES, PHOTOSY	W71-08687
, INHIBITION, *PLANT PHYSIOLOGY,	*PLANT GROWTH, PLANT GROWTH SUBST	W70-07842
TLES, *POLLUTANTS, *MAINTENANCE,	*PLANT MORPHOLOGY, *PHOSPHATES, *	W72-10162
CYTOLOGIC/ *ALGAE, *RHODOPHYTA,	*PLANT MORPHOLOGY, *PLANT GROWTH,	W73-01622
ALINITY, *ENVIRONMENTAL EFFECTS,	*PLANT MORPHOLOGY, *GENETICS, *VA	W73-00369
LIFE CYCL/ *ALGAE, *SYSTEMATICS,	*PLANT MORPHOLOGY, REPRODUCTION,	W72-07141
LANT PIGMENTS, PLANT PHYSIOLOGY,	*PLANT PATHOLOGY, PLANT GROWTH RE	W70-09429
HYTA, *GROWTH RATES, INHIBITION,	*PLANT PHYSIOLOGY, *PLANT GROWTH,	W70-07842
*CARBOHYDRATES, *ORGANICS ACIDS,	*PLANT PHYSIOLOGY, *GROWTH RATES,	W72-07710
DESCRIPTORS:*DDT, *RESISTANCE,	*PLANT PHYSIOLOGY, CULTURES, PHOT	W73-00281
*BIOINDICATORS, *PHOTOSYNTHESIS,	*PLANT PHYSIOLOGY, POLLUTANTS, AN	W71-10553
OXICITY, WAT/ *ALGAE, *BIOASSAY,	*PLANT PHYSIOLOGY, POLLUTANTS, *T	W72-12576
TENCE, *WATER POLLUTION EFFECTS,	*PLANT PHYSIOLOGY, *DIATOMS, PEST	W73-02280
VITY, *PRIMARY PR/ *CHLOROPHYLL,	*PLANT PIGMENTS, AQUATIC PRODUCTI	W72-12192
TEBRA/ *WATER POLLUTION EFFECTS,	*PLANTS, *ANIMALS, *ALGAE, *INVER	W71-13233
OONS, *CHANNELS, *HEAT TRANSFER,	*PLASTICS, DOMESTIC WASTES, BACTE	W72-10401
1D, PHYCOVIRUS, MITO/ *LYSOGENY,	*PLECTONEMA BORYANUM, LPP-1, LPP-	W72-03188
IRUS, *BLUE-GREEN ALGAL VIRUSES,	*PLECTONEMA BORYANIUM, SUSQUEHANN	W72-04736
LOGICAL SAMPLES, SARGASSO WEED,/	*PLUTONIUM, DETECTION LIMITS, BIO	W72-05965
*POLONIUM,	*PLUTONIUM.:	W71-09193
.:	*PLUTONIUM, *SILVER RADIOISOTOPES	W72-07826
:	*PLUTONIUM, CESIUM RADIOISOTOPES.	W73-00790
NUISANCE ALGAE, / *ALGAL TOXINS,	*POISONOUS PLANTS, *CYANOPHYTA, *	W72-14056
R, ALGAE, PHOTO/ *COPPER, *IONS,	*POISONS, *SEA WATER, *FRESH WATE	W71-03027
UTROPHICATION, *CHEMICAL WASTES,	*POLITICAL ASPECTS, LAKES, GREAT	W72-14782
ARED RADIATION, *REMOTE SENSING,	*POLLUTANT IDENTIFICATION, AQUATI	W72-14728
*BIOINDICATORS, *EUTROPHICATION,	*POLLUTANT IDENTIFICATION, *LAKE	W73-00432
*BIOINDICATORS, *PHYTOPLANKTON,	*POLLUTANT IDENTIFICATION, PROTOZ	W73-00918
*MARINE ALGAE, CULTURES, PHYTOP/	*POLLUTANTS, *SEA WATER, *ALGAE,	W70-07280
STREAMS, ALGAE, PESTICIDES, DDT/	*POLLUTANTS, *RECREATION, LAKES,	W71-06444
*MICROORGANISMS, ALGAL CONTROL,	*POLLUTANT IDENTIFICATION, ANALYT	W72-00801
TS, *ORGANIC WASTES, *NUTRIENTS,	*POLLUTANT IDENTIFICATION, ALGAE,	W72-01472
GEN, ELECTROCHEMISTRY, TRACE EL/	*POLLUTANT IDENTIFICATION, *NITRO	W72-01693
IC PLANTS, *AQUEOUS SOLUTIONS, /	*POLLUTANT IDENTIFICATION, *AQUAT	W72-01473
MORPHOLOGY, *PHO/ *SEA NETTLES,	*POLLUTANTS, *MAINTENANCE, *PLANT	W72-10162
DICATORS, *ALGAE, GROWTH RATES,/	*POLLUTANT IDENTIFICATION, *BIOIN	W71-09795
SAY, *INDICATORS, *WATER POLLUT/	*POLLUTANT IDENTIFICATION, *BIOAS	W71-10553
L PHOTOGRAPHY, *MATHEMATICAL ME/	*POLLUTANT IDENTIFICATION, *AERIA	W71-11685
FOOD CHAINS, *SPECTROPHOTOMETRY,	*POLLUTANT IDENTIFICATION, ONTARI	W71-11529
POLLUTION, *MO/ *REMOTE SENSING,	*POLLUTANT IDENTIFICATION, WATER	W72-12487
Y RIVERS.: *ENGLAND,	*POLLUTED ESTUARIES, DURHAM COUNT	W72-08804
ATION, *PHOSPHATES, *DETERGENTS,	*POLLUTION ABATEMENT, WATER POLLU	W71-09429
	*POLLUTION INDEX.:	W72-01780
UTION EFFECTS, *WATER POLLUTION/	*POLLUTION ABATEMENT, *WATER POLL	W71-13723
ARM WASTES, *HOGS, *ENVIRONMENT,	*POLLUTION ABATEMENT, WATER QUALI	W71-08214
TS, *WATER REUSE, *WATER SUPPLY,	*POLLUTION ABATEMENT, *WATER POLL	W70-09193
HATES, *WATER POLLUTION SOURCES,	*POLLUTION ABATEMENT, EUTROPHICAT	W72-06655
	*POLONIUM, PLUTONIUM.:	W73-00833
	*POLONIUM, *PLUTONIUM.:	W71-09193
*MULTIVALENT METALLIC CATIONS,	*POLYACRYLAMIDE, *POLYETHYLENEMIN	W72-14840
RNATED HYDROCARBON PESTICIDES, /	*POLYCHLORINATED BIPHENYLS, *CHLO	W72-08436
OR 1254, ESCAMBIA RIVER, PINFIS/	*POLYCHLORINATED BIPHENYLS, AROCL	W71-07412
E, *WATER POLLUTION EFFECTS, *P/	*POLYCHLORINATED BIPHENYLS, *ALGA	W72-05614
TALLIC CATIONS, *POLYACRYLAMIDE,	*POLYETHYLENEMINE, *POLYSTYRENE S	W72-14840
I, *IONS, *WASTEWATER TREATMENT,	*POLYMERS, CATIONS, ANIONS, CHLOR	W72-14840
R POLLUTION TREATMENT, COLLOIDS,	*POLYMERS, ANIONS, CATIONS, ORGAN	W72-11833
PTICA, *CYCLOTELLA MENEGHINIANA,	*POLYMORPHISM, *VALVE MORPHOLOGY,	W73-02548
QUECIRRHA, *MEDUSAE MAINTENANCE,	*POLYP MAINTENANCE, *POLYP METABO	W72-10162
MAINTENANCE, *POLYP MAINTENANCE,	*POLYP METABOLISM, TETRACYCLINE,	W72-10162
LYACRYLAMIDE, *POLYETHYLENEMINE,	*POLYSTYRENE SULFONATE, CHLORELLA	W72-14840
CAL OXY/ *WASTE WATER TREATMENT,	*PONDS, *ALGAE, LAGOONS, BIOCHEMI	W70-07838
RATURE, SEASONAL, ION/ *ECOLOGY,	*PONDS, *CHEMICAL ANALYSIS, TEMPE	W72-01363
, ALGAE, PROTEIN, *OXIDATION LA/	*PONDS, *WATER REUSE, RECLAMATION	W71-08951
C ALGAE, *RADIOACTIVITY EFFECTS,	*PONDS, WATER POLLUTION EFFECTS,	W71-09256
SKA, *CYANOPHYTA, ALGAL CONTROL,	*PONDWEEDS, PONDS, *THERMAL STRAT	W73-02349
S/ *MATHEMATICAL MODELS, *ALGAE,	*POPULATION, DYNAMICS, LAKES, DIS	W71-03034
ION, *ENZYME/ *AQUATIC BACTERIA,	*POPULATION, *ALGAE, *STRATIFICAT	W72-03224
ON, *SPATIAL DISTRIBUTI/ *ALGAE,	*POPULATION, *TEMPORAL DISTRIBUTI	W72-08560
ECOSYSTE/ *CARBON, *PHOSPHORUS,	*POPULATION, AQUATIC ENVIRONMENT,	W72-09159
*DEPTH, ON-SITE INVESTIGATIONS,	*POPULATION, TEMPORAL DISTRIBUTIO	W72-11719
, *FILTERS, *DIATOMACEOUS EARTH,	*POROUS MEDIA, *WASTE WATER TREAT	W72-00974
MEDIA, HUMMOCK CHANNEL, CHLOROP/	*PORTERINEMA FLUVIATILE, CULTURE	W72-12633
RVEYS, MICROBIOLOGY/ *LAKE ERIE,	*POTABLE WATER, *TASTE, *ODOR, SU	W72-10076
CONTROL, *NUCLEAR POWER PLANTS,	*POTABLE WATER, *BIOTA, SAMPLING,	W71-09018
ADSORPTION, FIL/ *TASTE, *ODOR,	*POTABLE WATER, ACTIVATED CARBON,	W72-08357
LKYLB/ *DETERGENTS, *PHOSPHATES,	*POTABLE WATER, EUTROPHICATION, A	W72-06837
ALGAL PONDS, ALGAL GROWTH RATES,	*POTAMOCYPRIS.: /ROWTH REACTORS,	W72-04787
*ION-SELECTIVE ELECTRODES, EDTA,	*POTENTIOMETRY, ELECTROANALYSIS,	W73-02112
	*POTOMAC ESTUARY.:	W72-09171

*SPECIES DIVERSITY,	*POTOMAC RIVER.:	W71-11583
	*POTOMAC RIVER ESTUARY.:	W71-08775
	*POTOMAC RIVER.:	W70-08248
*SURVEYS, *STATISTICAL METHODS,	*POTOMAC RIVER, DATA COLLECTIONS,	W72-07791
RIENTS, AQUATIC AL/ *FARM WASTE,	*POULTRY, *OXIDATION LAGOONS, NUT	W72-02850
POLLUTION, *ENVIRONM/ *NEW YORK,	*POWER PLANTS, *STREAM, *THERMAL	W72-14659
, *THERMAL STRATIFICATION, HYDR/	*POWER PLANTS, *THERMAL POLLUTION	W70-09614
, *REVIEWS, WATER TEMPERATURE, /	*POWER PLANTS, *THERMAL POLLUTION	W71-07417
TER CONSE/ *THERMAL POWERPLANTS,	*POWERPLANTS, *ENERGY LOSSES, *WA	W70-09193
SPECIFIC RESI/ *MICROSTRAINING,	*PRECOAT FILTRATION, BACKWASHING,	W72-09393
TEMS, FOOD CHAINS, *ZOOPLANKTON,	*PREDATION, *BIOASSAY, BIOLOGICAL	W72-07132
*MARINE BACTERIA, *ATTRACTANTS,	*PREDATION, ORGANIC COMPOUNDS, AM	W73-00942
TS, *SAMPLING, *ECOSY/ *ECOLOGY,	*PREIMPOUNDMENTS, *AQUATIC HABITA	W72-06526
O/ *CHEMORECEPTION, *CHEMOTAXIS,	*PREY, PYTHIUM DEBARYANUM, SKELET	W73-00942
ALS, *TR/ *ALGAE, *GROWTH RATES,	*PRIMARY PRODUCTIVITY, *HEAVY MET	W73-01434
*WATER ANALYSIS, *STANDING CROP,	*PRIMARY PRODUCTIVITY, LAKES, SAM	W73-01435
LGAE, *COMPARATIVE PRO/ *STRESS,	*PRIMARY PRODUCTIVITY, *AQUATIC A	W73-01080
R, *MARSHES, SALINE WATER, BIOM/	*PRIMARY PRODUCTIVITY, *FRESHWATE	W73-01089
*MARINE ALGAE, *BENTHIC.FLORA,	*PRIMARY PRODUCTIVITY, *ENERGY CO	W72-06283
S, *ARCTIC, *ALGAE.:	*PRIMARY PRODUCTIVITY, *OIL WASTE	W72-07922
TES, *PLANKT/ *ALGAE, *PROTOZOA,	*PRIMARY PRODUCTIVITY, INVERTEBRA	W72-07899
LORA, MICROORGANISMS, ALGAE, LA/	*PRIMARY PRODUCTIVITY, *BENTHIC F	W72-07501
N/ *PHYTOPLANKTON, *THERMOCLINE,	*PRIMARY PRODUCTIVITY, *NERITIC,	W73-00431
BENTHIC/ *ESTUARINE ENVIRONMENT,	*PRIMARY PRODUCTIVITY, *SHOALS, *	W72-14317
ATION, *BIOMASS/ *PHYTOPLANKTON,	*PRIMARY PRODUCTIVITY, *EUTROPHIC	W72-13871
SCENEDESMUS, GROWTH RATES, CULT/	*PRIMARY PRODUCTIVITY, BACTERIA,	W72-14314
GET, *MARINE ALG/ *GROWTH RATES,	*PRIMARY PRODUCTIVITY, ENERGY BUD	W73-00366
*MARINE ALGAE, DDT, *PESTICIDES,	*PRIMARY PRODUCTIVITY, WATER POLL	W72-09653
, CHLOR/ *ALGAE, *PHYTOPLANKTON,	*PRIMARY PRODUCTIVITY, MICROSCOPY	W72-09239
ORDS, SURFACE / *CARBON DIOXIDE,	*PRIMARY PRODUCTIVITY, ALASKA, FI	W72-09122
VITY TECHNIQU/ *LIGHT INTENSITY,	*PRIMARY PRODUCTIVITY, *RADIOACTI	W72-09102
AL MODELS, *ESSENTIAL NUTRIENTS,	*PRIMARY PRODUCTIVITY, WATER QUAL	W72-08986
ATION, *NUTRIENTS, *BIOLOGICAL /	*PRIMARY PRODUCTIVITY, *EUTROPHIC	W72-03216
KTON, *SEA WATER, COASTS, BIORT/	*PRIMARY PRODUCTIVITY, *PHYTOPLAN	W72-03228
D HYDROC/ *INSECTICIDES, *ALGAE,	*PRIMARY PRODUCTIVITY, CHLORINATE	W71-10096
CROP, *ALGAE, DIATOMS, MEASUREM/	*PRIMARY PRODUCTIVITY, *STANDING	W71-12855
*WATER QUALITY, *EUTROPHICATION,	*PRIMARY PRODUCTIVITY, *ALGAE, NI	W71-13183
PHYTOPLANKTON, *ESTUARIES, *STA/	*PRIMARY PRODUCTIVITY, *COASTS, *	W71-07875
GAE, *AQUATIC ANIMALS, *BIOMASS,	*PRIMARY PRODUCTIVITY, *SECONDARY	W72-05968
S, *NEBRASKA, *EUTROPHICATION, /	*PRIMARY PRODUCTIVITY, *RESERVOIR	W72-04270
, *PHOTOSYNTHESIS,/ *RESERVOIRS,	*PRIMARY PRODUCTIVITY, *TURBIDITY	W72-12393
PIGMENTS, AQUATIC PRODUCTIVITY,	*PRIMARY PRODUCTIVITY, *ALGAE, *D	W72-12192
NANNOPLANKTON, LAKES, *SEASONAL,	*PRIMARY PRODUCTIVITY, AQUATIC AL	W72-12738
MENT, *DOMESTIC WA/ *ECOSYSTEMS,	*PRODUCTIVITY, *ESTUARINE ENVIRON	W72-12567
*PHOTOSYNTHESIS, *PHYTOPLANKTON,	*PRODUCTIVITY, ALGAE, NITROGEN, P	W71-08670
TANCES, / *ALGAE, *GROWTH RATES,	*PRODUCTIVITY, *PLANT GROWTH SUBS	W71-08687
ONT/ *ECOSYSTEMS, *SALT MARSHES,	*PRODUCTIVITY, RADIOISOTOPES, STR	W71-09254
OMASS, / *PERIPHYTON, *PLANKTON,	*PRODUCTIVITY, PHOTOSYNTHESIS, BI	W71-09165
FOOD/ *MARINE ALGAE, *ECONOMICS,	*PRODUCTIVITY, ALGAE, RESOURCES,	W71-13246
ESTRIAL HABITATS, *HEAVY METALS,	*PRODUCTIVITY, *ECOSYSTEMS, POTAS	W72-14303
YTOPLANKTON, *LAKES, *TURBIDITY,	*PROFILES, PHOTOMETER, HYPOLIMNIO	W72-07145
OOCYSTIS, OSCILLATORIA, SPIRULI/	*PROTEASE ENZYMES, CHLOROGONIUM,	W72-07661
AL PROPERTIES, ANTICANCER, ANTI/	*PROTHROMBIN TIME, *PHARMACOLOGIC	W72-07360
INVERTEBRATES, *PLANKT/ *ALGAE,	*PROTOZOA, *PRIMARY PRODUCTIVITY,	W72-07899
S, *ENVIRONMENTA/ *SALT MARSHES,	*PROTOZOA, *BIOLOGICAL COMMUNITIE	W73-00367
UTROPHICATION, PONDS,/ *GRAZING,	*PROTOZOA, *ALGAE, SCENEDESMUS, E	W72-14796
D/ *MARINE ALGAE, *SALT MARSHES,	*PROTOZOA, *BACTERIA, FLORIDA, RA	W73-00796
ARSHE/ *WATER POLLUTION EFFECTS,	*PROTOZOA, *MARINE ALGAE, *SALT M	W72-00948
WATER POLLUTION EFFECTS, *ALGAE,	*PROTOZOA, *BIOLOGICAL COMMUNITIE	W72-08579
CHLOROPHYTA, HABITATS, AQUATIC /	*PROTOZOA, *SYSTEMATICS, *ALGAE,	W72-09119
TS, *DISPERSION, CADDIS/ *ALGAE,	*PROTOZOA, *FUNGI, *AQUATIC INSEC	W70-06655
PHICATION, *ECONOMIC EFFICIENCY,	*PUBLIC HEALTH, WATER POLLUTION,	W72-04266
QUES, / *AQUATIC PLANTS, *ALGAE,	*PUBLIC HEALTH, ANALYTICAL TECHNI	W72-04275
ATHOGENIC BACTERIA, CONFERENCES,	*PUBLIC HEALTH, RIVERS, *SEWAGE,	W72-09675
POLLUTION EFFECTS, T/ *MERCURY,	*PUBLIC HEALTH, *SHELLFISH, WATER	W72-12257
POLLUTION, *PATH OF POLLUTANTS,	*PUBLIC HEALTH, ENZYMES, ANALYTIC	W72-13347
ION, *WATER POLLUTION TREATMENT,	*PUBLIC HEALTH, ANALYTICAL TECHNI	W73-02144
ORNE DISEASES, *TOXIC CHEMICALS,	*PUBLIC HEALTH STANDARDS, TOXICOL	W73-02144
N, COAGULATION, BENTONITE, LIME,	*PULP WASTES, TEOTIARY TREATMENT,	W72-03753
*MODEL S/ *ECOSYSTEMS, *ECOLOGY,	*PUMPED STORAGE, *WATER QUALITY,	W72-05282
MOUNTAIN RESERVOIR(VA).:	*PUMPED STORAGE OPERATION, SMITH	W72-14405
PROP/ *WATER LEVEL FLUCTUATIONS,	*PUMPED STORAGE, PHYSICOCHEMICAL	W72-14280
HYTIN, TUNICATES, FORAMINIFERA,/	*PYCNOCLINE, CHLOROPHYLL A, PHEOP	W72-14313
AE, *CHLOROPHYTA, *EUGLENOPHYTA,	*PYRROPHYTA, *SYSTEMATICS, DINOFL	W72-07683
ECONOMIC JUSTIFICATION, *COBALT,	*RADIATION, COSTS, SEWAGE BACTERI	W72-00383
TON, PR/ *ANALYTICAL TECHNIQUES,	*RADIOACTIVITY TECHNIQUES, *PLANK	W72-00161
POWERPLANTS, MONI/ *FOOD CHAINS,	*RADIOACTIVITY EFFECTS, *NUCLEAR	W72-00959
MS, *RAIN FORESTS, CYCLING NUTR/	*RADIOACTIVITY EFFECTS, *ECOSYSTE	W72-00949
ENTAL EFFECTS, *NUCLEAR EXPLOSI/	*RADIOACTIVITY EFFECTS, *ENVIRONM	W72-00960
POLLUTION, CON/ *NUCLEAR WASTES,	*RADIOACTIVITY EFFECTS, *THERMAL	W72-00939
ANIMALS, *ALGAE, *INVERTEBRATES,	*RADIOACTIVITY EFFECTS, *RADIOISO	W71-13233
SIONS, *WATER POLLUTION EFFECTS,	*RADIOACTIVITY EFFECTS, UNDERGROU	W72-03347
NTENSITY, *PRIMARY PRODUCTIVITY,	*RADIOACTIVITY TECHNIQUES, *PHOTO	W72-09102
LOGY, *RESERVOI/ *RADIOISOTOPES,	*RADIOACTIVITY EFFECTS, *RADIOECO	W72-10694
IC FLORA, *GROWTH RATE, *AQUATI/	*RADIOACTIVITY TECHNIQUES, *BENTH	W71-11486
ODEGRADATION, *METABOLISM, *DDT,	*RADIOACTIVITY TECHNIQUES, PESTIC	W72-07703
NG, *MARINE ALGAE, RHODOPHYTA, /	*RADIOACTIVITY EFFECTS, *MONITORI	W72-07826
S, *ENVIRONMENT, *RADIOACTIVITY,	*RADIOACTIVITY EFFECTS, RIVERS, E	W72-07841
*DATA COLLECTIONS, *ENVIRONMENT,	*RADIOACTIVITY, *RADIOACTIVITY EF	W72-07841
*NUCLEAR WASTES,/ *FOOD CHAINS,	*RADIOACTIVITY EFFECTS, *TRITIUM,	W72-06342

Context	Keyword	Accession No.
S, *ABSORPTION, *CARBON DIOXIDE,	*RADIOACTIVITY TECHNIQUES, SOIL A	W72-14301
OLLUTANTS, *NUCLEAR POWERPLANTS,	*RADIOACTIVITY EFFECTS, *MATHEMAT	W73-00831
*MARINE/ *BACKGROUND RADIATION,	*RADIOACTIVITY EFFECTS, *FALLOUT,	W73-00828
WASTES, *FALLOUT, *OCEANS, WATE/	*RADIOACTIVITY EFFECTS, *NUCLEAR	W72-04463
*FOOD CHAINS, *AQUATIC LIFE, R/	*RADIOACTIVITY EFFECTS, *FALLOUT,	W72-04461
ATER POLLUTION / *AQUATIC ALGAE,	*RADIOACTIVITY EFFECTS, *PONDS, W	W71-09256
ADIOISOTOPES, *RADIOSENSITIVITY,	*RADIOACTIVITY EFFECTS, *ALGAE, C	W72-11836
ER, *SEDIMENTS, *MARINE ANIMALS/	*RADIOCHEMICAL ANALYSIS, *SEA WAT	W72-05965
ATIC ENVIRONMENT, *RADIOECOLOGY,	*RADIOCHEMICAL ANALYSIS, MEASUREM	W73-01673
*SURVEYS, *AQUATIC ENVIRONMENT,	*RADIOECOLOGY, *RADIOCHEMICAL ANA	W73-01673
SOTOPES, *RADIOACTIVITY EFFECTS,	*RADIOECOLOGY, *RESERVOIRS, *FRES	W72-10694
IRONMENTAL EFFECTS, NUCLEAR WAS/	*RADIOISOTOPES, *ABSORPTION, *ENV	W72-10955
ION, *PATH OF POLLUTANTS, *WATE/	*RADIOISOTOPES, *RADIUM, *ABSORPT	W72-10686
FECTS, *RADIOECOLOGY, *RESERVOI/	*RADIOISOTOPES, *RADIOACTIVITY EF	W72-10694
R, *FOOD CHAINS, *ABSORPTION, B/	*RADIOISOTOPES, *ALGAE, *SEA WATE	W71-13243
EBRATES, *RADIOACTIVITY EFFECTS,	*RADIOISOTOPES.: /*ALGAE, *INVERT	W71-13233
*MARINE ALGAE, SEDIME/ *REVIEWS,	*RADIOISOTOPES, *MARINE ANIMALS,	W73-00817
SPHATE, RIO GRAN/ *ALGAL GROWTH,	*RADIOISOTOPE P32, SODIUM PYROPHO	W72-13992
NETICS, / *LABORATORY EQUIPMENT,	*RADIOISOTOPES, *MEASUREMENT, *KI	W72-14326
*ABSORPTION, AQUARIA, MOLLUSKS,/	*RADIOISOTOPES, *MARINE ANIMALS,	W71-09182
S, *MARINE ALGAE, PHYTOPLANKTON/	*RADIOISOTOPES, *ORGANIC COMPOUND	W71-09188
RATURE, STRONTI/ *AQUATIC ALGAE,	*RADIOISOTOPES, ABSORPTION, TEMPE	W71-09250
*MA/ *ABSORPTION, *DISTRIBUTION,	*RADIOISOTOPES, *MARINE ANIMALS,	W71-09168
METABOLISM, *ABSORPTION, MARINE/	*RADIOISOTOPES, *PHYTOPLANKTON, *	W71-09193
ON, MONITORING, ANALYTICAL TECH/	*RADIOISOTOPES, *ALGAE, *ABSORPTI	W71-08032
IC FAUNA, *INVERTEBRATES, *MICR/	*RADIOISOTOPES, *DETRITUS, *BENTH	W70-06668
BSORPTION, *MINE WASTES, RADIOE/	*RADIOISOTOPES, *MARINE ALGAE, *A	W72-12204
EFFECTS/ *COBALT RADIOISOTOPES,	*RADIOSENSITIVITY, *RADIOACTIVITY	W72-11836
LLUTANTS, *WATE/ *RADIOISOTOPES,	*RADIUM, *ABSORPTION, *PATH OF PO	W72-10686
IOACTIVITY EFFECTS, *ECOSYSTEMS,	*RAIN FORESTS, CYCLING NUTRIENTS,	W72-00949
TION, *NITROGEN FIXING BACTERIA,	*RAINFALL, *ARID LANDS, *ARIZONA,	W72-12505
REGION).:	*RECIRCULATION RATIO, *ISRAEL(DAN	W72-11642
HICATION, *NUTRIENTS, *PLANKTON,	*RECREATION, *WATER QUALITY, CYAN	W72-10431
TORM RUN-OFF, *SUSPENDED SOLIDS,	*RECREATION, *TREATMENT FACILITIE	W72-00042
E, PESTICIDES, DDT/ *POLLUTANTS,	*RECREATION, LAKES, STREAMS, ALGA	W71-06444
TION EFFECTS, *LAKES, *COLORADO,	*RECREATION FACILITIES, WATER TEM	W72-04791
*WATER SUPPLY, *WASTE DISPOSAL,	*RECREATION WASTES, *MICROORGANIS	W72-14363
, SEWAGE TREAT/ *EUTROPHICATION,	*RECYCLING, WATER QUALITY CONTROL	W72-09378
LGAE, DIATOMS, /*DINOFLAGELLATES,	*RED TIDE, WATER POLLUTION SOURCE	W72-01329
CITY, *CYANOPHYT/ *ALGAL TOXINS,	*RED TIDE, *BEGGIATOA, *PHYTOTOXI	W72-14542
*ALGAE, *AQUATIC WEEDS, *WEEDS,	*RED TIDE, COASTS, LAKES, PONDS,	W72-14673
*GRIFFITHSIA PACIFICA,	*REGENERATION.:	W73-01622
WTH RATES, PHOSPHORUS, NITROGEN,	*REGRESSION ANALYSIS, VARIABILITY	W72-12966
IFICATION, WATER POLLUTION, *MO/	*REMOTE SENSING, *POLLUTANT IDENT	W72-12487
AKES, ALGAE, INFRARED RADIATION,	*REMOTE SENSING, *POLLUTANT IDENT	W72-14728
QUES, MEASUREMENT, CURR/ *ALGAE,	*REMOTE SENSING, *TRACKING TECHNI	W72-01095
PHY, *INFRARED RADIATION, *AQUA/	*REMOTE SENSING, *AERIAL PHOTOGRA	W71-12034
CONTROL, *SEA WATER, *OILS, *N/	*REMOTE SENSING, *WATER POLLUTION	W72-04742
NS, *MASSACHUSETTS, SYSTEMATICS,	*REPRODUCTION, RHODOPHYTA, PHAEOP	W73-01449
L EFFECTS, ESTUARIES,/ *DIATOMS,	*RESEARCH EQUIPMENT, ENVIRONMENTA	W72-08141
POLLUTION, PHOSPHORUS, THERMAL/	*RESEARCH AND DEVELOPMENT, *WATER	W70-09947
, *FUTURE TRENDS, CLADOPHORA, R/	*RESEARCH POLICY, *UNITED KINGDOM	W71-13172
ATER QUALITY, *IMPOUNDED WATERS,	*RESERVOIRS, *HYDROLOGIC DATA, *K	W72-02274
ACTIVITY EFFECTS, *RADIOECOLOGY,	*RESERVOIRS, *FRESHWATER, HYDROBI	W72-10694
EMATICAL MODELS, *ALGAL CONTROL,	*RESERVOIRS, DESTRATIFICATION, MI	W72-08559
ENTS, RECREATI/ *EUTROPHICATION,	*RESERVOIRS, *GREAT PLAINS, NUTRI	W72-04759
AE, *PHYTOPLANKTON, *RESERVOIRS,	*RESERVOIR STAGES, EARLY IMPOUNDM	W72-04855
QUATIC ANIMALS, *AQUATIC PLANTS,	*RESERVOIR STAGES, EARLY IMPOUNDM	W72-04853
*AQUATIC ALGAE, *PHYTOPLANKTON,	*RESERVOIRS, *RESERVOIR STAGES, E	W72-04855
ATIC ANIMALS, *BIOMA/ *PLANKTON,	*RESERVOIRS, *AQUATIC ALGAE, *AQU	W72-05968
CATION, / *PRIMARY PRODUCTIVITY,	*RESERVOIRS, *NEBRASKA, *EUTROPHI	W72-04270
Y, *TURBIDITY, *PHOTOSYNTHESIS,/	*RESERVOIRS, *PRIMARY PRODUCTIVIT	W72-12393
	*RESIDUAL, MOST PROBABLE NUMBER.:	W71-07096
FLY LARVAE, CLADOPHORA, CYPRINU/	*RESIDUES, *DDD, DDE, TDE, BLACK-	W70-08098
ULTURES, PHOT/ DESCRIPTORS:*DDT,	*RESISTANCE, *PLANT PHYSIOLOGY, C	W73-00281
NTS, CULTURE/ *LABORATORY TESTS,	*RESPIRATION, *CHLORELLA, *NUTRIE	W73-01626
*SYMBIOSIS, *ALGAE, *BACTERIA,	*RESPIRATION, ACTIVATED SLUDGE, W	W71-04079
TION, COPPER, F/ *ALGAL CONTROL,	*RESPIRATION, *CHLORELLA, *INHIBI	W72-11572
LLUTION, DESALINATION, ALGAE, D/	*RETURN FLOW, *NITRATES, WATER PO	W71-08223
REATMENT, WATER QUALITY CONTROL,	*REVIEWS.: /NANCE, *WASTE WATER T	W72-09386
ATIC ENVIRONMENT, SURFACE WATER/	*REVIEWS, *BOTTOM SEDIMENTS, *AQU	W72-00847
CATION, *ALGAE, *SURFACE WATERS,	*REVIEWS, *WATER POLLUTION SOURCE	W71-10580
OWER PLANTS, *THERMAL POLLUTION,	*REVIEWS, WATER TEMPERATURE, DISS	W71-07417
URFACE WATERS, *LAKES, *STREAMS,	*REVIEWS, *ILLINOIS, RECREATION,	W72-05869
ANIMALS, *MARINE ALGAE, SEDIME/	*REVIEWS, RADIOISOTOPES, *MARINE	W73-00817
ENTS, *WATER POLLUTION CONTROL,/	*REVIEWS, *EUTROPHICATION, *NUTRI	W73-00840
DICATORS, *MARINE ANIMALS, ASIA/	*REVIEWS, *NUCLEAR WASTES, *BIOIN	W73-00811
*AQUATIC ENVIRONMENT, *TOXICITY/	*REVIEWS, *HEAVY METALS, *ALGAE,	W72-14694
INE ALGAE, *SEDIMENTATION RATES,	*REVIEWS, BIOCHEMISTRY, GEOCHEMIS	W72-06313
NS, *CHLOROPHYTA, *MARINE ALGAE,	*RHODOPHYTA, *SYSTEMATICS, PHOTOG	W73-00284
PLANT GROWTH, CYTOLOGIC/ *ALGAE,	*RHODOPHYTA, *PLANT MORPHOLOGY, *	W73-01622
GAE.:	*RHODOPHYTA, *LAKE ERIE, OHIO, AL	W71-05630
ECOSYSTEM ANALYSIS, MACROCONSU/	*RIFFLE BENTHOS, WOODLAND STREAM,	W71-10064
	*RIFFLES, *CALORIC CONTENT.:	W71-07097
NET(ENGLAND),/ *CENTRIC DIATOMS,	*RIVER THAMES(ENGLAND), RIVER KEN	W72-14797
ENTHIC / *ALGAE, *PHYTOPLANKTON,	*RIVERS, *DIATOMS, CHLOROPHYTA, B	W72-14797
LOGICAL PROPERTIES, *ECOSYSTEMS,	*RIVERS, *VIRGINIA, *WATER POLLUT	W73-01972
YSICOCHEMICAL PROPERTIES, *ALGA/	*RIVERS, *BACTERIA, IPLANKTON, PH	W72-07896
	*ROADSIDE REST AREAS.:	W72-14363
PENDED SOLIDS, SODIUM HYDROXIDE,	*ROCK FILTERS.: *SUS	W71-13341

HYADESIA FUSCA, *SARCOPTIFORMES,	*ROCKPOOLS, *ENTEROMORPHA, SWEDEN	W71-08026
	*RUHR VALLEY.:	W71-12410
Y, HOGS, ANIMAL DISEASE/ *FEEDS,	*RUMINANTS, ALGAE, CATTLE, POULTR	W71-07544
*AQUATIC ENVIRONMENT, WATER PO/	*RUNNING WATERS, *AQUATIC PLANTS,	W72-14690
LVED OXYGEN, DIVE/ *FARM WASTES,	*RUNOFF, *FISHKILL, CATTLE, DISSO	W72-08401
	*RUTHENIUM RADIOISOTOPES.:	W71-09182
UM RADIO/ *CESIUM RADIOISOTOPES,	*RUTHENIUM RADIOISOTOPES, ZIRCONI	W71-09225
SOLUBLE FORMS, LABORATORY EXPER/	*RUTHENIUM 106, SOLUBLE FORMS, IN	W71-13233
RIBUTION PATTERNS, *SYSTEMATICS,	*SALINE LAKES, *EUTROPHICATION, W	W72-09111
, *PLANT MORPHOLOG/ *PHAEOPHYTA,	*SALINITY, *ENVIRONMENTAL EFFECTS	W73-00369
Y, *HALOPHILIC ANIMALS, *HALOPH/	*SALINITY, *LAGOONS, *HYDROBIOLOG	W73-00855
SH WATER, FRESHWATER,/ *DIATOMS,	*SALINITY, *BIOINDICATORS, BRACKI	W73-02548
LOGY, EPIDEMIOLOGY, CULTURES, M/	*SALMONELLA, *BIBLIOGRAPHIES, BIO	W72-13800
UTION SOURCES, *FISH HATCHERIES,	*SALMONIDS, WATER REQUIREMENTS, T	W71-11006
FECTS, *PROTOZOA, *MARINE ALGAE,	*SALT MARSHES, NEW YORK, NEMATODE	W72-00948
E ENVIRONMENT, *DOMESTIC WASTES,	*SALT MARSHES, *CARBON CYCLE, *PH	W72-12567
IA, FLORIDA, RAD/ *MARINE ALGAE,	*SALT MARSHES, *PROTOZOA, *BACTER	W73-00796
ICAL COMMUNITIES, *ENVIRONMENTA/	*SALT MARSHES, *PROTOZOA, *BIOLOG	W73-00367
IOISOTOPES, STRONT/ *ECOSYSTEMS,	*SALT MARSHES, *PRODUCTIVITY, RAD	W71-09254
S, NITRATE NITROGEN, ORTHOPHOSP/	*SALT RIVER(KENT), VASCULAR PLANT	W72-06526
, *SODIUM CHLORIDE, *INHIBITION,	*SALT TOLERANCE, WATER POLLUTION	W72-13809
RA PLANKTON, C-14.:	*SALT VALLEY RESERVOIRS(NEB), ULT	W72-04270
NAGEMENT.:	*SALTON SEA(CALIF), ECOLOGICAL MA	W71-10577
SION SPECTROPHOTOME/ *STRONTIUM,	*SAMPLE PRESERVATION, *FLAME EMIS	W71-11529
MATERIALS, *BIOLOGICAL SAMPLES,	*SAMPLE PRESERVATION, CHEMICAL PR	W73-01676
IMPOUNDMENTS, *AQUATIC HABITATS,	*SAMPLING, *ECOSYSTEMS, *ALGAE, *	W72-06526
HY, *PACIF/ *PLANKTON, *BIOMASS,	*SAMPLING, *NUTRIENTS, *HYDROGRAP	W72-07892
, *ARTIFICIAL SUBSTRATES, DEPTH,	*SAMPLING, *PERIPHYTON, EQUIPMENT	W73-00374
*LAKE MICHIGAN, *EUTROPHICATION/	*SAMPLING, *NUTRIENTS, *DIATOMS,	W73-00384
ACTIVATION ANALYSIS, RADIOACTI/	*SAMPLING, *ENVIRONMENT, *NEUTRON	W71-11036
*EUTROPHICATION, *MEASUREMENT,	*SAMPLING, WATER RESOURCES, AREA	W72-03215
, *PHOSPHA/ *PLANKTON, *DIATOMS,	*SAMPLING, *DISTRIBUTION PATTERNS	W72-03567
TEBRATES, ST. LAWRENC/ *BENTHOS,	*SAMPLING, *SESSILE ALGAE, *INVER	W72-05432
	*SAN FRANCISCO BAY-DELTA.:	W71-09581
ALGAL GROWTH, ALGAL HARVESTING,	*SAN JOAQUIN VALLEY(CALIF).: /US,	W72-02975
	*SAN JOAQUIN RIVER.:	W71-07098
NUOUS FLOW TURBIDOSTAT, *GROWTH/	*SAN JOAQUIN DELTA(CALIF), *CONTI	W70-10403
	*SANTA BARBARA CHANNEL.:	W71-08793
	*SANTA BARBARA CHANNEL.:	W71-08792
EROMORPHA, SWE/ *HYADESIA FUSCA,	*SARCOPTIFORMES, *ROCKPOOLS, *ENT	W71-08026
FLAGEL/ *SUBSTRATE UTILIZATION,	*SARGASSO SEA, COCCOLITHOPHORIDS,	W73-00380
*NITZSCHIA A/ *CHLORELLA FUSCA,	*SCENEDEMUS ACUTUS VAR ALTERNANS,	W72-12739
TOXINS, *INHIBITION, *CHLORELLA,	*SCENEDESMUS, AQUATIC MICROORGANI	W72-12255
*AEROBIC CONDITIONS, *KINETICS,	*SCENEDESMUS, *DISSOLVED SOLIDS,	W73-00379
MS, ENVIRONMENTAL E/ *CHLORELLA,	*SCENEDESMUS, ALGAE, MICROORGANIS	W72-10861
COLLEC/ *ON-SITE INVESTIGATIONS,	*SCUBA DIVING, *SEA WATER, *DATA	W72-11925
OUNDE/ *SEA WEEDS, *MACROPHYTES,	*SEA GRASSES, TELEVISION, DEPTH S	W72-09120
ENANCE, *PLANT MORPHOLOGY, *PHO/	*SEA NETTLES, *POLLUTANTS, *MAINT	W72-10162
RY PRODUCTIVITY, *PHYTOPLANKTON,	*SEA WATER, COASTS, BIORTHYMS, ON	W72-03228
TION, B/ *RADIOISOTOPES, *ALGAE,	*SEA WATER, *FOOD CHAINS, *ABSORP	W71-13243
OGY, *CALIFORN/ *SURFACE WATERS,	*SEA WATER, *WATER QUALITY, *ECOL	W71-10577
, CH/ *BIOMASS, *ORGANIC MATTER,	*SEA WATER, WATER ANALYSIS, ALGAE	W71-11234
POLLUTION EFFECTS, *INDICATORS,	*SEA WATER, MARINE PLANTS, ASSESS	W71-12867
E INVESTIGATIONS, *SCUBA DIVING,	*SEA WATER, *DATA COLLECTIONS, CO	W72-11925
HYDRATES, MARINE PLANTS, MARINE/	*SEA WATER, *MARINE ALGAE, *CARBO	W73-02100
PHOTO/ *COPPER, *IONS, *POISONS,	*SEA WATER, *FRESH WATER, ALGAE,	W71-03027
, CULTURES, PHYTOP/ *POLLUTANTS,	*SEA WATER, *ALGAE, *MARINE ALGAE	W70-07280
IA, NITRI/ *BACTERIA, *CULTURES,	*SEA WATER, *NITRIFICATION, AMMON	W72-05421
NIMALS/ *RADIOCHEMICAL ANALYSIS,	*SEA WATER, *SEDIMENTS, *MARINE A	W72-05965
NSING, *WATER POLLUTION CONTROL,	*SEA WATER, *OILS, *NUTRIENTS, AI	W72-04742
RADIOCHEMICAL ANALYSIS, RESINS,	*SEA WATER, BIOASSAY, COPEPODS, O	W72-04708
ASSES, TELEVISION, DEPTH SOUNDE/	*SEA WEEDS, *MACROPHYTES, *SEA GR	W72-09120
, *MASSACHUSETTS, SYSTEMATICS, /	*SEASONAL, *DISTRIBUTION PATTERNS	W73-01449
RONMENTAL EFFECTS, *SYSTEMATICS,	*SEASONAL, SEA WATER, SAMPLING, B	W73-00442
METHODS, MARINE ALGAE, *OREGON,	*SEASONAL, CHRYSOPHYTA, SYSTEMATI	W72-08141
PLANKTON, *NANNOPLANKTON, LAKES,	*SEASONAL, *PRIMARY PRODUCTIVITY,	W72-12738
URE, *PHOTOPERIODISM, G/ *ALGAE,	*SEASONAL, *CLADOPHORA, *TEMPERAT	W72-13651
*CHEMICAL WASTE, *GROWTH RATES,	*SEAWATER, *TOXICITY, *NUTRIENTS,	W72-12239
BIOMASS,/ *CARBON, *CHLOROPHYLL,	*SEAWATER, *AQUATIC PRODUCTION, *	W71-11562
LAMINARIA DIGITATA, AGARUM CRI/	*SEAWEEDS, LAMINARIA LONGICRURIS,	W73-00366
*BIOMASS, *PRIMARY PRODUCTIVITY,	*SECONDARY PRODUCTIVITY, PHYTOPLA	W72-05968
AL/ *SOIL ALGAE, *SOIL SURFACES,	*SEDIMENT YIELD, *SIMULATED RAINF	W72-02218
PTION, LAKES, NUTR/ *PHOSPHATES,	*SEDIMENT-WATER INTERFACES, ADSOR	W72-01108
, *AQUATIC MICROBIOLOGY, OREGON,	*SEDIMENT-WATER INTERFACES, SEDIM	W72-08989
ERIA, *SULFUR BACTERIA, *OXYGEN,	*SEDIMENT-WATER INTERFACES, *LAKE	W72-06690
D OXYGEN, *LAKE ERIE, *BACTERIA,	*SEDIMENT-WATER INTERFACES, *HYPO	W72-12996
*PHYTOPLANKTON, *MARINE ALGAE,	*SEDIMENTATION RATES, *REVIEWS, B	W72-06313
AE, *OKLAHOMA, *IODINE RADIOISO/	*SEDIMENTARY ROCKS, *BRINES, *ALG	W72-02073
GRAPHY, CHLO/ *PIGMENTS, *LAKES,	*SEDIMENTS, *INDICATORS, CHROMATO	W72-01784
S, *NEUTRON ACTIVATION ANALYSIS/	*SEDIMENTS, *ANALYTICAL TECHNIQUE	W72-01101
D OXYGEN, OXYGEN DEMAND, HYPOLI/	*SEDIMENTS, *LAKE ERIE, *DISSOLVE	W72-12994
IOCHEMICAL ANALYSIS, *SEA WATER,	*SEDIMENTS, *MARINE ANIMALS, *MAR	W72-05965
NUTRIENTS, *CHEMICAL PROPERTIES,	*SEDIMENTS, CLAYS, CORES, SEASONA	W72-05469
ATIO/ *PHOSPHATES, *GREAT LAKES,	*SEDIMENTS, *NUTRIENTS, EUTROPHIC	W72-04263
ON, LAKES, AQUATIC/ *PHOSPHORUS,	*SEDIMENTS, *ALGAE, *EUTROPHICATI	W72-03742
MORPHOLOGY, SEDIME/ *NUTRIENTS,	*SEDIMENTS, *WATER ANALYSIS, LAKE	W71-05626
*ADENOSINE TRIPHOSPHATE,	*SELENASTRUM CAPRICORNUTUM.:	W71-13794
L GROWTH, *LAKE GEORGE(NY).:	*SELENASTRUM CAPRICORNUTUM, *ALGA	W72-08376
*ANACYSTIS NIDULANS, GROWTH INH/	*SELENIUM, *ANABAENA VARIABILIS,	W72-00153
AMS, LABORATORY TESTS, DISSOLVE/	*SELF-PURIFICATION, *ALGAE, *STRE	W71-04518

, *ANACYSTIS NIDULANS, *MUTANTS,	*SENSITIVITY, CHLCRAMPHENICOL, RE	W72-14327
OOPLANKTON, *LARVAE, *FISH EGGS,	*SEPARATION TECHNIQUES, *MARINE F	W73-00371
BACTERIA, *FUNGI, *BACTERIOPHAG/	*SEPARATION TECHNIQUES, *ALGAE, *	W72-06124
N LAGOONS, *HARVESTING OF ALGAE,	*SEPARATION TECHNIQUES, CHEMICAL	W72-10237
*NUTRIENTS, *MUNICIPAL WASTES,	*SEPARATION TECHNIQUES, FERMENTAT	W72-10390
KTON, *CHLOROPHYLL, *EVALUATION,	*SEPARATION TECHNIQUES, SPECTROPH	W72-13673
MS, *PERIPHYTON, *PHYTOPLANKTON,	*SESSILE ALGAE, *LAKE SUPERIOR, L	W72-12192
T. LAWRENC/ *BENTHOS, *SAMPLING,	*SESSILE ALGAE, *INVERTEBRATES, S	W72-05432
TIDAL EFF/ *ESTUARIES, *DIATOMS,	*SESSILE ALGAE, LIGHT INTENSITY,	W73-00853
, *NITROGEN, PHOSPHORUS, CARBON,	*SEWAGE BACTERIA, SEWAGE, ALGAE,	W72-04309
EATMENT, *SEWAGE LAGOONS, *COLD/	*SEWAGE BACTERIA, *WASTE WATER TR	W72-12565
TION, BIOCHEMICAL OXYGEN DEMAND,	*SEWAGE EFFLUENTS, *SEWAGE TREATM	W72-06612
WATER TREATMENT, A/ *PHOSPHORUS,	*SEWAGE EFFLUENTS, *ALGAE, WASTE	W72-09176
ENT, *ECONOMIC JUSTIFICATION, */	*SEWAGE EFFLUENTS, *SEWAGE TREATM	W72-00383
OCLINE/ *EUTROPHICATION, *ALGAE,	*SEWAGE EFFLUENTS, GEORGIA, THERM	W71-11914
TANS, ZOOGLOEA/ *UNITED KINGDOM,	*SEWAGE FUNGUS, SPHAEROTCTILUS NA	W72-08573
, *OXIDATION LAGOONS, *DESIGN C/	*SEWAGE LAGOONS, *AERATED LAGOONS	W72-09386
S, ALGAE, PRODUCTIVITY/ *CARBON,	*SEWAGE LAGOONS, *LIMITING FACTOR	W72-09162
ERGENTS, *ALGAE, *PHYTOPLANKTON,	*SEWAGE LAGOONS, BIODEGRADATION,	W72-07661
ACTERIA, *WASTE WATER TREATMENT,	*SEWAGE LAGOONS, *COLD REGIONS, A	W72-12565
*METABOLISM, *SEWAGE TREATMENT,	*SEWAGE LAGOONS, OHIO, OXYGEN, DI	W72-05459
ROPHICATION, *OXIDATION LAGOONS,	*SEWAGE LAGOONS, BIODEGRADATION,	W71-07973
MENT, BIOCHEMI/ *SURFACE RUNOFF,	*SEWAGE TREATMENT, TERTIARY TREAT	W71-06445
S,/ *PHYTOPLANKTON, *METABOLISM,	*SEWAGE TREATMENT, *SEWAGE LAGOON	W72-05459
ES, CHEMIC/ *NITROGEN COMPOUNDS,	*SEWAGE TREATMENT, *DOMESTIC WAST	W70-09907
REATMENT, *TERTIARY TREATMENT, /	*SEWAGE TREATMENT, *WASTE WATER T	W70-10381
XYGEN DEMAND, *SEWAGE EFFLUENTS,	*SEWAGE TREATMENT, CULTURE, *WAST	W72-06612
TE WATER TREATMENT, *FILTRATION,	*SEWAGE TREATMENT, ALGAE, *AERATI	W72-14469
*EUTROPHICATION, *LAKES, *WATE/	*SEWAGE TREATMENT, *EFFICIENCIES,	W72-11702
EUTROPHICATION, *WATER POLLUTIO/	*SEWAGE TREATMENT, *PHOSPHATES, *	W71-11507
IFICATION, */ *SEWAGE EFFLUENTS,	*SEWAGE TREATMENT, *ECONOMIC JUST	W72-00383
ES, *EUTROPHICATION, *CHELATION,	*SEWAGE, *ORGANIC COMPOUNDS, WATE	W71-09674
AE, *ESTUARIES, *WASTE DISPOSAL,	*SEWAGE, *INDUSTRIAL WASTES, WATE	W72-08804
CAL COMMUNITIES, RIVERS, BACTER/	*SEWAGE, *FUNGI, *SLIME, *BIOLOGI	W72-08573
AQUATIC ENVIRONMENT, CHEMICAL /	*SEWAGE, *EUTROPHICATION, *LAKES,	W71-00117
MIDGES, LAGOONS/ *ENERGY BUDGET,	*SEWAGE, *AQUATIC PRODUCTIVITY, *	W72-03556
ERENCES, *PUBLIC HEALTH, RIVERS,	*SEWAGE, BIOCHEMICAL OXYGEN DEMAN	W72-09675
TS, T/ *MERCURY, *PUBLIC HEALTH,	*SHELLFISH, WATER POLLUTION EFFEC	W72-12257
IRONMENT, *PRIMARY PRODUCTIVITY,	*SHOALS, *BENTHIC FLORA, SEA WATE	W72-14317
ION, SOUTH WALES, MICROCYSTIS A/	*SHOCK-WAVES, *GAS-VACUOLE DEFLAT	W72-11564
T, *MARINE FISH, *PHYTOPLANKTON,	*SHRIMP, *BIOASSAY, *ABSORPTION,	W73-02105
STAGE, *DETERMINATION, *ORIGIN,	*SIGNIFICANCE, *EXTRACTION, LUCIF	W72-04292
ECTROPHORESIS, SETTLING, *ALGAE,	*SILICA, *BACTERIA, WATER POLLUTI	W71-01899
ATER TRE/ *FLOCCULATION, *ALGAE,	*SILICA, *E. COLI, *IONS, *WASTEW	W72-14840
*PLUTONIUM,	*SILVER RADIOISOTOPES.:	W72-07826
ARCINA, ANKISTRODESMUS BRAUNII,/	*SIMAZINE, *DETOXICATION, CHLOROS	W71-07675
*SOIL SURFACES, *SEDIMENT YIELD,	*SIMULATED RAINFALL, *SOIL TYPES,	W72-02218
AHALINE, HYPERHALINE, EURYHALIN/	*SINAI PENINSULA, CHLORINITY, MET	W73-00855
DISPERSANTS, *COLLECTING AGENTS,	*SINKING AGENTS, EFFECTIVENESS, T	W71-09789
DIFFUS/ *ANTIMICROBIAL ACTIVITY,	*SKELLYSOLVE, *FILTER PAPER DISC	W72-08586
RIVERS, BACTER/ *SEWAGE, *FUNGI,	*SLIME, *BIOLOGICAL COMMUNITIES,	W72-08573
	*SLIMICIDES.:	W72-08683
*EUTROPHICATION, *BIOCHEMICAL /	*SLUDGE WORMS, *DISSOLVED OXYGEN,	W70-09189
ROBIC CONDI/ *OXIDATION LAGOONS,	*SLUDGE, ANAEROBIC CONDITIONS, AE	W71-07102
NMENT, WATERSHEDS(BASINS), *DDT,	*SMALL WATERSHEDS, ECOSYSTEMS, PE	W72-12930
TION, WATER POLLUTION EFFECTS, /	*SNAILS, *FROGS, *TRITIUM, ABSORP	W72-03348
TIUM RADIOISOTOPES, *ABSORPTION,	*SNAILS, GASTROPODS, PATH OF POLL	W72-10978
RGANIC CARBON, *AMERI/ *GLUCOSE,	*SNAKE RIVER(IDAHO), *DISSOLVED O	W72-13300
	*SNAKE RIVER(WASH).:	W72-14728
LUTION CONTROL/ *EUTROPHICATION,	*SOCIAL PARTICIPATION, *WATER POL	W73-02479
ALT TOLERANCE, WATER POLL/ *DDT,	*SODIUM CHLORIDE, *INHIBITION, *S	W72-13809
IMENT YIELD, *SIMULATED RAINFAL/	*SOIL ALGAE, *SOIL SURFACES, *SED	W72-02218
S, *SYSTEMATICS, *AQUATIC ALGAE,	*SOIL BACTERIA, SAMPLING, CULTURE	W73-00854
SPHORUS, *EUTROPHICATION, LAKES/	*SOIL BACTERIA, *LAKE SOILS, *PHO	W72-04292
, *CARBON DIOXIDE, *RADIOACTIVI/	*SOIL MICROORGANISMS, *ABSORPTION	W72-14301
*SIMULATED RAINFAL/ *SOIL ALGAE,	*SOIL SURFACES, *SEDIMENT YIELD,	W72-02218
MENT YIELD, *SIMULATED RAINFALL,	*SOIL TYPES, SOIL TEXTURE, CLAYS,	W72-02218
*FALLOUT, *CYCLING NUTRIENTS, E/	*SOIL-WATER-PLANT RELATIONSHIPS,	W72-03342
S, / *ALGAE, *NITROGEN FIXATION,	*SOILS, *TEMPERATE, CLIMATIC ZONE	W73-02471
.:	*SOLAR INSOLATION, *STOICHIOMETRY	W72-02363
CULTURES.: *PHOSPHATES,	*SOLUBILITY, *CYANOPHYTA, ALGAE,	W71-10070
OLOGY, RESERVOIR SILTING, LAKES,	*SOUTH DAKOTA, PLANKTON, *ALGAE,	W73-00938
	*SOUTH RIVER(VA).:	W73-01972
A COSTATUM, TH/ *COASTAL WATERS,	*SOUTHERN LONG ISLAND, SKELETONEM	W71-07875
T, *CENTRAL HEATING AND COOLING,	*SPACE HEATING, *FARM USES OF WAS	W70-09193
*DEPTH, MARINE AN/ *CHLOROPHYLL,	*SPATIAL DISTRIBUTION, *DENSITY,	W72-14313
ATICS, *ECOLOGICAL DISTRIBUTION,	*SPATIAL DISTRIBUTION, GEOGRAPHIC	W73-00428
*LAKE MICHIGAN, *EUTROPHICATION,	*SPATIAL DISTRIBUTION, PHOSPHORUS	W73-00384
ULATION, *TEMPORAL DISTRIBUTION,	*SPATIAL DISTRIBUTION, ENVIRONMEN	W72-08560
.:	*SPECIES DIVERSITY, CRYSTAL RIVER	W72-11252
R.:	*SPECIES DIVERSITY, *POTOMAC RIVE	W71-11583
ENTIFIC/ *CALCIUM, *FOOD CHAINS,	*SPECTROPHOTOMETRY, *POLLUTANT ID	W71-11529
ANABAENA VARIABILIS,/ *LYSOZYME,	*SPHEROPLASTS, BLUE-GREEN ALGAE,	W71-08683
*MET/ *CLADOPHORA, *ACHNANTHES,	*SPIROGYRA, CLODOPHORA GLOMERATA,	W72-07901
A, *AMPHIPODA, *F/ *CHLOROPHYTA,	*SPRINGS, DIATOMS, BIOTA, PROTOZO	W72-12744
ON SOURCES, WATE/ *INSECTICIDES,	*ST LAWRENCE RIVER, WATER POLLUTI	W70-08098
OXYGEN DEMAND, HYDROGEN ION CON/	*ST. LAWRENCE RIVER, BIOCHEMICAL	W70-10427
AGARUM CRIBROSUM, *NOVA SCOTIA,	*ST. MARGARET'S BAY.: / DIGITATA,	W73-00366
REATMENT, B/ *OXIDATION LAGOONS,	*STABILIZATION, *ALGAE, AEROBIC T	W71-03896
ERRIC CHLORIDE, *FERRIC SULFATE,	*STABILIZATION PONDS, ALUM, POST-	W71-13356

OEIC BACTERIA, *TEXAS, / *ALGAE,	*STABLE ISOTOPES, *CARBON, *ANAER	W71-04873
OTOPES, ENVIR/ *ISOTOPE STUDIES,	*STABLE ISOTOPES, *CARBON RADIOIS	W72-12709
EASONAL, / *TEMPERATURE CONTROL,	*STANDARDS, *WATER TEMPERATURE, S	W70-09595
STS, *PHYTOPLANKTON, *ESTUARIES,	*STANDING CROP, SEASONAL, DIATOMS	W71-07875
ENERGY BUDGET, PRODUC/ *STREAMS,	*STANDING CROP, *TROPHIC LEVEL, *	W71-10064
MEASUREM/ *PRIMARY PRODUCTIVITY,	*STANDING CROP, *ALGAE, DIATOMS,	W71-12855
UNITIES, *ENVIRONMENTAL EFFECTS,	*STANDING CROPS, TIDAL EFFECTS, M	W73-00367
TRIENTS, *FRESHWATER, *BIOASSAY,	*STANDING CROP, CARBON, NITROGEN,	W72-07648
*PHYTOPLANKTON, *WATER ANALYSIS,	*STANDING CROP, *PRIMARY PRODUCTI	W73-01435
ION, POYELECTROLYTES.:	*STARCH REMOVAL, STARCH SUBSTITUT	W72-03753
VER, DATA COLLECTIONS/ *SURVEYS,	*STATISTICAL METHODS, *POTOMAC RI	W72-07791
OLOGICAL STUDIE/ *MODEL STUDIES,	*STATISTICAL METHODS, *ALGAE, CYT	W71-12868
M LONGIPES VAR CONTRACTUM, ARTH/	*STAURASTRUM, DESMIDS, STAURASTRU	W73-01094
NANS, *NITZSCHIA ACTINASTROIDES,	*STAURASTRUM PINGUE, SCENEDESMUS	W72-12739
LLUTION CONTROL, *WATER COOLING,	*STEAM TURBINES, *THERMAL POLLUTI	W70-09193
SIRA BINDERANA, SCANNING ELECTR/	*STEPHANODISCUS BINDERANUS, *MELO	W73-00432
*SOLAR INSOLATION,	*STOICHIOMETRY.:	W72-02363
, *RECREATION, *TREATMENT FACIL/	*STORM RUN-OFF, *SUSPENDED SOLIDS	W72-00042
C BACTERIA, *POPULATION, *ALGAE,	*STRATIFICATION, *ENZYMES, THERMO	W72-03224
IRONM/ *NEW YORK, *POWER PLANTS,	*STREAM, *THERMAL POLLUTION, *ENV	W72-14659
*ALGAE, *BIOLOGICAL COMMUNITIES,	*STREAMS, LAKE ONTARIO, PLANKTON,	W73-01627
LLUTION EFFECTS, *MODEL STUDIES,	*STREAMS, *ECOSYSTEMS, AQUATIC HA	W72-06340
C LEVEL, *ENERGY BUDGET, PRODUC/	*STREAMS, *STANDING CROP, *TROPHI	W71-10064
CREATI/ *SURFACE WATERS, *LAKES,	*STREAMS, *REVIEWS, *ILLINOIS, RE	W72-05869
LVE/ *SELF-PURIFICATION, *ALGAE,	*STREAMS, LABORATORY TESTS, DISSO	W71-04518
AQUATIC ALGAE, *COMPARATIVE PRO/	*STRESS, *PRIMARY PRODUCTIVITY, *	W73-01080
ICROORGANISMS, *ACID MINE WATER,	*STRIP MINES, *WATER POLLUTION EF	W71-07942
POLLUTION EFFECTS, NUCLEAR WAST/	*STRONTIUM RADIOISOTOPES, *WATER	W71-09225
T, *FOOD CHAINS, / *GREAT LAKES,	*STRONTIUM RADIOISOTOPES, *FALLOU	W73-00790
ANIMALS, *MARINE ALGAE, *FOOD /	*STRONTIUM RADIOISOTOPES, *MARINE	W73-00819
*FLAME EMISSION SPECTROPHOTOME/	*STRONTIUM, *SAMPLE PRESERVATION,	W71-11529
TION, *SNAILS, GASTROPODS, PATH/	*STRONTIUM RADIOISOTOPES, *ABSORP	W72-10978
ADIOISOTOPES, AB/ *MARINE ALGAE,	*STRONTIUM RADIOISOTOPES, *ZINC R	W72-12203
RIA DE ALDAN(SPAIN), LAMINARIA/	*SUBLITTORAL BENTHIC MACROPHYTES,	W72-04766
ODES, ALGAE, OLIGO/ *PERIPHYTON,	*SUBMERGED PLANTS, BIOMASS, NEMAT	W71-11012
MA, ACHNANTHES / *LAKE WINNIPEG,	*SUBSTRATES, ACHNANTHES MINUTISSI	W72-13142
SEA, COCCOLITHOPHORIDS, FLAGEL/	*SUBSTRATE UTILIZATION, *SARGASSO	W73-00380
OROPHYLL, PL/ *LAKES, *PIGMENTS,	*SUCCESSION, *EUTROPHICATION, CHL	W73-00916
YSTEMS, *BIOLOGICAL COMMUNITIES,	*SUCCESSION, AQUATIC ENVIRONMENT,	W72-11854
RANEAN SE/ *CERATIUM EGYPTIACUM,	*SUEZ CANAL, FLAGELLATES, MEDITER	W73-00296
ENT-WATER IN/ *AQUATIC BACTERIA,	*SULFUR BACTERIA, *OXYGEN, *SEDIM	W72-06690
SPENDED SOLIDS.:	*SURFACE CHARGE, REGENERATION, SU	W71-00138
T, TERTIARY TREATMENT, BIOCHEMI/	*SURFACE RUNOFF, *SEWAGE TREATMEN	W71-06445
ES, MINERALIZATION, NITROGEN SO/	*SURFACE WATERS, GEOLOGICAL SOURC	W71-06435
C WEED CONTROL, *DOMESTIC WATER,	*SURFACE WATERS, WATER SUPPLY, EU	W71-05083
T, *FILTRATION, *MICROORGANISMS,	*SURFACE WATERS, *PLANKTON, *ALGA	W72-06191
, *REVIEWS, *ILLINOIS, RECREATI/	*SURFACE WATERS, *LAKES, *STREAMS	W72-05869
, PHOSPHATES, CLAY, TASTE, ODOR,	*SURFACE WATERS, *GROUNDWATERS, S	W72-06536
RARED RADIATION, *AQUATIC ALGAE,	*SURFACE WATERS, ANALYTICAL TECHN	W71-12034
POLLU/ *EUTROPHICATION, *ALGAE,	*SURFACE WATERS, *REVIEWS, *WATER	W71-10580
ER QUALITY, *ECOLOGY, *CALIFORN/	*SURFACE WATERS, *SEA WATER, *WAT	W71-10577
*LA/ *ASSAY, *ZINC, *ADSORPTION,	*SURFACES, LABORATORY EQUIPMENT,	W72-12210
EANING, *DISPERSION, *CHEMICALS,	*SURFACTANTS, *TOXICITY, *BIOASSA	W71-09789
S, *ALGAL CONTROL, *FORMULATION,	*SURFACTANTS, *EUTROPHICATION, LI	W71-06247
ROLOGIC DATA, *MEASUREMENT, *AS/	*SURVEYS, *DATA COLLECTIONS, *HYD	W72-10678
LECTIONS, *ENVIRON/ *MONITORING,	*SURVEYS, *MEASUREMENT, *DATA COL	W72-07841
POTOMAC RIVER, DATA COLLECTIONS/	*SURVEYS, *STATISTICAL METHODS, *	W72-07791
RADIOECOLOGY, *RADIOCHEMICAL AN/	*SURVEYS, *AQUATIC ENVIRONMENT, *	W73-01673
YING, REFRIGERATION, INCUBATION/	*SURVEYS, *METHODOLOGY, FREEZE DR	W73-01676
TREATMENT FACIL/ *STORM RUN-OFF,	*SUSPENDED SOLIDS, *RECREATION, *	W72-00042
DE, *ROCK FILTERS.:	*SUSPENDED SOLIDS, SODIUM HYDROXI	W71-13341
MOUNT CLEMENS/ *COMBINED SEWERS,	*SUSPENDED SOLIDS, BACKWASHING, *	W72-00042
BLUE-GREEN ALGAE.:	*SUSPENDED SEDIMENT PRODUCTION, *	W72-02218
LECTROCHEMISTRY, *WATER PURIFIC/	*SUSPENSION, *ELECTROPHORESIS, *E	W73-01362
TREATMENT, *FILTRATION, *ALGAE,	*SUSPENSION, HARVESTING, PONDS, C	W70-10174
ERLABORATORY STUDIES, PRECISION,	*SWEDEN, LAKE MALAREN, MICROCYSTI	W73-00286
, / *CHLOROPHYLL A, MACROPHYTES,	*SWEDEN, MICROCYSIS, OSCILLATORIA	W72-12729
	*SWEDEN, SPECIES COMPOSITION.:	W73-01632
ESPIRATION, ACTIVATED SLUDGE, W/	*SYMBIOSIS, *ALGAE, *BACTERIA, *R	W71-04079
TION, RESPIR/ *ALGAE, *BACTERIA,	*SYMBIOSIS, PHOTOSYNTHESIS, OXIDA	W72-04983
, *NUTRIENTS, *ALGAE, *BACTERIA,	*SYMBIOSIS, ACTIVATED SLUDGE, NIT	W72-11100
S SOLI, SCENEDESMUS QUADRICAUDA/	*SYNCHRONOUS CULTURES, SCENEDESMU	W72-07144
*DIVISION PATTERNS, *AUTOSPORES,	*SYNCHRONOUS CULTURES, SYNCHRONY	W72-12724
, VARIABILITY, BIAS, S/ *ERRORS,	*SYNCHRONOUS CULTURES, AUTOSPORES	W72-11920
EUTROPHICATION, *LAKES, *ALPINE,	*SYNOPTIC ANALYSIS, COLD REGIONS,	W72-12549
OXYGEN, BI/ *NITROGEN FIXATION,	*SYNTHESIS, *CULTURES, *NITROGEN,	W72-07941
THYL MERCURIC CHLORID/ *MERCURY,	*SYNTHESIS, MERCURIC CHLORIDE, ME	W72-06037
REPRODUCTION, LIFE CYCL/ *ALGAE,	*SYSTEMATICS, *PLANT MORPHOLOGY,	W72-07141
YTA, *EUGLENOPHYTA, *PYRROPHYTA,	*SYSTEMATICS, DINOFLAGELLATES, SP	W72-07683
, SEASONAL, FLU/ *PHYTOPLANKTON,	*SYSTEMATICS, *DIATOMS, SEA WATER	W72-07663
ES, *GRO/ *ALGAE, *GROWTH RATES,	*SYSTEMATICS, *LIFE HISTORY STUDI	W73-02099
*PHYTOPLANKTON, *AQUATIC ALGAE,	*SYSTEMATICS, *DISTRIBUTION PATTE	W73-01094
L DISTRIBUTION, *MARINE ANIMALS,	*SYSTEMATICS, *ECOLOGY, *MARINE A	W73-00932
TARCTIC, *DISTRIBUTION PATTERNS,	*SYSTEMATICS, *AQUATIC ALGAE, *SO	W73-00854
YTA, *MARINE ALGAE, *RHODOPHYTA,	*SYSTEMATICS, PHOTOGRAPHY, SEA WA	W73-00284
UTION, / *DISTRIBUTION PATTERNS,	*SYSTEMATICS, *ECOLOGICAL DISTRIB	W73-00428
LANKTON, *ENVIRONMENTAL EFFECTS,	*SYSTEMATICS, *SEASONAL, SEA WATE	W73-00442
ION EFFECTS, WATER PURIFICATION,	*SYSTEMATICS.: /PHY, WATER POLLUT	W73-00432
, PHAEOPH/ *CYTOLOGICAL STUDIES,	*SYSTEMATICS, *ALGAE, CHRYSOPHYTA	W72-14808
TURES, CHLOROPHYTA, PROTOZOA, */	*SYSTEMATICS, *CHLAMYDOMONAS, CUL	W72-12632

UATIC A/ *CYANOPHYTA, *CULTURES, *SYSTEMATICS, CLASSIFICATION, *AQ W72-12172
NOPHYTA, *CHLOROPHYTA, PHAEOPHY/ *SYSTEMATICS, *MARINE ALGAE, *CYA W72-11800
LANKTON, *DISTRIBUTION PATTERNS, *SYSTEMATICS, *SALINE LAKES, *EUT W72-09111
, HABITATS, AQUATIC / *PROTOZOA, *SYSTEMATICS, *ALGAE, CHLOROPHYTA W72-09119
ARINE PLANTS, *ALASKA, SAMPLING, *SYSTEMATICS, BENTHOS, SCUBA DIVI W72-09120
ORA, RHODOPHYTA,/ *MARINE ALGAE, *SYSTEMATICS, *COASTS, BENTHIC FL W71-10066
, *EUTROPHICATION, *MATHEMATICA/ *SYSTEMS ANALYSIS, *WATER QUALITY W71-03021
, ANALYTICAL TECHNIQUES, SEPARA/ *TASTE, *ODOR, *ORGANIC COMPOUNDS W72-11604
OGY/ *LAKE ERIE, *POTABLE WATER, *TASTE, *ODOR, SURVEYS, MICROBIOL W72-10076
ALGAE,/ *THERMAL STRATIFICATION, *TASTE, *ODOR, DISSOLVED SOLIDS, W73-02138
TIVATED CARBON, ADSORPTION, FIL/ *TASTE, *ODOR, *POTABLE WATER, AC W72-08357
GAE, *NITROGEN FIXATION, *SOILS, *TEMPERATE, CLIMATIC ZONES, CYANO W73-02471
*ALGAE, *SEASONAL, *CLADOPHORA, *TEMPERATURE, *PHOTOPERICDISM, GR W72-13651
BOR, *TINTINNIDS, *TYCHOPELAGIC, *TEMPERATURE TOLERANCES, SAMPLE P W72-07899
L COMMUNITIES, SYSTEMAT/ *LIGHT, *TEMPERATURE, *DIATCMS, BIOLOGICA W73-00242
HERMAL POLLUTION, WATER/ *BIOTA, *TEMPERATURE, *CANADA, *LAKES, *T W73-01704
*WATER TEMPERATURE, SEASONAL, / *TEMPERATURE CONTROL, *STANDARDS, W70-09595
DISTRIBUTI/ *ALGAE, *POPULATION, *TEMPORAL DISTRIBUTION, *SPATIAL W72-08560
ALS, *PRODUCTIVITY, *ECOSYSTEMS/ *TERRESTRIAL HABITATS, *HEAVY MET W72-14303
PIRATION, PHOTOSYNTHETIC OXYGEN, *TERTIARY TREATMENT.: /CMASS, RES W72-11100
*ALGAE, *WASTE WATER TREATMENT, *TERTIARY TREATMENT, *COAGULATION W71-09629
TY CONTROL, AM/ *EUTROPHICATION, *TERTIARY TREATMENT, *WATER QUALI W71-09450
ION, / *PHOSPHORUS, *DETERGENTS, *TERTIARY TREATMENT, *EUTROPHICAT W70-09325
EATMENT, *WASTE WATER TREATMENT, *TERTIARY TREATMENT, *WATER REUSE W70-10381
ERIA, TRICK/ *OXIDATION LAGOONS, *TERTIARY TREATMENT, *DESIGN CRIT W71-07113
ALGAE, / *ANALYTICAL TECHNIQUES, *TESTING PROCEDURES, *BIOASSAY, * W72-10619
S, *CARBON, *ANAEROBIC BACTERIA, *TEXAS, BIODEGRADATION, TRACERS, W71-04873
KES, NITROGEN, PHOSPHORUS, IRON, *TEXAS, LABORATORY TESTS, CARBON W72-13692
ALANCING TANKS.: *TEXTILE MILL WASTES, MONTREAL, B W70-10427
N FLOS-AQUAE, *PHARMACOLOGICAL / *THERAPEUTIC VALUE, *APHANIZOMENO W72-04259
*THERMAL DISCHARGES.: W71-12092

CTIONS, *WATER QUALITY, *CHEMIC/ *THERMAL POLLUTION, *CHEMICAL REA W71-12092
LUTION, *BIOLOGICAL COMMUNITIES, *THERMAL POWER PLANTS, *DISCHARGE W71-11517
OMMUNITIES, *THERMAL POWER PLAN/ *THERMAL POLLUTION, *BIOLOGICAL C W71-11517
F MEXICO, *ELECTRIC POWERPLANTS, *THERMAL POLLUTION, *AQUATIC ENVI W72-11252
WASTES, *RADIOACTIVITY EFFECTS, *THERMAL POLLUTION, CONNECTICUT, W72-00939
ATURE, *CYANOPHYTA, GEYSERS, PH/ *THERMAL POLLUTION, *WATER TEMPER W72-03557
TIFICATION, HYDR/ *POWER PLANTS, *THERMAL POLLUTION, *THERMAL STRA W70-09614
R PLANTS, *COOLING WATER, *WATE/ *THERMAL POLLUTION, *NUCLEAR POWE W70-09612
*WATER COOLING, *STEAM TURBINES, *THERMAL POLLUTION, LONG TUBE VER W70-09193
S, *ENERGY LOSSES, *WATER CONSE/ *THERMAL POWERPLANTS, *PCWERPLANT W70-09193
ONDITIONING, *HEATING, *COOLING, *THERMAL POWER, *MULTIPLE-PURPOSE W70-09193
WATER TEMPERATURE, POWERPLANTS/ *THERMAL POLLUTION, *GREENHOUSES, W70-08832
ER TEMPERATURE, / *POWER PLANTS, *THERMAL POLLUTION, *REVIEWS, WAT W71-07417
QUALITY STANDARDS, *ESTURARIES, *THERMAL POLLUTION, *NORTH CAROLI W72-13636
INE ANIMALS, *THERMAL POLLUTICN, *THERMAL POWER, ATLANTIC OCEAN, M W72-11876
*MARINE ALGAE, *MARINE ANIMALS, *THERMAL POLLUTION, *THERMAL POWE W72-11876
EW YORK, *POWER PLANTS, *STREAM, *THERMAL POLLUTION, *ENVIRONMENTA W72-14659
R POLLUTION/ *INDUSTRIAL WASTES, *THERMAL POLLUTION, *COASTS, WATE W73-01632
, *TEMPERATURE, *CANADA, *LAKES, *THERMAL POLLUTION, WATER POLLUTI W73-01704
*FISH, HEATED WATER, NUCLEAR P/ *THERMAL POLLUTION, *CRUSTACEANS, W72-07907
MS, *FLUORESCE/ *THERMAL STRESS, *THERMAL POLLUTION, *MICRCORGANIS W72-07526
ION EFFECTS, *AQUATIC MICROORGA/ *THERMAL POLLUTION, *WATER POLLUT W72-07225
ON, *MICROORGANISMS, *FLUORESCE/ *THERMAL STRESS, *THERMAL PCLLUTI W72-07526
LGAL CONTROL, *PONDWEEDS, PONDS, *THERMAL STRATIFICATION, DIATOMS, W73-02349
*ODOR, DISSOLVED SOLIDS, ALGAE,/ *THERMAL STRATIFICATION, *TASTE, W73-02138
OWER PLANTS, *THERMAL POLLUTION, *THERMAL STRATIFICATION, HYDRAULI W70-09614
*ALGAE, *DISTRIBUTION PATTERNS, *THERMAL SPRINGS, CYANOPHYTA, ALK W72-05610
N FIXATION, *ALGAE, *CYANOPHYTA, *THERMAL WATER, *WYOMING, NATIONA W72-08508
TY, *NERITIC, N/ *PHYTOPLANKTON, *THERMOCLINE, *PRIMARY PRODUCTIVI W73-00431
IDATION, PHOTOSYNTHESIS, OXYGEN, *THERMODYNAMICS, ANALYTICAL TECHN W72-02363
LTURES, SALINITY, / *PHAEOPHYTA, *TIDAL MARSHES, *MARINE ALGAE, CU W72-12633
, *FERTILIZATION, *CULTIVATED L/ *TILE DRAINAGE, *WATER PROPERTIES W71-12252
BAY, *JAMAICA, FALMOUTH HARBCR, *TINTINNIDS, *TYCHOPELAGIC, *TEMP W72-07899
ICAN SHAD, / MACROINVERTEBRATES, *TOCKS ISLAND RESERVOIR(NJ), AMER W72-14280
OSIRA, TOLYPOTHRIX, ANACYSTIS N/ *TOLERANCE, ANABAENA, NCSTOC, AUL W72-14807
NUIS, AULOSIRA FERTILISSIMA, AN/ *TOLERANCE LEVELS, TOLYPCTHRIX TE W71-12071
LABCRATORY SCALE, *TOTAL ORGANIC CARBON.: W71-07102
*DIURON, *CARBARYL, *TOXAPHENE.: W71-10566
STANDARDS/ *WATERBORNE DISEASES, *TOXIC CHEMICALS, *PUBLIC HEALTH W73-02144
LOGY, ICHTH/ *USSR, *TOXICOLOGY, *TOXICANTS, *HYDROBIONTS, ICHTHYO W72-09646
SHWATER, *AQUATI/ *HYDROBIOLOGY, *TOXICITY, *WATER POLLUTICN, *FRE W72-09649
, *W/ *HYDROBIOLOGY, *PATHOLOGY, *TOXICITY, *WATER QUALITY CONTROL W72-09646
TESTS, FISH, / *COOLING TOWERS, *TOXICITY, *BIOASSAYS, LABORATORY W72-08683
RSION, *CHEMICALS, *SURFACTANTS, *TOXICITY, *BIOASSAY, FISH, SHELL W71-09789
TRIENTS, N/ *ALGAE, *DETERGENTS, *TOXICITY, *ACTIVATED SLUDGE, *NU W71-11972
E, *DIATOMS, CHEMICAL ANALYSIS,/ *TOXICITY, *AMMONIA, *MARINE ALGA W71-11008
LGAE, GRANULES, INHIBITION, MET/ *TOXICITY, *CYANOPHYTA, SULFUR, A W72-00153
RIN, *GROWTH RATES, *CYANOPHYTA, *TOXICITY, PESTICIDES, PESTICIDE W72-03545
, *PLANT PHYSIOLOGY, POLLUTANTS, *TOXICITY, WATER POLLUTICN EFFECT W72-12576
AQUAT/ *PHYTOPLANKTON, *MERCURY, *TOXICITY, CULTURES, *CHLORELLA, W72-11727
WASTE, *GROWTH RATES, *SEAWATER, *TOXICITY, *NUTRIENTS, LETHAL LIM W72-12239
HOSPHATES, *INHIBITIONS, *MICRO/ *TOXICITY, *ARSENIC COMPOUNDS, *P W72-13813
ER POLLUTION EFFECTS, *BIOASSAY, *TOXICITY, *CHLORINATED HYDROCARB W71-07412
DICATORS, *ALGAE, *HEAVY METALS, *TOXICITY, COPPER, CHLORCPHYTA, W W71-05991
HERBICIDES, *PESTICIDES, *ALGAE, *TOXICITY, SOILS, TRIAGINE PESTIC W71-07675
PHYTA/ *FOOD WEBS, *FOOD CHAINS, *TOXICITY, *ALGAL TOXINS, *CHLORO W71-08597
ERGE/ *ALGAE, *ACTIVATED SLUDGE, *TOXICITY, *LABORATORY TESTS, DET W70-08976
S, *ALGAE, *AQUATIC ENVIRONMENT, *TOXICITY, *WATER POLLUTICN EFFEC W72-14694
LU/ *AQUATIC PLANTS, *MINNESOTA, *TOXICITY, COAGULATION, WATER POL W72-07360

Y METALS, *ALGAE, *GROWTH RATES,	*TOXICITY, *FLUORESCENCE, TRACE E	W72-07660
ER, CHLOROPHYTA, CHL/ *BACTERIA,	*TOXICITY, *MARINE ALGAE, SEA WAT	W72-08051
NDUSTRIAL WASTES, *PESTICIDES, /	*TOXICITY, REVIEWS, *BIOASSAY, *I	W73-01976
*PHOTOSYNTHESIS, *OIL POLLUTION,	*TOXICITY, CHEMICAL REACTIONS, CA	W73-01074
OGY.:	*TOXICOLOGY, BIOTOXINS, PHARMACOL	W73-00243
HYTES, SAPROPHYTES.: *USSR,	*TOXICOLOGY, *HYDROBIONTS, MACROP	W72-09650
ONTS, ICHTHYOLOGY, ICHTH/ *USSR,	*TOXICOLOGY, *TOXICANTS, *HYDROBI	W72-09646
FEDERAL GOVERNMEN/ *LEGISLATION,	*TOXINS, *ENVIRONMENTAL EFFECTS,	W73-00703
ARY PRODUCTIVITY, *HEAVY METALS,	*TRACE ELEMENTS, NUTRIENTS, MERCU	W73-01434
PES, HEAVY METALS/ *FOOD CHAINS,	*TRACE ELEMENTS, *LEAD RADIOISOTO	W72-00940
E, *NEUTRON ACTIVATION ANALYSIS,	*TRACE ELEMENTS, COPPER, MOLYBDEN	W71-11515
N, CHROMATOGRAPHY, HEAVY METALS,	*TRACE ELEMENTS, RADIOCHEMICAL AN	W72-04708
BI/ *TRACE ORGANIC CONTAMINANTS,	*TRACE MATERIALS, BIOINHIBITION,	W71-11793
CE MATERIALS, BIOINHIBITION, BI/	*TRACE ORGANIC CONTAMINANTS, *TRA	W71-11793
, CURR/ *ALGAE, *REMOTE SENSING,	*TRACKING TECHNIQUES, MEASUREMENT	W72-01095
*SUSPENDED SOLIDS, *RECREATION,	*TREATMENT FACILITIES, SAMPLING,	W72-00042
ATMENT, WATER POLLUTION EFFECTS,	*TREATMENT FACILITIES, *INFLUENT	W73-01669
NOSTOC, TOLYPOTHRIX TENUIS, AU/	*TRICALCIUM PHOSPHATES, ANABAENA,	W71-10070
RING TECHNIQUES, CULTURE MEDIA,/	*TRICHODESMIUM ERYTHRAEUM, *CULTU	W73-00430
NOTONECTIDAE, BODO/ *HEMIPTERA,	*TRICHOPTERA, CORIXIDS, EUDORINA,	W70-06655
N, / *OXIDATION LAGOONS, *ALGAE,	*TRICKLING FILTERS, EUTROPHICATIO	W71-07097
LORELLA PYRENOIDOSA, SOURCE OF /	*TRISODIUM NITRILOTRIACETATE, *CH	W70-08976
*CHLORELLA PYRENOIDOSA,	*TRISODIUM NITRILOTRIACETATE.:	W71-11972
TION EFFECTS, / *SNAILS, *FROGS,	*TRITIUM, ABSORPTION, WATER POLLU	W72-03348
CHAINS, *RADIOACTIVITY EFFECTS,	*TRITIUM, *NUCLEAR WASTES, WATER	W72-06342
HORUS, NUTRIENTS, MEASUREMENT, /	*TROPHIC LEVELS, *BIOASSAY, PHOSP	W71-13794
RODUC/ *STREAMS, *STANDING CROP,	*TROPHIC LEVEL, *ENERGY BUDGET, P	W71-10064
S, *CHEMICAL PROPERTIES, *SEDIM/	*TROPHIC LEVEL, *LAKES, *NUTRIENT	W72-05469
MATICAL STUDIES, ATLANTIC OCEAN,	*TROPICAL REGIONS, MARINE ALGAE,	W72-11719
TA,/ *ALGAE, *NITROGEN FIXATION,	*TROPICAL REGIONS, RICE, CYANOPHY	W73-02473
OF P/ *PHOSPHORUS RADIOISOTOPES,	*TUBIFICIDS, *FOOD CHAINS, *PATH	W72-09668
SERVOIRS, *PRIMARY PRODUCTIVITY,	*TURBIDITY, *PHOTOSYNTHESIS, LIMN	W72-12393
*WATER TREATMENT, *PHOSPHATES,	*TURBIDITY, *FLOCCULATION, *FILTR	W72-06201
MAND, D/ *WASTE WATER TREATMENT,	*TURBIDITY, BIOCHEMICAL OXYGEN DE	W72-03753
, HYPOL/ *PHYTOPLANKTON, *LAKES,	*TURBIDITY, *PROFILES, PHOTOMETER	W72-07145
MAR/ *HYDRODYNAMICS, *FRICTION,	*TURBULENT FLOW, ALGAE, BACTERIA,	W72-03220
AT PLAIN/ *PHOTOSYNTHETIC MODEL,	*TUTTLE CREEK RESERVOIR(KAN), GRE	W72-12393
A, FALMOUTH HARBOR, *TINTINNICS,	*TYCHOPELAGIC, *TEMPERATURE TOLER	W72-07899
N TECHNIQUES, CYANO/ *ISOLATION,	*ULTRAVIOLET RADIATION, SEPARATIO	W72-14327
N, ALGAE, ANALYTICAL TECHNIQUES,	*ULTRAVIOLET RADIATION, DISSOLVED	W72-01780
*LAMINARIA SP.,	*ULVA SP., MACROCYSTIS SP.:	W71-10553
ENGLAND), IRISH SEA.:	*ULVA, *LAMINARIA, LIVERPOOL BAY(	W71-12867
ENGLAND), LAKE WINDERMERE(ENGLA/	*UNITED KINGDOM, ESTHWAITE WATER(	W71-13185
ENGLAND), BLELHAM TARN(ENGLAND)/	*UNITED KINGDOM, ESTHWAITE WATER(	W71-13183
CLADOPHORA, R/ *RESEARCH POLICY,	*UNITED KINGDOM, *FUTURE TRENDS,	W71-13172
SPHAEROTOTILUS NATANS, ZOOGLOEA/	*UNITED KINGDOM, *SEWAGE FUNGUS,	W72-08573
OSPHATES, *WATER POLLUTION CONT/	*UNITED STATES, *DETERGETNTS, *PH	W72-01085
URCES RESEARCH ACT, CONNECTICUT,	*UNIVERSITIES, *ABSTRACTS, WATER	W71-01192
INCOLN(NEBR).:	*URBAN RUNOFF, URBAN HYDROLOGY, L	W71-13466
WATERS, CONTINE/ *DISTRIBUTION,	*UREAS, *COASTS, *OCEANS, SURFACE	W72-08056
TING AND COOLING, / *WASTE HEAT,	*USES OF WASTE HEAT, *CENTRAL HEA	W70-09193
ICMETRY.:	*USSR, *HYDROBIONTS, *CERIUM, RAD	W72-09649
ESERVOIR, *HYDROBIOLOGY, METALI/	*USSR, *IRKUTSK OBLAST, *BRATSK R	W72-04855
ESERVOIR, ANGARA RIVER, OKA RIV/	*USSR, *IRKUTSK OBLAST, *BRATSK R	W72-04853
MACROPHYTES, SAPROPHYTES.:	*USSR, *TOXICOLOGY, *HYDROBIONTS,	W72-09650
HYDROBIONTS, ICHTHYOLOGY, ICHTH/	*USSR, *TOXICOLOGY, *TOXICANTS, *	W72-09646
OAM, NATURAL WATERS.:	*USSR, MOSCOW RIVER, OKA RIVER, F	W72-05506
LLA MENEGHINIANA, *POLYMORPHISM,	*VALVE MORPHOLOGY, *VALVE PATTERN	W73-02548
POLYMORPHISM, *VALVE MORPHOLOGY,	*VALVE PATTERN CHARACTERISTICS, C	W73-02548
WASH), PORTLAND(ORE), WATER QUA/	*VANCOUVER LAKE(WASH), VANCOUVER(	W72-03215
PTH, ON-SITE INVESTIGATIONS, *P/	*VARIABILITY, *PHYTOPLANKTON, *DE	W72-11719
S, *PLANT MORPHOLOGY, *GENETICS,	*VARIABILITY, WAVES(WATER), MAINE	W73-00369
COLLECTORS, BRYOZOANS, FORAMIN/	*VERTICAL COLLECTORS, *HORIZONTAL	W72-05432
DIFFUSION COEFF/ *ALGAL SINKING,	*VERTICAL MIXING, EUPHOTIC ZONE,	W71-03034
ECTS, ENVIRONMENTAL ENGINEERING,	*VIRGINIA, PROJECTIONS, MODEL STU	W72-14405
ROPERTIES, *ECOSYSTEMS, *RIVERS,	*VIRGINIA, *WATER POLLUTION EFFEC	W73-01972
NITY, SALT MARSHES, SYSTEMATICS,	*VIRGINIA, *ATLANTIC OCEAN, NORTH	W72-12633
*CYANOPHYTA, ALGAE,	*VIRUSES.:	W72-03188
*ALGAL CONTROL, *INFECTION,	*VIRUSES, CYANOPHYTA.:	W73-00246
ACTERIA, *FUNGI, *BACTERIOPHAGE,	*VIRUSES, ABSORPTION, ELECTROPHOR	W72-06124
ION PATTERNS, *WATER ANALYSIS, /	*VIRUSES, *CYANOPHYTA, *DISTRIBUT	W72-04736
ATIC MICROBIOLOGY, OREGON, *SED/	*VITAMIN B, *EUTROPHICATION, *AQU	W72-08989
RACILIMUM, CRYPTOMONAS EROSA, P/	*VOLTA LAKE(GHANA), ACTINASTRUM G	W73-00256
MULTI-LAYER FILTERS, ANTHRACITE,	*WAHNBACH RESERVOIR(GERMANY), OSC	W72-06201
T IDENTIFICATION, AQUATIC WEEDS,	*WASHINGTON, *CHLOROPHYLL, WATER	W72-14728
REMENTS, AMMONIA, *GROWTH RATES,	*WASTE ASSIMILATIVE CAPACITY, OXI	W72-04787
GANIC / *WASTE WATER(POLLUTION),	*WASTE ASSIMILATION CAPACITY, *OR	W71-11017
IS NIDULANS, *MUTANTS, *SENSITI/	*WASTE CHARACTERIZATION, *ANACYST	W72-14327
CONTROL, *NUISANCE ALGAE, *EUTR/	*WASTE DILUTION, WATER POLLUTION	W72-00799
MATHEMATICAL MODELS, *BIOASSAY,	*WASTE DILUTION, LAGOONS, *WASTE	W72-04787
TION LAGOONS, AER/ *FARM WASTES,	*WASTE DISPOSAL, *LAGOONS, *OXIDA	W71-07551
RIAL WASTES/ *ALGAE, *ESTUARIES,	*WASTE DISPOSAL, *SEWAGE, *INDUST	W72-08804
ES, *MICROORGANI/ *WATER SUPPLY,	*WASTE DISPOSAL, *RECREATION WAST	W72-14363
HEAT CONSUMPTION.:	*WASTE HEAT, THERMAL EFFICIENCY,	W70-08832
*CENTRAL HEATING AND COOLING, /	*WASTE HEAT, *USES OF WASTE HEAT,	W70-09193
IXING ZONES, MODELING.:	*WASTE HEAT, SUBLETHEL EFFECTS, M	W70-09612
DEPTH, SLOPE, *BOD LOADING,	*WASTE STABILIZATION.:	W71-03896
EMICAL OXYGEN DEMAND, NUTRIENTS,	*WASTE TREATMENT, *ALGAE, COSTS.:	W72-08396
CAL OXYGEN DEMAND, PILOT PLANTS,	*WASTE WATER TREATMENT, WYOMING,	W72-06838
ZATION, *WASTE WATER(POLLUTION),	*WASTE WATER DISPOSAL, *WATER POL	W72-14878

AT/ *IRRIGATION, *FERTILIZATION,	*WASTE WATER(POLLUTION), *WASTE W	W72-14878
ON, *SEWAGE TREATMENT/ *PATENTS,	*WASTE WATER TREATMENT, *FILTRATI	W72-14469
ION CONCEN/ *PHOSPHORUS, *ALGAE,	*WASTE WATER TREATMENT, HYDROGEN	W73-01367
TREATMENT, / *SEWAGE TREATMENT,	*WASTE WATER TREATMENT, *TERTIARY	W70-10381
SIS, DETER/ *PHOSPHATES, *LAKES,	*WASTE WATER(POLLUTION), *HYDROLY	W71-00116
ON, *ALGAE, *SUSPENSION, HARVES/	*WASTE WATER TREATMENT, *FILTRATI	W70-10174
ALGAE, LAGOONS, BIOCHEMICAL OXY/	*WASTE WATER TREATMENT, *PONDS, *	W70-07838
, PHOTOSYNTHESIS, STABILIZATICN,	*WASTE WATER TREATMENT, *FARM WAS	W70-06619
-USE, BIOCH/ *OXIDATION LAGOONS,	*WASTE WATER TREATMENT, *WATER RE	W71-07123
OADING, BIO/ *OXIDATION LAGOONS,	*WASTE WATER TREATMENT, ORGANIC L	W71-07124
TECHNIQUES, TERTIARY TREATMENT,	*WASTE WATER TREATMENT.: /LYTICAL	W72-04983
ASTES, SEWAGE TREATME/ *PATENTS,	*WASTE WATER TREATMENT, *LIQUID W	W72-05315
REQUIREMENT/ *ALGAE, *KINETICS,	*WASTE WATER TREATMENT, *NUTRIENT	W72-04789
ASSAY, *WASTE DILUTION, LAGOONS,	*WASTE WATER TREATMENT, ALGAL CON	W72-04787
TION, ALGAL CONTROL, *CHLORELLA,	*WASTE WATER TREATMENT.: /RCPHICA	W72-04788
ATION, *WATER POLLUTION CONTROL,	*WASTE WATER(POLLUTION), VIRGINIA	W72-05473
HATES, *WATER POLLUTION CONTROL,	*WASTE WATER TREATMENT, *COST-BEN	W72-04734
ICATICN, WATER REUSE, *NITROGEN/	*WASTE WATER TREATMENT, *DENITRIF	W72-04309
Y, BIOCHEMICAL OXYGEN DEMAND, D/	*WASTE WATER TREATMENT, *TURBIDIT	W72-03753
NTROL, *AQUATIC AL/ *PHOSPHORUS,	*WASTE WATER TREATMENT, *ALGAL CO	W72-03970
RIENTS, SLUDGE, ALGAE, BACTERIA,	*WASTE WATER TREATMENT, EUTROPHIC	W72-04086
, CALCIUM, MAGNESIUM, NUTRIENTS,	*WASTE WATER TREATMENT, TERTIARY	W72-03613
S, *ALGAE, *BACTERIA, *SYMBIOSI/	*WASTE WATER TREATMENT, *NUTRIENT	W72-11100
COST ANALYSIS, NITROGEN, CLAYS,	*WASTE WATER TREATMENT.: /SLUDGE,	W72-11642
YTICAL TECHNIQUES, MICROBIOLOGY,	*WASTE WATER TREATMENT.: /S, ANAL	W72-09590
R TREATMENT, TERTIARY TREATMENT,	*WASTE WATER TREATMENT.: /, *WATE	W72-09393
STES, OPERATION AND MAINTENANCE,	*WASTE WATER TREATMENT, WATER QUA	W72-09386
ATOMACEOUS EARTH, *POROUS MEDIA,	*WASTE WATER TREATMENT, *DENITRIF	W72-00974
USION, *ALGAL CONTROL, AERATION,	*WASTE WATER TREATMENT.: /N, DIFF	W72-00800
ION, DISINFECTION, FLOCCULATICN,	*WASTE WATER TREATMENT.: /FILTRAT	W72-00411
CHNIQUES, WATER QUALITY CONTROL,	*WASTE WATER TREATMENT.: /ICAL TE	W72-00422
ES, TREATMENT FACILITIES, ALGAE,	*WASTE WATER TREATMENT, *ALGAL CO	W72-02975
KS, METABOLISM, DESIGN CRITERIA,	*WASTE WATER TREATMENT, EFFLUENTS	W71-13341
FORMS, BACTERIA, CENTRIFUGATION,	*WASTE WATER TREATMENT.: /N, COLI	W71-13326
ATION, *DENITRIFICATION, *NITRA/	*WASTE WATER TREATMENT, *NITRIFIC	W71-13939
L TECHNIQUES, COA/ *COAGULATICN,	*WASTE WATER TREATMENT, ANALYTICA	W71-13356
QUES, DESIGN CRITERIA, KINETICS,	*WASTE WATER TREATMENT.: / TECHNI	W72-02363
SSIMILATION CAPACITY, *ORGANIC /	*WASTE WATER(POLLUTION), *WASTE A	W71-11017
CTIONATION, LIME, / *PHOSPHATES,	*WASTE WATER TREATMENT, *FOAM FRA	W71-12200
LOGICAL TREATMENT, *PHOSPHOROUS,	*WASTE WATER TREATMENT, NUTRIENTS	W71-12188
AL TREATMENT, *ACTIVATED SLUDGE/	*WASTE WATER TREATMENT, *BIOLOGIC	W71-12223
TREATMENT, *COAGULATIO/ *ALGAE,	*WASTE WATER TREATMENT, *TERTIARY	W71-09629
L OXYGEN DEMAND, COOLING TOWERS,	*WASTE WATER TREATMENT.: /CHEMICA	W71-09524
NITRIFICATION, DENITRIFICATION,	*WASTE WATER TREATMENT.: /TATION,	W71-09450
ICA/ *OXIDATION LAGOONS, *ALGAE,	*WASTE WATER TREATMENT, *MATHEMAT	W72-12289
AGOONS, *COLD/ *SEWAGE BACTERIA,	*WASTE WATER TREATMENT, *SEWAGE L	W72-12565
TS, M/ *OXIDATION PONDS, *ALGAE,	*WASTE WATER TREATMENT, *BYPRODUC	W72-13508
IES, HYDROGEN ION CONCENTRATICN,	*WASTE WATER TREATMENT.: /PROPERT	W72-11833
NTS, *SEWAGE TREATMENT, CULTURE,	*WASTE-WATER TREATMENT, *CARBON D	W72-06612
NISMS, *WATER POLLUTION CONTROL,	*WASTES, *PATHOGENIC BACTERIA, CO	W72-09675
PLANTS, DEPTH, CALIF/ *ECOLOGY,	*WASTES, *MARINE ANIMALS, *MARINE	W71-01475
ALGAE, *SILICA, *E. COLI, *IONS,	*WASTEWATER TREATMENT, *POLYMERS,	W72-14840
ER QUALITY CONTROL, *CYANOPHYTA,	*WATER ANALYSIS, LAKES, SAMPLING,	W72-07890
*PR/ *LIMNOLOGY, *PHYTOPLANKTCN,	*WATER ANALYSIS, *STANDING CROP,	W73-01435
TRODES,/ *ANALYTICAL TECHNIQUES,	*WATER ANALYSIS, *NITRATES, *ELEC	W71-00474
NOPHYTA, *DISTRIBUTION PATTERNS,	*WATER ANALYSIS, *BIOINDICATORS,	W72-04736
SEDIME/ *NUTRIENTS, *SEDIMENTS,	*WATER ANALYSIS, LAKE MORPHOLOGY,	W71-05626
N, *FOAM FRACTIONATION, *ORGANI/	*WATER CHEMISTRY, *FOAM SEPARATIO	W72-05506
S, *POWERPLANTS, *ENERGY LOSSES/	*WATER CONSERVATION, *AIR CONDITI	W70-09193
EMENT, *WATER POLLUTION CONTROL,	*WATER COOLING, *STEAM TURBINES,	W70-09193
PHORUS, *DETENTION RESERVOIRS, /	*WATER DILUTION, *NITROGEN, *PHOS	W72-02412
D STORAGE, PHYSICOCHEMICAL PROP/	*WATER LEVEL FLUCTUATIONS, *PUMPE	W72-14280
RCURY, *MUNICIPAL WASTES, *INDU/	*WATER POLLUTION, *LAKE ERIE, *ME	W72-14282
REATION WASTES, *MICROORGANISMS,	*WATER POLLUTION SOURCES, REVIEWS	W72-14363
*AQUATIC ENVIRONMENT, *TOXICITY,	*WATER POLLUTION EFFECTS, ZINC, L	W72-14694
LLUTICN), *WASTE WATER DISPOSAL,	*WATER POLLUTION CONTROL, ECONOMI	W72-14878
YNTHESIS, *OIL P/ *MARINE ALGAE,	*WATER POLLUTION EFFECTS, *PHOTOS	W73-01074
BLIOGRAPHIES, *DATA COLLECTIONS,	*WATER POLLUTION EFFECTS, DOCUMEN	W73-01976
*ECOSYSTEMS, *RIVERS, *VIRGINIA,	*WATER POLLUTION EFFECTS, SURVEYS	W73-01972
WS, *EUTROPHICATION, *NUTRIENTS,	*WATER POLLUTION CONTROL, ORGANIC	W73-00840
*INORGANIC COMPOUNDS, *NUTRIEN/	*WATER POLLUTION, MICROCRGANISMS,	W72-07933
STES, *CRUSTACEANS, OILY WATER,/	*WATER POLLUTION EFFECTS, *OIL WA	W72-07911
STUDIES, *STREAMS, *ECOSYSTEMS,/	*WATER POLLUTION EFFECTS, *MODEL	W72-06340
*PHOSPHATES, *LOCAL GOVERNMENTS,	*WATER POLLUTION CONTROL, WATER Q	W72-06638
ION A/ *DETERGENTS, *PHOSPHATES,	*WATER POLLUTION SOURCES, *POLLUT	W72-06655
CONFINEMENT PENS, IMPOUNDMENTS,	*WATER POLLUTION SOURCES, *AGRICU	W72-08401
Y, PESTICI/ *PESTICIDE TOXICITY,	*WATER POLLUTION EFFECTS, BIOASSA	W72-08440
N SOURCES, *AGRICULTURAL RUNOFF,	*WATER POLLUTION EFFECTS, *CATTLE	W72-08401
C MICROORGA/ *THERMAL POLLUTICN,	*WATER POLLUTION EFFECTS, *AQUATI	W72-07225
	*WATER POLLUTION CONTROL ACT.:	W72-01685
C WASTES, *NUTRIENTS, *POLLUTAN/	*WATER POLLUTION EFFECTS, *ORGANI	W72-01472
QUALITY CONTROL, *FEDER/ *LAKES,	*WATER POLLUTION CONTROL, *WATER	W72-01659
	*WATER POLLUTION CONTROL ACT.:	W72-01093
NMENTAL SANITATION, *LE/ *LAKES,	*WATER POLLUTION CONTROL, *ENVIRO	W72-01660
ATES, *DETERGENTS, *LEGISLATION/	*WATER POLLUTION CONTROL, *PHOSPH	W72-01685
POLLUTION/ *POLLUTION ABATEMENT,	*WATER POLLUTION EFFECTS, *WATER	W71-13723
EMENT, *WATER POLLUTION EFFECTS,	*WATER POLLUTION SOURCES, *FISHER	W71-13723
OLLECTIONS, ALGAE, ECOTYPES, IN/	*WATER POLLUTION EFFECTS, *DATA C	W71-13746
ATES, *NITRATES, *NEBRASKA, URB/	*WATER POLLUTION SOURCES, *PHOSPH	W71-13466
, *ANIMALS, *ALGAE, *INVERTEBRA/	*WATER POLLUTION EFFECTS, *PLANTS	W71-13233

CTIVITY EF/ *NUCLEAR EXPLOSIONS, *WATER POLLUTION EFFECTS, *RADIOA W72-03347
TROPHICATION, *LABORATORY TESTS, *WATER POLLUTION EFFECTS, *ALGAE, W72-03218
ENT, *ENVIRONMENT, *AGRICULTURE, *WATER POLLUTION CONTROL, SEDIMEN W72-00846
ATES, *DETERGETNTS, *PHOSPHATES, *WATER POLLUTION CONTROL, *WATER W72-01085
ATES, *FEDERAL GOV/ *DETERGENTS, *WATER POLLUTION CONTROL, *PHOSPH W72-01093
OA, *MARINE ALGAE, *SALT MARSHE/ *WATER POLLUTION EFFECTS, *PROTOZ W72-00948
HATES, *WATER POLLUTION CONTROL, *WATER POLLUTION SOURCES, EUTROPH W72-01085
, *PATHOGENIC / *MICROORGANISMS, *WATER POLLUTION CONTROL, *WASTES W72-09675
QUATI/ *HYDROBIOLOGY, *TOXICITY, *WATER POLLUTION, *FRESHWATER, *A W72-09649
OXICITY, *WATER QUALITY CONTROL, *WATER POLLUTION EFFECTS, WATER P W72-09646
ABSORPTION, *PATH OF POLLUTANTS, *WATER POLLUTION, WATER POLLUTION W72-10686
ENCIES, *EUTROPHICATION, *LAKES, *WATER POLLUTION EFFECTS, RECLAIM W72-11702
HICATION, *DETERGE/ *PHOSPHATES, *WATER POLLUTION SOURCES, *EUTROP W72-11186
*PROT/ *AQUATIC MICROORGANISMS, *WATER POLLUTION EFFECTS, *ALGAE, W72-08579
RAPHIES, *ANALYTICAL TECHNIQUES, *WATER POLLUTION CONTROL, *INSTRU W72-08790
TS, CALIFOR/ *INDUSTRIAL WASTES, *WATER POLLUTION EFFECTS, EFFLUEN W72-09092
EN/ *ESTUARIES, *EUTROPHICATION, *WATER POLLUTION CONTROL, MANAGEM W72-09171
LGAE, *SURFACE WATERS, *REVIEWS, *WATER POLLUTION SOURCES, NUTRIEN W71-10580
4-D, *DDT, *DIELDRIN, *DIAZINON, *WATER POLLUTION EFFECTS, *PHOTOS W71-10566
*DELTAS, *BAYS, *WATER QUALITY, *WATER POLLUTION CONTROL, *DIVERS W71-09581
ES, *NUTRIENTS, *ALGAE, *BACTER/ *WATER POLLUTION EFFECTS, *NITRAT W71-09958
ICATION, *BIOASSAY, *INDICATORS, *WATER POLLUTION EFFECTS, *ALGAE, W71-10553
C WASTES, *EUTROPHICATION, *CHE/ *WATER POLLUTION EFFECTS, *ORGANI W71-09674
LUTION CONTROL, *EUTROPHICATION, *WATER POLLUTION SOURCES, POPULAT W71-12091
IDES, *DDT, *PESTICIDE TOXICITY, *WATER POLLUTION EFFECTS, PESTICI W71-12303
*FILTRATION, *ELECTROCHEMISTRY, *WATER POLLUTION TREATMENT, ANALY W71-12467
HICATION, *WATER POLLUTION SOUR/ *WATER POLLUTION CONTROL, *EUTROP W71-12091
TORS, *SEA WATER, MARIN/ *ALGAE, *WATER POLLUTION EFFECTS, *INDICA W71-12867
T, *PHOSPHATES, *EUTROPHICATION, *WATER POLLUTION CONTROL, SEWAGE W71-11507
ATCHERIES, *SALMONIDS, WATER RE/ *WATER POLLUTION SOURCES, *FISH H W71-11006
UTRIEN/ *EUTROPHICATION, *LAKES, *WATER POLLUTION EFFECTS, OHIO, N W71-11949
S, WATER POLLUTION SOURCES, IND/ *WATER POLLUTION EFFECTS, *DIATOM W71-11583
HICATION, *DETERGENTS, *PHOSPHA/ *WATER POLLUTION CONTROL, *EUTROP W71-11516
DEMAND, *CHEMICAL OXYGEN DEMAND, *WATER POLLUTION SOURCES, *WATER W70-09189
EMAND, *WATER POLLUTION SOURCES, *WATER POLLUTION EFFECTS, *ORGANI W70-09189
ER SUPPLY, *POLLUTION ABATEMENT, *WATER POLLUTION CONTROL, *WATER W70-09193
ATES, *NITRATES, *NUTRIENTS, *E/ *WATER POLLUTION EFFECTS, *PHOSPH W70-09388
CT OF COLUMBIA, TID/ *ESTUARIES, *WATER POLLUTION EFFECTS, *DISTRI W70-08249
CT OF COLUMBIA, TID/ *ESTUARIES, *WATER POLLUTION EFFECTS, *DISTRI W70-08248
RMAL/ *RESEARCH AND DEVELOPMENT, *WATER POLLUTION, PHOSPHORUS, THE W70-09947
OLLUTION SOURCES, INDUSTRIAL WA/ *WATER POLLUTION EFFECTS, WATER P W71-03095
R QUAL/ *EUTROPHICATION, *LAKES, *WATER POLLUTION, MINNESOTA, WATE W71-03012
OLLUTION EFFEC/ *EUTROPHICATION, *WATER POLLUTION CONTROL, WATER P W71-01488
IE/ *NUTRIENTS, *EUTROPHICATION, *WATER POLLUTION SOURCES, *ESTUAR W70-06509
WATER(POLLUTIO/ *EUTROPHICATION, *WATER POLLUTION CONTROL, *WASTE W72-05473
ECOSYSTEMS, ABSORPTION, FISH, I/ *WATER POLLUTION, *FOOD CHAINS, * W72-05952
LYCHLORINATED BIPHENYLS, *ALGAE, *WATER POLLUTION EFFECTS, *PESTIC W72-05614
TER, *OILS, *N/ *REMOTE SENSING, *WATER POLLUTION CONTROL, *SEA WA W72-04742
OLLUT/ *EUTROPHICATION, *CARBON, *WATER POLLUTION CONTROL, WATER P W72-04769
W/ *EUTROPHICATION, *PHOSPHATES, *WATER POLLUTION CONTROL, *WASTE W72-04734
*COLORADO, *RE/ *WATER QUALITY, *WATER POLLUTION EFFECTS, *LAKES, W72-04791
ASTES, OKLA/ *OXIDATION LAGOONS, *WATER POLLUTION EFFECTS, *FARM W W72-04773
RESOURCES DEVELOPME/ *ESTUARIES, *WATER POLLUTION EFFECTS, *WATER W71-05553
AY, *TOXICITY, *CHLORINATED HYD/ *WATER POLLUTION EFFECTS, *BIOASS W71-07412
POLLU/ *PHOSPHATES, *DETERGENTS, *WATER POLLUTION EFFECTS, *WATER W71-08768
*ACID MINE WATER, *STRIP MINES, *WATER POLLUTION EFFECTS, *KENTUC W71-07942
GENTS, *WATER POLLUTION EFFECTS, *WATER POLLUTION CONTROL, *BIODEG W71-08768
WAST/ *STRONTIUM RADIOISOTOPES, *WATER POLLUTION EFFECTS, NUCLEAR W71-09225
R POWER PLANTS, *POTABLE WATER,/ *WATER POLLUTION CONTROL, *NUCLEA W71-09018
WATER, *WATER TEMPERATURE, *WA/ *WATER POLLUTION EFFECTS, *HEATED W72-13636
ATIC LIFE, *AQUATIC ENVIRONMENT, *WATER POLLUTION, *PATH OF POLLUT W72-13347
OGY, WATER POLLUTION, W/ *LAKES, *WATER POLLUTION EFFECTS, *LIMNOL W72-12729
HICATION, *LAKES, SEWAGE EFFLUE/ *WATER POLLUTION SOURCES, *EUTROP W72-12955
ESTICIDE KINETICS, *PERSISTENCE, *WATER POLLUTION EFFECTS, *PLANT W73-02280
IC HEALTH,/ *WATER PURIFICATION, *WATER POLLUTION TREATMENT, *PUBL W73-02144
HICATION, *SOCIAL PARTICIPATION, *WATER POLLUTION CONTROL, WISCONS W73-02479
ORUS, *EUTROPHICATION, CHEMICAL/ *WATER POLLUTION CONTROL, *PHOSPH W73-02478
*INSTRUMENTATION, *MONITORING, *WATER PROPERTIES, *ON-SITE DATA W72-12941
, *CULTIVATED L/ *TILE DRAINAGE, *WATER PROPERTIES, *FERTILIZATION W71-12252
*ELECTROCHEMISTRY, *WATER POLL/ *WATER PURIFICATION, *FILTRATION, W71-12467
TION TREATMENT, *PUBLIC HEALTH,/ *WATER PURIFICATION, *WATER POLLU W73-02144
ESERVOIRS, RESERVOIRS, PHOTOSYN/ *WATER PURIFICATION, *DETENTION R W71-01491
CTROPHORESIS, *ELECTROCHEMISTRY, *WATER PURIFICATION, CLAY, ALGAE, W73-01362
*CRUSTACEANS, *MARINE BACTERIA, *WATER QUALITY, BENTHOS, CRABS, C W72-08142
VITY, *AQUATIC MICROORGANISMS, / *WATER QUALITY, *AQUATIC PRODUCTI W72-06532
ICATION, *BIOLOGICAL PROPERTIES, *WATER QUALITY, *BIOINDICATORS, * W73-00002
ANTS, *MATHEMATICAL MODELS, THE/ *WATER QUALITY, *HYDROELECTRIC PL W72-14405
POLLUTION, NUTRIENTS, BIOASSAY, *WATER QUALITY, ALGAE, PESTICIDES W72-14282
*MATHEMATICA/ *SYSTEMS ANALYSIS, *WATER QUALITY, *EUTROPHICATION, W71-03021
CATION, PHYTOPLANKTON, PLANKTON/ *WATER QUALITY, *LAKES, *EUTROPHI W71-02880
, STREAMS, FLOW, SE/ *NUTRIENTS, *WATER QUALITY, *IMPOUNDED WATERS W71-03028
C/ *NUTRIENTS, *EUTROPHICATION, *WATER QUALITY CONTROL, *NITROGEN W70-04080
HORUS, PULP AND PAPER INDUSTRY,/ *WATER QUALITY, *WISCONSIN, PHOSP W70-05103
ERTIARY TREATMENT, *WATER REUSE, *WATER QUALITY, WATER SUPPLY, ALG W70-10381
NTROL, *FORAGE FISH, BIOCONTROL/ *WATER QUALITY CONTROL, *ALGAE CO W71-09163
ATION LAGOONS, *DESIGN CRITERIA, *WATER QUALITY, ALGAE, HARVESTING W71-07100
WEED CONTROL, *DOME/ *ALGICIDES, *WATER QUALITY CONTROL, *AQUATIC W71-05083
TEMS, *ECOLOGY, *PUMPED STORAGE, *WATER QUALITY, *MODEL STUDIES, Z W72-05282
EFFECTS, *LAKES, *COLORADO, *RE/ *WATER QUALITY, *WATER POLLUTION W72-04791
N, *ALGAE, TURBIDITY, *NEW YORK, *WATER QUALITY,: /ATERS, *PLANKTO W72-06191

ELOPMENT, *DISTRIBUTION SYSTEMS,	*WATER QUALITY, WATER CONSERVATIO	W72-05594
CTURAL/ *MAINE, *EUTROPHICATION,	*WATER QUALITY CONTROL, *NON-STRU	W72-04279
MENTATION, WATER QUALITY, RESER/	*WATER QUALITY CONTROL, *FLOW AUG	W72-04260
EATED WATER, *WATER TEMPERATURE,	*WATER QUALITY STANDARDS, *ESTURA	W72-13636
*WATER QUALITY, *CHLORINATION,	*WATER QUALITY CONTROL, *ODOR, AL	W71-13187
POLLUTION, *CHEMICAL REACTIONS,	*WATER QUALITY, *CHEMICAL PROPERT	W71-12092
*PRIMARY PRODUCTIVITY, *ALGAE, /	*WATER QUALITY, *EUTROPHICATION,	W71-13183
ATER QUALITY CONTROL, *ODOR, AL/	*WATER QUALITY, *CHLORINATION, *W	W71-13187
, *PHYTOPLA/ *INDUSTRIAL WASTES,	*WATER QUALITY, *DISSOLVED OXYGEN	W71-11876
TOXICITY, EUT/ *OCEANS, *COASTS,	*WATER QUALITY CONTROL, BIOMASS,	W71-11793
CONTROL, *DIVER/ *DELTAS, *BAYS,	*WATER QUALITY, *WATER POLLUTION	W71-09581
DEVELOPMENT, *WATER/ *ESTUARIES,	*WATER QUALITY, *WATER RESOURCES	W71-09788
OPHICATION, *TERTIARY TREATMENT,	*WATER QUALITY CONTROL, AMMONIA,	W71-09450
RN/ *SURFACE WATERS, *SEA WATER,	*WATER QUALITY, *ECOLOGY, *CALIFO	W71-10577
TROPHICATION, *DESTRATIFICATION,	*WATER QUALITY CONTROL, *LAKE MOR	W72-09304
TRIENTS, *PLANKTON, *RECREATION,	*WATER QUALITY, CYANOPHYTA, LAKES	W72-10431
ALGAE, INDU/ *ENTERIC BACTERIA,	*WATER QUALITY, *DOMESTIC WASTES,	W72-09590
OBIOLOGY, *PATHOLOGY, *TOXICITY,	*WATER QUALITY CONTROL, *WATER PO	W72-09646
EUTROPHICATION, AQUATIC ALGAE, /	*WATER QUALITY CONTROL, *LAKES, *	W71-13256
, *RESERVOIRS, *HYDROLOGIC DATA/	*WATER QUALITY, *IMPOUNDED WATERS	W72-02274
LAKES, *WATER POLLUTION CONTROL,	*WATER QUALITY CONTROL, *FEDERAL	W72-01659
LAGOONS, *WASTE WATER TREATMENT,	*WATER RE-USE, BIOCHEMICAL OXYGEN	W71-07123
OMPOSITION.:	*WATER RECLAMATION, VARIATIONS, C	W70-07838
NNECTICUT, *UNIVERSITIES, *ABST/	*WATER RESOURCES RESEARCH ACT, CO	W71-01192
ARIES, *WATER POLLUTION EFFECTS,	*WATER RESOURCES DEVELOPMENT, *MA	W71-05553
STRIBUTION SYSTE/ *WATER SUPPLY,	*WATER RESOURCES DEVELOPMENT, *DI	W72-05594
TER/ *ESTUARIES, *WATER QUALITY,	*WATER RESOURCES DEVELOPMENT, *WA	W71-09788
INDUSTRIAL WASTES, ACTIVATED SL/	*WATER REUSE, *DOMESTIC WASTES, *	W71-09524
TION LAGOONS, *MUNICIPAL WASTES,	*WATER REUSE, ALGAE, PHOTOSYNTHES	W72-11642
PROTEIN, *OXIDATION LA/ *PONDS,	*WATER REUSE, RECLAMATION, ALGAE,	W71-08951
TREATMENT, *TERTIARY TREATMENT,	*WATER REUSE, *WATER QUALITY, WAT	W70-10381
WER, *MULTIPLE-PURPOSE PROJECTS,	*WATER REUSE, *WATER SUPPLY, *POL	W70-09193
, *IMPAIRED WATER / *CALIFORNIA,	*WATER RIGHTS, *WATER UTILIZATION	W70-07100
-PURPOSE PROJECTS, *WATER REUSE,	*WATER SUPPLY, *POLLUTION ABATEME	W70-09193
EVELOPMENT, *DISTRIBUTION SYSTE/	*WATER SUPPLY, *WATER RESOURCES D	W72-05594
Y, *WATER RESOURCES DEVELOPMENT,	*WATER SUPPLY, WASTE WATER, RUNOF	W71-09788
FECTS, NUTRIEN/ *EUTROPHICATION,	*WATER SUPPLY, WATER POLLUTION EF	W71-13185
RECREATION WASTES, *MICROORGANI/	*WATER SUPPLY, *WASTE DISPOSAL, *	W72-14363
WATERS, *GROUNDWATERS, SURVEYS,	*WATER SUPPLY.: /, ODOR, *SURFACE	W72-06536
IVITY, INVERTEBRATES, *PLANKTON,	*WATER TEMPERATURE, SYSTEMATICS,	W72-07899
GEYSERS, PH/ *THERMAL POLLUTION,	*WATER TEMPERATURE, *CYANOPHYTA,	W72-03557
TEMPERATURE CONTROL, *STANDARDS,	*WATER TEMPERATURE, SEASONAL, MON	W70-09595
AR POWER PLANTS, *COOLING WATER,	*WATER TEMPERATURE, CONDENSERS, D	W70-09612
OLLUTION EFFECTS, *HEATED WATER,	*WATER TEMPERATURE, *WATER QUALIT	W72-13636
*ALGAL CONTROL, RIVERS, WATER P/	*WATER TREATMENT, *CHLORINATION,	W71-00099
ICROORGANISMS, *SURFACE WATERS,/	*WATER TREATMENT, *FILTRATION, *M	W72-06191
URBIDITY, *FLOCCULATION, *FILTR/	*WATER TREATMENT, *PHOSPHATES, *T	W72-06201
, *ALGAE, *ALGICIDES, MICROSTRA/	*WATER TREATMENT, *MICROORGANISMS	W72-06536
IDES, OPERATION AND MAINTENANCE,	*WATER TREATMENT, *MICHIGAN.: /IC	W72-08357
EFFECTS, *TREATMENT FACILITIES/	*WATER TREATMENT, WATER POLLUTION	W73-01669
CED INFILTRATION, *PIT RECHARGE,	*WATER TREATMENT, *FILTATION, ALL	W71-12410
, ALGAE, FLOCCULATION, CLEANING,	*WATER TREATMENT, TERTIARY TREATM	W72-09393
ER / *CALIFORNIA, *WATER RIGHTS,	*WATER UTILIZATION, *IMPAIRED WAT	W70-07100
ICALS, *PUBLIC HEALTH STANDARDS/	*WATERBORNE DISEASES, *TOXIC CHEM	W73-02144
ICIDES, *NUTRIENTS, *ECOSYSTEMS,	*WATERSHEDS(BASINS), STREAMFLOW,	W71-01489
*ALGIC/ *ALGAE, *AQUATIC WEEDS,	*WEED CONTROL, GROWTH, *CULTURES,	W71-06102
LANKTON, *ALGAE, *AQUATIC WEEDS,	*WEEDS, *RED TIDE, COASTS, LAKES,	W72-14673
LAKE SIDNEY LANIER(GEORGIA), L/	*WHITE ALGAL BLOOMS, HYPNODINIUM,	W71-09173
ORIUM, ANTIMONY, MACRONUTRIENTS/	*WILLIAMS LAKE(WASH), URANIUM, TH	W73-01434
, BENTHIC BLUE-GREEN ALGAE.:	*WINDERMERE(ENGLAND), HETEROCYSTS	W73-02472
IRCINALIS, BOSMINA LONGIROSTRIS/	*WINNISQUAM LAKE(N H), ANABAENA C	W71-03014
PAPER INDUSTRY,/ *WATER QUALITY,	*WISCONSIN, PHOSPHORUS, PULP AND	W70-05103
TOGRAPHY, SEPARATION TECHNIQUES,	*WISCONSIN, OLIGOTROPHY, EUTROPHI	W73-01456
, HYDROIDS, SPONGES, COELENTERA/	*WOOD BORERS, BARNACLES, BRYOZOAS	W73-00374
	*WOODS HOLE(MASS).:	W72-06046
AE, *CYANOPHYTA, *THERMAL WATER,	*WYOMING, NATIONAL PARKS, HOT SPR	W72-08508
SIRA NUMMULOIDES, CHLOROPHYLL A,	*YAQUINA BAY, AMPHORA SPP, MELOSI	W73-00853
Y, DATA INTERPRETATION, MACROAL/	*YAQUINA BAY(ORE), YAQUINA ESTUAR	W72-08141
GYRINA, KEMMERER SAMPLER, SECCHI	*YARDS CREEK(NJ), RELATIVE CONDIT	W72-14280
, RHODOPHYTA, PHAEOPHYTA, CHLOR/	*YEASTS, *CULTURES, *MARINE ALGAE	W72-04748
BLUE-GREEN ALGAE, MASTIGOCLADUS/	*YELLOWSTONE NATIONAL PARK(WYO),	W72-08508
YNGBYA MAJUSCULA, GOMONTIA POLY/	*YUCATAN, LYNGBYA CONFERVOIDES, L	W72-11800
ALGAE, *STRONTIUM RADIOISOTOPES,	*ZINC RADIOISOTOPES, ABSORPTION,	W72-12203
OISOTOPES, *MARINE ALGAE, HEAVY/	*ZINC RADIOISOTOPES, *COBALT RADI	W72-10957
*MARINE ALGAE, *PHYTOPLANKTON,/	*ZINC RADIOISOTOPES, *ABSORPTION,	W73-00802
*PHYTOPLANKTON, *MARINE ALGAE,/	*ZINC RADIOISOTOPES, *ABSORPTION,	W73-00823
BORATORY EQUIPMENT, *LA/ *ASSAY,	*ZINC, *ADSORPTION, *SURFACES, LA	W72-12210
, CLADOPHORA GLOMERATA, ULOTRIC/	*ZINC, *LEAD, STIGEOCLONIUM TENUE	W71-05991
UTROPHICATION, NUTRIENT REMOVAL,	*ZINC, CHELATION, FOOD CHAINS, PA	W73-00802
TER PAPER DISC DIFFUSION METHOD,	*ZONES OF INHIBITION, ANTIFUNGAL	W72-08586
ESIS, RAINFALL, DIURNAL, FLUCTU/	*ZOOPLANKTON, *ALGAE, *PHOTOSYNTH	W72-00150
*ECOLOGY, *AQUATIC PRODUCTIVITY,	*ZOOPLANKTON, *PHYTOPLANKTON, NER	W73-00935
, *SEPARATION TECHNIQUES, *MARI/	*ZOOPLANKTON, *LARVAE, *FISH EGGS	W73-00371
PEPODS, ALGAE, NUTRIENT REQUIRE/	*ZOOPLANKTON, *FOODS, DAPHNIA, CO	W72-07940
AY, B/ *ECOSYSTEMS, FOOD CHAINS,	*ZOOPLANKTON, *PREDATION, *BIOASS	W72-07132
HAMPSH/ *COPPER SULFATE, *ALGAE,	*ZOOPLANKTERS, *CHEMCONTROL, NEW	W71-03014
, *ORGANOPHOSPHOROUS PESTICIDES,	*2-4-D, *DDT, *DIELDRIN, *DIAZINO	W71-10566

GENERAL INDEX

The middle rank of this index lists in alphabetical order each word of each descriptor except those included in the *Significant Descriptor Index,* which should be consulted first. The numbers at the right are the accession numbers appearing at the end of the abstract.

PHORUS, EXCHANGEABLE PHOSPHORUS,	A*DENOSINE TRIPHOSPHATE(ATP).: /S	W72-03742
IN, LINDANE, DIADEMA ANTILLARUM,	AANDOMENE, SIPHONALES, *HALIMEDA	W72-14294
EMENT, CRITICAL NUCLIDE PATHWAY,	ABALONE.: ENVIRONMENTAL STAT	W72-00959
EDRO(CALIF), EPIBENTHIC SPECIES,	ABALONE, TUNICATES, ARTHROPODS, C	W71-01475
A, DUNALIELLA, SKELETONEMA.:	ABATE, BAYTEX, BAYGONE, CYCLOTELL	W71-10096
RINE FISHERIES, LAKES, POLLUTION	ABATEMENT.: /GAE, FOOD CHAINS, MA	W72-00960
IOECOLOGY, MONITORING, POLLUTION	ABATEMENT, ENVIRONMENTAL EFFECTS,	W72-00939
L ASPECTS, REGULATION, POLLUTION	ABATEMENT, WASTE DISPOSAL, STANDA	W72-01085
TS, *WATER POLLUTION/ *POLLUTION	ABATEMENT, *WATER POLLUTION EFFEC	W71-13723
, MATHEMATICAL MODELS, NITROGEN,	ABATEMENT, COSTS, ANALYTICAL TECH	W72-09171
REUSE, *WATER SUPPLY, *POLLUTION	ABATEMENT, *WATER POLLUTION CONTR	W70-09193
NITRATES, FILTRATION, POLLUTION	ABATEMENT, ALGAE, NUTRIENT, *BIOD	W72-05315
SPHATES, *DETERGENTS, *POLLUTION	ABATEMENT, WATER POLLUTION EFFECT	W71-09429
*HOGS, *ENVIRONMENT, *POLLUTION	ABATEMENT, WATER QUALITY, STANDAR	W71-08214
LORINATION, OXIDATION, POLLUTION	ABATEMENT, WATER POLLUTION TREATM	W72-14469
ER POLLUTION SOURCES, *POLLUTION	ABATEMENT, EUTROPHICATION, LEGISL	W72-06655
ZONE, DENITRIFICATION, POLLUTION	ABATEMENT, WATER POLLUTION CONTRO	W73-02187
MULATION, METAL COMPLEXES, PICEA	ABIES, VACCINIUM VITIS IDAEA, VAC	W72-14303
PHOSPHATES, CHLORIDES, BORATES,	ABSORBANCE.: /NITRATES, SULFATES,	W72-06274
RRY, COLLAR LAKE, CHLOROPHYLL A,	ABSORBANCE.: /SCN, CLEARWATER QUA	W72-13673
ION CONCENTRATION, VERTEBRATES,	ABSORBANCE, MAGNESIUM ION CONCENT	W72-09092
BATION, RUNOFF, SALINITY, DEPTH,	ABSORPTION, ALGAE, SEASONAL, WEAT	W72-09122
ATER, HYDROBIOLOGY, ALGAE, FISH,	ABSORPTION, BIOASSAY, METABOLISM.	W72-10694
ES, *MARINE ALGAE, HEAVY METALS,	ABSORPTION, BIOINDICATORS, PATH O	W72-10957
NT RELATIONSHIPS, RADIOISOTOPES,	ABSORPTION, RADIOACTIVITY TECHNIQ	W72-00949
OECOLOGY, RADIOACTIVITY EFFECTS,	ABSORPTION, FOOD WEBS, STRONTIUM	W72-00948
ANTS, MARINE ALGAE, MARINE FISH,	ABSORPTION, WATER POLLUTION SOURC	W72-00940
TS, / *SNAILS, *FROGS, *TRITIUM,	ABSORPTION, WATER POLLUTION EFFEC	W72-03348
OPODS, AMPHIPODA, RADIOISOTOPES,	ABSORPTION, PACIFIC OCEAN, FOOD C	W72-03347
ATION, MARINE ANIMALS, BACTERIA,	ABSORPTION, CULTURES.: / CONCENTR	W71-10083
N, SEDIMENTS, WATER CIRCULATION,	ABSORPTION, NUTRIENTS, PESTICIDES	W71-09581
S, PHOTOSYNTHESIS, CARBON CYCLE,	ABSORPTION, DECOMPOSING ORGANIC M	W71-12068
NC, ANTIMONY, NORWAY, A/ *ATOMIC	ABSORPTION, ARSENIC, SELENIUM, ZI	W71-11515
MY, OXYGEN PLASMA ASHING, ATOMIC	ABSORPTION, TITRATION.: /URCANATO	W71-11036
IOISOTOPES, *ZINC RADIOISOTOPES,	ABSORPTION, SEDIMENTATION, ANALYT	W72-12203
ESTICIDES, *CYTOLOGICAL STUDIES,	ABSORPTION, DDT, PHYTOPLANKTON, P	W72-12998
S, MARINE FISH, INSTRUMENTATION,	ABSORPTION, SEDIMENT DISCHARGE, R	W72-06283
, CARP, CRAYFISH, AQUATIC ALGAE,	ABSORPTION, AQUARIA, ANIMAL PHYSI	W72-06342
NMENT, RADIOACTIVITY TECHNIQUES,	ABSORPTION.: /S, ESTUARINE ENVIRO	W72-07826
RY, EQUIPMENT, INHIBITION, IONS,	ABSORPTION, BIOASSAY, *MERCURY.: /	W72-07660
ALGAE, PHYTOPLANKTON, NITROGEN,	ABSORPTION, SEA WATER.: / *MARINE	W72-07497
TOSYNTHESIS, OPTICAL PROPERTIES,	ABSORPTION, MATHEMATICAL STUDIES,	W72-14800
OPHYTA, WATER POLLUTION EFFECTS,	ABSORPTION, PESTICIDE TOXICITY, L	W73-00281
OECOLOGY, ENVIRONMENTAL EFFECTS,	ABSORPTION, SORPTION, WASTE DILUT	W73-00817
*MARINE FISH, MUSSLES, LOBSTERS,	ABSORPTION, COBALT RADIOISOTOPES,	W73-00832
LS, *MARINE ALGAE, *FOOD CHAINS,	ABSORPTION, CALCIUM, PUBLIC HEALT	W73-00819
CE ELEMENTS, PATH OF POLLUTANTS,	ABSORPTION, RADIOISOTOPES, STABLE	W73-00811
C COASTAL PLAIN, NORTH CAROLINA,	ABSORPTION, SOUNDS, ESTUARINE ENV	W73-00818
ALT RADIOISOTOPES, MARINE ALGAE,	ABSORPTION, SEDIMENTS, WASTE DILU	W73-00831
SOTOPES, *FALLOUT, *FOOD CHAINS,	ABSORPTION, SEDIMENTS, CORES, ANA	W73-00790
NE ALGAE, ESTUARINE ENVIRONMENT,	ABSORPTION, ANIMAL POPULATIONS, A	W73-00826
METHODOLOGY, NITROGEN, SILICON,	ABSORPTION, PHYTOPLANKTON.: /ION,	W73-01445
, ALKALI METALS, SEASONAL, ZINC,	ABSORPTION, LIGHT PENETRATION, AL	W73-01434
KTON, OREGON, ORGANIC COMPOUNDS,	ABSORPTION, KINETICS, WATER TEMPE	W73-01269
S, BACTERIA, POPULATION, HAWAII,	ABSORPTION, TROPICAL REGIONS, PAC	W71-08042
*AQUATIC ALGAE, *RADIOISOTOPES,	ABSORPTION, TEMPERATURE, STRONTIU	W71-09250
CHEMICAL ANALYSIS, ION EXCHANGE,	ABSORPTION, STRONTIUM RADIOISOTOP	W71-09188
OISOTOPES, COBALT RADIOISOTOPES,	ABSORPTION, WATER POLLUTION SOURC	W71-09018
PERATURE, PROTOZOA, CHLOROPHYTA,	ABSORPTION.: /, PLANT GROWTH, TEM	W71-09254
EFFECTS, NUCLEAR WASTES, RIVERS,	ABSORPTION, FISH, MOLLUSKS, MARIN	W71-09225
RELLA, TOXICITY, WATER ANALYSIS,	ABSORPTION, CULTURES, DDT, ANALYT	W72-04134
S, *AQUATIC LIFE, RADIOISOTOPES,	ABSORPTION, NUCLEAR WASTES, WATER	W72-04461
ETTLING VELOCITY, RADIOISOTOPES,	ABSORPTION, REVIEWS.: /DIMENTS, S	W72-04463

FUNGI, *BACTERIOPHAGE, *VIRUSES,	ABSORPTION, ELECTROPHORESIS, DETE	W72-06124
TION, *FOOD CHAINS, *ECOSYSTEMS,	ABSORPTION, FISH, INDUSTRIAL WAST	W72-05952
RES, METHODOLOGY, DOCUMENTATION,	ABSTRACTS, DATA COLLECTIONS, MICR	W72-13800
, NICKEL, RUTHEN/ CONCENTRATION,	ABUNDANCE, IRON, CHROMIUM, COPPER	W71-13243
ES, ORGANOPHOSPHORUS PESTICIDES,	ACARICIDES, FOOD CHAINS, WATER PO	W73-00238
CODS, HARPACTICOIDS, CIRRIPEDIA,	ACARINA.: /IDS, POLYCHAETA, OSTRA	W72-05432
ATION, SETTLING, MICROORGANISMS,	ACCLIMATION, CALCIUM, SEDIMENTATI	W71-13334
FULIGINOSA, BATOPHORA OERSTEDI,	ACETABULARIA CRENULATA, VALONIA V	W72-11800
ROSIN AMINE D	ACETATE(RADA).:	W71-05083
OL, GLYCEROL, LACTATE, PYRUVATE,	ACETATE, ETHANOL, BENZOATE, PROPI	W72-05421
YLUM TRICORNUTUM, PHENYLMERCURIC	ACETATE, MERCURY CHLORIDE, PHENYL	W72-11727
SCENEDESMUS QUADRICAUDA,	ACETATE, PYRUVATE, LACTATE.:	W72-00838
HETEROTROPHIC BACTERIA, GLUCOSE,	ACETATES, BIOTIN, NIACIN, COBALAM	W73-02095
ABAENA CYLINDRICA, ANABAENOPSIS,	ACETYLENE TEST, CYLINDROSPERMUM L	W71-12070
CHLOROPHYLL A, OPTICAL DENSITY,	ACETYLENE REDUCTION, SCHIZOTHRIX	W72-10883
SPIRATION, *ANABAENA CYLINDRICA,	ACETYLENE REDUCTION, WARBURG EFFE	W72-08584
(WIS), LAKE / *NUTRIENT SOURCES,	ACETYLENE REDUCTION, LAKE MENDOTA	W72-10608
REMENTS, CARBON-14 MEASUREMENTS,	ACETYLENE REDUCTION MEASUREMENTS,	W72-13692
OPHIC INDEX, MACROINVERTEBRATES,	ACETYLENE, ETHYLENE, NITROGEN RAD	W72-10883
T RIVER, HETEROTROPHIC BACTERIA,	ACETYLENE, ETHYLENE.: /ER, DETROI	W73-01456
MARK), COCCONEIS, RHOICOSPHENIA,	ACHANTHES, FRAGILARIA CONSTRUENTS	W72-07501
YTOCEPHALA, ACHNANTHES BREVIPES,	ACHNANTHES JAVANICA, LICMOPHORA J	W72-08141
ACENTULA, NAVICULA SRYTOCEPHALA,	ACHNANTHES BREVIPES, ACHNANTHES J	W72-08141
ESTERN ONTARIO, EPILITHIC ALGAE,	ACHNANTHES MINUTISSIMA, ACHNANTHE	W71-11029
C ALGAE, ACHNANTHES MINUTISSIMA,	ACHNANTHES FLEXELLA, EUNOTIA PECT	W71-11029
STRATES, ACHNANTHES MINUTISSIMA,	ACHNANTHES CF. BIRGIANI, AMPHIPLE	W72-13142
S / *LAKE WINNIPEG, *SUBSTRATES,	ACHNANTHES MINUTISSIMA, ACHNANTHE	W72-13142
OPHYCEAE, XANTHOPHYCEAE, VIBRIO,	ACHROMOBACTER, FLAVOBACTERIA.: /I	W72-08051
, PERIDINIUM AFRICANUM, NITZCHIA	ACICULARIS, SYNEDRA ACUS.: /EROSA	W73-00256
YLATI/ *GLUCOSE, *MALATE, CITRIC	ACID CYCLE, CYTOCHROMES, PHOSPHOR	W72-04309
TION PERIOD, FLUSHING OPERATION,	ACID INJECTION PUMP, ALUM INJECTI	W72-02850
QUATIC ALGAE, WATER TEMPERATURE,	ACID STREAMS, HYDROGEN ION CONCEN	W72-14690
, NITRATE REDUCTASE, RIBONUCLEIC	ACID.: CHLORELLA VULGARIS	W72-03222
ACID, 2,4-DICHLOROPHENOXYACETIC	ACID.: /-3,5,6-TRICHLOROPICOLINIC	W72-08440
ILS, AXENIC CULTURES, GIBBERELIC	ACID.: /IBER MEDIUM, ESTUARINE SO	W73-00430
VINEYARD SOUND, ORGANIC CARBON,	ACID-VOLATILE COMPOUNDS, WOODS HO	W72-07710
GREEN BAY(WIS), NITRILOTRIACETIC	ACID, ALGAL ASSAY PROCEDURES, ALG	W71-12072
ES, PHOTOASSIMILATION, GLYOXYLIC	ACID, ASSIMILATION.: /PERMATOPHYT	W73-02100
CAULERPICIN, CAULERPIN, PALMITIC	ACID, BETA-SITOSTEROL, THIN-LAYER	W71-08597
A, GLOEOTRICHIA, MICROC/ *FULVIC	ACID, GROWTH STIMULATION, ANABAEN	W72-01097
AND DETERGENT / NITRILOTRIACETIC	ACID, HOUSEHOLD DETERGENTS, SOAP	W72-04266
BENZENE SULFONATE, COCONUT FATTY	ACID, PLURONICS, DIETHANOLAMIDE.:	W72-12576
AKE TAHOE(CALIF), ZINC, ASCORBIC	ACID, PYRIDOXINE, INDAL COMPOUNDS	W72-07514
TRIPHOSPHATE, *DEOXYRIBONUCLEIC	ACID, SAMPLE PRESERVATION, COULTE	W71-11562
COSE, MANNOSE, RHAMNOSE, ACRYLIC	ACID, SODIUM, PYRIDINE, DEXTRAM,	W72-07710
ARABINOSE, GALACTOSE, HEXURONIC	ACID, VINEYARD SOUND, ORGANIC CAR	W72-07710
4-AMINO-3,5,6-TRICHLOROPICOLINIC	ACID, 2,4-DICHLOROPHENOXYACETIC A	W72-08440
URE, DEPTH, CHEMICAL PROPERTIES,	ACIDITY, ALKALINITY, CONDUCTIVITY	W72-04269
DIGESTION, AEROBIC, ALKALINITY,	ACIDITY, LABORATORY, DIGESTION TA	W72-02850
CARBON DIOXIDE, ALGAE, CULTURES,	ACIDITY, ORGANIC COMPOUNDS.: /M,	W72-00838
, OREGON, ORGANIC COMPOU/ *AMINO	ACIDS, *EUTROPHICATION, *PLANKTON	W73-01269
, METHYL ESTERS, ESTERS,/ *FATTY	ACIDS, *GAS LIQUID CHROMATOGRAPHY	W72-09365
NOPHYTA, *GROWTH RATES, *ORGANIC	ACIDS, *HYDROGEN ION CONCENTRATIO	W72-00838
ICAL TECHNIQUES, *CARBON, *AMINO	ACIDS, *MARINE ALGAE, PHYTOPLANKT	W72-07497
ALGAE, *CARBOHYDRATES, *ORGANICS	ACIDS, *PLANT PHYSIOLOGY, *GROWTH	W72-07710
SE, DIALYSIS, SEA WATER, ORGANIC	ACIDS, CHRYSOPHYTA, INHIBITION, M	W73-00942
RINE ALGAE, PHYTOPLANKTON, HUMIC	ACIDS, FULVIC ACIDS, TRACE ELEMEN	W71-09188
N EXCHANGE, METHODOLOGY, ORGANIC	ACIDS, FULVIC ACIDS, KELPS, WATER	W72-12166
OISOTOPES, CHROMATOGRAPHY, AMINO	ACIDS, GAS CHROMATOGRAPHY, PHAEOP	W72-07710
THODOLOGY, ORGANIC ACIDS, FULVIC	ACIDS, KELPS, WATER ANALYSIS, EXU	W72-12166
ROPHICATION/ *CYANOPHYTA, *HUMIC	ACIDS, ORGANIC MATTER, ALGAE, EUT	W72-01097
RET'S BAY, SCOUDOUC RIVER, HUMIC	ACIDS, OXIMATION, SAMPLE PREPARAT	W72-12166
ECTROSCOPY, CULTURE MEDIA, FATTY	ACIDS, SAMPLE PREPARATION, NATURA	W73-01066
ALYSIS, *ALGAE, COLUMNS, ORGANIC	ACIDS, SEPARATION TECHNIQUES, *CH	W72-09365
ECTROPHORESIS, DETERGENTS, AMINO	ACIDS, SPECTROPHOTOMETRY, MAGNESI	W72-06124
YTOPLANKTON, HUMIC ACIDS, FULVIC	ACIDS, TRACE ELEMENTS, CHEMICAL A	W71-09188
DATION, ORGANIC COMPOUNDS, AMINO	ACIDS, VITAMINS, CARBOHYDRATES, A	W73-00942
ECTUM, SCHOENOPLECTUS LACUSTRIS,	ACORUS CALAMUS, TYPHA ANGUSTIFOLI	W73-00286
ATIC ENVIRONMENT, AQUATIC FUNGI,	ACQUIFER CHARACTERISTICS, GENETIC	W72-11186
ONATE, ARSENATES, SODIUM CACODY/	ACREMONIELLA, DIMETHYL SODIUM ARS	W72-13813
A FLABELLUM, BRYOPSIS, CAULERPA,	ACROPORA PALMATA, PENICILLUS LAMO	W72-14294
ETI, GLUCOSE, MANNOSE, RHAMNOSE,	ACRYLIC ACID, SODIUM, PYRIDINE, D	W72-07710
, CELLS(BIOLOGY), *BIODETERIORA/	ACRYLIC RESINS, GLASS, THYMIDINES	W71-09261
*WATER POLLUTION CONTROL	ACT.:	W72-01685
*WATER POLLUTION CONTROL	ACT.:	W72-01093
WATER POLLUTION CONTROL	ACT.:	W72-01085
*ABST/ *WATER RESOURCES RESEARCH	ACT, CONNECTICUT, *UNIVERSITIES,	W71-01192
ROGEN, PHOSPHORUS, WATER QUALITY	ACT, WATER QUALITY.: /OXYGEN, NIT	W71-06444
G, TERRAMYCIN, CHLORAMPHENICOL,	ACTIDON, CULTURE MEDIA, CHLOROPHY	W73-01445
*CHEMOSTATS, ORTHOPH/ *NITZSCHIA	ACTINASTROIDES, *BATCH CULTURES,	W73-01445
ACUTUS VAR ALTERNANS, *NITZSCHIA	ACTINASTROIDES, *STAURASTRUM PING	W72-12739
AS EROSA, P/ *VOLTA LAKE(GHANA),	ACTINASTRUM GRACILIMUM, CRYPTOMON	W73-00256
, DIATOMA, MELOSIRA, TABELLARIA,	ACTINASTRUM, PHYTOCONIS, CYMBELLA	W72-10875
TS, SCANDIUM, RUBIDIUM, BROMINE,	ACTINIDES, RADIOACTIVE DECAY, GAM	W73-01434
GMENTS, CARBOHYDRATES, PROTEINS,	ACTINOMYCETES, CHLORELLA.: /Y, PI	W72-07951
MATOGRAPHY, SEPCTROSCOPY, ALGAE,	ACTINOMYCETES, FUNGI, LABORATORY	W72-11604
GEN, LIGHT INTENSITY, NUTRIENTS,	ACTINOMYCETES, TASTE, EUTROPHICAT	W72-10613
ON, TASTE, RESERVOIRS, SEASONAL,	ACTINOMYCETES, CYANOPHYTA, PLANKT	W71-13187
ING ALGAE, ODOR-PRODUCING ALGAE,	ACTINOMYCETES, PHOSPHATES.: /ODUC	W73-02144
E, PILIDIUM, CLADOCERA, LINGULA,	ACTINOTROCHA, VELIGER, PTEROPODS,	W73-00935

DETROIT LAKES(MINN), COMMUNITY	ACTION, LAKE SALLIE(MINN), LAKE M	W71-03012
TES, PHOTOSYNTHETIC OXYGENATION,	ACTIVATED ALGAE.: /RESPIRATORY RA	W71-04079
*ALGAE, *BACTERIA, *RESPIRATION,	ACTIVATED SLUDGE, WASTE WATER TRE	W71-04079
URE, HYDROGEN ION CONCENTRATION,	ACTIVATED SLUDGE, TREATMENT, EUTR	W71-00116
ER QUALITY, WATER SUPPLY, ALGAE,	ACTIVATED CARBON, ADSORPTION, FIL	W70-10381
YDROLYSIS, BIOLOGICAL TREATMENT,	ACTIVATED SLUDGE, TERTIARY TREATM	W70-09907
ILTRATION, *ALGAE, ELECTROLYTES,	ACTIVATED CARBON, HYDROGEN ION CO	W72-06201
CHLORINE, OXIDATION, ADSORPTION,	ACTIVATED CARBON, SLUDGE TREATMEN	W73-01669
*TASTE, *ODOR, *POTABLE WATER,	ACTIVATED CARBON, ADSORPTION, FIL	W72-08357
ENT, TASTE, ODOR, POTABLE WATER,	ACTIVATED CARBON, COLIFORMS, VIRU	W73-02144
RS, LAGOONS, TERTIARY TREATMENT,	ACTIVATED CARBON, FLOCCULATION, A	W71-09524
STIC WASTES, *INDUSTRIAL WASTES,	ACTIVATED SLUDGE, TRICKLING FILTE	W71-09524
LGAE, BACTERIA, DIATOMS, OCEANS,	ACTIVATED SLUDGE, SOIL BACTERIA,	W72-10883
, *ALGAE, *BACTERIA, *SYMBIOSIS,	ACTIVATED SLUDGE, NITROGEN, PHOSP	W72-11100
, *ALGAE, WASTE WATER TREATMENT,	ACTIVATED SLUDGE, LIME, AEROBIC T	W72-09176
VERS, BACTERIA, WATER POLLUTION,	ACTIVATED SLUDGE, ALGAE, PROTOZOA	W72-08573
QUALITY CONTROL, EUTROPHICATION,	ACTIVATED SLUDGE, DESIGN, NITROGE	W72-00800
, *PHOSPHOROUS, ALGAE, NITROGEN,	ACTIVATED SLUDGE, CHEMICAL OXYGEN	W71-13334
, *MOLYBDENUM, BIOASSAY, NEUTRON	ACTIVATION ANALYSIS.: /ANT GROWTH	W72-00801
NEUTRON	ACTIVATION ANALYSIS.:	W72-02073
*ANALYTICAL TECHNIQUES, *NEUTRON	ACTIVATION ANALYSIS, *CHEMICAL AN	W72-01101
NTS, C/ *AQUATIC ALGAE, *NEUTRON	ACTIVATION ANALYSIS, *TRACE ELEME	W71-11515
SAMPLING, *ENVIRONMENT, *NEUTRON	ACTIVATION ANALYSIS, RADIOACTIVIT	W71-11036
NTS, PATH OF POLLUTANTS, NEUTRON	ACTIVATION ANALYSIS, SPECTROSCOPY	W73-00832
*ZONES OF INHIBITION, ANTIFUNGAL	ACTIVITY.: /SC DIFFUSION METHOD,	W72-08586
CULTURE, CHEMOSTATS, PHOSPHATASE	ACTIVITY.: /S-AQUAE, *CONTINUOUS	W72-05476
, ALGISTATIC ACTIVITY, ALGICIDAL	ACTIVITY.: /SYNERGISM, FISH PONDS	W71-06189
YNERGISM, FISH PONDS, ALGISTATIC	ACTIVITY, ALGICIDAL ACTIVITY.: /S	W71-06189
GANIUM FLUCTUANS, ANTINEOPLASTIC	ACTIVITY, ALKALOIDS, FLAVONOIDS,	W72-04275
APER DISC DIFFUS/ *ANTIMICROBIAL	ACTIVITY, *SKELLYSOLVE, *FILTER P	W72-08586
ANABAENA FLOS-AQUA, *NITROGENASE	ACTIVITY, OXYGEN INACTIVATION.: /	W72-07941
NEYS, LIVER, BLOOD, GASTEROSTEUS	ACULEATUS, DAPHNIA MAGNA, DAPHNIA	W72-05952
TEROCYPRIS SALINUS, GASTEROSTEUS	ACULEATUS, BALTIC SEA, NITOCRA SP	W71-08026
VISCHERIA STELLATA, ELLIPSOIDION	ACUMINATUM.: /ISCHERIA PUNCTATA,	W72-14808
UM, NITZCHIA ACICULARIS, SYNEDRA	ACUS.: /EROSA, PERIDINIUM AFRICAN	W73-00256
INDIA, ORGANIC NITROGEN, EUGLENA	ACUS, EUGLENA CAUDATA, EUGLENA LI	W73-00918
OSCILLATORIA RUBESCENS, SYNEDRA	ACUS, RHODOMONAS, LAKE CONSTANCE,	W71-00117
NIZOMENON FLOS-AQUAE, CHROOMONAS	ACUTA, MELOSIRA GRANULATA, CERATI	W72-12738
A/ *CHLORELLA FUSCA, *SCENEDEMUS	ACUTUS VAR ALTERNANS, *NITZSCHIA	W72-12739
OBACTER, NAJAS FLEXILIS, SCIRPUS	ACUTUS, C-14, DISSOLVED ORGANIC M	W73-02095
SM, LAWRENCE LAKE(MICH), SCIRPUS	ACUTUS, NAJAS FLEXILIS.: /ETABOLI	W72-03216
OGIA HISTORY, PLEISTORENE EPOCH,	ADAPTATION, ENVIRONMENTAL EFFECTS	W73-00855
N-SELECTIVE ELECTRODES, STANDARD	ADDITION, POTENTIOMETRY, INTERFER	W71-00474
LEGAL ASPECTS, WATER POLLUTION,	ADJUDICATION PROCEDURE, ADMINISTR	W72-01093
LLUTION, ADJUDICATION PROCEDURE,	ADMINISTRATIVE AGENCIES, ADMINIST	W72-01093
CEDURE, ADMINISTRATIVE AGENCIES,	ADMINISTRATION, ALGAE, BIODEGRADA	W72-01093
VERNMENTS, TREATMENT FACILITIES,	ADMINISTRATIVE AGENCI: / STATE GO	W72-01659
JUDICIAL DECISIONS, LEGISLATION,	ADMINISTRATION, NON-CONSUMPTIVE U	W70-07100
SUPPLY, ALGAE, ACTIVATED CARBON,	ADSORPTION, FILTRATION, CHLORINAT	W70-10381
SNAILS, FISH, ALGAE, GASTROPODS,	ADSORPTION, TEMPERATURE, METABOLI	W71-09168
ICS, NUCLEAR WASTES, SOLUBILITY,	ADSORPTION, WATER POLLUTION EFFEC	W71-09182
M CARBONATE, DETRITUS, PLANKTON,	ADSORPTION, EUTROPHICATION, TEMPE	W72-05469
TES, *SEDIMENT-WATER INTERFACES,	ADSORPTION, LAKES, NUTRIENTS, LAK	W72-01108
DES, TROPHIC LEVEL, GASES, MUDS,	ADSORPTION, COPP: /RTIES, PESTICI	W72-00847
ION, DENITRIFICATION, OXIDATION,	ADSORPTION, COLIFORMS, BACTERIA,	W71-13326
ALGAE, FOOD CHAINS, MEASUREMENT,	ADSORPTION.: /N SOURCES, AQUATIC	W72-10686
IVATION ANALYSIS, RADIOACTIVITY,	ADSORPTION, FISH, LAKE ERIE, SEDI	W71-11036
S, PHENOLS, CHLORINE, OXIDATION,	ADSORPTION, ACTIVATED CARBON, SLU	W73-01669
DIFFUSION, CYTOLOGICAL STUDIES,	ADSORPTION, PENETRATION, CATION A	W73-00823
ADSORPTION, PENETRATION, CATION	ADSORPTION, CHELATION, MATHEMATIC	W73-00823
POTABLE WATER, ACTIVATED CARBON,	ADSORPTION, FILTRATION, ALGAE, PH	W72-08357
OTOPES, LAKE ERIE, LAKE ONTARIO,	ADSORPTION, AGRICULTURAL RUNOFF,	W72-13657
MARINE ALGAE, CHEMICAL ANALYSIS,	ADSORPTION, SEPARATION TECHNIQUES	W73-02105
MOSUM, LEERSIA ORYZOIDES, NUPHAR	ADVENA, EICHHORNIA CRASSIPES, JUS	W73-01089
ARII, AMMOTIUM SALSUM, ELPHIDIUM	ADVENUM, ELPHIDIUM CLAVATUM, ELPH	W73-00367
BAETIDAE, EPHYDRA, CHAETOGASTER,	AEOLOSOMA, CHAOBORUS, *GAMBUSIA A	W72-07132
QUADRICAUDA, MOUGEOTIA, ULOTHRIX	AEQUALIS, STEPHANOPYXIS TURRIS.: /	W72-07144
*SEWAGE LAGOONS, *COLD REGIONS,	AERATED LAGOONS, *NORTH DAKOTA, B	W72-12565
AND DISPOSAL, OXIDATION DITCHES,	AERATED LAGOONS, COMPOSTING, REFE	W71-07551
DE, BICARBONATES, ALGAL CONTROL,	AERATION.: /LORELLA, CARBON DIOXI	W72-09168
ALITY CONTROL, *LAKE MORPHOLOGY,	AERATION, MIXING, ALGAL BLOOMS, A	W72-09304
ORY TESTS, LABORATORY EQUIPMENT,	AERATION, DISSOLVED OXYGEN, AEROB	W72-11100
CARBON, CHLORELLA, BICARBONATES,	AERATION, LIMITING FACTORS, EUTRO	W72-10607
MAND, BIOCHEMICAL OXYGEN DEMAND,	AERATION, *OXIDATION, HYDROGEN IO	W71-12223
CHEMICAL OXYGEN DEMAND, OXYGEN,	AERATION, DISSOLVED OXYGEN, HYDRO	W71-13334
MONITORING, AUTOMATIC CONTROLS,	AERATION, SEDIMENTATION, FILTRATI	W72-00042
OGEN, DIFFUSION, *ALGAL CONTROL,	AERATION, *WASTE WATER TREATMENT.	W72-00800
TS, *DECOMPOSING ORGANIC MATTER,	AERATION, *ALGAE, *BIODEGRADATION	W72-01881
VOLUME, DRYING, ODOR, NITROGEN,	AERATION, HOGS, CATTLE, POULTRY,	W71-07551
TEMPERATURE, PHOTOSYNTHESIS, RE-	AERATION, LIGHT PENETRATION, NUTR	W71-07098
LIGHT INTENSITY, EVAPORATION, RE-	AERATION, DISSOLVED OXYGEN, ALGAE	W71-07099
, TEMPERATURE, DISSOLVED OXYGEN,	AERATION, OXYGENATION, NUTRIENTS,	W71-07113
N CONCENTRATION, CARBON DIOXIDE,	AERATION, MIXING, *EUTROPHICATION	W72-04433
INSTRUMENTATION, WASTE DISPOSAL,	AERATION, EROSION CONTROL, WATER	W71-01192
NUTIRIENTS, AEROBIC CONDITIONS,	AERATION, MICROBIOLOGY, BIOMASS,	W71-04079
GAE, *ALGICIDES, MICROSTRAINING,	AERATION, PHOSPHATES, CLAY, TASTE	W72-06536
DATION LAGOONS, ORGANIC LOADING,	AERATION, MIXING, ALGAE, TEMPERAT	W72-06838
UTRIENTS, SEWAGE, WATER QUALITY,	AERATION, WATER PROPERTIES, *NEW	W72-07890
*AERATED LAGOONS, SURFACE	AERATORS.:	W71-07109
G, DEPTH, DISTRIBUTION PATTERNS,	AERIAL PHOTOGRAPHY, REMOTE SENSIN	W72-09120

N, WATER POLLUTION, *MONITORING,	AERIAL PHOTOGRAPHY, *INFRARED RAD	W72-12487
S, INVERTEBRATES, AQUATIC ALGAE,	AERIAL PHOTOGRAPHY, PHOSPHORUS, N	W72-12729
CENCE, SELENASTRUM CAPRICONUTUM,	AEROBACTER AEROGENES.: /IOLUMINES	W72-04292
THERMOCLINE, ANAEROBIC BACTERIA,	AEROBIC BACTERIA, SULFUR BACTERIA	W72-12996
IFICATION, LABORATORY EQUIPMENT,	AEROBIC BACTERIA, SOLVENT EXTRACT	W73-01066
	AEROBIC BIOLOGICAL DEGRADATION.:	W72-05315
HOSPHATES, ANAEROBIC CONDITIONS,	AEROBIC CONDITIONS, METABOLISM, S	W72-04789
NUTRIENTS, ANAEROBIC CONDITIONS,	AEROBIC CONDITIONS, BIOCHEMICAL O	W71-07100
GRADATION, ANAEROBIC CONDITIONS,	AEROBIC CONDITIONS, SLUDGE, LIGHT	W71-07106
, *SLUDGE, ANAEROBIC CONDITIONS,	AEROBIC CONDITIONS, PILOT PLANTS,	W71-07102
EN, ALGAE, ANAEROBIC CONDITIONS,	AEROBIC CONDITIONS, BIODEGRADATIO	W71-07099
YGENATION, ANAEROBIC CONDITIONS,	AEROBIC CONDITIONS, BIOCHEMICAL O	W71-07109
SYNTHESIS, ANAEROBIC CONDITIONS,	AEROBIC CONDITIONS, SLUDGE, WASTE	W71-07084
L, *LAGOONS, *OXIDATION LAGOONS,	AEROBIC CONDITIONS, ANAEROBIC CON	W71-07551
STE WATER TREATMENT, NUTIRIENTS,	AEROBIC CONDITIONS, AERATION, MIC	W71-04079
OGEN, WATER QUALITY, PHOSPHORUS,	AEROBIC CONDITIONS, ANAEROBIC CON	W70-10405
SOURCES, ENVIRONMENTAL EFFECTS,	AEROBIC CONDITIONS, CHLOROPHYTA,	W73-01066
ATMENT, CHLOROPHYTA, CYANOPHYTA,	AEROBIC CONDITIONS, BIOASSAY, GRO	W72-07664
ENT, AERATION, DISSOLVED OXYGEN,	AEROBIC CONDITIONS, BIOMASS, RESP	W72-11100
HOTOSYNTHESIS, DISSOLVED OXYGEN,	AEROBIC CONDITIONS, ANAEROBIC CON	W72-11642
PONDS, *ALGAE, LABORATORY TESTS,	AEROBIC CONDITIONS, ANAEROBIC CON	W71-13326
EATMENT, ACTIVATED SLUDGE, LIME,	AEROBIC TREATMENT, DOMESTIC WASTE	W72-09176
COAGULATION, *OXIDATION LAGOONS,	AEROBIC TREATMENT, OXIDATION, ORG	W71-09629
TRY, NITROGEN, LAGOONS, PUMPING,	AEROBIC TREATMENT, BIOCHEMICAL OX	W72-08396
LAGOONS, *STABILIZATION, *ALGAE,	AEROBIC TREATMENT, BACTERIA, MUNI	W71-03896
, NITROGEN, ANAEROBIC DIGESTION,	AEROBIC, ALKALINITY, ACIDITY, LAB	W72-02850
NASTRUM CAPRICONUTUM, AEROBACTER	AEROGENES.: /IOLUMINESCENCE, SELE	W72-04292
CALIGENES FAECALIS, ENTEROBACTER	AEROGENES, PROTEUS VULGARIS, PSEU	W72-07664
, BACILLUS, VIBRIO, RHODOTORULA,	AEROMONAS.: / GROWTH, MICROCOCCUS	W72-10861
LATION, SOUTH WALES, MICROCYSTIS	AERUGINOSA, ANABAENA.: /CUOLE DEF	W72-11564
S, PROTEUS VULGARIS, PSEUDOMONAS	AERUGINOSA, SERRATIA: /R AEROGENE	W72-07664
MENON, ANABAENA SPP, MICROCYSTIS	AERUGINOSA, APHANIZOMENON HOLSATI	W72-07890
ROCYSTIS AERUGINOSA, PSEUDOMONAS	AERUGINOSA, LIMITING NUTRIENTS.: /	W73-01657
A, LIMITING NUTRIEN/ MICROCYSTIS	AERUGINOSA, PSEUDOMONAS AERUGINOS	W73-01657
BETA STREPTOCOCCUS, PSEUDOMONAS	AERUGINOSA, MYCOBACTERIUM FURTUIT	W72-04259
RYPTOMONAS PULSILLA, MICROCYSTIS	AERUGINOSA, CHROOMONAS CAUDATA, C	W72-13816
, *NUTRIENTS, *ALGAE, *BACTERIA,	AESTHETICS, COLOR, TASTE, CORROSI	W71-09958
BETATA F. SEMISPINA, CHAETOCEROS	AFFINIA,: /SSIMA, RHIZOSCLENIA HE	W72-03543
CARPIO, MOSQUITO FISH, GAMBUSIA	AFFINIS, CESIUM RADIOISOTOPES, CL	W73-01673
LENA, CRYPTOPHYCEAE, PONTOPOREIA	AFFINIS, ESOX LUCIUS, PERCA FLUVI	W72-12729
UNDIFOLIA, MILORGANITE, GAMBUSIA	AFFINIS, PRODUCTIVITY MEASUREMENT	W71-01488
EUCAMPIA ZOODIACUS, CHAETOCEROS	AFFINIS, SKELETONEMA COSTATUM, CH	W73-00442
AEOLOSOMA, CHAOBORUS, *GAMBUSIA	AFFINS, EPHEMEROPTERA, MENIDIA, B	W72-07132
WINDHOEK, SOUTH	AFRICA.:	W70-10381
UCTION, FUNGI, SOILS, SYMBIOSIS,	AFRICA, ASIA, ANABAENA.: /OP PROD	W73-02473
LYTICAL TECHNIQUES, FOOD CHAINS,	AFRICA, EFFLUENTS, PATH OF POLLUT	W73-00832
M, CRYPTOMONAS EROSA, PERIDINIUM	AFRICANUM, NITZCHIA ACICULARIS, S	W73-00256
*AXEMIC CULTURES, *DIEOFF RATES,	AFTERGROWTH, BLUE-GREEN ALGAE, FI	W70-08319
*MIXED CULTURES, DIEOFF RATES,	AFTERGROWTH.:	W71-05155
IODINE, ATOMIC EN/ FODDER, GLUE,	AGAR, POLYSIPHONIA, ANTIBIOTICS,	W71-09157
SITY, MICROAMMETER, OSCILLATORIA	AGARDHII VAR ISOTHRIX, LAKE ITASC	W72-07145
S, LAMINARIA DIGITATA, LAMINARIA	AGARDHII, ULVA LACTUCA, RHODYMENI	W72-04748
IA SP., GELIDIUM SP., CHLORELLA,	AGARS.: /ADOPHORA SP., POLYSIPHON	W71-11851
NGUILLARUM, ENRICHMENT CULTURES,	AGARS, CHEMOSTATS, NITROSOMONAS,	W72-05421
LONGICRURIS, LAMINARIA DIGITATA,	AGARUM CRIBROSUM, *NOVA SCOTIA, *	W73-00366
, GRAZING, GASTROPODS, LITTORAL,	AGE, BIOMASS, LIGHT QUALITY, LIGH	W72-04766
TMENT FACILITIES, ADMINISTRATIVE	AGENCI: / STATE GOVERNMENTS, TREA	W72-01659
CATION PROCEDURE, ADMINISTRATIVE	AGENCIES, ADMINISTRATION, ALGAE,	W72-01093
POSED), ENVIRONMENTAL PROTECTION	AGENCY.: *LEGISLATION(PRO	W73-00703
UTION, *ENVIRONMENTAL PROTECTION	AGENCY.: *MERCURY POLL	W72-14782
ENESS/ *DISPERSANTS, *COLLECTING	AGENTS, *SINKING AGENTS, EFFECTIV	W71-09789
TS, *COLLECTING AGENTS, *SINKING	AGENTS, EFFECTIVENESS, TESTING PR	W71-09789
YSIS, *ALGAL CULTURES, CHELATING	AGENTS, OOCYSTIS MARSSONII.: /NAL	W73-02112
NUTRIENTS, OXYGEN REQUIREMENTS,	AGING(BIOLOGICAL), AQUATIC ALGAE,	W72-01685
AKES, GREAT LAKES, ALGAL BLOOMS,	AGING(BIOLOGICAL), LAKE STAGES, M	W72-14782
L, *NON-STRUCTURAL ALTERNATIVES,	AGING(BIOLOGICAL), AQUATIC ALGAE,	W72-04279
UTANTS, METABOLITES, DUNALIELLA,	AGMENELLUM QUADRAPLICATUM, C-14,	W73-00361
OTODIELDRIN, ANACYSTIS NIDULANS,	AGMENELLUM QUADRUPLICATUM STRAIN	W72-03545
ION CONTROL, *BIOLOGICAL TREATM/	AGRICULTURAL WASTES, WATER POLLUT	W72-02975
N, DESALINATION, DRAINAGE WATER,	AGRICULTURAL CHEMICALS, FARM WAST	W71-11698
R POLLUTION EFFECTS, PESTICIDES,	AGRICULTURAL CHEMICALS, PESTICIDE	W71-12303
N, AQUATIC WEED CONTROL, RIVERS,	AGRICULTURE, ROOTED AQUATIC PLANT	W71-13172
TION, WATER QUALITY, STREAMFLOW,	AGRICULTURE, INDUSTRIES, WASTE TR	W71-11017
OURCES, WATER POLLUTION EFFECTS,	AGRICULTURAL RUNOFF, PHOSPHORUS,	W72-10625
GICIDES, HERBICIDES, IRRIGATION,	AGRICULTURE, RICE, CROP RESPONSE.	W72-14807
STICIDE TOXICITY, BIOINDICATORS,	AGRICULTURAL CHEMICALS, FISH, CHE	W73-01976
COLLECTIONS, INDUSTRIAL WASTES,	AGRICULTURAL RUNOFF, DOMESTIC WAS	W73-01972
OXYGEN DEMAND, ECONOMICS, ALGAE,	AGRICULTURAL: /NING, BIOCHEMICAL	W72-04734
NTATION, WATER QUALITY, SILTING,	AGRICULTURAL CHEMICALS, WATER POL	W72-05869
TS, COLIFORMS, MUNICIPAL WASTES,	AGRICULTURE, FALLOUT, DRAINAGE, M	W71-03012
WISCONSIN, LAKES, URBAN RUNOFF,	AGRICULTURAL RUNOFF, FARM WASTES,	W73-02479
ERIE, LAKE ONTARIO, ADSORPTION,	AGRICULTURAL RUNOFF, ALGAE, FOOD	W72-13657
ENERGY CONSUMPTION, MODEL TEST/	AGRO-INDUSTRIAL COMPLEX, ELECTRIC	W72-03327
WASTE M/ *ENVIRONMENTAL QUALITY,	AIR CONTAMINANT, STREAM QUALITY,	W71-08214
*KEZAR LAKE,	AIR INJECTION, COMPRESSED AIR.:	W72-09304
SH, BAYS, AIR POLLUTION EFFECTS,	AIR POLLUTION, WATER POLLUTION, L	W72-03347
LASKA, PONDS, WELLS, FISH, BAYS,	AIR POLLUTION EFFECTS, AIR POLLUT	W72-03347
EAD RADIOISOTOPES, HEAVY METALS,	AIR POLLUTION EFFECTS, SURFACE WA	W72-00940
MONITORING, PATH OF POLLUTANTS,	AIR POLLUTION EFFECTS, WATER POLL	W72-00959
STEMS ANALYSIS, WASTE TREATMENT,	AIR POLLUTION, LEGISLATION, BENTH	W71-13723

ITY, STANDARDS, WATER POLLUTICN, AIR POLLUTION, ODOR, CONFINEMENT W71-08214
L CONDUCTANCE, DISSOLVED OXYGEN, AIR POLLUTION, CHEMICAL ANALYSIS, W71-01489
TAL EFFECTS, FEDERAL GOVERNMENT, AIR POLLUTION, WATER POLLUTICN, T W73-00703
AKES, SOYBEANS, LIGHT INTENSITY, AIR TEMPERATURE, SAMPLING, AQUATI W73-01456
LAKE, AIR INJECTICN, COMPRESSED AIR.: *KEZAR W72-09304
PONDS, WATER POLLUTION CONTRCL, AIRCRAFT, COPPER SULFATE.: /AKES, W72-14673
, *SEA WATER, *OILS, *NUTRIENTS, AIRCRAFT, SATELLITES(ARTIFICIAL), W72-04742
UNINERVIS, CASSIOPEA ANDROMEDA, AKABARIA, STROMBUS TRICORNI: /ERA W73-00855
OBATA, TROITSKIELLA TRIANGULATA, AKIYAMAMONAS TERRESTRIS, BRASILOB W72-07683
NTENSITY, EUTROPHICATION, MARINE AL: /ERMOCLINE, NITRATES, LIGHT I W71-07875
TIVITY, ZINC, WATER TEMPERATURE, AL: /GEN, HARDNESS(WATER), CONDUC W72-07890
SA, SODIUM SILIC/ POTOMAC RIVER, ALAFIA RIVER, CHLORELLA PYRENOIDO W72-04734
TATE, METHIONINE, PHENYLALANINE, ALANINE, ISOLEUCINE, VALINE, LYSI W73-00942
RIBUTION, COLD BAY, BERMERS BAY, ALARIA, NEREOCYSTIS, LAMINARIA, T W72-09120
LINDRUS, PACIFIC OCE/ COOK INLET(ALASKA), CYCLOTELLA NANA, LEPTOCY W71-11008
TER TEMPERATURE, MARINE ANIMALS, ALASKA, BIOMASS, SALINITY, BENTHI W73-00374
DIOXIDE, *PRIMARY PRODUCTIVITY, ALASKA, FIORDS, SURFACE WATERS, N W72-09122
IOACTIVITY EFFECTS, UNDERGROUND, ALASKA, PONDS, WELLS, FISH, BAYS, W72-03347
WASHINGTON, OREGON, CALIFORNIA, ALASKA, SEA WATER.: /NTS, CANADA, W73-00428
COCCOLITHUS) HUXLEYI, AMPHIPRORA ALATA, CHRYSOCHROMULINA, SKELETON W73-00380
OSIRA, CHAETOCEROS, RHIZOSOLENIA ALATA, PERIDINIUM DEPRESSUM, PERI W71-07875
SKY, ZOOGLEA RAMIGERA, BEGGIATOA ALBA, BEGGIATOA, THIOTHRIX NIVEA, W72-07896
, STAPHYLOCOCCUS AUREUS, CANDIDA ALBICANS.: /COBACTERIUM FURTUITUM W72-04259
, COREGONUS LAVARETUS, COREGONUS ALBULA, WHITEFISH, BRANCHIURA: /T W72-12729
RUTHENIUM, FIXATION, TISSUES, ALBURNUS LUCIDUS, BELGIUM.: W71-09168
VULGARIS, SCENEDESMUS OBLIQUUS, ALCALIGENES FAECALIS, ENTEROBACTE W72-07664
ACIDS, VITAMINS, CARBOHYDRATES, ALCOHOLS, POLLUTANT IDENTIFICATIO W73-00942
ES, MERGUS MERGANSER, MEGACERYLE ALCYON, BIOACCUMULATION.: / SAMPL W72-12930
ORAL BENTHIC MACROPHYTES, RIA DE ALDAN(SPAIN), LAMINARIA OCHROLEUC W72-04766
, LEPTOCYLINDRUS, PACIFIC OCEAN, ALEUTIAN ISLANDS, PRYMNESIUM PARV W71-11008
ARIA, TABELLARIA/ *INTERFERENCE, ALEWIVES, DICHOTOMOSIPHON, FRAGIL W73-01669
HEMICAL OXYGEN DEMAND, BACTERIA, ALG: /CALS, CHEMICAL WASTES, BIOC W73-01976
STERS, MUSSELS, PLANKTON, MARINE ALG: /NT, REGULATION, FALLOUT, OY W72-00959
TAL EFFECTS, CHLOROPHYTA, MARINE ALGAE , SALT TOLERANCE, CRUSTACEA W73-00855
RMENTATION, ALKALINE HYDROLYSIS, ALGAE BLOOMS, UTAH, GREAT SALT LA W71-12223
STERS, DINOFLAGELLATES, DIATOMS, ALGAE CULTURES, COPPER, TOXICITY. W72-04708
IS NIDULANS, AQUATIC ECOSYSTEMS, ALGAE GROWTH, ALGAL POPULATIONS.: W71-00475
TION, FILTER / *NITRATE REMOVAL, ALGAE STRIPPING, POND DENITRIFICA W71-08223
ORY TESTS, *MEASUREMENT, *MARINE ALGAE TRACERS, CHEMICAL PRECIPITA W72-12210
MICROSCOPY, CHRYSOPHYTA, AQUATIC ALGAE.: *DIATOMS, * W72-11738
HETEROCYSTS, BENTHIC BLUE-GREEN ALGAE.: *WINDERMERE(ENGLAND), W73-02472
*RHODOPHYTA, *LAKE ERIE, OHIO, ALGAE.: W71-05630
SSAYS, ENZYMES, LIGHT INTENSITY, ALGAE.: *CHLORELLA, *BIOA W71-11916
SEDIMENT PRODUCTION, *BLUE-GREEN ALGAE.: *SUSPENDED W72-02218
*CARBON, *PHOSPHORUS, NITROGEN, ALGAE.: *EUTROPHICATION, W72-09155
*ODOR-PRODUCING ALGAE.: W72-11604
THODS, SEA WATER, COASTS, MARINE ALGAE.: /), MAINE, STATISTICAL ME W73-00369
ENOPHYTA, PYRROPHYTA, FRESHWATER ALGAE.: /, AQUATIC HABITATS, EUGL W72-09119
DISSOLVED OXYGEN, LAKE MICHIGAN, ALGAE.: /, PESTICIDES, TCXICITY, W71-12092
TORS, NITROGEN, PLANKTON, DEPTH, ALGAE.: /, NITRATES, LIMITING FAC W73-02472
ABITATS, CLASSIFICATION, AQUATIC ALGAE.: /, SCENEDESMUS, AQUATIC H W72-12744
, PRODUCT EVALUATION, BLUE-GREEN ALGAE.: /ATIC WEED STUDY FACILITY W71-06102
R POLLUTION, HAZARDS, NUTRIENTS, ALGAE.: /CY, *PUBLIC HEALTH, WATE W72-04266
BIOMASS, ON-SITE INVESTIGATIONS, ALGAE.: /CAL ANALYSIS, SEASCNAL, W73-00242
PLANKTON, ANALYTICAL TECHNIQUES, ALGAE.: /ERATURE, WATER QUALITY, W72-07663
CHNIQUES, SEDIMENTS, OILY WATER, ALGAE.: /ECHNIQUES, SEPARATION TE W71-12467
TIC PRODUCTIVITY, PHYTOPLANKTON, ALGAE.: /ERFACES, SEDIMENTS, AQUA W72-08989
S, LABORATORY TESTS, CYANOPHYTO, ALGAE.: /ES, DENSITY, GROWTH RATE W72-03217
NTS, STATIONS, DATA COLLECTIONS, ALGAE.: /FIC OCEAN, BOTTOM SEDIME W72-03567
PHYTOPLANKTON, FUNGI, BACTERIA, ALGAE.: /IOLOGY, WATER TREATMENT, W72-10076
*LAKE ERIE, HYPOLIMNION, AQUATIC ALGAE.: /IMENT-WATER INTERFACES, W72-06690
OLLUTION SOURCES, NITRIFICATION, ALGAE.: /LLUTION EFFECTS, WATER P W71-05390
ND, HYPOLIMNION, AQUATIC PLANTS, ALGAE.: /LVED OXYGEN, OXYGEN DEMA W72-12994
Y, CARBON DIOXIDE, SALTS, MARINE ALGAE.: /M, TRACERS, RADIOACTIVIT W72-12634
NDS, CHLORELLA, TRACERS, AQUATIC ALGAE.: /M, FUNGI, ORGANIC COMPOU W72-14326
PROPERTIES, SOIL BACTERIA, SOIL ALGAE.: /MPERATURE, SOIL CHEMICAL W72-03342
Y, INHIBITION, UREAS, SYNTHESIS, ALGAE.: /ORY TESTS, AMMONIA, ASSA W72-03222
A, TEMPERATURE, SIZE, FECUNDITY, ALGAE.: /ODS, DIATOMS, CHLOROPHYT W72-14805
NSECTS, AQUATIC ANIMALS, AQUATIC ALGAE.: /OMS, PROTOZOA, AQUATIC I W72-07791
AEOPHYTA, WATER QUALITY, *MARINE ALGAE.: /ODUCTION, RHODOPHYTA, PH W73-01449
TIC ALGAE, MARINE ALGAE, SESSILE ALGAE.: /PHYTA, CHLOROPHYTA, AQUA W73-00273
MATICS, CLASSIFICATION, *AQUATIC ALGAE.: /PHYTA, *CULTURES, *SYSTE W72-12172
RATION, INSECTS, EUTROPHICATION, ALGAE.: /PAGE, PERCOLATION, EVAPO W71-07124
SYNTHETIC OXYGENATION, ACTIVATED ALGAE.: /RESPIRATORY RATES, PHOTO W71-04079
S, NAVICULA, NITZSCHIA, EPIPYTIC ALGAE.: /S, FRAGILARIA CONSTRUENT W72-07501
RE, ORGANOPHOSPHOROUS COMPOUNDS, ALGAE.: /SOLVED OXYGEN, TEMPERATU W71-10645
ARBOHYDRATES, CYCLING NUTRIENTS, ALGAE.: /TY, ORGANIC COMPOUNDS, C W73-02100
LITY, HERBICIDES, ODOR-PRODUCING ALGAE.: /UTION EFFECTS, WATER QUA W72-14692
ASSAY, MICROORGANISMS, *EUGLENA, ALGAE.: /YSIOLOGICAL ECOLOGY, BIO W71-12303
(ALGAE), PHOSPHORUS REQUIREMENTS(ALGAE), NUTRIENT CHEMISTRY(AQUATI W70-04080
LGAE), NU/ NITROGEN REQUIREMENTS(ALGAE), PHOSPHORUS REQUIREMENTS(A W70-04080
RADIOE/ *RADIOISOTOPES, *MARINE ALGAE, *ABSORPTION, *MINE WASTES, W72-12204
*FILTRATION, *SEWAGE TREATMENT, ALGAE, *AERATION, SEDIMENTATION, W72-14469
EED C/ *EUTROPHICATION, *AQUATIC ALGAE, *ALGAL CONTROL, *AQUATIC W W72-14692
AL, CRUSTACEANS, ISOPODS, MARINE ALGAE, *ALASKA.: /NNELIDS, SEASON W73-00374
CLADOPHORA.: *FILAMENTCUS ALGAE, *AQUATIC ANGIOSPERM WEEDS, W72-14692
S PLANTS, *CYANOPHYTA, *NUISANCE ALGAE, *AQUATIC ALGAE, *CRUSTACEA W72-14056
HT, TURBIDIT/ *BIOMASS, *AQUATIC ALGAE, *AQUATIC PRODUCTIVITY, LIG W72-13142
CHANICAL CONTROL/ *HARVESTING OF ALGAE, *AQUATIC WEED CONTROL, *ME W71-06188
*PLANKTON, *RESERVOIRS, *AQUATIC ALGAE, *AQUATIC ANIMALS, *BIOMASS W72-05968

RODUCTIVITY, *ENERGY CO/ *MARINE ALGAE, *BENTHIC FLORA, *PRIMARY P W72-06283
NT GROWTH, *CHLOROPHYTA, *MARINE ALGAE, *BIOASSAY, REPRODUCTION, S W72-14294
LLUTION EFFECTS, *ALGAE, *MARINE ALGAE, *BIOINDICATORS, *PHOTOSYNT W71-10553
NTROL, CYANOPHYTA, CA/ *NUISANCE ALGAE, *CARBON DIOXIDE, *ALGAL CO W72-10607
, *DEFICIENT NUTRIENTS, *AQUATIC ALGAE, *CARBON DIOXIDE, NUISANCE W72-07514
NTS, MARINE/ *SEA WATER, *MARINE ALGAE, *CARBOHYDRATES, MARINE PLA W73-02100
BERS, *LIFE CYCLES, CU/ *AQUATIC ALGAE, *CHLOROPHYTA, *GROWTH CHAM W72-07144
E/ *OCEANS, *MATABOLISM, *MARINE ALGAE, *CHLORINATED HYDROCARBON P W72-12998
*GROWTH RATES, *INHIBI/ *AQUATIC ALGAE, *COMPETITION, *NUTRIENTS, W72-12734
*PRIMARY PRODUCTIVITY, *AQUATIC ALGAE, *COMPARATIVE PRODUCTIVITY, W73-01080
PHYTA, *NUISANCE ALGAE, *AQUATIC ALGAE, *CRUSTACEANS, *FISH TOXINS W72-14056
PHAEOPHY/ *SYSTEMATICS, *MARINE ALGAE, *CYANOPHYTA, *CHLOROPHYTA, W72-11800
COLOGICAL DISTRIBUTION, *AQUATIC ALGAE, *CYANOPHYTA, *BEGGIATOA, P W72-09061
L, LABORATORY TESTS, / *NUISANCE ALGAE, *CYANOPHYTA, *ALGAL CONTRO W71-13184
ONS, BIODEGRADATION, *NUTRIENTS, ALGAE, *DEGRADATION(STREAM), WATE W71-07973
S,/ *TOXICITY, *AMMONIA, *MARINE ALGAE, *DIATOMS, CHEMICAL ANALYSI W71-11008
ALGAE, RESOURCES, FOOD/ *MARINE ALGAE, *ECONOMICS, *PRODUCTIVITY, W71-13246
AND, HYDROGEN ION CONCENTRATION, ALGAE, *ENTROPHICATION, NEUTRALIZ W70-10427
OPHYTE-EPIPHTYE META/ *EPIPHYTIC ALGAE, *EPIPHYTIC BACTERIA, *MACR W72-03216
ER(ENGLAND), AMPHOR/ *EPIPSAMMIC ALGAE, *EPIPELIC ALGAE, SHEAR WAT W71-12855
C PLANTS, *DISTRIBUTION, *MARINE ALGAE, *ESTUARIES, PRODUCTIVITY, W72-04766
ATMENT, *ALGAL CONTROL, *AQUATIC ALGAE, *EUTROPHICATION, MUNICIPAL W72-03970
TRIENTS, *NITROGEN, *PHOSPHORUS, ALGAE, *EUTROPHICATION, AQUATIC E W71-08953
*CYANOPHYTA, *NUISANCE ALGAE, *EUTROPHICATION, REVIEWS.: W72-01358
TER POLLUTION CONTROL, *NUISANCE ALGAE, *EUTROPHICATION, LAKES, WA W72-00799
S, *ALGAL NUTRIENTS, *CHLORELLA, ALGAE, *EUTROPHICATION, NEW MEXIC W72-13992
FAR EXPLOSIONS, *ALASKA, *MARINE ALGAE, *FISHERIES, FISH POPULATIO W73-00764
OTOPES, *MARINE ANIMALS, *MARINE ALGAE, *FOOD CHAINS, ABSORPTION, W73-00819
POWERP/ *BENTHIC FLORA, *MARINE ALGAE, *GULF OF MEXICO, *ELECTRIC W72-11252
RATES, GROWTH AUTOSTIMUL/ *GREEN ALGAE, *HORMOTILA BLENNISTA, FILT W71-08687
C/ *BENTHOS, *SAMPLING, *SESSILE ALGAE, *INVERTEBRATES, ST. LAWREN W72-05432
OLIDS, F/ *ALGAE, *HARVESTING OF ALGAE, *ION EXCHANGE, DISSOLVED S W71-00138
*CULTURES, *CYANOPHYTA, *MARINE ALGAE, *ISOLATION, MARINE PLANTS, W73-00430
NDUSTRIAL WASTES, INVERTEBRATES, ALGAE, *KELP.: /, HEAVY METALS, I W72-11854
LOGY, PHAEOPHYTA, STAND/ *MARINE ALGAE, *KELPS, *PHYSIOLOGICAL ECO W72-12940
AL WASTE, *GROWTH RATES/ *MARINE ALGAE, *LABORATORY TESTS, *CHEMIC W72-12239
PHYTON, *PHYTOPLANKTON, *SESSILE ALGAE, *LAKE SUPERIOR, LIMNOLOGY, W72-12192
T LAKES, *LAKES, EUTROPHICATION, ALGAE, *LAKE SUPERIOR, *LAKE MICH W72-01094
ENSITY STRATIFICATION,/ *AQUATIC ALGAE, *LAKES, *PHYTOPLANKTON, *D W71-09852
TOPLANKTON, NATURAL RES/ *MARINE ALGAE, *LIGHT, *GROWTH RATES, PHY W71-13252
NSAS, MODEL STUDIES, ECOSYSTEMS, ALGAE, *LIGHT QUALITY, EUTROPHIC W72-12393
OPLANKTON, *ABSORPTION, *AQUATIC ALGAE, *LIGHT INTENSITY, *PHOSPHO W72-07166
ISH, M/ *NUCLEAR WASTES, *MARINE ALGAE, *MARINE ANIMALS, *MARINE F W73-00832
POLLUTION, *THERM/ BAYS, *MARINE ALGAE, *MARINE ANIMALS, *THERMAL W72-11876
*RADIOACTIVITY TECHNIQUES, SOIL ALGAE, *METHODOLOGY, MEASUREMENT, W72-14301
IS, *TRACE ELEMENTS, C/ *AQUATIC ALGAE, *NEUTRON ACTIVATION ANALYS W71-11515
GAE, STATISTICAL METHODS, MARINE ALGAE, *OREGON, *SEASONAL, CHRYSO W72-08141
IOISOTOPES, *ABSORPTION, *MARINE ALGAE, *PHYTOPLANKTON, BACTERIA, W73-00802
*EUTRO/ *PHYTOPLANKTON, *AQUATIC ALGAE, *PHOSPHORUS, *CYANOPHYTA, W71-10645
S, *R/ *AQUATIC PLANTS, *AQUATIC ALGAE, *PHYTOPLANKTON, *RESERVOIR W72-04855
ONDS, WATER POLLUTION / *AQUATIC ALGAE, *RADIOACTIVITY EFFECTS, *P W71-09256
, TEMPERATURE, STRONTI/ *AQUATIC ALGAE, *RADIOISOTOPES, ABSORPTION W71-09250
PATTERNS, *CHLOROPHYTA, *MARINE ALGAE, *RHODOPHYTA, *SYSTEMATICS, W73-00284
*BACTERIA, FLORIDA, RAD/ *MARINE ALGAE, *SALT MARSHES, *PROTOZOA, W73-00796
TION EFFECTS, *PROTOZOA, *MARINE ALGAE, *SALT MARSHES, NEW YORK, N W72-00948
XIDATION LAGOONS, *HARVESTING OF ALGAE, *SEPARATION TECHNIQUES, CH W72-10237
IEWS, B/ *PHYTOPLANKTON, *MARINE ALGAE, *SEDIMENTATION RATES, *REV W72-06313
PATTERNS, *SYSTEMATICS, *AQUATIC ALGAE, *SOIL BACTERIA, SAMPLING, W73-00854
YIELD, *SIMULATED RAINFAL/ *SOIL ALGAE, *SOIL SURFACES, *SEDIMENT W72-02218
*ZINC RADIOISOTOPES, AB/ *MARINE ALGAE, *STRONTIUM RADIOISOTOPES, W72-12203
Y, *INFRARED RADIATION, *AQUATIC ALGAE, *SURFACE WATERS, ANALYTICA W71-12034
THIC FLORA, RHODOPHYTA,/ *MARINE ALGAE, *SYSTEMATICS, *COASTS, BEN W71-10066
N PATT/ *PHYTOPLANKTON, *AQUATIC ALGAE, *SYSTEMATICS, *DISTRIBUTIO W73-01094
*CYANOPHYTA, ALGAE, *VIRUSES.: W72-03188
*NITRATES, TREATMENT FACILITIES, ALGAE, *WASTE WATER TREATMENT, *A W72-02975
*PHOTOSYNTHESIS, *OIL P/ *MARINE ALGAE, *WATER POLLUTION EFFECTS, W73-01074
NG, COBALT RADIOISOTOPES, MARINE ALGAE, ABSORPTION, SEDIMENTS, WAS W73-00831
, CLAMS, CARP, CRAYFISH, AQUATIC ALGAE, ABSORPTION, AQUARIA, ANIMA W72-06342
*NORTHWESTERN ONTARIO, EPILITHIC ALGAE, ACHNANTHES MINUTISSIMA, AC W71-11029
AS CHROMATOGRAPHY, SEPCTROSCOPY, ALGAE, ACTINOMYCETES, FUNGI, LABO W72-11604
-PRODUCING ALGAE, ODOR-PRODUCING ALGAE, ACTINOMYCETES, PHOSPHATES. W73-02144
E, *WATER QUALITY, WATER SUPPLY, ALGAE, ACTIVATED CARBON, ADSORPTI W70-10381
ANIMALS, INVERTEBRATES, AQUATIC ALGAE, AERIAL PHOTOGRAPHY, PHOSPH W72-12729
EMICAL OXYGEN DEMAND, ECONOMICS, ALGAE, AGRICULTURAL: /NING, BIOCH W72-04734
AL, INVERTEBRATES, ALGAE, MARINE ALGAE, ALGAL POISONING, CHROMATOG W71-08597
NISMS, AQUATIC BACTERIA, AQUATIC ALGAE, ALGAE, CHLOROPHYTA, PSEUDO W72-12255
, ENVIRONMENTAL EFFECTS, AQUATIC ALGAE, ALGAL CONTROL, ALGAL POISO W72-11186
*ENVIRONMENTAL EFFECTS, DIQUAT, ALGAE, ALGAL CONTROL, DIATOMS, PE W72-00155
ON, WATER TEMPERATURE, BACTERIA, ALGAE, ALKALINITY, IRON, PHOSPHAT W71-05626
OZYME, *SPHEROPLASTS, BLUE-GREEN ALGAE, ANABAENA VARIABILIS, ANACY W71-08683
, RE-AERATION, DISSOLVED OXYGEN, ALGAE, ANAEROBIC CONDITIONS, AERO W71-07099
ATION, POLLUTANT IDENTIFICATION, ALGAE, ANALYTICAL TECHNIQUES, *UL W72-01780
BUOYS, HARBORS, TUNNELS, MARINE ALGAE, ANIMAL POPULATIONS, LOBSTE W73-02029
*SYSTEMATICS, *ECOLOGY, *MARINE ALGAE, ANNELIDS, MOLLUSKS, RHODOP W73-00932
CONTROL, ASSAY, REMOTE SENSING, ALGAE, ANTIBIOTICS, MERCURY, ARSE W72-10162
ION, PHOSPHORUS, EUTROPHICATION, ALGAE, AQUATIC WEEDS, PHYTOPLANKT W72-08459
ENTS, *POLLUTANT IDENTIFICATION, ALGAE, AQUATIC BACTERIA, SEWAGE B W72-01472
LGAE, DDT, PHYTOPLANKTON, MARINE ALGAE, AQUATIC ALGAE, COLORIMETRY W72-03545
S, PHOSPHORUS COMPOUNDS, POTASH, ALGAE, AQUATIC PLANTS, CHEMICAL A W71-12252
HLOROPHYTA, SCENEDESMUS, AQUATIC ALGAE, AQUATIC PRODUCTIVITY, POLL W73-01066

DOMESTIC WASTES, ECOLOGY, FISH, ALGAE, AQUATIC PLANTS, PROTOZOA, W73-01972
MS, AQUATIC ENVIRONMENT, AQUATIC ALGAE, AQUATIC ANIMALS, AQUATIC I W72-07132
UATIC HABITATS, BIOMASS, AQUATIC ALGAE, AQUATIC ANIMALS, BALANCE O W72-06340
PLANKTON, DIATOMS, ALGAE, MARINE ALGAE, AQUATIC ALGAE, EUGLENA, CH W72-08436
2-4-D, PLANT GROWTH REGULATORS, ALGAE, AQUATIC ALGAE, CHLORELLA.: W72-08440
ER ANALYSIS, POLLUTANTS, AQUATIC ALGAE, AQUATIC MICROORGANISMS, AQ W72-07896
RE, DISSOLVED OXYGEN, COMPUTERS, ALGAE, AQUATIC LIFE, ENVIRONMENTA W72-14405
ICATION, *BIOMASS, *HYDROGRAPHY, ALGAE, AQUATIC ALGAE, OLIGOTROPHY W72-13871
*BIOLOGICAL COMMUNITIES, AQUATIC ALGAE, AQUATIC PRODUCTIVITY, BIOC W73-00002
BILITY, WATER POLLUTION EFFECTS, ALGAE, AQUATIC HABITATS, AQUATIC W71-07337
OLLUSKS, SNAILS, MUSSELS, MARINE ALGAE, AQUATIC ANIMALS, FRESH WAT W71-09182
, TURBIDITY, RUNOFF, PHOSPHORUS, ALGAE, AQUATIC PLANTS, CLAYS, ODO W72-04759
IVES, AGING(BIOLOGICAL), AQUATIC ALGAE, AQUATIC PRODUCTIVITY, FISH W72-04279
RPTION, FISH, INDUSTRIAL WASTES, ALGAE, AQUATIC PLANTS, SEA WATER, W72-05952
PHORUS, PULP AND PAPER INDUSTRY, ALGAE, AQUATIC WEEDS, RECREATION, W70-05103
TURE, POWERPLANTS, DISTILLATION, ALGAE, AQUICULTURE, FISH, SHELLFI W70-08832
PESTICIDES, SALINITY, NUTRIENTS, ALGAE, AQUATIC ANIMALS, RADIONUCL W72-12190
IRONMENTAL EFFECTS, METHODOLOGY, ALGAE, AQUATIC PLANTS, ISOTOPE FR W72-12709
ES, PESTICIDE DRIFT, METABOLISM, ALGAE, AQUATIC ALGAE, CHRYSOPHYTA W73-02280
GRICULTURAL RUNOFF, FARM WASTES, ALGAE, AQUATIC WEEDS, FERTILIZATI W73-02479
ENT, SLUDGE TREATMENT, TOXICITY, ALGAE, BACTERIA, FISH, SEWAGE SLU W70-06968
ITRIFICATION, NUTRIENTS, SLUDGE, ALGAE, BACTERIA, *WASTE WATER TRE W72-04086
*EUTROPHICATION, LAKES, AQUATIC ALGAE, BACTERIA, BIOCHEMISTRY, WI W72-04292
LOADING, ODORS, PHOTOSYNTHESIS, ALGAE, BACTERIA, CHLORINATION, GR W71-07123
TES, HYDROGEN ION CONCENTRATION, ALGAE, BACTERIA, WATER POLLUTION W71-07942
POUNDS, *NUTRIENTS, FRESH WATER, ALGAE, BACTERIA, NITROGEN, PHOSPH W72-07933
STRY, *WATER PURIFICATION, CLAY, ALGAE, BACTERIA, WASTE WATER TREA W73-01362
AL COMMUNITIES, *AQUATIC PLANTS, ALGAE, BACTERIA, LITTORAL, LAKES, W72-03216
ICS, *FRICTION, *TURBULENT FLOW, ALGAE, BACTERIA, MARINE PLANTS, E W72-03220
AKES, EUTROPHICATION, NUTRIENTS, ALGAE, BACTERIA, LAKE ERIE, ALUMI W72-01101
SALT BALANCE, DISSOLVED OXYGEN, ALGAE, BACTERIA, ODOR, INFLOW, DI W72-02274
ICAL), TROPHIC LEVELS, PROTOZOA, ALGAE, BACTERIA, DIATOMS, OCEANS, W72-10883
ATION, IRRIGATION WATER, AQUATIC ALGAE, BIOLOGICAL PROPERTIES, WAT W72-00974
RATIVE AGENCIES, ADMINISTRATION, ALGAE, BIODEGRADATION, LOCAL GOVE W72-01093
RCES, INDUSTRIAL WASTES, AQUATIC ALGAE, BIOINDICATORS, BIOASSAY, A W71-11583
TRY, DEFICIENT ELEMENTS, AQUATIC ALGAE, BIOINDI: /RIENTS, COLORIME W73-00916
TRIBUTION, PRIMARY PRODUCTIVITY, ALGAE, BIOMASS, OLIGOTROPHY, DIAT W71-09171
XIDATION LAGOONS, *PILOT PLANTS, ALGAE, BIOCHEMICAL OXYGEN DEMAND, W70-06619
ELDRIN, *ALDRIN, *ENDRIN, MARINE ALGAE, BOTTOM SEDIMENTS, PLANKTON W73-00361
TION EFFECTS, WATER TEMPERATURE, ALGAE, BULLHEADS, SALMON, SHELLFI W72-07907
IC HEALTH, AQUATIC LIFE, AQUATIC ALGAE, CALCIUM, RADIOACTIVITY EFF W72-10978
SSIE/ *CALCIFICATION, *CORALLINE ALGAE, CALCIUM RADIOISOTOPES, *BO W72-12634
, BIORTHYTHMS, LABORATORY TESTS, ALGAE, CARBON DIOXIDE, CYCLING NU W72-09159
IVITY, AQUATIC BACTERIA, AQUATIC ALGAE, CARBON CYCLE, ON-SITE TEST W73-02095
MAL DISEASE/ *FEEDS, *RUMINANTS, ALGAE, CATTLE, POULTRY, HOGS, ANI W71-07544
ES, *AM/ *PHYTOPLANKTON, *MARINE ALGAE, CHEMICAL ANALYSIS, *NITRAT W73-00380
OASSAYS, LABORATORY TESTS, FISH, ALGAE, CHEMICAL WASTES, ALGICIDES W72-08683
TER, *SEA WATER, WATER ANALYSIS, ALGAE, CHEMICAL ANALYSIS, SAMPLIN W71-11234
PERMEABILITY, PLANT POPULATIONS, ALGAE, CHEMICAL PROPERTIES, WATER W72-01473
OGEN, NUTRIENTS, SURFACE WATERS, ALGAE, CHEMICAL PROPERTIES, LANDF W72-12955
EA WATER, WATER ANALYSIS, MARINE ALGAE, CHEMICAL ANALYSIS, ADSORPT W73-02105
ROWTH RATES, *CULTURES, *AQUATIC ALGAE, CHLOROPHYTA, INHIBITORS, W W72-12736
AQUATIC BACTERIA, AQUATIC ALGAE, ALGAE, CHLOROPHYTA, PSEUDOMONAS, W72-12255
Y, CULTURES, *CHLORELLA, AQUATIC ALGAE, CHLOROPHYTA, HEAVY METALS, W72-11727
SENSING, SATELLITES(ARTIFICIAL), ALGAE, CHLOROPHYTA, PHAEOPHYTA, R W72-09120
ULTURES, PHOTOSYNTHESIS, AQUATIC ALGAE, CHLOROPHYLL, OXYGEN, RESPI W73-00281
OGY, TESTING PROCEDURES, AQUATIC ALGAE, CHLOROPHYTA, LABORATORY TE W72-14314
HY, NITRITES, NITRATES, AMMONIA, ALGAE, CHLORELLA.: /ON, HYDROGRAP W72-08056
ROWTH REGULATORS, ALGAE, AQUATIC ALGAE, CHLORELLA.: /-4-D, PLANT G W72-08440
CEAN, COLUMNS, BOTTOM SEDIMENTS, ALGAE, CHLORELLA, BENTHIC FLORA, W71-08042
PATH OF POLLUTANTS, *ABSORPTION, ALGAE, CHORELLA, TRACERS, PHOSPHA W72-09668
STES, PATH OF POLLUTANTS, MARINE ALGAE, CHROMOSOMES, EMBRYONIC GRO W72-10955
POLLUTANT IDENTIFICATION, MARINE ALGAE, CHRYSOPHYTA, AUTUMN, WINTE W73-00442
WATER POLLUTION EFFECTS, AQUATIC ALGAE, CHRYSOPHYTA, LABORATORY EQ W73-01445
RIFT, METABOLISM, ALGAE, AQUATIC ALGAE, CHRYSOPHYTA, PLANKTON, PHY W73-02280
YSOCHEMICAL PROPERTIES, DIATOMS, ALGAE, CLADOPHORA, PRODUCTIVITY, W72-13662
EATMENT, SEWAGE DISPOSAL, MARINE ALGAE, CLAMS, MUSSELS, OYSTERS, E W72-07715
ATER PROPERTIES, *NEW HAMPSHIRE, ALGAE, COLOR, COPPER SULFATE, NET W72-07890
DIATION, COSTS, SEWAGE BACTERIA, ALGAE, COLIFORMS, INSECTICIDES.: / W72-00383
OPLANKTON, MARINE ALGAE, AQUATIC ALGAE, COLORIMETRY, TRACES.: /HYT W72-03545
TANTS, SAMPLING, WATER ANALYSIS, ALGAE, COLIFORMS, DISSOLVED OXYGE W70-08249
TANTS, SAMPLING, WATER ANALYSIS, ALGAE, COLIFORMS, DISSOLVED OXYGE W70-08248
PHOTOSYNTHESIS, BIOMASS, OXYGEN, ALGAE, COMPARATIVE PRODUCTIVITY, W71-09165
ER QUALITY, CYANOPHYTA, NUISANCE ALGAE, COMPREHENSIVE PLANNING, AQ W72-09171
ON, NITRATES, NITROGEN FIXATION, ALGAE, CONDUCTIVITY, CHLOROPHYLL, W72-04773
TION, BIOCHEMICAL OXYGEN DEMAND, ALGAE, CONFINEMENT PENS, IMPOUNDM W72-08401
S, DESIGN, STANDARDS, FISH KILL, ALGAE, COOLING TOWERS, DENSITY, V W70-09612
VY METALS, ELECTRODES, NITRATES, ALGAE, COPPER, IONS, ION TRANSPOR W72-01693
ION, METABOLISM, CARBON DIOXIDE, ALGAE, CULTURES, ACIDITY, ORGANIC W72-00838
, CARBON DIOXIDE, PHYTOPLANKTON, ALGAE, CULTURES, *FLORIDA, SEA WA W72-09102
HATES, *SOLUBILITY, *CYANOPHYTA, ALGAE, CULTURES.: *PHOSP W71-10070
ENZYMES, LAKE ONTARIO, SEASONAL, ALGAE, CULTURES, MICROORGANISMS, W71-00116
NTS, *SEA WATER, *ALGAE, *MARINE ALGAE, CULTURES, PHYTOPLANKTON, C W70-07280
EMENTS, *PHYTOPLANKTON, *AQUATIC ALGAE, CULTURES, PHOSPHATES, NITR W72-12739
EOPHYTA, *TIDAL MARSHES, *MARINE ALGAE, CULTURES, SALINITY, SALT M W72-12633
EWAGE EFFLUENTS, EUTNOPHICATION, ALGAE, CYANOPHYTA, CHLOROPHYTA, M W72-13636
OR, OLIGOTROPHY, EUTROPHICATION, ALGAE, CYANOPHYTA, PHYTOPLANKTON, W72-04269
ER POLLUTION EFFECTS, NUTRIENTS, ALGAE, CYANOPHYTA, PHOSPHORUS, WA W72-04769
WATER TEMPERATURE, RESPIRATION, ALGAE, CYANOPHYTA, ANALYTICAL TEC W73-01269
, *DEPTH, MARINE ANIMALS, MARINE ALGAE, CYANOPHYTA, PYRROPHYTA, LA W72-14313

*AQUATIC MICROORGANISMS, SPORES,	ALGAE, CYTOLOGICAL STUDIES, MYCOB	W71-08027
IC ENVIRONMENT, FISH PHYSIOLOGY,	ALGAE, DAPHNIA.: /TIC LIFE, AQUAT	W72-12257
RATES, *PHOTOSYNTHESIS, *MARINE	ALGAE, DDT, *PESTICIDES, *PRIMARY	W72-09653
PESTICIDES, PESTICIDE RESIDUES,	ALGAE, DDT, PHYTOPLANKTON, MARINE	W72-03545
, TROPHIC LEVEL, EUTROPHICATION,	ALGAE, DECOMPOSING ORGANIC MATTER	W72-01784
ITY, *BIOASSAY, FISH, SHELLFISH,	ALGAE, DEGRADATION.: /NTS, *TOXIC	W71-09789
, DESALINATION, DRAIN/ NITRATES,	ALGAE, DENTRIFICATION, FILTRATION	W71-11698
, WATER POLLUTION, DESALINATION,	ALGAE, DENITRIFICATION, CALIFORNI	W71-08223
ERCHES, PIKES, SAMPLING, SNAILS,	ALGAE, DIATOMS, CLAMS, SHELLFISH,	W70-08098
RIVERS, WATER POLLUTION EFFECTS,	ALGAE, DIATOMS.: /ALGAL CONTROL,	W71-00099
CIAL DRIFT, NUTRIENTS, SALANITY,	ALGAE, DIATOMS.: /OTA, LAKES, GLA	W71-13149
RS, TOXINS, INORGANIC COMPOUNDS,	ALGAE, DIATOMS, CHLAMYDOMONAS, SC	W72-14706
CHRYSOPHYTA, INHIBITION, MARINE	ALGAE, DIATOMS, BEHAVIOR.: /CIDS,	W73-00942
AGOONS, BIOMASS, OREGON, SLUDGE,	ALGAE, DIPTERA, DISSOLVED OXYGEN,	W72-03556
ER POLLUTION SOURCES, WATER LAW,	ALGAE, DISSOLVED OXYGEN.: /L, WAT	W71-01192
TER RELATIONSHIPS, PERMEABILITY,	ALGAE, DISSOLVED OXYGEN.: /OUNDWA	W71-12410
TER POLLUTION SOURCES, BIOASSAY,	ALGAE, DOMESTIC WASTES, NUTRIENTS	W73-02478
SORPTION, NUTRIENTS, PESTICIDES,	ALGAE, DREDGING, ESTUARINE ENVIRO	W71-09581
HYSIOLOGY, LAKES, PONDS, AQUATIC	ALGAE, ECOLOGY, BIOLOGICAL PROPER	W71-10580
TION EFFECTS, *DATA COLLECTIONS,	ALGAE, ECOTYPES, INDUSTRIAL PLANT	W71-13746
ARVEST, PLANKTON, BENTHIC FAUNA,	ALGAE, ECOLOGY, MISSOURI.: /ISH H	W70-09596
GEMENT, WISCONSIN, PRODUCTIVITY,	ALGAE, ECONOMICS, POLITICAL ASPEC	W71-03012
*FRESHWATER	ALGAE, EDTA.:	W72-14694
HICATION, WASTE WATER TREATMENT,	ALGAE, EFFLUENTS, NUTRIENTS, SEWA	W70-09325
E MICHIGAN, CHRYSOPHYTA, AQUATIC	ALGAE, ELECTRON MICROSCOPY, PHOTO	W73-00432
ICAL REGIONS, PHAEOPHYTA, MARINE	ALGAE, ENVIRONMENTAL EFFECTS, MAR	W73-00428
CIDES), PHOTOACTIVATION, AQUATIC	ALGAE, ENVIRONMENTAL EFFECTS.: /I	W72-14327
ADIATION, HYDROGRAPHY, PLANKTON,	ALGAE, ENVIRONMENTAL EFFECTS, BIO	W73-01704
OD CHAINS, PHYTOPLANKTON, MARINE	ALGAE, ENVIRONMENTAL EFFECTS, TRA	W73-00818
OPHIC LEVEL, PROTOZOANS, AQUATIC	ALGAE, ENVIRONMENTAL EFFECTS, CHL	W72-13816
IFICATION, PHYTOPLANKTON, MARINE	ALGAE, ESSENTIAL NUTRIENTS.: /ENT	W73-00441
CALIFORNIA, RADIOECOLOGY, MARINE	ALGAE, ESTUARINE ENVIRONMENT, ABS	W73-00826
USTRIAL WASTES, FLORIDA, SESSILE	ALGAE, ESTUARIES, CHEMICAL WASTES	W71-07412
MS, ALGAE, MARINE ALGAE, AQUATIC	ALGAE, EUGLENA, CHLAMYDOMONAS.: /	W72-08436
ION EFFECTS, BIOMASS, NUTRIENTS,	ALGAE, EUGLENOPHYTA, CHRYSOPHYTA,	W72-09111
UTRIENTS, MEASUREMENT, BACTERIA,	ALGAE, EUTROPHICATION, DATA COLLE	W71-13794
A, *HUMIC ACIDS, ORGANIC MATTER,	ALGAE, EUTROPHICATION, CHELATION,	W72-01097
AL TURNOVER, ENERGY, CYANOPHYTA,	ALGAE, EUTROPHICATION, SUMMER, EN	W72-03228
, *WATER QUALITY CONTROL, *ODOR,	ALGAE, EUTROPHICATION, TASTE, RES	W71-13187
ROPHI/ *ALGAL CONTROL, *NUISANCE	ALGAE, EUTROPHICATION, REVIEWS, T	W72-07937
ALGAE, *CARBON DIOXIDE, NUISANCE	ALGAE, EUTROPHICATION, PHOSPHATES	W72-07514
ITRATES, PHOSPHORUS, PHOSPHATES,	ALGAE, EUTROPHICATION, WATER QUAL	W71-06445
ITY, OXYGEN DEMAND, HEAT BUDGET,	ALGAE, EUTROPHICATION.: / ALKALIN	W72-12997
NISTRATION, NON-CONSUMPTIVE USE,	ALGAE, EVAPORATION, REASONABLE US	W70-07100
ISSUES, FRESHWATER FISH, AQUATIC	ALGAE, EVALUATION, SAMPLING, CARP	W73-01673
ICATION, HYDRAULIC MODELS, FISH,	ALGAE, FATHOMETERS, BENTHIC FAUNA	W70-09614
, AMMONIA, NITROGEN, PHOSPHORUS,	ALGAE, FEASIBILITY STUDIES, TECHN	W71-09450
EPODS, CRUSTACEANS, ZOOPLANKTON,	ALGAE, FEEDING RATES, *LARVAL GRO	W72-09248
F RATES, AFTERGROWTH, BLUE-GREEN	ALGAE, FIELD STUDIES, GREEN ALGAE	W70-08319
NT EXTRACTIONS, DIATOMS, AQUATIC	ALGAE, FILTRATION, CYANOPHYTA, *E	W72-13673
QUALITY, TROPHIC LEVEL, AQUATIC	ALGAE, FISH POPULATIONS, FLORIDA,	W72-08986
OIRS, *FRESHWATER, HYDROBIOLOGY,	ALGAE, FISH, ABSORPTION, BIOASSAY	W72-10694
S, CHEMICALS, PLANKTON, BENTHOS,	ALGAE, FISH, WATER POLLUTION, WAT	W72-07841
, HEADLOSS, PRESSURE, TURBIDITY,	ALGAE, FLOCCULATION, CLEANING, *W	W72-09393
LLA, *ALGAE, *FOULING, *NUISANCE	ALGAE, FLOATING PLANTS, SEAWATER,	W71-09261
WATER POLLUTION SOURCES, AQUATIC	ALGAE, FOOD CHAINS, MEASUREMENT,	W72-10686
, PHYTOPLANKTON, MAMMALS, MARINE	ALGAE, FOOD CHAINS, MARINE FISHER	W72-00960
TAMIN B, PIGMENTS, GROWTH RATES,	ALGAE, FOOD WEBS, COPEPODS, ORGAN	W72-07892
ORPTION, *PHYTOPLANKTON, *MARINE	ALGAE, FOOD CHAINS, PUBLIC HEALTH	W73-00823
ADSORPTION, AGRICULTURAL RUNOFF,	ALGAE, FOOD CHAINS, CHLORINATED H	W72-13657
ICATION, NUTRIENTS, ASSESSMENTS,	ALGAE, FORMULATION, WASTE TREATME	W72-09169
ORUS, *PHOSPHORUS RADIOISOTOPES,	ALGAE, FRESH WATER, FILTRATION, L	W72-07143
TED WATER, CONDENSERS, PROTOZOA,	ALGAE, FUNGI, WATER POLLUTION EFF	W72-07526
BS, FOOD CHAINS, FROGS, LICHENS,	ALGAE, FUNGI, SMALL ANIMALS(MAMMA	W72-00949
, CYANOPHYTA, PYRROPHYTA, MARINE	ALGAE, GASTROPODS.: / CHRYSOPHYTA	W73-00935
NITRATES, MUSSELS, SNAILS, FISH,	ALGAE, GASTROPODS, ADSORPTION, TE	W71-09168
MICALS, WATER POLLUTION SOURCES,	ALGAE, GEOLOGY, DATA COLLECTIONS.	W72-05869
EALTH, COST COMPARISONS, AQUATIC	ALGAE, GRANTS, MODEL STUDIES.: /H	W71-09429
*TOXICITY, *CYANOPHYTA, SULFUR,	ALGAE, GRANULES, INHIBITION, META	W72-00153
TICAL MODELS, SELF-PURIFICATION,	ALGAE, GREAT LAKES, NUTRIENTS, SE	W71-12072
*BENTHIC FLORA, *PHYTOPLANKTON,	ALGAE, GROWTH RATES, BIOASSAY, DI	W72-06046
CULTURES, CHLOROPHYLL, PROTEINS,	ALGAE, GROWTH RATES, TRACERS.: /	W72-03557
RATURE, SORPTION, RADIOISOTOPES,	ALGAE, GROWTH RATES, THERMAL POLL	W71-02075
*METABOLISM, *CHLORELLA, *LIGHT,	ALGAE, GROWTH RATES, CHLOROPHYLL.	W73-00251
M, CULTURES, RESISTANCE, AQUATIC	ALGAE, GROWTH RATES, BIOASSAY, LA	W72-13809
DIES, *PHAEOPHYTA, EPIPHYTOLOGY,	ALGAE, HABITATS, CULTURES.: / STU	W72-01370
DESIGN CRITERIA, *WATER QUALITY,	ALGAE, HARVESTING, DEPTH, TEMPERA	W71-07100
, RADIOISOTOPES, TRACERS, MARINE	ALGAE, HAWAII, WATER TEMPERATURE,	W71-11486
, *COBALT RADIOISOTOPES, *MARINE	ALGAE, HEAVY METALS, ABSORPTION,	W72-10957
FFECTS, WATER POLLUTION EFFECTS,	ALGAE, HEAVY METALS, GROWTH, PSEU	W72-13813
OMPOSING ORGANIC MATTER, *MARINE	ALGAE, HUMUS, ORGANIC COMPOUNDS,	W72-12166
EROBIC CONDITIONS, PILOT PLANTS,	ALGAE, HYDROGEN ION CONCENTRATION	W71-07102
IES, FISH, MARINE PLANTS, MARINE	ALGAE, HYDROLYSIS, CHELATION, SED	W72-04463
ARBON, *SEWAGE BACTERIA, SEWAGE,	ALGAE, HYDROGEN ION CONCENTRATION	W72-04309
STORAGE, NUTRIENTS, TEMPERATURE,	ALGAE, HYDROLOGIC DATA, DISSOLVED	W72-05282
S, EQUATIONS, LIGHT PENETRATION,	ALGAE, HYDROGEN ION CONCENTRATION	W72-05459
T STREAMS, *GREAT LAKES, AQUATIC	ALGAE, ICE, FISH, MICROORGANISMS,	W73-01669
R, ATLANTIC OCEAN, GROWTH RATES,	ALGAE, INCUBATION, PHYTOPLANKTON.	W71-11562
WATER QUALITY, *DOMESTIC WASTES,	ALGAE, INDUSTRIAL WASTES, METABOL	W72-09590
EFFECTS, EUTROPHICATION, *LAKES,	ALGAE, INFRARED RADIATION, *REMOT	W72-14728

YORK, WATER QUALITY, NUTRIENTS, ALGAE, INFLOW, DISCHARGE(WATER), W72-13851
N, CARBON DIOXIDE, BIOCHEMISTRY, ALGAE, INHIBITION, LABORATORY TES W72-08584
HOSPHORUS, *LAKE ERIE, NITROGEN, ALGAE, INORGANIC COMPOUNDS, STAND W72-13653
DISPOSAL, MONITORING, NUTRIENTS, ALGAE, INSECTICIDES, HERBICIDES, W71-12091
E MICROORGANISMS, SEDIMENTATION, ALGAE, INVERTEBRATES, RADIATION, W71-13723
ES, *MARINE FISH, SILICA, MARINE ALGAE, INVERTEBRATES, ISOLATION, W73-00371
, PRODUCTIVITY, TOXICITY, CRABS, ALGAE, INVERTEBRATES, FISH, POLLU W71-01475
LATORS, PLANT GROWTH SUBSTANCES, ALGAE, IRON, BIOASSAY, BIOINDICAT W71-03095
REEN ALGAE, FIELD STUDIES, GREEN ALGAE, LABORATORY STUDIES.: /UE-G W70-08319
SYSTEMS, AQUATIC LIFE, SAMPLING, ALGAE, LABORATORY TESTS, DETRITUS W71-10064
NTROL, SEWAGE EFFLUENTS, AQUATIC ALGAE, LAKE ERIE, LAKE ONTARIO, N W71-11507
ION, *ON-SITE TESTS, *NUTRIENTS, ALGAE, LAKES, PHOSPHORUS, CARBON, W72-09165
ALYTICAL TECHNIQUES, PHOSPHORUS, ALGAE, LAKES, MICROORGANISMS, TER W71-03021
WATERS, SOIL EROSION, NUTRIENTS, ALGAE, LAKE SUPERIOR, WATER POLLU W71-04216
STRATIFICATION, CURRENTS(WATER), ALGAE, LAKES, RIVERS, SEWAGE DISP W70-09947
*BENTHIC FLORA, MICROORGANISMS, ALGAE, LAKES, EUTROPHICATION, LIT W72-07501
*ESTUARIES, *DIATOMS, *SESSILE ALGAE, LIGHT INTENSITY, TIDAL EFF W73-00853
AL DI/ *HEAT RESISTANCE, *MARINE ALGAE, LIGHT INTENSITY, *ECOLOGIC W73-00838
*MARINE PLANTS, *PHYTOPLANKTON, ALGAE, LIGHT INTENSITY, SALINITY, W71-10083
NION, *LAKE ERIE, BENTHIC FLORA, ALGAE, LIGHT PENETRATION, PLANKTO W72-13659
, DISTRIBUTION PATTERNS, AQUATIC ALGAE, LIMNOLOGY, CHRYSOPHYTA, EU W72-07145
TICIDES, PHYTOPLANKTON, DIATOMS, ALGAE, MARINE ALGAE, AQUATIC ALGA W72-08436
CTS, NUCLEAR POWERPLANTS, MARINE ALGAE, MARINE ANIMALS, MARINE PLA W73-00833
ECOLOGICAL DISTRIBUTION, MARINE ALGAE, MARINE ANIMALS.: /IBITION, W73-00348
UCTIVITY, ENERGY BUDGET, *MARINE ALGAE, MARINE PLANTS, BIOMASS, CA W73-00366
HRYSOPHYTA, CHLOROPHYTA, AQUATIC ALGAE, MARINE ALGAE, SESSILE ALGA W73-00273
ON, PHAEOPHYTA, MOLLUSKS, MARINE ALGAE, MARINE ANIMALS.: / POLLUTI W72-11925
TERS, PATH OF POLLUTANTS, MARINE ALGAE, MARINE FISH, ABSORPTION, W W72-00940
UT, TOXICITY, WASHINGTON, MARINE ALGAE, MARINE BACTERIA.: /ON, TRO W72-05421
US PLANTS, CORAL, INVERTEBRATES, ALGAE, MARINE ALGAE, ALGAL POISON W71-08597
E NATIONAL PARK(WYO), BLUE-GREEN ALGAE, MASTIGOCLADUS, CALOTHRIX.: W72-08508
DIOACTIVITY TECHNIQUES, AQUARIA, ALGAE, MATHEMATICAL MODELS, PATH W72-03348
ION EFFECTS, SLUDGE, RIVER BEDS, ALGAE, MATHEMATICAL MODELS.: /TAT W70-09189
N, DISSOLVED OXYGEN, SATURATION, ALGAE, METHANE BACTERIA, EUTROPHI W71-07106
, PHOSPHATES, CHEMICAL ANALYSIS, ALGAE, MICROORGANISMS, HYDROGEN I W71-01474
TAL E/ *CHLORELLA, *SCENEDESMUS, ALGAE, MICROORGANISMS, ENVIRONMEN W72-10861
ING FACTORS, RHODOPHYTA, AQUATIC ALGAE, MICROORGANISMS, CHLOROPHYT W73-02099
OLOGICAL STUDIES, PLANT TISSUES, ALGAE, MICROORGANISMS, BACTERIA, W73-01676
GAGES, GROUNDWATER, SAFE YIELD, ALGAE, MINERALOGY, TURBIDITY, COL W72-05594
ATER-PLANT RELATIONSHIPS, MARINE ALGAE, MOSSES, CRABS, ALGAE, SNAI W72-03347
OXYGEN, LAKE MORPHOLOGY, AQUATIC ALGAE, MOSSES, WATER TEMPERATURE, W71-11949
LLUTANT IDENTIFICATION, ONTARIO, ALGAE, MOSSES, AQUATIC PLANTS.: / W71-11529
DIMENTS, CLAYS, CORES, SEASONAL, ALGAE, MUD-WATER INTERFACES, IRON W72-05469
VERTEBRATES, ST. LAWRENCE RIVER, ALGAE, NEMATODES, GASTROPODS, MOL W72-05432
ASTES, *CRUSTACEANS, OILY WATER, ALGAE, NEMATODES, OYSTERS, MUSSEL W72-07911
RESINS, AQUATIC PLANTS, AQUATIC ALGAE, NITRATES, NITRITES, EUTROP W73-00286
SOPHYTA, PLANT PIGMENTS, AQUATIC ALGAE, NITROGEN.: /MICHIGAN, CHRY W73-00384
A, AQUATIC PRODUCTIVITY, AQUATIC ALGAE, NITROGEN, POLLUTANT IDENTI W73-00379
NITROGEN, POTASSIUM, DIVERSION, ALGAE, NITROGEN FIXATION, PLANKTO W71-00117
, *PHYTOPLANKTON, *PRODUCTIVITY, ALGAE, NITROGEN, PHOSPHORUS, MATH W71-08670
ITROGEN, *PHOSPHORUS, NUTRIENTS, ALGAE, NITRATES, SURFACE RUNOFF, W71-06443
RIO, *EUTROPHICATION, LAKE ERIE, ALGAE, NITROGEN, PHOSPHORUS, DRAI W71-11009
M WASTES, NUTRIENTS, PHOSPHORUS, ALGAE, NITRATES, WATER REUSE, SAL W72-00846
*EUTROPHICATION, *PHOSPHOROUS, ALGAE, NITROGEN, ACTIVATED SLUDGE W71-13334
TION LAGOONS, NUTRIENTS, AQUATIC ALGAE, NITROGEN, ANAEROBIC DIGEST W72-02850
WATER POLLUTION EFFECTS, AQUATIC ALGAE, NORTH CAROLINA, *ALGAE, DI W72-01329
E ANIMALS, MARINE PLANTS, MARINE ALGAE, NUCLEAR WASTES, PATH OF PO W73-00828
LOGY, CHEMICAL ANALYSIS, AQUATIC ALGAE, NUTRIENTS, WATER QUALITY, W73-00918
LAKE STAGES, LIMNOLOGY, NUISANCE ALGAE, NUTRIENT REMOVAL, NUTRIENT W72-14692
KTON, *FOODS, DAPHNIA, COPEPODS, ALGAE, NUTRIENT REQUIREMENTS, CYA W72-07940
ONMENT, WATER POLLUTION EFFECTS, ALGAE, NUTRIENTS, WATER POLLUTION W72-06638
NANCE, ENVIRONMENTAL SANITATION, ALGAE, NUTRIENTS, INDUSTRIAL WAST W72-01659
*LAKES, *EUTROPHICATION, AQUATIC ALGAE, NUTRIENTS, WATER RESOURCES W71-13256
INS, DETERGENTS, EUTROPHICATION, ALGAE, NUTRIENTS.: /TS, STORM DRA W71-13466
WATER, RUNOFF, DISSOLVED OXYGEN, ALGAE, NUTRIENTS, BACTERIA, VIRUS W71-09788
ION, WATER REUSE, POTABLE WATER, ALGAE, NUTRIENT REQUIREMENTS, CYA W71-13185
TICS, DOMESTIC WASTES, BACTERIA, ALGAE, NUTRIENTS, OXIDATION-REDUC W72-10401
S, WATER SUPPLY, EUTROPHICATION, ALGAE, NUTRIENTS, WATER POLLUTION W71-05083
OMPOUNDS, *PHOSPHORUS COMPOUNDS, ALGAE, NUTRIENT REQUIREMENTS, PRO W70-04080
FILTRATION, POLLUTION ABATEMENT, ALGAE, NUTRIENT, *BIODEGRADATION, W72-05315
TEMPERATURE, CHEMICAL ANALYSIS, ALGAE, NUTRIENTS, WATER ANALYSIS, W72-04791
AND DEVELOPMENT, TRACE ELEMENTS, ALGAE, NUTRIENTS, LIGHT INTENSITY W72-04433
FFECTS, WATER POLLUTION SOURCES, ALGAE, NUTRIENTS.: /R POLLUTION E W72-04280
DISSOLVED SOLIDS, PHYTOPLANKTON, ALGAE, NUTRIENTS.: /T INTENSITY, W72-04270
ALS, PESTICIDES, TASTE-PRODUCING ALGAE, ODOR-PRODUCING ALGAE, ACTI W73-02144
RGED PLANTS, BIOMASS, NEMATODES, ALGAE, OLIGOCHAETES, LIFE CYCLES, W71-11012
SS, *HYDROGRAPHY, ALGAE, AQUATIC ALGAE, OLIGOTROPHY, DISSOLVED OXY W72-13871
RPLANTS, SITES, MOLLUSKS, MARINE ALGAE, ON-SITE INVESTIGATIONS, CA W72-03324
KINETICS, CHEMOSTATS, FLAGELLATE ALGAE, OPTICAL DENSITY SENSOR.: / W70-10403
ONDITIONS, ANAEROBIC CONDITIONS, ALGAE, ORGANIC MATTER, VOLUME, DR W71-07551
IGN CRITERIA, TRICKLING FILTERS, ALGAE, ORGANIC LOADING, ODOR, COL W71-07113
WATERSHEDS(BASINS), PHOSPHORUS, ALGAE, ORGANIC MATTER, LAKE ERIE, W71-11017
, SYMBIOSIS, AZOTOBACTER, MARINE ALGAE, ORGANIC MATTER.: /XIDATION W72-11563
DISINFECTION, ULTRAVIOLET LIGHT, ALGAE, OXYGENATION, BIOCHEMICAL O W72-00042
GEN, PHOSPHORUS, EUTROPHICATION, ALGAE, PATHOGENIC BACTERIA, EFFLU W71-08214
PRODUCTION, LIFE CYCLES, AQUATIC ALGAE, PERIPHYTON.: /RPHOLOGY, RE W72-07141
TRIENTS, BIOLOGICAL COMMUNITIES, ALGAE, PERIPHYTON, PHYTOPLANKTON, W72-12549
IENTS, BIOASSAY, *WATER QUALITY, ALGAE, PESTICIDES, HEAVY METALS, W72-14282
TS, *RECREATION, LAKES, STREAMS, ALGAE, PESTICIDES, DDT, RETURN FL W71-06444
GANISMS, AQUATIC PLANTS, AQUATIC ALGAE, PESTICIDES, INSECTICIDES, W72-09646

BSORPTION, MARINE PLANTS, MARINE ALGAE, PHAEPHYTA, AQUARIA, PARTIC W71-09193
CARBON, ADSORPTION, FILTRATION, ALGAE, PHENOLS, DETERGENTS, PESTI W72-08357
TTS, *EUTROPHICATION, NUTRIENTS, ALGAE, PHOSPHATES, NITRATES, SURV W71-05553
RATURE, TURBULENCE, OXYGENATION, ALGAE, PHOTOSYNTHESIS, MUNICIPAL W71-06033
STRATIFICATION, SOLAR RADIATION, ALGAE, PHOTOSYNTHESIS, ANAEROBIC W71-07084
ISONS, *SEA WATER, *FRESH WATER, ALGAE, PHOTOSYNTHESIS, ORGANIC MA W71-03027
*MUNICIPAL WASTES, *WATER REUSE, ALGAE, PHOTOSYNTHESIS, DISSOLVED W72-11642
DESIGN CRITERIA, BIODEGRADATION, ALGAE, PHOTOSYNTHESIS, SUSPENDED W72-09386
S, EUTROPHICATION, PRODUCTIVITY, ALGAE, PHOSPHATES, NITROGEN, SEWA W72-09167
, LABORATORY TESTS, TEMPERATURE, ALGAE, PHOSPHORUS, ION EXCHANGE, W72-01108
MENT, SURFACE WATERS, NUTRIENTS, ALGAE, PHOSPHORUS, NITROGEN, EUTR W72-00847
, *PLANT GROWTH, *LAKE MICHIGAN, ALGAE, PHOSPHATES, HYDROLYSIS, ST W72-13644
ING, SECONDARY TREATMENT, MARINE ALGAE, PHYTOPLANKTON, BIOMASS, GR W72-09378
XICITY, *AQUATIC PLANTS, AQUATIC ALGAE, PHYTOPLANKTON, AQUATIC MIC W72-09650
UCTIVITY, SEASONAL, CHLOROPHYLL, ALGAE, PHYTOPLANKTON, WATER ANALY W72-09108
UTROPHICATION, RED TIDE, SILICA, ALGAE, PHYTOPLANKTON, FREEZING, T W71-09190
PES, *ORGANIC COMPOUNDS, *MARINE ALGAE, PHYTOPLANKTON, HUMIC ACIDS W71-09188
LY IMPOUNDMENT, BIOMASS, AQUATIC ALGAE, PHYTOPLANKTON, ZOOPLANKTON W72-04853
AL TECHNIQUES, MICHIGAN, SESTON, ALGAE, PHYTOPLANKTON, NANNOPLANKT W72-08048
, *CARBON, *AMINO ACIDS, *MARINE ALGAE, PHYTOPLANKTON, NITROGEN, A W72-07497
ANOPHYTA, *NEW HAMPSHIRE, MARINE ALGAE, PHYTOPLANKTON, AQUATIC PLA W72-14542
AMS, CHEMICAL ANALYSIS, BIOMASS, ALGAE, PHYSICOCHEMICAL PROPERTIES W73-01435
R TEMPERATURE, SAMPLING, AQUATIC ALGAE, PHYTOPLANKTON, TROPHIC LEV W73-01456
INO-3,5,6-TR/ TORDON, FRESHWATER ALGAE, PICLORAM, PEDIASTRUM, 4-AM W72-08440
CHEMISTRY, CYANOPHYTA, BACTERIA, ALGAE, PLANKTON, PLANTS, SPECTROS W73-01439
UTION), WATER POLLUTION SOURCES, ALGAE, PLANKTON, SEASONAL.: /POLL W72-05506
OTOPES, STRONTIUM RADIOISOTOPES, ALGAE, PLANT GROWTH, TEMPERATURE, W71-09254
NUTRIENTS, NITROGEN, PHOSPHORUS, ALGAE, PLANT GROWTH, WATER POLLUT W72-11702
NTHIC FAUNA, PERIPHYTON, AQUATIC ALGAE, PLANTS, ENVIRONMENTAL ENGI W72-00939
L CONTROL, *CYANPHYTA, *NUISANCE ALGAE, PLANKTON, WASHINGTON, DIAT W72-00798
RESERVOIRS, TERTIARY TREATMENT, ALGAE, PLANTS, TROUT, AQUATIC ENV W72-02412
CULTURE, *ECONOMICS, CYANOPHYTA, ALGAE, PLANTS, PROTEINS, NUTRIENT W72-03327
INFILTRATION, INDUSTRIAL WASTES, ALGAE, PONDS, FARM WASTES.: /GE, W71-06435
MENT, WATER POLLUTION TREATMENT, ALGAE, POTABLE WATER, TASTE, AMMO W71-01491
ALGAE, BIOINDICATORS, BIOASSAY, ALGAE, POWERPLANTS, EFFLUENTS, DI W71-11583
PERATURE EFFEC/ *BP 1002, *GREEN ALGAE, PRASINOCLADUS MARINUS, TEM W70-09429
ON, SECONDARY TREATMENT, AQUATIC ALGAE, PRIMARY PRODUCTIVITY, TRAC W71-09674
UTRIENTS, BIOMASS, HARVESTING OF ALGAE, PRODUCTIVITY.: /ROTEINS, N W72-03327
DOLOGY, FOREIGN RESEARCH, MARINE ALGAE, PRODUCTIVITY, POLLUTANTS, W72-08804
TER POLLUTION, ACTIVATED SLUDGE, ALGAE, PROTOZOA, DIATOMS, EUGLENA W72-08573
WAGE LAGOONS, *LIMITING FACTORS, ALGAE, PRODUCTIVITY, PHOTOSYNTHES W72-09162
ONDS, *WATER REUSE, RECLAMATION, ALGAE, PROTEIN, *OXIDATION LAGOON W71-08951
XYGEN, BIOCHEMISTRY, INHIBITORS, ALGAE, PROTEINS.: /, *NITROGEN, O W72-07941
AL TECHNIQUES, DATA COLLECTIONS, ALGAE, PROTOZA, FUNGI, EQUIPMENT, W72-07225
MOLLUSKS, MARINE PLANTS, MARINE ALGAE, PUBLIC HEALTH, ESTUARINE E W72-04461
WATER POLLUTION SOURCES, AQUATIC ALGAE, PUBLIC HEALTH, BIODEGRADAT W71-11516
IMENTS, *MARINE ANIMALS, *MARINE ALGAE, RADIOISOTOPES, FALLOUT, CH W72-05965
S, CESIUM, COBALT RADIOISOTOPES, ALGAE, RADIOACTIVITY, TROPHIC LEV W70-06668
SEDIMENTS, EUTROPHICATION, FISH, ALGAE, RECREATION, ENVIRONMENTAL W71-10577
ONS, HYDROGEN ION CONCENTRATION, ALGAE, REPRODUCTION, SAMPLING, SE W71-08026
LGAE, *ECONOMICS, *PRODUCTIVITY, ALGAE, RESOURCES, FOOD CHAINS, FI W71-13246
, SAMPLING, VOLUMETRIC ANALYSIS, ALGAE, RESERVOIR STORAGE, IMPOUND W72-14280
*TASTE, *ODOR, DISSOLVED SOLIDS, ALGAE, RESERVOIR STORAGE, *HYPOLI W73-02138
TY EFFECTS, *MONITORING, *MARINE ALGAE, RHODOPHYTA, PHAEOPHYTA, *C W72-07826
LOR/ *YEASTS, *CULTURES, *MARINE ALGAE, RHODOPHYTA, PHAEOPHYTA, CH W72-04748
IDES, *FERTILIZATION, NUTRIENTS, ALGAE, RICE, NITROGEN, NITROGEN F W71-12071
*WATER ANALYSIS, *BIOINDICATORS, ALGAE, RIVERS, STREAMS, DELAWARE W72-04736
RVOIRS, WATER POLLUTION SOURCES, ALGAE, RIVER FLOW, PHOSPHORUS, NI W72-04260
TIC WEED CONTROL, TILAPIA, CARP, ALGAE, ROOTED AQUATIC PLANTS, PLA W71-04528
IS, TEMPERATURE, SEASONAL, IONS, ALGAE, SALINITY, PHYTOPLANKTON, T W72-01363
, AQUATIC MICROORGANISMS, MARINE ALGAE, SALINITY, MIGRATION, TIDES W72-14317
LUS, CHLORELLA, SULFUR BACTERIA, ALGAE, SALMONELLA, E. COLI.: /CIL W72-12565
COLIFORMS, MATHEMATICAL MODELS, ALGAE, SAMPLIN: /HATES, NITROGEN, W71-13356
ASSIUM, IRON, MANGANESE, COPPER, ALGAE, SAMPLING, CHLOROPHYTA.: /T W71-11001
DAPHNIA, AQUATIC PLANTS, AQUATIC ALGAE, SCENEDESMUS, RADIOISOTOPES W72-09649
FF, SALINITY, DEPTH, ABSORPTION, ALGAE, SEASONAL, WEATHER, WATER A W72-09122
ING CROPS, TIDAL EFFECTS, MARINE ALGAE, SEA WATER, SALINITY, SAMPL W73-00367
L/ *BACTERIA, *TOXICITY, *MARINE ALGAE, SEA WATER, CHLOROPHYTA, CH W72-08051
TS, SYNTHESIS, SLUDGE, EFFLUENT, ALGAE, SECONDARY TREATMENT, BIOCH W71-12188
PATH OF POLLUTANTS, ASIA, MARINE ALGAE, SEDIMENTS, ESTUARINE ENVIR W72-10958
OTOPES, *MARINE ANIMALS, *MARINE ALGAE, SEDIMENTS, NUCLEAR WASTES, W73-00817
ORPTION, *MARINE ANIMALS, MARINE ALGAE, SEDIMENTS, PUERTO RICO, ES W73-00821
ENTS(WATER), WINDS, THERMOCLINE, ALGAE, SEDIMENTS.: /ERATURE, CURR W72-12992
D OXYGEN, *ANAEROBIC CONDITIONS, ALGAE, SEDIMENTS, OXYGEN DEMAND, W72-12990
ANTI/ *MARINE FUNGI, *ISOLATION, ALGAE, SEPARATION TECHNIQUES, ATL W71-11851
LOROPHYTA, AQUATIC ALGAE, MARINE ALGAE, SESSILE ALGAE.: /PHYTA, CH W73-00273
ES, NITROGEN COMPOUNDS, AMMONIA, ALGAE, SEWAGE TREATMENT, TERTIARY W71-13939
OR/ *EPIPSAMMIC ALGAE, *EPIPELIC ALGAE, SHEAR WATER(ENGLAND), AMPH W71-12855
BENTHOS, CRABS, COPEPODS, MARINE ALGAE, SHRIMP, FISH EGGS, FISH, B W72-08142
ITROGEN, PHOSPHORUS, METABOLISM, ALGAE, SLUDGE, ANALYTICAL TECHNIQ W71-11793
ACTIVATED CARBON, FLOCCULATION, ALGAE, SLUDGE, CHROMIUM, TOXICITY W71-09524
RE, NUTRIENTS, DISSOLVED OXYGEN, ALGAE, SLUDGE, OXYGENATION, ANAER W71-07109
PS, MARINE ALGAE, MOSSES, CRABS, ALGAE, SNAILS, MUSSELS, ISOPODS, W72-03347
TY, FOOD CHAINS, FISH, BIOASSAY, ALGAE, SNAILS, MOSQUITOES, MODEL W72-07703
BIOMASS, LICHENS, MOSSES, FUNGI, ALGAE, SOIL CONTAMINATION, TRACE W72-14303
EN ION CONCENTRATION, ECONOMICS, ALGAE, SOLUBILITY, SEWAGE, BIOCHE W71-12200
MENTAL EFFECTS, ALGAL POISONING, ALGAE, SOLVENTS, SURFACTANTS, WAT W70-09429
RBONATES, DIATONICS, CYANOPHYTA, ALGAE, SOUTH DAKOTA, WATER POLLUT W71-02880
HYTA, RESERVOIRS, ODOR-PRODUCING ALGAE, SOUTHWEST U.S., TEMPERATUR W72-10613
OCEAN, *TROPICAL REGIONS, MARINE ALGAE, SPATIAL DISTRIBUTION, FOUR W72-11719

ULFUR, SAMPLING, WATER ANALYSIS,	ALGAE, SPECTROPHOTOMETRY, PENNSYL	W72-04743
DGES, FISH, LAKES, CONDUCTIVITY,	ALGAE, STANDING CROP, CHEMICAL AN	W71-01488
*DISTRIBUTION PATTERNS, *AQUATIC	ALGAE, STATISTICAL METHODS, MARIN	W72-08141
TAMINS, YEASTS, SULFUR, FOULING,	ALGAE, SULFATES, PERCOLATION, HER	W72-05421
BIOINDICATORS, E. COLI, AQUATIC	ALGAE, SUNFISHES, SNAILS, EELS, C	W71-07417
ES, *NUTRIENTS, *EUTROPHICATION,	ALGAE, SURFACTANTS, WATER POLLUTI	W70-09388
*DISTRIBUTION, FLOATING, WINDS,	ALGAE, SURFACE WATERS, HYDROGEN I	W73-00236
*PHOSPHORUS, *CARBON, BACTERIA,	ALGAE, SYMBIOSIS, CARBON DIOXIDE,	W70-07283
TASSIUM, PHOSPHATES, METABOLISM,	ALGAE, SYNTHESIS, VITAMINS, SULFA	W73-01626
TENSITY, PHOTOSYNTHESIS, SESSILE	ALGAE, TEMPERATURE, DEPTH, LIGHT	W72-07501
GANIC LOADING, AERATION, MIXING,	ALGAE, TEMPERATURE, ALTITUDE, CLI	W72-06838
, *OXIDATION LAGOONS, COLIFORMS,	ALGAE, TEMPERATURE, ANALYTICAL TE	W72-00422
NEMA SPP., DUNALIELL/ FRESHWATER	ALGAE, THALASIOSIRA SPP., SKELETO	W72-08436
NG, DIATOMS, NITROGEN, BACTERIA,	ALGAE, THERMAL STRATIFICATION, DI	W71-03028
OXYGEN DEMAND, DISSOLVED OXYGEN,	ALGAE, THERMAL STRATIFICATION, CO	W72-03753
ES, *HYPOLIMNION, NITRIFICATION,	ALGAE, THERMOCLINE, ANAEROBIC BAC	W72-12996
LITY, STREAMS, SEWAGE EFFLUENTS,	ALGAE, TOURISM, RECREATION, FISHE	W72-14878
ES, IODINE RADIOISOTOPES, MARINE	ALGAE, TRACE ELEMENTS, PATH OF PO	W73-00811
RIE, SEDIMENTS, SILTS, PLANKTON,	ALGAE, TRACE ELEMENTS, VOLATILITY	W71-11036
, HARBORS, SPATIAL DISTRIBUTION,	ALGAE, TRIBUTARIES, CYCLING NUTRI	W72-04263
INE FISH, MARINE ANIMALS, MARINE	ALGAE, TRITIUM, IODINE RADIOISOTO	W71-09018
IDE, ECOSYSTEMS, SOUTH CAROLINA,	ALGAE, TROPHIC LEVEL.: /RBON DIOX	W71-00475
S, *ELECTRODES, IONS, FLUORIDES,	ALGAE, VOLUMETRIC ANALYSIS, WATER	W71-00474
RIA, OXIDATION, ORGANIC LOADING,	ALGAE, WASTE WATER TREATMENT, BIO	W71-07096
ENTS, AGING(BIOLOGICAL), AQUATIC	ALGAE, WASTE ASSIMILATIVE CAPACIT	W72-01685
LABORATORY TESTS, COST ANALYSIS,	ALGAE, WASTE WATER TREATMENT.: /	W72-10390
DIFFUSION, *NUTRIENTS, *NUISANCE	ALGAE, WASHINGTON, PHYTOPLANKTON,	W72-08049
, CORROSION, ENZYMES, CHLEATION,	ALGAE, WATER QUALITY, BIODEGRADAT	W72-06837
SAY, ECOSYSTEMS, BIODEGRADATION,	ALGAE, WATER POLLUTION EFFECTS, H	W73-00840
, ROOTED AQUATIC PLANTS, AQUATIC	ALGAE, WATER TEMPERATURE, ACID ST	W72-14690
S, STREAMS, SEPTIC TANKS, FUNGI,	ALGAE, WATER POLLUTION CONTROL.: /	W72-14363
CHNIQUES, EUTROPHICATION, MARINE	ALGAE, WATER POLLUTION SOURCES, S	W73-00431
WATER SOFTENING, SOAPS, AQUATIC	ALGAE, WATER POLLUTION EFFECTS, W	W72-01085
NIA, *EUTROPHICATION, *NITROGEN,	ALGAE, WATER POLLUTION SOURCES, C	W71-11496
NS, WASTES, CHLOROPHYTA, SESSILE	ALGAE, WATER ANALYSIS.: /ES, TOXI	W71-10553
ES, INDUSTRIAL WASTES, NEW YORK,	ALGAE, WATER POLLUTION.: /AL WAST	W71-11685
SPHATES, NITRATES, HUDSON RIVER,	ALGAE, WATER QUALITY, SEWAGE DISP	W70-06509
AGE EFFLUENTS, SEWAGE TREATMENT,	ALGAE, WATER POLLUTION SOURCES, C	W72-05473
FRARED RADIATION, WATER QUALITY,	ALGAE, WATER QUALITY CONTROL, EUT	W72-12487
REMENTS, TEMPERATURE, NUTRIENTS,	ALGAE, WEEDS, TASTE, ODOR, PATHOG	W71-11006
RRIGATION, FLOW PROFILES, MARINE	ALGAE, WINDS, SEA WATER, PHYSICOC	W71-09225
RITE, AMMONIA, SALINITY, AQUATIC	ALGAE, ZOOPLANKTON, BIOCHEMICAL O	W71-11876
ENT, BIOLOGICAL CHARACTERISTICS,	ALGAE, ZOOPLANKTON, BACTERIA, NUT	W72-03215
IMARY PRODUCTIVITY, FOOD HABITS,	ALGAE, ZOOPLANKTON, CARNIVORES, D	W72-00161
, *PRIMARY PRODUCTIVITY, AQUATIC	ALGAE, ZOOPLANKTON, CHRYSOPHYTA,	W72-12738
ER, PYRROPHYTA, PROTOZOA, MARINE	ALGAE, ZOOPLANKTON, PHYTOPLANKTON	W73-00296
PHORUS SOURCES.:	ALGAL ASSAY, ORTHOPHOSPHATE, PHOS	W73-02478
VERSITY INDEX.: *MICROCOSM	ALGAL ASSAY PROCEDURE, SPECIES DI	W72-03218
BAY(WIS), NITRILOTRIACETIC ACID,	ALGAL ASSAY PROCEDURES, ALGAL GRO	W71-12072
KE MORPHOLOGY, AERATION, MIXING,	ALGAL BLOOMS, ALGAL TOXINS, WATER	W72-09304
CAL ASPECTS, LAKES, GREAT LAKES,	ALGAL BLOOMS, AGING(BIOLOGICAL),	W72-14782
	ALGAL BLOOMS.:	W72-08401
IDNEY LANIER(GEORGIA), L/ *WHITE	ALGAL BLOOMS, HYPNODINIUM, LAKE S	W71-09173
ATION, FISHKILL, EUTROPHICATION,	ALGAL BLOOMS, WATER QUALITY.: /AD	W71-08768
NT DILUTION, LIMITING NUTRIENTS,	ALGAL CELL WASHOUT.: /SH), NUTRIE	W72-08049
ONTROL, *KINETICS, *CHLOROPHYTA,	ALGAL CONTROL, *NEW YORK.: /ITY C	W72-08376
C HEALTH, WATER QUALITY CONTROL,	ALGAL CONTROL, FEDERAL GOVERNMENT	W72-06655
SILVER, CHLOROPHYTA, CYANOPHYTA,	ALGAL CONTROL, ALGICIDES, COPPER	W72-14694
, CYANOPHYTA, IRRIGATION CANALS,	ALGAL CONTROL.: /ATURE, LIMNOLOGY	W71-06102
LAGOONS, *WASTE WATER TREATMENT,	ALGAL CONTROL, EUTROPHICATION, MO	W72-04787
QUALITY CONTROL, EUTROPHICATION,	ALGAL CONTROL, *CHLORELLA, *WASTE	W72-04788
R COOLING, ANAEROBIC CONDITIONS,	ALGAL CONTROL, SPAWNING, WARM-WAT	W70-09595
A, CARBON DIOXIDE, BICARBONATES,	ALGAL CONTROL, AERATION.: /LORELL	W72-09168
C ENVIRONMENT, WATER PROPERTIES,	ALGAL CONTROL, WATER ANALYSIS, CH	W72-11724
*PHOSPHORUS, *LAKE BEDS, *ALGAE,	ALGAL CONTROL, NUTRIENT REQUIREME	W72-11617
ONMENTAL EFFECTS, AQUATIC ALGAE,	ALGAL CONTROL, ALGAL POISONING, A	W72-11186
AE, BACTERIA, *LAKES, SYMBIOSIS,	ALGAL CONTROL, EUTROPHICATION, WA	W71-13257
ONMENTAL EFFECTS, DIQUAT, ALGAE,	ALGAL CONTROL, DIATOMS, PERSISTEN	W72-00155
ATOR/ *CHLOROPHYTA, *INHIBITORS,	ALGAL CONTROL, PLANT GROWTH REGUL	W71-13793
LUTION EFFECTS, *MICROORGANISMS,	ALGAL CONTROL, *POLLUTANT IDENTIF	W72-00801
TROPHICATION, LAKES, *NUTRIENTS,	ALGAL CONTROL, *CYANPHYTA, *NUISA	W72-00798
DERAL GOVERNMENT, UNITED STATES,	ALGAL CONTROL, STANDARDS, LEGISLA	W72-01093
TENSITY, *NEBRASKA, *CYANOPHYTA,	ALGAL CONTROL, *PONDWEEDS, PONDS,	W73-02349
-NJ), SAN JOAQUIN DELTA(CALIF),/	ALGAL CYCLE, DELAWARE ESTUARY(DEL	W71-05390
NT UPTAKE, CHEMICAL COMPOSITION,	ALGAL DECOMPOSITION.: /TH, NUTRIE	W70-05469
ULATION, EXTRACELLULAR PRODUCTS,	ALGAL EXCRETIONS.: /N, GROWTH REG	W71-08687
LLARIS, UNDISSOCIATED MOLECULES,	ALGAL GROWTH.: /STICHOCOCCUS BACI	W70-07255
	ALGAL GROWTH RATES.:	W72-04789
, *GROWTH REACTORS, ALGAL PONDS,	ALGAL GROWTH RATES, *POTAMOCYPRIS	W72-04787
PYRENOIDOSA, MONOD GROWTH MODEL,	ALGAL GROWTH RATES.: / CHLORELLA	W72-04788
*ORGANIC CARBON, *ALGAL MASS,	ALGAL GROWTH.:	W72-04431
FALLS RESERVOIR, *APHANIZOMENON,	ALGAL GROWTH.: /ARBON, *AMERICAN	W72-13300
*ALGAE STRIPPING, SCENEDESMUS,	ALGAL GROWTH, ALGAL HARVESTING, *	W72-02975
E(ENGLAND), GREAT OUSE(ENGLAND),	ALGAL GROWTH.: /), LAKE WINDERMER	W71-13185
IC ACID, ALGAL ASSAY PROCEDURES,	ALGAL GROWTH, IN-LAKE NUTRIENT CO	W71-12072
	ALGAL GROWTH.:	W71-12868
RIVER(ILL), DIAGNOSTIC CRITERIO/	ALGAL GROWTH POTENTIAL, ILLINOIS	W71-11033
PING, SCENEDESMUS, ALGAL GROWTH,	ALGAL HARVESTING, *SAN JOAQUIN VA	W72-02975
S, AQUATIC ALGAE, ALGAL CONTROL,	ALGAL POISONING, AQUATIC ENVIRONM	W72-11186
EFFECTS, ENVIRONMENTAL EFFECTS,	ALGAL POISONING, ALGAE, SOLVENTS,	W70-09429

ERTEBRATES, ALGAE, MARINE ALGAE,	ALGAL POISONING, CHROMATOGRAPHY,	W71-08597
ONIA-NITROGEN, *GROWTH REACTORS,	ALGAL PCNDS, ALGAL GROWTH RATES,	W72-04787
QUATIC ECOSYSTEMS, ALGAE GROWTH,	ALGAL POPULATIONS.: / NIDULANS, A	W71-00475
ENT REGENERATION, DEOXYGENATION,	ALGAL SEDIMENTATION.: /SIN, NUTRI	W72-12997
*CYANOPHYTA, *LABORATORY TESTS,	ALGAL TOXINS, HUMAN PATHOLOGY.:	W72-04259
PHYTA, CHLOROPHYTA, SCENEDESMUS,	ALGAL TOXINS, ANTIBIOTICS(PESTICI	W71-05155
AERATION, MIXING, ALGAL BLOOMS,	ALGAL TOXINS, WATER CIRCULATICN,	W72-09304
ENVIRONMENT, TOXICITY, POISONS,	ALGAL TOXINS, FISH TOXINS, SEDIME	W72-10678
OZOA, PATHOGENS, MICROORGANISMS,	ALGAL TOXINS.: /ERIA, FUNGI, PROT	W72-08586
TOXICITY, CHEMICALS, REGULATICN,	ALGAL TOXINS, MERCURY, DETERGENTS	W73-00703
IUM, SU/ *LPP VIRUS, *BLUE-GREEN	ALGAL VIRUSES, *PLECTONEMA BORYAN	W72-04736
ATION LAGOONS, SEWAGE TREATMENT,	ALGAL: /Y, DESIGN CRITERIA, *OXID	W72-04789
*NUTRIENT REQUIREMENTS, NUISANCE	ALGAS, *ENTROPHICATION, CYTOLOGIC	W72-03985
FISH PONDS, ALGISTATIC ACTIVITY,	ALGICIDAL ACTIVITY.: /SYNERGISM,	W71-06189
HYTA, CYANOPHYTA, ALGAL CONTROL,	ALGICIDES, COPPER SULFATE, SALINI	W72-14694
TICIDES, HERBICIDES, FUNGICIDES,	ALGICIDES, CARBAMATE PESTICIDES,	W73-00238
S, FISH, ALGAE, CHEMICAL WASTES,	ALGICIDES, BACTERICIDES, CHROMIUM	W72-08683
AENA, OSCILLATORIA, MICROCYSTIS,	ALGISTAT.: /LCNIL, ULOTHRIX, ANAB	W72-07508
FICATION, SYNERGISM, FISH PONDS,	ALGISTATIC ACTIVITY, ALGICIDAL AC	W71-06189
URING TECHNIQUES, CULTURE MEDIA,	ALGOLOGY, TAXON.: CULTUT	W72-12172
SPRINGS, JACKSON HOT SPRINGS, /	ALHAMBRA HOT SPRINGS, BOULDER HOT	W72-05610
DELAWARE BAY, SCOLOPLOS ARMIGER,	ALIMITOS BAY, LAKE PONTCHARTRAIN,	W73-02029
TADECANE, HYDROCARBONS, ALKANES,	ALIPHATIC HYDROCARBONS, MASS SPEC	W73-01439
PROPERTIES, GOLD, IRON, CESIUM,	ALKALI METALS, SEASONAL, ZINC, AB	W73-01434
N, MICROORGANISMS, FERMENTATION,	ALKALINE HYDROLYSIS, ALGAE BLOOMS	W71-12223
ORGANIC PHOSPHORUS, *EXCRETION,	ALKALINE PHOSPHATASE, REASSIMILAT	W71-10083
*EPIPHYTOLOGY, *LAKES, *ALGAE,	ALKALINE WATER, MAGNESIUM, SODIUM	W71-04526
GEN, HYDROGEN ION CONCENTRATION,	ALKALINITY, RESPIRATION, PHOTOSYN	W71-04518
ER TEMPERATURE, BACTERIA, ALGAE,	ALKALINITY, IRON, PHOSPHATE, PHOS	W71-05626
MENTS, VOLCANOES, BENTHIC FLORA,	ALKALINITY, PHOTOSYNTHESIS, HYDRO	W71-09190
S, *THERMAL SPRINGS, CYANOPHYTA,	ALKALINITY, NUTRIENTS, WATER QUAL	W72-05610
E, MINERALOGY, TURBIDITY, COLOR,	ALKALINITY, VOLUMETRIC ANALYSIS,	W72-05594
ION, DETERGENTS, CARBON DIOXIDE,	ALKALINITY, BURNING, DOMESTIC WAS	W72-05473
H, CHEMICAL PROPERTIES, ACIDITY,	ALKALINITY, CONDUCTIVITY, CALCIUM	W72-04269
HOSPHATES, NITRATES, *NUTRIENTS,	ALKALINITY, CHEMICAL OXYGEN DEMAN	W72-04983
OXYGEN DEMAND, LABORATORY TESTS,	ALKALINITY, CARBONATES, BICARBONA	W72-04431
ON, CHLORELLA, SODIUM COMPOUNDS,	ALKALINITY, MINING, BIOCHEMICAL O	W72-04734
ITY, HYDROGEN ION CONCENTRATION,	ALKALINITY, CARBONATES, HARDNE: /	W71-00117
OIS, HYDROGEN ION CONCENTRATION,	ALKALINITY, HARDNESS(WATER), SAMP	W71-03028
HYLL, DIVERSION, CARBON DIOXIDE,	ALKALINITY, EPILIMNION, WASHINGTO	W71-02681
AN, MEASUREMENT, SUSPENDED LOAD,	ALKALINITY, HARDNESS(WATER), ANAL	W71-11033
ION, DECOMPOSING ORGANIC MATTER,	ALKALINITY, HYDROGEN ION CONCENTR	W71-12068
NTHESIS, *ALGAE, CARBON DIOXIDE,	ALKALINITY, NUTRIENTS, AMMONIA, H	W71-11519
DUCTIVITY, PHOSPHATES, NITRATES,	ALKALINITY, DIATOMS, CYANOPHYTA,	W72-08560
NTS, EUTROPHICATION, CYANOPHYTA,	ALKALINITY, DIURNAL, NITROGEN.: /	W72-09162
GEN ICN CONCENTRATION, HARDNESS,	ALKALINITY, CONDUCTIVITY, S: /DRO	W72-10875
N, ANAEROBIC DIGESTION, AEROBIC,	ALKALINITY, ACIDITY, LABORATORY,	W72-02850
, ABSORPTION, LIGHT PENETRATION,	ALKALINITY, HALOGENS, HYDROGEN IO	W73-01434
N, TURBIDITY, WATER TEMPERATURE,	ALKALINITY, HYDROGEN ION CONCENTR	W73-00916
, BICARBONATES, HARDNESS(WATER),	ALKALINITY, ENZYMES, TEMPERATURE,	W72-07514
LOROPHYTA, DIATOMS, CHLOROPHYLL,	ALKALINITY, CARBON DIOXIDE, TEMPE	W72-08049
*CHLOROPHYLL, WATER TEMPERATURE,	ALKALINITY, DISSOLVED OXYGEN, CON	W72-14728
OGEN, HEAT TRANSFER, PHOSPHATES,	ALKALINITY, OXYGEN DEMAND, HEAT B	W72-12997
CTUANS, ANTINEOPLASTIC ACTIVITY,	ALKALOIDS, FLAVONOIDS, STEROIDS,	W72-04275
METHYLHEPTADECANE, HYDROCARBONS,	ALKANES, ALIPHATIC HYDROCARBONS,	W73-01439
ERCURIAL COMPOUNDS, ESOX LUCIUS,	ALKYL MERCURY, EAGLES, SALMO GAIR	W72-05952
ACTANTS, *EUTROPHICATION, LINEAR	ALKYLATE SULFONATES, CHELATION, P	W71-06247
SULFONATES, SURFACTANTS, LINEAR	ALKYLATE SULFONATES, SEWAGE, NUTR	W72-06837
*POTABLE WATER, EUTROPHICATION,	ALKYLBENZENE SULFONATES, SURFACTA	W72-06837
OLS, *DETERGENTS, *LETHAL LIMIT,	ALKYLBENZENE SULFONATES.: /, PHEN	W72-12576
NDARDS, WATER POLLUTION EFFECTS,	ALKYLBENZENE SULFONATES, WATER PO	W71-11516
MERS, TURBULANT FRICTIONAL DRAG,	ALKYLENE OXIDE, POLYMERIZATION, E	W70-06968
DE, NUTRIENTS, LIMITING FACTORS,	ALLOGENIC SUCCESSION, EUTROPHICAT	W72-14790
E, *WATER TREATMENT, *FILTATION,	ALLUVIA CHANNELS, SURFACE-GROUNDW	W71-12410
, ANACYSTIS NIDULANS, GLOEOCAPSA	ALPICOLA, OSCILLATORIA CHALYBIA,	W72-07664
S, PERCA FLUVIATILIS, SALVELINUS	ALPINUS, SALMO TRUTTA, VENDACE, S	W72-12729
LA FUSCA, *SCENEDEMUS ACUTUS VAR	ALTERNANS, *NITZSCHIA ACTINASTROI	W72-12739
A CRASSIPES, JUSTICIA AMERICANA,	ALTERNANTHERA PHILOXEROIDES, GLY:	W73-01089
QUALITY CONTROL, *NON-STRUCTURAL	ALTERNATIVES, AGING(BIOLOGICAL),	W72-04279
ION, MIXING, ALGAE, TEMPERATURE,	ALTITUDE, CLIMATES, METABOLISM, B	W72-06838
*ORGANIC NUTRIENTS,	ALUM COAGULATION.:	W73-02187
OPERATION, ACID INJECTION PUMP,	ALUM INJECTION PUMP, MANURE GRIND	W72-02850
(ISRAEL).:	ALUM, FERRIC CHLORIDE, *JERUSALEM	W72-12289
C SULFATE, *STABILIZATION PONDS,	ALUM, POST-TREATMENT.: /E, *FERRI	W71-13356
ASIA.:	ALUM, RECYCLING, TREATMENT COSTS,	W72-13508
NTS, ALGAE, BACTERIA, LAKE ERIE,	ALUMINUM, MANGANESE, SODIUM, CHRO	W72-01101
OISOTOPES, RADIUM RAIODISOTOPES,	ALUMINUM, IRON, COPPER, MAGNESIUM	W71-09188
AMPLE PRESERVATION, SALPA, LUDOX	AM, SUCROSE, DENSITY GRADIENTS, T	W73-00371
DICA, MELOSIRA ITALICA, MELOSIRA	AMBIGUA, SYNEDRA FILIFORMIS, FRAG	W73-00384
RUS NEBULOSUS, PERCA FLAVESCENS,	AMBLOPLITES RUPESTRIS, ESOX LUCIU	W70-08098
NIKIN TEST.:	AMCHITKA ISLAND, MILROW TEST, CAN	W73-00764
S CARPIO, CATOSTOMUS COMMERSONI,	AMEIURUS NEBULOSUS, PERCA FLAVESC	W70-08098
VIRGINIA, *ATLANTIC OCEAN, NORTH	AMERICA.: /ARSHES, SYSTEMATICS, *	W72-12633
N, CALCIUM, PUBLIC HEALTH, SOUTH	AMERICA, ATLANTIC COASTAL PLAIN,	W73-00819
ES, *TOCKS ISLAND RESERVOIR(NJ),	AMERICAN SHAD, EKMAN DREDGE, PUMP	W72-14280
, EICHHORNIA CRASSIPES, JUSTICIA	AMERICANA, ALTERNANTHERA PHILOXER	W73-01089
BIOLOGICAL SYSTEMS, BELEMNITELLA	AMERICANA, TERRESTRIAL PLANTS, CO	W72-12709
NIA SUBEQUICILIATA, UROGLENOPSIS	AMERICANA, CRYPTOMONAS PULSILLA,	W72-13816
ROSIN	AMINE D ACETATE(RADA).:	W71-05083
A, CADMIUM, DIMETHYLNITROSAMINE,	AMINES, CARCINOGENS, TORULOPSIS S	W72-09675

ON, ELECTROPHORESIS, DETERGENTS, AMINO ACIDS, SPECTROPHOTOMETRY, M W72-06124
, *PREDATION, ORGANIC COMPOUNDS, AMINO ACIDS, VITAMINS, CARBOHYDRA W73-00942
N RADIOISOTOPES, CHROMATOGRAPHY, AMINO ACIDS, GAS CHROMATOGRAPHY, W72-07710
R ALGAE, PICLORAM, PEDIASTRUM, 4- AMINO-3,5,6-TRICHLOROPICOLINIC AC W72-08440
OTROPHY, HARDNESS(WATER), SOILS, AMINO: /EEDS, PHYTOPLANKTON, OLIG W72-08459
IS, CODIUM, FUSCUS, MACROPHYTES, AMMOBACULITES DILATATUS, AMMONIA W73-00367
PHYTES, AMMOBACULITES DILATATUS, AMMONIA BECCARII, AMMOTIUM SALSUM W73-00367
ATMENT, PHYSICOCHEMICAL METHODS, AMMONIA STRIPPING, GUGGENHEIM PRO W70-09907
ITROGEN, *NUTRIENT REQUIREMENTS, AMMONIA, *GROWTH RATES, *WASTE AS W72-04787
HYDROGRAPHY, NITRITES, NITRATES, AMMONIA, ALGAE, CHLORELLA.: /ON, W72-08056
, *NITRATES, NITROGEN COMPOUNDS, AMMONIA, ALGAE, SEWAGE TREATMENT, W71-13939
MES, PROTEINS, LABORATORY TESTS, AMMONIA, ASSAY, INHIBITION, UREAS W72-03222
EN DEMAND, PHOSPHORUS, NITRATES, AMMONIA, COAGULATION, FLOCCULATIO W71-13341
METABOLISM, RESPIRATION, LIPIDS, AMMONIA, CARBOHYDRATES, CARBON DI W72-10233
PHOTOSYNTHESIS, LIGHT INTENSITY, AMMONIA, CARBON DIOXIDE, FLORIDA, W72-09103
N EFFECTS, PHOSPHORUS, NITROGEN, AMMONIA, CARBON, CHLORELLA.: /TIO W71-12857
EDIMENTS, SEA WATER, FRESHWATER, AMMONIA, EUTROPHICATION, OXIDATIO W72-11563
DIOXIDE, ALKALINITY, NUTRIENTS, AMMONIA, HYDROGEN ION CONCENTRATI W71-11519
N, LABORATORY TESTS, FILTRATION, AMMONIA, HYDROGEN ION CONCENTRATI W72-01881
MENTATION, NITROGEN, PHOSPHORUS, AMMONIA, METHANE, TEMPERATURE, HY W72-10390
CAL OXYGEN DEMAND, SOLID WASTES, AMMONIA, NITROGEN COMPOUNDS, NITR W72-09675
, INDUSTRIAL WASTES, DETERGENTS, AMMONIA, NITRITES, NITRATES.: /PY W71-11507
S(WATER), ANALYTICAL TECHNIQUES, AMMONIA, NITRATES, PHOSPHATES, WA W71-11033
E, PHYTOPLANKTON, NANNOPLANKTON, AMMONIA, NITRATES, NITRITES, HARD W72-08048
TROGEN FIXATION, NITROGEN CYCLE, AMMONIA, NITRATES, MODEL STUDIES, W72-08459
RES, *SEA WATER, *NITRIFICATION, AMMONIA, NITRITE, SPHAEROTILUS, N W72-05421
LS, *NITROGEN CYCLE, *OXIDATION, AMMONIA, NITRITES, OXYGEN, PHYTOP W71-05390
EATMENT, *WATER QUALITY CONTROL, AMMONIA, NITROGEN, PHOSPHORUS, AL W71-09450
LANKTON, PRODUCTIVITY, ANABAENA, AMMONIA, NITRATES, LIMITING FACTO W73-02472
ISSOLVED OXYGEN, CARBON DIOXIDE, AMMONIA, ORGANIC MATTER.: /ION, D W72-01363
, CYTOLOGICAL STUDIES, CULTURES, AMMONIA, POTASSIUM COMPOUNDS, NIT W72-03217
ITROGEN, *NITRATES, GROUNDWATER, AMMONIA, PRECIPITATION, SEDIMENTS W71-06435
TEMPERATURE, LIGHT PENETRATION, AMMONIA, PHOSPHATES, SALINE WATER W72-05469
ISSOLVED OXYGEN, CARBON DIOXIDE, AMMONIA, PHOSPHATE, ANALYTICAL TE W72-06046
TRATIFICATION, NITRATES, SILICA, AMMONIA, PHOSPHATES, RUNOFF, ILLI W71-03028
NT, ALGAE, POTABLE WATER, TASTE, AMMONIA, SILICA.: /LUTION TREATME W71-01491
OGEN ION CONCENTRATION, NITRITE, AMMONIA, SALINITY, AQUATIC ALGAE, W71-11876
COMPOSTING, REFEEDING, ANHYDROUS AMMONIA, WASTE CHARACTERIZATION.: W71-07551
L, LAKES, SOILS, EUTROPHICATION, AMMONIFICATION, NITRIFICATION, DE W72-07933
, PHOTOSYNTHESIS, PHYTOPLANKTON, AMMONIUM COMPOUNDS, NUTRIENTS, PR W71-11008
SSOCIATIONS, LA GORCE MOUNTAINS, AMMONIUM IONS.: /PHILIC, BIOTIC A W73-00854
TES DILATATUS, AMMONIA BECCARII, AMMOTIUM SALSUM, ELPHIDIUM ADVENU W73-00367
ENII, CHROMATIUM, MACROMONAS MO/ AMOEBOBACTER ROSEUS CHROMATIUM OK W72-07896
ILLATORIA PRINCEPS, OSCILLATORIA AMOENA, OSCILLATORIA TENUIS, OSCI W72-07896
ION, METHODOLOGY, INVERTEBRATES, AMPHIBIANS, TOXICITY, CHARA, SUBM W72-00155
NANA, THALASSIOSIRA FLUVIATILIS, AMPHIDINIUM CARTERI, COCCOLITHUS W72-12998
TORIA, STOMATOPODA, POLYCHAETES, AMPHIOXUS, SALPS, BALAN: /OSCILLA W73-00935
CULATA, PLAGIOGRAMMA BROCKMANNI, AMPHIPLEURA RUTILANS, THALASSIOSI W73-00853
ISSIMA, ACHNANTHES CF. BIRGIANI, AMPHIPLEURA PELLUCIDA, AMPHORA OV W72-13142
SEASONAL, BOTTOM SEDIMENTS, ICE, AMPHIPODA.: /ODUCTION, SAMPLING, W71-08026
LGAE, DIATOMS, CLAMS, SHELLFISH, AMPHIPODA, WATER BIRDS, WISCONSIN W70-08098
NEMATODES, GASTROPODS, MOLLUSKS, AMPHIPODA, MECHANICAL EQUIPMENT.: W72-05432
ALGAE, SNAILS, MUSSELS, ISOPODS, AMPHIPODA, RADIOISOTOPES, ABSORPT W72-03347
, EMILIANA(COCCOLITHUS) HUXLEYI, AMPHIPRORA ALATA, CHRYSOCHROMULIN W73-00380
JAVANICA, LICMOPHORA JURGENSII, AMPHORA OVALIS, SYNEDRA FASCIULAT W72-08141
LIC ALGAE, SHEAR WATER(ENGLAND), AMPHORA OVALIS, OPEPHORA MARTYI, W71-12855
LEURA PELLUCIDA, AMPHORA OVALIS, AMPHORA OVALIS V. PEDICULUS, COCC W72-13142
BIRGIANI, AMPHIPLEURA PELLUCIDA, AMPHORA OVALIS, AMPHORA OVALIS V. W72-13142
ES, CHLOROPHYLL A, *YAQUINA BAY, AMPHORA SPP, MELOSIRA SULCATA, NA W73-00853
HNIQUES, SALMONELLA TYPHIMURIUM, AMPICILLIN, VIBRIOS, SALMONELLA T W72-13800
DULANS, GL/ STABILIZATION PONDS, ANABAENA CYLINDRICA, ANACYSTIS NI W72-07664
ETYLENE REDUCTION, *HETEROCYSTS, ANABAENA CYLINDRICA, ANABAENOPSIS W71-12070
NDRIN, METABOLITES, SKELETONEMA, ANABAENA CYLINDRICA, DUNALIELLA.: W72-03545
INDRICA, APHANIZOMENON FLOS-AQU/ ANABAENA CIRCINALIS, ANABAENA CYL W72-10613
N FLOS-AQU/ ANABAENA CIRCINALIS, ANABAENA CYLINDRICA, APHANIZOMENO W72-10613
IROSTRIS/ *WINNISQUAM LAKE(N H), ANABAENA CIRCINALIS, BOSMINA LONG W71-03014
ON FLOW-AQUAE, DESTRATIFICATION, ANABAENA FLOW-AQUAE, HOMOGENIZING W72-07890
EEK, CHARA, ELODEA, CALLITRICHE, ANABAENA FLOSAQUAE, SPIRULINA, GY W73-00840
KEZAR LAKE(N.H.), APHANIZOMENON, ANABAENA SPP, MICROCYSTIS AERUGIN W72-07890
ORIA REDEKEI, LYNGBYA LIMNETICA, ANABAENA SPP., CHROMATIUM, MERISM W72-07145
DROCHLORIDE, MERISOMOPEDIA SPP., ANABAENA SPP., GOMPHONEMA SPP., N W72-14706
*SPHEROPLASTS, BLUE-GREEN ALGAE, ANABAENA VARIABILIS, ANACYSTIS NI W71-08683
ATMOSPHERIC CARBON, MICROCYSTIS, ANABAENA.: * W72-10607
H WALES, MICROCYSTIS AERUGINOSA, ANABAENA.: /CUOLE DEFLATION, SOUT W72-11564
SOILS, SYMBIOSIS, AFRICA, ASIA, ANABAENA.: /OP PRODUCTION, FUNGI, W73-02473
ON, PHYTOPLANKTON, PRODUCTIVITY, ANABAENA, AMMONIA, NITRATES, LIMI W73-02472
N, LIGHT INTENSITY, TEMPERATURE, ANABAENA, BACTERIA, PHOTOSYNTHESI W73-02471
KTON NETS, DIATOMS, FOOD CHAINS, ANABAENA, SCENEDESMUS, SEASONAL, W72-12738
TEMPERATURE, MUD, GROWTH RATES, ANABAENA, NITROGEN FIXATION, DENI W72-12544
ELLA, OSCILLATORIA, APHANOTHECE, ANABAENA, APHANIZOMENON, NAVICULA W72-10875
*LIMITING NUTRIENTS, ANABAENA, CARBON SOURCES.: W72-09168
FULVIC ACID, GROWTH STIMULATION, ANABAENA, GLOEOTRICHIA, MICROCYST W72-01097
UIS, AU/ *TRICALCIUM PHOSPHATES, ANABAENA, NOSTOC, TOLYPOTHRIX TEN W71-10070
*BAVARIA, SCHLIERSEE, TEGERNSEE, ANABAENA, COREGONS, LYNGBYA LIMNE W71-00117
MESOTROPHIC, MICROCYSTIS, ANABAENA, LYNGBYA.: W72-04269
OTHRIX, ANACYSTIS N/ *TOLERANCE, ANABAENA, NOSTOC, AULOSIRA, TOLYP W72-14807
IL, *CHLOROPHTHALONIL, ULOTHRIX, ANABAENA, OSCILLATORIA, MICROCYST W72-07508
MUS, ELAKOTOTHRIX, OSCILLATORIA, ANABAENA, MICROCYSTIS, TRIBONEMA, W72-07940
RIDINIUM, OSCILLATORIA RUBECENS, ANABAENA, CRUSSOSTREA, OSCILLATOR W73-00840
ETEROCYSTS, ANABAENA CYLINDRICA, ANABAENOPSIS, ACETYLENE TEST, CYL W71-12070

CUS, MICROCYSTIS, APHANIZOMENON,	ANABENA SPIROIDES.: / STEPHANODIS	W71-12855
X TENUIS, AULOSIRA FERTILISSIMA,	ANACYSTIS NIDULANS, CERASON, DIAT	W71-12071
OMONAS, FONTINALIS, ASTINASTRUM,	ANACYSTIS NIDULANS, MICROCYSTIS.:	W71-11519
X TENUIS, AULOSIRA FERTILISSIMA,	ANACYSTIS NIDULANS.: / TOLYPOTHRI	W71-10070
ADRUPLICATUM STR/ PHOTODIELDRIN,	ANACYSTIS NIDULANS, AGMENELLUM QU	W72-03545
TION PONDS, ANABAENA CYLINDRICA,	ANACYSTIS NIDULANS, GLOEOCAPSA AL	W72-07664
, NOSTOC, AULOSIRA, TOLYPOTHRIX,	ANACYSTIS NIDULANS.: /E, ANABAENA	W72-14807
, LAKE HARTWELL(SOUTH CAROLINA),	ANACYSTIS NIDULANS, AQUATIC ECOSY	W71-00475
REEN ALGAE, ANABAENA VARIABILIS,	ANACYSTIS NIDULANS.: /STS, BLUE-G	W71-08683
, EUGLENA GRACLIS, CALVIN CYCLE,	ANACYSTIS NIDULANS.: /BURG EFFECT	W71-09172
*ENUMERATION,	ANACYSTIS.:	W71-00139
AL BASIN, TRIBONEMA, OEDOGONIUM,	ANACYSTIS, CERATIUM, COSMARIUM.: /	W72-12995
MEMBRANACEA, BOODLEA COMPOSITA,	ANADYOMENE STELLATA, BR: /OROPSIS	W72-11800
LS, IRON, MANGANESE, PHOSPHORUS,	ANAEROBIC CONDITIONS, CHLOROPHYLL	W72-12997
TRIFICATION, ALGAE, THERMOCLINE,	ANAEROBIC BACTERIA, AEROBIC BACTE	W72-12996
, PRIMARY PRODUCTIVITY, BIOMASS,	ANAEROBIC CONDITIONS, WATER TEMPE	W72-12940
PHOSPHORUS, AEROBIC CONDITIONS,	ANAEROBIC CONDITIONS, CHLORELLA,	W70-10405
, MIXING, SAFETY, WATER COOLING,	ANAEROBIC CONDITIONS, ALGAL CONTR	W70-09595
IFORNIA, FILTERS, PARTICLE SIZE,	ANAEROBIC CONDITIONS, WASTE WATER	W71-08223
RATION, DISSOLVED OXYGEN, ALGAE,	ANAEROBIC CONDITIONS, AEROBIC CON	W71-07099
PHOSPHATES, NITRATES, NUTRIENTS,	ANAEROBIC CONDITIONS, AEROBIC CON	W71-07100
ION LAGOONS, AEROBIC CONDITIONS,	ANAEROBIC CONDITIONS, ALGAE, ORGA	W71-07551
TES, PHOSPHATES, BIODEGRADATION,	ANAEROBIC CONDITIONS, AEROBIC CON	W71-07106
ADIATION, ALGAE, PHOTOSYNTHESIS,	ANAEROBIC CONDITIONS, AEROBIC CON	W71-07084
GEN, ALGAE, SLUDGE, OXYGENATION,	ANAEROBIC CONDITIONS, AEROBIC CON	W71-07109
DI/ *OXIDATION LAGOONS, *SLUDGE,	ANAEROBIC CONDITIONS, AEROBIC CON	W71-07102
ERATURE, *NITROGEN, *PHOSPHATES,	ANAEROBIC CONDITIONS, AEROBIC CON	W72-04789
, E. COLI, BIOINDICATORS, PONDS,	ANAEROBIC CONDITIONS, WASTE WATER	W72-07664
ANOPHYTA, SYSTEMATICS, VITAMINS,	ANAEROBIC BACTERIA, ELECTRON MICR	W72-07951
RIENTS, AQUATIC ALGAE, NITROGEN,	ANAEROBIC DIGESTION, AEROBIC, ALK	W72-02850
ATORY TESTS, AEROBIC CONDITIONS,	ANAEROBIC CONDITIONS, ORGANIC LOA	W71-13326
TION, SEDIMENTATION, FILTRATION,	ANAEROBIC DIGESTION, CHLORINATION	W72-00042
EN CYCLE, ENVIRONMENTAL EFFECTS,	ANAEROBIC CONDITIONS, NITRATES, D	W72-11563
LVED OXYGEN, AEROBIC CONDITIONS,	ANAEROBIC CONDITIONS, ORGANIC LOA	W72-11642
VENUS, TIVELA STULTORUM, EMERITA	ANALOGA, DIASTYLOPSIS, OXYUROSTYL	W72-09092
TATION, DEWATERING, DRYING, COST	ANALYSES, WASTE WATER TREATMENT.:	W71-08951
SEPTIC TANKS, BIOASSAY, CHEMICAL	ANALYSES, NITROGEN, NUTRIENTS, SU	W72-12955
REGRESSION ANALYSIS, CORRELATION	ANALYSIS.: / NITRATES, SEASONAL,	W72-12393
ULTIVARIATE ANALYSIS, ORDINATION	ANALYSIS.: *LAKE WINGRA(WIS.), M	W73-02469
ORGANIC MATTER, BACTERIA, WATER	ANALYSIS.: /NS, RIVERS, CULTURES,	W71-08032
CHLORINATION, GROUNDWATER, COST	ANALYSIS.: /SIS, ALGAE, BACTERIA,	W71-07123
ETHODOLOGY, EVALUATION, CHEMICAL	ANALYSIS.: /E, WATER RESOURCES, M	W71-05083
ATHEMATICAL STUDIES, CORRELATION	ANALYSIS.: /, CULTURES, VOLUME, M	W70-05547
NEUTRON ACTIVATION	ANALYSIS.:	W72-02073
UM, BIOASSAY, NEUTRON ACTIVATION	ANALYSIS.: /ANT GROWTH, *MOLYBDEN	W72-00801
HLOROPHYTA, SESSILE ALGAE, WATER	ANALYSIS.: /ES, TOXINS, WASTES, C	W71-10553
: *CLUSTER	ANALYSIS(FISH), *DICKERSON PLANT.	W72-07791
ITY CONTROL, *CYANOPHYTA, *WATER	ANALYSIS, LAKES, SAMPLING, DIATOM	W72-07890
OOPLANKTON, PHYTOPLANKTON, WATER	ANALYSIS, POLLUTANTS, AQUATIC ALG	W72-07896
UTRIENTS, *FOOD CHAIN/ *CHEMICAL	ANALYSIS, *MONITORING, *CYCLING N	W72-07715
RATURE, CHEMICAL ANALYSIS, WATER	ANALYSIS, TIDAL EFFECTS, STABILIT	W72-08141
ITY, WATER TEMPERATURE, CHEMICAL	ANALYSIS, WATER ANALYSIS, TIDAL E	W72-08141
OLS, WATER TEMPERATURE, CHEMICAL	ANALYSIS, SAMPLING, ECOLOGY, PHYT	W72-07710
, SEDIMENT DISCHARGE, REGRESSION	ANALYSIS, BIOMASS, METHODOLOGY, R	W72-06283
PHORUS, RAIN, LAKES, CLADOPHORA,	ANALYSIS, PHYTOPLANKTON, CHEMICAL	W73-00232
NALYSIS, PHYTOPLANKTON, CHEMICAL	ANALYSIS, DIATOMS, CYANOPHYTA, TU	W73-00232
LANKTON, *MARINE ALGAE, CHEMICAL	ANALYSIS, *NITRATES, *AMMONIA, DI	W73-00380
SOURCES, ION EXCHANGE, *CHEMICAL	ANALYSIS, WATER POLLUTION, WATER	W73-00286
YSTEMATICS, CYANOPHYTA, CHEMICAL	ANALYSIS, SEASONAL, BIOMASS, ON-S	W73-00242
ANALYSIS, WATER POLLUTION, WATER	ANALYSIS, BIOASSAY, CYANOPHYTA, R	W73-00286
RIA, PRODUCTIVITY, RADIOCHEMICAL	ANALYSIS, THERMOPHILIC BACTERIA,	W72-14301
Y, FISH, FISH BEHAVIOR, CHEMICAL	ANALYSIS, SAMPLING, VOLUMETRIC AN	W72-14280
L ANALYSIS, SAMPLING, VOLUMETRIC	ANALYSIS, ALGAE, RESERVOIR STORAG	W72-14280
OF POLLUTANTS, SEA WATER, WATER	ANALYSIS, RADIOACTIVITY TECHNIQUE	W73-00817
NTS, MONITORING, SAMPLING, WATER	ANALYSIS, SEA WATER, STABLE ISOTO	W73-00831
TRACE ELEMENTS, SAMPLING, WATER	ANALYSIS, SEA WATER.: /CTROSCOPY,	W73-00832
F POLLUTANTS, NEUTRON ACTIVATION	ANALYSIS, SPECTROSCOPY, TRACE ELE	W73-00832
DS, LIMNOLOGY, ECOLOGY, CHEMICAL	ANALYSIS, AQUATIC ALGAE, NUTRIENT	W73-00918
NUTRIENTS, WATER QUALITY, WATER	ANALYSIS, FREQUENCY, FLUCTUATIONS	W73-00918
PIRATION, MARSH PLANTS, CHEMICAL	ANALYSIS, LEAVES, SAMPLING, PHYTO	W73-01089
, GRAZING, SYSTEMATICS, CHEMICAL	ANALYSIS, PHYSICAL PROPERTIES, CH	W73-01627
IRS, TRIBUTARIES, DAMS, CHEMICAL	ANALYSIS, BIOMASS, ALGAE, PHYSICO	W73-01435
IMNOLOGY, *PHYTOPLANKTON, *WATER	ANALYSIS, *STANDING CROP, *PRIMAR	W73-01435
S, STRONTIUM RADIOISOTOPES, SOIL	ANALYSIS, SOIL CONTAMINATION, CLA	W73-01673
ATURE, WATER CHEMISTRY, CHEMICAL	ANALYSIS, NUTRIENTS.: /TER TEMPER	W73-01972
TS, MARINE MICROORGANISMS, WATER	ANALYSIS, BENTHIC FLORA, ORGANIC	W73-02100
T, *RADIOECOLOGY, *RADIOCHEMICAL	ANALYSIS, MEASUREMENT, *NEVADA, A	W73-01673
ORNIA, HYDROLOGIC DATA, CHEMICAL	ANALYSIS, DATA COLLECTIONS, NUTRI	W71-10577
THOS, WOODLAND STREAM, ECOSYSTEM	ANALYSIS, MACROCONSUMERS, AUTUMNA	W71-10064
MISTRY, CHLOROPHYLL, HYPSOMETRIC	ANALYSIS, HYPOLIMNION, EPILIMNION	W71-11949
ALGAE, AQUATIC PLANTS, CHEMICAL	ANALYSIS, WATER POLLUTION SOURCES	W71-12252
MARINE ALGAE, *DIATOMS, CHEMICAL	ANALYSIS, HYDROGEN ION CONCENTRAT	W71-11008
WATER ANALYSIS, ALGAE, CHEMICAL	ANALYSIS, SAMPLING, ANALYTICAL TE	W71-11234
UATIC ALGAE, *NEUTRON ACTIVATION	ANALYSIS, *TRACE ELEMENTS, COPPER	W71-11515
ENVIRONMENT, *NEUTRON ACTIVATION	ANALYSIS, RADIOACTIVITY, ADSORPTI	W71-11036
.: *ATP	ANALYSIS, *ADENOSINE TRIPHOSPHATE	W71-11234
RGANIC MATTER, *SEA WATER, WATER	ANALYSIS, ALGAE, CHEMICAL ANALYSI	W71-11234
A, CYANOPHYTA, BIOMASS, CHEMICAL	ANALYSIS, STATISTICAL METHODS, NI	W71-11029
ION/ *ECOLOGY, *PONDS, *CHEMICAL	ANALYSIS, TEMPERATURE, SEASONAL,	W72-01363
TECHNIQUES, *NEUTRON ACTIVATION	ANALYSIS, *CHEMICAL ANALYSIS, GRE	W72-01101

N ACTIVATION ANALYSIS, *CHEMICAL ANALYSIS, GREAT LAKES, EUTROPHICA W72-01101
ANSAS, WATER CHEMISTRY, CHEMICAL ANALYSIS, EVAPORATION, SEEPAGE, T W72-02274
EMAND, WATER SPORTS, PARKS, COST ANALYSIS, MICHIGAN, WASTE WATER T W72-00042
REEFS, WATER RESOURCES, SYSTEMS ANALYSIS, WASTE TREATMENT, AIR PO W71-13723
ELS, LINEAR PROGRAMMING, SYSTEMS ANALYSIS, DECIDUOUS FORESTS, DECO W72-03342
COMPOUNDS, *CARBON, GRAVIMETRIC ANALYSIS, PRODUCTIVITY, SEASONAL, W72-09108
YLL, ALGAE, PHYTOPLANKTON, WATER ANALYSIS, CARBON CYCLE, ORGANIC M W72-09108
ENVIRONMENT, SAMPLING, CHEMICAL ANALYSIS, PLANKTON, COMPUTER PROG W72-08560
TION, MATHEMATICAL MODELS, WATER ANALYSIS, POLLUTANT IDENTIFICATIO W72-08790
NIQUES, CHEMICAL ANALYSIS, WATER ANALYSIS, EVALUATION, WATER QUALI W72-10883
ANIC LOADING, ODOR, SLUDGE, COST ANALYSIS, NITROGEN, CLAYS, *WASTE W72-11642
NALYSIS, BIORHYTHMS, CORRELATION ANALYSIS, PRIMARY PRODUCTIVITY.: / W72-11719
PROPERTIES, ALGAL CONTROL, WATER ANALYSIS, CHEMCONTROL, WATER POLL W72-11724
E, SPATIAL DISTRIBUTION, FOURIER ANALYSIS, BIORHYTHMS, CORRELATION W72-11719
*ANALYTICAL TECHNIQUES, CHEMICAL ANALYSIS, WATER ANALYSIS, EVALUAT W72-10883
OPERATION AND MAINTENANCE, *COST ANALYSIS, *FILTERS, HEADLOSS, PRE W72-09393
R POLLUTION CONTROL, MANAGEMENT, ANALYSIS, WATER QUALITY, CYANOPHY W72-09171
ALGAE, SEASONAL, WEATHER, WATER ANALYSIS, TRACERS, RADIOACTIVITY W72-09122
ROPS, LABORATORY TESTS, CHEMICAL ANALYSIS, BIOASSAY, CYTOLOGICAL S W72-09163
C ACIDS, SEPARATION TE/ CHEMICAL ANALYSIS, *ALGAE, COLUMNS, ORGANI W72-09365
PROTEIN, LABORATORY TESTS, COST ANALYSIS, ALGAE, WASTE WATER TREA W72-10390
, NITROGEN, PHOSPHORUS, CHEMICAL ANALYSIS, PROTEINS, CARBOHYDRATES W70-05469
H OF POLLUTANTS, SAMPLING, WATER ANALYSIS, ALGAE, COLIFORMS, DISSO W70-08249
H OF POLLUTANTS, SAMPLING, WATER ANALYSIS, ALGAE, COLIFORMS, DISSO W70-08248
IOACTIVITY, PHOSPHATES, CHEMICAL ANALYSIS, ALGAE, MICROORGANISMS, W71-01474
Y, CHLORELLA, NITRATES, CHEMICAL ANALYSIS, CHLOROPHYTA, DIATOMS, T W70-10403
NS, FLUORIDES, ALGAE, VOLUMETRIC ANALYSIS, WATER CHEMISTRY.: /, IO W71-00474
MENT, *DOMESTIC WASTES, CHEMICAL ANALYSIS, EFFLUENTS, TRICKLING FI W70-09907
*ANALYTICAL TECHNIQUES, *WATER ANALYSIS, *NITRATES, *ELECTRODES, W71-00474
, ALGAE, STANDING CROP, CHEMICAL ANALYSIS, MOSQUITOES, BACTERIA, F W71-01488
OXYGEN, AIR POLLUTION, CHEMICAL ANALYSIS, EUTROPHICATION, TRANSPI W71-01489
HICATION, *MATHEMATICA/ *SYSTEMS ANALYSIS, *WATER QUALITY, *EUTROP W71-03021
LIFE, IRRIGATION SYSTEMS, WATER ANALYSIS, TEMPERATURE, LIMNOLOGY, W71-06102
*NUTRIENTS, *SEDIMENTS, *WATER ANALYSIS, LAKE MORPHOLOGY, SEDIME W71-05626
LAGOONS, *CLIMATIC ZONES, *COST ANALYSIS, TEMPERATURE, NUTRIENTS, W71-07109
GRADATION, SEEPAGE, SLUDGE, COST ANALYSIS, *BIOCHEMICAL OXYGEN DEM W71-07099
SONING, CHROMATOGRAPHY, CHEMICAL ANALYSIS, BIOCHEMISTRY, PLANT PHY W71-08597
ACIDS, TRACE ELEMENTS, CHEMICAL ANALYSIS, ION EXCHANGE, ABSORPTIO W71-09188
ING, WATER QUALITY CONTROL, COST ANALYSIS, WASTE WATER TREATMENT, W71-08953
NALYSIS, ALGAE, NUTRIENTS, WATER ANALYSIS, DOMESTIC WASTES.: /AL A W72-04791
IES, WATER TEMPERATURE, CHEMICAL ANALYSIS, ALGAE, NUTRIENTS, WATER W72-04791
PACIFIC OCEAN, NITRATE, CHEMICAL ANALYSIS, SODIUM CHLORIDE, COLUMB W72-05421
HYTA, CHLORELLA, TOXICITY, WATER ANALYSIS, ABSORPTION, CULTURES, D W72-04134
CHING, VOLUMETRIC ANALYSIS, SOIL ANALYSIS, SPECTROMETERS, ZOOPLANK W72-05965
*MARINE ANIMALS/ *RADIOCHEMICAL ANALYSIS, *SEA WATER, *SEDIMENTS, W72-05965
, BIOASSAY, LEACHING, VOLUMETRIC ANALYSIS, SOIL ANALYSIS, SPECTROM W72-05965
Y, COLOR, ALKALINITY, VOLUMETRIC ANALYSIS, FISH, INTAKE STRUCTURES W72-05594
, *TRACE ELEMENTS, RADIOCHEMICAL ANALYSIS, RESINS, *SEA WATER, BIO W72-04708
, *DISTRIBUTION PATTERNS, *WATER ANALYSIS, *BIOINDICATORS, ALGAE, W72-04736
CARBON, SULFUR, SAMPLING, WATER ANALYSIS, ALGAE, SPECTROPHOTOMETR W72-04743
E WATER TREATMENT, *COST-BENEFIT ANALYSIS, DETERGENTS, PHOSPHORUS, W72-04734
HORUS, *EUTROPHICATION, CHEMICAL ANALYSIS, WATER POLLUTION SOURCES W73-02478
*LAKE WINGRA(WIS.), MULTIVARIATE ANALYSIS, ORDINATION ANALYSIS.: W73-02469
NT, *PHOSPHORUS, *NITROGEN, COST ANALYSIS, OZONE, DENITRIFICATION, W73-02187
ANALYSIS, MARINE ALGAE, CHEMICAL ANALYSIS, ADSORPTION, SEPARATION W73-02105
S, *CALIFORNIA, SEA WATER, WATER ANALYSIS, MARINE ALGAE, CHEMICAL W73-02105
COMPOUNDS, CHLAMYDOMONAS, WATER ANALYSIS, FLUOROMETRY, PIGMENTS.: W72-12637
TION, *LAKES, *ALPINE, *SYNOPTIC ANALYSIS, COLD REGIONS, NUTRIENTS W72-12549
ANALYTICAL TECHNIQUES, NUMERICAL ANALYSIS, MODEL STUDIES, COMPUTER W72-12724
, NITRATES, SEASONAL, REGRESSION ANALYSIS, CORRELATION ANALYSIS.: / W72-12393
HOSPHORUS, NITROGEN, *REGRESSION ANALYSIS, VARIABILITY.: /RATES, P W72-12966
ENVIRONMENT, SALINITY, CHEMICAL ANALYSIS, BIOMASS, DISSOLVED OXYG W72-12941
CIDS, FULVIC ACIDS, KELPS, WATER ANALYSIS, EXUDATION, WATER POLLUT W72-12166
TTER, *MARINE ALGAE, / *CHEMICAL ANALYSIS, *DECOMPOSING ORGANIC MA W72-12166
STUDIES, DIGITAL COMPUTERS, DATA ANALYSIS, EQUATIONS, REPRODUCTION W72-11920
OPES, ABSORPTION, SEDIMENTATION, ANALYTICAL TECHNIQUES, MOLECULAR W72-12203
*EUTROPHICATION, *ASSAY, *ALGAE, ANALYTICAL TECHNIQUES, WASTE WATE W72-12966
OGICAL PROPERTIES, REPRODUCTION, ANALYTICAL TECHNIQUES, NUMERICAL W72-12724
OMATOGRAPHY, GAS CHROMATOGRAPHY, ANALYTICAL TECHNIQUES: /USES, CHR W72-13800
TUDIES, NUTRIENTS, PLANT GROWTH, ANALYTICAL TECHNIQUES, WATER TEMP W72-13651
UTANTS, *PUBLIC HEALTH, ENZYMES, ANALYTICAL TECHNIQUES, PESTICIDES W72-13347
IC WASTES, NUTRIENTS, SEDIMENTS, ANALYTICAL TECHNIQUES, FERTILIZAT W73-02478
UTION TREATMENT, *PUBLIC HEALTH, ANALYTICAL TECHNIQUES, WASTE WATE W73-02144
BON DIOXIDE, AMMONIA, PHOSPHATE, ANALYTICAL TECHNIQUES, *PHOTOSYNT W72-06046
YSIS, ABSORPTION, CULTURES, DDT, ANALYTICAL TECHNIQUES, BIOMASS, G W72-04134
OPHICATION, CYTOLOGICAL STUDIES, ANALYTICAL TECHNIQUES, POLLUTANT W72-03985
STORAGE, FLOW RATES, INHIBITION, ANALYTICAL TECHNIQUES, ENZYMES, C W72-04070
PLANTS, *ALGAE, *PUBLIC HEALTH, ANALYTICAL TECHNIQUES, TOXICITY, W72-04275
ALINITY, CHEMICAL OXYGEN DEMAND, ANALYTICAL TECHNIQUES, TERTIARY T W72-04983
*ALGAE, *ABSORPTION, MONITORING, ANALYTICAL TECHNIQUES, LABORATORY W71-08032
MATHEMATICAL MODELS, SIMULATION, ANALYTICAL TECHNIQUES, PHOSPHORUS W71-03021
PHIPODA, WATER BIRDS, WISCONSIN, ANALYTICAL TECHNIQUES.: /FISH, AM W70-08098
OGIC DATA, *MEASUREMENT, *ASSAY, ANALYTICAL TECHNIQUES, EVALUATION W72-10678
N LAGOONS, COLIFORMS, NUTRIENTS, ANALYTICAL TECHNIQUES, MICROBIOLO W72-09590
ELS, NITROGEN, ABATEMENT, COSTS, ANALYTICAL TECHNIQUES, MUNICIPAL W72-09171
TOMS, BACTERIA, LAKES, NITROGEN, ANALYTICAL TECHNIQUES, CARBON, EU W72-09163
T RADIOISOTOPES, *BIOINDICATORS, ANALYTICAL TECHNIQUES, POLLUTANT W72-10958
ASTE, *ODOR, *ORGANIC COMPOUNDS, ANALYTICAL TECHNIQUES, SEPARATION W72-11604
OSTOC STREAMS, DATA COLLECTIONS, ANALYTICAL TECHNIQUES, NITROGEN F W72-08508
NMENTAL EFFECTS, TRACE ELEMENTS, ANALYTICAL TECHNIQUES, COLUMBIA R W72-03215

ION, OLIGOTROPHY, PHYTOPLANKTCN,	ANALYTICAL TECHNIQUES, TURBIDITY,	W72-03224
ULATION, *WASTE WATER TREATMENT,	ANALYTICAL TECHNIQUES, COALESCENC	W71-13356
T PROCEDURES, CYCLING NUTRIENTS,	ANALYTICAL TECHNIQUES, WATER POLL	W71-13794
ISOTOPES, URANIUM RADIOISOTOPES,	ANALYTICAL TECHNIQUES, PALEOZCIC	W72-02073
POLLUTANT IDENTIFICATION, ALGAE,	ANALYTICAL TECHNIQUES, *ULTRAVIOL	W72-01780
THESIS, OXYGEN, *THERMODYNAMICS,	ANALYTICAL TECHNIQUES, DESIGN CRI	W72-02363
TROL, *POLLUTANT IDENTIFICATION,	ANALYTICAL TECHNIQUES, *ALGAE, *N	W72-00801
, COLIFORMS, ALGAE, TEMPERATURE,	ANALYTICAL TECHNIQUES, WATER QUAL	W72-00422
AD, ALKALINITY, HARDNESS(WATER),	ANALYTICAL TECHNIQUES, AMMONIA, N	W71-11033
AE, CHEMICAL ANALYSIS, SAMPLING,	ANALYTICAL TECHNIQUES, NUTRIENTS,	W71-11234
, DIELDRIN, DDT, ORGANIC MATTER,	ANALYTICAL TECHNIQUES, SAMPLING,	W71-11498
TRY, *WATER POLLUTION TREATMENT,	ANALYTICAL TECHNIQUES, SEPARATION	W71-12467
HOSPHATES, DETERGENTS, BIOASSAY,	ANALYTICAL TECHNIQUES, FORECASTIN	W71-12072
*AQUATIC ALGAE, *SURFACE WATERS,	ANALYTICAL TECHNIQUES, PHOTOMETRY	W71-12034
ORUS, METABOLISM, ALGAE, SLUDGE,	ANALYTICAL TECHNIQUES, INDUSTRIAL	W71-11793
, *PLANT PHYSIOLOGY, POLLUTANTS,	ANALYTICAL TECHNIQUES, WATER POLL	W71-10553
RESPIRATION, ALGAE, CYANOPHYTA,	ANALYTICAL TECHNIQUES, WATER POLL	W73-01269
OLOGY, RADIOACTIVITY TECHNIQUES,	ANALYTICAL TECHNIQUES.: / RADIOEC	W73-00802
OLOGY, RADIOACTIVITY TECHNIQUES,	ANALYTICAL TECHNIQUES, CYCLING NU	W73-00821
OPES, MANGANESE, CHROMIUM, IRON,	ANALYTICAL TECHNIQUES, FOOD CHAIN	W73-00832
S, ABSORPTION, SEDIMENTS, CORES,	ANALYTICAL TECHNIQUES, RADIOACTIV	W73-00790
ESCENCE, FOOD CHAINS, EQUIPMENT,	ANALYTICAL TECHNIQUES, HEATED WAT	W72-07526
RATURE, WATER QUALITY, PLANKTON,	ANALYTICAL TECHNIQUES, ALGAE.: /E	W72-07663
SESTON, ALGA/ *NITROGEN, *LAKES,	ANALYTICAL TECHNIQUES, MICHIGAN,	W72-08048
PLANKTON, AUTOMATION, EQUIPMENT,	ANALYTICAL TECHNIQUES, SEWAGE, RE	W72-07715
H), PORTLAND(ORE), WATER QUALITY	ANALYZER.: /(WASH), VANCOUVER(WAS	W72-03215
AUKE BAY, GLACIAL WATER, BECKMAN	ANALYZER, VAN DORN SAMPLERS, C-14	W72-09122
ELEMENTAL	ANALYZERS, MACROINVERTEBRATES.:	W72-07715
A).:	ANDERSON-CUE LAKE, MELROSE(FLORID	W72-08986
DIPLANTHERA UNINERVIS, CASSIOPEA	ANDROMEDA, AKABARIA, STROMBUS TRI	W73-00855
, BARNACLES, EPIFAUNA, SAGARTIA,	ANEMONE, MOLGULA, TUNICATE, PISMO	W72-07907
BRATSK RESERVOIR(SIBERIA),	ANGARA RIVER(SIBERIA).:	W71-09165
KUTSK OBLAST, *BRATSK RESERVOIR,	ANGARA RIVER, OKA RIVER, HYDROBIO	W72-04853
ING, IMPERIAL VALLEY(CALIF), LOS	ANGELES(CALIF), TUCSON(ARIZ).: /L	W71-09157
*FILAMENTOUS ALGAE, *AQUATIC	ANGIOSPERM WEEDS, CLADOPHORA.:	W72-14692
Y, LICHENS, MOSSES, GYMNOSPERMS,	ANGIOSPERMS.: /ATIOS, SPECTROMETR	W72-12709
AGARS, / HYPHOMICROBIUM, VIBRIO	ANGUILLARUM, ENRICHMENT CULTURES,	W72-05421
IA), EICHHORNIA CRASSIPES, TYPHA	ANGUSTATA, IPOMOEA AQUATICA, TOTA	W72-01363
LACUSTRIS, ACORUS CALAMUS, TYPHA	ANGUSTIFOLIA.: /, SCHOENOPLECTUS	W73-00286
LAGOONS, COMPOSTING, REFEEDING,	ANHYDROUS AMMONIA, WASTE CHARACTE	W71-07551
S, ALGAE, CATTLE, POULTRY, HOGS,	ANIMAL DISEASES, CATFISH, WASTE T	W71-07544
ASS, GROWTH RATES, PRODUCTIVITY,	ANIMAL GROWTH, PLANT GROWTH, CYST	W72-09378
S, ANIMAL PHYSIOLOGY, SEDIMENTS,	ANIMAL METABOLISM.: /NMENT, COAST	W73-00811
, INVERTEBRATES, FUNGI, ENZYMES,	ANIMAL PARASITES, VIRUSES, MOLLUS	W73-01676
E ASSIMILATIVE CAPACITY, EUROPE,	ANIMAL PHYSIOLOGY, TRACE ELEMENTS	W73-00817
, ESTUARINE ENVIRONMENT, COASTS,	ANIMAL PHYSIOLOGY, SEDIMENTS, ANI	W73-00811
ABSORPTION, ANIMAL POPULATIONS,	ANIMAL PHYSIOLOGY, RADIOACTIVITY	W73-00826
ATIC ALGAE, ABSORPTION, AQUARIA,	ANIMAL PHYSIOLOGY, WATER BALANCE,	W72-06342
LS, PATH OF POLLUTANTS, TRACERS,	ANIMAL PHYSIOLOGY, NUCLEAR WASTES	W72-03348
DE RESIDUES, PESTICIDE KINETICS,	ANIMAL PHYSIOLOGY, PLANT PHYSIOLO	W71-12303
TUARINE ENVIRONMENT, ABSORPTION,	ANIMAL POPULATIONS, ANIMAL PHYSIO	W73-00826
HARBORS, TUNNELS, MARINE ALGAE,	ANIMAL POPULATIONS, LOBSTERS, CRU	W73-02029
AT LAKE, GAMMA RAY SPECTROMETRY,	ANIMAL TISSUES, MACROPHYTES, CYPR	W73-01673
HISTOCHEMISTRY, LYOPHILIZATION,	ANIMAL TISSUES, HISTOLOGICAL STUD	W73-01676
BEHAVIOR, BITUMINOUS MATERIALS,	ANIMAL WASTES, FISH, FLUORESCENCE	W72-08790
CIPITATION(ATMOSPHERIC), RUNOFF,	ANIMAL WASTES, SURFACE WATERS, SO	W71-04216
STRIBUTION, MARINE ALGAE, MARINE	ANIMALS.: /IBITION, ECOLOGICAL DI	W73-00348
, MOLLUSKS, MARINE ALGAE, MARINE	ANIMALS.: / POLLUTION, PHAEOPHYTA	W72-11925
GS, LICHENS, ALGAE, FUNGI, SMALL	ANIMALS(MAMMALS), CYTOLOGICAL STU	W72-00949
RM/ BAYS, *MARINE ALGAE, *MARINE	ANIMALS, *THERMAL POLLUTION, *THE	W72-11876
NKTON, *AQUATIC PLANTS, *AQUATIC	ANIMALS, *MODEL STUDIES, LIMNOLOG	W72-14280
REVIEWS, *RADIOISOTOPES, *MARINE	ANIMALS, *MARINE ALGAE, SEDIMENTS	W73-00817
R WASTES, *MARINE ALGAE, *MARINE	ANIMALS, *MARINE FISH, MUSSELS, L	W73-00832
UCLEAR WASTES, *COBALT / *MARINE	ANIMALS, *CLAMS, *FOOD CHAINS, *N	W73-00826
STRONTIUM RADIOISOTOPES, *MARINE	ANIMALS, *MARINE ALGAE, *FOOD CHA	W73-00819
ECOLOGICAL DISTRIBUTION, *MARINE	ANIMALS, *SYSTEMATICS, *ECOLOGY,	W73-00932
OONS, *HYDROBIOLOGY, *HALOPHILIC	ANIMALS, *HALOPHYTES, SEA WATER,	W73-00855
CRUSTACEANS, *MOLLUSKS, *AQUATIC	ANIMALS, *BIOTA, *AQUATIC PLANTS,	W72-06526
ALIF/ *ECOLOGY, *WASTES, *MARINE	ANIMALS, *MARINE PLANTS, DEPTH, C	W71-01475
ACARICIDES, BRACKISH W/ *AQUATIC	ANIMALS, *AQUATIC MICROBIOLOGY, *	W71-08026
IBUTION, *RADIOISOTOPES, *MARINE	ANIMALS, *MARINE PLANTS, AQUATIC	W71-09168
LLUSKS,/ *RADIOISOTOPES, *MARINE	ANIMALS, *ABSORPTION, AQUARIA, MO	W71-09182
IFE, *AQUATIC BACTERIA, *AQUATIC	ANIMALS, *AQUATIC PLANTS, *RESERV	W72-04853
RVOIRS, *AQUATIC ALGAE, *AQUATIC	ANIMALS, *BIOMASS, *PRIMARY PRODU	W72-05968
*SEA WATER, *SEDIMENTS, *MARINE	ANIMALS, *MARINE ALGAE, RADIOISOT	W72-05965
IRONMENT, AQUATIC ALGAE, AQUATIC	ANIMALS, AQUATIC INSECTS, DAPHNIA	W72-07132
OTOZOA, AQUATIC INSECTS, AQUATIC	ANIMALS, AQUATIC ALGAE.: /OMS, PR	W72-07791
SALMON, SHELLFISH, CLAMS, MARINE	ANIMALS, AQUATIC LIFE, ECOSYSTEMS	W72-07907
WASTES, *BIOINDICATORS, *MARINE	ANIMALS, ASIA, MOLLUSKS, COBALT R	W73-00811
ON CYCLE, ON-SITE TESTS, AQUATIC	ANIMALS, AQUATIC POPULATIONS, BIO	W73-02095
LUSKS, WATER TEMPERATURE, MARINE	ANIMALS, ALASKA, BIOMASS, SALINIT	W73-00374
N SOURCES, AQUATIC LIFE, AQUATIC	ANIMALS, AQUATIC MICROORGANISMS,	W72-09646
DROGEN ION CONCENTRATION, MARINE	ANIMALS, BACTERIA, ABSORPTION, CU	W71-10083
, *HALOPHYTES, SEA WATER, MARINE	ANIMALS, BIOLOGICAL COMMUNITIES,	W73-00855
BIOMASS, AQUATIC ALGAE, AQUATIC	ANIMALS, BALANCE OF NATURE, ECOLO	W72-06340
UTION EFFECTS, CATTAILS, AQUATIC	ANIMALS, CLAMS, CARP, CRAYFISH, A	W72-06342
ESHWATER, *AQUATIC LIFE, AQUATIC	ANIMALS, CARP, DAPHNIA, AQUATIC P	W72-09649
SORPTION, FISH, MOLLUSKS, MARINE	ANIMALS, COBALT RADIOISOTOPES, FO	W71-09225
AQUATIC MICROORGANISMS, AQUATIC	ANIMALS, DAPHNIA, APPLICATION MET	W72-09650
ESTS, DETRITUS, BENTHOS, AQUATIC	ANIMALS, DIATOMS, INSECTS, BACTER	W71-10064

, MUSSELS, MARINE ALGAE, AQUATIC	ANIMALS, FRESH WATER, FRESHWATER	W71-09182
ANOPHYTA, BENTHIC FAUNA, AQUATIC	ANIMALS, INVERTEBRATES, AQUATIC A	W72-12729
ANIMALS, *MARINE PLANTS, AQUATIC	ANIMALS, LABORATORY TESTS, NITRAT	W71-09168
RROPHYTA, INVERTEBRATES, AQUATIC	ANIMALS, LITTORAL, WAVES(WATER).:	W71-11012
A, SAMPLING, MARINE FISH, MARINE	ANIMALS, MARINE ALGAE, TRITIUM, I	W71-09018
LOGY, BIRD TYPES, BIRDS, AQUATIC	ANIMALS, MOSSES, INVERTEBRATES, D	W72-05952
S, *COBALT, *ABSORPTION, *MARINE	ANIMALS, MARINE ALGAE, SEDIMENTS,	W73-00821
OWERPLANTS, MARINE ALGAE, MARINE	ANIMALS, MARINE PLANTS, HUMAN POP	W73-00833
IVITY EFFECTS, *FALLOUT, *MARINE	ANIMALS, MARINE PLANTS, MARINE AL	W73-00828
, REFRACTIVITY, COPEPODS, MARINE	ANIMALS, MOLLUSKS, GASTROPODS, PH	W73-00371
BUTION, *DENSITY, *DEPTH, MARINE	ANIMALS, MARINE ALGAE, CYANOPHYTA	W72-14313
AQUATIC MICROORGANISMS, AQUATIC	ANIMALS, NUTRIENTS, WATER POLLUTI	W72-07896
EFFECTS, *MARINE PLANTS, *MARINE	ANIMALS, OFFSHORE PLATFORMS, NUCL	W73-02029
INITY, NUTRIENTS, ALGAE, AQUATIC	ANIMALS, RADIONUCLIDES.: /ES, SAL	W72-12190
Y, DISTRIBUTION PATTERNS, MARINE	ANIMALS, SALINITY, SAMPLING, SEA	W73-00296
TREATMENT, COLLOIDS, *POLYMERS,	ANIONS, CATIONS, ORGANIC COMPOUND	W72-11833
R TREATMENT, *POLYMERS, CATIONS,	ANIONS, CHLORELLA, FILTRATION, EL	W72-14840
LATION EFFICIENCES, CRYPTOMONAS,	ANKISTRODESMUS, ELAKOTOTHRIX, OSC	W72-07940
FORMOSA, PHORMIDIUM FAVEOLARUM,	ANKISTRODESMUS BRAUNII, CHLORELLA	W72-07664
FORMOSA, STEPHANODISCUS ASTREA,	ANKISTRODESMUS, CHRYSOCOCCUS MINU	W72-12738
ORIDE, METHYL MERCURIC CHLORIDE,	ANKISTRODESMUS BRAUNII, EUGLENA G	W72-06037
E, *DETOXICATION, CHLOROSARCINA,	ANKISTRODESMUS BRAUNII, CHLORELLA	W71-07675
A.: MOUGEOTIA GENUFLEXA,	ANKISTRODESMUS FALCATUS, PODOHEDR	W71-07360
CONIS, CYMBELLA, STEPHANODISCUS,	ANKISTRODESMUS, TRACHELOMONAS, SA	W72-10875
R(ME), CULTURAL EUTROPHIC/ *LAKE	ANNABESSACOOK(ME), *KENNEBEC RIVE	W72-04280
US, CYANOPHYTA, MOLLUSKS, MITES,	ANNE: /AS, CHLOROPHYTA, SCENEDESM	W72-00155
EACHES, SANDS, INTERTIDAL AREAS,	ANNELIDS, COPEPODS, CRABS, AQUATI	W72-07911
COMPOUNDS, WORMS, INVERTEBRATES,	ANNELIDS, SEASONAL, CRUSTACEANS,	W73-00374
S, DINOFLAGELLATES, CRUSTACEANS,	ANNELIDS, WATER POLLUTION EFFECTS	W73-00935
MATICS, *ECOLOGY, *MARINE ALGAE,	ANNELIDS, MOLLUSKS, RHODOPHYTA, P	W73-00932
ARUM, TRAMETES VERSICOLOR, FOMES	ANNOSUS, FUSCOPORIA CONTIGUA, TRA	W72-13813
, ECONOMICS, FEASIBILITY, COSTS,	ANNUAL COSTS, PILOT PLANTS.: /TES	W70-10174
CIPAL WASTES, INDUSTRIAL WASTES,	ANNUAL TURNOVER, ENERGY, CYANOPHY	W72-03228
S HOFFMEISTERI, STICTOCHIRONOMUS	ANNULIORUS, PHYSA HETEROSTROPHA,	W70-06668
RS.: *CHELATORS,	ANODIC STRIPPING, ORGANIC CHELATO	W71-09674
RGHEI, BORRELIA KANSAS, BORRELIA	ANSERINA.: / LIVER, PLASMODIUM BE	W73-01676
IRONOMUS ATTENTUATUS, CHIRONOMUS	ANTHRACINUS, LIMNODRILUS HOFFMEIS	W71-11009
C CHLORIDE, MULTI-LAYER FILTERS,	ANTHRACITE, *WAHNBACH RESERVOIR(G	W72-06201
OPHYTA, CHLOROPHYTA, INHIBITORS,	ANTIBIOTICS, SEA WATER, RHODE ISL	W72-04748
ODDER, GLUE, AGAR, POLYSIPHONIA,	ANTIBIOTICS, IODINE, ATOMIC ENERG	W71-09157
HYTA, SCENEDESMUS, ALGAL TOXINS,	ANTIBIOTICS(PESTICIDES), DATA PRO	W71-05155
HATES, ECOSYSTEMS, AQUATIC LIFE,	ANTIBIOTICS(PESTICIDES), ECOLOGY,	W72-09668
L, ASSAY, REMOTE SENSING, ALGAE,	ANTIBIOTICS, MERCURY, ARSENIC COM	W72-10162
MARY PRODUCTIVITY, CORAL, REEFS,	ANTIBIOTICS(PESTICIDES), CHLORINA	W72-14294
ECHNIQUES, CYANOPHYTA, CULTURES,	ANTIBIOTICS(PESTICIDES), PHOTOACT	W72-14327
FISH, BACTERIA, FISH PARASITES,	ANTIBIOTICS(PESTICIDES), BACTERIC	W72-08142
, DNA, MOTILITY, CYST FORMATION,	ANTIBIOTIC SENSITIVITY, CYTOPHAGA	W72-07951
NIC BACTERIA, COLIFORMS, TOXINS,	ANTIBIOTICS(PESTICIDES), ENTERIC	W72-13800
YPHIMURI/ DETECTION, FLUORESCENT	ANTIBODY TECHNIQUES, SALMONELLA T	W72-13800
ME, *PHARMACOLOGICAL PROPERTIES,	ANTICANCER, ANTINEOPLASTICS.: /TI	W72-07360
NUPHAR JAPONICUM, NUPHAR LUTEN,	ANTICANCER POTENTIAL, ANTICOAGULA	W72-04275
HAR LUTEN, ANTICANCER POTENTIAL,	ANTICOAGULANT POTENTIAL, SPARGANI	W72-04275
HIDLEY(LABRADOR), NEW BRUNSWICK,	ANTICOSTI ISLAND(QUEBEC), MAGDALE	W71-10066
ON METHOD, *ZONES OF INHIBITION,	ANTIFUNGAL ACTIVITY.: /SC DIFFUSI	W72-08586
LS, PENICILLIN, LINDANE, DIADEMA	ANTILLARUM, AANDOMENE, SIPHONALES	W72-14294
SELENOMETHIONINE, SELENOPURINE,	ANTIMETABOLITES.: /ATE, SELENITE,	W72-00153
PROPERTIES, PHYTOCHEMICAL SCRE/	ANTIMICROBIOTICS, PHARMACOLOGICAL	W72-04275
S-AQUAE, *PHARMACOLOGICAL TESTS,	ANTIMICROBIOTICS, ESCHERICHIA COL	W72-04259
RPTION, ARSENIC, SELENIUM, ZINC,	ANTIMONY, NORWAY, ASHING, BIOLOGI	W71-11515
OPE.:	ANTIMONY, CESIUM, *CAPE OF GOOD H	W73-00832
MS LAKE(WASH), URANIUM, THORIUM,	ANTIMONY, MACRONUTRIENTS, MICRONU	W73-01434
OLOGICAL PROPERTIES, ANTICANCER,	ANTINEOPLASTICS.: /TIME, *PHARMAC	W72-07360
POTENTIAL, SPARGANIUM FLUCTUANS,	ANTINEOPLASTIC ACTIVITY, ALKALOID	W72-04275
NABAENA, MICROCYSTIS, TRIBONEMA,	APHANIZOMENON, OOCYSTIS, GLOEOCYS	W72-07940
DISC, NEWFOUND LAKE, SQUAM LAKE,	APHANIZOMENON FLOW-AQUAE, DESTRAT	W72-07890
ENA SPP, MICROCYSTIS AERUGINOSA,	APHANIZOMENON HOLSATICUM, ICTALUR	W72-07890
OCYSTIS AERU/ *KEZAR LAKE(N.H.),	APHANIZOMENON, ANABAENA SPP, MICR	W72-07890
RIVER, SALT RIVER, MICROCYSTIS,	APHANIZOMENON, CYANIDES, COHO SAL	W72-14282
S HOFFMEISTERI, TUBIFEX TUBIFEX,	APHANIZOMENON, CLADOPHORA.: /RILU	W71-11009
IA, STEPHANODISCUS, MICROCYSTIS,	APHANIZOMENON, ANABENA SPIROIDES.	W71-12855
M,/ DOLLAR LAKE(OHIO), SPHAGNUM,	APHANIZOMENON FLOS-AQUAE, CERATIU	W71-11949
CIRCINALIS, ANABAENA CYLINDRICA,	APHANIZOMENON FLOS-AQUAE.: /AENA	W72-10613
LLATORIA, APHANOTHECE, ANABAENA,	APHANIZOMENON, NAVICULA, GLENODIN	W72-10875
ISM, AXENIC CULTURES, EXCRETION,	APHANIZOMENON GRACILE.: /OANTAGON	W72-12734
OMTA, STEPHANODISCUS HANTZSCHII,	APHANIZOMENON FLOS-AQUAE, CHROOMO	W72-12738
UYAHOGA RIVER(OHIO), CLADOPHORA,	APHANOCHAETE, RHIZOCLONIUM, VAUCH	W71-05629
NTING, CYCLOTELLA, OSCILLATORIA,	APHANOTHECE, ANABAENA, APHANIZOME	W72-10875
SEAWEEDS,	APHIDS, SPONGES, CARBON-14.:	W72-06283
TOCOCCUS PYRIFORMIS NAUTOCOCCUS,	APIOCOCCUS, OSCILLATORIA, NAUTOCO	W72-07141
TRIA), CHAMAESIPHON SUBAEQUALIS,	APISTONEMA EXPANSUM, PORPHYRIDIUM	W71-04526
AMONAS SAUPAULENSIS, ENEIDAMONAS	APPLANATA, BICUDOMONAS CYANOPH: /	W72-09119
C TANKS, OXIDATION LAGOONS, LAND	APPLICAT: /DENITRIFICATION, SEPTI	W70-09907
OIRS, SWIMMING POOLS, CHELATION,	APPLICATION METHODS, COOLING TOWE	W71-06189
NISMS, AQUATIC ANIMALS, DAPHNIA,	APPLICATION METHODS, GRANULES, PL	W72-09650
QUALITY, RIPARIAN RIGHTS, PRIOR	APPROPRIATION, GROUNDWATER, WATER	W70-07100
EN INACTIVATION./ *ANABAENA FLOS-	AQUA, *NITROGENASE ACTIVITY, OXYG	W72-07941
ISH, MUSSELS, SHELLFISH FARMING,	AQUACULTURE, COMMERCIAL SHELLFISH	W72-09378
A CYLINDRICA, APHANIZOMENON FLOS-	AQUAE.: /AENA CIRCINALIS, ANABAEN	W72-10613
LINE PHOSPHATASE, *ANABAENA FLOS-	AQUAE, *CONTINUOUS CULTURE, CHEMO	W72-05476
EUTIC VALUE, *APHANIZOMENON FLOS-	AQUAE, *PHARMACOLOGICAL TESTS, AN	W72-04259

OI, SPHAGNUM, APHANIZOMENON FLOS- AQUAE, CERATIUM, OSCILLATORIA, TR W71-11949
S HANTZSCHII, APHANIZOMENON FLOS- AQUAE, CHROOMONAS ACUTA, MELOSIRA W72-12738
, SQUAM LAKE, APHANIZOMENON FLOW- AQUAE, DESTRATIFICATION, ANABAENA W72-07890
DESTRATIFICATION, ANABAENA FLOW- AQUAE, HOMOGENIZING, TRA: /AQUAE, W72-07890
CYANOPHAGES, *APHANIZOMENON FLOS- AQUAE, LAKE ERKEN(SWEDEN).: * W73-00246
, LAKE MALAREN, MICROCYSTIS FLOS- AQUAE, PHRAGMITES COMMUNIS, *NORW W73-00286
FISH, AQUATIC ALGAE, ABSORPTION, AQUARIA, ANIMAL PHYSIOLOGY, WATER W72-06342
FECTS, RADIOACTIVITY TECHNIQUES, AQUARIA, ALGAE, MATHEMATICAL MODE W72-03348
ALGAE, *PHYTOPLANKTON, BACTERIA, AQUARIA, DECOMPOSING ORGANIC MATT W73-00802
SPECIATION, VERTICAL MIGRATION, AQUARIA, GROWTH RATES, SEDIMENTAT W71-09852
IDE TOXICITY, PLANT POPULATIONS, AQUARIA, LABORATORY EQUIPMENT, PL W72-14294
S, *MARINE ANIMALS, *ABSORPTION, AQUARIA, MOLLUSKS, SNAILS, MUSSEL W71-09182
PLANTS, MARINE ALGAE, PHAEPHYTA, AQUARIA, PARTICLE SIZE, COLLOIDS, W71-09193
TICAL DISTILLATION, EVAPORATORS, AQUAT: / POLLUTION, LONG TUBE VER W70-09193
PUBLIC HEALTH, COST COMPARISONS, AQUATIC ALGAE, GRANTS, MODEL STUD W71-09429
OXYGEN, BIOINDICATORS, E. COLI, AQUATIC ALGAE, SUNFISHES, SNAILS, W71-07417
SPHORUS, *EUTROPHICATION, LAKES, AQUATIC ALGAE, BACTERIA, BIOCHEMI W72-04292
ALTERNATIVES, AGING(BIOLOGICAL), AQUATIC ALGAE, AQUATIC PRODUCTIVI W72-04279
GES, EARLY IMPOUNDMENT, BIOMASS, AQUATIC ALGAE, PHYTOPLANKTON, ZOO W72-04853
C COMPOUNDS, CHLORELLA, TRACERS, AQUATIC ALGAE.: /M, FUNGI, ORGANI W72-14326
CS(PESTICIDES), PHOTOACTIVATION, AQUATIC ALGAE, ENVIRONMENTAL EFFE W72-14327
PHYSIOLOGY, TESTING PROCEDURES, AQUATIC ALGAE, CHLOROPHYTA, LABOR W72-14314
OLOGY, CULTURES, PHOTOSYNTHESIS, AQUATIC ALGAE, CHLOROPHYLL, OXYGE W73-00281
NOPHYTA, RESINS, AQUATIC PLANTS, AQUATIC ALGAE, NITRATES, NITRITES W73-00286
TOZOA, CHRYSOPHYTA, CHLOROPHYTA, AQUATIC ALGAE, MARINE ALGAE, SESS W73-00273
AN, CHRYSOPHYTA, PLANT PIGMENTS, AQUATIC ALGAE, NITROGEN.: /MICHIG W73-00384
LOROPHYTA, AQUATIC PRODUCTIVITY, AQUATIC ALGAE, NITROGEN, POLLUTAN W73-00379
ON, *LAKE MICHIGAN, CHRYSOPHYTA, AQUATIC ALGAE, ELECTRON MICROSCOP W73-00432
CATORS, *BIOLOGICAL COMMUNITIES, AQUATIC ALGAE, AQUATIC PRODUCTIVI W73-00002
, RIVERS, ROOTED AQUATIC PLANTS, AQUATIC ALGAE, WATER TEMPERATURE, W72-14690
SITY, AIR TEMPERATURE, SAMPLING, AQUATIC ALGAE, PHYTOPLANKTON, TRO W73-01456
EMENTS, WATER POLLUTION EFFECTS, AQUATIC ALGAE, CHRYSOPHYTA, LABOR W73-01445
S, LIMITING FACTORS, RHODOPHYTA, AQUATIC ALGAE, MICROORGANISMS, CH W73-02099
PRODUCTIVITY, AQUATIC BACTERIA, AQUATIC ALGAE, CARBON CYCLE, ON-S W73-02095
PLANT TISSUES, FRESHWATER FISH, AQUATIC ALGAE, EVALUATION, SAMPLI W73-01673
*INFLUENT STREAMS, *GREAT LAKES, AQUATIC ALGAE, ICE, FISH, MICROOR W73-01669
TIONS, CHLOROPHYTA, SCENEDESMUS, AQUATIC ALGAE, AQUATIC PRODUCTIVI W73-01066
OGY, ECOLOGY, CHEMICAL ANALYSIS, AQUATIC ALGAE, NUTRIENTS, WATER Q W73-00918
COLORIMETRY, DEFICIENT ELEMENTS, AQUATIC ALGAE, BIOINDI: /RIENTS, W73-00916
LOGY, REPRODUCTION, LIFE CYCLES, AQUATIC ALGAE, PERIPHYTON.: /RPHO W72-07141
ANIMALS, CLAMS, CARP, CRAYFISH, AQUATIC ALGAE, ABSORPTION, AQUARI W72-06342
ORGANISMS, AQUATIC ENVIRONMENT, AQUATIC ALGAE, AQUATIC ANIMALS, A W72-07132
TEMS, AQUATIC HABITATS, BIOMASS, AQUATIC ALGAE, AQUATIC ANIMALS, B W72-06340
PHOMETRY, DISTRIBUTION PATTERNS, AQUATIC ALGAE, LIMNOLOGY, CHRYSOP W72-07145
RFACES, *LAKE ERIE, HYPOLIMNION, AQUATIC ALGAE.: /IMENT-WATER INTE W72-06690
QUATIC INSECTS, AQUATIC ANIMALS, AQUATIC ALGAE.: /OMS, PROTOZOA, A W72-07791
TON, WATER ANALYSIS, POLLUTANTS, AQUATIC ALGAE, AQUATIC MICROORGAN W72-07896
N, DIATOMS, ALGAE, MARINE ALGAE, AQUATIC ALGAE, EUGLENA, CHLAMYDOM W72-08436
PLANT GROWTH REGULATORS, ALGAE, AQUATIC ALGAE, CHLORELLA.: /-4-D, W72-08440
SANITATION, SECONDARY TREATMENT, AQUATIC ALGAE, PRIMARY PRODUCTIVI W71-09674
PLANT PHYSIOLOGY, LAKES, PONDS, AQUATIC ALGAE, ECOLOGY, BIOLOGICA W71-10580
ONATES, WATER POLLUTION SOURCES, AQUATIC ALGAE, PUBLIC HEALTH, BIO W71-11516
TION SOURCES, INDUSTRIAL WASTES, AQUATIC ALGAE, BIOINDICATORS, BIO W71-11583
ION, NITRITE, AMMONIA, SALINITY, AQUATIC ALGAE, ZOOPLANKTON, BIOCH W71-11876
SSOLVED OXYGEN, LAKE MORPHOLOGY, AQUATIC ALGAE, MOSSES, WATER TEMP W71-11949
UTION CONTROL, SEWAGE EFFLUENTS, AQUATIC ALGAE, LAKE ERIE, LAKE ON W71-11507
DT, PHYTOPLANKTON, MARINE ALGAE, AQUATIC ALGAE, COLORIMETRY, TRACE W72-03545
ONTROL, *LAKES, *EUTROPHICATION, AQUATIC ALGAE, NUTRIENTS, WATER R W71-13256
PECTION, WATER SOFTENING, SOAPS, AQUATIC ALGAE, WATER POLLUTION EF W72-01085
KTON, BENTHIC FAUNA, PERIPHYTON, AQUATIC ALGAE, PLANTS, ENVIRONMEN W72-00939
UTROPHICATION, IRRIGATION WATER, AQUATIC ALGAE, BIOLOGICAL PROPERT W72-00974
, *OXIDATION LAGOONS, NUTRIENTS, AQUATIC ALGAE, NITROGEN, ANAEROBI W72-02850
REQUIREMENTS, AGING(BIOLOGICAL), AQUATIC ALGAE, WASTE ASSIMILATIVE W72-01685
RIENTS, WATER POLLUTION EFFECTS, AQUATIC ALGAE, NORTH CAROLINA, *A W72-01329
MICROORGANISMS, AQUATIC PLANTS, AQUATIC ALGAE, PESTICIDES, INSECT W72-09646
LUTION, WATER POLLUTION SOURCES, AQUATIC ALGAE, FOOD CHAINS, MEASU W72-10686
ICIDE TOXICITY, *AQUATIC PLANTS, AQUATIC ALGAE, PHYTOPLANKTON, AQU W72-09650
, CARP, DAPHNIA, AQUATIC PLANTS, AQUATIC ALGAE, SCENEDESMUS, RADIO W72-09649
NS, PUBLIC HEALTH, AQUATIC LIFE, AQUATIC ALGAE, CALCIUM, RADIOACTI W72-10978
OFTENING, ENVIRONMENTAL EFFECTS, AQUATIC ALGAE, ALGAL CONTROL, ALG W72-11186
*TOXICITY, CULTURES, *CHLORELLA, AQUATIC ALGAE, CHLOROPHYTA, HEAVY W72-11727
Y, WATER QUALITY, TROPHIC LEVEL, AQUATIC ALGAE, FISH POPULATIONS, W72-08986
QUATIC HABITATS, CLASSIFICATION, AQUATIC ALGAE.: /, SCENEDESMUS, A W72-12744
SEASONAL, *PRIMARY PRODUCTIVITY, AQUATIC ALGAE, ZOOPLANKTON, CHRYS W72-12738
POTASSIUM, CULTURES, RESISTANCE, AQUATIC ALGAE, GROWTH RATES, BIOA W72-13809
TION, TROPHIC LEVEL, PROTOZOANS, AQUATIC ALGAE, ENVIRONMENTAL EFFE W72-13816
, *BIOMASS, *HYDROGRAPHY, ALGAE, AQUATIC ALGAE, OLIGOTROPHY, DISSO W72-13871
A, SOLVENT EXTRACTIONS, DIATOMS, AQUATIC ALGAE, FILTRATION, CYANOP W72-13673
ICROORGANISMS, AQUATIC BACTERIA, AQUATIC ALGAE, ALGAE, CHLOROPHYTA W72-12255
ATOMS, *MICROSCOPY, CHRYSOPHYTA, AQUATIC ALGAE.: *DI W72-11738
AQUATIC ANIMALS, INVERTEBRATES, AQUATIC ALGAE, AERIAL PHOTOGRAPHY W72-12729
TICIDE DRIFT, METABOLISM, ALGAE, AQUATIC ALGAE, CHRYSOPHYTA, PLANK W73-02280
TOMS, CYANOPHYTA, BENTHIC FAUNA, AQUATIC ANIMALS, INVERTEBRATES, A W72-12729
DES, SALINITY, NUTRIENTS, ALGAE, AQUATIC ANIMALS, RADIONUCLIDES.: / W72-12190
ION, *FRESHWATER, *AQUATIC LIFE, AQUATIC ANIMALS, CARP, DAPHNIA, A W72-09649
LANKTON, AQUATIC MICROORGANISMS, AQUATIC ANIMALS, DAPHNIA, APPLICA W72-09650
POLLUTION SOURCES, AQUATIC LIFE, AQUATIC ANIMALS, AQUATIC MICROORG W72-09646
HYTA, PYRROPHYTA, INVERTEBRATES, AQUATIC ANIMALS, LITTORAL, WAVES(W71-11012
RATORY TESTS, DETRITUS, BENTHOS, AQUATIC ANIMALS, DIATOMS, INSECTS W71-10064
TOMS, PROTOZOA, AQUATIC INSECTS, AQUATIC ANIMALS, AQUATIC ALGAE.: / W72-07791

C ALGAE, AQUATIC MICROORGANISMS, AQUATIC ANIMALS, NUTRIENTS, WATER W72-07896
ABITATS, BIOMASS, AQUATIC ALGAE, AQUATIC ANIMALS, BALANCE OF NATUR W72-06340
ATIC ENVIRONMENT, AQUATIC ALGAE, AQUATIC ANIMALS, AQUATIC INSECTS, W72-07132
TER POLLUTION EFFECTS, CATTAILS, AQUATIC ANIMALS, CLAMS, CARP, CRA W72-06342
AE, CARBON CYCLE, ON-SITE TESTS, AQUATIC ANIMALS, AQUATIC POPULATI W73-02095
DES, ECOLOGY, BIRD TYPES, BIRDS, AQUATIC ANIMALS, MOSSES, INVERTEB W72-05952
, SNAILS, MUSSELS, MARINE ALGAE, AQUATIC ANIMALS, FRESH WATER, FRE W71-09182
*MARINE ANIMALS, *MARINE PLANTS, AQUATIC ANIMALS, LABORATORY TESTS W71-09168
CTERIA, PHOTOSYNTHETIC BACTERIA, AQUATIC BACTERIA, LAKES, FILTERS, W72-04743
AQUATIC HABITATS, PRODUCTIVITY, AQUATIC BACTERIA, AQUATIC ALGAE, W73-02095
POLLUTANT IDENTIFICATION, ALGAE, AQUATIC BACTERIA, SEWAGE BACTERIA W72-01472
EDESMUS, AQUATIC MICROORGANISMS, AQUATIC BACTERIA, AQUATIC ALGAE, W72-12255
AQUATIC BIOTOXINS, *BIOTOXINS.: W72-14056
EK RESERVOIR(KAN), GREAT PLAINS, AQUATIC ECOSYSTEM.: / *TUTTLE CRE W72-12393
, *ALGAE, *LIG/ *EUTROPHICATION, AQUATIC ECOSYSTEMS, WATER QUALITY W73-02349
H CAROLINA), ANACYSTIS NIDULANS, AQUATIC ECOSYSTEMS, ALGAE GROWTH, W71-00475
ARBON, *PHOSPHORUS, *CYANOPHYTA, AQUATIC ENVIRONMENT, CARBON DIOXI W71-00475
SEWAGE, *EUTROPHICATION, *LAKES, AQUATIC ENVIRONMENT, CHEMICAL PRO W71-00117
E ERIE, AQUATIC HABITATS, LAKES, AQUATIC ENVIRONMENT.: /LGAE, *LAK W71-09156
FFECTS, ALGAE, AQUATIC HABITATS, AQUATIC ENVIRONMENT.: /OLLUTION E W71-07337
SPHORUS, ALGAE, *EUTROPHICATION, AQUATIC ENVIRONMENTS, GREAT LAKES W71-08953
SH, RAINBOW TROUT, AQUATIC LIFE, AQUATIC ENVIRONMENT, FISH PHYSIOL W72-12257
OGICAL COMMUNITIES, *SUCCESSION, AQUATIC ENVIRONMENT, *ECOLOGICAL W72-11854
D WEBS, WATER POLLUTION EFFECTS, AQUATIC ENVIRONMENT, TURBIDITY, W W72-11925
TER QUALITY CONTROL, PHOSPHORUS, AQUATIC ENVIRONMENT, WATER POLLUT W72-06638
COMMUNITIES, DOMINANT ORGANISMS, AQUATIC ENVIRONMENT, AQUATIC ALGA W72-07132
TREATMENT, ALGAE, PLANTS, TROUT, AQUATIC ENVIRONMENT, EUTROPHICATI W72-02412
EN ION CONCENTRATION, NUTRIENTS, AQUATIC ENVIRONMENT, CARBON DIOXI W71-12068
ALGAL CONTROL, ALGAL POISONING, AQUATIC ENVIRONMENT, AQUATIC FUNG W72-11186
MATHEMATICAL STUDIES, EQUATIONS, AQUATIC ENVIRONMENT, WATER PROPER W72-11724
ARBON, *PHOSPHORUS, *POPULATION, AQUATIC ENVIRONMENT, ECOSYSTEMS, W72-09159
POISONING, AQUATIC ENVIRONMENT, AQUATIC FUNGI, ACQUIFER CHARACTER W72-11186
, *ALGAE, CHLOROPHYTA, HABITATS, AQUATIC HABITATS, EUGLENOPHYTA, P W72-09119
POWER PLANTS, *DISCHARGE(WATER), AQUATIC HABITATS, ENVIRONMENTAL E W71-11517
STUDIES, *STREAMS, *ECOSYSTEMS, AQUATIC HABITATS, BIOMASS, AQUATI W72-06340
OGICAL DISTRIBUTION, ECOSYSTEMS, AQUATIC HABITATS, PRODUCTIVITY, A W73-02095
MNOLOGY, WATER QUALITY, ECOLOGY, AQUATIC HABITATS, CLADOPHORA, LIM W72-13142
, STREAMS, ISOPODS, SCENEDESMUS, AQUATIC HABITATS, CLASSIFICATION, W72-12744
WATER POLLUTION EFFECTS, ALGAE, AQUATIC HABITATS, AQUATIC ENVIRON W71-07337
LAKE SHORES, *ALGAE, *LAKE ERIE, AQUATIC HABITATS, LAKES, AQUATIC W71-09156
XIDATION LAGOONS, PHYTOPLANKTON, AQUATIC I: /OLOGICAL TREATMENT, O W72-12567
BIRDS, STREAMS, AQUATIC PLANTS, AQUATIC INSECTS, CRUSTACEANS, MUS W72-12930
AQUATIC ALGAE, AQUATIC ANIMALS, AQUATIC INSECTS, DAPHNIA, AQUATIC W72-07132
URCES, *FISH, DIATOMS, PROTOZOA, AQUATIC INSECTS, AQUATIC ANIMALS, W72-07791
REAS, ANNELIDS, COPEPODS, CRABS, AQUATIC LIFE, BACTERIA.: /TIDAL A W72-07911
HELLFISH, CLAMS, MARINE ANIMALS, AQUATIC LIFE, ECOSYSTEMS, SALMONI W72-07907
COASTS, WATER POLLUTION EFFECTS, AQUATIC LIFE, HABITATS, ECOLOGY, W73-01632
SOLVED OXYGEN, COMPUTERS, ALGAE, AQUATIC LIFE, ENVIRONMENTAL EFFEC W72-14405
RBON RADIOISOTOPES, NEW ENGLAND, AQUATIC LIFE, AQUATIC MICROORGANI W72-14317
TROPHY, OLIGOTROPHY, OXYGEN SAG, AQUATIC LIFE, BIOASSAY, CLASSIFIC W73-00002
ESTUARINE ENVIRONMENT, HABITATS, AQUATIC LIFE, HEAVY METALS, INDUS W72-11854
, BIOASSAY, FISH, RAINBOW TROUT, AQUATIC LIFE, AQUATIC ENVIRONMENT W72-12257
CIDES, BIBLIOGRAPHIES, *ECOLOGY, AQUATIC LIFE, IRRIGATION SYSTEMS, W71-06102
TEMPERATURE, SEASONAL, MONTHLY, AQUATIC LIFE, INTERSTATE RIVERS, W70-09595
BIVORES, CARNIVORES, ECOSYSTEMS, AQUATIC LIFE, SAMPLING, ALGAE, LA W71-10064
DS, NUTRIENTS, DISSOLVED OXYGEN, AQUATIC LIFE, MUNICIPAL WASTES, P W71-11017
CTS, FOOD CHAINS, PUBLIC HEALTH, AQUATIC LIFE, AQUATIC ALGAE, CALC W72-10978
FFECTS, WATER POLLUTION SOURCES, AQUATIC LIFE, AQUATIC ANIMALS, AQ W72-09646
TRACERS, PHOSPHATES, ECOSYSTEMS, AQUATIC LIFE, ANTIBIOTICS(PESTICI W72-09668
BATEMENT, ENVIRONMENTAL EFFECTS, AQUATIC LIFE, FISH POPULATIONS, P W72-00939
, AQUATIC LIFE, AQUATIC ANIMALS, AQUATIC MICROORGANISMS, AQUATIC P W72-09646
S, AQUATIC ALGAE, PHYTOPLANKTON, AQUATIC MICROORGANISMS, AQUATIC A W72-09650
ITION, *CHLORELLA, *SCENEDESMUS, AQUATIC MICROORGANISMS, AQUATIC B W72-12255
OPES, NEW ENGLAND, AQUATIC LIFE, AQUATIC MICROORGANISMS, MARINE AL W72-14317
YSIS, POLLUTANTS, AQUATIC ALGAE, AQUATIC MICROORGANISMS, AQUATIC A W72-07896
UTROPHICATION, LARVAE, ROTIFERS, AQUATIC MICROORGANISMS, DIPTERA, W72-07132
ANCE, TRACERS, PLANT PHYSIOLOGY, AQUATIC PLANTS, RADIOECOLOGY.: /L W72-06342
EFFECTS, STREAMS, RIVERS, ROOTED AQUATIC PLANTS, AQUATIC ALGAE, WA W72-14690
RE, MARINE ALGAE, PHYTOPLANKTON, AQUATIC PLANTS, CHEMICAL PROPERTI W72-14542
S, BIOASSAY, CYANOPHYTA, RESINS, AQUATIC PLANTS, AQUATIC ALGAE, NI W73-00286
TICAL TECHNIQUES, *ALGAL TOXINS, AQUATIC PLANTS, TOXINS, RED TIDE, W73-00243
, *DISTRIBUTION PATTERNS, LAKES, AQUATIC PLANTS, AUSTRALIA, SAMPLI W73-01094
REVIEWS, BIOLOGICAL COMMUNITIES, AQUATIC PLANTS, LIMITING FACTORS, W73-01089
PLANTS, LIMITING FACTORS, ROOTED AQUATIC PLANTS, CYCLING NUTRIENTS W73-01089
ION, PHYSICOCHEMICAL PROPERTIES, AQUATIC PLANTS, INDUSTRIAL: /TRAT W73-01615
ANALYSIS, MEASUREMENT, *NEVADA, AQUATIC PLANTS, WATER POLLUTION S W73-01673
IC WASTES, ECOLOGY, FISH, ALGAE, AQUATIC PLANTS, PROTOZOA, BACTERI W73-01972
RTEBRATES, FISH, BIRDS, STREAMS, AQUATIC PLANTS, AQUATIC INSECTS, W72-12930
GEN, OXYGEN DEMAND, HYPOLIMNION, AQUATIC PLANTS, ALGAE.: /LVED OXY W72-12994
TAL EFFECTS, METHODOLOGY, ALGAE, AQUATIC PLANTS, ISOTOPE FRACTIONA W72-12709
TICAL TECHNIQUES, FERTILIZATION, AQUATIC PLANTS, NITROGEN, NUTRIEN W73-02478
AQUATIC ANIMALS, CARP, DAPHNIA, AQUATIC PLANTS, AQUATIC ALGAE, SC W72-09649
ANIMALS, AQUATIC MICROORGANISMS, AQUATIC PLANTS, AQUATIC ALGAE, PE W72-09646
ICATION, ONTARIO, ALGAE, MOSSES, AQUATIC PLANTS.: /LLUTANT IDENTIF W71-11529
*EUTROPHICATION, FERTILIZATION, AQUATIC PLANTS, MATHEMATICAL MODE W71-12072
ROL, RIVERS, AGRICULTURE, ROOTED AQUATIC PLANTS, DRAINAGE, RUNOFF, W71-13172
PHORUS COMPOUNDS, POTASH, ALGAE, AQUATIC PLANTS, CHEMICAL ANALYSIS W71-12252
OL, TILAPIA, CARP, ALGAE, ROOTED AQUATIC PLANTS, PLANKTON, ARKANSA W71-04528
DITY, RUNOFF, PHOSPHORUS, ALGAE, AQUATIC PLANTS, CLAYS, ODOR, WATE W72-04759
FISH, INDUSTRIAL WASTES, ALGAE, AQUATIC PLANTS, SEA WATER, TRACE W72-05952

*ALGAE, *EUTROPHICATION, LAKES,	AQUATIC PLANTS, IRON, MICROORGANI	W72-03742
ER, ICNS, NUTRIENT REQUIREMENTS,	AQUATIC POPULATIONS, PRIMARY PROD	W72-09103
IRONMENTAL EFFECTS, COMPETITICN,	AQUATIC POPULATIONS, NUTRIENTS, N	W72-12736
ON-SITE TESTS, AQUATIC ANIMALS,	AQUATIC POPULATICNS, BIOASSAY, EU	W73-02095
IMALS, AQUATIC INSECTS, DAPHNIA,	AQUATIC POPULATIONS, FISH DIETS,	W72-07132
RIVERS, EFFLUENT, RADIOECOLOGY,	AQUATIC POPULATIONS, HYDROLCGIC A	W72-07841
YTA, SCENEDESMUS, AQUATIC ALGAE,	AQUATIC PRODUCTIVITY, PCLLUTANT I	W73-01066
ICAL COMMUNITIES, AQUATIC ALGAE,	AQUATIC PRODUCTIVITY, BIOCHEMICAL	W73-00002
EDESMUS, GROWTH RATES, CULTURES,	AQUATIC PRODUCTIVITY, METABOLISM,	W72-14314
POLLUTION SOURCES, CHLOROPHYTA,	AQUATIC PRODUCTIVITY, AQUATIC ALG	W73-00379
*CHLOROPHYLL, *PLANT PIGMENTS,	AQUATIC PRODUCTIVITY, *PRIMARY PR	W72-12192
ENT-WATER INTERFACES, SEDIMENTS,	AQUATIC PRODUCTIVITY, PHYTOPLANKT	W72-08989
COMMUNITIES, DOMINANT ORGANISMS,	AQUATIC PRODUCTIVITY, BIOMASS, CA	W72-09239
ED OXYGEN, DIURNAL DISTRIBUTION,	AQUATIC PRODUCTIVITY, NITROGEN, P	W71-11519
GING(BIOLOGICAL), AQUATIC ALGAE,	AQUATIC PRODUCTIVITY, FISH, FISHK	W72-04279
ER POLLUTION SOURCES, SEDIMENTS,	AQUATIC SOILS, PLANT TISSUES, FRE	W73-01673
SING, *POLLUTANT IDENTIFICATION,	AQUATIC WEEDS, *WASHINGTCN, *CHLO	W72-14728
OSPHORUS, EUTROPHICATION, ALGAE,	AQUATIC WEEDS, PHYTOPLANKTON, OLI	W72-08459
UCT E/ *ALGAE CULTURE TECHNIQUE,	AQUATIC WEED STUDY FACILITY, PROD	W71-06102
PULP AND PAPER INDUSTRY, ALGAE,	AQUATIC WEEDS, RECREATION, GROUND	W70-05103
DEMAND, ECONOMIC JUSTIFICATION,	AQUATIC WEED CONTROL, RIVERS, AGR	W71-13172
E ALGAE, COMPREHENSIVE PLANNING,	AQUATIC WEEDS, WATER POLLUTICN SO	W72-09171
UTRIENTS, LAKES, MUD, WISCONSIN,	AQUATIC WEEDS, EUTROPHICATION.: /	W72-10608
URAL RUNOFF, FARM WASTES, ALGAE,	AQUATIC WEEDS, FERTILIZATION, LEA	W73-02479
MENTS(ALGAE), NUTRIENT CHEMISTRY(	AQUATIC).: /), PHOSPHORUS REQUIRE	W70-04080
SSIPES, TYPHA ANGUSTATA, IPOMOEA	AQUATICA, TOTAL SOLIDS.: /NIA CRA	W72-01363
BORATES, ARTHROBACTE/ NEOCHLORIS	AQUATICA, SCHIZOTHRIX CALCICOLA,	W73-00854
SCIRPUS SPP, TYPHA SPP, ZIZANIA	AQUATICA, CAREX SPP, BUTCMUS UMBE	W73-01089
MUSSELS, CLAMS, FISH, SHELLFISH,	AQUEOUS SOLUTIONS, PHAECPHYTA, AT	W72-05965
OUNDS, POLLUTANT IDENTIFICATICN,	AQUEOUS SOLUTIONS, ION EXCHANGE,	W72-12166
OWERPLANTS, DISTILLATION, ALGAE,	AQUICULTURE, FISH, SHELLFISH, CRU	W70-08832
ESCENT CULTURES, XYLOSE, RIBOSE,	ARABINOSE, GALACTCSE, HEXURONIC A	W72-07710
WAY, MACROPHYTES, FYRISAN RIVER,	ARBOGAAN RIVER, BOTORPSSTROMMEN R	W73-00286
KE, BIG ARBOR VITAE LAKE, LITTLE	ARBOR VITAE LAKE, GREEN BAY, FOX	W73-01456
A, CRYSTAL LAKE, TROUT LAKE, BIG	ARBOR VITAE LAKE, LITTLE ARBOR VI	W73-01456
LAKE MARY, LAKE JOSEPHINE, LAKE	ARCO, LAKE DEMING, LOWER LA SALLE	W72-07145
TOMONAS-SPP.: *CANADIAN	ARCTIC, *CHLAMYDOMONAS-SPP, *CRYP	W72-07922
I/ *ECOSYSTEMS, *ALASKA, *LAKES,	ARCTIC, PRODUCTIVITY, EUTROPHICAT	W71-09190
ENT, *SAMPLING, WATER RESOURCES,	AREA REDEVELOPMENT, BIOLOGICAL CH	W72-03215
ITHIC ALGAE/ *EXPERIMENTAL LAKES	AREA, *NORTHWESTERN ONTARIO, EPIL	W71-11029
*ROADSIDE REST	AREAS.:	W72-14363
R POLLUTION EFFECTS, *INTERTIDAL	AREAS, *INVERTEBRATES, *ALGAE, CA	W71-08792
IDES, BEACHES, SANDS, INTERTIDAL	AREAS, ANNELIDS, COPEPODS, CRABS,	W72-07911
L MARSHES, FOOD WEBS, INTERTIDAL	AREAS, CONNATE WATER.: /IES, TIDA	W73-00796
ESTICIDES, ESTUARIES, INTERTIDAL	AREAS, ESTUARINE FISHERIES.: /, P	W72-10678
, OTTERS, ROCKSLIDES, INTERTIDAL	AREAS, CN-SITE INVESTIGATIONS, FO	W73-00764
	ARGENTINA.:	W73-00819
NE, PROLINE, GLUTAMATE, LEUCINE,	ARGININE, SERINE, TYROSINE, TRYPT	W73-00942
ATUS, CERAMIUM NITENS, WRANGELIA	ARGUS, VALONIA VENTRICOSA, BRYOPS	W73-00838
OOMS, 2,4 DINITRO PHENOL, TUCSON(	ARIZ).: /ALCIUM RADIOISOTOPES, BL	W71-01474
LIF), LOS ANGELES(CALIF), TUCSON(	ARIZ).: /LING, IMPERIAL VALLEY(CA	W71-09157
R POLLUTION CONTROL, CALIFORNIA,	ARIZONA.: /OMIC FEASIBILITY, WATE	W71-09157
OGEN ICN CONCENTRATION, PHENOLS,	ARIZONA.: /, MICROORGANISMS, HYDR	W71-01474
ROOTED AQUATIC PLANTS, PLANKTON,	ARKANSAS, INSECTS, TURBIDITY, CHA	W71-04528
A, NEPHTYS, CAPITELLA, MAGELONA,	ARMANDIA, TELLINA BUTTONI, VENUS,	W72-09092
TES, PHACUS PYRUM, TRACHELOMONAS	ARMATA, TRACHELOMONAS HISPIDA VAR	W73-00918
RUCTICN, DELAWARE BAY, SCOLOPLOS	ARMIGER, ALIMITOS BAY, LAKE PONTC	W73-02029
E.:	ARNOLDIELLA-CONCHOPHILA, LAKE ERI	W71-12489
LGAL SPECIES, WESTERN LAKE ERIE,	ARNOLDIELLA CONCHCPHILA MILLER.: /	W71-09156
FIS/ *POLYCHLORINATED BIPHENYLS,	AROCLOR 1254, ESCAMBIA RIVER, PIN	W71-07412
ICROCYSIS, OSCILLATORIA, SYNURA,	AROGLENA, CRYPTOPHYCEAE, PONTOPCR	W72-12729
TER SUPPLY PROJECT, LOPEZ CREEK,	ARROYO GRANDE CREEK.: *LOPEZ WA	W72-05594
IELLA, DIMETHYL SODIUM ARSONATE,	ARSENATES, SODIUM CACODYLATE, CUL	W72-13813
NG, ALGAE, ANTIBIOTICS, MERCURY,	ARSENIC COMPOUNDS, CHESAPEAKE BAY	W72-10162
ILICATES, POLYCARBOXYLATES.:	ARSENIC, NITRILOTRIACETATE, POLYS	W72-06837
, NORWAY, A/ *ATOMIC ABSORPTION,	ARSENIC, SELENIUM, ZINC, ANTIMONY	W71-11515
Y/ ACREMONIELLA, DIMETHYL SODIUM	ARSONATE, ARSENATES, SODIUM CACOD	W72-13813
LA NANA.: CHELEX 100,	ARTEMIA SALINA, BIO-RAD, CYCLOTEL	W72-04708
A FLEXNERI, BIOCHEMICAL STUDIES,	ARTHOPODS, STREPTOM: /AE, SHIGELL	W72-13800
SCHIZOTHRIX CALCICOLA, BORATES,	ARTHROBACTER, NOSTOC COMMUNE, POR	W73-00854
CHAETES, CHIRONOMIDS, BARNACLES,	ARTHROPODS, SARGASSUM CRISPUM, DI	W73-00855
MS, SIPUNCULIDS,/ COELENTERATES,	ARTHROPCDS, NEMERTEANS, ECHINODER	W73-00932
CEPHALOCHORDATES, HEMICHORDATES,	ARTHROPODS, *BAY OF BENGAL, COELE	W73-00935
HIC SPECIES, ABALONE, TUNICATES,	ARTHROPCDS, COELENTERATES, BRYOZO	W71-01475
RASTRUM LONGIPES VAR CONTRACTUM,	ARTHURS LAKE, LAKE SORELL, LAKE C	W73-01094
KE, THOMAS L/ COOLING RESERVOIR,	ARTIFICIAL RESERVOIR, MCNTRCSE LA	W70-09596
*NUTRIENTS, AIRCRAFT, SATELLITES(	ARTIFICIAL), PHOSPHORUS, NITROGEN	W72-04742
APHY, REMOTE SENSING, SATELLITES(	ARTIFICIAL), ALGAE, CHLCROPHYTA,	W72-09120
RGENTS, WATER POLLUTION SOURCES,	ASBESTOS, EUTROPHICATION, POLYCHL	W73-00703
IOMASS, SALINITY, BENTHIC FAUNA,	ASBESTOS, INORGANIC COMPOUNDS, WO	W73-00374
DIGITATA, LAMINARIA HYPERBOREA,	ASCOPHYLLUM NODOSUM, GIGARTINA MA	W71-11515
RULA GLUTINIS, SARGASSUM NATANS,	ASCOPHYLLUM NODOSUM, FUCUS VESICU	W72-04748
COMPOU/ LAKE TAHOE(CALIF), ZINC,	ASCORBIC ACID, PYRIDOXINE, INDAL	W72-07514
ION, NEUROANATOMY, OXYGEN PLASMA	ASHING, ATOMIC ABSORPTION, TITRAT	W71-11036
ELENIUM, ZINC, ANTIMONY, NORWAY,	ASHING, BIOLOGICAL MAGNIFICATION,	W71-11515
PHOTOMETRY, PERCH LAKE(ONTARIO),	ASHING, BIOLOGICAL MAGNIFICATION.	W71-11529
RILIZATION, FIXATION, EMBEDDING,	ASHING, BONE, ORGANS, BRAIN, BLOO	W73-01676
IN, LCW-TEMPERATURE OVEN DRYING,	ASHING, EXTRACTIVE DIGESTION, NEU	W71-11036
ELLIPSOIDEA, RNA, CHLOROPHYLL B,	ASHING, LABORATORY TECHNIQUES.: /	W72-06274
LUM, RECYCLING, TREATMENT COSTS,	ASIA.: A	W72-13508

FUNGI, SOILS, SYMBIOSIS, AFRICA, ASIA, ANABAENA.: /OP PRODUCTION, W73-02473
LEAR WASTES, PATH OF POLLUTANTS, ASIA, MARINE ALGAE, SEDIMENTS, ES W72-10958
*BIOINDICATORS, *MARINE ANIMALS, ASIA, MOLLUSKS, COBALT RADIOISOTO W73-00811
INE, ISOLEUCINE, VALINE, LYSINE, ASPARAGINE, CYSTINE, CY: /E, ALAN W73-00942
TYROSINE, TRYPTOPHAN, HISTIDINE, ASPARTATE, METHIONINE, PHENYLALAN W73-00942
ITY, ALGAE, ECONOMICS, POLITICAL ASPECTS, GEOLOGY, TOPOGRAPHY, GRO W71-03012
SPHOROUS, WATER SOFTENING, LEGAL ASPECTS, HARD WATER, PUBLIC HEALT W71-09429
ULATED WATER, WATER REUSE, LEGAL ASPECTS, JUDICIAL DECISIONS, LEGI W70-07100
ON, *CHEMICAL WASTES, *POLITICAL ASPECTS, LAKES, GREAT LAKES, ALGA W72-14782
AQUATIC POPULATIONS, HYDROLOGIC ASPECTS, NUTRIENTS, CHEMICALS, PL W72-07841
ALTH, ENVIRONMENT CONTROL, LEGAL ASPECTS, REGULATION, PERMITS.: /E W73-00703
TROPHICATION, LEGISLATION, LEGAL ASPECTS, REGULATION, POLLUTION AB W72-01085
L, STANDARDS, LEGISLATICN, LEGAL ASPECTS, WATER POLLUTION, ADJUDIC W72-01093
S, CARCINOGENS, TORULOPSIS SPP., ASPERGILLUS, PENICILLIUM, FUSARIA W72-09675
NO-GILVA, XYLOBOLUS FRUSTULATUS, ASPERGILLU: /GUA, TRAMETES LILACI W72-13813
A, EUDORINA, CYCLOPS, KERATELLA, ASPLANCHNA, LITHODESMIUM, HARPEST W71-11876
ELEOCHARIS MONTEVIDENSES, CHARA ASPPRA, SCIRPUS SPP, GILLS, MUSCL W73-01673
S), NITRILOTRIACETIC ACID, ALGAL ASSAY PROCEDURES, ALGAL GROWTH, I W71-12072
Y INDEX.: *MICROCOSM ALGAL ASSAY PROCEDURE, SPECIES DIVERSIT W72-03218
OLVED PHOSPHORUS, *ALGAL ASSAYS, ASSAY PROCEDURE.: DISS W72-12966
RICOR/ *CHEMOSTAL ASSAYS, *ALGAL ASSAY PROCEDURES, SELENASTRUM CAP W72-03985
TRIENT REMOVAL, WATER TREATMENT, ASSAY, BACTERIA.: /, NITROGEN, NU W73-02478
EINS, LABORATORY TESTS, AMMONIA, ASSAY, INHIBITION, UREAS, SYNTHES W72-03222
ES, *ALGAE, EVALUATION, TESTING, ASSAY, NUTRIENTS, CULTURES, MEASU W73-00255
SOURCES.: ALGAL ASSAY, ORTHOPHOSPHATE, PHOSPHORUS W73-02478
EFFLUENTS, TEMPERATURE CONTROL, ASSAY, REMOTE SENSING, ALGAE, ANT W72-10162
CHLAMYDOMONAS, SCENEDESMUS, BIO- ASSAY, RESISTANCE, LETHAL LIMIT, W72-14706
SELENASTRUM CAPRICOR/ *CHEMOSTAL ASSAYS, *ALGAL ASSAY PROCEDURES, W72-03985
DISSOLVED PHOSPHORUS, *ALGAL ASSAYS, ASSAY PROCEDURE.: W72-12966
TES, *EUTROPHICATION, NUTRIENTS, ASSESSMENTS, ALGAE, FORMULATION, W72-09169
TORS, *SEA WATER, MARINE PLANTS, ASSESSMENTS, MONITORING, SEWAGE, W71-12867
*WASTE WATER(POLLUTION), *WASTE ASSIMILATION CAPACITY, *ORGANIC L W71-11017
IOLOGICAL), AQUATIC ALGAE, WASTE ASSIMILATIVE CAPACITY, WATER POLL W72-01685
UTRIENTS, WATER RESOURCES, WASTE ASSIMILATIVE CAPACITY, RESEARCH A W71-13256
REQUIREMENTS, OXYGEN SAG, WASTE ASSIMILATIVE CAPACITY, WATER PROP W72-04279
, AMMONIA, *GROWTH RATES, *WASTE ASSIMILATIVE CAPACITY, OXIDATION, W72-04787
H, ON-SITE INVESTIGATIONS, WASTE ASSIMILATIVE CAPACITY, WASTE DIL: W73-00790
TOGRAPHY, INFRARED SPECTROSCOPY, ASSIMILATION, KJEDAHL PROCEDURE, W73-00379
LLUTANTS, STEPHANOPYXIS COSTATA, ASSIMILATION, EMILIANA(COCCOLITHU W73-00380
OTOASSIMILATION, GLYOXYLIC ACID, ASSIMILATION.: /PERMATOPHYTES, PH W73-02100
SEDIMENTS, WASTE DILUTION, WASTE ASSIMILATIVE CAPACITY, CALIFORNIA W73-00831
SORPTION, WASTE DILUTION, WASTE ASSIMILATIVE CAPACITY, EUROPE, AN W73-00817
TOPES, PATH OF POLLUTANTS, WASTE ASSIMILATIVE CAPACITY, *CALIFORNI W73-00826
LOGY, MATHEMATICAL MODELS, WASTE ASSIMILATIVE CAPACITY, ESTUARINE W73-00828
ECOMPOSING ORGANIC MATTER, WASTE ASSIMILATIVE CAPACITY, EUTROPHICA W73-00802
CLOPS STRENUUS, INGESTION RATES, ASSIMILATION EFFICIENCES, CRYPTOM W72-07940
NOTARISII, PSYCHROPHILIC, BIOTIC ASSOCIATIONS, LA GORCE MOUNTAINS, W73-00854
CORPORATION, SOAP AND DETERGENT ASSOCIATION, CARBOY TRIALS, CANAD W70-07283
TELLA CAPITATA, MACOMA CALCAREA, ASTARTE BOREALIS,: /ORQUATA, CAPI W73-02029
AMPLES, SARGASSO WEED, STARFISH, ASTERIAS FORBESI, PLUTONIUM 236, W72-05965
AERIUM, CELL COUNTS, FRAGILARIA, ASTERIONELLA, CRYPTOMONAS, SPHAER W71-03014
, CHLAMIDOMONAS, B/ RHINE RIVER, ASTERIONELLA FORMOSA, CRYPTOMONAS W71-01491
OSIPHON, FRAGILARIA, TABELLARIA, ASTERIONELLA, SYNEDRA, DINOBRYON, W73-01669
OVATA, STEPHANODISCUS NIAGARAE, ASTERIONELLA FORMOSA, RHODOMONAS W73-01435
OSTOMUS COMMERSONII COMMERSONII, ASTERIONELLA SPP, ORTHOPHOSPHATES W72-07890
CYSTIS, GLOEOCYSTIS, COELASTRUM, ASTERIONELLA, CRYPTOMONAS.: /, OO W72-07940
N(ENGLAND), WINDERMERE(ENGLAND)/ ASTERIONELLA FORMOSA, BLELHAM TAR W72-10625
, LITHIUM, ASTERIONELLA FORMOSA, ASTERIONELLA JAPONICA, STAURASTRU W72-12739
DOSA, LITHIUM CHLORIDE, LITHIUM, ASTERIONELLA FORMOSA, ASTERIONELL W72-12739
MICROCYSTIS, RHODOMONAS MINUTA, ASTERIONELLA FORMOSA, STEPHANODIS W72-12738
Y NET, ASTEROLAMPRA MARYLANDICA, ASTEROLAMPRA HEPTACTIS, BACTERIAS W71-11527
AMPRA HEPTACTIS, BAC/ JUDAY NET, ASTEROLAMPRA MARYLANDICA, ASTEROL W71-11527
DOSA, CHLAMYDOMONAS, FONTINALIS, ASTINASTRUM, ANACYSTIS NIDULANS, W71-11519
RIONELLA FORMOSA, STEPHANODISCUS ASTREA, ANKISTRODESMUS, CHRYSOCOC W72-12738
M, PUBLIC HEALTH, SOUTH AMERICA, ATLANTIC COASTAL PLAIN, COASTS, C W73-00819
PHYTA, EUTROPHICATION, PLANKTON, ATLANTIC OCEAN, SEA WATER, SALINI W72-07710
PLANTS, BIOMASS, CANADA, COASTS, ATLANTIC OCEAN, SEASONAL, BAYS, S W73-00366
ER, *NORTH CAROLINA, MICROSCOPY, ATLANTIC OCEAN.: /GRAPHY, SEA WAT W73-00284
ERMAL POLLUTION, *THERMAL POWER, ATLANTIC OCEAN, MARINE BIOLOGY, B W72-11876
CTION, *BIOMASS, ORGANIC MATTER, ATLANTIC OCEAN, GROWTH RATES, ALG W71-11562
N, ALGAE, SEPARATION TECHNIQUES, ATLANTIC OCEAN, PACIFIC OCEAN, GU W71-11851
SAMPLING, MATHEMATICAL STUDIES, ATLANTIC OCEAN, *TROPICAL REGIONS W72-11719
URRENTS, *DISTRIBUTION PATTERNS, ATLANTIC OCEAN, PACIFIC OCEAN, TR W72-03543
, AQUEOUS SOLUTIONS, PHAEOPHYTA, ATLANTIC OCEAN, SHARKS, RADIOACTI W72-05965
ARIES, PRODUCTIVITY, POPULATION, ATLANTIC OCEAN, DEPTH, STANDING C W72-04766
TOCARPUS PAPENFUSSII, GLOIODERMA ATLANTICA, CODIUM CAROLINIANUM, * W73-00284
CTERIASTRUM COMOSUM, CHAETOCEROS ATLANTICUS VAR. NEAPOLITANA, CHAE W71-11527
CHEMISTRY OF THE ATMOSPHERE.: W72-00949
ENTS, FERTILIZERS, PRECIPITATION(ATMOSPHERIC), RUNOFF, ANIMAL WAST W71-04216
DRAINAGE, RUNOFF, PRECIPITATION(ATMOSPHERIC).: /LLS, PERCOLATION, W72-12955
CHMENT, LEUCOTHRIX MUCOR, MAJURO ATOLL, MARCHALL ISLAND, POLYSIPHO W72-08142
OTICS(PESTICIDES), BACTERICIDES, ATOLLS, CULTURES, SPHAEROTILUS, C W72-08142
ROANATOMY, OXYGEN PLASMA ASHING, ATOMIC ABSORPTION, TITRATION.: /U W71-11036
LYSIPHONIA, ANTIBIOTICS, IODINE, ATOMIC ENERGY, REACTOR COOLING, I W71-09157
SPHORUS, A*DENOSINE TRIPHOSPHATE(ATP).: /SPHORUS, EXCHANGEABLE PHO W72-03742
AN RIVER, BOTORPSSTROMMEN RIVER, ATRAN RIVER, MOTALA STROM RIVER, W73-00286
PPLE DAM, STONE VALLEY DAM(PA.), ATRAZINE, BIOTYPES, POUR PLATE ME W72-04743
RAL REGION, CULTURE EXPERIMENTS, ATRAZINE, MCPA, ENDOSULFAN, FAIRY W72-12576
PHRAGMITES COMMUNIS, JUNCUS SPP, ATRIPLEX HASTATA, SCIRPUS SPP, TY W73-01089
SALVELINUS FONTINALIS, SEMOTILUS ATROMACULATUS, CLEANUP, SAMPLE PR W72-12930

CHIRONOMUS PLUMOSUS, CHIRONOMUS	ATTENTUATUS, CHIRONOMUS ANTHRACIN	W71-11009
*FISHERIES, WATER QUALITY, FISH	ATTRACTANTS, FISH HARVEST, PLANKT	W70-09596
R/ *WATER QUALITY CONTROL, *FLOW	AUGMENTATION, WATER QUALITY, RESE	W72-04260
ANALYZER, VAN DORN SAMPLERS, C-/	AUKE BAY, GLACIAL WATER, BECKMAN	W72-09122
ANCE LEVELS, TOLYPOTHRIX TENUIS,	AULOSIRA FERTILISSIMA, ANACYSTIS	W71-12071
ENA, NOSTOC, TOLYPOTHRIX TENUIS,	AULOSIRA FERTILISSIMA, ANACYSTIS	W71-10070
N/ *TOLERANCE, ANABAENA, NOSTOC,	AULOSIRA, TOLYPOTHRIX, ANACYSTIS	W72-14807
TERIUM FURTUITUM, STAPHYLOCOCCUS	AUREUS, CANDIDA ALBICANS.: /COBAC	W72-04259
PATTERNS, LAKES, AQUATIC PLANTS,	AUSTRALIA, SAMPLING, MICROSCOPY,	W73-01094
S, APISTONEMA / *LAKE NEUSIEDLER(	AUSTRIA), CHAMAESIPHON SUBAEQUALI	W71-04526
ORINA MORUM, / HETEROINHIBITORS,	AUTOINHIBITORS, VOLVOCACEAE, PAND	W71-13793
RHODOMONAS, COCCOLITHUS HUXLEYI,	AUTOLYSIS, PHAEODACTYLUM.: /CUS,	W71-10083
ACILITIES, SAMPLING, MONITORING,	AUTOMATIC CONTROLS, AERATION, SED	W72-00042
INSTRUMENTATION, PHYTOPLANKTON,	AUTOMATION, EQUIPMENT, ANALYTICAL	W72-07715
UM, HYALOTHECA, LAKE LIVINGSTON,	AUTORADIOGRAPHY, SAMPLE PREPARATI	W72-09239
*ERRORS, *SYNCHRONOUS CULTURES,	AUTOSPORES, VARIABILITY, BIAS, SY	W72-11920
ILA BLENNISTA, FILTRATES, GROWTH	AUTOSTIMULATION, GROWTH REGULATIO	W71-08687
VORAX, PSEUDOMONAS FLUORESCENS,	AUTOTROPHIC BACTERIA, HETEROTROPH	W73-02095
HETEROTROPHS,	AUTOTROPHS.:	W71-03021
ASS, PYRROPHYTA, SPRING, SUMMER,	AUTUMN, SAMPLING, PLANKTON NETS,	W72-12738
TION, MARINE ALGAE, CHRYSOPHYTA,	AUTUMN, WINTER, SUMMER, PRIMARY P	W73-00442
SYSTEM ANALYSIS, MACROCONSUMERS,	AUTUMNAL COMMUNITY, CALORIES, DET	W71-10064
E PHOSPHORUS, A*DENOSINE TRIPHO/	AVAILABLE PHOSPHORUS, EXCHANGEABL	W72-03742
ATES, DIESEL FUEL, BUNKER C OIL,	AVICENNIA, RHIZOPHORA, BOSTRYCHIE	W72-07911
HREIBER MEDIUM, ESTUARINE SOILS,	AXENIC CULTURES, GIBBERELIC ACID.	W73-00430
TOANTAGONISM, *HETEROANTAGONISM,	AXENIC CULTURES, EXCRETION, APHAN	W72-12734
RY, MAGNESIUM, SULFATES, SPORES,	AZOTOBACTER, CYANOPHYTA, YEASTS,	W72-06124
NITROSOMONAS, NITROBACTER,	AZOTOBACTER.:	W72-11563
PHICATION, OXIDATION, SYMBIOSIS,	AZOTOBACTER, MARINE ALGAE, ORGANI	W72-11563
EFFECTS, GROWTH RATES, E. COLI,	AZOTOBACTER, PSEUDOMONAS, WATER P	W72-10861
STERA, PINAEUS DUORARUM, PINAEUS	AZTECUS, PENAEUS SETIFERUS, MERCE	W73-02029
ROGEN, THIAMINE, BIOTIN, VITAMIN	B 12, GONYAULAX POLYEDRA, PERIDIN	W72-07892
LASER, RHODAMINE	B.:	W72-01095
OBIOLOGY, OREGON, *SED/ *VITAMIN	B, *EUTROPHICATION, *AQUATIC MICR	W72-08989
LA ELLIPSOIDEA, RNA, CHLOROPHYLL	B, ASHING, LABORATORY TECHNIQUES.	W72-06274
, PSEUDOMONAS, CULTURES, VITAMIN	B, BIOASSAY, CELLULOSE, DIALYSIS,	W73-00942
RUS, PHOSPHATES, CARBON, VITAMIN	B, NITROGEN, CYANOPHYTA, LAKE MIC	W72-07937
OPHYLL, SOLAR RADIATION, VITAMIN	B, PIGMENTS, GROWTH RATES, ALGAE,	W72-07892
TECHNIQUES, CULTIVATION, VITAMIN	B, PLANT GROWTH REGULATORS, NITRA	W73-00430
ILAMIDE, PENICILLIN G, POLYMYXIN	B, PODOCYSTS.: /RACYCLINE, SULFAN	W72-10162
OPTICAL DENSITY, PRASINOPHYCEAE,	BACCILARIOPHYCEAE, XANTHOPHYCEAE,	W72-08051
RIDES, CRYPTOPHYCEA, DINOPHYCEA,	BACCILARIOPHYCEA, PRYMNESIUM PARV	W70-07280
RIPUNCTATA, GOMPHONEMA PARVULUM,	BACILLARI PARADOXA, SPHAEROTILUS	W72-10883
ROSIGMA FASCIOLA, NITZSCHIA SPP,	BACILLARIA PAXILLIFER,: /IGMA, GY	W73-00853
TOCRYPTOCHRYSIS OBOVATA, GUTTULA	BACILLARIOPHAGA, REFRACTOCYSTIS P	W72-09119
OIR, *HYDROBIOLOGY, METALIMNION,	BACILLARIOPHYTA, VOLVOCINEAE, PRO	W72-04855
ORELLA PYRENOIDOSA, STICHOCOCCUS	BACILLARIS, UNDISSOCIATED MOLECUL	W70-07255
*MICROBIAL GROWTH, MICROCOCCUS,	BACILLUS, VIBRIO, RHODOTORULA, AE	W72-10861
GI, POLYETHYLENE, POLYPROPYLENE,	BACILLUS, CORYNEBACTERIA, STREPTO	W72-09675
IVE BACTERIA.:	BACILLUS, MICROCOCCUS, GRAM NEGAT	W72-12255
OSTRAINING, *PRECOAT FILTRATION,	BACKWASHING, SPECIFIC RESISTANCE.	W72-09393
BINED SEWERS, *SUSPENDED SOLIDS,	BACKWASHING, *MOUNT CLEMENS(MICH)	W72-00042
VOLATILE SOLIDS, SANBORN FILTER,	BACKWASHING.: FIREBAUGH(CALIF),	W70-10174
*OXIDATION LAGOONS, *PATHOGENIC	BACTERIA.: /S, *ENTERIC BACTERIA,	W70-08319
WASHINGTON, MARINE ALGAE, MARINE	BACTERIA.: /ON, TROUT, TOXICITY,	W72-05421
, PHAEOPHYTA, RHODOPHYTA, MARINE	BACTERIA.: /, OCEANS, CHLOROPHYTA	W71-07878
IES, MYCOBACTERIUM, PSEUDOMONAS,	BACTERIA.: /GAE, CYTOLOGICAL STUD	W71-08027
ON, FLUORESCENCE, PHOTOSYNTHETIC	BACTERIA.: /ING, MUD, PHYTOPLANKT	W72-01784
SPHAEROTOTILUS NATANS, ZOOGLOEAL	BACTERIA.: /DOM, *SEWAGE FUNGUS,	W72-08573
HLOROPHYTA, PSEUDOMONAS, ENTERIC	BACTERIA.: /UATIC ALGAE, ALGAE, C	W72-12255
LLUS, MICROCOCCUS, GRAM NEGATIVE	BACTERIA.: BACI	W72-12255
REMOVAL, WATER TREATMENT, ASSAY,	BACTERIA.: /, NITROGEN, NUTRIENT	W73-02478
, COPEPODS, CRABS, AQUATIC LIFE,	BACTERIA.: /TIDAL AREAS, ANNELIDS	W72-07911
, ORGANIC MATTER, CULTURES, SOIL	BACTERIA.: /THERMOPHILIC BACTERIA	W72-14301
ISOTOPES, *ON-SITE TESTS, SULFUR	BACTERIA, PRODUCTIVITY, RADIOCHEM	W72-14301
DEMAND, *OXYGENATION, *ANAEROBIC	BACTERIA, BIOLOGICAL TREATMENT.: /	W72-14469
ES, CULT/ *PRIMARY PRODUCTIVITY,	BACTERIA, SCENEDESMUS, GROWTH RAT	W72-14314
OCHEMICAL ANALYSIS, THERMOPHILIC	BACTERIA, ORGANIC MATTER, CULTURE	W72-14301
INETICS, *SCENEDESMUS,/ *AEROBIC	BACTERIA, *AEROBIC CONDITIONS, *K	W73-00379
MICROSCOPY, DIATOMS, NUTRIENTS,	BACTERIA, CHRYSOPHYTA, PHYTOPLANK	W72-07899
*NUTRIENTS, FRESH WATER, ALGAE,	BACTERIA, NITROGEN, PHOSPHORUS, O	W72-07933
, *ALGAE, *HYDROBIOLOGY, ENTERIC	BACTERIA, SAMPLING, ZOOPLANKTON,	W72-07896
ICAL TECHNIQUES, NITROGEN FIXING	BACTERIA, THERMAL POLLUTION.: /YT	W72-08508
, CRABS, / *CRUSTACEANS, *MARINE	BACTERIA, *WATER QUALITY, BENTHOS	W72-08142
ALGAE, SHRIMP, FISH EGGS, FISH,	BACTERIA, FISH PARASITES, ANTIBIO	W72-08142
SYSTEMATICS, VITAMINS, ANAEROBIC	BACTERIA, ELECTRON MICROSCOPY, PI	W72-07951
, *ENTERIC BACTERIA, *PATHOGENIC	BACTERIA, CULTURES, E. COLI, BIOI	W72-07664
ULTURES, E. CO/ *ALGAE, *ENTERIC	BACTERIA, *PATHOGENIC BACTERIA, C	W72-07664
R IN/ *AQUATIC BACTERIA, *SULFUR	BACTERIA, *OXYGEN, *SEDIMENT-WATE	W72-06690
E.: *FILTER CLOGGING, IRON	BACTERIA, POND WEEDS, PERMANGANAT	W72-06536
EN, *SEDIMENT-WATER IN/ *AQUATIC	BACTERIA, *SULFUR BACTERIA, *OXYG	W72-06690
*KINETICS, *DISSOLVE/ *ANAEROBIC	BACTERIA, *ANAEROBIC CONDITIONS,	W73-01066
N, LABORATORY EQUIPMENT, AEROBIC	BACTERIA, SOLVENT EXTRACTIONS, FR	W73-01066
N, ORGANIC COMPOUNDS, A/ *MARINE	BACTERIA, *ATTRACTANTS, *PREDATIO	W73-00942
STEMATICS, *AQUATIC ALGAE, *SOIL	BACTERIA, SAMPLING, CULTURES, ECO	W73-00854
TISSUES, ALGAE, MICROORGANISMS,	BACTERIA, INVERTEBRATES, FUNGI, E	W73-01676
HABITATS, PRODUCTIVITY, AQUATIC	BACTERIA, AQUATIC ALGAE, CARBON C	W73-02095
OTROPHIC BACTERIA, HETEROTROPHIC	BACTERIA, GLUCOSE, ACETATES, BIOT	W73-02095
STES, BIOCHEMICAL OXYGEN DEMAND,	BACTERIA, ALG: /CALS, CHEMICAL WA	W73-01976
DOMONAS FLUORESCENS, AUTOTROPHIC	BACTERIA, HETEROTROPHIC BACTERIA,	W73-02095

ALGAE, AQUATIC PLANTS, PROTOZOA, BACTERIA, STREAMFLOW, FLOW RATES, W73-01972
ER, DETROIT RIVER, HETEROTROPHIC BACTERIA, ACETYLENE, ETHYLENE.: / W73-01456
UNDS, *GEOCHEMISTRY, CYANOPHYTA, BACTERIA, ALGAE, PLANKTON, PLANTS W73-01439
WATER PURIFICATION, CLAY, ALGAE, BACTERIA, WASTE WATER TREATMENT, W73-01362
, *MARINE ALGAE, *PHYTOPLANKTON, BACTERIA, AQUARIA, DECOMPOSING OR W73-00802
NTENSITY, TEMPERATURE, ANABAENA, BACTERIA, PHOTOSYNTHESIS.: /GHT I W73-02471
AQUATIC MICROORGANISMS, AQUATIC BACTERIA, AQUATIC ALGAE, ALGAE, C W72-12255
N, ALGAE, THERMOCLINE, ANAEROBIC BACTERIA, AEROBIC BACTERIA, SULFU W72-12996
CTERIA, AEROBIC BACTERIA, SULFUR BACTERIA, IRON COMPOUNDS.: /IC BA W72-12996
NTS, DECOMPOSING ORGANIC MATTER, BACTERIA, CHEMICAL PROPERTIES, NI W72-12997
INE, ANAEROBIC BACTERIA, AEROBIC BACTERIA, SULFUR BACTERIA, IRON C W72-12996
T PENETRATION, DISSOLVED OXYGEN, BACTERIA, THERMOCLINE, SEDIMENT-W W72-12995
*SEWAGE LAGOONS, *COLD/ *SEWAGE BACTERIA, *WASTE WATER TREATMENT, W72-12565
KOTA, BIOCHEMICAL OXYGEN DEMAND, BACTERIA, TEMPERATURE, COLIFORMS, W72-12565
GI, THERMAL SPRINGS, CYANOPHYTA, BACTERIA, ECOLOGY, WATER POLLUTIO W72-12734
ROGEN FIXATION, *NITROGEN FIXING BACTERIA, *RAINFALL, *ARID LANDS, W72-12505
LACTOBACILLUS, CHLORELLA, SULFUR BACTERIA, ALGAE, SALMONELLA, E. C W72-12565
HYTA, NITROGEN, CARBON, DIATOMS, BACTERIA, CARBON DIOXIDE, CHLAMYD W72-12544
TEMPERATURE, COLIFORMS, ENTERIC BACTERIA, NITROGEN, VITAMINS, SUS W72-12565
, *INHIBITIONS, *MICROORGANISMS, BACTERIA, FUNGI, METHODOLOGY, CUL W72-13813
ANTIBIOTICS(PESTICIDES), ENTERIC BACTERIA, DISEASES, E. COLI, SHIG W72-13800
TEMATICS, POLLUTANTS, PATHOGENIC BACTERIA, COLIFORMS, TOXINS, ANTI W72-13800
EN/ *AQUATIC PLANTS, *MINNESOTA, BACTERIA, FUNGI, PROTOZOA, PATHOG W72-08586
LS, THERMAL POLLUTION, SAMPLING, BACTERIA, TOXICITY, BEHAVIOR, BIT W72-08790
*BIOLOGICAL COMMUNITIES, RIVERS, BACTERIA, WATER POLLUTION, ACTIVA W72-08573
IC WASTES, ALGAE, INDU/ *ENTERIC BACTERIA, *WATER QUALITY, *DOMEST W72-09590
FACTORS, GROWTH RATES, DIATOMS, BACTERIA, LAKES, NITROGEN, ANALYT W72-09163
FER, *PLASTICS, DOMESTIC WASTES, BACTERIA, ALGAE, NUTRIENTS, OXIDA W72-10401
DGE TREATMENT, DISSOLVED OXYGEN, BACTERIA, FUNGI, PROTOZOA, FERMEN W72-09675
TREATMENT, PHYTOPLANKTON, FUNGI, BACTERIA, ALGAE.: /IOLOGY, WATER W72-10076
ON CONTROL, *WASTES, *PATHOGENIC BACTERIA, CONFERENCES, *PUBLIC HE W72-09675
TROPHIC LEVELS, PROTOZOA, ALGAE, BACTERIA, DIATOMS, OCEANS, ACTIVA W72-10883
, OCEANS, ACTIVATED SLUDGE, SOIL BACTERIA, PHO: /BACTERIA, DIATOMS W72-10883
*MICROBIOLOGY, *NITROGEN FIXING BACTERIA, *NITROGEN CYCLE, ENVIRO W72-11563
UTROPHICATION, NUTRIENTS, ALGAE, BACTERIA, LAKE ERIE, ALUMINUM, MA W72-01101
ALGAE, AQUATIC BACTERIA, SEWAGE BACTERIA, ENVIRONMENTAL EFFECTS, W72-01472
T IDENTIFICATION, ALGAE, AQUATIC BACTERIA, SEWAGE BACTERIA, ENVIRO W72-01472
ALANCE, DISSOLVED OXYGEN, ALGAE, BACTERIA, ODOR, INFLOW, DISSOLVED W72-02274
SPHORUS, NUTRIENTS, MEASUREMENT, BACTERIA, ALGAE, EUTROPHICATION, W71-13794
OBALT, *RADIATION, COSTS, SEWAGE BACTERIA, ALGAE, COLIFORMS, INSEC W72-00383
L CONTROL, EUTROPHICATI/ *ALGAE, BACTERIA, *LAKES, SYMBIOSIS, ALGA W71-13257
XIDATION, ADSORPTION, COLIFORMS, BACTERIA, CENTRIFUGATION, *WASTE W71-13326
L CONTROL, DIATOMS, PERSISTENCE, BACTERIA, BIRDS, FISH, ON-SITE IN W72-00155
STRONTIUM RADIOISOTOPES, MARINE BACTERIA, COASTAL MARSHES, MICROO W72-00948
ACTERISTICS, ALGAE, ZOOPLANKTON, BACTERIA, NUTRIENTS, SEDIMENTS, E W72-03215
UNITIES, *AQUATIC PLANTS, ALGAE, BACTERIA, LITTORAL, LAKES, PERIPH W72-03216
TA/ *EPIPHYTIC ALGAE, *EPIPHYTIC BACTERIA, *MACROPHYTE-EPIPHTYE ME W72-03216
TRATIFICATION, *ENZYME/ *AQUATIC BACTERIA, *POPULATION, *ALGAE, *S W72-03224
RICTION, *TURBULENT FLOW, ALGAE, BACTERIA, MARINE PLANTS, ENGINEER W72-03220
, SOIL CHEMICAL PROPERTIES, SOIL BACTERIA, SOIL ALGAE.: /MPERATURE W72-03342
UATIC ANIMALS, DIATOMS, INSECTS, BACTERIA, FUNGI, TURNOVERS.: / AQ W71-10064
SOLVED OXYGEN, ALGAE, NUTRIENTS, BACTERIA, VIRUSES, HEAVY METALS, W71-09788
N CONCENTRATION, MARINE ANIMALS, BACTERIA, ABSORPTION, CULTURES.: / W71-10083
AE, *NITROGEN FIXATION, ENZYMES, BACTERIA, SPORES, NITROGEN, METAB W71-12070
, WEEDS, TASTE, ODOR, PATHOGENIC BACTERIA, ORGANIC WASTES, SOLID W W71-11006
TEMPERATURE, COPEPODS, NITROGEN, BACTERIA, OXIDATION-REDUCTION PO: W71-11008
E MICROORGANISMS, MARINE PLANTS, BACTERIA, POPULATION, HAWAII, ABS W71-08042
CES, PHOTOSYNTHESIS, METABOLISM, BACTERIA, HYDROGEN ION CONCENTRAT W71-08687
CEAN, SAPROPHYTES, HETEROTROPHIC BACTERIA, C-14.: / ISLAND, OPEN O W71-08042
UTROPHICATION, ALGAE, PATHOGENIC BACTERIA, EFFLUENT, IRRIGATION, S W71-08214
DROGEN ION CONCENTRATION, ALGAE, BACTERIA, WATER POLLUTION SOURCES W71-07942
UTION EFFECTS, *KENTUCKY, SULFUR BACTERIA, WATER QUALITY, SULFATES W71-07942
IVERS, CULTURES, ORGANIC MATTER, BACTERIA, WATER ANALYSIS.: /NS, R W71-08032
ONCENTRATION, WATER TEMPERATURE, BACTERIA, ALGAE, ALKALINITY, IRON W71-05626
LE ISOTOPES, *CARBON, *ANAEROBIC BACTERIA, *TEXAS, BIODEGRADATION, W71-04873
TAGES, PATHOGE/ *ALGAE, *ENTERIC BACTERIA, *BACTERICIDES, GROWTH S W71-05155
CIDES, GROWTH STAGES, PATHOGENIC BACTERIA, CYANOPHYTA, CHLOROPHYTA W71-05155
G, ODORS, PHOTOSYNTHESIS, ALGAE, BACTERIA, CHLORINATION, GROUNDWAT W71-07123
EN ION CONCENTRATION, COLIFORMS, BACTERIA, PHOTOSYNTHESIS, ORGANIC W71-07100
YGEN, SATURATION, ALGAE, METHANE BACTERIA, EUTROPHICATION, NITRATE W71-07106
NTACT TIME, SAMPLING, COLIFORMS, BACTERIA, OXIDATION, ORGANIC LOAD W71-07096
DEL STUDIES, ZOOPLANKTON, BIOTA, BACTERIA, BENTHIC FAUNA, FISH, EN W72-05282
TIC PLA/ *AQUATIC LIFE, *AQUATIC BACTERIA, *AQUATIC ANIMALS, *AQUA W72-04853
ATER, / *BIOCONTROL, *PATHOGENIC BACTERIA, *ALGAL CONTROL, FRESH W W72-05457
BITION, HOSTS, MYXOPHYCEA/ LYTIC BACTERIA, ENGLAND, METABOLIC INHI W72-05457
S, *EUTROPHICATION, LAKES/ *SOIL BACTERIA, *LAKE SOILS, *PHOSPHORU W72-04292
GEN, PHOSPHORUS, CARBON, *SEWAGE BACTERIA, SEWAGE, ALGAE, HYDROGEN W72-04309
ATION, NUTRIENTS, SLUDGE, ALGAE, BACTERIA, *WASTE WATER TREATMENT, W72-04086
PHICATION, LAKES, AQUATIC ALGAE, BACTERIA, BIOCHEMISTRY, WISCONSIN W72-04292
*ISOLATION, *MEMBRANE PROCESSES, BACTERIA, SULFUR BACTERIA, PHOTOS W72-04743
PHOTOSYNTHETIC BACTERIA, AQUATIC BACTERIA, LAKES, FILTERS, COLIFOR W72-04743
DUSTRIAL WASTES, VIRUSES, ENERIC BACTERIA, TRITIUM, RADIOACTIVE WA W72-04742
SULFUR BACTERIA, PHOTOSYNTHETIC BACTERIA, AQUATIC BACTERIA, LAKES W72-04743
RANE PROCESSES, BACTERIA, SULFUR BACTERIA, PHOTOSYNTHETIC BACTERIA W72-04743
ROCESSES,/ *CULTURES, *ANAEROBIC BACTERIA, *ISOLATION, *MEMBRANE P W72-04743
THO/ *ALGAE, *CULTURES, *ENTERIC BACTERIA, *OXIDATION LAGOONS, *PA W70-08319
PHICATION, *PHOSPHORUS, *CARBON, BACTERIA, ALGAE, SYMBIOSIS, CARBO W70-07283
XATION, PLANKTON, PHYTOPLANKTON, BACTERIA, ZOOPLANKTON, SALMONIDS, W71-00117
UDGE TREATMENT, TOXICITY, ALGAE, BACTERIA, FISH, SEWAGE SLUDGE, FL W70-06968
R), SAMPLING, DIATOMS, NITROGEN, BACTERIA, ALGAE, THERMAL STRATIFI W71-03028

, CHEMICAL ANALYSIS, MOSQUITOES,	BACTERIA, FLOATING PLANTS, ON-SIT	W71-01488
TION, *ALGAE, AEROBIC TREATMENT,	BACTERIA, MUNICIPAL WASTES, BIOCH	W71-03896
PHOSPHORUS, FISH, WEEDS, CYCLES,	BACTERIA, CHEMIST: /YGEN DEMAND,	W71-03012
ILTER DENITRIFICATION, METHANOL,	BACTERIAL DENITRIFICATION, SAN JO	W71-08223
LANDICA, ASTEROLAMPRA HEPTACTIS,	BACTERIASTRUM COMOSUM, CHAETOCERO	W71-11527
GAE, CHEMICAL WASTES, ALGICIDES,	BACTERICIDES, CHROMIUM, ZINC, TEM	W72-08683
ASITES, ANTIBIOTICS(PESTICIDES),	BACTERICIDES, ATOLLS, CULTURES, S	W72-08142
NG, LABORATORY TESTS, CHEMICALS,	BACTERIOCIDES, CULTURES, TEST PRO	W71-06189
RIVER, OKA RIVER, HYDROBIOLOGY,	BACTERIOLOGY, SAPROPHYTES, STURGE	W72-04853
, FUNGI, PROTOZOA, FERMENTATION,	BACTERIOPHAGE, NUTRIENTS, FILTERS	W72-09675
RATION, TOTAL O/ FUSARIUM NO 83,	BACTERIUM NO 52, ENDOGENOUS RESPI	W71-12223
CYSTIS, SPIROGYRA, EPHEMEROPTERA	BAETIDAE, EPHYDRA, CHAETOGASTER,	W72-07132
	BAFFIN BAY(TEX).:	W71-04873
RIA REEF, BEAR BAY, NEVA ISLAND,	BAIRD ISLAND, SHOLIN ISLAND, WINI	W72-09120
ISCUS REX, ETHMODISCUS GAZELLAE,	BAJA CALIFORNIA, HEMIAULUS HAUCKI	W72-03567
, POLYCHAETES, AMPHIOXUS, SALPS,	BALAN: /OSCILLATORIA, STOMATOPODA	W73-00935
AQUATIC ALGAE, AQUATIC ANIMALS,	BALANCE OF NATURE, ECOLOGY, TROPH	W72-06340
*OXYGEN	BALANCE.:	W71-04518
INDIAN CREEK RESERVOIR, NUTRIENT	BALANCE.: / UTILITIES DISTRICT, *	W72-02412
DIATION, WATER TEMPERATURE, SALT	BALANCE, DISSOLVED OXYGEN, ALGAE,	W72-02274
INFLOW, DISCHARGE(WATER), WATER	BALANCE, SEDIMENTS, CHEMICAL PROP	W72-13851
QUARIA, ANIMAL PHYSIOLOGY, WATER	BALANCE, TRACERS, PLANT PHYSIOLOG	W72-06342
*TEXTILE MILL WASTES, MONTREAL,	BALANCING TANKS.:	W70-10427
EY POINT, BISCAYNE BAY, FLORIDA,	BALANUS, BARNACLES, EPIFAUNA, SAG	W72-07907
SALINUS, GASTEROSTEUS ACULEATUS,	BALTIC SEA, NITOCRA SPINIPES.: /	W71-08026
L, *CANADIAN PHOSPHATE DETERGENT	BAN, WYANDOTTE CHEMICAL CORPORATI	W70-07283
	BANGALORE(INDIA).:	W72-09176
), SEDIMENT / LAKE MENDOTA(WIS),	BANTAM LAKE(CONN), TROUT LAKE(WIS	W72-08459
ICITY, TRICHODESMIUM THIEBAUDII,	BARBADOS.: *PERIOD	W72-11719
*SANTA	BARBARA CHANNEL.:	W71-08793
*SANTA	BARBARA CHANNEL.:	W71-08792
GLYTOTENDIPES	BARBIPES, EKMAN DREDGE.:	W72-03556
ION CONCENTRATI/ *MOSS LANDING,	BARNACLES, LETHAL DOSAGE, CALCIUM	W72-09092
, CONNECTICUT RIVER, SAN JOAQUI/	BARNACLES, FLOUNDER, HUMBOLDT BAY	W71-07417
BISCAYNE BAY, FLORIDA, BALANUS,	BARNACLES, EPIFAUNA, SAGARTIA, AN	W72-07907
DERMS, POLYCHAETES, CHIRONOMIDS,	BARNACLES, ARTHROPODS, SARGASSUM	W73-00855
ONGES, COELENTERA/ *WOOD BORERS,	BARNACLES, BRYOZOAS, HYDROIDS, SP	W73-00374
X, CARBOWAX, FLOCCULANT DOSAGES,	BAROCO CLAY SOILS, URANIUM ORE.: /	W70-06968
, SPYRIDIA FILAMENTOSA, DICTYOTA	BARTAYRESII, LAURENCIA OBTUSA, LA	W73-00838
P, SAMPLE PRESERVATION, CAMBARUS	BARTONI, BIOLOGICAL SAMPLES, MERG	W72-12930
ALGAE, NITRATES, SURFACE RUNOFF,	BASE FLOW, PERCOLATION, LEACHING,	W71-06443
LAKE ERIE CENTRAL	BASIN.:	W72-12994
, PONDAGE, PILOT PLANT, SETTLING	BASIN.: /BORATORY, DIGESTION TANK	W72-02850
REGENERATION, LAKE ERIE CENTRAL	BASIN.: /DEOXYGENATION, *NUTRIENT	W72-12996
CE), CLADOPHORA, PHO/ RHINE-MAAS	BASIN(FRANCE), MEURTHE RIVER(FRAN	W71-00099
*BEAVER CREEK	BASIN(KY).:	W71-07942
LOADING GUIDELINES, *GRAND RIVER	BASIN(ONTARIO), CANADA, BUFFER CA	W71-11017
FOX RIVER	BASIN(WIS), FWPCA.:	W70-05103
*LAKE ERIE CENTRAL	BASIN, *OXYGEN DEPLETION.:	W72-13659
XYGENATION, / *LAKE ERIE CENTRAL	BASIN, NUTRIENT REGENERATION, DEO	W72-12997
GEN DEPLETIO/ *LAKE ERIE CENTRAL	BASIN, POINTE PELEE(ONTARIO), OXY	W72-12992
POTOMAC RIVER	BASIN, POTOMAC ESTUARY.:	W70-06509
CYSTIS, CERA/ *LAKE ERIE CENTRAL	BASIN, TRIBONEMA, OEDOGONIUM, ANA	W72-12995
, *LOTIC ENVIRONMENT, WATERSHEDS(	BASINS), *DDT, *SMALL WATERSHEDS,	W72-12930
IES, WASTE TREATMENT, WATERSHEDS(	BASINS), PHOSPHORUS, ALGAE, ORGAN	W71-11017
LUTION EFFECTS, BAYS, WATERSHEDS(	BASINS), SMALL WATERSHEDS, CROPS,	W72-00959
RIENTS, *ECOSYSTEMS, *WATERSHEDS(	BASINS), STREAMFLOW, IONS, NITRAT	W71-01489
S, GREAT LAKES, BAYS, WATERSHEDS(	BASINS), SURFACE DRAINAGE, MUNICI	W71-08953
POPULATION, TOURISM, WATERSHEDS(	BASINS), TROPHIC LEVELS, TURBIDIT	W71-11009
ROORGANISMS, EFFLUENTS, SETTLING	BASINS, WASTE WATER TREATMENT, LA	W71-00116
WATER POLLUTION EFFECTS, STRIPED	BASS, DISSOLVED OXYGEN, OIL, BRID	W73-02029
E, EGGS, CHINOOK SALMON, STRIPED	BASS, MINNOWS, FISHKILL, HERRING,	W71-11517
DDT, SILTS, SUCKERS, CATFISHES,	BASS, PERCHES, PIKES, SAMPLING, S	W70-08098
S, SNAILS, EELS, CHINOOK SALMON,	BASS, TROUT, MINNOWS, CRAYFISH, M	W71-07417
ENA SP L-31, CONTINUOUS CULTURE,	BATCH CULTURES.: /ROCYSTS, *ANABA	W72-03217
GAL GROWTH, CONTINUOUS CULTURES,	BATCH CULTURES.: *AL	W72-14790
SELENASTRUM CAPRICORNUTUM, PAAP	BATCH TEST, CULTURE MEDIA.: /DES,	W73-02099
TER), CHLOROPHYLL, FLUORESCENCE,	BATHYMETRY.: /REMENT, CURRENTS(WA	W72-01095
A RIGIDA, CLADOPHORA FULIGINOSA,	BATOPHORA OERSTEDI, ACETABULARIA	W72-11800
MARK,/ LAKE ESROM, TYSTRUP LAKE,	BAVELSE LAKE, CRYPTOPHYCEAE, *DEN	W72-12738
ELODEA, CALLITRICHE, ANABAENA /	BAY OF QUINTE, SALT CREEK, CHARA,	W73-00840
	BAY OF QUINTE, *LAKE ONTARIO.:	W71-08953
PLANT, SAN JOAQUIN RIVER, MORRO	BAY POWER PLANT, HUMBOLDT BAY, CO	W72-07907
M, *NOVA SCOTIA, *ST. MARGARET'S	BAY.: / DIGITATA, AGARUM CRIBROSU	W73-00366
Y, ARSENIC COMPOUNDS, CHESAPEAKE	BAY.: /ALGAE, ANTIBIOTICS, MERCUR	W72-10162
, BISCAYNE BAY(FLORIDA), HUMBOLT	BAY(CALIFORNIA), POTOMAC RIVER, C	W71-11517
*ULVA, *LAMINARIA, LIVERPOOL	BAY(ENGLAND), IRISH SEA.:	W71-12867
ORK), PATUXENT ESTUARY, BISCAYNE	BAY(FLORIDA), HUMBOLT BAY(CALIFOR	W71-11517
RKEY POINT(FLORIDA), / *BISCAYNE	BAY(FLORIDA), HEATED EFFLUENT, TU	W72-11876
OLOGICAL MAGNIFICATION, MINAMATA	BAY(JAPAN), ORGANIC MERCURY COMPO	W72-12257
CA, CODIUM CAROLINIANUM, *ONSLOW	BAY(NC).: /II, GLOIODERMA ATLANTI	W73-00284
*GEORGIAN	BAY(ONTARIO).:	W72-13662
NTERPRETATION, MACROAL/ *YAQUINA	BAY(ORE), YAQUINA ESTUARY, DATA I	W72-08141
BAFFIN	BAY(TEX).:	W71-04873
, RICHELIEU RIVER(QUEBEC), GREEN	BAY(WIS), MACKEREL, MONTREAL(CANA	W70-08098
ALGAL ASSAY PRO/ FLUSHING, GREEN	BAY(WIS), NITRILOTRIACETIC ACID,	W71-12072
*SAN FRANCISCO	BAY-DELTA.:	W71-09581
*HUMBOLDT	BAY, *CESIUM RADIOISOTOPES.:	W73-00826
ION, COASTA/ *FINLAND, *BOTHNIAN	BAY, *GULF OF FINLAND, DARK FIXAT	W72-03228
TINTINNIDS, *TYCHOPELAG/ *OYSTER	BAY, *JAMAICA, FALMOUTH HARBOR, *	W72-07899

DISTRIBUTION, COLD BAY, BERMERS BAY, ALARIA, NEREOCYSTIS, LAMINAR W72-09120
ULOIDES, CHLOROPHYLL A, *YAQUINA BAY, AMPHORA SPP, MELOSIRA SULCAT W73-00853
DER, VERTICAL DISTRIBUTION, COLD BAY, BERMERS BAY, ALARIA, NEREOCY W72-09120
*CAPE COD CANAL(MASS), BUZZARDS BAY, CAPE COD BAY, DATA INTERPRET W73-01449
MORRC BAY POWER PLANT, HUMBOLDT BAY, CONNETICUT YANKEE NUCLEAR PL W72-07907
I/ BARNACLES, FLOUNDER, HUMBOLDT BAY, CONNECTICUT RIVER, SAN JOAQU W71-07417
AL(MASS), BUZZARDS BAY, CAPE COD BAY, DATA INTERPRETATION, SPECIES W73-01449
UT RIVER, TURKEY POINT, BISCAYNE BAY, FLORIDA, BALANUS, BARNACLES, W72-07907
, LITTLE ARBOR VITAE LAKE, GREEN BAY, FOX RIVER, DETROIT RIVER, HE W73-01456
NCE, TRANSPARENCY, ORTHOP/ GREEN BAY, FOX RIVER, SPECIFIC CONDUCTA W71-10645
ZER, VAN DORN SAMPLERS, C-/ AUKE BAY, GLACIAL WATER, BECKMAN ANALY W72-09122
BAY, SCOLOPLOS ARMIGER, ALIMITOS BAY, LAKE PONTCHARTRAIN, PEARL RI W73-02029
ARMION ISLAND, MORRIA REEF, BEAR BAY, NEVA ISLAND, BAIRD ISLAND, S W72-09120
RODESMUS, TRACHELOMONAS, SAGINAW BAY, ORGANIC NITROGEN.: /, ANKIST W72-10875
CUS, CHA/ SEA OF JAPAN, *AMURSKY BAY, PERIDINIANS, EUCAMPIA ZOODIA W73-00442
*OFFSHORE CONSTRUCTION, DELAWARE BAY, SCOLOPLOS ARMIGER, ALIMITOS W73-02029
MINARIA DIGITATA, ST. MARGARET'S BAY, SCOUDOUC RIVER, HUMIC ACIDS, W72-12166
QUIN RIVER(CALIFORNIA), BISCAYNE BAY, WHITE RIVER(INDIANA), PATUXE W71-07417
SKELETONEMA.: ABATE, BAYTEX, BAYGONE, CYCLOTELLA, DUNALIELLA, W71-10096
ALS, *THERMAL POLLUTION, *THERM/ BAYS, *MARINE ALGAE, *MARINE ANIM W72-11876
UND, ALASKA, PONDS, WELLS, FISH, BAYS, AIR POLLUTION EFFECTS, AIR W72-03347
OASTS, ATLANTIC OCEAN, SEASONAL, BAYS, SEA WATER, ORGANIC MATTER, W73-00366
FFECTS, WATER POLLUTION EFFECTS, BAYS, WATERSHEDS(BASINS), SMALL W W72-00959
UATIC ENVIRONMENTS, GREAT LAKES, BAYS, WATERSHEDS(BASINS), SURFACE W71-08953
LIELLA, SKELETONEMA.: ABATE BAYTEX, BAYGONE, CYCLOTELLA, DUNA W71-10096
IRUSES, PLASTICIZERS, PHTHALATE/ BDELLOVIBRID, PELODICTYON, POLIOV W72-09675
CANDICA ZEYLANOIDES, CAMP VARNUM BEACH.: /A, RHODOTORULA LACTOSA, W72-04748
SWAMPS, LITTORAL, SHORES, TIDES, BEACHES, SANDS, INTERTIDAL AREAS, W72-07911
IA, MARMION ISLAND, MORRIA REEF, BEAR BAY, NEVA ISLAND, BAIRD ISLA W72-09120
AMMOBACULITES DILATATUS, AMMONIA BECCARII, AMMOTIUM SALSUM, ELPHID W73-00367
RS, C-/ AUKE BAY, GLACIAL WATER, BECKMAN ANALYZER, VAN DORN SAMPLE W72-09122
IENT REQUIRE/ *PHOSPHORUS, *LAKE BEDS, *ALGAE, ALGAL CONTROL, NUTR W72-11617
EGETATION EFFECTS, SLUDGE, RIVER BEDS, ALGAE, MATHEMATICAL MODELS. W70-09189
EXTINCTION COEFFICIENT, LAMBERT- BEER EQUATION, EDDY DIFFUSIVITY, W72-05459
A WINOGRADSKY, ZOOGLEA RAMIGERA, BEGGIATOA ALBA, BEGGIATOA, THIOTH W72-07896
OOGLEA RAMIGERA, BEGGIATOA ALBA, BEGGIATOA, THIOTHRIX NIVEA, THIOT W72-07896
LGAL POISONING, DINOFLAGELLATES, BEGGIATOA, PHYTOPLANKTON BLOOMS, W72-14056
HIBITION, MARINE ALGAE, DIATOMS, BEHAVIOR.: /CIDS, CHRYSOPHYTA, IN W73-00942
SH WATER, DISTRIBUTION PATTERNS, BEHAVIOR.: *ALGAE, *FRE W71-07360
TOWERS, DENSITY, VISCOSITY, FISH BEHAVIOR, FERTILITY, SEASONAL, DI W70-09612
OPERTIES, MINERALOGY, FISH, FISH BEHAVIOR, CHEMICAL ANALYSIS, SAMP W72-14280
ISHERIES, FISH POPULATIONS, FISH BEHAVIOR, INVERTEBRATES, OTTERS, W73-00764
ES, GROWTH STAGES, DISTRIBUTION, BEHAVIOR, SEASONAL, WATER POLLUTI W72-14659
N, SAMPLING, BACTERIA, TOXICITY, BEHAVIOR, BITUMINOUS MATERIALS, A W72-08790
C-13, *C-12, BIOLOGICAL SYSTEMS, BELEMNITELLA AMERICANA, TERRESTRI W72-12709
TION, TISSUES, ALBURNUS LUCIDUS, BELGIUM.: RUTHENIUM, FIXA W71-09168
VER, CLINTON RIVER, HURON RIVER, BELL RIVER, SALT RIVER, MICROCYST W72-14282
UTION SOURCES, COST COMPARISONS, BENEFICIAL USE, WATER QUALITY, CA W72-00974
L, *WASTE WATER TREATMENT, *COST- BENEFIT ANALYSIS, DETERGENTS, PHO W72-04734
TY, CALIFORNIA, BIOCONTROL, COST- BENEFIT THEORY.: /SE, WATER QUALI W72-00974
MICHORDATES, ARTHROPODS, *BAY OF BENGAL, COELENTERATES, CTENOPHORA W73-00935
CHNIQUES, CULTURE MEDIA, *BAY OF BENGAL, INDIAN OCEAN, SOIL EXTRAC W73-00430
INDERMERE(ENGLAND), HETEROCYSTS, BENTHIC BLUE-GREEN ALGAE.: *W W73-02472
HLOROPHYTA, DIATOMS, CYANOPHYTA, BENTHIC FAUNA, AQUATIC ANIMALS, I W72-12729
NTHESIS, INVERTEBRATES, BIOMASS, BENTHIC FAUNA, BENTHIC FLORA, HUM W72-12549
MALS, ALASKA, BIOMASS, SALINITY, BENTHIC FAUNA, ASBESTOS, INORGANI W73-00374
BUTION, ECOLOGICAL DISTRIBUTION, BENTHIC FAUNA, SIEVES, WATER TEMP W73-00367
S, SNAILS, CRABS, BENTHIC FLORA, BENTHIC FAUNA, DISTRIBUTIO: /OPOD W73-00932
TEMPERATURE, OXYGEN, TURBIDITY, BENTHIC FAUNA, WATER POLLUTION EF W73-02029
DATION(DECOMPOSITION), SAMPLING, BENTHIC FAUNA, SEDIMENTATION, SED W72-00847
FISH POPULATIONS, PHYTOPLANKTON, BENTHIC FAUNA, PERIPHYTON, AQUATI W72-00939
CHLOROPHYLL, SEDIMENT TRANSPORT, BENTHIC FAUNA, PLANKTON, BIOINDIC W72-09092
TON, WATER TEMPERATURE, E. COLI, BENTHIC FAUNA, CELLULOSE, ENZYMES W72-09668
S, ZOOPLANKTON, BIOTA, BACTERIA, BENTHIC FAUNA, FISH, ENVIRONMENTA W72-05282
ODELS, FISH, ALGAE, FATHOMETERS, BENTHIC FAUNA, BENTHIC FLORA, LAK W70-09614
ACTANTS, FISH HARVEST, PLANKTON, BENTHIC FAUNA, ALGAE, ECOLOGY, MI W70-09596
GAE, FATHOMETERS, BENTHIC FAUNA, BENTHIC FLORA, LAKE ONTARIO, HUDS W70-09614
ENTS, TRACE ELEMENTS, VOLCANOES, BENTHIC FLORA, ALKALINITY, PHOTOS W71-09190
TOM SEDIMENTS, ALGAE, CHLORELLA, BENTHIC FLORA, MUD, MICROORGANISM W71-08042
ENT, AIR POLLUTION, LEGISLATION, BENTHIC FLORA, F: /, WASTE TREATM W71-13723
NE ALGAE, *SYSTEMATICS, *COASTS, BENTHIC FLORA, RHODOPHYTA, PHAEOP W71-10066
E, *LITTORAL, LIGHT PENETRATION, BENTHIC FLORA, LAKES, DEPTH, DIAT W71-11029
MICROORGANISMS, WATER ANALYSIS, BENTHIC FLORA, ORGANIC MATTER, DI W73-02100
EANS, GASTROPODS, SNAILS, CRABS, BENTHIC FLORA, BENTHIC FAUNA, DIS W73-00932
S, *ALGAE, *LAKE ERIE, PLANKTON, BENTHIC FLORA, DIATOMS, PHYTOPLAN W73-01615
*RIVERS, *DIATOMS, CHLOROPHYTA, BENTHIC FLORA, EUTROPHICATION, DO W72-14797
EBRATES, BIOMASS, BENTHIC FAUNA, BENTHIC FLORA, HUMAN POPULATION, W72-12549
ATLANTIC OCEAN, MARINE BIOLOGY, BENTHIC FLORA, BIOASSAY, CYANOPHY W72-11876
XYGEN, *HYPOLIMNION, *LAKE ERIE, BENTHIC FLORA, ALGAE, LIGHT PENET W72-13659
(SPAIN), LAMINARIA/ *SUBLITTORAL BENTHIC MACROPHYTES, RIA DE ALDAN W72-04766
OXYGEN, CHEMICAL OXYGEN DEMAND, BENTHOS.: /ATION, ORGANIC MATTER, W71-04518
NUTRIENTS, CHEMICALS, PLANKTON, BENTHOS, ALGAE, FISH, WATER POLLU W72-07841
GAE, LABORATORY TESTS, DETRITUS, BENTHOS, AQUATIC ANIMALS, DIATOMS W71-10064
CHLOROPHYTA, METABOLISM, NEKTON, BENTHOS, BIOMASS.: / CYANOPHYTA, W72-13636
MARINE BACTERIA, *WATER QUALITY, BENTHOS, CRABS, COPEPODS, MARINE W72-08142
RGANIC MATTER, DISSOLVED SOLIDS, BENTHOS, CARBON, PRODUCTIVITY, OR W73-02100
OMS, SEDIMENTS, HARDNESS(WATER), BENTHOS, GRAZING, SYSTEMATICS, CH W73-01627
(WATER), WAVES(WATER), SAMPLING, BENTHOS, GRAZING, GASTROPODS, LIT W72-04766
NTIBIOTICS(PESTICIDES), ECOLOGY, BENTHOS, INVERTEBRATES, PHYTOPLAN W72-09668
CYANOPHYTA, GEOLOGIC FORMATIONS, BENTHOS, MAYFLIES, TUBIFICIDS, DI W71-11009

*ALASKA, SAMPLING, *SYSTEMATICS,	BENTHOS, SCUBA DIVING, DEPTH, DIS	W72-09120
PHICATION, PONDS, FEEDING RATES,	BENTHOS, STANDING CROPS.: / EUTRO	W72-14796
EM ANALYSIS, MACROCONSU/ *RIFFLE	BENTHOS, WOODLAND STREAM, ECOSYST	W71-10064
MAL STRATIFICATION, COAGULATION,	BENTONITE, LIME, *PULP WASTES, TE	W72-03753
, ETHER SULFATE, SODIUM DODECYL,	BENZENE SULFONATE, COCONUT FATTY	W72-12576
ATE, PYRUVATE, ACETATE, ETHANOL,	BENZOATE, PROPIANATE, PROPANOL, L	W72-05421
OSA, CRYPTOMONAS, CHLAMIDOMONAS,	BERENPLAAT RESERVOIR(ROTTERDAM),	W71-01491
LOOD, KIDNEYS, LIVER, PLASMODIUM	BERGHEI, BORRELIA KANSAS, BORRELI	W73-01676
VERTICAL DISTRIBUTION, COLD BAY,	BERMERS BAY, ALARIA, NEREOCYSTIS,	W72-09120
IMICROBIOTICS, ESCHERICHIA COLI,	BETA STREPTOCOCCUS, PSEUDOMONAS A	W72-04259
PICIN, CAULERPIN, PALMITIC ACID,	BETA-SITOSTEROL, THIN-LAYER CHROM	W71-08597
LTURES, AUTOSPORES, VARIABILITY,	BIAS, SYNCHRONY INDEX, COUNTING.:	W72-11920
, GROWTH, *CULTURES, *ALGICIDES,	BIBLIOGRAPHIES, *ECOLOGY, AQUATIC	W71-06102
Y TESTS, ALKALINITY, CARBONATES,	BICARBONATES, CARBON DIOXIDE, HYD	W72-04431
KES, IONS, SULFATES, CARBONATES,	BICARBONATES, DIATONICS, CYANOPHY	W71-02880
IDE, RESPIRATION, MARINE PLANTS,	BICARBONATES, MASS SPECTROMETRY,	W72-12709
, RESPIRATION, PLANT PHYSIOLOGY,	BICARBONATES, WATER POLLUTION EFF	W72-12940
RBON DIOXIDE, CYCLING NUTRIENTS,	BICARBONATES, PLANT GROWTH, DURNA	W72-09159
ATES, CHLORELLA, CARBON DIOXIDE,	BICARBONATES, ALGAL CONTROL, AERA	W72-09168
, CYANOPHYTA, CARBON, CHLORELLA,	BICARBONATES, AERATION, LIMITING	W72-10607
S, SEWAGE TREATMENT, CARBONATES,	BICARBONATES, HARDNESS(WATER), AL	W72-07514
AULENSIS, ENEIDAMONAS APPLANATA,	BICUDOMONAS CYANOPH: /AMONAS SAUP	W72-09119
TICOIDS, NOCTILUCA, CENTROPYXIS,	BIDDULPHIA, CHAETOCEROS, THALASSI	W71-11876
), MEUSE RIVER(THE NETHERLANDS),	BIESBOSCH RESERVOIR(THE NETHERLAN	W72-08559
NDOTA, CRYSTAL LAKE, TROUT LAKE,	BIG ARBOR VITAE LAKE, LITTLE ARBO	W73-01456
ITE CLAY RIVER, CHRISTINA RIVER,	BIG ELK RIVER, LITTLE ELK RIVER,	W72-04736
UT), LAKE MINNETONKA(MINNESOTA),	BIG STONE LAKE(MINNESOTA), LAKE C	W71-04216
LLOWTAIL D/ *BIGHORN LAKE(MONT),	BIGHORN RIVER, SHOSHONE RIVER, YE	W73-01435
*NEW YORK	BIGHT, COLLABORATIVE STUDIES.:	W72-12941
ES, MACROINVERTEBRATES, NEW YORK	BIGHT, EELGRASS, ZOSTERA, PINAEUS	W73-02029
HANODISCUS BINDERANUS, *MELOSIRA	BINDERANA, SCANNING ELECTRON MICR	W73-00432
SCANNING ELECTR/ *STEPHANODISCUS	BINDERANUS, *MELOSIRA BINDERANA,	W73-00432
OMS, CHLAMYDOMONAS, SCENEDESMUS,	BIO-ASSAY, RESISTANCE, LETHAL LIM	W72-14706
RATION, STORM RUNOFF, BIOASSAYS,	BIO-INDICATORS, STORAGE, FLOW RAT	W72-04070
CHELEX 100, ARTEMIA SALINA,	BIO-RAD, CYCLOTELLA NANA.:	W72-04708
PICEA AB/ VANADIUM, BRYOPHYTES,	BIOACCUMULATION, METAL COMPLEXES,	W72-14303
US MERGANSER, MEGACERYLE ALCYON,	BIOACCUMULATION.: / SAMPLES, MERG	W72-12930
FICATION, FATE OF POLL/ ENGLAND,	BIOACCUMULATION, BIOLOGICAL MAGNI	W72-11925
, MEASUREMENT, LABORATORY TESTS,	BIOASSAY.: /, NUTRIENTS, CULTURES	W73-00255
IDENTIFICATION, LIGHT INTENSITY,	BIOASSAY.: /, BIOLOGY, POLLUTANT	W73-02099
TY, MINERALOGY, LIGHT, CULTURES,	BIOASSAY.: /S, PRIMARY PRODUCTIVI	W72-07166
CYANOPHYTA, AEROBIC CONDITIONS,	BIOASSAY, GROWTH RATES, MORTALITY	W72-07664
PONDS, SCENDESOMUS, CYANOPHYTA,	BIOASSAY, SAMPLING, OXIDATION LAG	W72-07661
T, INHIBITION, IONS, ABSORPTION,	BIOASSAY, *MERCURY.: /Y, EQUIPMEN	W72-07660
IDE TOXICITY, FOOD CHAINS, FISH,	BIOASSAY, ALGAE, SNAILS, MOSQUITO	W72-07703
HYTA, INHIBITORS, PLANT TISSUES,	BIOASSAY, PHAEPHYTA.: /TH, RHODOP	W72-07496
ICITY, *WATER POLLUTION EFFECTS,	BIOASSAY, PESTICIDES, HERBICIDES,	W72-08440
IC ANIMALS, AQUATIC POPULATIONS,	BIOASSAY, EUTROPHICATION.: /AQUAT	W73-02095
SEUDOMONAS, CULTURES, VITAMIN B,	BIOASSAY, CELLULOSE, DIALYSIS, SE	W73-00942
WASTES, WASTE WATER(POLLUTION),	BIOASSAY, ECOSYSTEMS, BIODEGRADAT	W73-00840
INS, RED TIDE, PLANT PHYSIOLOGY,	BIOASSAY, CYTOLOGICAL STUDIES, DI	W73-00243
WATER POLLUTION, WATER ANALYSIS,	BIOASSAY, CYANOPHYTA, RESINS, AQU	W73-00286
S, THERMAL POLLUTION, NUTRIENTS,	BIOASSAY, *WATER QUALITY, ALGAE,	W72-14282
CATORS, *EUTROPHICATION, FJORDS,	BIOASSAY, CHLORELLA, NUTRIENTS, W	W73-00234
GAE, *AQUATIC WEEDS, *NUTRIENTS,	BIOASSAY, LIMITING FACTORS, NITRO	W73-00232
ROPHY, OXYGEN SAG, AQUATIC LIFE,	BIOASSAY, CLASSIFICATION.: /LIGOT	W73-00002
USTRIAL WASTES, DOMESTIC WASTES,	BIOASSAY, SULFATES, IRON COMPOUND	W72-12239
FOOD CHAINS, PATH OF POLLUTANTS,	BIOASSAY, FISH, RAINBOW TROUT, AQ	W72-12257
, MARINE BIOLOGY, BENTHIC FLORA,	BIOASSAY, CYANOPHYTA, *FLORIDA, C	W72-11876
EFFLUENTS, SEEPAGE, SEPTIC TANKS,	BIOASSAY, CHEMICAL ANALYSES, NITR	W72-12955
MUSSELS, COMMON MERGANSER DUCK,	BIOASSAY, WATER POLLUTION EFFECTS	W72-12930
A, FUNGI, METHODOLOGY, CULTURES,	BIOASSAY, ENVIRONMENTAL EFFECTS,	W72-13813
EFFECTS, FERTILIZERS, NITROGEN,	BIOASSAY, PHOSPHORUS, CYANOPHYTA,	W72-13816
CE, AQUATIC ALGAE, GROWTH RATES,	BIOASSAY, LABORATORY TESTS, CALCI	W72-13809
ALYSIS, WATER POLLUTION SOURCES,	BIOASSAY, ALGAE, DOMESTIC WASTES,	W73-02478
AL ANALYSIS, RESINS, *SEA WATER,	BIOASSAY, COPEPODS, OYSTERS, DINO	W72-04708
PHY, CENTRIFUGATION, PESTICIDES,	BIOASSAY, DISTRIBUTION PATTERNS,	W72-04134
T, CHEMICAL PRECIPITATION, IRON,	BIOASSAY, LEACHING, VOLUMETRIC AN	W72-05965
TOPLANKTON, ALGAE, GROWTH RATES,	BIOASSAY, DISSOLVED OXYGEN, CARBO	W72-06046
GROWTH SUBSTANCES, ALGAE, IRON,	BIOASSAY, BIOINDICATORS, FOOD CHA	W71-03095
DETERGENTS, OIL, *OIL WASTES, /	BIOASSAY, TOXICITY, *EMULSIFIERS,	W70-09429
D ORGANISMS, PATH OF POLLUTANTS,	BIOASSAY, TOXINS, POISONS, POISON	W71-08597
RATORY TESTS, CHEMICAL ANALYSIS,	BIOASSAY, CYTOLOGICAL STUDIES, NU	W72-09163
LUTANT IDENTIFICATION, SAMPLING,	BIOASSAY, ENVIRONMENT, TOXICITY,	W72-10678
IOLOGY, ALGAE, FISH, ABSORPTION,	BIOASSAY, METABOLISM.: /R, HYDROB	W72-10694
S, AQUATIC ALGAE, BIOINDICATORS,	BIOASSAY, ALGAE, POWERPLANTS, EFF	W71-11583
LEVELS, PHOSPHATES, DETERGENTS,	BIOASSAY, ANALYTICAL TECHNIQUES,	W71-12072
YSIOLOGY, PHYSIOLOGICAL ECOLOGY,	BIOASSAY, MICROORGANISMS, *EUGLEN	W71-12303
AKES, PLANT GROWTH, *MOLYBDENUM,	BIOASSAY, NEUTRON ACTIVATION ANAL	W72-00801
OGEN, EVAPORATION, STORM RUNOFF,	BIOASSAYS, BIO-INDICATORS, STORAG	W72-04070
ALYZER, VAN DORN SAMPLERS, C-14,	BIOCARBONATES.: /ATER, BECKMAN AN	W72-09122
PHORUS, WATER POLLUTION SOURCES,	BIOCHEMICAL OXYGEN DEMAND, WATER	W72-09167
*PUBLIC HEALTH, RIVERS, *SEWAGE,	BIOCHEMICAL OXYGEN DEMAND, SOLID	W72-09675
UBILITY, CHEMICAL OXYGEN DEMAND,	BIOCHEMICAL OXYGEN DEMAND, FILTRA	W72-00411
EUTROPHICATION, WATER POLLUTION,	BIOCHEMICAL OXYGEN DEMAND, ORGANI	W71-13257
IOLET LIGHT, ALGAE, OXYGENATION,	BIOCHEMICAL OXYGEN DEMAND, WATER	W72-00042
L WASTES, *FILTRATION, *FILTERS,	BIOCHEMICAL OXYGEN DEMAND, CHEMIC	W71-13341
USTRIAL WASTES, MUNICIPAL WASTE,	BIOCHEMICAL OXYGEN DEMAND, CHEMIC	W71-13356
IENTS, MICROORGANISMS, EFFLUENT,	BIOCHEMICAL OXYGEN DEMAND, SATURA	W71-13553
DIMANTATION, OXIDATION, STORAGE,	BIOCHEMICAL OXYGEN DEMAND, CHEMIC	W72-01881

ONOMICS, CHEMICAL OXYGEN DEMAND, BIOCHEMICAL OXYGEN DEMAND, AERATI W71-12223
MICS, ALGAE, SOLUBILITY, SEWAGE, BIOCHEMICAL OXYGEN DEMAND, ION EX W71-12200
ENT, ALGAE, SECONDARY TREATMENT, BIOCHEMICAL OXYGEN DEMAND, CHEMIC W71-12188
ITY, AQUATIC ALGAE, ZOOPLANKTON, BIOCHEMICAL OXYGEN DEMAND, COPEPO W71-11876
HINGTON, CHEMICAL OXYGEN DEMAND, BIOCHEMICAL OXYGEN DEMAND, DISSOL W71-11006
STE WATER TREATMENT, *TURBIDITY, BIOCHEMICAL OXYGEN DEMAND, DISSOL W72-03753
, OXIDATION-REDUCTION POTENTIAL, BIOCHEMICAL OXYGEN DEMAND, NITRIF W72-04086
US, CARBON, NITROGEN, OXIDATION, BIOCHEMICAL OXYGEN DEMAND, RADIOA W72-05506
M COMPOUNDS, ALKALINITY, MINING, BIOCHEMICAL OXYGEN DEMAND, ECONOM W72-04734
CONDITIONS, METABOLISM, SLUDGE, BIOCHEMICAL OXYGEN DEMAND, CHEMIC W72-04789
E TREATMENT, TERTIARY TREATMENT, BIOCHEMICAL OXYGEN DEMAND, INDUST W71-06445
NITRITES, OXYGEN, PHYTOPLANKTON, BIOCHEMICAL OXYGEN DEMAND, DISSOL W71-05390
HOTOSYNTHESIS, MUNICIPAL WASTES, BIOCHEMICAL OXYGEN DEMAND, WASTE W71-06033
ION CONCENTRATION, TEMPERATURE, BIOCHEMICAL OXYGEN DEMAND, CHEMIC W71-09524
ATER TREATMENT, ORGANIC LOADING, BIOCHEMICAL OXYGEN DEMAND, SOLAR W71-07124
CONDITIONS, AEROBIC CONDITIONS, BIOCHEMICAL OXYGEN DEMAND, WASTE W71-07109
MENTATION, CHLORINATION, SLOPES, BIOCHEMICAL OXYGEN DEMAND, HYDROG W71-07113
WATERS, CHEMICAL OXYGEN DEMAND, BIOCHEMICAL OXYGEN DEMAND, WASTE W71-07098
DGE, LIGHT INTENSITY, COLIFORMS, BIOCHEMICAL OXYGEN DEMAND, WASTE W71-07106
WATER TREATMENT, BIODEGRADATION, BIOCHEMICAL OXYGEN DEMAND.: /STE W71-07084
AMPLING, CHEMICAL OXYGEN DEMAND, BIOCHEMICAL OXYGEN DEMAND, STABIL W71-07097
CONDITIONS, AEROBIC CONDITIONS, BIOCHEMICAL OXYGEN DEMAND, WASTE W71-07100
SMS, SLUDGE, TERTIARY TREATMENT, BIOCHEMICAL OXYGEN DEMAND, WASTE W71-07382
WATER TREATMENT, *WATER RE-USE, BIOCHEMICAL OXYGEN DEMAND, ORGANI W71-07123
G, ALGAE, WASTE WATER TREATMENT, BIOCHEMICAL OXYGEN DEMAND.: /ADIN W71-07096
ATMENT, *PONDS, *ALGAE, LAGOONS, BIOCHEMICAL OXYGEN DEMAND, CHEMIC W70-07838
ENT, BACTERIA, MUNICIPAL WASTES, BIOCHEMICAL OXYGEN DEMAND, WASTE W71-03896
EOLOGY, TOPOGRAPHY, GROUNDWATER, BIOCHEMICAL OXYGEN DEMAND, PHOSPH W71-03012
EN ION CON/ *ST. LAWRENCE RIVER, BIOCHEMICAL OXYGEN DEMAND, HYDROG W70-10427
N LAGOONS, *PILOT PLANTS, ALGAE, BIOCHEMICAL OXYGEN DEMAND, BIOLOG W70-06619
DYSENTERIAE, SHIGELLA FLEXNERI, BIOCHEMICAL STUDIES, ARTHOPODS, S W72-13800
CONCENTRATION, DISSOLVED OXYGEN, BIOCHEMICAL OXYGEN DEMAND, PERFOR W72-13508
ENETRATION, PLANKTON, NUTRIENTS, BIOCHEMICAL OXYGEN DEMAND, INORGA W72-13659
LANS, *ADENOSINE TRIPHOSPHATASE, BIOCHEMICAL TESTS, OUABAIN, POTAS W72-13809
OPHYTA, CHLOROPHYTA, CYANOPHYTA, BIOCHEMICAL OXYGEN DEMAND.: /HRYS W72-12995
AERATED LAGOONS, *NORTH DAKOTA, BIOCHEMICAL OXYGEN DEMAND, BACTER W72-12565
CONTROL, *AQUATIC WEED CONTROL, BIOCHEMICAL OXYGEN DEMAND, LAKE S W72-14692
TIC ALGAE, AQUATIC PRODUCTIVITY, BIOCHEMICAL OXYGEN DEMAND, LAKE S W73-00002
, WATER QUALITY, STREAMS, LAKES, BIOCHEMICAL OXYGEN DEMAND, *OXYGE W72-14469
ISH, CHEMICALS, CHEMICAL WASTES, BIOCHEMICAL OXYGEN DEMAND, BACTER W73-01976
RSION STRUCTURES, SEDIMENTATION, BIOCHEMICAL OXYGEN DEMAND, ALGAE, W72-08401
ONS, PUMPING, AEROBIC TREATMENT, BIOCHEMICAL OXYGEN DEMAND, CHEMIC W72-08396
ALTITUDE, CLIMATES, METABOLISM, BIOCHEMICAL OXYGEN DEMAND, PILOT W72-06838
GEN, HYDROGEN ION CONCENTRATION, BIOCHEMICAL OXYGEN DEMAND, *SEWAG W72-06612
*SEDIMENTATION RATES, *REVIEWS, BIOCHEMISTRY, GEOCHEMISTRY, RADIO W72-06313
S, *CULTURES, *NITROGEN, OXYGEN, BIOCHEMISTRY, INHIBITORS, ALGAE, W72-07941
ROMATOGRAPHY, CHEMICAL ANALYSIS, BIOCHEMISTRY, PLANT PHYSIOLOGY, B W71-08597
LAKES, AQUATIC ALGAE, BACTERIA, BIOCHEMISTRY, WISCONSIN.: /ATION, W72-04292
ATER, *FOOD CHAINS, *ABSORPTION, BIOCHEMISTRY, PHYSIOLOGICAL ECOLO W71-13243
DISM, GROWTH RATES, TEMPERATURE, BIOCHEMISTRY, PHYSIOLOGY, LIGHT.: W72-00377
IXATION, OXYGEN, CARBON DIOXIDE, BIOCHEMISTRY, ALGAE, INHIBITION, W72-08584
RUNOFF, RADIOACTIVITY, FALLOUT, BIOCONTROL, INSECT CONTROL, IRRIG W72-00846
USE, WATER QUALITY, CALIFORNIA, BIOCONTROL, COST-BENEFIT THEORY.: W72-00974
ROPHICATION, NIRTROGEN FIXATION, BIOCONTROL, PLANT PHYSIOLOGY, MOV W71-13184
L, *ALGAE CONTROL, *FORAGE FISH, BIOCONTROL, REMEDIES, FOODS, FISH W71-09163
TING PLANTS, SEAWATER, PLANKTON, BIODEGRADATION, LIGHT, TEMPERATUR W71-09261
DATION LAGOONS, *SEWAGE LAGOONS, BIODEGRADATION, *NUTRIENTS, ALGAE W71-07973
, SLUDGE, WASTE WATER TREATMENT, BIODEGRADATION, BIOCHEMICAL OXYGE W71-07084
CONDITIONS, AEROBIC CONDITIONS, BIODEGRADATION, SEEPAGE, SLUDGE, W71-07099
PHICATION, NITRATES, PHOSPHATES, BIODEGRADATION, ANAEROBIC CONDITI W71-07106
ILIZATION PONDS, MICROORGANISMS, BIODEGRADATION, WASTE WATER TREAT W71-07097
ON, *ANAEROBIC BACTERIA, *TEXAS, BIODEGRADATION, TRACERS, LAGOONS, W71-04873
N DIOXIDE, PRIMARY PRODUCTIVITY, BIODEGRADATION, ORGANIC MATTER, O W71-04518
S, AQUATIC ALGAE, PUBLIC HEALTH, BIODEGRADATION, TOXICITY, CONTRAC W71-11516
AGENCIES, ADMINISTRATION, ALGAE, BIODEGRADATION, LOCAL GOVERNMENTS W72-01093
ATMENT, WATER POLLUTION CONTROL, BIODEGRADATION.: /T, TERTIARY TRE W71-13939
ATION LAGOONS, *DESIGN CRITERIA, BIODEGRADATION, ALGAE, PHOTOSYNTH W72-09386
SURFACTANTS, TOXICITY, HAZARDS, BIODEGRADATION.: /TY, TECHNOLOGY, W72-09169
CHLEATION, ALGAE, WATER QUALITY, BIODEGRADATION, SEWAGE TREATMENT W72-06837
ELOPMENT, WASTE WATER TREATMENT, BIODEGRADATION, LOCAL GOVERNMENTS W72-06655
*PHYTOPLANKTON, *SEWAGE LAGOONS, BIODEGRADATION, SEWAGE TREATMENT, W72-07661
ONMENT, WATER POLLUTION SOURCES, BIODEGRADATION, ECOSYSTEMS, WATER W73-00853
OLLUTION), BIOASSAY, ECOSYSTEMS, BIODEGRADATION, ALGAE, WATER POLL W73-00840
AE, *EUTROPHICATION, NEW MEXICO, BIODEGRADATION.: /*CHLORELLA, ALG W72-13992
NICKEL, TOXICITY, BIOINDICATORS, BIODEGRADATION, BIOMASS, LICHENS, W72-14303
TICIDES, SEA WATER, *METABOLISM, BIODEGRADATION, *MARINE MICROORGA W73-00361
E MEDIA, MACRO/ GROWTH KINETICS, BIOENERGETICS, ENRICHMENT, CULTUR W72-09378
D, PROCESSING WASTES, PATHOGENS, BIOENVIRONMENTAL CONTROLS.: / FOO W72-00846
FICIENT ELEMENTS, AQUATIC ALGAE, BIOINDI: /RIENTS, COLORIMETRY, DE W73-00916
UBLICATIONS, PESTICIDE TOXICITY, BIOINDICATORS, AGRICULTURAL CHEMI W73-01976
WASTES, LEAD, NICKEL, TOXICITY, BIOINDICATORS, BIODEGRADATION, BI W72-14303
NIC BACTERIA, CULTURES, E. COLI, BIOINDICATORS, PONDS, ANAEROBIC C W72-07664
ALITY, SEASONAL, SUSPENDED LOAD, BIOINDICATORS, WATER POLLUTION, M W72-07896
NITIES, ECOLOGICAL DISTRIBUTION, BIOINDICATORS, WATER POLLUTION SO W72-08579
NSPORT, BENTHIC FAUNA, PLANKTON, BIOINDICATORS, CURRENTS(WATER), T W72-09092
ALS, SECCHI DISKS, HEAVY METALS, BIOINDICATORS, SILICA, CHROMIUM, W72-09111
ALGAE, HEAVY METALS, ABSORPTION, BIOINDICATORS, PATH OF POLLUTANTS W72-10957
NDUSTRIAL WASTES, AQUATIC ALGAE, BIOINDICATORS, BIOASSAY, ALGAE, P W71-11583
BSTANCES, ALGAE, IRON, BIOASSAY, BIOINDICATORS, FOOD CHAINS, TITAN W71-03095
R TEMPERATURE, DISSOLVED OXYGEN, BIOINDICATORS, E. COLI, AQUATIC A W71-07417

RCES, SEWAGE, INDUSTRIAL WASTES,	BIOINDICATORS, CHEMICAL PROPERTIE	W72-12729
NEITY, CHLOROPHYTA, CHRYSOPHYTA,	BIOINDICATORS, PHYSICOCHEMICAL PR	W72-13142
UTRIENT REQUIREMENTS, NUTRIENTS,	BIOINDICATORS, WATER POLLUTION EF	W72-13816
CONTAMINANTS, *TRACE MATERIALS,	BIOINHIBITION, BIOSTIMULATION.: /	W71-11793
RY, PERCH LAKE(ONTARIO), ASHING,	BIOLOGICAL MAGNIFICATION.: /TOMET	W71-11529
ZINC, ANTIMONY, NORWAY, ASHING,	BIOLOGICAL MAGNIFICATION, PELVETI	W71-11515
SEWAGE, ESTUARIES, GROWTH RATES,	BIOLOGICAL COMMUNITIES.: /ORING,	W71-12867
, PONDS, AQUATIC ALGAE, ECOLOGY,	BIOLOGICAL PROPERTIES, FRESH WATE	W71-10580
URFACE WATERS, COASTS, SAMPLING,	BIOLOGICAL COMMUNITIES, LAKE MICH	W72-10875
ON, *PERIPHYTON, *PHYTOPLANKTON,	BIOLOGICAL COMMUNITIES, DOMINANT	W72-10883
UACULTURE, COMMERCIAL SHELLFISH,	BIOLOGICAL COMMUNITIES, FOOD: /AQ	W72-09378
EDESMUS, *CYANOPHYTA, CHLORELLA,	BIOLOGICAL COMMUNITIES, DOMINANT	W72-09239
MYDOMONAS, MATHEMATICAL STUDIES,	BIOLOGICAL COMMUNITIES, FOOD CHAI	W72-09248
FEX, UNIALGAL CULTURES, OUABAIN,	BIOLOGICAL MAGNIFICATION, PHOSPHO	W72-09668
IRRIGATION WATER, AQUATIC ALGAE,	BIOLOGICAL PROPERTIES, WATER POLL	W72-00974
GLAND), POLYGONUM, TYPHA, ELODE/	BIOLOGICAL EFFECTS, NOTTINGHAM(EN	W72-00155
R, SUSPENDED SOLIDS, / *CHEMICAL-	BIOLOGICAL TREATMENT, MIXED LIQUO	W71-13334
BACTERIA, ENVIRONMENTAL EFFECTS,	BIOLOGICAL COMMUNITIES, HEAVY MET	W72-01472
ENTS, OXYGEN REQUIREMENTS, AGING(	BIOLOGICAL), AQUATIC ALGAE, WASTE	W72-01685
R RESOURCES, AREA REDEVELOPMENT,	BIOLOGICAL CHARACTERISTICS, ALGAE	W72-03215
TIONS, LAKE MORPHOLOGY, ECOLOGY,	BIOLOGICAL PROPERTIES.: /A COLLEC	W72-13851
PRESERVATION, CAMBARUS BARTONI,	BIOLOGICAL SAMPLES, MERGUS MERGAN	W72-12930
IONS, NEW YORK, BIOLUMINESCENCE,	BIOLOGICAL PROPERTIES, CHLOROPHYL	W72-12941
ALYSIS, COLD REGIONS, NUTRIENTS,	BIOLOGICAL COMMUNITIES, ALGAE, PE	W72-12549
ICAL STUDIES, *ALGAE, EQUATIONS,	BIOLOGICAL PROPERTIES, REPRODUCTI	W72-12724
GE LAGOONS, BRACKISH WATER FISH,	BIOLOGICAL TREATMENT, OXIDATION L	W72-12567
AMERICANA, TERRES/ *C-13, *C-12,	BIOLOGICAL SYSTEMS, BELEMNITELLA	W72-12709
RY COMPOUNDS, *MINAMATA DISEASE,	BIOLOGICAL MAGNIFICATION, MINAMAT	W72-12257
POLL/ ENGLAND, BIOACCUMULATION,	BIOLOGICAL MAGNIFICATION, FATE OF	W72-11925
NON-	BIOLOGICAL UPTAKE.:	W72-12210
ROGRAMS, *ALGAE, *PHYTOPLANKTON,	BIOLOGICAL COMMUNITIES, STANDING	W73-02469
GAE, NUTRIENTS, WATER POLLUTION,	BIOLOGICAL TREATMENT, CHEMICALS,	W71-05083
BIOCHEMISTRY, PLANT PHYSIOLOGY,	BIOLOGICAL COMMUNITIES.: /ALYSIS,	W71-08597
LGAE, BIOCHEMICAL OXYGEN DEMAND,	BIOLOGICAL TREATMENT, PHOTOSYNTHE	W70-06619
, TRICKLING FILTERS, HYDROLYSIS,	BIOLOGICAL TREATMENT, ACTIVATED S	W70-09907
APHY, POLYCHLORINATED BIPHENYLS,	BIOLOGICAL MAGNIFICATION, CHLAMYD	W72-04134
N-STRUCTURAL ALTERNATIVES, AGING(	BIOLOGICAL), AQUATIC ALGAE, AQUAT	W72-04279
AEROBIC	BIOLOGICAL DEGRADATION.:	W72-05315
,/ *PLUTONIUM, DETECTION LIMITS,	BIOLOGICAL SAMPLES, SARGASSO WEED	W72-05965
*BENTHIC FLORA, *OREGON, RIVERS,	BIOLOGICAL COMMUNITIES, SEASONAL,	W72-07901
HLOR, *METHYLCHLOR, METABOLITES,	BIOLOGICAL MAGNIFICATION, BIOLOGI	W72-07703
LITES, BIOLOGICAL MAGNIFICATION,	BIOLOGICAL SAMPLES, LIVER, URINE,	W72-07703
PLANKTON, *PREDATION, *BIOASSAY,	BIOLOGICAL COMMUNITIES, DOMINANT	W72-07132
XYGENATION, *ANAEROBIC BACTERIA,	BIOLOGICAL TREATMENT.: /EMAND, *O	W72-14469
MALS, *MODEL STUDIES, LIMNOLOGY,	BIOLOGICAL PROPERTIES, MINERALOGY	W72-14280
*LIGHT, *TEMPERATURE, *DIATOMS,	BIOLOGICAL COMMUNITIES, SYSTEMATI	W73-00242
TANTS, META/ *BIOTRANSFORMATION,	BIOLOGICAL SAMPLES, FATE OF POLLU	W73-00361
TS, *ECOLOGY, *PLANKTON, *ALGAE,	BIOLOGICAL COMMUNITIES, PRODUCTIV	W72-14659
ROPHICATION, DOMINANT ORGANISMS,	BIOLOGICAL COMMUNITIES, NUTRIENTS	W72-14797
GREAT LAKES, ALGAL BLOOMS, AGING(	BIOLOGICAL), LAKE STAGES, MESOTRO	W72-14782
N, ALGAE, ENVIRONMENTAL EFFECTS,	BIOLOGICAL COMMUNITY, CANADA.: /O	W73-01704
*LAKE WABAMUN,	BIOLOGICAL STUDIES, MACROPHYTES.:	W73-01704
TURE DETECTOR, PRECONCENTRATION,	BIOLOGICAL SAMPLES, LIQUID-LIQUID	W73-02105
VED ORGANIC MATTER, MACROPHYTES,	BIOLOGICAL SAMPLES: /C-14, DISSOL	W73-02095
YTES, SEA WATER, MARINE ANIMALS,	BIOLOGICAL COMMUNITIES, CHLORINE,	W73-00855
ALYSIS, FREQUENCY, FLUCTUATIONS,	BIOLOGICAL COMMUNITIES, PHYSIOCHE	W73-00918
SEDIMENTATION, *EUTROPHICATION,	BIOLOGICAL COMMUNITIES, FOOD CHAI	W73-00938
HICATION, CHLOROPHYLL, PLANKTON,	BIOLOGICAL COMMUNITIES, ECOSYSTEM	W73-00916
CHAINS, GAMMA RAYS, RESISTANCE,	BIOLOGICAL COMMUNITIES, TIDAL MAR	W73-00796
IMARY PRODUCTIVITY, CHRYSOPHYTA,	BIOLOGICAL COMMUNITIES, ESTUARINE	W73-00853
RGY BUDGETS, NUTRIENTS, REVIEWS,	BIOLOGICAL COMMUNITIES, AQUATIC P	W73-01089
RESINS, GLASS, THYMIDINES, CELLS(	BIOLOGY), *BIODETERIORATION, DEMO	W71-09261
AL POWER, ATLANTIC OCEAN, MARINE	BIOLOGY, BENTHIC FLORA, BIOASSAY,	W72-11876
M/ *SALMONELLA, *BIBLIOGRAPHIES,	BIOLOGY, EPIDEMIOLOGY, CULTURES,	W72-13800
AE, MICROORGANISMS, CHLOROPHYTA,	BIOLOGY, POLLUTANT IDENTIFICATION	W73-02099
SITE DATA COLLECTIONS, NEW YORK,	BIOLUMINESCENCE, BIOLOGICAL PROPE	W72-12941
OD, SEDIMENT-WATER INTERACTIONS,	BIOLUMINESCENCE, SELENASTRUM CAPR	W72-04292
HNIQUES, NUTRIENTS, FOOD CHAINS,	BIOLUMINESCENCE, POLLUTANT IDENTI	W71-11234
TA, METABOLISM, NEKTON, BENTHOS,	BIOMASS.: / CYANOPHYTA, CHLOROPHY	W72-13636
ION CONCENTRATION, SCENEDESMUS,	BIOMASS.: /OLVED OXYGEN, HYDROGEN	W72-08049
GTON, *ALGAL CONTROL, NUTRIENTS,	BIOMASS, *CYANOPHYTA, PLANT GROWT	W72-00799
ENOSINE TRIPHOSPHATE, *MICROBIAL	BIOMASS, *GROWTH STAGE, *DETERMIN	W72-04292
RVOIR STAGES, EARLY IMPOUNDMENT,	BIOMASS, AQUATIC ALGAE, PHYTOPLAN	W72-04853
, *ECOSYSTEMS, AQUATIC HABITATS,	BIOMASS, AQUATIC ALGAE, AQUATIC A	W72-06340
TARIES, DAMS, CHEMICAL ANALYSIS,	BIOMASS, ALGAE, PHYSICOCHEMICAL P	W73-01435
ESISTANCE, PRIMARY PRODUCTIVITY,	BIOMASS, ANAEROBIC CONDITIONS, WA	W72-12940
, PHOTOSYNTHESIS, INVERTEBRATES,	BIOMASS, BENTHIC FAUNA, BENTHIC F	W72-12549
PRIMARY PRODUCTIVITY, CULTURES,	BIOMASS, CARBON RADIOISOTOPES, LI	W72-07648
T, *MARINE ALGAE, MARINE PLANTS,	BIOMASS, CANADA, COASTS, ATLANTIC	W73-00366
DITIONS, AERATION, MICROBIOLOGY,	BIOMASS, CHLORELLA.: /AEROBIC CON	W71-04079
IATOMS, CHLOROPHYTA, CYANOPHYTA,	BIOMASS, CHEMICAL ANALYSIS, STATI	W71-11029
ORGANISMS, AQUATIC PRODUCTIVITY,	BIOMASS, CARBON RADIOISOTOPES, SU	W72-09239
RVOIR STAGES, EARLY IMPOUNDMENT,	BIOMASS, DIATOMS, CHLOROPHYTA, CY	W72-04855
*SEASONAL, SEA WATER, SAMPLING,	BIOMASS, DIATOMS, POLLUTANT IDENT	W73-00442
NT, SALINITY, CHEMICAL ANALYSIS,	BIOMASS, DISSOLVED OXYGEN, THERMO	W72-12941
VITY, SYSTEMATICS, GROWTH RATES,	BIOMASS, DATA COLLECTIONS.: /UCTI	W73-02469
YTOLOGICAL STUDIES, SYSTEMATICS,	BIOMASS, FLUOROMETRY, MICROSCOPY,	W72-14313
RES, DDT, ANALYTICAL TECHNIQUES,	BIOMASS, GAS CHROMATOGRAPHY, CENT	W72-04134
PLANT PHYSIOLOGY, PLANT GROWTH,	BIOMASS, GROWTH RATES, LIGHT INTE	W72-05476
NT, MARINE ALGAE, PHYTOPLANKTON,	BIOMASS, GROWTH RATES, PRODUCTIVI	W72-09378

AE, PLANTS, PROTEINS, NUTRIENTS,	BIOMASS, HARVESTING OF ALGAE, PRO	W72-03327
PHYTA, EUGLENOPHYTA, PYRROPHYTA,	BIOMASS, HYDROGEN ICN CCNCENTRATI	W72-07145
, BIOINDICATORS, BIODEGRADATICN,	BIOMASS, LICHENS, MOSSES, FUNGI,	W72-14303
LGAE, *CULTURES, PHOTOSYNTHESIS,	BIOMASS, LIGHT INTENSITY, GROWTH	W72-06274
ZING, GASTROPODS, LITTORAL, AGE,	BIOMASS, LIGHT QUALITY, LIGHT PEN	W72-04766
DISCHARGE, REGRESSION ANALYSIS,	BIOMASS, METHODOLOGY, REEFS, SAMP	W72-06283
KES, PERIPHYTON, ORGANIC MATTER,	BIOMASS, METABOLISM, PHYTOPLANKTO	W72-03216
MPLING, WATER POLLUTION EFFECTS,	BIOMASS, NUTRIENTS, ALGAE, EUGLEN	W72-09111
*PERIPHYTON, *SUBMERGED PLANTS,	BICMASS, NEMATODES, ALGAE, OLIGOC	W71-11012
ON, FORECASTING, PHOTOSYNTHESIS,	BIOMASS, NITRATES, RESERVCIRS, LI	W71-13183
SHWATER, *MARSHES, SALINE WATER,	BICMASS, NUTRIENTS, FOOD WEBS, DE	W73-01089
TA, CHEMICAL ANALYSIS, SEASONAL,	BIOMASS, ON-SITE INVESTIGATIONS,	W73-00242
PRODUCTIVITY, *MIDGES, LAGOONS,	BIOMASS, OREGON, SLUDGE, ALGAE, D	W72-03556
ON, PRIMARY PRODUCTIVITY, ALGAE,	BICMASS, OLIGOTROPHY, DIATOMS, EU	W71-09171
, *PRODUCTIVITY, PHOTOSYNTHESIS,	BIOMASS, OXYGEN, ALGAE, CCMPARATI	W71-09165
ICITY, HERBICIDES, INSECTICIDES,	BICMASS, PRODUCTIVITY, SCENE: /OX	W71-10566
N SOURCES, SPATIAL DISTRIBUTION,	BIOMASS, PHOTOSYNTHESIS, NUTRIENT	W73-00431
YSTEMS, *BIOLOGICAL CCMMUNITIES,	BIOMASS, PRODUCTIVITY, METABOLISM	W73-01080
NKTON, CHRYSOPHYTA, CHLOROPHYTA,	BIOMASS, PYRRCPHYTA, SPRING, SUMM	W72-12738
LVED OXYGEN, AEROBIC CONDITIONS,	BIOMASS, RESPIRATION, PHOTOSYNTHE	W72-11100
ERATURE, MARINE ANIMALS, ALASKA,	BICMASS, SALINITY, BENTHIC FAUNA,	W73-00374
EFFECTS, SALINITY, HEATED WATER,	BIOMASS, THERMAL STRESS, GROWTH R	W73-00853
IS, PLANKTON, COMPUTER PROGRAMS,	BICMASS, TEMPERATURE, CONDUCTIVIT	W72-08560
*COASTS, *WATER QUALITY CONTROL,	BIOMASS, TOXICITY, EUTROPHICATION	W71-11793
S, *CYANOPHYTA, *EUTROPHICATICN,	BIOMASS, WATER QUALITY, CHLCROPHY	W71-10645
DISTRIBUTION, FOURIER ANALYSIS,	BIORHYTHMS, CORRELATION ANALYSIS,	W72-11719
ALYSIS, EQUATIONS, REPRODUCTICN,	BIORHYTHMS, MEASUREMENT, POPULATI	W72-11920
YTOPLANKTON, *SEA WATER, COASTS,	BIORTHYMS, ON-SITE TESTS, WATER P	W72-03228
NVIRONMENT, ECOSYSTEMS, GEORGIA,	BIORTHYTHMS, LABORATORY TESTS, AL	W72-09159
*TRACE MATERIALS, BIOINHIBITICN,	BIOSTIMULATION.: / CONTAMINANTS,	W71-11793
PAN), ORGANIC MERCURY COMPOUNDS,	BIOSYNTHESIS, GUPPIES.: /A BAY(JA	W72-12257
TY, *MODEL STUDIES, ZOOPLANKTON,	BIOTA, BACTERIA, BENTHIC FAUNA, F	W72-05282
*CHLOROPHYTA, *SPRINGS, DIATCMS,	BIOTA, PRCTOZCA, *AMPHIPODA, *FLO	W72-12744
	BIOTESTS.:	W73-00002
SIPHON NOTARISII, PSYCHROPHILIC,	BIOTIC ASSOCIATIONS, LA GORCE MOU	W73-00854
HIC BACTERIA, GLUCOSE, ACETATES,	BIOTIN, NIACIN, COBALAMINS, EUGLE	W73-02095
L A, CRGANIC NITROGEN, THIAMINE,	BIOTIN, VITAMIN B 12, GONYAULAX P	W72-07892
*TOXICOLOGY,	BIOTOXINS, PHARMACOLOGY.:	W73-00243
AQUATIC	BIOTOXINS, *BIOTOXINS.:	W72-14056
STONE VALLEY DAM(PA.), ATRAZINE,	BIOTYPES, POUR PLATE METHOD, MOST	W72-04743
ON EFFECTS, *P/ *POLYCHLORINATED	BIPHENYLS, *ALGAE, *WATER POLLUTI	W72-05614
CHROMATOGRAPHY, POLYCHLORINATED	BIPHENYLS, BIOLOGICAL MAGNIFICATI	W72-04134
RIVER, PINFIS/ *POLYCHLORINATED	BIPHENYLS, ARCCLOR 1254, ESCAMBIA	W71-07412
EUTROPHICATION, POLYCHLORINATED	BIPHENYLS, PUBLIC HEALTH, ENVIRON	W73-00703
N PESTICIDES, / *POLYCHLORINATED	BIPHENYLS, *CHLCRNATED HYDROCARBO	W72-08436
FUNGICIDES, PESTICIDES, ECOLOGY,	BIRD TYPES, BIRDS, AQUATIC ANIMAL	W72-05952
PESTICIDES, ECOLOGY, BIRD TYPES,	BIRDS, AQUATIC ANIMALS, MOSSES, I	W72-05952
DIATCMS, PERSISTENCE, BACTERIA,	BIRDS, FISH, ON-SITE INVESTIGATIO	W72-00155
AL EFFECTS, *NUCLEAR EXPLOSIONS,	BIRDS, SALMON, STICKLEBACKS, CRAB	W72-00960
AINE, MUDS, INVERTEBRATES, FISH,	BIRDS, STREAMS, AQUATIC PLANTS, A	W72-12930
AMS, SHELLFISH, AMPHIPODA, WATER	BIRDS, WISCONSIN, ANALYTICAL TECH	W70-08098
THES MINUTISSIMA, ACHNANTHES CF.	BIRGIANI, AMPHIPLEURA PELLUCIDA,	W72-13142
, SAN JOAQUIN RIVER(CALIFCRNIA),	BISCAYNE BAY, WHITE RIVER(INDIANA	W71-07417
AKE(NEW YORK), PATUXENT ESTUARY,	BISCAYNE BAY(FLORIDA), HUMBOLT BA	W71-11517
CONNECTICUT RIVER, TURKEY POINT,	BISCAYNE BAY, FLORIDA, BALANUS, B	W72-07907
G, BACTERIA, TOXICITY, BEHAVIOR,	BITUMINCUS MATERIALS, ANIMAL WAST	W72-08790
SHAD, EKMAN DREDGE, PUMPKINSEED,	BLACK CRAPPIE, CHAIN PICKEREL, TE	W72-14280
, LAKE ST. CLAIR, DETROIT RIVER,	BLACK RIVER, PINE RIVER, CLINTON	W72-14282
ILLINCIS RIVER, ST. LOUIS RIVER,	BLACK RIVER(MINNESOTA), LAKE WASH	W71-04216
	BLACK SEA.:	W72-10955
RINU/ *RESIDUES, *DDD, DDE, TDE,	BLACK-FLY LARVAE, CLADOPHORA, CYP	W70-08098
(ENGLAND)/ ASTERIONELLA FORMOSA,	BLELHAM TARN(ENGLANC), WINDERMERE	W72-10625
NGDOM, ESTHWAITE WATER(ENGLAND),	BLELHAM TARN(ENGLAND), LUXURY UPT	W71-13183
STIMUL/ *GREEN ALGAE, *HORMOTILA	BLENNISTA, FILTRATES, GROWTH AUTO	W71-08687
PARTRIDGE, DDE, KIDNEYS, LIVER,	BLOOD, GASTEROSTEUS ACULEATUS, DA	W72-05952
NG, ASHING, BONE, ORGANS, BRAIN,	BLOCD, KIDNEYS, LIVER, PLASMODIUM	W73-01676
E, CRAYFISH, MIDGES, GAMMA RAYS,	BLOODWORMS, SLUDGE WORMS.: /NESSE	W70-06668
TION.: LAKE	BLOOMINGTON(ILL), CHEMOSTRATIFICA	W71-03028
ALGAL	BLOOMS.:	W72-08401
*NITROGE/ EUTROPHICATION, *ALGAL	BLOOMS, *NUTRIENTS, *PHOSPHORUS,	W70-09669
TUARIES, *EUTROPHICATION, *ALGAL	BLOOMS, *PHYTOPLANKTON, *NUTRIENT	W72-01329
PECTS, LAKES, GREAT LAKES, ALGAL	BLOOMS, AGING(BIOLOGICAL), LAKE S	W72-14782
PHOLOGY, AERATION, MIXING, ALGAL	BLOOMS, ALGAL TOXINS, WATER CIRCU	W72-09304
LANIER(GEORGIA), L/ *WHITE ALGAL	BLOOMS, HYPNODINIUM, LAKE SIDNEY	W71-09173
LLATES, BEGGIATOA, PHYTOPLANKTON	BLOCMS, PHYSIOLOGICAL ECOLOGY, MO	W72-14056
TION, ALKALINE HYDROLYSIS, ALGAE	BLOCMS, UTAH, GREAT SALT LAKE, OR	W71-12223
FISHKILL, EUTROPHICATION, ALGAL	BLOOMS, WATER QUALITY.: /ADATION,	W71-08768
L UPTAKE, CALCIUM RADIOISOTOPES,	BLOOMS, 2,4 DINITRO PHENOL, TUCSO	W71-01474
UDACORUS, LYSIMACHIA NUMMULARIA,	BLUE HERON, STERNA HIRUNDO, RATS,	W72-05952
RES, *DIEOFF RATES, AFTERGROWTH,	BLUE-GREEN ALGAE, FIELD STUDIES,	W70-08319
ILIS,/ *LYSOZYME, *SPHEROPLASTS,	BLUE-GREEN ALGAE, ANABAENA VARIAB	W71-08683
DY FACILITY, PRODUCT EVALUATION,	BLUE-GREEN ALGAE.: /ATIC WEED STU	W71-06102
*YELLOWSTONE NATIONAL PARK(WYO),	BLUE-GREEN ALGAE, MASTIGOCLADUS,	W72-08508
E(ENGLAND), HETEROCYSTS, BENTHIC	BLUE-GREEN ALGAE.: *WINDERMER	W73-02472
MCPA, ENDOSULFAN, FAIRY LIQUID,	BLUSYL, SODIUM LAURYL, ETHER SULF	W72-12576
, RECREATION, FISHING, SWIMMING,	BOATING, DOCUMENTATION, WATER QUA	W72-05869
, SALINITY, SAMPLING, SEA WATER,	BODIES OF WATER, PYRROPHYTA, PROT	W73-00296
ORIXIDS, EUDORINA, NOTONECTIDAE,	BODO, HYDROPSYCHIDAE, PLECOPTERA,	W70-06655
L TISSUES, HISTOLOGICAL STUDIES,	BODY FLUIDS, VERTEBRATES, HEAT ST	W73-01676
RICHORIXA, RAMPHOCORIXA, SIGARA,	BOENOA, SALDULA, CRYNELLUS, CHEUM	W70-06655

*CRANBERRY	BOGS.:	W71-11498
MANGANESE RADIOISOTOPES,	BOGUE SOUND, CORE SOUND.:	W73-00818
CONCENTRATION, MUSCLE,	BONE PHYSIOLOGY/METABOLISM.:	W72-10694
NIUM 239, POLYMERIZATION, LIVER,	BONE, DETECTION LIMITS, CHEMICAL	W72-05965
ON, FIXATION, EMBEDDING, ASHING,	BONE, ORGANS, BRAIN, BLOOD, KIDNE	W73-01676
OSA, CLADOPHOROPSIS MEMBRANACEA,	BOODLEA COMPOSITA, ANADYOMENE STE	W72-11800
AQUATICA, SCHIZOTHRIX CALCICOLA,	BORATES, ARTHROBACTER, NOSTOC COM	W73-00854
SULFATES, PHOSPHATES, CHLORIDES,	BORATES, ABSORBANCE.: /NITRATES,	W72-06274
PITATA, MACOMA CALCAREA, ASTARTE	BOREALIS,: /ORQUATA, CAPITELLA CA	W73-02029
OIDS, SPONGES, COELENTERA/ *WOOD	BORERS, BARNACLES, BRYOZOAS, HYDR	W73-00374
PONTCHARTRAIN, PEARL RIVER, LAKE	BORGNE, SUBSTRATES, MACROINVERTEB	W73-02029
MODIUM BERGHEI, BORRELIA KANSAS,	BORRELIA ANSERINA.: / LIVER, PLAS	W73-01676
NEYS, LIVER, PLASMODIUM BERGHEI,	BORRELIA KANSAS, BORRELIA ANSERIN	W73-01676
ISTICHLIS SPICATA, FIMBRISTYLIS,	BORRICHIA, PHRAGMITES COMMUNIS, J	W73-01089
GREEN ALGAL VIRUSES, *PLECTONEMA	BORYANIUM, SUSQUEHANNA RIVER, BRA	W72-04736
US, MITO/ *LYSOGENY, *PLECTONEMA	BORYANUM, LPP-1, LPP-1D, PHYCOVIR	W72-03188
LAKE(N H), ANABAENA CIRCINALIS,	BOSMINA LONGIROSTRIS, SYNURA UVEL	W71-03014
S.:	BOSMINA, CERIODAPHNIA, EUDIAPTOMU	W72-00161
ER C OIL, AVICENNIA, RHIZOPHORA,	BOSTRYCHIETUM, FIDDLER CRABS, MES	W72-07911
, FYRISAN RIVER, ARBOGAAN RIVER,	BOTORPSSTROMMEN RIVER, ATRAN RIVE	W73-00286
*PAAP TEST, *ALGAL GROWTH,	BOTTLE TEST.:	W73-00255
LF STREAM, CAPE HATTERAS, NANSEN	BOTTLE, SPECIES DIVERSITY, SAMPLE	W72-07663
, LAKES, STRATIFICATION, MIXING,	BOTTOM SEDIMENTS, SUCCESSION, CHL	W72-08049
*ALDRIN, *ENDRIN, MARINE ALGAE,	BOTTOM SEDIMENTS, PLANKTON, RADIO	W73-00361
ERA, DISSOLVED OXYGEN, DREDGING,	BOTTOM SEDIMENTS, STABILIZATION,	W72-03556
IONS, NITRATES, DENITRIFICATION,	BOTTOM SEDIMENTS, SEA WATER, FRES	W72-11563
G, PHYTOPLANKTON, PACIFIC OCEAN,	BOTTOM SEDIMENTS, STATIONS, DATA	W72-03567
REGIONS, PACIFIC OCEAN, COLUMNS,	BOTTOM SEDIMENTS, ALGAE, CHLORELL	W71-08042
EPRODUCTION, SAMPLING, SEASONAL,	BOTTOM SEDIMENTS, ICE, AMPHIPODA.	W71-08026
SIUM ION CONCENTRATION, BRUCITE,	BOTTOM TOPOGRAPHY, PRIONOSPIO, LU	W72-09092
L).: *LAKE KINNERET(ISRAEL),	BOTTOM WATERS, RIVER JORDAN(ISRAE	W72-05469
SPRINGS, / ALHAMBRA HOT SPRINGS,	BOULDER HOT SPRINGS, JACKSON HOT	W72-05610
LFATE, COBALT SULFATE, POTASSIUM	BR: /ANGANESE CHLORIDE, NICKEL SU	W72-12739
COMPOSITA, ANADYOMENE STELLATA,	BR: /OROPSIS MEMBRANACEA, BOODLEA	W72-11800
AFFINS, EPHEMEROPTERA, MENIDIA,	BRACHYDEUTERA, CHYDORUS SPHAERICU	W72-07132
SEWAGE DISPOSAL, SEWAGE LAGOONS,	BRACKISH WATER FISH, BIOLOGICAL T	W72-12567
TOMS, *SALINITY, *BIOINDICATORS,	BRACKISH WATER, FRESHWATER, *ALGA	W73-02548
UATIC MICROBIOLOGY, *ACARICIDES,	BRACKISH WATER, MICROENVIRONMENT,	W71-08026
AE, EVAPORATION, REASONABLE USE,	BRACKISH WATER, D: /TIVE USE, ALG	W70-07100
EMBEDDING, ASHING, BONE, ORGANS,	BRAIN, BLOOD, KIDNEYS, LIVER, PLA	W73-01676
TISSUES, MERCURY ISOTOPES, HUMAN	BRAIN, LOW-TEMPERATURE OVEN DRYIN	W71-11036
ESOX/ *MERCURY, *METHYLMERCURY,	BRAIN, ORGANOMERCURIAL COMPOUNDS,	W72-05952
ESICULATION, SAMPLE PREPARATION,	BRANCHING.: /FUCUS VESICULOSUS, V	W73-00369
US, COREGONUS ALBULA, WHITEFISH,	BRANCHIURA: /T, COREGONUS LAVARET	W72-12729
MA BORYANIUM, SUSQUEHANNA RIVER,	BRANDYWINE RIVER, RED CLAY RIVER,	W72-04736
PROWSEMONAS TROPICA, STIGMOBODO	BRASILIANA, PAVLOVIAMONAS FRUTICO	W72-07683
ESIAMONAS STAGNALIS, REFRACTODES	BRASILIANA, OYEMONAS PULSULAE, RE	W72-09119
OYEMONAS PULSULAE, REFRACTOMONAS	BRASILIANA, HORTOBAGGIAMONAS PLIC	W72-09119
GULATA, AKIYAMAMONAS TERRESTRIS,	BRASILOBIA PISCIFORMIS, PROWSEMON	W72-07683
RIVER(SIBERIA).:	BRATSK RESERVOIR(SIBERIA), ANGARA	W71-09165
N, CHLOROSARCINA, ANKISTRODESMUS	BRAUNII, CHLORELLA VULGARIS.: /IO	W71-07675
IDIUM FAVEOLARUM, ANKISTRODESMUS	BRAUNII, CHLORELLA PYRENOIDOSA, C	W72-07664
ERCURIC CHLORIDE, ANKISTRODESMUS	BRAUNII, EUGLENA GRACILIS.: /YL M	W72-06037
LOSAQUAE, SPIRULINA, GYMNODINIUM	BREVE, PERIDINIUM, OSCILLATORIA R	W73-00840
LENZITES TRABEA, SCOPULARIOPSIS	BREVICAULIS, CLADOSPORIUM HERBARU	W72-13813
*ENTEROMORPHA, SWEDEN, TIGRIOPUS	BREVICORNIS, GAMMARUS DUEBENI, FU	W71-08026
AVICULA SRYTOCEPHALA, ACHNANTHES	BREVIPES, ACHNANTHES JAVANICA, LI	W72-08141
PED BASS, DISSOLVED OXYGEN, OIL,	BRIDGES, BUOYS, HARBORS, TUNNELS,	W73-02029
AT SALT LAKE, ORGANIC COMPOUNDS,	BRINES.: /ALGAE BLOOMS, UTAH, GRE	W71-12223
SINCLAIRII, *LAMINARIA LONGIPES,	BRITISH COLUMBIA, RUSSIA, LAMINAR	W73-00428
	BRITISH ISLES.:	W72-10619
YNEDRA FASCICULATA, PLAGIOGRAMMA	BROCKMANNI, AMPHIPLEURA RUTILANS,	W73-00853
ES, PHAEOCYSTIS, SORTING, SODIUM	BROMIDE, DEXTRAN, OIKOPLEURA, SAG	W73-00371
RONUTRIENTS, SCANDIUM, RUBIDIUM,	BROMINE, ACTINIDES, RADIOACTIVE D	W73-01434
ULATE MATTER, NITROSOM/ *HUBBARD	BROOK EXPERIMENTAL FOREST, PARTIC	W71-01489
ION EFFECTS, GAS CHROMATOGRAPHY,	BROOK TROUT, SAMPLING, SEPARATION	W72-12930
CE, MAGNESIUM ION CONCENTRATION,	BRUCITE, BOTTOM TOPOGRAPHY, PRION	W72-09092
ADA, CAPE CHIDLEY(LABRADOR), NEW	BRUNSWICK, ANTICOSTI ISLAND(QUEBE	W71-10066
L COMPLEXES, PICEA AB/ VANADIUM,	BRYOPHYTES, BIOACCUMULATION, META	W72-14303
GELIA ARGUS, VALONIA VENTRICOSA,	BRYOPSIS PENNATA, SPYRIDIA FILAME	W73-00838
HALUS PHOENIX, UDOTEA FLABELLUM,	BRYOPSIS, CAULERPA, ACROPORA PALM	W72-14294
ATES, ARTHROPODS, COELENTERATES,	BRYOZOA, ECHINODERMS.: /NE, TUNIC	W71-01475
TEANS, ECHINODERMS, SIPUNCULIDS,	BRYOZOA, INTOPROCTA, TUNICATES, S	W73-00932
TURBELLARIA, SPONGES, TUNICATES,	BRYOZOANS, SEA TURTLES.: /FAUNA,	W72-07911
LECTORS, *HORIZONTAL COLLECTORS,	BRYOZOANS, FORAMINIFERA, HYDROIDS	W72-05432
ENTERA/ *WOOD BORERS, BARNACLES,	BRYOZOAS, HYDROIDS, SPONGES, COEL	W73-00374
A), LAKE CARDINAL(GEORGIA), LAKE	BUCKHORN(GEORGIA).: /CKSON(GEORGI	W71-09173
A), LAKE CARDINAL(GEORGIA), LAKE	BUCKHORN(GEORGIA), HYPNODINIUM SP	W71-11914
AKE, LAKE LONG, LAKE SQUAW, LAKE	BUDD, OSCILLATORIA REDEKEI, LYNGB	W72-07145
S, *PRIMARY PRODUCTIVITY, ENERGY	BUDGET, *MARINE ALGAE, MARINE PLA	W73-00366
IVITY, *MIDGES, LAGOONS/ *ENERGY	BUDGET, *SEWAGE, *AQUATIC PRODUCT	W72-03556
ALKALINITY, OXYGEN DEMAND, HEAT	BUDGET, ALGAE, EUTROPHICATION.: /	W72-12997
NG CROP, *TROPHIC LEVEL, *ENERGY	BUDGET, PRODUCTIVITY, HERBIVORES,	W71-10064
, SEEPAGE, TURBIDITY, HYDROLOGIC	BUDGET, SOLAR RADIATION, WATER TE	W72-02274
NUTRIENT INPUTS, CHI/ *NUTRIENT	BUDGETS, *BAY OF QUINTE(ONTARIO),	W71-11009
BS, DETRITUS, ECOSYSTEMS, ENERGY	BUDGETS, NUTRIENTS, REVIEWS, BIOL	W73-01089
ND RIVER BASIN(ONTARIO), CANADA,	BUFFER CAPACITY.: /IDELINES, *GRA	W71-11017
TIFICATION, PHYSICAL PROPERTIES,	BULK DENSITY, POROSITY, COLOR, HY	W73-00854
FECTS, WATER TEMPERATURE, ALGAE,	BULLHEADS, SALMON, SHELLFISH, CLA	W72-07907
MACROINVERTEBRATES, DIESEL FUEL,	BUNKER C OIL, AVICENNIA, RHIZOPHO	W72-07911

IA, GROWTH RATES, SEDIMENTATION,	BUOYANCY, WIND VELOCITY, CONVECTI	W71-09852
ALGAL CONTROL, LABORATORY TESTS,	BUOYANCY, EUTROPHICATION, NIRTROG	W71-13184
DISSOLVED OXYGEN, OIL, BRIDGES,	BUOYS, HARBORS, TUNNELS, MARINE A	W73-02029
EN, CHAR, LUCIOPERCA LUCISPERCA,	BURBOT, COREGONUS LAVARETUS, CORE	W72-12729
NTS, CARBON DIOXIDE, ALKALINITY,	BURNING, DOMESTIC WASTES, INDUSTR	W72-05473
SES OF WASTE HEAT, STEAM POWERED	BUSES.: / *SPACE HEATING, *FARM U	W70-09193
UM, HYPENA MUSCIFORMIS, CERAMIUM	BUSSOIDEUM, CHONDRIA LI: /LAVULAT	W73-00838
, SUCCINATE, FUMARATE, BUTYRATE,	BUTANOL, GLYCEROL, LACTATE, PYRUV	W72-05421
PP, ZIZANIA AQUATICA, CAREX SPP,	BUTOMUS UMBELLATUS, SPARGANIUM RA	W73-01089
LLA, MAGELONA, ARMANDIA, TELLINA	BUTTONI, VENUS, TIVELA STULTORUM,	W72-09092
, SORBITAL, SUCCINATE, FUMARATE,	BUTYRATE, BUTANOL, GLYCEROL, LACT	W72-05421
INTERPRE/ *CAPE COD CANAL(MASS),	BUZZARDS BAY, CAPE COD BAY, DATA	W73-01449
VERTEBRATES, DIESEL FUEL, BUNKER	C OIL, AVICENNIA, RHIZOPHORA, BOS	W72-07911
IA, DNA, MOTILITY, CYST FO/ *G +	C RATIOS, *NONFRUITING MYXOBACTER	W72-07951
1, LPP-1D, PHYCOVIRUS, MITOMYCIN	C.: /, *PLECTONEMA BORYANUM, LPP-	W72-03188
RESERVOIRS(NEB), ULTRA PLANKTON,	C-14.: *SALT VALLEY	W72-04270
OPHYTES, HETEROTROPHIC BACTERIA,	C-14.: / ISLAND, OPEN OCEAN, SAPR	W71-08042
PORA CRASSA, SCINTILLATION COUN/	C-14, *CHLORELLA VULGARIS, NEUROS	W72-14326
MAN ANALYZER, VAN DORN SAMPLERS,	C-14, BIOCARBONATES.: /ATER, BECK	W72-09122
ELICATISSIMA, NITZSCHIA PUNGENS,	C-14, CHLOROPHYLL A, MEDITERRANEA	W73-00431
NAJAS FLEXILIS, SCIRPUS ACUTUS,	C-14, DISSOLVED ORGANIC MATTER, M	W73-02095
ELLA, AGMENELLUM QUADRAPLICATUM,	C-14, THIN LAYER CHROMATOGRAPHY,	W73-00361
ENSI/ CHLOROPHYLL A, CHLOROPHYLL	C, CAROTENOIDS, FRAGILARIA CROTON	W73-00384
OSTYLIS, HEMILAMPROPS, LAMPROPS,	C: / ANALOGA, DIASTYLOPSIS, OXYUR	W72-09092
ENCE, INFRARED RADIATION, RADAR,	C: /TRAVIOLET RADIATION, FLUORESC	W72-04742
OGRAPHY, SCINTILLATION COUNTING,	CA-45.: /RBIGNIANA, PAPER CHROMAT	W72-12634
HICATION, NITROGEN, PHOSPHOROUS,	CA: /MS, MUSSELS, OYSTERS, EUTROP	W72-07715
DIUM ARSONATE, ARSENATES, SODIUM	CACODYLATE, CULTURE MEDIA, PORIA	W72-13813
, *AQUATIC INSECTS, *DISPERSION,	CADDISFLIES, STONEFLIES, DOBSONFL	W70-06655
DISTRIBUTION, CARBON, SEASONAL,	CADDISFLIES, GASTROPODS, SANDS, C	W71-11029
DIPTERA, MAYFLIES, DRAGONFLIES,	CADDISFLIES, MOLLUSKS, CYANOPHYTA	W71-11012
IS.: POTENTIOMETRY,	CADMIUM IONS, COPPER IONS, OOCYST	W72-01693
MANGANESE, COBALT RADIOISOTOPES,	CADMIUM RADIOISOTOPES, METABOLISM	W71-13243
OLLUTANT IDENTIFICATION, COPPER,	CADMIUM, CHROMIUM, NICKEL,: /E, P	W72-14303
BACTERIA, STREPTOTHRIX HYORHINA,	CADMIUM, DIMETHYLNITROSAMINE, AMI	W72-09675
DOSA, PHAEODACTYLUM TRI/ SILVER,	CADMIUM, LEAD, CHLORELLA PHYRENOI	W72-07660
OECETIS, LEPTOCELLA, ISONYCHIA,	CAENIS, PERLESTA.: /ODES INUUSTA,	W70-06655
SCHOENOPLECTUS LACUSTRIS, ACORUS	CALAMUS, TYPHA ANGUSTIFOLIA.: /,	W73-00286
, PERIDINIUM DEPRESSUM, METAZOA,	CALANUS HELGOLANDICUS(PACIFICUS),	W72-07892
UATA, CAPITELLA CAPITATA, MACOMA	CALCAREA, ASTARTE BOREALIS,: /ORQ	W73-02029
, *PARTICULATE ORGANIC NITROGEN,	CALCAREOUS LAKES, ULTRA-VIOLET LA	W72-08048
CHLOROPHYLL A, SCHIZOTHRIX	CALCICOLA.:	W73-01080
NEOCHLORIS AQUATICA, SCHIZOTHRIX	CALCICOLA, BORATES, ARTHROBACTER,	W73-00854
E, OSCILLATORIACEAE, SCHIZOTHRIX	CALCICOLA, ELKTON POND, RISING SU	W72-04736
ACETYLENE REDUCTION, SCHIZOTHRIX	CALCIOCOLA, CLOSTERIUM MONILIFERU	W72-10883
HORUS, NITROGEN, ORGANIC MATTER,	CALCIUM CARBONATE, DETRITUS, PLAN	W72-05469
ESMUS, SUSPENDED LOAD, NITRATES,	CALCIUM CARBONATE, PROTEINS, FILT	W70-10174
TREATMENT, DIATOMS, COAGULATION,	CALCIUM CARBONATE, HYDROGEN ION C	W73-01669
ECHNIQUES, MET/ *PHOTOSYNTHESIS,	CALCIUM CHLORIDE, RADIOACTIVITY T	W72-12634
NDING, BARNACLES, LETHAL DOSAGE,	CALCIUM ION CONCENTRATION, VERTEB	W72-09092
CALCIFICATION, *CORALLINE ALGAE,	CALCIUM RADIOISOTOPES, *BOSSIELLA	W72-12634
4 DINITRO P/ *BIOLOGICAL UPTAKE,	CALCIUM RADIOISOTOPES, BLOOMS, 2,	W71-01474
TES, BIOASSAY, LABORATORY TESTS,	CALCIUM.: /UATIC ALGAE, GROWTH RA	W72-13809
N, COPPER, MAGNESIUM, MANAGNESE,	CALCIUM, CHROMIUM, POTASS: /, IRO	W71-09188
S, POTASSIUM, MAGNESIUM, SODIUM,	CALCIUM, HYDROGEN, CHEMICAL WASTE	W72-14303
N, NITROGEN, PHOSPHORUS, SULFUR,	CALCIUM, MANGANESE, POTASSIUM, SO	W73-00840
ATION, COLLOIDS, CARBON DIOXIDE,	CALCIUM, MAGNESIUM, NUTRIENTS, *W	W72-03613
ON, SULFATES, SODIUM, POTASSIUM,	CALCIUM, MAGNESIUM, CHLORIDES, HY	W72-05610
OMPOUNDS, SEPARATION TECHNIQUES,	CALCIUM, MAGNESIUM, E. COLI, SILI	W72-11833
PHOSPHORUS, NITRATES, CHLORIDES,	CALCIUM, MAGNESIUM, SODIUM, POTAS	W71-11001
ALGAE, *FOOD CHAINS, ABSORPTION,	CALCIUM, PUBLIC HEALTH, SOUTH AME	W73-00819
ATION, NITROGEN, CARBON, SODIUM,	CALCIUM, POT: /ROGEN ION CONCENTR	W73-00854
TH, AQUATIC LIFE, AQUATIC ALGAE,	CALCIUM, RADIOACTIVITY EFFECTS.: /	W72-10978
NG, MICROORGANISMS, ACCLIMATION,	CALCIUM, SEDIMENTATION, WASTE WAT	W71-13334
IDITY, ALKALINITY, CONDUCTIVITY,	CALCIUM, SCENEDESMUS, WATER HYACI	W72-04269
CINTILLATION COUNTING, CYANIDIUM	CALDARIUM, THIOBACILLUS, SULFOLOB	W72-14301
*LAKE TAHOE(	CALIF.).:	W72-12955
*ALGAL GROWTH, *LAKE TAHOE(	CALIF).:	W71-13553
HARVESTING, *SAN JOAQUIN VALLEY(	CALIF).: /US, ALGAL GROWTH, ALGAL	W72-02975
TAT, *GROWTH/ *SAN JOAQUIN DELTA(	CALIF), *CONTINUOUS FLOW TURBIDOS	W70-10403
NEOUS DISEQUILIBRIUM, LAKE TAHOE(	CALIF), CASTLE LAKE(CALIF), DIVER	W71-09171
, LAKE TAHOE(CALIF), CASTLE LAKE(	CALIF), DIVERSITY, PATCHES.: /IUM	W71-09171
*SALTON SEA(	CALIF), ECOLOGICAL MANAGEMENT.:	W71-10577
NE, TUNICATES, ARTHRO/ SAN PEDRO(	CALIF), EPIBENTHIC SPECIES, ABALO	W71-01475
REACTOR COOLING, IMPERIAL VALLEY(	CALIF), LOS ANGELES(CALIF), TUCSO	W71-09157
E LAKE(CALIF), SAN JOAQUIN RIVER(	CALIF), NITZSCHIA.: /), SEARSVILL	W70-10405
SHADFLY, CHIRONOMIDS, CLEAR LAKE(	CALIF), RICHELIEU RIVER(QUEBEC),	W70-08098
N RIVER(MICH), SAN JOAQUIN RIVER(	CALIF), RIFLE FALLS(COLO), FISH F	W71-11006
TO RIVER(CALIF), SEARSVILLE LAKE(	CALIF), SAN JOAQUIN RIVER(CALIF),	W70-10405
SEARSVILLE LA/ SAN PABLO ESTUARY(	CALIF), SACRAMENTO RIVER(CALIF),	W70-10405
ESTUARY(CALIF), SACRAMENTO RIVER(	CALIF), SEARSVILLE LAKE(CALIF), S	W70-10405
TUARY(DEL-NJ), SAN JOAQUIN DELTA(	CALIF), STEADY-STATE MODELS.: /ES	W71-05390
ERIAL VALLEY(CALIF), LOS ANGELES(	CALIF), TUCSON(ARIZ).: /LING, IMP	W71-09157
FILTER, BACKWASHING.: FIREBAUGH(	CALIF), VOLATILE SOLIDS, SANBORN	W70-10174
DOXINE, INDAL COMPOU/ LAKE TAHOE(	CALIF), ZINC, ASCORBIC ACID, PYRI	W72-07514
*LAKE TAHOE(	CALIF-NEV).:	W72-12549
NTS, CANADA, WASHINGTON, OREGON,	CALIFORNIA, ALASKA, SEA WATER.: /	W73-00428
ON, WASTE ASSIMILATIVE CAPACITY,	CALIFORNIA, TRITIUM, MARINE PLANT	W73-00831
ANIMALS, *MARINE PLANTS, DEPTH,	CALIFORNIA, SANDS, SEDIMENTS, PRO	W71-01475
*SUSPENSION, HARVESTING, PONDS,	CALIFORNIA, SCENEDESMUS, SUSPENDE	W70-10174

, HABITATS, NICHES, COMPETITION, CALIFORNIA.: /S, EUGLENA, DAPHNIA W71-09171
BILITY, WATER POLLUTION CONTROL, CALIFORNIA, ARIZONA.: /OMIC FEASI W71-09157
OTOPES, TEMPERATURE, ECOSYSTEMS, CALIFORNIA, IRRIGATION, ECONOMIC W71-09157
CURRENTS(WATER), WINDS, NEVADA, CALIFORNIA, NEW YORK, OREGON, WIS W71-05626
POPULATIONS, *MORTALITY, *ALGAE, CALIFORNIA.: /VERTEBRATES, *FISH W71-08793
INATION, ALGAE, DENITRIFICATION, CALIFORNIA, FILTERS, PARTICLE SIZ W71-08223
L AREAS, *INVERTEBRATES, *ALGAE, CALIFORNIA.: /EFFECTS, *INTERTIDA W71-08792
ECTICUT RIVER, SAN JOAQUIN RIVER(CALIFORNIA), BISCAYNE BAY, WHITE W71-07417
OYSTERS, CLAMS, COLUMBIA RIVER, CALIFORNIA, WASHINGTON, CONNECTIC W71-07417
REX, ETHMODISCUS GAZELLAE, BAJA CALIFORNIA, HEMIAULUS HAUCKII, ST W72-03567
LID WASTES, CHEMICALS, MICHIGAN, CALIFORNIA, TUBIFICIDS, COLORADO, W71-11006
DREDGING, ESTUARINE ENVIRONMENT, CALIFORNIA WATER CONVEYANCE, CANA W71-09581
M WASTES, WASTE WATER TREATMENT, CALIFORNIA.: /URAL CHEMICALS, FAR W71-11698
SCAYNE BAY(FLORIDA), HUMBOLT BAY(CALIFORNIA), POTOMAC RIVER, COLUM W71-11517
E ALGAE, ON-SITE INVESTIGATIONS, CALIFORNIA, FOOD CHAINS, PUBLIC H W72-03324
EN, PHOSPHOROUS, *ALGAL CONTROL, CALIFORNIA.: / METABOLISM, NITROG W71-13553
CONTROL, *AQUATIC WEED CONTROL, CALIFORNIA.: /R TREATMENT, *ALGAL W72-02975
, BENEFICIAL USE, WATER QUALITY, CALIFORNIA, BIOCONTROL, COST-BENE W72-00974
ER POLLUTION EFFECTS, EFFLUENTS, CALIFORNIA, SEDIMENTS, SAMPLING, W72-09092
M, PLATYDORINA CAUDATA, EUDORINA CALIFORNICA, EUDORINA ILLINOISENS W71-13793
E MAY(/ *COASTAL WATERS, PANAMA, CALLAO(PERU), CAPE COD(MASS), CAP W72-08056
INTE, SALT CREEK, CHARA, ELODEA, CALLITRICHE, ANABAENA FLOSAQUAE, W73-00840
HYROPSIS COCCINEA VAR. DAWSONII, CALLOPHYLLIS FIRMA, PLENOSPORIUM W73-02099
ROCONSUMERS, AUTUMNAL COMMUNITY, CALORIES, DETRIVORES, LINESVILLE W71-10064
BLUE-GREEN ALGAE, MASTIGOCLADUS, CALOTHRIX.: /NATIONAL PARK(WYO), W72-08508
WARBURG EFFECT, EUGLENA GRACLIS, CALVIN CYCLE, ANACYSTIS NIDULANS. W71-09172
S, CLEANUP, SAMPLE PRESERVATION, CAMBARUS BARTONI, BIOLOGICAL SAMP W72-12930
LA LACTOSA, CANDIDA ZEYLANOIDES, CAMP VARNUM BEACH.: /A, RHODOTORU W72-04748
LOPLITES RUPESTRIS, ESOX LUCIUS, CAMPELOMA, PSIDIUM, SHADFLY, CHIR W70-08098
CANADA.: W70-09388
*LIMITING NUTRIENTS, CANADA.: W71-12857
L EFFECTS, BIOLOGICAL COMMUNITY, CANADA.: /ON, ALGAE, ENVIRONMENTA W73-01704
EEN BAY(WIS), MACKEREL, MONTREAL(CANADA).: /LIEU RIVER(QUEBEC), GR W70-08098
IA), ST PIERRE(CANADA), MIQUELON(CANADA), MIKROSYPHAR PORPHYRAE, P W71-10066
E ISLAND(NOVA SCOTIA), ST PIERRE(CANADA), MIQUELON(CANADA), MIKROS W71-10066
E RIVER(INDIANA), PATUXENT RIVER(CANADA), ONTARIO.: /YNE BAY, WHIT W71-07417
ES, *GRAND RIVER BASIN(ONTARIO), CANADA, BUFFER CAPACITY.: /IDELIN W71-11017
EW BRUNSWI/ *CHECKLIST, *EASTERN CANADA, CAPE CHIDLEY(LABRADOR), N W71-10066
E ALGAE, MARINE PLANTS, BIOMASS, CANADA, COASTS, ATLANTIC OCEAN, S W73-00366
INTERNATIONAL JOINT COMMISSION, CANADA, FEDERAL GOV: / LIMNOLOGY, W72-14782
GENT ASSOCIATION, CARBOY TRIALS, CANADA, NITROGEN: PHOSPHORUS RATI W70-07283
PHAEROPHORA, MELANOSIP/ *EASTERN CANADA, NOVA SCOTIA, ISTHMOPLEA S W72-01370
ONMENTAL EFFECTS, MARINE PLANTS, CANADA, WASHINGTON, OREGON, CALIF W73-00428
NY).: *LIMITING NUTRIENTS, CANADARAGO LAKE(NY), LAKE GEORGE(W72-09163
S, MYRIOPHYLLUM SPECATUM, ELODEA CANADENSIS, PHRAGMITES, FONTINALI W71-11012
S, S.S. WITWATER, GALETA ISLAND, CANAL ZONE, MACROINVERTEBRATES, D W72-07911
OD BAY, DATA INTERPRE/ *CAPE COD CANAL(MASS), BUZZARDS BAY, CAPE C W73-01449
SE/ *CERATIUM EGYPTIACUM, *SUEZ CANAL, FLAGELLATES, MEDITERRANEAN W73-00296
OLOGICAL MAGNIFICATION, PELVETIA CANALICULATA, FUCUS SERRATUS, FUC W71-11515
ODOMONAS, LAKE CONSTANCE, SEWAGE CANALS.: /SCENS, SYNEDRA ACUS, RH W71-00117
IMNOLOGY, CYANOPHYTA, IRRIGATION CANALS, ALGAL CONTROL.: /ATURE, L W71-06102
NT, CALIFORNIA WATER CONVEYANCE, CANALS, HEAVY METALS.: /ENVIRONME W71-09581
URTUITUM, STAPHYLOCOCCUS AUREUS, CANDIDA ALBICANS.: /COBACTERIUM F W72-04259
ISPUS, RHODO/ *AGARS, EPIPHYTES, CANDIDA PARAPSILOSIS, CHONDRUS CR W72-04748
RULA RUBRA, RHODOTORULA LACTOSA, CANDIDA ZEYLANOIDES, CAMP VARNUM W72-04748
ND, IZEMBEK LAGOON, CHAETOMORPHA CANN: /OLIN ISLAND, WINIFRED ISLA W72-09120
AMCHITKA ISLAND, MILROW TEST, CANNIKIN TEST.: W73-00764
R BASIN(ONTARIO), CANADA, BUFFER CAPACITY.: /IDELINES, *GRAND RIVE W71-11017
(POLLUTION), *WASTE ASSIMILATION CAPACITY, *ORGANIC LOADS, NUTRIEN W71-11017
QUATIC ALGAE, WASTE ASSIMILATIVE CAPACITY, WATER POLLUTION EFFECTS W72-01685
ER RESOURCES, WASTE ASSIMILATIVE CAPACITY, RESEARCH AND DEVELOPMEN W71-13256
VESTIGATIONS, WASTE ASSIMILATIVE CAPACITY, WASTE DIL: / ON-SITE IN W73-00790
STE DILUTION, WASTE ASSIMILATIVE CAPACITY, EUROPE, ANIMAL PHYSIOLO W73-00817
GANIC MATTER, WASTE ASSIMILATIVE CAPACITY, EUTROPHICATION, NUTRIEN W73-00802
STE DILUTION, WASTE ASSIMILATIVE CAPACITY, CALIFORNIA, TRITIUM, MA W73-00831
TICAL MODELS, WASTE ASSIMILATIVE CAPACITY, ESTUARINE ENVIRONMENT, W73-00828
F POLLUTANTS, WASTE ASSIMILATIVE CAPACITY, *CALIFORNIA, RADIOECOLO W73-00826
ROWTH RATES, *WASTE ASSIMILATIVE CAPACITY, OXIDATION, DISSOLVED OX W72-04787
, OXYGEN SAG, WASTE ASSIMILATIVE CAPACITY, WATER PROPERTIES, WATER W72-04279
WI/ *CHECKLIST, *EASTERN CANADA, CAPE CHIDLEY(LABRADOR), NEW BRUNS W71-10066
E COD CANAL(MASS), BUZZARDS BAY, CAPE COD BAY, DATA INTERPRETATION W73-01449
AL WATERS, PANAMA, CALLAO(PERU), CAPE COD(MASS), CAPE MAY(N J), SA W72-08056
SEA, SHELF WATERS, GULF STREAM, CAPE HATTERAS, NANSEN BOTTLE, SPE W72-07663
A, CALLAO(PERU), CAPE COD(MASS), CAPE MAY(N J), SARGASSO SEA.: /AM W72-08056
, CLYMENELLA TORQUATA, CAPITELLA CAPITATA, MACOMA CALCAREA, ASTART W73-02029
HOROPSIS MEMBRANACEA, PENICILLUS CAPITATUS, CERAMIUM NITENS, WRANG W73-00838
ALES, *HALIMEDA SPP., PENICILLUS CAPITATUS, PENICILLUS DUMENTOSUS, W72-14294
MERCENARIA, CLYMENELLA TORQUATA, CAPITELLA CAPITATA, MACOMA CALCAR W73-02029
LUMBRINEREIS, GLYCERA, NEPHTYS, CAPITELLA, MAGELONA, ARMANDIA, TE W72-09092
NS, BIOLUMINESCENCE, SELENASTRUM CAPRICONUTUM, AEROBACTER AEROGENE W72-04292
AL ASSAY PROCEDURES, SELENASTRUM CAPRICORNUTUM.: /TAL ASSAYS, *ALG W72-03985
SELENASTRUM CAPRICORNUTUM.: W72-04433
OSINE TRIPHOSPHATE, *SELENASTRUM CAPRICORNUTUM.: *ADEN W71-13794
NOSPORIUM DASYOIDES, SELENASTRUM CAPRICORNUTUM, PAAP BATCH TEST, C W73-02099
KE GEORGE(NY).: *SELENASTRUM CAPRICORNUTUM, *ALGAL GROWTH, *LA W72-08376
LIQUID CHROMATOGRAPHY, ELECTRON CAPTURE DETECTOR, PRECONCENTRATIO W73-02105
GICAL MAGNIFICATION, / *ELECTRON CAPTURE GAS CHROMATOGRAPHY *BIOLO W73-02105
P P'-DDE, CLOPHEN A 50, ELECTRON CAPTURE GAS CHROMATOGRAPHY, POLYC W72-04134
, SYNEDRA FILIFORMIS, FRAGILARIA CAPUCINA, CYCLOTELLA STELLIGERA, W73-00384
RBICIDES, FUNGICIDES, ALGICIDES, CARBAMATE PESTICIDES, ORGANOPHOSP W73-00238

, ENDRIN, DIELDRIN, OXYGENATION,	CARBAMATE PESTICIDES.: / CULTRUES	W71-10096
ORINATED HYDROCARBON PESTICIDES,	CARBAMATE PESTICIDES, ORGANIC PES	W72-13347
POLYMERS, SEWAGE SETTLING, UNION	CARBIDE CORP, POLYOX, CARBOWAX, F	W70-06968
US, CHEMICAL ANALYSIS, PROTEINS,	CARBOHYDRATES, LIPIDS.: /PHOSPHOR	W70-05469
A, CULTURES, METABOLISM, SEWAGE,	CARBOHYDRATES, HYDROGEN ION CONCE	W72-05457
M, RESPIRATION, LIPIDS, AMMONIA,	CARBOHYDRATES, CARBON DIOXIDE, PO	W72-10233
PRODUCTIVITY, ORGANIC COMPOUNDS,	CARBOHYDRATES, CYCLING NUTRIENTS,	W73-02100
OMPOUNDS, AMINO ACIDS, VITAMINS,	CARBOHYDRATES, ALCOHOLS, POLLUTAN	W73-00942
, ELECTRON MICROSCOPY, PIGMENTS,	CARBOHYDRATES, PROTEINS, ACTINOMY	W72-07951
AQUATIC BACTERIA, AQUATIC ALGAE,	CARBON CYCLE, ON-SITE TESTS, AQUA	W73-02095
, PHYTOPLANKTON, WATER ANALYSIS,	CARBON CYCLE, ORGANIC MATTER, NUT	W72-09108
IMITING FACTORS, PHOTOSYNTHESIS,	CARBON CYCLE, ABSORPTION, DECOMPO	W71-12068
OTOSYNTHESIS, *ALGAE, PHYLOGENY,	CARBON CYCLE, ENZYMES, METABOLISM	W71-09172
LOROPHYLL, CARBON RADIOISOTOPES,	CARBON DIOXIDE, RHODOPHYTA.: / CH	W71-09172
, NITROGEN FIXATION, DETERGENTS,	CARBON DIOXIDE, ALKALINITY, BURNI	W72-05473
LVED OXYGEN, LIGHT, DIFFUSIVITY,	CARBON DIOXIDE, LIGHT INTENSITY,	W72-05459
DEMAND, CHEMICAL OXYGEN DEMAND,	CARBON DIOXIDE, LININGS, DISSOLVE	W72-04789
G, DISSOLVED OXYGEN, METABOLISM,	CARBON DIOXIDE, PHOSPHATES, NITRA	W72-04983
ENTRATION, FILTRATION, COLLOIDS,	CARBON DIOXIDE, CALCIUM, MAGNESIU	W72-03613
GAE, HYDROGEN ION CONCENTRATION,	CARBON DIOXIDE, NITRATES, NITRITE	W72-04309
INITY, CARBONATES, BICARBONATES,	CARBON DIOXIDE, HYDROGEN ION CONC	W72-04431
CES, HYDROGEN ION CONCENTRATION,	CARBON DIOXIDE, NITROGEN, CHLOROP	W72-04769
ITY, HYDROGEN ION CONCENTRATION,	CARBON DIOXIDE, AERATION, MIXING,	W72-04433
ALYSIS, FISH, INTAKE STRUCTURES,	CARBON DIOXIDE,: /, VOLUMETRIC AN	W72-05594
, GROWTH RATES, LIGHT INTENSITY,	CARBON DIOXIDE, NITROGEN, LABORAT	W72-05476
ATER TEMPERATURE, WATER QUALITY,	CARBON DIOXIDE, HYDROGEN ION CONC	W72-06274
AZOTOBACTER, CYANOPHYTA, YEASTS,	CARBON DIOXIDE, CHLORELLA, CLOSTR	W72-06124
TES, BIOASSAY, DISSOLVED OXYGEN,	CARBON DIOXIDE, AMMONIA, PHOSPHAT	W72-06046
CYANOPHYTA, AQUATIC ENVIRONMENT,	CARBON DIOXIDE, ECOSYSTEMS, SOUTH	W71-00475
BON, BACTERIA, ALGAE, SYMBIOSIS,	CARBON DIOXIDE, NITROGEN, LAKES,	W70-07283
TRATES, *CHLOROPHYLL, DIVERSION,	CARBON DIOXIDE, ALKALINITY, EPILI	W71-02681
TY, RESPIRATION, PHOTOSYNTHESIS,	CARBON DIOXIDE, PRIMARY PRODUCTIV	W71-04518
CARBON, *PHOTOSYNTHESIS, *ALGAE,	CARBON DIOXIDE, ALKALINITY, NUTRI	W71-11519
NUTRIENTS, AQUATIC ENVIRONMENT,	CARBON DIOXIDE.: / CONCENTRATION,	W71-12068
YLL, HYDROGEN ION CONCENTRATION,	CARBON DIOXIDE.: /LAKES, CHLOROPH	W71-12855
HYTA, NITROGEN FIXATION, OXYGEN,	CARBON DIOXIDE, BIOCHEMISTRY, ALG	W72-08584
HESIS, LIGHT INTENSITY, AMMONIA,	CARBON DIOXIDE, FLORIDA, CULTURES	W72-09103
RATIFICATION, CARBON, SEDIMENTS,	CARBON DIOXIDE, CHLORPHYLL, DC: /	W72-08560
ITY TECHNIQUES, *PHOTOSYNTHESIS,	CARBON DIOXIDE, PHYTOPLANKTON, AL	W72-09102
LIPIDS, AMMONIA, CARBOHYDRATES,	CARBON DIOXIDE, POLYMERS, HYDROGE	W72-10233
HYTHMS, LABORATORY TESTS, ALGAE,	CARBON DIOXIDE, CYCLING NUTRIENTS	W72-09159
ATION, *GROWTH RATES, CHLORELLA,	CARBON DIOXIDE, BICARBONATES, ALG	W72-09168
E, PHOSPHATES, NITROGEN, SEWAGE,	CARBON DIOXIDE, SEASONAL, PHOSPHO	W72-09167
FACTORS, *DIURNAL DISTRIBUTION,	CARBON DIOXIDE, NITROGEN, PHOSPHO	W72-11724
AND, ORGANIC MATTER, PHOSPHORUS,	CARBON DIOXIDE.: /ICAL OXYGEN DEM	W71-13257
CONCENTRATION, DISSOLVED OXYGEN,	CARBON DIOXIDE, AMMONIA, ORGANIC	W72-01363
N ION CONCENTRATION, METABOLISM,	CARBON DIOXIDE, ALGAE, CULTURES,	W72-00838
S, *GROWTH STAGES, REPRODUCTION,	CARBON DIOXIDE, CULTURES, LIMITIN	W73-02099
, *TOXICITY, CHEMICAL REACTIONS,	CARBON DIOXIDE, WATER POLLUTION S	W73-01074
M, MANGANESE, POTASSIUM, SODIUM,	CARBON DIOXIDE, CYANOPHYTA, FISH,	W73-00840
IATOMS, CHLOROPHYLL, ALKALINITY,	CARBON DIOXIDE, TEMPERATURE, DISS	W72-08049
ES, CYANOPHYTA, TOXICITY, COSTS,	CARBON DIOXIDE.: /IATOMS, SILICAT	W72-07508
OGEN, PHOSPHORUS, PHYTOPLANKTON,	CARBON DIOXIDE, SESTON, FERTILIZA	W72-07648
ES, NITROGEN FIXATION, CULTURES,	CARBON DIOXIDE, NUTRIENTS, LIMITI	W72-14790
OGEN, CARBON, DIATOMS, BACTERIA,	CARBON DIOXIDE, CHLAMYDOMONAS, SC	W72-12544
ABOLISM, TRACERS, RADIOACTIVITY,	CARBON DIOXIDE, SALTS, MARINE ALG	W72-12634
C PLANTS, ISOTOPE FRACTIONATION,	CARBON DIOXIDE, RESPIRATION, MARI	W72-12709
O/ PLATYMONAS, NITZSCHIA OVALIS,	CARBON EXTRUSION, CARBON-14, NITR	W72-07497
CAL TECHNIQUES, *PHOTOSYNTHESIS,	CARBON FIXATION, EUTROPHICATION,	W72-06046
NOPHYTA, RHODOPHYTA, PHAEOPHYTA,	CARBON RADIOISOTOPES, MARINE FISH	W72-06283
NIC MATTER, PLANKTON, CHELATION,	CARBON RADIOISOTOPES, CHLORELLA,	W71-03027
LOROPHYTA, PRIMARY PRODUCTIVITY,	CARBON RADIOISOTOPES, PHOTOSYNTHE	W71-03014
MONAS, RESPIRATION, CHLOROPHYLL,	CARBON RADIOISOTOPES, CARBON DIOX	W71-09172
TY, DISSOLVED OXYGEN, NUTRIENTS,	CARBON RADIOISOTOPES, PHOSPHATES,	W71-07875
PRODUCTIVITY, CULTURES, BIOMASS,	CARBON RADIOISOTOPES, LIMITING FA	W72-07648
OLOGY, CULTURES, PHOTOSYNTHESIS,	CARBON RADIOISOTOPES, CHROMATOGRA	W72-07710
NTS, LIMITING FACTORS, SALINITY,	CARBON RADIOISOTOPES, SESTON, ORG	W73-00431
MENTS, LIGHT INTENSITY, DIATOMS,	CARBON RADIOISOTOPES, NEW ENGLAND	W72-14317
ATION, RADIOACTIVITY TECHNIQUES,	CARBON RADIOISOTOPES, PRODUCTIVIT	W72-14326
ODOLOGY, MEASUREMENT, OXIDATION,	CARBON RADIOISOTOPES, *ON-SITE TE	W72-14301
IRON, *TEXAS, LABORATORY TESTS,	CARBON RADIOISOTOPES, *CYANOPHYTA	W72-13692
*DDT, *DIELDRIN, *PHYTOPLANKTON,	CARBON RADIOISOTOPES, LAKE ERIE,	W72-13657
, AQUATIC PRODUCTIVITY, BIOMASS,	CARBON RADIOISOTOPES, SUCCESSION,	W72-09239
*LIMITING NUTRIENTS, ANABAENA,	CARBON SOURCES.:	W72-09168
C CARBON, CARBON TRANSFORMATION,	CARBON SOURCES, ORGANIC CARBON, C	W71-12068
CHEMISTRY OF INORGANIC CARBON,	CARBON TRANSFORMATION, CARBON SOU	W71-12068
CARBON SOURCES, ORGANIC CARBON,	CARBON UTILIZATION.: /SFORMATION,	W71-12068
LABORATORY SCALE, *TOTAL ORGANIC	CARBON.:	W71-07102
OLLUTION EFFECTS, WATER QUALITY,	CARBON.: /S, CHEMCONTROL, WATER P	W72-11724
LIVINGSTON, FIELD MEASUREMENTS,	CARBON-14 MEASUREMENTS, ACETYLENE	W72-13692
SEAWEEDS, APHIDS, SPONGES,	CARBON-14.:	W72-06283
ICHOCOCCUS, CHLOROCOCCUM, HYALO/	CARBON-14, CARTERIA EUGAMETOS, ST	W72-09239
IELLA TERIOLECTA, PHAEODACTYLUM/	CARBON-14, *FLORIDA STRAIT, DUNAL	W72-09103
, PHAEODACTYLU/ *FLORIDA STRAIT,	CARBON-14, DUNALIELLA TERTIOLECTA	W72-09102
ZSCHIA OVALIS, CARBON EXTRUSION,	CARBON-14, NITROGEN SOURCE.: /NIT	W72-07497
.: *ORGANIC	CARBON, *ALGAL MASS, ALGAL GROWTH	W72-04431
RIVER(IDAHO), *DISSOLVED ORGANIC	CARBON, *AMERICAN FALLS RESERVOIR	W72-13300
*DISSOLVED ORGANIC	CARBON, *MENAI STRAIT(WALES).:	W72-09108
ER REUSE, *NITROGEN, PHOSPHORUS,	CARBON, *SEWAGE BACTERIA, SEWAGE,	W72-04309
IC ACID, VINEYARD SOUND, ORGANIC	CARBON, ACID-VOLATILE COMPOUNDS,	W72-07710

*ODOR, *POTABLE WATER, ACTIVATED CARBON, ADSORPTION, FILTRATION, A W72-08357
, WATER SUPPLY, ALGAE, ACTIVATED CARBON, ADSORPTION, FILTRATION, C W70-10381
RMATION, CARBON SOURCES, ORGANIC CARBON, CARBON UTILIZATION.: /SFO W71-12068
RBON SOU/ CHEMISTRY OF INORGANIC CARBON, CARBON TRANSFORMATION, CA W71-12068
, PHOSPHORUS, NITROGEN, AMMONIA, CARBON, CHLORELLA.: /TION EFFECTS W71-12857
IDE, *ALGAL CONTROL, CYANOPHYTA, CARBON, CHLORELLA, BICARBONATES, W72-10607
TROGEN, THIAMINE, BIOTI/ ORGANIC CARBON, CHLOROPHYLL A, ORGANIC NI W72-07892
, ODOR, POTABLE WATER, ACTIVATED CARBON, COLIFORMS, VIRUSES, CRUST W73-02144
ATION, ORGANIC NITROGEN, ORGANIC CARBON, CULTURE MEDIA, PAPER CHRO W73-00379
MODEL STUDIES, DISSOLVED OXYGEN, CARBON, DISTRIBUTION, PHOSPHORUS, W72-08459
LOROPHYTA, CYANOPHYTA, NITROGEN, CARBON, DIATOMS, BACTERIA, CARBON W72-12544
POLLUTION EFFECTS, TEMPERATURE, CARBON, DIATOMS, LIGHT, OXYGEN, M W71-08670
NITROGEN, ANALYTICAL TECHNIQUES, CARBON, EUTROPHICATION, STANDING W72-09163
DEMAND, CHEMICAL OXYGEN DEMAND, CARBON, FERTILIZATION, DISSOLVED W71-12188
S, TERTIARY TREATMENT, ACTIVATED CARBON, FLOCCULATION, ALGAE, SLUD W71-09524
ON, LAKES, NITROGEN, PHOSPHORUS, CARBON, GROWTH RATES, OLIGOTROPHY W72-09164
*ALGAE, ELECTROLYTES, ACTIVATED CARBON, HYDROGEN ION CONCENTRATIO W72-06201
TEMPERATURE, LIGHT PENETRATION, CARBON, LAKE ERIE, RUNOFF, EROSIO W72-04734
E, ODOR, RECREATION, PHOSPHORUS, CARBON, LAKE ERIE, DETERGENTS, LA W71-12091
EUGLENOPHYTA, DISSOLVED OXYGEN, CARBON, LAKE MORPHOMETRY, CHEMICA W72-08048
, SAMPLE PREP/ DISSOLVED ORGANIC CARBON, MACROPHYTES, SARGASSO SEA W73-02100
*ATMOSPHERIC CARBON, MICROCYSTIS, ANABAENA.: W72-10607
IENTS, ALGAE, LAKES, PHOSPHORUS, CARBON, NITROGEN.: / TESTS, *NUTR W72-09165
HEAVY METALS, CYCLING NUTRIENTS, CARBON, NITROGEN, PHOSPHORUS, SUL W73-00840
ATER, *BIOASSAY, *STANDING CROP, CARBON, NITROGEN, PHOSPHORUS, PHY W72-07648
DS, IRON, MANGANESE, PHOSPHORUS, CARBON, NITROGEN, OXIDATION, BIOC W72-05506
TY, IRON, PHOSPHATE, PHOSPHORUS, CARBON, NITROGEN, CURRENTS(WATER) W71-05626
ICATION, PHOSPHATES, PHOSPHORUS, CARBON, ORGANIC MATTER, MANGANESE W72-07514
TE UTILIZATION, PARTICU/ ORGANIC CARBON, ORGANIC NITROGEN, SUBSTRA W73-01066
SOLIDS, *MICROBIAL DEGRADATION, CARBON, ORGANIC MATTER, WATER POL W73-00379
IDE, HYDROGEN ION CONCENTRATION, CARBON, OXYGEN, NITROGEN, HYDROGE W72-06274
TTER, DISSOLVED SOLIDS, BENTHOS, CARBON, PRODUCTIVITY, ORGANIC COM W73-02100
TRIENTS, THERMAL STRATIFICATION, CARBON, SEDIMENTS, CARBON DIOXIDE W72-08560
ORUS, CHLOROPHYLL, DISTRIBUTION, CARBON, SEASONAL, CADDISFLIES, GA W71-11029
OXIDATION, ADSORPTION, ACTIVATED CARBON, SLUDGE TREATMENT, DIATOMS W73-01669
GEN ION CONCENTRATION, NITROGEN, CARBON, SODIUM, CALCIUM, POT: /RO W73-00854
FILTERS, COLIFORMS, METABOLISM, CARBON, SULFUR, SAMPLING, WATER A W72-04743
C LEVEL, PHOSPHORUS, PHOSPHATES, CARBON, VITAMIN B, NITROGEN, CYAN W72-07937
RGANIC COMPOUNDS, PHYTOPLANKTON, CARBON, WATER TEMPERATURE, GREAT W73-00384
E(MICH), WINTERGREEN LAKE(MICH), CARBONATE PARTICLES, PHOSPHORUS C W72-08048
T, DIATOMS, COAGULATION, CALCIUM CARBONATE, HYDROGEN ION CONCENTRA W73-01669
ITROGEN, ORGANIC MATTER, CALCIUM CARBONATE, DETRITUS, PLANKTON, AD W72-05469
USPENDED LOAD, NITRATES, CALCIUM CARBONATE, PROTEINS, FILTERS, ENG W70-10174
N ION CONCENTRATION, ALKALINITY, CARBONATES, HARDNE: /ITY, HYDROGE W71-00117
ITY, ICED LAKES, IONS, SULFATES, CARBONATES, BICARBONATES, DIATONI W71-02880
D, LABORATORY TESTS, ALKALINITY, CARBONATES, BICARBONATES, CARBON W72-04431
M, DETERGENTS, SEWAGE TREATMENT, CARBONATES, BICARBONATES, HARDNES W72-07514
ING, UNION CARBIDE CORP, POLYOX, CARBOWAX, FLOCCULANT DOSAGES, BAR W70-06968
SOAP AND DETERGENT ASSOCIATION, CARBOY TRIALS, CANADA, NITROGEN: W70-07283
ICORNUTUM, SKELETONEMA COSTATUM, CARBOY, OSCILLATORIA ERYTHRAEA.: / W72-09103
UM, DIMETHYLNITROSAMINE, AMINES, CARCINOGENS, TORULOPSIS SPP., ASP W72-09675
IA), LAKE JACKSON(GEORGIA), LAKE CARDINAL(GEORGIA), LAKE BUCKHORN(W71-11914
IA), LAKE JACKSON(GEORGIA), LAKE CARDINAL(GEORGIA), LAKE BUCKHORN(W71-09173
TIVELA STULTORUM, OSTREA LURIDA, CARDIUM: / TUNICATE, PISMO CLAM, W72-07907
PP, TYPHA SPP, ZIZANIA AQUATICA, CAREX SPP, BUTOMUS UMBELLATUS, SP W73-01089
UDGET, PRODUCTIVITY, HERBIVORES, CARNIVORES, ECOSYSTEMS, AQUATIC L W71-10064
FOOD HABITS, ALGAE, ZOOPLANKTON, CARNIVORES, DAPHNIA.: /UCTIVITY, W72-00161
D, WASTE WATER TREATMENT, *NORTH CAROLINA.: /CHEMICAL OXYGEN DEMAN W71-07382
, *PHOSPHORUS, *NITROGEN, *NORTH CAROLINA.: /AL BLOOMS, *NUTRIENTS W70-09669
POPULATIONS, LAKE HARTWELL(SOUTH CAROLINA), ANACYSTIS NIDULANS, AQ W71-00475
ARBON DIOXIDE, ECOSYSTEMS, SOUTH CAROLINA, ALGAE, TROPHIC LEVEL.: / W71-00475
, *DISTRIBUTION PATTERNS, *NORTH CAROLINA, *COMPUTER PROGRAMS, DAT W71-07337
ON EFFECTS, AQUATIC ALGAE, NORTH CAROLINA, *ALGAE, DIATOMS, *DINOF W72-01329
, *ATLANTIC COASTAL PLAIN, NORTH CAROLINA, ABSORPTION, SOUNDS, EST W73-00818
TRIBUTION, NORTH CAROLINA, SOUTH CAROLINA, SAMPLING, MICROSCOPY, E W72-07663
LF, VERTICAL DISTRIBUTION, NORTH CAROLINA, SOUTH CAROLINA, SAMPLIN W72-07663
, PHOTOGRAPHY, SEA WATER, *NORTH CAROLINA, MICROSCOPY, ATLANTIC OC W73-00284
RIES, *THERMAL POLLUTION, *NORTH CAROLINA, NUTRIENTS, SEWAGE EFFLU W72-13636
II, GLOIODERMA ATLANTICA, CODIUM CAROLINIANUM, *ONSLOW BAY(NC).: / W73-00284
I/ CHLOROPHYLL A, CHLOROPHYLL C, CAROTENOIDS, FRAGILARIA CROTONENS W73-00384
OTA LAKES, LUTEIN.: CAROTENOIDS, XANTHOPHYLLS, MINNES W72-01784
*AQUATIC WEED CONTROL, TILAPIA, CARP, ALGAE, ROOTED AQUATIC PLANT W71-04528
ATTAILS, AQUATIC ANIMALS, CLAMS, CARP, CRAYFISH, AQUATIC ALGAE, AB W72-06342
*AQUATIC LIFE, AQUATIC ANIMALS, CARP, DAPHNIA, AQUATIC PLANTS, AQ W72-09649
TIC ALGAE, EVALUATION, SAMPLING, CARP, POTASSIUM RADIOISOTOPES, ST W73-01673
FLY LARVAE, CLADOPHORA, CYPRINUS CARPIO, CATOSTOMUS COMMERSONI, AM W70-08098
, TILAPIA MELANOPLEURA, CYPRINUS CARPIO, CTENOPHARYNGODON IDELLUS. W71-04528
L TISSUES, MACROPHYTES, CYPRINUS CARPIO, MOSQUITO FISH, GAMBUSIA A W73-01673
DIVERSITY INDEX, CHLOROPHYLL A, CARTENOIDS, SAMPLE PREPARATION, M W73-00916
SIOSIRA FLUVIATILIS, AMPHIDINIUM CARTERI, COCCOLITHUS HUXLEYI, POR W72-12998
CHLOROCOCCUM, HYALO/ CARBON-14, CARTERIA EUGAMETOS, STICHOCOCCUS, W72-09239
S EPIPHYTICA, CHLAMYDOMONAS SPP, CARTERIA FRITSCHII, PANDORINA MOR W72-09111
IPULACEA, DIPLANTHERA UNINERVIS, CASSIOPEA ANDROMEDA, AKABARIA, ST W73-00855
SEQUILIBRIUM, LAKE TAHOE(CALIF), CASTLE LAKE(CALIF), DIVERSITY, PA W71-09171
POULTRY, HOGS, ANIMAL DISEASES, CATFISH, WASTE TREATMENT, FARM WA W71-07544
H, INSECTS, DDT, SILTS, SUCKERS, CATFISHES, BASS, PERCHES, PIKES, W70-08098
*ION-SELECTIVE CATHODES.: W72-00800
TUDIES, ADSORPTION, PENETRATION, CATION ADSORPTION, CHELATION, MAT W73-00823
TECHNIQUES, MOLECULAR STRUCTURE, CATION EXCHANGE, RADIOECOLOGY, *P W72-12203
HYLENEMIN/ *MULTIVALENT METALLIC CATIONS, *POLYACRYLAMIDE, *POLYET W72-14840

WASTEWATER TREATMENT, *POLYMERS,	CATIONS, ANIONS, CHLORELLA, FILTR	W72-14840
NT, COLLOIDS, *POLYMERS, ANIONS,	CATIONS, ORGANIC COMPOUNDS, SEPAR	W72-11833
ULOSUS, NOTEMIGONUS CRYSOLEUCAS,	CATOSTOMUS COMMERSONII COMMERSONI	W72-07890
AE, CLADOPHORA, CYPRINUS CARPIO,	CATOSTOMUS COMMERSONI, AMEIURUS N	W70-08098
WASTES, WATER POLLUTION EFFECTS,	CATTAILS, AQUATIC ANIMALS, CLAMS,	W72-06342
FARM WASTES, *RUNOFF, *FISHKILL,	CATTLE, DISSOLVED OXYGEN, DIVERSI	W72-08401
EASE/ *FEEDS, *RUMINANTS, ALGAE,	CATTLE, POULTRY, HOGS, ANIMAL DIS	W71-07544
ODOR, NITROGEN, AERATION, HOGS,	CATTLE, POULTRY, RESEARCH AND DEV	W71-07551
TER POLLUTION SOURCES, COLORADO,	CATTLE, URINE, PATH OF POLLUTANTS	W71-11496
CROCYSTIS AERUGINOSA, CHROOMONAS	CAUDATA, CHLAMYDOMONAS PERTUSA, P	W72-13816
AE, PANDORINA MORUM, PLATYDORINA	CAUDATA, EUDORINA CALIFORNICA, EU	W71-13793
NITROGEN, EUGLENA ACUS, EUGLENA	CAUDATA, EUGLENA LIMNOPHYLA, EUGL	W73-00918
OCCUS, OSCILLATORIA, NAUTOCOCCUS	CAUDATUS, NAUTOCOCCUS CONSTRICTUS	W72-07141
PIN, PALMITIC ACID,/ *CIGUATERA,	CAULERPA SP., CAULERPICIN, CAULER	W71-08597
NIX, UDOTEA FLABELLUM, BRYOPSIS,	CAULERPA, ACROPORA PALMATA, PENIC	W72-14294
ACID,/ *CIGUATERA, CAULERPA SP.,	CAULERPICIN, CAULERPIN, PALMITIC	W71-08597
TERA, CAULERPA SP., CAULERPICIN,	CAULERPIN, PALMITIC ACID, BETA-SI	W71-08597
EUGLENA GRACILIS, COLONIZATION,	CAULOBACTER, NAJAS FLEXILIS, SCIR	W73-02095
AERIA VANBOSSEAE, DICTYOSPHAERIA	CAVERNOSA, CLADOPHOROPSIS MEMBRAN	W72-11800
ECTICUT), GREEN RIVER(KENTUCKY),	CAYUGA LAKE(NEW YORK), PATUXENT E	W71-11517
UDIES.: LAKE	CAYUGA, CLADOPHORA, ECOLOGICAL ST	W70-09614
ORIA LIMNETICA, DICTYOSPHAERIUM,	CELL COUNTS, FRAGILARIA, ASTERION	W71-03014
AMPHORA OVALIS, OPEPHORA MARTYI,	CELL COUNTS, NAVICULA, NITZSCHIA,	W71-12855
UTION, LIMITING NUTRIENTS, ALGAL	CELL WASHOUT.: /SH), NUTRIENT DIL	W72-08049
RYLIC RESINS, GLASS, THYMIDINES,	CELLS(BIOLOGY), *BIODETERIORATION	W71-09261
*LYSIS, *CONTACT TIME,	CELLULAR METABOLITES.:	W72-00411
ERATURE, E. COLI, BENTHIC FAUNA,	CELLULOSE, ENZYMES, PULP WASTES,	W72-09668
ENTRATION, TEMPERATURE, ENZYMES,	CELLULOSE, VIRUSES, HUMAN PATHOLO	W72-05457
, CULTURES, VITAMIN B, BIOASSAY,	CELLULOSE, DIALYSIS, SEA WATER, O	W73-00942
: *LAKE ERIE	CENTRAL BASIN, *OXYGEN DEPLETION.	W72-13659
IC), OXYGEN DEPLETIO/ *LAKE ERIE	CENTRAL BASIN, POINTE PELEE(ONTAR	W72-12992
IUM, ANACYSTIS, CERA/ *LAKE ERIE	CENTRAL BASIN, TRIBONEMA, OEDOGON	W72-12995
ION, DEOXYGENATION, / *LAKE ERIE	CENTRAL BASIN, NUTRIENT REGENERAT	W72-12997
NUTRIENT REGENERATION, LAKE ERIE	CENTRAL BASIN.: /DEOXYGENATION, *	W72-12996
LAKE ERIE	CENTRAL BASIN.:	W72-12994
SAMPLING, SEPARATION TECHNIQUES,	CENTRIFUGATION, TURBIDITY, WATER	W73-00916
RE, SYSTEMATICS, SAMPLING, NETS,	CENTRIFUGATION, MICROSCOPY, DIATO	W72-07899
ADIENT CENTRIFUGATION, ISOPYCNIC	CENTRIFUGATION, SAMPLE PRESERVATI	W73-00371
UGATION, SAMP/ *DENSITY-GRADIENT	CENTRIFUGATION, ISOPYCNIC CENTRIF	W73-00371
ES, BIOMASS, GAS CHROMATOGRAPHY,	CENTRIFUGATION, PESTICIDES, BIOAS	W72-04134
IME, NEUTRALIZATION, FILTRATION,	CENTRIFUGATION, VACUUM DRYING, FL	W72-10237
A, CULTURES, *ALGAE, INCUBATION,	CENTRIFUGATION, RADIOACTIVITY TEC	W72-09103
ADSORPTION, COLIFORMS, BACTERIA,	CENTRIFUGATION, *WASTE WATER TREA	W71-13326
PILLOSA, COELOTHRIX IRREGULARIS,	CENTROCERAS CLAVULATUM, HYPENA MU	W73-00838
SMIUM, HARPESTICOIDS, NOCTILUCA,	CENTROPYXIS, BIDDULPHIA, CHAETOCE	W71-11876
ARTHROP/ SECONDARY PRODUCTIVITY,	CEPHALOCHORDATES, HEMICHORDATES,	W73-00935
CLAVULATUM, HYPENA MUSCIFORMIS,	CERAMIUM BUSSOIDEUM, CHONDRIA LI:	W73-00838
MBRANACEA, PENICILLUS CAPITATUS,	CERAMIUM NITENS, WRANGELIA ARGUS,	W73-00838
ERTILISSIMA, ANACYSTIS NIDULANS,	CERASON, DIATHENE, PROPAZINE, COT	W71-12071
M PALLIDUM, CERATIUM MASSILENSE,	CERATIUM FURCA, CERATIUM TRIPOS,	W71-07875
*NORTH SEA, PROROCENTRUM MICANS,	CERATIUM FURCA, TURBIDOSTAT CULTU	W71-03095
ULATA, MIKOLAJSKIE LAKE(POLAND),	CERATIUM HIRUNDINELLA.: /IA ECHIN	W72-14812
DEPRESSUM, PERIDINIUM PALLIDUM,	CERATIUM MASSILENSE, CERATIUM FUR	W71-07875
CERATIUM FURCA, CERATIUM TRIPOS,	CERATIUM MACROCEROS, NOCTILUCA MI	W71-07875
TIUM MASSILENSE, CERATIUM FURCA,	CERATIUM TRIPOS, CERATIUM MACROCE	W71-07875
AGNUM, APHANIZOMENON FLOS-AQUAE,	CERATIUM, OSCILLATORIA, TRANSPARE	W71-11949
ON, FRAGILARI/ *CLEVELAND(OHIO),	CERATIUM, COELOSPHAERIUM, DINOBRY	W72-10076
RIBONEMA, OEDOGONIUM, ANACYSTIS,	CERATIUM, COSMARIUM.: /L BASIN, T	W72-12995
MONAS ACUTA, MELOSIRA GRANULATA,	CERATIUM, MICROCYSTIS, RHODOMONAS	W72-12738
HRAGMITES, FONTINALIS, CHARALES,	CERATOPHYLLUM, LEMNA,: /DENSIS, P	W71-11012
BOSMINA,	CERIODAPHNIA, EUDIAPTOMUS.:	W72-00161
OTOPES, RUTHENIUM RADIOISOTOPES,	CERIUM RADIOISOTOPES, THORIUM RAD	W71-09188
FAUNA, MACROPHYTES, SAPROPHYTES,	CERIUM.: /S, ICHTHYOLOGY, ICHTHYO	W72-09646
MIUM, COPPER, NICKEL, RUTHENIUM,	CERIUM, LINCONIUM.: /, IRON, CHRO	W71-13243
INDIA,	CESIUM RADIOISOTOPES.:	W72-10958
ADIOIS/ MANGANESE RADIOISOTOPES,	CESIUM RADIOISOTOPES, RUTHENIUM R	W71-09188
MANGANESE RADIOISOTOPES,	CESIUM RADIOISOTOPES.:	W71-09018
CROINVERTEBRATES, RUTHENIUM-106,	CESIUM RADIOISOTOPES, SPIROGYRA,	W70-06668
*PLUTONIUM,	CESIUM RADIOISOTOPES.:	W73-00790
	CESIUM RADIOISOTOPES.:	W73-00831
	CESIUM RADIOISOTOPES.:	W73-00811
MOSQUITO FISH, GAMBUSIA AFFINIS,	CESIUM RADIOISOTOPES, CLADOPHORA	W73-01673
ANTIMONY,	CESIUM, *CAPE OF GOOD HOPE.:	W73-00832
CHEMICAL PROPERTIES, GOLD, IRON,	CESIUM, ALKALI METALS, SEASONAL,	W73-01434
*MICROENVIRONMENT, *ECOSYSTEMS,	CESIUM, COBALT RADIOISOTOPES, ALG	W70-06668
TOPES, PHOSPHORUS RADIOISOTOPES,	CESIUM, RADIOACTIVITY EFFECTS, LA	W72-09649
HEMISTRY, PHYSIOLOGICAL ECOLOGY,	CESIUM, STRONTIUM RADIOISOTOPES,	W71-13243
AL TECHNIQUES, LABORATORY TESTS,	CESIUM, STRONTIUM RADIOISOTOPES,	W71-08032
HNANTHES MINUTISSIMA, ACHNANTHES	CF. BIRGIANI, AMPHIPLEURA PELLUCI	W72-13142
COCCONEIS PLACENTULA, CYCLOTELLA	CF. SOCIALIS, CYCLOTELLA STELLIGE	W72-13142
ABSORPTION, PACIFIC OCEAN, FOOD	CHA: /, AMPHIPODA, RADIOISOTOPES,	W72-03347
ZOSOLENIA HEBETATA F. SEMISPINA,	CHAETOCEROS AFFINIA,: /SSIMA, RHI	W72-03543
ZIANUS, CHAETOCEROS MESSANENSIS,	CHAETOCEROS PERUVIANUS, CLADOGRAM	W71-11527
ROS ATLANTICUS VAR. NEAPOLITANA,	CHAETOCEROS COARCTATUS, CHAETOCER	W71-11527
EPTACTIS, BACTERIASTRUM COMOSUM,	CHAETOCEROS ATLANTICUS VAR. NEAPO	W71-11527
TILUCA, CENTROPYXIS, BIDDULPHIA,	CHAETOCEROS, THALASSISTHRIX, INDI	W71-11876
OLITANA, CHAETOCEROS COARCTATUS,	CHAETOCERCS LORENZIANUS, CHAETOCE	W71-11527
CTATUS, CHAETOCEROS LORENZIANUS,	CHAETOCEROS MESSANENSIS, CHAETOCE	W71-11527
RHIZOCLONIUM, COSMARIUM, PHACUS,	CHAETOCEROS, MOUGEOTIA, NAVICULA,	W72-09111
OROCENTRUM, NITZSCHIA, LAUDERIA,	CHAETOCEROS.: /IALIS, DITYLUM, PR	W72-09248

ETONEMA COSTATUM, THALASSIOSIRA, CHAETOCEROS, RHIZOSOLENIA ALATA, W71-07875
S AFFINIS, SKELETONEMA COSTATUM, CHAETOCERCS SOCIALIS.: /HAETOCERO W73-00442
, CHRYSOCHROMULINA, SKELETONEMA, CHAETOCEROS SIMPLEX.: /RORA ALATA W73-00380
PERIDINIANS, EUCAMPIA ZOODIACUS, CHAETOCEROS AFFINIS, SKELETONEMA W73-00442
AX POLYEDRA, PLATYMONAS VIRIDIS, CHAETOCEROS CURVISETUS, SKELETONE W72-07166
UM TRICORNUTUM, CYCLOTELLA NANA, CHAETOCEROS GALVESTONENSIS, MERCU W72-07660
EPHEMEROPTERA BAETIDAE, EPHYDRA, CHAETOGASTER, AEOLOSOMA, CHAOBORU W72-07132
NGAL, COELENTERATES, CTENOPHORA, CHAETOGNATHS, CYPHONAUTES, ECHINO W73-00935
WINIFRED ISLAND, IZEMBEK LAGOCN, CHAETOMORPHA CANN: /OLIN ISLAND, W72-09120
DGE, PUMPKINSEED, BLACK CRAPPIE, CHAIN PICKEREL, TENDIPES TENTANS, W72-14280
ES, BIOLOGICAL COMMUNITIES, FOOD CHAIN, DIATOMS, LARVAE, DIETS.: / W72-09248
NOWS, CRAYFISH, MOSQUITOES, FOOD CHAIN, HERRING, OYSTERS, CLAMS, C W71-07417
S, WATER POLLUTION EFFECTS, FOOD CHAINES, PUBLIC HEALTH, ESTUARINE W72-04463
, RADIOACTIVITY TECHNIQUES, FOOD CHAINS.: /ICROORGANISMS, PROTEINS W71-08042
N, WATER POLLUTION EFFECTS, FOOD CHAINS.: /, SOLUBILITY, ADSORPTIO W71-09182
CTIVITY EFFECTS, *FALLOUT, *FOOD CHAINS, *AQUATIC LIFE, RADICISOTO W72-04461
TOPES, *ALGAE, *SEA WATER, *FOOD CHAINS, *ABSORPTION, BIOCHEMISTRY W71-13243
FISH, I/ *WATER POLLUTION, *FOOD CHAINS, *ECOSYSTEMS, ABSORPTION, W72-05952
*MARINE ANIMALS, *CLAMS, *FOOD CHAINS, *NUCLEAR WASTES, *COBALT W73-00826
ORING, *CYCLING NUTRIENTS, *FOOD CHAINS, *ORGANIC COMPOUNDS, SHELL W72-07715
ADIOISOTOPES, *TUBIFICIDS, *FOOD CHAINS, *PATH OF POLLUTANTS, *ABS W72-09668
NUCLEAR POWERPLANTS, MONI/ *FOOD CHAINS, *RADIOACTIVITY EFFECTS, * W72-00959
TRITIUM, *NUCLEAR WASTES,/ *FOOD CHAINS, *RADIOACTIVITY EFFECTS, * W72-06342
UTANT IDENTIFIC/ *CALCIUM, *FOOD CHAINS, *SPECTROPHOTOMETRY, *POLL W71-11529
DIOISOTOPES, HEAVY METALS/ *FOOD CHAINS, *TRACE ELEMENTS, *LEAD RA W72-00940
*CHLOROPHYTA/ *FOOD WEBS, *FOOD CHAINS, *TOXICITY, *ALGAL TOXINS, W71-08597
*BIOASSAY, B/ *ECOSYSTEMS, FOOD CHAINS, *ZOOPLANKTON, *PREDATION, W72-07132
NE ANIMALS, *MARINE ALGAE, *FOOD CHAINS, ABSORPTION, CALCIUM, PUBL W73-00819
M RADIOISOTOPES, *FALLOUT, *FOOD CHAINS, ABSORPTION, SEDIMENTS, CO W73-00790
RON, ANALYTICAL TECHNIQUES, FOOD CHAINS, AFRICA, EFFLUENTS, PATH O W73-00832
NG, PLANKTON NETS, DIATOMS, FOOD CHAINS, ANABAENA, SCENEDESMUS, SE W72-12738
ICAL TECHNIQUES, NUTRIENTS, FOOD CHAINS, BIOLUMINESCENCE, POLLUTAN W71-11234
AGRICULTURAL RUNOFF, ALGAE, FOOD CHAINS, CHLORINATED HYDROCARBON P W72-13657
ION, SEPARATION TECHNIQUES, FOOD CHAINS, CULTURES, FISH DIET, GAS W73-02105
CLAYS, ODOR, WATER QUALITY, FOOD CHAINS, DISSOLVED OXYGEN, COPPER W72-04759
ROORGANISMS, *FLUORESCENCE, FOOD CHAINS, EQUIPMENT, ANALYTICAL TEC W72-07526
, *CRABS, *FISHES, ECOLOGY, FOOD CHAINS, ESTUARINE FISHERIES, MARI W72-12567
OURCES, PESTICIDE TOXICITY, FOOD CHAINS, FISH, BIOASSAY, ALGAE, SN W72-07703
UCTIVITY, ALGAE, RESOURCES, FOOD CHAINS, FISH FOOD ORGANISMS, OXYG W71-13246
ATH OF POLLUTANTS, REVIEWS, FOOD CHAINS, FORECASTING, CONSTRAINTS, W72-06340
LPS, STARFISH, CRUSTACEANS, FOOD CHAINS, FOOD WEBS, WATER POLLUTIO W72-11925
INVESTIGATIONS, FOOD WEBS, FOOD CHAINS, FROGS, LICHENS, ALGAE, FU W72-00949
ONS, ULTRAVIOLET RADIATION, FOOD CHAINS, GAMMA RAYS, RESISTANCE, B W73-00796
MALS, COBALT RADIOISOTOPES, FOOD CHAINS, IRRIGATION, FLOW PROFILES W71-09225
ON, BIOLOGICAL COMMUNITIES, FOOD CHAINS, LAKE MORPHOLOGY, RESERVOI W73-00938
TON, MAMMALS, MARINE ALGAE, FOOD CHAINS, MARINE FISHERIES, LAKES, W72-00960
ION SOURCES, AQUATIC ALGAE, FOOD CHAINS, MEASUREMENT, ADSORPTION.: W72-10686
CATORS, PATH OF POLLUTANTS, FOOD CHAINS, NUCLEAR WASTES, ENVIRONME W72-10957
LIZERS, DIATOMACEOUS EARTH, FOOD CHAINS, OXYGEN, ECOLOGY, RHODOPHY W71-09157
REMOVAL, *ZINC, CHELATION, FOOD CHAINS, PATH OF POLLUTANTS, PUBLI W73-00802
TOXICITY, WATER POLLUTION, FOOD CHAINS, PATH OF POLLUTANTS, BIOAS W72-12257
ONMENT, PATH OF POLLUTANTS, FOOD CHAINS, PHYTOPLANKTON, MARINE ALG W73-00818
ELATION, ORGANIC COMPOUNDS, FOOD CHAINS, PUBLIC HEALTH, PHYTOPLANK W73-00821
ITY, ESTUARINE ENVIRONMENT, FOOD CHAINS, PUBLIC HEALTH, EUROPE, RA W73-00828
YTOPLANKTON, *MARINE ALGAE, FOOD CHAINS, PUBLIC HEALTH, NUCLEAR WA W73-00823
CLEAR WASTES, *ABSORPTION, *FOOD CHAINS, PUBLIC HEALTH, PATH OF PO W73-00833
S, WATER POLLUTION EFFECTS, FOOD CHAINS, PUBLIC HEALTH, AQUATIC LI W72-10978
INVESTIGATIONS, CALIFORNIA, FOOD CHAINS, PUBLIC HEALTH, WATER POLL W72-03324
AS, ON-SITE INVESTIGATIONS, FOOD CHAINS, SPECIATION, REPRODUCTION. W73-00764
N, BIOASSAY, BIOINDICATORS, FOOD CHAINS, TITANIUM, PLANT POPULATIO W71-03095
RUS PESTICIDES, ACARICIDES, FOOD CHAINS, WATER POLLUTION SOURCES.: W73-00238
XENT RIVER, CONTRA COSTA POWER / CHALK POINT, COLUMBIA RIVER, PATU W72-07907
LOEOCAPSA ALPICOLA, OSCILLATORIA CHALYBIA, OSCILLATORIA FORMOSA, P W72-07664
EMA / *LAKE NEUSIEDLER(AUSTRIA), CHAMAESIPHON SUBAEQUALIS, APISTON W71-04526
S, SAMPLE PRESERVATION, COUNTING CHAMBERS, NUTRIENT CYCLING.: /NCE W72-07899
TIC ALGAE, *CHLOROPHYTA, *GROWTH CHAMBERS, *LIFE CYCLES, CULTURES, W72-07144
*SANTA BARBARA CHANNEL.: W71-08792
*SANTA BARBARA CHANNEL.: W71-08793
UVIATILE, CULTURE MEDIA, HUMMOCK CHANNEL, CHLOROPLASTS, MORPHOLOGY W72-12633
R TREATMENT, *FILTATION, ALLUVIA CHANNELS, SURFACE-GROUNDWATER REL W71-12410
PHYDRA, CHAETOGASTER, AEOLOSOMA, CHAOBORUS, *GAMBUSIA AFFINS, EPHE W72-07132
LOTA LOTA, LAKE MALOGEN, SWEDEN, CHAR, LUCIOPERCA LUCISPERCA, BURB W72-12729
SOIL CONTAMINATION, CLADOPHORA, CHAR: /IOISOTOPES, SOIL ANALYSIS, W73-01673
NATUS, ELEOCHARIS MONTEVIDENSES, CHARA ASPPRA, SCIRPUS SPP, GILLS, W73-01673
NDS, ON-SITE TESTS, METHODOLOGY, CHARA.: / ECOLOGY, ORGANIC COMPOU W72-03216
ENA / BAY OF QUINTE, SALT CREEK, CHARA, ELODEA, CALLITRICHE, ANABA W73-00840
N, ARKANSAS, INSECTS, TURBIDITY, CHARA, PONDWEEDS.: /ANTS, PLANKTO W71-04528
RTEBRATES, AMPHIBIANS, TOXICITY, CHARA, SUBMERGED PLANTS, PROTOZOA W72-00155
LE THEORY, ELECTRONIC POTENTIAL, CHARACEAE.: *CAB W73-01630
EINHARDTII, TETR/ MINERAL CYCLE, CHARACTERIZATION, CHLAMYDOMONAS R W73-02095
LANS, *MUTANTS, *SENSITI/ *WASTE CHARACTERIZATION, *ANACYSTIS NIDU W72-14327
, AREA REDEVELOPMENT, BIOLOGICAL CHARACTERISTICS, ALGAE, ZOOPLANKT W72-03215
SOLI/ ORGANIC NITROGEN, SETTLING CHARACTERISTICS, SOLIDS RECYCLE, W72-11100
RONMENT, AQUATIC FUNGI, ACQUIFER CHARACTERISTICS, GENETICS, DISSOL W72-11186
EEDING, ANHYDROUS AMMONIA, WASTE CHARACTERIZATION.: /MPOSTING, REF W71-07551
VALVE MORPHOLOGY, *VALVE PATTERN CHARACTERISTICS, CLONES.: /ISM, * W73-02548
ADENSIS, PHRAGMITES, FONTINALIS, CHARALES, CERATOPHYLLUM, LEMNA.: / W71-11012
OLIDS.: *SURFACE CHARGE, REGENERATION, SUSPENDED S W71-00138
LECTROANALYSIS, *ALGAL CULTURES, CHELATING AGENTS, OOCYSTIS MARSSO W73-02112
EFFECTS, EFFLUENTS, MONITORING, CHELATION.: /ASTES, ENVIRONMENTAL W72-10957

C MATTER, ALGAE, EUTROPHICATION, CHELATION, DECOMPOSING ORGANIC MA W72-01097
CATION, NUTRIENT REMOVAL, *ZINC, CHELATION, FOOD CHAINS, PATH OF P W73-00802
CLING NUTRIENTS, TRACE ELEMENTS, CHELATION, ORGANIC COMPOUNDS, FOO W73-00821
PENETRATION, CATION ADSORPTION, CHELATION, MATHEMATICAL MODELS.: / W73-00823
FATE, SALINITY, HARDNESS(WATER), CHELATION, MINE DRAINAGE, WATER P W72-14694
HESIS, ORGANIC MATTER, PLANKTON, CHELATION, CARBON RADIOISOTOPES, W71-03027
ION, LINEAR ALKYLATE SULFONATES, CHELATION, PHOSPHATES, ORGANIC CO W71-06247
KES, RESERVOIRS, SWIMMING POOLS, CHELATION, APPLICATION METHODS, C W71-06189
PLANKTON, FREEZING, TEMPERATURE, CHELATION, HYDROGEN ION CONCENTRA W71-09190
LANTS, MARINE ALGAE, HYDROLYSIS, CHELATION, SEDIMENTS, SETTLING VE W72-04463
ATORS, ANODIC STRIPPING, ORGANIC CHELATORS.: *CHEL W71-09674
AD, CYCLOTELLA NANA.: CHELEX 100, ARTEMIA SALINA, BIO-R W72-04708
, ALGAL CONTROL, WATER ANALYSIS, CHEMCONTROL, WATER POLLUTION EFFE W72-11724
, EVALU/ *ANALYTICAL TECHNIQUES, CHEMICAL ANALYSIS, WATER ANALYSIS W72-10883
S, ORGANIC ACIDS, SEPARATION TE/ CHEMICAL ANALYSIS, *ALGAE, COLUMN W72-09365
TANDING CROPS, LABORATORY TESTS, CHEMICAL ANALYSIS, BIOASSAY, CYTO W72-09163
RIBUTION, ENVIRONMENT, SAMPLING, CHEMICAL ANALYSIS, PLANKTON, COMP W72-08560
Y, *CALIFORNIA, HYDROLOGIC DATA, CHEMICAL ANALYSIS, DATA COLLECTIO W71-10577
MMONIA, *MARINE ALGAE, *DIATOMS, CHEMICAL ANALYSIS, HYDROGEN ION C W71-11008
EA WATER, WATER ANALYSIS, ALGAE, CHEMICAL ANALYSIS, SAMPLING, ANAL W71-11234
HLOROPHYTA, CYANOPHYTA, BIOMASS, CHEMICAL ANALYSIS, STATISTICAL ME W71-11029
, POTASH, ALGAE, AQUATIC PLANTS, CHEMICAL ANALYSIS, WATER POLLUTIO W71-12252
DATA, *KANSAS, WATER CHEMISTRY, CHEMICAL ANALYSIS, EVAPORATION, S W72-02274
TRIENTS, PACIFIC OCEAN, NITRATE, CHEMICAL ANALYSIS, SODIUM CHLORID W72-05421
N FACILITIES, WATER TEMPERATURE, CHEMICAL ANALYSIS, ALGAE, NUTRIEN W72-04791
S, FULVIC ACIDS, TRACE ELEMENTS, CHEMICAL ANALYSIS, ION EXCHANGE, W71-09188
OURCES, METHODOLOGY, EVALUATION, CHEMICAL ANALYSIS.: /E, WATER RES W71-05083
ALGAL POISONING, CHROMATOGRAPHY, CHEMICAL ANALYSIS, BIOCHEMISTRY, W71-08597
DISSOLVED OXYGEN, AIR POLLUTION, CHEMICAL ANALYSIS, EUTROPHICATION W71-01489
WAGE, RADIOACTIVITY, PHOSPHATES, CHEMICAL ANALYSIS, ALGAE, MICROOR W71-01474
DUCTIVITY, ALGAE, STANDING CROP, CHEMICAL ANALYSIS, MOSQUITOES, BA W71-01488
AE, *ASSAY, CHLORELLA, NITRATES, CHEMICAL ANALYSIS, CHLOROPHYTA, D W70-10403
AGE TREATMENT, *DOMESTIC WASTES, CHEMICAL ANALYSIS, EFFLUENTS, TRI W70-09907
NUTRIENTS, NITROGEN, PHOSPHORUS, CHEMICAL ANALYSIS, PROTEINS, CARB W70-05469
OPHORA, ANALYSIS, PHYTOPLANKTON, CHEMICAL ANALYSIS, DIATOMS, CYANO W73-00232
MINERALOGY, FISH, FISH BEHAVIOR, CHEMICAL ANALYSIS, SAMPLING, VOLU W72-14280
NITIES, SYSTEMATICS, CYANOPHYTA, CHEMICAL ANALYSIS, SEASONAL, BIOM W73-00242
*PHYTOPLANKTON, *MARINE ALGAE, CHEMICAL ANALYSIS, *NITRATES, *AM W73-00380
ER TEMPERATURE, WATER CHEMISTRY, CHEMICAL ANALYSIS, NUTRIENTS.: /T W73-01972
R, WATER ANALYSIS, MARINE ALGAE, CHEMICAL ANALYSIS, ADSORPTION, SE W73-02105
ENTS, RESPIRATION, MARSH PLANTS, CHEMICAL ANALYSIS, LEAVES, SAMPLI W73-01089
, RESERVOIRS, TRIBUTARIES, DAMS, CHEMICAL ANALYSIS, BIOMASS, ALGAE W73-01435
, BENTHOS, GRAZING, SYSTEMATICS, CHEMICAL ANALYSIS, PHYSICAL PROPE W73-01627
OZOA, PONDS, LIMNOLOGY, ECOLOGY, CHEMICAL ANALYSIS, AQUATIC ALGAE, W73-00918
ITY, PHENOLS, WATER TEMPERATURE, CHEMICAL ANALYSIS, SAMPLING, ECOL W72-07710
ER, SALINITY, WATER TEMPERATURE, CHEMICAL ANALYSIS, WATER ANALYSIS W72-08141
L, *PHOSPHORUS, *EUTROPHICATION, CHEMICAL ANALYSIS, WATER POLLUTIO W73-02478
SEEPAGE, SEPTIC TANKS, BIOASSAY, CHEMICAL ANALYSES, NITROGEN, NUTR W72-12955
ESTUARINE ENVIRONMENT, SALINITY, CHEMICAL ANALYSIS, BIOMASS, DISSO W72-12941
*ALGAL GROWTH, NUTRIENT UPTAKE, CHEMICAL COMPOSITION, ALGAL DECOM W70-05469
OSPHATE DETERGENT BAN, WYANDOTTE CHEMICAL CORPORATION, FMC CORPORA W70-07283
OLLUTION EFFECTS, WATER QUALITY, CHEMICAL DEGRADATION.: /, WATER P W72-01085
ATORY STUDI/ SAMPLE PREPARATION, CHEMICAL INTERFERENCE, INTERLABOR W73-00286
ION COUNTING, CYANIDIUM CALDARI/ CHEMICAL INTERFERENCE, SCINTILLAT W72-14301
MENT, BIOCHEMICAL OXYGEN DEMAND, CHEMICAL OXYGEN DEMAND, NUTRIENTS W72-08396
ION, *BIOCHEMICAL OXYGEN DEMAND, CHEMICAL OXYGEN DEMAND.: /EGRADAT W72-06612
RAGE, BIOCHEMICAL OXYGEN DEMAND, CHEMICAL OXYGEN DEMAND, PHOSPHORU W72-01881
TERS, BIOCHEMICAL OXYGEN DEMAND, CHEMICAL OXYGEN DEMAND, PHOSPHORU W71-13341
AE, *PHOTOSYNTHESIS, SOLUBILITY, CHEMICAL OXYGEN DEMAND, BIOCHEMIC W72-00411
ASTE, BIOCHEMICAL OXYGEN DEMAND, CHEMICAL OXYGEN DEMAND, *PHOSPHAT W71-13356
GAE, NITROGEN, ACTIVATED SLUDGE, CHEMICAL OXYGEN DEMAND, OXYGEN, A W71-13334
MENT, BIOCHEMICAL OXYGEN DEMAND, CHEMICAL OXYGEN DEMAND, CARBON, F W71-12188
ENT, *ORGANIC WASTES, ECONOMICS, CHEMICAL OXYGEN DEMAND, BIOCHEMIC W71-12223
CAL OXYGEN DEMAND, ION EXCHANGE, CHEMICAL OXYGEN DEMAND, FERTILIZE W71-12200
R POLLUTION CONTROL, WASHINGTON, CHEMICAL OXYGEN DEMAND, BIOCHEMIC W71-11006
ED SLUDGE, NITROGEN, PHOSPHORUS, CHEMICAL OXYGEN DEMAND, CHLORELLA W72-11100
OONS, BIOCHEMICAL OXYGEN DEMAND, CHEMICAL OXYGEN DEMAND, PHOSPHATE W70-07838
ADATION, ORGANIC MATTER, OXYGEN, CHEMICAL OXYGEN DEMAND, BENTHOS.: W71-04518
TURE, BIOCHEMICAL OXYGEN DEMAND, CHEMICAL OXYGEN DEMAND, COOLING T W71-09524
SPHORUS, NITROGEN, TIDAL WATERS, CHEMICAL OXYGEN DEMAND, BIOCHEMIC W71-07098
GAE, HYDROGEN ION CONCENTRATION, CHEMICAL OXYGEN DEMAND, *BIOCHEMI W71-07102
OXYGEN, SEDIMENTATION, SAMPLING, CHEMICAL OXYGEN DEMAND, BIOCHEMIC W71-07097
ITRATES, *NUTRIENTS, ALKALINITY, CHEMICAL OXYGEN DEMAND, ANALYTICA W72-04983
UDGE, BIOCHEMICAL OXYGEN DEMAND, CHEMICAL OXYGEN DEMAND, CARBON DI W72-04789
HICATION, PHOSPHOROUS, NITROGEN, CHEMICAL OXYGEN DEMAND, LABORATOR W72-04431
IO/ *PHOSPHATES, *CLAMS, *ALGAE, CHEMICAL PRECIPITATION, HYDROGEN W72-03613
ANALYTICAL TECHNIQUES, TOXICITY, CHEMICAL PROPERTIES, MINNESOTA, L W72-04275
, LIMESTONE, TEMPERATURE, DEPTH, CHEMICAL PROPERTIES, ACIDITY, ALK W72-04269
E ALGAE, RADIOISOTOPES, FALLOUT, CHEMICAL PRECIPITATION, IRON, BIO W72-05965
H, TEMPERATURE, LIGHT INTENSITY, CHEMICAL PRECIPITATION, HYDROGEN W71-07100
NTS, ION EXCHANGE, ELECTROLYSIS, CHEMICAL PRECIPITATION, NITRIFICA W71-09450
ON, *LAKES, AQUATIC ENVIRONMENT, CHEMICAL PROPERTIES, PHOSPHORUS, W71-00117
PTION, FILTRATION, CHLORINATION, CHEMICAL PRECIPITATION, FLOCCULAT W70-10381
TS, DETERGENTS, SODIUM CHLORIDE, CHEMICAL PROPERTIES, SECCHI DISKS W72-09167
F ALGAE, *SEPARATION TECHNIQUES, CHEMICAL PRECIPITATION, COAGULATI W72-10237
ILITY, PLANT POPULATIONS, ALGAE, CHEMICAL PROPERTIES, WATER POLLUT W72-01473
IRONMENT, SOIL TEMPERATURE, SOIL CHEMICAL PROPERTIES, SOIL BACTERI W72-03342
XYGEN, CARBON, LAKE MORPHOMETRY, CHEMICAL PROPERTIES, O: /SOLVED O W72-08048
, PHYTOPLANKTON, AQUATIC PLANTS, CHEMICAL PROPERTIES, WATER QUALIT W72-14542
YTA, HYDROGEN ION CONCENTRATION, CHEMICAL PORPERTIES, NITROGEN FIX W72-14790

TER, LAKES, PHYSICAL PROPERTIES,	CHEMICAL PROPERTIES, GOLD, IRON,	W73-01434
L SAMPLES, *SAMPLE PRESERVATION,	CHEMICAL PRESERVATION, HISTOCHEMI	W73-01676
POSING ORGANIC MATTER, BACTERIA,	CHEMICAL PROPERTIES, NITROGEN, HE	W72-12997
YDROGEN ION CONCENTRATION, FLOW,	CHEMICAL PROPERTIES, PHYSICAL: /H	W72-12941
UTRIENTS, SURFACE WATERS, ALGAE,	CHEMICAL PROPERTIES, LANDFILLS, P	W72-12955
NDUSTRIAL WASTES, BIOINDICATORS,	CHEMICAL PROPERTIES, PHYSICAL PRO	W72-12729
CAL PROPERTIES, ELECTROPHORESIS,	CHEMICAL PROPERTIES, HYDROGEN ION	W72-11833
SUREMENT, *MARINE ALGAE TRACERS,	CHEMICAL PRECIPITATION, HYDROGEN	W72-12210
ATER), WATER BALANCE, SEDIMENTS,	CHEMICAL PROPERTIES, DISSOLVED SO	W72-13851
ESIS, *OIL POLLUTION, *TOXICITY,	CHEMICAL REACTIONS, CARBON DIOXID	W73-01074
ROENVIRONMENT, LITTORAL, OXYGEN,	CHEMICAL REACTIONS, HYDROGEN ION	W71-08026
, LIVER, BONE, DETECTION LIMITS,	CHEMICAL RECOVERY, SAMPLE PREPARA	W72-05965
ORIDA, SESSILE ALGAE, ESTUARIES,	CHEMICAL WASTES, GAS CHROMATOGRAP	W71-07412
URAL CHEMICALS, FISH, CHEMICALS,	CHEMICAL WASTES, BIOCHEMICAL OXYG	W73-01976
ISPOSAL, PHOSPHATES, PHOSPHORUS,	CHEMICAL WASTES, LIMNOLOGY, INTER	W72-14782
SIUM, SODIUM, CALCIUM, HYDROGEN,	CHEMICAL WASTES, LEAD, NICKEL, TO	W72-14303
CTS, RESPIRATION, *HEAVY METALS,	CHEMICAL WASTES, MERCURY, COPPER,	W72-12576
RGENTS, *LEGISLATION, CHEMICALS,	CHEMICAL WASTES, CLEANING, DOMEST	W72-01685
ALGAE, PRODUCTIVITY, POLLUTANTS,	CHEMICAL WASTES, SEWAGE EFFLUENTS	W72-08804
, LABORATORY TESTS, FISH, ALGAE,	CHEMICAL WASTES, ALGICIDES, BACTE	W72-08683
S, DOMINANT ORGANISMS, REDUCTION(	CHEMICAL), TROPHIC LEVELS, PROTOZ	W72-10883
YTA, PLANTS, NITROGEN, REDUCTION(	CHEMICAL), GAS CHROMATOGRAPHY, SE	W73-01456
IGERATION, INCUBATION, DIALYSIS,	CHEMICALS, DRYING, FREEZING, RADI	W73-01676
S, AGRICULTURAL CHEMICALS, FISH,	CHEMICALS, CHEMICAL WASTES, BIOCH	W73-01976
ITY, BIOINDICATORS, AGRICULTURAL	CHEMICALS, FISH, CHEMICALS, CHEMI	W73-01976
RGANISMS, ODOR, TASTE, PLANKTON,	CHEMICALS, PHENOLS, CHLORINE, OXI	W73-01669
TION, WATER POLLUTION, TOXICITY,	CHEMICALS, REGULATION, ALGAL TOXI	W73-00703
, HYDROLOGIC ASPECTS, NUTRIENTS,	CHEMICALS, PLANKTON, BENTHOS, ALG	W72-07841
ATES, *DETERGENTS, *LEGISLATION,	CHEMICALS, CHEMICAL WASTES, CLEAN	W72-01685
FFECTS, PESTICIDES, AGRICULTURAL	CHEMICALS, PESTICIDE RESIDUES, PE	W71-12303
A, ORGANIC WASTES, SOLID WASTES,	CHEMICALS, MICHIGAN, CALIFORNIA,	W71-11006
ON, DRAINAGE WATER, AGRICULTURAL	CHEMICALS, FARM WASTES, WASTE WAT	W71-11698
DS/ *WATERBORNE DISEASES, *TOXIC	CHEMICALS, *PUBLIC HEALTH STANDAR	W73-02144
TROL, TESTING, LABORATORY TESTS,	CHEMICALS, BACTERIOCIDES, CULTURE	W71-06189
POLLUTION, BIOLOGICAL TREATMENT,	CHEMICALS, COPPER SULFATE, CHLORI	W71-05083
R QUALITY, SILTING, AGRICULTURAL	CHEMICALS, WATER POLLUTION SOURCE	W72-05869
, FISH, WEEDS, CYCLES, BACTERIA,	CHEMIST: /YGEN DEMAND, PHOSPHORUS	W71-03012
RBON TRANSFORMATION, CARBON SCU/	CHEMISTRY OF INORGANIC CARBON, CA	W71-12068
	CHEMISTRY OF THE ATMOSPHERE.:	W72-00949
TECHNIQUES, PALEOZOIC ERA, WATER	CHEMISTRY.: /SOTOPES, ANALYTICAL	W72-02073
LGAE, VOLUMETRIC ANALYSIS, WATER	CHEMISTRY.: /, IONS, FLUORIDES, A	W71-00474
US REQUIREMENTS(ALGAE), NUTRIENT	CHEMISTRY(AQUATIC).: /), PHOSPHOR	W70-04080
NUTRIENTS, WATER QUALITY, WATER	CHEMISTRY, CYCLING NUTRIENTS.: /L	W70-04080
M FRACTIONATION, *ORGANI/ *WATER	CHEMISTRY, *FOAM SEPARATION, *FOA	W72-05506
NIC MATTER, WATER QUALITY, WATER	CHEMISTRY, WATER POLLUTION SOURCE	W72-04853
*HYDROLOGIC DATA, *KANSAS, WATER	CHEMISTRY, CHEMICAL ANALYSIS, EVA	W72-02274
MALS), CYTOLOGICAL STUDIES, SOIL	CHEMISTRY, RADIOSENSITIV: /LS(MAM	W72-00949
MOSSES, WATER TEMPERATURE, WATER	CHEMISTRY, CHLOROPHYLL, HYPSOMETR	W71-11949
ION, PRIMARY PRODUCTIVITY, WATER	CHEMISTRY, *CALIFORNIA, *NEVADA.:	W72-12549
UTROPHICATION, EVALUATION, WATER	CHEMISTRY, WATER TEMPERATURE, DIS	W72-06526
RATES, WATER TEMPERATURE, WATER	CHEMISTRY, CHEMICAL ANALYSIS, NUT	W73-01972
RUM, ENRICHMENT CULTURES, AGARS,	CHEMOSTATS, NITROSOMONAS, NITROBA	W72-05421
FLOS-AQUAE, *CONTINUOUS CULTURE,	CHEMOSTATS, PHOSPHATASE ACTIVITY.	W72-05476
W TURBIDOSTAT, *GROWTH KINETICS,	CHEMOSTATS, FLAGELLATE ALGAE, OPT	W70-10403
LAKE BLOOMINGTON(ILL),	CHEMOSTRATIFICATION.:	W71-03028
ICS, MERCURY, ARSENIC COMPOUNDS,	CHESAPEAKE BAY.: /ALGAE, ANTIBIOT	W72-10162
ARA, BOENOA, SALDULA, CRYNELLUS,	CHEUMATOPSYCHE, TRIAENODES INUUST	W70-06655
*PHOSPHORUS REMOVAL,	CHICAGO(ILLINOIS).:	W72-03970
CHECKLIST, *EASTERN CANADA, CAPE	CHIDLEY(LABRADOR), NEW BRUNSWICK,	W71-10066
PIPESTONE HOT SPRINGS, SLEEPING	CHILD HOT SPRINGS.: /HOT SPRINGS,	W72-05610
RUBBLE	CHIMNEY, SAN JOAQUIN VALLEY.:	W72-00974
KTON, ZOOPLANKTON, LARVAE, EGGS,	CHINOOK SALMON, STRIPED BASS, MIN	W71-11517
ALGAE, SUNFISHES, SNAILS, EELS,	CHINOOK SALMON, BASS, TROUT, MINN	W71-07417
US, CAMPELOMA, PSIDIUM, SHADFLY,	CHIRONOMIDS, CLEAR LAKE(CALIF), R	W70-08098
RATES, ECHINODERMS, POLYCHAETES,	CHIRONOMIDS, BARNACLES, ARTHROPOD	W73-00855
LUMOSUS, CHIRONOMUS ATTENTUATUS,	CHIRONOMUS ANTHRACINUS, LIMNODRIL	W71-11009
ENT INPUTS, CHIRONOMUS PLUMOSUS,	CHIRONOMUS ATTENTUATUS, CHIRONOMU	W71-11009
UINTE(ONTARIO), NUTRIENT INPUTS,	CHIRONOMUS PLUMOSUS, CHIRONOMUS A	W71-11009
RIFERA, SEA CUCUMBERS, STARFISH,	CHITONS, RHABDODERMELLA, LOPHOPAN	W73-00932
TERIONELLA FORMOSA, CRYPTOMONAS,	CHLAMIDOMONAS, BERENPLAAT RESERVO	W71-01491
A TERTIOLECTA, EUGLENA GRACILIS,	CHLAMYDOMANAS REINHARDTII, LETHAL	W72-05614
LIGOTROPHY, PROTOZOA, CHLORELLA,	CHLAMYDOMONAS, CYANOPHYTA, CHLORO	W72-05968
ENYLS, BIOLOGICAL MAGNIFICATION,	CHLAMYDOMONAS, METABOLITES, HAEMA	W72-04134
ORINATED HERBICIDES, MCPB, MCPA,	CHLAMYDOMONAS GLOBUSA, CHLORELLA	W70-07255
HRYSOPHYTA, CHLORELLA, ROTIFERS,	CHLAMYDOMONAS, PHYTOPLANKTON, ZOO	W70-06655
ANAEROBIC CONDITIONS, CHLORELLA,	CHLAMYDOMONAS, SCENEDESMUS, EFFLU	W70-10405
EUGLENA, CHLORELLA, SCENEDESMUS,	CHLAMYDOMONAS, CYANOPHYTA, CHLORO	W71-09172
MINERAL CYCLE, CHARACTERIZATION,	CHLAMYDOMONAS REINHARDTII, TETRAH	W73-02095
PHYTA, CHRYSOPHYTA, SCENEDESMUS,	CHLAMYDOMONAS.: /LENOPHYTA, CYANO	W73-01627
E ALGAE, SEA WATER, CHLOROPHYTA,	CHLAMYDOMONAS, LABORATORY TESTS,	W72-08051
E ALGAE, AQUATIC ALGAE, EUGLENA,	CHLAMYDOMONAS.: /MS, ALGAE, MARIN	W72-08436
PHICATION, NUTRIENTS, CHLORELLA,	CHLAMYDOMONAS, DIATOMS, SILICATES	W72-07508
ING, OXIDATION LAGOONS, EUGLENA,	CHLAMYDOMONAS, EUGLENOPHYTA, PROT	W72-07661
GANIC COMPOUNDS, ALGAE, DIATOMS,	CHLAMYDOMONAS, SCENEDESMUS, BIO-A	W72-14706
ONMENTAL EFFECTS, PHYTOPLANKTON,	CHLAMYDOMONAS, CHLOROPHYTA, WATER	W73-00281
ORMIDIUM, CHLORELLA PYRENOIDOSA,	CHLAMYDOMONAS, FONTINALIS, ASTINA	W71-11519
TRIENT REQUIREMENTS, CYANOPHYTA,	CHLAMYDOMONAS.: /WATER, ALGAE, NU	W71-13185
ERGED PLANTS, PROTOZOA, EUGLENA,	CHLAMYDOMONAS, CHLOROPHYTA, SCENE	W72-00155
DORN SAMPLER, KEMMERER SAMPLER,	CHLAMYDOMONAS EPIPHYTICA, CHLAMYD	W72-09111
MPLER, CHLAMYDOMONAS EPIPHYTICA,	CHLAMYDOMONAS SPP, CARTERIA FRITS	W72-09111

ENT REQUIREMENTS, INVERTEBRATES,	CHLAMYDOMONAS, MATHEMATICAL STUDI	W72-09248
CHL/ CHLAMYDOMONAS ISABELIENSIS,	CHLAMYDOMONAS PALLIDOSTIGMATICA,	W72-12632
LAKE, PINE LA/ *ORTHOPHOSPHATES,	CHLAMYDOMONAS REINHARDTII, MOSES	W72-12637
ORESCENCE, PHOSPHORUS COMPOUNDS,	CHLAMYDOMONAS, WATER ANALYSIS, FL	W72-12637
ATOMS, BACTERIA, CARBON DIOXIDE,	CHLAMYDOMONAS, SCENEDESMUS, CHLOR	W72-12544
CHLAMYDOMONAS PALLIDOSTIGMATICA,	CHLAMYDOMONAS FOTTII, CHLAMYDOMON	W72-12632
TIGMATICA, CHLAMYDOMONAS FOTTII,	CHLAMYDOMONAS TEXENSIS, CHLAMYDOM	W72-12632
YDOMONAS PALLIDOSTIGMATICA, CHL/	CHLAMYDOMONAS ISABELIENSIS, CHLAM	W72-12632
FOTTII, CHLAMYDOMONAS TEXENSIS,	CHLAMYDOMONAS PSEUDOMICROSPHAERA,	W72-12632
PHORUS, CYANOPHYTA, OLIGOTROPHY,	CHLAMYDOMONAS, DEFICIENT ELEMENTS	W72-13816
AERUGINOSA, CHROOMONAS CAUDATA,	CHLAMYDOMONAS PERTUSA, PANDORINA	W72-13816
TASTE, ODOR, CORROSION, ENZYMES,	CHLEATION, ALGAE, WATER QUALITY,	W72-06837
IDULANS, *MUTANTS, *SENSITIVITY,	CHLORAMPHENICOL, RECOVERY, SURVIV	W72-14327
ATION, PENICILLIN G, TERRAMYCIN,	CHLORAMPHENICOL, ACTIDON, CULTURE	W73-01445
LIN, DRUGS, CORTISONE, HORMONES,	CHLORAMPHENICOL, SALMONELLA ENTER	W72-13800
O-60, GAMMA RADIATION, SURVIVAL,	CHLORELLA PYRENOIDOSA.: /NIANA, C	W72-11836
PINGUE, SCENEDESMUS QUADRICAUDA,	CHLORELLA PYRENOIDOSA, LITHIUM CH	W72-12739
ENEMINE, *POLYSTYRENE SULFONATE,	CHLORELLA ELLIPSOIDIA.: /OLYETHYL	W72-14840
	CHLORELLA VULGARIS.:	W72-14800
EOLARUM, ANKISTRODESMUS BRAUNII,	CHLORELLA PYRENOIDOSA, CHLORELLA	W72-07664
BRAUNII, CHLORELLA PYRENOIDOSA,	CHLORELLA VULGARIS, SCENEDESMUS O	W72-07664
YLUM TRI/ SILVER, CADMIUM, LEAD,	CHLORELLA PHYRENOIDOSA, PHAEODACT	W72-07660
TOGRAPHY, METHYL ESTERS, ESTERS,	CHLORELLA VULGARIS, SAMPLE PREPAR	W72-09365
	CHLORELLA PYRENOIDOSA.:	W72-11572
TASE, RIBONUCLEIC ACID.:	CHLORELLA VULGARIS, NITRATE REDUC	W72-03222
NAS, FONTINALIS, AS/ PHORMIDIUM,	CHLORELLA PYRENOIDOSA, CHLAMYDOMO	W71-11519
E ENZYME SYSTEM, PLANT EXTRACTS,	CHLORELLA SOROKINIANA.: /UCIFERAS	W71-11916
SARCINA, ANKISTRODESMUS BRAUNII,	CHLORELLA VULGARIS.: /ION, CHLORO	W71-07675
PB, MCPA, CHLAMYDOMONAS GLOBUSA,	CHLORELLA PYRENOIDOSA, STICHOCOCC	W70-07255
PALEA, SKELETONEMA COSTATUM.:	CHLORELLA PYRENOIDOSA, NITZSCHIA	W71-03027
CHLORELLA / DNA, CHLOROPHYLL A,	CHLORELLA PYRENOIDOSA, SPIROGIRA,	W72-06274
HLORELLA PYRENOIDOSA, SPIROGIRA,	CHLORELLA ELLIPSOIDEA, RNA, CHLOR	W72-06274
, SATURATION CONSTANT, DILUTION,	CHLORELLA PYRENOIDOSA, MONOD GROW	W72-04788
IC/ POTOMAC RIVER, ALAFIA RIVER,	CHLORELLA PYRENOIDOSA, SODIUM SIL	W72-04734
AERATION, MICROBIOLOGY, BIOMASS,	CHLORELLA.: /AEROBIC CONDITIONS,	W71-04079
ORUS, NITROGEN, AMMONIA, CARBON,	CHLORELLA.: /TION EFFECTS, PHOSPH	W71-12857
RITES, NITRATES, AMMONIA, ALGAE,	CHLORELLA.: /ON, HYDROGRAPHY, NIT	W72-08056
EGULATORS, ALGAE, AQUATIC ALGAE,	CHLORELLA.: /-4-D, PLANT GROWTH R	W72-08440
DRATES, PROTEINS, ACTINOMYCETES,	CHLORELLA.: /Y, PIGMENTS, CARBOHY	W72-07951
ANT IDENTIFICATION, CHLOROPHYTA,	CHLORELLA.: /UM COMPOUNDS, POLLUT	W72-13813
S, LIGHT INTENSITY, SCENEDESMUS,	CHLORELLA, GROWTH RATES, RESISTAN	W72-12739
*RADIOACTIVITY EFFECTS, *ALGAE,	CHLORELLA, IRRADIATION, RADIOACTI	W72-11836
IDE, CHLAMYDOMONAS, SCENEDESMUS,	CHLORELLA, SILICA, LIGHT INTENSIT	W72-12544
SUSPENDED SOLIDS, LACTOBACILLUS,	CHLORELLA, SULFUR BACTERIA, ALGAE	W72-12565
TION, EUTROPHICATION, NUTRIENTS,	CHLORELLA, CHLAMYDOMONAS, DIATOMS	W72-07508
ESCENCE, TRACE ELEMENTS, COPPER,	CHLORELLA, CULTURES, NUTRIENTS, R	W72-07660
UTROPHICATION, FJORDS, BIOASSAY,	CHLORELLA, NUTRIENTS, WATER SAMPL	W73-00234
ENT, *POLYMERS, CATIONS, ANIONS,	CHLORELLA, FILTRATION, ELECTROPHO	W72-14840
OLISM, FUNGI, ORGANIC COMPOUNDS,	CHLORELLA, TRACERS, AQUATIC ALGAE	W72-14326
OMASS, PRODUCTIVITY, METABOLISM,	CHLORELLA, RESPIRATION, CHLOROPHY	W73-01080
POLYSIPHONIA SP., GELIDIUM SP.,	CHLORELLA, AGARS.: /ADOPHORA SP.,	W71-11851
N ION CONCENTRATION, CYANOPHYTA,	CHLORELLA, CHLOROPHYTA, SCENEDESM	W71-11519
SALINITY, NUTRIENT REQUIREMENTS,	CHLORELLA, GROWTH RATES, TEMPERAT	W71-10083
RHODOPHYTA, DIATOMS, PYRROPHYTA,	CHLORELLA, PSEUDOMONAS, TESTING,	W72-03220
SPHORUS, CHEMICAL OXYGEN DEMAND,	CHLORELLA, PERFORMANCE, EFFICIENC	W72-11100
PHYTA, SCENEDESMUS, *CYANOPHYTA,	CHLORELLA, BIOLOGICAL COMMUNITIES	W72-09239
*EUTROPHICATION, *GROWTH RATES,	CHLORELLA, CARBON DIOXIDE, BICARB	W72-09168
GAL CONTROL, CYANOPHYTA, CARBON,	CHLORELLA, BICARBONATES, AERATION	W72-10607
CHELATION, CARBON RADIOISOTOPES,	CHLORELLA, GROWTH RATES, DIATOMS,	W71-03027
*ALGAE, *HERBICIDES, PESTICIDES,	CHLORELLA, RESPIRATION, PHOSPHORU	W70-07255
PHYTA, CHLOROPHYTA, CHRYSOPHYTA,	CHLORELLA, ROTIFERS, CHLAMYDOMONA	W70-06655
OSS, HYDROGEN ION CONCENTRATION,	CHLORELLA, SCENEDESMUS, WASTE WAT	W71-00138
ONDITIONS, ANAEROBIC CONDITIONS,	CHLORELLA, CHLAMYDOMONAS, SCENEDE	W70-10405
SSIFICATION, *OXIDATION LAGOONS,	CHLORELLA, OXYGEN DEMAND, PHOTOSY	W71-00139
ICAL TECHNIQUES, *ALGAE, *ASSAY,	CHLORELLA, NITRATES, CHEMICAL ANA	W70-10403
RM WASTES, OKLAHOMA, CYANOPHYTA,	CHLORELLA, IONS, PHYTOPLANKTON, N	W72-04773
BON, LAKE ERIE, RUNOFF, EROSION,	CHLORELLA, SODIUM COMPOUNDS, ALKA	W72-04734
NOPHYTA, YEASTS, CARBON DIOXIDE,	CHLORELLA, CLOSTRIDIUM, E. COLI,	W72-06124
HICATION, OLIGOTROPHY, PROTOZOA,	CHLORELLA, CHLAMYDOMONAS, CYANOPH	W72-05968
E RESIDUES, *ALGAE, CHLOROPHYTA,	CHLORELLA, TOXICITY, WATER ANALYS	W72-04134
, METABOLISM, HYDROGEN, EUGLENA,	CHLORELLA, SCENEDESMUS, CHLAMYDOM	W71-09172
ECOLOGY, RHODOPHYTA, PHAEOPHYTA,	CHLORELLA, PLANT PATHOLOGY, INDUS	W71-09157
OLUMNS, BOTTOM SEDIMENTS, ALGAE,	CHLORELLA, BENTHIC FLORA, MUD, MI	W71-08042
EMICAL TESTS, OUABAIN, POTASSIUM	CHLORIDE.: /TRIPHOSPHATASE, BIOCH	W72-13809
RANCE, WATER POLL/ *DDT, *SODIUM	CHLORIDE, *INHIBITION, *SALT TOLE	W72-13809
S, MET/ *PHOTOSYNTHESIS, CALCIUM	CHLORIDE, RADIOACTIVITY TECHNIQUE	W72-12634
ALUM, FERRIC	CHLORIDE, *JERUSALEM(ISRAEL).:	W72-12289
, CHLORELLA PYRENOIDOSA, LITHIUM	CHLORIDE, LITHIUM, ASTERIONELLA F	W72-12739
E MEDIA, ZINC SULFATE, MANGANESE	CHLORIDE, NICKEL SULFATE, COBALT	W72-12739
*MERCURY, *SYNTHESIS, MERCURIC	CHLORIDE, METHYL MERCURIC CHLORID	W72-06037
RCURIC CHLORIDE, METHYL MERCURIC	CHLORIDE, ANKISTRODESMUS BRAUNII,	W72-06037
THRACITE, *WAHNBACH RES/ *FERRIC	CHLORIDE, MULTI-LAYER FILTERS, AN	W72-06201
TRATE, CHEMICAL ANALYSIS, SODIUM	CHLORIDE, COLUMBIA RIVER, ORGANIC	W72-05421
*LIMITING NUTRIENTS, POTASSIUM	CHLORIDE, SHRINER'S POND(GA).:	W72-09159
S, EFFLUENTS, DETERGENTS, SODIUM	CHLORIDE, CHEMICAL PROPERTIES, SE	W72-09167
PHENYLMERCURIC ACETATE, MERCURY	CHLORIDE, PHENYLACETATE, MERCURY	W72-11727
LIZA/ *ALUMINUM SULFATE, *FERRIC	CHLORIDE, *FERRIC SULFATE, *STABI	W71-13356
ROCHLORIDE, MERI/ *CHLOROCHOLINE	CHLORIDE, *CCC, *DIETHYLAMINE HYD	W72-14706
TOCEROS GALVESTONENSIS, MERCURIC	CHLORIDE, DIMETHYL MERCURY, FLUOR	W72-07660
IMPOUNDMENTS, DISSOLVED OXYGEN,	CHLORIDES, INVERTEBRATES, HA: /E,	W72-14280

MPERATURE, PHOSPHORUS, NITRATES,	CHLORIDES, CALCIUM, MAGNESIUM, SO	W71-11001
, POTASSIUM, CALCIUM, MAGNESIUM,	CHLORIDES, HYDROGEN ION CONCENTRA	W72-05610
NITRATES, SULFATES, PHOSPHATES,	CHLORIDES, BORATES, ABSORBANCE.: /	W72-06274
C CONDITIONS, WATER TEMPERATURE,	CHLORIDES, PHOTOSYNTHESIS, LITTOR	W72-12940
SCILLATORIA TENUIS, OSCILLATORIA	CHLORINA, OSCILLATORI: /AMOENA, O	W72-07896
REEFS, ANTIBIOTICS(PESTICIDES),	CHLORINATED HYDROCARBON PESTICIDE	W72-14294
ONAS REINHARDTII, CULTURE MEDIA,	CHLORINATED HYDROCARBONS.: /MYDOM	W73-00281
HLOROPHYLL, OXYGEN, RESPIRATION,	CHLORINATED HYDROCARBON PESTICIDE	W73-00281
ALGAE, *INDICATORS, *PESTICIDES,	CHLORINATED HYDROCARBON PESTICIDE	W73-00238
S, ORGANOPHOSPHOROUS PESTICIDES,	CHLORINATED HYDROCARBON PESTICIDE	W72-13347
URAL RUNOFF, ALGAE, FOOD CHAINS,	CHLORINATED HYDROCARBON PESTICIDE	W72-13657
ERANCE, WATER POLLUTION EFFECTS,	CHLORINATED HYDROCARBON PESTICIDE	W72-13809
POLLUTANTS, PESTICIDE RESIDUES,	CHLORINATED HYDROCARBON PESTICIDE	W73-02280
, *ALGAE, *PRIMARY PRODUCTIVITY,	CHLORINATED HYDROCARBON PESTICIDE	W71-10096
FILTRATION, ANAEROBIC DIGESTION,	CHLORINATION, DISINFECTION, ULTRA	W72-00042
MICAL WASTES, INDUSTRIAL WASTES,	CHLORINATION, PHOSPHATES, NITRATE	W72-05315
ION CONCENTRATION, TEMPERATURE,	CHLORINATION, FLOATATION, DEWATER	W71-08951
ATION, NUTRIENTS, SEDIMENTATION,	CHLORINATION, SLOPES, BIOCHEMICAL	W71-07113
PHOTOSYNTHESIS, ALGAE, BACTERIA,	CHLORINATION, GROUNDWATER, COST A	W71-07123
TRIENTS, TEMPERATURE, COLIFORMS,	CHLORINATION, SEEPAGE, PERCOLATIO	W71-07124
CARBON, ADSORPTION, FILTRATION,	CHLORINATION, CHEMICAL PRECIPITAT	W70-10381
ONS, *DISINFECTION, DEGRADATION,	CHLORINE, CONTACT TIME, SAMPLING,	W71-07096
MENT, CHEMICALS, COPPER SULFATE,	CHLORINE, WATER RESOURCES, METHOD	W71-05083
ANIMALS, BIOLOGICAL COMMUNITIES,	CHLORINE, MOLLUSKS, COPEPODS, DIP	W73-00855
E, PLANKTON, CHEMICALS, PHENOLS,	CHLORINE, OXIDATION, ADSORPTION,	W73-01669
NE, EURYHALIN/ *SINAI PENINSULA,	CHLORINITY, METAHALINE, HYPERHALI	W73-00855
MONAS GELATINOSA, THIORHODACEAE,	CHLOROBACTERIACEAE.: /RHODOPSEUDO	W72-04743
RA, EPHEMEROPTERA, NANNOCHLORIS,	CHLOROCOCCUM, MICROCYSTIS, FUSAR	W70-06655
ARTERIA EUGAMETOS, STICHOCOCCUS,	CHLOROCOCCUM, HYALOTHECA, LAKE LI	W72-09239
RIA, SPIRULI/ *PROTEASE ENZYMES,	CHLOROGONIUM, OOCYSTIS, OSCILLATO	W72-07661
TURES, NUTRIENTS, RADIOISOTOPES,	CHLOROPHYLL, PIGMENT, MAGNESIUM,	W72-07660
THIAMINE, BIOTI/ ORGANIC CARBON,	CHLOROPHYLL A, ORGANIC NITROGEN,	W72-07892
URE, PHYTOPLANKTON, ZOOPLANKTON,	CHLOROPHYLL, SOLAR RADIATION, VIT	W72-07892
UCCESSION, CHLOROPHYTA, DIATOMS,	CHLOROPHYLL, ALKALINITY, CARBON D	W72-08049
S, *SUCCESSION, *EUTROPHICATION,	CHLOROPHYLL, PLANKTON, BIOLOGICAL	W73-00916
RATIC, PIGMENT DIVERSITY INDEX,	CHLOROPHYLL A, CARTENOIDS, SAMPLE	W73-00916
ABOLISM, CHLORELLA, RESPIRATION,	CHLOROPHYLL, LABORATORY TESTS, IR	W73-01080
OLA.:	CHLOROPHYLL A, SCHIZOTHRIX CALCIC	W73-01080
TRIS, COUNTING, ORTHOPHOSPHATES,	CHLOROPHYLL A.: /RHODOMONAS LACUS	W73-01435
HENICOL, ACTIDON, CULTURE MEDIA,	CHLOROPHYLL A, TERATOLOGY, NITZSC	W73-01445
NITROGEN, CHRYSOPHYTA, NITRATES,	CHLOROPHYLL, PHO: /HT INTENSITY,	W73-01435
DIVERSITY, MELOSIRA NUMMULOIDES,	CHLOROPHYLL A, *YAQUINA BAY, AMPH	W73-00853
LA, *LIGHT, ALGAE, GROWTH RATES,	CHLOROPHYLL.: /TABOLISM, *CHLOREL	W73-00251
, PHOTOSYNTHESIS, AQUATIC ALGAE,	CHLOROPHYLL, OXYGEN, RESPIRATION,	W73-00281
TES, FORAMINIFERA,/ *PYCNOCLINE,	CHLOROPHYLL A, PHEOPHYTIN, TUNICA	W72-14313
BSORPTION, MATHEMATICAL STUDIES,	CHLOROPHYLL, OXIDATION REDUCTION	W72-14800
OTENOIDS, FRAGILARIA CROTONENSI/	CHLOROPHYLL A, CHLOROPHYLL C, CAR	W73-00384
ISSIMA, NITZSCHIA PUNGENS, C-14,	CHLOROPHYLL A, MEDITERRANEAN SEA,	W73-00431
LARIA CROTONENSI/ CHLOROPHYLL A,	CHLOROPHYLL C, CAROTENOIDS, FRAGI	W73-00384
ES, N-15, LUCIFERIN, LUCIFERASE,	CHLOROPHYLL A, OPTICAL DENSITY, A	W72-10883
PULATION, TEMPORAL DISTRIBUTION,	CHLOROPHYLL, SAMPLING, MATHEMATIC	W72-11719
TY CONTROL, *BIOASSAY, *BIOMASS,	CHLOROPHYLL, NITROGEN FIXATION, *	W72-10883
NALYSIS, PRODUCTIVITY, SEASONAL,	CHLOROPHYLL, ALGAE, PHYTOPLANKTON	W72-09108
ATION, RADIOACTIVITY TECHNIQUES,	CHLOROPHYLL, SEA WATER, IONS, NUT	W72-09103
CALIFORNIA, SEDIMENTS, SAMPLING,	CHLOROPHYLL, SEDIMENT TRANSPORT,	W72-09092
, CULTURES, *FLORIDA, SEA WATER,	CHLOROPHYLL, NITRATES, NITRITES,	W72-09102
E(ENGLAND), EXOENZYME PRODUCERS,	CHLOROPHYLL A.: /, LAKE WINDERMER	W72-03224
YSERS, PHOTOSYNTHESIS, CULTURES,	CHLOROPHYLL, PROTEINS, ALGAE, GRO	W72-03557
S, MEASUREMENT, CURRENTS(WATER),	CHLOROPHYLL, FLUORESCENCE, BATHYM	W72-01095
TS, *INDICATORS, CHROMATOGRAPHY,	CHLOROPHYLL, TROPHIC LEVEL, EUTRO	W72-01784
ICATION, BIOMASS, WATER QUALITY,	CHLOROPHYLL, NITROGEN, WISCONSIN,	W71-10645
ER TEMPERATURE, WATER CHEMISTRY,	CHLOROPHYLL, HYPSOMETRIC ANALYSIS	W71-11949
, METHODOLOGY, SEDIMENTS, LAKES,	CHLOROPHYLL, HYDROGEN ION CONCENT	W71-12855
L METHODS, NITROGEN, PHOSPHORUS,	CHLOROPHYLL, DISTRIBUTION, CARBON	W71-11029
N FIXATION, ALGAE, CONDUCTIVITY,	CHLOROPHYLL, PIGMENTS, DISPERSION	W72-04773
IRA, CHLORELLA ELLIPSOIDEA, RNA,	CHLOROPHYLL B, ASHING, LABORATORY	W72-06274
OSA, SPIROGIRA, CHLORELLA / DNA,	CHLOROPHYLL A, CHLORELLA PYRENOID	W72-06274
ABORATORY TESTS, PHOTOSYNTHESIS,	CHLOROPHYLL, ENZYMES, PERMEABILIT	W72-06037
ASKA, *EUTROPHICATION, PLANKTON,	CHLOROPHYLL, RECREATION, RUNOFF,	W72-04270
, RIVERS, IOWA, CORN, DISCHARGE,	CHLOROPHYLL.: /ION, WATER QUALITY	W71-06445
SONAL, DIATOMS, DINOFLAGELLATES,	CHLOROPHYLL, PHOTOSYNTHESIS, TEMP	W71-07875
IFERS, PROTOZOA, PHOTOSYNTHESIS,	CHLOROPHYLL, TEMPERATURE, HYDROGE	W71-07382
RIENTS, MEASUREMENT, POPULATION,	CHLOROPHYLL, WATER POLLUTION EFFE	W71-08670
OPHYTA, OCHROMONAS, RESPIRATION,	CHLOROPHYLL, CARBON RADIOISOTOPES	W71-09172
CLEARWATER QUARRY, DOLLAR LAKE,	CHLOROPHYLL A, ABSORBANCE.: /SON,	W73-13673
HOSPHORUS, ANAEROBIC CONDITIONS,	CHLOROPHYLL, EPILIMNION, SEDIMENT	W72-12997
NESCENCE, BIOLOGICAL PROPERTIES,	CHLOROPHYLL, TURBIDITY, SAMPLING,	W72-12941
DT, PHYTOPLANKTON, PLANT GROWTH,	CHLOROPHYLL, DDE, INHIBITION.: /D	W72-12998
*LIGHT QUALITY, EUTROPHIC ZONE,	CHLOROPHYLL, PHOSPHATES, NITRATES	W72-12393
AKE, PINE LAKE, LAKE WASHINGTON,	CHLOROPHYLL A, *DNA, *DISSOLVED O	W72-12637
DISTILLATION, SPECTROPHOTOMETRY,	CHLOROPHYLL, HYDROLYSIS, PHYTOPLA	W72-12637
, *LIMITING FACTORS, PHOSPHATES,	CHLOROPHYTA, CYANOPHYTA, NITROGEN	W72-12544
IES, PHYTOPLANKTON, ZOOPLANKTON,	CHLOROPHYTA, DIATOMS, CYANOPHYTA,	W72-12729
ATICS, *CHLAMYDOMONAS, CULTURES,	CHLOROPHYTA, PROTOZOA, *ALGAE.: /	W72-12632
CULTURES, PHOSPHATES, NUTRIENTS,	CHLOROPHYTA, *ORGANOPHOSPHORUS CO	W72-12637
ALGAE, ZOOPLANKTON, CHRYSOPHYTA,	CHLOROPHYTA, BIOMASS, PYRROPHYTA,	W72-12738
ATES, *CULTURES, *AQUATIC ALGAE,	CHLOROPHYTA, INHIBITORS, WATER PO	W72-12736
BACTERIA, AQUATIC ALGAE, ALGAE,	CHLOROPHYTA, PSEUDOMONAS, ENTERIC	W72-12255
ATION, CLASSIFICATION, SAMPLING,	CHLOROPHYTA, *CHRYSOPHYTA, CYANOP	W72-12192
T-WATER INTERFACES, CHRYSOPHYTA,	CHLOROPHYTA, CYANOPHYTA, BIOCHEMI	W72-12995

IDITY, NUTRIENTS, HETEROGENEITY, CHLOROPHYTA, CHRYSOPHYTA, BIOINDI W72-13142
OUNDS, POLLUTANT IDENTIFICATION, CHLOROPHYTA, CHLORELLA.: /UM COMP W72-13813
IC ALGAE, ENVIRONMENTAL EFFECTS, CHLOROPHYTA, PRIMARY PRODUCTIVITY W72-13816
LAKES, PLANT PIGMENTS, DETRITUS, CHLOROPHYTA, CHRYSOPHYTA, SOLVENT W72-13673
TNOPHICATION, ALGAE, CYANOPHYTA, CHLOROPHYTA, METABOLISM, NEKTON, W72-13636
T GROWTH, TEMPERATURE, PROTOZOA, CHLOROPHYTA, ABSORPTION.: /, PLAN W71-09254
SMUS, CHLAMYDOMONAS, CYANOPHYTA, CHLOROPHYTA, OXYGEN, LIGHT, CHRYS W71-09172
ID MECHANICS, EXUDATION, OCEANS, CHLOROPHYTA, PHAEOPHYTA, RHODOPHY W71-07878
HEAVY METALS, *TOXICITY, COPPER, CHLOROPHYTA, WATER POLLUTION, NUT W71-05991
YTA, CYANOPHYTA, / *ALGAE, OHIO, CHLOROPHYTA, CHRYSOPHYTA, RHODOPH W71-05629
PATHOGENIC BACTERIA, CYANOPHYTA, CHLOROPHYTA, SCENEDESMUS, ALGAL T W71-05155
ES, *PESTICIDE RESIDUES, *ALGAE, CHLOROPHYTA, CHLORELLA, TOXICITY, W72-04134
CONTROL, WASTE WATER TREATMENT, CHLOROPHYTA, GROWTH RATES.: /LITY W72-04431
ELLA, CHLAMYDOMONAS, CYANOPHYTA, CHLOROPHYTA, CHRYSOPHYTA, DIATOMS W72-05968
TRIENTS, WATER QUALITY, DIATOMS, CHLOROPHYTA, THERMAL WATER, *MONT W72-05610
NERGY CONVERSION, STANDING CROP, CHLOROPHYTA, CYANOPHYTA, RHODOPHY W72-06283
E. COLI, EUGLENA, EUGLENOPHYTA, CHLOROPHYTA, FERROBACILLUS, GAS C W72-06124
ATION, CARBON DIOXIDE, NITROGEN, CHLOROPHYTA, NITROGEN FIXATION, L W72-04769
E ALGAE, RHODOPHYTA, PHAEOPHYTA, CHLOROPHYTA, INHIBITORS, ANTIBIOT W72-04748
Y IMPOUNDMENT, BIOMASS, DIATOMS, CHLOROPHYTA, CYANOPHYTA, PYRROPHY W72-04855
ALGAE, WATER POLLUTION SOURCES, CHLOROPHYTA, CYANOPHYTA, WASHINGT W72-05473
CYTOLOGICAL STUDIES, CYANOPHYTA, CHLOROPHYTA, CHRYSOPHYTA, RHODOPH W70-07280
BSONFLIES, MAYFLIES, CYANOPHYTA, CHLOROPHYTA, CHRYSOPHYTA, CHLOREL W70-06655
*ALGAE, *NUTRIENT REQUIREMENTS, CHLOROPHYTA, CYANOPHYTA, CYCLING W70-05469
LA, NITRATES, CHEMICAL ANALYSIS, CHLOROPHYTA, DIATOMS, TEMPERATURE W70-10403
DEPTH, CYANOPHYTA, CHRYSOPHYTA, CHLOROPHYTA, PRIMARY PRODUCTIVITY W71-03014
OPLANKTON, DIATOMS, CHRYSOPHYTA, CHLOROPHYTA, *GROWTH RATES, INHIB W70-07842
IC FLORA, LAKES, DEPTH, DIATOMS, CHLOROPHYTA, CYANOPHYTA, BIOMASS, W71-11029
LLUSKS, CYANOPHYTA, CHRYSOPHYTA, CHLOROPHYTA, PYRROPHYTA, INVERTEB W71-11012
NTRATION, CYANOPHYTA, CHLORELLA, CHLOROPHYTA, SCENEDESMUS, EUGLENA W71-11519
C FLORA, RHODOPHYTA, PHAEOPHYTA, CHLOROPHYTA, MAINE, SPECIATION.: / W71-10066
LLUTION SOURCES, TOXINS, WASTES, CHLOROPHYTA, SESSILE ALGAE, WATER W71-10553
GANESE, COPPER, ALGAE, SAMPLING, CHLOROPHYTA.: /TASSIUM, IRON, MAN W71-11001
, PLANKTON, WASHINGTON, DIATOMS, CHLOROPHYTA.: /A, *NUISANCE ALGAE W72-00798
WATER PROPERTIES, PHYTOPLANKTON, CHLOROPHYTA, PHAEOPHYTA, RHODOPHY W72-03220
RAINFALL, DIURNAL, FLUCTUATIONS, CHLOROPHYTA, EUGLENOPHYTA, CYANOP W72-00150
ROTOZOA, EUGLENA, CHLAMYDOMONAS, CHLOROPHYTA, SCENEDESMUS, CYANOPH W72-00155
ALKALINITY, DIATOMS, CYANOPHYTA, CHLOROPHYTA, SEASONAL, NUTRIENTS, W72-08560
, SATELLITES(ARTIFICIAL), ALGAE, CHLOROPHYTA, PHAEOPHYTA, RHODOPHY W72-09120
*PROTOZOA, *SYSTEMATICS, *ALGAE, CHLOROPHYTA, HABITATS, AQUATIC HA W72-09119
LGAE, EUGLENOPHYTA, CHRYSOPHYTA, CHLOROPHYTA, PYRROPHYTA, CYANOPHY W72-09111
ERATURE, RHODOPHYTA, PHAEOPHYTA, CHLOROPHYTA, COOLING WATER, HYDRO W72-11252
URES, *CHLORELLA, AQUATIC ALGAE, CHLOROPHYTA, HEAVY METALS, *CHLAM W72-11727
RIMARY PRODUCTIVITY, MICROSCOPY, CHLOROPHYTA, SCENEDESMUS, *CYANOP W72-09239
MATTER, WATER POLLUTION SOURCES, CHLOROPHYTA, AQUATIC PRODUCTIVITY W73-00379
HYTOPLANKTON, *RIVERS, *DIATOMS, CHLOROPHYTA, BENTHIC FLORA, EUTRO W72-14797
LEAD, COPPER, CHROMIUM, SILVER, CHLOROPHYTA, CYANOPHYTA, ALGAL CO W72-14694
TORY STUDIES, COPEPODS, DIATOMS, CHLOROPHYTA, TEMPERATURE, SIZE, F W72-14805
FLUENTS, REGULATION, CYANOPHYTA, CHLOROPHYTA, HYDROGEN ION CONCENT W72-14790
STING PROCEDURES, AQUATIC ALGAE, CHLOROPHYTA, LABORATORY TESTS, OX W72-14314
S, PHYTOPLANKTON, CHLAMYDOMONAS, CHLOROPHYTA, WATER POLLUTION EFFE W73-00281
FICATION, PROTOZOA, CHRYSOPHYTA, CHLOROPHYTA, AQUATIC ALGAE, MARIN W73-00273
BUTION, SURFACE WATERS, LAGOONS, CHLOROPHYTA, RHODOPHYTA, PHYSICAL W73-00838
L ANALYSIS, PHYSICAL PROPERTIES, CHLOROPHYTA, EUGLENOPHYTA, CYANOP W73-01627
CAL DISTRIBUTION, PLANKTON NETS, CHLOROPHYTA, PLANKTON.: / ECOLOGI W73-01094
, LABORATORY TESTS, IRRADIATION, CHLOROPHYTA, INHIBITION, GAMMA RA W73-01080
TAL EFFECTS, AEROBIC CONDITIONS, CHLOROPHYTA, SCENEDESMUS, AQUATIC W73-01066
PHOSPHORUS COMPOUNDS, CYANOPYTA, CHLOROPHYTA.: /LANKTON, *ALGAE, * W73-00938
APTATION, ENVIRONMENTAL EFFECTS, CHLOROPHYTA, MARINE ALGAE , SALT W73-00855
, AQUATIC ALGAE, MICROORGANISMS, CHLOROPHYTA, BIOLOGY, POLLUTANT I W73-02099
G, BOTTOM SEDIMENTS, SUCCESSION, CHLOROPHYTA, DIATOMS, CHLOROPHYLL W72-08049
SIS, EUTROPHICATION, PHOSPHORUS, CHLOROPHYTA, CYANOPHYTA, CHRYSOPH W72-08048
ICITY, *MARINE ALGAE, SEA WATER, CHLOROPHYTA, CHLAMYDOMONAS, LABOR W72-08051
TRIENT REQUIREMENTS, CYANOPHYTA, CHLOROPHYTA, PHYTOPLANKTON, GRAZI W72-07940
TA, DINOFLAGELLATES, CYANOPHYTA, CHLOROPHYTA, EUGLENOPHYTA, GRAZIN W72-07899
I/ *ALGAL TOXINS, *MYXOBACTERIA, CHLOROPHYTA, CYANOPHYTA, SYSTEMAT W72-07951
NDITIONS, WASTE WATER TREATMENT, CHLOROPHYTA, CYANOPHYTA, AEROBIC W72-07664
NTENSITY, *PHOSPHORUS, PLANKTON, CHLOROPHYTA, PYRROPHYTA, DINOFLAG W72-07166
CULTURE MEDIA, HUMMOCK CHANNEL, CHLOROPLASTS, MORPHOLOGY, EPIPHYT W72-12633
UNII,/ *SIMAZINE, *DETOXICATION, CHLOROSARCINA, ANKISTRODESMUS BRA W71-07675
RBON, SEDIMENTS, CARBON DIOXIDE, CHLORPHYLL, DO: /RATIFICATION, CA W72-08560
USCIFORMIS, CERAMIUM BUSSOIDEUM, CHONDRIA LI: /LAVULATUM, HYPENA M W73-00838
CEA, PENI/ SURVIVAL, PENICILLUS, CHONDRIA, CLADOPHOROPSIS MEMBRANA W73-00838
ALL ISLAND, POLYSIPHONIA LANOSA, CHONDRUS CRISPUS, COD, FLOUNDER, W72-08142
EPIPHYTES, CANDIDA PARAPSILOSIS, CHONDRUS CRISPUS, RHODOTORULA GLU W72-04748
SEA OTTER, MACROCYSTIS PYRIFERA, CHORDATES, POLYCHAETES: /NEREIS, W73-00932
POLLUTANTS, *ABSORPTION, ALGAE, CHORELLA, TRACERS, PHOSPHATES, EC W72-09668
ED CLAY RIVER, WHITE CLAY RIVER, CHRISTINA RIVER, BIG ELK RIVER, L W72-04736
YNGBYA LIMNETICA, ANABAENA SPP., CHROMATIUM, MERISMOPEDIA TROLLERI W72-07145
CROMONAS MO/ AMOEBOBACTER ROSEUS CHROMATIUM OKENII, CHROMATIUM, MA W72-07896
BACTER ROSEUS CHROMATIUM OKENII, CHROMATIUM, MACROMONAS MOBILIS, T W72-07896
SYNTHESIS, CARBON RADIOISOTOPES, CHROMATOGRAPHY, AMINO ACIDS, GAS W72-07710
CHROMATOGRAPHY, AMINO ACIDS, GAS CHROMATOGRAPHY, PHAEOPHYTA, EUTRO W72-07710
ATTER, FATE OF POLLUTANTS, PAPER CHROMATOGRAPHY, INFRARED SPECTROS W73-01066
OGICAL MAGNIFICATION, GAS LIQUID CHROMATOGRAPHY, ELECTRON CAPTURE W73-02105
CHAINS, CULTURES, FISH DIET, GAS CHROMATOGRAPHY, PARTICULATE MATTE W73-02105
ICATION, / *ELECTRON CAPTURE GAS CHROMATOGRAPHY *BIOLOGICAL MAGNIF W73-02105
PECTROSCOPY, CHROMATOGRAPHY, GAS CHROMATOGRAPHY.: /KTON, PLANTS, S W73-01439
PLANKTON, PLANTS, SPECTROSCOPY, CHROMATOGRAPHY, GAS CHROMATOGRAPH W73-01439
TROGEN, REDUCTION(CHEMICAL), GAS CHROMATOGRAPHY, SEPARATION TECHNI W73-01456

QUADRAPLICATUM, C-14, THIN LAYER CHROMATOGRAPHY, CLEANUP, RADIOAUT W73-00361
NIC CARBON, CULTURE MEDIA, PAPER CHROMATOGRAPHY, INFRARED SPECTROS W73-00379
ACE ELEMENTS, RADIO/ *CHELATION, CHROMATOGRAPHY, HEAVY METALS, *TR W72-04708
CHLOROPHYTA, FERROBACILLUS, GAS CHROMATOGRAPHY, IONS,: /ENOPHYTA, W72-06124
BIOASSAY, DISTRIBUTION PATTERNS, CHROMATOGRAPHY, GROWTH RATES, POL W72-04134
OPHEN A 50, ELECTRON CAPTURE GAS CHROMATOGRAPHY, POLYCHLORINATED B W72-04134
LYTICAL TECHNIQUES, BIOMASS, GAS CHROMATOGRAPHY, CENTRIFUGATION, P W72-04134
ESTUARIES, CHEMICAL WASTES, GAS CHROMATOGRAPHY, SEDIMENTS, OYSTER W71-07412
CID, BETA-SITOSTEROL, THIN-LAYER CHROMATOGRAPHY.: /PIN, PALMITIC A W71-08597
, MARINE ALGAE, ALGAL POISONING, CHROMATOGRAPHY, CHEMICAL ANALYSIS W71-08597
TERS,/ *FATTY ACIDS, *GAS LIQUID CHROMATOGRAPHY, METHYL ESTERS, ES W72-09365
QUES, SEPARATION TECHNIQUES, GAS CHROMATOGRAPHY, SEPCTROSCOPY, ALG W72-11604
*LAKES, *SEDIMENTS, *INDICATORS, CHROMATOGRAPHY, CHLOROPHYLL, TROP W72-01784
ES, *BOSSIELLA ORBIGNIANA, PAPER CHROMATOGRAPHY, SCINTILLATION COU W72-12634
LA, VIRUSES, CHROMATOGRAPHY, GAS CHROMATOGRAPHY, ANALYTICAL TECHNI W72-13800
MERCURIAL PESTICIDES, THIN LAYER CHROMATOGRAPHY, GROWTH INHIBITION W72-13347
SES, E. COLI, SHIGELLA, VIRUSES, CHROMATOGRAPHY, GAS CHROMATOGRAPH W72-13800
AY, WATER POLLUTION EFFECTS, GAS CHROMATOGRAPHY, BROOK TROUT, SAMP W72-12930
, LAMI/ *HUMIC COMPOUNDS, COLUMN CHROMATOGRAPHY, FUCUS VESICULOSUS W72-12166
IE, ALUMINUM, MANGANESE, SODIUM, CHROMIUM, IRON.: /CTERIA, LAKE ER W72-01101
Y METALS, BIOINDICATORS, SILICA, CHROMIUM, COPPER, *NEW YORK, DIAT W72-09111
WASTES, ALGICIDES, BACTERICIDES, CHROMIUM, ZINC, TEMPERATURE, HYDR W72-08683
CONCENTRATION, ABUNDANCE, IRON, CHROMIUM, COPPER, NICKEL, RUTHENI W71-13243
, MAGNESIUM, MANAGNESE, CALCIUM, CHROMIUM, POTASS: /, IRON, COPPER W71-09188
ON, FLOCCULATION, ALGAE, SLUDGE, CHROMIUM, TOXICITY, NEUTRALIZATIO W71-09524
IDENTIFICATION, COPPER, CADMIUM, CHROMIUM, NICKEL,: /E, POLLUTANT W72-14303
ION EFFECTS, ZINC, LEAD, COPPER, CHROMIUM, SILVER, CHLOROPHYTA, CY W72-14694
CE ELEMENTS, NUTRIENTS, MERCURY, CHROMIUM, COBALT, FRESH WATER, LA W73-01434
, ZINC RADIOISOTOPES, MANGANESE, CHROMIUM, IRON, ANALYTICAL TECHNI W73-00832
WATER, TRACE ELEMENTS, TOXICITY, CHROMOSOMES, FUNGICIDES, PESTICID W72-05952
ATH OF POLLUTANTS, MARINE ALGAE, CHROMOSOMES, EMBRYONIC GROWTH STA W72-10955
CILLATORIA, RADON RADIOISOTOPES, CHROMULINA, CRYPTOMONAS, MALLOMON W72-07648
KINIA SUBEQUICILIAT/ ENRICHMENT, CHROMULINA, CRYPTOMONAS EROSA, ER W72-13816
ULSILLA, MICROCYSTIS AERUGINOSA, CHROOMONAS CAUDATA, CHLAMYDOMONAS W72-13816
SCHII, APHANIZOMENON FLOS-AQUAE, CHROOMONAS ACUTA, MELOSIRA GRANUL W72-12738
PHYTOPLANKTON(PRYMNESIUM PARVUM)(CHROOMONAS SALINA), GLYCEROL POLL W70-07842
MNESIUM PARVUM, PHOTOAUTOTROPHS, CHROOMONAS SALINA.: /OPHYCEA, PRY W70-07280
THUS) HUXLEYI, AMPHIPRORA ALATA, CHRYSOCHROMULINA, SKELETONEMA, CH W73-00380
NODISCUS ASTREA, ANKISTRODESMUS, CHRYSOCOCCUS MINUTUS, DICTYOSPHAE W72-12738
LINE, SEDIMENT-WATER INTERFACES, CHRYSOPHYTA, CHLOROPHYTA, CYANOPH W72-12995
NTS, HETEROGENEITY, CHLOROPHYTA, CHRYSOPHYTA, BIOINDICATORS, PHYSI W72-13142
PIGMENTS, DETRITUS, CHLOROPHYTA, CHRYSOPHYTA, SOLVENT EXTRACTIONS, W72-13673
LOROPHYTA, PRIMARY PRODUCTIVITY, CHRYSOPHYTA.: /MENTAL EFFECTS, CH W72-13816
ITY, AQUATIC ALGAE, ZOOPLANKTON, CHRYSOPHYTA, CHLOROPHYTA, BIOMASS W72-12738
ETABOLISM, ALGAE, AQUATIC ALGAE, CHRYSOPHYTA, PLANKTON, PHYTOPLANK W73-02280
NT IDENTIFICATION, MARINE ALGAE, CHRYSOPHYTA, AUTUMN, WINTER, SUMM W73-00442
IDENTIFICATION, *LAKE MICHIGAN, CHRYSOPHYTA, AQUATIC ALGAE, ELECT W73-00432
KES REGION, WISCONSIN, MICHIGAN, CHRYSOPHYTA, PLANT PIGMENTS, AQUA W73-00384
NITRATES, PHOSPHORUS, NUTRIENTS, CHRYSOPHYTA, COMPETITION, WATER P W73-00441
Y, METABOLISM, PLANT PHYSIOLOGY, CHRYSOPHYTA, CULTURES, SEA WATER, W73-00380
UNITIES, NUTRIENTS, SCENEDESMUS, CHRYSOPHYTA, SEASONAL, SILICA, DI W72-14797
L STUDIES, *SYSTEMATICS, *ALGAE, CHRYSOPHYTA, PHAEOPHYTA.: /LOGICA W72-14808
IGRATION, TIDES, LOW WATER MARK, CHRYSOPHYTA.: /ALGAE, SALINITY, M W72-14317
LUTANT IDENTIFICATION, PROTOZOA, CHRYSOPHYTA, CHLOROPHYTA, AQUATIC W73-00273
Y, DIATOMS, NUTRIENTS, BACTERIA, CHRYSOPHYTA, PHYTOPLANKTON, ZOOPL W72-07899
TERNS, AQUATIC ALGAE, LIMNOLOGY, CHRYSOPHYTA, EUGLENOPHYTA, PYRROP W72-07145
ARINE ALGAE, *OREGON, *SEASONAL, CHRYSOPHYTA, SYSTEMATICS, *ESTUAR W72-08141
PHORUS, CHLOROPHYTA, CYANOPHYTA, CHRYSOPHYTA, EUGLENOPHYTA, DISSOL W72-08048
CHLAMYDOMONAS, LABORATORY TESTS, CHRYSOPHYTA, OPTICAL PROPERTIES.: W72-08051
TRIBUTION, PRIMARY PRODUCTIVITY, CHRYSOPHYTA, BIOLOGICAL COMMUNITI W73-00853
EAMS, LIGHT INTENSITY, NITROGEN, CHRYSOPHYTA, NITRATES, CHLOROPHYL W73-01435
PHYTA, EUGLENOPHYTA, CYANOPHYTA, CHRYSOPHYTA, SCENEDESMUS, CHLAMYD W73-01627
OLLUTION EFFECTS, AQUATIC ALGAE, CHRYSOPHYTA, LABORATORY EQUIPMENT W73-01445
LYSIS, SEA WATER, ORGANIC ACIDS, CHRYSOPHYTA, INHIBITION, MARINE A W73-00942
TANDING CROPS, MOLLUSKS, LARVAE, CHRYSOPHYTA, CYANOPHYTA, PYRROPHY W73-00935
YFLIES, CYANOPHYTA, CHLOROPHYTA, CHRYSOPHYTA, CHLORELLA, ROTIFERS, W70-06655
TUDIES, CYANOPHYTA, CHLOROPHYTA, CHRYSOPHYTA, RHODOPHYTA, TESTING, W70-07280
*ALGAE, PHYTOPLANKTON, DIATOMS, CHRYSOPHYTA, CHLOROPHYTA, *GROWTH W70-07842
IA, SAMPLING, DEPTH, CYANOPHYTA, CHRYSOPHYTA, CHLOROPHYTA, PRIMARY W71-03014
OMONAS, CYANOPHYTA, CHLOROPHYTA, CHRYSOPHYTA, DIATOMS, DINOFLAGELL W72-05968
ROPHYTA, CYANOPHYTA, PYRROPHYTA, CHRYSOPHYTA, EUGLENOPHYTA, WATER W72-04855
YTA, CHLOROPHYTA, OXYGEN, LIGHT, CHRYSOPHYTA, OCHROMONAS, RESPIRAT W71-09172
TA, / *ALGAE, OHIO, CHLOROPHYTA, CHRYSOPHYTA, RHODOPHYTA, CYANOPHY W71-05629
*DIATOMS, *MICROSCOPY, CHRYSOPHYTA, AQUATIC ALGAE.: W72-11738
NUTRIENTS, ALGAE, EUGLENOPHYTA, CHRYSOPHYTA, CHLOROPHYTA, PYRROPH W72-09111
, LIGHT PENETRATION, CYANOPHYTA, CHRYSOPHYTA.: /TRATES, RESERVOIRS W71-13183
DDISFLIES, MOLLUSKS, CYANOPHYTA, CHRYSOPHYTA, CHLOROPHYTA, PYRROPH W71-11012
CADDISFLIES, GASTROPODS, SANDS, CHRYSOPHYTA.: / CARBON, SEASONAL, W71-11029
PHYTA, EUGLENOPHYTA, CYANOPHYTA, CHRYSOPHYTA, COPEPODS, CRUSTACEAN W72-00150
IONS, POLAR REGIONS, CYANOPHYTA, CHRYSOPHYTA, PYRROPHYTA, SAMPLING W72-03543
EXPANSUM, PORPHYRIDIUM GRISEUM, CHRYSOTRICHALAE.: /IS, APISTONEMA W71-04526
TINALIS, SEMOTILUS ATROMACULATU/ CHUBS, KINGFISHES, SALVELINUS FON W72-12930
ANIDES, COHO SALMON, SHEEPSHEAD, CHUBS, NORTHERN PIKE, GIZZARD SHA W72-14282
ROPTERA, MENIDIA, BRACHYDEUTERA, CHYDORUS SPHAERICUS, CHYDORUS, MO W72-07132
CHYDEUTERA, CHYDORUS SPHAERICUS, CHYDORUS, MONOSTYLA, LEPADELLA, T W72-07132
: RESEARCH, CHYTRIDIACEOUS FUNGI, HEAT SHOCK. W72-07225
RIA, TRICHODESMIUM, FLAGELLATES, CILIATES, NAUPLII, COPEPODIDS.: / W72-14313
*WINNISQUAM LAKE(N H), ANABAENA CIRCINALIS, BOSMINA LONGIROSTRIS, W71-03014
APHANIZOMENON FLOS-AQU/ ANABAENA CIRCINALIS, ANABAENA CYLINDRICA, W72-10613
*ARTIFICIAL CIRCULATION, *KEZAR LAKE(NH).: W72-09061

LGAL BLOOMS, ALGAL TOXINS, WATER	CIRCULATION, WATER POLLUTION CONT	W72-09304
OL, *DIVERSION, SEDIMENTS, WATER	CIRCULATION, ABSORPTION, NUTRIENT	W71-09581
H RATES, PLANT MORPHOLOGY, WATER	CIRCULATION.: /COMPETITION, GROWT	W72-04766
ERMAL POWERPLANTS, HEATED WATER,	CIRCULATION, INFRARED RADIATION,	W73-01704
*PHOSPHORUS	CIRCULATION, SMALL LAKES.:	W72-12543
ROGYRA, LEPOCINCLIS OVUM, PHACUS	CIRCUMFLEXUS, PHACUS ONYX, PHACUS	W73-00918
HAETA, OSTRACODS, HARPACTICOIDS,	CIRRIPEDIA, ACARINA.: /IDS, POLYC	W72-05432
UTROPHICATION, MUNICIPAL WASTES,	CITIES, PHOSPHATES, ILLINOIS, ECO	W72-03970
TRATES, *NEBRASKA, URBANIZATION,	CITIES, STORM RUNOFF, SEWAGE, WAS	W71-13466
SODIUM NITRILOTRIACETATE, SODIUM	CITRATE.: /E-FREE DETERGENTS, TRI	W71-06247
HOSPHORYLATI/ *GLUCOSE, *MALATE,	CITRIC ACID CYCLE, CYTOCHROMES, P	W72-04309
IPHONOPHORES, MEDUSAE, PILIDIUM,	CLADOCERA, LINGULA, ACTINOTROCHA,	W73-00935
LADOGRAMMA LINEATUS-EXCENTRICUS,	CLADOGRAMMA NODULIFER, DACTYLIOSO	W71-11527
PERUVIANUS, CLADOGRAMMA KOLBEI,	CLADOGRAMMA CRENULATUS, CLADOGRAM	W71-11527
KOLBEI, CLADOGRAMMA CRENULATUS,	CLADOGRAMMA LINEATUS-EXCENTRICUS,	W71-11527
ANENSIS, CHAETOCEROS PERUVIANUS,	CLADOGRAMMA KOLBEI, CLADOGRAMMA C	W71-11527
YLLOSPADEX SP., MICROCLADIA SP.,	CLADOPHORA SP., POLYSIPHONIA SP.,	W71-11851
TUBIFEX TUBIFEX, APHANIZOMENON,	CLADOPHORA.: /RILUS HOFFMEISTERI,	W71-11009
*UNITED KINGDOM, *FUTURE TRENDS,	CLADOPHORA, RESEARCH STRATEGY, TH	W71-13172
GOMONTIA POLYRHIZA, ULVA RIGIDA,	CLADOPHORA FULIGINOSA, BATOPHORA	W72-11800
ROPHYTA, PHAEOPHYTA, RHODOPHYTA,	CLADOPHORA.: / *CYANOPHYTA, *CHLO	W72-11800
IL ANALYSIS, SOIL CONTAMINATION,	CLADOPHORA, CHAR: /IOISOTOPES, SO	W73-01673
A AFFINIS, CESIUM RADIOISOTOPES,	CLADOPHORA SPP, PASPALUM DISTICHU	W73-01673
TROGEN, PHOSPHORUS, RAIN, LAKES,	CLADOPHORA, ANALYSIS, PHYTOPLANKT	W73-00232
LGAE, *AQUATIC ANGIOSPERM WEEDS,	CLADOPHORA.: *FILAMENTOUS A	W72-14692
ATOLLS, CULTURES, SPHAEROTILUS,	CLADOPHORA.: /DES), BACTERICIDES,	W72-08142
INC, *LEAD, STIGEOCLONIUM TENUE,	CLADOPHORA GLOMERATA, ULOTRICHALE	W71-05991
ONIUM, V/ *CUYAHOGA RIVER(OHIO),	CLADOPHORA, APHANOCHAETE, RHIZOCL	W71-05629
CUS SERRATUS, FUCUS VESICULOSUS,	CLADOPHORA GLOMERATA, HETEROCYPRI	W71-08026
DDD, DDE, TDE, BLACK-FLY LARVAE,	CLADOPHORA, CYPRINUS CARPIO, CATO	W70-08098
(FRANCE), MEURTHE RIVER(FRANCE),	CLADOPHORA, PHORMIDIUM, OSCILLATO	W71-00099
LAKE CAYUGA,	CLADOPHORA, ECOLOGICAL STUDIES.:	W70-09614
LITY, ECOLOGY, AQUATIC HABITATS,	CLADOPHORA, LIMITING FACTORS, ECO	W72-13142
ARBOR(WISCONSIN), POLYPHOSPHATE/	CLADOPHORA GLOMERATA, MILWAUKEE H	W72-13644
ICAL PROPERTIES, DIATOMS, ALGAE,	CLADOPHORA, PRODUCTIVITY, WATER P	W72-13662
SSEAE, DICTYOSPHAERIA CAVERNOSA,	CLADOPHOROPSIS MEMBRANACEA, BOODL	W72-11800
SURVIVAL, PENICILLUS, CHONDRIA,	CLADOPHOROPSIS MEMBRANACEA, PENIC	W73-00838
BEA, SCOPULARIOPSIS BREVICAULIS,	CLADOSPORIUM HERBARUM, TRAMETES V	W72-13813
OIT RIVER, BL/ MAUMEE RIVER, ST.	CLAIR RIVER, LAKE ST. CLAIR, DETR	W72-14282
RIVER, ST. CLAIR RIVER, LAKE ST.	CLAIR, DETROIT RIVER, BLACK RIVER	W72-14282
NEMONE, MOLGULA, TUNICATE, PISMO	CLAM, TIVELA STULTORUM, OSTREA LU	W72-07907
ECTS, CATTAILS, AQUATIC ANIMALS,	CLAMS, CARP, CRAYFISH, AQUATIC AL	W72-06342
S, FOOD CHAIN, HERRING, OYSTERS,	CLAMS, COLUMBIA RIVER, CALIFORNIA	W71-07417
TER), TIDES, WIND, LIGHT, WAVES,	CLAMS, CRUSTACEANS, WORMS, MOLLUS	W72-09092
TROMETERS, ZOOPLANKTON, MUSSELS,	CLAMS, FISH, SHELLFISH, AQUEOUS S	W72-05965
E, BULLHEADS, SALMON, SHELLFISH,	CLAMS, MARINE ANIMALS, AQUATIC LI	W72-07907
, SEWAGE DISPOSAL, MARINE ALGAE,	CLAMS, MUSSELS, OYSTERS, EUTROPHI	W72-07715
AMPLING, SNAILS, ALGAE, DIATOMS,	CLAMS, SHELLFISH, AMPHIPODA, WATE	W70-08098
LIGASAPROBIC STREAMS, FINGERNAIL	CLAMS, UNIONIDS, ORCONECTES RUSTI	W72-06526
ERIZATION, ETHYLENE OXIDE, WATER	CLARIFICATION, SLUDGE DRYING, POL	W70-06968
REVIEWS, MICROBIAL DEGRADATION,	CLASSIFICATION, SEWAGE, SOIL PROP	W72-14363
GEN SAG, AQUATIC LIFE, BIOASSAY,	CLASSIFICATION.: /LIGOTROPHY, OXY	W73-00002
S, TESTING PROCEDURES, CRITERIA,	CLASSIFICATIONS.: /, EFFECTIVENES	W71-09789
UTROPHICATION, PLANT POPULATION,	CLASSIFICATION, SAMPLING, CHLOROP	W72-12192
OPHYTA, *CULTURES, *SYSTEMATICS,	CLASSIFICATION, *AQUATIC ALGAE.: /	W72-12172
, SCENEDESMUS, AQUATIC HABITATS,	CLASSIFICATION, AQUATIC ALGAE.: /	W72-12744
UM, ELPHIDIUM ADVENUM, ELPHIDIUM	CLAVATUM, ELPHIDIUM GALVESTONENSE	W73-00367
LOTHRIX IRREGULARIS, CENTROCERAS	CLAVULATUM, HYPENA MUSCIFORMIS, C	W73-00838
GOMPHONEMA OLIVACEOIDES, WHITE	CLAY CREEK(PA.).:	W73-00242
INE RIVER, RED CLAY RIVER, WHITE	CLAY RIVER, CHRISTINA RIVER, BIG	W72-04736
NNA RIVER, BRANDYWINE RIVER, RED	CLAY RIVER, WHITE CLAY RIVER, CHR	W72-04736
OWAX, FLOCCULANT DOSAGES, BAROCO	CLAY SOILS, URANIUM ORE.: /, CARB	W70-06968
OCHEMISTRY, *WATER PURIFICATION,	CLAY, ALGAE, BACTERIA, WASTE WATE	W73-01362
STRAINING, AERATION, PHOSPHATES,	CLAY, TASTE, ODOR, *SURFACE WATER	W72-06536
SLUDGE, COST ANALYSIS, NITROGEN,	CLAYS, *WASTE WATER TREATMENT.: /	W72-11642
CHEMICAL PROPERTIES, *SEDIMENTS,	CLAYS, CORES, SEASONAL, ALGAE, MU	W72-05469
FALL, *SOIL TYPES, SOIL TEXTURE,	CLAYS, LABORATORY TESTS, STATISTI	W72-02218
OSPHORUS, ALGAE, AQUATIC PLANTS,	CLAYS, ODOR, WATER QUALITY, FOOD	W72-04759
ION, CHEMICALS, CHEMICAL WASTES,	CLEANING, DOMESTIC WASTES, EUTROP	W72-01685
TURBIDITY, ALGAE, FLOCCULATION,	CLEANING, *WATER TREATMENT, TERTI	W72-09393
C-14, THIN LAYER CHROMATOGRAPHY,	CLEANUP, RADIOAUTOGRAPHY, SUBSTRA	W73-00361
INALIS, SEMOTILUS ATROMACULATUS,	CLEANUP, SAMPLE PRESERVATION, CAM	W72-12930
, PSIDIUM, SHADFLY, CHIRONOMIDS,	CLEAR LAKE(CALIF), RICHELIEU RIVE	W70-08098
KE, LAKE ROCKWELL, LAKE HODGSON,	CLEARWATER QUARRY, DOLLAR LAKE, C	W72-13673
REGENERATION, *BACKWASHING, *MT	CLEMENS(MICH).: *CARBON	W72-08357
NDED SOLIDS, BACKWASHING, *MOUNT	CLEMENS(MICH).: /D SEWERS, *SUSPE	W72-00042
G, ALGAE, TEMPERATURE, ALTITUDE,	CLIMATES, METABOLISM, BIOCHEMICAL	W72-06838
EN FIXATION, *SOILS, *TEMPERATE,	CLIMATIC ZONES, CYANOPHYTA, SYMBI	W73-02471
AE, PROTEIN, *OXIDATION LAGOONS,	CLIMATIC ZONES, PRECIPITATION, CO	W71-08951
T, *ANAEROBIC CONDITIONS, PONDS,	CLIMATIC ZONES, TEMPERATURE, TURB	W71-06033
RIVER, BLACK RIVER, PINE RIVER,	CLINTON RIVER, HURON RIVER, BELL	W72-14282
OPHORA, *ACHNANTHES, *SPIROGYRA,	CLODOPHORA GLOMERATA, *METOLIUS R	W72-07901
DS, PERMANGANATE.: *FILTER	CLOGGING, IRON BACTERIA, POND WEE	W72-06536
*VALVE PATTERN CHARACTERISTICS,	CLONES.: /ISM, *VALVE MORPHOLOGY,	W73-02548
NOSACCHARIDES, OLIGOSACCHARIDES,	CLONES, PHAEOCYSTIS POUCHETI, GLU	W72-07710
PYRENOIDOSA, LINDANE, P P'-DDE,	CLOPHEN A 50, ELECTRON CAPTURE GA	W72-04134
WTH MEDIA, COCCONEIS PLACENTULA,	CLOSTERIUM, GOMPHONEMA LONGICEPS,	W72-12744
M, MICRASTERIAS CRUX-MELITENSIS,	CLOSTERIUM STRIGOSUM, CULTURE MED	W72-12736
DUCTION, SCHIZOTHRIX CALCIOCOLA,	CLOSTERIUM MONILIFERUM, MELOSIRA	W72-10883
ASTS, CARBON DIOXIDE, CHLORELLA,	CLOSTRIDIUM, E. COLI, EUGLENA, EU	W72-06124

E, CYTOCHROMES, PHOSPHORYLATION,	CLYCOLYSIS.: /E, CITRIC ACID CYCL	W72-04309
ETIFERUS, MERCENARIA MERCENARIA,	CLYMENELLA TORQUATA, CAPITELLA CA	W73-02029
ELLA, LOPHOPANOPEUS, EUPENTACTA,	CNEMIDOCARPA, ENHYDRA LUTRIS NERE	W73-00932
CHLORE/ *CHLORELLA SOROKINIANA,	CO-60, GAMMA RADIATION, SURVIVAL,	W72-11836
*ORGANIC NUTRIENTS, ALUM	COAGULATION.:	W73-02187
TEMENT, WATER POLLUTION CONTROL,	COAGULATION.: /ION, POLLUTION ABA	W73-02187
RBON, SLUDGE TREATMENT, DIATOMS,	COAGULATION, CALCIUM CARBONATE, H	W73-01669
C PLANTS, *MINNESOTA, *TOXICITY,	COAGULATION, WATER POLLUTION EFFE	W72-07360
E SULFONATES, SEWAGE, NUTRIENTS,	COAGULATION, SEDIMENTATION, TOXIC	W72-06837
, ALGAE, THERMAL STRATIFICATION,	COAGULATION, BENTONITE, LIME, *PU	W72-03753
, CLIMATIC ZONES, PRECIPITATION,	COAGULATION, LIME, HYDROGEN ION C	W71-08951
T, COSTS, SEPARATION TECHNIQUES,	COAGULATION, WASTE WATER TREATMEN	W70-08904
WAGE SLUDGE, FLOW, FLOCCULATION,	COAGULATION, FILTRATION, VISCOSIT	W70-06968
HNIQUES, CHEMICAL PRECIPITATION,	COAGULATION, SEDIMENTATION, POLYE	W72-10237
, PHOSPHORUS, NITRATES, AMMONIA,	COAGULATION, FLOCCULATION, SEDIME	W71-13341
REATMENT, ANALYTICAL TECHNIQUES,	COALESCENCE, FLOCCULATION, LIQUID	W71-13356
US VAR. NEAPOLITANA, CHAETOCEROS	COARCTATUS, CHAETOCEROS LORENZIAN	W71-11527
AR WASTES, *MONITORING, *PACIFIC	COAST REGION, TRITIUM, OCEAN CURR	W72-03324
OF THANET, MEDWELL ESTUA/ *KENT	COAST, INDUSTRIAL EXPANSION, ISLE	W71-13746
RADIOISOTOPES, MARINE BACTERIA,	COASTAL MARSHES, MICROORGANISMS.:	W72-00948
HEALTH, SOUTH AMERICA, ATLANTIC	COASTAL PLAIN, COASTS, CRUSTACEAN	W73-00819
, *FALLOUT, *MOLLUSKS, *ATLANTIC	COASTAL PLAIN, NORTH CAROLINA, AB	W73-00818
*GULF OF FINLAND, DARK FIXATION,	COASTAL WATERS, TROPHOGENIC LAYER	W72-03228
ISOTOPES, ESTUARINE ENVIRONMENT,	COASTS, ANIMAL PHYSIOLOGY, SEDIME	W73-00811
MARINE PLANTS, BIOMASS, CANADA,	COASTS, ATLANTIC OCEAN, SEASONAL,	W73-00366
ITY, *PHYTOPLANKTON, *SEA WATER,	COASTS, BIORTHYMS, ON-SITE TESTS,	W72-03228
AMERICA, ATLANTIC COASTAL PLAIN,	COASTS, CRUSTACEANS, MARINE FISH,	W73-00819
QUATIC WEEDS, *WEEDS, *RED TIDE,	COASTS, LAKES, PONDS, WATER POLLU	W72-14673
STATISTICAL METHODS, SEA WATER,	COASTS, MARINE ALGAE.: /), MAINE,	W73-00369
ALITY, MICHIGAN, SURFACE WATERS,	COASTS, SAMPLING, BIOLOGICAL COMM	W72-10875
UCOSE, ACETATES, BIOTIN, NIACIN,	COBALAMINS, EUGLENA GRACILIS, COL	W73-02095
*MARINE ANIMALS, ASIA, MOLLUSKS,	COBALT RADIOISOTOPES, IODINE RADI	W73-00811
ATHEMATICAL MODELS, FORECASTING,	COBALT RADIOISOTOPES, MARINE ALGA	W73-00831
, MUSSLES, LOBSTERS, ABSORPTION,	COBALT RADIOISOTOPES, ZINC RADIOI	W73-00832
REGION, TRITIUM, OCEAN CURRENTS,	COBALT RADIOISOTOPES, NUCLEAR POW	W72-03324
, LINC RADIOISOTOPES, MANGANESE,	COBALT RADIOISOTOPES, CADMIUM RAD	W71-13243
NVIRONMENT, *ECOSYSTEMS, CESIUM,	COBALT RADIOISOTOPES, ALGAE, RADI	W70-06668
DIOISOTOPES, ZINC RADIOISOTOPES,	COBALT RADIOISOTOPES, ABSORPTION,	W71-09018
DIOISOTOPES, ZINC RADIOISOTOPES,	COBALT RADIOISOTOPES, WATER POLLU	W71-09250
, DYSTROPHY, ZINC RADIOISOTOPES,	COBALT RADIOISOTOPES, PHOSPHORUS	W71-09190
FISH, MOLLUSKS, MARINE ANIMALS,	COBALT RADIOISOTOPES, FOOD CHAINS	W71-09225
DIOISOTOPES, ZINC RADIOISOTOPES,	COBALT RADIOISOTOPES, URANIUM RAD	W71-09188
GANESE CHLORIDE, NICKEL SULFATE,	COBALT SULFATE, POTASSIUM BR: /AN	W72-12739
ADIATION SYSTEM, MODULAR SYSTEM,	COBALT-60.: *BIOLOGICAL--GAMMA-R	W72-00383
, IRON, MANGANESE, ZINC, NICKEL,	COBALT, COPPER SULFATE, HEAVY MET	W72-12739
S, NUTRIENTS, MERCURY, CHROMIUM,	COBALT, FRESH WATER, LAKES, PHYSI	W73-01434
, COPPER, MOLYBDENUM, MANGANESE,	COBALT, IRON, INDUSTRIAL WASTES,	W71-11515
LIS FIRMA, PLENOSP/ PORPHYROPSIS	COCCINEA VAR. DAWSONII, CALLOPHYL	W73-02099
RA,/ *GAMBUSIA, CYCLOPS NAUPLII,	COCCOCHLORIS PENIOCYSTIS, SPIROGY	W72-07132
ES, SARGASSO SEA, SHELF WATERS,/	COCCOLITHOPHORES, SILICOFLAGELLAT	W72-07663
RATE UTILIZATION, *SARGASSO SEA,	COCCOLITHOPHORIDS, FLAGELLATES, F	W73-00380
COSTATA, ASSIMILATION, EMILIANA(	COCCOLITHUS) HUXLEYI, AMPHIPRORA	W73-00380
ECTA, SYNECHOCOCCUS, RHODOMONAS,	COCCOLITHUS HUXLEYI, AUTOLYSIS, P	W71-10083
LUVIATILIS, AMPHIDINIUM CARTERI,	COCCOLITHUS HUXLEYI, PORPHYRIDIUM	W72-12998
IS, AMPHORA OVALIS V. PEDICULUS,	COCCONEIS PEDICULUS, COCCONEIS PL	W72-13142
ROSUS, *EPIPHYTES, GROWTH MEDIA,	COCCONEIS PLACENTULA, CLOSTERIUM,	W72-12744
PEDICULUS, COCCONEIS PEDICULUS,	COCCONEIS PLACENTULA, CYCLOTELLA	W72-13142
TY INDEX, FRAGILARIA CONSTRUENS,	COCCONEIS PLACENTULA, NAVICULA SR	W72-08141
ES, FRAG/ LAKE FURESO (DENMARK),	COCCONEIS, RHOICOSPHENIA, ACHANTH	W72-07501
VER, COLUMBIA RIVER(WASHINGTON),	COCKLES, GOLDFISH, MERRIMACK RIVE	W71-11517
DIUM DODECYL, BENZENE SULFONATE,	COCONUT FATTY ACID, PLURONICS, DI	W72-12576
RE MEDIA, PORIA MONTICOLA, PORIA	COCOS, PORIA VAILLANTII, CONIOPHO	W72-13813
CANAL(MASS), BUZZARDS BAY, CAPE	COD BAY, DATA INTERPRETATION, SPE	W73-01449
PE COD BAY, DATA INTERPRE/ *CAPE	COD CANAL(MASS), BUZZARDS BAY, CA	W73-01449
TERS, PANAMA, CALLAO(PERU), CAPE	COD(MASS), CAPE MAY(N J), SARGASS	W72-08056
PHONIA LANOSA, CHONDRUS CRISPUS,	COD, FLOUNDER, CRAB EGGS, PELAGIC	W72-08142
PENFUSSII, GLOIODERMA ATLANTICA,	CODIUM CAROLINIANUM, *ONSLOW BAY(	W73-00284
LACTUCA, ZANICHELLIA PALUSTRIS,	CODIUM, FUSCUS, MACROPHYTES, AMMO	W73-00367
MIXING, EUPHOTIC ZONE, DIFFUSION	COEFFICIENT, LAKE SAMMAMISH(WASH)	W71-03034
N, EDDY DIFFUSIVITY,/ EXTINCTION	COEFFICIENT, LAMBERT-BEER EQUATIO	W72-05459
IZOMENON, OOCYSTIS, GLOEOCYSTIS,	COELASTRUM, ASTERIONELLA, CRYPTOM	W72-07940
ES, BRYOZOAS, HYDROIDS, SPONGES,	COELENTERATES, PORIFERA, POLYCHAE	W73-00374
, COLEOPTERA, MEDITERRANEAN SEA,	COELENTERATES, ECHINODERMS, POLYC	W73-00855
TES, ARTHROPODS, *BAY OF BENGAL,	COELENTERATES, CTENOPHORA, CHAETO	W73-00935
EANS, ECHINODERMS, SIPUNCULIDS,/	COELENTERATES, ARTHROPODS, NEMERT	W73-00932
ABALONE, TUNICATES, ARTHROPODS,	COELENTERATES, BRYOZOA, ECHINODER	W71-01475
ARI/ *CLEVELAND(OHIO), CERATIUM,	COELOSPHAERIUM, DINOBRYON, FRAGIL	W72-10076
CIA OBTUSA, LAURENCIA PAPILLOSA,	COELOTHRIX IRREGULARIS, CENTROCER	W73-00838
CYSTIS, APHANIZOMENON, CYANIDES,	COHO SALMON, SHEEPSHEAD, CHUBS, N	W72-14282
SOUNDER, VERTICAL DISTRIBUTION,	COLD BAY, BERMERS BAY, ALARIA, NE	W72-09120
ES, *ALPINE, *SYNOPTIC ANALYSIS,	COLD REGIONS, NUTRIENTS, BIOLOGIC	W72-12549
RUTICOSA, NODAMASTIX SPIROGYRAE,	COLEMANIA V: /NA, PAVLOVIAMONAS F	W72-07683
RED SEA, GULF OF SUEZ, DECAPODS,	COLEOPTERA, MEDITERRANEAN SEA, CO	W73-00855
BACTERIA, ALGAE, SALMONELLA, E.	COLI.: /CILLUS, CHLORELLA, SULFUR	W72-12565
OCCULATION, *ALGAE, *SILICA, *E.	COLI, *IONS, *WASTEWATER TREATMEN	W72-14840
SOLVED OXYGEN, BIOINDICATORS, E.	COLI, AQUATIC ALGAE, SUNFISHES, S	W71-07417
MENTAL EFFECTS, GROWTH RATES, E.	COLI, AZOTOBACTER, PSEUDOMONAS, W	W72-10861
PLANKTON, WATER TEMPERATURE, E.	COLI, BENTHIC FAUNA, CELLULOSE, E	W72-09668
S, ANTIMICROBIOTICS, ESCHERICHIA	COLI, BETA STREPTOCOCCUS, PSEUDOM	W72-04259
ATHOGENIC BACTERIA, CULTURES, E.	COLI, BIOINDICATORS, PONDS, ANAER	W72-07664

XIDE, CHLORELLA, CLOSTRIDIUM, E.	COLI, EUGLENA, EUGLENOPHYTA, CHLO	W72-06124
OA, YEASTS, INSECTS, MILDEWS, E.	COLI, FISH, FOODS, GRASSES, WORMS	W73-01676
, ENTERIC BACTERIA, DISEASES, E.	COLI, SHIGELLA, VIRUSES, CHROMATO	W72-13800
CHNIQUES, CALCIUM, MAGNESIUM, E.	COLI, SILICA, *FILTRATION, LIGHT	W72-11833
POLLUTANTS, PATHOGENIC BACTERIA,	COLIFORMS, TOXINS, ANTIBIOTICS(PE	W72-13800
N DEMAND, BACTERIA, TEMPERATURE,	COLIFORMS, ENTERIC BACTERIA, NITR	W72-12565
POTABLE WATER, ACTIVATED CARBON,	COLIFORMS, VIRUSES, CRUSTACEANS,	W73-02144
QUATIC BACTERIA, LAKES, FILTERS,	COLIFORMS, METABOLISM, CARBON, SU	W72-04743
DIATION, NUTRIENTS, TEMPERATURE,	COLIFORMS, CHLORINATION, SEEPAGE,	W71-07124
S, ALGAE, ORGANIC LOADING, ODOR,	COLIFORMS, TEMPERATURE, DISSOLVED	W71-07113
ION, HYDROGEN ION CONCENTRATION,	COLIFORMS, BACTERIA, PHOTOSYNTHES	W71-07100
ITIONS, SLUDGE, LIGHT INTENSITY,	COLIFORMS, BIOCHEMICAL OXYGEN DEM	W71-07106
LLUTION SOURCES, EUTROPHICATION,	COLIFORMS, SPORES, LIFE CYCLES, O	W71-09173
HLORINE, CONTACT TIME, SAMPLING,	COLIFORMS, BACTERIA, OXIDATION, O	W71-07096
CS(PESTICIDES), DATA PROCESSING,	COLIFORMS, MICROORGANISMS, HYDROG	W71-05155
R QUALITY, NUTRIENTS, SEDIMENTS,	COLIFORMS, MUNICIPAL WASTES, AGRI	W71-03012
SAMPLING, WATER ANALYSIS, ALGAE,	COLIFORMS, DISSOLVED OXYGEN.: /,	W70-08249
SAMPLING, WATER ANALYSIS, ALGAE,	COLIFORMS, DISSOLVED OXYGEN.: /,	W70-08248
GROWTH RATES, OXIDATION LAGOONS,	COLIFORMS, NUTRIENTS, ANALYTICAL	W72-09590
R TEMPERATURE, DISSOLVED OXYGEN,	COLIFORMS, RECREATION, ODOR.: /TE	W71-11914
ISINFECTION, *OXIDATION LAGOONS,	COLIFORMS, ALGAE, TEMPERATURE, AN	W72-00422
N DEMAND, *PHOSPHATES, NITROGEN,	COLIFORMS, MATHEMATICAL MODELS, A	W71-13356
, COSTS, SEWAGE BACTERIA, ALGAE,	COLIFORMS, INSECTICIDES.: /IATION	W72-00383
FICATION, OXIDATION, ADSORPTION,	COLIFORMS, BACTERIA, CENTRIFUGATI	W71-13326
OS, SALMONELLA TYPHOSA, TYPHOID,	COLIPHAGE, IMMUNOLOGY, PASTEURELL	W72-13800
*NEW YORK BIGHT,	COLLABORATIVE STUDIES.:	W72-12941
*WATER PROPERTIES, *ON-SITE DATA	COLLECTIONS, NEW YORK, BIOLUMINES	W72-12941
ITROGEN, NUTRIENTS, ON-SITE DATA	COLLECTIONS, LAKES.: /OSPHORUS, N	W72-12990
ED SOLIDS, HYDROLOGIC DATA, DATA	COLLECTIONS, LAKE MORPHOLOGY, ECO	W72-13851
, DOCUMENTATION, ABSTRACTS, DATA	COLLECTIONS, MICROBIOLOGY, SYSTEM	W72-13800
RIZONA, GRASSLANDS, ON-SITE DATA	COLLECTIONS, PLANT GROWTH.: /, *A	W72-12505
HERMAL POLLUTION, REVIEWS, *DATA	COLLECTIONS, TOXICITY, PATH OF PO	W72-12190
*SCUBA DIVING, *SEA WATER, *DATA	COLLECTIONS, COPPER, LEAD, HEAVY	W72-11925
ICS, GROWTH RATES, BIOMASS, DATA	COLLECTIONS.: /UCTIVITY, SYSTEMAT	W73-02469
RIA, ALGAE, EUTROPHICATION, DATA	COLLECTIONS, TEST PROCEDURES, CYC	W71-13794
*WATER POLLUTION EFFECTS, *DATA	COLLECTIONS, ALGAE, ECOTYPES, IND	W71-13746
BOTTOM SEDIMENTS, STATIONS, DATA	COLLECTIONS, ALGAE.: /FIC OCEAN,	W72-03567
, INFLOW, DISSOLVED SOLIDS, DATA	COLLECTIONS.: /AE, BACTERIA, ODOR	W72-02274
IC DATA, CHEMICAL ANALYSIS, DATA	COLLECTIONS, NUTRIENTS, SALINITY,	W71-10577
EASUREMENT, *AS/ *SURVEYS, *DATA	COLLECTIONS, *HYDROLOGIC DATA, *M	W72-10678
PROGRAMS, DATA PROCESSING, DATA	COLLECTIONS, VARIABILITY, PROBABI	W71-07337
UALITY, WASTE MANAGEMENT, MANURE	COLLECTION, MANURE TRANSPORT.: /Q	W71-08214
ON SOURCES, ALGAE, GEOLOGY, DATA	COLLECTIONS.: /ALS, WATER POLLUTI	W72-05869
EFFECTS, SURVEYS, SAMPLING, DATA	COLLECTIONS, INDUSTRIAL WASTES, A	W73-01972
MCONTROL, *BIBLIOGRAPHIES, *DATA	COLLECTIONS, *WATER POLLUTION EFF	W73-01976
E FISH, PERIPHYTON, ON-SITE DATA	COLLECTIONS, CRUSTACEANS, GASTROP	W73-00932
PERTIES, WATER TEMPERATURE, DATA	COLLECTIONS, MARINE PLANTS, ENVIR	W73-00838
AL METHODS, *POTOMAC RIVER, DATA	COLLECTIONS, EVALUATION, ON-SITE	W72-07791
CTIONS, EVALUATION, ON-SITE DATA	COLLECTIONS, RIVERS, *ELECTRIC PO	W72-07791
G, *SURVEYS, *MEASUREMENT, *DATA	COLLECTIONS, *ENVIRONMENT, *RADIO	W72-07841
MPERATURES, NOSTOC STREAMS, DATA	COLLECTIONS, ANALYTICAL TECHNIQUE	W72-08508
NS, *ANALYTICAL TECHNIQUES, DATA	COLLECTIONS, ALGAE, PROTOZA, FUNG	W72-07225
VERTICAL COLLECTORS, *HORIZONTAL	COLLECTORS, BRYOZOANS, FORAMINIFE	W72-05432
S, BRYOZOANS, FORAMIN/ *VERTICAL	COLLECTORS, *HORIZONTAL COLLECTOR	W72-05432
SWEDEN,	COLLEMA, PELTIGERA, NOSTOC.:	W73-02471
TION, WATER POLLUTION TREATMENT,	COLLOIDS, *POLYMERS, ANIONS, CATI	W72-11833
N ION CONCENTRATION, FILTRATION,	COLLOIDS, CARBON DIOXIDE, CALCIUM	W72-03613
AEPHYTA, AQUARIA, PARTICLE SIZE,	COLLOIDS, WATER POLLUTION EFFECTS	W71-09193
NMENTAL EFFECTS, TRACE ELEMENTS,	COLLOIDS, NUCLEAR WASTES, RADIOEC	W73-00818
MOUNTAIN LAKE(COLO), LAKE GRANBY(	COLO).: /RAND LAKE(COLO), SHADOW	W72-04791
OAQUIN RIVER(CALIF), RIFLE FALLS(	COLO), FISH FECAL WASTES, RESIDUA	W71-11006
LAKE(COLO), SHADOW MOUNTAIN LAKE(	COLO), LAKE GRANBY(COLO).: /RAND	W72-04791
, LAKE GRANBY(COLO)./ GRAND LAKE(	COLO), SHADOW MOUNTAIN LAKE(COLO)	W72-04791
YTES, *MIKOLAJSKIE LAKE(POLAND),	COLONIZATION, POTAMOGETON LUCENS,	W71-11012
N, COBALAMINS, EUGLENA GRACILIS,	COLONIZATION, CAULOBACTER, NAJAS	W73-02095
D, ALGAE, MINERALOGY, TURBIDITY,	COLOR, ALKALINITY, VOLUMETRIC ANA	W72-05594
OPERTIES, *NEW HAMPSHIRE, ALGAE,	COLOR, COPPER SULFATE, NETS, HYDR	W72-07890
PERTIES, BULK DENSITY, POROSITY,	COLOR, HYDROGEN ION CONCENTRATION	W73-00854
TRATION, IMPOUNDMENTS, SEASONAL,	COLOR, IRON, PHYSICOCHEMICAL PROP	W73-00256
STRATIFICATION, HARDNESS(WATER),	COLOR, OLIGOTROPHY, EUTROPHICATIO	W72-04269
, *ALGAE, *BACTERIA, AESTHETICS,	COLOR, TASTE, CORROSION, DISEASES	W71-09958
ANALYTICAL TECHNIQUES, ENZYMES,	COLOR, WATER QUALITY CONTROL, RES	W72-04070
ICHIGAN, CALIFORNIA, TUBIFICIDS,	COLORADO, WATER POLLUTION CONTROL	W71-11006
ALGAE, WATER POLLUTION SOURCES,	COLORADO, CATTLE, URINE, PATH OF	W71-11496
RAYS, POLLUTANT IDENTIFICATION,	COLORIMETRY, TRACERS.: /MS, GAMMA	W71-11036
ON, MARINE ALGAE, AQUATIC ALGAE,	COLORIMETRY, TRACES.: /HYTOPLANKT	W72-03545
IC PLANTS, IRON, MICROORGANISMS,	COLORIMETRY, WISCONSIN.: /, AQUAT	W72-03742
ATION, WATER QUALITY, NUTRIENTS,	COLORIMETRY, DEFICIENT ELEMENTS,	W73-00916
ONTRA COSTA POWER / CHALK POINT,	COLUMBIA RIVER, PATUXENT RIVER, C	W72-07907
MICAL ANALYSIS, SODIUM CHLORIDE,	COLUMBIA RIVER, ORGANIC COMPOUNDS	W72-05421
CHAIN, HERRING, OYSTERS, CLAMS,	COLUMBIA RIVER, CALIFORNIA, WASHI	W71-07417
ELEMENTS, ANALYTICAL TECHNIQUES,	COLUMBIA RIVER, WASHINGTON, OREGO	W72-03215
BAY(CALIFORNIA), POTOMAC RIVER,	COLUMBIA RIVER(WASHINGTON), COCKL	W71-11517
POLLUTION EFFECTS, *DISTRICT OF	COLUMBIA, TIDAL EFFECTS, NUTRIENT	W70-08248
POLLUTION EFFECTS, *DISTRICT OF	COLUMBIA, TIDAL EFFECTS, NUTRIENT	W70-08249
II, *LAMINARIA LONGIPES, BRITISH	COLUMBIA, RUSSIA, LAMINARIA RODRI	W73-00428
CULOSUS, LAMI/ *HUMIC COMPOUNDS,	COLUMN CHROMATOGRAPHY, FUCUS VESI	W72-12166
TROPICAL REGIONS, PACIFIC OCEAN,	COLUMNS, BOTTOM SEDIMENTS, ALGAE,	W71-08042
N TE/ CHEMICAL ANALYSIS, *ALGAE,	COLUMNS, ORGANIC ACIDS, SEPARATIO	W72-09365
A AMERICANA, TERRESTRIAL PLANTS,	COMBUSTION, ISOTOPE RATIOS, SPECT	W72-12709

SHELLFISH FARMING, AQUACULTURE, COMMERCIAL SHELLFISH, BIOLOGICAL W72-09378
ORA, CYPRINUS CARPIO, CATOSTOMUS COMMERSONI, AMEIURUS NEBULOSUS, P W70-08098
SOLEUCAS, CATOSTOMUS COMMERSONII COMMERSONII, ASTERIONELLA SPP, OR W72-07890
EMIGONUS CRYSOLEUCAS, CATOSTOMUS COMMERSONII COMMERSONII, ASTERION W72-07890
, LIMNOLOGY, INTERNATIONAL JOINT COMMISSION, CANADA, FEDERAL GOV: / W72-14782
RENCE RIVER, INTERNATIONAL JOINT COMMISSION.: /AKE ONTARIO, ST LAW W70-07283
Y TREATMENT, INTERNATIONAL JOINT COMMISSION.: /KE ONTARIO, TERTIAR W71-12091
C INSECTS, CRUSTACEANS, MUSSELS, COMMON MERGANSER DUCK, BIOASSAY, W72-12930
A, BORATES, ARTHROBACTER, NOSTOC COMMUNE, PORPHYROSIPHON NOTARISII W73-00854
BRISTYLIS, BORRICHIA, PHRAGMITES COMMUNIS, JUNCUS SPP, ATRIPLEX HA W73-01089
CROCYSTIS FLOS-AQUAE, PHRAGMITES COMMUNIS, *NORWAY, MACROPHYTES, F W73-00286
EMPERATURE, *DIATOMS, BIOLOGICAL COMMUNITIES, SYSTEMATICS, CYANOPH W73-00242
MARSHES, *PROTOZOA, *BIOLOGICAL COMMUNITIES, *ENVIRONMENTAL EFFEC W73-00367
ITY, *BICINDICATORS, *BIOLOGICAL COMMUNITIES, AQUATIC ALGAE, AQUAT W73-00002
, DOMINANT ORGANISMS, BIOLOGICAL COMMUNITIES, NUTRIENTS, SCENEDESM W72-14797
MMA RAYS, RESISTANCE, BIOLOGICAL COMMUNITIES, TIDAL MARSHES, FOOD W73-00796
Y, *PLANKTON, *ALGAE, BIOLOGICAL COMMUNITIES, PRODUCTIVITY, LIFE C W72-14659
, NUTRIENTS, REVIEWS, BIOLOGICAL COMMUNITIES, AQUATIC PLANTS, LIMI W73-01089
ION, *EUTROPHICATION, BIOLOGICAL COMMUNITIES, FOOD CHAINS, LAKE MO W73-00938
TIVITY, *ECOSYSTEMS, *BIOLOGICAL COMMUNITIES, BIOMASS, PRODUCTIVIT W73-01080
ATER, MARINE ANIMALS, BIOLOGICAL COMMUNITIES, CHLORINE, MOLLUSKS, W73-00855
QUENCY, FLUCTUATIONS, BIOLOGICAL COMMUNITIES, PHYSIOCHEMICAL PROPE W73-00918
HLOROPHYLL, PLANKTON, BIOLOGICAL COMMUNITIES, ECOSYSTEMS, SPECTROP W73-00916
CTIVITY, CHRYSOPHYTA, BIOLOGICAL COMMUNITIES, ESTUARINE ENVIRONMEN W73-00853
AQUATIC ENVIRONMENT, *BIOLOGICAL COMMUNITIES, *AQUATIC MICROORGANI W73-02095
IO, PLANKTO/ *ALGAE, *BIOLOGICAL COMMUNITIES, *STREAMS, LAKE ONTAR W73-01627
ORA, *OREGON, RIVERS, BIOLOGICAL COMMUNITIES, SEASONAL, DIATOMS, * W72-07901
PREDATION, *BIOASSAY, BIOLOGICAL COMMUNITIES, DOMINANT ORGANISMS, W72-07132
D REGIONS, NUTRIENTS, BIOLOGICAL COMMUNITIES, ALGAE, PERIPHYTON, P W72-12549
ECOLOGY, ECOSYSTEMS, *BIOLOGICAL COMMUNITIES, *SUCCESSION, AQUATIC W72-11854
LGAE, *PHYTOPLANKTON, BIOLOGICAL COMMUNITIES, STANDING CROPS, WISC W73-02469
*THERMAL POLLUTION, *BIOLOGICAL COMMUNITIES, *THERMAL POWER PLANT W71-11517
UARIES, GROWTH RATES, BIOLOGICAL COMMUNITIES.: /ORING, SEWAGE, EST W71-12867
YANOPHYTA, CHLORELLA, BIOLOGICAL COMMUNITIES, DOMINANT ORGANISMS, W72-09239
MATHEMATICAL STUDIES, BIOLOGICAL COMMUNITIES, FOOD CHAIN, DIATOMS, W72-09248
COMMERCIAL SHELLFISH, BIOLOGICAL COMMUNITIES, FOOD: /AQUACULTURE, W72-09378
RS, COASTS, SAMPLING, BIOLOGICAL COMMUNITIES, LAKE MICHIGAN, LAKE W72-10875
AGE, *FUNGI, *SLIME, *BIOLOGICAL COMMUNITIES, RIVERS, BACTERIA, WA W72-08573
, *ALGAE, *PROTOZOA, *BIOLOGICAL COMMUNITIES, ECOLOGICAL DISTRIBUT W72-08579
YTON, *PHYTOPLANKTON, BIOLOGICAL COMMUNITIES, DOMINANT ORGANISMS, W72-10883
ICATION, *NUTRIENTS, *BIOLOGICAL COMMUNITIES, *AQUATIC PLANTS, ALG W72-03216
NVIRONMENTAL EFFECTS, BIOLOGICAL COMMUNITIES, HEAVY METALS, HERBIC W72-01472
RY, PLANT PHYSIOLOGY, BIOLOGICAL COMMUNITIES.: /ALYSIS, BIOCHEMIST W71-08597
N), LAKE M/ DETROIT LAKES(MINN), COMMUNITY ACTION, LAKE SALLIE(MIN W71-03012
, RECREATION DEMAND, POPULATION, COMMUNITY DEVELOPMENT, ENVIRONMEN W72-01660
ALYSIS, MACROCONSUMERS, AUTUMNAL COMMUNITY, CALORIES, DETRIVORES, W71-10064
NVIRONMENTAL EFFECTS, BIOLOGICAL COMMUNITY, CANADA.: /ON, ALGAE, E W73-01704
DRIELLA HELVETICA, PLEUROCHLORIS COMMUTATA, PLEUROCHLORIS MAGNA, V W72-14808
OLAMPRA HEPTACTIS, BACTERIASTRUM COMOSUM, CHAETOCEROS ATLANTICUS V W71-11527
CONN), TROUT LAKE(WIS), SEDIMENT COMPACTION.: /(WIS), BANTAM LAKE(W72-08459
POLLUTION EFFECTS, FLUCTUATION, COMPARATIVE PRODUCTIVITY, MUNICIP W72-03228
NTHESIS, BIOMASS, OXYGEN, ALGAE, COMPARATIVE PRODUCTIVITY, EUTROPH W71-09165
HARD WATER, PUBLIC HEALTH, COST COMPARISONS, AQUATIC ALGAE, GRANT W71-09429
T, WATER POLLUTION SOURCES, COST COMPARISONS, BENEFICIAL USE, WATE W72-00974
ION, GROUNDWATER, WATER QUALITY, COMPETING USES, DIVERSION, RECIRC W70-07100
LENA, DAPHNIA, HABITATS, NICHES, COMPETITION, CALIFORNIA.: /S, EUG W71-09171
OLLUTION EFFECTS, RADIOISOTOPES, COMPETITION, SPECIATION.: /ATER P W71-09256
LIGHT PENETRATION, EFFICIENCES, COMPETITION, GROWTH RATES, PLANT W72-04766
SPHORUS, NUTRIENTS, CHRYSOPHYTA, COMPETITION, WATER POLLUTION SOUR W73-00441
EFFECTS, ENVIRONMENTAL EFFECTS, COMPETITION, AQUATIC POPULATIONS, W72-12736
ION, MODEL TEST/ AGRO-INDUSTRIAL COMPLEX, ELECTRIC ENERGY CONSUMPT W72-03327
YOPHYTES, BIOACCUMULATION, METAL COMPLEXES, PICEA ABIES, VACCINIUM W72-14303
DOPHOROPSIS MEMBRANACEA, BOODLEA COMPOSITA, ANADYOMENE STELLATA, B W72-11800
*SWEDEN, SPECIES COMPOSITION.: W73-01632
NDLAND, PERIDINIANS, QUALITATIVE COMPOSITION, COSCINOSIRA POLYCHOR W72-03543
ROWTH, NUTRIENT UPTAKE, CHEMICAL COMPOSITION, ALGAL DECOMPOSITION. W70-05469
*WATER RECLAMATION, VARIATIONS, COMPOSITION.: W70-07838
DATION DITCHES, AERATED LAGOONS, COMPOSTING, REFEEDING, ANHYDROUS W71-07551
LING FOOD, PROCESSING/ FEEDLOTS, COMPOSTING, PLANT RESIDUES, RECYC W72-00846
LGAE, CULTURES, ACIDITY, ORGANIC COMPOUNDS.: /M, CARBON DIOXIDE, A W72-00838
CHLORIDE, PHENYLACETATE, MERCURY COMPOUNDS.: /IC ACETATE, MERCURY W72-11727
ASCORBIC ACID, PYRIDOXINE, INDAL COMPOUNDS.: /TAHOE(CALIF), ZINC, W72-07514
BACTERIA, SULFUR BACTERIA, IRON COMPOUNDS.: /IC BACTERIA, AEROBIC W72-12996
R, *MARINE ALGAE, HUMUS, ORGANIC COMPOUNDS, POLLUTANT IDENTIFICATI W72-12166
MATA BAY(JAPAN), ORGANIC MERCURY COMPOUNDS, BIOSYNTHESIS, GUPPIES. W72-12257
LOGICAL MAGNIFICATION,/ *MERCURY COMPOUNDS, *MINAMATA DISEASE, BIO W72-12257
LYMERS, ANIONS, CATIONS, ORGANIC COMPOUNDS, SEPARATION TECHNIQUES, W72-11833
WASTES, BIOASSAY, SULFATES, IRON COMPOUNDS, SEWAGE, NITRATES, PHOS W72-12239
FUCUS VESICULOSUS, LAMI/ *HUMIC COMPOUNDS, COLUMN CHROMATOGRAPHY, W72-12166
SHES, *CARBON CYCLE, *PHOSPHORUS COMPOUNDS, *NITROGEN CYCLE, *CRAB W72-12567
ANKTON, FLUORESCENCE, PHOSPHORUS COMPOUNDS, CHLAMYDOMONAS, WATER A W72-12637
DNA, *DISSOLVED ORGANOPHOSPHORUS COMPOUNDS, NATURAL ORGANICS.: / * W72-12637
, CHLOROPHYTA, *ORGANOPHOSPHORUS COMPOUNDS, WASHINGTON, SEPARATION W72-12637
HEMICAL OXYGEN DEMAND, INORGANIC COMPOUNDS, THERMAL STRATIFICATION W72-13659
ERIE, NITROGEN, ALGAE, INORGANIC COMPOUNDS, STANDING CROPS, LIMITI W72-13653
ONS, *MICRO/ *TOXICITY, *ARSENIC COMPOUNDS, *PHOSPHATES, *INHIBITI W72-13813
, GROWTH, PSEUDOMONAS, POTASSIUM COMPOUNDS, POLLUTANT IDENTIFICATI W72-13813
UTRIENTS, *FOOD CHAINS, *ORGANIC COMPOUNDS, SHELLFISH, INSTRUMENTA W72-07715
D, ORGANIC CARBON, ACID-VOLATILE COMPOUNDS, WOODS HOLE, HAPTAPHYCE W72-07710
WEBS, COPEPODS, ORGANOPHOSPHORUS COMPOUNDS, PRIMARY PRODUCTIVITY, W72-07892

TION, MICROORGANISMS, *INORGANIC	COMPOUNDS, *NUTRIENTS, FRESH WATE	W72-07933
TION, *PLANKTON, OREGON, ORGANIC	COMPOUNDS, ABSORPTION, KINETICS,	W73-01269
YTA, BACTERIA/ *COASTS, *ORGANIC	COMPOUNDS, *GEOCHEMISTRY, CYANOPH	W73-01439
CULATE MATTER, DETRITUS, ORGANIC	COMPOUNDS, PATH OF POLL: /, PARTI	W73-02105
S, CARBON, PRODUCTIVITY, ORGANIC	COMPOUNDS, CARBOHYDRATES, CYCLING	W73-02100
ACE ELEMENTS, CHELATION, ORGANIC	COMPOUNDS, FOOD CHAINS, PUBLIC HE	W73-00821
A, PLANKTON, *ALGAE, *PHOSPHORUS	COMPOUNDS, CYANOPYTA, CHLOROPHYTA	W73-00938
ATTRACTANTS, *PREDATION, ORGANIC	COMPOUNDS, AMINO ACIDS, VITAMINS,	W73-00942
VITY, METABOLISM, FUNGI, ORGANIC	COMPOUNDS, CHLORELLA, TRACERS, AQ	W72-14326
NTHIC FAUNA, ASBESTOS, INORGANIC	COMPOUNDS, WORMS, INVERTEBRATES,	W73-00374
ICA, PATH OF POLLUTANTS, ORGANIC	COMPOUNDS, PHYTOPLANKTON, CARBON,	W73-00384
Y, INHIBITORS, TOXINS, INORGANIC	COMPOUNDS, ALGAE, DIATOMS, CHLAMY	W72-14706
SEPARA/ *TASTE, *ODOR, *ORGANIC	COMPOUNDS, ANALYTICAL TECHNIQUES,	W72-11604
*ANALYTICAL TECHNIQUES, *ORGANIC	COMPOUNDS, *CARBON, GRAVIMETRIC A	W72-09108
SOLID WASTES, AMMONIA, NITROGEN	COMPOUNDS, NITRATES, PHOSPHATES,	W72-09675
PHOSPHATES, *NITRATES, *AMMONIUM	COMPOUNDS, SEWAGE EFFLUENTS, TEMP	W72-10162
, PHOSPHORUS COMPOUNDS, NITROGEN	COMPOUNDS, CYCLING NUTRIENTS, ESS	W72-10608
E, ANTIBIOTICS, MERCURY, ARSENIC	COMPOUNDS, CHESAPEAKE BAY.: /ALGA	W72-10162
S, *ALGAE, *BIOASSAY, PHOSPHORUS	COMPOUNDS, NITROGEN COMPOUNDS, CY	W72-10608
, PHYSIOLOGICAL ECOLOGY, ORGANIC	COMPOUNDS, ON-SITE TESTS, METHODO	W72-03216
ES, CULTURES, AMMONIA, POTASSIUM	COMPOUNDS, NITRATES, DENSITY, GRO	W72-03217
TRIFICATION, *NITRATES, NITROGEN	COMPOUNDS, AMMONIA, ALGAE, SEWAGE	W71-13939
N, *CHELATION, *SEWAGE, *ORGANIC	COMPOUNDS, WATER POLLUTION TREATM	W71-09674
LANTS, *PHYTOPLANKTON,/ *ORGANIC	COMPOUNDS, *PHOSPHORUS, *MARINE P	W71-10083
, TEMPERATURE, ORGANOPHOSPHOROUS	COMPOUNDS, ALGAE.: /SOLVED OXYGEN	W71-10645
, UTAH, GREAT SALT LAKE, ORGANIC	COMPOUNDS, BRINES.: /ALGAE BLOOMS	W71-12223
FERTILIZERS, NUTRIENTS, NITROGEN	COMPOUNDS, PHOSPHORUS COMPOUNDS,	W71-12252
, NITROGEN COMPOUNDS, PHOSPHORUS	COMPOUNDS, POTASH, ALGAE, AQUATIC	W71-12252
NTHESIS, PHYTOPLANKTON, AMMONIUM	COMPOUNDS, NUTRIENTS, PRIMARY PRO	W71-11008
ACTERIA, PHOTOSYNTHESIS, ORGANIC	COMPOUNDS, PHOSPHATES, NITRATES,	W71-07100
ANKTON/ *RADIOISOTOPES, *ORGANIC	COMPOUNDS, *MARINE ALGAE, PHYTOPL	W71-09188
L EXUDAT/ *HIGH MOLECULAR WEIGHT	COMPOUNDS, *MOLECULAR SIZE, *ALGA	W71-07878
, CHELATION, PHOSPHATES, ORGANIC	COMPOUNDS, WATER POLLUTION CONTRO	W71-06247
*NITROGEN COMPOUNDS, *PHOSPHORUS	COMPOUNDS, ALGAE, NUTRIENT REQUIR	W70-04080
WATER QUALITY CONTROL, *NITROGEN	COMPOUNDS, *PHOSPHORUS COMPOUNDS,	W70-04080
MESTIC WASTES, CHEMIC/ *NITROGEN	COMPOUNDS, *SEWAGE TREATMENT, *DO	W70-09907
NOFF, EROSION, CHLORELLA, SODIUM	COMPOUNDS, ALKALINITY, MINING, BI	W72-04734
HLORIDE, COLUMBIA RIVER, ORGANIC	COMPOUNDS, VITAMINS, YEASTS, SULF	W72-05421
ION, *ORGANIC MATTER, *INORGANIC	COMPOUNDS, IRON, MANGANESE, PHOSP	W72-05506
LMERCURY, BRAIN, ORGANOMERCURIAL	COMPOUNDS, ESOX LUCIUS, ALKYL MER	W72-05952
ITY, CYANOPHYTA, NUISANCE ALGAE,	COMPREHENSIVE PLANNING, AQUATIC W	W72-09171
*KEZAR LAKE, AIR INJECTION,	COMPRESSED AIR.:	W72-09304
NG, CHEMICAL ANALYSIS, PLANKTON,	COMPUTER PROGRAMS, BIOMASS, TEMPE	W72-08560
ZE DRYING, X-RAYS, ELECTROLYSIS,	COMPUTER PROGRAMS, GAMMA RAYS, PO	W71-11036
, TEMPERATURE, DISSOLVED OXYGEN,	COMPUTERS, ALGAE, AQUATIC LIFE, E	W72-14405
MERICAL ANALYSIS, MODEL STUDIES,	COMPUTERS, CULTURES.: /NIQUES, NU	W72-12724
S, MATHEMATICAL STUDIES, DIGITAL	COMPUTERS, DATA ANALYSIS, EQUATIO	W72-11920
PTOPHYCEAE, *DENMARK, CYCLOTELLA	COMTA, STEPHANODISCUS HANTZSCHII,	W72-12738
XYGEN, THERMOCLINE, HYDROGEN ION	CONCENTRATION, FLOW, CHEMICAL PRO	W72-12941
ICAL PRECIPITATION, HYDROGEN ION	CONCENTRATION.: /AE TRACERS, CHEM	W72-12210
HEMICAL PROPERTIES, HYDROGEN ION	CONCENTRATION, *WASTE WATER TREAT	W72-11833
ORTHOPHOSPHATE	CONCENTRATIONS, GULF OF FINLAND.:	W72-13871
WATER TEMPERATURE, HYDROGEN ION	CONCENTRATION, STRATIFICATION.: /	W72-13871
URNAL DISTRIBUTION, HYDROGEN ION	CONCENTRATION, DISSOLVED OXYGEN,	W72-13508
TURE, ACID STREAMS, HYDROGEN ION	CONCENTRATION, HEAVY METALS, DISS	W72-14690
AE, SURFACE WATERS, HYDROGEN ION	CONCENTRATION, LIGHT PENETRATION,	W73-00236
PHYTA, CHLOROPHYTA, HYDROGEN ION	CONCENTRATION, CHEMICAL PORPERTIE	W72-14790
YGEN, CONDUCTIVITY, HYDROGEN ION	CONCENTRATION.: /TY, DISSOLVED OX	W72-14728
RATURE, ALKALINITY, HYDROGEN ION	CONCENTRATION, WATER QUALITY, NUT	W73-00916
Y, POROSITY, COLOR, HYDROGEN ION	CONCENTRATION, NITROGEN, CARBON,	W73-00854
TE WATER TREATMENT, HYDROGEN ION	CONCENTRATION, CYANOPHYTA.: /*WAS	W73-01367
KALINITY, HALOGENS, HYDROGEN ION	CONCENTRATION, PHOSPHATES, NITRAT	W73-01434
LTURES, CYANOPHYTA, HYDROGEN ION	CONCENTRATION, LIGHT INTENSITY, N	W73-01657
CALCIUM CARBONATE, HYDROGEN ION	CONCENTRATION, FILTRATION, PLA: /	W73-01669
RADIATION CONTROL.:	CONCENTRATION, RADIATION SAFETY,	W72-07841
PPER SULFATE, NETS, HYDROGEN ION	CONCENTRATION, PHOSPHATES, NITROG	W72-07890
, DISSOLVED OXYGEN, HYDROGEN ION	CONCENTRATION, SCENEDESMUS, BIOMA	W72-08049
*ALGAE, *NITROGEN, HYDROGEN ION	CONCENTRATION, BIOCHEMICAL OXYGEN	W72-06612
YRROPHYTA, BIOMASS, HYDROGEN ION	CONCENTRATION, MINNESOTA, WATER T	W72-07145
CHEMICAL ANALYSIS, HYDROGEN ION	CONCENTRATION, SEA WATER, PHOTOSY	W71-11008
RATION, *OXIDATION, HYDROGEN ION	CONCENTRATION, MICROORGANISMS, FE	W71-12223
LAKES, CHLOROPHYLL, HYDROGEN ION	CONCENTRATION, CARBON DIOXIDE.: /	W71-12855
VED OXYGEN, SEWAGE, HYDROGEN ION	CONCENTRATION.: /LIZATION, DISSOL	W71-12188
HROMIUM, COPPER, NICKEL, RUTHEN/	CONCENTRATION, ABUNDANCE, IRON, C	W71-13243
RACTIONATION, LIME, HYDROGEN ION	CONCENTRATION, ECONOMICS, ALGAE,	W71-12200
OISOTOPES, DIATOMS, HYDROGEN ION	CONCENTRATION, MARINE ANIMALS, BA	W71-10083
ORGIA, THERMOCLINE, HYDROGEN ION	CONCENTRATION, WATER TEMPERATURE,	W71-11914
NUTRIENTS, AMMONIA, HYDROGEN ION	CONCENTRATION, CYANOPHYTA, CHLORE	W71-11519
TY, PHOTOSYNTHESIS, HYDROGEN ION	CONCENTRATION, NITRITE, AMMONIA,	W71-11876
MATTER, ALKALINITY, HYDROGEN ION	CONCENTRATION, NUTRIENTS, AQUATIC	W71-12068
RATES, ABSORBANCE, MAGNESIUM ION	CONCENTRATION, BRUCITE, BOTTOM TO	W72-09092
ZINC, TEMPERATURE, HYDROGEN ION	CONCENTRATION, CORROSION, WATER P	W72-08683
CLES, LETHAL DOSAGE, CALCIUM ION	CONCENTRATION, VERTEBRATES, ABSOR	W72-09092
ITY, ON-SITE TESTS, HYDROGEN ION	CONCENTRAT: /EA WATER, WATER QUAL	W72-09092
DIOXIDE, POLYMERS, HYDROGEN ION	CONCENTRATION, PROTOZOA, DISSOLVE	W72-10233
WATER TEMPERATURE, HYDROGEN ION	CONCENTRATION, HARDNESS, ALKALINI	W72-10875
THANE, TEMPERATURE, HYDROGEN ION	CONCENTRATION, PROTEIN, LABORATOR	W72-10390
	CONCENTRATION, UPTAKE.:	W72-10686
OLOGY/METABOLISM.:	CONCENTRATION, MUSCLE, BONE PHYSI	W72-10694
TES, *LIMITING FA/ *HYDROGEN ION	CONCENTRATION, *ALGAE, *GROWTH RA	W72-11724

, DISSOLVED OXYGEN, HYDROGEN ION CONCENTRATION, SETTLING, MICROORG W71-13334
HEMICAL PROPERTIES, HYDROGEN ION CONCENTRATION, TEMPERATURE, DISSO W72-03224
S, *ORGANIC ACIDS, *HYDROGEN ION CONCENTRATION, METABOLISM, CARBON W72-00838
THERMAL PROPERTIES, HYDROGEN ION CONCENTRATION, DISSOLVED OXYGEN, W72-01363
EDUCTION POTENTIAL, HYDROGEN ION CONCENTRATION.: /NGE, OXIDATION-R W72-01108
ILTRATION, AMMONIA, HYDROGEN ION CONCENTRATION, TEMPERATURE, LIGHT W72-01881
GNESIUM, CHLORIDES, HYDROGEN ION CONCENTRATION, PERIPHYTON, WATER W72-05610
TY, CARBON DIOXIDE, HYDROGEN ION CONCENTRATION, CARBON, OXYGEN, NI W72-06274
, ACTIVATED CARBON, HYDROGEN ION CONCENTRATION, NUTRIENTS.: /LYTES W72-06201
PENETRATION, ALGAE, HYDROGEN ION CONCENTRATION.: /QUATIONS, LIGHT W72-05459
MENTS, TEMPERATURE, HYDROGEN ION CONCENTRATION, EFFICIENCIES, WATE W72-04788
AGE, CARBOHYDRATES, HYDROGEN ION CONCENTRATION, TEMPERATURE, ENZYM W72-05457
, DISSOLVED OXYGEN, HYDROGEN ION CONCENTRATION, DESIGN CRITERIA, M W72-04787
POLLUTION SOURCES, HYDROGEN ION CONCENTRATION, CARBON DIOXIDE, NI W72-04769
S, LIGHT INTENSITY, HYDROGEN ION CONCENTRATION, CARBON DIOXIDE, AE W72-04433
ICAL PRECIPITATION, HYDROGEN ION CONCENTRATION, FILTRATION, COLLOI W72-03613
RIA, SEWAGE, ALGAE, HYDROGEN ION CONCENTRATION, CARBON DIOXIDE, NI W72-04309
ES, CARBON DIOXIDE, HYDROGEN ION CONCENTRATION, WATER QUALITY CONT W72-04431
ICAL OXYGEN DEMAND, HYDROGEN ION CONCENTRATION, ALGAE, *EUTROPHICA W70-10427
LAKES, TEMPERATURE, HYDROGEN ION CONCENTRATION, ACTIVATED SLUDGE, W71-00116
LANKTON, TURBIDITY, HYDROGEN ION CONCENTRATION, ALKALINITY, CARBON W71-00117
ULATION, HEAD LOSS, HYDROGEN ION CONCENTRATION, CHLORELLA, SCENEDE W71-00138
AE, MICROORGANISMS, HYDROGEN ION CONCENTRATION, PHENOLS, ARIZONA.: W71-01474
OLOGICAL LIFE, *DISSOLVED OXYGEN CONCENTRATION.: /ARE ESTUARY, *BI W70-09189
ICITY, IRON, LIGHT, HYDROGEN ION CONCENTRATION.: /YTOPLANKTON, TOX W71-03027
DS, NITROGEN CYCLE, HYDROGEN ION CONCENTRATION, TEMPERATURE, ELECT W71-01489
, DISSOLVED OXYGEN, HYDROGEN ION CONCENTRATION, ALKALINITY, RESPIR W71-04518
, RUNOFF, ILLINOIS, HYDROGEN ION CONCENTRATION, ALKALINITY, HARDNE W71-03028
IGRATION, PROTOZOA, HYDROGEN ION CONCENTRATION, WATER TEMPERATURE, W71-05626
MS, MICROORGANISMS, HYDROGEN ION CONCENTRATION, WASTE WATER TREATM W71-05155
QUALITY, SULFATES, HYDROGEN ION CONCENTRATION, ALGAE, BACTERIA, W W71-07942
TABOLISM, BACTERIA, HYDROGEN ION CONCENTRATION, LABORATORY TESTS.: W71-08687
CHEMICAL REACTIONS, HYDROGEN ION CONCENTRATION, ALGAE, REPRODUCTIO W71-08026
COAGULATION, LIME, HYDROGEN ION CONCENTRATION, TEMPERATURE, CHLOR W71-08951
TY, NEUTRALIZATION, HYDROGEN ION CONCENTRATION, TEMPERATURE, BIOCH W71-09524
ERATURE, CHELATION, HYDROGEN ION CONCENTRATION, DYSTROPHY, ZINC RA W71-09190
ICAL PRECIPITATION, HYDROGEN ION CONCENTRATION, COLIFORMS, BACTERI W71-07100
ILOT PLANTS, ALGAE, HYDROGEN ION CONCENTRATION, CHEMICAL OXYGEN DE W71-07102
ICAL OXYGEN DEMAND, HYDROGEN ION CONCENTRATION, WASTE WATER TREATM W71-07113
PHYLL, TEMPERATURE, HYDROGEN ION CONCENTRATION, DISSOLVED OXYGEN, W71-07382
, WESTERN LAKE ERIE, ARNOLDIELLA CONCHOPHILA MILLER.: /GAL SPECIES W71-09156
ARNOLDIELLA- CONCHOPHILA, LAKE ERIE.: W71-12489
OLING WATER, *WATER TEMPERATURE, CONDENSERS, DESIGN, STANDARDS, FI W70-09612
YTICAL TECHNIQUES, HEATED WATER, CONDENSERS, PROTOZOA, ALGAE, FUNG W72-07526
ECCHI *YARDS CREEK(NJ), RELATIVE CONDITION FACTOR.: /ER SAMPLER, S W72-14280
OSSES, *WATER CONSERVATION, *AIR CONDITIONING, *HEATING, *COOLING, W70-09193
SAFETY, WATER COOLING, ANAEROBIC CONDITIONS, ALGAL CONTROL, SPAWNI W70-09595
R TREATMENT, NUTIRIENTS, AEROBIC CONDITIONS, AERATION, MICROBIOLOG W71-04079
S, AEROBIC CONDITIONS, ANAEROBIC CONDITIONS, CHLORELLA, CHLAMYDOMO W70-10405
TER QUALITY, PHOSPHORUS, AEROBIC CONDITIONS, ANAEROBIC CONDITIONS, W70-10405
ILTERS, PARTICLE SIZE, ANAEROBIC CONDITIONS, WASTE WATER TREATMENT W71-08223
E, ANAEROBIC CONDITIONS, AEROBIC CONDITIONS, BIODEGRADATION, SEEPA W71-07099
, SLUDGE, OXYGENATION, ANAEROBIC CONDITIONS, AEROBIC CONDITIONS, B W71-07109
S, ANAEROBIC CONDITIONS, AEROBIC CONDITIONS, BIOCHEMICAL OXYGEN DE W71-07100
ONS, *OXIDATION LAGOONS, AEROBIC CONDITIONS, ANAEROBIC CONDITIONS, W71-07551
N, ANAEROBIC CONDITIONS, AEROBIC CONDITIONS, SLUDGE, LIGHT INTENSI W71-07106
SSOLVED OXYGEN, ALGAE, ANAEROBIC CONDITIONS, AEROBIC CONDITIONS, B W71-07099
HATES, BIODEGRADATION, ANAEROBIC CONDITIONS, AEROBIC CONDITIONS, S W71-07106
E, ANAEROBIC CONDITIONS, AEROBIC CONDITIONS, PILOT PLANTS, ALGAE, W71-07102
, NITRATES, NUTRIENTS, ANAEROBIC CONDITIONS, AEROBIC CONDITIONS, B W71-07100
S, AEROBIC CONDITIONS, ANAEROBIC CONDITIONS, ALGAE, ORGANIC MATTER W71-07551
TION LAGOONS, *SLUDGE, ANAEROBIC CONDITIONS, AEROBIC CONDITIONS, P W71-07102
N, ANAEROBIC CONDITIONS, AEROBIC CONDITIONS, BIOCHEMICAL OXYGEN DE W71-07109
S, ANAEROBIC CONDITIONS, AEROBIC CONDITIONS, SLUDGE, WASTE WATER T W71-07084
ALGAE, PHOTOSYNTHESIS, ANAEROBIC CONDITIONS, AEROBIC CONDITIONS, S W71-07084
, *AEROBIC TREATMENT, *ANAEROBIC CONDITIONS, PONDS, CLIMATIC ZONES W71-06033
NITROGEN, *PHOSPHATES, ANAEROBIC CONDITIONS, AEROBIC CONDITIONS, M W72-04789
S, ANAEROBIC CONDITIONS, AEROBIC CONDITIONS, METABOLISM, SLUDGE, B W72-04789
US,/ *AEROBIC BACTERIA, *AEROBIC CONDITIONS, *KINETICS, *SCENEDESM W73-00379
CHLOROPHYTA, CYANOPHYTA, AEROBIC CONDITIONS, BIOASSAY, GROWTH RATE W72-07664
BIOINDICATORS, PONDS, ANAEROBIC CONDITIONS, WASTE WATER TREATMENT W72-07664
DIOXIDE, *GROWTH RATES, *AEROBIC CONDITIONS, *GROWTH STAGES, *BIOD W72-06612
, ENVIRONMENTAL EFFECTS, AEROBIC CONDITIONS, CHLOROPHYTA, SCENEDES W73-01066
*ANAEROBIC BACTERIA, *ANAEROBIC CONDITIONS, *KINETICS, *DISSOLVED W73-01066
TROGEN, EUTROPHICATION, *AEROBIC CONDITIONS, *AEROBIC: /PHORUS, NI W72-01881
ALGAE, LABORATORY TESTS, AEROBIC CONDITIONS, ANAEROBIC CONDITIONS, W71-13326
S, AEROBIC CONDITIONS, ANAEROBIC CONDITIONS, ORGANIC LOADING, *BIO W71-13326
ATION, DISSOLVED OXYGEN, AEROBIC CONDITIONS, BIOMASS, RESPIRATION, W72-11100
HESIS, DISSOLVED OXYGEN, AEROBIC CONDITIONS, ANAEROBIC CONDITIONS, W72-11642
N, AEROBIC CONDITIONS, ANAEROBIC CONDITIONS, ORGANIC LOADING, ODOR W72-11642
ENVIRONMENTAL EFFECTS, ANAEROBIC CONDITIONS, NITRATES, DENITRIFICA W72-11563
S, ESSENTIAL NUTRIENTS, *AEROBIC CONDITIONS, WATER POLLUTION CONTR W72-11617
BENT/ *OXYGEN DEMAND, *ANAEROBIC CONDITIONS, *LAKE ERIE, *ALGAE, * W72-12995
N, *DISSOLVED OXYGEN, *ANAEROBIC CONDITIONS, ALGAE, SEDIMENTS, OXY W72-12990
PRODUCTIVITY, BIOMASS, ANAEROBIC CONDITIONS, WATER TEMPERATURE, CH W72-12940
MANGANESE, PHOSPHORUS, ANAEROBIC CONDITIONS, CHLOROPHYLL, EPILIMNI W72-12997
GREEN BAY, FOX RIVER, SPECIFIC CONDUCTANCE, TRANSPARENCY, ORTHOP W71-10645
TRATION, TEMPERATURE, ELECTRICAL CONDUCTANCE, DISSOLVED OXYGEN, AI W71-01489
NUTRIENTS, MIDGES, FISH, LAKES, CONDUCTIVITY, ALGAE, STANDING CRO W71-01488
PROPERTIES, ACIDITY, ALKALINITY, CONDUCTIVITY, CALCIUM, SCENEDESMU W72-04269

RATES, NITROGEN FIXATION, ALGAE, CONDUCTIVITY, CHLOROPHYLL, PIGMEN W72-04773
ENTRATION, HARDNESS, ALKALINITY, CONDUCTIVITY, S: /DROGEN ION CONC W72-10875
PROGRAMS, BIOMASS, TEMPERATURE, CONDUCTIVITY, PHOSPHATES, NITRATE W72-08560
ATES, NITROGEN, HARDNESS(WATER), CONDUCTIVITY, ZINC, WATER TEMPERA W72-07890
E, ALKALINITY, DISSOLVED OXYGEN, CONDUCTIVITY, HYDROGEN ICN CONCEN W72-14728
, *WASTES, *PATHOGENIC BACTERIA, CONFERENCES, *PUBLIC HEALTH, RIVE W72-09675
, POLLUTANTS, INDUSTRIAL WASTES, CONFERENCES.: /S, PHENOLS, METALS W72-09646
GOMONTIA POLY/ *YUCATAN, LYNGBYA CONFERVOIDES, LYNGBYA MAJUSCULA, W72-11800
IOCHEMICAL OXYGEN DEMAND, ALGAE, CONFINEMENT PENS, IMPOUNDMENTS, * W72-08401
POLLUTION, AIR POLLUTION, ODOR, CONFINEMENT PENS, ORGANIC MATTER, W71-08214
, PORIA COCOS, PORIA VAILLANTII, CONIOPHORA OLIVACEA, LENZITES TRA W72-13813
LAKE MENDOTA(WIS), BANTAM LAKE(CONN), TROUT LAKE(WIS), SEDIMENT W72-08459
ES, FOOD WEBS, INTERTIDAL AREAS, CONNATE WATER.: /IES, TIDAL MARSH W73-00796
CONNETICUT YANKEE NUCLEAR PLANT, CONNECTICUT RIVER, TURKEY POINT, W72-07907
A RIVER, CALIFORNIA, WASHINGTON, CONNECTICUT, FLORIDA, DELAWARE RI W71-07417
RNACLES, FLOUNDER, HUMBOLDT BAY, CONNECTICUT RIVER, SAN JOAQUIN RI W71-07417
*WATER RESOURCES RESEARCH ACT, CONNECTICUT, *UNIVERSITIES, *ABST W71-01192
LEY POND(CONNECTICUT), LAKE ZOAR(CONNECTICUT), LAKE MINNETONKA(MIN W71-04216
INGTON(WASHINGTON), LINSLEY POND(CONNECTICUT), LAKE ZOAR(CONNECTIC W71-04216
SOURCES, MINNESOTA, WASHINGTON, CONNECTICUT.: /R, WATER POLLUTION W71-04216
IVER(INDIANA), CONNECTICUT RIVER(CONNECTICUT), GREEN RIVER(KENTUCK W71-11517
TRAINMENT, WHITE RIVER(INDIANA), CONNECTICUT RIVER(CONNECTICUT), G W71-11517
ITY EFFECTS, *THERMAL POLLUTION, CONNECTICUT, RIVERS, POST-IMPOUND W72-00939
CONNECTICUT RIVER.: W72-00939
O BAY POWER PLANT, HUMBOLDT BAY, CONNETICUT YANKEE NUCLEAR PLANT, W72-07907
ERPLANTS, *ENERGY LOSSES, *WATER CONSERVATION, *AIR CONDITIONING, W70-09193
TY, WATER CONSERVATION, WILDLIFE CONSERVATION, FLOOD CONTROL, RECR W72-05594
N SYSTEMS, *WATER QUALITY, WATER CONSERVATION, WILDLIFE CONSERVATI W72-05594
, SYNEDRA ACUS, RHODOMONAS, LAKE CONSTANCE, SEWAGE CANALS.: /SCENS W71-00117
SPECIFIC GROWTH RATE, SATURATION CONSTANT, DILUTION, CHLORELLA PYR W72-04788
VIEWS, FOOD CHAINS, FORECASTING, CONSTRAINTS, RESEARCH FACILITIES. W72-06340
AUTOCOCCUS CAUDATUS, NAUTOCOCCUS CONSTRICTUS, NAUTOCOCCUS EMERSUS. W72-07141
PLOS ARMIGER, ALIMITO/ *OFFSHORE CONSTRUCTION, DELAWARE BAY, SCOLO W73-02029
CIES DIVERSITY INDEX, FRAGILARIA CONSTRUENS, COCCONEIS PLACENTULA, W72-08141
COSPHENIA, ACHANTHES, FRAGILARIA CONSTRUENTS, NAVICULA, NITZSCHIA, W72-07501
E HEAT, THERMAL EFFICIENCY, HEAT CONSUMPTION.: *WAST W70-08832
USTRIAL COMPLEX, ELECTRIC ENERGY CONSUMPTION, MODEL TESTING.: /IND W72-03327
LEGISLATION, ADMINISTRATION, NON- CONSUMPTIVE USE, ALGAE, EVAPORATI W70-07100
HYDROELECTRIC POWER, NAVIGATION, CONSUMPTIVE USE, WATER POLLUTION, W72-14878
NFECTION, DEGRADATION, CHLORINE, CONTACT TIME, SAMPLING, COLIFORMS W71-07096
E M/ *ENVIRONMENTAL QUALITY, AIR CONTAMINANT, STREAM QUALITY, WAST W71-08214
IOINHIBITION, BI/ *TRACE ORGANIC CONTAMINANTS, *TRACE MATERIALS, B W71-11793
ARCH AND DEVELOPMENT, SOIL, SOIL CONTAMINATION.: /E, POULTRY, RESE W71-07551
HENS, MOSSES, FUNGI, ALGAE, SOIL CONTAMINATION, TRACE ELEMENTS, IO W72-14303
DIOISOTOPES, SOIL ANALYSIS, SOIL CONTAMINATION, CLADOPHORA, CHAR: / W73-01673
*RIFFLES, *CALORIC CONTENT.: W71-07097
COLOR, FOMES ANNOSUS, FUSCOPORIA CONTIGUA, TRAMETES LILACINO-GILVA W72-13813
A WATER, SEASONAL, FLUCTUATIONS, CONTINENTAL SHELF, VERTICAL DISTR W72-07663
COASTS, *OCEANS, SURFACE WATERS, CONTINENTAL SHELF, DEPTH, NITROGE W72-08056
ES.: *ALGAL GROWTH, CONTINUOUS CULTURES, BATCH CULTUR W72-14790
IFIC GROWTH RATE, / *CHEMOSTATIC CONTINUOUS CULTURES, MAXIMUM SPEC W72-04788
*HETEROCYSTS, *ANABAENA SP L-31, CONTINUOUS CULTURE, BATCH CULTURE W72-03217
COLUMBIA RIVER, PATUXENT RIVER, CONTRA COSTA POWER PLANT, SAN JOA W72-07907
EALTH, BIODEGRADATION, TOXICITY, CONTRACTS, GRANTS, RESEARCH AND D W71-11516
ESMIDS, STAURASTRUM LONGIPES VAR CONTRACTUM, ARTHURS LAKE, LAKE SO W73-01094
WATER POLLUTION CONTROL ACT.: W72-01085
*WATER POLLUTION CONTROL ACT.: W72-01093
*WATER POLLUTION CONTROL ACT.: W72-01685
NUTRIENT CONTROL METHODS.: W71-13256
, ALGAL GROWTH, IN-LAKE NUTRIENT CONTROL.: /ALGAL ASSAY PROCEDURES W71-12072
E WATER TREATMENT, WATER QUALITY CONTROL.: /HUMAN POPULATION, WAST W71-09788
LABORATORY TESTS, WATER QUALITY CONTROL.: / ACTINOMYCETES, FUNGI, W72-11604
ERISTICS, SOLIDS RECYCLE, SOLIDS CONTROL.: /OGEN, SETTLING CHARACT W72-11100
OBIC CONDITIONS, WATER POLLUTION CONTROL.: /ENTIAL NUTRIENTS, *AER W72-11617
AGE, *HYPOLIMNION, WATER QUALITY CONTROL.: / ALGAE, RESERVOIR STOR W73-02138
ION, RADIATION SAFETY, RADIATION CONTROL.: CONCENTRAT W72-07841
S, FUNGI, ALGAE, WATER POLLUTION CONTROL.: /, STREAMS, SEPTIC TANK W72-14363
, *ANALYTICAL TECHNIQUES, *ALGAL CONTROL.: /IXING, *EUTROPHICATION W72-04433
LLUTION SOURCES, WATER POLLUTION CONTROL.: /TION EFFECTS, WATER PO W72-04759
GANIC COMPOUNDS, WATER POLLUTION CONTROL.: /LATION, PHOSPHATES, OR W71-06247
OPHYTA, IRRIGATION CANALS, ALGAL CONTROL.: /ATURE, LIMNOLOGY, CYAN W71-06102
LLUTION EFFECTS, WATER POLLUTION CONTROL.: /ITRIFICATION, WATER PO W72-12544
IENTS, *CHLORELLA, ALGAE/ *ALGAL CONTROL, *PHOSPHATES, *ALGAL NUTR W72-13992
*IDAHO, *CHLOROPHYLL, *A/ *ALGAL CONTROL, *PHOSPHORUS, *NITROGEN, W72-13300
ION, *MACROPHYTES, *AQUATIC WEED CONTROL, *LIGHT INTENSITY, *NEBRA W73-02349
Y, *NEBRASKA, *CYANOPHYTA, ALGAL CONTROL, *PONDWEEDS, PONDS, *THER W73-02349
TION, CHEMICAL/ *WATER POLLUTION CONTROL, *PHOSPHORUS, *EUTROPHICA W73-02478
QUATIC WEED CONTROL, *MECHANICAL CONTROL, *DEWATERING.: /ALGAE, *A W71-06188
DOME/ *ALGICIDES, *WATER QUALITY CONTROL, *AQUATIC WEED CONTROL, * W71-05083
R QUALITY CONTROL, *AQUATIC WEED CONTROL, *DOMESTIC WATER, *SURFAC W71-05083
TS, *EUTROP/ *DETERGENTS, *ALGAL CONTROL, *FORMULATION, *SURFACTAN W71-06247
RVESTING OF ALGAE, *AQUATIC WEED CONTROL, *MECHANICAL CONTROL, *DE W71-06188
POTABLE WATER,/ *WATER POLLUTION CONTROL, *NUCLEAR POWER PLANTS, * W71-09018
ARIES, *NUTRIENTS, WATER QUALITY CONTROL, *ECOLOGY, *ALGAE, *EUTRO W71-08775
LUTION EFFECTS, *WATER POLLUTION CONTROL, *BIODEGRADATION, FISHKIL W71-08768
FISH, BIOCONTROL/ *WATER QUALITY CONTROL, *ALGAE CONTROL, *FORAGE W71-09163
*WATER QUALITY, *WATER POLLUTION CONTROL, *DIVERSION, SEDIMENTS, W W71-09581
*WATER QUALITY CONTROL, *ALGAE CONTROL, *FORAGE FISH, BIOCONTROL W71-09163
REMOTE SENSING, *WATER POLLUTION CONTROL, *SEA WATER, *OILS, *NUTR W72-04742
N, *PHOSPHATES, *WATER POLLUTION CONTROL, *WASTE WATER TREATMENT, W72-04734

EUTROPHICATION, *WATER POLLUTION CONTROL, *WASTE WATER(POLLUTION), W72-05473
Y CONTROL, EUTROPHICATION, ALGAL CONTROL, *CHLORELLA, *WASTE WATER W72-04788
, *WASTE WATER TREATMENT, *ALGAL CONTROL, *AQUATIC ALGAE, *EUTROPH W72-03970
R QUALITY, RESER/ *WATER QUALITY CONTROL, *FLOW AUGMENTATION, WATE W72-04260
*EUTROPHICATION, *WATER QUALITY CONTROL, *NON-STRUCTURAL ALTERNAT W72-04279
*EUTROPHICATION, *WATER QUALITY CONTROL, *NITROGEN COMPOUNDS, *PH W70-04080
RATURE, SEASONAL, / *TEMPERATURE CONTROL, *STANDARDS, *WATER TEMPE W70-09595
TION ABATEMENT, *WATER POLLUTION CONTROL, *WATER COOLING, *STEAM T W70-09193
HICATION, *AQUATIC ALGAE, *ALGAL CONTROL, *AQUATIC WEED CONTROL, B W72-14692
ANOPHYTA.: *ALGAL CONTROL, *INFECTION, *VIRUSES, CY W73-00246
YSIS, LAKES, SAMP/ WATER QUALITY CONTROL, *CYANOPHYTA, *WATER ANAL W72-07890
ICATION, REVIEWS, TROPHI/ *ALGAL CONTROL, *NUISANCE ALGAE, EUTROPH W72-07937
ON, *MATHEMATICAL MODELS, *ALGAL CONTROL, *RESERVOIRS, DESTRATIFIC W72-08559
LABORATORY TESTS, WATER QUALITY CONTROL, *KINETICS, *CHLOROPHYTA, W72-08376
, *KINETICS, *CHLOROPHYTA, ALGAL CONTROL, *NEW YORK.: /ITY CONTROL W72-08376
*NUISANCE ALGAE, WASHING/ *ALGAL CONTROL, *DIFFUSION, *NUTRIENTS, W72-08049
EFICIENT NUTRIENTS, *AQU/ *ALGAL CONTROL, *ESSENTIAL NUTRIENTS, *D W72-07514
CIDES, *AQUATIC ORGANISMS, *PEST CONTROL, *CHEMCONTROL, *BIBLIOGRA W73-01976
, *INHIBITION, COPPER, F/ *ALGAL CONTROL, *RESPIRATION, *CHLORELLA W72-11572
LYSIS, EVALUATION, WATER QUALITY CONTROL, *BIOASSAY, *BIOMASS, CHL W72-10883
COSTS, FISH, FISHKILL, /.*ALGAL CONTROL, *CYANOPHYTA, EXPLOSIVES, W72-11564
OLOGY, *TOXICITY, *WATER QUALITY CONTROL, *WATER POLLUTION EFFECTS W72-09646
E WATER TREATMENT, WATER QUALITY CONTROL, *REVIEWS.: /NANCE, *WAST W72-09386
TIFICATION, *WATER QUALI/ *ALGAL CONTROL, *EUTROPHICATION, *DESTRA W72-09304
DESTRATIFICATION, *WATER QUALITY CONTROL, *LAKE MORPHOLOGY, AERATI W72-09304
MICROORGANISMS, *WATER POLLUTION CONTROL, *WASTES, *PATHOGENIC BAC W72-09675
CAL TECHNIQUES, *WATER POLLUTION CONTROL, *INSTRUMENTATION, MATHEM W72-08790
POLLUTION SOUR/ *WATER POLLUTION CONTROL, *EUTROPHICATION, *WATER W71-12091
ENTS, *PHOSPHA/ *WATER POLLUTION CONTROL, *EUTROPHICATION, *DETERG W71-11516
Y, *CHLORINATION, *WATER QUALITY CONTROL, *ODOR, ALGAE, EUTROPHICA W71-13187
AQUATIC ALGAE, / *WATER QUALITY CONTROL, *LAKES, *EUTROPHICATION, W71-13256
CATION, LAKES, *NUTRIENTS, ALGAL CONTROL, *CYANPHYTA, *NUISANCE AL W72-00798
YTICAL TECHNIQUES, WATER QUALITY CONTROL, *WASTE WATER TREATMENT.: W72-00422
L EF/ *HERBICIDES, *AQUATIC WEED CONTROL, *PARAQUAT, *ENVIRONMENTA W72-00155
*FEDER/ *LAKES, *WATER POLLUTION CONTROL, *WATER QUALITY CONTROL, W72-01659
CULTURAL WASTES, WATER POLLUTION CONTROL, *BIOLOGICAL TREATMENT, * W72-02975
, *LEGISLATION/ *WATER POLLUTION CONTROL, *PHOSPHATES, *DETERGENTS W72-01685
N, *LE/ *LAKES, *WATER POLLUTION CONTROL, *ENVIRONMENTAL SANITATIO W72-01660
, *WASTE WATER TREATMENT, *ALGAL CONTROL, *AQUATIC WEED CONTROL, C W72-02975
OLLUTION CONTROL, *WATER QUALITY CONTROL, *FEDERAL GOVERNMENT, *GR W72-01659
V/ *DETERGENTS, *WATER POLLUTION CONTROL, *PHOSPHATES, *FEDERAL GO W72-01093
EFFECTS, *MICROORGANISMS, ALGAL CONTROL, *POLLUTANT IDENTIFICATIO W72-00801
*WASTE DILUTION, WATER POLLUTION CONTROL, *NUISANCE ALGAE, *EUTROP W72-00799
S, *PHOSPHATES, *WATER POLLUTION CONTROL, *WATER POLLUTION SOURCES W72-01085
IGN, NITROGEN, DIFFUSION, *ALGAL CONTROL, AERATION, *WASTE WATER T W72-00800
S, SEWAGE EFFLUENTS, TEMPERATURE CONTROL, ASSAY, REMOTE SENSING, A W72-10162
BON DIOXIDE, BICARBONATES, ALGAL CONTROL, AERATION.: /LORELLA, CAR W72-09168
AL EFFECTS, AQUATIC ALGAE, ALGAL CONTROL, ALGAL POISONING, AQUATIC W72-11186
ON, PUBLIC HEALTH, WATER QUALITY CONTROL, ALGAL CONTROL, FEDERAL G W72-06655
, CHLOROPHYTA, CYANOPHYTA, ALGAL CONTROL, ALGICIDES, COPPER SULFAT W72-14694
S, LAKES, PONDS, WATER POLLUTION CONTROL, AIRCRAFT, COPPER SULFATE W72-14673
RTIARY TREATMENT, *WATER QUALITY CONTROL, AMMONIA, NITROGEN, PHOSP W71-09450
E, *ALGAL CONTROL, *AQUATIC WEED CONTROL, BIOCHEMICAL OXYGEN DEMAN W72-14692
TIARY TREATMENT, WATER POLLUTION CONTROL, BIODEGRADATION.: /T, TER W71-13939
*OCEANS, *COASTS, *WATER QUALITY CONTROL, BIOMASS, TOXICITY, EUTRO W71-11793
M, NITROGEN, PHOSPHOROUS, *ALGAL CONTROL, CALIFORNIA.: / METABOLIS W71-13553
T, *ALGAL CONTROL, *AQUATIC WEED CONTROL, CALIFORNIA.: /R TREATMEN W72-02975
E ALGAE, *CARBON DIOXIDE, *ALGAL CONTROL, CYANOPHYTA, CARBON, CHLO W72-10607
MIC FEASIBILITY, WATER POLLUTION CONTROL, CALIFORNIA, ARIZONA.: /O W71-09157
WASTES, SAMPLING, WATER QUALITY CONTROL, COST ANALYSIS, WASTE WAT W71-08953
UTION ABATEMENT, WATER POLLUTION CONTROL, COAGULATION.: /ION, POLL W73-02187
TION, *DETERGENTS, WATER QUALITY CONTROL, DOMESTIC WASTES, RUNOFF, W72-11186
AL EFFECTS, DIQUAT, ALGAE, ALGAL CONTROL, DIATOMS, PERSISTENCE, BA W72-00155
CTERIA, *LAKES, SYMBIOSIS, ALGAL CONTROL, EUTROPHICATION, WATER PO W71-13257
*DENITRIFICATION, WATER QUALITY CONTROL, EUTROPHICATION, ACTIVATE W72-00800
LLUTION EFFECTS, WATER POLLUTION CONTROL, ENERGY, DIFFUSION, THERM W71-12092
ER QUALITY, ALGAE, WATER QUALITY CONTROL, EUTROPHICATION, WASTE WA W72-12487
ION, EFFICIENCIES, WATER QUALITY CONTROL, EUTROPHICATION, ALGAL CO W72-04788
S, *WASTE WATER TREATMENT, ALGAL CONTROL, EUTROPHICATION, MODEL ST W72-04787
WATER DISPOSAL, *WATER POLLUTION CONTROL, ECONOMIC FEASIBILITY, ST W72-14878
TH, WATER QUALITY CONTROL, ALGAL CONTROL, FEDERAL GOVERNMENT, NUTR W72-06655
OL, *PATHOGENIC BACTERIA, *ALGAL CONTROL, FRESH WATER, CYANOPHYTA, W72-05457
TOXICI/ *EUTROPHICATION, *ALGAL CONTROL, FINANCING, PLANT GROWTH, W71-13172
TER CIRCULATION, WATER POLLUTION CONTROL, FISHKILL, NEW HAMPSHIRE, W72-09304
C/ *ALGAE, *AQUATIC WEEDS, *WEED CONTROL, GROWTH, *CULTURES, *ALGI W71-06102
ITY, FALLOUT, BIOCONTROL, INSECT CONTROL, IRRIGATION, PREDATION, P W72-00846
SANCE ALGAE, *CYANOPHYTA, *ALGAL CONTROL, LABORATORY TESTS, BUOYAN W71-13184
NYLS, PUBLIC HEALTH, ENVIRONMENT CONTROL, LEGAL ASPECTS, REGULATIO W73-00703
LGAE, *PLANKTON, WATER POLLUTION CONTROL, METHODOLOGY, MICROENVIRO W72-03218
EUTROPHICATION, *WATER POLLUTION CONTROL, MANAGEMENT, ANALYSIS, WA W72-09171
N, EUTROPHICATION, WATER QUALITY CONTROL, MASSACHUSETTS, *NUTRIENT W72-06046
L OXYGEN DEMAND, WATER POLLUTION CONTROL, NITRATES, EFFLUENTS, DET W72-09167
HORUS, *LAKE BEDS, *ALGAE, ALGAL CONTROL, NUTRIENT REQUIREMENTS, S W72-11617
ATION, LAKES, WASHINGTON, *ALGAL CONTROL, NUTRIENTS, BIOMASS, *CYA W72-00799
ON, *NUTRIENTS, *WATER POLLUTION CONTROL, ORGANIC WASTES, INDUSTRI W73-00840
POLLUTION CONTROL, WATER QUALITY CONTROL, PHOSPHORUS, AQUATIC ENVI W72-06638
*CHLOROPHYTA, *INHIBITORS, ALGAL CONTROL, PLANT GROWTH REGULATORS, W71-13793
OMIC JUSTIFICATION, AQUATIC WEED CONTROL, RIVERS, AGRICULTURE, ROO W71-13172
ON, WILDLIFE CONSERVATION, FLOOD CONTROL, RECREATION, HYDROLOGIC D W72-05594
S, ENZYMES, COLOR, WATER QUALITY CONTROL, RESEARCH AND DEVELOPMENT W72-04070

.TREATMENT, *CHLORINATION, *ALGAL	CONTROL, RIVERS, WATER POLLUTION	W71-00099
ING, ANAEROBIC CONDITIONS, ALGAL	CONTROL, SPAWNING, WARM-WATER FIS	W70-09595
, WATER QUALITY, WATER POLLUTION	CONTROL, SURVEYS, INVESTIGATIONS,	W70-08249
, WATER QUALITY, WATER POLLUTION	CONTROL, SURVEYS, INVESTIGATIONS,	W70-08248
EUTROPHICATION, *WATER POLLUTION	CONTROL, SEWAGE EFFLUENTS, AQUATI	W71-11507
GOVERNMENT, UNITED STATES, ALGAL	CONTROL, STANDARDS, LEGISLATION,	W72-01093
, *AGRICULTURE, *WATER POLLUTION	CONTROL, SEDIMENTS, FARM WASTES,	W72-00846
, WATER TREATMENT, WATER QUALITY	CONTROL, STANDARDS, WATER POLLUTI	W72-01660
ATION, *RECYCLING, WATER QUALITY	CONTROL, SEWAGE TREATMENT, SEWAGE	W72-09378
LLUTION SOURCES, WATER POLLUTION	CONTROL, TROPHIC LEVELS, PHOSPHAT	W71-12072
FISH, *HERBIVORES, *AQUATIC WEED	CONTROL, TILAPIA, CARP, ALGAE, RO	W71-04528
S, C/ *ALGICIDES, *ALGAE, *ALGAL	CONTROL, TESTING, LABORATORY TEST	W71-06189
LICA, *BACTERIA, WATER POLLUTION	CONTROL, WASTE WATER TREATMENT.: /	W71-01899
ASTE DISPOSAL, AERATION, EROSION	CONTROL, WATER POLLUTION SOURCES,	W71-01192
EUTROPHICATION, *WATER POLLUTION	CONTROL, WATER POLLUTION EFFECTS,	W71-01488
ION CONCENTRATION, WATER QUALITY	CONTROL, WASTE WATER TREATMENT, C	W72-04431
MES, PERMEABILITY, WATER QUALITY	CONTROL, WATER POLLUTION EFFECTS.	W72-06037
ATION, *CARBON, *WATER POLLUTION	CONTROL, WATER POLLUTION EFFECTS,	W72-04769
ICIDS, COLORADO, WATER POLLUTION	CONTROL, WASHINGTON, CHEMICAL OXY	W71-11006
RONMENT, WATER PROPERTIES, ALGAL	CONTROL, WATER ANALYSIS, CHEMCONT	W72-11724
AL GOVERNMENTS, *WATER POLLUTION	CONTROL, WATER QUALITY CONTROL, P	W72-06638
ICATION, NUT/ *ALGICIDES, *ALGAL	CONTROL, WATER POLLUTION, EUTROPH	W72-07508
DUSTRIAL WASTES, WATER POLLUTION	CONTROL, WATER POLLUTION SOURCES,	W72-14282
PARTICIPATION, *WATER POLLUTION	CONTROL, WISCONSIN, LAKES, URBAN	W73-02479
TES, PATHOGENS, BIOENVIRONMENTAL	CONTROLS.: / FOOD, PROCESSING WAS	W72-00846
SAMPLING, MONITORING, AUTOMATIC	CONTROLS, AERATION, SEDIMENTATION	W72-00042
TATION, BUOYANCY, WIND VELOCITY,	CONVECTION, WATTER POLLUTION EFFE	W71-09852
, *PRIMARY PRODUCTIVITY, *ENERGY	CONVERSION, STANDING CROP, CHLORO	W72-06283
NE ENVIRONMENT, CALIFORNIA WATER	CONVEYANCE, CANALS, HEAVY METALS.	W71-09581
NA, LEPTOCYLINDRUS, PACIFIC OCE/	COOK INLET(ALASKA), CYCLOTELLA NA	W71-11008
ERVOIR, MONTROSE LAKE, THOMAS L/	COOLING RESERVOIR, ARTIFICIAL RES	W70-09596
GN, STANDARDS, FISH KILL, ALGAE,	COOLING TOWERS, DENSITY, VISCOSIT	W70-09612
DEMAND, CHEMICAL OXYGEN DEMAND,	COOLING TOWERS, *WASTE WATER TREA	W71-09524
CHELATION, APPLICATION METHODS,	COOLING TOWERS, WATER POLLUTION E	W71-06189
ANTS, ENVIRONMENTAL ENGINEERING,	COOLING TOWER: /AQUATIC ALGAE, PL	W72-00939
OPHYTA, PHAEOPHYTA, CHLOROPHYTA,	COOLING WATER, HYDROLOGIC DATA.: /	W72-11252
*WATER POLLUTION CONTROL, *WATER	COOLING, *STEAM TURBINES, *THERMA	W70-09193
WASTE HEAT, *CENTRAL HEATING AND	COOLING, *SPACE HEATING, *FARM US	W70-09193
N, VOLUME, MIXING, SAFETY, WATER	COOLING, ANAEROBIC CONDITIONS, AL	W70-09595
, IODINE, ATOMIC ENERGY, REACTOR	COOLING, IMPERIAL VALLEY(CALIF),	W71-09157
RMAL POLLUTION, WATER POLLUTION,	COOLING, THERMAL POWERPLANTS, HEA	W73-01704
COMMUNITIES, CHLORINE, MOLLUSKS,	COPEPODS, DIPTERA, GULFS, GEOLOGI	W73-00855
FLAGELLATES, CILIATES, NAUPLII,	COPEPODIDS.: /RIA, TRICHODESMIUM,	W72-14313
US HELGOLANDICUS, COSCINODISCUS,	COPEPODIDS, *CALANUS GLACIALIS, D	W72-09248
KTON, BIOCHEMICAL OXYGEN DEMAND,	COPEPODS.: /QUATIC ALGAE, ZOOPLAN	W71-11876
EFFECTS, SAMPLING, TEMPERATURE,	COPEPODS, NITROGEN, BACTERIA, OXI	W71-11008
ASS, MINNOWS, FISHKILL, HERRING,	COPEPODS, CRUSTACEANS, DIATOMS, C	W71-11517
OPHYTA, CYANOPHYTA, CHRYSOPHYTA,	COPEPODS, CRUSTACEANS, LABORATORY	W72-00150
, *FOODS, *LIFE HISTORY STUDIES,	COPEPODS, DIATOMS, CHLOROPHYTA, T	W72-14805
BRATES, ISOLATION, REFRACTIVITY,	COPEPODS, MARINE ANIMALS, MOLLUSK	W73-00371
ENVIRONMENT, SEA WATER, DIATOMS,	COPEPODS, DINOFLAGELLATES, CRUSTA	W73-00935
*WATER QUALITY, BENTHOS, CRABS,	COPEPODS, MARINE ALGAE, SHRIMP, F	W72-08142
*ZOOPLANKTON, *FOODS, DAPHNIA,	COPEPODS, ALGAE, NUTRIENT REQUIRE	W72-07940
GROWTH RATES, ALGAE, FOOD WEBS,	COPEPODS, ORGANOPHOSPHORUS COMPOU	W72-07892
NDS, INTERTIDAL AREAS, ANNELIDS,	COPEPODS, CRABS, AQUATIC LIFE, BA	W72-07911
RS, *CHEMCONTROL, NEW HAMPSHIRE,	COPEPODS, DAPHNIA, SAMPLING, DEPT	W71-03014
NEDESMUS, ROTIFERS, CRUSTACEANS,	COPEPODS, NUTRIENTS, NITROGE: /CE	W72-05968
S, RESINS, *SEA WATER, BIOASSAY,	COPEPODS, OYSTERS, DINOFLAGELLATE	W72-04708
LEVEL, GASES, MUDS, ADSORPTION,	COPP: /RTIES, PESTICIDES, TROPHIC	W72-00847
POTENTIOMETRY, CADMIUM IONS,	COPPER IONS, OOCYSTIS.:	W72-01693
PSHIRE, WATER POLLUTION EFFECTS,	COPPER SULFATE.: /SHKILL, NEW HAM	W72-09304
, FOOD CHAINS, DISSOLVED OXYGEN,	COPPER SULFATE, HERBICIDES, WATER	W72-04759
BIOLOGICAL TREATMENT, CHEMICALS,	COPPER SULFATE, CHLORINE, WATER R	W71-05083
S, *NEW HAMPSHIRE, ALGAE, COLOR,	COPPER SULFATE, NETS, HYDROGEN IO	W72-07890
PHYTA, ALGAL CONTROL, ALGICIDES,	COPPER SULFATE, SALINITY, HARDNES	W72-14694
TER POLLUTION CONTROL, AIRCRAFT,	COPPER SULFATE.: /AKES, PONDS, WA	W72-14673
MANGANESE, ZINC, NICKEL, COBALT,	COPPER SULFATE, HEAVY METALS, MET	W72-12739
BIOINDICATORS, SILICA, CHROMIUM,	COPPER, *NEW YORK, DIATOMS, PHOSP	W72-09111
IUM, POTASSIUM, IRON, MANGANESE,	COPPER, ALGAE, SAMPLING, CHLOROPH	W71-11001
HANGE, POLLUTANT IDENTIFICATION,	COPPER, CADMIUM, CHROMIUM, NICKEL	W72-14303
R POLLUTION EFFECTS, ZINC, LEAD,	COPPER, CHROMIUM, SILVER, CHLOROP	W72-14694
, *FLUORESCENCE, TRACE ELEMENTS,	COPPER, CHLORELLA, CULTURES, NUTR	W72-07660
ALGAE, *HEAVY METALS, *TOXICITY,	COPPER, CHLOROPHYTA, WATER POLLUT	W71-05991
RATION, *CHLORELLA, *INHIBITION,	COPPER, FLUORIDES, IONS, CYTOLOGI	W72-11572
LS, ELECTRODES, NITRATES, ALGAE,	COPPER, IONS, ION TRANSPORT.: /TA	W72-01693
, *SEA WATER, *DATA COLLECTIONS,	COPPER, LEAD, HEAVY METALS, MUSSE	W72-11925
M RAIODISOTOPES, ALUMINUM, IRON,	COPPER, MAGNESIUM, MANAGNESE, CAL	W71-09188
ATION ANALYSIS, *TRACE ELEMENTS,	COPPER, MOLYBDENUM, MANGANESE, CO	W71-11515
TION, ABUNDANCE, IRON, CHROMIUM,	COPPER, NICKEL, RUTHENIUM, CERIUM	W71-13243
LLATES, DIATOMS, ALGAE CULTURES,	COPPER, TOXICITY.: /RS, DINOFLAGE	W72-04708
ETALS, CHEMICAL WASTES, MERCURY,	COPPER, ZINC, PESTICIDES, PESTICI	W72-12576
XINS, POISONS, POISONOUS PLANTS,	CORAL, INVERTEBRATES, ALGAE, MARI	W71-08597
NTS, TURBIDITY, ORGANIC LOADING,	CORAL, MOLLUSKS, WORMS.: /POLLUTA	W71-01475
ADIATION, ECOSYSTEMS, ESTUARIES,	CORAL, REEFS, WATER RESOURCES, SY	W71-13723
NE PLANTS, PRIMARY PRODUCTIVITY,	CORAL, REEFS, ANTIBIOTICS(PESTICI	W72-14294
BIOASSAY, CYANOPHYTA, *FLORIDA,	CORAL, SALINITY, CURRENTS(WATER),	W72-11876
NESE RADIOISOTOPES, BOGUE SOUND,	CORE SOUND.: MANGA	W73-00818
SCHLIERSEE, TEGERNSEE, ANABAENA,	COREGONS, LYNGBYA LIMNETICA, OSCI	W71-00117
, LUCIOPERCA LUCISPERCA, BURBOT,	COREGONUS LAVARETUS, COREGONUS AL	W72-12729
CA, BURBOT, COREGONUS LAVARETUS,	COREGONUS ALBULA, WHITEFISH, BRAN	W72-12729

D CHAINS, ABSORPTION, SEDIMENTS,	CORES, ANALYTICAL TECHNIQUES, RAD	W73-00790
, SEA WATER, SALINITY, SAMPLING,	CORES, DISTRIBUTION PATTERNS, CUR	W73-00367
L PROPERTIES, *SEDIMENTS, CLAYS,	CORES, SEASONAL, ALGAE, MUD-WATER	W72-05469
BODO/ *HEMIPTERA, *TRICHOPTERA,	CORIXIDS, EUDORINA, NOTONECTIDAE,	W70-06655
ON, WATER QUALITY, RIVERS, IOWA,	CORN, DISCHARGE, CHLOROPHYLL.: /I	W71-06445
, SEWAGE SETTLING, UNION CARBIDE	CORP, POLYOX, CARBOWAX, FLOCCULAN	W70-06968
ETERGENT BAN, WYANDOTTE CHEMICAL	CORPORATION, FMC CORPORATION, SOA	W70-07283
NDOTTE CHEMICAL CORPORATION, FMC	CORPORATION, SOAP AND DETERGENT A	W70-07283
S, VOLUME, MATHEMATICAL STUDIES,	CORRELATION ANALYSIS.: /, CULTURE	W70-05547
, SEASONAL, REGRESSION ANALYSIS,	CORRELATION ANALYSIS.: / NITRATES	W72-12393
N, FOURIER ANALYSIS, BIORHYTHMS,	CORRELATION ANALYSIS, PRIMARY PRO	W72-11719
URE, HYDROGEN ION CONCENTRATION,	CORROSION, WATER POLLUTION EFFECT	W72-08683
TERIA, AESTHETICS, COLOR, TASTE,	CORROSION, DISEASES, EUTROPHICATI	W71-09958
A, NITRATES, NITRITES, TOXICITY,	CORROSION, PUBLIC HEALTH, ODORS,	W72-04086
ENTATION, TOXICITY, TASTE, ODOR,	CORROSION, ENZYMES, CHLEATION, AL	W72-06837
TUBERCULOSIS, PENICILLIN, DRUGS,	CORTISONE, HORMONES, CHLORAMPHENI	W72-13800
HYLENE, POLYPROPYLENE, BACILLUS,	CORYNEBACTERIA, STREPTOTHRIX HYOR	W72-09675
ATTERNS, *CALANUS HELGOLANDICUS,	COSCINODISCUS, COPEPODIDS, *CALAN	W72-09248
TORIA, PLEUROSIGMA, NAUPLII, LA/	COSCINODISCUS, NITZSCHIA, OSCILLA	W71-11876
INIANS, QUALITATIVE COMPOSITION,	COSCINOSIRA POLYCHORDA, THALASSIO	W72-03543
OEDOGONIUM, ANACYSTIS, CERATIUM,	COSMARIUM.: /L BASIN, TRIBONEMA,	W72-12995
LAKE, OEDOGONIUM, RHIZOCLONIUM,	COSMARIUM, PHACUS, CHAETOCEROS, M	W72-09111
TION, PROTEIN, LABORATORY TESTS,	COST ANALYSIS, ALGAE, WASTE WATER	W72-10390
, ORGANIC LOADING, ODOR, SLUDGE,	COST ANALYSIS, NITROGEN, CLAYS, *	W72-11642
GEN DEMAND, WATER SPORTS, PARKS,	COST ANALYSIS, MICHIGAN, WASTE WA	W72-00042
EATMENT, *PHOSPHORUS, *NITROGEN,	COST ANALYSIS, OZONE, DENITRIFICA	W73-02187
SAMPLING, WATER QUALITY CONTROL,	COST ANALYSIS, WASTE WATER TREATM	W71-08953
FLOATATION, DEWATERING, DRYING,	COST ANALYSES, WASTE WATER TREATM	W71-08951
ERIA, CHLORINATION, GROUNDWATER,	COST ANALYSIS.: /SIS, ALGAE, BACT	W71-07123
BIODEGRADATION, SEEPAGE, SLUDGE,	COST ANALYSIS, *BIOCHEMICAL OXYGE	W71-07099
ECTS, HARD WATER, PUBLIC HEALTH,	COST COMPARISONS, AQUATIC ALGAE,	W71-09429
EFFECT, WATER POLLUTION SOURCES,	COST COMPARISONS, BENEFICIAL USE,	W72-00974
QUALITY, CALIFORNIA, BIOCONTROL,	COST-BENEFIT THEORY.: /SE, WATER	W72-00974
IA RIVER, PATUXENT RIVER, CONTRA	COSTA POWER PLANT, SAN JOAQUIN RI	W72-07907
ATE OF POLLUTANTS, STEPHANOPYXIS	COSTATA, ASSIMILATION, EMILIANA(C	W73-00380
SA, NITZSCHIA PALEA, SKELETONEMA	COSTATUM.: CHLORELLA PYRENOIDO	W71-03027
XLEYI, PORPHYRIDIUM, SKELETONEMA	COSTATUM.: /RTERI, COCCOLITHUS HU	W72-12998
OUTHERN LONG ISLAND, SKELETONEMA	COSTATUM, THALASSIOSIRA, CHAETOCE	W71-07875
SSIOSIRA PSEUDONANA, SKELETONEMA	COSTATUM, DUNALIELLA TERTIOLECTA,	W72-05614
I/ *INTERREGULATION, SKELETONEMA	COSTATUM, THALASSIONEMA NITZSCHIO	W73-00441
CHAETOCEROS AFFINIS, SKELETONEMA	COSTATUM, CHAETOCEROS SOCIALIS.: /	W73-00442
ETOCEROS CURVISETUS, SKELETONEMA	COSTATUM, GLENODINIUM FOLIACUM, G	W72-07166
PYTHIUM DEBARYANUM, SKELETONEMA	COSTATUM, NUCLEOTIDES, ETHANOL, M	W73-00942
ACTYLUM TRICORNUTUM, SKELETONEMA	COSTATUM, CARBOY, OSCILLATORIA ER	W72-09103
IC FEASIBILITY, COSTS, ESTIMATED	COSTS.: /ATION, FLOTATION, ECONOM	W72-13508
IENTS, *WASTE TREATMENT, *ALGAE,	COSTS.: /ICAL OXYGEN DEMAND, NUTR	W72-08396
CAL MODELS, NITROGEN, ABATEMENT,	COSTS, ANALYTICAL TECHNIQUES, MUN	W72-09171
TIMATES, ECONOMICS, FEASIBILITY,	COSTS, ANNUAL COSTS, PILOT PLANTS	W70-10174
ALUM, RECYCLING, TREATMENT	COSTS, ASIA.:	W72-13508
SILICATES, CYANOPHYTA, TOXICITY,	COSTS, CARBON DIOXIDE.: /IATOMS,	W72-07508
FLOTATION, ECONOMIC FEASIBILITY,	COSTS, ESTIMATED COSTS.: /ATION,	W72-13508
ONTROL, *CYANOPHYTA, EXPLOSIVES,	COSTS, FISH, FISHKILL, RESERVOIRS	W72-11564
MICS, FEASIBILITY, COSTS, ANNUAL	COSTS, PILOT PLANTS.: /TES, ECONO	W70-10174
ALGAE, *FLOCCULATION, EQUIPMENT,	COSTS, SEPARATION TECHNIQUES, COA	W70-08904
TIFICATION, *COBALT, *RADIATION,	COSTS, SEWAGE BACTERIA, ALGAE, CO	W72-00383
RVOIRS, FISHERIES, WATER SUPPLY,	COSTS, WATER DEMAND, ECONOMIC JUS	W71-13172
S, CERASON, DIATHENE, PROPAZINE,	COTORON, DIURON, LINURON.: /DULAN	W71-12071
CLEIC ACID, SAMPLE PRESERVATION,	COULTER COUNTER.: /, *DEOXYRIBONU	W71-11562
ID, SAMPLE PRESERVATION, COULTER	COUNTER.: /, *DEOXYRIBONUCLEIC AC	W71-11562
TOLERANCES, SAMPLE PRESERVATION,	COUNTING CHAMBERS, NUTRIENT CYCLI	W72-07899
HERMISTORS, DATA INTERPRETATION,	COUNTING.: /ANDICUS(PACIFICUS), T	W72-07892
DIVERSITY, SAMPLE PRESERVATION,	COUNTING.: /ANSEN BOTTLE, SPECIES	W72-07663
E DECAY, GAMMA-RAY SPECTROMETRY,	COUNTING.: /ACTINIDES, RADIOACTIV	W73-01434
AMPLE PREPARATION, SCINTILLATION	COUNTING.: /N, AUTORADIOGRAPHY, S	W72-09239
IABILITY, BIAS, SYNCHRONY INDEX,	COUNTING.: /URES, AUTOSPORES, VAR	W72-11920
ER CHROMATOGRAPHY, SCINTILLATION	COUNTING, CA-45.: /RBIGNIANA, PAP	W72-12634
SAMPLE PRESERVATION, FORMALIN,	COUNTING, CYCLOTELLA, OSCILLATORI	W72-10875
A FORMOSA, RHODOMONAS LACUSTRIS,	COUNTING, ORTHOPHOSPHATES, CHLORO	W73-01435
TARALDEHYDE, SAMPLE PREPARATION,	COUNTING, SAMPLE PRESERVATION, FL	W73-00273
NEUROSPORA CRASSA, SCINTILLATION	COUNTING, GLUCOSE, MEMBRANE FILTE	W72-14326
ICAL INTERFERENCE, SCINTILLATION	COUNTING, CYANIDIUM CALDARIUM, TH	W72-14301
*MICROSCOPIC	COUNTS, *AMORPHOUS MATTER.:	W72-06191
LIMNETICA, DICTYOSPHAERIUM, CELL	COUNTS, FRAGILARIA, ASTERIONELLA,	W71-03014
RA OVALIS, OPEPHORA MARTYI, CELL	COUNTS, NAVICULA, NITZSCHIA, STEP	W71-12855
AND, *POLLUTED ESTUARIES, DURHAM	COUNTY RIVERS.: *ENGL	W72-08804
*GREENE	COUNTY(ILL).:	W72-05869
ESIDUAL FOOD, GREEN RIVER(WASH),	COWLITZ TROUT HATCHERY(WASH).: /R	W71-11006
CHONDRUS CRISPUS, COD, FLOUNDER,	CRAB EGGS, PELAGIC EGGS, GADUS MO	W72-08142
ADUS MORHUA, GRASS SHRIMP, GREEN	CRAB, PENICILLIUM, STREPTOMYCIN,	W72-08142
TIONSHIPS, MARINE ALGAE, MOSSES,	CRABS, ALGAE, SNAILS, MUSSELS, IS	W72-03347
DIMENTS, PRODUCTIVITY, TOXICITY,	CRABS, ALGAE, INVERTEBRATES, FISH	W71-01475
TIDAL AREAS, ANNELIDS, COPEPODS,	CRABS, AQUATIC LIFE, BACTERIA.: /	W72-07911
CRUSTACEANS, GASTROPODS, SNAILS,	CRABS, BENTHIC FLORA, BENTHIC FAU	W73-00932
CTERIA, *WATER QUALITY, BENTHOS,	CRABS, COPEPODS, MARINE ALGAE, SH	W72-08142
IZOPHORA, BOSTRYCHIETUM, FIDDLER	CRABS, MESOFAUNA, TURBELLARIA, SP	W72-07911
NS, BIRDS, SALMON, STICKLEBACKS,	CRABS, OTTERS, ZOOPLANKTON, PHYTO	W72-00960
TUARINE ENVIRONMENT, MONITORING,	CRABS, SHRIMP.: /E, SEDIMENTS, ES	W72-10958
PHY, SEDIMENTS, OYSTERS, SHRIMP,	CRABS, WATER POLLUTION SOURCES, S	W71-07412
EKMAN DREDGE, PUMPKINSEED, BLACK	CRAPPIE, CHAIN PICKEREL, TENDIPES	W72-14280
*CHLORELLA VULGARIS, NEUROSPORA	CRASSA, SCINTILLATION COUNTING, G	W72-14326

OIDES, NUPHAR ADVENA, EICHHORNIA	CRASSIPES, JUSTICIA AMERICANA, AL	W73-01089
CHIRONOMIDS, SAWDUST, EICHHORNIA	CRASSIPES, SALVINIA RCTUNDIFOLIA,	W71-01488
E/ *HYDERABAD(INDIA), EICHHORNIA	CRASSIPES, TYPHA ANGUSTATA, IPOMO	W72-01363
, RADIOACTIVE WASTES, TENNESSEE,	CRAYFISH, MIDGES, GAMMA RAYS, BLO	W70-06668
OK SALMON, BASS, TROUT, MINNOWS,	CRAYFISH, MOSQUITOES, FOOD CHAIN,	W71-07417
S, AQUATIC ANIMALS, CLAMS, CARP,	CRAYFISH, AQUATIC ALGAE, ABSORPTI	W72-06342
*BEAVER	CREEK BASIN(KY).:	W71-07942
BLIC UTILITIES DISTRICT, *INDIAN	CREEK RESERVOIR, NUTRIENT BALANCE	W72-02412
*PHOTOSYNTHETIC MODEL, *TUTTLE	CREEK RESERVOIR(KAN), GREAT PLAIN	W72-12393
JECT, LOPEZ CREEK, ARROYO GRANDE	CREEK.: *LOPEZ WATER SUPPLY PRO	W72-05594
KEMMERER SAMPLER, SECCHI *YARDS	CREEK(NJ), RELATIVE CONDITION FAC	W72-14280
SHELTER VALLEY	CREEK(ONTARIO).:	W73-01627
PHONEMA OLIVACEOIDES, WHITE CLAY	CREEK(PA.).: GOM	W73-00242
CALORIES, DETRIVORES, LINESVILLE	CREEK(PA).: /AUTUMNAL COMMUNITY,	W71-10064
IES DIVERSITY, DIRECT EF/ RIDLEY	CREEK(PA), LITITZ CREEK(PA), SPEC	W72-08579
ECT EF/ RIDLEY CREEK(PA), LITITZ	CREEK(PA), SPECIES DIVERSITY, DIR	W72-08579
OPEZ WATER SUPPLY PROJECT, LOPEZ	CREEK, ARROYO GRANDE CREEK.: *L	W72-05594
, ANABAENA / BAY OF QUINTE, SALT	CREEK, CHARA, ELODEA, CALLITRICHE	W73-00840
RIDINIUM, SENECA RIVER, NINEMILE	CREEK, VAN DORN SAMPLER, KEMMERER	W72-09111
BATOPHORA OERSTEDI, ACETABULARIA	CRENULATA, VALONIA VENTRICOSA, VA	W72-11800
RMATA, TRACHELOMONAS HISPIDA VAR	CRENULATOCOLLIS.: /RACHELOMONAS A	W73-00918
CLADOGRAMMA KOLBEI, CLADOGRAMMA	CRENULATUS, CLADOGRAMMA LINEATUS-	W71-11527
ARTHURS LAKE, LAKE SORELL, LAKE	CRESCENT, TOOMS LAKE, WOODS LAKE,	W73-01094
URIS, LAMINARIA DIGITATA, AGARUM	CRIBROSUM, *NOVA SCOTIA, *ST. MAR	W73-00366
BARNACLES, ARTHROPODS, SARGASSUM	CRISPUM, DIGENIA SIMPLEX, HALOPHI	W73-00855
D, POLYSIPHONIA LANOSA, CHONDRUS	CRISPUS, COD, FLOUNDER, CRAB EGGS	W72-08142
, CANDIDA PARAPSILOSIS, CHONDRUS	CRISPUS, RHODOTORULA GLUTINIS, SA	W72-04748
DROGEN ION CONCENTRATION, DESIGN	CRITERIA, MATHEMATICAL MODELS, *B	W72-04787
OXYGEN, LIGHT INTENSITY, DESIGN	CRITERIA, *OXIDATION LAGOONS, SEW	W72-04789
NS, *TERTIARY TREATMENT, *DESIGN	CRITERIA, TRICKLING FILTERS, ALGA	W71-07113
HAR/ *OXIDATION LAGOONS, *DESIGN	CRITERIA, *WATER QUALITY, ALGAE,	W71-07100
FECTIVENESS, TESTING PROCEDURES,	CRITERIA, CLASSIFICATIONS.: /, EF	W71-09789
ONS, *OXIDATION LAGOONS, *DESIGN	CRITERIA, BIODEGRADATION, ALGAE,	W72-09386
S, ANALYTICAL TECHNIQUES, DESIGN	CRITERIA, KINETICS, *WASTE WATER	W72-02363
LIME, ROCKS, METABOLISM, DESIGN	CRITERIA, *WASTE WATER TREATMENT,	W71-13341
ILLINOIS RIVER(ILL), DIAGNOSTIC	CRITERION.: /AL GROWTH POTENTIAL,	W71-11033
.: ENVIRONMENTAL STATEMENT,	CRITICAL NUCLIDE PATHWAY, ABALONE	W72-00959
CAMBIA RIVER, PINFISH, FLOUNDER,	CROAKER, MENHADEN.: /LOR 1254, ES	W71-07412
ION, 2-4-D, DALAPON, HERBICIDES,	CROP PRODUCTION.: /NITROGEN FIXAT	W71-12071
YTA, FERTILIZATION, TEMPERATURE,	CROP PRODUCTION, FUNGI, SOILS, SY	W73-02473
, IRRIGATION, AGRICULTURE, RICE,	CROP RESPONSE.: /IDES, HERBICIDES	W72-14807
*PRIMARY PRODUCTIVITY, *STANDING	CROP, *ALGAE, DIATOMS, MEASUREMEN	W71-12855
KTON, *WATER ANALYSIS, *STANDING	CROP, *PRIMARY PRODUCTIVITY, LAKE	W73-01435
GET, PRODUC/ *STREAMS, *STANDING	CROP, *TROPHIC LEVEL, *ENERGY BUD	W71-10064
FRESHWATER, *BIOASSAY, *STANDING	CROP, CARBON, NITROGEN, PHOSPHORU	W72-07648
S, CONDUCTIVITY, ALGAE, STANDING	CROP, CHEMICAL ANALYSIS, MOSQUITO	W71-01488
TY, *ENERGY CONVERSION, STANDING	CROP, CHLOROPHYTA, CYANOPHYTA, RH	W72-06283
ATLANTIC OCEAN, DEPTH, STANDING	CROP, CURRENTS(WATER), WAVES(WATE	W72-04766
RIENTS, PHOTOSYNTHESIS, STANDING	CROP, ECOLOGICAL DISTRIBUTION: /T	W72-14313
PRODUCTIVITY, PLANKTON, STANDING	CROP, LIMNOLOGY, EUTROPHICATION.:	W72-08559
OPLANKTON, *ESTUARIES, *STANDING	CROP, SEASONAL, DIATOMS, DINOFLAG	W71-07875
FEEDING RATES, BENTHOS, STANDING	CROPS.: / EUTROPHICATION, PONDS,	W72-14796
IVITY, WATER POLLUTION, STANDING	CROPS.: /GAE, CLADOPHORA, PRODUCT	W72-13662
, PRIMARY PRODUCTIVITY, STANDING	CROPS, *PHOTOSYNTHESIS.: /OXICITY	W72-14690
SHEDS(BASINS), SMALL WATERSHEDS,	CROPS, EGGS, FISHERIES, ON-SITE I	W72-00959
CARBON, EUTROPHICATION, STANDING	CROPS, LABORATORY TESTS, CHEMICAL	W72-09163
E, INORGANIC COMPOUNDS, STANDING	CROPS, LIMITING FACTORS, EUTROPHI	W72-13653
ATER POLLUTION EFFECTS, STANDING	CROPS, MOLLUSKS, LARVAE, CHRYSOPH	W73-00935
SCUBA DIVING, ECOLOGY, STANDING	CROPS, PATH OF POLLUTANTS.: /ING,	W73-00366
AL ECOLOGY, PHAEOPHYTA, STANDING	CROPS, RESISTANCE, PRIMARY PRODUC	W72-12940
ENVIRONMENTAL EFFECTS, *STANDING	CROPS, TIDAL EFFECTS, MARINE ALGA	W73-00367
BIOLOGICAL COMMUNITIES, STANDING	CROPS, WISCONSIN, PRIMARY PRODUCT	W73-02469
PHYLL C, CAROTENOIDS, FRAGILARIA	CROTONENSIS, TABELLARIA FLOCCULOS	W73-00384
IVER, YELLOWTAIL DAM, FRAGILARIA	CROTONENSIS, CRYPTOMONAS OVATA, S	W73-01435
UMBILICALIS, SAMPLE PREPARATION,	CRUDE OIL.: / DIGITATA, PORPHYRA	W73-01074
FRICTION, SEAWEED, PORPHYRIDIUM	CRUENTUM, PORPHYRA, GIGARTINA.: /	W71-07878
OSCILLATORIA RUBECENS, ANABAENA,	CRUSSOSTREA, OSCILLATORIA.: /UM,	W73-00840
E, ANIMAL POPULATIONS, LOBSTERS,	CRUSTACE: /, TUNNELS, MARINE ALGA	W73-02029
ATED CARBON, COLIFORMS, VIRUSES,	CRUSTACEANS, NEMATODES, EPIDEMIOL	W73-02144
ATLANTIC COASTAL PLAIN, COASTS,	CRUSTACEANS, MARINE FISH, MOLLUSK	W73-00819
TOMS, COPEPODS, DINOFLAGELLATES,	CRUSTACEANS, ANNELIDS, WATER POLL	W73-00935
HYTON, ON-SITE DATA COLLECTIONS,	CRUSTACEANS, GASTROPODS, SNAILS,	W73-00932
, MARINE ALGAE , SALT TOLERANCE,	CRUSTACEANS, MARINE FISH, MARSH P	W73-00855
VERTEBRATES, ANNELIDS, SEASONAL,	CRUSTACEANS, ISOPODS, MARINE ALGA	W73-00374
GELLATES, SCENEDESMUS, ROTIFERS,	CRUSTACEANS, COPEPODS, NUTRIENTS,	W72-05968
E, AQUICULTURE, FISH, SHELLFISH,	CRUSTACEANS, OYSTERS, SEASONAL.: /	W70-08832
AQUATIC PLANTS, AQUATIC INSECTS,	CRUSTACEANS, MUSSELS, COMMON MERG	W72-12930
TALS, MUSSELS, *KELPS, STARFISH,	CRUSTACEANS, FOOD CHAINS, FOOD WE	W72-11925
FEEDING RATES, *LARV/ *COPEPODS,	CRUSTACEANS, ZOOPLANKTON, ALGAE,	W72-09248
MOSOMES, EMBRYONIC GROWTH STAGE,	CRUSTACEANS, MONITORING, WATER PO	W72-10955
IDES, WIND, LIGHT, WAVES, CLAMS,	CRUSTACEANS, WORMS, MOLLUSKS, WAT	W72-09092
ANOPHYTA, CHRYSOPHYTA, COPEPODS,	CRUSTACEANS, LABORATORY TESTS, VI	W72-00150
TOMS, LAKES, PROTOZOA, ROTIFERS,	CRUSTACEANS, DIPTERA, MAYFLIES, D	W71-11012
WS, FISHKILL, HERRING, COPEPODS,	CRUSTACEANS, DIATOMS, CYANOPHYTA,	W71-11517
*ELECTRA	CRUSTULENTA, *BRYOZOA.:	W73-00348
ALDII VAR. ORNATUM, MICRASTERIAS	CRUX-MELITENSIS, CLOSTERIUM STRIG	W72-12736
CORIXA, SIGARA, BOENOA, SALDULA,	CRYNELLUS, CHEUMATOPSYCHE, TRIAEN	W70-06655
A, *POLYMORPHISM, */ *CYCLOTELLA	CRYPTICA, *CYCLOTELLA MENEGHINIAN	W73-02548
ON, GREEN FLAGELLATE, CYCLOTELLA	CRYPTICA, THALASSIOSIRA FLUVIATIL	W71-10083
ICILIAT/ ENRICHMENT, CHROMULINA,	CRYPTOMONAS EROSA, ERKINIA SUBEQU	W72-13816

CILIATA, UROGLENOPSIS AMERICANA, CRYPTOMONAS PULSILLA, MICROCYSTIS W72-13816
INE RIVER, ASTERIONELLA FORMOSA, CRYPTOMONAS, CHLAMIDOMONAS, BEREN W71-01491
OUNTS, FRAGILARIA, ASTERIONELLA, CRYPTOMONAS, SPHAEROCYSTIS, DAPHN W71-03014
(GHANA), ACTINASTRUM GRACILIMUM, CRYPTOMONAS EROSA, PERIDINIUM AFR W73-00256
AND), STEPHANODISCUS HANTZSCHII, CRYPTOMONAS, RHODOMONAS.: /T(ENGL W72-14797
AIL DAM, FRAGILARIA CROTONENSIS, CRYPTOMONAS OVATA, STEPHANODISCUS W73-01435
RADON RADIOISOTOPES, CHROMULINA, CRYPTOMONAS, MALLOMONAS, STAURAST W72-07648
ROMATIUM, MERISMOPEDIA TROLLERI, CRYPTOMONADS, STICHOGLOEA D: / CH W72-07145
YSTIS, COELASTRUM, ASTERIONELLA, CRYPTOMONAS.: /, OOCYSTIS, GLOEOC W72-07940
RATES, ASSIMILATION EFFICIENCES, CRYPTOMONAS, ANKISTRODESMUS, ELAK W72-07940
LUMINARIN-TYPE POLYSACCHARIDES, CRYPTOPHYCEA, DINOPHYCEA, BACCILA W70-07280
ROM, TYSTRUP LAKE, BAVELSE LAKE, CRYPTOPHYCEAE, *DENMARK, CYCLOTEL W72-12738
OSCILLATORIA, SYNURA, AROGLENA, CRYPTOPHYCEAE, PONTOPOREIA AFFINI W72-12729
ICTALURUS NEBULOSUS, NOTEMIGONUS CRYSOLEUCAS, CATOSTOMUS COMMERSON W72-07890
IZOBIUM JAPONICUM, LAKE MENDOTA, CRYSTAL LAKE, TROUT LAKE, BIG ARB W73-01456
*SPECIES DIVERSITY, CRYSTAL RIVER.: W72-11252
BIG STONE LAKE(MINNESOTA), LAKE CRYSTAL(MINNESOTA).: /MINNESOTA), W71-04216
BIOLOGICAL SAMPLES, SR-90, K-40, CS-137, *PAHRANAGAT LAKE, GAMMA R W73-01673
A MELANOPLEURA, CYPRINUS CARPIO, CTENOPHARYNGODON IDELLUS.: /ILAPI W71-04528
, *BAY OF BENGAL, COELENTERATES, CTENOPHORA, CHAETOGNATHS, CYPHONA W73-00935
SPECIES DIVERSITY, PORIFERA, SEA CUCUMBERS, STARFISH, CHITONS, RHA W73-00932
ANDERSON- CUE LAKE, MELROSE(FLORIDA).: W72-08986
LIVER, URINE, OEDOGONIUM, PHYSA, CULEX, EAMBUSIA, FATE OF POLLUTAN W72-07703
SAMPLING, SEPARATION TECHNIQUES, CULTIVATION, VITAMIN B, PLANT GRO W73-00430
FFECTS, WATER POLLUTION SOURCES, CULTIVATION.: / WATER POLLUTION E W71-02880
TON, TOXICITY, DISSOLVED OXYGEN, CULTRUES, ENDRIN, DIELDRIN, OXYGE W71-10096
SACOOK(ME), *KENNEBEC RIVER(ME), CULTURAL EUTROPHICATION.: /NNABES W72-04280
*LAMINARIA, SUBLITTORAL REGION, CULTURE EXPERIMENTS, ATRAZINE, MC W72-12576
HLAMYDOMONAS PSEUDOMICROSPHAERA, CULTURE MEDIA, CULTURING TECHNIQU W72-12632
ELITENSIS, CLOSTERIUM STRIGOSUM, CULTURE MEDIA.: /RASTERIAS CRUX-M W72-12736
HLOROP/ *PORTERINEMA FLUVIATILE, CULTURE MEDIA, HUMMOCK CHANNEL, C W72-12633
E, ARSENATES, SODIUM CACODYLATE, CULTURE MEDIA, PORIA MONTICOLA, P W72-13813
CULTUTURING TECHNIQUES, CULTURE MEDIA, ALGOLOGY, TAXON.: W72-12172
APONICA, STAURASTRUM PARADOXIUM, CULTURE MEDIA, ZINC SULFATE, MANG W72-12739
L EFFECTS/ SAVANNAH RIVER PLANT, CULTURE MEDIA, DRY WEIGHT, THERMA W71-02075
TICS, BIOENERGETICS, ENRICHMENT, CULTURE MEDIA, MACROINVERTEBRATES W72-09378
YTHRAEUM, *CULTURING TECHNIQUES, CULTURE MEDIA, *BAY OF BENGAL, IN W73-00430
RGANIC NITROGEN, ORGANIC CARBON, CULTURE MEDIA, PAPER CHROMATOGRAP W73-00379
ARB/ *CHLAMYDOMONAS REINHARDTII, CULTURE MEDIA, CHLORINATED HYDROC W73-00281
TOGRAPHY, INFRARED SPECTROSCOPY, CULTURE MEDIA, FATTY ACIDS, SAMPL W73-01066
TIDES, ETHANOL, MONOSACCHARIDES, CULTURE MEDIA, DISACCHARIDES, THR W73-00942
CAPRICORNUTUM, PAAP BATCH TEST, CULTURE MEDIA.: /DES, SELENASTRUM W73-02099
MYCIN, CHLORAMPHENICOL, ACTIDON, CULTURE MEDIA, CHLOROPHYLL A, TER W73-01445
*CHLORELLA ELLIPSOIDEA, CULTURE MEDIA.: W73-01626
TUDY FACILITY, PRODUCT E/ *ALGAE CULTURE TECHNIQUE, AQUATIC WEED S W71-06102
GEN PRODUCTION, LIGHT INTENSITY, CULTURE THICKNESS, CULTURE VOLUME W70-05547
HT INTENSITY, CULTURE THICKNESS, CULTURE VOLUME.: /PRODUCTION, LIG W70-05547
GE EFFLUENTS, *SEWAGE TREATMENT, CULTURE, *WASTE-WATER TREATMENT, W72-06612
S, *ANABAENA SP L-31, CONTINUOUS CULTURE, BATCH CULTURES.: /ROCYST W72-03217
ANABAENA FLOS-AQUAE, *CONTINUOUS CULTURE, CHEMOSTATS, PHOSPHATASE W72-05476
ANS, CERATIUM FURCA, TURBIDOSTAT CULTURES.: /SEA, PROROCENTRUM MIC W71-03095
L-31, CONTINUOUS CULTURE, BATCH CULTURES.: /ROCYSTS, *ANABAENA SP W72-03217
, EPIPHYTOLOGY, ALGAE, HABITATS, CULTURES.: / STUDIES, *PHAEOPHYTA W72-01370
SSION, RADIOACTIVITY TECHNIQUES, CULTURES.: / RADIOISOTOPES, SUCCE W72-09239
*SOLUBILITY, *CYANOPHYTA, ALGAE, CULTURES.: *PHOSPHATES, W71-10070
E ANIMALS, BACTERIA, ABSORPTION, CULTURES.: / CONCENTRATION, MARIN W71-10083
OWTH, CONTINUOUS CULTURES, BATCH CULTURES.: *ALGAL GR W72-14790
AL STUDIES, STATISTICAL METHODS, CULTURES.: /OPULATIONS, CYTOLOGIC W72-11920
LYSIS, MODEL STUDIES, COMPUTERS, CULTURES.: /NIQUES, NUMERICAL ANA W72-12724
TAURASTRUM PA/ *DESMIDS, *AXENIC CULTURES, CULTURING TECHNIQUES, S W72-12736
ONISM, *HETEROANTAGONISM, AXENIC CULTURES, EXCRETION, APHANIZOMENO W72-12734
CH/ *ALGAE, *PHOSPHORUS, *LAKES, CULTURES, PHOSPHATES, NUTRIENTS, W72-12637
, *TIDAL MARSHES, *MARINE ALGAE, CULTURES, SALINITY, SALT MARSHES, W72-12633
TERNS, *AUTOSPORES, *SYNCHRONOUS CULTURES, SYNCHRONY INDEX.: / PAT W72-12724
*/ *SYSTEMATICS, *CHLAMYDOMONAS, CULTURES, CHLOROPHYTA, PROTOZOA, W72-12632
ACTIVITY TECHNIQUES, RESISTANCE, CULTURES, *CHLOROPHYTA.: /, RADIO W72-11836
ORTH SEA, PHAEOCYSTIS SPP., SOIL CULTURES, TITANIUM DIOXIDE WASTES W72-12239
, BIAS, S/ *ERRORS, *SYNCHRONOUS CULTURES, AUTOSPORES, VARIABILITY W72-11920
*PHYTOPLANKTON, *AQUATIC ALGAE, CULTURES, PHOSPHATES, NITRATES, N W72-12739
GRAPHIES, BIOLOGY, EPIDEMIOLOGY, CULTURES, METHODOLOGY, DOCUMENTAT W72-13800
*CYANOPHYTA, SODIUM, POTASSIUM, CULTURES, RESISTANCE, AQUATIC ALG W72-13809
S, BACTERIA, FUNGI, METHODOLOGY, CULTURES, BIOASSAY, ENVIRONMENTAL W72-13813
*ALGAL GROWTH, CONTINUOUS CULTURES, BATCH CULTURES.: W72-14790
L PORPERTIES, NITROGEN FIXATION, CULTURES, CARBON DIOXIDE, NUTRIEN W72-14790
F POLLUTANTS, OCEANS, ESTUARIES, CULTURES, SAMPLING, HAWAII, S: /O W73-00361
TION, TESTING, ASSAY, NUTRIENTS, CULTURES, MEASUREMENT, LABORATORY W73-00255
*RESISTANCE, *PLANT PHYSIOLOGY, CULTURES, PHOTOSYNTHESIS, AQUATIC W73-00281
ALASSIONEMA NITZSCHIOIDES, MIXED CULTURES, METABOLITES, SUBSTRATE W73-00441
MEDIUM, ESTUARINE SOILS, AXENIC CULTURES, GIBBERELIC ACID.: /IBER W73-00430
, PLANT PHYSIOLOGY, CHRYSOPHYTA, CULTURES, SEA WATER, NUTRIENT REQ W73-00380
PHILIC BACTERIA, ORGANIC MATTER, CULTURES, SOIL BACTERIA.: /THERMO W72-14301
PARATION TECHNIQUES, CYANOPHYTA, CULTURES, ANTIBIOTICS(PESTICIDES) W72-14327
ERIA, SCENEDESMUS, GROWTH RATES, CULTURES, AQUATIC PRODUCTIVITY, M W72-14314
DESMUS QUADRICAUDA/ *SYNCHRONOUS CULTURES, SCENEDESMUS SOLI, SCENE W72-07144
PRODUCTIVITY, MINERALOGY, LIGHT, CULTURES, BIOASSAY.: /S, PRIMARY W72-07166
*GROWTH CHAMBERS, *LIFE CYCLES, CULTURES, WATER QUALITY, SCENEDES W72-07144
OISOTOPES, PRIMARY PRODUCTIVITY, CULTURES, BIOMASS, CARBON RADIOIS W72-07648
M, RAFFINOSE, SUCROSE, SENESCENT CULTURES, XYLOSE, RIBOSE, ARABINO W72-07710
*GROWTH RATES, PLANT MORPHOLOGY, CULTURES, PHOTOSYNTHESIS, CARBON W72-07710
BACTERIA, *PATHOGENIC BACTERIA, CULTURES, E. COLI, BIOINDICATORS, W72-07664

ACE ELEMENTS, COPPER, CHLORELLA,	CULTURES, NUTRIENTS, RADIOISOTOPE	W72-07660
LGAE, *GROWTH RATES, *NUTRIENTS,	CULTURES, INHIBITION, OLIGOTROPHY	W72-08376
STICIDES), BACTERICIDES, ATOLLS,	CULTURES, SPHAEROTILUS, CLADOPHOR	W72-08142
NITZSCHIA ACTINASTROIDES, *BATCH	CULTURES, *CHEMOSTATS, ORTHOPHOSP	W73-01445
IRATION, *CHLORELLA, *NUTRIENTS,	CULTURES, NITROGEN, POTASSIUM, PH	W73-01626
IOMETRY, ELECTROANALYSIS, *ALGAL	CULTURES, CHELATING AGENTS, COCYS	W73-02112
ORT, *HEAVY METALS, EU/ *COPPER,	CULTURES, *CHELATION, *ION TRANSP	W73-02112
S, REPRODUCTION, CARBON DIOXIDE,	CULTURES, LIMITING FACTORS, RHODO	W73-02099
ARATION TECHNIQUES, FOOD CHAINS,	CULTURES, FISH DIET, GAS CHROMATO	W73-02105
EMATICS, NUTRIENTS, PSEUDOMONAS,	CULTURES, VITAMIN B, BIOASSAY, CE	W73-00942
ALGAE, *SOIL BACTERIA, SAMPLING,	CULTURES, ECOLOGY, SOIL MICROBIOL	W73-00854
NDICATORS, *ALGAE, GROWTH RATES,	CULTURES, SUSPENDED LOAD, SEWAGE,	W71-09795
MMONIA, CARBON DIOXIDE, FLORIDA,	CULTURES, *ALGAE, INCUBATION, CEN	W72-09103
N DIOXIDE, PHYTOPLANKTON, ALGAE,	CULTURES, *FLORIDA, SEA WATER, CH	W72-09102
TOPLANKTON, *MERCURY, *TOXICITY,	CULTURES, *CHLORELLA, AQUATIC ALG	W72-11727
NODRILUS SPP., TUBIFEX, UNIALGAL	CULTURES, OUABAIN, BIOLOGICAL MAG	W72-09668
OPHYTA, GEYSERS, PHOTOSYNTHESIS,	CULTURES, CHLOROPHYLL, PROTEINS,	W72-03557
PHYSIOLOGY, CYTOLOGICAL STUDIES,	CULTURES, AMMONIA, POTASSIUM COMP	W72-03217
RANULES, INHIBITION, METABOLISM,	CULTURES, ENVIRONMENTAL EFFECTS,	W72-00153
ANT GROWTH REGULATORS, PLANKTON,	CULTURES; WATER POLLUTION EFFECTS	W71-13793
TABOLISM, CARBON DIOXIDE, ALGAE,	CULTURES, ACIDITY, ORGANIC COMPOU	W72-00838
OFLAGELLATES, *LABORATORY TESTS,	CULTURES, *GROWTH RATES, PLANT GR	W71-03095
*PHOTOSYNTHETIC OXYGEN, *ALGAE,	CULTURES, VOLUME, MATHEMATICAL ST	W70-05547
EA WATER, *ALGAE, *MARINE ALGAE,	CULTURES, PHYTOPLANKTON, CYTOLOGI	W70-07280
*AUTOGOMISTIC EFFECTS, *AXEMIC	CULTURES, *DIEOFF RATES, AFTERGRO	W70-08319
, LAKE ONTARIO, SEASONAL, ALGAE,	CULTURES, MICROORGANISMS, EFFLUEN	W71-00116
WATER, CYANOPHYTA, MYXOBACTERIA,	CULTURES, METABOLISM, SEWAGE, CAR	W72-05457
, VIBRIO ANGUILLARUM, ENRICHMENT	CULTURES, AGARS, CHEMOSTATS, NITR	W72-05421
RATE, / *CHEMOSTATIC CONTINUOUS	CULTURES, MAXIMUM SPECIFIC GROWTH	W72-04788
UTRIENT REQUIREMENTS/ *BIOASSAY,	CULTURES, *BIOMASS, NUTRIENTS, *N	W72-03985
ITY, WATER ANALYSIS, ABSORPTION,	CULTURES, DDT, ANALYTICAL TECHNIQ	W72-04134
IATOMS, PHYTOPLANKTON, TOXICITY,	CULTURES, DDT, GROWTH RATES.: / D	W72-05614
DINOFLAGELLATES, DIATOMS, ALGAE	CULTURES, COPPER, TOXICITY.: /RS,	W72-04708
TH.: *MIXED	CULTURES, DIEOFF RATES, AFTERGROW	W71-05155
TESTS, CHEMICALS, BACTERIOCIDES,	CULTURES, TEST PROCEDURES, PESTIC	W71-06189
IUM RADIOISOTOPES, IONS, RIVERS,	CULTURES, ORGANIC MATTER, BACTERI	W71-08032
TRIAGINE PESTICIDES, RESISTANCE,	CULTURES, METABOLISM.: /, SOILS,	W71-07675
NDANE, DIADEMA ANTIL/ EPIPHYTES,	CULTURING VESSELS, PENICILLIN, LI	W72-14294
PA/ *DESMIDS, *AXENIC CULTURES,	CULTURING TECHNIQUES, STAURASTRUM	W72-12736
EUDOMICROSPHAERA, CULTURE MEDIA,	CULTURING TECHNIQUES, MORPHOLOGY,	W72-12632
EDIA, ALGOLOGY, TAXON.:	CULTUTURING TECHNIQUES, CULTURE M	W72-12172
XUOSA, PARMELIA PHYSODES, HYPNUM	CUPRESSIFORME.: / DESCHAMPSIA FLE	W72-14303
G, CORES, DISTRIBUTION PATTERNS,	CURRENTS(WATER), PARTICLE SIZE, S	W73-00367
ANKTON, STRATIFICATION, DIATOMS,	CURRENTS(WATER), FISH FARMING.: /	W73-00236
HYTA, *FLORIDA, CORAL, SALINITY,	CURRENTS(WATER), MOLLUSKS, ENVIRO	W72-11876
*DISSOLVED OXYGEN, TEMPERATURE,	CURRENTS(WATER), WINDS, THERMOCLI	W72-12992
E, PHOSPHORUS, CARBON, NITROGEN,	CURRENTS(WATER), WINDS, NEVADA, C	W71-05626
TIC OCEAN, DEPTH, STANDING CROP,	CURRENTS(WATER), WAVES(WATER), SA	W72-04766
SPHORUS, THERMAL STRATIFICATION,	CURRENTS(WATER), ALGAE, LAKES, RI	W70-09947
RACKING TECHNIQUES, MEASUREMENT,	CURRENTS(WATER), CHLOROPHYLL, FLU	W72-01095
FAUNA, PLANKTON, BIOINDICATORS,	CURRENTS(WATER), TIDES, WIND, LIG	W72-09092
FIC COAST REGION, TRITIUM, OCEAN	CURRENTS, COBALT RADIOISOTOPES, N	W72-03324
*CLADOPHORA GLOMERATA, GROWTH	CURVES.:	W72-13651
PLATYMONAS VIRIDIS, CHAETOCEROS	CURVISETUS, SKELETONEMA COSTATUM,	W72-07166
*NUTRIENT LEVELS,	CUYAHOGA RIVER(OHIO).:	W72-04263
NE, LYSINE, ASPARAGINE, CYSTINE,	CY: /E, ALANINE, ISOLEUCINE, VALI	W73-00942
	CYANIDES, PHOSPHORUS-32.:	W72-07143
VER, MICROCYSTIS, APHANIZOMENON,	CYANIDES, COHO SALMON, SHEEPSHEAD	W72-14282
FERENCE, SCINTILLATION COUNTING,	CYANIDIUM CALDARIUM, THIOBACILLUS	W72-14301
EIDAMONAS APPLANATA, BICUDOMONAS	CYANOPH: /AMONAS SAUPAULENSIS, EN	W72-09119
*GAS VACUOLES, MOTILE FORMS,	CYANOPHAGES.:	W71-13184
ARVAE, SHAD, MACROINVERTEBRATES,	CYANOPHYCEAE.: /IFLORA, WALLEYE L	W72-08142
PLECTONEMA, LYNGBYA, PHORMIDIUM,	CYANOPHYCEAE, OSCILLATORIACEAE, S	W72-04736
EFFECTS, *FARM WASTES, OKLAHOMA,	CYANOPHYTA, CHLORELLA, IONS, PHYT	W72-04773
UTION EFFECTS, NUTRIENTS, ALGAE,	CYANOPHYTA, PHOSPHORUS, WATER POL	W72-04769
GOTROPHY, EUTROPHICATION, ALGAE,	CYANOPHYTA, PHYTOPLANKTON, LAKE M	W72-04269
ION, STANDING CROP, CHLOROPHYTA,	CYANOPHYTA, RHODOPHYTA, PHAEOPHYT	W72-06283
TOZOA, CHLORELLA, CHLAMYDOMONAS,	CYANOPHYTA, CHLOROPHYTA, CHRYSOPH	W72-05968
, SULFATES, SPORES, AZOTOBACTER,	CYANOPHYTA, YEASTS, CARBON DIOXID	W72-06124
TION PATTERNS, *THERMAL SPRINGS,	CYANOPHYTA, ALKALINITY, NUTRIENTS	W72-05610
IDE, NITROGEN, LABORATORY TESTS,	CYANOPHYTA, CYTOLOGICAL STUDIES.:	W72-05476
IA, *ALGAL CONTROL, FRESH WATER,	CYANOPHYTA, MYXOBACTERIA, CULTURE	W72-05457
POLLUTION SOURCES, CHLOROPHYTA,	CYANOPHYTA, WASHINGTON, DIVERSION	W72-05473
, BIOMASS, DIATOMS, CHLOROPHYTA,	CYANOPHYTA, PYRROPHYTA, CHRYSOPHY	W72-04855
TOPLANKTON, CYTOLOGICAL STUDIES,	CYANOPHYTA, CHLOROPHYTA, CHRYSOPH	W70-07280
ONEFLIES, DOBSONFLIES, MAYFLIES,	CYANOPHYTA, CHLOROPHYTA, CHRYSOPH	W70-06655
RIENT REQUIREMENTS, CHLOROPHYTA,	CYANOPHYTA, CYCLING NUTRIENTS, NI	W70-05469
EPODS, DAPHNIA, SAMPLING, DEPTH,	CYANOPHYTA, CHRYSOPHYTA, CHLOROPH	W71-03014
ES, DISSOLVED OXYGEN, NUTRIENTS,	CYANOPHYTA, MIXING, PHYTOPLANKTON	W71-03034
TER, MAGNESIUM, SODIUM SULFATES,	CYANOPHYTA, RHODOPHYTA.: /LINE WA	W71-04526
IODEGRADATION, TRACERS, LAGOONS,	CYANOPHYTA.: /BACTERIA, *TEXAS, B	W71-04873
ONATES, BICARBONATES, DIATONICS,	CYANOPHYTA, ALGAE, SOUTH DAKOTA,	W71-02880
WTH STAGES, PATHOGENIC BACTERIA,	CYANOPHYTA, CHLOROPHYTA, SCENEDES	W71-05155
NALYSIS, TEMPERATURE, LIMNOLOGY,	CYANOPHYTA, IRRIGATION CANALS, AL	W71-06102
OPHYTA, CHRYSOPHYTA, RHODOPHYTA,	CYANOPHYTA, EUGLENOPHYTA.: /CHLOR	W71-05629
ANT PATHOLOGY, INDUSTRIES, IONS,	CYANOPHYTA, RADIOISOTOPES, TEMPER	W71-09157
LLA, SCENEDESMUS, CHLAMYDOMONAS,	CYANOPHYTA, CHLOROPHYTA, OXYGEN,	W71-09172
LGAE, WASHINGTON, PHYTOPLANKTON,	CYANOPHYTA, NITRATES, PHOSPHATES,	W72-08049
N FIXATION, *ALGAE, MEASUREMENT,	CYANOPHYTA, PHYTOPLANKTON, PHOTOS	W72-08054
, NITRATES, ALKALINITY, DIATOMS,	CYANOPHYTA, CHLOROPHYTA, SEASONAL	W72-08560

R TREATMENT, PONDS, SCENDESOMUS, CYANOPHYTA, BIOASSAY, SAMPLING, O W72-07661
LAMYDOMONAS, DIATOMS, SILICATES, CYANOPHYTA, TOXICITY, COSTS, CARB W72-07508
TE WATER TREATMENT, CHLOROPHYTA, CYANOPHYTA, AEROBIC CONDITIONS, B W72-07664
CATION, PHOSPHORUS, CHLOROPHYTA, CYANOPHYTA, CHRYSOPHYTA, EUGLENOP W72-08048
ES, CARBON, VITAMIN B, NITROGEN, CYANOPHYTA, LAKE MICHIGAN, NUTRIE W72-07937
INS, *MYXOBACTERIA, CHLOROPHYTA, CYANOPHYTA, SYSTEMATICS, VITAMINS W72-07951
ON, PYRROPHYTA, DINOFLAGELLATES, CYANOPHYTA, CHLOROPHYTA, EUGLENOP W72-07899
S, ALGAE, NUTRIENT REQUIREMENTS, CYANOPHYTA, CHLOROPHYTA, PHYTOPLA W72-07940
H, MARINE ANIMALS, MARINE ALGAE, CYANOPHYTA, PYRROPHYTA, LARVAE, S W72-14313
ADIATION, SEPARATION TECHNIQUES, CYANOPHYTA, CULTURES, ANTIBIOTICS W72-14327
TON, CHEMICAL ANALYSIS, DIATOMS, CYANOPHYTA, TURNOVERS, EUTROPHICA W73-00232
, CHROMIUM, SILVER, CHLOROPHYTA, CYANOPHYTA, ALGAL CONTROL, ALGICI W72-14694
H, SEWAGE EFFLUENTS, REGULATION, CYANOPHYTA, CHLOROPHYTA, HYDROGEN W72-14790
UTION, WATER ANALYSIS, BIOASSAY, CYANOPHYTA, RESINS, AQUATIC PLANT W73-00286
OGICAL COMMUNITIES, SYSTEMATICS, CYANOPHYTA, CHEMICAL ANALYSIS, SE W73-00242
OGICAL STUDIES, DINOFLAGELLATES, CYANOPHYTA.: /GY, BIOASSAY, CYTOL W73-00243
L CONTROL, *INFECTION, *VIRUSES, CYANOPHYTA.: *ALGA W73-00246
, MOLLUSKS, LARVAE, CHRYSOPHYTA, CYANOPHYTA, PYRROPHYTA, MARINE AL W73-00935
RES, ECOLOGY, SOIL MICROBIOLOGY, CYANOPHYTA, FROZEN SOILS, IONS, N W73-00854
TASSIUM, SODIUM, CARBON DIOXIDE, CYANOPHYTA, FISH,: /MANGANESE, PO W73-00840
S, *BACTERIA, *ALGAE, *CULTURES, CYANOPHYTA, HYDROGEN ION CONCENTR W73-01657
ENT, HYDROGEN ION CONCENTRATION, CYANOPHYTA.: /*WASTE WATER TREATM W73-01367
TIES, CHLOROPHYTA, EUGLENOPHYTA, CYANOPHYTA, CHRYSOPHYTA, SCENEDES W73-01627
RGANIC COMPOUNDS, *GEOCHEMISTRY, CYANOPHYTA, BACTERIA, ALGAE, PLAN W73-01439
TEMPERATURE, RESPIRATION, ALGAE, CYANOPHYTA, ANALYTICAL TECHNIQUES W73-01269
RVOIRS, SEASONAL, ACTINOMYCETES, CYANOPHYTA, PLANKTON, ON-SITE TES W71-13187
R, ALGAE, NUTRIENT REQUIREMENTS, CYANOPHYTA, CHLAMYDOMONAS.: /WATE W71-13185
, RESERVOIRS, LIGHT PENETRATION, CYANOPHYTA, CHRYSOPHYTA.: /TRATES W71-13183
ES, DEPTH, DIATOMS, CHLOROPHYTA, CYANOPHYTA, BIOMASS, CHEMICAL ANA W71-11029
COPEPODS, CRUSTACEANS, DIATOMS, CYANOPHYTA, PHOTOSYNTHESI: /RING, W71-11517
GONFLIES, CADDISFLIES, MOLLUSKS, CYANOPHYTA, CHRYSOPHYTA, CHLOROPH W71-11012
CREATION, INVERTEBRATES, RIVERS, CYANOPHYTA, GEOLOGIC FORMATIONS, W71-11009
ALYSIS, HYPOLIMNION, EPILIMNION, CYANOPHYTA, DINOFLOGELLATES, SEAS W71-11949
NIA, HYDROGEN ION CONCENTRATION, CYANOPHYTA, CHLORELLA, CHLOROPHYT W71-11519
OPHYTA, CHLOROPHYTA, PYRROPHYTA, CYANOPHYTA, MANGANESE, ZINC, META W72-09111
ON, *RECREATION, *WATER QUALITY, CYANOPHYTA, LAKES, MISSOURI, NITR W72-10431
*CARBON DIOXIDE, *ALGAL CONTROL, CYANOPHYTA, CARBON, CHLORELLA, BI W72-10607
GEMENT, ANALYSIS, WATER QUALITY, CYANOPHYTA, NUISANCE ALGAE, COMPR W72-09171
ESIS, NUTRIENTS, EUTROPHICATION, CYANOPHYTA, ALKALINITY, DIURNAL, W72-09162
WASTES, ANNUAL TURNOVER, ENERGY, CYANOPHYTA, ALGAE, EUTROPHICATION W72-03228
TROPICAL REGIONS, POLAR REGIONS, CYANOPHYTA, CHRYSOPHYTA, PYRROPHY W72-03543
GRAMS, *AGRICULTURE, *ECONOMICS, CYANOPHYTA, ALGAE, PLANTS, PROTEI W72-03327
IONS, CHLOROPHYTA, EUGLENOPHYTA, CYANOPHYTA, CHRYSOPHYTA, COPEPODS W72-00150
MONAS, CHLOROPHYTA, SCENEDESMUS, CYANOPHYTA, MOLLUSKS, MITES, ANNE W72-00155
OSING ORGANIC MATTER, MINNESOTA, CYANOPHYTA, TREES, GRASSES, LEAVE W72-01784
TOMS, AQUATIC ALGAE, FILTRATION, CYANOPHYTA, *EUTROPHICATION, *OHI W72-13673
NITROGEN, BIOASSAY, PHOSPHORUS, CYANOPHYTA, OLIGOTROPHY, CHLAMYDO W72-13816
FFLUENTS, EUTNOPHICATION, ALGAE, CYANOPHYTA, CHLOROPHYTA, METABOLI W72-13636
FACES, CHRYSOPHYTA, CHLOROPHYTA, CYANOPHYTA, BIOCHEMICAL OXYGEN DE W72-12995
IOLOGY, BENTHIC FLORA, BIOASSAY, CYANOPHYTA, *FLORIDA, CORAL, SALI W72-11876
LING, CHLOROPHYTA, *CHRYSOPHYTA, CYANOPHYTA.: /LASSIFICATION, SAMP W72-12192
OPLANKTON, CHLOROPHYTA, DIATOMS, CYANOPHYTA, BENTHIC FAUNA, AQUATI W72-12729
IBITION, FUNGI, THERMAL SPRINGS, CYANOPHYTA, BACTERIA, ECOLOGY, WA W72-12734
ACTORS, PHOSPHATES, CHLOROPHYTA, CYANOPHYTA, NITROGEN, CARBON, DIA W72-12544
XATION, *TROPICAL REGIONS, RICE, CYANOPHYTA, FERTILIZATION, TEMPER W73-02473
ILS, *TEMPERATE, CLIMATIC ZONES, CYANOPHYTA, SYMBIOSIS, LICHENS, N W73-02471
OPH/ *NITROGEN FIXATION, *LAKES, CYANOPHYTA, LIGHT INTENSITY, EUTR W73-02472
EFFECTS, WASTE WATER TREATMENT, CYANOPHYTE, ENGLENOPHYTA, ORGANIC W71-07973
GROWTH RATES, LABORATORY TESTS, CYANOPHYTO, ALGAE.: /ES, DENSITY, W72-03217
, *ALGAE, *PHOSPHORUS COMPOUNDS, CYANOPYTA, CHLOROPHYTA.: /LANKTON W73-00938
CARBONATE PARTICLES, PHOSPHORUS CYCLE.: / WINTERGREEN LAKE(MICH), W72-08048
ALITY, WATER TREATMENT, NITROGEN CYCLE.: /SAG, TURBIDITY, WATER QU W71-09958
*PHOSPHORUS COMPOUNDS, *NITROGEN CYCLE, *CRABS, *FISHES, ECOLOGY, W72-12567
*MATHEMATICAL MODELS, *NITROGEN CYCLE, *OXIDATION, AMMONIA, NITRI W71-05390
C WASTES, *SALT MARSHES, *CARBON CYCLE, *PHOSPHORUS COMPOUNDS, *NI W72-12567
FACTORS, PHOTOSYNTHESIS, CARBON CYCLE, ABSORPTION, DECOMPOSING OR W71-12068
ION, NITROGEN FIXATION, NITROGEN CYCLE, AMMONIA, NITRATES, MODEL S W72-08459
EFFECT, EUGLENA GRACLIS, CALVIN CYCLE, ANACYSTIS NIDULANS.: /BURG W71-09172
MONAS REINHARDTII, TETR/ MINERAL CYCLE, CHARACTERIZATION, CHLAMYDO W73-02095
*GLUCOSE, *MALATE, CITRIC ACID CYCLE, CYTOCHROMES, PHOSPHORYLATI W72-04309
SAN JOAQUIN DELTA(CALIF),/ ALGAL CYCLE, DELAWARE ESTUARY(DEL-NJ), W71-05390
ROGEN FIXING BACTERIA, *NITROGEN CYCLE, ENVIRONMENTAL EFFECTS, ANA W72-11563
HESIS, *ALGAE, PHYLOGENY, CARBON CYCLE, ENZYMES, METABOLISM, HYDRO W71-09172
ATER, DISSOLVED SOLIDS, NITROGEN CYCLE, HYDROGEN ION CONCENTRATION W71-01489
BACTERIA, AQUATIC ALGAE, CARBON CYCLE, ON-SITE TESTS, AQUATIC ANI W73-02095
RITES, HARDNESS(WATER), NITROGEN CYCLE, ORGANIC MATTER, PHOTOSYNTH W72-08048
PLANKTON, WATER ANALYSIS, CARBON CYCLE, ORGANIC MATTER, NUTRIENTS. W72-09108
T MORPHOLOGY, REPRODUCTION, LIFE CYCLES, AQUATIC ALGAE, PERIPHYTON W72-07141
DEMAND, PHOSPHORUS, FISH, WEEDS, CYCLES, BACTERIA, CHEMIST: /YGEN W71-03012
ROPHYTA, *GROWTH CHAMBERS, *LIFE CYCLES, CULTURES, WATER QUALITY, W72-07144
COMMUNITIES, PRODUCTIVITY, LIFE CYCLES, GROWTH STAGES, DISTRIBUTI W72-14659
ICATION, COLIFORMS, SPORES, LIFE CYCLES, ODOR, RECREATION, SYSTEMA W71-09173
TODES, ALGAE, OLIGOCHAETES, LIFE CYCLES, SEASONAL, DENSITY, DIATOM W71-11012
RY TESTS, ALGAE, CARBON DIOXIDE, CYCLING NUTRIENTS, BICARBONATES, W72-09159
S COMPOUNDS, NITROGEN COMPOUNDS, CYCLING NUTRIENTS, ESSENTIAL NUTR W72-10608
TA COLLECTIONS, TEST PROCEDURES, CYCLING NUTRIENTS, ANALYTICAL TEC W71-13794
ROPHICATION, SEDIMENT TRANSPORT, CYCLING NUTRIENTS, DEGRADATION(DE W72-00847
CTS, *ECOSYSTEMS, *RAIN FORESTS, CYCLING NUTRIENTS, SOIL-WATER-PLA W72-00949
PHORUS, *CYANOPHYTA, AQUATIC EN/ CYCLING NUTRIENTS, *CARBON, *PHOS W71-00475
EMENTS, CHLOROPHYTA, CYANOPHYTA, CYCLING NUTRIENTS, NITROGEN, PHOS W70-05469

WATER QUALITY, WATER CHEMISTRY,	CYCLING NUTRIENTS.: /L NUTRIENTS,	W70-04080
ISTRIBUTION, ALGAE, TRIBUTARIES,	CYCLING NUTRIENTS.: /S, SPATIAL D	W72-04263
SITIVITY, RADIOECOLOGY, ECOLOGY,	CYCLING NUTRIENTS, THERMAL POLLUT	W73-00796
NITIES, *AQUATIC MICROORGANISMS,	CYCLING NUTRIENTS, ECOLOGICAL DIS	W73-02095
RGANIC COMPOUNDS, CARBOHYDRATES,	CYCLING NUTRIENTS, ALGAE.: /TY, O	W73-02100
FACTORS, ROOTED AQUATIC PLANTS,	CYCLING NUTRIENTS, RESPIRATION, M	W73-01089
CHNIQUES, ANALYTICAL TECHNIQUES,	CYCLING NUTRIENTS, TRACE ELEMENTS	W73-00821
OLOGY, RADIOACTIVITY TECHNIQUES,	CYCLING NUTRIENTS, TRACE ELEMENTS	W73-00826
POLLUTION EFFECTS, HEAVY METALS,	CYCLING NUTRIENTS, CARBON, NITROG	W73-00840
OSPHORUS, *MODEL STUDIES, LAKES,	CYCLING NUTRIENTS, TRACERS, LITTO	W72-12543
ION, COUNTING CHAMBERS, NUTRIENT	CYCLING.: /NCES, SAMPLE PRESERVAT	W72-07899
IOCYSTIS, SPIROGYRA,/ *GAMBUSIA,	CYCLOPS NAUPLII, COCCOCHLORIS PEN	W72-07132
LONGISPINA, DIAPTOMUS GRACILIS,	CYCLOPS STRENUUS, INGESTION RATES	W72-07940
ELLIBRANCH, SPIROGYRA, EUDORINA,	CYCLOPS, KERATELLA, ASPLANCHNA, L	W71-11876
LOSA, STEPHANODISCUS HANTZSCHII,	CYCLOT MICHIGANIANA, RHIZOSOLENIA	W73-00384
FILIFORMIS, FRAGILARIA CAPUCINA,	CYCLOTELLA STELLIGERA, STEPHANODI	W73-00384
DOSA, PHAEODACTYLUM TRICORNUTUM,	CYCLOTELLA NANA, CHAETOCEROS GALV	W72-07660
PACIFIC OCE/ COOK INLET(ALASKA),	CYCLOTELLA NANA, LEPTOCYLINDRUS,	W71-11008
MA.: ABATE, BAYTEX, BAYGONE,	CYCLOTELLA, DUNALIELLA, SKELETONE	W71-10096
EASSIMILATION, GREEN FLAGELLATE,	CYCLOTELLA CRYPTICA, THALASSIOSIR	W71-10083
RESERVATION, FORMALIN, COUNTING,	CYCLOTELLA, OSCILLATORIA, APHANOT	W72-10875
E LAKE, CRYPTOPHYCEAE, *DENMARK,	CYCLOTELLA COMTA, STEPHANODISCUS	W72-12738
PEDICULUS, COCCONEIS PLACENTULA,	CYCLOTELLA CF. SOCIALIS, CYCLOTEL	W72-13142
ENTULA, CYCLOTELLA CF. SOCIALIS,	CYCLOTELLA STELLIGERA, CYMBELLA S	W72-13142
ANSPORT, DUNALIELLA TERTIOLECTA,	CYCLOTELLA NANA, THALASSIOSIRA FL	W72-12998
EX 100, ARTEMIA SALINA, BIO-RAD,	CYCLOTELLA NANA,: CHEL	W72-04708
U/ ANABAENA CIRCINALIS, ANABAENA	CYLINDRICA, APHANIZOMENON FLOS-AQ	W72-10613
WA/ *PHOTORESPIRATION, *ANABAENA	CYLINDRICA, ACETYLENE REDUCTION,	W72-08584
EDUCTION, *HETEROCYSTS, ANABAENA	CYLINDRICA, ANABAENOPSIS, ACETYLE	W71-12070
TABOLITES, SKELETONEMA, ANABAENA	CYLINDRICA, DUNALIELLA.: /RIN, ME	W72-03545
L/ STABILIZATION PONDS, ANABAENA	CYLINDRICA, ANACYSTIS NIDULANS, G	W72-07664
A, ANABAENOPSIS, ACETYLENE TEST,	CYLINDROSPERMUM LICHENEFORME, MAS	W71-12070
NOSTOC,	CYLINDROTHECA.:	W71-12858
SOCIALIS, CYCLOTELLA STELLIGERA,	CYMBELLA SPP, DIATOMA SSP.: /CF.	W72-13142
LOSTERIUM, GOMPHONEMA LONGICEPS,	CYMBELLA, VAUCHERIA.: /CENTULA, C	W72-12744
LLARIA, ACTINASTRUM, PHYTOCONIS,	CYMBELLA, STEPHANODISCUS, ANKISTR	W72-10875
RATES, CTENOPHORA, CHAETOGNATHS,	CYPHONAUTES, ECHINODERMS, TUNICAT	W73-00935
RY, ANIMAL TISSUES, MACROPHYTES,	CYPRINUS CARPIO, MOSQUITO FISH, G	W73-01673
OSSAMBICA, TILAPIA MELANOPLEURA,	CYPRINUS CARPIO, CTENOPHARYNGODON	W71-04528
E, BLACK-FLY LARVAE, CLADOPHORA,	CYPRINUS CARPIO, CATOSTOMUS COMME	W70-08098
ING MYXOBACTERIA, DNA, MOTILITY,	CYST FORMATION, ANTIBIOTIC SENSIT	W72-07951
INE, VALINE, LYSINE, ASPARAGINE,	CYSTINE, CY: /E, ALANINE, ISOLEUC	W73-00942
OSE, *MALATE, CITRIC ACID CYCLE,	CYTOCHROMES, PHOSPHORYLATION, CLY	W72-04309
NUISANCE ALGAS, *ENTROPHICATION,	CYTOLOGICAL STUDIES, ANALYTICAL T	W72-03985
N, LABORATORY TESTS, CYANOPHYTA,	CYTOLOGICAL STUDIES.: /E, NITROGE	W72-05476
ALGAE, CULTURES, PHYTOPLANKTON,	CYTOLOGICAL STUDIES, CYANOPHYTA,	W70-07280
C MICROORGANISMS, SPORES, ALGAE,	CYTOLOGICAL STUDIES, MYCOBACTERIU	W71-08027
LS, DRYING, FREEZING, RADIATION,	CYTOLOGICAL STUDIES, PLANT TISSUE	W73-01676
, PATH OF POLLUTANTS, DIFFUSION,	CYTOLOGICAL STUDIES, ADSORPTION,	W73-00823
PLANT MORPHOLOGY, *PLANT GROWTH,	CYTOLOGICAL STUDIES.: /OOPHYTA, *	W73-01622
*GULF OF MEXICO, PHYTOPLANKTON,	CYTOLOGICAL STUDIES, SYSTEMATICS,	W72-14313
IDE, PLANT PHYSIOLOGY, BIOASSAY,	CYTOLOGICAL STUDIES, DINOFLAGELLA	W73-00243
TS, CHEMICAL ANALYSIS, BIOASSAY,	CYTOLOGICAL STUDIES, NUTRIENTS, E	W72-09163
BITION, COPPER, FLUORIDES, IONS,	CYTOLOGICAL STUDIES, WATER POLLUT	W72-11572
S, *STATISTICAL METHODS, *ALGAE,	CYTOLOGICAL STUDIES, SELECTIVITY,	W71-12868
RELLA, *BORON, LABORATORY TESTS,	CYTOLOGICAL STUDIES, PLANT GROWTH	W71-12858
, *CYANOPHYTA, PLANT PHYSIOLOGY,	CYTOLOGICAL STUDIES, CULTURES, AM	W72-03217
, FUNGI, SMALL ANIMALS(MAMMALS),	CYTOLOGICAL STUDIES, SOIL CHEMIST	W72-00949
OWTH, SULFATE, LABORATORY TESTS,	CYTOLOGICAL STUDIES.: /, PLANT GR	W72-00153
YTHMS, MEASUREMENT, POPULATIONS,	CYTOLOGICAL STUDIES, STATISTICAL	W72-11920
RMATION, ANTIBIOTIC SENSITIVITY,	CYTOPHAGA JOHNSONII, SPOROCYTOPHA	W72-07951
ROSIN AMINE	D ACETATE(RADA).:	W71-05083
GANOPHOSPHOROUS PESTICIDES, *2-4-	D, *DDT, *DIELDRIN, *DIAZINON, *W	W71-10566
NITROGEN, NITROGEN FIXATION, 2-4-	D, DALAPON, HERBICIDES, CROP PROD	W71-12071
ICITY, *HERBICIDES, DALAPON, 2-4-	D, INSECTICIDES, PHENOLS, *DETERG	W72-12576
SAY, PESTICIDES, HERBICIDES, 2-4-	D, PLANT GROWTH REGULATORS, ALGAE	W72-08440
LLERI, CRYPTOMONADS, STICHOGLOEA	D: / CHROMATIUM, MERISMOPEDIA TRO	W72-07145
REASONABLE USE, BRACKISH WATER,	D: /TIVE USE, ALGAE, EVAPORATION,	W70-07100
ENTRICUS, CLADOGRAMMA NODULIFER,	DACTYLIOSOLEN: /AMMA LINEATUS-EXC	W71-11527
ROTULA, SCHRODERELLA DELICATULA,	DACTYLIOSOLEN MEDITERRANEUS, LEPT	W72-03543
RIX TENUIS, MERISMOPEDIA GLAUCA,	DACTYLOCOCCOPSIS SMITHII, OSCILLA	W72-07896
RMAN(S. DAK), ENEMY SWIM LAKE(S.	DAK).: LAKE HE	W71-02880
LAKE HERMAN(S.	DAK), ENEMY SWIM LAKE(S. DAK).:	W71-02880
PRAIRIE LAKES, *LAKE HERMAN(SO.	DAKOTA), *ORTHOPHOSPHATES.:	W73-00938
REGIONS, AERATED LAGOONS, *NORTH	DAKOTA, BIOCHEMICAL OXYGEN DEMAND	W72-12565
RESERVOIR SILTING, LAKES, *SOUTH	DAKOTA, PLANKTON, *ALGAE, *PHOSPH	W73-00938
TONICS, CYANOPHYTA, ALGAE, SOUTH	DAKOTA, WATER POLLUTION EFFECTS,	W71-02880
ROGEN, NITROGEN FIXATION, 2-4-D,	DALAPON, HERBICIDES, CROP PRODUCT	W71-12071
PESTICIDE TOXICITY, *HERBICIDES,	DALAPON, 2-4-D, INSECTICIDES, PHE	W72-12576
ACEAE, WHIPPLE DAM, STONE VALLEY	DAM(PA.), ATRAZINE, BIOTYPES, POU	W72-04743
IVER, SHOSHONE RIVER, YELLOWTAIL	DAM, FRAGILARIA CROTONENSIS, CRYP	W73-01435
INE, B/ *ATHIORHODACEAE, WHIPPLE	DAM, STONE VALLEY DAM(PA.), ATRAZ	W72-04743
MPLING, RESERVOIRS, TRIBUTARIES,	DAMS, CHEMICAL ANALYSIS, BIOMASS,	W73-01435
*RECIRCULATION RATIO, *ISRAEL(	DAN REGION).:	W72-11642
EN MEDITERRANEUS, LEPTOCYLINDRUS	DANICUS, RHIZOSOLENIA FRAGILISSIM	W72-03543
LLA, CRYPTOMONAS, SPHAEROCYSTIS,	DAPHNIA GALEATA MENDOTAE, DAPHNIA	W71-03014
CILIS, CYCLOPS STRENUUS, INGEST/	DAPHNIA LONGISPINA, DIAPTOMUS GRA	W72-07940
YSTIS, DAPHNIA GALEATA MENDOTAE,	DAPHNIA MAGNA.: /OMONAS, SPHAEROC	W71-03014
, BLOOD, GASTEROSTEUS ACULEATUS,	DAPHNIA MAGNA, DAPHNIA, GASTEROST	W72-05952
ALGAE, ZOOPLANKTON, CARNIVORES,	DAPHNIA.: /UCTIVITY, FOOD HABITS,	W72-00161

RONMENT, FISH PHYSIOLOGY, ALGAE, DAPHNIA.: /TIC LIFE, AQUATIC ENVI W72-12257
TIC LIFE, AQUATIC ANIMALS, CARP, DAPHNIA, AQUATIC PLANTS, AQUATIC W72-09649
MICROORGANISMS, AQUATIC ANIMALS, DAPHNIA, APPLICATION METHODS, GRA W72-09650
QUATIC ANIMALS, AQUATIC INSECTS, DAPHNIA, AQUATIC POPULATIONS, FIS W72-07132
T REQUIRE/ *ZOOPLANKTON, *FOODS, DAPHNIA, COPEPODS, ALGAE, NUTRIEN W72-07940
OSTEUS ACULEATUS, DAPHNIA MAGNA, DAPHNIA, GASTEROSTEUS.: /, GASTER W72-05952
, OLIGOTROPHY, DIATOMS, EUGLENA, DAPHNIA, HABITATS, NICHES, COMPET W71-09171
ONTROL, NEW HAMPSHIRE, COPEPODS, DAPHNIA, SAMPLING, DEPTH, CYANOPH W71-03014
*BOTHNIAN BAY, *GULF OF FINLAND, DARK FIXATION, COASTAL WATERS, TR W72-03228
CALLOPHYLLIS FIRMA, PLENOSPORIUM DASYOIDES, SELENASTRUM CAPRICORNU W73-02099
ICAL STUDIES, DIGITAL COMPUTERS, DATA ANALYSIS, EQUATIONS, REPRODU W72-11920
S, *ARIZONA, GRASSLANDS, ON-SITE DATA COLLECTIONS, PLANT GROWTH.: / W72-12505
ING, *WATER PROPERTIES, *ON-SITE DATA COLLECTIONS, NEW YORK, BIOLU W72-12941
US, NITROGEN, NUTRIENTS, ON-SITE DATA COLLECTIONS, LAKES.: /OSPHOR W72-12990
SSOLVED SOLIDS, HYDROLOGIC DATA, DATA COLLECTIONS, LAKE MORPHOLOGY W72-13851
OLOGY, DOCUMENTATION, ABSTRACTS, DATA COLLECTIONS, MICROBIOLOGY, S W72-13800
TEMATICS, GROWTH RATES, BIOMASS, DATA COLLECTIONS.: /UCTIVITY, SYS W73-02469
TION EFFECTS, SURVEYS, SAMPLING, DATA COLLECTIONS, INDUSTRIAL WAST W73-01972
MARINE FISH, PERIPHYTON, ON-SITE DATA COLLECTIONS, CRUSTACEANS, GA W73-00932
L PROPERTIES, WATER TEMPERATURE, DATA COLLECTIONS, MARINE PLANTS, W73-00838
GATIONS, *ANALYTICAL TECHNIQUES, DATA COLLECTIONS, ALGAE, PROTOZA, W72-07225
ER TEMPERATURES, NOSTOC STREAMS, DATA COLLECTIONS, ANALYTICAL TECH W72-08508
COLLECTIONS, EVALUATION, ON-SITE DATA COLLECTIONS, RIVERS, *ELECTR W72-07791
ISTICAL METHODS, *POTOMAC RIVER, DATA COLLECTIONS, EVALUATION, ON- W72-07791
EAN, BOTTOM SEDIMENTS, STATIONS, DATA COLLECTIONS, ALGAE.: /FIC OC W72-03567
BACTERIA, ALGAE, EUTROPHICATION, DATA COLLECTIONS, TEST PROCEDURES W71-13794
ODOR, INFLOW, DISSOLVED SOLIDS, DATA COLLECTIONS.: /AE, BACTERIA, W72-02274
ROLOGIC DATA, CHEMICAL ANALYSIS, DATA COLLECTIONS, NUTRIENTS, SALI W71-10577
PUTER PROGRAMS, DATA PROCESSING, DATA COLLECTIONS, VARIABILITY, PR W71-07337
LLUTION SOURCES, ALGAE, GEOLOGY, DATA COLLECTIONS.: /ALS, WATER PO W72-05869
LOSIUM, LYNGBYA, PSEUDOANABAENA, DATA INTERPRETATION.: /XA, SPONDY W72-07648
QUINA BAY(ORE), YAQUINA ESTUARY, DATA INTERPRETATION, MACROALGAE, W72-08141
ANDICUS(PACIFICUS), THERMISTORS, DATA INTERPRETATION, COUNTING.: / W72-07892
SS), BUZZARDS BAY, CAPE COD BAY, DATA INTERPRETATION, SPECIES DIVE W73-01449
TH CAROLINA, *COMPUTER PROGRAMS, DATA PROCESSING, DATA COLLECTIONS W71-07337
TOXINS, ANTIBIOTICS(PESTICIDES), DATA PROCESSING, COLIFORMS, MICRO W71-05155
SUBMARINE LIGHT DATA SPHERE.: W71-13252
PHYTA, COOLING WATER, HYDROLOGIC DATA.: /PHYTA, PHAEOPHYTA, CHLORO W72-11252
MICROORGANISMS, RUNOFF, HYDROLOGIC DATA.: /TATISTICAL METHODS, SOIL W72-02218
WATERS, *RESERVOIRS, *HYDROLOGIC DATA, *KANSAS, WATER CHEMISTRY, C W72-02274
, *DATA COLLECTIONS, *HYDROLOGIC DATA, *MEASUREMENT, *ASSAY, ANALY W72-10678
ECOLOGY, *CALIFORNIA, HYDROLOGIC DATA, CHEMICAL ANALYSIS, DATA COL W71-10577
ES, DISSOLVED SOLIDS, HYDROLOGIC DATA, DATA COLLECTIONS, LAKE MORP W72-13851
, TEMPERATURE, ALGAE, HYDROLOGIC DATA, DISSOLVED OXYGEN, WEATHER, W72-05282
OLLUTION SOURCES, METEOROLOGICAL DATA, SEASONAL.: /MISTRY, WATER P W72-04853
CONTROL, RECREATION, HYDROLOGIC DATA, WATER TREATMENT, STREAM GAG W72-05594
NOSP/ PORPHYROPSIS COCCINEA VAR. DAWSONII, CALLOPHYLLIS FIRMA, PLE W73-02099
HYSIOLOGY, *DIATOMS, PESTICIDES, DDD, PESTICIDE TOXICITY, PATH OF W73-02280
RELLA PYRENOIDOSA, LINDANE, P P'- DDE, CLOPHEN A 50, ELECTRON CAPTU W72-04134
KTON, PLANT GROWTH, CHLOROPHYLL, DDE, INHIBITION.: /DDT, PHYTOPLAN W72-12998
STERNA HIRUNDO, RATS, PARTRIDGE, DDE, KIDNEYS, LIVER, BLOOD, GASTE W72-05952
PHORA, CYPRINU/ *RESIDUES, *DDD, DDE, TDE, BLACK-FLY LARVAE, CLADO W70-08098
*PHOTOSYNTHESIS, *MARINE ALGAE, DDT, *PESTICIDES, *PRIMARY PRODUC W72-09653
ANALYSIS, ABSORPTION, CULTURES, DDT, ANALYTICAL TECHNIQUES, BIOMA W72-04134
YTOPLANKTON, TOXICITY, CULTURES, DDT, GROWTH RATES.: / DIATOMS, PH W72-05614
ORINATED HYDROCARBON PESTICIDES, DDT, ORGANOPHOSPHORUS PESTICIDES, W71-10096
, *PATH OF POLLUTANTS, DIELDRIN, DDT, ORGANIC MATTER, ANALYTICAL T W71-11498
IDES, PESTICIDE RESIDUES, ALGAE, DDT, PHYTOPLANKTON, MARINE ALGAE, W72-03545
CYTOLOGICAL STUDIES, ABSORPTION, DDT, PHYTOPLANKTON, PLANT GROWTH, W72-12998
KES, STREAMS, ALGAE, PESTICIDES, DDT, RETURN FLOW, SEDIMENTS, SOIL W71-06444
S, MUD, MOLLUSKS, FISH, INSECTS, DDT, SILTS, SUCKERS, CATFISHES, B W70-08098
ITTORAL BENTHIC MACROPHYTES, RIA DE ALDAN(SPAIN), LAMINARIA OCHROL W72-04766
ION, *CHEMOTAXIS, *PREY, PYTHIUM DEBARYANUM, SKELETONEMA COSTATUM, W73-00942
RYHALINE, RED SEA, GULF OF SUEZ, DECAPODS, COLEOPTERA, MEDITERRANE W73-00855
BROMINE, ACTINIDES, RADIOACTIVE DECAY, GAMMA-RAY SPECTROMETRY, CO W73-01434
R PROGRAMMING, SYSTEMS ANALYSIS, DECIDUOUS FORESTS, DECOMPOSING OR W72-03342
NOSIRA POLYCHORDA, THALASSIOSIRA DECIPIENS, THALASSIOSIRA GRAVIDA, W72-03543
R REUSE, LEGAL ASPECTS, JUDICIAL DECISIONS, LEGISLATION, ADMINISTR W70-07100
SYNTH/ *NUTRIENT REMOVAL, OXYGEN DECLINE, RESPIRATORY RATES, PHOTO W71-04079
EMS ANALYSIS, DECIDUOUS FORESTS, DECOMPOSING ORGANIC MATERIAL, FER W72-03342
IC LEVEL, EUTROPHICATION, ALGAE, DECOMPOSING ORGANIC MATTER, MINNE W72-01784
LGAE, EUTROPHICATION, CHELATION, DECOMPOSING ORGANIC MATTER, PHYTO W72-01097
HESIS, CARBON CYCLE, ABSORPTION, DECOMPOSING ORGANIC MATTER, ALKAL W71-12068
HYTOPLANKTON, BACTERIA, AQUARIA, DECOMPOSING ORGANIC MATTER, WASTE W73-00802
OROPHYLL, EPILIMNION, SEDIMENTS, DECOMPOSING ORGANIC MATTER, BACTE W72-12997
, CYCLING NUTRIENTS, DEGRADATION(DECOMPOSITION), SAMPLING, BENTHIC W72-00847
AKE, CHEMICAL COMPOSITION, ALGAL DECOMPOSITION.: /TH, NUTRIENT UPT W70-05469
MENTS, SORPTION, EUTROPHICATION, DEFICIENT ELEMENTS, ESSENTIAL NUT W72-11617
YTA, OLIGOTROPHY, CHLAMYDOMONAS, DEFICIENT ELEMENTS, MESOTROPHY, E W72-13816
NT IDENTIFICATION, GROWTH RATES, DEFICIENT ELEMENTS, WATER POLLUTI W73-01445
QUALITY, NUTRIENTS, COLORIMETRY, DEFICIENT ELEMENTS, AQUATIC ALGAE W73-00916
IS A/ *SHOCK-WAVES, *GAS-VACUOLE DEFLATION, SOUTH WALES, MICROCYST W72-11564
NT TRANSPORT, CYCLING NUTRIENTS, DEGRADATION(DECOMPOSITION), SAMPL W72-00847
EFFECTS, WATER QUALITY, CHEMICAL DEGRADATION.: /, WATER POLLUTION W72-01085
IOASSAY, FISH, SHELLFISH, ALGAE, DEGRADATION.: /NTS, *TOXICITY, *B W71-09789
S, *DISSOLVED SOLIDS, *MICROBIAL DEGRADATION, CARBON, ORGANIC MATT W73-00379
ARINE MICROORGANISMS, *MICROBIAL DEGRADATION, *DDT, *DIELDRIN, *AL W73-00361
TION SOURCES, REVIEWS, MICROBIAL DEGRADATION, CLASSIFICATION, SEWA W72-14363
AEROBIC BIOLOGICAL DEGRADATION.: W72-05315
XIDATION LAGOONS, *DISINFECTION, DEGRADATION, CHLORINE, CONTACT TI W71-07096

H, WASTE TREATMENT, FARM WASTES, DEHYDRATION, FEASIBILITY.: /ATFIS W71-07544
STE DISPOSAL, OXIDATION LAGOONS, DEHYDRATION, RUNOFF, RADIOACTIVIT W72-00846
,/ ALGAL CYCLE, DELAWARE ESTUARY(DEL-NJ), SAN JOAQUIN DELTA(CALIF) W71-05390
ALIMITO/ *OFFSHORE CONSTRUCTION, DELAWARE BAY, SCOLOPLOS ARMIGER, W73-02029
QUIN DELTA(CALIF),/ ALGAL CYCLE, DELAWARE ESTUARY(DEL-NJ), SAN JOA W71-05390
ASHINGTON, CONNECTICUT, FLORIDA, DELAWARE RIVER, INDIANA.: /NIA, W W71-07417
ICATORS, ALGAE, RIVERS, STREAMS, DELAWARE RIVER, OXIDATION LAGOONS W72-04736
MERRIMACK RIVER(NEW HAMPSHIRE), DELAWARE RIVER, ILLINOIS RIVER, H W71-11517
RIVER, HUDSON RIVER, INDICATORS, DELAWARE, PATH OF POLLUTANTS, HOS W72-04736
C-14, CHLOROPHYLL A,/ NITZSCHIA DELICATISSIMA, NITZSCHIA PUNGENS, W73-00431
ALASSIOSIRA ROTULA, SCHRODERELLA DELICATULA, DACTYLIOSOLEN MEDITER W72-03543
*SAN FRANCISCO BAY- DELTA.: W71-09581
ARE ESTUARY(DEL-NJ), SAN JOAQUIN DELTA(CALIF), STEADY-STATE MODELS W71-05390
RBIDOSTAT, *GROWTH/ *SAN JOAQUIN DELTA(CALIF), *CONTINUOUS FLOW TU W70-10403
ODEGRADATION, BIOCHEMICAL OXYGEN DEMAND.: /STE WATER TREATMENT, BI W71-07084
ER TREATMENT, BIOCHEMICAL OXYGEN DEMAND.: /ADING, ALGAE, WASTE WAT W71-07096
L OXYGEN DEMAND, CHEMICAL OXYGEN DEMAND.: /EGRADATION, *BIOCHEMICA W72-06612
, CYANOPHYTA, BIOCHEMICAL OXYGEN DEMAND.: /HRYSOPHYTA, CHLOROPHYTA W72-12995
AKE ERIE, *ALGAE, *BENT/ *OXYGEN DEMAND, *ANAEROBIC CONDITIONS, *L W72-12995
N CONCENTRATION, CHEMICAL OXYGEN DEMAND, *BIOCHEMICAL OXYGEN DEMAN W71-07102
ROPHICATION, *BIOCHEMICAL OXYGEN DEMAND, *CHEMICAL OXYGEN DEMAND, W70-09189
OUS/ *ALGAE, *BIOCHEMICAL OXYGEN DEMAND, *EUTROPHICATION, PHOSPHOR W72-04431
REAMS, LAKES, BIOCHEMICAL OXYGEN DEMAND, *OXYGENATION, *ANAEROBIC W72-14469
L OXYGEN DEMAND, CHEMICAL OXYGEN DEMAND, *PHOSPHATES, NITROGEN, CO W71-13356
ONCENTRATION, BIOCHEMICAL OXYGEN DEMAND, *SEWAGE EFFLUENTS, *SEWAG W72-06612
OXYGEN DEMAND, *CHEMICAL OXYGEN DEMAND, *WATER POLLUTION SOURCES, W70-09189
XYGEN DEMAND, BIOCHEMICAL OXYGEN DEMAND, AERATION, *OXIDATION, HYD W71-12223
EDIMENTATION, BIOCHEMICAL OXYGEN DEMAND, ALGAE, CONFINEMENT PENS, W72-08401
NTS, ALKALINITY, CHEMICAL OXYGEN DEMAND, ANALYTICAL TECHNIQUES, TE W72-04983
MICAL WASTES, BIOCHEMICAL OXYGEN DEMAND, BACTERIA, ALG: /CALS, CHE W73-01976
NORTH DAKOTA, BIOCHEMICAL OXYGEN DEMAND, BACTERIA, TEMPERATURE, CO W72-12565
MATTER, OXYGEN, CHEMICAL OXYGEN DEMAND, BENTHOS.: /ATION, ORGANIC W71-04518
LANTS, ALGAE, BIOCHEMICAL OXYGEN DEMAND, BIOLOGICAL TREATMENT, PHO W70-06619
N, TIDAL WATERS, CHEMICAL OXYGEN DEMAND, BIOCHEMICAL OXYGEN DEMAND W71-07098
ATION, SAMPLING, CHEMICAL OXYGEN DEMAND, BIOCHEMICAL OXYGEN DEMAND W71-07097
STES, ECONOMICS, CHEMICAL OXYGEN DEMAND, BIOCHEMICAL OXYGEN DEMAND W71-12223
ROL, WASHINGTON, CHEMICAL OXYGEN DEMAND, BIOCHEMICAL OXYGEN DEMAND W71-11006
SIS, SOLUBILITY, CHEMICAL OXYGEN DEMAND, BIOCHEMICAL OXYGEN DEMAND W72-00411
L OXYGEN DEMAND, CHEMICAL OXYGEN DEMAND, CARBON, FERTILIZATION, DI W71-12188
L OXYGEN DEMAND, CHEMICAL OXYGEN DEMAND, CARBON DIOXIDE, LININGS, W72-04789
LISM, SLUDGE, BIOCHEMICAL OXYGEN DEMAND, CHEMICAL OXYGEN DEMAND, C W72-04789
TEMPERATURE, BIOCHEMICAL OXYGEN DEMAND, CHEMICAL OXYGEN DEMAND, C W71-09524
GAE, LAGOONS, BIOCHEMICAL OXYGEN DEMAND, CHEMICAL OXYGEN DEMAND, P W70-07838
RY TREATMENT, BIOCHEMICAL OXYGEN DEMAND, CHEMICAL OXYGEN DEMAND, C W71-12188
ON, *FILTERS, BIOCHEMICAL OXYGEN DEMAND, CHEMICAL OXYGEN DEMAND, P W71-13341
ICIPAL WASTE, BIOCHEMICAL OXYGEN DEMAND, CHEMICAL OXYGEN DEMAND, * W71-13356
ION, STORAGE, BIOCHEMICAL OXYGEN DEMAND, CHEMICAL OXYGEN DEMAND, P W72-01881
GEN, PHOSPHORUS, CHEMICAL OXYGEN DEMAND, CHLORELLA, PERFORMANCE, E W72-11100
IC TREATMENT, BIOCHEMICAL OXYGEN DEMAND, CHEMICAL OXYGEN DEMAND, N W72-08396
DEGRADATION, *BIOCHEMICAL OXYGEN DEMAND, CHEMICAL OXYGEN DEMAND.: / W72-06612
ZOOPLANKTON, BIOCHEMICAL OXYGEN DEMAND, COPEPODS.: /QUATIC ALGAE, W71-11876
L OXYGEN DEMAND, CHEMICAL OXYGEN DEMAND, COOLING TOWERS, *WASTE WA W71-09524
HYTOPLANKTON, BIOCHEMICAL OXYGEN DEMAND, DISSOLVED OXYGEN, ESTUARI W71-05390
, *TURBIDITY, BIOCHEMICAL OXYGEN DEMAND, DISSOLVED OXYGEN, ALGAE, W72-03753
XYGEN DEMAND, BIOCHEMICAL OXYGEN DEMAND, DISSOLVED OXYGEN: /ICAL O W71-11006
RIES, WATER SUPPLY, COSTS, WATER DEMAND, ECONOMIC JUSTIFICATION, A W71-13172
NITY, MINING, BIOCHEMICAL OXYGEN DEMAND, ECONOMICS, ALGAE, AGRICUL W72-04734
D, ION EXCHANGE, CHEMICAL OXYGEN DEMAND, FERTILIZER, TURBIDITY.: / W71-12200
XYGEN DEMAND, BIOCHEMICAL OXYGEN DEMAND, FILTRATION, DISINFECTION, W72-00411
GENOUS RESPIRATION, TOTAL OXYGEN DEMAND, GLYCOLS.: /UM NO 52, ENDO W71-12223
, PHOSPHATES, ALKALINITY, OXYGEN DEMAND, HEAT BUDGET, ALGAE, EUTRO W72-12997
ERIE, *DISSOLVED OXYGEN, OXYGEN DEMAND, HYPOLIMNION, AQUATIC PLAN W72-12994
TION, SLOPES, BIOCHEMICAL OXYGEN DEMAND, HYDROGEN ION CONCENTRATIO W71-07113
WRENCE RIVER, BIOCHEMICAL OXYGEN DEMAND, HYDROGEN ION CONCENTRATIO W70-10427
RY TREATMENT, BIOCHEMICAL OXYGEN DEMAND, INDUSTRIAL WASTES, FARM W W71-06445
N, NUTRIENTS, BIOCHEMICAL OXYGEN DEMAND, INORGANIC COMPOUNDS, THER W72-13659
LITY, SEWAGE, BIOCHEMICAL OXYGEN DEMAND, ION EXCHANGE, CHEMICAL OX W71-12200
OROUS, NITROGEN, CHEMICAL OXYGEN DEMAND, LABORATORY TESTS, ALKALIN W72-04431
WEED CONTROL, BIOCHEMICAL OXYGEN DEMAND, LAKE STAGES, LIMNOLOGY, N W72-14692
PRODUCTIVITY, BIOCHEMICAL OXYGEN DEMAND, LAKE STAGES, NUTRIENTS, M W73-00002
YGEN DEMAND, *BIOCHEMICAL OXYGEN DEMAND, MIXING, SAMPLING, ORGANIC W71-07102
ON POTENTIAL, BIOCHEMICAL OXYGEN DEMAND, NITRIFICATION, NUTRIENTS, W72-04086
NIC LOADING, *BIOCHEMICAL OXYGEN DEMAND, NITROGEN, PHOSPHOROUS, PH W71-13326
L OXYGEN DEMAND, CHEMICAL OXYGEN DEMAND, NUTRIENTS, *WASTE TREATME W72-08396
ER POLLUTION, BIOCHEMICAL OXYGEN DEMAND, ORGANIC MATTER, PHOSPHORU W71-13257
WATER RE-USE, BIOCHEMICAL OXYGEN DEMAND, ORGANIC LOADING, ODORS, P W71-07123
CTIVATED SLUDGE, CHEMICAL OXYGEN DEMAND, OXYGEN, AERATION, DISSOLV W71-13334
OLVED OXYGEN, BIOCHEMICAL OXYGEN DEMAND, PERFORMANCE, FLOCCULATION W72-13508
ITIONS, ALGAE, SEDIMENTS, OXYGEN DEMAND, PHOSPHORUS, NITROGEN, NUT W72-12990
L OXYGEN DEMAND, CHEMICAL OXYGEN DEMAND, PHOSPHORUS, NITRATES, AMM W71-13341
L OXYGEN DEMAND, CHEMICAL OXYGEN DEMAND, PHOSPHORUS, NITROGEN, EUT W72-01881
ATION LAGOONS, CHLORELLA, OXYGEN DEMAND, PHOTOSYNTHESIS, RESPIRATI W71-00139
L OXYGEN DEMAND, CHEMICAL OXYGEN DEMAND, PHOSPHATES, NITROGEN, STA W70-07838
GROUNDWATER, BIOCHEMICAL OXYGEN DEMAND, PHOSPHORUS, FISH, WEEDS, W71-03012
, METABOLISM, BIOCHEMICAL OXYGEN DEMAND, PILOT PLANTS, *WASTE WATE W72-06838
ER POLLUTION EFFECTS, RECREATION DEMAND, POPULATION, COMMUNITY DEV W72-01660
N, OXIDATION, BIOCHEMICAL OXYGEN DEMAND, RADIOACTIVITY, RADIOACTIV W72-05506
MS, EFFLUENT, BIOCHEMICAL OXYGEN DEMAND, SATURATION, OXYGEN, METAB W71-13553
ERS, *SEWAGE, BIOCHEMICAL OXYGEN DEMAND, SOLID WASTES, AMMONIA, NI W72-09675
ANIC LOADING, BIOCHEMICAL OXYGEN DEMAND, SOLAR RADIATION, NUTRIENT W71-07124

XYGEN DEMAND, BIOCHEMICAL OXYGEN	DEMAND, STABILIZATION PONDS, MICR	W71-07097
CIPAL WASTES, BIOCHEMICAL OXYGEN	DEMAND, WASTE WATER TREATMENT, MI	W71-06033
C CONDITIONS, BIOCHEMICAL OXYGEN	DEMAND, WASTE WATER TREAT: /EROBI	W71-07100
ST ANALYSIS, *BIOCHEMICAL OXYGEN	DEMAND, WASTE WATER TREATMENT.: /	W71-07099
RY TREATMENT, BIOCHEMICAL OXYGEN	DEMAND, WASTE WATER TREATMENT, *N	W71-07382
C CONDITIONS, BIOCHEMICAL OXYGEN	DEMAND, WASTE WATER TREATMENT.: /	W71-07109
XYGEN DEMAND, BIOCHEMICAL OXYGEN	DEMAND, WASTE WATER TREATMENT, WA	W71-07098
Y, COLIFORMS, BIOCHEMICAL OXYGEN	DEMAND, WASTE WATER TREATMENT.: /	W71-07106
CIPAL WASTES, BIOCHEMICAL OXYGEN	DEMAND, WASTE WATER TREATMENT.: /	W71-03896
TION SOURCES, BIOCHEMICAL OXYGEN	DEMAND, WATER POLLUTION CONTROL,	W72-09167
OXYGENATION, BIOCHEMICAL OXYGEN	DEMAND, WATER SPORTS, PARKS, COST	W72-00042
LAKE JOSEPHINE, LAKE ARCO, LAKE	DEMING, LOWER LA SALLE LAKE, LAKE	W72-07145
LLS(BIOLOGY), *BIODETERIORATION,	DEMOCOLCINE.: /SS, THYMIDINES, CE	W71-09261
AL PRECIPITATION, NITRIFICATION,	DENITRIFICATION, *WASTE WATER TRE	W71-09450
MONIA, PRECIPITATION, SEDIMENTS,	DENITRIFICATION, RUNOFF, UREAS, F	W71-06435
POLLUTION, DESALINATION, ALGAE,	DENITRIFICATION, CALIFORNIA, FILT	W71-08223
TRIFICATION, METHANOL, BACTERIAL	DENITRIFICATION, SAN JOAQUIN VALL	W71-08223
NG, POND DENITRIFICATION, FILTER	DENITRIFICATION, METHANOL, BACTER	W71-08223
E REMOVAL, ALGAE STRIPPING, POND	DENITRIFICATION, FILTER DENITRIFI	W71-08223
IS, DISTILLATION, NITRIFICATION,	DENITRIFICATION, SEPTIC TANKS, OX	W70-09907
, OXIDATION-REDUCTION POTENTIAL,	DENITRIFICATION, NITROGEN FIXATIO	W72-08459
, AMMONIFICATION, NITRIFICATION,	DENITRIFICATION, NITROGEN FIXATIO	W72-07933
S, LIGHT INTENSITY, OXYGENATION,	DENITRIFICATION, OXIDATION, ADSOR	W71-13326
ANAEROBIC CONDITIONS, NITRATES,	DENITRIFICATION, BOTTOM SEDIMENTS	W72-11563
ES, ANABAENA, NITROGEN FIXATION,	DENITRIFICATION, WATER POLLUTION	W72-12544
*NITROGEN, COST ANALYSIS, OZONE,	DENITRIFICATION, POLLUTION ABATEM	W73-02187
A, ACHANTHES, FRAG/ LAKE FURESO (	DENMARK), COCCONEIS, RHOICOSPHENI	W72-07501
ATION, SALPA, LUDOX AM, SUCROSE,	DENSITY GRADIENTS, TUNICATES, PHA	W73-00371
STATS, FLAGELLATE ALGAE, OPTICAL	DENSITY SENSOR.: /KINETICS, CHEMO	W70-10403
CIFERASE, CHLOROPHYLL A, OPTICAL	DENSITY, ACETYLENE REDUCTION, SCH	W72-10883
OCHAETES, LIFE CYCLES, SEASONAL,	DENSITY, DIATOMS, LAKES, PROTOZOA	W71-11012
, POTASSIUM COMPOUNDS, NITRATES,	DENSITY, GROWTH RATES, LABORATORY	W72-03217
*MICROSTRATIFICATION, *OPTICAL	DENSITY, MICROAMMETER, OSCILLATOR	W72-07145
IOPHYCEAE,/ *ANTIBIOSIS, OPTICAL	DENSITY, PRASINOPHYCEAE, BACCILAR	W72-08051
ATION, PHYSICAL PROPERTIES, BULK	DENSITY, POROSITY, COLOR, HYDROGE	W73-00854
ISH KILL, ALGAE, COOLING TOWERS,	DENSITY, VISCOSITY, FISH BEHAVIOR	W70-09612
INATION, DRAIN/ NITRATES, ALGAE,	DENTRIFICATION, FILTRATION, DESAL	W71-11698
AL BASIN, NUTRIENT REGENERATION,	DEOXYGENATION, ALGAL SEDIMENTATIO	W72-12997
N, POINTE PELEE(ONTARIO), OXYGEN	DEPLETION.: /KE ERIE CENTRAL BASI	W72-12992
LAKE ERIE CENTRAL BASIN, *OXYGEN	DEPLETION.: *	W72-13659
, RHIZOSOLENIA ALATA, PERIDINIUM	DEPRESSUM, PERIDINIUM PALLIDUM, C	W71-07875
, GONYAULAX POLYEDRA, PERIDINIUM	DEPRESSUM, METAZOA, CALANUS HELGO	W72-07892
HYTES, *SEA GRASSES, TELEVISION,	DEPTH SOUNDER, VERTICAL DISTRIBUT	W72-09120
FOULING, *ARTIFICIAL SUBSTRATES,	DEPTH, *SAMPLING, *PERIPHYTON, EQ	W73-00374
G, INCUBATION, RUNOFF, SALINITY,	DEPTH, ABSORPTION, ALGAE, SEASONA	W72-09122
ING FACTORS, NITROGEN, PLANKTON,	DEPTH, ALGAE.: /, NITRATES, LIMIT	W73-02472
*MARINE ANIMALS, *MARINE PLANTS,	DEPTH, CALIFORNIA, SANDS, SEDIMEN	W71-01475
GEOLOGY, LIMESTONE, TEMPERATURE,	DEPTH, CHEMICAL PROPERTIES, ACIDI	W72-04269
RE, COPEPODS, DAPHNIA, SAMPLING,	DEPTH, CYANOPHYTA, CHRYSOPHYTA, C	W71-03014
NETRATION, BENTHIC FLORA, LAKES,	DEPTH, DIATOMS, CHLOROPHYTA, CYAN	W71-11029
HYTOPLANKTON, NATURAL RESOURCES,	DEPTH, DISTRIBUTION PATTERNS, MEA	W71-13252
TEMATICS, BENTHOS, SCUBA DIVING,	DEPTH, DISTRIBUTION PATTERNS, AER	W72-09120
, SOLAR RADIATION, DISTRIBUTION,	DEPTH, DIURNAL, SEASONAL.: /ATION	W72-08054
GEN, TURBULENCE, PHOTOSYNTHESIS,	DEPTH, LIGHT PENTRATION, RESPRIAT	W72-08559
SIS, SESSILE ALGAE, TEMPERATURE,	DEPTH, LIGHT PENETRATION.: /YNTHE	W72-07501
RFACE WATERS, CONTINENTAL SHELF,	DEPTH, NITROGEN, PHYTOPLANKTON, H	W72-08056
F STABILIZATION.:	DEPTH, SLOPE, *BOD LOADING, *WAST	W71-03896
ITY, POPULATION, ATLANTIC OCEAN,	DEPTH, STANDING CROP, CURRENTS(WA	W72-04766
N LAGOONS, *OXYGENATION, *ALGAE,	DEPTH, TEMPERATURE, PHOTOSYNTHESI	W71-07098
ATER QUALITY, ALGAE, HARVESTING,	DEPTH, TEMPERATURE, LIGHT INTENSI	W71-07100
TES, SILICATES, SOLAR RADIATION,	DEPTH, THERMOCLINE, NITRATES, LIG	W71-07875
: INDUSTRIAL WATER POLLUTION,	DES MOINES RIVER, PACKING PLANTS.	W71-06445
LOW, *NITRATES, WATER POLLUTION,	DESALINATION, ALGAE, DENITRIFICAT	W71-08223
GAE, DENTRIFICATION, FILTRATION,	DESALINATION, DRAINAGE WATER, AGR	W71-11698
ITIS IDAEA, VACCINIUM MYRTILLUS,	DESCHAMPSIA FLEXUOSA, PARMELIA PH	W72-14303
LANT PHYSIOLOGY, CULTURES, PHOT/	DESCRIPTORS:*DDT, *RESISTANCE, *P	W73-00281
BEER EQUATION, EDDY DIFFUSIVITY,	DESHLER(OHIO).: /ICIENT, LAMBERT-	W72-05459
SSOLVED OXYGEN, LIGHT INTENSITY,	DESIGN CRITERIA, *OXIDATION LAGOO	W72-04789
GEN, HYDROGEN ION CONCENTRATION,	DESIGN CRITERIA, MATHEMATICAL MOD	W72-04787
RATION, LIME, ROCKS, METABOLISM,	DESIGN CRITERIA, *WASTE WATER TRE	W71-13341
DYNAMICS, ANALYTICAL TECHNIQUES,	DESIGN CRITERIA, KINETICS, *WASTE	W72-02363
LUM, POLYELECTROLYTES, FACTORIAL	DESIGN.: *A	W70-08904
UTROPHICATION, ACTIVATED SLUDGE,	DESIGN, NITROGEN, DIFFUSION, *ALG	W72-00800
*WATER TEMPERATURE, CONDENSERS,	DESIGN, STANDARDS, FISH KILL, ALG	W70-09612
HICATION/ *NUTRIENTS, *DREDGING,	DESILTING, SEDIMENTATION, *EUTROP	W73-00938
, PROTOCOCCINEAE, ULOTRICHINEAE,	DESMIDIALES.: /PHYTA, VOLVOCINEAE	W72-04855
	DESMIDS.:	W73-01615
ES FLEXELLA, EUNOTIA PECTINALIS,	DESMIDS.: / MINUTISSIMA, ACHNANTH	W71-11029
CONTRACTUM, ARTH/ *STAURASTRUM,	DESMIDS, STAURASTRUM LONGIPES VAR	W73-01094
ELOSIRA NUMMULOIDES, CHLOROPHYL/	DESSICATION, SPECIES DIVERSITY, M	W73-00853
LS, *ALGAL CONTROL, *RESERVOIRS,	DESTRATIFICATION, MIXING, PHOSPHO	W72-08559
LAKE, APHANIZOMENON FLOW-AQUAE,	DESTRATIFICATION, ANABAENA FLOW-A	W72-07890
ACTER, THIOBACILLUS THIOOXIDANS,	DESULFOVIBRIO, ULOTHRIZ ZONATA.: /	W71-01489
39, POLYMERIZATION, LIVER, BONE,	DETECTION LIMITS, CHEMICAL RECOVE	W72-05965
LES, SARGASSO WEED,/ *PLUTONIUM,	DETECTION LIMITS, BIOLOGICAL SAMP	W72-05965
ECHNIQUES, SALMONELLA TYPHIMURI/	DETECTION, FLUORESCENT ANTIBODY T	W72-13800
CHROMATOGRAPHY, ELECTRON CAPTURE	DETECTOR, PRECONCENTRATION, BIOLO	W73-02105
, HOUSEHOLD DETERGENTS, SOAP AND	DETERGENT INDUSTRY, NTA.: /C ACID	W72-04266
ATION, FMC CORPORATION, SOAP AND	DETERGENT ASSOCIATION, CARBOY TRI	W70-07283
US MATERIAL, *CANADIAN PHOSPHATE	DETERGENT BAN, WYANDOTTE CHEMICAL	W70-07283

EN, LAKES, ESTUARIES, NUTRIENTS, DETERGENTS, WISCONSIN, LAKE ERIE, W70-07283
IOASSAY, TOXICITY, *EMULSIFIERS, DETERGENTS, OIL, *OIL WASTES, OIL W70-09429
E, *TOXICITY, *LABORATORY TESTS, DETERGENTS, WASTE WATER TREATMENT W70-08976
E WATER(POLLUTION), *HYDROLYSIS, DETERGENTS, GREAT LAKES, TEMPERAT W71-00116
NITRILOTRIACETIC ACID, HOUSEHOLD DETERGENTS, SOAP AND DETERGENT IN W72-04266
ES, ABSORPTION, ELECTROPHORESIS, DETERGENTS, AMINO ACIDS, SPECTROP W72-06124
N, LAKE ERIE, NITROGEN FIXATION, DETERGENTS, CARBON DIOXIDE, ALKAL W72-05473
EATMENT, *COST-BENEFIT ANALYSIS, DETERGENTS, PHOSPHORUS, NITROGEN, W72-04734
ENZYME PRESOAKS, DISHWATER DETERGENTS, NTA.: W71-08768
CETATE, SODIUM C/ PHOSPHATE-FREE DETERGENTS, TRISODIUM NITRILOTRIA W71-06247
A, NUTRIENT REQUIREMENT, *ALGAE, DETERGENTS, PHOSPHATES.: /HLORELL W73-02112
ION, FILTRATION, ALGAE, PHENOLS, DETERGENTS, PESTICIDES, OPERATION W72-08357
, SULFUR, POTASSIUM, MOLYBDENUM, DETERGENTS, SEWAGE TREATMENT, CAR W72-07514
GULATION, ALGAL TOXINS, MERCURY, DETERGENTS, WATER POLLUTION SOURC W73-00703
, MICROSCOPY, INDUSTRIAL WASTES, DETERGENTS, AMMONIA, NITRITES, NI W71-11507
R, LAKE ERIE, EFFLUENTS, SEWAGE, DETERGENTS, PLANTS, LA: /IC MATTE W71-11017
, PHOSPHORUS, CARBON, LAKE ERIE, DETERGENTS, LAKE ONTARIO, TERTIAR W71-12091
ROL, TROPHIC LEVELS, PHOSPHATES, DETERGENTS, BIOASSAY, ANALYTICAL W71-12072
ATH OF POLLUTANTS, STORM DRAINS, DETERGENTS, EUTROPHICATION, ALGAE W71-13466
LYSIS, POLLUTANT IDENTIFICATION, DETERGENTS, NUTRIENTS, PESTICIDES W72-08790
ON CONTROL, NITRATES, EFFLUENTS, DETERGENTS, SODIUM CHLORIDE, CHEM W72-09167
HOSPHORUS, NITROGEN, PHOSPHATES, DETERGENTS, POTABLE WATER.: /F, P W72-10625
ONDS, ALGISTATIC ACTIVITY, ALGI/ DETOXIFICATION, SYNERGISM, FISH P W71-06189
GANIC MATTER, CALCIUM CARBONATE, DETRITUS, PLANKTON, ADSORPTION, E W72-05469
P, *ALGAE, DIATOMS, MEASUREMENT, DETRITUS, METHODOLOGY, SEDIMENTS, W71-12855
MPLING, ALGAE, LABORATORY TESTS, DETRITUS, BENTHOS, AQUATIC ANIMAL W71-10064
OMATOGRAPHY, PARTICULATE MATTER, DETRITUS, ORGANIC COMPOUNDS, PATH W73-02105
, BIOMASS, NUTRIENTS, FOOD WEBS, DETRITUS, ECOSYSTEMS, ENERGY BUDG W73-01089
OTOMETRY, LAKES, PLANT PIGMENTS, DETRITUS, CHLOROPHYTA, CHRYSOPHYT W72-13673
S, AUTUMNAL COMMUNITY, CALORIES, DETRIVORES, LINESVILLE CREEK(PA). W71-10064
TION, LAKE SALLIE(MINN), LAKE M/ DETROIT LAKES(MINN), COMMUNITY AC W71-03012
ITAE LAKE, GREEN BAY, FOX RIVER, DETROIT RIVER, HETEROTROPHIC BACT W73-01456
ST. CLAIR RIVER, LAKE ST. CLAIR, DETROIT RIVER, BLACK RIVER, PINE W72-14282
POLLUTION EFFECTS, RESEARCH AND DEVELOPMENT, WASTE WATER TREATMEN W72-06655
TE WATER TREATMENT, RESEARCH AND DEVELOPMENT, MUNICIPAL WASTES, LE W72-06638
OSPHORUS, THERMAL/ *RESEARCH AND DEVELOPMENT, *WATER POLLUTION, PH W70-09947
*WATER SUPPLY, *WATER RESOURCES DEVELOPMENT, *DISTRIBUTION SYSTEM W72-05594
*LABORATORY TESTS, RESEARCH AND DEVELOPMENT, TRACE ELEMENTS, ALGA W72-04433
ER QUALITY CONTROL, RESEARCH AND DEVELOPMENT.: /NZYMES, COLOR, WAT W72-04070
LUTION EFFECTS, *WATER RESOURCES DEVELOPMENT, *MASSACHUSETTS, *EUT W71-05553
S, CATTLE, POULTRY, RESEARCH AND DEVELOPMENT, SOIL, SOIL CONTAMINA W71-07551
, INVERTEBRATES, WATER RESOURCES DEVELOPMENT.: /IRONMENTAL EFFECTS W71-10577
*WATER QUALITY, *WATER RESOURCES DEVELOPMENT, *WATER SUPPLY, WASTE W71-09788
IMILATIVE CAPACITY, RESEARCH AND DEVELOPMENT, WASTE WATER TREATMEN W71-13256
CONTRACTS, GRANTS, RESEARCH AND DEVELOPMENT, INSPECTION, FEDERAL: W71-11516
, SEWAGE EFFLUENTS, RESEARCH AND DEVELOPMENT, NUTRIENTS, NITROGEN, W72-11702
ON DEMAND, POPULATION, COMMUNITY DEVELOPMENT, ENVIRONMENTAL EFFECT W72-01660
ATURE, CHLORINATION, FLOATATION, DEWATERING, DRYING, COST ANALYSES W71-08951
ACRYLIC ACID, SODIUM, PYRIDINE, DEXTRAM, RAFFINOSE, SUCROSE, SENE W72-07710
CYSTIS, SORTING, SODIUM BROMIDE, DEXTRAN, OIKOPLEURA, SAGITTA, PTE W73-00371
NG VESSELS, PENICILLIN, LINDANE, DIADEMA ANTILLARUM, AANDOMENE, SI W72-14294
POTENTIAL, ILLINOIS RIVER(ILL), DIAGNOSTIC CRITERION.: /AL GROWTH W71-11033
VITAMIN B, BIOASSAY, CELLULOSE, DIALYSIS, SEA WATER, ORGANIC ACID W73-00942
YING, REFRIGERATION, INCUBATION, DIALYSIS, CHEMICALS, DRYING, FREE W73-01676
UUS, INGEST/ DAPHNIA LONGISPINA, DIAPTOMUS GRACILIS, CYCLOPS STREN W72-07940
VELA STULTORUM, EMERITA ANALOGA, DIASTYLOPSIS, OXYUROSTYLIS, HEMIL W72-09092
, BENTHOS, MAYFLIES, TUBIFICIDS, DIAT: /PHYTA, GEOLOGIC FORMATIONS W71-11009
MA, ANACYSTIS NIDULANS, CERASON, DIATHENE, PROPAZINE, COTORON, DIU W71-12071
OTELLA STELLIGERA, CYMBELLA SPP, DIATOMA SSP.: /CF. SOCIALIS, CYCL W72-13142
NOBRYON, OEDOGONIUM, FRAGILARIA, DIATOMA, MELOSIRA, TABELLARIA, AC W72-10875
*HUMAN POPULATION, FERTILIZERS, DIATOMACEOUS EARTH, FOOD CHAINS, W71-09157
WATER POLLUTION EFFECTS, ALGAE, DIATOMS.: /ALGAL CONTROL, RIVERS, W71-00099
IFT, NUTRIENTS, SALANITY, ALGAE, DIATOMS.: /OTA, LAKES, GLACIAL DR W71-13149
MENTATION, HYDROSTATIC PRESSURE, DIATOMS.: /PHOTOSYNTHESIS, INSTRU W71-10791
C ALGAE, NORTH CAROLINA, *ALGAE, DIATOMS, *DINOFLAGELLATES, *RED T W72-01329
PONDS, *THERMAL STRATIFICATION, DIATOMS, *CHARA.: /L, *PONDWEEDS, W73-02349
IOLOGICAL COMMUNITIES, SEASONAL, DIATOMS, *LIGHT, SOLAR RADIATION, W72-07901
RIVER KENNET(ENGLAND),/ *CENTRIC DIATOMS, *RIVER THAMES(ENGLAND), W72-14797
*DDT, PESTICIDES, PHYTOPLANKTON, DIATOMS, ALGAE, MARINE ALGAE, AQU W72-08436
ROGEN, PHYSOCHEMICAL PROPERTIES, DIATOMS, ALGAE, CLADOPHORA, PRODU W72-13662
HRYSOPHYTA, SOLVENT EXTRACTIONS, DIATOMS, AQUATIC ALGAE, FILTRATIO W72-13673
EPODS, OYSTERS, DINOFLAGELLATES, DIATOMS, ALGAE CULTURES, COPPER, W72-04708
ODA, *F/ *CHLOROPHYTA, *SPRINGS, DIATOMS, BIOTA, PROTOZOA, *AMPHIP W72-12744
A, CYANOPHYTA, NITROGEN, CARBON, DIATOMS, BACTERIA, CARBON DIOXIDE W72-12544
PHYTA, INHIBITION, MARINE ALGAE, DIATOMS, BEHAVIOR.: /CIDS, CHRYSO W73-00942
*LIMITING FACTORS, GROWTH RATES, DIATOMS, BACTERIA, LAKES, NITROGE W72-09163
NCE ALGAE, PLANKTON, WASHINGTON, DIATOMS, CHLOROPHYTA.: /A, *NUISA W72-00798
HERRING, COPEPODS, CRUSTACEANS, DIATOMS, CYANOPHYTA, PHOTOSYNTHES W71-11517
ON, BENTHIC FLORA, LAKES, DEPTH, DIATOMS, CHLOROPHYTA, CYANOPHYTA, W71-11029
STUARINE ENVIRONMENT, SEA WATER, DIATOMS, COPEPODS, DINOFLAGELLATE W73-00935
IVATED CARBON, SLUDGE TREATMENT, DIATOMS, COAGULATION, CALCIUM CAR W73-01669
IMENTS, SUCCESSION, CHLOROPHYTA, DIATOMS, CHLOROPHYLL, ALKALINITY, W72-08049
HOSPHATES, NITRATES, ALKALINITY, DIATOMS, CYANOPHYTA, CHLOROPHYTA, W72-08560
*LIFE HISTORY STUDIES, COPEPODS, DIATOMS, CHLOROPHYTA, TEMPERATURE W72-14805
, PHYTOPLANKTON, STRATIFICATION, DIATOMS, CURRENTS(WATER), FISH FA W73-00236
INS, INORGANIC COMPOUNDS, ALGAE, DIATOMS, CHLAMYDOMONAS, SCENEDESM W72-14706
HYTOPLANKTON, CHEMICAL ANALYSIS, DIATOMS, CYANOPHYTA, TURNOVERS, E W73-00232
TER, SEDIMENTS, LIGHT INTENSITY, DIATOMS, CARBON RADIOISOTOPES, NE W72-14317
NKTON, ZOOPLANKTON, CHLOROPHYTA, DIATOMS, CYANOPHYTA, BENTHIC FAUN W72-12729
INITY, NUTRIENTS, WATER QUALITY, DIATOMS, CHLOROPHYTA, THERMAL WAT W72-05610

GES, EARLY IMPOUNDMENT, BIOMASS,	DIATOMS, CHLOROPHYTA, CYANOPHYTA,	W72-04855
EFFECTS, *ALGAE, PHYTOPLANKTON,	DIATOMS, CHRYSOPHYTA, CHLOROPHYTA	W70-07842
PIKES, SAMPLING, SNAILS, ALGAE,	DIATOMS, CLAMS, SHELLFISH, AMPHIP	W70-08098
PHYTA, CHLOROPHYTA, CHRYSOPHYTA,	DIATOMS, DINOFLAGELLATES, SCENEDE	W72-05968
ARIES, *STANDING CROP, SEASONAL,	DIATOMS, DINOFLAGELLATES, CHLOROP	W71-07875
PHOSPHORUS, IRON, PHYTOPLANKTON,	DIATOMS, DINOFLAGELLATES.: /GEN,	W73-00234
RON, PHYSICOCHEMICAL PROPERTIES,	DIATOMS, DOMINANT ORGANISMS.: / I	W73-00256
TIVATED SLUDGE, ALGAE, PROTOZOA,	DIATOMS, EUGLENA, MATHEMATICAL ST	W72-08573
TY, ALGAE, BIOMASS, OLIGOTROPHY,	DIATOMS, EUGLENA, DAPHNIA, HABITA	W71-09171
MPOUNDS, THERMAL STRATIFICATION,	DIATOMS, EPILIMNION, EUTROPHICATI	W72-13659
*ALGAE, FRESHWATER, MEASUREMENT,	DIATOMS, EUTROPHICATION.: /SSAY,	W72-10619
AUTUMN, SAMPLING, PLANKTON NETS,	DIATOMS, FOOD CHAINS, ANABAENA, S	W72-12738
L ANALYSIS, *NITRATES, *AMMONIA,	DIATOMS, GROWTH RATES, NITROGEN,	W73-00380
HATES, PHOSPHORUS RADIOISOTOPES,	DIATOMS, HYDROGEN ION CONCENTRATI	W71-10083
RITUS, BENTHOS, AQUATIC ANIMALS,	DIATOMS, INSECTS, BACTERIA, FUNGI	W71-10064
LIFE CYCLES, SEASONAL, DENSITY,	DIATOMS, LAKES, PROTOZOA, ROTIFER	W71-11012
LOGICAL COMMUNITIES, FOOD CHAIN,	DIATOMS, LARVAE, DIETS.: /ES, BIO	W72-09248
OW WATER, PERIPHYTON, SEDIMENTS,	DIATOMS, LIGHT INTENSITY, PHOTOSY	W72-07501
ON EFFECTS, TEMPERATURE, CARBON,	DIATOMS, LIGHT, OXYGEN, MINNESOTA	W71-08670
CTIVITY, *STANDING CROP, *ALGAE,	DIATOMS, MEASUREMENT, DETRITUS, M	W71-12855
, THERMAL POLLUTION, MICROSCOPY,	DIATOMS, NUTRIENTS, PHYTOPLANKTON	W72-10875
NITY, HARDNESS(WATER), SAMPLING,	DIATOMS, NITROGEN, BACTERIA, ALGA	W71-03028
ETS, CENTRIFUGATION, MICROSCOPY,	DIATOMS, NUTRIENTS, BACTERIA, CHR	W72-07899
VELS, PROTOZOA, ALGAE, BACTERIA,	DIATOMS, OCEANS, ACTIVATED SLUDGE	W72-10883
CA, CHROMIUM, COPPER, *NEW YORK,	DIATOMS, PHOSP: /INDICATORS, SILI	W72-09111
S, DIQUAT, ALGAE, ALGAL CONTROL,	DIATOMS, PERSISTENCE, BACTERIA, B	W72-00155
ROPHYTA, PHAEOPHYTA, RHODOPHYTA,	DIATOMS, PYRROPHYTA, CHLORELLA, P	W72-03220
WATER POLLUTION SOURCES, *FISH,	DIATOMS, PROTOZOA, AQUATIC INSECT	W72-07791
DATION, SEWAGE TREATMENT PLANTS,	DIATOMS, PHYTOPLANKTON, HYDORGEN	W72-06837
L, SEA WATER, SAMPLING, BIOMASS,	DIATOMS, POLLUTANT IDENTIFICATION	W73-00442
*LIMITING FACTORS, *PHOSPHATES,	DIATOMS, POLLUTANT IDENTIFICATION	W73-01445
E ERIE, PLANKTON, BENTHIC FLORA,	DIATOMS, PHYTOPLANKTON, PHOTOSYNT	W73-01615
OTOPES, CHLORELLA, GROWTH RATES,	DIATOMS, PHYTOPLANKTON, TOXICITY,	W71-03027
POLLUTION EFFECTS, *PESTICIDES,	DIATOMS, PHYTOPLANKTON, TOXICITY,	W72-05614
ANIMALS, MOSSES, INVERTEBRATES,	DIATOMS, PHYTOPLANKTON, PHOTOSYNT	W72-05952
GAE, PHYSICOCHEMICAL PROPERTIES,	DIATOMS, SUCCESSION, NUTRIENTS, R	W73-01435
STREAMS, LAKE ONTARIO, PLANKTON,	DIATOMS, SEDIMENTS, HARDNESS(WATE	W73-01627
IENTS, CHLORELLA, CHLAMYDOMONAS,	DIATOMS, SILICATES, CYANOPHYTA, T	W72-07508
WATER ANALYSIS, LAKES, SAMPLING,	DIATOMS, TURBIDITY, WATER POLLUTI	W72-07890
CHEMICAL ANALYSIS, CHLOROPHYTA,	DIATOMS, TEMPERATURE, OPTICAL PRO	W70-10403
FATES, CARBONATES, BICARBONATES,	DIATONICS, CYANOPHYTA, ALGAE, SOU	W71-02880
5,6-TRICHLOROPICOLINIC ACID, 2,4-	DICHLOROPHENOXYACETIC ACID.: /-3,	W72-08440
LLARIA/ *INTERFERENCE, ALEWIVES,	DICHOTOMOSIPHON, FRAGILARIA, TABE	W73-01669
NOBRYON, OSCILLATORIA LIMNETICA,	DICTYOSPHAERIUM, CELL COUNTS, FRA	W71-03014
ELLATA, ERNODESMIS VERTICILLATA,	DICTYOSPHAERIA VANBOSSEAE, DICTYO	W72-11800
LATA, DICTYOSPHAERIA VANBOSSEAE,	DICTYOSPHAERIA CAVERNOSA, CLADOPH	W72-11800
TRODESMUS, CHRYSOCOCCUS MINUTUS,	DICTYOSPHAERIUM, LAGERHEIMIA.: /S	W72-12738
S PENNATA, SPYRIDIA FILAMENTOSA,	DICTYOTA BARTAYRESII, LAURENCIA O	W73-00838
HOLOGY, SEDIMENTATION, PLANKTON,	DIEL MIGRATION, PROTOZOA, HYDROGE	W71-05626
EAMS, PULP WASTES, PRODUCTIVITY,	DIELDRIN, PATH OF POLLUTANTS, REV	W72-06340
FLOODWATER, *PATH OF POLLUTANTS,	DIELDRIN, DDT, ORGANIC MATTER, AN	W71-11498
SOLVED OXYGEN, CULTRUES, ENDRIN,	DIELDRIN, OXYGENATION, CARBAMATE	W71-10096
*MIXED CULTURES,	DIEOFF RATES, AFTERGROWTH.:	W71-05155
CANAL ZONE, MACROINVERTEBRATES,	DIESEL FUEL, BUNKER C OIL, AVICEN	W72-07911
UES, FOOD CHAINS, CULTURES, FISH	DIET, GAS CHROMATOGRAPHY, PARTICU	W73-02105
, COCONUT FATTY ACID, PLURONICS,	DIETHANOLAMIDE.: /NZENE SULFONATE	W72-12576
ES, FOOD CHAIN, DIATOMS, LARVAE,	DIETS.: /ES, BIOLOGICAL COMMUNITI	W72-09248
PHNIA, AQUATIC POPULATIONS, FISH	DIETS, FRESHWATER FISH, PONDS, LA	W72-07132
*SKELLYSOLVE, *FILTER PAPER DISC	DIFFUSION METHOD, *ZONES OF INHIB	W72-08586
*VERTICAL MIXING, EUPHOTIC ZONE,	DIFFUSION COEFFICIENT, LAKE SAMMA	W71-03034
WATER POLLUTION CONTROL, ENERGY,	DIFFUSION, THERMODYNAMICS, SOLUBI	W71-12092
IVATED SLUDGE, DESIGN, NITROGEN,	DIFFUSION, *ALGAL CONTROL, AERATI	W72-00800
LEAR WASTES, PATH OF POLLUTANTS,	DIFFUSION, CYTOLOGICAL STUDIES, A	W73-00823
OPHYTA, OILY WATER, EVAPORATION,	DIFFUSION, RHODOPHYTA.: /TS, PHAE	W73-01074
H BEHAVIOR, FERTILITY, SEASONAL,	DIFFUSIVITY.: /TY, VISCOSITY, FIS	W70-09612
ENT, LAMBERT-BEER EQUATION, EDDY	DIFFUSIVITY, DESHLER(OHIO).: /ICI	W72-05459
OXYGEN, DISSOLVED OXYGEN, LIGHT,	DIFFUSIVITY, CARBON DIOXIDE, LIGH	W72-05459
, ARTHROPODS, SARGASSUM CRISPUM,	DIGENIA SIMPLEX, HALOPHILA STIPUL	W73-00855
ALKALINITY, ACIDITY, LABORATORY,	DIGESTION TANK, PONDAGE, PILOT PL	W72-02850
UATIC ALGAE, NITROGEN, ANAEROBIC	DIGESTION, AEROBIC, ALKALINITY, A	W72-02850
MENTATION, FILTRATION, ANAEROBIC	DIGESTION, CHLORINATION, DISINFEC	W72-00042
OVEN DRYING, ASHING, EXTRACTIVE	DIGESTION, NEUROANATOMY, OXYGEN P	W71-11036
L STUDIES, MATHEMATICAL STUDIES,	DIGITAL COMPUTERS, DATA ANALYSIS,	W72-11920
HY, FUCUS VESICULOSUS, LAMINARIA	DIGITATA, ST. MARGARET'S BAY, SCO	W72-12166
IS, FUCUS VESICULOSUS, LAMINARIA	DIGITATA, LAMINARIA HYPERBOREA, A	W71-11515
AM/ FUCUS VESICULOSUS, LAMINARIA	DIGITATA, PORPHYRA UMBILICALIS, S	W73-01074
LAMINARIA LONGICRURIS, LAMINARIA	DIGITATA, AGARUM CRIBROSUM, *NOVA	W73-00366
UM, FUCUS VESICULOSUS, LAMINARIA	DIGITATA, LAMINARIA AGARDHII, ULV	W72-04748
STE ASSIMILATIVE CAPACITY, WASTE	DIL: / ON-SITE INVESTIGATIONS, WA	W73-00790
SCUS, MACROPHYTES, AMMOBACULITES	DILATATUS, AMMONIA BECCARII, AMMO	W73-00367
AE, ABSORPTION, SEDIMENTS, WASTE	DILUTION, WASTE ASSIMILATIVE CAPA	W73-00831
CTS, ABSORPTION, SORPTION, WASTE	DILUTION, WASTE ASSIMILATIVE CAPA	W73-00817
AL / *MOSES LAKE(WASH), NUTRIENT	DILUTION, LIMITING NUTRIENTS, ALG	W72-08049
ATICAL MODELS, *BIOASSAY, *WASTE	DILUTION, LAGOONS, *WASTE WATER T	W72-04787
ROWTH RATE, SATURATION CONSTANT,	DILUTION, CHLORELLA PYRENOIDOSA,	W72-04788
, *NUISANCE ALGAE, *EUTR/ *WASTE	DILUTION, WATER POLLUTION CONTROL	W72-00799
*DETENTION RESERVOIRS, / *WATER	DILUTION, *NITROGEN, *PHOSPHORUS,	W72-02412
ETALS, HERBICIDES, TOXINS, WASTE	DILUTION, NITROGEN, PHOSPHORUS, E	W72-01472
LVESTONENSIS, MERCURIC CHLORIDE,	DIMETHYL MERCURY, FLUOROMICROPHOT	W72-07660
ES, SODIUM CACODY/ ACREMONIELLA,	DIMETHYL SODIUM ARSONATE, ARSENAT	W72-13813

STREPTOTHRIX HYORHINA, CADMIUM, DIMETHYLNITROSAMINE, AMINES, CARC W72-09675
LCIUM RADIOISOTOPES, BLOOMS, 2,4 DINITRO PHENOL, TUCSON(ARIZ).: /A W71-01474
INA LONGIROSTRIS, SYNURA UVELLA, DINOBRYON, OSCILLATORIA LIMNETICA W71-03014
LENODINIUM, MICROSPORA, SYNEDRA, DINOBRYON, OEDOGONIUM, FRAGILARIA W72-10875
OHIO), CERATIUM, COELOSPHAERIUM, DINOBRYON, FRAGILARIA, PEDIASTRUM W72-10076
BELLARIA, ASTERIONELLA, SYNEDRA, DINOBRYON, MELOSIRA.: /ILARIA, TA W73-01669
T, SEA WATER, DIATOMS, COPEPODS, DINOFLAGELLATES, CRUSTACEANS, ANN W73-00935
HYTA, *PYRROPHYTA, *SYSTEMATICS, DINOFLAGELLATES, SPECIATION.: /OP W72-07683
ANKTON, CHLOROPHYTA, PYRROPHYTA, DINOFLAGELLATES, PRIMARY PRODUCTI W72-07166
ANKTON, ZOOPLANKTON, PYRROPHYTA, DINOFLAGELLATES, CYANOPHYTA, CHLO W72-07899
, BIOASSAY, CYTOLOGICAL STUDIES, DINOFLAGELLATES, CYANOPHYTA.: /GY W73-00243
R, SUMMER, PRIMARY PRODUCTIVITY, DINOFLAGELLATES, PYRROPHYTA.: /TE W73-00442
S, IRON, PHYTOPLANKTON, DIATOMS, DINOFLAGELLATES.: /GEN, PHOSPHORU W73-00234
USTRIAL WASTES, DOMESTIC WASTES, DINOFLAGELLATES, *LABORATORY TEST W71-03095
ER, BIOASSAY, COPEPODS, OYSTERS, DINOFLAGELLATES, DIATOMS, ALGAE C W72-04708
LOROPHYTA, CHRYSOPHYTA, DIATOMS, DINOFLAGELLATES, SCENEDESMUS, ROT W72-05968
TANDING CROP, SEASONAL, DIATOMS, DINOFLAGELLATES, CHLOROPHYLL, PHO W71-07875
*FISH TOXINS, *ALGAL POISONING, DINOFLAGELLATES, BEGGIATOA, PHYTO W72-14056
ANABAENA, SCENEDESMUS, SEASONAL, DINOFLAGELLATES.: / FOOD CHAINS, W72-12738
LIMNION, EPILIMNION, CYANOPHYTA, DINOFLOGELLATES, SEASONAL, ZOOPLA W71-11949
E POLYSACCHARIDES, CRYPTOPHYCEA, DINOPHYCEA, BACCILARIOPHYCEA, PRY W70-07280
IS SPP., SOIL CULTURES, TITANIUM DIOXIDE WASTES, HELGOLAND.: /CYST W72-12239
NTS, AQUATIC ENVIRONMENT, CARBON DIOXIDE.: / CONCENTRATION, NUTRIE W71-12068
DROGEN ION CONCENTRATION, CARBON DIOXIDE.: /LAKES, CHLOROPHYLL, HY W71-12855
GANIC MATTER, PHOSPHORUS, CARBON DIOXIDE.: /ICAL OXYGEN DEMAND, OR W71-13257
NOPHYTA, TOXICITY, COSTS, CARBON DIOXIDE.: /IATOMS, SILICATES, CYA W72-07508
*WASTE-WATER TREATMENT, *CARBON DIOXIDE, *GROWTH RATES, *AEROBIC W72-06612
OORGANISMS, *ABSORPTION, *CARBON DIOXIDE, *RADIOACTIVITY TECHNIQUE W72-14301
TA, CA/ *NUISANCE ALGAE, *CARBON DIOXIDE, *ALGAL CONTROL, CYANOPHY W72-10607
LASKA, FIORDS, SURFACE / *CARBON DIOXIDE, *PRIMARY PRODUCTIVITY, A W72-09122
RATION, DISSOLVED OXYGEN, CARBON DIOXIDE, AMMONIA, ORGANIC MATTER. W72-01363
ONCENTRATION, METABOLISM, CARBON DIOXIDE, ALGAE, CULTURES, ACIDITY W72-00838
*PHOTOSYNTHESIS, *ALGAE, CARBON DIOXIDE, ALKALINITY, NUTRIENTS, A W71-11519
*CHLOROPHYLL, DIVERSION, CARBON DIOXIDE, ALKALINITY, EPILIMNION, W71-02681
OASSAY, DISSOLVED OXYGEN, CARBON DIOXIDE, AMMONIA, PHOSPHATE, ANAL W72-06046
GEN FIXATION, DETERGENTS, CARBON DIOXIDE, ALKALINITY, BURNING, DOM W72-05473
DROGEN ION CONCENTRATION, CARBON DIOXIDE, AERATION, MIXING, *EUTRO W72-04433
ITROGEN FIXATION, OXYGEN, CARBON DIOXIDE, BIOCHEMISTRY, ALGAE, INH W72-08584
*GROWTH RATES, CHLORELLA, CARBON DIOXIDE, BICARBONATES, ALGAL CONT W72-09168
LABORATORY TESTS, ALGAE, CARBON DIOXIDE, CYCLING NUTRIENTS, BICAR W72-09159
ON, FILTRATION, COLLOIDS, CARBON DIOXIDE, CALCIUM, MAGNESIUM, NUTR W72-03613
CTER, CYANOPHYTA, YEASTS, CARBON DIOXIDE, CHLORELLA, CLOSTRIDIUM, W72-06124
ATION, CARBON, SEDIMENTS, CARBON DIOXIDE, CHLORPHYLL, DO: /RATIFIC W72-08560
ANESE, POTASSIUM, SODIUM, CARBON DIOXIDE, CYANOPHYTA, FISH,: /MANG W73-00840
WTH STAGES, REPRODUCTION, CARBON DIOXIDE, CULTURES, LIMITING FACTO W73-02099
ARBON, DIATOMS, BACTERIA, CARBON DIOXIDE, CHLAMYDOMONAS, SCENEDESM W72-12544
YTA, AQUATIC ENVIRONMENT, CARBON DIOXIDE, ECOSYSTEMS, SOUTH CAROLI W71-00475
LIGHT INTENSITY, AMMONIA, CARBON DIOXIDE, FLORIDA, CULTURES, *ALGA W72-09103
MPERATURE, WATER QUALITY, CARBON DIOXIDE, HYDROGEN ION CONCENTRATI W72-06274
CARBONATES, BICARBONATES, CARBON DIOXIDE, HYDROGEN ION CONCENTRATI W72-04431
YGEN, LIGHT, DIFFUSIVITY, CARBON DIOXIDE, LIGHT INTENSITY, PHOTOSY W72-05459
, CHEMICAL OXYGEN DEMAND, CARBON DIOXIDE, LININGS, DISSOLVED OXYGE W72-04789
H RATES, LIGHT INTENSITY, CARBON DIOXIDE, NITROGEN, LABORATORY TES W72-05476
DROGEN ION CONCENTRATION, CARBON DIOXIDE, NITRATES, NITRITES, *ENZ W72-04309
DROGEN ION CONCENTRATION, CARBON DIOXIDE, NITROGEN, CHLOROPHYTA, N W72-04769
CTERIA, ALGAE, SYMBIOSIS, CARBON DIOXIDE, NITROGEN, LAKES, ESTUARI W70-07283
S, *DIURNAL DISTRIBUTION, CARBON DIOXIDE, NITROGEN, PHOSPHORUS, LI W72-11724
TRIENTS, *AQUATIC ALGAE, *CARBON DIOXIDE, NUISANCE ALGAE, EUTROPHI W72-07514
ROGEN FIXATION, CULTURES, CARBON DIOXIDE, NUTRIENTS, LIMITING FACT W72-14790
HNIQUES, *PHOTOSYNTHESIS, CARBON DIOXIDE, PHYTOPLANKTON, ALGAE, CU W72-09102
, AMMONIA, CARBOHYDRATES, CARBON DIOXIDE, POLYMERS, HYDROGEN ION C W72-10233
PIRATION, PHOTOSYNTHESIS, CARBON DIOXIDE, PRIMARY PRODUCTIVITY, BI W71-04518
OLVED OXYGEN, METABOLISM, CARBON DIOXIDE, PHOSPHATES, NITRATES, *N W72-04983
LL, CARBON RADIOISOTOPES, CARBON DIOXIDE, RHODOPHYTA.: / CHLOROPHY W71-09172
S, ISOTOPE FRACTIONATION, CARBON DIOXIDE, RESPIRATION, MARINE PLAN W72-12709
, TRACERS, RADIOACTIVITY, CARBON DIOXIDE, SALTS, MARINE ALGAE.: /M W72-12634
PHATES, NITROGEN, SEWAGE, CARBON DIOXIDE, SEASONAL, PHOSPHORUS, WA W72-09167
HOSPHORUS, PHYTOPLANKTON, CARBON DIOXIDE, SESTON, FERTILIZATION, W W72-07648
CHLOROPHYLL, ALKALINITY, CARBON DIOXIDE, TEMPERATURE, DISSOLVED O W72-08049
CITY, CHEMICAL REACTIONS, CARBON DIOXIDE, WATER POLLUTION SOURCES, W73-01074
FISH, INTAKE STRUCTURES, CARBON DIOXIDE,: /, VOLUMETRIC ANALYSIS, W72-05594
A SIMPLEX, HALOPHILA STIPULACEA, DIPLANTHERA UNINERVIS, CASSIOPEA W73-00855
BIOMASS, OREGON, SLUDGE, ALGAE, DIPTERA, DISSOLVED OXYGEN, DREDGI W72-03556
S, CHLORINE, MOLLUSKS, COPEODS, DIPTERA, GULFS, GEOLOGIA HISTORY, W73-00855
PROTOZOA, ROTIFERS, CRUSTACEANS, DIPTERA, MAYFLIES, DRAGONFLIES, C W71-11012
OTIFERS, AQUATIC MICROORGANISMS, DIPTERA, PHOSPHO: /ION, LARVAE, R W72-07132
ARAQUAT, *ENVIRONMENTAL EFFECTS, DIQUAT, ALGAE, ALGAL CONTROL, DIA W72-00155
TZ CREEK(PA), SPECIES DIVERSITY, DIRECT EFFECTS, INDIRECT EFFECTS. W72-08579
ER CHEMISTRY, WATER TEMPERATURE, DIS: /OPHICATION, EVALUATION, WAT W72-06526
LIGOSACCHARIDES, CL/ *EXCRETION, DISACCHARIDES, MONOSACCHARIDES, O W72-07710
MONOSACCHARIDES, CULTURE MEDIA, DISACCHARIDES, THREONINE, PROLINE W73-00942
ITY, *SKELLYSOLVE, *FILTER PAPER DISC DIFFUSION METHOD, *ZONES OF W72-08586
ED PHOSPHORUS, OSCILLATO/ SECCHI DISC, DISSOLVED NITROGEN, DISSOLV W72-07648
LLA SPP, ORTHOPHOSPHATES, SECCHI DISC, NEWFOUND LAKE, SQUAM LAKE, W72-07890
*CONTINUOUS DISCHARGE LAGOON.: W72-12565
ALITY, NUTRIENTS, ALGAE, INFLOW, DISCHARGE(WATER), WATER BALANCE, W72-13851
, CHRYSOPHYTA, SEASONAL, SILICA, DISCHARGE(WATER).: /, SCENEDESMUS W72-14797
, ALGAE, POWERPLANTS, EFFLUENTS, DISCHARGE(WATER), STATISTICAL MOD W71-11583
, EPILIMNION, INTAKE STRUCTURES, DISCHARGE(WATER), WATER TEMPERATU W70-09614
UMENTATION, ABSORPTION, SEDIMENT DISCHARGE, REGRESSION ANALYSIS, B W72-06283

TER QUALITY, RIVERS, IOWA, CORN,	DISCHARGE, CHLOROPHYLL.: /ION, WA	W71-06445
*THERMAL	DISCHARGES.:	W71-12092
,/ *MERCURY COMPOUNDS, *MINAMATA	DISEASE, BIOLOGICAL MAGNIFICATION	W72-12257
NESOTA, LIPIDS, HUMAN PATHOLOGY,	DISEASES.: /MICAL PROPERTIES, MIN	W72-04275
E, CATTLE, POULTRY, HOGS, ANIMAL	DISEASES, CATFISH, WASTE TREATMEN	W71-07544
S(PESTICIDES), ENTERIC BACTERIA,	DISEASES, E. COLI, SHIGELLA, VIRU	W72-13800
HETICS, COLOR, TASTE, CORROSION,	DISEASES, EUTROPHICATION, FOULING	W71-09958
IC HEALTH STANDARDS/ *WATERBORNE	DISEASES, *TOXIC CHEMICALS, *PUBL	W73-02144
, CASTLE LAKE(/ *CONTEMPORANEOUS	DISEQUILIBRIUM, LAKE TAHOE(CALIF)	W71-09171
ENZYME PRESOAKS,	DISHWATER DETERGENTS, NTA.:	W71-08768
MICAL OXYGEN DEMAND, FILTRATION,	DISINFECTION, FLOCCULATION, *WAST	W72-00411
AEROBIC DIGESTION, CHLORINATION,	DISINFECTION, ULTRAVIOLET LIGHT,	W72-00042
MANGANESE, ZINC, METALS, SECCHI	DISKS, HEAVY METALS, BIOINDICATOR	W72-09111
IDE, CHEMICAL PROPERTIES, SECCHI	DISKS, METHODOLOGY.: /ODIUM CHLOR	W72-09167
UCTIVITY, CHLOROPHYLL, PIGMENTS,	DISPERSION, RUNOFF.: /ALGAE, COND	W72-04773
UENT, IRRIGATION, STORAGE, WASTE	DISPOSAL.: /OGENIC BACTERIA, EFFL	W71-08214
GOONS, AER/ *FARM WASTES, *WASTE	DISPOSAL, *LAGOONS, *OXIDATION LA	W71-07551
TED LA/ *DISPOSAL SYSTEMS, *LAND	DISPOSAL, OXIDATION DITCHES, AERA	W71-07551
ATES, SURVEYS, PESTICIDES, WASTE	DISPOSAL, PATH OF POLLUTANTS, WAT	W71-05553
TY, SEWAGE DISPOSAL, WASTE WATER	DISPOSAL, SEWAGE EFFLUENTS, FARM	W70-06509
ER, ALGAE, WATER QUALITY, SEWAGE	DISPOSAL, WASTE WATER DISPOSAL, S	W70-06509
REATMENT, INSTRUMENTATION, WASTE	DISPOSAL, AERATION, EROSION CONTR	W71-01192
R), ALGAE, LAKES, RIVERS, SEWAGE	DISPOSAL, SEWAGE TREATMENT, WASTE	W70-09947
*MARINE	DISPOSAL, *GOTHENBRUG, SWEDEN.:	W70-09947
STES/ *ALGAE, *ESTUARIES, *WASTE	DISPOSAL, *SEWAGE, *INDUSTRIAL WA	W72-08804
WASTES, FISH REPRODUCTION, WASTE	DISPOSAL, MARINE MICROORGANISMS,	W71-13723
TORM RUNOFF, SEWAGE, WASTE WATER	DISPOSAL, PATH OF POLLUTANTS, STO	W71-13466
DATION, LOCAL GOVERNMENTS, WASTE	DISPOSAL, WATER POLLUTION SOURCES	W72-01093
TION, POLLUTION ABATEMENT, WASTE	DISPOSAL, STANDARDS, PUBLIC HEALT	W72-01085
TY, PESTICIDES, LIVESTOCK, WASTE	DISPOSAL, OXIDATION LAGOONS, DEHY	W72-00846
CIPAL WASTES, FARM WASTES, WASTE	DISPOSAL, PESTICIDE REMOVAL, OXYG	W72-01659
S, POPULATION, INDUSTRIES, WASTE	DISPOSAL, MONITORING, NUTRIENTS,	W71-12091
TES, METALS, HEAVY METALS, WASTE	DISPOSAL, WATER POLLUTION EFFECTS	W71-11793
E WATER(POLLUTION), *WASTE WATER	DISPOSAL, *WATER POLLUTION CONTRO	W72-14878
LLUTION EFFECTS, MERCURY, SEWAGE	DISPOSAL, PHOSPHATES, PHOSPHORUS,	W72-14782
CROORGANI/ *WATER SUPPLY, *WASTE	DISPOSAL, *RECREATION WASTES, *MI	W72-14363
LUENTS, SEWAGE TREATMENT, SEWAGE	DISPOSAL, MARINE ALGAE, CLAMS, MU	W72-07715
HERIES, MARINE FISHERIES, SEWAGE	DISPOSAL, SEWAGE LAGOONS, BRACKIS	W72-12567
ENTRATION, PHOSPHATES, NITRATES,	DISSO: /LOGENS, HYDROGEN ION CONC	W73-01434
YTES, SARGASSO SEA, SAMPLE PREP/	DISSOLVED ORGANIC CARBON, MACROPH	W73-02100
FLEXILIS, SCIRPUS ACUTUS, C-14,	DISSOLVED ORGANIC MATTER, MACROPH	W73-02095
L STRATIFICATION, *TASTE, *ODOR,	DISSOLVED SOLIDS, ALGAE, RESERVOI	W73-02138
, BENTHIC FLORA, ORGANIC MATTER,	DISSOLVED SOLIDS, BENTHOS, CARBON	W73-02100
POLLUTION EFFECTS, STRIPED BASS,	DISSOLVED OXYGEN, OIL, BRIDGES, B	W73-02029
SPHORUS, OSCILLATO/ SECCHI DISC,	DISSOLVED NITROGEN, DISSOLVED PHO	W72-07648
SECCHI DISC, DISSOLVED NITROGEN,	DISSOLVED PHOSPHORUS, OSCILLATORI	W72-07648
HYTA, CHRYSOPHYTA, EUGLENOPHYTA,	DISSOLVED OXYGEN, CARBON, LAKE MO	W72-08048
N, MINNESOTA, WATER TEMPERATURE,	DISSOLVED OXYGEN.: / CONCENTRATIO	W72-07145
TES, *RUNOFF, *FISHKILL, CATTLE,	DISSOLVED OXYGEN, DIVERSION STRUC	W72-08401
TY, CARBON DIOXIDE, TEMPERATURE,	DISSOLVED OXYGEN, HYDROGEN ION CO	W72-08049
MMONIA, NITRATES, MODEL STUDIES,	DISSOLVED OXYGEN, CARBON, DISTRIB	W72-08459
MAL STRATIFICATION, TEMPERATURE,	DISSOLVED OXYGEN, COMPUTERS, ALGA	W72-14405
ION CONCENTRATION, HEAVY METALS,	DISSOLVED OXYGEN, *PLANT GROWTH,	W72-14690
, WATER TEMPERATURE, ALKALINITY,	DISSOLVED OXYGEN, CONDUCTIVITY, H	W72-14728
SEDIMENTS, CHEMICAL PROPERTIES,	DISSOLVED SOLIDS, HYDROLOGIC DATA	W72-13851
GAE, AQUATIC ALGAE, OLIGOTROPHY,	DISSOLVED OXYGEN, WATER TEMPERATU	W72-13871
RESERVOIR STORAGE, IMPOUNDMENTS,	DISSOLVED OXYGEN, CHLORIDES, INVE	W72-14280
HYTOPLANKTON, LIGHT PENETRATION,	DISSOLVED OXYGEN, BACTERIA, THERM	W72-12995
ITY, CHEMICAL ANALYSIS, BIOMASS,	DISSOLVED OXYGEN, THERMOCLINE, HY	W72-12941
ION, HYDROGEN ION CONCENTRATION,	DISSOLVED OXYGEN, BIOCHEMICAL OXY	W72-13508
YS, ASSAY PROCEDURE.:	DISSOLVED PHOSPHORUS, *ALGAL ASSA	W72-12966
LOROPHYLL, *ANABAENA, NUTRIENTS,	DISSOLVED SOLIDS, PHYTOPLANKON, W	W72-13300
SORPTION, PESTICIDES, TOXICITY,	DISSOLVED OXYGEN, LAKE MICHIGAN,	W71-12092
LOROPHYTA, SCENEDESMUS, EUGLENA,	DISSOLVED OXYGEN, DIURNAL DISTRIB	W71-11519
ONCENTRATION, WATER TEMPERATURE,	DISSOLVED OXYGEN, COLIFORMS, RECR	W71-11914
RIENTS, WATER POLLUTION SOURCES,	DISSOLVED OXYGEN, LAKE MORPHOLOGY	W71-11949
R POLLUTION EFFECTS, INTERFACES,	DISSOLVED OXYGEN, TEMPERATURE, OR	W71-10645
MAND, BIOCHEMICAL OXYGEN DEMAND,	DISSOLVED OXYGEN: /ICAL OXYGEN DE	W71-11006
ICIDES, PHYTOPLANKTON, TOXICITY,	DISSOLVED OXYGEN, CULTRUES, ENDRI	W71-10096
TER SUPPLY, WASTE WATER, RUNOFF,	DISSOLVED OXYGEN, ALGAE, NUTRIENT	W71-09788
CITY, *ORGANIC LOADS, NUTRIENTS,	DISSOLVED OXYGEN, AQUATIC LIFE, M	W71-11017
ATIONSHIPS, PERMEABILITY, ALGAE,	DISSOLVED OXYGEN.: /OUNDWATER REL	W71-12410
N DEMAND, CARBON, FERTILIZATION,	DISSOLVED OXYGEN, SEWAGE, HYDROGE	W71-12188
IES, HYDROGEN ION CONCENTRATION,	DISSOLVED OXYGEN, CARBON DIOXIDE,	W72-01363
WATER TEMPERATURE, SALT BALANCE,	DISSOLVED OXYGEN, ALGAE, BACTERIA	W72-02274
, ALGAE, BACTERIA, ODOR, INFLOW,	DISSOLVED SOLIDS, DATA COLLECTION	W72-02274
HNIQUES, *ULTRAVIOLET RADIATION,	DISSOLVED OXYGEN, LAKES, PHYTOPLA	W72-01780
OXYGEN DEMAND, OXYGEN, AERATION,	DISSOLVED OXYGEN, HYDROGEN ION CO	W71-13334
ION CONCENTRATION, TEMPERATURE,	DISSOLVED OXYGEN, EUTROPHICATION,	W72-03224
OREGON, SLUDGE, ALGAE, DIPTERA,	DISSOLVED OXYGEN, DREDGING, BOTTO	W72-03556
GEN ION CONCENTRATION, PROTOZOA,	DISSOLVED OXYGEN, SLUDGE, WASTE W	W72-10233
S, PHOSPHATES, SLUDGE TREATMENT,	DISSOLVED OXYGEN, BACTERIA, FUNGI	W72-09675
UIFER CHARACTERISTICS, GENETICS,	DISSOLVED OXYGEN.: /IC FUNGI, ACQ	W72-11186
LABORATORY EQUIPMENT, AERATION,	DISSOLVED OXYGEN, AEROBIC CONDITI	W72-11100
ER REUSE, ALGAE, PHOTOSYNTHESIS,	DISSOLVED OXYGEN, AEROBIC CONDITI	W72-11642
RATES, SULFATES, DRAINAGE WATER,	DISSOLVED SOLIDS, NITROGEN CYCLE,	W71-01489
UTION SOURCES, WATER LAW, ALGAE,	DISSOLVED OXYGEN.: /L, WATER POLL	W71-01192
VESTING OF ALGAE, *ION EXCHANGE,	DISSOLVED SOLIDS, FLOCCULATION, H	W71-00138
ERATURE, ELECTRICAL CONDUCTANCE,	DISSOLVED OXYGEN, AIR POLLUTION,	W71-01489
ATER ANALYSIS, ALGAE, COLIFORMS,	DISSOLVED OXYGEN.: /, SAMPLING, W	W70-08248

ATER ANALYSIS, ALGAE, COLIFORMS,	DISSOLVED OXYGEN.: /, SAMPLING, W	W70-08249
, ALGAE, THERMAL STRATIFICATION,	DISSOLVED OXYGEN, HYPOLIMNION.: /	W71-03028
IRS, RESERVOIRS, PHOTOSYNTHESIS,	DISSOLVED OXYGEN, PHOSPHATES, WAT	W71-01491
E, *POPULATION, DYNAMICS, LAKES,	DISSOLVED OXYGEN, NUTRIENTS, CYAN	W71-03034
GAE, *STREAMS, LABORATORY TESTS,	DISSOLVED OXYGEN, HYDROGEN ION CO	W71-04518
KTON, BIOCHEMICAL OXYGEN DEMAND,	DISSOLVED OXYGEN, ESTUARIES, WATE	W71-05390
EUTROPHICATION, PHOTOSYNTHESIS,	DISSOLVED OXYGEN, SEDIMENTATION,	W71-07097
FISH EGGS, TURBIDITY, NUTRIENTS,	DISSOLVED OXYGEN, NITROGEN, PHOSP	W71-06444
NALYSIS, TEMPERATURE, NUTRIENTS,	DISSOLVED OXYGEN, ALGAE, SLUDGE,	W71-07109
URE, HYDROGEN ION CONCENTRATION,	DISSOLVED OXYGEN, STRATIFICATION,	W71-07382
TEMPERATURE, SAMPLING, SALINITY,	DISSOLVED OXYGEN, NUTRIENTS, CARB	W71-07875
G, ODOR, COLIFORMS, TEMPERATURE,	DISSOLVED OXYGEN, AERATION, OXYGE	W71-07113
OXIDATION LAGOONS, *OXYGENATION,	DISSOLVED OXYGEN, SATURATION, ALG	W71-07106
ON, *REVIEWS, WATER TEMPERATURE,	DISSOLVED OXYGEN, BIOINDICATORS,	W71-07417
NSITY, EVAPORATION, RE-AERATION,	DISSOLVED OXYGEN, ALGAE, ANAEROBI	W71-07099
SSIMILATIVE CAPACITY, OXIDATION,	DISSOLVED OXYGEN, HYDROGEN ION CO	W72-04787
DOR, WATER QUALITY, FOOD CHAINS,	DISSOLVED OXYGEN, COPPER SULFATE,	W72-04759
DITY, BIOCHEMICAL OXYGEN DEMAND,	DISSOLVED OXYGEN, ALGAE, THERMAL	W72-03753
PHOSPHORUS, NITRATES, POTASSIUM,	DISSOLVED OXYGEN, FORECASTING, FL	W72-04260
HT PENETRATION, LIGHT INTENSITY,	DISSOLVED SOLIDS, PHYTOPLANKTON,	W72-04270
, ALGAE, GROWTH RATES, BIOASSAY,	DISSOLVED OXYGEN, CARBON DIOXIDE,	W72-06046
RATURE, LIGHT INTENSITY, MIXING,	DISSOLVED OXYGEN, METABOLISM, CAR	W72-04983
, *SEWAGE LAGOONS, OHIO, OXYGEN,	DISSOLVED OXYGEN, LIGHT, DIFFUSIV	W72-05459
DEMAND, CARBON DIOXIDE, LININGS,	DISSOLVED OXYGEN, LIGHT INTENSITY	W72-04789
ERATURE, ALGAE, HYDROLOGIC DATA,	DISSOLVED OXYGEN, WEATHER, RESERV	W72-05282
, SWEDEN, GERMANY, SPARTINA SPP,	DISTICHLIS SPICATA, FIMBRISTYLIS,	W73-01089
OTOPES, CLADOPHORA SPP, PASPALUM	DISTICHUM, POTAMOGETON PECTINATUS	W73-01673
AL POLLUTION, LONG TUBE VERTICAL	DISTILLATION, EVAPORATORS, AQUAT:	W70-09193
WATER TEMPERATURE, POWERPLANTS,	DISTILLATION, ALGAE, AQUICULTURE,	W70-08832
EVERSE OSMOSIS, ELECTRODIALYSIS,	DISTILLATION, NITRIFICATION, DENI	W70-09907
SHINGTON, SEPARATION TECHNIQUES,	DISTILLATION, SPECTROPHOTOMETRY,	W72-12637
BICARBONATES, MASS SPECTROMETRY,	DISTRIBUTION, SEPARATION TECHNIQU	W72-12709
IATION, ORGANIC LOADING, DIURNAL	DISTRIBUTION, HYDROGEN ION CONCEN	W72-13508
RA, LIMITING FACTORS, ECOLOGICAL	DISTRIBUTION.: /ABITATS, CLADOPHO	W72-13142
AQUATIC ENVIRONMENT, *ECOLOGICAL	DISTRIBUTION, *ESTUARINE ENVIRONM	W72-11854
EASON, OXYGEN, HYDROGEN SULFIDE,	DISTRIBUTION, NANNOPLANKTON, TURB	W71-00117
OGEN, LITTORAL, HARBORS, SPATIAL	DISTRIBUTION, ALGAE, TRIBUTARIES,	W72-04263
IFUGATION, PESTICIDES, BIOASSAY,	DISTRIBUTION PATTERNS, CHROMATOGR	W72-04134
*ALGAE, *FRESH WATER,	DISTRIBUTION PATTERNS, BEHAVIOR.:	W71-07360
*LAKES, EPILIMNION, ZOOPLANKTON,	DISTRIBUTION, PRIMARY PRODUCTIVIT	W71-09171
S, CYCLING NUTRIENTS, ECOLOGICAL	DISTRIBUTION, ECOSYSTEMS, AQUATIC	W73-02095
S, BENTHIC FLORA, BENTHIC FAUNA,	DISTRIBUTIO: /OPODS, SNAILS, CRAB	W73-00932
SAMPLING, MICROSCOPY, ECOLOGICAL	DISTRIBUTION, PLANKTON NETS, CHLO	W73-01094
YSTEMATICS,/ *KELPS, *ECOLOGICAL	DISTRIBUTION, *MARINE ANIMALS, *S	W73-00932
ISTRIBUTION PATTERNS, ECOLOGICAL	DISTRIBUTION, WATER POLLUTION EFF	W73-00918
TIES, PHYSIOCHEMICAL PROPERTIES,	DISTRIBUTION PATTERNS, ECOLOGICAL	W73-00918
AE, LIGHT INTENSITY, *ECOLOGICAL	DISTRIBUTION, SURFACE WATERS, LAG	W73-00838
AL STRESS, GROWTH RATES, SPATIAL	DISTRIBUTION, PRIMARY PRODUCTIVIT	W73-00853
HESIS, STANDING CROP, ECOLOGICAL	DISTRIBUTION: /TRIENTS, PHOTOSYNT	W72-14313
ARINE AN/ *CHLOROPHYLL, *SPATIAL	DISTRIBUTION, *DENSITY, *DEPTH, M	W72-14313
ITY, LIFE CYCLES, GROWTH STAGES,	DISTRIBUTION, BEHAVIOR, SEASONAL,	W72-14659
HIGAN, *EUTROPHICATION, *SPATIAL	DISTRIBUTION, PHOSPHORUS, NITRATE	W73-00384
SOTOPES, SESTON, ORGANIC MATTER,	DISTRIBUTION PATTERNS, THERMAL PR	W73-00431
WATER POLLUTION SOURCES, SPATIAL	DISTRIBUTION, BIOMASS, PHOTOSYNTH	W73-00431
TERNS, *SYSTEMATICS, *ECOLOGICAL	DISTRIBUTION, *SPATIAL DISTRIBUTI	W73-00428
COLOGICAL DISTRIBUTION, *SPATIAL	DISTRIBUTION, GEOGRAPHICAL REGION	W73-00428
SPATIAL DISTRIBUTION, ECOLOGICAL	DISTRIBUTION, BENTHIC FAUNA, SIEV	W73-00367
DIN/ *BIOINDICATORS, *ECOLOGICAL	DISTRIBUTION, *CURRENTS(WATER), *	W73-00296
TER), *DINOFLAGELLATES, ECOLOGY,	DISTRIBUTION PATTERNS, MARINE ANI	W73-00296
ATER, SALINITY, SAMPLING, CORES,	DISTRIBUTION PATTERNS, CURRENTS(W	W73-00367
, SEPARATION TECHNIQUES, SPATIAL	DISTRIBUTION, ECOLOGICAL DISTRIBU	W73-00367
QUIPMENT, INHIBITION, ECOLOGICAL	DISTRIBUTION, MARINE ALGAE, MARIN	W73-00348
*TEMPORAL DISTRIBUTION, *SPATIAL	DISTRIBUTION, ENVIRONMENT, SAMPLI	W72-08560
*ALGAE, *POPULATION, *TEMPORAL	DISTRIBUTION, *SPATIAL DISTRIBUTI	W72-08560
UDIES, DISSOLVED OXYGEN, CARBON,	DISTRIBUTION, PHOSPHORUS, EUTROPH	W72-08459
EUTROPHICATION, SOLAR RADIATION,	DISTRIBUTION, DEPTH, DIURNAL, SEA	W72-08054
N, EPILIMNION, LAKE MORPHOMETRY,	DISTRIBUTION PATTERNS, AQUATIC AL	W72-07145
ELECTRON MICROSCOPY, PYRROPHYTA,	DISTRIBUTION PATTERNS, WATER TEMP	W72-07663
ONS, CONTINENTAL SHELF, VERTICAL	DISTRIBUTION, NORTH CAROLINA, SOU	W72-07663
TES, *LIMITING FACTORS, *DIURNAL	DISTRIBUTION, CARBON DIOXIDE, NIT	W72-11724
L REGIONS, MARINE ALGAE, SPATIAL	DISTRIBUTION, FOURIER ANALYSIS, B	W72-11719
TIGATIONS, *POPULATION, TEMPORAL	DISTRIBUTION, CHLOROPHYLL, SAMPLI	W72-11719
OLOGICAL COMMUNITIES, ECOLOGICAL	DISTRIBUTION, BIOINDICATORS, WATE	W72-08579
S, BENTHOS, SCUBA DIVING, DEPTH,	DISTRIBUTION PATTERNS, AERIAL PHO	W72-09120
ON, *EUTROPHICATION, *ECOLOGICAL	DISTRIBUTION, *AQUATIC ALGAE, *CY	W72-09061
EVISION, DEPTH SOUNDER, VERTICAL	DISTRIBUTION, COLD BAY, BERMERS B	W72-09120
C FAUNA, SEDIMENTATION, SEDIMENT	DISTRIBUTION, PHYSICOCHEMICAL PRO	W72-00847
NKTON, NATURAL RESOURCES, DEPTH,	DISTRIBUTION PATTERNS, MEASUREMEN	W71-13252
TROGEN, PHOSPHORUS, CHLOROPHYLL,	DISTRIBUTION, CARBON, SEASONAL, C	W71-11029
GLENA, DISSOLVED OXYGEN, DIURNAL	DISTRIBUTION, AQUATIC PRODUCTIVIT	W71-11519
, / SOUTH TAHOE PUBLIC UTILITIES	DISTRICT, *INDIAN CREEK RESERVOIR	W72-02412
: *OXIDATION	DITCH, HYDRAULIC MANURE HANDLING.	W72-08396
STEMS, *LAND DISPOSAL, OXIDATION	DITCHES, AERATED LAGOONS, COMPOST	W71-07551
COPEPODIDS, *CALANUS GLACIALIS,	DITYLUM, PROROCENTRUM, NITZSCHIA,	W72-09248
SMUS, EUGLENA, DISSOLVED OXYGEN,	DIURNAL DISTRIBUTION, AQUATIC PRO	W71-11519
OLAR RADIATION, ORGANIC LOADING,	DIURNAL DISTRIBUTION, HYDROGEN IO	W72-13508
LGAE, *PHOTOSYNTHESIS, RAINFALL,	DIURNAL, FLUCTUATIONS, CHLOROPHYT	W72-00150
ICATION, CYANOPHYTA, ALKALINITY,	DIURNAL, NITROGEN.: /NTS, EUTROPH	W72-09162
RADIATION, DISTRIBUTION, DEPTH,	DIURNAL, SEASONAL.: /ATION, SOLAR	W72-08054
N, DIATHENE, PROPAZINE, COTORON,	DIURON, LINURON.: /DULANS, CERASO	W71-12071

HKILL, CATTLE, DISSOLVED OXYGEN, DIVERSION STRUCTURES, SEDIMENTATI W72-08401
ROPHYTA, CYANOPHYTA, WASHINGTON, DIVERSION, LAKE ERIE, NITROGEN FI W72-05473
PHOSPHORUS, NITROGEN, POTASSIUM, DIVERSION, ALGAE, NITROGEN FIXATI W71-00117
PHATES, *NITRATES, *CHLOROPHYLL, DIVERSION, CARBON DIOXIDE, ALKALI W71-02681
, WATER QUALITY, COMPETING USES, DIVERSION, RECIRCULATED WATER, WA W70-07100
ERPRETATION, MACROALGAE, SPECIES DIVERSITY INDEX, FRAGILARIA CONST W72-08141
DIVERSITY INDICES.: W72-14690
MARGALEF PIGMENT RATIO, PIGMENT DIVERSITY INDEX, CHLOROPHYLL A, C W73-00916
M ALGAL ASSAY PROCEDURE, SPECIES DIVERSITY INDEX.: *MICROCOS W72-03218
AY, DATA INTERPRETATION, SPECIES DIVERSITY.: /ARDS BAY, CAPE COD B W73-01449
TES, MACROINVERTEBRATES, SPECIES DIVERSITY, PORIFERA, SEA CUCUMBER W73-00932
CHLOROPHYL/ DESSICATION, SPECIES DIVERSITY, MELOSIRA NUMMULOIDES, W73-00853
HATTERAS, NANSEN BOTTLE, SPECIES DIVERSITY, SAMPLE PRESERVATION, C W72-07663
*SPECIES DIVERSITY, *POTOMAC RIVER.: W71-11583
K(PA), LITITZ CREEK(PA), SPECIES DIVERSITY, DIRECT EFFECTS, INDIRE W72-08579
*SPECIES DIVERSITY, CRYSTAL RIVER.: W72-11252
AHOE(CALIF), CASTLE LAKE(CALIF), DIVERSITY, PATCHES.: /IUM, LAKE T W71-09171
*ON-SITE INVESTIGATIONS, *SCUBA DIVING, *SEA WATER, *DATA COLLECT W72-11925
NG, *SYSTEMATICS, BENTHOS, SCUBA DIVING, DEPTH, DISTRIBUTION PATTE W72-09120
URE, PHAEOPHYTA, SAMPLING, SCUBA DIVING, ECOLOGY, STANDING CROPS, W73-00366
, PHAEOPHYTA, PHOTOGRAPHY, SCUBA DIVING, SEA WATER, SAMPLING, ENVI W73-00932
ENOIDOSA, SPIROGIRA, CHLORELLA / DNA, CHLOROPHYLL A, CHLORELLA PYR W72-06274
TIOS, *NONFRUITING MYXOBACTERIA, DNA, MOTILITY, CYST FORMATION, AN W72-07951
NTS, CARBON DIOXIDE, CHLORPHYLL, DO: /RATIFICATION, CARBON, SEDIME W72-08560
ERSION, CADDISFLIES, STONEFLIES, DOBSONFLIES, MAYFLIES, CYANOPHYTA W70-06655
ION, FISHING, SWIMMING, BOATING, DOCUMENTATION, WATER QUALITY, SIL W72-05869
TIONS, *WATER POLLUTION EFFECTS, DOCUMENTATION, PUBLICATIONS, PEST W73-01976
EMIOLOGY, CULTURES, METHODOLOGY, DOCUMENTATION, ABSTRACTS, DATA CO W72-13800
UM LAURYL, ETHER SULFATE, SODIUM DODECYL, BENZENE SULFONATE, COCON W72-12576
LAKE HODGSON, CLEARWATER QUARRY, DOLLAR LAKE, CHLOROPHYLL A, ABSOR W72-13673
NIZOMENON FLOS-AQUAE, CERATIUM,/ DOLLAR LAKE(OHIO), SPHAGNUM, APHA W71-11949
TERGENTS, WATER QUALITY CONTROL, DOMESTIC WASTES, RUNOFF, WATER SO W72-11186
SLUDGE, LIME, AEROBIC TREATMENT, DOMESTIC WASTES.: /NT, ACTIVATED W72-09176
NELS, *HEAT TRANSFER, *PLASTICS, DOMESTIC WASTES, BACTERIA, ALGAE, W72-10401
CALS, CHEMICAL WASTES, CLEANING, DOMESTIC WASTES, EUTROPHICATION, W72-01685
METHODOLOGY, INDUSTRIAL WASTES, DOMESTIC WASTES, BIOASSAY, SULFAT W72-12239
LUTION SOURCES, BIOASSAY, ALGAE, DOMESTIC WASTES, NUTRIENTS, SEDIM W73-02478
IAL WASTES, AGRICULTURAL RUNOFF, DOMESTIC WASTES, ECOLOGY, FISH, A W73-01972
LGAE, NUTRIENTS, WATER ANALYSIS, DOMESTIC WASTES.: /AL ANALYSIS, A W72-04791
ON DIOXIDE, ALKALINITY, BURNING, DOMESTIC WASTES, INDUSTRIAL WAS: / W72-05473
TION SOURCES, INDUSTRIAL WASTES, DOMESTIC WASTES, DINOFLAGELLATES, W71-03095
SURFACE WATERS, WATER POLLUTION, DOMESTIC WASTES, MUNICIPAL WASTES W70-09388
IOASSAY, BIOLOGICAL COMMUNITIES, DOMINANT ORGANISMS, AQUATIC ENVIR W72-07132
ICOCHEMICAL PROPERTIES, DIATOMS, DOMINANT ORGANISMS.: / IRON, PHYS W73-00256
, BENTHIC FLORA, EUTROPHICATION, DOMINANT ORGANISMS, BIOLOGICAL CO W72-14797
LORELLA, BIOLOGICAL COMMUNITIES, DOMINANT ORGANISMS, AQUATIC PRODU W72-09239
LANKTON, BIOLOGICAL COMMUNITIES, DOMINANT ORGANISMS, REDUCTION(CHE W72-10883
IAL WATER, BECKMAN ANALYZER, VAN DORN SAMPLERS, C-14, BICARBONATE W72-09122
ENECA RIVER, NINEMILE CREEK, VAN DORN SAMPLER, KEMMERER SAMPLER, C W72-09111
HLAMYDOMANAS REINHARDTII, LETHAL DOSAGE.: /TA, EUGLENA GRACILIS, C W72-05614
*MOSS LANDING, BARNACLES, LETHAL DOSAGE, CALCIUM ION CONCENTRATION W72-09092
RP, POLYOX, CARBOWAX, FLOCCULANT DOSAGES, BAROCO CLAY SOILS, URANI W70-06968
TION REDUCTION, POLYSACCHARIDES, DRAG REDUCTION.: /POLYMERS, *FRIC W72-03220
T POLYMERS, TURBULANT FRICTIONAL DRAG, ALKYLENE OXIDE, POLYMERIZAT W70-06968
CRUSTACEANS, DIPTERA, MAYFLIES, DRAGONFLIES, CADDISFLIES, MOLLUSK W71-11012
ATION, FILTRATION, DESALINATION, DRAINAGE WATER, AGRICULTURAL CHEM W71-11698
MFLOW, IONS, NITRATES, SULFATES, DRAINAGE WATER, DISSOLVED SOLIDS, W71-01489
ION, RUNOFF, UREAS, FERTILIZERS, DRAINAGE WATER, IRRIGATION, RETUR W71-06435
ASE FLOW, PERCOLATION, LEACHING, DRAINAGE, FARM WASTES, SOIL MANAG W71-06443
TY, PARTICULATE F/ *AGRICULTURAL DRAINAGE, LAKE METABOLISM, MOBILI W71-06443
AYS, WATERSHEDS(BASINS), SURFACE DRAINAGE, MUNICIPAL WASTES, INDUS W71-08953
AL WASTES, AGRICULTURE, FALLOUT, DRAINAGE, MANAGEMENT, WISCONSIN, W71-03012
IE, ALGAE, NITROGEN, PHOSPHORUS, DRAINAGE, INDUSTRIES, MUNICIPAL W W71-11009
TILIZATION, *CULTIVATED L/ *TILE DRAINAGE, *WATER PROPERTIES, *FER W71-12252
ICULTURE, ROOTED AQUATIC PLANTS, DRAINAGE, RUNOFF, FERTILIZERS, LI W71-13172
HARDNESS(WATER), CHELATION, MINE DRAINAGE, WATER PROPERTIES.: /Y, W72-14694
PERTIES, LANDFILLS, PERCOLATION, DRAINAGE, RUNOFF, PRECIPITATION(A W72-12955
POSAL, PATH OF POLLUTANTS, STORM DRAINS, DETERGENTS, EUTROPHICATIO W71-13466
GLYTOTENDIPES BARBIPES, EKMAN DREDGE.: W72-03556
ERVOIR(NJ), AMERICAN SHAD, EKMAN DREDGE, PUMPKINSEED, BLACK CRAPPI W72-14280
LGAE, DIPTERA, DISSOLVED OXYGEN, DREDGING, BOTTOM SEDIMENTS, STABI W72-03556
, PESTICIDE REMOVAL, OXYGEN SAG, DREDGING, STATE GOVERNMENTS, TREA W72-01659
N, NUTRIENTS, PESTICIDES, ALGAE, DREDGING, ESTUARINE ENVIRONMENT, W71-09581
TES, MEDITERRANEAN SEA, RED SEA, DRIFT ORGANISMS.: /ANAL, FLAGELLA W73-00296
YDROCARBON PESTICIDES, PESTICIDE DRIFT, METABOLISM, ALGAE, AQUATIC W73-02280
ROPHY, MINNESOTA, LAKES, GLACIAL DRIFT, NUTRIENTS, SALANITY, ALGAE W71-13149
PSEUDOTUBERCULOSIS, PENICILLIN, DRUGS, CORTISONE, HORMONES, CHLOR W72-13800
, TRANSPARENCY, ORTHOPHOSPHATES, DRY WEIGHT, PHOSPHATE UPTAKE.: /E W71-10645
NNAH RIVER PLANT, CULTURE MEDIA, DRY WEIGHT, THERMAL EFFECTS.: /VA W71-02075
UMAN BRAIN, LOW-TEMPERATURE OVEN DRYING, ASHING, EXTRACTIVE DIGEST W71-11036
INATION, FLOATATION, DEWATERING, DRYING, COST ANALYSES, WASTE WATE W71-08951
LTRATION, CENTRIFUGATION, VACUUM DRYING, FLOTATION, WASTEWATER TRE W72-10237
INCUBATION, DIALYSIS, CHEMICALS, DRYING, FREEZING, RADIATION, CYTO W73-01676
, ALGAE, ORGANIC MATTER, VOLUME, DRYING, ODOR, NITROGEN, AERATION, W71-07551
IDE, WATER CLARIFICATION, SLUDGE DRYING, POLYETHYLENE OXIDE POLYME W70-06968
*SURVEYS, *METHODOLOGY, FREEZE DRYING, REFRIGERATION, INCUBATION W73-01676
TS, LABORATORY EQUIPMENT, FREEZE DRYING, SEPARATION TECHNIQUES, SO W73-00379
RIA, SOLVENT EXTRACTIONS, FREEZE DRYING, SYST: /ENT, AEROBIC BACTE W73-01066
ACE ELEMENTS, VOLATILITY, FREEZE DRYING, X-RAYS, ELECTROLYSIS, COM W71-11036
CEANS, MUSSELS, COMMON MERGANSER DUCK, BIOASSAY, WATER POLLUTION E W72-12930

TIGRIOPUS BREVICORNIS, GAMMARUS DUEBENI, FUCUS SERRATUS, FUCUS VE W71-08026
PENICILLUS CAPITATUS, PENICILLUS DUMENTOSUS, RHIPOCEPHALUS PHOENIX W72-14294
FATE CF POLLUTANTS, METABOLITES, DUNALIELLA, AGMENELLUM QUADRAPLIC W73-00361
ASIOSIRA SPP., SKELETONEMA SPP., DUNALIELLA SPP.: /TER ALGAE, THAL W72-08436
SEUDONANA, SKELETONEMA COSTATUM, DUNALIELLA TERTIOLECTA, EUGLENA G W72-05614
A NANA, TH/ *ELECTRON TRANSPORT, DUNALIELLA TERTIOLECTA, CYCLOTELL W72-12998
TE, BAYTEX, BAYGONE, CYCLOTELLA, DUNALIELLA, SKELETONEMA.: ABA W71-10096
TICA, THALASSIOSIRA FLUVIATILIS, DUNALIELLA TERTIOLECTA, SYNECHOCO W71-10083
LUM/ CARBON-14, *FLORIDA STRAIT, DUNALIELLA TERIOLECTA, PHAEODACTY W72-09103
YLU/ *FLORIDA STRAIT, CARBON-14, DUNALIELLA TERTIOLECTA, PHAEODACT W72-09102
KELETCNEMA, ANABAENA CYLINDRICA, DUNALIELLA.: /RIN, METABOLITES, S W72-03545
IGHT, EELGRASS, ZOSTERA, PINAEUS DUORARUM, PINAEUS AZTECUS, PENAEU W73-02029
*ENGLAND, *POLLUTED ESTUARIES, DURHAM COUNTY RIVERS.: W72-08804
DURHAM(NC).: W71-07382
NTS, BICARBONATES, PLANT GROWTH, DURNAL.: /DIOXIDE, CYCLING NUTRIE W72-09159
*ANALINE DYE.: W73-02349
SON(WIS), LAKE WASHINGTON(WASH), DYEING WATER.: /RVOIRS(NEB), MADI W72-04759
CAL MODELS, *ALGAE, *POPULATION, DYNAMICS, LAKES, DISSOLVED OXYGEN W71-03034
LMONELLA ENTEROCOLITIS, SHIGELLA DYSENTERIAE, SHIGELLA FLEXNERI, B W72-13800
ION, HYDROGEN ION CONCENTRATION, DYSTROPHY, ZINC RADIOISOTOPES, CO W71-09190
FUR BACTERIA, ALGAE, SALMONELLA, E. COLI.: /CILLUS, CHLORELLA, SUL W72-12565
DISSOLVED OXYGEN, BIOINDICATORS, E. COLI, AQUATIC ALGAE, SUNFISHES W71-07417
RONMENTAL EFFECTS, GROWTH RATES, E. COLI, AZOTOBACTER, PSEUDOMONAS W72-10861
ON, PLANKTON, WATER TEMPERATURE, E. COLI, BENTHIC FAUNA, CELLULOSE W72-09668
*PATHOGENIC BACTERIA, CULTURES, E. COLI, BIOINDICATORS, PONDS, AN W72-07664
DIOXIDE, CHLORELLA, CLOSTRIDIUM, E. COLI, EUGLENA, EUGLENOPHYTA, C W72-06124
TOZOA, YEASTS, INSECTS, MILDEWS, E. COLI, FISH, FOODS, GRASSES, WO W73-01676
TECHNIQUES, CALCIUM, MAGNESIUM, E. COLI, SILICA, *FILTRATION, LIG W72-11833
ES), ENTERIC BACTERIA, DISEASES, E. COLI, SHIGELLA, VIRUSES, CHROM W72-13800
AVATUM, ELPHIDIUM GALVESTONENSE, E: /PHIDIUM ADVENUM, ELPHIDIUM CL W73-00367
NDS, ESOX LUCIUS, ALKYL MERCURY, EAGLES, SALMO GAIRDNERII, IRIS PS W72-05952
URINE, OEDOGONIUM, PHYSA, CULEX, EAMBUSIA, FATE OF POLLUTANTS.: / W72-07703
UATIC PLANTS, *RESERVOIR STAGES, EARLY IMPOUNDMENT, BIOMASS, AQUAT W72-04853
*RESERVOIRS, *RESERVOIR STAGES, EARLY IMPOUNDMENT, BIOMASS, DIATO W72-04855
PLOSION, *FILTERS, *DIATOMACEOUS EARTH, *POROUS MEDIA, *WASTE WATE W72-00974
ATION, FERTILIZERS, DIATOMACEOUS EARTH, FOOD CHAINS, OXYGEN, ECOLO W71-09157
*MEMBRANE FILTERS, GRANGE LAKE, EAST TWIN LAKE, LAKE ROCKWELL, LA W72-13673
HROPODS, COELENTERATES, BRYOZOA, ECHINODERMS.: /NE, TUNICATES, ART W71-01475
TERATES, ARTHROPODS, NEMERTEANS, ECHINODERMS, SIPUNCULIDS, BRYOZOA W73-00932
HORA, CHAETOGNATHS, CYPHONAUTES, ECHINODERMS, TUNICATES, NOCTILUCA W73-00935
EDITERRANEAN SEA, COELENTERATES, ECHINODERMS, POLYCHAETES, CHIRONO W73-00855
ND), CERATIUM HIR/ *GLOEOTRICHIA ECHINULATA, MIKOLAJSKIE LAKE(POLA W72-14812
USCIFORMIS, GRACILARIA VERUCOSA, ECKLONIA RADIATA, GROWTH REGULATI W72-07496
ECHNIQUES, SPATIAL DISTRIBUTION, ECOLOGICAL DISTRIBUTION, BENTHIC W73-00367
ABORATORY EQUIPMENT, INHIBITION, ECOLOGICAL DISTRIBUTION, MARINE A W73-00348
, PHOTOSYNTHESIS, STANDING CROP, ECOLOGICAL DISTRIBUTION: /TRIENTS W72-14313
OPERTIES, DISTRIBUTION PATTERNS, ECOLOGICAL DISTRIBUTION, WATER PO W73-00918
AUSTRALIA, SAMPLING, MICROSCOPY, ECOLOGICAL DISTRIBUTION, PLANKTON W73-01094
CROORGANISMS, CYCLING NUTRIENTS, ECOLOGICAL DISTRIBUTION, ECOSYSTE W73-02095
LAKE CAYUGA, CLADOPHORA, ECOLOGICAL STUDIES.: W70-09614
S, CLADOPHORA, LIMITING FACTORS, ECOLOGICAL DISTRIBUTION.: /ABITAT W72-13142
OTOZOA, *BIOLOGICAL COMMUNITIES, ECOLOGICAL DISTRIBUTION, BIOINDIC W72-08579
*SALTON SEA(CALIF), ECOLOGICAL MANAGEMENT.: W71-10577
HRYSOPHYTA, RHODOPHYTA, TESTING, ECOLOGY.: /OPHYTA, CHLOROPHYTA, C W70-07280
EFFECTS, AQUATIC LIFE, HABITATS, ECOLOGY, *ECOSYSTEM, *ALGAE, *PHY W73-01632
SCOPY, LIMNOLOGY, WATER QUALITY, ECOLOGY, AQUATIC HABITATS, CLADOP W72-13142
TA COLLECTIONS, LAKE MORPHOLOGY, ECOLOGY, BIOLOGICAL PROPERTIES.: / W72-13851
MOSOMES, FUNGICIDES, PESTICIDES, ECOLOGY, BIRD TYPES, BIRDS, AQUAT W72-05952
GY, LAKES, PONDS, AQUATIC ALGAE, ECOLOGY, BIOLOGICAL PROPERTIES, F W71-10580
PLANT PHYSIOLOGY, PHYSIOLOGICAL ECOLOGY, BIOASSAY, MICROORGANISMS W71-12303
C LIFE, ANTIBIOTICS(PESTICIDES), ECOLOGY, BENTHOS, INVERTEBRATES, W72-09668
ION, BIOCHEMISTRY, PHYSIOLOGICAL ECOLOGY, CESIUM, STRONTIUM RADIOI W71-13243
ION, PROTOZOA, PONDS, LIMNOLOGY, ECOLOGY, CHEMICAL ANALYSIS, AQUAT W73-00918
ATION SENSITIVITY, RADIOECOLOGY, ECOLOGY, CYCLING NUTRIENTS, THERM W73-00796
RRENTS(WATER), *DINOFLAGELLATES, ECOLOGY, DISTRIBUTION PATTERNS, M W73-00296
ATH OF POLLUTANTS, PHYSIOLOGICAL ECOLOGY, ENVIRONMENTAL EFFECTS, P W72-06313
METRY, PENNSYLVANIA, INHIBITORS, ECOLOGY, ECOTYPES, FRESH WATER.: / W72-04743
COMMUN/ WATER POLLUTION EFFECTS, ECOLOGY, ECOSYSTEMS, *BIOLOGICAL W72-11854
NITROGEN CYCLE, *CRABS, *FISHES, ECOLOGY, FOOD CHAINS, ESTUARINE F W72-12567
INS, *CHLOROPHYTA, *PHYTOTOXINS, ECOLOGY, FISH FOOD ORGANISMS, PAT W71-08597
SUPPLY, SYSTEMATICS, LIMNOLOGY, ECOLOGY, FISH, WATER POLLUTION, P W73-01615
ULTURAL RUNOFF, DOMESTIC WASTES, ECOLOGY, FISH, ALGAE, AQUATIC PLA W73-01972
PLANKTON, BENTHIC FAUNA, ALGAE, ECOLOGY, MISSOURI.: /ISH HARVEST, W70-09596
TOPLANKTON BLOOMS, PHYSIOLOGICAL ECOLOGY, MORTALITY, RED TIDE, TOX W72-14056
PLANT PHYSIOLOGY, PHYSIOLOGICAL ECOLOGY, ORGANIC COMPOUNDS, ON-SI W72-03216
ER, NUCLEAR POWER PLANTS, STEAM, ECOLOGY, PHYTOPLANKTON, ZOOPLANKT W71-11517
TOXINS, SEDIMENTS, SALT MARSHES, ECOLOGY, PESTICIDES, ESTUARIES, I W72-10678
NE ALGAE, *KELPS, *PHYSIOLOGICAL ECOLOGY, PHAEOPHYTA, STANDING CRO W72-12940
RE, CHEMICAL ANALYSIS, SAMPLING, ECOLOGY, PHYTOP: /WATER TEMPERATU W72-07710
MPOUNDMENT, NUCLEAR POWERPLANTS, ECOLOGY, RADIOECOLOGY, MONITORING W72-00939
MARY PRODUCTIVITY, PHYSIOLOGICAL ECOLOGY, RADIOECOLOGY, RADIOACTIV W72-00948
EOUS EARTH, FOOD CHAINS, OXYGEN, ECOLOGY, RHODOPHYTA, PHAEOPHYTA, W71-09157
IL BACTERIA, SAMPLING, CULTURES, ECOLOGY, SOIL MICROBIOLOGY, CYANO W73-00854
EOPHYTA, SAMPLING, SCUBA DIVING, ECOLOGY, STANDING CROPS, PATH OF W73-00366
BRATES, WATER POLLUTION EFFECTS, ECOLOGY, SEA WATER, LABORATORY EQ W73-00348
ATIC ANIMALS, BALANCE OF NATURE, ECOLOGY, TROPHIC LEVEL, HERBIVORE W72-06340
S, CITIES, PHOSPHATES, ILLINOIS, ECOLOGY, TERTIARY TREATMENT, SEWA W72-03970
S, *HYDROGRAPHY, *PACIFIC OCEAN, ECOLOGY, WATER TEMPERATURE, PHYTO W72-07892
RYSOPHYTA, PYRROPHYTA, SAMPLING, ECOLOGY, WATER TEMPERATURE.: / CH W72-03543
L SPRINGS, CYANOPHYTA, BACTERIA, ECOLOGY, WATER POLLUTION EFFECTS. W72-12734

RMANCE, FLOCCULATION, FLOTATION, ECONOMIC FEASIBILITY, COSTS, ESTI W72-13508
POSAL, *WATER POLLUTION CONTROL, ECONOMIC FEASIBILITY, STREAMS, SE W72-14878
SYSTEMS, CALIFORNIA, IRRIGATION, ECONOMIC FEASIBILITY, WATER POLLU W71-09157
STUDIES, TECHNICAL FEASIBILITY, ECONOMIC FEASIBILITY, PILOT PLANT W71-09450
TER SUPPLY, COSTS, WATER DEMAND, ECONOMIC JUSTIFICATION, AQUATIC W W71-13172
RUNOFF, FERTILIZERS, LIVESTOCK, ECONOMICS,: /IC PLANTS, DRAINAGE, W71-13172
UDGE, EFFLUENT, *ORGANIC WASTES, ECONOMICS, CHEMICAL OXYGEN DEMAND W71-12223
ATION, WATER SUPPLY, SUCCESSION, ECONOMICS, NUTRIENTS, NITROGEN FI W71-13183
IME, HYDROGEN ION CONCENTRATION, ECONOMICS, ALGAE, SOLUBILITY, SEW W71-12200
NING, BIOCHEMICAL OXYGEN DEMAND, ECONOMICS, ALGAE, AGRICULTURAL: / W72-04734
S, FILTERS, ENGINEERS ESTIMATES, ECONOMICS, FEASIBILITY, COSTS, AN W70-10174
WISCONSIN, PRODUCTIVITY, ALGAE, ECONOMICS, POLITICAL ASPECTS, GEO W71-03012
RIFFLE BENTHOS, WOODLAND STREAM, ECOSYSTEM ANALYSIS, MACRCCONSUMER W71-10064
VOIR(KAN), GREAT PLAINS, AQUATIC ECOSYSTEM.: / *TUTTLE CREEK RESER W72-12393
ATER POLLUTION EFFECTS, ECOLOGY, ECOSYSTEMS, *BIOLOGICAL CCMMUNITI W72-11854
ICATICN, *KANSAS, MODEL STUDIES, ECOSYSTEMS, ALGAE, *LIGHT QUALITY W72-12393
ASINS), *DDT, *SMALL WATERSHEDS, ECOSYSTEMS, PESTICIDE KINETICS, * W72-12930
, *LIG/ *EUTROPHICATION, AQUATIC ECOSYSTEMS, WATER QUALITY, *ALGAE W73-02349
CTIVITY, HERBIVORES, CARNIVORES, ECOSYSTEMS, AQUATIC LIFE, SAMPLIN W71-10064
NS, FISH FOOD ORGANISMS, OXYGEN, ECOSYSTEMS.: /ESOURCES, FOOD CHAI W71-13246
S, *FALLOUT, *CYCLING NUTRIENTS, ECOSYSTEMS, RAIN FORESTS, MATHEMA W72-03342
ALGAE, INVERTEBRATES, RADIATION, ECOSYSTEMS, ESTUARIES, CORAL, REE W71-13723
, CHORELLA, TRACERS, PHOSPHATES, ECOSYSTEMS, AQUATIC LIFE, ANTIBIO W72-09668
POPULATION, AQUATIC ENVIRONMENT, ECOSYSTEMS, GEORGIA, BIORTHYTHMS, W72-09159
TIC ENVIRONMENT, CARBON DIOXIDE, ECOSYSTEMS, SOUTH CAROLINA, ALGAE W71-00475
NA), ANACYSTIS NIDULANS, AQUATIC ECOSYSTEMS, ALGAE GROWTH, ALGAL P W71-00475
YTA, RADIOISOTOPES, TEMPERATURE, ECOSYSTEMS, CALIFORNIA, IRRIGATIO W71-09157
S, MARINE ANIMALS, AQUATIC LIFE, ECOSYSTEMS, SALMCNIDS.: /SH, CLAM W72-07907
ACTIVITY TECHNIQUES, PESTICIDES, ECOSYSTEMS, POLLUTICN SOURCES, PE W72-07703
ASTE WATER(POLLUTION), BIOASSAY, ECOSYSTEMS, BIODEGRADATICN, ALGAE W73-00840
LLUTICN SOURCES, BIODEGRADATION, ECOSYSTEMS, WATER POLLUTION.: /PO W73-00853
RIENTS, ECOLOGICAL DISTRIBUTION, ECOSYSTEMS, AQUATIC HABITATS, PRO W73-02095
NUTRIENTS, FOOD WEBS, DETRITUS, ECOSYSTEMS, ENERGY BUDGETS, NUTRI W73-01089
LANKTON, BIOLOGICAL COMMUNITIES, ECOSYSTEMS, SPECTROPHOTCMETRY, VI W73-00916
NNSYLVANIA, INHIBITORS, ECOLOGY, ECOTYPES, FRESH WATER.: /ETRY, PE W72-04743
FECTS, *DATA COLLECTIONS, ALGAE, ECOTYPES, INDUSTRIAL PLANTS, INDU W71-13746
FFICIENT, LAMBERT-BEER EQUATION, EDDY DIFFUSIVITY, DESHLER(OHIO).: W72-05459
*FRESHWATER ALGAE, EDTA.: W72-14694
YSIS/ *ION-SELECTIVE ELECTRODES, EDTA, *POTENTIOMETRY, ELECTROANAL W73-02112
ROINVERTEBRATES, NEW YORK BIGHT, EELGRASS, ZOSTERA, PINAEUS DUCRAR W73-02029
QUATIC ALGAE, SUNFISHES, SNAILS, EELS, CHINOOK SALMON, BASS, TROUT W71-07417
TH MOUNTAIN PROJECT(VA), WEATHER EFFECT.: SMI W72-05282
CA, ACETYLENE REDUCTION, WARBURG EFFECT.: /ION, *ANABAENA CYLINDRI W72-08584
YCLE, ANACYSTIS NIDULAN/ WARBURG EFFECT, EUGLENA GRACLIS, CALVIN C W71-09172
ICAL PROPERTIES, WATER POLLUTION EFFECT, WATER POLLUTION SCURCES, W72-00974
LECTING AGENTS, *SINKING AGENTS, EFFECTIVENESS, TESTING PROCEDURES W71-09789
TY, CONVECTION, WATTER POLLUTICN EFFECTS.: / BUOYANCY, WIND VELOCI W71-09852
L WASTES, SILTS, WATER POLLUTION EFFECTS.: /OAC, SEWAGE, INDUSTRIA W71-09795
WASTE DISPOSAL, WATER POLLUTICN EFFECTS.: / METALS, HEAVY METALS, W71-11793
NKTON, CULTURES, WATER POLLUTION EFFECTS.: /GROWTH REGULATCRS, PLA W71-13793
TION, CORROSICN, WATER POLLUTION EFFECTS.: /HYDROGEN ICN CONCENTRA W72-08683
ERSITY, DIRECT EFFECTS, INDIRECT EFFECTS.: /CREEK(PA), SPECIES DIV W72-08579
EWAGE EFFLUENTS, WATER POLLUTION EFFECTS.: /TS, CHEMICAL WASTES, S W72-08804
HAMPSHIRE, MIXING, ENVIRONMENTAL EFFECTS.: /, PHYTOPLANKTCN, *NEW W72-09061
ER, PSEUDOMONAS, WATER POLLUTION EFFECTS.: /ES, E. COLI, AZOTOBACT W72-10861
IC ALGAE, CALCIUM, RADIOACTIVITY EFFECTS.: /H, AQUATIC LIFE, AQUAT W72-10978
LOGICAL STUDIES, WATER POLLUTICN EFFECTS.: / FLUORIDES, ICNS, CYTO W72-11572
SIZE, COLLOIDS, WATER POLLUTION EFFECTS.: /YTA, AQUARIA, PARTICLE W71-09193
WATER TREATMENT, WATER POLLUTICN EFFECTS.: / OXYGEN DEMAND, WASTE W71-07098
WATER TREATMENT, WATER POLLUTION EFFECTS.: /BIODEGRADATICN, WASTE W71-07097
WATER TREATMENT, WATER POLLUTION EFFECTS.: / CONCENTRATION, WASTE W71-05155
COOLING TOWERS, WATER POLLUTION EFFECTS.: /, APPLICATION METHODS, W71-06189
QUALITY CONTROL, WATER POLLUTION EFFECTS.: /, PERMEABILITY, WATER W72-06037
LTURE MEDIA, DRY WEIGHT, THERMAL EFFECTS.: /VANNAH RIVER PLANT, CU W71-02075
ASINOCLADUS MARINUS, TEMPERATURE EFFECTS.: /1002, *GREEN ALGAE, PR W70-09429
LIMIT, POISONS, WATER POLLUTION EFFECTS.: /AY, RESISTANCE, LETHAL W72-14706
Y TESTS, OXYGEN, WATER POLLUTION EFFECTS.: /CHLOROPHYTA, LABCRATOR W72-14314
ON, AQUATIC ALGAE, ENVIRONMENTAL EFFECTS.: /ICIDES), PHOTCACTIVATI W72-14327
AVIOR, SEASONAL, WATER POLLUTION EFFECTS.: /GES, DISTRIBUTION, BEH W72-14659
A, ALGAE, FUNGI, WATER POLLUTION EFFECTS.: /R, CONDENSERS, PROTOZO W72-07526
OZOA, EQUIPMENT, WATER POLLUTION EFFECTS.: /AS, EUGLENOPHYTA, PROT W72-07661
SHWATER, *ALGAE, WATER POLLUTION EFFECTS.: /S, BRACKISH WATER, FRE W73-02548
Y, BICARBONATES, WATER POLLUTION EFFECTS.: /ATION, PLANT PHYSIOLOG W72-12940
S, PHYTOPLANKON, WATER POLLUTION EFFECTS.: /IENTS, DISSOLVED SCLID W72-13300
(WATER), MOLLUSKS, ENVIRONMENTAL EFFECTS.: /AL, SALINITY, CURRENTS W72-11876
CTERIA, ECOLOGY, WATER POLLUTION EFFECTS.: /PRINGS, CYANOPHYTA, BA W72-12734
EUTROPHICATION, WATER POLLUTION EFFECTS.: /ISOTOPES, *CYANOPHYTA, W72-13692
ION, W/ *LAKES, *WATER POLLUTION EFFECTS, *LIMNOLOGY, WATER POLLUT W72-12729
MPERATURE, *WA/ *WATER POLLUTION EFFECTS, *HEATED WATER, *WATER TE W72-13636
, *PERSISTENCE, *WATER POLLUTION EFFECTS, *PLANT PHYSIOLOGY, *DIAT W73-02280
AE, RHODOPHYTA, / *RADIOACTIVITY EFFECTS, *MONITORING, *MARINE ALG W72-07826
S, OILY WATER,/ *WATER POLLUTION EFFECTS, *OIL WASTES, *CRUSTACEAN W72-07911
TY, COAGULATICN, WATER POLLUTION EFFECTS, *ALGAL TOXINS.: /*TOXICI W72-07360
RMAL POLLUTION, *WATER POLLUTION EFFECTS, *AQUATIC MICROCRGANISMS, W72-07225
MICROORGANISMS, *WATER POLLUTION EFFECTS, *ALGAE, *PROTOZCA, *BIOL W72-08579
ULTURAL RUNOFF, *WATER POLLUTION EFFECTS, *CATTLE, *FEED LCTS.: /C W72-08401
HERMAL POLLUTION, *ENVIRONMENTAL EFFECTS, *ECOLOGY, *PLANKTCN, *AL W72-14659
OSING ORGANIC M/ WATER POLLUTION EFFECTS, *ORGANIC MATTER, *DECOMP W72-14673
ICAL CCMMUNITIES, *ENVIRONMENTAL EFFECTS, *STANDING CRCPS, TIDAL E W73-00367
PHYTA, *SALINITY, *ENVIRONMENTAL EFFECTS. *PLANT MORPHOLOGY, *GENE W73-00369

*PHYTOPLANKTON, *ENVIRONMENTAL EFFECTS, *SYSTEMATICS, *SEASONAL, W73-00442
ANIMAL/ *ECOLOGY, *ENVIRONMENTAL EFFECTS, *MARINE PLANTS, *MARINE W73-02029
TICIDE RESIDUES, WATER POLLUTION EFFECTS, *CALIFORNIA, SEA WATER, W73-02105
*MARINE ALGAE, *WATER POLLUTION EFFECTS, *PHOTOSYNTHESIS, *OIL PO W73-01074
GROUND RADIATION, *RADIOACTIVITY EFFECTS, *FALLOUT, *MARINE ANIMAL W73-00828
LEAR POWERPLANTS, *RADIOACTIVITY EFFECTS, *MATHEMATICAL MODELS, FO W73-00831
WATER TREATMENT, WATER POLLUTION EFFECTS, *TREATMENT FACILITIES, * W73-01669
ID/ *ESTUARIES, *WATER POLLUTION EFFECTS, *DISTRICT OF COLUMBIA, T W70-08248
LUTION SOURCES, *WATER POLLUTION EFFECTS, *ORGANIC MATTER, *OXYGEN W70-09189
F RATES, AFTERGRO/ *AUTOGOMISTIC EFFECTS, *AXEMIC CULTURES, *DIEOF W70-08319
ID/ *ESTUARIES, *WATER POLLUTION EFFECTS, *DISTRICT OF COLUMBIA, T W70-08249
*NUTRIENTS, *E/ *WATER POLLUTION EFFECTS, *PHOSPHATES, *NITRATES, W70-09388
ORGANIC WASTES, WATER POLLUTION EFFECTS, *ALGAE, PHYTOPLANKTON, D W70-07842
LLUTION CONTROL, WATER POLLUTION EFFECTS, *AQUATIC PLANTS, HARVEST W71-01488
S,/ *FOOD CHAINS, *RADIOACTIVITY EFFECTS, *TRITIUM, *NUCLEAR WASTE W72-06342
HENYLS, *ALGAE, *WATER POLLUTION EFFECTS, *PESTICIDES, DIATOMS, PH W72-05614
, *ECOSYSTEMS,/ *WATER POLLUTION EFFECTS, *MODEL STUDIES, *STREAMS W72-06340
*WATER QUALITY, *WATER POLLUTION EFFECTS, *LAKES, *COLORADO, *RECR W72-04791
DATION LAGOONS, *WATER POLLUTION EFFECTS, *FARM WASTES, OKLAHOMA, W72-04773
T, *OCEANS, WATE/ *RADIOACTIVITY EFFECTS, *NUCLEAR WASTES, *FALLOU W72-04463
*AQUATIC LIFE, R/ *RADIOACTIVITY EFFECTS, *FALLOUT, *FOOD CHAINS, W72-04461
ME/ *ESTUARIES, *WATER POLLUTION EFFECTS, *WATER RESOURCES DEVELOP W71-05553
HLORINATED HYD/ *WATER POLLUTION EFFECTS, *BIOASSAY, *TOXICITY, *C W71-07412
HICATION, *CHE/ *WATER POLLUTION EFFECTS, *ORGANIC WASTES, *EUTROP W71-09674
*AQUATIC ALGAE, *RADIOACTIVITY EFFECTS, *PONDS, WATER POLLUTION W71-09256
, *STRIP MINES, *WATER POLLUTION EFFECTS, *KENTUCKY, SULFUR BACTER W71-07942
POLLUTION EFFECTS, ENVIRONMENTAL EFFECTS, *BENTHIC FAUNA, *INVERTE W71-08793
S, *DETERGENTS, *WATER POLLUTION EFFECTS, *WATER POLLUTION CONTROL W71-08768
TS, *OILY WATER, WATER POLLUTION EFFECTS, *INTERTIDAL AREAS, *INVE W71-08792
RADIOSENSITIVITY, *RADIOACTIVITY EFFECTS, *ALGAE, CHLORELLA, IRRAD W72-11836
*RADIOISOTOPES, *RADIOACTIVITY EFFECTS, *RADIOECOLOGY, *RESERVOI W72-10694
TION ABATEMENT, *WATER POLLUTION EFFECTS, *WATER POLLUTION SOURCES W71-13723
*NUTRIENTS, ALG/ WATER POLLUTION EFFECTS, *EUTROPHICATION, LAKES, W72-00798
, ECOTYPES, IN/ *WATER POLLUTION EFFECTS, *DATA COLLECTIONS, ALGAE W71-13746
NI/ *FOOD CHAINS, *RADIOACTIVITY EFFECTS, *NUCLEAR POWERPLANTS, MO W72-00959
TS, CYCLING NUTR/ *RADIOACTIVITY EFFECTS, *ECOSYSTEMS, *RAIN FORES W72-00949
ACTIVITY EFFECTS, *ENVIRONMENTAL EFFECTS, *NUCLEAR EXPLOSIONS, BIR W72-00960
, *SALT MARSHE/ *WATER POLLUTION EFFECTS, *PROTOZOA, *MARINE ALGAE W72-00948
ONTROL, *POLLUT/ WATER POLLUTION EFFECTS, *MICROORGANISMS, ALGAL C W72-00801
*NUCLEAR WASTES, *RADIOACTIVITY EFFECTS, *THERMAL POLLUTION, CONN W72-00939
*NUCLEAR EXPLOSI/ *RADIOACTIVITY EFFECTS, *ENVIRONMENTAL EFFECTS, W72-00960
EAR EXPLOSIONS, *WATER POLLUTION EFFECTS, *RADIOACTIVITY EFFECTS, W72-03347
BORATORY TESTS, *WATER POLLUTION EFFECTS, *ALGAE, *PLANKTON, WATER W72-03218
NTS, *POLLUTAN/ *WATER POLLUTION EFFECTS, *ORGANIC WASTES, *NUTRIE W72-01472
N SOURCES, IND/ *WATER POLLUTION EFFECTS, *DIATOMS, WATER POLLUTIO W71-11583
Y, *INDICATORS, *WATER POLLUTION EFFECTS, *ALGAE, *MARINE ALGAE, * W71-10553
ALGAE, *BACTER/ *WATER POLLUTION EFFECTS, *NITRATES, *NUTRIENTS, * W71-09958
RIN, *DIAZINON, *WATER POLLUTION EFFECTS, *PHOTOSYNTHESIS, PESTICI W71-10566
MARIN/ *ALGAE, *WATER POLLUTION EFFECTS, *INDICATORS, *SEA WATER, W71-12867
, *INVERTEBRATES, *RADIOACTIVITY EFFECTS, *RADIOISOTOPES.: /*ALGAE W71-13233
E, *INVERTEBRA/ *WATER POLLUTION EFFECTS, *PLANTS, *ANIMALS, *ALGA W71-13233
TION, STANDARDS, WATER POLLUTION EFFECTS, ALKYLBENZENE SULFONATES, W71-11516
WELLS, FISH, BAYS, AIR POLLUTION EFFECTS, AIR POLLUTION, WATER POL W72-03347
LLUTION ABATEMENT, ENVIRONMENTAL EFFECTS, AQUATIC LIFE, FISH POPUL W72-00939
TON, *NUTRIENTS, WATER POLLUTION EFFECTS, AQUATIC ALGAE, NORTH CAR W72-01329
OGY, RADIOECOLOGY, RADIOACTIVITY EFFECTS, ABSORPTION, FOOD WEBS, S W72-00948
LLUTION SOURCES, WATER POLLUTION EFFECTS, AGRICULTURAL RUNOFF, PHO W72-10625
, *NITROGEN CYCLE, ENVIRONMENTAL EFFECTS, ANAEROBIC CONDITIONS, NI W72-11563
, WATER SOFTENING, ENVIRONMENTAL EFFECTS, AQUATIC ALGAE, ALGAL CON W72-11186
TY, PROBABILITY, WATER POLLUTION EFFECTS, ALGAE, AQUATIC HABITATS, W71-07337
CONTROL, RIVERS, WATER POLLUTION EFFECTS, ALGAE, DIATOMS.: /ALGAL W71-00099
POLLUTION EFFECTS, ENVIRONMENTAL EFFECTS, ALGAL POISONING, ALGAE, W70-09429
LUTION, *COASTS, WATER POLLUTION EFFECTS, AQUATIC LIFE, HABITATS, W73-01632
ICIENT ELEMENTS, WATER POLLUTION EFFECTS, AQUATIC ALGAE, CHRYSOPHY W73-01445
UES, RADIOECOLOGY, ENVIRONMENTAL EFFECTS, ABSORPTION, SORPTION, WA W73-00817
POLLUTION SOURCES, ENVIRONMENTAL EFFECTS, AEROBIC CONDITIONS, CHLO W73-01066
AS, CHLOROPHYTA, WATER POLLUTION EFFECTS, ABSORPTION, PESTICIDE TO W73-00281
TIC ENVIRONMENT, WATER POLLUTION EFFECTS, ALGAE, NUTRIENTS, WATER W72-06638
NMENTAL EFFECTS, WATER POLLUTION EFFECTS, ALGAE, HEAVY METALS, GRO W72-13813
AINS, FOOD WEBS, WATER POLLUTION EFFECTS, AQUATIC ENVIRONMENT, TUR W72-11925
ICIDE TOXICITY, *WATER POLLUTION EFFECTS, BIOASSAY, PESTICIDES, HE W72-08440
, PLANKTON, ALGAE, ENVIRONMENTAL EFFECTS, BIOLOGICAL COMMUNITY, CA W73-01704
WATER SAMPLING, WATER POLLUTION EFFECTS, BIOMASS, NUTRIENTS, ALGA W72-09111
LLUTION EFFECTS, WATER POLLUTION EFFECTS, BAYS, WATERSHEDS(BASINS) W72-00959
, SEWAGE BACTERIA, ENVIRONMENTAL EFFECTS, BIOLOGICAL COMMUNITIES, W72-01472
, NEW HAMPSHIRE, WATER POLLUTION EFFECTS, COPPER SULFATE.: /SHKILL W72-09304
EPOCH, ADAPTATION, ENVIRONMENTAL EFFECTS, CHLOROPHYTA, MARINE ALGA W73-00855
*SALT TOLERANCE, WATER POLLUTION EFFECTS, CHLORINATED HYDROCARBON W72-13809
NS, AQUATIC ALGAE, ENVIRONMENTAL EFFECTS, CHLOROPHYTA, PRIMARY PRO W72-13816
POLLUTION EFFECTS, ENVIRONMENTAL EFFECTS, COMPETITION, AQUATIC POP W72-12736
*NUCLEAR WASTES, WATER POLLUTION EFFECTS, CATTAILS, AQUATIC ANIMAL W72-06342
TA COLLECTIONS, *WATER POLLUTION EFFECTS, DOCUMENTATION, PUBLICATI W73-01976
NTROL, *PARAQUAT, *ENVIRONMENTAL EFFECTS, DIQUAT, ALGAE, ALGAL CON W72-00155
TE INVESTIGATIONS, ENVIRONMENTAL EFFECTS, ESTUARINE ENVIRONMENT, R W72-00959
*PHYTOPLANKTON, WATER POLLUTION EFFECTS, EUTROPHICATION, PRODUCTI W72-09167
USTRIAL WASTES, *WATER POLLUTION EFFECTS, EFFLUENTS, CALIFORNIA, S W72-09092
S, NUCLEAR WASTES, ENVIRONMENTAL EFFECTS, EFFLUENTS, MONITORING, C W72-10957
RY PRODUCTIVITY, WATER POLLUTION EFFECTS, ESTUARINE ENVIRONMENT.: / W72-09653
ESEARCH EQUIPMENT, ENVIRONMENTAL EFFECTS, ESTUARIES, *DISTRIBUTION W72-08141
, INVERTEBRATES, WATER POLLUTION EFFECTS, ECOLOGY, SEA WATER, LABO W73-00348

GAE, AQUATIC LIFE, ENVIRONMENTAL EFFECTS, ENVIRONMENTAL ENGINEERIN W72-14405
IAL PHOTOGRAPHY, WATER POLLUTION EFFECTS, EUTROPHICATION, *LAKES, W72-14728
UTION TREATMENT, WATER POLLUTION EFFECTS, ENVIRONMENTAL EFFECTS, A W70-09429
*OIL, SEEPAGE, WATER POLLUTION EFFECTS, ENVIRONMENTAL EFFECTS, * W71-08793
YTA, INHIBITORS, WATER POLLUTION EFFECTS, ENVIRONMENTAL EFFECTS, C W72-12736
OLOGICAL COMMUN/ WATER POLLUTION EFFECTS, ECOLOGY, ECOSYSTEMS, *BI W72-11854
, BIOINDICATORS, WATER POLLUTION EFFECTS, FERTILIZERS, NITROGEN, B W72-13816
ITY, ADSORPTION, WATER POLLUTION EFFECTS, FOOD CHAINS.: /, SOLUBIL W71-09182
T RADIOISOTOPES, WATER POLLUTION EFFECTS, FRESH WATER,: /ES, COBAL W71-09250
ALLOUT, *OCEANS, WATER POLLUTION EFFECTS, FOOD CHAINES, PUBLIC HEA W72-04463
SLATION, *TOXINS, *ENVIRONMENTAL EFFECTS, FEDERAL GOVERNMENT, AIR W73-00703
AL DISTRIBUTION, WATER POLLUTION EFFECTS, F: /N PATTERNS, ECOLOGIC W73-00918
NUCLEAR WASTES, WATER POLLUTION EFFECTS, FOOD CHAINS, PUBLIC HEAL W72-10978
LLUTION SOURCES, WATER POLLUTION EFFECTS, FLUCTUATION, COMPARATIVE W72-03228
UNITY DEVELOPMENT, ENVIRONMENTAL EFFECTS, GOVERNMENT FINANCE,: /MM W72-01660
E, MICROORGANISMS, ENVIRONMENTAL EFFECTS, GROWTH RATES, E. COLI, A W72-10861
DUCK, BIOASSAY, WATER POLLUTION EFFECTS, GAS CHROMATOGRAPHY, BROO W72-12930
AQUATIC HABITATS, ENVIRONMENTAL EFFECTS, HEATED WATER, NUCLEAR PO W71-11517
RADATION, ALGAE, WATER POLLUTION EFFECTS, HEAVY METALS, CYCLING NU W73-00840
NSIN, NUTRIENTS, WATER POLLUTION EFFECTS, INTERFACES, DISSOLVED OX W71-10645
ALGAE, RECREATION, ENVIRONMENTAL EFFECTS, INVERTEBRATES, WATER RES W71-10577
K(PA), SPECIES DIVERSITY, DIRECT EFFECTS, INDIRECT EFFECTS.: /CREE W72-08579
OISOTOPES, CESIUM, RADIOACTIVITY EFFECTS, LABORATORY TESTS.: /RADI W72-09649
NT IDENTIFICATION, ENVIRONMENTAL EFFECTS, LABORATORY EQUIPMENT, FR W73-00379
T, *GROWTH RATES, *ENVIRONMENTAL EFFECTS, LARVAE, ON-SITE TESTS, I W73-00348
TS, WATER QUALITY, ENVIRONMENTAL EFFECTS, LIMNOLOGY, PLANNING, EUT W72-06526
EFFECTS, *STANDING CROPS, TIDAL EFFECTS, MARINE ALGAE, SEA WATER, W73-00367
YTA, MARINE ALGAE, ENVIRONMENTAL EFFECTS, MARINE PLANTS, CANADA, W W73-00428
LLUTION SOURCES, WATER POLLUTION EFFECTS, MERCURY, SEWAGE DISPOSAL W72-14782
A WATER, SAMPLING, ENVIRONMENTAL EFFECTS, MARINE FISH, PERIPHYTON, W73-00932
ANS, MONITORING, WATER POLLUTION EFFECTS, MOLLUSKS.: /GE, CRUSTACE W72-10955
BON RADIOISOTOPES, ENVIRONMENTAL EFFECTS, METHODOLOGY, ALGAE, AQUA W72-12709
NUCLEAR WASTES, WATER POLLUTION EFFECTS, MOLLUSKS, MARINE PLANTS, W72-04461
*WASTE HEAT, SUBLETHEL EFFECTS, MIXING ZONES, MODELING.: W70-09612
LLUTION SOURCES, WATER POLLUTION EFFECTS, MUD, MOLLUSKS, FISH, INS W70-08098
TS, *DISTRICT OF COLUMBIA, TIDAL EFFECTS, NUTRIENTS, EUTROPHICATIO W70-08248
TS, *DISTRICT OF COLUMBIA, TIDAL EFFECTS, NUTRIENTS, EUTROPHICATIO W70-08249
LLUTION CONTROL, WATER POLLUTION EFFECTS, NUTRIENTS, ALGAE, CYANOP W72-04769
UTION ABATEMENT, WATER POLLUTION EFFECTS, NUTRIENTS, PHOSPHORUS, W71-09429
OLISM, KINETICS, WATER POLLUTION EFFECTS, NUCLEAR WASTES.: / METAB W71-09168
RADIOISOTOPES, *WATER POLLUTION EFFECTS, NUCLEAR WASTES, RIVERS, W71-09225
PES, *ABSORPTION, *ENVIRONMENTAL EFFECTS, NUCLEAR WASTES, PATH OF W72-10955
ATIONS, FLORIDA, WATER POLLUTION EFFECTS, NITROGEN, PHOSPHORUS.: / W72-08986
, *WATER SUPPLY, WATER POLLUTION EFFECTS, NUTRIENTS, RESERVOIR OPE W71-13185
YGONUM, TYPHA, ELODE/ BIOLOGICAL EFFECTS, NOTTINGHAM(ENGLAND), POL W72-00155
NTS, RADIOECOLOGY, RADIOACTIVITY EFFECTS, NUCLEAR POWERPLANTS, MAR W73-00833
ICAL TECHNIQUES, WATER POLLUTION EFFECTS, NUTRIENT REQUIREMENTS.: / W73-01269
CATION, *LAKES, *WATER POLLUTION EFFECTS, OHIO, NUTRIENTS, WATER P W71-11949
E, PRODUCTIVITY, WATER POLLUTION EFFECTS, PHOSPHORUS, NITROGEN, AM W71-12857
ICIDE TOXICITY, *WATER POLLUTION EFFECTS, PESTICIDES, AGRICULTURAL W71-12303
ABOLISM, CULTURES, ENVIRONMENTAL EFFECTS, PLANT GROWTH, SULFATE, L W72-00153
ICAL TECHNIQUES, WATER POLLUTION EFFECTS, POLLUTANT IDENTIFICATION W71-13794
LATIVE CAPACITY, WATER POLLUTION EFFECTS, POLLUTANTS,: /STE ASSIMI W72-01685
ILS, IONS, NOSTOC, ENVIRONMENTAL EFFECTS, POLLUTANT IDENTIFICATION W73-00854
DES, GROWTH RATES, ENVIRONMENTAL EFFECTS, PHYTOPLANKTON, CHLAMYDOM W73-00281
IOLOGICAL ECOLOGY, ENVIRONMENTAL EFFECTS, PRODUCTIVITY, RADIOECOLO W72-06313
NTHIC FAUNA, FISH, ENVIRONMENTAL EFFECTS, RESERVOIR STORAGE, NUTRI W72-05282
EFFECTS, *PONDS, WATER POLLUTION EFFECTS, RADIOISOTOPES, COMPETITI W71-09256
RATES, NITROGEN, WATER POLLUTION EFFECTS, RADIOACTIVITY, METABOLIS W73-00380
NS, MARINE PLANTS, ENVIRONMENTAL EFFECTS, REEFS, LABORATORY TESTS, W73-00838
ATES, NUTRIENTS, WATER POLLUTION EFFECTS, RADIOACTIVITY TECHNIQUES W72-07143
WATER POLLUTION, WATER POLLUTION EFFECTS, RESEARCH AND DEVELOPMENT W72-06655
, *RADIOACTIVITY, *RADIOACTIVITY EFFECTS, RIVERS, EFFLUENT, RADIOE W72-07841
LLUTION SOURCES, WATER POLLUTION EFFECTS, RECREATION DEMAND, POPUL W72-01660
POLLUTION EFFECTS, RADIOACTIVITY EFFECTS, RADIOACTIVITY TECHNIQUES W72-03348
IUM, ABSORPTION, WATER POLLUTION EFFECTS, RADIOACTIVITY EFFECTS, R W72-03348
POLLUTION SOURCES, RADIOACTIVITY EFFECTS, RADIOACTIVITY TECHNIQUES W72-00940
CATION, *LAKES, *WATER POLLUTION EFFECTS, RECLAIMED WATER, SEWAGE W72-11702
ANTS, *TOXICITY, WATER POLLUTION EFFECTS, RESPIRATION, *HEAVY META W72-12576
PES, HEAVY METALS, AIR POLLUTION EFFECTS, SURFACE WATERS, PATH OF W72-00940
LLUTION SOURCES, WATER POLLUTION EFFECTS, SAMPLING, TEMPERATURE, C W71-11008
ANALYSIS, WATER ANALYSIS, TIDAL EFFECTS, STABILITY, SOLAR RADIATI W72-08141
LE ALGAE, LIGHT INTENSITY, TIDAL EFFECTS, SALINITY, HEATED WATER, W73-00853
CEANS, ANNELIDS, WATER POLLUTION EFFECTS, STANDING CROPS, MOLLUSKS W73-00935
, BENTHIC FAUNA, WATER POLLUTION EFFECTS, STRIPED BASS, DISSOLVED W73-02029
ERS, *VIRGINIA, *WATER POLLUTION EFFECTS, SURVEYS, SAMPLING, DATA W73-01972
ORGANIC MATTER, WATER POLLUTION EFFECTS, SEWAGE, NITROGEN, PHOSPH W73-00234
TIC ENVIRONMENT, WATER POLLUTION EFFECTS, STREAMS, RIVERS, ROOTED W72-14690
IES, *DELAWARE RIVER, VEGETATION EFFECTS, SLUDGE, RIVER BEDS, ALGA W70-09189
ON, CHLOROPHYLL, WATER POLLUTION EFFECTS, TEMPERATURE, CARBON, DIA W71-08670
TON, MARINE ALGAE, ENVIRONMENTAL EFFECTS, TRACE ELEMENTS, COLLOIDS W73-00818
RIENTS, SEDIMENTS, ENVIRONMENTAL EFFECTS, TRACE ELEMENTS, ANALYTIC W72-03215
LTH, *SHELLFISH, WATER POLLUTION EFFECTS, TOXICITY, WATER POLLUTIO W72-12257
OLLUTION EFFECTS, *RADIOACTIVITY EFFECTS, UNDERGROUND, ALASKA, PON W72-03347
RIO, *LAKE ERIE, WATER POLLUTION EFFECTS, WATER POLLUTION SOURCES, W72-01094
, AQUATIC ALGAE, WATER POLLUTION EFFECTS, WATER QUALITY, CHEMICAL W72-01085
ATH OF POLLUTANTS, AIR POLLUTION EFFECTS, WATER POLLUTION EFFECTS, W72-00959
TER TEMPERATURE, WATER POLLUTION EFFECTS, WATER POLLUTION CONTROL, W71-12092
IENTS, NITROGEN, WATER POLLUTION EFFECTS, WATER POLLUTION SOURCES. W71-11972

, *LETHAL LIMIT, WATER POLLUTION EFFECTS, WATER POLLUTION SOURCES, W72-11727
IS, CHEMCONTROL, WATER POLLUTION EFFECTS, WATER QUALITY, CARBON.: / W72-11724
UALITY CONTROL, *WATER POLLUTION EFFECTS, WATER POLLUTION SOURCES, W72-09646
DENITRIFICATION, WATER POLLUTION EFFECTS, WATER POLLUTION CONTROL. W72-12544
ULTURES, BIOASSAY, ENVIRONMENTAL EFFECTS, WATER POLLUTION EFFECTS, W72-13813
OVAL, NUTRIENTS, WATER POLLUTION EFFECTS, WATER QUALITY, HERBICIDE W72-14692
PY, PHOTOGRAPHY, WATER POLLUTION EFFECTS, WATER PURIFICATION, *SYS W73-00432
RIC POWERPLANTS, WATER POLLUTION EFFECTS, WATER POLLUTION SOURCES, W72-07791
MALS, NUTRIENTS, WATER POLLUTION EFFECTS, WATER QUALITY, SEASONAL, W72-07896
, HEAT TRANSFER, WATER POLLUTION EFFECTS, WATER TEMPERATURE, ALGAE W72-07907
ADATION(STREAM), WATER POLLUTION EFFECTS, WASTE WATER TREATMENT, C W71-07973
EWAGE EFFLUENTS, WATER POLLUTION EFFECTS, WATER POLLUTION SOURCES, W71-09173
YGEN, ESTUARIES, WATER POLLUTION EFFECTS, WATER POLLUTION SOURCES, W71-05390
E, SOUTH DAKOTA, WATER POLLUTION EFFECTS, WATER POLLUTION SOURCES, W71-02880
INDUSTRIAL WA/ *WATER POLLUTION EFFECTS, WATER POLLUTION SOURCES, W71-03095
ATE, HERBICIDES, WATER POLLUTION EFFECTS, WATER POLLUTION SOURCES, W72-04759
LLUTION SOURCES, WATER POLLUTION EFFECTS, WATER POL: /ON, WATER PO W72-04279
N, WATER SUPPLY, WATER POLLUTION EFFECTS, WATER POLLUTION SOURCES, W72-04280
ENT, *TOXICITY, *WATER POLLUTION EFFECTS, ZINC, LEAD, COPPER, CHRO W72-14694
LLUTION SOURCES, WATER POLLUTION EFFECTS,: /OTCSYNTHESIS, WATER PO W72-05952
IGHT QUALITY, LIGHT PENETRATION, EFFICIENCES, COMPETITION, GROWTH W72-04766
S, INGESTION RATES, ASSIMILATION EFFICIENCES, CRYPTOMONAS, ANKISTR W72-07940
URE, HYDROGEN ION CONCENTRATION, EFFICIENCIES, WATER QUALITY CONTR W72-04788
TION, FLOCCULATION, PERFORMANCE, EFFICIENCIES, *BIOMASS, ORGANIC M W72-12289
DEMAND, CHLORELLA, PERFORMANCE, EFFICIENCIES, LABORATORY TESTS, L W72-11100
ENTS, *EUTROPHICATION, *ECONOMIC EFFICIENCY, *PUBLIC HEALTH, WATER W72-04266
*WASTE HEAT, THERMAL EFFICIENCY, HEAT CONSUMPTION.: W70-08832
IENTS, RIVERS, INFLUENT STREAMS, EFFLUENT STREAMS, LIGHT INTENSITY W73-01435
*RADIOACTIVITY EFFECTS, RIVERS, EFFLUENT, RADIOECOLOGY, AQUATIC P W72-07841
ION, ALGAE, PATHOGENIC BACTERIA, EFFLUENT, IRRIGATION, STORAGE, WA W71-08214
T, NUTRIENTS, SYNTHESIS, SLUDGE, EFFLUENT, ALGAE, SECONDARY TREATM W71-12188
AL TREATMENT, *ACTIVATED SLUDGE, EFFLUENT, *ORGANIC WASTES, ECONOM W71-12223
ION, *NUTRIENTS, MICROORGANISMS, EFFLUENT, BIOCHEMICAL OXYGEN DEMA W71-13553
*BISCAYNE BAY(FLORIDA), HEATED EFFLUENT, TURKEY POINT(FLORIDA), W72-11876
RITERIA, *WASTE WATER TREATMENT, EFFLUENTS.: /METABOLISM, DESIGN C W71-13341
PULP WASTES, TEOTIARY TREATMENT, EFFLUENTS.: /, BENTONITE, LIME, * W72-03753
C HEALTH, EUROPE, RADIOISOTOPES, EFFLUENTS.: /, FOOD CHAINS, PUBLI W73-00828
TECHNIQUES, FOOD CHAINS, AFRICA, EFFLUENTS, PATH OF POLLUTANTS, NE W73-00832
WAGE, RESEARCH EQUIPMENT, SEWAGE EFFLUENTS, SEWAGE TREATMENT, SEWA W72-07715
OCHEMICAL OXYGEN DEMAND, *SEWAGE EFFLUENTS, *SEWAGE TREATMENT, CUL W72-06612
MIC FEASIBILITY, STREAMS, SEWAGE EFFLUENTS, ALGAE, TOURISM, RECREA W72-14878
EN, *ALGAE, PLANT GROWTH, SEWAGE EFFLUENTS, REGULATION, CYANOPHYTA W72-14790
WATER FISH, YELLOW PERCH, SEWAGE EFFLUENTS, OIL SPILLS, THERMAL PO W72-14282
LOGY, TERTIARY TREATMENT, SEWAGE EFFLUENTS, WATER REUSE.: /IS, ECO W72-03970
TS, NITROGEN, PHOSPHORUS, SEWAGE EFFLUENTS, SEWAGE TREATMENT, ALGA W72-05473
*ALGAE, *LAKES, *GEORGIA, SEWAGE EFFLUENTS, WATER POLLUTION EFFECT W71-09173
ESTIC WASTES, CHEMICAL ANALYSIS, EFFLUENTS, TRICKLING FILTERS, HYD W70-09907
N, WASTE WATER TREATMENT, ALGAE, EFFLUENTS, NUTRIENTS, SEWAGE TREA W70-09325
ROORGANISMS, TERTIARY TREATMENT, EFFLUENTS, MODEL STUDIES.: /, MIC W71-03021
, *AQUATIC PRODUCTIVITY,/ SEWAGE EFFLUENTS, *ALGAE, *PHYTOPLANKTON W71-02681
LLA, CHLAMYDOMONAS, SCENEDESMUS, EFFLUENTS, OXIDATION LAGOONS.: /E W70-10405
ALGAE, CULTURES, MICROORGANISMS, EFFLUENTS, SETTLING BASINS, WASTE W71-00116
AL, WASTE WATER DISPOSAL, SEWAGE EFFLUENTS, FARM WASTES.: / DISPOS W70-06509
ONOMIC JUSTIFICATION, */ *SEWAGE EFFLUENTS, *SEWAGE TREATMENT, *EC W72-00383
*EUTROPHICATION, *ALGAE, *SEWAGE EFFLUENTS, GEORGIA, THERMOCLINE, W71-11914
S, BIOASSAY, ALGAE, POWERPLANTS, EFFLUENTS, DISCHARGE(WATER), STAT W71-11583
R(M/ SETTLEABLE SOLIDS, HATCHERY EFFLUENTS, PARASITES, JORDAN RIVE W71-11006
IMARY PRODUCTIVITY, FERTILIZERS, EFFLUENTS, WATER POLLUTION SOURCE W71-11008
IRON, INDUSTRIAL WASTES, SEWAGE EFFLUENTS, POLLUTANT IDENTIFICATI W71-11515
LGAE, ORGANIC MATTER, LAKE ERIE, EFFLUENTS, SEWAGE, DETERGENTS, PL W71-11017
*WATER POLLUTION CONTROL, SEWAGE EFFLUENTS, AQUATIC ALGAE, LAKE ER W71-11507
R WASTES, ENVIRONMENTAL EFFECTS, EFFLUENTS, MONITORING, CHELATION. W72-10957
TIFICATION, NUCLEAR POWERPLANTS, EFFLUENTS, NUCLEAR WASTES, PATH O W72-10958
EFFECTS, RECLAIMED WATER, SEWAGE EFFLUENTS, RESEARCH AND DEVELOPME W72-11702
ONTROL, SEWAGE TREATMENT, SEWAGE EFFLUENTS, SANITARY ENGINEERING, W72-09378
EATMENT, A/ *PHOSPHORUS, *SEWAGE EFFLUENTS, *ALGAE, WASTE WATER TR W72-09176
TER POLLUTION CONTROL, NITRATES, EFFLUENTS, DETERGENTS, SODIUM CHL W72-09167
LUTANTS, CHEMICAL WASTES, SEWAGE EFFLUENTS, WATER POLLUTION EFFECT W72-08804
ASTES, *WATER POLLUTION EFFECTS, EFFLUENTS, CALIFORNIA, SEDIMENTS, W72-09092
TES, *AMMONIUM COMPOUNDS, SEWAGE EFFLUENTS, TEMPERATURE CONTROL, A W72-10162
ICATION, *NUTRIENTS, LAKE HURON, EFFLUENTS, WATER QUALITY, PHOSPHO W72-13662
ORTH CAROLINA, NUTRIENTS, SEWAGE EFFLUENTS, EUTNOPHICATION, ALGAE, W72-13636
*EUTROPHICATION, *LAKES, SEWAGE EFFLUENTS, SEEPAGE, SEPTIC TANKS, W72-12955
RI/ *ZOOPLANKTON, *LARVAE, *FISH EGGS, *SEPARATION TECHNIQUES, *MA W73-00371
TOPLANKTON, ZOOPLANKTON, LARVAE, EGGS, CHINOOK SALMON, STRIPED BAS W71-11517
ASINS), SMALL WATERSHEDS, CROPS, EGGS, FISHERIES, ON-SITE INVESTIG W72-00959
PODS, MARINE ALGAE, SHRIMP, FISH EGGS, FISH, BACTERIA, FISH PARASI W72-08142
OD, FLOUNDER, CRAB EGGS, PELAGIC EGGS, GADUS MORHUA, GRASS SHRIMP, W72-08142
RUS CRISPUS, COD, FLOUNDER, CRAB EGGS, PELAGIC EGGS, GADUS MORHUA, W72-08142
ILIZERS, FARM WASTES, FISH, FISH EGGS, TURBIDITY, NUTRIENTS, DISSO W71-06444
C SOLIDS.: *ALEXANDRIA(EGYPT), PANDORINA, SOLUBLE ORGANI W70-06619
TES, MEDITERRANEAN SE/ *CERATIUM EGYPTIACUM, *SUEZ CANAL, FLAGELLA W73-00296
EERSIA ORYZOIDES, NUPHAR ADVENA, EICHHORNIA CRASSIPES, JUSTICIA AM W73-01089
TUNDIFOL/ *CHIRONOMIDS, SAWDUST, EICHHORNIA CRASSIPES, SALVINIA RO W71-01488
TATA, IPOMOE/ *HYDERABAD(INDIA), EICHHORNIA CRASSIPES, TYPHA ANGUS W72-01363
GLYTOTENDIPES BARBIPES, EKMAN DREDGE.: W72-03556
ND RESERVOIR(NJ), AMERICAN SHAD, EKMAN DREDGE, PUMPKINSEED, BLACK W72-14280
TESTINALIS, HECATONEMA MACULANS, ELACHISTA LUBRICA, RALFSIA VERRUC W72-01370
ES, CRYPTOMONAS, ANKISTRODESMUS, ELAKOTOTHRIX, OSCILLATORIA, ANABA W72-07940
L TEST/ AGRO-INDUSTRIAL COMPLEX, ELECTRIC ENERGY CONSUMPTION, MODE W72-03327

, *AQUATIC ENVIRONMENT, FLORIDA, ELECTRIC GENERATORS, WATER TEMPER W72-11252
TED WATER, NUCLEAR POWER PLANTS, ELECTRICAL POWER PLANTS, HEAT TRA W72-07907
RPHOLOGY, EQUATIONS, RESISTANCE, ELECTRICAL PROPERTIES.: /PLANT MO W73-01630
ION CONCENTRATION, TEMPERATURE, ELECTRICAL CONDUCTANCE, DISSOLVED W71-01489
LECTRODES, EDTA, *POTENTIOMETRY, ELECTROANALYSIS, *ALGAL CULTURES, W73-02112
CTIVITY, METABOLISM, ELECTRODES, ELECTROCHEMISTRY, PHOTOSYNTHESIS, W72-14314
ERTIARY TREATMENT, ION EXCHANGE, ELECTROCHEMISTRY, REVERSE OSMOSIS W70-09907
UTANT IDENTIFICATION, *NITROGEN, ELECTROCHEMISTRY, TRACE ELEMENTS, W72-01693
ETRY, RADIOGRAPHY, ION SELECTIVE ELECTRODES, LASERS, ORTHOPHOSPHAT W72-08790
TENTIOMETRY, INTE/ ION-SELECTIVE ELECTRODES, STANDARD ADDITION, PO W71-00474
QUATIC PRODUCTIVITY, METABOLISM, ELECTRODES, ELECTROCHEMISTRY, PHO W72-14314
ELECTROANALYSIS/ *ION-SELECTIVE ELECTRODES, EDTA, *POTENTIOMETRY, W73-02112
ECTROCHEMISTRY, REVERSE OSMOSIS, ELECTRODIALYSIS, DISTILLATION, NI W70-09907
ITY, PILOT PLANTS, ION EXCHANGE, ELECTROLYSIS, CHEMICAL PRECIPITAT W71-09450
LATILITY, FREEZE DRYING, X-RAYS, ELECTROLYSIS, COMPUTER PROGRAMS, W71-11036
OCCULATION, *FILTRATION, *ALGAE, ELECTROLYTES, ACTIVATED CARBON, H W72-06201
LINDANE, P P'-DDE, CLOPHEN A 50, ELECTRON CAPTURE GAS CHROMATOGRAP W72-04134
TION, GAS LIQUID CHROMATOGRAPHY, ELECTRON CAPTURE DETECTOR, PRECON W73-02105
LECTRON MICROSCOPY, TRANSMISSION ELECTRON MICROSCOPY, LIGHT MICROS W73-00432
GAN, CHRYSOPHYTA, AQUATIC ALGAE, ELECTRON MICROSCOPY, PHOTOGRAPHY, W73-00432
S, *MELOSIRA BINDERANA, SCANNING ELECTRON MICROSCOPY, TRANSMISSION W73-00432
S, VITAMINS, ANAEROBIC BACTERIA, ELECTRON MICROSCOPY, PIGMENTS, CA W72-07951
CAROLINA, SAMPLING, MICROSCOPY, ELECTRON MICROSCOPY, PYRROPHYTA, W72-07663
*CABLE THEORY, ELECTRONIC POTENTIAL, CHARACEAE.: W73-01630
, ANIONS, CHLORELLA, FILTRATION, ELECTROPHORESIS, TRACERS.: /TIONS W72-14840
ERIOPHAGE, *VIRUSES, ABSORPTION, ELECTROPHORESIS, DETERGENTS, AMIN W72-06124
, *S/ *FLOCCULATION, FILTRATION, ELECTROPHORESIS, SETTLING, *ALGAE W71-01899
PENTRATION, PHYSICAL PROPERTIES, ELECTROPHORESIS, CHEMICAL PROPERT W72-11833
BRATES.: ELEMENTAL ANALYZERS, MACROINVERTE W72-07715
*TOXICITY, *FLUORESCENCE, TRACE ELEMENTS, COPPER, CHLORELLA, CULT W72-07660
DIOISOTOPES, MARINE ALGAE, TRACE ELEMENTS, PATH OF POLLUTANTS, ABS W73-00811
ALGAE, SOIL CONTAMINATION, TRACE ELEMENTS, ION EXCHANGE, POLLUTANT W72-14303
ICATION, GROWTH RATES, DEFICIENT ELEMENTS, WATER POLLUTION EFFECTS W73-01445
DUCTIVITY, *HEAVY METALS, *TRACE ELEMENTS, NUTRIENTS, MERCURY, CHR W73-01434
UTRIENTS, COLORIMETRY, DEFICIENT ELEMENTS, AQUATIC ALGAE, BICINDI: W73-00916
EUROPE, ANIMAL PHYSIOLOGY, TRACE ELEMENTS, HEAVY METALS, PHYSICOCH W73-00817
ON ANALYSIS, SPECTROSCOPY, TRACE ELEMENTS, SAMPLING, WATER ANALYSI W73-00832
AE, ENVIRONMENTAL EFFECTS, TRACE ELEMENTS, COLLOIDS, NUCLEAR WASTE W73-00818
NIQUES, CYCLING NUTRIENTS, TRACE ELEMENTS, PUBLIC HEALTH, RADIOECO W73-00826
NIQUES, CYCLING NUTRIENTS, TRACE ELEMENTS, CHELATION, ORGANIC COMP W73-00821
PTION, EUTROPHICATION, DEFICIENT ELEMENTS, ESSENTIAL NUTRIENTS, *A W72-11617
TRON ACTIVATION ANALYSIS, *TRACE ELEMENTS, COPPER, MOLYBDENUM, MAN W71-11515
S, SILTS, PLANKTON, ALGAE, TRACE ELEMENTS, VOLATILITY, FREEZE DRYI W71-11036
TOXICITY, EUTROPHICATION, TRACE ELEMENTS, NITROGEN, PHOSPHORUS, M W71-11793
TS, ENVIRONMENTAL EFFECTS, TRACE ELEMENTS, ANALYTICAL TECHNIQUES, W72-03215
ITROGEN, ELECTROCHEMISTRY, TRACE ELEMENTS, HEAVY METALS, ELECTRODE W72-01693
AVY METALS/ *FOOD CHAINS, *TRACE ELEMENTS, *LEAD RADIOISOTOPES, HE W72-00940
AQUATIC PLANTS, SEA WATER, TRACE ELEMENTS, TOXICITY, CHROMOSOMES, W72-05952
RESEARCH AND DEVELOPMENT, TRACE ELEMENTS, ALGAE, NUTRIENTS, LIGHT W72-04433
MATOGRAPHY, HEAVY METALS, *TRACE ELEMENTS, RADIOCHEMICAL ANALYSIS, W72-04708
RADIOISOTOPES, SEDIMENTS, TRACE ELEMENTS, VOLCANOES, BENTHIC FLOR W71-09190
GAE, PRIMARY PRODUCTIVITY, TRACE ELEMENTS, PHOSPHATES, NITRATES, L W71-09674
HUMIC ACIDS, FULVIC ACIDS, TRACE ELEMENTS, CHEMICAL ANALYSIS, ION W71-09188
TROPHY, CHLAMYDOMONAS, DEFICIENT ELEMENTS, MESOTROPHY, EUTROPHICAT W72-13816
TH OF POLLUTANTS, *METALS, TRACE ELEMENTS, PESTICIDES, SALINITY, N W72-12190
ITRATES, NUTRIENTS, SALTS, TRACE ELEMENTS, LIGHT INTENSITY, SCENED W72-12739
STICHUM, POTAMOGETON PECTINATUS, ELEOCHARIS MONTEVIDENSES, CHARA A W73-01673
*PHOSPHORUS ELIMINATION.: W73-01367
CLAY RIVER, CHRISTINA RIVER, BIG ELK RIVER, LITTLE ELK RIVER, OCTO W72-04736
INA RIVER, BIG ELK RIVER, LITTLE ELK RIVER, OCTORARA RIVER, PLECTO W72-04736
VAR ISOTHRIX, LAKE ITASCA, LAKE ELK, LAKE MARY, LAKE JOSEPHINE, L W72-07145
ORIACEAE, SCHIZOTHRIX CALCICOLA, ELKTON POND, RISING SUN POND, IND W72-04736
YRENOIDOSA, SPIROGIRA, CHLORELLA ELLIPSOIDEA, RNA, CHLOROPHYLL B, W72-06274
*CHLORELLA ELLIPSOIDEA, CULTURE MEDIA.: W73-01626
IA PUNCTATA, VISCHERIA STELLATA, ELLIPSOIDION ACUMINATUM.: /ISCHER W72-14808
POLYSTYRENE SULFONATE, CHLORELLA ELLIPSOIDIA.: /OLYETHYLENEMINE, * W72-14840
E, POLYETHYLENEIMINE, *CHLORELLA ELLIPSOIDIA.: /LYSTYRENE SULFONAT W72-11833
CTINATUS, MYRIOPHYLLUM SPECATUM, ELODEA CANADENSIS, PHRAGMITES, FO W71-11012
GHAM(ENGLAND), POLYGONUM, TYPHA, ELODEA.: /LOGICAL EFFECTS, NOTTIN W72-00155
AY OF QUINTE, SALT CREEK, CHARA, ELODEA, CALLITRICHE, ANABAENA FLO W73-00840
IUM ADVENUM, ELPHIDIUM CLAVATUM, ELPHIDIUM GALVESTONENSE, E: /PHID W73-00367
MONIA BECCARII, AMMOTIUM SALSUM, ELPHIDIUM ADVENUM, ELPHIDIUM CLAV W73-00367
OTIUM SALSUM, ELPHIDIUM ADVENUM, ELPHIDIUM CLAVATUM, ELPHIDIUM GAL W73-00367
S, HEAT STERILIZATION, FIXATION, EMBEDDING, ASHING, BONE, ORGANS, W73-01676
ANTS, MARINE ALGAE, CHROMOSOMES, EMBRYONIC GROWTH STAGE, CRUSTACEA W72-10955
UTTONI, VENUS, TIVELA STULTORUM, EMERITA ANALOGA, DIASTYLOPSIS, OX W72-09092
OCOCCUS CONSTRICTUS, NAUTOCOCCUS EMERSUS,: /OCOCCUS CAUDATUS, NAUT W72-07141
HANOPYXIS COSTATA, ASSIMILATION, EMILIANA(COCCOLITHUS) HUXLEYI, AM W73-00380
UM, *SAMPLE PRESERVATION, *FLAME EMISSION SPECTROPHOTOMETRY, PERCH W71-11529
CHLORELLA MARINA, RHODOMELA SP., ENDOCLADIA SP., PHYLLOSPADEX SP., W71-11851
FUSARIUM NO 83, BACTERIUM NO 52, ENDOGENOUS RESPIRATION, TOTAL OXY W71-12223
URE EXPERIMENTS, ATRAZINE, MCPA, ENDOSULFAN, FAIRY LIQUID, BLUSYL, W72-12576
ITY, DISSOLVED OXYGEN, CULTRUES, ENDRIN, DIELDRIN, OXYGENATION, CA W71-10096
TANS, MARINIAMONAS SAUPAULENSIS, ENEIDAMONAS APPLANATA, BICUDOMONA W72-09119
LAKE HERMAN(S. DAK), ENEMY SWIM LAKE(S. DAK).: W71-02880
TH RATES, *PRIMARY PRODUCTIVITY, ENERGY BUDGET, *MARINE ALGAE, MAR W73-00366
FOOD WEBS, DETRITUS, ECOSYSTEMS, ENERGY BUDGETS, NUTRIENTS, REVIEW W73-01089
GRO-INDUSTRIAL COMPLEX, ELECTRIC ENERGY CONSUMPTION, MODEL TESTING W72-03327
USTRIAL WASTES, ANNUAL TURNOVER, ENERGY, CYANOPHYTA, ALGAE, EUTROP W72-03228
FFECTS, WATER POLLUTION CONTROL, ENERGY, DIFFUSION, THERMODYNAMICS W71-12092

NIA, ANTIBIOTICS, IODINE, ATOMIC ENERGY, REACTOR COOLING, IMPERIAL W71-09157
, ALGAE, EUTROPHICATION, SUMMER, ENERGY, SEASONAL.: /Y, CYANOPHYTA W72-03228
DES, INDUSTRIAL WASTES, VIRUSES, ENERIC BACTERIA, TRITIUM, RADIOAC W72-04742
ALGAE, BACTERIA, MARINE PLANTS, ENGINEERING, WATER PROPERTIES, PH W72-03220
TIC ALGAE, PLANTS, ENVIRONMENTAL ENGINEERING, COOLING TOWER: /AQUA W72-00939
T, EUTROPHICATION, ENVIRONMENTAL ENGINEERING, WATER TREATMENT, WAT W72-01660
MENT, SEWAGE EFFLUENTS, SANITARY ENGINEERING, SECONDARY TREATMENT, W72-09378
RONMENTAL EFFECTS, ENVIRONMENTAL ENGINEERING, *VIRGINIA, PROJECTIO W72-14405
UM CARBONATE, PROTEINS, FILTERS, ENGINEERS ESTIMATES, ECONOMICS, F W70-10174
LELHAM TARN(ENGLAND), WINDERMERE(ENGLAND).: /TERIONELLA FORMOSA, B W72-10625
LEVEN(SCOTLAND), LAKE WINDERMERE(ENGLAND).: / LUXURY UPTAKE, LOCH W71-13183
THAMES RIVER(ENGLAND), LEE RIVER(ENGLAND).: /, RESEARCH STRATEGY, W71-13172
RESEARCH STRATEGY, THAMES RIVER(ENGLAND), LEE RIVER(ENGLAND).: /, W71-13172
*ULVA, *LAMINARIA, LIVERPOOL BAY(ENGLAND), IRISH SEA.: W71-12867
*UNITED KINGDOM, ESTHWAITE WATER(ENGLAND), BLELHAM TARN(ENGLAND), W71-13183
WATER(ENGLAND), LAKE WINDERMERE(ENGLAND), GREAT OUSE(ENGLAND), AL W71-13185
AE, *EPIPELIC ALGAE, SHEAR WATER(ENGLAND), AMPHORA OVALIS, OPEPHOR W71-12855
WINDERMERE(ENGLAND), GREAT OUSE(ENGLAND), ALGAL GROWTH.: /), LAKE W71-13185
*UNITED KINGDOM, ESTHWAITE WATER(ENGLAND), LAKE WINDERMERE(ENGLAND W71-13185
ITE WATER(ENGLAND), BLELHAM TARN(ENGLAND), LUXURY UPTAKE, LOCH LEV W71-13183
TERIONELLA FORMOSA, BLELHAM TARN(ENGLAND), WINDERMERE(ENGLAND).: / W72-10625
), EXOENZYME PR/ ESTHWAITE WATER(ENGLAND), LAKE WINDERMERE(ENGLAND W72-03224
WATER(ENGLAND), LAKE WINDERMERE(ENGLAND), EXOENZYME PRODUCERS, CH W72-03224
BIOLOGICAL EFFECTS, NOTTINGHAM(ENGLAND), POLYGONUM, TYPHA, ELODE W72-00155
ER THAMES(ENGLAND), RIVER KENNET(ENGLAND), STEPHANODISCUS HANTZSCH W72-14797
*CENTRIC DIATOMS, *RIVER THAMES(ENGLAND), RIVER KENNET(ENGLAND), W72-14797
UE-GREEN ALGAE.: *WINDERMERE(ENGLAND), HETEROCYSTS, BENTHIC BL W73-02472
ATOMS, CARBON RADIOISOTOPES, NEW ENGLAND, AQUATIC LIFE, AQUATIC MI W72-14317
CAL MAGNIFICATION, FATE OF POLL/ ENGLAND, BIOACCUMULATION, BIOLOGI W72-11925
STS, MYXOPHYCEA/ LYTIC BACTERIA, ENGLAND, METABOLIC INHIBITION, HO W72-05457
A SPP, DISTICHLIS SPICATA, FIMB/ ENGLAND, SWEDEN, GERMANY, SPARTIN W73-01089
STE WATER TREATMENT, CYANOPHYTE, ENGLENOPHYTA, ORGANIC LOADING.: / W71-07973
S MEXICANUS, EUPHAUSIA PACIFICA, ENGRAULIS MORDAX.: /N, TRIPHOTURU W73-02105
OPEUS, EUPENTACTA, CNEMIDOCARPA, ENHYDRA LUTRIS NEREIS, SEA OTTER, W73-00932
ANDS.: PHILIPPINES, ENIWETOK, MICRONESIA, FANNING ISL W71-11486
GROWTH KINETICS, BIOENERGETICS, ENRICHMENT, CULTURE MEDIA, MACROI W72-09378
URO ATOLL, M/ PLEOPODS, UROPODS, ENRICHMENT, LEUCOTHRIX MUCOR, MAJ W72-08142
HOMICROBIUM, VIBRIO ANGUILLARUM, ENRICHMENT CULTURES, AGARS, CHEMO W72-05421
REDUCTION MEASUREMENTS, NUTRIENT ENRICHMENT TESTS.: /S, ACETYLENE W72-13692
AS EROSA, ERKINIA SUBEQUICILIAT/ ENRICHMENT, CHROMULINA, CRYPTOMON W72-13816
TOXINS, ANTIBIOTICS(PESTICIDES), ENTERIC BACTERIA, DISEASES, E. CO W72-13800
ALGAE, CHLOROPHYTA, PSEUDOMONAS, ENTERIC BACTERIA.: /UATIC ALGAE, W72-12255
ACTERIA, TEMPERATURE, COLIFORMS, ENTERIC BACTERIA, NITROGEN, VITAM W72-12565
OPERTIES, *ALGAE, *HYDROBIOLOGY, ENTERIC BACTERIA, SAMPLING, ZOOPL W72-07896
OBLIQUUS, ALCALIGENES FAECALIS, ENTEROBACTER AEROGENES, PROTEUS V W72-07664
NES, CHLORAMPHENICOL, SALMONELLA ENTEROCOLITIS, SHIGELLA DYSENTERI W72-13800
EUNOTIA PECTINALIS, LITHOPHYTES, ENTEROMORPHA, INSOLATION, TEXTURE W72-08141
ATION, SUBLITTORAL, LONG ISLAND, ENTEROMORPHA INTESTINALIS, POLYSI W73-00367
ROGEN, POLLUTANT IDENTIFICATION, ENVIRONMENTAL EFFECTS, LABORATORY W73-00379
CARBON PESTICIDES, GROWTH RATES, ENVIRONMENTAL EFFECTS, PHYTOPLANK W73-00281
TIC LIFE, ENVIRONMENTAL EFFECTS, ENVIRONMENTAL ENGINEERING, *VIRGI W72-14405
COMPUTERS, ALGAE, AQUATIC LIFE, ENVIRONMENTAL EFFECTS, ENVIRONMEN W72-14405
PHOTOACTIVATION, AQUATIC ALGAE, ENVIRONMENTAL EFFECTS.: /ICIDES), W72-14327
ATERS, *AQUATIC PLANTS, *AQUATIC ENVIRONMENT, WATER POLLUTION EFFE W72-14690
Y, *SHOALS, *BENTHIC/ *ESTUARINE ENVIRONMENT, *PRIMARY PRODUCTIVIT W72-14317
*HEAVY METALS, *ALGAE, *AQUATIC ENVIRONMENT, *TOXICITY, *WATER PO W72-14694
GIONS, PHAEOPHYTA, MARINE ALGAE, ENVIRONMENTAL EFFECTS, MARINE PLA W73-00428
*LEGISLATION(PROPOSED), ENVIRONMENTAL PROTECTION AGENCY.: W73-00703
OPES, STABLE ISOTOPES, ESTUARINE ENVIRONMENT, COASTS, ANIMAL PHYSI W73-00811
INATED BIPHENYLS, PUBLIC HEALTH, ENVIRONMENT CONTROL, LEGAL ASPECT W73-00703
SOPHYTA, SYSTEMATICS, *ESTUARINE ENVIRONMENT, SEA WATER, SALINITY, W72-08141
RIBUTION, *SPATIAL DISTRIBUTION, ENVIRONMENT, SAMPLING, CHEMICAL A W72-08560
NTS, NITRIF/ *NITROGEN, *AQUATIC ENVIRONMENT, PRODUCTIVITY, SEDIME W72-08459
*DIATOMS, *RESEARCH EQUIPMENT, ENVIRONMENTAL EFFECTS, ESTUARIES, W72-08141
S, PATH OF POLLUTANTS, ESTUARINE ENVIRONMENT, RADIOACTIVITY TECHNI W72-07826
ITY CONTROL, PHOSPHORUS, AQUATIC ENVIRONMENT, WATER POLLUTION EFFE W72-06638
IES, DOMINANT ORGANISMS, AQUATIC ENVIRONMENT, AQUATIC ALGAE, AQUAT W72-07132
UBA DIVING, SEA WATER, SAMPLING, ENVIRONMENTAL EFFECTS, MARINE FIS W73-00932
TROGEN, WATER POLLUTION SOURCES, ENVIRONMENTAL EFFECTS, AEROBIC CO W73-01066
HYTOPLANKTON, NERITIC, ESTUARINE ENVIRONMENT, SEA WATER, DIATOMS, W73-00935
N, HYDROGRAPHY, PLANKTON, ALGAE, ENVIRONMENTAL EFFECTS, BIOLOGICAL W73-01704
PRODUCTIVITY, *ECOLOGY, *AQUATIC ENVIRONMENT, *BIOLOGICAL COMMUNIT W73-02095
IOLOGICAL COMMUNITIES, ESTUARINE ENVIRONMENT, WATER POLLUTION SOUR W73-00853
ASSIMILATIVE CAPACITY, ESTUARINE ENVIRONMENT, FOOD CHAINS, PUBLIC W73-00828
A, ABSORPTION, SOUNDS, ESTUARINE ENVIRONMENT, PATH OF POLLUTANTS, W73-00818
, PLEISTOCENE EPOCH, ADAPTATION, ENVIRONMENTAL EFFECTS, CHLOROPHYT W73-00855
NS, PHYTOPLANKTON, MARINE ALGAE, ENVIRONMENTAL EFFECTS, TRACE ELEM W73-00818
EDIMENTS, PUERTO RICO, ESTUARINE ENVIRONMENT, PATH OF POLLUTANTS, W73-00821
DATA COLLECTIONS, MARINE PLANTS, ENVIRONMENTAL EFFECTS, REEFS, LAB W73-00838
YTA, FROZEN SOILS, IONS, NOSTOC, ENVIRONMENTAL EFFECTS, POLLUTANT W73-00854
ECOLOGY, MARINE ALGAE, ESTUARINE ENVIRONMENT, ABSORPTION, ANIMAL P W73-00826
TIVITY TECHNIQUES, RADIOECOLOGY, ENVIRONMENTAL EFFECTS, ABSORPTION W73-00817
OCHEMICAL AN/ *SURVEYS, *AQUATIC ENVIRONMENT, *RADIOECOLOGY, *RADI W73-01673
EVEL, PROTOZOANS, AQUATIC ALGAE, ENVIRONMENTAL EFFECTS, CHLOROPHYT W72-13816
METHODOLOGY, CULTURES, BIOASSAY, ENVIRONMENTAL EFFECTS, WATER POLL W72-13813
ISOTOPES, *CARBON RADIOISOTOPES, ENVIRONMENTAL EFFECTS, METHODOLOG W72-12709
STEMS, *PRODUCTIVITY, *ESTUARINE ENVIRONMENT, *DOMESTIC WASTES, *S W72-12567
BITORS, WATER POLLUTION EFFECTS, ENVIRONMENTAL EFFECTS, COMPETITIO W72-12736
WATER POLLUTION EFFECTS, AQUATIC ENVIRONMENT, TURBIDITY, WATER POL W72-11925
BOW TROUT, AQUATIC LIFE, AQUATIC ENVIRONMENT, FISH PHYSIOLOGY, ALG W72-12257

NITY, CURRENTS(WATER), MOLLUSKS, ENVIRONMENTAL EFFECTS.: /AL, SALI W72-11876
ESIDUES, *AQUATIC LIFE, *AQUATIC ENVIRONMENT, *WATER POLLUTION, *P W72-13347
, MATHEMATICAL MODELS, ESTUARINE ENVIRONMENT, SALINITY, CHEMICAL A W72-12941
*DD/ *PESTICIDE RESIDUES, *LOTIC ENVIRONMENT, WATERSHEDS(BASINS), W72-12930
, BACTERIA, BENTHIC FAUNA, FISH, ENVIRONMENTAL EFFECTS, RESERVOIR W72-05282
HAINES, PUBLIC HEALTH, ESTUARINE ENVIRONMENT, ZINC RADIOISOTOPES, W72-04463
LLUTANTS, PHYSIOLOGICAL ECOLOGY, ENVIRONMENTAL EFFECTS, PRODUCTIVI W72-06313
*AQUATIC PLANTS, WATER QUALITY, ENVIRONMENTAL EFFECTS, LIMNOLOGY, W72-06526
ALGAE, PUBLIC HEALTH, ESTUARINE ENVIRONMENT, ON-SITE INVESTIGATIO W72-04461
AQUATIC HABITATS, LAKES, AQUATIC ENVIRONMENT.: /LGAE, *LAKE ERIE, W71-09156
EEPAGE, WATER POLLUTION EFFECTS, ENVIRONMENTAL EFFECTS, *BENTHIC F W71-08793
ALGAE, *EUTROPHICATION, AQUATIC ENVIRONMENTS, GREAT LAKES, BAYS, W71-08953
IDES, ALGAE, DREDGING, ESTUARINE ENVIRONMENT, CALIFORNIA WATER CON W71-09581
ATMENT, WATER POLLUTION SOURCES, ENVIRONMENTAL SANITATION, SECONDA W71-09674
ALGAE, AQUATIC HABITATS, AQUATIC ENVIRONMENT.: /OLLUTION EFFECTS, W71-07337
PHOSPHORUS, *CYANOPHYTA, AQUATIC ENVIRONMENT, CARBON DIOXIDE, ECOS W71-00475
*EUTROPHICATION, *LAKES, AQUATIC ENVIRONMENT, CHEMICAL PROPERTIES, W71-00117
ATMENT, WATER POLLUTION EFFECTS, ENVIRONMENTAL EFFECTS, ALGAL POIS W70-09429
PHOSPHORUS, *POPULATION, AQUATIC ENVIRONMENT, ECOSYSTEMS, GEORGIA, W72-09159
TER POLLUTION EFFECTS, ESTUARINE ENVIRONMENT.: /Y PRODUCTIVITY, WA W72-09653
NEDESMUS, ALGAE, MICROORGANISMS, ENVIRONMENTAL EFFECTS, GROWTH RAT W72-10861
NTIFICATION, SAMPLING, BIOASSAY, ENVIRONMENT, TOXICITY, POISONS, A W72-10678
LANKTON, *NEW HAMPSHIRE, MIXING, ENVIRONMENTAL EFFECTS.: /, PHYTOP W72-09061
ONTROL, ALGAL POISONING, AQUATIC ENVIRONMENT, AQUATIC FUNGI, ACQUI W72-11186
RINE ALGAE, SEDIMENTS, ESTUARINE ENVIRONMENT, MONITORING, CRABS, S W72-10958
OMMUNITIES, *SUCCESSION, AQUATIC ENVIRONMENT, *ECOLOGICAL DISTRIBU W72-11854
WASTES, RUNOFF, WATER SOFTENING, ENVIRONMENTAL EFFECTS, AQUATIC AL W72-11186
TS, *THERMAL POLLUTION, *AQUATIC ENVIRONMENT, FLORIDA, ELECTRIC GE W72-11252
ICAL STUDIES, EQUATIONS, AQUATIC ENVIRONMENT, WATER PROPERTIES, AL W72-11724
LOGICAL DISTRIBUTION, *ESTUARINE ENVIRONMENT, HABITATS, AQUATIC LI W72-11854
IXING BACTERIA, *NITROGEN CYCLE, ENVIRONMENTAL EFFECTS, ANAEROBIC W72-11563
TS, FOOD CHAINS, NUCLEAR WASTES, ENVIRONMENTAL EFFECTS, EFFLUENTS, W72-10957
CHARGE(WATER), AQUATIC HABITATS, ENVIRONMENTAL EFFECTS, HEATED WAT W71-11517
ONCENTRATION, NUTRIENTS, AQUATIC ENVIRONMENT, CARBON DIOXIDE.: / C W71-12068
CATION, FISH, ALGAE, RECREATION, ENVIRONMENTAL EFFECTS, INVERTEBRA W71-10577
NHIBITION, METABOLISM, CULTURES, ENVIRONMENTAL EFFECTS, PLANT GROW W72-00153
TH RATES, TEM/ *ALGAE, *ECOLOGY, ENVIRONMENT, PHOTOPERIODISM, GROW W72-00377
, ROOT SYSTEMS, SOIL FUNGI, SOIL ENVIRONMENT, SOIL TEMPERATURE, SO W72-03342
BACTERIA, NUTRIENTS, SEDIMENTS, ENVIRONMENTAL EFFECTS, TRACE ELEM W72-03215
ROPHICATION, GOVERNMENT FINANCE, ENVIRONMENTAL SANITATION, ALGAE, W72-01659
PULATION, COMMUNITY DEVELOPMENT, ENVIRONMENTAL EFFECTS, GOVERNMENT W72-01660
ERAL GOVERNMENT, EUTROPHICATION, ENVIRONMENTAL ENGINEERING, WATER W72-01660
T, ALGAE, PLANTS, TROUT, AQUATIC ENVIRONMENT, EUTROPHICATION, NITR W72-02412
UATIC BACTERIA, SEWAGE BACTERIA, ENVIRONMENTAL EFFECTS, BIOLOGICAL W72-01472
ENVIRONMENTAL EFFECTS, ESTUARINE ENVIRONMENT, REGULATION, FALLOUT, W72-00959
EWS, *BOTTOM SEDIMENTS, *AQUATIC ENVIRONMENT, SURFACE WATERS, NUTR W72-00847
SHERIES, ON-SITE INVESTIGATIONS, ENVIRONMENTAL EFFECTS, ESTUARINE W72-00959
MONITORING, POLLUTION ABATEMENT, ENVIRONMENTAL EFFECTS, AQUATIC LI W72-00939
NUCLIDE PATHWAY, ABALONE.: ENVIRONMENTAL STATEMENT, CRITICAL W72-00959
RIPHYTON, AQUATIC ALGAE, PLANTS, ENVIRONMENTAL ENGINEERING, COOLIN W72-00939
ENTS, NTA.: ENZYME PRESOAKS, DISHWATER DETERG W71-08768
PHOSPHATE, *LUCIFERIN-LUCIFERASE ENZYME SYSTEM, PLANT EXTRACTS, CH W71-11916
CYTOLOGICAL STUDIES, NUTRIENTS, ENZYMES.: /AL ANALYSIS, BIOASSAY, W72-09163
H OF POLLUTANTS, *PUBLIC HEALTH, ENZYMES, ANALYTICAL TECHNIQUES, P W72-13347
BACTERIA, INVERTEBRATES, FUNGI, ENZYMES, ANIMAL PARASITES, VIRUSE W73-01676
YTA, *ALGAE, *NITROGEN FIXATION, ENZYMES, BACTERIA, SPORES, NITROG W71-12070
OXICITY, TASTE, ODOR, CORROSION, ENZYMES, CHLEATION, ALGAE, WATER W72-06837
OSCILLATORIA, SPIRULI/ *PROTEASE ENZYMES, CHLOROGONIUM, OOCYSTIS, W72-07661
HIBITION, ANALYTICAL TECHNIQUES, ENZYMES, COLOR, WATER QUALITY CON W72-04070
ION CONCENTRATION, TEMPERATURE, ENZYMES, CELLULOSE, VIRUSES, HUMA W72-05457
NT, EUTROPHICATION, SURFACTANTS, ENZYMES, LAKE ONTARIO, SEASONAL, W71-00116
*CHLORELLA, *BIOASSAYS, ENZYMES, LIGHT INTENSITY, ALGAE.: W71-11916
*ALGAE, PHYLOGENY, CARBON CYCLE, ENZYMES, METABOLISM, HYDROGEN, EU W71-09172
TS, PHOTOSYNTHESIS, CHLOROPHYLL, ENZYMES, PERMEABILITY, WATER QUAL W72-06037
COLI, BENTHIC FAUNA, CELLULOSE, ENZYMES, PULP WASTES, RADIOECOLOG W72-09668
HEMISTRY, *NITRATES, *CHLORELLA, ENZYMES, PROTEINS, LABORATORY TES W72-03222
ES, HARDNESS(WATER), ALKALINITY, ENZYMES, TEMPERATURE, PHOTOSYNT: / W72-07514
MA, CHAOBORUS, *GAMBUSIA AFFINS, EPHEMEROPTERA, MENIDIA, BRACHYDEU W72-07132
OCHLORIS PENIOCYSTIS, SPIROGYRA, EPHEMEROPTERA BAETIDAE, EPHYDRA, W72-07132
CHIDAE, PLECOPTERA, MAGALOPTERA, EPHEMEROPTERA, NANNOCHLORIS, CHLO W70-06655
IROGYRA, EPHEMEROPTERA BAETIDAE, EPHYDRA, CHAETOGASTER, AEOLOSOMA, W72-07132
CATES, ARTHRO/ SAN PEDRO(CALIF), EPIBENTHIC SPECIES, ABALONE, TUNI W71-01475
VIRUSES, CRUSTACEANS, NEMATODES, EPIDEMIOLOGY, PHOSPHATES, HEAVY M W73-02144
NELLA, *BIBLIOGRAPHIES, BIOLOGY, EPIDEMIOLOGY, CULTURES, METHODOLO W72-13800
AMINAREA SPP., SEWAGE POLLUTION, EPIFAUNA.: *NORTH SEA, *L W72-11854
AY, FLORIDA, BALANUS, BARNACLES, EPIFAUNA, SAGARTIA, ANEMONE, MOLG W72-07907
CRAB, PENICILLIUM, STREPTOMYCIN, EPIFLORA, WALLEYE LARVAE, SHAD, M W72-08142
OFILES, PHOTOMETER, HYPOLIMNION, EPILIMNION, LAKE MORPHOMETRY, DIS W72-07145
N, PRODUCTIVITY, PHOTOSYNTHESIS, EPILIMNION, LIGHT INTENSITY, RESP W72-14812
THERMAL STRATIFICATION, DIATOMS, EPILIMNION, EUTROPHICATION.: /S, W72-13659
AEROBIC CONDITIONS, CHLOROPHYLL, EPILIMNION, SEDIMENTS, DECOMPOSIN W72-12997
IVER, STANDARDS, SURFACE WATERS, EPILIMNION, INTAKE STRUCTURES, DI W70-09614
ION, CARBON DIOXIDE, ALKALINITY, EPILIMNION, WASHINGTON, PHOSPHORU W71-02681
TION, P/ *PHYTOPLANKTON, *LAKES, EPILIMNION, ZOOPLANKTON, DISTRIBU W71-09171
FICATION, *ENZYMES, THERMOCLINE, EPILIMNION, MUD, PHYSICOCHEMICAL W72-03224
PSOMETRIC ANALYSIS, HYPOLIMNION, EPILIMNION, CYANOPHYTA, DINOFLOGE W71-11949
KES AREA, *NORTHWESTERN ONTARIO, EPILITHIC ALGAE, ACHNANTHES MINUT W71-11029
US, NAUTOCOCCUS PYRIFORMIS NAUT/ EPINEUSTONT, NAUTOCOCCUS MAMMILAT W72-07141
*EPIPHYTIC BACTERIA, *MACROPHYTE- EPIPHTYE METABOLISM, LAWRENCE LAK W72-03216
ANNEL, CHLOROPLASTS, MORPHOLOGY, EPIPHYTES.: /RE MEDIA, HUMMOCK CH W72-12633

ICILLIN, LINDANE, DIADEMA ANTIL/ EPIPHYTES, CULTURING VESSELS, PEN W72-14294
AMPL/ *FORAMINIFERA, SUBSTRATES, EPIPHYTES, SAMPLE PRESERVATION, S W73-00367
CHONDRUS CRISPUS, RHODO/ *AGARS, EPIPHYTES, CANDIDA PARAPSILOSIS, W72-04748
KEMMERER SAMPLER, CHLAMYDOMONAS EPIPHYTICA, CHLAMYDOMONAS SPP, CA W72-09111
LT MARSHES, NEW YORK, NEMATODES, EPIPHYTOLOGY, PRIMARY PRODUCTIVIT W72-00948
FE HISTORY STUDIES, *PHAEOPHYTA, EPIPHYTOLOGY, ALGAE, HABITATS, CU W72-01370
ONSTRUENTS, NAVICULA, NITZSCHIA, EPIPYTIC ALGAE.: /S, FRAGILARIA C W72-07501
S, GEOLOGIA HISTORY, PLEISTORENE EPOCH, ADAPTATION, ENVIRONMENTAL W73-00855
ERVOIR(THE NETHERLANDS), KINETIC EQUATION.: /LANDS), BIESBOSCH RES W72-08559
ES, SUBSTRATE UTILIZATION, MONOD EQUATION, PENICILLIN G, TERRAMYCI W73-01445
NCTION COEFFICIENT, LAMBERT-BEER EQUATION, EDDY DIFFUSIVITY, DESHL W72-05459
LIGHT INTENSITY, PHOTOSYNTHESIS, EQUATIONS, LIGHT PENETRATION, ALG W72-05459
GICAL STUDIES, PLANT MORPHOLOGY, EQUATIONS, RESISTANCE, ELECTRICAL W73-01630
NUTRIENTS, MATHEMATICAL STUDIES, EQUATIONS, AQUATIC ENVIRONMENT, W W72-11724
*MATHEMATICAL STUDIES, *ALGAE, EQUATIONS, BIOLOGICAL PROPERTIES, W72-12724
IGITAL COMPUTERS, DATA ANALYSIS, EQUATIONS, REPRODUCTION, BIORHYTH W72-11920
MOLLUSKS, AMPHIPODA, MECHANICAL EQUIPMENT.: /MATODES, GASTROPODS, W72-05432
NIQUES, / *ALGAE, *FLOCCULATION, EQUIPMENT, COSTS, SEPARATION TECH W70-08904
DSORPTION, *SURFACES, LABORATORY EQUIPMENT, *LABORATORY TESTS, *ME W72-12210
ES, LABORATORY TESTS, LABORATORY EQUIPMENT, AERATION, DISSOLVED OX W72-11100
ACTERIOPHAGE, NUTRIENTS, FILTERS EQUIPMENT, OXI: / FERMENTATION, B W72-09675
C ALGAE, CHRYSOPHYTA, LABORATORY EQUIPMENT, FRESHWATER, WATER POLL W73-01445
UTANT IDENTIFICATION, LABORATORY EQUIPMENT, AEROBIC BACTERIA, SOLV W73-01066
ESTUARIES,/ *DIATOMS, *RESEARCH EQUIPMENT, ENVIRONMENTAL EFFECTS, W72-08141
QUALITY, SCENEDESMUS, LABORATORY EQUIPMENT, PHOTOGRAPHY.: / WATER W72-07144
LECTIONS, ALGAE, PROTOZA, FUNGI, EQUIPMENT, INSTRUMENTATION, FLUOR W72-07225
SMS, *FLUORESCENCE, FOOD CHAINS, EQUIPMENT, ANALYTICAL TECHNIQUES, W72-07526
ITROGEN, SEA WATER, FLUOROMETRY, EQUIPMENT, INHIBITION, IONS, ABSO W72-07660
DOMONAS, EUGLENOPHYTA, PROTOZOA, EQUIPMENT, WATER POLLUTION EFFECT W72-07661
TION, PHYTOPLANKTON, AUTOMATION, EQUIPMENT, ANALYTICAL TECHNIQUES, W72-07715
CAL TECHNIQUES, SEWAGE, RESEARCH EQUIPMENT, SEWAGE EFFLUENTS, SEWA W72-07715
NVIRONMENTAL EFFECTS, LABORATORY EQUIPMENT, FREEZE DRYING, SEPARAT W73-00379
, ECOLOGY, SEA WATER, LABORATORY EQUIPMENT, INHIBITION, ECOLOGICAL W73-00348
, DEPTH, *SAMPLING, *PERIPHYTON, EQUIPMENT, TUBIFICIDS, MOLLUSKS, W73-00374
LOGY/ *PHYTOPLANKTON, LABORATORY EQUIPMENT, SEDIMENTATION, METHODO W73-00273
REMENT, *KINETICS, / *LABORATORY EQUIPMENT, *RADIOISOTOPES, *MEASU W72-14326
POPULATIONS, AQUARIA, LABORATORY EQUIPMENT, PLANT PHYSIOLOGY.: /T W72-14294
ANALYTICAL TECHNIQUES, PALEOZOIC ERA, WATER CHEMISTRY.: /SOTOPES, W72-02073
AL, INDIAN OCEAN, SOIL EXTRACTS, ERDSCHREIBER MEDIUM, ESTUARINE SO W73-00430
, MOTALA STROM RIVER, SPARGANIUM ERECTUM, SCHOENOPLECTUS LACUSTRIS W73-00286
ON, *NUTRIENT REGENERATION, LAKE ERIE CENTRAL BASIN.: /DEOXYGENATI W72-12996
DOGONIUM, ANACYSTIS, CERA/ *LAKE ERIE CENTRAL BASIN, TRIBONEMA, OE W72-12995
NERATION, DEOXYGENATION, / *LAKE ERIE CENTRAL BASIN, NUTRIENT REGE W72-12997
LAKE ERIE CENTRAL BASIN.: W72-12994
ONTARIO), OXYGEN DEPLETIO/ *LAKE ERIE CENTRAL BASIN, POINTE PELEE(W72-12992
TION.: *LAKE ERIE CENTRAL BASIN, *OXYGEN DEPLE W72-13659
ARNOLDIELLA-CONCHOPHILA, LAKE ERIE.: W71-12489
ND, *ANAEROBIC CONDITIONS, *LAKE ERIE, *ALGAE, *BENTHIC FAUNA, *BO W72-12995
INTERF/ *DISSOLVED OXYGEN, *LAKE ERIE, *BACTERIA, *SEDIMENT-WATER W72-12996
URE, CURREN/ *HYPOLIMNION, *LAKE ERIE, *DISSOLVED OXYGEN, TEMPERAT W72-12992
EMAND, HYPOLI/ *SEDIMENTS, *LAKE ERIE, *DISSOLVED OXYGEN, OXYGEN D W72-12994
ON, *DISSOLVED OXYGEN, *A/ *LAKE ERIE, *EUTROPHICATION, *HYPOLIMNI W72-12990
WASTES, *POLITICAL ASPECT/ *LAKE ERIE, *EUTROPHICATION, *CHEMICAL W72-14782
SOLVED OXYGEN, *NUTRIENTS, *LAKE ERIE, *HYPOLIMNION, MATHEMATICAL W72-12997
, *INDU/ *WATER POLLUTION, *LAKE ERIE, *MERCURY, *MUNICIPAL WASTES W72-14282
OR, SURVEYS, MICROBIOLOGY/ *LAKE ERIE, *POTABLE WATER, *TASTE, *OD W72-10076
E ONTARIO, *EUTROPHICATION, LAKE ERIE, ALGAE, NITROGEN, PHOSPHORUS W71-11009
NUTRIENTS, ALGAE, BACTERIA, LAKE ERIE, ALUMINUM, MANGANESE, SODIUM W72-01101
UAT/ *LAKE SHORES, *ALGAE, *LAKE ERIE, AQUATIC HABITATS, LAKES, AQ W71-09156
LE/ *ALGAL SPECIES, WESTERN LAKE ERIE, ARNOLDIELLA CONCHOPHILA MIL W71-09156
LVED OXYGEN, *HYPOLIMNION, *LAKE ERIE, BENTHIC FLORA, ALGAE, LIGHT W72-13659
EATION, PHOSPHORUS, CARBON, LAKE ERIE, DETERGENTS, LAKE ONTARIO, T W71-12091
RUS, ALGAE, ORGANIC MATTER, LAKE ERIE, EFFLUENTS, SEWAGE, DETERGEN W71-11017
LAKE MICHIGAN, LAKE HURON, LAKE ERIE, GREAT LAKES, SOYBEANS, LIGH W73-01456
SEDIMENT-WATER INTERFACES, *LAKE ERIE, HYPOLIMNION, AQUATIC ALGAE. W72-06690
HOTOSYNTHESIS, GREAT LAKES, LAKE ERIE, LAKE ONTARIO, HYDRO: /RS, P W72-07648
E EFFLUENTS, AQUATIC ALGAE, LAKE ERIE, LAKE ONTARIO, NUTRIENTS, NI W71-11507
SORPTION, LAKES, NUTRIENTS, LAKE ERIE, LAKE SUPERIOR, LABORATORY T W72-01108
COMMUNITIES, LAKE MICHIGAN, LAKE ERIE, LAKE HURON, LAKE SUPERIOR, W72-10875
KTON, CARBON RADIOISOTOPES, LAKE ERIE, LAKE ONTARIO, ADSORPTION, A W72-13657
NTS, DETERGENTS, WISCONSIN, LAKE ERIE, LAKE ONTARIO, ST LAWRENCE R W70-07283
YTA, WASHINGTON, DIVERSION, LAKE ERIE, NITROGEN FIXATION, DETERGEN W72-05473
*CHLOROPHYLL, *PHOSPHORUS, *LAKE ERIE, NITROGEN, ALGAE, INORGANIC W72-13653
*RHODOPHYTA, *LAKE ERIE, OHIO, ALGAE.: W71-05630
*BIBLIOGRAPHIES, *ALGAE, *LAKE ERIE, PLANKTON, BENTHIC FLORA, DI W73-01615
LIGHT PENETRATION, CARBON, LAKE ERIE, RUNOFF, EROSION, CHLORELLA, W72-04734
ACTIVITY, ADSORPTION, FISH, LAKE ERIE, SEDIMENTS, SILTS, PLANKTON, W71-11036
LAKE HURON, *LAKE ONTARIO, *LAKE ERIE, WATER POLLUTION EFFECTS, WA W72-01094
YCLOT MICHIGANIANA, RHIZOSOLENIA ERIENSIS, MELOSIRA ISLANDICA, MEL W73-00384
*APHANIZOMENON FLOS-AQUAE, LAKE ERKEN(SWEDEN).: *CYANOPHAGES, W73-00246
ATION.: *LAKE ERKEN(SWEDEN), *DARK NITROGEN FIX W72-08054
, CHROMULINA, CRYPTOMONAS EROSA, ERKINIA SUBEQUICILIATA, UROGLENOP W72-13816
IA VENTRICOSA, VALONIA OCELLATA, ERNODESMIS VERTICILLATA, DICTYOSP W72-11800
ICHMENT, CHROMULINA, CRYPTOMONAS EROSA, ERKINIA SUBEQUICILIATA, UR W72-13816
INASTRUM GRACILIMUM, CRYPTOMONAS EROSA, PERIDINIUM AFRICANUM, NITZ W73-00256
ATION, WASTE DISPOSAL, AERATION, EROSION CONTROL, WATER POLLUTION W71-01192
TION, CARBON, LAKE ERIE, RUNOFF, EROSION, CHLORELLA, SODIUM COMPOU W72-04734
DT, RETURN FLOW, SEDIMENTS, SOIL EROSION, FERTILIZERS, FARM WASTES W71-06444
MAL WASTES, SURFACE WATERS, SOIL EROSION, NUTRIENTS, ALGAE, LAKE S W71-04216
HORES, MONITORING, MARSHES, SOIL EROSION, WASTE WATER TREATMENT.: / W73-02479

A COSTATUM, CARBOY, OSCILLATORIA	ERYTHRAEA,: /CORNUTUM, SKELETCNEM	W72-09103
, CULTURE MEDIA,/ *TRICHODESMIUM	ERYTHRAEUM, *CULTURING TECHNIQUES	W73-00430
RINATED BIPHENYLS, AROCLOR 1254,	ESCAMBIA RIVER, PINFISH, FLCUNDER	W71-07412
LOGICAL TESTS, ANTIMICROBIOTICS,	ESCHERICHIA COLI, BETA STREPTOCOC	W72-04259
RAIN, ORGANOMERCURIAL COMPOUNDS,	ESOX LUCIUS, ALKYL MERCURY, EAGLE	W72-05952
AVESCENS, AMBLOPLITES RUPESTRIS,	ESOX LUCIUS, CAMPELCMA, PSIDIUM,	W70-08098
PTOPHYCEAE, PONTOPOREIA AFFINIS,	ESOX LUCIUS, PERCA FLUVIATILIS, S	W72-12729
, CRYPTOPHYCEAE, *DENMARK,/ LAKE	ESROM, TYSTRUP LAKE, BAVELSE LAKE	W72-12738
IENT REQUIREMENTS, PRODUCTIVITY,	ESSENTIAL NUTRIENTS, WATER QUALIT	W70-04080
ON, PHYTOPLANKTON, MARINE ALGAE,	ESSENTIAL NUTRIENTS.: /ENTIFICATI	W73-00441
OSPHATES, NUTRIENT REQUIREMENTS,	ESSENTIAL NUTRIENTS.: /TRATES, PH	W73-00430
ROPHICATION, DEFICIENT ELEMENTS,	ESSENTIAL NUTRIENTS, *AEROBIC CON	W72-11617
EN COMPOUNDS, CYCLING NUTRIENTS,	ESSENTIAL NUTRIENTS, LAKES, MUD,	W72-10608
D CHRCMATOGRAPHY, METHYL ESTERS,	ESTERS, CHLORELLA VULGARIS, SAMPL	W72-09365
AS LIQUID CHROMATOGRAPHY, METHYL	ESTERS, ESTERS, CHLORELLA VULGARI	W72-09365
NDERMERE(ENGLAND), EXOENZYME PR/	ESTHWAITE WATER(ENGLAND), LAKE WI	W72-03224
TARN(ENGLAND)/ *UNITED KINGDOM,	ESTHWAITE WATER(ENGLAND), BLELHAM	W71-13183
NDERMERE(ENGLA/ *UNITED KINGDOM,	ESTHWAITE WATER(ENGLAND), LAKE WI	W71-13185
ON, ECONOMIC FEASIBILITY, COSTS,	ESTIMATED COSTS.: /ATION, FLOTATI	W72-13508
TE, PROTEINS, FILTERS, ENGINEERS	ESTIMATES, ECONOMICS, FEASIBILITY	W70-10174
CARBON DIOXIDE, NITROGEN, LAKES,	ESTUARIES, NUTRIENTS, DETERGENTS,	W70-07283
WASTES, FLORIDA, SESSILE ALGAE,	ESTUARIES, CHEMICAL WASTES, GAS C	W71-07412
OXYGEN DEMAND, DISSOLVED OXYGEN,	ESTUARIES, WATER POLLUTICN EFFECT	W71-05390
ASSESSMENTS, MONITORING, SEWAGE,	ESTUARIES, GROWTH RATES, BIOLOGIC	W71-12867
ISSOLVED OXYGEN, *PHYTOPLANKTON,	ESTUARIES, WATER TEMPERATURE, TUR	W71-11876
, INDUSTRIAL WASTES, OIL WASTES,	ESTUARIES, MONITORING.: /L PLANTS	W71-13746
TEBRATES, RADIATION, ECOSYSTEMS,	ESTUARIES, CORAL, REEFS, WATER RE	W71-13723
LT MARSHES, ECOLOGY, PESTICIDES,	ESTUARIES, INTERTIDAL AREAS, ESTU	W72-10678
*ENGLAND, *POLLUTED	ESTUARIES, DURHAM COUNTY RIVERS.:	W72-08804
NCE, PATH OF POLLUTANTS, OCEANS,	ESTUARIES, CULTURES, SAMPLING, HA	W73-00361
QUIPMENT, ENVIRONMENTAL EFFECTS,	ESTUARIES, *DISTRIBUTION PATTERNS	W72-08141
LEAR WASTES, PATH OF POLLUTANTS,	ESTUARINE ENVIRONMENT, RADIOACTIV	W72-07826
L EXTRACTS, ERDSCHREIBER MEDIUM,	ESTUARINE SOILS, AXENIC CULTURES,	W73-00430
RADIOISOTOPES, STABLE ISOTOPES,	ESTUARINE ENVIRONMENT, COASTS, AN	W73-00811
ANKTON, *PHYTOPLANKTON, NERITIC,	ESTUARINE ENVIRONMENT, SEA WATER,	W73-00935
NIA, RADIOECOLOGY, MARINE ALGAE,	ESTUARINE ENVIRONMENT, ABSORPTION	W73-00826
E ALGAE, SEDIMENTS, PUERTO RICO,	ESTUARINE ENVIRONMENT, PATH OF PO	W73-00821
TH CAROLINA, ABSORPTION, SOUNDS,	ESTUARINE ENVIRONMENT, PATH OF PO	W73-00818
LS, WASTE ASSIMILATIVE CAPACITY,	ESTUARINE ENVIRONMENT, FCOD CHAIN	W73-00828
SOPHYTA, BIOLOGICAL CCMMUNITIES,	ESTUARINE ENVIRONMENT, WATER POLL	W73-00853
ES, ESTUARIES, INTERTIDAL AREAS,	ESTUARINE FISHERIES.: /, PESTICID	W72-10678
TIVITY, WATER POLLUTION EFFECTS,	ESTUARINE ENVIRONMENT.: /Y PRODUC	W72-09653
, ASIA, MARINE ALGAE, SEDIMENTS,	ESTUARINE ENVIRONMENT, MCNITCRING	W72-10958
IGATICNS, ENVIRONMENTAL EFFECTS,	ESTUARINE ENVIRONMENT, REGULATION	W72-00959
TS, PESTICIDES, ALGAE, DREDGING,	ESTUARINE ENVIRONMENT, CALIFORNIA	W71-09581
TS, FOOD CHAINES, PUBLIC HEALTH,	ESTUARINE ENVIRONMENT, ZINC RADIO	W72-04463
TS, MARINE ALGAE, PUBLIC HEALTH,	ESTUARINE ENVIRONMENT, ON-SITE IN	W72-04461
, SAMPLING, MATHEMATICAL MODELS,	ESTUARINE ENVIRONMENT, SALINITY,	W72-12941
, *FISHES, ECOLOGY, FOOD CHAINS,	ESTUARINE FISHERIES, MARINE FISHE	W72-12567
*POTOMAC RIVER	ESTUARY.:	W71-08775
PCTOMAC RIVER BASIN, POTOMAC	ESTUARY.:	W70-06509
*PAMLICO RIVER	ESTUARY.:	W72-01329
PANSICN, ISLE OF THANET, MEDWELL	ESTUARY.: /T COAST, INDUSTRIAL EX	W71-13746
*POTOMAC	ESTUARY.:	W72-09171
POTOMAC	ESTUARY.:	W71-09788
CALIF), SEARSVILLE LA/ SAN PABLO	ESTUARY(CALIF), SACRAMENTO RIVER(	W70-10405
A(CALIF),/ ALGAL CYCLE, DELAWARE	ESTUARY(DEL-NJ), SAN JOAQUIN DELT	W71-05390
LVED CXYGEN CONCENTRA/ *DELAWARE	ESTUARY, *BIOLOGICAL LIFE, *DISSO	W70-09189
CAYUGA LAKE(NEW YORK), PATUXENT	ESTUARY, BISCAYNE BAY(FLORIDA), H	W71-11517
ROAL/ *YAQUINA BAY(ORE), YAQUINA	ESTUARY, DATA INTERPRETATICN, MAC	W72-08141
ON RATIOS.: *PAMLICO RIVER	ESTUARY, PHOTOSYNTHESIS/RESPIRATI	W72-13636
ROL, LACTATE, PYRUVATE, ACETATE,	ETHANOL, BENZOATE, PROPIANATE, PR	W72-05421
ELETCNEMA COSTATUM, NUCLEOTIDES,	ETHANOL, MONOSACCHARIDES, CULTURE	W73-00942
Y LIQUID, BLUSYL, SODIUM LAURYL,	ETHER SULFATE, SODIUM DODECYL, BE	W72-12576
S HAUCKII, STIGMAPHORA ROSTRATA,	ETHMODISCUS OOZES.: /IA, HEMIAULU	W72-03567
RNIA, HEMIAUL/ *ETHMODISCUS REX,	ETHMODISCUS GAZELLAE, BAJA CALIFO	W72-03567
ALKYLENE OXIDE, POLYMERIZATION,	ETHYLENE OXIDE, WATER CLARIFICATI	W70-06968
TEROTROPHIC BACTERIA, ACETYLENE,	ETHYLENE.: /ER, DETROIT RIVER, HE	W73-01456
, MACRCINVERTEBRATES, ACETYLENE,	ETHYLENE, NITROGEN RADIOISOTOPES,	W72-10883
APAN, *AMURSKY BAY, PERIDINIANS,	EUCAMPIA ZOODIACUS, CHAETCCEROS A	W73-00442
BOSMINA, CERIODAPHNIA,	EUDIAPTOMUS.:	W72-00161
RINA MORUM, PLATYDORINA CAUDATA,	EUDORINA CALIFORNICA, EUDCRINA IL	W71-13793
A CAUDATA, EUDORINA CALIFORNICA,	EUDORINA ILLINOISENSIS.: /TYDORIN	W71-13793
UPLII, LAMELLIBRANCH, SPIROGYRA,	EUDORINA, CYCLOPS, KERATELLA, ASP	W71-11876
MIPTERA, *TRICHOPTERA, CORIXIDS,	EUDORINA, NOTONECTIDAE, BODO, HYD	W70-06655
CCUM, HYALO/ CARBON-14, CARTERIA	EUGAMETOS, STICHOCOCCUS, CHLOROCO	W72-09239
GLENA/ *INDIA, ORGANIC NITROGEN,	EUGLENA ACUS, EUGLENA CAUDATA, EU	W73-00918
ORGANIC NITROGEN, EUGLENA ACUS,	EUGLENA CAUDATA, EUGLENA LIMNOPHY	W73-00918
TES, BIOTIN, NIACIN, COBALAMINS,	EUGLENA GRACILIS, COLONIZATION, C	W73-02095
HLORIDE, ANKISTRODESMUS BRAUNII,	EUGLENA GRACILIS.: /YL MERCURIC C	W72-06037
OSTATUM, DUNALIELLA TERTIOLECTA,	EUGLENA GRACILIS, CHLAMYDOMANAS R	W72-05614
ACYSTIS NIDULAN/ WARBURG EFFECT,	EUGLENA GRACLIS, CALVIN CYCLE, AN	W71-09172
, EUGLENA ACUS, EUGLENA CAUDATA,	EUGLENA LIMNOPHYLA, EUGLENA OBLON	W73-00918
ENA CAUDATA, EUGLENA LIMNOPHYLA,	EUGLENA OBLONGA, EUGLENA OXYURIS,	W73-00918
ENA LIMNOPHYLA, EUGLENA OBLONGA,	EUGLENA OXYURIS, EUGLENA POLYMORP	W73-00918
UGLENA OBLONGA, EUGLENA OXYURIS,	EUGLENA POLYMORPHA, EUGLENA SPIRO	W73-00918
ENA OXYURIS, EUGLENA POLYMORPHA,	EUGLENA SPIROGYRA, LEPOCINCLIS OV	W73-00918
AE, MARINE ALGAE, AQUATIC ALGAE,	EUGLENA, CHLAMYDOMONAS.: /MS, ALG	W72-08436
AY, SAMPLING, OXIDATION LAGOONS,	EUGLENA, CHLAMYDOMONAS, EUGLENOPH	W72-07661
, ENZYMES, METABOLISM, HYDROGEN,	EUGLENA, CHLORELLA, SCENEDESMUS,	W71-09172

ARA, SUBMERGED PLANTS, PROTOZOA, EUGLENA, CHLAMYDOMONAS, CHLOROPHY W72-00155
RELLA, CHLOROPHYTA, SCENEDESMUS, EUGLENA, DISSOLVED OXYGEN, DIURNA W71-11519
, BIOMASS, OLIGOTROPHY, DIATOMS, EUGLENA, DAPHNIA, HABITATS, NICHE W71-09171
CHLORELLA, CLOSTRIDIUM, E. COLI, EUGLENA, EUGLENOPHYTA, CHLOROPHYT W72-06124
LUDGE, ALGAE, PROTOZOA, DIATOMS, EUGLENA, MATHEMATICAL STUDIES.: / W72-08573
LAGOONS, EUGLENA, CHLAMYDOMONAS, EUGLENOPHYTA, PROTOZOA, EQUIPMENT W72-07661
C ALGAE, LIMNOLOGY, CHRYSOPHYTA, EUGLENOPHYTA, PYRROPHYTA, BIOMASS W72-07145
LLATES, CYANOPHYTA, CHLOROPHYTA, EUGLENOPHYTA, GRAZING, WATER QUAL W72-07899
OPHYTA, CYANOPHYTA, CHRYSOPHYTA, EUGLENOPHYTA, DISSOLVED OXYGEN, C W72-08048
HYSICAL PROPERTIES, CHLOROPHYTA, EUGLENOPHYTA, CYANOPHYTA, CHRYSOP W73-01627
, CLOSTRIDIUM, E. COLI, EUGLENA, EUGLENOPHYTA, CHLOROPHYTA, FERROB W72-06124
OPHYTA, PYRROPHYTA, CHRYSOPHYTA, EUGLENOPHYTA, WATER TEMPERATURE, W72-04855
SOPHYTA, RHODOPHYTA, CYANOPHYTA, EUGLENOPHYTA.: /CHLOROPHYTA, CHRY W71-05629
RNAL, FLUCTUATIONS, CHLOROPHYTA, EUGLENOPHYTA, CYANOPHYTA, CHRYSOP W72-00150
YTA, HABITATS, AQUATIC HABITATS, EUGLENOPHYTA, PYRROPHYTA, FRESHWA W72-09119
ECTS, BIOMASS, NUTRIENTS, ALGAE, EUGLENOPHYTA, CHRYSOPHYTA, CHLORO W72-09111
INUTISSIMA, ACHNANTHES FLEXELLA, EUNOTIA PECTINALIS, DESMIDS.: / M W71-11029
HORA OVALIS, SYNEDRA FASCIULATA, EUNOTIA PECTINALIS, LITHOPHYTES, W72-08141
, RHABDODERMELLA, LOPHOPANOPEUS, EUPENTACTA, CNEMIDOCARPA, ENHYDRA W73-00932
TRACTION, TRIPHOTURUS MEXICANUS, EUPHAUSIA PACIFICA, ENGRAULIS MOR W73-02105
OIKOPLEURA, SAGITTA, PTEROPODS, EUPHAUSIIDS, GADUS MORHUA, UROPHY W73-00371
ALGAL SINKING, *VERTICAL MIXING, EUPHOTIC ZONE, DIFFUSION COEFFICI W71-03034
ON, WASTE ASSIMILATIVE CAPACITY, EUROPE, ANIMAL PHYSIOLOGY, TRACE W73-00817
ENT, FOOD CHAINS, PUBLIC HEALTH, EUROPE, RADIOISOTOPES, EFFLUENTS. W73-00828
RINITY, METAHALINE, HYPERHALINE, EURYHALINE, RED SEA, GULF OF SUEZ W73-00855
ATER TEMPERATURE, PHYTOPLANKTON, EUTHROPHICATION.: /R OPERATION, W W72-05282
NA, NUTRIENTS, SEWAGE EFFLUENTS, EUTNOPHICATION, ALGAE, CYANOPHYTA W72-13636
IUM SPLENDENS, EXUVIELLA MARINA, EUTREPTIA, NITZSCHIA SERIATA, THA W71-07875
OSYSTEMS, ALGAE, *LIGHT QUALITY, EUTROPHIC ZONE, CHLOROPHYLL, PHOS W72-12393
TY, PHOTOSYNTHESIS, RESPIRATION, EUTROPHICATION, PLANT POPULATION, W72-12192
Y, ALGAE, WATER QUALITY CONTROL, EUTROPHICATION, WASTE WATER TREAT W72-12487
YGEN DEMAND, HEAT BUDGET, ALGAE, EUTROPHICATION.: / ALKALINITY, OX W72-12997
AINFALL, WASTE WATER(POLLUTION), EUTROPHICATION, WATER POLLUTION S W72-13644
TIFICATION, DIATOMS, EPILIMNION, EUTROPHICATION.: /S, THERMAL STRA W72-13659
RBON RADIOISOTOPES, *CYANOPHYTA, EUTROPHICATION, WATER POLLUTION E W72-13692
DEFICIENT ELEMENTS, MESOTROPHY, EUTROPHICATION, TROPHIC LEVEL, PR W72-13816
TANDING CROPS, LIMITING FACTORS, EUTROPHICATION.: /IC COMPOUNDS, S W72-13653
ES, CYANOPHYTA, LIGHT INTENSITY, EUTROPHICATION, PHYTOPLANKTON, PR W73-02472
RATION, ALGAE, METHANE BACTERIA, EUTROPHICATION, NITRATES, PHOSPHA W71-07106
LINE, NITRATES, LIGHT INTENSITY, EUTROPHICATION, MARINE AL: /ERMOC W71-07875
RCOLATION, EVAPORATION, INSECTS, EUTROPHICATION, ALGAE.: /PAGE, PE W71-07124
ONS, *ALGAE, *TRICKLING FILTERS, EUTROPHICATION, PHOTOSYNTHESIS, D W71-07097
, PHOSPHORUS, PHOSPHATES, ALGAE, EUTROPHICATION, WATER QUALITY, RI W71-06445
ALGAE, COMPARATIVE PRODUCTIVITY, EUTROPHICATION.: /OMASS, OXYGEN, W71-09165
FFECTS, WATER POLLUTION SOURCES, EUTROPHICATION, COLIFORMS, SPORES W71-09173
A, *LAKES, ARCTIC, PRODUCTIVITY, EUTROPHICATION, RED TIDE, SILICA, W71-09190
NUTRIENTS, NITROGEN, PHOSPHORUS, EUTROPHICATION, ALGAE, PATHOGENIC W71-08214
TROL, *BIODEGRADATION, FISHKILL, EUTROPHICATION, ALGAL BLOOMS, WAT W71-08768
RE, THERMAL STRATIFICATION, ICE, EUTROPHICATION, OXYGEN, SEASONAL. W72-04855
DETRITUS, PLANKTON, ADSORPTION, EUTROPHICATION, TEMPERATURE, LIGH W72-05469
ITY, PHYTOPLANKTON, ZOOPLANKTON, EUTROPHICATION, OLIGOTROPHY, PROT W72-05968
PHOTOSYNTHESIS, CARBON FIXATION, EUTROPHICATION, WATER QUALITY CON W72-06046
AL EFFECTS, LIMNOLOGY, PLANNING, EUTROPHICATION, EVALUATION, WATER W72-06526
ACTERIA, *WASTE WATER TREATMENT, EUTROPHICATION.: /LUDGE, ALGAE, B W72-04086
T LAKES, *SEDIMENTS, *NUTRIENTS, EUTROPHICATION, NITROGEN, LITTORA W72-04263
), *KENNEBEC RIVER(ME), CULTURAL EUTROPHICATION.: /NNABESSACOOK(ME W72-04280
NESS(WATER), COLOR, OLIGOTROPHY, EUTROPHICATION, ALGAE, CYANOPHYTA W72-04269
CIENCIES, WATER QUALITY CONTROL, EUTROPHICATION, ALGAL CONTROL, *C W72-04788
WATER TREATMENT, ALGAL CONTROL, EUTROPHICATION, MODEL STUDIES, MA W72-04787
, *SURFACE WATERS, WATER SUPPLY, EUTROPHICATION, ALGAE, NUTRIENTS, W71-05083
UTRIENTS, *PHOSPHORUS, *NITROGE/ EUTROPHICATION, *ALGAL BLOOMS, *N W70-09669
UMBIA, TIDAL EFFECTS, NUTRIENTS, EUTROPHICATION, WATER QUALITY, WA W70-08249
UMBIA, TIDAL EFFECTS, NUTRIENTS, EUTROPHICATION, WATER QUALITY, WA W70-08248
ON, ACTIVATED SLUDGE, TREATMENT, EUTROPHICATION, SURFACTANTS, ENZY W71-00116
IR POLLUTION, CHEMICAL ANALYSIS, EUTROPHICATION, TRANSPIRATION, PR W71-01489
, *ION TRANSPORT, *HEAVY METALS, EUTROPHICATION, POLLUTANT IDENTIF W73-02112
, AQUATIC POPULATIONS, BIOASSAY, EUTROPHICATION.: /AQUATIC ANIMALS W73-02095
WATER POLLUTION, PROTOZOA, OHIO, EUTROPHICATION, LAKE ONTARIO, LIG W73-01615
NIQUES, *WISCONSIN, OLIGOTROPHY, EUTROPHICATION, LAKE SUPERIOR, LA W73-01456
UATIC ALGAE, NITRATES, NITRITES, EUTROPHICATION.: /ATIC PLANTS, AQ W73-00286
LGAE, *ISOLATION, MARINE PLANTS, EUTROPHICATION, OCEANS, SAMPLING, W73-00430
TER POLLUTION SOURCES, ASBESTOS, EUTROPHICATION, POLYCHLORINATED B W73-00703
RATES, RADIOACTIVITY TECHNIQUES, EUTROPHICATION, MARINE ALGAE, WAT W73-00431
ER, WASTE ASSIMILATIVE CAPACITY, EUTROPHICATION, NUTRIENT REMOVAL, W73-00802
GRAPHY, WATER POLLUTION EFFECTS, EUTROPHICATION, *LAKES, ALGAE, IN W72-14728
DIATOMS, CYANOPHYTA, TURNOVERS, EUTROPHICATION.: /MICAL ANALYSIS, W73-00232
G FACTORS, ALLOGENIC SUCCESSION, EUTROPHICATION.: /RIENTS, LIMITIN W72-14790
OMS, CHLOROPHYTA, BENTHIC FLORA, EUTROPHICATION, DOMINANT ORGANISM W72-14797
*PROTOZOA, *ALGAE, SCENEDESMUS, EUTROPHICATION, PONDS, FEEDING RA W72-14796
SYNTHESIS, PRIMARY PRODUCTIVITY, EUTROPHICATION, SOLAR RADIATION, W72-08054
NKTON, STANDING CROP, LIMNOLOGY, EUTROPHICATION.: /ODUCTIVITY, PLA W72-08559
ARBON, DISTRIBUTION, PHOSPHORUS, EUTROPHICATION, ALGAE, AQUATIC WE W72-08459
ORGANIC MATTER, PHOTOSYNTHESIS, EUTROPHICATION, PHOSPHORUS, CHLOR W72-08048
*ALGAL CONTROL, *NUISANCE ALGAE, EUTROPHICATION, REVIEWS, TROPHIC W72-07937
DUCTION POTENTIAL, LAKES, SOILS, EUTROPHICATION, AMMONIFICATION, N W72-07933
TS, *PHOSPHATES, *POTABLE WATER, EUTROPHICATION, ALKYLBENZENE SULF W72-06837
*ALGAL CONTROL, WATER POLLUTION, EUTROPHICATION, NUTRIENTS, CHLORE W72-07508
A, MICROORGANISMS, ALGAE, LAKES, EUTROPHICATION, LITTORAL, SANDS, W72-07501
FISH, PONDS, LAKES, FOOD HABITS, EUTROPHICATION, LARVAE, ROTIFERS, W72-07132
N SOURCES, *POLLUTION ABATEMENT, EUTROPHICATION, LEGISLATION, PUBL W72-06655

*CARBON DIOXIDE, NUISANCE ALGAE, EUTROPHICATION, PHOSPHATES, PHOSP W72-07514
ALGAE, CLAMS, MUSSELS, OYSTERS, EUTROPHICATION, NITROGEN, PHOSPHO W72-07715
GAS CHROMATOGRAPHY, PHAEOPHYTA, EUTROPHICATION, PLANKTON, ATLANTI W72-07710
R QUALITY, *CHEMICAL PROPERTIES, EUTROPHICATION, WATER TEMPERATURE W71-12092
LITY CONTROL, BIOMASS, TOXICITY, EUTROPHICATION, TRACE ELEMENTS, N W71-11793
NUTRIENTS, SALINITY, SEDIMENTS, EUTROPHICATION, FISH, ALGAE, RECR W71-10577
LOR, TASTE, CORROSION, DISEASES, EUTROPHICATION, FOULING, TOXICITY W71-09958
LAKES, SYMBIOSIS, ALGAL CONTROL, EUTROPHICATION, WATER POLLUTION, W71-13257
RCL, LABORATORY TESTS, BUOYANCY, EUTROPHICATION, NIRTROGEN FIXATIO W71-13184
R QUALITY CONTROL, *ODOR, ALGAE, EUTROPHICATION, TASTE, RESERVOIRS W71-13187
, ANALYTICAL TECHNIQUES, CARBON, EUTROPHICATION, STANDING CROPS, L W72-09163
ANKTON, WATER POLLUTION EFFECTS, EUTROPHICATION, PRODUCTIVITY, ALG W72-09167
VITY, PHOTOSYNTHESIS, NUTRIENTS, EUTROPHICATION, CYANOPHYTA, ALKAL W72-09162
RBON, GROWTH RATES, OLIGOTROPHY, EUTROPHICATION.: / PHOSPHORUS, CA W72-09164
TES, AERATION, LIMITING FACTORS, EUTROPHICATION.: /ELLA, BICARBONA W72-10607
RESHWATER, MEASUREMENT, DIATOMS, EUTROPHICATION.: /SSAY, *ALGAE, F W72-10619
, MUD, WISCONSIN, AQUATIC WEEDS, EUTROPHICATION.: /UTRIENTS, LAKES W72-10608
NUTRIENTS, ACTINOMYCETES, TASTE, EUTROPHICATION.: /GHT INTENSITY, W72-10613
NUTRIENT REQUIREMENTS, SORPTION, EUTROPHICATION, DEFICIENT ELEMENT W72-11617
SEA WATER, FRESHWATER, AMMONIA, EUTROPHICATION, OXIDATION, SYMBIO W72-11563
FICATION, WATER QUALITY CONTROL, EUTROPHICATION, ACTIVATED SLUDGE, W72-00800
S, MEASUREMENT, BACTERIA, ALGAE, EUTROPHICATION, DATA COLLECTIONS, W71-13794
TANTS, STORM DRAINS, DETERGENTS, EUTROPHICATION, ALGAE, NUTRIENTS. W71-13466
R TREATMENT, TERTIARY TREATMENT, EUTROPHICATION.: /TS, *WASTE WATE W72-03613
, TEMPERATURE, DISSOLVED OXYGEN, EUTROPHICATION, OLIGOTROPHY, PHYT W72-03224
OVER, ENERGY, CYANOPHYTA, ALGAE, EUTROPHICATION, SUMMER, ENERGY, S W72-03228
*CHEMICAL ANALYSIS, GREAT LAKES, EUTROPHICATION, NUTRIENTS, ALGAE, W72-01101
NT, *DENITRIFICATION, *NITRATES, EUTROPHICATION, IRRIGATION WATER, W72-00974
TS, ALGAE, PHOSPHORUS, NITROGEN, EUTROPHICATION, SEDIMENT TRANSPOR W72-00847
RIOR, *LA/ *GREAT LAKES, *LAKES, EUTROPHICATION, ALGAE, *LAKE SUPE W72-01094
NTROL, *WATER POLLUTION SOURCES, EUTROPHICATION, LEGISLATION, LEGA W72-01085
IC ACIDS, ORGANIC MATTER, ALGAE, EUTROPHICATION, CHELATION, DECOMP W72-01097
NTS, TROUT, AQUATIC ENVIRONMENT, EUTROPHICATION, NITRIFICATION, RE W72-02412
EGISLATION, *FEDERAL GOVERNMENT, EUTROPHICATION, ENVIRONMENTAL ENG W72-01660
PHY, CHLOROPHYLL, TROPHIC LEVEL, EUTROPHICATION, ALGAE, DECOMPOSIN W72-01784
STES, CLEANING, DOMESTIC WASTES, EUTROPHICATION, WATER POLLUTION, W72-01685
L, *FEDERAL GOVERNMENT, *GRANTS, EUTROPHICATION, GOVERNMENT FINANC W72-01659
EN DEMAND, PHOSPHORUS, NITROGEN, EUTROPHICATION, *AEROBIC CONDITIO W72-01881
DILUTION, NITROGEN, PHOSPHORUS, EUTROPHICATION, SURFACE WATERS.: / W72-01472
OWTH, WATER POLLUTION TREATMENT, EVALUATION, WATER QUALITY.: /T GR W72-11702
, *ASSAY, ANALYTICAL TECHNIQUES, EVALUATION, WATER POLLUTION, POLL W72-10678
EMICAL ANALYSIS, WATER ANALYSIS, EVALUATION, WATER QUALITY CONTROL W72-10883
POTOMAC RIVER, DATA COLLECTIONS, EVALUATION, ON-SITE DATA COLLECTI W72-07791
*ANALYTICAL TECHNIQUES, *ALGAE, EVALUATION, TESTING, ASSAY, NUTRI W73-00255
FRESHWATER FISH, AQUATIC ALGAE, EVALUATION, SAMPLING, CARP, POTAS W73-01673
E, WATER RESOURCES, METHODOLOGY, EVALUATION, CHEMICAL ANALYSIS.: / W71-05083
OLOGY, PLANNING, EUTROPHICATION, EVALUATION, WATER CHEMISTRY, WATE W72-06526
TIC WEED STUDY FACILITY, PRODUCT EVALUATION, BLUE-GREEN ALGAE.: /A W71-06102
, WATER RE-USE, LIGHT INTENSITY, EVAPORATION, RE-AERATION, DISSOLV W71-07099
ORINATION, SEEPAGE, PERCOLATION, EVAPORATION, INSECTS, EUTROPHICAT W71-07124
ICATION, *PHOSPHORUS, *NITROGEN, EVAPORATION, STORM RUNOFF, BIOASS W72-04070
ION, NON-CONSUMPTIVE USE, ALGAE, EVAPORATION, REASONABLE USE, BRAC W70-07100
PLANTS, PHAEOPHYTA, OILY WATER, EVAPORATION, DIFFUSION, RHODOPHYT W73-01074
ER CHEMISTRY, CHEMICAL ANALYSIS, EVAPORATION, SEEPAGE, TURBIDITY, W72-02274
CATION, NITRIFICATION, RESERVOIR EVAPORATION, RESERVOIRS, NUTRIENT W72-02412
LONG TUBE VERTICAL DISTILLATION, EVAPORATORS, AQUAT: / POLLUTION, W70-09193
*CHEMICAL EVOLUTION, *MACROMOLECULES.: W72-01473
CRENULATUS, CLADOGRAMMA LINEATUS- EXCENTRICUS, CLADOGRAMMA NODULIFE W71-11527
SEDIMENT-WATER NUTRIENT- EXCHANGE.: W70-08944
SLUDGE, TERTIARY TREATMENT, ION EXCHANGE, ELECTROCHEMISTRY, REVER W70-09907
LGAE, *HARVESTING OF ALGAE, *ION EXCHANGE, DISSOLVED SOLIDS, FLOCC W71-00138
C FEASIBILITY, PILOT PLANTS, ION EXCHANGE, ELECTROLYSIS, CHEMICAL W71-09450
ELEMENTS, CHEMICAL ANALYSIS, ION EXCHANGE, ABSORPTION, STRONTIUM R W71-09188
, BIOCHEMICAL OXYGEN DEMAND, ION EXCHANGE, CHEMICAL OXYGEN DEMAND, W71-12200
PERATURE, ALGAE, PHOSPHORUS, ION EXCHANGE, OXIDATION-REDUCTION POT W72-01108
RS, WATER POLLUTION SOURCES, ION EXCHANGE, *CHEMICAL ANALYSIS, WAT W73-00286
NTAMINATION, TRACE ELEMENTS, ION EXCHANGE, POLLUTANT IDENTIFICATIO W72-14303
FICATION, AQUEOUS SOLUTIONS, ION EXCHANGE, METHODOLOGY, ORGANIC AC W72-12166
UES, MOLECULAR STRUCTURE, CATION EXCHANGE, RADIOECOLOGY, *PATH OF W72-12203
NE TRIPHO/ AVAILABLE PHOSPHORUS, EXCHANGEABLE PHOSPHORUS, A*DENOSI W72-03742
TEROANTAGONISM, AXENIC CULTURES, EXCRETION, APHANIZOMENON GRACILE. W72-12734
N, EXTRACELLULAR PRODUCTS, ALGAL EXCRETIONS.: /N, GROWTH REGULATIO W71-08687
LAND), LAKE WINDERMERE(ENGLAND), EXOENZYME PRODUCERS, CHLOROPHYLL W72-03224
L ESTUA/ *KENT COAST, INDUSTRIAL EXPANSION, ISLE OF THANET, MEDWEL W71-13746
AESIPHON SUBAEQUALIS, APISTONEMA EXPANSUM, PORPHYRIDIUM GRISEUM, C W71-04526
MATTER, NITROSOM/ *HUBBARD BROOK EXPERIMENTAL FOREST, PARTICULATE W71-01489
RMS, INSOLUBLE FORMS, LABORATORY EXPERIMENTS.: /UM 106, SOLUBLE FO W71-13233
RIA, SUBLITTORAL REGION, CULTURE EXPERIMENTS, ATRAZINE, MCPA, ENDO W72-12576
S EARTH, *POROUS MEDIA/ *NUCLEAR EXPLOSION, *FILTERS, *DIATOMACEOU W72-00974
*ENVIRONMENTAL EFFECTS, *NUCLEAR EXPLOSIONS, BIRDS, SALMON, STICKL W72-00960
CTS, *RADIOACTIVITY EF/ *NUCLEAR EXPLOSIONS, *WATER POLLUTION EFFE W72-03347
E, *FISHERIES, FISH PO/ *NUCLEAR EXPLOSIONS, *ALASKA, *MARINE ALGA W73-00764
, / *ALGAL CONTROL, *CYANOPHYTA, EXPLOSIVES, COSTS, FISH, FISHKILL W72-11564
, SEPARATION TECHNIQUES, SOLVENT EXT: /RY EQUIPMENT, FREEZE DRYING W73-00379
EER EQUATION, EDDY DIFFUSIVITY,/ EXTINCTION COEFFICIENT, LAMBERT-B W72-05459
OSTIMULATION, GROWTH REGULATION, EXTRACELLULAR PRODUCTS, ALGAL EXC W71-08687
MOLECULAR SIZE, *ALGAL EXUDATES, EXTRACELLULAR MATERIAL, TURBULENT W71-07878
PMENT, AEROBIC BACTERIA, SOLVENT EXTRACTIONS, FREEZE DRYING, SYST: W73-01066
IOLOGICAL SAMPLES, LIQUID-LIQUID EXTRACTION, TRIPHOTURUS MEXICANUS W73-02105
HLOROPHYTA, CHRYSOPHYTA, SOLVENT EXTRACTIONS, DIATOMS, AQUATIC ALG W72-13673

TEMPERATURE OVEN DRYING, ASHING, EXTRACTIVE DIGESTION, NEUROANATOM W71-11036
-LUCIFERASE ENZYME SYSTEM, PLANT EXTRACTS, CHLORELLA SOROKINIANA.: W71-11916
AY OF BENGAL, INDIAN OCEAN, SOIL EXTRACTS, ERDSCHREIBER MEDIUM, ES W73-00430
YMONAS, NITZSCHIA OVALIS, CARBON EXTRUSION, CARBON-14, NITROGEN SO W72-07497
MPOUNDS, *MOLECULAR SIZE, *ALGAL EXUDATES, EXTRACELLULAR MATERIAL, W71-07878
PHYTOPLANKTON, FLUID MECHANICS, EXUDATION, OCEANS, CHLOROPHYTA, P W71-07878
IC ACIDS, KELPS, WATER ANALYSIS, EXUDATION, WATER POLLUTION, PHYTO W72-12166
MILIARIS, GYMNODINIUM SPLENDENS, EXUVIELLA MARINA, EUTREPTIA, NITZ W71-07875
PUS SPP, GILLS, MUSCLE, VISCERA, EYES, SCALES(FI: /RA ASPPRA, SCIR W73-01673
GILISSIMA, RHIZOSOLENIA HEBETATA F. SEMISPINA, CHAETOCEROS AFFINIA W72-03543
ION, LEGISLATION, BENTHIC FLORA, F: /, WASTE TREATMENT, AIR POLLUT W71-13723
BUTION, WATER POLLUTION EFFECTS, F: /N PATTERNS, ECOLOGICAL DISTRI W73-00918
ER POLLUTION EFFECTS, *TREATMENT FACILITIES, *INFLUENT STREAMS, *G W73-01669
SOLIDS, *RECREATION, *TREATMENT FACILITIES, SAMPLING, MONITORING, W72-00042
NG, STATE GOVERNMENTS, TREATMENT FACILITIES, ADMINISTRATIVE AGENCI W72-01659
TREATMENT, *NITRATES, TREATMENT FACILITIES, ALGAE, *WASTE WATER T W72-02975
, *LAKES, *COLORADO, *RECREATION FACILITIES, WATER TEMPERATURE, CH W72-04791
RECASTING, CONSTRAINTS, RESEARCH FACILITIES.: /WS, FOOD CHAINS, FO W72-06340
RE TECHNIQUE, AQUATIC WEED STUDY FACILITY, PRODUCT EVALUATION, BLU W71-06102
DS CREEK(NJ), RELATIVE CONDITION FACTOR.: /ER SAMPLER, SECCHI *YAR W72-14280
*ALUM, POLYELECTROLYTES, FACTORIAL DESIGN.: W70-08904
INTENSITY, RESPIRATION, LIMITING FACTORS.: /IS, EPILIMNION, LIGHT W72-14812
NUTRIENT REQUIREMENTS, *LIMITING FACTORS, *PHOSPHATES, DIATOMS, PO W73-01445
*ALGAE, *GROWTH RATES, *LIMITING FACTORS, *DIURNAL DISTRIBUTION, C W72-11724
RBON, *SEWAGE LAGOONS, *LIMITING FACTORS, ALGAE, PRODUCTIVITY, PHO W72-09162
BON DIOXIDE, NUTRIENTS, LIMITING FACTORS, ALLOGENIC SUCCESSION, EU W72-14790
BICARBONATES, AERATION, LIMITING FACTORS, EUTROPHICATION.: /ELLA, W72-10607
POUNDS, STANDING CROPS, LIMITING FACTORS, EUTROPHICATION.: /IC COM W72-13653
C HABITATS, CLADOPHORA, LIMITING FACTORS, ECOLOGICAL DISTRIBUTION. W72-13142
US, *PLANKTON, *ALGAE, *LIMITING FACTORS, GROWTH RATES, DIATOMS, B W72-09163
ENA, AMMONIA, NITRATES, LIMITING FACTORS, NITROGEN, PLANKTON, DEPT W73-02472
, *NUTRIENTS, BIOASSAY, LIMITING FACTORS, NITROGEN, PHOSPHORUS, RA W73-00232
, CARBON RADIOISOTOPES, LIMITING FACTORS, PHOTOSYNTHESIS, GREAT LA W72-07648
N, *NUTRIENTS, *ALGAE, *LIMITING FACTORS, PHOSPHATES, CHLOROPHYTA, W72-12544
LGAE, *EUTROPHICATION, *LIMITING FACTORS, PHOTOSYNTHESIS, CARBON C W71-12068
NITIES, AQUATIC PLANTS, LIMITING FACTORS, ROOTED AQUATIC PLANTS, C W73-01089
RBON DIOXIDE, CULTURES, LIMITING FACTORS, RHODOPHYTA, AQUATIC ALGA W73-02099
*LIMITING FACTORS, SILICON.: W72-07937
TOSYNTHESIS, NUTRIENTS, LIMITING FACTORS, SALINITY, CARBON RADIOIS W73-00431
.: FACULTATIVE MICROORGANISMS, INDIA W71-06033
CENEDESMUS OBLIQUUS, ALCALIGENES FAECALIS, ENTEROBACTER AEROGENES, W72-07664
NTS, ATRAZINE, MCPA, ENDOSULFAN, FAIRY LIQUID, BLUSYL, SODIUM LAUR W72-12576
GEOTIA GENUFLEXA, ANKISTRODESMUS FALCATUS, PODOHEDRA.: MCU W71-07360
YDRATION, RUNOFF, RADIOACTIVITY, FALLOUT, BIOCONTROL, INSECT CONTR W72-00846
S, *MARINE ALGAE, RADIOISOTOPES, FALLOUT, CHEMICAL PRECIPITATION, W72-05965
, MUNICIPAL WASTES, AGRICULTURE, FALLOUT, DRAINAGE, MANAGEMENT, WI W71-03012
ENVIRONMENT, PATH OF POLLUTANTS, FALLOUT, NUCLEAR WASTES, RADIOECO W73-00821
TUARINE ENVIRONMENT, REGULATION, FALLOUT, OYSTERS, MUSSELS, PLANKT W72-00959
SOLVED ORGANIC CARBON, *AMERICAN FALLS RESERVOIR, *APHANIZOMENON, W72-13300
SAN JOAQUIN RIVER(CALIF), RIFLE FALLS(COLO), FISH FECAL WASTES, R W71-11006
CHOPELAG/ *OYSTER BAY, *JAMAICA, FALMOUTH HARBOR, *TINTINNIDS, *TY W72-07899
ILIPPINES, ENIWETOK, MICRONESIA, FANNING ISLANDS.: PH W71-11486
ERY, UTILIZATION, MODIFICATION./ FARM MACHINERY, INDUSTRIAL MACHIN W71-06188
UTION, NUTRIENTS, RIVERS, LAKES, FARM PONDS.: /ROPHYTA, WATER POLL W71-05991
RIVER, OXIDATION LAGOONS, PONDS, FARM PONDS, MARYLAND, QUARRIES, O W72-04736
N, *IONS, *AQUATIC PRODUCTIVITY, FARM PONDS, NUTRIENTS, SURFACES, W71-11001
TTLE, URINE, PATH OF POLLUTANTS, FARM WASTES.: /RCES, COLORADO, CA W71-11496
E WATER, AGRICULTURAL CHEMICALS, FARM WASTES, WASTE WATER TREATMEN W71-11698
ER POLLUTION CONTROL, SEDIMENTS, FARM WASTES, NUTRIENTS, PHOSPHORU W72-00846
STRIAL WASTES, MUNICIPAL WASTES, FARM WASTES, WASTE DISPOSAL, PEST W72-01659
ENTS, SOIL EROSION, FERTILIZERS, FARM WASTES, FISH, FISH EGGS, TUR W71-06444
XYGEN DEMAND, INDUSTRIAL WASTES, FARM WASTES, POLLUTANTS, FERTILIZ W71-06445
INDUSTRIAL WASTES, ALGAE, PONDS, FARM WASTES.: /GE, INFILTRATION, W71-06435
PERCOLATION, LEACHING, DRAINAGE, FARM WASTES, SOIL MANAGEMENT, GRO W71-06443
EASES, CATFISH, WASTE TREATMENT, FARM WASTES, DEHYDRATION, FEASIBI W71-07544
ATER DISPOSAL, SEWAGE EFFLUENTS, FARM WASTES.: / DISPOSAL, WASTE W W70-06509
BAN RUNOFF, AGRICULTURAL RUNOFF, FARM WASTES, ALGAE, AQUATIC WEEDS W73-02479
, DIATOMS, CURRENTS(WATER), FISH FARMING.: /ANKTON, STRATIFICATION W73-00236
S, SHELLFISH, MUSSELS, SHELLFISH FARMING, AQUACULTURE, COMMERCIAL W72-09378
A SULCATA, NAVICULA SPP, SYNEDRA FASCICULATA, PLAGIOGRAMMA BROCKMA W73-00853
HEURCKII, PLEUROSIGMA, GYROSIGMA FASCIOLA, NITZSCHIA SPP, BACILLAR W73-00853
RGENSII, AMPHORA OVALIS, SYNEDRA FASCIULATA, EUNOTIA PECTINALIS, L W72-08141
OGONIUM, PHYSA, CULEX, GAMBUSIA, FATE OF POLLUTANTS.: / URINE, OED W72-07703
UTILIZATION, PARTICULATE MATTER, FATE OF POLLUTANTS, PAPER CHROMAT W73-01066
COCCOLITHOPHORIDS, FLAGELLATES, FATE OF POLLUTANTS, STEPHANOPYXIS W73-00380
NSFORMATION, BIOLOGICAL SAMPLES, FATE OF POLLUTANTS, METABOLITES, W73-00361
LIZATION, ORGANIC NITROGEN, ORG/ FATE OF POLLUTANTS, SUBSTRATE UTI W73-00379
FATE OF POLLUTANTS.: W72-11702
ATION, BIOLOGICAL MAGNIFICATION, FATE OF POLLUTANTS.: / BIOACCUMUL W72-11925
, HYDRAULIC MODELS, FISH, ALGAE, FATHOMETERS, BENTHIC FAUNA, BENTH W70-09614
ECYL, BENZENE SULFONATE, COCONUT FATTY ACID, PLURONICS, DIETHANOLA W72-12576
RED SPECTROSCOPY, CULTURE MEDIA, FATTY ACIDS, SAMPLE PREPARATION, W73-01066
KTON, ALGA/ *NUTRIENTS, *BENTHIC FAUNA, *BENTHIC FLORA, *PHYTOPLAN W72-06046
NS, *LAKE ERIE, *ALGAE, *BENTHIC FAUNA, *BOTTOM SEDIMENTS, *HYPOLI W72-12995
DIOISOTOPES, *DETRITUS, *BENTHIC FAUNA, *INVERTEBRATES, *MICROENVI W70-06668
ENVIRONMENTAL EFFECTS, *BENTHIC FAUNA, *INVERTEBRATES, *FISH POPU W71-08793
FISH HARVEST, PLANKTON, BENTHIC FAUNA, ALGAE, ECOLOGY, MISSOURI.: W70-09596
TA, DIATOMS, CYANOPHYTA, BENTHIC FAUNA, AQUATIC ANIMALS, INVERTEBR W72-12729
ASKA, BIOMASS, SALINITY, BENTHIC FAUNA, ASBESTOS, INORGANIC COMPOU W73-00374
INVERTEBRATES, BIOMASS, BENTHIC FAUNA, BENTHIC FLORA, HUMAN POPUL W72-12549

ISH, ALGAE, FATHOMETERS, BENTHIC FAUNA, BENTHIC FLORA, LAKE CNTARI W70-09614
ER TEMPERATURE, E, COLI, BENTHIC FAUNA, CELLULCSE, ENZYMES, PULP W W72-09668
S, CRABS, BENTHIC FLORA, BENTHIC FAUNA, DISTRIBUTIO: /OPODS, SNAIL W73-00932
ANKTON, BIOTA, BACTERIA, BENTHIC FAUNA, FISH, ENVIRONMENTAL EFFECT W72-05282
ULATIONS, PHYTOPLANKTON, BENTHIC FAUNA, PERIPHYTON, AQUATIC ALGAE, W72-00939
YLL, SEDIMENT TRANSPORT, BENTHIC FAUNA, PLANKTON, BIOINDICATORS, C W72-09092
ECOMPOSITION), SAMPLING, BENTHIC FAUNA, SEDIMENTATION, SEDIMENT DI W72-00847
ECOLOGICAL DISTRIBUTION, BENTHIC FAUNA, SIEVES, WATER TEMPERATURE, W73-00367
TURE, OXYGEN, TURBIDITY, BENTHIC FAUNA, WATER POLLUTION EFFECTS, S W73-02029
OSCILLATORIA FORMOSA, PHORMIDIUM FAVEOLARUM, ANKISTRODESMUS BRAUNI W72-07664
TAHOE, LAKE SEBASTICOOK(MAINE), FAYETTEVILLE GREEN LAKE(NEW YORK) W71-05626
CALIFORNIA, IRRIGATION, ECONOMIC FEASIBILITY, WATER POLLUTION CONT W71-09157
TMENT, FARM WASTES, DEHYDRATION, FEASIBILITY.: /ATFISH, WASTE TREA W71-07544
TECHNICAL FEASIBILITY, ECONOMIC FEASIBILITY, PILOT PLANTS, ION EX W71-09450
, FEASIBILITY STUDIES, TECHNICAL FEASIBILITY, ECONOMIC FEASIBILITY W71-09450
IA, NITROGEN, PHOSPHORUS, ALGAE, FEASIBILITY STUDIES, TECHNICAL FE W71-09450
ENGINEERS ESTIMATES, ECONOMICS, FEASIBILITY, COSTS, ANNUAL COSTS, W70-10174
ATER POLLUTION CONTROL, ECONOMIC FEASIBILITY, STREAMS, SEWAGE EFFL W72-14878
LOCCULATION, FLOTATION, ECONOMIC FEASIBILITY, COSTS, ESTIMATED COS W72-13508
SEBASTICOOK(MAINE), FAYETTEVILL/ FECAL MATERIAL, LAKE TAHOE, LAKE W71-05626
(CALIF), RIFLE FALLS(COLO), FISH FECAL WASTES, RESIDUAL FOOD, GREE W71-11006
INGTON, NITRILOTRIACETATE, HUMAN FECES.: / METASILICATE, LAKE WASH W72-04734
CHLOROPHYTA, TEMPERATURE, SIZE, FECUNDITY, ALGAE,: /ODS, DIATOMS, W72-14805
TIONAL JOINT COMMISSION, CANADA, FEDERAL GOV: / LIMNOLOGY, INTERNA W72-14782
*TOXINS, *ENVIRONMENTAL EFFECTS, FEDERAL GOVERNMENT, AIR POLLUTION W73-00703
QUALITY CONTROL, ALGAL CONTROL, FEDERAL GOVERNMENT, NUTRIENTS, SE W72-06655
SPHATES, LEGISLATION, WATER LAW, FEDERAL GOVERNMENT, WATER POLLUTI W71-11516
RCH AND DEVELOPMENT, INSPECTION, FEDERAL: /ONTRACTS, GRANTS, RESEA W71-11516
CRUSTACEANS, ZOOPLANKTON, ALGAE, FEEDING RATES, *LARVAL GROWTH STA W72-09248
NEDESMUS, EUTROPHICATION, PONDS, FEEDING RATES, BENTHOS, STANDING W72-14796
IMAL MANURE, *LITERATURE REVIEW, FEEDING VALUE.: *AN W71-07544
N, NITROGEN SOURCES, WELL WATER, FEEDLOTS.: /OURCES, MINERALIZATIO W71-06435
FEEDLOTS.: W71-11496
UES, RECYCLING FOOD, PROCESSING/ FEEDLOTS, COMPOSTING, PLANT RESID W72-00846
ISM, MOBILITY, PARTICULATE FORM, FEEDLOTS, NUTRIENT SOURCES,: /BOL W71-06443
N CONCENTRATION, MICROORGANISMS, FERMENTATION, ALKALINE HYDROLYSIS W71-12223
WASTES, *SEPARATION TECHNIQUES, FERMENTATION, NITROGEN, PHOSPHORU W72-10390
YGEN, BACTERIA, FUNGI, PROTOZOA, FERMENTATION, BACTERIOPHAGE, NUTR W72-09675
L).: ALUM, FERRIC CHLORIDE, *JERUSALEM(ISRAE W72-12289
LENA, EUGLENOPHYTA, CHLOROPHYTA, FERROBACILLUS, GAS CHROMATOGRAPHY W72-06124
OC, TOLYPOTHRIX TENUIS, AULOSIRA FERTILISSIMA, ANACYSTIS NIDULANS. W71-10070
LS, TOLYPOTHRIX TENUIS, AULOSIRA FERTILISSIMA, ANACYSTIS NIDULANS, W71-12071
S, DECOMPOSING ORGANIC MATERIAL, FERTILITY, RHIZOSPHERE, ROOT SYST W72-03342
NSITY, VISCOSITY, FISH BEHAVIOR, FERTILITY, SEASONAL, DIFFUSIVITY. W70-09612
LGAE, *LAKE SUPERIOR, LIMNOLOGY, FERTILITY, PHOTOSYNTHESIS, RESPIR W72-12192
EDIMENTS, ANALYTICAL TECHNIQUES, FERTILIZATION, AQUATIC PLANTS, NI W73-02478
RM WASTES, ALGAE, AQUATIC WEEDS, FERTILIZATION, LEAVES, LEGISLATIO W73-02479
PICAL REGIONS, RICE, CYANOPHYTA, FERTILIZATION, TEMPERATURE, CROP W73-02473
ROGEN FIXATION, LAND MANAGEMENT, FERTILIZATION, PRIMARY PRODUCTIVI W72-04769
CHEMICAL OXYGEN DEMAND, CARBON, FERTILIZATION, DISSOLVED OXYGEN, W71-12188
T/ *PHOSPHORUS, *EUTROPHICATION, FERTILIZATION, AQUATIC PLANTS, MA W71-12072
LANKTON, CARBON DIOXIDE, SESTON, FERTILIZATION, WATER QUALITY, RAD W72-07648
XCHANGE, CHEMICAL OXYGEN DEMAND, FERTILIZER, TURBIDITY.: /D, ION E W71-12200
NUTRIENTS, PRIMARY PRODUCTIVITY, FERTILIZERS, EFFLUENTS, WATER POL W71-11008
QUATIC PLANTS, DRAINAGE, RUNOFF, FERTILIZERS, LIVESTOCK, ECONOMICS W71-13172
TED LANDS, *IOWA, WATER QUALITY, FERTILIZERS, NUTRIENTS, NITROGEN W71-12252
OPHICATION, *LEACHING NUTRIENTS, FERTILIZERS, PRECIPITATION(ATMOSP W71-04216
N FLOW, SEDIMENTS, SOIL EROSION, FERTILIZERS, FARM WASTES, FISH, F W71-06444
WASTES, FARM WASTES, POLLUTANTS, FERTILIZERS, NITROGEN, NITRATES, W71-06445
DENITRIFICATION, RUNOFF, UREAS, FERTILIZERS, DRAINAGE WATER, IRRI W71-06435
FOOD/ *ALGAE, *HUMAN POPULATION, FERTILIZERS, DIATOMACEOUS EARTH, W71-09157
CATORS, WATER POLLUTION EFFECTS, FERTILIZERS, NITROGEN, BIOASSAY, W72-13816
S, MUSCLE, VISCERA, EYES, SCALES(FI: /RA ASPPRA, SCIRPUS SPP, GILL W73-01673
GRANGE LAKE, EAST TWIN / *GLASS FIBER FILTERS, *MEMBRANE FILTERS, W72-13673
NNIA, RHIZOPHORA, BOSTRYCHIETUM, FIDDLER CRABS, MESOFAUNA, TURBELL W72-07911
*LAKE TRAVIS, *LAKE LIVINGSTON, FIELD MEASUREMENTS, CARBON-14 MEA W72-13692
, AFTERGROWTH, BLUE-GREEN ALGAE, FIELD STUDIES, GREEN ALGAE, LABOR W70-08319
GREEN MANURE, INDIA, JAPAN, RICE FIELDS.: * W73-02473
COSA, BRYOPSIS PENNATA, SPYRIDIA FILAMENTOSA, DICTYOTA BARTAYRESII W73-00838
ALICA, MELOSIRA AMBIGUA, SYNEDRA FILIFORMIS, FRAGILARIA CAPUCINA, W73-00384
STRIPPING, POND DENITRIFICATION, FILTER DENITRIFICATION, METHANOL, W71-08223
CALIF), VOLATILE SOLIDS, SANBORN FILTER, BACKWASHING.: FIREBAUGH(W70-10174
ATION, BACTERIOPHAGE, NUTRIENTS, FILTERS EQUIPMENT, OXI: / FERMENT W72-09675
SOLIDS, SODIUM HYDROXIDE, *ROCK FILTERS.: *SUSPENDED W71-13341
E LAKE, EAST TWIN / *GLASS FIBER FILTERS, *MEMBRANE FILTERS, GRANG W72-13673
ENT, *DESIGN CRITERIA, TRICKLING FILTERS, ALGAE, ORGANIC LOADING, W71-07113
S/ *FERRIC CHLORIDE, MULTI-LAYER FILTERS, ANTHRACITE, *WAHNBACH RE W72-06201
CTERIA, AQUATIC BACTERIA, LAKES, FILTERS, COLIFORMS, METABOLISM, C W72-04743
TION LAGOONS, *ALGAE, *TRICKLING FILTERS, EUTROPHICATION, PHOTOSYN W71-07097
ES, CALCIUM CARBONATE, PROTEINS, FILTERS, ENGINEERS ESTIMATES, ECO W70-10174
*GLASS FIBER FILTERS, *MEMBRANE FILTERS, GRANGE LAKE, EAST TWIN L W72-13673
L ANALYSIS, EFFLUENTS, TRICKLING FILTERS, HYDROLYSIS, BIOLOGICAL T W70-09907
TES, ACTIVATED SLUDGE, TRICKLING FILTERS, LAGOONS, TERTIARY TREATM W71-09524
TION COUNTING, GLUCOSE, MEMBRANE FILTERS, MONOSACCHARIDES, MEMBRAN W72-14326
AE, DENITRIFICATION, CALIFORNIA, FILTERS, PARTICLE SIZE, ANAEROBIC W71-08223
ZATION, SEDIMENTATION, TRICKLING FILTERS, WASTE WATER TREATMENT.: / W70-10427
EEN ALGAE, *HORMOTILA BLENNISTA, FILTRATES, GROWTH AUTOSTIMULATION W71-08687
E, ACTIVATED CARBON, ADSORPTION, FILTRATION, CHLORINATION, CHEMICA W70-10381
LING, *ALGAE, *S/ *FLOCCULATION, FILTRATION, ELECTROPHORESIS, SETT W71-01899
FLOW, FLOCCULATION, COAGULATION, FILTRATION, VISCOSITY.: /SLUDGE, W70-06968

T PROBABLE-NUMBER TEST, MEMBRANE FILTRATION, RHODOPSEUDOMONAS GELA W72-04743
ORINATION, PHOSPHATES, NITRATES, FILTRATION, POLLUTION ABATEMENT, W72-05315
LTERS, MONOSACCHARIDES, MEMBRANE FILTRATION.: /LUCOSE, MEMBRANE FI W72-14326
ERS, CATIONS, ANIONS, CHLORELLA, FILTRATION, ELECTROPHORESIS, TRAC W72-14840
ATE, HYDROGEN ION CONCENTRATION, FILTRATION, PLA: / CALCIUM CARBON W73-01669
DIOISOTOPES, ALGAE, FRESH WATER, FILTRATION, LEACHING, FUNGI, LIGH W72-07143
R, ACTIVATED CARBON, ADSORPTION, FILTRATION, ALGAE, PHENOLS, DETER W72-08357
ACTIONS, DIATOMS, AQUATIC ALGAE, FILTRATION, CYANOPHYTA, *EUTROPHI W72-13673
NTROLS, AERATION, SEDIMENTATION, FILTRATION, ANAEROBIC DIGESTION, W72-00042
MAND, BIOCHEMICAL OXYGEN DEMAND, FILTRATION, DISINFECTION, FLOCCUL W72-00411
ON, FLOCCULATION, SEDIMENTATION, FILTRATION, LIME, ROCKS, METABOLI W71-13341
ION, HYDROGEN ION CONCENTRATION, FILTRATION, COLLOIDS, CARBON DIOX W72-03613
IODEGRADATION, LABORATORY TESTS, FILTRATION, AMMONIA, HYDROGEN ION W72-01881
CTROLYTES, LIME, NEUTRALIZATION, FILTRATION, CENTRIFUGATION, VACUU W72-10237
D SOLIDS, SEPARATION TECHNIQUES, FILTRATION, MUNICIPAL WASTES, IND W72-09386
RESI/ *MICROSTRAINING, *PRECOAT FILTRATION, BACKWASHING, SPECIFIC W72-09393
NITRATES, ALGAE, DENTRIFICATION, FILTRATION, DESALINATION, DRAINAG W71-11698
KE ONTARIO, NUTRIENTS, NITROGEN, FILTRATION, MICROSCOPY, INDUSTRIA W71-11507
PARTINA SPP, DISTICHLIS SPICATA, FIMBRISTYLIS, BORRICHIA, PHRAGMIT W73-01089
ANTS, EUTROPHICATION, GOVERNMENT FINANCE, ENVIRONMENTAL SANITATION W72-01659
NVIRONMENTAL EFFECTS, GOVERNMENT FINANCE,: /MMUNITY DEVELOPMENT, E W72-01660
*EUTROPHICATION, *ALGAL CONTROL, FINANCING, PLANT GROWTH, TOXICITY W71-13172
SPHATES, *OLIGASAPROBIC STREAMS, FINGERNAIL CLAMS, UNIONIDS, ORCON W72-06526
HOSPHATE CONCENTRATIONS, GULF OF FINLAND.: ORTHOP W72-13871
FINLAND, *BOTHNIAN BAY, *GULF OF FINLAND, DARK FIXATION, COASTAL W W72-03228
, *PRIMARY PRODUCTIVITY, ALASKA, FIORDS, SURFACE WATERS, NITROGEN, W72-09122
, SANBORN FILTER, BACKWASHING.: FIREBAUGH(CALIF), VOLATILE SOLIDS W70-10174
INEA VAR. DAWSONII, CALLOPHYLLIS FIRMA, PLENOSPORIUM DASYOIDES, SE W73-02099
ATER, *FISHERIES, WATER QUALITY, FISH ATTRACTANTS, FISH HARVEST, P W70-09596
LING TOWERS, DENSITY, VISCOSITY, FISH BEHAVIOR, FERTILITY, SEASONA W70-09612
E, *FISHERIES, FISH POPULATIONS, FISH BEHAVIOR, INVERTEBRATES, OTT W73-00764
AL PROPERTIES, MINERALOGY, FISH, FISH BEHAVIOR, CHEMICAL ANALYSIS, W72-14280
CHNIQUES, FOOD CHAINS, CULTURES, FISH DIET, GAS CHROMATOGRAPHY, PA W73-02105
S, DAPHNIA, AQUATIC POPULATIONS, FISH DIETS, FRESHWATER FISH, POND W72-07132
COPEPODS, MARINE ALGAE, SHRIMP, FISH EGGS, FISH, BACTERIA, FISH P W72-08142
FERTILIZERS, FARM WASTES, FISH, FISH EGGS, TURBIDITY, NUTRIENTS, W71-06444
ATION, DIATOMS, CURRENTS(WATER), FISH FARMING.: /ANKTON, STRATIFIC W73-00236
RIVER(CALIF), RIFLE FALLS(COLO), FISH FECAL WASTES, RESIDUAL FOOD, W71-11006
, ALGAE, RESOURCES, FOOD CHAINS, FISH FOOD ORGANISMS, OXYGEN, ECOS W71-13246
OROPHYTA, *PHYTOTOXINS, ECOLOGY, FISH FOOD ORGANISMS, PATH OF POLL W71-08597
WATER QUALITY, FISH ATTRACTANTS, FISH HARVEST, PLANKTON, BENTHIC F W70-09596
, CONDENSERS, DESIGN, STANDARDS, FISH KILL, ALGAE, COOLING TOWERS, W70-09612
MEDIES, FOODS, FISH, HERBIVORES, FISH MANAGEMENT.: /BIOCONTROL, RE W71-09163
RIMP, FISH EGGS, FISH, BACTERIA, FISH PARASITES, ANTIBIOTICS(PESTI W72-08142
UATIC LIFE, AQUATIC ENVIRONMENT, FISH PHYSIOLOGY, ALGAE, DAPHNIA.: W72-12257
ALGI/ DETOXIFICATION, SYNERGISM, FISH PONDS, ALGISTATIC ACTIVITY, W71-06189
GAE, PHYTOPLANKTON, ZOOPLANKTON, FISH POPULATIONS, WATER TEMPERATU W72-04853
ASKA, *MARINE ALGAE, *FISHERIES, FISH POPULATIONS, FISH BEHAVIOR, W73-00764
Y, TROPHIC LEVEL, AQUATIC ALGAE, FISH POPULATIONS, FLORIDA, WATER W72-08986
RONMENTAL EFFECTS, AQUATIC LIFE, FISH POPULATIONS, PHYTOPLANKTON, W72-00939
SOURCES, *FISHERIES, OIL WASTES, FISH REPRODUCTION, WASTE DISPOSAL W71-13723
TOXICITY, POISONS, ALGAL TOXINS, FISH TOXINS, SEDIMENTS, SALT MARS W72-10678
AL CONTROL, SPAWNING, WARM-WATER FISH.: /ANAEROBIC CONDITIONS, ALG W70-09595
*CLUSTER ANALYSIS(FISH), *DICKERSON PLANT.: W72-07791
IOASSAY, *ABSORPT/ *DDT, *MARINE FISH, *PHYTOPLANKTON, *SHRIMP, *B W73-02105
FRESHWATER, HYDROBIOLOGY, ALGAE, FISH, ABSORPTION, BIOASSAY, METAB W72-10694
POLLUTANTS, MARINE ALGAE, MARINE FISH, ABSORPTION, WATER POLLUTION W72-00940
Y, *BIOASSAYS, LABORATORY TESTS, FISH, ALGAE, CHEMICAL WASTES, ALG W72-08683
NITY, SEDIMENTS, EUTROPHICATION, FISH, ALGAE, RECREATION, ENVIRONM W71-10577
UNOFF, DOMESTIC WASTES, ECOLOGY, FISH, ALGAE, AQUATIC PLANTS, PROT W73-01972
TRATIFICATION, HYDRAULIC MODELS, FISH, ALGAE, FATHOMETERS, BENTHIC W70-09614
ESTS, NITRATES, MUSSELS, SNAILS, FISH, ALGAE, GASTROPODS, ADSORPTI W71-09168
SOILS, PLANT TISSUES, FRESHWATER FISH, AQUATIC ALGAE, EVALUATION, W73-01673
MARINE ALGAE, SHRIMP, FISH EGGS, FISH, BACTERIA, FISH PARASITES, A W72-08142
DERGROUND, ALASKA, PONDS, WELLS, FISH, BAYS, AIR POLLUTION EFFECTS W72-03347
PESTICIDE TOXICITY, FOOD CHAINS, FISH, BIOASSAY, ALGAE, SNAILS, MO W72-07703
CONTROL, *ALGAE CONTROL, *FORAGE FISH, BIOCONTROL, REMEDIES, FOODS W71-09163
, SEWAGE LAGOONS, BRACKISH WATER FISH, BIOLOGICAL TREATMENT, OXIDA W72-12567
CS, *MAINE, MUDS, INVERTEBRATES, FISH, BIRDS, STREAMS, AQUATIC PLA W72-12930
ICATORS, AGRICULTURAL CHEMICALS, FISH, CHEMICALS, CHEMICAL WASTES, W73-01976
BIOTA, BACTERIA, BENTHIC FAUNA, FISH, ENVIRONMENTAL EFFECTS, RESE W72-05282
TIC ALGAE, AQUATIC PRODUCTIVITY, FISH, FISHKILL, LIMNOLOGY, NUTRIE W72-04279
OSION, FERTILIZERS, FARM WASTES, FISH, FISH EGGS, TURBIDITY, NUTRI W71-06444
OLOGICAL PROPERTIES, MINERALOGY, FISH, FISH BEHAVIOR, CHEMICAL ANA W72-14280
*CYANOPHYTA, EXPLOSIVES, COSTS, FISH, FISHKILL, RESERVOIRS, FISHE W72-11564
WEEDS, RECREATION, GROUNDWATER, FISH, FLOOD PLAINS, IRRIGATION, P W70-05103
MINOUS MATERIALS, ANIMAL WASTES, FISH, FLUORESCENCE, INFRARED RADI W72-08790
ASTS, INSECTS, MILDEWS, E. COLI, FISH, FOODS, GRASSES, WORMS.: /YE W73-01676
HYTES, CYPRINUS CARPIO, MOSQUITO FISH, GAMBUSIA AFFINIS, CESIUM RA W73-01673
SH, BIOCONTROL, REMEDIES, FOODS, FISH, HERBIVORES, FISH MANAGEMENT W71-09163
CHAINS, *ECOSYSTEMS, ABSORPTION, FISH, INDUSTRIAL WASTES, ALGAE, A W72-05952
OLLUTION EFFECTS, MUD, MOLLUSKS, FISH, INSECTS, DDT, SILTS, SUCKER W70-08098
TA, CARBON RADIOISOTOPES, MARINE FISH, INSTRUMENTATION, ABSORPTION W72-06283
ALKALINITY, VOLUMETRIC ANALYSIS, FISH, INTAKE STRUCTURES, CARBON D W72-05594
TER HYACINTH, NUTRIENTS, MIDGES, FISH, LAKES, CONDUCTIVITY, ALGAE, W71-01488
YSIS, RADIOACTIVITY, ADSORPTION, FISH, LAKE ERIE, SEDIMENTS, SILTS W71-11036
PES, PHYSICOCHEMICAL PROPERTIES, FISH, MARINE PLANTS, MARINE ALGAE W72-04463
WATER, *BIOTA, SAMPLING, MARINE FISH, MARINE ANIMALS, MARINE ALGA W71-09018
T TOLERANCE, CRUSTACEANS, MARINE FISH, MARSH PLANTS.: /ALGAE , SAL W73-00855
GREAT LAKES, AQUATIC ALGAE, ICE, FISH, MICROORGANISMS, ODOR, TASTE W73-01669

AIN, COASTS, CRUSTACEANS, MARINE FISH, MOLLUSKS, POLLUTANT IDENTIF W73-00819
LEAR WASTES, RIVERS, ABSORPTION, FISH, MOLLUSKS, MARINE ANIMALS, C W71-09225
ALGAE, *MARINE ANIMALS, *MARINE FISH, MUSSLES, LOBSTERS, ABSORPTI W73-00832
S, PERSISTENCE, BACTERIA, BIRDS, FISH, ON-SITE INVESTIGATION, METH W72-00155
G, ENVIRONMENTAL EFFECTS, MARINE FISH, PERIPHYTON, ON-SITE DATA CO W73-00932
TY, CRABS, ALGAE, INVERTEBRATES, FISH, POLLUTANTS, TURBIDITY, ORGA W71-01475
ULATIONS, FISH DIETS, FRESHWATER FISH, PONDS, LAKES, FOOD HABITS, W72-07132
S, PATH OF POLLUTANTS, BIOASSAY, FISH, RAINBOW TROUT, AQUATIC LIFE W72-12257
OPHIC LEVELS, TURBIDITY, OXYGEN, FISH, RECREATION, INVERTEBRATES, W71-11009
MENT, TOXICITY, ALGAE, BACTERIA, FISH, SEWAGE SLUDGE, FLOW, FLOCCU W70-06968
ISTILLATION, ALGAE, AQUICULTURE, FISH, SHELLFISH, CRUSTACEANS, OYS W70-08832
RS, ZOOPLANKTON, MUSSELS, CLAMS, FISH, SHELLFISH, AQUEOUS SOLUTION W72-05965
REACTANTS, *TOXICITY, *BIOASSAY, FISH, SHELLFISH, ALGAE, DEGRADATI W71-09789
*SEPARATION TECHNIQUES, *MARINE FISH, SILICA, MARINE ALGAE, INVER W73-00371
ANIMALS, FRESH WATER, FRESHWATER FISH, TEMPERATURE, KINETICS, NUCL W71-09182
ICALS, PLANKTON, BENTHOS, ALGAE, FISH, WATER POLLUTION, WATER POLL W72-07841
SYSTEMATICS, LIMNOLOGY, ECOLOGY, FISH, WATER POLLUTION, PROTOZOA, W73-01615
MICAL OXYGEN DEMAND, PHOSPHORUS, FISH, WEEDS, CYCLES, BACTERIA, CH W71-03012
SOURCES, TRIBUTARIES, FRESHWATER FISH, YELLOW PERCH, SEWAGE EFFLUE W72-14282
IUM, CARBON DIOXIDE, CYANOPHYTA, FISH,: /MANGANESE, POTASSIUM, SOD W73-00840
ORINATED HYDROCARBON PESTICIDES, FISHERIES.: /AE, FOOD CHAINS, CHL W72-13657
IES, INTERTIDAL AREAS, ESTUARINE FISHERIES.: /, PESTICIDES, ESTUAR W72-10678
STS, FISH, FISHKILL, RESERVOIRS, FISHERIES, PRESSURE.: /CSIVES, CO W72-11564
ITROGEN, PHOSPHORUS, RESERVOIRS, FISHERIES, WATER SUPPLY, COSTS, W W71-13172
ARINE ALGAE, FOOD CHAINS, MARINE FISHERIES, LAKES, POLLUTION ABATE W72-00960
, SMALL WATERSHEDS, CROPS, EGGS, FISHERIES, ON-SITE INVESTIGATIONS W72-00959
ECOLOGY, FOOD CHAINS, ESTUARINE FISHERIES, MARINE FISHERIES, SEWA W72-12567
INS, ESTUARINE FISHERIES, MARINE FISHERIES, SEWAGE DISPOSAL, SEWAG W72-12567
NTS, ALGAE, TOURISM, RECREATION, FISHERIES, HYDROELECTRIC POWER, N W72-14878
*REVIEWS, *ILLINOIS, RECREATION, FISHING, SWIMMING, BOATING, DOCUM W72-05869
GAE, AQUATIC PRODUCTIVITY, FISH, FISHKILL, LIMNOLOGY, NUTRIENTS, O W72-04279
LUTION CONTROL, *BIODEGRADATION, FISHKILL, EUTROPHICATION, ALGAL B W71-08768
K SALMON, STRIPED BASS, MINNOWS, FISHKILL, HERRING, COPEPODS, CRUS W71-11517
OPHYTA, EXPLOSIVES, COSTS, FISH, FISHKILL, RESERVOIRS, FISHERIES, W72-11564
LATION, WATER POLLUTION CONTROL, FISHKILL, NEW HAMPSHIRE, WATER PO W72-09304
GLENODINIUM FOLIACUM, GYRODINIUM FISUM.: /, SKELETONEMA COSTATUM, W72-07166
KE ERKEN(SWEDEN), *DARK NITROGEN FIXATION.: *LA W72-08054
HERMAL WATER, *WYOMIN/ *NITROGEN FIXATION, *ALGAE, *CYANOPHYTA, *T W72-08508
NTIAL, DENITRIFICATION, NITROGEN FIXATION, NITROGEN CYCLE, AMMONIA W72-08459
SYNTHESIS, *CYANOPHYTA, NITROGEN FIXATION, OXYGEN, CARBON DIOXIDE, W72-08584
ANOPHYTA, PHYTOPLANKT/ *NITROGEN FIXATION, *ALGAE, MEASUREMENT, CY W72-08054
*NITROGEN, OXYGEN, BI/ *NITROGEN FIXATION, *SYNTHESIS, *CULTURES, W72-07941
ATION, DENITRIFICATION, NITROGEN FIXATION, SEDIMENTS, NITRATES, WI W72-07933
N, CHEMICAL PORPERTIES, NITROGEN FIXATION, CULTURES, CARBON DIOXID W72-14790
TRO/ *ALGAE, *LEGUMES, *NITROGEN FIXATION, *CYANOPHYTA, PLANTS, NI W73-01456
VERTEBRATES, HEAT STERILIZATION, FIXATION, EMBEDDING, ASHING, BONE W73-01676
*BIOMASS, CHLOROPHYLL, NITROGEN FIXATION, *PERIPHYTON, *PHYTOPLAN W72-10883
, ECONOMICS, NUTRIENTS, NITROGEN FIXATION, FORECASTING, PHOTOSYNTH W71-13183
YANCY, EUTROPHICATION, NIRTROGEN FIXATION, BIOCONTROL, PLANT PHYSI W71-13184
ALGAE, RICE, NITROGEN, NITROGEN FIXATION, 2-4-D, DALAPON, HERBICI W71-12071
*CYANOPHYTA, *ALGAE, *NITROGEN FIXATION, ENZYMES, BACTERIA, SPOR W71-12070
NIAN BAY, *GULF OF FINLAND, DARK FIXATION, COASTAL WATERS, TROPHOG W72-03228
IOLOGY, CYTOLOGICAL S/ *NITROGEN FIXATION, *CYANOPHYTA, PLANT PHYS W72-03217
US, BELGIUM.: RUTHENIUM, FIXATION, TISSUES, ALBURNUS LUCID W71-09168
HNIQUES, *PHOTOSYNTHESIS, CARBON FIXATION, EUTROPHICATION, WATER Q W72-06046
, DIVERSION, LAKE ERIE, NITROGEN FIXATION, DETERGENTS, CARBON DIOX W72-05473
NITROGEN, CHLOROPHYTA, NITROGEN FIXATION, LAND MANAGEMENT, FERTIL W72-04769
HYTOPLANKTON, NITRATES, NITROGEN FIXATION, ALGAE, CONDUCTIVITY, CH W72-04773
SIUM, DIVERSION, ALGAE, NITROGEN FIXATION, PLANKTON, PHYTOPLANKTON W71-00117
N, *ALGAE, *NUTRIENTS, *NITROGEN FIXATION, LAKES, NITROGEN, PHOSPH W72-13692
IA, *RAINFALL/ *ALGAE, *NITROGEN FIXATION, *NITROGEN FIXING BACTER W72-12505
GROWTH RATES, ANABAENA, NITROGEN FIXATION, DENITRIFICATION, WATER W72-12544
, CYANOPHYTA,/ *ALGAE, *NITROGEN FIXATION, *TROPICAL REGIONS, RICE W73-02473
HT INTENSITY, EUTROPH/ *NITROGEN FIXATION, *LAKES, CYANOPHYTA, LIG W73-02472
MATIC ZONES, / *ALGAE, *NITROGEN FIXATION, *SOILS, *TEMPERATE, CLI W73-02471
E, *NITROGEN FIXATION, *NITROGEN FIXING BACTERIA, *RAINFALL, *ARID W72-12505
ENVIR/ *MICROBIOLOGY, *NITROGEN FIXING BACTERIA, *NITROGEN CYCLE, W72-11563
ANALYTICAL TECHNIQUES, NITROGEN FIXING BACTERIA, THERMAL POLLUTIO W72-08508
S, *INDICATORS, *EUTROPHICATION, FJORDS, BIOASSAY, CHLORELLA, NUTR W73-00234
S, RHIPOCEPHALUS PHOENIX, UDOTEA FLABELLUM, BRYOPSIS, CAULERPA, AC W72-14294
FERA, RADIOLARIA, TRICHODESMIUM, FLAGELLATES, CILIATES, NAUPLII, C W72-14313
, COUNTING, SAMPLE PRESERVATION, FLAGELLATES, RHODOMONAS, IKROAVIC W73-00273
ERATIUM EGYPTIACUM, *SUEZ CANAL, FLAGELLATES, MEDITERRANEAN SEA, R W73-00296
SARGASSO SEA, COCCOLITHOPHORIDS, FLAGELLATES, FATE OF POLLUTANTS, W73-00380
OSPHATASE, REASSIMILATION, GREEN FLAGELLATE, CYCLOTELLA CRYPTICA, W71-10083
ULTURING TECHNIQUES, MORPHOLOGY, FLAGELLATES.: /, CULTURE MEDIA, C W72-12632
T, *GROWTH KINETICS, CHEMOSTATS, FLAGELLATE ALGAE, OPTICAL DENSITY W70-10403
RSONI, AMEIURUS NEBULOSUS, PERCA FLAVESCENS, AMBLOPLITES RUPESTRIS W70-08098
PERFUSION UNIT, FLAVOBACTERIUM, MICROCOCCUS.: W71-08027
OPHYCEAE, VIBRIO, ACHROMOBACTER, FLAVOBACTERIA.: /IOPHYCEAE, XANTH W72-08051
INEOPLASTIC ACTIVITY, ALKALOIDS, FLAVONOIDS, STEROIDS, TANNINS, SA W72-04275
Y, TRACE ELEMENTS, HEAVY METALS, FLECTRODES, NITRATES, ALGAE, COPP W72-01693
HNANTHES MINUTISSIMA, ACHNANTHES FLEXELLA, EUNOTIA PECTINALIS, DES W71-11029
YELLOWSTONE PARK, SYNECHOCOCCUS, FLEXIBACTERIA, PHORMIDIUM, MASTIG W72-03557
AKE(MICH), SCIRPUS ACUTUS, NAJAS FLEXILIS.: /ETABOLISM, LAWRENCE L W72-03216
COLONIZATION, CAULOBACTER, NAJAS FLEXILIS, SCIRPUS ACUTUS, C-14, D W73-02095
, SHIGELLA DYSENTERIAE, SHIGELLA FLEXNERI, BIOCHEMICAL STUDIES, AR W72-13800
VACCINIUM MYRTILLUS, DESCHAMPSIA FLEXUOSA, PARMELIA PHYSODES, HYPN W72-14303
TION, TEMPERATURE, CHLORINATION, FLOATATION, DEWATERING, DRYING, C W71-08951
LGAE, *FOULING, *NUISANCE ALGAE, FLOATING PLANTS, SEAWATER, PLANKT W71-09261

ANALYSIS, MOSQUITOES, BACTERIA,	FLOATING PLANTS, ON-SITE INVESTIG	W71-01488
*EUTROPHICATION, *DISTRIBUTION,	FLOATING, WINDS, ALGAE, SURFACE W	W73-00236
CARBIDE CORP, POLYOX, CARBOWAX,	FLOCCULANT DOSAGES, BARCCO CLAY S	W70-06968
ERIA, FISH, SEWAGE SLUDGE, FLOW,	FLOCCULATION, COAGULATION, FILTRA	W70-06968
*ION EXCHANGE, DISSOLVED SOLIDS,	FLOCCULATION, HEAD LOSS, HYDROGEN	W71-00138
INATION, CHEMICAL PRECIPITATION,	FLOCCULATION, FLOTATION.: / CHLOR	W70-10381
ARY TREATMENT, ACTIVATED CARBON,	FLOCCULATION, ALGAE, SLUDGE, CHRO	W71-09524
TES, SOLAR RADIATION, FLOTATION,	FLOCCULATION, PERFORMANCE, EFFICI	W72-12289
ICAL OXYGEN DEMAND, PERFORMANCE,	FLOCCULATION, FLOTATION, ECONOMIC	W72-13508
EMAND, FILTRATION, DISINFECTION,	FLOCCULATION, *WASTE WATER TREATM	W72-00411
NITRATES, AMMONIA, COAGULATION,	FLOCCULATION, SEDIMENTATION, FILT	W71-13341
LYTICAL TECHNIQUES, COALESCENCE,	FLOCCULATION, LIQUID WASTES, SEPA	W71-13356
OSS, PRESSURE, TURBIDITY, ALGAE,	FLOCCULATION, CLEANING, *WATER TR	W72-09393
AGILARIA CROTONENSIS, TABELLARIA	FLOCCULOSA, STEPHANODISCUS HANTZS	W73-00384
ERVATION, WILDLIFE CONSERVATION,	FLOOD CONTROL, RECREATION, HYDROL	W72-05594
, RECREATION, GROUNDWATER, FISH,	FLOOD PLAINS, IRRIGATION, PESTICI	W70-05103
DIOACTIVITY TECHNIQUES, *BENTHIC	FLORA, *GROWTH RATE, *AQUATIC PRO	W71-11486
XICO, *ELECTRIC POWERP/ *BENTHIC	FLORA, *MARINE ALGAE, *GULF OF ME	W72-11252
L COMMUNITIES,/ *ALGAE, *BENTHIC	FLORA, *OREGON, RIVERS, BIOLOGICA	W72-07901
RIENTS, *BENTHIC FAUNA, *BENTHIC	FLORA, *PHYTOPLANKTON, ALGAE, GRO	W72-06046
ERGY CO/ *MARINE ALGAE, *BENTHIC	FLORA, *PRIMARY PRODUCTIVITY, *EN	W72-06283
HYPOLIMNION, *LAKE ERIE, BENTHIC	FLORA, ALGAE, LIGHT PENETRATION,	W72-13659
ACE ELEMENTS, VOLCANOES, BENTHIC	FLORA, ALKALINITY, PHOTOSYNTHESIS	W71-09190
STROPODS, SNAILS, CRABS, BENTHIC	FLORA, BENTHIC FAUNA, DISTRIBUTIO	W73-00932
C OCEAN, MARINE BIOLOGY, BENTHIC	FLORA, BIOASSAY, CYANOPHYTA, *FLO	W72-11876
E, *LAKE ERIE, PLANKTON, BENTHIC	FLORA, DIATOMS, PHYTOPLANKTON, PH	W73-01615
, *DIATOMS, CHLOROPHYTA, BENTHIC	FLORA, EUTROPHICATION, DOMINANT O	W72-14797
POLLUTION, LEGISLATION, BENTHIC	FLORA, F: /, WASTE TREATMENT, AIR	W71-13723
BIOMASS, BENTHIC FAUNA, BENTHIC	FLORA, HUMAN POPULATION, PRIMARY	W72-12549
ORAL, LIGHT PENETRATION, BENTHIC	FLORA, LAKES, DEPTH, DIATOMS, CHL	W71-11029
HOMETERS, BENTHIC FAUNA, BENTHIC	FLORA, LAKE ONTARIO, HUDSON RIVER	W70-09614
*PRIMARY PRODUCTIVITY, *BENTHIC	FLORA, MICROORGANISMS, ALGAE, LAK	W72-07501
MENTS, ALGAE, CHLORELLA, BENTHIC	FLORA, MUD, MICROORGANISMS, PROTE	W71-08042
GANISMS, WATER ANALYSIS, BENTHIC	FLORA, ORGANIC MATTER, DISSOLVED	W73-02100
, *SYSTEMATICS, *COASTS, BENTHIC	FLORA, RHODOPHYTA, PHAEOPHYTA, CH	W71-10066
PRODUCTIVITY, *SHOALS, *BENTHIC	FLORA, SEA WATER, SEDIMENTS, LIGH	W72-14317
ANDERSON-CUE LAKE, MELROSE(	FLORIDA).:	W72-08986
, PATUXENT ESTUARY, BISCAYNE BAY(	FLORIDA), HUMBOLT BAY(CALIFORNIA)	W71-11517
), HEATED EFFLUENT, TURKEY POINT(	FLORIDA), TURTLE GRASS, THALASSIA	W72-11876
POINT(FLORIDA), / *BISCAYNE BAY(	FLORIDA), HEATED EFFLUENT, TURKEY	W72-11876
VER, TURKEY POINT, BISCAYNE BAY,	FLORIDA, BALANUS, BARNACLES, EPIF	W72-07907
ENSITY, AMMONIA, CARBON DIOXIDE,	FLORIDA, CULTURES, *ALGAE, INCUBA	W72-09103
FORNIA, WASHINGTON, CONNECTICUT,	FLORIDA, DELAWARE RIVER, INDIANA.	W71-07417
POLLUTION, *AQUATIC ENVIRONMENT,	FLORIDA, ELECTRIC GENERATORS, WAT	W72-11252
T MARSHES, *PROTOZOA, *BACTERIA,	FLORIDA, RADIATION SENSITIVITY, R	W73-00796
IOINDICATORS, INDUSTRIAL WASTES,	FLORIDA, SESSILE ALGAE, ESTUARIES	W71-07412
AQUATIC ALGAE, FISH POPULATIONS,	FLORIDA, WATER POLLUTION EFFECTS,	W72-08986
OXYGEN INACTIVATION./ *ANABAENA	FLOS-AQUA, *NITROGENASE ACTIVITY,	W72-07941
*CYANOPHAGES, *APHANIZOMENON	FLOS-AQUAE, LAKE ERKEN(SWEDEN).:	W73-00246
WEDEN, LAKE MALAREN, MICROCYSTIS	FLOS-AQUAE, PHRAGMITES COMMUNIS,	W73-00286
ABAENA CYLINDRICA, APHANIZOMENON	FLOS-AQUAE.: /AENA CIRCINALIS, AN	W72-10613
E(OHIO), SPHAGNUM, APHANIZOMENON	FLOS-AQUAE, CERATIUM, OSCILLATORI	W71-11949
*ALKALINE PHOSPHATASE, *ANABAENA	FLOS-AQUAE, *CONTINUOUS CULTURE,	W72-05476
HERAPEUTIC VALUE, *APHANIZOMENON	FLOS-AQUAE, *PHARMACOLOGICAL TEST	W72-04259
DISCUS HANTZSCHII, APHANIZOMENON	FLOS-AQUAE, CHROOMONAS ACUTA, MEL	W72-12738
A, ELODEA, CALLITRICHE, ANABAENA	FLOSAQUAE, SPIRULINA, GYMNODINIUM	W73-00840
*FRACTIONATION-	FLOTATION PROCESS.:	W71-12200
CAL PRECIPITATION, FLOCCULATION,	FLOTATION.: / CHLORINATION, CHEMI	W70-10381
, CENTRIFUGATION, VACUUM DRYING,	FLOTATION, WASTEWATER TREATMENT.:	W72-10237
, GROWTH RATES, SOLAR RADIATION,	FLOTATION, FLOCCULATION, PERFORMA	W72-12289
MAND, PERFORMANCE, FLOCCULATION,	FLOTATION, ECONOMIC FEASIBILITY,	W72-13508
UT RIVER, SAN JOAQUI/ BARNACLES,	FLOUNDER, HUMBOLDT BAY, CONNECTIC	W71-07417
R 1254, ESCAMBIA RIVER, PINFISH,	FLOUNDER, CROAKER, MENHADEN.: /LO	W71-07412
A LANOSA, CHONDRUS CRISPUS, COD,	FLOUNDER, CRAB EGGS, PELAGIC EGGS	W72-08142
OTOPES, FOOD CHAINS, IRRIGATION,	FLOW PROFILES, MARINE ALGAE, WIND	W71-09225
ASSAYS, BIO-INDICATORS, STORAGE,	FLOW RATES, INHIBITION, ANALYTICA	W72-04070
PROTOZOA, BACTERIA, STREAMFLOW,	FLOW RATES, WATER TEMPERATURE, WA	W73-01972
OAQUIN DELTA(CALIF), *CONTINUOUS	FLOW TURBIDOSTAT, *GROWTH KINETIC	W70-10403
QUAE, DESTRATIFICATION, ANABAENA	FLOW-AQUAE, HOMOGENIZING, TRA: /A	W72-07890
LAKE, SQUAM LAKE, APHANIZOMENON	FLOW-AQUAE, DESTRATIFICATION, ANA	W72-07890
DESALINATION, ALGAE, D/ *RETURN	FLOW, *NITRATES, WATER POLLUTION,	W71-08223
ODYNAMICS, *FRICTION, *TURBULENT	FLOW, ALGAE, BACTERIA, MARINE PLA	W72-03220
INE, HYDROGEN ION CONCENTRATION,	FLOW, CHEMICAL PROPERTIES, PHYSIC	W72-12941
, BACTERIA, FISH, SEWAGE SLUDGE,	FLOW, FLOCCULATION, COAGULATION,	W70-06968
, NITRATES, SURFACE RUNOFF, BASE	FLOW, PERCOLATION, LEACHING, DRAI	W71-06443
POLLUTION SOURCES, ALGAE, RIVER	FLOW, PHOSPHORUS, NITRATES, POTAS	W72-04260
ITY, *IMPOUNDED WATERS, STREAMS,	FLOW, SEASONAL, STRATIFICATION, N	W71-03028
, ALGAE, PESTICIDES, DDT, RETURN	FLOW, SEDIMENTS, SOIL EROSION, FE	W71-06444
AINAGE WATER, IRRIGATION, RETURN	FLOW, WATER SUPPLY, LIVESTOCK, SE	W71-06435
ICOAGULANT POTENTIAL, SPARGANIUM	FLUCTUANS, ANTINEOPLASTIC ACTIVIT	W72-04275
YSICCHEMICAL PROP/ *WATER LEVEL	FLUCTUATIONS, *PUMPED STORAGE, PH	W72-14280
OURCES, WATER POLLUTION EFFECTS,	FLUCTUATION, COMPARATIVE PRODUCTI	W72-03228
OTOSYNTHESIS, RAINFALL, DIURNAL,	FLUCTUATIONS, CHLOROPHYTA, EUGLEN	W72-00150
, *DIATOMS, SEA WATER, SEASONAL,	FLUCTUATIONS, CONTINENTAL SHELF,	W72-07663
LITY, WATER ANALYSIS, FREQUENCY,	FLUCTUATIONS, BIOLOGICAL COMMUNIT	W73-00918
ROWTH SUBSTANCES, PHYTOPLANKTON,	FLUID MECHANICS, EXUDATION, OCEAN	W71-07878
SUES, HISTOLOGICAL STUDIES, BODY	FLUIDS, VERTEBRATES, HEAT STERILI	W73-01676
, TETRAHYMENA VORAX, PSEUDOMONAS	FLUORESCENS, AUTOTROPHIC BACTERIA	W73-02095
NGI, EQUIPMENT, INSTRUMENTATION,	FLUORESCENCE, WATER TEMPERATURES,	W72-07225
TOGRAPHY, ULTRAVIOLET RADIATION,	FLUORESCENCE, INFRARED RADIATION,	W72-04742

S, SAMPLING, MUD, PHYTOPLANKTON,	FLUORESCENCE, PHOTOSYNTHETIC BACT	W72-01784
T, CURRENTS(WATER), CHLOROPHYLL,	FLUORESCENCE, BATHYMETRY.: /REMEN	W72-01095
MATERIALS, ANIMAL WASTES, FISH,	FLUORESCENCE, INFRARED RADIATION,	W72-08790
SALMONELLA TYPHIMURI/ DETECTION,	FLUORESCENT ANTIBODY TECHNIQUES,	W72-13800
HYLL, HYDROLYSIS, PHYTOPLANKTON,	FLUORESCENCE, PHOSPHORUS COMPOUND	W72-12637
*CHLORELLA, *INHIBITION, COPPER,	FLUORIDES, IONS, CYTOLOGICAL STUD	W72-11572
, DISSOLVED OXYGEN, FORECASTING,	FLUORIDES, OHIO.: /TES, POTASSIUM	W72-04260
S, *NITRATES, *ELECTRODES, IONS,	FLUORIDES, ALGAE, VOLUMETRIC ANAL	W71-00474
, CHLAMYDOMONAS, WATER ANALYSIS,	FLUOROMETRY, PIGMENTS.: /OMPOUNDS	W72-12637
PHOSPHATES, NITROGEN, SEA WATER,	FLUOROMETRY, EQUIPMENT, INHIBITIO	W72-07660
L STUDIES, SYSTEMATICS, BIOMASS,	FLUOROMETRY, MICROSCOPY, ZOOPLANK	W72-14313
URIC CHLORIDE, DIMETHYL MERCURY,	FLUOROMICROPHOTOMETER.: /IS, MERC	W72-07660
N PUMP, ALUM/ *DETENTION PERIOD,	FLUSHING OPERATION, ACID INJECTIO	W72-02850
TRIACETIC ACID, ALGAL ASSAY PRO/	FLUSHING, GREEN BAY(WIS), NITRILO	W71-12072
K CHANNEL, CHLOROP/ *PORTERINEMA	FLUVIATILE, CULTURE MEDIA, HUMMOC	W72-12633
REIA AFFINIS, ESOX LUCIUS, PERCA	FLUVIATILIS, SALVELINUS ALPINUS,	W72-12729
, CYCLOTELLA NANA, THALASSIOSIRA	FLUVIATILIS, AMPHIDINIUM CARTERI,	W72-12998
CLOTELLA CRYPTICA, THALASSIOSIRA	FLUVIATILIS, DUNALIELLA TERTIOLEC	W71-10083
*RESIDUES, *DDD, DDE, TDE, BLACK-	FLY LARVAE, CLADOPHORA, CYPRINUS	W70-08098
WYANDOTTE CHEMICAL CORPORATION,	FMC CORPORATION, SOAP AND DETERGE	W70-07283
*USSR, MOSCOW RIVER, OKA RIVER,	FOAM, NATURAL WATERS.:	W72-05506
ANTIBIOTICS, IODINE, ATOMIC EN/	FODDER, GLUE, AGAR, POLYSIPHONIA,	W71-09157
KELETONEMA COSTATUM, GLENODINIUM	FOLIACUM, GYRODINIUM FISUM.: /, S	W72-07166
M HERBARUM, TRAMETES VERSICOLOR,	FOMES ANNOSUS, FUSCOPORIA CONTIGU	W72-13813
U/ CHUBS, KINGFISHES, SALVELINUS	FONTINALIS, SEMOTILUS ATROMACULAT	W72-12930
ELLA PYRENOIDOSA, CHLAMYDOMONAS,	FONTINALIS, ASTINASTRUM, ANACYSTI	W71-11519
, ELODEA CANADENSIS, PHRAGMITES,	FONTINALIS, CHARALES, CERATOPHYLL	W71-11012
OPES, ABSORPTION, PACIFIC OCEAN,	FOOD CHA: /, AMPHIPODA, RADIOISOT	W72-03347
SITE INVESTIGATIONS, CALIFORNIA,	FOOD CHAINS, PUBLIC HEALTH, WATER	W72-03324
PLANKTON, MAMMALS, MARINE ALGAE,	FOOD CHAINS, MARINE FISHERIES, LA	W72-00960
-SITE INVESTIGATIONS, FOOD WEBS,	FOOD CHAINS, FROGS, LICHENS, ALGA	W72-00949
NALYTICAL TECHNIQUES, NUTRIENTS,	FOOD CHAINS, BIOLUMINESCENCE, POL	W71-11234
*PRODUCTIVITY, ALGAE, RESOURCES,	FOOD CHAINS, FISH FOOD ORGANISMS,	W71-13246
WASTES, WATER POLLUTION EFFECTS,	FOOD CHAINS, PUBLIC HEALTH, AQUAT	W72-10978
OINDICATORS, PATH OF POLLUTANTS,	FOOD CHAINS, NUCLEAR WASTES, ENVI	W72-10957
OLLUTION SOURCES, AQUATIC ALGAE,	FOOD CHAINS, MEASUREMENT, ADSORPT	W72-10686
STUDIES, BIOLOGICAL COMMUNITIES,	FOOD CHAIN, DIATOMS, LARVAE, DIET	W72-09248
ION, AGRICULTURAL RUNOFF, ALGAE,	FOOD CHAINS, CHLORINATED HYDROCAR	W72-13657
CYCLE, *CRABS, *FISHES, ECOLOGY,	FOOD CHAINS, ESTUARINE FISHERIES,	W72-12567
AMPLING, PLANKTON NETS, DIATOMS,	FOOD CHAINS, ANABAENA, SCENEDESMU	W72-12738
ECTS, TOXICITY, WATER POLLUTION,	FOOD CHAINS, PATH OF POLLUTANTS,	W72-12257
, *KELPS, STARFISH, CRUSTACEANS,	FOOD CHAINS, FOOD WEBS, WATER POL	W72-11925
TION, *BIOASSAY, B/ *ECOSYSTEMS,	FOOD CHAINS, *ZOOPLANKTON, *PREDA	W72-07132
*MICROORGANISMS, *FLUORESCENCE,	FOOD CHAINS, EQUIPMENT, ANALYTICA	W72-07526
ION SOURCES, PESTICIDE TOXICITY,	FOOD CHAINS, FISH, BIOASSAY, ALGA	W72-07703
OSPHORUS PESTICIDES, ACARICIDES,	FOOD CHAINS, WATER POLLUTION SOUR	W73-00238
REGIONS, ULTRAVIOLET RADIATION,	FOOD CHAINS, GAMMA RAYS, RESISTAN	W73-00796
L AREAS, ON-SITE INVESTIGATIONS,	FOOD CHAINS, SPECIATION, REPRODUC	W73-00764
RIENT REMOVAL, *ZINC, CHELATION,	FOOD CHAINS, PATH OF POLLUTANTS,	W73-00802
SORPTION, SEPARATION TECHNIQUES,	FOOD CHAINS, CULTURES, FISH DIET,	W73-02105
ICATION, BIOLOGICAL COMMUNITIES,	FOOD CHAINS, LAKE MORPHOLOGY, RES	W73-00938
CAPACITY, ESTUARINE ENVIRONMENT,	FOOD CHAINS, PUBLIC HEALTH, EUROP	W73-00828
S, CHELATION, ORGANIC COMPOUNDS,	FOOD CHAINS, PUBLIC HEALTH, PHYTO	W73-00821
ENVIRONMENT, PATH OF POLLUTANTS,	FOOD CHAINS, PHYTOPLANKTON, MARIN	W73-00818
, *PHYTOPLANKTON, *MARINE ALGAE,	FOOD CHAINS, PUBLIC HEALTH, NUCLE	W73-00823
UM, IRON, ANALYTICAL TECHNIQUES,	FOOD CHAINS, AFRICA, EFFLUENTS, P	W73-00832
TEINS, RADIOACTIVITY TECHNIQUES,	FOOD CHAINS.: /ICROORGANISMS, PRO	W71-08042
FERTILIZERS, DIATOMACEOUS EARTH,	FOOD CHAINS, OXYGEN, ECOLOGY, RHO	W71-09157
, MINNOWS, CRAYFISH, MOSQUITOES,	FOOD CHAIN, HERRING, OYSTERS, CLA	W71-07417
E ANIMALS, COBALT RADIOISOTOPES,	FOOD CHAINS, IRRIGATION, FLOW PRO	W71-09225
RPTION, WATER POLLUTION EFFECTS,	FOOD CHAINS.: /, SOLUBILITY, ADSO	W71-09182
OCEANS, WATER POLLUTION EFFECTS,	FOOD CHAINES, PUBLIC HEALTH, ESTU	W72-04463
NTS, CLAYS, ODOR, WATER QUALITY,	FOOD CHAINS, DISSOLVED OXYGEN, CO	W72-04759
IN, PATH OF POLLUTANTS, REVIEWS,	FOOD CHAINS, FORECASTING, CONSTRA	W72-06340
, IRON, BIOASSAY, BIOINDICATORS,	FOOD CHAINS, TITANIUM, PLANT POPU	W71-03095
, FRESHWATER FISH, PONDS, LAKES,	FOOD HABITS, EUTROPHICATION, LARV	W72-07132
*PLANKTON, PRIMARY PRODUCTIVITY,	FOOD HABITS, ALGAE, ZOOPLANKTON,	W72-00161
AE, RESOURCES, FOOD CHAINS, FISH	FOOD ORGANISMS, OXYGEN, ECOSYSTEM	W71-13246
YTA, *PHYTOTOXINS, ECOLOGY, FISH	FOOD ORGANISMS, PATH OF POLLUTANT	W71-08597
DIOACTIVITY EFFECTS, ABSORPTION,	FOOD WEBS, STRONTIUM RADIOISOTOPE	W72-00948
LUTANTS, ON-SITE INVESTIGATIONS,	FOOD WEBS, FOOD CHAINS, FROGS, LI	W72-00949
, PIGMENTS, GROWTH RATES, ALGAE,	FOOD WEBS, COPEPODS, ORGANOPHOSPH	W72-07892
ALINE WATER, BIOMASS, NUTRIENTS,	FOOD WEBS, DETRITUS, ECOSYSTEMS,	W73-01089
ICAL COMMUNITIES, TIDAL MARSHES,	FOOD WEBS, INTERTIDAL AREAS, CONN	W73-00796
RFISH, CRUSTACEANS, FOOD CHAINS,	FOOD WEBS, WATER POLLUTION EFFECT	W72-11925
LO), FISH FECAL WASTES, RESIDUAL	FOOD, GREEN RIVER(WASH), COWLITZ	W71-11006
STING, PLANT RESIDUES, RECYCLING	FOOD, PROCESSING WASTES, PATHOGEN	W72-00846
ELLFISH, BIOLOGICAL COMMUNITIES,	FOOD: /AQUACULTURE, COMMERCIAL SH	W72-09378
RAGE FISH, BIOCONTROL, REMEDIES,	FOODS, FISH, HERBIVORES, FISH MAN	W71-09163
INSECTS, MILDEWS, E. COLI, FISH,	FOODS, GRASSES, WORMS.: /YEASTS,	W73-01676
	FORAMINIFERA.:	W73-00796
OPHYLL A, PHEOPHYTIN, TUNICATES,	FORAMINIFERA, RADIOLARIA, TRICHOD	W72-14313
ORIZONTAL COLLECTORS, BRYOZOANS,	FORAMINIFERA, HYDROIDS, POLYCHAET	W72-05432
MEIOFAUNA,	FORAMINIFERA.:	W72-12567
ARGASSO WEED, STARFISH, ASTERIAS	FORBESI, PLUTONIUM 236, PLUTONIUM	W72-05965
OLLUTANTS, REVIEWS, FOOD CHAINS,	FORECASTING, CONSTRAINTS, RESEARC	W72-06340
ES, POTASSIUM, DISSOLVED OXYGEN,	FORECASTING, FLUORIDES, OHIO.: /T	W72-04260
Y EFFECTS, *MATHEMATICAL MODELS,	FORECASTING, COBALT RADIOISOTOPES	W73-00831
S, NUTRIENTS, NITROGEN FIXATION,	FORECASTING, PHOTOSYNTHESIS, BIOM	W71-13183
BIOASSAY, ANALYTICAL TECHNIQUES,	FORECASTING, SILICA.: /TERGENTS,	W71-12072

ITE INVESTIGATIONS, METHODOLOGY,	FOREIGN RESEARCH, MARINE ALGAE, P	W72-08804
NOPHYTA, TREES, GRASSES, LEAVES,	FOREST SOILS, SAMPLING, MUD, PHYT	W72-01784
SOM/ *HUBBARD BROOK EXPERIMENTAL	FOREST, PARTICULATE MATTER, NITRO	W71-01489
VITY EFFECTS, *ECOSYSTEMS, *RAIN	FORESTS, CYCLING NUTRIENTS, SOIL-	W72-00949
ING, SYSTEMS ANALYSIS, DECIDUCUS	FORESTS, DECOMPOSING ORGANIC MATE	W72-03342
LING NUTRIENTS, ECOSYSTEMS, RAIN	FORESTS, MATHEMATICAL MODELS, LIN	W72-03342
ETABOLISM, MOBILITY, PARTICULATE	FORM, FEEDLOTS, NUTRIENT SOURCES.	W71-06443
SCILLATORI/ SAMPLE PRESERVATION,	FORMALIN, COUNTING, CYCLOTELLA, O	W72-10875
YXOBACTERIA, DNA, MOTILITY, CYST	FORMATION, ANTIBIOTIC SENSITIVITY	W72-07951
ES, RIVERS, CYANOPHYTA, GEOLOGIC	FORMATIONS, BENTHOS, MAYFLIES, TU	W71-11009
CHLORIDE, LITHIUM, ASTERIONELLA	FORMOSA, ASTERIONELLA JAPONICA, S	W72-12739
INDERMERE(ENGLAND)/ ASTERIONELLA	FORMOSA, BLELHAM TARN(ENGLAND), W	W72-10625
AS, B/ RHINE RIVER, ASTERIONELLA	FORMOSA, CRYPTOMONAS, CHLAMIDOMON	W71-01491
ILLATORIA CHALYBIA, OSCILLATORIA	FORMOSA, PHORMIDIUM FAVEOLARUM, A	W72-07664
ANODISCUS NIAGARAE, ASTERIONELLA	FORMOSA, RHODOMONAS LACUSTRIS, CO	W73-01435
RHODOMONAS MINUTA, ASTERIONELLA	FORMOSA, STEPHANODISCUS ASTREA, A	W72-12738
*GAS VACUOLES, MOTILE	FORMS, CYANOPHAGES.:	W71-13184
Y EXPER/ *RUTHENIUM 106, SOLUBLE	FORMS, INSOLUBLE FORMS, LABORATOR	W71-13233
UM 106, SOLUBLE FORMS, INSOLUBLE	FORMS, LABORATORY EXPERIMENTS.: /	W71-13233
, NUTRIENTS, ASSESSMENTS, ALGAE,	FORMULATION, WASTE TREATMENT, IND	W72-09169
PALLIDOSTIGMATICA, CHLAMYDOMONAS	FOTTII, CHLAMYDOMONAS TEXENSIS, C	W72-12632
OUNDS, VITAMINS, YEASTS, SULFUR,	FOULING, ALGAE, SULFATES, PERCOLA	W72-05421
OSION, DISEASES, EUTROPHICATION,	FOULING, TOXICITY, ODOR, OXYGEN S	W71-09958
INE ALGAE, SPATIAL DISTRIBUTION,	FOURIER ANALYSIS, BIORHYTHMS, COR	W72-11719
	FOX RIVER BASIN(WIS), FWPCA.:	W70-05103
TRANSPARENCY, ORTHOP/ GREEN BAY,	FOX RIVER, SPECIFIC CONDUCTANCE,	W71-10645
TLE ARBOR VITAE LAKE, GREEN BAY,	FOX RIVER, DETROIT RIVER, HETEROT	W73-01456
S, *WASTE WATER TREATMENT, *FOAM	FRACTIONATION, LIME, HYDROGEN ION	W71-12200
EMISTRY, *FOAM SEPARATION, *FOAM	FRACTIONATION, *ORGANIC MATTER, *	W72-05506
, ALGAE, AQUATIC PLANTS, ISOTOPE	FRACTIONATION, CARBON DIOXIDE, RE	W72-12709
A, DICTYOSPHAERIUM, CELL COUNTS,	FRAGILARIA, ASTERIONELLA, CRYPTOM	W71-03014
TIUM, COELOSPHAERIUM, DINOBRYON,	FRAGILARIA, PEDIASTRUM, STAURASTR	W72-10076
SYNEDRA, DINOBRYON, OEDOGONIUM,	FRAGILARIA, DIATOMA, MELOSIRA, TA	W72-10875
ENCE, ALEWIVES, DICHOTOMOSIPHON,	FRAGILARIA, TABELLARIA, ASTERIONE	W73-01669
SHOSHONE RIVER, YELLOWTAIL DAM,	FRAGILARIA CROTONENSIS, CRYPTOMON	W73-01435
ONEIS, RHOICOSPHENIA, ACHANTHES,	FRAGILARIA CONSTRUENTS, NAVICULA,	W72-07501
OALGAE, SPECIES DIVERSITY INDEX,	FRAGILARIA CONSTRUENS, COCCONEIS	W72-08141
L A, CHLOROPHYLL C, CAROTENOIDS,	FRAGILARIA CROTONENSIS, TABELLARI	W73-00384
IRA AMBIGUA, SYNEDRA FILIFORMIS,	FRAGILARIA CAPUCINA, CYCLOTELLA S	W73-00384
OCYLINDRUS DANICUS, RHIZOSOLENIA	FRAGILISSIMA, RHIZOSOLENIA HEBETA	W72-03543
AAS BASIN(FRANCE), MEURTHE RIVER(	FRANCE), CLADOPHORA, PHORMIDIUM,	W71-00099
LADOPHORA, PHO/ RHINE-MAAS BASIN(	FRANCE), MEURTHE RIVER(FRANCE), C	W71-00099
HLOROPHYLL A, MEDITERRANEAN SEA,	FRANCE, PYCNOCLINE.: /NS, C-14, C	W73-00431
*SAN	FRANCISCO BAY-DELTA.:	W71-09581
OTRIACETATE, SODIUM C/ PHOSPHATE-	FREE DETERGENTS, TRISODIUM NITRIL	W71-06247
L EFFECTS, LABORATORY EQUIPMENT,	FREEZE DRYING, SEPARATION TECHNIQ	W73-00379
UBATION/ *SURVEYS, *METHODOLOGY,	FREEZE DRYING, REFRIGERATION, INC	W73-01676
C BACTERIA, SOLVENT EXTRACTIONS,	FREEZE DRYING, SYST: /ENT, AEROBI	W73-01066
GAE, TRACE ELEMENTS, VOLATILITY,	FREEZE DRYING, X-RAYS, ELECTROLYS	W71-11036
ON, DIALYSIS, CHEMICALS, DRYING,	FREEZING, RADIATION, CYTOLOGICAL	W73-01676
E, SILICA, ALGAE, PHYTOPLANKTON,	FREEZING, TEMPERATURE, CHELATION,	W71-09190
, WATER QUALITY, WATER ANALYSIS,	FREQUENCY, FLUCTUATIONS, BIOLOGIC	W73-00918
ENTS, MERCURY, CHROMIUM, COBALT,	FRESH WATER, LAKES, PHYSICAL PROP	W73-01434
PHOSPHORUS RADIOISOTOPES, ALGAE,	FRESH WATER, FILTRATION, LEACHING	W72-07143
INORGANIC COMPOUNDS, *NUTRIENTS,	FRESH WATER, ALGAE, BACTERIA, NIT	W72-07933
OTOPES, WATER POLLUTION EFFECTS,	FRESH WATER.: /ES, COBALT RADIOIS	W71-09250
, MARINE ALGAE, AQUATIC ANIMALS,	FRESH WATER, FRESHWATER FISH, TEM	W71-09182
OGENIC BACTERIA, *ALGAL CONTROL,	FRESH WATER, CYANOPHYTA, MYXOBACT	W72-05457
, INHIBITORS, ECOLOGY, ECOTYPES,	FRESH WATER.: /ETRY, PENNSYLVANIA	W72-04743
ECOLOGY, BIOLOGICAL PROPERTIES,	FRESH WATER, WATER QUALITY.: /AE,	W71-10580
G PROCEDURES, *BIOASSAY, *ALGAE,	FRESHWATER, MEASUREMENT, DIATOMS,	W72-10619
ON, BOTTOM SEDIMENTS, SEA WATER,	FRESHWATER, AMMONIA, EUTROPHICATI	W72-11563
ITATS, EUGLENOPHYTA, PYRROPHYTA,	FRESHWATER ALGAE.: /, AQUATIC HAB	W72-09119
E, AQUATIC ANIMALS, FRESH WATER,	FRESHWATER FISH, TEMPERATURE, KIN	W71-09182
AQUATIC POPULATIONS, FISH DIETS,	FRESHWATER FISH, PONDS, LAKES, FO	W72-07132
STRUM, 4-AMINO-3,5,6-TR/ TORDON,	FRESHWATER ALGAE, PICLORAM, PEDIA	W72-08440
P., SKELETONEMA SPP., DUNALIELL/	FRESHWATER ALGAE, THALASSIOSIRA SP	W72-08436
S, AQUATIC SOILS, PLANT TISSUES,	FRESHWATER FISH, AQUATIC ALGAE, E	W73-01673
RYSOPHYTA, LABORATORY EQUIPMENT,	FRESHWATER, WATER POLLUTION SOURC	W73-01445
UTANTS, METHODOLOGY, RIVERS, WA/	FRESHWATER, *BORON, *PATH OF POLL	W73-00286
POLLUTION SOURCES, TRIBUTARIES,	FRESHWATER FISH, YELLOW PERCH, SE	W72-14282
*BIOINDICATORS, BRACKISH WATER,	FRESHWATER, *ALGAE, WATER POLLUTI	W73-02548
CULAR STRUCTURES, *ALGAE, *FLUID	FRICTION, *LITTORAL, PLANT GROWTH	W71-07878
XTRACELLULAR MATERIAL, TURBULENT	FRICTION, SEAWEED, PORPHYRIDIUM C	W71-07878
WATER TREATMEN/ *OXIDES, *FLUID	FRICTION, SEWERS, SLURRIES, WASTE	W70-06968
CULAR-WEIGHT POLYMERS, TURBULANT	FRICTIONAL DRAG, ALKYLENE OXIDE,	W70-06968
ICA, CHLAMYDOMONAS SPP, CARTERIA	FRITSCHII, PANDORINA MORUM, SPHAE	W72-09111
GATIONS, FOOD WEBS, FOOD CHAINS,	FROGS, LICHENS, ALGAE, FUNGI, SMA	W72-00949
, SOIL MICROBIOLOGY, CYANOPHYTA,	FROZEN SOILS, IONS, NOSTOC, ENVIR	W73-00854
AMETES LILACINO-GILVA, XYLOBOLUS	FRUSTULATUS, ASPERGILLU: /GUA, TR	W72-13813
MOBODO BRASILIANA, PAVLOVIAMONAS	FRUTICOSA, NODAMASTIX SPIROGYRAE,	W72-07683
FICATION, PELVETIA CANALICULATA,	FUCUS SERRATUS, FUCUS SPIRALIS, F	W71-11515
S BREVICORNIS, GAMMARUS DUEBENI,	FUCUS SERRATUS, FUCUS VESICULOSUS	W71-08026
IA CANALICULATA, FUCUS SERRATUS,	FUCUS SPIRALIS, FUCUS VESICULOSUS	W71-11515
.: *FUCUS VIRSOIDES,	FUCUS SPIRALIS, MEDITERRANEAN SEA	W72-12940
OMPOUNDS, COLUMN CHROMATOGRAPHY,	FUCUS VESICULOSUS, LAMINARIA DIGI	W72-12166
FUCUS SERRATUS, FUCUS SPIRALIS,	FUCUS VESICULOSUS, LAMINARIA DIGI	W71-11515
AMMARUS DUEBENI, FUCUS SERRATUS,	FUCUS VESICULOSUS, CLADOPHORA GLO	W71-08026
SUM NATANS, ASCOPHYLLUM NODOSUM,	FUCUS VESICULOSUS, LAMINARIA DIGI	W72-04748
TATA, PORPHYRA UMBILICALIS, SAM/	FUCUS VESICULOSUS, LAMINARIA DIGI	W73-01074

IS, LAMINARIA, THALASSIOPHYLLUM, FUCUS, SCHIZYMENIA, RHODOMELA, RH W72-09120
RHODYMENIA, FUCUS, ULVA.: W72-09653
ZONE, MACROINVERTEBRATES, DIESEL FUEL, BUNKER C OIL, AVICENNIA, RH W72-07911
LYRHIZA, ULVA RIGIDA, CLADOPHORA FULIGINOSA, BATOPHORA OERSTEDI, A W72-11800
GAE, PHYTOPLANKTON, HUMIC ACIDS, FULVIC ACIDS, TRACE ELEMENTS, CHE W71-09188
NGE, METHODOLOGY, ORGANIC ACIDS, FULVIC ACIDS, KELPS, WATER ANALYS W72-12166
, MANNITAL, SORBITAL, SUCCINATE, FUMARATE, BUTYRATE, BUTANOL, GLYC W72-05421
ION TECHNIQUES, ATLANTI/ *MARINE FUNGI, *ISOLATION, ALGAE, SEPARAT W71-11851
NG, AQUATIC ENVIRONMENT, AQUATIC FUNGI, ACQUIFER CHARACTERISTICS, W72-11186
SERVERS, STREAMS, SEPTIC TANKS, FUNGI, ALGAE, WATER POLLUTION CON W72-14363
ATION, BIOMASS, LICHENS, MOSSES, FUNGI, ALGAE, SOIL CONTAMINATION, W72-14303
WATER TREATMENT, PHYTOPLANKTON, FUNGI, BACTERIA, ALGAE.: /IOLOGY, W72-10076
ANISMS, BACTERIA, INVERTEBRATES, FUNGI, ENZYMES, ANIMAL PARASITES, W73-01676
ATA COLLECTIONS, ALGAE, PROTOZA, FUNGI, EQUIPMENT, INSTRUMENTATION W72-07225
RESEARCH, CHYTRIDIACECUS FUNGI, HEAT SHOCK.: W72-07225
CTROSCOPY, ALGAE, ACTINOMYCETES, FUNGI, LABORATORY TESTS, WATER QU W72-11604
ESH WATER, FILTRATION, LEACHING, FUNGI, LIGHT, PHOSPHATES, NUTRIEN W72-07143
IONS, *MICROORGANISMS, BACTERIA, FUNGI, METHODOLOGY, CULTURES, BIO W72-13813
TOPES, PRODUCTIVITY, METABOLISM, FUNGI, ORGANIC COMPOUNDS, CHLOREL W72-14326
ICIZERS, PHTHALATE, THERMOPHILIC FUNGI, POLYETHYLENE, POLYPROPYLEN W72-09675
ENT, DISSOLVED OXYGEN, BACTERIA, FUNGI, PROTOZOA, FERMENTATION, BA W72-09675
IC PLANTS, *MINNESOTA, BACTERIA, FUNGI, PROTOZOA, PATHOGENS, MICRO W72-08586
D CHAINS, FROGS, LICHENS, ALGAE, FUNGI, SMALL ANIMALS(MAMMALS), CY W72-00949
RHIZOSPHERE, ROOT SYSTEMS, SOIL FUNGI, SOIL ENVIRONMENT, SOIL TEM W72-03342
N, TEMPERATURE, CROP PRODUCTION, FUNGI, SOILS, SYMBIOSIS, AFRICA, W73-02473
NTS, *GROWTH RATES, *INHIBITION, FUNGI, THERMAL SPRINGS, CYANOPHYT W72-12734
ALS, DIATOMS, INSECTS, BACTERIA, FUNGI, TURNOVERS.: / AQUATIC ANIM W71-10064
ER, CONDENSERS, PROTOZOA, ALGAE, FUNGI, WATER POLLUTION EFFECTS.: / W72-07526
ALGAE, *CYANOPHYTA, *PESTICIDES, FUNGICIDES, HERBICIDES, IRRIGATIO W72-14807
CIDES, RODENTICIDES, HERBICIDES, FUNGICIDES, ALGICIDES, CARBAMATE W73-00238
ALYTICAL TECHNIQUES, PESTICIDES, FUNGICIDES, ORGANOPHOSPHOROUS PES W72-13347
ELEMENTS, TOXICITY, CHROMOSOMES, FUNGICIDES, PESTICIDES, ECOLOGY, W72-05952
OGLOEA/ *UNITED KINGDOM, *SEWAGE FUNGUS, SPHAEROTOTILUS NATANS, ZO W72-08573
M, CERATIUM MASSILENSE, CERATIUM FURCA, CERATIUM TRIPOS, CERATIUM W71-07875
A, PRROCENTRUM MICANS, CERATIUM FURCA, TURBIDOSTAT CULTURES.: /SE W71-03095
COSPHENIA, ACHANTHES, FRAG/ LAKE FURESO (DENMARK), COCCONEIS, RHOI W72-07501
OMONAS AERUGINOSA, MYCOBACTERIUM FURTUITUM, STAPHYLOCOCCUS AUREUS, W72-04259
SPP., ASPERGILLUS, PENICILLIUM, FUSARIA, VIBRIS.: /NS, TORULOPSIS W72-09675
ENDOGENOUS RESPIRATION, TOTAL O/ FUSARIUM NO 83, BACTERIUM NO 52, W71-12223
RIS, CHLOROCOCCUM, MICROCYSTIS, FUSARIUM, TRICHORIXA, RAMPHOCORIX W70-06655
S, *ENTEROMORPHA, SWE/ *HYADESIA FUSCA, *SARCOPTIFORMES, *ROCKPOOL W71-08026
ERNANS, *NITZSCHIA A/ *CHLORELLA FUSCA, *SCENEDEMUS ACUTUS VAR ALT W72-12739
METES VERSICOLOR, FOMES ANNOSUS, FUSCOPORIA CONTIGUA, TRAMETES LIL W72-13813
, ZANICHELLIA PALUSTRIS, CODIUM, FUSCUS, MACROPHYTES, AMMOBACULITE W73-00367
FOX RIVER BASIN(WIS), FWPCA.: W70-05103
COMMUNIS, *NORWAY, MACROPHYTES, FYRISAN RIVER, ARBOGAAN RIVER, BO W73-00286
CLINE, SULFANILAMIDE, PENICILLIN G, POLYMYXIN B, PODOCYSTS.: /RACY W72-10162
TION, MONOD EQUATION, PENICILLIN G, TERRAMYCIN, CHLORAMPHENICOL, A W73-01445
TASSIUM CHLORIDE, SHRINER'S POND(GA).: *LIMITING NUTRIENTS, PO W72-09159
SAGITTA, PTEROPODS, EUPHAUSIIDS, GADUS MORHUA, UROPHYCIS.: /EURA, W73-00371
OUNDER, CRAB EGGS, PELAGIC EGGS, GADUS MORHUA, GRASS SHRIMP, GREEN W72-08142
IC DATA, WATER TREATMENT, STREAM GAGES, GROUNDWATER, SAFE YIELD, A W72-05594
US, ALKYL MERCURY, EAGLES, SALMO GAIRDNERII, IRIS PSEUDACORUS, LYS W72-05952
S, SPHAEROTILUS NATANS, GLUCOSE, GALACTOSE, SUCROSE, MALTOSE, MANN W72-05421
URES, XYLOSE, RIBOSE, ARABINOSE, GALACTOSE, HEXURONIC ACID, VINEYA W72-07710
PTOMONAS, SPHAEROCYSTIS, DAPHNIA GALEATA MENDOTAE, DAPHNIA MAGNA.: W71-03014
NVE/ *OIL SPILLS, S.S. WITWATER, GALETA ISLAND, CANAL ZONE, MACROI W72-07911
UM, CYCLOTELLA NANA, CHAETOCEROS GALVESTONENSIS, MERCURIC CHLORIDE W72-07660
M, ELPHIDIUM CLAVATUM, ELPHIDIUM GALVESTONENSE, E: /PHIDIUM ADVENU W73-00367
CYPRINUS CARPIO, MOSQUITO FISH, GAMBUSIA AFFINIS, CESIUM RADIOISO W73-01673
VINIA ROTUNDIFOLIA, MILORGANITE, GAMBUSIA AFFINIS, PRODUCTIVITY ME W71-01488
*CHLORELLA SOROKINIANA, CO-60, GAMMA RADIATION, SURVIVAL, CHLORE W72-11836
K-40, CS-137, *PAHRANAGAT LAKE, GAMMA RAY SPECTROMETRY, ANIMAL TI W73-01673
IATION, CHLOROPHYTA, INHIBITION, GAMMA RAYS.: /RATORY TESTS, IRRAD W73-01080
RAVIOLET RADIATION, FOOD CHAINS, GAMMA RAYS, RESISTANCE, BIOLOGICA W73-00796
ELECTROLYSIS, COMPUTER PROGRAMS, GAMMA RAYS, POLLUTANT IDENTIFICAT W71-11036
ES, TENNESSEE, CRAYFISH, MIDGES, GAMMA RAYS, BLOODWORMS, SLUDGE WO W70-06668
YSTEM, COBALT-60.: *BIOLOGICAL-- GAMMA-RADIATION SYSTEM, MODULAR S W72-00383
E, ACTINIDES, RADIOACTIVE DECAY, GAMMA-RAY SPECTROMETRY, COUNTING. W73-01434
, SWEDEN, TIGRIOPUS BREVICORNIS, GAMMARUS DUEBENI, FUCUS SERRATUS, W71-08026
CTES RUSTICUS, LIRCEUS LINEATUS, GAMMARUS.: /AMS, UNIONIDS, ORCONE W72-06526
YTA, CHLOROPHYTA, FERROBACILLUS, GAS CHROMATOGRAPHY, IONS,: /ENOPH W72-06124
, CLOPHEN A 50, ELECTRON CAPTURE GAS CHROMATOGRAPHY, POLYCHLORINAT W72-04134
ANALYTICAL TECHNIQUES, BIOMASS, GAS CHROMATOGRAPHY, CENTRIFUGATIO W72-04134
GAE, ESTUARIES, CHEMICAL WASTES, GAS CHROMATOGRAPHY, SEDIMENTS, OY W71-07412
, NITROGEN, REDUCTION(CHEMICAL), GAS CHROMATOGRAPHY, SEPARATION TE W73-01456
S, SPECTROSCOPY, CHROMATOGRAPHY, GAS CHROMATOGRAPHY.: /KTON, PLANT W73-01439
GNIFICATION, / *ELECTRON CAPTURE GAS CHROMATOGRAPHY *BIOLOGICAL MA W73-02105
OOD CHAINS, CULTURES, FISH DIET, GAS CHROMATOGRAPHY, PARTICULATE M W73-02105
ES, CHROMATOGRAPHY, AMINO ACIDS, GAS CHROMATOGRAPHY, PHAEOPHYTA, E W72-07710
CHNIQUES, SEPARATION TECHNIQUES, GAS CHROMATOGRAPHY, SEPCTROSCOPY, W72-11604
IGELLA, VIRUSES, CHROMATOGRAPHY, GAS CHROMATOGRAPHY, ANALYTICAL TE W72-13800
OASSAY, WATER POLLUTION EFFECTS, GAS CHROMATOGRAPHY, BROCK TROUT, W72-12930
RAPHY *BIOLOGICAL MAGNIFICATION, GAS LIQUID CHROMATOGRAPHY, ELECTR W73-02105
TIES, PESTICIDES, TROPHIC LEVEL, GASES, MUDS, ADSORPTION, COPP: /R W72-00847
GLOMERATA, HETEROCYPRIS SALINUS, GASTEROSTEUS ACULEATUS, BALTIC SE W71-08026
ULEATUS, DAPHNIA MAGNA, DAPHNIA, GASTEROSTEUS.: /, GASTEROSTEUS AC W72-05952
DGE, DDE, KIDNEYS, LIVER, BLOOD, GASTEROSTEUS ACULEATUS, DAPHNIA M W72-05952
AWRENCE RIVER, ALGAE, NEMATODES, GASTROPODS, MOLLUSKS, AMPHIPODA, W72-05432
ER), SAMPLING, BENTHOS, GRAZING, GASTROPODS, LITTORAL, AGE, BIOMAS W72-04766

S, MUSSELS, SNAILS, FISH, ALGAE, GASTROPODS, ADSORPTION, TEMPERATU W71-09168
OISOTOPES, *ABSORPTION, *SNAILS, GASTROPODS, PATH OF POLLUTANTS, N W72-10978
, CARBON, SEASONAL, CADDISFLIES, GASTROPODS, SANDS, CHRYSOPHYTA.: / W71-11029
PHYTA, PYRROPHYTA, MARINE ALGAE, GASTROPODS.: / CHRYSOPHYTA, CYANO W73-00935
E DATA COLLECTIONS, CRUSTACEANS, GASTROPODS, SNAILS, CRABS, BENTHI W73-00932
-EPODS, MARINE ANIMALS, MOLLUSKS, GASTROPODS, PHAEOPHYTA.: /TY, COP W73-00371
L/ *ETHMODISCUS REX, ETHMODISCUS GAZELLAE, BAJA CALIFORNIA, HEMIAU W72-03567
ANE FILTRATION, RHODOPSEUDOMONAS GELATINOSA, THIORHODACEAE, CHLORO W72-04743
LADOPHORA SP., POLYSIPHONIA SP., GELIDIUM SP., CHLORELLA, AGARS.: / W71-11851
C ENVIRONMENT, FLORIDA, ELECTRIC GENERATORS, WATER TEMPERATURE, RH W72-11252
FUNGI, ACQUIFER CHARACTERISTICS, GENETICS, DISSOLVED OXYGEN.: /IC W72-11186
S, PODOHEDRA.: MOUGEOTIA GENUFLEXA, ANKISTRODESMUS FALCATU W71-07360
N RATES, *REVIEWS, BIOCHEMISTRY, GEOCHEMISTRY, RADIOISOTOPES, PATH W72-06313
RIBUTION, *SPATIAL DISTRIBUTION, GEOGRAPHICAL REGIONS, PHAEOPHYTA, W73-00428
LUSKS, COPEODDS, DIPTERA, GULFS, GEOLOGIA HISTORY, PLEISTCRENE EPO W73-00855
VERTEBRATES, RIVERS, CYANOPHYTA, GEOLOGIC FORMATIONS, BENTHOS, MAY W71-11009
N, NITROGEN SO/ *SURFACE WATERS, GEOLOGICAL SOURCES, MINERALIZATIO W71-06435
WATER POLLUTION SOURCES, ALGAE, GEOLOGY, DATA COLLECTIONS.: /ALS, W72-05869
PHYTOPLANKTON, LAKE MORPHOMETRY, GEOLOGY, LIMESTONE, TEMPERATURE, W72-04269
E, ECONOMICS, POLITICAL ASPECTS, GEOLOGY, TOPOGRAPHY, GROUNDWATER, W71-03012
IENTS, CANADARAGO LAKE(NY), LAKE GEORGE(NY).: *LIMITING NUTR W72-09163
RICORNUTUM, *ALGAL GROWTH, *LAKE GEORGE(NY).: *SELENASTRUM CAP W72-08376
CARDINAL(GEORGIA), LAKE BUCKHORN(GEORGIA).: /CKSON(GEORGIA), LAKE W71-09173
JACKSON(GEORGIA), LAKE CARDINAL(GEORGIA), LAKE BUCKHORN(GEORGIA). W71-09173
HYPNODINIUM, LAKE SIDNEY LANIER(GEORGIA), LAKE JACKSON(GEORGIA), W71-09173
EY LANIER(GEORGIA), LAKE JACKSON(GEORGIA), LAKE CARDINAL(GEORGIA), W71-09173
EY LANIER(GEORGIA), LAKE JACKSON(GEORGIA), LAKE CARDINAL(GEORGIA), W71-11914
JACKSON(GEORGIA), LAKE CARDINAL(GEORGIA), LAKE BUCKHORN(GEORGIA), W71-11914
LAKE CARDIN/ *LAKE SIDNEY LANIER(GEORGIA), LAKE JACKSON(GEORGIA), W71-11914
CARDINAL(GEORGIA), LAKE BUCKHORN(GEORGIA), HYPNODINIUM SPHAERICUM. W71-11914
AQUATIC ENVIRONMENT, ECOSYSTEMS, GEORGIA, BIORTHYTHMS, LABORATORY W72-09159
TION, *ALGAE, *SEWAGE EFFLUENTS, GEORGIA, THERMOCLINE, HYDROGEN IO W71-11914
ANTHRACITE, *WAHNBACH RESERVOIR(GERMANY), OSCILLATORA RUBESCENS.: W72-06201
SPICATA, FIMB/ ENGLAND, SWEDEN, GERMANY, SPARTINA SPP, DISTICHLIS W73-01089
SAALE RIVER, GERMANY, SCENEDESMUS OBLIQUUS.: W72-14314
*WATER TEMPERATURE, *CYANOPHYTA, GEYSERS, PHOTOSYNTHESIS, CULTURES W72-03557
RYPTOMONAS EROSA, P/ *VOLTA LAKE(GHANA), ACTINASTRUM GRACILIMUM, C W73-00256
STUARINE SOILS, AXENIC CULTURES, GIBBERELIC ACID.: /IBER MEDIUM, E W73-00430
HYPERBOREA, ASCOPHYLLUM NODOSUM, GIGARTINA MAMILLOSA, RHODYMENIA P W71-11515
PORPHYRIDIUM CRUENTUM, PORPHYRA, GIGARTINA.: / FRICTION, SEAWEED, W71-07878
NSES, CHARA ASPPRA, SCIRPUS SPP, GILLS, MUSCLE, VISCERA, EYES, SCA W73-01673
ORIA CONTIGUA, TRAMETES LILACINO- GILVA, XYLOBOLUS FRUSTULATUS, ASP W72-13813
HEEPSHEAD, CHUBS, NORTHERN PIKE, GIZZARD SHAD.: /S, COHO SALMON, S W72-14282
*OLIGOTROPHY, MINNESOTA, LAKES, GLACIAL DRIFT, NUTRIENTS, SALANIT W71-13149
VAN DORN SAMPLERS, C-/ AUKE BAY, GLACIAL WATER, BECKMAN ANALYZER, W72-09122
CINODISCUS, COPEPODIDS, *CALANUS GLACIALIS, DITYLUM, PRORCCENTRUM, W72-09248
, *BIODETERIORA/ ACRYLIC RESINS, GLASS, THYMIDINES, CELLS(BIOLOGY) W71-09261
, THIOTHRIX TENUIS, MERISMOPEDIA GLAUCA, DACTYLOCOCCOPSIS SMITHII, W72-07896
URVISETUS, SKELETONEMA COSTATUM, GLENODINIUM FOLIACUM, GYRODINIUM W72-07166
ABAENA, APHANIZOMENON, NAVICULA, GLENODINIUM, MICROSPORA, SYNEDRA, W72-10875
CIDES, MCPB, MCPA, CHLAMYDOMONAS GLOBUSA, CHLORELLA PYRENOIDOSA, S W70-07255
CYLINDRICA, ANACYSTIS NIDULANS, GLOEOCAPSA ALPICOLA, OSCILLATORIA W72-07664
BONEMA, APHANIZOMENON, OOCYSTIS, GLOEOCYSTIS, COELASTRUM, ASTERION W72-07940
D, GROWTH STIMULATION, ANABAENA, GLOEOTRICHIA, MICROCYSTIS, NOSTOC W72-01097
LINI/ TREMATOCARPUS PAPENFUSSII, GLOIODERMA ATLANTICA, CODIUM CARO W73-00284
HNANTHES, *SPIROGYRA, CLODOPHORA GLOMERATA, *METOLIUS RIVER(ORE).: W72-07901
STIGEOCLONIUM TENUE, CLADOPHORA GLOMERATA, ULOTRICHALES, ZYGNEMAL W71-05991
: *CLADOPHORA GLOMERATA, *PITHOPHORA OEDOGONIA. W71-08032
S, FUCUS VESICULOSUS, CLADOPHORA GLOMERATA, HETEROCYPRIS SALINUS, W71-08026
*CLADOPHORA GLOMERATA, GROWTH CURVES.: W72-13651
NSIN), POLYPHOSPHATE/ CLADOPHORA GLOMERATA, MILWAUKEE HARBOR(WISCO W72-13644
NTANS, TRICHOPTERA, LIMNEPHILUS, GLOSSOSOMA, PHYSA GYRINA, KEMMERE W72-14280
ACTERIA, HETEROTROPHIC BACTERIA, GLUCOSE, ACETATES, BIOTIN, NIACIN W73-02095
IS OCEANUS, SPHAEROTILUS NATANS, GLUCOSE, GALACTOSE, SUCROSE, MALT W72-05421
S, CLONES, PHAEOCYSTIS POUCHETI, GLUCOSE, MANNOSE, RHAMNOSE, ACRYL W72-07710
CRASSA, SCINTILLATION COUNTING, GLUCOSE, MEMBRANE FILTERS, MONOSA W72-14326
ARIUM, THIOBACILLUS, SULFOLOBUS, GLUCOSE, URACIL.: /CYANIDIUM CALD W72-14301
TICS, IODINE, ATOMIC EN/ FODDER, GLUE, AGAR, POLYSIPHONIA, ANTIBIO W71-09157
SACCHARIDES, THREONINE, PROLINE, GLUTAMATE, LEUCINE, ARGININE, SER W73-00942
NT MICROSCOPY, *AUTORADIOGRAPHY, GLUTARALDEHYDE, SAMPLE PREPARATIO W73-00273
S, CHONDRUS CRISPUS, RHODOTORULA GLUTINIS, SARGASSUM NATANS, ASCOP W72-04748
NA, ALTERNANTHERA PHILOXEROIDES, GLY: /CRASSIPES, JUSTICIA AMERICA W73-01089
RAPHY, PRIONOSPIO, LUMBRINEREIS, GLYCERA, NEPHTYS, CAPITELLA, MAGE W72-09092
SIUM PARVUM)(CHROOMONAS SALINA), GLYCEROL POLLUTION.: /KTON(PRYMNE W70-07842
TE, FUMARATE, BUTYRATE, BUTANOL, GLYCEROL, LACTATE, PYRUVATE, ACET W72-05421
ESPIRATION, TOTAL OXYGEN DEMAND, GLYCOLS.: /UM NO 52, ENDOGENOUS R W71-12223
ERMATOPHYTES, PHOTOASSIMILATION, GLYOXYLIC ACID, ASSIMILATION.: /P W73-02100
DGE.: GLYTOTENDIPES BARBIPES, EKMAN DRE W72-03556
PROPERTIES, CHEMICAL PROPERTIES, GOLD, IRON, CESIUM, ALKALI METALS W73-01434
MBIA RIVER(WASHINGTON), COCKLES, GOLDFISH, MERRIMACK RIVER(NEW HAM W71-11517
PHAGA, REFRACTOCYSTIS PLUVIALIS, GOMESIAMONAS STAGNALIS, REFRACTOD W72-09119
CONFERVOIDES, LYNGBYA MAJUSCULA, GOMONTIA POLYRHIZA, ULVA RIGIDA, W72-11800
A VARIANS, NAVICULA TRIPUNCTATA, GOMPHONEMA PARVULUM, BACILLARI PA W72-10883
RISOMOPEDIA SPP., ANABAENA SPP., GOMPHONEMA SPP., NITZCHIA SPP., S W72-14706
AY CREEK(PA.).: GOMPHONEMA OLIVACEOIDES, WHITE CL W73-00242
OCCONEIS PLACENTULA, CLOSTERIUM, GOMPHONEMA LONGICEPS, CYMBELLA, V W72-12744
OUGOTIA, SPOROTETRAS PYRIFORMIS, GONGROSIRA, MICROSPORA, SPIROGYRA W71-05991
THIAMINE, BIOTIN, VITAMIN B 12, GONYAULAX POLYEDRA, PERIDINIUM DE W72-07892
RIDI/ PHAEODACTYLUM TRICORNUTUM, GONYAULAX POLYEDRA, PLATYMONAS VI W72-07166
ANTIMONY, CESIUM, *CAPE OF GOOD HOPE.: W73-00832

OPHILIC, BIOTIC ASSOCIATIONS, LA	GORCE MOUNTAINS, AMMONIUM IONS.: /	W73-00854
OINT COMMISSION, CANACA, FEDERAL	GOV: / LIMNOLOGY, INTERNATIONAL J	W72-14782
*ENVIRONMENTAL EFFECTS, FEDERAL	GOVERNMENT, AIR POLLUTION, WATER	W73-00703
TREATMENT, BIODEGRADATION, LOCAL	GOVERNMENTS, MUNICI: /ASTE WATER	W72-06655
CONTROL, ALGAL CONTROL, FEDERAL	GOVERNMENT, NUTRIENTS, SEWAGE TRE	W72-06655
*DETERGENTS, *PHOSPHATES, *LOCAL	GOVERNMENTS, *WATER POLLUTION CON	W72-06638
LEGISLATION, WATER LAW, FEDERAL	GOVERNMENT, WATER POLLUTION, STAN	W71-11516
N CONTROL, *PHOSPHATES, *FEDERAL	GOVERNMENT, UNITED STATES, ALGAL	W72-01093
ON, ALGAE, BIODEGRADATION, LOCAL	GOVERNMENTS, WASTE DISPOSAL, WATE	W72-01093
VAL, OXYGEN SAG, DREDGING, STATE	GOVERNMENTS, TREATMENT FACILITIES	W72-01659
*WATER QUALITY CONTROL, *FEDERAL	GOVERNMENT, *GRANTS, EUTROPHICATI	W72-01659
RNMENT, *GRANTS, EUTROPHICATION,	GOVERNMENT FINANCE, ENVIRONMENTAL	W72-01659
NITATION, *LEGISLATION, *FEDERAL	GOVERNMENT, EUTROPHICATION, ENVIR	W72-01660
ELOPMENT, ENVIRONMENTAL EFFECTS,	GOVERNMENT FINANCE,: /MMUNITY DEV	W72-01660
GIBBERELLIN, HYPNEA MUSCIFORMIS,	GRACILARIA VERUCOSA, ECKLONIA RAD	W72-07496
LTURES, EXCRETION, APHANIZOMENON	GRACILE.: /ANTAGONISM, AXENIC CU	W72-12734
*VOLTA LAKE(GHANA), ACTINASTRUM	GRACILIMUM, CRYPTOMONAS EROSA, PE	W73-00256
ANKISTRODESMUS BRAUNII, EUGLENA	GRACILIS.: /YL MERCURIC CHLORIDE,	W72-06037
DUNALIELLA TERTIOLECTA, EUGLENA	GRACILIS, CHLAMYDOMONAS REINHARDT	W72-05614
T/ DAPHNIA LONGISPINA, DIAPTOMUS	GRACILIS, CYCLOPS STRENUUS, INGES	W72-07940
TIN, NIACIN, COBALAMINS, EUGLENA	GRACILIS, COLONIZATION, CAULOBACT	W73-02095
NIDULAN/ WARBURG EFFECT, EUGLENA	GRACLIS, CALVIN CYCLE, ANACYSTIS	W71-09172
C CENTRIFUGATION, SAMP/ *DENSITY-	GRADIENT CENTRIFUGATION, ISOPYCNI	W73-00371
ALPA, LUDOX AM, SUCROSE, DENSITY	GRADIENTS, TUNICATES, PHAEOCYSTIS	W73-00371
BACILLUS, MICROCOCCUS,	GRAM NEGATIVE BACTERIA.:	W72-12255
SHADOW MOUNTAIN LAKE(COLO), LAKE	GRANBY(COLO).: /RAND LAKE(COLO),	W72-04791
LAKE(COLO), LAKE GRANBY(COLO)./	GRAND LAKE(COLO), SHADOW MOUNTAIN	W72-04791
PLY PROJECT, LOPEZ CREEK, ARROYO	GRANDE CREEK.: *LOPEZ WATER SUP	W72-05594
E P32, SODIUM PYROPHOSPHATE, RIO	GRANDE VALLEY.: /TH, *RADIOISOTOP	W72-13992
IBER FILTERS, *MEMBRANE FILTERS,	GRANGE LAKE, EAST TWIN LAKE, LAKE	W72-13673
COST COMPARISONS, AQUATIC ALGAE,	GRANTS, MODEL STUDIES.: /HEALTH,	W71-09429
EGRADATION, TOXICITY, CONTRACTS,	GRANTS, RESEARCH AND DEVELOPMENT,	W71-11516
QUAE, CHROOMONAS ACUTA, MELOSIRA	GRANULATA, CERATIUM, MICROCYSTIS,	W72-12738
ITY, *CYANOPHYTA, SULFUR, ALGAE,	GRANULES, INHIBITION, METABOLISM,	W72-00153
S, DAPHNIA, APPLICATION METHODS,	GRANULES, PLANT GROWTH REGULATORS	W72-09650
GGS, PELAGIC EGGS, GADUS MORHUA,	GRASS SHRIMP, GREEN CRAB, PENICIL	W72-08142
T, TURKEY POINT(FLORIDA), TURTLE	GRASS, THALASSIA SPP., UDOTEA SPP	W72-11876
R, MINNESOTA, CYANOPHYTA, TREES,	GRASSES, LEAVES, FOREST SOILS, SA	W72-01784
*SEA WEEDS, *MACROPHYTES, *SEA	GRASSES, TELEVISION, DEPTH SOUNDE	W72-09120
, MILDEWS, E. COLI, FISH, FOODS,	GRASSES, WORMS.: /YEASTS, INSECTS	W73-01676
DIUM, MASTIGOCLADUS, PHEOPHYTIN,	GRASSLAND SPRING.: /TERIA, PHORMI	W72-03557
RAINFALL, *ARID LANDS, *ARIZONA,	GRASSLANDS, ON-SITE DATA COLLECTI	W72-12505
SIOSIRA DECIPIENS, THALASSIOSIRA	GRAVIDA, THALASSIOSIRA NORDENSKIO	W72-03543
ES, *ORGANIC COMPOUNDS, *CARBON,	GRAVIMETRIC ANALYSIS, PRODUCTIVIT	W72-09108
YTA, CHLOROPHYTA, PHYTOPLANKTON,	GRAZING.: / REQUIREMENTS, CYANOPH	W72-07940
WAVES(WATER), SAMPLING, BENTHOS,	GRAZING, GASTROPODS, LITTORAL, AG	W72-04766
MENTS, HARDNESS(WATER), BENTHOS,	GRAZING, SYSTEMATICS, CHEMICAL AN	W73-01627
HYTA, CHLOROPHYTA, EUGLENOPHYTA,	GRAZING, WATER QUALITY.: / CYANOP	W72-07899
IMITING FACTORS, PHOTOSYNTHESIS,	GREAT LAKES, LAKE ERIE, LAKE ONTA	W72-07648
SYNTHESIS, PRIMARY PRODUCTIVITY,	GREAT LAKES, NEW YORK, WATER SUPP	W73-01615
MICHIGAN, LAKE HURON, LAKE ERIE,	GREAT LAKES, SOYBEANS, LIGHT INTE	W73-01456
KTON, CARBON, WATER TEMPERATURE,	GREAT LAKES REGION, WISCONSIN, MI	W73-00384
STES, *POLITICAL ASPECTS, LAKES,	GREAT LAKES, ALGAL BLOOMS, AGING(	W72-14782
PHICATION, AQUATIC ENVIRONMENTS,	GREAT LAKES, BAYS, WATERSHEDS(BAS	W71-08953
UTION), *HYDROLYSIS, DETERGENTS,	GREAT LAKES, TEMPERATURE, HYDROGE	W71-00116
ON ANALYSIS, *CHEMICAL ANALYSIS,	GREAT LAKES, EUTROPHICATION, NUTR	W72-01101
ODELS, SELF-PURIFICATION, ALGAE,	GREAT LAKES, NUTRIENTS, SEDIMENTS	W71-12072
, *TEMPERATURE, *PHOTOPERIODISM,	GREAT LAKES, MODEL STUDIES, NUTRI	W72-13651
LAND), LAKE WINDERMERE(ENGLAND),	GREAT OUSE(ENGLAND), ALGAL GROWTH	W71-13185
L, *TUTTLE CREEK RESERVOIR(KAN),	GREAT PLAINS, AQUATIC ECOSYSTEM.:	W72-12393
HYDROLYSIS, ALGAE BLOOMS, UTAH,	GREAT SALT LAKE, ORGANIC COMPOUND	W71-12223
ENDED SEDIMENT PRODUCTION, *BLUE-	GREEN ALGAE.: *SUSP	W72-02218
LAND), HETEROCYSTS, BENTHIC BLUE-	GREEN ALGAE.: *WINDERMERE(ENG	W73-02472
BLUE-GREEN ALGAE, FIELD STUDIES,	GREEN ALGAE, LABORATORY STUDIES.:	W70-08319
*DIEOFF RATES, AFTERGROWTH, BLUE-	GREEN ALGAE, FIELD STUDIES, GREEN	W70-08319
*LYSOZYME, *SPHEROPLASTS, BLUE-	GREEN ALGAE, ANABAENA VARIABILIS,	W71-08683
CILITY, PRODUCT EVALUATION, BLUE-	GREEN ALGAE.: /ATIC WEED STUDY FA	W71-06102
BORYANIUM, SU/ *LPP VIRUS, *BLUE-	GREEN ALGAL VIRUSES, *PLECTONEMA	W72-04736
OWSTONE NATIONAL PARK(WYO), BLUE-	GREEN ALGAE, MASTIGOCLADUS, CALOT	W72-08508
CALIF), RICHELIEU RIVER(QUEBEC),	GREEN BAY(WIS), MACKEREL, MONTREA	W70-08098
ACID, ALGAL ASSAY PRO/ FLUSHING,	GREEN BAY(WIS), NITRILOTRIACETIC	W71-12072
NDUCTANCE, TRANSPARENCY, ORTHOP/	GREEN BAY, FOX RIVER, SPECIFIC CO	W71-10645
E LAKE, LITTLE ARBOR VITAE LAKE,	GREEN BAY, FOX RIVER, DETROIT RIV	W73-01456
GGS, GADUS MORHUA, GRASS SHRIMP,	GREEN CRAB, PENICILLIUM, STREPTOM	W72-08142
INE PHOSPHATASE, REASSIMILATION,	GREEN FLAGELLATE, CYCLOTELLA CRYP	W71-10083
NSA(WIS), LAKE WASHINGTON(WASH),	GREEN LAKE(WASH).: /S), LAKE KEGO	W72-05473
SEBASTICOOK(MAINE), FAYETTEVILLE	GREEN LAKE(NEW YORK), UPPER KLAMA	W71-05626
ISH FECAL WASTES, RESIDUAL FOOD,	GREEN RIVER(WASH), COWLITZ TROUT	W71-11006
CONNECTICUT RIVER(CONNECTICUT),	GREEN RIVER(KENTUCKY), CAYUGA LAK	W71-11517
UMP, ALUM INJECTION PUMP, MANURE	GRINDER.: /TION, ACID INJECTION P	W72-02850
PISTONEMA EXPANSUM, PORPHYRIDIUM	GRISEUM, CHRYSOTRICHALAE.: /IS, A	W71-04526
AL ASPECTS, GEOLOGY, TOPOGRAPHY,	GROUNDWATER, BIOCHEMICAL OXYGEN D	W71-03012
IAN RIGHTS, PRIOR APPROPRIATION,	GROUNDWATER, WATER QUALITY, COMPE	W70-07100
LGAE, AQUATIC WEEDS, RECREATION,	GROUNDWATER, FISH, FLOOD PLAINS,	W70-05103
E, FARM WASTES, SOIL MANAGEMENT,	GROUNDWATER, WISCONSIN, WATER SUP	W71-06443
ON, SEDIM/ *NITROGEN, *NITRATES,	GROUNDWATER, AMMONIA, PRECIPITATI	W71-06435
, ALGAE, BACTERIA, CHLORINATION,	GROUNDWATER, COST ANALYSIS.: /SIS	W71-07123
, WATER TREATMENT, STREAM GAGES,	GROUNDWATER, SAFE YIELD, ALGAE, M	W72-05594
ATION, ALLUVIA CHANNELS, SURFACE-	GROUNDWATER RELATIONSHIPS, PERMEA	W71-12410
CATION, SEWAGE, SOIL PROPERTIES,	GROUNDWATER MOVEMENT, OUTFALL SER	W72-14363

*HORMOTILA BLENNISTA, FILTRATES,	GROWTH AUTOSTIMULATION, GROWTH RE	W71-08687
*CLADOPHORA GLOMERATA,	GROWTH CURVES.:	W72-13651
IDES, THIN LAYER CHROMATOGRAPHY,	GROWTH INHIBITION.: /URIAL PESTIC	W72-13347
VARIABILIS, *ANACYSTIS NIDULANS,	GROWTH INHIBITION, SELENATE, SELE	W72-00153
NRICHMENT, CULTURE MEDIA, MACRO/	GROWTH KINETICS, BIOENERGETICS, E	W72-09378
OMOSIPHON TUBEROSUS, *EPIPHYTES,	GROWTH MEDIA, COCCONEIS PLACENTUL	W72-12744
ON, CHLORELLA PYRENOIDOSA, MONOD	GROWTH MODEL, ALGAL GROWTH RATES.	W72-04788
ILL), DIAGNOSTIC CRITERIO/ ALGAL	GROWTH POTENTIAL, ILLINOIS RIVER(	W71-11033
*PLANT GROWTH, *EUTROPHICATION,	GROWTH RATES, SAMPLING, LAKE MICH	W71-11033
, MONITORING, SEWAGE, ESTUARIES,	GROWTH RATES, BIOLOGICAL COMMUNIT	W71-12867
UTRIENT REQUIREMENTS, CHLORELLA,	GROWTH RATES, TEMPERATURE, SEA WA	W71-10083
ON, VERTICAL MIGRATION, AQUARIA,	GROWTH RATES, SEDIMENTATION, BUOY	W71-09852
ICATION, *BIOINDICATORS, *ALGAE,	GROWTH RATES, CULTURES, SUSPENDED	W71-09795
ORGANIC MATTER, ATLANTIC OCEAN,	GROWTH RATES, ALGAE, INCUBATION,	W71-11562
, INDUSTRIAL WASTES, METABOLISM,	GROWTH RATES, OXIDATION LAGOONS,	W72-09590
S, NITROGEN, PHOSPHORUS, CARBON,	GROWTH RATES, OLIGOTROPHY, EUTROP	W72-09164
KTON, *ALGAE, *LIMITING FACTORS,	GROWTH RATES, DIATOMS, BACTERIA,	W72-09163
E ALGAE, PHYTOPLANKTON, BIOMASS,	GROWTH RATES, PRODUCTIVITY, ANIMA	W72-09378
RGANISMS, ENVIRONMENTAL EFFECTS,	GROWTH RATES, E. COLI, AZOTOBACTE	W72-10861
GY, ENVIRONMENT, PHOTOPERIODISM,	GROWTH RATES, TEMPERATURE, BIOCHE	W72-00377
UM COMPOUNDS, NITRATES, DENSITY,	GROWTH RATES, LABORATORY TESTS, C	W72-03217
S, CHLOROPHYLL, PROTEINS, ALGAE,	GROWTH RATES, TRACERS.: / CULTURE	W72-03557
ILIZATION, PRIMARY PRODUCTIVITY,	GROWTH RATES, RESPIRATION.: /STAB	W72-03556
ATION, EFFICIENCES, COMPETITION,	GROWTH RATES, PLANT MORPHOLOGY, W	W72-04766
NUOUS CULTURES, MAXIMUM SPECIFIC	GROWTH RATE, SATURATION CONSTANT,	W72-04788
WTH REACTORS, ALGAL PONDS, ALGAL	GROWTH RATES, *POTAMOCYPRIS.: /RO	W72-04787
IDOSA, MONOD GROWTH MODEL, ALGAL	GROWTH RATES.: / CHLORELLA PYRENO	W72-04788
ALGAL	GROWTH RATES.:	W72-04789
YSIOLOGY, PLANT GROWTH, BIOMASS,	GROWTH RATES, LIGHT INTENSITY, CA	W72-05476
IC FLORA, *PHYTOPLANKTON, ALGAE,	GROWTH RATES, BIOASSAY, DISSOLVED	W72-06046
ANKTON, TOXICITY, CULTURES, DDT,	GROWTH RATES.: / DIATOMS, PHYTOPL	W72-05614
HESIS, BIOMASS, LIGHT INTENSITY,	GROWTH RATES, MATHEMATICAL MODELS	W72-06274
TE WATER TREATMENT, CHLOROPHYTA,	GROWTH RATES.: /LITY CONTROL, WAS	W72-04431
BUTION PATTERNS, CHROMATOGRAPHY,	GROWTH RATES, POLLUTANT IDENTIFIC	W72-04134
SORPTION, RADIOISOTOPES, ALGAE,	GROWTH RATES, THERMAL POLLUTION.:	W71-02075
CARBON RADIOISOTOPES, CHLORELLA,	GROWTH RATES, DIATOMS, PHYTOPLANK	W71-03027
TEMPERATURE, OPTICAL PROPERTIES,	GROWTH RATES.: /OPHYTA, DIATOMS,	W70-10403
UTRIENTS, NUTRIENT REQUIREMENTS,	GROWTH RATES, PHOSPHORUS, NITROGE	W72-12966
URES, RESISTANCE, AQUATIC ALGAE,	GROWTH RATES, BIOASSAY, LABORATOR	W72-13809
GHT INTENSITY, TEMPERATURE, MUD,	GROWTH RATES, ANABAENA, NITROGEN	W72-12544
TREATMENT, *MATHEMATICAL MODELS,	GROWTH RATES, SOLAR RADIATION, FL	W72-12289
TENSITY, SCENEDESMUS, CHLORELLA,	GROWTH RATES, RESISTANCE, PHOSPHO	W72-12739
IMARY PRODUCTIVITY, SYSTEMATICS,	GROWTH RATES, BIOMASS, DATA COLLE	W73-02469
UCTIVITY, BACTERIA, SCENEDESMUS,	GROWTH RATES, CULTURES, AQUATIC P	W72-14314
LISM, *CHLORELLA, *LIGHT, ALGAE,	GROWTH RATES, CHLOROPHYLL.: /TABO	W73-00251
S, *NITRATES, *AMMONIA, DIATOMS,	GROWTH RATES, NITROGEN, WATER POL	W73-00380
ORINATED HYDROCARBON PESTICIDES,	GROWTH RATES, ENVIRONMENTAL EFFEC	W73-00281
A, AEROBIC CONDITIONS, BIOASSAY,	GROWTH RATES, MORTALITY.: /NOPHYT	W72-07664
RADIATION, VITAMIN B, PIGMENTS,	GROWTH RATES, ALGAE, FOOD WEBS, C	W72-07892
ATION, SUCCESSION, PRODUCTIVITY,	GROWTH RATES.: /SYNTHESIS, RESPIR	W73-01632
ATOMS, POLLUTANT IDENTIFICATION,	GROWTH RATES, DEFICIENT ELEMENTS,	W73-01445
WATER, BIOMASS, THERMAL STRESS,	GROWTH RATES, SPATIAL DISTRIBUTIO	W73-00853
ICIDES, HERBICIDES, 2-4-D, PLANT	GROWTH REGULATORS, ALGAE, AQUATIC	W72-08440
ARIA VERUCOSA, ECKLONIA RADIATA,	GROWTH REGULATION.: /RMIS, GRACIL	W72-07496
YANOPHYTA, / *INHIBITION, *PLANT	GROWTH REGULATORS, *CHLORIDES, *C	W72-14706
S, CULTIVATION, VITAMIN B, PLANT	GROWTH REGULATORS, NITRATES, PHOS	W73-00430
, CULTURES, *GROWTH RATES, PLANT	GROWTH REGULATORS, PLANT GROWTH S	W71-03095
SIOLOGY, *PLANT PATHOLOGY, PLANT	GROWTH REGULATORS, SALINITY, PLAN	W70-09429
LTRATES, GROWTH AUTOSTIMULATION,	GROWTH REGULATION, EXTRACELLULAR	W71-08687
INHIBITORS, ALGAL CONTROL, PLANT	GROWTH REGULATORS, PLANKTON, CULT	W71-13793
ICATION METHODS, GRANULES, PLANT	GROWTH REGULATORS, INHIBITORS, HE	W72-09650
HESIS, PESTICIDE TOXICITY, PLANT	GROWTH REGULATORS, INHIBITION, IN	W71-10566
N, ALGAE, FEEDING RATES, *LARVAL	GROWTH STAGE, PHYTOPLANKTON, NUTR	W72-09248
NE ALGAE, CHROMOSOMES, EMBRYONIC	GROWTH STAGE, CRUSTACEANS, MONITO	W72-10955
ENTERIC BACTERIA, *BACTERICIDES,	GROWTH STAGES, PATHOGENIC BACTERI	W71-05155
TIES, PRODUCTIVITY, LIFE CYCLES,	GROWTH STAGES, DISTRIBUTION, BEHA	W72-14659
EOTRICHIA, MICROC/ *FULVIC ACID,	GROWTH STIMULATION, ANABAENA, GLO	W72-01097
, PLANT GROWTH REGULATORS, PLANT	GROWTH SUBSTANCES, ALGAE, IRON, B	W71-03095
PHYSIOLOGY, *PLANT GROWTH, PLANT	GROWTH SUBSTANCES, PLANT MORPHOLO	W70-07842
WTH RATES, *PRODUCTIVITY, *PLANT	GROWTH SUBSTANCES, PHOTOSYNTHESIS	W71-08687
FLUID FRICTION, *LITTORAL, PLANT	GROWTH SUBSTANCES, PHYTOPLANKTON,	W71-07878
GANIC CARBON, *ALGAL MASS, ALGAL	GROWTH.: *OR	W72-04431
*MICRO-NUTRIENTS, *ALGAL	GROWTH.:	W72-00801
ALGAL	GROWTH.:	W71-12868
AND), GREAT OUSE(ENGLAND), ALGAL	GROWTH.: /), LAKE WINDERMERE(ENGL	W71-13185
, UNDISSOCIATED MOLECULES, ALGAL	GROWTH.: /STICHOCOCCUS BACILLARIS	W70-07255
ABOLITES, SUBSTRATE UTILIZATION,	GROWTH.: /ES, MIXED CULTURES, MET	W73-00441
ON-SITE DATA COLLECTIONS, PLANT	GROWTH.: /, *ARIZONA, GRASSLANDS,	W72-12505
RESERVOIR, *APHANIZOMENON, ALGAL	GROWTH.: /ARBON, *AMERICAN FALLS	W72-13300
AE, *BIOASSAY/ *CULTURES, *PLANT	GROWTH, *CHLOROPHYTA, *MARINE ALG	W72-14294
*ALGAE	GROWTH, *CHEMOSTATS.:	W72-06612
, *AQUATIC WEEDS, *WEED CONTROL,	GROWTH, *CULTURES, *ALGICIDES, BI	W71-06102
AT/ *FORECASTING, *ALGAE, *PLANT	GROWTH, *EUTROPHICATION, GROWTH R	W71-11033
*ALGAL	GROWTH, *LAKE TAHOE(CALIF).:	W71-13553
ELENASTRUM CAPRICORNUTUM, *ALGAL	GROWTH, *LAKE GEORGE(NY).: *S	W72-08376
*PHOSPHORUS, *CLADOPHORA, *PLANT	GROWTH, *LAKE MICHIGAN, ALGAE, PH	W72-13644
*ALGAE, *NUTRIENTS, LAKES, PLANT	GROWTH, *MOLYBDENUM, BIOASSAY, NE	W72-00801
MNESIU/ *GLYCEROL, *PHOTOTROPHIC	GROWTH, *MARINE PHYTOPLANKTON(PRY	W70-07842
PYROPHOSPHATE, RIO GRAN/ *ALGAL	GROWTH, *RADIOISOTOPE P32, SODIUM	W72-13992
ULANS, AQUATIC ECOSYSTEMS, ALGAE	GROWTH, ALGAL POPULATIONS.: / NID	W71-00475

AE STRIPPING, SCENEDESMUS, ALGAL	GROWTH, ALGAL HARVESTING, *SAN JO	W72-02975
MODEL STUDIES, NUTRIENTS, PLANT	GROWTH, ANALYTICAL TECHNIQUES, WA	W72-13651
ITRATES, PLANT PHYSIOLOGY, PLANT	GROWTH, BIOMASS, GROWTH RATES, LI	W72-05476
*PAAP TEST, *ALGAL	GROWTH, BOTTLE TEST.:	W73-00255
PTION, DDT, PHYTOPLANKTON, PLANT	GROWTH, CHLOROPHYLL, DDE, INHIBIT	W72-12998
H CULTURES.: *ALGAL	GROWTH, CONTINUOUS CULTURES, BATC	W72-14790
PHYTA, *PLANT MORPHOLOGY, *PLANT	GROWTH, CYTOLOGICAL STUDIES.: /DO	W73-01622
G NUTRIENTS, BICARBONATES, PLANT	GROWTH, DURNAL.: /DIOXIDE, CYCLIN	W72-09159
D, ALGAL ASSAY PROCEDURES, ALGAL	GROWTH, IN-LAKE NUTRIENT CONTROL.	W71-12072
METALS, DISSOLVED OXYGEN, *PLANT	GROWTH, LIGHT INTENSITY, TOXICITY	W72-14690
R/ *GLYCEROL, PHOTOHETEROTROPHIC	GROWTH, LUMINARIN-TYPE POLYSACCHA	W70-07280
BRIO, RHODOTORULA, A/ *MICROBIAL	GROWTH, MICROCOCCUS, BACILLUS, VI	W72-10861
COMPOSITION, ALGAL DECO/ *ALGAL	GROWTH, NUTRIENT UPTAKE, CHEMICAL	W70-05469
	GROWTH, OXYGEN, MONITORS.:	W71-10791
ODUCTIVITY, ANIMAL GROWTH, PLANT	GROWTH, OYSTERS, SHELLFISH, MUSSE	W72-09378
OWTH RATES, PRODUCTIVITY, ANIMAL	GROWTH, PLANT GROWTH, OYSTERS, SH	W72-09378
NTS, BIOMASS, *CYANOPHYTA, PLANT	GROWTH, PLANKTON.: /NTROL, NUTRIE	W72-00799
ITION, *PLANT PHYSIOLOGY, *PLANT	GROWTH, PLANT GROWTH SUBSTANCES,	W70-07842
ON EFFECTS, ALGAE, HEAVY METALS,	GROWTH, PSEUDOMONAS, POTASSIUM CO	W72-13813
LANT TISSUES, BI/ *ALGAE, *PLANT	GROWTH, RHODOPHYTA, INHIBITORS, P	W72-07496
ESTS, CYTOLOGICAL STUDIES, PLANT	GROWTH, SCENEDESMUS.: /BORATORY T	W71-12858
CARBON, *NITROGEN, *ALGAE, PLANT	GROWTH, SEWAGE EFFLUENTS, REGULAT	W72-14790
ES, ENVIRONMENTAL EFFECTS, PLANT	GROWTH, SULFATE, LABORATORY TESTS	W72-00153
TIUM RADIOISOTOPES, ALGAE, PLANT	GROWTH, TEMPERATURE, PROTOZOA, CH	W71-09254
*ALGAL CONTROL, FINANCING, PLANT	GROWTH, TOXICITY, NUTRIENTS, WATE	W71-13172
TROGEN, PHOSPHORUS, ALGAE, PLANT	GROWTH, WATER POLLUTION TREATMENT	W72-11702
ICAL METHODS, AMMONIA STRIPPING,	GUGGENHEIM PROCESS.: /PHYSICOCHEM	W70-09907
TARIO), CANADA, BUFFER/ *LOADING	GUIDELINES, *GRAND RIVER BASIN(ON	W71-11017
ORTHOPHOSPHATE CONCENTRATIONS,	GULF OF FINLAND.:	W72-13871
, ATLANTIC OCEAN, PACIFIC OCEAN,	GULF OF MEXICO.: /TION TECHNIQUES	W71-11851
YPERHALINE, EURYHALINE, RED SEA,	GULF OF SUEZ, DECAPODS, COLEOPTER	W73-00855
TES, SARGASSO SEA, SHELF WATERS,	GULF STREAM, CAPE HATTERAS, NANSE	W72-07663
NE, MOLLUSKS, COPEPODS, DIPTERA,	GULFS, GEOLOGIA HISTORY, PLEISTOR	W73-00855
MERCURY COMPOUNDS, BIOSYNTHESIS,	GUPPIES.: /A BAY(JAPAN), ORGANIC	W72-12257
CYS/ PROTOCRYPTOCHRYSIS OBOVATA,	GUTTULA BACILLARIOPHAGA, REFRACTO	W72-09119
, ANABAENA FLOSAQUAE, SPIRULINA,	GYMNODINIUM BREVE, PERIDINIUM, OS	W73-00840
MACROCEROS, NOCTILUCA MILIARIS,	GYMNODINIUM SPLENDENS, EXUVIELLA	W71-07875
, SPECTROMETRY, LICHENS, MOSSES,	GYMNOSPERMS, ANGIOSPERMS.: /ATIOS	W72-12709
, LIMNEPHILUS, GLOSSOSOMA, PHYSA	GYRINA, KEMMERER SAMPLER, SECCHI	W72-14280
COSTATUM, GLENODINIUM FOLIACUM,	GYRODINIUM FISUM.: /, SKELETONEMA	W72-07166
GRAMMA VANHEURCKII, PLEUROSIGMA,	GYROSIGMA FASCIOLA, NITZSCHIA SPP	W73-00853
MICROCYSTIS AERU/ *KEZAR LAKE(N.	H.), APHANIZOMENON, ANABAENA SPP,	W72-07890
LONGIROSTRIS/ *WINNISQUAM LAKE(N	H), ANABAENA CIRCINALIS, BOSMINA	W71-03014
XYGEN, CHLORIDES, INVERTEBRATES,	HA: /E, IMPOUNDMENTS, DISSOLVED O	W72-14280
WATER QUALITY, ECOLOGY, AQUATIC	HABITATS, CLADOPHORA, LIMITING FA	W72-13142
S, ISOPODS, SCENEDESMUS, AQUATIC	HABITATS, CLASSIFICATION, AQUATIC	W72-12744
OLLUTION EFFECTS, ALGAE, AQUATIC	HABITATS, AQUATIC ENVIRONMENT.: /	W71-07337
OPHY, DIATOMS, EUGLENA, DAPHNIA,	HABITATS, NICHES, COMPETITION, CA	W71-09171
RES, *ALGAE, *LAKE ERIE, AQUATIC	HABITATS, LAKES, AQUATIC ENVIRONM	W71-09156
, *STREAMS, *ECOSYSTEMS, AQUATIC	HABITATS, BIOMASS, AQUATIC ALGAE,	W72-06340
LOGY, *PREIMPOUNDMENTS, *AQUATIC	HABITATS, *SAMPLING, *ECOSYSTEMS,	W72-06526
POLLUTION EFFECTS, AQUATIC LIFE,	HABITATS, ECOLOGY, *ECOSYSTEM, *A	W73-01632
ISTRIBUTION, ECOSYSTEMS, AQUATIC	HABITATS, PRODUCTIVITY, AQUATIC B	W73-02095
IVITY, *ECOSYSTEMS/ *TERRESTRIAL	HABITATS, *HEAVY METALS, *PRODUCT	W72-14303
YSTEMATICS, *ALGAE, CHLOROPHYTA,	HABITATS, AQUATIC HABITATS, EUGLE	W72-09119
, CHLOROPHYTA, HABITATS, AQUATIC	HABITATS, EUGLENOPHYTA, PYRROPHYT	W72-09119
IBUTION, *ESTUARINE ENVIRONMENT,	HABITATS, AQUATIC LIFE, HEAVY MET	W72-11854
ANTS, *DISCHARGE(WATER), AQUATIC	HABITATS, ENVIRONMENTAL EFFECTS,	W71-11517
PHAEOPHYTA, EPIPHYTOLOGY, ALGAE,	HABITATS, CULTURES.: / STUDIES, *	W72-01370
KTON, PRIMARY PRODUCTIVITY, FOOD	HABITS, ALGAE, ZOOPLANKTON, CARNI	W72-00161
SHWATER FISH, PONDS, LAKES, FOOD	HABITS, EUTROPHICATION, LARVAE, R	W72-07132
NOSTYLA, LEPADELLA, TRICHOCERCA,	HAEMATOCOCCUS LACUSTRIS.: /US, MO	W72-07132
ION, CHLAMYDOMONAS, METABOLITES,	HAEMATOCRIT TUBE.: /AL MAGNIFICAT	W72-04134
, LIGHT PENETRATION, ALKALINITY,	HALOGENS, HYDROGEN ION CONCENTRAT	W73-01434
GASSUM CRISPUM, DIGENIA SIMPLEX,	HALOPHILA STIPULACEA, DIPLANTHERA	W73-00855
S, GOLDFISH, MERRIMACK RIVER(NEW	HAMPSHIRE), DELAWARE RIVER, ILLIN	W71-11517
*BEGGIATOA, PHYTOPLANKTON, *NEW	HAMPSHIRE, MIXING, ENVIRONMENTAL	W72-09061
POLLUTION CONTROL, FISHKILL, NEW	HAMPSHIRE, WATER POLLUTION EFFECT	W72-09304
AERATION, WATER PROPERTIES, *NEW	HAMPSHIRE, ALGAE, COLOR, COPPER S	W72-07890
PHYTOTOXICITY, *CYANOPHYTA, *NEW	HAMPSHIRE, MARINE ALGAE, PHYTOPLA	W72-14542
STS, PENNSYLVANIA, NEW YORK, NEW	HAMPSHIRE, OHIO, MICHIGAN.: /, HO	W72-04736
*ZOOPLANKTERS, *CHEMCONTROL, NEW	HAMPSHIRE, COPEPODS, DAPHNIA, SAM	W71-03014
XIDATION DITCH, HYDRAULIC MANURE	HANDLING.: *O	W72-08396
KENNET(ENGLAND), STEPHANODISCUS	HANTZSCHII, CRYPTOMONAS, RHODOMON	W72-14797
LARIA FLOCCULOSA, STEPHANODISCUS	HANTZSCHII, CYCLOT MICHIGANIANA,	W73-00384
CYCLOTELLA COMTA, STEPHANODISCUS	HANTZSCHII, APHANIZOMENON FLOS-AQ	W72-12738
-VOLATILE COMPOUNDS, WOODS HOLE,	HAPTAPHYCEAE,: /ANIC CARBON, ACID	W72-07710
*BOSTON(MASS), LYNN	HARBOR(MASS).:	W71-05553
CLADOPHORA GLOMERATA, MILWAUKEE	HARBOR(WISCONSIN), POLYPHOSPHATES	W72-13644
*OYSTER BAY, *JAMAICA, FALMOUTH	HARBOR, *TINTINNIDS, *TYCHOPELAGI	W72-07899
ROPHICATION, NITROGEN, LITTORAL,	HARBORS, SPATIAL DISTRIBUTION, AL	W72-04263
VED OXYGEN, OIL, BRIDGES, BUOYS,	HARBORS, TUNNELS, MARINE ALGAE, A	W73-02029
WATER SOFTENING, LEGAL ASPECTS,	HARD WATER, PUBLIC HEALTH, COST C	W71-09429
TRATION, ALKALINITY, CARBONATES,	HARDNE: /ITY, HYDROGEN ION CONCEN	W71-00117
N ION CONCENTRATION, ALKALINITY,	HARDNESS(WATER), SAMPLING, DIATOM	W71-03028
AMPLING, THERMAL STRATIFICATION,	HARDNESS(WATER), COLOR, OLIGOTROP	W72-04269
O, PLANKTON, DIATOMS, SEDIMENTS,	HARDNESS(WATER), BENTHOS, GRAZING	W73-01627
ON, AMMONIA, NITRATES, NITRITES,	HARDNESS(WATER), NITROGEN CYCLE,	W72-08048
TMENT, CARBONATES, BICARBONATES,	HARDNESS(WATER), ALKALINITY, ENZY	W72-07514
ENTRATION, PHOSPHATES, NITROGEN,	HARDNESS(WATER), CONDUCTIVITY, ZI	W72-07890

EDS, PHYTOPLANKTON, OLIGOTROPHY, HARDNESS(WATER), SOILS, AMINO: /E W72-08459
CIDES, COPPER SULFATE, SALINITY, HARDNESS(WATER), CHELATION, MINE W72-14694
ENT, SUSPENDED LOAD, ALKALINITY, HARDNESS(WATER), ANALYTICAL TECHN W71-11033
URE, HYDROGEN ION CONCENTRATION, HARDNESS, ALKALINITY, CONDUCTIVIT W72-10875
HYDROIDS, POLYCHAETA, OSTRACODS, HARPACTICOIDS, CIRRIPEDIA, ACARIN W72-05432
TELLA, ASPLANCHNA, LITHODESMIUM, HARPESTICOIDS, NOCTILUCA, CENTROP W71-11876
*HETEROTROPHIC POPULATIONS, LAKE HARTWELL(SOUTH CAROLINA), ANACYST W71-00475
QUALITY, FISH ATTRACTANTS, FISH HARVEST, PLANKTON, BENTHIC FAUNA, W70-09596
LUTION EFFECTS, *AQUATIC PLANTS, HARVESTING, PONDS, WATER HYACINTH W71-01488
FILTRATION, *ALGAE, *SUSPENSION, HARVESTING, PONDS, CALIFORNIA, SC W70-10174
*ALGAE HARVESTING.: W71-08951
CRITERIA, *WATER QUALITY, ALGAE, HARVESTING, DEPTH, TEMPERATURE, L W71-07100
SCENEDESMUS, ALGAL GROWTH, ALGAL HARVESTING, *SAN JOAQUIN VALLEY(C W72-02975
S, PROTEINS, NUTRIENTS, BIOMASS, HARVESTING OF ALGAE, PRODUCTIVITY W72-03327
OLYSIPHONIA LANOSA, POLYSIPHONIA HARVEYI, RHODOTORULA RUBRA, RHODO W72-04748
S COMMUNIS, JUNCUS SPP, ATRIPLEX HASTATA, SCIRPUS SPP, TYPHA SPP, W73-01089
*WATER POLLUTION SOURCES, *FISH HATCHERIES, *SALMONIDS, WATER REQ W71-11006
RDAN RIVER(M/ SETTLEABLE SOLIDS, HATCHERY EFFLUENTS, PARASITES, JO W71-11006
GREEN RIVER(WASH), COWLITZ TROUT HATCHERY(WASH).: /RESIDUAL FOOD, W71-11006
SHELF WATERS, GULF STREAM, CAPE HATTERAS, NANSEN BOTTLE, SPECIES W72-07663
LLAE, BAJA CALIFORNIA, HEMIAULUS HAUCKII, STIGMAPHORA ROSTRATA, ET W72-03567
NUTRIENT INTERCHANGE, HAW RIVER, NEW HOPE RIVER.: W70-09669
NE PLANTS, BACTERIA, POPULATION, HAWAII, ABSORPTION, TROPICAL REGI W71-08042
, ESTUARIES, CULTURES, SAMPLING, HAWAII, S: /OF POLLUTANTS, OCEANS W73-00361
ISOTOPES, TRACERS, MARINE ALGAE, HAWAII, WATER TEMPERATURE, WAVES, W71-11486
CHNOLOGY, SURFACTANTS, TOXICITY, HAZARDS, BIODEGRADATION.: /TY, TE W72-09169
*PUBLIC HEALTH, WATER POLLUTION, HAZARDS, NUTRIENTS, ALGAE.: /CY, W72-04266
DISSOLVED SOLIDS, FLOCCULATION, HEAD LOSS, HYDROGEN ION CONCENTRA W71-00138
NANCE, *COST ANALYSIS, *FILTERS, HEADLOSS, PRESSURE, TURBIDITY, AL W72-09393
EASES, *TOXIC CHEMICALS, *PUBLIC HEALTH STANDARDS, TOXICOLOGY, NTA W73-02144
ON EFFECTS, T/ *MERCURY, *PUBLIC HEALTH, *SHELLFISH, WATER POLLUTI W72-12257
TER POLLUTION TREATMENT, *PUBLIC HEALTH, ANALYTICAL TECHNIQUES, WA W73-02144
*AQUATIC PLANTS, *ALGAE, *PUBLIC HEALTH, ANALYTICAL TECHNIQUES, TO W72-04275
ION EFFECTS, FOOD CHAINS, PUBLIC HEALTH, AQUATIC LIFE, AQUATIC ALG W72-10978
N SOURCES, AQUATIC ALGAE, PUBLIC HEALTH, BIODEGRADATION, TOXICITY, W71-11516
EGAL ASPECTS, HARD WATER, PUBLIC HEALTH, COST COMPARISONS, AQUATIC W71-09429
OLYCHLORINATED BIPHENYLS, PUBLIC HEALTH, ENVIRONMENT CONTROL, LEGA W73-00703
ON, *PATH OF POLLUTANTS, *PUBLIC HEALTH, ENZYMES, ANALYTICAL TECHN W72-13347
INE PLANTS, MARINE ALGAE, PUBLIC HEALTH, ESTUARINE ENVIRONMENT, ON W72-04461
ON EFFECTS, FOOD CHAINES, PUBLIC HEALTH, ESTUARINE ENVIRONMENT, ZI W72-04463
ENVIRONMENT, FOOD CHAINS, PUBLIC HEALTH, EUROPE, RADIOISOTOPES, EF W73-00828
ASTE DISPOSAL, STANDARDS, PUBLIC HEALTH, INSPECTION, WATER SOFTENI W72-01085
ARINE ALGAE, FOOD CHAINS, PUBLIC HEALTH, NUCLEAR WASTES, PATH OF P W73-00823
TES, TOXICITY, CORROSION, PUBLIC HEALTH, ODORS, OXIDATION-REDUCTIO W72-04086
WATER POLLUTION SOURCES, PUBLIC HEALTH, ON-SITE INVESTIGATIONS, W W73-00790
ABSORPTION, *FOOD CHAINS, PUBLIC HEALTH, PATH OF POLLUTANTS, RADIO W73-00833
C COMPOUNDS, FOOD CHAINS, PUBLIC HEALTH, PHYTOPLANKTON: /N, ORGANI W73-00821
UTRIENTS, TRACE ELEMENTS, PUBLIC HEALTH, RADIOECOLOGY.: /CYCLING N W73-00826
C BACTERIA, CONFERENCES, *PUBLIC HEALTH, RIVERS, *SEWAGE, BIOCHEMI W72-09675
INS, ABSORPTION, CALCIUM, PUBLIC HEALTH, SOUTH AMERICA, ATLANTIC C W73-00819
AINS, PATH OF POLLUTANTS, PUBLIC HEALTH, WATER POLLUTION EFFECTS, W73-00802
ROPHICATION, LEGISLATION, PUBLIC HEALTH, WATER QUALITY CONTROL, AL W72-06655
OURCES, SEWAGE TREATMENT, PUBLIC HEALTH, WASTE WATER TREATMENT, RE W72-06638
CALIFORNIA, FOOD CHAINS, PUBLIC HEALTH, WATER POLLUTION SOURCES.: W72-03324
N, *ECONOMIC EFFICIENCY, *PUBLIC HEALTH, WATER POLLUTION, HAZARDS, W72-04266
SOURCES, SOAPS, TOXICITY, PUBLIC HEALTH: /SPOSAL, WATER POLLUTION W72-01093
ATES, ALKALINITY, OXYGEN DEMAND, HEAT BUDGET, ALGAE, EUTROPHICATIO W72-12997
*WASTE HEAT, THERMAL EFFICIENCY, HEAT CONSUMPTION.: W70-08832
RESEARCH, CHYTRIDIACEOUS FUNGI, HEAT SHOCK.: W72-07225
UDIES, BODY FLUIDS, VERTEBRATES, HEAT STERILIZATION, FIXATION, EMB W73-01676
PLANTS, ELECTRICAL POWER PLANTS, HEAT TRANSFER, WATER POLLUTION EF W72-07907
, CHEMICAL PROPERTIES, NITROGEN, HEAT TRANSFER, PHOSPHATES, ALKALI W72-12997
G, / *WASTE HEAT, *USES OF WASTE HEAT, *CENTRAL HEATING AND COOLIN W70-09193
AL HEATING AND COOLING, / *WASTE HEAT, *USES OF WASTE HEAT, *CENTR W70-09193
ACE HEATING, *FARM USES OF WASTE HEAT, STEAM POWERED BUSES.: / *SP W70-09193
ONES, MODELING.: *WASTE HEAT, SUBLETHEL EFFECTS, MIXING Z W70-09612
NSUMPTION.: *WASTE HEAT, THERMAL EFFICIENCY, HEAT CO W70-08832
RIDA), / *BISCAYNE BAY(FLORIDA), HEATED EFFLUENT, TURKEY POINT(FLO W72-11876
POLLUTION, *CRUSTACEANS, *FISH, HEATED WATER, NUCLEAR POWER PLANT W72-07907
QUIPMENT, ANALYTICAL TECHNIQUES, HEATED WATER, CONDENSERS, PROTOZO W72-07526
N, COOLING, THERMAL POWERPLANTS, HEATED WATER, CIRCULATION, INFRAR W73-01704
ENSITY, TIDAL EFFECTS, SALINITY, HEATED WATER, BIOMASS, THERMAL ST W73-00853
HABITATS, ENVIRONMENTAL EFFECTS, HEATED WATER, NUCLEAR POWER PLANT W71-11517
T, *USES OF WASTE HEAT, *CENTRAL HEATING AND COOLING, *SPACE HEATI W70-09193
TRAL HEATING AND COOLING, *SPACE HEATING, *FARM USES OF WASTE HEAT W70-09193
DIO/ *CHELATION, CHROMATOGRAPHY, HEAVY METALS, *TRACE ELEMENTS, RA W72-04708
TIFICIAL), PHOSPHORUS, NITROGEN, HEAVY METALS, ORGANOPHOSPHORUS PE W72-04742
*INHIBITION, *LIPIDS, TOXICITY, HEAVY METALS, LABORATORY TESTS, P W72-06037
E, NUTRIENTS, BACTERIA, VIRUSES, HEAVY METALS, HUMAN POPULATION, W W71-09788
FORNIA WATER CONVEYANCE, CANALS, HEAVY METALS.: /ENVIRONMENT, CALI W71-09581
ASTES, MUNICIPAL WASTES, METALS, HEAVY METALS, WASTE DISPOSAL, WAT W71-11793
ALGAE, INSECTICIDES, HERBICIDES, HEAVY METALS, TASTE, ODOR, RECREA W71-12091
E ELEMENTS, *LEAD RADIOISOTOPES, HEAVY METALS, AIR POLLUTION EFFEC W72-00940
LECTROCHEMISTRY, TRACE ELEMENTS, HEAVY METALS, ELECTRODES, NITRATE W72-01693
EFFECTS, BIOLOGICAL COMMUNITIES, HEAVY METALS, HERBICIDES, TOXINS, W72-01472
RONMENT, HABITATS, AQUATIC LIFE, HEAVY METALS, INDUSTRIAL WASTES, W72-11854
LLA, AQUATIC ALGAE, CHLOROPHYTA, HEAVY METALS, *CHLAMYDOMONAS, INH W72-11727
LT RADIOISOTOPES, *MARINE ALGAE, HEAVY METALS, ABSORPTION, BICINDI W72-10957
ESE, ZINC, METALS, SECCHI DISKS, HEAVY METALS, BIOINDICATORS, SILI W72-09111
ALGAE, WATER POLLUTION EFFECTS, HEAVY METALS, CYCLING NUTRIENTS, W73-00840

TODES, EPIDEMIOLOGY, PHOSPHATES,	HEAVY METALS, PESTICIDES, TASTE-P	W73-02144
IMAL PHYSIOLOGY, TRACE ELEMENTS,	HEAVY METALS, PHYSICOCHEMICAL PR:	W73-00817
AMS, HYDROGEN ION CONCENTRATICN,	HEAVY METALS, DISSOLVED OXYGEN, *	W72-14690
*DATA COLLECTIONS, COPPER, LEAD,	HEAVY METALS, MUSSELS, *KELPS, ST	W72-11925
NICKEL, COBALT, COPPER SULFATE,	HEAVY METALS, METALS, SULFATES.: /	W72-12739
ATER QUALITY, ALGAE, PESTICIDES,	HEAVY METALS, WASTEWATER TREATMEN	W72-14282
WATER POLLUTION EFFECTS, ALGAE,	HEAVY METALS, GROWTH, PSEUDOMCNAS	W72-13813
LENIA FRAGILISSIMA, RHIZOSCLENIA	HEBETATA F. SEMISPINA, CHAETOCERO	W72-03543
HORA, MELANOSIPHON INTESTINALIS,	HECATONEMA MACULANS, ELACHISTA LU	W72-01370
LTURES, TITANIUM DIOXIDE WASTES,	HELGOLAND.: /CYSTIS SPP., SOIL CU	W72-12239
EPO/ *FEEDING PATTERNS, *CALANUS	HELGOLANDICUS, COSCINCDISCUS, COP	W72-09248
NIUM DEPRESSUM, METAZOA, CALANUS	HELGOLANDICUS(PACIFICUS), THERMIS	W72-07892
EAE, XANTHOPHYCEAE, POLYEDRIELLA	HELVETICA, PLEUROCHLORIS COMMUTAT	W72-14808
ISCUS GAZELLAE, BAJA CALIFORNIA,	HEMIAULUS HAUCKII, STIGMAPHCRA RO	W72-03567
PRODUCTIVITY, CEPHALCCHORDATES,	HEMICHORDATES, ARTHROPODS, *BAY O	W73-00935
OGA, DIASTYLOPSIS, OXYUROSTYLIS,	HEMILAMPROPS, LAMPROPS, C: / ANAL	W72-09092
LAMPRA MARYLANDICA, ASTEROLAMPRA	HEPTACTIS, BACTERIASTRUM COMOSUM,	W71-11527
IOPSIS BREVICAULIS, CLADOSPORIUM	HERBARUM, TRAMETES VERSICOLOR, FO	W72-13813
PESTICIDES, ORGANIC PESTICIDES,	HERBICIDES, WATER POLLUTICN SCURC	W72-13347
NUTRIENTS, ALGAE, INSECTICIDES,	HERBICIDES, HEAVY METALS, TASTE,	W71-12091
TROGEN FIXATION, 2-4-D, DALAPCN,	HERBICIDES, CROP PRODUCTION.: /NI	W71-12071
, PHYTOPLANKTON, PHYTO-TOXICITY,	HERBICIDES, INSECTICIDES, BIOMASS	W71-10566
ALGAE, PESTICIDES, INSECTICIDES,	HERBICIDES, PHOSPHORUS, PHENOLS,	W72-09646
T GROWTH REGULATORS, INHIBITORS,	HERBICIDES, MCNURCN, PHOTOSYNTHES	W72-09650
GICAL COMMUNITIES, HEAVY METALS,	HERBICIDES, TOXINS, WASTE DILUTIO	W72-01472
DES, INSECTICIDES, RODENTICIDES,	HERBICIDES, FUNGICIDES, ALGICIDES	W73-00238
OPHYTA, *PESTICIDES, FUNGICIDES,	HERBICIDES, IRRIGATION, AGRICULTU	W72-14807
OLLUTION EFFECTS, WATER QUALITY,	HERBICIDES, OCOR-PRODUCING ALGAE.	W72-14692
N EFFECTS, BIOASSAY, PESTICIDES,	HERBICIDES, 2-4-D, PLANT GROWTH R	W72-08440
ISSOLVED OXYGEN, COPPER SULFATE,	HERBICIDES, WATER POLLUTION EFFEC	W72-04759
ONAS GLOBUSA, CHLO/ *CHLORINATED	HERBICIDES, MCPB, MCPA, CHLAMYDOM	W70-07255
NATURE, ECOLOGY, TROPHIC LEVEL,	HERBIVORES, NATURAL STREAMS, PULP	W72-06340
OCONTROL, REMEDIES, FOODS, FISH,	HERBIVORES, FISH MANAGEMENT.: /BI	W71-09163
L, *ENERGY BUDGET, PRODUCTIVITY,	HERBIVORES, CARNIVORES, ECOSYSTEM	W71-10064
. DAK).: LAKE	HERMAN(S. DAK), ENEMY SWIM LAKE(S	W71-02880
TES.: PRAIRIE LAKES, *LAKE	HERMAN(SO. DAKOTA), *ORTHOPHOSPHA	W73-00938
RUS, LYSIMACHIA NUMMULARIA, BLUE	HERON, STERNA HIRUNDO, RATS, PART	W72-05952
STRIPED BASS, MINNOWS, FISHKILL,	HERRING, COPEPODS, CRUSTACEANS, D	W71-11517
RAYFISH, MOSQUITOES, FOOD CHAIN,	HERRING, OYSTERS, CLAMS, COLUMBIA	W71-07417
G, ALGAE, SULFATES, PERCOLATICN,	HERRING, SALMON, TROUT, TOXICITY,	W72-05421
SICULOSUS, CLADOPHORA GLOMERATA,	HETEROCYPRIS SALINUS, GASTEROSTEU	W71-08026
LGAE.: *WINDERMERE(ENGLAND),	HETEROCYSTS, BENTHIC BLUE-GREEN A	W73-02472
TY, LIGHT, TURBIDITY, NUTRIENTS,	HETEROGENEITY, CHLOROPHYTA, CHRYS	W72-13142
VOLVOCACEAE, PANDORINA MORUM, /	HETEROINHIBITORS, AUTOINHIBITORS,	W71-13793
CTINOTROCHA, VELIGER, PTEROPODS,	HETEROPODS, OSCILLATORIA, STOMATO	W73-00935
ICTOCHIRONOMUS ANNULIORUS, PHYSA	HETEROSTROPHA, PROCLADIUS, WHITE	W70-06668
	HETEROTROPHS, AUTOTROPHS.:	W71-03021
ISLAND, OPEN OCEAN, SAPROPHYTES,	HETEROTROPHIC BACTERIA, C-14.: /	W71-08042
UORESCENS, AUTOTROPHIC BACTERIA,	HETEROTROPHIC BACTERIA, GLUCOSE,	W73-02095
N BAY, FOX RIVER, DETROIT RIVER,	HETEROTROPHIC BACTERIA, ACETYLENE	W73-01456
E, RIBOSE, ARABINOSE, GALACTOSE,	HEXURONIC ACID, VINEYARD SOUND, O	W72-07710
IFLAGELLATA, TSUMURAIA NUMEROSA,	HIROSEIA QUINQUELOBATA, TROITSKIE	W72-07683
KOLAJSKIE LAKE(POLAND), CERATIUM	HIRUNDINELLA.: /IA ECHINULATA, MI	W72-14812
A NUMMULARIA, BLUE HERON, STERNA	HIRUNDO, RATS, PARTRIDGE, DDE, KI	W72-05952
CHELOMONAS ARMATA, TRACHELOMONAS	HISPIDA VAR CRENULATOCOLLIS.: /RA	W73-00918
E, SERINE, TYROSINE, TRYPTOPHAN,	HISTIDINE, ASPARTATE, METHIONINE,	W73-00942
ERVATION, CHEMICAL PRESERVATION,	HISTOCHEMISTRY, LYOPHILIZATION, A	W73-01676
LYOPHILIZATION, ANIMAL TISSUES,	HISTOLOGICAL STUDIES, BODY FLUIDS	W73-01676
ROWTH RATES, *SYSTEMATICS, *LIFE	HISTORY STUDIES, *GROWTH STAGES,	W73-02099
S, / *CRUSTACEANS, *FOODS, *LIFE	HISTORY STUDIES, COPEPODS, DIATOM	W72-14805
DIOACTIVITY, TROPHIC LEVEL, LIFE	HISTORY STUDIES, TIME, SNAILS, RA	W70-06668
PHYTOLOGY, ALGAE, HABITAT/ *LIFE	HISTORY STUDIES, *PHAEOPHYTA, EPI	W72-01370
PEOODS, DIPTERA, GULFS, GEOLOGIA	HISTORY, PLEISTOCENE EPOCH, ADAPT	W73-00855
T TWIN LAKE, LAKE ROCKWELL, LAKE	HODGSON, CLEARWATER QUARRY, DOLLA	W72-13673
RONOMUS ANTHRACINUS, LIMNODRILUS	HOFFMEISTERI, TUBIFEX TUBIFEX, AP	W71-11009
ISOTOPES, SPIROGYRA, LIMNODRILUS	HOFFMEISTERI, STICTOCHIRONOMUS AN	W70-06668
MINANTS, ALGAE, CATTLE, POULTRY,	HOGS, ANIMAL DISEASES, CATFISH, W	W71-07544
RYING, ODOR, NITROGEN, AERATION,	HOGS, CATTLE, POULTRY, RESEARCH A	W71-07551
*WOODS	HOLE(MASS).:	W72-06046
, ACID-VOLATILE COMPOUNDS, WOODS	HOLE, HAPTAPHYCEAE,: /ANIC CARBON	W72-07710
CYSTIS AERUGINOSA, APHANIZOMENON	HOLSATICUM, ICTALURUS NEBULOSUS,	W72-07890
DELAWARE RIVER, ILLINOIS RIVER,	HOLSTON RIVER(TENNESSEE).: /IRE),	W71-11517
TIFICATION, ANABAENA FLOW-AQUAE,	HOMOGENIZING, TRA: /AQUAE, DESTRA	W72-07890
, TONGA ISLANDS, TUTUILA ISLAND,	HONSHU ISLAND, OPEN OCEAN, SAPROP	W71-08042
IENT INTERCHANGE, HAW RIVER, NEW	HOPE RIVER.: NUTR	W70-09669
ANTIMONY, CESIUM, *CAPE OF GOOD	HOPE.:	W73-00832
S, PENICILLIN, DRUGS, CORTISONE,	HORMONES, CHLORAMPHENICOL, SALMON	W72-13800
SULAE, REFRACTOMONAS BRASILIANA,	HORTOBAGGIAMONAS PLICATA, TRICHOC	W72-09119
, ENGLAND, METABOLIC INHIBITION,	HOSTS, MYXOPHYCEAE, SCOTLAND.: /A	W72-05457
S, DELAWARE, PATH OF POLLUTANTS,	HOSTS, PENNSYLVANIA, NEW YORK, NE	W72-04736
RINGS, JACKSON HOT SPRINGS, LOLO	HOT SPRINGS, PIPESTONE HOT SPRING	W72-05610
TONE HOT SPRINGS, SLEEPING CHILD	HOT SPRINGS.: /HOT SPRINGS, PIPES	W72-05610
/ ALHAMBRA HOT SPRINGS, BOULDER	HOT SPRINGS, JACKSON HOT SPRINGS,	W72-05610
NGS, LOLO HOT SPRINGS, PIPESTONE	HOT SPRINGS, SLEEPING CHILD HOT S	W72-05610
GS, BOULDER HOT SPRINGS, JACKSON	HOT SPRINGS, LOLO HOT SPRINGS, PI	W72-05610
JACKSON HOT SPRINGS, / ALHAMBRA	HOT SPRINGS, BOULDER HOT SPRINGS,	W72-05610
WATER, *WYOMING, NATIONAL PARKS,	HOT SPRINGS, WATER TEMPERATURES,	W72-08508
TERGENT / NITRILOTRIACETIC ACID,	HOUSEHOLD DETERGENTS, SOAP AND DE	W72-04266
MARYLAND, QUARRIES, OHIO RIVER,	HUDSON RIVER, INDICATORS, DELAWAR	W72<04736
NA, BENTHIC FLORA, LAKE ONTARIO,	HUDSON RIVER, STANDARDS, SURFACE	W¯C-09614

S, RIVERS, PHOSPHATES, NITRATES, HUDSON RIVER, ALGAE, WATER QUALIT W70-06509
GICAL TISSUES, MERCURY ISOTOPES, HUMAN BRAIN, LOW-TEMPERATURE OVEN W71-11036
E WASHINGTON, NITRILOTRIACETATE, HUMAN FECES.: / METASILICATE, LAK W72-04734
L PROPERTIES, MINNESOTA, LIPIDS, HUMAN PATHOLOGY, DISEASES.: /MICA W72-04275
*LABORATORY TESTS, ALGAL TOXINS, HUMAN PATHOLOGY.: *CYANOPHYTA, W72-04259
RE, ENZYMES, CELLULOSE, VIRUSES, HUMAN PATHOLOGY, SYSTEMATICS.: /U W72-05457
BACTERIA, VIRUSES, HEAVY METALS, HUMAN POPULATION, WASTE WATER TRE W71-09788
AL WASTES, TURNOVERS, SEDIMENTS, HUMAN POPULATION, TOURISM, WATERS W71-11009
, MARINE ANIMALS, MARINE PLANTS, HUMAN POPULATION, RADIOACTIVITY T W73-00833
S, BENTHIC FAUNA, BENTHIC FLORA, HUMAN POPULATION, PRIMARY PRODUCT W72-12549
IN RIVER, MORRO BAY POWER PLANT, HUMBOLDT BAY, CONNETICUT YANKEE N W72-07907
SAN JOAQUI/ BARNACLES, FLOUNDER, HUMBOLDT BAY, CONNECTICUT RIVER, W71-07417
ESTUARY, BISCAYNE BAY(FLORIDA), HUMBOLT BAY(CALIFORNIA), POTOMAC W71-11517
S, *MARINE ALGAE, PHYTOPLANKTON, HUMIC ACIDS, FULVIC ACIDS, TRACE W71-09188
MARGARET'S BAY, SCOUDOUC RIVER, HUMIC ACIDS, OXIMATION, SAMPLE PR W72-12166
INEMA FLUVIATILE, CULTURE MEDIA, HUMMOCK CHANNEL, CHLOROPLASTS, MO W72-12633
G ORGANIC MATTER, *MARINE ALGAE, HUMUS, ORGANIC COMPOUNDS, POLLUTA W72-12166
IVER, PINE RIVER, CLINTON RIVER, HURON RIVER, BELL RIVER, SALT RIV W72-14282
SUPERIOR, *LAKE MICHIGAN, *LAKE HURON, *LAKE ONTARIO, *LAKE ERIE, W72-01094
EUTROPHICATION, *NUTRIENTS, LAKE HURON, EFFLUENTS, WATER QUALITY, W72-13662
, LAKE MICHIGAN, LAKE ERIE, LAKE HURON, LAKE SUPERIOR, WATER TEMPE W72-10875
KE SUPERIOR, LAKE MICHIGAN, LAKE HURON, LAKE ERIE, GREAT LAKES, SO W73-01456
IMILATION, EMILIANA(COCCOLITHUS) HUXLEYI, AMPHIPRORA ALATA, CHRYSO W73-00380
OCOCCUS, RHODOMONAS, COCCOLITHUS HUXLEYI, AUTOLYSIS, PHAEODACTYLUM W71-10083
AMPHIDINIUM CARTERI, COCCOLITHUS HUXLEYI, PORPHYRIDIUM, SKELETONEM W72-12998
ITY, CALCIUM, SCENEDESMUS, WATER HYACINTH, LIGHT PENETRATION, OXYG W72-04269
PLANTS, HARVESTING, PONDS, WATER HYACINTH, NUTRIENTS, MIDGES, FISH W71-01488
TOS, STICHOCOCCUS, CHLOROCOCCUM, HYALOTHECA, LAKE LIVINGSTON, AUTO W72-09239
PLANTS, DIATOMS, PHYTOPLANKTON, HYDORGEN ION TEMPERATURE,: /TMENT W72-06837
*OXIDATION DITCH, HYDRAULIC MANURE HANDLING.: W72-08396
LUTION, *THERMAL STRATIFICATION, HYDRAULIC MODELS, FISH, ALGAE, FA W70-09614
LAKES, LAKE ERIE, LAKE ONTARIO, HYDRO: /RS, PHOTOSYNTHESIS, GREAT W72-07648
ERVOIR, ANGARA RIVER, OKA RIVER, HYDROBIOLOGY, BACTERIOLOGY, SAPRO W72-04853
OLOGY, *RESERVOIRS, *FRESHWATER, HYDROBIOLOGY, ALGAE, FISH, ABSORP W72-10694
RIMARY PRODUCTIVITY, CHLORINATED HYDROCARBON PESTICIDES, DDT, ORGA W71-10096
E PES/ *PESTICIDES, *CHLORINATED HYDROCARBON PESTICIDES, *CARBAMAT W71-10566
STICIDE TOXICITY, / *CHLORINATED HYDROCARBON PESTICIDES, *DDT, *PE W71-12303
*DIELDRIN, *ENDRIN/ *CHLORINATED HYDROCARBON PESTICIDES, *ALDRIN, W72-03545
E RESIDUES, *ALGAE/ *CHLORINATED HYDROCARBON PESTICIDES, *PESTICID W72-04134
IOASSAY, *TOXICITY, *CHLORINATED HYDROCARBONS, *BIOINDICATORS, IND W71-07412
LORINATED BIPHENYLS, *CHLORNATED HYDROCARBON PESTICIDES, *PESTICID W72-08436
CATORS, *PESTICIDES, CHLORINATED HYDROCARBON PESTICIDES, INSECTICI W73-00238
BIOTICS(PESTICIDES), CHLORINATED HYDROCARBON PESTICIDES, PESTICIDE W72-14294
, *METABOLISM, BIO/ *CHLORINATED HYDROCARBON PESTICIDES, SEA WATER W73-00361
DTII, CULTURE MEDIA, CHLORINATED HYDROCARBONS.: /MYDOMONAS REINHAR W73-00281
OXYGEN, RESPIRATION, CHLORINATED HYDROCARBON PESTICIDES, GROWTH RA W73-00281
EPTADECANE, 8-METHYLHEPTADECANE, HYDROCARBONS, ALKANES, ALIPHATIC W73-01439
HYDROCARBONS, ALKANES, ALIPHATIC HYDROCARBONS, MASS SPECTROMETRY, W73-01439
PESTICIDE RESIDUES, CHLORINATED HYDROCARBON PESTICIDES, PESTICIDE W73-02280
SPHORCUS PESTICIDES, CHLORINATED HYDROCARBON PESTICIDES, CARBAMATE W72-13347
ISM, *MARINE ALGAE, *CHLORINATED HYDROCARBON PESTICIDES, *CYTOLOGI W72-12998
R POLLUTION EFFECTS, CHLORINATED HYDROCARBON PESTICIDES, *CYANOPHY W72-13809
ALGAE, FOOD CHAINS, CHLORINATED HYDROCARBON PESTICIDES, FISHERIES W72-13657
NE CHLORIDE, *CCC, *DIETHYLAMINE HYDROCHLORIDE, MERISOMOPEDIA SPP. W72-14706
*INORGANIC IONS, ZINC, VOLVOX, HYDRODICTYON, SPODYLOSIUM, SPIROG W71-11001
TOURISM, RECREATION, FISHERIES, HYDROELECTRIC POWER, NAVIGATION, W72-14878
LATION, CYANOPHYTA, CHLOROPHYTA, HYDROGEN ION CONCENTRATION, CHEMI W72-14790
G, WINDS, ALGAE, SURFACE WATERS, HYDROGEN ION CONCENTRATION, LIGHT W73-00236
DISSOLVED OXYGEN, CONDUCTIVITY, HYDROGEN ION CONCENTRATION.: /TY, W72-14728
WATER TEMPERATURE, ACID STREAMS, HYDROGEN ION CONCENTRATION, HEAVY W72-14690
NETRATION, ALKALINITY, HALOGENS, HYDROGEN ION CONCENTRATION, PHOSP W73-01434
, *ALGAE, *CULTURES, CYANOPHYTA, HYDROGEN ION CONCENTRATION, LIGHT W73-01657
COAGULATION, CALCIUM CARBONATE, HYDROGEN ION CONCENTRATION, FILTR W73-01669
, BULK DENSITY, POROSITY, COLOR, HYDROGEN ION CONCENTRATION, NITRO W73-00854
, WATER TEMPERATURE, ALKALINITY, HYDROGEN ION CONCENTRATION, WATER W73-00916
*ALGAE, *WASTE WATER TREATMENT, HYDROGEN ION CONCENTRATION, CYANO W73-01367
AE, COLOR, COPPER SULFATE, NETS, HYDROGEN ION CONCENTRATION, PHOSP W72-07890
EMICAL OXYGE/ *ALGAE, *NITROGEN, HYDROGEN ION CONCENTRATION, BIOCH W72-06612
GLENOPHYTA, PYRROPHYTA, BIOMASS, HYDROGEN ION CONCENTRATION, MINNE W72-07145
, TEMPERATURE, DISSOLVED OXYGEN, HYDROGEN ION CONCENTRATION, SCENE W72-08049
OSPHORUS RADIOISOTOPES, DIATOMS, HYDROGEN ION CONCENTRATION, MARIN W71-10083
AE, *DIATOMS, CHEMICAL ANALYSIS, HYDROGEN ION CONCENTRATION, SEA W W71-11008
, SEDIMENTS, LAKES, CHLOROPHYLL, HYDROGEN ION CONCENTRATION, CARBO W71-12855
GEN, AERATION, DISSOLVED OXYGEN, HYDROGEN ION CONCENTRATION, SETTL W71-13334
ALKALINITY, NUTRIENTS, AMMONIA, HYDROGEN ION CONCENTRATION, CYANO W71-11519
TURE, TURBIDITY, PHOTOSYNTHESIS, HYDROGEN ION CONCENTRATION, NITRI W71-11876
EN DEMAND, AERATION, *OXIDATION, HYDROGEN ION CONCENTRATION, MICRO W71-12223
MENT, *FOAM FRACTIONATION, LIME, HYDROGEN ION CONCENTRATION, ECONO W71-12200
SING ORGANIC MATTER, ALKALINITY, HYDROGEN ION CONCENTRATION, NUTRI W71-12068
ATION, DISSOLVED OXYGEN, SEWAGE, HYDROGEN ION CONCENTRATION.: /LIZ W71-12188
EFFLUENTS, GEORGIA, THERMOCLINE, HYDROGEN ION CONCENTRATION, WATER W71-11914
MUD, PHYSICOCHEMICAL PROPERTIES, HYDROGEN ION CONCENTRATION, TEMPE W72-03224
*ALGAE, CHEMICAL PRECIPITATION, HYDROGEN ION CONCENTRATION, FILTR W72-03613
YTOPLANKTON, THERMAL PROPERTIES, HYDROGEN ION CONCENTRATION, DISSO W72-01363
, OXIDATION-REDUCTION POTENTIAL, HYDROGEN ION CONCENTRATION.: /NGE W72-01108
TORY TESTS, FILTRATION, AMMONIA, HYDROGEN ION CONCENTRATION, TEMPE W72-01881
, AMMONIA, METHANE, TEMPERATURE, HYDROGEN ION CONCENTRATION, PROTE W72-10390
AKE SUPERIOR, WATER TEMPERATURE, HYDROGEN ION CONCENTRATION, HARDN W72-10875
RATES, CARBON DIOXIDE, POLYMERS, HYDROGEN ION CONCENTRATION, PROTO W72-10233
ES, CHROMIUM, ZINC, TEMPERATURE, HYDROGEN ION CONCENTRATION, CORRO W72-08683

R, WATER QUALITY, ON-SITE TESTS,	HYDROGEN ION CONCENTRAT: /EA WATE	W72-09092
ROPHORESIS, CHEMICAL PROPERTIES,	HYDROGEN ION CONCENTRATION, *WAST	W72-11833
OLVED OXYGEN, WATER TEMPERATURE,	HYDROGEN ION CONCENTRATICN, STRAT	W72-13871
C LOADING, DIURNAL DISTRIBUTION,	HYDROGEN ION CONCENTRATION, DISSO	W72-13508
, DISSOLVED OXYGEN, THERMOCLINE,	HYDROGEN ION CONCENTRATICN, FLOW,	W72-12941
TRACERS, CHEMICAL PRECIPITATION,	HYDROGEN ION CONCENTRATICN.: /AE	W72-12210
CTERIA, WATER QUALITY, SULFATES,	HYDROGEN ION CONCENTRATION, ALGAE	W71-07942
OPES, BIOCHEMICAL OXYGEN DEMAND,	HYDROGEN ION CONCENTRATICN, WASTE	W71-07113
CONDITIONS, PILOT PLANTS, ALGAE,	HYDROGEN ION CONCENTRATICN, CHEMI	W71-07102
TENSITY, CHEMICAL PRECIPITATION,	HYDROGEN ION CONCENTRATICN, CCLIF	W71-07100
HESIS, CHLOROPHYLL, TEMPERATURE,	HYDROGEN ION CONCENTRATICN, DISSO	W71-07382
OMIUM, TOXICITY, NEUTRALIZATION,	HYDROGEN ION CONCENTRATICN, TEMPE	W71-09524
REEZING, TEMPERATURE, CHELATICN,	HYDROGEN ION CONCENTRATICN, DYSTR	W71-09190
SYNTHESIS, METABOLISM, BACTERIA,	HYDROGEN ION CONCENTRATICN, LABOR	W71-08687
RECIPITATION, COAGULATION, LIME,	HYDROGEN ION CONCENTRATICN, TEMPE	W71-08951
RAL, CXYGEN, CHEMICAL REACTIONS,	HYDROGEN ION CONCENTRATION, ALGAE	W71-08026
NKTON, DIEL MIGRATION, PROTOZCA,	HYDROGEN ION CONCENTRATION, WATER	W71-05626
S, BICARBONATES, CARBON DIOXIDE,	HYDROGEN ION CONCENTRATICN, WATER	W72-04431
GAE, NUTRIENTS, LIGHT INTENSITY,	HYDROGEN ION CONCENTRATICN, CARBO	W72-04433
*SEWAGE BACTERIA, SEWAGE, ALGAE,	HYDROGEN ION CONCENTRATICN, CARBO	W72-04309
TIONS, LIGHT PENETRATION, ALGAE,	HYDROGEN ION CONCENTRATICN.: /QUA	W72-05459
TABOLISM, SEWAGE, CARBOHYDRATES,	HYDROGEN ION CONCENTRATICN, TEMPE	W72-05457
RIENT REQUIREMENTS, TEMPERATURE,	HYDROGEN ION CONCENTRATICN, EFFIC	W72-04788
TY, OXIDATION, DISSOLVED OXYGEN,	HYDROGEN ION CONCENTRATION, CESIG	W72-04787
PHORUS, WATER POLLUTION SOURCES,	HYDROGEN ION CONCENTRATICN, CARBO	W72-04769
, WATER QUALITY, CARBON DIOXIDE,	HYDROGEN ION CONCENTRATION, CARBO	W72-06274
, CALCIUM, MAGNESIUM, CHLORIDES,	HYDROGEN ION CONCENTRATION, PERIP	W72-05610
ELECTROLYTES, ACTIVATED CARBON,	HYDROGEN ION CONCENTRATICN, NUTRI	W72-06201
GENTS, GREAT LAKES, TEMPERATURE,	HYDROGEN ION CONCENTRATICN, ACTIV	W71-00116
IVER, BIOCHEMICAL OXYGEN DEMAND,	HYDROGEN ION CONCENTRATICN, ALGAE	W70-10427
UTION, NANNOPLANKTON, TURBIDITY,	HYDROGEN ION CONCENTRATION, ALKAL	W71-00117
ISSOLVED SOLIDS, NITROGEN CYCLE,	HYDROGEN ION CONCENTRATICN, TEMPE	W71-01489
ANALYSIS, ALGAE, MICROORGANISMS,	HYDROGEN ION CONCENTRATICN, PHENO	W71-01474
SOLIDS, FLOCCULATION, HEAD LOSS,	HYDROGEN ION CONCENTRATICN, CHLOR	W71-00138
ORATORY TESTS, DISSOLVED OXYGEN,	HYDROGEN ION CONCENTRATICN, ALKAL	W71-04518
A, PHCSPHATES, RUNOFF, ILLINOIS,	HYDROGEN ION CONCENTRATION, ALKAL	W71-03028
SING, CCLIFORMS, MICROORGANISMS,	HYDROGEN ION CONCENTRATICN, WASTE	W71-05155
PLANKTON, TOXICITY, IRON, LIGHT,	HYDROGEN ION CONCENTRATICN.: /YTO	W71-03027
KTON, SALMCNIDS, SEASON, OXYGEN,	HYDROGEN SULFIDE, DISTRIBUTION, N	W71-00117
ATION, CARBON, OXYGEN, NITROGEN,	HYDROGEN, PRIMARY PRODUCTIVITY, R	W72-06274
RBON CYCLE, ENZYMES, METABOLISM,	HYDROGEN, EUGLENA, CHLORELLA, SCE	W71-09172
IUM, MAGNESIUM, SODIUM, CALCIUM,	HYDROGEN, CHEMICAL WASTES, LEAD,	W72-14303
PHOSPHATES, NITRATES, SILICATES,	HYDROGRAPHS: /ALINITY, PRCFILES,	W72-07892
DEPTH, NITROGEN, PHYTOPLANKTON,	HYDROGRAPHY, NITRITES, NITRATES,	W72-08056
CIRCULATION, INFRARED RADIATION,	HYDROGRAPHY, PLANKTON, ALGAE, ENV	W73-01704
OOD BORERS, BARNACLES, BRYOZOAS,	HYDROIDS, SPONGES, COELENTERATES,	W73-00374
ECTORS, BRYOZOANS, FORAMINIFERA,	HYDROIDS, POLYCHAETA, OSTRACODS,	W72-05432
DS, SCIL MICROORGANISMS, RUNOFF,	HYDROLIC DATA.: /TATISTICAL METHO	W72-02218
ORA, ALKALINITY, PHOTOSYNTHESIS,	HYDROLO: /, VOLCANOES, BENTHIC FL	W71-09190
TION, FLOOD CONTROL, RECREATION,	HYDROLOGIC DATA, WATER TREATMENT,	W72-05594
, NUTRIENTS, TEMPERATURE, ALGAE,	HYDROLOGIC DATA, DISSOLVED OXYGEN	W72-05282
EVAPORATION, SEEPAGE, TURBIDITY,	HYDROLCGIC BUDGET, SOLAR RADIATIO	W72-02274
YTA, CHLOROPHYTA, COOLING WATER,	HYDROLOGIC DATA.: /PHYTA, PHAEOPH	W72-11252
QUALITY, *ECOLOGY, *CALIFORNIA,	HYDROLOGIC DATA, CHEMICAL ANALYSI	W71-10577
DIOECOLOGY, AQUATIC POPULATIONS,	HYDROLOGIC ASPECTS, NUTRIENTS, CH	W72-07841
AL PRCPERTIES, DISSOLVED SOLIDS,	HYDROLOGIC DATA, DATA COLLECTIONS	W72-13851
*URBAN RUNOFF, URBAN	HYDROLOGY, LINCCLN(NEBR).:	W71-13466
LA ISLAND, HONSHU ISLAND/ *ALGAL	HYDROLYSATE, TONGA ISLANDS, TUTUI	W71-08042
SH, MARINE PLANTS, MARINE ALGAE,	HYDROLYSIS, CHELATION, SEDIMENTS,	W72-04463
S, EFFLUENTS, TRICKLING FILTERS,	HYDROLYSIS, BIOLOGICAL TREATMENT,	W70-09907
RGANISMS, FERMENTATION, ALKALINE	HYDROLYSIS, ALGAE BLOOMS, UTAH, G	W71-12223
AKE MICHIGAN, ALGAE, PHOSPHATES,	HYDROLYSIS, STORM WATER, RAINFALL	W72-13644
SPECTROPHOTOMETRY, CHLOROPHYLL,	HYDROLYSIS, PHYTOPLANKTON, FLUORE	W72-12637
S, EUDORINA, NOTONECTIDAE, BODC,	HYDROPSYCHIDAE, PLECOPTERA, MAGAL	W70-06655
PHOTOSYNTHESIS, INSTRUMENTATION,	HYDROSTATIC PRESSURE, DIATOMS.: /	W71-10791
*SUSPENDED SOLIDS, SODIUM	HYDROXIDE, *ROCK FILTERS.:	W71-13341
US, CORYNEBACTERIA, STREPTOTHRIX	HYORHINA, CADMIUM, DIMETHYLNITROS	W72-09675
GULARIS, CENTROCERAS CLAVULATUM,	HYPENA MUSCIFORMIS, CERAMIUM BUSS	W73-00838
S, LAMINARIA DIGITATA, LAMINARIA	HYPERBOREA, ASCOPHYLLUM NODOSUM,	W71-11515
(WASH).:	HYPEREUTROPHIC LAKES, *MOSES LAKE	W72-00798
*MOSES LAKE(WASH),	HYPEREUTROPHIC LAKES.:	W72-00799
NINSULA, CHLORINITY, METAHALINE,	HYPERHALINE, EURYHALINE, RED SEA,	W73-00855
M, ENRICHMENT CULTURES, AGARS, /	HYPHOMICROBIUM, VIBRIC ANGUILLARU	W72-05421
RU/ *BIOSYNTHESIS, *GIBBERELLIN,	HYPNEA MUSCIFORMIS, GRACILARIA VE	W72-07496
EORGIA), L/ *WHITE ALGAL BLOOMS,	HYPNODINIUM, LAKE SIDNEY LANIER(G	W71-09173
EORGIA), LAKE BUCKHORN(GEORGIA),	HYPNODINIUM SPHAERICUM.: /DINAL(G	W71-11914
SIA FLEXUOSA, PARMELIA PHYSODES,	HYPNUM CUPRESSIFORME.: / DESCHAMP	W72-14303
NT-WATER INTERFACES, *LAKE ERIE,	HYPCLIMNION, AQUATIC ALGAE.: /IME	W72-06690
URBIDITY, *PROFILES, PHOTOMETER,	HYPCLIMNION, EPILIMNION, LAKE MOR	W72-07145
LOROPHYLL, HYPSOMETRIC ANALYSIS,	HYPCLIMNION, EPILIMNION, CYANOPHY	W71-11949
AQUATIC LIFE, INTERSTATE RIVERS,	HYPOLIMNION, VOLUME, MIXING, SAFE	W70-09595
TRATIFICATION, DISSOLVED OXYGEN,	HYPCLIMNION.: /, ALGAE, THERMAL S	W71-03028
DISSOLVED OXYGEN, OXYGEN DEMAND,	HYPOLIMNION, AQUATIC PLANTS, ALGA	W72-12994
E, WATER CHEMISTRY, CHLOROPHYLL,	HYPSOMETRIC ANALYSIS, HYPOLIMNION	W71-11949
LAGOONS, PHYTOPLANKTON, AQUATIC	I: /OLOGICAL TREATMENT, OXIDATION	W72-12567
ING, SEASONAL, BOTTOM SEDIMENTS,	ICE, AMPHIPODA.: /ODUCTICN, SAMPL	W71-08026
ERATURE, THERMAL STRATIFICATICN,	ICE, EUTROPHICATION, OXYGEN, SEAS	W72-04855
MS, *GREAT LAKES, AQUATIC ALGAE,	ICE, FISH, MICROORGANISMS, CDCR,	W73-01669
DUCTIVITY, PRIMARY PRODUCTIVITY,	ICED LAKES, IONS, SULFATES, CARBO	W71-02880
ANTS, *HYDROBIONTS, ICHTHYOLOGY,	ICHTHYCFAUNA, MACROPHYTES, SAPROP	W72-09646

OLOGY, *TOXICANTS, *HYDROBIONTS, ICHTHYOLOGY, ICHTHYOFAUNA, MACROP W72-09646
INOSA, APHANIZOMENON HOLSATICUM, ICTALURUS NEBULOSUS, NOTEMIGONUS W72-07890
ES, PICEA ABIES, VACCINIUM VITIS IDAEA, VACCINIUM MYRTILLUS, DESCH W72-14303
LUTION, OXYGEN, SOIL PROPERTIES, IDAHO.: /NSUMPTIVE USE, WATER POL W72-14878
, *AMERI/ *GLUCOSE, *SNAKE RIVER(IDAHO), *DISSOLVED ORGANIC CARBON W72-13300
YPRINUS CARPIO, CTENOPHARYNGODON IDELLUS.: /ILAPIA MELANOPLEURA, C W71-04528
OGRAPHY, GROWTH RATES, POLLUTANT IDENTIFICATION.: /TTERNS, CHROMAT W72-04134
ANALYTICAL TECHNIQUES, POLLUTANT IDENTIFICATION.: /GICAL STUDIES, W72-03985
ALGAE, GROWTH RATES,/ *POLLUTANT IDENTIFICATION, *BIOINDICATORS, * W71-09795
, POTASSIUM COMPOUNDS, POLLUTANT IDENTIFICATION, CHLOROPHYTA, CHLO W72-13813
US, ORGANIC COMPOUNDS, POLLUTANT IDENTIFICATION, AQUEOUS SOLUTIONS W72-12166
*MO/ *REMOTE SENSING, *POLLUTANT IDENTIFICATION, WATER POLLUTION, W72-12487
ION, *REMOTE SENSING, *POLLUTANT IDENTIFICATION, AQUATIC WEEDS, *W W72-14728
LEMENTS, ION EXCHANGE, POLLUTANT IDENTIFICATION, COPPER, CADMIUM, W72-14303
ENTATION, METHODOLOGY, POLLUTANT IDENTIFICATION, PROTOZOA, CHRYSOP W73-00273
UATIC ALGAE, NITROGEN, POLLUTANT IDENTIFICATION, ENVIRONMENTAL EFF W73-00379
ORS, *EUTROPHICATION, *POLLUTANT IDENTIFICATION, *LAKE MICHIGAN, C W73-00432
TER POLLUTION SOURCES, POLLUTANT IDENTIFICATION, PHYTOPLANKTON, MA W73-00441
ING, BIOMASS, DIATOMS, POLLUTANT IDENTIFICATION, MARINE ALGAE, CHR W73-00442
*PHOSPHATES, DIATOMS, POLLUTANT IDENTIFICATION, GROWTH RATES, DEF W73-01445
ENVIRONMENTAL EFFECTS, POLLUTANT IDENTIFICATION, PHYSICAL PROPERTI W73-00854
MARINE FISH, MOLLUSKS, POLLUTANT IDENTIFICATION.: /, CRUSTACEANS, W73-00819
ETALS, EUTROPHICATION, POLLUTANT IDENTIFICATION, TOXICITY, *IRON, W73-02112
CHLOROPHYTA, BIOLOGY, POLLUTANT IDENTIFICATION, LIGHT INTENSITY, W73-02099
AQUATIC PRODUCTIVITY, POLLUTANT IDENTIFICATION, LABORATORY EQUIPM W73-01066
RBOHYDRATES, ALCOHOLS, POLLUTANT IDENTIFICATION, SYSTEMATICS, NUTR W73-00942
TORS, *PHYTOPLANKTON, *POLLUTANT IDENTIFICATION, PROTOZOA, PONDS, W73-00918
TION, WATER POLLUTION, POLLUTANT IDENTIFICATION, SAMPLING, BIOASSA W72-10678
ANALYTICAL TECHNIQUES, POLLUTANT IDENTIFICATION, NUCLEAR POWERPLAN W72-10958
ODELS, WATER ANALYSIS, POLLUTANT IDENTIFICATION, DETERGENTS, NUTRI W72-08790
HY, *MATHEMATICAL ME/ *POLLUTANT IDENTIFICATION, *AERIAL PHOTOGRAP W71-11685
PROGRAMS, GAMMA RAYS, POLLUTANT IDENTIFICATION, COLORIMETRY, TRAC W71-11036
, *SPECTROPHOTOMETRY, *POLLUTANT IDENTIFICATION, ONTARIO, ALGAE, M W71-11529
MPLING, MASSACHUSETTS, POLLUTANT IDENTIFICATION.: / TECHNIQUES, SA W71-11498
AINS, BIOLUMINESCENCE, POLLUTANT IDENTIFICATION.: /RIENTS, FOOD CH W71-11234
TES, SEWAGE EFFLUENTS, POLLUTANT IDENTIFICATION.: / INDUSTRIAL WAS W71-11515
ATORS, *WATER POLLUT/ *POLLUTANT IDENTIFICATION, *BIOASSAY, *INDIC W71-10553
NISMS, ALGAL CONTROL, *POLLUTANT IDENTIFICATION, ANALYTICAL TECHNI W72-00801
TER POLLUTION EFFECTS, POLLUTANT IDENTIFICATION.: / TECHNIQUES, WA W71-13794
C WASTES, *NUTRIENTS, *POLLUTANT IDENTIFICATION, ALGAE, AQUATIC BA W72-01472
KTON, *EUTROPHICATION, POLLUTANT IDENTIFICATION, ALGAE, ANALYTICAL W72-01780
OCHEMISTRY, TRACE EL/ *POLLUTANT IDENTIFICATION, *NITROGEN, ELECTR W72-01693
*AQUEOUS SOLUTIONS, / *POLLUTANT IDENTIFICATION, *AQUATIC PLANTS, W72-01473
VATION, FLAGELLATES, RHODOMONAS, IKROAVICH LAKE, UTERMOHL METHOD.: W73-00273
*GREENE COUNTY(ILL).: W72-05869
LAKE BLOOMINGTON(ILL), CHEMOSTRATIFICATION.: W71-03028
GROWTH POTENTIAL, ILLINOIS RIVER(ILL), DIAGNOSTIC CRITERION.: /AL W71-11033
RITERIO/ ALGAL GROWTH POTENTIAL, ILLINOIS RIVER(ILL), DIAGNOSTIC C W71-11033
(NEW HAMPSHIRE), DELAWARE RIVER, ILLINOIS RIVER, HOLSTON RIVER(TEN W71-11517
BLACK RIVER(MINNESOTA), LAKE WA/ ILLINOIS RIVER, ST. LOUIS RIVER, W71-04216
*PHOSPHORUS REMOVAL, CHICAGO(ILLINOIS).: W72-03970
IPAL WASTES, CITIES, PHOSPHATES, ILLINOIS, ECOLOGY, TERTIARY TREAT W72-03970
CA, AMMONIA, PHOSPHATES, RUNOFF, ILLINOIS, HYDROGEN ION CONCENTRAT W71-03028
, EUDORINA CALIFORNICA, EUDORINA ILLINOISENSIS.: /TYDORINA CAUDATA W71-13793
LLA TYPHOSA, TYPHOID, COLIPHAGE, IMMUNOLOGY, PASTEURELLA PSEUDOTUB W72-13800
ATOMIC ENERGY, REACTOR COOLING, IMPERIAL VALLEY(CALIF), LOS ANGEL W71-09157
RVOIRS, *RESERVOIR STAGES, EARLY IMPOUNDMENT, BIOMASS, DIATOMS, CH W72-04855
PLANTS, *RESERVOIR STAGES, EARLY IMPOUNDMENT, BIOMASS, AQUATIC ALG W72-04853
LYSIS, ALGAE, RESERVOIR STORAGE, IMPOUNDMENTS, DISSOLVED OXYGEN, C W72-14280
UTION, CONNECTICUT, RIVERS, POST- IMPOUNDMENT, NUCLEAR POWERPLANTS, W72-00939
S, TURBIDITY, LIGHT PENETRATION, IMPOUNDMENTS, SEASONAL, COLOR, IR W73-00256
DEMAND, ALGAE, CONFINEMENT PENS, IMPOUNDMENTS, *WATER POLLUTION SO W72-08401
OXYGEN, PHOSPHATES, WATER YIELD IMPROVEMENT, WATER POLLUTION TREA W71-01491
A, *NITROGENASE ACTIVITY, OXYGEN INACTIVATION.: /ANABAENA FLOS-AQU W72-07941
Y, FREEZE DRYING, REFRIGERATION, INCUBATION, DIALYSIS, CHEMICALS, W73-01676
NTIC OCEAN, GROWTH RATES, ALGAE, INCUBATION, PHYTOPLANKTON.: /ATLA W71-11562
XIDE, FLORIDA, CULTURES, *ALGAE, INCUBATION, CENTRIFUGATION, RADIO W72-09103
GEN, SEA WATER, WINDS, SAMPLING, INCUBATION, RUNOFF, SALINITY, DEP W72-09122
ZINC, ASCORBIC ACID, PYRIDOXINE, INDAL COMPOUNDS.: /TAHOE(CALIF), W72-07514
SAY PROCEDURE, SPECIES DIVERSITY INDEX.: *MICROCOSM ALGAL AS W72-03218
*POLLUTION INDEX.: W72-01780
*SYNCHRONOUS CULTURES, SYNCHRONY INDEX.: / PATTERNS, *AUTOSPORES, W72-12724
PIGMENT RATIO, PIGMENT DIVERSITY INDEX, CHLOROPHYLL A, CARTENOIDS, W73-00916
ES, VARIABILITY, BIAS, SYNCHRONY INDEX, COUNTING.: /URES, AUTOSPOR W72-11920
N, MACROALGAE, SPECIES DIVERSITY INDEX, FRAGILARIA CONSTRUENS, COC W72-08141
OSINE TRIPHOSPHATE, *AUTOTROPHIC INDEX, MACROINVERTEBRATES, ACETYL W72-10883
FACULTATIVE MICROORGANISMS, INDIA.: W71-06033
IA, CHAETOCEROS, THALASSISTHRIX, INDIA.: /A, CENTROPYXIS, BIDDULPH W71-11876
BANGALORE(INDIA).: W72-09176
HA ANGUSTATA, IPOMOE/ *HYDERABAD(INDIA), EICHHORNIA CRASSIPES, TYP W72-01363
INDIA, CESIUM RADIOISOTOPES.: W72-10958
*GREEN MANURE, INDIA, JAPAN, RICE FIELDS.: W73-02473
, CULTURE MEDIA, *BAY OF BENGAL, INDIAN OCEAN, SOIL EXTRACTS, ERDS W73-00430
CTICUT, FLORIDA, DELAWARE RIVER, INDIANA.: /NIA, WASHINGTON, CONNE W71-07417
RNIA), BISCAYNE BAY, WHITE RIVER(INDIANA), PATUXENT RIVER(CANADA), W71-07417
CTICU/ *ENTRAINMENT, WHITE RIVER(INDIANA), CONNECTICUT RIVER(CONNE W71-11517
ON, STORM RUNOFF, BIOASSAYS, BIO- INDICATORS, STORAGE, FLOW RATES, W72-04070
RRIES, OHIO RIVER, HUDSON RIVER, INDICATORS, DELAWARE, PATH OF POL W72-04736
DIVERSITY INDICES.: W72-14690
ECIES DIVERSITY, DIRECT EFFECTS, INDIRECT EFFECTS.: /CREEK(PA), SP W72-08579

ICAL PROPERTIES, AQUATIC PLANTS,	INDUSTRIAL: /TRATION, PHYSICOCHEM	W73-01615
EYS, SAMPLING, DATA COLLECTIONS,	INDUSTRIAL WASTES, AGRICULTURAL R	W73-01972
LLUTICN CONTROL, ORGANIC WASTES,	INDUSTRIAL WASTES, WASTE WATER(PO	W73-00840
LS, ORGANOPHOSPHORUS PESTICIDES,	INDUSTRIAL WASTES, VIRUSES, ENERI	W72-04742
A, ELKTON POND, RISING SUN POND,	INDUSTRIAL STORAGE T: /X CALCICCL	W72-04736
ON TECHNIQUES, *CHEMICAL WASTES,	INDUSTRIAL WASTES, CHLORINATION,	W72-05315
INITY, BURNING, DOMESTIC WASTES,	INDUSTRIAL WAS: /N DIOXIDE, ALKAL	W72-05473
, *ECOSYSTEMS, ABSORPTION, FISH,	INDUSTRIAL WASTES, ALGAE, AQUATIC	W72-05952
ED HYDROCARBONS, *BIOINDICATORS,	INDUSTRIAL WASTES, FLORIDA, SESSI	W71-07412
MENT, BIOCHEMICAL OXYGEN DEMAND,	INDUSTRIAL WASTES, FARM WASTES, P	W71-06445
LIVESTOCK, SEWAGE, INFILTRATICN,	INDUSTRIAL WASTES, ALGAE, PONDS,	W71-06435
OINES RIVER, PACKING PLANTS.:	INDUSTRIAL WATER POLLUTICN, DES M	W71-06445
, MODIFICATION./ FARM MACHINERY,	INDUSTRIAL MACHINERY, UTILIZATION	W71-06188
FACE DRAINAGE, MUNICIPAL WASTES,	INDUSTRIAL WASTES, SAMPLING, WATE	W71-08953
ULTURES, SUSPENDED LOAD, SEWAGE,	INDUSTRIAL WASTES, SILTS, WATER P	W71-09795
FFECTS, WATER POLLUTION SOURCES,	INDUSTRIAL WASTES, DOMESTIC WASTE	W71-03095
ITROGEN, FILTRATION, MICROSCOPY,	INDUSTRIAL WASTES, DETERGENTS, AM	W71-11507
BDENUM, MANGANESE, COBALT, IRON,	INDUSTRIAL WASTES, SEWAGE EFFLUEN	W71-11515
TICAL METHODS, MUNICIPAL WASTES,	INDUSTRIAL WASTES, NEW YORK, ALGA	W71-11685
IATOMS, WATER POLLUTION SOURCES,	INDUSTRIAL WASTES, AQUATIC ALGAE,	W71-11583
, SLUDGE, ANALYTICAL TECHNIQUES,	INDUSTRIAL WASTES, MUNICIPAL WAST	W71-11793
ATS, AQUATIC LIFE, HEAVY METALS,	INDUSTRIAL WASTES, INVERTEBRATES,	W72-11854
UALITY, *DOMESTIC WASTES, ALGAE,	INDUSTRIAL WASTES, METABOLISM, GR	W72-09590
US, PHENOLS, METALS, POLLUTANTS,	INDUSTRIAL WASTES, CONFERENCES.: /	W72-09646
S, FILTRATION, MUNICIPAL WASTES,	INDUSTRIAL WASTES, OPERATION AND	W72-09386
AL SANITATION, ALGAE, NUTRIENTS,	INDUSTRIAL WASTES, MUNICIPAL WAST	W72-01659
PRODUCTIVITY, MUNICIPAL WASTES,	INDUSTRIAL WASTES, ANNUAL TURNOVE	W72-03228
GY CONSUMPTION, MODEL TEST/ AGRO-	INDUSTRIAL COMPLEX, ELECTRIC ENER	W72-03327
AE, ECOTYPES, INDUSTRIAL PLANTS,	INDUSTRIAL WASTES, OIL WASTES, ES	W71-13746
NET, MEDWELL ESTUA/ *KENT COAST,	INDUSTRIAL EXPANSION, ISLE OF THA	W71-13746
TA COLLECTIONS, ALGAE, ECOTYPES,	INDUSTRIAL PLANTS, INDUSTRIAL WAS	W71-13746
XIDATION LAGOONS, STABILIZATION,	INDUSTRIAL WASTES, MUNICIPAL WAST	W71-13356
NTS, LETHAL LIMITS, METHODOLOGY,	INDUSTRIAL WASTES, DOMESTIC WASTE	W72-12239
WATER POLLUTION SOURCES, SEWAGE,	INDUSTRIAL WASTES, BIOINDICATORS,	W72-12729
E, FORMULATION, WASTE TREATMENT,	INDUSTRIES, WATER QUALITY, TECHNO	W72-09169
R POLLUTION SOURCES, POPULATION,	INDUSTRIES, WASTE DISPOSAL, MONIT	W71-12091
UALITY, STREAMFLOW, AGRICULTURE,	INDUSTRIES, WASTE TREATMENT, WATE	W71-11017
NITROGEN, PHOSPHORUS, DRAINAGE,	INDUSTRIES, MUNICIPAL WASTES, TUR	W71-11009
YTA, CHLORELLA, PLANT PATHOLOGY,	INDUSTRIES, IONS, CYANOPHYTA, RAD	W71-09157
LOT PLANTS, ALGAE, BIOCH/ *DAIRY	INDUSTRY, *OXIDATION LAGOONS, *PI	W70-06619
NSIN, PHOSPHORUS, PULP AND PAPER	INDUSTRY, ALGAE, AQUATIC WEEDS, R	W70-05103
D DETERGENTS, SOAP AND DETERGENT	INDUSTRY, NTA.: /C ACID, HOUSEHOL	W72-04266
WATER SUPPLY, LIVESTOCK, SEWAGE,	INFILTRATION, INDUSTRIAL WASTES,	W71-06435
*ARTIFICIAL RECHARGE, *INDUCED	INFILTRATION, *PIT RECHARGE, *WAT	W71-12410
D OXYGEN, ALGAE, BACTERIA, ODOR,	INFLOW, DISSOLVED SOLIDS, DATA CO	W72-02274
WATER QUALITY, NUTRIENTS, ALGAE,	INFLOW, DISCHARGE(WATER), WATER B	W72-13851
, SUCCESSION, NUTRIENTS, RIVERS,	INFLUENT STREAMS, EFFLUENT STREAM	W73-01435
*COLOR	INFRARED PHOTOGRAPHY.:	W72-12487
ANTS, HEATED WATER, CIRCULATION,	INFRARED RADIATION, HYDROGRAPHY,	W73-01704
, EUTROPHICATION, *LAKES, ALGAE,	INFRARED RADIATION, *REMOTE SENSI	W72-14728
IMAL WASTES, FISH, FLUORESCENCE,	INFRARED RADIATION, IRRAD: /S, AN	W72-08790
AVIOLET RADIATION, FLUORESCENCE,	INFRARED RADIATION, RADAR, C: /TR	W72-04742
URE MEDIA, PAPER CHROMATOGRAPHY,	INFRARED SPECTROSCOPY, ASSIMILATI	W73-00379
HYDROCARBONS, MASS SPECTROMETRY,	INFRARED SPECTROSCOPY.: /IPHATIC	W73-01439
OLLUTANTS, PAPER CHROMATOGRAPHY,	INFRARED SPECTROSCOPY, CULTURE ME	W73-01066
OMUS GRACILIS, CYCLOPS STRENUUS,	INGESTION RATES, ASSIMILATION EFF	W72-07940
DISC DIFFUSION METHOD, *ZONES OF	INHIBITION, ANTIFUNGAL ACTIVITY.:	W72-08586
WTH RATES, *NUTRIENTS, CULTURES,	INHIBITION, OLIGOTROPHY, PHOSPHOR	W72-08376
ON DIOXIDE, BIOCHEMISTRY, ALGAE,	INHIBITION, LABORATORY TESTS.: /B	W72-08584
A WATER, FLUOROMETRY, EQUIPMENT,	INHIBITION, IONS, ABSORPTION, BIO	W72-07660
TESTS, IRRADIATION, CHLOROPHYTA,	INHIBITION, GAMMA RAYS.: /RATORY	W73-01080
TER, ORGANIC ACIDS, CHRYSOPHYTA,	INHIBITION, MARINE ALGAE, DIATOMS	W73-00942
SEA WATER, LABORATORY EQUIPMENT,	INHIBITION, ECOLOGICAL DISTRIBUTI	W73-00348
INDICATORS, STORAGE, FLOW RATES,	INHIBITION, ANALYTICAL TECHNIQUES	W72-04070
TIC BACTERIA, ENGLAND, METABOLIC	INHIBITION, HOSTS, MYXOPHYCEAE, S	W72-05457
PHORUS, TOXICITY, RADIOISOTOPES,	INHIBITION.: /, RESPIRATION, PHOS	W70-07255
YTA, CHLOROPHYTA, *GROWTH RATES,	INHIBITION, *PLANT PHYSIOLOGY, *P	W70-07842
FFECTS, WATER POLLUTION SOURCES,	INHIBITION.: /, WATER POLLUTION E	W72-11727
LIS, *ANACYSTIS NIDULANS, GROWTH	INHIBITION, SELENATE, SELENITE, S	W72-00153
OPHYTA, SULFUR, ALGAE, GRANULES,	INHIBITION, METABOLISM, CULTURES,	W72-00153
ABORATORY TESTS, AMMONIA, ASSAY,	INHIBITION, UREAS, SYNTHESIS, ALG	W72-03222
XICITY, PLANT GROWTH REGULATORS,	INHIBITION, INHIBITORS, PHYTOPLAN	W71-10566
PLANT GROWTH, CHLOROPHYLL, DDE,	INHIBITION.: /DDT, PHYTOPLANKTON,	W72-12998
HIN LAYER CHROMATOGRAPHY, GROWTH	INHIBITION.: /URIAL PESTICIDES, T	W72-13347
ES, *AQUATIC ALGAE, CHLOROPHYTA,	INHIBITORS, WATER POLLUTION EFFEC	W72-12736
T GROWTH REGULATORS, INHIBITION,	INHIBITORS, PHYTOPLANKTON, PHYTO-	W71-10566
A, HEAVY METALS, *CHLAMYDOMONAS,	INHIBITORS, *LETHAL LIMIT, WATER	W72-11727
ANULES, PLANT GROWTH REGULATORS,	INHIBITORS, HERBICIDES, MONURON,	W72-09650
SPECTROPHOTOMETRY, PENNSYLVANIA,	INHIBITORS, ECOLOGY, ECOTYPES, FR	W72-04743
OPHYTA, PHAEOPHYTA, CHLOROPHYTA,	INHIBITORS, ANTIBIOTICS, SEA WATE	W72-04748
ROPHYTA, *CHRYSOPHYTA, TOXICITY,	INHIBITORS, TOXINS, INORGANIC COM	W72-14706
*NITROGEN, OXYGEN, BIOCHEMISTRY,	INHIBITORS, ALGAE, PROTEINS.: /,	W72-07941
LGAE, *PLANT GROWTH, RHODOPHYTA,	INHIBITORS, PLANT TISSUES, BIOASS	W72-07496
ATION, ACID INJECTION PUMP, ALUM	INJECTION PUMP, MANURE GRINDER.: /	W72-02850
PERIOD, FLUSHING OPERATION, ACID	INJECTION PUMP, ALUM INJECTION PU	W72-02850
*KEZAR LAKE, AIR	INJECTION, COMPRESSED AIR.:	W72-09304
EPTOCYLINDRUS, PACIFIC OCE/ COOK	INLET(ALASKA), CYCLOTELLA NANA, L	W71-11008
MATION, CARBON SOU/ CHEMISTRY OF	INORGANIC CARBON, CARBON TRANSFOR	W71-12068
A, TOXICITY, INHIBITORS, TOXINS,	INORGANIC COMPOUNDS, ALGAE, DIATO	W72-14706
LINITY, BENTHIC FAUNA, ASBESTOS,	INORGANIC COMPOUNDS, WORMS, INVER	W73-00374

US, *LAKE ERIE, NITROGEN, ALGAE, INORGANIC COMPOUNDS, STANDING CRO W72-13653
ENTS, BIOCHEMICAL OXYGEN DEMAND, INORGANIC COMPOUNDS, THERMAL STRA W72-13659
BAY OF QUINTE(ONTARIO), NUTRIENT INPUTS, CHIRONOMUS PLUMOSUS, CHIR W71-11009
IOACTIVITY, FALLOUT, BIOCONTROL, INSECT CONTROL, IRRIGATION, PREDA W72-00846
RIGATION, PREDATION, PARASITISM, INSECT RESISTANCE: /T CONTROL, IR W72-00846
WAGE BACTERIA, ALGAE, COLIFORMS, INSECTICIDES.: /IATION, COSTS, SE W72-00383
L, MONITORING, NUTRIENTS, ALGAE, INSECTICIDES, HERBICIDES, HEAVY M W71-12091
TON, PHYTO-TOXICITY, HERBICIDES, INSECTICIDES, BIOMASS, PRODUCTIVI W71-10566
ANTS, AQUATIC ALGAE, PESTICIDES, INSECTICIDES, HERBICIDES, PHOSPHO W72-09646
TY, *HERBICIDES, DALAPON, 2-4-D, INSECTICIDES, PHENOLS, *DETERGENT W72-12576
ORINATED HYDROCARBON PESTICIDES, INSECTICIDES, RODENTICIDES, HERBI W73-00238
OR, METABOLITES, BIOLOGIC/ MICE, INSECTS, *ETHOXYCHLOR, *METHYLCHL W72-07703
, *ALGAE, *CHLORAPHYTA, *AQUATIC INSECTS, *CHLORAPHYTA, *AQUATIC I W72-06526
INSECTS, *CHLORAPHYTA, *AQUATIC INSECTS, *CRUSTACEANS, *MOLLUSKS, W72-06526
GAE, *PROTOZOA, *FUNGI, *AQUATIC INSECTS, *DISPERSION, CADDISFLIES W70-06655
FISH, DIATOMS, PROTOZOA, AQUATIC INSECTS, AQUATIC ANIMALS, AQUATIC W72-07791
NTHOS, AQUATIC ANIMALS, DIATOMS, INSECTS, BACTERIA, FUNGI, TURNOVE W71-10064
STREAMS, AQUATIC PLANTS, AQUATIC INSECTS, CRUSTACEANS, MUSSELS, CO W72-12930
ALGAE, AQUATIC ANIMALS, AQUATIC INSECTS, DAPHNIA, AQUATIC POPULAT W72-07132
ON EFFECTS, MUD, MOLLUSKS, FISH, INSECTS, DDT, SILTS, SUCKERS, CAT W70-08098
EPAGE, PERCOLATION, EVAPORATION, INSECTS, EUTROPHICATION, ALGAE.: / W71-07124
SES, MOLLUSKS, PROTOZOA, YEASTS, INSECTS, MILDEWS, E. COLI, FISH, W73-01676
ATIC PLANTS, PLANKTON, ARKANSAS, INSECTS, TURBIDITY, CHARA, PONDWE W71-04528
ALIS, LITHOPHYTES, ENTEROMORPHA, INSOLATION, TEXTURE, SUBSTRA: /IN W72-08141
*SOLAR INSOLATION, *STOICHIOMETRY.: W72-02363
*RUTHENIUM 106, SOLUBLE FORMS, INSOLUBLE FORMS, LABORATORY EXPER W71-13233
RANTS, RESEARCH AND DEVELOPMENT, INSPECTION, FEDERAL: /ONTRACTS, G W71-11516
POSAL, STANDARDS, PUBLIC HEALTH, INSPECTION, WATER SOFTENING, SOAP W72-01085
NTHETIC OXYGEN, *PHOTOSYNTHESIS, INSTRUMENTATION, HYDROSTATIC PRES W71-10791
LGAE, PROTOZA, FUNGI, EQUIPMENT, INSTRUMENTATION, FLUORESCENCE, WA W72-07225
, *ORGANIC COMPOUNDS, SHELLFISH, INSTRUMENTATION, PHYTOPLANKTON, A W72-07715
HWATER, WATER POLLUTION SOURCES, INSTRUMENTATION, METHODOLOGY, NIT W73-01445
UREMENT, *KINETICS, *ABSORPTION, INSTRUMENTATION, RADIOACTIVITY TE W72-14326
ACTS, WATER POLLUTION TREATMENT, INSTRUMENTATION, WASTE DISPOSAL, W71-01192
RBON RADIOISOTOPES, MARINE FISH, INSTRUMENTATION, ABSORPTION, SEDI W72-06283
NITY, VOLUMETRIC ANALYSIS, FISH, INTAKE STRUCTURES, CARBON DIOXIDE W72-05594
RDS, SURFACE WATERS, EPILIMNION, INTAKE STRUCTURES, DISCHARGE(WATE W70-09614
WATER TEMPERATURE, WAVES, LIGHT INTENSITY.: /ARINE ALGAE, HAWAII, W71-11486
HOTOSYNTHESIS, TURBULENCE, LIGHT INTENSITY.: /MOVEMENT, VIRUSES, P W71-13184
OSPHOROUS, PHOTOSYNTHESIS, LIGHT INTENSITY, OXYGENATION, DENITRIFI W71-13326
TS, *PHYTOPLANKTON, ALGAE, LIGHT INTENSITY, SALINITY, NUTRIENT REQ W71-10083
ELLA, *BIOASSAYS, ENZYMES, LIGHT INTENSITY, ALGAE.: *CHLOR W71-11916
ONCENTRATION, TEMPERATURE, LIGHT INTENSITY, SEDIMANTATION, OXIDATI W72-01881
UTRIENTS, *PHOTOSYNTHESIS, LIGHT INTENSITY, AMMONIA, CARBON DIOXID W72-09103
*RADIOACTIVITY TECHNIQU/ *LIGHT INTENSITY, *PRIMARY PRODUCTIVITY, W72-09102
U.S., TEMPERATURE, OXYGEN, LIGHT INTENSITY, NUTRIENTS, ACTINOMYCET W72-10613
OIDOSA, OXYGEN PRODUCTION, LIGHT INTENSITY, CULTURE THICKNESS, CUL W70-05547
FFUSIVITY, CARBON DIOXIDE, LIGHT INTENSITY, PHOTOSYNTHESIS, EQUATI W72-05459
TH, BIOMASS, GROWTH RATES, LIGHT INTENSITY, CARBON DIOXIDE, NITROG W72-05476
OXYGENATION, TEMPERATURE, LIGHT INTENSITY, MIXING, DISSOLVED OXYG W72-04983
, PHOTOSYNTHESIS, BIOMASS, LIGHT INTENSITY, GROWTH RATES, MATHEMAT W72-06274
LININGS, DISSOLVED OXYGEN, LIGHT INTENSITY, DESIGN CRITERIA, *OXID W72-04789
LEMENTS, ALGAE, NUTRIENTS, LIGHT INTENSITY, HYDROGEN ION CONCENTRA W72-04433
ASONAL, LIGHT PENETRATION, LIGHT INTENSITY, DISSOLVED SOLIDS, PHYT W72-04270
LGAE, *OXIDATION LAGOONS, *LIGHT INTENSITY, PONDS, ROTIFERS, PROTO W71-07382
STING, DEPTH, TEMPERATURE, LIGHT INTENSITY, CHEMICAL PRECIPITATION W71-07100
TH, THERMOCLINE, NITRATES, LIGHT INTENSITY, EUTROPHICATION, MARINE W71-07875
EROBIC CONDITIONS, SLUDGE, LIGHT INTENSITY, COLIFORMS, BIOCHEMICAL W71-07106
ELS, MIXING, WATER RE-USE, LIGHT INTENSITY, EVAPORATION, RE-AERATI W71-07099
VED OXYGEN, *PLANT GROWTH, LIGHT INTENSITY, TOXICITY, PRIMARY PROD W72-14690
ORA, SEA WATER, SEDIMENTS, LIGHT INTENSITY, DIATOMS, CARBON RADIOI W72-14317
HOTOSYNTHESIS, EPILIMNION, LIGHT INTENSITY, RESPIRATION, LIMITING W72-14812
YDROGEN ION CONCENTRATION, LIGHT INTENSITY, NUTRIENTS, PSEUDOMONAS W73-01657
IE, GREAT LAKES, SOYBEANS, LIGHT INTENSITY, AIR TEMPERATURE, SAMPL W73-01456
STREAMS, EFFLUENT STREAMS, LIGHT INTENSITY, NITROGEN, CHRYSOPHYTA, W73-01435
S, *AQUATIC WEED CONTROL, *LIGHT INTENSITY, *NEBRASKA, *CYANOPHYTA W73-02349
POLLUTANT IDENTIFICATION, LIGHT INTENSITY, BIOASSAY.: /, BIOLOGY, W73-02099
*DIATOMS, *SESSILE ALGAE, LIGHT INTENSITY, TIDAL EFFECTS, SALINIT W73-00853
RESISTANCE, *MARINE ALGAE, LIGHT INTENSITY, *ECOLOGICAL DISTRIBUTI W73-00838
SORPTION, *AQUATIC ALGAE, *LIGHT INTENSITY, *PHOSPHORUS, PLANKTON, W72-07166
HYTON, SEDIMENTS, DIATOMS, LIGHT INTENSITY, PHOTOSYNTHESIS, SESSIL W72-07501
TS, SALTS, TRACE ELEMENTS, LIGHT INTENSITY, SCENEDESMUS, CHLORELLA W72-12739
DESMUS, CHLORELLA, SILICA, LIGHT INTENSITY, TEMPERATURE, MUD, GROW W72-12544
ATION, *LAKES, CYANOPHYTA, LIGHT INTENSITY, EUTROPHICATION, PHYTOP W73-02472
BIOSIS, LICHENS, NITROGEN, LIGHT INTENSITY, TEMPERATURE, ANABAENA, W73-02471
UCIFERASE METHOD, SEDIMENT-WATER INTERACTIONS, BIOLUMINESCENCE, SE W72-04292
RIVER.: NUTRIENT INTERCHANGE, HAW RIVER, NEW HOPE W70-09669
IL WASTES, OILY WATER, OIL-WATER INTERFACES, WATER POLLUTION TREAT W70-09429
ORES, SEASONAL, ALGAE, MUD-WATER INTERFACES, IRON, MANGANESE, PHOS W72-05469
ERIE, *BACTERIA, *SEDIMENT-WATER INTERFACES, *HYPOLIMNION, NITRIFI W72-12996
RIA, THERMOCLINE, SEDIMENT-WATER INTERFACES, CHRYSOPHYTA, CHLOROPH W72-12995
CTERIA, *OXYGEN, *SEDIMENT-WATER INTERFACES, *LAKE ERIE, HYPOLIMNI W72-06690
BIOLOGY, OREGON, *SEDIMENT-WATER INTERFACES, SEDIMENTS, AQUATIC PR W72-08989
TR/ *PHOSPHATES, *SEDIMENT-WATER INTERFACES, ADSORPTION, LAKES, NU W72-01108
RIENTS, WATER POLLUTION EFFECTS, INTERFACES, DISSOLVED OXYGEN, TEM W71-10645
ING, CYANIDIUM CALDARI/ CHEMICAL INTERFERENCE, SCINTILLATION COUNT W72-14301
DI/ SAMPLE PREPARATION, CHEMICAL INTERFERENCE, INTERLABORATORY STU W73-00286
TANDARD ADDITION, POTENTIOMETRY, INTERFERING IONS.: /ELECTRODES, S W71-00474
PARATION, CHEMICAL INTERFERENCE, INTERLABORATORY STUDIES, PRECISIO W73-00286
RUS, CHEMICAL WASTES, LIMNOLOGY, INTERNATIONAL JOINT COMMISSION, C W72-14782

LAKE ONTARIO, ST LAWRENCE RIVER,	INTERNATIONAL JOINT COMMISSION.: /	W70-07283
AKE ONTARIO, TERTIARY TREATMENT,	INTERNATIONAL JOINT COMMISSION.: /	W71-12091
M, LYNGBYA, PSEUDOANABAENA, DATA	INTERPRETATION.: /XA, SPCNDYLOSIU	W72-07648
BAY(ORE), YAQUINA ESTUARY, DATA	INTERPRETATION, MACROALGAE, SPECI	W72-08141
US(PACIFICUS), THERMISTORS, DATA	INTERPRETATION, COUNTING.: /ANDIC	W72-07892
BUZZARDS BAY, CAPE COD BAY, DATA	INTERPRETATION, SPECIES DIVERSITY	W73-01449
SEASONAL, MONTHLY, AQUATIC LIFE,	INTERSTATE RIVERS, HYPOLIMNION, V	W70-09595
, SHORES, TIDES, BEACHES, SANDS,	INTERTIDAL AREAS, ANNELIDS, COPEP	W72-07911
ITIES, TIDAL MARSHES, FOOD WEBS,	INTERTIDAL AREAS, CONNATE WATER.:	W73-00796
VERTEBRATES, OTTERS, ROCKSLIDES,	INTERTIDAL AREAS, ON-SITE INVESTI	W73-00764
ECOLOGY, PESTICIDES, ESTUARIES,	INTERTIDAL AREAS, ESTUARINE FISHE	W72-10678
OPLEA SPHAEROPHORA, MELANOSIPHON	INTESTINALIS, HECATONEMA MACULANS	W72-01370
TORAL, LONG ISLAND, ENTEROMORPHA	INTESTINALIS, POLYSIPHONIA, ULVA	W73-00367
HINODERMS, SIPUNCULIDS, BRYOZOA,	INTOPROCTA, TUNICATES, SPONGES, V	W73-00932
LLUS, CHEUMATOPSYCHE, TRIAENODES	INUUSTA, OECETIS, LEPTOCELLA, ISO	W70-06655
CTIVITY, TOXICITY, CRABS, ALGAE,	INVERTEBRATES, FISH, POLLUTANTS,	W71-01475
BIRDS, AQUATIC ANIMALS, MOSSES,	INVERTEBRATES, DIATOMS, PHYTOPLAN	W72-05952
OISONS, POISONOUS PLANTS, CORAL,	INVERTEBRATES, ALGAE, MARINE ALGA	W71-08597
ALGAE, MICROORGANISMS, BACTERIA,	INVERTEBRATES, FUNGI, ENZYMES, AN	W73-01676
EFFECTS, LARVAE, ON-SITE TESTS,	INVERTEBRATES, WATER POLLUTION EF	W73-00348
RINE FISH, SILICA, MARINE ALGAE,	INVERTEBRATES, ISOLATION, REFRACT	W73-00371
TOS, INORGANIC COMPOUNDS, WORMS,	INVERTEBRATES, ANNELIDS, SEASONAL	W73-00374
FISH POPULATIONS, FISH BEHAVIOR,	INVERTEBRATES, OTTERS, ROCKSLIDES	W73-00764
A, PYRROPHYTA, LARVAE, SAMPLING,	INVERTEBRATES, *ATLANTIC OCEAN, *	W72-14313
PROTOZOA, *PRIMARY PRODUCTIVITY,	INVERTEBRATES, *PLANKTON, *WATER	W72-07899
ORGANISMS, SEDIMENTATION, ALGAE,	INVERTEBRATES, RADIATION, ECOSYST	W71-13723
SITE INVESTIGATION, METHODOLOGY,	INVERTEBRATES, AMPHIBIANS, TOXICI	W72-00155
S(PESTICIDES), ECOLOGY, BENTHOS,	INVERTEBRATES, PHYTOPLANKTON, PLA	W72-09668
PLANKTON, NUTRIENT REQUIREMENTS,	INVERTEBRATES, CHLAMYDOMONAS, MAT	W72-09248
HEAVY METALS, INDUSTRIAL WASTES,	INVERTEBRATES, ALGAE, *KELP.: /,	W72-11854
IDITY, OXYGEN, FISH, RECREATION,	INVERTEBRATES, RIVERS, CYANOPHYTA	W71-11009
CREATION, ENVIRONMENTAL EFFECTS,	INVERTEBRATES, WATER RESOURCES DE	W71-10577
OPHYTA, CHLOROPHYTA, PYRROPHYTA,	INVERTEBRATES, AQUATIC ANIMALS, L	W71-11012
ESTICIDE KINETICS, *MAINE, MUDS,	INVERTEBRATES, FISH, BIRDS, STREA	W72-12930
, PHYTOPLANKTON, PHOTOSYNTHESIS,	INVERTEBRATES, BIOMASS, BENTHIC F	W72-12549
BENTHIC FAUNA, AQUATIC ANIMALS,	INVERTEBRATES, AQUATIC ALGAE, AER	W72-12729
TS, DISSOLVED OXYGEN, CHLORIDES,	INVERTEBRATES, HA: /E, IMPOUNDMEN	W72-14280
EA WATER, *DATA COLLEC/ *ON-SITE	INVESTIGATIONS, *SCUBA DIVING, *S	W72-11925
*PHYTOPLANKTON, *DEPTH, ON-SITE	INVESTIGATIONS, *POPULATION, TEMP	W72-11719
, / *OXIDATION LAGOONS, *ON-SITE	INVESTIGATIONS, *BACTERIA, *ALGAE	W72-10233
WATER POLLUTION SOURCES, ON-SITE	INVESTIGATIONS, METHODOLOGY, FORE	W72-08804
, BACTERIA, BIRDS, FISH, ON-SITE	INVESTIGATION, METHODOLOGY, INVER	W72-00155
CROPS, EGGS, FISHERIES, ON-SITE	INVESTIGATIONS, ENVIRONMENTAL EFF	W72-00959
ERS, PATH OF POLLUTANTS, ON-SITE	INVESTIGATIONS, FOOD WEBS, FOOD C	W72-00949
MOLLUSKS, MARINE ALGAE, ON-SITE	INVESTIGATIONS, CALIFORNIA, FOOD	W72-03324
IDATION LAGOONS, ORGAN/ *ON-SITE	INVESTIGATION, *COLD REGIONS, *OX	W72-06838
LIDES, INTERTIDAL AREAS, ON-SITE	INVESTIGATIONS, FOOD CHAINS, SPEC	W73-00764
SOURCES, PUBLIC HEALTH, ON-SITE	INVESTIGATIONS, WASTE ASSIMILATIV	W73-00790
YSIS, SEASONAL, BIOMASS, ON-SITE	INVESTIGATIONS, ALGAE.: /CAL ANAL	W73-00242
, ESTUARINE ENVIRONMENT, ON-SITE	INVESTIGATIONS, REVIEWS.: /HEALTH	W72-04461
CTERIA, FLOATING PLANTS, ON-SITE	INVESTIGATIONS, TURTLES.: /ES, BA	W71-01488
ATER POLLUTION CONTROL, SURVEYS,	INVESTIGATIONS, PATH OF POLLUTANT	W70-08248
ATER POLLUTION CONTROL, SURVEYS,	INVESTIGATIONS, PATH OF POLLUTANT	W70-08249
DE P/ *MOUNTING MEDIA, METHYLENE	IODIDE, TOLUENE, POLYSTYRENE, SLI	W72-11738
ANIMALS, MARINE ALGAE, TRITIUM,	IODINE RADIOISOTOPES, STRONTIUM R	W71-09018
MOLLUSKS, COBALT RADIOISOTOPES,	IODINE RADIOISOTOPES, MARINE ALGA	W73-00811
AGAR, POLYSIPHONIA, ANTIBIOTICS,	IODINE, ATOMIC ENERGY, REACTOR CO	W71-09157
ION, COAGULATION, LIME, HYDROGEN	ION CONCENTRATION, TEMPERATURE, C	W71-08951
, METABOLISM, BACTERIA, HYDROGEN	ION CONCENTRATION, LABORATORY TES	W71-08687
EL MIGRATION, PROTOZOA, HYDROGEN	ION CONCENTRATION, WATER TEMPERAT	W71-05626
LOROPHYLL, TEMPERATURE, HYDROGEN	ION CONCENTRATION, DISSOLVED OXYG	W71-07382
ATER QUALITY, SULFATES, HYDROGEN	ION CONCENTRATION, ALGAE, BACTERI	W71-07942
EN, CHEMICAL REACTIONS, HYDROGEN	ION CONCENTRATION, ALGAE, REPRODU	W71-08026
CHEMICAL OXYGEN DEMAND, HYDROGEN	ION CONCENTRATION, WASTE WATER TR	W71-07113
S, PILOT PLANTS, ALGAE, HYDROGEN	ION CONCENTRATION, CHEMICAL OXYGE	W71-07102
CHEMICAL PRECIPITATION, HYDROGEN	ION CONCENTRATION, COLIFORMS, BAC	W71-07100
TEMPERATURE, CHELATION, HYDROGEN	ION CONCENTRATION, DYSTROPHY, ZIN	W71-09190
XICITY, NEUTRALIZATION, HYDROGEN	ION CONCENTRATION, TEMPERATURE, B	W71-09524
ALGAE, MICROORGANISMS, HYDROGEN	ION CONCENTRATION, PHENOLS, ARIZO	W71-01474
LOCCULATION, HEAD LOSS, HYDROGEN	ION CONCENTRATION, CHLORELLA, SCE	W71-00138
NNOPLANKTON, TURBIDITY, HYDROGEN	ION CONCENTRATION, ALKALINITY, CA	W71-00117
EAT LAKES, TEMPERATURE, HYDROGEN	ION CONCENTRATION, ACTIVATED SLUD	W71-00116
CHEMICAL OXYGEN DEMAND, HYDROGEN	ION CONCENTRATION, ALGAE, *ENTROP	W70-10427
SOLIDS, NITROGEN CYCLE, HYDROGEN	ION CONCENTRATION, TEMPERATURE, E	W71-01489
ESTS, DISSOLVED OXYGEN, HYDROGEN	ION CONCENTRATION, ALKALINITY, RE	W71-04518
IFORMS, MICROORGANISMS, HYDROGEN	ION CONCENTRATION, WASTE WATER TR	W71-05155
ATES, RUNOFF, ILLINOIS, HYDROGEN	ION CONCENTRATION, ALKALINITY, HA	W71-03028
TOXICITY, IRON, LIGHT, HYDROGEN	ION CONCENTRATION.: /YTOPLANKTON,	W71-03027
IENTS, LIGHT INTENSITY, HYDROGEN	ION CONCENTRATION, CARBON DIOXIDE	W72-04433
ACTERIA, SEWAGE, ALGAE, HYDROGEN	ION CONCENTRATION, CARBON DIOXIDE	W72-04309
ONATES, CARBON DIOXIDE, HYDROGEN	ION CONCENTRATION, WATER QUALITY	W72-04431
UALITY, CARBON DIOXIDE, HYDROGEN	ION CONCENTRATION, CARBON, OXYGEN	W72-06274
YTES, ACTIVATED CARBON, HYDROGEN	ION CONCENTRATION, NUTRIENTS.: /L	W72-06201
YGE/ *ALGAE, *NITROGEN, HYDROGEN	ION CONCENTRATION, BIOCHEMICAL OX	W72-06612
GHT PENETRATION, ALGAE, HYDROGEN	ION CONCENTRATION.: /QUATIONS, LI	W72-05459
SEWAGE, CARBOHYDRATES, HYDROGEN	ION CONCENTRATION, TEMPERATURE, E	W72-05457
, MAGNESIUM, CHLORIDES, HYDROGEN	ION CONCENTRATION, PERIPHYTON, WA	W72-05610
UIREMENTS, TEMPERATURE, HYDROGEN	ION CONCENTRATION, EFFICIENCIES,	W72-04788
ATER POLLUTION SOURCES, HYDROGEN	ION CONCENTRATION, CARBON DIOXIDE	W72-04769
TION, DISSOLVED OXYGEN, HYDROGEN	ION CONCENTRATION, DESIGN CRITERI	W72-04787

YANOPHYTA, CHLOROPHYTA, HYDROGEN ION CONCENTRATION, CHEMICAL PORPE W72-14790
ALGAE, SURFACE WATERS, HYDROGEN ICN CONCENTRATION, LIGHT PENETRAT W73-00236
D OXYGEN, CONDUCTIVITY, HYDROGEN ICN CONCENTRATION.: /TY, DISSOLVE W72-14728
PERATURE, ACID STREAMS, HYDROGEN ICN CONCENTRATION, HEAVY METALS, W72-14690
A, PYRROPHYTA, BIOMASS, HYDROGEN ION CONCENTRATION, MINNESOTA, WAT W72-07145
TURE, DISSOLVED OXYGEN, HYDROGEN ICN CONCENTRATION, SCENEDESMUS, B W72-08049
, COPPER SULFATE, NETS, HYDROGEN ICN CONCENTRATION, PHOSPHATES, NI W72-07890
*CULTURES, CYANOPHYTA, HYDROGEN ION CONCENTRATION, LIGHT INTENSIT W73-01657
, ALKALINITY, HALOGENS, HYDROGEN ICN CONCENTRATION, PHCSPHATES, NI W73-01434
ION, CALCIUM CARBONATE, HYDROGEN ION CONCENTRATION, FILTRATION, PL W73-01669
*WASTE WATER TREATMENT, HYDROGEN ION CONCENTRATION, CYANOPHYTA.: / W73-01367
NSITY, POROSITY, COLOR, HYDROGEN ICN CONCENTRATION, NITROGEN, CARB W73-00854
EMPERATURE, ALKALINITY, HYDROGEN ION CONCENTRATION, WATER QUALITY, W73-00916
S, CHEMICAL PROPERTIES, HYDROGEN ICN CONCENTRATION, *WASTE WATER T W72-11833
H RATES, *LIMITING FA/ *HYDROGEN ION CONCENTRATION, *ALGAE, *GROWT W72-11724
IUM, ZINC, TEMPERATURE, HYDROGEN ION CONCENTRATION, CORROSION, WAT W72-08683
RTEBRATES, ABSORBANCE, MAGNESIUM ICN CONCENTRATION, BRUCITE, BOTTO W72-09092
QUALITY, ON-SITE TESTS, HYDROGEN ICN CONCENTRAT: /EA WATER, WATER W72-09092
ARNACLES, LETHAL DOSAGE, CALCIUM ION CONCENTRATION, VERTEBRATES, A W72-09092
RBON DIOXIDE, POLYMERS, HYDROGEN ICN CONCENTRATION, PROTOZOA, DISS W72-10233
IOR, WATER TEMPERATURE, HYDROGEN ICN CONCENTRATION, HARDNESS, ALKA W72-10875
, METHANE, TEMPERATURE, HYDROGEN ICN CONCENTRATION, PROTEIN, LABOR W72-10390
ICOCHEMICAL PROPERTIES, HYDROGEN ION CONCENTRATION, TEMPERATURE, D W72-03224
CHEMICAL PRECIPITATION, HYDROGEN ION CONCENTRATION, FILTRATION, CO W72-03613
ON, THERMAL PROPERTIES, HYDROGEN ION CONCENTRATION, DISSOLVED OXYG W72-01363
ON-REDUCTION POTENTIAL, HYDROGEN ION CONCENTRATION.: /NGE, OXIDATI W72-01108
RATES, *ORGANIC ACIDS, *HYDROGEN ION CONCENTRATION, METABOLISM, CA W72-00838
S, FILTRATION, AMMONIA, HYDROGEN ION CONCENTRATION, TEMPERATURE, L W72-01881
TY, NUTRIENTS, AMMONIA, HYDROGEN ION CONCENTRATION, CYANOPHYTA, CH W71-11519
RADIOISOTOPES, DIATOMS, HYDROGEN ION CONCENTRATION, MARINE ANIMALS W71-10083
OMS, CHEMICAL ANALYSIS, HYDROGEN ION CONCENTRATION, SEA WATER, PHO W71-11008
, AERATION, *OXIDATION, HYDROGEN ION CONCENTRATION, MICROORGANISMS W71-12223
NIC MATTER, ALKALINITY, HYDROGEN ION CONCENTRATION, NUTRIENTS, AQU W71-12068
SSOLVED OXYGEN, SEWAGE, HYDROGEN ICN CONCENTRATION.: /LIZATICN, DI W71-12188
, GEORGIA, THERMOCLINE, HYDROGEN ION CONCENTRATION, WATER TEMPERAT W71-11914
BIDITY, PHOTOSYNTHESIS, HYDROGEN ION CONCENTRATION, NITRITE, AMMON W71-11876
AM FRACTIONATION, LIME, HYDROGEN ION CONCENTRATION, ECONOMICS, ALG W71-12200
TION, DISSOLVED OXYGEN, HYDROGEN ION CONCENTRATION, SETTLING, MICR W71-13334
TS, LAKES, CHLOROPHYLL, HYDROGEN ICN CONCENTRATION, CARBON DIOXIDE W71-12855
CHEMICAL PRECIPITATION, HYDROGEN ION CONCENTRATION.: /AE TRACERS, W72-12210
GEN, WATER TEMPERATURE, HYDROGEN ION CONCENTRATION, STRATIFICATION W72-13871
ED OXYGEN, THERMOCLINE, HYDROGEN ION CONCENTRATION, FLOW, CHEMICAL W72-12941
, DIURNAL DISTRIBUTION, HYDROGEN ION CONCENTRATION, DISSOLVED OXYG W72-13508
ENTIFICATION, AQUEOUS SOLUTIONS, ION EXCHANGE, METHODOLOGY, ORGANI W72-12166
WAGE, BIOCHEMICAL OXYGEN DEMAND, ION EXCHANGE, CHEMICAL OXYGEN DEM W71-12200
TEMPERATURE, ALGAE, PHOSPHORUS, ION EXCHANGE, OXIDATION-REDUCTION W72-01108
L CONTAMINATION, TRACE ELEMENTS, ION EXCHANGE, POLLUTANT IDENTIFIC W72-14303
RIVERS, WATER POLLUTION SOURCES, ION EXCHANGE, *CHEMICAL ANALYSIS, W73-00286
ATED SLUDGE, TERTIARY TREATMENT, ION EXCHANGE, ELECTROCHEMISTRY, R W70-09907
ACE ELEMENTS, CHEMICAL ANALYSIS, ION EXCHANGE, ABSORPTION, STRONTI W71-09188
NOMIC FEASIBILITY, PILOT PLANTS, ION EXCHANGE, ELECTROLYSIS, CHEMI W71-09450
ORTH/ VOLTAMMETRY, RADIOGRAPHY, ION SELECTIVE ELECTRODES, LASERS, W72-08790
DIATOMS, PHYTOPLANKTON, HYDORGEN ICN TEMPERATURE,: /TMENT PLANTS, W72-06837
, NITRATES, ALGAE, COPPER, IONS, ICN TRANSPORT.: /TALS, ELECTRODES W72-01693
D ADDITION, POTENTIOMETRY, INTE/ ICN-SELECTIVE ELECTRODES, STANDAR W71-00474
TION, POTENTIOMETRY, INTERFERING IONS.: /ELECTRODES, STANDARD ADDI W71-00474
NS, LA GORCE MOUNTAINS, AMMONIUM IONS.: /PHILIC, BIOTIC ASSOCIATIO W73-00854
, SYNTHESIS, VITAMINS, SULFATES, IONS.: /PHATES, METABOLISM, ALGAE W73-01626
OROMETRY, EQUIPMENT, INHIBITION, IONS, ABSORPTION, BIOASSAY, *MERC W72-07660
ANALYSIS, TEMPERATURE, SEASONAL, IONS, ALGAE, SALINITY, PHYTOPLANK W72-01363
POTENTIOMETRY, CADMIUM IONS, COPPER IONS, OOCYSTIS.: W72-01693
LA, PLANT PATHOLOGY, INDUSTRIES, IONS, CYANOPHYTA, RADIOISOTOPES, W71-09157
*INHIBITION, COPPER, FLUORIDES, ICNS, CYTOLOGICAL STUDIES, WATER W72-11572
NALYSIS, *NITRATES, *ELECTRODES, ICNS, FLUORIDES, ALGAE, VOLUMETRI W71-00474
TRODES, NITRATES, ALGAE, COPPER, IONS, ION TRANSPORT.: /TALS, FLEC W72-01693
*WATERSHEDS(BASINS), STREAMFLOW, IONS, NITRATES, SULFATES, DRAINAG W71-01489
OLOGY, CYANOPHYTA, FROZEN SOILS, IONS, NOSTOC, ENVIRONMENTAL EFFEC W73-00854
HNIQUES, CHLOROPHYLL, SEA WATER, ICNS, NUTRIENT REQUIREMENTS, AQUA W72-09103
ENTIOMETRY, CADMIUM IONS, COPPER IONS, OOCYSTIS.: POT W72-01693
OKLAHOMA, CYANOPHYTA, CHLORELLA, IONS, PHYTOPLANKTON, NITRATES, NI W72-04773
CESIUM, STRONTIUM RADIOISOTOPES, IONS, RIVERS, CULTURES, ORGANIC M W71-08032
RIMARY PRODUCTIVITY, ICED LAKES, IONS, SULFATES, CARBONATES, BICAR W71-02880
SPODYLOSIUM, SPIROG/ *INORGANIC ICNS, ZINC, VOLVOX, HYDRODICTYON, W71-11001
RROBACILLUS, GAS CHROMATOGRAPHY, ICNS,: /ENOPHYTA, CHLOROPHYTA, FE W72-06124
HICATION, WATER QUALITY, RIVERS, IOWA, CORN, DISCHARGE, CHLOROPHYL W71-06445
TIES, *ALGA/ *RIVERS, *BACTERIA, IPLANKTON, PHYSICOCHEMICAL PROPER W72-07896
RNIA CRASSIPES, TYPHA ANGUSTATA, IPOMOEA AQUATICA, TOTAL SOLIDS.: / W72-01363
RCURY, EAGLES, SALMO GAIRDNERII, IRIS PSEUDACORUS, LYSIMACHIA NUMM W72-05952
MINARIA, LIVERPOOL BAY(ENGLAND), IRISH SEA.: *ULVA, *LA W71-12867
GANATE.: *FILTER CLOGGING, IRON BACTERIA, POND WEEDS, PERMAN W72-06536
STIC WASTES, BIOASSAY, SULFATES, IRON COMPOUNDS, SEWAGE, NITRATES, W72-12239
ROBIC BACTERIA, SULFUR BACTERIA, IRON COMPOUNDS.: /IC BACTERIA, AE W72-12996
OTOPES, RUTHENIUM RADIOISOTOPES, IRON RADIOISOTOPES.: /IUM RADIOIS W72-04463
ISOTOPES, THORIUM RADIOISOTOPES, IRON RADIOISOTOPES, ZIRCONIUM RAD W71-09188
IOISOTOP/ *CESIUM RADIOISOTOPES, IRON RADIOISOTOPES, MANGANESE RAD W71-09250
ER, PHYTOPLANKTON, PRODUCTIVITY, IRON.: / DECOMPOSING ORGANIC MATT W72-01097
UM, MANGANESE, SODIUM, CHROMIUM, IRON.: /CTERIA, LAKE ERIE, ALUMIN W72-01101
ON, LAKES, NITROGEN, PHOSPHORUS, IRON, *TEXAS, LABORATORY TESTS, C W72-13692
IOISOTOPES, MANGANESE, CHROMIUM, IRON, ANALYTICAL TECHNIQUES, FOOD W73-00832
FALLOUT, CHEMICAL PRECIPITATION, IRON, BIOASSAY, LEACHING, VOLUMET W72-05965

PLANT GROWTH SUBSTANCES, ALGAE,	IRON, BIOASSAY, BIOINDICATORS, FO	W71-03095
TIES, CHEMICAL PROPERTIES, GOLD,	IRON, CESIUM, ALKALI METALS, SEAS	W73-01434
UTHEN/ CONCENTRATION, ABUNDANCE,	IRON, CHROMIUM, COPPER, NICKEL, R	W71-13243
RADIUM RAIODISOTOPES, ALUMINUM,	IRON, COPPER, MAGNESIUM, MANAGNES	W71-09188
, MOLYBDENUM, MANGANESE, COBALT,	IRON, INDUSTRIAL WASTES, SEWAGE E	W71-11515
IATOMS, PHYTOPLANKTON, TOXICITY,	IRON, LIGHT, HYDROGEN ION CONCENT	W71-03027
LEAVES, SAMPLING, PHYTOPLANKTON,	IRON, MAN: /, CHEMICAL ANALYSIS,	W73-01089
AL, ALGAE, MUD-WATER INTERFACES,	IRON, MANGANESE, PHOSPHORUS, NITR	W72-05469
IC MATTER, *INORGANIC COMPOUNDS,	IRON, MANGANESE, PHOSPHORUS, CARB	W72-05506
M, MAGNESIUM, SODIUM, POTASSIUM,	IRON, MANGANESE, COPPER, ALGAE, S	W71-11001
YPOLIMNION, MATHEMATICAL MODELS,	IRON, MANGANESE, PHOSPHORUS, ANAE	W72-12997
H RATES, RESISTANCE, PHOSPHORUS,	IRON, MANGANESE, ZINC, NICKEL, CO	W72-12739
HICATION, LAKES, AQUATIC PLANTS,	IRON, MICROORGANISMS, COLORIMETRY	W72-03742
RE, BACTERIA, ALGAE, ALKALINITY,	IRON, PHOSPHATE, PHOSPHORUS, CARB	W71-05626
, IMPOUNDMENTS, SEASONAL, COLOR,	IRON, PHYSICOCHEMICAL PROPERTIES,	W73-00256
S, SEWAGE, NITROGEN, PHOSPHORUS,	IRON, PHYTOPLANKTON, DIATOMS, DIN	W73-00234
PHOSPHATES, NITRATES, NITRITES,	IRON, SILICON, SULFATES, SODIUM,	W72-05610
RBON, ORGANIC MATTER, MANGANESE,	IRON, SULFUR, POTASSIUM, MOLYBDEN	W72-07514
LUORESCENCE, INFRARED RADIATION,	IRRAD: /S, ANIMAL WASTES, FISH, F	W72-08790
VITY EFFECTS, *ALGAE, CHLORELLA,	IRRADIATION, RADIOACTIVITY TECHNI	W72-11836
, CHLOROPHYLL, LABORATORY TESTS,	IRRADIATION, CHLOROPHYTA, INHIBIT	W73-01080
LAURENCIA PAPILLOSA, COELOTHRIX	IRREGULARIS, CENTROCERAS CLAVULAT	W73-00838
TICIDES, FUNGICIDES, HERBICIDES,	IRRIGATION, AGRICULTURE, RICE, CR	W72-14807
TION, *NITRATES, EUTROPHICATION,	IRRIGATION WATER, AQUATIC ALGAE,	W72-00974
OUT, BIOCONTROL, INSECT CONTROL,	IRRIGATION, PREDATION, PARASITISM	W72-00846
AS, FERTILIZERS, DRAINAGE WATER,	IRRIGATION, RETURN FLOW, WATER SU	W71-06435
RAPHIES, *ECOLOGY, AQUATIC LIFE,	IRRIGATION SYSTEMS, WATER ANALYSI	W71-06102
PERATURE, LIMNOLOGY, CYANOPHYTA,	IRRIGATION CANALS, ALGAL CONTROL.	W71-06102
BALT RADIOISOTOPES, FOOD CHAINS,	IRRIGATION, FLOW PROFILES, MARINE	W71-09225
ERATURE, ECOSYSTEMS, CALIFORNIA,	IRRIGATION, ECONOMIC FEASIBILITY,	W71-09157
, PATHOGENIC BACTERIA, EFFLUENT,	IRRIGATION, STORAGE, WASTE DISPOS	W71-08214
GROUNDWATER, FISH, FLOOD PLAINS,	IRRIGATION, PESTICIDES.: /ATION,	W70-05103
DOSTIGMATICA, CHL/ CHLAMYDOMONAS	ISABELIENSIS, CHLAMYDOMONAS PALLI	W72-12632
AD, / MACROINVERTEBRATES, *TOCKS	ISLAND RESERVOIR(NJ), AMERICAN SH	W72-14280
S, ANTIBIOTICS, SEA WATER, RHODE	ISLAND.: / CHLOROPHYTA, INHIBITOR	W72-04748
, MAGDALEN ISLAND(QUEBEC), SABLE	ISLAND(NOVA SCOTIA), ST PIERRE(CA	W71-10066
RADOR), NEW BRUNSWICK, ANTICOSTI	ISLAND(QUEBEC), MAGDALEN ISLAND(Q	W71-10066
TICOSTI ISLAND(QUEBEC), MAGDALEN	ISLAND(QUEBEC), SABLE ISLAND(NOVA	W71-10066
*NORTHPORT PLANT(LONG	ISLAND).:	W72-14659
AND, MORRIA REEF, BEAR BAY, NEVA	ISLAND, BAIRD ISLAND, SHOLIN ISLA	W72-09120
IL SPILLS, S.S. WITWATER, GALETA	ISLAND, CANAL ZONE, MACROINVERTEB	W72-07911
E PREPARATION, SUBLITTORAL, LONG	ISLAND, ENTEROMORPHA INTESTINALIS	W73-00367
ROLYSATE, TONGA ISLANDS, TUTUILA	ISLAND, HONSHU ISLAND, OPEN OCEAN	W71-08042
ISLAND, SHOLIN ISLAND, WINIFRED	ISLAND, IZEMBEK LAGOON, CHAETOMOR	W72-09120
T.: AMCHITKA	ISLAND, MILROW TEST, CANNIKIN TES	W73-00764
, RHODOMELA, RHODOMENIA, MARMION	ISLAND, MORRIA REEF, BEAR BAY, NE	W72-09120
ISLANDS, TUTUILA ISLAND, HONSHU	ISLAND, OPEN OCEAN, SAPROPHYTES,	W71-08042
IX MUCOR, MAJURO ATOLL, MARCHALL	ISLAND, POLYSIPHONIA LANOSA, CHON	W72-08142
EF, BEAR BAY, NEVA ISLAND, BAIRD	ISLAND, SHOLIN ISLAND, WINIFRED I	W72-09120
*COASTAL WATERS, *SOUTHERN LONG	ISLAND, SKELETONEMA COSTATUM, THA	W71-07875
EVA ISLAND, BAIRD ISLAND, SHOLIN	ISLAND, WINIFRED ISLAND, IZEMBEK	W72-09120
RHIZOSOLENIA ERIENSIS, MELOSIRA	ISLANDICA, MELOSIRA ITALICA, MELO	W73-00384
S, ENIWETOK, MICRONESIA, FANNING	ISLANDS.: PHILIPPINE	W71-11486
LINDRUS, PACIFIC OCEAN, ALEUTIAN	ISLANDS, PRYMNESIUM PARVUM.: /OCY	W71-11008
OOMS LAKE, WOODS LAKE, LAGOON OF	ISLANDS, STAURASTRUM SPP.: /NT, T	W73-01094
SLAND/ *ALGAL HYDROLYSATE, TONGA	ISLANDS, TUTUILA ISLAND, HONSHU I	W71-08042
ENT COAST, INDUSTRIAL EXPANSION,	ISLE OF THANET, MEDWELL ESTUARY.:	W71-13746
BRITISH	ISLES.:	W72-10619
CA, MARINE ALGAE, INVERTEBRATES,	ISOLATION, REFRACTIVITY, COPEPODS	W73-00371
HIONINE, PHENYLALANINE, ALANINE,	ISOLEUCINE, VALINE, LYSINE, ASPAR	W73-00942
ES INUUSTA, OECETIS, LEPTOCELLA,	ISONYCHIA, CAENIS, PERLESTA.: /OD	W70-06655
, CRABS, ALGAE, SNAILS, MUSSELS,	ISOPODS, AMPHIPODA, RADIOISOTOPES	W72-03347
ANNELIDS, SEASONAL, CRUSTACEANS,	ISOPODS, MARINE ALGAE, *ALASKA.: /	W73-00374
, *AMPHIPODA, *FLORIDA, STREAMS,	ISOPODS, SCENEDESMUS, AQUATIC HAB	W72-12744
DENSITY-GRADIENT CENTRIFUGATION,	ISOPYCNIC CENTRIFUGATION, SAMPLE	W73-00371
METER, OSCILLATORIA AGARDHII VAR	ISOTHRIX, LAKE ITASCA, LAKE ELK,	W72-07145
HODOLOGY, ALGAE, AQUATIC PLANTS,	ISOTOPE FRACTIONATION, CARBON DIO	W72-12709
TERRESTRIAL PLANTS, COMBUSTION,	ISOTOPE RATIOS, SPECTROMETRY, LIC	W72-12709
ENVIR/ *ISOTOPE STUDIES, *STABLE	ISOTOPES, *CARBON RADIOISOTOPES,	W72-12709
BSORPTION, RADIOISOTOPES, STABLE	ISOTOPES, ESTUARINE ENVIRONMENT,	W73-00811
ATER ANALYSIS, SEA WATER, STABLE	ISOTOPES, RADIOISOTOPES, NUCLEAR	W73-00831
RY, *BIOLOGICAL TISSUES, MERCURY	ISOTOPES, HUMAN BRAIN, LOW-TEMPER	W71-11036
TERIA, *TEXAS, / *ALGAE, *STABLE	ISOTOPES, *CARBON, *ANAEROBIC BAC	W71-04873
*CYSTS,	ISRAEL.:	W72-00422
EL), BOTTOM WATERS, RIVER JORDAN(	ISRAEL).: *LAKE KINNERET(ISRA	W72-05469
LUM, FERRIC CHLORIDE, *JERUSALEM(	ISRAEL).: A	W72-12289
DAN(ISRAEL).: *LAKE KINNERET(	ISRAEL), BOTTOM WATERS, RIVER JOR	W72-05469
ATORIA / *ISRAEL, *LAKE KINNERET(	ISRAEL), PERIDINIUM WESTI, OSCILL	W71-13187
P/ *EASTERN CANADA, NOVA SCOTIA,	ISTHMOPLEA SPHAEROPHORA, MELANOSI	W72-01370
IS, MELOSIRA ISLANDICA, MELOSIRA	ITALICA, MELOSIRA AMBIGUA, SYNEDR	W73-00384
ORIA AGARDHII VAR ISOTHRIX, LAKE	ITASCA, LAKE ELK, LAKE MARY, LAKE	W72-07145
SHOLIN ISLAND, WINIFRED ISLAND,	IZEMBEK LAGOON, CHAETOMORPHA CANN	W72-09120
ERU), CAPE COD(MASS), CAPE MAY(N	J), SARGASSO SEA.: /AMA, CALLAO(P	W72-08056
OT SPRINGS, BOULDER HOT SPRINGS,	JACKSON HOT SPRINGS, LOLO HOT SPR	W72-05610
AKE SIDNEY LANIER(GEORGIA), LAKE	JACKSON(GEORGIA), LAKE CARDINAL(G	W71-09173
AKE SIDNEY LANIER(GEORGIA), LAKE	JACKSON(GEORGIA), LAKE CARDINAL(G	W71-11914
LAKE	JACOMO(MO).:	W72-10431
*MICROCYSTIS, *LAKE SUWA(	JAPAN).:	W73-00236
ICAL MAGNIFICATION, MINAMATA BAY(	JAPAN), ORGANIC MERCURY COMPOUNDS	W72-12257
EUCAMPIA ZOODIACUS, CHA/ SEA OF	JAPAN, *AMURSKY BAY, PERIDINIANS,	W73-00442

*GREEN MANURE, INDIA, JAPAN, RICE FIELDS.: W73-02473
TERIONELLA FORMOSA, ASTERIONELLA JAPONICA, STAURASTRUM PARADOXIUM, W72-12739
*ACETYLENE REDUCTION, RHIZOBIUM JAPONICUM, LAKE MENDOTA, CRYSTAL W73-01456
EROSA, NUPHAR VARIEGATUM, NUPHAR JAPONICUM, NUPHAR LUTEN, ANTICANC W72-04275
ACHNANTHES BREVIPES, ACHNANTHES JAVANICA, LICMOPHORA JURGENSII, A W72-08141
E, DELAWARE ESTUARY(DEL-NJ), SAN JOACUIN DELTA(CALIF), STEADY-STAT W71-05390
FLOW TURBIDOSTAT, *GROWTH/ *SAN JOAQUIN DELTA(CALIF), *CONTINUOUS W70-10403
IF), SEARSVILLE LAKE(CALIF), SAN JOAQUIN RIVER(CALIF), NITZSCHIA.: W70-10405
*SAN JOAQUIN RIVER.: W71-07098
OLDT BAY, CONNECTICUT RIVER, SAN JOAQUIN RIVER(CALIFORNIA), BISCAY W71-07417
R, CONTRA COSTA POWER PLANT, SAN JOACUIN RIVER, MORRO BAY POWER PL W72-07907
RASITES, JORDAN RIVER(MICH), SAN JOAQUIN RIVER(CALIF), RIFLE FALLS W71-11006
RUBBLE CHIMNEY, SAN JOAQUIN VALLEY.: W72-00974
L GROWTH, ALGAL HARVESTING, *SAN JOAQUIN VALLEY(CALIF).: /US, ALGA W72-02975
, BACTERIAL DENITRIFICATION, SAN JOAQUIN VALLEY.: /ATION, METHANOL W71-08223
NTIBIOTIC SENSITIVITY, CYTOPHAGA JOHNSONII, SPOROCYTOPHAGA MYXOCOC W72-07951
WASTES, LIMNOLOGY, INTERNATIONAL JOINT COMMISSION, CANADA, FEDERAL W72-14782
ST LAWRENCE RIVER, INTERNATIONAL JOINT COMMISSION.: /AKE ONTARIO, W70-07283
ERTIARY TREATMENT, INTERNATIONAL JOINT COMMISSION.: /KE ONTARIO, T W71-12091
, HATCHERY EFFLUENTS, PARASITES, JORDAN RIVER(MICH), SAN JOAQUIN R W71-11006
ET(ISRAEL), BOTTOM WATERS, RIVER JORDAN(ISRAEL).: *LAKE KINNER W72-05469
TASCA, LAKE ELK, LAKE MARY, LAKE JOSEPHINE, LAKE ARCO, LAKE DEMING W72-07145
CA, ASTEROLAMPRA HEPTACTIS, BAC/ JUDAY NET, ASTEROLAMPRA MARYLANDI W71-11527
TER, WATER REUSE, LEGAL ASPECTS, JUDICIAL DECISIONS, LEGISLATION, W70-07100
BORRICHIA, PHRAGMITES COMMUNIS, JUNCUS SPP, ATRIPLEX HASTATA, SCI W73-01089
ACHNANTHES JAVANICA, LICMOPHORA JURGENSII, AMPHORA OVALIS, SYNEDR W72-08141
AR ADVENA, EICHHORNIA CRASSIPES, JUSTICIA AMERICANA, ALTERNANTHERA W73-01089
Y, COSTS, WATER DEMAND, ECONOMIC JUSTIFICATION, AQUATIC WEED CONTR W71-13172
TS, *SEWAGE TREATMENT, *ECONOMIC JUSTIFICATION, *COBALT, *RADIATIO W72-00383
AMM/ *BIOLOGICAL SAMPLES, SR-90, K-40, CS-137, *PAHRANAGAT LAKE, G W73-01673
C MODEL, *TUTTLE CREEK RESERVOIR(KAN), GREAT PLAINS, AQUATIC ECOSY W72-12393
NTRATION, WASTE WATER TREATMENT, KANSAS.: /AND, HYDROGEN ION CONCE W71-07113
*CHENEY RESERVOIR(KANSAS).: W72-02274
ER, PLASMODIUM BERGHEI, BORRELIA KANSAS, BORRELIA ANSERINA.: / LIV W73-01676
OTA(WIS), LAKE MONONA(WIS), LAKE KEGONSA(WIS), LAKE WASHINGTON(WAS W72-05473
GY, ORGANIC ACIDS, FULVIC ACIDS, KELPS, WATER ANALYSIS, EXUDATION, W72-12166
HILUS, GLOSSOSOMA, PHYSA GYRINA, KEMMERER SAMPLER, SECCHI *YARDS C W72-14280
INEMILE CREEK, VAN DORN SAMPLER, KEMMERER SAMPLER, CHLAMYDOMONAS E W72-09111
S, *RIVER THAMES(ENGLAND), RIVER KENNET(ENGLAND), STEPHANODISCUS H W72-14797
ITROGEN, ORTHOPHOSP/ *SALT RIVER(KENT), VASCULAR PLANTS, NITRATE N W72-06526
RIVER(CONNECTICUT), GREEN RIVER(KENTUCKY), CAYUGA LAKE(NEW YORK), W71-11517
H, SPIROGYRA, EUDORINA, CYCLOPS, KERATELLA, ASPLANCHNA, LITHODESMI W71-11876
ICATUM STRAIN PR-6, PHOTOALDRIN, KETOENDRIN, METABOLITES, SKELETON W72-03545
A HIRUNDO, RATS, PARTRIDGE, DDE, KIDNEYS, LIVER, BLOOD, GASTEROSTE W72-05952
ING, BONE, ORGANS, BRAIN, BLOOD, KIDNEYS, LIVER, PLASMODIUM BERGHE W73-01676
DENSERS, DESIGN, STANDARDS, FISH KILL, ALGAE, COOLING TOWERS, DENS W70-09612
OSCH RESERVOIR(THE NETHERLANDS), KINETIC EQUATION.: /LANDS), BIESB W72-08559
LLUTION / *DDT, *DDE, *PESTICIDE KINETICS, *PERSISTENCE, *WATER PO W73-02280
, ORGANIC COMPOUNDS, ABSORPTION, KINETICS, WATER TEMPERATURE, RESP W73-01269
INUOUS FLOW TURBIDOSTAT, *GROWTH KINETICS, CHEMOSTATS, FLAGELLATE W70-10403
RPTION, TEMPERATURE, METABOLISM, KINETICS, WATER POLLUTION EFFECTS W71-09168
R, FRESHWATER FISH, TEMPERATURE, KINETICS, NUCLEAR WASTES, SOLUBIL W71-09182
CAL TECHNIQUES, DESIGN CRITERIA, KINETICS, *WASTE WATER TREATMENT. W72-02363
PLANKTON, *EUTROPHICATION, *NUT/ KINETICS, *BIOLOGICAL TREATMENT, W71-13553
S, PESTICIDE RESIDUES, PESTICIDE KINETICS, ANIMAL PHYSIOLOGY, PLAN W71-12303
NT, CULTURE MEDIA, MACRO/ GROWTH KINETICS, BIOENERGETICS, ENRICHME W72-09378
NG NUTRIENTS, TRACERS, LITTORAL, KINETICS, PHOSPHATES.: /ES, CYCLI W72-12543
ATERSHEDS, ECOSYSTEMS, PESTICIDE KINETICS, *MAINE, MUDS, INVERTEBR W72-12930
RA, R/ *RESEARCH POLICY, *UNITED KINGDOM, *FUTURE TRENDS, CLADOPHO W71-13172
OTILUS NATANS, ZOOGLOEA/ *UNITED KINGDOM, *SEWAGE FUNGUS, SPHAEROT W72-08573
, LAKE WINDERMERE(ENGLA/ *UNITED KINGDOM, ESTHWAITE WATER(ENGLAND) W71-13185
, BLELHAM TARN(ENGLAND)/ *UNITED KINGDOM, ESTHWAITE WATER(ENGLAND) W71-13183
, SEMOTILUS ATROMACULATU/ CHUBS, KINGFISHES, SALVELINUS FONTINALIS W72-12930
I, OSCILLATORIA / *ISRAEL, *LAKE KINNERET(ISRAEL), PERIDINIUM WEST W71-13187
RIVER JORDAN(ISRAEL).: *LAKE KINNERET(ISRAEL), BOTTOM WATERS, W72-05469
ARED SPECTROSCOPY, ASSIMILATION, KJEDAHL PROCEDURE, SAMPLE PREPARA W73-00379
ON.: UPPER KLAMATH LAKE, SUBSTRATE UTILIZATI W73-01269
ILLE GREEN LAKE(NEW YORK), UPPER KLAMATH LAKE(OREGON), LAKE MENDOT W71-05626
UPPER KLAMATH LAKE(OREGON).: W72-08989
UM SINICUM, SINAMONAS STAGNALIS, KOFOIDIELLA UNIFLAGELLATA, TSUMUR W72-07683
LANCTONICA, MANCHUDINIUM SINICU/ KOLBEANA OVOIDEA, PALMERIAMONAS P W72-07683
ETOCEROS PERUVIANUS, CLADOGRAMMA KOLBEI, CLADOGRAMMA CRENULATUS, C W71-11527
ERI, ULOTHRIX SPP, MICROTHAMNION KUETZ: /UM, SPHAEROCYSTIS SCHROET W72-09111
*BEAVER CREEK BASIN(KY).: W71-07942
ULTU/ *HETEROCYSTS, *ANABAENA SP L-31, CONTINUOUS CULTURE, BATCH C W72-03217
CHROPHILIC, BIOTIC ASSOCIATIONS, LA GORCE MOUNTAINS, AMMONIUM IONS W73-00854
E, LAKE ARCO, LAKE DEMING, LOWER LA SALLE LAKE, LAKE LONG, LAKE SQ W72-07145
NTS, SEWAGE, DETERGENTS, PLANTS, LA: /IC MATTER, LAKE ERIE, EFFLUE W71-11017
, CALCAREOUS LAKES, ULTRA-VIOLET LABILE NITROGEN, ULTRA-VIOLET REF W72-08048
RES, WATER QUALITY, SCENEDESMUS, LABORATORY EQUIPMENT, PHOTOGRAPHY W72-07144
TER, CHLOROPHYTA, CHLAMYDOMONAS, LABORATORY TESTS, CHRYSOPHYTA, OP W72-08051
ORUS, NITROGEN, *EUTROPHICATION, LABORATORY TESTS, WATER QUALITY C W72-08376
G TOWERS, *TOXICITY, *BIOASSAYS, LABORATORY TESTS, FISH, ALGAE, CH W72-08683
BIOCHEMISTRY, ALGAE, INHIBITION, LABORATORY TESTS.: /BON DIOXIDE, W72-08584
S, ENVIRONMENTAL EFFECTS, REEFS, LABORATORY TESTS, SAMPLING, PHAEO W73-00838
RELLA, RESPIRATION, CHLOROPHYLL, LABORATORY TESTS, IRRADIATION, CH W73-01080
IVITY, POLLUTANT IDENTIFICATION, LABORATORY EQUIPMENT, AEROBIC BAC W73-01066
CTS, AQUATIC ALGAE, CHRYSOPHYTA, LABORATORY EQUIPMENT, FRESHWATER, W73-01445
ON, METHODOLOGY/ *PHYTOPLANKTON, LABORATORY EQUIPMENT, SEDIMENTATI W73-00273
ION EFFECTS, ECOLOGY, SEA WATER, LABORATORY EQUIPMENT, INHIBITION, W73-00348

FICATION, ENVIRONMENTAL EFFECTS,	LABORATORY EQUIPMENT, FREEZE DRYI	W73-00379
ABSORPTION, PESTICIDE TOXICITY,	LABCRATORY TESTS.: /TION EFFECTS,	W73-00281
UTRIENTS, CULTURES, MEASUREMENT,	LABCRATORY TESTS, BIOASSAY.: /, N	W73-00255
RES, AQUATIC ALGAE, CHLOROPHYTA,	LABORATORY TESTS, OXYGEN, WATER P	W72-14314
ITY, PLANT POPULATIONS, AQUARIA,	LABORATORY EQUIPMENT, PLANT PHYSI	W72-14294
AE, *CYANOPHYTA, *ALGAL CONTROL,	LABORATORY TESTS, BUOYANCY, EUTRO	W71-13184
OPHYTA, PLANKTON, ON-SITE TESTS,	LABCRATORY TESTS.: /MYCETES, CYAN	W71-13187
ONS, / *OXIDATION PONDS, *ALGAE,	LABORATORY TESTS, AEROBIC CONDITI	W71-13326
DIE/ *ALGAE, *CHLORELLA, *BORCN,	LABORATORY TESTS, CYTOLCGICAL STU	W71-12858
SOLUBLE FORMS, INSOLUBLE FORMS,	LABCRATORY EXPERIMENTS.: /UM 106,	W71-13233
, AQUATIC LIFE, SAMPLING, ALGAE,	LABORATORY TESTS, DETRITUS, BENTH	W71-10064
RATION, *ALGAE, *BIODEGRADATICN,	LABCRATORY TESTS, FILTRATION, AMM	W72-01881
SOIL TYPES, SOIL TEXTURE, CLAYS,	LABCRATORY TESTS, STATISTICAL MET	W72-02218
NITRATES, DENSITY, GROWTH RATES,	LABORATORY TESTS, CYANOPHYTO, ALG	W72-03217
N, AEROBIC, ALKALINITY, ACIDITY,	LABCRATORY, DIGESTION TANK, PCNDA	W72-02850
EFFECTS, PLANT GROWTH, SULFATE,	LABCRATORY TESTS, CYTOLOGICAL STU	W72-00153
YSOPHYTA, COPEPODS, CRUSTACEANS,	LABCRATORY TESTS, VIRGINIA.: /CHR	W72-00150
, *CHLORELLA, ENZYMES, PROTEINS,	LABCRATORY TESTS, AMMONIA, ASSAY,	W72-03222
IENTS, LAKE ERIE, LAKE SUPERIOR,	LABORATORY TESTS, TEMPERATURE, AL	W72-01108
EUTROPHICATION, STANDING CROPS,	LABCRATORY TESTS, CHEMICAL ANALYS	W72-09163
COSYSTEMS, GEORGIA, BIORTHYTHMS,	LABCRATORY TESTS, ALGAE, CARBON D	W72-09159
, CESIUM, RADIOACTIVITY EFFECTS,	LABORATORY TESTS.: /RADICISOTOPES	W72-09649
OGEN ION CONCENTRATION, PROTEIN,	LABORATORY TESTS, COST ANALYSIS,	W72-10390
EFFICIENCIES, LABORATORY TESTS,	LABCRATORY EQUIPMENT, AERATION, D	W72-11100
PY, ALGAE, ACTINOMYCETES, FUNGI,	LABORATORY TESTS, WATER QUALITY C	W72-11604
ELLA, PERFORMANCE, EFFICIENCIES,	LABORATORY TESTS, LABORATORY EQUI	W72-11100
CARBON.:	LABORATORY SCALE, *TOTAL CRGANIC	W71-07102
*ALGAE, *ALGAL CONTROL, TESTING,	LABCRATORY TESTS, CHEMICALS, BACT	W71-06189
ELEMENTS, PHOSPHATES, NITRATES,	LABORATORY TESTS.: /TIVITY, TRACE	W71-09674
RIA, HYDROGEN ION CONCENTRATION,	LABORATORY TESTS.: /BOLISM, BACTE	W71-08687
*MARINE PLANTS, AQUATIC ANIMALS,	LABORATORY TESTS, NITRATES, MUSSE	W71-09168
NITORING, ANALYTICAL TECHNIQUES,	LABCRATORY TESTS, CESIUM, STRCNTI	W71-08032
NSITY, CARBON DIOXIDE, NITROGEN,	LABORATORY TESTS, CYANOPHYTA, CYT	W72-05476
DEA, RNA, CHLOROPHYLL B, ASHING,	LABORATORY TECHNIQUES.: /ELLIPSOI	W72-06274
*LIPIDS, TOXICITY, HEAVY METALS,	LABORATORY TESTS, PHOTOSYNTHESIS,	W72-06037
ITROGEN, CHEMICAL OXYGEN DEMAND,	LABORATORY TESTS, ALKALINITY, CAR	W72-04431
G BASINS, WASTE WATER TREATMENT,	LABORATORY TESTS.: /ENTS, SETTLIN	W71-00116
GAE, FIELD STUDIES, GREEN ALGAE,	LABCRATORY STUDIES.: /UE-GREEN AL	W70-08319
-PURIFICATION, *ALGAE, *STREAMS,	LABORATORY TESTS, DISSOLVED OXYGE	W71-04518
, *ZINC, *ADSORPTION, *SURFACES,	LABCRATORY EQUIPMENT, *LABORATORY	W72-12210
C ALGAE, GROWTH RATES, BIOASSAY,	LABCRATORY TESTS, CALCIUM.: /UATI	W72-13809
ROGEN, PHOSPHORUS, IRON, *TEXAS,	LABORATORY TESTS, CARBON RADIOISO	W72-13692
T, *EASTERN CANADA, CAPE CHIDLEY(	LABRADOR), NEW BRUNSWICK, ANTICOS	W71-10066
QUADRICAUDA, ACETATE, PYRUVATE,	LACTATE.: SCENEDESMUS	W72-00838
TE, BUTYRATE, BUTANOL, GLYCEROL,	LACTATE, PYRUVATE, ACETATE, ETHAN	W72-05421
GEN, VITAMINS, SUSPENDED SOLIDS,	LACTOBACILLUS, CHLORELLA, SULFUR	W72-12565
, RHODOTORULA RUBRA, RHODOTORULA	LACTOSA, CANDICA ZEYLANCIDES, CAM	W72-04748
GITATA, LAMINARIA AGARDHII, ULVA	LACTUCA, RHODYMENIS PALMATA, POLY	W72-04748
INTESTINALIS, POLYSIPHONIA, ULVA	LACTUCA, ZANICHELLIA PALUSTRIS, C	W73-00367
ELLA, TRICHOCERCA, HAEMATOCOCCUS	LACUSTRIS.: /US, MONOSTYLA, LEPAD	W72-07132
ARGANIUM ERECTUM, SCHOENOPLECTUS	LACUSTRIS, ACORUS CALAMUS, TYPHA	W73-00286
ASTERIONELLA FORMOSA, RHODCMONAS	LACUSTRIS, COUNTING, ORTHCPHOSPHA	W73-01435
8-MET/ NOSTOC MUSCORUM, LYNGBYA	LAGERHAIMII, 7-METHYLHEPTADECANE,	W73-01439
COCCUS MINUTUS, DICTYOSPHAERIUM,	LAGERHEIMIA.: /STRODESMUS, CHRYSO	W72-12738
RESCENT, TOOMS LAKE, WOODS LAKE,	LAGCON OF ISLANDS, STAURASTRUM SP	W73-01094
*CONTINUOUS DISCHARGE	LAGOON.:	W72-12565
ISLAND, WINIFRED ISLAND, IZEMBEK	LAGOON, CHAETOMORPHA CANN: /OLIN	W72-09120
*FACULTATIVE LAGOONS, *ANAEROBIC	LAGOONS.:	W72-09386
TER TREATMENT, WYOMING, *AERATED	LAGOONS.: /ILOT PLANTS, *WASTE WA	W72-06838
, *AEROBIC TREATMENT, *OXIDATION	LAGOONS.: /ON, WATER PURIFICATION	W72-05315
CENEDESMUS, EFFLUENTS, OXIDATION	LAGOONS.: /ELLA, CHLAMYDOMONAS, S	W70-10405
AEROBIC TREATMENT, B/ *OXIDATION	LAGOONS, *STABILIZATION, *ALGAE,	W71-03896
S, *ENTERIC BACTERIA, *OXIDATION	LAGOONS, *PATHOGENIC BACTERIA.: /	W70-08319
OCH/ *DAIRY INDUSTRY, *OXIDATION	LAGOONS, *PILOT PLANTS, ALGAE, BI	W70-06619
ELS, *BIOASSAY, *WASTE DILUTICN,	LAGOONS, *WASTE WATER TREATMENT,	W72-04787
, *FARM WASTES, OKLA/ *OXIDATION	LAGOONS, *WATER POLLUTICN EFFECTS	W72-04773
EROBIC CONDITIONS, P/ *OXIDATION	LAGOONS, *AEROBIC TREATMENT, *ANA	W71-06033
OMPUTER SIMULATION, / *OXIDATION	LAGOONS, *MATHEMATICAL MCDELS, *C	W71-07084
RS, EUTROPHICATION, / *OXIDATION	LAGOONS, *ALGAE, *TRICKLING FILTE	W71-07097
QUALITY, ALGAE, HAR/ *OXIDATION	LAGOONS, *DESIGN CRITERIA, *WATER	W71-07100
PTH, TEMPERATURE, PH/ *OXIDATION	LAGOONS, *OXYGENATION, *ALGAE, DE	W71-07098
ON, CHLORINE, CONTAC/ *OXIDATION	LAGOONS, *DISINFECTION, DEGRADATI	W71-07096
XING, WATER RE-USE, / *OXIDATION	LAGOONS, *MATHEMATICAL MODELS, MI	W71-07099
OXYGEN, SATURATION, / *OXIDATION	LAGCONS, *OXYGENATION, DISSOLVED	W71-07106
ORGANIC LOADING, BIO/ *OXIDATION	LAGCONS, *WASTE WATER TREATMENT,	W71-07124
SIGN CRITERIA, TRICK/ *OXIDATION	LAGOONS, *TERTIARY TREATMENT, *DE	W71-07113
*WATER RE-USE, BIOCH/ *OXIDATION	LAGCONS, *WASTE WATER TREATMENT,	W71-07123
TIONS, AEROBIC CONDI/ *OXIDATION	LAGCONS, *SLUDGE, ANAERCBIC CCNDI	W71-07102
ROTIFERS, P/ *ALGAE, *OXIDATION	LAGOONS, *LIGHT INTENSITY, PONDS,	W71-07382
NALYSIS, TEMPERATURE/ *OXIDATION	LAGOONS, *CLIMATIC ZONES, *COST A	W71-07109
ADA/ *EUTROPHICATION, *OXIDATION	LAGOONS, *SEWAGE LAGOONS, BIODEGR	W71-07973
*FACULTATIVE	LAGOONS, *ANAEROBIC LAGCNS.:	W72-09386
IGN C/ *SEWAGE LAGOONS, *AERATED	LAGOONS, *OXIDATION LAGOCNS, *DES	W72-09386
NS, *AERATED LAGOONS, *OXIDATION	LAGOONS, *DESIGN CRITERIA, BIODEG	W72-09386
TION LAGOONS, *DESIGN C/ *SEWAGE	LAGCONS, *AERATED LAGOONS, *OXIDA	W72-09386
, PRODUCTIVITY/ *CARBON, *SEWAGE	LAGOONS, *LIMITING FACTORS, ALGAE	W72-09162
R REUSE, ALGAE, PHOT/ *OXIDATION	LAGOONS, *MUNICIPAL WASTES, *WATE	W72-11642
EPARATION TECHNIQUES/ *OXIDATION	LAGCONS, *HARVESTING CF ALGAE, *S	W72-10237
R, *PLASTICS, DOMEST/ *OXIDATION	LAGOONS, *CHANNELS, *HEAT TRANSFE	W72-10401
*BACTERIA, *ALGAE, / *OXIDATION	LAGCONS, *ON-SITE INVESTIGATIONS,	W72-10233

OSYNTHESIS, OXYGEN, / *OXIDATION	LAGOONS, *ALGAE, *OXIDATION, PHOT	W72-02363
, *FILTRATION, *FILT/ *OXIDATION	LAGOONS, *ALGAE, MUNICIPAL WASTES	W71-13341
*WASTE WATER TREATMENT, *SEWAGE	LAGOONS, *COLD REGIONS, AERATED L	W72-12565
LAGOONS, *COLD REGIONS, AERATED	LAGOONS, *NORTH DAKOTA, BIOCHEMIC	W72-12565
ATMENT, *MATHEMATICA/ *OXIDATION	LAGOONS, *ALGAE, *WASTE WATER TRE	W72-12289
E DISPOSAL, *LAGOONS, *OXIDATION	LAGOONS, AEROBIC CONDITIONS, ANAE	W71-07551
ATMENT, *COAGULATION, *OXIDATION	LAGOONS, AEROBIC TREATMENT, OXIDA	W71-09629
ION, *OXIDATION LAGOONS, *SEWAGE	LAGOONS, BIODEGRADATION, *NUTRIEN	W71-07973
WATER TREATMENT, *PONDS, *ALGAE,	LAGOONS, BIOCHEMICAL OXYGEN DEMAN	W70-07838
SHERIES, SEWAGE DISPOSAL, SEWAGE	LAGOONS, BRACKISH WATER FISH, BIO	W72-12567
*AQUATIC PRODUCTIVITY, *MIDGES,	LAGOONS, BIOMASS, OREGON, SLUDGE,	W72-03556
*ALGAE, *PHYTOPLANKTON, *SEWAGE	LAGOONS, BIODEGRADATION, SEWAGE T	W72-07661
AL DISTRIBUTION, SURFACE WATERS,	LAGOONS, CHLOROPHYTA, RHODOPHYTA,	W73-00838
ATION, *DISINFECTION, *OXIDATION	LAGOONS, COLIFORMS, ALGAE, TEMPER	W72-00422
ABOLISM, GROWTH RATES, OXIDATION	LAGOONS, COLIFORMS, NUTRIENTS, AN	W72-09590
*TEXAS, BIODEGRADATION, TRACERS,	LAGOONS, CYANOPHYTA.: /BACTERIA,	W71-04873
YLL, *CLASSIFICATION, *OXIDATION	LAGOONS, CHLORELLA, OXYGEN DEMAND	W71-00139
OSAL, OXIDATION DITCHES, AERATED	LAGOONS, COMPOSTING, REFEEDING, A	W71-07551
TION, ALGAE, PROTEIN, *OXIDATION	LAGOONS, CLIMATIC ZONES, PRECIPIT	W71-08951
STOCK, WASTE DISPOSAL, OXIDATION	LAGOONS, DEHYDRATION, RUNOFF, RAD	W72-00846
A, BIOASSAY, SAMPLING, OXIDATION	LAGOONS, EUGLENA, CHLAMYDOMONAS,	W72-07661
ICATION, SEPTIC TANKS, OXIDATION	LAGOONS, LAND APPLICAT: /DENITRIF	W70-09907
FARM WASTE, *POULTRY, *OXIDATION	LAGOONS, NUTRIENTS, AQUATIC ALGAE	W72-02850
LISM, *SEWAGE TREATMENT, *SEWAGE	LAGOONS, OHIO, OXYGEN, DISSOLVED	W72-05459
ATION, *COLD REGIONS, *OXIDATION	LAGOONS, ORGANIC LOADING, AERATIO	W72-06838
*FARM WASTES, POULTRY, NITROGEN,	LAGOONS, PUMPING, AEROBIC TREATME	W72-08396
REAMS, DELAWARE RIVER, OXIDATION	LAGOONS, PONDS, FARM PONDS, MARYL	W72-04736
BIOLOGICAL TREATMENT, OXIDATION	LAGOONS, PHYTOPLANKTON, AQUATIC I	W72-12567
ITY, DESIGN CRITERIA, *OXIDATION	LAGOONS, SEWAGE TREATMENT, ALGAL:	W72-04789
*AERATED	LAGOONS, SUSPENDED SOLIDS.:	W71-07106
*AERATED	LAGOONS, SUSPENDED SOLIDS.:	W71-07113
*AERATED	LAGOONS, SURFACE AERATORS.:	W71-07109
, WATER PURIFICATION, *OXIDATION	LAGOONS, STABILIZATION, INDUSTRIA	W71-13356
VATED SLUDGE, TRICKLING FILTERS,	LAGOONS, TERTIARY TREATMENT, ACTI	W71-09524
WESTHAMPTON	LAKE (VA).:	W72-00150
ELK, LAKE MARY, LAKE JOSEPHINE,	LAKE ARCO, LAKE DEMING, LOWER LA	W72-07145
IFICATION.:	LAKE BLOOMINGTON(ILL), CHEMOSTRAT	W71-03028
LAKE PONTCHARTRAIN, PEARL RIVER,	LAKE BORGNE, SUBSTRATES, MACROINV	W73-02029
EORGIA), LAKE CARDINAL(GEORGIA),	LAKE BUCKHORN(GEORGIA).: /CKSON(G	W71-09173
EORGIA), LAKE CARDINAL(GEORGIA),	LAKE BUCKHORN(GEORGIA), HYPNODINI	W71-11914
LLE LAKE, LAKE LONG, LAKE SQUAW,	LAKE BUDD, OSCILLATORIA REDEKEI,	W72-07145
GEORGIA), LAKE JACKSON(GEORGIA),	LAKE CARDINAL(GEORGIA), LAKE BUCK	W71-11914
GEORGIA), LAKE JACKSON(GEORGIA),	LAKE CARDINAL(GEORGIA), LAKE BUCK	W71-09173
AL STUDIES.:	LAKE CAYUGA, CLADOPHORA, ECOLOGIC	W70-09614
SCENS, SYNEDRA ACUS, RHODOMONAS,	LAKE CONSTANCE, SEWAGE CANALS.: /	W71-00117
CTUM, ARTHURS LAKE, LAKE SORELL,	LAKE CRESCENT, TOOMS LAKE, WOODS	W73-01094
OTA), BIG STONE LAKE(MINNESOTA),	LAKE CRYSTAL(MINNESOTA).: /MINNES	W71-04216
MARY, LAKE JOSEPHINE, LAKE ARCO,	LAKE DEMING, LOWER LA SALLE LAKE,	W72-07145
RDHII VAR ISOTHRIX, LAKE ITASCA,	LAKE ELK, LAKE MARY, LAKE JOSEPHI	W72-07145
	LAKE ERIE CENTRAL BASIN.:	W72-12994
ENATION, *NUTRIENT REGENERATION,	LAKE ERIE CENTRAL BASIN.: /DEOXYG	W72-12996
ARNOLDIELLA-CONCHOPHILA,	LAKE ERIE.:	W71-12489
RECREATION, PHOSPHORUS, CARBON,	LAKE ERIE, DETERGENTS, LAKE ONTAR	W71-12091
*LAKE ONTARIO, *EUTROPHICATION,	LAKE ERIE, ALGAE, NITROGEN, PHOSP	W71-11009
RADIOACTIVITY, ADSORPTION, FISH,	LAKE ERIE, SEDIMENTS, SILTS, PLAN	W71-11036
OSPHORUS, ALGAE, ORGANIC MATTER,	LAKE ERIE, EFFLUENTS, SEWAGE, DET	W71-11017
SEWAGE EFFLUENTS, AQUATIC ALGAE,	LAKE ERIE, LAKE ONTARIO, NUTRIENT	W71-11507
S, ADSORPTION, LAKES, NUTRIENTS,	LAKE ERIE, LAKE SUPERIOR, LABORAT	W72-01108
ION, NUTRIENTS, ALGAE, BACTERIA,	LAKE ERIE, ALUMINUM, MANGANESE, S	W72-01101
ICAL COMMUNITIES, LAKE MICHIGAN,	LAKE ERIE, LAKE HURON, LAKE SUPER	W72-10875
OPLANKTON, CARBON RADIOISOTOPES,	LAKE ERIE, LAKE ONTARIO, ADSORPTI	W72-13657
RS, PHOTOSYNTHESIS, GREAT LAKES,	LAKE ERIE, LAKE ONTARIO, HYDRO: /	W72-07648
RIOR, LAKE MICHIGAN, LAKE HURON,	LAKE ERIE, GREAT LAKES, SOYBEANS,	W73-01456
UTRIENTS, DETERGENTS, WISCONSIN,	LAKE ERIE, LAKE ONTARIO, ST LAWRE	W70-07283
A MILLE/ *ALGAL SPECIES, WESTERN	LAKE ERIE, ARNOLDIELLA CONCHOPHIL	W71-09156
TURE, LIGHT PENETRATION, CARBON,	LAKE ERIE, RUNOFF, EROSION, CHLOR	W72-04734
ANOPHYTA, WASHINGTON, DIVERSION,	LAKE ERIE, NITROGEN FIXATION, DET	W72-05473
AGES, *APHANIZOMENON FLOS-AQUAE,	LAKE ERKEN(SWEDEN).: *CYANOPH	W73-00246
LAKE, CRYPTOPHYCEAE, *DENMARK,/	LAKE ESROM, TYSTRUP LAKE, BAVELSE	W72-12738
RHOICOSPHENIA, ACHANTHES, FRAG/	LAKE FURESO (DENMARK), COCCONEIS,	W72-07501
NUTRIENTS, CANADARAGO LAKE(NY),	LAKE GEORGE(NY).: *LIMITING	W72-09163
LO), SHADOW MOUNTAIN LAKE(COLO),	LAKE GRANBY(COLO).: /RAND LAKE(CO	W72-04791
ONS, *HETEROTROPHIC POPULATIONS,	LAKE HARTWELL(SOUTH CAROLINA), AN	W71-00475
AKE(S. DAK).:	LAKE HERMAN(S. DAK), ENEMY SWIM L	W71-02880
, EAST TWIN LAKE, LAKE ROCKWELL,	LAKE HODGSON, CLEARWATER QUARRY,	W72-13673
IT/ *EUTROPHICATION, *NUTRIENTS,	LAKE HURON, EFFLUENTS, WATER QUAL	W72-13662
ITIES, LAKE MICHIGAN, LAKE ERIE,	LAKE HURON, LAKE SUPERIOR, WATER	W72-10875
N, LAKE SUPERIOR, LAKE MICHIGAN,	LAKE HURON, LAKE ERIE, GREAT LAKE	W73-01456
ILLATORIA AGARDHII VAR ISOTHRIX,	LAKE ITASCA, LAKE ELK, LAKE MARY,	W72-07145
N/ *LAKE SIDNEY LANIER(GEORGIA),	LAKE JACKSON(GEORGIA), LAKE CARDI	W71-11914
UM, LAKE SIDNEY LANIER(GEORGIA),	LAKE JACKSON(GEORGIA), LAKE CARDI	W71-09173
	LAKE JACOMO(MO).:	W72-10431
AKE ITASCA, LAKE ELK, LAKE MARY,	LAKE JOSEPHINE, LAKE ARCO, LAKE D	W72-07145
MENDOTA(WIS), LAKE MONONA(WIS),	LAKE KEGONSA(WIS), LAKE WASHINGTO	W72-05473
OCCUS, CHLOROCOCCUM, HYALOTHECA,	LAKE LIVINGSTON, AUTORADIOGRAPHY,	W72-09239
AKE DEMING, LOWER LA SALLE LAKE,	LAKE LONG, LAKE SQUAW, LAKE BUDD,	W72-07145
ORY STUDIES, PRECISION, *SWEDEN,	LAKE MALAREN, MICROCYSTIS FLOS-AQ	W73-00286
ENDACE, SALMON TROUT, LOTA LOTA,	LAKE MALOGEN, SWEDEN, CHAR, LUCIO	W72-12729
ISOTHRIX, LAKE ITASCA, LAKE ELK,	LAKE MARY, LAKE JOSEPHINE, LAKE A	W72-07145
UNITY ACTION, LAKE SALLIE(MINN),	LAKE MELISSA(MINN), NUTRIENT REMO	W71-03012

UAN RESERVOIR(VA), MADISON(WIS),	LAKE MENDOTA(WIS), LAKE MONONA(WI	W72-05473
RK), UPPER KLAMATH LAKE(OREGON),	LAKE MENDOTA(WISCONSIN).: /NEW YO	W71-05626
NN), TROUT LAKE(WIS), SEDIMENT /	LAKE MENDOTA(WIS), BANTAM LAKE(CO	W72-08459
*LIMITING NUTRIENTS,	LAKE MENDOTA(WIS).:	W72-07933
IS.), LAKE WINGRA(WIS.), LIMITI/	LAKE MENDOTA(WIS.), LAKE MONONA(W	W73-00232
REDUCTION, RHIZOBIUM JAPONICUM,	LAKE MENDOTA, CRYSTAL LAKE, TROUT	W73-01456
CE.: *MADISON(WIS),	LAKE MENDOTA(WIS), LAWN MAINTENAN	W73-02479
NT SOURCES, ACETYLENE REDUCTION,	LAKE MENDOTA(WIS), LAKE MONONA(WI	W72-10608
ULATE F/ *AGRICULTURAL DRAINAGE,	LAKE METABOLISM, MOBILITY, PARTIC	W71-06443
AMPLING, BIOLOGICAL COMMUNITIES,	LAKE MICHIGAN, LAKE ERIE, LAKE HU	W72-10875
EAT LAKES, NUTRIENTS, SEDIMENTS,	LAKE MICHIGAN, WATER POLLUTION SO	W71-12072
DES, TOXICITY, DISSOLVED OXYGEN,	LAKE MICHIGAN, ALGAE.: /, PESTICI	W71-12092
ICATION, GROWTH RATES, SAMPLING,	LAKE MICHIGAN, MEASUREMENT, SUSPE	W71-11033
, EUTROPHICATION, LAKE SUPERIOR,	LAKE MICHIGAN, LAKE HURON, LAKE E	W73-01456
VITAMIN B, NITROGEN, CYANOPHYTA,	LAKE MICHIGAN, NUTRIENTS.: /BON,	W72-07937
CTICUT), LAKE ZOAR(CONNECTICUT),	LAKE MINNETONKA(MINNESOTA), BIG S	W71-04216
MADISON(WIS), LAKE MENDOTA(WIS),	LAKE MONONA(WIS), LAKE KEGONSA(WI	W72-05473
S.), LIMITI/ LAKE MENDOTA(WIS.),	LAKE MONONA(WIS.), LAKE WINGRA(WI	W73-00232
NE REDUCTION, LAKE MENDOTA(WIS),	LAKE MONONA(WIS).: /RCES, ACETYLE	W72-10608
UTION SOURCES, DISSOLVED OXYGEN,	LAKE MORPHOLOGY, AQUATIC ALGAE, M	W71-11949
PHYTA, DISSOLVED OXYGEN, CARBON,	LAKE MORPHOMETRY, CHEMICAL PROPER	W72-08048
OMETER, HYPOLIMNION, EPILIMNION,	LAKE MORPHOMETRY, DISTRIBUTION PA	W72-07145
OGICAL COMMUNITIES, FOOD CHAINS,	LAKE MORPHOLOGY, RESERVOIR SILTIN	W73-00938
LGAE, CYANOPHYTA, PHYTOPLANKTON,	LAKE MORPHOMETRY, GEOLOGY, LIMEST	W72-04269
TS, *SEDIMENTS, *WATER ANALYSIS,	LAKE MORPHOLOGY, SEDIMENTATION, P	W71-05626
DROLOGIC DATA, DATA COLLECTIONS,	LAKE MORPHOLOGY, ECOLOGY, BIOLOGI	W72-13851
OSPHORUS REMOVAL, RATE OF SORPT/	LAKE MUD, PHOSPHORUS SORPTION, PH	W72-11617
SAY PROCEDURES, ALGAL GROWTH, IN-	LAKE NUTRIENT CONTROL.: /ALGAL AS	W71-12072
, CARBON, LAKE ERIE, DETERGENTS,	LAKE ONTARIO, TERTIARY TREATMENT,	W71-12091
UENTS, AQUATIC ALGAE, LAKE ERIE,	LAKE ONTARIO, NUTRIENTS, NITROGEN	W71-11507
CARBON RADIOISOTOPES, LAKE ERIE,	LAKE ONTARIO, ADSORPTION, AGRICUL	W72-13657
PHICATION, SURFACTANTS, ENZYMES,	LAKE ONTARIO, SEASONAL, ALGAE, CU	W71-00116
ETERGENTS, WISCONSIN, LAKE ERIE,	LAKE ONTARIO, ST LAWRENCE RIVER,	W70-07283
S, BENTHIC FAUNA, BENTHIC FLORA,	LAKE ONTARIO, HUDSON RIVER, STAND	W70-09614
PROTOZOA, OHIO, EUTROPHICATION,	LAKE ONTARIO, LIGHT PENETRATION,	W73-01615
IOLOGICAL COMMUNITIES, *STREAMS,	LAKE ONTARIO, PLANKTON, DIATOMS,	W73-01627
NTHESIS, GREAT LAKES, LAKE ERIE,	LAKE ONTARIO, HYDRO: /RS, PHOTOSY	W72-07648
SCOLOPLOS ARMIGER, ALIMITOS BAY,	LAKE PONTCHARTRAIN, PEARL RIVER,	W73-02029
RS, GRANGE LAKE, EAST TWIN LAKE,	LAKE ROCKWELL, LAKE HODGSON, CLEA	W72-13673
T LAKES(MINN), COMMUNITY ACTION,	LAKE SALLIE(MINN), LAKE MELISSA(M	W71-03012
TIC ZONE, DIFFUSION COEFFICIENT,	LAKE SAMMAMISH(WASH).: /NG, EUPHO	W71-03034
ILL/ FECAL MATERIAL, LAKE TAHOE,	LAKE SEBASTICOOK(MAINE), FAYETTEV	W71-05626
WHITE ALGAL BLOOMS, HYPNODINIUM,	LAKE SIDNEY LANIER(GEORGIA), LAKE	W71-09173
ES VAR CONTRACTUM, ARTHURS LAKE,	LAKE SORELL, LAKE CRESCENT, TOOMS	W73-01094
LOWER LA SALLE LAKE, LAKE LONG,	LAKE SQUAW, LAKE BUDD, OSCILLATOR	W72-07145
MAUMEE RIVER, ST. CLAIR RIVER,	LAKE ST. CLAIR, DETROIT RIVER, BL	W72-14282
ALGAL BLOOMS, AGING(BIOLOGICAL),	LAKE STAGES, MESOTROPHY, OLIGOTRO	W72-14782
VITY, BIOCHEMICAL OXYGEN DEMAND,	LAKE STAGES, NUTRIENTS, MESTROPHY	W73-00002
TROL, BIOCHEMICAL OXYGEN DEMAND,	LAKE STAGES, LIMNOLOGY, NUISANCE	W72-14692
IN, OLIGOTROPHY, EUTROPHICATION,	LAKE SUPERIOR, LAKE MICHIGAN, LAK	W73-01456
SOIL EROSION, NUTRIENTS, ALGAE,	LAKE SUPERIOR, WATER POLLUTION SO	W71-04216
MICHIGAN, LAKE ERIE, LAKE HURON,	LAKE SUPERIOR, WATER TEMPERATURE,	W72-10875
ON, LAKES, NUTRIENTS, LAKE ERIE,	LAKE SUPERIOR, LABORATORY TESTS,	W72-01108
E), FAYETTEVILL/ FECAL MATERIAL,	LAKE TAHOE, LAKE SEBASTICOOK(MAIN	W71-05626
*CONTEMPORANEOUS DISEQUILIBRIUM,	LAKE TAHOE(CALIF), CASTLE LAKE(CA	W71-09171
ACID, PYRIDOXINE, INDAL COMPOU/	LAKE TAHOE(CALIF), ZINC, ASCORBIC	W72-07514
S RIVER, BLACK RIVER(MINNESOTA),	LAKE WASHINGTON(WASHINGTON), LINS	W71-04216
	LAKE WASHINGTON(WASHINGTON).:	W71-02681
MONONA(WIS), LAKE KEGONSA(WIS),	LAKE WASHINGTON(WASH), GREEN LAKE	W72-05473
M SILICATE, SODIUM METASILICATE,	LAKE WASHINGTON, NITRILOTRIACETAT	W72-04734
Y RESERVOIRS(NEB), MADISON(WIS),	LAKE WASHINGTON(WASH), DYEING WAT	W72-04759
MERCURY, PHOSPHORUS REMOVAL,	LAKE WASHINGTON(WASH).:	W71-12091
NHARDTII, MOSES LAKE, PINE LAKE,	LAKE WASHINGTON, CHLOROPHYLL A, *	W72-12637
VOI/ *NITROGEN:PHOSPHORUS RATIO,	LAKE WAUBESA(WIS), OCCOQUAN RESER	W72-05473
NGDOM, ESTHWAITE WATER(ENGLAND),	LAKE WINDERMERE(ENGLAND), GREAT O	W71-13185
RY UPTAKE, LOCH LEVEN(SCOTLAND),	LAKE WINDERMERE(ENGLAND).: / LUXU	W71-13183
ME PR/ ESTHWAITE WATER(ENGLAND),	LAKE WINDERMERE(ENGLAND), EXOENZY	W72-03224
ENDOTA(WIS.), LAKE MONONA(WIS.),	LAKE WINGRA(WIS.), LIMITING NUTRI	W73-00232
TON), LINSLEY POND(CONNECTICUT),	LAKE ZOAR(CONNECTICUT), LAKE MINN	W71-04216
DS, SAMPLE PREPARATION, MOUNTAIN	LAKE.: /, CHLOROPHYLL A, CARTENOI	W73-00916
RESERVOIR, MONTROSE LAKE, THOMAS	LAKE.: /NG RESERVOIR, ARTIFICIAL	W70-09596
IUM, SHADFLY, CHIRONOMIDS, CLEAR	LAKE(CALIF), RICHELIEU RIVER(QUEB	W70-08098
RAMENTO RIVER(CALIF), SEARSVILLE	LAKE(CALIF), SAN JOAQUIN RIVER(CA	W70-10405
BRIUM, LAKE TAHOE(CALIF), CASTLE	LAKE(CALIF), DIVERSITY, PATCHES.:	W71-09171
RAND LAKE(COLO), SHADOW MOUNTAIN	LAKE(COLO), LAKE GRANBY(COLO).: /	W72-04791
COLO), LAKE GRANBY(COLO)./ GRAND	LAKE(COLO), SHADOW MOUNTAIN LAKE(	W72-04791
MENT / LAKE MENDOTA(WIS), BANTAM	LAKE(CONN), TROUT LAKE(WIS), SEDI	W72-08459
UM, CRYPTOMONAS EROSA, P/ *VOLTA	LAKE(GHANA), ACTINASTRUM GRACILIM	W73-00256
LAWRENCE LAKE(MICH), WINTERGREEN	LAKE(MICH), CARBONATE PARTICLES,	W72-08048
NITROGEN, NETPLANKTON, LAWRENCE	LAKE(MICH), WINTERGREEN LAKE(MICH	W72-08048
TE-EPIPHYTE METABOLISM, LAWRENCE	LAKE(MICH), SCIRPUS ACUTUS, NAJAS	W72-03216
*LIMITING NUTRIENTS, SHAGAWA	LAKE(MINN), WALDO LAKE(ORE).:	W72-09165
MINNETONKA(MINNESOTA), BIG STONE	LAKE(MINNESOTA), LAKE CRYSTAL(MIN	W71-04216
NE RIVER, YELLOWTAIL D/ *BIGHORN	LAKE(MONT), BIGHORN RIVER, SHOSHO	W73-01435
OSMINA LONGIROSTRIS/ *WINNISQUAM	LAKE(N H), ANABAENA CIRCINALIS, B	W71-03014
NA SPP, MICROCYSTIS AERU/ *KEZAR	LAKE(N.H.), APHANIZOMENON, ANABAE	W72-07890
ICOOK(MAINE), FAYETTEVILLE GREEN	LAKE(NEW YORK), UPPER KLAMATH LAK	W71-05626
), GREEN RIVER(KENTUCKY), CAYUGA	LAKE(NEW YORK), PATUXENT ESTUARY,	W71-11517
*ARTIFICIAL CIRCULATION, *KEZAR	LAKE(NH).:	W72-09061
*KEZAR	LAKE(NH).:	W72-14542

*ONEIDA	LAKE(NY).:	W72-13851
*LIMITING NUTRIENTS, CANADARAGO	LAKE(NY), LAKE GEORGE(NY).:	W72-09163
ON FLCS-AQUAE, CERATIUM,/ DOLLAR	LAKE(OHIO), SPHAGNUM, APHANIZOMEN	W71-11949
MISSICN SPECTROPHOTOMETRY, PERCH	LAKE(CNTARIO), ASHING, BICLCGICAL	W71-11529
IENTS, SHAGAWA LAKE(MINN), WALDO	LAKE(ORE).: *LIMITING NUTR	W72-09165
UPPER KLAMATH	LAKE(OREGON).:	W72-08989
EN LAKE(NEW YORK), UPPER KLAMATH	LAKE(OREGON), LAKE MENDCTA(WISCON	W71-05626
OGET/ *MACROPHYTES, *MIKOLAJSKIE	LAKE(POLAND), COLONIZATICN, POTAM	W71-11012
OTRICHIA ECHINULATA, MIKOLAJSKIE	LAKE(POLAND), CERATIUM HIRUNDINEL	W72-14812
LAKE HERMAN(S. DAK), ENEMY SWIM	LAKE(S. DAK).:	W71-02880
ROSTROPHA, PROCLADIUS, WHITE OAK	LAKE(TENN), UPTAKE.: / PHYSA HETE	W70-06668
S), LAKE WASHINGTON(WASH), GREEN	LAKE(WASH).: /S), LAKE KEGONSA(WI	W72-05473
MITING NUTRIENTS, ALGAL / *MOSES	LAKE(WASH), NUTRIENT DILUTION, LI	W72-08049
*MOSES	LAKE(WASH).:	W72-08560
IMONY, MACRONUTRIENTS/ *WILLIAMS	LAKE(WASH), URANIUM, THORIUM, ANT	W73-01434
LAND(CRE), WATER QUA/ *VANCOUVER	LAKE(WASH), VANCOUVER(WASH), PORT	W72-03215
: *MOSES	LAKE(WASH), HYPEREUTROPHIC LAKES.	W72-00799
HYPEREUTROPHIC LAKES, *MOSES	LAKE(WASH).:	W72-00798
A(WIS), BANTAM LAKE(CONN), TROUT	LAKE(WIS), SEDIMENT COMPACTION.: /	W72-08459
IR.: *KEZAR	LAKE, AIR INJECTION, COMPRESSED A	W72-09304
ECCHI DISC, NEWFOUND LAKE, SQUAM	LAKE, APHANIZCMENON FLOW-AQUAE, D	W72-07890
, *DENMARK,/ LAKE ESROM, TYSTRUP	LAKE, BAVELSE LAKE, CRYPTCPHYCEAE	W72-12738
AKE MENDOTA, CRYSTAL LAKE, TROUT	LAKE, BIG ARBCR VITAE LAKE, LITTL	W73-01456
DGSON, CLEARWATER QUARRY, DOLLAR	LAKE, CHLOROPHYLL A, ABSCRBANCE.:	W72-13673
AKE ESROM, TYSTRUP LAKE, BAVELSE	LAKE, CRYPTOPHYCEAE, *DENMARK, CY	W72-12738
LTERS, *MEMBRANE FILTERS, GRANGE	LAKE, EAST TWIN LAKE, LAKE ROCKWE	W72-13673
SR-90, K-40, CS-137, *PAHRANAGAT	LAKE, GAMMA RAY SPECTROMETRY, ANI	W73-01673
R VITAE LAKE, LITTLE ARBOR VITAE	LAKE, GREEN BAY, FOX RIVER, DETRO	W73-01456
LAKE CRESCENT, TOOMS LAKE, WOODS	LAKE, LAGCON OF ISLANDS, STAURAST	W73-01094
LONGIPES VAR CONTRACTUM, ARTHURS	LAKE, LAKE SORELL, LAKE CRESCENT,	W73-01094
RCO, LAKE DEMING, LOWER LA SALLE	LAKE, LAKE LCNG, LAKE SQUAW, LAKE	W72-07145
FILTERS, GRANGE LAKE, EAST TWIN	LAKE, LAKE ROCKWELL, LAKE HODGSON	W72-13673
AS REINHARDTII, MOSES LAKE, PINE	LAKE, LAKE WASHINGTON, CHLOROPHYL	W72-12637
AKE, TROUT LAKE, BIG ARBOR VITAE	LAKE, LITTLE ARBOR VITAE LAKE, GR	W73-01456
ANDERSON-CUE	LAKE, MELROSE(FLORIDA).:	W72-08986
OSMARIUM, PHACUS, CHA/ *ONONDAGA	LAKE, OEDOGONIUM, RHIZOCLONIUM, C	W72-09111
, ALGAE BLOOMS, UTAH, GREAT SALT	LAKE, ORGANIC COMPOUNDS, BRINES.:	W71-12223
CHLAMYDOMONAS REINHARDTII, MOSES	LAKE, PINE LAKE, LAKE WASHINGTON,	W72-12637
HOSPHATES, SECCHI DISC, NEWFOUND	LAKE, SQUAM LAKE, APHANIZCMENON F	W72-07890
UPPER KLAMATH	LAKE, SUBSTRATE UTILIZATION.:	W73-01269
, ARTIFICIAL RESERVOIR, MONTROSE	LAKE, THOMAS LAKE.: /NG RESERVOIR	W70-09596
JAPONICUM, LAKE MENDOTA, CRYSTAL	LAKE, TROUT LAKE, BIG ARBOR VITAE	W73-01456
AGELLATES, RHODOMONAS, IKROAVICH	LAKE, UTERMOHL METHOD.: /TICN, FL	W73-00273
AKE SORELL, LAKE CRESCENT, TOOMS	LAKE, WOODS LAKE, LAGOON OF ISLAN	W73-01094
, EPILITHIC ALGAE/ *EXPERIMENTAL	LAKES AREA, *NORTHWESTERN ONTARIO	W71-11029
CARBON, WATER TEMPERATURE, GREAT	LAKES REGION, WISCONSIN, MICHIGAN	W73-00384
MOSES LAKE(WASH), HYPEREUTROPHIC	LAKES.: *	W72-00799
*PHOSPHORUS CIRCULATION, SMALL	LAKES.:	W72-12543
DUCTIVITY, NITROGEN, PHOSPHORUS,	LAKES.: /ISTRIBUTION, AQUATIC PRO	W71-11519
IENTS, ON-SITE DATA COLLECTIONS,	LAKES.: /OSPHORUS, NITRCGEN, NUTR	W72-12990
TS, NUTRIENTS, SEWAGE TREATMENT,	LAKES.: /REATMENT, ALGAE, EFFLUEN	W70-09325
KE SALLIE(MINN), LAKE M/ DETROIT	LAKES(MINN), COMMUNITY ACTION, LA	W71-03012
PATTERNS, *SYSTEMATICS, *SALINE	LAKES, *EUTROPHICATION, WATER SAM	W72-09111
GAE, *LAKE SUPERIOR, *LA/ *GREAT	LAKES, *LAKES, EUTROPHICATION, AL	W72-01094
*ORTHOPHOSPHATES.: PRAIRIE	LAKES, *LAKE HERMAN(SO. DAKOTA),	W73-00938
HYPEREUTROPHIC	LAKES, *MOSES LAKE(WASH).:	W72-00798
S, THERMAL POLLU/ *ALGAE, *GREAT	LAKES, *MONITORING, *BIOINDICATOR	W72-10875
LUTION EFFECTS, *EUTROPHICATICN,	LAKES, *NUTRIENTS, ALGAL CONTROL,	W72-00798
TROPHICATIO/ *PHOSPHATES, *GREAT	LAKES, *SEDIMENTS, *NUTRIENTS, EU	W72-04263
*PHYTOPLANKTON, *NANNOPLANKTCN,	LAKES, *SEASONAL, *PRIMARY PRODUC	W72-12738
E MORPHOLOGY, RESERVOIR SILTING,	LAKES, *SOUTH DAKOTA, PLANKTCN, *	W73-00938
*FALLCUT, *FOOD CHAINS, / *GREAT	LAKES, *STRONTIUM RADIOISOTOPES,	W73-00790
*POLITICAL ASPECTS, LAKES, GREAT	LAKES, ALGAL BLOOMS, AGING(BIOLOG	W72-14782
EMATICS, *DISTRIBUTION PATTERNS,	LAKES, AQUATIC PLANTS, AUSTRALIA,	W73-01094
ITIES, *INFLUENT STREAMS, *GREAT	LAKES, AQUATIC ALGAE, ICE, FISH,	W73-01669
S, *PHOSPHORUS, *EUTROPHICATION,	LAKES, AQUATIC ALGAE, BACTERIA, B	W72-04292
E, *LAKE ERIE, AQUATIC HABITATS,	LAKES, AQUATIC ENVIRONMENT.: /LGA	W71-09156
IMENTS, *ALGAE, *EUTROPHICATICN,	LAKES, AQUATIC PLANTS, IRCN, MICR	W72-03742
ION, AQUATIC ENVIRONMENTS, GREAT	LAKES, BAYS, WATERSHEDS(BASINS),	W71-08953
EATMENT, WATER QUALITY, STREAMS,	LAKES, BIOCHEMICAL OXYGEN DEMAND,	W72-14469
ETRITUS, METHODOLOGY, SEDIMENTS,	LAKES, CHLOROPHYLL, HYDROGEN ION	W71-12855
ORS, NITROGEN, PHOSPHORUS, RAIN,	LAKES, CLADOPHORA, ANALYSIS, PHYT	W73-00232
ACINTH, NUTRIENTS, MIDGES, FISH,	LAKES, CONCUCTIVITY, ALGAE, STAND	W71-01488
AE, *PHOSPHORUS, *MODEL STUDIES,	LAKES, CYCLING NUTRIENTS, TRACERS	W72-12543
IGHT PENETRATION, BENTHIC FLCRA,	LAKES, CEPTH, DIATOMS, CHLOROPHYT	W71-11029
, *ALGAE, *POPULATION, DYNAMICS,	LAKES, DISSOLVED OXYGEN, NUTRIENT	W71-03034
IOSIS, CARBON DIOXIDE, NITROGEN,	LAKES, ESTUARIES, NUTRIENTS, DETE	W70-07283
LYSIS, *CHEMICAL ANALYSIS, GREAT	LAKES, EUTROPHICATION, NUTRIENTS,	W72-01101
IC FLCRA, MICROORGANISMS, ALGAE,	LAKES, EUTROPHICATION, LITTORAL,	W72-07501
ER POLLUTION, NUTRIENTS, RIVERS,	LAKES, FARM PONDS.: /ROPHYTA, WAT	W71-05991
ETIC BACTERIA, AQUATIC BACTERIA,	LAKES, FILTERS, COLIFORMS, METABO	W72-04743
H DIETS, FRESHWATER FISH, PONDS,	LAKES, FOOD HABITS, EUTROPHICATIO	W72-07132
CATION, *OLIGOTROPHY, MINNESOTA,	LAKES, GLACIAL DRIFT, NUTRIENTS,	W71-13149
ICAL WASTES, *POLITICAL ASPECTS,	LAKES, GREAT LAKES, ALGAL BLOOMS,	W72-14782
VITY, PRIMARY PRODUCTIVITY, ICED	LAKES, IONS, SULFATES, CARBONATES	W71-02880
G FACTORS, PHOTOSYNTHESIS, GREAT	LAKES, LAKE ERIE, LAKE CNTARIC, H	W72-07648
TENOIDS, XANTHOPHYLLS, MINNESOTA	LAKES, LUTEIN.: CARO	W72-01784
L TECHNIQUES, PHOSPHORUS, ALGAE,	LAKES, MICROORGANISMS, TERTIARY T	W71-03021
ION, *WATER QUALITY, CYANOPHYTA,	LAKES, MISSOURI, NITROGEN, PHOSPH	W72-10431
PERATURE, *PHOTOPERIODISM, GREAT	LAKES, MODEL STUDIES, NUTRIENTS,	W72-13651

NUTRIENTS, ESSENTIAL NUTRIENTS,	LAKES, MUD, WISCONSIN, AQUATIC WE	W72-10608
SIS, PRIMARY PRODUCTIVITY, GREAT	LAKES, NEW YORK, WATER SUPPLY, SY	W73-01615
GROWTH RATES, DIATOMS, BACTERIA,	LAKES, NITROGEN, ANALYTICAL TECHN	W72-09163
RATORY TESTS, *BIOASSAY, OREGON,	LAKES, NITROGEN, PHOSPHORUS, CARB	W72-09164
*NUTRIENTS, *NITROGEN FIXATION,	LAKES, NITROGEN, PHOSPHORUS, IRON	W72-13692
NT-WATER INTERFACES, ADSORPTION,	LAKES, NUTRIENTS, LAKE ERIE, LAKE	W72-01108
SELF-PURIFICATION, ALGAE, GREAT	LAKES, NUTRIENTS, SEDIMENTS, LAKE	W71-12072
ANTS, ALGAE, BACTERIA, LITTORAL,	LAKES, PERIPHYTON, ORGANIC MATTER	W72-03216
N-SITE TESTS, *NUTRIENTS, ALGAE,	LAKES, PHOSPHORUS, CARBON, NITROG	W72-09165
LET RADIATION, DISSOLVED OXYGEN,	LAKES, PHYTOPLANKTON, NUTRIENTS.:	W72-01780
, CHROMIUM, COBALT, FRESH WATER,	LAKES, PHYSICAL PROPERTIES, CHEMI	W73-01434
TECHNIQUES, *ALGAE, *NUTRIENTS,	LAKES, PLANT GROWTH, *MOLYBDENUM,	W72-00801
N TECHNIQUES, SPECTROPHOTOMETRY,	LAKES, PLANT PIGMENTS, DETRITUS,	W72-13673
, FOOD CHAINS, MARINE FISHERIES,	LAKES, POLLUTION ABATEMENT.: /GAE	W72-00960
ES, NUTRIENTS, PLANT PHYSIOLOGY,	LAKES, PONDS, AQUATIC ALGAE, ECOL	W71-10580
EEDS, *WEEDS, *RED TIDE, COASTS,	LAKES, PONDS, WATER POLLUTION CON	W72-14673
LES, SEASONAL, DENSITY, DIATOMS,	LAKES, PROTOZOA, ROTIFERS, CRUSTA	W71-11012
PESTICIDE TOXICITY, RESISTANCE,	LAKES, RESERVOIRS, SWIMMING POOLS	W71-06189
ICATION, CURRENTS(WATER), ALGAE,	LAKES, RIVERS, SEWAGE DISPOSAL, S	W70-09947
ING CROP, *PRIMARY PRODUCTIVITY,	LAKES, SAMPLING, RESERVOIRS, TRIB	W73-01435
L, *CYANOPHYTA, *WATER ANALYSIS,	LAKES, SAMPLING, DIATOMS, TURBIDI	W72-07890
, OXIDATION-REDUCTION POTENTIAL,	LAKES, SOILS, EUTROPHICATION, AMM	W72-07933
AN, LAKE HURON, LAKE ERIE, GREAT	LAKES, SOYBEANS, LIGHT INTENSITY,	W73-01456
YANOPHYTA, NITRATES, PHOSPHATES,	LAKES, STRATIFICATION, MIXING, BO	W72-08049
, DDT/ *POLLUTANTS, *RECREATION,	LAKES, STREAMS, ALGAE, PESTICIDES	W71-06444
, *HYDROLYSIS, DETERGENTS, GREAT	LAKES, TEMPERATURE, HYDROGEN ION	W71-00116
ATE ORGANIC NITROGEN, CALCAREOUS	LAKES, ULTRA-VIOLET LABILE NITROG	W72-08048
ER POLLUTION CONTROL, WISCONSIN,	LAKES, URBAN RUNOFF, AGRICULTURAL	W73-02479
NUISANCE ALGAE, *EUTROPHICATION,	LAKES, WASHINGTON, *ALGAL CONTROL	W72-00799
NOLOGY, *ALGAE, *EUTROPHICATION,	LAKES, WATER QUALITY, PHOSPHORUS,	W70-08944
SIVITY,/ EXTINCTION COEFFICIENT,	LAMBERT-BEER EQUATION, EDDY DIFFU	W72-05459
ILLATORIA, PLEUROSIGMA, NAUPLII,	LAMELLIBRANCH, SPIROGYRA, EUDORIN	W71-11876
CUS SPIRALIS, FUCUS VESICULOSUS,	LAMINARIA DIGITATA, LAMINARIA HYP	W71-11515
VESICULOSUS, LAMINARIA DIGITATA,	LAMINARIA HYPERBOREA, ASCOPHYLLUM	W71-11515
ACROPHYTES, RIA DE ALDAN(SPAIN),	LAMINARIA OCHROLEUCA, SACCORHIZA	W72-04766
VESICULOSUS, LAMINARIA DIGITATA,	LAMINARIA AGARDHII, ULVA LACTUCA,	W72-04748
LLUM NODOSUM, FUCUS VESICULOSUS,	LAMINARIA DIGITATA, LAMINARIA AGA	W72-04748
ROMATOGRAPHY, FUCUS VESICULOSUS,	LAMINARIA DIGITATA, ST. MARGARET'	W72-12166
LICALIS, SAM/ FUCUS VESICULOSUS,	LAMINARIA DIGITATA, PORPHYRA UMBI	W73-01074
GIPES, BRITISH COLUMBIA, RUSSIA,	LAMINARIA RODRIGUEZII.: /ARIA LON	W73-00428
SEAWEEDS, LAMINARIA LONGICRURIS,	LAMINARIA DIGITATA, AGARUM CRIBRO	W73-00366
DIGITATA, AGARUM CRI/ *SEAWEEDS,	LAMINARIA LONGICRURIS, LAMINARIA	W73-00366
ERMERS BAY, ALARIA, NEREOCYSTIS,	LAMINARIA, THALASSIOPHYLLUM, FUCU	W72-09120
RMUM LICHENEFORME, MASTIGOCLADUS	LAMINOSUS.: /NE TEST, CYLINDROSPE	W71-12070
PA, ACROPORA PALMATA, PENICILLUS	LAMOUROUXII.: /, BRYOPSIS, CAULER	W72-14294
SIS, OXYUROSTYLIS, HEMILAMPROPS,	LAMPROPS, C: / ANALOGA, DIASTYLOP	W72-09092
SEPTIC TANKS, OXIDATION LAGOONS,	LAND APPLICAT: /DENITRIFICATION,	W70-09907
CHLOROPHYTA, NITROGEN FIXATION,	LAND MANAGEMENT, FERTILIZATION, P	W72-04769
TILIZATION, LEAVES, LEGISLATION,	LAND USE, SHORES, MONITORING, MAR	W73-02479
ERS, ALGAE, CHEMICAL PROPERTIES,	LANDFILLS, PERCOLATION, DRAINAGE,	W72-12955
, CALCIUM ION CONCENTRATI/ *MOSS	LANDING, BARNACLES, LETHAL DOSAGE	W72-09092
IXING BACTERIA, *RAINFALL, *ARID	LANDS, *ARIZONA, GRASSLANDS, ON-S	W72-12505
IES, *FERTILIZATION, *CULTIVATED	LANDS, *IOWA, WATER QUALITY, FERT	W71-12252
RGIA), LAKE CARDIN/ *LAKE SIDNEY	LANIER(GEORGIA), LAKE JACKSON(GEO	W71-11914
BLOOMS, HYPNODINIUM, LAKE SIDNEY	LANIER(GEORGIA), LAKE JACKSON(GEO	W71-09173
L, MARCHALL ISLAND, POLYSIPHONIA	LANOSA, CHONDRUS CRISPUS, COD, FL	W72-08142
RHODYMENIS PALMATA, POLYSIPHONIA	LANOSA, POLYSIPHONIA HARVEYI, RHO	W72-04748
VANADIUM,	LANTHANUM, SCANDIUM.:	W72-01101
FECTS, STANDING CROPS, MOLLUSKS,	LARVAE, CHRYSOPHYTA, CYANOPHYTA,	W73-00935
IDUES, *DDD, DDE, TDE, BLACK-FLY	LARVAE, CLADOPHORA, CYPRINUS CARP	W70-08098
OMMUNITIES, FOOD CHAIN, DIATOMS,	LARVAE, DIETS.: /ES, BIOLOGICAL C	W72-09248
OGY, PHYTOPLANKTON, ZOOPLANKTON,	LARVAE, EGGS, CHINOOK SALMON, STR	W71-11517
H RATES, *ENVIRONMENTAL EFFECTS,	LARVAE, ON-SITE TESTS, INVERTEBRA	W73-00348
ES, FOOD HABITS, EUTROPHICATION,	LARVAE, ROTIFERS, AQUATIC MICROOR	W72-07132
E ALGAE, CYANOPHYTA, PYRROPHYTA,	LARVAE, SAMPLING, INVERTEBRATES,	W72-14313
STREPTOMYCIN, EPIFLORA, WALLEYE	LARVAE, SHAD, MACROINVERTEBRATES,	W72-08142
	LASER, RHODAMINE B.:	W72-01095
SPACE PLATFORMS, LEAD, MERCURY,	LASERS, MICROWAVE RADIOMETRY, VID	W72-04742
RAPHY, ION SELECTIVE ELECTRODES,	LASERS, ORTHOPHOSPHATES.: /RADIOG	W72-08790
ITYLUM, PROROCENTRUM, NITZSCHIA,	LAUDERIA, CHAETOCEROS.: /IALIS, D	W72-09248
BENZOATE, PROPIANATE, PROPANOL,	LAURATO, PALM: /ACETATE, ETHANOL,	W72-05421
., UDOTEA SPP., PENICILLUS SPP.,	LAURENCIA SPP., SPONGES.: /IA SPP	W72-11876
A BARTAYRESII, LAURENCIA OBTUSA,	LAURENCIA PAPILLOSA, COELOTHRIX I	W73-00838
LAMENTOSA, DICTYOTA BARTAYRESII,	LAURENCIA OBTUSA, LAURENCIA PAPIL	W73-00838
AN, FAIRY LIQUID, BLUSYL, SODIUM	LAURYL, ETHER SULFATE, SODIUM DOD	W72-12576
CA LUCISPERCA, BURBOT, COREGONUS	LAVARETUS, COREGONUS ALBULA, WHIT	W72-12729
, WATER POLLUTION SOURCES, WATER	LAW, ALGAE, DISSOLVED OXYGEN.: /L	W71-01192
*PHOSPHATES, LEGISLATION, WATER	LAW, FEDERAL GOVERNMENT, WATER PO	W71-11516
MADISON(WIS), LAKE MENDOTA(WIS),	LAWN MAINTENANCE.: *	W73-02479
*MACROPHYTE-EPIPHTYE METABOLISM,	LAWRENCE LAKE(MICH), SCIRPUS ACUT	W72-03216
EFRACTORY NITROGEN, NETPLANKTON,	LAWRENCE LAKE(MICH), WINTERGREEN	W72-08048
N DEMAND, HYDROGEN ION CON/ *ST.	LAWRENCE RIVER, BIOCHEMICAL OXYGE	W70-10427
OURCES, WATE/ *INSECTICIDES, *ST	LAWRENCE RIVER, WATER POLLUTION S	W70-08098
SIN, LAKE ERIE, LAKE ONTARIO, ST	LAWRENCE RIVER, INTERNATIONAL JOI	W70-07283
SSILE ALGAE, *INVERTEBRATES, ST.	LAWRENCE RIVER, ALGAE, NEMATODES,	W72-05432
ITIC ACID, BETA-SITOSTEROL, THIN-	LAYER CHROMATOGRAPHY.: /PIN, PALM	W71-08597
ELLUM QUADRAPLICATUM, C-14, THIN	LAYER CHROMATOGRAPHY, CLEANUP, RA	W73-00361
BITI/ MERCURIAL PESTICIDES, THIN	LAYER CHROMATOGRAPHY, GROWTH INHI	W72-13347
ACH RES/ *FERRIC CHLORIDE, MULTI-	LAYER FILTERS, ANTHRACITE, *WAHNB	W72-06201
ION, COASTAL WATERS, TROPHOGENIC	LAYER.: /F OF FINLAND, DARK FIXAT	W72-03228

HORUS, PHOSPHATES, PRODUCTIVITY,	LEACHING.: / WATER QUALITY, PHOSP	W70-08944
L PRECIPITATION, IRON, BIOASSAY,	LEACHING, VOLUMETRIC ANALYSIS, SO	W72-05965
RUNOFF, BASE FLOW, PERCOLATION,	LEACHING, DRAINAGE, FARM WASTES,	W71-06443
ALGAE, FRESH WATER, FILTRATION,	LEACHING, FUNGI, LIGHT, PHOSPHATE	W72-07143
EODACTYLUM TRI/ SILVER, CADMIUM,	LEAD, CHLORELLA PHYRENOIDOSA, PHA	W72-07660
*WATER POLLUTION EFFECTS, ZINC,	LEAD, COPPER, CHROMIUM, SILVER, C	W72-14694
ATER, *DATA COLLECTIONS, COPPER,	LEAD, HEAVY METALS, MUSSELS, *KEL	W72-11925
RADIOMETRY, VI/ SPACE PLATFORMS,	LEAD, MERCURY, LASERS, MICROWAVE	W72-04742
CIUM, HYDROGEN, CHEMICAL WASTES,	LEAD, NICKEL, TOXICITY, BIOINDICA	W72-14303
OTA, CYANOPHYTA, TREES, GRASSES,	LEAVES, FOREST SOILS, SAMPLING, M	W72-01784
E, AQUATIC WEEDS, FERTILIZATION,	LEAVES, LEGISLATION, LAND USE, SH	W73-02479
MARSH PLANTS, CHEMICAL ANALYSIS,	LEAVES, SAMPLING, PHYTOPLANKTON,	W73-01089
STRATEGY, THAMES RIVER(ENGLAND),	LEE RIVER(ENGLAND).: /, RESEARCH	W71-13172
UMBELLATUS, SPARGANIUM RAMOSUM,	LEERSIA ORYZOIDES, NUPHAR ADVENA,	W73-01089
LIC HEALTH, ENVIRONMENT CONTROL,	LEGAL ASPECTS, REGULATION, PERMIT	W73-00703
CONTROL, STANDARDS, LEGISLATION,	LEGAL ASPECTS, WATER POLLUTION, A	W72-01093
ES, EUTROPHICATION, LEGISLATION,	LEGAL ASPECTS, REGULATION, POLLUT	W72-01085
S, PHOSPHOROUS, WATER SOFTENING,	LEGAL ASPECTS, HARD WATER, PUBLIC	W71-09429
RECIRCULATED WATER, WATER REUSE,	LEGAL ASPECTS, JUDICIAL DECISIONS	W70-07100
GAL ASPECTS, JUDICIAL DECISIONS,	LEGISLATION, ADMINISTRATION, NON-	W70-07100
TATES, ALGAL CONTROL, STANDARDS,	LEGISLATION, LEGAL ASPECTS, WATER	W72-01093
LLUTION SOURCES, EUTROPHICATION,	LEGISLATION, LEGAL ASPECTS, REGUL	W72-01085
WASTE TREATMENT, AIR POLLUTION,	LEGISLATION, BENTHIC FLORA, F: /,	W71-13723
ATION, *DETERGENTS, *PHOSPHATES,	LEGISLATION, WATER LAW, FEDERAL G	W71-11516
D DEVELOPMENT, MUNICIPAL WASTES,	LEGISLATION, REGULATION,: /RCH AN	W72-06638
UTION ABATEMENT, EUTROPHICATION,	LEGISLATION, PUBLIC HEALTH, WATER	W72-06655
IC WEEDS, FERTILIZATION, LEAVES,	LEGISLATION, LAND USE, SHORES, MO	W73-02479
*FEEDLOTS,	LEMNA.:	W72-04773
INALIS, CHARALES, CERATOPHYLLUM,	LEMNA.: /DENSIS, PHRAGMITES, FONT	W71-11012
VAILLANTII, CONIOPHORA OLIVACEA,	LENZITES TRABEA, SCOPULARIOPSIS B	W72-13813
SPHAERICUS, CHYDORUS, MONOSTYLA,	LEPADELLA, TRICHOCERCA, HAEMATOCO	W72-07132
A POLYMORPHA, EUGLENA SPIROGYRA,	LEPOCINCLIS OVUM, PHACUS CIRCUMFL	W73-00918
HE, TRIAENODES INUUSTA, OECETIS,	LEPTOCELLA, ISONYCHIA, CAENIS, PE	W70-06655
INLET(ALASKA), CYCLOTELLA NANA,	LEPTOCYLINDRUS, PACIFIC OCEAN, AL	W71-11008
LA, DACTYLIOSOLEN MEDITERRANEUS,	LEPTOCYLINDRUS DANICUS, RHIZOSOLE	W72-03543
TRATI/ *MOSS LANDING, BARNACLES,	LETHAL DOSAGE, CALCIUM ION CONCEN	W72-09092
ILIS, CHLAMYDOMANAS REINHARDTII,	LETHAL DOSAGE.: /TA, EUGLENA GRAC	W72-05614
NEDESMUS, BIO-ASSAY, RESISTANCE,	LETHAL LIMIT, POISONS, WATER POLL	W72-14706
SEAWATER, *TOXICITY, *NUTRIENTS,	LETHAL LIMITS, METHODOLOGY, INDUS	W72-12239
, THREONINE, PROLINE, GLUTAMATE,	LEUCINE, ARGININE, SERINE, TYROSI	W73-00942
PLECPODS, UROPODS, ENRICHMENT,	LEUCOTHRIX MUCOR, MAJURO ATOLL, M	W72-08142
GE, PHYSICOCHEMICAL PROP/ *WATER	LEVEL FLUCTUATIONS, *PUMPED STORA	W72-14280
TY, NITROGEN, NUTRIENTS, TROPHIC	LEVEL.: / PHOSPHATES, WATER QUALI	W71-11033
, SOUTH CAROLINA, ALGAE, TROPHIC	LEVEL.: /RBON DIOXIDE, ECOSYSTEMS	W71-00475
TREAMS, *STANDING CROP, *TROPHIC	LEVEL, *ENERGY BUDGET, PRODUCTIVI	W71-10064
CAL PROPERTIES, *SEDIM/ *TROPHIC	LEVEL, *LAKES, *NUTRIENTS, *CHEMI	W72-05469
UCTIVITY, WATER QUALITY, TROPHIC	LEVEL, AQUATIC ALGAE, FISH POPULA	W72-08986
MATOGRAPHY, CHLOROPHYLL, TROPHIC	LEVEL, EUTROPHICATION, ALGAE, DEC	W72-01784
PROPERTIES, PESTICIDES, TROPHIC	LEVEL, GASES, MUDS, ADSORPTION, C	W72-00847
ANCE OF NATURE, ECOLOGY, TROPHIC	LEVEL, HERBIVORES, NATURAL STREAM	W72-06340
S, ALGAE, RADIOACTIVITY, TROPHIC	LEVEL, LIFE HISTORY STUDIES, TIME	W70-06668
EUTROPHICATION, REVIEWS, TROPHIC	LEVEL, PHOSPHORUS, PHOSPHATES, CA	W72-07937
OTROPHY, EUTROPHICATION, TROPHIC	LEVEL, PROTOZOANS, AQUATIC ALGAE,	W72-13816
ENTS, THERMAL POLLUTION, TROPHIC	LEVEL, TROPICAL REGIONS, ULTRAVIO	W73-00796
IC ALGAE, PHYTOPLANKTON, TROPHIC	LEVEL: /PERATURE, SAMPLING, AQUAT	W73-01456
DIOISOTOPES, METABOLISM, TROPHIC	LEVELS.: /DIOISOTOPES, CADMIUM RA	W71-13243
TRIENTS, MEASUREMENT, / *TROPHIC	LEVELS, *BIOASSAY, PHOSPHORUS, NU	W71-13794
*NUTRIENT	LEVELS, CUYAHOGA RIVER(OHIO).:	W72-04263
WATER POLLUTION CONTROL, TROPHIC	LEVELS, PHOSPHATES, DETERGENTS, B	W71-12072
MS, REDUCTION(CHEMICAL), TROPHIC	LEVELS, PROTOZOA, ALGAE, BACTERIA	W72-10883
IRA FERTILISSIMA, AN/ *TOLERANCE	LEVELS, TOLYPOTHRIX TENUIS, AULOS	W71-12071
ISM, WATERSHEDS(BASINS), TROPHIC	LEVELS, TURBIDITY, OXYGEN, FISH,	W71-11009
RN(ENGLAND), LUXURY UPTAKE, LOCH	LEVEN(SCOTLAND), LAKE WINDERMERE(	W71-13183
S, CERAMIUM BUSSOIDEUM, CHONDRIA	LI: /LAVULATUM, HYPENA MUSCIFORMI	W73-00838
ACETYLENE TEST, CYLINDROSPERMUM	LICHENEFORME, MASTIGOCLADUS LAMIN	W71-12070
, FOOD WEBS, FOOD CHAINS, FROGS,	LICHENS, ALGAE, FUNGI, SMALL ANIM	W72-00949
CATORS, BIODEGRADATION, BIOMASS,	LICHENS, MOSSES, FUNGI, ALGAE, SO	W72-14303
N, ISOTOPE RATIOS, SPECTROMETRY,	LICHENS, MOSSES, GYMNOSPERMS, ANG	W72-12709
IC ZONES, CYANOPHYTA, SYMBIOSIS,	LICHENS, NITROGEN, LIGHT INTENSIT	W73-02471
AIR POLLUTION, WATER POLLUTION,	LICHENS, SOIL-WATER-PLANT RELATIO	W72-03347
S BREVIPES, ACHNANTHES JAVANICA,	LICMOPHORA JURGENSII, AMPHORA OVA	W72-08141
*PLANT MORPHOLOGY, REPRODUCTION,	LIFE CYCLES, AQUATIC ALGAE, PERIP	W72-07141
GICAL COMMUNITIES, PRODUCTIVITY,	LIFE CYCLES, GROWTH STAGES, DISTR	W72-14659
NEMATODES, ALGAE, OLIGOCHAETES,	LIFE CYCLES, SEASONAL, DENSITY, D	W71-11012
TROPHICATION, COLIFORMS, SPORES,	LIFE CYCLES, ODOR, RECREATION, SY	W71-09173
E, RADIOACTIVITY, TROPHIC LEVEL,	LIFE HISTORY STUDIES, TIME, SNAIL	W70-06668
ANIMALS, *AQUATIC PLA/ *AQUATIC	LIFE, *AQUATIC BACTERIA, *AQUATIC	W72-04853
R/ *PESTICIDE RESIDUES, *AQUATIC	LIFE, *AQUATIC ENVIRONMENT, *WATE	W72-13347
*DELAWARE ESTUARY, *BIOLOGICAL	LIFE, *DISSOLVED OXYGEN CONCENTRA	W70-09189
PHOSPHATES, ECOSYSTEMS, AQUATIC	LIFE, ANTIBIOTICS(PESTICIDES), EC	W72-09668
WATER POLLUTION SOURCES, AQUATIC	LIFE, AQUATIC ANIMALS, AQUATIC MI	W72-09646
POLLUTION, *FRESHWATER, *AQUATIC	LIFE, AQUATIC ANIMALS, CARP, DAPH	W72-09649
D CHAINS, PUBLIC HEALTH, AQUATIC	LIFE, AQUATIC ALGAE, CALCIUM, RAD	W72-10978
AY, FISH, RAINBOW TROUT, AQUATIC	LIFE, AQUATIC ENVIRONMENT, FISH P	W72-12257
IOISOTOPES, NEW ENGLAND, AQUATIC	LIFE, AQUATIC MICROORGANISMS, MAR	W72-14317
NELIDS, COPEPODS, CRABS, AQUATIC	LIFE, BACTERIA.: /TIDAL AREAS, AN	W72-07911
OLIGOTROPHY, OXYGEN SAG, AQUATIC	LIFE, BIOASSAY, CLASSIFICATION.: /	W73-00002
, CLAMS, MARINE ANIMALS, AQUATIC	LIFE, ECOSYSTEMS, SALMONIDS.: /SH	W72-07907
XYGEN, COMPUTERS, ALGAE, AQUATIC	LIFE, ENVIRONMENTAL EFFECTS, ENVI	W72-14405
, ENVIRONMENTAL EFFECTS, AQUATIC	LIFE, FISH POPULATIONS, PHYTOPLAN	W72-00939

WATER POLLUTION EFFECTS, AQUATIC LIFE, HABITATS, ECOLOGY, *ECOSYST W73-01632
E ENVIRONMENT, HABITATS, AQUATIC LIFE, HEAVY METALS, INDUSTRIAL WA W72-11854
TURE, SEASONAL, MONTHLY, AQUATIC LIFE, INTERSTATE RIVERS, HYPOLIMN W70-09595
IBLIOGRAPHIES, *ECOLOGY, AQUATIC LIFE, IRRIGATION SYSTEMS, WATER A W71-06102
IENTS, DISSOLVED OXYGEN, AQUATIC LIFE, MUNICIPAL WASTES, PHOTCSYNT W71-11017
*FALLOUT, *FOOD CHAINS, *AQUATIC LIFE, RADIOISOTOPES, ABSCRPTION, W72-04461
CARNIVORES, ECOSYSTEMS, AQUATIC LIFE, SAMPLING, ALGAE, LABORATORY W71-10064
SUBMARINE LIGHT DATA SPHERE.: W71-13252
SES, PHOTOSYNTHESIS, TURBULENCE, LIGHT INTENSITY.: /MOVEMENT, VIRU W71-13184
EN, PHOSPHOROUS, PHOTOSYNTHESIS, LIGHT INTENSITY, CXYGENATION, DEN W71-13326
E PLANTS, *PHYTOPLANKTON, ALGAE, LIGHT INTENSITY, SALINITY, NUTRIE W71-10083
AWAII, WATER TEMPERATURE, WAVES, LIGHT INTENSITY.: /ARINE ALGAE, H W71-11486
*CHLORELLA, *BIOASSAYS, ENZYMES, LIGHT INTENSITY, ALGAE.: W71-11916
HWEST U.S., TEMPERATURE, OXYGEN, LIGHT INTENSITY, NUTRIENTS, ACTIN W72-10613
EN, *NUTRIENTS, *PHOTOSYNTHESIS, LIGHT INTENSITY, AMMONIA, CARBON W72-09103
ION CONCENTRATION, TEMPERATURE, LIGHT INTENSITY, SEDIMANTATION, O W72-01881
IS, SEASONAL, LIGHT PENETRATION, LIGHT INTENSITY, DISSCLVED SOLIDS W72-04270
RACE ELEMENTS, ALGAE, NUTRIENTS, LIGHT INTENSITY, HYDROGEN ION CON W72-04433
XIDE, LININGS, DISSOLVED OXYGEN, LIGHT INTENSITY, CESIGN CRITERIA, W72-04789
LTURES, PHOTOSYNTHESIS, BIOMASS, LIGHT INTENSITY, GROWTH RATES, MA W72-06274
ATION, OXYGENATION, TEMPERATURE, LIGHT INTENSITY, MIXING, DISSOLVE W72-04983
T GROWTH, BIOMASS, GROWTH RATES, LIGHT INTENSITY, CARBCN DIOXIDE, W72-05476
HT, DIFFUSIVITY, CARBON DIOXIDE, LIGHT INTENSITY, PHOTOSYNTHESIS, W72-05459
AL MODELS, MIXING, WATER RE-USE, LIGHT INTENSITY, EVAPORATION, RE- W71-07099
HARVESTING, DEPTH, TEMPERATURE, LIGHT INTENSITY, CHEMICAL PRECIPI W71-07100
ONS, AEROBIC CONDITIONS, SLUDGE, LIGHT INTENSITY, COLIFORMS, BIOCH W71-07106
N, DEPTH, THERMOCLINE, NITRATES, LIGHT INTENSITY, EUTROPHICATION, W71-07875
PYRENOIDOSA, OXYGEN PRODUCTION, LIGHT INTENSITY, CULTURE THICKNES W70-05547
LUENT STREAMS, EFFLUENT STREAMS, LIGHT INTENSITY, NITROGEN, CHRYSO W73-01435
AKE ERIE, GREAT LAKES, SOYBEANS, LIGHT INTENSITY, AIR TEMPERATURE, W73-01456
YTA, HYDROGEN ION CONCENTRATICN, LIGHT INTENSITY, NUTRIENTS, PSEUD W73-01657
ARIES, *DIATOMS, *SESSILE ALGAE, LIGHT INTENSITY, TIDAL EFFECTS, S W73-00853
*HEAT RESISTANCE, *MARINE ALGAE, LIGHT INTENSITY, *ECOLOGICAL DIST W73-00838
OLOGY, POLLUTANT IDENTIFICATICN, LIGHT INTENSITY, BIOASSAY.: /, BI W73-02099
HIC FLORA, SEA WATER, SEDIMENTS, LIGHT INTENSITY, DIATCMS, CARBON W72-14317
DISSOLVED OXYGEN, *PLANT GROWTH, LIGHT INTENSITY, TOXICITY, PRIMAR W72-14690
ITY, PHOTOSYNTHESIS, EPILIMNICN, LIGHT INTENSITY, RESPIRATICN, LIM W72-14812
PERIPHYTON, SEDIMENTS, DIATOMS, LIGHT INTENSITY, PHOTOSYNTHESIS, W72-07501
SCENEDESMUS, CHLORELLA, SILICA, LIGHT INTENSITY, TEMPERATURE, MUD W72-12544
UTRIENTS, SALTS, TRACE ELEMENTS, LIGHT INTENSITY, SCENEDESMUS, CHL W72-12739
A, SYMBIOSIS, LICHENS, NITROGEN, LIGHT INTENSITY, TEMPERATURE, ANA W73-02471
EN FIXATION, *LAKES, CYANOPHYTA, LIGHT INTENSITY, EUTROPHICATION, W73-02472
RANSMISSION ELECTRON MICROSCOPY, LIGHT MICROSCOPY.: /MICROSCOPY, T W73-00432
YGEN, PHOTOSYNTHESIS, TURBIDITY, LIGHT PENETRATION, IMPOUNDMENTS, W73-00256
ERS, HYDROGEN ION CONCENTRATION, LIGHT PENETRATION, PHOTCSYNTHESIS W73-00236
SSILE ALGAE, TEMPERATURE, DEPTH, LIGHT PENETRATION.: /YNTHESIS, SE W72-07501
O, EUTROPHICATION, LAKE ONTARIO, LIGHT PENETRATION, PHYSICOCHEMICA W73-01615
ALS, SEASONAL, ZINC, ABSORPTION, LIGHT PENETRATION, ALKALINITY, HA W73-01434
TS, *HYPOLIMNION, PHYTOPLANKTON, LIGHT PENETRATION, DISSCLVED OXYG W72-12995
LAKE ERIE, BENTHIC FLORA, ALGAE, LIGHT PENETRATION, PLANKTCN, NUTR W72-13659
RE, PHOTOSYNTHESIS, RE-AERATION, LIGHT PENETRATION, NUTRIENTS, PHO W71-07098
SITY, PHOTOSYNTHESIS, EQUATICNS, LIGHT PENETRATION, ALGAE, HYDROGE W72-05459
ON, EUTROPHICATION, TEMPERATURE, LIGHT PENETRATION, AMMONIA, PHOSP W72-05469
AL, AGE, BIOMASS, LIGHT QUALITY, LIGHT PENETRATION, EFFICIENCES, C W72-04766
T PLAINS, NUTRIENTS, RECREATICN, LIGHT PENETRATION, NEBRASKA, WAST W72-04759
ENTS, SEWAGE, WATER TEMPERATURE, LIGHT PENETRATION, CARBON, LAKE E W72-04734
UM, SCENEDESMUS, WATER HYACINTH, LIGHT PENETRATION, OXYGEN,: /ALCI W72-04269
DURES, PHOTOSYNTHESIS, SEASONAL, LIGHT PENETRATION, LIGHT INTENSIT W72-04270
*PERIPHYTON, *ALGAE, *LITTORAL, LIGHT PENETRATION, BENTHIC FLORA, W71-11029
, BIOMASS, NITRATES, RESERVOIRS, LIGHT PENETRATION, CYANOPHYTA, CH W71-13183
M, E. COLI, SILICA, *FILTRATION, LIGHT PENTRATION, PHYSICAL PROPER W72-11833
RBULENCE, PHOTOSYNTHESIS, DEPTH, LIGHT PENTRATICN, RESPRIATION, NU W72-08559
TROPODS, LITTORAL, AGE, BIOMASS, LIGHT QUALITY, LIGHT PENETRATION, W72-04766
ATURE, BIOCHEMISTRY, PHYSIOLOGY, LIGHT.: /SM, GROWTH RATES, TEMPER W72-00377
ATION, DISINFECTION, ULTRAVIOLET LIGHT, ALGAE, OXYGENATICN, BIOCHE W72-00042
CYANOPHYTA, CHLOROPHYTA, OXYGEN, LIGHT, CHRYSOPHYTA, OCHRCMONAS, R W71-09172
RIMARY PRODUCTIVITY, MINERALOGY, LIGHT, CULTURES, BIOASSAY.: /S, P W72-07166
OHIO, OXYGEN, DISSOLVED OXYGEN, LIGHT, DIFFUSIVITY, CARBCN DIOXID W72-05459
, PHYTOPLANKTON, TOXICITY, IRON, LIGHT, HYDROGEN ION CONCENTRATION W71-03027
N DIOXIDE, NITROGEN, PHOSPHORUS, LIGHT, NUTRIENTS, MATHEMATICAL ST W72-11724
S, TEMPERATURE, CARBON, DIATCMS, LIGHT, OXYGEN, MINNESOTA.: /FFECT W71-08670
ER, FILTRATION, LEACHING, FUNGI, LIGHT, PHOSPHATES, NUTRIENTS, WAT W72-07143
ELLA, *PHOTOACTIVATION, *OXYGEN, LIGHT, PHOTOSYNTHESIS, CPTICAL PR W72-14800
WATER, PLANKTON, BIODEGRADATION, LIGHT, TEMPERATURE.: /PLANTS, SEA W71-09261
IC ALGAE, *AQUATIC PRODUCTIVITY, LIGHT, TURBICITY, NUTRIENTS, HETE W72-13142
, SAMPLING, OXYGEN, TEMPERATURE, LIGHT, WAVES(WATER), PHYTCPLANKTO W72-06283
S, CURRENTS(WATER), TIDES, WIND, LIGHT, WAVES, CLAMS, CRUSTACEANS, W72-09092
S, FUSCOPORIA CONTIGUA, TRAMETES LILACINO-GILVA, XYLOBOLUS FRUSTUL W72-13813
HIA, MICROCYSTIS, NOSTOC, SLAKED LIME.: /TION, ANABAENA, GLOEOTRIC W72-01097
ICATICN, COAGULATION, BENTONITE, LIME, *PULP WASTES, TEOTIARY TREA W72-03753
TER TREATMENT, ACTIVATED SLUDGE, LIME, AEROBIC TREATMENT, DOMESTIC W72-09176
TREATMENT, *FOAM FRACTIONATION, LIME, HYDROGEN ION CONCENTRATION, W71-12200
NES, PRECIPITATION, COAGULATION, LIME, HYDROGEN ION CONCENTRATION, W71-08951
SEDIMENTATION, POLYELECTROLYTES, LIME, NEUTRALIZATION, FILTRATION, W72-10237
TION, SEDIMENTATION, FILTRATION, LIME, ROCKS, METABOLISM, DESIGN C W71-13341
KTON, LAKE MORPHOMETRY, GEOLOGY, LIMESTONE, TEMPERATURE, DEPTH, CH W72-04269
S, PHENOLS, *DETERGENTS, *LETHAL LIMIT, ALKYLBENZENE SULFONATES.: / W72-12576
S, BIO-ASSAY, RESISTANCE, LETHAL LIMIT, POISONS, WATER PCLLUTICN E W72-14706
LAMYDOMONAS, INHIBITORS, *LETHAL LIMIT, WATER POLLUTION EFFECTS, W W72-11727
LORELLA, BICARBONATES, AERATION, LIMITING FACTORS, EUTROPHICATION. W72-10607

N, LIGHT INTENSITY, RESPIRATION,   LIMITING FACTORS.: /IS, EPILIMNIO   W72-14812
TIC WEEDS, *NUTRIENTS, BIOASSAY,   LIMITING FACTORS, NITROGEN, PHOSP   W73-00232
URES, CARBON DIOXIDE, NUTRIENTS,   LIMITING FACTORS, ALLCGENIC SUCCE   W72-14790
MASS, PHOTOSYNTHESIS, NUTRIENTS,   LIMITING FACTORS, SALINITY, CARBO   W73-00431
, BIOMASS, CARBON RADIOISOTOPES,   LIMITING FACTORS, PHOTOSYNTHESIS,   W72-07648
CAL CCMMUNITIES, AQUATIC PLANTS,   LIMITING FACTORS, ROOTED AQUATIC   W73-01089
CTION, CARBON DIOXIDE, CULTURES,   LIMITING FACTORS, RHODOPHYTA, AQU   W73-02099
GANIC COMPOUNDS, STANDING CROPS,   LIMITING FACTORS, EUTROPHICATION.   W72-13653
Y, AQUATIC HABITATS, CLADOPHORA,   LIMITING FACTORS, ECOLOGICAL DIST   W72-13142
TY, ANABAENA, AMMONIA, NITRATES,   LIMITING FACTORS, NITROGEN, PLANK   W73-02472
NETHERLANDS,   LIMITING NUTRIENTS.:   W72-12544
UGINOSA, PSEUDOMONAS AERUGINCSA,   LIMITING NUTRIENTS.: /OCYSTIS AER   W73-01657
S LAKE(WASH), NUTRIENT DILUTION,   LIMITING NUTRIENTS, ALGAL CELL WA   W72-08049
MONONA(WIS.), LAKE WINGRA(WIS.),   LIMITING NUTRIENTS, SURPLUS NUTRI   W73-00232
SSO WEED,/ *PLUTONIUM, DETECTION   LIMITS, BIOLOGICAL SAMPLES, SARGA   W72-05965
RIZATICN, LIVER, BONE, DETECTION   LIMITS, CHEMICAL RECOVERY, SAMPLE   W72-05965
R, *TCXICITY, *NUTRIENTS, LETHAL   LIMITS, METHODOLOGY, INCUSTRIAL W   W72-12239
, TENDIPES TENTANS, TRICHOPTERA,   LIMNEPHILUS, GLOSSOSOMA, PHYSA GY   W72-14280
UVELLA, DINOBRYON, OSCILLATORIA   LIMNETICA, DICTYOSPHAERIUM, CELL   W71-03014
SEE, ANABAENA, COREGONS, LYNGBYA   LIMNETICA, OSCILLATORIA REDECKEI,   W71-00117
D, OSCILLATORIA REDEKEI, LYNGBYA   LIMNETICA, ANABAENA SPP., CHROMAT   W72-07145
CESIUM RADIOISOTOPES, SPIROGYRA,   LIMNODRILUS HOFFMEISTERI, STICTOC   W70-06668
A PYRENOIDOSA, *CHLORAMPHENICAL,   LIMNODRILUS SPP., TUBIFEX, UNIALG   W72-09668
AL MAGNIFICATION, PHOSPHORUS-32,   LIMNODRILUS.: / OUABAIN, BIOLOGIC   W72-09668
NTUATUS, CHIRONOMUS ANTHRACINUS,   LIMNODRILUS HOFFMEISTERI, TUBIFEX   W71-11009
FFECTS, WATER POLLUTION SOURCES,   LIMNOLOGY.: /E, WATER POLLUTICN E   W72-01094
QUALITY, ENVIRONMENTAL EFFECTS,   LIMNOLOGY, PLANNING, EUTROPHICATI   W72-06526
IC PRODUCTIVITY, FISH, FISHKILL,   LIMNOLOGY, NUTRIENTS, OXYGEN, OXY   W72-04279
MS, WATER ANALYSIS, TEMPERATURE,   LIMNOLOGY, CYANOPHYTA, IRRIGATION   W71-06102
IBUTICN PATTERNS, AQUATIC ALGAE,   LIMNOLOGY, CHRYSOPHYTA, EUGLENOPH   W72-07145
TIVITY, PLANKTON, STANDING CROP,   LIMNOLOGY, EUTROPHICATICN.: /ODUC   W72-08559
ES, PHOSPHORUS, CHEMICAL WASTES,   LIMNOLOGY, INTERNATIONAL JOINT CO   W72-14782
ICAL OXYGEN DEMAND, LAKE STAGES,   LIMNOLOGY, NUISANCE ALGAE, NUTRIE   W72-14692
YORK, WATER SUPPLY, SYSTEMATICS,   LIMNOLOGY, ECOLOGY, FISH, WATER P   W73-01615
IDENTIFICATION, PROTOZOA, PONDS,   LIMNOLOGY, ECOLOGY, CHEMICAL ANAL   W73-00918
AQUATIC ANIMALS, *MODEL STUDIES,   LIMNOLOGY, BIOLOGICAL PROPERTIES,   W72-14280
*SESSILE ALGAE, *LAKE SUPERIOR,   LIMNOLOGY, FERTILITY, PHOTOSYNTHE   W72-12192
TY, *TURBIDITY, *PHOTOSYNTHESIS,   LIMNOLOGY, *EUTROPHICATICN, *KANS   W72-12393
CHEMICAL PROPERTIES, MICROSCOPY,   LIMNOLOGY, WATER QUALITY, ECOLOGY   W72-13142
A ACUS, EUGLENA CAUDATA, EUGLENA   LIMNOPHYLA, EUGLENA OBLCNGA, EUGL   W73-00918
CESIUM, STRONTIUM RADIOISOTOPES,   LINC RADIOISOTOPES, MANGANESE, CO   W71-13243
*URBAN RUNOFF, URBAN HYDROLOGY,   LINCOLN(NEBR).:   W71-13466
PPER, NICKEL, RUTHENIUM, CERIUM,   LINCONIUM.: /, IRON, CHRCMIUM, CO   W71-13243
, CULTURING VESSELS, PENICILLIN,   LINDANE, DIADEMA ANTILLARUM, AAND   W72-14294
ELECTRO/ *CHLORELLA PYRENOIDOSA,   LINDANE, P P'-DDE, CLCPHEN A 50,   W72-04134
, *SURFACTANTS, *EUTROPHICATION,   LINEAR ALKYLATE SULFONATES, CHELA   W71-06247
BENZENE SULFONATES, SURFACTANTS,   LINEAR ALKYLATE SULFONATES, SEWAG   W72-06837
IN FORESTS, MATHEMATICAL MODELS,   LINEAR PROGRAMMING, SYSTEMS ANALY   W72-03342
DOGRAMMA CRENULATUS, CLADOGRAMMA   LINEATUS-EXCENTRICUS, CLADOGRAMMA   W71-11527
DS, ORCONECTES RUSTICUS, LIRCEUS   LINEATUS, GAMMARUS.: /AMS, UNIONI   W72-06526
COMMUNITY, CALORIES, DETRIVORES,   LINESVILLE CREEK(PA).: /AUTUMNAL   W71-10064
S, MEDUSAE, PILIDIUM, CLADOCERA,   LINGULA, ACTINOTROCHA, VELIGER, P   W73-00935
L OXYGEN DEMAND, CARBON DIOXIDE,   LININGS, DISSOLVED OXYGEN, LIGHT   W72-04789
A), LAKE WASHINGTON(WASHINGTON),   LINSLEY POND(CONNECTICUT), LAKE Z   W71-04216
ENE, PROPAZINE, COTORON, DIURON,   LINURON.: /DULANS, CERASCN, DIATH   W71-12071
ALYSIS, PROTEINS, CARBOHYDRATES,   LIPIDS.: /PHOSPHORUS, CHEMICAL AN   W70-05469
DATION, METABOLISM, RESPIRATION,   LIPIDS, AMMONIA, CARBCHYCRATES, C   W72-10233
CHEMICAL PROPERTIES, MINNESOTA,   LIPIDS, HUMAN PATHOLOGY, DISEASES   W72-04275
ERS, ESTERS,/ *FATTY ACIDS, *GAS   LIQUID CHROMATOGRAPHY, METHYL EST   W72-09365
Y *BIOLOGICAL MAGNIFICATION, GAS   LIQUID CHROMATOGRAPHY, ELECTRON C   W73-02105
TION, BIOLOGICAL SAMPLES, LIQUID-   LIQUID EXTRACTION, TRIPHOTURUS ME   W73-02105
ARATION, ZO/ *C-14 PRODUCTIVITY,   LIQUID SCINTILLATICN, SAMPLE PREP   W72-14317
QUES, COALESCENCE, FLOCCULATICN,   LIQUID WASTES, SEPARATICN TECHNIQ   W71-13356
NCENTRATION, BIOLOGICAL SAMPLES,   LIQUID-LIQUID EXTRACTION, TRIPHOT   W73-02105
TRAZINE, MCPA, ENDOSULFAN, FAIRY   LIQUID, BLUSYL, SCDIUM LAURYL, ET   W72-12576
ICAL-BIOLOGICAL TREATMENT, MIXED   LIQUOR, SUSPENDED SOLIDS, UPTAKE.   W71-13334
, UNICNIDS, ORCONECTES RUSTICUS,   LIRCEUS LINEATUS, GAMMARUS.: /AMS   W72-06526
DRICAUDA, CHLORELLA PYRENOIDOSA,   LITHIUM CHLORIDE, LITHIUM, ASTERI   W72-12739
A PYRENOIDOSA, LITHIUM CHLORIDE,   LITHIUM, ASTERICNELLA FCRMOSA, AS   W72-12739
CYCLOPS, KERATELLA, ASPLANCHNA,   LITHODESMIUM, HARPESTICCIDS, NOCT   W71-11876
FASCIULATA, EUNOTIA PECTINALIS,   LITHOPHYTES, ENTEROMORPHA, INSOLA   W72-08141
TY, DIRECT EF/ RIDLEY CREEK(PA),   LITITZ CREEK(PA), SPECIES DIVERSI   W72-08579
ROUT LAKE, BIG ARBOR VITAE LAKE,   LITTLE ARBOR VITAE LAKE, GREEN BA   W73-01456
CHRISTINA RIVER, BIG ELK RIVER,   LITTLE ELK RIVER, OCTORARA RIVER,   W72-04736
G, BENTHOS, GRAZING, GASTROPODS,   LITTORAL, AGE, BIOMASS, LIGHT QUA   W72-04766
IENTS, EUTROPHICATION, NITROGEN,   LITTORAL, HARBORS, SPATIAL DISTRI   W72-04263
RACKISH WATER, MICROENVIRONMENT,   LITTORAL, OXYGEN, CHEMICAL REACTI   W71-08026
S, ALGAE, LAKES, EUTROPHICATICN,   LITTORAL, SANDS, SHALLOW WATER, P   W72-07501
SELS, PROTOZOA, MANGROVE SWAMPS,   LITTORAL, SHORES, TIDES, BEACHES,   W72-07911
INVERTEBRATES, AQUATIC ANIMALS,   LITTORAL, WAVES(WATER).: /OPHYTA,   W71-11012
AQUATIC PLANTS, ALGAE, BACTERIA,   LITTORAL, LAKES, PERIPHYTON, ORGA   W72-03216
TURE, CHLORIDES, PHOTOSYNTHESIS,   LITTORAL, RESPIRATICN, PLANT PHYS   W72-12940
KES, CYCLING NUTRIENTS, TRACERS,   LITTORAL, KINETICS, PHOSPHATES.: /   W72-12543
, RATS, PARTRIDGE, DDE, KIDNEYS,   LIVER, BLCOD, GASTEROSTEUS ACULEA   W72-05952
, PLUTONIUM 239, POLYMERIZATION,   LIVER, BONE, DETECTION LIMITS, CH   W72-05965
, ORGANS, BRAIN, BLOOC, KIDNEYS,   LIVER, PLASMOCIUM BERGHEI, BORREL   W73-01676
GNIFICATION, BIOLOGICAL SAMPLES,   LIVER, URINE, OEDCGONIUM, PHYSA,   W72-07703
.:   *ULVA, *LAMINARIA,   LIVERPOOL BAY(ENGLAND), IRISH SEA   W71-12867
, DRAINAGE, RUNOFF, FERTILIZERS,   LIVESTOCK, ECONOMICS,: /IC PLANTS   W71-13172
TER REUSE, SALINITY, PESTICIDES,   LIVESTOCK, WASTE DISPOSAL, CXIDAT   W72-00846

TION, RETURN FLOW, WATER SUPPLY, LIVESTOCK, SEWAGE, INFILTRATION, W71-06435
, CHLCROCOCCUM, HYALOTHECA, LAKE LIVINGSTON, AUTORADIOGRAPHY, SAMP W72-09239
ARBON-14 ME/ *LAKE TRAVIS, *LAKE LIVINGSTON, FIELD MEASUREMENTS, C W72-13692
MICHIGAN, MEASUREMENT, SUSPENDED LCAD, ALKALINITY, HARDNESS(WATER) W71-11033
TER QUALITY, SEASONAL, SUSPENDED LOAD, BIOINDICATORS, WATER POLLUT W72-07896
LIFORNIA, SCENEDESMUS, SUSPENDED LCAD, NITRATES, CALCIUM CARBONATE W70-10174
ROWTH RATES, CULTURES, SUSPENDED LOAD, SEWAGE, INDUSTRIAL WASTES, W71-09795
EMAND, MIXING, SAMPLING, ORGANIC LOADING.: / *BIOCHEMICAL OXYGEN D W71-07102
YANOPHYTE, ENGLENOPHYTA, ORGANIC LOADING.: /STE WATER TREATMENT, C W71-07973
DEPTH, SLOPE, *BOD LOADING, *WASTE STABILIZATION.: W71-03896
S, ANAEROBIC CONDITIONS, ORGANIC LOADING, *BIOCHEMICAL OXYGEN DEMA W71-13326
MS, BACTERIA, OXIDATION, ORGANIC LOACING, ALGAE, WASTE WATER TREAT W71-07096
ONS, *OXIDATION LAGOONS, ORGANIC LOADING, AERATION, MIXING, ALGAE, W72-06838
*WASTE WATER TREATMENT, ORGANIC LOADING, BIOCHEMICAL OXYGEN DEMAN W71-07124
, POLLUTANTS, TURBIDITY, ORGANIC LOADING, CORAL, MOLLUSKS, WORMS.: W71-01475
MIXING, SOLAR RADIATION, ORGANIC LOADING, DIURNAL DISTRIBUTION, HY W72-13508
RICKLING FILTERS, ALGAE, ORGANIC LOADING, ODOR, COLIFORMS, TEMPERA W71-07113
OCHEMICAL OXYGEN DEMAND, ORGANIC LOADING, ODORS, PHOTOSYNTHESIS, A W71-07123
S, ANAEROBIC CONDITIONS, ORGANIC LOADING, ODOR, SLUDGE, COST ANALY W72-11642
ASSIMILATION CAPACITY, *ORGANIC LOADS, NUTRIENTS, DISSOLVED OXYGE W71-11017
ARINE ALGAE, ANIMAL POPULATIONS, LOBSTERS, CRUSTACE: /, TUNNELS, M W73-02029
ANIMALS, *MARINE FISH, MUSSLES, LOBSTERS, ABSORPTION, COBALT RADI W73-00832
WATER TREATMENT, BIODEGRADATION, LOCAL GOVERNMENTS, MUNICI: /ASTE W72-06655
STRATION, ALGAE, BIODEGRADATION, LOCAL GOVERNMENTS, WASTE DISPOSAL W72-01093
AM TARN(ENGLAND), LUXURY UPTAKE, LOCH LEVEN(SCOTLAND), LAKE WINDER W71-13183
OT SPRINGS, JACKSON HOT SPRINGS, LOLO HOT SPRINGS, PIPESTONE HOT S W72-05610
, TH/ *COASTAL WATERS, *SOUTHERN LONG ISLAND, SKELETONEMA COSTATUM W71-07875
*NORTHPORT PLANT(LONG ISLAND).: W72-14659
SAMPLE PREPARATION, SUBLITTORAL, LONG ISLAND, ENTEROMORPHA INTESTI W73-00367
AM TURBINES, *THERMAL POLLUTION, LONG TUBE VERTICAL DISTILLATION, W70-09193
EMING, LOWER LA SALLE LAKE, LAKE LONG, LAKE SQUAW, LAKE BUDD, OSCI W72-07145
ACENTULA, CLOSTERIUM, GOMPHONEMA LONGICEPS, CYMBELLA, VAUCHERIA.: / W72-12744
AGARUM CRI/ *SEAWEEDS, LAMINARIA LONGICRURIS, LAMINARIA DIGITATA, W73-00366
TAURASTRUM, DESMIDS, STAURASTRUM LONGIPES VAR CONTRACTUM, ARTHURS W73-01094
LAMINARIA SINCLAIRII, *LAMINARIA LONGIPES, BRITISH COLUMBIA, RUSSI W73-00428
H), ANABAENA CIRCINALIS, BOSMINA LONGIROSTRIS, SYNURA UVELLA, DINO W71-03014
YCLOPS STRENUUS, INGEST/ DAPHNIA LONGISPINA, DIAPTOMUS GRACILIS, C W72-07940
: *LOPEZ WATER SUPPLY PROJECT, LOPEZ CREEK, ARROYO GRANDE CREEK. W72-05594
ARFISH, CHITONS, RHABDODERMELLA, LOPHOPANOPEUS, EUPENTACTA, CNEMID W73-00932
ETOCEROS COARCTATUS, CHAETOCEROS LORENZIANUS, CHAETOCEROS MESSANEN W71-11527
COOLING, IMPERIAL VALLEY(CALIF), LOS ANGELES(CALIF), TUCSON(ARIZ). W71-09157
OLVED SOLIDS, FLOCCULATION, HEAD LOSS, HYDROGEN ION CONCENTRATION, W71-00138
WERPLANTS, *POWERPLANTS, *ENERGY LOSSES, *WATER CONSERVATION, *AIR W70-09193
O TRUTTA, VENDACE, SALMON TROUT, LOTA LOTA, LAKE MALOGEN, SWEDEN, W72-12729
TTA, VENDACE, SALMON TROUT, LOTA LOTA, LAKE MALOGEN, SWEDEN, CHAR, W72-12729
OLLUTION EFFECTS, *CATTLE, *FEED LOTS.: /CULTURAL RUNOFF, *WATER P W72-08401
A), LAKE WA/ ILLINOIS RIVER, ST. LOUIS RIVER, BLACK RIVER(MINNESOT W71-04216
GAE, SALINITY, MIGRATION, TIDES, LOW WATER MARK, CHRYSOPHYTA.: /AL W72-14317
, MERCURY ISOTOPES, HUMAN BRAIN, LOW-TEMPERATURE OVEN DRYING, ASHI W71-11036
SEPHINE, LAKE ARCO, LAKE DEMING, LOWER LA SALLE LAKE, LAKE LONG, L W72-07145
*LYSOGENY, *PLECTONEMA BORYANUM, LPP-1, LPP-1D, PHYCOVIRUS, MITOMY W72-03188
NY, *PLECTONEMA BORYANUM, LPP-1, LPP-1D, PHYCOVIRUS, MITOMYCIN C.: W72-03188
, HECATONEMA MACULANS, ELACHISTA LUBRICA, RALFSIA VERRUCOSA.: /LIS W72-01370
LAND), COLONIZATION, POTAMOGETON LUCENS, POTAMOGETON PERFOLIATUS, W71-11012
IUM, FIXATION, TISSUES, ALBURNUS LUCIDUS, BELGIUM.: RUTHEN W71-09168
IFICANCE, *EXTRACTION, LUCIFERIN- LUCIFERASE METHOD, SEDIMENT-WATER W72-04292
ENOSINE TRIPHOSPHATE, *LUCIFERIN- LUCIFERASE ENZYME SYSTEM, PLANT E W71-11916
RADIOISOTOPES, N-15, LUCIFERIN, LUCIFERASE, CHLOROPHYLL A, OPTICA W72-10883
GIN, *SIGNIFICANCE, *EXTRACTION, LUCIFERIN-LUCIFERASE METHOD, SEDI W72-04292
E, NITROGEN RADIOISOTOPES, N-15, LUCIFERIN, LUCIFERASE, CHLOROPHYL W72-10883
OTA, LAKE MALOGEN, SWEDEN, CHAR, LUCIOPERCA LUCISPERCA, BURBOT, CO W72-12729
ALOGEN, SWEDEN, CHAR, LUCIOPERCA LUCISPERCA, BURBOT, COREGONUS LAV W72-12729
ORGANOMERCURIAL COMPOUNDS, ESOX LUCIUS, ALKYL MERCURY, EAGLES, SA W72-05952
ENS, AMBLOPLITES RUPESTRIS, ESOX LUCIUS, CAMPELOMA, PSIDIUM, SHADF W70-08098
YCEAE, PONTOPOREIA AFFINIS, ESOX LUCIUS, PERCA FLUVIATILIS, SALVEL W72-12729
ION, SAMPLE PRESERVATION, SALPA, LUDOX AM, SUCROSE, DENSITY GRADIE W73-00371
, BOTTOM TOPOGRAPHY, PRIONOSPIO, LUMBRINEREIS, GLYCERA, NEPHTYS, C W72-09092
EROL, PHOTOHETEROTROPHIC GROWTH, LUMINARIN-TYPE POLYSACCHARIDES, C W70-07280
O CLAM, TIVELA STULTORUM, OSTREA LURIDA, CARDIUM: / TUNICATE, PISM W72-07907
, XANTHOPHYLLS, MINNESOTA LAKES, LUTEIN.: CAROTENOIDS W72-01784
EGATUM, NUPHAR JAPONICUM, NUPHAR LUTEN, ANTICANCER POTENTIAL, ANTI W72-04275
UPENTACTA, CNEMIDOCARPA, ENHYDRA LUTRIS NEREIS, SEA OTTER, MACROCY W73-00932
ENGLAND), BLELHAM TARN(ENGLAND), LUXURY UPTAKE, LOCH LEVEN(SCOTLAN W71-13183
USCULA, GOMONTIA POLY/ *YUCATAN, LYNGBYA CONFERVOIDES, LYNGBYA MAJ W72-11800
ADECANE, 8-MET/ NOSTOC MUSCORUM, LYNGBYA LAGERHAIMII, 7-METHYLHEPT W73-01439
LAKE BUDD, OSCILLATORIA REDEKEI, LYNGBYA LIMNETICA, ANABAENA SPP., W72-07145
, TEGERNSEE, ANABAENA, COREGONS, LYNGBYA LIMNETICA, OSCILLATORIA R W71-00117
*YUCATAN, LYNGBYA CONFERVOIDES, LYNGBYA MAJUSCULA, GOMONTIA POLYR W72-11800
OTROPHIC, MICROCYSTIS, ANABAENA, LYNGBYA.: MES W72-04269
VER, OCTORARA RIVER, PLECTONEMA, LYNGBYA, PHORMIDIUM, CYANOPHYCEAE W72-04736
ASTRUM, PHACOMYXA, SPONDYLOSIUM, LYNGBYA, PSEUDOANABAENA, DATA INT W72-07648
*BOSTON(MASS), LYNN HARBOR(MASS).: W71-05553
AL PRESERVATION, HISTOCHEMISTRY, LYOPHILIZATION, ANIMAL TISSUES, H W73-01676
MO GAIRDNERII, IRIS PSEUDACORUS, LYSIMACHIA NUMMULARIA, BLUE HERON W72-05952
NE, ALANINE, ISOLEUCINE, VALINE, LYSINE, ASPARAGINE, CYSTINE, CY: / W73-00942
C INHIBITION, HOSTS, MYXOPHYCEA/ LYTIC BACTERIA, ENGLAND, METABOLI W72-05457
(FRANCE), CLADOPHORA, PHO/ RHINE- MAAS BASIN(FRANCE), MEURTHE RIVER W71-00099
UTILIZATION, MODIFICATION./ FARM MACHINERY, INDUSTRIAL MACHINERY, W71-06188
ION./ FARM MACHINERY, INDUSTRIAL MACHINERY, UTILIZATION, MODIFICAT W71-06188
U RIVER(QUEBEC), GREEN BAY(WIS), MACKEREL, MONTREAL(CANADA).: /LIE W70-08098

LA TORQUATA, CAPITELLA CAPITATA, MACOMA CALCAREA, ASTARTE BOREALIS W73-02029
NA ESTUARY, DATA INTERPRETATION, MACROALGAE, SPECIES DIVERSITY IND W72-08141
FURCA, CERATIUM TRIPOS, CERATIUM MACROCEROS, NOCTILUCA MILIARIS, G W71-07875
LAND STREAM, ECOSYSTEM ANALYSIS, MACROCONSUMERS, AUTUMNAL COMMUNIT W71-10064
*LAMINARIA SP., *ULVA SP., MACROCYSTIS SP.: W71-10553
NHYDRA LUTRIS NEREIS, SEA OTTER, MACROCYSTIS PYRIFERA, CHORDATES, W73-00932
TUNICATES, SPONGES, VERTEBRATES, MACROINVERTEBRATES, SPECIES DIVER W73-00932
RIVER, LAKE BORGNE, SUBSTRATES, MACROINVERTEBRATES, NEW YORK BIGH W73-02029
EPIFLORA, WALLEYE LARVAE, SHAD, MACROINVERTEBRATES, CYANOPHYCEAE. W72-08142
ELEMENTAL ANALYZERS, MACROINVERTEBRATES.: W72-07715
ATER, GALETA ISLAND, CANAL ZONE, MACROINVERTEBRATES, DIESEL FUEL, W72-07911
RIPHOSPHATE, *AUTOTROPHIC INDEX, MACROINVERTEBRATES, ACETYLENE, ET W72-10883
TICS, ENRICHMENT, CULTURE MEDIA, MACROINVERTEBRATES.: /, BIOENERGE W72-09378
RESERVOIR(NJ), AMERICAN SHAD, / MACROINVERTEBRATES, *TOCKS ISLAND W72-14280
S CHROMATIUM OKENII, CHROMATIUM, MACROMONAS MOBILIS, THIOVULUM MAJ W72-07896
SH), URANIUM, THORIUM, ANTIMONY, MACRONUTRIENTS, MICRONUTRIENTS, S W73-01434
PREP/ DISSOLVED ORGANIC CARBON, MACROPHYTES, SARGASSO SEA, SAMPLE W73-02100
C-14, DISSOLVED ORGANIC MATTER, MACROPHYTES, BIOLOGICAL SAMPLES: / W73-02095
AKE WABAMUN, BIOLOGICAL STUDIES, MACROPHYTES.: *L W73-01704
AY SPECTROMETRY, ANIMAL TISSUES, MACROPHYTES, CYPRINUS CARPIO, MOS W73-01673
ELLIA PALUSTRIS, CODIUM, FUSCUS, MACROPHYTES, AMMOBACULITES DILATA W73-00367
E, PHRAGMITES COMMUNIS, *NORWAY, MACROPHYTES, FYRISAN RIVER, ARBOG W73-00286
OSCILLATORIA, / *CHLOROPHYLL A, MACROPHYTES, *SWEDEN, MICROCYSIS, W72-12729
USSR, *TOXICOLOGY, *HYDROBIONTS, MACROPHYTES, SAPROPHYTES.: * W72-09650
ONTS, ICHTHYOLOGY, ICHTHYOFAUNA, MACROPHYTES, SAPROPHYTES, CERIUM. W72-09646
LAMINARIA/ *SUBLITTORAL BENTHIC MACROPHYTES, RIA DE ALDAN(SPAIN), W72-04766
OSIPHON INTESTINALIS, HECATONEMA MACULANS, ELACHISTA LUBRICA, RALF W72-01370
H)/ SALT VALLEY RESERVOIRS(NEB), MADISON(WIS), LAKE WASHINGTON(WAS W72-04759
SA(WIS), OCCOQUAN RESERVOIR(VA), MADISON(WIS), LAKE MENDOTA(WIS), W72-05473
ODO, HYDROPSYCHIDAE, PLECOPTERA, MAGALOPTERA, EPHEMEROPTERA, NANNO W70-06655
SWICK, ANTICOSTI ISLAND(QUEBEC), MAGDALEN ISLAND(QUEBEC), SABLE IS W71-10066
IS, GLYCERA, NEPHTYS, CAPITELLA, MAGELONA, ARMANDIA, TELLINA BUTTO W72-09092
APHNIA GALEATA MENDOTAE, DAPHNIA MAGNA.: /OMONAS, SPHAEROCYSTIS, D W71-03014
GASTEROSTEUS ACULEATUS, DAPHNIA MAGNA, DAPHNIA, GASTEROSTEUS.: /, W72-05952
CHLORIS COMMUTATA, PLEUROCHLORIS MAGNA, VISCHERIA PUNCTATA, VISCHE W72-14808
RATION, VERTEBRATES, ABSORBANCE, MAGNESIUM ION CONCENTRATION, BRUC W72-09092
SEPARATION TECHNIQUES, CALCIUM, MAGNESIUM, E. COLI, SILICA, *FILT W72-11833
S, NITRATES, CHLORIDES, CALCIUM, MAGNESIUM, SODIUM, POTASSIUM, IRO W71-11001
LLOIDS, CARBON DIOXIDE, CALCIUM, MAGNESIUM, NUTRIENTS, *WASTE WATE W72-03613
CTIVITY, *ECOSYSTEMS, POTASSIUM, MAGNESIUM, SODIUM, CALCIUM, HYDRO W72-14303
OISOTOPES, CHLOROPHYLL, PIGMENT, MAGNESIUM, PHOSPHATES, NITROGEN, W72-07660
AMINO ACIDS, SPECTROPHOTOMETRY, MAGNESIUM, SULFATES, SPORES, AZOT W72-06124
TES, SODIUM, POTASSIUM, CALCIUM, MAGNESIUM, CHLORIDES, HYDROGEN IO W72-05610
*LAKES, *ALGAE, ALKALINE WATER, MAGNESIUM, SODIUM SULFATES, CYANO W71-04526
SOTOPES, ALUMINUM, IRON, COPPER, MAGNESIUM, MANAGNESE, CALCIUM, CH W71-09188
HLORINATED BIPHENYLS, BIOLOGICAL MAGNIFICATION, CHLAMYDOMONAS, MET W72-04134
YLCHLOR, METABOLITES, BIOLOGICAL MAGNIFICATION, BIOLOGICAL SAMPLES W72-07703
E GAS CHROMATOGRAPHY *BIOLOGICAL MAGNIFICATION, GAS LIQUID CHROMAT W73-02105
AKE(ONTARIO), ASHING, BIOLOGICAL MAGNIFICATION.: /TOMETRY, PERCH L W71-11529
MONY, NORWAY, ASHING, BIOLOGICAL MAGNIFICATION, PELVETIA CANALICUL W71-11515
AL CULTURES, OUABAIN, BIOLOGICAL MAGNIFICATION, PHOSPHORUS-32, LIM W72-09668
S, *MINAMATA DISEASE, BIOLOGICAL MAGNIFICATION, MINAMATA BAY(JAPAN W72-12257
AND, BIOACCUMULATION, BIOLOGICAL MAGNIFICATION, FATE OF POLLUTANTS W72-11925
*LOXODES MAGNUS.: W72-14796
IA, NEW YORK, OREGON, WISCONSIN, MAINE.: / WINDS, NEVADA, CALIFORN W71-05626
AL, LAKE TAHOE, LAKE SEBASTICOOK(MAINE), FAYETTEVILLE GREEN LAKE(N W71-05626
OPHYTA, PHAEOPHYTA, CHLOROPHYTA, MAINE, SPECIATION.: / FLORA, RHOD W71-10066
ICS, *VARIABILITY, WAVES(WATER), MAINE, STATISTICAL METHODS, SEA W W73-00369
GENTS, PESTICIDES, OPERATION AND MAINTENANCE, *WATER TREATMENT, *M W72-08357
TER/ *FILTRATION, *OPERATION AND MAINTENANCE, *COST ANALYSIS, *FIL W72-09393
INDUSTRIAL WASTES, OPERATION AND MAINTENANCE, *WASTE WATER TREATME W72-09386
HA, *MEDUSAE MAINTENANCE, *POLYP MAINTENANCE, *POLYP METABOLISM, T W72-10162
HRYSAORA QUINQUECIRRHA, *MEDUSAE MAINTENANCE, *POLYP MAINTENANCE, W72-10162
POTENTIAL, ODORS, OPERATION AND MAINTENANCE, WASTE WATER TREATMEN W72-10401
ON(WIS), LAKE MENDOTA(WIS), LAWN MAINTENANCE.: *MADIS W73-02479
S, ENRICHMENT, LEUCOTHRIX MUCOR, MAJURO ATOLL, MARCHALL ISLAND, PO W72-08142
M, MACROMONAS MOBILIS, THIOVULUM MAJUS, THIOSPIRA WINOGRADSKY, ZOO W72-07896
N, LYNGBYA CONFERVOIDES, LYNGBYA MAJUSCULA, GOMONTIA POLYRHIZA, UL W72-11800
TUDIES, PRECISION, *SWEDEN, LAKE MALAREN, MICROCYSTIS FLOS-AQUAE, W73-00286
OTOPES, CHROMULINA, CRYPTOMONAS, MALLOMONAS, STAURASTRUM, PHACOMYX W72-07648
E, SALMON TROUT, LOTA LOTA, LAKE MALOGEN, SWEDEN, CHAR, LUCIOPERCA W72-12729
NS, GLUCOSE, GALACTOSE, SUCROSE, MALTOSE, MANNITAL, SORBITAL, SUCC W72-05421
, ASCOPHYLLUM NODOSUM, GIGARTINA MAMILLOSA, RHODYMENIA PALMATA.: / W71-11515
ENS, ALGAE, FUNGI, SMALL ANIMALS(MAMMALS), CYTOLOGICAL STUDIES, SO W72-00949
ERS, ZOOPLANKTON, PHYTOPLANKTON, MAMMALS, MARINE ALGAE, FOOD CHAIN W72-00960
S NAUT/ EPINEUSTONT, NAUTOCOCCUS MAMMILATUS, NAUTOCOCCUS PYRIFORMI W72-07141
, SAMPLING, PHYTOPLANKTON, IRON, MAN: /, CHEMICAL ANALYSIS, LEAVES W73-01089
*SALTON SEA(CALIF), ECOLOGICAL MANAGEMENT.: W71-10577
ATION, *WATER POLLUTION CONTROL, MANAGEMENT, ANALYSIS, WATER QUALI W72-09171
ROPHYTA, NITROGEN FIXATION, LAND MANAGEMENT, FERTILIZATION, PRIMAR W72-04769
ING, DRAINAGE, FARM WASTES, SOIL MANAGEMENT, GROUNDWATER, WISCONSI W71-06443
S, FOODS, FISH, HERBIVORES, FISH MANAGEMENT.: /BIOCONTROL, REMEDIE W71-09163
NTAMINANT, STREAM QUALITY, WASTE MANAGEMENT, MANURE COLLECTION, MA W71-08214
AGRICULTURE, FALLOUT, DRAINAGE, MANAGEMENT, WISCONSIN, PRODUCTIVI W71-03012
TS, *ECOSYST/ *FORESTS, *CUTTING MANAGEMENT, *HERBICIDES, *NUTRIEN W71-01489
UMINUM, IRON, COPPER, MAGNESIUM, MANAGNESE, CALCIUM, CHROMIUM, POT W71-09188
IDEA, PALMERIAMONAS PLANCTONICA, MANCHUDINIUM SINICUM, SINAMONAS S W72-07683
UND, CORE SOUND.: MANGANESE RADIOISOTOPES, BOGUE SO W73-00818
ADIOISOTOPES, RUTHENIUM RADIOIS/ MANGANESE RADIOISOTOPES, CESIUM R W71-09188
DIOISOTOPES, IRON RADIOISOTOPES, MANGANESE RADIOISOTOPES.: /IUM RA W71-09250

ZINC,	MANGANESE RADIOISOTOPES.:	W71-09190
ADIOISOTOPES,:	MANGANESE RADIOISOTOPES, CESIUM R	W71-09018
UM, CULTURE MEDIA, ZINC SULFATE,	MANGANESE CHLORIDE, NICKEL SULFAT	W72-12739
S, RESISTANCE, PHOSPHORUS, IRON,	MANGANESE, ZINC, NICKEL, COBALT,	W72-12739
NION, MATHEMATICAL MODELS, IRON,	MANGANESE, PHOSPHORUS, ANAEROBIC	W72-12997
TER, *INORGANIC COMPOUNDS, IRON,	MANGANESE, PHOSPHORUS, CARBON, NI	W72-05506
GAE, MUD-WATER INTERFACES, IRON,	MANGANESE, PHOSPHORUS, NITROGEN,	W72-05469
SPHORUS, CARBON, ORGANIC MATTER,	MANGANESE, IRON, SULFUR, POTASSIU	W72-07514
EN, PHOSPHORUS, SULFUR, CALCIUM,	MANGANESE, POTASSIUM, SODIUM, CAR	W73-00840
DIOISOTOPES, ZINC RADIOISOTOPES,	MANGANESE, CHROMIUM, IRON, ANALYT	W73-00832
ROPHYTA, PYRROPHYTA, CYANOPHYTA,	MANGANESE, ZINC, METALS, SECCHI D	W72-09111
NESIUM, SODIUM, POTASSIUM, IRON,	MANGANESE, COPPER, ALGAE, SAMPLIN	W71-11001
CE ELEMENTS, COPPER, MOLYBDENUM,	MANGANESE, COBALT, IRON, INDUSTRI	W71-11515
DIOISOTOPES, LINC RADIOISOTOPES,	MANGANESE, COBALT RADIOISOTOPES,	W71-13243
, BACTERIA, LAKE ERIE, ALUMINUM,	MANGANESE, SODIUM, CHROMIUM, IRON	W72-01101
DES, OYSTERS, MUSSELS, PROTOZOA,	MANGROVE SWAMPS, LITTORAL, SHORES	W72-07911
SE, GALACTOSE, SUCROSE, MALTOSE,	MANNITAL, SORBITAL, SUCCINATE, FU	W72-05421
, PHAEOCYSTIS POUCHETI, GLUCOSE,	MANNOSE, RHAMNOSE, ACRYLIC ACID,	W72-07710
TREAM QUALITY, WASTE MANAGEMENT,	MANURE COLLECTION, MANURE TRANSPO	W71-08214
CTION PUMP, ALUM INJECTION PUMP,	MANURE GRINDER.: /TION, ACID INJE	W72-02850
*OXIDATION DITCH, HYDRAULIC	MANURE HANDLING.:	W72-08396
E MANAGEMENT, MANURE COLLECTION,	MANURE TRANSPORT.: /QUALITY, WAST	W71-08214
NG VALUE.: *ANIMAL	MANURE, *LITERATURE REVIEW, FEEDI	W71-07544
.: *GREEN	MANURE, INDIA, JAPAN, RICE FIELDS	W73-02473
TES, *AQUATIC POPULATIONS, MAPS,	MAPPING, PHYTOPLANKTON, PACIFIC O	W72-03567
HOSPHATES, *AQUATIC POPULATIONS,	MAPS, MAPPING, PHYTOPLANKTON, PAC	W72-03567
LEUCOTHRIX MUCOR, MAJURO ATOLL,	MARCHALL ISLAND, POLYSIPHONIA LAN	W72-08142
IVERSITY INDEX, CHLOROPHYLL A, /	MARGALEF PIGMENT RATIO, PIGMENT D	W73-00916
UM CRIBROSUM, *NOVA SCOTIA, *ST.	MARGARET'S BAY.: / DIGITATA, AGAR	W73-00366
CULOSUS, LAMINARIA DIGITATA, ST.	MARGARET'S BAY, SCOUDOUC RIVER, H	W72-12166
GYMNODINIUM SPLENDENS, EXUVIELLA	MARINA, EUTREPTIA, NITZSCHIA SERI	W71-07875
SP., PHYLLOSPA/ *HYALOCHLORELLA	MARINA, RHODOMELA SP., ENDOCLADIA	W71-11851
LIGHT INTENSITY, EUTROPHICATION,	MARINE AL: /ERMOCLINE, NITRATES,	W71-07875
NG, MARINE FISH, MARINE ANIMALS,	MARINE ALGAE, TRITIUM, IODINE RAD	W71-09018
TS, CORAL, INVERTEBRATES, ALGAE,	MARINE ALGAE, ALGAL POISONING, CH	W71-08597
AINS, IRRIGATION, FLOW PROFILES,	MARINE ALGAE, WINDS, SEA WATER, P	W71-09225
ISM, *ABSORPTION, MARINE PLANTS,	MARINE ALGAE, PHAEPHYTA, AQUARIA,	W71-09193
ARIA, MOLLUSKS, SNAILS, MUSSELS,	MARINE ALGAE, AQUATIC ANIMALS, FR	W71-09182
ON, TROUT, TOXICITY, WASHINGTON,	MARINE ALGAE, MARINE BACTERIA.: /	W72-05421
PROPERTIES, FISH, MARINE PLANTS,	MARINE ALGAE, HYDROLYSIS, CHELATI	W72-04463
FFECTS, MOLLUSKS, MARINE PLANTS,	MARINE ALGAE, PUBLIC HEALTH, ESTU	W72-04461
CTIVITY, RADIOISOTOPES, TRACERS,	MARINE ALGAE, HAWAII, WATER TEMPE	W71-11486
DUES, ALGAE, DDT, PHYTOPLANKTON,	MARINE ALGAE, AQUATIC ALGAE, COLO	W72-03545
AR POWERPLANTS, SITES, MOLLUSKS,	MARINE ALGAE, ON-SITE INVESTIGATI	W72-03324
SOIL-WATER-PLANT RELATIONSHIPS,	MARINE ALGAE, MOSSES, CRABS, ALGA	W72-03347
LANKTON, PHYTOPLANKTON, MAMMALS,	MARINE ALGAE, FOOD CHAINS, MARINE	W72-00960
FACE WATERS, PATH OF POLLUTANTS,	MARINE ALGAE, MARINE FISH, ABSORP	W72-00940
OUT, OYSTERS, MUSSELS, PLANKTON,	MARINE ALG: /NT, REGULATION, FALL	W72-00959
, METHODOLOGY, FOREIGN RESEARCH,	MARINE ALGAE, PRODUCTIVITY, POLLU	W72-08804
NGINEERING, SECONDARY TREATMENT,	MARINE ALGAE, PHYTOPLANKTON, BIOM	W72-09378
LANTIC OCEAN, *TROPICAL REGIONS,	MARINE ALGAE, SPATIAL DISTRIBUTIO	W72-11719
IDATION, SYMBIOSIS, AZOTOBACTER,	MARINE ALGAE, ORGANIC MATTER.: /X	W72-11563
ASTES, PATH OF POLLUTANTS, ASIA,	MARINE ALGAE, SEDIMENTS, ESTUARIN	W72-10958
LEAR WASTES, PATH OF POLLUTANTS,	MARINE ALGAE, CHROMOSOMES, EMBRYO	W72-10955
POLLUTION, PHAEOPHYTA, MOLLUSKS,	MARINE ALGAE, MARINE ANIMALS.: /	W72-11925
ACTIVITY, CARBON DIOXIDE, SALTS,	MARINE ALGAE.: /M, TRACERS, RADIO	W72-12634
DT, *DIELDRIN, *ALDRIN, *ENDRIN,	MARINE ALGAE, BOTTOM SEDIMENTS, P	W73-00361
ECHNIQUES, *MARINE FISH, SILICA,	MARINE ALGAE, INVERTEBRATES, ISOL	W73-00371
*STANDING CROPS, TIDAL EFFECTS,	MARINE ALGAE, SEA WATER, SALINITY	W73-00367
ICAL METHODS, SEA WATER, COASTS,	MARINE ALGAE.: /), MAINE, STATIST	W73-00369
SEASONAL, CRUSTACEANS, ISOPODS,	MARINE ALGAE, *ALASKA.: /NNELIDS,	W73-00374
BITION, ECOLOGICAL DISTRIBUTION,	MARINE ALGAE, MARINE ANIMALS.: /I	W73-00348
OF WATER, PYRROPHYTA, PROTOZOA,	MARINE ALGAE, ZOOPLANKTON, PHYTOP	W73-00296
EOGRAPHICAL REGIONS, PHAEOPHYTA,	MARINE ALGAE, ENVIRONMENTAL EFFEC	W73-00428
VITY TECHNIQUES, EUTROPHICATION,	MARINE ALGAE, WATER POLLUTION SOU	W73-00431
ATOMS, POLLUTANT IDENTIFICATION,	MARINE ALGAE, CHRYSOPHYTA, AUTUMN	W73-00442
NTS, FOOD CHAINS, PHYTOPLANKTON,	MARINE ALGAE, ENVIRONMENTAL EFFEC	W73-00818
OISOTOPES, IODINE RADIOISOTOPES,	MARINE ALGAE, TRACE ELEMENTS, PAT	W73-00811
T IDENTIFICATION, PHYTOPLANKTON,	MARINE ALGAE, ESSENTIAL NUTRIENTS	W73-00441
YTA, CHLOROPHYTA, AQUATIC ALGAE,	MARINE ALGAE, SESSILE ALGAE.: /PH	W73-00273
TY, *CYANOPHYTA, *NEW HAMPSHIRE,	MARINE ALGAE, PHYTOPLANKTON, AQUA	W72-14542
DENSITY, *DEPTH, MARINE ANIMALS,	MARINE ALGAE, CYANOPHYTA, PYRROPH	W72-14313
IC LIFE, AQUATIC MICROORGANISMS,	MARINE ALGAE, SALINITY, MIGRATION	W72-14317
*MARINE ANIMALS, MARINE PLANTS,	MARINE ALGAE, NUCLEAR WASTES, PAT	W73-00828
IRONMENTAL EFFECTS, CHLOROPHYTA,	MARINE ALGAE , SALT TOLERANCE, CR	W73-00855
TY EFFECTS, NUCLEAR POWERPLANTS,	MARINE ALGAE, MARINE ANIMALS, MAR	W73-00833
T, *ABSORPTION, *MARINE ANIMALS,	MARINE ALGAE, SEDIMENTS, PUERTO R	W73-00821
RECASTING, COBALT RADIOISOTOPES,	MARINE ALGAE, ABSORPTION, SEDIMEN	W73-00831
CITY, *CALIFORNIA, RADIOECOLOGY,	MARINE ALGAE, ESTUARINE ENVIRONME	W73-00826
ACIDS, CHRYSOPHYTA, INHIBITION,	MARINE ALGAE, DIATOMS, BEHAVIOR.:	W73-00942
SOPHYTA, CYANOPHYTA, PYRROPHYTA,	MARINE ALGAE, GASTROPODS.: / CHRY	W73-00935
RNIA, SEA WATER, WATER ANALYSIS,	MARINE ALGAE, CHEMICAL ANALYSIS,	W73-02105
RIDGES, BUOYS, HARBORS, TUNNELS,	MARINE ALGAE, ANIMAL POPULATIONS,	W73-02029
ALITY, BENTHOS, CRABS, COPEPODS,	MARINE ALGAE, SHRIMP, FISH EGGS,	W72-08142
, PHYTOPLANKTON, DIATOMS, ALGAE,	MARINE ALGAE, AQUATIC ALGAE, EUGL	W72-08436
ATIC ALGAE, STATISTICAL METHODS,	MARINE ALGAE, *OREGON, *SEASONAL,	W72-08141
WAGE TREATMENT, SEWAGE DISPOSAL,	MARINE ALGAE, CLAMS, MUSSELS, OYS	W72-07715
HEADS, SALMON, SHELLFISH, CLAMS,	MARINE ANIMALS, AQUATIC LIFE, ECO	W72-07907
CLEAR POWERPLANTS, MARINE ALGAE,	MARINE ANIMALS, MARINE PLANTS, HU	W73-00833
ANIMALS, *HALOPHYTES, SEA WATER,	MARINE ANIMALS, BIOLOGICAL COMMUN	W73-00855

DISTRIBUTION, *DENSITY, *DEPTH,	MARINE ANIMALS, MARINE ALGAE, CYA	W72-14313
DS, MOLLUSKS, WATER TEMPERATURE,	MARINE ANIMALS, ALASKA, BIOMASS,	W73-00374
ICAL DISTRIBUTION, MARINE ALGAE,	MARINE ANIMALS.: /IBITION, ECOLOG	W73-00348
ECOLOGY, DISTRIBUTION PATTERNS,	MARINE ANIMALS, SALINITY, SAMPLIN	W73-00296
OLATION, REFRACTIVITY, COPEPODS,	MARINE ANIMALS, MOLLUSKS, GASTROP	W73-00371
EOPHYTA, MOLLUSKS, MARINE ALGAE,	MARINE ANIMALS.: / POLLUTION, PHA	W72-11925
OMS, HYDROGEN ION CONCENTRATION,	MARINE ANIMALS, BACTERIA, ABSORPT	W71-10083
ERS, ABSORPTION, FISH, MOLLUSKS,	MARINE ANIMALS, COBALT RADIOISOTO	W71-09225
, *BIOTA, SAMPLING, MARINE FISH,	MARINE ANIMALS, MARINE ALGAE, TRI	W71-09018
ROPHYTA, PHAEOPHYTA, RHODOPHYTA,	MARINE BACTERIA.: /, OCEANS, CHLO	W71-07878
ICITY, WASHINGTON, MARINE ALGAE,	MARINE BACTERIA.: /ON, TROUT, TOX	W72-05421
D WEBS, STRONTIUM RADIOISOTOPES,	MARINE BACTERIA, COASTAL MARSHES,	W72-00948
*THERMAL POWER, ATLANTIC OCEAN,	MARINE BIOLOGY, BENTHIC FLORA, BI	W72-11876
MALS, MARINE ALGAE, FOOD CHAINS,	MARINE FISHERIES, LAKES, POLLUTIO	W72-00960
ATH OF POLLUTANTS, MARINE ALGAE,	MARINE FISH, ABSORPTION, WATER PO	W72-00940
HAEOPHYTA, CARBON RADIOISOTOPES,	MARINE FISH, INSTRUMENTATION, ABS	W72-06283
POTABLE WATER, *BIOTA, SAMPLING,	MARINE FISH, MARINE ANIMALS, MARI	W71-09018
OOD CHAINS, ESTUARINE FISHERIES,	MARINE FISHERIES, SEWAGE DISPOSAL	W72-12567
STAL PLAIN, COASTS, CRUSTACEANS,	MARINE FISH, MOLLUSKS, POLLUTANT	W73-00819
E , SALT TOLERANCE, CRUSTACEANS,	MARINE FISH, MARSH PLANTS.: /ALGA	W73-00855
SAMPLING, ENVIRONMENTAL EFFECTS,	MARINE FISH, PERIPHYTON, ON-SITE	W73-00932
, *CARBOHYDRATES, MARINE PLANTS,	MARINE MICROORGANISMS, WATER ANAL	W73-02100
SH REPRODUCTION, WASTE DISPOSAL,	MARINE MICROORGANISMS, SEDIMENTAT	W71-13723
TURBULENT FLOW, ALGAE, BACTERIA,	MARINE PLANTS, ENGINEERING, WATER	W72-03220
FFECTS, *INDICATORS, *SEA WATER,	MARINE PLANTS, ASSESSMENTS, MONIT	W71-12867
, *MARINE ALGAE, *CARBOHYDRATES,	MARINE PLANTS, MARINE MICROORGANI	W73-02100
SOURCES, OIL SPILLS, SEA WATER,	MARINE PLANTS, PHAEOPHYTA, OILY W	W73-01074
ECTS, *FALLOUT, *MARINE ANIMALS,	MARINE PLANTS, MARINE ALGAE, NUCL	W73-00828
E CAPACITY, CALIFORNIA, TRITIUM,	MARINE PLANTS, MONITORING, SAMPLI	W73-00831
R TEMPERATURE, DATA COLLECTIONS,	MARINE PLANTS, ENVIRONMENTAL EFFE	W73-00838
S, MARINE ALGAE, MARINE ANIMALS,	MARINE PLANTS, HUMAN POPULATION,	W73-00833
NE ALGAE, ENVIRONMENTAL EFFECTS,	MARINE PLANTS, CANADA, WASHINGTON	W73-00428
HYTA, *MARINE ALGAE, *ISOLATION,	MARINE PLANTS, EUTROPHICATION, OC	W73-00430
Y, ENERGY BUDGET, *MARINE ALGAE,	MARINE PLANTS, BIOMASS, CANADA, C	W73-00366
ON, CARBON DIOXIDE, RESPIRATION,	MARINE PLANTS, BICARBONATES, MASS	W72-12709
OASSAY, REPRODUCTION, SEA WATER,	MARINE PLANTS, PRIMARY PRODUCTIVI	W72-14294
HNIQUES, *MARINE MICROORGANISMS,	MARINE PLANTS, BACTERIA, POPULATI	W71-08042
NKTON, *METABOLISM, *ABSORPTION,	MARINE PLANTS, MARINE ALGAE, PHAE	W71-09193
HYSICOCHEMICAL PROPERTIES, FISH,	MARINE PLANTS, MARINE ALGAE, HYDR	W72-04463
TER POLLUTION EFFECTS, MOLLUSKS,	MARINE PLANTS, MARINE ALGAE, PUBL	W72-04461
PHAERICA, TETRACULAMONAS NATANS,	MARINIAMONAS SAUPAULENSIS, ENEIDA	W72-09119
002, *GREEN ALGAE, PRASINOCLADUS	MARINUS, TEMPERATURE EFFECTS.: /1	W70-09429
ITY, MIGRATION, TIDES, LOW WATER	MARK, CHRYSOPHYTA.: /ALGAE, SALIN	W72-14317
IZYMENIA, RHODOMELA, RHODOMENIA,	MARMION ISLAND, MORRIA REEF, BEAR	W72-09120
RANCE, CRUSTACEANS, MARINE FISH,	MARSH PLANTS.: /ALGAE , SALT TOLE	W73-00855
CYCLING NUTRIENTS, RESPIRATION,	MARSH PLANTS, CHEMICAL ANALYSIS,	W73-01089
OMMUNITIES, *ENVIRONMENTA/ *SALT	MARSHES, *PROTOZOA, *BIOLOGICAL C	W73-00367
ORIDA, RAD/ *MARINE ALGAE, *SALT	MARSHES, *PROTOZOA, *BACTERIA, FL	W73-00796
OPES, STRONT/ *ECOSYSTEMS, *SALT	MARSHES, *PRODUCTIVITY, RADIOISOT	W71-09254
SALINITY, / *PHAEOPHYTA, *TIDAL	MARSHES, *MARINE ALGAE, CULTURES,	W72-12633
RONMENT, *DOMESTIC WASTES, *SALT	MARSHES, *CARBON CYCLE, *PHOSPHOR	W72-12567
NS, FISH TOXINS, SEDIMENTS, SALT	MARSHES, ECOLOGY, PESTICIDES, EST	W72-10678
E, BIOLOGICAL COMMUNITIES, TIDAL	MARSHES, FOOD WEBS, INTERTIDAL AR	W73-00796
OTOPES, MARINE BACTERIA, COASTAL	MARSHES, MICROORGANISMS.: /ADIOIS	W72-00948
*PROTOZOA, *MARINE ALGAE, *SALT	MARSHES, NEW YORK, NEMATODES, EPI	W72-00948
ALGAE, CULTURES, SALINITY, SALT	MARSHES, SYSTEMATICS, *VIRGINIA,	W72-12633
N, LAND USE, SHORES, MONITORING,	MARSHES, SOIL EROSION, WASTE WATE	W73-02479
URES, CHELATING AGENTS, OOCYSTIS	MARSSONII.: /NALYSIS, *ALGAL CULT	W73-02112
GLAND), AMPHORA OVALIS, OPEPHORA	MARTYI, CELL COUNTS, NAVICULA, NI	W71-12855
RIX, LAKE ITASCA, LAKE ELK, LAKE	MARY, LAKE JOSEPHINE, LAKE ARCO,	W72-07145
TION LAGOONS, PONDS, FARM PONDS,	MARYLAND, QUARRIES, OHIO RIVER, H	W72-04736
IS, BAC/ JUDAY NET, ASTEROLAMPRA	MARYLANDICA, ASTEROLAMPRA HEPTACT	W71-11527
ALKANES, ALIPHATIC HYDROCARBONS,	MASS SPECTROMETRY, INFRARED SPECT	W73-01439
ON, MARINE PLANTS, BICARBONATES,	MASS SPECTROMETRY, DISTRIBUTION,	W72-12709
*WOODS HOLE(	MASS).:	W72-06046
*BOSTON(MASS), LYNN HARBOR(	MASS).:	W71-05553
, DATA INTERPRE/ *CAPE COD CANAL(	MASS), BUZZARDS BAY, CAPE COD BAY	W73-01449
, PANAMA, CALLAO(PERU), CAPE COD(	MASS), CAPE MAY(N J), SARGASSO SE	W72-08056
*BOSTON(	MASS), LYNN HARBOR(MASS).:	W71-05553
*ORGANIC CARBON, *ALGAL	MASS, ALGAL GROWTH.:	W72-04431
HICATION, WATER QUALITY CONTROL,	MASSACHUSETTS, *NUTRIENTS, OUTLET	W72-06046
ANALYTICAL TECHNIQUES, SAMPLING,	MASSACHUSETTS, POLLUTANT IDENTIFI	W71-11498
M, PERIDINIUM PALLIDUM, CERATIUM	MASSILENSE, CERATIUM FURCA, CERAT	W71-07875
T, CYLINDROSPERMUM LICHENIFORME,	MASTIGOCLADUS LAMINOSUS.: /NE TES	W71-12070
CCUS, FLEXIBACTERIA, PHORMIDIUM,	MASTIGOCLADUS, PHEOPHYTIN, GRASSL	W72-03557
NAL PARK(WYO), BLUE-GREEN ALGAE,	MASTIGOCLADUS, CALOTHRIX.: /NATIO	W72-08508
OUS FORESTS, DECOMPOSING ORGANIC	MATERIAL, FERTILITY, RHIZOSPHERE,	W72-03342
, *ALGAL EXUDATES, EXTRACELLULAR	MATERIAL, TURBULENT FRICTION, SEA	W71-07878
ICOOK(MAINE), FAYETTEVILL/ FECAL	MATERIAL, LAKE TAHOE, LAKE SEBAST	W71-05626
ERGENT BAN, WYAND/ *CARBONACEOUS	MATERIAL, *CANADIAN PHOSPHATE DET	W70-07283
REFRACTORY	MATERIALS.:	W72-01881
ACE ORGANIC CONTAMINANTS, *TRACE	MATERIALS, BIOINHIBITION, BIOSTIM	W71-11793
, TOXICITY, BEHAVIOR, BITUMINOUS	MATERIALS, ANIMAL WASTES, FISH, F	W72-08790
SAMPLE PRESERVATION/ *BIOLOGICAL	MATERIALS, *BIOLOGICAL SAMPLES, *	W73-01676
N, CATION ADSORPTION, CHELATION,	MATHEMATICAL MODELS.: /PENETRATIO	W73-00823
ATH OF POLLUTANTS, RADIOECOLOGY,	MATHEMATICAL MODELS, WASTE ASSIMI	W73-00828
UTION CONTROL, *INSTRUMENTATION,	MATHEMATICAL MODELS, WATER ANALYS	W72-08790
GAE, PROTOZOA, DIATOMS, EUGLENA,	MATHEMATICAL STUDIES.: /LUDGE, AL	W72-08573
OPTICAL PROPERTIES, ABSORPTION,	MATHEMATICAL STUDIES, CHLOROPHYLL	W72-14800
RIBUTION, CHLOROPHYLL, SAMPLING,	MATHEMATICAL STUDIES, ATLANTIC OC	W72-11719

PUTERS,/ *ALGAE, *MODEL STUDIES, MATHEMATICAL STUDIES, DIGITAL COM W72-11920
N, PHOSPHORUS, LIGHT, NUTRIENTS, MATHEMATICAL STUDIES, EQUATIONS, W72-11724
S, INVERTEBRATES, CHLAMYDOMONAS, MATHEMATICAL STUDIES, BIOLOGICAL W72-09248
WEEDS, WATER POLLUTION SOURCES, MATHEMATICAL MODELS, NITROGEN, AB W72-09171
, FERTILIZATION, AQUATIC PLANTS, MATHEMATICAL MODELS, SELF-PURIFIC W71-12072
VITY TECHNIQUES, AQUARIA, ALGAE, MATHEMATICAL MODELS, PATH OF POLL W72-03348
IENTS, ECOSYSTEMS, RAIN FORESTS, MATHEMATICAL MODELS, LINEAR PROGR W72-03342
PHOSPHATES, NITROGEN, COLIFORMS, MATHEMATICAL MODELS, ALGAE, SAMPL W71-13356
XYGEN, *ALGAE, CULTURES, VOLUME, MATHEMATICAL STUDIES, CORRELATION W70-05547
ECTS, SLUDGE, RIVER BEDS, ALGAE, MATHEMATICAL MODELS.: /TATION EFF W70-09189
TY, ALGAE, NITROGEN, PHOSPHORUS, MATHEMATICAL MODELS, NUTRIENTS, M W71-08670
, LIGHT INTENSITY, GROWTH RATES, MATHEMATICAL MODELS, WATER TEMPER W72-06274
CONCENTRATION, DESIGN CRITERIA, MATHEMATICAL MODELS, *BIOASSAY, * W72-04787
, EUTROPHICATION, MODEL STUDIES, MATHEMATICAL MODELS.: /AL CONTROL W72-04787
IENTS, *LAKE ERIE, *HYPOLIMNION, MATHEMATICAL MODELS, IRON, MANGAN W72-12997
HLOROPHYLL, TURBIDITY, SAMPLING, MATHEMATICAL MODELS, ESTUARINE EN W72-12941
*MICROSCOPIC COUNTS, *AMORPHOUS MATTER.: W72-06191
IC TREATMENT, OXIDATION, ORGANIC MATTER.: /XIDATION LAGOONS, AEROB W71-09629
CARBON DIOXIDE, AMMONIA, ORGANIC MATTER.: /ION, DISSOLVED OXYGEN, W72-01363
OTOBACTER, MARINE ALGAE, ORGANIC MATTER.: /XIDATION, SYMBIOSIS, AZ W72-11563
NALYTICAL TEC/ *OCEANS, *ORGANIC MATTER, *CARBON RADIOISOTOPES, *A W71-08042
ATER POLLUTION EFFECTS, *ORGANIC MATTER, *DECOMPOSING ORGANIC MATT W72-14673
NIC MATTER, *DECOMPOSING ORGANIC MATTER, *EUTROPHICATION, *PHYTOPL W72-14673
N, *FOAM FRACTIONATION, *ORGANIC MATTER, *INORGANIC COMPOUNDS, IRO W72-05506
L ANALYSIS, *DECOMPOSING ORGANIC MATTER, *MARINE ALGAE, HUMUS, ORG W72-12166
ATER POLLUTION EFFECTS, *ORGANIC MATTER, *OXYGEN SAG, *IMPAIRED WA W70-09189
S, ALGAE, CH/ *BIOMASS, *ORGANIC MATTER, *SEA WATER, WATER ANALYSI W71-11234
QUIREMENTS, *DECOMPOSING ORGANIC MATTER, AERATION, *ALGAE, *BIODEG W72-01881
YANOPHYTA, *HUMIC ACIDS, ORGANIC MATTER, ALGAE, EUTROPHICATION, CH W72-01097
ABSORPTION, DECOMPOSING ORGANIC MATTER, ALKALINITY, HYDROGEN ION W71-12068
LLUTANTS, DIELDRIN, DDT, ORGANIC MATTER, ANALYTICAL TECHNIQUES, SA W71-11498
IC PRODUCTION, *BIOMASS, ORGANIC MATTER, ATLANTIC OCEAN, GROWTH RA W71-11562
IONS, RIVERS, CULTURES, ORGANIC MATTER, BACTERIA, WATER ANALYSIS. W71-08032
, SEDIMENTS, DECOMPOSING ORGANIC MATTER, BACTERIA, CHEMICAL PROPER W72-12997
ORAL, LAKES, PERIPHYTON, ORGANIC MATTER, BIOMASS, METABOLISM, PHYT W72-03216
E, PHOSPHORUS, NITROGEN, ORGANIC MATTER, CALCIUM CARBONATE, DETRIT W72-05469
, THERMOPHILIC BACTERIA, ORGANIC MATTER, CULTURES, SOIL BACTERIA.: W72-14301
GAS CHROMATOGRAPHY, PARTICULATE MATTER, DETRITUS, ORGANIC COMPOUN W73-02105
ANALYSIS, BENTHIC FLORA, ORGANIC MATTER, DISSOLVED SOLIDS, BENTHOS W73-02100
N RADIOISOTOPES, SESTON, ORGANIC MATTER, DISTRIBUTION PATTERNS, TH W73-00431
BSTRATE UTILIZATION, PARTICULATE MATTER, FATE OF POLLUTANTS, PAPER W73-01066
INS), PHOSPHORUS, ALGAE, ORGANIC MATTER, LAKE ERIE, EFFLUENTS, SEW W71-11017
ACUTUS, C-14, DISSOLVED ORGANIC MATTER, MACROPHYTES, BIOLOGICAL S W73-02095
TES, PHOSPHORUS, CARBON, ORGANIC MATTER, MANGANESE, IRON, SULFUR, W72-07514
TION, ALGAE, DECOMPOSING ORGANIC MATTER, MINNESOTA, CYANOPHYTA, TR W72-01784
ASONAL, BAYS, SEA WATER, ORGANIC MATTER, MORTALITY, WATER TEMPERAT W73-00366
TICS, *DISSOLVED SOLIDS, ORGANIC MATTER, NITROGEN, WATER POLLUTION W73-01066
EXPERIMENTAL FOREST, PARTICULATE MATTER, NITROSOMONAS, NITROBACTER W71-01489
ON, *ALGAE, *DECOMPOSING ORGANIC MATTER, NUTRIENTS, SURFACE WATERS W70-10405
ODOR, CONFINEMENT PENS, ORGANIC MATTER, NUTRIENTS, NITROGEN, PHOS W71-08214
ANALYSIS, CARBON CYCLE, ORGANIC MATTER, NUTRIENTS.: /NKTON, WATER W72-09108
EFFICIENCIES, *BIOMASS, ORGANIC MATTER, NUTRIENTS, PHOTOSYNTHESIS W72-12289
CTIVITY, BIODEGRADATION, ORGANIC MATTER, OXYGEN, CHEMICAL OXYGEN D W71-04518
, CHELATION, DECOMPOSING ORGANIC MATTER, PHYTOPLANKTON, PRODUCTIVI W72-01097
OCHEMICAL OXYGEN DEMAND, ORGANIC MATTER, PHOSPHORUS, CARBON DIOXID W71-13257
(WATER), NITROGEN CYCLE, ORGANIC MATTER, PHOTOSYNTHESIS, EUTROPHIC W72-08048
, ALGAE, PHOTOSYNTHESIS, ORGANIC MATTER, PLANKTON, CHELATION, CARB W71-03027
ROBIC CONDITIONS, ALGAE, ORGANIC MATTER, VOLUME, DRYING, ODOR, NIT W71-07551
RATIFICATION, OXIDATION, ORGANIC MATTER, WATER QUALITY, WATER CHEM W72-04853
IAL DEGRADATION, CARBON, ORGANIC MATTER, WATER POLLUTION SOURCES, W73-00379
IA, AQUARIA, DECOMPOSING ORGANIC MATTER, WASTE ASSIMILATIVE CAPACI W73-00802
TRIENTS, WATER SAMPLING, ORGANIC MATTER, WATER POLLUTION EFFECTS, W73-00234
KE ST. CLAIR, DETROIT RIVER, BL/ MAUMEE RIVER, ST. CLAIR RIVER, LA W72-14282
CHEMOSTATIC CONTINUOUS CULTURES, MAXIMUM SPECIFIC GROWTH RATE, SAT W72-04788
LLAO(PERU), CAPE COD(MASS), CAPE MAY(N J), SARGASSO SEA.: /AMA, CA W72-08056
SFLIES, STONEFLIES, DOBSONFLIES, MAYFLIES, CYANOPHYTA, CHLOROPHYTA W70-06655
A, GEOLOGIC FORMATIONS, BENTHOS, MAYFLIES, TUBIFICIDS, DIAT: /PHYT W71-11009
ROTIFERS, CRUSTACEANS, DIPTERA, MAYFLIES, DRAGONFLIES, CADDISFLIE W71-11012
*CHLORINATED HERBICIDES, MCPB, MCPA, CHLAMYDOMONAS GLOBUSA, CHLO W70-07255
, CULTURE EXPERIMENTS, ATRAZINE, MCPA, ENDOSULFAN, FAIRY LIQUID, B W72-12576
, CHLO/ *CHLORINATED HERBICIDES, MCPB, MCPA, CHLAMYDOMONAS GLOBUSA W70-07255
L EUTROPHIC/ *LAKE ANNABESSACOOK(ME), *KENNEBEC RIVER(ME), CULTURA W72-04280
NABESSACOOK(ME), *KENNEBEC RIVER(ME), CULTURAL EUTROPHICATION.: /N W72-04280
CHLOROPHYLL, RECREATION, RUNOFF, MEASUREMENT, TEST PROCEDURES, PHO W72-04270
, GAMBUSIA AFFINIS, PRODUCTIVITY MEASUREMENTS, NUTRIENT REMOVAL.: / W71-01488
MATHEMATICAL MODELS, NUTRIENTS, MEASUREMENT, POPULATION, CHLOROPH W71-08670
N, FIELD MEASUREMENTS, CARBON-14 MEASUREMENTS, ACETYLENE REDUCTION W72-13692
EASUREMENTS, ACETYLENE REDUCTION MEASUREMENTS, NUTRIENT ENRICHMENT W72-13692
TRAVIS, *LAKE LIVINGSTON, FIELD MEASUREMENTS, CARBON-14 MEASUREME W72-13692
S, DEPTH, DISTRIBUTION PATTERNS, MEASUREMENT, PHOTOSYNTHESIS, PLAN W71-13252
*STANDING CROP, *ALGAE, DIATOMS, MEASUREMENT, DETRITUS, METHODOLOG W71-12855
RATES, SAMPLING, LAKE MICHIGAN, MEASUREMENT, SUSPENDED LOAD, ALKA W71-11033
E SENSING, *TRACKING TECHNIQUES, MEASUREMENT, CURRENTS(WATER), CHL W72-01095
CHLORELLA, PSEUDOMONAS, TESTING, MEASUREMENT, PHEOLOGY.: /OPHYTA, W72-03220
, METHODOLOGY, MICROENVIRONMENT, MEASUREMENT.: / POLLUTION CONTROL W72-03218
BIOASSAY, PHOSPHORUS, NUTRIENTS, MEASUREMENT, BACTERIA, ALGAE, EUT W71-13794
TIONS, REPRODUCTION, BIORHYTHMS, MEASUREMENT, POPULATIONS, CYTOLOG W72-11920
CES, AQUATIC ALGAE, FOOD CHAINS, MEASUREMENT, ADSORPTION.: /N SOUR W72-10686
, *BIOASSAY, *ALGAE, FRESHWATER, MEASUREMENT, DIATOMS, EUTROPHICAT W72-10619
NKT/ *NITROGEN FIXATION, *ALGAE, MEASUREMENT, CYANOPHYTA, PHYTOPLA W72-08054

ING, ASSAY, NUTRIENTS, CULTURES, MEASUREMENT, LABORATORY TESTS, BI W73-00255
IQUES, SOIL ALGAE, *METHODOLOGY, MEASUREMENT, OXIDATION, CARBON RA W72-14301
COLOGY, *RADIOCHEMICAL ANALYSIS, MEASUREMENT, *NEVADA, AQUATIC PLA W73-01673
GASTROPODS, MOLLUSKS, AMPHIPODA, MECHANICAL EQUIPMENT.: /MATODES, W72-05432
SUBSTANCES, PHYTOPLANKTON, FLUID MECHANICS, EXUDATION, OCEANS, CHL W71-07878
*CHLORELLA ELLIPSOIDEA, CULTURE MEDIA.: W73-01626
RNUTUM, PAAP BATCH TEST, CULTURE MEDIA.: /DES, SELENASTRUM CAPRICO W73-02099
S, CLOSTERIUM STRIGOSUM, CULTURE MEDIA.: /RASTERIAS CRUX-MELITENSI W72-12736
, *CULTURING TECHNIQUES, CULTURE MEDIA, *BAY OF BENGAL, INDIAN OCE W73-00430
RS, *DIATOMACEOUS EARTH, *PORCUS MEDIA, *WASTE WATER TREATMENT, *D W72-00974
CULTUTURING TECHNIQUES, CULTURE MEDIA, ALGOLOGY, TAXON.: W72-12172
LAMYDOMONAS REINHARDTII, CULTURE MEDIA, CHLORINATED HYDROCARBONS.: W73-00281
HLORAMPHENICOL, ACTIDON, CULTURE MEDIA, CHLOROPHYLL A, TERATOLOGY, W73-01445
ON TUBEROSUS, *EPIPHYTES, GROWTH MEDIA, COCCONEIS PLACENTULA, CLOS W72-12744
ONAS PSEUDOMICROSPHAERA, CULTURE MEDIA, CULTURING TECHNIQUES, MORP W72-12632
THANOL, MONOSACCHARIDES, CULTURE MEDIA, DISACCHARIDES, THREONINE, W73-00942
S/ SAVANNAH RIVER PLANT, CULTURE MEDIA, DRY WEIGHT, THERMAL EFFECT W71-02075
, INFRARED SPECTROSCOPY, CULTURE MEDIA, FATTY ACIDS, SAMPLE PREPAR W73-01066
*PORTERINEMA FLUVIATILE, CULTURE MEDIA, HUMMOCK CHANNEL, CHLOROPLA W72-12633
OENERGETICS, ENRICHMENT, CULTURE MEDIA, MACROINVERTEBRATES.: /, BI W72-09378
POLYSTYRENE, SLIDE P/ *MOUNTING MEDIA, METHYLENE IODIDE, TOLUENE, W72-11738
ITROGEN, ORGANIC CARBON, CULTURE MEDIA, PAPER CHROMATOGRAPHY, INFR W73-00379
ATES, SODIUM CACODYLATE, CULTURE MEDIA, PORIA MONTICOLA, PORIA COC W72-13813
STAURASTRUM PARADOXIUM, CULTURE MEDIA, ZINC SULFATE, MANGANESE CH W72-12739
FUCUS VIRSOIDES, FUCUS SPIRALIS, MEDITERRANEAN SEA.: * W72-12940
IACUM, *SUEZ CANAL, FLAGELLATES, MEDITERRANEAN SEA, RED SEA, DRIFT W73-00296
IA PUNGENS, C-14, CHLOROPHYLL A, MEDITERRANEAN SEA, FRANCE, PYCNOC W73-00431
F OF SUEZ, DECAPODS, COLEOPTERA, MEDITERRANEAN SEA, COELENTERATES, W73-00855
ERELLA DELICATULA, DACTYLIOSOLEN MEDITERRANEUS, LEPTOCYLINDRUS DAN W72-03543
EAN, SOIL EXTRACTS, ERDSCHREIBER MEDIUM, ESTUARINE SOILS, AXENIC C W73-00430
CATES, NOCTILUCA, SIPHONOPHORES, MEDUSAE, PILIDIUM, CLADOCERA, LIN W73-00935
TRIAL EXPANSION, ISLE OF THANET, MEDWELL ESTUARY.: /T COAST, INDUS W71-13746
GICAL SAMPLES, MERGUS MERGANSER, MEGACERYLE ALCYON, BIOACCUMULATIO W72-12930
MEIOFAUNA, FORAMINIFERA.: W72-12567
ICA, TILAPIA MOSSAMBICA, TILAPIA MELANOPLEURA, CYPRINUS CARPIO, CT W71-04528
SCOTIA, ISTHMOPLEA SPHAEROPHORA, MELANOSIPHON INTESTINALIS, HECATO W72-01370
ACTION, LAKE SALLIE(MINN), LAKE MELISSA(MINN), NUTRIENT REMOVAL.: W71-03012
VAR. ORNATUM, MICRASTERIAS CRUX- MELITENSIS, CLOSTERIUM STRIGOSUM, W72-12736
IRA ISLANDICA, MELOSIRA ITALICA, MELOSIRA AMBIGUA, SYNEDRA FILIFOR W73-00384
ON FLOS-AQUAE, CHROOMONAS ACUTA, MELOSIRA GRANULATA, CERATIUM, MIC W72-12738
IA ERIENSIS, MELOSIRA ISLANDICA, MELOSIRA ITALICA, MELOSIRA AMBIGU W73-00384
IGANIANA, RHIZOSOLENIA ERIENSIS, MELOSIRA ISLANDICA, MELOSIRA ITAL W73-00384
DESSICATION, SPECIES DIVERSITY, MELOSIRA NUMMULOIDES, CHLOROPHYLL W73-00853
LL A, *YAQUINA BAY, AMPHORA SPP, MELOSIRA SULCATA, NAVICULA SPP, S W73-00853
CIOCOLA, CLOSTERIUM MONILIFERUM, MELOSIRA VARIANS, NAVICULA TRIPUN W72-10883
STERIONELLA, SYNEDRA, DINOBRYON, MELOSIRA.: /ILARIA, TABELLARIA, A W73-01669
OEDOGONIUM, FRAGILARIA, DIATOMA, MELOSIRA, TABELLARIA, ACTINASTRUM W72-10875
ANDERSON-CUE LAKE, MELROSE(FLORIDA).: W72-08986
HAERIA CAVERNOSA, CLADOPHOROPSIS MEMBRANACEA, BOODLEA COMPOSITA, A W72-11800
CILLUS, CHONDRIA, CLADOPHOROPSIS MEMBRANACEA, PENICILLUS CAPITATUS W73-00838
MBRANE FILTERS, MONOSACCHARIDES, MEMBRANE FILTRATION.: /LUCOSE, ME W72-14326
SCINTILLATION COUNTING, GLUCOSE, MEMBRANE FILTERS, MONOSACCHARIDES W72-14326
THOD, MOST PROBABLE-NUMBER TEST, MEMBRANE FILTRATION, RHODOPSEUDOM W72-04743
TUDIES, PLANT MORPH/ *BIOLOGICAL MEMBRANES, *ALGAE, *CYTOLOGICAL S W73-01630
LAKE WINGRA(WIS.), LIMITI/ LAKE MENDOTA(WIS.), LAKE MONONA(WIS.), W73-00232
*LIMITING NUTRIENTS, LAKE MENDOTA(WIS).: W72-07933
TROUT LAKE(WIS), SEDIMENT / LAKE MENDOTA(WIS), BANTAM LAKE(CONN), W72-08459
ESERVOIR(VA), MADISON(WIS), LAKE MENDOTA(WIS), LAKE MONONA(WIS), L W72-05473
UPPER KLAMATH LAKE(OREGON), LAKE MENDOTA(WISCONSIN).: /NEW YORK), W71-05626
URCES, ACETYLENE REDUCTION, LAKE MENDOTA(WIS), LAKE MONONA(WIS).: / W72-10608
*MADISON(WIS), LAKE MENDOTA(WIS), LAWN MAINTENANCE.: W73-02479
CTION, RHIZOBIUM JAPONICUM, LAKE MENDOTA, CRYSTAL LAKE, TROUT LAKE W73-01456
, SPHAEROCYSTIS, DAPHNIA GALEATA MENDOTAE, DAPHNIA MAGNA.: /OMONAS W71-03014
CYCLOTELLA CRYPTICA, *CYCLOTELLA MENEGHINIANA, *POLYMORPHISM, *VAL W73-02548
VER, PINFISH, FLOUNDER, CROAKER, MENHADEN.: /LOR 1254, ESCAMBIA RI W71-07412
*GAMBUSIA AFFINS, EPHEMEROPTERA, MENIDIA, BRACHYCEUTERA, CHYDORUS W72-07132
S, PENAEUS SETIFERUS, MERCENARIA MERCENARIA, CLYMENELLA TORQUATA, W73-02029
AEUS AZTECUS, PENAEUS SETIFERUS, MERCENARIA MERCENARIA, CLYMENELLA W73-02029
CHROMATOGRAPHY, GROWTH INHIBITI/ MERCURIAL PESTICIDES, THIN LAYER W72-13347
ANA, CHAETOCEROS GALVESTONENSIS, MERCURIC CHLORIDE, DIMETHYL MERCU W72-07660
C CHLORID/ *MERCURY, *SYNTHESIS, MERCURIC CHLORIDE, METHYL MERCURI W72-06037
HESIS, MERCURIC CHLORIDE, METHYL MERCURIC CHLORIDE, ANKISTRODESMUS W72-06037
ORNUTUM, PHENYLMERCURIC ACETATE, MERCURY CHLORIDE, PHENYLACETATE, W72-11727
MERCURY CHLORIDE, PHENYLACETATE, MERCURY COMPOUNDS.: /IC ACETATE, W72-11727
ON, MINAMATA BAY(JAPAN), ORGANIC MERCURY COMPOUNDS, BIOSYNTHESIS, W72-12257
*MERCURY, *BIOLOGICAL TISSUES, MERCURY ISOTOPES, HUMAN BRAIN, LO W71-11036
OTE SENSING, ALGAE, ANTIBIOTICS, MERCURY, ARSENIC COMPOUNDS, CHESA W72-10162
*HEAVY METALS, CHEMICAL WASTES, MERCURY, COPPER, ZINC, PESTICIDES W72-12576
ALS, *TRACE ELEMENTS, NUTRIENTS, MERCURY, CHROMIUM, COBALT, FRESH W73-01434
ICALS, REGULATION, ALGAL TOXINS, MERCURY, DETERGENTS, WATER POLLUT W73-00703
AL COMPOUNDS, ESOX LUCIUS, ALKYL MERCURY, EAGLES, SALMO GAIRDNERII W72-05952
SIS, MERCURIC CHLORIDE, DIMETHYL MERCURY, FLUOROMICROPHOTOMETER.: / W72-07660
ETRY, VI/ SPACE PLATFORMS, LEAD, MERCURY, LASERS, MICROWAVE RADIOM W72-04742
WASHINGTON(WASH).: MERCURY, PHOSPHORUS REMOVAL, LAKE W71-12091
OURCES, WATER POLLUTION EFFECTS, MERCURY, SEWAGE DISPOSAL, PHOSPHA W72-14782
TS, CRUSTACEANS, MUSSELS, COMMON MERGANSER DUCK, BIOASSAY, WATER P W72-12930
TONI, BIOLOGICAL SAMPLES, MERGUS MERGANSER, MEGACERYLE ALCYON, BIO W72-12930
RUS BARTONI, BIOLOGICAL SAMPLES, MERGUS MERGANSER, MEGACERYLE ALCY W72-12930
TICA, ANABAENA SPP., CHROMATIUM, MERISMOPEDIA TROLLERI, CRYPTOMONA W72-07145
IOTHRIX NIVEA, THIOTHRIX TENUIS, MERISMOPEDIA GLAUCA, DACTYLOCOCCO W72-07896

CC, *DIETHYLAMINE HYDROCHLORIDE, MERISOMOPEDIA SPP., ANABAENA SPP. W72-14706
(WASHINGTON), COCKLES, GOLDFISH, MERRIMACK RIVER(NEW HAMPSHIRE), D W71-11517
A, BOSTRYCHIETUM, FIDDLER CRABS, MESCFAUNA, TURBELLARIA, SPONGES, W72-07911
A, LYNGBYA.: MESOTROPHIC, MICROCYSTIS, ANABAEN W72-04269
AGING(BIOLOGICAL), LAKE STAGES, MESOTROPHY, OLIGOTROPHY, WATER PO W72-14782
LAMYDOMONAS, DEFICIENT ELEMENTS, MESOTROPHY, EUTROPHICATICN, TROPH W72-13816
TOCERCS LORENZIANUS, CHAETOCEROS MESSANENSIS, CHAETOCEROS PERUVIAN W71-11527
DEMAND, LAKE STAGES, NUTRIENTS, MESTROPHY, OLIGCTROPHY, CXYGEN SA W73-00002
PHYCEA/ LYTIC BACTERIA, ENGLAND, METABOLIC INHIBITION, HOSTS, MYXO W72-05457
NOPHYTA, MYXOBACTERIA, CULTURES, METABOLISM, SEWAGE, CARBOHYDRATES W72-05457
NSITY, MIXING, DISSOLVED OXYGEN, METABOLISM, CARBON DIOXIDE, PHOSP W72-04983
CONDITIONS, AEROBIC CONDITIONS, METABOLISM, SLUDGE, BIOCHEMICAL O W72-04789
ERIA, LAKES, FILTERS, COLIFORMS, METABOLISM, CARBON, SULFUR, SAMPL W72-04743
ESTICIDES, RESISTANCE, CULTURES, METABOLISM.: /, SOILS, TRIAGINE P W71-07675
F/ *AGRICULTURAL DRAINAGE, LAKE METABOLISM, MOBILITY, PARTICULATE W71-06443
HYLOGENY, CARBON CYCLE, ENZYMES, METABOLISM, HYDROGEN, EUGLENA, CH W71-09172
OWTH SUBSTANCES, PHOTOSYNTHESIS, METABOLISM, BACTERIA, HYDROGEN IO W71-08687
ROPODS, ADSORPTION, TEMPERATURE, METABOLISM, KINETICS, WATER POLLU W71-09168
AL PHYSIOLOGY, SEDIMENTS, ANIMAL METABOLISM.: /NMENT, COASTS, ANIM W73-00811
BON RADIOISOTOPES, PRODUCTIVITY, METABOLISM, FUNGI, ORGANIC COMPOU W72-14326
CULTURES, AQUATIC PRODUCTIVITY, METABOLISM, ELECTRODES, ELECTROCH W72-14314
OLLUTION EFFECTS, RADIOACTIVITY, METABOLISM, PLANT PHYSIOLOGY, CHR W73-00380
TROGEN, *PARTICULATE ORGA/ *LAKE METABOLISM, *DISSOLVED ORGANIC NI W72-08048
TEMPERATURE, ALTITUDE, CLIMATES, METABOLISM, BIOCHEMICAL OXYGEN DE W72-06838
MUNITIES, BIOMASS, PRODUCTIVITY, METABOLISM, CHLORELLA, RESPIRATIO W73-01080
BON PESTICIDES, PESTICIDE DRIFT, METABOLISM, ALGAE, AQUATIC ALGAE, W73-02280
NITROGEN, POTASSIUM, PHOSPHATES, METABOLISM, ALGAE, SYNTHESIS, VIT W73-01626
ELEMENTS, NITROGEN, PHOSPHORUS, METABOLISM, ALGAE, SLUDGE, ANALYT W71-11793
MES, BACTERIA, SPORES, NITROGEN, METABOLISM.: /OGEN FIXATION, ENZY W71-12070
ISOTOPES, CADMIUM RADIOISOTOPES, METABOLISM, TROPHIC LEVELS.: /DIO W71-13243
TATION, FILTRATION, LIME, ROCKS, METABOLISM, DESIGN CRITERIA, *WAS W71-13341
S, *BACTERIA, *ALGAE, OXIDATION, METABOLISM, RESPIRATION, LIPIDS, W72-10233
ANCE, *POLYP MAINTENANCE, *POLYP METABOLISM, TETRACYCLINE, SULFANI W72-10162
GAE, FISH, ABSORPTION, BIOASSAY, METABOLISM.: /R, HYDROBIOLOGY, AL W72-10694
ASTES, ALGAE, INDUSTRIAL WASTES, METABOLISM, GROWTH RATES, OXIDATI W72-09590
YGEN DEMAND, SATURATION, OXYGEN, METABOLISM, NITROGEN, PHOSPHOROUS W71-13553
DS, *HYDROGEN ION CONCENTRATION, METABOLISM, CARBON DIOXIDE, ALGAE W72-00838
UR, ALGAE, GRANULES, INHIBITION, METABOLISM, CULTURES, ENVIRONMENT W72-00153
C BACTERIA, *MACROPHYTE-EPIPHTYE METABOLISM, LAWRENCE LAKE(MICH), W72-03216
PHYTON, ORGANIC MATTER, BIOMASS, METABOLISM, PHYTOPLANKTON, PLANT W72-03216
ALGAE, CYANOPHYTA, CHLOROPHYTA, METABOLISM, NEKTON, BENTHOS, BIOM W72-13636
ORIDE, RADIOACTIVITY TECHNIQUES, METABOLISM, TRACERS, RADIOACTIVIT W72-12634
*LYSIS, *CONTACT TIME, CELLULAR METABOLITES.: W72-00411
N PR-6, PHOTOALDRIN, KETOENDRIN, METABOLITES, SKELETONEMA, ANABAEN W72-03545
CTS, *ETHOXYCHLOR, *METHYLCHLOR, METABOLITES, BIOLOGICAL MAGNIFICA W72-07703
CAL SAMPLES, FATE OF POLLUTANTS, METABOLITES, DUNALIELLA, AGMENELL W73-00361
A NITZSCHIOIDES, MIXED CULTURES, METABOLITES, SUBSTRATE UTILIZATIO W73-00441
AL MAGNIFICATION, CHLAMYDOMONAS, METABOLITES, HAEMATOCRIT TUBE.: / W72-04134
N/ *SINAI PENINSULA, CHLORINITY, METAHALINE, HYPERHALINE, EURYHALI W73-00855
UM, BRYOPHYTES, BIOACCUMULATION, METAL COMPLEXES, PICEA ABIES, VAC W72-14303
BRATSK RESERVOIR, *HYDROBIOLOGY, METALIMNION, BACILLARIOPHYTA, VOL W72-04855
ERIOLOGY, SAPROPHYTES, STURGEON, METALIMNION.: /HYDROBIOLOGY, BACT W72-04853
, *POLYETHYLENEMIN/ *MULTIVALENT METALLIC CATIONS, *POLYACRYLAMIDE W72-14840
WATER CONVEYANCE, CANALS, HEAVY METALS.: /ENVIRONMENT, CALIFORNIA W71-09581
ENT, *TOXICITY/ *REVIEWS, *HEAVY METALS, *ALGAE, *AQUATIC ENVIRONM W72-14694
OXICITY, *FLUORESCENCE, / *HEAVY METALS, *ALGAE, *GROWTH RATES, *T W72-07660
QUATIC ALGAE, CHLOROPHYTA, HEAVY METALS, *CHLAMYDOMONAS, INHIBITOR W72-11727
S/ *TERRESTRIAL HABITATS, *HEAVY METALS, *PRODUCTIVITY, *ECOSYSTEM W72-14303
S, *PRIMARY PRODUCTIVITY, *HEAVY METALS, *TRACE ELEMENTS, NUTRIENT W73-01434
PHY/ *INDICATORS, *ALGAE, *HEAVY METALS, *TOXICITY, COPPER, CHLORO W71-05991
CHELATION, CHROMATOGRAPHY, HEAVY METALS, *TRACE ELEMENTS, RADIOCHE W72-04708
IOISOTOPES, *MARINE ALGAE, HEAVY METALS, ABSORPTION, BIOINDICATORS W72-10957
ENTS, *LEAD RADIOISOTOPES, HEAVY METALS, AIR POLLUTION EFFECTS, SU W72-00940
INC, METALS, SECCHI DISKS, HEAVY METALS, BIOINDICATORS, SILICA, CH W72-09111
ION EFFECTS, RESPIRATION, *HEAVY METALS, CHEMICAL WASTES, MERCURY, W72-12576
, WATER POLLUTION EFFECTS, HEAVY METALS, CYCLING NUTRIENTS, CARBON W73-00840
YDROGEN ION CONCENTRATION, HEAVY METALS, DISSOLVED OXYGEN, *PLANT W72-14690
HELATION, *ION TRANSPORT, *HEAVY METALS, EUTROPHICATION, POLLUTANT W73-02112
CHEMISTRY, TRACE ELEMENTS, HEAVY METALS, ELECTRODES, NITRATES, ALG W72-01693
POLLUTION EFFECTS, ALGAE, HEAVY METALS, GROWTH, PSEUDOMONAS, POTA W72-13813
S, BIOLOGICAL COMMUNITIES, HEAVY METALS, HERBICIDES, TOXINS, WASTE W72-01472
STRIAL WASTES, MUNICIPAL WASTES, METALS, HEAVY METALS, WASTE DISPO W71-11793
RIENTS, BACTERIA, VIRUSES, HEAVY METALS, HUMAN POPULATION, WASTE W W71-09788
T, HABITATS, AQUATIC LIFE, HEAVY METALS, INDUSTRIAL WASTES, INVERT W72-11854
BITION, *LIPIDS, TOXICITY, HEAVY METALS, LABORATORY TESTS, PHOTOSY W72-06037
L, COBALT, COPPER SULFATE, HEAVY METALS, METALS, SULFATES.: /NICKE W72-12739
COLLECTIONS, COPPER, LEAD, HEAVY METALS, MUSSELS, *KELPS, STARFISH W72-11925
AL), PHOSPHORUS, NITROGEN, HEAVY METALS, ORGANOPHOSPHORUS PESTICID W72-04742
EPIDEMIOLOGY, PHOSPHATES, HEAVY METALS, PESTICIDES, TASTE-PRODUCI W72-02144
HYSIOLOGY, TRACE ELEMENTS, HEAVY METALS, PHYSICOCHEMICAL PR: /AL P W73-00817
HERBICIDES, PHOSPHORUS, PHENOLS, METALS, POLLUTANTS, INDUSTRIAL WA W72-09646
TA, CYANOPHYTA, MANGANESE, ZINC, METALS, SECCHI DISKS, HEAVY METAL W72-09111
TIES, GOLD, IRON, CESIUM, ALKALI METALS, SEASONAL, ZINC, ABSORPTIO W73-01434
T, COPPER SULFATE, HEAVY METALS, METALS, SULFATES.: /NICKEL, COBAL W72-12739
INSECTICIDES, HERBICIDES, HEAVY METALS, TASTE, ODOR, RECREATION, W71-12091
NTS, NUTRIENTS, PESTICIDES, OIL, METALS, THERMAL POLLUTION, SAMPLI W72-08790
MUNICIPAL WASTES, METALS, HEAVY METALS, WASTE DISPOSAL, WATER POL W71-11793
UALITY, ALGAE, PESTICIDES, HEAVY METALS, WASTEWATER TREATMENT, SME W72-14282
NOIDOSA, SODIUM SILICATE, SODIUM METASILICATE, LAKE WASHINGTON, NI W72-04734
POLYEDRA, PERIDINIUM DEPRESSUM, METAZOA, CALANUS HELGOLANDICUS(PA W72-07892

MISTRY, WATER POLLUTION SOURCES, METEOROLOGICAL DATA, SEASONAL.: / W72-04853
OLVED OXYGEN, SATURATION, ALGAE, METHANE BACTERIA, EUTROPHICATION, W71-07106
, NITROGEN, PHOSPHORUS, AMMONIA, METHANE, TEMPERATURE, HYDROGEN IO W72-10390
ICATION, FILTER DENITRIFICATION, METHANOL, BACTERIAL DENITRIFICATI W71-08223
RYPTOPHAN, HISTIDINE, ASPARTATE, METHIONINE, PHENYLALANINE, ALANIN W73-00942
OMONAS, IKROAVICH LAKE, UTERMOHL METHOD.: /TION, FLAGELLATES, RHOD W73-00273
VE, *FILTER PAPER DISC DIFFUSION METHOD, *ZONES OF INHIBITION, ANT W72-08586
, ATRAZINE, BIOTYPES, POUR PLATE METHOD, MOST PROBABLE-NUMBER TEST W72-04743
EXTRACTION, LUCIFERIN-LUCIFERASE METHOD, SEDIMENT-WATER INTERACTIO W72-04292
E, REGRESSION ANALYSIS, BIOMASS, METHODOLOGY, REEFS, SAMPLING, OXY W72-06283
FATE, CHLORINE, WATER RESOURCES, METHODOLOGY, EVALUATION, CHEMICAL W71-05083
RATORY EQUIPMENT, SEDIMENTATION, METHODOLOGY, POLLUTANT IDENTIFICA W73-00273
ER, *BORON, *PATH OF POLLUTANTS, METHODOLOGY, RIVERS, WATER POLLUT W73-00286
LUTION SOURCES, INSTRUMENTATION, METHODOLOGY, NITROGEN, SILICON, A W73-01445
SOURCES, ON-SITE INVESTIGATIONS, METHODOLOGY, FOREIGN RESEARCH, MA W72-08804
EMICAL PROPERTIES, SECCHI DISKS, METHODOLOGY.: /ODIUM CHLORIDE, CH W72-09167
DIATOMS, MEASUREMENT, DETRITUS, METHODOLOGY, SEDIMENTS, LAKES, CH W71-12855
ANKTON, WATER POLLUTION CONTROL, METHODOLOGY, MICROENVIRONMENT, ME W72-03218
RGANIC COMPOUNDS, ON-SITE TESTS, METHODOLOGY, CHARA.: / ECOLOGY, O W72-03216
DS, FISH, ON-SITE INVESTIGATION, METHODOLOGY, INVERTEBRATES, AMPHI W72-00155
BIOLOGY, EPIDEMIOLOGY, CULTURES, METHODOLOGY, DOCUMENTATION, ABSTR W72-13800
MICROORGANISMS, BACTERIA, FUNGI, METHODOLOGY, CULTURES, BIOASSAY, W72-13813
ISOTOPES, ENVIRONMENTAL EFFECTS, METHODOLOGY, ALGAE, AQUATIC PLANT W72-12709
CITY, *NUTRIENTS, LETHAL LIMITS, METHODOLOGY, INDUSTRIAL WASTES, D W72-12239
AQUEOUS SOLUTIONS, ION EXCHANGE, METHODOLOGY, ORGANIC ACIDS, FULVI W72-12166
NUTRIENT CONTROL METHODS.: W71-13256
IE/ *MODEL STUDIES, *STATISTICAL METHODS, *ALGAE, CYTOLOGICAL STUD W71-12868
LECTIONS/ *SURVEYS, *STATISTICAL METHODS, *POTOMAC RIVER, DATA COL W72-07791
EWAGE TREATMENT, PHYSICOCHEMICAL METHODS, AMMONIA STRIPPING, GUGGE W70-09907
NG POOLS, CHELATION, APPLICATION METHODS, COOLING TOWERS, WATER PO W71-06189
CYTOLOGICAL STUDIES, STATISTICAL METHODS, CULTURES.: /OPULATIONS, W72-11920
IC ANIMALS, DAPHNIA, APPLICATION METHODS, GRANULES, PLANT GROWTH R W72-09650
ERIAL PHOTOGRAPHY, *MATHEMATICAL METHODS, MUNICIPAL WASTES, INDUST W71-11685
RNS, *AQUATIC ALGAE, STATISTICAL METHODS, MARINE ALGAE, *OREGON, * W72-08141
, CHEMICAL ANALYSIS, STATISTICAL METHODS, NITROGEN, PHOSPHORUS, CH W71-11029
S, LABORATORY TESTS, STATISTICAL METHODS, SOIL MICROORGANISMS, RUN W72-02218
WAVES(WATER), MAINE, STATISTICAL METHODS, SEA WATER, COASTS, MARIN W73-00369
IDS, *GAS LIQUID CHROMATOGRAPHY, METHYL ESTERS, ESTERS, CHLORELLA W72-09365
, *SYNTHESIS, MERCURIC CHLORIDE, METHYL MERCURIC CHLORIDE, ANKISTR W72-06037
YRENE, SLIDE P/ *MOUNTING MEDIA, METHYLENE IODIDE, TOLUENE, POLYST W72-11738
MUSCORUM, LYNGBYA LAGERHAIMII, 7- METHYLHEPTADECANE, 8-METHYLHEPTAD W73-01439
ERHAIMII, 7-METHYLHEPTADECANE, 8- METHYLHEPTADECANE, HYDROCARBONS, W73-01439
, PHO/ RHINE-MAAS BASIN(FRANCE), MEURTHE RIVER(FRANCE), CLADOPHORA W71-00099
S/ RHINE RIVER(THE NETHERLANDS), MEUSE RIVER(THE NETHERLANDS), BIE W72-08559
D-LIQUID EXTRACTION, TRIPHOTURUS MEXICANUS, EUPHAUSIA PACIFICA, EN W73-02105
IC OCEAN, PACIFIC OCEAN, GULF OF MEXICO.: /TION TECHNIQUES, ATLANT W71-11851
C FLORA, *MARINE ALGAE, *GULF OF MEXICO, *ELECTRIC POWERPLANTS, *T W72-11252
LLA, ALGAE, *EUTROPHICATION, NEW MEXICO, BIODEGRADATION.: /*CHLORE W72-13992
RATES, *ATLANTIC OCEAN, *GULF OF MEXICO, PHYTOPLANKTON, CYTOLOGICA W72-14313
AT CUL/ *NORTH SEA, PROROCENTRUM MICANS, CERATIUM FURCA, TURBIDOST W71-03095
HYLCHLOR, METABOLITES, BIOLOGIC/ MICE, INSECTS, *ETHOXYCHLOR, *MET W72-07703
ATION, *BACKWASHING, *MT CLEMENS(MICH).: *CARBON REGENER W72-08357
IDS, BACKWASHING, *MOUNT CLEMENS(MICH).: /D SEWERS, *SUSPENDED SOL W72-00042
NCE LAKE(MICH), WINTERGREEN LAKE(MICH), CARBONATE PARTICLES, PHOSP W72-08048
FLUENTS, PARASITES, JORDAN RIVER(MICH), SAN JOAQUIN RIVER(CALIF), W71-11006
IPHTYE METABOLISM, LAWRENCE LAKE(MICH), SCIRPUS ACUTUS, NAJAS FLEX W72-03216
OGEN, NETPLANKTON, LAWRENCE LAKE(MICH), WINTERGREEN LAKE(MICH), CA W72-08048
, NEW YORK, NEW HAMPSHIRE, OHIO, MICHIGAN.: /, HOSTS, PENNSYLVANIA W72-04736
IN B, NITROGEN, CYANOPHYTA, LAKE MICHIGAN, NUTRIENTS.: /BON, VITAM W72-07937
, *LAKES, ANALYTICAL TECHNIQUES, MICHIGAN, SESTON, ALGAE, PHYTOPLA W72-08048
ING, *NUTRIENTS, *DIATOMS, *LAKE MICHIGAN, *EUTROPHICATION, *SPATI W73-00384
, GREAT LAKES REGION, WISCONSIN, MICHIGAN, CHRYSOPHYTA, PLANT PIGM W73-00384
*POLLUTANT IDENTIFICATION, *LAKE MICHIGAN, CHRYSOPHYTA, AQUATIC AL W73-00432
ROPHICATION, LAKE SUPERIOR, LAKE MICHIGAN, LAKE HURON, LAKE ERIE, W73-01456
ER SPORTS, PARKS, COST ANALYSIS, MICHIGAN, WASTE WATER TREATMENT.: W72-00042
ON, ALGAE, *LAKE SUPERIOR, *LAKE MICHIGAN, *LAKE HURON, *LAKE ONTA W72-01094
WASTES, SOLID WASTES, CHEMICALS, MICHIGAN, CALIFORNIA, TUBIFICIDS, W71-11006
TOXICITY, DISSOLVED OXYGEN, LAKE MICHIGAN, ALGAE.: /, PESTICIDES, W71-12092
AKES, NUTRIENTS, SEDIMENTS, LAKE MICHIGAN, WATER POLLUTION SOURCES W71-12072
ON, GROWTH RATES, SAMPLING, LAKE MICHIGAN, MEASUREMENT, SUSPENDED W71-11033
S, PHYTOPLANKTON, WATER QUALITY, MICHIGAN, SURFACE WATERS, COASTS, W72-10875
NG, BIOLOGICAL COMMUNITIES, LAKE MICHIGAN, LAKE ERIE, LAKE HURON, W72-10875
S, WASTEWATER TREATMENT, SMELTS, MICHIGAN, OHIO,: /ES, HEAVY METAL W72-14282
CLADOPHORA, *PLANT GROWTH, *LAKE MICHIGAN, ALGAE, PHOSPHATES, HYDR W72-13644
TEPHANODISCUS HANTZSCHII, CYCLOT MICHIGANIANA, RHIZOSOLENIA ERIENS W73-00384
AURASTRUM SEBALDII VAR. ORNATUM, MICRASTERIAS CRUX-MELITENSIS, CLO W72-12736
TRATIFICATION, *OPTICAL DENSITY, MICROAMMETER, OSCILLATORIA AGARDH W72-07145
BIOINDICATORS, WATER POLLUTION, MICROB: /EASONAL, SUSPENDED LOAD, W72-07896
ATER POLLUTION SOURCES, REVIEWS, MICROBIAL DEGRADATION, CLASSIFICA W72-14363
AMPLING, CULTURES, ECOLOGY, SOIL MICROBIOLOGY, CYANOPHYTA, FROZEN W73-00854
ON, ABSTRACTS, DATA COLLECTIONS, MICROBIOLOGY, SYSTEMATICS, POLLUT W72-13800
E WATER, *TASTE, *ODOR, SURVEYS, MICROBIOLOGY, WATER TREATMENT, PH W72-10076
UTRIENTS, ANALYTICAL TECHNIQUES, MICROBIOLOGY, *WASTE WATER TREATM W72-09590
MIN B, *EUTROPHICATION, *AQUATIC MICROBIOLOGY, OREGON, *SEDIMENT-W W72-08989
S, AEROBIC CONDITIONS, AERATION, MICROBIOLOGY, BIOMASS, CHLORELLA. W71-04079
SH W/ *AQUATIC ANIMALS, *AQUATIC MICROBIOLOGY, *ACARICIDES, BRACKI W71-08026
NDOCLADIA SP., PHYLLOSPADEX SP., MICROCLADIA SP., CLADOPHORA SP., W71-11851
ODOTORULA, A/ *MICROBIAL GROWTH, MICROCOCCUS, BACILLUS, VIBRIO, RH W72-10861
PERFUSION UNIT, FLAVOBACTERIUM, MICROCOCCUS.: W71-08027
IA.: BACILLUS, MICROCOCCUS, GRAM NEGATIVE BACTER W72-12255

ROPHYLL A, MACROPHYTES, *SWEDEN,	MICROCYSIS, OSCILLATORIA, SYNURA,	W72-12729
A, MELOSIRA GRANULATA, CERATIUM,	MICROCYSTIS, RHODOMONAS MINUTA, A	W72-12738
AMERICANA, CRYPTOMONAS PULSILLA,	MICROCYSTIS AERUGINOSA, CHRCOMONA	W72-13816
N RIVER, BELL RIVER, SALT RIVER,	MICROCYSTIS, APHANIZOMENON, CYANI	W72-14282
RA, NANNOCHLORIS, CHLOROCCOCUM,	MICROCYSTIS, FUSARIUM, TRICHORIXA	W70-06655
MESOTROPHIC,	MICROCYSTIS, ANABAENA, LYNGBYA.:	W72-04269
*ATMOSPHERIC CARBON,	MICROCYSTIS, ANABAENA.:	W72-10607
-VACUOLE DEFLATION, SOUTH WALES,	MICROCYSTIS AERUGINOSA, ANABAENA.	W72-11564
ASTINASTRUM, ANACYSTIS NIDULANS,	MICROCYSTIS.: /ONAS, FONTINALIS,	W71-11519
CULA, NITZSCHIA, STEPHANODISCUS,	MICROCYSTIS, APHANIZOMENON, ANABE	W71-12855
ULATION, ANABAENA, GLOEOTRICHIA,	MICROCYSTIS, NOSTOC, SLAKED LIME.	W72-01097
AS AERUGINOSA, LIMITING NUTRIEN/	MICROCYSTIS AERUGINOSA, PSEUDOMON	W73-01657
RECISION, *SWEDEN, LAKE MALAREN,	MICROCYSTIS FLOS-AQUAE, PHRAGMITE	W73-00286
TOTHRIX, OSCILLATORIA, ANABAENA,	MICROCYSTIS, TRIBONEMA, APHANIZOM	W72-07940
LOTHRIX, ANABAENA, OSCILLATORIA,	MICROCYSTIS, ALGISTAT.: /LONIL, U	W72-07508
.), APHANIZOMENON, ANABAENA SPP,	MICROCYSTIS AERUGINOSA, APHANIZOM	W72-07890
POLLUTION CONTROL, METHODOLOGY,	MICROENVIRONMENT, MEASUREMENT.: /	W72-03218
GY, *ACARICIDES, BRACKISH WATER,	MICROENVIRONMENT, LITTORAL, OXYGE	W71-08026
PHILIPPINES, ENIWETOK,	MICRONESIA, FANNING ISLANDS.:	W71-11486
ORIUM, ANTIMONY, MACRONUTRIENTS,	MICRONUTRIENTS, SCANDIUM, RUBIDIU	W73-01434
L STUDIES, PLANT TISSUES, ALGAE,	MICROORGANISMS, BACTERIA, INVERTE	W73-01676
LAKES, AQUATIC ALGAE, ICE, FISH,	MICROORGANISMS, ODOR, TASTE, PLAN	W73-01669
TORS, RHODOPHYTA, AQUATIC ALGAE,	MICROORGANISMS, CHLOROPHYTA, BIOL	W73-02099
OHYDRATES, MARINE PLANTS, MARINE	MICROORGANISMS, WATER ANALYSIS, B	W73-02100
BIOLOGICAL COMMUNITIES, *AQUATIC	MICROORGANISMS, CYCLING NUTRIENTS	W73-02095
RY PRODUCTIVITY, *BENTHIC FLORA,	MICROORGANISMS, ALGAE, LAKES, EUT	W72-07501
ATION, LARVAE, ROTIFERS, AQUATIC	MICROORGANISMS, DIPTERA, PHOSPHO:	W72-07132
ATER POLLUTION EFFECTS, *AQUATIC	MICROORGANISMS, *INVESTIGATIONS,	W72-07225
LLUTANTS, AQUATIC ALGAE, AQUATIC	MICROORGANISMS, AQUATIC ANIMALS,	W72-07896
NDS, *NUTRIEN/ *WATER POLLUTION,	MICROORGANISMS, *INORGANIC COMPOU	W72-07933
EFFECTS, *ALGAE, *PROT/ *AQUATIC	MICROORGANISMS, *WATER POLLUTION	W72-08579
RIA, FUNGI, PROTOZOA, PATHOGENS,	MICROORGANISMS, ALGAL TOXINS.: /E	W72-08586
ABOLISM, BIODEGRADATION, *MARINE	MICROORGANISMS, *MICROBIAL DEGRAD	W73-00361
W ENGLAND, AQUATIC LIFE, AQUATIC	MICROORGANISMS, MARINE ALGAE, SAL	W72-14317
BON DIOXIDE, *RADIOACTIVI/ *SOIL	MICROORGANISMS, *ABSORPTION, *CAR	W72-14301
GEN ION CONCENTRATION, SETTLING,	MICROORGANISMS, ACCLIMATION, CALC	W71-13334
PHYSIOLOGICAL ECOLOGY, BIOASSAY,	MICROORGANISMS, *EUGLENA, ALGAE.:	W71-12303
ION, HYDROGEN ION CONCENTRATION,	MICROORGANISMS, FERMENTATION, ALK	W71-12223
TESTS, STATISTICAL METHODS, SOIL	MICROORGANISMS, RUNOFF, HYDROLIC	W72-02218
ARINE BACTERIA, COASTAL MARSHES,	MICROORGANISMS.: /ADIOISOTOPES, M	W72-00948
ON, *EUTROPHICATION, *NUTRIENTS,	MICROORGANISMS, EFFLUENT, BIOCHEM	W71-13553
ODUCTION, WASTE DISPOSAL, MARINE	MICROORGANISMS, SEDIMENTATION, AL	W71-13723
ON, LAKES, AQUATIC PLANTS, IRON,	MICROORGANISMS, COLORIMETRY, WISC	W72-03742
*CHLORELLA, *SCENEDESMUS, ALGAE,	MICROORGANISMS, ENVIRONMENTAL EFF	W72-10861
IC ALGAE, PHYTOPLANKTON, AQUATIC	MICROORGANISMS, AQUATIC ANIMALS,	W72-09650
C LIFE, AQUATIC ANIMALS, AQUATIC	MICROORGANISMS, AQUATIC PLANTS, A	W72-09646
*CHLORELLA, *CULTURES, *AQUATIC	MICROORGANISMS, SPORES, ALGAE, CY	W71-08027
*STRIP MINES, *WATER / *AQUATIC	MICROORGANISMS, *ACID MINE WATER,	W71-07942
XYGEN, STRATIFICATION, SAMPLING,	MICROORGANISMS, SLUDGE, TERTIARY	W71-07382
*ANALYTICAL TECHNIQUES, *MARINE	MICROORGANISMS, MARINE PLANTS, BA	W71-08042
, CHLORELLA, BENTHIC FLORA, MUD,	MICROORGANISMS, PROTEINS, RADIOAC	W71-08042
FACULTATIVE	MICROORGANISMS, INDIA.:	W71-06033
GEN DEMAND, STABILIZATION PONDS,	MICROORGANISMS, BIODEGRADATION, W	W71-07097
N DEMAND, WASTE WATER TREATMENT,	MICROORGANISMS.: /OCHEMICAL OXYGE	W71-06033
*AQUATIC PRODUCTIVITY, *AQUATIC	MICROORGANISMS, *ALGAE, *NITROGEN	W72-06532
ES), DATA PROCESSING, COLIFORMS,	MICROORGANISMS, HYDROGEN ION CONC	W71-05155
IQUES, PHOSPHORUS, ALGAE, LAKES,	MICROORGANISMS, TERTIARY TREATMEN	W71-03021
HATES, CHEMICAL ANALYSIS, ALGAE,	MICROORGANISMS, HYDROGEN ION CONC	W71-01474
ARIO, SEASONAL, ALGAE, CULTURES,	MICROORGANISMS, EFFLUENTS, SETTLI	W71-00116
CHLORELLA, *SCENEDESMUS, AQUATIC	MICROORGANISMS, AQUATIC BACTERIA,	W72-12255
ORS, PHYSICOCHEMICAL PROPERTIES,	MICROSCOPY, LIMNOLOGY, WATER QUAL	W72-13142
PLANKTON, *PRIMARY PRODUCTIVITY,	MICROSCOPY, CHLOROPHYTA, SCENEDES	W72-09239
IOINDICATORS, THERMAL POLLUTION,	MICROSCOPY, DIATOMS, NUTRIENTS, P	W72-10875
NUTRIENTS, NITROGEN, FILTRATION,	MICROSCOPY, INDUSTRIAL WASTES, DE	W71-11507
STEMATICS, BIOMASS, FLUOROMETRY,	MICROSCOPY, ZOOPLANKTON, NUTRIENT	W72-14313
PHY, SEA WATER, *NORTH CAROLINA,	MICROSCOPY, ATLANTIC OCEAN.: /GRA	W73-00284
SOPHYTA, AQUATIC ALGAE, ELECTRON	MICROSCOPY, PHOTOGRAPHY, WATER PO	W73-00432
IRA BINDERANA, SCANNING ELECTRON	MICROSCOPY, TRANSMISSION ELECTRON	W73-00432
SSION ELECTRON MICROSCOPY, LIGHT	MICROSCOPY.: /MICROSCOPY, TRANSMI	W73-00432
ICROSCOPY, TRANSMISSION ELECTRON	MICROSCOPY, LIGHT MICROSCOPY.: /M	W73-00432
TARALDEHYDE, SAMPL/ *FLUORESCENT	MICROSCOPY, *AUTORADIOGRAPHY, GLU	W73-00273
NS, ANAEROBIC BACTERIA, ELECTRON	MICROSCOPY, PIGMENTS, CARBOHYDRAT	W72-07951
SAMPLING, NETS, CENTRIFUGATION,	MICROSCOPY, DIATOMS, NUTRIENTS, B	W72-07899
, SAMPLING, MICROSCOPY, ELECTRON	MICROSCOPY, PYRROPHYTA, DISTRIBUT	W72-07663
OLINA, SOUTH CAROLINA, SAMPLING,	MICROSCOPY, ELECTRON MICROSCOPY,	W72-07663
TIC PLANTS, AUSTRALIA, SAMPLING,	MICROSCOPY, ECOLOGICAL DISTRIBUTI	W73-01094
IZOMENON, NAVICULA, GLENODINIUM,	MICROSPORA, SYNEDRA, DINOBRYON, O	W72-10875
ROTETRAS PYRIFORMIS, GONGROSIRA,	MICROSPORA, SPIROGYRA, ULOTHRIX.:	W71-05991
ROORGANISMS, *ALGAE, *ALGICIDES,	MICROSTRAINING, AERATION, PHOSPHA	W72-06536
CYSTIS SCHROETERI, ULOTHRIX SPP,	MICROTHAMNION KUETZ: /UM, SPHAERO	W72-09111
LATFORMS, LEAD, MERCURY, LASERS,	MICROWAVE RADIOMETRY, VIDEO, SCAN	W72-04742
ONDS, WATER HYACINTH, NUTRIENTS,	MIDGES, FISH, LAKES, CONDUCTIVITY	W71-01488
IVE WASTES, TENNESSEE, CRAYFISH,	MIDGES, GAMMA RAYS, BLOODWORMS, S	W70-06668
Y, SEDIMENTATION, PLANKTON, DIEL	MIGRATION, PROTOZOA, HYDROGEN ION	W71-05626
TIFICATION, SPECIATION, VERTICAL	MIGRATION, AQUARIA, GROWTH RATES,	W71-09852
GANISMS, MARINE ALGAE, SALINITY,	MIGRATION, TIDES, LOW WATER MARK,	W72-14317
M HIR/ *GLOEOTRICHIA ECHINULATA,	MIKOLAJSKIE LAKE(POLAND), CERATIU	W72-14812
IERRE(CANADA), MIQUELON(CANADA),	MIKROSYPHAR PORPHYRAE, PROTECTOCA	W71-10066
USKS, PROTOZOA, YEASTS, INSECTS,	MILDEWS, E. COLI, FISH, FOODS, GR	W73-01676
, CERATIUM MACROCEROS, NOCTILUCA	MILIARIS, GYMNODINIUM SPLENDENS,	W71-07875

TANKS.: *TEXTILE MILL WASTES, MONTREAL, BALANCING W70-10427
KE ERIE, ARNOLDIELLA CONCHOPHILA MILLER.: /GAL SPECIES, WESTERN LA W71-09156
RASSIPES, SALVINIA ROTUNDIFOLIA, MILORGANITE, GAMBUSIA AFFINIS, PR W71-01488
AMCHITKA ISLAND, MILROW TEST, CANNIKIN TEST.: W73-00764
PHOSPHATE/ CLADOPHORA GLOMERATA, MILWAUKEE HARBOR(WISCONSIN), POLY W72-13644
SEASE, BIOLOGICAL MAGNIFICATION, MINAMATA BAY(JAPAN), ORGANIC MERC W72-12257
ITY, HARDNESS(WATER), CHELATICN, MINE DRAINAGE, WATER PROPERTIES.: W72-14694
*AQUATIC MICROORGANISMS, *ACID MINE WATER, *STRIP MINES, *WATER W71-07942
CHLAMYDOMONAS REINHARDTII, TETR/ MINERAL CYCLE, CHARACTERIZATION, W73-02095
FACE WATERS, GEOLOGICAL SOURCES, MINERALIZATION, NITROGEN SOURCES, W71-06435
GROUNDWATER, SAFE YIELD, ALGAE, MINERALOGY, TURBIDITY, COLOR, ALK W72-05594
AGELLATES, PRIMARY PRODUCTIVITY, MINERALOGY, LIGHT, CULTURES, BIOA W72-07166
IMNOLOGY, BIOLOGICAL PROPERTIES, MINERALOGY, FISH, FISH BEHAVIOR, W72-14280
ANISMS, *ACID MINE WATER, *STRIP MINES, *WATER POLLUTION EFFECTS, W71-07942
A, SODIUM COMPOUNDS, ALKALINITY, MINING, BIOCHEMICAL OXYGEN DEMAND W72-04734
*LAKE MINNETONKA(MINN).: W71-08670
LIE(MINN), LAKE M/ DETROIT LAKES(MINN), COMMUNITY ACTION, LAKE SAL W71-03012
), COMMUNITY ACTION, LAKE SALLIE(MINN), LAKE MELISSA(MINN), NUTRIE W71-03012
LAKE SALLIE(MINN), LAKE MELISSA(MINN), NUTRIENT REMOVAL.: /CTION, W71-03012
LIMITING NUTRIENTS, SHAGAWA LAKE(MINN), WALDO LAKE(ORE).: * W72-09165
CAROTENOIDS, XANTHOPHYLLS, MINNESOTA LAKES, LUTEIN.: W72-01784
CARBON, DIATOMS, LIGHT, OXYGEN, MINNESOTA.: /FFECTS, TEMPERATURE, W71-08670
NE LAKE(MINNESOTA), LAKE CRYSTAL(MINNESOTA).: /MINNESOTA), BIG STO W71-04216
TONKA(MINNESOTA), BIG STONE LAKE(MINNESOTA), LAKE CRYSTAL(MINNESOT W71-04216
AR(CONNECTICUT), LAKE MINNETONKA(MINNESOTA), BIG STONE LAKE(MINNES W71-04216
ER, ST. LOUIS RIVER, BLACK RIVER(MINNESOTA), LAKE WASHINGTON(WASHI W71-04216
ATION, *LAKES, *WATER POLLUTION, MINNESOTA, WATER QUALITY, NUTRIEN W71-03012
PERIOR, WATER POLLUTION SOURCES, MINNESOTA, WASHINGTON, CONNECTICU W71-04216
, TOXICITY, CHEMICAL PROPERTIES, MINNESOTA, LIPIDS, HUMAN PATHOLOG W72-04275
GAE, DECOMPOSING ORGANIC MATTER, MINNESOTA, CYANOPHYTA, TREES, GRA W72-01784
, *EUTROPHICATION, *OLIGOTROPHY, MINNESOTA, LAKES, GLACIAL DRIFT, W71-13149
ASS, HYDROGEN ION CONCENTRATION, MINNESOTA, WATER TEMPERATURE, DIS W72-07145
T), LAKE ZOAR(CONNECTICUT), LAKE MINNETONKA(MINNESOTA), BIG STONE W71-04216
*LAKE MINNETONKA(MINN).: W71-08670
LS, CHINOOK SALMON, BASS, TROUT, MINNOWS, CRAYFISH, MOSQUITOES, FO W71-07417
S, CHINOOK SALMON, STRIPED BASS, MINNOWS, FISHKILL, HERRING, COPEP W71-11517
ERATIUM, MICROCYSTIS, RHODOMONAS MINUTA, ASTERIONELLA FORMOSA, STE W72-12738
INNIPEG, *SUBSTRATES, ACHNANTHES MINUTISSIMA, ACHNANTHES CF. BIRGI W72-13142
RIO, EPILITHIC ALGAE, ACHNANTHES MINUTISSIMA, ACHNANTHES FLEXELLA, W71-11029
EA, ANKISTRODESMUS, CHRYSOCOCCUS MINUTUS, DICTYOSPHAERIUM, LAGERHE W72-12738
NOVA SCOTIA), ST PIERRE(CANADA), MIQUELON(CANADA), MIKROSYPHAR POR W71-10066
, BENTHIC FAUNA, ALGAE, ECOLOGY, MISSOURI.: /ISH HARVEST, PLANKTON W70-09596
ATER QUALITY, CYANOPHYTA, LAKES, MISSOURI, NITROGEN, PHOSPHORUS.: / W72-10431
ENEDESMUS, CYANOPHYTA, MOLLUSKS, MITES, ANNE: /AS, CHLOROPHYTA, SC W72-00155
ANUM, LPP-1, LPP-1D, PHYCOVIRUS, MITOMYCIN C.: /, *PLECTONEMA BORY W72-03188
UM, THALASSIONEMA NITZSCHIOIDES, MIXED CULTURES, METABOLITES, SUBS W73-00441
*CHEMICAL-BIOLOGICAL TREATMENT, MIXED LIQUOR, SUSPENDED SOLIDS, U W71-13334
*WASTE HEAT, SUBLETHEL EFFECTS, MIXING ZONES, MODELING.: W70-09612
ATION, CARBON DIOXIDE, AERATION, MIXING, *EUTROPHICATION, *ANALYTI W72-04433
ROL, *LAKE MORPHOLOGY, AERATION, MIXING, ALGAL BLOOMS, ALGAL TOXIN W72-09304
OONS, ORGANIC LOADING, AERATION, MIXING, ALGAE, TEMPERATURE, ALTIT W72-06838
OSPHATES, LAKES, STRATIFICATION, MIXING, BOTTOM SEDIMENTS, SUCCESS W72-08049
N, TEMPERATURE, LIGHT INTENSITY, MIXING, DISSOLVED OXYGEN, METABOL W72-04983
, PHYTOPLANKTON, *NEW HAMPSHIRE, MIXING, ENVIRONMENTAL EFFECTS.: / W72-09061
COEFF/ *ALGAL SINKING, *VERTICAL MIXING, EUPHOTIC ZONE, DIFFUSION W71-03034
D OXYGEN, NUTRIENTS, CYANOPHYTA, MIXING, PHYTOPLANKTON.: /DISSOLVE W71-03034
, *RESERVOIRS, DESTRATIFICATION, MIXING, PHOSPHORUS, NITROGEN, TUR W72-08559
ATE RIVERS, HYPOLIMNION, VOLUME, MIXING, SAFETY, WATER COOLING, AN W70-09595
AND, *BIOCHEMICAL OXYGEN DEMAND, MIXING, SAMPLING, ORGANIC LOADING W71-07102
TE WATER TREATMENT, *BYPRODUCTS, MIXING, SOLAR RADIATION, ORGANIC W72-13508
AL MODELS, *COMPUTER SIMULATION, MIXING, TURBULENCE, TEMPERATURE, W71-07084
N LAGOONS, *MATHEMATICAL MODELS, MIXING, WATER RE-USE, LIGHT INTEN W71-07099
LAKE JACOMO(MO).: W72-10431
M OKENII, CHROMATIUM, MACROMONAS MOBILIS, THIOVULUM MAJUS, THIOSPI W72-07896
TURAL DRAINAGE, LAKE METABOLISM, MOBILITY, PARTICULATE FORM, FEEDL W71-06443
PARISONS, AQUATIC ALGAE, GRANTS, MODEL STUDIES.: /HEALTH, COST COM W71-09429
, TERTIARY TREATMENT, EFFLUENTS, MODEL STUDIES.: /, MICROORGANISMS W71-03021
, ALGAL CONTROL, EUTROPHICATION, MODEL STUDIES, MATHEMATICAL MODEL W72-04787
TROGEN CYCLE, AMMONIA, NITRATES, MODEL STUDIES, DISSOLVED OXYGEN, W72-08459
SSAY, ALGAE, SNAILS, MOSQUITOES, MODEL STUDIES.: /AINS, FISH, BIOA W72-07703
NEERING, *VIRGINIA, PROJECTIONS, MODEL STUDIES.: /VIRONMENTAL ENGI W72-14405
ATER TREATMENT, WATER TREATMENT, MODEL STUDIES.: /ACTERIA, WASTE W W73-01362
E, *PHOTOPERIODISM, GREAT LAKES, MODEL STUDIES, NUTRIENTS, PLANT G W72-13651
TECHNIQUES, NUMERICAL ANALYSIS, MODEL STUDIES, COMPUTERS, CULTURE W72-12724
OLOGY, *EUTROPHICATION, *KANSAS, MODEL STUDIES, ECOSYSTEMS, ALGAE, W72-12393
EX, ELECTRIC ENERGY CONSUMPTION, MODEL TESTING.: /INDUSTRIAL COMPL W72-03327
*PARALLEL PLATE MODEL.: W73-01362
N), GREAT PLAIN/ *PHOTOSYNTHETIC MODEL, *TUTTLE CREEK RESERVOIR(KA W72-12393
ORELLA PYRENOIDOSA, MONOD GROWTH MODEL, ALGAL GROWTH RATES.: / CHL W72-04788
SUBLETHEL EFFECTS, MIXING ZONES, MODELING.: *WASTE HEAT, W70-09612
RIVER BEDS, ALGAE, MATHEMATICAL MODELS.: /TATION EFFECTS, SLUDGE, W70-09189
AQUIN DELTA(CALIF), STEADY-STATE MODELS.: /ESTUARY(DEL-NJ), SAN JO W71-05390
ION, MODEL STUDIES, MATHEMATICAL MODELS.: /AL CONTROL, EUTROPHICAT W72-04787
ORPTION, CHELATION, MATHEMATICAL MODELS.: /PENETRATION, CATION ADS W73-00823
S, DISCHARGE(WATER), STATISTICAL MODELS.: /, POWERPLANTS, EFFLUENT W71-11583
R/ *PHYTOPLANKTON, *MATHEMATICAL MODELS, *ALGAL CONTROL, *RESERVOI W72-08559
MICS, LAKES, DISS/ *MATHEMATICAL MODELS, *ALGAE, *POPULATION, DYNA W71-03034
N, DESIGN CRITERIA, MATHEMATICAL MODELS, *BIOASSAY, *WASTE DILUTIO W72-04787
OXIDATION LAGOONS, *MATHEMATICAL MODELS, *COMPUTER SIMULATION, MIX W71-07084
*EUTROPHICATION, *MATHEMATICAL MODELS, *ESSENTIAL NUTRIENTS, *PR W72-08986

ON, AMMONIA, NITR/ *MATHEMATICAL MODELS, *NITROGEN CYCLE, *OXIDATI W71-05390
ITROGEN, COLIFORMS, MATHEMATICAL MODELS, ALGAE, SAMPLIN: /HATES, N W71-13356
URBIDITY, SAMPLING, MATHEMATICAL MODELS, ESTUARINE ENVIRONMENT, SA W72-12941
HERMAL STRATIFICATION, HYDRAULIC MODELS, FISH, ALGAE, FATHOMETERS, W70-09614
OACTIVITY EFFECTS, *MATHEMATICAL MODELS, FORECASTING, COBALT RADIO W73-00831
E WATER TREATMENT, *MATHEMATICAL MODELS, GROWTH RATES, SOLAR RADIA W72-12289
ERIE, *HYPOLIMNION, MATHEMATICAL MODELS, IRON, MANGANESE, PHOSPHOR W72-12997
TEMS, RAIN FORESTS, MATHEMATICAL MODELS, LINEAR PROGRAMMING, SYSTE W72-03342
OXIDATION LAGOONS, *MATHEMATICAL MODELS, MIXING, WATER RE-USE, LIG W71-07099
POLLUTION SOURCES, MATHEMATICAL MODELS, NITROGEN, ABATEMENT, COST W72-09171
TROGEN, PHOSPHORUS, MATHEMATICAL MODELS, NUTRIENTS, MEASUREMENT, P W71-08670
ES, AQUARIA, ALGAE, MATHEMATICAL MODELS, PATH OF POLLUTANTS, TRACE W72-03348
ON, AQUATIC PLANTS, MATHEMATICAL MODELS, SELF-PURIFICATION, ALGAE, W71-12072
, *EUTROPHICATION, *MATHEMATICAL MODELS, SIMULATION, ANALYTICAL TE W71-03021
ROELECTRIC PLANTS, *MATHEMATICAL MODELS, THERMAL STRATIFICATION, T W72-14405
ANTS, RADIOECOLOGY, MATHEMATICAL MODELS, WASTE ASSIMILATIVE CAPACI W73-00828
, *INSTRUMENTATION, MATHEMATICAL MODELS, WATER ANALYSIS, POLLUTANT W72-08790
SITY, GROWTH RATES, MATHEMATICAL MODELS, WATER TEMPERATURE, WATER W72-06274
DUSTRIAL MACHINERY, UTILIZATION, MODIFICATION.: /ARM MACHINERY, IN W71-06188
LOGICAL--GAMMA-RADIATION SYSTEM, MODULAR SYSTEM, COBALT-60.: *BIO W72-00383
INDUSTRIAL WATER POLLUTION, DES MOINES RIVER, PACKING PLANTS.: W71-06445
CULAR SIZE, *ALGAL EXUDAT/ *HIGH MOLECULAR WEIGHT COMPOUNDS, *MOLE W71-07878
LUTIONS, *BIOLOGICAL PROPERTIES, MOLECULAR STRUCTURE, PERMEABILITY W72-01473
ENTATION, ANALYTICAL TECHNIQUES, MOLECULAR STRUCTURE, CATION EXCHA W72-12203
ANT FRICTIONAL DRAG, ALKY/ *HIGH- MOLECULAR-WEIGHT POLYMERS, TURBUL W70-06968
COCCUS BACILLARIS, UNDISSOCIATED MOLECULES, ALGAL GROWTH.: /STICHO W70-07255
ES, EPIFAUNA, SAGARTIA, ANEMONE, MOLGULA, TUNICATE, PISMO CLAM, TI W72-07907
TORING, WATER POLLUTION EFFECTS, MOLLUSKS.: /GE, CRUSTACEANS, MONI W72-10955
AVES, CLAMS, CRUSTACEANS, WORMS, MOLLUSKS, WATER TEMPERATURE, SALI W72-09092
ORAL, SALINITY, CURRENTS(WATER), MOLLUSKS, ENVIRONMENTAL EFFECTS.: W72-11876
OPHYTA, SCENEDESMUS, CYANOPHYTA, MOLLUSKS, MITES, ANNE: /AS, CHLOR W72-00155
PES, NUCLEAR POWERPLANTS, SITES, MOLLUSKS, MARINE ALGAE, ON-SITE I W72-03324
FLIES, DRAGONFLIES, CADDISFLIES, MOLLUSKS, CYANOPHYTA, CHRYSOPHYTA W71-11012
IOLOGICAL COMMUNITIES, CHLORINE, MOLLUSKS, COPEODS, DIPTERA, GULF W73-00855
COLOGY, *MARINE ALGAE, ANNELIDS, MOLLUSKS, RHODOPHYTA, PHAEOPHYTA, W73-00932
LLUTION EFFECTS, STANDING CROPS, MOLLUSKS, LARVAE, CHRYSOPHYTA, CY W73-00935
YMES, ANIMAL PARASITES, VIRUSES, MOLLUSKS, PROTOZOA, YEASTS, INSEC W73-01676
OASTS, CRUSTACEANS, MARINE FISH, MOLLUSKS, POLLUTANT IDENTIFICATIO W73-00819
DICATORS, *MARINE ANIMALS, ASIA, MOLLUSKS, COBALT RADIOISOTOPES, I W73-00811
RIPHYTON, EQUIPMENT, TUBIFICIDS, MOLLUSKS, WATER TEMPERATURE, MARI W73-00374
IVITY, COPEPODS, MARINE ANIMALS, MOLLUSKS, GASTROPODS, PHAEOPHYTA. W73-00371
S, WATER POLLUTION EFFECTS, MUD, MOLLUSKS, FISH, INSECTS, DDT, SIL W70-08098
RBIDITY, ORGANIC LOADING, CORAL, MOLLUSKS, WORMS.: /POLLUTANTS, TU W71-01475
ASTES, RIVERS, ABSORPTION, FISH, MOLLUSKS, MARINE ANIMALS, COBALT W71-09225
E ANIMALS, *ABSORPTION, AQUARIA, MOLLUSKS, SNAILS, MUSSELS, MARINE W71-09182
R, ALGAE, NEMATODES, GASTROPODS, MOLLUSKS, AMPHIPODA, MECHANICAL E W72-05432
WASTES, WATER POLLUTION EFFECTS, MOLLUSKS, MARINE PLANTS, MARINE A W72-04461
TY, WATER POLLUTION, PHAEOPHYTA, MOLLUSKS, MARINE ALGAE, MARINE AN W72-11925
GANESE, IRON, SULFUR, POTASSIUM, MOLYBDENUM, DETERGENTS, SEWAGE TR W72-07514
ALYSIS, *TRACE ELEMENTS, COPPER, MOLYBDENUM, MANGANESE, COBALT, IR W71-11515
HIZOTHRIX CALCIOCOLA, CLOSTERIUM MONILIFERUM, MELOSIRA VARIANS, NA W72-10883
NVIRONMENTAL EFFECTS, EFFLUENTS, MONITORING, CHELATION.: /ASTES, E W72-10957
EDIMENTS, ESTUARINE ENVIRONMENT, MONITORING, CRABS, SHRIMP.: /E, S W72-10958
YONIC GROWTH STAGE, CRUSTACEANS, MONITORING, WATER POLLUTION EFFEC W72-10955
ION, INDUSTRIES, WASTE DISPOSAL, MONITORING, NUTRIENTS, ALGAE, INS W71-12091
TER, MARINE PLANTS, ASSESSMENTS, MONITORING, SEWAGE, ESTUARIES, GR W71-12867
L WASTES, OIL WASTES, ESTUARIES, MONITORING.: /L PLANTS, INDUSTRIA W71-13746
*TREATMENT FACILITIES, SAMPLING, MONITORING, AUTOMATIC CONTROLS, A W72-00042
ERPLANTS, ECOLOGY, RADIOECOLOGY, MONITORING, POLLUTION ABATEMENT, W72-00939
Y EFFECTS, *NUCLEAR POWERPLANTS, MONITORING, PATH OF POLLUTANTS, A W72-00959
NKTON, RADIOACTIVITY TECHNIQUES, MONITORING, SURFACE WATERS, PERSI W73-00361
CHNIQUES, RADIOECOLOGY, TRACERS, MONITORING, NUCLEAR POWERPLANTS, W73-00790
IFORNIA, TRITIUM, MARINE PLANTS, MONITORING, SAMPLING, WATER ANALY W73-00831
, LEGISLATION, LAND USE, SHORES, MONITORING, MARSHES, SOIL EROSION W73-02479
IA, TRITIUM, RADIOACTIVE WASTES, MONITORING, PHOTOGRAPHY, ULTRAVIO W72-04742
IOISOTOPES, *ALGAE, *ABSORPTION, MONITORING, ANALYTICAL TECHNIQUES W71-08032
GROWTH, OXYGEN, MONITORS.: W71-10791
OSPHATES, SUBSTRATE UTILIZATION, MONOD EQUATION, PENICILLIN G, TER W73-01445
DILUTION, CHLORELLA PYRENOIDOSA, MONOD GROWTH MODEL, ALGAL GROWTH W72-04788
ON(WIS), LAKE MENDOTA(WIS), LAKE MONONA(WIS), LAKE KEGONSA(WIS), L W72-05473
LIMITI/ LAKE MENDOTA(WIS.), LAKE MONONA(WIS.), LAKE WINGRA(WIS.), W73-00232
DUCTION, LAKE MENDOTA(WIS), LAKE MONONA(WIS).: /RCES, ACETYLENE RE W72-10608
TING, GLUCOSE, MEMBRANE FILTERS, MONOSACCHARIDES, MEMBRANE FILTRAT W72-14326
COSTATUM, NUCLEOTIDES, ETHANOL, MONOSACCHARIDES, CULTURE MEDIA, D W73-00942
, CL/ *EXCRETION, DISACCHARIDES, MONOSACCHARIDES, OLIGOSACCHARIDES W72-07710
, CHYDORUS SPHAERICUS, CHYDORUS, MONOSTYLA, LEPADELLA, TRICHOCERCA W72-07132
VER, YELLOWTAIL D/ *BIGHORN LAKE(MONT), BIGHORN RIVER, SHOSHONE RI W73-01435
TAMOGETON PECTINATUS, ELEOCHARIS MONTEVIDENSES, CHARA ASPPRA, SCIR W73-01673
S, *WATER TEMPERATURE, SEASONAL, MONTHLY, AQUATIC LIFE, INTERSTATE W70-09595
CACODYLATE, CULTURE MEDIA, PORIA MONTICOLA, PORIA COCOS, PORIA VAI W72-13813
EBEC), GREEN BAY(WIS), MACKEREL, MONTREAL(CANADA).: /LIEU RIVER(QU W70-08098
*TEXTILE MILL WASTES, MONTREAL, BALANCING TANKS.: W70-10427
RESERVOIR, ARTIFICIAL RESERVOIR, MONTROSE LAKE, THOMAS LAKE.: /NG W70-09596
ULATORS, INHIBITORS, HERBICIDES, MONURON, PHOTOSYNTHESIS.: /TH REG W72-09650
S, EUPHAUSIA PACIFICA, ENGRAULIS MORDAX.: /N, TRIPHOTURUS MEXICANU W73-02105
, CRAB EGGS, PELAGIC EGGS, GADUS MORHUA, GRASS SHRIMP, GREEN CRAB, W72-08142
A, PTEROPODS, EUPHAUSIIDS, GADUS MORHUA, UROPHYCIS.: /EURA, SAGITT W73-00371
, *ENVIRONMENTAL EFFECTS, *PLANT MORPHOLOGY, *GENETICS, *VARIABILI W73-00369
CL/ *ALGAE, *SYSTEMATICS, *PLANT MORPHOLOGY, REPRODUCTION, LIFE CY W72-07141
PHYSIOLOGY, *GROWTH RATES, PLANT MORPHOLOGY, CULTURES, PHOTOSYNTHE W72-07710

GAE, *CYTYLOGICAL STUDIES, PLANT MORPHOLOGY, EQUATIONS, RESISTANCE W73-01630
GIC/ *ALGAE, *RHODOPHYTA, *PLANT MORPHOLOGY, *PLANT GROWTH, CYTOLO W73-01622
L COMMUNITIES, FOOD CHAINS, LAKE MORPHOLOGY, RESERVOIR SILTING, LA W73-00938
N, *WATER QUALITY CONTROL, *LAKE MORPHOLOGY, AERATION, MIXING, ALG W72-09304
POLLUTANTS, *MAINTENANCE, *PLANT MORPHOLOGY, *PHOSPHATES, *NITRATE W72-10162
SOURCES, DISSOLVED OXYGEN, LAKE MORPHOLOGY, AQUATIC ALGAE, MOSSES W71-11949
, PLANT GROWTH SUBSTANCES, PLANT MORPHOLOGY.: /LOGY, *PLANT GROWTH W70-07842
SEDIMENTS, *WATER ANALYSIS, LAKE MORPHOLOGY, SEDIMENTATION, PLANKT W71-05626
COMPETITION, GROWTH RATES, PLANT MORPHOLOGY, WATER CIRCULATION.: / W72-04766
GIC DATA, DATA COLLECTIONS, LAKE MORPHOLOGY, ECOLOGY, BIOLOGICAL P W72-13851
, HUMMOCK CHANNEL, CHLOROPLASTS, MORPHOLOGY, EPIPHYTES.: /RE MEDIA W72-12633
URE MEDIA, CULTURING TECHNIQUES, MORPHOLOGY, FLAGELLATES.: /, CULT W72-12632
EGHINIANA, *POLYMORPHISM, *VALVE MORPHOLOGY, *VALVE PATTERN CHARAC W73-02548
CYANOPHYTA, PHYTOPLANKTON, LAKE MORPHOMETRY, GEOLOGY, LIMESTONE, W72-04269
R, HYPOLIMNION, EPILIMNION, LAKE MORPHOMETRY, DISTRIBUTION PATTERN W72-07145
, DISSOLVED OXYGEN, CARBON, LAKE MORPHOMETRY, CHEMICAL PROPERTIES, W72-08048
ELA, RHODOMENIA, MARMION ISLAND, MORRIA REEF, BEAR BAY, NEVA ISLAN W72-09120
POWER PLANT, SAN JOAQUIN RIVER, MORRO BAY POWER PLANT, HUMBOLDT B W72-07907
DITIONS, BIOASSAY, GROWTH RATES, MORTALITY.: /NOPHYTA, AEROBIC CON W72-07664
BAYS, SEA WATER, ORGANIC MATTER, MORTALITY, WATER TEMPERATURE, PHA W73-00366
N BLOOMS, PHYSIOLOGICAL ECOLOGY, MORTALITY, RED TIDE, TOXICITY.: / W72-14056
CHLAMYDOMONAS PERTUSA, PANDORINA MORUM.: /SA, CHROOMONAS CAUDATA, W72-13816
HIBITORS, VOLVOCACEAE, PANDORINA MORUM, PLATYDORINA CAUDATA, EUDOR W71-13793
P, CARTERIA FRITSCHII, PANDORINA MORUM, SPHAEROCYSTIS SCHROETERI, W72-09111
TURAL WATERS.: *USSR, MOSCOW RIVER, OKA RIVER, FOAM, NA W72-05506
ATES, CHLAMYDOMONAS REINHARDTII, MOSES LAKE, PINE LAKE, LAKE WASHI W72-12637
S, MACROPHYTES, CYPRINUS CARPIO, MOSQUITO FISH, GAMBUSIA AFFINIS, W73-01673
, FISH, BIOASSAY, ALGAE, SNAILS, MOSQUITOES, MODEL STUDIES.: /AINS W72-07703
TANDING CROP, CHEMICAL ANALYSIS, MOSQUITOES, BACTERIA, FLOATING PL W71-01488
BASS, TROUT, MINNOWS, CRAYFISH, MOSQUITOES, FOOD CHAIN, HERRING, W71-07417
CYPR/ TILAPIA NILOTICA, TILAPIA MOSSAMBICA, TILAPIA MELANOPLEURA, W71-04528
IDENTIFICATION, ONTARIO, ALGAE, MOSSES, AQUATIC PLANTS.: /LLUTANT W71-11529
ANT RELATIONSHIPS, MARINE ALGAE, MOSSES, CRABS, ALGAE, SNAILS, MUS W72-03347
IODEGRADATION, BIOMASS, LICHENS, MOSSES, FUNGI, ALGAE, SOIL CONTAM W72-14303
E RATIOS, SPECTROMETRY, LICHENS, MOSSES, GYMNOSPERMS, ANGIOSPERMS. W72-12709
D TYPES, BIRDS, AQUATIC ANIMALS, MOSSES, INVERTEBRATES, DIATOMS, P W72-05952
LAKE MORPHOLOGY, AQUATIC ALGAE, MOSSES, WATER TEMPERATURE, WATER W71-11949
NE, BIOTYPES, POUR PLATE METHOD, MOST PROBABLE-NUMBER TEST, MEMBRA W72-04743
*RESIDUAL, MOST PROBABLE NUMBER.: W71-07096
ORPSSTROMMEN RIVER, ATRAN RIVER, MOTALA STROM RIVER, SPARGANIUM ER W73-00286
*GAS VACUOLES, MOTILE FORMS, CYANOPHAGES.: W71-13184
*NONFRUITING MYXOBACTERIA, DNA, MOTILITY, CYST FORMATION, ANTIBIC W72-07951
US FALCATUS, PODOHEDRA.: MOUGEOTIA GENUFLEXA, ANKISTRODESM W71-07360
ASTRUM, STAURASTRUM, TABELLARIA, MOUGEOTIA.: /ON, FRAGILARIA, PEDI W72-10076
COSMARIUM, PHACUS, CHAETOCEROS, MOUGEOTIA, NAVICULA, PINNULARIA, W72-09111
S SOLI, SCENEDESMUS QUADRICAUDA, MOUGEOTIA, ULOTHRIX AEQUALIS, STE W72-07144
ICHALES, ZYGNEMALES, OEDOGONIUM, MOUGOTIA, SPOROTETRAS PYRIFORMIS, W71-05991
COLO).: GRAND LAKE(COLO), SHADOW MOUNTAIN LAKE(COLO), LAKE GRANBY(W72-04791
CARTENOIDS, SAMPLE PREPARATION, MOUNTAIN LAKE.: /, CHLOROPHYLL A, W73-00916
ECT.: SMITH MOUNTAIN PROJECT(VA), WEATHER EFF W72-05282
*PUMPED STORAGE OPERATION, SMITH MOUNTAIN RESERVOIR(VA).: W72-14405
C, BIOTIC ASSOCIATIONS, LA GORCE MOUNTAINS, AMMONIUM IONS.: /PHILI W73-00854
GE, SOIL PROPERTIES, GROUNDWATER MOVEMENT, OUTFALL SERVERS, STREAM W72-14363
N, BIOCONTROL, PLANT PHYSIOLOGY, MOVEMENT, VIRUSES, PHOTOSYNTHESIS W71-13184
UROPODS, ENRICHMENT, LEUCOTHRIX MUCOR, MAJURO ATOLL, MARCHALL ISL W72-08142
, CLAYS, CORES, SEASONAL, ALGAE, MUD-WATER INTERFACES, IRON, MANGA W72-05469
A, LIGHT INTENSITY, TEMPERATURE, MUD, GROWTH RATES, ANABAENA, NITR W72-12544
ALGAE, CHLORELLA, BENTHIC FLORA, MUD, MICROORGANISMS, PROTEINS, RA W71-08042
OURCES, WATER POLLUTION EFFECTS, MUD, MOLLUSKS, FISH, INSECTS, DDT W70-08098
RUS REMOVAL, RATE OF SORPT/ LAKE MUD, PHOSPHORUS SORPTION, PHOSPHO W72-11617
NZYMES, THERMOCLINE, EPILIMNION, MUD, PHYSICOCHEMICAL PROPERTIES, W72-03224
LEAVES, FOREST SOILS, SAMPLING, MUD, PHYTOPLANKTON, FLUORESCENCE, W72-01784
NTS, ESSENTIAL NUTRIENTS, LAKES, MUD, WISCONSIN, AQUATIC WEEDS, EU W72-10608
ESTICIDES, TROPHIC LEVEL, GASES, MUDS, ADSORPTION, COPP: /RTIES, P W72-00847
EMS, PESTICIDE KINETICS, *MAINE, MUDS, INVERTEBRATES, FISH, BIRDS, W72-12930
*WAHNBACH RES/ *FERRIC CHLORIDE, MULTI-LAYER FILTERS, ANTHRACITE, W72-06201
ANALYSIS.: *LAKE WINGRA(WIS.), MULTIVARIATE ANALYSIS, ORDINATION W73-02469
ODEGRADATION, LOCAL GOVERNMENTS, MUNICI: /ASTE WATER TREATMENT, BI W72-06655
TMENT, RESEARCH AND DEVELOPMENT, MUNICIPAL WASTES, LEGISLATION, RE W72-06638
AE, AEROBIC TREATMENT, BACTERIA, MUNICIPAL WASTES, BIOCHEMICAL OXY W71-03896
NUTRIENTS, SEDIMENTS, COLIFORMS, MUNICIPAL WASTES, AGRICULTURE, FA W71-03012
ATER POLLUTION, DOMESTIC WASTES, MUNICIPAL WASTES.: /ACE WATERS, W W70-09388
HICATION, NUTRIENT REQUIREMENTS, MUNICIPAL WASTES.: /LGAE, *EUTROP W71-08775
SHEDS(BASINS), SURFACE DRAINAGE, MUNICIPAL WASTES, INDUSTRIAL WAST W71-08953
GENATION, ALGAE, PHOTOSYNTHESIS, MUNICIPAL WASTES, BIOCHEMICAL OXY W71-06033
E, NUTRIENTS, INDUSTRIAL WASTES, MUNICIPAL WASTES, FARM WASTES, WA W72-01659
*AQUATIC ALGAE, *EUTROPHICATION, MUNICIPAL WASTES, CITIES, PHOSPHA W72-03970
ATION, COMPARATIVE PRODUCTIVITY, MUNICIPAL WASTES, INDUSTRIAL WAST W72-03228
TABILIZATION, INDUSTRIAL WASTES, MUNICIPAL WASTE, BIOCHEMICAL OXYG W71-13356
T, COSTS, ANALYTICAL TECHNIQUES, MUNICIPAL WATER, PHOSPHORUS.: /EN W72-09171
PARATION TECHNIQUES, FILTRATION, MUNICIPAL WASTES, INDUSTRIAL WAST W72-09386
ILT/ *OXIDATION LAGOONS, *ALGAE, MUNICIPAL WASTES, *FILTRATION, *F W71-13341
L TECHNIQUES, INDUSTRIAL WASTES, MUNICIPAL WASTES, METALS, HEAVY M W71-11793
TOGRAPHY, *MATHEMATICAL METHODS, MUNICIPAL WASTES, INDUSTRIAL WAST W71-11685
DISSOLVED OXYGEN, AQUATIC LIFE, MUNICIPAL WASTES, PHOTOSYNTHESIS, W71-11017
HOSPHORUS, DRAINAGE, INDUSTRIES, MUNICIPAL WASTES, TURNOVERS, SEDI W71-11009
OSYNTHESIS, *GIBBERELLIN, HYPNEA MUSCIFORMIS, GRACILARIA VERUCOSA, W72-07496
, CENTROCERAS CLAVULATUM, HYPENA MUSCIFORMIS, CERAMIUM BUSSOIDEUM, W73-00838
M.: CONCENTRATION, MUSCLE, BONE PHYSIOLOGY/METABOLIS W72-10694
HARA ASPPRA, SCIRPUS SPP, GILLS, MUSCLE, VISCERA, EYES, SCALES(FI: W73-01673

METHYLHEPTADECANE, 8-MET/ NOSTOC MUSCORUM, LYNGBYA LAGERHAIMII, 7- W73-01439
ONS, COPPER, LEAD, HEAVY METALS, MUSSELS, *KELPS, STARFISH, CRUSTA W72-11925
S, AQUATIC INSECTS, CRUSTACEANS, MUSSELS, COMMON MERGANSER DUCK, B W72-12930
SIS, SPECTROMETERS, ZOOPLANKTON, MUSSELS, CLAMS, FISH, SHELLFISH, W72-05965
E, MOSSES, CRABS, ALGAE, SNAILS, MUSSELS, ISOPODS, AMPHIPODA, RADI W72-03347
TION, AQUARIA, MOLLUSKS, SNAILS, MUSSELS, MARINE ALGAE, AQUATIC AN W71-09182
E DISPOSAL, MARINE ALGAE, CLAMS, MUSSELS, OYSTERS, EUTROPHICATION, W72-07715
ATER, ALGAE, NEMATODES, OYSTERS, MUSSELS, PROTOZOA, MANGROVE SWAMP W72-07911
T, REGULATION, FALLOUT, OYSTERS, MUSSELS, PLANKTON, MARINE ALG: /N W72-00959
LANT GROWTH, OYSTERS, SHELLFISH, MUSSELS, SHELLFISH FARMING, AQUAC W72-09378
ALS, LABORATORY TESTS, NITRATES, MUSSELS, SNAILS, FISH, ALGAE, GAS W71-09168
, *MARINE ANIMALS, *MARINE FISH, MUSSLES, LOBSTERS, ABSORPTION, CO W73-00832
RES, ALGAE, CYTOLOGICAL STUDIES, MYCOBACTERIUM, PSEUDOMONAS, BACTE W71-08027
OCOCCUS, PSEUDOMONAS AERUGINOSA, MYCOBACTERIUM FURTUITUM, STAPHYLO W72-04259
OLIATUS, POTAMOGETON PECTINATUS, MYRIOPHYLLUM SPECATUM, ELODEA CAN W71-11012
VACCINIUM VITIS IDAEA, VACCINIUM MYRTILLUS, DESCHAMPSIA FLEXUOSA, W72-14303
FO/ *G + C RATIOS, *NONFRUITING MYXOBACTERIA, DNA, MOTILITY, CYST W72-07951
ONTROL, FRESH WATER, CYANOPHYTA, MYXOBACTERIA, CULTURES, METABOLIS W72-05457
OPHAGA JOHNSONII, SPOROCYTOPHAGA MYXOCOCCOIDES.: /SENSITIVITY, CYT W72-07951
ND, METABOLIC INHIBITION, HOSTS, MYXOPHYCEAE, SCOTLAND.: /A, ENGLA W72-05457
A LONGIROSTRIS/ *WINNISQUAM LAKE(N H), ANABAENA CIRCINALIS, BOSMIN W71-03014
(PERU), CAPE COD(MASS), CAPE MAY(N J), SARGASSO SEA.: /AMA, CALLAO W72-08056
P, MICROCYSTIS AERU/ *KEZAR LAKE(N.H.), APHANIZOMENON, ANABAENA SP W72-07890
THYLENE, NITROGEN RADIOISOTOPES, N-15, LUCIFERIN, LUCIFERASE, CHLO W72-10883
ENCE LAKE(MICH), SCIRPUS ACUTUS, NAJAS FLEXILIS.: /ETABOLISM, LAWR W72-03216
ILIS, COLONIZATION, CAULOBACTER, NAJAS FLEXILIS, SCIRPUS ACUTUS, C W73-02095
EMIA SALINA, BIO-RAD, CYCLOTELLA NANA.: CHELEX 100, ART W72-04708
DACTYLUM TRICORNUTUM, CYCLOTELLA NANA, CHAETOCEROS GALVESTONENSIS, W72-07660
COOK INLET(ALASKA), CYCLOTELLA NANA, LEPTOCYLINDRUS, PACIFIC OCE W71-11008
NALIELLA TERTIOLECTA, CYCLOTELLA NANA, THALASSIOSIRA FLUVIATILIS, W72-12998
ERA, MAGALOPTERA, EPHEMEROPTERA, NANNOCHLORIS, CHLOROCCOCCUM, MICR W70-06655
HYDROGEN SULFIDE, DISTRIBUTION, NANNOPLANKTON, TURBIDITY, HYDROGE W71-00117
N, SESTON, ALGAE, PHYTOPLANKTON, NANNOPLANKTON, AMMONIA, NITRATES, W72-08048
ERS, GULF STREAM, CAPE HATTERAS, NANSEN BOTTLE, SPECIES DIVERSITY, W72-07663
BACILLARI PARADOXA, SPHAEROTILUS NAT: /TATA, GOMPHONEMA PARVULUM, W72-10883
RHODOTORULA GLUTINIS, SARGASSUM NATANS, ASCOPHYLLUM NODOSUM, FUCU W72-04748
ROSOCYSTIS OCEANUS, SPHAEROTILUS NATANS, GLUCOSE, GALACTOSE, SUCRO W72-05421
MIA SUBSPHAERICA, TETRACULAMONAS NATANS, MARINIAMONAS SAUPAULENSIS W72-09119
, *SEWAGE FUNGUS, SPHAEROTOTILUS NATANS, ZOOGLOEAL BACTERIA.: /DOM W72-08573
GAE, MASTIGOCLADUS/ *YELLOWSTONE NATIONAL PARK(WYO), BLUE-GREEN AL W72-08508
PHYTA, *THERMAL WATER, *WYOMING, NATIONAL PARKS, HOT SPRINGS, WATE W72-08508
FATTY ACIDS, SAMPLE PREPARATION, NATURAL ORGANICS.: /LTURE MEDIA, W73-01066
LVED ORGANOPHOSPHORUS COMPOUNDS, NATURAL ORGANICS.: / *DNA, *DISSO W72-12637
T, *GROWTH RATES, PHYTOPLANKTON, NATURAL RESOURCES, DEPTH, DISTRIB W71-13252
LOGY, TROPHIC LEVEL, HERBIVORES, NATURAL STREAMS, PULP WASTES, PRO W72-06340
, MOSCOW RIVER, OKA RIVER, FOAM, NATURAL WATERS.: *USSR W72-05506
GAE, AQUATIC ANIMALS, BALANCE OF NATURE, ECOLOGY, TROPHIC LEVEL, H W72-06340
, SPIROGYRA,/ *GAMBUSIA, CYCLOPS NAUPLII, COCCOCHLORIS PENIOCYSTIS W72-07132
ODESMIUM, FLAGELLATES, CILIATES, NAUPLII, COPEPODIDS.: /RIA, TRICH W72-14313
CHIA, OSCILLATORIA, PLEUROSIGMA, NAUPLII, LAMELLIBRANCH, SPIROGYRA W71-11876
EUSTONT, NAUTOCOCCUS MAMMILATUS, NAUTOCOCCUS PYRIFORMIS NAUTOCOCCU W72-07141
UDATUS, NAUTOCOCCUS CONSTRICTUS, NAUTOCOCCUS EMERSUS.: /COCCUS CA W72-07141
OCCUS, APIOCOCCUS, OSCILLATORIA, NAUTOCOCCUS CAUDATUS, NAUTOCOCCUS W72-07141
ILLATORIA, NAUTOCOCCUS CAUDATUS, NAUTOCOCCUS CONSTRICTUS, NAUTOCOC W72-07141
MMILATUS, NAUTOCOCCUS PYRIFORMIS NAUTOCOCCUS, APIOCOCCUS, OSCILLAT W72-07141
US PYRIFORMIS NAUT/.EPINEUSTONT, NAUTOCOCCUS MAMMILATUS, NAUTOCOCC W72-07141
ONSTRUENS, COCCONEIS PLACENTULA, NAVICULA SRYTOCEPHALA, ACHNANTHES W72-08141
, AMPHORA SPP, MELOSIRA SULCATA, NAVICULA SPP, SYNEDRA FASCICULATA W73-00853
M MONILIFERUM, MELOSIRA VARIANS, NAVICULA TRIPUNCTATA, GOMPHONEMA W72-10883
OTHECE, ANABAENA, APHANIZOMENON, NAVICULA, GLENODINIUM, MICROSPORA W72-10875
PHACUS, CHAETOCEROS, MOUGEOTIA, NAVICULA, PINNULARIA, SURIRELLA, W72-09111
S, OPEPHORA MARTYI, CELL COUNTS, NAVICULA, NITZSCHIA, STEPHANODISC W71-12855
HANTHES, FRAGILARIA CONSTRUENTS, NAVICULA, NITZSCHIA, EPIPYTIC ALG W72-07501
FISHERIES, HYDROELECTRIC POWER, NAVIGATION, CONSUMPTIVE USE, WATE W72-14878
*PAMLICO RIVER(NC).: W71-07337
DURHAM(NC).: W71-07382
CODIUM CAROLINIANUM, *ONSLOW BAY(NC).: /II, GLOIODERMA ATLANTICA, W73-00284
SUM, CHAETOCEROS ATLANTICUS VAR. NEAPOLITANA, CHAETOCEROS COARCTAT W71-11527
ON(WASH)/ SALT VALLEY RESERVOIRS(NEB), MADISON(WIS), LAKE WASHINGT W72-04759
*SALT VALLEY RESERVOIRS(NEB), ULTRA PLANKTON, C-14.: W72-04270
RUNOFF, URBAN HYDROLOGY, LINCOLN(NEBR).: *URBAN W71-13466
, RECREATION, LIGHT PENETRATION, NEBRASKA, WASTEWATER TREATMENT, T W72-04759
CATOSTOMUS COMMERSONI, AMEIURUS NEBULOSUS, PERCA FLAVESCENS, AMBL W70-08098
ANIZOMENON HOLSATICUM, ICTALURUS NEBULOSUS, NOTEMIGONUS CRYSOLEUCA W72-07890
BACILLUS, MICROCOCCUS, GRAM NEGATIVE BACTERIA.: W72-12255
OPHYTA, CHLOROPHYTA, METABOLISM, NEKTON, BENTHOS, BIOMASS.: / CYAN W72-13636
*CRUSTACEANS, OILY WATER, ALGAE, NEMATODES, OYSTERS, MUSSELS, PROT W72-07911
COLIFORMS, VIRUSES, CRUSTACEANS, NEMATODES, EPIDEMIOLOGY, PHOSPHAT W73-02144
ATES, ST. LAWRENCE RIVER, ALGAE, NEMATODES, GASTROPODS, MOLLUSKS, W73-05432
ALGAE, *SALT MARSHES, NEW YORK, NEMATODES, EPIPHYTOLOGY, PRIMARY W72-00948
TON, *SUBMERGED PLANTS, BIOMASS, NEMATODES, ALGAE, OLIGOCHAETES, L W71-11012
IDS,/ COELENTERATES, ARTHROPODS, NEMERTEANS, ECHINODERMS, SIPUNCUL W73-00932
CALCICOLA, BORATES, ARTHROBACTE/ NEOCHLORIS AQUATICA, SCHIZOTHRIX W73-00854
IONOSPIO, LUMBRINEREIS, GLYCERA, NEPHTYS, CAPITELLA, MAGELONA, ARM W72-09092
TA, CNEMIDOCARPA, ENHYDRA LUTRIS NEREIS, SEA OTTER, MACROCYSTIS PY W73-00932
, COLD BAY, BERMERS BAY, ALARIA, NEREOCYSTIS, LAMINARIA, THALASSIO W72-09120
Y, *ZOOPLANKTON, *PHYTOPLANKTON, NERITIC, ESTUARINE ENVIRONMENT, S W73-00935
TEROLAMPRA HEPTACTIS, BAC/ JUDAY NET, ASTEROLAMPRA MARYLANDICA, AS W71-11527
RLANDS), BIESBOSCH RESERVOIR(THE NETHERLANDS), KINETIC EQUATION.: / W72-08559
HE NETHERLANDS), MEUSE RIVER(THE NETHERLANDS), BIESBOSCH RESERVOIR W72-08559

HERLANDS), BIES/ RHINE RIVER(THE	NETHERLANDS), MEUSE RIVER(THE NET	W72-08559
NPLAAT RESERVOIR(ROTTERDAM), THE	NETHERLANDS.: /HLAMIDOMONAS, BERE	W71-01491
	NETHERLANDS, LIMITING NUTRIENTS.:	W72-12544
LTRA-VIOLET REFRACTORY NITROGEN,	NETPLANKTON, LAWRENCE LAKE(MICH),	W72-08048
PERATURE, SYSTEMATICS, SAMPLING,	NETS, CENTRIFUGATION, MICROSCOPY,	W72-07899
COLOGICAL DISTRIBUTION, PLANKTON	NETS, CHLOROPHYTA, PLANKTON.: / E	W73-01094
MMER, AUTUMN, SAMPLING, PLANKTON	NETS, DIATOMS, FOOD CHAINS, ANABA	W72-12738
E, ALGAE, COLOR, COPPER SULFATE,	NETS, HYDROGEN ION CONCENTRATION,	W72-07890
E, *PLANT MORPHOLOGY, *PHO/ *SEA	NETTLES, *POLLUTANTS, *MAINTENANC	W72-10162
G, ASHING, EXTRACTIVE DIGESTION,	NEUROANATOMY, OXYGEN PLASMA ASHIN	W71-11036
COUN/ C-14, *CHLORELLA VULGARIS,	NEUROSPORA CRASSA, SCINTILLATION	W72-14326
SUBAEQUALIS, APISTONEMA / *LAKE	NEUSIEDLER(AUSTRIA), CHAMAESIPHON	W71-04526
TRATION, ALGAE, *ENTROPHICATION,	NEUTRALIZATION, SEDIMENTATION, TR	W70-10427
GAE, SLUDGE, CHROMIUM, TOXICITY,	NEUTRALIZATION, HYDROGEN ION CONC	W71-09524
NTATION, POLYELECTROLYTES, LIME,	NEUTRALIZATION, FILTRATION, CENTR	W72-10237
T GROWTH, *MOLYBDENUM, BIOASSAY,	NEUTRON ACTIVATION ANALYSIS.: /AN	W72-00801
	NEUTRON ACTIVATION ANALYSIS.:	W72-02073
, EFFLUENTS, PATH OF POLLUTANTS,	NEUTRON ACTIVATION ANALYSIS, SPEC	W73-00832
*LAKE TAHOE(CALIF-	NEV).:	W72-12549
N ISLAND, MORRIA REEF, BEAR BAY,	NEVA ISLAND, BAIRD ISLAND, SHOLIN	W72-09120
ITROGEN, CURRENTS(WATER), WINDS,	NEVADA, CALIFORNIA, NEW YORK, ORE	W71-05626
CANADA, CAPE CHIDLEY(LABRADOR),	NEW BRUNSWICK, ANTICOSTI ISLAND(Q	W71-10066
, DIATOMS, CARBON RADIOISOTOPES,	NEW ENGLAND, AQUATIC LIFE, AQUATI	W72-14317
CKLES, GOLDFISH, MERRIMACK RIVER(	NEW HAMPSHIRE), DELAWARE RIVER, I	W71-11517
TER POLLUTION CONTROL, FISHKILL,	NEW HAMPSHIRE, WATER POLLUTION EF	W72-09304
AE, *ZOOPLANKTERS, *CHEMCONTROL,	NEW HAMPSHIRE, COPEPODS, DAPHNIA,	W71-03014
, HOSTS, PENNSYLVANIA, NEW YORK,	NEW HAMPSHIRE, OHIO, MICHIGAN.: /	W72-04736
NUTRIENT INTERCHANGE, HAW RIVER,	NEW HOPE RIVER.:	W70-09669
LORELLA, ALGAE, *EUTROPHICATION,	NEW MEXICO, BIODEGRADATION.: /*CH	W72-13992
SUBSTRATES, MACROINVERTEBRATES,	NEW YORK BIGHT, EELGRASS, ZOSTERA	W73-02029
(MAINE), FAYETTEVILLE GREEN LAKE(	NEW YORK), UPPER KLAMATH LAKE(ORE	W71-05626
EEN RIVER(KENTUCKY), CAYUGA LAKE(	NEW YORK), PATUXENT ESTUARY, BISC	W71-11517
CIPAL WASTES, INDUSTRIAL WASTES,	NEW YORK, ALGAE, WATER POLLUTION,	W71-11685
A, *MARINE ALGAE, *SALT MARSHES,	NEW YORK, NEMATODES, EPIPHYTOLOGY	W72-00948
TER), WINDS, NEVADA, CALIFORNIA,	NEW YORK, OREGON, WISCONSIN, MAIN	W71-05626
POLLUTANTS, HOSTS, PENNSYLVANIA,	NEW YORK, NEW HAMPSHIRE, OHIO, MI	W72-04736
IMARY PRODUCTIVITY, GREAT LAKES,	NEW YORK, WATER SUPPLY, SYSTEMATI	W73-01615
TIES, *ON-SITE DATA COLLECTIONS,	NEW YORK, BIOLUMINESCENCE, BIOLOG	W72-12941
P, ORTHOPHOSPHATES, SECCHI DISC,	NEWFOUND LAKE, SQUAM LAKE, APHANI	W72-07890
ATIVE COMPOSITION, COSCINOSIRA /	NEWFOUNDLAND, PERIDINIANS, QUALIT	W72-03543
IFICIAL CIRCULATION, *KEZAR LAKE(	NH).: *ART	W72-09061
*KEZAR LAKE(	NH).:	W72-14542
ERIA, GLUCOSE, ACETATES, BIOTIN,	NIACIN, COBALAMINS, EUGLENA GRACI	W73-02095
	NIAGARA RIVER.:	W71-11685
RYPTOMONAS OVATA, STEPHANODISCUS	NIAGARAE, ASTERIONELLA FORMOSA, R	W73-01435
OMS, EUGLENA, DAPHNIA, HABITATS,	NICHES, COMPETITION, CALIFORNIA.:	W71-09171
INC SULFATE, MANGANESE CHLORIDE,	NICKEL SULFATE, COBALT SULFATE, P	W72-12739
OSPHORUS, IRON, MANGANESE, ZINC,	NICKEL, COBALT, COPPER SULFATE, H	W72-12739
UNDANCE, IRON, CHROMIUM, COPPER,	NICKEL, RUTHENIUM, CERIUM, LINCON	W71-13243
HYDROGEN, CHEMICAL WASTES, LEAD,	NICKEL, TOXICITY, BIOINDICATORS,	W72-14303
TION, COPPER, CADMIUM, CHROMIUM,	NICKEL,: /E, POLLUTANT IDENTIFICA	W72-14303
AULOSIRA, TOLYPOTHRIX, ANACYSTIS	NIDULANS.: /E, ANABAENA, NOSTOC,	W72-14807
AULOSIRA FERTILISSIMA, ANACYSTIS	NIDULANS.: / TOLYPOTHRIX TENUIS,	W71-10070
GRACLIS, CALVIN CYCLE, ANACYSTIS	NIDULANS.: /BURG EFFECT, EUGLENA	W71-09172
, ANABAENA VARIABILIS, ANACYSTIS	NIDULANS.: /STS, BLUE-GREEN ALGAE	W71-08683
TWELL(SOUTH CAROLINA), ANACYSTIS	NIDULANS, AQUATIC ECOSYSTEMS, ALG	W71-00475
NTINALIS, ASTINASTRUM, ANACYSTIS	NIDULANS, MICROCYSTIS.: /ONAS, FO	W71-11519
AULOSIRA FERTILISSIMA, ANACYSTIS	NIDULANS, CERASON, DIATHENE, PROP	W71-12071
UM STR/ PHOTODIELDRIN, ANACYSTIS	NIDULANS, AGMENELLUM QUADRUPLICAT	W72-03545
*ANABAENA VARIABILIS, *ANACYSTIS	NIDULANS, GROWTH INHIBITION, SELE	W72-00153
STE CHARACTERIZATION, *ANACYSTIS	NIDULANS, *MUTANTS, *SENSITIVITY,	W72-14327
, ANABAENA CYLINDRICA, ANACYSTIS	NIDULANS, GLOEOCAPSA ALPICOLA, OS	W72-07664
SE, BIOCHEMICAL TEST/ *ANACYSTIS	NIDULANS, *ADENOSINE TRIPHOSPHATA	W72-13809
APIA MELANOPLEURA, CYPR/ TILAPIA	NILOTICA, TILAPIA MOSSAMBICA, TIL	W71-04528
RELLA, PERIDINIUM, SENECA RIVER,	NINEMILE CREEK, VAN DORN SAMPLER,	W72-09111
TESTS, BUOYANCY, EUTROPHICATION,	NIRTROGEN FIXATION, BIOCONTROL, P	W71-13184
, PENICILLUS CAPITATUS, CERAMIUM	NITENS, WRANGELIA ARGUS, VALONIA	W73-00838
TEROSTEUS ACULEATUS, BALTIC SEA,	NITOCRA SPINIPES.: / SALINUS, GAS	W71-08026
LT RIVER(KENT), VASCULAR PLANTS,	NITRATE NITROGEN, ORTHOPHOSPHATES	W72-06526
ID.: CHLORELLA VULGARIS,	NITRATE REDUCTASE, RIBONUCLEIC AC	W72-03222
TILUS, NUTRIENTS, PACIFIC OCEAN,	NITRATE, CHEMICAL ANALYSIS, SODIU	W72-05421
, DETERGENTS, AMMONIA, NITRITES,	NITRATES.: /PY, INDUSTRIAL WASTES	W71-11507
FILTRATION, DESALINATION, DRAIN/	NITRATES, ALGAE, CENTRIFICATION,	W71-11698
ANALYTICAL TECHNIQUES, AMMONIA,	NITRATES, PHOSPHATES, WATER QUALI	W71-11033
ICITY, NUTRIENTS, WATER QUALITY,	NITRATES, PLANNING, NITROGEN, PHO	W71-13172
ASTING, PHOTOSYNTHESIS, BIOMASS,	NITRATES, RESERVOIRS, LIGHT PENET	W71-13183
MICAL OXYGEN DEMAND, PHOSPHORUS,	NITRATES, AMMONIA, COAGULATION, F	W71-13341
RFACES, TEMPERATURE, PHOSPHORUS,	NITRATES, CHLORIDES, CALCIUM, MAG	W71-11001
ROSION, PU/ *NITROGEN, *AMMONIA,	NITRATES, NITRITES, TOXICITY, COR	W72-04086
NALYTICAL TECHNIQUES, TURBIDITY,	NITRATES, PHOSPHATES, WIND VELOCI	W72-03224
S, NUTRIENTS, PHOSPHORUS, ALGAE,	NITRATES, WATER REUSE, SALINITY,	W72-00846
MENTS, HEAVY METALS, ELECTRODES,	NITRATES, ALGAE, COPPER, IONS, IO	W72-01693
S, AMMONIA, POTASSIUM COMPOUNDS,	NITRATES, DENSITY, GROWTH RATES,	W72-03217
FLORIDA, SEA WATER, CHLOROPHYLL,	NITRATES, NITRITES, PHOSPHATES, N	W72-09102
DEMAND, WATER POLLUTION CONTROL,	NITRATES, EFFLUENTS, DETERGENTS,	W72-09167
ES, AMMONIA, NITROGEN COMPOUNDS,	NITRATES, PHOSPHATES, SLUDGE TREA	W72-09675
L EFFECTS, ANAEROBIC CONDITIONS,	NITRATES, DENITRIFICATION, BOTTOM	W72-11563
PHOSPHATES, *ENZYMES, *CULTURES,	NITRATES, PLANT PHYSIOLOGY, PLANT	W72-05476
ISM, CARBON DIOXIDE, PHOSPHATES,	NITRATES, *NUTRIENTS, ALKALINITY,	W72-04983
MAL WATER, *MONTANA, PHOSPHATES,	NITRATES, NITRITES, IRON, SILICON	W72-05610

ASTES, CHLORINATION, PHOSPHATES, NITRATES, FILTRATION, POLLUTION A W72-05315
ON, PLANT PHYSIOLOGY, NUTRIENTS, NITRATES, SULFATES, PHOSPHATES, C W72-06274
CHLORELLA, IONS, PHYTOPLANKTON, NITRATES, NITROGEN FIXATION, ALGA W72-04773
N CONCENTRATION, CARBON DIOXIDE, NITRATES, NITRITES, *ENZYMES.: /O W72-04309
, ALGAE, RIVER FLOW, PHOSPHORUS, NITRATES, POTASSIUM, DISSOLVED OX W72-04260
R RADIATION, DEPTH, THERMOCLINE, NITRATES, LIGHT INTENSITY, EUTROP W71-07875
UATIC ANIMALS, LABORATORY TESTS, NITRATES, MUSSELS, SNAILS, FISH, W71-09168
, ORGANIC COMPOUNDS, PHOSPHATES, NITRATES, NUTRIENTS, ANAEROBIC CO W71-07100
LLUTANTS, FERTILIZERS, NITROGEN, NITRATES, PHOSPHORUS, PHOSPHATES, W71-06445
, *PHOSPHORUS, NUTRIENTS, ALGAE, NITRATES, SURFACE RUNOFF, BASE FL W71-06443
ETHANE BACTERIA, EUTROPHICATION, NITRATES, PHOSPHATES, BIODEGRADAT W71-07106
ITY, TRACE ELEMENTS, PHOSPHATES, NITRATES, LABORATORY TESTS.: /TIV W71-09674
N, NUTRIENTS, ALGAE, PHOSPHATES, NITRATES, SURVEYS, PESTICIDES, WA W71-05553
FLOW, SEASONAL, STRATIFICATION, NITRATES, SILICA, AMMONIA, PHOSPH W71-03028
SHEDS(BASINS), STREAMFLOW, IONS, NITRATES, SULFATES, DRAINAGE WATE W71-01489
QUES, *ALGAE, *ASSAY, CHLORELLA, NITRATES, CHEMICAL ANALYSIS, CHLO W70-10403
IA, SCENEDESMUS, SUSPENDED LOAD, NITRATES, CALCIUM CARBONATE, PROT W70-10174
*ESTUARIES, RIVERS, PHOSPHATES, NITRATES, HUDSON RIVER, ALGAE, WA W70-06509
NTENSITY, NITROGEN, CHRYSOPHYTA, NITRATES, CHLOROPHYLL, PHO: /HT I W73-01435
N ION CONCENTRATION, PHOSPHATES, NITRATES, DISSO: /LOGENS, HYDROGE W73-01434
SALINITY, PROFILES, PHOSPHATES, NITRATES, SILICATES, HYDROGRAPHS: W72-07892
GTON, PHYTOPLANKTON, CYANOPHYTA, NITRATES, PHOSPHATES, LAKES, STRA W72-08049
PLANKTON, HYDROGRAPHY, NITRITES, NITRATES, AMMONIA, ALGAE, CHLOREL W72-08056
LANKTON, NANNOPLANKTON, AMMONIA, NITRATES, NITRITES, HARDNESS(WATE W72-08048
N, NITROGEN FIXATION, SEDIMENTS, NITRATES, WISCONSIN.: /TRIFICATIO W72-07933
ATURE, CONDUCTIVITY, PHOSPHATES, NITRATES, ALKALINITY, DIATOMS, CY W72-08560
XATION, NITROGEN CYCLE, AMMONIA, NITRATES, MODEL STUDIES, DISSOLVE W72-08459
, AQUATIC PLANTS, AQUATIC ALGAE, NITRATES, NITRITES, EUTROPHICATIO W73-00286
PATIAL DISTRIBUTION, PHOSPHORUS, NITRATES, SILICA, PATH OF POLLUTA W73-00384
*PRIMARY PRODUCTIVITY, *NERITIC, NITRATES, RADIOACTIVITY TECHNIQUE W73-00431
AMIN B, PLANT GROWTH REGULATORS, NITRATES, PHOSPHATES, NUTRIENT RE W73-00430
UTRIENT REQUIREMENTS, SEA WATER, NITRATES, PHOSPHORUS, NUTRIENTS, W73-00441
TIC ALGAE, CULTURES, PHOSPHATES, NITRATES, NUTRIENTS, SALTS, TRACE W72-12739
ULFATES, IRON COMPOUNDS, SEWAGE, NITRATES, PHOSPHATES.: /OASSAY, S W72-12239
C ZONE, CHLOROPHYLL, PHOSPHATES, NITRATES, SEASONAL, REGRESSION AN W72-12393
PRODUCTIVITY, ANABAENA, AMMONIA, NITRATES, LIMITING FACTORS, NITRO W73-02472
AERIAL PHOTOGRAPHY, PHOSPHORUS, NITRATES,: /RATES, AQUATIC ALGAE, W72-12729
-WATER INTERFACES, *HYPOLIMNION, NITRIFICATION, ALGAE, THERMOCLINE W72-12996
ONMENT, PRODUCTIVITY, SEDIMENTS, NITRIFICATION, OXIDATION-REDUCTIO W72-08459
EUTROPHICATION, AMMONIFICATION, NITRIFICATION, DENITRIFICATION, N W72-07933
, ELECTRODIALYSIS, DISTILLATION, NITRIFICATION, DENITRIFICATION, S W70-09907
FFECTS, WATER POLLUTION SOURCES, NITRIFICATION, ALGAE.: /LLUTION E W71-05390
ROLYSIS, CHEMICAL PRECIPITATION, NITRIFICATION, DENITRIFICATION, * W71-09450
TIAL, BIOCHEMICAL OXYGEN DEMAND, NITRIFICATION, NUTRIENTS, SLUDGE, W72-04086
TIC ENVIRONMENT, EUTROPHICATION, NITRIFICATION, RESERVOIR EVAPORAT W72-02412
NITRILOTRIACETATE.: W72-09169
HLORELLA PYRENOIDOSA, *TRISODIUM NITRILOTRIACETATE.: *C W71-11972
Y PRO/ FLUSHING, GREEN BAY(WIS), NITRILOTRIACETIC ACID, ALGAL ASSA W71-12072
DETERGENTS, SOAP AND DETERGENT / NITRILOTRIACETIC ACID, HOUSEHOLD W72-04266
M METASILICATE, LAKE WASHINGTON, NITRILOTRIACETATE, HUMAN FECES.: / W72-04734
PHATE-FREE DETERGENTS, TRISODIUM NITRILOTRIACETATE, SODIUM CITRATE W71-06247
ENOIDOSA, SOURCE OF / *TRISODIUM NITRILOTRIACETATE, *CHLORELLA PYR W70-08976
POLYCARBOXYLATES.: ARSENIC, NITRILOTRIACETATE, POLYSILICATES, W72-06837
SIS, HYDROGEN ION CONCENTRATION, NITRITE, AMMONIA, SALINITY, AQUAT W71-11876
WATER, *NITRIFICATION, AMMONIA, NITRITE, SPHAEROTILUS, NUTRIENTS, W72-05421
*MONTANA, PHOSPHATES, NITRATES, NITRITES, IRON, SILICON, SULFATES W72-05610
ATION, CARBON DIOXIDE, NITRATES, NITRITES, *ENZYMES.: /ON CONCENTR W72-04309
OGEN CYCLE, *OXIDATION, AMMONIA, NITRITES, OXYGEN, PHYTOPLANKTON, W71-05390
IAL WASTES, DETERGENTS, AMMONIA, NITRITES, NITRATES.: /PY, INDUSTR W71-11507
EA WATER, CHLOROPHYLL, NITRATES, NITRITES, PHOSPHATES, NUTRIENTS.: W72-09102
*NITROGEN, *AMMONIA, NITRATES, NITRITES, TOXICITY, CORROSION, PU W72-04086
GEN, PHYTOPLANKTON, HYDROGRAPHY, NITRITES, NITRATES, AMMONIA, ALGA W72-08056
ANNOPLANKTON, AMMONIA, NITRATES, NITRITES, HARDNESS(WATER), NITROG W72-08048
PLANTS, AQUATIC ALGAE, NITRATES, NITRITES, EUTROPHICATION.: /ATIC W73-00286
NITROSOMONAS, NITROBACTER, AZOTOBACTER.: W72-11563
ARTICULATE MATTER, NITROSOMONAS, NITROBACTER, THIOBACILLUS THIOOXI W71-01489
AGARS, CHEMOSTATS, NITROSOMONAS, NITROBACTER, NITROSOCYSTIS OCEANU W72-05421
RUSTACEANS, COPEPODS, NUTRIENTS, NITROGE: /CENEDESMUS, ROTIFERS, C W72-05968
RAINAGE WATER, DISSOLVED SOLIDS, NITROGEN CYCLE, HYDROGEN ION CONC W71-01489
N DEMAND, SOLID WASTES, AMMONIA, NITROGEN COMPOUNDS, NITRATES, PHO W72-09675
*BIOASSAY, PHOSPHORUS COMPOUNDS, NITROGEN COMPOUNDS, CYCLING NUTRI W72-10608
ON, *DENITRIFICATION, *NITRATES, NITROGEN COMPOUNDS, AMMONIA, ALGA W71-13939
QUALITY, FERTILIZERS, NUTRIENTS, NITROGEN COMPOUNDS, PHOSPHORUS CO W71-12252
WATER QUALITY, WATER TREATMENT, NITROGEN CYCLE.: /SAG, TURBIDITY, W71-09958
ATES, NITRITES, HARDNESS(WATER), NITROGEN CYCLE, ORGANIC MATTER, P W72-08048
ITRIFICATION, NITROGEN FIXATION, NITROGEN CYCLE, AMMONIA, NITRATES W72-08459
LECTIONS, ANALYTICAL TECHNIQUES, NITROGEN FIXING BACTERIA, THERMAL W72-08508
T, *PHOTOSYNTHESIS, *CYANOPHYTA, NITROGEN FIXATION, OXYGEN, CARBON W72-08584
TION POTENTIAL, DENITRIFICATION, NITROGEN FIXATION, NITROGEN CYCLE W72-08459
NITRIFICATION, DENITRIFICATION, NITROGEN FIXATION, SEDIMENTS, NIT W72-07933
*LAKE ERKEN(SWEDEN), *DARK NITROGEN FIXATION.: W72-08054
CENTRATION, CHEMICAL PORPERTIES, NITROGEN FIXATION, CULTURES, CARB W72-14790
UTRIENTS, ALGAE, RICE, NITROGEN, NITROGEN FIXATION, 2-4-D, DALAPON W71-12071
UCCESSION, ECONOMICS, NUTRIENTS, NITROGEN FIXATION, FORECASTING, P W71-13183
BIOASSAY, *BIOMASS, CHLOROPHYLL, NITROGEN FIXATION, *PERIPHYTON, * W72-10883
EN, POTASSIUM, DIVERSION, ALGAE, NITROGEN FIXATION, PLANKTON, PHYT W71-00117
ASHINGTON, DIVERSION, LAKE ERIE, NITROGEN FIXATION, DETERGENTS, CA W72-05473
, IONS, PHYTOPLANKTON, NITRATES, NITROGEN FIXATION, ALGAE, CONDUCT W72-04773
DIOXIDE, NITROGEN, CHLOROPHYTA, NITROGEN FIXATION, LAND MANAGEMEN W72-04769
RE, MUD, GROWTH RATES, ANABAENA, NITROGEN FIXATION, DENITRIFICATIO W72-12544

SPHORUS REQUIREMENTS(ALGAE), NU/	NITROGEN REQUIREMENTS(ALGAE), PHO	W70-04080
ERTEBRATES, ACETYLENE, ETHYLENE,	NITROGEN RADIOISOTOPES, N-15, LUC	W72-10883
LOGICAL SOURCES, MINERALIZATION,	NITROGEN SOURCES, WELL WATER, FEE	W71-06435
IS, CARBON EXTRUSION, CARBON-14,	NITROGEN SOURCE.: /NITZSCHIA OVAL	W72-07497
, PLANT PIGMENTS, AQUATIC ALGAE,	NITROGEN.: /MICHIGAN, CHRYSOPHYTA	W73-00384
CHELOMONAS, SAGINAW BAY, ORGANIC	NITROGEN.: /, ANKISTRODESMUS, TRA	W72-10875
CYANOPHYTA, ALKALINITY, DIURNAL,	NITROGEN.: /NTS, EUTROPHICATION,	W72-09162
LGAE, LAKES, PHOSPHORUS, CARBON,	NITROGEN.: / TESTS, *NUTRIENTS, A	W72-09165
RATES, DIATOMS, BACTERIA, LAKES,	NITROGEN, ANALYTICAL TECHNIQUES,	W72-09163
LORIDA, WATER POLLUTION EFFECTS,	NITROGEN, PHOSPHORUS.: /ATIONS, F	W72-08986
PHICATION, *CARBON, *PHOSPHORUS,	NITROGEN, ALGAE.: *EUTRO	W72-09155
TESTS, *BIOASSAY, OREGON, LAKES,	NITROGEN, PHOSPHORUS, CARBON, GRO	W72-09164
ALASKA, FIORDS, SURFACE WATERS,	NITROGEN, SEA WATER, WINDS, SAMPL	W72-09122
TY, CYANOPHYTA, LAKES, MISSOURI,	NITROGEN, PHOSPHORUS.: /TER QUALI	W72-10431
RATION TECHNIQUES, FERMENTATION,	NITROGEN, PHOSPHORUS, AMMONIA, ME	W72-10390
AGRICULTURAL RUNOFF, PHOSPHORUS,	NITROGEN, PHOSPHATES, DETERGENTS,	W72-10625
PRODUCTIVITY, ALGAE, PHOSPHATES,	NITROGEN, SEWAGE, CARBON DIOXIDE,	W72-09167
ON SOURCES, MATHEMATICAL MODELS,	NITROGEN, ABATEMENT, COSTS, ANALY	W72-09171
ARCH AND DEVELOPMENT, NUTRIENTS,	NITROGEN, PHOSPHORUS, ALGAE, PLAN	W72-11702
AL DISTRIBUTION, CARBON DIOXIDE,	NITROGEN, PHOSPHORUS, LIGHT, NUTR	W72-11724
S, SOLIDS RECYCLE, SOLI/ ORGANIC	NITROGEN, SETTLING CHARACTERISTIC	W72-11100
A, *SYMBIOSIS, ACTIVATED SLUDGE,	NITROGEN, PHOSPHORUS, CHEMICAL OX	W72-11100
NG, ODOR, SLUDGE, COST ANALYSIS,	NITROGEN, CLAYS, *WASTE WATER TRE	W72-11642
OPHICATION, *PHOSPHOROUS, ALGAE,	NITROGEN, ACTIVATED SLUDGE, CHEMI	W71-13334
R POLLUTION EFFECTS, PHOSPHORUS,	NITROGEN, AMMONIA, CARBON, CHLORE	W71-12857
ICAL OXYGEN DEMAND, *PHOSPHATES,	NITROGEN, COLIFORMS, MATHEMATICAL	W71-13356
, *PRIMARY PRODUCTIVITY, *ALGAE,	NITROGEN, PHOSPHORUS, THERMAL STR	W71-13183
TER QUALITY, NITRATES, PLANNING,	NITROGEN, PHOSPHORUS, RESERVOIRS,	W71-13172
ING, *BIOCHEMICAL OXYGEN DEMAND,	NITROGEN, PHOSPHOROUS, PHOTOSYNTH	W71-13326
TION, ENZYMES, BACTERIA, SPORES,	NITROGEN, METABOLISM.: /OGEN FIXA	W71-12070
, *ACTIVATED SLUDGE, *NUTRIENTS,	NITROGEN, WATER POLLUTION EFFECTS	W71-11972
IZATION, NUTRIENTS, ALGAE, RICE,	NITROGEN, NITROGEN FIXATION, 2-4-	W71-12071
ASS, WATER QUALITY, CHLOROPHYLL,	NITROGEN, WISCONSIN, NUTRIENTS, W	W71-10645
SAMPLING, TEMPERATURE, COPEPODS,	NITROGEN, BACTERIA, OXIDATION-RED	W71-11008
UTROPHICATION, LAKE ERIE, ALGAE,	NITROGEN, PHOSPHORUS, DRAINAGE, I	W71-11009
E ERIE, LAKE ONTARIO, NUTRIENTS,	NITROGEN, FILTRATION, MICROSCOPY,	W71-11507
L ANALYSIS, STATISTICAL METHODS,	NITROGEN, PHOSPHORUS, CHLOROPHYLL	W71-11029
EUTROPHICATION, TRACE ELEMENTS,	NITROGEN, PHOSPHORUS, METABOLISM,	W71-11793
TRIBUTION, AQUATIC PRODUCTIVITY,	NITROGEN, PHOSPHORUS, LAKES.: /IS	W71-11519
ATES, PHOSPHATES, WATER QUALITY,	NITROGEN, NUTRIENTS, TROPHIC LEVE	W71-11033
SATURATION, OXYGEN, METABOLISM,	NITROGEN, PHOSPHOROUS, *ALGAL CON	W71-13553
S, NUTRIENTS, ALGAE, PHOSPHORUS,	NITROGEN, EUTROPHICATION, SEDIMEN	W72-00847
ATION, ACTIVATED SLUDGE, DESIGN,	NITROGEN, DIFFUSION, *ALGAL CONTR	W72-00800
MICAL OXYGEN DEMAND, PHOSPHORUS,	NITROGEN, EUTROPHICATION, *AEROBI	W72-01881
GOONS, NUTRIENTS, AQUATIC ALGAE,	NITROGEN, ANAEROBIC DIGESTION, AE	W72-02850
BICIDES, TOXINS, WASTE DILUTION,	NITROGEN, PHOSPHORUS, EUTROPHICAT	W72-01472
, SUBSTRATE UTILIZATION, ORGANIC	NITROGEN, ORGANIC CARBON, CULTURE	W73-00379
TIC PRODUCTIVITY, AQUATIC ALGAE,	NITROGEN, POLLUTANT IDENTIFICATIO	W73-00379
*AMMONIA, DIATOMS, GROWTH RATES,	NITROGEN, WATER POLLUTION EFFECTS	W73-00380
NTS, BIOASSAY, LIMITING FACTORS,	NITROGEN, PHOSPHORUS, RAIN, LAKES	W73-00232
WATER POLLUTION EFFECTS, SEWAGE,	NITROGEN, PHOSPHORUS, IRON, PHYTO	W73-00234
SCILLATO/ SECCHI DISC, DISSOLVED	NITROGEN, DISSOLVED PHOSPHORUS, O	W72-07648
IOASSAY, *STANDING CROP, CARBON,	NITROGEN, PHOSPHORUS, PHYTOPLANKT	W72-07648
S, *MARINE ALGAE, PHYTOPLANKTON,	NITROGEN, ABSORPTION, SEA WATER.:	W72-07497
C NITROGEN, *PARTICULATE ORGANIC	NITROGEN, CALCAREOUS LAKES, ULTRA	W72-08048
REOUS LAKES, ULTRA-VIOLET LABILE	NITROGEN, ULTRA-VIOLET REFRACTORY	W72-08048
, PHOSPHATES, CARBON, VITAMIN B,	NITROGEN, CYANOPHYTA, LAKE MICHIG	W72-07937
ITROGEN, ULTRA-VIOLET REFRACTORY	NITROGEN, NETPLANKTON, LAWRENCE L	W72-08048
ATERS, CONTINENTAL SHELF, DEPTH,	NITROGEN, PHYTOPLANKTON, HYDROGRA	W72-08056
E METABOLISM, *DISSOLVED ORGANIC	NITROGEN, *PARTICULATE ORGANIC NI	W72-08048
S, FRESH WATER, ALGAE, BACTERIA,	NITROGEN, PHOSPHORUS, OXIDATION-R	W72-07933
ATIFICATION, MIXING, PHOSPHORUS,	NITROGEN, TURBULENCE, PHOTOSYNTHE	W72-08559
BITION, OLIGOTROPHY, PHOSPHORUS,	NITROGEN, *EUTROPHICATION, LABORA	W72-08376
IC TREAT/ *FARM WASTES, POULTRY,	NITROGEN, LAGOONS, PUMPING, AEROB	W72-08396
N ION CONCENTRATION, PHOSPHATES,	NITROGEN, HARDNESS(WATER), CONDUC	W72-07890
C CARBON, CHLOROPHYLL A, ORGANIC	NITROGEN, THIAMINE, BIOTIN, VITAM	W72-07892
USSELS, OYSTERS, EUTROPHICATION,	NITROGEN, PHOSPHOROUS, CA: /MS, M	W72-07715
PIGMENT, MAGNESIUM, PHOSPHATES,	NITROGEN, SEA WATER, FLUOROMETRY,	W72-07660
PARTICU/ ORGANIC CARBON, ORGANIC	NITROGEN, SUBSTRATE UTILIZATION,	W73-01066
FLUENT STREAMS, LIGHT INTENSITY,	NITROGEN, CHRYSOPHYTA, NITRATES,	W73-01435
ISSOLVED SOLIDS, ORGANIC MATTER,	NITROGEN, WATER POLLUTION SOURCES	W73-01066
CHLORELLA, *NUTRIENTS, CULTURES,	NITROGEN, POTASSIUM, PHOSPHATES,	W73-01626
N FIXATION, *CYANOPHYTA, PLANTS,	NITROGEN, REDUCTION(CHEMICAL), GA	W73-01456
S, INSTRUMENTATION, METHODOLOGY,	NITROGEN, SILICON, ABSORPTION, PH	W73-01445
AUDATA, EUGLENA/ *INDIA, ORGANIC	NITROGEN, EUGLENA ACUS, EUGLENA C	W73-00918
LOR, HYDROGEN ION CONCENTRATION,	NITROGEN, CARBON, SODIUM, CALCIUM	W73-00854
TALS, CYCLING NUTRIENTS, CARBON,	NITROGEN, PHOSPHORUS, SULFUR, CAL	W73-00840
CYANOPHYTA, SYMBIOSIS, LICHENS,	NITROGEN, LIGHT INTENSITY, TEMPER	W73-02471
TY, NUTRIENTS, DISSOLVED OXYGEN,	NITROGEN, PHOSPHORUS, WATER QUALI	W71-06444
WASTES, POLLUTANTS, FERTILIZERS,	NITROGEN, NITRATES, PHOSPHORUS, P	W71-06445
ETRATION, NUTRIENTS, PHOSPHORUS,	NITROGEN, TIDAL WATERS, CHEMICAL	W71-07098
*WATER QUALITY CONTROL, AMMONIA,	NITROGEN, PHOSPHORUS, ALGAE, FEAS	W71-09450
PENS, ORGANIC MATTER, NUTRIENTS,	NITROGEN, PHOSPHORUS, EUTROPHICAT	W71-08214
OPLANKTON, *PRODUCTIVITY, ALGAE,	NITROGEN, PHOSPHORUS, MATHEMATICA	W71-08670
IC MATTER, VOLUME, DRYING, ODOR,	NITROGEN, AERATION, HOGS, CATTLE,	W71-07551
MICAL OXYGEN DEMAND, PHOSPHATES,	NITROGEN, STABILIZATION.: /D, CHE	W70-07838
LGAE, SYMBIOSIS, CARBON DIOXIDE,	NITROGEN, LAKES, ESTUARIES, NUTRI	W70-07283
, CYANOPHYTA, CYCLING NUTRIENTS,	NITROGEN, PHOSPHORUS, CHEMICAL AN	W70-05469
TTER, NUTRIENTS, SURFACE WATERS,	NITROGEN, WATER QUALITY, PHOSPHOR	W70-10405
LIMNION, WASHINGTON, PHOSPHORUS,	NITROGEN, NUTRIENTS.: /INITY, EPI	W71-02681

CHEMICAL PROPERTIES, PHOSPHORUS,	NITROGEN, POTASSIUM, DIVERSION, A	W71-00117
, PHOSPHATE, PHOSPHORUS, CARBON,	NITROGEN, CURRENTS(WATER), WINDS,	W71-05626
DNESS(WATER), SAMPLING, DIATOMS,	NITROGEN, BACTERIA, ALGAE, THERMA	W71-03028
CHLORELLA PYRENOIDOSA, SOURCE OF	NITROGEN, NTA.: /ILOTRIACETATE, *	W70-08976
TOR-PREY RELATIONSHIPS, *AMMONIA-	NITROGEN, *GROWTH REACTORS, ALGAL	W72-04787
ELLITES(ARTIFICIAL), PHOSPHORUS,	NITROGEN, HEAVY METALS, ORGANOPHO	W72-04742
N CONCENTRATION, CARBON DIOXIDE,	NITROGEN, CHLOROPHYTA, NITROGEN F	W72-04769
, MANGANESE, PHOSPHORUS, CARBON,	NITROGEN, OXIDATION, BIOCHEMICAL	W72-05506
ES, IRON, MANGANESE, PHOSPHORUS,	NITROGEN, ORGANIC MATTER, CALCIUM	W72-05469
LIGHT INTENSITY, CARBON DIOXIDE,	NITROGEN, LABORATORY TESTS, CYANO	W72-05476
POLLUTION), VIRGINIA, NUTRIENTS,	NITROGEN, PHOSPHORUS, SEWAGE EFFL	W72-05473
N CONCENTRATION, CARBON, OXYGEN,	NITROGEN, HYDROGEN, PRIMARY PRODU	W72-06274
(KENT), VASCULAR PLANTS, NITRATE	NITROGEN, ORTHOPHOSPHATES, *OLIGA	W72-06526
NTS, *NUTRIENTS, EUTROPHICATION,	NITROGEN, LITTORAL, HARBORS, SPAT	W72-04263
D, *EUTROPHICATION, PHOSPHOROUS,	NITROGEN, CHEMICAL OXYGEN DEMAND,	W72-04431
NALYSIS, DETERGENTS, PHOSPHORUS,	NITROGEN, NUTRIENTS, SEWAGE, WATE	W72-04734
RE, COLIFORMS, ENTERIC BACTERIA,	NITROGEN, VITAMINS, SUSPENDED SOL	W72-12565
PHATES, CHLOROPHYTA, CYANOPHYTA,	NITROGEN, CARBON, DIATOMS, BACTER	W72-12544
OPHYLL, *PHOSPHORUS, *LAKE ERIE,	NITROGEN, ALGAE, INORGANIC COMPOU	W72-13653
KS, BIOASSAY, CHEMICAL ANALYSES,	NITROGEN, NUTRIENTS, SURFACE WATE	W72-12955
MENTS, GROWTH RATES, PHOSPHORUS,	NITROGEN, *REGRESSION ANALYSIS, V	W72-12966
, BACTERIA, CHEMICAL PROPERTIES,	NITROGEN, HEAT TRANSFER, PHOSPHAT	W72-12997
ENTS, OXYGEN DEMAND, PHOSPHORUS,	NITROGEN, NUTRIENTS, ON-SITE DATA	W72-12990
ENTS, *NITROGEN FIXATION, LAKES,	NITROGEN, PHOSPHORUS, IRON, *TEXA	W72-13692
POLLUTION EFFECTS, FERTILIZERS,	NITROGEN, BIOASSAY, PHOSPHORUS, C	W72-13816
PHYTOPLANKTON, SEWAGE TREATMENT,	NITROGEN, PHYSOCHEMICAL PROPERTIE	W72-13662
NIA, NITRATES, LIMITING FACTORS,	NITROGEN, PLANKTON, DEPTH, ALGAE.	W73-02472
, FERTILIZATION, AQUATIC PLANTS,	NITROGEN, NUTRIENT REMOVAL, WATER	W73-02478
OCIATION, CARBOY TRIALS, CANADA,	NITROGEN: PHOSPHORUS RATIO.: /ASS	W70-07283
TATS, NITROSOMONAS, NITROBACTER,	NITROSOCYSTIS OCEANUS, SPHAEROTIL	W72-05421
ENT CULTURES, AGARS, CHEMOSTATS,	NITROSOMONAS, NITROBACTER, NITROS	W72-05421
NTAL FOREST, PARTICULATE MATTER,	NITROSOMONAS, NITROBACTER, THIOBA	W71-01489
ACTER.:	NITROSOMONAS, NITROBACTER, AZOTOB	W72-11563
NITZSCHIA SERIATA, THALASSIONEMA	NITZ: /VIELLA MARINA, EUTREPTIA,	W71-07875
NAS EROSA, PERIDINIUM AFRICANUM,	NITZCHIA ACICULARIS, SYNEDRA ACUS	W73-00256
ANABAENA SPP., GOMPHONEMA SPP.,	NITZCHIA SPP., SENSITIVITY.: /P.,	W72-14706
A PUNGENS, C-14, CHLOROPHYLL A,/	NITZSCHIA DELICATISSIMA, NITZSCHI	W73-00431
YLL A,/ NITZSCHIA DELICATISSIMA,	NITZSCHIA PUNGENS, C-14, CHLOROPH	W73-00431
PLEUROSIGMA, GYROSIGMA FASCIOLA,	NITZSCHIA SPP, BACILLARIA PAXILLI	W73-00853
N, CARBON-14, NITRO/ PLATYMONAS,	NITZSCHIA OVALIS, CARBON EXTRUSIO	W72-07497
NS, EXUVIELLA MARINA, EUTREPTIA,	NITZSCHIA SERIATA, THALASSIONEMA	W71-07875
ATUM.: CHLORELLA PYRENOIDOSA,	NITZSCHIA PALEA, SKELETONEMA COST	W71-03027
ALIF), SAN JOAQUIN RIVER(CALIF),	NITZSCHIA.: /), SEARSVILLE LAKE(C	W70-10405
EDIA, CHLOROPHYLL A, TERATOLOGY,	NITZSCHIA.: /, ACTIDON, CULTURE M	W73-01445
RAGILARIA CONSTRUENTS, NAVICULA,	NITZSCHIA, EPIPYTIC ALGAE.: /S, F	W72-07501
LACIALIS, DITYLUM, PROROCENTRUM,	NITZSCHIA, LAUDERIA, CHAETOCEROS.	W72-09248
GMA, NAUPLII, LA/ COSCINODISCUS,	NITZSCHIA, OSCILLATORIA, PLEUROSI	W71-11876
A MARTYI, CELL COUNTS, NAVICULA,	NITZSCHIA, STEPHANODISCUS, MICROC	W71-12855
NS, THALASSIOSIRA, THALASSIONEMA	NITZSCHIOIDES, PLAGIOGRAMMA VANHE	W73-00853
LETONEMA COSTATUM, THALASSIONEMA	NITZSCHIOIDES, MIXED CULTURES, ME	W73-00441
IATOA ALBA, BEGGIATOA, THIOTHRIX	NIVEA, THIOTHRIX TENUIS, MERISMOP	W72-07896
EBRATES, *TOCKS ISLAND RESERVOIR(	NJ), AMERICAN SHAD, EKMAN DREDGE,	W72-14280
RER SAMPLER, SECCHI *YARDS CREEK(	NJ), RELATIVE CONDITION FACTOR.: /	W72-14280
LGAL CYCLE, DELAWARE ESTUARY(DEL-	NJ), SAN JOAQUIN DELTA(CALIF), ST	W71-05390
TAL O/ FUSARIUM NO 83, BACTERIUM	NO 52, ENDOGENOUS RESPIRATION, TO	W71-12223
S RESPIRATION, TOTAL O/ FUSARIUM	NO 83, BACTERIUM NO 52, ENDOGENOU	W71-12223
IUM TRIPOS, CERATIUM MACROCEROS,	NOCTILUCA MILIARIS, GYMNODINIUM S	W71-07875
NA, LITHODESMIUM, HARPESTICOIDS,	NOCTILUCA, CENTROPYXIS, BIDDULPHI	W71-11876
ONAUTES, ECHINODERMS, TUNICATES,	NOCTILUCA, SIPHONOPHORES, MEDUSAE	W73-00935
ILIANA, PAVLOVIAMONAS FRUTICOSA,	NODAMASTIX SPIROGYRAE, COLEMANIA	W72-07683
S, SARGASSUM NATANS, ASCOPHYLLUM	NODOSUM, FUCUS VESICULOSUS, LAMIN	W72-04748
AMINARIA HYPERBOREA, ASCOPHYLLUM	NODOSUM, GIGARTINA MAMILLOSA, RHO	W71-11515
INEATUS-EXCENTRICUS, CLADOGRAMMA	NODULIFER, DACTYLIOSOLEN: /AMMA L	W71-11527
	NON-BIOLOGICAL UPTAKE.:	W72-12210
NS, LEGISLATION, ADMINISTRATION,	NON-CONSUMPTIVE USE, ALGAE, EVAPO	W70-07100
ASSIOSIRA GRAVIDA, THALASSIOSIRA	NORDENSKIOLDII, THALASSIOSIRA ROT	W72-03543
ICS, *VIRGINIA, *ATLANTIC OCEAN,	NORTH AMERICA.: /ARSHES, SYSTEMAT	W72-12633
OLLUTION EFFECTS, AQUATIC ALGAE,	NORTH CAROLINA, *ALGAE, DIATOMS,	W72-01329
AL SHELF, VERTICAL DISTRIBUTION,	NORTH CAROLINA, SOUTH CAROLINA, S	W72-07663
LLUSKS, *ATLANTIC COASTAL PLAIN,	NORTH CAROLINA, ABSORPTION, SOUND	W73-00818
SILT-WATER-PLANT RELATIONSHIPS,	NORTH SEA, PHAEOCYSTIS SPP., SOIL	W72-12239
COHO SALMON, SHEEPSHEAD, CHUBS,	NORTHERN PIKE, GIZZARD SHAD.: /S,	W72-14282
GRAPHIE/ *OCEANOGRAPHY, *PACIFIC	NORTHWEST, *COAST REVIEW, *BIBLIO	W72-12190
*OSLOFJORD(	NORWAY).:	W73-00234
SENIC, SELENIUM, ZINC, ANTIMONY,	NORWAY, ASHING, BIOLOGICAL MAGNIF	W71-11515
ALCICOLA, BORATES, ARTHROBACTER,	NOSTOC COMMUNE, PORPHYROSIPHON NO	W73-00854
MII, 7-METHYLHEPTADECANE, 8-MET/	NOSTOC MUSCORUM, LYNGBYA LAGERHAI	W73-01439
HOT SPRINGS, WATER TEMPERATURES,	NOSTOC STREAMS, DATA COLLECTIONS,	W72-08508
SWEDEN, COLLEMA, PELTIGERA,	NOSTOC.:	W73-02471
ACYSTIS N/ *TOLERANCE, ANABAENA,	NOSTOC, AULOSIRA, TOLYPOTHRIX, AN	W72-14807
	NOSTOC, CYLINDROTHECA.:	W71-12858
CYANOPHYTA, FROZEN SOILS, IONS,	NOSTOC, ENVIRONMENTAL EFFECTS, PO	W73-00854
AENA, GLOEOTRICHIA, MICROCYSTIS,	NOSTOC, SLAKED LIME.: /TION, ANAB	W72-01097
TRICALCIUM PHOSPHATES, ANABAENA,	NOSTOC, TOLYPOTHRIX TENUIS, AULOS	W71-10070
, NOSTOC COMMUNE, PORPHYROSIPHON	NOTARISII, PSYCHROPHILIC, BIOTIC	W73-00854
HOLSATICUM, ICTALURUS NEBULOSUS,	NOTEMIGONUS CRYSOLEUCAS, CATOSTOM	W72-07890
TRICHOPTERA, CORIXIDS, EUDORINA,	NOTONECTIDAE, BODO, HYDROPSYCHIDA	W70-06655
YPHA, ELODE/ BIOLOGICAL EFFECTS,	NOTTINGHAM(ENGLAND), POLYGONUM, T	W72-00155
ORA, MELANOSIP/ *EASTERN CANADA,	NOVA SCOTIA, ISTHMOPLEA SPHAEROPH	W72-01370
LEN ISLAND(QUEBEC), SABLE ISLAND(	NOVA SCOTIA), ST PIERRE(CANADA),	W71-10066

PRESOAKS, DISHWATER DETERGENTS, NTA.: ENZYME W71-08768
TS, SOAP AND DETERGENT INDUSTRY, NTA.: /C ACID, HOUSEHCLD DETERGEN W72-04266
PYRENOIDOSA, SOURCE OF NITROGEN, NTA.: /ILOTRIACETATE, *CHLORELLA W70-08976
IC HEALTH STANDARDS, TOXICOLOGY, NTA.: /S, *TOXIC CHEMICALS, *PUBL W73-02144
HUMBCLDT BAY, CONNETICUT YANKEE NUCLEAR PLANT, CONNECTICUT RIVER, W72-07907
RUSTACEANS, *FISH, HEATED WATER, NUCLEAR POWER PLANTS, ELECTRICAL W72-07907
INE ANIMALS, OFFSHORE PLATFORMS, NUCLEAR POWERPLANTS, WATER PROPER W73-02029
OECOLOGY, RADIOACTIVITY EFFECTS, NUCLEAR POWERPLANTS, MARINE ALGAE W73-00833
DIOECCLOGY, TRACERS, MONITORING, NUCLEAR POWERPLANTS, NUCLEAR WAST W73-00790
RONMENTAL EFFECTS, HEATED WATER, NUCLEAR POWER PLANTS, STEAM, ECOL W71-11517
TICUT, RIVERS, POST-IMPOUNDMENT, NUCLEAR POWERPLANTS, ECOLOGY, RAD W72-00939
CURRENTS, COBALT RADIOISOTOPES, NUCLEAR POWERPLANTS, SITES, MOLLU W72-03324
IQUES, POLLUTANT IDENTIFICATION, NUCLEAR POWERPLANTS, EFFLUENTS, N W72-10958
GASTROPODS, PATH OF POLLUTANTS, NUCLEAR WASTES, WATER POLLUTION E W72-10978
PATH OF POLLUTANTS, FOOD CHAINS, NUCLEAR WASTES, ENVIRONMENTAL EFF W72-10957
NUCLEAR POWERPLANTS, EFFLUENTS, NUCLEAR WASTES, PATH OF POLLUTANT W72-10958
ORPTICN, *ENVIRONMENTAL EFFECTS, NUCLEAR WASTES, PATH OF POLLUTANT W72-10955
NTS, TRACERS, ANIMAL PHYSIOLOGY, NUCLEAR WASTES.: /PATH OF POLLUTA W72-03348
MONITORING, NUCLEAR POWERPLANTS, NUCLEAR WASTES, PATH OF POLLUTANT W73-00790
IMALS, *MARINE ALGAE, SEDIMENTS, NUCLEAR WASTES, PATH OF POLLUTANT W73-00817
FECTS, TRACE ELEMENTS, COLLOIDS, NUCLEAR WASTES, RADIOECOLOGY.: /F W73-00818
NT, PATH OF POLLUTANTS, FALLOUT, NUCLEAR WASTES, RADIOECOLOGY, RAD W73-00821
LS, MARINE PLANTS, MARINE ALGAE, NUCLEAR WASTES, PATH OF POLLUTANT W73-00828
STABLE ISOTOPES, RADIOISOTOPES, NUCLEAR WASTES: /YSIS, SEA WATER, W73-00831
GAE, FOOD CHAINS, PUBLIC HEALTH, NUCLEAR WASTES, PATH OF POLLUTANT W73-00823
AEOPHYTA, *COBALT RADIOISOTOPES, NUCLEAR WASTES, PATH OF POLLUTANT W72-07826
LIFE, RADIOISOTOPES, ABSORPTION, NUCLEAR WASTES, WATER POLLUTION E W72-04461
NETICS, WATER POLLUTION EFFECTS, NUCLEAR WASTES.: / METABOLISM, KI W71-09168
TOPES, *WATER POLLUTION EFFECTS, NUCLEAR WASTES, RIVERS, ABSORPTIO W71-09225
TER FISH, TEMPERATURE, KINETICS, NUCLEAR WASTES, SOLUBILITY, ADSOR W71-09182
EBARYANUM, SKELETONEMA COSTATUM, NUCLEOTIDES, ETHANOL, MONOSACCHAR W73-00942
NVIRONMENTAL STATEMENT, CRITICAL NUCLIDE PATHWAY, ABALONE.: E W72-00959
TRIENTS, *NUTRIENT REQUIREMENTS, NUISANCE ALGAS, *ENTROPHICATION, W72-03985
YSIS, WATER QUALITY, CYANOPHYTA, NUISANCE ALGAE, COMPREHENSIVE PLA W72-09171
*AQUATIC ALGAE, *CARBON DIOXIDE, NUISANCE ALGAE, EUTROPHICATION, P W72-07514
DEMAND, LAKE STAGES, LIMNOLOGY, NUISANCE ALGAE, NUTRIENT REMOVAL, W72-14692
POUR PLATE METHOD, MOST PROBABLE- NUMBER TEST, MEMBRANE FILTRATION, W72-04743
*RESIDUAL, MOST PROBABLE NUMBER.: W71-07096
ATORIA PROLIFICA, THRESHOLD ODOR NUMBER.: /ERICINIUM WESTI, OSCILL W71-13187
ODUCTION, ANALYTICAL TECHNIQUES, NUMERICAL ANALYSIS, MODEL STUDIES W72-12724
IDIELLA UNIFLAGELLATA, TSUMURAIA NUMEROSA, HIROSEIA QUINCUELCBATA, W72-07683
II, IRIS PSEUDACORUS, LYSIMACHIA NUMMULARIA, BLUE HERON, STERNA HI W72-05952
ION, SPECIES DIVERSITY, MELOSIRA NUMMULOIDES, CHLOROPHYLL A, *YAQU W73-00853
NIUM RAMOSUM, LEERSIA ORYZOIDES, NUPHAR ADVENA, EICHHORNIA CRASSIP W73-01089
AEA TUBEROSA, NUPHAR VARIEGATUM, NUPHAR JAPONICUM, NUPHAR LUTEN, A W72-04275
AR VARIEGATUM, NUPHAR JAPONICUM, NUPHAR LUTEN, ANTICANCER POTENTIA W72-04275
L SCREENINGS, NYMPHAEA TUBEROSA, NUPHAR VARIEGATUM, NUPHAR JAPONIC W72-04275
D SLUDGE, WASTE WATER TREATMENT, NUTIRIENTS, AEROBIC CONDITIONS, A W71-04079
STRICT, *INDIAN CREEK RESERVOIR, NUTRIENT BALANCE.: / UTILITIES DI W72-02412
NUTRIENT CONTROL METHODS.: W71-13256
ROCEDURES, ALGAL GROWTH, IN-LAKE NUTRIENT CONTROL.: /ALGAL ASSAY P W71-12072
PHOSPHORUS REQUIREMENTS(ALGAE), NUTRIENT CHEMISTRY(AQUATIC).: /), W70-04080
PRESERVATION, COUNTING CHAMBERS, NUTRIENT CYCLING.: /NCES, SAMPLE W72-07899
ENTS, ALGAL / *MOSES LAKE(WASH), NUTRIENT DILUTION, LIMITING NUTRI W72-08049
CETYLENE REDUCTION MEASUREMENTS, NUTRIENT ENRICHMENT TESTS.: /S, A W72-13692
NEW HOPE RIVER.: NUTRIENT INTERCHANGE, HAW RIVER, W70-09669
UDGETS, *BAY OF QUINTE(ONTARIO), NUTRIENT INPUTS, CHIRONOMUS PLUMO W71-11009
LGAE, LIGHT INTENSITY, SALINITY, NUTRIENT REQUIREMENTS, CHLORELLA, W71-10083
TER REUSE, POTABLE WATER, ALGAE, NUTRIENT REQUIREMENTS, CYANOPHYTA W71-13185
VAL GROWTH STAGE, PHYTOPLANKTON, NUTRIENT REQUIREMENTS, INVERTEBRA W72-09248
AKE BEDS, *ALGAE, ALGAL CONTROL, NUTRIENT REQUIREMENTS, SORPTION, W72-11617
S, CHLOROPHYLL, SEA WATER, IONS, NUTRIENT REQUIREMENTS, AQUATIC PO W72-09103
S, *PHOSPHORUS COMPOUNDS, ALGAE, NUTRIENT REQUIREMENTS, PRODUCTIVI W70-04080
ALLIE(MINN), LAKE MELISSA(MINN), NUTRIENT REMOVAL.: /CTION, LAKE S W71-03012
INIS, PRODUCTIVITY MEASUREMENTS, NUTRIENT REMOVAL.: / GAMBUSIA AFF W71-01488
NUTRIENT REMOVAL.: W71-09629
ANALYSIS, WASTE WATER TREATMENT, NUTRIENT REQUIREMENTS.: /L, COST W71-08953
COLOGY, *ALGAE, *EUTROPHICATION, NUTRIENT REQUIREMENTS, MUNICIPAL W71-08775
AQUATIC POPULATIONS, NUTRIENTS, NUTRIENT REQUIREMENTS.: /ETITION, W72-12736
ASTE WATER TREATMENT, NUTRIENTS, NUTRIENT REQUIREMENTS, GROWTH RAT W72-12966
ION, / *LAKE ERIE CENTRAL BASIN, NUTRIENT REGENERATION, DEOXYGENAT W72-12997
ATION, AQUATIC PLANTS, NITROGEN, NUTRIENT REMOVAL, WATER TREATMENT W73-02478
FOODS, DAPHNIA, COPEPODS, ALGAE, NUTRIENT REQUIREMENTS, CYANOPHYTA W72-07940
NIQUES, WATER POLLUTION EFFECTS, NUTRIENT REQUIREMENTS.: /CAL TECH W73-01269
ON, TOXICITY, *IRON, *CHLORELLA, NUTRIENT REQUIREMENT, *ALGAE, DET W73-02112
AGES, LIMNOLOGY, NUISANCE ALGAE, NUTRIENT REMOVAL, NUTRIENTS, WATE W72-14692
LATIVE CAPACITY, EUTROPHICATION, NUTRIENT REMOVAL, *ZINC, CHELATIO W73-00802
EGULATORS, NITRATES, PHOSPHATES, NUTRIENT REQUIREMENTS, ESSENTIAL W73-00430
HRYSOPHYTA, CULTURES, SEA WATER, NUTRIENT REQUIREMENTS, WATER POLL W73-00380
NG NUTRIENTS, SURPLUS NUTRIENTS, NUTRIENT SOURCES.: /WIS.), LIMITI W73-00232
ITY, PARTICULATE FORM, FEEDLOTS, NUTRIENT SOURCES.: /BOLISM, MOBIL W71-06443
TION, ALGAL DECO/ *ALGAL GROWTH, NUTRIENT UPTAKE, CHEMICAL COMPOSI W70-05469
SEDIMENT-WATER NUTRIENT-EXCHANGE.: W70-08944
ION, POLLUTION ABATEMENT, ALGAE, NUTRIENT, *BIODEGRADATION, WATER W72-05315
ED SOLIDS, PHYTOPLANKTON, ALGAE, NUTRIENTS.: /T INTENSITY, DISSOLV W72-04270
ION, ALGAE, TRIBUTARIES, CYCLING NUTRIENTS.: /S, SPATIAL DISTRIBUT W72-04263
WATER POLLUTION SOURCES, ALGAE, NUTRIENTS.: /R POLLUTION EFFECTS, W72-04280
BON, HYDROGEN ION CONCENTRATION, NUTRIENTS.: /LYTES, ACTIVATED CAR W72-06201
UALITY, WATER CHEMISTRY, CYCLING NUTRIENTS.: /L NUTRIENTS, WATER Q W70-04080
ASHINGTON, PHOSPHORUS, NITROGEN, NUTRIENTS.: /INITY, EPILIMNION, W W71-02681

NUTRIENT REQUIREMENTS, ESSENTIAL	NUTRIENTS.: /TRATES, PHOSPHATES,	W73-00430
LANKTON, MARINE ALGAE, ESSENTIAL	NUTRIENTS.: /ENTIFICATION, PHYTOP	W73-00441
ER CHEMISTRY, CHEMICAL ANALYSIS,	NUTRIENTS.: /TER TEMPERATURE, WAT	W73-01972
PSEUDOMONAS AERUGINOSA, LIMITING	NUTRIENTS.: /OCYSTIS AERUGINOSA,	W73-01657
OGEN, CYANOPHYTA, LAKE MICHIGAN,	NUTRIENTS.: /BON, VITAMIN B, NITR	W72-07937
CATION, WATER POLLUTION SOURCES,	NUTRIENTS.: /POLLUTION), EUTROPHI	W72-13644
NETHERLANDS, LIMITING	NUTRIENTS.:	W72-12544
NITRATES, NITRITES, PHOSPHATES,	NUTRIENTS.: / WATER, CHLOROPHYLL,	W72-09102
S, CARBON CYCLE, ORGANIC MATTER,	NUTRIENTS.: /NKTON, WATER ANALYSI	W72-09108
*LIMITING	NUTRIENTS.:	W72-09155
*LIMITING	NUTRIENTS.:	W72-09164
*LIMITING	NUTRIENTS.:	W72-09162
ED OXYGEN, LAKES, PHYTOPLANKTON,	NUTRIENTS.: /T RADIATION, DISSOLV	W72-01780
TERGENTS, EUTROPHICATION, ALGAE,	NUTRIENTS.: /TS, STORM DRAINS, DE	W71-13466
ION, SEDIMENT TRANSPORT, CYCLING	NUTRIENTS, DEGRADATION(DECOMPOSIT	W72-00847
TIC ENVIRONMENT, SURFACE WATERS,	NUTRIENTS, ALGAE, PHOSPHORUS, NIT	W72-00847
C LEVELS, *BIOASSAY, PHOSPHORUS,	NUTRIENTS, MEASUREMENT, BACTERIA,	W71-13794
CONTROL, SEDIMENTS, FARM WASTES,	NUTRIENTS, PHOSPHORUS, ALGAE, NIT	W72-00846
*MICRO-	NUTRIENTS, *ALGAL GROWTH.:	W72-00801
CTIONS, TEST PROCEDURES, CYCLING	NUTRIENTS, ANALYTICAL TECHNIQUES,	W71-13794
KES, WASHINGTON, *ALGAL CONTROL,	NUTRIENTS, BIOMASS, *CYANOPHYTA,	W72-00799
E, *POULTRY, *OXIDATION LAGOONS,	NUTRIENTS, AQUATIC ALGAE, NITROGE	W72-02850
RCES, WATER POLLUTION TREATMENT,	NUTRIENTS, OXYGEN REQUIREMENTS, A	W72-01685
SERVOIR EVAPORATION, RESERVOIRS,	NUTRIENTS, WATER REUSE.: /ION, RE	W72-02412
S, ALGAE, ZOOPLANKTON, BACTERIA,	NUTRIENTS, SEDIMENTS, ENVIRONMENT	W72-03215
AL OXYGEN DEMAND, NITRIFICATION,	NUTRIENTS, SLUDGE, ALGAE, BACTERI	W72-04086
*BIOASSAY, CULTURES, *BIOMASS,	NUTRIENTS, *NUTRIENT REQUIREMENTS	W72-03985
BON DIOXIDE, CALCIUM, MAGNESIUM,	NUTRIENTS, *WASTE WATER TREATMENT	W72-03613
ELATIONSHIPS, *FALLOUT, *CYCLING	NUTRIENTS, ECOSYSTEMS, RAIN FORES	W72-03342
OPHYTA, ALGAE, PLANTS, PROTEINS,	NUTRIENTS, BIOMASS, HARVESTING OF	W72-03327
IS, GREAT LAKES, EUTROPHICATION,	NUTRIENTS, ALGAE, BACTERIA, LAKE	W72-01101
OSYSTEMS, *RAIN FORESTS, CYCLING	NUTRIENTS, SOIL-WATER-PLANT RELAT	W72-00949
R INTERFACES, ADSORPTION, LAKES,	NUTRIENTS, LAKE ERIE, LAKE SUPERI	W72-01108
ENVIRONMENTAL SANITATION, ALGAE,	NUTRIENTS, INDUSTRIAL WASTES, MUN	W72-01659
LDO LAKE(ORE).: *LIMITING	NUTRIENTS, SHAGAWA LAKE(MINN), WA	W72-09165
E, PRODUCTIVITY, PHOTOSYNTHESIS,	NUTRIENTS, EUTROPHICATION, CYANOP	W72-09162
, ALGAE, CARBON DIOXIDE, CYCLING	NUTRIENTS, BICARBONATES, PLANT GR	W72-09159
, BIOASSAY, CYTOLOGICAL STUDIES,	NUTRIENTS, ENZYMES.: /AL ANALYSIS	W72-09163
ATER POLLUTION EFFECTS, BIOMASS,	NUTRIENTS, ALGAE, EUGLENOPHYTA, C	W72-09111
AKE GEORGE(NY).: *LIMITING	NUTRIENTS, CANADARAGO LAKE(NY), L	W72-09163
RINER'S POND(GA).: *LIMITING	NUTRIENTS, POTASSIUM CHLORIDE, SH	W72-09159
*MATHEMATICAL MODELS, *ESSENTIAL	NUTRIENTS, *PRIMARY PRODUCTIVITY,	W72-08986
UENTS, RESEARCH AND DEVELOPMENT,	NUTRIENTS, NITROGEN, PHOSPHORUS,	W72-11702
N, DEFICIENT ELEMENTS, ESSENTIAL	NUTRIENTS, *AEROBIC CONDITIONS, W	W72-11617
DE, NITROGEN, PHOSPHORUS, LIGHT,	NUTRIENTS, MATHEMATICAL STUDIES,	W72-11724
OA, FERMENTATION, BACTERIOPHAGE,	NUTRIENTS, FILTERS EQUIPMENT, OXI	W72-09675
ES.: *LIMITING	NUTRIENTS, ANABAENA, CARBON SOURC	W72-09168
S, OXIDATION LAGOONS, COLIFORMS,	NUTRIENTS, ANALYTICAL TECHNIQUES,	W72-09590
S, *PHOSPHATES, *EUTROPHICATION,	NUTRIENTS, ASSESSMENTS, ALGAE, FO	W72-09169
.: *LIMITING	NUTRIENTS, *LAKE WASHINGTON(WASH)	W72-09167
OMESTIC WASTES, BACTERIA, ALGAE,	NUTRIENTS, OXIDATION-REDUCTION PO	W72-10401
NDS, NITROGEN COMPOUNDS, CYCLING	NUTRIENTS, ESSENTIAL NUTRIENTS, L	W72-10608
RATURE, OXYGEN, LIGHT INTENSITY,	NUTRIENTS, ACTINOMYCETES, TASTE,	W72-10613
POLLUTION, MICROSCOPY, DIATOMS,	NUTRIENTS, PHYTOPLANKTON, WATER Q	W72-10875
DS, CYCLING NUTRIENTS, ESSENTIAL	NUTRIENTS, LAKES, MUD, WISCONSIN,	W72-10608
SUPPLY, WATER POLLUTION EFFECTS,	NUTRIENTS, RESERVOIR OPERATION, W	W71-13185
*LIMITING	NUTRIENTS, CANADA.:	W71-12857
NANCING, PLANT GROWTH, TOXICITY,	NUTRIENTS, WATER QUALITY, NITRATE	W71-13172
*EUTROPHICATION, AQUATIC ALGAE,	NUTRIENTS, WATER RESOURCES, WASTE	W71-13256
MINNESOTA, LAKES, GLACIAL DRIFT,	NUTRIENTS, SALANITY, ALGAE, DIATO	W71-13149
R SUPPLY, SUCCESSION, ECONOMICS,	NUTRIENTS, NITROGEN FIXATION, FOR	W71-13183
ICAL ANALYSIS, DATA COLLECTIONS,	NUTRIENTS, SALINITY, SEDIMENTS, E	W71-10577
WATER REQUIREMENTS, TEMPERATURE,	NUTRIENTS, ALGAE, WEEDS, TASTE, O	W71-11006
HLOROPHYLL, NITROGEN, WISCONSIN,	NUTRIENTS, WATER POLLUTION EFFECT	W71-10645
QUATIC PRODUCTIVITY, FARM PONDS,	NUTRIENTS, SURFACES, TEMPERATURE,	W71-11001
VIEWS, *WATER POLLUTION SOURCES,	NUTRIENTS, PLANT PHYSIOLOGY, LAKE	W71-10580
YTOPLANKTON, AMMONIUM COMPOUNDS,	NUTRIENTS, PRIMARY PRODUCTIVITY,	W71-11008
TA, *PESTICIDES, *FERTILIZATION,	NUTRIENTS, ALGAE, RICE, NITROGEN,	W71-12071
ITY, HYDROGEN ION CONCENTRATION,	NUTRIENTS, AQUATIC ENVIRONMENT, C	W71-12068
PHOROUS, *WASTE WATER TREATMENT,	NUTRIENTS, SYNTHESIS, SLUDGE, EFF	W71-12188
OWA, WATER QUALITY, FERTILIZERS,	NUTRIENTS, NITROGEN COMPOUNDS, PH	W71-12252
URIFICATION, ALGAE, GREAT LAKES,	NUTRIENTS, SEDIMENTS, LAKE MICHIG	W71-12072
IES, WASTE DISPOSAL, MONITORING,	NUTRIENTS, ALGAE, INSECTICIDES, H	W71-12091
*WATER POLLUTION EFFECTS, OHIO,	NUTRIENTS, WATER POLLUTION SOURCE	W71-11949
ALGAE, LAKE ERIE, LAKE ONTARIO,	NUTRIENTS, NITROGEN, FILTRATION,	W71-11507
PHATES, WATER QUALITY, NITROGEN,	NUTRIENTS, TROPHIC LEVEL.: / PHOS	W71-11033
SAMPLING, ANALYTICAL TECHNIQUES,	NUTRIENTS, FOOD CHAINS, BIOLUMINE	W71-11234
LATION CAPACITY, *ORGANIC LOADS,	NUTRIENTS, DISSOLVED OXYGEN, AQUA	W71-11017
GAE, CARBON DIOXIDE, ALKALINITY,	NUTRIENTS, AMMONIA, HYDROGEN ION	W71-11519
, *MODEL STUDIES, LAKES, CYCLING	NUTRIENTS, TRACERS, LITTORAL, KIN	W72-12543
SYNOPTIC ANALYSIS, COLD REGIONS,	NUTRIENTS, BIOLOGICAL COMMUNITIES	W72-12549
ELEMENTS, PESTICIDES, SALINITY,	NUTRIENTS, ALGAE, AQUATIC ANIMALS	W72-12190
NCIES, *BIOMASS, ORGANIC MATTER,	NUTRIENTS, PHOTOSYNTHESIS.: /ICIE	W72-12289
AY, CHEMICAL ANALYSES, NITROGEN,	NUTRIENTS, SURFACE WATERS, ALGAE,	W72-12955
PRODUCTIVITY, LIGHT, TURBIDITY,	NUTRIENTS, HETEROGENEITY, CHLOROP	W72-13142
*IDAHO, *CHLOROPHYLL, *ANABAENA,	NUTRIENTS, DISSOLVED SOLIDS, PHYT	W72-13300
ISM, GREAT LAKES, MODEL STUDIES,	NUTRIENTS, PLANT GROWTH, ANALYTIC	W72-13651
RMAL POLLUTION, *NORTH CAROLINA,	NUTRIENTS, SEWAGE EFFLUENTS, EUTN	W72-13636
CHNIQUES, WASTE WATER TREATMENT,	NUTRIENTS, NUTRIENT REQUIREMENTS,	W72-12966
AE, LIGHT PENETRATION, PLANKTON,	NUTRIENTS, BIOCHEMICAL OXYGEN DEM	W72-13659

EN DEMAND, PHOSPHORUS, NITROGEN, NUTRIENTS, ON-SITE CATA CCLLECTIO W72-12990
OMPETITION, AQUATIC POPULATIONS, NUTRIENTS, NUTRIENT REQUIREMENTS. W72-12736
S, *LAKES, CULTURES, PHOSPHATES, NUTRIENTS, CHLOROPHYTA, *CRGANOPH W72-12637
CULTURES, PHOSPHATES, NITRATES, NUTRIENTS, SALTS, TRACE ELEMENTS, W72-12739
OLOGY, *NEW YORK, WATER QUALITY, NUTRIENTS, ALGAE, INFLOW, DISCHAR W72-13851
GAL CCNTROL, *PHOSPHATES, *ALGAL NUTRIENTS, *CHLORELLA, ALGAE, *EU W72-13992
, OIL SPILLS, THERMAL POLLUTION, NUTRIENTS, BIOASSAY, *WATER QUALI W72-14282
LANKTCN, *NUTRIENT REQUIREMENTS, NUTRIENTS, BIOINDICATORS, WATER P W72-13816
IOASSAY, ALGAE, DOMESTIC WASTES, NUTRIENTS, SEDIMENTS, ANALYTICAL W73-02478
SH), NUTRIENT DILUTION, LIMITING NUTRIENTS, ALGAL CELL WASHOUT,: / W72-08049
*LIMITING NUTRIENTS, LAKE MENDOTA(WIS).: W72-07933
RIFUGATION, MICROSCOPY, DIATOMS, NUTRIENTS, BACTERIA, CHRYSOPHYTA, W72-07899
MICROCRGANISMS, AQUATIC ANIMALS, NUTRIENTS, WATER POLLUTICN EFFECT W72-07896
BIDITY, WATER POLLUTION SOURCES, NUTRIENTS, SEWAGE, WATER QUALITY, W72-07890
TS, CCPPER, CHLORELLA, CULTURES, NUTRIENTS, RADIOISOTOPES, CHLOROP W72-07660
ANALYSIS, *MONITORING, *CYCLING NUTRIENTS, *FCOD CHAINS, *ORGANIC W72-07715
POPULATIONS, HYDROLOGIC ASPECTS, NUTRIENTS, CHEMICALS, PLANKTCN, B W72-07841
EAR ALKYLATE SULFONATES, SEWAGE, NUTRIENTS, COAGULATION, SEDIMENTA W72-06837
GAL CONTROL, FEDERAL GOVERNMENT, NUTRIENTS, SEWAGE TREATMENT, WATE W72-06655
WATER POLLUTION, EUTROPHICATICN, NUTRIENTS, CHLORELLA, CHLAMYDOMON W72-07508
CHING, FUNGI, LIGHT, PHOSPHATES, NUTRIENTS, WATER POLLUTICN EFFECT W72-07143
*AQU/ *ALGAL CONTROL, *ESSENTIAL NUTRIENTS, *DEFICIENT NUTRIENTS, W72-07514
*ESSENTIAL NUTRIENTS, *DEFICIENT NUTRIENTS, *AQUATIC ALGAE, *CARBO W72-07514
ANOPHYTA, CHLOROPHYTA, SEASONAL, NUTRIENTS, THERMAL STRATIFICATION W72-08560
, LIGHT PENTRATION, RESPRIATION, NUTRIENTS, TURBIDITY, PRCDUCTIVIT W72-08559
TANT IDENTIFICATION, DETERGENTS, NUTRIENTS, PESTICIDES, OIL, METAL W72-08790
DEMAND, CHEMICAL OXYGEN DEMAND, NUTRIENTS, *WASTE TREATMENT, *ALG W72-08396
CONCENTRATION, LIGHT INTENSITY, NUTRIENTS, PSEUDOMONAS, SYMBIOSIS W73-01657
OMPOUNDS, CARBOHYDRATES, CYCLING NUTRIENTS, ALGAE.: /TY, CRGANIC C W73-02100
*AQUATIC MICROORGANISMS, CYCLING NUTRIENTS, ECOLOGICAL DISTRIBUTIO W73-02095
*ORGANIC NUTRIENTS, ALUM CCAGULATICN.: W73-02187
TUS, ECOSYSTEMS, ENERGY BUDGETS, NUTRIENTS, REVIEWS, BIOLCGICAL CO W73-01089
*HEAVY METALS, *TRACE ELEMENTS, NUTRIENTS, MERCURY, CHRCMIUM, COB W73-01434
ANT IDENTIFICATION, SYSTEMATICS, NUTRIENTS, PSEUDOMONAS, CULTURES, W73-00942
PROPERTIES, DIATOMS, SUCCESSICN, NUTRIENTS, RIVERS, INFLUENT STREA W73-01435
*MARSHES, SALINE WATER, BIOMASS, NUTRIENTS, FOOD WEBS, DETRITUS, E W73-01089
, ROOTED AQUATIC PLANTS, CYCLING NUTRIENTS, RESPIRATION, MARSH PLA W73-01089
ON CONCENTRATION, WATER QUALITY, NUTRIENTS, COLORIMETRY, DEFICIENT W73-00916
HEMICAL ANALYSIS, AQUATIC ALGAE, NUTRIENTS, WATER QUALITY, WATER A W73-00918
N EFFECTS, HEAVY METALS, CYCLING NUTRIENTS, CARBON, NITRCGEN, PHOS W73-00840
ADIOACTIVITY TECHNIQUES, CYCLING NUTRIENTS, TRACE ELEMENTS, PUBLIC W73-00826
BUTION, BIOMASS, PHOTOSYNTHESIS, NUTRIENTS, LIMITING FACTORS, SALI W73-00431
SEA WATER, NITRATES, PHOSPHORUS, NUTRIENTS, CHRYSOPHYTA, CCMPETITI W73-00441
, RADIOECOLOGY, ECOLOGY, CYCLING NUTRIENTS, THERMAL POLLUTION, TRO W73-00796
, ANALYTICAL TECHNIQUES, CYCLING NUTRIENTS, TRACE ELEMENTS, CHELAT W73-00821
ICAL OXYGEN DEMAND, LAKE STAGES, NUTRIENTS, MESTROPHY, OLIGOTROPHY W73-00002
ATION, CULTURES, CARBON DIOXIDE, NUTRIENTS, LIMITING FACTORS, ALLO W72-14790
S.), LAKE WINGRA(WIS.), LIMITING NUTRIENTS, SURPLUS NUTRIENTS, NUT W73-00232
ON, FJORDS, BIOASSAY, CHLORELLA, NUTRIENTS, WATER SAMPLING, ORGANI W73-00234
GANISMS, BIOLOGICAL COMMUNITIES, NUTRIENTS, SCENEDESMUS, CHRYSCPHY W72-14797
GAE, EVALUATION, TESTING, ASSAY, NUTRIENTS, CULTURES, MEASUREMENT, W73-00255
S.), LIMITING NUTRIENTS, SURPLUS NUTRIENTS, NUTRIENT SOURCES,: /WI W73-00232
UISANCE ALGAE, NUTRIENT REMOVAL, NUTRIENTS, WATER POLLUTICN EFFECT W72-14692
OMETRY, MICROSCOPY, ZOOPLANKTON, NUTRIENTS, PHOTOSYNTHESIS, STANDI W72-14313
AE, *DECOMPOSING ORGANIC MATTER, NUTRIENTS, SURFACE WATERS, NITROG W70-10405
UTION, MINNESOTA, WATER QUALITY, NUTRIENTS, SEDIMENTS, COLIFCRMS, W71-03012
*CYANOPHYTA, AQUATIC EN/ CYCLING NUTRIENTS, *CARBON, *PHOSPHORUS, W71-00475
RVESTING, PONDS, WATER HYACINTH, NUTRIENTS, MIDGES, FISH, LAKES, C W71-01488
REMENTS, PRODUCTIVITY, ESSENTIAL NUTRIENTS, WATER QUALITY, WATER C W70-04080
IDE, NITROGEN, LAKES, ESTUARIES, NUTRIENTS, DETERGENTS, WISCCNSIN, W70-07283
CHLORCPHYTA, CYANOPHYTA, CYCLING NUTRIENTS, NITROGEN, PHCSPHCRUS, W70-05469
RICT CF COLUMBIA, TIDAL EFFECTS, NUTRIENTS, EUTROPHICATION, WATER W70-08248
TER TREATMENT, ALGAE, EFFLUENTS, NUTRIENTS, SEWAGE TREATMENT, LAKE W70-09325
RICT CF COLUMBIA, TIDAL EFFECTS, NUTRIENTS, EUTROPHICATICN, WATER W70-08249
*MASSACHUSETTS, *EUTROPHICATICN, NUTRIENTS, ALGAE, PHOSPHATES, NIT W71-05553
ANTS, *EUTROPHICATION, *LEACHING NUTRIENTS, FERTILIZERS, PRECIPITA W71-04216
NAMICS, LAKES, DISSOLVED OXYGEN, NUTRIENTS, CYANOPHYTA, MIXING, PH W71-03034
S, SURFACE WATERS, SOIL EROSICN, NUTRIENTS, ALGAE, LAKE SUPERIOR, W71-04216
R SUPPLY, EUTROPHICATION, ALGAE, NUTRIENTS, WATER POLLUTICN, BIOLO W71-05083
ROTIFERS, CRUSTACEANS, COPEPODS, NUTRIENTS, NITROGE: /CENEDESMUS, W72-05968
WATER POLLUTION EFFECTS, ALGAE, NUTRIENTS, WATER POLLUTICN SOURCE W72-06638
, RESPIRATION, PLANT PHYSIOLOGY, NUTRIENTS, NITRATES, SULFATES, PH W72-06274
VITY, FISH, FISHKILL, LIMNOLOGY, NUTRIENTS, OXYGEN, OXYGEN REQUIRE W72-04279
ELOPMENT, TRACE ELEMENTS, ALGAE, NUTRIENTS, LIGHT INTENSITY, HYDRO W72-04433
EALTH, WATER POLLUTION, HAZARDS, NUTRIENTS, ALGAE.: /CY, *PUBLIC H W72-04266
ETERGENTS, PHOSPHORUS, NITROGEN, NUTRIENTS, SEWAGE, WATER TEMPERAT W72-04734
NTAL EFFECTS, RESERVOIR STORAGE, NUTRIENTS, TEMPERATURE, ALGAE, HY W72-05282
SPRINGS, CYANOPHYTA, ALKALINITY, NUTRIENTS, WATER QUALITY, DIATOMS W72-05610
ASTE WATER(POLLUTION), VIRGINIA, NUTRIENTS, NITROGEN, PHOSPHCRUS, W72-05473
AMMONIA, NITRITE, SPHAEROTILUS, NUTRIENTS, PACIFIC OCEAN, NITRATE W72-05421
ATURE, CHEMICAL ANALYSIS, ALGAE, NUTRIENTS, WATER ANALYSIS, DCMEST W72-04791
ONTRCL, WATER POLLUTION EFFECTS, NUTRIENTS, ALGAE, CYANOPHYTA, PHO W72-04769
ION, *RESERVOIRS, *GREAT PLAINS, NUTRIENTS, RECREATION, LIGHT PENE W72-04759
COMPOUNDS, PHOSPHATES, NITRATES, NUTRIENTS, ANAEROBIC CONDITIONS, W71-07100
RE-AERATION, LIGHT PENETRATICN, NUTRIENTS, PHOSPHCRUS, NITROGEN, W71-07098
ICATICN, *NITROGEN, *PHOSPHORUS, NUTRIENTS, ALGAE, NITRATES, SURFA W71-06443
TES, FISH, FISH EGGS, TURBIDITY, NUTRIENTS, DISSOLVED OXYGEN, NITR W71-06444
R, CHLOROPHYTA, WATER PCLLUTICN, NUTRIENTS, RIVERS, LAKES, FARM PO W71-05991
PHOSPHORUS, MATHEMATICAL MODELS, NUTRIENTS, MEASUREMENT, PCPULATIO W71-08670
ONFINEMENT PENS, ORGANIC MATTER, NUTRIENTS, NITROGEN, PHCSPHCRUS, W71-08214

RUNOFF, DISSOLVED OXYGEN, ALGAE, NUTRIENTS, BACTERIA, VIRUSES, HEA W71-09788
, WATER CIRCULATION, ABSORPTION, NUTRIENTS, PESTICIDES, ALGAE, DRE W71-09581
TEMENT, WATER POLLUTION EFFECTS, NUTRIENTS, PHOSPHOROUS, WATER SOF W71-09429
ING, SALINITY, DISSOLVED OXYGEN, NUTRIENTS, CARBON RADIOISOTOPES, W71-07875
OXYGEN DEMAND, SOLAR RADIATION, NUTRIENTS, TEMPERATURE, COLIFORMS W71-07124
ES, *COST ANALYSIS, TEMPERATURE, NUTRIENTS, DISSOLVED OXYGEN, ALGA W71-07109
D OXYGEN, AERATION, OXYGENATION, NUTRIENTS, SEDIMENTATION, CHLORIN W71-07113
TUM, *ALGAL GROWTH, *LAKE GEORGE(NY).: *SELENASTRUM CAPRICORNU W72-08376
*ONEIDA LAKE(NY).: W72-13851
CANADARAGO LAKE(NY), LAKE GEORGE(NY).: *LIMITING NUTRIENTS, W72-09163
ITING NUTRIENTS, CANADARAGO LAKE(NY), LAKE GEORGE(NY).: *LIM W72-09163
RTIES, PHYTOCHEMICAL SCREENINGS, NYMPHAEA TUBEROSA, NUPHAR VARIEGA W72-04275
ORPHOMETRY, CHEMICAL PROPERTIES, O: /SOLVED OXYGEN, CARBON, LAKE M W72-08048
HETERCSTROPHA, PROCLADIUS, WHITE OAK LAKE(TENN), UPTAKE.: / PHYSA W70-06668
RIVER, OKA RIV/ *USSR, *IRKUTSK OBLAST, *BRATSK RESERVOIR, ANGARA W72-04853
BIOLOGY, METALI/ *USSR, *IRKUTSK OBLAST, *BRATSK RESERVOIR, *HYDRO W72-04855
AALE RIVER, GERMANY, SCENEDESMUS OBLIQUUS.: S W72-14314
CHLORELLA VULGARIS, SCENEDESMUS OBLIQUUS, ALCALIGENES FAECALIS, E W72-07664
ATA, EUGLENA LIMNOPHYLA, EUGLENA OBLONGA, EUGLENA OXYURIS, EUGLENA W73-00918
REFRACTOCYS/ PROTOCRYPTOCHRYSIS OBOVATA, GUTTULA BACILLARIOPHAGA, W72-09119
DICTYOTA BARTAYRESII, LAURENCIA OBTUSA, LAURENCIA PAPILLOSA, COEL W73-00838
PHORUS RATIO, LAKE WAUBESA(WIS), OCCOQUAN RESERVOIR(VA), MADISON(W W72-05473
*PACIFIC COAST REGION, TRITIUM, OCEAN CURRENTS, COBALT RADIOISOTO W72-03324
H CAROLINA, MICROSCOPY, ATLANTIC OCEAN.: /GRAPHY, SEA WATER, *NORT W73-00284
IATOMS, *PHYTOPLANKTON, *PACIFIC OCEAN, *ALGAE.: *D W71-11527
MPLING, INVERTEBRATES, *ATLANTIC OCEAN, *GULF OF MEXICO, PHYTOPLAN W72-14313
, MATHEMATICAL STUDIES, ATLANTIC OCEAN, *TROPICAL REGIONS, MARINE W72-11719
LA NANA, LEPTOCYLINDRUS, PACIFIC OCEAN, ALEUTIAN ISLANDS, PRYMNESI W71-11008
MAPPING, PHYTOPLANKTON, PACIFIC OCEAN, BOTTOM SEDIMENTS, STATIONS W72-03567
PTION, TROPICAL REGIONS, PACIFIC OCEAN, COLUMNS, BOTTOM SEDIMENTS, W71-08042
ODUCTIVITY, POPULATION, ATLANTIC OCEAN, DEPTH, STANDING CROP, CURR W72-04766
UTRIENTS, *HYDROGRAPHY, *PACIFIC OCEAN, ECOLOGY, WATER TEMPERATURE W72-07892
DIOISOTOPES, ABSORPTION, PACIFIC OCEAN, FOOD CHA: /, AMPHIPODA, RA W72-03347
IOMASS, ORGANIC MATTER, ATLANTIC OCEAN, GROWTH RATES, ALGAE, INCUB W71-11562
HNIQUES, ATLANTIC OCEAN, PACIFIC OCEAN, GULF OF MEXICO.: /TION TEC W71-11851
LUTION, *THERMAL POWER, ATLANTIC OCEAN, MARINE BIOLOGY, BENTHIC FL W72-11876
SPHAEROTILUS, NUTRIENTS, PACIFIC OCEAN, NITRATE, CHEMICAL ANALYSIS W72-05421
YSTEMATICS, *VIRGINIA, *ATLANTIC OCEAN, NORTH AMERICA.: /ARSHES, S W72-12633
SEPARATION TECHNIQUES, ATLANTIC OCEAN, PACIFIC OCEAN, GULF OF MEX W71-11851
*DISTRIBUTION PATTERNS, ATLANTIC OCEAN, PACIFIC OCEAN, TROPICAL RE W72-03543
UILA ISLAND, HONSHU ISLAND, OPEN OCEAN, SAPROPHYTES, HETEROTROPHIC W71-08042
TROPHICATION, PLANKTON, ATLANTIC OCEAN, SEA WATER, SALINITY, PHENO W72-07710
IOMASS, CANADA, COASTS, ATLANTIC OCEAN, SEASONAL, BAYS, SEA WATER, W73-00366
SOLUTIONS, PHAEOPHYTA, ATLANTIC OCEAN, SHARKS, RADIOACTIVITY TECH W72-05965
RE MEDIA, *BAY OF BENGAL, INDIAN OCEAN, SOIL EXTRACTS, ERDSCHREIBE W73-00430
ATTERNS, ATLANTIC OCEAN, PACIFIC OCEAN, TROPICAL REGIONS, POLAR RE W72-03543
CTS, PRODUCTIVITY, RADIOECOLOGY, OCEANOGRAPHY.: /NVIRONMENTAL EFFE W72-06313
TOZOA, ALGAE, BACTERIA, DIATOMS, OCEANS, ACTIVATED SLUDGE, SOIL BA W72-10883
TON, FLUID MECHANICS, EXUDATION, OCEANS, CHLOROPHYTA, PHAEOPHYTA, W71-07878
PERSISTENCE, PATH OF POLLUTANTS, OCEANS, ESTUARIES, CULTURES, SAMP W73-00361
, MARINE PLANTS, EUTROPHICATION, OCEANS, SAMPLING, SEPARATION TECH W73-00430
ONAS, NITROBACTER, NITROSOCYSTIS OCEANUS, SPHAEROTILUS NATANS, GLU W72-05421
ATA, VALONIA VENTRICOSA, VALONIA OCELLATA, ERNODESMIS VERTICILLATA W72-11800
, RIA DE ALDAN(SPAIN), LAMINARIA OCHROLEUCA, SACCORHIZA POLYSCHIDE W72-04766
YTA, OXYGEN, LIGHT, CHRYSOPHYTA, OCHROMONAS, RESPIRATION, CHLOROPH W71-09172
BIG ELK RIVER, LITTLE ELK RIVER, OCTORARA RIVER, PLECTONEMA, LYNGB W72-04736
SCILLATORIA PROLIFICA, THRESHOLD ODOR NUMBER.: /ERIDINIUM WESTI, O W71-13187
D OXYGEN, COLIFORMS, RECREATION, ODOR.: /TER TEMPERATURE, DISSOLVE W71-11914
WATER, *CYANOPHYTA, RESERVOIRS, ODOR-PRODUCING ALGAE, SOUTHWEST U W72-10613
ECTS, WATER QUALITY, HERBICIDES, ODOR-PRODUCING ALGAE.: /UTION EFF W72-14692
STICIDES, TASTE-PRODUCING ALGAE, ODOR-PRODUCING ALGAE, ACTINOMYCET W73-02144
RATION, PHOSPHATES, CLAY, TASTE, ODOR, *SURFACE WATERS, *GROUNDWAT W72-06536
FILTERS, ALGAE, ORGANIC LOADING, ODOR, COLIFORMS, TEMPERATURE, DIS W71-07113
WATER POLLUTION, AIR POLLUTION, ODOR, CONFINEMENT PENS, ORGANIC M W71-08214
SEDIMENTATION, TOXICITY, TASTE, ODOR, CORROSION, ENZYMES, CHLEATI W72-06837
SSOLVED OXYGEN, ALGAE, BACTERIA, ODOR, INFLOW, DISSOLVED SOLIDS, D W72-02274
ORGANIC MATTER, VOLUME, DRYING, ODOR, NITROGEN, AERATION, HOGS, C W71-07551
TROPHICATION, FOULING, TOXICITY, ODOR, OXYGEN SAG, TURBIDITY, WATE W71-09958
NUTRIENTS, ALGAE, WEEDS, TASTE, ODOR, PATHOGENIC BACTERIA, ORGANI W71-11006
S, WASTE WATER TREATMENT, TASTE, ODOR, POTABLE WATER, ACTIVATED CA W73-02144
HERBICIDES, HEAVY METALS, TASTE, ODOR, RECREATION, PHOSPHORUS, CAR W71-12091
COLIFORMS, SPORES, LIFE CYCLES, ODOR, RECREATION, SYSTEMATICS.: / W71-09173
BIC CONDITIONS, ORGANIC LOADING, ODOR, SLUDGE, COST ANALYSIS, NITR W72-11642
LGAE, ICE, FISH, MICROORGANISMS, ODOR, TASTE, PLANKTON, CHEMICALS, W73-01669
S, ALGAE, AQUATIC PLANTS, CLAYS, ODOR, WATER QUALITY, FOOD CHAINS, W72-04759
, OXIDATION-REDUCTION POTENTIAL, ODORS, OPERATION AND MAINTENANCE, W72-10401
ICITY, CORROSION, PUBLIC HEALTH, ODORS, OXIDATION-REDUCTION POTENT W72-04086
OXYGEN DEMAND, ORGANIC LOADING, ODORS, PHOTOSYNTHESIS, ALGAE, BAC W71-07123
UMATOPSYCHE, TRIAENODES INUUSTA, OECETIS, LEPTOCELLA, ISONYCHIA, C W70-06655
LADOPHORA GLOMERATA, *PITHOPHORA OEDOGONIA.: *C W71-08032
ERATA, ULOTRICHALES, ZYGNEMALES, OEDOGONIUM, MOUGOTIA, SPOROTETRAS W71-05991
PHORA, PHORMIDIUM, OSCILLATORIA, OEDOGONIUM.: /IVER(FRANCE), CLADO W71-00099
MICROSPORA, SYNEDRA, DINOBRYON, OEDOGONIUM, FRAGILARIA, DIATOMA, W72-10875
UM, PHACUS, CHA/ *ONONDAGA LAKE, OEDOGONIUM, RHIZOCLONIUM, COSMARI W72-09111
IOLOGICAL SAMPLES, LIVER, URINE, OEDOGONIUM, PHYSA, CULEX, EAMBUSI W72-07703
E ERIE CENTRAL BASIN, TRIBONEMA, OEDOGONIUM, ANACYSTIS, CERATIUM, W72-12995
CLADOPHORA FULIGINOSA, BATOPHORA OERSTEDI, ACETABULARIA CRENULATA, W72-11800
ON, *TREATMENT FACIL/ *STORM RUN- OFF, *SUSPENDED SOLIDS, *RECREATI W72-00042
*MARINE PLANTS, *MARINE ANIMALS, OFFSHORE PLATFORMS, NUCLEAR POWER W73-02029
FARM PONDS, MARYLAND, QUARRIES, OHIO RIVER, HUDSON RIVER, INDICAT W72-04736

OXYGEN, FORECASTING, FLUORIDES,	OHIO.: /TES, POTASSIUM, DISSOLVED	W72-04260
*NUTRIENT LEVELS, CUYAHOGA RIVER(	OHIO).:	W72-04263
SANDUSKY RIVER(	OHIO).:	W72-04260
ATION, EDDY DIFFUSIVITY, DESHLER(	OHIO).: /ICIENT, LAMBERT-BEER EQU	W72-05459
DINOBRYON, FRAGILARI/ *CLEVELAND(	OHIO), CERATIUM, COELOSPHAERIUM,	W72-10076
RHIZOCLONIUM, V/ *CUYAHOGA RIVER(	CHIO), CLADOPHORA, APHANOCHAETE,	W71-05629
OS-AQUAE, CERATIUM,/ DOLLAR LAKE(	OHIO), SPHAGNUM, APHANIZOMENON FL	W71-11949
*RHODOPHYTA, *LAKE ERIE,	OHIO, ALGAE.:	W71-05630
HODOPHYTA, CYANOPHYTA, / *ALGAE,	OHIC, CHLOROPHYTA, CHRYSOPHYTA, R	W71-05629
FISH, WATER POLLUTION, PROTOZOA,	CHIO, EUTROPHICATION, LAKE ONTARI	W73-01615
LVANIA, NEW YORK, NEW HAMPSHIRE,	OHIO, MICHIGAN.: /, HOSTS, PENNSY	W72-04736
LAKES, *WATER POLLUTION EFFECTS,	OHIO, NUTRIENTS, WATER POLLUTION	W71-11949
WAGE TREATMENT, *SEWAGE LAGOONS,	OHIO, OXYGEN, DISSOLVED OXYGEN, L	W72-05459
TER TREATMENT, SMELTS, MICHIGAN,	OHIO,: /ES, HEAVY METALS, WASTEWA	W72-14282
ORTING, SODIUM BROMIDE, DEXTRAN,	OIKOPLEURA, SAGITTA, PTEROPODS, E	W73-00371
IOXIDE, WATER POLLUTION SOURCES,	OIL SPILLS, SEA WATER, MARINE PLA	W73-01074
YELLOW PERCH, SEWAGE EFFLUENTS,	OIL SPILLS, THERMAL POLLUTION, NU	W72-14282
TREATMENT, SEDIMENTATION RATES,	OIL SPILLS, WATER POLLUTION SOURC	W72-12487
TRIAL PLANTS, INDUSTRIAL WASTES,	OIL WASTES, ESTUARIES, MONITORING	W71-13746
R POLLUTION SOURCES, *FISHERIES,	OIL WASTES, FISH REPRODUCTION, WA	W71-13723
CALIS, SAMPLE PREPARATION, CRUDE	OIL.: / DIGITATA, PORPHYRA UMBILI	W73-01074
S, OIL, *OIL WASTES, OILY WATER,	OIL-WATER INTERFACES, WATER POLLU	W70-09429
ICITY, *EMULSIFIERS, DETERGENTS,	OIL, *OIL WASTES, OILY WATER, OIL	W70-09429
RTEBRATES, DIESEL FUEL, BUNKER C	OIL, AVICENNIA, RHIZOPHORA, BOSTR	W72-07911
STRIPED BASS, DISSOLVED OXYGEN,	OIL, BRIDGES, BUOYS, HARBORS, TUN	W73-02029
TERGENTS, NUTRIENTS, PESTICIDES,	OIL, METALS, THERMAL POLLUTION, S	W72-08790
ECTS, *OIL WASTES, *CRUSTACEANS,	OILY WATER, ALGAE, NEMATODES, OYS	W72-07911
ATER, MARINE PLANTS, PHAEOPHYTA,	OILY WATER, EVAPORATION, DIFFUSIO	W73-01074
S, DETERGENTS, OIL, *OIL WASTES,	OILY WATER, OIL-WATER INTERFACES,	W70-09429
EPARATION TECHNIQUES, SEDIMENTS,	OILY WATER, ALGAE.: /ECHNIQUES, S	W71-12467
*USSR, MOSCOW RIVER,	OKA RIVER, FOAM, NATURAL WATERS.:	W72-05506
*BRATSK RESERVOIR, ANGARA RIVER,	OKA RIVER, HYDROBIOLOGY, BACTERIO	W72-04853
AMOEBOBACTER ROSEUS CHROMATIUM	OKENII, CHROMATIUM, MACROMONAS MO	W72-07896
POLLUTION EFFECTS, *FARM WASTES,	OKLAHOMA, CYANOPHYTA, CHLORELLA,	W72-04773
ANTS, BIOMASS, NEMATODES, ALGAE,	OLIGOCHAETES, LIFE CYCLES, SEASON	W71-11012
DISACCHARIDES, MONOSACCHARIDES,	OLIGOSACCHARIDES, CLONES, PHAEOCY	W72-07710
NUTRIENTS, CULTURES, INHIBITION,	OLIGOTROPHY, PHOSPHORUS, NITROGEN	W72-08376
E, AQUATIC WEEDS, PHYTOPLANKTON,	OLIGOTROPHY, HARDNESS(WATER), SOI	W72-08459
PARATION TECHNIQUES, *WISCONSIN,	OLIGOTROPHY, EUTROPHICATION, LAKE	W73-01456
KE STAGES, NUTRIENTS, MESTROPHY,	OLIGOTROPHY, OXYGEN SAG, AQUATIC	W73-00002
GICAL), LAKE STAGES, MESOTROPHY,	OLIGOTROPHY, WATER POLLUTION SOUR	W72-14782
ISSOLVED OXYGEN, EUTROPHICATION,	OLIGOTROPHY, PHYTOPLANKTON, ANALY	W72-03224
HOSPHORUS, CARBON, GROWTH RATES,	OLIGOTROPHY, EUTROPHICATION.: / P	W72-09164
ICATION, HARDNESS(WATER), COLOR,	OLIGOTROPHY, EUTROPHICATION, ALGA	W72-04269
ON, ZOOPLANKTON, EUTROPHICATION,	OLIGOTROPHY, PROTOZOA, CHLORELLA,	W72-05968
RY PRODUCTIVITY, ALGAE, BIOMASS,	OLIGOTROPHY, DIATOMS, EUGLENA, DA	W71-09171
DROGRAPHY, ALGAE, AQUATIC ALGAE,	OLIGOTROPHY, DISSOLVED OXYGEN, WA	W72-13871
IOASSAY, PHOSPHORUS, CYANOPHYTA,	OLIGOTROPHY, CHLAMYDOMONAS, DEFIC	W72-13816
OS, PORIA VAILLANTII, CONIOPHORA	OLIVACEA, LENZITES TRABEA, SCOPUL	W72-13813
.).: GOMPHONEMA	OLIVACEOIDES, WHITE CLAY CREEK(PA	W73-00242
BAY OF QUINTE, *LAKE	ONTARIO.:	W71-08953
NDIANA), PATUXENT RIVER(CANADA),	ONTARIO.: /YNE BAY, WHITE RIVER(I	W71-07417
SHELTER VALLEY CREEK(	ONTARIO).:	W73-01627
*GEORGIAN BAY(	ONTARIO).:	W72-13662
ERIE CENTRAL BASIN, POINTE PELEE(	ONTARIO), OXYGEN DEPLETION.: /KE	W72-12992
NUTRIENT BUDGETS, *BAY OF QUINTE(	ONTARIO), NUTRIENT INPUTS, CHIRON	W71-11009
G GUIDELINES, *GRAND RIVER BASIN(	ONTARIO), CANADA, BUFFER CAPACITY	W71-11017
ON SPECTROPHOTOMETRY, PERCH LAKE(	ONTARIO), ASHING, BIOLOGICAL MAGN	W71-11529
IE, ALGAE, NI/ *NUTRIENTS, *LAKE	ONTARIO, *EUTROPHICATION, LAKE ER	W71-11009
AKE MICHIGAN, *LAKE HURON, *LAKE	ONTARIO, *LAKE ERIE, WATER POLLUT	W72-01094
ETRY, *POLLUTANT IDENTIFICATION,	ONTARIO, ALGAE, MOSSES, AQUATIC P	W71-11529
N RADIOISOTOPES, LAKE ERIE, LAKE	ONTARIO, ADSORPTION, AGRICULTURAL	W72-13657
MENTAL LAKES AREA, *NORTHWESTERN	ONTARIO, EPILITHIC ALGAE, ACHNANT	W71-11029
IS, GREAT LAKES, LAKE ERIE, LAKE	ONTARIO, HYDRO: /RS, PHOTOSYNTHES	W72-07648
NTHIC FAUNA, BENTHIC FLORA, LAKE	ONTARIO, HUDSON RIVER, STANDARDS,	W70-09614
OZOA, OHIO, EUTROPHICATION, LAKE	ONTARIO, LIGHT PENETRATION, PHYSI	W73-01615
, AQUATIC ALGAE, LAKE ERIE, LAKE	ONTARIO, NUTRIENTS, NITROGEN, FIL	W71-11507
ICAL COMMUNITIES, *STREAMS, LAKE	ONTARIO, PLANKTON, DIATOMS, SEDIM	W73-01627
TION, SURFACTANTS, ENZYMES, LAKE	ONTARIO, SEASONAL, ALGAE, CULTURE	W71-00116
ENTS, WISCONSIN, LAKE ERIE, LAKE	ONTARIO, ST LAWRENCE RIVER, INTER	W70-07283
RON, LAKE ERIE, DETERGENTS, LAKE	ONTARIO, TERTIARY TREATMENT, INTE	W71-12091
VUM, PHACUS CIRCUMFLEXUS, PHACUS	ONYX, PHACUS PLEURONECTES, PHACUS	W73-00918
LGAL CULTURES, CHELATING AGENTS,	OOCYSTIS MARSSONII.: /NALYSIS, *A	W73-02112
ETRY, CADMIUM IONS, COPPER IONS,	OOCYSTIS.: POTENTIOM	W72-01693
*PROTEASE ENZYMES, CHLOROGONIUM,	OOCYSTIS, OSCILLATORIA, SPIRULINA	W72-07661
YSTIS, TRIBONEMA, APHANIZOMENON,	OOCYSTIS, GLOEOCYSTIS, COELASTRUM	W72-07940
TIGMAPHORA ROSTRATA, ETHMODISCUS	OOZES.: /IA, HEMIAULUS HAUCKII, S	W72-03567
, TUTUILA ISLAND, HONSHU ISLAND,	OPEN OCEAN, SAPROPHYTES, HETEROTR	W71-08042
WATER(ENGLAND), AMPHORA OVALIS,	OPEPHORA MARTYI, CELL COUNTS, NAV	W71-12855
TION-REDUCTION POTENTIAL, ODORS,	OPERATION AND MAINTENANCE, WASTE	W72-10401
CIPAL WASTES, INDUSTRIAL WASTES,	OPERATION AND MAINTENANCE, *WASTE	W72-09386
PHENOLS, DETERGENTS, PESTICIDES,	OPERATION AND MAINTENANCE, *WATER	W72-08357
IR(VA).: *PUMPED STORAGE	OPERATION, SMITH MOUNTAIN RESERVO	W72-14405
ON EFFECTS, NUTRIENTS, RESERVOIR	OPERATION, WATER REUSE, POTABLE W	W71-13185
LUM/ *DETENTION PERIOD, FLUSHING	OPERATION, ACID INJECTION PUMP, A	W72-02850
OLVED OXYGEN, WEATHER, RESERVOIR	OPERATION, WATER TEMPERATURE, PHY	W72-05282
S, CHEMOSTATS, FLAGELLATE ALGAE,	OPTICAL DENSITY SENSOR.: /KINETIC	W70-10403
ERIN, LUCIFERASE, CHLOROPHYLL A,	OPTICAL DENSITY, ACETYLENE REDUCT	W72-10883
BACCILARIOPHYCEAE,/ *ANTIBIOSIS,	OPTICAL DENSITY, PRASINOPHYCEAE,	W72-08051
, LABORATORY TESTS, CHRYSOPHYTA,	OPTICAL PROPERTIES.: /LAMYDOMONAS	W72-08051

*OXYGEN, LIGHT, PHOTOSYNTHESIS, OPTICAL PROPERTIES, ABSORPTION, M W72-14800
LOROPHYTA, DIATOMS, TEMPERATURE, OPTICAL PROPERTIES, GROWTH RATES. W70-10403
ALCIUM RADIOISOTOPES, *BOSSIELLA ORBIGNIANA, PAPER CHROMATOGRAPHY, W72-12634
AMS, FINGERNAIL CLAMS, UNIONIDS, ORCONECTES RUSTICUS, LIRCEUS LINE W72-06526
RA(WIS.), MULTIVARIATE ANALYSIS, ORDINATION ANALYSIS.: *LAKE WING W73-02469
AGES, BAROCO CLAY SOILS, URANIUM ORE.: /, CARBOWAX, FLOCCULANT DOS W70-06968
, SHAGAWA LAKE(MINN), WALDO LAKE(ORE).: *LIMITING NUTRIENTS W72-09165
PHORA GLOMERATA, *METOLIUS RIVER(ORE).: /ANTHES, *SPIROGYRA, CLODO W72-07901
WASH), VANCOUVER(WASH), PORTLAND(ORE), WATER QUALITY ANALYZER.: /(W72-03215
PRETATION, MACROAL/ *YAQUINA BAY(ORE), YAQUINA ESTUARY, DATA INTER W72-08141
UES, COLUMBIA RIVER, WASHINGTON, OREGON.: /NTS, ANALYTICAL TECHNIQ W72-03215
UPPER KLAMATH LAKE(OREGON).: W72-08989
KE(NEW YORK), UPPER KLAMATH LAKE(OREGON), LAKE MENDOTA(WISCONSIN). W71-05626
HICATION, *AQUATIC MICROBIOLOGY, OREGON, *SEDIMENT-WATER INTERFACE W72-08989
RINE PLANTS, CANADA, WASHINGTON, OREGON, CALIFORNIA, ALASKA, SEA W W73-00428
S, *LABORATORY TESTS, *BIOASSAY, OREGON, LAKES, NITROGEN, PHOSPHOR W72-09164
IDS, *EUTROPHICATION, *PLANKTON, OREGON, ORGANIC COMPOUNDS, ABSORP W73-01269
VITY, *MIDGES, LAGOONS, BIOMASS, OREGON, SLUDGE, ALGAE, DIPTERA, D W72-03556
S, NEVADA, CALIFORNIA, NEW YORK, OREGON, WISCONSIN, MAINE.: / WIND W71-05626
MICAL ANALYSIS, *ALGAE, COLUMNS, ORGANIC ACIDS, SEPARATION TECHNIQ W72-09365
CELLULOSE, DIALYSIS, SEA WATER, ORGANIC ACIDS, CHRYSOPHYTA, INHIB W73-00942
IONS, ION EXCHANGE, METHODOLOGY, ORGANIC ACIDS, FULVIC ACIDS, KELP W72-12166
*SNAKE RIVER(IDAHO), *DISSOLVED ORGANIC CARBON, *AMERICAN FALLS R W72-13300
SUBSTRATE UTILIZATION, PARTICU/ ORGANIC CARBON, ORGANIC NITROGEN, W73-01066
ASSO SEA, SAMPLE PREP/ DISSOLVED ORGANIC CARBON, MACROPHYTES, SARG W73-02100
E UTILIZATION, ORGANIC NITROGEN, ORGANIC CARBON, CULTURE MEDIA, PA W73-00379
HEXURONIC ACID, VINEYARD SOUND, ORGANIC CARBON, ACID-VOLATILE COM W72-07710
GANIC NITROGEN, THIAMINE, BIOTI/ ORGANIC CARBON, CHLOROPHYLL A, OR W72-07892
ES).: *DISSOLVED ORGANIC CARBON, *MENAI STRAIT(WAL W72-09108
TRANSFORMATION, CARBON SOURCES, ORGANIC CARBON, CARBON UTILIZATIO W71-12068
LABORATORY SCALE, *TOTAL ORGANIC CARBON.: W71-07102
*CHELATORS, ANODIC STRIPPING, ORGANIC CHELATORS.: W71-09674
FORMS, BACTERIA, PHOTOSYNTHESIS, ORGANIC COMPOUNDS, PHOSPHATES, NI W71-07100
LFONATES, CHELATION, PHOSPHATES, ORGANIC COMPOUNDS, WATER POLLUTIO W71-06247
SODIUM CHLORIDE, COLUMBIA RIVER, ORGANIC COMPOUNDS, VITAMINS, YEAS W72-05421
RIALS, BIOINHIBITION, BI/ *TRACE ORGANIC CONTAMINANTS, *TRACE MATE W71-11793
E BLOOMS, UTAH, GREAT SALT LAKE, ORGANIC COMPOUNDS, BRINES.: /ALGA W71-12223
IDS, *POLYMERS, ANIONS, CATIONS, ORGANIC COMPOUNDS, SEPARATION TEC W72-11833
YSIOLOGY, PHYSIOLOGICAL ECOLOGY, ORGANIC COMPOUNDS, ON-SITE TESTS, W72-03216
OXIDE, ALGAE, CULTURES, ACIDITY, ORGANIC COMPOUNDS.: /M, CARBON DI W72-00838
TES, SILICA, PATH OF POLLUTANTS, ORGANIC COMPOUNDS, PHYTOPLANKTON, W73-00384
ENTS, TRACE ELEMENTS, CHELATION, ORGANIC COMPOUNDS, FOOD CHAINS, P W73-00821
PRODUCTIVITY, METABOLISM, FUNGI, ORGANIC COMPOUNDS, CHLORELLA, TRA W72-14326
Y, PARTICULATE MATTER, DETRITUS, ORGANIC COMPOUNDS, PATH OF POLL: / W73-02105
, BENTHOS, CARBON, PRODUCTIVITY, ORGANIC COMPOUNDS, CARBOHYDRATES, W73-02100
TERIA, *ATTRACTANTS, *PREDATION, ORGANIC COMPOUNDS, AMINO ACIDS, V W73-00942
TROPHICATION, *PLANKTON, OREGON, ORGANIC COMPOUNDS, ABSORPTION, KI W73-01269
IC MATTER, *MARINE ALGAE, HUMUS, ORGANIC COMPOUNDS, POLLUTANT IDEN W72-12166
ODUCTS, MIXING, SOLAR RADIATION, ORGANIC LOADING, DIURNAL DISTRIBU W72-13508
OLD REGIONS, *OXIDATION LAGOONS, ORGANIC LOADING, AERATION, MIXING W72-06838
ONDITIONS, ANAEROBIC CONDITIONS, ORGANIC LOADING, ODOR, SLUDGE, CO W72-11642
ONDITIONS, ANAEROBIC CONDITIONS, ORGANIC LOADING, *BIOCHEMICAL OXY W71-13326
OXYGEN DEMAND, MIXING, SAMPLING, ORGANIC LOADING.: / *BIOCHEMICAL W71-07102
COLIFORMS, BACTERIA, OXIDATION, ORGANIC LOADING, ALGAE, WASTE WAT W71-07096
LAGOONS, *WASTE WATER TREATMENT, ORGANIC LOADING, BIOCHEMICAL OXYG W71-07124
TERIA, TRICKLING FILTERS, ALGAE, ORGANIC LOADING, ODOR, COLIFORMS, W71-07113
TMENT, CYANOPHYTE, ENGLENOPHYTA, ORGANIC LOADING.: /STE WATER TREA W71-07973
-USE, BIOCHEMICAL OXYGEN DEMAND, ORGANIC LOADING, ODORS, PHOTOSYNT W71-07123
ES, FISH, POLLUTANTS, TURBIDITY, ORGANIC LOADING, CORAL, MOLLUSKS, W71-01475
OPHICATION, *ALGAE, *DECOMPOSING ORGANIC MATTER, NUTRIENTS, SURFAC W70-10405
RY PRODUCTIVITY, BIODEGRADATION, ORGANIC MATTER, OXYGEN, CHEMICAL W71-04518
SH WATER, ALGAE, PHOTOSYNTHESIS, ORGANIC MATTER, PLANKTON, CHELATI W71-03027
NS, ANAEROBIC CONDITIONS, ALGAE, ORGANIC MATTER, VOLUME, DRYING, C W71-07551
SOTOPES, IONS, RIVERS, CULTURES, ORGANIC MATTER, BACTERIA, WATER A W71-08032
S, AEROBIC TREATMENT, OXIDATION, ORGANIC MATTER.: /XIDATION LAGOON W71-09629
LLUTION, ODOR, CONFINEMENT PENS, ORGANIC MATTER, NUTRIENTS, NITROG W71-08214
MANGANESE, PHOSPHORUS, NITROGEN, ORGANIC MATTER, CALCIUM CARBONATE W72-05469
ERMAL STRATIFICATION, OXIDATION, ORGANIC MATTER, WATER QUALITY, WA W72-04853
TION, BIOCHEMICAL OXYGEN DEMAND, ORGANIC MATTER, PHOSPHORUS, CARBO W71-13257
N CYCLE, ABSORPTION, DECOMPOSING ORGANIC MATTER, ALKALINITY, HYDRO W71-12068
HEDS(BASINS), PHOSPHORUS, ALGAE, ORGANIC MATTER, LAKE ERIE, EFFLUE W71-11017
, *AQUATIC PRODUCTION, *BIOMASS, ORGANIC MATTER, ATLANTIC OCEAN, G W71-11562
TH OF POLLUTANTS, DIELDRIN, DDT, ORGANIC MATTER, ANALYTICAL TECHNI W71-11498
OSIS, AZOTOBACTER, MARINE ALGAE, ORGANIC MATTER.: /XIDATION, SYMBI W72-11563
N, WATER ANALYSIS, CARBON CYCLE, ORGANIC MATTER, NUTRIENTS.: /NKTO W72-09108
IA, LITTORAL, LAKES, PERIPHYTON, ORGANIC MATTER, BIOMASS, METABOLI W72-03216
XYGEN REQUIREMENTS, *DECOMPOSING ORGANIC MATTER, AERATION, *ALGAE, W72-01881
TROPHICATION, ALGAE, DECOMPOSING ORGANIC MATTER, MINNESOTA, CYANOP W72-01784
, DECIDUOUS FORESTS, DECOMPOSING ORGANIC MATERIAL, FERTILITY, RHIZ W72-03342
TION/ *CYANOPHYTA, *HUMIC ACIDS, ORGANIC MATTER, ALGAE, EUTROPHICA W72-01097
OXYGEN, CARBON DIOXIDE, AMMONIA, ORGANIC MATTER.: /ION, DISSOLVED W72-01363
HICATION, CHELATION, DECOMPOSING ORGANIC MATTER, PHYTOPLANKTON, PR W72-01097
PHOSPHATES, PHOSPHORUS, CARBON, ORGANIC MATTER, MANGANESE, IRON, W72-07514
HARDNESS(WATER), NITROGEN CYCLE, ORGANIC MATTER, PHOTOSYNTHESIS, E W72-08048
S, *KINETICS, *DISSOLVED SOLIDS, ORGANIC MATTER, NITROGEN, WATER P W73-01066
SCIRPUS ACUTUS, C-14, DISSOLVED ORGANIC MATTER, MACROPHYTES, BIOL W73-02095
, WATER ANALYSIS, BENTHIC FLORA, ORGANIC MATTER, DISSOLVED SOLIDS, W73-02100
S, *ORGANIC MATTER, *DECOMPOSING ORGANIC MATTER, *EUTROPHICATION, W72-14673
ANALYSIS, THERMOPHILIC BACTERIA, ORGANIC MATTER, CULTURES, SOIL BA W72-14301
Y, CARBON RADIOISOTOPES, SESTON, ORGANIC MATTER, DISTRIBUTION PATT W73-00431

, BACTERIA, AQUARIA, DECOMPOSING	ORGANIC MATTER, WASTE ASSIMILATIV	W73-00802
*MICROBIAL DEGRADATION, CARBON,	ORGANIC MATTER, WATER POLLUTION S	W73-00379
CEAN, SEASONAL, BAYS, SEA WATER,	ORGANIC MATTER, MORTALITY, WATER	W73-00366
ELLA, NUTRIENTS, WATER SAMPLING,	ORGANIC MATTER, WATER POLLUTION E	W73-00234
ILIMNION, SEDIMENTS, DECOMPOSING	ORGANIC MATTER, BACTERIA, CHEMICA	W72-12997
ORMANCE, EFFICIENCIES, *BIOMASS,	ORGANIC MATTER, NUTRIENTS, PHOTOS	W72-12289
*CHEMICAL ANALYSIS, *DECOMPOSING	ORGANIC MATTER, *MARINE ALGAE, HU	W72-12166
NIFICATION, MINAMATA BAY(JAPAN),	ORGANIC MERCURY COMPOUNDS, BIOSYN	W72-12257
LLUTANTS, SUBSTRATE UTILIZATION,	ORGANIC NITROGEN, ORGANIC CARBON,	W73-00379
ZATION, PARTICU/ ORGANIC CARBON,	ORGANIC NITROGEN, SUBSTRATE UTILI	W73-01066
UGLENA CAUDATA, EUGLENA/ *INDIA,	ORGANIC NITROGEN, EUGLENA ACUS, E	W73-00918
GA/ *LAKE METABOLISM, *DISSOLVED	ORGANIC NITROGEN, *PARTICULATE OR	W72-08048
D ORGANIC NITROGEN, *PARTICULATE	ORGANIC NITROGEN, CALCAREOUS LAKE	W72-08048
ORGANIC CARBON, CHLOROPHYLL A,	ORGANIC NITROGEN, THIAMINE, BIOTI	W72-07892
TERISTICS, SOLIDS RECYCLE, SOLI/	ORGANIC NITROGEN, SETTLING CHARAC	W72-11100
MUS, TRACHELOMONAS, SAGINAW BAY,	ORGANIC NITROGEN.: /, ANKISTRODES	W72-10875
ESTICIDES, CARBAMATE PESTICIDES,	ORGANIC PESTICIDES, HERBICIDES, W	W72-13347
LKALINE PHOSPHATASE,/ *DISSOLVED	ORGANIC PHOSPHORUS, *EXCRETION, A	W71-10083
NDRIA(EGYPT), PANDORINA, SOLUBLE	ORGANIC SOLIDS.: *ALEXA	W70-06619
FFECTS, *ALGAE, PHYTOPLANKTON, /	ORGANIC WASTES, WATER POLLUTION E	W70-07842
ITROGEN, *PHOSPHORUS, SEDIMENTS,	ORGANIC WASTES.: /SMS, *ALGAE, *N	W72-06532
ASTE, ODOR, PATHOGENIC BACTERIA,	ORGANIC WASTES, SOLID WASTES, CHE	W71-11006
IENTS, *WATER POLLUTION CONTROL,	ORGANIC WASTES, INDUSTRIAL WASTES	W73-00840
IDS, SAMPLE PREPARATION, NATURAL	ORGANICS.: /LTURE MEDIA, FATTY AC	W73-01066
ANOPHOSPHORUS COMPOUNDS, NATURAL	ORGANICS.: / *DNA, *DISSOLVED ORG	W72-12637
EDITERRANEAN SEA, RED SEA, DRIFT	ORGANISMS.: /ANAL, FLAGELLATES, M	W73-00296
AL PROPERTIES, DIATOMS, DOMINANT	ORGANISMS.: / IRON, PHYSICOCHEMIC	W73-00256
FLORA, EUTROPHICATION, DOMINANT	ORGANISMS, BIOLOGICAL COMMUNITIES	W72-14797
AL WASTES, *PESTICIDES, *AQUATIC	ORGANISMS, *PEST CONTROL, *CHEMCO	W73-01976
BIOLOGICAL COMMUNITIES, DOMINANT	ORGANISMS, AQUATIC ENVIRONMENT, A	W72-07132
ESOURCES, FOOD CHAINS, FISH FOOD	ORGANISMS, OXYGEN, ECOSYSTEMS.: /	W71-13246
BIOLOGICAL COMMUNITIES, DOMINANT	ORGANISMS, REDUCTION(CHEMICAL), T	W72-10883
BIOLOGICAL COMMUNITIES, DOMINANT	ORGANISMS, AQUATIC PRODUCTIVITY,	W72-09239
*PHYTOTOXINS, ECOLOGY, FISH FOOD	ORGANISMS, PATH OF POLLUTANTS, BI	W71-08597
*MERCURY, *METHYLMERCURY, BRAIN,	ORGANOMERCURIAL COMPOUNDS, ESOX L	W72-05952
SPHORUS, NITROGEN, HEAVY METALS,	ORGANOPHOSPHORUS PESTICIDES, INDU	W72-04742
TED HYDROCARBON PESTICIDES, DDT,	ORGANOPHOSPHORUS PESTICIDES, PHYT	W71-10096
, DISSOLVED OXYGEN, TEMPERATURE,	ORGANOPHOSPHOROUS COMPOUNDS, ALGA	W71-10645
TES, ALGAE, FOOD WEBS, COPEPODS,	ORGANOPHOSPHORUS COMPOUNDS, PRIMA	W72-07892
ALGICIDES, CARBAMATE PESTICIDES,	ORGANOPHOSPHORUS PESTICIDES, ACAR	W73-00238
CHLOROPHYLL A, *DNA, *DISSOLVED	ORGANOPHOSPHORUS COMPOUNDS, NATUR	W72-12637
HNIQUES, PESTICIDES, FUNGICIDES,	ORGANOPHOSPHOROUS PESTICIDES, CHL	W72-13347
XATION, EMBEDDING, ASHING, BONE,	ORGANS, BRAIN, BLOOD, KIDNEYS, LI	W73-01676
DOXUM, STAURASTRUM SEBALDII VAR.	ORNATUM, MICRASTERIAS CRUX-MELITE	W72-12736
RBOR(WISCONSIN), POLYPHOSPHATES,	ORTHOPHOSPHATES.: /, MILWAUKEE HA	W72-13644
LF OF FINLAND.:	ORTHOPHOSPHATE CONCENTRATIONS, GU	W72-13871
S.: ALGAL ASSAY,	ORTHOPHOSPHATE, PHOSPHORUS SOURCE	W73-02478
S, *BATCH CULTURES, *CHEMOSTATS,	ORTHOPHOSPHATES, SUBSTRATE UTILIZ	W73-01445
RHODOMONAS LACUSTRIS, COUNTING,	ORTHOPHOSPHATES, CHLOROPHYLL A.: /	W73-01435
I COMMERSONII, ASTERIONELLA SPP,	ORTHOPHOSPHATES, SECCHI DISC, NEW	W72-07890
ON SELECTIVE ELECTRODES, LASERS,	ORTHOPHOSPHATES.: /RADIOGRAPHY, I	W72-08790
CIFIC CONDUCTANCE, TRANSPARENCY,	ORTHOPHOSPHATES, DRY WEIGHT, PHOS	W71-10645
	ORTHOPHOSPHATES.:	W71-11507
SCULAR PLANTS, NITRATE NITROGEN,	ORTHOPHOSPHATES, *OLIGASAPROBIC S	W72-06526
TUS, SPARGANIUM RAMOSUM, LEERSIA	ORYZOIDES, NUPHAR ADVENA, EICHHOR	W73-01089
VELIGER, PTEROPODS, HETEROPODS,	OSCILLATORIA, STOMATOPODA, POLYCH	W73-00935
, GYMNODINIUM BREVE, PERIDINIUM,	OSCILLATORIA RUBECENS, ANABAENA,	W73-00840
RUBECENS, ANABAENA, CRUSSOSTREA,	OSCILLATORIA.: /UM, OSCILLATORIA	W73-00840
S NIDULANS, GLOEOCAPSA ALPICOLA,	OSCILLATORIA CHALYBIA, OSCILLATOR	W72-07664
ENZYMES, CHLOROGONIUM, OOCYSTIS,	OSCILLATORIA, SPIRULINA, PANDORIN	W72-07661
ALPICOLA, OSCILLATORIA CHALYBIA,	OSCILLATORIA FORMOSA, PHORMIDIUM	W72-07664
NITROGEN, DISSOLVED PHOSPHORUS,	OSCILLATORIA, RADON RADIOISOTOPES	W72-07648
*OPTICAL DENSITY, MICROAMMETER,	OSCILLATORIA AGARDHII VAR ISOTHRI	W72-07145
AKE LONG, LAKE SQUAW, LAKE BUDD,	OSCILLATORIA REDEKEI, LYNGBYA LIM	W72-07145
OPHTHALONIL, ULOTHRIX, ANABAENA,	OSCILLATORIA, MICROCYSTIS, ALGIST	W72-07508
IFORMIS NAUTOCOCCUS, APIOCOCCUS,	OSCILLATORIA, NAUTOCOCCUS CAUDATU	W72-07141
SMITHII, OSCILLATORIA PRINCEPS,	OSCILLATORIA AMOENA, OSCILLATORIA	W72-07896
A TENUIS, OSCILLATORIA CHLORINA,	OSCILLATORI: /AMOENA, OSCILLATORI	W72-07896
LAUCA, DACTYLOCOCCOPSIS SMITHII,	OSCILLATORIA PRINCEPS, OSCILLATOR	W72-07896
S, ANKISTRODESMUS, ELAKOTOTHRIX,	OSCILLATORIA, ANABAENA, MICROCYST	W72-07940
RIA AMOENA, OSCILLATORIA TENUIS,	OSCILLATORIA CHLORINA, OSCILLATOR	W72-07896
A PRINCEPS, OSCILLATORIA AMOENA,	OSCILLATORIA TENUIS, OSCILLATORIA	W72-07896
E, *WAHNBACH RESERVOIR(GERMANY),	OSCILLATORA RUBESCENS.: /NTHRACIT	W72-06201
NGBYA, PHORMIDIUM, CYANOPHYCEAE,	OSCILLATORIACEAE, SCHIZOTHRIX CAL	W72-04736
ZOCLONIUM, VAUCHERIA, TRIBONEMA,	OSCILLATORIA.: /APHANOCHAETE, RHI	W71-05629
STRIS, SYNURA UVELLA, DINOBRYON,	OSCILLATORIA LIMNETICA, DICTYOSPH	W71-03014
NA, COREGONS, LYNGBYA LIMNETICA,	OSCILLATORIA REDECKEI, OSCILLATOR	W71-00117
IMNETICA, OSCILLATORIA REDECKEI,	OSCILLATORIA RUBESCENS, SYNEDRA A	W71-00117
FRANCE), CLADOPHORA, PHORMIDIUM,	OSCILLATORIA, OEDOGONIUM.: /IVER(	W71-00099
NERET(ISRAEL), PERIDINIUM WESTI,	OSCILLATORIA PROLIFICA, THRESHOLD	W71-13187
ANIZOMENON FLOS-AQUAE, CERATIUM,	OSCILLATORIA, TRANSPARENCY.: /APH	W71-11949
I, LA/ COSCINODISCUS, NITZSCHIA,	OSCILLATORIA, PLEUROSIGMA, NAUPLI	W71-11876
FORMALIN, COUNTING, CYCLOTELLA,	OSCILLATORIA, APHANOTHECE, ANABAE	W72-10875
M, SKELETONEMA COSTATUM, CARBOY,	OSCILLATORIA ERYTHRAEA.: /CORNUTU	W72-09103
ACROPHYTES, *SWEDEN, MICROCYSIS,	OSCILLATORIA, SYNURA, ARCGLENA, C	W72-12729
HANGE, ELECTROCHEMISTRY, REVERSE	OSMOSIS, ELECTRODIALYSIS, DISTILL	W70-09907
AMINIFERA, HYDROIDS, POLYCHAETA,	OSTRACODS, HARPACTICOIDS, CIRRIPE	W72-05432
E, PISMO CLAM, TIVELA STULTORUM,	OSTREA LURIDA, CARDIUM: / TUNICAT	W72-07907
ARPA, ENHYDRA LUTRIS NEREIS, SEA	OTTER, MACROCYSTIS PYRIFERA, CHOR	W73-00932
S, FISH BEHAVIOR, INVERTEBRATES,	OTTERS, ROCKSLIDES, INTERTIDAL AR	W73-00764

DS, SALMON, STICKLEBACKS, CRABS, OTTERS, ZOOPLANKTON, PHYTOPLANKTO W72-00960
PP., TUBIFEX, UNIALGAL CULTURES, OUABAIN, BIOLOGICAL MAGNIFICATION W72-09668
IPHOSPHATASE, BIOCHEMICAL TESTS, OUABAIN, POTASSIUM CHLORIDE.: /TR W72-13809
LAKE WINDERMERE(ENGLAND), GREAT OUSE(ENGLAND), ALGAL GROWTH.: /), W71-13185
ROPERTIES, GROUNDWATER MOVEMENT, CUTFALL SERVERS, STREAMS, SEPTIC W72-14363
TROL, MASSACHUSETTS, *NUTRIENTS, OUTLETS.: /ION, WATER QUALITY CON W72-06046
LLUCIDA, AMPHORA OVALIS, AMPHORA OVALIS V. PEDICULUS, COCCONEIS PE W72-13142
, AMPHIPLEURA PELLUCIDA, AMPHCRA OVALIS, AMPHCRA OVALIS V. PEDICUL W72-13142
14, NITRO/ PLATYMONAS, NITZSCHIA OVALIS, CARBON EXTRUSION, CARBON- W72-07497
E, SHEAR WATER(ENGLAND), AMPHORA OVALIS, OPEPHORA MARTYI, CELL CCU W71-12855
A, LICMOPHORA JURGENSII, AMPHORA OVALIS, SYNEDRA FASCIULATA, EUNOT W72-08141
GILARIA CROTONENSIS, CRYPTOMONAS OVATA, STEPHANODISCUS NIAGARAE, A W73-01435
ES, HUMAN BRAIN, LOW-TEMPERATURE OVEN DRYING, ASHING, EXTRACTIVE D W71-11036
A, MANCHUDINIUM SINICU/ KOLBEANA OVOIDEA, PALMERIAMONAS PLANCTONIC W72-07683
, EUGLENA SPIROGYRA, LEPOCINCLIS OVUM, PHACUS CIRCUMFLEXUS, PHACUS W73-00918
E, NUTRIENTS, FILTERS EQUIPMENT, OXI: / FERMENTATICN, BACTERIOPHAG W72-09675
ASTES, METABOLISM, GROWTH RATES, OXIDATION LAGOONS, COLIFCRMS, NUT W72-09590
IDES, LIVESTOCK, WASTE DISPOSAL, CXIDATION LAGOONS, DEHYDRATION, R W72-00846
CYANOPHYTA, BIOASSAY, SAMPLING, OXIDATION LAGOONS, EUGLENA, CHLAM W72-07661
THEMATICAL STUDIES, CHLOROPHYLL, OXIDATION REDUCTION POTENTIAL.: / W72-14800
ATER FISH, BIOLOGICAL TREATMENT, OXIDATION LAGOONS, PHYTCPLANKTON, W72-12567
RIVERS, STREAMS, DELAWARE RIVER, OXIDATION LAGOONS, PONDS, FARM PO W72-04736
, DENITRIFICATION, SEPTIC TANKS, OXIDATION LAGOONS, LAND APPLICAT: W70-09907
DOMONAS, SCENEDESMUS, EFFLUENTS, OXIDATION LAGOONS.: /ELLA, CHLAMY W70-10405
ISPOSAL SYSTEMS, *LAND DISPOSAL, CXIDATION DITCHES, AERATED LAGOON W71-07551
R POLLUTION SOURCES, PHAEOPHYTA, OXIDATION.: / PHYTOPLANKTON, WATE W72-12166
IVITY, SEDIMENTS, NITRIFICATICN, OXIDATION-REDUCTION POTENTIAL, DE W72-08459
BACTERIA, NITROGEN, PHOSPHORUS, OXIDATION-REDUCTION POTENTIAL, LA W72-07933
ALGAE, PHOSPHORUS, ION EXCHANGE, OXIDATION-REDUCTICN POTENTIAL, HY W72-01108
CORROSION, PUBLIC HEALTH, ODORS, OXIDATION-REDUCTION POTENTIAL, BI W72-04086
TES, BACTERIA, ALGAE, NUTRIENTS, OXIDATION-REDUCTICN PCTENTIAL, OD W72-10401
E, COPEPODS, NITROGEN, BACTERIA, OXIDATION-REDUCTION PO: /MPERATUR W71-11008
Y, OXYGENATION, DENITRIFICATION, CXIDATION, ADSORPTION, CCLIFORMS, W71-13326
VESTIGATIONS, *BACTERIA, *ALGAE, OXIDATION, METABOLISM, RESPIRATIO W72-10233
HWATER, AMMONIA, EUTROPHICATION, OXIDATION, SYMBIOSIS, AZOTOBACTER W72-11563
LIGHT INTENSITY, SEDIMANTATICN, OXIDATICN, STCRAGE, BIOCHEMICAL O W72-01881
N, SEDIMENTATION, *CHLORINATICN, OXIDATION, POLLUTION ABATEMENT, W W72-14469
N, CHEMICALS, PHENOLS, CHLORINE, CXIDATION, ADSORPTION, ACTIVATED W73-01669
LGAE, *METHODOLOGY, MEASUREMENT, OXIDATION, CARBON RADICISOTOPES, W72-14301
, SAMPLING, COLIFORMS, BACTERIA, OXIDATION, ORGANIC LOADING, ALGAE W71-07096
TION LAGOONS, AEROBIC TREATMENT, OXIDATION, ORGANIC MATTER.: /XIDA W71-09629
RIA, *SYMBIOSIS, PHOTOSYNTHESIS, OXIDATION, RESPIRATION, CXYGENATI W72-04983
E, PHOSPHORUS, CARBON, NITROGEN, CXIDATION, BIOCHEMICAL OXYGEN DEM W72-05506
S, *WASTE ASSIMILATIVE CAPACITY, OXIDATION, DISSOLVED OXYGEN, HYDR W72-04787
ERATURE, THERMAL STRATIFICATION, OXIDATION, ORGANIC MATTER, WATER W72-04853
ION, SLUDGE DRYING, POLYETHYLENE OXIDE POLYMERS, SEWAGE SETTLING, W70-06968
BULANT FRICTIONAL DRAG, ALKYLENE OXIDE, POLYMERIZATION, ETHYLENE O W70-06968
OXIDE, POLYMERIZATION, ETHYLENE OXIDE, WATER CLARIFICATICN, SLUDG W70-06968
AY, SCOUDOUC RIVER, HUMIC ACIDS, OXIMATION, SAMPLE PREPARATION.: / W72-12166
RY, *BIOLOGICAL LIFE, *DISSOLVED OXYGEN CONCENTRATION.: /ARE ESTUA W70-09189
, PHOTOSYNTH/ *NUTRIENT REMOVAL, OXYGEN DECLINE, RESPIRATCRY RATES W71-04079
YGEN, PHYTOPLANKTON, BICCHEMICAL OXYGEN DEMAND, DISSOLVED OXYGEN, W71-05390
A, MUNICIPAL WASTES, BIOCHEMICAL OXYGEN DEMAND, WASTE WATER TREATM W71-03896
ORGANIC MATTER, OXYGEN, CHEMICAL OXYGEN DEMANC, BENTHOS.: /ATICN, W71-04518
GRAPHY, GROUNDWATER, BIOCHEMICAL OXYGEN DEMAND, PHOSPHORUS, FISH, W71-03012
HEMICAL OXYGEN DEMAND, *CHEMICAL OXYGEN DEMAND, *WATER PCLLUTION S W70-09189
N, *EUTROPHICATION, *BICCHEMICAL OXYGEN DEMAND, *CHEMICAL CXYGEN D W70-09189
CHEMICAL OXYGEN DEMAND, CHEMICAL CXYGEN DEMAND, PHOSPHATES, NITROG W70-07838
DS, *ALGAE, LAGOONS, BIOCHEMICAL OXYGEN DEMAND, CHEMICAL CXYGEN DE W7C-07838
PILOT PLANTS, ALGAE, BIOCHEMICAL OXYGEN DEMAND, BIOLOGICAL TREATME W70-06619
, *OXIDATION LAGOONS, CHLORELLA, OXYGEN DEMAND, PHOTOSYNTHESIS, RE W71-00139
*ST. LAWRENCE RIVER, BIOCHEMICAL OXYGEN DEMAND, HYDROGEN ICN CCNCE W70-10427
METABOLISM, SLUDGE, BIOCHEMICAL OXYGEN DEMAND, CHEMICAL CXYGEN DE W72-04789
CHEMICAL OXYGEN DEMAND, CHEMICAL OXYGEN DEMAND, CARBON DIOXIDE, LI W72-04789
NITROGEN, OXIDATION, BIOCHEMICAL CXYGEN DEMAND, RADIOACTIVITY, RAD W72-05506
*NUTRIENTS, ALKALINITY, CHEMICAL OXYGEN DEMAND, ANALYTICAL TECHNIQ W72-04983
ALKALINITY, MINING, BIOCHEMICAL OXYGEN DEMAND, ECONOMICS, ALGAE, W72-04734
HOSPHOROUS/ *ALGAE, *BIOCHEMICAL OXYGEN DEMANC, *EUTROPHICATICN, P W72-04431
PHOSPHOROUS, NITROGEN, CHEMICAL OXYGEN DEMAND, LABORATORY TESTS, W72-04431
CHEMICAL OXYGEN DEMAND, CHEMICAL CXYGEN DEMANC.: /EGRADATION, *BIO W72-06612
S, *BICDEGRADATION, *BIOCHEMICAL OXYGEN DEMAND, CHEMICAL CXYGEN DE W72-06612
N ION CONCENTRATION, BIOCHEMICAL OXYGEN DEMANC, *SEWAGE EFFLUENTS, W72-06612
CHEMICAL OXYGEN DEMAND, CHEMICAL OXYGEN DEMAND, COOLING TCWERS, *W W71-09524
RATION, TEMPERATURE, BIOCHEMICAL OXYGEN DEMAND, CHEMICAL CXYGEN DE W71-09524
TERTIARY TREATMENT, BIOCHEMICAL OXYGEN DEMANC, INDUSTRIAL WASTES, W71-06445
S, MUNICIPAL WASTES, BIOCHEMICAL OXYGEN DEMAND, WASTE WATER TREATM W71-06033
OGEN ICN CONCENTRATION, CHEMICAL OXYGEN DEMAND, *BIOCHEMICAL OXYGE W71-07102
ENT, BIODEGRADATION, BIOCHEMICAL OXYGEN DEMAND.: /STE WATER TREATM W71-07084
ICAL CXYGEN DEMAND, *BIOCHEMICAL OXYGEN DEMAND, MIXING, SAMPLING, W71-07102
MICAL OXYGEN DEMAND, BIOCHEMICAL OXYGEN CEMANC, WASTE WATER TREATM W71-07098
MICAL OXYGEN DEMANC, BIOCHEMICAL OXYGEN DEMAND, STABILIZATION POND W71-07097
AEROBIC CONDITIONS, BICCHEMICAL OXYGEN CEMANC, WASTE WATER TREATM W71-07109
DGE, COST ANALYSIS, *BIOCHEMICAL OXYGEN DEMAND, WASTE WATER TREATM W71-07099
NTENSITY, COLIFORMS, BIOCHEMICAL OXYGEN DEMAND, WASTE WATER TREATM W71-07106
NITROGEN, TIDAL WATERS, CHEMICAL CXYGEN DEMAND, BIOCHEMICAL CXYGEN W71-07098
AEROBIC CONDITIONS, BIOCHEMICAL OXYGEN DEMAND, WASTE WATER TREAT: W71-07100
STE WATER TREATMENT, BIOCHEMICAL OXYGEN CEMANC.: /ADING, ALGAE, WA W71-07096
EDIMENTATION, SAMPLING, CHEMICAL OXYGEN DEMAND, BIOCHEMICAL CXYGEN W71-07097
HLORINATION, SLOPES, BIOCHEMICAL OXYGEN DEMAND, HYDROGEN ION CONCE W71-07113
NT, ORGANIC LOADING, BIOCHEMICAL OXYGEN DEMAND, SOLAR RADIATION, N W71-07124

MENT, *WATER RE-USE, BIOCHEMICAL OXYGEN DEMAND, ORGANIC LOADING, O W71-07123
TERTIARY TREATMENT, BIOCHEMICAL OXYGEN DEMAND, WASTE WATER TREATM W71-07382
OCNS, *NORTH DAKOTA, BIOCHEMICAL OXYGEN DEMAND, BACTERIA, TEMPERAT W72-12565
PLANKTON, NUTRIENTS, BIOCHEMICAL OXYGEN DEMAND, INORGANIC COMPOUND W72-13659
IC CONDITIONS, ALGAE, SEDIMENTS, OXYGEN DEMAND, PHOSPHORUS, NITROG W72-12990
, *LAKE ERIE, *DISSOLVED OXYGEN, OXYGEN DEMAND, HYPOLIMNION, AQUAT W72-12994
ROPHYTA, CYANOPHYTA, BIOCHEMICAL OXYGEN DEMAND.: /HRYSOPHYTA, CHLO W72-12995
N, DISSOLVED OXYGEN, BIOCHEMICAL OXYGEN DEMAND, PERFORMANCE, FLOCC W72-13508
RANSFER, PHOSPHATES, ALKALINITY, OXYGEN DEMAND, HEAT BUDGET, ALGAE W72-12997
LS, CHEMICAL WASTES, BIOCHEMICAL OXYGEN DEMAND, BACTERIA, ALG: /CA W73-01976
ITY, STREAMS, LAKES, BIOCHEMICAL OXYGEN DEMAND, *OXYGENATION, *ANA W72-14469
QUATIC WEED CONTROL, BIOCHEMICAL OXYGEN DEMAND, LAKE STAGES, LIMNO W72-14692
QUATIC PRODUCTIVITY, BIOCHEMICAL OXYGEN DEMAND, LAKE STAGES, NUTRI W73-00002
CHEMICAL OXYGEN DEMAND, CHEMICAL OXYGEN DEMAND, NUTRIENTS, *WASTE W72-08396
, AEROBIC TREATMENT, BIOCHEMICAL OXYGEN DEMAND, CHEMICAL OXYGEN DE W72-08396
URES, SEDIMENTATION, BIOCHEMICAL OXYGEN DEMAND, ALGAE, CONFINEMENT W72-08401
LIMATES, METABOLISM, BIOCHEMICAL OXYGEN DEMAND, PILOT PLANTS, *WAS W72-06838
CHEMICAL OXYGEN DEMAND, CHEMICAL OXYGEN DEMAND, PHOSPHORUS, NITROG W72-01881
OXIDATION, STORAGE, BIOCHEMICAL OXYGEN DEMAND, CHEMICAL OXYGEN DE W72-01881
EATMENT, *TURBIDITY, BIOCHEMICAL OXYGEN DEMAND, DISSOLVED OXYGEN, W72-03753
REDUCTION POTENTIAL, BIOCHEMICAL OXYGEN DEMAND, NITRIFICATION, NUT W72-04086
ORGANISMS, EFFLUENT, BIOCHEMICAL OXYGEN DEMAND, SATURATION, OXYGEN W71-13553
ALGAE, OXYGENATION, BIOCHEMICAL OXYGEN DEMAND, WATER SPORTS, PARK W72-00042
MICAL OXYGEN DEMAND, BIOCHEMICAL OXYGEN DEMAND, FILTRATION, DISINF W72-00411
OSYNTHESIS, SOLUBILITY, CHEMICAL OXYGEN DEMAND, BIOCHEMICAL OXYGEN W72-00411
, NITROGEN, PHOSPHORUS, CHEMICAL OXYGEN DEMAND, CHLORELLA, PERFORM W72-11100
TH, RIVERS, *SEWAGE, BIOCHEMICAL OXYGEN DEMAND, SOLID WASTES, AMMO W72-09675
R POLLUTION SOURCES, BIOCHEMICAL OXYGEN DEMAND, WATER POLLUTION CO W72-09167
OGEN, ACTIVATED SLUDGE, CHEMICAL OXYGEN DEMAND, OXYGEN, AERATION, W71-13334
CHEMICAL OXYGEN DEMAND, CHEMICAL OXYGEN DEMAND, PHOSPHORUS, NITRAT W71-13341
S, ORGANIC LOADING, *BIOCHEMICAL OXYGEN DEMAND, NITROGEN, PHOSPHOR W71-13326
CHEMICAL OXYGEN DEMAND, CHEMICAL OXYGEN DEMAND, *PHOSPHATES, NITRO W71-13356
ON, WATER POLLUTION, BIOCHEMICAL OXYGEN DEMAND, ORGANIC MATTER, PH W71-13257
ES, MUNICIPAL WASTE, BIOCHEMICAL OXYGEN DEMAND, CHEMICAL OXYGEN DE W71-13356
ILTRATION, *FILTERS, BIOCHEMICAL OXYGEN DEMAND, CHEMICAL OXYGEN DE W71-13341
ON CONTROL, WASHINGTON, CHEMICAL OXYGEN DEMAND, BIOCHEMICAL OXYGEN W71-11006
MICAL OXYGEN DEMAND, BIOCHEMICAL OXYGEN DEMAND, DISSOLVED OXYGEN: / W71-11006
2, ENDOGENOUS RESPIRATION, TOTAL OXYGEN DEMAND, GLYCOLS.: /UM NO 5 W71-12223
SECONDARY TREATMENT, BIOCHEMICAL OXYGEN DEMAND, CHEMICAL OXYGEN DE W71-12188
ALGAE, ZOOPLANKTON, BIOCHEMICAL OXYGEN DEMAND, COPEPODS.: /QUATIC W71-11876
ANIC WASTES, ECONOMICS, CHEMICAL OXYGEN DEMAND, BIOCHEMICAL OXYGEN W71-12223
CHEMICAL OXYGEN DEMAND, CHEMICAL OXYGEN DEMAND, CARBON, FERTILIZAT W71-12188
MICAL OXYGEN DEMAND, BIOCHEMICAL OXYGEN DEMAND, AERATION, *OXIDATI W71-12223
SOLUBILITY, SEWAGE, BIOCHEMICAL OXYGEN DEMAND, ION EXCHANGE, CHEM W71-12200
N DEMAND, ION EXCHANGE, CHEMICAL OXYGEN DEMAND, FERTILIZER, TURBID W71-12200
AL BASIN, POINTE PELEE(ONTARIO), OXYGEN DEPLETION.: /KE ERIE CENTR W72-12992
LOS-AQUA, *NITROGENASE ACTIVITY, OXYGEN INACTIVATION.: /ANABAENA F W72-07941
RACTIVE DIGESTION, NEUROANATOMY, OXYGEN PLASMA ASHING, ATOMIC ABSO W71-11036
Y, CULT/ *CHLORELLA PYRENOIDOSA, OXYGEN PRODUCTION, LIGHT INTENSIT W70-05547
L, LIMNOLOGY, NUTRIENTS, OXYGEN, OXYGEN REQUIREMENTS, OXYGEN SAG, W72-04279
POLLUTION TREATMENT, NUTRIENTS, OXYGEN REQUIREMENTS, AGING(BIOLOG W72-01685
STE DISPOSAL, PESTICIDE REMOVAL, OXYGEN SAG, DREDGING, STATE GOVER W72-01659
TS, OXYGEN, OXYGEN REQUIREMENTS, OXYGEN SAG, WASTE ASSIMILATIVE CA W72-04279
CATION, FOULING, TOXICITY, ODOR, OXYGEN SAG, TURBIDITY, WATER QUAL W71-09958
TRIENTS, MESTROPHY, OLIGOTROPHY, OXYGEN SAG, AQUATIC LIFE, BIOASSA W73-00002
TA, WATER TEMPERATURE, DISSOLVED OXYGEN.: / CONCENTRATION, MINNESO W72-07145
CES, WATER LAW, ALGAE, DISSOLVED OXYGEN.: /L, WATER POLLUTION SOUR W71-01192
SIS, ALGAE, COLIFORMS, DISSOLVED OXYGEN.: /, SAMPLING, WATER ANALY W70-08248
SIS, ALGAE, COLIFORMS, DISSOLVED OXYGEN.: /, SAMPLING, WATER ANALY W70-08249
, PERMEABILITY, ALGAE, DISSOLVED OXYGEN.: /OUNDWATER RELATIONSHIPS W71-12410
ACTERISTICS, GENETICS, DISSOLVED OXYGEN.: /IC FUNGI, ACQUIFER CHAR W72-11186
LORELLA, *LIGHT, *PHOTOSYNTHETIC OXYGEN, *ALGAE, CULTURES, VOLUME, W70-05547
CATION, *HYPOLIMNION, *DISSOLVED OXYGEN, *ANAEROBIC CONDITIONS, AL W72-12990
ICAL / *SLUDGE WORMS, *DISSOLVED OXYGEN, *EUTROPHICATION, *BIOCHEM W70-09189
UTION, / *CYANOPHYTA, *DISSOLVED OXYGEN, *EUTROPHICATION, *DISTRIB W73-00236
BENTHIC FLORA, ALGA/ *DISSOLVED OXYGEN, *HYPOLIMNION, *LAKE ERIE, W72-13659
EDIMENT-WATER INTERF/ *DISSOLVED OXYGEN, *LAKE ERIE, *BACTERIA, *S W72-12996
HYPOLIMNION, MATHEMA/ *DISSOLVED OXYGEN, *NUTRIENTS, *LAKE ERIE, * W72-12997
TRATION, HEAVY METALS, DISSOLVED OXYGEN, *PLANT GROWTH, LIGHT INTE W72-14690
STES, *WATER QUALITY, *DISSOLVED OXYGEN, *PHYTOPLANKTON, ESTUARIES W71-11876
NTATION/ *ALGAE, *PHOTOSYNTHETIC OXYGEN, *PHOTOSYNTHESIS, INSTRUME W71-10791
ASS, RESPIRATION, PHOTOSYNTHETIC OXYGEN, *TERTIARY TREATMENT.: /OM W72-11100
GAE, *OXIDATION, PHOTOSYNTHESIS, OXYGEN, *THERMODYNAMICS, ANALYTIC W72-02363
ALGAE, PHOTOSYNTHESIS, DISSOLVED OXYGEN, AEROBIC CONDITIONS, ANAER W72-11642
Y EQUIPMENT, AERATION, DISSOLVED OXYGEN, AEROBIC CONDITIONS, BIOMA W72-11100
SLUDGE, CHEMICAL OXYGEN DEMAND, OXYGEN, AERATION, DISSOLVED OXYGE W71-13334
OLIFORMS, TEMPERATURE, DISSOLVED OXYGEN, AERATION, OXYGENATION, NU W71-07113
LECTRICAL CONDUCTANCE, DISSOLVED OXYGEN, AIR POLLUTION, CHEMICAL A W71-01489
, WASTE WATER, RUNOFF, DISSOLVED OXYGEN, ALGAE, NUTRIENTS, BACTERI W71-09788
PORATION, RE-AERATION, DISSOLVED OXYGEN, ALGAE, ANAEROBIC CONDITIO W71-07099
EMPERATURE, NUTRIENTS, DISSOLVED OXYGEN, ALGAE, SLUDGE, OXYGENATIO W71-07109
TIVITY, PHOTOSYNTHESIS, BIOMASS, OXYGEN, ALGAE, COMPARATIVE PRODUC W71-09165
ERATURE, SALT BALANCE, DISSOLVED OXYGEN, ALGAE, BACTERIA, ODOR, IN W72-02274
HEMICAL OXYGEN DEMAND, DISSOLVED OXYGEN, ALGAE, THERMAL STRATIFICA W72-03753
ANIC LOADS, NUTRIENTS, DISSOLVED OXYGEN, AQUATIC LIFE, MUNICIPAL W W71-11017
TES, SLUDGE TREATMENT, DISSOLVED OXYGEN, BACTERIA, FUNGI, PROTOZOA W72-09675
ON, LIGHT PENETRATION, DISSOLVED OXYGEN, BACTERIA, THERMOCLINE, SE W72-12995
GEN ION CONCENTRATION, DISSOLVED OXYGEN, BIOCHEMICAL OXYGEN DEMAND W72-13508
WS, WATER TEMPERATURE, DISSOLVED OXYGEN, BIOINDICATORS, E. COLI, A W71-07417
SYNTHESIS, *CULTURES, *NITROGEN, OXYGEN, BIOCHEMISTRY, INHIBITORS, W72-07941

SOPHYTA, EUGLENOPHYTA, DISSOLVED OXYGEN, CARBON, LAKE MORPHOMETRY, W72-08048
*CYANOPHYTA, NITROGEN FIXATION, OXYGEN, CARBON DIOXIDE, BIOCHEMIS W72-08584
TRATES, MODEL STUDIES, DISSOLVED OXYGEN, CARBON, DISTRIBUTION, PHO W72-08459
ROWTH RATES, BIOASSAY, DISSOLVED OXYGEN, CARBON DIOXIDE, AMMONIA, W72-06046
GEN ION CONCENTRATION, DISSOLVED OXYGEN, CARBON DIOXIDE, AMMONIA, W72-01363
TER, MICROENVIRONMENT, LITTORAL, OXYGEN, CHEMICAL REACTIONS, HYDRO W71-08026
BIODEGRADATION, ORGANIC MATTER, OXYGEN, CHEMICAL OXYGEN DEMAND, B W71-04518
STORAGE, IMPOUNDMENTS, DISSOLVED OXYGEN, CHLORIDES, INVERTEBRATES, W72-14280
QUALITY, FOOD CHAINS, DISSOLVED OXYGEN, COPPER SULFATE, HERBICIDE W72-04759
ON, WATER TEMPERATURE, DISSOLVED OXYGEN, COLIFORMS, RECREATION, OD W71-11914
FICATION, TEMPERATURE, DISSOLVED OXYGEN, COMPUTERS, ALGAE, AQUATIC W72-14405
MPERATURE, ALKALINITY, DISSOLVED OXYGEN, CONDUCTIVITY, HYDROGEN IO W72-14728
YTOPLANKTON, TOXICITY, DISSOLVED OXYGEN, CULTRUES, ENDRIN, DIELDRI W71-10096
SCENEDESMUS, EUGLENA, DISSOLVED OXYGEN, DIURNAL DISTRIBUTION, AQU W71-11519
FF, *FISHKILL, CATTLE, DISSOLVED OXYGEN, DIVERSION STRUCTURES, SED W72-08401
REATMENT, *SEWAGE LAGOONS, OHIO, OXYGEN, DISSOLVED OXYGEN, LIGHT, W72-05459
LUDGE, ALGAE, DIPTERA, DISSOLVED OXYGEN, DREDGING, BOTTOM SEDIMENT W72-03556
OOD CHAINS, FISH FOOD ORGANISMS, OXYGEN, ECOSYSTEMS.: /ESOURCES, F W71-13246
DIATOMACEOUS EARTH, FOOD CHAINS, OXYGEN, ECOLOGY, RHODOPHYTA, PHAE W71-09157
HEMICAL OXYGEN DEMAND, DISSOLVED OXYGEN, ESTUARIES, WATER POLLUTIO W71-05390
NTRATION, TEMPERATURE, DISSOLVED OXYGEN, EUTROPHICATION, OLIGOTROP W72-03224
INS), TROPHIC LEVELS, TURBIDITY, OXYGEN, FISH, RECREATION, INVERTE W71-11009
, NITRATES, POTASSIUM, DISSOLVED OXYGEN, FORECASTING, FLUORIDES, O W72-04260
E CAPACITY, OXIDATION, DISSOLVED OXYGEN, HYDROGEN ION CONCENTRATIO W72-04787
AMS, LABORATORY TESTS, DISSOLVED OXYGEN, HYDROGEN ION CONCENTRATIO W71-04518
HERMAL STRATIFICATION, DISSOLVED OXYGEN, HYPOLIMNION.: /, ALGAE, T W71-03028
ZOOPLANKTON, SALMONIDS, SEASON, OXYGEN, HYDROGEN SULFIDE, DISTRIB W71-00117
AND, OXYGEN, AERATION, DISSOLVED OXYGEN, HYDROGEN ION CONCENTRATIO W71-13334
DIOXIDE, TEMPERATURE, DISSOLVED OXYGEN, HYDROGEN ION CONCENTRATIO W72-08049
TER POLLUTION SOURCES, DISSOLVED OXYGEN, LAKE MORPHOLOGY, AQUATIC W71-11949
PESTICIDES, TOXICITY, DISSOLVED OXYGEN, LAKE MICHIGAN, ALGAE.: /, W71-12092
ULTRAVIOLET RADIATION, DISSOLVED OXYGEN, LAKES, PHYTOPLANKTON, NUT W72-01780
AE, SOUTHWEST U.S., TEMPERATURE, OXYGEN, LIGHT INTENSITY, NUTRIENT W72-10613
RBON DIOXIDE, LININGS, DISSOLVED OXYGEN, LIGHT INTENSITY, DESIGN C W72-04789
LAGOONS, OHIO, OXYGEN, DISSOLVED OXYGEN, LIGHT, DIFFUSIVITY, CARBO W72-05459
OMONAS, CYANOPHYTA, CHLOROPHYTA, OXYGEN, LIGHT, CHRYSOPHYTA, OCHRO W71-09172
GHT INTENSITY, MIXING, DISSOLVED OXYGEN, METABOLISM, CARBON DIOXID W72-04983
MICAL OXYGEN DEMAND, SATURATION, OXYGEN, METABOLISM, NITROGEN, PHO W71-13553
ERATURE, CARBON, DIATOMS, LIGHT, OXYGEN, MINNESOTA.: /FFECTS, TEMP W71-08670
GROWTH, OXYGEN, MONITORS.: W71-10791
TURBIDITY, NUTRIENTS, DISSOLVED OXYGEN, NITROGEN, PHOSPHORUS, WAT W71-06444
ROGEN ION CONCENTRATION, CARBON, OXYGEN, NITROGEN, HYDROGEN, PRIMA W72-06274
E, SAMPLING, SALINITY, DISSOLVED OXYGEN, NUTRIENTS, CARBON RADIOIS W71-07875
TION, DYNAMICS, LAKES, DISSOLVED OXYGEN, NUTRIENTS, CYANOPHYTA, MI W71-03034
EFFECTS, STRIPED BASS, DISSOLVED OXYGEN, OIL, BRIDGES, BUOYS, HARB W73-02029
FISHKILL, LIMNOLOGY, NUTRIENTS, OXYGEN, OXYGEN REQUIREMENTS, OXYG W72-04279
EDIMENTS, *LAKE ERIE, *DISSOLVED OXYGEN, OXYGEN DEMAND, HYPOLIMNIO W72-12994
, *OXIDATION, AMMONIA, NITRITES, OXYGEN, PHYTOPLANKTON, BIOCHEMICA W71-05390
VOIRS, PHOTOSYNTHESIS, DISSOLVED OXYGEN, PHOSPHATES, WATER YIELD I W71-01491
OSYSTEM, *ALGAE, *PHYTOPLANKTON, OXYGEN, PHOTOSYNTHESIS, RESPIRATI W73-01632
HYTOPLANKTON, *ALGAE, *DISSOLVED OXYGEN, PHOTOSYNTHESIS, TURBIDITY W73-00256
SIS, AQUATIC ALGAE, CHLOROPHYLL, OXYGEN, RESPIRATION, CHLORINATED W73-00281
LAGOONS, *OXYGENATION, DISSOLVED OXYGEN, SATURATION, ALGAE, METHAN W71-07106
ATION, PHOTOSYNTHESIS, DISSOLVED OXYGEN, SEDIMENTATION, SAMPLING, W71-07097
TIFICATION, ICE, EUTROPHICATION, OXYGEN, SEASONAL.: / THERMAL STRA W72-04855
CARBON, FERTILIZATION, DISSOLVED OXYGEN, SEWAGE, HYDROGEN ION CONC W71-12188
NCENTRATION, PROTOZOA, DISSOLVED OXYGEN, SLUDGE, WASTE WATER TREAT W72-10233
ONSUMPTIVE USE, WATER POLLUTION, OXYGEN, SOIL PROPERTIES, IDAHO.: / W72-14878
GEN ION CONCENTRATION, DISSOLVED OXYGEN, STRATIFICATION, SAMPLING, W71-07382
S, METHODOLOGY, REEFS, SAMPLING, OXYGEN, TEMPERATURE, LIGHT, WAVES W72-06283
N EFFECTS, INTERFACES, DISSOLVED OXYGEN, TEMPERATURE, ORGANOPHOSPH W71-10645
OLIMNION, *LAKE ERIE, *DISSOLVED OXYGEN, TEMPERATURE, CURRENTS(WAT W72-12992
CAL ANALYSIS, BIOMASS, DISSOLVED OXYGEN, THERMOCLINE, HYDROGEN ION W72-12941
ES, SALINITY, WATER TEMPERATURE, OXYGEN, TURBIDITY, BENTHIC FAUNA, W73-02029
, CHLOROPHYTA, LABORATORY TESTS, OXYGEN, WATER POLLUTION EFFECTS.: W72-14314
IC ALGAE, OLIGOTROPHY, DISSOLVED OXYGEN, WATER TEMPERATURE, HYDROG W72-13871
LGAE, HYDROLOGIC DATA, DISSOLVED OXYGEN, WEATHER, RESERVOIR OPERAT W72-05282
TER HYACINTH, LIGHT PENETRATION, OXYGEN,: /ALCIUM, SCENEDESMUS, WA W72-04269
HEMICAL OXYGEN DEMAND, DISSOLVED OXYGEN: /ICAL OXYGEN DEMAND, BIOC W71-11006
GEN, CULTRUES, ENDRIN, DIELDRIN, OXYGENATION, CARBAMATE PESTICIDES W71-10096
PHOTOSYNTHESIS, LIGHT INTENSITY, OXYGENATION, DENITRIFICATION, OXI W71-13326
CTION, ULTRAVIOLET LIGHT, ALGAE, OXYGENATION, BIOCHEMICAL OXYGEN D W72-00042
NTHESIS, OXIDATION, RESPIRATION, OXYGENATION, TEMPERATURE, LIGHT I W72-04983
URE, DISSOLVED OXYGEN, AERATION, OXYGENATION, NUTRIENTS, SEDIMENTA W71-07113
DISSOLVED OXYGEN, ALGAE, SLUDGE, OXYGENATION, ANAEROBIC CONDITIONS W71-07109
ZONES, TEMPERATURE, TURBULENCE, OXYGENATION, ALGAE, PHOTOSYNTHESI W71-06033
ESPIRATORY RATES, PHOTOSYNTHETIC OXYGENATION, ACTIVATED ALGAE.: /R W71-04079
OPHYLA, EUGLENA OBLONGA, EUGLENA OXYURIS, EUGLENA POLYMORPHA, EUGL W73-00918
, EMERITA ANALOGA, DIASTYLOPSIS, OXYUROSTYLIS, HEMILAMPROPS, LAMPR W72-09092
AGNALIS, REFRACTODES BRASILIANA, OYEMONAS PULSULAE, REFRACTOMONAS W72-09119
MOSQUITOES, FOOD CHAIN, HERRING, OYSTERS, CLAMS, COLUMBIA RIVER, C W71-07417
*SEA WATER, BIOASSAY, COPEPODS, OYSTERS, DINOFLAGELLATES, DIATOMS W72-04708
L, MARINE ALGAE, CLAMS, MUSSELS, OYSTERS, EUTROPHICATION, NITROGEN W72-07715
S, OILY WATER, ALGAE, NEMATODES, OYSTERS, MUSSELS, PROTOZOA, MANGR W72-07911
NVIRONMENT, REGULATION, FALLOUT, OYSTERS, MUSSELS, PLANKTON, MARIN W72-00959
TY, ANIMAL GROWTH, PLANT GROWTH, OYSTERS, SHELLFISH, MUSSELS, SHEL W72-09378
, GAS CHROMATOGRAPHY, SEDIMENTS, OYSTERS, SHRIMP, CRABS, WATER POL W71-07412
E, FISH, SHELLFISH, CRUSTACEANS, OYSTERS, SEASONAL.: /, AQUICULTUR W70-08832
TER POLLUTION SOURCES, PLANKTON, OYSTERS, SHEL: /S, HERBICIDES, WA W72-13347
HORUS, *NITROGEN, COST ANALYSIS, OZONE, DENITRIFICATION, POLLUTION W73-02187

*CHLORELLA PYRENOIDOSA, LINDANE,	P P'-DDE, CLOPHEN A 50, ELECTRON	W72-04134
HLORELLA PYRENOIDOSA, LINDANE, P	P'-DDE, CLOPHEN A 50, ELECTRON CA	W72-04134
A OLIVACEOIDES, WHITE CLAY CREEK(	PA.).: GOMPHONEM	W73-00242
E, WHIPPLE DAM, STONE VALLEY DAM(	PA.), ATRAZINE, BIOTYPES, POUR PL	W72-04743
ES, DETRIVORES, LINESVILLE CREEK(	PA).: /AUTUMNAL COMMUNITY, CALORI	W71-10064
VERSITY, DIRECT EF/ RIDLEY CREEK(	PA), LITITZ CREEK(PA), SPECIES DI	W72-08579
RIDLEY CREEK(PA), LITITZ CREEK(	PA), SPECIES DIVERSITY, DIRECT EF	W72-08579
IDES, SELENASTRUM CAPRICORNUTUM,	PAAP BATCH TEST, CULTURE MEDIA.: /	W73-02099
RIVER(CALIF), SEARSVILLE LA/ SAN	PABLO ESTUARY(CALIF), SACRAMENTO	W70-10405
ITRITE, SPHAEROTILUS, NUTRIENTS,	PACIFIC OCEAN, NITRATE, CHEMICAL	W72-05421
I, ABSORPTION, TROPICAL REGIONS,	PACIFIC OCEAN, COLUMNS, BOTTOM SE	W71-08042
CYCLOTELLA NANA, LEPTOCYLINDRUS,	PACIFIC OCEAN, ALEUTIAN ISLANDS,	W71-11008
TION TECHNIQUES, ATLANTIC OCEAN,	PACIFIC OCEAN, GULF OF MEXICO.: /	W71-11851
S, MAPS, MAPPING, PHYTOPLANKTON,	PACIFIC OCEAN, BOTTOM SEDIMENTS,	W72-03567
PODA, RADIOISOTOPES, ABSORPTION,	PACIFIC OCEAN, FOOD CHA: /, AMPHI	W72-03347
BUTION PATTERNS, ATLANTIC OCEAN,	PACIFIC OCEAN, TROPICAL REGIONS,	W72-03543
TRIPHOTURUS MEXICANUS, EUPHAUSIA	PACIFICA, ENGRAULIS MORDAX.: /N,	W73-02105
*GRIFFITHSIA	PACIFICA, *REGENERATION.:	W73-01622
, METAZOA, CALANUS HELGOLANDICUS(	PACIFICUS), THERMISTORS, DATA INT	W72-07892
TER POLLUTION, DES MOINES RIVER,	PACKING PLANTS.: INDUSTRIAL WA	W71-06445
CHLORELLA PYRENOIDOSA, NITZSCHIA	PALEA, SKELETONEMA COSTATUM.:	W71-03027
ISOTOPES, ANALYTICAL TECHNIQUES,	PALEOZOIC ERA, WATER CHEMISTRY.: /	W72-02073
ONAS ISABELIENSIS, CHLAMYDOMONAS	PALLIDOSTIGMATICA, CHLAMYDOMONAS	W72-12632
PERIDINIUM DEPRESSUM, PERIDINIUM	PALLIDUM, CERATIUM MASSILENSE, CE	W71-07875
, PROPIANATE, PROPANOL, LAURATE,	PALM: /ACETATE, ETHANOL, BENZOATE	W72-05421
GIGARTINA MAMILLOSA, RHODYMENIA	PALMATA.: /, ASCOPHYLLUM NODOSUM,	W71-11515
ARDHII, ULVA LACTUCA, RHODYMENIS	PALMATA, POLYSIPHONIA LANOSA, POL	W72-04748
UM, BRYOPSIS, CAULERPA, ACROPORA	PALMATA, PENICILLUS LAMOUROUXII.:	W72-14294
DINIUM SINICU/ KOLBEANA OVOIDEA,	PALMERIAMONAS PLANCTONICA, MANCHU	W72-07683
RPA SP., CAULERPICIN, CAULERPIN,	PALMITIC ACID, BETA-SITOSTEROL, T	W71-08597
HONIA, ULVA LACTUCA, ZANICHELLIA	PALUSTRIS, CODIUM, FUSCUS, MACROP	W73-00367
SS), CAPE MAY(/ *COASTAL WATERS,	PANAMA, CALLAO(PERU), CAPE COD(MA	W72-08056
CAUDATA, CHLAMYDOMONAS PERTUSA,	PANDORINA MORUM.: /SA, CHROOMONAS	W72-13816
RS, AUTOINHIBITORS, VOLVOCACEAE,	PANDORINA MORUM, PLATYDORINA CAUD	W71-13793
DOMONAS SPP, CARTERIA FRITSCHII,	PANDORINA MORUM, SPHAEROCYSTIS SC	W72-09111
CYSTIS, OSCILLATORIA, SPIRULINA,	PANDORINA, PHOCUS, STAURASTRUM, T	W72-07661
.: *ALEXANDRIA(EGYPT),	PANDORINA, SOLUBLE ORGANIC SOLIDS	W70-06619
, CODIUM CAROLINI/ TREMATOCARPUS	PAPENFUSSII, GLOIODERMA ATLANTICA	W73-00284
, ORGANIC CARBON, CULTURE MEDIA,	PAPER CHROMATOGRAPHY, INFRARED SP	W73-00379
LATE MATTER, FATE OF POLLUTANTS,	PAPER CHROMATOGRAPHY, INFRARED SP	W73-01066
ISOTOPES, *BOSSIELLA ORBIGNIANA,	PAPER CHROMATOGRAPHY, SCINTILLATI	W72-12634
ACTIVITY, *SKELLYSOLVE, *FILTER	PAPER DISC DIFFUSION METHOD, *ZON	W72-08586
*WISCONSIN, PHOSPHORUS, PULP AND	PAPER INDUSTRY, ALGAE, AQUATIC WE	W70-05103
SII, LAURENCIA OBTUSA, LAURENCIA	PAPILLOSA, COELOTHRIX IRREGULARIS	W73-00838
, GOMPHONEMA PARVULUM, BACILLARI	PARADOXA, SPHAEROTILUS NAT: /TATA	W72-10883
TERIONELLA JAPONICA, STAURASTRUM	PARADOXIUM, CULTURE MEDIA, ZINC S	W72-12739
ULTURING TECHNIQUES, STAURASTRUM	PARADOXUM, STAURASTRUM SEBALDII V	W72-12736
HODO/ *AGARS, EPIPHYTES, CANDIDA	PARAPSILOSIS, CHONDRUS CRISPUS, R	W72-04748
ABLE SOLIDS, HATCHERY EFFLUENTS,	PARASITES, JORDAN RIVER(MICH), SA	W71-11006
TEBRATES, FUNGI, ENZYMES, ANIMAL	PARASITES, VIRUSES, MOLLUSKS, PRO	W73-01676
FISH EGGS, FISH, BACTERIA, FISH	PARASITES, ANTIBIOTICS(PESTICIDES	W72-08142
CONTROL, IRRIGATION, PREDATION,	PARASITISM, INSECT RESISTANCE: /T	W72-00846
IGOCLADUS/ *YELLOWSTONE NATIONAL	PARK(WYO), BLUE-GREEN ALGAE, MAST	W72-08508
A, PHORMIDIUM, MAST/ YELLOWSTONE	PARK, SYNECHOCOCCUS, FLEXIBACTERI	W72-03557
CAL OXYGEN DEMAND, WATER SPORTS,	PARKS, COST ANALYSIS, MICHIGAN, W	W72-00042
HERMAL WATER, *WYOMING, NATIONAL	PARKS, HOT SPRINGS, WATER TEMPERA	W72-08508
MYRTILLUS, DESCHAMPSIA FLEXUOSA,	PARMELIA PHYSODES, HYPNUM CUPRESS	W72-14303
ONTROL/ *EUTROPHICATION, *SOCIAL	PARTICIPATION, *WATER POLLUTION C	W73-02479
UTION PATTERNS, CURRENTS(WATER),	PARTICLE SIZE, SEDIMENTS, SEPARAT	W73-00367
RIFICATION, CALIFORNIA, FILTERS,	PARTICLE SIZE, ANAEROBIC CONDITIO	W71-08223
ARINE ALGAE, PHAEPHYTA, AQUARIA,	PARTICLE SIZE, COLLOIDS, WATER PO	W71-09193
INTERGREEN LAKE(MICH), CARBONATE	PARTICLES, PHOSPHORUS CYCLE.: / W	W72-08048
NITROGEN, SUBSTRATE UTILIZATION,	PARTICULATE MATTER, FATE OF POLLU	W73-01066
, FISH DIET, GAS CHROMATOGRAPHY,	PARTICULATE MATTER, DETRITUS, ORG	W73-02105
NAGE, LAKE METABOLISM, MOBILITY,	PARTICULATE FORM, FEEDLOTS, NUTRI	W71-06443
BBARD BROOK EXPERIMENTAL FOREST,	PARTICULATE MATTER, NITROSOMONAS,	W71-01489
LUE HERON, STERNA HIRUNDO, RATS,	PARTRIDGE, DDE, KIDNEYS, LIVER, B	W72-05952
NAVICULA TRIPUNCTATA, GOMPHONEMA	PARVULUM, BACILLARI PARADOXA, SPH	W72-10883
AN, ALEUTIAN ISLANDS, PRYMNESIUM	PARVUM.: /OCYLINDRUS, PACIFIC OCE	W71-11008
*MARINE PHYTOPLANKTON(PRYMNESIUM	PARVUM)(CHROOMONAS SALINA), GLYCE	W70-07842
EA, BACCILARIOPHYCEA, PRYMNESIUM	PARVUM, PHOTOAUTOTROPHS, CHROOMON	W70-07280
M RADIOISOTOPES, CLADOPHORA SPP,	PASPALUM DISTICHUM, POTAMOGETON P	W73-01673
TYPHOID, COLIPHAGE, IMMUNOLOGY,	PASTEURELLA PSEUDOTUBERCULOSIS, P	W72-13800
, CASTLE LAKE(CALIF), DIVERSITY,	PATCHES.: /IUM, LAKE TAHOE(CALIF)	W71-09171
S, ECOLOGY, FISH FOOD ORGANISMS,	PATH OF POLLUTANTS, BIOASSAY, TOX	W71-08597
EYS, PESTICIDES, WASTE DISPOSAL,	PATH OF POLLUTANTS, WATER QUALITY	W71-05553
ONTROL, SURVEYS, INVESTIGATIONS,	PATH OF POLLUTANTS, SAMPLING, WAT	W70-08249
ONTROL, SURVEYS, INVESTIGATIONS,	PATH OF POLLUTANTS, SAMPLING, WAT	W70-08248
WASTES, PRODUCTIVITY, DIELDRIN,	PATH OF POLLUTANTS, REVIEWS, FOOD	W72-06340
RY, GEOCHEMISTRY, RADIOISOTOPES,	PATH OF POLLUTANTS, PHYSIOLOGICAL	W72-06313
SON RIVER, INDICATORS, DELAWARE,	PATH OF POLLUTANTS, HOSTS, PENNSY	W72-04736
WS, *DATA COLLECTIONS, TOXICITY,	PATH OF POLLUTANTS, *METALS, TRAC	W72-12190
Y, WATER POLLUTION, FOOD CHAINS,	PATH OF POLLUTANTS, BIOASSAY, FIS	W72-12257
ER, DETRITUS, ORGANIC COMPOUNDS,	PATH OF POLL: /, PARTICULATE MATT	W73-02105
ICIDES, DDD, PESTICIDE TOXICITY,	PATH OF POLLUTANTS, PESTICIDE RES	W73-02280
S, MARINE ALGAE, NUCLEAR WASTES,	PATH OF POLLUTANTS, RADIOECOLOGY,	W73-00828
FOOD CHAINS, AFRICA, EFFLUENTS,	PATH OF POLLUTANTS, NEUTRON ACTIV	W73-00832
R WASTES, *COBALT RADIOISOTOPES,	PATH OF POLLUTANTS, WASTE ASSIMIL	W73-00826
ON, *FOOD CHAINS, PUBLIC HEALTH,	PATH OF POLLUTANTS, RADIOECOLOGY,	W73-00833
T RADIOISOTOPES, NUCLEAR WASTES,	PATH OF POLLUTANTS, ESTUARINE ENV	W72-07826

NG, SURFACE WATERS, PERSISTENCE,	PATH OF POLLUTANTS, OCEANS, ESTUA	W73-00361
N, PHOSPHORUS, NITRATES, SILICA,	PATH OF POLLUTANTS, ORGANIC COMPO	W73-00384
DIVING, ECOLOGY, STANDING CROPS,	PATH OF POLLUTANTS.: /ING, SCUBA	W73-00366
RTO RICO, ESTUARINE ENVIRONMENT,	PATH OF POLLUTANTS, FALLOUT, NUCL	W73-00821
EAR POWERPLANTS, NUCLEAR WASTES,	PATH OF POLLUTANTS, WATER POLLUTI	W73-00790
, PUBLIC HEALTH, NUCLEAR WASTES,	PATH OF POLLUTANTS, DIFFUSION, CY	W73-00823
, *ZINC, CHELATION, FOOD CHAINS,	PATH OF POLLUTANTS, PUBLIC HEALTH	W73-00802
S, MARINE ALGAE, TRACE ELEMENTS,	PATH OF POLLUTANTS, ABSORPTION, R	W73-00811
LGAE, SEDIMENTS, NUCLEAR WASTES,	PATH OF POLLUTANTS, SEA WATER, WA	W73-00817
, SOUNDS, ESTUARINE ENVIRONMENT,	PATH OF POLLUTANTS, FOOD CHAINS,	W73-00818
OURCES, COLORADO, CATTLE, URINE,	PATH OF POLLUTANTS, FARM WASTES.:	W71-11496
F, SEWAGE, WASTE WATER DISPOSAL,	PATH OF POLLUTANTS, STORM DRAINS,	W71-13466
ABSORPTION, *SNAILS, GASTROPODS,	PATH OF POLLUTANTS, NUCLEAR WASTE	W72-10978
TALS, ABSORPTION, BIOINDICATORS,	PATH OF POLLUTANTS, FOOD CHAINS,	W72-10957
ANTS, EFFLUENTS, NUCLEAR WASTES,	PATH OF POLLUTANTS, ASIA, MARINE	W72-10958
NMENTAL EFFECTS, NUCLEAR WASTES,	PATH OF POLLUTANTS, MARINE ALGAE,	W72-10955
RIA, ALGAE, MATHEMATICAL MODELS,	PATH OF POLLUTANTS, TRACERS, ANIM	W72-03348
DIOACTIVITY TECHNIQUES, TRACERS,	PATH OF POLLUTANTS, ON-SITE INVES	W72-00949
NUCLEAR POWERPLANTS, MONITORING,	PATH OF POLLUTANTS, AIR POLLUTION	W72-00959
LLUTION EFFECTS, SURFACE WATERS,	PATH OF POLLUTANTS, MARINE ALGAE,	W72-00940
ENTS, ALGAE, WEEDS, TASTE, ODOR,	PATHOGENIC BACTERIA, ORGANIC WAST	W71-11006
IOLOGY, SYSTEMATICS, POLLUTANTS,	PATHOGENIC BACTERIA, COLIFORMS, T	W72-13800
A, *BACTERICIDES, GROWTH STAGES,	PATHOGENIC BACTERIA, CYANOPHYTA,	W71-05155
OSPHORUS, EUTROPHICATION, ALGAE,	PATHOGENIC BACTERIA, EFFLUENT, IR	W71-08214
CYCLING FOOD, PROCESSING WASTES,	PATHOGENS, BIOENVIRONMENTAL CONTR	W72-00846
SOTA, BACTERIA, FUNGI, PROTOZOA,	PATHOGENS, MICROORGANISMS, ALGAL	W72-08586
ATORY TESTS, ALGAL TOXINS, HUMAN	PATHOLOGY.: *CYANOPHYTA, *LABOR	W72-04259
ERTIES, MINNESOTA, LIPIDS, HUMAN	PATHOLOGY, DISEASES.: /MICAL PROP	W72-04275
ZYMES, CELLULOSE, VIRUSES, HUMAN	PATHOLOGY, SYSTEMATICS.: /URE, EN	W72-05457
TA, PHAEOPHYTA, CHLORELLA, PLANT	PATHOLOGY, INDUSTRIES, IONS, CYAN	W71-09157
GMENTS, PLANT PHYSIOLOGY, *PLANT	PATHOLOGY, PLANT GROWTH REGULATOR	W70-09429
NTAL STATEMENT, CRITICAL NUCLIDE	PATHWAY, ABALONE.: ENVIRONME	W72-00959
PHISM, *VALVE MORPHOLOGY, *VALVE	PATTERN CHARACTERISTICS, CLONES.:	W73-02548
US CULTURES, SYNCHRON/ *DIVISION	PATTERNS, *AUTOSPORES, *SYNCHRONO	W72-12724
IATOMS, *CURRENTS, *DISTRIBUTION	PATTERNS, ATLANTIC OCEAN, PACIFIC	W72-03543
IATOMS, *SAMPLING, *DISTRIBUTION	PATTERNS, *PHOSPHATES, *AQUATIC P	W72-03567
L RESOURCES, DEPTH, DISTRIBUTION	PATTERNS, MEASUREMENT, PHOTOSYNTH	W71-13252
A/ *PHYTOPLANKTON, *DISTRIBUTION	PATTERNS, *SYSTEMATICS, *SALINE L	W72-09111
CUBA DIVING, DEPTH, DISTRIBUTION	PATTERNS, AERIAL PHOTOGRAPHY, REM	W72-09120
COSCINODISCUS, COPEPO/ *FEEDING	PATTERNS, *CALANUS HELGOLANDICUS,	W72-09248
LGAE, *FRESH WATER, DISTRIBUTION	PATTERNS, BEHAVIOR.: *A	W71-07360
S, *PHYTOPLANKTON, *DISTRIBUTION	PATTERNS, *NORTH CAROLINA, *COMPU	W71-07337
PHYTA, AL/ *ALGAE, *DISTRIBUTION	PATTERNS, *THERMAL SPRINGS, CYANO	W72-05610
STICIDES, BIOASSAY, DISTRIBUTION	PATTERNS, CHROMATOGRAPHY, GROWTH	W72-04134
USES, *CYANOPHYTA, *DISTRIBUTION	PATTERNS, *WATER ANALYSIS, *BIOIN	W72-04736
OSCOPY, PYRROPHYTA, DISTRIBUTION	PATTERNS, WATER TEMPERATURE, WATE	W72-07663
FFECTS, ESTUARIES, *DISTRIBUTION	PATTERNS, *AQUATIC ALGAE, STATIST	W72-08141
, LAKE MORPHOMETRY, DISTRIBUTION	PATTERNS, AQUATIC ALGAE, LIMNOLOG	W72-07145
ON, ORGANIC MATTER, DISTRIBUTION	PATTERNS, THERMAL PROPERTIES, THE	W73-00431
AL DISTRIBUTION, / *DISTRIBUTION	PATTERNS, *SYSTEMATICS, *ECOLOGIC	W73-00428
AGELLATES, ECOLOGY, DISTRIBUTION	PATTERNS, MARINE ANIMALS, SALINIT	W73-00296
Y, SAMPLING, CORES, DISTRIBUTION	PATTERNS, CURRENTS(WATER), PARTIC	W73-00367
LGAE, *RHODOPHYTA/ *DISTRIBUTION	PATTERNS, *CHLOROPHYTA, *MARINE A	W73-00284
ALGAE/ *ANTARCTIC, *DISTRIBUTION	PATTERNS, *SYSTEMATICS, *AQUATIC	W73-00854
HEMICAL PROPERTIES, DISTRIBUTION	PATTERNS, ECOLOGICAL DISTRIBUTION	W73-00918
TICS, / *SEASONAL, *DISTRIBUTION	PATTERNS, *MASSACHUSETTS, SYSTEMA	W73-01449
GAE, *SYSTEMATICS, *DISTRIBUTION	PATTERNS, LAKES, AQUATIC PLANTS,	W73-01094
ENTUCKY), CAYUGA LAKE(NEW YORK),	PATUXENT ESTUARY, BISCAYNE BAY(FL	W71-11517
R / CHALK POINT, COLUMBIA RIVER,	PATUXENT RIVER, CONTRA COSTA POWE	W72-07907
CAYNE BAY, WHITE RIVER(INDIANA),	PATUXENT RIVER(CANADA), ONTARIO.:	W71-07417
TROPICA, STIGMOBODO BRASILIANA,	PAVLOVIAMONAS FRUTICOSA, NODAMAST	W72-07683
CIOLA, NITZSCHIA SPP, BACILLARIA	PAXILLIFER,: /IGMA, GYROSIGMA FAS	W73-00853
LIMITOS BAY, LAKE PONTCHARTRAIN,	PEARL RIVER, LAKE BORGNE, SUBSTRA	W73-02029
LIS, SYNEDRA FASCIULATA, EUNOTIA	PECTINALIS, LITHOPHYTES, ENTEROMO	W72-08141
MA, ACHNANTHES FLEXELLA, EUNOTIA	PECTINALIS, DESMIDS.: / MINUTISSI	W71-11029
MOGETON PERFOLIATUS, POTAMOGETON	PECTINATUS, MYRIOPHYLLUM SPECATUM	W71-11012
PASPALUM DISTICHUM, POTAMOGETON	PECTINATUS, ELEOCHARIS MONTEVIDEN	W73-01673
DON, FRESHWATER ALGAE, PICLORAM,	PEDIASTRUM, 4-AMINO-3,5,6-TRICHLO	W72-08440
PHAERIUM, DINOBRYON, FRAGILARIA,	PEDIASTRUM, STAURASTRUM, TABELLAR	W72-10076
MPHORA OVALIS, AMPHORA OVALIS V.	PEDICULUS, COCCONEIS PEDICULUS, C	W72-13142
A OVALIS V. PEDICULUS, COCCONEIS	PEDICULUS, COCCONEIS PLACENTULA,	W72-13142
ABALONE, TUNICATES, ARTHRO/ SAN	PEDRO(CALIF), EPIBENTHIC SPECIES,	W71-01475
ISPUS, COD, FLOUNDER, CRAB EGGS,	PELAGIC EGGS, GADUS MORHUA, GRASS	W72-08142
*LAKE ERIE CENTRAL BASIN, POINTE	PELEE(ONTARIO), OXYGEN DEPLETION.	W72-12992
ANTHES CF. BIRGIANI, AMPHIPLEURA	PELLUCIDA, AMPHORA OVALIS, AMPHOR	W72-13142
CIZERS, PHTHALATE/ BDELLOVIBRIO,	PELODICTYON, POLIOVIRUSES, PLASTI	W72-09675
SWEDEN, COLLEMA,	PELTIGERA, NOSTOC.:	W73-02471
SHING, BIOLOGICAL MAGNIFICATION,	PELVETIA CANALICULATA, FUCUS SERR	W71-11515
NAEUS DUORARUM, PINAEUS AZTECUS,	PENAEUS SETIFERUS, MERCENARIA MER	W73-02029
S, WATER QUALITY, *ALGAE, *LIGHT	PENETRATION, *MACROPHYTES, *AQUAT	W73-02349
ROPHICATION, LAKE ONTARIO, LIGHT	PENETRATION, PHYSICOCHEMICAL PROP	W73-01615
EASONAL, ZINC, ABSORPTION, LIGHT	PENETRATION, ALKALINITY, HALOGENS	W73-01434
ALGAE, TEMPERATURE, DEPTH, LIGHT	PENETRATION.: /YNTHESIS, SESSILE	W72-07501
YDROGEN ION CONCENTRATION, LIGHT	PENETRATION, PHOTOSYNTHESIS, PHYT	W73-00236
PHOTOSYNTHESIS, TURBIDITY, LIGHT	PENETRATION, IMPOUNDMENTS, SEASON	W73-00256
CYTOLOGICAL STUDIES, ADSORPTION,	PENETRATION, CATION ADSORPTION, C	W73-00823
PHYTON, *ALGAE, *LITTORAL, LIGHT	PENETRATION, BENTHIC FLORA, LAKES	W71-11029
ASS, NITRATES, RESERVOIRS, LIGHT	PENETRATION, CYANOPHYTA, CHRYSOPH	W71-13183
YPOLIMNION, PHYTOPLANKTON, LIGHT	PENETRATION, DISSOLVED OXYGEN, BA	W72-12995
RIE, BENTHIC FLORA, ALGAE, LIGHT	PENETRATION, PLANKTON, NUTRIENTS,	W72-13659

OTOSYNTHESIS, RE-AERATION, LIGHT PENETRATION, NUTRIENTS, PHOSPHORU W71-07098
ENEDESMUS, WATER HYACINTH, LIGHT PENETRATION, OXYGEN,: /ALCIUM, SC W72-04269
SEWAGE, WATER TEMPERATURE, LIGHT PENETRATION, CARBON, LAKE ERIE, R W72-04734
PHOTOSYNTHESIS, SEASONAL, LIGHT PENETRATION, LIGHT INTENSITY, DIS W72-04270
TROPHICATION, TEMPERATURE, LIGHT PENETRATION, AMMONIA, PHOSPHATES, W72-05469
PHOTOSYNTHESIS, EQUATIONS, LIGHT PENETRATION, ALGAE, HYDROGEN ION W72-05459
NS, NUTRIENTS, RECREATION, LIGHT PENETRATION, NEBRASKA, WASTEWATER W72-04759
E, BIOMASS, LIGHT QUALITY, LIGHT PENETRATION, EFFICIENCES, COMPETI W72-04766
L/ EPIPHYTES, CULTURING VESSELS, PENICILLIN, LINDANE, DIADEMA ANTI W72-14294
PASTEURELLA PSEUDOTUBERCULOSIS, PENICILLIN, DRUGS, CORTISONE, HOR W72-13800
SM, TETRACYCLINE, SULFANILAMIDE, PENICILLIN G, POLYMYXIN B, PODOCY W72-10162
ATE UTILIZATION, MONOD EQUATION, PENICILLIN G, TERRAMYCIN, CHLORAM W73-01445
ORHUA, GRASS SHRIMP, GREEN CRAB, PENICILLIUM, STREPTOMYCIN, EPIFLO W72-08142
S, TORULOPSIS SPP., ASPERGILLUS, PENICILLIUM, FUSARIA, VIBRIS.: /N W72-09675
SS, THALASSIA SPP., UDOTEA SPP., PENICILLUS SPP., LAURENCIA SPP., W72-11876
SIS MEMBRANACEA, PENI/ SURVIVAL, PENICILLUS, CHONDRIA, CLADOPHOROP W73-00838
RIA, CLADOPHOROPSIS MEMBRANACEA, PENICILLUS CAPITATUS, CERAMIUM NI W73-00838
MEDA SPP., PENICILLUS CAPITATUS, PENICILLUS DUMENTOSUS, RHIPOCEPHA W72-14294
SIS, CAULERPA, ACROPORA PALMATA, PENICILLUS LAMOUROUXII.: /, BRYOP W72-14294
ENE, SIPHONALES, *HALIMEDA SPP., PENICILLUS CAPITATUS, PENICILLUS W72-14294
, HYPERHALINE, EURYHALIN/ *SINAI PENINSULA, CHLORINITY, METAHALINE W73-00855
A, CYCLOPS NAUPLII, COCCOCHLORIS PENIOCYSTIS, SPIROGYRA, EPHEMEROP W72-07132
US, VALONIA VENTRICOSA, BRYOPSIS PENNATA, SPYRIDIA FILAMENTOSA, DI W73-00838
LYSIS, ALGAE, SPECTROPHOTOMETRY, PENNSYLVANIA, INHIBITORS, ECOLOGY W72-04743
WARE, PATH OF POLLUTANTS, HOSTS, PENNSYLVANIA, NEW YORK, NEW HAMPS W72-04736
XYGEN DEMAND, ALGAE, CONFINEMENT PENS, IMPOUNDMENTS, *WATER POLLUT W72-08401
AIR POLLUTION, ODOR, CONFINEMENT PENS, ORGANIC MATTER, NUTRIENTS, W71-08214
CE, PHOTOSYNTHESIS, DEPTH, LIGHT PENTRATION, RESPRIATION, NUTRIENT W72-08559
COLI, SILICA, *FILTRATION, LIGHT PENTRATION, PHYSICAL PROPERTIES, W72-11833
COMMERSONI, AMEIURUS NEBULOSUS, PERCA FLAVESCENS, AMBLOPLITES RUP W70-08098
ONTOPOREIA AFFINIS, ESOX LUCIUS, PERCA FLUVIATILIS, SALVELINUS ALP W72-12729
LAME EMISSION SPECTROPHOTOMETRY, PERCH LAKE(ONTARIO), ASHING, BIOL W71-11529
UTARIES, FRESHWATER FISH, YELLOW PERCH, SEWAGE EFFLUENTS, OIL SPIL W72-14282
SILTS, SUCKERS, CATFISHES, BASS, PERCHES, PIKES, SAMPLING, SNAILS, W70-08098
ATES, SURFACE RUNOFF, BASE FLOW, PERCOLATION, LEACHING, DRAINAGE, W71-06443
OLIFORMS, CHLORINATION, SEEPAGE, PERCOLATION, EVAPORATION, INSECTS W71-07124
ULFUR, FOULING, ALGAE, SULFATES, PERCOLATION, HERRING, SALMON, TRO W72-05421
CHEMICAL PROPERTIES, LANDFILLS, PERCOLATION, DRAINAGE, RUNOFF, PR W72-12955
POTAMOGETON LUCENS, POTAMOGETON PERFOLIATUS, POTAMOGETON PECTINAT W71-11012
EMICAL OXYGEN DEMAND, CHLORELLA, PERFORMANCE, EFFICIENCIES, LABORA W72-11100
YGEN, BIOCHEMICAL OXYGEN DEMAND, PERFORMANCE, FLOCCULATION, FLOTAT W72-13508
IATION, FLOTATION, FLOCCULATION, PERFORMANCE, EFFICIENCIES, *BIOMA W72-12289
ICROCOCCUS.: PERFUSION UNIT, FLAVOBACTERIUM, M W71-08027
ION, COSCINOSIRA / NEWFOUNDLAND, PERIDINIANS, QUALITATIVE COMPOSIT W72-03543
CHA/ SEA OF JAPAN, *AMURSKY BAY, PERIDINIANS, EUCAMPIA ZOODIACUS, W73-00442
M GRACILIMUM, CRYPTOMONAS EROSA, PERIDINIUM AFRICANUM, NITZCHIA AC W73-00256
ITAMIN B 12, GONYAULAX POLYEDRA, PERIDINIUM DEPRESSUM, METAZOA, CA W72-07892
E, SPIRULINA, GYMNODINIUM BREVE, PERIDINIUM, OSCILLATORIA RUBECENS W73-00840
NAVICULA, PINNULARIA, SURIRELLA, PERIDINIUM, SENECA RIVER, NINEMIL W72-09111
*ISRAEL, *LAKE KINNERET(ISRAEL), PERIDINIUM WESTI, OSCILLATORIA PR W71-13187
NIA ALATA, PERIDINIUM DEPRESSUM, PERIDINIUM PALLIDUM, CERATIUM MAS W71-07875
CHAETOCEROS, RHIZOSOLENIA ALATA, PERIDINIUM DEPRESSUM, PERIDINIUM W71-07875
INJECTION PUMP, ALUM/ *DETENTION PERIOD, FLUSHING OPERATION, ACID W72-02850
LGAE, BACTERIA, LITTORAL, LAKES, PERIPHYTON, ORGANIC MATTER, BIOMA W72-03216
S, PHYTOPLANKTON, BENTHIC FAUNA, PERIPHYTON, AQUATIC ALGAE, PLANTS W72-00939
DES, HYDROGEN ION CONCENTRATION, PERIPHYTON, WATER TEMPERATURE.: / W72-05610
NAS, PHYTOPLANKTON, ZOOPLANKTON, PERIPHYTON.: /OTIFERS, CHLAMYDOMO W70-06655
IRONMENTAL EFFECTS, MARINE FISH, PERIPHYTON, ON-SITE DATA COLLECTI W73-00932
ION, LIFE CYCLES, AQUATIC ALGAE, PERIPHYTON.: /RPHOLOGY, REPRODUCT W72-07141
LITTORAL, SANDS, SHALLOW WATER, PERIPHYTON, SEDIMENTS, DIATOMS, L W72-07501
OLAR RADIATION, *PHOTOPERIODISM, PERIPHYTON, *BENTHOS.: /*LIGHT, S W72-07901
, BIOLOGICAL COMMUNITIES, ALGAE, PERIPHYTON, PHYTOPLANKTON, PHOTOS W72-12549
, LEPTOCELLA, ISONYCHIA, CAENIS, PERLESTA.: /ODES INUUSTA, OECETIS W70-06655
GING, IRON BACTERIA, POND WEEDS, PERMANGANATE.: *FILTER CLOG W72-06536
SYNTHESIS, CHLOROPHYLL, ENZYMES, PERMEABILITY, WATER QUALITY CONTR W72-06037
PROPERTIES, MOLECULAR STRUCTURE, PERMEABILITY, PLANT POPULATIONS, W72-01473
RFACE-GROUNDWATER RELATIONSHIPS, PERMEABILITY, ALGAE, DISSOLVED OX W71-12410
TROL, LEGAL ASPECTS, REGULATION, PERMITS.: /EALTH, ENVIRONMENT CON W73-00703
UES, MONITORING, SURFACE WATERS, PERSISTENCE, PATH OF POLLUTANTS, W73-00361
, ALGAE, ALGAL CONTROL, DIATOMS, PERSISTENCE, BACTERIA, BIRDS, FIS W72-00155
HROOMONAS CAUDATA, CHLAMYDOMONAS PERTUSA, PANDORINA MORUM.: /SA, C W72-13816
*COASTAL WATERS, PANAMA, CALLAO(PERU), CAPE COD(MASS), CAPE MAY(N W72-08056
TOCEROS MESSANENSIS, CHAETOCEROS PERUVIANUS, CLADOGRAMMA KOLBEI, C W71-11527
TICIDES, AGRICULTURAL CHEMICALS, PESTICIDE RESIDUES, PESTICIDE KIN W71-12303
L CHEMICALS, PESTICIDE RESIDUES, PESTICIDE KINETICS, ANIMAL PHYSIO W71-12303
LUTION EFFECTS, *PHOTOSYNTHESIS, PESTICIDE TOXICITY, PLANT GROWTH W71-10566
ES, FARM WASTES, WASTE DISPOSAL, PESTICIDE REMOVAL, OXYGEN SAG, DR W72-01659
ANOPHYTA, *TOXICITY, PESTICIDES, PESTICIDE RESIDUES, ALGAE, DDT, P W72-03545
, ECOSYSTEMS, POLLUTION SOURCES, PESTICIDE TOXICITY, FOOD CHAINS, W72-07703
R POLLUTION EFFECTS, ABSORPTION, PESTICIDE TOXICITY, LABORATORY TE W73-00281
TS, DOCUMENTATION, PUBLICATIONS, PESTICIDE TOXICITY, BIOINDICATORS W73-01976
*SHRIMP, *BIOASSAY, *ABSORPTION, PESTICIDE RESIDUES, WATER POLLUTI W73-02105
DE TOXICITY, PATH OF POLLUTANTS, PESTICIDE RESIDUES, CHLORINATED H W73-02280
ORINATED HYDROCARBON PESTICIDES, PESTICIDE DRIFT, METABOLISM, ALGA W73-02280
LOGY, *DIATOMS, PESTICIDES, DDD, PESTICIDE TOXICITY, PATH OF POLLU W73-02280
ORINATED HYDROCARBON PESTICIDES, PESTICIDE TOXICITY, PLANT POPULAT W72-14294
, *SMALL WATERSHEDS, ECOSYSTEMS, PESTICIDE KINETICS, *MAINE, MUDS, W72-12930
RCURY, COPPER, ZINC, PESTICIDES, PESTICIDE TOXICITY, *HERBICIDES, W72-12576
IDES, CULTURES, TEST PROCEDURES, PESTICIDE TOXICITY, RESISTANCE, L W71-06189
CREATION, LAKES, STREAMS, ALGAE, PESTICIDES, DDT, RETURN FLOW, SED W71-06444

LGAE, *TOXICITY, SOILS, TRIAGINE PESTICIDES, RESISTANCE, CULTURES, W71-07675
CULATION, ABSORPTION, NUTRIENTS, PESTICIDES, ALGAE, DREDGING, ESTU W71-09581
XICITY, CHROMOSOMES, FUNGICIDES, PESTICIDES, ECOLOGY, BIRD TYPES, W72-05952
, HEAVY METALS, ORGANOPHOSPHORUS PESTICIDES, INDUSTRIAL WASTES, VI W72-04742
AMYDOMONAS, *ALGAE, *HERBICIDES, PESTICIDES, CHLORELLA, RESPIRATIO W70-07255
FISH, FLOOD PLAINS, IRRIGATION, PESTICIDES.: /ATION, GROUNDWATER, W70-05103
, PHOSPHATES, NITRATES, SURVEYS, PESTICIDES, WASTE DISPOSAL, PATH W71-05553
ESMUS, ALGAL TOXINS, ANTIBIOTICS(PESTICIDES), DATA PROCESSING, COL W71-05155
L WASTES, MERCURY, COPPER, ZINC, PESTICIDES, PESTICIDE TOXICITY, * W72-12576
IVITY, CORAL, REEFS, ANTIBIOTICS(PESTICIDES), CHLORINATED HYDROCAR W72-14294
, COLIFORMS, TOXINS, ANTIBIOTICS(PESTICIDES), ENTERIC BACTERIA, DI W72-13800
BIOASSAY, *WATER QUALITY, ALGAE, PESTICIDES, HEAVY METALS, WASTEWA W72-14282
ICIDES), CHLORINATED HYDROCARBON PESTICIDES, PESTICIDE TOXICITY, P W72-14294
EFFECTS, CHLORINATED HYDROCARBON PESTICIDES, *CYANOPHYTA, SODIUM, W72-13809
UTANTS, *METALS, TRACE ELEMENTS, PESTICIDES, SALINITY, NUTRIENTS, W72-12190
YDROCARBON PESTICIDES, CARBAMATE PESTICIDES, ORGANIC PESTICIDES, H W72-13347
ENZYMES, ANALYTICAL TECHNIQUES, PESTICIDES, FUNGICIDES, ORGANOPHO W72-13347
ALGAE, *CHLORINATED HYDROCARBON PESTICIDES, *CYTOLOGICAL STUDIES, W72-12998
TICIDES, CHLORINATED HYDROCARBON PESTICIDES, CARBAMATE PESTICIDES, W72-13347
CHAINS, CHLORINATED HYDROCARBON PESTICIDES, FISHERIES.: /AE, FOOD W72-13657
APHY, GROWTH INHIBITI/ MERCURIAL PESTICIDES, THIN LAYER CHROMATOGR W72-13347
S, CARBAMATE PESTICIDES, ORGANIC PESTICIDES, HERBICIDES, WATER POL W72-13347
S, FUNGICIDES, ORGANOPHOSPHORUS PESTICIDES, CHLORINATED HYDROCARB W72-13347
TS, *PLANT PHYSIOLOGY, *DIATOMS, PESTICIDES, DDD, PESTICIDE TOXICI W73-02280
OLOGY, PHOSPHATES, HEAVY METALS, PESTICIDES, TASTE-PRODUCING ALGAE W73-02144
ESIDUES, CHLORINATED HYDROCARBON PESTICIDES, PESTICIDE DRIFT, META W73-02280
IRATION, CHLORINATED HYDROCARBON PESTICIDES, GROWTH RATES, ENVIRON W73-00281
ATE PESTICIDES, ORGANOPHOSPHORUS PESTICIDES, ACARICIDES, FOOD CHAI W73-00238
TICIDES, CHLORINATED HYDROCARBON PESTICIDES, INSECTICIDES, RODENTI W73-00238
FUNGICIDES, ALGICIDES, CARBAMATE PESTICIDES, ORGANOPHOSPHORUS PEST W73-00238
M, BIO/ *CHLORINATED HYDROCARBON PESTICIDES, SEA WATER, *METABOLIS W73-00361
YANOPHYTA, CULTURES, ANTIBIOTICS(PESTICIDES), PHOTOACTIVATION, AQU W72-14327
*DDT, *RADIOACTIVITY TECHNIQUES, PESTICIDES, ECOSYSTEMS, POLLUTION W72-07703
ION, ALGAE, PHENOLS, DETERGENTS, PESTICIDES, OPERATION AND MAINTEN W72-08357
IDES, *PESTICIDE TOXICITY, *DDT, PESTICIDES, PHYTOPLANKTON, DIATOM W72-08436
RIA, FISH PARASITES, ANTIBIOTICS(PESTICIDES), BACTERICIDES, ATOLLS W72-08142
TER POLLUTION EFFECTS, BIOASSAY, PESTICIDES, HERBICIDES, 2-4-D, PL W72-08440
PHENYLS, *CHLORNATED HYDROCARBON PESTICIDES, *PESTICIDE TOXICITY, W72-08436
FICATION, DETERGENTS, NUTRIENTS, PESTICIDES, OIL, METALS, THERMAL W72-08790
H RATES, *CYANOPHYTA, *TOXICITY, PESTICIDES, PESTICIDE RESIDUES, A W72-03545
*ALGAE/ *CHLORINATED HYDROCARBON PESTICIDES, *PESTICIDE RESIDUES, W72-04134
CHROMATOGRAPHY, CENTRIFUGATION, PESTICIDES, BIOASSAY, DISTRIBUTIO W72-04134
ENDRIN/ *CHLORINATED HYDROCARBON PESTICIDES, *ALDRIN, *DIELDRIN, * W72-03545
NITRATES, WATER REUSE, SALINITY, PESTICIDES, LIVESTOCK, WASTE DISP W72-00846
ION, PHYSICOCHEMICAL PROPERTIES, PESTICIDES, TROPHIC LEVEL, GASES, W72-00847
DROCARBON PESTICIDES, *CARBAMATE PESTICIDES, *ORGANOPHOSPHOROUS PE W71-10566
E PESTICIDES, *ORGANOPHOSPHOROUS PESTICIDES, *2-4-D, *DDT, *DIELDR W71-10566
ESTICIDES, DDT, ORGANOPHOSPHORUS PESTICIDES, PHYTOPLANKTON, TOXICI W71-10096
DIELDRIN, OXYGENATION, CARBAMATE PESTICIDES.: / CULTRUES, ENDRIN, W71-10096
CTIVITY, CHLORINATED HYDROCARBON PESTICIDES, DDT, ORGANOPHOSPHORUS W71-10096
ICIDES, *CHLORINATED HYDROCARBON PESTICIDES, *CARBAMATE PESTICIDES W71-10566
ICITY, *WATER POLLUTION EFFECTS, PESTICIDES, AGRICULTURAL CHEMICAL W71-12303
ODYNAMICS, SOLUBILITY, SORPTION, PESTICIDES, TOXICITY, DISSOLVED O W71-12092
CITY, / *CHLORINATED HYDROCARBON PESTICIDES, *DDT, *PESTICIDE TOXI W71-12303
EDIMENTS, SALT MARSHES, ECOLOGY, PESTICIDES, ESTUARIES, INTERTIDAL W72-10678
, AQUATIC PLANTS, AQUATIC ALGAE, PESTICIDES, INSECTICIDES, HERBICI W72-09646
STEMS, AQUATIC LIFE, ANTIBIOTICS(PESTICIDES), ECOLOGY, BENTHOS, IN W72-09668
OMONAS, MALLOMONAS, STAURASTRUM, PHACOMYXA, SPONDYLOSIUM, LYNGBYA, W72-07648
ENA SPIROGYRA, LEPOCINCLIS OVUM, PHACUS CIRCUMFLEXUS, PHACUS ONYX, W73-00918
NCLIS OVUM, PHACUS CIRCUMFLEXUS, PHACUS ONYX, PHACUS PLEURONECTES, W73-00918
HACUS CIRCUMFLEXUS, PHACUS ONYX, PHACUS PLEURONECTES, PHACUS PYRUM W73-00918
HACUS ONYX, PHACUS PLEURONECTES, PHACUS PYRUM, TRACHELOMONAS ARMAT W73-00918
GONIUM, RHIZOCLONIUM, COSMARIUM, PHACUS, CHAETOCEROS, MOUGEOTIA, N W72-09111
RIDES, OLIGOSACCHARIDES, CLONES, PHAEOCYSTIS POUCHETI, GLUCOSE, MA W72-07710
E, DENSITY GRADIENTS, TUNICATES, PHAEOCYSTIS, SORTING, SODIUM BROM W73-00371
-PLANT RELATIONSHIPS, NORTH SEA, PHAEOCYSTIS SPP., SOIL CULTURES, W72-12239
M, LEAD, CHLORELLA PHYRENOIDOSA, PHAEODACTYLUM TRICORNUTUM, CYCLOT W72-07660
LAX POLYEDRA, PLATYMONAS VIRIDI/ PHAEODACTYLUM TRICORNUTUM, GONYAU W72-07166
RBON-14, DUNALIELLA TERTIOLECTA, PHAEODACTYLUM TRICORNUTUM.: /, CA W72-09102
A STRAIT, DUNALIELLA TERIOLECTA, PHAEODACTYLUM TRICORNUTUM, SKELET W72-09103
MERCURIC ACETATE, MERCURY CHLOR/ PHAEODACTYLUM TRICORNUTUM, PHENYL W72-11727
COCCOLITHUS HUXLEYI, AUTOLYSIS, PHAEODACTYLUM.: /CUS, RHODOMONAS, W71-10083
ASTS, BENTHIC FLORA, RHODOPHYTA, PHAEOPHYTA, CHLOROPHYTA, MAINE, S W71-10066
, WATER TEMPERATURE, RHODOPHYTA, PHAEOPHYTA, CHLOROPHYTA, COOLING W72-11252
ENT, TURBIDITY, WATER POLLUTION, PHAEOPHYTA, MOLLUSKS, MARINE ALGA W72-11925
LGAE, *CYANOPHYTA, *CHLOROPHYTA, PHAEOPHYTA, RHODOPHYTA, CLADOPHOR W72-11800
ARTIFICIAL), ALGAE, CHLOROPHYTA, PHAEOPHYTA, RHODOPHYTA.: /LLITES(W72-09120
IES, PHYTOPLANKTON, CHLOROPHYTA, PHAEOPHYTA, RHODOPHYTA, DIATOMS, W72-03220
RING, *MARINE ALGAE, RHODOPHYTA, PHAEOPHYTA, *COBALT RADIOISOTOPES W72-07826
AMINO ACIDS, GAS CHROMATOGRAPHY, PHAEOPHYTA, EUTROPHICATION, PLANK W72-07710
R, MORTALITY, WATER TEMPERATURE, PHAEOPHYTA, SAMPLING, SCUBA DIVIN W73-00366
E ANIMALS, MOLLUSKS, GASTROPODS, PHAEOPHYTA.: /TY, COPEPODS, MARIN W73-00371
TRIBUTION, GEOGRAPHICAL REGIONS, PHAEOPHYTA, MARINE ALGAE, ENVIRON W73-00428
YSTEMATICS, *ALGAE, CHRYSOPHYTA, PHAEOPHYTA.: /LOGICAL STUDIES, *S W72-14808
ANNELIDS, MOLLUSKS, RHODOPHYTA, PHAEOPHYTA, PHOTOGRAPHY, SCUBA DI W73-00932
EFS, LABORATORY TESTS, SAMPLING, PHAEOPHYTA.: /NMENTAL EFFECTS, RE W73-00838
TICS, *REPRODUCTION, RHODOPHYTA, PHAEOPHYTA, WATER QUALITY, *MARIN W73-01449
PILLS, SEA WATER, MARINE PLANTS, PHAEOPHYTA, OILY WATER, EVAPORATI W73-01074
ANKTON, WATER POLLUTION SOURCES, PHAEOPHYTA, OXIDATION.: / PHYTOPL W72-12166
*KELPS, *PHYSIOLOGICAL ECOLOGY, PHAEOPHYTA, STANDING CROPS, RESIS W72-12940

URES, *MARINE ALGAE, RHODOPHYTA, PHAEOPHYTA, CHLOROPHYTA, INHIBITO W72-04748
ROPHYTA, CYANOPHYTA, RHODOPHYTA, PHAEOPHYTA, CARBON RADIOISOTOPES, W72-06283
H, SHELLFISH, AQUEOUS SOLUTIONS, PHAEOPHYTA, ATLANTIC OCEAN, SHARK W72-05965
EXUDATION, OCEANS, CHLOROPHYTA, PHAEOPHYTA, RHODOPHYTA, MARINE BA W71-07878
NS, OXYGEN, ECOLOGY, RHODOPHYTA, PHAEOPHYTA, CHLORELLA, PLANT PATH W71-09157
BITORS, PLANT TISSUES, BIOASSAY, PHAEPHYTA.: /TH, RHODOPHYTA, INHI W72-07496
ON, MARINE PLANTS, MARINE ALGAE, PHAEPHYTA, AQUARIA, PARTICLE SIZE W71-09193
CHEMICAL SCRE/ ANTIMICROBIOTICS, PHARMACOLOGICAL PROPERTIES, PHYTO W72-04275
*TOXICOLOGY, BIOTOXINS, PHARMACOLOGY.: W73-00243
DIOISOTOPES, BLOOMS, 2,4 DINITRO PHENOL, TUCSON(ARIZ).: /ALCIUM RA W71-01474
S, DALAPON, 2-4-D, INSECTICIDES, PHENOLS, *DETERGENTS, *LETHAL LIM W72-12576
SMS, HYDROGEN ION CONCENTRATION, PHENOLS, ARIZONA.: /, MICROORGANI W71-01474
DOR, TASTE, PLANKTON, CHEMICALS, PHENOLS, CHLORINE, OXIDATION, ADS W73-01669
, ADSORPTION, FILTRATION, ALGAE, PHENOLS, DETERGENTS, PESTICIDES, W72-08357
TICIDES, HERBICIDES, PHOSPHORUS, PHENOLS, METALS, POLLUTANTS, INDU W72-09646
NTIC OCEAN, SEA WATER, SALINITY, PHENOLS, WATER TEMPERATURE, CHEMI W72-07710
CURIC ACETATE, MERCURY CHLORIDE, PHENYLACETATE, MERCURY COMPOUNDS. W72-11727
ISTIDINE, ASPARTATE, METHIONINE, PHENYLALANINE, ALANINE, ISOLEUCIN W73-00942
HLOR/ PHAEODACTYLUM TRICORNUTUM, PHENYLMERCURIC ACETATE, MERCURY C W72-11727
EUDOMONAS, TESTING, MEASUREMENT, PHEOLOGY.: /OPHYTA, CHLORELLA, PS W72-03220
ERIA, PHORMIDIUM, MASTIGOCLADUS, PHEOPHYTIN, GRASSLAND SPRING.: /T W72-03557
RA,/ *PYCNOCLINE, CHLOROPHYLL A, PHEOPHYTIN, TUNICATES, FORAMINIFE W72-14313
, FANNING ISLANDS.: PHILIPPINES, ENIWETOK, MICRONESIA W71-11486
USTICIA AMERICANA, ALTERNANTHERA PHILOXEROIDES, GLY: /CRASSIPES, J W73-01089
ACTIVATED SLUDGE, SOIL BACTERIA, PHO: /BACTERIA, DIATOMS, OCEANS, W72-10883
YSOPHYTA, NITRATES, CHLOROPHYLL, PHO: /HT INTENSITY, NITROGEN, CHR W73-01435
ILLATORIA, SPIRULINA, PANDORINA, PHOCUS, STAURASTRUM, TRACHELOMONA W72-07661
CILLUS DUMENTOSUS, RHIPOCEPHALUS PHOENIX, UDOTEA FLABELLUM, BRYOPS W72-14294
CHALYBIA, OSCILLATORIA FORMOSA, PHORMIDIUM FAVEOLARUM, ANKISTRODE W72-07664
, CHLAMYDOMONAS, FONTINALIS, AS/ PHORMIDIUM, CHLORELLA PYRENOIDOSA W71-11519
K, SYNECHOCOCCUS, FLEXIBACTERIA, PHORMIDIUM, MASTIGOCLADUS, PHEOPH W72-03557
URTHE RIVER(FRANCE), CLADOPHORA, PHORMIDIUM, OSCILLATORIA, OEDOGON W71-00099
RARA RIVER, PLECTONEMA, LYNGBYA, PHORMIDIUM, CYANOPHYCEAE, OSCILLA W72-04736
IUM, COPPER, *NEW YORK, DIATOMS, PHOSP: /INDICATORS, SILICA, CHROM W72-09111
PHOSPHORUS, *EXCRETION, ALKALINE PHOSPHATASE, REASSIMILATION, GREE W71-10083
, *CONTINUOUS CULTURE/ *ALKALINE PHOSPHATASE, *ANABAENA FLOS-AQUAE W72-05476
*CONTINUOUS CULTURE, CHEMOSTATS, PHOSPHATASE ACTIVITY.: /S-AQUAE, W72-05476
CARBONACEOUS MATERIAL, *CANADIAN PHOSPHATE DETERGENT BAN, WYANDOTT W70-07283
CY, ORTHOPHOSPHATES, DRY WEIGHT, PHOSPHATE UPTAKE.: /E, TRANSPAREN W71-10645
IUM NITRILOTRIACETATE, SODIUM C/ PHOSPHATE-FREE DETERGENTS, TRISOD W71-06247
CTERIA, ALGAE, ALKALINITY, IRON, PHOSPHATE, PHOSPHORUS, CARBON, NI W71-05626
OXYGEN, CARBON DIOXIDE, AMMONIA, PHOSPHATE, ANALYTICAL TECHNIQUES, W72-06046
CIDES, MICROSTRAINING, AERATION, PHOSPHATES, CLAY, TASTE, ODOR, *S W72-06536
, NUTRIENTS, NITRATES, SULFATES, PHOSPHATES, CHLORIDES, BORATES, A W72-06274
GEN, METABOLISM, CARBON DIOXIDE, PHOSPHATES, NITRATES, *NUTRIENTS, W72-04983
INDUSTRIAL WASTES, CHLORINATION, PHOSPHATES, NITRATES, FILTRATION, W72-05315
OPHYTA, THERMAL WATER, *MONTANA, PHOSPHATES, NITRATES, NITRITES, I W72-05610
URE, LIGHT PENETRATION, AMMONIA, PHOSPHATES, SALINE WATER.: /PERAT W72-05469
UTROPHICATION, NUTRIENTS, ALGAE, PHOSPHATES, NITRATES, SURVEYS, PE W71-05553
TION, NITRATES, SILICA, AMMONIA, PHOSPHATES, RUNOFF, ILLINOIS, HYD W71-03028
DEMAND, CHEMICAL OXYGEN DEMAND, PHOSPHATES, NITROGEN, STABILIZATI W70-07838
ION SOURCES, *ESTUARIES, RIVERS, PHOSPHATES, NITRATES, HUDSON RIVE W70-06509
ISOTOPES, SEWAGE, RADIOACTIVITY, PHOSPHATES, CHEMICAL ANALYSIS, AL W71-01474
HOTOSYNTHESIS, DISSOLVED OXYGEN, PHOSPHATES, WATER YIELD IMPROVEME W71-01491
TE, SODIUM PYROPHOSP/ *CONDENSED PHOSPHATES, SODIUM TRIPOLYPHOSPHA W71-00116
AKES, WATER QUALITY, PHOSPHORUS, PHOSPHATES, PRODUCTIVITY, LEACHIN W70-08944
NITROGEN, NITRATES, PHOSPHORUS, PHOSPHATES, ALGAE, EUTROPHICATION W71-06445
ALKYLATE SULFONATES, CHELATION, PHOSPHATES, ORGANIC COMPOUNDS, WA W71-06247
OTOSYNTHESIS, ORGANIC COMPOUNDS, PHOSPHATES, NITRATES, NUTRIENTS, W71-07100
TERIA, EUTROPHICATION, NITRATES, PHOSPHATES, BIODEGRADATION, ANAER W71-07106
RY PRODUCTIVITY, TRACE ELEMENTS, PHOSPHATES, NITRATES, LABORATORY W71-09674
NUTRIENTS, CARBON RADIOISOTOPES, PHOSPHATES, SILICATES, SOLAR RADI W71-07875
YPOTHRIX TENUIS, AU/ *TRICALCIUM PHOSPHATES, ANABAENA, NOSTOC, TOL W71-10070
H RATES, TEMPERATURE, SEA WATER, PHOSPHATES, PHOSPHORUS RADIOISOTO W71-10083
L TECHNIQUES, AMMONIA, NITRATES, PHOSPHATES, WATER QUALITY, NITROG W71-11033
LLUTION CONTROL, TROPHIC LEVELS, PHOSPHATES, DETERGENTS, BIOASSAY, W71-12072
CHLOROPHYLL, NITRATES, NITRITES, PHOSPHATES, NUTRIENTS.: / WATER, W72-09102
OPHICATION, PRODUCTIVITY, ALGAE, PHOSPHATES, NITROGEN, SEWAGE, CAR W72-09167
AL RUNOFF, PHOSPHORUS, NITROGEN, PHOSPHATES, DETERGENTS, POTABLE W W72-10625
A, NITROGEN COMPOUNDS, NITRATES, PHOSPHATES, SLUDGE TREATMENT, DIS W72-09675
PTION, ALGAE, CHORELLA, TRACERS, PHOSPHATES, ECOSYSTEMS, AQUATIC L W72-09668
ATION, MUNICIPAL WASTES, CITIES, PHOSPHATES, ILLINOIS, ECOLOGY, TE W72-03970
TECHNIQUES, TURBIDITY, NITRATES, PHOSPHATES, WIND VELOCITY, RAINFA W72-03224
ETS, HYDROGEN ION CONCENTRATION, PHOSPHATES, NITROGEN, HARDNESS(WA W72-07890
RODUCTIVITY, SALINITY, PROFILES, PHOSPHATES, NITRATES, SILICATES, W72-07892
MASS, TEMPERATURE, CONDUCTIVITY, PHOSPHATES, NITRATES, ALKALINITY, W72-08560
CHLOROPHYLL, PIGMENT, MAGNESIUM, PHOSPHATES, NITROGEN, SEA WATER, W72-07660
NUISANCE ALGAE, EUTROPHICATION, PHOSPHATES, PHOSPHORUS, CARBON, O W72-07514
TRATION, LEACHING, FUNGI, LIGHT, PHOSPHATES, NUTRIENTS, WATER POLL W72-07143
IEWS, TROPHIC LEVEL, PHOSPHORUS, PHOSPHATES, CARBON, VITAMIN B, NI W72-07937
OPLANKTON, CYANOPHYTA, NITRATES, PHOSPHATES, LAKES, STRATIFICATION W72-08049
ENS, HYDROGEN ION CONCENTRATION, PHOSPHATES, NITRATES, DISSO: /LOG W73-01434
, CULTURES, NITROGEN, POTASSIUM, PHOSPHATES, METABOLISM, ALGAE, SY W73-01626
-PRODUCING ALGAE, ACTINOMYCETES, PHOSPHATES.: /ODUCING ALGAE, ODOR W73-02144
ACEANS, NEMATODES, EPIDEMIOLOGY, PHOSPHATES, HEAVY METALS, PESTICI W73-02144
REQUIREMENT, *ALGAE, DETERGENTS, PHOSPHATES.: /HLORELLA, NUTRIENT W73-02112
FECTS, MERCURY, SEWAGE DISPOSAL, PHOSPHATES, PHOSPHORUS, CHEMICAL W72-14782
ANT GROWTH REGULATORS, NITRATES, PHOSPHATES, NUTRIENT REQUIREMENTS W73-00430
NKTON, *AQUATIC ALGAE, CULTURES, PHOSPHATES, NITRATES, NUTRIENTS, W72-12739
, *PHOSPHORUS, *LAKES, CULTURES, PHOSPHATES, NUTRIENTS, CHLOROPHYT W72-12637

TS, TRACERS, LITTORAL, KINETICS,	PHOSPHATES.: /ES, CYCLING NUTRIEN	W72-12543
RON COMPOUNDS, SEWAGE, NITRATES,	PHOSPHATES.: /OASSAY, SULFATES, I	W72-12239
ENTS, *ALGAE, *LIMITING FACTORS,	PHOSPHATES, CHLOROPHYTA, CYANOPHY	W72-12544
TY, EUTROPHIC ZONE, CHLOROPHYLL,	PHOSPHATES, NITRATES, SEASONAL, R	W72-12393
ERTIES, NITROGEN, HEAT TRANSFER,	PHOSPHATES, ALKALINITY, OXYGEN DE	W72-12997
T GROWTH, *LAKE MICHIGAN, ALGAE,	PHOSPHATES, HYDROLYSIS, STORM WAT	W72-13644
AQUATIC MICROORGANISMS, DIPTERA,	PHOSPHO: /ION, LARVAE, ROTIFERS,	W72-07132
STERS, EUTROPHICATION, NITROGEN,	PHOSPHOROUS, CA: /MS, MUSSELS, OY	W72-07715
N, OXYGEN, METABOLISM, NITROGEN,	PHOSPHOROUS, *ALGAL CONTROL, CALI	W71-13553
HEMICAL OXYGEN DEMAND, NITROGEN,	PHOSPHOROUS, PHOTOSYNTHESIS, LIGH	W71-13326
ER POLLUTION EFFECTS, NUTRIENTS,	PHOSPHOROUS, WATER SOFTENING, LEG	W71-09429
OXYGEN DEMAND, *EUTROPHICATION,	PHOSPHOROUS, NITROGEN, CHEMICAL O	W72-04431
ST-BENEFIT ANALYSIS, DETERGENTS,	PHOSPHORUS, NITROGEN, NUTRIENTS,	W72-04734
ICATION, WATER REUSE, *NITROGEN,	PHOSPHORUS, CARBON, *SEWAGE BACTE	W72-04309
TION SOURCES, ALGAE, RIVER FLOW,	PHOSPHORUS, NITRATES, POTASSIUM,	W72-04260
, VIRGINIA, NUTRIENTS, NITROGEN,	PHOSPHORUS, SEWAGE EFFLUENTS, SEW	W72-05473
TER INTERFACES, IRON, MANGANESE,	PHOSPHORUS, NITROGEN, ORGANIC MAT	W72-05469
S), OCCOQUAN RESERVOI/ *NITROGEN:	PHOSPHORUS RATIO, LAKE WAUBESA(WI	W72-05473
ANIC COMPOUNDS, IRON, MANGANESE,	PHOSPHORUS, CARBON, NITROGEN, OXI	W72-05506
CONTROL, WATER QUALITY CONTROL,	PHOSPHORUS, AQUATIC ENVIRONMENT,	W72-06638
S, NUTRIENTS, ALGAE, CYANOPHYTA,	PHOSPHORUS, WATER POLLUTION SOURC	W72-04769
IRCRAFT, SATELLITES(ARTIFICIAL),	PHOSPHORUS, NITROGEN, HEAVY METAL	W72-04742
ER TREATMENT, TURBIDITY, RUNOFF,	PHOSPHORUS, ALGAE, AQUATIC PLANTS	W72-04759
OISOTOPES, COBALT RADIOISOTOPES,	PHOSPHORUS RADIOISOTOPES, SEDIMEN	W71-09190
LITY CONTROL, AMMONIA, NITROGEN,	PHOSPHORUS, ALGAE, FEASIBILITY ST	W71-09450
FERTILIZERS, NITROGEN, NITRATES,	PHOSPHORUS, PHOSPHATES, ALGAE, EU	W71-06445
N, LIGHT PENETRATION, NUTRIENTS,	PHOSPHORUS, NITROGEN, TIDAL WATER	W71-07098
NTS, DISSOLVED OXYGEN, NITROGEN,	PHOSPHORUS, WATER QUALITY ACT, WA	W71-06444
*PRODUCTIVITY, ALGAE, NITROGEN,	PHOSPHORUS, MATHEMATICAL MODELS,	W71-08670
NIC MATTER, NUTRIENTS, NITROGEN,	PHOSPHORUS, EUTROPHICATION, ALGAE	W71-08214
PHICATION, LAKES, WATER QUALITY,	PHOSPHORUS, PHOSPHATES, PRODUCTIV	W70-08944
D DEVELOPMENT, *WATER POLLUTION,	PHOSPHORUS, THERMAL STRATIFICATIO	W70-09947
*PHOSPHORUS, *ACTIVATED SLUDGE,	PHOSPHORUS RADIOISOTOPES, SEWAGE,	W71-01474
ALINITY, EPILIMNION, WASHINGTON,	PHOSPHORUS, NITROGEN, NUTRIENTS.:	W71-02681
NVIRONMENT, CHEMICAL PROPERTIES,	PHOSPHORUS, NITROGEN, POTASSIUM,	W71-00117
ATER, BIOCHEMICAL OXYGEN DEMAND,	PHOSPHORUS, FISH, WEEDS, CYCLES,	W71-03012
WATERS, NITROGEN, WATER QUALITY,	PHOSPHORUS, AEROBIC CONDITIONS, A	W70-10405
CARBOY TRIALS, CANADA, NITROGEN:	PHOSPHORUS RATIO.: /ASSOCIATION,	W70-07283
TA, CYCLING NUTRIENTS, NITROGEN,	PHOSPHORUS, CHEMICAL ANALYSIS, PR	W70-05469
U/ NITROGEN REQUIREMENTS(ALGAE),	PHOSPHORUS REQUIREMENTS(ALGAE), N	W70-04080
TICIDES, CHLORELLA, RESPIRATION,	PHOSPHORUS, TOXICITY, RADIOISOTOP	W70-07255
RY,/ *WATER QUALITY, *WISCONSIN,	PHOSPHORUS, PULP AND PAPER INDUST	W70-05103
MULATION, ANALYTICAL TECHNIQUES,	PHOSPHORUS, ALGAE, LAKES, MICROOR	W71-03021
AE, ALKALINITY, IRON, PHOSPHATE,	PHOSPHORUS, CARBON, NITROGEN, CUR	W71-05626
L OXYGEN DEMAND, ORGANIC MATTER,	PHOSPHORUS, CARBON DIOXIDE.: /ICA	W71-13257
PRODUCTIVITY, *ALGAE, NITROGEN,	PHOSPHORUS, THERMAL STRATIFICATIO	W71-13183
Y, NITRATES, PLANNING, NITROGEN,	PHOSPHORUS, RESERVOIRS, FISHERIES	W71-13172
TIVITY, WATER POLLUTION EFFECTS,	PHOSPHORUS, NITROGEN, AMMONIA, CA	W71-12857
DEMAND, CHEMICAL OXYGEN DEMAND,	PHOSPHORUS, NITRATES, AMMONIA, CO	W71-13341
, NUTRIENTS, NITROGEN COMPOUNDS,	PHOSPHORUS COMPOUNDS, POTASH, ALG	W71-12252
METALS, TASTE, ODOR, RECREATION,	PHOSPHORUS, CARBON, LAKE ERIE, DE	W71-12091
ON(WASH).: MERCURY,	PHOSPHORUS REMOVAL, LAKE WASHINGT	W71-12091
ATION, TRACE ELEMENTS, NITROGEN,	PHOSPHORUS, METABOLISM, ALGAE, SL	W71-11793
AQUATIC PRODUCTIVITY, NITROGEN,	PHOSPHORUS, LAKES.: /ISTRIBUTION,	W71-11519
, STATISTICAL METHODS, NITROGEN,	PHOSPHORUS, CHLOROPHYLL, DISTRIBU	W71-11029
E TREATMENT, WATERSHEDS(BASINS),	PHOSPHORUS, ALGAE, ORGANIC MATTER	W71-11017
ION, LAKE ERIE, ALGAE, NITROGEN,	PHOSPHORUS, DRAINAGE, INDUSTRIES,	W71-11009
UTRIENTS, SURFACES, TEMPERATURE,	PHOSPHORUS, NITRATES, CHLORIDES,	W71-11001
PHOSPHATASE,/ *DISSOLVED ORGANIC	PHOSPHORUS, *EXCRETION, ALKALINE	W71-10083
PERATURE, SEA WATER, PHOSPHATES,	PHOSPHORUS RADIOISOTOPES, DIATOMS	W71-10083
T, / *TROPHIC LEVELS, *BIOASSAY,	PHOSPHORUS, NUTRIENTS, MEASUREMEN	W71-13794
DIMENTS, FARM WASTES, NUTRIENTS,	PHOSPHORUS, ALGAE, NITRATES, WATE	W72-00846
URFACE WATERS, NUTRIENTS, ALGAE,	PHOSPHORUS, NITROGEN, EUTROPHICAT	W72-00847
DEMAND, CHEMICAL OXYGEN DEMAND,	PHOSPHORUS, NITROGEN, EUTROPHICAT	W72-01881
AILABLE PHOSPHORUS, EXCHANGEABLE	PHOSPHORUS, A*DENOSINE TRIPHOSPHA	W72-03742
US, A*DENOSINE TRIPHO/ AVAILABLE	PHOSPHORUS, EXCHANGEABLE PHOSPHOR	W72-03742
SEDIMENTARY	PHOSPHORUS RELEASE.:	W72-01108
ATORY TESTS, TEMPERATURE, ALGAE,	PHOSPHORUS, ION EXCHANGE, OXIDATI	W72-01108
OXINS, WASTE DILUTION, NITROGEN,	PHOSPHORUS, EUTROPHICATION, SURFA	W72-01472
OTOPES, STRONTIUM RADIOISOTOPES,	PHOSPHORUS RADIOISOTOPES, CESIUM,	W72-09649
CAL TECHNIQUES, MUNICIPAL WATER,	PHOSPHORUS.: /ENT, COSTS, ANALYTI	W72-09171
ABAIN, BIOLOGICAL MAGNIFICATION,	PHOSPHORUS-32, LIMNODRILUS,: / OU	W72-09668
CIDES, INSECTICIDES, HERBICIDES,	PHOSPHORUS, PHENOLS, METALS, POLL	W72-09646
ON EFFECTS, AGRICULTURAL RUNOFF,	PHOSPHORUS, NITROGEN, PHOSPHATES,	W72-10625
HYTA, LAKES, MISSOURI, NITROGEN,	PHOSPHORUS.: /TER QUALITY, CYANOP	W72-10431
, *NUTRIENTS, *ALGAE, *BIOASSAY,	PHOSPHORUS COMPOUNDS, NITROGEN CO	W72-10608
HNIQUES, FERMENTATION, NITROGEN,	PHOSPHORUS, AMMONIA, METHANE, TEM	W72-10390
SIS, ACTIVATED SLUDGE, NITROGEN,	PHOSPHORUS, CHEMICAL OXYGEN DEMAN	W72-11100
TESTS, *NUTRIENTS, ALGAE, LAKES,	PHOSPHORUS, CARBON, NITROGEN.: /	W72-09165
OASSAY, OREGON, LAKES, NITROGEN,	PHOSPHORUS, CARBON, GROWTH RATES,	W72-09164
EWAGE, CARBON DIOXIDE, SEASONAL,	PHOSPHORUS, WATER POLLUTION SOURC	W72-09167
EVELOPMENT, NUTRIENTS, NITROGEN,	PHOSPHORUS, ALGAE, PLANT GROWTH,	W72-11702
LAKE MUD, PHOSPHORUS SORPTION,	PHOSPHORUS REMOVAL, RATE OF SORPT	W72-11617
UTION, CARBON DIOXIDE, NITROGEN,	PHOSPHORUS, LIGHT, NUTRIENTS, MAT	W72-11724
EMOVAL, RATE OF SORPT/ LAKE MUD,	PHOSPHORUS SORPTION, PHOSPHORUS R	W72-11617
CYANIDES,	PHOSPHORUS-32.:	W72-07143
STANDING CROP, CARBON, NITROGEN,	PHOSPHORUS, PHYTOPLANKTON, CARBON	W72-07648
C, DISSOLVED NITROGEN, DISSOLVED	PHOSPHORUS, OSCILLATORIA, RADON R	W72-07648
GAE, EUTROPHICATION, PHOSPHATES,	PHOSPHORUS, CARBON, ORGANIC MATTE	W72-07514
PHOTOSYNTHESIS, EUTROPHICATION,	PHOSPHORUS, CHLOROPHYTA, CYANOPHY	W72-08048

ICATICN, REVIEWS, TROPHIC LEVEL,	PHOSPHORUS, PHUSPHATES, CARBON, V	W72-07937
LAKE(MICH), CARBONATE PARTICLES,	PHOSPHORUS CYCLE.: / WINTERGREEN	W72-08048
ATER, ALGAE, BACTERIA, NITROGEN,	PHOSPHORUS, OXICATION-REDUCTION P	W72-07933
ED OXYGEN, CARBON, DISTRIBUTION,	PHOSPHCRUS, EUTROPHICATION, ALGAE	W72-08459
LTURES, INHIBITION, OLIGOTROPHY,	PHOSPHORUS, NITROGEN, *EUTRCPHICA	W72-08376
TER POLLUTION EFFECTS, NITROGEN,	PHOSPHORUS.: /ATIONS, FLORIDA, WA	W72-08986
VOIRS, DESTRATIFICATION, MIXING,	PHOSPHORUS, NITROGEN, TURBULENCE,	W72-08559
QUIREMENTS, SEA WATER, NITRATES,	PHOSPHORUS, NUTRIENTS, CHRYSCPHYT	W73-00441
RY, SEWAGE DISPOSAL, PHOSPHATES,	PHOSPHORUS, CHEMICAL WASTES, LIMN	W72-14782
UTION EFFECTS, SEWAGE, NITROGEN,	PHOSPHORUS, IRON, PHYTOPLANKTCN,	W73-00234
SAY, LIMITING FACTORS, NITROGEN,	PHOSPHORUS, RAIN, LAKES, CLADOPHO	W73-00232
HICATION, *SPATIAL DISTRIBUTICN,	PHOSPHORUS, NITRATES, SILICA, PAT	W73-00384
ING NUTRIENTS, CARBON, NITROGEN,	PHOSPHORUS, SULFUR, CALCIUM, MANG	W73-00840
MATICAL MODELS, IRON, MANGANESE,	PHOSPHORUS, ANAEROBIC CONDITIONS,	W72-12997
IENT REQUIREMENTS, GROWTH RATES,	PHCSPHORUS, NITROGEN, *REGRESSION	W72-12966
PROCEDURE.: DISSOLVED	PHOSPHORUS, *ALGAL ASSAYS, ASSAY	W72-12966
HURON, EFFLUENTS, WATER QUALITY,	PHOSPHORUS, PHYTOPLANKTON, SEWAGE	W72-13662
ALGAE, SEDIMENTS, OXYGEN DEMAND,	PHOSPHORUS, NITROGEN, NUTRIENTS,	W72-12990
RELLA, GROWTH RATES, RESISTANCE,	PHOSPHORUS, IRON, MANGANESE, ZINC	W72-12739
UATIC ALGAE, AERIAL PHOTOGRAPHY,	PHOSPHORUS, NITRATES,: /RATES, AQ	W72-12729
IS, PHYTOPLANKTON, FLUORESCENCE,	PHOSPHORUS COMPOUNDS, CHLAMYDOMON	W72-12637
ROGEN FIXATION, LAKES, NITROGEN,	PHOSPHORUS, IRON, *TEXAS, LABORAT	W72-13692
FERTILIZERS, NITROGEN, BIOASSAY,	PHOSPHORUS, CYANOPHYTA, CLIGCTROP	W72-13816
ALGAL ASSAY, ORTHOPHOSPHATE,	PHOSPHORUS SOURCES.:	W73-02478
CITRIC ACID CYCLE, CYTOCHROMES,	PHOSPHORYLATION, GLYCOLYSIS.: /E,	W72-04309
LTURES, ANTIBIOTICS(PESTICIDES),	PHOTOACTIVATION, AQUATIC ALGAE, E	W72-14327
LLUM QUADRUPLICATUM STRAIN PR-6,	PHOTCALDRIN, KETOENDRIN, METABOLI	W72-03545
PLE PREPARATION, SPERMATOPHYTES,	PHOTOASSIMILATION, GLYOXYLIC ACID	W73-02100
ILARICPHYCEA, PRYMNESIUM PARVUM,	PHOTOAUTOTROPHS, CHROOMONAS SALIN	W70-07280
, AGMENELLUM QUADRUPLICATUM STR/	PHOTODIELDRIN, ANACYSTIS NIDULANS	W72-03545
H, DISTRIBUTION PATTERNS, AERIAL	PHOTOGRAPHY, REMOTE SENSING, SATE	W72-09120
OLLUTANT IDENTIFICATION, *AERIAL	PHOTOGRAPHY, *MATHEMATICAL METHOD	W71-11685
*AQUA/ *REMOTE SENSING, *AERIAL	PHOTOGRAPHY, *INFRARED RADIATION,	W71-12034
: *INFRARED	PHOTOGRAPHY, *AQUATIC VEGETATION.	W71-12034
RADIOACTIVE WASTES, MONITORING,	PHOTOGRAPHY, ULTRAVIOLET RADIATIO	W72-04742
OLLUSKS, RHODOPHYTA, PHAEOPHYTA,	PHOTOGRAPHY, SCUBA DIVING, SEA WA	W73-00932
CTS, EUTROPHIC/ SURVEYS, *AERIAL	PHOTOGRAPHY, WATER POLLUTION EFFE	W72-14728
LGAE, *RHODOPHYTA, *SYSTEMATICS,	PHOTOGRAPHY, SEA WATER, *NORTH CA	W73-00284
ATIC ALGAE, ELECTRON MICROSCOPY,	PHOTOGRAPHY, WATER POLLUTION EFFE	W73-00432
ENEDESMUS, LABORATORY EQUIPMENT,	PHOTOGRAPHY.: / WATER QUALITY, SC	W72-07144
RTEBRATES, AQUATIC ALGAE, AERIAL	PHOTOGRAPHY, PHOSPHORUS, NITRATES	W72-12729
R POLLUTION, *MONITORING, AERIAL	PHOTOGRAPHY, *INFRARED RADIATION,	W72-12487
*COLOR INFRARED	PHOTOGRAPHY.:	W72-12487
RIN-TYPE POLYSACCHAR/ *GLYCEROL,	PHOTOHETEROTROPHIC GROWTH, LUMINA	W70-07280
, *LAKES, *TURBIDITY, *PROFILES,	PHOTOMETER, HYPOLIMNION, EPILIMNI	W72-07145
E WATERS, ANALYTICAL TECHNIQUES,	PHOTOMETRY, WATER POLLUTION.: /AC	W71-12034
CTS, STABILITY, SOLAR RADIATION,	PHOTOPERI: / ANALYSIS, TIDAL EFFE	W72-08141
*ALGAE, *ECOLOGY, ENVIRONMENT,	PHOTOPERIODISM, GROWTH RATES, TEM	W72-00377
	PHOTOSENSITIVITY,:	W72-00377
LKALINITY, ENZYMES, TEMPERATURE,	PHOTOSYNT: /S, HARDNESS(WATER), A	W72-07514
MENTS, DIATOMS, LIGHT INTENSITY,	PHOTOSYNTHESIS, SESSILE ALGAE, TE	W72-07501
RADIOISOTOPES, LIMITING FACTORS,	PHOTOSYNTHESIS, GREAT LAKES, LAKE	W72-07648
NITROGEN CYCLE, ORGANIC MATTER,	PHOTOSYNTHESIS, EUTROPHICATION, P	W72-08048
MENT, CYANOPHYTA, PHYTOPLANKTON,	PHOTOSYNTHESIS, PRIMARY PRODUCTIV	W72-08054
HOSPHORUS, NITROGEN, TURBULENCE,	PHOTOSYNTHESIS, DEPTH, LIGHT PENT	W72-08559
TES, PLANT MORPHOLOGY, CULTURES,	PHOTOSYNTHESIS, CARBON RADIOISOTO	W72-07710
, SPATIAL DISTRIBUTION, BIOMASS,	PHOTOSYNTHESIS, NUTRIENTS, LIMITI	W73-00431
PHOTOACTIVATION, *OXYGEN, LIGHT,	PHOTOSYNTHESIS, OPTICAL PROPERTIE	W72-14800
CE, *PLANT PHYSIOLOGY, CULTURES,	PHOTOSYNTHESIS, AQUATIC ALGAE, CH	W73-00281
KTON, *ALGAE, *DISSOLVED OXYGEN,	PHOTOSYNTHESIS, TURBIDITY, LIGHT	W73-00256
TA, PHYTOPLANKTON, PRODUCTIVITY,	PHOTOSYNTHESIS, EPILIMNION, LIGHT	W72-14812
ONCENTRATION, LIGHT PENETRATION,	PHOTOSYNTHESIS, PHYTOPLANKTON, ST	W73-00236
ROSCOPY, ZOOPLANKTON, NUTRIENTS,	PHOTOSYNTHESIS, STANDING CROP, EC	W72-14313
M, ELECTRODES, ELECTROCHEMISTRY,	PHOTOSYNTHESIS, RIVERS, PLANT PHY	W72-14314
TEMPERATURE, ANABAENA, BACTERIA,	PHOTOSYNTHESIS.: /GHT INTENSITY,	W73-02471
C FLORA, DIATOMS, PHYTOPLANKTON,	PHOTOSYNTHESIS, PRIMARY PRODUCTIV	W73-01615
*ALGAE, *PHYTOPLANKTON, OXYGEN,	PHOTOSYNTHESIS, RESPIRATION, SUCC	W73-01632
MPERATURE, *CYANOPHYTA, GEYSERS,	PHOTOSYNTHESIS, CULTURES, CHLOROP	W72-03557
ION LAGOONS, *ALGAE, *OXIDATION,	PHOTOSYNTHESIS, OXYGEN, *THERMODY	W72-02363
UD, PHYTOPLANKTON, FLUORESCENCE,	PHOTOSYNTHETIC BACTERIA.: /ING, M	W72-01784
TROPHICATION, *LIMITING FACTORS,	PHOTOSYNTHESIS, CARBON CYCLE, ABS	W71-12068
S, WATER TEMPERATURE, TURBIDITY,	PHOTOSYNTHESIS, HYDROGEN ION CONC	W71-11876
RUSTACEANS, DIATOMS, CYANOPHYTA,	PHOTOSYNTHESI: /RING, COPEPODS, C	W71-11517
AQUATIC LIFE, MUNICIPAL WASTES,	PHOTOSYNTHESIS, RESPIRATION, WATE	W71-11017
EN ION CONCENTRATION, SEA WATER,	PHOTOSYNTHESIS, PHYTOPLANKTON, AM	W71-11008
T PHYSIOLOGY, MOVEMENT, VIRUSES,	PHOTOSYNTHESIS, TURBULENCE, LIGHT	W71-13184
N DEMAND, NITROGEN, PHOSPHOROUS,	PHOTOSYNTHESIS, LIGHT INTENSITY,	W71-13326
NITROGEN FIXATION, FORECASTING,	PHOTOSYNTHESIS, BIOMASS, NITRATES	W71-13183
TRIBUTION PATTERNS, MEASUREMENT,	PHOTOSYNTHESIS, PLANT PHYSIOLOGY,	W71-13252
NG FACTORS, ALGAE, PRODUCTIVITY,	PHOTOSYNTHESIS, NUTRIENTS, EUTROP	W72-09162
PAL WASTES, *WATER REUSE, ALGAE,	PHOTOSYNTHESIS, DISSOLVED OXYGEN,	W72-11642
ONDITIONS, BIOMASS, RESPIRATION,	PHOTOSYNTHETIC OXYGEN, *TERTIARY	W72-11100
CRITERIA, BIODEGRADATION, ALGAE,	PHOTOSYNTHESIS, SUSPENDED SOLIDS,	W72-09386
INHIBITORS, HERBICIDES, MONURON,	PHOTOSYNTHESIS.: /TH REGULATORS,	W72-09650
EN DEMAND, BIOLOGICAL TREATMENT,	PHOTOSYNTHESIS, STABILIZATION, *W	W70-06619
RATION, ALKALINITY, RESPIRATION,	PHOTOSYNTHESIS, CARBON DIOXIDE, P	W71-04518
*SEA WATER, *FRESH WATER, ALGAE,	PHOTOSYNTHESIS, ORGANIC MATTER, P	W71-03027
DUCTIVITY, CARBON RADIOISOTOPES,	PHOTOSYNTHESIS.: /TA, PRIMARY PRO	W71-03014
YGEN DECLINE, RESPIRATORY RATES,	PHOTOSYNTHETIC OXYGENATION, ACTIV	W71-04079
GOONS, CHLORELLA, OXYGEN DEMAND,	PHOTOSYNTHESIS, RESPIRATICN, SCEN	W71-00139

ETENTION RESERVOIRS, RESERVOIRS, PHOTOSYNTHESIS, DISSOLVED OXYGEN, W71-01491
SSES, BACTERIA, SULFUR BACTERIA, PHOTOSYNTHETIC BACTERIA, AQUATIC W72-04743
*ALGAE, *BACTERIA, *SYMBIOSIS, PHOTOSYNTHESIS, OXIDATION, RESPIR W72-04983
F, MEASUREMENT, TEST PROCEDURES, PHOTOSYNTHESIS, SEASONAL, LIGHT P W72-04270
TENSITY, GRO/ *ALGAE, *CULTURES, PHOTOSYNTHESIS, BIOMASS, LIGHT IN W72-06274
HEAVY METALS, LABORATORY TESTS, PHOTOSYNTHESIS, CHLOROPHYLL, ENZY W72-06037
CARBON DIOXIDE, LIGHT INTENSITY, PHOTOSYNTHESIS, EQUATIONS, LIGHT W72-05459
EBRATES, DIATOMS, PHYTOPLANKTON, PHOTOSYNTHESIS, WATER POLLUTION S W72-05952
IVITY, *PLANT GROWTH SUBSTANCES, PHOTOSYNTHESIS, METABOLISM, BACTE W71-08687
ETABOLISM, *CYTOLOGICAL STUDIES, PHOTOSYNTHESIS, RESPIRATION.: /*M W71-08683
HYTON, *PLANKTON, *PRODUCTIVITY, PHOTOSYNTHESIS, BIOMASS, OXYGEN, W71-09165
TURBULENCE, OXYGENATION, ALGAE, PHOTOSYNTHESIS, MUNICIPAL WASTES, W71-06033
CENTRATION, COLIFORMS, BACTERIA, PHOTOSYNTHESIS, ORGANIC COMPOUNDS W71-07100
ICKLING FILTERS, EUTROPHICATION, PHOTOSYNTHESIS, DISSOLVED OXYGEN, W71-07097
ICATION, SOLAR RADIATION, ALGAE, PHOTOSYNTHESIS, ANAEROBIC CONDITI W71-07084
ION, *ALGAE, DEPTH, TEMPERATURE, PHOTOSYNTHESIS, RE-AERATION, LIGH W71-07098
NOES, BENTHIC FLORA, ALKALINITY, PHOTOSYNTHESIS, HYDROLO: /, VOLCA W71-09190
S, DINOFLAGELLATES, CHLOROPHYLL, PHOTOSYNTHESIS, TEMPERATURE, SAMP W71-07875
SITY, PONDS, ROTIFERS, PROTOZOA, PHOTOSYNTHESIS, CHLOROPHYLL, TEMP W71-07382
DEMAND, ORGANIC LOADING, ODORS, PHOTOSYNTHESIS, ALGAE, BACTERIA, W71-07123
SUPERIOR, LIMNOLOGY, FERTILITY, PHOTOSYNTHESIS, RESPIRATION, EUTR W72-12192
MASS, ORGANIC MATTER, NUTRIENTS, PHOTOSYNTHESIS.: /ICIENCIES, *BIO W72-12289
LGAE, PERIPHYTON, PHYTOPLANKTON, PHOTOSYNTHESIS, INVERTEBRATES, BI W72-12549
S, WATER TEMPERATURE, CHLORIDES, PHOTOSYNTHESIS, LITTORAL, RESPIRA W72-12940
.: *PAMLICO RIVER ESTUARY, PHOTOSYNTHESIS/RESPIRATION RATIOS W72-13636
LUM SPECATUM, ELODEA CANADENSIS, PHRAGMITES, FONTINALIS, CHARALES, W71-11012
PICATA, FIMBRISTYLIS, BORRICHIA, PHRAGMITES COMMUNIS, JUNCUS SPP, W73-01089
MALAREN, MICROCYSTIS FLOS-AQUAE, PHRAGMITES COMMUNIS, *NORWAY, MAC W73-00286
YON, POLIOVIRUSES, PLASTICIZERS, PHTHALATE, THERMOPHILIC FUNGI, PO W72-09675
CTONEMA BORYANUM, LPP-1, LPP-1D, PHYCOVIRUS, MITOMYCIN C.: /, *PLE W72-03188
, RHODOMELA SP., ENDOCLADIA SP., PHYLLOSPADEX SP., MICROCLADIA SP. W71-11851
MISTRY, *PHOTOSYNTHESIS, *ALGAE, PHYLOGENY, CARBON CYCLE, ENZYMES, W71-09172
SILVER, CADMIUM, LEAD, CHLORELLA PHYRENOIDOSA, PHAEODACTYLUM TRICO W72-07660
OPTERA, LIMNEPHILUS, GLOSSOSOMA, PHYSA GYRINA, KEMMERER SAMPLER, S W72-14280
RI, STICTOCHIRONOMUS ANNULIORUS, PHYSA HETEROSTROPHA, PROCLADIUS, W70-06668
MPLES, LIVER, URINE, OEDOGONIUM, PHYSA, CULEX, GAMBUSIA, FATE OF P W72-07703
IUM, COBALT, FRESH WATER, LAKES, PHYSICAL PROPERTIES, CHEMICAL PRO W73-01434
SYSTEMATICS, CHEMICAL ANALYSIS, PHYSICAL PROPERTIES, CHLOROPHYTA, W73-01627
AGOONS, CHLOROPHYTA, RHODOPHYTA, PHYSICAL PROPERTIES, WATER TEMPER W73-00838
FECTS, POLLUTANT IDENTIFICATION, PHYSICAL PROPERTIES, BULK DENSITY W73-00854
INDICATORS, CHEMICAL PROPERTIES, PHYSICAL PROPERTIES, PHYTOPLANKTO W72-12729
, *FILTRATION, LIGHT PENTRATION, PHYSICAL PROPERTIES, ELECTROPHORE W72-11833
TION, FLOW, CHEMICAL PROPERTIES, PHYSICAL: /HYDROGEN ION CONCENTRA W72-12941
YTA, CHRYSOPHYTA, BIOINDICATORS, PHYSICOCHEMICAL PROPERTIES, MICRO W72-13142
L FLUCTUATIONS, *PUMPED STORAGE, PHYSICOCHEMICAL PROPERTIES, *PLAN W72-14280
DIOECOLOGY, *PATH OF POLLUTANTS, PHYSICOCHEMICAL PROPERTIES.: / RA W72-12204
S, THERMOCLINE, EPILIMNION, MUD, PHYSICOCHEMICAL PROPERTIES, HYDRO W72-03224
ENTATION, SEDIMENT DISTRIBUTION, PHYSICOCHEMICAL PROPERTIES, PESTI W72-00847
LAKE ONTARIO, LIGHT PENETRATION, PHYSICOCHEMICAL PROPERTIES, AQUAT W73-01615
EMICAL ANALYSIS, BIOMASS, ALGAE, PHYSICOCHEMICAL PROPERTIES, DIATO W73-01435
*RIVERS, *BACTERIA, IPLANKTON, PHYSICOCHEMICAL PROPERTIES, *ALGA W72-07896
UNDMENTS, SEASONAL, COLOR, IRON, PHYSICOCHEMICAL PROPERTIES, DIATO W73-00256
Y, TRACE ELEMENTS, HEAVY METALS, PHYSICOCHEMICAL PR: /AL PHYSIOLOG W73-00817
ENT, SECONDARY SEWAGE TREATMENT, PHYSICOCHEMICAL METHODS, AMMONIA W70-09907
MARINE ALGAE, WINDS, SEA WATER, PHYSICOCHEMICAL PROPERTIES.: /ES, W71-09225
ENVIRONMENT, ZINC RADIOISOTOPES, PHYSICOCHEMICAL PROPERTIES, FISH, W72-04463
UATIONS, BIOLOGICAL COMMUNITIES, PHYSIOCHEMICAL PROPERTIES, DISTRI W73-00918
DIOISOTOPES, PATH OF POLLUTANTS, PHYSIOLOGICAL ECOLOGY, ENVIRONMEN W72-06313
PHYTOPLANKTON, PLANT PHYSIOLOGY, PHYSIOLOGICAL ECOLOGY, ORGANIC CO W72-03216
PHYTOLOGY, PRIMARY PRODUCTIVITY, PHYSIOLOGICAL ECOLOGY, RADIOECOLO W72-00948
AL PHYSIOLOGY, PLANT PHYSIOLOGY, PHYSIOLOGICAL ECOLOGY, BIOASSAY, W71-12303
AINS, *ABSORPTION, BIOCHEMISTRY, PHYSIOLOGICAL ECOLOGY, CESIUM, ST W71-13243
BEGGIATOA, PHYTOPLANKTON BLOOMS, PHYSIOLOGICAL ECOLOGY, MORTALITY, W72-14056
RIA, LABORATORY EQUIPMENT, PLANT PHYSIOLOGY.: /T POPULATIONS, AQUA W72-14294
, WAT/ *ALGAE, *BIOASSAY, *PLANT PHYSIOLOGY, POLLUTANTS, *TOXICITY W72-12576
LIFE, AQUATIC ENVIRONMENT, FISH PHYSIOLOGY, ALGAE, DAPHNIA,: /TIC W72-12257
IS, LITTORAL, RESPIRATION, PLANT PHYSIOLOGY, BICARBONATES, WATER P W72-12940
OGEN FIXATION, BIOCONTROL, PLANT PHYSIOLOGY, MOVEMENT, VIRUSES, PH W71-13184
ASUREMENT, PHOTOSYNTHESIS, PLANT PHYSIOLOGY, PIGMENTS.: /TERNS, ME W71-13252
NETICS, ANIMAL PHYSIOLOGY, PLANT PHYSIOLOGY, PHYSIOLOGICAL ECOLOGY W71-12303
DUES, PESTICIDE KINETICS, ANIMAL PHYSIOLOGY, PLANT PHYSIOLOGY, PHY W71-12303
LUTION SOURCES, NUTRIENTS, PLANT PHYSIOLOGY, LAKES, PONDS, AQUATIC W71-10580
ICATORS, *PHOTOSYNTHESIS, *PLANT PHYSIOLOGY, POLLUTANTS, ANALYTICA W71-10553
METABOLISM, PHYTOPLANKTON, PLANT PHYSIOLOGY, PHYSIOLOGICAL ECOLOGY W72-03216
GEN FIXATION, *CYANOPHYTA, PLANT PHYSIOLOGY, CYTOLOGICAL STUDIES, W72-03217
ATES, TEMPERATURE, BIOCHEMISTRY, PHYSIOLOGY, LIGHT.: /SM, GROWTH R W72-00377
H OF POLLUTANTS, TRACERS, ANIMAL PHYSIOLOGY, NUCLEAR WASTES.: /PAT W72-03348
CONCENTRATION, MUSCLE, BONE PHYSIOLOGY/METABOLISM.: W72-10694
GAE, ABSORPTION, AQUARIA, ANIMAL PHYSIOLOGY, WATER BALANCE, TRACER W72-06342
Y, WATER BALANCE, TRACERS, PLANT PHYSIOLOGY, AQUATIC PLANTS, RADIO W72-06342
PRODUCTIVITY, RESPIRATION, PLANT PHYSIOLOGY, NUTRIENTS, NITRATES, W72-06274
YMES, *CULTURES, NITRATES, PLANT PHYSIOLOGY, PLANT GROWTH, BIOMASS W72-05476
AL ANALYSIS, BIOCHEMISTRY, PLANT PHYSIOLOGY, BIOLOGICAL COMMUNITIE W71-08597
MPERATURE, PLANT PIGMENTS, PLANT PHYSIOLOGY, *PLANT PATHOLOGY, PLA W70-09429
GROWTH RATES, INHIBITION, *PLANT PHYSIOLOGY, *PLANT GROWTH, PLANT W70-07842
TION, ANIMAL POPULATIONS, ANIMAL PHYSIOLOGY, RADIOACTIVITY TECHNIQ W73-00826
*WATER POLLUTION EFFECTS, *PLANT PHYSIOLOGY, *DIATOMS, PESTICIDES, W73-02280
RINE ENVIRONMENT, COASTS, ANIMAL PHYSIOLOGY, SEDIMENTS, ANIMAL MET W73-00811
ILATIVE CAPACITY, EUROPE, ANIMAL PHYSIOLOGY, TRACE ELEMENTS, HEAVY W73-00817
IPTORS:*DDT, *RESISTANCE, *PLANT PHYSIOLOGY, CULTURES, PHOTOSYNTHE W73-00281

PLANTS, TOXINS, RED TIDE, PLANT PHYSIOLOGY, BIOASSAY, CYTCLCGICAL W73-00243
RADIOACTIVITY, METABOLISM, PLANT PHYSIOLOGY, CHRYSOPHYTA, CULTURES W73-00380
Y, PHOTOSYNTHESIS, RIVERS, PLANT PHYSIOLOGY, TESTING PROCEDURES, A W72-14314
YDRATES, *ORGANICS ACIDS, *PLANT PHYSIOLOGY, *GROWTH RATES, PLANT W72-07710
TON, SEWAGE TREATMENT, NITROGEN, PHYSOCHEMICAL PROPERTIES, DIATOMS W72-13662
, DESCHAMPSIA FLEXUOSA, PARMELIA PHYSODES, HYPNUM CUPRESSIFORME.: / W72-14303
TION, INHIBITORS, PHYTOPLANKTCN, PHYTO-TOXICITY, HERBICIDES, INSEC W71-10566
ICS, PHARMACOLOGICAL PROPERTIES, PHYTOCHEMICAL SCREENINGS, NYMPHAE W72-04275
LOSIRA, TABELLARIA, ACTINASTRUM, PHYTOCONIS, CYMBELLA, STEPHANODIS W72-10875
CAL ANALYSIS, SAMPLING, ECOLOGY, PHYTOP: /WATER TEMPERATURE, CHEMI W72-07710
EAN, ECOLOGY, WATER TEMPERATURE, PHYTOPLANKTON, ZOOPLANKTCN, CHLOR W72-07892
BACTERIA, SAMPLING, ZOOPLANKTON, PHYTOPLANKTON, WATER ANALYSIS, PO W72-07896
NDS, SHELLFISH, INSTRUMENTATION, PHYTOPLANKTON, AUTOMATICN, EQUIPM W72-07715
ON, *AMINO ACIDS, *MARINE ALGAE, PHYTOPLANKTON, NITRCGEN, ABSORPTI W72-07497
P, CARBON, NITROGEN, PHOSPHORUS, PHYTOPLANKTON, CARBON DICXIDE, SE W72-07648
EWAGE TREATMENT PLANTS, DIATOMS, PHYTOPLANKTON, HYDORGEN ICN TEMPE W72-06837
PHICATION, ALGAE, AQUATIC WEEDS, PHYTOPLANKTON, OLIGOTROPHY, HARDN W72-08459
CIDE TOXICITY, *DDT, PESTICIDES, PHYTOPLANKTON, DIATCMS, ALGAE, MA W72-08436
UTRIENTS, BACTERIA, CHRYSOPHYTA, PHYTOPLANKTON, ZOOPLANKTCN, PYRRO W72-07899
NIQUES, MICHIGAN, SESTON, ALGAE, PHYTOPLANKTON, NANNOPLANKTON, AMM W72-08048
TS, *NUISANCE ALGAE, WASHINGTON, PHYTOPLANKTON, CYANCPHYTA, NITRAT W72-08049
*ALGAE, MEASUREMENT, CYANOPHYTA, PHYTOPLANKTON, PHOTOSYNTHESIS, PR W72-08054
TINENTAL SHELF, DEPTH, NITROGEN, PHYTOPLANKTON, HYDRCGRAPHY, NITRI W72-08056
EMENTS, CYANOPHYTA, CHLOROPHYTA, PHYTOPLANKTON, GRAZING.: / REQUIR W72-07940
A, *NEW HAMPSHIRE, MARINE ALGAE, PHYTOPLANKTON, AQUATIC PLANTS, CH W72-14542
ATLANTIC OCEAN, *GULF OF MEXICO, PHYTOPLANKTON, CYTOLOGICAL STUDIE W72-14313
F POLLUTANTS, ORGANIC COMPOUNDS, PHYTOPLANKTON, CARBON, WATER TEMP W73-00384
OZOA, MARINE ALGAE, ZOOPLANKTON, PHYTOPLANKTON.: /PYRROPHYTA, PROT W73-00296
TH RATES, ENVIRONMENTAL EFFECTS, PHYTOPLANKTON, CHLAMYDOMCNAS, CHL W73-00281
AGE, NITROGEN, PHOSPHORUS, IRON, PHYTOPLANKTON, DIATOMS, DINOFLAGE W73-00234
*ALGAE, *BIOMASS, *CYANOPHYTA, PHYTOPLANKTON, PRODUCTIVITY, PHOT W72-14812
GHT PENETRATION, PHOTOSYNTHESIS, PHYTOPLANKTON, STRATIFICATION, DI W73-00236
IN, LAKES, CLADOPHORA, ANALYSIS, PHYTOPLANKTON, CHEMICAL ANALYSIS, W73-00232
URCES, POLLUTANT IDENTIFICATION, PHYTOPLANKTON, MARINE ALGAE, ESSE W73-00441
PATH OF POLLUTANTS, FOOD CHAINS, PHYTOPLANKTON, MARINE ALGAE, ENVI W73-00818
NDS, FOOD CHAINS, PUBLIC HEALTH, PHYTOPLANKTON: /N, ORGANIC COMPOU W73-00821
IC ALGAE, CHRYSOPHYTA, PLANKTON, PHYTOPLANKTON.: /SM, ALGAE, AQUAT W73-02280
LIGHT INTENSITY, EUTROPHICATION, PHYTOPLANKTON, PRODUCTIVITY, ANAB W73-02472
ICAL ANALYSIS, LEAVES, SAMPLING, PHYTOPLANKTON, IRON, MAN: /, CHEM W73-01089
LANKTCN, BENTHIC FLORA, DIATCMS, PHYTOPLANKTON, PHOTOSYNTHESIS, PR W73-01615
RATURE, SAMPLING, AQUATIC ALGAE, PHYTOPLANKTON, TROPHIC LEVEL: /PE W73-01456
, NITROGEN, SILICON, ABSORPTION, PHYTOPLANKTON.: /ION, METHODOLOGY W73-01445
MICRCSCOPY, DIATOMS, NUTRIENTS, PHYTOPLANKTON, WATER QUALITY, MIC W72-10875
, MICROBIOLOGY, WATER TREATMENT, PHYTOPLANKTON, FUNGI, BACTERIA, A W72-10076
ING RATES, *LARVAL GROWTH STAGE, PHYTOPLANKTON, NUTRIENT REQUIREME W72-09248
*AQUATIC PLANTS, AQUATIC ALGAE, PHYTOPLANKTON, AQUATIC MICROORGAN W72-09650
ECOLOGY, BENTHOS, INVERTEBRATES, PHYTOPLANKTON, PLANKTCN, WATER TE W72-09668
CONDARY TREATMENT, MARINE ALGAE, PHYTOPLANKTON, BIOMASS, GROWTH RA W72-09378
SEDIMENTS, AQUATIC PRODUCTIVITY, PHYTOPLANKTON, ALGAE.: /ERFACES, W72-08989
Y, SEASONAL, CHLOROPHYLL, ALGAE, PHYTOPLANKTON, WATER ANALYSIS, CA W72-09108
*PHOTOSYNTHESIS, CARBON DIOXIDE, PHYTOPLANKTON, ALGAE, CULTURES, * W72-09102
ALGAE, *CYANOPHYTA, *BEGGIATOA, PHYTOPLANKTON, *NEW HAMPSHIRE, MI W72-09061
ULATORS, INHIBITION, INHIBITORS, PHYTOPLANKTON, PHYTO-TOXICITY, HE W71-10566
TION, SEA WATER, PHOTOSYNTHESIS, PHYTOPLANKTON, AMMONIUM COMPOUNDS W71-11008
DT, ORGANOPHOSPHORUS PESTICIDES, PHYTOPLANKTON, TOXICITY, DISSOLVE W71-10096
GELLATES, SEASONAL, ZOOPLANKTON, PHYTOPLANKTON.: /NOPHYTA, DINOFLO W71-11949
NE ALGAE, *LIGHT, *GROWTH RATES, PHYTOPLANKTON, NATURAL RESOURCES, W71-13252
GROWTH RATES, ALGAE, INCUBATION, PHYTOPLANKTON.: /ATLANTIC OCEAN, W71-11562
AR POWER PLANTS, STEAM, ECOLOGY, PHYTOPLANKTON, ZOOPLANKTON, LARVA W71-11517
PESTICIDE RESIDUES, ALGAE, DDT, PHYTOPLANKTON, MARINE ALGAE, AQUA W72-03545
ATIC POPULATIONS, MAPS, MAPPING, PHYTOPLANKTON, PACIFIC OCEAN, BOT W72-03567
AQUATIC LIFE, FISH POPULATIONS, PHYTOPLANKTON, BENTHIC FAUNA, PER W72-00939
NIC MATTER, BIOMASS, METABOLISM, PHYTOPLANKTON, PLANT PHYSIOLOGY, W72-03216
IATION, DISSOLVED OXYGEN, LAKES, PHYTOPLANKTON, NUTRIENTS,: /T RAD W72-01780
, ENGINEERING, WATER PROPERTIES, PHYTOPLANKTON, CHLOROPHYTA, PHAEO W72-03220
EN, EUTROPHICATION, OLIGOTROPHY, PHYTOPLANKTON, ANALYTICAL TECHNIQ W72-03224
ES, FOREST SOILS, SAMPLING, MUD, PHYTOPLANKTON, FLUORESCENCE, PHOT W72-01784
CKS, CRABS, OTTERS, ZOOPLANKTCN, PHYTOPLANKTON, MAMMALS, MARINE AL W72-00960
ION, DECOMPOSING ORGANIC MATTER, PHYTOPLANKTON, PRODUCTIVITY, IRON W72-01097
SEASONAL, IONS, ALGAE, SALINITY, PHYTOPLANKTON, THERMAL PROPERTIES W72-01363
TROPHICATION, ALGAE, CYANOPHYTA, PHYTOPLANKTON, LAKE MORPHOMETRY, W72-04269
GHT INTENSITY, DISSOLVED SOLIDS, PHYTOPLANKTON, ALGAE, NUTRIENTS.: W72-04270
IR OPERATION, WATER TEMPERATURE, PHYTOPLANKTON, EUTHROPHICATICN.: / W72-05282
MOSSES, INVERTEBRATES, DIATOMS, PHYTOPLANKTON, PHOTOSYNTHESIS, WA W72-05952
N EFFECTS, *PESTICIDES, DIATOMS, PHYTOPLANKTON, TOXICITY, CULTURES W72-05614
EMPERATURE, LIGHT, WAVES(WATER), PHYTOPLANKTON: /MPLING, OXYGEN, T W72-06283
TIVITY, *SECONDARY PRODUCTIVITY, PHYTOPLANKTON, ZOOPLANKTCN, EUTRO W72-05968
MA, CYANOPHYTA, CHLORELLA, IONS, PHYTOPLANKTON, NITRATES, NITRCGEN W72-04773
UNDMENT, BIOMASS, AQUATIC ALGAE, PHYTOPLANKTON, ZOOPLANKTCN, FISH W72-04853
*ALGAE, *MARINE ALGAE, CULTURES, PHYTOPLANKTON, CYTOLOGICAL STUDIE W70-07280
L, *PHOTOTROPHIC GROWTH, *MARINE PHYTOPLANKTON(PRYMNESIUM PARVUM)(W70-07842
WATER POLLUTION EFFECTS, *ALGAE, PHYTOPLANKTON, DIATOMS, CHRYSOPHY W70-07842
ORELLA, ROTIFERS, CHLAMYDOMONAS, PHYTOPLANKTON, ZOOPLANKTCN, PERIP W70-06655
UALITY, *LAKES, *EUTROPHICATION, PHYTOPLANKTON, PLANKTON, PRODUCTI W71-02880
AE, NITROGEN FIXATION, PLANKTON, PHYTOPLANKTON, BACTERIA, ZOOPLANK W71-00117
TION, AMMONIA, NITRITES, OXYGEN, PHYTOPLANKTON, BIOCHEMICAL OXYGEN W71-05390
HLORELLA, GROWTH RATES, DIATOMS, PHYTOPLANKTON, TOXICITY, IRON, LI W71-03027
, NUTRIENTS, CYANOPHYTA, MIXING, PHYTOPLANKTON.: /DISSOLVED OXYGEN W71-03034
RGANIC COMPOUNDS, *MARINE ALGAE, PHYTOPLANKTON, HUMIC ACIDS, FULVI W71-09188
CATION, RED TIDE, SILICA, ALGAE, PHYTOPLANKTON, FREEZING, TEMPERAT W71-09190

TTORAL, PLANT GROWTH SUBSTANCES, PHYTOPLANKTON, FLUID MECHANICS, E W71-07878
NA, NUTRIENTS, DISSOLVED SOLIDS, PHYTOPLANKON, WATER POLLUTION EFF W72-13300
*BOTTOM SEDIMENTS, *HYPOLIMNION, PHYTOPLANKTON, LIGHT PENETRATION, W72-12995
OGICAL STUDIES, ABSORPTION, DDT, PHYTOPLANKTON, PLANT GROWTH, CHLO W72-12998
ENTS, WATER QUALITY, PHOSPHORUS, PHYTOPLANKTON, SEWAGE TREATMENT, W72-13662
PROPERTIES, PHYSICAL PROPERTIES, PHYTOPLANKTON, ZOOPLANKTON, CHLOR W72-12729
OMETRY, CHLOROPHYLL, HYDROLYSIS, PHYTOPLANKTON, FLUORESCENCE, PHOS W72-12637
COMMUNITIES, ALGAE, PERIPHYTON, PHYTOPLANKTON, PHOTOSYNTHESIS, IN W72-12549
SIS, EXUDATION, WATER POLLUTION, PHYTOPLANKTON, WATER POLLUTION SO W72-12166
AL TREATMENT, OXIDATION LAGOONS, PHYTOPLANKTON, AQUATIC I: /OLOGIC W72-12567
ING, DINOFLAGELLATES, BEGGIATOA, PHYTOPLANKTON BLOOMS, PHYSIOLOGIC W72-14056
IOACCUMULATION, METAL COMPLEXES, PICEA ABIES, VACCINIUM VITIS IDAE W72-14303
UMPKINSEED, BLACK CRAPPIE, CHAIN PICKEREL, TENDIPES TENTANS, TRICH W72-14280
,6-TR/ TORDON, FRESHWATER ALGAE, PICLORAM, PEDIASTRUM, 4-AMINO-3,5 W72-08440
), SABLE ISLAND(NOVA SCOTIA), ST PIERRE(CANADA), MIQUELON(CANADA), W71-10066
YLL A, / MARGALEF PIGMENT RATIO, PIGMENT DIVERSITY INDEX, CHLOROPH W73-00916
INDEX, CHLOROPHYLL A, / MARGALEF PIGMENT RATIO, PIGMENT DIVERSITY W73-00916
NTS, RADIOISOTOPES, CHLOROPHYLL, PIGMENT, MAGNESIUM, PHOSPHATES, N W72-07660
HOTOSYNTHESIS, PLANT PHYSIOLOGY, PIGMENTS.: /TERNS, MEASUREMENT, P W71-13252
AS, WATER ANALYSIS, FLUOROMETRY, PIGMENTS.: /OMPOUNDS, CHLAMYDOMON W72-12637
PRIMARY PR/ *CHLOROPHYLL, *PLANT PIGMENTS, AQUATIC PRODUCTIVITY, * W72-12192
SPECTROPHOTOMETRY, LAKES, PLANT PIGMENTS, DETRITUS, CHLOROPHYTA, W72-13673
C BACTERIA, ELECTRON MICROSCOPY, PIGMENTS, CARBOHYDRATES, PROTEINS W72-07951
YLL, SOLAR RADIATION, VITAMIN B, PIGMENTS, GROWTH RATES, ALGAE, FO W72-07892
IN, MICHIGAN, CHRYSOPHYTA, PLANT PIGMENTS, AQUATIC ALGAE, NITROGEN W73-00384
CTANTS, WATER TEMPERATURE, PLANT PIGMENTS, PLANT PHYSIOLOGY, *PLAN W70-09429
LGAE, CONDUCTIVITY, CHLOROPHYLL, PIGMENTS, DISPERSION, RUNOFF.: /A W72-04773
MON, SHEEPSHEAD, CHUBS, NORTHERN PIKE, GIZZARD SHAD.: /S, COHO SAL W72-14282
CKERS, CATFISHES, BASS, PERCHES, PIKES, SAMPLING, SNAILS, ALGAE, D W70-08098
CTILUCA, SIPHONOPHORES, MEDUSAE, PILIDIUM, CLADOCERA, LINGULA, ACT W73-00935
LISM, BIOCHEMICAL OXYGEN DEMAND, PILOT PLANTS, *WASTE WATER TREATM W72-06838
EASIBILITY, COSTS, ANNUAL COSTS, PILOT PLANTS.: /TES, ECONOMICS, F W70-10174
ASIBILITY, ECONOMIC FEASIBILITY, PILOT PLANTS, ION EXCHANGE, ELECT W71-09450
CONDITIONS, AEROBIC CONDITIONS, PILOT PLANTS, ALGAE, HYDROGEN ION W71-07102
RATORY, DIGESTION TANK, PONDAGE, PILOT PLANT, SETTLING BASIN.: /BO W72-02850
RASS, ZOSTERA, PINAEUS DUORARUM, PINAEUS AZTECUS, PENAEUS SETIFERU W73-02029
W YORK BIGHT, EELGRASS, ZOSTERA, PINAEUS DUORARUM, PINAEUS AZTECUS W73-02029
DOMONAS REINHARDTII, MOSES LAKE, PINE LAKE, LAKE WASHINGTON, CHLOR W72-12637
AIR, DETROIT RIVER, BLACK RIVER, PINE RIVER, CLINTON RIVER, HURON W72-14282
S, AROCLOR 1254, ESCAMBIA RIVER, PINFISH, FLOUNDER, CROAKER, MENHA W71-07412
HIA ACTINASTROIDES, *STAURASTRUM PINGUE, SCENEDESMUS QUADRICAUDA, W72-12739
HAETOCEROS, MOUGEOTIA, NAVICULA, PINNULARIA, SURIRELLA, PERIDINIUM W72-09111
N HOT SPRINGS, LOLO HOT SPRINGS, PIPESTONE HOT SPRINGS, SLEEPING C W72-05610
YAMAMONAS TERRESTRIS, BRASILOBIA PISCIFORMIS, PROWSEMONAS TROPICA, W72-07683
TIA, ANEMONE, MOLGULA, TUNICATE, PISMO CLAM, TIVELA STULTORUM, OST W72-07907
N ION CONCENTRATION, FILTRATION, PLA: / CALCIUM CARBONATE, HYDROGE W73-01669
FRAGILARIA CONSTRUENS, COCCONEIS PLACENTULA, NAVICULA SRYTOCEPHALA W72-08141
IPHYTES, GROWTH MEDIA, COCCONEIS PLACENTULA, CLOSTERIUM, GOMPHONEM W72-12744
, COCCONEIS PEDICULUS, COCCONEIS PLACENTULA, CYCLOTELLA CF. SOCIAL W72-13142
RA, THALASSIONEMA NITZSCHIOIDES, PLAGIOGRAMMA VANHEURCKII, PLEUROS W73-00853
VICULA SPP, SYNEDRA FASCICULATA, PLAGIOGRAMMA BROCKMANNI, AMPHIPLE W73-00853
SOUTH AMERICA, ATLANTIC COASTAL PLAIN, COASTS, CRUSTACEANS, MARIN W73-00819
UT, *MOLLUSKS, *ATLANTIC COASTAL PLAIN, NORTH CAROLINA, ABSORPTION W73-00818
TTLE CREEK RESERVOIR(KAN), GREAT PLAINS, AQUATIC ECOSYSTEM.: / *TU W72-12393
EATION, GROUNDWATER, FISH, FLOOD PLAINS, IRRIGATION, PESTICIDES.: / W70-05103
ROPHICATION, *RESERVOIRS, *GREAT PLAINS, NUTRIENTS, RECREATION, LI W72-04759
KOLBEANA OVOIDEA, PALMERIAMONAS PLANCTONICA, MANCHUDINIUM SINICUM W72-07683
OSCOPY, ECOLOGICAL DISTRIBUTION, PLANKTON NETS, CHLOROPHYTA, PLANK W73-01094
PRING, SUMMER, AUTUMN, SAMPLING, PLANKTON NETS, DIATOMS, FOOD CHAI W72-12738
ION, PLANKTON NETS, CHLOROPHYTA, PLANKTON.: / ECOLOGICAL DISTRIBUT W73-01094
MASS, *CYANOPHYTA, PLANT GROWTH, PLANKTON.: /NTROL, NUTRIENTS, BIO W72-00799
OL, *CYANPHYTA, *NUISANCE ALGAE, PLANKTON, WASHINGTON, DIATOMS, CH W72-00798
ONTROL, PLANT GROWTH REGULATORS, PLANKTON, CULTURES, WATER POLLUTI W71-13793
ANT IDENTIFICATION, / *NITROGEN, PLANKTON, *EUTROPHICATION, POLLUT W72-01780
TION, FALLOUT, OYSTERS, MUSSELS, PLANKTON, MARINE ALG: /NT, REGULA W72-00959
DIMENT TRANSPORT, BENTHIC FAUNA, PLANKTON, BIOINDICATORS, CURRENTS W72-09092
S, INVERTEBRATES, PHYTOPLANKTON, PLANKTON, WATER TEMPERATURE, E. C W72-09668
KINETICS, *BIOLOGICAL TREATMENT, PLANKTON, *EUTROPHICATION, *NUTRI W71-13553
ONAL, ACTINOMYCETES, CYANOPHYTA, PLANKTON, ON-SITE TESTS, LABORATO W71-13187
SH, LAKE ERIE, SEDIMENTS, SILTS, PLANKTON, ALGAE, TRACE ELEMENTS, W71-11036
RY, CYANOPHYTA, BACTERIA, ALGAE, PLANKTON, PLANTS, SPECTROSCOPY, C W73-01439
R SILTING, LAKES, *SOUTH DAKOTA, PLANKTON, *ALGAE, *PHOSPHORUS COM W73-00938
N, *EUTROPHICATION, CHLOROPHYLL, PLANKTON, BIOLOGICAL COMMUNITIES, W73-00916
UNITIES, *STREAMS, LAKE ONTARIO, PLANKTON, DIATOMS, SEDIMENTS, HAR W73-01627
INFRARED RADIATION, HYDROGRAPHY, PLANKTON, ALGAE, ENVIRONMENTAL EF W73-01704
LIOGRAPHIES, *ALGAE, *LAKE ERIE, PLANKTON, BENTHIC FLORA, DIATOMS, W73-01615
SH, MICROORGANISMS, ODOR, TASTE, PLANKTON, CHEMICALS, PHENOLS, CHL W73-01669
TES, LIMITING FACTORS, NITROGEN, PLANKTON, DEPTH, ALGAE.: /, NITRA W73-02472
GAE, AQUATIC ALGAE, CHRYSOPHYTA, PLANKTON, PHYTOPLANKTON.: /SM, AL W73-02280
ATER TEMPERATURE, WATER QUALITY, PLANKTON, ANALYTICAL TECHNIQUES, W72-07663
PHY, PHAEOPHYTA, EUTROPHICATION, PLANKTON, ATLANTIC OCEAN, SEA WAT W72-07710
C ASPECTS, NUTRIENTS, CHEMICALS, PLANKTON, BENTHOS, ALGAE, FISH, W W72-07841
, *LIGHT INTENSITY, *PHOSPHORUS, PLANKTON, CHLOROPHYTA, PYRROPHYTA W72-07166
RIENTS, TURBIDITY, PRODUCTIVITY, PLANKTON, STANDING CROP, LIMNOLOG W72-08559
NT, SAMPLING, CHEMICAL ANALYSIS, PLANKTON, COMPUTER PROGRAMS, BIOM W72-08560
MARINE ALGAE, BOTTOM SEDIMENTS, PLANKTON, RADIOACTIVITY TECHNIQUE W73-00361
FLORA, ALGAE, LIGHT PENETRATION, PLANKTON, NUTRIENTS, BIOCHEMICAL W72-13659
ICIDES, WATER POLLUTION SOURCES, PLANKTON, OYSTERS, SHEL: /S, HERB W72-13347
WATER POLLUTION SOURCES, ALGAE, PLANKTON, SEASONAL.: /POLLUTION), W72-05506
ER, CALCIUM CARBONATE, DETRITUS, PLANKTON, ADSORPTION, EUTROPHICAT W72-05469

IRS, *NEBRASKA, *EUTROPHICATION, PLANKTON, CHLOROPHYLL, RECREATION W72-04270
LT VALLEY RESERVOIRS(NEB), ULTRA PLANKTON, C-14.: *SA W72-04270
FISH ATTRACTANTS, FISH HARVEST, PLANKTON, BENTHIC FAUNA, ALGAE, E W70-09596
PHOTOSYNTHESIS, ORGANIC MATTER, PLANKTON, CHELATION, CARBON RADIO W71-03027
LAKE MORPHOLOGY, SEDIMENTATION, PLANKTON, DIEL MIGRATION, PROTOZO W71-05626
P, ALGAE, ROOTED AQUATIC PLANTS, PLANKTON, ARKANSAS, INSECTS, TURB W71-04528
RSION, ALGAE, NITROGEN FIXATION, PLANKTON, PHYTOPLANKTON, BACTERIA W71-00117
*EUTROPHICATION, PHYTOPLANKTON, PLANKTON, PRODUCTIVITY, PRIMARY P W71-02880
LGAE, FLOATING PLANTS, SEAWATER, PLANKTON, BIODEGRADATION, LIGHT, W71-09261
NVIRONMENTAL EFFECTS, LIMNOLOGY, PLANNING, EUTROPHICATION, EVALUAT W72-06526
RIENTS, WATER QUALITY, NITRATES, PLANNING, NITROGEN, PHOSPHORUS, R W71-13172
A, NUISANCE ALGAE, COMPREHENSIVE PLANNING, AQUATIC WEEDS, WATER PO W72-09171
IFERIN-LUCIFERASE ENZYME SYSTEM, PLANT EXTRACTS, CHLORELLA SOROKIN W71-11916
TION, *ALGAL CONTROL, FINANCING, PLANT GROWTH, TOXICITY, NUTRIENTS W71-13172
TORY TESTS, CYTOLOGICAL STUDIES, PLANT GROWTH, SCENEDESMUS.: /BORA W71-12858
TOSYNTHESIS, PESTICIDE TOXICITY, PLANT GROWTH REGULATORS, INHIBITI W71-10566
, APPLICATION METHODS, GRANULES, PLANT GROWTH REGULATORS, INHIBITO W72-09650
ES, PRODUCTIVITY, ANIMAL GROWTH, PLANT GROWTH, OYSTERS, SHELLFISH, W72-09378
CYCLING NUTRIENTS, BICARBONATES, PLANT GROWTH, DURNAL,: /DIOXIDE, W72-09159
TS, NITROGEN, PHOSPHORUS, ALGAE, PLANT GROWTH, WATER POLLUTION TRE W72-11702
QUES, *ALGAE, *NUTRIENTS, LAKES, PLANT GROWTH, *MOLYBDENUM, BIOASS W72-00801
YTA, *INHIBITORS, ALGAL CONTROL, PLANT GROWTH REGULATORS, PLANKTON W71-13793
CULTURES, ENVIRONMENTAL EFFECTS, PLANT GROWTH, SULFATE, LABORATORY W72-00153
NUTRIENTS, BIOMASS, *CYANOPHYTA, PLANT GROWTH, PLANKTON.: /NTROL, W72-00799
RES, NITRATES, PLANT PHYSIOLOGY, PLANT GROWTH, BIOMASS, GROWTH RAT W72-05476
STRONTIUM RADIOISOTOPES, ALGAE, PLANT GROWTH, TEMPERATURE, PROTOZ W71-09254
GAE, *FLUID FRICTION, *LITTORAL, PLANT GROWTH SUBSTANCES, PHYTOPLA W71-07878
TESTS, CULTURES, *GROWTH RATES, PLANT GROWTH REGULATORS, PLANT GR W71-03095
RATES, PLANT GROWTH REGULATORS, PLANT GROWTH SUBSTANCES, ALGAE, I W71-03095
NT PHYSIOLOGY, *PLANT PATHOLOGY, PLANT GROWTH REGULATORS, SALINITY W70-09429
PLANT PHYSIOLOGY, *PLANT GROWTH, PLANT GROWTH SUBSTANCES, PLANT MO W70-07842
ABSORPTION, DDT, PHYTOPLANKTON, PLANT GROWTH, CHLOROPHYLL, DDE, I W72-12998
LAKES, MODEL STUDIES, NUTRIENTS, PLANT GROWTH, ANALYTICAL TECHNIQU W72-13651
LANDS, ON-SITE DATA COLLECTIONS, PLANT GROWTH.: /, *ARIZONA, GRASS W72-12505
HNIQUES, CULTIVATION, VITAMIN B, PLANT GROWTH REGULATORS, NITRATES W73-00430
EGU/ *CARBON, *NITROGEN, *ALGAE, PLANT GROWTH, SEWAGE EFFLUENTS, R W72-14790
, PESTICIDES, HERBICIDES, 2-4-D, PLANT GROWTH REGULATORS, ALGAE, A W72-08440
PLANT PHYSIOLOGY, *GROWTH RATES, PLANT MORPHOLOGY, CULTURES, PHOTO W72-07710
S, *ALGAE, *CYTYLOGICAL STUDIES, PLANT MORPHOLOGY, EQUATIONS, RESI W73-01630
GROWTH, PLANT GROWTH SUBSTANCES, PLANT MORPHOLOGY.: /LOGY, *PLANT W70-07842
NCES, COMPETITION, GROWTH RATES, PLANT MORPHOLOGY, WATER CIRCULATI W72-04766
ODOPHYTA, PHAEOPHYTA, CHLORELLA, PLANT PATHOLOGY, INDUSTRIES, IONS W71-09157
CHEMICAL ANALYSIS, BIOCHEMISTRY, PLANT PHYSIOLOGY, BIOLOGICAL COMM W71-08597
, *ENZYMES, *CULTURES, NITRATES, PLANT PHYSIOLOGY, PLANT GROWTH, B W72-05476
SIOLOGY, WATER BALANCE, TRACERS, PLANT PHYSIOLOGY, AQUATIC PLANTS, W72-06342
IMARY PRODUCTIVITY, RESPIRATION, PLANT PHYSIOLOGY, NUTRIENTS, NITR W72-06274
TER TEMPERATURE, PLANT PIGMENTS, PLANT PHYSIOLOGY, *PLANT PATHOLOG W70-09429
QUATIC PLANTS, TOXINS, RED TIDE, PLANT PHYSIOLOGY, BIOASSAY, CYTOL W73-00243
ECTS, RADIOACTIVITY, METABOLISM, PLANT PHYSIOLOGY, CHRYSOPHYTA, CU W73-00380
EMISTRY, PHOTOSYNTHESIS, RIVERS, PLANT PHYSIOLOGY, TESTING PROCEDU W72-14314
YNTHESIS, LITTORAL, RESPIRATION, PLANT PHYSIOLOGY, BICARBONATES, W W72-12940
, AQUARIA, LABORATORY EQUIPMENT, PLANT PHYSIOLOGY.: /T POPULATIONS W72-14294
MASS, METABOLISM, PHYTOPLANKTON, PLANT PHYSIOLOGY, PHYSIOLOGICAL E W72-03216
*NITROGEN FIXATION, *CYANOPHYTA, PLANT PHYSIOLOGY, CYTOLOGICAL STU W72-03217
ER POLLUTION SOURCES, NUTRIENTS, PLANT PHYSIOLOGY, LAKES, PONDS, A W71-10580
NIRTROGEN FIXATION, BIOCONTROL, PLANT PHYSIOLOGY, MOVEMENT, VIRUS W71-13184
NS, MEASUREMENT, PHOTOSYNTHESIS, PLANT PHYSIOLOGY, PIGMENTS.: /TER W71-13252
IDE KINETICS, ANIMAL PHYSIOLOGY, PLANT PHYSIOLOGY, PHYSIOLOGICAL E W71-12303
IQUES, SPECTROPHOTOMETRY, LAKES, PLANT PIGMENTS, DETRITUS, CHLOROP W72-13673
ISCONSIN, MICHIGAN, CHRYSOPHYTA, PLANT PIGMENTS, AQUATIC ALGAE, NI W73-00384
SURFACTANTS, WATER TEMPERATURE, PLANT PIGMENTS, PLANT PHYSIOLOGY, W70-09429
ANT GROWTH REGULATORS, SALINITY, PLANT POPUL: /PLANT PATHOLOGY, PL W70-09429
DICATORS, FOOD CHAINS, TITANIUM, PLANT POPULATIONS, TOXICITY.: /IN W71-03095
PESTICIDES, PESTICIDE TOXICITY, PLANT POPULATIONS, AQUARIA, LABOR W72-14294
IS, RESPIRATION, EUTROPHICATION, PLANT POPULATION, CLASSIFICATION, W72-12192
LECULAR STRUCTURE, PERMEABILITY, PLANT POPULATIONS, ALGAE, CHEMICA W72-01473
S, CYCLING NUTRIENTS, SOIL-WATER- PLANT RELATIONSHIPS, RADIOISOTOPE W72-00949
R POLLUTION, LICHENS, SOIL-WATER- PLANT RELATIONSHIPS, MARINE ALGAE W72-03347
YCLING NUTRIENTS, E/ *SOIL-WATER- PLANT RELATIONSHIPS, *FALLOUT, *C W72-03342
HAEOCYSTIS SPP., SOI/ SILT-WATER- PLANT RELATIONSHIPS, NORTH SEA, P W72-12239
ROCESSING/ FEEDLOTS, COMPOSTING, PLANT RESIDUES, RECYCLING FOOD, P W72-00846
URCES, SEDIMENTS, AQUATIC SOILS, PLANT TISSUES, FRESHWATER FISH, A W73-01673
RADIATION, CYTOLOGICAL STUDIES, PLANT TISSUES, ALGAE, MICROORGANI W73-01676
GROWTH, RHODOPHYTA, INHIBITORS, PLANT TISSUES, BIOASSAY, PHAEPHYT W72-07496
USTER ANALYSIS(FISH), *DICKERSON PLANT.: *CL W72-07791
WINDSCALE PROCESSING PLANT.: W73-00828
*NORTHPORT PLANT(LONG ISLAND).: W72-14659
T BAY, CONNETICUT YANKEE NUCLEAR PLANT, CONNECTICUT RIVER, TURKEY W72-07907
THERMAL EFFECTS/ SAVANNAH RIVER PLANT, CULTURE MEDIA, DRY WEIGHT, W71-02075
N JOAQUIN RIVER, MORRO BAY POWER PLANT, HUMBOLDT BAY, CONNETICUT Y W72-07907
TUXENT RIVER, CONTRA COSTA POWER PLANT, SAN JOAQUIN RIVER, MORRO B W72-07907
, DIGESTION TANK, PONDAGE, PILOT PLANT, SETTLING BASIN.: /BORATORY W72-02850
UTION, DES MOINES RIVER, PACKING PLANTS.: INDUSTRIAL WATER POLL W71-06445
LITY, COSTS, ANNUAL COSTS, PILOT PLANTS.: /TES, ECONOMICS, FEASIBI W70-10174
ONTARIO, ALGAE, MOSSES, AQUATIC PLANTS.: /LLUTANT IDENTIFICATION, W71-11529
CRUSTACEANS, MARINE FISH, MARSH PLANTS.: /ALGAE , SALT TOLERANCE, W73-00855
ER PO/ *RUNNING WATERS, *AQUATIC PLANTS, *AQUATIC ENVIRONMENT, WAT W72-14690
LLUTANT IDENTIFICATION, *AQUATIC PLANTS, *AQUEOUS SOLUTIONS, *BIOL W72-01473
MATICS, BENTHOS, SCUBA / *MARINE PLANTS, *ALASKA, SAMPLING, *SYSTE W72-09120
NKTON, *RESERVOIRS, *R/ *AQUATIC PLANTS, *AQUATIC ALGAE, *PHYTOPLA W72-04855

NALYTICAL TECHNIQUES, / *AQUATIC	PLANTS, *ALGAE, *PUBLIC HEALTH, A	W72-04275
PROPERTIES, *PLANKTON, *AQUATIC	PLANTS, *AQUATIC ANIMALS, *MODEL	W72-14280
GAE, / *ALGAL TOXINS, *POISONOUS	PLANTS, *CYANOPHYTA, *NUISANCE AL	W72-14056
HERMAL POLLUTION, *NUCLEAR POWER	PLANTS, *COOLING WATER, *WATER TE	W70-09612
GAE, *ESTUARIES, PRODU/ *AQUATIC	PLANTS, *DISTRIBUTION, *MARINE AL	W72-04766
ICAL COMMUNITIES, *THERMAL POWER	PLANTS, *DISCHARGE(WATER), AQUATI	W71-11517
CLEAR POWERPLANTS, *DESALINATION	PLANTS, *ELECTRIC POWER PRODUCTIO	W72-03327
*WATER QUALITY, *HYDROELECTRIC	PLANTS, *MATHEMATICAL MODELS, THE	W72-14405
*ENVIRONMENTAL EFFECTS, *MARINE	PLANTS, *MARINE ANIMALS, OFFSHORE	W73-02029
AGULATION, WATER POLLU/ *AQUATIC	PLANTS, *MINNESOTA, *TOXICITY, CO	W72-07360
GI, PROTOZOA, PATHOGEN/ *AQUATIC	PLANTS, *MINNESOTA, BACTERIA, FUN	W72-08586
ION EFFECTS, *INTERTIDA/ *MARINE	PLANTS, *OILY WATER, WATER POLLUT	W71-08792
OLLUTION CONTROL, *NUCLEAR POWER	PLANTS, *POTABLE WATER, *BIOTA, S	W71-09018
COMPOUNDS, *PHOSPHORUS, *MARINE	PLANTS, *PHYTOPLANKTON, ALGAE, LI	W71-10083
ERIA, *AQUATIC ANIMALS, *AQUATIC	PLANTS, *RESERVOIR STAGES, EARLY	W72-04853
ON, *ENVIRONM/ *NEW YORK, *POWER	PLANTS, *STREAM, *THERMAL POLLUTI	W72-14659
EWS, WATER TEMPERATURE, / *POWER	PLANTS, *THERMAL POLLUTION, *REVI	W71-07417
MAL STRATIFICATION, HYDR/ *POWER	PLANTS, *THERMAL POLLUTION, *THER	W70-09614
BIOCHEMICAL OXYGEN DEMAND, PILOT	PLANTS, *WASTE WATER TREATMENT, W	W72-06838
STRY, *OXIDATION LAGOONS, *PILOT	PLANTS, ALGAE, BIOCHEMICAL OXYGEN	W70-06619
TIONS, AEROBIC CONDITIONS, PILOT	PLANTS, ALGAE, HYDROGEN ION CONCE	W71-07102
BIOLOGICAL COMMUNITIES, *AQUATIC	PLANTS, ALGAE, BACTERIA, LITTORAL	W72-03216
GEN DEMAND, HYPOLIMNION, AQUATIC	PLANTS, ALGAE.: /LVED OXYGEN, OXY	W72-12994
S, FISH, BIRDS, STREAMS, AQUATIC	PLANTS, AQUATIC INSECTS, CRUSTACE	W72-12930
S, *PESTICIDE TOXICITY, *AQUATIC	PLANTS, AQUATIC ALGAE, PHYTOPLANK	W72-09650
ANIMALS, CARP, DAPHNIA, AQUATIC	PLANTS, AQUATIC ALGAE, SCENEDESMU	W72-09649
AQUATIC MICROORGANISMS, AQUATIC	PLANTS, AQUATIC ALGAE, PESTICIDES	W72-09646
OTOPES, *MARINE ANIMALS, *MARINE	PLANTS, AQUATIC ANIMALS, LABORATO	W71-09168
STREAMS, RIVERS, ROOTED AQUATIC	PLANTS, AQUATIC ALGAE, WATER TEMP	W72-14690
SAY, CYANOPHYTA, RESINS, AQUATIC	PLANTS, AQUATIC ALGAE, NITRATES,	W73-00286
*INDICATORS, *SEA WATER, MARINE	PLANTS, ASSESSMENTS, MONITORING,	W71-12867
IBUTION PATTERNS, LAKES, AQUATIC	PLANTS, AUSTRALIA, SAMPLING, MICR	W73-01094
, *MARINE MICROORGANISMS, MARINE	PLANTS, BACTERIA, POPULATION, HAW	W71-08042
GY BUDGET, *MARINE ALGAE, MARINE	PLANTS, BIOMASS, CANADA, COASTS,	W73-00366
, OLIGO/ *PERIPHYTON, *SUBMERGED	PLANTS, BIOMASS, NEMATODES, ALGAE	W71-11012
BON DIOXIDE, RESPIRATION, MARINE	PLANTS, BICARBONATES, MASS SPECTR	W72-12709
E, ENVIRONMENTAL EFFECTS, MARINE	PLANTS, CANADA, WASHINGTON, OREGO	W73-00428
NE ALGAE, PHYTOPLANKTON, AQUATIC	PLANTS, CHEMICAL PROPERTIES, WATE	W72-14542
NG NUTRIENTS, RESPIRATION, MARSH	PLANTS, CHEMICAL ANALYSIS, LEAVES	W73-01089
OMPOUNDS, POTASH, ALGAE, AQUATIC	PLANTS, CHEMICAL ANALYSIS, WATER	W71-12252
NOFF, PHOSPHORUS, ALGAE, AQUATIC	PLANTS, CLAYS, ODOR, WATER QUALIT	W72-04759
SSAY, TOXINS, POISONS, POISONOUS	PLANTS, CORAL, INVERTEBRATES, ALG	W71-08597
EMNITELLA AMERICANA, TERRESTRIAL	PLANTS, COMBUSTION, ISOTOPE RATIO	W72-12709
LIMITING FACTORS, ROOTED AQUATIC	PLANTS, CYCLING NUTRIENTS, RESPIR	W73-01089
WASTES, *MARINE ANIMALS, *MARINE	PLANTS, DEPTH, CALIFORNIA, SANDS,	W71-01475
BIODEGRADATION, SEWAGE TREATMENT	PLANTS, DIATOMS, PHYTOPLANKTON, H	W72-06837
ERS, AGRICULTURE, ROOTED AQUATIC	PLANTS, DRAINAGE, RUNOFF, FERTILI	W71-13172
ISH, HEATED WATER, NUCLEAR POWER	PLANTS, ELECTRICAL POWER PLANTS,	W72-07907
RATURE, DATA COLLECTIONS, MARINE	PLANTS, ENVIRONMENTAL EFFECTS, RE	W73-00838
NT FLOW, ALGAE, BACTERIA, MARINE	PLANTS, ENGINEERING, WATER PROPER	W72-03220
AUNA, PERIPHYTON, AQUATIC ALGAE,	PLANTS, ENVIRONMENTAL ENGINEERING	W72-00939
MARINE ALGAE, *ISOLATION, MARINE	PLANTS, EUTROPHICATION, OCEANS, S	W73-00430
ATER POLLUTION EFFECTS, *AQUATIC	PLANTS, HARVESTING, PONDS, WATER	W71-01488
R POWER PLANTS, ELECTRICAL POWER	PLANTS, HEAT TRANSFER, WATER POLL	W72-07907
NE ALGAE, MARINE ANIMALS, MARINE	PLANTS, HUMAN POPULATION, RADIOAC	W73-00833
SICOCHEMICAL PROPERTIES, AQUATIC	PLANTS, INDUSTRIAL: /TRATION, PHY	W73-01615
ONS, ALGAE, ECOTYPES, INDUSTRIAL	PLANTS, INDUSTRIAL WASTES, OIL WA	W71-13746
ITY, ECONOMIC FEASIBILITY, PILOT	PLANTS, ION EXCHANGE, ELECTROLYSI	W71-09450
*EUTROPHICATION, LAKES, AQUATIC	PLANTS, IRON, MICROORGANISMS, COL	W72-03742
CTS, METHODOLOGY, ALGAE, AQUATIC	PLANTS, ISOTOPE FRACTIONATION, CA	W72-12709
, EFFLUENTS, SEWAGE, DETERGENTS,	PLANTS, LA: /IC MATTER, LAKE ERIE	W71-11017
BIOLOGICAL COMMUNITIES, AQUATIC	PLANTS, LIMITING FACTORS, ROOTED	W73-01089
FALLOUT, *MARINE ANIMALS, MARINE	PLANTS, MARINE ALGAE, NUCLEAR WAS	W73-00828
NE ALGAE, *CARBOHYDRATES, MARINE	PLANTS, MARINE MICROORGANISMS, WA	W73-02100
HICATION, FERTILIZATION, AQUATIC	PLANTS, MATHEMATICAL MODELS, SELF	W71-12072
*METABOLISM, *ABSORPTION, MARINE	PLANTS, MARINE ALGAE, PHAEPHYTA,	W71-09193
LUTION EFFECTS, MOLLUSKS, MARINE	PLANTS, MARINE ALGAE, PUBLIC HEAL	W72-04461
HEMICAL PROPERTIES, FISH, MARINE	PLANTS, MARINE ALGAE, HYDROLYSIS,	W72-04463
ITY, CALIFORNIA, TRITIUM, MARINE	PLANTS, MONITORING, SAMPLING, WAT	W73-00831
*NITROGEN FIXATION, *CYANOPHYTA,	PLANTS, NITROGEN, REDUCTION(CHEMI	W73-01456
OSP/ *SALT RIVER(KENT), VASCULAR	PLANTS, NITRATE NITROGEN, ORTHOPH	W72-06526
CHNIQUES, FERTILIZATION, AQUATIC	PLANTS, NITROGEN, NUTRIENT REMOVA	W73-02478
, MOSQUITOES, BACTERIA, FLOATING	PLANTS, ON-SITE INVESTIGATIONS, T	W71-01488
S, OIL SPILLS, SEA WATER, MARINE	PLANTS, PHAEOPHYTA, OILY WATER, E	W73-01074
PIA, CARP, ALGAE, ROOTED AQUATIC	PLANTS, PLANKTON, ARKANSAS, INSEC	W71-04528
S, ECOLOGY, FISH, ALGAE, AQUATIC	PLANTS, PROTOZOA, BACTERIA, STREA	W73-01972
REPRODUCTION, SEA WATER, MARINE	PLANTS, PRIMARY PRODUCTIVITY, COR	W72-14294
, *ECONOMICS, CYANOPHYTA, ALGAE,	PLANTS, PROTEINS, NUTRIENTS, BIOM	W72-03327
IANS, TOXICITY, CHARA, SUBMERGED	PLANTS, PROTOZOA, EUGLENA, CHLAMY	W72-00155
ACERS, PLANT PHYSIOLOGY, AQUATIC	PLANTS, RADIOECOLOGY.: /LANCE, TR	W72-06342
NDUSTRIAL WASTES, ALGAE, AQUATIC	PLANTS, SEA WATER, TRACE ELEMENTS	W72-05952
ULING, *NUISANCE ALGAE, FLOATING	PLANTS, SEAWATER, PLANKTON, BIODE	W71-09261
HYTA, BACTERIA, ALGAE, PLANKTON,	PLANTS, SPECTROSCOPY, CHROMATOGRA	W73-01439
CTS, HEATED WATER, NUCLEAR POWER	PLANTS, STEAM, ECOLOGY, PHYTOPLAN	W71-11517
CHNIQUES, *ALGAL TOXINS, AQUATIC	PLANTS, TOXINS, RED TIDE, PLANT P	W73-00243
OIRS, TERTIARY TREATMENT, ALGAE,	PLANTS, TROUT, AQUATIC ENVIRONMEN	W72-02412
S, MEASUREMENT, *NEVADA, AQUATIC	PLANTS, WATER POLLUTION SOURCES,	W73-01673
QUATIC ANIMALS, *BIOTA, *AQUATIC	PLANTS, WATER QUALITY, ENVIRONMEN	W72-06526
DIGESTION, NEUROANATOMY, OXYGEN	PLASMA ASHING, ATOMIC ABSORPTION,	W71-11036
S, BRAIN, BLOOD, KIDNEYS, LIVER,	PLASMODIUM BERGHEI, BORRELIA KANS	W73-01676

BRID, PELODICTYON, POLIOVIRUSES,	PLASTICIZERS, PHTHALATE, THERMOPH	W72-09675
M(PA.), ATRAZINE, BIOTYPES, POUR	PLATE METHOD, MOST PROBABLE-NUMBE	W72-04743
*PARALLEL	PLATE MODEL.:	W73-01362
LANTS, *MARINE ANIMALS, OFFSHORE	PLATFORMS, NUCLEAR POWERPLANTS, W	W73-02029
MICROWAVE RADIOMETRY, VI/ SPACE	PLATFORMS, LEAD, MERCURY, LASERS,	W72-04742
S, VOLVOCACEAE, PANDORINA MORUM,	PLATYDORINA CAUDATA, EUDORINA CAL	W71-13793
TRICORNUTUM, GONYAULAX POLYEDRA,	PLATYMONAS VIRIDIS, CHAETOCEROS C	W72-07166
BON EXTRUSION, CARBON-14, NITRO/	PLATYMONAS, NITZSCHIA OVALIS, CAR	W72-07497
ONECTIDAE, BODO, HYDROPSYCHIDAE,	PLECOPTERA, MAGALOPTERA, EPHEMERO	W70-06655
ITTLE ELK RIVER, OCTORARA RIVER,	PLECTONEMA, LYNGBYA, PHORMIDIUM,	W72-04736
IPTERA, GULFS, GEOLOGIA HISTORY,	PLEISTORENE EPOCH, ADAPTATION, EN	W73-00855
R. DAWSONII, CALLOPHYLLIS FIRMA,	PLENOSPORIUM DASYOIDES, SELENASTR	W73-02099
UCOTHRIX MUCOR, MAJURO ATOLL, M/	PLEOPODS, UROPODS, ENRICHMENT, LE	W72-08142
PHYCEAE, POLYEDRIELLA HELVETICA,	PLEUROCHLORIS COMMUTATA, PLEUROCH	W72-14808
VETICA, PLEUROCHLORIS COMMUTATA,	PLEUROCHLORIS MAGNA, VISCHERIA PU	W72-14808
IRCUMFLEXUS, PHACUS ONYX, PHACUS	PLEURONECTES, PHACUS PYRUM, TRACH	W73-00918
OIDES, PLAGIOGRAMMA VANHEURCKII,	PLEUROSIGMA, GYROSIGMA FASCIOLA,	W73-00853
DISCUS, NITZSCHIA, OSCILLATORIA,	PLEUROSIGMA, NAUPLII, LAMELLIBRAN	W71-11876
NAS BRASILIANA, HORTOBAGGIAMONAS	PLICATA, TRICHOCYANELLA SPIRALIS,	W72-09119
IO), NUTRIENT INPUTS, CHIRONOMUS	PLUMOSUS, CHIRONOMUS ATTENTUATUS,	W71-11009
E SULFONATE, COCONUT FATTY ACID,	PLURONICS, DIETHANOLAMIDE.: /NZEN	W72-12576
EED, STARFISH, ASTERIAS FORBESI,	PLUTONIUM 236, PLUTONIUM 238, PLU	W72-05965
ASTERIAS FORBESI, PLUTONIUM 236,	PLUTONIUM 238, PLUTONIUM 239, PCL	W72-05965
I, PLUTONIUM 236, PLUTONIUM 238,	PLUTONIUM 239, POLYMERIZATION, LI	W72-05965
M RADIOISOTOPES, IRON RADIOISOT/	PLUTONIUM RADIOISOTOPES, RUTHENIU	W72-04463
	PLUTONIUM.:	W72-04461
*POLONIUM,	PLUTONIUM.:	W73-00833
BACILLARIOPHAGA, REFRACTOCYSTIS	PLUVIALIS, GOMESIAMONAS STAGNALIS	W72-09119
N, BACTERIA, OXIDATION-REDUCTION	PO: /MPERATURE, COPEPODS, NITROGE	W71-11008
MIDE, PENICILLIN G, POLYMYXIN B,	PODOCYSTS.: /RACYCLINE, SULFANILA	W72-10162
UFLEXA, ANKISTRODESMUS FALCATUS,	PODOHEDRA.: MOUGEOTIA GEN	W71-07360
LORIDA), HEATED EFFLUENT, TURKEY	POINT(FLORIDA), TURTLE GRASS, THA	W72-11876
PLANT, CONNECTICUT RIVER, TURKEY	POINT, BISCAYNE BAY, FLORIDA, BAL	W72-07907
IVER, CONTRA COSTA POWER / CHALK	POINT, COLUMBIA RIVER, PATUXENT R	W72-07907
LETIO/ *LAKE ERIE CENTRAL BASIN,	POINTE PELEE(ONTARIO), OXYGEN DEP	W72-12992
RUSTACEANS, *FISH TOXINS, *ALGAL	POISONING, DINOFLAGELLATES, BEGGI	W72-14056
ATIC ALGAE, ALGAL CONTROL, ALGAL	POISONING, AQUATIC ENVIRONMENT, A	W72-11186
ATES, ALGAE, MARINE ALGAE, ALGAL	POISONING, CHROMATOGRAPHY, CHEMIC	W71-08597
TS, ENVIRONMENTAL EFFECTS, ALGAL	POISONING, ALGAE, SOLVENTS, SURFA	W70-09429
ANTS, BIOASSAY, TOXINS, POISONS,	POISONOUS PLANTS, CORAL, INVERTEB	W71-08597
BIOASSAY, ENVIRONMENT, TOXICITY,	POISONS, ALGAL TOXINS, FISH TOXIN	W72-10678
OF POLLUTANTS, BIOASSAY, TOXINS,	POISONS, POISONOUS PLANTS, CORAL,	W71-08597
ASSAY, RESISTANCE, LETHAL LIMIT,	POISONS, WATER POLLUTION EFFECTS.	W72-14706
, WATER POLLUTION EFFECTS, WATER	POL: /ON, WATER POLLUTION SOURCES	W72-04279
HIA ECHINULATA, MIKOLAJSKIE LAKE(	POLAND), CERATIUM HIRUNDINELLA.: /	W72-14812
*MACROPHYTES, *MIKOLAJSKIE LAKE(	POLAND), COLONIZATION, POTAMOGETO	W71-11012
PACIFIC OCEAN, TROPICAL REGIONS,	POLAR REGIONS, CYANOPHYTA, CHRYSO	W72-03543
TRENDS, CLADOPHORA, R/ *RESEARCH	POLICY, *UNITED KINGDOM, *FUTURE	W71-13172
LATE/ BDELLOVIBRID, PELODICTYON,	POLIOVIRUSES, PLASTICIZERS, PHTHA	W72-09675
PRODUCTIVITY, ALGAE, ECONOMICS,	POLITICAL ASPECTS, GEOLOGY, TOPOG	W71-03012
ITUS, ORGANIC COMPOUNDS, PATH OF	POLL: /, PARTICULATE MATTER, DETR	W73-02105
PULP WASTES, RADIOECOLOGY, WATER	POLLUT: /NA, CELLULOSE, ENZYMES,	W72-09668
ES, EVALUATION, WATER POLLUTION,	POLLUTANT IDENTIFICATION, SAMPLIN	W72-10678
DICATORS, ANALYTICAL TECHNIQUES,	POLLUTANT IDENTIFICATION, NUCLEAR	W72-10958
STRIAL WASTES, SEWAGE EFFLUENTS,	POLLUTANT IDENTIFICATION.: / INDU	W71-11515
S, FOOD CHAINS, BIOLUMINESCENCE,	POLLUTANT IDENTIFICATION.: /RIENT	W71-11234
NIQUES, SAMPLING, MASSACHUSETTS,	POLLUTANT IDENTIFICATION.: / TECH	W71-11498
, COMPUTER PROGRAMS, GAMMA RAYS,	POLLUTANT IDENTIFICATION, COLORIM	W71-11036
STUDIES, ANALYTICAL TECHNIQUES,	POLLUTANT IDENTIFICATION.: /GICAL	W72-03985
S, CHROMATOGRAPHY, GROWTH RATES,	POLLUTANT IDENTIFICATION.: /TTERN	W72-04134
NIQUES, WATER POLLUTION EFFECTS,	POLLUTANT IDENTIFICATION.: / TECH	W71-13794
OGEN, PLANKTON, *EUTROPHICATION,	POLLUTANT IDENTIFICATION, ALGAE,	W72-01780
ORGANISMS, CHLOROPHYTA, BIOLOGY,	POLLUTANT IDENTIFICATION, LIGHT I	W73-02099
, *HEAVY METALS, EUTROPHICATION,	POLLUTANT IDENTIFICATION, TOXICIT	W73-02112
, NOSTOC, ENVIRONMENTAL EFFECTS,	POLLUTANT IDENTIFICATION, PHYSICA	W73-00854
TIC ALGAE, AQUATIC PRODUCTIVITY,	POLLUTANT IDENTIFICATION, LABORAT	W73-01066
TAMINS, CARBOHYDRATES, ALCOHOLS,	POLLUTANT IDENTIFICATION, SYSTEMA	W73-00942
G FACTORS, *PHOSPHATES, DIATOMS,	POLLUTANT IDENTIFICATION, GROWTH	W73-01445
ENT, SEDIMENTATION, METHODOLOGY,	POLLUTANT IDENTIFICATION, PROTOZO	W73-00273
N, TRACE ELEMENTS, ION EXCHANGE,	POLLUTANT IDENTIFICATION, COPPER,	W72-14303
TIVITY, AQUATIC ALGAE, NITROGEN,	POLLUTANT IDENTIFICATION, ENVIRON	W73-00379
STACEANS, MARINE FISH, MOLLUSKS,	POLLUTANT IDENTIFICATION.: /, CRU	W73-00819
TER, SAMPLING, BIOMASS, DIATOMS,	POLLUTANT IDENTIFICATION, MARINE	W73-00442
TITION, WATER POLLUTION SOURCES,	POLLUTANT IDENTIFICATION, PHYTOPL	W73-00441
EMATICAL MODELS, WATER ANALYSIS,	POLLUTANT IDENTIFICATION, DETERGE	W72-08790
SEUDOMONAS, POTASSIUM COMPOUNDS,	POLLUTANT IDENTIFICATION, CHLOROP	W72-13813
ALGAE, HUMUS, ORGANIC COMPOUNDS,	POLLUTANT IDENTIFICATION, AQUEOUS	W72-12166
POLLUTION, FOOD CHAINS, PATH OF	POLLUTANTS, BIOASSAY, FISH, RAINB	W72-12257
E, *BIOASSAY, *PLANT PHYSIOLOGY,	POLLUTANTS, *TOXICITY, WATER POLL	W72-12576
E WASTES, RADIOECOLOGY, *PATH OF	POLLUTANTS, PHYSICOCHEMICAL PROPE	W72-12204
A COLLECTIONS, TOXICITY, PATH OF	POLLUTANTS, *METALS, TRACE ELEMEN	W72-12190
EXCHANGE, RADIOECOLOGY, *PATH OF	POLLUTANTS.: / STRUCTURE, CATION	W72-12203
IONS, MICROBIOLOGY, SYSTEMATICS,	POLLUTANTS, PATHOGENIC BACTERIA,	W72-13800
MENT, *WATER POLLUTION, *PATH OF	POLLUTANTS, *PUBLIC HEALTH, ENZYM	W72-13347
RCH, MARINE ALGAE, PRODUCTIVITY,	POLLUTANTS, CHEMICAL WASTES, SEWA	W72-08804
PHYSA, CULEX, GAMBUSIA, FATE OF	POLLUTANTS.: / URINE, OEDOGONIUM,	W72-07703
, PHYTOPLANKTON, WATER ANALYSIS,	POLLUTANTS, AQUATIC ALGAE, AQUATI	W72-07896
SOTOPES, NUCLEAR WASTES, PATH OF	POLLUTANTS, ESTUARINE ENVIRONMENT	W72-07826
RPLANTS, NUCLEAR WASTES, PATH OF	POLLUTANTS, WATER POLLUTION SOURC	W73-00790
DIMENTS, NUCLEAR WASTES, PATH OF	POLLUTANTS, SEA WATER, WATER ANAL	W73-00817

CHELATION, FOOD CHAINS, PATH OF POLLUTANTS, PUBLIC HEALTH, WATER W73-00802
, *COBALT RADIOISOTOPES, PATH OF POLLUTANTS, WASTE ASSIMILATIVE CA W73-00826
, ESTUARINE ENVIRONMENT, PATH OF POLLUTANTS, FALLOUT, NUCLEAR WAST W73-00821
HEALTH, NUCLEAR WASTES, PATH OF POLLUTANTS, DIFFUSION, CYTOLOGICA W73-00823
, ESTUARINE ENVIRONMENT, PATH OF POLLUTANTS, FOOD CHAINS, PHYTOPLA W73-00818
E ALGAE, TRACE ELEMENTS, PATH OF POLLUTANTS, ABSORPTION, RADIOISOT W73-00811
ION, BIOLOGICAL SAMPLES, FATE OF POLLUTANTS, METABOLITES, DUNALIEL W73-00361
HORUS, NITRATES, SILICA, PATH OF POLLUTANTS, ORGANIC COMPOUNDS, PH W73-00384
ACE WATERS, PERSISTENCE, PATH OF POLLUTANTS, OCEANS, ESTUARIES, CU W73-00361
THOPHORIDS, FLAGELLATES, FATE OF POLLUTANTS, STEPHANOPYXIS COSTATA W73-00380
, ORGANIC NITROGEN, ORG/ FATE OF POLLUTANTS, SUBSTRATE UTILIZATION W73-00379
ECOLOGY, STANDING CROPS, PATH OF POLLUTANTS.: /ING, SCUBA DIVING, W73-00366
WA/ FRESHWATER, *BORON, *PATH OF POLLUTANTS, METHODOLOGY, RIVERS, W73-00286
ION, PARTICULATE MATTER, FATE OF POLLUTANTS, PAPER CHROMATOGRAPHY, W73-01066
AINS, AFRICA, EFFLUENTS, PATH OF POLLUTANTS, NEUTRON ACTIVATION AN W73-00832
E ALGAE, NUCLEAR WASTES, PATH OF POLLUTANTS, RADIOECOLOGY, MATHEMA W73-00828
D CHAINS, PUBLIC HEALTH, PATH OF POLLUTANTS, RADIOECOLOGY, RADIOAC W73-00833
*RADIOACTIVITY EFFECT/ *PATH OF POLLUTANTS, *NUCLEAR POWERPLANTS, W73-00831
DDD, PESTICIDE TOXICITY, PATH OF POLLUTANTS, PESTICIDE RESIDUES, C W73-02280
PACITY, WATER POLLUTION EFFECTS, POLLUTANTS,: /STE ASSIMILATIVE CA W72-01685
AE, MATHEMATICAL MODELS, PATH OF POLLUTANTS, TRACERS, ANIMAL PHYSI W72-03348
POWERPLANTS, MONITORING, PATH OF POLLUTANTS, AIR POLLUTION EFFECTS W72-00959
ITY TECHNIQUES, TRACERS, PATH OF POLLUTANTS, ON-SITE INVESTIGATION W72-00949
EFFECTS, SURFACE WATERS, PATH OF POLLUTANTS, MARINE ALGAE, MARINE W72-00940
COLORADO, CATTLE, URINE, PATH OF POLLUTANTS, FARM WASTES.: /RCES, W71-11496
E, *ALGAE, *FLOODWATER, *PATH OF POLLUTANTS, DIELDRIN, DDT, ORGANI W71-11498
E, WASTE WATER DISPOSAL, PATH OF POLLUTANTS, STORM DRAINS, DETERGE W71-13466
OTOSYNTHESIS, *PLANT PHYSIOLOGY, POLLUTANTS, ANALYTICAL TECHNIQUES W71-10553
SORPTION, BIOINDICATORS, PATH OF POLLUTANTS, FOOD CHAINS, NUCLEAR W72-10957
ON, *SNAILS, GASTROPODS, PATH OF POLLUTANTS, NUCLEAR WASTES, WATER W72-10978
EFFECTS, NUCLEAR WASTES, PATH OF POLLUTANTS, MARINE ALGAE, CHROMOS W72-10955
FLUENTS, NUCLEAR WASTES, PATH OF POLLUTANTS, ASIA, MARINE ALGAE, S W72-10958
, *RADIUM, *ABSORPTION, *PATH OF POLLUTANTS, *WATER POLLUTION, WAT W72-10686
BIFICIDS, *FOOD CHAINS, *PATH OF POLLUTANTS, *ABSORPTION, ALGAE, C W72-09668
ES, PHOSPHORUS, PHENOLS, METALS, POLLUTANTS, INDUSTRIAL WASTES, CO W72-09646
IOLOGICAL MAGNIFICATION, FATE OF POLLUTANTS.: / BIOACCUMULATION, B W72-11925
FATE OF POLLUTANTS.: W72-11702
ABS, ALGAE, INVERTEBRATES, FISH, POLLUTANTS, TURBIDITY, ORGANIC LO W71-01475
SURVEYS, INVESTIGATIONS, PATH OF POLLUTANTS, SAMPLING, WATER ANALY W70-08249
SURVEYS, INVESTIGATIONS, PATH OF POLLUTANTS, SAMPLING, WATER ANALY W70-08248
TICIDES, WASTE DISPOSAL, PATH OF POLLUTANTS, WATER QUALITY.: / PES W71-05553
CHING NUT/ *PHOSPHATES, *PATH OF POLLUTANTS, *EUTROPHICATION, *LEA W71-04216
R, INDICATORS, DELAWARE, PATH OF POLLUTANTS, HOSTS, PENNSYLVANIA, W72-04736
HEMISTRY, RADIOISOTOPES, PATH OF POLLUTANTS, PHYSIOLOGICAL ECOLOGY W72-06313
PRODUCTIVITY, DIELDRIN, PATH OF POLLUTANTS, REVIEWS, FOOD CHAINS, W72-06340
GY, FISH FOOD ORGANISMS, PATH OF POLLUTANTS, BIOASSAY, TOXINS, POI W71-08597
INDUSTRIAL WASTES, FARM WASTES, POLLUTANTS, FERTILIZERS, NITROGEN W71-06445
ND, WASTE WATER TREATMENT, WATER POLLUTION EFFECTS.: / OXYGEN DEMA W71-07098
N METHODS, COOLING TOWERS, WATER POLLUTION EFFECTS.: /, APPLICATIO W71-06189
ON, WASTE WATER TREATMENT, WATER POLLUTION EFFECTS.: /BIODEGRADATI W71-07097
PHATES, ORGANIC COMPOUNDS, WATER POLLUTION CONTROL.: /LATION, PHOS W71-06247
*WATER POLLUTION EFFECTS, *WATER POLLUTION CONTROL, *BIODEGRADATIO W71-08768
*PHOSPHATES, *DETERGENTS, *WATER POLLUTION EFFECTS, *WATER POLLUTI W71-08768
ARINE PLANTS, *OILY WATER, WATER POLLUTION EFFECTS, *INTERTIDAL AR W71-08792
, WATER POLLUTION EFFECTS, WATER POLLUTION SOURCES, EUTROPHICATION W71-09173
GEORGIA, SEWAGE EFFLUENTS, WATER POLLUTION EFFECTS, WATER POLLUTIO W71-09173
EFFECTS, */ *OIL, SEEPAGE, WATER POLLUTION EFFECTS, ENVIRONMENTAL W71-08793
ION, ECONOMIC FEASIBILITY, WATER POLLUTION CONTROL, CALIFORNIA, AR W71-09157
PLANTS, *POTABLE WATER,/ *WATER POLLUTION CONTROL, *NUCLEAR POWER W71-09018
, POPULATION, CHLOROPHYLL, WATER POLLUTION EFFECTS. TEMPERATURE, C W71-08670
URE, METABOLISM, KINETICS, WATER POLLUTION EFFECTS, NUCLEAR WASTES W71-09168
S, SOLUBILITY, ADSORPTION, WATER POLLUTION EFFECTS, FOOD CHAINS.: / W71-09182
RADIOISOTOPES, ABSORPTION, WATER POLLUTION SOURCES.: /PES, COBALT W71-09018
MINE WATER, *STRIP MINES, *WATER POLLUTION EFFECTS, *KENTUCKY, SUL W71-07942
GAE, *DEGRADATION(STREAM), WATER POLLUTION EFFECTS, WASTE WATER TR W71-07973
VARIABILITY, PROBABILITY, WATER POLLUTION EFFECTS, ALGAE, AQUATIC W71-07337
S, OYSTERS, SHRIMP, CRABS, WATER POLLUTION SOURCES, SEA WATER, TRO W71-07412
NTRATION, ALGAE, BACTERIA, WATER POLLUTION SOURCES.: /EN ION CONCE W71-07942
XICITY, *CHLORINATED HYD/ *WATER POLLUTION EFFECTS, *BIOASSAY, *TO W71-07412
OACTIVITY EFFECTS, *PONDS, WATER POLLUTION EFFECTS, RADIOISOTOPES, W71-09256
EWAGE, *ORGANIC COMPOUNDS, WATER POLLUTION TREATMENT, WATER POLLUT W71-09674
S, *BAYS, *WATER QUALITY, *WATER POLLUTION CONTROL, *DIVERSION, SE W71-09581
, PARTICLE SIZE, COLLOIDS, WATER POLLUTION EFFECTS.: /YTA, AQUARIA W71-09193
INDUSTRIAL WASTES, SILTS, WATER POLLUTION EFFECTS.: /CAD, SEWAGE, W71-09795
WATER POLLUTION TREATMENT, WATER POLLUTION SOURCES, ENVIRONMENTAL W71-09674
NTS, *POLLUTION ABATEMENT, WATER POLLUTION EFFECTS, NUTRIENTS, PHO W71-09429
TRIENTS, *ALGAE, *BACTER/ *WATER POLLUTION EFFECTS, *NITRATES, *NU W71-09958
S, *EUTROPHICATION, *CHE/ *WATER POLLUTION EFFECTS, *ORGANIC WASTE W71-09674
PES, COBALT RADIOISOTOPES, WATER POLLUTION EFFECTS, FRESH WATER.: / W71-09250
IND VELOCITY, CONVECTION, WATTER POLLUTION EFFECTS.: / BUOYANCY, W W71-09852
*STRONTIUM RADIOISOTOPES, *WATER POLLUTION EFFECTS, NUCLEAR WASTES W71-09225
*DETERGENTS, *PHOSPHATES, *WATER POLLUTION SOURCES, *POLLUTION ABA W72-06655
*TRITIUM, *NUCLEAR WASTES, WATER POLLUTION EFFECTS, CATTAILS, AQUA W72-06342
EFFECTS, ALGAE, NUTRIENTS, WATER POLLUTION SOURCES, SEWAGE TREATME W72-06638
TY, WATER QUALITY CONTROL, WATER POLLUTION EFFECTS.: /, PERMEABILI W72-06037
, *STREAMS, *ECOSYSTEMS,/ *WATER POLLUTION EFFECTS, *MODEL STUDIES W72-06340
ATES, *LOCAL GOVERNMENTS, *WATER POLLUTION CONTROL, WATER QUALITY W72-06638
REATMENT, WATER POLLUTION, WATER POLLUTION EFFECTS, RESEARCH AND D W72-06655
ORUS, AQUATIC ENVIRONMENT, WATER POLLUTION EFFECTS, ALGAE, NUTRIEN W72-06638
, WATER POLLUTION EFFECTS, WATER POLLUTION SOURCES, ALGAE, NUTRIEN W72-04280

RECREATION, WATER SUPPLY, WATER POLLUTION EFFECTS, WATER POLLUTIO W72-04280
ROPHICATION, *PHOSPHATES, *WATER POLLUTION CONTROL, *WASTE WATER T W72-04734
, WATER POLLUTION SOURCES, WATER POLLUTION EFFECTS, WATER POL: /ON W72-04279
WATER QUALITY, RESERVOIRS, WATER POLLUTION SOURCES, ALGAE, RIVER F W72-04260
WASTES, *FALLOUT, *OCEANS, WATER POLLUTION EFFECTS, FOOD CHAINES, W72-04463
BSORPTION, NUCLEAR WASTES, WATER POLLUTION EFFECTS, MOLLUSKS, MARI W72-04461
OPERTIES, WATER POLLUTION, WATER POLLUTION SOURCES, WATER POLLUTIO W72-04279
ADO, *RE/ *WATER QUALITY, *WATER POLLUTION EFFECTS, *LAKES, *COLOR W72-04791
OPPER SULFATE, HERBICIDES, WATER POLLUTION EFFECTS, WATER POLLUTIO W72-04759
, WATER POLLUTION SOURCES, WATER POLLUTION CONTROL.: /TION EFFECTS W72-04759
OKLA/ *OXIDATION LAGOONS, *WATER POLLUTION EFFECTS, *FARM WASTES, W72-04773
*WATER POLLUTION CONTROL, WATER POLLUTION EFFECTS, NUTRIENTS, ALG W72-04769
QUALITY, WATER CHEMISTRY, WATER POLLUTION SOURCES, METEOROLOGICAL W72-04853
, WATER POLLUTION EFFECTS, WATER POLLUTION SOURCES, WATER POLLUTIO W72-04759
E, CYANOPHYTA, PHOSPHORUS, WATER POLLUTION SOURCES, HYDROGEN ION C W72-04769
*EUTROPHICATION, *CARBON, *WATER POLLUTION CONTROL, WATER POLLUTIO W72-04769
ILS, *N/ *REMOTE SENSING, *WATER POLLUTION CONTROL, *SEA WATER, *O W72-04742
S, WASTE WATER(POLLUTION), WATER POLLUTION SOURCES, ALGAE, PLANKTO W72-05506
OLLUTIO/ *EUTROPHICATION, *WATER POLLUTION CONTROL, *WASTE WATER(P W72-05473
, SEWAGE TREATMENT, ALGAE, WATER POLLUTION SOURCES, CHLOROPHYTA, C W72-05473
INATED BIPHENYLS, *ALGAE, *WATER POLLUTION EFFECTS, *PESTICIDES, D W72-05614
HOSPHATES, NITRATES, FILTRATION, POLLUTION ABATEMENT, ALGAE, NUTRI W72-05315
, WATER POLLUTION SOURCES, WATER POLLUTION EFFECTS,: /OTOSYNTHESIS W72-05952
OPLANKTON, PHOTOSYNTHESIS, WATER POLLUTION SOURCES, WATER POLLUTIO W72-05952
G, AGRICULTURAL CHEMICALS, WATER POLLUTION SOURCES, ALGAE, GEOLOGY W72-05869
ON, WASTE WATER TREATMENT, WATER POLLUTION EFFECTS.: / CONCENTRATI W71-05155
, WATER POLLUTION EFFECTS, WATER POLLUTION SOURCES, NITRIFICATION, W71-05390
N SOURCES, INDUSTRIAL WA/ *WATER POLLUTION EFFECTS, WATER POLLUTIO W71-03095
NTS, ALGAE, LAKE SUPERIOR, WATER POLLUTION SOURCES, MINNESOTA, WAS W71-04216
ES DEVELOPME/ *ESTUARIES, *WATER POLLUTION EFFECTS, *WATER RESOURC W71-05553
SSOLVED OXYGEN, ESTUARIES, WATER POLLUTION EFFECTS, WATER POLLUTIO W71-05390
*WATER POLLUTION EFFECTS, WATER POLLUTION SOURCES, INDUSTRIAL WAS W71-03095
TRIENTS, *EUTROPHICATION, *WATER POLLUTION SOURCES, *ESTUARIES, RI W70-06509
OLUMBIA, TID/ *ESTUARIES, *WATER POLLUTION EFFECTS, *DISTRICT OF C W70-08249
OLUMBIA, TID/ *ESTUARIES, *WATER POLLUTION EFFECTS, *DISTRICT OF C W70-08248
CIDES, *ST LAWRENCE RIVER, WATER POLLUTION SOURCES, WATER POLLUTIO W70-08098
LANKTON, / ORGANIC WASTES, WATER POLLUTION EFFECTS, *ALGAE, PHYTOP W70-07842
OPHICATION, WATER QUALITY, WATER POLLUTION CONTROL, SURVEYS, INVES W70-08248
, WATER POLLUTION SOURCES, WATER POLLUTION EFFECTS, MUD, MOLLUSKS, W70-08098
NITRATES, *NUTRIENTS, *E/ *WATER POLLUTION EFFECTS, *PHOSPHATES, * W70-09388
*WATER POLLUTION SOURCES, *WATER POLLUTION EFFECTS, *ORGANIC MATTE W70-09189
WATER POLLUTION TREATMENT, WATER POLLUTION EFFECTS, ENVIRONMENTAL W70-09429
TER, OIL-WATER INTERFACES, WATER POLLUTION TREATMENT, WATER POLLUT W70-09429
*CHEMICAL OXYGEN DEMAND, *WATER POLLUTION SOURCES, *WATER POLLUTI W70-09189
LY, *POLLUTION ABATEMENT, *WATER POLLUTION CONTROL, *WATER COOLING W70-09193
OPHICATION, WATER QUALITY, WATER POLLUTION CONTROL, SURVEYS, INVES W70-08249
ATION, ALGAE, SURFACTANTS, WATER POLLUTION SOURCES, SURFACE WATERS W70-09388
HYTA, ALGAE, SOUTH DAKOTA, WATER POLLUTION EFFECTS, WATER POLLUTIO W71-02880
N EFFEC/ *EUTROPHICATION, *WATER POLLUTION CONTROL, WATER POLLUTIO W71-01488
, WATER YIELD IMPROVEMENT, WATER POLLUTION TREATMENT, ALGAE, POTAB W71-01491
AERATION, EROSION CONTROL, WATER POLLUTION SOURCES, WATER LAW, ALG W71-01192
, WATER POLLUTION EFFECTS, WATER POLLUTION SOURCES, CULTIVATION.: / W71-02880
*WATER POLLUTION CONTROL, WATER POLLUTION EFFECTS, *AQUATIC PLANT W71-01488
ALGAE, *SILICA, *BACTERIA, WATER POLLUTION CONTROL, WASTE WATER TR W71-01899
N, *ALGAL CONTROL, RIVERS, WATER POLLUTION EFFECTS, ALGAE, DIATOMS W71-00099
*UNIVERSITIES, *ABSTRACTS, WATER POLLUTION TREATMENT, INSTRUMENTAT W71-01192
TER ANALYSIS, CHEMCONTROL, WATER POLLUTION EFFECTS, WATER QUALITY, W72-11724
ENTS, *AEROBIC CONDITIONS, WATER POLLUTION CONTROL.: /ENTIAL NUTRI W72-11617
IONS, CYTOLOGICAL STUDIES, WATER POLLUTION EFFECTS.: / FLUORIDES, W72-11572
*EUTROPHICATION, *LAKES, *WATER POLLUTION EFFECTS, RECLAIMED WATE W72-11702
STEMS, *BIOLOGICAL COMMUN/ WATER POLLUTION EFFECTS, ECOLOGY, ECOSY W72-11854
ORUS, ALGAE, PLANT GROWTH, WATER POLLUTION TREATMENT, EVALUATION, W72-11702
S, FOOD CHAINS, FOOD WEBS, WATER POLLUTION EFFECTS, AQUATIC ENVIRO W72-11925
OL/ *ALGAE, *FLOCCULATION, WATER POLLUTION TREATMENT, COLLOIDS, *P W72-11833
, WATER POLLUTION EFFECTS, WATER POLLUTION SOURCES, INHIBITION.: / W72-11727
INHIBITORS, *LETHAL LIMIT, WATER POLLUTION EFFECTS, WATER POLLUTIO W72-11727
TUARIES, *EUTROPHICATION, *WATER POLLUTION CONTROL, MANAGEMENT, AN W72-09171
, *WATER QUALITY CONTROL, *WATER POLLUTION EFFECTS, WATER POLLUTIO W72-09646
OGENIC / *MICROORGANISMS, *WATER POLLUTION CONTROL, *WASTES, *PATH W72-09675
, FISHKILL, NEW HAMPSHIRE, WATER POLLUTION EFFECTS, COPPER SULFATE W72-09304
TOXINS, WATER CIRCULATION, WATER POLLUTION CONTROL, FISHKILL, NEW W72-09304
E PLANNING, AQUATIC WEEDS, WATER POLLUTION SOURCES, MATHEMATICAL M W72-09171
*WATER POLLUTION EFFECTS, WATER POLLUTION SOURCES, AQUATIC LIFE, W72-09646
ES, *PRIMARY PRODUCTIVITY, WATER POLLUTION EFFECTS, ESTUARINE ENVI W72-09653
AZOTOBACTER, PSEUDOMONAS, WATER POLLUTION EFFECTS.: /ES, E. COLI, W72-10861
*ALGAE, SEWAGE TREATMENT, WATER POLLUTION SOURCES, WATER POLLUTIO W72-10625
LUTANTS, *WATER POLLUTION, WATER POLLUTION SOURCES, AQUATIC ALGAE, W72-10686
OLLUTANTS, NUCLEAR WASTES, WATER POLLUTION EFFECTS, FOOD CHAINS, P W72-10978
, WATER POLLUTION SOURCES, WATER POLLUTION EFFECTS, AGRICULTURAL R W72-10625
N, *DETERGE/ *PHOSPHATES, *WATER POLLUTION SOURCES, *EUTROPHICATIO W72-11186
, CRUSTACEANS, MONITORING, WATER POLLUTION EFFECTS, MOLLUSKS.: /GE W72-10955
NUTRIENTS, *PHYTOPLANKTON, WATER POLLUTION EFFECTS, EUTROPHICATION W72-09167
IFOR/ *INDUSTRIAL WASTES, *WATER POLLUTION EFFECTS, EFFLUENTS, CAL W72-09092
IDE, SEASONAL, PHOSPHORUS, WATER POLLUTION SOURCES, BIOCHEMICAL OX W72-09167
PHICATION, WATER SAMPLING, WATER POLLUTION EFFECTS, BIOMASS, NUTRI W72-09111
BIOCHEMICAL OXYGEN DEMAND, WATER POLLUTION CONTROL, NITRATES, EFFL W72-09167
GEN, WISCONSIN, NUTRIENTS, WATER POLLUTION EFFECTS, INTERFACES, DI W71-10645
TS, ANALYTICAL TECHNIQUES, WATER POLLUTION SOURCES, TOXINS, WASTES W71-10553
DT, *DIELDRIN, *DIAZINON, *WATER POLLUTION EFFECTS, *PHOTOSYNTHESI W71-10566
ES, *SALMONIDS, WATER RE/ *WATER POLLUTION SOURCES, *FISH HATCHERI W71-11006

Y, FERTILIZERS, EFFLUENTS, WATER	POLLUTION SOURCES, WATER POLLUTIO	W71-11008
, *BIOASSAY, *INDICATORS, *WATER	POLLUTION EFFECTS, *ALGAE, *MARIN	W71-10553
SURFACE WATERS, *REVIEWS, *WATER	POLLUTION SOURCES, NUTRIENTS, PLA	W71-10580
NIA, TUBIFICIDS, COLORADO, WATER	POLLUTION CONTROL, WASHINGTON, CH	W71-11006
, WATER POLLUTION SOURCES, WATER	POLLUTION EFFECTS, SAMPLING, TEMP	W71-11008
ALS, *ALGAE, *INVERTEBRA/ *WATER	POLLUTION EFFECTS, *PLANTS, *ANIM	W71-13233
GAE, SEWAGE, PRODUCTIVITY, WATER	POLLUTION EFFECTS, PHOSPHORUS, NI	W71-12857
OPHICATION, *WATER SUPPLY, WATER	POLLUTION EFFECTS, NUTRIENTS, RES	W71-13185
SEA WATER, MARIN/ *ALGAE, *WATER	POLLUTION EFFECTS, *INDICATORS, *	W71-12867
NITRATES, *NEBRASKA, URB/ *WATER	POLLUTION SOURCES, *PHOSPHATES, *	W71-13466
ON/ *POLLUTION ABATEMENT, *WATER	POLLUTION EFFECTS, *WATER POLLUTI	W71-13723
ICATION, *NITROGEN, ALGAE, WATER	POLLUTION SOURCES, COLORADO, CATT	W71-11496
N, *DETERGENTS, *PHOSPHA/ *WATER	POLLUTION CONTROL, *EUTROPHICATIO	W71-11516
, ALKYLBENZENE SULFONATES, WATER	POLLUTION SOURCES, AQUATIC ALGAE,	W71-11516
LLUTION EFFECTS, *DIATOMS, WATER	POLLUTION SOURCES, INDUSTRIAL WAS	W71-11583
SPHATES, *EUTROPHICATION, *WATER	POLLUTION CONTROL, SEWAGE EFFLUEN	W71-11507
ATER POLLUTION, STANDARDS, WATER	POLLUTION EFFECTS, ALKYLBENZENE S	W71-11516
R POLLUTION SOURCES, IND/ *WATER	POLLUTION EFFECTS, *DIATOMS, WATE	W71-11583
VY METALS, WASTE DISPOSAL, WATER	POLLUTION EFFECTS.: / METALS, HEA	W71-11793
N, *WATER POLLUTION SOUR/ *WATER	POLLUTION CONTROL, *EUTROPHICATIO	W71-12091
PLANTS, CHEMICAL ANALYSIS, WATER	POLLUTION SOURCES.: /AE, AQUATIC	W71-12252
CATION, WATER TEMPERATURE, WATER	POLLUTION EFFECTS, WATER POLLUTIO	W71-12092
, WATER POLLUTION EFFECTS, WATER	POLLUTION CONTROL, ENERGY, DIFFUS	W71-12092
, WATER POLLUTION SOURCES, WATER	POLLUTION CONTROL, TROPHIC LEVELS	W71-12072
EFFECTS, OHIO, NUTRIENTS, WATER	POLLUTION SOURCES, DISSOLVED OXYG	W71-11949
DDT, *PESTICIDE TOXICITY, *WATER	POLLUTION EFFECTS, PESTICIDES, AG	W71-12303
ATION, *ELECTROCHEMISTRY, *WATER	POLLUTION TREATMENT, ANALYTICAL T	W71-12467
DGE, *NUTRIENTS, NITROGEN, WATER	POLLUTION EFFECTS, WATER POLLUTIO	W71-11972
CONTROL, *EUTROPHICATION, *WATER	POLLUTION SOURCES, POPULATION, IN	W71-12091
, WATER POLLUTION EFFECTS, WATER	POLLUTION SOURCES.: /TS, NITROGEN	W71-11972
SEDIMENTS, LAKE MICHIGAN, WATER	POLLUTION SOURCES, WATER POLLUTIO	W71-12072
*EUTROPHICATION, *LAKES, *WATER	POLLUTION EFFECTS, OHIO, NUTRIENT	W71-11949
INOFLAGELLATES, *RED TIDE, WATER	POLLUTION SOURCES.: / DIATOMS, *D	W72-01329
UALITY CONTROL, STANDARDS, WATER	POLLUTION SOURCES, WATER POLLUTIO	W72-01660
S, *NUTRIENTS, *POLLUTAN/ *WATER	POLLUTION EFFECTS, *ORGANIC WASTE	W72-01472
*LAKE ONTARIO, *LAKE ERIE, WATER	POLLUTION EFFECTS, WATER POLLUTIO	W72-01094
, WATER POLLUTION SOURCES, WATER	POLLUTION EFFECTS, RECREATION DEM	W72-01660
ING, SOAPS, AQUATIC ALGAE, WATER	POLLUTION EFFECTS, WATER QUALITY,	W72-01085
RINE ALGAE, *SALT MARSHE/ *WATER	POLLUTION EFFECTS, *PROTOZOA, *MA	W72-00948
WATER	POLLUTION CONTROL ACT.:	W72-01085
FEDERAL GOV/ *DETERGENTS, *WATER	POLLUTION CONTROL, *PHOSPHATES, *	W72-01093
CONTROL, *FEDER/ *LAKES, *WATER	POLLUTION CONTROL, *WATER QUALITY	W72-01659
*WATER	POLLUTION CONTROL ACT.:	W72-01093
TION, LEGAL ASPECTS, REGULATION,	POLLUTION ABATEMENT, WASTE DISPOS	W72-01085
, WATER POLLUTION EFFECTS, WATER	POLLUTION SOURCES, LIMNOLOGY.: /E	W72-01094
DETERGETNTS, *PHOSPHATES, *WATER	POLLUTION CONTROL, *WATER POLLUTI	W72-01085
SANITATION, *LE/ *LAKES, *WATER	POLLUTION CONTROL, *ENVIRONMENTAL	W72-01660
*WATER POLLUTION CONTROL, *WATER	POLLUTION SOURCES, EUTROPHICATION	W72-01085
ITORING, PATH OF POLLUTANTS, AIR	POLLUTION EFFECTS, WATER POLLUTIO	W72-00959
S, WATER POLLUTION EFFECT, WATER	POLLUTION SOURCES, COST COMPARISO	W72-00974
DETERGENTS, *LEGISLATION/ *WATER	POLLUTION CONTROL, *PHOSPHATES, *	W72-01685
CHAINS, MARINE FISHERIES, LAKES,	POLLUTION ABATEMENT.: /GAE, FOOD	W72-00960
TS, AIR POLLUTION EFFECTS, WATER	POLLUTION EFFECTS, BAYS, WATERSHE	W72-00959
PHYTOPLANKTON, *NUTRIENTS, WATER	POLLUTION EFFECTS, AQUATIC ALGAE,	W72-01329
VERNMENTS, WASTE DISPOSAL, WATER	POLLUTION SOURCES, SOAPS, TOXICIT	W72-01093
AE, BIOLOGICAL PROPERTIES, WATER	POLLUTION EFFECT, WATER POLLUTION	W72-00974
OGS, *TRITIUM, ABSORPTION, WATER	POLLUTION EFFECTS, RADIOACTIVITY	W72-03348
EF/ *NUCLEAR EXPLOSIONS, *WATER	POLLUTION EFFECTS, *RADIOACTIVITY	W72-03347
A, PONDS, WELLS, FISH, BAYS, AIR	POLLUTION EFFECTS, AIR POLLUTION,	W72-03347
OOD CHAINS, PUBLIC HEALTH, WATER	POLLUTION SOURCES.: /ALIFORNIA, F	W72-03324
, WATER POLLUTION SOURCES, WATER	POLLUTION EFFECTS, FLUCTUATION, C	W72-03228
*WATER	POLLUTION CONTROL ACT.:	W72-01685
EATM/ AGRICULTURAL WASTES, WATER	POLLUTION CONTROL, *BIOLOGICAL TR	W72-02975
BIORTHYMS, ON-SITE TESTS, WATER	POLLUTION SOURCES, WATER POLLUTIO	W72-03228
, WATER POLLUTION SOURCES, WATER	POLLUTION TREATMENT, NUTRIENTS, O	W72-01685
FFECTS, *ALGAE, *PLANKTON, WATER	POLLUTION CONTROL, METHODOLOGY, M	W72-03218
STE ASSIMILATIVE CAPACITY, WATER	POLLUTION EFFECTS, POLLUTANTS,: /	W72-01685
HICATION, WATER POLLUTION, WATER	POLLUTION SOURCES, WATER POLLUTIO	W72-01685
ATION, *LABORATORY TESTS, *WATER	POLLUTION EFFECTS, *ALGAE, *PLANK	W72-03218
NVIRONMENT, *AGRICULTURE, *WATER	POLLUTION CONTROL, SEDIMENTS, FAR	W72-00846
*WATER POLLUTION EFFECTS, *WATER	POLLUTION SOURCES, *FISHERIES, OI	W71-13723
TMENT, TERTIARY TREATMENT, WATER	POLLUTION CONTROL, BIODEGRADATION	W71-13939
S, ALGAL CONTROL, *POLLUT/ WATER	POLLUTION EFFECTS, *MICROORGANISM	W72-00801
N, LAKES, *NUTRIENTS, ALG/ WATER	POLLUTION EFFECTS, *EUTROPHICATIO	W72-00798
RADIOISOTOPES, HEAVY METALS, AIR	POLLUTION EFFECTS, SURFACE WATERS	W72-00940
E, *EUTR/ *WASTE DILUTION, WATER	POLLUTION CONTROL, *NUISANCE ALGA	W72-00799
TS, ANALYTICAL TECHNIQUES, WATER	POLLUTION EFFECTS, POLLUTANT IDEN	W71-13794
OLOGY, RADIOECOLOGY, MONITORING,	POLLUTION ABATEMENT, ENVIRONMENTA	W72-00939
ATORS, PLANKTON, CULTURES, WATER	POLLUTION EFFECTS.: /GROWTH REGUL	W71-13793
, MARINE FISH, ABSORPTION, WATER	POLLUTION SOURCES, RADIOACTIVITY	W72-00940
ONS, ALGAE, ECOTYPES, IN/ *WATER	POLLUTION EFFECTS, *DATA COLLECTI	W71-13746
PTION, PESTICIDE RESIDUES, WATER	POLLUTION EFFECTS, *CALIFORNIA, S	W73-02105
TURBIDITY, BENTHIC FAUNA, WATER	POLLUTION EFFECTS, STRIPED BASS,	W73-02029
E KINETICS, *PERSISTENCE, *WATER	POLLUTION EFFECTS, *PLANT PHYSIOL	W73-02280
EUTROPHICATION, CHEMICAL/ *WATER	POLLUTION CONTROL, *PHOSPHORUS, *	W73-02478
NALYSIS, OZONE, DENITRIFICATION,	POLLUTION ABATEMENT, WATER POLLUT	W73-02187
TION, POLLUTION ABATEMENT, WATER	POLLUTION CONTROL, COAGULATION.: /	W73-02187
CATION, CHEMICAL ANALYSIS, WATER	POLLUTION SOURCES, BIOASSAY, ALGA	W73-02478
TH,/ *WATER PURIFICATION, *WATER	POLLUTION TREATMENT, *PUBLIC HEAL	W73-02144
ES, ESTUARINE ENVIRONMENT, WATER	POLLUTION SOURCES, BIODEGRADATION	W73-00853

MS, BIODEGRADATION, ALGAE, WATER POLLUTION EFFECTS, HEAVY METALS, W73-00840
TROPHICATION, *NUTRIENTS, *WATER POLLUTION CONTROL, ORGANIC WASTES W73-00840
, ECOLOGICAL DISTRIBUTION, WATER POLLUTION EFFECTS, F: /N PATTERNS W73-00918
REACTIONS, CARBON DIOXIDE, WATER POLLUTION SOURCES, OIL SPILLS, SE W73-01074
ES, CRUSTACEANS, ANNELIDS, WATER POLLUTION EFFECTS, STANDING CROPS W73-00935
ORGANIC MATTER, NITROGEN, WATER POLLUTION SOURCES, ENVIRONMENTAL W73-01066
TA, ANALYTICAL TECHNIQUES, WATER POLLUTION EFFECTS, NUTRIENT REQUI W73-01269
S, *OIL P/ *MARINE ALGAE, *WATER POLLUTION EFFECTS, *PHOTOSYNTHESI W73-01074
RATES, DEFICIENT ELEMENTS, WATER POLLUTION EFFECTS, AQUATIC ALGAE, W73-01445
ORY EQUIPMENT, FRESHWATER, WATER POLLUTION SOURCES, INSTRUMENTATIO W73-01445
, *NEVADA, AQUATIC PLANTS, WATER POLLUTION SOURCES, SEDIMENTS, AQU W73-01673
TEMS, *RIVERS, *VIRGINIA, *WATER POLLUTION EFFECTS, SURVEYS, SAMPL W73-01972
HERMAL POLLUTION, *COASTS, WATER POLLUTION EFFECTS, AQUATIC LIFE, W73-01632
PHIES, *DATA COLLECTIONS, *WATER POLLUTION EFFECTS, DOCUMENTATION, W73-01976
ILITIES/ *WATER TREATMENT, WATER POLLUTION EFFECTS, *TREATMENT FAC W73-01669
HLAMYDOMONAS, CHLOROPHYTA, WATER POLLUTION EFFECTS, ABSORPTION, PE W73-00281
, WATER POLLUTION SOURCES, WATER POLLUTION EFFECTS, MERCURY, SEWAG W72-14782
, ACARICIDES, FOOD CHAINS, WATER POLLUTION SOURCES.: /S PESTICIDES W73-00238
, MESOTROPHY, OLIGOTROPHY, WATER POLLUTION SOURCES, WATER POLLUTIO W72-14782
ANTS, METHODOLOGY, RIVERS, WATER POLLUTION SOURCES, ION EXCHANGE, W73-00286
), *WASTE WATER DISPOSAL, *WATER POLLUTION CONTROL, ECONOMIC FEASI W72-14878
SAMPLING, ORGANIC MATTER, WATER POLLUTION EFFECTS, SEWAGE, NITROG W73-00234
N, CARBON, ORGANIC MATTER, WATER POLLUTION SOURCES, CHLOROPHYTA, A W73-00379
S, GROWTH RATES, NITROGEN, WATER POLLUTION EFFECTS, RADIOACTIVITY, W73-00380
SITE TESTS, INVERTEBRATES, WATER POLLUTION EFFECTS, ECOLOGY, SEA W W73-00348
XINS, MERCURY, DETERGENTS, WATER POLLUTION SOURCES, ASBESTOS, EUTR W73-00703
CHRYSOPHYTA, COMPETITION, WATER POLLUTION SOURCES, POLLUTANT IDEN W73-00441
ROPHICATION, MARINE ALGAE, WATER POLLUTION SOURCES, SPATIAL DISTRI W73-00431
N MICROSCOPY, PHOTOGRAPHY, WATER POLLUTION EFFECTS, WATER PURIFICA W73-00432
POLLUTANTS, PUBLIC HEALTH, WATER POLLUTION EFFECTS, RADIOECOLOGY, W73-00802
ASTES, PATH OF POLLUTANTS, WATER POLLUTION SOURCES, PUBLIC HEALTH, W73-00790
LABORATORY TESTS, OXYGEN, WATER POLLUTION EFFECTS.: /CHLOROPHYTA, W72-14314
TRIENT REMOVAL, NUTRIENTS, WATER POLLUTION EFFECTS, WATER QUALITY, W72-14692
TION, POLLUTION ABATEMENT, WATER POLLUTION TREATMENT, WATER QUALIT W72-14469
R, *DECOMPOSING ORGANIC M/ WATER POLLUTION EFFECTS, *ORGANIC MATTE W72-14673
EPTIC TANKS, FUNGI, ALGAE, WATER POLLUTION CONTROL.: /, STREAMS, S W72-14363
NTS, *AQUATIC ENVIRONMENT, WATER POLLUTION EFFECTS, STREAMS, RIVER W72-14690
CE, LETHAL LIMIT, POISONS, WATER POLLUTION EFFECTS.: /AY, RESISTAN W72-14706
WASTES, *MICROORGANISMS, *WATER POLLUTION SOURCES, REVIEWS, MICRO W72-14363
IDE, COASTS, LAKES, PONDS, WATER POLLUTION CONTROL, AIRCRAFT, COPP W72-14673
VEYS, *AERIAL PHOTOGRAPHY, WATER POLLUTION EFFECTS, EUTROPHICATION W72-14728
C ENVIRONMENT, *TOXICITY, *WATER POLLUTION EFFECTS, ZINC, LEAD, CO W72-14694
ATION, *CHLORINATION, OXIDATION, POLLUTION ABATEMENT, WATER POLLUT W72-14469
UTION, BEHAVIOR, SEASONAL, WATER POLLUTION EFFECTS.: /GES, DISTRIB W72-14659
HYTA, PROTOZOA, EQUIPMENT, WATER POLLUTION EFFECTS.: /AS, EUGLENOP W72-07661
QUATIC ANIMALS, NUTRIENTS, WATER POLLUTION EFFECTS, WATER QUALITY, W72-07896
, WATER POLLUTION EFFECTS, WATER POLLUTION SOURCES, *FISH, DIATOMS W72-07791
PLING, DIATOMS, TURBIDITY, WATER POLLUTION SOURCES, NUTRIENTS, SEW W72-07890
AE, FISH, WATER POLLUTION, WATER POLLUTION SOURCES.: /BENTHOS, ALG W72-07841
RS, *ELECTRIC POWERPLANTS, WATER POLLUTION EFFECTS, WATER POLLUTIO W72-07791
HNIQUES, PESTICIDES, ECOSYSTEMS, POLLUTION SOURCES, PESTICIDE TOXI W72-07703
EMENT PENS, IMPOUNDMENTS, *WATER POLLUTION SOURCES, *AGRICULTURAL W72-08401
CONCENTRATION, CORROSION, WATER POLLUTION EFFECTS.: /HYDROGEN ION W72-08683
, *ANALYTICAL TECHNIQUES, *WATER POLLUTION CONTROL, *INSTRUMENTATI W72-08790
FISH POPULATIONS, FLORIDA, WATER POLLUTION EFFECTS, NITROGEN, PHOS W72-08986
ES, *AGRICULTURAL RUNOFF, *WATER POLLUTION EFFECTS, *CATTLE, *FEED W72-08401
STRIBUTION, BIOINDICATORS, WATER POLLUTION SOURCES.: /COLOGICAL DI W72-08579
EWAGE, *INDUSTRIAL WASTES, WATER POLLUTION SOURCES, ON-SITE INVEST W72-08804
*AQUATIC MICROORGANISMS, *WATER POLLUTION EFFECTS, *ALGAE, *PROTO W72-08579
ICI/ *PESTICIDE TOXICITY, *WATER POLLUTION EFFECTS, BIOASSAY, PEST W72-08440
WASTES, SEWAGE EFFLUENTS, WATER POLLUTION EFFECTS.: /TS, CHEMICAL W72-08804
WER PLANTS, HEAT TRANSFER, WATER POLLUTION EFFECTS, WATER TEMPERAT W72-07907
CRUSTACEANS, OILY WATER,/ *WATER POLLUTION EFFECTS, *OIL WASTES, * W72-07911
ORGA/ *THERMAL POLLUTION, *WATER POLLUTION EFFECTS, *AQUATIC MICRO W72-07225
HT, PHOSPHATES, NUTRIENTS, WATER POLLUTION EFFECTS, RADIOACTIVITY W72-07143
S, PROTOZOA, ALGAE, FUNGI, WATER POLLUTION EFFECTS.: /R, CONDENSER W72-07526
A, *TOXICITY, COAGULATION, WATER POLLUTION EFFECTS, *ALGAL TOXINS. W72-07360
N, *LAKES, SEWAGE EFFLUE/ *WATER POLLUTION SOURCES, *EUTROPHICATIO W72-12955
YANOPHYTA, EUTROPHICATION, WATER POLLUTION EFFECTS.: /ISOTOPES, *C W72-13692
*WATER TEMPERATURE, *WA/ *WATER POLLUTION EFFECTS, *HEATED WATER, W72-13636
IC PESTICIDES, HERBICIDES, WATER POLLUTION SOURCES, PLANKTON, CYST W72-13347
OLLUTION), EUTROPHICATION, WATER POLLUTION SOURCES, NUTRIENTS.: /P W72-13644
LVED SOLIDS, PHYTOPLANKON, WATER POLLUTION EFFECTS.: /IENTS, DISSO W72-13300
NUTRIENTS, BIOINDICATORS, WATER POLLUTION EFFECTS, FERTILIZERS, N W72-13816
AY, ENVIRONMENTAL EFFECTS, WATER POLLUTION EFFECTS, ALGAE, HEAVY M W72-13813
, WATER POLLUTION CONTROL, WATER POLLUTION SOURCES, TRIBUTARIES, F W72-14282
HIBITION, *SALT TOLERANCE, WATER POLLUTION EFFECTS, CHLORINATED HY W72-13809
ASTES, *INDUSTRIAL WASTES, WATER POLLUTION CONTROL, WATER POLLUTIO W72-14282
T REVIEW, *BIBLIOGRAPHIES, WATER POLLUTION SOURCES, THERMAL POLLUT W72-12190
GY, POLLUTANTS, *TOXICITY, WATER POLLUTION EFFECTS, RESPIRATION, * W72-12576
NTATION RATES, OIL SPILLS, WATER POLLUTION SOURCES.: /MENT, SEDIME W72-12487
POLLUTION, PHYTOPLANKTON, WATER POLLUTION SOURCES, PHAEOPHYTA, OX W72-12166
FIXATION, DENITRIFICATION, WATER POLLUTION EFFECTS, WATER POLLUTIO W72-12544
PUBLIC HEALTH, *SHELLFISH, WATER POLLUTION EFFECTS, TOXICITY, WATE W72-12257
, WATER POLLUTION EFFECTS, WATER POLLUTION CONTROL.: /ITRIFICATION W72-12544
OPHYTA, BACTERIA, ECOLOGY, WATER POLLUTION EFFECTS.: /PRINGS, CYAN W72-12734
MERGANSER DUCK, BIOASSAY, WATER POLLUTION EFFECTS, GAS CHROMATOGR W72-12930
TER POLLUTION, W/ *LAKES, *WATER POLLUTION EFFECTS, *LIMNOLOGY, WA W72-12729
PHYSIOLOGY, BICARBONATES, WATER POLLUTION EFFECTS.: /ATION, PLANT W72-12940
IMNOLOGY, WATER POLLUTION, WATER POLLUTION SOURCES, SEWAGE, INDUST W72-12729

, CHLOROPHYTA, INHIBITORS, WATER	POLLUTION EFFECTS, ENVIRCNMENTAL	W72-12736
WATER, FRESHWATER, *ALGAE, WATER	POLLUTICN EFFECTS.: /S, BRACKISH	W73-02548
N, *SOCIAL PARTICIPATION, *WATER	POLLUTICN CONTROL, WISCCNSIN, LAK	W73-02479
ITROGEN FIXING BACTERIA, THERMAL	POLLUTION.: /YTICAL TECHNIQUES, N	W72-08508
ER, NUTRIENT REQUIREMENTS, WATER	POLLUTICN.: /A, CULTURES, SEA WAT	W73-00380
IODEGRADATION, ECOSYSTEMS, WATER	POLLUTION.: /POLLUTION SOURCES, B	W73-00853
AL TECHNIQUES, PHOTOMETRY, WATER	POLLUTICN.: /ACE WATERS, ANALYTIC	W71-12034
L WASTES, NEW YORK, ALGAE, WATER	POLLUTION.: /AL WASTES, INDUSTRIA	W71-11685
ES, ALGAE, GROWTH RATES, THERMAL	POLLUTION.: /ORPTION, RADIOISOTOP	W71-02075
UM)(CHROOMONAS SALINA), GLYCEROL	POLLUTION.: /KTON(PRYMNESIUM PARV	W70-07842
PHOSPHATES, *LAKES, *WASTE WATER(	POLLUTION), *HYDROLYSIS, DETERGEN	W71-00116
RADIOACTIVE WASTES, WASTE WATER(	POLLUTICN), WATER POLLUTICN SOURC	W72-05506
POLLUTION CONTROL, *WASTE WATER(	POLLUTION), VIRGINIA, NUTRIENTS,	W72-05473
APACITY, *ORGANIC / *WASTE WATER(	POLLUTION), *WASTE ASSIMILATION C	W71-11017
, INDUSTRIAL WASTES, WASTE WATER(	POLLUTION), BIOASSAY, ECCSYSTEMS,	W73-00840
ON, *FERTILIZATION, *WASTE WATER(	POLLUTION), *WASTE WATER DISPOSAL	W72-14878
ORM WATER, RAINFALL, WASTE WATER(	POLLUTICN), EUTROPHICATICN, WATER	W72-13644
FE, *AQUATIC ENVIRONMENT, *WATER	POLLUTION, *PATH OF POLLUTANTS, *	W72-13347
CLADCPHORA, PRODUCTIVITY, WATER	POLLUTIÇN, STANDING CROPS.: /GAE,	W72-13662
STANDARDS, *ESTUARIES, *THERMAL	POLLUTION, *NORTH CAROLINA, NUTRI	W72-13636
UTION EFFECTS, *LIMNOLOGY, WATER	POLLUTION, WATER POLLUTICN SOURCE	W72-12729
LLUTION EFFECTS, TOXICITY, WATER	POLLUTICN, FOOD CHAINS, PATH OF P	W72-12257
WATER POLLUTION SOURCES, THERMAL	POLLUTION, REVIEWS, *DATA COLLECT	W72-12190
*POLLUTANT IDENTIFICATION, WATER	POLLUTICN, *MCNITORING, AERIAL PH	W72-12487
WATER ANALYSIS, EXUDATION, WATER	POLLUTION, PHYTOPLANKTON, WATER P	W72-12166
E EFFLUENTS, OIL SPILLS, THERMAL	POLLUTION, NUTRIENTS, BIOASSAY, *	W72-14282
*MUNICIPAL WASTES, *INDU/ *WATER	POLLUTICN, *LAKE ERIE, *MERCURY,	W72-14282
HANGE, *CHEMICAL ANALYSIS, WATER	POLLUTION, WATER ANALYSIS, BIOASS	W73-00286
ION AGENCY.: *MERCURY	POLLUTION, *ENVIRONMENTAL PROTECT	W72-14782
VIGATION, CONSUMPTIVE USE, WATER	POLLUTICN, OXYGEN, SOIL PROPERTIE	W72-14878
*POWER PLANTS, *STREAM, *THERMAL	POLLUTION, *ENVIRONMENTAL EFFECTS	W72-14659
LOGY, CYCLING NUTRIENTS, THERMAL	POLLUTICN, TROPHIC LEVEL, TROPICA	W73-00796
GOVERNMENT, AIR POLLUTION, WATER	POLLUTION, TOXICITY, CHEMICALS, R	W73-00703
EFFECTS, FEDERAL GOVERNMENT, AIR	POLLUTION, WATER POLLUTICN, TOXIC	W73-00703
ATURE, *CANADA, *LAKES, *THERMAL	POLLUTICN, WATER POLLUTICN, COOLI	W73-01704
LIMNCLOGY, ECOLOGY, FISH, WATER	POLLUTICN, PROTOZOA, OHIO, EUTROP	W73-01615
LAKES, *THERMAL POLLUTION, WATER	POLLUTICN, COOLING, THERMAL POWER	W73-01704
ON/ *INDUSTRIAL WASTES, *THERMAL	POLLUTION, *COASTS, WATER POLLUTI	W73-01632
N EFFECTS, *PHOTOSYNTHESIS, *OIL	POLLUTION, *TOXICITY, CHEMICAL RE	W73-01074
PESTICIDES, OIL, METALS, THERMAL	POLLUTICN, SAMPLING, BACTERIA, TO	W72-08790
UNITIES, RIVERS, BACTERIA, WATER	POLLUTION, ACTIVATED SLUDGE, ALGA	W72-08573
ALGICIDES, *ALGAL CONTROL, WATER	POLLUTICN, EUTROPHICATICN, NUTRIE	W72-07508
RESCE/ *THERMAL STRESS, *THERMAL	POLLUTION, *MICROORGANISMS, *FLUO	W72-07526
TS, *AQUATIC MICROORGA/ *THERMAL	POLLUTICN, *WATER POLLUTION EFFEC	W72-07225
EATED WATER, NUCLEAR P/ *THERMAL	POLLUTICN, *CRUSTACEANS, *FISH, H	W72-07907
ANIC COMPOUNDS, *NUTRIEN/ *WATER	POLLUTION, MICROORGANISMS, *INORG	W72-07933
TON, BENTHOS, ALGAE, FISH, WATER	POLLUTION, WATER POLLUTICN SOURCE	W72-07841
ENDED LOAD, BIOINDICATORS, WATER	POLLUTICN, MICROB: /EASONAL, SUSP	W72-07896
S, *THERMAL POWER PLAN/ *THERMAL	POLLUTION, *BIOLOGICAL CCMMUNITIE	W71-11517
R LAW, FEDERAL GOVERNMENT, WATER	POLLUTICN, STANDARDS, WATER POLLU	W71-11516
WATER QUALITY, *CHEMIC/ *THERMAL	POLLUTION, *CHEMICAL REACTIONS, *	W71-12092
L CONTROL, EUTROPHICATION, WATER	POLLUTICN, BIOCHEMICAL CXYGEN DEM	W71-13257
*RADIOACTIVITY EFFECTS, *THERMAL	POLLUTICN, CONNECTICUT, RIVERS, P	W72-00939
S ANALYSIS, WASTE TREATMENT, AIR	POLLUTICN, LEGISLATION, BENTHIC F	W71-13723
IC WASTES, EUTROPHICATION, WATER	POLLUTICN, WATER POLLUTICN SOURCE	W72-01685
ON EFFECTS, AIR POLLUTION, WATER	POLLUTICN, LICHENS, SOIL-WATER-PL	W72-03347
BAYS, AIR POLLUTION EFFECTS, AIR	POLLUTION, WATER POLLUTION, LICHE	W72-03347
YANOPHYTA, GEYSERS, PH/ *THERMAL	POLLUTION, *WATER TEMPERATURE, *C	W72-03557
LGAE, CHEMICAL PROPERTIES, WATER	POLLUTICN, PROTEIN, PRIMARY PRODU	W72-01473
EGISLATION, LEGAL ASPECTS, WATER	POLLUTICN, ADJUDICATION PROCEDURE	W72-01093
AL TECHNIQUES, EVALUATION, WATER	POLLUTION, POLLUTANT IDENTIFICATI	W72-10678
ION, *PATH OF POLLUTANTS, *WATER	POLLUTICN, WATER POLLUTICN SOURCE	W72-10686
ITORING, *BIOINDICATORS, THERMAL	POLLUTICN, MICROSCOPY, DIATOMS, N	W72-10875
*HYDROBIOLOGY, *TOXICITY, *WATER	POLLUTION, *FRESHWATER, *AQUATIC	W72-09649
*ELECTRIC POWERPLANTS, *THERMAL	POLLUTICN, *AQUATIC ENVIRONMENT,	W72-11252
ALGAE, *MARINE ANIMALS, *THERMAL	POLLUTION, *THERMAL POWER, ATLANT	W72-11876
IC ENVIRONMENT, TURBIDITY, WATER	POLLUTICN, PHAEOPHYTA, MOLLUSKS,	W72-11925
RTH SEA, *LAMINAREA SPP., SEWAGE	POLLUTION, EPIFAUNA.: *NO	W72-11854
EMS, ABSORPTION, FISH, I/ *WATER	POLLUTION, *FOOD CHAINS, *ECOSYST	W72-05952
APACITY, WATER PROPERTIES, WATER	POLLUTION, WATER POLLUTION SOURCE	W72-04279
FFICIENCY, *PUBLIC HEALTH, WATER	POLLUTION, HAZARDS, NUTRIENTS, AL	W72-04266
TRIENTS, SEWAGE TREATMENT, WATER	POLLUTICN, WATER POLLUTICN EFFECT	W72-06655
*EUTROPHICATION, *LAKES, *WATER	POLLUTICN, MINNESOTA, WATER QUALI	W71-03012
NDUCTANCE, DISSOLVED OXYGEN, AIR	POLLUTION, CHEMICAL ANALYSIS, EUT	W71-01489
N, HYDR/ *POWER PLANTS, *THERMAL	POLLUTION, *THERMAL STRATIFICATIO	W70-09614
MPERATURE, POWERPLANTS/ *THERMAL	POLLUTION, *GREENHOUSES, WATER TE	W70-08832
RESEARCH AND DEVELOPMENT, *WATER	POLLUTICN, PHOSPHORUS, THERMAL ST	W70-09947
OLING, *STEAM TURBINES, *THERMAL	POLLUTICN, LONG TUBE VERTICAL DIS	W70-09193
N SOURCES, SURFACE WATERS, WATER	POLLUTION, DOMESTIC WASTES, MUNIC	W70-09388
*COOLING WATER, *WATE/ *THERMAL	POLLUTION, *NUCLEAR POWER PLANTS,	W70-09612
CITY, COPPER, CHLOROPHYTA, WATER	POLLUTION, NUTRIENTS, RIVERS, LAK	W71-05991
ICATION, ALGAE, NUTRIENTS, WATER	POLLUTION, BIOLOGICAL TREATMENT,	W71-05083
ATURE, / *POWER PLANTS, *THERMAL	POLLUTICN, *REVIEWS, WATER TEMPER	W71-07417
WATER QUALITY, STANDARDS, WATER	POLLUTION, AIR POLLUTICN, ODOR, C	W71-08214
STANDARDS, WATER POLLUTION, AIR	POLLUTION, ODOR, CONFINEMENT PENS	W71-08214
*RETURN FLOW, *NITRATES, WATER	POLLUTION, DESALINATION, ALGAE, D	W71-08223
ING PLANTS.: INDUSTRIAL WATER	POLLUTICN, DES MOINES RIVER, PACK	W71-06445
:	POLYACRYLAMIDE, POLYETHYLENIMINE.	W71-01899
ONATE, POLYETHYLENEIMINE, *CHLO/	POLYACRYLAMIDES, POLYSTYRENE SULF	W72-11833
ITRILOTRIACETATE, POLYSILICATES,	POLYCARBOXYLATES.: ARSENIC, N	W72-06837

YCZOANS, FORAMINIFERA, HYDROIDS, POLYCHAETA, OSTRACODS, HARPACTICO W72-05432
PODS, OSCILLATORIA, STOMATOPODA, POLYCHAETES, AMPHIOXUS, SALPS, BA W73-00935
SEA, COELENTERATES, ECHINODERMS, POLYCHAETES, CHIRONOMIDS, BARNACL W73-00855
MACROCYSTIS PYRIFERA, CHORDATES, POLYCHAETES: /NEREIS, SEA OTTER, W73-00932
PONGES, COELENTERATES, PORIFERA, POLYCHAETES, SERPULIDS.: /OIDS, S W73-00374
URCES, ASBESTOS, EUTROPHICATION, POLYCHLORINATED BIPHENYLS, PUBLIC W73-00703
TRON CAPTURE GAS CHROMATOGRAPHY, POLYCHLORINATED BIPHENYLS, BIOLOG W72-04134
ITATIVE COMPOSITION, COSCINOSIRA POLYCHORDA, THALASSIOSIRA DECIPIE W72-03543
ODACTYLUM TRICORNUTUM, GONYAULAX POLYEDRA, PLATYMONAS VIRIDIS, CHA W72-07166
BIOTIN, VITAMIN B 12, GONYAULAX POLYEDRA, PERIDINIUM DEPRESSUM, M W72-07892
USTIGMATOPHYCEAE, XANTHOPHYCEAE, POLYEDRIELLA HELVETICA, PLEUROCHL W72-14808
ION, COAGULATION, SEDIMENTATION, POLYELECTROLYTES, LIME, NEUTRALIZ W72-10237
N.: *ALUM, POLYELECTROLYTES, FACTORIAL DESIG W70-08904
POLYACRYLAMIDE, POLYETHYLENIMINE.: W71-01899
ER CLARIFICATION, SLUDGE DRYING, POLYETHYLENE OXIDE POLYMERS, SEWA W70-06968
YLAMIDES, POLYSTYRENE SULFONATE, POLYETHYLENEIMINE, *CHLORELLA ELL W72-11833
, PHTHALATE, THERMOPHILIC FUNGI, POLYETHYLENE, POLYPROPYLENE, BACI W72-09675
AL EFFECTS, NOTTINGHAM(ENGLAND), POLYGONUM, TYPHA, ELODEA.: /LOGIC W72-00155
FRICTIONAL DRAG, ALKYLENE OXIDE, POLYMERIZATION, ETHYLENE OXIDE, W W70-06968
6, PLUTONIUM 238, PLUTONIUM 239, POLYMERIZATION, LIVER, BONE, DETE W72-05965
AG, ALKY/ *HIGH-MOLECULAR-WEIGHT POLYMERS, TURBULANT FRICTIONAL DR W70-06968
LUDGE DRYING, POLYETHYLENE OXIDE POLYMERS, SEWAGE SETTLING, UNION W70-06968
LYSACCHARIDES, DRAG/ *BIOLOGICAL POLYMERS, *FRICTION REDUCTION, PO W72-03220
, CARBOHYDRATES, CARBON DIOXIDE, POLYMERS, HYDROGEN ION CONCENTRAT W72-10233
BLONGA, EUGLENA OXYURIS, EUGLENA POLYMORPHA, EUGLENA SPIROGYRA, LE W73-00918
NE, SULFANILAMIDE, PENICILLIN G, POLYMYXIN B, PODOCYSTS.: /RACYCLI W72-10162
GE SETTLING, UNION CARBIDE CORP, POLYOX, CARBOWAX, FLOCCULANT DOSA W70-06968
TA, MILWAUKEE HARBOR(WISCONSIN), POLYPHOSPHATES, ORTHOPHOSPHATES.: W72-13644
HERMOPHILIC FUNGI, POLYETHYLENE, POLYPROPYLENE, BACILLUS, CORYNEBA W72-09675
DES, LYNGBYA MAJUSCULA, GOMONTIA POLYRHIZA, ULVA RIGIDA, CLADOPHOR W72-11800
L POLYMERS, *FRICTION REDUCTION, POLYSACCHARIDES, DRAG REDUCTION.: W72-03220
ROTROPHIC GROWTH, LUMINARIN-TYPE POLYSACCHARIDES, CRYPTOPHYCEA, DI W70-07280
LAMINARIA OCHROLEUCA, SACCORHIZA POLYSCHIDES.: / DE ALDAN(SPAIN), W72-04766
ARSENIC, NITRILOTRIACETATE, POLYSILICATES, POLYCARBOXYLATES.: W72-06837
, MAJURO ATOLL, MARCHALL ISLAND, POLYSIPHONIA LANOSA, CHONDRUS CRI W72-08142
LAND, ENTEROMORPHA INTESTINALIS, POLYSIPHONIA, ULVA LACTUCA, ZANIC W73-00367
LVA LACTUCA, RHODYMENIS PALMATA, POLYSIPHONIA LANOSA, POLYSIPHONIA W72-04748
IS PALMATA, POLYSIPHONIA LANOSA, POLYSIPHONIA HARVEYI, RHODOTORULA W72-04748
, ATOMIC EN/ FODDER, GLUE, AGAR, POLYSIPHONIA, ANTIBIOTICS, IODINE W71-09157
MICROCLADIA SP., CLADOPHORA SP., POLYSIPHONIA SP., GELIDIUM SP., C W71-11851
NEIMINE, *CHLO/ POLYACRYLAMIDES, POLYSTYRENE SULFONATE, POLYETHYLE W72-11833
EDIA, METHYLENE IODIDE, TOLUENE, POLYSTYRENE, SLIDE PREPARATION, S W72-11738
ITRATE REMOVAL, ALGAE STRIPPING, POND DENITRIFICATION, FILTER DENI W71-08223
*FILTER CLOGGING, IRON BACTERIA, POND WEEDS, PERMANGANATE.: W72-06536
WASHINGTON(WASHINGTON), LINSLEY POND(CONNECTICUT), LAKE ZOAR(CONN W71-04216
S, POTASSIUM CHLORIDE, SHRINER'S POND(GA).: *LIMITING NUTRIENT W72-09159
LCICOLA, ELKTON POND, RISING SUN POND, INDUSTRIAL STORAGE T: /X CA W72-04736
E, SCHIZOTHRIX CALCICOLA, ELKTON POND, RISING SUN POND, INDUSTRIAL W72-04736
ITY, LABORATORY, DIGESTION TANK, PONDAGE, PILOT PLANT, SETTLING BA W72-02850
, NUTRIENTS, RIVERS, LAKES, FARM PONDS.: /ROPHYTA, WATER POLLUTION W71-05991
AEROBIC CONDITIONS, / *OXIDATION PONDS, *ALGAE, LABORATORY TESTS, W71-13326
MENT, *BYPRODUCTS, M/ *OXIDATION PONDS, *ALGAE, *WASTE WATER TREAT W72-13508
HYTA, ALGAL CONTROL, *PONDWEEDS, PONDS, *THERMAL STRATIFICATION, D W73-02349
ITROGEN, *GROWTH REACTORS, ALGAL PONDS, ALGAL GROWTH RATES, *POTAM W72-04787
DETOXIFICATION, SYNERGISM, FISH PONDS, ALGISTATIC ACTIVITY, ALGIC W71-06189
*FERRIC SULFATE, *STABILIZATION PONDS, ALUM, POST-TREATMENT.: /E, W71-13356
ULTURES, E. COLI, BIOINDICATORS, PONDS, ANAEROBIC CONDITIONS, WAST W72-07664
STIS NIDULANS, GL/ STABILIZATION PONDS, ANABAENA CYLINDRICA, ANACY W72-07664
RIENTS, PLANT PHYSIOLOGY, LAKES, PONDS, AQUATIC ALGAE, ECOLOGY, BI W71-10580
*ALGAE, *SUSPENSION, HARVESTING, PONDS, CALIFORNIA, SCENEDESMUS, S W70-10174
REATMENT, *ANAEROBIC CONDITIONS, PONDS, CLIMATIC ZONES, TEMPERATUR W71-06033
ATION, INDUSTRIAL WASTES, ALGAE, PONDS, FARM WASTES.: /GE, INFILTR W71-06435
LAWARE RIVER, OXIDATION LAGOONS, PONDS, FARM PONDS, MARYLAND, QUAR W72-04736
AE, SCENEDESMUS, EUTROPHICATION, PONDS, FEEDING RATES, BENTHOS, ST W72-14796
NS, FISH DIETS, FRESHWATER FISH, PONDS, LAKES, FOOD HABITS, EUTROP W72-07132
LUTANT IDENTIFICATION, PROTOZOA, PONDS, LIMNOLOGY, ECOLOGY, CHEMIC W73-00918
, OXIDATION LAGOONS, PONDS, FARM PONDS, MARYLAND, QUARRIES, OHIO R W72-04736
CAL OXYGEN DEMAND, STABILIZATION PONDS, MICROORGANISMS, BIODEGRADA W71-07097
ONS, *AQUATIC PRODUCTIVITY, FARM PONDS, NUTRIENTS, SURFACES, TEMPE W71-11001
ATION LAGOONS, *LIGHT INTENSITY, PONDS, ROTIFERS, PROTOZOA, PHOTOS W71-07382
C/ *LIMNOLOGY, *FLORIDA, *LAKES, PONDS, SAMPLING, THERMAL STRATIFI W72-04269
REATMENT, WASTE WATER TREATMENT, PONDS, SCENDESOMUS, CYANOPHYTA, B W72-07661
WEEDS, *RED TIDE, COASTS, LAKES, PONDS, WATER POLLUTION CONTROL, A W72-14673
TS, *AQUATIC PLANTS, HARVESTING, PONDS, WATER HYACINTH, NUTRIENTS, W71-01488
TY EFFECTS, UNDERGROUND, ALASKA, PONDS, WELLS, FISH, BAYS, AIR POL W72-03347
NSAS, INSECTS, TURBIDITY, CHARA, PONDWEEDS.: /ANTS, PLANKTON, ARKA W71-04528
PLOS ARMIGER, ALIMITOS BAY, LAKE PONTCHARTRAIN, PEARL RIVER, LAKE W73-02029
SYNURA, AROGLENA, CRYPTOPHYCEAE, PONTOPOREIA AFFINIS, ESOX LUCIUS, W72-12729
NCE, LAKES, RESERVOIRS, SWIMMING POOLS, CHELATION, APPLICATION MET W71-06189
OWTH REGULATORS, SALINITY, PLANT POPUL: /PLANT PATHOLOGY, PLANT GR W70-09429
RS, FOOD CHAINS, TITANIUM, PLANT POPULATIONS, TOXICITY.: /INDICATO W71-03095
PHIC POPULATIONS, *HETEROTROPHIC POPULATIONS, LAKE HARTWELL(SOUTH W71-00475
ATIONS, LAKE HARTW/ *ANTOTROPHIC POPULATIONS, *HETEROTROPHIC POPUL W71-00475
ECOSYSTEMS, ALGAE GROWTH, ALGAL POPULATIONS.: / NIDULANS, AQUATIC W71-00475
ANISMS, MARINE PLANTS, BACTERIA, POPULATION, HAWAII, ABSORPTION, T W71-08042
HIC FAUNA, *INVERTEBRATES, *FISH POPULATIONS, *MORTALITY, *ALGAE, W71-08793
EOUS EARTH, FOOD/ *ALGAE, *HUMAN POPULATION, FERTILIZERS, DIATOMAC W71-09157
MODELS, NUTRIENTS, MEASUREMENT, POPULATION, CHLOROPHYLL, WATER PO W71-08670
IA, VIRUSES, HEAVY METALS, HUMAN POPULATION, WASTE WATER TREATMENT W71-09788
ALGAE, *ESTUARIES, PRODUCTIVITY, POPULATION, ATLANTIC OCEAN, DEPTH W72-04766

PHYTOPLANKTON, ZOOPLANKTON, FISH	POPULATIONS, WATER TEMPERATURE, T	W72-04853
AL EFFECTS, COMPETITION, AQUATIC	POPULATIONS, NUTRIENTS, NUTRIENT	W72-12736
CIDES, PESTICIDE TOXICITY, PLANT	POPULATIONS, AQUARIA, LABORATORY	W72-14294
THIC FAUNA, BENTHIC FLORA, HUMAN	POPULATION, PRIMARY PRODUCTIVITY,	W72-12549
SPIRATION, EUTROPHICATION, PLANT	POPULATION, CLASSIFICATION, SAMPL	W72-12192
TESTS, AQUATIC ANIMALS, AQUATIC	POPULATIONS, BIOASSAY, EUTROPHICA	W73-02095
S, TUNNELS, MARINE ALGAE, ANIMAL	POPULATIONS, LOBSTERS, CRUSTACE: /	W73-02029
NE ANIMALS, MARINE PLANTS, HUMAN	POPULATION, RADIOACTIVITY TECHNIQ	W73-00833
ENVIRONMENT, ABSORPTION, ANIMAL	POPULATIONS, ANIMAL PHYSIOLOGY, R	W73-00826
*MARINE ALGAE, *FISHERIES, FISH	POPULATIONS, FISH BEHAVIOR, INVER	W73-00764
QUATIC INSECTS, DAPHNIA, AQUATIC	POPULATIONS, FISH DIETS, FRESHWAT	W72-07132
EFFLUENT, RADIOECOLOGY, AQUATIC	POPULATIONS, HYDROLOGIC ASPECTS,	W72-07841
OPHIC LEVEL, AQUATIC ALGAE, FISH	POPULATIONS, FLORIDA, WATER POLLU	W72-08986
PATTERNS, *PHOSPHATES, *AQUATIC	POPULATIONS, MAPS, MAPPING, PHYTO	W72-03567
NTAL EFFECTS, AQUATIC LIFE, FISH	POPULATIONS, PHYTOPLANKTON, BENTH	W72-00939
R STRUCTURE, PERMEABILITY, PLANT	POPULATIONS, ALGAE, CHEMICAL PROP	W72-01473
TION EFFECTS, RECREATION DEMAND,	POPULATION, COMMUNITY DEVELOPMENT	W72-01660
TES, TURNOVERS, SEDIMENTS, HUMAN	POPULATION, TOURISM, WATERSHEDS(B	W71-11009
ATION, *WATER POLLUTION SOURCES,	POPULATION, INDUSTRIES, WASTE DIS	W71-12091
, NUTRIENT REQUIREMENTS, AQUATIC	POPULATIONS, PRIMARY PRODUCTIVITY	W72-09103
UCTION, BIORHYTHMS, MEASUREMENT,	POPULATIONS, CYTOLOGICAL STUDIES,	W72-11920
CULTURE MEDIA, PORIA MONTICOLA,	PORIA COCOS, PORIA VAILLANTII, CO	W72-13813
ODIUM CACODYLATE, CULTURE MEDIA,	PORIA MONTICOLA, PORIA COCOS, POR	W72-13813
A, PORIA MONTICOLA, PORIA COCOS,	PORIA VAILLANTII, CONIOPHORA OLIV	W72-13813
YDROIDS, SPONGES, COELENTERATES,	PORIFERA, POLYCHAETES, SERPULIDS.	W73-00374
NVERTEBRATES, SPECIES DIVERSITY,	PORIFERA, SEA CUCUMBERS, STARFISH	W73-00932
YSICAL PROPERTIES, BULK DENSITY,	POROSITY, COLOR, HYDROGEN ION CON	W73-00854
OGEN ION CONCENTRATION, CHEMICAL	PORPERTIES, NITROGEN FIXATION, CU	W72-14790
VESICULOSUS, LAMINARIA DIGITATA,	PORPHYRA UMBILICALIS, SAMPLE PREP	W73-01074
SEAWEED, PORPHYRIDIUM CRUENTUM,	PORPHYRA, GIGARTINA.: / FRICTION,	W71-07878
), MIQUELON(CANADA), MIKROSYPHAR	PORPHYRAE, PROTECTOCARPUS SPECIOS	W71-10066
AL, TURBULENT FRICTION, SEAWEED,	PORPHYRIDIUM CRUENTUM, PORPHYRA,	W71-07878
UBAEQUALIS, APISTONEMA EXPANSUM,	PORPHYRIDIUM GRISEUM, CHRYSOTRICH	W71-04526
UM CARTERI, COCCOLITHUS HUXLEYI,	PORPHYRIDIUM, SKELETONEMA COSTATU	W72-12998
II, CALLOPHYLLIS FIRMA, PLENOSP/	PORPHYROPSIS COCCINEA VAR. DAWSON	W73-02099
S, ARTHROBACTER, NOSTOC COMMUNE,	PORPHYROSIPHON NOTARISII, PSYCHRO	W73-00854
VER LAKE(WASH), VANCOUVER(WASH),	PORTLAND(ORE), WATER QUALITY ANAL	W72-03215
POLLUTION, CONNECTICUT, RIVERS,	POST-IMPOUNDMENT, NUCLEAR POWERPL	W72-00939
ATE, *STABILIZATION PONDS, ALUM,	POST-TREATMENT.: /E, *FERRIC SULF	W71-13356
TROGEN, CARBON, SODIUM, CALCIUM,	POT: /ROGEN ION CONCENTRATION, NI	W73-00854
TE WATER TREATMENT, TASTE, ODOR,	POTABLE WATER, ACTIVATED CARBON,	W73-02144
ESERVOIR OPERATION, WATER REUSE,	POTABLE WATER, ALGAE, NUTRIENT RE	W71-13185
ITROGEN, PHOSPHATES, DETERGENTS,	POTABLE WATER.: /F, PHOSPHORUS, N	W72-10625
ATER POLLUTION TREATMENT, ALGAE,	POTABLE WATER, TASTE, AMMONIA, SI	W71-01491
OLONIZATION, POTAMOGETON LUCENS,	POTAMOGETON PERFOLIATUS, POTAMOGE	W71-11012
SKIE LAKE(POLAND), COLONIZATION,	POTAMOGETON LUCENS, POTAMOGETON P	W71-11012
LUCENS, POTAMOGETON PERFOLIATUS,	POTAMOGETON PECTINATUS, MYRIOPHYL	W71-11012
DOPHORA SPP, PASPALUM DISTICHUM,	POTAMOGETON PECTINATUS, ELEOCHARI	W73-01673
COMPOUNDS, PHOSPHORUS COMPOUNDS,	POTASH, ALGAE, AQUATIC PLANTS, CH	W71-12252
M, MANAGNESE, CALCIUM, CHROMIUM,	POTASS: /, IRON, COPPER, MAGNESIU	W71-09188
D(GA).: *LIMITING NUTRIENTS,	POTASSIUM CHLORIDE, SHRINER'S PON	W72-09159
ICAL STUDIES, CULTURES, AMMONIA,	POTASSIUM COMPOUNDS, NITRATES, DE	W72-03217
GAE, EVALUATION, SAMPLING, CARP,	POTASSIUM RADIOISOTOPES, STRONTIU	W73-01673
AVY METALS, GROWTH, PSEUDOMONAS,	POTASSIUM COMPOUNDS, POLLUTANT ID	W72-13813
ASE, BIOCHEMICAL TESTS, OUABAIN,	POTASSIUM CHLORIDE.: /TRIPHOSPHAT	W72-13809
NICKEL SULFATE, COBALT SULFATE,	POTASSIUM BR: /ANGANESE CHLORIDE,	W72-12739
PESTICIDES, *CYANOPHYTA, SODIUM,	POTASSIUM, CULTURES, RESISTANCE,	W72-13809
ALS, *PRODUCTIVITY, *ECOSYSTEMS,	POTASSIUM, MAGNESIUM, SODIUM, CAL	W72-14303
*NUTRIENTS, CULTURES, NITROGEN,	POTASSIUM, PHOSPHATES, METABOLISM	W73-01626
RUS, SULFUR, CALCIUM, MANGANESE,	POTASSIUM, SODIUM, CARBON DIOXIDE	W73-00840
MATTER, MANGANESE, IRON, SULFUR,	POTASSIUM, MOLYBDENUM, DETERGENTS	W72-07514
DES, CALCIUM, MAGNESIUM, SODIUM,	POTASSIUM, IRON, MANGANESE, COPPE	W71-11001
ROPERTIES, PHOSPHORUS, NITROGEN,	POTASSIUM, DIVERSION, ALGAE, NITR	W71-00117
IVER FLOW, PHOSPHORUS, NITRATES,	POTASSIUM, DISSOLVED OXYGEN, FORE	W72-04260
IRON, SILICON, SULFATES, SODIUM,	POTASSIUM, CALCIUM, MAGNESIUM, CH	W72-05610
CHLOROPHYLL, OXIDATION REDUCTION	POTENTIAL.: /THEMATICAL STUDIES,	W72-14800
TRIFICATION, OXIDATION-REDUCTION	POTENTIAL, DENITRIFICATION, NITRO	W72-08459
PHOSPHORUS, OXIDATION-REDUCTION	POTENTIAL, LAKES, SOILS, EUTROPHI	W72-07933
*CABLE THEORY, ELECTRONIC	POTENTIAL, CHARACEAE.:	W73-01630
ONICUM, NUPHAR LUTEN, ANTICANCER	POTENTIAL, ANTICOAGULANT POTENTIA	W72-04275
ICANCER POTENTIAL, ANTICOAGULANT	POTENTIAL, SPARGANIUM FLUCTUANS,	W72-04275
IAGNOSTIC CRITERIO/ ALGAL GROWTH	POTENTIAL, ILLINOIS RIVER(ILL), D	W71-11033
ON EXCHANGE, OXIDATION-REDUCTION	POTENTIAL, HYDROGEN ION CONCENTRA	W72-01108
ALTH, ODORS, OXIDATION-REDUCTION	POTENTIAL, BIOCHEMICAL OXYGEN DEM	W72-04086
, NUTRIENTS, OXIDATION-REDUCTION	POTENTIAL, ODORS, OPERATION AND M	W72-10401
ER IONS, OOCYSTIS.:	POTENTIOMETRY, CADMIUM IONS, COPP	W72-01693
E ELECTRODES, STANDARD ADDITION,	POTENTIOMETRY, INTERFERING IONS.:	W71-00474
POTOMAC RIVER BASIN,	POTOMAC ESTUARY.:	W70-06509
	POTOMAC ESTUARY.:	W71-09788
ARY.:	POTOMAC RIVER BASIN, POTOMAC ESTU	W70-06509
RELLA PYRENOIDOSA, SODIUM SILIC/	POTOMAC RIVER, ALAFIA RIVER, CHLO	W72-04734
ORIDA), HUMBOLT BAY(CALIFORNIA),	POTOMAC RIVER, COLUMBIA RIVER(WAS	W71-11517
SACCHARIDES, CLONES, PHAEOCYSTIS	POUCHETI, GLUCOSE, MANNOSE, RHAMN	W72-07710
EEDS, *RUMINANTS, ALGAE, CATTLE,	POULTRY, HOGS, ANIMAL DISEASES, C	W71-07544
NG, AEROBIC TREAT/ *FARM WASTES,	POULTRY, NITROGEN, LAGOONS, PUMPI	W72-08396
ITROGEN, AERATION, HOGS, CATTLE,	POULTRY, RESEARCH AND DEVELOPMENT	W71-07551
EY DAM(PA.), ATRAZINE, BIOTYPES,	POUR PLATE METHOD, MOST PROBABLE-	W72-04743
ATER POLLUTION CONTROL, *NUCLEAR	POWER PLANTS, *POTABLE WATER, *BI	W71-09018
TE/ *THERMAL POLLUTION, *NUCLEAR	POWER PLANTS, *COOLING WATER, *WA	W70-09612
NS, *FISH, HEATED WATER, NUCLEAR	POWER PLANTS, ELECTRICAL POWER PL	W72-07907

ER, PATUXENT RIVER, CONTRA COSTA POWER PLANT, SAN JOAQUIN RIVER, M W72-07907
NUCLEAR POWER PLANTS, ELECTRICAL POWER PLANTS, HEAT TRANSFER, WATE W72-07907
NT, SAN JOAQUIN RIVER, MORRO BAY POWER PLANT, HUMBOLDT BAY, CONNET W72-07907
L EFFECTS, HEATED WATER, NUCLEAR POWER PLANTS, STEAM, ECOLOGY, PHY W71-11517
BIOLOGICAL COMMUNITIES, *THERMAL POWER PLANTS, *DISCHARGE(WATER), W71-11517
*DESALINATION PLANTS, *ELECTRIC POWER PRODUCTION, *IRRIGATION PRO W72-03327
NG, *HEATING, *COOLING, *THERMAL POWER, *MULTIPLE-PURPOSE PROJECTS W70-09193
LS, *THERMAL POLLUTION, *THERMAL POWER, ATLANTIC OCEAN, MARINE BIO W72-11876
EATION, FISHERIES, HYDROELECTRIC POWER, NAVIGATION, CONSUMPTIVE US W72-14878
*FARM USES OF WASTE HEAT, STEAM POWERED BUSES.: / *SPACE HEATING, W70-09193
*GREENHOUSES, WATER TEMPERATURE, POWERPLANTS, DISTILLATION, ALGAE, W70-08832
Y LOSSES, *WATER CONSE/ *THERMAL POWERPLANTS, *POWERPLANTS, *ENERG W70-09193
GY, TRACERS, MONITORING, NUCLEAR POWERPLANTS, NUCLEAR WASTES, PATH W73-00790
A COLLECTIONS, RIVERS, *ELECTRIC POWERPLANTS, WATER POLLUTION EFFE W72-07791
ATER POLLUTION, COOLING, THERMAL POWERPLANTS, HEATED WATER, CIRCUL W73-01704
T/ *PATH OF POLLUTANTS, *NUCLEAR POWERPLANTS, *RADIOACTIVITY EFFEC W73-00831
, RADIOACTIVITY EFFECTS, NUCLEAR POWERPLANTS, MARINE ALGAE, MARINE W73-00833
ALS, OFFSHORE PLATFORMS, NUCLEAR POWERPLANTS, WATER PROPERTIES, SA W73-02029
LGAE, *GULF OF MEXICO, *ELECTRIC POWERPLANTS, *THERMAL POLLUTION, W72-11252
OLLUTANT IDENTIFICATION, NUCLEAR POWERPLANTS, EFFLUENTS, NUCLEAR W W72-10958
, *ELECTRIC POWER PROD/ *NUCLEAR POWERPLANTS, *DESALINATION PLANTS W72-03327
S, COBALT RADIOISOTOPES, NUCLEAR POWERPLANTS, SITES, MOLLUSKS, MAR W72-03324
*RADIOACTIVITY EFFECTS, *NUCLEAR POWERPLANTS, MONITORING, PATH OF W72-00959
IVERS, POST-IMPOUNDMENT, NUCLEAR POWERPLANTS, ECOLOGY, RADIOECOLOG W72-00939
BIOINDICATORS, BIOASSAY, ALGAE, POWERPLANTS, EFFLUENTS, DISCHARGE W71-11583
CH REMOVAL, STARCH SUBSTITUTION, POYELECTROLYTES.: *STAR W72-03753
AGMENELLUM QUADRUPLICATUM STRAIN PR-6, PHOTOALDRIN, KETOENDRIN, ME W72-03545
S, HEAVY METALS, PHYSICOCHEMICAL PR: /AL PHYSIOLOGY, TRACE ELEMENT W73-00817
AKOTA), *ORTHOPHOSPHATES.: PRAIRIE LAKES, *LAKE HERMAN(SO. D W73-00938
E EFFEC/ *BP 1002, *GREEN ALGAE, PRASINOCLADUS MARINUS, TEMPERATUR W70-09429
,/ *ANTIBIOSIS, OPTICAL DENSITY, PRASINOPHYCEAE, BACCILARIOPHYCEAE W72-08051
LTRATION, CHLORINATION, CHEMICAL PRECIPITATION, FLOCCULATION, FLOT W70-10381
, EUTROPHICATION, TRANSPIRATION, PRECIPITATI: /, CHEMICAL ANALYSIS W71-01489
LEACHING NUTRIENTS, FERTILIZERS, PRECIPITATION(ATMOSPHERIC), RUNOF W71-04216
IDATION LAGOONS, CLIMATIC ZONES, PRECIPITATION, COAGULATION, LIME, W71-08951
EXCHANGE, ELECTROLYSIS, CHEMICAL PRECIPITATION, NITRIFICATION, DEN W71-09450
ATURE, LIGHT INTENSITY, CHEMICAL PRECIPITATION, HYDROGEN ION CONCE W71-07100
*NITRATES, GROUNDWATER, AMMONIA, PRECIPITATION, SEDIMENTS, DENITRI W71-06435
RADIOISOTOPES, FALLOUT, CHEMICAL PRECIPITATION, IRON, BIOASSAY, LE W72-05965
PHATES, *CLAMS, *ALGAE, CHEMICAL PRECIPITATION, HYDROGEN ION CONCE W72-03613
*SEPARATION TECHNIQUES, CHEMICAL PRECIPITATION, COAGULATION, SEDIM W72-10237
, PERCOLATION, DRAINAGE, RUNOFF, PRECIPITATION(ATMOSPHERIC).: /LLS W72-12955
*MARINE ALGAE TRACERS, CHEMICAL PRECIPITATION, HYDROGEN ION CONCE W72-12210
ERENCE, INTERLABORATORY STUDIES, PRECISION, *SWEDEN, LAKE MALAREN, W73-00286
APHY, ELECTRON CAPTURE DETECTOR, PRECONCENTRATION, BIOLOGICAL SAMP W73-02105
ROL, INSECT CONTROL, IRRIGATION, PREDATION, PARASITISM, INSECT RES W72-00846
ONIA-NITROGEN, *GRO/ *OSTRACODS, PREDATOR-PREY RELATIONSHIPS, *AMM W72-04787
IMITS, CHEMICAL RECOVERY, SAMPLE PREPARATION.: / BONE, DETECTION L W72-05965
YRENE, SLIDE PREPARATION, SAMPLE PREPARATION.: /E, TOLUENE, POLYST W72-11738
IDE, TOLUENE, POLYSTYRENE, SLIDE PREPARATION, SAMPLE PREPARATION.: W72-11738
INGSTON, AUTORADIOGRAPHY, SAMPLE PREPARATION, SCINTILLATION COUNTI W72-09239
TERS, CHLORELLA VULGARIS, SAMPLE PREPARATION.: / METHYL ESTERS, ES W72-09365
ACROPHYTES, SARGASSO SEA, SAMPLE PREPARATION, SPERMATOPHYTES, PHOT W73-02100
TA, PORPHYRA UMBILICALIS, SAMPLE PREPARATION, CRUDE OIL.: / DIGITA W73-01074
LTURE MEDIA, FATTY ACIDS, SAMPLE PREPARATION, NATURAL ORGANICS.: / W73-01066
HLOROPHYLL A, CARTENOIDS, SAMPLE PREPARATION, MOUNTAIN LAKE.: /, C W73-00916
IOGRAPHY, GLUTARALDEHYDE, SAMPLE PREPARATION, COUNTING, SAMPLE PRE W73-00273
E, INTERLABORATORY STUDI/ SAMPLE PREPARATION, CHEMICAL INTERFERENC W73-00286
TES, SAMPLE PRESERVATION, SAMPLE PREPARATION, SUBLITTORAL, LONG IS W73-00367
ATION, KJEDAHL PROCEDURE, SAMPLE PREPARATION.: /CTROSCOPY, ASSIMIL W73-00379
ESICULOSUS, VESICULATION, SAMPLE PREPARATION, BRANCHING.: /FUCUS V W73-00369
TY, LIQUID SCINTILLATION, SAMPLE PREPARATION, ZOSTREA, ULVA.: /IVI W72-14317
, HUMIC ACIDS, OXIMATION, SAMPLE PREPARATION.: /AY, SCOUDOUC RIVER W72-12166
S ATROMACULATUS, CLEANUP, SAMPLE PRESERVATION, CAMBARUS BARTONI, B W72-12930
ISOPYCNIC CENTRIFUGATION, SAMPLE PRESERVATION, SALPA, LUDOX AM, SU W73-00371
A, SUBSTRATES, EPIPHYTES, SAMPLE PRESERVATION, SAMPLE PREPARATION, W73-00367
LE PREPARATION, COUNTING, SAMPLE PRESERVATION, FLAGELLATES, RHODOM W73-00273
, *SAMPLE PRESERVATION, CHEMICAL PRESERVATION, HISTOCHEMISTRY, LYO W73-01676
LS, *BIOLOGICAL SAMPLES, *SAMPLE PRESERVATION, CHEMICAL PRESERVATI W73-01676
*TEMPERATURE TOLERANCES, SAMPLE PRESERVATION, COUNTING CHAMBERS, W72-07899
OTTLE, SPECIES DIVERSITY, SAMPLE PRESERVATION, COUNTING.: /ANSEN B W72-07663
CYCLOTELLA, OSCILLATORI/ SAMPLE PRESERVATION, FORMALIN, COUNTING, W72-10875
CTROPHOTOME/ *STRONTIUM, *SAMPLE PRESERVATION, *FLAME EMISSION SPE W71-11529
, *DEOXYRIBONUCLEIC ACID, SAMPLE PRESERVATION, COULTER COUNTER.: / W71-11562
TA.: ENZYME PRESOAKS, DISHWATER DETERGENTS, N W71-08768
FISHKILL, RESERVOIRS, FISHERIES, PRESSURE.: /OSIVES, COSTS, FISH, W72-11564
ST ANALYSIS, *FILTERS, HEADLOSS, PRESSURE, TURBIDITY, ALGAE, FLOCC W72-09393
IS, INSTRUMENTATION, HYDROSTATIC PRESSURE, DIATOMS.: /PHOTOSYNTHES W71-10791
OGEN, *GRO/ *OSTRACODS, PREDATOR- PREY RELATIONSHIPS, *AMMONIA-NITR W72-04787
LAND MANAGEMENT, FERTILIZATION, PRIMARY PRODUCTIVITY.: /FIXATION, W72-04769
BON, OXYGEN, NITROGEN, HYDROGEN, PRIMARY PRODUCTIVITY, RESPIRATION W72-06274
NION, ZOOPLANKTON, DISTRIBUTION, PRIMARY PRODUCTIVITY, ALGAE, BIOM W71-09171
ONDARY TREATMENT, AQUATIC ALGAE, PRIMARY PRODUCTIVITY, TRACE ELEME W71-09674
PHOTOSYNTHESIS, CARBON DIOXIDE, PRIMARY PRODUCTIVITY, BIODEGRADAT W71-04518
PHYTA, CHRYSOPHYTA, CHLOROPHYTA, PRIMARY PRODUCTIVITY, CARBON RADI W71-03014
LANKTON, PLANKTON, PRODUCTIVITY, PRIMARY PRODUCTIVITY, ICED LAKES, W71-02880
, AMMONIUM COMPOUNDS, NUTRIENTS, PRIMARY PRODUCTIVITY, FERTILIZERS W71-11008
IORHYTHMS, CORRELATION ANALYSIS, PRIMARY PRODUCTIVITY.: /ALYSIS, B W72-11719
QUIREMENTS, AQUATIC POPULATIONS, PRIMARY PRODUCTIVITY.: /TRIENT RE W72-09103
OACTIVITY TECHNIQUES, *PLANKTON, PRIMARY PRODUCTIVITY, FOOD HABITS W72-00161

BOTTOM SEDIMENTS, STABILIZATION, PRIMARY PRODUCTIVITY, GROWTH RATE W72-03556
RTIES, WATER POLLUTION, PROTEIN, PRIMARY PRODUCTIVITY.: /CAL PROPE W72-01473
W YORK, NEMATODES, EPIPHYTOLOGY, PRIMARY PRODUCTIVITY, PHYSIOLOGIC W72-00948
ODS, ORGANOPHOSPHORUS COMPOUNDS, PRIMARY PRODUCTIVITY, SALINITY, P W72-07892
, PHYTOPLANKTON, PHOTOSYNTHESIS, PRIMARY PRODUCTIVITY, EUTROPHICAT W72-08054
ER QUALITY, RADON RADIOISOTOPES, PRIMARY PRODUCTIVITY, CULTURES, B W72-07648
TA, PYRROPHYTA, DINOFLAGELLATES, PRIMARY PRODUCTIVITY, MINERALOGY, W72-07166
, PHYTOPLANKTON, PHOTOSYNTHESIS, PRIMARY PRODUCTIVITY, GREAT LAKES W73-01615
WTH RATES, SPATIAL DISTRIBUTION, PRIMARY PRODUCTIVITY, CHRYSOPHYTA W73-00853
TIES, STANDING CROPS, WISCONSIN, PRIMARY PRODUCTIVITY, SYSTEMATICS W73-02469
OWTH, LIGHT INTENSITY, TOXICITY, PRIMARY PRODUCTIVITY, STANDING CR W72-14690
SOPHYTA, AUTUMN, WINTER, SUMMER, PRIMARY PRODUCTIVITY, DINOFLAGELL W73-00442
YTA, STANDING CROPS, RESISTANCE, PRIMARY PRODUCTIVITY, BIOMASS, AN W72-12940
BENTHIC FLORA, HUMAN POPULATION, PRIMARY PRODUCTIVITY, WATER CHEMI W72-12549
IRONMENTAL EFFECTS, CHLOROPHYTA, PRIMARY PRODUCTIVITY, CHRYSOPHYTA W72-13816
CTION, SEA WATER, MARINE PLANTS, PRIMARY PRODUCTIVITY, CORAL, REEF W72-14294
RY SEWAGE T/ *LITERATURE SURVEY, PRIMARY SEWAGE TREATMENT, SECONDA W70-09907
OCOCCOPSIS SMITHII, OSCILLATORIA PRINCEPS, OSCILLATORIA AMOENA, OS W72-07896
ION, BRUCITE, BOTTOM TOPOGRAPHY, PRIONOSPIO, LUMBRINEREIS, GLYCERA W72-09092
WATER QUALITY, RIPARIAN RIGHTS, PRIOR APPROPRIATION, GROUNDWATER, W70-07100
, DATA COLLECTIONS, VARIABILITY, PROBABILITY, WATER POLLUTION EFFE W71-07337
*RESIDUAL, MOST PROBABLE NUMBER.: W71-07096
IOTYPES, POUR PLATE METHOD, MOST PROBABLE-NUMBER TEST, MEMBRANE FI W72-04743
PHOSPHORUS, *ALGAL ASSAYS, ASSAY PROCEDURE.: DISSOLVED W72-12966
S, WATER POLLUTION, ADJUDICATION PROCEDURE, ADMINISTRATIVE AGENCIE W72-01093
X.: *MICROCOSM ALGAL ASSAY PROCEDURE, SPECIES DIVERSITY INDE W72-03218
CTROSCOPY, ASSIMILATION, KJEDAHL PROCEDURE, SAMPLE PREPARATION.: / W73-00379
IVERS, PLANT PHYSIOLOGY, TESTING PROCEDURES, AQUATIC ALGAE, CHLORO W72-14314
*CHEMOSTAL ASSAYS, *ALGAL ASSAY PROCEDURES, SELENASTRUM CAPRICORN W72-03985
HICATION, DATA COLLECTIONS, TEST PROCEDURES, CYCLING NUTRIENTS, AN W71-13794
*ANALYTICAL TECHNIQUES, *TESTING PROCEDURES, *BIOASSAY, *ALGAE, FR W72-10619
TRILOTRIACETIC ACID, ALGAL ASSAY PROCEDURES, ALGAL GROWTH, IN-LAKE W71-12072
ATION, RUNOFF, MEASUREMENT, TEST PROCEDURES, PHOTOSYNTHESIS, SEASO W72-04270
S, BACTERIOCIDES, CULTURES, TEST PROCEDURES, PESTICIDE TOXICITY, R W71-06189
G AGENTS, EFFECTIVENESS, TESTING PROCEDURES, CRITERIA, CLASSIFICAT W71-09789
S, AMMONIA STRIPPING, GUGGENHEIM PROCESS.: /PHYSICOCHEMICAL METHOD W70-09907
*FRACTIONATION-FLOTATION PROCESS.: W71-12200
BACTERIA, *ISOLATION, *MEMBRANE PROCESSES, BACTERIA, SULFUR BACTE W72-04743
S, ANTIBIOTICS(PESTICIDES), DATA PROCESSING, COLIFORMS, MICROORGAN W71-05155
ROLINA, *COMPUTER PROGRAMS, DATA PROCESSING, DATA COLLECTIONS, VAR W71-07337
PLANT RESIDUES, RECYCLING FOOD, PROCESSING WASTES, PATHOGENS, BIO W72-00846
WINDSCALE PROCESSING PLANT.: W73-00828
ANNULIORUS, PHYSA HETEROSTROPHA, PROCLADIUS, WHITE OAK LAKE(TENN), W70-06668
E WINDERMERE(ENGLAND), EXOENZYME PRODUCERS, CHLOROPHYLL A.: /, LAK W72-03224
R, *CYANOPHYTA, RESERVOIRS, ODOR- PRODUCING ALGAE, SOUTHWEST U.S., W72-10613
*ODOR- PRODUCING ALGAE.: W72-11604
DES, TASTE-PRODUCING ALGAE, ODOR- PRODUCING ALGAE, ACTINOMYCETES, P W73-02144
HEAVY METALS, PESTICIDES, TASTE- PRODUCING ALGAE, ODOR-PRODUCING A W73-02144
WATER QUALITY, HERBICIDES, ODOR- PRODUCING ALGAE.: /UTION EFFECTS, W72-14692
UE, AQUATIC WEED STUDY FACILITY, PRODUCT EVALUATION, BLUE-GREEN AL W71-06102
*CHLORELLA PYRENOIDOSA, OXYGEN PRODUCTION, LIGHT INTENSITY, CULT W70-05547
FERTILIZATION, TEMPERATURE, CROP PRODUCTION, FUNGI, SOILS, SYMBIOS W73-02473
*SUSPENDED SEDIMENT PRODUCTION, *BLUE-GREEN ALGAE.: W72-02218
LINATION PLANTS, *ELECTRIC POWER PRODUCTION, *IRRIGATION PROGRAMS, W72-03327
2-4-D, DALAPON, HERBICIDES, CROP PRODUCTION.: /NITROGEN FIXATION, W71-12071
CHLOROPHYLL, *SEAWATER, *AQUATIC PRODUCTION, *BIOMASS, ORGANIC MAT W71-11562
IC FLORA, *GROWTH RATE, *AQUATIC PRODUCTIVITY, RADIOISOTOPES, TRAC W71-11486
N, DIURNAL DISTRIBUTION, AQUATIC PRODUCTIVITY, NITROGEN, PHOSPHORU W71-11519
ION, *NUTRIENTS, *ALGAE, SEWAGE, PRODUCTIVITY, WATER POLLUTION EFF W71-12857
GAE, DIATOMS, MEASUREM/ *PRIMARY PRODUCTIVITY, *STANDING CROP, *AL W71-12855
*EUTROPHICATION, *IONS, *AQUATIC PRODUCTIVITY, FARM PONDS, NUTRIEN W71-11001
RBICIDES, INSECTICIDES, BIOMASS, PRODUCTIVITY, SCENE: /OXICITY, HE W71-10566
*INSECTICIDES, *ALGAE, *PRIMARY PRODUCTIVITY, CHLORINATED HYDROCA W71-10096
UM COMPOUNDS, NUTRIENTS, PRIMARY PRODUCTIVITY, FERTILIZERS, EFFLUE W71-11008
ALITY, *EUTROPHICATION, *PRIMARY PRODUCTIVITY, *ALGAE, NITROGEN, P W71-13183
YTOLOGICAL STUDIES, SELECTIVITY, PRODUCTIVITY.: /ETHODS, *ALGAE, C W71-12868
ENERGY BUDGET, *SEWAGE, *AQUATIC PRODUCTIVITY, *MIDGES, LAGOONS, B W72-03556
EDIMENTS, STABILIZATION, PRIMARY PRODUCTIVITY, GROWTH RATES, RESPI W72-03556
S, BIOMASS, HARVESTING OF ALGAE, PRODUCTIVITY.: /ROTEINS, NUTRIENT W72-03327
FFECTS, FLUCTUATION, COMPARATIVE PRODUCTIVITY, MUNICIPAL WASTES, I W72-03228
A WATER, COASTS, BIORT/ *PRIMARY PRODUCTIVITY, *PHYTOPLANKTON, *SE W72-03228
UTRIENTS, *BIOLOGICAL / *PRIMARY PRODUCTIVITY, *EUTROPHICATION, *N W72-03216
Y TECHNIQUES, *PLANKTON, PRIMARY PRODUCTIVITY, FOOD HABITS, ALGAE, W72-00161
ATER POLLUTION, PROTEIN, PRIMARY PRODUCTIVITY.: /CAL PROPERTIES, W W72-01473
NEMATODES, EPIPHYTOLOGY, PRIMARY PRODUCTIVITY, PHYSIOLOGICAL ECOLO W72-00948
G ORGANIC MATTER, PHYTOPLANKTON, PRODUCTIVITY, IRON.: / DECOMPOSIN W72-01097
S, CORRELATION ANALYSIS, PRIMARY PRODUCTIVITY.: /ALYSIS, BIORHYTHM W72-11719
FACE / *CARBON DIOXIDE, *PRIMARY PRODUCTIVITY, ALASKA, FIORDS, SUR W72-09122
LLUTION EFFECTS, EUTROPHICATION, PRODUCTIVITY, ALGAE, PHOSPHATES, W72-09167
TS, AQUATIC POPULATIONS, PRIMARY PRODUCTIVITY.: /TRIENT REQUIREMEN W72-09103
, *CARBON, GRAVIMETRIC ANALYSIS, PRODUCTIVITY, SEASONAL, CHLOROPHY W72-09108
GOONS, *LIMITING FACTORS, ALGAE, PRODUCTIVITY, PHOTOSYNTHESIS, NUT W72-09162
NIQU/ *LIGHT INTENSITY, *PRIMARY PRODUCTIVITY, *RADIOACTIVITY TECH W72-09102
IES, DOMINANT ORGANISMS, AQUATIC PRODUCTIVITY, BIOMASS, CARBON RAD W72-09239
PLANKTON, BIOMASS, GROWTH RATES, PRODUCTIVITY, ANIMAL GROWTH, PLAN W72-09378
*ALGAE, *PHYTOPLANKTON, *PRIMARY PRODUCTIVITY, MICROSCOPY, CHLOROP W72-09239
LGAE, DDT, *PESTICIDES, *PRIMARY PRODUCTIVITY, WATER POLLUTION EFF W72-09653
ANDING CROPS, WISCONSIN, PRIMARY PRODUCTIVITY, SYSTEMATICS, GROWTH W73-02469
SSOLVED SOLIDS, BENTHOS, CARBON, PRODUCTIVITY, ORGANIC COMPOUNDS, W73-02100
, EUTROPHICATION, PHYTOPLANKTON, PRODUCTIVITY, ANABAENA, AMMONIA, W73-02472

ENVIRONMENT, *BIOLOGIC/ *AQUATIC	PRODUCTIVITY, *ECOLOGY, *AQUATIC	W73-02095
N, ECOSYSTEMS, AQUATIC HABITATS,	PRODUCTIVITY, AQUATIC BACTERIA, A	W73-02095
OPLANKTON, N/ *ECOLOGY, *AQUATIC	PRODUCTIVITY, *ZOOPLANKTON, *PHYT	W73-00935
S, SPATIAL DISTRIBUTION, PRIMARY	PRODUCTIVITY, CHRYSOPHYTA, BIOLOG	W73-00853
THESIS, RESPIRATION, SUCCESSION,	PRODUCTIVITY, GROWTH RATES.: /SYN	W73-01632
LANKTON, PHOTOSYNTHESIS, PRIMARY	PRODUCTIVITY, GREAT LAKES, NEW YO	W73-01615
*ALGAE, *GROWTH RATES, *PRIMARY	PRODUCTIVITY, *HEAVY METALS, *TRA	W73-01434
BIOLOGICAL COMMUNITIES, BIOMASS,	PRODUCTIVITY, METABOLISM, CHLOREL	W73-01080
TY, *AQUATIC ALGAE, *COMPARATIVE	PRODUCTIVITY, *ECOSYSTEMS, *BIOLO	W73-01080
NEDESMUS, AQUATIC ALGAE, AQUATIC	PRODUCTIVITY, POLLUTANT IDENTIFIC	W73-01066
ALYSIS, *STANDING CROP, *PRIMARY	PRODUCTIVITY, LAKES, SAMPLING, RE	W73-01435
ES, SALINE WATER, BIOM/ *PRIMARY	PRODUCTIVITY, *FRESHWATER, *MARSH	W73-01089
EMICHORDATES, ARTHROP/ SECONDARY	PRODUCTIVITY, CEPHALOCHORDATES, H	W73-00935
MPARATIVE PRO/ *STRESS, *PRIMARY	PRODUCTIVITY, *AQUATIC ALGAE, *CO	W73-01080
GROWTH RATES, CULTURES, AQUATIC	PRODUCTIVITY, METABOLISM, ELECTRO	W72-14314
GHT INTENSITY, TOXICITY, PRIMARY	PRODUCTIVITY, STANDING CROPS, *PH	W72-14690
*ESTUARINE ENVIRONMENT, *PRIMARY	PRODUCTIVITY, *SHOALS, *BENTHIC F	W72-14317
US, GROWTH RATES, CULT/ *PRIMARY	PRODUCTIVITY, BACTERIA, SCENEDESM	W72-14314
*ALGAE, BIOLOGICAL COMMUNITIES,	PRODUCTIVITY, LIFE CYCLES, GROWTH	W72-14659
N, SAMPLE PREPARATION, ZO/ *C-14	PRODUCTIVITY, LIQUID SCINTILLATIO	W72-14317
ECHNIQUES, CARBON RADIOISOTOPES,	PRODUCTIVITY, METABOLISM, FUNGI,	W72-14326
INE ALG/ *GROWTH RATES, *PRIMARY	PRODUCTIVITY, ENERGY BUDGET, *MAR	W73-00366
ON SOURCES, CHLOROPHYTA, AQUATIC	PRODUCTIVITY, AQUATIC ALGAE, NITR	W73-00379
PLANKTON, *THERMOCLINE, *PRIMARY	PRODUCTIVITY, *NERITIC, NITRATES,	W73-00431
AUTUMN, WINTER, SUMMER, PRIMARY	PRODUCTIVITY, DINOFLAGELLATES, PY	W73-00442
MUNITIES, AQUATIC ALGAE, AQUATIC	PRODUCTIVITY, BIOCHEMICAL OXYGEN	W73-00002
ASS, *CYANOPHYTA, PHYTOPLANKTON,	PRODUCTIVITY, PHOTOSYNTHESIS, EPI	W72-14812
ANOPHOSPHORUS COMPOUNDS, PRIMARY	PRODUCTIVITY, SALINITY, PROFILES,	W72-07892
OPHYTA, DINOFLAGELLATES, PRIMARY	PRODUCTIVITY, MINERALOGY, LIGHT,	W72-07166
ROORGANISMS, ALGAE, LA/ *PRIMARY	PRODUCTIVITY, *BENTHIC FLORA, MIC	W72-07501
TY, RADON RADIOISOTOPES, PRIMARY	PRODUCTIVITY, CULTURES, BIOMASS,	W72-07648
LANKTON, PHOTOSYNTHESIS, PRIMARY	PRODUCTIVITY, EUTROPHICATION, SOL	W72-08054
C, *ALGAE.: *PRIMARY	PRODUCTIVITY, *OIL WASTES, *ARCTI	W72-07922
NKT/ *ALGAE, *PROTOZOA, *PRIMARY	PRODUCTIVITY, INVERTEBRATES, *PLA	W72-07899
SPRIATION, NUTRIENTS, TURBIDITY,	PRODUCTIVITY, PLANKTON, STANDING	W72-08559
FOREIGN RESEARCH, MARINE ALGAE,	PRODUCTIVITY, POLLUTANTS, CHEMICA	W72-08804
*NITROGEN, *AQUATIC ENVIRONMENT,	PRODUCTIVITY, SEDIMENTS, NITRIFIC	W72-08459
R INTERFACES, SEDIMENTS, AQUATIC	PRODUCTIVITY, PHYTOPLANKTON, ALGA	W72-08989
, *ESSENTIAL NUTRIENTS, *PRIMARY	PRODUCTIVITY, WATER QUALITY, TROP	W72-08986
S, ALGAE, NUTRIENT REQUIREMENTS,	PRODUCTIVITY, ESSENTIAL NUTRIENTS	W70-04080
NTHESIS, CARBON DIOXIDE, PRIMARY	PRODUCTIVITY, BIODEGRADATION, ORG	W71-04518
QUALITY, PHOSPHORUS, PHOSPHATES,	PRODUCTIVITY, LEACHING.: / WATER	W70-08944
, MILORGANITE, GAMBUSIA AFFINIS,	PRODUCTIVITY MEASUREMENTS, NUTRIE	W71-01488
*ALGAE, *PHYTOPLANKTON, *AQUATIC	PRODUCTIVITY, *PHOSPHATES, *NITRA	W71-02681
HRYSOPHYTA, CHLOROPHYTA, PRIMARY	PRODUCTIVITY, CARBON RADIOISOTOPE	W71-03014
DRAINAGE, MANAGEMENT, WISCONSIN,	PRODUCTIVITY, ALGAE, ECONOMICS, P	W71-03012
H, CALIFORNIA, SANDS, SEDIMENTS,	PRODUCTIVITY, TOXICITY, CRABS, AL	W71-01475
PLANKTON, PRODUCTIVITY, PRIMARY	PRODUCTIVITY, ICED LAKES, IONS, S	W71-02880
CATION, PHYTOPLANKTON, PLANKTON,	PRODUCTIVITY, PRIMARY PRODUCTIVIT	W71-02880
KTON, *ESTUARIES, *STA/ *PRIMARY	PRODUCTIVITY, *COASTS, *PHYTOPLAN	W71-07875
REATMENT, AQUATIC ALGAE, PRIMARY	PRODUCTIVITY, TRACE ELEMENTS, PHO	W71-09674
YSTEMS, *ALASKA, *LAKES, ARCTIC,	PRODUCTIVITY, EUTROPHICATION, RED	W71-09190
*TROPHIC LEVEL, *ENERGY BUDGET,	PRODUCTIVITY, HERBIVORES, CARNIVO	W71-10064
MASS, OXYGEN, ALGAE, COMPARATIVE	PRODUCTIVITY, EUTROPHICATION.: /O	W71-09165
OPLANKTON, DISTRIBUTION, PRIMARY	PRODUCTIVITY, ALGAE, BIOMASS, OLI	W71-09171
TION, *MARINE ALGAE, *ESTUARIES,	PRODUCTIVITY, POPULATION, ATLANTI	W72-04766
NAGEMENT, FERTILIZATION, PRIMARY	PRODUCTIVITY.: /FIXATION, LAND MA	W72-04769
SKA, *EUTROPHICATION, / *PRIMARY	PRODUCTIVITY, *RESERVOIRS, *NEBRA	W72-04270
LOGICAL), AQUATIC ALGAE, AQUATIC	PRODUCTIVITY, FISH, FISHKILL, LIM	W72-04279
ECOLOGY, ENVIRONMENTAL EFFECTS,	PRODUCTIVITY, RADIOECOLOGY, OCEAN	W72-06313
S, NATURAL STREAMS, PULP WASTES,	PRODUCTIVITY, DIELDRIN, PATH OF P	W72-06340
ISMS, / *WATER QUALITY, *AQUATIC	PRODUCTIVITY, *AQUATIC MICROORGAN	W72-06532
ALGAE, *BENTHIC FLORA, *PRIMARY	PRODUCTIVITY, *ENERGY CONVERSION,	W72-06283
GEN, NITROGEN, HYDROGEN, PRIMARY	PRODUCTIVITY, RESPIRATION, PLANT	W72-06274
PRIMARY PRODUCTIVITY, *SECONDARY	PRODUCTIVITY, PHYTOPLANKTON, ZOOP	W72-05968
ATIC ANIMALS, *BIOMASS, *PRIMARY	PRODUCTIVITY, *SECONDARY PRODUCTI	W72-05968
IES, DIATOMS, ALGAE, CLADOPHORA,	PRODUCTIVITY, WATER POLLUTION, ST	W72-13662
IOMASS, *AQUATIC ALGAE, *AQUATIC	PRODUCTIVITY, LIGHT, TURBIDITY, N	W72-13142
EA WATER, MARINE PLANTS, PRIMARY	PRODUCTIVITY, CORAL, REEFS, ANTIB	W72-14294
*ON-SITE TESTS, SULFUR BACTERIA,	PRODUCTIVITY, RADIOCHEMICAL ANALY	W72-14301
AL EFFECTS, CHLOROPHYTA, PRIMARY	PRODUCTIVITY, CHRYSOPHYTA.: /MENT	W72-13816
IOMASS/ *PHYTOPLANKTON, *PRIMARY	PRODUCTIVITY, *EUTROPHICATION, *B	W72-13871
FLORA, HUMAN POPULATION, PRIMARY	PRODUCTIVITY, WATER CHEMISTRY, *C	W72-12549
OPHYLL, *PLANT PIGMENTS, AQUATIC	PRODUCTIVITY, *PRIMARY PRODUCTIVI	W72-12192
YNTHESIS,/ *RESERVOIRS, *PRIMARY	PRODUCTIVITY, *TURBIDITY, *PHOTOS	W72-12393
, AQUATIC PRODUCTIVITY, *PRIMARY	PRODUCTIVITY, *ALGAE, *DIATOMS, *	W72-12192
NDING CROPS, RESISTANCE, PRIMARY	PRODUCTIVITY, BIOMASS, ANAEROBIC	W72-12940
KTON, LAKES, *SEASONAL, *PRIMARY	PRODUCTIVITY, AQUATIC ALGAE, ZOOP	W72-12738
OANTAGONISM, AXENIC / *EXCRETORY	PRODUCTS, *AUTOANTAGONISM, *HETER	W72-12734
GROWTH REGULATION, EXTRACELLULAR	PRODUCTS, ALGAL EXCRETIONS.: /N,	W71-08687
S, FOOD CHAINS, IRRIGATION, FLOW	PROFILES, MARINE ALGAE, WINDS, SE	W71-09225
PRIMARY PRODUCTIVITY, SALINITY,	PROFILES, PHOSPHATES, NITRATES, S	W72-07892
STS, MATHEMATICAL MODELS, LINEAR	PROGRAMMING, SYSTEMS ANALYSIS, DE	W72-03342
IC POWER PRODUCTION, *IRRIGATION	PROGRAMS, *AGRICULTURE, *ECONOMIC	W72-03327
, X-RAYS, ELECTROLYSIS, COMPUTER	PROGRAMS, GAMMA RAYS, POLLUTANT I	W71-11036
CAL ANALYSIS, PLANKTON, COMPUTER	PROGRAMS, BIOMASS, TEMPERATURE, C	W72-08560
ANALYTICAL TECHNIQUES, *COMPUTER	PROGRAMS, *ALGAE, *PHYTOPLANKTON,	W73-02469
ERNS, *NORTH CAROLINA, *COMPUTER	PROGRAMS, DATA PROCESSING, DATA C	W71-07337
SMITH MOUNTAIN	PROJECT(VA), WEATHER EFFECT.:	W72-05282
DE CREEK.: *LOPEZ WATER SUPPLY	PROJECT, LOPEZ CREEK, ARROYO GRAN	W72-05594

ONMENTAL ENGINEERING, *VIRGINIA, PROJECTIONS, MODEL STUDIES.: /VIR W72-14405
THERMAL POWER, *MULTIPLE-PURPOSE PROJECTS, *WATER REUSE, *WATER SU W70-09193
, PERIDINIUM WESTI, OSCILLATORIA PROLIFICA, THRESHOLD ODOR NUMBER. W71-13187
MEDIA, DISACCHARIDES, THREONINE, PROLINE, GLUTAMATE, LEUCINE, ARGI W73-00942
, ETHANOL, BENZOATE, PROPIANATE, PROPANOL, LAURATO, PALM: /ACETATE W72-05421
TIS NIDULANS, CERASON, DIATHENE, PROPAZINE, COTORON, DIURON, LINUR W71-12071
TIONS, *WATER QUALITY, *CHEMICAL PROPERTIES, EUTROPHICATION, WATER W71-12092
IVATED L/ *TILE DRAINAGE, *WATER PROPERTIES, *FERTILIZATION, *CULT W71-12252
UATIC ALGAE, ECOLOGY, BIOLOGICAL PROPERTIES, FRESH WATER, WATER QU W71-10580
SOIL TEMPERATURE, SOIL CHEMICAL PROPERTIES, SOIL BACTERIA, SOIL A W72-03342
*AQUEOUS SOLUTIONS, *BIOLOGICAL PROPERTIES, MOLECULAR STRUCTURE, W72-01473
ANT POPULATIONS, ALGAE, CHEMICAL PROPERTIES, WATER POLLUTION, PROT W72-01473
SALINITY, PHYTOPLANKTON, THERMAL PROPERTIES, HYDROGEN ION CONCENTR W72-01363
WATER, AQUATIC ALGAE, BIOLOGICAL PROPERTIES, WATER POLLUTION EFFEC W72-00974
NT DISTRIBUTION, PHYSICOCHEMICAL PROPERTIES, PESTICIDES, TROPHIC L W72-00847
EPILIMNION, MUD, PHYSICOCHEMICAL PROPERTIES, HYDROGEN ION CONCENTR W72-03224
ARINE PLANTS, ENGINEERING, WATER PROPERTIES, PHYTOPLANKTON, CHLORO W72-03220
GENTS, SODIUM CHLORIDE, CHEMICAL PROPERTIES, SECCHI DISKS, METHODO W72-09167
RTIES, ELECTROPHORESIS, CHEMICAL PROPERTIES, HYDROGEN ION CONCENTR W72-11833
TION, LIGHT PENTRATION, PHYSICAL PROPERTIES, ELECTROPHORESIS, CHEM W72-11833
IONS, AQUATIC ENVIRONMENT, WATER PROPERTIES, ALGAL CONTROL, WATER W72-11724
L, *LAKES, *NUTRIENTS, *CHEMICAL PROPERTIES, *SEDIMENTS, CLAYS, CO W72-05469
NE, TEMPERATURE, DEPTH, CHEMICAL PROPERTIES, ACIDITY, ALKALINITY, W72-04269
NTIMICROBIOTICS, PHARMACOLOGICAL PROPERTIES, PHYTOCHEMICAL SCREENI W72-04275
STE ASSIMILATIVE CAPACITY, WATER PROPERTIES, WATER POLLUTION, WATE W72-04279
L TECHNIQUES, TOXICITY, CHEMICAL PROPERTIES, MINNESOTA, LIPIDS, HU W72-04275
C RADIOISOTOPES, PHYSICOCHEMICAL PROPERTIES, FISH, MARINE PLANTS, W72-04463
A, DIATOMS, TEMPERATURE, OPTICAL PROPERTIES, GROWTH RATES.: /OPHYT W70-10403
S, AQUATIC ENVIRONMENT, CHEMICAL PROPERTIES, PHOSPHORUS, NITROGEN, W71-00117
INDS, SEA WATER, PHYSICOCHEMICAL PROPERTIES.: /ES, MARINE ALGAE, W W71-09225
BIOMASS, ALGAE, PHYSICOCHEMICAL PROPERTIES, DIATOMS, SUCCESSION, W73-01435
LT, FRESH WATER, LAKES, PHYSICAL PROPERTIES, CHEMICAL PROPERTIES, W73-01434
S, PHYSICAL PROPERTIES, CHEMICAL PROPERTIES, GOLD, IRON, CESIUM, A W73-01434
ORMS, NUCLEAR POWERPLANTS, WATER PROPERTIES, SALINITY, WATER TEMPE W73-02029
*VIRGINIA, *WATER / *BIOLOGICAL PROPERTIES, *ECOSYSTEMS, *RIVERS, W73-01972
ICS, CHEMICAL ANALYSIS, PHYSICAL PROPERTIES, CHLOROPHYTA, EUGLENOP W73-01627
QUATIONS, RESISTANCE, ELECTRICAL PROPERTIES.: /PLANT MORPHOLOGY, E W73-01630
GHT PENETRATION, PHYSICOCHEMICAL PROPERTIES, AQUATIC PLANTS, INDUS W73-01615
ICAL COMMUNITIES, PHYSIOCHEMICAL PROPERTIES, DISTRIBUTION PATTERNS W73-00918
LLUTANT IDENTIFICATION, PHYSICAL PROPERTIES, BULK DENSITY, POROSIT W73-00854
HLOROPHYTA, RHODOPHYTA, PHYSICAL PROPERTIES, WATER TEMPERATURE, DA W73-00838
ON, CLASSIFICATION, SEWAGE, SOIL PROPERTIES, GROUNDWATER MOVEMENT, W72-14363
CHELATION, MINE DRAINAGE, WATER PROPERTIES.: /Y, HARDNESS(WATER), W72-14694
ANKTON, AQUATIC PLANTS, CHEMICAL PROPERTIES, WATER QUALITY.: /TOPL W72-14542
E, WATER POLLUTION, OXYGEN, SOIL PROPERTIES, IDAHO.: /NSUMPTIVE US W72-14878
ND/ *EUTROPHICATION, *BIOLOGICAL PROPERTIES, *WATER QUALITY, *BIOI W73-00002
, LIGHT, PHOTOSYNTHESIS, OPTICAL PROPERTIES, ABSORPTION, MATHEMATI W72-14800
AL, COLOR, IRON, PHYSICOCHEMICAL PROPERTIES, DIATOMS, DOMINANT ORG W73-00256
, DISTRIBUTION PATTERNS, THERMAL PROPERTIES, THERMAL: /ANIC MATTER W73-00431
, WATER QUALITY, AERATION, WATER PROPERTIES, *NEW HAMPSHIRE, ALGAE W72-07890
ERIA, IPLANKTON, PHYSICOCHEMICAL PROPERTIES, *ALGAE, *HYDROBIOLOGY W72-07896
TORY TESTS, CHRYSOPHYTA, OPTICAL PROPERTIES.: /LAMYDOMONAS, LABORA W72-08051
RBON, LAKE MORPHOMETRY, CHEMICAL PROPERTIES, O: /SOLVED OXYGEN, CA W72-08048
OTHROMBIN TIME, *PHARMACOLOGICAL PROPERTIES, ANTICANCER, ANTINEOPL W72-07360
WASTES, BIOINDICATORS, CHEMICAL PROPERTIES, PHYSICAL PROPERTIES, W72-12729
S, CHEMICAL PROPERTIES, PHYSICAL PROPERTIES, PHYTOPLANKTON, ZOOPLA W72-12729
RUMENTATION, *MONITORING, *WATER PROPERTIES, *ON-SITE DATA COLLECT W72-12941
ON CONCENTRATION, FLOW, CHEMICAL PROPERTIES, PHYSICAL: /HYDROGEN I W72-12941
S, *ALGAE, EQUATIONS, BIOLOGICAL PROPERTIES, REPRODUCTION, ANALYTI W72-12724
ORK, BIOLUMINESCENCE, BIOLOGICAL PROPERTIES, CHLOROPHYLL, TURBIDIT W72-12941
H OF POLLUTANTS, PHYSICOCHEMICAL PROPERTIES.: / RADIOECOLOGY, *PAT W72-12204
*PUMPED STORAGE, PHYSICOCHEMICAL PROPERTIES, *PLANKTON, *AQUATIC P W72-14280
L STUDIES, LIMNOLOGY, BIOLOGICAL PROPERTIES, MINERALOGY, FISH, FIS W72-14280
MORPHOLOGY, ECOLOGY, BIOLOGICAL PROPERTIES.: /A COLLECTIONS, LAKE W72-13851
TER BALANCE, SEDIMENTS, CHEMICAL PROPERTIES, DISSOLVED SOLIDS, HYD W72-13851
SURFACE WATERS, ALGAE, CHEMICAL PROPERTIES, LANDFILLS, PERCOLATIO W72-12955
, BIOINDICATORS, PHYSICOCHEMICAL PROPERTIES, MICROSCOPY, LIMNOLOGY W72-13142
EATMENT, NITROGEN, PHYSOCHEMICAL PROPERTIES, DIATOMS, ALGAE, CLADO W72-13662
GANIC MATTER, BACTERIA, CHEMICAL PROPERTIES, NITROGEN, HEAT TRANSF W72-12997
ATE, ACETATE, ETHANOL, BENZOATE, PROPIANATE, PROPANOL, LAURATO, PA W72-05421
ON AGENCY.: *LEGISLATION(PROPOSED), ENVIRONMENTAL PROTECTI W73-00703
CA, TURBIDOSTAT CUL/ *NORTH SEA, PROROCENTRUM MICANS, CERATIUM FUR W71-03095
DS, *CALANUS GLACIALIS, DITYLUM, PROROCENTRUM, NITZSCHIA, LAUDERIA W72-09248
SLATION(PROPOSED), ENVIRONMENTAL PROTECTION AGENCY.: *LEGI W73-00703
ERCURY POLLUTION, *ENVIRONMENTAL PROTECTION AGENCY.: *M W72-14782
(CANADA), MIKROSYPHAR PORPHYRAE, PROTECTOCARPUS SPECIOSUS.: /UELON W71-10066
WATER REUSE, RECLAMATION, ALGAE, PROTEIN, *OXIDATION LAGOONS, CLIM W71-08951
URE, HYDROGEN ION CONCENTRATION, PROTEIN, LABORATORY TESTS, COST A W72-10390
CAL PROPERTIES, WATER POLLUTION, PROTEIN, PRIMARY PRODUCTIVITY.: / W72-01473
BIOCHEMISTRY, INHIBITORS, ALGAE, PROTEINS.: /, *NITROGEN, OXYGEN, W72-07941
OSCOPY, PIGMENTS, CARBOHYDRATES, PROTEINS, ACTINOMYCETES, CHLORELL W72-07951
*NITRATES, *CHLORELLA, ENZYMES, PROTEINS, LABORATORY TESTS, AMMON W72-03222
MICS, CYANOPHYTA, ALGAE, PLANTS, PROTEINS, NUTRIENTS, BIOMASS, HAR W72-03327
YNTHESIS, CULTURES, CHLOROPHYLL, PROTEINS, ALGAE, GROWTH RATES, TR W72-03557
THIC FLORA, MUD, MICROORGANISMS, PROTEINS, RADIOACTIVITY TECHNIQUE W71-08042
AD, NITRATES, CALCIUM CARBONATE, PROTEINS, FILTERS, ENGINEERS ESTI W70-10174
, PHOSPHORUS, CHEMICAL ANALYSIS, PROTEINS, CARBOHYDRATES, LIPIDS.: W70-05469
AECALIS, ENTEROBACTER AEROGENES, PROTEUS VULGARIS, PSEUDOMONAS AER W72-07664
N, BACILLARIOPHYTA, VOLVOCINEAE, PROTOCOCCINEAE, ULOTRICHINEAE, DE W72-04855
LA BACILLARIOPHAGA, REFRACTOCYS/ PROTOCRYPTOCHRYSIS OBOVATA, GUTTU W72-09119

NIQUES, DATA COLLECTIONS, ALGAE, PROTOZOA, FUNGI, EQUIPMENT, INSTRU W72-07225
IQUES, HEATED WATER, CONDENSERS, PROTOZOA, ALGAE, FUNGI, WATER POL W72-07526
NA, CHLAMYDOMONAS, EUGLENOPHYTA, PROTOZOA, EQUIPMENT, WATER POLLUT W72-07661
LLUTICN SOURCES, *FISH, DIATCMS, PROTOZOA, AQUATIC INSECTS, AQUATI W72-07791
AE, NEMATODES, OYSTERS, MUSSELS, PROTOZOA, MANGROVE SWAMPS, LITTOR W72-07911
TS, *MINNESOTA, BACTERIA, FUNGI, PROTOZOA, PATHOGENS, MICROORGANIS W72-08586
LUTION, ACTIVATED SLUDGE, ALGAE, PROTOZOA, DIATOMS, EUGLENA, MATHE W72-08573
OLOGY, POLLUTANT IDENTIFICATION, PROTOZOA, CHRYSOPHYTA, CHLOROPHYT W73-00273
ER, BODIES OF WATER, PYRRCPHYTA, PROTOZOA, MARINE ALGAE, ZOOPLANKT W73-00296
KTON, *POLLUTANT IDENTIFICATION, PROTOZOA, PONDS, LIMNOLCGY, ECOLO W73-00918
GY, FISH, ALGAE, AQUATIC PLANTS, PROTOZOA, BACTERIA, STREAMFLOW, F W73-01972
ECOLOGY, FISH, WATER POLLUTION, PROTOZCA, OHIO, EUTROPHICATION, L W73-01615
AL PARASITES, VIRUSES, MOLLUSKS, PROTOZOA, YEASTS, INSECTS, MILDEW W73-01676
CTION(CHEMICAL), TROPHIC LEVELS, PROTOZOA, ALGAE, BACTERIA, DIATOM W72-10883
ERS, HYDROGEN ION CONCENTRATION, PROTOZOA, DISSOLVED OXYGEN, SLUDG W72-10233
SSOLVED OXYGEN, BACTERIA, FUNGI, PROTOZOA, FERMENTATION, BACTERIOP W72-09675
XICITY, CHARA, SUBMERGED PLANTS, PROTOZOA, EUGLENA, CHLAMYDOMONAS, W72-00155
ASONAL, DENSITY, DIATOMS, LAKES, PROTOZOA, ROTIFERS, CRUSTACEANS, W71-11012
ON, EUTROPHICATION, OLIGOTROPHY, PROTOZOA, CHLORELLA, CHLAMYDOMONA W72-05968
ATION, PLANKTON, DIEL MIGRATION, PROTOZOA, HYDROGEN ION CONCENTRAT W71-05626
IGHT INTENSITY, PONDS, ROTIFERS, PROTOZOA, PHOTOSYNTHESIS, CHLOROP W71-07382
LGAE, PLANT GROWTH, TEMPERATURE, PROTOZOA, CHLOROPHYTA, ABSORPTION W71-09254
YDOMONAS, CULTURES, CHLOROPHYTA, PROTOZOA, *ALGAE.: /ATICS, *CHLAM W72-12632
PHYTA, *SPRINGS, DIATOMS, BIOTA, PROTOZOA, *AMPHIPODA, *FLORIDA, S W72-12744
, EUTROPHICATION, TROPHIC LEVEL, PROTOZOANS, AQUATIC ALGAE, ENVIRO W72-13816
RESTRIS, BRASILOBIA PISCIFORMIS, PROWSEMONAS TROPICA, STIGMOBODO B W72-07683
A, DINOPHYCEA, BACCILARIOPHYCEA, PRYMNESIUM PARVUM, PHOTOAUTOTROPH W70-07280
IC GROWTH, *MARINE PHYTOPLANKTON(PRYMNESIUM PARVUM)(CHROOMONAS SAL W70-07842
PACIFIC OCEAN, ALEUTIAN ISLANDS, PRYMNESIUM PARVUM.: /CCYLINDRUS, W71-11008
, EAGLES, SALMO GAIRDNERII, IRIS PSEUDACORUS, LYSIMACHIA NUMMULARI W72-05952
HACOMYXA, SPONDYLOSIUM, LYNGBYA, PSEUDOANABAENA, DATA INTERPRETATI W72-07648
YDOMONAS TEXENSIS, CHLAMYDOMONAS PSEUDOMICROSPHAERA, CULTURE MEDIA W72-12632
TS, ALGAE, HEAVY METALS, GROWTH, PSEUDOMONAS, POTASSIUM COMPOUNDS, W72-13813
UATIC ALGAE, ALGAE, CHLOROPHYTA, PSEUDOMONAS, ENTERIC BACTERIA.: / W72-12255
TER AEROGENES, PROTEUS VULGARIS, PSEUDOMONAS AERUGINOSA, SERRATIA: W72-07664
ION, LIGHT INTENSITY, NUTRIENTS, PSEUDOMONAS, SYMBIOSIS.: /CENTRAT W73-01657
NUTRIEN/ MICROCYSTIS AERUGINOSA, PSEUDOMONAS AERUGINOSA, LIMITING W73-01657
REINHARDTII, TETRAHYMENA VORAX, PSEUDOMONAS FLUORESCENS, AUTOTROP W73-02095
ICATION, SYSTEMATICS, NUTRIENTS, PSEUDOMONAS, CULTURES, VITAMIN B, W73-00942
OLOGICAL STUDIES, MYCOBACTERIUM, PSEUDOMONAS, BACTERIA.: /GAE, CYT W71-08027
RICHIA COLI, BETA STREPTOCOCCUS, PSEUDOMONAS AERUGINOSA, MYCOBACTE W72-04259
DIATOMS, PYRROPHYTA, CHLORELLA, PSEUDOMONAS, TESTING, MEASUREMENT W72-03220
WTH RATES, E. COLI, AZOTOBACTER, PSEUDOMONAS, WATER POLLUTION EFFE W72-10861
DUNALIELLA TERTI/ THALASSIOSIRA PSEUDONANA, SKELETONEMA COSTATUM, W72-05614
LIPHAGE, IMMUNOLOGY, PASTEURELLA PSEUDOTUBERCULOSIS, PENICILLIN, D W72-13800
PESTRIS, ESOX LUCIUS, CAMPELOMA, PSIDIUM, SHADFLY, CHIRONOMIDS, CL W70-08098
MMUNE, PORPHYROSIPHON NOTARISII, PSYCHROPHILIC, BIOTIC ASSOCIATION W73-00854
LINGULA, ACTINOTROCHA, VELIGER, PTEROPODS, HETEROPODS, OSCILLATOR W73-00935
E, DEXTRAN, OIKOPLEURA, SAGITTA, PTEROPODS, EUPHAUSIIDS, GADUS MOR W73-00371
FOOD CHAINS, PATH OF POLLUTANTS, PUBLIC HEALTH, WATER POLLUTION EF W73-00802
OOD CHAINS, ABSORPTION, CALCIUM, PUBLIC HEALTH, SOUTH AMERICA, ATL W73-00819
ORGANIC COMPOUNDS, FOOD CHAINS, PUBLIC HEALTH, PHYTOPLANKTON: /N, W73-00821
CLING NUTRIENTS, TRACE ELEMENTS, PUBLIC HEALTH, RADIOECOLOGY.: /CY W73-00826
TON, *MARINE ALGAE, FOOD CHAINS, PUBLIC HEALTH, NUCLEAR WASTES, PA W73-00823
UTANTS, WATER POLLUTION SOURCES, PUBLIC HEALTH, ON-SITE INVESTIGAT W73-00790
TION, POLYCHLORINATED BIPHENYLS, PUBLIC HEALTH, ENVIRONMENT CONTRO W73-00703
UARINE ENVIRONMENT, FOOD CHAINS, PUBLIC HEALTH, EUROPE, RADIOISOTO W73-00828
STES, *ABSORPTION, *FOOD CHAINS, PUBLIC HEALTH, PATH OF POLLUTANTS W73-00833
NT, EUTROPHICATION, LEGISLATION, PUBLIC HEALTH, WATER QUALITY CONT W72-06655
UTION SOURCES, SEWAGE TREATMENT, PUBLIC HEALTH, WASTE WATER TREATM W72-06638
KS, MARINE PLANTS, MARINE ALGAE, PUBLIC HEALTH, ESTUARINE ENVIRONM W72-04461
POLLUTION EFFECTS, FOOD CHAINES, PUBLIC HEALTH, ESTUARINE ENVIRONM W72-04463
NING, LEGAL ASPECTS, HARD WATER, PUBLIC HEALTH, COST COMPARISONS, W71-09429
POLLUTION EFFECTS, FOOD CHAINS, PUBLIC HEALTH, AQUATIC LIFE, AQUA W72-10978
, NITRITES, TOXICITY, CORROSION, PUBLIC HEALTH, ODORS, OXIDATION-R W72-04086
ATIONS, CALIFORNIA, FOOD CHAINS, PUBLIC HEALTH, WATER POLLUTION SO W72-03324
MENT, WASTE DISPOSAL, STANDARDS, PUBLIC HEALTH, INSPECTION, WATER W72-01085
LUTION SOURCES, SOAPS, TOXICITY, PUBLIC HEALTH: /SPOSAL, WATER POL W72-01093
OLLUTION SOURCES, AQUATIC ALGAE, PUBLIC HEALTH, BIODEGRADATION, TO W71-11516
N CREEK RESERVOIR, / SOUTH TAHOE PUBLIC UTILITIES DISTRICT, *INDIA W72-02412
OLLUTION EFFECTS, DOCUMENTATION, PUBLICATIONS, PESTICIDE TOXICITY, W73-01976
NIMALS, MARINE ALGAE, SEDIMENTS, PUERTO RICO, ESTUARINE ENVIRONMEN W73-00821
QUALITY, *WISCONSIN, PHOSPHORUS, PULP AND PAPER INDUSTRY, ALGAE, A W70-05103
EL, HERBIVORES, NATURAL STREAMS, PULP WASTES, PRODUCTIVITY, DIELDR W72-06340
NTHIC FAUNA, CELLULOSE, ENZYMES, PULP WASTES, RADIOECOLOGY, WATER W72-09668
GLENOPSIS AMERICANA, CRYPTOMONAS PULSILLA, MICROCYSTIS AERUGINOSA, W72-13816
REFRACTODES BRASILIANA, CYEMONAS PULSULAE, REFRACTOMONAS BRASILIAN W72-09119
USHING OPERATION, ACID INJECTION PUMP, ALUM INJECTION PUMP, MANURE W72-02850
D INJECTION PUMP, ALUM INJECTION PUMP, MANURE GRINDER.: /TION, ACI W72-02850
TES, POULTRY, NITROGEN, LAGOONS, PUMPING, AEROBIC TREATMENT, BIOCH W72-08396
J), AMERICAN SHAD, EKMAN DREDGE, PUMPKINSEED, BLACK CRAPPIE, CHAIN W72-14280
, PLEUROCHLORIS MAGNA, VISCHERIA PUNCTATA, VISCHERIA STELLATA, ELL W72-14808
TZSCHIA DELICATISSIMA, NITZSCHIA PUNGENS, C-14, CHLOROPHYLL A, MED W73-00431
, WATER POLLUTION EFFECTS, WATER PURIFICATION, *SYSTEMATICS.: /PHY W73-00432
RESIS, *ELECTROCHEMISTRY, *WATER PURIFICATION, CLAY, ALGAE, BACTER W73-01362
EATMENT, *PUBLIC HEALTH,/ *WATER PURIFICATION, *WATER POLLUTION TR W73-02144
ROCHEMISTRY, *WATER POLL/ *WATER PURIFICATION, *FILTRATION, *ELECT W71-12467
LANTS, MATHEMATICAL MODELS, SELF- PURIFICATION, ALGAE, GREAT LAKES, W71-12072
ES, SEPARATION TECHNIQUES, WATER PURIFICATION, *OXIDATION LAGOONS, W71-13356
NUTRIENT, *BIODEGRADATION, WATER PURIFICATION, *AEROBIC TREATMENT, W72-05315

ABORATORY TESTS, DISSOLVE/ *SELF-	PURIFICATION, *ALGAE, *STREAMS, L	W71-04518
RS, RESERVOIRS, PHOTOSYN/ *WATER	PURIFICATION, *DETENTION RESERVOI	W71-01491
OLING, *THERMAL POWER, *MULTIPLE-	PURPOSE PROJECTS, *WATER REUSE, *	W70-09193
LL A, MEDITERRANEAN SEA, FRANCE,	PYCNOCLINE.: /NS, C-14, CHLOROPHY	W73-00431
*CHLORELLA	PYRENOIDOSA.:	W73-00251
NKISTRODESMUS BRAUNII, CHLORELLA	PYRENOIDOSA, CHLORELLA VULGARIS,	W72-07664
UM NITRILOTRIACETATE, *CHLORELLA	PYRENOIDOSA, SOURCE OF NITROGEN,	W70-08976
LETONEMA COSTATUM.: CHLORELLA	PYRENOIDOSA, NITZSCHIA PALEA, SKE	W71-03027
IGHT INTENSITY, CULT/ *CHLORELLA	PYRENOIDOSA, OXYGEN PRODUCTION, L	W70-05547
CHLAMYDOMONAS GLOBUSA, CHLORELLA	PYRENOIDOSA, STICHOCOCCUS BACILLA	W70-07255
/ DNA, CHLOROPHYLL A, CHLORELLA	PYRENOIDOSA, SPIROGIRA, CHLORELLA	W72-06274
C RIVER, ALAFIA RIVER, CHLORELLA	PYRENOIDOSA, SODIUM SILICATE, SOD	W72-04734
ON CONSTANT, DILUTION, CHLORELLA	PYRENOIDOSA, MONOD GROWTH MODEL,	W72-04788
IACETATE.: *CHLORELLA	PYRENOIDOSA, *TRISODIUM NITRILOTR	W71-11972
NALIS, AS/ PHORMIDIUM, CHLORELLA	PYRENOIDOSA, CHLAMYDOMONAS, FONTI	W71-11519
LOPHEN A 50, ELECTRO/ *CHLORELLA	PYRENOIDOSA, LINDANE, P P'-DDE, C	W72-04134
MNODRILUS SPP., TUBI/ *CHLORELLA	PYRENOIDOSA, *CHLORAMPHENICAL, LI	W72-09668
A RADIATION, SURVIVAL, CHLORELLA	PYRENOIDOSA.: /NIANA, CO-60, GAMM	W72-11836
CHLORELLA	PYRENOIDOSA.:	W72-11572
ENEDESMUS QUADRICAUDA, CHLORELLA	PYRENOIDOSA, LITHIUM CHLORIDE, LI	W72-12739
RHAMNOSE, ACRYLIC ACID, SODIUM,	PYRIDINE, DEXTRAM, RAFFINOSE, SUC	W72-07710
HOE(CALIF), ZINC, ASCORBIC ACID,	PYRIDOXINE, INDAL COMPOUNDS.: /TA	W72-07514
S NEREIS, SEA OTTER, MACROCYSTIS	PYRIFERA, CHORDATES, POLYCHAETES:	W73-00932
TOCOCCUS MAMMILATUS, NAUTOCOCCUS	PYRIFORMIS NAUTOCOCCUS, APIOCOCCU	W72-07141
EDOGONIUM, MOUGOTIA, SPOROTETRAS	PYRIFORMIS, GONGROSIRA, MICROSPOR	W71-05991
SODIUM TRIPOLYPHOSPHATE, SODIUM	PYROPHOSPHATE.: /NSED PHOSPHATES,	W71-00116
ROWTH, *RADIOISOTOPE P32, SODIUM	PYROPHOSPHATE, RIO GRANDE VALLEY.	W72-13992
RYSOPHYTA, CHLOROPHYTA, BIOMASS,	PYRROPHYTA, SPRING, SUMMER, AUTUM	W72-12738
IATOMS, CHLOROPHYTA, CYANOPHYTA,	PYRROPHYTA, CHRYSOPHYTA, EUGLENOP	W72-04855
LOGY, CHRYSOPHYTA, EUGLENOPHYTA,	PYRROPHYTA, BIOMASS, HYDROGEN ION	W72-07145
OSPHORUS, PLANKTON, CHLOROPHYTA,	PYRROPHYTA, DINOFLAGELLATES, PRIM	W72-07166
MICROSCOPY, ELECTRON MICROSCOPY,	PYRROPHYTA, DISTRIBUTION PATTERNS	W72-07663
YTA, PHYTOPLANKTON, ZOOPLANKTON,	PYRROPHYTA, DINOFLAGELLATES, CYAN	W72-07899
LARVAE, CHRYSOPHYTA, CYANOPHYTA,	PYRROPHYTA, MARINE ALGAE, GASTROP	W73-00935
Y PRODUCTIVITY, DINOFLAGELLATES,	PYRROPHYTA.: /TER, SUMMER, PRIMAR	W73-00442
ING, SEA WATER, BODIES OF WATER,	PYRROPHYTA, PROTOZOA, MARINE ALGA	W73-00296
IMALS, MARINE ALGAE, CYANOPHYTA,	PYRROPHYTA, LARVAE, SAMPLING, INV	W72-14313
AQUATIC HABITATS, EUGLENOPHYTA,	PYRROPHYTA, FRESHWATER ALGAE.: /,	W72-09119
PHYTA, CHRYSOPHYTA, CHLOROPHYTA,	PYRROPHYTA, CYANOPHYTA, MANGANESE	W72-09111
EGIONS, CYANOPHYTA, CHRYSOPHYTA,	PYRROPHYTA, SAMPLING, ECOLOGY, WA	W72-03543
PHAEOPHYTA, RHODOPHYTA, DIATOMS,	PYRROPHYTA, CHLORELLA, PSEUDOMONA	W72-03220
PHYTA, CHRYSOPHYTA, CHLOROPHYTA,	PYRROPHYTA, INVERTEBRATES, AQUATI	W71-11012
NYX, PHACUS PLEURONECTES, PHACUS	PYRUM, TRACHELOMONAS ARMATA, TRAC	W73-00918
CENEDESMUS QUADRICAUDA, ACETATE,	PYRUVATE, LACTATE.: S	W72-00838
ATE, BUTANOL, GLYCEROL, LACTATE,	PYRUVATE, ACETATE, ETHANOL, BENZO	W72-05421
MORECEPTION, *CHEMOTAXIS, *PREY,	PYTHIUM DEBARYANUM, SKELETONEMA C	W73-00942
AN/ *ALGAL GROWTH, *RADIOISOTOPE	P32, SODIUM PYROPHOSPHATE, RIO GR	W72-13992
ABOLITES, DUNALIELLA, AGMENELLUM	QUADRAPLICATUM, C-14, THIN LAYER	W73-00361
S, SCENEDESMUS SOLI, SCENEDESMUS	QUADRICAUDA, MOUGEOTIA, ULOTHRIX	W72-07144
*STAURASTRUM PINGUE, SCENEDESMUS	QUADRICAUDA, CHLORELLA PYRENOIDOS	W72-12739
ACTATE.: SCENEDESMUS	QUADRICAUDA, ACETATE, PYRUVATE, L	W72-00838
, ANACYSTIS NIDULANS, AGMENELLUM	QUADRUPLICATUM STRAIN PR-6, PHOTO	W72-03545
IRA / NEWFOUNDLAND, PERIDINIANS,	QUALITATIVE COMPOSITION, COSCINOS	W72-03543
GEN, NITROGEN, PHOSPHORUS, WATER	QUALITY ACT, WATER QUALITY.: /OXY	W71-06444
UVER(WASH), PORTLAND(ORE), WATER	QUALITY ANALYZER.: /(WASH), VANCO	W72-03215
ON, WATER QUALITY, RESER/ *WATER	QUALITY CONTROL, *FLOW AUGMENTATI	W72-04260
ECHNIQUES, ENZYMES, COLOR, WATER	QUALITY CONTROL, RESEARCH AND DEV	W72-04070
RE, ANALYTICAL TECHNIQUES, WATER	QUALITY CONTROL, *WASTE WATER TRE	W72-00422
ICATION, *DENITRIFICATION, WATER	QUALITY CONTROL, EUTROPHICATION,	W72-00800
INEERING, WATER TREATMENT, WATER	QUALITY CONTROL, STANDARDS, WATER	W72-01660
*WATER POLLUTION CONTROL, *WATER	QUALITY CONTROL, *FEDERAL GOVERNM	W72-01659
Y, EUT/ *OCEANS, *COASTS, *WATER	QUALITY CONTROL, BIOMASS, TOXICIT	W71-11793
R QUALITY, *CHLORINATION, *WATER	QUALITY CONTROL, *ODOR, ALGAE, EU	W71-13187
ICATION, AQUATIC ALGAE, / *WATER	QUALITY CONTROL, *LAKES, *EUTROPH	W71-13256
, FUNGI, LABORATORY TESTS, WATER	QUALITY CONTROL.: / ACTINOMYCETES	W72-11604
UTROPHICATION, *RECYCLING, WATER	QUALITY CONTROL, SEWAGE TREATMENT	W72-09378
E, *WASTE WATER TREATMENT, WATER	QUALITY CONTROL, *REVIEWS.: /NANC	W72-09386
ATION, *DESTRATIFICATION, *WATER	QUALITY CONTROL, *LAKE MORPHOLOGY	W72-09304
Y, *PATHOLOGY, *TOXICITY, *WATER	QUALITY CONTROL, *WATER POLLUTION	W72-09646
ATER ANALYSIS, EVALUATION, WATER	QUALITY CONTROL, *BIOASSAY, *BIOM	W72-10883
TROPHICATION, *DETERGENTS, WATER	QUALITY CONTROL, DOMESTIC WASTES,	W72-11186
ON, WASTE WATER TREATMENT, WATER	QUALITY CONTROL.: /HUMAN POPULATI	W71-09788
ION, *TERTIARY TREATMENT, *WATER	QUALITY CONTROL, AMMONIA, NITROGE	W71-09450
DUSTRIAL WASTES, SAMPLING, WATER	QUALITY CONTROL, COST ANALYSIS, W	W71-08953
*FORAGE FISH, BIOCONTROL/ *WATER	QUALITY CONTROL, *ALGAE CONTROL,	W71-09163
,/ *ESTUARIES, *NUTRIENTS, WATER	QUALITY CONTROL, *ECOLOGY, *ALGAE	W71-08775
NCENTRATION, EFFICIENCIES, WATER	QUALITY CONTROL, EUTROPHICATION,	W72-04788
YDROGEN ION CONCENTRATION, WATER	QUALITY CONTROL, WASTE WATER TREA	W72-04431
*MAINE, *EUTROPHICATION, *WATER	QUALITY CONTROL, *NON-STRUCTURAL	W72-04279
EGISLATION, PUBLIC HEALTH, WATER	QUALITY CONTROL, ALGAL CONTROL, F	W72-06655
LL, ENZYMES, PERMEABILITY, WATER	QUALITY CONTROL, WATER POLLUTION	W72-06037
FIXATION, EUTROPHICATION, WATER	QUALITY CONTROL, MASSACHUSETTS, *	W72-06046
*WATER POLLUTION CONTROL, WATER	QUALITY CONTROL, PHOSPHORUS, AQUA	W72-06638
NTROL, *DOME/ *ALGICIDES, *WATER	QUALITY CONTROL, *AQUATIC WEED CO	W71-05083
TRIENTS, *EUTROPHICATION, *WATER	QUALITY CONTROL, *NITROGEN COMPOU	W70-04080
ION, WATER QUALITY, ALGAE, WATER	QUALITY CONTROL, EUTROPHICATION,	W72-12487
TER ANALYSIS, LAKES, SAMP/ WATER	QUALITY CONTROL, *CYANOPHYTA, *WA	W72-07890
ICATION, LABORATORY TESTS, WATER	QUALITY CONTROL, *KINETICS, *CHLO	W72-08376
OIR STORAGE, *HYPOLIMNION, WATER	QUALITY CONTROL.: / ALGAE, RESERV	W73-02138
ATER, *WATER TEMPERATURE, *WATER	QUALITY STANDARDS, *ESTURARIES, *	W72-13636

TA, EUGLENOPHYTA, GRAZING, WATER QUALITY.: / CYANOPHYTA, CHLOROPHY W72-07899
ANTS, CHEMICAL PROPERTIES, WATER QUALITY.: /TOPLANKTON, AQUATIC PL W72-14542
POSAL, PATH OF POLLUTANTS, WATER QUALITY.: / PESTICIDES, WASTE DIS W71-05553
AE, TURBIDITY, *NEW YORK, *WATER QUALITY.: /ATERS, *PLANKTON, *ALG W72-06191
ROPHICATION, ALGAL BLOOMS, WATER QUALITY.: /ADATION, FISHKILL, EUT W71-08768
PHORUS, WATER QUALITY ACT, WATER QUALITY.: /OXYGEN, NITROGEN, PHOS W71-06444
ION TREATMENT, EVALUATION, WATER QUALITY.: /T GROWTH, WATER POLLUT W72-11702
L PROPERTIES, FRESH WATER, WATER QUALITY.: /AE, ECOLOGY, BIOLOGICA W71-10580
RFACE WATERS, *SEA WATER, *WATER QUALITY, *ECOLOGY, *CALIFORNIA, H W71-10577
ALITY CONTROL, *ODOR, AL/ *WATER QUALITY, *CHLORINATION, *WATER QU W71-13187
Y PRODUCTIVITY, *ALGAE, / *WATER QUALITY, *EUTROPHICATION, *PRIMAR W71-13183
OPLA/ *INDUSTRIAL WASTES, *WATER QUALITY, *DISSOLVED OXYGEN, *PHYT W71-11876
ION, *CHEMICAL REACTIONS, *WATER QUALITY, *CHEMICAL PROPERTIES, EU W71-12092
INDU/ *ENTERIC BACTERIA, *WATER QUALITY, *DOMESTIC WASTES, ALGAE, W72-09590
RVOIRS, *HYDROLOGIC DATA/ *WATER QUALITY, *IMPOUNDED WATERS, *RESE W72-02274
, *DIVER/ *DELTAS, *BAYS, *WATER QUALITY, *WATER POLLUTION CONTROL W71-09581
MENT, *WATER/ *ESTUARIES, *WATER QUALITY, *WATER RESOURCES DEVELOP W71-09788
AQUATIC MICROORGANISMS, / *WATER QUALITY, *AQUATIC PRODUCTIVITY, * W72-06532
ECOLOGY, *PUMPED STORAGE, *WATER QUALITY, *MODEL STUDIES, ZOOPLANK W72-05282
, *LAKES, *COLORADO, *RE/ *WATER QUALITY, *WATER POLLUTION EFFECTS W72-04791
MS, FLOW, SE/ *NUTRIENTS, *WATER QUALITY, *IMPOUNDED WATERS, STREA W71-03028
ATICA/ *SYSTEMS ANALYSIS, *WATER QUALITY, *EUTROPHICATION, *MATHEM W71-03021
PULP AND PAPER INDUSTRY,/ *WATER QUALITY, *WISCONSIN, PHOSPHORUS, W70-05103
PHYTOPLANKTON, PLANKTON/ *WATER QUALITY, *LAKES, *EUTROPHICATION, W71-02880
ER, *OXYGEN SAG, *IMPAIRED WATER QUALITY, *ESTUARIES, *DELAWARE RI W70-09189
MATHEMATICAL MODELS, THE/ *WATER QUALITY, *HYDROELECTRIC PLANTS, * W72-14405
, *BIOLOGICAL PROPERTIES, *WATER QUALITY, *BIOINDICATORS, *BIOLOGI W73-00002
ATION, AQUATIC ECOSYSTEMS, WATER QUALITY, *ALGAE, *LIGHT PENETRATI W73-02349
N, RHODOPHYTA, PHAEOPHYTA, WATER QUALITY, *MARINE ALGAE.: /ODUCTIO W73-01449
OURCES, NUTRIENTS, SEWAGE, WATER QUALITY, AERATION, WATER PROPERTI W72-07890
AGOONS, *DESIGN CRITERIA, *WATER QUALITY, ALGAE, HARVESTING, DEPTH W71-07100
QUALITY, WASTE M/ *ENVIRONMENTAL QUALITY, AIR CONTAMINANT, STREAM W71-08214
APHY, *INFRARED RADIATION, WATER QUALITY, ALGAE, WATER QUALITY CON W72-12487
ION, NUTRIENTS, BIOASSAY, *WATER QUALITY, ALGAE, PESTICIDES, HEAVY W72-14282
ENZYMES, CHLEATION, ALGAE, WATER QUALITY, BIODEGRADATION, SEWAGE T W72-06837
ACEANS, *MARINE BACTERIA, *WATER QUALITY, BENTHOS, CRABS, COPEPODS W72-08142
MODELS, WATER TEMPERATURE, WATER QUALITY, CARBON DIOXIDE, HYDROGEN W72-06274
PPROPRIATION, GROUNDWATER, WATER QUALITY, COMPETING USES, DIVERSIO W70-07100
, WATER POLLUTION EFFECTS, WATER QUALITY, CHEMICAL DEGRADATION.: / W72-01085
MPARISONS, BENEFICIAL USE, WATER QUALITY, CALIFORNIA, BIOCONTROL, W72-00974
ROL, MANAGEMENT, ANALYSIS, WATER QUALITY, CYANOPHYTA, NUISANCE ALG W72-09171
, WATER POLLUTION EFFECTS, WATER QUALITY, CARBON.: /S, CHEMCONTROL W72-11724
, *PLANKTON, *RECREATION, *WATER QUALITY, CYANOPHYTA, LAKES, MISSO W72-10431
*EUTROPHICATION, BIOMASS, WATER QUALITY, CHLOROPHYLL, NITROGEN, W W71-10645
TA, ALKALINITY, NUTRIENTS, WATER QUALITY, DIATOMS, CHLOROPHYTA, TH W72-05610
, *BIOTA, *AQUATIC PLANTS, WATER QUALITY, ENVIRONMENTAL EFFECTS, L W72-06526
UDIES, ECOSYSTEMS, ALGAE, *LIGHT QUALITY, EUTROPHIC ZONE, CHLOROPH W72-12393
ES, MICROSCOPY, LIMNOLOGY, WATER QUALITY, ECOLOGY, AQUATIC HABITAT W72-13142
UATIC PLANTS, CLAYS, ODOR, WATER QUALITY, FOOD CHAINS, DISSOLVED O W72-04759
COOLING WATER, *FISHERIES, WATER QUALITY, FISH ATTRACTANTS, FISH H W70-09596
*CULTIVATED LANDS, *IOWA, WATER QUALITY, FERTILIZERS, NUTRIENTS, W71-12252
, WATER POLLUTION EFFECTS, WATER QUALITY, HERBICIDES, ODOR-PRODUCI W72-14692
S, LITTORAL, AGE, BIOMASS, LIGHT QUALITY, LIGHT PENETRATION, EFFIC W72-04766
NUTRIENTS, PHYTOPLANKTON, WATER QUALITY, MICHIGAN, SURFACE WATERS W72-10875
NIA, NITRATES, PHOSPHATES, WATER QUALITY, NITROGEN, NUTRIENTS, TRO W71-11033
OWTH, TOXICITY, NUTRIENTS, WATER QUALITY, NITRATES, PLANNING, NITR W71-13172
ATER POLLUTION, MINNESOTA, WATER QUALITY, NUTRIENTS, SEDIMENTS, CO W71-03012
YDROGEN ION CONCENTRATION, WATER QUALITY, NUTRIENTS, COLORIMETRY, W73-00916
ON, *HYDROLOGY, *NEW YORK, WATER QUALITY, NUTRIENTS, ALGAE, INFLOW W72-13851
TURE, SALINITY, SEA WATER, WATER QUALITY, ON-SITE TESTS, HYDROGEN W72-09092
TS, LAKE HURON, EFFLUENTS, WATER QUALITY, PHOSPHORUS, PHYTOPLANKTO W72-13662
TTERNS, WATER TEMPERATURE, WATER QUALITY, PLANKTON, ANALYTICAL TEC W72-07663
SURFACE WATERS, NITROGEN, WATER QUALITY, PHOSPHORUS, AEROBIC COND W70-10405
E, *EUTROPHICATION, LAKES, WATER QUALITY, PHOSPHORUS, PHOSPHATES, W70-08944
TER UTILIZATION, *IMPAIRED WATER QUALITY, RIPARIAN RIGHTS, PRIOR A W70-07100
ES, ALGAE, EUTROPHICATION, WATER QUALITY, RIVERS, IOWA, CORN, DISC W71-06445
DE, SESTON, FERTILIZATION, WATER QUALITY, RADON RADIOISOTOPES, PRI W72-07648
NTROL, *FLOW AUGMENTATION, WATER QUALITY, RESERVOIRS, WATER POLLUT W72-04260
OTOSYNTHESIS, RESPIRATION, WATER QUALITY, STREAMFLOW, AGRICULTURE, W71-11017
S, *LIFE CYCLES, CULTURES, WATER QUALITY, SCENEDESMUS, LABORATORY W72-07144
, WATER POLLUTION EFFECTS, WATER QUALITY, SEASONAL, SUSPENDED LOAD W72-07896
WATER POLLUTION TREATMENT, WATER QUALITY, STREAMS, LAKES, BIOCHEMI W72-14469
ENT, *POLLUTION ABATEMENT, WATER QUALITY, STANDARDS, WATER POLLUTI W71-08214
KENTUCKY, SULFUR BACTERIA, WATER QUALITY, SULFATES, HYDROGEN ION C W71-07942
ATES, HUDSON RIVER, ALGAE, WATER QUALITY, SEWAGE DISPOSAL, WASTE W W70-06509
G, BOATING, DOCUMENTATION, WATER QUALITY, SILTING, AGRICULTURAL CH W72-05869
TS, *PRIMARY PRODUCTIVITY, WATER QUALITY, TROPHIC LEVEL, AQUATIC A W72-08986
STE TREATMENT, INDUSTRIES, WATER QUALITY, TECHNOLOGY, SURFACTANTS, W72-09169
AQUATIC ALGAE, NUTRIENTS, WATER QUALITY, WATER ANALYSIS, FREQUENC W73-00918
T, *DISTRIBUTION SYSTEMS, *WATER QUALITY, WATER CONSERVATION, WILD W72-05594
OXIDATION, ORGANIC MATTER, WATER QUALITY, WATER CHEMISTRY, WATER P W72-04853
NUTRIENTS, EUTROPHICATION, WATER QUALITY, WATER POLLUTION CONTROL, W70-08248
VITY, ESSENTIAL NUTRIENTS, WATER QUALITY, WATER CHEMISTRY, CYCLING W70-04080
NUTRIENTS, EUTROPHICATION, WATER QUALITY, WATER POLLUTION CONTROL, W70-08249
TREATMENT, *WATER REUSE, *WATER QUALITY, WATER SUPPLY, ALGAE, ACT W70-10381
QUALITY, AIR CONTAMINANT, STREAM QUALITY, WASTE MANAGEMENT, MANURE W71-08214
OR, OXYGEN SAG, TURBIDITY, WATER QUALITY, WATER TREATMENT, NITROGE W71-09958
NS, PONDS, FARM PONDS, MARYLAND, QUARRIES, OHIO RIVER, HUDSON RIVE W72-04736
CKWELL, LAKE HODGSON, CLEARWATER QUARRY, DOLLAR LAKE, CHLOROPHYLL W72-13673
EAR LAKE(CALIF), RICHELIEU RIVER(QUEBEC), GREEN BAY(WIS), MACKEREL W70-08098

NEW BRUNSWICK, ANTICOSTI ISLAND(	QUEBEC), MAGDALEN ISLAND(QUEBEC),	W71-10066
ISLAND(QUEBEC), MAGDALEN ISLAND(	QUEBEC), SABLE ISLAND(NOVA SCOTIA	W71-10066
CE, *POLYP MAINTENAN/ *CHRYSAORA	QUINQUECIRRHA, *MEDUSAE MAINTENAN	W72-10162
TA, TSUMURAIA NUMEROSA, HIROSEIA	QUINQUELOBATA, TROITSKIELLA TRIAN	W72-07683
CHI/ *NUTRIENT BUDGETS, *BAY OF	QUINTE(ONTARIO), NUTRIENT INPUTS,	W71-11009
BAY OF	QUINTE, *LAKE ONTARIO.:	W71-08953
, CALLITRICHE, ANABAENA / BAY OF	QUINTE, SALT CREEK, CHARA, ELODEA	W73-00840
CHELEX 100, ARTEMIA SALINA, BIO-	RAD, CYCLOTELLA NANA.:	W72-04708
ROSIN AMINE D ACETATE(	RADA).:	W71-05083
LUORESCENCE, INFRARED RADIATION,	RADAR, C: /TRAVIOLET RADIATION, F	W72-04742
S, GRACILARIA VERUCOSA, ECKLONIA	RADIATA, GROWTH REGULATION.: /RMI	W72-07496
CONCENTRATION, RADIATION SAFETY,	RADIATION CONTROL.:	W72-07841
OL.: CONCENTRATION,	RADIATION SAFETY, RADIATION CONTR	W72-07841
, *PROTOZOA, *BACTERIA, FLORIDA,	RADIATION SENSITIVITY, RADIOECOLO	W73-00796
COBALT-60.: *BIOLOGICAL--GAMMA-	RADIATION SYSTEM, MODULAR SYSTEM,	W72-00383
MENTATION, ALGAE, INVERTEBRATES,	RADIATION, ECOSYSTEMS, ESTUARIES,	W71-13723
BIDITY, HYDROLOGIC BUDGET, SOLAR	RADIATION, WATER TEMPERATURE, SAL	W72-02274
LYTICAL TECHNIQUES, *ULTRAVIOLET	RADIATION, DISSOLVED OXYGEN, LAKE	W72-01780
, *AERIAL PHOTOGRAPHY, *INFRARED	RADIATION, *AQUATIC ALGAE, *SURFA	W71-12034
ORELLA SOROKINIANA, CO-60, GAMMA	RADIATION, SURVIVAL, CHLORELLA PY	W72-11836
L, TROPICAL REGIONS, ULTRAVIOLET	RADIATION, FOOD CHAINS, GAMMA RAY	W73-00796
, *FALLOUT, *MARINE/ *BACKGROUND	RADIATION, *RADIOACTIVITY EFFECTS	W73-00828
ICATION, *LAKES, ALGAE, INFRARED	RADIATION, *REMOTE SENSING, *POLL	W72-14728
CYANO/ *ISOLATION, *ULTRAVIOLET	RADIATION, SEPARATION TECHNIQUES,	W72-14327
ZOOPLANKTON, CHLOROPHYLL, SOLAR	RADIATION, VITAMIN B, PIGMENTS, G	W72-07892
ES, FISH, FLUORESCENCE, INFRARED	RADIATION, IRRAD: /S, ANIMAL WAST	W72-08790
SEASONAL, DIATOMS, *LIGHT, SOLAR	RADIATION, *PHOTOPERIODISM, PERIP	W72-07901
DUCTIVITY, EUTROPHICATION, SOLAR	RADIATION, DISTRIBUTION, DEPTH, D	W72-08054
TIDAL EFFECTS, STABILITY, SOLAR	RADIATION, PHOTOPERI: / ANALYSIS,	W72-08141
RPTION, *FOOD CHAIN/ *BACKGROUND	RADIATION, *NUCLEAR WASTES, *ABSO	W73-00833
TED WATER, CIRCULATION, INFRARED	RADIATION, HYDROGRAPHY, PLANKTON,	W73-01704
IS, CHEMICALS, DRYING, FREEZING,	RADIATION, CYTOLOGICAL STUDIES, P	W73-01676
TORING, PHOTOGRAPHY, ULTRAVIOLET	RADIATION, FLUORESCENCE, INFRARED	W72-04742
ADIATION, FLUORESCENCE, INFRARED	RADIATION, RADAR, C: /TRAVIOLET R	W72-04742
BIOCHEMICAL OXYGEN DEMAND, SOLAR	RADIATION, NUTRIENTS, TEMPERATURE	W71-07124
ES, PHOSPHATES, SILICATES, SOLAR	RADIATION, DEPTH, THERMOCLINE, NI	W71-07875
MPERATURE, STRATIFICATION, SOLAR	RADIATION, ALGAE, PHOTOSYNTHESIS,	W71-07084
MENT, *BYPRODUCTS, MIXING, SOLAR	RADIATION, ORGANIC LOADING, DIURN	W72-13508
ICAL MODELS, GROWTH RATES, SOLAR	RADIATION, FLOTATION, FLOCCULATIO	W72-12289
G, AERIAL PHOTOGRAPHY, *INFRARED	RADIATION, WATER QUALITY, ALGAE,	W72-12487
TECHNIQUES, METABOLISM, TRACERS,	RADIOACTIVITY, CARBON DIOXIDE, SA	W72-12634
HOTOSYNTHESIS, CALCIUM CHLORIDE,	RADIOACTIVITY TECHNIQUES, METABOL	W72-12634
, MUD, MICROORGANISMS, PROTEINS,	RADIOACTIVITY TECHNIQUES, FOOD CH	W71-08042
RUSES, ENERIC BACTERIA, TRITIUM,	RADIOACTIVE WASTES, MONITORING, P	W72-04742
AL OXYGEN DEMAND, RADIOACTIVITY,	RADIOACTIVE WASTES, WASTE WATER(P	W72-05506
TION, BIOCHEMICAL OXYGEN DEMAND,	RADIOACTIVITY, RADIOACTIVE WASTES	W72-05506
EOPHYTA, ATLANTIC OCEAN, SHARKS,	RADIOACTIVITY TECHNIQUES.: /, PHA	W72-05965
UM, COBALT RADIOISOTOPES, ALGAE,	RADIOACTIVITY, TROPHIC LEVEL, LIF	W70-06668
E HISTORY STUDIES, TIME, SNAILS,	RADIOACTIVE WASTES, TENNESSEE, CR	W70-06668
HOSPHORUS RADIOISOTOPES, SEWAGE,	RADIOACTIVITY, PHOSPHATES, CHEMIC	W71-01474
MARINE PLANTS, HUMAN POPULATION,	RADIOACTIVITY TECHNIQUES, RADIUM	W73-00833
ATH OF POLLUTANTS, RADIOECOLOGY,	RADIOACTIVITY EFFECTS, NUCLEAR PO	W73-00833
M, RUBIDIUM, BROMINE, ACTINIDES,	RADIOACTIVE DECAY, GAMMA-RAY SPEC	W73-01434
LLUTANTS, ESTUARINE ENVIRONMENT,	RADIOACTIVITY TECHNIQUES, ABSORPT	W72-07826
RIENTS, WATER POLLUTION EFFECTS,	RADIOACTIVITY TECHNIQUES.: /, NUT	W72-07143
S, *ABSORPTION, INSTRUMENTATION,	RADIOACTIVITY TECHNIQUES, CARBON	W72-14326
S, CORES, ANALYTICAL TECHNIQUES,	RADIOACTIVITY TECHNIQUES, RADIOEC	W73-00790
ANTS, SEA WATER, WATER ANALYSIS,	RADIOACTIVITY TECHNIQUES, RADIOEC	W73-00817
POLLUTION EFFECTS, RADIOECOLOGY,	RADIOACTIVITY TECHNIQUES, ANALYTI	W73-00802
POPULATIONS, ANIMAL PHYSIOLOGY,	RADIOACTIVITY TECHNIQUES, CYCLING	W73-00826
T, NUCLEAR WASTES, RADIOECOLOGY,	RADIOACTIVITY TECHNIQUES, ANALYTI	W73-00821
RODUCTIVITY, *NERITIC, NITRATES,	RADIOACTIVITY TECHNIQUES, EUTROPH	W73-00431
GAE, BOTTOM SEDIMENTS, PLANKTON,	RADIOACTIVITY TECHNIQUES, MONITOR	W73-00361
TROGEN, WATER POLLUTION EFFECTS,	RADIOACTIVITY, METABOLISM, PLANT	W73-00380
*ALGAE, CHLORELLA, IRRADIATION,	RADIOACTIVITY TECHNIQUES, RESISTA	W72-11836
ARBON RADIOISOTOPES, SUCCESSION,	RADIOACTIVITY TECHNIQUES, CULTURE	W72-09239
HOSPHORUS RADIOISOTOPES, CESIUM,	RADIOACTIVITY EFFECTS, LABORATORY	W72-09649
EATHER, WATER ANALYSIS, TRACERS,	RADIOACTIVITY TECHNIQUES.: /AL, W	W72-09122
GAE, INCUBATION, CENTRIFUGATION,	RADIOACTIVITY TECHNIQUES, CHLOROP	W72-09103
IC LIFE, AQUATIC ALGAE, CALCIUM,	RADIOACTIVITY EFFECTS.: /H, AQUAT	W72-10978
T, *NEUTRON ACTIVATION ANALYSIS,	RADIOACTIVITY, ADSORPTION, FISH,	W71-11036
RPTION, WATER POLLUTION SOURCES,	RADIOACTIVITY EFFECTS, RADIOACTIV	W72-00940
ON LAGOONS, DEHYDRATION, RUNOFF,	RADIOACTIVITY, FALLOUT, BIOCONTRO	W72-00846
IOLOGICAL ECOLOGY, RADIOECOLOGY,	RADIOACTIVITY EFFECTS, ABSORPTION	W72-00948
SOURCES, RADIOACTIVITY EFFECTS,	RADIOACTIVITY TECHNIQUES.: /UTION	W72-00940
EFFECTS, RADIOACTIVITY EFFECTS,	RADIOACTIVITY TECHNIQUES, AQUARIA	W72-03348
RPTION, WATER POLLUTION EFFECTS,	RADIOACTIVITY EFFECTS, RADIOACTIV	W72-03348
HIPS, RADIOISOTOPES, ABSORPTION,	RADIOACTIVITY TECHNIQUES, TRACERS	W72-00949
N LAYER CHROMATOGRAPHY, CLEANUP,	RADIOAUTOGRAPHY, SUBSTRATE UTILIZ	W73-00361
, HEAVY METALS, *TRACE ELEMENTS,	RADIOCHEMICAL ANALYSIS, RESINS, *	W72-04708
, SULFUR BACTERIA, PRODUCTIVITY,	RADIOCHEMICAL ANALYSIS, THERMOPHI	W72-14301
LGAE, *ABSORPTION, *MINE WASTES,	RADIOECOLOGY, *PATH OF POLLUTANTS	W72-12204
ULAR STRUCTURE, CATION EXCHANGE,	RADIOECOLOGY, *PATH OF POLLUTANTS	W72-12203
LANT PHYSIOLOGY, AQUATIC PLANTS,	RADIOECOLOGY.: /LANCE, TRACERS, P	W72-06342
RONMENTAL EFFECTS, PRODUCTIVITY,	RADIOECOLOGY, OCEANOGRAPHY.: /NVI	W72-06313
IMILATIVE CAPACITY, *CALIFORNIA,	RADIOECOLOGY, MARINE ALGAE, ESTUA	W73-00826
LEAR WASTES, PATH OF POLLUTANTS,	RADIOECOLOGY, MATHEMATICAL MODELS	W73-00828
LYSIS, RADIOACTIVITY TECHNIQUES,	RADIOECOLOGY, ENVIRONMENTAL EFFEC	W73-00817
HEALTH, WATER POLLUTION EFFECTS,	RADIOECOLOGY, RADIOACTIVITY TECHN	W73-00802
UTANTS, FALLOUT, NUCLEAR WASTES,	RADIOECOLOGY, RADIOACTIVITY TECHN	W73-00821

FLORIDA, RADIATION SENSITIVITY,	RADIOECOLOGY, ECOLOGY, CYCLING NU	W73-00796
IQUES, RADIOACTIVITY TECHNIQUES,	RADIOECOLOGY, TRACERS, MONITORING	W73-00790
, TRACE ELEMENTS, PUBLIC HEALTH,	RADIOECOLOGY.: /CYCLING NUTRIENTS	W73-00826
MENTS, COLLOIDS, NUCLEAR WASTES,	RADIOECOLOGY.: /FFECTS, TRACE ELE	W73-00818
IVITY EFFECTS, RIVERS, EFFLUENT,	RADIOECOLOGY, AQUATIC POPULATIONS	W72-07841
BLIC HEALTH, PATH OF POLLUTANTS,	RADIOECOLOGY, RADIOACTIVITY EFFEC	W73-00833
UCTIVITY, PHYSIOLOGICAL ECOLOGY,	RADIOECOLOGY, RADIOACTIVITY EFFEC	W72-00948
T, NUCLEAR POWERPLANTS, ECOLOGY,	RADIOECOLOGY, MONITORING, POLLUTI	W72-00939
CELLULOSE, ENZYMES, PULP WASTES,	RADIOECOLOGY, WATER POLLUT: /NA,	W72-09668
ODES, LASERS, ORTH/ VOLTAMMETRY,	RADIOGRAPHY, ION SELECTIVE ELECTR	W72-08790
*PLUTONIUM, *SILVER	RADIOISOTOPES.:	W72-07826
CULTURES, PHOTOSYNTHESIS, CARBON	RADIOISOTOPES, CHROMATOGRAPHY, AM	W72-07710
RHODOPHYTA, PHAEOPHYTA, *COBALT	RADIOISOTOPES, NUCLEAR WASTES, PA	W72-07826
IVITY, CULTURES, BIOMASS, CARBON	RADIOISOTOPES, LIMITING FACTORS,	W72-07648
CHLORELLA, CULTURES, NUTRIENTS,	RADIOISOTOPES, CHLOROPHYLL, PIGME	W72-07660
PHOSPHORUS, OSCILLATORIA, RADON	RADIOISOTOPES, CHROMULINA, CRYPTO	W72-07648
TILIZATION, WATER QUALITY, RADON	RADIOISOTOPES, PRIMARY PRODUCTIVI	W72-07648
RPTION, *PHOSPHORUS, *PHOSPHORUS	RADIOISOTOPES, ALGAE, FRESH WATER	W72-07143
TION, COBALT RADIOISOTOPES, ZINC	RADIOISOTOPES, MANGANESE, CHROMIU	W73-00832
RADIOACTIVITY TECHNIQUES, RADIUM	RADIOISOTOPES.: /MAN POPULATION,	W73-00833
SIS, SEA WATER, STABLE ISOTOPES,	RADIOISOTOPES, NUCLEAR WASTES: /Y	W73-00831
ES, LOBSTERS, ABSORPTION, COBALT	RADIOISOTOPES, ZINC RADIOISOTOPES	W73-00832
CESIUM	RADIOISOTOPES.:	W73-00831
TASSIUM RADIOISOTOPES, STRONTIUM	RADIOISOTOPES, SOIL ANALYSIS, SOI	W73-01673
ATION, SAMPLING, CARP, POTASSIUM	RADIOISOTOPES, STRONTIUM RADIOISO	W73-01673
O FISH, GAMBUSIA AFFINIS, CESIUM	RADIOISOTOPES, CLADOPHORA SPP, PA	W73-01673
MARINE ALGAE, *FOOD / *STRONTIUM	RADIOISOTOPES, *MARINE ANIMALS, *	W73-00819
PATH OF POLLUTANTS, ABSORPTION,	RADIOISOTOPES, STABLE ISOTOPES, E	W73-00811
OPLANKTON, *MARINE ALGAE,/ *ZINC	RADIOISOTOPES, *ABSORPTION, *PHYT	W73-00823
MITING FACTORS, SALINITY, CARBON	RADIOISOTOPES, SESTON, ORGANIC MA	W73-00431
*HUMBOLDT BAY, *CESIUM	RADIOISOTOPES.:	W73-00826
ICAL MODELS, FORECASTING, COBALT	RADIOISOTOPES, MARINE ALGAE, ABSO	W73-00831
ON, *MARINE ANIMALS, MA/ *COBALT	RADIOISOTOPES, *COBALT, *ABSORPTI	W73-00821
NE ALGAE, *PHYTOPLANKTON,/ *ZINC	RADIOISOTOPES, *ABSORPTION, *MARI	W73-00802
ANIMALS, ASIA, MOLLUSKS, COBALT	RADIOISOTOPES, IODINE RADIOISOTOP	W73-00811
CHAINS, *NUCLEAR WASTES, *COBALT	RADIOISOTOPES, PATH OF POLLUTANTS	W73-00826
*PLUTONIUM, CESIUM	RADIOISOTOPES.:	W73-00790
SOUND.: MANGANESE	RADIOISOTOPES, BOGUE SOUND, CORE	W73-00818
D CHAINS, PUBLIC HEALTH, EUROPE,	RADIOISOTOPES, EFFLUENTS.: /, FOO	W73-00828
CESIUM	RADIOISOTOPES.:	W73-00811
AINS, / *GREAT LAKES, *STRONTIUM	RADIOISOTOPES, *FALLOUT, *FOOD CH	W73-00790
KS, COBALT RADIOISOTOPES, IODINE	RADIOISOTOPES, MARINE ALGAE, TRAC	W73-00811
LIGHT INTENSITY, DIATOMS, CARBON	RADIOISOTOPES, NEW ENGLAND, AQUAT	W72-14317
RADIOACTIVITY TECHNIQUES, CARBON	RADIOISOTOPES, PRODUCTIVITY, META	W72-14326
NTS, AQUATIC ALGAE, SCENEDESMUS,	RADIOISOTOPES, STRONTIUM RADIOISO	W72-09649
CHAINS, *PATH OF P/ *PHOSPHORUS	RADIOISOTOPES, *TUBIFICIDS, *FOOD	W72-09668
IC PRODUCTIVITY, BIOMASS, CARBON	RADIOISOTOPES, SUCCESSION, RADIOA	W72-09239
DESMUS, RADIOISOTOPES, STRONTIUM	RADIOISOTOPES, PHOSPHORUS RADIOIS	W72-09649
ONTIUM RADIOISOTOPES, PHOSPHORUS	RADIOISOTOPES, CESIUM, RADIOACTIV	W72-09649
INDIA, CESIUM	RADIOISOTOPES.:	W72-10958
PES, *COBALT RADIOISOTO/ *IODINE	RADIOISOTOPES, *CESIUM RADIOISOTO	W72-10958
*IODINE RADIOISOTOPES, *CESIUM	RADIOISOTOPES, *COBALT RADIOISOTO	W72-10958
PES, *MARINE ALGAE, HEAVY/ *ZINC	RADIOISOTOPES, *COBALT RADIOISOTO	W72-10957
S, ACETYLENE, ETHYLENE, NITROGEN	RADIOISOTOPES, N-15, LUCIFERIN, L	W72-10883
VY/ *ZINC RADIOISOTOPES, *COBALT	RADIOISOTOPES, *MARINE ALGAE, HEA	W72-10957
, *CESIUM RADIOISOTOPES, *COBALT	RADIOISOTOPES, *BIOINDICATORS, AN	W72-10958
LS, GASTROPODS, PATH/ *STRONTIUM	RADIOISOTOPES, *ABSORPTION, *SNAI	W72-10978
*RADIOACTIVITY EFFECTS/ *COBALT	RADIOISOTOPES, *RADIOSENSITIVITY,	W72-11836
D CHAINS, *TRACE ELEMENTS, *LEAD	RADIOISOTOPES, HEAVY METALS, AIR	W72-00940
ABSORPTION, FOOD WEBS, STRONTIUM	RADIOISOTOPES, MARINE BACTERIA, C	W72-00948
SOIL-WATER-PLANT RELATIONSHIPS,	RADIOISOTOPES, ABSORPTION, RADIOA	W72-00949
LS, MUSSELS, ISOPODS, AMPHIPODA,	RADIOISOTOPES, ABSORPTION, PACIFI	W72-03347
TRITIUM, OCEAN CURRENTS, COBALT	RADIOISOTOPES, NUCLEAR POWERPLANT	W72-03324
INES, *ALGAE, *OKLAHOMA, *IODINE	RADIOISOTOPES, URANIUM RADIOISOTO	W72-02073
, *IODINE RADIOISOTOPES, URANIUM	RADIOISOTOPES, ANALYTICAL TECHNIQ	W72-02073
WTH RATE, *AQUATIC PRODUCTIVITY,	RADIOISOTOPES, TRACERS, MARINE AL	W71-11486
EA WATER, PHOSPHATES, PHOSPHORUS	RADIOISOTOPES, DIATOMS, HYDROGEN	W71-10083
E, COBALT RADIOISOTOPES, CADMIUM	RADIOISOTOPES, METABOLISM, TROPHI	W71-13243
GICAL ECOLOGY, CESIUM, STRONTIUM	RADIOISOTOPES, LINC RADIOISOTOPES	W71-13243
RADIOISOTOPES, MANGANESE, COBALT	RADIOISOTOPES, CADMIUM RADIOISOTO	W71-13243
M, STRONTIUM RADIOISOTOPES, LINC	RADIOISOTOPES, MANGANESE, COBALT	W71-13243
, RHODOPHYTA, PHAEOPHYTA, CARBON	RADIOISOTOPES, MARINE FISH, INSTR	W72-06283
EWS, BIOCHEMISTRY, GEOCHEMISTRY,	RADIOISOTOPES, PATH OF POLLUTANTS	W72-06313
N, SEDIMENTS, SETTLING VELOCITY,	RADIOISOTOPES, ABSORPTION, REVIEW	W72-04463
UT, *FOOD CHAINS, *AQUATIC LIFE,	RADIOISOTOPES, ABSORPTION, NUCLEA	W72-04461
UTONIUM RADIOISOTOPES, RUTHENIUM	RADIOISOTOPES, IRON RADIOISOTOPES	W72-04463
TOPES, IRON RADIOISOT/ PLUTONIUM	RADIOISOTOPES, RUTHENIUM RADIOISO	W72-04463
S, RUTHENIUM RADIOISOTOPES, IRON	RADIOISOTOPES.: /IUM RADIOISOTOPE	W72-04463
LTH, ESTUARINE ENVIRONMENT, ZINC	RADIOISOTOPES, PHYSICOCHEMICAL PR	W72-04463
*MARINE ANIMALS, *MARINE ALGAE,	RADIOISOTOPES, FALLOUT, CHEMICAL	W72-05965
TA, PRIMARY PRODUCTIVITY, CARBON	RADIOISOTOPES, PHOTOSYNTHESIS.: /	W71-03014
O P/ *BIOLOGICAL UPTAKE, CALCIUM	RADIOISOTOPES, BLOOMS, 2,4 DINITR	W71-01474
S, THERM/ TEMPERATURE, SORPTION,	RADIOISOTOPES, ALGAE, GROWTH RATE	W71-02075
S, *ACTIVATED SLUDGE, PHOSPHORUS	RADIOISOTOPES, SEWAGE, RADIOACTIV	W71-01474
ENT, *ECOSYSTEMS, CESIUM, COBALT	RADIOISOTOPES, ALGAE, RADIOACTIVI	W70-06668
RTEBRATES, RUTHENIUM-106, CESIUM	RADIOISOTOPES, SPIROGYRA, LIMNODR	W70-06668
SPIRATION, PHOSPHORUS, TOXICITY,	RADIOISOTOPES, INHIBITION.: /, RE	W70-07255
TER, PLANKTON, CHELATION, CARBON	RADIOISOTOPES, CHLORELLA, GROWTH	W71-03027
ORATORY TESTS, CESIUM, STRONTIUM	RADIOISOTOPES, IONS, RIVERS, CULT	W71-08032
SOLVED OXYGEN, NUTRIENTS, CARBON	RADIOISOTOPES, PHOSPHATES, SILICA	W71-07875

OCEANS, *ORGANIC MATTER, *CARBON	RADIOISOTOPES, *ANALYTICAL TECHNI	W71-08042
EXCHANGE, ABSORPTION, STRONTIUM	RADIOISOTOPES, ZINC RADIOISOTOPES	W71-09188
N, STRONTIUM RADIOISOTOPES, ZINC	RADIOISOTOPES, COBALT RADIOISOTOP	W71-09188
ES.: MANGANESE	RADIOISOTOPES, CESIUM RADIOISOTOP	W71-09018
OPES, ZINC RADIOISOTOPES, COBALT	RADIOISOTOPES, URANIUM RADIOISOTO	W71-09188
IODINE RADIOISOTOPES, STRONTIUM	RADIOISOTOPES, ZINC RADIOISOTOPES	W71-09018
S, COBALT RADIOISOTOPES, URANIUM	RADIOISOTOPES, RADIUM RAIODISOTOP	W71-09188
MANGANESE RADIOISOTOPES, CESIUM	RADIOISOTOPES.:	W71-09018
Y, INDUSTRIES, IONS, CYANOPHYTA,	RADIOISOTOPES, TEMPERATURE, ECOSY	W71-09157
S, STRONTIUM RADIOISOTOPES, ZINC	RADIOISOTOPES, COBALT RADIOISOTOP	W71-09018
OPES, ZINC RADIOISOTOPES, COBALT	RADIOISOTOPES, ABSORPTION, WATER	W71-09018
S, MARINE ALGAE, TRITIUM, IODINE	RADIOISOTOPES, STRONTIUM RADIOISO	W71-09018
RESPIRATION, CHLOROPHYLL, CARBON	RADIOISOTOPES, CARBON DIOXIDE, RH	W71-09172
*RUTHENIUM	RADIOISOTOPES.:	W71-09182
COBALT RADIOISOTOPES, PHOSPHORUS	RADIOISOTOPES, SEDIMENTS, TRACE E	W71-09190
THENIUM RADIOISOTOPES, ZIRCONIUM	RADIOISOTOPES.: /DIOISOTOPES, *RU	W71-09225
SORPTION, TEMPERATURE, STRONTIUM	RADIOISOTOPES, ZINC RADIOISOTOPES	W71-09250
*CESIUM	RADIOISOTOPES.:	W71-09256
ES, RUTHENIUM RADIOIS/ MANGANESE	RADIOISOTOPES, CESIUM RADIOISOTOP	W71-09188
CESIUM RADIOISOTOPES, *RUTHENIUM	RADIOISOTOPES, ZIRCONIUM RADIOISO	W71-09225
MOLLUSKS, MARINE ANIMALS, COBALT	RADIOISOTOPES, FOOD CHAINS, IRRIG	W71-09225
, MANGANESE RADIOISOTOP/ *CESIUM	RADIOISOTOPES, IRON RADIOISOTOPES	W71-09250
OTOPES, ZIRCONIUM RADIO/ *CESIUM	RADIOISOTOPES, *RUTHENIUM RADIOIS	W71-09225
N CONCENTRATION, DYSTROPHY, ZINC	RADIOISOTOPES, COBALT RADIOISOTOP	W71-09190
OPES, ZINC RADIOISOTOPES, COBALT	RADIOISOTOPES, WATER POLLUTION EF	W71-09250
S, *SALT MARSHES, *PRODUCTIVITY,	RADIOISOTOPES, STRONTIUM RADIOISO	W71-09254
TOP/ *CESIUM RADIOISOTOPES, IRON	RADIOISOTOPES, MANGANESE RADIOISO	W71-09250
PES, THORIUM RADIOISOTOPES, IRON	RADIOISOTOPES, ZIRCONIUM RADIOISO	W71-09188
S, IRON RADIOISOTOPES, MANGANESE	RADIOISOTOPES.: /IUM RADIOISOTOPE	W71-09250
ZINC, MANGANESE	RADIOISOTOPES.:	W71-09190
RUTHENIUM RADIOISOTOPES, CERIUM	RADIOISOTOPES, THORIUM RADIOISOTO	W71-09188
CESIUM RADIOISOTOPES, RUTHENIUM	RADIOISOTOPES, CERIUM RADIOISOTOP	W71-09188
FFECTS, NUCLEAR WAST/ *STRONTIUM	RADIOISOTOPES, *WATER POLLUTION E	W71-09225
TIVITY, RADIOISOTOPES, STRONTIUM	RADIOISOTOPES, ALGAE, PLANT GROWT	W71-09254
S, CERIUM RADIOISOTOPES, THORIUM	RADIOISOTOPES, IRON RADIOISOTOPES	W71-09188
E, STRONTIUM RADIOISOTOPES, ZINC	RADIOISOTOPES, COBALT RADIOISOTOP	W71-09250
MANGANESE RADIOISOTOPES, CESIUM	RADIOISOTOPES, RUTHENIUM RADIOISO	W71-09188
OPHY, ZINC RADIOISOTOPES, COBALT	RADIOISOTOPES, PHOSPHORUS RADIOIS	W71-09190
*PONDS, WATER POLLUTION EFFECTS,	RADIOISOTOPES, COMPETITION, SPECI	W71-09256
S, IRON RADIOISOTOPES, ZIRCONIUM	RADIOISOTOPES, ZINC.: /DIOISOTOPE	W71-09188
S, AB/ *MARINE ALGAE, *STRONTIUM	RADIOISOTOPES, *ZINC RADIOISOTOPE	W72-12203
THORIUM	RADIOISOTOPES.:	W72-12204
*STRONTIUM RADIOISOTOPES, *ZINC	RADIOISOTOPES, ABSORPTION, SEDIME	W72-12203
*CERIUM	RADIOISOTOPES.:	W72-12203
, MEASUREMENT, OXIDATION, CARBON	RADIOISOTOPES, *ON-SITE TESTS, SU	W72-14301
ATION, *CORALLINE ALGAE, CALCIUM	RADIOISOTOPES, *BOSSIELLA ORBIGNI	W72-12634
UDIES, *STABLE ISOTOPES, *CARBON	RADIOISOTOPES, ENVIRONMENTAL EFFE	W72-12709
*TEXAS, LABORATORY TESTS, CARBON	RADIOISOTOPES, *CYANOPHYTA, EUTRO	W72-13692
DIELDRIN, *PHYTOPLANKTON, CARBON	RADIOISOTOPES, LAKE ERIE, LAKE ON	W72-13657
PHYTIN, TUNICATES, FORAMINIFERA,	RADIOLARIA, TRICHODESMIUM, FLAGEL	W72-14313
LEAD, MERCURY, LASERS, MICROWAVE	RADIOMETRY, VIDEO, SCANNERS.: /,	W72-04742
*USSR, *HYDROBIONTS, *CERIUM,	RADIOMETRY.:	W72-09649
	RADIONUCLIDE UPTAKE.:	W72-03347
	RADIONUCLIDE UPTAKE.:	W72-00960
TRIENTS, ALGAE, AQUATIC ANIMALS,	RADIONUCLIDES.: /ES, SALINITY, NU	W72-12190
LOGICAL STUDIES, SOIL CHEMISTRY,	RADIOSENSITIV: /LS(MAMMALS), CYTO	W72-00949
ATION, RADIOACTIVITY TECHNIQUES,	RADIUM RADIOISOTOPES.: /MAN POPUL	W73-00833
ISOTOPES, URANIUM RADIOISOTOPES,	RADIUM RAIODISOTOPES, ALUMINUM, I	W71-09188
N, FERTILIZATION, WATER QUALITY,	RADON RADIOISOTOPES, PRIMARY PROD	W72-07648
SOLVED PHOSPHORUS, OSCILLATORIA,	RADON RADIOISOTOPES, CHROMULINA,	W72-07648
ACID, SODIUM, PYRIDINE, DEXTRAM,	RAFFINOSE, SUCROSE, SENESCENT CUL	W72-07710
*CYCLING NUTRIENTS, ECOSYSTEMS,	RAIN FORESTS, MATHEMATICAL MODELS	W72-03342
G FACTORS, NITROGEN, PHOSPHORUS,	RAIN, LAKES, CLADOPHORA, ANALYSIS	W73-00232
H OF POLLUTANTS, BIOASSAY, FISH,	RAINBOW TROUT, AQUATIC LIFE, AQUA	W72-12257
ATES, PHOSPHATES, WIND VELOCITY,	RAINFALL.: /QUES, TURBIDITY, NITR	W72-03224
CES, *SEDIMENT YIELD, *SIMULATED	RAINFALL, *SOIL TYPES, SOIL TEXTU	W72-02218
ANKTON, *ALGAE, *PHOTOSYNTHESIS,	RAINFALL, DIURNAL, FLUCTUATIONS,	W72-00150
PHATES, HYDROLYSIS, STORM WATER,	RAINFALL, WASTE WATER(POLLUTION),	W72-13644
S, URANIUM RADIOISOTOPES, RADIUM	RAIODISOTOPES, ALUMINUM, IRON, CO	W71-09188
EMA MACULANS, ELACHISTA LUBRICA,	RALFSIA VERRUCOSA.: /LIS, HECATON	W72-01370
, THIOSPIRA WINOGRADSKY, ZOOGLEA	RAMIGERA, BEGGIATOA ALBA, BEGGIAT	W72-07896
, BUTOMUS UMBELLATUS, SPARGANIUM	RAMOSUM, LEERSIA ORYZOIDES, NUPHA	W73-01089
CROCYSTIS, FUSARIUM, TRICHORIXA,	RAMPHOCORIXA, SIGARA, BOENOA, SAL	W70-06655
US SORPTION, PHOSPHORUS REMOVAL,	RATE OF SORPTION.: /MUD, PHOSPHOR	W72-11617
HNIQUES, *BENTHIC FLORA, *GROWTH	RATE, *AQUATIC PRODUCTIVITY, RADI	W71-11486
ULTURES, MAXIMUM SPECIFIC GROWTH	RATE, SATURATION CONSTANT, DILUTI	W72-04788
ALGAL GROWTH	RATES.:	W72-04789
MONOD GROWTH MODEL, ALGAL GROWTH	RATES.: / CHLORELLA PYRENOIDOSA,	W72-04788
TOXICITY, CULTURES, DDT, GROWTH	RATES.: / DIATOMS, PHYTOPLANKTON,	W72-05614
R TREATMENT, CHLOROPHYTA, GROWTH	RATES.: /LITY CONTROL, WASTE WATE	W72-04431
TURE, OPTICAL PROPERTIES, GROWTH	RATES.: /OPHYTA, DIATOMS, TEMPERA	W70-10403
SUCCESSION, PRODUCTIVITY, GROWTH	RATES.: /SYNTHESIS, RESPIRATION,	W73-01632
ATMENT, *CARBON DIOXIDE, *GROWTH	RATES, *AEROBIC CONDITIONS, *GROW	W72-06612
CHLOROPHYTA,/ *CALCIUM, *GROWTH	RATES, *CULTURES, *AQUATIC ALGAE,	W72-12736
RIN, *DIELDRIN, *ENDRIN, *GROWTH	RATES, *CYANOPHYTA, *TOXICITY, PE	W72-03545
RVAE, ON-SITE T/ *LIGHT, *GROWTH	RATES, *ENVIRONMENTAL EFFECTS, LA	W73-00348
COMPETITION, *NUTRIENTS, *GROWTH	RATES, *INHIBITION, FUNGI, THERMA	W72-12734
UTRIENT REQUIRE/ *ALGAE, *GROWTH	RATES, *KINETICS, *PHOSPHORUS, *N	W72-04788
ANS, ZOOPLANKTON, ALGAE, FEEDING	RATES, *LARVAL GROWTH STAGE, PHYT	W72-09248
N CONCENTRATION, *ALGAE, *GROWTH	RATES, *LIMITING FACTORS, *DIURNA	W72-11724

BITION, OLIGUIR/ *ALGAE, *GROWTH RATES, *NUTRIENTS, CULTURES, INHI W72-08376
ION CONCEN/ *CYANOPHYTA, *GROWTH RATES, *ORGANIC ACIDS, *HYDROGEN W72-00838
LGAE, D/ *ANALYTICAL TECHNIQUES, RATES, *PHOTOSYNTHESIS, *MARINE A W72-09653
CTORS, ALGAL PONDS, ALGAL GROWTH RATES, *POTAMOCYPRIS.: /ROWTH REA W72-04787
TH SUBSTANCES, / *ALGAE, *GROWTH RATES, *PRODUCTIVITY, *PLANT GROW W71-08687
RGY BUDGET, *MARINE ALG/ *GROWTH RATES, *PRIMARY PRODUCTIVITY, ENE W73-00366
AVY METALS, *TR/ *ALGAE, *GROWTH RATES, *PRIMARY PRODUCTIVITY, *HE W73-01434
N, *MARINE ALGAE, *SEDIMENTATION RATES, *REVIEWS, BIOCHEMISTRY, GE W72-06313
TESTS, *CHEMICAL WASTE, *GROWTH RATES, *SEAWATER, *TOXICITY, *NUT W72-12239
Y STUDIES, *GRO/ *ALGAE, *GROWTH RATES, *SYSTEMATICS, *LIFE HISTOR W73-02099
*HEAVY METALS, *ALGAE, *GROWTH RATES, *TOXICITY, *FLUORESCENCE, W72-07660
T REQUIREMENTS, AMMONIA, *GROWTH RATES, *WASTE ASSIMILATIVE CAPACI W72-04787
FECTS, *AXEMIC CULTURES, *DIEOFF RATES, AFTERGROWTH, BLUE-GREEN AL W70-08319
*MIXED CULTURES, DIEOFF RATES, AFTERGROWTH.: W71-05155
ION, VITAMIN B, PIGMENTS, GROWTH RATES, ALGAE, FOOD WEBS, COPEPODS W72-07892
C MATTER, ATLANTIC OCEAN, GROWTH RATES, ALGAE, INCUBATION, PHYTOPL W71-11562
ENSITY, TEMPERATURE, MUD, GROWTH RATES, ANABAENA, NITROGEN FIXATIO W72-12544
LIS, CYCLOPS STRENUUS, INGESTION RATES, ASSIMILATION EFFICIENCES, W72-07940
, EUTROPHICATION, PONDS, FEEDING RATES, BENTHOS, STANDING CROPS.: / W72-14796
RODUCTIVITY, SYSTEMATICS, GROWTH RATES, BIOMASS, DATA COLLECTIONS. W73-02469
ESISTANCE, AQUATIC ALGAE, GROWTH RATES, BIOASSAY, LABORATORY TESTS W72-13809
ORING, SEWAGE, ESTUARIES, GROWTH RATES, BIOLOGICAL COMMUNITIES.: / W71-12867
A, *PHYTOPLANKTON, ALGAE, GROWTH RATES, BIOASSAY, DISSOLVED OXYGEN W72-06046
*ALGAE, *EUTROPHICATION, *GROWTH RATES, CHLORELLA, CARBON DIOXIDE, W72-09168
CHLORELLA, *LIGHT, ALGAE, GROWTH RATES, CHLOROPHYLL.: /TABOLISM, * W73-00251
Y, BACTERIA, SCENEDESMUS, GROWTH RATES, CULTURES, AQUATIC PRODUCTI W72-14314
, *BIOINDICATORS, *ALGAE, GROWTH RATES, CULTURES, SUSPENDED LOAD, W71-09795
POLLUTANT IDENTIFICATION, GROWTH RATES, DEFICIENT ELEMENTS, WATER W73-01445
RADIOISOTOPES, CHLORELLA, GROWTH RATES, DIATOMS, PHYTOPLANKTON, TO W71-03027
ALGAE, *LIMITING FACTORS, GROWTH RATES, DIATOMS, BACTERIA, LAKES, W72-09163
S, ENVIRONMENTAL EFFECTS, GROWTH RATES, E. COLI, AZOTOBACTER, PSEU W72-10861
D HYDROCARBON PESTICIDES, GROWTH RATES, ENVIRONMENTAL EFFECTS, PHY W73-00281
S, BIO-INDICATORS, STORAGE, FLOW RATES, INHIBITION, ANALYTICAL TEC W72-04070
HRYSOPHYTA, CHLOROPHYTA, *GROWTH RATES, INHIBITION, *PLANT PHYSIOL W70-07842
OUNDS, NITRATES, DENSITY, GROWTH RATES, LABORATORY TESTS, CYANOPHY W72-03217
Y, PLANT GROWTH, BIOMASS, GROWTH RATES, LIGHT INTENSITY, CARBON DI W72-05476
BIOMASS, LIGHT INTENSITY, GROWTH RATES, MATHEMATICAL MODELS, WATER W72-06274
BIC CONDITIONS, BIOASSAY, GROWTH RATES, MORTALITY.: /NOPHYTA, AERO W72-07664
RATES, *AMMONIA, DIATOMS, GROWTH RATES, NITROGEN, WATER POLLUTION W73-00380
E WATER TREATMENT, SEDIMENTATION RATES, OIL SPILLS, WATER POLLUTIO W72-12487
OGEN, PHOSPHORUS, CARBON, GROWTH RATES, OLIGOTROPHY, EUTROPHICATIO W72-09164
TRIAL WASTES, METABOLISM, GROWTH RATES, OXIDATION LAGOONS, COLIFOR W72-09590
S, NUTRIENT REQUIREMENTS, GROWTH RATES, PHOSPHORUS, NITROGEN, *REG W72-12966
VAL, OXYGEN DECLINE, RESPIRATORY RATES, PHOTOSYNTHETIC OXYGENATION W71-04079
*MARINE ALGAE, *LIGHT, *GROWTH RATES, PHYTOPLANKTON, NATURAL RES W71-13252
ORATORY TESTS, CULTURES, *GROWTH RATES, PLANT GROWTH REGULATORS, P W71-03095
EFFICIENCES, COMPETITION, GROWTH RATES, PLANT MORPHOLOGY, WATER CI W72-04766
CIDS, *PLANT PHYSIOLOGY, *GROWTH RATES, PLANT MORPHOLOGY, CULTURES W72-07710
PATTERNS, CHROMATOGRAPHY, GROWTH RATES, POLLUTANT IDENTIFICATION.: W72-04134
, PHYTOPLANKTON, BIOMASS, GROWTH RATES, PRODUCTIVITY, ANIMAL GROWT W72-09378
ON, PRIMARY PRODUCTIVITY, GROWTH RATES, RESPIRATION.: /STABILIZATI W72-03556
, SCENEDESMUS, CHLORELLA, GROWTH RATES, RESISTANCE, PHOSPHORUS, IR W72-12739
GROWTH, *EUTROPHICATION, GROWTH RATES, SAMPLING, LAKE MICHIGAN, M W71-11033
TICAL MIGRATION, AQUARIA, GROWTH RATES, SEDIMENTATION, BUOYANCY, W W71-09852
NT, *MATHEMATICAL MODELS, GROWTH RATES, SOLAR RADIATION, FLOTATION W72-12289
BIOMASS, THERMAL STRESS, GROWTH RATES, SPATIAL DISTRIBUTION, PRIM W73-00853
REQUIREMENTS, CHLORELLA, GROWTH RATES, TEMPERATURE, SEA WATER, PH W71-10083
IRONMENT, PHOTOPERIODISM, GROWTH RATES, TEMPERATURE, BIOCHEMISTRY, W72-00377
ON, RADIOISOTOPES, ALGAE, GROWTH RATES, THERMAL POLLUTION.: /ORPTI W71-02075
ROPHYLL, PROTEINS, ALGAE, GROWTH RATES, TRACERS.: / CULTURES, CHLO W72-03557
OZOA, BACTERIA, STREAMFLOW, FLOW RATES, WATER TEMPERATURE, WATER C W73-01972
LS, CANADA, NITROGEN: PHOSPHORUS RATIO.: /ASSOCIATION, CARBOY TRIA W70-07283
*RECIRCULATION RATIO, *ISRAEL(DAN REGION).: W72-11642
N RESERVOI/ *NITROGEN:PHOSPHORUS RATIO, LAKE WAUBESA(WIS), OCCOQUA W72-05473
HLOROPHYLL A, / MARGALEF PIGMENT RATIO, PIGMENT DIVERSITY INDEX, C W73-00916
UARY, PHOTOSYNTHESIS/RESPIRATION RATIOS.: *PAMLICO RIVER EST W72-13636
, DNA, MOTILITY, CYST FO/ *G + C RATIOS, *NONFRUITING MYXOBACTERIA W72-07951
RIAL PLANTS, COMBUSTION, ISOTOPE RATIOS, SPECTROMETRY, LICHENS, MO W72-12709
RIA, BLUE HERON, STERNA HIRUNDO, RATS, PARTRIDGE, DDE, KIDNEYS, LI W72-05952
CS-137, *PAHRANAGAT LAKE, GAMMA RAY SPECTROMETRY, ANIMAL TISSUES, W73-01673
INIDES, RADIOACTIVE DECAY, GAMMA- RAY SPECTROMETRY, COUNTING.: /ACT W73-01434
, CHLOROPHYTA, INHIBITION, GAMMA RAYS.: /RATORY TESTS, IRRADIATION W73-01080
NNESSEE, CRAYFISH, MIDGES, GAMMA RAYS, BLOODWORMS, SLUDGE WORMS.: / W70-06668
TS, VOLATILITY, FREEZE DRYING, X- RAYS, ELECTROLYSIS, COMPUTER PROG W71-11036
OLYSIS, COMPUTER PROGRAMS, GAMMA RAYS, POLLUTANT IDENTIFICATION, C W71-11036
ET RADIATION, FOOD CHAINS, GAMMA RAYS, RESISTANCE, BIOLOGICAL COMM W73-00796
TH, TEMPERATURE, PHOTOSYNTHESIS, RE-AERATION, LIGHT PENETRATION, N W71-07098
E, LIGHT INTENSITY, EVAPORATION, RE-AERATION, DISSOLVED OXYGEN, AL W71-07099
, *WASTE WATER TREATMENT, *WATER RE-USE, BIOCHEMICAL OXYGEN DEMAND W71-07123
THEMATICAL MODELS, MIXING, WATER RE-USE, LIGHT INTENSITY, EVAPORAT W71-07099
MENT, LITTORAL, OXYGEN, CHEMICAL REACTIONS, HYDROGEN ION CONCENTRA W71-08026
L POLLUTION, *TOXICITY, CHEMICAL REACTIONS, CARBON DIOXIDE, WATER W73-01074
C/ *THERMAL POLLUTION, *CHEMICAL REACTIONS, *WATER QUALITY, *CHEMI W71-12092
IBIOTICS, IODINE, ATOMIC ENERGY, REACTOR COOLING, IMPERIAL VALLEY(W71-09157
HIPS, *AMMONIA-NITROGEN, *GROWTH REACTORS, ALGAL PONDS, ALGAL GROW W72-04787
UMPTIVE USE, ALGAE, EVAPORATION, REASONABLE USE, BRACKISH WATER, D W70-07100
EXCRETION, ALKALINE PHOSPHATASE, REASSIMILATION, GREEN FLAGELLATE, W71-10083
*PIT RECHARGE, *WAT/ *ARTIFICIAL RECHARGE, *INDUCED INFILTRATION, W71-12410
RGE, *INDUCED INFILTRATION, *PIT RECHARGE, *WATER TREATMENT, *FILT W71-12410
LITY, COMPETING USES, DIVERSION, RECIRCULATED WATER, WATER REUSE, W70-07100

LAKES, *WATER POLLUTION EFFECTS, RECLAIMED WATER, SEWAGE EFFLUENTS W72-11702
ION.: *WATER RECLAMATION, VARIATIONS, COMPOSIT W70-07838
DATION LA/ *PONDS, *WATER REUSE, RECLAMATION, ALGAE, PROTEIN, *OXI W71-08951
BONE, DETECTION LIMITS, CHEMICAL RECOVERY, SAMPLE PREPARATION.: / W72-05965
, *SENSITIVITY, CHLORAMPHENICOL, RECOVERY, SURVIVAL.: /S, *MUTANTS W72-14327
EWAGE EFFLUENTS, ALGAE, TOURISM, RECREATION, FISHERIES, HYDROELECT W72-14878
VOIRS, *GREAT PLAINS, NUTRIENTS, RECREATION, LIGHT PENETRATION, NE W72-04759
IFE CONSERVATION, FLOOD CONTROL, RECREATION, HYDROLOGIC DATA, WATE W72-05594
, *STREAMS, *REVIEWS, *ILLINOIS, RECREATION, FISHING, SWIMMING, BO W72-05869
HICATION, PLANKTON, CHLOROPHYLL, RECREATION, RUNOFF, MEASUREMENT, W72-04270
*EUTROPHICATION, *LAKES, *MAINE, RECREATION, WATER SUPPLY, WATER P W72-04280
ORMS, SPORES, LIFE CYCLES, ODOR, RECREATION, SYSTEMATICS.: / COLIF W71-09173
INDUSTRY, ALGAE, AQUATIC WEEDS, RECREATION, GROUNDWATER, FISH, FL W70-05103
RE, DISSOLVED OXYGEN, COLIFORMS, RECREATION, ODOR.: /TER TEMPERATU W71-11914
IDES, HEAVY METALS, TASTE, ODOR, RECREATION, PHOSPHORUS, CARBON, L W71-12091
TS, EUTROPHICATION, FISH, ALGAE, RECREATION, ENVIRONMENTAL EFFECTS W71-10577
LEVELS, TURBIDITY, OXYGEN, FISH, RECREATION, INVERTEBRATES, RIVERS W71-11009
OURCES, WATER POLLUTION EFFECTS, RECREATION DEMAND, POPULATION, CO W72-01660
SETTLING CHARACTERISTICS, SOLIDS RECYCLE, SOLIDS CONTROL.: /OGEN, W72-11100
OTS, COMPOSTING, PLANT RESIDUES, RECYCLING FOOD, PROCESSING WASTES W72-00846
: ALUM, RECYCLING, TREATMENT COSTS, ASIA. W72-13508
UEHANNA RIVER, BRANDYWINE RIVER, RED CLAY RIVER, WHITE CLAY RIVER, W72-04736
FLAGELLATES, MEDITERRANEAN SEA, RED SEA, DRIFT ORGANISMS.: /ANAL, W73-00296
HALINE, HYPERHALINE, EURYHALINE, RED SEA, GULF OF SUEZ, DECAPODS, W73-00855
TOXINS, AQUATIC PLANTS, TOXINS, RED TIDE, PLANT PHYSIOLOGY, BIOAS W73-00243
C, PRODUCTIVITY, EUTROPHICATION, RED TIDE, SILICA, ALGAE, PHYTOPLA W71-09190
HYSIOLOGICAL ECOLOGY, MORTALITY, RED TIDE, TOXICITY.: /N BLOOMS, P W72-14056
LYNGBYA LIMNETICA, OSCILLATORIA REDECKEI, OSCILLATORIA RUBESCENS, W71-00117
F SQUAW, LAKE BUDD, OSCILLATORIA REDEKEI, LYNGBYA LIMNETICA, ANABA W72-07145
*SAMPLING, WATER RESOURCES, AREA REDEVELOPMENT, BIOLOGICAL CHARACT W72-03215
CHLORELLA VULGARIS, NITRATE REDUCTASE, RIBONUCLEIC ACID.: W72-03222
SPHORUS, ION EXCHANGE, OXIDATION- REDUCTION POTENTIAL, HYDROGEN ION W72-01108
PUBLIC HEALTH, ODORS, OXIDATION- REDUCTION POTENTIAL, BIOCHEMICAL W72-04086
RIA, ALGAE, NUTRIENTS, OXIDATION- REDUCTION POTENTIAL, ODORS, OPERA W72-10401
S, NITROGEN, BACTERIA, OXIDATION- REDUCTION PO: /MPERATURE, COPEPOD W71-11008
NITROGEN, PHOSPHORUS, OXIDATION- REDUCTION POTENTIAL, LAKES, SOILS W72-07933
IMENTS, NITRIFICATION, OXIDATION- REDUCTION POTENTIAL, DENITRIFICAT W72-08459
STUDIES, CHLOROPHYLL, OXIDATION REDUCTION POTENTIAL.: /THEMATICAL W72-14800
ARBON-14 MEASUREMENTS, ACETYLENE REDUCTION MEASUREMENTS, NUTRIENT W72-13692
REDUCTION, POLYSACCHARIDES, DRAG REDUCTION.: /POLYMERS, *FRICTION W72-03220
COMMUNITIES, DOMINANT ORGANISMS, REDUCTION(CHEMICAL), TROPHIC LEVE W72-10883
, *CYANOPHYTA, PLANTS, NITROGEN, REDUCTION(CHEMICAL), GAS CHROMATO W73-01456
AKE MENDOTA, CRYSTAL/ *ACETYLENE REDUCTION, RHIZOBIUM JAPONICUM, L W73-01456
*ANABAENA CYLINDRICA, ACETYLENE REDUCTION, WARBURG EFFECT.: /ION, W72-08584
E / *NUTRIENT SOURCES, ACETYLENE REDUCTION, LAKE MENDOTA(WIS), LAK W72-10608
LL A, OPTICAL DENSITY, ACETYLENE REDUCTION, SCHIZOTHRIX CALCIOCOLA W72-10883
*BIOLOGICAL POLYMERS, *FRICTION REDUCTION, POLYSACCHARIDES, DRAG W72-03220
CYLINDRICA, ANABAEN/ *ACETYLENE REDUCTION, *HETEROCYSTS, ANABAENA W71-12070
ODOMENIA, MARMION ISLAND, MORRIA REEF, BEAR BAY, NEVA ISLAND, BAIR W72-09120
TS, PRIMARY PRODUCTIVITY, CORAL, REEFS, ANTIBIOTICS(PESTICIDES), C W72-14294
E PLANTS, ENVIRONMENTAL EFFECTS, REEFS, LABORATORY TESTS, SAMPLING W73-00838
ANALYSIS, BIOMASS, METHODOLOGY, REEFS, SAMPLING, OXYGEN, TEMPERAT W72-06283
N, ECOSYSTEMS, ESTUARIES, CORAL, REEFS, WATER RESOURCES, SYSTEMS A W71-13723
ES, AERATED LAGOONS, COMPOSTING, REFEEDING, ANHYDROUS AMMONIA, WAS W71-07551
ALGAE, INVERTEBRATES, ISOLATION, REFRACTIVITY, COPEPODS, MARINE AN W73-00371
BOVATA, GUTTULA BACILLARIOPHAGA, REFRACTOCYSTIS PLUVIALIS, GOMESIA W72-09119
UVIALIS, GOMESIAMONAS STAGNALIS, REFRACTODES BRASILIANA, OYEMONAS W72-09119
S BRASILIANA, OYEMONAS PULSULAE, REFRACTOMONAS BRASILIANA, HORTOBA W72-09119
REFRACTORY MATERIALS.: W72-01881
ET LABILE NITROGEN, ULTRA-VIOLET REFRACTORY NITROGEN, NETPLANKTON, W72-08048
YS, *METHODOLOGY, FREEZE DRYING, REFRIGERATION, INCUBATION, DIALYS W73-01676
LEMENS(MICH).: *CARBON REGENERATION, *BACKWASHING, *MT C W72-08357
*SURFACE CHARGE, REGENERATION, SUSPENDED SOLIDS.: W71-00138
AKE ERIE CENTRAL BASIN, NUTRIENT REGENERATION, DEOXYGENATION, ALGA W72-12997
ASIN./ *DEOXYGENATION, *NUTRIENT REGENERATION, LAKE ERIE CENTRAL B W72-12996
RECIRCULATION RATIO, *ISRAEL(DAN REGION).: * W72-11642
ZINE, M/ *LAMINARIA, SUBLITTORAL REGION, CULTURE EXPERIMENTS, ATRA W72-12576
TES, *MONITORING, *PACIFIC COAST REGION, TRITIUM, OCEAN CURRENTS, W72-03324
, WATER TEMPERATURE, GREAT LAKES REGION, WISCONSIN, MICHIGAN, CHRY W73-00384
N/ *ON-SITE INVESTIGATION, *COLD REGIONS, *OXIDATION LAGOONS, ORGA W72-06838
REATMENT, *SEWAGE LAGOONS, *COLD REGIONS, AERATED LAGOONS, *NORTH W72-12565
C OCEAN, TROPICAL REGIONS, POLAR REGIONS, CYANOPHYTA, CHRYSOPHYTA, W72-03543
UDIES, ATLANTIC OCEAN, *TROPICAL REGIONS, MARINE ALGAE, SPATIAL DI W72-11719
ALPINE, *SYNOPTIC ANALYSIS, COLD REGIONS, NUTRIENTS, BIOLOGICAL CO W72-12549
C OCEAN, PACIFIC OCEAN, TROPICAL REGIONS, POLAR REGIONS, CYANOPHYT W72-03543
ATIAL DISTRIBUTION, GEOGRAPHICAL REGIONS, PHAEOPHYTA, MARINE ALGAE W73-00428
ON, HAWAII, ABSORPTION, TROPICAL REGIONS, PACIFIC OCEAN, COLUMNS, W71-08042
E, *NITROGEN FIXATION, *TROPICAL REGIONS, RICE, CYANOPHYTA, FERTIL W73-02473
LLUTION, TROPHIC LEVEL, TROPICAL REGIONS, ULTRAVIOLET RADIATION, F W73-00796
ABSORPTION, SEDIMENT DISCHARGE, REGRESSION ANALYSIS, BIOMASS, MET W72-06283
PHOSPHATES, NITRATES, SEASONAL, REGRESSION ANALYSIS, CORRELATION W72-12393
, MUNICIPAL WASTES, LEGISLATION, REGULATION,: /RCH AND DEVELOPMENT W72-06638
, GROWTH AUTOSTIMULATION, GROWTH REGULATION, EXTRACELLULAR PRODUCT W71-08687
IRONMENT CONTROL, LEGAL ASPECTS, REGULATION, PERMITS.: /EALTH, ENV W73-00703
POLLUTION, TOXICITY, CHEMICALS, REGULATION, ALGAL TOXINS, MERCURY W73-00703
PLANT GROWTH, SEWAGE EFFLUENTS, REGULATION, CYANOPHYTA, CHLOROPHY W72-14790
RUCOSA, ECKLONIA RADIATA, GROWTH REGULATION.: /RMIS, GRACILARIA VE W72-07496
ION, LEGISLATION, LEGAL ASPECTS, REGULATION, POLLUTION ABATEMENT, W72-01085
EFFECTS, ESTUARINE ENVIRONMENT, REGULATION, FALLOUT, OYSTERS, MUS W72-00959
ORS, ALGAL CONTROL, PLANT GROWTH REGULATORS, PLANKTON, CULTURES, W W71-13793

METHODS, GRANULES, PLANT GROWTH	REGULATORS, INHIBITORS, HERBICIDE	W72-09650
PESTICIDE TOXICITY, PLANT GROWTH	REGULATORS, INHIBITION, INHIBITOR	W71-10566
HERBICIDES, 2-4-D, PLANT GROWTH	REGULATORS, ALGAE, AQUATIC ALGAE,	W72-08440
IVATION, VITAMIN B, PLANT GROWTH	REGULATORS, NITRATES, PHOSPHATES,	W73-00430
TA, / *INHIBITION, *PLANT GROWTH	REGULATORS, *CHLORIDES, *CYANOPHY	W72-14706
RES, *GROWTH RATES, PLANT GROWTH	REGULATORS, PLANT GROWTH SUBSTANC	W71-03095
, *PLANT PATHOLOGY, PLANT GROWTH	REGULATORS, SALINITY, PLANT POPUL	W70-09429
EUGLENA GRACILIS, CHLAMYDOMANAS	REINHARDTII, LETHAL DOSAGE.: /TA,	W72-05614
INATED HYDROCARB/ *CHLAMYDOMONAS	REINHARDTII, CULTURE MEDIA, CHLOR	W73-00281
CHARACTERIZATION, CHLAMYDOMONAS	REINHARDTII, TETRAHYMENA VORAX, P	W73-02095
*ORTHOPHOSPHATES, CHLAMYDOMONAS	REINHARDTII, MOSES LAKE, PINE LAK	W72-12637
STIS SPP., SOI/ SILT-WATER-PLANT	RELATIONSHIPS, NORTH SEA, PHAEOCY	W72-12239
*GRO/ *OSTRACODS, PREDATOR-PREY	RELATIONSHIPS, *AMMONIA-NITROGEN,	W72-04787
IA CHANNELS, SURFACE-GROUNDWATER	RELATIONSHIPS, PERMEABILITY, ALGA	W71-12410
LING NUTRIENTS, SOIL-WATER-PLANT	RELATIONSHIPS, RADIOISOTOPES, ABS	W72-00949
UTION, LICHENS, SOIL-WATER-PLANT	RELATIONSHIPS, MARINE ALGAE, MOSS	W72-03347
NUTRIENTS, E/ *SOIL-WATER-PLANT	RELATIONSHIPS, *FALLOUT, *CYCLING	W72-03342
AMPLER, SECCHI *YARDS CREEK(NJ),	RELATIVE CONDITION FACTOR.: /ER S	W72-14280
SEDIMENTARY PHOSPHORUS	RELEASE.:	W72-01108
NTROL, *FORAGE FISH, BIOCONTROL,	REMEDIES, FOODS, FISH, HERBIVORES	W71-09163
NTS, TEMPERATURE CONTROL, ASSAY,	REMOTE SENSING, ALGAE, ANTIBIOTIC	W72-10162
ON PATTERNS, AERIAL PHOTOGRAPHY,	REMOTE SENSING, SATELLITES(ARTIFI	W72-09120
*NITROGEN	REMOVAL.:	W71-13939
NUTRIENT	REMOVAL.:	W71-09629
DUCTIVITY MEASUREMENTS, NUTRIENT	REMOVAL.: / GAMBUSIA AFFINIS, PRO	W71-01488
N), LAKE MELISSA(MINN), NUTRIENT	REMOVAL.: /CTION, LAKE SALLIE(MIN	W71-03012
PACITY, EUTROPHICATION, NUTRIENT	REMOVAL, *ZINC, CHELATION, FOOD C	W73-00802
NITRIFICATION, FILTER / *NITRATE	REMOVAL, ALGAE STRIPPING, POND DE	W71-08223
*PHOSPHORUS	REMOVAL, CHICAGO(ILLINOIS).:	W72-03970
MERCURY, PHOSPHORUS	REMOVAL, LAKE WASHINGTON(WASH).:	W71-12091
NOLOGY, NUISANCE ALGAE, NUTRIENT	REMOVAL, NUTRIENTS, WATER POLLUTI	W72-14692
ASTES, WASTE DISPOSAL, PESTICIDE	REMOVAL, OXYGEN SAG, DREDGING, ST	W72-01659
ORY RATES, PHOTOSYNTH/ *NUTRIENT	REMOVAL, OXYGEN DECLINE, RESPIRAT	W71-04079
PHOSPHORUS SORPTION, PHOSPHORUS	REMOVAL, RATE OF SORPTION.: /MUD,	W72-11617
ELECTROLYTES.: *STARCH	REMOVAL, STARCH SUBSTITUTION, POY	W72-03753
UATIC PLANTS, NITROGEN, NUTRIENT	REMOVAL, WATER TREATMENT, ASSAY,	W73-02478
HISTORY STUDIES, *GROWTH STAGES,	REPRODUCTION, CARBON DIOXIDE, CUL	W73-02099
ATIONS, FOOD CHAINS, SPECIATION,	REPRODUCTION.: / ON-SITE INVESTIG	W73-00764
*SYSTEMATICS, *PLANT MORPHOLOGY,	REPRODUCTION, LIFE CYCLES, AQUATI	W72-07141
UTERS, DATA ANALYSIS, EQUATIONS,	REPRODUCTION, BIORHYTHMS, MEASURE	W72-11920
ES, *FISHERIES, OIL WASTES, FISH	REPRODUCTION, WASTE DISPOSAL, MAR	W71-13723
DROGEN ION CONCENTRATION, ALGAE,	REPRODUCTION, SAMPLING, SEASONAL,	W71-08026
PHYTA, *MARINE ALGAE, *BIOASSAY,	REPRODUCTION, SEA WATER, MARINE P	W72-14294
QUATIONS, BIOLOGICAL PROPERTIES,	REPRODUCTION, ANALYTICAL TECHNIQU	W72-12724
POPULATIONS, NUTRIENTS, NUTRIENT	REQUIREMENTS.: /ETITION, AQUATIC	W72-12736
UATIC AL/ *MINERALOGY, *NUTRIENT	REQUIREMENTS, *PHYTOPLANKTON, *AQ	W72-12739
ATORS/ *PHYTOPLANKTON, *NUTRIENT	REQUIREMENTS, NUTRIENTS, BICINDIC	W72-13816
R TREATMENT, NUTRIENTS, NUTRIENT	REQUIREMENTS, GROWTH RATES, PHOSP	W72-12966
ALGAE, *EUTROPHICATION, NUTRIENT	REQUIREMENTS, MUNICIPAL WASTES.: /	W71-08775
WASTE WATER TREATMENT, NUTRIENT	REQUIREMENTS.: /L, COST ANALYSIS,	W71-08953
REQUIREMENTS(ALGAE), PHOSPHORUS	REQUIREMENTS(ALGAE), NUTRIENT CHE	W70-04080
EQUIREMENTS(ALGAE), NU/ NITROGEN	REQUIREMENTS(ALGAE), PHOSPHORUS R	W70-04080
HORUS COMPOUNDS, ALGAE, NUTRIENT	REQUIREMENTS, PRODUCTIVITY, ESSEN	W70-04080
UTROPHICATION, *ALGAE, *NUTRIENT	REQUIREMENTS, CHLOROPHYTA, CYANOP	W70-05469
WASTE WATER TREATMENT, *NUTRIENT	REQUIREMENTS, TEMPERATURE, *NITRO	W72-04789
*KINETICS, *NITROGEN, *NUTRIENT	REQUIREMENTS, AMMONIA, *GROWTH RA	W72-04787
KINETICS, *PHOSPHORUS, *NUTRIENT	REQUIREMENTS, TEMPERATURE, HYDROG	W72-04788
OLOGY, NUTRIENTS, OXYGEN, OXYGEN	REQUIREMENTS, OXYGEN SAG, WASTE A	W72-04279
, POTABLE WATER, ALGAE, NUTRIENT	REQUIREMENTS, CYANOPHYTA, CHLAMYD	W71-13185
HT INTENSITY, SALINITY, NUTRIENT	REQUIREMENTS, CHLORELLA, GROWTH R	W71-10083
SH HATCHERIES, *SALMONIDS, WATER	REQUIREMENTS, TEMPERATURE, NUTRIE	W71-11006
*ALGAE, ALGAL CONTROL, NUTRIENT	REQUIREMENTS, SORPTION, EUTROPHIC	W72-11617
PHYLL, SEA WATER, IONS, NUTRIENT	REQUIREMENTS, AQUATIC POPULATIONS	W72-09103
H STAGE, PHYTOPLANKTON, NUTRIENT	REQUIREMENTS, INVERTEBRATES, CHLA	W72-09248
, *BIOMASS, NUTRIENTS, *NUTRIENT	REQUIREMENTS, NUISANCE ALGAS, *EN	W72-03985
ION TREATMENT, NUTRIENTS, OXYGEN	REQUIREMENTS, AGING(BIOLOGICAL),	W72-01685
C MATTER, AERATION, *AL/ *OXYGEN	REQUIREMENTS, *DECOMPOSING ORGANI	W72-01881
PHNIA, COPEPODS, ALGAE, NUTRIENT	REQUIREMENTS, CYANOPHYTA, CHLOROP	W72-07940
*DIATOMS, *CULTURES, *NUTRIENT	REQUIREMENTS, SEA WATER, NITRATES	W73-00441
, NITRATES, PHOSPHATES, NUTRIENT	REQUIREMENTS, ESSENTIAL NUTRIENTS	W73-00430
A, CULTURES, SEA WATER, NUTRIENT	REQUIREMENTS, WATER POLLUTION.: /	W73-00380
ITY, *IRON, *CHLORELLA, NUTRIENT	REQUIREMENT, *ALGAE, DETERGENTS,	W73-02112
*PHOSPHATES, DIATOMS,/ *NUTRIENT	REQUIREMENTS, *LIMITING FACTORS,	W73-01445
ATER POLLUTION EFFECTS, NUTRIENT	REQUIREMENTS.: /CAL TECHNIQUES, W	W73-01269
S, COLOR, WATER QUALITY CONTROL,	RESEARCH AND DEVELOPMENT.: /NZYME	W72-04070
CLAIMED WATER, SEWAGE EFFLUENTS,	RESEARCH AND DEVELOPMENT, NUTRIEN	W72-11702
ES, WASTE ASSIMILATIVE CAPACITY,	RESEARCH AND DEVELOPMENT, WASTE W	W71-13256
ON, TOXICITY, CONTRACTS, GRANTS,	RESEARCH AND DEVELOPMENT, INSPECT	W71-11516
L/ *BIOASSAY, *LABORATORY TESTS,	RESEARCH AND DEVELOPMENT, TRACE E	W72-04433
C HEALTH, WASTE WATER TREATMENT,	RESEARCH AND DEVELOPMENT, MUNICIP	W72-06638
LUTION, WATER POLLUTION EFFECTS,	RESEARCH AND DEVELOPMENT, WASTE W	W72-06655
RSITIES, *ABST/ *WATER RESOURCES	RESEARCH ACT, CONNECTICUT, *UNIVE	W71-01192
AERATION, HOGS, CATTLE, POULTRY,	RESEARCH AND DEVELOPMENT, SOIL, S	W71-07551
, ANALYTICAL TECHNIQUES, SEWAGE,	RESEARCH EQUIPMENT, SEWAGE EFFLUE	W72-07715
HAINS, FORECASTING, CONSTRAINTS,	RESEARCH FACILITIES.: /WS, FOOD C	W72-06340
DOM, *FUTURE TRENDS, CLADOPHORA,	RESEARCH STRATEGY, THAMES RIVER(E	W71-13172
WATER RESOURCES	RESEARCH.:	W71-01192
EAT SHOCK.:	RESEARCH, CHYTRIDIACEOUS FUNGI, H	W72-07225
STIGATIONS, METHODOLOGY, FOREIGN	RESEARCH, MARINE ALGAE, PRODUCTIV	W72-08804
S, FOOD CHAINS, LAKE MORPHOLOGY,	RESERVOIR SILTING, LAKES, *SOUTH	W73-00938

*ODOR, DISSOLVED SOLIDS, ALGAE, RESERVOIR STORAGE, *HYPOLIMNION, W73-02138
NA, FISH, ENVIRONMENTAL EFFECTS, RESERVOIR STORAGE, NUTRIENTS, TEM W72-05282
DATA, DISSOLVED OXYGEN, WEATHER, RESERVOIR OPERATION, WATER TEMPER W72-05282
ER POLLUTION EFFECTS, NUTRIENTS, RESERVOIR OPERATION, WATER REUSE, W71-13185
, EUTROPHICATION, NITRIFICATION, RESERVOIR EVAPORATION, RESERVOIRS W72-02412
ING, VOLUMETRIC ANALYSIS, ALGAE, RESERVOIR STORAGE, IMPOUNDMENTS, W72-14280
ACROINVERTEBRATES, *TOCKS ISLAND RESERVOIR(NJ), AMERICAN SHAD, EKM W72-14280
TOSYNTHETIC MODEL, *TUTTLE CREEK RESERVOIR(KAN), GREAT PLAINS, AQU W72-12393
*CHENEY RESERVOIR(KANSAS).: W72-02274
R FILTERS, ANTHRACITE, *WAHNBACH RESERVOIR(GERMANY), OSCILLATORA R W72-06201
TIO, LAKE WAUBESA(WIS), OCCOQUAN RESERVOIR(VA), MADISON(WIS), LAKE W72-05473
MONAS, CHLAMIDOMONAS, BERENPLAAT RESERVOIR(ROTTERDAM), THE NETHERL W71-01491
SIBERIA).: BRATSK RESERVOIR(SIBERIA), ANGARA RIVER(W71-09165
IVER(THE NETHERLANDS), BIESBOSCH RESERVOIR(THE NETHERLANDS), KINET W72-08559
TORAGE OPERATION, SMITH MOUNTAIN RESERVOIR(VA).: *PUMPED S W72-14405
MONTROSE LAKE, THOMAS L/ COOLING RESERVOIR, ARTIFICIAL RESERVOIR, W70-09596
L/ COOLING RESERVOIR, ARTIFICIAL RESERVOIR, MONTROSE LAKE, THOMAS W70-09596
*USSR, *IRKUTSK OBLAST, *BRATSK RESERVOIR, ANGARA RIVER, CKA RIVE W72-04853
*USSR, *IRKUTSK OBLAST, *BRATSK RESERVOIR, *HYDROBIOLOGY, METALIM W72-04855
TILITIES DISTRICT, *INDIAN CREEK RESERVOIR, NUTRIENT BALANCE.: / U W72-02412
ORGANIC CARBON, *AMERICAN FALLS RESERVOIR, *APHANIZOMENON, ALGAL W72-13300
FICATION, RESERVOIR EVAPORATION, RESERVOIRS, NUTRIENTS, WATER REUS W72-02412
ITROGEN, *PHOSPHORUS, *DETENTION RESERVOIRS, TERTIARY TREATMENT, A W72-02412
LOW AUGMENTATION, WATER QUALITY, RESERVOIRS, WATER POLLUTION SOURC W72-04260
R, ALGAE, EUTROPHICATION, TASTE, RESERVOIRS, SEASONAL, ACTINOMYCET W71-13187
PLANNING, NITROGEN, PHOSPHORUS, RESERVOIRS, FISHERIES, WATER SUPP W71-13172
OTOSYNTHESIS, BIOMASS, NITRATES, RESERVOIRS, LIGHT PENETRATION, CY W71-13183
PLOSIVES, COSTS, FISH, FISHKILL, RESERVOIRS, FISHERIES, PRESSURE.: W72-11564
/ *DOMESTIC WATER, *CYANOPHYTA, RESERVOIRS, ODOR-PRODUCING ALGAE, W72-10613
KE WASHINGTON(WASH)/ SALT VALLEY RESERVOIRS(NEB), MADISON(WIS), LA W72-04759
C-14.: *SALT VALLEY RESERVOIRS(NEB), ULTRA PLANKTON, W72-04270
FICATION, *DETENTION RESERVOIRS, RESERVOIRS, PHOTOSYNTHESIS, DISSO W71-01491
*WATER PURIFICATION, *DETENTION RESERVOIRS, RESERVOIRS, PHOTOSYNT W71-01491
IDE TOXICITY, RESISTANCE, LAKES, RESERVOIRS, SWIMMING POOLS, CHELA W71-06189
Y PRODUCTIVITY, LAKES, SAMPLING, RESERVOIRS, TRIBUTARIES, DAMS, CH W73-01435
FALLS(COLO), FISH FECAL WASTES, RESIDUAL FOOD, GREEN RIVER(WASH), W71-11006
GRICULTURAL CHEMICALS, PESTICIDE RESIDUES, PESTICIDE KINETICS, ANI W71-12303
BIOLOGY, *PESTICIDES, *PESTICIDE RESIDUES, *PESTICIDE TOXICITY, *A W72-09650
DROCARBON PESTICIDES, *PESTICIDE RESIDUES, *ALGAE, CHLOROPHYTA, CH W72-04134
*TOXICITY, PESTICIDES, PESTICIDE RESIDUES, ALGAE, DDT, PHYTOPLANKT W72-03545
ING/ FEEDLOTS, COMPOSTING, PLANT RESIDUES, RECYCLING FOOD, PROCESS W72-00846
Y, PATH OF POLLUTANTS, PESTICIDE RESIDUES, CHLORINATED HYDROCARBON W73-02280
BIOASSAY, *ABSORPTION, PESTICIDE RESIDUES, WATER POLLUTION EFFECTS W73-02105
ENVIRONMENT, *WATER/ *PESTICIDE RESIDUES, *AQUATIC LIFE, *AQUATIC W72-13347
ERSHEDS(BASINS), *DD/ *PESTICIDE RESIDUES, *LOTIC ENVIRONMENT, WAT W72-12930
LEMENTS, RADIOCHEMICAL ANALYSIS, RESINS, *SEA WATER, BIOASSAY, COP W72-04708
ANALYSIS, BIOASSAY, CYANOPHYTA, RESINS, AQUATIC PLANTS, AQUATIC A W73-00286
BIOLOGY), *BIODETERIORA/ ACRYLIC RESINS, GLASS, THYMIDINES, CELLS(W71-09261
PROCEDURES, PESTICIDE TOXICITY, RESISTANCE, LAKES, RESERVOIRS, SW W71-06189
ITY, SOILS, TRIAGINE PESTICIDES, RESISTANCE, CULTURES, METABOLISM. W71-07675
DOMONAS, SCENEDESMUS, BIO-ASSAY, RESISTANCE, LETHAL LIMIT, POISONS W72-14706
IATION, FOOD CHAINS, GAMMA RAYS, RESISTANCE, BIOLOGICAL COMMUNITIE W73-00796
ES, PLANT MORPHOLOGY, EQUATIONS, RESISTANCE, ELECTRICAL PROPERTIES W73-01630
INTENSITY, *ECOLOGICAL DI/ *HEAT RESISTANCE, *MARINE ALGAE, LIGHT W73-00838
DESMUS, CHLORELLA, GROWTH RATES, RESISTANCE, PHOSPHORUS, IRON, MAN W72-12739
OGY, PHAEOPHYTA, STANDING CROPS, RESISTANCE, PRIMARY PRODUCTIVITY, W72-12940
TA, SODIUM, POTASSIUM, CULTURES, RESISTANCE, AQUATIC ALGAE, GROWTH W72-13809
N, PREDATION, PARASITISM, INSECT RESISTANCE: /T CONTROL, IRRIGATIO W72-00846
ILTRATION, BACKWASHING, SPECIFIC RESISTANCE.: /RAINING, *PRECOAT F W72-09393
ATION, RADIOACTIVITY TECHNIQUES, RESISTANCE, CULTURES, *CHLOROPHYT W72-11836
AL EFFECTS, INVERTEBRATES, WATER RESOURCES DEVELOPMENT.: /IRONMENT W71-10577
STUARIES, *WATER QUALITY, *WATER RESOURCES DEVELOPMENT, *WATER SUP W71-09788
ION SYSTE/ *WATER SUPPLY, *WATER RESOURCES DEVELOPMENT, *DISTRIBUT W72-05594
WATER RESOURCES RESEARCH.: W71-01192
UT, *UNIVERSITIES, *ABST/ *WATER RESOURCES RESEARCH ACT, CONNECTIC W71-01192
*WATER POLLUTION EFFECTS, *WATER RESOURCES DEVELOPMENT, *MASSACHUS W71-05553
COPPER SULFATE, CHLORINE, WATER RESOURCES, METHODOLOGY, EVALUATIO W71-05083
, ESTUARIES, CORAL, REEFS, WATER RESOURCES, SYSTEMS ANALYSIS, WAST W71-13723
TH RATES, PHYTOPLANKTON, NATURAL RESOURCES, DEPTH, DISTRIBUTION PA W71-13252
AQUATIC ALGAE, NUTRIENTS, WATER RESOURCES, WASTE ASSIMILATIVE CAP W71-13256
ECONOMICS, *PRODUCTIVITY, ALGAE, RESOURCES, FOOD CHAINS, FISH FOOD W71-13246
, *MEASUREMENT, *SAMPLING, WATER RESOURCES, AREA REDEVELOPMENT, BI W72-03215
MARY PRODUCTIVITY, GROWTH RATES, RESPIRATION.: /STABILIZATION, PRI W72-03556
UNICIPAL WASTES, PHOTOSYNTHESIS, RESPIRATION, WATER QUALITY, STREA W71-11017
83, BACTERIUM NO 52, ENDOGENOUS RESPIRATION, TOTAL OXYGEN DEMAND, W71-12223
, *ALGAE, OXIDATION, METABOLISM, RESPIRATION, LIPIDS, AMMONIA, CAR W72-10233
EN, AEROBIC CONDITIONS, BIOMASS, RESPIRATION, PHOTOSYNTHETIC OXYGE W72-11100
N ION CONCENTRATION, ALKALINITY, RESPIRATION, PHOTOSYNTHESIS, CARB W71-04518
, OXYGEN DEMAND, PHOTOSYNTHESIS, RESPIRATION, SCENEDESMUS, WASTE W W71-00139
RBICIDES, PESTICIDES, CHLORELLA, RESPIRATION, PHOSPHORUS, TOXICITY W70-07255
OSIS, PHOTOSYNTHESIS, OXIDATION, RESPIRATION, OXYGENATION, TEMPERA W72-04983
HYDROGEN, PRIMARY PRODUCTIVITY, RESPIRATION, PLANT PHYSIOLOGY, NU W72-06274
LOGICAL STUDIES, PHOTOSYNTHESIS, RESPIRATION.: /*METABOLISM, *CYTO W71-08683
LIGHT, CHRYSOPHYTA, OCHROMONAS, RESPIRATION, CHLOROPHYLL, CARBON W71-09172
RIDES, PHOTOSYNTHESIS, LITTORAL, RESPIRATION, PLANT PHYSIOLOGY, BI W72-12940
E FRACTIONATION, CARBON DIOXIDE, RESPIRATION, MARINE PLANTS, BICAR W72-12709
XICITY, WATER POLLUTION EFFECTS, RESPIRATION, *HEAVY METALS, CHEMI W72-12576
LOGY, FERTILITY, PHOTOSYNTHESIS, RESPIRATION, EUTROPHICATION, PLAN W72-12192
LANKTON, OXYGEN, PHOTOSYNTHESIS, RESPIRATION, SUCCESSION, PRODUCTI W73-01632
ON, KINETICS, WATER TEMPERATURE, RESPIRATION, ALGAE, CYANOPHYTA, A W73-01269

UATIC PLANTS, CYCLING NUTRIENTS, RESPIRATICN, MARSH PLANTS, CHEMIC W73-01089
UCTIVITY, METABOLISM, CHLORELLA, RESPIRATION, CHLOROPHYLL, LABORAT W73-01080
IS, EPILIMNION, LIGHT INTENSITY, RESPIRATION, LIMITING FACTORS.: / W72-14812
ATIC ALGAE, CHLOROPHYLL, OXYGEN, RESPIRATICN, CHLORINATED HYDROCAR W73-00281
UTRIENT REMOVAL, OXYGEN DECLINE, RESPIRATORY RATES, PHOTCSYNTHETIC W71-04079
IGATICN, AGRICULTURE, RICE, CROP RESPONSE.: /ICES, HERBICIDES, IRR W72-14807
THESIS, DEPTH, LIGHT PENTRATION, RESPRIATION, NUTRIENTS, TURBIDITY W72-08559
*ROADSIDE REST AREAS.: W72-14363
ERS, DRAINAGE WATER, IRRIGATION, RETURN FLOW, WATER SUPPLY, LIVEST W71-06435
STREAMS, ALGAE, PESTICIDES, DDT, RETURN FLOW, SEDIMENTS, SCIL EROS W71-06444
EATMENT, SEWAGE EFFLUENTS, WATER REUSE.: /IS, ECOLOGY, TERTIARY TR W72-03970
ON, RESERVOIRS, NUTRIENTS, WATER REUSE.: /ION, RESERVOIR EVAPORATI W72-02412
IAL WASTES, ACTIVATED SL/ *WATER REUSE, *DOMESTIC WASTES, *INDUSTR W71-09524
EATMENT, *DENITRIFICATION, WATER REUSE, *NITROGEN, PHOSPHORUS, CAR W72-04309
ENT, *TERTIARY TREATMENT, *WATER REUSE, *WATER QUALITY, WATER SUPP W70-10381
ULTIPLE-PURPOSE PROJECTS, *WATER REUSE, *WATER SUPPLY, *PCLLUTION W70-09193
GOONS, *MUNICIPAL WASTES, *WATER REUSE, ALGAE, PHOTOSYNTHESIS, DIS W72-11642
RSION, RECIRCULATED WATER, WATER REUSE, LEGAL ASPECTS, JUDICIAL DE W70-07100
ENTS, RESERVOIR OPERATION, WATER REUSE, POTABLE WATER, ALGAE, NUTR W71-13185
N, *OXIDATION LA/ *PONDS, *WATER REUSE, RECLAMATION, ALGAE, PROTEI W71-08951
OSPHORUS, ALGAE, NITRATES, WATER REUSE, SALINITY, PESTICIDES, LIVE W72-00846
ION EXCHANGE, ELECTROCHEMISTRY, REVERSE OSMOSIS, ELECTRODIALYSIS, W70-09907
APHY, *PACIFIC NORTHWEST, *COAST REVIEW, *BIBLIOGRAPHIES, WATER PO W72-12190
*ANIMAL MANURE, *LITERATURE REVIEW, FEEDING VALUE.: W71-07544
RONMENT, ON-SITE INVESTIGATIONS, REVIEWS.: /HEALTH, ESTUARINE ENVI W72-04461
CITY, RADIOISOTOPES, ABSORPTION, REVIEWS.: /DIMENTS, SETTLING VELO W72-04463
*LITERATURE REVIEWS.: W72-12190
NUISANCE ALGAE, *EUTROPHICATION, REVIEWS.: *CYANOPHYTA, * W72-01358
TION SOURCES, THERMAL POLLUTION, REVIEWS, *DATA COLLECTIONS, TCXIC W72-12190
ASTES, *PESTICIDES, / *TOXICITY, REVIEWS, *BIOASSAY, *INDUSTRIAL W W73-01976
TEMS, ENERGY BUDGETS, NUTRIENTS, REVIEWS, BIOLOGICAL COMMUNITIES, W73-01089
Y, DIELDRIN, PATH OF POLLUTANTS, REVIEWS, FOOD CHAINS, FORECASTING W72-06340
NISMS, *WATER POLLUTION SOURCES, REVIEWS, MICROBIAL DEGRADATION, C W72-14363
*NUISANCE ALGAE, EUTROPHICATICN, REVIEWS, TROPHIC LEVEL, PHOSPHORU W72-07937
ALIFORNIA, HEMIAUL/ *ETHMODISCUS REX, ETHMODISCUS GAZELLAE, BAJA C W72-03567
EA CUCUMBERS, STARFISH, CHITONS, RHABDODERMELLA, LOPHOPANOPEUS, EU W73-00932
STIS POUCHETI, GLUCOSE, MANNOSE, RHAMNOSE, ACRYLIC ACID, SODIUM, P W72-07710
SE RIVER(THE NETHERLANDS), BIES/ RHINE RIVER(THE NETHERLANDS), MEU W72-08559
, CRYPTOMONAS, CHLAMIDOMONAS, B/ RHINE RIVER, ASTERIONELLA FORMOSA W71-01491
RIVER(FRANCE), CLADOPHORA, PHO/ RHINE-MAAS BASIN(FRANCE), MEURTHE W71-00099
APITATUS, PENICILLUS DUMENTOSUS, RHIPOCEPHALUS PHOENIX, UDOTEA FLA W72-14294
, CRYSTAL/ *ACETYLENE REDUCTION, RHIZOBIUM JAPONICUM, LAKE MENDOTA W73-01456
OHIO), CLADOPHORA, APHANOCHAETE, RHIZOCLONIUM, VAUCHERIA, TRIBONEM W71-05629
CHA/ *ONONDAGA LAKE, OEDOGONIUM, RHIZOCLONIUM, COSMARIUM, PHACUS, W72-09111
L FUEL, BUNKER C OIL, AVICENNIA, RHIZOPHORA, BOSTRYCHIETUM, FIDDLE W72-07911
HANTZSCHII, CYCLOT MICHIGANIANA, RHIZOSOLENIA ERIENSIS, MELOSIRA I W73-00384
ICUS, RHIZOSOLENIA FRAGILISSIMA, RHIZOSOLENIA HEBETATA F. SEMISPIN W72-03543
RRANEUS, LEPTOCYLINDRUS DANICUS, RHIZOSOLENIA FRAGILISSIMA, RHIZOS W72-03543
TUM, THALASSIOSIRA, CHAETOCEROS, RHIZOSOLENIA ALATA, PERIDINIUM DE W71-07875
ING ORGANIC MATERIAL, FERTILITY, RHIZOSPHERE, ROOT SYSTEMS, SOIL F W72-03342
LASER, RHODAMINE B.: W72-01095
IBITORS, ANTIBIOTICS, SEA WATER, RHODE ISLAND.: / CHLOROPHYTA, INH W72-04748
YLLOSPA/ *HYALOCHLORELLA MARINA, RHODOMELA SP., ENDOCLADIA SP., PH W71-11851
SSIOPHYLLUM, FUCUS, SCHIZYMENIA, RHODOMELA, RHODOMENIA, MARMION IS W72-09120
, FUCUS, SCHIZYMENIA, RHODOMELA, RHODOMENIA, MARMION ISLAND, MORRI W72-09120
ELLA TERTIOLECTA, SYNECHOCOCCUS, RHODOMONAS, COCCOLITHUS HUXLEYI, W71-10083
LATORIA RUBESCENS, SYNEDRA ACUS, RHODOMONAS, LAKE CONSTANCE, SEWAG W71-00117
AMPLE PRESERVATION, FLAGELLATES, RHODOMONAS, IKROAVICH LAKE, UTERM W73-00273
ODISCUS HANTZSCHII, CRYPTOMONAS, RHODOMONAS.: /T(ENGLAND), STEPHAN W72-14797
NIAGARAE, ASTERIONELLA FORMOSA, RHODOMONAS LACUSTRIS, COUNTING, O W73-01435
RANULATA, CERATIUM, MICROCYSTIS, RHODOMONAS MINUTA, ASTERIONELLA F W72-12738
Y WATER, EVAPORATION, DIFFUSION, RHODOPHYTA.: /TS, PHAEOPHYTA, OIL W73-01074
TTS, SYSTEMATICS, *REPRODUCTICN, RHODOPHYTA, PHAEOPHYTA, WATER QUA W73-01449
CE WATERS, LAGOONS, CHLOROPHYTA, RHODOPHYTA, PHYSICAL PROPERTIES, W73-00838
ARINE ALGAE, ANNELIDS, MOLLUSKS, RHODOPHYTA, PHAEOPHYTA, PHOTOGRAP W73-00932
IDE, CULTURES, LIMITING FACTORS, RHODOPHYTA, AQUATIC ALGAE, MICROO W73-02099
CTS, *MONITORING, *MARINE ALGAE, RHODOPHYTA, PHAEOPHYTA, *COBALT R W72-07826
SUES, BI/ *ALGAE, *PLANT GROWTH, RHODOPHYTA, INHIBITORS, PLANT TIS W72-07496
UM, SODIUM SULFATES, CYANOPHYTA, RHODOPHYTA.: /LINE WATER, MAGNESI W71-04526
OHIO, CHLOROPHYTA, CHRYSOPHYTA, RHODOPHYTA, CYANOPHYTA, EUGLENOPH W71-05629
PHYTA, CHLOROPHYTA, CHRYSOPHYTA, RHODOPHYTA, TESTING, ECOLOGY.: /C W70-07280
EASTS, *CULTURES, *MARINE ALGAE, RHODOPHYTA, PHAEOPHYTA, CHLOROPHY W72-04748
G CROP, CHLOROPHYTA, CYANOPHYTA, RHODOPHYTA, PHAEOPHYTA, CARBON RA W72-06283
OCEANS, CHLOROPHYTA, PHAEOPHYTA, RHODOPHYTA, MARINE BACTERIA.: /, W71-07878
N RADIOISOTOPES, CARBON DIOXIDE, RHODOPHYTA.: / CHLOROPHYLL, CARBO W71-09172
H, FOOD CHAINS, OXYGEN, ECOLOGY, RHODOPHYTA, PHAEOPHYTA, CHLORELLA W71-09157
EMATICS, *COASTS, BENTHIC FLORA, RHODOPHYTA, PHAEOPHYTA, CHLOROPHY W71-10066
HLOROPHYTA, *CYANOPHYTA, *LAKES, RHODOPHYTA, SHORES.: *ALGAE, *C W71-12489
ALGAE, CHLOROPHYTA, PHAEOPHYTA, RHODOPHYTA.: /LLITES(ARTIFICIAL), W72-09120
PHYTA, *CHLOROPHYTA, PHAEOPHYTA, RHODOPHYTA, CLADOPHORA.: / *CYANO W72-11800
C GENERATORS, WATER TEMPERATURE, RHODOPHYTA, PHAEOPHYTA, CHLOROPHY W72-11252
ANKTON, CHLOROPHYTA, PHAEOPHYTA, RHODOPHYTA, DIATOMS, PYRROPHYTA, W72-03220
UMBER TEST, MEMBRANE FILTRATICN, RHODOPSEUDOMONAS GELATINOSA, THIO W72-04743
PARAPSILOSIS, CHONDRUS CRISPUS, RHODOTORULA GLUTINIS, SARGASSUM N W72-04748
IA LANOSA, POLYSIPHONIA HARVEYI, RHODOTORULA RUBRA, RHODOTORULA LA W72-04748
ONIA HARVEYI, RHODOTORULA RUBRA, RHODOTORULA LACTOSA, CANDIDA ZEYL W72-04748
, MICROCOCCUS, BACILLUS, VIBRIO, RHODOTORULA, AEROMONAS.: / GROWTH W72-10861
RHODYMENIA, FUCUS, ULVA.: W72-09653
UM NODOSUM, GIGARTINA MAMILLOSA, RHODYMENIA PALMATA.: /, ASCOPHYLL W71-11515
AMINARIA AGARDHII, ULVA LACTUCA, RHODYMENIS PALMATA, POLYSIPHONIA W72-04748

AKE FURESO (DENMARK), COCCONEIS,	RHOICOSPHENIA, ACHANTHES, FRAGILA	W72-07501
SUBLITTORAL BENTHIC MACROPHYTES,	RIA DE ALDAN(SPAIN), LAMINARIA OC	W72-04766
LLA VULGARIS, NITRATE REDUCTASE,	RIBONUCLEIC ACID.: CHLORE	W72-03222
OSE, SENESCENT CULTURES, XYLOSE,	RIBOSE, ARABINOSE, GALACTOSE, HEX	W72-07710
*GREEN MANURE, INDIA, JAPAN,	RICE FIELDS.:	W73-02473
ICIDES, IRRIGATION, AGRICULTURE,	RICE, CROP RESPONSE.: /ICES, HERB	W72-14807
GEN FIXATION, *TROPICAL REGIONS,	RICE, CYANOPHYTA, FERTILIZATION,	W73-02473
FERTILIZATION, NUTRIENTS, ALGAE,	RICE, NITROGEN, NITROGEN FIXATION	W71-12071
CHIRONOMIDS, CLEAR LAKE(CALIF),	RICHELIEU RIVER(QUEBEC), GREEN BA	W70-08098
MARINE ALGAE, SEDIMENTS, PUERTO	RICO, ESTUARINE ENVIRONMENT, PATH	W73-00821
), SPECIES DIVERSITY, DIRECT EF/	RIDLEY CREEK(PA), LITITZ CREEK(PA	W72-08579
MICH), SAN JOAQUIN RIVER(CALIF),	RIFLE FALLS(COLO), FISH FECAL WAS	W71-11006
IRED WATER / *CALIFORNIA, *WATER	RIGHTS, *WATER UTILIZATION, *IMPA	W70-07100
IMPAIRED WATER QUALITY, RIPARIAN	RIGHTS, PRIOR APPROPRIATION, GROU	W70-07100
USCULA, GOMONTIA POLYRHIZA, ULVA	RIGIDA, CLADOPHORA FULIGINOSA, BA	W72-11800
OTOPE P32, SODIUM PYROPHOSPHATE,	RIO GRANDE VALLEY.: /TH, *RADIOIS	W72-13992
ZATION, *IMPAIRED WATER QUALITY,	RIPARIAN RIGHTS, PRIOR APPROPRIAT	W70-07100
IZOTHRIX CALCICOLA, ELKTON POND,	RISING SUN POND, INDUSTRIAL STORA	W72-04736
POTOMAC	RIVER BASIN, POTOMAC ESTUARY.:	W70-06509
FOX	RIVER BASIN(WIS), FWPCA.:	W70-05103
FER/ *LOADING GUIDELINES, *GRAND	RIVER BASIN(ONTARIO), CANADA, BUF	W71-11017
VER, VEGETATION EFFECTS, SLUDGE,	RIVER BEDS, ALGAE, MATHEMATICAL M	W70-09189
*POTOMAC	RIVER ESTUARY.:	W71-08775
*PAMLICO	RIVER ESTUARY.:	W72-01329
PIRATION RATIOS.: *PAMLICO	RIVER ESTUARY, PHOTOSYNTHESIS/RES	W72-13636
WATER POLLUTION SOURCES, ALGAE,	RIVER FLOW, PHOSPHORUS, NITRATES,	W72-04260
KINNERET(ISRAEL), BOTTOM WATERS,	RIVER JORDAN(ISRAEL).: *LAKE	W72-05469
DIATOMS, *RIVER THAMES(ENGLAND),	RIVER KENNET(ENGLAND), STEPHANODI	W72-14797
EIGHT, THERMAL EFFECTS/ SAVANNAH	RIVER PLANT, CULTURE MEDIA, DRY W	W71-02075
INTERCHANGE, HAW RIVER, NEW HOPE	RIVER.: NUTRIENT	W70-09669
*POTOMAC	RIVER.:	W70-08248
*ANACOSTIA	RIVER.:	W70-08249
*SAN JOAQUIN	RIVER.:	W71-07098
CONNECTICUT	RIVER.:	W72-00939
NIAGARA	RIVER.:	W71-11685
*SPECIES DIVERSITY, *POTOMAC	RIVER.:	W71-11583
*SPECIES DIVERSITY, CRYSTAL	RIVER.:	W72-11252
JORDAN RIVER(MICH), SAN JOAQUIN	RIVER(CALIF), RIFLE FALLS(COLO),	W71-11006
, CONNECTICUT RIVER, SAN JOAQUIN	RIVER(CALIFORNIA), BISCAYNE BAY,	W71-07417
RSVILLE LAKE(CALIF), SAN JOAQUIN	RIVER(CALIF), NITZSCHIA.: /), SEA	W70-10405
PABLO ESTUARY(CALIF), SACRAMENTO	RIVER(CALIF), SEARSVILLE LAKE(CAL	W70-10405
, WHITE RIVER(INDIANA), PATUXENT	RIVER(CANADA), ONTARIO.: /YNE BAY	W71-07417
HITE RIVER(INDIANA), CONNECTICUT	RIVER(CONNECTICUT), GREEN RIVER(K	W71-11517
PHORA, RESEARCH STRATEGY, THAMES	RIVER(ENGLAND), LEE RIVER(ENGLAND	W71-13172
TEGY, THAMES RIVER(ENGLAND), LEE	RIVER(ENGLAND).: /, RESEARCH STRA	W71-13172
HINE-MAAS BASIN(FRANCE), MEURTHE	RIVER(FRANCE), CLADOPHORA, PHORMI	W71-00099
CARBON, *AMERI/ *GLUCOSE, *SNAKE	RIVER(IDAHO), *DISSOLVED ORGANIC	W72-13300
ALGAL GROWTH POTENTIAL, ILLINOIS	RIVER(ILL), DIAGNOSTIC CRITERION.	W71-11033
(CONNECTICU/ *ENTRAINMENT, WHITE	RIVER(INDIANA), CONNECTICUT RIVER	W71-11517
CALIFORNIA), BISCAYNE BAY, WHITE	RIVER(INDIANA), PATUXENT RIVER(CA	W71-07417
RATE NITROGEN, ORTHOPHOSP/ *SALT	RIVER(KENT), VASCULAR PLANTS, NIT	W72-06526
CTICUT RIVER(CONNECTICUT), GREEN	RIVER(KENTUCKY), CAYUGA LAKE(NEW	W71-11517
AKE ANNABESSACOOK(ME), *KENNEBEC	RIVER(ME), CULTURAL EUTROPHICATIO	W72-04280
ERY EFFLUENTS, PARASITES, JORDAN	RIVER(MICH), SAN JOAQUIN RIVER(CA	W71-11006
IS RIVER, ST. LOUIS RIVER, BLACK	RIVER(MINNESOTA), LAKE WASHINGTON	W71-04216
*PAMLICO	RIVER(NC).:	W71-07337
N), COCKLES, GOLDFISH, MERRIMACK	RIVER(NEW HAMPSHIRE), DELAWARE RI	W71-11517
SANDUSKY	RIVER(OHIO).:	W72-04260
AETE, RHIZOCLONIUM, V/ *CUYAHOGA	RIVER(OHIO), CLADOPHORA, APHANOCH	W71-05629
*NUTRIENT LEVELS, CUYAHOGA	RIVER(OHIO).:	W72-04263
CLODOPHORA GLOMERATA, *METOLIUS	RIVER(ORE).: /ANTHES, *SPIROGYRA,	W72-07901
DS, CLEAR LAKE(CALIF), RICHELIEU	RIVER(QUEBEC), GREEN BAY(WIS), MA	W70-08098
RATSK RESERVOIR(SIBERIA), ANGARA	RIVER(SIBERIA).: B	W71-09165
E RIVER, ILLINOIS RIVER, HOLSTON	RIVER(TENNESSEE).: /IRE), DELAWAR	W71-11517
ER(THE NETHERLANDS), BIES/ RHINE	RIVER(THE NETHERLANDS), MEUSE RIV	W72-08559
NE RIVER(THE NETHERLANDS), MEUSE	RIVER(THE NETHERLANDS), BIESBOSCH	W72-08559
*SOUTH	RIVER(VA).:	W73-01972
*SNAKE	RIVER(WASH).:	W72-14728
FORNIA), POTOMAC RIVER, COLUMBIA	RIVER(WASHINGTON), COCKLES, GOLDF	W71-11517
CAL WASTES, RESIDUAL FOOD, GREEN	RIVER(WASH), COWLITZ TROUT HATCHE	W71-11006
RENOIDOSA, SODIUM SILIC/ POTOMAC	RIVER, ALAFIA RIVER, CHLORELLA PY	W72-04734
AE, *INVERTEBRATES, ST. LAWRENCE	RIVER, ALGAE, NEMATODES, GASTROPO	W72-05432
RS, PHOSPHATES, NITRATES, HUDSON	RIVER, ALGAE, WATER QUALITY, SEWA	W70-06509
S, *NORWAY, MACROPHYTES, FYRISAN	RIVER, ARBOGAAN RIVER, BOTORPSSTR	W73-00286
TOMONAS, CHLAMIDOMONAS, B/ RHINE	RIVER, ASTERIONELLA FORMOSA, CRYP	W71-01491
ARBOGAAN RIVER, BOTORPSSTROMMEN	RIVER, ATRAN RIVER, MOTALA STROM	W73-00286
PINE RIVER, CLINTON RIVER, HURON	RIVER, BELL RIVER, SALT RIVER, MI	W72-14282
VER, WHITE CLAY RIVER, CHRISTINA	RIVER, BIG ELK RIVER, LITTLE ELK	W72-04736
HYDROGEN ION CON/ *ST. LAWRENCE	RIVER, BIOCHEMICAL OXYGEN DEMAND,	W70-10427
KE WA/ ILLINOIS RIVER, ST. LOUIS	RIVER, BLACK RIVER(MINNESOTA), LA	W71-04216
R RIVER, LAKE ST. CLAIR, DETROIT	RIVER, BLACK RIVER, PINE RIVER, C	W72-14282
OPHYTES, FYRISAN RIVER, ARBOGAAN	RIVER, BOTORPSSTROMMEN RIVER, ATR	W73-00286
LECTONEMA BORYANIUM, SUSQUEHANNA	RIVER, BRANDYWINE RIVER, RED CLAY	W72-04736
ERRING, OYSTERS, CLAMS, COLUMBIA	RIVER, CALIFORNIA, WASHINGTON, CO	W71-07417
IUM SILIC/ POTOMAC RIVER, ALAFIA	RIVER, CHLORELLA PYRENOIDOSA, SOD	W72-04734
IVER, RED CLAY RIVER, WHITE CLAY	RIVER, CHRISTINA RIVER, BIG ELK R	W72-04736
DETROIT RIVER, BLACK RIVER, PINE	RIVER, CLINTON RIVER, HURON RIVER	W72-14282
HUMBOLT BAY(CALIFORNIA), POTOMAC	RIVER, COLUMBIA RIVER(WASHINGTON)	W71-11517
POINT, COLUMBIA RIVER, PATUXENT	RIVER, CONTRA COSTA POWER PLANT,	W72-07907
, *STATISTICAL METHODS, *POTOMAC	RIVER, DATA COLLECTIONS, EVALUATI	W72-07791

ARBOR VITAE LAKE, GREEN BAY, FOX RIVER, DETROIT RIVER, HETEROTROPH W73-01456
*USSR, MOSCOW RIVER, OKA RIVER, FOAM, NATURAL WATERS.: W72-05506
UUS.: SAALE RIVER, GERMANY, SCENEDESMUS OBLIQ W72-14314
E, GREEN BAY, FOX RIVER, DETROIT RIVER, HETEROTROPHIC BACTERIA, AC W73-01456
SHIRE), DELAWARE RIVER, ILLINOIS RIVER, HOLSTON RIVER(TENNESSEE).: W71-11517
PONDS, MARYLAND, QUARRIES, OHIO RIVER, HUDSON RIVER, INDICATORS, W72-04736
TA, ST. MARGARET'S BAY, SCOUDOUC RIVER, HUMIC ACIDS, OXIMATION, SA W72-12166
BLACK RIVER, PINE RIVER, CLINTON RIVER, HURON RIVER, BELL RIVER, S W72-14282
TSK RESERVOIR, ANGARA RIVER, OKA RIVER, HYDROBIOLOGY, BACTERIOLOGY W72-04853
K RIVER(NEW HAMPSHIRE), DELAWARE RIVER, ILLINOIS RIVER, HOLSTON RI W71-11517
ND, QUARRIES, OHIO RIVER, HUDSON RIVER, INDICATORS, DELAWARE, PATH W72-04736
, CONNECTICUT, FLORIDA, DELAWARE RIVER, INDIANA.: /NIA, WASHINGTON W71-07417
ERIE, LAKE ONTARIO, ST LAWRENCE RIVER, INTERNATIONAL JOINT COMMIS W70-07283
VER, BL/ MAUMEE RIVER, ST. CLAIR RIVER, LAKE ST. CLAIR, DETROIT RI W72-14282
S BAY, LAKE PONTCHARTRAIN, PEARL RIVER, LAKE BORGNE, SUBSTRATES, M W73-02029
RIVER, CHRISTINA RIVER, BIG ELK RIVER, LITTLE ELK RIVER, OCTORARA W72-04736
R, HURON RIVER, BELL RIVER, SALT RIVER, MICROCYSTIS, APHANIZOMENON W72-14282
A COSTA POWER PLANT, SAN JOAQUIN RIVER, MORRO BAY POWER PLANT, HUM W72-07907
ER, BOTORPSSTROMMEN RIVER, ATRAN RIVER, MOTALA STROM RIVER, SPARGA W73-00286
NUTRIENT INTERCHANGE, HAW RIVER, NEW HOPE RIVER.: W70-09669
A, SURIRELLA, PERIDINIUM, SENECA RIVER, NINEMILE CREEK, VAN DORN S W72-09111
RIVER, BIG ELK RIVER, LITTLE ELK RIVER, OCTORARA RIVER, PLECTONEMA W72-04736
BLAST, *BRATSK RESERVOIR, ANGARA RIVER, OKA RIVER, HYDROBIOLOGY, B W72-04853
ATERS.: *USSR, MOSCOW RIVER, OKA RIVER, FOAM, NATURAL W W72-05506
LYSIS, SODIUM CHLORIDE, COLUMBIA RIVER, ORGANIC COMPOUNDS, VITAMIN W72-05421
ALGAE, RIVERS, STREAMS, DELAWARE RIVER, OXIDATION LAGOONS, PONDS, W72-04736
RIAL WATER POLLUTION, DES MOINES RIVER, PACKING PLANTS.: INDUST W71-06445
TA POWER / CHALK POINT, COLUMBIA RIVER, PATUXENT RIVER, CONTRA COS W72-07907
IPHENYLS, AROCLOR 1254, ESCAMBIA RIVER, PINFISH, FLOUNDER, CROAKER W71-07412
ST. CLAIR, DETROIT RIVER, BLACK RIVER, PINE RIVER, CLINTON RIVER, W72-14282
IVER, LITTLE ELK RIVER, OCTORARA RIVER, PLECTONEMA, LYNGBYA, PHORM W72-04736
M, SUSQUEHANNA RIVER, BRANDYWINE RIVER, RED CLAY RIVER, WHITE CLAY W72-04736
CLINTON RIVER, HURON RIVER, BELL RIVER, SALT RIVER, MICROCYSTIS, A W72-14282
UNDER, HUMBOLDT BAY, CONNECTICUT RIVER, SAN JOAQUIN RIVER(CALIFORN W71-07417
D/ *BIGHORN LAKE(MONT), BIGHORN RIVER, SHOSHONE RIVER, YELLOWTAIL W73-01435
RIVER, ATRAN RIVER, MOTALA STROM RIVER, SPARGANIUM ERECTUM, SCHOEN W73-00286
SPARENCY, ORTHOP/ GREEN BAY, FOX RIVER, SPECIFIC CONDUCTANCE, TRAN W71-10645
ER(MINNESOTA), LAKE WA/ ILLINOIS RIVER, ST. LOUIS RIVER, BLACK RIV W71-04216
CLAIR, DETROIT RIVER, BL/ MAUMEE RIVER, ST. CLAIR RIVER, LAKE ST. W72-14282
THIC FLORA, LAKE ONTARIO, HUDSON RIVER, STANDARDS, SURFACE WATERS, W70-09614
ANKEE NUCLEAR PLANT, CONNECTICUT RIVER, TURKEY POINT, BISCAYNE BAY W72-07907
R QUALITY, *ESTUARIES, *DELAWARE RIVER, VEGETATION EFFECTS, SLUDGE W70-09189
ANALYTICAL TECHNIQUES, COLUMBIA RIVER, WASHINGTON, OREGON.: /NTS, W72-03215
ATE/ *INSECTICIDES, *ST LAWRENCE RIVER, WATER POLLUTION SOURCES, W W70-08098
IVER, BRANDYWINE RIVER, RED CLAY RIVER, WHITE CLAY RIVER, CHRISTIN W72-04736
E(MONT), BIGHORN RIVER, SHOSHONE RIVER, YELLOWTAIL DAM, FRAGILARIA W73-01435
OLLUTED ESTUARIES, DURHAM COUNTY RIVERS.: *ENGLAND, *P W72-08804
ATION, ON-SITE DATA COLLECTIONS, RIVERS, *ELECTRIC POWERPLANTS, WA W72-07791
IA, CONFERENCES, *PUBLIC HEALTH, RIVERS, *SEWAGE, BIOCHEMICAL OXYG W72-09675
LLUTION EFFECTS, NUCLEAR WASTES, RIVERS, ABSORPTION, FISH, MOLLUSK W71-09225
IFICATION, AQUATIC WEED CONTROL, RIVERS, AGRICULTURE, ROOTED AQUAT W71-13172
*SLIME, *BIOLOGICAL COMMUNITIES, RIVERS, BACTERIA, WATER POLLUTION W72-08573
*ALGAE, *BENTHIC FLORA, *OREGON, RIVERS, BIOLOGICAL COMMUNITIES, S W72-07901
, STRONTIUM RADIOISOTOPES, IONS, RIVERS, CULTURES, ORGANIC MATTER, W71-08032
FISH, RECREATION, INVERTEBRATES, RIVERS, CYANOPHYTA, GEOLOGIC FORM W71-11009
CTIVITY, *RADIOACTIVITY EFFECTS, RIVERS, EFFLUENT, RADIOECOLOGY, A W72-07841
ONTHLY, AQUATIC LIFE, INTERSTATE RIVERS, HYPOLIMNION, VOLUME, MIXI W70-09595
DIATOMS, SUCCESSION, NUTRIENTS, RIVERS, INFLUENT STREAMS, EFFLUEN W73-01435
, EUTROPHICATION, WATER QUALITY, RIVERS, IOWA, CORN, DISCHARGE, CH W71-06445
YTA, WATER POLLUTION, NUTRIENTS, RIVERS, LAKES, FARM PONDS.: /ROPH W71-05991
R POLLUTION SOURCES, *ESTUARIES, RIVERS, PHOSPHATES, NITRATES, HUD W70-06509
LECTROCHEMISTRY, PHOTOSYNTHESIS, RIVERS, PLANT PHYSIOLOGY, TESTING W72-14314
*THERMAL POLLUTION, CONNECTICUT, RIVERS, POST-IMPOUNDMENT, NUCLEAR W72-00939
ATER POLLUTION EFFECTS, STREAMS, RIVERS, ROOTED AQUATIC PLANTS, AQ W72-14690
, CURRENTS(WATER), ALGAE, LAKES, RIVERS, SEWAGE DISPOSAL, SEWAGE T W70-09947
ANALYSIS, *BIOINDICATORS, ALGAE, RIVERS, STREAMS, DELAWARE RIVER, W72-04736
, *CHLORINATION, *ALGAL CONTROL, RIVERS, WATER POLLUTION EFFECTS, W71-00099
PATH OF POLLUTANTS, METHODOLOGY, RIVERS, WATER POLLUTION SOURCES, W73-00286
PIROGIRA, CHLORELLA ELLIPSOIDEA, RNA, CHLOROPHYLL B, ASHING, LABOR W72-06274
, *IODINE RADIOISO/ *SEDIMENTARY ROCKS, *BRINES, *ALGAE, *OKLAHOMA W72-02073
SEDIMENTATION, FILTRATION, LIME, ROCKS, METABOLISM, DESIGN CRITERI W71-13341
BEHAVIOR, INVERTEBRATES, OTTERS, ROCKSLIDES, INTERTIDAL AREAS, ON- W73-00764
RANGE LAKE, EAST TWIN LAKE, LAKE ROCKWELL, LAKE HODGSON, CLEARWATE W72-13673
CARBON PESTICIDES, INSECTICIDES, RODENTICIDES, HERBICIDES, FUNGICI W73-00238
TISH COLUMBIA, RUSSIA, LAMINARIA RODRIGUEZII.: /ARIA LONGIPES, BRI W73-00428
ATERIAL, FERTILITY, RHIZOSPHERE, ROOT SYSTEMS, SOIL FUNGI, SOIL EN W72-03342
ED CONTROL, RIVERS, AGRICULTURE, ROOTED AQUATIC PLANTS, DRAINAGE, W71-13172
LUTION EFFECTS, STREAMS, RIVERS, ROOTED AQUATIC PLANTS, AQUATIC AL W72-14690
QUATIC PLANTS, LIMITING FACTORS, ROOTED AQUATIC PLANTS, CYCLING NU W73-01089
D CONTROL, TILAPIA, CARP, ALGAE, ROOTED AQUATIC PLANTS, PLANKTON, W71-04528
IUM, MACROMONAS MO/ AMOEBOBACTER ROSEUS CHROMATIUM OKENII, CHROMAT W72-07896
ROSIN AMINE D ACETATE(RADA).: W71-05083
, HEMIAULUS HAUCKII, STIGMAPHORA ROSTRATA, ETHMODISCUS OOZES.: /IA W72-03567
NSITY, DIATOMS, LAKES, PROTOZOA, ROTIFERS, CRUSTACEANS, DIPTERA, M W71-11012
ROPHYTA, CHRYSOPHYTA, CHLORELLA, ROTIFERS, CHLAMYDOMONAS, PHYTOPLA W70-06655
S, DINOFLAGELLATES, SCENEDESMUS, ROTIFERS, CRUSTACEANS, COPEPODS, W72-05968
AGOONS, *LIGHT INTENSITY, PONDS, ROTIFERS, PROTOZOA, PHOTOSYNTHESI W71-07382
HABITS, EUTROPHICATION, LARVAE, ROTIFERS, AQUATIC MICROORGANISMS, W72-07132
AMIDOMONAS, BERENPLAAT RESERVOIR(ROTTERDAM), THE NETHERLANDS.: /HL W71-01491
RA NORDENSKIOLDII, THALASSIOSIRA ROTULA, SCHRODERELLA DELICATULA, W72-03543

, EICHHORNIA CRASSIPES, SALVINIA RCTUNDIFOLIA, MILORGANITE, GAMBUS W71-01488
Y.: RUBBLE CHIMNEY, SAN JOAQUIN VALLE W72-00974
BREVE, PERIDINIUM, OSCILLATORIA RUBECENS, ANABAENA, CRUSSOSTREA, W73-00840
RESERVOIR(GERMANY), OSCILLATCRA RUBESCENS.: /NTHRACITE, *WAHNBACH W72-06201
ILLATORIA REDECKEI, OSCILLATORIA RUBESCENS, SYNEDRA ACUS, RHODOMCN W71-00117
IENTS, MICRONUTRIENTS, SCANDIUM, RUBIDIUM, BROMINE, ACTINIDES, RAD W73-01434
OLYSIPHONIA HARVEYI, RHODOTORULA RUBRA, RHODOTCRULA LACTCSA, CANDI W72-04748
EATION, *TREATMENT FACIL/ *STORM RUN-OFF, *SUSPENDED SOLIDS, *RECR W72-00042
LOROPHYLL, PIGMENTS, DISPERSION, RUNOFF.: /ALGAE, CONDUCTIVITY, CH W72-04773
RY TREATMENT, BIOCHEMI/ *SURFACE RUNOFF, *SEWAGE TREATMENT, TERTIA W71-06445
POLLUTION SOURCES, *AGRICULTURAL RUNOFF, *WATER POLLUTION EFFECTS, W72-08401
CONTROL, WISCONSIN, LAKES, URBAN RUNOFF, AGRICULTURAL RUNOFF, FARM W73-02479
NTARIO, ADSORPTION, AGRICULTURAL RUNOFF, ALGAE, FOOD CHAINS, CHLOR W72-13657
ERS, PRECIPITATION(ATMOSPHERIC), RUNOFF, ANIMAL WASTES, SURFACE WA W71-04216
RIENTS, ALGAE, NITRATES, SURFACE RUNOFF, BASE FLOW, PERCOLATION, L W71-06443
S, *NITROGEN, EVAPORATION, STORM RUNOFF, BIOASSAYS, BIO-INDICATORS W72-04070
ENT, *WATER SUPPLY, WASTE WATER, RUNOFF, DISSOLVED OXYGEN, ALGAE, W71-09788
INDUSTRIAL WASTES, AGRICULTURAL RUNOFF, DOMESTIC WASTES, ECOLOGY, W73-01972
PENETRATION, CARBON, LAKE ERIE, RUNOFF, EROSION, CHLORELLA, SODIU W72-04734
AKES, URBAN RUNOFF, AGRICULTURAL RUNOFF, FARM WASTES, ALGAE, AQUAT W73-02479
ROOTED AQUATIC PLANTS, DRAINAGE, RUNOFF, FERTILIZERS, LIVESTOCK, E W71-13172
AL METHODS, SOIL MICROORGANISMS, RUNOFF, HYDROLIC DATA.: /TATISTIC W72-02218
ES, SILICA, AMMONIA, PHOSPHATES, RUNOFF, ILLINOIS, HYDROGEN ION CO W71-03028
ANKTON, CHLOROPHYLL, RECREATION, RUNOFF, MEASUREMENT, TEST PROCEDU W72-04270
WASTEWATER TREATMENT, TURBIDITY, RUNOFF, PHOSPHORUS, ALGAE, AQUATI W72-04759
POLLUTION EFFECTS, AGRICULTURAL RUNOFF, PHOSPHORUS, NITROGEN, PHO W72-10625
ANDFILLS, PERCOLATION, DRAINAGE, RUNOFF, PRECIPITATION(ATMOSPHERIC W72-12955
OXIDATION LAGOONS, DEHYDRATION, RUNOFF, RADIOACTIVITY, FALLOUT, B W72-00846
ER, WINDS, SAMPLING, INCUBATION, RUNOFF, SALINITY, DEPTH, ABSORPTI W72-09122
SKA, URBANIZATION, CITIES, STORM RUNOFF, SEWAGE, WASTE WATER DISPO W71-13466
NEBR).: *URBAN RUNOFF, URBAN HYDROLOGY, LINCOLN(W71-13466
ION, SEDIMENTS, DENITRIFICATION, RUNOFF, UREAS, FERTILIZERS, DRAIN W71-06435
UALITY CONTROL, DOMESTIC WASTES, RUNOFF, WATER SOFTENING, ENVIRONM W72-11186
S, PERCA FLAVESCENS, AMBLOPLITES RUPESTRIS, ESOX LUCIUS, CAMPELOMA W70-08098
ARIA LONGIPES, BRITISH COLUMBIA, RUSSIA, LAMINARIA RODRIGUEZII.: / W73-00428
NAIL CLAMS, UNIONIDS, ORCONECTES RUSTICUS, LIRCEUS LINEATUS, GAMMA W72-06526
IOISOT/ PLUTONIUM RADIOISOTOPES, RUTHENIUM RADIOISOTOPES, IRON RAD W72-04463
OISOTOPES, CESIUM RADIOISOTOPES, RUTHENIUM RADIOISOTOPES, CERIUM R W71-09188
ES, SPIROG/ *MACROINVERTEBRATES, RUTHENIUM-106, CESIUM RADIOISOTOP W70-06668
URNUS LUCIDUS, BELGIUM.: RUTHENIUM, FIXATION, TISSUES, ALB W71-09168
IRON, CHROMIUM, COPPER, NICKEL, RUTHENIUM, CERIUM, LINCONIUM.: /, W71-13243
IOGRAMMA BROCKMANNI, AMPHIPLEURA RUTILANS, THALASSIOSIRA, THALASSI W73-00853
SUM, *NOVA SCOTIA, *ST. MARGARET' S BAY.: / DIGITATA, AGARUM CRIBRO W73-00366
LAMINARIA DIGITATA, ST. MARGARET' S BAY, SCOUDOUC RIVER, HUMIC ACID W72-12166
NTS, POTASSIUM CHLORIDE, SHRINER' S POND(GA).: *LIMITING NUTRIE W72-09159
HERMAN(S. DAK), ENEMY SWIM LAKE(S. DAK).: LAKE W71-02880
: LAKE HERMAN(S. DAK), ENEMY SWIM LAKE(S. DAK). W71-02880
ZONE, MACROINVE/ *OIL SPILLS, S. S. WITWATER, GALETA ISLAND, CANAL W72-07911
ODOR-PRODUCING ALGAE, SOUTHWEST U. S., TEMPERATURE, OXYGEN, LIGHT IN W72-10613
AL ZONE, MACROINVE/ *OIL SPILLS, S.S. WITWATER, GALETA ISLAND, CAN W72-07911
DNESS, ALKALINITY, CONDUCTIVITY, S: /DROGEN ION CONCENTRATION, HAR W72-10875
IES, CULTURES, SAMPLING, HAWAII, S: /OF POLLUTANTS, OCEANS, ESTUAR W73-00361
OBLIQUUS.: SAALE RIVER, GERMANY, SCENEDESMUS W72-14314
UEBEC), MAGDALEN ISLAND(QUEBEC), SABLE ISLAND(NOVA SCOTIA), ST PIE W71-10066
*ENGLAND, *LAMINARIA SACCHARINA.: W71-09795
AN(SPAIN), LAMINARIA OCHROLEUCA, SACCORHIZA POLYSCHIDES.: / DE ALD W72-04766
LE LA/ SAN PABLO ESTUARY(CALIF), SACRAMENTO RIVER(CALIF), SEARSVIL W70-10405
MENT, STREAM GAGES, GROUNDWATER, SAFE YIELD, ALGAE, MINERALOGY, TU W72-05594
CONCENTRATION, RADIATION SAFETY, RADIATION CONTROL.: W72-07841
RS, HYPOLIMNION, VOLUME, MIXING, SAFETY, WATER COOLING, ANAEROBIC W70-09595
FFECTS, *ORGANIC MATTER, *OXYGEN SAG, *IMPAIRED WATER QUALITY, *ES W70-09189
, MESTROPHY, OLIGOTROPHY, OXYGEN SAG, AQUATIC LIFE, BIOASSAY, CLAS W73-00002
POSAL, PESTICIDE REMOVAL, OXYGEN SAG, DREDGING, STATE GOVERNMENTS, W72-01659
FOULING, TOXICITY, ODOR, OXYGEN SAG, TURBIDITY, WATER QUALITY, WA W71-09958
GEN, OXYGEN REQUIREMENTS, OXYGEN SAG, WASTE ASSIMILATIVE CAPACITY, W72-04279
A, BALANUS, BARNACLES, EPIFAUNA, SAGARTIA, ANEMONE, MOLGULA, TUNIC W72-07907
, ANKISTRODESMUS, TRACHELOMONAS, SAGINAW BAY, ORGANIC NITROGEN.: / W72-10875
UM BROMIDE, DEXTRAN, OIKOPLEURA, SAGITTA, PTEROPODS, EUPHAUSIIDS, W73-00371
LAKES, GLACIAL DRIFT, NUTRIENTS, SALANITY, ALGAE, DIATOMS.: /OTA, W71-13149
A, RAMPHOCORIXA, SIGARA, BOENOA, SALDULA, CRYNELLUS, CHEUMATOPSYCH W70-06655
VUM, PHOTOAUTOTROPHS, CHROMONAS SALINA.: /OPHYCEA, PRYMNESIUM PAR W70-07280
ON(PRYMNESIUM PARVUM)(CHROOMONAS SALINA), GLYCEROL POLLUTION.: /KT W70-07842
: CHELEX 100, ARTEMIA SALINA, BIO-RAD, CYCLOTELLA NANA. W72-04708
ENETRATION, AMMONIA, PHOSPHATES, SALINE WATER.: /PERATURE, LIGHT P W72-05469
UCTIVITY, *FRESHWATER, *MARSHES, SALINE WATER, BIOMASS, NUTRIENTS, W73-01089
LIGHT INTENSITY, TIDAL EFFECTS, SALINITY, HEATED WATER, BIOMASS, W73-00853
R POWERPLANTS, WATER PROPERTIES, SALINITY, WATER TEMPERATURE, OXYG W73-02029
MARINE ANIMALS, ALASKA, BIOMASS, SALINITY, BENTHIC FAUNA, ASBESTOS W73-00374
FFECTS, MARINE ALGAE, SEA WATER, SALINITY, SAMPLING, CORES, DISTRI W73-00367
BUTION PATTERNS, MARINE ANIMALS, SALINITY, SAMPLING, SEA WATER, BO W73-00296
IC MICROORGANISMS, MARINE ALGAE, SALINITY, MIGRATION, TIDES, LOW W W72-14317
TROL, ALGICIDES, COPPER SULFATE, SALINITY, HARDNESS(WATER), CHELAT W72-14694
IS, NUTRIENTS, LIMITING FACTORS, SALINITY, CARBON RADIOISOTOPES, S W73-00431
STUARINE ENVIRONMENT, SEA WATER, SALINITY, WATER TEMPERATURE, CHEM W72-08141
KTON, ATLANTIC OCEAN, SEA WATER, SALINITY, PHENOLS, WATER TEMPERAT W72-07710
COMPOUNDS, PRIMARY PRODUCTIVITY, SALINITY, PROFILES, PHOSPHATES, N W72-07892
HOLOGY, PLANT GROWTH REGULATORS, SALINITY, PLANT POPUL: /PLANT PAT W70-09429
YNTHESIS, TEMPERATURE, SAMPLING, SALINITY, DISSOLVED OXYGEN, NUTRI W71-07875
IS, DATA COLLECTIONS, NUTRIENTS, SALINITY, SEDIMENTS, EUTROPHICATI W71-10577
LANKTON, ALGAE, LIGHT INTENSITY, SALINITY, NUTRIENT REQUIREMENTS, W71-10083

CONCENTRATION, NITRITE, AMMONIA, SALINITY, AQUATIC ALGAE, ZOOPLANK W71-11876
S, SAMPLING, INCUBATION, RUNOFF, SALINITY, DEPTH, ABSORPTION, ALGA W72-09122
MS, MOLLUSKS, WATER TEMPERATURE, SALINITY, SEA WATER, WATER QUALIT W72-09092
AY, CYANOPHYTA, *FLORIDA, CORAL, SALINITY, CURRENTS(WATER), MOLLUS W72-11876
PERATURE, SEASONAL, IONS, ALGAE, SALINITY, PHYTOPLANKTON, THERMAL W72-01363
S, ALGAE, NITRATES, WATER REUSE, SALINITY, PESTICIDES, LIVESTOCK, W72-00846
L MODELS, ESTUARINE ENVIRONMENT, SALINITY, CHEMICAL ANALYSIS, BIOM W72-12941
ALS, TRACE ELEMENTS, PESTICIDES, SALINITY, NUTRIENTS, ALGAE, AQUAT W72-12190
ARSHES, *MARINE ALGAE, CULTURES, SALINITY, SALT MARSHES, SYSTEMATI W72-12633
ADOPHORA GLOMERATA, HETEROCYPRIS SALINUS, GASTEROSTEUS ACULEATUS, W71-08026
LAKE ARCO, LAKE DEMING, LOWER LA SALLE LAKE, LAKE LONG, LAKE SQUAW W72-07145
ES(MINN), COMMUNITY ACTION, LAKE SALLIE(MINN), LAKE MELISSA(MINN), W71-03012
IMALS, AQUATIC LIFE, ECOSYSTEMS, SALMONIDS,: /SH, CLAMS, MARINE AN W72-07907
X LUCIUS, ALKYL MERCURY, EAGLES, SALMO GAIRDNERII, IRIS PSEUDACORU W72-05952
FLUVIATILIS, SALVELINUS ALPINUS, SALMO TRUTTA, VENDACE, SALMON TRO W72-12729
ALPINUS, SALMO TRUTTA, VENDACE, SALMON TROUT, LOTA LOTA, LAKE MAL W72-12729
SUNFISHES, SNAILS, EELS, CHINOOK SALMON, BASS, TROUT, MINNOWS, CRA W71-07417
S, APHANIZOMENON, CYANIDES, COHO SALMON, SHEEPSHEAD, CHUBS, NORTHE W72-14282
R TEMPERATURE, ALGAE, BULLHEADS, SALMON, SHELLFISH, CLAMS, MARINE W72-07907
CTS, *NUCLEAR EXPLOSIONS, BIRDS, SALMON, STICKLEBACKS, CRABS, OTTE W72-00960
OPLANKTON, LARVAE, EGGS, CHINOOK SALMON, STRIPED BASS, MINNOWS, FI W71-11517
SULFATES, PERCOLATION, HERRING, SALMON, TROUT, TOXICITY, WASHINGT W72-05421
SONE, HORMONES, CHLORAMPHENICOL, SALMONELLA ENTEROCOLITIS, SHIGELL W72-13800
FLUORESCENT ANTIBODY TECHNIQUES, SALMONELLA TYPHIMURIUM, AMPICILLI W72-13800
YPHIMURIUM, AMPICILLIN, VIBRIOS, SALMONELLA TYPHOSA, TYPHOID, COLI W72-13800
LORELLA, SULFUR BACTERIA, ALGAE, SALMONELLA, E. COLI.: /CILLUS, CH W72-12565
PLANKTON, BACTERIA, ZOOPLANKTON, SALMONIDS, SEASON, OXYGEN, HYDROG W71-00117
RIFUGATION, SAMPLE PRESERVATION, SALPA, LUDOX AM, SUCROSE, DENSITY W73-00371
ATOPODA, POLYCHAETES, AMPHIOXUS, SALPS, BALAN: /OSCILLATORIA, STOM W73-00935
ATUS, AMMONIA BECCARII, AMMOTIUM SALSUM, ELPHIDIUM ADVENUM, ELPHID W73-00367
AR RADIATION, WATER TEMPERATURE, SALT BALANCE, DISSOLVED OXYGEN, A W72-02274
RICHE, ANABAENA / BAY OF QUINTE, SALT CREEK, CHARA, ELODEA, CALLIT W73-00840
LYSIS, ALGAE BLOOMS, UTAH, GREAT SALT LAKE, ORGANIC COMPOUNDS, BRI W71-12223
TOXINS, FISH TOXINS, SEDIMENTS, SALT MARSHES, ECOLOGY, PESTICIDES W72-10678
ARINE ALGAE, CULTURES, SALINITY, SALT MARSHES, SYSTEMATICS, *VIRGI W72-12633
RIVER, HURON RIVER, BELL RIVER, SALT RIVER, MICROCYSTIS, APHANIZO W72-14282
CTS, CHLOROPHYTA, MARINE ALGAE , SALT TOLERANCE, CRUSTACEANS, MARI W73-00855
SON(WIS), LAKE WASHINGTON(WASH)/ SALT VALLEY RESERVOIRS(NEB), MADI W72-04759
, RADIOACTIVITY, CARBON DIOXIDE, SALTS, MARINE ALGAE.: /M, TRACERS W72-12634
PHOSPHATES, NITRATES, NUTRIENTS, SALTS, TRACE ELEMENTS, LIGHT INTE W72-12739
ESOX LUCIUS, PERCA FLUVIATILIS, SALVELINUS ALPINUS, SALMO TRUTTA, W72-12729
ATROMACULATU/ CHUBS, KINGFISHES, SALVELINUS FONTINALIS, SEMOTILUS W72-12930
, SAWDUST, EICHHORNIA CRASSIPES, SALVINIA ROTUNDIFOLIA, MILORGANIT W71-01488
ONE, DIFFUSION COEFFICIENT, LAKE SAMMAMISH(WASH).: /NG, EUPHOTIC Z W71-03034
CTION LIMITS, CHEMICAL RECOVERY, SAMPLE PREPARATION.: / BONE, DETE W72-05965
EMOTILUS ATROMACULATUS, CLEANUP, SAMPLE PRESERVATION, CAMBARUS BAR W72-12930
NDEX, CHLOROPHYLL A, CARTENOIDS, SAMPLE PREPARATION, MOUNTAIN LAKE W73-00916
DIGITATA, PORPHYRA UMBILICALIS, SAMPLE PREPARATION, CRUDE OIL.: / W73-01074
OPY, CULTURE MEDIA, FATTY ACIDS, SAMPLE PREPARATION, NATURAL ORGAN W73-01066
RBON, MACROPHYTES, SARGASSO SEA, SAMPLE PREPARATION, SPERMATOPHYTE W73-02100
EPIPHYTES, SAMPLE PRESERVATION, SAMPLE PREPARATION, SUBLITTORAL, W73-00367
MINIFERA, SUBSTRATES, EPIPHYTES, SAMPLE PRESERVATION, SAMPLE PREPA W73-00367
ASSIMILATION, KJEDAHL PROCEDURE, SAMPLE PREPARATION.: /CTROSCOPY, W73-00379
FUCUS VESICULOSUS, VESICULATION, SAMPLE PREPARATION, BRANCHING.: / W73-00369
ATION, ISOPYCNIC CENTRIFUGATION, SAMPLE PRESERVATION, SALPA, LUDOX W73-00371
DUCTIVITY, LIQUID SCINTILLATION, SAMPLE PREPARATION, ZOSTREA, ULVA W72-14317
AUTORADIOGRAPHY, GLUTARALDEHYDE, SAMPLE PREPARATION, COUNTING, SAM W73-00273
E, SAMPLE PREPARATION, COUNTING, SAMPLE PRESERVATION, FLAGELLATES, W73-00273
RFERENCE, INTERLABORATORY STUDI/ SAMPLE PREPARATION, CHEMICAL INTE W73-00286
FLAGIC, *TEMPERATURE TOLERANCES, SAMPLE PRESERVATION, COUNTING CHA W72-07899
ANSEN BOTTLE, SPECIES DIVERSITY, SAMPLE PRESERVATION, COUNTING.: / W72-07663
UNTING, CYCLOTELLA, OSCILLATORI/ SAMPLE PRESERVATION, FORMALIN, CO W72-10875
C RIVER, HUMIC ACIDS, OXIMATION, SAMPLE PREPARATION.: /AY, SCCUDOU W72-12166
POLYSTYRENE, SLIDE PREPARATION, SAMPLE PREPARATION.: /E, TOLUENE, W72-11738
AKE LIVINGSTON, AUTORADIOGRAPHY, SAMPLE PREPARATION, SCINTILLATION W72-09239
ERS, ESTERS, CHLORELLA VULGARIS, SAMPLE PREPARATION.: / METHYL EST W72-09365
OSPHATE, *DEOXYRIBONUCLEIC ACID, SAMPLE PRESERVATION, COULTER COUN W71-11562
REEK, VAN DORN SAMPLER, KEMMERER SAMPLER, CHLAMYDOMONAS EPIPHYTICA W72-09111
RIVER, NINEMILE CREEK, VAN DORN SAMPLER, KEMMERER SAMPLER, CHLAMY W72-09111
OSSOSOMA, PHYSA GYRINA, KEMMERER SAMPLER, SECCHI *YARDS CREEK(NJ), W72-14280
ATER, BECKMAN ANALYZER, VAN DORN SAMPLERS, C-14, BIOCARBONATES.: / W72-09122
IOLOGICAL MATERIALS, *BIOLOGICAL SAMPLES, *SAMPLE PRESERVATION, CH W73-01676
*BIOTRANSFORMATION, BIOLOGICAL SAMPLES, FATE OF POLLUTANTS, META W73-00361
OR, PRECONCENTRATION, BIOLOGICAL SAMPLES, LIQUID-LIQUID EXTRACTION W73-02105
OGICAL MAGNIFICATION, BIOLOGICAL SAMPLES, LIVER, URINE, OEDOGONIUM W72-07703
ON, CAMBARUS BARTONI, BIOLOGICAL SAMPLES, MERGUS MERGANSER, MEGACE W72-12930
HRANAGAT LAKE, GAMM/ *BIOLOGICAL SAMPLES, SR-90, K-40, CS-137, *PA W73-01673
UM, DETECTION LIMITS, BIOLOGICAL SAMPLES, SARGASSO WEED, STARFISH, W72-05965
MATTER, MACROPHYTES, BIOLOGICAL SAMPLES: /C-14, DISSOLVED ORGANIC W73-02095
RMS, MATHEMATICAL MODELS, ALGAE, SAMPLIN: /HATES, NITROGEN, COLIFO W71-13356
C MATTER, ANALYTICAL TECHNIQUES, SAMPLING, MASSACHUSETTS, POLLUTAN W71-11498
LYSIS, ALGAE, CHEMICAL ANALYSIS, SAMPLING, ANALYTICAL TECHNIQUES, W71-11234
OURCES, WATER POLLUTION EFFECTS, SAMPLING, TEMPERATURE, COPEPODS, W71-11008
, *EUTROPHICATION, GROWTH RATES, SAMPLING, LAKE MICHIGAN, MEASUREM W71-11033
IRON, MANGANESE, COPPER, ALGAE, SAMPLING, CHLOROPHYTA.: /TASSIUM, W71-11001
ERS, NITROGEN, SEA WATER, WINDS, SAMPLING, INCUBATION, RUNOFF, SAL W72-09122
NE LAKES, *EUTROPHICATION, WATER SAMPLING, WATER POLLUTION EFFECTS W72-09111
SCUBA / *MARINE PLANTS, *ALASKA, SAMPLING, *SYSTEMATICS, BENTHOS, W72-09120
PORAL DISTRIBUTION, CHLOROPHYLL, SAMPLING, MATHEMATICAL STUDIES, A W72-11719
ICHIGAN, SURFACE WATERS, COASTS, SAMPLING, BIOLOGICAL COMMUNITIES, W72-10875

UTION, POLLUTANT IDENTIFICATION, SAMPLING, BIOASSAY, ENVIRONMENT, W72-10678
, GRASSES, LEAVES, FOREST SOILS, SAMPLING, MUD, PHYTOPLANKTON, FLU W72-01784
CREATION, *TREATMENT FACILITIES, SAMPLING, MONITORING, AUTOMATIC C W72-00042
NTS, DEGRADATION(DECOMPOSITION), SAMPLING, BENTHIC FAUNA, SEDIMENT W72-00847
OPHYTA, CHRYSOPHYTA, PYRROPHYTA, SAMPLING, ECOLOGY, WATER TEMPERAT W72-03543
ATER POLLUTION EFFECTS, SURVEYS, SAMPLING, DATA COLLECTIONS, INDUS W73-01972
IGHT INTENSITY, AIR TEMPERATURE, SAMPLING, AQUATIC ALGAE, PHYTOPLA W73-01456
FISH, AQUATIC ALGAE, EVALUATION, SAMPLING, CARP, POTASSIUM RADIOIS W73-01673
P, *PRIMARY PRODUCTIVITY, LAKES, SAMPLING, RESERVOIRS, TRIBUTARIES W73-01435
ANTS, CHEMICAL ANALYSIS, LEAVES, SAMPLING, PHYTOPLANKTON, IRON, MA W73-01089
AKES, AQUATIC PLANTS, AUSTRALIA, SAMPLING, MICROSCOPY, ECOLOGICAL W73-01094
GRAPHY, SCUBA DIVING, SEA WATER, SAMPLING, ENVIRONMENTAL EFFECTS, W73-00932
S, SPECTROSCOPY, TRACE ELEMENTS, SAMPLING, WATER ANALYSIS, SEA WAT W73-00832
*AQUATIC ALGAE, *SOIL BACTERIA, SAMPLING, CULTURES, ECOLOGY, SOIL W73-00854
MS, SPECTROPHOTOMETRY, VIRGINIA, SAMPLING, SEPARATION TECHNIQUES, W73-00916
FFECTS, REEFS, LABORATORY TESTS, SAMPLING, PHAEOPHYTA.: /NMENTAL E W73-00838
TEMPERATURE, CHEMICAL ANALYSIS, SAMPLING, ECOLOGY, PHYTOP: /WATER W72-07710
*HYDROBIOLOGY, ENTERIC BACTERIA, SAMPLING, ZOOPLANKTON, PHYTOPLANK W72-07896
*WATER TEMPERATURE, SYSTEMATICS, SAMPLING, NETS, CENTRIFUGATION, M W72-07899
NOPHYTA, *WATER ANALYSIS, LAKES, SAMPLING, DIATOMS, TURBIDITY, WAT W72-07890
NORTH CAROLINA, SOUTH CAROLINA, SAMPLING, MICROSCOPY, ELECTRON MI W72-07663
ENDESCMUS, CYANOPHYTA, BIOASSAY, SAMPLING, OXIDATION LAGOONS, EUGL W72-07661
ATIAL DISTRIBUTION, ENVIRONMENT, SAMPLING, CHEMICAL ANALYSIS, PLAN W72-08560
OIL, METALS, THERMAL POLLUTION, SAMPLING, BACTERIA, TOXICITY, BEH W72-08790
FFLUENTS, CALIFORNIA, SEDIMENTS, SAMPLING, CHLOROPHYLL, SEDIMENT T W72-09092
TS, OCEANS, ESTUARIES, CULTURES, SAMPLING, HAWAII, S: /OF POLLUTAN W73-00361
, WATER TEMPERATURE, PHAEOPHYTA, SAMPLING, SCUBA DIVING, ECOLOGY, W73-00366
RINE ALGAE, SEA WATER, SALINITY, SAMPLING, CORES, DISTRIBUTION PAT W73-00367
PLANTS, EUTROPHICATION, OCEANS, SAMPLING, SEPARATION TECHNIQUES, W73-00430
TERNS, MARINE ANIMALS, SALINITY, SAMPLING, SEA WATER, BODIES OF WA W73-00296
SAY, CHLORELLA, NUTRIENTS, WATER SAMPLING, ORGANIC MATTER, WATER P W73-00234
TIUM, MARINE PLANTS, MONITORING, SAMPLING, WATER ANALYSIS, SEA WAT W73-00831
STEMATICS, *SEASONAL, SEA WATER, SAMPLING, BIOMASS, DIATOMS, POLLU W73-00442
IS, BIOMASS, METHODOLOGY, REEFS, SAMPLING, OXYGEN, TEMPERATURE, LI W72-06283
, CURRENTS(WATER), WAVES(WATER), SAMPLING, BENTHOS, GRAZING, GASTR W72-04766
RMS, METABOLISM, CARBON, SULFUR, SAMPLING, WATER ANALYSIS, ALGAE, W72-04743
NOLOGY, *FLORIDA, *LAKES, PONDS, SAMPLING, THERMAL STRATIFICATION, W72-04269
ON, ALKALINITY, HARDNESS(WATER), SAMPLING, DIATOMS, NITROGEN, BACT W71-03028
EW HAMPSHIRE, COPEPODS, DAPHNIA, SAMPLING, DEPTH, CYANOPHYTA, CHRY W71-03014
ESTIGATIONS, PATH OF POLLUTANTS, SAMPLING, WATER ANALYSIS, ALGAE, W70-08248
CATFISHES, BASS, PERCHES, PIKES, SAMPLING, SNAILS, ALGAE, DIATOMS, W70-08098
ESTIGATIONS, PATH OF POLLUTANTS, SAMPLING, WATER ANALYSIS, ALGAE, W70-08249
ISSOLVED OXYGEN, STRATIFICATION, SAMPLING, MICROORGANISMS, SLUDGE, W71-07382
CENTRATION, ALGAE, REPRODUCTION, SAMPLING, SEASONAL, BOTTOM SEDIME W71-08026
LL, PHOTOSYNTHESIS, TEMPERATURE, SAMPLING, SALINITY, DISSOLVED OXY W71-07875
VORES, ECOSYSTEMS, AQUATIC LIFE, SAMPLING, ALGAE, LABORATORY TESTS W71-10064
CIPAL WASTES, INDUSTRIAL WASTES, SAMPLING, WATER QUALITY CONTROL, W71-08953
PLANTS, *POTABLE WATER, *BIOTA, SAMPLING, MARINE FISH, MARINE ANI W71-09018
ADATION, CHLORINE, CONTACT TIME, SAMPLING, COLIFORMS, BACTERIA, OX W71-07096
OCHEMICAL OXYGEN DEMAND, MIXING, SAMPLING, ORGANIC LOADING.: / *BI W71-07102
DISSOLVED OXYGEN, SEDIMENTATION, SAMPLING, CHEMICAL OXYGEN DEMAND, W71-07097
GAS CHROMATOGRAPHY, BROOK TROUT, SAMPLING, SEPARATION T: /FFECTS, W72-12930
ROPHYTA, SPRING, SUMMER, AUTUMN, SAMPLING, PLANKTON NETS, DIATOMS, W72-12738
PERTIES, CHLOROPHYLL, TURBIDITY, SAMPLING, MATHEMATICAL MODELS, ES W72-12941
ISH BEHAVIOR, CHEMICAL ANALYSIS, SAMPLING, VOLUMETRIC ANALYSIS, AL W72-14280
CYANOPHYTA, PYRROPHYTA, LARVAE, SAMPLING, INVERTEBRATES, *ATLANTI W72-14313
LANT POPULATION, CLASSIFICATION, SAMPLING, CHLOROPHYTA, *CHRYSOPHY W72-12192
ANOL, BACTERIAL DENITRIFICATION, SAN JOAQUIN VALLEY.: /ATION, METH W71-08223
HUMBOLDT BAY, CONNECTICUT RIVER, SAN JOAQUIN RIVER(CALIFORNIA), BI W71-07417
CYCLE, DELAWARE ESTUARY(DEL-NJ), SAN JOAQUIN DELTA(CALIF), STEADY- W71-05390
(CALIF), SEARSVILLE LAKE(CALIF), SAN JOAQUIN RIVER(CALIF), NITZSCH W70-10405
RIVER, CONTRA COSTA POWER PLANT, SAN JOAQUIN RIVER, MORRO BAY POWE W72-07907
RUBBLE CHIMNEY, SAN JOAQUIN VALLEY.: W72-00974
, PARASITES, JORDAN RIVER(MICH), SAN JOAQUIN RIVER(CALIF), RIFLE F W71-11006
NTO RIVER(CALIF), SEARSVILLE LA/ SAN PABLO ESTUARY(CALIF), SACRAME W70-10405
IES, ABALONE, TUNICATES, ARTHRO/ SAN PEDRO(CALIF), EPIBENTHIC SPEC W71-01475
REBAUGH(CALIF), VOLATILE SOLIDS, SANBORN FILTER, BACKWASHING.: FI W70-10174
ASONAL, CADDISFLIES, GASTROPODS, SANDS, CHRYSOPHYTA.: / CARBON, SE W71-11029
ITTORAL, SHORES, TIDES, BEACHES, SANDS, INTERTIDAL AREAS, ANNELIDS W72-07911
ARINE PLANTS, DEPTH, CALIFORNIA, SANDS, SEDIMENTS, PRODUCTIVITY, T W71-01475
LAKES, EUTROPHICATION, LITTORAL, SANDS, SHALLOW WATER, PERIPHYTON, W72-07501
SANDUSKY RIVER(OHIO).: W72-04260
AGE TREATMENT, SEWAGE EFFLUENTS, SANITARY ENGINEERING, SECONDARY T W72-09378
OLLUTION CONTROL, *ENVIRONMENTAL SANITATION, *LEGISLATION, *FEDERA W72-01660
OVERNMENT FINANCE, ENVIRONMENTAL SANITATION, ALGAE, NUTRIENTS, IND W72-01659
POLLUTION SOURCES, ENVIRONMENTAL SANITATION, SECONDARY TREATMENT, W71-09674
, FLAVONOIDS, STEROIDS, TANNINS, SAPONINS.: /C ACTIVITY, ALKALOIDS W72-04275
VER, HYDROBIOLOGY, BACTERIOLOGY, SAPROPHYTES, STURGEON, METALIMNIO W72-04853
LAND, HONSHU ISLAND, OPEN OCEAN, SAPROPHYTES, HETEROTROPHIC BACTER W71-08042
LOGY, *HYDROBIONTS, MACROPHYTES, SAPROPHYTES.: *USSR, *TOXICO W72-09650
LOGY, ICHTHYOFAUNA, MACROPHYTES, SAPROPHYTES, CERIUM.: /S, ICHTHYO W72-09646
, CAPE COD(MASS), CAPE MAY(N J), SARGASSO SEA.: /AMA, CALLAO(PERU) W72-08056
OLITHOPHORES, SILICOFLAGELLATES, SARGASSO SEA, SHELF WATERS, GULF W72-07663
VED ORGANIC CARBON, MACROPHYTES, SARGASSO SEA, SAMPLE PREPARATION, W73-02100
TION LIMITS, BIOLOGICAL SAMPLES, SARGASSO WEED, STARFISH, ASTERIAS W72-05965
S CRISPUS, RHODOTORULA GLUTINIS, SARGASSUM NATANS, ASCOPHYLLUM NOD W72-04748
RONOMIDS, BARNACLES, ARTHROPODS, SARGASSUM CRISPUM, DIGENIA SIMPLE W73-00855
ER, *OILS, *NUTRIENTS, AIRCRAFT, SATELLITES(ARTIFICIAL), PHOSPHORU W72-04742
IAL PHOTOGRAPHY, REMOTE SENSING, SATELLITES(ARTIFICIAL), ALGAE, CH W72-09120
UENT, BIOCHEMICAL OXYGEN DEMAND, SATURATION, OXYGEN, METABOLISM, N W71-13553

S, MAXIMUM SPECIFIC GROWTH RATE,	SATURATION CONSTANT, DILUTION, CH	W72-04788
*OXYGENATION, DISSOLVED OXYGEN,	SATURATION, ALGAE, METHANE BACTER	W71-07106
RACULAMONAS NATANS, MARINIAMONAS	SAUPAULENSIS, ENEIDAMONAS APPLANA	W72-09119
IA, DRY WEIGHT, THERMAL EFFECTS/	SAVANNAH RIVER PLANT, CULTURE MED	W71-02075
LVINIA ROTUNDIFOL/ *CHIRONOMIDS,	SAWDUST, EICHHORNIA CRASSIPES, SA	W71-01488
LABORATORY	SCALE, *TOTAL ORGANIC CARBON.:	W71-07102
P, GILLS, MUSCLE, VISCERA, EYES,	SCALES(FI: /RA ASPPRA, SCIRPUS SP	W73-01673
VANADIUM, LANTHANUM,	SCANDIUM.:	W72-01101
MACRONUTRIENTS, MICRONUTRIENTS,	SCANDIUM, RUBIDIUM, BROMINE, ACTI	W73-01434
RS, MICROWAVE RADIOMETRY, VIDEO,	SCANNERS.: /, LEAD, MERCURY, LASE	W72-04742
BINDERANUS, *MELOSIRA BINDERANA,	SCANNING ELECTRON MICROSCOPY, TRA	W73-00432
T, WASTE WATER TREATMENT, PONDS,	SCENDESOMUS, CYANOPHYTA, BIOASSAY	W72-07661
CTICIDES, BIOMASS, PRODUCTIVITY,	SCENE: /OXICITY, HERBICIDES, INSE	W71-10566
NOPHYTA, CHLORELLA, CHLOROPHYTA,	SCENEDESMUS, EUGLENA, DISSOLVED O	W71-11519
TOLOGICAL STUDIES, PLANT GROWTH,	SCENEDESMUS.: /BORATORY TESTS, CY	W71-12858
ENA, CHLAMYDOMONAS, CHLOROPHYTA,	SCENEDESMUS, CYANOPHYTA, MOLLUSKS	W72-00155
PYRUVATE, LACTATE.:	SCENEDESMUS QUADRICAUDA, ACETATE,	W72-00838
HARVESTING, */ *ALGAE STRIPPING,	SCENEDESMUS, ALGAL GROWTH, ALGAL	W72-02975
, AQUATIC PLANTS, AQUATIC ALGAE,	SCENEDESMUS, RADIOISOTOPES, STRON	W72-09649
TIVITY, MICROSCOPY, CHLOROPHYTA,	SCENEDESMUS, *CYANOPHYTA, CHLOREL	W72-09239
NOUS CULTURES, SCENEDESMUS SOLI,	SCENEDESMUS QUADRICAUDA, MOUGEOTI	W72-07144
DRICAUDA/ *SYNCHRONOUS CULTURES,	SCENEDESMUS SOLI, SCENEDESMUS QUA	W72-07144
CYCLES, CULTURES, WATER QUALITY,	SCENEDESMUS, LABORATORY EQUIPMENT	W72-07144
PYRENOIDOSA, CHLORELLA VULGARIS,	SCENEDESMUS OBLIQUUS, ALCALIGENES	W72-07664
GEN, HYDROGEN ION CONCENTRATION,	SCENEDESMUS, BIOMASS.: /OLVED OXY	W72-08049
OLOGICAL COMMUNITIES, NUTRIENTS,	SCENEDESMUS, CHRYSOPHYTA, SEASONA	W72-14797
S,/ *GRAZING, *PROTOZOA, *ALGAE,	SCENEDESMUS, EUTROPHICATION, POND	W72-14796
SAALE RIVER, GERMANY,	SCENEDESMUS OBLIQUUS.:	W72-14314
*PRIMARY PRODUCTIVITY, BACTERIA,	SCENEDESMUS, GROWTH RATES, CULTUR	W72-14314
, ALGAE, DIATOMS, CHLAMYDOMONAS,	SCENEDESMUS, BIO-ASSAY, RESISTANC	W72-14706
AEROBIC CONDITIONS, CHLOROPHYTA,	SCENEDESMUS, AQUATIC ALGAE, AQUAT	W73-01066
OPHYTA, CYANOPHYTA, CHRYSOPHYTA,	SCENEDESMUS, CHLAMYDOMONAS.: /LEN	W73-01627
KALINITY, CONDUCTIVITY, CALCIUM,	SCENEDESMUS, WATER HYACINTH, LIGH	W72-04269
PHYTA, DIATOMS, DINOFLAGELLATES,	SCENEDESMUS, ROTIFERS, CRUSTACEAN	W72-05968
M, HYDROGEN, EUGLENA, CHLORELLA,	SCENEDESMUS, CHLAMYDOMONAS, CYANO	W71-09172
ND, PHOTOSYNTHESIS, RESPIRATION,	SCENEDESMUS, WASTE WATER TREATMEN	W71-00139
EN ION CONCENTRATION, CHLORELLA,	SCENEDESMUS, WASTE WATER TREATMEN	W71-00138
, HARVESTING, PONDS, CALIFORNIA,	SCENEDESMUS, SUSPENDED LOAD, NITR	W70-10174
TIONS, CHLORELLA, CHLAMYDOMONAS,	SCENEDESMUS, EFFLUENTS, OXIDATION	W70-10405
CTERIA, CYANOPHYTA, CHLOROPHYTA,	SCENEDESMUS, ALGAL TOXINS, ANTIBI	W71-05155
, CARBON DIOXIDE, CHLAMYDOMONAS,	SCENEDESMUS, CHLORELLA, SILICA, L	W72-12544
ODA, *FLORIDA, STREAMS, ISOPODS,	SCENEDESMUS, AQUATIC HABITATS, CL	W72-12744
DIATOMS, FOOD CHAINS, ANABAENA,	SCENEDESMUS, SEASONAL, DINOFLAGEL	W72-12738
TRACE ELEMENTS, LIGHT INTENSITY,	SCENEDESMUS, CHLORELLA, GROWTH RA	W72-12739
NASTROIDES, *STAURASTRUM PINGUE,	SCENEDESMUS QUADRICAUDA, CHLORELL	W72-12739
CYANOPHYCEAE, OSCILLATORIACEAE,	SCHIZOTHRIX CALCICOLA, ELKTON PON	W72-04736
CHLOROPHYLL A,	SCHIZOTHRIX CALCICOLA.:	W73-01080
RTHROBACTE/ NEOCHLORIS AQUATICA,	SCHIZOTHRIX CALCICOLA, BORATES, A	W73-00854
AL DENSITY, ACETYLENE REDUCTION,	SCHIZOTHRIX CALCIOCOLA, CLOSTERIU	W72-10883
INARIA, THALASSIOPHYLLUM, FUCUS,	SCHIZYMENIA, RHODOMELA, RHODOMENI	W72-09120
COREGONS, LYNGBYA LIM/ *BAVARIA,	SCHLIERSEE, TEGERNSEE, ANABAENA,	W71-00117
STROM RIVER, SPARGANIUM ERECTUM,	SCHOENOPLECTUS LACUSTRIS, ACORUS	W73-00286
NSKIOLDII, THALASSIOSIRA ROTULA,	SCHRODERELLA DELICATULA, DACTYLIO	W72-03543
, PANDORINA MORUM, SPHAEROCYSTIS	SCHROETERI, ULOTHRIX SPP, MICROTH	W72-09111
RADIOGRAPHY, SAMPLE PREPARATION,	SCINTILLATION COUNTING.: /N, AUTO	W72-09239
LLA VULGARIS, NEUROSPORA CRASSA,	SCINTILLATION COUNTING, GLUCOSE,	W72-14326
, ZO/ *C-14 PRODUCTIVITY, LIQUID	SCINTILLATION, SAMPLE PREPARATION	W72-14317
RBIGNIANA, PAPER CHROMATOGRAPHY,	SCINTILLATION COUNTING, CA-45.: /	W72-12634
CALDARI/ CHEMICAL INTERFERENCE,	SCINTILLATION COUNTING, CYANIDIUM	W72-14301
ON, CAULOBACTER, NAJAS FLEXILIS,	SCIRPUS ACUTUS, C-14, DISSOLVED O	W73-02095
METABOLISM, LAWRENCE LAKE(MICH),	SCIRPUS ACUTUS, NAJAS FLEXILIS.: /	W72-03216
S, JUNCUS SPP, ATRIPLEX HASTATA,	SCIRPUS SPP, TYPHA SPP, ZIZANIA A	W73-01089
RIS MONTEVIDENSES, CHARA ASPPRA,	SCIRPUS SPP, GILLS, MUSCLE, VISCE	W73-01673
HORE CONSTRUCTION, DELAWARE BAY,	SCOLOPLOS ARMIGER, ALIMITOS BAY,	W73-02029
PHORA OLIVACEA, LENZITES TRABEA,	SCOPULARIOPSIS BREVICAULIS, CLADO	W72-13813
SLAND(QUEBEC), SABLE ISLAND(NOVA	SCOTIA), ST PIERRE(CANADA), MIQUE	W71-10066
IGITATA, AGARUM CRIBROSUM, *NOVA	SCOTIA, *ST. MARGARET'S BAY.: / D	W73-00366
MELANOSIP/ *EASTERN CANADA, NOVA	SCOTIA, ISTHMOPLEA SPHAEROPHORA,	W72-01370
INHIBITION, HOSTS, MYXOPHYCEAE,	SCOTLAND.: /A, ENGLAND, METABOLIC	W72-05457
LAND), LUXURY UPTAKE, LOCH LEVEN(	SCOTLAND), LAKE WINDERMERE(ENGLAN	W71-13183
IA DIGITATA, ST. MARGARET'S BAY,	SCOUDOUC RIVER, HUMIC ACIDS, OXIM	W72-12166
*MICRO	SCREENING.:	W71-09524
OGICAL PROPERTIES, PHYTOCHEMICAL	SCREENINGS, NYMPHAEA TUBEROSA, NU	W72-04275
SAMPLING, *SYSTEMATICS, BENTHOS,	SCUBA DIVING, DEPTH, DISTRIBUTION	W72-09120
MPERATURE, PHAEOPHYTA, SAMPLING,	SCUBA DIVING, ECOLOGY, STANDING C	W73-00366
OPHYTA, PHAEOPHYTA, PHOTOGRAPHY,	SCUBA DIVING, SEA WATER, SAMPLING	W73-00932
ES, SPECIES DIVERSITY, PORIFERA,	SEA CUCUMBERS, STARFISH, CHITONS,	W73-00932
INIANS, EUCAMPIA ZOODIACUS, CHA/	SEA OF JAPAN, *AMURSKY BAY, PERID	W73-00442
IDOCARPA, ENHYDRA LUTRIS NEREIS,	SEA OTTER, MACROCYSTIS PYRIFERA,	W73-00932
, SPONGES, TUNICATES, BRYOZOANS,	SEA TURTLES.: /FAUNA, TURBELLARIA	W72-07911
OPLANKTON, NITROGEN, ABSORPTION,	SEA WATER.: / *MARINE ALGAE, PHYT	W72-07497
MENTS, SAMPLING, WATER ANALYSIS,	SEA WATER.: /CTROSCOPY, TRACE ELE	W73-00832
TON, OREGON, CALIFORNIA, ALASKA,	SEA WATER.: /NTS, CANADA, WASHING	W73-00428
ORINATED HYDROCARBON PESTICIDES,	SEA WATER, *METABOLISM, BIODEGRAD	W73-00361
ER), MAINE, STATISTICAL METHODS,	SEA WATER, COASTS, MARINE ALGAE.:	W73-00369
ATLANTIC OCEAN, SEASONAL, BAYS,	SEA WATER, ORGANIC MATTER, MORTAL	W73-00366
YSIOLOGY, CHRYSOPHYTA, CULTURES,	SEA WATER, NUTRIENT REQUIREMENTS,	W73-00380
PS, TIDAL EFFECTS, MARINE ALGAE,	SEA WATER, SALINITY, SAMPLING, CO	W73-00367
ATER POLLUTION EFFECTS, ECOLOGY,	SEA WATER, LABORATORY EQUIPMENT,	W73-00348
ORING, SAMPLING, WATER ANALYSIS,	SEA WATER, STABLE ISOTOPES, RADIO	W73-00831

LEAR WASTES, PATH OF POLLUTANTS, SEA WATER, WATER ANALYSIS, RADIOA W73-00817
ULTURES, *NUTRIENT REQUIREMENTS, SEA WATER, NITRATES, PHOSPHORUS, W73-00441
FFECTS, *SYSTEMATICS, *SEASONAL, SEA WATER, SAMPLING, BIOMASS, DIA W73-00442
TIVITY, *SHOALS, *BENTHIC FLORA, SEA WATER, SEDIMENTS, LIGHT INTEN W72-14317
HYTA, *SYSTEMATICS, PHOTOGRAPHY, SEA WATER, *NORTH CAROLINA, MICRO W73-00284
INE ANIMALS, SALINITY, SAMPLING, SEA WATER, BODIES OF WATER, PYRRO W73-00296
HYTA, PHOTOGRAPHY, SCUBA DIVING, SEA WATER, SAMPLING, ENVIRONMENTA W73-00932
NERITIC, ESTUARINE ENVIRONMENT, SEA WATER, DIATOMS, COPEPODS, DIN W73-00935
HALOPHILIC ANIMALS, *HALOPHYTES, SEA WATER, MARINE ANIMALS, BIOLOG W73-00855
POLLUTION EFFECTS, *CALIFORNIA, SEA WATER, WATER ANALYSIS, MARINE W73-02105
, BIOASSAY, CELLULOSE, DIALYSIS, SEA WATER, ORGANIC ACIDS, CHRYSOP W73-00942
R POLLUTION SOURCES, OIL SPILLS, SEA WATER, MARINE PLANTS, PHAEOPH W73-01074
MAGNESIUM, PHOSPHATES, NITROGEN, SEA WATER, FLUOROMETRY, EQUIPMENT W72-07660
LANKTON, *SYSTEMATICS, *DIATOMS, SEA WATER, SEASONAL, FLUCTUATIONS W72-07663
TERIA, *TOXICITY, *MARINE ALGAE, SEA WATER, CHLOROPHYTA, CHLAMYDOM W72-08051
EMATICS, *ESTUARINE ENVIRONMENT, SEA WATER, SALINITY, WATER TEMPER W72-08141
ATION, PLANKTON, ATLANTIC OCEAN, SEA WATER, SALINITY, PHENOLS, WAT W72-07710
KTON, ALGAE, CULTURES, *FLORIDA, SEA WATER, CHLOROPHYLL, NITRATES, W72-09102
KS, WATER TEMPERATURE, SALINITY, SEA WATER, WATER QUALITY, ON-SITE W72-09092
IORDS, SURFACE WATERS, NITROGEN, SEA WATER, WINDS, SAMPLING, INCUB W72-09122
CTIVITY TECHNIQUES, CHLOROPHYLL, SEA WATER, IONS, NUTRIENT REQUIRE W72-09103
NITRIFICATION, BOTTOM SEDIMENTS, SEA WATER, FRESHWATER, AMMONIA, E W72-11563
ELLA, GROWTH RATES, TEMPERATURE, SEA WATER, PHOSPHATES, PHOSPHORUS W71-10083
SIS, HYDROGEN ION CONCENTRATION, SEA WATER, PHOTOSYNTHESIS, PHYTOP W71-11008
L WASTES, ALGAE, AQUATIC PLANTS, SEA WATER, TRACE ELEMENTS, TOXICI W72-05952
OPHYTA, INHIBITORS, ANTIBIOTICS, SEA WATER, RHODE ISLAND.: / CHLOR W72-04748
W PROFILES, MARINE ALGAE, WINDS, SEA WATER, PHYSICOCHEMICAL PROPER W71-09225
CRABS, WATER POLLUTION SOURCES, SEA WATER, TROUT, TOXINS.: /RIMP, W71-07412
ALGAE, *BIOASSAY, REPRODUCTION, SEA WATER, MARINE PLANTS, PRIMARY W72-14294
S, FUCUS SPIRALIS, MEDITERRANEAN SEA.: *FUCUS VIRSOIDE W72-12940
A, LIVERPOOL BAY(ENGLAND), IRISH SEA.: *ULVA, *LAMINARI W71-12867
BLACK SEA.: W72-10955
D(MASS), CAPE MAY(N J), SARGASSO SEA.: /AMA, CALLAO(PERU), CAPE CO W72-08056
.: *SALTON SEA(CALIF), ECOLOGICAL MANAGEMENT W71-10577
UTION, EPIFAUNA.: *NORTH SEA, *LAMINAREA SPP., SEWAGE POLL W72-11854
SUBSTRATE UTILIZATION, *SARGASSO SEA, COCCOLITHOPHORIDS, FLAGELLAT W73-00380
APODS, COLEOPTERA, MEDITERRANEAN SEA, COELENTERATES, ECHINODERMS, W73-00855
GELLATES, MEDITERRANEAN SEA, RED SEA, DRIFT ORGANISMS.: /ANAL, FLA W73-00296
14, CHLOROPHYLL A, MEDITERRANEAN SEA, FRANCE, PYCNOCLINE.: /NS, C- W73-00431
NE, HYPERHALINE, EURYHALINE, RED SEA, GULF OF SUEZ, DECAPODS, COLE W73-00855
, GASTEROSTEUS ACULEATUS, BALTIC SEA, NITOCRA SPINIPES.: / SALINUS W71-08026
WATER-PLANT RELATIONSHIPS, NORTH SEA, PHAEOCYSTIS SPP., SOIL CULTU W72-12239
M FURCA, TURBIDOSTAT CUL/ *NORTH SEA, PROROCENTRUM MICANS, CERATIU W71-03095
ANAL, FLAGELLATES, MEDITERRANEAN SEA, RED SEA, DRIFT ORGANISMS.: / W73-00296
IC CARBON, MACROPHYTES, SARGASSO SEA, SAMPLE PREPARATION, SPERMATO W73-02100
RES, SILICOFLAGELLATES, SARGASSO SEA, SHELF WATERS, GULF STREAM, C W72-07663
CALIF), SACRAMENTO RIVER(CALIF), SEARSVILLE LAKE(CALIF), SAN JOAQU W70-10405
ACTERIA, ZOOPLANKTON, SALMONIDS, SEASON, OXYGEN, HYDROGEN SULFIDE, W71-00117
SHELLFISH, CRUSTACEANS, OYSTERS, SEASONAL.: /, AQUICULTURE, FISH, W70-08832
ON SOURCES, METEOROLOGICAL DATA, SEASONAL.: /MISTRY, WATER POLLUTI W72-04853
ON, ICE, EUTROPHICATION, OXYGEN, SEASONAL.: / THERMAL STRATIFICATI W72-04855
LUTION SOURCES, ALGAE, PLANKTON, SEASONAL.: /POLLUTION), WATER POL W72-05506
N, DISTRIBUTION, DEPTH, DIURNAL, SEASONAL.: /ATION, SOLAR RADIATIO W72-08054
EUTROPHICATION, SUMMER, ENERGY, SEASONAL.: /Y, CYANOPHYTA, ALGAE, W72-03228
*CHEMICAL ANALYSIS, TEMPERATURE, SEASONAL, IONS, ALGAE, SALINITY, W72-01363
INITY, DEPTH, ABSORPTION, ALGAE, SEASONAL, WEATHER, WATER ANALYSIS W72-09122
VIMETRIC ANALYSIS, PRODUCTIVITY, SEASONAL, CHLOROPHYLL, ALGAE, PHY W72-09108
ITROGEN, SEWAGE, CARBON DIOXIDE, SEASONAL, PHOSPHORUS, WATER POLLU W72-09167
LOROPHYLL, DISTRIBUTION, CARBON, SEASONAL, CADDISFLIES, GASTROPODS W71-11029
LGAE, OLIGOCHAETES, LIFE CYCLES, SEASONAL, DENSITY, DIATOMS, LAKES W71-11012
TROPHICATION, TASTE, RESERVOIRS, SEASONAL, ACTINOMYCETES, CYANOPHY W71-13187
ON, CYANOPHYTA, DINOFLOGELLATES, SEASONAL, ZOOPLANKTON, PHYTOPLANK W71-11949
YSTEMATICS, *DIATOMS, SEA WATER, SEASONAL, FLUCTUATIONS, CONTINENT W72-07663
RIVERS, BIOLOGICAL COMMUNITIES, SEASONAL, DIATOMS, *LIGHT, SOLAR W72-07901
OLLUTION EFFECTS, WATER QUALITY, SEASONAL, SUSPENDED LOAD, BIOINDI W72-07896
IATOMS, CYANOPHYTA, CHLOROPHYTA, SEASONAL, NUTRIENTS, THERMAL STRA W72-08560
LD, IRON, CESIUM, ALKALI METALS, SEASONAL, ZINC, ABSORPTION, LIGHT W73-01434
CANADA, COASTS, ATLANTIC OCEAN, SEASONAL, BAYS, SEA WATER, ORGANI W73-00366
WORMS, INVERTEBRATES, ANNELIDS, SEASONAL, CRUSTACEANS, ISOPODS, M W73-00374
, CYANOPHYTA, CHEMICAL ANALYSIS, SEASONAL, BIOMASS, ON-SITE INVEST W73-00242
IENTS, SCENEDESMUS, CHRYSOPHYTA, SEASONAL, SILICA, DISCHARGE(WATER W72-14797
LIGHT PENETRATION, IMPOUNDMENTS, SEASONAL, COLOR, IRON, PHYSICOCHE W73-00256
STAGES, DISTRIBUTION, BEHAVIOR, SEASONAL, WATER POLLUTION EFFECTS W72-14659
RTIES, *SEDIMENTS, CLAYS, CORES, SEASONAL, ALGAE, MUD-WATER INTERF W72-05469
TEST PROCEDURES, PHOTOSYNTHESIS, SEASONAL, LIGHT PENETRATION, LIGH W72-04270
FACTANTS, ENZYMES, LAKE ONTARIO, SEASONAL, ALGAE, CULTURES, MICROO W71-00116
*STANDARDS, *WATER TEMPERATURE, SEASONAL, MONTHLY, AQUATIC LIFE, W70-09595
OSITY, FISH BEHAVIOR, FERTILITY, SEASONAL, DIFFUSIVITY.: /TY, VISC W70-09612
IMPOUNDED WATERS, STREAMS, FLOW, SEASONAL, STRATIFICATION, NITRATE W71-03028
TON, *ESTUARIES, *STANDING CROP, SEASONAL, DIATOMS, DINOFLAGELLATE W71-07875
, ALGAE, REPRODUCTION, SAMPLING, SEASONAL, BOTTOM SEDIMENTS, ICE, W71-08026
LOROPHYLL, PHOSPHATES, NITRATES, SEASONAL, REGRESSION ANALYSIS, CO W72-12393
D CHAINS, ANABAENA, SCENEDESMUS, SEASONAL, DINOFLAGELLATES.: / FOO W72-12738
NUISANCE ALGAE, FLOATING PLANTS, SEAWATER, PLANKTON, BIODEGRADATIO W71-09261
AR MATERIAL, TURBULENT FRICTION, SEAWEED, PORPHYRIDIUM CRUENTUM, P W71-07878
-14.: SEAWEEDS, APHIDS, SPONGES, CARBON W72-06283
AURASTRUM PARADOXUM, STAURASTRUM SEBALDII VAR. ORNATUM, MICRASTERI W72-12736
FECAL MATERIAL, LAKE TAHOE, LAKE SEBASTICOOK(MAINE), FAYETTEVILLE W71-05626
PHYSA GYRINA, KEMMERER SAMPLER, SECCHI *YARDS CREEK(NJ), RELATIVE W72-14280
TERIONELLA SPP, ORTHOPHOSPHATES, SECCHI DISC, NEWFOUND LAKE, SQUAM W72-07890

DISSOLVED PHOSPHORUS, OSCILLATO/ SECCHI DISC, DISSOLVED NITROGEN, W72-07648
M CHLORIDE, CHEMICAL PROPERTIES, SECCHI DISKS, METHODOLOGY.: /ODIU W72-09167
OPHYTA, MANGANESE, ZINC, METALS, SECCHI DISKS, HEAVY METALS, BIOIN W72-09111
EFFLUENTS, SANITARY ENGINEERING, SECONDARY TREATMENT, MARINE ALGAE W72-09378
THESIS, SLUDGE, EFFLUENT, ALGAE, SECONDARY TREATMENT, BIOCHEMICAL W71-12188
ORDATES, HEMICHORDATES, ARTHROP/ SECONDARY PRODUCTIVITY, CEPHALOCH W73-00935
URVEY, PRIMARY SEWAGE TREATMENT, SECONDARY SEWAGE TREATMENT, PHYSI W70-09907
URCES, ENVIRONMENTAL SANITATION, SECONDARY TREATMENT, AQUATIC ALGA W71-09674
N, TEMPERATURE, LIGHT INTENSITY, SEDIMANTATION, OXIDATION, STORAGE W72-01881
TAM LAKE(CONN), TROUT LAKE(WIS), SEDIMENT COMPACTION.: /(WIS), BAN W72-08459
G, BENTHIC FAUNA, SEDIMENTATION, SEDIMENT DISTRIBUTION, PHYSICOCHE W72-00847
SH, INSTRUMENTATION, ABSORPTION, SEDIMENT DISCHARGE, REGRESSION AN W72-06283
ALGAE.: *SUSPENDED SEDIMENT PRODUCTION, *BLUE-GREEN W72-02218
HORUS, NITROGEN, EUTROPHICATION, SEDIMENT TRANSPORT, CYCLING NUTRI W72-00347
EDIMENTS, SAMPLING, CHLOROPHYLL, SEDIMENT TRANSPORT, BENTHIC FAUNA W72-09092
ON, LUCIFERIN-LUCIFERASE METHOD, SEDIMENT-WATER INTERACTIONS, BIOL W72-04292
: SEDIMENT-WATER NUTRIENT-EXCHANGE. W70-08944
D OXYGEN, BACTERIA, THERMOCLINE, SEDIMENT-WATER INTERFACES, CHRYSO W72-12995
SEDIMENTARY PHOSPHORUS RELEASE.: W72-01108
ITION), SAMPLING, BENTHIC FAUNA, SEDIMENTATION, SEDIMENT DISTRIBUT W72-00847
G, AUTOMATIC CONTROLS, AERATION, SEDIMENTATION, FILTRATION, ANAERO W72-00042
ONIA, COAGULATION, FLOCCULATION, SEDIMENTATION, FILTRATION, LIME, W71-13341
ORGANISMS, ACCLIMATION, CALCIUM, SEDIMENTATION, WASTE WATER TREATM W71-13334
DISPOSAL, MARINE MICROORGANISMS, SEDIMENTATION, ALGAE, INVERTEBRAT W71-13723
ICAL PRECIPITATION, COAGULATION, SEDIMENTATION, POLYELECTROLYTES, W72-10237
GENERATION, DEOXYGENATION, ALGAL SEDIMENTATION.: /SIN, NUTRIENT RE W72-12997
*ZINC RADIOISOTOPES, ABSORPTION, SEDIMENTATION, ANALYTICAL TECHNIQ W72-12203
HICATION, WASTE WATER TREATMENT, SEDIMENTATION RATES, OIL SPILLS, W72-12487
*ENTROPHICATION, NEUTRALIZATION, SEDIMENTATION, TRICKLING FILTERS, W70-10427
WATER ANALYSIS, LAKE MORPHOLOGY, SEDIMENTATION, PLANKTON, DIEL MIG W71-05626
SEWAGE, NUTRIENTS, COAGULATION, SEDIMENTATION, TOXICITY, TASTE, O W72-06837
IGRATION, AQUARIA, GROWTH RATES, SEDIMENTATION, BUOYANCY, WIND VEL W71-09852
ERATION, OXYGENATION, NUTRIENTS, SEDIMENTATION, CHLORINATION, SLOP W71-07113
HOTOSYNTHESIS, DISSOLVED OXYGEN, SEDIMENTATION, SAMPLING, CHEMICAL W71-07097
ED OXYGEN, DIVERSION STRUCTURES, SEDIMENTATION, BIOCHEMICAL OXYGEN W72-08401
NUTRIENTS, *DREDGING, DESILTING, SEDIMENTATION, *EUTROPHICATION, B W73-00938
AGE TREATMENT, ALGAE, *AERATION, SEDIMENTATION, *CHLORINATION, OXI W72-14469
OPLANKTON, LABORATORY EQUIPMENT, SEDIMENTATION, METHODOLOGY, POLLU W73-00273
TER), WINDS, THERMOCLINE, ALGAE, SEDIMENTS.: /ERATURE, CURRENTS(WA W72-12992
N, *ANAEROBIC CONDITIONS, ALGAE, SEDIMENTS, OXYGEN DEMAND, PHOSPHO W72-12990
*ALGAE, *BENTHIC FAUNA, *BOTTOM SEDIMENTS, *HYPOLIMNION, PHYTOPLA W72-12995
ITIONS, CHLOROPHYLL, EPILIMNION, SEDIMENTS, DECOMPOSING ORGANIC MA W72-12997
DISCHARGE(WATER), WATER BALANCE, SEDIMENTS, CHEMICAL PROPERTIES, D W72-13851
OALS, *BENTHIC FLORA, SEA WATER, SEDIMENTS, LIGHT INTENSITY, DIATO W72-14317
CURRENTS(WATER), PARTICLE SIZE, SEDIMENTS, SEPARATION TECHNIQUES, W73-00367
N, *ENDRIN, MARINE ALGAE, BOTTOM SEDIMENTS, PLANKTON, RADIOACTIVIT W73-00361
LLOUT, *FOOD CHAINS, ABSORPTION, SEDIMENTS, CORES, ANALYTICAL TECH W73-00790
MENT, COASTS, ANIMAL PHYSIOLOGY, SEDIMENTS, ANIMAL METABOLISM.: /N W73-00811
TOPES, MARINE ALGAE, ABSORPTION, SEDIMENTS, WASTE DILUTION, WASTE W73-00831
, *MARINE ANIMALS, MARINE ALGAE, SEDIMENTS, PUERTO RICO, ESTUARINE W73-00821
*MARINE ANIMALS, *MARINE ALGAE, SEDIMENTS, NUCLEAR WASTES, PATH O W73-00817
GAE, DOMESTIC WASTES, NUTRIENTS, SEDIMENTS, ANALYTICAL TECHNIQUES, W73-02478
LAKE ONTARIO, PLANKTON, DIATOMS, SEDIMENTS, HARDNESS(WATER), BENTH W73-01627
PLANTS, WATER POLLUTION SOURCES, SEDIMENTS, AQUATIC SOILS, PLANT T W73-01673
THERMAL STRATIFICATION, CARBON, SEDIMENTS, CARBON DIOXIDE, CHLORP W72-08560
GON, *SEDIMENT-WATER INTERFACES, SEDIMENTS, AQUATIC PRODUCTIVITY, W72-08989
UATIC ENVIRONMENT, PRODUCTIVITY, SEDIMENTS, NITRIFICATION, OXIDATI W72-08459
EFFECTS, EFFLUENTS, CALIFORNIA, SEDIMENTS, SAMPLING, CHLOROPHYLL, W72-09092
ANDS, SHALLOW WATER, PERIPHYTON, SEDIMENTS, DIATOMS, LIGHT INTENSI W72-07501
ITRIFICATION, NITROGEN FIXATION, SEDIMENTS, NITRATES, WISCONSIN.: / W72-07933
, STRATIFICATION, MIXING, BOTTOM SEDIMENTS, SUCCESSION, CHLOROPHYT W72-08049
E, PESTICIDES, DDT, RETURN FLOW, SEDIMENTS, SOIL EROSION, FERTILIZ W71-06444
NDWATER, AMMONIA, PRECIPITATION, SEDIMENTS, DENITRIFICATION, RUNOF W71-06435
TOPES, PHOSPHORUS RADIOISOTOPES, SEDIMENTS, TRACE ELEMENTS, VOLCAN W71-09190
R POLLUTION CONTROL, *DIVERSION, SEDIMENTS, WATER CIRCULATION, ABS W71-09581
TION, SAMPLING, SEASONAL, BOTTOM SEDIMENTS, ICE, AMPHIPODA.: /ODUC W71-08026
, PACIFIC OCEAN, COLUMNS, BOTTOM SEDIMENTS, ALGAE, CHLORELLA, BENT W71-08042
ICAL WASTES, GAS CHROMATOGRAPHY, SEDIMENTS, OYSTERS, SHRIMP, CRABS W71-07412
*ALGAE, *NITROGEN, *PHOSPHORUS, SEDIMENTS, ORGANIC WASTES.: /SMS, W72-06532
NE ALGAE, HYDROLYSIS, CHELATION, SEDIMENTS, SETTLING VELOCITY, RAD W72-04463
UTROPHICATI/ *NUTRIENTS, *BOTTOM SEDIMENTS, *LIMNOLOGY, *ALGAE, *E W70-08944
LANTS, DEPTH, CALIFORNIA, SANDS, SEDIMENTS, PRODUCTIVITY, TOXICITY W71-01475
ESOTA, WATER QUALITY, NUTRIENTS, SEDIMENTS, COLIFORMS, MUNICIPAL W W71-03012
POLLUTANTS, ASIA, MARINE ALGAE, SEDIMENTS, ESTUARINE ENVIRONMENT, W72-10958
SONS, ALGAL TOXINS, FISH TOXINS, SEDIMENTS, SALT MARSHES, ECOLOGY, W72-10678
ITRATES, DENITRIFICATION, BOTTOM SEDIMENTS, SEA WATER, FRESHWATER, W72-11563
CHNIQUES, SEPARATION TECHNIQUES, SEDIMENTS, OILY WATER, ALGAE.: /E W71-12467
, ALGAE, GREAT LAKES, NUTRIENTS, SEDIMENTS, LAKE MICHIGAN, WATER P W71-12072
SUREMENT, DETRITUS, METHODOLOGY, SEDIMENTS, LAKES, CHLOROPHYLL, HY W71-12855
ES, MUNICIPAL WASTES, TURNOVERS, SEDIMENTS, HUMAN POPULATION, TOUR W71-11009
OLLECTIONS, NUTRIENTS, SALINITY, SEDIMENTS, EUTROPHICATION, FISH, W71-10577
TY, ADSORPTION, FISH, LAKE ERIE, SEDIMENTS, SILTS, PLANKTON, ALGAE W71-11036
LTURE, *WATER POLLUTION CONTROL, SEDIMENTS, FARM WASTES, NUTRIENTS W72-00846
SURFACE WATER/ *REVIEWS, *BOTTOM SEDIMENTS, *AQUATIC ENVIRONMENT, W72-00847
OOPLANKTON, BACTERIA, NUTRIENTS, SEDIMENTS, ENVIRONMENTAL EFFECTS, W72-03215
OPLANKTON, PACIFIC OCEAN, BOTTOM SEDIMENTS, STATIONS, DATA COLLECT W72-03567
SSOLVED OXYGEN, DREDGING, BOTTOM SEDIMENTS, STABILIZATION, PRIMARY W72-03556
RATURE, COLIFORMS, CHLORINATION, SEEPAGE, PERCOLATION, EVAPORATION W71-07124
OBIC CONDITIONS, BIODEGRADATION, SEEPAGE, SLUDGE, COST ANALYSIS, * W71-07099
ATION, *LAKES, SEWAGE EFFLUENTS, SEEPAGE, SEPTIC TANKS, BIOASSAY, W72-12955

CHEMICAL ANALYSIS, EVAPORATION, SEEPAGE, TURBIDITY, HYDROLOGIC BU W72-02274
ENVIRONMENTAL EFFECTS, */ *OIL, SEEPAGE, WATER POLLUTION EFFECTS, W71-08793
DITION, POTENTIOMETRY, INTE/ ION- SELECTIVE ELECTRODES, STANDARD AD W71-00474
*ION- SELECTIVE CATHODES.: W72-00800
H/ VOLTAMMETRY, RADIOGRAPHY, ION SELECTIVE ELECTRODES, LASERS, ORT W72-08790
NTIOMETRY, ELECTROANALYSIS/ *ION- SELECTIVE ELECTRODES, EDTA, *POTE W73-02112
DS, *ALGAE, CYTOLOGICAL STUDIES, SELECTIVITY, PRODUCTIVITY.: /ETHO W71-12868
ASSAYS, *ALGAL ASSAY PROCEDURES, SELENASTRUM CAPRICORNUTUM.: /TAL W72-03985
S FIRMA, PLENOSPORIUM DASYOIDES, SELENASTRUM CAPRICORNUTUM, PAAP B W73-02099
SELENASTRUM CAPRICORNUTUM.: W72-04433
R INTERACTIONS, BIOLUMINESCENCE, SELENASTRUM CAPRICONUTUM, AEROBAC W72-04292
TIS NIDULANS, GROWTH INHIBITION, SELENATE, SELENITE, SELENOMETHION W72-00153
NS, GROWTH INHIBITION, SELENATE, SELENITE, SELENOMETHIONINE, SELEN W72-00153
A/ *ATOMIC ABSORPTION, ARSENIC, SELENIUM, ZINC, ANTIMONY, NORWAY, W71-11515
INHIBITION, SELENATE, SELENITE, SELENOMETHIONINE, SELENOPURINE, A W72-00153
ATE, SELENITE, SELENOMETHIONINE, SELENOPURINE, ANTIMETABOLITES.: / W72-00153
TIC PLANTS, MATHEMATICAL MODELS, SELF-PURIFICATION, ALGAE, GREAT L W71-12072
ISSIMA, RHIZOSOLENIA HEBETATA F. SEMISPINA, CHAETOCEROS AFFINIA,: / W72-03543
NGFISHES, SALVELINUS FONTINALIS, SEMOTILUS ATROMACULATUS, CLEANUP, W72-12930
NNULARIA, SURIRELLA, PERIDINIUM, SENECA RIVER, NINEMILE CREEK, VAN W72-09111
NE, DEXTRAM, RAFFINOSE, SUCROSE, SENESCENT CULTURES, XYLOSE, RIBOS W72-07710
GAE, INFRARED RADIATION, *REMOTE SENSING, *POLLUTANT IDENTIFICATIO W72-14728
ASUREMENT, CURR/ *ALGAE, *REMOTE SENSING, *TRACKING TECHNIQUES, ME W72-01095
FRARED RADIATION, *AQUA/ *REMOTE SENSING, *AERIAL PHOTOGRAPHY, *IN W71-12034
N, WATER POLLUTION, *MO/ *REMOTE SENSING, *POLLUTANT IDENTIFICATIO W72-12487
, *SEA WATER, *OILS, *N/ *REMOTE SENSING, *WATER POLLUTION CONTROL W72-04742
MPERATURE CONTROL, ASSAY, REMOTE SENSING, ALGAE, ANTIBIOTICS, MERC W72-10162
ERNS, AERIAL PHOTOGRAPHY, REMOTE SENSING, SATELLITES(ARTIFICIAL), W72-09120
GOMPHONEMA SPP., NITZCHIA SPP., SENSITIVITY.: /P., ANABAENA SPP., W72-14706
A, *BACTERIA, FLORIDA, RADIATION SENSITIVITY, RADIOECOLOGY, ECOLOG W73-00796
LITY, CYST FORMATION, ANTIBIOTIC SENSITIVITY, CYTOPHAGA JOHNSONII, W72-07951
LAGELLATE ALGAE, OPTICAL DENSITY SENSOR.: /KINETICS, CHEMOSTATS, F W70-10403
*FLOCCULATION, EQUIPMENT, COSTS, SEPARATION TECHNIQUES, COAGULATIO W70-08904
*ORGANI/ *WATER CHEMISTRY, *FOAM SEPARATION, *FOAM FRACTIONATION, W72-05506
LIQUID WASTES, SEWAGE TREATMENT, SEPARATION TECHNIQUES, *CHEMICAL W72-05315
OLATION, *ULTRAVIOLET RADIATION, SEPARATION TECHNIQUES, CYANOPHYTA W72-14327
ATER), PARTICLE SIZE, SEDIMENTS, SEPARATION TECHNIQUES, SPATIAL DI W73-00367
UTROPHICATION, OCEANS, SAMPLING, SEPARATION TECHNIQUES, CULTIVATIO W73-00430
RATORY EQUIPMENT, FREEZE DRYING, SEPARATION TECHNIQUES, SOLVENT EX W73-00379
, CHEMICAL ANALYSIS, ADSORPTION, SEPARATION TECHNIQUES, FOOD CHAIN W73-02105
N(CHEMICAL), GAS CHROMATOGRAPHY, SEPARATION TECHNIQUES, *WISCONSIN W73-01456
OPHOTOMETRY, VIRGINIA, SAMPLING, SEPARATION TECHNIQUES, CENTRIFUGA W73-00916
*ALGAE, COLUMNS, ORGANIC ACIDS, SEPARATION TECHNIQUES, *CHROMATOG W72-09365
HOTOSYNTHESIS, SUSPENDED SOLIDS, SEPARATION TECHNIQUES, FILTRATION W72-09386
OMPOUNDS, ANALYTICAL TECHNIQUES, SEPARATION TECHNIQUES, GAS CHROMA W72-11604
ONS, CATIONS, ORGANIC COMPOUNDS, SEPARATION TECHNIQUES, CALCIUM, M W72-11833
REATMENT, ANALYTICAL TECHNIQUES, SEPARATION TECHNIQUES, SEDIMENTS, W71-12467
MARINE FUNGI, *ISOLATION, ALGAE, SEPARATION TECHNIQUES, ATLANTIC O W71-11851
CE, FLOCCULATION, LIQUID WASTES, SEPARATION TECHNIQUES, WATER PURI W71-13356
MASS SPECTROMETRY, DISTRIBUTION, SEPARATION TECHNIQUES.: /ONATES, W72-12709
TOGRAPHY, BROOK TROUT, SAMPLING, SEPARATION T: /EFFECTS, GAS CHROMA W72-12930
HOSPHORUS COMPOUNDS, WASHINGTON, SEPARATION TECHNIQUES, DISTILLATI W72-12637
TECHNIQUES, GAS CHROMATOGRAPHY, SEPCTROSCOPY, ALGAE, ACTINOMYCETE W72-11604
AKES, SEWAGE EFFLUENTS, SEEPAGE, SEPTIC TANKS, BIOASSAY, CHEMICAL W72-12955
EMENT, OUTFALL SERVERS, STREAMS, SEPTIC TANKS, FUNGI, ALGAE, WATER W72-14363
NITRIFICATION, DENITRIFICATION, SEPTIC TANKS, OXIDATION LAGOONS, W70-09907
LLA MARINA, EUTREPTIA, NITZSCHIA SERIATA, THALASSIONEMA NITZ: /VIE W71-07875
E, GLUTAMATE, LEUCINE, ARGININE, SERINE, TYROSINE, TRYPTOPHAN, HIS W73-00942
NTERATES, PORIFERA, POLYCHAETES, SERPULIDS.: /OIDS, SPONGES, COELE W73-00374
ULGARIS, PSEUDOMONAS AERUGINOSA, SERRATIA: /R AEROGENES, PROTEUS V W72-07664
ICORNIS, GAMMARUS DUEBENI, FUCUS SERRATUS, FUCUS VESICULOSUS, CLAD W71-08026
ON, PELVETIA CANALICULATA, FUCUS SERRATUS, FUCUS SPIRALIS, FUCUS V W71-11515
S, GROUNDWATER MOVEMENT, OUTFALL SERVERS, STREAMS, SEPTIC TANKS, F W72-14363
TA, AQUATIC ALGAE, MARINE ALGAE, SESSILE ALGAE.: /PHYTA, CHLOROPHY W73-00273
LIGHT INTENSITY, PHOTOSYNTHESIS, SESSILE ALGAE, TEMPERATURE, DEPTH W72-07501
ES, TOXINS, WASTES, CHLOROPHYTA, SESSILE ALGAE, WATER ANALYSIS.: / W71-10553
ORS, INDUSTRIAL WASTES, FLORIDA, SESSILE ALGAE, ESTUARIES, CHEMICA W71-07412
ANALYTICAL TECHNIQUES, MICHIGAN, SESTON, ALGAE, PHYTOPLANKTON, NAN W72-08048
, PHYTOPLANKTON, CARBON DIOXIDE, SESTON, FERTILIZATION, WATER QUAL W72-07648
SALINITY, CARBON RADIOISOTOPES, SESTON, ORGANIC MATTER, DISTRIBUT W73-00431
ORARUM, PINAEUS AZTECUS, PENAEUS SETIFERUS, MERCENARIA MERCENARIA, W73-02029
ENTS, PARASITES, JORDAN RIVER(M/ SETTLEABLE SOLIDS, HATCHERY EFFLU W71-11006
TION TANK, PONDAGE, PILOT PLANT, SETTLING BASIN.: /BORATORY, DIGES W72-02850
URES, MICROORGANISMS, EFFLUENTS, SETTLING BASINS, WASTE WATER TREA W71-00116
RECYCLE, SOLI/ ORGANIC NITROGEN, SETTLING CHARACTERISTICS, SOLIDS W72-11100
YDROLYSIS, CHELATION, SEDIMENTS, SETTLING VELOCITY, RADIOISOTOPES, W72-04463
ON, FILTRATION, ELECTROPHORESIS, SETTLING, *ALGAE, *SILICA, *BACTE W71-01899
YETHYLENE OXIDE POLYMERS, SEWAGE SETTLING, UNION CARBIDE CORP, POL W70-06968
GEN, HYDROGEN ION CONCENTRATION, SETTLING, MICROORGANISMS, ACCLIMA W71-13334
CATION, ALGAE, AQUATIC BACTERIA, SEWAGE BACTERIA, ENVIRONMENTAL EF W72-01472
ION, *COBALT, *RADIATION, COSTS, SEWAGE BACTERIA, ALGAE, COLIFORMS W72-00383
CUS, RHODOMONAS, LAKE CONSTANCE, SEWAGE CANALS.: /SCENS, SYNEDRA A W71-00117
SON RIVER, ALGAE, WATER QUALITY, SEWAGE DISPOSAL, WASTE WATER DISP W70-06509
TS(WATER), ALGAE, LAKES, RIVERS, SEWAGE DISPOSAL, SEWAGE TREATMENT W70-09947
ATER POLLUTION EFFECTS, MERCURY, SEWAGE DISPOSAL, PHOSPHATES, PHOS W72-14782
AGE EFFLUENTS, SEWAGE TREATMENT, SEWAGE DISPOSAL, MARINE ALGAE, CL W72-07715
INE FISHERIES, MARINE FISHERIES, SEWAGE DISPOSAL, SEWAGE LAGOONS, W72-12567
OURCES, *EUTROPHICATION, *LAKES, SEWAGE EFFLUENTS, SEEPAGE, SEPTIC W72-12955
, FRESHWATER FISH, YELLOW PERCH, SEWAGE EFFLUENTS, OIL SPILLS, THE W72-14282
ION, *NORTH CAROLINA, NUTRIENTS, SEWAGE EFFLUENTS, EUTNOPHICATION, W72-13636

UES, SEWAGE, RESEARCH EQUIPMENT,	SEWAGE EFFLUENTS, SEWAGE TREATMEN	W72-07715
TY, POLLUTANTS, CHEMICAL WASTES,	SEWAGE EFFLUENTS, WATER POLLUTION	W72-08804
*NITROGEN, *ALGAE, PLANT GROWTH,	SEWAGE EFFLUENTS, REGULATION, CYA	W72-14790
, ECONOMIC FEASIBILITY, STREAMS,	SEWAGE EFFLUENTS, ALGAE, TOURISM,	W72-14878
DISPOSAL, WASTE WATER DISPOSAL,	SEWAGE EFFLUENTS, FARM WASTES.: /	W70-06509
LANKTON, *AQUATIC PRODUCTIVITY,/	SEWAGE EFFLUENTS, *ALGAE, *PHYTOP	W71-02681
NUTRIENTS, NITROGEN, PHOSPHORUS,	SEWAGE EFFLUENTS, SEWAGE TREATMEN	W72-05473
EFFE/ *ALGAE, *LAKES, *GEORGIA,	SEWAGE EFFLUENTS, WATER POLLUTION	W71-09173
IS, ECOLOGY, TERTIARY TREATMENT,	SEWAGE EFFLUENTS, WATER REUSE.: /	W72-03970
ATION, *WATER POLLUTION CONTROL,	SEWAGE EFFLUENTS, AQUATIC ALGAE,	W71-11507
COBALT, IRON, INDUSTRIAL WASTES,	SEWAGE EFFLUENTS, POLLUTANT IDENT	W71-11515
LUTION EFFECTS, RECLAIMED WATER,	SEWAGE EFFLUENTS, RESEARCH AND DE	W72-11702
ALITY CONTROL, SEWAGE TREATMENT,	SEWAGE EFFLUENTS, SANITARY ENGINE	W72-09378
*NITRATES, *AMMONIUM COMPOUNDS,	SEWAGE EFFLUENTS, TEMPERATURE CON	W72-10162
RINE FISHERIES, SEWAGE DISPOSAL,	SEWAGE LAGOONS, BRACKISH WATER FI	W72-12567
*NORTH SEA, *LAMINAREA SPP.,	SEWAGE POLLUTION, EPIFAUNA.:	W72-11854
NG, POLYETHYLENE OXIDE POLYMERS,	SEWAGE SETTLING, UNION CARBIDE CO	W70-06968
TOXICITY, ALGAE, BACTERIA, FISH,	SEWAGE SLUDGE, FLOW, FLOCCULATION	W70-06968
MARY SEWAGE TREATMENT, SECONDARY	SEWAGE TREATMENT, PHYSICOCHEMICAL	W70-09907
E T/ *LITERATURE SURVEY, PRIMARY	SEWAGE TREATMENT, SECONDARY SEWAG	W70-09907
NT, ALGAE, EFFLUENTS, NUTRIENTS,	SEWAGE TREATMENT, LAKES,: /REATME	W70-09325
LAKES, RIVERS, SEWAGE DISPOSAL,	SEWAGE TREATMENT, WASTE WATER TRE	W70-09947
N, PHOSPHORUS, SEWAGE EFFLUENTS,	SEWAGE TREATMENT, ALGAE, WATER PO	W72-05473
GN CRITERIA, *OXIDATION LAGOONS,	SEWAGE TREATMENT, ALGAL: /Y, DESI	W72-04789
WATER TREATMENT, *LIQUID WASTES,	SEWAGE TREATMENT, SEPARATION TECH	W72-05315
, FEDERAL GOVERNMENT, NUTRIENTS,	SEWAGE TREATMENT, WATER POLLUTION	W72-06655
, WATER QUALITY, BIODEGRADATION,	SEWAGE TREATMENT PLANTS, DIATOMS,	W72-06837
RIENTS, WATER POLLUTION SOURCES,	SEWAGE TREATMENT, PUBLIC HEALTH,	W72-06638
ECYCLING, WATER QUALITY CONTROL,	SEWAGE TREATMENT, SEWAGE EFFLUENT	W72-09378
ROPHICATION, *NUTRIENTS, *ALGAE,	SEWAGE TREATMENT, WATER POLLUTION	W72-10625
ROGEN COMPOUNDS, AMMONIA, ALGAE,	SEWAGE TREATMENT, TERTIARY TREATM	W71-13939
LITY, PHOSPHORUS, PHYTOPLANKTON,	SEWAGE TREATMENT, NITROGEN, PHYSO	W72-13662
RCH EQUIPMENT, SEWAGE EFFLUENTS,	SEWAGE TREATMENT, SEWAGE DISPOSAL	W72-07715
*SEWAGE LAGOONS, BIODEGRADATION,	SEWAGE TREATMENT, WASTE WATER TRE	W72-07661
TASSIUM, MOLYBDENUM, DETERGENTS,	SEWAGE TREATMENT, CARBONATES, BIC	W72-07514
HORUS, CARBON, *SEWAGE BACTERIA,	SEWAGE, ALGAE, HYDROGEN ION CONCE	W72-04309
N, ECONOMICS, ALGAE, SOLUBILITY,	SEWAGE, BIOCHEMICAL OXYGEN DEMAND	W71-12200
TY, ALGAE, PHOSPHATES, NITROGEN,	SEWAGE, CARBON DIOXIDE, SEASONAL,	W72-09167
OBACTERIA, CULTURES, METABOLISM,	SEWAGE, CARBOHYDRATES, HYDROGEN I	W72-05457
IC MATTER, LAKE ERIE, EFFLUENTS,	SEWAGE, DETERGENTS, PLANTS, LA: /	W71-11017
PLANTS, ASSESSMENTS, MONITORING,	SEWAGE, ESTUARIES, GROWTH RATES,	W71-12867
FERTILIZATION, DISSOLVED OXYGEN,	SEWAGE, HYDROGEN ION CONCENTRATIO	W71-12188
N FLOW, WATER SUPPLY, LIVESTOCK,	SEWAGE, INFILTRATION, INDUSTRIAL	W71-06435
RATES, CULTURES, SUSPENDED LOAD,	SEWAGE, INDUSTRIAL WASTES, SILTS,	W71-09795
LUTION, WATER POLLUTION SOURCES,	SEWAGE, INDUSTRIAL WASTES, BIOIND	W72-12729
ASSAY, SULFATES, IRON COMPOUNDS,	SEWAGE, NITRATES, PHOSPHATES.: /O	W72-12239
MATTER, WATER POLLUTION EFFECTS,	SEWAGE, NITROGEN, PHOSPHORUS, IRO	W73-00234
NTS, LINEAR ALKYLATE SULFONATES,	SEWAGE, NUTRIENTS, COAGULATION, S	W72-06837
ROPHICATION, *NUTRIENTS, *ALGAE,	SEWAGE, PRODUCTIVITY, WATER POLLU	W71-12857
LUDGE, PHOSPHORUS RADIOISOTOPES,	SEWAGE, RADIOACTIVITY, PHOSPHATES	W71-01474
QUIPMENT, ANALYTICAL TECHNIQUES,	SEWAGE, RESEARCH EQUIPMENT, SEWAG	W72-07715
IAL DEGRADATION, CLASSIFICATION,	SEWAGE, SOIL PROPERTIES, GROUNDWA	W72-14363
ER POLLUTION SOURCES, NUTRIENTS,	SEWAGE, WATER QUALITY, AERATION,	W72-07890
PHOSPHORUS, NITROGEN, NUTRIENTS,	SEWAGE, WATER TEMPERATURE, LIGHT	W72-04734
ANIZATION, CITIES, STORM RUNOFF,	SEWAGE, WASTE WATER DISPOSAL, PAT	W71-13466
SHING, *MOUNT CLEMENS/ *COMBINED	SEWERS, *SUSPENDED SOLIDS, BACKWA	W72-00042
ATMEN/ *OXIDES, *FLUID FRICTION,	SEWERS, SLURRIES, WASTE WATER TRE	W70-06968
D, CHUBS, NORTHERN PIKE, GIZZARD	SHAD.: /S, COHO SALMON, SHEEPSHEA	W72-14282
S ISLAND RESERVOIR(NJ), AMERICAN	SHAD, EKMAN DREDGE, PUMPKINSEED,	W72-14280
MYCIN, EPIFLORA, WALLEYE LARVAE,	SHAD, MACROINVERTEBRATES, CYANOPH	W72-08142
ESOX LUCIUS, CAMPELOMA, PSIDIUM,	SHADFLY, CHIRONOMIDS, CLEAR LAKE(	W70-08098
GRANBY(COLO),/ GRAND LAKE(COLO),	SHADOW MOUNTAIN LAKE(COLO), LAKE	W72-04791
E).: *LIMITING NUTRIENTS,	SHAGAWA LAKE(MINN), WALDO LAKE(OR	W72-09165
EUTROPHICATION, LITTORAL, SANDS,	SHALLOW WATER, PERIPHYTON, SEDIME	W72-07501
ONS, PHAEOPHYTA, ATLANTIC OCEAN,	SHARKS, RADIOACTIVITY TECHNIQUES.	W72-05965
IPSAMMIC ALGAE, *EPIPELIC ALGAE,	SHEAR WATER(ENGLAND), AMPHORA OVA	W71-12855
IZOMENON, CYANIDES, COHO SALMON,	SHEEPSHEAD, CHUBS, NORTHERN PIKE,	W72-14282
TION SOURCES, PLANKTON, OYSTERS,	SHEL: /S, HERBICIDES, WATER POLLU	W72-13347
SILICOFLAGELLATES, SARGASSO SEA,	SHELF WATERS, GULF STREAM, CAPE H	W72-07663
ANS, SURFACE WATERS, CONTINENTAL	SHELF, DEPTH, NITROGEN, PHYTOPLAN	W72-08056
SONAL, FLUCTUATIONS, CONTINENTAL	SHELF, VERTICAL DISTRIBUTION, NOR	W72-07663
TH, OYSTERS, SHELLFISH, MUSSELS,	SHELLFISH FARMING, AQUACULTURE, C	W72-09378
FARMING, AQUACULTURE, COMMERCIAL	SHELLFISH, BIOLOGICAL COMMUNITIES	W72-09378
L GROWTH, PLANT GROWTH, OYSTERS,	SHELLFISH, MUSSELS, SHELLFISH FAR	W72-09378
ATURE, ALGAE, BULLHEADS, SALMON,	SHELLFISH, CLAMS, MARINE ANIMALS,	W72-07907
FOOD CHAINS, *ORGANIC COMPOUNDS,	SHELLFISH, INSTRUMENTATION, PHYTO	W72-07715
OPLANKTON, MUSSELS, CLAMS, FISH,	SHELLFISH, AQUEOUS SOLUTIONS, PHA	W72-05965
ATION, ALGAE, AQUICULTURE, FISH,	SHELLFISH, CRUSTACEANS, OYSTERS,	W70-08832
, SNAILS, ALGAE, DIATOMS, CLAMS,	SHELLFISH, AMPHIPODA, WATER BIRDS	W70-08098
NTS, *TOXICITY, *BIOASSAY, FISH,	SHELLFISH, ALGAE, DEGRADATION.: /	W71-09789
	SHELTER VALLEY CREEK(ONTARIO).:	W73-01627
NICOL, SALMONELLA ENTEROCOLITIS,	SHIGELLA DYSENTERIAE, SHIGELLA FL	W72-13800
ROCOLITIS, SHIGELLA DYSENTERIAE,	SHIGELLA FLEXNERI, BIOCHEMICAL ST	W72-13800
RIC BACTERIA, DISEASES, E. COLI,	SHIGELLA, VIRUSES, CHROMATOGRAPHY	W72-13800
ARCH, CHYTRIDIACEOUS FUNGI, HEAT	SHOCK.: RESE	W72-07225
BAY, NEVA ISLAND, BAIRD ISLAND,	SHOLIN ISLAND, WINIFRED ISLAND, I	W72-09120
*CYANOPHYTA, *LAKES, RHODOPHYTA,	SHORES.: *ALGAE, *CHLOROPHYTA,	W71-12489
IC HABITATS, LAKES, AQUAT/ *LAKE	SHORES, *ALGAE, *LAKE ERIE, AQUAT	W71-09156
, LEAVES, LEGISLATION, LAND USE,	SHORES, MONITORING, MARSHES, SOIL	W73-02479
OZOA, MANGROVE SWAMPS, LITTORAL,	SHORES, TIDES, BEACHES, SANDS, IN	W72-07911

GHORN LAKE(MONT), BIGHORN RIVER, SHOSHONE RIVER, YELLOWTAIL DAM, F W73-01435
ENVIRONMENT, MONITORING, CRABS, SHRIMP.: /E, SEDIMENTS, ESTUARINE W72-10958
OMATOGRAPHY, SEDIMENTS, OYSTERS, SHRIMP, CRABS, WATER POLLUTION SO W71-07412
, CRABS, COPEPODS, MARINE ALGAE, SHRIMP, FISH EGGS, FISH, BACTERIA W72-08142
ELAGIC EGGS, GADUS MORHUA, GRASS SHRIMP, GREEN CRAB, PENICILLIUM, W72-08142
G NUTRIENTS, POTASSIUM CHLORIDE, SHRINER'S POND(GA).: *LIMITIN W72-09159
RESERVOIR(SIBERIA), ANGARA RIVER(SIBERIA).: BRATSK W71-09165
BRATSK RESERVOIR(SIBERIA), ANGARA RIVER(SIBERIA).: W71-09165
ALGAL BLOOMS, HYPNODINIUM, LAKE SIDNEY LANIER(GEORGIA), LAKE JACK W71-09173
SON(GEORGIA), LAKE CARDIN/ *LAKE SIDNEY LANIER(GEORGIA), LAKE JACK W71-11914
CAL DISTRIBUTION, BENTHIC FAUNA, SIEVES, WATER TEMPERATURE,: /LOGI W73-00367
ARIUM, TRICHORIXA, RAMPHOCORIXA, SIGARA, BOENOA, SALDULA, CRYNELLU W70-06655
, POTABLE WATER, TASTE, AMMONIA, SILICA.: /LUTION TREATMENT, ALGAE W71-01491
LYTICAL TECHNIQUES, FORECASTING, SILICA.: /TERGENTS, BIOASSAY, ANA W71-12072
ES, CALCIUM, MAGNESIUM, E. COLI, SILICA, *FILTRATION, LIGHT PENTRA W72-11833
IVITY, EUTROPHICATION, RED TIDE, SILICA, ALGAE, PHYTOPLANKTON, FRE W71-09190
SONAL, STRATIFICATION, NITRATES, SILICA, AMMONIA, PHOSPHATES, RUNO W71-03028
KS, HEAVY METALS, BIOINDICATORS, SILICA, CHROMIUM, COPPER, *NEW YO W72-09111
NEDESMUS, CHRYSOPHYTA, SEASONAL, SILICA, DISCHARGE(WATER).: /, SCE W72-14797
DOMONAS, SCENEDESMUS, CHLORELLA, SILICA, LIGHT INTENSITY, TEMPERAT W72-12544
RATION TECHNIQUES, *MARINE FISH, SILICA, MARINE ALGAE, INVERTEBRAT W73-00371
TRIBUTION, PHOSPHORUS, NITRATES, SILICA, PATH OF POLLUTANTS, ORGAN W73-00384
R, CHLORELLA PYRENOIDOSA, SODIUM SILICATE, SODIUM METASILICATE, LA W72-04734
ARBON RADIOISOTOPES, PHOSPHATES, SILICATES, SOLAR RADIATION, DEPTH W71-07875
LORELLA, CHLAMYDOMONAS, DIATOMS, SILICATES, CYANOPHYTA, TOXICITY, W72-07508
PROFILES, PHOSPHATES, NITRATES, SILICATES, HYDROGRAPHS: /ALINITY, W72-07892
SHELF WATERS,/ COCCOLITHOPHORES, SILICOFLAGELLATES, SARGASSO SEA, W72-07663
*LIMITING FACTORS, SILICON.: W72-07937
ENTATION, METHODOLOGY, NITROGEN, SILICON, ABSORPTION, PHYTOPLANKTO W73-01445
HATES, NITRATES, NITRITES, IRON, SILICON, SULFATES, SODIUM, POTASS W72-05610
ORTH SEA, PHAEOCYSTIS SPP., SOI/ SILT-WATER-PLANT RELATIONSHIPS, N W72-12239
G, DOCUMENTATION, WATER QUALITY, SILTING, AGRICULTURAL CHEMICALS, W72-05869
AINS, LAKE MORPHOLOGY, RESERVOIR SILTING, LAKES, *SOUTH DAKOTA, PL W73-00938
ION, FISH, LAKE ERIE, SEDIMENTS, SILTS, PLANKTON, ALGAE, TRACE ELE W71-11036
D, MOLLUSKS, FISH, INSECTS, DDT, SILTS, SUCKERS, CATFISHES, BASS, W70-08098
LOAD, SEWAGE, INDUSTRIAL WASTES, SILTS, WATER POLLUTION EFFECTS.: / W71-09795
PHYRENOIDOSA, PHAEODACTYLUM TRI/ SILVER, CADMIUM, LEAD, CHLORELLA W72-07660
S, ZINC, LEAD, COPPER, CHROMIUM, SILVER, CHLOROPHYTA, CYANOPHYTA, W72-14694
MULINA, SKELETONEMA, CHAETOCEROS SIMPLEX.: /RORA ALATA, CHRYSOCHRO W73-00380
PODS, SARGASSUM CRISPUM, DIGENIA SIMPLEX, HALOPHILA STIPULACEA, DI W73-00855
*MATHEMATICAL MODELS, *COMPUTER SIMULATION, MIXING, TURBULENCE, T W71-07084
PHICATION, *MATHEMATICAL MODELS, SIMULATION, ANALYTICAL TECHNIQUES W71-03021
ANCTONICA, MANCHUDINIUM SINICUM, SINAMONAS STAGNALIS, KOFOIDIELLA W72-07683
BRITISH COLUMBIA, RU/ *LAMINARIA SINCLAIRII, *LAMINARIA LONGIPES, W73-00428
AMONAS PLANCTONICA, MANCHUDINIUM SINICUM, SINAMONAS STAGNALIS, KOF W72-07683
IC ZONE, DIFFUSION COEFF/ *ALGAL SINKING, *VERTICAL MIXING, EUPHOT W71-03034
, DIADEMA ANTILLARUM, AANDOMENE, SIPHONALES, *HALIMEDA SPP., PENIC W72-14294
HINODERMS, TUNICATES, NOCTILUCA, SIPHONOPHORES, MEDUSAE, PILIDIUM, W73-00935
ROPODS, NEMERTEANS, ECHINODERMS, SIPUNCULIDS, BRYOZOA, INTOPROCTA, W73-00932
CTS, MARINE FISH, PERIPHYTON, ON- SITE DATA COLLECTIONS, CRUSTACEAN W73-00932
DATA COLLECTIONS, EVALUATION, ON- SITE DATA COLLECTIONS, RIVERS, *E W72-07791
LANDS, *ARIZONA, GRASSLANDS, ON- SITE DATA COLLECTIONS, PLANT GROW W72-12505
SPHORUS, NITROGEN, NUTRIENTS, ON- SITE DATA COLLECTIONS, LAKES.: /O W72-12990
NITORING, *WATER PROPERTIES, *ON- SITE DATA COLLECTIONS, NEW YORK, W72-12941
TES, WATER POLLUTION SOURCES, ON- SITE INVESTIGATIONS, METHODOLOGY, W72-08804
ANALYSIS, SEASONAL, BIOMASS, ON- SITE INVESTIGATIONS, ALGAE.: /CAL W73-00242
UTION SOURCES, PUBLIC HEALTH, ON- SITE INVESTIGATIONS, WASTE ASSIMI W73-00790
ROCKSLIDES, INTERTIDAL AREAS, ON- SITE INVESTIGATIONS, FOOD CHAINS, W73-00764
S, BACTERIA, FLOATING PLANTS, ON- SITE INVESTIGATIONS, TURTLES.: /E W71-01488
EALTH, ESTUARINE ENVIRONMENT, ON- SITE INVESTIGATIONS, REVIEWS.: /H W72-04461
, *OXIDATION LAGOONS, ORGAN/ *ON- SITE INVESTIGATION, *COLD REGIONS W72-06838
G, *SEA WATER, *DATA COLLEC/ *ON- SITE INVESTIGATIONS, *SCUBA DIVIN W72-11925
LITY, *PHYTOPLANKTON, *DEPTH, ON- SITE INVESTIGATIONS, *POPULATION, W72-11719
ALGAE, / *OXIDATION LAGOONS, *ON- SITE INVESTIGATIONS, *BACTERIA, * W72-10233
TENCE, BACTERIA, BIRDS, FISH, ON- SITE INVESTIGATION, METHODOLOGY, W72-00155
TRACERS, PATH OF POLLUTANTS, ON- SITE INVESTIGATIONS, FOOD WEBS, F W72-00949
HEDS, CROPS, EGGS, FISHERIES, ON- SITE INVESTIGATIONS, ENVIRONMENTA W72-00959
ITES, MOLLUSKS, MARINE ALGAE, ON- SITE INVESTIGATIONS, CALIFORNIA, W72-03324
SEA WATER, COASTS, BIORTHYMS, ON- SITE TESTS, WATER POLLUTION SOURC W72-03228
L ECOLOGY, ORGANIC COMPOUNDS, ON- SITE TESTS, METHODOLOGY, CHARA.: / W72-03216
KES, PHOSP/ *EUTROPHICATION, *ON- SITE TESTS, *NUTRIENTS, ALGAE, LA W72-09165
TY, SEA WATER, WATER QUALITY, ON- SITE TESTS, HYDROGEN ION CONCENTR W72-09092
YCETES, CYANOPHYTA, PLANKTON, ON- SITE TESTS, LABORATORY TESTS.: /M W71-13187
NVIRONMENTAL EFFECTS, LARVAE, ON- SITE TESTS, INVERTEBRATES, WATER W73-00348
AQUATIC ALGAE, CARBON CYCLE, ON- SITE TESTS, AQUATIC ANIMALS, AQUA W73-02095
ATION, CARBON RADIOISOTOPES, *ON- SITE TESTS, SULFUR BACTERIA, PROD W72-14301
IOISOTOPES, NUCLEAR POWERPLANTS, SITES, MOLLUSKS, MARINE ALGAE, ON W72-03324
, CAULERPIN, PALMITIC ACID, BETA- SITOSTEROL, THIN-LAYER CHROMATOGR W71-08597
LAR WEIGHT COMPOUNDS, *MOLECULAR SIZE, *ALGAL EXUDATES, EXTRACELLU W71-07878
N, CALIFORNIA, FILTERS, PARTICLE SIZE, ANAEROBIC CONDITIONS, WASTE W71-08223
AE, PHAEPHYTA, AQUARIA, PARTICLE SIZE, COLLOIDS, WATER POLLUTION E W71-09193
ATOMS, CHLOROPHYTA, TEMPERATURE, SIZE, FECUNDITY, ALGAE.: /ODS, DI W72-14805
TERNS, CURRENTS(WATER), PARTICLE SIZE, SEDIMENTS, SEPARATION TECHN W73-00367
HIPRORA ALATA, CHRYSOCHROMULINA, SKELETONEMA, CHAETOCEROS SIMPLEX. W73-00380
MA NITZSCHIOI/ *INTERREGULATION, SKELETONEMA COSTATUM, THALASSIONE W73-00441
ZOODIACUS, CHAETOCEROS AFFINIS, SKELETONEMA COSTATUM, CHAETOCEROS W73-00442
AXIS, *PREY, PYTHIUM DEBARYANUM, SKELETONEMA COSTATUM, NUCLEOTIDES W73-00942
HWATER ALGAE, THALASIOSIRA SPP., SKELETONEMA SPP., DUNALIELLA SPP. W72-08436
VIRIDIS, CHAETOCEROS CURVISETUS, SKELETONEMA COSTATUM, GLENODINIUM W72-07166
L WATERS, *SOUTHERN LONG ISLAND, SKELETONEMA COSTATUM, THALASSIOSI W71-07875

TERTI/ THALASSIOSIRA PSEUDONANA, SKELETONEMA COSTATUM, DUNALIELLA W72-05614
LA PYRENOIDOSA, NITZSCHIA PALEA, SKELETONEMA COSTATUM.: CHLOREL W71-03027
ALDRIN, KETOENDRIN, METABOLITES, SKELETONEMA, ANABAENA CYLINDRICA, W72-03545
BAYGONE, CYCLOTELLA, DUNALIELLA, SKELETONEMA.: ABATE, BAYTEX, W71-10096
ECTA, PHAEODACTYLUM TRICORNUTUM, SKELETONEMA COSTATUM, CARBOY, OSC W72-09103
COOLITHUS HUXLEYI, PORPHYRIDIUM, SKELETONEMA COSTATUM.: /RTERI, CO W72-12998
OECTRICHIA, MICROCYSTIS, NOSTOC, SLAKED LIME.: /TION, ANABAENA, GL W72-01097
SPRINGS, PIPESTONE HOT SPRINGS, SLEEPING CHILD HOT SPRINGS.: /HOT W72-05610
NE IODIDE, TOLUENE, POLYSTYRENE, SLIDE PREPARATION, SAMPLE PREPARA W72-11738
LIZATION.: DEPTH, SLOPE, *BOD LOADING, *WASTE STABI W71-03896
TS, SEDIMENTATION, CHLORINATION, SLOPES, BIOCHEMICAL OXYGEN DEMAND W71-07113
LENE OXIDE, WATER CLARIFICATION, SLUDGE DRYING, POLYETHYLENE OXIDE W70-06968
SLURRIES, WASTE WATER TREATMENT, SLUDGE TREATMENT, TOXICITY, ALGAE W70-06968
COMPOUNDS, NITRATES, PHOSPHATES, SLUDGE TREATMENT, DISSOLVED OXYGE W72-09675
N, ADSORPTION, ACTIVATED CARBON, SLUDGE TREATMENT, DIATOMS, COAGUL W73-01669
MIDGES, GAMMA RAYS, BLOODWORMS, SLUDGE WORMS.: /NESSEE, CRAYFISH, W70-06668
ETERGENTS, *TOXICITY, *ACTIVATED SLUDGE, *NUTRIENTS, NITROGEN, WAT W71-11972
STS, DETERGE/ *ALGAE, *ACTIVATED SLUDGE, *TOXICITY, *LABORATORY TE W70-08976
IDGES, LAGOONS, BIOMASS, OREGON, SLUDGE, ALGAE, DIPTERA, DISSOLVED W72-03556
EMAND, NITRIFICATION, NUTRIENTS, SLUDGE, ALGAE, BACTERIA, *WASTE W W72-04086
ERIA, WATER POLLUTION, ACTIVATED SLUDGE, ALGAE, PROTOZOA, DIATOMS, W72-08573
, PHOSPHORUS, METABOLISM, ALGAE, SLUDGE, ANALYTICAL TECHNIQUES, IN W71-11793
AEROBIC CONDITIONS, METABOLISM, SLUDGE, BIOCHEMICAL OXYGEN DEMAND W72-04789
TED CARBON, FLOCCULATION, ALGAE, SLUDGE, CHROMIUM, TOXICITY, NEUTR W71-09524
ROUS, ALGAE, NITROGEN, ACTIVATED SLUDGE, CHEMICAL OXYGEN DEMAND, O W71-13334
NDITIONS, ORGANIC LOADING, ODOR, SLUDGE, COST ANALYSIS, NITROGEN, W72-11642
ITIONS, BIODEGRADATION, SEEPAGE, SLUDGE, COST ANALYSIS, *BIOCHEMIC W71-07099
NTROL, EUTROPHICATION, ACTIVATED SLUDGE, DESIGN, NITROGEN, DIFFUSI W72-00800
BIOLOGICAL TREATMENT, *ACTIVATED SLUDGE, EFFLUENT, *ORGANIC WASTES W71-12223
TREATMENT, NUTRIENTS, SYNTHESIS, SLUDGE, EFFLUENT, ALGAE, SECONDAR W71-12188
Y, ALGAE, BACTERIA, FISH, SEWAGE SLUDGE, FLOW, FLOCCULATION, COAGU W70-06968
CONDITIONS, AEROBIC CONDITIONS, SLUDGE, LIGHT INTENSITY, COLIFORM W71-07106
WASTE WATER TREATMENT, ACTIVATED SLUDGE, LIME, AEROBIC TREATMENT, W72-09176
*BACTERIA, *SYMBIOSIS, ACTIVATED SLUDGE, NITROGEN, PHOSPHORUS, CHE W72-11100
RIENTS, DISSOLVED OXYGEN, ALGAE, SLUDGE, OXYGENATION, ANAEROBIC CO W71-07109
SEWAGE/ *PHOSPHORUS, *ACTIVATED SLUDGE, PHOSPHORUS RADIOISOTOPES, W71-01474
AWARE RIVER, VEGETATION EFFECTS, SLUDGE, RIVER BEDS, ALGAE, MATHEM W70-09189
ERIA, DIATOMS, OCEANS, ACTIVATED SLUDGE, SOIL BACTERIA, PHO: /BACT W72-10883
BIOLOGICAL TREATMENT, ACTIVATED SLUDGE, TERTIARY TREATMENT, ION E W70-09907
ATION, SAMPLING, MICROORGANISMS, SLUDGE, TERTIARY TREATMENT, BIOCH W71-07382
S, *INDUSTRIAL WASTES, ACTIVATED SLUDGE, TRICKLING FILTERS, LAGOON W71-09524
GEN ION CONCENTRATION, ACTIVATED SLUDGE, TREATMENT, EUTROPHICATION W71-00116
ACTERIA, *RESPIRATION, ACTIVATED SLUDGE, WASTE WATER TREATMENT, NU W71-04079
CONDITIONS, AEROBIC CONDITIONS, SLUDGE, WASTE WATER TREATMENT, BI W71-07084
ION, PROTOZOA, DISSOLVED OXYGEN, SLUDGE, WASTE WATER TREATMENT.: / W72-10233
OXIDES, *FLUID FRICTION, SEWERS, SLURRIES, WASTE WATER TREATMENT, W70-06968
S, FROGS, LICHENS, ALGAE, FUNGI, SMALL ANIMALS(MAMMALS), CYTOLOGIC W72-00949
*PHOSPHORUS CIRCULATION, SMALL LAKES.: W72-12543
FECTS, BAYS, WATERSHEDS(BASINS), SMALL WATERSHEDS, CROPS, EGGS, FI W72-00959
VY METALS, WASTEWATER TREATMENT, SMELTS, MICHIGAN, OHIO,: /ES, HEA W72-14282
ER EFFECT.: SMITH MOUNTAIN PROJECT(VA), WEATH W72-05282
*PUMPED STORAGE OPERATION, SMITH MOUNTAIN RESERVOIR(VA).: W72-14405
MOPEDIA GLAUCA, DACTYLOCOCCOPSIS SMITHII, OSCILLATORIA PRINCEPS, O W72-07896
BASS, PERCHES, PIKES, SAMPLING, SNAILS, ALGAE, DIATOMS, CLAMS, SH W70-08098
CTIONS, CRUSTACEANS, GASTROPODS, SNAILS, CRABS, BENTHIC FLORA, BEN W73-00932
COLI, AQUATIC ALGAE, SUNFISHES, SNAILS, EELS, CHINOOK SALMON, BAS W71-07417
RATORY TESTS, NITRATES, MUSSELS, SNAILS, FISH, ALGAE, GASTROPODS, W71-09168
D CHAINS, FISH, BIOASSAY, ALGAE, SNAILS, MOSQUITOES, MODEL STUDIES W72-07703
*ABSORPTION, AQUARIA, MOLLUSKS, SNAILS, MUSSELS, MARINE ALGAE, AQ W71-09182
INE ALGAE, MOSSES, CRABS, ALGAE, SNAILS, MUSSELS, ISOPODS, AMPHIPO W72-03347
VEL, LIFE HISTORY STUDIES, TIME, SNAILS, RADIOACTIVE WASTES, TENNE W70-06668
PRAIRIE LAKES, *LAKE HERMAN(SO. DAKOTA), *ORTHOPHOSPHATES.: W73-00938
AL CORPORATION, FMC CORPORATION, SOAP AND DETERGENT ASSOCIATION, C W70-07283
ETIC ACID, HOUSEHOLD DETERGENTS, SOAP AND DETERGENT INDUSTRY, NTA. W72-04266
TH, INSPECTION, WATER SOFTENING, SOAPS, AQUATIC ALGAE, WATER POLLU W72-01085
SPOSAL, WATER POLLUTION SOURCES, SOAPS, TOXICITY, PUBLIC HEALTH: / W72-01093
KELETONEMA COSTATUM, CHAETOCEROS SOCIALIS.: /HAETOCEROS AFFINIS, S W73-00442
ONEIS PLACENTULA, CYCLOTELLA CF. SOCIALIS, CYCLOTELLA STELLIGERA, W72-13142
M CACODY/ ACREMONIELLA, DIMETHYL SODIUM ARSONATE, ARSENATES, SODIU W72-13813
TUNICATES, PHAEOCYSTIS, SORTING, SODIUM BROMIDE, DEXTRAN, OIKOPLEU W73-00371
THYL SODIUM ARSONATE, ARSENATES, SODIUM CACODYLATE, CULTURE MEDIA, W72-13813
NITRATES, EFFLUENTS, DETERGENTS, SODIUM CHLORIDE, CHEMICAL PROPERT W72-09167
EAN, NITRATE, CHEMICAL ANALYSIS, SODIUM CHLORIDE, COLUMBIA RIVER, W72-05421
TS, TRISODIUM NITRILOTRIACETATE, SODIUM CITRATE.: /E-FREE DETERGEN W71-06247
RIE, RUNOFF, EROSION, CHLORELLA, SODIUM COMPOUNDS, ALKALINITY, MIN W72-04734
L, SODIUM LAURYL, ETHER SULFATE, SODIUM DODECYL, BENZENE SULFONATE W72-12576
*SUSPENDED SOLIDS, SODIUM HYDROXIDE, *ROCK FILTERS.: W71-13341
NDOSULFAN, FAIRY LIQUID, BLUSYL, SODIUM LAURYL, ETHER SULFATE, SOD W72-12576
LA PYRENOIDOSA, SODIUM SILICATE, SODIUM METASILICATE, LAKE WASHING W72-04734
PHATES, SODIUM TRIPOLYPHOSPHATE, SODIUM PYROPHOSPHATE.: /NSED PHOS W71-00116
ALGAL GROWTH, *RADIOISOTOPE P32, SODIUM PYROPHOSPHATE, RIO GRANDE W72-13992
IA RIVER, CHLORELLA PYRENOIDOSA, SODIUM SILICATE, SODIUM METASILIC W72-04734
LGAE, ALKALINE WATER, MAGNESIUM, SODIUM SULFATES, CYANOPHYTA, RHOD W71-04526
YROPHOSP/ *CONDENSED PHOSPHATES, SODIUM TRIPOLYPHOSPHATE, SODIUM P W71-00116
COSYSTEMS, POTASSIUM, MAGNESIUM, SODIUM, CALCIUM, HYDROGEN, CHEMIC W72-14303
, CALCIUM, MANGANESE, POTASSIUM, SODIUM, CARBON DIOXIDE, CYANOPHYT W73-00840
CONCENTRATION, NITROGEN, CARBON, SODIUM, CALCIUM, POT: /ROGEN ION W73-00854
LAKE ERIE, ALUMINUM, MANGANESE, SODIUM, CHROMIUM, IRON.: /CTERIA, W72-01101
, CHLORIDES, CALCIUM, MAGNESIUM, SODIUM, POTASSIUM, IRON, MANGANES W71-11001
OCARBON PESTICIDES, *CYANOPHYTA, SODIUM, POTASSIUM, CULTURES, RESI W72-13809

TRITES, IRON, SILICON, SULFATES, SODIUM, POTASSIUM, CALCIUM, MAGNE W72-05610
MANNOSE, RHAMNOSE, ACRYLIC ACID, SODIUM, PYRIDINE, DEXTRAM, RAFFIN W72-07710
S, NUTRIENTS, PHOSPHOROUS, WATER SOFTENING, LEGAL ASPECTS, HARD WA W71-09429
PUBLIC HEALTH, INSPECTION, WATER SOFTENING, SOAPS, AQUATIC ALGAE, W72-01085
, DOMESTIC WASTES, RUNOFF, WATER SOFTENING, ENVIRONMENTAL EFFECTS, W72-11186
MICAL PROPERTIES, SOIL BACTERIA, SOIL ALGAE.: /MPERATURE, SOIL CHE W72-03342
XIDE, *RADIOACTIVITY TECHNIQUES, SOIL ALGAE, *METHODOLOGY, MEASURE W72-14301
, LEACHING, VOLUMETRIC ANALYSIS, SOIL ANALYSIS, SPECTROMETERS, ZOO W72-05965
OTOPES, STRONTIUM RADIOISOTOPES, SOIL ANALYSIS, SOIL CONTAMINATION W73-01673
TERIA, ORGANIC MATTER, CULTURES, SOIL BACTERIA.: /THERMOPHILIC BAC W72-14301
ATURE, SOIL CHEMICAL PROPERTIES, SOIL BACTERIA, SOIL ALGAE.: /MPER W72-03342
ATOMS, OCEANS, ACTIVATED SLUDGE, SOIL BACTERIA, PHO: /BACTERIA, DI W72-10883
L ENVIRONMENT, SOIL TEMPERATURE, SOIL CHEMICAL PROPERTIES, SOIL BA W72-03342
S(MAMMALS), CYTOLOGICAL STUDIES, SOIL CHEMISTRY, RADIOSENSITIV: /L W72-00949
, LICHENS, MOSSES, FUNGI, ALGAE, SOIL CONTAMINATION, TRACE ELEMENT W72-14303
UM RADIOISOTOPES, SOIL ANALYSIS, SOIL CONTAMINATION, CLADOPHORA, C W73-01673
RESEARCH AND DEVELOPMENT, SOIL, SOIL CONTAMINATION.: /E, POULTRY, W71-07551
PS, NORTH SEA, PHAEOCYSTIS SPP., SOIL CULTURES, TITANIUM DIOXIDE W W72-12239
PHERE, ROOT SYSTEMS, SOIL FUNGI, SOIL ENVIRONMENT, SOIL TEMPERATUR W72-03342
ES, DDT, RETURN FLOW, SEDIMENTS, SOIL EROSION, FERTILIZERS, FARM W W71-06444
, ANIMAL WASTES, SURFACE WATERS, SOIL EROSION, NUTRIENTS, ALGAE, L W71-04216
SE, SHORES, MONITORING, MARSHES, SOIL EROSION, WASTE WATER TREATME W73-02479
A, *BAY OF BENGAL, INDIAN OCEAN, SOIL EXTRACTS, ERDSCHREIBER MEDIU W73-00430
LITY, RHIZOSPHERE, ROOT SYSTEMS, SOIL FUNGI, SOIL ENVIRONMENT, SOI W72-03342
LEACHING, DRAINAGE, FARM WASTES, SOIL MANAGEMENT, GROUNDWATER, WIS W71-06443
TORY TESTS, STATISTICAL METHODS, SOIL MICROORGANISMS, RUNOFF, HYDR W72-02218
IA, SAMPLING, CULTURES, ECOLOGY, SOIL MICROBIOLOGY, CYANOPHYTA, FR W73-00854
ADATION, CLASSIFICATION, SEWAGE, SOIL PROPERTIES, GROUNDWATER MOVE W72-14363
VE USE, WATER POLLUTION, OXYGEN, SOIL PROPERTIES, IDAHO.: /NSUMPTI W72-14878
S, SOIL FUNGI, SOIL ENVIRONMENT, SOIL TEMPERATURE, SOIL CHEMICAL P W72-03342
SIMULATED RAINFALL, *SOIL TYPES, SOIL TEXTURE, CLAYS, LABORATORY T W72-02218
UTION, WATER POLLUTION, LICHENS, SOIL-WATER-PLANT RELATIONSHIPS, M W72-03347
RAIN FORESTS, CYCLING NUTRIENTS, SOIL-WATER-PLANT RELATIONSHIPS, R W72-00949
ULTRY, RESEARCH AND DEVELOPMENT, SOIL, SOIL CONTAMINATION.: /E, PO W71-07551
ON, LAKES/ *SOIL BACTERIA, *LAKE SOILS, *PHOSPHORUS, *EUTROPHICATI W72-04292
N, OLIGOTROPHY, HARDNESS(WATER), SOILS, AMINO: /EEDS, PHYTOPLANKTO W72-08459
, ERDSCHREIBER MEDIUM, ESTUARINE SOILS, AXENIC CULTURES, GIBBERELI W73-00430
TION-REDUCTION POTENTIAL, LAKES, SOILS, EUTROPHICATION, AMMONIFICA W72-07933
MICROBIOLOGY, CYANOPHYTA, FROZEN SOILS, IONS, NOSTOC, ENVIRONMENTA W73-00854
TION SOURCES, SEDIMENTS, AQUATIC SOILS, PLANT TISSUES, FRESHWATER W73-01673
, TREES, GRASSES, LEAVES, FOREST SOILS, SAMPLING, MUD, PHYTOPLANKT W72-01784
ERATURE, CROP PRODUCTION, FUNGI, SOILS, SYMBIOSIS, AFRICA, ASIA, A W73-02473
*PESTICIDES, *ALGAE, *TOXICITY, SOILS, TRIAGINE PESTICIDES, RESIS W71-07675
FLOCCULANT DOSAGES, BAROCO CLAY SOILS, URANIUM ORE.: /, CARBOWAX, W70-06968
ISOTOPES, PHOSPHATES, SILICATES, SOLAR RADIATION, DEPTH, THERMOCLI W71-07875
DING, BIOCHEMICAL OXYGEN DEMAND, SOLAR RADIATION, NUTRIENTS, TEMPE W71-07124
CE, TEMPERATURE, STRATIFICATION, SOLAR RADIATION, ALGAE, PHOTOSYNT W71-07084
LYSIS, TIDAL EFFECTS, STABILITY, SOLAR RADIATION, PHOTOPERI: / ANA W72-08141
RY PRODUCTIVITY, EUTROPHICATION, SOLAR RADIATION, DISTRIBUTION, DE W72-08054
NKTON, ZOOPLANKTON, CHLOROPHYLL, SOLAR RADIATION, VITAMIN B, PIGME W72-07892
TIES, SEASONAL, DIATOMS, *LIGHT, SOLAR RADIATION, *PHOTOPERIODISM, W72-07901
E, TURBIDITY, HYDROLOGIC BUDGET, SOLAR RADIATION, WATER TEMPERATUR W72-02274
THEMATICAL MODELS, GROWTH RATES, SOLAR RADIATION, FLOTATION, FLOCC W72-12289
TREATMENT, *BYPRODUCTS, MIXING, SOLAR RADIATION, ORGANIC LOADING, W72-13508
YNCHRONOUS CULTURES, SCENEDESMUS SOLI, SCENEDESMUS QUADRICAUDA, MO W72-07144
WAGE, BIOCHEMICAL OXYGEN DEMAND, SOLID WASTES, AMMONIA, NITROGEN C W72-09675
OGENIC BACTERIA, ORGANIC WASTES, SOLID WASTES, CHEMICALS, MICHIGAN W71-11006
CHARACTERISTICS, SOLIDS RECYCLE, SOLIDS CONTROL.: /OGEN, SETTLING W72-11100
ROGEN, SETTLING CHARACTERISTICS, SOLIDS RECYCLE, SOLIDS CONTROL.: / W72-11100
SUSPENDED SOLIDS.: W71-12188
SUSPENDED SOLIDS.: W71-07123
*AERATED LAGOONS, SUSPENDED SOLIDS.: W71-07106
*AERATED LAGOONS, SUSPENDED SOLIDS.: W71-07113
YPT), PANDORINA, SOLUBLE ORGANIC SOLIDS.: *ALEXANDRIA(EG W70-06619
CHARGE, REGENERATION, SUSPENDED SOLIDS.: *SURFACE W71-00138
GUSTATA, IPOMOEA AQUATICA, TOTAL SOLIDS.: /NIA CRASSIPES, TYPHA AN W72-01363
NETICS, *SCENEDESMUS, *DISSOLVED SOLIDS, *MICROBIAL DEGRADATION, C W73-00379
ACIL/ *STORM RUN-OFF, *SUSPENDED SOLIDS, *RECREATION, *TREATMENT F W72-00042
CATION, *TASTE, *ODOR, DISSOLVED SOLIDS, ALGAE, RESERVOIR STORAGE, W73-02138
NS/ *COMBINED SEWERS, *SUSPENDED SOLIDS, BACKWASHING, *MOUNT CLEME W72-00042
FLORA, ORGANIC MATTER, DISSOLVED SOLIDS, BENTHOS, CARBON, PRODUCTI W73-02100
ACTERIA, ODOR, INFLOW, DISSOLVED SOLIDS, DATA COLLECTIONS.: /AE, B W72-02274
ALGAE, *ION EXCHANGE, DISSOLVED SOLIDS, FLOCCULATION, HEAD LOSS, W71-00138
ITES, JORDAN RIVER(M/ SETTLEABLE SOLIDS, HATCHERY EFFLUENTS, PARAS W71-11006
, CHEMICAL PROPERTIES, DISSOLVED SOLIDS, HYDROLOGIC DATA, DATA COL W72-13851
A, NITROGEN, VITAMINS, SUSPENDED SOLIDS, LACTOBACILLUS, CHLORELLA, W72-12565
FATES, DRAINAGE WATER, DISSOLVED SOLIDS, NITROGEN CYCLE, HYDROGEN W71-01489
ONDITIONS, *KINETICS, *DISSOLVED SOLIDS, ORGANIC MATTER, NITROGEN, W73-01066
TION, LIGHT INTENSITY, DISSOLVED SOLIDS, PHYTOPLANKTON, ALGAE, NUT W72-04270
*ANABAENA, NUTRIENTS, DISSOLVED SOLIDS, PHYTOPLANKON, WATER POLLU W72-13300
NG.: FIREBAUGH(CALIF), VOLATILE SOLIDS, SANBORN FILTER, BACKWASHI W70-10174
ALGAE, PHOTOSYNTHESIS, SUSPENDED SOLIDS, SEPARATION TECHNIQUES, FI W72-09386
ILTERS.: *SUSPENDED SOLIDS, SODIUM HYDROXIDE, *ROCK F W71-13341
EATMENT, MIXED LIQUOR, SUSPENDED SOLIDS, UPTAKE.: /L-BIOLOGICAL TR W71-13334
ERGY, DIFFUSION, THERMODYNAMICS, SOLUBILITY, SORPTION, PESTICIDES, W71-12092
CONCENTRATION, ECONOMICS, ALGAE, SOLUBILITY, SEWAGE, BIOCHEMICAL O W71-12200
NATION, *ALGAE, *PHOTOSYNTHESIS, SOLUBILITY, CHEMICAL OXYGEN DEMAN W72-00411
ATURE, KINETICS, NUCLEAR WASTES, SOLUBILITY, ADSORPTION, WATER POL W71-09182
ABORATORY EXPER/ *RUTHENIUM 106, SOLUBLE FORMS, INSOLUBLE FORMS, L W71-13233
*ALEXANDRIA(EGYPT), PANDORINA, SOLUBLE ORGANIC SOLIDS.: W70-06619

CLAMS, FISH, SHELLFISH, AQUEOUS SOLUTIONS, PHAEOPHYTA, ATLANTIC O W72-05965
ATION, *AQUATIC PLANTS, *AQUEOUS SOLUTIONS, *BIOLOGICAL PROPERTIES W72-01473
OLLUTANT IDENTIFICATION, AQUEOUS SOLUTIONS, ION EXCHANGE, METHODOL W72-12166
RITUS, CHLOROPHYTA, CHRYSOPHYTA, SOLVENT EXTRACTIONS, DIATOMS, AQU W72-13673
ORY EQUIPMENT, AEROBIC BACTERIA, SOLVENT EXTRACTIONS, FREEZE DRYIN W73-01066
E DRYING, SEPARATION TECHNIQUES, SOLVENT EXT: /RY EQUIPMENT, FREEZ W73-00379
EFFECTS, ALGAL POISONING, ALGAE, SOLVENTS, SURFACTANTS, WATER TEMP W70-09429
OSE, SUCROSE, MALTOSE, MANNITAL, SORBITAL, SUCCINATE, FUMARATE, BU W72-05421
R CONTRACTUM, ARTHURS LAKE, LAKE SORELL, LAKE CRESCENT, TOOMS LAKE W73-01094
ON, SURVIVAL, CHLORE/ *CHLORELLA SOROKINIANA, CO-60, GAMMA RADIATI W72-11836
YSTEM, PLANT EXTRACTS, CHLORELLA SOROKINIANA.: /UCIFERASE ENZYME S W71-11916
ION, PHOSPHORUS REMOVAL, RATE OF SORPTION.: /MUD, PHOSPHORUS SORPT W72-11617
CONTROL, NUTRIENT REQUIREMENTS, SORPTION, EUTROPHICATION, DEFICIE W72-11617
E OF SORPT/ LAKE MUD, PHOSPHORUS SORPTION, PHOSPHORUS REMOVAL, RAT W72-11617
ION, THERMODYNAMICS, SOLUBILITY, SORPTION, PESTICIDES, TOXICITY, D W71-12092
VIRONMENTAL EFFECTS, ABSORPTION, SORPTION, WASTE DILUTION, WASTE A W73-00817
ROWTH RATES, THERM/ TEMPERATURE, SORPTION, RADIOISOTOPES, ALGAE, G W71-02075
ADIENTS, TUNICATES, PHAEOCYSTIS, SORTING, SODIUM BROMIDE, DEXTRAN, W73-00371
RADIOISOTOPES, BOGUE SOUND, CORE SOUND.: MANGANESE W73-00818
MANGANESE RADIOISOTOPES, BOGUE SOUND, CORE SOUND.: W73-00818
ACTOSE, HEXURONIC ACID, VINEYARD SOUND, ORGANIC CARBON, ACID-VOLAT W72-07710
*SEA GRASSES, TELEVISION, DEPTH SOUNDER, VERTICAL DISTRIBUTION, C W72-09120
AIN, NORTH CAROLINA, ABSORPTION, SOUNDS, ESTUARINE ENVIRONMENT, PA W73-00818
ACETATE, *CHLORELLA PYRENOIDOSA, SOURCE OF NITROGEN, NTA.: /ILOTRI W70-08976
N EXTRUSION, CARBON-14, NITROGEN SOURCE.: /NITZSCHIA OVALIS, CARBO W72-07497
WATER POLLUTION, WATER POLLUTION SOURCES.: /BENTHOS, ALGAE, FISH, W72-07841
, BIOINDICATORS, WATER POLLUTION SOURCES.: /COLOGICAL DISTRIBUTION W72-08579
ES, FOOD CHAINS, WATER POLLUTION SOURCES.: /S PESTICIDES, ACARICID W73-00238
NTS, SURPLUS NUTRIENTS, NUTRIENT SOURCES.: /WIS.), LIMITING NUTRIE W73-00232
SSAY, ORTHOPHOSPHATE, PHOSPHORUS SOURCES.: ALGAL A W73-02478
PES, ABSORPTION, WATER POLLUTION SOURCES.: /PES, COBALT RADIOISOTO W71-09018
ICULATE FORM, FEEDLOTS, NUTRIENT SOURCES.: /BOLISM, MOBILITY, PART W71-06443
ALGAE, BACTERIA, WATER POLLUTION SOURCES.: /EN ION CONCENTRATION, W71-07942
TING NUTRIENTS, ANABAENA, CARBON SOURCES.: *LIMI W72-09168
LLUTION EFFECTS, WATER POLLUTION SOURCES.: /TS, NITROGEN, WATER PO W71-11972
EMICAL ANALYSIS, WATER POLLUTION SOURCES.: /AE, AQUATIC PLANTS, CH W71-12252
ATES, *RED TIDE, WATER POLLUTION SOURCES.: / DIATOMS, *DINOFLAGELL W72-01329
, PUBLIC HEALTH, WATER POLLUTION SOURCES.: /ALIFORNIA, FOOD CHAINS W72-03324
TES, OIL SPILLS, WATER POLLUTION SOURCES.: /MENT, SEDIMENTATION RA W72-12487
SEWAGE EFFLUE/ *WATER POLLUTION SOURCES, *EUTROPHICATION, *LAKES, W72-12955
*NEBRASKA, URB/ *WATER POLLUTION SOURCES, *PHOSPHATES, *NITRATES, W71-13466
LUTION EFFECTS, *WATER POLLUTION SOURCES, *FISHERIES, OIL WASTES, W71-13723
NIDS, WATER RE/ *WATER POLLUTION SOURCES, *FISH HATCHERIES, *SALMO W71-11006
E/ *PHOSPHATES, *WATER POLLUTION SOURCES, *EUTROPHICATION, *DETERG W72-11186
OXYGEN DEMAND, *WATER POLLUTION SOURCES, *WATER POLLUTION EFFECTS W70-09189
EUTROPHICATION, *WATER POLLUTION SOURCES, *ESTUARIES, RIVERS, PHOS W70-06509
S, *PHOSPHATES, *WATER POLLUTION SOURCES, *POLLUTION ABATEMENT, EU W72-06655
, IMPOUNDMENTS, *WATER POLLUTION SOURCES, *AGRICULTURAL RUNOFF, *W W72-08401
LLUTION EFFECTS, WATER POLLUTION SOURCES, *FISH, DIATOMS, PROTOZOA W72-07791
URY, DETERGENTS, WATER POLLUTION SOURCES, ASBESTOS, EUTROPHICATION W73-00703
TURAL CHEMICALS, WATER POLLUTION SOURCES, ALGAE, GEOLOGY, DATA COL W72-05869
ATER(POLLUTION), WATER POLLUTION SOURCES, ALGAE, PLANKTON, SEASONA W72-05506
LLUTION EFFECTS, WATER POLLUTION SOURCES, ALGAE, NUTRIENTS.: /R PO W72-04280
WATER POLLUTION, WATER POLLUTION SOURCES, AQUATIC ALGAE, FOOD CHAI W72-10686
E MENDOTA(WIS), LAKE / *NUTRIENT SOURCES, ACETYLENE REDUCTION, LAK W72-10608
LLUTION EFFECTS, WATER POLLUTION SOURCES, AQUATIC LIFE, AQUATIC AN W72-09646
ZENE SULFONATES, WATER POLLUTION SOURCES, AQUATIC ALGAE, PUBLIC HE W71-11516
ITY, RESERVOIRS, WATER POLLUTION SOURCES, ALGAE, RIVER FLOW, PHOSP W72-04260
NAL, PHOSPHORUS, WATER POLLUTION SOURCES, BIOCHEMICAL OXYGEN DEMAN W72-09167
EMICAL ANALYSIS, WATER POLLUTION SOURCES, BIOASSAY, ALGAE, DOMESTI W73-02478
INE ENVIRONMENT, WATER POLLUTION SOURCES, BIODEGRADATION, ECOSYSTE W73-00853
ORGANIC MATTER, WATER POLLUTION SOURCES, CHLOROPHYTA, AQUATIC PRO W73-00379
OLLUTION EFFECT, WATER POLLUTION SOURCES, COST COMPARISONS, BENEFI W72-00974
NITROGEN, ALGAE, WATER POLLUTION SOURCES, COLORADO, CATTLE, URINE, W71-11496
REATMENT, ALGAE, WATER POLLUTION SOURCES, CHLOROPHYTA, CYANOPHYTA, W72-05473
LLUTION EFFECTS, WATER POLLUTION SOURCES, CULTIVATION.: / WATER PO W71-02880
OHIO, NUTRIENTS, WATER POLLUTION SOURCES, DISSOLVED OXYGEN, LAKE M W71-11949
LUTION CONTROL, *WATER POLLUTION SOURCES, EUTROPHICATION, LEGISLAT W72-01085
LLUTION EFFECTS, WATER POLLUTION SOURCES, EUTROPHICATION, COLIFORM W71-09173
UTION TREATMENT, WATER POLLUTION SOURCES, ENVIRONMENTAL SANITATION W71-09674
ATTER, NITROGEN, WATER POLLUTION SOURCES, ENVIRONMENTAL EFFECTS, A W73-01066
YTA, PHOSPHORUS, WATER POLLUTION SOURCES, HYDROGEN ION CONCENTRATI W72-04769
LLUTION EFFECTS, WATER POLLUTION SOURCES, INDUSTRIAL WASTES, DOMES W71-03095
ENT, FRESHWATER, WATER POLLUTION SOURCES, INSTRUMENTATION, METHODO W73-01445
ODOLOGY, RIVERS, WATER POLLUTION SOURCES, ION EXCHANGE, *CHEMICAL W73-00286
FECTS, *DIATOMS, WATER POLLUTION SOURCES, INDUSTRIAL WASTES, AQUAT W71-11583
LLUTION EFFECTS, WATER POLLUTION SOURCES, INHIBITION.: /, WATER PO W72-11727
LLUTION EFFECTS, WATER POLLUTION SOURCES, LIMNOLOGY.: /E, WATER PO W72-01094
, AQUATIC WEEDS, WATER POLLUTION SOURCES, MATHEMATICAL MODELS, NIT W72-09171
, LAKE SUPERIOR, WATER POLLUTION SOURCES, MINNESOTA, WASHINGTON, C W71-04216
WATER CHEMISTRY, WATER POLLUTION SOURCES, METEOROLOGICAL DATA, SEA W72-04853
SO/ *SURFACE WATERS, GEOLOGICAL SOURCES, MINERALIZATION, NITROGEN W71-06435
LLUTION EFFECTS, WATER POLLUTION SOURCES, NITRIFICATION, ALGAE.: / W71-05390
TERS, *REVIEWS, *WATER POLLUTION SOURCES, NUTRIENTS, PLANT PHYSIOL W71-10580
TOMS, TURBIDITY, WATER POLLUTION SOURCES, NUTRIENTS, SEWAGE, WATER W72-07890
EUTROPHICATION, WATER POLLUTION SOURCES, NUTRIENTS.: /POLLUTION), W72-13644
DUSTRIAL WASTES, WATER POLLUTION SOURCES, ON-SITE INVESTIGATIONS, W72-08804
CARBON DIOXIDE, WATER POLLUTION SOURCES, OIL SPILLS, SEA WATER, M W73-01074
N, CARBON TRANSFORMATION, CARBON SOURCES, ORGANIC CARBON, CARBON U W71-12068
EUTROPHICATION, *WATER POLLUTION SOURCES, POPULATION, INDUSTRIES, W71-12091

, PHYTOPLANKTON, WATER POLLUTION SOURCES, PHAEOPHYTA, OXIDATION.: / W72-12166
ESTICIDES, ECOSYSTEMS, POLLUTION SOURCES, PESTICIDE TOXICITY, FOOD W72-07703
TA, COMPETITION, WATER POLLUTION SOURCES, POLLUTANT IDENTIFICATION W73-00441
H OF POLLUTANTS, WATER POLLUTION SOURCES, PUBLIC HEALTH, ON-SITE I W73-00790
DES, HERBICIDES, WATER POLLUTION SOURCES, PLANKTON, OYSTERS, SHEL: W72-13347
MICROORGANISMS, *WATER POLLUTION SOURCES, REVIEWS, MICROBIAL DEGRA W72-14363
ISH, ABSORPTION, WATER POLLUTION SOURCES, RADIOACTIVITY EFFECTS, R W72-00940
WASTE DISPOSAL, WATER POLLUTION SOURCES, SOAPS, TOXICITY, PUBLIC W72-01093
N, MARINE ALGAE, WATER POLLUTION SOURCES, SPATIAL DISTRIBUTION, BI W73-00431
AQUATIC PLANTS, WATER POLLUTION SOURCES, SEDIMENTS, AQUATIC SOILS W73-01673
WATER POLLUTION, WATER POLLUTION SOURCES, SEWAGE, INDUSTRIAL WASTE W72-12729
AE, SURFACTANTS, WATER POLLUTION SOURCES, SURFACE WATERS, WATER PO W70-09388
, SHRIMP, CRABS, WATER POLLUTION SOURCES, SEA WATER, TROUT, TOXINS W71-07412
LGAE, NUTRIENTS, WATER POLLUTION SOURCES, SEWAGE TREATMENT, PUBLIC W72-06638
LLUTION CONTROL, WATER POLLUTION SOURCES, TRIBUTARIES, FRESHWATER W72-14282
*BIBLIOGRAPHIES, WATER POLLUTION SOURCES, THERMAL POLLUTION, REVIE W72-12190
ICAL TECHNIQUES, WATER POLLUTION SOURCES, TOXINS, WASTES, CHLOROPH W71-10553
ZERS, EFFLUENTS, WATER POLLUTION SOURCES, WATER POLLUTION EFFECTS, W71-11008
, LAKE MICHIGAN, WATER POLLUTION SOURCES, WATER POLLUTION CONTROL, W71-12072
EWAGE TREATMENT, WATER POLLUTION SOURCES, WATER POLLUTION EFFECTS, W72-10625
TROL, STANDARDS, WATER POLLUTION SOURCES, WATER POLLUTION EFFECTS, W72-01660
WATER POLLUTION, WATER POLLUTION SOURCES, WATER POLLUTION TREATMEN W72-01685
, ON-SITE TESTS, WATER POLLUTION SOURCES, WATER POLLUTION EFFECTS, W72-03228
LLUTION EFFECTS, WATER POLLUTION SOURCES, WATER POLLUTION CONTROL. W72-04759
PHOTOSYNTHESIS, WATER POLLUTION SOURCES, WATER POLLUTION EFFECTS, W72-05952
WATER POLLUTION, WATER POLLUTION SOURCES, WATER POLLUTION EFFECTS, W72-04279
OURCES, MINERALIZATION, NITROGEN SOURCES, WELL WATER, FEEDLOTS.: / W71-06435
EROSION CONTROL, WATER POLLUTION SOURCES, WATER LAW, ALGAE, DISSOL W71-01192
LAWRENCE RIVER, WATER POLLUTION SOURCES, WATER POLLUTION EFFECTS, W70-08098
HY, OLIGOTROPHY, WATER POLLUTION SOURCES, WATER POLLUTION EFFECTS, W72-14782
WINDHOEK, SOUTH AFRICA.: W70-10381
ORPTION, CALCIUM, PUBLIC HEALTH, SOUTH AMERICA, ATLANTIC COASTAL P W73-00819
AL DISTRIBUTION, NORTH CAROLINA, SOUTH CAROLINA, SAMPLING, MICROSC W72-07663
OPHIC POPULATIONS, LAKE HARTWELL(SOUTH CAROLINA), ANACYSTIS NIDULA W71-00475
ENT, CARBON DIOXIDE, ECOSYSTEMS, SOUTH CAROLINA, ALGAE, TROPHIC LE W71-00475
S, DIATONICS, CYANOPHYTA, ALGAE, SOUTH DAKOTA, WATER POLLUTION EFF W71-02880
RICT, *INDIAN CREEK RESERVOIR, / SOUTH TAHOE PUBLIC UTILITIES DIST W72-02412
K-WAVES, *GAS-VACUOLE DEFLATION, SOUTH WALES, MICROCYSTIS AERUGINO W72-11564
ESERVOIRS, ODOR-PRODUCING ALGAE, SOUTHWEST U.S., TEMPERATURE, OXYG W72-10613
E HURON, LAKE ERIE, GREAT LAKES, SOYBEANS, LIGHT INTENSITY, AIR TE W73-01456
H CULTU/ *HETEROCYSTS, *ANABAENA SP L-31, CONTINUOUS CULTURE, BATC W72-03217
*LAMINARIA SP., *ULVA SP., MACROCYSTIS SP.: W71-10553
ITIC ACID,/ *CIGUATERA, CAULERPA SP., CAULERPICIN, CAULERPIN, PALM W71-08597
SP., POLYSIPHONIA SP., GELIDIUM SP., CHLORELLA, AGARS.: /ADOPHORA W71-11851
., PHYLLOSPADEX SP., MICROCLADIA SP., CLADOPHORA SP., POLYSIPHONIA W71-11851
HYALOCHLORELLA MARINA, RHODOMELA SP., ENDOCLADIA SP., PHYLLOSPADEX W71-11851
P., CLADOPHORA SP., POLYSIPHONIA SP., GELIDIUM SP., CHLORELLA, AGA W71-11851
*LAMINARIA SP., *ULVA SP., MACROCYSTIS SP.: W71-10553
P., ENDOCLADIA SP., PHYLLOSPADEX SP., MICROCLADIA SP., CLADOPHORA W71-11851
ARINA, RHODOMELA SP., ENDOCLADIA SP., PHYLLOSPADEX SP., MICROCLADI W71-11851
SP., MICROCLADIA SP., CLADOPHORA SP., POLYSIPHONIA SP., GELIDIUM S W71-11851
ARIA SP., *ULVA SP., MACROCYSTIS SP.: *LAMIN W71-10553
ASERS, MICROWAVE RADIOMETRY, VI/ SPACE PLATFORMS, LEAD, MERCURY, L W72-04742
ENTHIC MACROPHYTES, RIA DE ALDAN(SPAIN), LAMINARIA OCHROLEUCA, SAC W72-04766
ENTIAL, ANTICOAGULANT POTENTIAL, SPARGANIUM FLUCTUANS, ANTINEOPLAS W72-04275
, CAREX SPP, BUTOMUS UMBELLATUS, SPARGANIUM RAMOSUM, LEERSIA ORYZO W73-01089
ATRAN RIVER, MOTALA STROM RIVER, SPARGANIUM ERECTUM, SCHOENOPLECTU W73-00286
FIMB/ ENGLAND, SWEDEN, GERMANY, SPARTINA SPP, DISTICHLIS SPICATA, W73-01089
S, THERMAL STRESS, GROWTH RATES, SPATIAL DISTRIBUTION, PRIMARY PRO W73-00853
ALGAE, WATER POLLUTION SOURCES, SPATIAL DISTRIBUTION, BIOMASS, PH W73-00431
EDIMENTS, SEPARATION TECHNIQUES, SPATIAL DISTRIBUTION, ECOLOGICAL W73-00367
ON, NITROGEN, LITTORAL, HARBORS, SPATIAL DISTRIBUTION, ALGAE, TRIB W72-04263
*TROPICAL REGIONS, MARINE ALGAE, SPATIAL DISTRIBUTION, FOURIER ANA W72-11719
ROBIC CONDITIONS, ALGAL CONTROL, SPAWNING, WARM-WATER FISH.: /ANAE W70-09595
MOGETON PECTINATUS, MYRIOPHYLLUM SPECATUM, ELODEA CANADENSIS, PHRA W71-11012
CTS, RADIOISOTOPES, COMPETITION, SPECIATION.: /ATER POLLUTION EFFE W71-09256
ANKTON, *DENSITY STRATIFICATION, SPECIATION, VERTICAL MIGRATION, A W71-09852
PHAEOPHYTA, CHLOROPHYTA, MAINE, SPECIATION.: / FLORA, RHODOPHYTA, W71-10066
ITE INVESTIGATIONS, FOOD CHAINS, SPECIATION, REPRODUCTION.: / ON-S W73-00764
, *SYSTEMATICS, DINOFLAGELLATES, SPECIATION.: /OPHYTA, *PYRROPHYTA W72-07683
*SWEDEN, SPECIES COMPOSITION.: W73-01632
LOIDES, CHLOROPHYL/ DESSICATION, SPECIES DIVERSITY, MELOSIRA NUMMU W73-00853
VERTEBRATES, MACROINVERTEBRATES, SPECIES DIVERSITY, PORIFERA, SEA W73-00932
PE COD BAY, DATA INTERPRETATION, SPECIES DIVERSITY.: /ARDS BAY, CA W73-01449
M, CAPE HATTERAS, NANSEN BOTTLE, SPECIES DIVERSITY, SAMPLE PRESERV W72-07663
LEY CREEK(PA), LITITZ CREEK(PA), SPECIES DIVERSITY, DIRECT EFFECTS W72-08579
DATA INTERPRETATION, MACROALGAE, SPECIES DIVERSITY INDEX, FRAGILAR W72-08141
MICROCOSM ALGAL ASSAY PROCEDURE, SPECIES DIVERSITY INDEX.: * W72-03218
RO/ SAN PEDRO(CALIF), EPIBENTHIC SPECIES, ABALONE, TUNICATES, ARTH W71-01475
DIELLA CONCHOPHILA MILLE/ *ALGAL SPECIES, WESTERN LAKE ERIE, ARNOL W71-09156
Y, ORTHOP/ GREEN BAY, FOX RIVER, SPECIFIC CONDUCTANCE, TRANSPARENC W71-10645
TIC CONTINUOUS CULTURES, MAXIMUM SPECIFIC GROWTH RATE, SATURATION W72-04788
PRECOAT FILTRATION, BACKWASHING, SPECIFIC RESISTANCE.: /RAINING, * W72-09393
SYPHAR PORPHYRAE, PROTECTOCARPUS SPECIOSUS.: /UELON(CANADA), MIKRO W71-10066
UMETRIC ANALYSIS, SOIL ANALYSIS, SPECTROMETERS, ZOOPLANKTON, MUSSE W72-05965
ES, ALIPHATIC HYDROCARBONS, MASS SPECTROMETRY, INFRARED SPECTROSCO W73-01439
ES, RADIOACTIVE DECAY, GAMMA-RAY SPECTROMETRY, COUNTING.: /ACTINID W73-01434
137, *PAHRANAGAT LAKE, GAMMA RAY SPECTROMETRY, ANIMAL TISSUES, MAC W73-01673
NTS, COMBUSTION, ISOTOPE RATIOS, SPECTROMETRY, LICHENS, MOSSES, GY W72-12709
ARINE PLANTS, BICARBONATES, MASS SPECTROMETRY, DISTRIBUTION, SEPAR W72-12709

RATION TECHNIQUES, DISTILLATION,	SPECTROPHOTOMETRY, CHLORCPHYLL, H	W72-12637
LUATION, *SEPARATION TECHNIQUES,	SPECTROPHOTOMETRY, LAKES, PLANT P	W72-13673
LOGICAL COMMUNITIES, ECOSYSTEMS,	SPECTROPHOTOMETRY, VIRGINIA, SAMP	W73-00916
SAMPLING, WATER ANALYSIS, ALGAE,	SPECTROPHOTOMETRY, PENNSYLVANIA,	W72-04743
ORESIS, DETERGENTS, AMINO ACIDS,	SPECTROPHOTOMETRY, MAGNESIUM, SUL	W72-06124
LE PRESERVATION, *FLAME EMISSION	SPECTROPHOTOMETRY, PERCH LAKE(ONT	W71-11529
TS, NEUTRON ACTIVATION ANALYSIS,	SPECTROSCOPY, TRACE ELEMENTS, SAM	W73-00832
CTERIA, ALGAE, PLANKTON, PLANTS,	SPECTROSCOPY, CHROMATOGRAPHY, GAS	W73-01439
ONS, MASS SPECTROMETRY, INFRARED	SPECTROSCOPY.: /IPHATIC HYDROCARB	W73-01439
, PAPER CHROMATOGRAPHY, INFRARED	SPECTROSCOPY, CULTURE MEDIA, FATT	W73-01066
, PAPER CHROMATOGRAPHY, INFRARED	SPECTROSCOPY, ASSIMILATION, KJEDA	W73-00379
ARGASSO SEA, SAMPLE PREPARATION,	SPERMATOPHYTES, PHOTOASSIMILATION	W73-02100
E BUCKHORN(GEORGIA), HYPNODINIUM	SPHAERICUM.: /DINAL(GEORGIA), LAK	W71-11914
MENIDIA, BRACHYDEUTERA, CHYDORUS	SPHAERICUS, CHYDORUS, MONOSTYLA,	W72-07132
ERIA FRITSCHII, PANDORINA MORUM,	SPHAEROCYSTIS SCHROETERI, ULOTHRI	W72-09111
ARIA, ASTERIONELLA, CRYPTOMONAS,	SPHAEROCYSTIS, DAPHNIA GALEATA ME	W71-03014
CANADA, NOVA SCOTIA, ISTHMOPLEA	SPHAEROPHORA, MELANOSIPHON INTEST	W72-01370
MA PARVULUM, BACILLARI PARADOXA,	SPHAEROTILUS NAT: /TATA, GOMPHONE	W72-10883
ROBACTER, NITROSOCYSTIS OCEANUS,	SPHAEROTILUS NATANS, GLUCOSE, GAL	W72-05421
NITRIFICATION, AMMONIA, NITRITE,	SPHAEROTILUS, NUTRIENTS, PACIFIC	W72-05421
BACTERICIDES, ATOLLS, CULTURES,	SPHAEROTILUS, CLADOPHORA.: /DES),	W72-08142
*UNITED KINGDOM, *SEWAGE FUNGUS,	SPHAEROTOTILUS NATANS, ZOOGLOEAL	W72-08573
E, CERATIUM,/ DOLLAR LAKE(OHIO),	SPHAGNUM, APHANIZOMENON FLOS-AQUA	W71-11949
SUBMARINE LIGHT DATA	SPHERE.:	W71-13252
ERMANY, SPARTINA SPP, DISTICHLIS	SPICATA, FIMBRISTYLIS, BORRICHIA,	W73-01089
AND, CANAL ZONE, MACROINVE/ *OIL	SPILLS, S.S. WITWATER, GALETA ISL	W72-07911
DE, WATER POLLUTION SOURCES, OIL	SPILLS, SEA WATER, MARINE PLANTS,	W73-01074
LOW PERCH, SEWAGE EFFLUENTS, OIL	SPILLS, THERMAL POLLUTION, NUTRIE	W72-14282
ATMENT, SEDIMENTATION RATES, OIL	SPILLS, WATER POLLUTION SOURCES.:	W72-12487
S ACULEATUS, BALTIC SEA, NITOCRA	SPINIPES.: / SALINUS, GASTEROSTEU	W71-08026
*FUCUS VIRSOIDES, FUCUS	SPIRALIS, MEDITERRANEAN SEA.:	W72-12940
ALICULATA, FUCUS SERRATUS, FUCUS	SPIRALIS, FUCUS VESICULOSUS, LAMI	W71-11515
GIAMONAS PLICATA, TRICHOCYANELLA	SPIRALIS, STROMIA SUBSPHAERICA, T	W72-09119
OPHYLL A, CHLORELLA PYRENOIDOSA,	SPIROGIRA, CHLORELLA ELLIPSOIDEA,	W72-06274
LVOX, HYDRODICTYON, SPODYLOSIUM,	SPIROGYRA.: /GANIC IONS, ZINC, VO	W71-11001
ROSIGMA, NAUPLII, LAMELLIBRANCH,	SPIROGYRA, EUDORINA, CYCLOPS, KER	W71-11876
ENIUM-106, CESIUM RADIOISOTOPES,	SPIROGYRA, LIMNODRILUS HOFFMEISTE	W70-06668
IFORMIS, GONGROSIRA, MICROSPORA,	SPIROGYRA, ULOTHRIX.: /TETRAS PYR	W71-05991
RIS, EUGLENA POLYMORPHA, EUGLENA	SPIROGYRA, LEPOCINCLIS OVUM, PHAC	W73-00918
UPLII, COCCOCHLORIS PENIOCYSTIS,	SPIROGYRA, EPHEMEROPTERA BAETIDAE	W72-07132
LOVIAMONAS FRUTICOSA, NODAMASTIX	SPIROGYRAE, COLEMANIA V: /NA, PAV	W72-07683
ROCYSTIS, APHANIZOMENON, ANABENA	SPIROIDES.: / STEPHANODISCUS, MIC	W71-12855
OGONIUM, OOCYSTIS, OSCILLATORIA,	SPIRULINA, PANDORINA, PHCCUS, STA	W72-07661
CALLITRICHE, ANABAENA FLOSAQUAE,	SPIRULINA, GYMNODINIUM BREVE, PER	W73-00840
NOCTILUCA MILIARIS, GYMNODINIUM	SPLENDENS, EXUVIELLA MARINA, EUTR	W71-07875
ONS, ZINC, VOLVOX, HYDRODICTYON,	SPODYLOSIUM, SPIROGYRA.: /GANIC I	W71-11001
LOMONAS, STAURASTRUM, PHACOMYXA,	SPONDYLOSIUM, LYNGBYA, PSEUDOANAB	W72-07648
PENICILLUS SPP., LAURENCIA SPP.,	SPONGES.: /IA SPP., UDOTEA SPP.,	W72-11876
, BARNACLES, BRYOZOAS, HYDROIDS,	SPONGES, COELENTERATES, PORIFERA,	W73-00374
SEAWEEDS, APHIDS,	SPONGES, CARBON-14.:	W72-06283
R CRABS, MESOFAUNA, TURBELLARIA,	SPONGES, TUNICATES, BRYOZOANS, SE	W72-07911
BRYOZOA, INTOPROCTA, TUNICATES,	SPONGES, VERTEBRATES, MACROINVERT	W73-00932
LTURES, *AQUATIC MICROORGANISMS,	SPORES, ALGAE, CYTOLOGICAL STUDIE	W71-08027
PHOTOMETRY, MAGNESIUM, SULFATES,	SPORES, AZOTOBACTER, CYANOPHYTA,	W72-06124
RCES, EUTROPHICATION, COLIFORMS,	SPORES, LIFE CYCLES, DDCR, RECREA	W71-09173
GEN FIXATION, ENZYMES, BACTERIA,	SPORES, NITROGEN, METABOLISM.: /O	W71-12070
ENSITIVITY, CYTOPHAGA JOHNSONII,	SPOROCYTOPHAGA MYXOCOCCOIDES.: /S	W72-07951
YGNEMALES, OEDOGONIUM, MOUGOTIA,	SPOROTETRAS PYRIFORMIS, GONGROSIR	W71-05991
BIOCHEMICAL OXYGEN DEMAND, WATER	SPORTS, PARKS, COST ANALYSIS, MIC	W72-00042
INE HYDROCHLORIDE, MERISOMOPEDIA	SPP., ANABAENA SPP., GOMPHONEMA S	W72-14706
AMINES, CARCINOGENS, TORULOPSIS	SPP., ASPERGILLUS, PENICILLIUM, F	W72-09675
KEI, LYNGBYA LIMNETICA, ANABAENA	SPP., CHROMATIUM, MERISMOPEDIA TR	W72-07145
, THALASIOSIRA SPP., SKELETONEMA	SPP., DUNALIELLA SPP.: /TER ALGAE	W72-08436
DE, MERISOMOPEDIA SPP., ANABAENA	SPP., GOMPHONEMA SPP., NITZCHIA S	W72-14706
IA SPP., UDOTEA SPP., PENICILLUS	SPP., LAURENCIA SPP., SPONGES.: /	W72-11876
SPP., ANABAENA SPP., GOMPHONEMA	SPP., NITZCHIA SPP., SENSITIVITY.	W72-14706
LE GRASS, THALASSIA SPP., UDOTEA	SPP., PENICILLUS SPP., LAURENCIA	W72-11876
AANDOMENE, SIPHONALES, *HALIMEDA	SPP., PENICILLUS CAPITATUS, PENIC	W72-14294
SPP., GOMPHONEMA SPP., NITZCHIA	SPP., SENSITIVITY.: /P., ANABAENA	W72-14706
: *NORTH SEA, *LAMINAREA	SPP., SEWAGE POLLUTION, EPIFAUNA,	W72-11354
FRESHWATER ALGAE, THALASIOSIRA	SPP., SKELETONEMA SPP., DUNALIELL	W72-08436
IONSHIPS, NORTH SEA, PHAEOCYSTIS	SPP., SOIL CULTURES, TITANIUM DIO	W72-12239
SPP., PENICILLUS SPP., LAURENCIA	SPP., SPONGES.: /IA SPP., UDOTEA	W72-11876
A, *CHLORAMPHENICAL, LIMNODRILUS	SPP., TUBIFEX, UNIALGAL CULTURES,	W72-09668
LORIDA), TURTLE GRASS, THALASSIA	SPP., UDOTEA SPP., PENICILLUS SPP	W72-11876
*CHLAMYDOMONAS-SPP, *CRYPTOMONAS-	SPP.: *CANADIAN ARCTIC,	W72-07922
, LAGOON OF ISLANDS, STAURASTRUM	SPP.: /NT, TOOMS LAKE, WOODS LAKE	W73-01094
P., SKELETONEMA SPP., DUNALIELLA	SPP.: /TER ALGAE, THALASICSIRA SP	W72-08436
*CANADIAN ARCTIC, *CHLAMYDOMONAS-	SPP, *CRYPTOMONAS-SPP.:	W72-07922
HIA, PHRAGMITES COMMUNIS, JUNCUS	SPP, ATRIPLEX HASTATA, SCIRPUS SP	W73-01089
A, GYROSIGMA FASCIOLA, NITZSCHIA	SPP, BACILLARIA PAXILLIFER,: /IGM	W73-00853
PHA SPP, ZIZANIA AQUATICA, CAREX	SPP, BUTOMUS UMBELLATUS, SPARGANI	W73-01089
OMONAS EPIPHYTICA, CHLAMYDOMONAS	SPP, CARTERIA FRITSCHII, PANDORIN	W72-09111
CYCLOTELLA STELLIGERA, CYMBELLA	SPP, DIATOMA SSP.: /CF. SOCIALIS,	W72-13142
GLAND, SWEDEN, GERMANY, SPARTINA	SPP, DISTICHLIS SPICATA, FIMBRIST	W73-01089
EVIDENSES, CHARA ASPPRA, SCIRPUS	SPP, GILLS, MUSCLE, VISCERA, EYES	W73-01673
ROPHYLL A, *YAQUINA BAY, AMPHORA	SPP, MELOSIRA SULCATA, NAVICULA S	W73-00853
E(N.H.), APHANIZOMENON, ANABAENA	SPP, MICROCYSTIS AERUGINOSA, APHA	W72-07890
HAEROCYSTIS SCHROETERI, ULOTHRIX	SPP, MICROTHAMNION KUETZ: /UM, SP	W72-09111
RSONII COMMERSONII, ASTERIONELLA	SPP, ORTHOPHOSPHATES, SECCHI DISC	W72-07890

CESIUM RADIOISOTOPES, CLADOPHORA SPP, PASPALUM DISTICHUM, POTAMOGE W73-01673
SPP, MELOSIRA SULCATA, NAVICULA SPP, SYNEDRA FASCICULATA, PLAGIOG W73-00853
S SPP, ATRIPLEX HASTATA, SCIRPUS SPP, TYPHA SPP, ZIZANIA AQUATICA, W73-01089
PLEX HASTATA, SCIRPUS SPP, TYPHA SPP, ZIZANIA AQUATICA, CAREX SPP, W73-01089
IGOCLADUS, PHEOPHYTIN, GRASSLAND SPRING.: /TERIA, PHORMIDIUM, MAST W72-03557
HLOROPHYTA, BIOMASS, PYRROPHYTA, SPRING, SUMMER, AUTUMN, SAMPLING, W72-12738
HOT SPRINGS, SLEEPING CHILD HOT SPRINGS.: /HOT SPRINGS, PIPESTONE W72-05610
KSON HOT SPRINGS, / ALHAMBRA HOT SPRINGS, BOULDER HOT SPRINGS, JAC W72-05610
*DISTRIBUTION PATTERNS, *THERMAL SPRINGS, CYANOPHYTA, ALKALINITY, W72-05610
TES, *INHIBITION, FUNGI, THERMAL SPRINGS, CYANOPHYTA, BACTERIA, EC W72-12734
LHAMBRA HOT SPRINGS, BOULDER HOT SPRINGS, JACKSON HOT SPRINGS, LOL W72-05610
BOULDER HOT SPRINGS, JACKSON HOT SPRINGS, LOLO HOT SPRINGS, PIPEST W72-05610
S, JACKSON HOT SPRINGS, LOLO HOT SPRINGS, PIPESTONE HOT SPRINGS, S W72-05610
LOLO HOT SPRINGS, PIPESTONE HOT SPRINGS, SLEEPING CHILD HOT SPRIN W72-05610
R, *WYOMING, NATIONAL PARKS, HOT SPRINGS, WATER TEMPERATURES, NOST W72-08508
IA VENTRICOSA, BRYOPSIS PENNATA, SPYRIDIA FILAMENTOSA, DICTYOTA BA W73-00838
TES, SECCHI DISC, NEWFOUND LAKE, SQUAM LAKE, APHANIZOMENON FLOW-AQ W72-07890
R LA SALLE LAKE, LAKE LONG, LAKE SQUAW, LAKE BUDD, OSCILLATORIA RE W72-07145
LAKE, GAMM/ *BIOLOGICAL SAMPLES, SR-90, K-40, CS-137, *PAHRANAGAT W73-01673
, COCCONEIS PLACENTULA, NAVICULA SRYTOCEPHALA, ACHNANTHES BREVIPES W72-08141
TELLIGERA, CYMBELLA SPP, DIATOMA SSP.: /CF. SOCIALIS, CYCLOTELLA S W72-13142
CONSIN, LAKE ERIE, LAKE ONTARIO, ST LAWRENCE RIVER, INTERNATIONAL W70-07283
BEC), SABLE ISLAND(NOVA SCOTIA), ST PIERRE(CANADA), MIQUELON(CANAD W71-10066
DETROIT RIVER, BL/ MAUMEE RIVER, ST. CLAIR RIVER, LAKE ST. CLAIR, W72-14282
MEE RIVER, ST. CLAIR RIVER, LAKE ST. CLAIR, DETROIT RIVER, BLACK R W72-14282
*SESSILE ALGAE, *INVERTEBRATES, ST. LAWRENCE RIVER, ALGAE, NEMATO W72-05432
ESOTA), LAKE WA/ ILLINOIS RIVER, ST. LOUIS RIVER, BLACK RIVER(MINN W71-04216
VESICULOSUS, LAMINARIA DIGITATA, ST. MARGARET'S BAY, SCOUDOUC RIVE W72-12166
, WATER ANALYSIS, TIDAL EFFECTS, STABILITY, SOLAR RADIATION, PHOTO W72-08141
INDRICA, ANACYSTIS NIDULANS, GL/ STABILIZATION PONDS, ANABAENA CYL W72-07664
GEN, DREDGING, BOTTOM SEDIMENTS, STABILIZATION, PRIMARY PRODUCTIVI W72-03556
URIFICATION, *OXIDATION LAGOONS, STABILIZATION, INDUSTRIAL WASTES, W71-13356
PTH, SLOPE, *BOD LOADING, *WASTE STABILIZATION.: DE W71-03896
GICAL TREATMENT, PHOTOSYNTHESIS, STABILIZATION, *WASTE WATER TREAT W70-06619
EN DEMAND, PHOSPHATES, NITROGEN, STABILIZATION.: /D, CHEMICAL OXYG W70-07838
MAND, BIOCHEMICAL OXYGEN DEMAND, STABILIZATION PONDS, MICROORGANIS W71-07097
LING, WATER ANALYSIS, SEA WATER, STABLE ISOTOPES, RADIOISOTOPES, N W73-00831
ANTS, ABSORPTION, RADIOISOTOPES, STABLE ISOTOPES, ESTUARINE ENVIRO W73-00811
ATE, *MICROBIAL BIOMASS, *GROWTH STAGE, *DETERMINATION, *ORIGIN, * W72-04292
E, CHROMOSOMES, EMBRYONIC GROWTH STAGE, CRUSTACEANS, MONITORING, W W72-10955
E, FEEDING RATES, *LARVAL GROWTH STAGE, PHYTOPLANKTON, NUTRIENT RE W72-09248
ES, *AEROBIC CONDITIONS, *GROWTH STAGES, *BIODEGRADATION, *BIOCHEM W72-06612
RODUCTIVITY, LIFE CYCLES, GROWTH STAGES, DISTRIBUTION, BEHAVIOR, S W72-14659
ALS, *AQUATIC PLANTS, *RESERVOIR STAGES, EARLY IMPOUNDMENT, BIOMAS W72-04853
LANKTON, *RESERVOIRS, *RESERVOIR STAGES, EARLY IMPOUNDMENT, BIOMAS W72-04855
BIOCHEMICAL OXYGEN DEMAND, LAKE STAGES, LIMNOLOGY, NUISANCE ALGAE W72-14692
BLOOMS, AGING(BIOLOGICAL), LAKE STAGES, MESOTROPHY, OLIGOTROPHY, W72-14782
BIOCHEMICAL OXYGEN DEMAND, LAKE STAGES, NUTRIENTS, MESTROPHY, OLI W73-00002
BACTERIA, *BACTERICIDES, GROWTH STAGES, PATHOGENIC BACTERIA, CYAN W71-05155
, *LIFE HISTORY STUDIES, *GROWTH STAGES, REPRODUCTION, CARBON DIOX W73-02099
MANCHUDINIUM SINICUM, SINAMONAS STAGNALIS, KOFOIDIELLA UNIFLAGELL W72-07683
TOCYSTIS PLUVIALIS, GOMESIAMONAS STAGNALIS, REFRACTOCES BRASILIANA W72-09119
INTE/ ION-SELECTIVE ELECTRODES, STANDARD ADDITION, POTENTIOMETRY, W71-00474
ORA, LAKE ONTARIO, HUDSON RIVER, STANDARDS, SURFACE WATERS, EPILIM W70-09614
TEMPERATURE, CONDENSERS, DESIGN, STANDARDS, FISH KILL, ALGAE, COOL W70-09612
LUTION ABATEMENT, WATER QUALITY, STANDARDS, WATER POLLUTION, AIR P W71-08214
RAL GOVERNMENT, WATER POLLUTION, STANDARDS, WATER POLLUTION EFFECT W71-11516
T, UNITED STATES, ALGAL CONTROL, STANDARDS, LEGISLATION, LEGAL ASP W72-01093
REATMENT, WATER QUALITY CONTROL, STANDARDS, WATER POLLUTION SOURCE W72-01660
UTION ABATEMENT, WASTE DISPOSAL, STANDARDS, PUBLIC HEALTH, INSPECT W72-01085
*TOXIC CHEMICALS, *PUBLIC HEALTH STANDARDS, TOXICOLOGY, NTA.: /S, W73-02144
ATER TEMPERATURE, *WATER QUALITY STANDARDS, *ESTURARIES, *THERMAL W72-13636
GEN, ALGAE, INORGANIC COMPOUNDS, STANDING CROPS, LIMITING FACTORS, W72-13653
, PRODUCTIVITY, WATER POLLUTION, STANDING CROPS.: /GAE, CLADOPHORA W72-13662
KTON, NUTRIENTS, PHOTOSYNTHESIS, STANDING CROP, ECOLOGICAL DISTRIB W72-14313
YSIOLOGICAL ECOLOGY, PHAEOPHYTA, STANDING CROPS, RESISTANCE, PRIMA W72-12940
LANKTON, BIOLOGICAL COMMUNITIES, STANDING CROPS, WISCONSIN, PRIMAR W73-02469
NELIDS, WATER POLLUTION EFFECTS, STANDING CROPS, MOLLUSKS, LARVAE, W73-00935
RBIDITY, PRODUCTIVITY, PLANKTON, STANDING CROP, LIMNOLOGY, EUTROPH W72-08559
TOXICITY, PRIMARY PRODUCTIVITY, STANDING CROPS, *PHOTOSYNTHESIS.: W72-14690
, PONDS, FEEDING RATES, BENTHOS, STANDING CROPS.: / EUTROPHICATION W72-14796
SAMPLING, SCUBA DIVING, ECOLOGY, STANDING CROPS, PATH OF POLLUTANT W73-00366
HNIQUES, CARBON, EUTROPHICATION, STANDING CROPS, LABORATORY TESTS, W72-09163
ISH, LAKES, CONDUCTIVITY, ALGAE, STANDING CROP, CHEMICAL ANALYSIS, W71-01488
PULATION, ATLANTIC OCEAN, DEPTH, STANDING CROP, CURRENTS(WATER), W W72-04766
RODUCTIVITY, *ENERGY CONVERSION, STANDING CROP, CHLOROPHYTA, CYANO W72-06283
GINOSA, MYCOBACTERIUM FURTUITUM, STAPHYLOCOCCUS AUREUS, CANDIDA AL W72-04259
TES.: *STARCH REMOVAL, STARCH SUBSTITUTION, POYELECTROLY W72-03753
, HEAVY METALS, MUSSELS, *KELPS, STARFISH, CRUSTACEANS, FOOD CHAIN W72-11925
OLOGICAL SAMPLES, SARGASSO WEED, STARFISH, ASTERIAS FORBESI, PLUTO W72-05965
ERSITY, PORIFERA, SEA CUCUMBERS, STARFISH, CHITONS, RHABDODERMELLA W73-00932
E REMOVAL, OXYGEN SAG, DREDGING, STATE GOVERNMENTS, TREATMENT FACI W72-01659
SAN JOAQUIN DELTA(CALIF), STEADY- STATE MODELS.: /ESTUARY(DEL-NJ), W71-05390
AY, ABALONE.: ENVIRONMENTAL STATEMENT, CRITICAL NUCLIDE PATHW W72-00959
, *WATER POLLUTION CONT/ *UNITED STATES, *DETERGETNTS, *PHOSPHATES W72-01085
TES, *FEDERAL GOVERNMENT, UNITED STATES, ALGAL CONTROL, STANDARDS, W72-01093
PACIFIC OCEAN, BOTTOM SEDIMENTS, STATIONS, DATA COLLECTIONS, ALGAE W72-03567
EXTURE, CLAYS, LABORATORY TESTS, STATISTICAL METHODS, SOIL MICROOR W72-02218
OPULATIONS, CYTOLOGICAL STUDIES, STATISTICAL METHODS, CULTURES.: / W72-11920
TS, EFFLUENTS, DISCHARGE(WATER), STATISTICAL MODELS.: /, POWERPLAN W71-11583

YTA, BIOMASS, CHEMICAL ANALYSIS, STATISTICAL METHODS, NITROGEN, PH W71-11029
ARIABILITY, WAVES(WATER), MAINE, STATISTICAL METHODS, SEA WATER, C W73-00369
BUTICN PATTERNS, *AQUATIC ALGAE, STATISTICAL METHODS, MARINE ALGAE W72-08141
A, SPIRULINA, PANDORINA, PHOCUS, STAURASTRUM, TRACHELOMONAS.: /ORI W72-07661
MULINA, CRYPTOMONAS, MALLOMONAS, STAURASTRUM, PHACOMYXA, SPONDYLOS W72-07648
UM, ARTH/ *STAURASTRUM, DESMIDS, STAURASTRUM LONGIPES VAR CONTRACT W73-01094
, WOODS LAKE, LAGOON OF ISLANDS, STAURASTRUM SPP.: /NT, TOOMS LAKE W73-01094
NOBRYON, FRAGILARIA, PEDIASTRUM, STAURASTRUM, TABELLARIA, MOUGEOTI W72-10076
CULTURES, CULTURING TECHNIQUES, STAURASTRUM PARADOXUM, STAURASTRU W72-12736
FORMOSA, ASTERIONELLA JAPONICA, STAURASTRUM PARADOXIUM, CULTURE M W72-12739
CHNIQUES, STAURASTRUM PARADOXUM, STAURASTRUM SEBALDII VAR. ORNATUM W72-12736
L-NJ), SAN JOAQUIN DELTA(CALIF), STEADY-STATE MODELS.: /ESTUARY(DE W71-05390
ATING, *FARM USES OF WASTE HEAT, STEAM POWERED BUSES.: / *SPACE HE W70-09193
TED WATER, NUCLEAR POWER PLANTS, STEAM, ECOLOGY, PHYTOPLANKTON, ZO W71-11517
A, BOODLEA COMPOSITA, ANADYOMENE STELLATA, BR: /OROPSIS MEMBRANACE W72-11800
A, VISCHERIA PUNCTATA, VISCHERIA STELLATA, ELLIPSOIDION ACUMINATUM W72-14808
FRAGILARIA CAPUCINA, CYCLOTELLA STELLIGERA, STEPHANODISCUS.: /IS, W73-00384
LOTELLA CF. SOCIALIS, CYCLOTELLA STELLIGERA, CYMBELLA SPP, DIATOMA W72-13142
AS MINUTA, ASTERIONELLA FORMOSA, STEPHANODISCUS ASTREA, ANKISTRODE W72-12738
EAE, *DENMARK, CYCLOTELLA COMTA, STEPHANODISCUS HANTZSCHII, APHANI W72-12738
TONENSIS, TABELLARIA FLOCCULOSA, STEPHANODISCUS HANTZSCHII, CYCLOT W73-00384
CAPUCINA, CYCLOTELLA STELLIGERA, STEPHANODISCUS.: /IS, FRAGILARIA W73-00384
ENGLAND), RIVER KENNET(ENGLAND), STEPHANODISCUS HANTZSCHII, CRYPTO W72-14797
CROTONENSIS, CRYPTOMONAS OVATA, STEPHANODISCUS NIAGARAE, ASTERION W73-01435
TINASTRUM, PHYTOCONIS, CYMBELLA, STEPHANODISCUS, ANKISTRODESMUS, T W72-10875
ELL COUNTS, NAVICULA, NITZSCHIA, STEPHANODISCUS, MICROCYSTIS, APHA W71-12855
FLAGELLATES, FATE OF POLLUTANTS, STEPHANOPYXIS COSTATA, ASSIMILATI W73-00380
A, MOUGEOTIA, ULOTHRIX AEQUALIS, STEPHANOPYXIS TURRIS.: /UADRICAUD W72-07144
, BODY FLUIDS, VERTEBRATES, HEAT STERILIZATION, FIXATION, EMBEDDIN W73-01676
SIMACHIA NUMMULARIA, BLUE HERON, STERNA HIRUNDO, RATS, PARTRIDGE, W72-05952
ACTIVITY, ALKALOIDS, FLAVONOIDS, STEROIDS, TANNINS, SAPONINS.: /C W72-04275
GLOBUSA, CHLORELLA PYRENOIDOSA, STICHOCOCCUS BACILLARIS, UNDISSOC W70-07255
CARBON-14, CARTERIA EUGAMETOS, STICHOCOCCUS, CHLOROCOCCUM, HYALO W72-09239
SMOPEDIA TROLLERI, CRYPTOMONADS, STICHOGLOEA D: / CHROMATIUM, MERI W72-07145
CLEAR EXPLOSIONS, BIRDS, SALMON, STICKLEBACKS, CRABS, OTTERS, ZOOP W72-00960
OGYRA, LIMNODRILUS HOFFMEISTERI, STICTOCHIRONOMUS ANNULICRUS, PHYS W70-06668
LOMERATA, ULOTRIC/ *ZINC, *LEAD, STIGEOCLONIUM TENUE, CLADOPHORA G W71-05991
A CALIFORNIA, HEMIAULUS HAUCKII, STIGMAPHORA ROSTRATA, ETHMODISCUS W72-03567
ISCIFORMIS, PROWSEMONAS TROPICA, STIGMOBODO BRASILIANA, PAVLOVIAMO W72-07683
IA, MICROC/ *FULVIC ACID, GROWTH STIMULATION, ANABAENA, GLOEOTRICH W72-01097
SPUM, DIGENIA SIMPLEX, HALOPHILA STIPULACEA, DIPLANTHERA UNINERVIS W73-00855
OPODS, HETEROPODS, OSCILLATORIA, STOMATOPODA, POLYCHAETES, AMPHIOX W73-00935
LAKE MINNETONKA(MINNESOTA), BIG STONE LAKE(MINNESOTA), LAKE CRYST W71-04216
B/ *ATHIORHODACEAE, WHIPPLE DAM, STONE VALLEY DAM(PA.), ATRAZINE, W72-04743
SECTS, *DISPERSION, CADDISFLIES, STONEFLIES, DOBSONFLIES, MAYFLIES W70-06655
RESERVOIR(VA).: *PUMPED STORAGE OPERATION, SMITH MOUNTAIN W72-14405
OND, RISING SUN POND, INDUSTRIAL STORAGE T: /X CALCICOLA, ELKTON P W72-04736
*ECOSYSTEMS, *ECOLOGY, *PUMPED STORAGE, *WATER QUALITY, *MODEL S W72-05282
SSOLVED SOLIDS, ALGAE, RESERVOIR STORAGE, *HYPOLIMNION, WATER QUAL W73-02138
NSITY, SEDIMANTATION, OXIDATION, STORAGE, BIOCHEMICAL OXYGEN DEMAN W72-01881
NOFF, BIOASSAYS, BIO-INDICATORS, STORAGE, FLOW RATES, INHIBITION, W72-04070
ETRIC ANALYSIS, ALGAE, RESERVOIR STORAGE, IMPOUNDMENTS, DISSOLVED W72-14280
ENVIRONMENTAL EFFECTS, RESERVOIR STORAGE, NUTRIENTS, TEMPERATURE, W72-05282
ATER LEVEL FLUCTUATIONS, *PUMPED STORAGE, PHYSICOCHEMICAL PROPERTI W72-14280
BACTERIA, EFFLUENT, IRRIGATION, STORAGE, WASTE DISPOSAL.: /OGENIC W71-08214
ER DISPOSAL, PATH OF POLLUTANTS, STORM DRAINS, DETERGENTS, EUTROPH W71-13466
*NEBRASKA, URBANIZATION, CITIES, STORM RUNOFF, SEWAGE, WASTE WATER W71-13466
SPHORUS, *NITROGEN, EVAPORATION, STORM RUNOFF, BIOASSAYS, BIO-INDI W72-04070
, ALGAE, PHOSPHATES, HYDROLYSIS, STORM WATER, RAINFALL, WASTE WATE W72-13644
ULANS, AGMENELLUM QUADRUPLICATUM STRAIN PR-6, PHOTODALDRIN, KETOEND W72-03545
DISSOLVED ORGANIC CARBON, *MENAI STRAIT(WALES).: * W72-09108
TIOLECTA, PHAEODACTYLU/ *FLORIDA STRAIT, CARBON-14, DUNALIELLA TER W72-09102
AEODACTYLUM/ CARBON-14, *FLORIDA STRAIT, DUNALIELLA TERIOLECTA, PH W72-09103
URE TRENDS, CLADOPHORA, RESEARCH STRATEGY, THAMES RIVER(ENGLAND), W71-13172
E, NITROGEN, PHOSPHORUS, THERMAL STRATIFICATION, WATER SUPPLY, SUC W71-13183
*LAKES, PONDS, SAMPLING, THERMAL STRATIFICATION, HARDNESS(WATER), W72-04269
DISSOLVED OXYGEN, ALGAE, THERMAL STRATIFICATION, COAGULATION, BENT W72-03753
ND, INORGANIC COMPOUNDS, THERMAL STRATIFICATION, DIATOMS, EPILIMNI W72-13659
URE, HYDROGEN ION CONCENTRATION, STRATIFICATION.: / WATER TEMPERAT W72-13871
CONCENTRATION, DISSOLVED OXYGEN, STRATIFICATION, SAMPLING, MICROOR W71-07382
MIXING, TURBULENCE, TEMPERATURE, STRATIFICATION, SOLAR RADIATION, W71-07084
*LAKES, *PHYTOPLANKTON, *DENSITY STRATIFICATION, SPECIATION, VERTI W71-09852
HYTA, WATER TEMPERATURE, THERMAL STRATIFICATION, ICE, EUTROPHICATI W72-04855
IONS, WATER TEMPERATURE, THERMAL STRATIFICATION, OXIDATION, ORGANI W72-04853
TROGEN, BACTERIA, ALGAE, THERMAL STRATIFICATION, DISSOLVED OXYGEN, W71-03028
TS, *THERMAL POLLUTION, *THERMAL STRATIFICATION, HYDRAULIC MODELS, W70-09614
R POLLUTION, PHOSPHORUS, THERMAL STRATIFICATION, CURRENTS(WATER), W70-09947
WATERS, STREAMS, FLOW, SEASONAL, STRATIFICATION, NITRATES, SILICA, W71-03028
ROL, *PONDWEEDS, PONDS, *THERMAL STRATIFICATION, DIATOMS, *CHARA.: W73-02349
SSOLVED SOLIDS, ALGAE,/ *THERMAL STRATIFICATION, *TASTE, *ODOR, DI W73-02138
S, *MATHEMATICAL MODELS, THERMAL STRATIFICATION, TEMPERATURE, DISS W72-14405
, PHOTOSYNTHESIS, PHYTOPLANKTON, STRATIFICATION, DIATOMS, CURRENTS W73-00236
TA, NITRATES, PHOSPHATES, LAKES, STRATIFICATION, MIXING, BOTTOM SE W72-08049
TA, SEASONAL, NUTRIENTS, THERMAL STRATIFICATION, CARBON, SEDIMENTS W72-08560
YDROLOGIC DATA, WATER TREATMENT, STREAM GAGES, GROUNDWATER, SAFE Y W72-05594
MENTAL QUALITY, AIR CONTAMINANT, STREAM QUALITY, WASTE MANAGEMENT, W71-08214
*NUTRIENTS, ALGAE, *DEGRADATION(STREAM), WATER POLLUTION EFFECTS, W71-07973
SARGASSO SEA, SHELF WATERS, GULF STREAM, CAPE HATTERAS, NANSEN BOT W72-07663
CONSU/ *RIFFLE BENTHOS, WOODLAND STREAM, ECOSYSTEM ANALYSIS, MACRO W71-10064
ECOSYSTEMS, *WATERSHEDS(BASINS), STREAMFLOW, IONS, NITRATES, SULFA W71-01489

ATIC PLANTS, PROTOZOA, BACTERIA,	STREAMFLOW, FLOW RATES, WATER TEM	W73-01972
SIS, RESPIRATION, WATER QUALITY,	STREAMFLOW, AGRICULTURE, INDUSTRI	W71-11017
*TREATMENT FACILITIES, *INFLUENT	STREAMS, *GREAT LAKES, AQUATIC AL	W73-01669
*POLLUTANTS, *RECREATION, LAKES,	STREAMS, ALGAE, PESTICIDES, DDT,	W71-06444
UDS, INVERTEBRATES, FISH, BIRDS,	STREAMS, AQUATIC PLANTS, AQUATIC	W72-12930
, *BIOINDICATORS, ALGAE, RIVERS,	STREAMS, DELAWARE RIVER, OXIDATIO	W72-04736
INGS, WATER TEMPERATURES, NOSTOC	STREAMS, DATA COLLECTIONS, ANALYT	W72-08508
ION, NUTRIENTS, RIVERS, INFLUENT	STREAMS, EFFLUENT STREAMS, LIGHT	W73-01435
ORTHOPHOSPHATES, *OLIGASAPROBIC	STREAMS, FINGERNAIL CLAMS, UNIONI	W72-06526
ATER QUALITY, *IMPOUNDED WATERS,	STREAMS, FLOW, SEASONAL, STRATIFI	W71-03028
C ALGAE, WATER TEMPERATURE, ACID	STREAMS, HYDROGEN ION CONCENTRATI	W72-14690
PROTOZOA, *AMPHIPODA, *FLORIDA,	STREAMS, ISOPODS, SCENEDESMUS, AQ	W72-12744
LUTION TREATMENT, WATER QUALITY,	STREAMS, LAKES, BIOCHEMICAL OXYGE	W72-14469
VERS, INFLUENT STREAMS, EFFLUENT	STREAMS, LIGHT INTENSITY, NITROGE	W73-01435
OPHIC LEVEL, HERBIVORES, NATURAL	STREAMS, PULP WASTES, PRODUCTIVIT	W72-06340
ONMENT, WATER POLLUTION EFFECTS,	STREAMS, RIVERS, ROOTED AQUATIC P	W72-14690
WATER MOVEMENT, OUTFALL SERVERS,	STREAMS, SEPTIC TANKS, FUNGI, ALG	W72-14363
N CONTROL, ECONOMIC FEASIBILITY,	STREAMS, SEWAGE EFFLUENTS, ALGAE,	W72-14878
INA, DIAPTOMUS GRACILIS, CYCLOPS	STRENUUS, INGESTICN RATES, ASSIMI	W72-07940
OBIOTICS, ESCHERICHIA COLI, BETA	STREPTOCOCCUS, PSEUDOMONAS AERUGI	W72-04259
BIOCHEMICAL STUDIES, ARTHOPODS,	STREPTOM: /AE, SHIGELLA FLEXNERI,	W72-13800
SHRIMP, GREEN CRAB, PENICILLIUM,	STREPTOMYCIN, EPIFLORA, WALLEYE L	W72-08142
YLENE, BACILLUS, CORYNEBACTERIA,	STREPTOTHRIX HYORHINA, CADMIUM, D	W72-09675
NCE, WATER TEMPERATURES, THERMAL	STRESS.: /TRUMENTATION, FLUORESCE	W72-07225
OORGANISMS, *FLUORESCE/ *THERMAL	STRESS, *THERMAL POLLUTION, *MICR	W72-07526
, HEATED WATER, BIOMASS, THERMAL	STRESS, GROWTH RATES, SPATIAL DIS	W73-00853
RIAS CRUX-MELITENSIS, CLOSTERIUM	STRIGOSUM, CULTURE MEDIA.: /RASTE	W72-12736
FAUNA, WATER POLLUTION EFFECTS,	STRIPED BASS, DISSOLVED OXYGEN, O	W73-02029
N, LARVAE, EGGS, CHINOOK SALMON,	STRIPED BASS, MINNOWS, FISHKILL,	W71-11517
WTH, ALGAL HARVESTING, */ *ALGAE	STRIPPING, SCENEDESMUS, ALGAL GRO	W72-02975
PHYSICOCHEMICAL METHODS, AMMONIA	STRIPPING, GUGGENHEIM PROCESS.: /	W70-09907
*CHELATORS, ANODIC	STRIPPING, ORGANIC CHELATORS.:	W71-09674
FILTER / *NITRATE REMOVAL, ALGAE	STRIPPING, POND DENITRIFICATION,	W71-08223
OMMEN RIVER, ATRAN RIVER, MOTALA	STROM RIVER, SPARGANIUM ERECTUM,	W73-00286
, CASSIOPEA ANDROMEDA, AKABARIA,	STROMBUS TRICORNI: /ERA UNINERVIS	W73-00855
LICATA, TRICHOCYANELLA SPIRALIS,	STROMIA SUBSPHAERICA, TETRACULAMO	W72-09119
GAE, SCENEDESMUS, RADIOISOTOPES,	STRONTIUM RADIOISOTOPES, PHOSPHOR	W72-09649
EFFECTS, ABSORPTION, FOOD WEBS,	STRONTIUM RADIOISOTOPES, MARINE B	W72-00948
, PHYSIOLOGICAL ECOLOGY, CESIUM,	STRONTIUM RADIOISOTOPES, LINC RAD	W71-13243
, CARP, POTASSIUM RADIOISOTOPES,	STRONTIUM RADIOISOTOPES, SOIL ANA	W73-01673
IQUES, LABORATORY TESTS, CESIUM,	STRONTIUM RADIOISOTOPES, IONS, RI	W71-08032
S, *PRODUCTIVITY, RADIOISOTOPES,	STRONTIUM RADIOISOTOPES, ALGAE, P	W71-09254
OTOPES, ABSORPTION, TEMPERATURE,	STRONTIUM RADIOISOTOPES, ZINC RAD	W71-09250
LYSIS, ION EXCHANGE, ABSORPTION,	STRONTIUM RADIOISOTOPES, ZINC RAD	W71-09188
, TRITIUM, IODINE RADIOISOTOPES,	STRONTIUM RADIOISOTOPES, ZINC RAD	W71-09018
ON, *WATER QUALITY CONTROL, *NON-	STRUCTURAL ALTERNATIVES, AGING(BI	W72-04279
BIOLOGICAL PROPERTIES, MOLECULAR	STRUCTURE, PERMEABILITY, PLANT PO	W72-01473
ANALYTICAL TECHNIQUES, MOLECULAR	STRUCTURE, CATION EXCHANGE, RADIO	W72-12203
OLUMETRIC ANALYSIS, FISH, INTAKE	STRUCTURES, CARBON DIOXIDE,: /, V	W72-05594
ON, *LITTORAL, PLANT/ *MOLECULAR	STRUCTURES, *ALGAE, *FLUID FRICTI	W71-07878
RFACE WATERS, EPILIMNION, INTAKE	STRUCTURES, DISCHARGE(WATER), WAT	W70-09614
TLE, DISSOLVED OXYGEN, DIVERSION	STRUCTURES, SEDIMENTATION, BIOCHE	W72-08401
, DIATOMS, EUGLENA, MATHEMATICAL	STUDIES.: /LUDGE, ALGAE, PROTOZOA	W72-08573
ALGAE, SNAILS, MOSQUITOES, MODEL	STUDIES.: /AINS, FISH, BIOASSAY,	W72-07703
LOGY, *PLANT GROWTH, CYTOLOGICAL	STUDIES.: /OOPHYTA, *PLANT MORPHO	W73-01622
REATMENT, WATER TREATMENT, MODEL	STUDIES.: /ACTERIA, WASTE WATER T	W73-01362
G, *VIRGINIA, PROJECTIONS, MODEL	STUDIES.: /VIRONMENTAL ENGINEERIN	W72-14405
E CAYUGA, CLADOPHORA, ECOLOGICAL	STUDIES.: LAK	W70-09614
IARY TREATMENT, EFFLUENTS, MODEL	STUDIES.: /, MICROORGANISMS, TERT	W71-03021
STUDIES, GREEN ALGAE, LABORATORY	STUDIES.: /UE-GREEN ALGAE, FIELD	W70-08319
NS, AQUATIC ALGAE, GRANTS, MODEL	STUDIES.: /HEALTH, COST COMPARISO	W71-09429
Y TESTS, CYANOPHYTA, CYTOLOGICAL	STUDIES.: /E, NITROGEN, LABORATOR	W72-05476
*NEW YORK BIGHT, COLLABORATIVE	STUDIES.:	W72-12941
E, LABORATORY TESTS, CYTOLOGICAL	STUDIES.: /, PLANT GROWTH, SULFAT	W72-00153
Y, ALGAE, HABITAT/ *LIFE HISTORY	STUDIES, *PHAEOPHYTA, EPIPHYTOLOG	W72-01370
LGAE, CYTOLOGICAL STUDIE/ *MODEL	STUDIES, *STATISTICAL METHODS, *A	W71-12868
GICAL PROPERTIES,/ *MATHEMATICAL	STUDIES, *ALGAE, EQUATIONS, BIOLO	W72-12724
N RADIOISOTOPES, ENVIR/ *ISOTOPE	STUDIES, *STABLE ISOTOPES, *CARBO	W72-12709
*WATER POLLUTION EFFECTS, *MODEL	STUDIES, *STREAMS, *ECOSYSTEMS, A	W72-06340
RYSOPHYTA, PHAEOPH/ *CYTOLOGICAL	STUDIES, *SYSTEMATICS, *ALGAE, CH	W72-14808
TES, *SYSTEMATICS, *LIFE HISTORY	STUDIES, *GROWTH STAGES, REPRODUC	W73-02099
LLUTANTS, DIFFUSION, CYTOLOGICAL	STUDIES, ADSORPTION, PENETRATION,	W73-00823
, SHIGELLA FLEXNERI, BIOCHEMICAL	STUDIES, ARTHOPODS, STREPTOM: /AE	W72-13800
OCARBON PESTICIDES, *CYTOLOGICAL	STUDIES, ABSORPTION, DDT, PHYTOPL	W72-12998
AS, *ENTROPHICATION, CYTOLOGICAL	STUDIES, ANALYTICAL TECHNIQUES, P	W72-03985
OROPHYLL, SAMPLING, MATHEMATICAL	STUDIES, ATLANTIC OCEAN, *TROPICA	W72-11719
TES, CHLAMYDOMONAS, MATHEMATICAL	STUDIES, BIOLOGICAL COMMUNITIES,	W72-09248
ON, ANIMAL TISSUES, HISTOLOGICAL	STUDIES, BODY FLUIDS, VERTEBRATES	W73-01676
ERTIES, ABSORPTION, MATHEMATICAL	STUDIES, CHLOROPHYLL, OXIDATION R	W72-14800
USTACEANS, *FOODS, *LIFE HISTORY	STUDIES, COPEPODS, DIATOMS, CHLOR	W72-14805
A, PLANT PHYSIOLOGY, CYTOLOGICAL	STUDIES, CULTURES, AMMONIA, POTAS	W72-03217
IQUES, NUMERICAL ANALYSIS, MODEL	STUDIES, COMPUTERS, CULTURES.: /N	W72-12724
URES, PHYTOPLANKTON, CYTOLOGICAL	STUDIES, CYANOPHYTA, CHLOROPHYTA,	W70-07280
, CULTURES, VOLUME, MATHEMATICAL	STUDIES, CORRELATION ANALYSIS.: /	W70-05547
AE, *MODEL STUDIES, MATHEMATICAL	STUDIES, DIGITAL COMPUTERS, DATA	W72-11920
HYSIOLOGY, BIOASSAY, CYTOLOGICAL	STUDIES, DINOFLAGELLATES, CYANOPH	W73-00243
CYCLE, AMMONIA, NITRATES, MODEL	STUDIES, DISSOLVED OXYGEN, CARBON	W72-08459
, LIGHT, NUTRIENTS, MATHEMATICAL	STUDIES, EQUATIONS, AQUATIC ENVIR	W72-11724
*EUTROPHICATION, *KANSAS, MODEL	STUDIES, ECOSYSTEMS, ALGAE, *LIGH	W72-12393
RGROWTH, BLUE-GREEN ALGAE, FIELD	STUDIES, GREEN ALGAE, LABORATORY	W70-08319

TON, *ALGAE, *PHOSPHORUS, *MODEL STUDIES, LAKES, CYCLING NUTRIENTS W72-12543
PLANTS, *AQUATIC ANIMALS, *MODEL STUDIES, LIMNOLOGY, BIOLCGICAL PR W72-14280
L CONTROL, EUTROPHICATION, MODEL STUDIES, MATHEMATICAL MODELS.: /A W72-04787
ISMS, SPORES, ALGAE, CYTCLCGICAL STUDIES, MYCOBACTERIUM, PSEUDOMON W71-08027
GITAL COMPUTERS,/ *ALGAE, *MODEL STUDIES, MATHEMATICAL STUDIES, DI W72-11920
*LAKE WABAMUN, BIOLOGICAL STUDIES, MACRCPHYTES.: W73-01704
ANALYSIS, BIOASSAY, CYTOLOGICAL STUDIES, NUTRIENTS, ENZYMES.: /AL W72-09163
OTOPERIODISM, GREAT LAKES, MODEL STUDIES, NUTRIENTS, PLANT GROWTH, W72-13651
N, LABORATORY TESTS, CYTOLCGICAL STUDIES, PLANT GRCWTH, SCENEDESMU W71-12858
FREEZING, RADIATION, CYTOLOGICAL STUDIES, PLANT TISSUES, ALGAE, MI W73-01676
MEMBRANES, *ALGAE, *CYTYLOGICAL STUDIES, PLANT MORPHOLOGY, EQUATI W73-01630
AL INTERFERENCE, INTERLABORATORY STUDIES, PRECISION, *SWEDEN, LAKE W73-00286
ALGAE, *METABOLISM, *CYTOLOGICAL STUDIES, PHOTOSYNTHESIS, RESPIRAT W71-08683
CAL METHODS, *ALGAE, CYTCLCGICAL STUCIES, SELECTIVITY, PRODUCTIVIT W71-12868
REMENT, POPULATIONS, CYTOLOGICAL STUDIES, STATISTICAL METHODS, CUL W72-11920
LL ANIMALS(MAMMALS), CYTOLOGICAL STUDIES, SOIL CHEMISTRY, RACIOSEN W72-00949
XICO, PHYTOPLANKTON, CYTOLOGICAL STUDIES, SYSTEMATICS, BICMASS, FL W72-14313
, PHOSPHORUS, ALGAE, FEASIBILITY STUDIES, TECHNICAL FEASIBILITY, E W71-09450
ITY, TROPHIC LEVEL, LIFE HISTORY STUDIES, TIME, SNAILS, RADIOACTIV W70-06668
ER, FLUORIDES, IONS, CYTOLOGICAL STUDIES, WATER POLLUTION EFFECTS. W72-11572
STORAGE, *WATER QUALITY, *MODEL STUCIES, ZOOPLANKTON, BICTA, BACT W72-05282
CULTURE TECHNIQUE, AQUATIC WEED STUDY FACILITY, PRODUCT EVALUATIO W71-06102
, TELLINA BUTTONI, VENUS, TIVELA STULTORUM, EMERITA ANALOGA, DIAST W72-09092
LA, TUNICATE, PISMO CLAM, TIVELA STULTORUM, OSTREA LURIDA, CARDIUM W72-07907
LOGY, BACTERIOLOGY, SAPROPHYTES, STURGEON, METALIMNION.: /HYDROBIO W72-04853
EUSIEDLER(AUSTRIA), CHAMAESIPHON SUBAEQUALIS, APISTONEMA EXPANSUM, W71-04526
LINA, CRYPTOMONAS EROSA, ERKINIA SUBEQUICILIATA, UROGLENCPSIS AMER W72-13816
MODELING.: *WASTE HEAT, SUBLETHEL EFFECTS, MIXING ZONES, W70-09612
IMENTS, ATRAZINE, M/ *LAMINARIA, SUBLITTORAL REGION, CULTURE EXPER W72-12576
RESERVATION, SAMPLE PREPARATION, SUBLITTORAL, LONG ISLAND, ENTERCM W73-00367
SUBMARINE LIGHT DATA SPHERE.: W71-13252
ES, AMPHIBIANS, TOXICITY, CHARA, SUBMERGED PLANTS, PROTOZCA, EUGLE W72-00155
TRICHOCYANELLA SPIRALIS, STROMIA SUBSPHAERICA, TETRACULAMCNAS NATA W72-09119
GROWTH REGULATORS, PLANT GROWTH SUBSTANCES, ALGAE, IRCN, BICASSAY W71-03095
OGY, *PLANT GROWTH, PLANT GROWTH SUBSTANCES, PLANT MORPHOLOGY.: /L W70-07842
ES, *PRODUCTIVITY, *PLANT GROWTH SUBSTANCES, PHOTOSYNTHESIS, METAB W71-08687
RICTICN, *LITTORAL, PLANT GROWTH SUBSTANCES, PHYTOPLANKTCN, FLUID W71-07878
*STARCH REMOVAL, STARCH SUBSTITUTION, POYELECTRCLYTES.: W72-03753
TEROMORPHA, INSOLATION, TEXTURE, SUBSTRA: /INALIS, LITHOPHYTES, EN W72-08141
TROGEN, ORG/ FATE OF POLLUTANTS, SUBSTRATE UTILIZATION, CRGANIC NI W73-00379
RAPHY, CLEANUP, RADIOAUTOGRAPHY, SUBSTRATE UTILIZATION.: /FRCMATOG W73-00361
ES, MIXED CULTURES, METABOLITES, SUBSTRATE UTILIZATION, GROWTH.: / W73-00441
UPPER KLAMATH LAKE, SUBSTRATE UTILIZATION.: W73-01269
RGANIC CARBON, ORGANIC NITROGEN, SUBSTRATE UTILIZATION, PARTICULAT W73-01066
S, *CHEMOSTATS, ORTHOPHOSPHATES, SUBSTRATE UTILIZATION, MONOD EQUA W73-01445
TRAIN, PEARL RIVER, LAKE BORGNE, SUBSTRATES, MACROINVERTEBRATES, N W73-02029
RIPHYTON,/ *FOULING, *ARTIFICIAL SUBSTRATES, DEPTH, *SAMPLING, *PE W73-00374
SERVATION, SAMPL/ *FORAMINIFERA, SUBSTRATES, EPIPHYTES, SAMPLE PRE W73-00367
NTS, LIMITING FACTCRS, ALLOGENIC SUCCESSION, EUTROPHICATION.: /RIE W72-14790
EN, PHOTOSYNTHESIS, RESPIRATICN, SUCCESSION, PRODUCTIVITY, GROWTH W73-01632
ICOCHEMICAL PROPERTIES, DIATOMS, SUCCESSION, NUTRIENTS, RIVERS, IN W73-01435
ATION, MIXING, BOTTOM SEDIMENTS, SUCCESSION, CHLOROPHYTA, DIATOMS, W72-08049
, BIOMASS, CARBON RADIOISOTOPES, SUCCESSION, RADIOACTIVITY TECHNIQ W72-09239
AL STRATIFICATION, WATER SUPPLY, SUCCESSION, ECONOMICS, NUTRIENTS, W71-13183
SE, MALTOSE, MANNITAL, SORBITAL, SUCCINATE, FUMARATE, BUTYRATE, BU W72-05421
USKS, FISH, INSECTS, DDT, SILTS, SUCKERS, CATFISHES, BASS, PERCHES W70-08098
E PRESERVATION, SALPA, LUDOX AM, SUCROSE, DENSITY GRADIENTS, TUNIC W73-00371
ILUS NATANS, GLUCOSE, GALACTOSE, SUCROSE, MALTOSE, MANNITAL, SORBI W72-05421
M, PYRIDINE, DEXTRAM, RAFFINOSE, SUCROSE, SENESCENT CULTURES, XYLO W72-07710
NE, EURYHALINE, RED SEA, GULF OF SUEZ, DECAPODS, COLEOPTERA, MEDIT W73-00855
QUINA BAY, AMPHORA SPP, MELOSIRA SULCATA, NAVICULA SPP, SYNEDRA FA W73-00853
*POLYP METABOLISM, TETRACYCLINE, SULFANILAMIDE, PENICILLIN G, POLY W72-10162
WATER POLLUTION EFFECTS, COPPER SULFATE.: /SHKILL, NEW HAMPSHIRE, W72-09304
LUTICN CONTROL, AIRCRAFT, COPPER SULFATE.: /AKES, PONDS, WATER POL W72-14673
C SULFATE, *STABILIZA/ *ALUMINUM SULFATE, *FERRIC CHLORIDE, *FERRI W71-13356
LFATE, *FERRIC CHLORIDE, *FERRIC SULFATE, *STABILIZATION PCNDS, AL W71-13356
TH OF POLLUTANTS, DIELD/ *COPPER SULFATE, *ALGAE, *FLOODWATER, *PA W71-11498
CHEMCCNTROL, NEW HAMPSH/ *COPPER SULFATE, *ALGAE, *ZOOPLANKTERS, * W71-03014
CAL TREATMENT, CHEMICALS, COPPER SULFATE, CHLORINE, WATER RESOURCE W71-05083
FATE, MANGANESE CHLORIDE, NICKEL SULFATE, COBALT SULFATE, POTASSIU W72-12739
SE, ZINC, NICKEL, COBALT, COPPER SULFATE, HEAVY METALS, METALS, SU W72-12739
CHAINS, DISSOLVED OXYGEN, COPPER SULFATE, HERBICIDES, WATER POLLUT W72-04759
RONMENTAL EFFECTS, PLANT GROWTH, SULFATE, LABORATORY TESTS, CYTOLO W72-00153
PARADOXIUM, CULTURE MEDIA, ZINC SULFATE, MANGANESE CHLORIDE, NICK W72-12739
HAMPSHIRE, ALGAE, COLOR, COPPER SULFATE, NETS, HYDROGEN ION CONCE W72-07890
CHLORIDE, NICKEL SULFATE, COBALT SULFATE, POTASSIUM BR: /ANGANESE W72-12739
ID, BLUSYL, SODIUM LAURYL, ETHER SULFATE, SODIUM DODECYL, BENZENE W72-12576
ALGAL CONTROL, ALGICIDES, COPPER SULFATE, SALINITY, HARDNESS(WATER W72-14694
R SULFATE, HEAVY METALS, METALS, SULFATES.: /NICKEL, CCBALT, COPPE W72-12739
STES, DOMESTIC WASTES, BIOASSAY, SULFATES, IRON COMPOUNDS, SEWAGE, W72-12239
ISM, ALGAE, SYNTHESIS, VITAMINS, SULFATES, IONS.: /PHATES, METABOL W73-01626
YEASTS, SULFUR, FOULING, ALGAE, SULFATES, PERCOLATICN, HERRING, S W72-05421
TRATES, NITRITES, IRON, SILICON, SULFATES, SODIUM, POTASSIUM, CALC W72-05610
PHYSIOLOGY, NUTRIENTS, NITRATES, SULFATES, PHOSPHATES, CHLORIDES, W72-06274
S, SPECTROPHOTOMETRY, MAGNESIUM, SULFATES, SPORES, AZOTOBACTER, CY W72-06124
LKALINE WATER, MAGNESIUM, SODIUM SULFATES, CYANOPHYTA, RHODOPHYTA. W71-04526
NS), STREAMFLOW, IONS, NITRATES, SULFATES, DRAINAGE WATER, DISSOLV W71-01489
PRODUCTIVITY, ICED LAKES, IONS, SULFATES, CARBONATES, BICARBONATE W71-02880
SULFUR BACTERIA, WATER QUALITY, SULFATES, HYDROGEN ION CONCENTRAT W71-07942
MONIDS, SEASON, OXYGEN, HYDROGEN SULFIDE, DISTRIBUTION, NANNOPLANK W71-00117

ANIDIUM CALDARIUM, THIOBACILLUS,	SULFOLOBUS, GLUCOSE, URACIL.: /CY	W72-14301
SULFATE, SODIUM DODECYL, BENZENE	SULFONATE, COCONUT FATTY ACID, PL	W72-12576
*POLYETHYLENEMINE, *POLYSTYRENE	SULFONATE, CHLORELLA ELLIPSCIDIA.	W72-14840
LO/ POLYACRYLAMIDES, POLYSTYRENE	SULFONATE, POLYETHYLENEIMINE, *CH	W72-11833
POLLUTION EFFECTS, ALKYLBENZENE	SULFONATES, WATER POLLUTION SOURC	W71-11516
NTS, *LETHAL LIMIT, ALKYLBENZENE	SULFONATES.: /, PHENOLS, *DETERGE	W72-12576
*EUTROPHICATION, LINEAR ALKYLATE	SULFONATES, CHELATION, PHOSPHATES	W71-06247
ER, EUTROPHICATION, ALKYLBENZENE	SULFONATES, SURFACTANTS, LINEAR A	W72-06837
ES, SURFACTANTS, LINEAR ALKYLATE	SULFONATES, SEWAGE, NUTRIENTS, CO	W72-06837
, *MEMBRANE PROCESSES, BACTERIA,	SULFUR BACTERIA, PHOTOSYNTHETIC B	W72-04743
ER POLLUTION EFFECTS, *KENTUCKY,	SULFUR BACTERIA, WATER QUALITY, S	W71-07942
OLIDS, LACTOBACILLUS, CHLORELLA,	SULFUR BACTERIA, ALGAE, SALMONELL	W72-12565
N RADIOISOTOPES, *ON-SITE TESTS,	SULFUR BACTERIA, PRODUCTIVITY, RA	W72-14301
OBIC BACTERIA, AEROBIC BACTERIA,	SULFUR BACTERIA, IRON COMPOUNDS.:	W72-12996
ON, MET/ *TOXICITY, *CYANOPHYTA,	SULFUR, ALGAE, GRANULES, INHIBITI	W72-00153
S, CARBON, NITROGEN, PHOSPHORUS,	SULFUR, CALCIUM, MANGANESE, POTAS	W73-00840
NIC COMPOUNDS, VITAMINS, YEASTS,	SULFUR, FOULING, ALGAE, SULFATES,	W72-05421
ORGANIC MATTER, MANGANESE, IRON,	SULFUR, POTASSIUM, MOLYBDENUM, DE	W72-07514
, COLIFORMS, METABOLISM, CARBON,	SULFUR, SAMPLING, WATER ANALYSIS,	W72-04743
	SUMMARIES.:	W73-01976
TA, BIOMASS, PYRROPHYTA, SPRING,	SUMMER, AUTUMN, SAMPLING, PLANKTO	W72-12738
ANOPHYTA, ALGAE, EUTROPHICATION,	SUMMER, ENERGY, SEASONAL.: /Y, CY	W73-03228
AE, CHRYSOPHYTA, AUTUMN, WINTER,	SUMMER, PRIMARY PRODUCTIVITY, DIN	W73-00442
X CALCICOLA, ELKTON POND, RISING	SUN POND, INDUSTRIAL STORAGE T: /	W72-04736
ICATORS, E. COLI, AQUATIC ALGAE,	SUNFISHES, SNAILS, EELS, CHINOOK	W71-07417
EROSION, NUTRIENTS, ALGAE, LAKE	SUPERIOR, WATER POLLUTION SOURCES	W71-04216
LIGOTROPHY, EUTROPHICATION, LAKE	SUPERIOR, LAKE MICHIGAN, LAKE HUR	W73-01456
ES, EUTROPHICATION, ALGAE, *LAKE	SUPERIOR, *LAKE MICHIGAN, *LAKE H	W72-01094
AKES, NUTRIENTS, LAKE ERIE, LAKE	SUPERIOR, LABORATORY TESTS, TEMPE	W72-01108
GAN, LAKE ERIE, LAKE HURON, LAKE	SUPERIOR, WATER TEMPERATURE, HYDR	W72-10875
OPLANKTON, *SESSILE ALGAE, *LAKE	SUPERIOR, LIMNOLOGY, FERTILITY, P	W72-12192
YO GRANDE CREEK.: *LOPEZ WATER	SUPPLY PROJECT, LOPEZ CREEK, ARRO	W72-05594
, *GROUNDWATERS, SURVEYS, *WATER	SUPPLY.: /, ODOR, *SURFACE WATERS	W72-06536
T, GROUNDWATER, WISCONSIN, WATER	SUPPLY.: / WASTES, SOIL MANAGEMEN	W71-06443
E PROJECTS, *WATER REUSE, *WATER	SUPPLY, *POLLUTION ABATEMENT, *WA	W70-09193
ENT, *DISTRIBUTION SYSTE/ *WATER	SUPPLY, *WATER RESOURCES DEVELOPM	W72-05594
ION WASTES, *MICROORGANI/ *WATER	SUPPLY, *WASTE DISPOSAL, *RECREAT	W72-14363
TER REUSE, *WATER QUALITY, WATER	SUPPLY, ALGAE, ACTIVATED CARBON,	W70-10381
US, RESERVOIRS, FISHERIES, WATER	SUPPLY, COSTS, WATER DEMAND, ECON	W71-13172
IC WATER, *SURFACE WATERS, WATER	SUPPLY, EUTROPHICATION, ALGAE, NU	W71-05083
, IRRIGATION, RETURN FLOW, WATER	SUPPLY, LIVESTOCK, SEWAGE, INFILT	W71-06435
S, THERMAL STRATIFICATION, WATER	SUPPLY, SUCCESSION, ECONOMICS, NU	W71-13183
TY, GREAT LAKES, NEW YORK, WATER	SUPPLY, SYSTEMATICS, LIMNOLOGY, E	W73-01615
NUTRIEN/ *EUTROPHICATION, *WATER	SUPPLY, WATER POLLUTION EFFECTS,	W71-13185
ER RESOURCES DEVELOPMENT, *WATER	SUPPLY, WASTE WATER, RUNOFF, DISS	W71-09788
LAKES, *MAINE, RECREATION, WATER	SUPPLY, WATER POLLUTION EFFECTS,	W72-04280
*AERATED LAGOONS,	SURFACE AERATORS.:	W71-07109
LAKES, BAYS, WATERSHEDS(BASINS),	SURFACE DRAINAGE, MUNICIPAL WASTE	W71-08953
RUS, NUTRIENTS, ALGAE, NITRATES,	SURFACE RUNOFF, BASE FLOW, PERCOL	W71-06443
SPHERIC), RUNOFF, ANIMAL WASTES,	SURFACE WATERS, SOIL EROSION, NUT	W71-04216
CTANTS, WATER POLLUTION SOURCES,	SURFACE WATERS, WATER POLLUTION,	W70-09388
NTARIO, HUDSON RIVER, STANDARDS,	SURFACE WATERS, EPILIMNION, INTAK	W70-09614
OSING ORGANIC MATTER, NUTRIENTS,	SURFACE WATERS, NITROGEN, WATER Q	W70-10405
ANKTON, WATER QUALITY, MICHIGAN,	SURFACE WATERS, COASTS, SAMPLING,	W72-10875
RY PRODUCTIVITY, ALASKA, FIORDS,	SURFACE WATERS, NITROGEN, SEA WAT	W72-09122
GEN, PHOSPHORUS, EUTROPHICATION,	SURFACE WATERS.: /DILUTION, NITRO	W72-01472
SEDIMENTS, *AQUATIC ENVIRONMENT,	SURFACE WATERS, NUTRIENTS, ALGAE,	W72-00847
Y METALS, AIR POLLUTION EFFECTS,	SURFACE WATERS, PATH OF POLLUTANT	W72-00940
NSITY, *ECOLOGICAL DISTRIBUTION,	SURFACE WATERS, LAGOONS, CHLOROPH	W73-00838
IBUTION, FLOATING, WINDS, ALGAE,	SURFACE WATERS, HYDROGEN ION CONC	W73-00236
ACTIVITY TECHNIQUES, MONITORING,	SURFACE WATERS, PERSISTENCE, PATH	W73-00361
UTION, *UREAS, *COASTS, *OCEANS,	SURFACE WATERS, CONTINENTAL SHELF	W72-08056
L ANALYSES, NITROGEN, NUTRIENTS,	SURFACE WATERS, ALGAE, CHEMICAL P	W72-12955
T, *FILTATION, ALLUVIA CHANNELS,	SURFACE-GROUNDWATER RELATIONSHIPS	W71-12410
UCTIVITY, FARM PONDS, NUTRIENTS,	SURFACES, TEMPERATURE, PHOSPHORUS	W71-11001
ATED RAINFAL/ *SOIL ALGAE, *SOIL	SURFACES, *SEDIMENT YIELD, *SIMUL	W72-02218
RIES, WATER QUALITY, TECHNOLOGY,	SURFACTANTS, TOXICITY, HAZARDS, B	W72-09169
UDGE, TREATMENT, EUTROPHICATION,	SURFACTANTS, ENZYMES, LAKE ONTARI	W71-00116
LGAL POISONING, ALGAE, SOLVENTS,	SURFACTANTS, WATER TEMPERATURE, P	W70-09429
TRIENTS, *EUTROPHICATION, ALGAE,	SURFACTANTS, WATER POLLUTION SOUR	W70-09388
CATION, ALKYLBENZENE SULFONATES,	SURFACTANTS, LINEAR ALKYLATE SULF	W72-06837
MOUGEOTIA, NAVICULA, PINNULARIA,	SURIRELLA, PERIDINIUM, SENECA RIV	W72-09111
INGRA(WIS.), LIMITING NUTRIENTS,	SURPLUS NUTRIENTS, NUTRIENT SOURC	W73-00232
SECONDARY SEWAGE T/ *LITERATURE	SURVEY, PRIMARY SEWAGE TREATMENT,	W70-09907
*SURFACE WATERS, *GROUNDWATERS,	SURVEYS, *WATER SUPPLY.: /, ODOR,	W72-06536
ER POLLUTION EFFECTS, EUTROPHIC/	SURVEYS, *AERIAL PHOTOGRAPHY, WAT	W72-14728
UALITY, WATER POLLUTION CONTROL,	SURVEYS, INVESTIGATIONS, PATH OF	W70-08248
UALITY, WATER POLLUTION CONTROL,	SURVEYS, INVESTIGATIONS, PATH OF	W70-08249
, *POTABLE WATER, *TASTE, *ODOR,	SURVEYS, MICROBIOLOGY, WATER TREA	W72-10076
TS, ALGAE, PHOSPHATES, NITRATES,	SURVEYS, PESTICIDES, WASTE DISPOS	W71-05553
GINIA, *WATER POLLUTION EFFECTS,	SURVEYS, SAMPLING, DATA COLLECTIO	W73-01972
VITY, CHLORAMPHENICOL, RECOVERY,	SURVIVAL.: /S, *MUTANTS, *SENSITI	W72-14327
LADOPHOROPSIS MEMBRANACEA, PENI/	SURVIVAL, PENICILLUS, CHONDRIA, C	W73-00838
KINIANA, CO-60, GAMMA RADIATION,	SURVIVAL, CHLORELLA PYRENOIDOSA.:	W72-11836
RADATION, ALGAE, PHOTOSYNTHESIS,	SUSPENDED SOLIDS, SEPARATION TECH	W72-09386
ING, LAKE MICHIGAN, MEASUREMENT,	SUSPENDED LOAD, ALKALINITY, HARDN	W71-11033
	SUSPENDED SOLIDS.:	W71-12188
LOGICAL TREATMENT, MIXED LIQUOR,	SUSPENDED SOLIDS, UPTAKE.: /L-BIO	W71-13334
FFECTS, WATER QUALITY, SEASONAL,	SUSPENDED LOAD, BIOINDICATORS, WA	W72-07896
PONDS, CALIFORNIA, SCENEDESMUS,	SUSPENDED LOAD, NITRATES, CALCIUM	W70-10174

*SURFACE CHARGE, REGENERATICN,	SUSPENDED SOLIDS.:	W71-00138
*AERATED LAGOONS,	SUSPENDED SOLIDS.:	W71-07113
*AERATED LAGOONS,	SUSPENDEC SOLIDS.:	W71-07106
	SUSPENDED SOLIDS.:	W71-07123
*ALGAE, GROWTH RATES, CULTURES,	SUSPENDED LOAC, SEWAGE, INDUSTRIA	W71-09795
IC BACTERIA, NITROGEN, VITAMINS,	SUSPENDED SOLICS, LACTOBACILLUS,	W72-12565
VIRUSES, *PLECTCNEMA BORYANIUM,	SUSQUEHANNA RIVER, BRANDYWINE RIV	W72-04736
*MICROCYSTIS, *LAKE	SUWA(JAPAN).:	W73-00236
ERS, MUSSELS, PROTOZOA, MANGROVE	SWAMPS, LITTORAL, SHORES, TIDES,	W72-07911
*MARINE DISPOSAL, *GOTHENBRUG,	SWEDEN.:	W70-09947
NIZOMENON FLOS-AQUAE, LAKE ERKEN(	SWEDEN).: *CYANOPHAGES, *APHA	W73-00246
: *LAKE ERKEN(	SWEDEN), *DARK NITROGEN FIXATION.	W72-08054
TROUT, LOTA LOTA, LAKE MALOGEN,	SWEDEN, CHAR, LUCIOPERCA LUCISPER	W72-12729
C.:	SWEDEN, COLLEMA, PELTIGERA, NOSTO	W73-02471
STICHLIS SPICATA, FIMB/ ENGLAND,	SWEDEN, GERMANY, SPARTINA SPP, DI	W73-01089
RMES, *ROCKPOOLS, *ENTEROMORPHA,	SWEDEN, TIGRIOPUS BREVICCRNIS, GA	W71-08026
LAKE HERMAN(S. DAK), ENEMY	SWIM LAKE(S. DAK).:	W71-02880
, RESISTANCE, LAKES, RESERVOIRS,	SWIMMING POOLS, CHELATICN, APPLIC	W71-06189
*ILLINOIS, RECREATION, FISHING,	SWIMMING, BOATING, DOCUMENTATION,	W72-05869
TENSITY, NUTRIENTS, PSEUDOMONAS,	SYMBIOSIS.: /CENTRATION, LIGHT IN	W73-01657
, CROP PRODUCTION, FUNGI, SCILS,	SYMBIOSIS, AFRICA, ASIA, ANABAENA	W73-02473
ATE, CLIMATIC ZONES, CYANOPHYTA,	SYMBIOSIS, LICHENS, NITRCGEN, LIG	W73-02471
HORUS, *CARBON, BACTERIA, ALGAE,	SYMBIOSIS, CARBON DIOXIDE, NITROG	W70-07283
ICATI/ *ALGAE, BACTERIA, *LAKES,	SYMBIOSIS, ALGAL CONTROL, EUTROPH	W71-13257
ONIA, EUTROPHICATICN, OXIDATION,	SYMBIOSIS, AZOTOBACTER, MARINE AL	W72-11563
, AUTOSPORES, VARIABILITY, BIAS,	SYNCHRONY INDEX, COUNTING.: /URES	W72-11920
TOSPORES, *SYNCHRONOUS CULTURES,	SYNCHRONY INDEX.: / PATTERNS, *AU	W72-12724
IATILIS, DUNALIELLA TERTIOLECTA,	SYNECHOCOCCUS, RHODOMCNAS, COCCOL	W71-10083
RMIDIUM, MAST/ YELLOWSTONE PARK,	SYNECHOCOCCUS, FLEXIBACTERIA, PHC	W72-03557
EDECKEI, OSCILLATORIA RUBESCENS,	SYNEDRA ACUS, RHOCOMONAS, LAKE CO	W71-00117
AFRICANUM, NITZCHIA ACICULARIS,	SYNEDRA ACUS.: /EROSA, PERIDINIUM	W73-00256
MELOSIRA SULCATA, NAVICULA SPP,	SYNEDRA FASCICULATA, PLAGIOGRAMMA	W73-00853
PHORA JURGENSII, AMPHORA OVALIS,	SYNEDRA FASCIULATA, EUNCTIA PECTI	W72-08141
OSIRA ITALICA, MELOSIRA AMBIGUA,	SYNEDRA FILIFORMIS, FRAGILARIA CA	W73-00384
LARIA, TABELLARIA, ASTERIONELLA,	SYNEDRA, DINOBRYON, MELCSIRA.: /I	W73-01669
VICULA, GLENODINIUM, MICROSPORA,	SYNEDRA, DINOBRYON, OEDCGONIUM, F	W72-10875
ACTIVITY, ALGI/ DETOXIFICATION,	SYNERGISM, FISH PONDS, ALGISTATIC	W71-06189
MONIA, ASSAY, INHIBITION, UREAS,	SYNTHESIS, ALGAE.: /ORY TESTS, AM	W72-03222
ASTE WATER TREATMENT, NUTRIENTS,	SYNTHESIS, SLUDGE, EFFLUENT, ALGA	W71-12188
, PHOSPHATES, METABOLISM, ALGAE,	SYNTHESIS, VITAMINS, SULFATES, IC	W73-01626
IRCINALIS, BOSMINA LONGIROSTRIS,	SYNURA UVELLA, DINOBRYCN, OSCILLA	W71-03014
WEDEN, MICROCYSIS, OSCILLATORIA,	SYNURA, AROGLENA, CRYPTOPHYCEAE,	W72-12729
VENT EXTRACTIONS, FREEZE DRYING,	SYST: /ENT, AEROBIC BACTERIA, SOL	W73-01066
-GAMMA-RADIATION SYSTEM, MODULAR	SYSTEM, COBALT-60.: *BIOLOGICAL-	W72-00383
.: *BIOLOGICAL--GAMMA-RADIATION	SYSTEM, MODULAR SYSTEM, COBALT-60	W72-00383
TE, *LUCIFERIN-LUCIFERASE ENZYME	SYSTEM, PLANT EXTRACTS, CHLORELLA	W71-11916
BUTION PATTERNS, *MASSACHUSETTS,	SYSTEMATICS, *REPRODUCTICN, RHODO	W73-01449
RDNESS(WATER), BENTHOS, GRAZING,	SYSTEMATICS, CHEMICAL ANALYSIS, P	W73-01627
T LAKES, NEW YORK, WATER SUPPLY,	SYSTEMATICS, LIMNOLOGY, ECOLOGY,	W73-01615
OHOLS, POLLUTANT IDENTIFICATION,	SYSTEMATICS, NUTRIENTS, PSEUDOMON	W73-00942
WISCONSIN, PRIMARY PRODUCTIVITY,	SYSTEMATICS, GROWTH RATES, BICMAS	W73-02469
DIATOMS, BIOLOGICAL COMMUNITIES,	SYSTEMATICS, CYANOPHYTA, CHEMICAL	W73-00242
*OREGON, *SEASONAL, CHRYSOPHYTA,	SYSTEMATICS, *ESTUARINE ENVIRONME	W72-08141
CTERIA, CHLOROPHYTA, CYANOPHYTA,	SYSTEMATICS, VITAMINS, ANAEROBIC	W72-07951
, *PLANKTON, *WATER TEMPERATURE,	SYSTEMATICS, SAMPLING, NETS, CENT	W72-07899
ULTURES, SALINITY, SALT MARSHES,	SYSTEMATICS, *VIRGINIA, *ATLANTIC	W72-12633
DATA COLLECTIONS, MICROBIOLOGY,	SYSTEMATICS, POLLUTANTS, PATHOGEN	W72-13800
TOPLANKTON, CYTOLOGICAL STUDIES,	SYSTEMATICS, BIOMASS, FLUORCMETRY	W72-14313
, LIFE CYCLES, ODOR, RECREATION,	SYSTEMATICS.: / COLIFORMS, SPORES	W71-09173
ULOSE, VIRUSES, HUMAN PATHOLOGY,	SYSTEMATICS.: /URE, ENZYMES, CELL	W72-05457
, CORAL, REEFS, WATER RESOURCES,	SYSTEMS ANALYSIS, WASTE TREATMENT	W71-13723
ICAL MODELS, LINEAR PROGRAMMING,	SYSTEMS ANALYSIS, DECIDUCUS FCRES	W72-03342
URCES DEVELOPMENT, *DISTRIBUTION	SYSTEMS, *WATER QUALITY, WATER CO	W72-05594
N DITCHES, AERATED LA/ *DISPOSAL	SYSTEMS, *LANC CISPOSAL, CXIDATIO	W71-07551
TERRES/ *C-13, *C-12, BIOLOGICAL	SYSTEMS, BELEMNITELLA AMERICANA,	W72-12709
AL, FERTILITY, RHIZOSPHERE, RCOT	SYSTEMS, SOIL FUNGI, SOIL ENVIRON	W72-03342
COLOGY, AQUATIC LIFE, IRRIGATION	SYSTEMS, WATER ANALYSIS, TEMPERAT	W71-06102
ROOK TROUT, SAMPLING, SEPARATION	T: /EFFECTS, GAS CHRCMATCGRAPHY, B	W72-12930
ING SUN POND, INDUSTRIAL STORAGE	T: /X CALCICOLA, ELKTCN PCND, RIS	W72-04736
, FRAGILARIA, DIATOMA, MELOSIRA,	TABELLARIA, ACTINASTRUM, PHYTOCON	W72-10875
ILARIA, PEDIASTRUM, STAURASTRUM,	TABELLARIA, MCUGECTIA.: /ON, FRAG	W72-10076
TENOIDS, FRAGILARIA CROTONENSIS,	TABELLARIA FLOCCULOSA, STEPHANODI	W73-00384
ES, DICHOTCMOSIPHON, FRAGILARIA,	TABELLARIA, ASTERIONELLA, SYNEDRA	W73-01669
*INDIAN CREEK RESERVOIR, / SOUTH	TAHOE PUBLIC UTILITIES DISTRICT,	W72-02412
*LAKE	TAHOE.:	W72-01660
*ALGAL GROWTH, *LAKE	TAHOE(CALIF).:	W71-13553
, PYRIDOXINE, INDAL COMPOU/ LAKE	TAHOE(CALIF), ZINC, ASCCRBIC ACID	W72-07514
EMPORANEOUS DISEQUILIBRIUM, LAKE	TAHOE(CALIF), CASTLE LAKE(CALIF),	W71-09171
*LAKE	TAHOE(CALIF.).:	W72-12955
*LAKE	TAHOE(CALIF-NEV).:	W72-12549
AYETTEVILL/ FECAL MATERIAL, LAKE	TAHOE, LAKE SEBASTICOOK(MAINE), F	W71-05626
, ACIDITY, LABORATORY, DIGESTION	TANK, PONDAGE, PILOT PLANT, SETTL	W72-02850
MILL WASTES, MONTREAL, BALANCING	TANKS.: *TEXTILE	W70-10427
EWAGE EFFLUENTS, SEEPAGE, SEPTIC	TANKS, BIOASSAY, CHEMICAL ANALYSE	W72-12955
OUTFALL SERVERS, STREAMS, SEPTIC	TANKS, FUNGI, ALGAE, WATER POLLUT	W72-14363
ICATION, DENITRIFICATION, SEPTIC	TANKS, CXIDATION LAGOCNS, LAND AP	W70-09907
ALKALCIDS, FLAVONOIDS, STEROIDS,	TANNINS, SAPONINS.: /C ACTIVITY,	W72-04275
STHWAITE WATER(ENGLAND), BLELHAM	TARN(ENGLAND), LUXURY UPTAKE, LOC	W71-13183
)/ ASTERIONELLA FORMOSA, BLELHAM	TARN(ENGLAND), WINDERMERE(ENGLAND	W72-10625
HATES, HEAVY METALS, PESTICIDES,	TASTE-PRODUCING ALGAE, ODCR-PRODU	W73-02144

TREATMENT, ALGAE, POTABLE WATER,	TASTE, AMMONIA, SILICA.: /LUTION	W71-01491
E, *BACTERIA, AESTHETICS, COLOR,	TASTE, CORROSION, DISEASES, EUTRO	W71-09958
NSITY, NUTRIENTS, ACTINOMYCETES,	TASTE, EUTROPHICATION.: /GHT INTE	W72-10613
CIDES, HERBICIDES, HEAVY METALS,	TASTE, ODOR, RECREATION, PHOSPHOR	W71-12091
RATURE, NUTRIENTS, ALGAE, WEEDS,	TASTE, ODOR, PATHOGENIC BACTERIA,	W71-11006
LATION, SEDIMENTATION, TOXICITY,	TASTE, ODOR, CORROSION, ENZYMES,	W72-06837
ING, AERATION, PHOSPHATES, CLAY,	TASTE, ODOR, *SURFACE WATERS, *GR	W72-06536
CHNIQUES, WASTE WATER TREATMENT,	TASTE, ODOR, POTABLE WATER, ACTIV	W73-02144
ICE, FISH, MICROORGANISMS, ODOR,	TASTE, PLANKTON, CHEMICALS, PHENO	W73-01669
L, *ODOR, ALGAE, EUTROPHICATION,	TASTE, RESERVOIRS, SEASONAL, ACTI	W71-13187
NIQUES, CULTURE MEDIA, ALGOLOGY,	TAXON.: CULTUTURING TECH	W72-12172
, CYPRINU/ *RESIDUES, *DDD, DDE,	TDE, BLACK-FLY LARVAE, CLADOPHORA	W70-08098
RUS, ALGAE, FEASIBILITY STUDIES,	TECHNICAL FEASIBILITY, ECONOMIC F	W71-09450
ILITY, PRODUCT E/ *ALGAE CULTURE	TECHNIQUE, AQUATIC WEED STUDY FAC	W71-06102
ON, EQUIPMENT, COSTS, SEPARATION	TECHNIQUES, COAGULATION, WASTE WA	W70-08904
TER BIRDS, WISCONSIN, ANALYTICAL	TECHNIQUES.: /FISH, AMPHIPODA, WA	W70-08098
RATES, *ELECTRODES,/ *ANALYTICAL	TECHNIQUES, *WATER ANALYSIS, *NIT	W71-00474
L MODELS, SIMULATION, ANALYTICAL	TECHNIQUES, PHOSPHORUS, ALGAE, LA	W71-03021
ELLA, NITRATES, CHE/ *ANALYTICAL	TECHNIQUES, *ALGAE, *ASSAY, CHLOR	W70-10403
SORPTION, MONITORING, ANALYTICAL	TECHNIQUES, LABORATORY TESTS, CES	W71-08032
GANISMS, PROTEINS, RADIOACTIVITY	TECHNIQUES, FOOD CHAINS.: /ICROOR	W71-08042
ARBON RADIOISOTOPES, *ANALYTICAL	TECHNIQUES, *MARINE MICROORGANISM	W71-08042
HLOROPHYLL B, ASHING, LABORATORY	TECHNIQUES.: /ELLIPSOIDEA, RNA, C	W72-06274
, AMMONIA, PHOSPHATE, ANALYTICAL	TECHNIQUES, *PHOTOSYNTHESIS, CARB	W72-06046
UNGI, *BACTERIOPHAG/ *SEPARATION	TECHNIQUES, *ALGAE, *BACTERIA, *F	W72-06124
NG, *EUTROPHICATION, *ANALYTICAL	TECHNIQUES, *ALGAL CONTROL.: /IXI	W72-04433
LGAE, *PUBLIC HEALTH, ANALYTICAL	TECHNIQUES, TOXICITY, CHEMICAL PR	W72-04275
TIC OCEAN, SHARKS, RADIOACTIVITY	TECHNIQUES.: /, PHAEOPHYTA, ATLAN	W72-05965
EMICAL OXYGEN DEMAND, ANALYTICAL	TECHNIQUES, TERTIARY TREATMENT, *	W72-04983
ES, SEWAGE TREATMENT, SEPARATION	TECHNIQUES, *CHEMICAL WASTES, IND	W72-05315
S, ORGANIC COMPOUNDS, SEPARATION	TECHNIQUES, CALCIUM, MAGNESIUM, E	W72-11833
ELLA, IRRADIATION, RADIOACTIVITY	TECHNIQUES, RESISTANCE, CULTURES,	W72-11836
NALYTICAL TECHNIQUES, SEPARATION	TECHNIQUES, GAS CHROMATOGRAPHY, S	W72-11604
, *ORGANIC COMPOUNDS, ANALYTICAL	TECHNIQUES, SEPARATION TECHNIQUES	W72-11604
GY, TAXON.: CULTUTURING	TECHNIQUES, CULTURE MEDIA, ALGOLO	W72-12172
*BIOASSAY, *ALGAE, / *ANALYTICAL	TECHNIQUES, *TESTING PROCEDURES,	W72-10619
OPES, *BIOINDICATORS, ANALYTICAL	TECHNIQUES, POLLUTANT IDENTIFICAT	W72-10958
BIOASSAY, PHOSPHORU/ *ANALYTICAL	TECHNIQUES, *NUTRIENTS, *ALGAE, *	W72-10608
*MEASUREMENT, *ASSAY, ANALYTICAL	TECHNIQUES, EVALUATION, WATER POL	W72-10678
TER ANALYSIS, EVALU/ *ANALYTICAL	TECHNIQUES, CHEMICAL ANALYSIS, WA	W72-10883
IS, SUSPENDED SOLIDS, SEPARATION	TECHNIQUES, FILTRATION, MUNICIPAL	W72-09386
S, *MARINE ALGAE, D/ *ANALYTICAL	TECHNIQUES, RATES, *PHOTOSYNTHESI	W72-09653
, *MUNICIPAL WASTES, *SEPARATION	TECHNIQUES, FERMENTATION, NITROGE	W72-10390
HARVESTING OF ALGAE, *SEPARATION	TECHNIQUES, CHEMICAL PRECIPITATIO	W72-10237
LUMNS, ORGANIC ACIDS, SEPARATION	TECHNIQUES, *CHROMATOGRAPHY.: /CO	W72-09365
COLIFORMS, NUTRIENTS, ANALYTICAL	TECHNIQUES, MICROBIOLOGY, *WASTE	W72-09590
ARY PRODUCTIVITY, *RADIOACTIVITY	TECHNIQUES, *PHOTOSYNTHESIS, CARB	W72-09102
EN, ABATEMENT, COSTS, ANALYTICAL	TECHNIQUES, MUNICIPAL WATER, PHOS	W72-09171
TOPES, SUCCESSION, RADIOACTIVITY	TECHNIQUES, CULTURES.: / RADIOISO	W72-09239
ANALYSIS, TRACERS, RADIOACTIVITY	TECHNIQUES.: /AL, WEATHER, WATER	W72-09122
RIA, LAKES, NITROGEN, ANALYTICAL	TECHNIQUES, CARBON, EUTROPHICATIO	W72-09163
N, CENTRIFUGATION, RADIOACTIVITY	TECHNIQUES, CHLOROPHYLL, SEA WATE	W72-09103
CARBON, GRAVIMETRIC/ *ANALYTICAL	TECHNIQUES, *ORGANIC COMPOUNDS, *	W72-09108
ASTE WATER TREATMENT, ANALYTICAL	TECHNIQUES, COALESCENCE, FLOCCULA	W71-13356
ATION, LIQUID WASTES, SEPARATION	TECHNIQUES, WATER PURIFICATION, *	W71-13356
YSIOLOGY, POLLUTANTS, ANALYTICAL	TECHNIQUES, WATER POLLUTION SOURC	W71-10553
ITY, HARDNESS(WATER), ANALYTICAL	TECHNIQUES, AMMONIA, NITRATES, PH	W71-11033
GAE, *SURFACE WATERS, ANALYTICAL	TECHNIQUES, PHOTOMETRY, WATER POL	W71-12034
NALYTICAL TECHNIQUES, SEPARATION	TECHNIQUES, SEDIMENTS, OILY WATER	W71-12467
POLLUTION TREATMENT, ANALYTICAL	TECHNIQUES, SEPARATION TECHNIQUES	W71-12467
DETERGENTS, BIOASSAY, ANALYTICAL	TECHNIQUES, FORECASTING, SILICA.:	W71-12072
OLISM, ALGAE, SLUDGE, ANALYTICAL	TECHNIQUES, INDUSTRIAL WASTES, MU	W71-11793
I, *ISOLATION, ALGAE, SEPARATION	TECHNIQUES, ATLANTIC OCEAN, PACIF	W71-11851
L ANALYSIS, SAMPLING, ANALYTICAL	TECHNIQUES, NUTRIENTS, FOOD CHAIN	W71-11234
TH RATE, *AQUATI/ *RADIOACTIVITY	TECHNIQUES, *BENTHIC FLORA, *GROW	W71-11486
DDT, ORGANIC MATTER, ANALYTICAL	TECHNIQUES, SAMPLING, MASSACHUSET	W71-11498
RANIUM RADIOISOTOPES, ANALYTICAL	TECHNIQUES, PALEOZOIC ERA, WATER	W72-02073
ROPHY, PHYTOPLANKTON, ANALYTICAL	TECHNIQUES, TURBIDITY, NITRATES,	W72-03224
ECTS, TRACE ELEMENTS, ANALYTICAL	TECHNIQUES, COLUMBIA RIVER, WASHI	W72-03215
GEN, *THERMODYNAMICS, ANALYTICAL	TECHNIQUES, DESIGN CRITERIA, KINE	W72-02363
NALYSIS/ *SEDIMENTS, *ANALYTICAL	TECHNIQUES, *NEUTRON ACTIVATION A	W72-01101
DENTIFICATION, ALGAE, ANALYTICAL	TECHNIQUES, *ULTRAVIOLET RADIATIO	W72-01780
LGAE, *REMOTE SENSING, *TRACKING	TECHNIQUES, MEASUREMENT, CURRENTS	W72-01095
PTION, CULTURES, DDT, ANALYTICAL	TECHNIQUES, BIOMASS, GAS CHROMATO	W72-04134
OW RATES, INHIBITION, ANALYTICAL	TECHNIQUES, ENZYMES, COLOR, WATER	W72-04070
CYTOLOGICAL STUDIES, ANALYTICAL	TECHNIQUES, POLLUTANT IDENTIFICAT	W72-03985
OACTIVITY EFFECTS, RADIOACTIVITY	TECHNIQUES, AQUARIA, ALGAE, MATHE	W72-03348
UTANT IDENTIFICATION, ANALYTICAL	TECHNIQUES, *ALGAE, *NUTRIENTS, L	W72-00801
S, CYCLING NUTRIENTS, ANALYTICAL	TECHNIQUES, WATER POLLUTION EFFEC	W71-13794
OACTIVITY EFFECTS, RADIOACTIVITY	TECHNIQUES.: /UTION SOURCES, RADI	W72-00940
, ALGAE, TEMPERATURE, ANALYTICAL	TECHNIQUES, WATER QUALITY CONTROL	W72-00422
QUES, *PLANKTON, PR/ *ANALYTICAL	TECHNIQUES, *RADIOACTIVITY TECHNI	W72-00161
TOPES, ABSORPTION, RADIOACTIVITY	TECHNIQUES, TRACERS, PATH OF POLL	W72-00949
TICAL TECHNIQUES, *RADIOACTIVITY	TECHNIQUES, *PLANKTON, PRIMARY PR	W72-00161
MENT, *PUBLIC HEALTH, ANALYTICAL	TECHNIQUES, WASTE WATER TREATMENT	W73-02144
ALGAE, *PHYTOPLANKT/ *ANALYTICAL	TECHNIQUES, *COMPUTER PROGRAMS, *	W73-02469
NUTRIENTS, SEDIMENTS, ANALYTICAL	TECHNIQUES, FERTILIZATION, AQUATI	W73-02478
ANALYSIS, ADSORPTION, SEPARATION	TECHNIQUES, FOOD CHAINS, CULTURES	W73-02105
, VIRGINIA, SAMPLING, SEPARATION	TECHNIQUES, CENTRIFUGATION, TURBI	W73-00916
HUMAN POPULATION, RADIOACTIVITY	TECHNIQUES, RADIUM RADIOISOTOPES.	W73-00833

N, ALGAE, CYANOPHYTA, ANALYTICAL TECHNIQUES, WATER POLLUTION EFFEC W73-01269
, GAS CHROMATOGRAPHY, SEPARATION TECHNIQUES, *WISCONSIN, OLIGOTROP W73-01456
, INSTRUMENTATION, RADIOACTIVITY TECHNIQUES, CARBON RADIOISOTOPES, W72-14326
LTRAVIOLET RADIATION, SEPARATION TECHNIQUES, CYANOPHYTA, CULTURES, W72-14327
ICLE SIZE, SEDIMENTS, SEPARATION TECHNIQUES, SPATIAL DISTRIBUTION, W73-00367
ON, OCEANS, SAMPLING, SEPARATION TECHNIQUES, CULTIVATION, VITAMIN W73-00430
DIMENTS, PLANKTON, RADIOACTIVITY TECHNIQUES, MONITORING, SURFACE W W73-00361
NERITIC, NITRATES, RADIOACTIVITY TECHNIQUES, EUTROPHICATION, MARIN W73-00431
*LARVAE, *FISH EGGS, *SEPARATION TECHNIQUES, *MARINE FISH, SILICA, W73-00371
HODESMIUM ERYTHRAEUM, *CULTURING TECHNIQUES, CULTURE MEDIA, *BAY O W73-00430
PMENT, FREEZE DRYING, SEPARATION TECHNIQUES, SOLVENT EXT: /RY EQUI W73-00379
ESTING, ASSAY, NUTR/ *ANALYTICAL TECHNIQUES, *ALGAE, EVALUATION, T W73-00255
C PLANTS, TOXINS, R/ *ANALYTICAL TECHNIQUES, *ALGAL TOXINS, AQUATI W73-00243
YTICAL TECHNIQUES, RADIOACTIVITY TECHNIQUES, RADIOECOLOGY, TRACERS W73-00790
OACTIVITY TECHNIQUES, ANALYTICAL TECHNIQUES, CYCLING NUTRIENTS, TR W73-00821
NESE, CHROMIUM, IRON, ANALYTICAL TECHNIQUES, FOOD CHAINS, AFRICA, W73-00832
OACTIVITY TECHNIQUES, ANALYTICAL TECHNIQUES.: / RADIOECOLOGY, RADI W73-00802
ON, SEDIMENTS, CORES, ANALYTICAL TECHNIQUES, RADIOACTIVITY TECHNIQ W73-00790
CTS, RADIOECOLOGY, RADIOACTIVITY TECHNIQUES, ANALYTICAL TECHNIQUES W73-00802
ANIMAL PHYSIOLOGY, RADIOACTIVITY TECHNIQUES, CYCLING NUTRIENTS, TR W73-00826
R, WATER ANALYSIS, RADIOACTIVITY TECHNIQUES, RADIOECOLOGY, ENVIRON W73-00817
TES, RADIOECOLOGY, RADIOACTIVITY TECHNIQUES, ANALYTICAL TECHNIQUES W73-00821
OD CHAINS, EQUIPMENT, ANALYTICAL TECHNIQUES, HEATED WATER, CONDENS W72-07526
MS, *INVESTIGATIONS, *ANALYTICAL TECHNIQUES, DATA COLLECTIONS, ALG W72-07225
ER QUALITY, PLANKTON, ANALYTICAL TECHNIQUES, ALGAE.: /ERATURE, WAT W72-07663
POLLUTION EFFECTS, RADIOACTIVITY TECHNIQUES.: /, NUTRIENTS, WATER W72-07143
, *MARINE ALGAE, PH/ *ANALYTICAL TECHNIQUES, *CARBON, *AMINO ACIDS W72-07497
UTOMATION, EQUIPMENT, ANALYTICAL TECHNIQUES, SEWAGE, RESEARCH EQUI W72-07715
METABOLISM, *DDT, *RADIOACTIVITY TECHNIQUES, PESTICIDES, ECOSYSTEM W72-07703
ARINE ENVIRONMENT, RADIOACTIVITY TECHNIQUES, ABSORPTION.: /S, ESTU W72-07826
A/ *NITROGEN, *LAKES, ANALYTICAL TECHNIQUES, MICHIGAN, SESTON, ALG W72-08048
MS, DATA COLLECTIONS, ANALYTICAL TECHNIQUES, NITROGEN FIXING BACTE W72-08508
RO/ *BIBLIOGRAPHIES, *ANALYTICAL TECHNIQUES, *WATER POLLUTION CONT W72-08790
OMETRY, DISTRIBUTION, SEPARATION TECHNIQUES.: /ONATES, MASS SPECTR W72-12709
ERTIES, REPRODUCTION, ANALYTICAL TECHNIQUES, NUMERICAL ANALYSIS, M W72-12724
TION, *ASSAY, *ALGAE, ANALYTICAL TECHNIQUES, WASTE WATER TREATMENT W72-12966
IDS, *AXENIC CULTURES, CULTURING TECHNIQUES, STAURASTRUM PARADOXUM W72-12736
OMPOUNDS, WASHINGTON, SEPARATION TECHNIQUES, DISTILLATION, SPECTRO W72-12637
PTION, SEDIMENTATION, ANALYTICAL TECHNIQUES, MOLECULAR STRUCTURE, W72-12203
PHAERA, CULTURE MEDIA, CULTURING TECHNIQUES, MORPHOLOGY, FLAGELLAT W72-12632
CALCIUM CHLORIDE, RADIOACTIVITY TECHNIQUES, METABOLISM, TRACERS, W72-12634
*CARBON DIOXIDE, *RADIOACTIVITY TECHNIQUES, SOIL ALGAE, *METHODOL W72-14301
DETECTION, FLUORESCENT ANTIBODY TECHNIQUES, SALMONELLA TYPHIMURIU W72-13800
OPHYLL, *EVALUATION, *SEPARATION TECHNIQUES, SPECTROPHOTOMETRY, LA W72-13673
RIENTS, PLANT GROWTH, ANALYTICAL TECHNIQUES, WATER TEMPERATURE.: / W72-13651
, GAS CHROMATOGRAPHY, ANALYTICAL TECHNIQUES: /USES, CHROMATOGRAPHY W72-13800
BLIC HEALTH, ENZYMES, ANALYTICAL TECHNIQUES, PESTICIDES, FUNGICIDE W72-13347
MENT, INDUSTRIES, WATER QUALITY, TECHNOLOGY, SURFACTANTS, TOXICITY W72-09169
NGBYA LIM/ *BAVARIA, SCHLIERSEE, TEGERNSEE, ANABAENA, COREGONS, LY W71-00117
EDS, *MACROPHYTES, *SEA GRASSES, TELEVISION, DEPTH SOUNDER, VERTIC W72-09120
, CAPITELLA, MAGELONA, ARMANDIA, TELLINA BUTTONI, VENUS, TIVELA ST W72-09092
IUM COMPOUNDS, SEWAGE EFFLUENTS, TEMPERATURE CONTROL, ASSAY, REMOT W72-10162
N, PHOSPHORUS, AMMONIA, METHANE, TEMPERATURE, HYDROGEN ION CONCENT W72-10390
, PHYTOPLANKTON, PLANKTON, WATER TEMPERATURE, E. COLI, BENTHIC FAU W72-09668
LAKE HURON, LAKE SUPERIOR, WATER TEMPERATURE, HYDROGEN ION CONCENT W72-10875
RIDA, ELECTRIC GENERATORS, WATER TEMPERATURE, RHODOPHYTA, PHAEOPHY W72-11252
PRODUCING ALGAE, SOUTHWEST U.S., TEMPERATURE, OXYGEN, LIGHT INTENS W72-10613
T, PHOTOPERIODISM, GROWTH RATES, TEMPERATURE, BIOCHEMISTRY, PHYSIO W72-00377
ATION LAGOONS, COLIFORMS, ALGAE, TEMPERATURE, ANALYTICAL TECHNIQUE W72-00422
MORPHOMETRY, GEOLOGY, LIMESTONE, TEMPERATURE, DEPTH, CHEMICAL PROP W72-04269
, PH/ *THERMAL POLLUTION, *WATER TEMPERATURE, *CYANOPHYTA, GEYSERS W72-03557
IL FUNGI, SOIL ENVIRONMENT, SOIL TEMPERATURE, SOIL CHEMICAL PROPER W72-03342
OPHYTA, SAMPLING, ECOLOGY, WATER TEMPERATURE.: / CHRYSOPHYTA, PYRR W72-03543
LAKE SUPERIOR, LABORATORY TESTS, TEMPERATURE, ALGAE, PHOSPHORUS, I W72-01108
OGY, *PONDS, *CHEMICAL ANALYSIS, TEMPERATURE, SEASONAL, IONS, ALGA W72-01363
NIA, HYDROGEN ION CONCENTRATION, TEMPERATURE, LIGHT INTENSITY, SED W72-01881
C BUDGET, SOLAR RADIATION, WATER TEMPERATURE, SALT BALANCE, DISSOL W72-02274
IES, HYDROGEN ION CONCENTRATION, TEMPERATURE, DISSOLVED OXYGEN, EU W72-03224
*PHYTOPLANKTON, ESTUARIES, WATER TEMPERATURE, TURBIDITY, PHOTOSYNT W71-11876
ERS, MARINE ALGAE, HAWAII, WATER TEMPERATURE, WAVES, LIGHT INTENSI W71-11486
YDROGEN ION CONCENTRATION, WATER TEMPERATURE, DISSOLVED OXYGEN, CO W71-11914
ROPERTIES, EUTROPHICATION, WATER TEMPERATURE, WATER POLLUTION EFFE W71-12092
GY, AQUATIC ALGAE, MOSSES, WATER TEMPERATURE, WATER CHEMISTRY, CHL W71-11949
*SALMONIDS, WATER REQUIREMENTS, TEMPERATURE, NUTRIENTS, ALGAE, WE W71-11006
FARM PONDS, NUTRIENTS, SURFACES, TEMPERATURE, PHOSPHORUS, NITRATES W71-11001
S, INTERFACES, DISSOLVED OXYGEN, TEMPERATURE, ORGANOPHOSPHOROUS CO W71-10645
RCURY ISOTOPES, HUMAN BRAIN, LOW- TEMPERATURE OVEN DRYING, ASHING, W71-11036
TER POLLUTION EFFECTS, SAMPLING, TEMPERATURE, COPEPODS, NITROGEN, W71-11008
CLE, HYDROGEN ION CONCENTRATION, TEMPERATURE, ELECTRICAL CONDUCTAN W71-01489
PES, ALGAE, GROWTH RATES, THERM/ TEMPERATURE, SORPTION, RADIOISOTO W71-02075
ANALYSIS, CHLOROPHYTA, DIATOMS, TEMPERATURE, OPTICAL PROPERTIES, W70-10403
EN ALGAE, PRASINOCLADUS MARINUS, TEMPERATURE EFFECTS.: /1002, *GRE W70-09429
OLYSIS, DETERGENTS, GREAT LAKES, TEMPERATURE, HYDROGEN ION CONCENT W71-00116
R PLANTS, *COOLING WATER, *WATER TEMPERATURE, CONDENSERS, DESIGN, W70-09612
TURE CONTROL, *STANDARDS, *WATER TEMPERATURE, SEASONAL, MONTHLY, A W70-09595
UCTURES, DISCHARGE(WATER), WATER TEMPERATURE.: /IMNION, INTAKE STR W70-09614
AE, SOLVENTS, SURFACTANTS, WATER TEMPERATURE, PLANT PIGMENTS, PLAN W70-09429
L POLLUTION, *GREENHOUSES, WATER TEMPERATURE, POWERPLANTS, DISTILL W70-08832
YDROGEN ION CONCENTRATION, WATER TEMPERATURE, BACTERIA, ALGAE, ALK W71-05626
NDITIONS, PONDS, CLIMATIC ZONES, TEMPERATURE, TURBULENCE, OXYGENAT W71-06033

IGATICN SYSTEMS, WATER ANALYSIS,	TEMPERATURE, LIMNOLOGY, CYANOPHYT	W71-06102
EATMENT, *NUTRIENT REQUIREMENTS,	TEMPERATURE, *NITROGEN, *PHOSPHAT	W72-04789
S, RESERVOIR STORAGE, NUTRIENTS,	TEMPERATURE, ALGAE, HYDRCLOGIC DA	W72-05282
LANKTON, FISH POPULATIONS, WATER	TEMPERATURE, THERMAL STRATIFICATI	W72-04853
O, *RECREATION FACILITIES, WATER	TEMPERATURE, CHEMICAL ANALYSIS, A	W72-04791
SPHORUS, *NUTRIENT REQUIREMENTS,	TEMPERATURE, HYDROGEN ION CONCENT	W72-04788
ATION, RESPIRATION, OXYGENATION,	TEMPERATURE, LIGHT INTENSITY, MIX	W72-04983
CHRYSOPHYTA, EUGLENOPHYTA, WATER	TEMPERATURE, THERMAL STRATIFICATI	W72-04855
THER, RESERVOIR OPERATION, WATER	TEMPERATURE, PHYTOPLANKTON, EUTHR	W72-05282
CONCENTRATION, PERIPHYTON, WATER	TEMPERATURE.: /DES, HYDROGEN ION	W72-05610
TES, HYDROGEN ION CONCENTRATION,	TEMPERATURE, ENZYMES, CELLULOSE,	W72-05457
TON, ADSORPTION, EUTROPHICATION,	TEMPERATURE, LIGHT PENETRATION, A	W72-05469
TROGEN, NUTRIENTS, SEWAGE, WATER	TEMPERATURE, LIGHT PENETRATION, C	W72-04734
OADING, AERATION, MIXING, ALGAE,	TEMPERATURE, ALTITUDE, CLIMATES,	W72-06838
ATES, MATHEMATICAL MODELS, WATER	TEMPERATURE, WATER QUALITY, CARBO	W72-06274
DOLOGY, REEFS, SAMPLING, OXYGEN,	TEMPERATURE, LIGHT, WAVES(WATER),	W72-06283
OMS, PHYTOPLANKTON, HYDORGEN ION	TEMPERATURE,: /TMENT PLANTS, DIAT	W72-06837
ALUATION, WATER CHEMISTRY, WATER	TEMPERATURE, CIS: /OPHICATION, EV	W72-06526
ES, CHLOROPHYLL, PHOTOSYNTHESIS,	TEMPERATURE, SAMPLING, SALINITY,	W71-07875
OA, PHOTOSYNTHESIS, CHLOROPHYLL,	TEMPERATURE, HYDROGEN ION CONCENT	W71-07382
ERMAL POLLUTION, *REVIEWS, WATER	TEMPERATURE, DISSOLVED OXYGEN, BI	W71-07417
IOISOTOPES, ALGAE, PLANT GROWTH,	TEMPERATURE, PROTOZOA, CHLOROPHYT	W71-09254
ION, HYDROGEN ION CONCENTRATION,	TEMPERATURE, BIOCHEMICAL OXYGEN D	W71-09524
EMENTS, CHLORELLA, GROWTH RATES,	TEMPERATURE, SEA WATER, PHOSPHATE	W71-10083
PLANKTON, BIODEGRADATION, LIGHT,	TEMPERATURE.: /PLANTS, SEAWATER,	W71-09261
GAE, *RADIOISOTOPES, ABSORPTION,	TEMPERATURE, STRONTIUM RADIOISOTO	W71-09250
OPHYLL, WATER POLLUTION EFFECTS,	TEMPERATURE, CARBON, DIATOMS, LIG	W71-08670
ALGAE, PHYTOPLANKTON, FREEZING,	TEMPERATURE, CHELATION, HYDROGEN	W71-09190
IONS, CYANOPHYTA, RADIOISOTOPES,	TEMPERATURE, ECOSYSTEMS, CALIFORN	W71-09157
IME, HYDROGEN ION CONCENTRATICN,	TEMPERATURE, CHLORINATION, FLOATA	W71-08951
S, FRESH WATER, FRESHWATER FISH,	TEMPERATURE, KINETICS, NUCLEAR WA	W71-09182
, ALGAE, GASTROPODS, ADSORPTION,	TEMPERATURE, METABOLISM, KINETICS	W71-09168
NS, *OXYGENATION, *ALGAE, DEPTH,	TEMPERATURE, PHOTOSYNTHESIS, RE-A	W71-07098
*CLIMATIC ZONES, *COST ANALYSIS,	TEMPERATURE, NUTRIENTS, DISSOLVED	W71-07109
AND, SOLAR RADIATION, NUTRIENTS,	TEMPERATURE, COLIFORMS, CHLORINAT	W71-07124
RGANIC LOADING, ODOR, COLIFORMS,	TEMPERATURE, DISSOLVED OXYGEN, AE	W71-07113
SIMULATION, MIXING, TURBULENCE,	TEMPERATURE, STRATIFICATION, SOLA	W71-07084
ALITY, ALGAE, HARVESTING, DEPTH,	TEMPERATURE, LIGHT INTENSITY, CHE	W71-07100
N EFFECTS, *HEATED WATER, *WATER	TEMPERATURE, *WATER QUALITY STAND	W72-13636
TH, ANALYTICAL TECHNIQUES, WATER	TEMPERATURE.: /RIENTS, PLANT GROW	W72-13651
OTROPHY, DISSOLVED OXYGEN, WATER	TEMPERATURE, HYDROGEN ION CONCENT	W72-13871
HEMICAL OXYGEN DEMAND, BACTERIA,	TEMPERATURE, COLIFORMS, ENTERIC B	W72-12565
ORELLA, SILICA, LIGHT INTENSITY,	TEMPERATURE, MUD, GROWTH RATES, A	W72-12544
, *LAKE ERIE, *DISSOLVED OXYGEN,	TEMPERATURE, CURRENTS(WATER), WIN	W72-12992
ASS, ANAEROBIC CONDITIONS, WATER	TEMPERATURE, CHLORIDES, PHOTOSYNT	W72-12940
TIONAL PARKS, HOT SPRINGS, WATER	TEMPERATURES, NOSTOC STREAMS, DAT	W72-08508
S, BACTERICIDES, CHROMIUM, ZINC,	TEMPERATURE, HYDROGEN ION CONCENT	W72-08683
TON, COMPUTER PROGRAMS, BIOMASS,	TEMPERATURE, CONDUCTIVITY, PHOSPH	W72-08560
STACEANS, WORMS, MOLLUSKS, WATER	TEMPERATURE, SALINITY, SEA WATER,	W72-09092
YLL, ALKALINITY, CARBON DIOXIDE,	TEMPERATURE, DISSOLVED OXYGEN, HY	W72-08049
MENT, SEA WATER, SALINITY, WATER	TEMPERATURE, CHEMICAL ANALYSIS, W	W72-08141
ATER), CONDUCTIVITY, ZINC, WATER	TEMPERATURE, AL: /GEN, HARDNESS(W	W72-07890
, WATER POLLUTION EFFECTS, WATER	TEMPERATURE, ALGAE, BULLHEADS, SA	W72-07907
INVERTEBRATES, *PLANKTON, *WATER	TEMPERATURE, SYSTEMATICS, SAMPLIN	W72-07899
WATER, SALINITY, PHENOLS, WATER	TEMPERATURE, CHEMICAL ANALYSIS, S	W72-07710
, *PACIFIC OCEAN, ECOLOGY, WATER	TEMPERATURE, PHYTOPLANKTON, ZOOPL	W72-07892
ESS(WATER), ALKALINITY, ENZYMES,	TEMPERATURE, PHOTOSYNT: /S, HARDN	W72-07514
, PHOTOSYNTHESIS, SESSILE ALGAE,	TEMPERATURE, DEPTH, LIGHT PENETRA	W72-07501
TA, DISTRIBUTION PATTERNS, WATER	TEMPERATURE, WATER QUALITY, PLANK	W72-07663
CONCENTRATION, MINNESOTA, WATER	TEMPERATURE, DISSOLVED OXYGEN.: /	W72-07145
RUMENTATION, FLUORESCENCE, WATER	TEMPERATURES, THERMAL STRESS.: /T	W72-07225
COPEPODS, DIATOMS, CHLOROPHYTA,	TEMPERATURE, SIZE, FECUNDITY, ALG	W72-14805
DS, PHYTOPLANKTON, CARBON, WATER	TEMPERATURE, GREAT LAKES REGION,	W73-00384
ENT, TUBIFICIDS, MOLLUSKS, WATER	TEMPERATURE, MARINE ANIMALS, ALAS	W73-00374
ORGANIC MATTER, MORTALITY, WATER	TEMPERATURE, PHAEOPHYTA, SAMPLING	W73-00366
ON, BENTHIC FAUNA, SIEVES, WATER	TEMPERATURE,: /LOGICAL DISTRIBUTI	W73-00367
MODELS, THERMAL STRATIFICATION,	TEMPERATURE, DISSOLVED OXYGEN, CO	W72-14405
TIC PLANTS, AQUATIC ALGAE, WATER	TEMPERATURE, ACID STREAMS, HYDROG	W72-14690
*WASHINGTON, *CHLOROPHYLL, WATER	TEMPERATURE, ALKALINITY, DISSOLVE	W72-14728
, SOYBEANS, LIGHT INTENSITY, AIR	TEMPERATURE, SAMPLING, AQUATIC AL	W73-01456
NDS, ABSORPTION, KINETICS, WATER	TEMPERATURE, RESPIRATION, ALGAE,	W73-01269
CENTRIFUGATION, TURBIDITY, WATER	TEMPERATURE, ALKALINITY, HYDROGEN	W73-00916
HYTA, PHYSICAL PROPERTIES, WATER	TEMPERATURE, DATA COLLECTIONS, MA	W73-00838
HENS, NITROGEN, LIGHT INTENSITY,	TEMPERATURE, ANABAENA, BACTERIA,	W73-02471
RICE, CYANOPHYTA, FERTILIZATION,	TEMPERATURE, CROP PRODUCTION, FUN	W73-02473
ATER PROPERTIES, SALINITY, WATER	TEMPERATURE, OXYGEN, TURBIDITY, B	W73-02029
A, STREAMFLOW, FLOW RATES, WATER	TEMPERATURE, WATER CHEMISTRY, CHE	W73-01972
ITE INVESTIGATIONS, *POPULATION,	TEMPORAL DISTRIBUTION, CHLOROPHYL	W72-11719
, BLACK CRAPPIE, CHAIN PICKEREL,	TENDIPES TENTANS, TRICHOPTERA, LI	W72-14280
OPHA, PROCLADIUS, WHITE OAK LAKE(	TENN), UPTAKE.: / PHYSA HETEROSTR	W70-06668
R, ILLINOIS RIVER, HOLSTON RIVER(	TENNESSEE).: /IRE), DELAWARE RIVE	W71-11517
IME, SNAILS, RADIOACTIVE WASTES,	TENNESSEE, CRAYFISH, MIDGES, GAMM	W70-06668
RAPPIE, CHAIN PICKEREL, TENDIPES	TENTANS, TRICHOPTERA, LIMNEPHILUS	W72-14280
RIC/ *ZINC, *LEAD, STIGEOCLONIUM	TENUE, CLADOPHORA GLOMERATA, ULOT	W71-05991
S, ANABAENA, NOSTOC, TOLYPOTHRIX	TENUIS, AULOSIRA FERTILISSIMA, AN	W71-10070
*TOLERANCE LEVELS, TOLYPOTHRIX	TENUIS, AULOSIRA FERTILISSIMA, AN	W71-12071
ATOA, THIOTHRIX NIVEA, THIOTHRIX	TENUIS, MERISMOPEDIA GLAUCA, DACT	W72-07896
SCILLATORIA AMOENA, OSCILLATORIA	TENUIS, OSCILLATORIA CHLORINA, OS	W72-07896
, BENTONITE, LIME, *PULP WASTES,	TECTIARY TREATMENT, EFFLUENTS.: /	W72-03753
N, CULTURE MEDIA, CHLOROPHYLL A,	TERATOLOGY, NITZSCHIA.: /, ACTIDO	W73-01445

-14, *FLORIDA STRAIT, DUNALIELLA TERIOLECTA, PHAEODACTYLUM TRICORN W72-09103
N, MONOD EQUATION, PENICILLIN G, TERRAMYCIN, CHLORAMPHENICOL, ACTI W73-01445
SYSTEMS, BELEMNITELLA AMERICANA, TERRESTRIAL PLANTS, COMBUSTION, I W72-12709
KIELLA TRIANGULATA, AKIYAMAMONAS TERRESTRIS, BRASILOBIA PISCIFORMI W72-07683
ION, CLEANING, *WATER TREATMENT, TERTIARY TREATMENT, *WASTE WATER W72-09393
TRIENTS, *WASTE WATER TREATMENT, TERTIARY TREATMENT, EUTROPHICATIO W72-03613
, PHOSPHATES, ILLINOIS, ECOLOGY, TERTIARY TREATMENT, SEWAGE EFFLUE W72-03970
OSPHORUS, *DETENTION RESERVOIRS, TERTIARY TREATMENT, ALGAE, PLANTS W72-02412
MMONIA, ALGAE, SEWAGE TREATMENT, TERTIARY TREATMENT, WATER POLLUTI W71-13939
ERIE, DETERGENTS, LAKE ONTARIO, TERTIARY TREATMENT, INTERNATIONAL W71-12091
DGE, TRICKLING FILTERS, LAGOONS, TERTIARY TREATMENT, ACTIVATED CAR W71-09524
RFACE RUNOFF, *SEWAGE TREATMENT, TERTIARY TREATMENT, BIOCHEMICAL O W71-06445
AMPLING, MICROORGANISMS, SLUDGE, TERTIARY TREATMENT, BIOCHEMICAL O W71-07382
CAL TREATMENT, ACTIVATED SLUDGE, TERTIARY TREATMENT, ION EXCHANGE, W70-09907
S, ALGAE, LAKES, MICROORGANISMS, TERTIARY TREATMENT, EFFLUENTS, MO W71-03021
N DEMAND, ANALYTICAL TECHNIQUES, TERTIARY TREATMENT, *WASTE WATER W72-04983
SKELETONEMA COSTATUM, DUNALIELLA TERTIOLECTA, EUGLENA GRACILIS, CH W72-05614
SSIOSIRA FLUVIATILIS, DUNALIELLA TERTIOLECTA, SYNECHOCOCCUS, RHODO W71-10083
DA STRAIT, CARBON-14, DUNALIELLA TERTIOLECTA, PHAEODACTYLUM TRICOR W72-09102
*ELECTRON TRANSPORT, DUNALIELLA TERTIOLECTA, CYCLOTELLA NANA, THA W72-12998
UTROPHICATION, DATA COLLECTIONS, TEST PROCEDURES, CYCLING NUTRIENT W71-13794
MICALS, BACTERIOCIDES, CULTURES, TEST PROCEDURES, PESTICIDE TOXICI W71-06189
RECREATION, RUNOFF, MEASUREMENT, TEST PROCEDURES, PHOTOSYNTHESIS, W72-04270
PAAP TEST, *ALGAL GROWTH, BOTTLE TEST.: * W73-00255
KA ISLAND, MILROW TEST, CANNIKIN TEST.: AMCHIT W73-00764
: *PAAP TEST, *ALGAL GROWTH, BOTTLE TEST. W73-00255
AMCHITKA ISLAND, MILROW TEST, CANNIKIN TEST.: W73-00764
ASTRUM CAPRICORNUTUM, PAAP BATCH TEST, CULTURE MEDIA.: /DES, SELEN W73-02099
INDRICA, ANABAENOPSIS, ACETYLENE TEST, CYLINDROSPERMUM LICHENEFORM W71-12070
ATE METHOD, MOST PROBABLE-NUMBER TEST, MEMBRANE FILTRATION, RHODOP W72-04743
*SINKING AGENTS, EFFECTIVENESS, TESTING PROCEDURES, CRITERIA, CLA W71-09789
HESIS, RIVERS, PLANT PHYSIOLOGY, TESTING PROCEDURES, AQUATIC ALGAE W72-14314
ECTRIC ENERGY CONSUMPTION, MODEL TESTING.: /INDUSTRIAL COMPLEX, EL W72-03327
TECHNIQUES, *ALGAE, EVALUATION, TESTING, ASSAY, NUTRIENTS, CULTUR W73-00255
OPHYTA, CHRYSOPHYTA, RHODOPHYTA, TESTING, ECOLOGY.: /OPHYTA, CHLOR W70-07280
GICIDES, *ALGAE, *ALGAL CONTROL, TESTING, LABORATORY TESTS, CHEMIC W71-06189
ROPHYTA, CHLORELLA, PSEUDOMONAS, TESTING, MEASUREMENT, PHEOLOGY.: / W72-03220
EN ION CONCENTRATION, LABORATORY TESTS.: /BOLISM, BACTERIA, HYDROG W71-08687
Y, ALGAE, INHIBITION, LABORATORY TESTS.: /BON DIOXIDE, BIOCHEMISTR W72-08584
ASTE WATER TREATMENT, LABORATORY TESTS.: /ENTS, SETTLING BASINS, W W71-00116
NKTON, ON-SITE TESTS, LABORATORY TESTS.: /MYCETES, CYANOPHYTA, PLA W71-13187
ADIOACTIVITY EFFECTS, LABORATORY TESTS.: /RADIOISOTOPES, CESIUM, R W72-09649
EASUREMENTS, NUTRIENT ENRICHMENT TESTS.: /S, ACETYLENE REDUCTION M W72-13692
PHOSPHATES, NITRATES, LABORATORY TESTS.: /TIVITY, TRACE ELEMENTS, W71-09674
, PESTICIDE TOXICITY, LABORATORY TESTS.: /TION EFFECTS, ABSORPTION W73-00281
ES, CYANOPHYTA, HYD/ *LABORATORY TESTS, *BACTERIA, *ALGAE, *CULTUR W73-01657
*ALGAE, *NUTRIENTS, *LABORATORY TESTS, *BIOASSAY, OREGON, LAKES, W72-09164
UATIC MICROORGANISM/ *LABORATORY TESTS, *CHLORELLA, *CULTURES, *AQ W71-08027
ATES/ *MARINE ALGAE, *LABORATORY TESTS, *CHEMICAL WASTE, *GROWTH R W72-12239
S, *ALGAE, SEWAGE, / *LABORATORY TESTS, *EUTROPHICATION, *NUTRIENT W71-12857
ABORATORY EQUIPMENT, *LABORATORY TESTS, *MEASUREMENT, *MARINE ALGA W72-12210
PHOSP/ *EUTROPHICATION, *ON-SITE TESTS, *NUTRIENTS, ALGAE, LAKES, W72-09165
*NUTRIENTS, CULTURE/ *LABORATORY TESTS, *RESPIRATION, *CHLORELLA, W73-01626
*A/ *EUTROPHICATION, *LABORATORY TESTS, *WATER POLLUTION EFFECTS, W72-03218
DATION PONDS, *ALGAE, LABORATORY TESTS, AEROBIC CONDITIONS, ANAERO W71-13326
OGY.: *CYANOPHYTA, *LABORATORY TESTS, ALGAL TOXINS, HUMAN PATHOL W72-04259
GEORGIA, BIORTHYTHMS, LABORATORY TESTS, ALGAE, CARBON DIOXIDE, CYC W72-09159
EMICAL OXYGEN DEMAND, LABORATORY TESTS, ALKALINITY, CARBONATES, BI W72-04431
A, ENZYMES, PROTEINS, LABORATORY TESTS, AMMONIA, ASSAY, INHIBITION W72-03222
NON FLOS-AQUAE, *PHARMACOLOGICAL TESTS, ANTIMICROBIOTICS, ESCHERIC W72-04259
TIC ALGAE, CARBON CYCLE, ON-SITE TESTS, AQUATIC ANIMALS, AQUATIC P W73-02095
ULTURES, MEASUREMENT, LABORATORY TESTS, BIOASSAY.: /, NUTRIENTS, C W73-00255
HYTA, *ALGAL CONTROL, LABORATORY TESTS, BUOYANCY, EUTROPHICATION, W71-13184
OWTH RATES, BIOASSAY, LABORATORY TESTS, CALCIUM.: /UATIC ALGAE, GR W72-13809
PHORUS, IRON, *TEXAS, LABORATORY TESTS, CARBON RADIOISOTOPES, *CYA W72-13692
NALYTICAL TECHNIQUES, LABORATORY TESTS, CESIUM, STRONTIUM RADIOISO W71-08032
GAL CONTROL, TESTING, LABORATORY TESTS, CHEMICALS, BACTERICCIDES, W71-06189
TION, STANDING CROPS, LABORATORY TESTS, CHEMICAL ANALYSIS, BIOASSA W72-09163
PHYTA, CHLAMYDOMONAS, LABORATORY TESTS, CHRYSOPHYTA, OPTICAL PROPE W72-08051
NCENTRATION, PROTEIN, LABORATORY TESTS, COST ANALYSIS, ALGAE, WAST W72-10390
ES, DINOFLAGELLATES, *LABORATORY TESTS, CULTURES, *GROWTH RATES, P W71-03095
ON DIOXIDE, NITROGEN, LABORATORY TESTS, CYANOPHYTA, CYTOLOGICAL ST W72-05476
ENSITY, GROWTH RATES, LABORATORY TESTS, CYANOPHYTO, ALGAE.: /ES, D W72-03217
LANT GROWTH, SULFATE, LABORATORY TESTS, CYTOLOGICAL STUDIES.: /, P W72-00153
, *CHLORELLA, *BORON, LABORATORY TESTS, CYTOLOGICAL STUDIES, PLANT W71-12858
D SLUDGE, *TOXICITY, *LABORATORY TESTS, DETERGENTS, WASTE WATER TR W70-08976
IFE, SAMPLING, ALGAE, LABORATORY TESTS, DETRITUS, BENTHOS, AQUATIC W71-10064
ON, *ALGAE, *STREAMS, LABORATORY TESTS, DISSOLVED OXYGEN, HYDROGEN W71-04518
GAE, *BIODEGRADATION, LABORATORY TESTS, FILTRATION, AMMONIA, HYDRO W72-01881
TOXICITY, *BIOASSAYS, LABORATORY TESTS, FISH, ALGAE, CHEMICAL WAST W72-08683
EA WATER, WATER QUALITY, ON-SITE TESTS, HYDROGEN ION CONCENTRAT: / W72-09092
NMENTAL EFFECTS, LARVAE, ON-SITE TESTS, INVERTEBRATES, WATER POLLU W73-00348
IRATION, CHLOROPHYLL, LABORATORY TESTS, IRRADIATION, CHLOROPHYTA, W73-01080
S, CYANOPHYTA, PLANKTON, ON-SITE TESTS, LABORATORY TESTS.: /MYCETE W71-13187
RMANCE, EFFICIENCIES, LABORATORY TESTS, LABORATORY EQUIPMENT, AERA W72-11100
LOGY, ORGANIC COMPOUNDS, ON-SITE TESTS, METHODOLOGY, CHARA.: / ECO W72-03216
NTS, AQUATIC ANIMALS, LABORATORY TESTS, NITRATES, MUSSELS, SNAILS, W71-09168
SINE TRIPHOSPHATASE, BIOCHEMICAL TESTS, OUABAIN, POTASSIUM CHLORID W72-13809
C ALGAE, CHLOROPHYTA, LABORATORY TESTS, OXYGEN, WATER POLLUTION EF W72-14314
XICITY, HEAVY METALS, LABORATORY TESTS, PHOTOSYNTHESIS, CHLOROPHYL W72-06037

TRACE EL/ *BIOASSAY, *LABORATORY	TESTS, RESEARCH AND DEVELOPMENT,	W72-04433
ENTAL EFFECTS, REEFS, LABORATORY	TESTS, SAMPLING, PHAEOPHYTA.: /NM	W73-00838
SOIL TEXTURE, CLAYS, LABORATORY	TESTS, STATISTICAL METHODS, SOIL	W72-02218
, CARBON RADIOISOTOPES, *ON-SITE	TESTS, SULFUR BACTERIA, PRODUCTIV	W72-14301
ERIE, LAKE SUPERIOR, LABORATORY	TESTS, TEMPERATURE, ALGAE, PHOSPH	W72-01108
OPEPODS, CRUSTACEANS, LABORATORY	TESTS, VIRGINIA.: /CHRYSOPHYTA, C	W72-00150
ATER, COASTS, BIORTHYMS, ON-SITE	TESTS, WATER POLLUTION SOURCES, W	W72-03228
ACTINOMYCETES, FUNGI, LABORATORY	TESTS, WATER QUALITY CONTROL.: /	W72-11604
GEN, *EUTROPHICATION, LABORATORY	TESTS, WATER QUALITY CONTROL, *KI	W72-08376
SPIRALIS, STROMIA SUBSPHAERICA,	TETRACULAMONAS NATANS, MARINIAMON	W72-09119
MAINTENANCE, *POLYP METABOLISM,	TETRACYCLINE, SULFANILAMIDE, PENI	W72-10162
TION, CHLAMYDOMONAS REINHARDTII,	TETRAHYMENA VORAX, PSEUDOMONAS FL	W73-02095
BAFFIN BAY(	TEX).:	W71-04873
AMYDOMONAS FOTTII, CHLAMYDOMONAS	TEXENSIS, CHLAMYDOMONAS PSEUDOMIC	W72-12632
ATED RAINFALL, *SOIL TYPES, SOIL	TEXTURE, CLAYS, LABORATORY TESTS,	W72-02218
HYTES, ENTEROMORPHA, INSOLATION,	TEXTURE, SUBSTRA: /INALIS, LITHOP	W72-08141
P., DUNALIELL/ FRESHWATER ALGAE,	THALASIOSIRA SPP., SKELETONEMA SP	W72-08436
EY POINT(FLORIDA), TURTLE GRASS,	THALASSIA SPP., UDOTEA SPP., PENI	W72-11876
IPLEURA RUTILANS, THALASSIOSIRA,	THALASSIONEMA NITZSCHIOIDES, PLAG	W73-00853
EGULATION, SKELETONEMA COSTATUM,	THALASSIONEMA NITZSCHIOIDES, MIXE	W73-00441
A, EUTREPTIA, NITZSCHIA SERIATA,	THALASSIONEMA NITZ: /VIELLA MARIN	W71-07875
ALARIA, NEREOCYSTIS, LAMINARIA,	THALASSIOPHYLLUM, FUCUS, SCHIZYME	W72-09120
A, THALASSIOSIRA NORDENSKIOLDII,	THALASSIOSIRA ROTULA, SCHRODERELL	W72-03543
ECIPIENS, THALASSIOSIRA GRAVIDA,	THALASSIOSIRA NORDENSKIOLDII, THA	W72-03543
CHORDA, THALASSIOSIRA DECIPIENS,	THALASSIOSIRA GRAVIDA, THALASSIOS	W72-03543
OSITION, COSCINOSIRA POLYCHORDA,	THALASSIOSIRA DECIPIENS, THALASSI	W72-03543
NG ISLAND, SKELETONEMA COSTATUM,	THALASSIOSIRA, CHAETOCEROS, RHIZO	W71-07875
FLAGELLATE, CYCLOTELLA CRYPTICA,	THALASSIOSIRA FLUVIATILIS, DUNALI	W71-10083
NEMA COSTATUM, DUNALIELLA TERTI/	THALASSIOSIRA PSEUDONANA, SKELETO	W72-05614
ROCKMANNI, AMPHIPLEURA RUTILANS,	THALASSIOSIRA, THALASSIONEMA NITZ	W73-00853
LA TERTIOLECTA, CYCLOTELLA NANA,	THALASSIOSIRA FLUVIATILIS, AMPHID	W72-12998
OPYXIS, BIDDULPHIA, CHAETOCEROS,	THALASSISTHRIX, INDIA.: /A, CENTR	W71-11876
, CLADOPHORA, RESEARCH STRATEGY,	THAMES RIVER(ENGLAND), LEE RIVER(	W71-13172
LAND),/ *CENTRIC DIATOMS, *RIVER	THAMES(ENGLAND), RIVER KENNET(ENG	W72-14797
T, INDUSTRIAL EXPANSION, ISLE OF	THANET, MEDWELL ESTUARY.: /T COAS	W71-13746
FORNIA, BIOCONTROL, COST-BENEFIT	THEORY.: /SE, WATER QUALITY, CALI	W72-00974
RACEAE.: *CABLE	THEORY, ELECTRONIC POTENTIAL, CHA	W73-01630
ION.: *WASTE HEAT,	THERMAL EFFICIENCY, HEAT CONSUMPT	W70-08832
LANT, CULTURE MEDIA, DRY WEIGHT,	THERMAL EFFECTS.: /VANNAH RIVER P	W71-02075
IOISOTOPES, ALGAE, GROWTH RATES,	THERMAL POLLUTION.: /ORPTION, RAD	W71-02075
UTION, WATER POLLUTION, COOLING,	THERMAL POWERPLANTS, HEATED WATER	W73-01704
OGY, ECOLOGY, CYCLING NUTRIENTS,	THERMAL POLLUTION, TROPHIC LEVEL,	W73-00796
IQUES, NITROGEN FIXING BACTERIA,	THERMAL POLLUTION.: /YTICAL TECHN	W72-08508
RIENTS, PESTICIDES, OIL, METALS,	THERMAL POLLUTION, SAMPLING, BACT	W72-08790
APHIES, WATER POLLUTION SOURCES,	THERMAL POLLUTION, REVIEWS, *DATA	W72-12190
ES, *MONITORING, *BIOINDICATORS,	THERMAL POLLUTION, MICROSCOPY, DI	W72-10875
H, SEWAGE EFFLUENTS, OIL SPILLS,	THERMAL POLLUTION, NUTRIENTS, BIO	W72-14282
ALGAE, SALINITY, PHYTOPLANKTON,	THERMAL PROPERTIES, HYDROGEN ION	W72-01363
C MATTER, DISTRIBUTION PATTERNS,	THERMAL PROPERTIES, THERMAL: /ANI	W73-00431
ROWTH RATES, *INHIBITION, FUNGI,	THERMAL SPRINGS, CYANOPHYTA, BACT	W72-12734
GEN DEMAND, INORGANIC COMPOUNDS,	THERMAL STRATIFICATION, DIATOMS,	W72-13659
IC PLANTS, *MATHEMATICAL MODELS,	THERMAL STRATIFICATION, TEMPERATU	W72-14405
HLOROPHYTA, SEASONAL, NUTRIENTS,	THERMAL STRATIFICATION, CARBON, S	W72-08560
LUORESCENCE, WATER TEMPERATURES,	THERMAL STRESS.: /TRUMENTATION, F	W72-07225
SALINITY, HEATED WATER, BIOMASS,	THERMAL STRESS, GROWTH RATES, SPA	W73-00853
LORIDA, *LAKES, PONDS, SAMPLING,	THERMAL STRATIFICATION, HARDNESS(	W72-04269
DEMAND, DISSOLVED OXYGEN, ALGAE,	THERMAL STRATIFICATION, COAGULATI	W72-03753
Y, *ALGAE, NITROGEN, PHOSPHORUS,	THERMAL STRATIFICATION, WATER SUP	W71-13183
TOMS, NITROGEN, BACTERIA, ALGAE,	THERMAL STRATIFICATION, DISSOLVED	W71-03028
T, *WATER POLLUTION, PHOSPHORUS,	THERMAL STRATIFICATION, CURRENTS(	W70-09947
EUGLENOPHYTA, WATER TEMPERATURE,	THERMAL STRATIFICATION, ICE, EUTR	W72-04855
POPULATIONS, WATER TEMPERATURE,	THERMAL STRATIFICATION, OXIDATION	W72-04853
R QUALITY, DIATOMS, CHLOROPHYTA,	THERMAL WATER, *MONTANA, PHOSPHAT	W72-05610
ON PATTERNS, THERMAL PROPERTIES,	THERMAL: /ANIC MATTER, DISTRIBUTI	W73-00431
ALANUS HELGOLANDICUS(PACIFICUS),	THERMISTORS, DATA INTERPRETATION,	W72-07892
LICATES, SOLAR RADIATION, DEPTH,	THERMOCLINE, NITRATES, LIGHT INTE	W71-07875
GAE, *SEWAGE EFFLUENTS, GEORGIA,	THERMOCLINE, HYDROGEN ION CONCENT	W71-11914
LGAE, *STRATIFICATION, *ENZYMES,	THERMOCLINE, EPILIMNION, MUD, PHY	W72-03224
ION, DISSOLVED OXYGEN, BACTERIA,	THERMOCLINE, SEDIMENT-WATER INTER	W72-12995
POLIMNION, NITRIFICATION, ALGAE,	THERMOCLINE, ANAEROBIC BACTERIA,	W72-12996
ERATURE, CURRENTS(WATER), WINDS,	THERMOCLINE, ALGAE, SEDIMENTS.: /	W72-12992
YSIS, BIOMASS, DISSOLVED OXYGEN,	THERMOCLINE, HYDROGEN ION CONCENT	W72-12941
TION CONTROL, ENERGY, DIFFUSION,	THERMODYNAMICS, SOLUBILITY, SORPT	W71-12092
IRUSES, PLASTICIZERS, PHTHALATE,	THERMOPHILIC FUNGI, POLYETHYLENE,	W72-09675
CTIVITY, RADIOCHEMICAL ANALYSIS,	THERMOPHILIC BACTERIA, ORGANIC MA	W72-14301
CHLOROPHYLL A, ORGANIC NITROGEN,	THIAMINE, BIOTIN, VITAMIN B 12, G	W72-07892
UCTION, LIGHT INTENSITY, CULTURE	THICKNESS, CULTURE VOLUME.: /PROD	W70-05547
*PERIODICITY, TRICHODESMIUM	THIEBAUDII, BARBADOS.:	W72-11719
AGMENELLUM QUADRAPLICATUM, C-14,	THIN LAYER CHROMATOGRAPHY, CLEANU	W73-00361
INHIBITI/ MERCURIAL PESTICIDES,	THIN LAYER CHROMATOGRAPHY, GROWTH	W72-13347
PALMITIC ACID, BETA-SITOSTEROL,	THIN-LAYER CHROMATOGRAPHY.: /PIN,	W71-08597
TTER, NITROSOMONAS, NITROBACTER,	THIOBACILLUS THIOOXIDANS, DESULFO	W71-01489
N COUNTING, CYANIDIUM CALDARIUM,	THIOBACILLUS, SULFOLOBUS, GLUCOSE	W72-14301
MONAS, NITROBACTER, THIOBACILLUS	THIOOXIDANS, DESULFOVIBRIO, ULOTH	W71-01489
ON, RHODOPSEUDOMONAS GELATINOSA,	THIORHODACEAE, CHLOROBACTERIACEAE	W72-04743
OMONAS MOBILIS, THIOVULUM MAJUS,	THIOSPIRA WINOGRADSKY, ZOOGLEA RA	W72-07896
GERA, BEGGIATOA ALBA, BEGGIATOA,	THIOTHRIX NIVEA, THIOTHRIX TENUIS	W72-07896
LBA, BEGGIATOA, THIOTHRIX NIVEA,	THIOTHRIX TENUIS, MERISMOPEDIA GL	W72-07896
CHROMATIUM, MACROMONAS MOBILIS,	THIOVULUM MAJUS, THIOSPIRA WINOGR	W72-07896
FICIAL RESERVOIR, MONTROSE LAKE,	THOMAS LAKE.: /NG RESERVOIR, ARTI	W70-09596

OISOTCPES, CERIUM RADIOISOTOPES,	THORIUM RADIOISOTOPES, IRCN RADIO	W71-09188
	THORIUM RADIOISOTOPES.:	W72-12204
*WILLIAMS LAKE(WASH), URANIUM,	THORIUM, ANTIMONY, MACRCNUTRIENTS	W73-01434
S, CULTURE MEDIA, DISACCHARIDES,	THREONINE, PROLINE, GLUTAMATE, LE	W73-00942
M WESTI, OSCILLATORIA PROLIFICA,	THRESHOLD ODOR NUMBER.: /ERIDINIU	W71-13187
ETERICRA/ ACRYLIC RESINS, GLASS,	THYMIDINES, CELLS(BIOLOGY), *BIOD	W71-09261
EFFECTS, *DISTRICT OF COLUMBIA,	TIDAL EFFECTS, NUTRIENTS, EUTROPH	W70-08249
EFFECTS, *DISTRICT OF COLUMBIA,	TIDAL EFFECTS, NUTRIENTS, EUTROPH	W70-08248
*SESSILE ALGAE, LIGHT INTENSITY,	TIDAL EFFECTS, SALINITY, HEATED W	W73-00853
EMICAL ANALYSIS, WATER ANALYSIS,	TIDAL EFFECTS, STABILITY, SOLAR R	W72-08141
MENTAL EFFECTS, *STANDING CROPS,	TIDAL EFFECTS, MARINE ALGAE, SEA	W73-00367
ISTANCE, BIOLOGICAL COMMUNITIES,	TIDAL MARSHES, FOOD WEBS, INTERTI	W73-00796
NUTRIENTS, PHOSPHORUS, NITROGEN,	TIDAL WATERS, CHEMICAL CXYGEN DEM	W71-07098
*CYANOPHYT/ *ALGAL TOXINS, *RED	TIDE, *BEGGIATOA, *PHYTOTOXICITY,	W72-14542
AE, *AQUATIC WEEDS, *WEEDS, *RED	TIDE, COASTS, LAKES, PONDS, WATER	W72-14673
INS, AQUATIC PLANTS, TOXINS, RED	TIDE, PLANT PHYSIOLOGY, BIOASSAY,	W73-00243
RODUCTIVITY, EUTROPHICATION, RED	TIDE, SILICA, ALGAE, PHYTOPLANKTO	W71-09190
OLOGICAL ECOLOGY, MORTALITY, RED	TIDE, TCXICITY.: /N BLOOMS, PHYSI	W72-14056
DIATOMS, *DINOFLAGELLATES, *RED	TIDE, WATER POLLUTICN SOURCES.: /	W72-01329
NGROVE SWAMPS, LITTORAL, SHORES,	TIDES, BEACHES, SANDS, INTERTIDAL	W72-07911
RINE ALGAE, SALINITY, MIGRATION,	TIDES, LOW WATER MARK, CHRYSOPHYT	W72-14317
BIOINDICATORS, CURRENTS(WATER),	TIDES, WIND, LIGHT, WAVES, CLAMS,	W72-09092
OCKPOCLS, *ENTEROMORPHA, SWEDEN,	TIGRIOPUS BREVICORNIS, GAMMARUS D	W71-08026
IA NILOTICA, TILAPIA MOSSAMBICA,	TILAPIA MELANOPLEURA, CYPRINUS CA	W71-04528
OPLEURA, CYPR/ TILAPIA NILOTICA,	TILAPIA MOSSAMBICA, TILAPIA MELAN	W71-04528
ICA, TILAPIA MELANOPLEURA, CYPR/	TILAPIA NILOTICA, TILAPIA MOSSAMB	W71-04528
RBIVORES, *AQUATIC WEED CONTROL,	TILAPIA, CARP, ALGAE, ROOTED AQUA	W71-04528
*DETENTION	TIME.:	W71-13326
, ANTICANCER, ANTI/ *PROTHROMBIN	TIME, *PHARMACOLOGICAL PROPERTIES	W72-07360
*LYSIS, *CONTACT	TIME, CELLULAR METABOLITES.:	W72-00411
, DEGRADATION, CHLORINE, CONTACT	TIME, SAMPLING, COLIFORMS, BACTER	W71-07096
HIC LEVEL, LIFE HISTORY STUDIES,	TIME, SNAILS, RADIOACTIVE WASTES,	W70-06669
M.: RUTHENIUM, FIXATION,	TISSUES, ALBURNUS LUCIDUS, BELGIU	W71-09168
TION, CYTOLOGICAL STUDIES, PLANT	TISSUES, ALGAE, MICROORGANISMS, B	W73-01676
H, RHODOPHYTA, INHIBITORS, PLANT	TISSUES, BIOASSAY, PHAEPHYTA.: /T	W72-07496
SEDIMENTS, AQUATIC SOILS, PLANT	TISSUES, FRESHWATER FISH, AQUATIC	W73-01673
HEMISTRY, LYOPHILIZATION, ANIMAL	TISSUES, HISTCLOGICAL STUDIES, BC	W73-01676
, GAMMA RAY SPECTROMETRY, ANIMAL	TISSUES, MACROPHYTES, CYPRINUS CA	W73-01673
BRAIN, LO/ *MERCURY, *BIOLOGICAL	TISSUES, MERCURY ISOTOPES, HUMAN	W71-11036
PHAEOCYSTIS SPP., SOIL CULTURES,	TITANIUM DIOXIDE WASTES, HELGOLAN	W72-12239
SAY, BIOINDICATORS, FOOD CHAINS,	TITANIUM, PLANT POPULATICNS, TOXI	W71-03095
LASMA ASHING, ATOMIC ABSORPTION,	TITRATION.: /UROANATOMY, CXYGEN P	W71-11036
RMANDIA, TELLINA BUTTONI, VENUS,	TIVELA STULTORUM, EMERITA ANALOGA	W72-09092
, MOLGULA, TUNICATE, PISMO CLAM,	TIVELA STULTORUM, OSTREA LURIDA,	W72-07907
CHLOROPHYTA, MARINE ALGAE , SALT	TOLERANCE, CRUSTACEANS, MARINE FI	W73-00855
IUM CHLORIDE, *INHIBITION, *SALT	TOLERANCE, WATER POLLUTICN EFFECT	W72-13809
IDS, *TYCHOPELAGIC, *TEMPERATURE	TOLERANCES, SAMPLE PRESERVATION,	W72-07899
OUNTING MEDIA, METHYLENE IODIDE,	TOLUENE, POLYSTYRENE, SLIDE PREPA	W72-11738
ILISSIMA, AN/ *TOLERANCE LEVELS,	TOLYPOTHRIX TENUIS, AULOSIRA FERT	W71-12071
NCE, ANABAENA, NOSTOC, AULOSIRA,	TOLYPOTHRIX, ANACYSTIS NIDULANS.:	W72-14807
UM PHOSPHATES, ANABAENA, NOSTCC,	TOLYPOTHRIX TENUIS, AULOSIRA FERT	W71-10070
NSHU ISLAND/ *ALGAL HYDROLYSATE,	TONGA ISLANDS, TUTUILA ISLAND, HO	W71-08042
AKE, LAKE SORELL, LAKE CRESCENT,	TOOMS LAKE, WOODS LAKE, LAGOON OF	W73-01094
N CONCENTRATION, BRUCITE, BOTTOM	TOPOGRAPHY, PRIONOSPIO, LUMBRINER	W72-09092
ICS, POLITICAL ASPECTS, GEOLOGY,	TOPOGRAPHY, GROUNDWATER, BIOCHEMI	W71-03012
M, PEDIASTRUM, 4-AMINO-3,5,6-TR/	TORDON, FRESHWATER ALGAE, PICLORA	W72-08440
ERCENARIA MERCENARIA, CLYMENELLA	TORQUATA, CAPITELLA CAPITATA, MAC	W73-02029
ITROSAMINE, AMINES, CARCINOGENS,	TORULOPSIS SPP., ASPERGILLUS, PEN	W72-09675
M NO 52, ENDOGENOUS RESPIRATION,	TOTAL OXYGEN DEMAND, GLYCOLS.: /U	W71-12223
PHA ANGUSTATA, IPOMOEA AQUATICA,	TOTAL SOLIDS.: /NIA CRASSIPES, TY	W72-01363
TREAMS, SEWAGE EFFLUENTS, ALGAE,	TOURISM, RECREATICN, FISHERIES, H	W72-14878
RS, SEDIMENTS, HUMAN POPULATION,	TOURISM, WATERSHEDS(BASINS), TROP	W71-11009
VIRONMENTAL ENGINEERING, COOLING	TOWER: /AQUATIC ALGAE, PLANTS, EN	W72-00939
BORATORY TESTS, FISH, / *COOLING	TOWERS, *TOXICITY, *BIOASSAYS, LA	W72-08683
CHEMICAL OXYGEN DEMAND, COOLING	TOWERS, *WASTE WATER TREATMENT.: /	W71-09524
DARDS, FISH KILL, ALGAE, COOLING	TOWERS, DENSITY, VISCOSITY, FISH	W70-09612
ON, APPLICATION METHODS, COOLING	TOWERS, WATER POLLUTION EFFECTS.:	W71-06189
NS, TITANIUM, PLANT POPULATIONS,	TOXICITY.: /INDICATORS, FOOD CHAI	W71-03095
DIATOMS, ALGAE CULTURES, COPPER,	TOXICITY.: /RS, DINOFLAGELLATES,	W72-04708
AL ECOLOGY, MORTALITY, RED TIDE,	TOXICITY.: /N BLOOMS, PHYSIOLOGIC	W72-14056
, CHEMICAL WASTES, LEAD, NICKEL,	TOXICITY, BIOINDICATORS, BIODEGRA	W72-14303
YDROCARBON PESTICIDES, PESTICIDE	TOXICITY, PLANT POPULATIONS, AQUA	W72-14294
LLFISH, WATER POLLUTION EFFECTS,	TOXICITY, WATER POLLUTION, FOOD C	W72-12257
PER, ZINC, PESTICIDES, PESTICIDE	TOXICITY, *HERBICIDES, DALAPON, 2	W72-12576
C HEALTH, ANALYTICAL TECHNIQUES,	TOXICITY, CHEMICAL PROPERTIES, MI	W72-04275
Y/ *ALGAE, *INHIBITION, *LIPIDS,	TOXICITY, HEAVY METALS, LABORATOR	W72-06037
ANTS, SEA WATER, TRACE ELEMENTS,	TOXICITY, CHROMOSOMES, FUNGICIDES	W72-05952
TICIDES, DIATOMS, PHYTOPLANKTON,	TOXICITY, CULTURES, DDT, GROWTH R	W72-05614
OLATION, HERRING, SALMON, TROUT,	TOXICITY, WASHINGTON, MARINE ALGA	W72-05421
NTS, COAGULATION, SEDIMENTATION,	TOXICITY, TASTE, ODOR, CORROSION,	W72-06837
S, OIL, *OIL WASTES, / BIOASSAY,	TOXICITY, *EMULSIFIERS, DETERGENT	W70-09429
H RATES, DIATOMS, PHYTOPLANKTON,	TOXICITY, IRON, LIGHT, HYDROGEN I	W71-03027
SANDS, SEDIMENTS, PRODUCTIVITY,	TOXICITY, CRABS, ALGAE, INVERTEBR	W71-01475
TER TREATMENT, SLUDGE TREATMENT,	TOXICITY, ALGAE, BACTERIA, FISH,	W70-06968
ORELLA, RESPIRATION, PHOSPHORUS,	TOXICITY, RADIOISOTOPES, INHIBITI	W70-07255
URES, TEST PROCEDURES, PESTICIDE	TOXICITY, RESISTANCE, LAKES, RESE	W71-06189
LATION, ALGAE, SLUDGE, CHROMIUM,	TOXICITY, NEUTRALIZATION, HYDROGE	W71-09524
SEASES, EUTROPHICATION, FOULING,	TOXICITY, ODOR, OXYGEN SAG, TURBI	W71-09958
S, BIOASSAY, PESTICI/ *PESTICIDE	TOXICITY, *WATER POLLUTION EFFECT	W72-08440
ORDCARBON PESTICIDES, *PESTICIDE	TOXICITY, *DDT, PESTICIDES, PHYTO	W72-08436

L POLLUTION, SAMPLING, BACTERIA, TOXICITY, BEHAVIOR, BITUMINOUS MA W72-08790
MS, POLLUTION SOURCES, PESTICIDE TOXICITY, FOOD CHAINS, FISH, BIOA W72-07703
DIATOMS, SILICATES, CYANOPHYTA, TOXICITY, COSTS, CARBON DIOXIDE.: W72-07508
N EFFECTS, ABSORPTION, PESTICIDE TOXICITY, LABORATORY TESTS.: /TIO W73-00281
YTA, *CHLOROPHYTA, *CHRYSOPHYTA, TOXICITY, INHIBITORS, TOXINS, INO W72-14706
*PLANT GROWTH, LIGHT INTENSITY, TOXICITY, PRIMARY PRODUCTIVITY, S W72-14690
AIR POLLUTION, WATER POLLUTION, TOXICITY, CHEMICALS, REGULATION, W73-00703
NTATION, PUBLICATIONS, PESTICIDE TOXICITY, BIOINDICATORS, AGRICULT W73-01976
ATION, POLLUTANT IDENTIFICATION, TOXICITY, *IRON, *CHLORELLA, NUTR W73-02112
TOMS, PESTICIDES, DDD, PESTICIDE TOXICITY, PATH OF POLLUTANTS, PES W73-02280
LOGY, INVERTEBRATES, AMPHIBIANS, TOXICITY, CHARA, SUBMERGED PLANTS W72-00155
WATER POLLUTION SOURCES, SOAPS, TOXICITY, PUBLIC HEALTH: /SPOSAL, W72-01093
N, *AMMONIA, NITRATES, NITRITES, TOXICITY, CORROSION, PUBLIC HEALT W72-04086
*ALGAE, CHLOROPHYTA, CHLORELLA, TOXICITY, WATER ANALYSIS, ABSORPT W72-04134
ECTS, *PHOTOSYNTHESIS, PESTICIDE TOXICITY, PLANT GROWTH REGULATORS W71-10566
HORUS PESTICIDES, PHYTOPLANKTON, TOXICITY, DISSOLVED OXYGEN, CULTR W71-10096
INHIBITORS, PHYTOPLANKTON, PHYTO- TOXICITY, HERBICIDES, INSECTICIDE W71-10566
BON PESTICIDES, *DDT, *PESTICIDE TOXICITY, *WATER POLLUTION EFFECT W71-12303
OLUBILITY, SORPTION, PESTICIDES, TOXICITY, DISSOLVED OXYGEN, LAKE W71-12092
CNTROL, FINANCING, PLANT GROWTH, TOXICITY, NUTRIENTS, WATER QUALIT W71-13172
, PUBLIC HEALTH, BIODEGRADATION, TOXICITY, CONTRACTS, GRANTS, RESE W71-11516
*WATER QUALITY CONTROL, BIOMASS, TOXICITY, EUTROPHICATION, TRACE E W71-11793
*PESTICIDE RESIDUES, *PESTICIDE TOXICITY, *AQUATIC PLANTS, AQUATI W72-09650
ION, REVIEWS, *DATA COLLECTIONS, TOXICITY, PATH OF POLLUTANTS, *ME W72-12190
UALITY, TECHNOLOGY, SURFACTANTS, TOXICITY, HAZARDS, BIODEGRADATION W72-09169
SAMPLING, BIOASSAY, ENVIRONMENT, TOXICITY, POISONS, ALGAL TOXINS, W72-10678
ICALS, *PUBLIC HEALTH STANDARDS, TOXICOLOGY, NTA.: /S, *TOXIC CHEM W73-02144
WATER POLLUTION EFFECTS, *ALGAL TOXINS.: /*TOXICITY, COAGULATION, W72-07360
PATHOGENS, MICROORGANISMS, ALGAL TOXINS.: /ERIA, FUNGI, PROTOZOA, W72-08586
UTION SOURCES, SEA WATER, TROUT, TOXINS.: /RIMP, CRABS, WATER POLL W71-07412
UATIC ALGAE, *CRUSTACEANS, *FISH TOXINS, *ALGAL POISONING, DINOFLA W72-14056
*FOOD CHAINS, *TOXICITY, *ALGAL TOXINS, *CHLOROPHYTA, *PHYTOTOXIN W71-08597
*SCENEDESMUS, AQUATIC MI/ *ALGAL TOXINS, *INHIBITION, *CHLORELLA, W72-12255
A, CYANOPHYTA, SYSTEMATI/ *ALGAL TOXINS, *MYXOBACTERIA, CHLOROPHYT W72-07951
PHYTA, *NUISANCE ALGAE, / *ALGAL TOXINS, *POISONOUS PLANTS, *CYANO W72-14056
HYTOTOXICITY, *CYANOPHYT/ *ALGAL TOXINS, *RED TIDE, *BEGGIATOA, *P W72-14542
PATHOGENIC BACTERIA, COLIFORMS, TOXINS, ANTIBIOTICS(PESTICIDES), W72-13800
CHLOROPHYTA, SCENEDESMUS, ALGAL TOXINS, ANTIBIOTICS(PESTICIDES), W71-05155
*ANALYTICAL TECHNIQUES, *ALGAL TOXINS, AQUATIC PLANTS, TOXINS, R W73-00243
ONMENT, TOXICITY, POISONS, ALGAL TOXINS, FISH TOXINS, SEDIMENTS, S W72-10678
OPHYTA, *LABORATORY TESTS, ALGAL TOXINS, HUMAN PATHOLOGY.: *CYAN W72-04259
RYSOPHYTA, TOXICITY, INHIBITORS, TOXINS, INORGANIC COMPOUNDS, ALGA W72-14706
TY, CHEMICALS, REGULATION, ALGAL TOXINS, MERCURY, DETERGENTS, WATE W73-00703
S, PATH OF POLLUTANTS, BIOASSAY, TOXINS, POISONS, POISONOUS PLANTS W71-08597
, *ALGAL TOXINS, AQUATIC PLANTS, TOXINS, RED TIDE, PLANT PHYSIOLOG W73-00243
ITY, POISONS, ALGAL TOXINS, FISH TOXINS, SEDIMENTS, SALT MARSHES, W72-10678
ION, MIXING, ALGAL BLOOMS, ALGAL TOXINS, WATER CIRCULATION, WATER W72-09304
ITIES, HEAVY METALS, HERBICIDES, TOXINS, WASTE DILUTION, NITROGEN, W72-01472
NIQUES, WATER POLLUTION SOURCES, TOXINS, WASTES, CHLOROPHYTA, SESS W71-10553
ABAENA FLOW-AQUAE, HOMOGENIZING, TRA: /AQUAE, DESTRATIFICATION, AN W72-07890
I, CONIOPHORA OLIVACEA, LENZITES TRABEA, SCOPULARIOPSIS BREVICAULI W72-13813
UNGI, ALGAE, SOIL CONTAMINATION, TRACE ELEMENTS, ION EXCHANGE, POL W72-14303
TES, NITRATES, NUTRIENTS, SALTS, TRACE ELEMENTS, LIGHT INTENSITY, W72-12739
RATES, *TOXICITY, *FLUORESCENCE, TRACE ELEMENTS, COPPER, CHLORELLA W72-07660
TIVATION ANALYSIS, SPECTROSCOPY, TRACE ELEMENTS, SAMPLING, WATER A W73-00832
Y TECHNIQUES, CYCLING NUTRIENTS, TRACE ELEMENTS, PUBLIC HEALTH, RA W73-00826
CITY, EUROPE, ANIMAL PHYSIOLOGY, TRACE ELEMENTS, HEAVY METALS, PHY W73-00817
L TECHNIQUES, CYCLING NUTRIENTS, TRACE ELEMENTS, CHELATION, ORGANI W73-00821
INE RADIOISOTOPES, MARINE ALGAE, TRACE ELEMENTS, PATH OF POLLUTANT W73-00811
NE ALGAE, ENVIRONMENTAL EFFECTS, TRACE ELEMENTS, COLLOIDS, NUCLEAR W73-00818
DIMENTS, SILTS, PLANKTON, ALGAE, TRACE ELEMENTS, VOLATILITY, FREEZ W71-11036
OMASS, TOXICITY, EUTROPHICATION, TRACE ELEMENTS, NITROGEN, PHOSPHO W71-11793
ON, *NITROGEN, ELECTROCHEMISTRY, TRACE ELEMENTS, HEAVY METALS, FLE W72-01693
EDIMENTS, ENVIRONMENTAL EFFECTS, TRACE ELEMENTS, ANALYTICAL TECHNI W72-03215
TY, PATH OF POLLUTANTS, *METALS, TRACE ELEMENTS, PESTICIDES, SALIN W72-12190
TIC ALGAE, PRIMARY PRODUCTIVITY, TRACE ELEMENTS, PHOSPHATES, NITRA W71-09674
PHORUS RADIOISOTOPES, SEDIMENTS, TRACE ELEMENTS, VOLCANOES, BENTHI W71-09190
KTON, HUMIC ACIDS, FULVIC ACIDS, TRACE ELEMENTS, CHEMICAL ANALYSIS W71-09188
LGAE, AQUATIC PLANTS, SEA WATER, TRACE ELEMENTS, TOXICITY, CHROMOS W72-05952
TESTS, RESEARCH AND DEVELOPMENT, TRACE ELEMENTS, ALGAE, NUTRIENTS, W72-04433
, PROTEINS, ALGAE, GROWTH RATES, TRACERS.: / CULTURES, CHLOROPHYLL W72-03557
ANT IDENTIFICATION, COLORIMETRY, TRACERS.: /MS, GAMMA RAYS, POLLUT W71-11036
LA, FILTRATION, ELECTROPHORESIS, TRACERS.: /TIONS, ANIONS, CHLOREL W72-14840
I, ORGANIC COMPOUNDS, CHLORELLA, TRACERS, AQUATIC ALGAE.: /M, FUNG W72-14326
ICAL MODELS, PATH OF POLLUTANTS, TRACERS, ANIMAL PHYSIOLOGY, NUCLE W72-03348
STS, *MEASUREMENT, *MARINE ALGAE TRACERS, CHEMICAL PRECIPITATION, W72-12210
UDIES, LAKES, CYCLING NUTRIENTS, TRACERS, LITTORAL, KINETICS, PHOS W72-12543
ACTERIA, *TEXAS, BIODEGRADATION, TRACERS, LAGOONS, CYANOPHYTA.: /B W71-04873
TIC PRODUCTIVITY, RADIOISOTOPES, TRACERS, MARINE ALGAE, HAWAII, WA W71-11486
TIVITY TECHNIQUES, RADIOECOLOGY, TRACERS, MONITORING, NUCLEAR POWE W73-00790
PTION, RADIOACTIVITY TECHNIQUES, TRACERS, PATH OF POLLUTANTS, ON-S W72-00949
S, *ABSORPTION, ALGAE, CHORELLA, TRACERS, PHOSPHATES, ECOSYSTEMS, W72-09668
NIMAL PHYSIOLOGY, WATER BALANCE, TRACERS, PLANT PHYSIOLOGY, AQUATI W72-06342
ASONAL, WEATHER, WATER ANALYSIS, TRACERS, RADIOACTIVITY TECHNIQUES W72-09122
ACTIVITY TECHNIQUES, METABOLISM, TRACERS, RADIOACTIVITY, CARBON DI W72-12634
GAE, AQUATIC ALGAE, COLORIMETRY, TRACES.: /HYTOPLANKTON, MARINE AL W72-03545
STEPHANODISCUS, ANKISTRODESMUS, TRACHELOMONAS, SAGINAW BAY, ORGAN W72-10875
PANDORINA, PHOCUS, STAURASTRUM, TRACHELOMONAS.: /ORIA, SPIRULINA, W72-07661
CUS PYRUM, TRACHELOMONAS ARMATA, TRACHELOMONAS HISPIDA VAR CRENULA W73-00918
ACUS PLEURONECTES, PHACUS PYRUM, TRACHELOMONAS ARMATA, TRACHELOMON W73-00918

ES ANNOSUS, FUSCOPORIA CONTIGUA, TRAMETES LILACINO-GILVA, XYLOBOLU W72-13813
VICAULIS, CLADOSPORIUM HERBARUM, TRAMETES VERSICOLOR, FOMES ANNOSU W72-13813
MICAL PROPERTIES, NITROGEN, HEAT TRANSFER, PHOSPHATES, ALKALINITY, W72-12997
S, ELECTRICAL POWER PLANTS, HEAT TRANSFER, WATER POLLUTION EFFECTS W72-07907
DATION LAGOONS, *CHANNELS, *HEAT TRANSFER, *PLASTICS, DOMESTIC WAS W72-10401
STRY OF INORGANIC CARBON, CARBON TRANSFORMATION, CARBON SOURCES, O W71-12068
A, SCANNING ELECTRON MICROSCOPY, TRANSMISSION ELECTRON MICROSCOPY, W73-00432
S-AQUAE, CERATIUM, OSCILLATORIA, TRANSPARENCY.: /APHANIZOMENON FLO W71-11949
FOX RIVER, SPECIFIC CONDUCTANCE, TRANSPARENCY, ORTHOPHOSPHATES, DR W71-10645
EMICAL ANALYSIS, EUTROPHICATION, TRANSPIRATION, PRECIPITATI: /, CH W71-01489
EMENT, MANURE COLLECTION, MANURE TRANSPORT.: /QUALITY, WASTE MANAG W71-08214
TRATES, ALGAE, COPPER, IONS, ION TRANSPORT.: /TALS, ELECTRODES, NI W72-01693
TROGEN, EUTROPHICATION, SEDIMENT TRANSPORT, CYCLING NUTRIENTS, DEG W72-00847
US, *PHOSP/ *PHYTOPLANKTON, *ION TRANSPORT, *ABSORPTION, *PHOSPHOR W72-07143
SAMPLING, CHLOROPHYLL, SEDIMENT TRANSPORT, BENTHIC FAUNA, PLANKTO W72-09092
PPER, CULTURES, *CHELATION, *ION TRANSPORT, *HEAVY METALS, EUTROPH W73-02112
, CYCLOTELLA NANA, TH/ *ELECTRON TRANSPORT, DUNALIELLA TERTIOLECTA W72-12998
EASUREMENTS, CARBON-14 ME/ *LAKE TRAVIS, *LAKE LIVINGSTON, FIELD M W72-13692
MICAL OXYGEN DEMAND, WASTE WATER TREAT: /EROBIC CONDITIONS, BIOCHE W71-07100
QUALITY, BIODEGRADATION, SEWAGE TREATMENT PLANTS, DIATOMS, PHYTOP W72-06837
ALUM, RECYCLING, TREATMENT COSTS, ASIA.: W72-13508
AG, DREDGING, STATE GOVERNMENTS, TREATMENT FACILITIES, ADMINISTRAT W72-01659
BIOLOGICAL TREATMENT, *NITRATES, TREATMENT FACILITIES, ALGAE, *WAS W72-02975
CRITERIA, KINETICS, *WASTE WATER TREATMENT.: / TECHNIQUES, DESIGN W72-02363
ANALYSIS, MICHIGAN, WASTE WATER TREATMENT.: / SPORTS, PARKS, COST W72-00042
ER QUALITY CONTROL, *WASTE WATER TREATMENT.: /ICAL TECHNIQUES, WAT W72-00422
CONTROL, AERATION, *WASTE WATER TREATMENT.: /N, DIFFUSION, *ALGAL W72-00800
TION, FLOCCULATION, *WASTE WATER TREATMENT.: /FILTRATION, DISINFEC W72-00411
*STABILIZATION PONDS, ALUM, POST- TREATMENT.: /F, *FERRIC SULFATE, W71-13356
IA, CENTRIFUGATION, *WASTE WATER TREATMENT.: /N, COLIFORMS, BACTER W71-13326
RCH AND DEVELOPMENT, WASTE WATER TREATMENT.: /TIVE CAPACITY, RESEA W71-13256
CIUM, SEDIMENTATION, WASTE WATER TREATMENT.: /MS, ACCLIMATION, CAL W71-13334
UM DRYING, FLOTATION, WASTEWATER TREATMENT.: /CENTRIFUGATION, VACU W72-10237
OST ANALYSIS, ALGAE, WASTE WATER TREATMENT.: / LABORATORY TESTS, C W72-10390
TERTIARY TREATMENT, *WASTE WATER TREATMENT.: /, *WATER TREATMENT, W72-09393
QUES, MICROBIOLOGY, *WASTE WATER TREATMENT.: /S, ANALYTICAL TECHNI W72-09590
LVED OXYGEN, SLUDGE, WASTE WATER TREATMENT.: /ION, PROTOZOA, DISSO W72-10233
ION AND MAINTENANCE, WASTE WATER TREATMENT.: /NTIAL, ODORS, OPERAT W72-10401
PHOTOSYNTHETIC OXYGEN, *TERTIARY TREATMENT.: /OMASS, RESPIRATION, W72-11100
ION CONCENTRATION, *WASTE WATER TREATMENT.: /PROPERTIES, HYDROGEN W72-11933
S, NITROGEN, CLAYS, *WASTE WATER TREATMENT.: /SLUDGE, COST ANALYSI W72-11642
TERTIARY TREATMENT, *WASTE WATER TREATMENT.: /LYTICAL TECHNIQUES, W72-04983
ONTROL, *CHLORELLA, *WASTE WATER TREATMENT.: /ROPHICATION, ALGAL C W72-04788
MICAL OXYGEN DEMAND, WASTE WATER TREATMENT.: /ST ANALYSIS, *BIOCHE W71-07099
MICAL OXYGEN DEMAND, WASTE WATER TREATMENT.: /Y, COLIFORMS, BIOCHE W71-07106
MICAL OXYGEN DEMAND, WASTE WATER TREATMENT.: /C CONDITIONS, BIOCHE W71-07109
NAEROBIC CONDITIONS, WASTE WATER TREATMENT.: /RS, PARTICLE SIZE, A W71-08223
YING, COST ANALYSES, WASTE WATER TREATMENT.: /TION, DEWATERING, DR W71-08951
ND, COOLING TOWERS, *WASTE WATER TREATMENT.: /CHEMICAL OXYGEN DEMA W71-09524
N, DENITRIFICATION, *WASTE WATER TREATMENT.: /TATION, NITRIFICATIO W71-09450
ORELLA, SCENEDESMUS, WASTE WATER TREATMENT.: /N CONCENTRATION, CHL W71-00138
RATION, SCENEDESMUS, WASTE WATER TREATMENT.: /HOTOSYNTHESIS, RESPI W71-00139
R POLLUTION CONTROL, WASTE WATER TREATMENT.: /ICA, *BACTERIA, WATE W71-01899
MICAL OXYGEN DEMAND, WASTE WATER TREATMENT.: /CIPAL WASTES, BIOCHE W71-03896
Y TESTS, DETERGENTS, WASTE WATER TREATMENT.: /TOXICITY, *LABORATOR W70-08976
NIQUES, COAGULATION, WASTE WATER TREATMENT.: /STS, SEPARATION TECH W70-08904
L, SEWAGE TREATMENT, WASTE WATER TREATMENT.: /VERS, SEWAGE DISPOSA W70-09947
, TRICKLING FILTERS, WASTE WATER TREATMENT.: /ATION, SEDIMENTATION W70-10427
RSHES, SOIL EROSION, WASTE WATER TREATMENT.: /ORES, MONITORING, MA W73-02479
*ANAEROBIC BACTERIA, BIOLOGICAL TREATMENT.: /EMAND, *OXYGENATION, W72-14469
TREATMENT, *FILTRATION, *SEWAGE TREATMENT, ALGAE, *AERATION, SEDI W72-14469
REATMENT/ *PATENTS, *WASTE WATER TREATMENT, *FILTRATION, *SEWAGE T W72-14469
UTION ABATEMENT, WATER POLLUTION TREATMENT, WATER QUALITY, STREAMS W72-14469
CA, *E. COLI, *IONS, *WASTEWATER TREATMENT, *POLYMERS, CATIONS, AN W72-14840
ALYTICAL TECHNIQUES, WASTE WATER TREATMENT, TASTE, ODOR, POTABLE W W73-02144
,/ *ALGAE, *NITRIFICATION, WASTE TREATMENT, *PHOSPHORUS, *NITROGEN W73-02187
R PURIFICATION, *WATER POLLUTION TREATMENT, *PUBLIC HEALTH, ANALYT W73-02144
ITROGEN, NUTRIENT REMOVAL, WATER TREATMENT, ASSAY, BACTERIA.: /, N W73-02478
RPTION, ACTIVATED CARBON, SLUDGE TREATMENT, DIATOMS, COAGULATION, W73-01669
S, *TREATMENT FACILITIES/ *WATER TREATMENT, WATER POLLUTION EFFECT W73-01669
PHOSPHORUS, *ALGAE, *WASTE WATER TREATMENT, HYDROGEN ION CONCENTRA W73-01367
IA, WASTE WATER TREATMENT, WATER TREATMENT, MODEL STUDIES.: /ACTER W73-01362
AY, ALGAE, BACTERIA, WASTE WATER TREATMENT, WATER TREATMENT, MODEL W73-01362
ROGEN, LAGOONS, PUMPING, AEROBIC TREATMENT, BIOCHEMICAL OXYGEN DEM W72-08396
OXYGEN DEMAND, NUTRIENTS, *WASTE TREATMENT, *ALGAE, COSTS.: /ICAL W72-08396
N, SEWAGE TREATMENT, WASTE WATER TREATMENT, PONDS, SCENDESOMUS, CY W72-07661
NAEROBIC CONDITIONS, WASTE WATER TREATMENT, CHLOROPHYTA, CYANOPHYT W72-07664
LAGOONS, BIODEGRADATION, SEWAGE TREATMENT, WASTE WATER TREATMENT, W72-07661
, MOLYBDENUM, DETERGENTS, SEWAGE TREATMENT, CARBONATES, BICARBONAT W72-07514
IPMENT, SEWAGE EFFLUENTS, SEWAGE TREATMENT, SEWAGE DISPOSAL, MARIN W72-07715
PERATION AND MAINTENANCE, *WATER TREATMENT, *MICHIGAN.: /ICIDES, O W72-08357
, *TERTIARY TREATMENT, / *SEWAGE TREATMENT, *WASTE WATER TREATMENT W70-10381
NT, *EUTROPHICATION, WASTE WATER TREATMENT, ALGAE, EFFLUENTS, NUTR W70-09325
*SEWAGE TREATMENT, *WASTE WATER TREATMENT, *TERTIARY TREATMENT, * W70-10381
RIVERS, SEWAGE DISPOSAL, SEWAGE TREATMENT, WASTE WATER TREATMENT. W70-09947
AE, EFFLUENTS, NUTRIENTS, SEWAGE TREATMENT, LAKES.: /REATMENT, ALG W70-09325
CONCENTRATION, ACTIVATED SLUDGE, TREATMENT, EUTROPHICATION, SURFAC W71-00116
ITERATURE SURVEY, PRIMARY SEWAGE TREATMENT, SECONDARY SEWAGE TREAT W70-09907
MENT, ACTIVATED SLUDGE, TERTIARY TREATMENT, ION EXCHANGE, ELECTROC W70-09907
ATER INTERFACES, WATER POLLUTION TREATMENT, WATER POLLUTION EFFECT W70-09429

TS, SETTLING BASINS, WASTE WATER TREATMENT, LABORATORY TESTS.: /EN W71-00116
WAGE TREATMENT, SECONDARY SEWAGE TREATMENT, PHYSICOCHEMICAL METHOD W70-09907
OSPHORUS, *DETERGENTS, *TERTIARY TREATMENT, *EUTROPHICATICN, WASTE W70-09325
FILTERS, HYDROLYSIS, BIOLOGICAL TREATMENT, ACTIVATED SLUDGE, TERT W70-09907
WASTE WATER TREATMENT, *TERTIARY TREATMENT, *WATER REUSE, *WATER Q W70-10381
CONTRCL, RIVERS, WATER P/ *WATER TREATMENT, *CHLORINATION, *ALGAL W71-00099
SUSPENSION, HARVES/ *WASTE WATER TREATMENT, *FILTRATION, *ALGAE, * W70-10174
IC/ *NITROGEN COMPOUNDS, *SEWAGE TREATMENT, *DOMESTIC WASTES, CHEM W70-09907
EMICAL OXYGEN DEMAND, BIOLOGICAL TREATMENT, PHOTOSYNTHESIS, STABIL W70-06619
SIS, STABILIZATION, *WASTE WATER TREATMENT, *FARM WASTES.: /SYNTHE W70-06619
S, WASTE WATER TREATMENT, SLUDGE TREATMENT, TOXICITY, ALGAE, BACTE W70-06968
N, SEWERS, SLURRIES, WASTE WATER TREATMENT, SLUDGE TREATMENT, TOXI W70-06968
S, BICCHEMICAL OXY/ *WASTE WATER TREATMENT, *PONDS, *ALGAE, LAGOON W70-07838
N, ACTIVATED SLUDGE, WASTE WATER TREATMENT, NUTIRIENTS, AEROBIC CO W71-04079
*STABILIZATION, *ALGAE, AEROBIC TREATMENT, BACTERIA, MUNICIPAL WA W71-03896
P/ *OXIDATION LAGOONS, *AEROBIC TREATMENT, *ANAEROBIC CONDITIONS, W71-06033
NTS, WATER POLLUTION, BIOLOGICAL TREATMENT, CHEMICALS, CCPPER SULF W71-05083
N ION CONCENTRATION, WASTE WATER TREATMENT, WATER POLLUTICN EFFECT W71-05155
MICAL OXYGEN DEMAND, WASTE WATER TREATMENT, MICROORGANISMS.: /OCHE W71-06033
ELD IMPROVEMENT, WATER POLLUTION TREATMENT, ALGAE, POTABLE WATER, W71-01491
LAKES, MICROORGANISMS, TERTIARY TREATMENT, EFFLUENTS, MODEL STUDI W71-03021
IES, *ABSTRACTS, WATER POLLUTION TREATMENT, INSTRUMENTATICN, WASTE W71-01192
IRONMENTAL SANITATION, SECONDARY TREATMENT, AQUATIC ALGAE, PRIMARY W71-09674
KLING FILTERS, LAGOONS, TERTIARY TREATMENT, ACTIVATED CARBON, FLOC W71-09524
, AM/ *EUTROPHICATION, *TERTIARY TREATMENT, *WATER QUALITY CONTROL W71-09450
TURBIDITY, WATER QUALITY, WATER TREATMENT, NITROGEN CYCLE.: /SAG, W71-09958
WASTE WATER TREATMENT, *TERTIARY TREATMENT, *COAGULATICN, *OXIDATI W71-09629
GANIC COMPOUNDS, WATER POLLUTION TREATMENT, WATER POLLUTICN SOURCE W71-09674
S, HUMAN POPULATION, WASTE WATER TREATMENT, WATER QUALITY CONTROL. W71-09788
ION, *OXIDATION LAGOONS, AEROBIC TREATMENT, OXIDATION, ORGANIC MAT W71-09629
COAGULATIO/ *ALGAE, *WASTE WATER TREATMENT, *TERTIARY TREATMENT, * W71-09629
TROL, COST ANALYSIS, WASTE WATER TREATMENT, NUTRIENT REQUIREMENTS. W71-08953
MICAL OXYGEN DEMAND, WASTE WATER TREATMENT, *NORTH CAROLINA.: /CHE W71-07382
MICROORGANISMS, SLUDGE, TERTIARY TREATMENT, BIOCHEMICAL CXYGEN DEM W71-07382
R POLLUTION EFFECTS, WASTE WATER TREATMENT, CYANOPHYTE, ENGLENOPHY W71-07973
ANIMAL DISEASES, CATFISH, WASTE TREATMENT, FARM WASTES, DEHYDRATI W71-07544
OCHEMI/ *SURFACE RUNOFF, *SEWAGE TREATMENT, TERTIARY TREATMENT, BI W71-06445
*OXIDATION LAGOONS, *WASTE WATER TREATMENT, ORGANIC LOADING, BIOCH W71-07124
*OXIDATION LAGOONS, *WASTE WATER TREATMENT, *WATER RE-USE, BIOCHEM W71-07123
OFF, *SEWAGE TREATMENT, TERTIARY TREATMENT, BIOCHEMICAL CXYGEN DEM W71-06445
CONDITIONS, SLUDGE, WASTE WATER TREATMENT, BIODEGRADATICN, BIOCHE W71-07084
SMS, BIODEGRADATION, WASTE WATER TREATMENT, WATER POLLUTICN EFFECT W71-07097
ANIC LOADING, ALGAE, WASTE WATER TREATMENT, BIOCHEMICAL CXYGEN DEM W71-07096
K/ *OXIDATION LAGOONS, *TERTIARY TREATMENT, *DESIGN CRITERIA, TRIC W71-07113
MICAL OXYGEN DEMAND, WASTE WATER TREATMENT, WATER POLLUTICN EFFECT W71-07098
N ION CONCENTRATION, WASTE WATER TREATMENT, KANSAS.: /ANC, HYDROGE W71-07113
REATMENT, *LIQUID WASTES, SEWAGE TREATMENT, SEPARATION TECHNIQUES, W72-05315
DILUTION, LAGOONS, *WASTE WATER TREATMENT, ALGAL CONTROL, EUTROPH W72-04787
ANALYTICAL TECHNIQUES, TERTIARY TREATMENT, *WASTE WATER TREATMENT W72-04983
ERIA, *OXIDATION LAGOONS, SEWAGE TREATMENT, ALGAL: /Y, DESIGN CRIT W72-04789
ON, WATER PURIFICATION, *AEROBIC TREATMENT, *OXIDATION LAGOONS.: / W72-05315
*ALGAE, *KINETICS, *WASTE WATER TREATMENT, *NUTRIENT REQUIREMENTS W72-04789
ENETRATION, NEBRASKA, WASTEWATER TREATMENT, TURBIDITY, RUNOFF, PHO W72-04759
TREATME/ *PATENTS, *WASTE WATER TREATMENT, *LIQUID WASTES, SEWAGE W72-05315
WATER POLLUTION SOURCES, SEWAGE TREATMENT, PUBLIC HEALTH, WASTE W W72-06638
ANISMS, *SURFACE WATERS,/ *WATER TREATMENT, *FILTRATION, *MICROORG W72-06191
MENT, PUBLIC HEALTH, WASTE WATER TREATMENT, RESEARCH AND DEVELOPME W72-06638
TREATMENT, CULTURE, *WASTE-WATER TREATMENT, *CARBON DIOXIDE, *GROW W72-06612
AL GOVERNMENT, NUTRIENTS, SEWAGE TREATMENT, WATER POLLUTION, WATER W72-06655
MAND, *SEWAGE EFFLUENTS, *SEWAGE TREATMENT, CULTURE, *WASTE-WATER W72-06612
RCH AND DEVELOPMENT, WASTE WATER TREATMENT, BIODEGRADATICN, LOCAL W72-06655
E, *ALGICIDES, MICROSTRA/ *WATER TREATMENT, *MICROORGANISMS, *ALGA W72-06536
Y, *FLOCCULATION, *FILTR/ *WATER TREATMENT, *PHOSPHATES, *TURBIDIT W72-06201
MAND, PILOT PLANTS, *WASTE WATER TREATMENT, WYOMING, *AERATED LAGO W72-06838
POLLUTION CONTROL, *WASTE WATER TREATMENT, *COST-BENEFIT ANALYSIS W72-04734
TER QUALITY CONTROL, WASTE WATER TREATMENT, CHLOROPHYTA, GROWTH RA W72-04431
R REUSE, *NITROGEN/ *WASTE WATER TREATMENT, *DENITRIFICATICN, WATE W72-04309
TOPLANKTON, *METABOLISM, *SEWAGE TREATMENT, *SEWAGE LAGOCNS, OHIO, W72-05459
PHORUS, SEWAGE EFFLUENTS, SEWAGE TREATMENT, ALGAE, WATER POLLUTICN W72-05473
CREATION, HYDROLOGIC DATA, WATER TREATMENT, STREAM GAGES, GRCUNDWA W72-05594
E, PLANT GROWTH, WATER POLLUTION TREATMENT, EVALUATION, WATER QUAL W72-11702
HICATION, *LAKES, *WATE/ *SEWAGE TREATMENT, *EFFICIENCIES, *EUTROP W72-11702
, *FLOCCULATION, WATER POLLUTION TREATMENT, COLLOIDS, *PCLYMERS, A W72-11833
TION, *NUTRIENTS, *ALGAE, SEWAGE TREATMENT, WATER POLLUTICN SOURCE W72-10625
ACTERIA, *SYMBIOSI/ *WASTE WATER TREATMENT, *NUTRIENTS, *ALGAE, *B W72-11100
G, WATER QUALITY CONTROL, SEWAGE TREATMENT, SEWAGE EFFLUENTS, SANI W72-09378
OR, SURVEYS, MICROBIOLOGY, WATER TREATMENT, PHYTOPLANKTON, FUNGI, W72-10076
NING, *WATER TREATMENT, TERTIARY TREATMENT, *WASTE WATER TREATMENT W72-09393
, FLOCCULATION, CLEANING, *WATER TREATMENT, TERTIARY TREATMENT, *W W72-09393
SANITARY ENGINEERING, SECONDARY TREATMENT, MARINE ALGAE, PHYTOPLA W72-09378
DS, NITRATES, PHOSPHATES, SLUDGE TREATMENT, DISSOLVED OXYGEN, BACT W72-09675
ON AND MAINTENANCE, *WASTE WATER TREATMENT, WATER QUALITY CONTROL, W72-09386
MENTS, ALGAE, FORMULATION, WASTE TREATMENT, INDUSTRIES, WATER QUAL W72-09169
ACTIVATED SLUDGE, LIME, AEROBIC TREATMENT, DOMESTIC WASTES.: /NT, W72-09176
E EFFLUENTS, *ALGAE, WASTE WATER TREATMENT, ACTIVATED SLUDGE, LIME W72-09176
ION, *NUT/ KINETICS, *BIOLOGICAL TREATMENT, PLANKTON, *EUTROPHICAT W71-13553
M, DESIGN CRITERIA, *WASTE WATER TREATMENT, EFFLUENTS.: /METABOLIS W71-13341
D SOLIDS, / *CHEMICAL-BIOLOGICAL TREATMENT, MIXED LIQUOR, SUSPENDE W71-13334
COA/ *COAGULATION, *WASTE WATER TREATMENT, ANALYTICAL TECHNIQUES, W71-13356
SOURCES, SYSTEMS ANALYSIS, WASTE TREATMENT, AIR PCLLUTICN, LEGISLA W71-13723

, AGRICULTURE, INDUSTRIES, WASTE TREATMENT, WATERSHEDS(BASINS), PH W71-11017
UDGE, EFFLUENT, ALGAE, SECONDARY TREATMENT, BIOCHEMICAL OXYGEN DEM W71-12188
STE WATER TREATMENT, *BIOLOGICAL TREATMENT, *ACTIVATED SLUDGE, EFF W71-12223
MENT, *PHOSPHOROUS, *WASTE WATER TREATMENT, NUTRIENTS, SYNTHESIS, W71-12188
*ACTIVATED SLUDGE/ *WASTE WATER TREATMENT, *BIOLOGICAL TREATMENT, W71-12223
TERGENTS, LAKE ONTARIO, TERTIARY TREATMENT, INTERNATIONAL JOINT CO W71-12091
ILTRATION, *PIT RECHARGE, *WATER TREATMENT, *FILTATION, ALLUVIA CH W71-12410
ECTROCHEMISTRY, *WATER POLLUTION TREATMENT, ANALYTICAL TECHNIQUES, W71-12467
ATER TREATMENT, NUT/ *BIOLOGICAL TREATMENT, *PHOSPHOROUS, *WASTE W W71-12188
IME, / *PHOSPHATES, *WASTE WATER TREATMENT, *FOAM FRACTIONATION, L W71-12200
MICALS, FARM WASTES, WASTE WATER TREATMENT, CALIFORNIA.: /URAL CHE W71-11698
CATION, *WATER POLLUTIO/ *SEWAGE TREATMENT, *PHOSPHATES, *EUTROPHI W71-11507
RIFICATION, *NITRA/ *WASTE WATER TREATMENT, *NITRIFICATION, *DENIT W71-13939
LGAE, SEWAGE TREATMENT, TERTIARY TREATMENT, WATER POLLUTION CONTRO W71-13939
N, */ *SEWAGE EFFLUENTS, *SEWAGE TREATMENT, *ECONOMIC JUSTIFICATIO W72-00383
OMPOUNDS, AMMONIA, ALGAE, SEWAGE TREATMENT, TERTIARY TREATMENT, WA W71-13939
*DETENTION RESERVOIRS, TERTIARY TREATMENT, ALGAE, PLANTS, TROUT, W72-02412
FACILITIES, ALGAE, *WASTE WATER TREATMENT, *ALGAL CONTROL, *AQUAT W72-02975
R POLLUTION CONTROL, *BIOLOGICAL TREATMENT, *NITRATES, TREATMENT F W72-02975
ENVIRONMENTAL ENGINEERING, WATER TREATMENT, WATER QUALITY CONTROL, W72-01660
RTH, *POROUS MEDIA, *WASTE WATER TREATMENT, *DENITRIFICATION, *NIT W72-00974
LLUTION SOURCES, WATER POLLUTION TREATMENT, NUTRIENTS, OXYGEN REQU W72-01685
*WASTE WATER TREATMENT, TERTIARY TREATMENT, EUTROPHICATION.: /TS, W72-03613
E, ALGAE, BACTERIA, *WASTE WATER TREATMENT, EUTROPHICATION.: /LUDG W72-04086
TE, LIME, *PULP WASTES, TEOTIARY TREATMENT, EFFLUENTS.: /, BENTONI W72-03753
GNESIUM, NUTRIENTS, *WASTE WATER TREATMENT, TERTIARY TREATMENT, EU W72-03613
IC AL/ *PHOSPHORUS, *WASTE WATER TREATMENT, *ALGAL CONTROL, *AQUAT W72-03970
L OXYGEN DEMAND, D/ *WASTE WATER TREATMENT, *TURBIDITY, BIOCHEMICA W72-03753
TES, ILLINOIS, ECOLOGY, TERTIARY TREATMENT, SEWAGE EFFLUENTS, WATE W72-03970
TION PONDS, *ALGAE, *WASTE WATER TREATMENT, *BYPRODUCTS, MIXING, S W72-13508
HOSPHORUS, PHYTOPLANKTON, SEWAGE TREATMENT, NITROGEN, PHYSCCHEMICA W72-13662
ICIDES, HEAVY METALS, WASTEWATER TREATMENT, SMELTS, MICHIGAN, OHIO W72-14282
ON LAGOONS, *ALGAE, *WASTE WATER TREATMENT, *MATHEMATICAL MODELS, W72-12289
*SEWAGE BACTERIA, *WASTE WATER TREATMENT, *SEWAGE LAGOONS, *COLD W72-12565
BRACKISH WATER FISH, BIOLOGICAL TREATMENT, OXIDATION LAGOONS, PHY W72-12567
ROL, EUTROPHICATION, WASTE WATER TREATMENT, SEDIMENTATION RATES, O W72-12487
ALYTICAL TECHNIQUES, WASTE WATER TREATMENT, NUTRIENTS, NUTRIENT RE W72-12966
C MATTER, MINNESOTA, CYANOPHYTA, TREES, GRASSES, LEAVES, FOREST SO W72-01784
ERMA ATLANTICA, CODIUM CAROLINI/ TREMATOCARPUS PAPENFUSSII, GLOIOD W73-00284
POLICY, *UNITED KINGDOM, *FUTURE TRENDS, CLADOPHORA, RESEARCH STRA W71-13172
DULA, CRYNELLUS, CHEUMATOPSYCHE, TRIAENODES INUUSTA, OECETIS, LEPT W70-06655
CIDES, *ALGAE, *TOXICITY, SOILS, TRIAGINE PESTICIDES, RESISTANCE, W71-07675
ND DETERGENT ASSOCIATION, CARBOY TRIALS, CANADA, NITROGEN: PHOSPHO W70-07283
SEIA QUINQUELOBATA, TROITSKIELLA TRIANGULATA, AKIYAMAMONAS TERREST W72-07683
LLATORIA, ANABAENA, MICROCYSTIS, TRIBONEMA, APHANIZOMENON, OOCYSTI W72-07940
CHAETE, RHIZOCLONIUM, VAUCHERIA, TRIBONEMA, OSCILLATORIA.: /APHANO W71-05629
CERA/ *LAKE ERIE CENTRAL BASIN, TRIBONEMA, OEDOGONIUM, ANACYSTIS, W72-12995
ONTROL, WATER POLLUTION SOURCES, TRIBUTARIES, FRESHWATER FISH, YEL W72-14282
TY, LAKES, SAMPLING, RESERVOIRS, TRIBUTARIES, DAMS, CHEMICAL ANALY W73-01435
RS, SPATIAL DISTRIBUTION, ALGAE, TRIBUTARIES, CYCLING NUTRIENTS.: / W72-04263
LORAM, PEDIASTRUM, 4-AMINO-3,5,6- TRICHLOROPICOLINIC ACID, 2,4-DICH W72-08440
CHYDORUS, MONOSTYLA, LEPADELLA, TRICHOCERCA, HAEMATOCOCCUS LACUST W72-07132
LIANA, HORTOBAGGIAMONAS PLICATA, TRICHOCYANELLA SPIRALIS, STROMIA W72-09119
S.: *PERIODICITY, TRICHODESMIUM THIEBAUDII, BARBADO W72-11719
CATES, FORAMINIFERA, RADIOLARIA, TRICHODESMIUM, FLAGELLATES, CILIA W72-14313
HAIN PICKEREL, TENDIPES TENTANS, TRICHOPTERA, LIMNEPHILUS, GLOSSOS W72-14280
OCCOCCUM, MICROCYSTIS, FUSARIUM, TRICHORIXA, RAMPHOCORIXA, SIGARA, W70-06655
, NEUTRALIZATION, SEDIMENTATION, TRICKLING FILTERS, WASTE WATER TR W70-10427
S, CHEMICAL ANALYSIS, EFFLUENTS, TRICKLING FILTERS, HYDROLYSIS, BI W70-09907
ARY TREATMENT, *DESIGN CRITERIA, TRICKLING FILTERS, ALGAE, ORGANIC W71-07113
STRIAL WASTES, ACTIVATED SLUDGE, TRICKLING FILTERS, LAGOONS, TERTI W71-09524
EA ANDROMEDA, AKABARIA, STROMBUS TRICORNI: /ERA UNINERVIS, CASSIOP W73-00855
PLATYMONAS VIRIDI/ PHAEODACTYLUM TRICORNUTUM, GONYAULAX POLYEDRA, W72-07166
ELLA PHYRENOIDOSA, PHAEODACTYLUM TRICORNUTUM, CYCLOTELLA NANA, CHA W72-07660
TE, MERCURY CHLOR/ PHAEODACTYLUM TRICORNUTUM, PHENYLMERCURIC ACETA W72-11727
IELLA TERTIOLECTA, PHAEODACTYLUM TRICORNUTUM.: /, CARBON-14, DUNAL W72-09102
LIELLA TERIOLECTA, PHAEODACTYLUM TRICORNUTUM, SKELETONEMA COSTATUM W72-09103
MACROINVERTEBRATES,/ *ADENOSINE TRIPHOSPHATE, *AUTOTROPHIC INDEX, W72-10883
HANGEABLE PHOSPHORUS, A*DENOSINE TRIPHOSPHATE(ATP).: /SPHORUS, EXC W72-03742
ORNUTUM.: *ADENOSINE TRIPHOSPHATE, *SELENASTRUM CAPRIC W71-13794
CID, SAMPLE PRESERVA/ *ADENOSINE TRIPHOSPHATE, *DEOXYRIBONUCLEIC A W71-11562
SE ENZYME SYSTEM, PL/ *ADENOSINE TRIPHOSPHATE, *LUCIFERIN-LUCIFERA W71-11916
*ATP ANALYSIS, *ADENOSINE TRIPHOSPHATE.: W71-11234
*GROWTH STAGE, *DET/ *ADENOSINE TRIPHOSPHATE, *MICROBIAL BIOMASS, W72-04292
*ANACYSTIS NIDULANS, *ADENOSINE TRIPHOSPHATASE, BIOCHEMICAL TESTS W72-13809
MPLES, LIQUID-LIQUID EXTRACTION, TRIPHOTURUS MEXICANUS, EUPHAUSIA W73-02105
P/ *CONDENSED PHOSPHATES, SODIUM TRIPOLYPHOSPHATE, SODIUM PYROPHOS W71-00116
ILENSE, CERATIUM FURCA, CERATIUM TRIPOS, CERATIUM MACROCEROS, NOCT W71-07875
ERUM, MELOSIRA VARIANS, NAVICULA TRIPUNCTATA, GOMPHONEMA PARVULUM, W72-10883
UM C/ PHOSPHATE-FREE DETERGENTS, TRISODIUM NITRILOTRIACETATE, SODI W71-06247
H, MARINE ANIMALS, MARINE ALGAE, TRITIUM, IODINE RADIOISOTOPES, ST W71-09018
SIMILATIVE CAPACITY, CALIFORNIA, TRITIUM, MARINE PLANTS, MONITORIN W73-00831
NITORING, *PACIFIC COAST REGION, TRITIUM, OCEAN CURRENTS, COBALT R W72-03324
ASTES, VIRUSES, ENERIC BACTERIA, TRITIUM, RADIOACTIVE WASTES, MONI W72-04742
UMEROSA, HIROSEIA QUINQUELOBATA, TROITSKIELLA TRIANGULATA, AKIYAMA W72-07683
A SPP., CHROMATIUM, MERISMOPEDIA TROLLERI, CRYPTOMONADS, STICHOGLO W72-07145
ARY PRODUCTIVITY, WATER QUALITY, TROPHIC LEVEL, AQUATIC ALGAE, FIS W72-08986
ALGAE, EUTROPHICATION, REVIEWS, TROPHIC LEVEL, PHOSPHORUS, PHOSPH W72-07937
NG NUTRIENTS, THERMAL POLLUTION, TROPHIC LEVEL, TROPICAL REGIONS, W73-00796
G, AQUATIC ALGAE, PHYTOPLANKTON, TROPHIC LEVEL: /PERATURE, SAMPLIN W73-01456

ALS, BALANCE OF NATURE, ECOLOGY, TROPHIC LEVEL, HERBIVORES, NATURA W72-06340
DISOTOPES, ALGAE, RADIOACTIVITY, TROPHIC LEVEL, LIFE HISTORY STUDI W70-06668
OSYSTEMS, SOUTH CAROLINA, ALGAE, TROPHIC LEVEL,: /RBON DIOXIDE, EC W71-00475
RS, CHROMATOGRAPHY, CHLOROPHYLL, TROPHIC LEVEL, EUTROPHICATION, AL W72-01784
CHEMICAL PROPERTIES, PESTICIDES, TROPHIC LEVEL, GASES, MUDS, ADSOR W72-00847
ORGANISMS, REDUCTION(CHEMICAL), TROPHIC LEVELS, PROTOZOA, ALGAE, W72-10883
DMIUM RADIOISOTOPES, METABOLISM, TROPHIC LEVELS.: /DIOISOTOPES, CA W71-13243
OURCES, WATER POLLUTION CONTROL, TROPHIC LEVELS, PHOSPHATES, DETER W71-12072
ER QUALITY, NITROGEN, NUTRIENTS, TROPHIC LEVEL.: / PHOSPHATES, WAT W71-11033
ON, TOURISM, WATERSHEDS(BASINS), TROPHIC LEVELS, TURBIDITY, OXYGEN W71-11009
NTS, MESOTROPHY, EUTROPHICATION, TROPHIC LEVEL, PROTOZOANS, AQUATI W72-13816
, DARK FIXATION, COASTAL WATERS, TROPHOGENIC LAYER.: /F OF FINLAND W72-03228
SILOBIA PISCIFORMIS, PROWSEMONAS TROPICA, STIGMOBODO BRASILIANA, P W72-07683
HERMAL POLLUTION, TROPHIC LEVEL, TROPICAL REGIONS, ULTRAVIOLET RAD W73-00796
, ATLANTIC OCEAN, PACIFIC OCEAN, TROPICAL REGIONS, POLAR REGIONS, W72-03543
POPULATION, HAWAII, ABSORPTION, TROPICAL REGIONS, PACIFIC OCEAN, W71-08042
FOOD, GREEN RIVER(WASH), COWLITZ TROUT HATCHERY(WASH).: /RESIDUAL W71-11006
MENDOTA(WIS), BANTAM LAKE(CONN), TROUT LAKE(WIS), SEDIMENT COMPACT W72-08459
CUM, LAKE MENDOTA, CRYSTAL LAKE, TROUT LAKE, BIG ARBOR VITAE LAKE, W73-01456
RTIARY TREATMENT, ALGAE, PLANTS, TROUT, AQUATIC ENVIRONMENT, EUTRO W72-02412
LUTANTS, BIOASSAY, FISH, RAINBOW TROUT, AQUATIC LIFE, AQUATIC ENVI W72-12257
S, SALMO TRUTTA, VENDACE, SALMON TROUT, LOTA LOTA, LAKE MALOGEN, S W72-12729
ILS, EELS, CHINOOK SALMON, BASS, TROUT, MINNOWS, CRAYFISH, MOSQUIT W71-07417
FECTS, GAS CHROMATOGRAPHY, BROOK TROUT, SAMPLING, SEPARATION T: /F W72-12930
ER POLLUTION SOURCES, SEA WATER, TROUT, TOXINS.: /RIMP, CRABS, WAT W71-07412
S, PERCOLATION, HERRING, SALMON, TROUT, TOXICITY, WASHINGTON, MARI W72-05421
TILIS, SALVELINUS ALPINUS, SALMO TRUTTA, VENDACE, SALMON TROUT, LO W72-12729
INE, ARGININE, SERINE, TYROSINE, TRYPTOPHAN, HISTIDINE, ASPARTATE, W73-00942
ALIS, KOFOIDIELLA UNIFLAGELLATA, TSUMURAIA NUMEROSA, HIRCSEIA QUIN W72-07683
RBINES, *THERMAL POLLUTION, LONG TUBE VERTICAL DISTILLATION, EVAPO W70-09193
OMONAS, METABOLITES, HAEMATOCRIT TUBE.: /AL MAGNIFICATION, CHLAMYD W72-04134
YTOCHEMICAL SCREENINGS, NYMPHAEA TUBEROSA, NUPHAR VARIEGATUM, NUPH W72-04275
IA, COCCONEIS / *DICHOTOMOSIPHON TUBEROSUS, *EPIPHYTES, GROWTH MED W72-12744
CINUS, LIMNODRILUS HOFFMEISTERI, TUBIFEX TUBIFEX, APHANIZOMENON, C W71-11009
IMNODRILUS HOFFMEISTERI, TUBIFEX TUBIFEX, APHANIZOMENON, CLADOPHOR W71-11009
LORAMPHENICAL, LIMNODRILUS SPP., TUBIFEX, UNIALGAL CULTURES, OUABA W72-09668
CHEMICALS, MICHIGAN, CALIFORNIA, TUBIFICIDS, COLORADO, WATER POLLU W71-11006
C FORMATIONS, BENTHOS, MAYFLIES, TUBIFICIDS, DIAT: /PHYTA, GEOLOGI W71-11009
AMPLING, *PERIPHYTON, EQUIPMENT, TUBIFICIDS, MOLLUSKS, WATER TEMPE W73-00374
PES, BLOOMS, 2,4 DINITRO PHENOL, TUCSON(ARIZ),: /ALCIUM RADIOISOTO W71-01474
LLEY(CALIF), LOS ANGELES(CALIF), TUCSON(ARIZ).: /LING, IMPERIAL VA W71-09157
UNA, SAGARTIA, ANEMONE, MOLGULA, TUNICATE, PISMO CLAM, TIVELA STUL W72-07907
MESOFAUNA, TURBELLARIA, SPONGES, TUNICATES, BRYOZOANS, SEA TURTLES W72-07911
AM, SUCROSE, DENSITY GRADIENTS, TUNICATES, PHAEOCYSTIS, SORTING, W73-00371
IPUNCULIDS, BRYOZOA, INTOPROCTA, TUNICATES, SPONGES, VERTEBRATES, W73-00932
NATHS, CYPHONAUTES, ECHINODERMS, TUNICATES, NOCTILUCA, SIPHONOPHOR W73-00935
F), EPIBENTHIC SPECIES, ABALONE, TUNICATES, ARTHROPODS, COELENTERA W71-01475
LINE, CHLOROPHYLL A, PHEOPHYTIN, TUNICATES, FORAMINIFERA, RADIOLAR W72-14313
N, OIL, BRIDGES, BUOYS, HARBORS, TUNNELS, MARINE ALGAE, ANIMAL POP W73-02029
IETUM, FIDDLER CRABS, MESOFAUNA, TURBELLARIA, SPONGES, TUNICATES, W72-07911
MICAL OXYGEN DEMAND, FERTILIZER, TURBIDITY.: /C, ION EXCHANGE, CHE W71-12200
ERSHEDS(BASINS), TROPHIC LEVELS, TURBIDITY, OXYGEN, FISH, RECREATI W71-11009
N, ESTUARIES, WATER TEMPERATURE, TURBIDITY, PHOTOSYNTHESIS, HYDROG W71-11876
S, *FILTERS, HEADLOSS, PRESSURE, TURBIDITY, ALGAE, FLOCCULATION, C W72-09393
ON EFFECTS, AQUATIC ENVIRONMENT, TURBIDITY, WATER POLLUTION, PHAEO W72-11925
PLANKTON, ANALYTICAL TECHNIQUES, TURBIDITY, NITRATES, PHOSPHATES, W72-03224
ANALYSIS, EVAPORATION, SEEPAGE, TURBIDITY, HYDROLOGIC BUDGET, SOL W72-02274
LYSIS, LAKES, SAMPLING, DIATOMS, TURBIDITY, WATER POLLUTION SOURCE W72-07890
TRATION, RESPRIATION, NUTRIENTS, TURBIDITY, PRODUCTIVITY, PLANKTON W72-08559
NITY, WATER TEMPERATURE, OXYGEN, TURBIDITY, BENTHIC FAUNA, WATER P W73-02029
TION TECHNIQUES, CENTRIFUGATION, TURBIDITY, WATER TEMPERATURE, ALK W73-00916
ISSOLVED OXYGEN, PHOTOSYNTHESIS, TURBIDITY, LIGHT PENETRATION, IMP W73-00256
LOGICAL PROPERTIES, CHLOROPHYLL, TURBIDITY, SAMPLING, MATHEMATICAL W72-12941
E, *AQUATIC PRODUCTIVITY, LIGHT, TURBIDITY, NUTRIENTS, HETEROGENEI W72-13142
INVERTEBRATES, FISH, POLLUTANTS, TURBIDITY, ORGANIC LOADING, CORAL W71-01475
DE, DISTRIBUTION, NANNOPLANKTON, TURBIDITY, HYDROGEN ION CONCENTRA W71-00117
TS, PLANKTON, ARKANSAS, INSECTS, TURBIDITY, CHARA, PONDWEEDS.: /AN W71-04528
S, FARM WASTES, FISH, FISH EGGS, TURBIDITY, NUTRIENTS, DISSOLVED O W71-06444
ING, TOXICITY, ODOR, OXYGEN SAG, TURBIDITY, WATER QUALITY, WATER T W71-09958
NEBRASKA, WASTEWATER TREATMENT, TURBIDITY, RUNOFF, PHOSPHORUS, AL W72-04759
RFACE WATERS, *PLANKTON, *ALGAE, TURBIDITY, *NEW YORK, *WATER QUAL W72-06191
, SAFE YIELD, ALGAE, MINERALOGY, TURBIDITY, COLOR, ALKALINITY, VOL W72-05594
OCENTRUM MICANS, CERATIUM FURCA, TURBIDOSTAT CULTURES.: /SEA, PROR W71-03095
N DELTA(CALIF), *CONTINUOUS FLOW TURBIDOSTAT, *GROWTH KINETICS, CH W70-10403
CONTROL, *WATER COOLING, *STEAM TURBINES, *THERMAL POLLUTION, LON W70-09193
*HIGH-MOLECULAR-WEIGHT POLYMERS, TURBULANT FRICTIONAL DRAG, ALKYLE W70-06968
DS, CLIMATIC ZONES, TEMPERATURE, TURBULENCE, OXYGENATION, ALGAE, P W71-06033
S, *COMPUTER SIMULATION, MIXING, TURBULENCE, TEMPERATURE, STRATIFI W71-07084
N, MIXING, PHOSPHORUS, NITROGEN, TURBULENCE, PHOTOSYNTHESIS, DEPTH W72-08559
VEMENT, VIRUSES, PHOTOSYNTHESIS, TURBULENCE, LIGHT INTENSITY.: /MO W71-13184
XUDATES, EXTRACELLULAR MATERIAL, TURBULENT FRICTION, SEAWEED, PORP W71-07878
E BAY(FLORIDA), HEATED EFFLUENT, TURKEY POINT(FLORIDA), TURTLE GRA W72-11876
UCLEAR PLANT, CONNECTICUT RIVER, TURKEY POINT, BISCAYNE BAY, FLORI W72-07907
ASTES, INDUSTRIAL WASTES, ANNUAL TURNOVER, ENERGY, CYANOPHYTA, ALG W72-03228
ATOMS, INSECTS, BACTERIA, FUNGI, TURNOVERS.: / AQUATIC ANIMALS, DI W71-10064
E, INDUSTRIES, MUNICIPAL WASTES, TURNOVERS, SEDIMENTS, HUMAN POPUL W71-11009
L ANALYSIS, DIATOMS, CYANOPHYTA, TURNOVERS, EUTROPHICATION.: /MICA W73-00232
ULOTHRIX AEQUALIS, STEPHANOPYXIS TURRIS.: /UADRICAUDA, MOUGEOTIA, W72-07144
EFFLUENT, TURKEY POINT(FLORIDA), TURTLE GRASS, THALASSIA SPP., UDO W72-11876
ONGES, TUNICATES, BRYOZOANS, SEA TURTLES.: /FAUNA, TURBELLARIA, SP W72-07911

PLANTS, ON-SITE INVESTIGATIONS,	TURTLES.: /ES, BACTERIA, FLOATING	W71-01488
LGAL HYDROLYSATE, TONGA ISLANDS,	TUTUILA ISLANC, HONSHU ISLAND, OP	W71-08042
BRANE FILTERS, GRANGE LAKE, EAST	TWIN LAKE, LAKE ROCKWELL, LAKE HO	W72-13673
OHETEROTROPHIC GROWTH, LUMINARIN-	TYPE POLYSACCHARIDES, CRYPTOPHYCE	W70-07280
CIDES, PESTICIDES, ECOLOGY, BIRD	TYPES, BIRDS, AQUATIC ANIMALS, MO	W72-05952
IELD, *SIMULATED RAINFALL, *SOIL	TYPES, SOIL TEXTURE, CLAYS, LABOR	W72-02218
AD(INDIA), EICHHORNIA CRASSIPES,	TYPHA ANGUSTATA, IPOMOEA AQUATICA	W72-01363
ECTUS LACUSTRIS, ACORUS CALAMUS,	TYPHA ANGUSTIFOLIA.: /, SCHOENOPL	W73-00286
, ATRIPLEX HASTATA, SCIRPUS SPP,	TYPHA SPP, ZIZANIA AQUATICA, CARE	W73-01089
NOTTINGHAM(ENGLAND), POLYGONUM,	TYPHA, ELODEA.: /LOGICAL EFFECTS,	W72-00155
ANTIBODY TECHNIQUES, SALMONELLA	TYPHIMURIUM, AMPICILLIN, VIBRIOS,	W72-13800
IN, VIBRIOS, SALMONELLA TYPHOSA,	TYPHOID, COLIPHAGE, IMMUNOLOGY, P	W72-13800
AMPICILLIN, VIBRIOS, SALMONELLA	TYPHOSA, TYPHOID, COLIPHAGE, IMMU	W72-13800
MATE, LEUCINE, ARGININE, SERINE,	TYROSINE, TRYPTOPHAN, HISTIDINE,	W73-00942
OPHYCEAE, *DENMARK,/ LAKE ESROM,	TYSTRUP LAKE, BAVELSE LAKE, CRYPT	W72-12738
ODOR-PRODUCING ALGAE, SOUTHWEST	U.S., TEMPERATURE, OXYGEN, LIGHT	W72-10613
MENTOSUS, RHIPOCEPHALUS PHOENIX,	UDOTEA FLABELLUM, BRYOPSIS, CAULE	W72-14294
), TURTLE GRASS, THALASSIA SPP.,	UDOTEA SPP., PENICILLUS SPP., LAU	W72-11876
NEDESMUS QUADRICAUDA, MOUGEOTIA,	ULOTHRIX AEQUALIS, STEPHANOPYXIS	W72-07144
MORUM, SPHAEROCYSTIS SCHROETERI,	ULOTHRIX SPP, MICROTHAMNION KUETZ	W72-09111
NGROSIRA, MICROSPORA, SPIROGYRA,	ULOTHRIX.: /TETRAS PYRIFORMIS, GO	W71-05991
M/ *DACONIL, *CHLOROPHTHALONIL,	ULOTHRIX, ANABAENA, OSCILLATORIA,	W72-07508
LLUS THIOOXIDANS, DESULFOVIBRIO,	ULOTHRIZ ZONATA.: /CTER, THIOBACI	W71-01489
IUM TENUE, CLADOPHORA GLOMERATA,	ULOTRICHALES, ZYGNEMALES, OEDOGON	W71-05991
TA, VOLVOCINEAE, PROTOCOCCINEAE,	ULOTRICHINEAE, DESMIDIALES.: /PHY	W72-04855
*SALT VALLEY RESERVOIRS(NEB),	ULTRA PLANKTON, C-14.:	W72-04270
ANIC NITROGEN, CALCAREOUS LAKES,	ULTRA-VIOLET LABILE NITROGEN, ULT	W72-08048
S, ULTRA-VIOLET LABILE NITROGEN,	ULTRA-VIOLET REFRACTORY NITROGEN,	W72-08048
TROPHIC LEVEL, TROPICAL REGIONS,	ULTRAVIOLET RADIATION, FOOD CHAIN	W73-00796
WASTES, MONITORING, PHOTOGRAPHY,	ULTRAVIOLET RADIATION, FLUORESCEN	W72-04742
ION, CHLORINATION, DISINFECTION,	ULTRAVIOLET LIGHT, ALGAE, OXYGENA	W72-00042
IA DIGITATA, LAMINARIA AGARDHII,	ULVA LACTUCA, RHODYMENIS PALMATA,	W72-04748
RPHA INTESTINALIS, POLYSIPHONIA,	ULVA LACTUCA, ZANICHELLIA PALUSTR	W73-00367
A MAJUSCULA, GOMONTIA POLYRHIZA,	ULVA RIGIDA, CLADOPHORA FULIGINOS	W72-11800
RHODYMENIA, FUCUS,	ULVA.:	W72-09653
ON, SAMPLE PREPARATION, ZOSTREA,	ULVA.: /IVITY, LIQUID SCINTILLATI	W72-14317
NIA AQUATICA, CAREX SPP, BUTOMUS	UMBELLATUS, SPARGANIUM RAMOSUM, L	W73-01089
US, LAMINARIA DIGITATA, PORPHYRA	UMBILICALIS, SAMPLE PREPARATION,	W73-01074
EFFECTS, *RADIOACTIVITY EFFECTS,	UNDERGROUND, ALASKA, PONDS, WELLS	W72-03347
OIDOSA, STICHOCOCCUS BACILLARIS,	UNDISSOCIATED MOLECULES, ALGAL GR	W70-07255
ICAL, LIMNODRILUS SPP., TUBIFEX,	UNIALGAL CULTURES, OUABAIN, BIOLO	W72-09668
SINAMONAS STAGNALIS, KOFOIDIELLA	UNIFLAGELLATA, TSUMURAIA NUMEROSA	W72-07683
ALOPHILA STIPULACEA, DIPLANTHERA	UNINERVIS, CASSIOPEA ANDROMEDA, A	W73-00855
OXIDE POLYMERS, SEWAGE SETTLING,	UNION CARBIDE CORP, POLYOX, CARBO	W70-06968
ROBIC STREAMS, FINGERNAIL CLAMS,	UNIONIDS, ORCONECTES RUSTICUS, LI	W72-06526
.: PERFUSION	UNIT, FLAVOBACTERIUM, MICROCOCCUS	W71-08027
PHOSPHATES, *FEDERAL GOVERNMENT,	UNITED STATES, ALGAL CONTROL, STA	W72-01093
YETTEVILLE GREEN LAKE(NEW YORK),	UPPER KLAMATH LAKE(OREGON), LAKE	W71-05626
LIZATION.:	UPPER KLAMATH LAKE, SUBSTRATE UTI	W73-01269
	UPPER KLAMATH LAKE(OREGON).:	W72-08989
RADIONUCLIDE	UPTAKE.:	W72-00960
RADIONUCLIDE	UPTAKE.:	W72-03347
CONCENTRATION,	UPTAKE.:	W72-10686
NON-BIOLOGICAL	UPTAKE.:	W72-12210
HOSPHATES, DRY WEIGHT, PHOSPHATE	UPTAKE.: /E, TRANSPARENCY, ORTHOP	W71-10645
MIXED LIQUOR, SUSPENDED SOLIDS,	UPTAKE.: /L-BIOLOGICAL TREATMENT,	W71-13334
ROCLADIUS, WHITE OAK LAKE(TENN),	UPTAKE.: / PHYSA HETEROSTROPHA, P	W70-06668
OOMS, 2,4 DINITRO P/ *BIOLOGICAL	UPTAKE, CALCIUM RADIOISOTOPES, BL	W71-01474
AL DECO/ *ALGAL GROWTH, NUTRIENT	UPTAKE, CHEMICAL COMPOSITION, ALG	W70-05469
), BLELHAM TARN(ENGLAND), LUXURY	UPTAKE, LOCH LEVEN(SCOTLAND), LAK	W71-13183
IOBACILLUS, SULFOLOBUS, GLUCOSE,	URACIL.: /CYANIDIUM CALDARIUM, TH	W72-14301
LANT DOSAGES, BAROCO CLAY SOILS,	URANIUM ORE.: /, CARBOWAX, FLOCCU	W70-06968
OISOTOPES, COBALT RADIOISOTOPES,	URANIUM RADIOISOTOPES, RADIUM RAI	W71-09188
OKLAHOMA, *IODINE RADIOISOTOPES,	URANIUM RADIOISOTOPES, ANALYTICAL	W72-02073
NUTRIENTS/ *WILLIAMS LAKE(WASH),	URANIUM, THORIUM, ANTIMONY, MACRO	W73-01434
*URBAN RUNOFF,	URBAN HYDROLOGY, LINCOLN(NEBR).:	W71-13466
UTION CONTROL, WISCONSIN, LAKES,	URBAN RUNOFF, AGRICULTURAL RUNOFF	W73-02479
HOSPHATES, *NITRATES, *NEBRASKA,	URBANIZATION, CITIES, STORM RUNOF	W71-13466
IMENTS, DENITRIFICATION, RUNOFF,	UREAS, FERTILIZERS, DRAINAGE WATE	W71-06435
STS, AMMONIA, ASSAY, INHIBITION,	UREAS, SYNTHESIS, ALGAE.: /ORY TE	W72-03222
TION, BIOLOGICAL SAMPLES, LIVER,	URINE, OEDOGONIUM, PHYSA, CULEX,	W72-07703
UTION SOURCES, COLORADO, CATTLE,	URINE, PATH OF POLLUTANTS, FARM W	W71-11496
S EROSA, ERKINIA SUBEQUICILIATA,	UROGLENOPSIS AMERICANA, CRYPTOMON	W72-13816
PODS, EUPHAUSIIDS, GADUS MORHUA,	UROPHYCIS.: /EURA, SAGITTA, PTERO	W73-00371
UCOR, MAJURO ATOLL, M/ PLEOPODS,	UROPODS, ENRICHMENT, LEUCOTHRIX M	W72-08142
WATE/ *HEATED WATER, *BENEFICIAL	USE, *COOLING WATER, *FISHERIES,	W70-09596
ADMINISTRATION, NON-CONSUMPTIVE	USE, ALGAE, EVAPORATION, REASONAB	W70-07100
WASTE WATER TREATMENT, *WATER RE-	USE, BIOCHEMICAL OXYGEN DEMAND, O	W71-07123
, ALGAE, EVAPORATION, REASONABLE	USE, BRACKISH WATER, D: /TIVE USE	W70-07100
MATICAL MODELS, MIXING, WATER RE-	USE, LIGHT INTENSITY, EVAPORATION	W71-07099
ATION, LEAVES, LEGISLATION, LAND	USE, SHORES, MONITORING, MARSHES,	W73-02479
C POWER, NAVIGATION, CONSUMPTIVE	USE, WATER POLLUTION, OXYGEN, SOI	W72-14878
ES, COST COMPARISONS, BENEFICIAL	USE, WATER QUALITY, CALIFORNIA, B	W72-00974
D COOLING, *SPACE HEATING, *FARM	USES OF WASTE HEAT, STEAM POWERED	W70-09193
DWATER, WATER QUALITY, COMPETING	USES, DIVERSION, RECIRCULATED WAT	W70-07100
	USSR.:	W71-09163
	USSR.:	W72-01358
KALINE HYDROLYSIS, ALGAE BLOOMS,	UTAH, GREAT SALT LAKE, ORGANIC CO	W71-12223
TES, RHODOMONAS, IKROAVICH LAKE,	UTERMOHL METHOD.: /TION, FLAGELLA	W73-00273
RESERVOIR, / SOUTH TAHOE PUBLIC	UTILITIES DISTRICT, *INDIAN CREEK	W72-02412

SOURCES, ORGANIC CARBON, CARBON	UTILIZATION.: /SFORMATION, CARBON	W71-12068
ANUP, RADIOAUTOGRAPHY, SUBSTRATE	UTILIZATION.: /HROMATOGRAPHY, CLE	W73-00361
G/ FATE OF POLLUTANTS, SUBSTRATE	UTILIZATION, ORGANIC NITROGEN, OR	W73-00379
LITHOPHORIDS, FLAGEL/ *SUBSTRATE	UTILIZATION, *SARGASSO SEA, COCCO	W73-00380
CULTURES, METABOLITES, SUBSTRATE	UTILIZATION, GROWTH.: /ES, MIXED	W73-00441
BON, ORGANIC NITROGEN, SUBSTRATE	UTILIZATION, PARTICULATE MATTER,	W73-01066
TATS, ORTHOPHOSPHATES, SUBSTRATE	UTILIZATION, MONOD EQUATION, PENI	W73-01445
UPPER KLAMATH LAKE, SUBSTRATE	UTILIZATION.:	W73-01269
ALIFORNIA, *WATER RIGHTS, *WATER	UTILIZATION, *IMPAIRED WATER QUAL	W70-07100
MACHINERY, INDUSTRIAL MACHINERY,	UTILIZATION, MODIFICATION.: /ARM	W71-06188
IS, BOSMINA LONGIROSTRIS, SYNURA	UVELLA, DINOBRYON, OSCILLATORIA L	W71-03014
, AMPHORA OVALIS, AMPHORA OVALIS	V. PEDICULUS, COCCONEIS PEDICULUS	W72-13142
NODAMASTIX SPIROGYRAE, COLEMANIA	V: /NA, PAVLOVIAMONAS FRUTICOSA,	W72-07683
*SOUTH RIVER(	VA).:	W73-01972
RATION, SMITH MOUNTAIN RESERVOIR(	VA).: *PUMPED STORAGE OPE	W72-14405
WESTHAMPTON LAKE (	VA).:	W72-00150
WAUBESA(WIS), OCCOQUAN RESERVOIR(	VA), MADISON(WIS), LAKE MENDOTA(W	W72-05473
SMITH MOUNTAIN PROJECT(	VA), WEATHER EFFECT.:	W72-05282
N, METAL COMPLEXES, PICEA ABIES,	VACCINIUM VITIS IDAEA, VACCINIUM	W72-14303
EA ABIES, VACCINIUM VITIS IDAEA,	VACCINIUM MYRTILLUS, DESCHAMPSIA	W72-14303
ICROCYSTIS A/ *SHOCK-WAVES, *GAS-	VACUOLE DEFLATION, SOUTH WALES, M	W72-11564
ES.: *GAS	VACUOLES, MOTILE FORMS, CYANOPHAG	W71-13184
ION, FILTRATION, CENTRIFUGATION,	VACUUM DRYING, FLOTATION, WASTEWA	W72-10237
IA MONTICOLA, PORIA COCOS, PORIA	VAILLANTII, CONIOPHORA CLIVACEA,	W72-13813
NYLALANINE, ALANINE, ISOLEUCINE,	VALINE, LYSINE, ASPARAGINE, CYSTI	W73-00942
SHELTER	VALLEY CREEK(ONTARIO).:	W73-01627
HIORHODACEAE, WHIPPLE DAM, STONE	VALLEY DAM(PA.), ATRAZINE, BIOTYP	W72-04743
NKTON, C-14.: *SALT	VALLEY RESERVOIRS(NEB), ULTRA PLA	W72-04270
IS), LAKE WASHINGTON(WASH)/ SALT	VALLEY RESERVOIRS(NEB), MADISON(W	W72-04759
*RUHR	VALLEY.:	W71-12410
RUBBLE CHIMNEY, SAN JOAQUIN	VALLEY.:	W72-00974
IAL DENITRIFICATION, SAN JOAQUIN	VALLEY.: /ATION, METHANOL, BACTER	W71-08223
SODIUM PYROPHOSPHATE, RIO GRANDE	VALLEY.: /TH, *RADIOISOTOPE P32,	W72-13992
NERGY, REACTOR COOLING, IMPERIAL	VALLEY(CALIF), LOS ANGELES(CALIF)	W71-09157
, ALGAL HARVESTING, *SAN JOAQUIN	VALLEY(CALIF).: /US, ALGAL GROWTH	W72-02975
A CRENULATA, VALONIA VENTRICOSA,	VALONIA OCELLATA, ERNODESMIS VERT	W72-11800
ERSTEDI, ACETABULARIA CRENULATA,	VALONIA VENTRICOSA, VALONIA OCELL	W72-11800
ERAMIUM NITENS, WRANGELIA ARGUS,	VALONIA VENTRICOSA, BRYOPSIS PENN	W73-00838
URE, *LITERATURE REVIEW, FEEDING	VALUE.: *ANIMAL MAN	W71-07544
*PHARMACOLOGICAL / *THERAPEUTIC	VALUE, *APHANIZOMENON FLOS-AQUAE,	W72-04259
GLACIAL WATER, BECKMAN ANALYZER,	VAN DORN SAMPLERS, C-14, BICCARBO	W72-09122
M, SENECA RIVER, NINEMILE CREEK,	VAN DORN SAMPLER, KEMMERER SAMPLE	W72-09111
	VANADIUM, LANTHANUM, SCANDIUM.:	W72-01101
TION, METAL COMPLEXES, PICEA AB/	VANADIUM, BRYOPHYTES, BIOACCUMULA	W72-14303
MIS VERTICILLATA, DICTYOSPHAERIA	VANBOSSEAE, DICTYOSPHAERIA CAVERN	W72-11800
ATER QUA/ *VANCOUVER LAKE(WASH),	VANCOUVER(WASH), PORTLAND(ORE), W	W72-03215
NEMA NITZSCHIOIDES, PLAGIOGRAMMA	VANHEURCKII, PLEUROSIGMA, GYROSIG	W73-00853
ORELLA FUSCA, *SCENEDEMUS ACUTUS	VAR ALTERNANS, *NITZSCHIA ACTINAS	W72-12739
M, DESMIDS, STAURASTRUM LONGIPES	VAR CONTRACTUM, ARTHURS LAKE, LAK	W73-01094
AS ARMATA, TRACHELOMONAS HISPIDA	VAR CRENULATOCOLLIS.: /RACHELOMON	W73-00918
ROAMMETER, OSCILLATORIA AGARDHII	VAR ISOTHRIX, LAKE ITASCA, LAKE E	W72-07145
, PLENOSP/ PORPHYROPSIS COCCINEA	VAR. DAWSONII, CALLOPHYLLIS FIRMA	W73-02099
COMOSUM, CHAETOCEROS ATLANTICUS	VAR. NEAPOLITANA, CHAETOCEROS COA	W71-11527
PARADOXUM, STAURASTRUM SEBALDII	VAR. ORNATUM, MICRASTERIAS CRUX-M	W72-12736
GROWTH INH/ *SELENIUM, *ANABAENA	VARIABILIS, *ANACYSTIS NIDULANS,	W72-00153
ASTS, BLUE-GREEN ALGAE, ANABAENA	VARIABILIS, ANACYSTIS NIDULANS.: /	W71-08683
TA PROCESSING, DATA COLLECTIONS,	VARIABILITY, PROBABILITY, WATER P	W71-07337
YNCHRONOUS CULTURES, AUTOSPORES,	VARIABILITY, BIAS, SYNCHRONY INDE	W72-11920
NITROGEN, *REGRESSION ANALYSIS,	VARIABILITY.: /RATES, PHOSPHORUS,	W72-12966
CLOSTERIUM MONILIFERUM, MELOSIRA	VARIANS, NAVICULA TRIPUNCTATA, GO	W72-10883
*WATER RECLAMATION,	VARIATIONS, COMPOSITION.:	W70-07838
NINGS, NYMPHAEA TUBEROSA, NUPHAR	VARIEGATUM, NUPHAR JAPONICUM, NUP	W72-04275
CTOSA, CANDIDA ZEYLANOIDES, CAMP	VARNUM BEACH.: /A, RHODOTORULA LA	W72-04748
, ORTHOPHOSP/ *SALT RIVER(KENT),	VASCULAR PLANTS, NITRATE NITROGEN	W72-06526
GOMPHONEMA LONGICEPS, CYMBELLA,	VAUCHERIA.: /CENTULA, CLOSTERIUM,	W72-12744
ORA, APHANOCHAETE, RHIZOCLONIUM,	VAUCHERIA, TRIBONEMA, OSCILLATORI	W71-05629
TY, *ESTUARIES, *DELAWARE RIVER,	VEGETATION EFFECTS, SLUDGE, RIVER	W70-09189
*INFRARED PHOTOGRAPHY, *AQUATIC	VEGETATION.:	W71-12034
LADOCERA, LINGULA, ACTINOTROCHA,	VELIGER, PTEROPODS, HETEROPODS, O	W73-00935
DITY, NITRATES, PHOSPHATES, WIND	VELOCITY, RAINFALL.: /QUES, TURBI	W72-03224
, CHELATION, SEDIMENTS, SETTLING	VELOCITY, RADIOISOTOPES, ABSORPTI	W72-04463
S, SEDIMENTATION, BUOYANCY, WIND	VELOCITY, CONVECTION, WATTER POLL	W71-09852
ALVELINUS ALPINUS, SALMO TRUTTA,	VENDACE, SALMON TROUT, LOTA LOTA,	W72-12729
ACETABULARIA CRENULATA, VALONIA	VENTRICOSA, VALONIA OCELLATA, ERN	W72-11800
NITENS, WRANGELIA ARGUS, VALONIA	VENTRICOSA, BRYOPSIS PENNATA, SPY	W73-00838
LONA, ARMANDIA, TELLINA BUTTONI,	VENUS, TIVELA STULTORUM, EMERITA	W72-09092
LANS, ELACHISTA LUBRICA, RALFSIA	VERRUCOSA.: /LIS, HECATONEMA MACU	W72-01370
CLADOSPORIUM HERBARUM, TRAMETES	VERSICOLOR, FOMES ANNOSUS, FUSCOP	W72-13813
SAGE, CALCIUM ION CONCENTRATION,	VERTEBRATES, ABSORBANCE, MAGNESIU	W72-09092
INTOPROCTA, TUNICATES, SPONGES,	VERTEBRATES, MACROINVERTEBRATES,	W73-00932
STOLOGICAL STUDIES, BODY FLUIDS,	VERTEBRATES, HEAT STERILIZATION,	W73-01676
FLUCTUATIONS, CONTINENTAL SHELF,	VERTICAL DISTRIBUTION, NORTH CARO	W72-07663
SSES, TELEVISION, DEPTH SOUNDER,	VERTICAL DISTRIBUTION, COLD BAY,	W72-09120
S, *THERMAL POLLUTION, LONG TUBE	VERTICAL DISTILLATION, EVAPORATOR	W70-09193
SITY STRATIFICATION, SPECIATION,	VERTICAL MIGRATION, AQUARIA, GROW	W71-09852
SA, VALONIA OCELLATA, ERNODESMIS	VERTICILLATA, DICTYOSPHAERIA VANB	W72-11800
, HYPNEA MUSCIFORMIS, GRACILARIA	VERUCOSA, ECKLONIA RADIATA, GROWT	W72-07496
BRANCHING./ *FUCUS VESICULOSUS,	VESICULATION, SAMPLE PREPARATION,	W73-00369
PREPARATION, BRANCHING./ *FUCUS	VESICULOSUS, VESICULATION, SAMPLE	W73-00369
PORPHYRA UMBILICALIS, SAM/ FUCUS	VESICULOSUS, LAMINARIA DIGITATA,	W73-01074

DS, COLUMN CHROMATOGRAPHY, FUCUS	VESICULOSUS, LAMINARIA DIGITATA,	W72-12166
SERRATUS, FUCUS SPIRALIS, FUCUS	VESICULOSUS, LAMINARIA DIGITATA,	W71-11515
S DUEBENI, FUCUS SERRATUS, FUCUS	VESICULOSUS, CLADOPHORA GLOMERATA	W71-08026
TANS, ASCOPHYLLUM NODOSUM, FUCUS	VESICULOSUS, LAMINARIA DIGITATA,	W72-04748
DEMA ANTIL/ EPIPHYTES, CULTURING	VESSELS, PENICILLIN, LINDANE, DIA	W72-14294
LTURES, AGARS, / HYPHOMICROBIUM,	VIBRIO ANGUILLARUM, ENRICHMENT CU	W72-05421
ACCILARIOPHYCEAE, XANTHOPHYCEAE,	VIBRIO, ACHROMOBACTER, FLAVOBACTE	W72-08051
L GROWTH, MICROCOCCUS, BACILLUS,	VIBRIO, RHODOTORULA, AEROMONAS.: /	W72-10861
MONELLA TYPHIMURIUM, AMPICILLIN,	VIBRIOS, SALMONELLA TYPHOSA, TYPH	W72-13800
PERGILLUS, PENICILLIUM, FUSARIA,	VIBRIS.: /NS, TORULOPSIS SPP., AS	W72-09675
Y, LASERS, MICROWAVE RADIOMETRY,	VIDEO, SCANNERS.: /, LEAD, MERCUR	W72-04742
NOSE, GALACTOSE, HEXURONIC ACID,	VINEYARD SOUND, ORGANIC CARBON, A	W72-07710
ITROGEN, CALCAREOUS LAKES, ULTRA-	VIOLET LABILE NITROGEN, ULTRA-VIO	W72-08048
RA-VIOLET LABILE NITROGEN, ULTRA-	VIOLET REFRACTORY NITROGEN, NETPL	W72-08048
, CRUSTACEANS, LABORATORY TESTS,	VIRGINIA.: /CHRYSOPHYTA, COPEPODS	W72-00150
, ECOSYSTEMS, SPECTROPHOTOMETRY,	VIRGINIA, SAMPLING, SEPARATION TE	W73-00916
ONTROL, *WASTE WATER(POLLUTION),	VIRGINIA, NUTRIENTS, NITROGEN, PH	W72-05473
, GONYAULAX POLYEDRA, PLATYMONAS	VIRIDIS, CHAETOCEROS CURVISETUS,	W72-07166
RRANEAN SEA.: *FUCUS	VIRSOIDES, FUCUS SPIRALIS, MEDITE	W72-12940
*PLECTONEMA BORYANIUM, SU/ *LPP	VIRUS, *BLUE-GREEN ALGAL VIRUSES,	W72-04736
U/ *LPP VIRUS, *BLUE-GREEN ALGAL	VIRUSES, *PLECTONEMA BORYANIUM, S	W72-04736
IA, DISEASES, E. COLI, SHIGELLA,	VIRUSES, CHROMATOGRAPHY, GAS CHRO	W72-13800
ER, ACTIVATED CARBON, COLIFORMS,	VIRUSES, CRUSTACEANS, NEMATODES,	W73-02144
S PESTICIDES, INDUSTRIAL WASTES,	VIRUSES, ENERIC BACTERIA, TRITIUM	W72-04742
TEMPERATURE, ENZYMES, CELLULOSE,	VIRUSES, HUMAN PATHOLOGY, SYSTEMA	W72-05457
GEN, ALGAE, NUTRIENTS, BACTERIA,	VIRUSES, HEAVY METALS, HUMAN POPU	W71-09788
UNGI, ENZYMES, ANIMAL PARASITES,	VIRUSES, MOLLUSKS, PROTOZOA, YEAS	W73-01676
ROL, PLANT PHYSIOLOGY, MOVEMENT,	VIRUSES, PHOTOSYNTHESIS, TURBULEN	W71-13184
PRA, SCIRPUS SPP, GILLS, MUSCLE,	VISCERA, EYES, SCALES(FI: /RA ASP	W73-01673
COMMUTATA, PLEUROCHLORIS MAGNA,	VISCHERIA PUNCTATA, VISCHERIA STE	W72-14808
LORIS MAGNA, VISCHERIA PUNCTATA,	VISCHERIA STELLATA, ELLIPSOIDION	W72-14808
LATION, COAGULATION, FILTRATION,	VISCOSITY.: /SLUDGE, FLOW, FLOCCU	W70-06968
ALGAE, COOLING TOWERS, DENSITY,	VISCOSITY, FISH BEHAVIOR, FERTILI	W70-09612
G ARBOR VITAE LAKE, LITTLE ARBOR	VITAE LAKE, GREEN BAY, FOX RIVER,	W73-01456
STAL LAKE, TROUT LAKE, BIG ARBOR	VITAE LAKE, LITTLE ARBOR VITAE LA	W73-01456
ANIC NITROGEN, THIAMINE, BIOTIN,	VITAMIN B 12, GONYAULAX POLYEDRA,	W72-07892
N, CHLOROPHYLL, SOLAR RADIATION,	VITAMIN B, PIGMENTS, GROWTH RATES	W72-07892
PHOSPHORUS, PHOSPHATES, CARBON,	VITAMIN B, NITROGEN, CYANOPHYTA,	W72-07937
UTRIENTS, PSEUDOMONAS, CULTURES,	VITAMIN B, BIOASSAY, CELLULOSE, D	W73-00942
ARATION TECHNIQUES, CULTIVATION,	VITAMIN B, PLANT GROWTH REGULATOR	W73-00430
ORGANIC COMPOUNDS, AMINO ACIDS,	VITAMINS, CARBOHYDRATES, ALCOHOLS	W73-00942
S, METABOLISM, ALGAE, SYNTHESIS,	VITAMINS, SULFATES, IONS.: /PHATE	W73-01626
OPHYTA, CYANOPHYTA, SYSTEMATICS,	VITAMINS, ANAEROBIC BACTERIA, ELE	W72-07951
LUMBIA RIVER, ORGANIC COMPOUNDS,	VITAMINS, YEASTS, SULFUR, FOULING	W72-05421
RMS, ENTERIC BACTERIA, NITROGEN,	VITAMINS, SUSPENDED SOLIDS, LACTO	W72-12565
OMPLEXES, PICEA ABIES, VACCINIUM	VITIS IDAEA, VACCINIUM MYRTILLUS,	W72-14303
YARD SOUND, ORGANIC CARBON, ACID-	VOLATILE COMPOUNDS, WOODS HOLE, H	W72-07710
BACKWASHING.: FIREBAUGH(CALIF),	VOLATILE SOLIDS, SANBORN FILTER,	W70-10174
PLANKTON, ALGAE, TRACE ELEMENTS,	VOLATILITY, FREEZE DRYING, X-RAYS	W71-11036
OPES, SEDIMENTS, TRACE ELEMENTS,	VOLCANOES, BENTHIC FLORA, ALKALIN	W71-09190
ECTIVE ELECTRODES, LASERS, ORTH/	VOLTAMMETRY, RADIOGRAPHY, ION SEL	W72-08790
SITY, CULTURE THICKNESS, CULTURE	VOLUME.: /PRODUCTION, LIGHT INTEN	W70-05547
NDITIONS, ALGAE, ORGANIC MATTER,	VOLUME, DRYING, ODOR, NITROGEN, A	W71-07551
THETIC OXYGEN, *ALGAE, CULTURES,	VOLUME, MATHEMATICAL STUDIES, COR	W70-05547
INTERSTATE RIVERS, HYPOLIMNION,	VOLUME, MIXING, SAFETY, WATER COO	W70-09595
CTRODES, IONS, FLUORIDES, ALGAE,	VOLUMETRIC ANALYSIS, WATER CHEMIS	W71-00474
Y, TURBIDITY, COLOR, ALKALINITY,	VOLUMETRIC ANALYSIS, FISH, INTAKE	W72-05594
ATION, IRON, BIOASSAY, LEACHING,	VOLUMETRIC ANALYSIS, SOIL ANALYSI	W72-05965
OR, CHEMICAL ANALYSIS, SAMPLING,	VOLUMETRIC ANALYSIS, ALGAE, RESER	W72-14280
ETEROINHIBITORS, AUTOINHIBITORS,	VOLVOCACEAE, PANDORINA MORUM, PLA	W71-13793
Y, METALIMNION, BACILLARIOPHYTA,	VOLVOCINEAE, PROTOCOCCINEAE, ULOT	W72-04855
, SPIROG/ *INORGANIC IONS, ZINC,	VOLVOX, HYDRODICTYON, SPODYLOSIUM	W71-11001
DOMONAS REINHARDTII, TETRAHYMENA	VORAX, PSEUDOMONAS FLUORESCENS, A	W73-02095
CHLORELLA	VULGARIS.:	W72-14800
NKISTRODESMUS BRAUNII, CHLORELLA	VULGARIS.: /ION, CHLOROSARCINA, A	W71-07675
TILLATION COUN/ C-14, *CHLORELLA	VULGARIS, NEUROSPORA CRASSA, SCIN	W72-14326
ENTEROBACTER AEROGENES, PROTEUS	VULGARIS, PSEUDOMONAS AERUGINOSA,	W72-07664
CHLORELLA PYRENOIDOSA, CHLORELLA	VULGARIS, SCENEDESMUS OBLIQUUS, A	W72-07664
NUCLEIC ACID.: CHLORELLA	VULGARIS, NITRATE REDUCTASE, RIBO	W72-03222
METHYL ESTERS, ESTERS, CHLORELLA	VULGARIS, SAMPLE PREPARATION.: /	W72-09365
OPHYTES.: *LAKE	WABAMUN, BIOLOGICAL STUDIES, MACR	W73-01704
G NUTRIENTS, SHAGAWA LAKE(MINN),	WALDO LAKE(ORE).: *LIMITIN	W72-09165
ED ORGANIC CARBON, *MENAI STRAIT(	WALES).: *DISSOLV	W72-09108
S, *GAS-VACUOLE DEFLATION, SOUTH	WALES, MICROCYSTIS AERUGINOSA, AN	W72-11564
CILLIUM, STREPTOMYCIN, EPIFLORA,	WALLEYE LARVAE, SHAD, MACROINVERT	W72-08142
CYLINDRICA, ACETYLENE REDUCTION,	WARBURG EFFECT.: /ION, *ANABAENA	W72-08584
CALVIN CYCLE, ANACYSTIS NIDULAN/	WARBURG EFFECT, EUGLENA GRACLIS,	W71-09172
ITIONS, ALGAL CONTROL, SPAWNING,	WARM-WATER FISH.: /ANAEROBIC COND	W70-09595
ING, DOMESTIC WASTES, INDUSTRIAL	WAS: /N DIOXIDE, ALKALINITY, BURN	W72-05473
*MOSES LAKE(	WASH).:	W72-08560
*SNAKE RIVER(	WASH).:	W72-14728
TING NUTRIENTS, *LAKE WASHINGTON(	WASH).: *LIMI	W72-09167
YPEREUTROPHIC LAKES, *MOSES LAKE(	WASH).: H	W72-00798
SPHORUS REMOVAL, LAKE WASHINGTON(	WASH).: MERCURY, PHO	W71-12091
SION COEFFICIENT, LAKE SAMMAMISH(	WASH).: /NG, EUPHOTIC ZONE, DIFFU	W71-03034
ER(WASH), COWLITZ TROUT HATCHERY(	WASH).: /RESIDUAL FOOD, GREEN RIV	W71-11006
AKE WASHINGTON(WASH), GREEN LAKE(	WASH).: /S), LAKE KEGONSA(WIS), L	W72-05473
STES, RESIDUAL FOOD, GREEN RIVER(	WASH), COWLITZ TROUT HATCHERY(WAS	W71-11006
), MADISON(WIS), LAKE WASHINGTON(	WASH), DYEING WATER.: /RVOIRS(NEB	W72-04759
KE KEGONSA(WIS), LAKE WASHINGTON(	WASH), GREEN LAKE(WASH).: /S), LA	W72-05473

*MOSES LAKE(WASH), HYPEREUTROPHIC LAKES.: W72-00799
G NUTRIENTS, ALGAL / *MOSES LAKE(WASH), NUTRIENT DILUTION, LIMITIN W72-08049
*VANCOUVER LAKE(WASH), VANCOUVER(WASH), PORTLAND(ORE), WATER QUALI W72-03215
, MACRONUTRIENTS/ *WILLIAMS LAKE(WASH), URANIUM, THORIUM, ANTIMONY W73-01434
ORE), WATER QUA/ *VANCOUVER LAKE(WASH), VANCOUVER(WASH), PORTLAND(W72-03215
ICAL TECHNIQUES, COLUMBIA RIVER, WASHINGTON, OREGON.: /NTS, ANALYT W72-03215
E ALGAE, *EUTROPHICATION, LAKES, WASHINGTON, *ALGAL CONTROL, NUTRI W72-00799
HYTA, *NUISANCE ALGAE, PLANKTON, WASHINGTON, DIATOMS, CHLOROPHYTA. W72-00798
LORADO, WATER POLLUTION CONTROL, WASHINGTON, CHEMICAL OXYGEN DEMAN W71-11006
ERCURY, PHOSPHORUS REMOVAL, LAKE WASHINGTON(WASH).: M W71-12091
), POTOMAC RIVER, COLUMBIA RIVER(WASHINGTON), COCKLES, GOLDFISH, M W71-11517
*LIMITING NUTRIENTS, *LAKE WASHINGTON(WASH).: W72-09167
ON, *NUTRIENTS, *NUISANCE ALGAE, WASHINGTON, PHYTOPLANKTON, CYANOP W72-08049
EFFECTS, MARINE PLANTS, CANADA, WASHINGTON, OREGON, CALIFORNIA, A W73-00428
NA(WIS), LAKE KEGONSA(WIS), LAKE WASHINGTON(WASH), GREEN LAKE(WASH W72-05473
OURCES, CHLOROPHYTA, CYANOPHYTA, WASHINGTON, DIVERSION, LAKE ERIE, W72-05473
ERVOIRS(NEB), MADISON(WIS), LAKE WASHINGTON(WASH), DYEING WATER.: / W72-04759
ERRING, SALMON, TROUT, TOXICITY, WASHINGTON, MARINE ALGAE, MARINE W72-05421
ICATE, SODIUM METASILICATE, LAKE WASHINGTON, NITRILOTRIACETATE, HU W72-04734
LAKE WASHINGTON(WASHINGTON).: W71-02681
DIOXIDE, ALKALINITY, EPILIMNION, WASHINGTON, PHOSPHORUS, NITROGEN, W71-02681
LAKE WASHINGTON(WASHINGTON).: W71-02681
ER POLLUTION SOURCES, MINNESOTA, WASHINGTON, CONNECTICUT.: /R, WAT W71-04216
ER, BLACK RIVER(MINNESOTA), LAKE WASHINGTON(WASHINGTON), LINSLEY P W71-04216
IVER(MINNESOTA), LAKE WASHINGTON(WASHINGTON), LINSLEY POND(CONNECT W71-04216
AMS, COLUMBIA RIVER, CALIFORNIA, WASHINGTON, CONNECTICUT, FLORIDA, W71-07417
TA, *ORGANOPHOSPHORUS COMPOUNDS, WASHINGTON, SEPARATION TECHNIQUES W72-12637
TII, MOSES LAKE, PINE LAKE, LAKE WASHINGTON, CHLOROPHYLL A, *DNA, W72-12637
, LIMITING NUTRIENTS, ALGAL CELL WASHOUT.: /SH), NUTRIENT DILUTION W72-08049
TION, SEDIMENTS, WASTE DILUTION, WASTE ASSIMILATIVE CAPACITY, CALI W73-00831
PTION, SORPTION, WASTE DILUTION, WASTE ASSIMILATIVE CAPACITY, EURO W73-00817
DIOISOTOPES, PATH OF POLLUTANTS, WASTE ASSIMILATIVE CAPACITY, *CAL W73-00826
RIA, DECOMPOSING ORGANIC MATTER, WASTE ASSIMILATIVE CAPACITY, EUTR W73-00802
DIOECOLOGY, MATHEMATICAL MODELS, WASTE ASSIMILATIVE CAPACITY, ESTU W73-00828
HEALTH, ON-SITE INVESTIGATIONS, WASTE ASSIMILATIVE CAPACITY, WAST W73-00790
OXYGEN REQUIREMENTS, OXYGEN SAG, WASTE ASSIMILATIVE CAPACITY, WATE W72-04279
GAE, NUTRIENTS, WATER RESOURCES, WASTE ASSIMILATIVE CAPACITY, RESE W71-13256
GING(BIOLOGICAL), AQUATIC ALGAE, WASTE ASSIMILATIVE CAPACITY, WATE W72-01685
G, REFEEDING, ANHYDROUS AMMONIA, WASTE CHARACTERIZATION.: /MPOSTIN W71-07551
NS, WASTE ASSIMILATIVE CAPACITY, WASTE DIL: / ON-SITE INVESTIGATIO W73-00790
NE ALGAE, ABSORPTION, SEDIMENTS, WASTE DILUTION, WASTE ASSIMILATIV W73-00831
L EFFECTS, ABSORPTION, SORPTION, WASTE DILUTION, WASTE ASSIMILATIV W73-00817
EAVY METALS, HERBICIDES, TOXINS, WASTE DILUTION, NITROGEN, PHOSPHO W72-01472
ODEGRADATION, LOCAL GOVERNMENTS, WASTE DISPOSAL, WATER POLLUTION S W72-01093
, MUNICIPAL WASTES, FARM WASTES, WASTE DISPOSAL, PESTICIDE REMOVAL W72-01659
REGULATION, POLLUTION ABATEMENT, WASTE DISPOSAL, STANDARDS, PUBLIC W72-01085
SALINITY, PESTICIDES, LIVESTOCK, WASTE DISPOSAL, OXIDATION LAGOONS W72-00846
, OIL WASTES, FISH REPRODUCTION, WASTE DISPOSAL, MARINE MICROORGAN W71-13723
AL WASTES, METALS, HEAVY METALS, WASTE DISPOSAL, WATER POLLUTION E W71-11793
SOURCES, POPULATION, INDUSTRIES, WASTE DISPOSAL, MONITORING, NUTRI W71-12091
, EFFLUENT, IRRIGATION, STORAGE, WASTE DISPOSAL.: /OGENIC BACTERIA W71-08214
, NITRATES, SURVEYS, PESTICIDES, WASTE DISPOSAL, PATH OF POLLUTANT W71-05553
TION TREATMENT, INSTRUMENTATION, WASTE DISPOSAL, AERATION, EROSION W71-01192
G, *SPACE HEATING, *FARM USES OF WASTE HEAT, STEAM POWERED BUSES.: W70-09193
COOLING, / *WASTE HEAT, *USES OF WASTE HEAT, *CENTRAL HEATING AND W70-09193
AIR CONTAMINANT, STREAM QUALITY, WASTE MANAGEMENT, MANURE COLLECTI W71-08214
HOGS, ANIMAL DISEASES, CATFISH, WASTE TREATMENT, FARM WASTES, DEH W71-07544
TER RESOURCES, SYSTEMS ANALYSIS, WASTE TREATMENT, AIR POLLUTION, L W71-13723
AMFLOW, AGRICULTURE, INDUSTRIES, WASTE TREATMENT, WATERSHEDS(BASIN W71-11017
ASSESSMENTS, ALGAE, FORMULATION, WASTE TREATMENT, INDUSTRIES, WATE W72-09169
TROGEN,/ *ALGAE, *NITRIFICATION, WASTE TREATMENT, *PHOSPHORUS, *NI W73-02187
NITORING, MARSHES, SOIL EROSION, WASTE WATER TREATMENT.: /ORES, MO W73-02479
C HEALTH, ANALYTICAL TECHNIQUES, WASTE WATER TREATMENT, TASTE, ODO W73-02144
FICATION, CLAY, ALGAE, BACTERIA, WASTE WATER TREATMENT, WATER TREA W73-01362
GANIC WASTES, INDUSTRIAL WASTES, WASTE WATER(POLLUTION), BIOASSAY, W73-00840
IODEGRADATION, SEWAGE TREATMENT, WASTE WATER TREATMENT, PONDS, SCE W72-07661
RS, PONDS, ANAEROBIC CONDITIONS, WASTE WATER TREATMENT, CHLOROPHYT W72-07664
ORUS, *SEWAGE EFFLUENTS, *ALGAE, WASTE WATER TREATMENT, ACTIVATED W72-09176
DORS, OPERATION AND MAINTENANCE, WASTE WATER TREATMENT.: /NTIAL, O W72-10401
ORY TESTS, COST ANALYSIS, ALGAE, WASTE WATER TREATMENT.: / LABORAT W72-10390
TOZOA, DISSOLVED OXYGEN, SLUDGE, WASTE WATER TREATMENT.: /ION, PRO W72-10233
IMATION, CALCIUM, SEDIMENTATION, WASTE WATER TREATMENT.: /MS, ACCL W71-13334
ACITY, RESEARCH AND DEVELOPMENT, WASTE WATER TREATMENT.: /TIVE CAP W71-13256
N, CITIES, STORM RUNOFF, SEWAGE, WASTE WATER DISPOSAL, PATH OF POL W71-13466
CULTURAL CHEMICALS, FARM WASTES, WASTE WATER TREATMENT, CALIFORNIA W71-11698
PARKS, COST ANALYSIS, MICHIGAN, WASTE WATER TREATMENT.: / SPORTS, W72-00042
ICLE SIZE, ANAEROBIC CONDITIONS, WASTE WATER TREATMENT.: /RS, PART W71-08223
TREAM), WATER POLLUTION EFFECTS, WASTE WATER TREATMENT, CYANOPHYTE W71-07973
WATERING, DRYING, COST ANALYSES, WASTE WATER TREATMENT.: /TION, DE W71-08951
QUALITY CONTROL, COST ANALYSIS, WASTE WATER TREATMENT, NUTRIENT R W71-08953
RCES DEVELOPMENT, *WATER SUPPLY, WASTE WATER, RUNOFF, DISSOLVED OX W71-09788
HEAVY METALS, HUMAN POPULATION, WASTE WATER TREATMENT, WATER QUAL W71-09788
ONS, AEROBIC CONDITIONS, SLUDGE, WASTE WATER TREATMENT, BIODEGRADA W71-07084
ORMS, BIOCHEMICAL OXYGEN DEMAND, WASTE WATER TREATMENT.: /Y, COLIF W71-07106
MENT, BIOCHEMICAL OXYGEN DEMAND, WASTE WATER TREATMENT, *NORTH CAR W71-07382
IONS, BIOCHEMICAL OXYGEN DEMAND, WASTE WATER TREATMENT.: /C CONDIT W71-07109
MICROORGANISMS, BIODEGRADATION, WASTE WATER TREATMENT, WATER POLL W71-07097
AND, HYDROGEN ION CONCENTRATION, WASTE WATER TREATMENT, KANSAS.: / W71-07113
SIS, *BIOCHEMICAL OXYGEN DEMAND, WASTE WATER TREATMENT.: /ST ANALY W71-07099
MAND, BIOCHEMICAL OXYGEN DEMAND, WASTE WATER TREATMENT, WATER POLL W71-07098

IDATICN, ORGANIC LOADING, ALGAE, WASTE WATER TREATMENT, BIOCHEMICA W71-07096
IONS, BIOCHEMICAL OXYGEN DEMAND, WASTE WATER TREAT: /EROBIC CONDIT W71-07100
EDIMENTATION, TRICKLING FILTERS, WASTE WATER TREATMENT.: /ATION, S W70-10427
WAGE DISPOSAL, SEWAGE TREATMENT, WASTE WATER TREATMENT.: /VERS, SE W70-09947
SMS, EFFLUENTS, SETTLING BASINS, WASTE WATER TREATMENT, LABORATORY W71-00116
IARY TREATMENT, *EUTROPHICATION, WASTE WATER TREATMENT, ALGAE, EFF W70-09325
TRATICN, CHLORELLA, SCENEDESMUS, WASTE WATER TREATMENT.: /N CONCEN W71-00138
HESIS, RESPIRATION, SCENEDESMUS, WASTE WATER TREATMENT.: /PHOTOSYNT W71-00139
CTERIA, WATER POLLUTICN CONTROL, WASTE WATER TREATMENT.: /ICA, *BA W71-01899
*RESPIRATION, ACTIVATED SLUDGE, WASTE WATER TREATMENT, NUTIRIENTS W71-04079
SMS, HYDROGEN ION CONCENTRATION, WASTE WATER TREATMENT, WATER POLL W71-05155
STES, BIOCHEMICAL OXYGEN DEMAND, WASTE WATER TREATMENT, MICRCORGAN W71-06033
STES, BIOCHEMICAL OXYGEN DEMAND, WASTE WATER TREATMENT.: /CIPAL WA W71-03896
WATER QUALITY, SEWAGE DISPOSAL, WASTE WATER DISPOSAL, SEWAGE EFFL W70-06509
LUID FRICTION, SEWERS, SLURRIES, WASTE WATER TREATMENT, SLUDGE TRE W70-06968
ARATION TECHNIQUES, COAGULATION, WASTE WATER TREATMENT.: /STS, SEP W70-08904
, *LABCRATORY TESTS, DETERGENTS, WASTE WATER TREATMENT.: /TOXICITY W70-08976
NTRATION, WATER QUALITY CONTROL, WASTE WATER TREATMENT, CHLOROPHYT W72-04431
DIOACTIVITY, RADIOACTIVE WASTES, WASTE WATER(POLLUTION), WATER POL W72-05506
FECTS, RESEARCH AND DEVELOPMENT, WASTE WATER TREATMENT, BIODEGRADA W72-06655
SEWAGE TREATMENT, PUBLIC HEALTH, WASTE WATER TREATMENT, RESEARCH A W72-06638
, *ALGAE, ANALYTICAL TECHNIQUES, WASTE WATER TREATMENT, NUTRIENTS, W72-12966
QUALITY CONTROL, EUTROPHICATION, WASTE WATER TREATMENT, SEDIMENTAT W72-12487
DROLYSIS, STORM WATER, RAINFALL, WASTE WATER(POLLUTICN), EUTROPHIC W72-13644
AE, *LABORATORY TESTS, *CHEMICAL WASTE, *GROWTH RATES, *SEAWATER, W72-12239
NS, NUTRIENTS, AQUATIC AL/ *FARM WASTE, *PCULTRY, *OXICATION LAGOO W72-02850
ON, INDUSTRIAL WASTES, MUNICIPAL WASTE, BIOCHEMICAL OXYGEN DEMAND, W71-13356
URINE, PATH OF POLLUTANTS, FARM WASTES.: /RCES, COLORADO, CATTLE, W71-11496
CERS, ANIMAL PHYSIOLOGY, NUCLEAR WASTES.: /PATH OF POLLUTANTS, TRA W72-03348
IME, AEROBIC TREATMENT, DOMESTIC WASTES.: /NT, ACTIVATED SLUDGE, L W72-09176
*PHOSPHORUS, SEDIMENTS, ORGANIC WASTES.: /SMS, *ALGAE, *NITROGEN, W72-06532
RIENTS, WATER ANALYSIS, DOMESTIC WASTES.: /AL ANALYSIS, ALGAE, NUT W72-04791
N, *WASTE WATER TREATMENT, *FARM WASTES.: /SYNTHESIS, STABILIZATIO W70-06619
DISPOSAL, SEWAGE EFFLUENTS, FARM WASTES.: / DISPOSAL, WASTE WATER W70-06509
TION, DOMESTIC WASTES, MUNICIPAL WASTES.: /ACE WATERS, WATER POLLU W70-09388
TRIAL WASTES, ALGAE, PONDS, FARM WASTES.: /GE, INFILTRATION, INDUS W71-06435
NUTRIENT REQUIREMENTS, MUNICIPAL WASTES.: /LGAE, *EUTROPHICATION, W71-08775
WATER POLLUTION EFFECTS, NUCLEAR WASTES.: / METABOLISM, KINETICS, W71-09168
*PRIMARY PRODUCTIVITY, *OIL WASTES, *ARCTIC, *ALGAE.: W72-07922
*BACKGROUND RADIATION, *NUCLEAR WASTES, *ABSORPTION, *FOOD CHAINS W73-00833
NIMALS, ASIA/ *REVIEWS, *NUCLEAR WASTES, *BIOINDICATORS, *MARINE A W73-00811
, *CLAMS, *FOOD CHAINS, *NUCLEAR WASTES, *COBALT RADIOISOTOPES, PA W73-00826
*WATER POLLUTION EFFECTS, *OIL WASTES, *CRUSTACEANS, OILY WATER, W72-07911
CHEMICALS, *SURFACTANTS, */ *OIL WASTES, *CLEANING, *DISPERSION, * W71-09789
ATER POLLUTION EFFECTS, *ORGANIC WASTES, *EUTROPHICATION, *CHELATI W71-09674
*RADIOACTIVITY EFFECTS, *NUCLEAR WASTES, *FALLOUT, *OCEANS, WATER W72-04463
ATION LAGOONS, *ALGAE, MUNICIPAL WASTES, *FILTRATION, *FILTERS, BI W71-13341
LUTION ABATEMENT, WATER Q/ *FARM WASTES, *HOGS, *ENVIRCNMENT, *POL W71-08214
ATED SL/ *WATER REUSE, *DOMESTIC WASTES, *INDUSTRIAL WASTES, ACTIV W71-09524
*LAKE ERIE, *MERCURY, *MUNICIPAL WASTES, *INDUSTRIAL WASTES, WATER W72-14282
ST REGION, TRITIUM, OC/ *NUCLEAR WASTES, *MONITORING, *PACIFIC COA W72-03324
IMALS, *MARINE FISH, M/ *NUCLEAR WASTES, *MARINE ALGAE, *MARINE AN W73-00832
LY, *WASTE DISPOSAL, *RECREATION WASTES, *MICRCORGANISMS, *WATER P W72-14363
ATER POLLUTION EFFECTS, *ORGANIC WASTES, *NUTRIENTS, *POLLUTANT ID W72-01472
ERIE, *EUTROPHICATION, *CHEMICAL WASTES, *POLITICAL ASPECTS, LAKES W72-14782
REVIEWS, *BIOASSAY, *INDUSTRIAL WASTES, *PESTICIDES, *AQUATIC ORG W73-01976
E, DISSOLVED OXYGEN, DIVE/ *FARM WASTES, *RUNOFF, *FISHKILL, CATTL W72-08401
THERMAL POLLUTION, CON/ *NUCLEAR WASTES, *RADIOACTIVITY EFFECTS, * W72-00939
ERMENTAT/ *NUTRIENTS, *MUNICIPAL WASTES, *SEPARATION TECHNIQUES, F W72-10390
ESTUARINE ENVIRONMENT, *DOMESTIC WASTES, *SALT MARSHES, *CARBON CY W72-12567
TS, WATER POLLUTION/ *INDUSTRIAL WASTES, *THERMAL POLLUTICN, *COAS W73-01632
EFFLUENTS, CALIFOR/ *INDUSTRIAL WASTES, *WATER POLLUTION EFFECTS, W72-09092
*OXIDATION LAGOONS, *MUNICIPAL WASTES, *WATER REUSE, ALGAE, PHOT W72-11642
D OXYGEN, *PHYTOPLA/ *INDUSTRIAL WASTES, *WATER QUALITY, *DISSOLVE W71-11876
, *OXIDATION LAGOONS, AER/ *FARM WASTES, *WASTE DISPOSAL, *LAGOONS W71-07551
E, *DOMESTIC WASTES, *INDUSTRIAL WASTES, ACTIVATED SLUDGE, TRICKLI W71-09524
SEDIMENTS, COLIFORMS, MUNICIPAL WASTES, AGRICULTURE, FALLOUT, DRA W71-03012
NG, DATA COLLECTIONS, INDUSTRIAL WASTES, AGRICULTURAL RUNOFF, DOME W73-01972
UNOFF, AGRICULTURAL RUNOFF, FARM WASTES, ALGAE, AQUATIC WEEDS, FER W73-02479
ORY TESTS, FISH, ALGAE, CHEMICAL WASTES, ALGICIDES, BACTERICIDES, W72-08683
SEWAGE, INFILTRATION, INDUSTRIAL WASTES, ALGAE, PONDS, FARM WASTES W71-06435
MS, ABSORPTION, FISH, INDUSTRIAL WASTES, ALGAE, AQUATIC PLANTS, SE W72-05952
TERIA, *WATER QUALITY, *DOMESTIC WASTES, ALGAE, INDUSTRIAL WASTES, W72-09590
BIOCHEMICAL OXYGEN DEMAND, SOLID WASTES, AMMONIA, NITROGEN COMPOUN W72-09675
TY, MUNICIPAL WASTES, INDUSTRIAL WASTES, ANNUAL TURNOVER, ENERGY, W72-03228
ER POLLUTION SOURCES, INDUSTRIAL WASTES, AQUATIC ALGAE, BICINDICAT W71-11583
AT TRANSFER, *PLASTICS, DOMESTIC WASTES, BACTERIA, ALGAE, NUTRIENT W72-10401
ALGAE, PHOTOSYNTHESIS, MUNICIPAL WASTES, BIOCHEMICAL OXYGEN DEMAND W71-06033
C TREATMENT, BACTERIA, MUNICIPAL WASTES, BIOCHEMICAL OXYGEN DEMAND W71-03896
ICALS, FISH, CHEMICALS, CHEMICAL WASTES, BIOCHEMICAL OXYGEN DEMAND W73-01976
OGY, INDUSTRIAL WASTES, DOMESTIC WASTES, BIOASSAY, SULFATES, IRON W72-12239
TION SOURCES, SEWAGE, INDUSTRIAL WASTES, BIOINDICATORS, CHEMICAL P W72-12729
DS, *SEWAGE TREATMENT, *DOMESTIC WASTES, CHEMICAL ANALYSIS, EFFLUE W70-09907
ES, *CHEMICAL WASTES, INDUSTRIAL WASTES, CHLORINATION, PHOSPHATES, W72-05315
WATER POLLUTION SOURCES, TOXINS, WASTES, CHLOROPHYTA, SESSILE ALGA W71-10553
BACTERIA, ORGANIC WASTES, SOLID WASTES, CHEMICALS, MICHIGAN, CALI W71-11006
LGAE, *EUTROPHICATION, MUNICIPAL WASTES, CITIES, PHOSPHATES, ILLIN W72-03970
LEGISLATION, CHEMICALS, CHEMICAL WASTES, CLEANING, DOMESTIC WASTES W72-01685
, METALS, POLLUTANTS, INDUSTRIAL WASTES, CONFERENCES.: /S, PHENOLS W72-09646
LTRATION, MICROSCOPY, INDUSTRIAL WASTES, DETERGENTS, AMMONIA, NITR W71-11507

, CATFISH, WASTE TREATMENT, FARM WASTES, DEHYDRATION, FEASIBILITY. W71-07544
CES, INDUSTRIAL WASTES, DOMESTIC WASTES, DINOFLAGELLATES, *LABORAT W71-03095
ER POLLUTION SOURCES, INDUSTRIAL WASTES, DOMESTIC WASTES, DINOFLAG W71-03095
LIMITS, METHODOLOGY, INDUSTRIAL WASTES, DOMESTIC WASTES, BIOASSAY W72-12239
VATED SLUDGE, EFFLUENT, *ORGANIC WASTES, ECONOMICS, CHEMICAL OXYGE W71-12223
S, AGRICULTURAL RUNOFF, DOMESTIC WASTES, ECOLOGY, FISH, ALGAE, AQU W73-01972
POLLUTANTS, FOOD CHAINS, NUCLEAR WASTES, ENVIRONMENTAL EFFECTS, EF W72-10957
L PLANTS, INDUSTRIAL WASTES, OIL WASTES, ESTUARIES, MONITORING.: / W71-13746
MICAL WASTES, CLEANING, DOMESTIC WASTES, EUTROPHICATION, WATER POL W72-01685
TS, INDUSTRIAL WASTES, MUNICIPAL WASTES, FARM WASTES, WASTE DISPOS W72-01659
EMICAL OXYGEN DEMAND, INDUSTRIAL WASTES, FARM WASTES, POLLUTANTS, W71-06445
SOIL EROSION, FERTILIZERS, FARM WASTES, FISH, FISH EGGS, TURBIDIT W71-06444
LLUTION SOURCES, *FISHERIES, OIL WASTES, FISH REPRODUCTION, WASTE W71-13723
OR, BITUMINOUS MATERIALS, ANIMAL WASTES, FISH, FLUORESCENCE, INFRA W72-08790
BONS, *BIOINDICATORS, INDUSTRIAL WASTES, FLORIDA, SESSILE ALGAE, E W71-07412
SSILE ALGAE, ESTUARIES, CHEMICAL WASTES, GAS CHROMATOGRAPHY, SEDIM W71-07412
SOIL CULTURES, TITANIUM DIOXIDE WASTES, HELGOLAND.: /CYSTIS SPP., W72-12239
NS), SURFACE DRAINAGE, MUNICIPAL WASTES, INDUSTRIAL WASTES, SAMPLI W71-08953
SEPARATION TECHNIQUES, *CHEMICAL WASTES, INDUSTRIAL WASTES, CHLORI W72-05315
E, ALKALINITY, BURNING, DOMESTIC WASTES, INDUSTRIAL WAS: /N DIOXID W72-05473
WATER POLLUTION CONTROL, ORGANIC WASTES, INDUSTRIAL WASTES, WASTE W73-00840
*MATHEMATICAL METHODS, MUNICIPAL WASTES, INDUSTRIAL WASTES, NEW YO W71-11685
PARATIVE PRODUCTIVITY, MUNICIPAL WASTES, INDUSTRIAL WASTES, ANNUAL W72-03228
ECHNIQUES, FILTRATION, MUNICIPAL WASTES, INDUSTRIAL WASTES, OPERAT W72-09386
C LIFE, HEAVY METALS, INDUSTRIAL WASTES, INVERTEBRATES, ALGAE, *KE W72-11854
EARCH AND DEVELOPMENT, MUNICIPAL WASTES, LEGISLATION, REGULATION,: W72-06638
IUM, CALCIUM, HYDROGEN, CHEMICAL WASTES, LEAD, NICKEL, TOXICITY, B W72-14303
PHOSPHATES, PHOSPHORUS, CHEMICAL WASTES, LIMNOLOGY, INTERNATIONAL W72-14782
IRATION, *HEAVY METALS, CHEMICAL WASTES, MERCURY, COPPER, ZINC, PE W72-12576
MESTIC WASTES, ALGAE, INDUSTRIAL WASTES, METABOLISM, GROWTH RATES, W72-09590
ES, INDUSTRIAL WASTES, MUNICIPAL WASTES, METALS, HEAVY METALS, WAS W71-11793
C BACTERIA, TRITIUM, RADIOACTIVE WASTES, MONITORING, PHOTOGRAPHY, W72-04742
.: *TEXTILE MILL WASTES, MONTREAL, BALANCING TANKS W70-10427
ATERS, WATER POLLUTION, DOMESTIC WASTES, MUNICIPAL WASTES.: /ACE W W70-09388
NALYTICAL TECHNIQUES, INDUSTRIAL WASTES, MUNICIPAL WASTES, METALS, W71-11793
GOONS, STABILIZATION, INDUSTRIAL WASTES, MUNICIPAL WASTE, BIOCHEMI W71-13356
ON, ALGAE, NUTRIENTS, INDUSTRIAL WASTES, MUNICIPAL WASTES, FARM WA W72-01659
DS, MUNICIPAL WASTES, INDUSTRIAL WASTES, NEW YORK, ALGAE, WATER PO W71-11685
LLUTION CONTROL, SEDIMENTS, FARM WASTES, NUTRIENTS, PHOSPHORUS, AL W72-00846
URCES, BIOASSAY, ALGAE, DOMESTIC WASTES, NUTRIENTS, SEDIMENTS, ANA W73-02478
S, INDUSTRIAL PLANTS, INDUSTRIAL WASTES, OIL WASTES, ESTUARIES, MO W71-13746
ULSIFIERS, DETERGENTS, OIL, *OIL WASTES, OILY WATER, OIL-WATER INT W70-09429
*WATER POLLUTION EFFECTS, *FARM WASTES, OKLAHOMA, CYANOPHYTA, CHL W72-04773
ON, MUNICIPAL WASTES, INDUSTRIAL WASTES, OPERATION AND MAINTENANCE W72-09386
*ENVIRONMENTAL EFFECTS, NUCLEAR WASTES, PATH OF POLLUTANTS, MARIN W72-10955
POWERPLANTS, EFFLUENTS, NUCLEAR WASTES, PATH OF POLLUTANTS, ASIA, W72-10958
DUES, RECYCLING FOOD, PROCESSING WASTES, PATHOGENS, BIOENVIRONMENT W72-00846
MARINE ALGAE, SEDIMENTS, NUCLEAR WASTES, PATH OF POLLUTANTS, SEA W W73-00817
NG, NUCLEAR POWERPLANTS, NUCLEAR WASTES, PATH OF POLLUTANTS, WATER W73-00790
NE PLANTS, MARINE ALGAE, NUCLEAR WASTES, PATH OF POLLUTANTS, RADIO W73-00828
D CHAINS, PUBLIC HEALTH, NUCLEAR WASTES, PATH OF POLLUTANTS, DIFFU W73-00823
, *COBALT RADIOISOTOPES, NUCLEAR WASTES, PATH OF POLLUTANTS, ESTUA W72-07826
OXYGEN, AQUATIC LIFE, MUNICIPAL WASTES, PHOTOSYNTHESIS, RESPIRATI W71-11017
S, PUMPING, AEROBIC TREAT/ *FARM WASTES, POULTRY, NITROGEN, LAGOON W72-08396
DEMAND, INDUSTRIAL WASTES, FARM WASTES, POLLUTANTS, FERTILIZERS, W71-06445
ERBIVORES, NATURAL STREAMS, PULP WASTES, PRODUCTIVITY, DIELDRIN, P W72-06340
RACE ELEMENTS, COLLOIDS, NUCLEAR WASTES, RADIOECOLOGY.: /FFECTS, T W73-00818
OF POLLUTANTS, FALLOUT, NUCLEAR WASTES, RADIOECOLOGY, RADIOACTIVI W73-00821
FAUNA, CELLULOSE, ENZYMES, PULP WASTES, RADIOECOLOGY, WATER POLLU W72-09668
MARINE ALGAE, *ABSORPTION, *MINE WASTES, RADIOECOLOGY, *PATH OF PO W72-12204
), RIFLE FALLS(COLO), FISH FECAL WASTES, RESIDUAL FOOD, GREEN RIVE W71-11006
WATER POLLUTION EFFECTS, NUCLEAR WASTES, RIVERS, ABSORPTION, FISH, W71-09225
WATER QUALITY CONTROL, DOMESTIC WASTES, RUNOFF, WATER SOFTENING, W72-11186
GE, MUNICIPAL WASTES, INDUSTRIAL WASTES, SAMPLING, WATER QUALITY C W71-08953
*WASTE WATER TREATMENT, *LIQUID WASTES, SEWAGE TREATMENT, SEPARAT W72-05315
OALESCENCE, FLOCCULATION, LIQUID WASTES, SEPARATION TECHNIQUES, WA W71-13356
GANESE, COBALT, IRON, INDUSTRIAL WASTES, SEWAGE EFFLUENTS, POLLUTA W71-11515
ODUCTIVITY, POLLUTANTS, CHEMICAL WASTES, SEWAGE EFFLUENTS, WATER P W72-08804
SPENDED LOAD, SEWAGE, INDUSTRIAL WASTES, SILTS, WATER POLLUTION EF W71-09795
, TEMPERATURE, KINETICS, NUCLEAR WASTES, SOLUBILITY, ADSORPTION, W W71-09182
LATION, LEACHING, DRAINAGE, FARM WASTES, SOIL MANAGEMENT, GROUNDWA W71-06443
OR, PATHOGENIC BACTERIA, ORGANIC WASTES, SOLID WASTES, CHEMICALS, W71-11006
ION(ATMOSPHERIC), RUNOFF, ANIMAL WASTES, SURFACE WATERS, SOIL EROS W71-04216
UDIES, TIME, SNAILS, RADIOACTIVE WASTES, TENNESSEE, CRAYFISH, MIDG W70-06668
GULATION, BENTONITE, LIME, *PULP WASTES, TEOTIARY TREATMENT, EFFLU W72-03753
DRAINAGE, INDUSTRIES, MUNICIPAL WASTES, TURNOVERS, SEDIMENTS, HUM W71-11009
HOSPHORUS PESTICIDES, INDUSTRIAL WASTES, VIRUSES, ENERIC BACTERIA, W72-04742
DIOISOTOPES, ABSORPTION, NUCLEAR WASTES, WATER POLLUTION EFFECTS, W72-04461
VITY EFFECTS, *TRITIUM, *NUCLEAR WASTES, WATER POLLUTION EFFECTS, W72-06342
MAND, RADIOACTIVITY, RADIOACTIVE WASTES, WASTE WATER(POLLUTION), W W72-05506
*ALGAE, PHYTOPLANKTON, / ORGANIC WASTES, WATER POLLUTION EFFECTS, W70-07842
ER, AGRICULTURAL CHEMICALS, FARM WASTES, WASTE WATER TREATMENT, CA W71-11698
L WASTES, MUNICIPAL WASTES, FARM WASTES, WASTE DISPOSAL, PESTICIDE W72-01659
*BIOLOGICAL TREATM/ AGRICULTURAL WASTES, WATER POLLUTION CONTROL, W72-02975
ODS, PATH OF POLLUTANTS, NUCLEAR WASTES, WATER POLLUTION EFFECTS, W72-10978
E DISPOSAL, *SEWAGE, *INDUSTRIAL WASTES, WATER POLLUTION SOURCES, W72-08804
TROL, ORGANIC WASTES, INDUSTRIAL WASTES, WASTE WATER(POLLUTION), B W73-00840
, *MUNICIPAL WASTES, *INDUSTRIAL WASTES, WATER POLLUTION CONTROL, W72-14282
ISOTOPES, RADIOISOTOPES, NUCLEAR WASTES: /YSIS, SEA WATER, STABLE W73-00831
ALGAE, PESTICIDES, HEAVY METALS, WASTEWATER TREATMENT, SMELTS, MIC W72-14282

ATION, VACUUM DRYING, FLOTATION, WASTEWATER TREATMENT,: /CENTRIFUG W72-10237
ON, LIGHT PENETRATION, NEBRASKA, WASTEWATER TREATMENT, TURBIDITY, W72-04759
ICAL ANALYSIS, ALGAE, NUTRIENTS, WATER ANALYSIS, DOMESTIC WASTES.: W72-04791
OLISM, CARBON, SULFUR, SAMPLING, WATER ANALYSIS, ALGAE, SPECTROPHO W72-04743
S, PATH OF POLLUTANTS, SAMPLING, WATER ANALYSIS, ALGAE, COLIFORMS, W70-08248
S, PATH OF POLLUTANTS, SAMPLING, WATER ANALYSIS, ALGAE, COLIFORMS, W70-08249
QUATIC LIFE, IRRIGATION SYSTEMS, WATER ANALYSIS, TEMPERATURE, LIMN W71-06102
TURES, ORGANIC MATTER, BACTERIA, WATER ANALYSIS.: /NS, RIVERS, CUL W71-08032
L TECHNIQUES, CHEMICAL ANALYSIS, WATER ANALYSIS, EVALUATION, WATER W72-10883
ANIC ACIDS, FULVIC ACIDS, KELPS, WATER ANALYSIS, EXUDATION, WATER W72-12166
WATER PROPERTIES, ALGAL CONTROL, WATER ANALYSIS, CHEMCONTROL, WATE W72-11724
PTION, ALGAE, SEASONAL, WEATHER, WATER ANALYSIS, TRACERS, RADIOACT W72-09122
LOROPHYLL, ALGAE, PHYTOPLANKTON, WATER ANALYSIS, CARBON CYCLE, ORG W72-09108
HLOROPHYTA, CHLORELLA, TOXICITY, WATER ANALYSIS, ABSORPTION, CULTU W72-04134
SS, *ORGANIC MATTER, *SEA WATER, WATER ANALYSIS, ALGAE, CHEMICAL A W71-11234
TES, CHLOROPHYTA, SESSILE ALGAE, WATER ANALYSIS.: /ES, TOXINS, WAS W71-10553
PHORUS COMPOUNDS, CHLAMYDOMONAS, WATER ANALYSIS, FLUOROMETRY, PIGM W72-12637
NE PLANTS, MONITORING, SAMPLING, WATER ANALYSIS, SEA WATER, STABLE W73-00831
, PATH OF POLLUTANTS, SEA WATER, WATER ANALYSIS, RADIOACTIVITY TEC W73-00817
SCOPY, TRACE ELEMENTS, SAMPLING, WATER ANALYSIS, SEA WATER.: /CTRO W73-00832
MICAL ANALYSIS, WATER POLLUTION, WATER ANALYSIS, BIOASSAY, CYANOPH W73-00286
ALGAE, NUTRIENTS, WATER QUALITY, WATER ANALYSIS, FREQUENCY, FLUCTU W73-00918
EFFECTS, *CALIFORNIA, SEA WATER, WATER ANALYSIS, MARINE ALGAE, CHE W73-02105
E PLANTS, MARINE MICROORGANISMS, WATER ANALYSIS, BENTHIC FLORA, OR W73-02100
UMENTATION, MATHEMATICAL MODELS, WATER ANALYSIS, POLLUTANT IDENTIF W72-08790
TEMPERATURE, CHEMICAL ANALYSIS, WATER ANALYSIS, TIDAL EFFECTS, ST W72-08141
ING, ZOOPLANKTON, PHYTOPLANKTON, WATER ANALYSIS, POLLUTANTS, AQUAT W72-07896
ALGAE, INFLOW, DISCHARGE(WATER), WATER BALANCE, SEDIMENTS, CHEMICA W72-13851
ION, AQUARIA, ANIMAL PHYSIOLOGY, WATER BALANCE, TRACERS, PLANT PHY W72-06342
MS, CLAMS, SHELLFISH, AMPHIPODA, WATER BIRDS, WISCONSIN, ANALYTICA W70-08098
ENTIAL NUTRIENTS, WATER QUALITY, WATER CHEMISTRY, CYCLING NUTRIENT W70-04080
DES, ALGAE, VOLUMETRIC ANALYSIS, WATER CHEMISTRY.: /, IONS, FLUORI W71-00474
ING, EUTROPHICATION, EVALUATION, WATER CHEMISTRY, WATER TEMPERATUR W72-06526
, ORGANIC MATTER, WATER QUALITY, WATER CHEMISTRY, WATER POLLUTION W72-04853
OPULATION, PRIMARY PRODUCTIVITY, WATER CHEMISTRY, *CALIFORNIA, *NE W72-12549
, FLOW RATES, WATER TEMPERATURE, WATER CHEMISTRY, CHEMICAL ANALYSI W73-01972
LGAE, MOSSES, WATER TEMPERATURE, WATER CHEMISTRY, CHLOROPHYLL, HYP W71-11949
OIRS, *HYDROLOGIC DATA, *KANSAS, WATER CHEMISTRY, CHEMICAL ANALYSI W72-02274
TICAL TECHNIQUES, PALEOZOIC ERA, WATER CHEMISTRY.: /SOTOPES, ANALY W72-02073
ING, ALGAL BLOOMS, ALGAL TOXINS, WATER CIRCULATION, WATER POLLUTIO W72-09304
GROWTH RATES, PLANT MORPHOLOGY, WATER CIRCULATION.: /COMPETITION, W72-04766
CONTROL, *DIVERSION, SEDIMENTS, WATER CIRCULATION, ABSORPTION, NU W71-09581
POLYMERIZATION, ETHYLENE OXIDE, WATER CLARIFICATION, SLUDGE DRYIN W70-06968
IBUTION SYSTEMS, *WATER QUALITY, WATER CONSERVATION, WILDLIFE CONS W72-05594
STUARINE ENVIRONMENT, CALIFORNIA WATER CONVEYANCE, CANALS, HEAVY M W71-09581
LIMNION, VOLUME, MIXING, SAFETY, WATER COOLING, ANAEROBIC CONDITIO W70-09595
FISHERIES, WATER SUPPLY, COSTS, WATER DEMAND, ECONOMIC JUSTIFICAT W71-13172
IES, STORM RUNOFF, SEWAGE, WASTE WATER DISPOSAL, PATH OF POLLUTANT W71-13466
QUALITY, SEWAGE DISPOSAL, WASTE WATER DISPOSAL, SEWAGE EFFLUENTS, W70-06509
*WASTE WATER(POLLUTION), *WASTE WATER DISPOSAL, *WATER POLLUTION W72-14878
S, ALGAL CONTROL, SPAWNING, WARM- WATER FISH.: /ANAEROBIC CONDITION W70-09595
SPOSAL, SEWAGE LAGOONS, BRACKISH WATER FISH, BIOLOGICAL TREATMENT, W72-12567
UATIC PLANTS, HARVESTING, PONDS, WATER HYACINTH, NUTRIENTS, MIDGES W71-01488
DUCTIVITY, CALCIUM, SCENEDESMUS, WATER HYACINTH, LIGHT PENETRATION W72-04269
ES, NUTR/ *PHOSPHATES, *SEDIMENT- WATER INTERFACES, ADSORPTION, LAK W72-01108
IL, *OIL WASTES, OILY WATER, OIL- WATER INTERFACES, WATER POLLUTION W70-09429
AYS, CORES, SEASONAL, ALGAE, MUD- WATER INTERFACES, IRON, MANGANESE W72-05469
FUR BACTERIA, *OXYGEN, *SEDIMENT- WATER INTERFACES, *LAKE ERIE, HYP W72-06690
ERIN-LUCIFERASE METHOD, SEDIMENT- WATER INTERACTIONS, BIOLUMINESCEN W72-04292
*LAKE ERIE, *BACTERIA, *SEDIMENT- WATER INTERFACES, *HYPOLIMNION, N W72-12996
BACTERIA, THERMOCLINE, SEDIMENT- WATER INTERFACES, CHRYSOPHYTA, CH W72-12995
MICROBIOLOGY, OREGON, *SEDIMENT- WATER INTERFACES, SEDIMENTS, AQUA W72-08989
ONTROL, WATER POLLUTION SOURCES, WATER LAW, ALGAE, DISSOLVED OXYGE W71-01192
GENTS, *PHOSPHATES, LEGISLATION, WATER LAW, FEDERAL GOVERNMENT, WA W71-11516
SALINITY, MIGRATION, TIDES, LOW WATER MARK, CHRYSOPHYTA.: /ALGAE, W72-14317
SEDIMENT- WATER NUTRIENT-EXCHANGE.: W70-08944
OURCES, WATER POLLUTION EFFECTS, WATER POL: /ON, WATER POLLUTION S W72-04279
CLEAR WASTES, *FALLOUT, *OCEANS, WATER POLLUTION EFFECTS, FOOD CHA W72-04463
TIVE CAPACITY, WATER PROPERTIES, WATER POLLUTION, WATER POLLUTION W72-04279
MAINE, RECREATION, WATER SUPPLY, WATER POLLUTION EFFECTS, WATER PO W72-04280
PES, ABSORPTION, NUCLEAR WASTES, WATER POLLUTION EFFECTS, MOLLUSKS W72-04461
LUTION, WATER POLLUTION SOURCES, WATER POLLUTION EFFECTS, WATER PO W72-04279
TER PROPERTIES, WATER POLLUTION, WATER POLLUTION SOURCES, WATER PO W72-04279
SUPPLY, WATER POLLUTION EFFECTS, WATER POLLUTION SOURCES, ALGAE, N W72-04280
ECTS, *TRITIUM, *NUCLEAR WASTES, WATER POLLUTION EFFECTS, CATTAILS W72-06342
NT, NUTRIENTS, SEWAGE TREATMENT, WATER POLLUTION, WATER POLLUTION W72-06655
PHOSPHORUS, AQUATIC ENVIRONMENT, WATER POLLUTION EFFECTS, ALGAE, N W72-06638
UTION EFFECTS, ALGAE, NUTRIENTS, WATER POLLUTION SOURCES, SEWAGE T W72-06638
WAGE TREATMENT, WATER POLLUTION, WATER POLLUTION EFFECTS, RESEARCH W72-06655
THESIS, WATER POLLUTION SOURCES, WATER POLLUTION EFFECTS,: /OTOSYN W72-05952
, PHYTOPLANKTON, PHOTOSYNTHESIS, WATER POLLUTION SOURCES, WATER PO W72-05952
WASTES, WASTE WATER(POLLUTION), WATER POLLUTION SOURCES, ALGAE, P W72-05506
EABILITY, WATER QUALITY CONTROL, WATER POLLUTION EFFECTS.: /, PERM W72-06037
SILTING, AGRICULTURAL CHEMICALS, WATER POLLUTION SOURCES, ALGAE, G W72-05869
LUENTS, SEWAGE TREATMENT, ALGAE, WATER POLLUTION SOURCES, CHLOROPH W72-05473
ARBON, *WATER POLLUTION CONTROL, WATER POLLUTION EFFECTS, NUTRIENT W72-04769
, ALGAE, CYANOPHYTA, PHOSPHORUS, WATER POLLUTION SOURCES, HYDROGEN W72-04769
ICIDES, WATER POLLUTION EFFECTS, WATER POLLUTION SOURCES, WATER PO W72-04759
GEN, COPPER SULFATE, HERBICIDES, WATER POLLUTION EFFECTS, WATER PO W72-04759
FFECTS, WATER POLLUTION SOURCES, WATER POLLUTION CONTROL.: /TION E W72-04759

WATER QUALITY, WATER CHEMISTRY, WATER POLLUTION SOURCES, METEOROL W72-04853
NSECTICIDES, *ST LAWRENCE RIVER, WATER POLLUTION SOURCES, WATER PO W70-08098
PHYTOPLANKTON, / ORGANIC WASTES, WATER POLLUTION EFFECTS, *ALGAE, W70-07842
, EUTROPHICATION, WATER QUALITY, WATER POLLUTION CONTROL, SURVEYS, W70-08249
RIVER, WATER POLLUTION SOURCES, WATER POLLUTION EFFECTS, MUD, MOL W70-08098
, EUTROPHICATION, WATER QUALITY, WATER POLLUTION CONTROL, SURVEYS, W70-08248
CYANOPHYTA, ALGAE, SOUTH DAKOTA, WATER POLLUTION EFFECTS, WATER PO W71-02880
AL WA/ *WATER POLLUTION EFFECTS, WATER POLLUTION SOURCES, INDUSTRI W71-03095
ATION, *WATER POLLUTION CONTROL, WATER POLLUTION EFFECTS, *AQUATIC W71-01488
DAKOTA, WATER POLLUTION EFFECTS, WATER POLLUTION SOURCES, CULTIVAT W71-02880
ING, *ALGAE, *SILICA, *BACTERIA, WATER POLLUTION CONTROL, WASTE WA W71-01899
OSAL, AERATION, EROSION CONTROL, WATER POLLUTION SOURCES, WATER LA W71-01192
PHATES, WATER YIELD IMPROVEMENT, WATER POLLUTION TREATMENT, ALGAE, W71-01491
ICUT, *UNIVERSITIES, *ABSTRACTS, WATER POLLUTION TREATMENT, INSTRU W71-01192
ROPHICATION, ALGAE, SURFACTANTS, WATER POLLUTION SOURCES, SURFACE W70-09388
INATION, *ALGAL CONTROL, RIVERS, WATER POLLUTION EFFECTS, ALGAE, D W71-00099
ILY WATER, OIL-WATER INTERFACES, WATER POLLUTION TREATMENT, WATER W70-09429
LLUTION SOURCES, SURFACE WATERS, WATER POLLUTION, DOMESTIC WASTES, W70-09388
ACES, WATER POLLUTION TREATMENT, WATER POLLUTION EFFECTS, ENVIRONM W70-09429
UTROPHICATION, ALGAE, NUTRIENTS, WATER POLLUTION, BIOLOGICAL TREAT W71-05083
ICATION METHODS, COOLING TOWERS, WATER POLLUTION EFFECTS.: /, APPL W71-06189
NUTRIENTS, ALGAE, LAKE SUPERIOR, WATER POLLUTION SOURCES, MINNESOT W71-04216
NTRATION, WASTE WATER TREATMENT, WATER POLLUTION EFFECTS.: / CONCE W71-05155
UARIES, WATER POLLUTION EFFECTS, WATER POLLUTION SOURCES, NITRIFIC W71-05390
ND, DISSOLVED OXYGEN, ESTUARIES, WATER POLLUTION EFFECTS, WATER PO W71-05390
*TOXICITY, COPPER, CHLOROPHYTA, WATER POLLUTION, NUTRIENTS, RIVER W71-05991
ON, *SEWAGE, *ORGANIC COMPOUNDS, WATER POLLUTION TREATMENT, WATER W71-09674
ETERGENTS, *POLLUTION ABATEMENT, WATER POLLUTION EFFECTS, NUTRIENT W71-09429
OISOTOPES, COBALT RADIOISOTOPES, WATER POLLUTION EFFECTS, FRESH WA W71-09250
EWAGE, INDUSTRIAL WASTES, SILTS, WATER POLLUTION EFFECTS.: /OAD, S W71-09795
*RADIOACTIVITY EFFECTS, *PONDS, WATER POLLUTION EFFECTS, RADIOISO W71-09256
UNDS, WATER POLLUTION TREATMENT, WATER POLLUTION SOURCES, ENVIRONM W71-09674
GAE, D/ *RETURN FLOW, *NITRATES, WATER POLLUTION, DESALINATION, AL W71-08223
REMENT, POPULATION, CHLOROPHYLL, WATER POLLUTION EFFECTS, TEMPERAT W71-08670
TS, ALGAE, *DEGRADATION(STREAM), WATER POLLUTION SOURCES, WASTE WA W71-07973
DIMENTS, OYSTERS, SHRIMP, CRABS, WATER POLLUTION SOURCES, SEA WATE W71-07412
EMENT, WATER QUALITY, STANDARDS, WATER POLLUTION, AIR POLLUTION, O W71-08214
CONCENTRATION, ALGAE, BACTERIA, WATER POLLUTION SOURCES.: /EN ION W71-07942
N DEMAND, WASTE WATER TREATMENT, WATER POLLUTION EFFECTS.: / OXYGE W71-07098
, PACKING PLANTS.: INDUSTRIAL WATER POLLUTION, DES MOINES RIVER W71-06445
RADATION, WASTE WATER TREATMENT, WATER POLLUTION EFFECTS.: /BIODEG W71-07097
TIONS, VARIABILITY, PROBABILITY, WATER POLLUTION EFFECTS, ALGAE, A W71-07337
, PHOSPHATES, ORGANIC COMPOUNDS, WATER POLLUTION CONTROL.: /LATION W71-06247
LUENTS, WATER POLLUTION EFFECTS, WATER POLLUTION SOURCES, EUTROPHI W71-09173
KES, *GEORGIA, SEWAGE EFFLUENTS, WATER POLLUTION EFFECTS, WATER PO W71-09173
RRIGATION, ECONOMIC FEASIBILITY, WATER POLLUTION CONTROL, CALIFORN W71-09157
DA/ *MARINE PLANTS, *OILY WATER, WATER POLLUTION EFFECTS, *INTERTI W71-08792
ENTAL EFFECTS, */ *OIL, SEEPAGE, WATER POLLUTION EFFECTS, ENVIRONM W71-08793
QUARIA, PARTICLE SIZE, COLLOIDS, WATER POLLUTION EFFECTS.: /YTA, A W71-09193
WASTES, SOLUBILITY, ADSORPTION, WATER POLLUTION EFFECTS, FOOD CHA W71-09182
OBALT RADIOISOTOPES, ABSORPTION, WATER POLLUTION SOURCES.: /PES, C W71-09018
MPERATURE, METABOLISM, KINETICS, WATER POLLUTION EFFECTS, NUCLEAR W71-09168
PHYTA, LABORATORY TESTS, OXYGEN, WATER POLLUTION EFFECTS.: /CHLORO W72-14314
BIOASSAY, ENVIRONMENTAL EFFECTS, WATER POLLUTION EFFECTS, ALGAE, H W72-13813
WASTES, WATER POLLUTION CONTROL, WATER POLLUTION SOURCES, TRIBUTAR W72-14282
MENTS, NUTRIENTS, BIOINDICATORS, WATER POLLUTION EFFECTS, FERTILIZ W72-13816
IPAL WASTES, *INDUSTRIAL WASTES, WATER POLLUTION CONTROL, WATER PO W72-14282
E, *INHIBITION, *SALT TOLERANCE, WATER POLLUTION EFFECTS, CHLORINA W72-13809
ORGANIC PESTICIDES, HERBICIDES, WATER POLLUTION SOURCES, PLANKTON W72-13347
DISSOLVED SOLIDS, PHYTOPLANKON, WATER POLLUTION EFFECTS.: /IENTS, W72-13300
ALGAE, CLADOPHORA, PRODUCTIVITY, WATER POLLUTION, STANDING CROPS.: W72-13662
ATER(POLLUTION), EUTROPHICATION, WATER POLLUTION SOURCES, NUTRIENT W72-13644
ES, *CYANOPHYTA, EUTROPHICATION, WATER POLLUTION EFFECTS.: /ISOTOP W72-13692
TER POLLUTION EFFECTS, TOXICITY, WATER POLLUTION, FOOD CHAINS, PAT W72-12257
SING, *POLLUTANT IDENTIFICATION, WATER POLLUTION, *MONITORING, AER W72-12487
YSIOLOGY, POLLUTANTS, *TOXICITY, WATER POLLUTION EFFECTS, RESPIRAT W72-12576
CATION, WATER POLLUTION EFFECTS, WATER POLLUTION CONTROL.: /ITRIFI W72-12544
URY, *PUBLIC HEALTH, *SHELLFISH, WATER POLLUTION EFFECTS, TOXICITY W72-12257
SEDIMENTATION RATES, OIL SPILLS, WATER POLLUTION SOURCES.: /MENT, W72-12487
ROGEN FIXATION, DENITRIFICATION, WATER POLLUTION EFFECTS, WATER PO W72-12544
TS, *LIMNOLOGY, WATER POLLUTION, WATER POLLUTION SOURCES, SEWAGE, W72-12729
ALGAE, CHLOROPHYTA, INHIBITORS, WATER POLLUTION EFFECTS, ENVIRONM W72-12736
COMMON MERGANSER DUCK, BIOASSAY, WATER POLLUTION EFFECTS, GAS CHRO W72-12930
PLANT PHYSIOLOGY, BICARBONATES, WATER POLLUTION EFFECTS.: /ATION, W72-12940
R POLLUTION EFFECTS, *LIMNOLOGY, WATER POLLUTION, WATER POLLUTION W72-12729
, CYANOPHYTA, BACTERIA, ECOLOGY, WATER POLLUTION EFFECTS.: /PRINGS W72-12734
TER POLLUTION EFFECTS, *DIATOMS, WATER POLLUTION SOURCES, INDUSTRI W71-11583
USTRIAL WASTES, NEW YORK, ALGAE, WATER POLLUTION.: /AL WASTES, IND W71-11685
ENT, WATER POLLUTION, STANDARDS, WATER POLLUTION EFFECTS, ALKYLBEN W71-11516
LUTION EFFECTS, OHIO, NUTRIENTS, WATER POLLUTION SOURCES, DISSOLVE W71-11949
UTROPHICATION, *NITROGEN, ALGAE, WATER POLLUTION SOURCES, COLORADO W71-11496
S, HEAVY METALS, WASTE DISPOSAL, WATER POLLUTION EFFECTS.: / METAL W71-11793
FFECTS, ALKYLBENZENE SULFONATES, WATER POLLUTION SOURCES, AQUATIC W71-11516
, WATER LAW, FEDERAL GOVERNMENT, WATER POLLUTION, STANDARDS, WATER W71-11516
, ALGAL CONTROL, EUTROPHICATION, WATER POLLUTION, BIOCHEMICAL OXYG W71-13257
E TREATMENT, TERTIARY TREATMENT, WATER POLLUTION CONTROL, BIODEGRA W71-13939
*EUTROPHICATION, *WATER SUPPLY, WATER POLLUTION EFFECTS, NUTRIENT W71-13185
REGULATORS, PLANKTON, CULTURES, WATER POLLUTION EFFECTS.: /GROWTH W71-13793
UTRIENTS, ANALYTICAL TECHNIQUES, WATER POLLUTION EFFECTS, POLLUTAN W71-13794
ALYTICAL TECHNIQUES, PHOTOMETRY, WATER POLLUTION.: /ACE WATERS, AN W71-12034

S, *ALGAE, SEWAGE, PRODUCTIVITY, WATER POLLUTION EFFECTS, PHOSPHOR W71-12857
TROPHICATION, WATER TEMPERATURE, WATER POLLUTION EFFECTS, WATER PO W71-12092
CHIGAN, WATER POLLUTION SOURCES, WATER POLLUTION CONTROL, TROPHIC W71-12072
ED SLUDGE, *NUTRIENTS, NITROGEN, WATER POLLUTION EFFECTS, WATER PO W71-11972
RATURE, WATER POLLUTION EFFECTS, WATER POLLUTION CONTROL, ENERGY, W71-12092
UATIC PLANTS, CHEMICAL ANALYSIS, WATER POLLUTION SOURCES.: /AE, AQ W71-12252
IENTS, SEDIMENTS, LAKE MICHIGAN, WATER POLLUTION SOURCES, WATER PO W71-12072
TROGEN, WATER POLLUTION EFFECTS, WATER POLLUTION SOURCES.: /TS, NI W71-11972
LLUTANTS, ANALYTICAL TECHNIQUES, WATER POLLUTION SOURCES, TOXINS, W71-10553
ALIFORNIA, TUBIFICIDS, COLORADO, WATER POLLUTION CONTROL, WASHINGT W71-11006
LUENTS, WATER POLLUTION SOURCES, WATER POLLUTION EFFECTS, SAMPLING W71-11008
CTIVITY, FERTILIZERS, EFFLUENTS, WATER POLLUTION SOURCES, WATER PO W71-11008
NITROGEN, WISCONSIN, NUTRIENTS, WATER POLLUTION EFFECTS, INTERFAC W71-10645
MS, *DINOFLAGELLATES, *RED TIDE, WATER POLLUTION SOURCES.: / DIATO W72-01329
SOFTENING, SOAPS, AQUATIC ALGAE, WATER POLLUTION EFFECTS, WATER QU W72-01085
ONS, ALGAE, CHEMICAL PROPERTIES, WATER POLLUTION, PROTEIN, PRIMARY W72-01473
E ERIE, WATER POLLUTION EFFECTS, WATER POLLUTION SOURCES, LIMNOLOG W72-01094
LUTION, WATER POLLUTION SOURCES, WATER POLLUTION TREATMENT, NUTRIE W72-01685
AE, WASTE ASSIMILATIVE CAPACITY, WATER POLLUTION EFFECTS, POLLUTAN W72-01685
WATER POLLUTION CONTROL ACT.: W72-01085
PERTIES, WATER POLLUTION EFFECT, WATER POLLUTION SOURCES, COST COM W72-00974
IC ALGAE, BIOLOGICAL PROPERTIES, WATER POLLUTION EFFECT, WATER POL W72-00974
OMS, *PHYTOPLANKTON, *NUTRIENTS, WATER POLLUTION EFFECTS, AQUATIC W72-01329
CAL GOVERNMENTS, WASTE DISPOSAL, WATER POLLUTION SOURCES, SOAPS, T W72-01093
EUTROPHICATION, WATER POLLUTION, WATER POLLUTION SOURCES, WATER PO W72-01685
DOMESTIC WASTES, EUTROPHICATION, WATER POLLUTION, WATER POLLUTION W72-01685
ATER QUALITY CONTROL, STANDARDS, WATER POLLUTION SOURCES, WATER PO W72-01660
RDS, LEGISLATION, LEGAL ASPECTS, WATER POLLUTION, ADJUDICATION PRO W72-01093
URON, *LAKE ONTARIO, *LAKE ERIE, WATER POLLUTION EFFECTS, WATER PO W72-01094
NDARDS, WATER POLLUTION SOURCES, WATER POLLUTION EFFECTS, RECREATI W72-01660
TION, WATER QUALITY, RESERVOIRS, WATER POLLUTION SOURCES, ALGAE, R W72-04260
S, *FROGS, *TRITIUM, ABSORPTION, WATER POLLUTION EFFECTS, RADIOACT W72-03348
OMIC EFFICIENCY, *PUBLIC HEALTH, WATER POLLUTION, HAZARDS, NUTRIEN W72-04266
TESTS, WATER POLLUTION SOURCES, WATER POLLUTION EFFECTS, FLUCTUAT W72-03228
CAL TREATM/ AGRICULTURAL WASTES, WATER POLLUTION CONTROL, *BIOLOGI W72-02975
OASTS, BIORTHYMS, ON-SITE TESTS, WATER POLLUTION SOURCES, WATER PO W72-03228
NIA, FOOD CHAINS, PUBLIC HEALTH, WATER POLLUTION SOURCES.: /ALIFOR W72-03324
TION EFFECTS, *ALGAE, *PLANKTON, WATER POLLUTION CONTROL, METHODOL W72-03218
OLLUTION EFFECTS, AIR POLLUTION, WATER POLLUTION, LICHENS, SOIL-WA W72-03347
ALGAE, MARINE FISH, ABSORPTION, WATER POLLUTION SOURCES, RADIOACT W72-00940
E ALGAE, *EUTR/ *WASTE DILUTION, WATER POLLUTION CONTROL, *NUISANC W72-00799
GANISMS, ALGAL CONTROL, *POLLUT/ WATER POLLUTION EFFECTS, *MICROOR W72-00801
LLUTANTS, AIR POLLUTION EFFECTS, WATER POLLUTION EFFECTS, BAYS, WA W72-00959
ICATION, LAKES, *NUTRIENTS, ALG/ WATER POLLUTION EFFECTS, *EUTROPH W72-00798
NTROL, *WATER POLLUTION EFFECTS, WATER POLLUTION SOURCES, AQUATIC W72-09646
YMES, PULP WASTES, RADIOECOLOGY, WATER POLLUT: /NA, CELLULOSE, ENZ W72-09668
ALGAL TOXINS, WATER CIRCULATION, WATER POLLUTION CONTROL, FISHKILL W72-09304
STICIDES, *PRIMARY PRODUCTIVITY, WATER POLLUTION EFFECTS, ESTUARIN W72-09653
ONTROL, FISHKILL, NEW HAMPSHIRE, WATER POLLUTION EFFECTS, COPPER S W72-09304
*EUTROPHICATION, WATER SAMPLING, WATER POLLUTION EFFECTS, BIOMASS, W72-09111
HENSIVE PLANNING, AQUATIC WEEDS, WATER POLLUTION SOURCES, MATHEMAT W72-09171
CAT/ *NUTRIENTS, *PHYTOPLANKTON, WATER POLLUTION EFFECTS, EUTROPHI W72-09167
RCES, BIOCHEMICAL OXYGEN DEMAND, WATER POLLUTION CONTROL, NITRATES W72-09167
N DIOXIDE, SEASONAL, PHOSPHORUS, WATER POLLUTION SOURCES, BIOCHEMI W72-09167
DS, *POL/ *ALGAE, *FLOCCULATION, WATER POLLUTION TREATMENT, COLLOI W72-11833
LIMIT, WATER POLLUTION EFFECTS, WATER POLLUTION SOURCES, INHIBITI W72-11727
OL, WATER ANALYSIS, CHEMCONTROL, WATER POLLUTION EFFECTS, WATER QU W72-11724
ELPS, WATER ANALYSIS, EXUDATION, WATER POLLUTION, PHYTOPLANKTON, W W72-12166
WATER POLLUTION, PHYTOPLANKTON, WATER POLLUTION SOURCES, PHAEOPHY W72-12166
*COAST REVIEW, *BIBLIOGRAPHIES, WATER POLLUTION SOURCES, THERMAL W72-12190
NUTRIENTS, *AEROBIC CONDITIONS, WATER POLLUTION CONTROL.: /ENTIAL W72-11617
AQUATIC ENVIRONMENT, TURBIDITY, WATER POLLUTION, PHAEOPHYTA, MOLL W72-11925
ECOSYSTEMS, *BIOLOGICAL COMMUN/ WATER POLLUTION EFFECTS, ECOLOGY, W72-11854
TACEANS, FOOD CHAINS, FOOD WEBS, WATER POLLUTION EFFECTS, AQUATIC W72-11925
PHOSPHORUS, ALGAE, PLANT GROWTH, WATER POLLUTION TREATMENT, EVALUA W72-11702
ONAS, INHIBITORS, *LETHAL LIMIT, WATER POLLUTION EFFECTS, WATER PO W72-11727
STAGE, CRUSTACEANS, MONITORING, WATER POLLUTION EFFECTS, MOLLUSKS W72-10955
IENTS, *ALGAE, SEWAGE TREATMENT, WATER POLLUTION SOURCES, WATER PO W72-10625
ALYTICAL TECHNIQUES, EVALUATION, WATER POLLUTION, POLLUTANT IDENTI W72-10678
IDES, IONS, CYTOLOGICAL STUDIES, WATER POLLUTION EFFECTS.: / FLUOR W72-11572
H OF POLLUTANTS, NUCLEAR WASTES, WATER POLLUTION EFFECTS, FOOD CHA W72-10978
OF POLLUTANTS, *WATER POLLUTION, WATER POLLUTION SOURCES, AQUATIC W72-10686
ATMENT, WATER POLLUTION SOURCES, WATER POLLUTION EFFECTS, AGRICULT W72-10625
COLI, AZOTOBACTER, PSEUDOMONAS, WATER POLLUTION EFFECTS.: /ES, E. W72-10861
LGAE, FISH POPULATIONS, FLORIDA, WATER POLLUTION EFFECTS, NITROGEN W72-08986
CAL DISTRIBUTION, BIOINDICATORS, WATER POLLUTION SOURCES.: /COLOGI W72-08579
L COMMUNITIES, RIVERS, BACTERIA, WATER POLLUTION, ACTIVATED SLUDGE W72-08573
EN ION CONCENTRATION, CORROSION, WATER POLLUTION EFFECTS.: /HYDROG W72-08683
AL, *SEWAGE, *INDUSTRIAL WASTES, WATER POLLUTION SOURCES, ON-SITE W72-08804
EMICAL WASTES, SEWAGE EFFLUENTS, WATER POLLUTION EFFECTS.: /TS, CH W72-08804
, RIVERS, *ELECTRIC POWERPLANTS, WATER POLLUTION EFFECTS, WATER PO W72-07791
SMS, AQUATIC ANIMALS, NUTRIENTS, WATER POLLUTION EFFECTS, WATER QU W72-07896
S, SAMPLING, DIATOMS, TURBIDITY, WATER POLLUTION SOURCES, NUTRIENT W72-07890
PLANTS, WATER POLLUTION EFFECTS, WATER POLLUTION SOURCES, *FISH, D W72-07791
PLANKTON, BENTHOS, ALGAE, FISH, WATER POLLUTION, WATER POLLUTION W72-07841
, SUSPENDED LOAD, BIOINDICATORS, WATER POLLUTION, MICROB: /EASONAL W72-07896
CAL POWER PLANTS, HEAT TRANSFER, WATER POLLUTION EFFECTS, WATER TE W72-07907
S, ALGAE, FISH, WATER POLLUTION, WATER POLLUTION SOURCES.: /BENTHO W72-07841
GLENOPHYTA, PROTOZOA, EQUIPMENT, WATER POLLUTION EFFECTS.: /AS, EU W72-07661
DENSERS, PROTOZOA, ALGAE, FUNGI, WATER POLLUTION EFFECTS.: /R, CON W72-07526

NUT/ *ALGICIDES, *ALGAL CONTROL,	WATER POLLUTION, EUTROPHICATION,	W72-07508
I, LIGHT, PHOSPHATES, NUTRIENTS,	WATER POLLUTION EFFECTS, RADICACT	W72-07143
NNESOTA, *TOXICITY, COAGULATION,	WATER POLLUTION EFFECTS, *ALGAL T	W72-07360
E, ON-SITE TESTS, INVERTEBRATES,	WATER POLLUTION EFFECTS, ECOLOGY,	W73-00348
ER, NAVIGATION, CONSUMPTIVE USE,	WATER POLLUTION, OXYGEN, SOIL PRO	W72-14878
POLLUTANTS, METHODOLOGY, RIVERS,	WATER POLLUTION SOURCES, ION EXCH	W73-00286
ICIDES, ACARICIDES, FOOD CHAINS,	WATER POLLUTION SOURCES.: /S PEST	W73-00238
TON, CHLAMYDOMONAS, CHLOROPHYTA,	WATER POLLUTION EFFECTS, ABSORPTI	W73-00281
ON EXCHANGE, *CHEMICAL ANALYSIS,	WATER POLLUTION, WATER ANALYSIS,	W73-00286
WATER SAMPLING, ORGANIC MATTER,	WATER POLLUTION EFFECTS, SEWAGE,	W73-00234
LEAR WASTES, PATH OF POLLUTANTS,	WATER POLLUTION SOURCES, PUBLIC H	W73-00790
TH OF POLLUTANTS, PUBLIC HEALTH,	WATER POLLUTION EFFECTS, RADIOECO	W73-00802
GAL TOXINS, MERCURY, DETERGENTS,	WATER POLLUTION SOURCES, ASBESTOS	W73-00703
DERAL GOVERNMENT, AIR POLLUTION,	WATER POLLUTION, TOXICITY, CHEMIC	W73-00703
IC PLANTS, *AQUATIC ENVIRONMENT,	WATER POLLUTION EFFECTS, STREAMS,	W72-14690
AMS, SEPTIC TANKS, FUNGI, ALGAE,	WATER POLLUTION CONTROL.: /, STRE	W72-14363
TROPHY, WATER POLLUTION SOURCES,	WATER POLLUTION EFFECTS, MERCURY,	W72-14782
OXIDATION, POLLUTION ABATEMENT,	WATER POLLUTION TREATMENT, WATER	W72-14469
ISTRIBUTION, BEHAVIOR, SEASONAL,	WATER POLLUTION EFFECTS.: /GES, D	W72-14659
SISTANCE, LETHAL LIMIT, POISONS,	WATER POLLUTION EFFECTS.: /AY, RE	W72-14706
MATTER, *DECOMPOSING ORGANIC M/	WATER POLLUTION EFFECTS, *ORGANIC	W72-14673
*RED TIDE, COASTS, LAKES, PONDS,	WATER POLLUTION CONTROL, AIRCRAFT	W72-14673
C/ SURVEYS, *AERIAL PHOTOGRAPHY,	WATER POLLUTION EFFECTS, EUTROPHI	W72-14728
AE, NUTRIENT REMOVAL, NUTRIENTS,	WATER POLLUTION EFFECTS, WATER QU	W72-14692
STAGES, MESOTROPHY, OLIGOTROPHY,	WATER POLLUTION SOURCES, WATER PO	W72-14782
DIATOMS, GROWTH RATES, NITROGEN,	WATER POLLUTION EFFECTS, RADIOACT	W73-00380
EA WATER, NUTRIENT REQUIREMENTS,	WATER POLLUTION.: /A, CULTURES, S	W73-00380
S, EUTROPHICATION, MARINE ALGAE,	WATER POLLUTION SOURCES, SPATIAL	W73-00431
ADATION, CARBON, ORGANIC MATTER,	WATER POLLUTION SOURCES, CHLOROPH	W73-00379
IENTS, CHRYSOPHYTA, COMPETITION,	WATER POLLUTION SOURCES, POLLUTAN	W73-00441
LECTRON MICROSCOPY, PHOTOGRAPHY,	WATER POLLUTION EFFECTS, WATER PU	W73-00432
ES, *THERMAL POLLUTION, *COASTS,	WATER POLLUTION EFFECTS, AQUATIC	W73-01632
ADA, *LAKES, *THERMAL POLLUTION,	WATER POLLUTION, COOLING, THERMAL	W73-01704
REMENT, *NEVADA, AQUATIC PLANTS,	WATER POLLUTION SOURCES, SEDIMENT	W73-01673
NT FACILITIES/ *WATER TREATMENT,	WATER POLLUTION EFFECTS, *TREATME	W73-01669
ATICS, LIMNOLOGY, ECOLOGY, FISH,	WATER POLLUTION, PROTOZOA, OHIO,	W73-01615
XYGEN, TURBIDITY, BENTHIC FAUNA,	WATER POLLUTION EFFECTS, STRIPED	W73-02029
CKISH WATER, FRESHWATER, *ALGAE,	WATER POLLUTION EFFECTS.: /S, BRA	W73-02548
RIFICATION, POLLUTION ABATEMENT,	WATER POLLUTION CONTROL, COAGULAT	W73-02187
TROPHICATION, CHEMICAL ANALYSIS,	WATER POLLUTION SOURCES, BIOASSAY	W73-02478
*ABSORPTION, PESTICIDE RESIDUES,	WATER POLLUTION EFFECTS, *CALIFOR	W73-02105
CES, BIODEGRADATION, ECOSYSTEMS,	WATER POLLUTION.: /POLLUTION SOUR	W73-00853
GELLATES, CRUSTACEANS, ANNELIDS,	WATER POLLUTION EFFECTS, STANDING	W73-00935
OSYSTEMS, BIODEGRADATION, ALGAE,	WATER POLLUTION EFFECTS, HEAVY ME	W73-00840
MUNITIES, ESTUARINE ENVIRONMENT,	WATER POLLUTION SOURCES, BIODEGRA	W73-00853
TTERNS, ECOLOGICAL DISTRIBUTION,	WATER POLLUTION EFFECTS, F: /N PA	W73-00918
ANOPHYTA, ANALYTICAL TECHNIQUES,	WATER POLLUTION EFFECTS, NUTRIENT	W73-01269
ABORATORY EQUIPMENT, FRESHWATER,	WATER POLLUTION SOURCES, INSTRUME	W73-01445
ROWTH RATES, DEFICIENT ELEMENTS,	WATER POLLUTION EFFECTS, AQUATIC	W73-01445
MICAL REACTIONS, CARBON DIOXIDE,	WATER POLLUTION SOURCES, OIL SPIL	W73-01074
OLIDS, ORGANIC MATTER, NITROGEN,	WATER POLLUTION SOURCES, ENVIRONM	W73-01066
PLATFORMS, NUCLEAR POWERPLANTS,	WATER PROPERTIES, SALINITY, WATER	W73-02029
ATER), CHELATION, MINE DRAINAGE,	WATER PROPERTIES.: /Y, HARDNESS(W	W72-14694
SEWAGE, WATER QUALITY, AERATION,	WATER PROPERTIES, *NEW HAMPSHIRE,	W72-07890
EQUATIONS, AQUATIC ENVIRONMENT,	WATER PROPERTIES, ALGAL CONTROL,	W72-11724
RIA, MARINE PLANTS, ENGINEERING,	WATER PROPERTIES, PHYTOPLANKTON,	W72-03220
AG, WASTE ASSIMILATIVE CAPACITY,	WATER PROPERTIES, WATER POLLUTION	W72-04279
LGAE, NUTRIENT, *BIODEGRADATION,	WATER PURIFICATION, *AEROBIC TREA	W72-05315
D WASTES, SEPARATION TECHNIQUES,	WATER PURIFICATION, *OXIDATION LA	W71-13356
GRAPHY, WATER POLLUTION EFFECTS,	WATER PURIFICATION, *SYSTEMATICS.	W73-00432
TIC PLANTS, CHEMICAL PROPERTIES,	WATER QUALITY.: /TOPLANKTON, AQUA	W72-14542
MENT, WATER POLLUTION TREATMENT,	WATER QUALITY, STREAMS, LAKES, BI	W72-14469
RIENTS, WATER POLLUTION EFFECTS,	WATER QUALITY, HERBICIDES, ODOR-P	W72-14692
TION SOURCES, NUTRIENTS, SEWAGE,	WATER QUALITY, AERATION, WATER PR	W72-07890
RIENTS, WATER POLLUTION EFFECTS,	WATER QUALITY, SEASONAL, SUSPENDE	W72-07896
OROPHYTA, EUGLENOPHYTA, GRAZING,	WATER QUALITY.: / CYANOPHYTA, CHL	W72-07899
A, *WATER ANALYSIS, LAKES, SAMP/	WATER QUALITY CONTROL, *CYANOPHYT	W72-07890
DIOXIDE, SESTON, FERTILIZATION,	WATER QUALITY, RADON RADIOISOTOPE	W72-07648
ION PATTERNS, WATER TEMPERATURE,	WATER QUALITY, PLANKTON, ANALYTIC	W72-07663
HAMBERS, *LIFE CYCLES, CULTURES,	WATER QUALITY, SCENEDESMUS, LABOR	W72-07144
EMPERATURE, SALINITY, SEA WATER,	WATER QUALITY, ON-SITE TESTS, HYD	W72-09092
UTRIENTS, *PRIMARY PRODUCTIVITY,	WATER QUALITY, TROPHIC LEVEL, AQU	W72-08986
UTROPHICATION, LABORATORY TESTS,	WATER QUALITY CONTROL, *KINETICS,	W72-08376
DUCTION, RHODOPHYTA, PHAEOPHYTA,	WATER QUALITY, *MARINE ALGAE.: /O	W73-01449
ITY, HYDROGEN ION CONCENTRATION,	WATER QUALITY, NUTRIENTS, COLORIM	W73-00916
LYSIS, AQUATIC ALGAE, NUTRIENTS,	WATER QUALITY, WATER ANALYSIS, FR	W73-00918
RESERVOIR STORAGE, *HYPOLIMNION,	WATER QUALITY CONTROL.: / ALGAE,	W73-02138
ROPHICATION, AQUATIC ECOSYSTEMS,	WATER QUALITY, *ALGAE, *LIGHT PEN	W73-02349
LOGICAL PROPERTIES, FRESH WATER,	WATER QUALITY.: /AE, ECOLOGY, BIO	W71-10580
ES, PHOTOSYNTHESIS, RESPIRATION,	WATER QUALITY, STREAMFLOW, AGRICU	W71-11017
PHYTA, *EUTROPHICATION, BIOMASS,	WATER QUALITY, CHLOROPHYLL, NITRO	W71-10645
, AMMONIA, NITRATES, PHOSPHATES,	WATER QUALITY, NITROGEN, NUTRIENT	W71-11033
ANT GROWTH, TOXICITY, NUTRIENTS,	WATER QUALITY, NITRATES, PLANNING	W71-13172
ATION, *CULTIVATED LANDS, *IOWA,	WATER QUALITY, FERTILIZERS, NUTRI	W71-12252
VANCOUVER(WASH), PORTLAND(ORE),	WATER QUALITY ANALYZER.: /(WASH),	W72-03215
NITRIFICATION, *DENITRIFICATION,	WATER QUALITY CONTROL, EUTROPHICA	W72-00800
PERATURE, ANALYTICAL TECHNIQUES,	WATER QUALITY CONTROL, *WASTE WAT	W72-00422
ICAL TECHNIQUES, ENZYMES, COLOR,	WATER QUALITY CONTROL, RESEARCH A	W72-04070
ITY CONTROL, *FLOW AUGMENTATION,	WATER QUALITY, RESERVOIRS, WATER	W72-04260
ALGAE, WATER POLLUTION EFFECTS,	WATER QUALITY, CHEMICAL DEGRADATI	W72-01085

OST COMPARISONS, BENEFICIAL USE, WATER QUALITY, CALIFORNIA, BIOCON W72-00974
AL ENGINEERING, WATER TREATMENT, WATER QUALITY CONTROL, STANDARDS, W72-01660
POLLUTION TREATMENT, EVALUATION, WATER QUALITY.: /T GROWTH, WATER W72-11702
ONTROL, WATER POLLUTION EFFECTS, WATER QUALITY, CARBON.: /S, CHEMC W72-11724
ATOMS, NUTRIENTS, PHYTOPLANKTON, WATER QUALITY, MICHIGAN, SURFACE W72-10875
SIS, WATER ANALYSIS, EVALUATION, WATER QUALITY CONTROL, *BIOASSAY, W72-10883
S, *EUTROPHICATION, *DETERGENTS, WATER QUALITY CONTROL, DOMESTIC W W72-11186
YCETES, FUNGI, LABORATORY TESTS, WATER QUALITY CONTROL.: / ACTINOM W72-11604
N CONTROL, MANAGEMENT, ANALYSIS, WATER QUALITY, CYANOPHYTA, NUISAN W72-09171
ON, WASTE TREATMENT, INDUSTRIES, WATER QUALITY, TECHNOLOGY, SURFAC W72-09169
TENANCE, *WASTE WATER TREATMENT, WATER QUALITY CONTROL, *REVIEWS.: W72-09386
AT/ *EUTROPHICATION, *RECYCLING, WATER QUALITY CONTROL, SEWAGE TRE W72-09378
ION CONCENTRATION, EFFICIENCIES, WATER QUALITY CONTROL, EUTROPHICA W72-04788
TION, OXIDATION, ORGANIC MATTER, WATER QUALITY, WATER CHEMISTRY, W W72-04853
AE, AQUATIC PLANTS, CLAYS, ODOR, WATER QUALITY, FOOD CHAINS, DISSO W72-04759
IDE, HYDROGEN ION CONCENTRATION, WATER QUALITY CONTROL, WASTE WATE W72-04431
ANOPHYTA, ALKALINITY, NUTRIENTS, WATER QUALITY, DIATOMS, CHLOROPHY W72-05610
OROPHYLL, ENZYMES, PERMEABILITY, WATER QUALITY CONTROL, WATER POLL W72-06037
WIMMING, BOATING, DOCUMENTATION, WATER QUALITY, SILTING, AGRICULTU W72-05869
ION, LEGISLATION, PUBLIC HEALTH, WATER QUALITY CONTROL, ALGAL CONT W72-06655
MENTS, *WATER POLLUTION CONTROL, WATER QUALITY CONTROL, PHOSPHORUS W72-06638
NIMALS, *BIOTA, *AQUATIC PLANTS, WATER QUALITY, ENVIRONMENTAL EFFE W72-06526
SION, ENZYMES, CHLEATION, ALGAE, WATER QUALITY, BIODEGRADATION, SE W72-06837
TICAL MODELS, WATER TEMPERATURE, WATER QUALITY, CARBON DIOXIDE, HY W72-06274
CARBON FIXATION, EUTROPHICATION, WATER QUALITY CONTROL, MASSACHUSE W72-06046
L, EUTROPHICATION, ALGAL BLOOMS, WATER QUALITY.: /ADATION, FISHKIL W71-08768
ES, INDUSTRIAL WASTES, SAMPLING, WATER QUALITY CONTROL, COST ANALY W71-08953
*ALGAE,/ *ESTUARIES, *NUTRIENTS, WATER QUALITY CONTROL, *ECOLOGY, W71-08775
, PHOSPHORUS, WATER QUALITY ACT, WATER QUALITY.: /OXYGEN, NITROGEN W71-06444
ED OXYGEN, NITROGEN, PHOSPHORUS, WATER QUALITY ACT, WATER QUALITY. W71-06444
OSPHATES, ALGAE, EUTROPHICATION, WATER QUALITY, RIVERS, IOWA, CORN W71-06445
VIRONMENT, *POLLUTION ABATEMENT, WATER QUALITY, STANDARDS, WATER P W71-08214
CTS, *KENTUCKY, SULFUR BACTERIA, WATER QUALITY, SULFATES, HYDROGEN W71-07942
TY, ODOR, OXYGEN SAG, TURBIDITY, WATER QUALITY, WATER TREATMENT, N W71-09958
PULATION, WASTE WATER TREATMENT, WATER QUALITY CONTROL.: /HUMAN PO W71-09788
TE DISPOSAL, PATH OF POLLUTANTS, WATER QUALITY.: / PESTICIDES, WAS W71-05553
USE, *COOLING WATER, *FISHERIES, WATER QUALITY, FISH ATTRACTANTS, W70-09596
IENTS, SURFACE WATERS, NITROGEN, WATER QUALITY, PHOSPHORUS, AEROBI W70-10405
ES, *WATER POLLUTION, MINNESOTA, WATER QUALITY, NUTRIENTS, SEDIMEN W71-03012
ODUCTIVITY, ESSENTIAL NUTRIENTS, WATER QUALITY, WATER CHEMISTRY, C W70-04080
ECTS, NUTRIENTS, EUTROPHICATION, WATER QUALITY, WATER POLLUTION CO W70-08249
*ALGAE, *EUTROPHICATION, LAKES, WATER QUALITY, PHOSPHORUS, PHOSPH W70-08944
, NITRATES, HUDSON RIVER, ALGAE, WATER QUALITY, SEWAGE DISPOSAL, W W70-06509
RIOR APPROPRIATION, GROUNDWATER, WATER QUALITY, COMPETING USES, DI W70-07100
S, *WATER UTILIZATION, *IMPAIRED WATER QUALITY, RIPARIAN RIGHTS, P W70-07100
ECTS, NUTRIENTS, EUTROPHICATION, WATER QUALITY, WATER POLLUTION CO W70-08248
C MATTER, *OXYGEN SAG, *IMPAIRED WATER QUALITY, *ESTUARIES, *DELAW W70-09189
RADIATION, WATER QUALITY, ALGAE, WATER QUALITY CONTROL, EUTROPHICA W72-12487
HOTOGRAPHY, *INFRARED RADIATION, WATER QUALITY, ALGAE, WATER QUALI W72-12487
OPERTIES, MICROSCOPY, LIMNOLOGY, WATER QUALITY, ECOLOGY, AQUATIC H W72-13142
UTRIENTS, LAKE HURON, EFFLUENTS, WATER QUALITY, PHOSPHORUS, PHYTOP W72-13662
HICATION, *HYDROLOGY, *NEW YORK, WATER QUALITY, NUTRIENTS, ALGAE, W72-13851
S, *MATHEMATICAL MODELS, MIXING, WATER RE-USE, LIGHT INTENSITY, EV W71-07099
S, *FISH HATCHERIES, *SALMONIDS, WATER REQUIREMENTS, TEMPERATURE, W71-11006
ONMENTAL EFFECTS, INVERTEBRATES, WATER RESOURCES DEVELOPMENT.: /IR W71-10577
YSTEMS, ESTUARIES, CORAL, REEFS, WATER RESOURCES, SYSTEMS ANALYSIS W71-13723
ATION, AQUATIC ALGAE, NUTRIENTS, WATER RESOURCES, WASTE ASSIMILATI W71-13256
CATION, *MEASUREMENT, *SAMPLING, WATER RESOURCES, AREA REDEVELOPME W72-03215
WATER RESOURCES RESEARCH.: W71-01192
ICALS, COPPER SULFATE, CHLORINE, WATER RESOURCES, METHODOLOGY, EVA W71-05083
, DIVERSION, RECIRCULATED WATER, WATER REUSE, LEGAL ASPECTS, JUDIC W70-07100
TER TREATMENT, *DENITRIFICATION, WATER REUSE, *NITROGEN, PHOSPHORU W72-04309
PORATION, RESERVOIRS, NUTRIENTS, WATER REUSE.: /ION, RESERVOIR EVA W72-02412
ARY TREATMENT, SEWAGE EFFLUENTS, WATER REUSE.: /IS, ECOLOGY, TERTI W72-03970
TS, PHOSPHORUS, ALGAE, NITRATES, WATER REUSE, SALINITY, PESTICIDES W72-00846
NUTRIENTS, RESERVOIR OPERATION, WATER REUSE, POTABLE WATER, ALGAE W71-13185
*SALINE LAKES, *EUTROPHICATION, WATER SAMPLING, WATER POLLUTION E W72-09111
BIOASSAY, CHLORELLA, NUTRIENTS, WATER SAMPLING, ORGANIC MATTER, W W73-00234
ONTROL, DOMESTIC WASTES, RUNOFF, WATER SOFTENING, ENVIRONMENTAL EF W72-11186
ARDS, PUBLIC HEALTH, INSPECTION, WATER SOFTENING, SOAPS, AQUATIC A W72-01085
EFFECTS, NUTRIENTS, PHOSPHOROUS, WATER SOFTENING, LEGAL ASPECTS, H W71-09429
TION, BIOCHEMICAL OXYGEN DEMAND, WATER SPORTS, PARKS, COST ANALYSI W72-00042
SPHORUS, THERMAL STRATIFICATION, WATER SUPPLY, SUCCESSION, ECONOMI W71-13183
OSPHORUS, RESERVOIRS, FISHERIES, WATER SUPPLY, COSTS, WATER DEMAND W71-13172
WATER, IRRIGATION, RETURN FLOW, WATER SUPPLY, LIVESTOCK, SEWAGE, W71-06435
AGEMENT, GROUNDWATER, WISCONSIN, WATER SUPPLY.: / WASTES, SOIL MAN W71-06443
ION, *LAKES, *MAINE, RECREATION, WATER SUPPLY, WATER POLLUTION EFF W72-04280
, ARROYO GRANDE CREEK.: *LOPEZ WATER SUPPLY PROJECT, LOPEZ CREEK W72-05594
DOMESTIC WATER, *SURFACE WATERS, WATER SUPPLY, EUTROPHICATION, ALG W71-05083
T, *WATER REUSE, *WATER QUALITY, WATER SUPPLY, ALGAE, ACTIVATED CA W70-10381
UCTIVITY, GREAT LAKES, NEW YORK, WATER SUPPLY, SYSTEMATICS, LIMNOL W73-01615
NTS, WATER PROPERTIES, SALINITY, WATER TEMPERATURE, OXYGEN, TURBID W73-02029
ACTERIA, STREAMFLOW, FLOW RATES, WATER TEMPERATURE, WATER CHEMISTR W73-01972
QUES, CENTRIFUGATION, TURBIDITY, WATER TEMPERATURE, ALKALINITY, HY W73-00916
RHODOPHYTA, PHYSICAL PROPERTIES, WATER TEMPERATURE, DATA COLLECTIO W73-00838
COMPOUNDS, ABSORPTION, KINETICS, WATER TEMPERATURE, RESPIRATION, A W73-01269
D AQUATIC PLANTS, AQUATIC ALGAE, WATER TEMPERATURE, ACID STREAMS, W72-14690
EEDS, *WASHINGTON, *CHLOROPHYLL, WATER TEMPERATURE, ALKALINITY, DI W72-14728
OMPOUNDS, PHYTOPLANKTON, CARBON, WATER TEMPERATURE, GREAT LAKES RE W73-00384
EQUIPMENT, TUBIFICIDS, MOLLUSKS, WATER TEMPERATURE, MARINE ANIMALS W73-00374

RIBUTION, BENTHIC FAUNA, SIEVES,	WATER TEMPERATURE,: /LOGICAL DIST	W73-00367
ATER, ORGANIC MATTER, MORTALITY,	WATER TEMPERATURE, PHAEOPHYTA, SA	W73-00366
NVIRONMENT, SEA WATER, SALINITY,	WATER TEMPERATURE, CHEMICAL ANALY	W72-08141
NG, NATIONAL PARKS, HOT SPRINGS,	WATER TEMPERATURES, NOSTOC STREAM	W72-08508
S, CRUSTACEANS, WORMS, MOLLUSKS,	WATER TEMPERATURE, SALINITY, SEA	W72-09092
RROPHYTA, DISTRIBUTION PATTERNS,	WATER TEMPERATURE, WATER QUALITY,	W72-07663
EN ION CONCENTRATION, MINNESOTA,	WATER TEMPERATURE, DISSOLVED OXYG	W72-07145
, INSTRUMENTATION, FLUORESCENCE,	WATER TEMPERATURES, THERMAL STRES	W72-07225
NESS(WATER), CONDUCTIVITY, ZINC,	WATER TEMPERATURE, AL: /GEN, HARD	W72-07890
N, SEA WATER, SALINITY, PHENOLS,	WATER TEMPERATURE, CHEMICAL ANALY	W72-07710
ANSFER, WATER POLLUTION EFFECTS,	WATER TEMPERATURE, ALGAE, BULLHEA	W72-07907
GRAPHY, *PACIFIC OCEAN, ECOLOGY,	WATER TEMPERATURE, PHYTOPLANKTON,	W72-07892
KE STRUCTURES, DISCHARGE(WATER),	WATER TEMPERATURE.: /IMNION, INTA	W70-09614
G, ALGAE, SOLVENTS, SURFACTANTS,	WATER TEMPERATURE, PLANT PIGMENTS	W70-09429
ZOA, HYDROGEN ION CONCENTRATION,	WATER TEMPERATURE, BACTERIA, ALGA	W71-05626
THERMAL POLLUTION, *GREENHOUSES,	WATER TEMPERATURE, POWERPLANTS, D	W70-08832
N ION CONCENTRATION, PERIPHYTON,	WATER TEMPERATURE.: /DES, HYDROGE	W72-05610
US, NITROGEN, NUTRIENTS, SEWAGE,	WATER TEMPERATURE, LIGHT PENETRAT	W72-04734
ON, EVALUATION, WATER CHEMISTRY,	WATER TEMPERATURE, DIS: /OPHICATI	W72-06526
OWTH RATES, MATHEMATICAL MODELS,	WATER TEMPERATURE, WATER QUALITY,	W72-06274
HYTA, CHRYSOPHYTA, EUGLENOPHYTA,	WATER TEMPERATURE, THERMAL STRATI	W72-04855
N, WEATHER, RESERVOIR OPERATION,	WATER TEMPERATURE, PHYTOPLANKTON,	W72-05282
OLORADO, *RECREATION FACILITIES,	WATER TEMPERATURE, CHEMICAL ANALY	W72-04791
, ZOOPLANKTON, FISH POPULATIONS,	WATER TEMPERATURE, THERMAL STRATI	W72-04853
S, *THERMAL POLLUTION, *REVIEWS,	WATER TEMPERATURE, DISSOLVED OXYG	W71-07417
ICAL PROPERTIES, EUTROPHICATICN,	WATER TEMPERATURE, WATER POLLUTIO	W71-12092
RPHOLOGY, AQUATIC ALGAE, MOSSES,	WATER TEMPERATURE, WATER CHEMISTR	W71-11949
YGEN, *PHYTOPLANKTON, ESTUARIES,	WATER TEMPERATURE, TURBIDITY, PHO	W71-11876
INE, HYDROGEN ION CONCENTRATION,	WATER TEMPERATURE, DISSOLVED OXYG	W71-11914
, TRACERS, MARINE ALGAE, HAWAII,	WATER TEMPERATURE, WAVES, LIGHT I	W71-11486
, PYRROPHYTA, SAMPLING, ECOLOGY,	WATER TEMPERATURE.: / CHRYSOPHYTA	W72-03543
ROLOGIC BUDGET, SOLAR RADIATION,	WATER TEMPERATURE, SALT BALANCE,	W72-02274
T, FLORIDA, ELECTRIC GENERATORS,	WATER TEMPERATURE, RHODOPHYTA, PH	W72-11252
ERIE, LAKE HURON, LAKE SUPERIOR,	WATER TEMPERATURE, HYDROGEN ION C	W72-10875
BRATES, PHYTOPLANKTON, PLANKTON,	WATER TEMPERATURE, E. COLI, BENTH	W72-09668
, OLIGOTROPHY, DISSOLVED OXYGEN,	WATER TEMPERATURE, HYDROGEN ION C	W72-13871
T GROWTH, ANALYTICAL TECHNIQUES,	WATER TEMPERATURE.: /RIENTS, PLAN	W72-13651
, BIOMASS, ANAEROBIC CONDITIONS,	WATER TEMPERATURE, CHLORIDES, PHO	W72-12940
AE, ANALYTICAL TECHNIQUES, WASTE	WATER TREATMENT, NUTRIENTS, NUTRI	W72-12966
*OXIDATION PONDS, *ALGAE, *WASTE	WATER TREATMENT, *BYPRODUCTS, MIX	W72-13508
Y CONTROL, EUTROPHICATION, WASTE	WATER TREATMENT, SEDIMENTATION RA	W72-12487
*COLD/ *SEWAGE BACTERIA, *WASTE	WATER TREATMENT, *SEWAGE LAGOONS,	W72-12565
XIDATION LAGOONS, *ALGAE, *WASTE	WATER TREATMENT, *MATHEMATICAL MO	W72-12289
PERATION AND MAINTENANCE, *WASTE	WATER TREATMENT, WATER QUALITY CO	W72-09386
OPERATION AND MAINTENANCE, WASTE	WATER TREATMENT.: /NTIAL, ODORS,	W72-10401
MENT, TERTIARY TREATMENT, *WASTE	WATER TREATMENT.: /, *WATER TREAT	W72-09393
TECHNIQUES, MICROBIOLOGY, *WASTE	WATER TREATMENT.: /S, ANALYTICAL	W72-09590
E, *ODOR, SURVEYS, MICROBIOLOGY,	WATER TREATMENT, PHYTOPLANKTON, F	W72-10076
STS, COST ANALYSIS, ALGAE, WASTE	WATER TREATMENT.: / LABORATORY TE	W72-10390
DISSOLVED OXYGEN, SLUDGE, WASTE	WATER TREATMENT.: /ION, PROTOZOA,	W72-10233
AE, *BACTERIA, *SYMBIOSI/ *WASTE	WATER TREATMENT, *NUTRIENTS, *ALG	W72-11100
*SEWAGE EFFLUENTS, *ALGAE, WASTE	WATER TREATMENT, ACTIVATED SLUDGE	W72-09176
NALYSIS, NITROGEN, CLAYS, *WASTE	WATER TREATMENT.: /SLUDGE, COST A	W72-11642
DROGEN ION CONCENTRATION, *WASTE	WATER TREATMENT.: /PROPERTIES, HY	W72-11833
ATMENT FACILITIES, ALGAE, *WASTE	WATER TREATMENT, *ALGAL CONTROL,	W72-02975
ESIGN CRITERIA, KINETICS, *WASTE	WATER TREATMENT.: / TECHNIQUES, D	W72-02363
*AQUATIC AL/ *PHOSPHORUS, *WASTE	WATER TREATMENT, *ALGAL CONTROL,	W72-03970
UM, MAGNESIUM, NUTRIENTS, *WASTE	WATER TREATMENT, TERTIARY TREATME	W72-03613
HEMICAL OXYGEN DEMAND, D/ *WASTE	WATER TREATMENT, *TURBIDITY, BIOC	W72-03753
SLUDGE, ALGAE, BACTERIA, *WASTE	WATER TREATMENT, EUTROPHICATION.:	W72-04086
*ALGAL CONTROL, AERATION, *WASTE	WATER TREATMENT.: /N, DIFFUSION,	W72-00800
S, WATER QUALITY CONTROL, *WASTE	WATER TREATMENT.: /ICAL TECHNIQUE	W72-00422
SINFECTION, FLOCCULATION, *WASTE	WATER TREATMENT.: /FILTRATION, DI	W72-00411
, COST ANALYSIS, MICHIGAN, WASTE	WATER TREATMENT.: / SPORTS, PARKS	W72-00042
OUS EARTH, *POROUS MEDIA, *WASTE	WATER TREATMENT, *DENITRIFICATION	W72-00974
TION, ENVIRONMENTAL ENGINEERING,	WATER TREATMENT, WATER QUALITY CO	W72-01660
AL CHEMICALS, FARM WASTES, WASTE	WATER TREATMENT, CALIFORNIA.: /UR	W71-11698
ION, LIME, / *PHOSPHATES, *WASTE	WATER TREATMENT, *FOAM FRACTIONAT	W71-12200
TREATMENT, *PHOSPHOROUS, *WASTE	WATER TREATMENT, NUTRIENTS, SYNTH	W71-12188
TMENT, *ACTIVATED SLUDGE/ *WASTE	WATER TREATMENT, *BIOLOGICAL TREA	W71-12223
ABOLISM, DESIGN CRITERIA, *WASTE	WATER TREATMENT, EFFLUENTS.: /MET	W71-13341
N, CALCIUM, SEDIMENTATION, WASTE	WATER TREATMENT.: /MS, ACCLIMATIO	W71-13334
RESEARCH AND DEVELOPMENT, WASTE	WATER TREATMENT.: /TIVE CAPACITY,	W71-13256
BACTERIA, CENTRIFUGATION, *WASTE	WATER TREATMENT.: /N, COLIFORMS,	W71-13326
*DENITRIFICATION, *NITRA/ *WASTE	WATER TREATMENT, *NITRIFICATION,	W71-13939
IQUES, COA/ *COAGULATION, *WASTE	WATER TREATMENT, ANALYTICAL TECHN	W71-13356
, WATER POLLUTION EFFECTS, WASTE	WATER TREATMENT, CYANOPHYTE, ENGL	W71-07973
IZE, ANAEROBIC CONDITIONS, WASTE	WATER TREATMENT.: /RS, PARTICLE S	W71-08223
IOCH/ *OXIDATION LAGOONS, *WASTE	WATER TREATMENT, *WATER RE-USE, B	W71-07123
ORGANISMS, BIODEGRADATION, WASTE	WATER TREATMENT, WATER POLLUTION	W71-07097
EROBIC CONDITIONS, SLUDGE, WASTE	WATER TREATMENT, BIODEGRADATION,	W71-07084
N, ORGANIC LOADING, ALGAE, WASTE	WATER TREATMENT, BIOCHEMICAL OXYG	W71-07096
BIOCHEMICAL OXYGEN DEMAND, WASTE	WATER TREATMENT.: /Y, COLIFORMS,	W71-07106
BIO/ *OXIDATION LAGOONS, *WASTE	WATER TREATMENT, ORGANIC LOADING,	W71-07124
YDROGEN ION CONCENTRATION, WASTE	WATER TREATMENT, KANSAS.: /AND, H	W71-07113
BIOCHEMICAL OXYGEN DEMAND, WASTE	WATER TREATMENT.: /ST ANALYSIS, *	W71-07099
BIOCHEMICAL OXYGEN DEMAND, WASTE	WATER TREAT: /EROBIC CONDITIONS,	W71-07100
BIOCHEMICAL OXYGEN DEMAND, WASTE	WATER TREATMENT, WATER POLLUTION	W71-07098
BIOCHEMICAL OXYGEN DEMAND, WASTE	WATER TREATMENT, *NORTH CAROLINA.	W71-07382
BIOCHEMICAL OXYGEN DEMAND, WASTE	WATER TREATMENT.: /C CONDITIONS,	W71-07109

ENT, *COAGULATIO/ *ALGAE, *WASTE WATER TREATMENT, *TERTIARY TREATM W71-09629
METALS, HUMAN POPULATION, WASTE WATER TREATMENT, WATER QUALITY CO W71-09788
N SAG, TURBIDITY, WATER QUALITY, WATER TREATMENT, NITROGEN CYCLE.: W71-09958
ICATICN, DENITRIFICATION, *WASTE WATER TREATMENT.: /TATION, NITRIF W71-09450
N DEMAND, COOLING TOWERS, *WASTE WATER TREATMENT.: /CHEMICAL OXYGE W71-09524
TY CONTROL, COST ANALYSIS, WASTE WATER TREATMENT, NUTRIENT REQUIRE W71-08953
NG, DRYING, COST ANALYSES, WASTE WATER TREATMENT.: /TION, DEWATERI W71-08951
LGAL CONTROL, *CHLORELLA, *WASTE WATER TREATMENT.: /ROPHICATION, A W72-04788
*WASTE DILUTION, LAGOONS, *WASTE WATER TREATMENT, ALGAL CONTROL, E W72-04787
QUES, TERTIARY TREATMENT, *WASTE WATER TREATMENT.: /LYTICAL TECHNI W72-04983
EMENT/ *ALGAE, *KINETICS, *WASTE WATER TREATMENT, *NUTRIENT REQUIR W72-04789
SEWAGE TREATME/ *PATENTS, *WASTE WATER TREATMENT, *LIQUID WASTES, W72-05315
RESEARCH AND DEVELOPMENT, WASTE WATER TREATMENT, BIODEGRADATION, W72-06655
EWAGE TREATMENT, CULTURE, *WASTE- WATER TREATMENT, *CARBON DIOXIDE, W72-06612
GEN DEMAND, PILOT PLANTS, *WASTE WATER TREATMENT, WYOMING, *AERATE W72-06838
TREATMENT, PUBLIC HEALTH, WASTE WATER TREATMENT, RESEARCH AND DEV W72-06638
ON, WATER QUALITY CONTROL, WASTE WATER TREATMENT, CHLOROPHYTA, GRO W72-04431
*WATER POLLUTION CONTROL, *WASTE WATER TREATMENT, *COST-BENEFIT AN W72-04734
, WATER REUSE, *NITROGEN/ *WASTE WATER TREATMENT, *DENITRIFICATION W72-04309
OL, RECREATION, HYDROLOGIC DATA, WATER TREATMENT, STREAM GAGES, GR W72-05594
ORATORY TESTS, DETERGENTS, WASTE WATER TREATMENT.: /TOXICITY, *LAB W70-08976
RICTICN, SEWERS, SLURRIES, WASTE WATER TREATMENT, SLUDGE TREATMENT W70-06968
SYNTHESIS, STABILIZATION, *WASTE WATER TREATMENT, *FARM WASTES.: / W70-06619
LAGOONS, BIOCHEMICAL OXY/ *WASTE WATER TREATMENT, *PONDS, *ALGAE, W70-07838
N TECHNIQUES, COAGULATION, WASTE WATER TREATMENT.: /STS, SEPARATIO W70-08904
BIOCHEMICAL OXYGEN DEMAND, WASTE WATER TREATMENT, MICROORGANISMS.: W71-06033
IRATION, ACTIVATED SLUDGE, WASTE WATER TREATMENT, NUTIRIENTS, AERO W71-04079
YDROGEN ION CONCENTRATION, WASTE WATER TREATMENT, WATER POLLUTION W71-05155
BIOCHEMICAL OXYGEN DEMAND, WASTE WATER TREATMENT.: /CIPAL WASTES, W71-03896
REATMENT, *EUTROPHICATION, WASTE WATER TREATMENT, ALGAE, EFFLUENTS W70-09325
ISPOSAL, SEWAGE TREATMENT, WASTE WATER TREATMENT.: /VERS, SEWAGE D W70-09947
TATION, TRICKLING FILTERS, WASTE WATER TREATMENT.: /ATION, SEDIMEN W70-10427
GAE, *SUSPENSION, HARVES/ *WASTE WATER TREATMENT, *FILTRATION, *AL W70-10174
ENT, / *SEWAGE TREATMENT, *WASTE WATER TREATMENT, *TERTIARY TREATM W70-10381
N, CHLORELLA, SCENEDESMUS, WASTE WATER TREATMENT.: /N CONCENTRATIO W71-00138
EFLUENTS, SETTLING BASINS, WASTE WATER TREATMENT, LABORATORY TESTS W71-00116
RESPIRATION, SCENEDESMUS, WASTE WATER TREATMENT.: /HOTOSYNTHESIS, W71-00139
, WATER POLLUTION CONTROL, WASTE WATER TREATMENT.: /ICA, *BACTERIA W71-01899
ADATICN, SEWAGE TREATMENT, WASTE WATER TREATMENT, PONDS, SCENDESOM W72-07661
NDS, ANAEROBIC CONDITIONS, WASTE WATER TREATMENT, CHLOROPHYTA, CYA W72-07664
WAGE TREATMENT/ *PATENTS, *WASTE WATER TREATMENT, *FILTRATION, *SE W72-14469
BACTERIA, WASTE WATER TREATMENT, WATER TREATMENT, MODEL STUDIES.: / W73-01362
CEN/ *PHOSPHORUS, *ALGAE, *WASTE WATER TREATMENT, HYDROGEN ION CON W73-01367
ON, CLAY, ALGAE, BACTERIA, WASTE WATER TREATMENT, WATER TREATMENT, W73-01362
NTS, NITROGEN, NUTRIENT REMOVAL, WATER TREATMENT, ASSAY, BACTERIA. W73-02478
TH, ANALYTICAL TECHNIQUES, WASTE WATER TREATMENT, TASTE, ODOR, POT W73-02144
NG, MARSHES, SOIL EROSION, WASTE WATER TREATMENT.: /ORES, MONITORI W73-02479
S, DISSOLVED OXYGEN, PHOSPHATES, WATER YIELD IMPROVEMENT, WATER PO W71-01491
NKTON, NITROGEN, ABSORPTION, SEA WATER.: / *MARINE ALGAE, PHYTOPLA W72-07497
S, SAMPLING, WATER ANALYSIS, SEA WATER.: /CTROSCOPY, TRACE ELEMENT W73-00832
BITORS, ECOLOGY, ECOTYPES, FRESH WATER.: /ETRY, PENNSYLVANIA, INHI W72-04743
, WATER POLLUTION EFFECTS, FRESH WATER.: /ES, COBALT RADICISOTCPES W71-09250
PHOSPHATES, DETERGENTS, POTABLE WATER.: /F, PHOSPHORUS, NITROGEN, W72-10625
WEBS, INTERTIDAL AREAS, CONNATE WATER.: /IES, TIDAL MARSHES, FOOD W73-00796
OREGCN, CALIFORNIA, ALASKA, SEA WATER.: /NTS, CANADA, WASHINGTON, W73-00428
ION, AMMONIA, PHOSPHATES, SALINE WATER.: /PERATURE, LIGHT PENETRAT W72-05469
), LAKE WASHINGTON(WASH), DYEING WATER.: /RVOIRS(NEB), MADISCN(WIS W72-04759
AND)/ *UNITED KINGDOM, ESTHWAITE WATER(ENGLAND), BLELHAM TARN(ENGL W71-13183
NGLA/ *UNITED KINGDOM, ESTHWAITE WATER(ENGLAND), LAKE WINDERMERE(E W71-13185
IC ALGAE, *EPIPELIC ALGAE, SHEAR WATER(ENGLAND), AMPHORA OVALIS, O W71-12855
NGLAND), EXOENZYME PR/ ESTHWAITE WATER(ENGLAND), LAKE WINDERMERE(E W72-03224
TION CAPACITY, *ORGANIC / *WASTE WATER(POLLUTION), *WASTE ASSIMILA W71-11017
*WATER POLLUTION CONTROL, *WASTE WATER(POLLUTION), VIRGINIA, NUTRI W72-05473
IVITY, RADIOACTIVE WASTES, WASTE WATER(POLLUTION), WATER POLLUTION W72-05506
TER/ *PHOSPHATES, *LAKES, *WASTE WATER(POLLUTION), *HYDROLYSIS, DE W71-00116
RIGATION, *FERTILIZATION, *WASTE WATER(POLLUTION), *WASTE WATER DI W72-14878
WASTES, INDUSTRIAL WASTES, WASTE WATER(POLLUTION), BIOASSAY, ECOSY W73-00840
IS, STORM WATER, RAINFALL, WASTE WATER(POLLUTION), EUTROPHICATION, W72-13644
YTA, SEASONAL, SILICA, DISCHARGE(WATER).: /, SCENEDESMUS, CHRYSOPH W72-14797
AQUATIC ANIMALS, LITTORAL, WAVES(WATER).: /OPHYTA, INVERTEBRATES, W71-11012
OLOGICAL DISTRIBUTION, *CURRENTS(WATER), *DINOFLAGELLATES, ECOLOGY W73-00296
RBONATES, BICARBONATES, HARDNESS(WATER), ALKALINITY, ENZYMES, TEMP W72-07514
THERMAL STRATIFICATION, CURRENTS(WATER), ALGAE, LAKES, RIVERS, SEW W70-09947
ENDED LOAD, ALKALINITY, HARDNESS(WATER), ANALYTICAL TECHNIQUES, AM W71-11033
THERMAL POWER PLANTS, *DISCHARGE(WATER), AQUATIC HABITATS, ENVIRON W71-11517
ON, DIATOMS, SEDIMENTS, HARDNESS(WATER), BENTHOS, GRAZING, SYSTEMA W73-01627
PPER SULFATE, SALINITY, HARDNESS(WATER), CHELATION, MINE DRAINAGE, W72-14694
ECHNIQUES, MEASUREMENT, CURRENTS(WATER), CHLOROPHYLL, FLUORESCENCE W72-01095
THERMAL STRATIFICATION, HARDNESS(WATER), COLOR, OLIGOTROPHY, EUTRO W72-04269
, PHOSPHATES, NITROGEN, HARDNESS(WATER), CONDUCTIVITY, ZINC, WATER W72-07890
TRATIFICATION, DIATOMS, CURRENTS(WATER), FISH FARMING.: /ANKTON, S W73-00236
, *GENETICS, *VARIABILITY, WAVES(WATER), MAINE, STATISTICAL METHOD W73-00369
ORIDA, CORAL, SALINITY, CURRENTS(WATER), MOLLUSKS, ENVIRONMENTAL E W72-11876
IA, NITRATES, NITRITES, HARDNESS(WATER), NITROGEN CYCLE, ORGANIC M W72-08048
DISTRIBUTION PATTERNS, CURRENTS(WATER), PARTICLE SIZE, SEDIMENTS, W73-00367
XYGEN, TEMPERATURE, LIGHT, WAVES(WATER), PHYTOPLANKTON: /MPLING, O W72-06283
ING CROP, CURRENTS(WATER), WAVES(WATER), SAMPLING, BENTHOS, GRAZIN W72-04766
CENTRATION, ALKALINITY, HARDNESS(WATER), SAMPLING, DIATOMS, NITROG W71-03028
OPLANKTON, OLIGOTROPHY, HARDNESS(WATER), SOILS, AMINO: /EEDS, PHYT W72-08459
OWERPLANTS, EFFLUENTS, DISCHARGE(WATER), STATISTICAL MODELS.: /, P W71-11583

LANKTON, BIOINDICATORS, CURRENTS(WATER), TIDES, WIND, LIGHT, WAVES W72-09092
ON, INTAKE STRUCTURES, DISCHARGE(WATER), WATER TEMPERATURE.: /IMNI W70-09614
, DEPTH, STANDING CROP, CURRENTS(WATER), WAVES(WATER), SAMPLING, B W72-04766
RIENTS, ALGAE, INFLOW, DISCHARGE(WATER), WATER BALANCE, SEDIMENTS, W72-13851
ED OXYGEN, TEMPERATURE, CURRENTS(WATER), WINDS, THERMOCLINE, ALGAE W72-12992
ORUS, CARBON, NITROGEN, CURRENTS(WATER), WINDS, NEVADA, CALIFORNIA W71-05626
SEA, PHAEOCYSTIS SPP., SOI/ SILT- WATER-PLANT RELATIONSHIPS, NORTH W72-12239
, WATER POLLUTION, LICHENS, SOIL- WATER-PLANT RELATIONSHIPS, MARINE W72-03347
UT, *CYCLING NUTRIENTS, E/ *SOIL- WATER-PLANT RELATIONSHIPS, *FALLO W72-03342
FORESTS, CYCLING NUTRIENTS, SOIL- WATER-PLANT RELATIONSHIPS, RADIOI W72-00949
TURES, PHYTOP/ *POLLUTANTS, *SEA WATER, *ALGAE, *MARINE ALGAE, CUL W70-07280
WATER, *FISHERIES, WATE/ *HEATED WATER, *BENEFICIAL USE, *COOLING W70-09596
*NUCLEAR POWER PLANTS, *POTABLE WATER, *BIOTA, SAMPLING, MARINE F W71-09018
DOR-PRODUCING ALGAE, / *DOMESTIC WATER, *CYANOPHYTA, RESERVOIRS, O W72-10613
ESTIGATIONS, *SCUBA DIVING, *SEA WATER, *DATA COLLECTIONS, COPPER, W72-11925
WATER, *BENEFICIAL USE, *COOLING WATER, *FISHERIES, WATER QUALITY, W70-09596
B/ *RADIOISOTOPES, *ALGAE, *SEA WATER, *FOOD CHAINS, *ABSORPTION, W71-13243
*COPPER, *IONS, *POISONS, *SEA WATER, *FRESH WATER, ALGAE, PHOTO W71-03027
TES, MARINE PLANTS, MARINE/ *SEA WATER, *MARINE ALGAE, *CARBOHYDRA W73-02100
ATED HYDROCARBON PESTICIDES, SEA WATER, *METABOLISM, BIODEGRADATIO W73-00361
Y, DIATOMS, CHLOROPHYTA, THERMAL WATER, *MONTANA, PHOSPHATES, NITR W72-05610
ITRI/ *BACTERIA, *CULTURES, *SEA WATER, *NITRIFICATION, AMMONIA, N W72-05421
, *SYSTEMATICS, PHOTOGRAPHY, SEA WATER, *NORTH CAROLINA, MICROSCOP W73-00284
, *WATER POLLUTION CONTROL, *SEA WATER, *OILS, *NUTRIENTS, AIRCRAF W72-04742
S/ *RADIOCHEMICAL ANALYSIS, *SEA WATER, *SEDIMENTS, *MARINE ANIMAL W72-05965
UATIC MICROORGANISMS, *ACID MINE WATER, *STRIP MINES, *WATER POLLU W71-07942
*AQUATIC WEED CONTROL, *DOMESTIC WATER, *SURFACE WATERS, WATER SUP W71-05083
CROBIOLOGY/ *LAKE ERIE, *POTABLE WATER, *TASTE, *ODOR, SURVEYS, MI W72-10076
*CALIFORN/ *SURFACE WATERS, *SEA WATER, *WATER QUALITY, *ECOLOGY, W71-10577
*NUCLEAR POWER PLANTS, *COOLING WATER, *WATER TEMPERATURE, CONDEN W70-09612
WATER POLLUTION EFFECTS, *HEATED WATER, *WATER TEMPERATURE, *WATER W72-13636
N, *ALGAE, *CYANOPHYTA, *THERMAL WATER, *WYOMING, NATIONAL PARKS, W72-08508
ON, FIL/ *TASTE, *ODOR, *POTABLE WATER, ACTIVATED CARBON, ADSORPTI W72-08357
TREATMENT, TASTE, ODOR, POTABLE WATER, ACTIVATED CARBON, COLIFORM W73-02144
LTRATION, DESALINATION, DRAINAGE WATER, AGRICULTURAL CHEMICALS, FA W71-11698
OPERATION, WATER REUSE, POTABLE WATER, ALGAE, NUTRIENT REQUIREMEN W71-13185
TION TECHNIQUES, SEDIMENTS, OILY WATER, ALGAE.: /ECHNIQUES, SEPARA W71-12467
NIC COMPOUNDS, *NUTRIENTS, FRESH WATER, ALGAE, BACTERIA, NITROGEN, W72-07933
*OIL WASTES, *CRUSTACEANS, OILY WATER, ALGAE, NEMATODES, OYSTERS, W72-07911
NS, *POISONS, *SEA WATER, *FRESH WATER, ALGAE, PHOTOSYNTHESIS, ORG W71-03027
ATES, EUTROPHICATION, IRRIGATION WATER, AQUATIC ALGAE, BIOLOGICAL W72-00974
SAMPLERS, C-/ AUKE BAY, GLACIAL WATER, BECKMAN ANALYZER, VAN DORN W72-09122
OCHEMICAL ANALYSIS, RESINS, *SEA WATER, BIOASSAY, COPEPODS, OYSTER W72-04708
TIDAL EFFECTS, SALINITY, HEATED WATER, BIOMASS, THERMAL STRESS, G W73-00853
Y, *FRESHWATER, *MARSHES, SALINE WATER, BIOMASS, NUTRIENTS, FOOD W W73-01089
ANIMALS, SALINITY, SAMPLING, SEA WATER, BODIES OF WATER, PYRROPHYT W73-00296
A, *TOXICITY, *MARINE ALGAE, SEA WATER, CHLOROPHYTA, CHLAMYDOMONAS W72-08051
, ALGAE, CULTURES, *FLORIDA, SEA WATER, CHLOROPHYLL, NITRATES, NIT W72-09102
ING, THERMAL POWERPLANTS, HEATED WATER, CIRCULATION, INFRARED RADI W73-01704
MAINE, STATISTICAL METHODS, SEA WATER, COASTS, MARINE ALGAE.: /), W73-00369
ODUCTIVITY, *PHYTOPLANKTON, *SEA WATER, COASTS, BIORTHYMS, ON-SITE W72-03228
T, ANALYTICAL TECHNIQUES, HEATED WATER, CONDENSERS, PROTOZOA, ALGA W72-07526
BACTERIA, *ALGAL CONTROL, FRESH WATER, CYANOPHYTA, MYXOBACTERIA, W72-05457
RATION, REASONABLE USE, BRACKISH WATER, D: /TIVE USE, ALGAE, EVAPO W70-07100
ITIC, ESTUARINE ENVIRONMENT, SEA WATER, DIATOMS, COPEPODS, DINOFLA W73-00935
NS, NITRATES, SULFATES, DRAINAGE WATER, DISSOLVED SOLIDS, NITROGEN W71-01489
AVIOR.: *ALGAE, *FRESH WATER, DISTRIBUTION PATTERNS, BEH W71-07360
ETERGENTS, *PHOSPHATES, *POTABLE WATER, EUTROPHICATION, ALKYLBENZE W72-06837
MARINE PLANTS, PHAEOPHYTA, OILY WATER, EVAPORATION, DIFFUSION, RH W73-01074
LIZATION, NITROGEN SOURCES, WELL WATER, FEEDLOTS.: /OURCES, MINERA W71-06435
ORUS RADIOISOTOPES, ALGAE, FRESH WATER, FILTRATION, LEACHING, FUNG W72-07143
ESIUM, PHOSPHATES, NITROGEN, SEA WATER, FLUOROMETRY, EQUIPMENT, IN W72-07660
LINITY, *BIOINDICATORS, BRACKISH WATER, FRESHWATER, *ALGAE, WATER W73-02548
NE ALGAE, AQUATIC ANIMALS, FRESH WATER, FRESHWATER FISH, TEMPERATU W71-09182
IFICATION, BOTTOM SEDIMENTS, SEA WATER, FRESHWATER, AMMONIA, EUTRO W72-11563
PHAEOPHYTA, CHLOROPHYTA, COOLING WATER, HYDROLOGIC DATA.: /PHYTA, W72-11252
ITY TECHNIQUES, CHLOROPHYLL, SEA WATER, IONS, NUTRIENT REQUIREMENT W72-09103
FF, UREAS, FERTILIZERS, DRAINAGE WATER, IRRIGATION, RETURN FLOW, W W71-06435
POLLUTION EFFECTS, ECOLOGY, SEA WATER, LABORATORY EQUIPMENT, INHI W73-00348
MERCURY, CHROMIUM, COBALT, FRESH WATER, LAKES, PHYSICAL PROPERTIES W73-01434
TOLOGY, *LAKES, *ALGAE, ALKALINE WATER, MAGNESIUM, SODIUM SULFATES W71-04526
LLUTION SOURCES, OIL SPILLS, SEA WATER, MARINE PLANTS, PHAEOPHYTA, W73-01074
PHILIC ANIMALS, *HALOPHYTES, SEA WATER, MARINE ANIMALS, BIOLOGICAL W73-00855
UTION EFFECTS, *INDICATORS, *SEA WATER, MARINE PLANTS, ASSESSMENTS W71-12867
AE, *BIOASSAY, REPRODUCTION, SEA WATER, MARINE PLANTS, PRIMARY PRO W72-14294
ROBIOLOGY, *ACARICIDES, BRACKISH WATER, MICROENVIRONMENT, LITTORAL W71-08026
RES, *NUTRIENT REQUIREMENTS, SEA WATER, NITRATES, PHOSPHORUS, NUTR W73-00441
ION, *CRUSTACEANS, *FISH, HEATED WATER, NUCLEAR POWER PLANTS, ELEC W72-07907
S, ENVIRONMENTAL EFFECTS, HEATED WATER, NUCLEAR POWER PLANTS, STEA W71-11517
LOGY, CHRYSOPHYTA, CULTURES, SEA WATER, NUTRIENT REQUIREMENTS, WAT W73-00380
TERGENTS, OIL, *OIL WASTES, OILY WATER, OIL-WATER INTERFACES, WATE W70-09429
ANTIC OCEAN, SEASONAL, BAYS, SEA WATER, ORGANIC MATTER, MORTALITY, W73-00366
OASSAY, CELLULOSE, DIALYSIS, SEA WATER, ORGANIC ACIDS, CHRYSOPHYTA W73-00942
CATION, LITTORAL, SANDS, SHALLOW WATER, PERIPHYTON, SEDIMENTS, DIA W72-07501
, GROWTH RATES, TEMPERATURE, SEA WATER, PHOSPHATES, PHOSPHORUS RAD W71-10083
HYDROGEN ION CONCENTRATION, SEA WATER, PHOTOSYNTHESIS, PHYTOPLANK W71-11008
ANALYTICAL TECHNIQUES, MUNICIPAL WATER, PHOSPHORUS.: /ENT, COSTS, W72-09171
OFILES, MARINE ALGAE, WINDS, SEA WATER, PHYSICOCHEMICAL PROPERTIES W71-09225
R SOFTENING, LEGAL ASPECTS, HARD WATER, PUBLIC HEALTH, COST COMPAR W71-09429
, SAMPLING, SEA WATER, BODIES OF WATER, PYRROPHYTA, PROTOZOA, MARI W73-00296

E, PHOSPHATES, HYDROLYSIS, STORM WATER, RAINFALL, WASTE WATER(POLL W72-13644
TA, INHIBITORS, ANTIBIOTICS, SEA WATER, RHODE ISLAND.: / CHLOROPHY W72-04748
EVELOPMENT, *WATER SUPPLY, WASTE WATER, RUNOFF, DISSOLVED OXYGEN, W71-09788
TIDAL EFFECTS, MARINE ALGAE, SEA WATER, SALINITY, SAMPLING, CORES, W73-00367
N, PLANKTON, ATLANTIC OCEAN, SEA WATER, SALINITY, PHENOLS, WATER T W72-07710
ICS, *ESTUARINE ENVIRONMENT, SEA WATER, SALINITY, WATER TEMPERATUR W72-08141
TS, *SYSTEMATICS, *SEASONAL, SEA WATER, SAMPLING, BIOMASS, DIATOMS W73-00442
, PHOTOGRAPHY, SCUBA DIVING, SEA WATER, SAMPLING, ENVIRONMENTAL EF W73-00932
TON, *SYSTEMATICS, *DIATOMS, SEA WATER, SEASONAL, FLUCTUATIONS, CO W72-07663
TY, *SHOALS, *BENTHIC FLORA, SEA WATER, SEDIMENTS, LIGHT INTENSITY W72-14317
TER POLLUTION EFFECTS, RECLAIMED WATER, SEWAGE EFFLUENTS, RESEARCH W72-11702
G, SAMPLING, WATER ANALYSIS, SEA WATER, STABLE ISOTOPES, RADIOISOT W73-00831
LUTION TREATMENT, ALGAE, POTABLE WATER, TASTE, AMMONIA, SILICA.: / W71-01491
STES, ALGAE, AQUATIC PLANTS, SEA WATER, TRACE ELEMENTS, TOXICITY, W72-05952
BS, WATER POLLUTION SOURCES, SEA WATER, TROUT, TOXINS.: /RIMP, CRA W71-07412
INTERTIDA/ *MARINE PLANTS, *OILY WATER, WATER POLLUTION EFFECTS, * W71-08792
NG USES, DIVERSION, RECIRCULATED WATER, WATER REUSE, LEGAL ASPECTS W70-07100
WASTES, PATH OF POLLUTANTS, SEA WATER, WATER ANALYSIS, RADIOACTIV W73-00817
WATER TEMPERATURE, SALINITY, SEA WATER, WATER QUALITY, ON-SITE TES W72-09092
LUTION EFFECTS, *CALIFORNIA, SEA WATER, WATER ANALYSIS, MARINE ALG W73-02105
*BIOMASS, *ORGANIC MATTER, *SEA WATER, WATER ANALYSIS, ALGAE, CHE W71-11234
GY, BIOLOGICAL PROPERTIES, FRESH WATER, WATER QUALITY.: /AE, ECOLO W71-10580
S, SURFACE WATERS, NITROGEN, SEA WATER, WINDS, SAMPLING, INCUBATIO W72-09122
RIVER, OKA RIVER, FOAM, NATURAL WATERS.: *USSR, MOSCOW W72-05506
SPHORUS, EUTROPHICATION, SURFACE WATERS.: /DILUTION, NITROGEN, PHO W72-01472
ENVIRONMENT, WATER PO/ *RUNNING WATERS, *AQUATIC PLANTS, *AQUATIC W72-14690
TES, CLAY, TASTE, ODOR, *SURFACE WATERS, *GROUNDWATERS, SURVEYS, * W72-06536
S, *ILLINOIS, RECREATI/ *SURFACE WATERS, *LAKES, *STREAMS, *REVIEW W72-05869
ATION, *MICROORGANISMS, *SURFACE WATERS, *PLANKTON, *ALGAE, TURBID W72-06191
DATA/ *WATER QUALITY, *IMPOUNDED WATERS, *RESERVOIRS, *HYDROLOGIC W72-02274
EUTROPHICATION, *ALGAE, *SURFACE WATERS, *REVIEWS, *WATER POLLUTIO W71-10580
Y, *ECOLOGY, *CALIFORN/ *SURFACE WATERS, *SEA WATER, *WATER QUALIT W71-10577
ELETONEMA COSTATUM, TH/ *COASTAL WATERS, *SOUTHERN LONG ISLAND, SK W71-07875
ES, NITROGEN, NUTRIENTS, SURFACE WATERS, ALGAE, CHEMICAL PROPERTIE W72-12955
IATION, *AQUATIC ALGAE, *SURFACE WATERS, ANALYTICAL TECHNIQUES, PH W71-12034
NTS, PHOSPHORUS, NITROGEN, TIDAL WATERS, CHEMICAL OXYGEN DEMAND, B W71-07098
WATER QUALITY, MICHIGAN, SURFACE WATERS, COASTS, SAMPLING, BIOLOGI W72-10875
UREAS, *COASTS, *OCEANS, SURFACE WATERS, CONTINENTAL SHELF, DEPTH, W72-08056
HUDSON RIVER, STANDARDS, SURFACE WATERS, EPILIMNION, INTAKE STRUCT W70-09614
ALIZATION, NITROGEN SO/ *SURFACE WATERS, GEOLOGICAL SOURCES, MINER W71-06435
FLAGELLATES, SARGASSO SEA, SHELF WATERS, GULF STREAM, CAPE HATTERA W72-07663
FLOATING, WINDS, ALGAE, SURFACE WATERS, HYDROGEN ION CONCENTRATIO W73-00236
ECOLOGICAL DISTRIBUTION, SURFACE WATERS, LAGOONS, CHLOROPHYTA, RHO W73-00838
GANIC MATTER, NUTRIENTS, SURFACE WATERS, NITROGEN, WATER QUALITY, W70-10405
CTIVITY, ALASKA, FIORDS, SURFACE WATERS, NITROGEN, SEA WATER, WIND W72-09122
S, *AQUATIC ENVIRONMENT, SURFACE WATERS, NUTRIENTS, ALGAE, PHOSPHO W72-00847
, AIR POLLUTION EFFECTS, SURFACE WATERS, PATH OF POLLUTANTS, MARIN W72-00940
E COD(MASS), CAPE MAY(/ *COASTAL WATERS, PANAMA, CALLAO(PERU), CAP W72-08056
TECHNIQUES, MONITORING, SURFACE WATERS, PERSISTENCE, PATH OF POLL W73-00361
*LAKE KINNERET(ISRAEL), BOTTOM WATERS, RIVER JORDAN(ISRAEL).: W72-05469
, RUNOFF, ANIMAL WASTES, SURFACE WATERS, SOIL EROSION, NUTRIENTS, W71-04216
ENTS, *WATER QUALITY, *IMPOUNDED WATERS, STREAMS, FLOW, SEASONAL, W71-03028
FINLAND, DARK FIXATION, COASTAL WATERS, TROPHOGENIC LAYER.: /F OF W72-03228
NTROL, *DOMESTIC WATER, *SURFACE WATERS, WATER SUPPLY, EUTROPHICAT W71-05083
WATER POLLUTION SOURCES, SURFACE WATERS, WATER POLLUTION, DOMESTIC W70-09388
ENVIRONMENTS, GREAT LAKES, BAYS, WATERSHEDS(BASINS), SURFACE DRAIN W71-08953
BAYS, WATERSHEDS(BASINS), SMALL WATERSHEDS, CROPS, EGGS, FISHERIE W72-00959
, WATER POLLUTION EFFECTS, BAYS, WATERSHEDS(BASINS), SMALL WATERSH W72-00959
ENTS, HUMAN POPULATION, TOURISM, WATERSHEDS(BASINS), TROPHIC LEVEL W71-11009
RE, INDUSTRIES, WASTE TREATMENT, WATERSHEDS(BASINS), PHOSPHORUS, A W71-11017
WATERSHEDS(BASINS), *DDT, *SMALL WATERSHEDS, ECOSYSTEMS, PESTICIDE W72-12930
DE RESIDUES, *LOTIC ENVIRONMENT, WATERSHEDS(BASINS), *DDT, *SMALL W72-12930
ANCY, WIND VELOCITY, CONVECTION, WATTER POLLUTION EFFECTS.: / BUOY W71-09852
*NITROGEN:PHOSPHORUS RATIO, LAKE WAUBESA(WIS), OCCOQUAN RESERVOIR(W72-05473
ING, OXYGEN, TEMPERATURE, LIGHT, WAVES(WATER), PHYTOPLANKTON: /MPL W72-06283
STANDING CROP, CURRENTS(WATER), WAVES(WATER), SAMPLING, BENTHOS, W72-04766
ATES, AQUATIC ANIMALS, LITTORAL, WAVES(WATER).: /OPHYTA, INVERTEBR W71-11012
HOLOGY, *GENETICS, *VARIABILITY, WAVES(WATER), MAINE, STATISTICAL W73-00369
UTH WALES, MICROCYSTIS A/ *SHOCK- WAVES, *GAS-VACUOLE DEFLATION, SO W72-11564
ENTS(WATER), TIDES, WIND, LIGHT, WAVES, CLAMS, CRUSTACEANS, WORMS, W72-09092
LGAE, HAWAII, WATER TEMPERATURE, WAVES, LIGHT INTENSITY.: /ARINE A W71-11486
SMITH MOUNTAIN PROJECT(VA), WEATHER EFFECT.: W72-05282
DROLOGIC DATA, DISSOLVED OXYGEN, WEATHER, RESERVOIR OPERATION, WAT W72-05282
TH, ABSORPTION, ALGAE, SEASONAL, WEATHER, WATER ANALYSIS, TRACERS, W72-09122
LGAL TOXINS, *CHLOROPHYTA/ *FOOD WEBS, *FOOD CHAINS, *TOXICITY, *A W71-08597
MENTS, GROWTH RATES, ALGAE, FOOD WEBS, COPEPODS, ORGANOPHOSPHORUS W72-07892
WATER, BIOMASS, NUTRIENTS, FOOD WEBS, DETRITUS, ECOSYSTEMS, ENERG W73-01089
TS, ON-SITE INVESTIGATIONS, FOOD WEBS, FOOD CHAINS, FROGS, LICHENS W72-00949
COMMUNITIES, TIDAL MARSHES, FOOD WEBS, INTERTIDAL AREAS, CONNATE W W73-00796
TIVITY EFFECTS, ABSORPTION, FOOD WEBS, STRONTIUM RADIOISOTOPES, MA W72-00948
, CRUSTACEANS, FOOD CHAINS, FOOD WEBS, WATER POLLUTION EFFECTS, AQ W72-11925
MENTAL EF/ *HERBICIDES, *AQUATIC WEED CONTROL, *PARAQUAT, *ENVIRON W72-00155
ATMENT, *ALGAL CONTROL, *AQUATIC WEED CONTROL, CALIFORNIA.: /R TRE W72-02975
ECONOMIC JUSTIFICATION, AQUATIC WEED CONTROL, RIVERS, AGRICULTURE W71-13172
ALGAE, *ALGAL CONTROL, *AQUATIC WEED CONTROL, BIOCHEMICAL OXYGEN W72-14692
ETRATION, *MACROPHYTES, *AQUATIC WEED CONTROL, *LIGHT INTENSITY, * W73-02349
E,/ *FISH, *HERBIVORES, *AQUATIC WEED CONTROL, TILAPIA, CARP, ALGA W71-04528
*WATER QUALITY CONTROL, *AQUATIC WEED CONTROL, *DOMESTIC WATER, *S W71-05083
*HARVESTING OF ALGAE, *AQUATIC WEED CONTROL, *MECHANICAL CONTROL W71-06188
ALGAE CULTURE TECHNIQUE, AQUATIC WEED STUDY FACILITY, PRODUCT EVAL W71-06102

TS, BIOLOGICAL SAMPLES, SARGASSO	WEED, STARFISH, ASTERIAS FORBESI,	W72-05965
, TELEVISION, DEPTH SOUNDE/ *SEA	WEEDS, *MACROPHYTES, *SEA GRASSES	W72-09120
TING FACTORS, / *ALGAE, *AQUATIC	WEEDS, *NUTRIENTS, BIOASSAY, LIMI	W73-00232
OLLUTANT IDENTIFICATION, AQUATIC	WEEDS, *WASHINGTON, *CHLOROPHYLL,	W72-14728
*PHYTOPLANKTON, *ALGAE, *AQUATIC	WEEDS, *WEEDS, *RED TIDE, COASTS,	W72-14673
LTURES, *ALGIC/ *ALGAE, *AQUATIC	WEEDS, *WEED CONTROL, GROWTH, *CU	W71-06102
NTOUS ALGAE, *AQUATIC ANGIOSPERM	WEEDS, CLADOPHORA.: *FILAME	W72-14692
OXYGEN DEMAND, PHOSPHORUS, FISH,	WEEDS, CYCLES, BACTERIA, CHEMIST:	W71-03012
, LAKES, MUD, WISCONSIN, AQUATIC	WEEDS, EUTROPHICATION.: /UTRIENTS	W72-10608
OFF, FARM WASTES, ALGAE, AQUATIC	WEEDS, FERTILIZATION, LEAVES, LEG	W73-02479
ER CLOGGING, IRON BACTERIA, POND	WEEDS, PERMANGANATE.: *FILT	W72-06536
, EUTROPHICATION, ALGAE, AQUATIC	WEEDS, PHYTOPLANKTON, OLIGOTROPHY	W72-08459
D PAPER INDUSTRY, ALGAE, AQUATIC	WEEDS, RECREATION, GROUNDWATER, F	W70-05103
, TEMPERATURE, NUTRIENTS, ALGAE,	WEEDS, TASTE, ODOR, PATHOGENIC BA	W71-11006
COMPREHENSIVE PLANNING, AQUATIC	WEEDS, WATER POLLUTION SOURCES, M	W72-09171
, *ALGAL EXUDAT/ *HIGH MOLECULAR	WEIGHT COMPOUNDS, *MOLECULAR SIZE	W71-07878
ONAL DRAG, ALKY/ *HIGH-MOLECULAR-	WEIGHT POLYMERS, TURBULANT FRICTI	W70-06968
ANSPARENCY, ORTHOPHOSPHATES, DRY	WEIGHT, PHOSPHATE UPTAKE.: /E, TR	W71-10645
RIVER PLANT, CULTURE MEDIA, DRY	WEIGHT, THERMAL EFFECTS.: /VANNAH	W71-02075
INERALIZATION, NITROGEN SOURCES,	WELL WATER, FEEDLOTS.: /OURCES, M	W71-06435
CTS, UNDERGROUND, ALASKA, PONDS,	WELLS, FISH, BAYS, AIR POLLUTION	W72-03347
NCHOPHILA MILLE/ *ALGAL SPECIES,	WESTERN LAKE ERIE, ARNOLDIELLA CO	W71-09156
	WESTHAMPTON LAKE (VA).:	W72-00150
AKE KINNERET(ISRAEL), PERIDINIUM	WESTI, OSCILLATORIA PROLIFICA, TH	W71-13187
), ATRAZINE, B/ *ATHIORHODACEAE,	WHIPPLE DAM, STONE VALLEY DAM(PA.	W72-04743
RANDYWINE RIVER, RED CLAY RIVER,	WHITE CLAY RIVER, CHRISTINA RIVER	W72-04736
GOMPHONEMA OLIVACEOIDES,	WHITE CLAY CREEK(PA.).:	W73-00242
PHYSA HETEROSTROPHA, PROCLADIUS,	WHITE OAK LAKE(TENN), UPTAKE.: /	W70-06668
RIVER(CALIFORNIA), BISCAYNE BAY,	WHITE RIVER(INDIANA), PATUXENT RI	W71-07417
RIVER(CONNECTICU/ *ENTRAINMENT,	WHITE RIVER(INDIANA), CONNECTICUT	W71-11517
NUS LAVARETUS, COREGONUS ALBULA,	WHITEFISH, BRANCHIURA: /T, COREGO	W72-12729
TER QUALITY, WATER CONSERVATION,	WILDLIFE CONSERVATION, FLOOD CONT	W72-05594
RATES, SEDIMENTATION, BUOYANCY,	WIND VELOCITY, CONVECTION, WATTER	W71-09852
TURBIDITY, NITRATES, PHOSPHATES,	WIND VELOCITY, RAINFALL.: /QUES,	W72-03224
ICATORS, CURRENTS(WATER), TIDES,	WIND, LIGHT, WAVES, CLAMS, CRUSTA	W72-09092
ESTHWAITE WATER(ENGLAND), LAKE	WINDERMERE(ENGLAND), EXCENZYME PR	W72-03224
TAKE, LOCH LEVEN(SCOTLAND), LAKE	WINDERMERE(ENGLAND).: / LUXURY UP	W71-13183
, ESTHWAITE WATER(ENGLAND), LAKE	WINDERMERE(ENGLAND), GREAT OUSE(E	W71-13185
FORMOSA, BLELHAM TARN(ENGLAND),	WINDERMERE(ENGLAND).: /TERIONELLA	W72-10625
	WINDHOEK, SOUTH AFRICA.:	W70-10381
CATION, *DISTRIBUTION, FLOATING,	WINDS, ALGAE, SURFACE WATERS, HYD	W73-00236
RBON, NITROGEN, CURRENTS(WATER),	WINDS, NEVADA, CALIFORNIA, NEW YO	W71-05626
ACE WATERS, NITROGEN, SEA WATER,	WINDS, SAMPLING, INCUBATION, RUNO	W72-09122
ON, FLOW PROFILES, MARINE ALGAE,	WINDS, SEA WATER, PHYSICOCHEMICAL	W71-09225
N, TEMPERATURE, CURRENTS(WATER),	WINDS, THERMOCLINE, ALGAE, SEDIME	W72-12992
	WINDSCALE PROCESSING PLANT.:	W73-00828
A(WIS.), LAKE MONONA(WIS.), LAKE	WINGRA(WIS.), LIMITING NUTRIENTS,	W73-00232
IS, ORDINATION ANALYSIS.: *LAKE	WINGRA(WIS.), MULTIVARIATE ANALYS	W73-02469
ND, BAIRD ISLAND, SHOLIN ISLAND,	WINIFRED ISLAND, IZEMBEK LAGOON,	W72-09120
MINUTISSIMA, ACHNANTHES / *LAKE	WINNIPEG, *SUBSTRATES, ACHNANTHES	W72-13142
ILIS, THIOVULUM MAJUS, THIOSPIRA	WINOGRADSKY, ZOOGLEA RAMIGERA, BE	W72-07896
RINE ALGAE, CHRYSOPHYTA, AUTUMN,	WINTER, SUMMER, PRIMARY PRODUCTIV	W73-00442
ETPLANKTON, LAWRENCE LAKE(MICH),	WINTERGREEN LAKE(MICH), CARBONATE	W72-08048
LAKE MENDOTA(WIS.), LAKE MONONA(	WIS.), LAKE WINGRA(WIS.), LIMITIN	W73-00232
NGRA(WIS.), LIMITI/ LAKE MENDOTA(	WIS.), LAKE MONONA(WIS.), LAKE WI	W73-00232
, LAKE MONONA(WIS.), LAKE WINGRA(	WIS.), LIMITING NUTRIENTS, SURPLU	W73-00232
INATION ANALYSIS.: *LAKE WINGRA(	WIS.), MULTIVARIATE ANALYSIS, ORD	W73-02469
LIMITING NUTRIENTS, LAKE MENDOTA(	WIS).: *	W72-07933
, LAKE MENDOTA(WIS), LAKE MONONA(	WIS).: /RCES, ACETYLENE REDUCTION	W72-10608
KE(WIS), SEDIMENT / LAKE MENDOTA(	WIS), BANTAM LAKE(CONN), TROUT LA	W72-08459
FOX RIVER BASIN(	WIS), FWPCA.:	W70-05103
(VA), MADISON(WIS), LAKE MENDOTA(	WIS), LAKE MONONA(WIS), LAKE KEGO	W72-05473
, LAKE MONONA(WIS), LAKE KEGONSA(	WIS), LAKE WASHINGTON(WASH), GREE	W72-05473
OCCOQUAN RESERVOIR(VA), MADISON(	WIS), LAKE MENDOTA(WIS), LAKE MON	W72-05473
, LAKE MENDOTA(WIS), LAKE MONONA(	WIS), LAKE KEGONSA(WIS), LAKE WAS	W72-05473
VALLEY RESERVOIRS(NEB), MADISON(	WIS), LAKE WASHINGTON(WASH), DYEI	W72-04759
NTENANCE.: *MADISON(	WIS), LAKE MENDOTA(WIS), LAWN MAI	W73-02479
CETYLENE REDUCTION, LAKE MENDOTA(	WIS), LAKE MONONA(WIS).: /RCES, A	W72-10608
*MADISON(WIS), LAKE MENDOTA(	WIS), LAWN MAINTENANCE.:	W73-02479
CHELIEU RIVER(QUEBEC), GREEN BAY(	WIS), MACKEREL, MONTREAL(CANADA).	W70-08098
L ASSAY PRO/ FLUSHING, GREEN BAY(	WIS), NITRILOTRIACETIC ACID, ALGA	W71-12072
N:PHOSPHORUS RATIO, LAKE WAUBESA(	WIS), OCCOQUAN RESERVOIR(VA), MAD	W72-05473
), BANTAM LAKE(CONN), TROUT LAKE(	WIS), SEDIMENT COMPACTION.: /(WIS	W72-08459
N FIXATION, SEDIMENTS, NITRATES,	WISCONSIN.: /TRIFICATION, NITROGE	W72-07933
C ALGAE, BACTERIA, BIOCHEMISTRY,	WISCONSIN.: /ATION, LAKES, AQUATI	W72-04292
ON, MICROORGANISMS, COLORIMETRY,	WISCONSIN.: /, AQUATIC PLANTS, IR	W72-03742
AMATH LAKE(OREGON), LAKE MENDOTA(	WISCONSIN).: /NEW YORK), UPPER KL	W71-05626
HORA GLOMERATA, MILWAUKEE HARBOR(	WISCONSIN), POLYPHOSPHATES, ORTHO	W72-13644
A, CALIFORNIA, NEW YORK, OREGON,	WISCONSIN, MAINE.: / WINDS, NEVAD	W71-05626
ELLFISH, AMPHIPODA, WATER BIRDS,	WISCONSIN, ANALYTICAL TECHNIQUES.	W70-08098
STUARIES, NUTRIENTS, DETERGENTS,	WISCONSIN, LAKE ERIE, LAKE ONTARI	W70-07283
, FALLOUT, DRAINAGE, MANAGEMENT,	WISCONSIN, PRODUCTIVITY, ALGAE, E	W71-03012
S, SOIL MANAGEMENT, GROUNDWATER,	WISCONSIN, WATER SUPPLY.: / WASTE	W71-06443
QUALITY, CHLOROPHYLL, NITROGEN,	WISCONSIN, NUTRIENTS, WATER POLLU	W71-10645
ESSENTIAL NUTRIENTS, LAKES, MUD,	WISCONSIN, AQUATIC WEEDS, EUTROPH	W72-10608
ATION, *WATER POLLUTION CONTROL,	WISCONSIN, LAKES, URBAN RUNOFF, A	W73-02479
CAL COMMUNITIES, STANDING CROPS,	WISCONSIN, PRIMARY PRODUCTIVITY,	W73-02469
TEMPERATURE, GREAT LAKES REGION,	WISCONSIN, MICHIGAN, CHRYSOPHYTA,	W73-00384
NE, MACROINVE/ *OIL SPILLS, S.S.	WITWATER, GALETA ISLAND, CANAL ZO	W72-07911
IS, MACROCONSU/ *RIFFLE BENTHOS,	WOODLAND STREAM, ECOSYSTEM ANALYS	W71-10064

CARBON, ACID-VOLATILE COMPOUNDS, WOODS HOLE, HAPTAPHYCEAE,: /ANIC W72-07710
RELL, LAKE CRESCENT, TOOMS LAKE, WOODS LAKE, LAGOON OF ISLANDS, ST W73-01094
, GAMMA RAYS, BLOODWORMS, SLUDGE WORMS.: /NESSEE, CRAYFISH, MIDGES W70-06668
RGANIC LOADING, CORAL, MOLLUSKS, WORMS.: /POLLUTANTS, TURBIDITY, C W71-01475
, E. COLI, FISH, FOODS, GRASSES, WORMS.: /YEASTS, INSECTS, MILDEWS W73-01676
HICATION, *BIOCHEMICAL / *SLUDGE WORMS, *DISSOLVED OXYGEN, *EUTROP W70-09189
, ASBESTOS, INORGANIC COMPOUNDS, WORMS, INVERTEBRATES, ANNELIDS, S W73-00374
IGHT, WAVES, CLAMS, CRUSTACEANS, WORMS, MOLLUSKS, WATER TEMPERATUR W72-09092
LLUS CAPITATUS, CERAMIUM NITENS, WRANGELIA ARGUS, VALONIA VENTRICO W73-00838
ANADIAN PHOSPHATE DETERGENT BAN, WYANDOTTE CHEMICAL CORPORATION, F W70-07283
ADUS/ *YELLOWSTONE NATIONAL PARK(WYO), BLUE-GREEN ALGAE, MASTIGOCL W72-08508
PLANTS, *WASTE WATER TREATMENT, WYOMING, *AERATED LAGOONS.: /ILOT W72-06838
ENTS, VOLATILITY, FREEZE DRYING, X-RAYS, ELECTROLYSIS, COMPUTER PR W71-11036
ASINOPHYCEAE, BACCILARIOPHYCEAE, XANTHOPHYCEAE, VIBRIO, ACHROMOBAC W72-08051
TICA, PLEUR/ *EUSTIGMATOPHYCEAE, XANTHOPHYCEAE, POLYEDRIELLA HELVE W72-14808
TEIN.: CAROTENOIDS, XANTHOPHYLLS, MINNESOTA LAKES, LU W72-01784
NTIGUA, TRAMETES LILACINO-GILVA, XYLOBOLUS FRUSTULATUS, ASPERGILLU W72-13813
SE, SUCROSE, SENESCENT CULTURES, XYLOSE, RIBOSE, ARABINOSE, GALACT W72-07710
PLANT, HUMBOLDT BAY, CONNETICUT YANKEE NUCLEAR PLANT, CONNECTICUT W72-07907
ION, MACROAL/ *YAQUINA BAY(ORE), YAQUINA ESTUARY, DATA INTERPRETAT W72-08141
SPORES, AZOTOBACTER, CYANOPHYTA, YEASTS, CARBON DIOXIDE, CHLORELLA W72-06124
ES, VIRUSES, MOLLUSKS, PROTOZOA, YEASTS, INSECTS, MILDEWS, E. COLI W73-01676
ER, ORGANIC COMPOUNDS, VITAMINS, YEASTS, SULFUR, FOULING, ALGAE, S W72-05421
S, TRIBUTARIES, FRESHWATER FISH, YELLOW PERCH, SEWAGE EFFLUENTS, O W72-14282
FLEXIBACTERIA, PHORMIDIUM, MAST/ YELLOWSTONE PARK, SYNECHOCOCCUS, W72-03557
, BIGHORN RIVER, SHOSHONE RIVER, YELLOWTAIL DAM, FRAGILARIA CROTON W73-01435
SOLVED OXYGEN, PHOSPHATES, WATER YIELD IMPROVEMENT, WATER POLLUTIO W71-01491
ALGAE, *SOIL SURFACES, *SEDIMENT YIELD, *SIMULATED RAINFALL, *SOIL W72-02218
STREAM GAGES, GROUNDWATER, SAFE YIELD, ALGAE, MINERALOGY, TURBIDI W72-05594
STRATES, MACROINVERTEBRATES, NEW YORK BIGHT, EELGRASS, ZOSTERA, PI W73-02029
.: *NEW YORK BIGHT, COLLABORATIVE STUDIES W72-12941
CHLOROPHYTA, ALGAL CONTROL, *NEW YORK.: /ITY CONTROL, *KINETICS, * W72-08376
RIVER(KENTUCKY), CAYUGA LAKE(NEW YORK), PATUXENT ESTUARY, BISCAYNE W71-11517
NE), FAYETTEVILLE GREEN LAKE(NEW YORK), UPPER KLAMATH LAKE(OREGON) W71-05626
ERMAL POLLUTION, *ENVIRONM/ *NEW YORK, *POWER PLANTS, *STREAM, *TH W72-14659
LANKTON, *ALGAE, TURBIDITY, *NEW YORK, *WATER QUALITY.: /ATERS, *P W72-06191
L WASTES, INDUSTRIAL WASTES, NEW YORK, ALGAE, WATER POLLUTION.: /A W71-11685
, *ON-SITE DATA COLLECTIONS, NEW YORK, BIOLUMINESCENCE, BIOLOGICAL W72-12941
, SILICA, CHROMIUM, COPPER, *NEW YORK, DIATOMS, PHOSP: /INDICATORS W72-09111
MARINE ALGAE, *SALT MARSHES, NEW YORK, NEMATODES, EPIPHYTOLOGY, PR W72-00948
UTANTS, HOSTS, PENNSYLVANIA, NEW YORK, NEW HAMPSHIRE, CHIC, MICHIG W72-04736
, WINDS, NEVADA, CALIFORNIA, NEW YORK, OREGON, WISCONSIN, MAINE.: / W71-05626
USTROPHICATION, *HYDROLOGY, *NEW YORK, WATER QUALITY, NUTRIENTS, A W72-13851
Y PRODUCTIVITY, GREAT LAKES, NEW YORK, WATER SUPPLY, SYSTEMATICS, W73-01615
*EUDIAPTOMUS ZACHARIASI.: W72-14805
LIS, POLYSIPHONIA, ULVA LACTUCA, ZANICHELLIA PALUSTRIS, CODIUM, FU W73-00367
RA, RHODOTORULA LACTOSA, CANDIDA ZEYLANOIDES, CAMP VARNUM BEACH.: / W72-04748
C HEALTH, ESTUARINE ENVIRONMENT, ZINC RADIOISOTOPES, PHYSICOCHEMIC W72-04463
RATURE, STRONTIUM RADIOISOTOPES, ZINC RADIOISOTOPES, COBALT RADIOI W71-09250
EN ION CONCENTRATION, DYSTROPHY, ZINC RADIOISOTOPES, COBALT RADIOI W71-09190
OTOPES, STRONTIUM RADIOISOTOPES, ZINC RADIOISOTOPES, COBALT RADIOI W71-09018
RPTION, STRONTIUM RADIOISOTOPES, ZINC RADIOISOTOPES, COBALT RADIOI W71-09188
BSORPTION, COBALT RADIOISOTOPES, ZINC RADIOISOTOPES, MANGANESE, CH W73-00832
STRUM PARADOXIUM, CULTURE MEDIA, ZINC SULFATE, MANGANESE CHLORIDE, W72-12739
OTOPES, ZIRCONIUM RADIOISOTOPES, ZINC.: /DIOISOTOPES, IRON RADIOIS W71-09188
CESIUM, ALKALI METALS, SEASONAL, ZINC, ABSORPTION, LIGHT PENETRATI W73-01434
C ABSORPTION, ARSENIC, SELENIUM, ZINC, ANTIMONY, NORWAY, ASHING, B W71-11515
INDAL COMPOU/ LAKE TAHOE(CALIF), ZINC, ASCORBIC ACID, PYRIDOXINE, W72-07514
ICITY, *WATER POLLUTION EFFECTS, ZINC, LEAD, COPPER, CHROMIUM, SIL W72-14694
ZINC, MANGANESE RADIOISOTOPES.: W71-09190
RROPHYTA, CYANOPHYTA, MANGANESE, ZINC, METALS, SECCHI DISKS, HEAVY W72-09111
CE, PHOSPHORUS, IRON, MANGANESE, ZINC, NICKEL, COBALT, COPPER SULF W72-12739
HEMICAL WASTES, MERCURY, COPPER, ZINC, PESTICIDES, PESTICIDE TOXIC W72-12576
GICIDES, BACTERICIDES, CHROMIUM, ZINC, TEMPERATURE, HYDROGEN ION C W72-08683
LOSIUM, SPIROG/ *INORGANIC IONS, ZINC, VOLVOX, HYDRODICTYON, SPODY W71-11001
, HARDNESS(WATER), CONDUCTIVITY, ZINC, WATER TEMPERATURE, AL: /GEN W72-07890
TOPES, *RUTHENIUM RADIOISOTOPES, ZIRCONIUM RADIOISOTOPES.: /DIOISO W71-09225
DIOISOTOPES, IRON RADIOISOTOPES, ZIRCONIUM RADIOISOTOPES, ZINC.: / W71-09188
HASTATA, SCIRPUS SPP, TYPHA SPP, ZIZANIA AQUATICA, CAREX SPP, BUTO W73-01089
LINSLEY POND(CONNECTICUT), LAKE ZOAR(CONNECTICUT), LAKE MINNETONK W71-04216
OXIDANS, DESULFOVIBRIO, ULOTHRIZ ZONATA.: /CTER, THIOBACILLUS THIO W71-01489
ALGAE, *LIGHT QUALITY, EUTROPHIC ZONE, CHLOROPHYLL, PHOSPHATES, NI W72-12393
KING, *VERTICAL MIXING, EUPHOTIC ZONE, DIFFUSION COEFFICIENT, LAKE W71-03034
. WITWATER, GALETA ISLAND, CANAL ZONE, MACROINVERTEBRATES, DIESEL W72-07911
E/ *OXIDATION LAGOONS, *CLIMATIC ZONES, *COST ANALYSIS, TEMPERATUR W71-07109
ON, *SOILS, *TEMPERATE, CLIMATIC ZONES, CYANOPHYTA, SYMBIOSIS, LIC W73-02471
HEAT, SUBLETHEL EFFECTS, MIXING ZONES, MODELING.: *WASTE W70-09612
IN, *OXIDATION LAGOONS, CLIMATIC ZONES, PRECIPITATION, COAGULATION W71-08951
OBIC CONDITIONS, PONDS, CLIMATIC ZONES, TEMPERATURE, TURBULENCE, O W71-06033
URSKY BAY, PERIDINIANS, EUCAMPIA ZOODIACUS, CHAETOCEROS AFFINIS, S W73-00442
UM MAJUS, THIOSPIRA WINOGRADSKY, ZOOGLEA RAMIGERA, BEGGIATOA ALBA, W72-07896
E FUNGUS, SPHAEROTOTILUS NATANS, ZOOGLOEAL BACTERIA.: /DOM, *SEWAG W72-08573
OGY, ENTERIC BACTERIA, SAMPLING, ZOOPLANKTON, PHYTOPLANKTON, WATER W72-07896
ATER TEMPERATURE, PHYTOPLANKTON, ZOOPLANKTON, CHLOROPHYLL, SOLAR R W72-07892
RIA, CHRYSOPHYTA, PHYTOPLANKTON, ZOOPLANKTON, PYRROPHYTA, DINOFLAG W72-07899
ROPHYTA, PROTOZOA, MARINE ALGAE, ZOOPLANKTON, PHYTOPLANKTON.: /PYR W73-00296
ANKTON, PHYTOPLANKTON, BACTERIA, ZOOPLANKTON, SALMONIDS, SEASON, O W71-00117
S, CHLAMYDOMONAS, PHYTOPLANKTON, ZOOPLANKTON, PERIPHYTON.: /OTIFER W70-06655
YTOPLANKTON, *LAKES, EPILIMNION, ZOOPLANKTON, DISTRIBUTION, PRIMAR W71-09171
ARY PRODUCTIVITY, PHYTOPLANKTON, ZOOPLANKTON, EUTROPHICATION, OLIG W72-05968

S, SOIL ANALYSIS, SPECTROMETERS, ZOOPLANKTON, MUSSELS, CLAMS, FISH W72-05965
S, AQUATIC ALGAE, PHYTOPLANKTON, ZOOPLANKTON, FISH POPULATIONS, WA W72-04853
*WATER QUALITY, *MODEL STUDIES, ZOOPLANKTON, BIOTA, BACTERIA, BEN W72-05282
ARY PRODUCTIVITY, AQUATIC ALGAE, ZOOPLANKTON, CHRYSOPHYTA, CHLOROP W72-12738
SICAL PROPERTIES, PHYTOPLANKTON, ZOOPLANKTON, CHLOROPHYTA, DIATOMS W72-12729
IOMASS, FLUOROMETRY, MICROSCOPY, ZOOPLANKTON, NUTRIENTS, PHOTOSYNT W72-14313
HYTA, DINOFLOGELLATES, SEASONAL, ZOOPLANKTON, PHYTOPLANKTON.: /NOP W71-11949
, STEAM, ECOLOGY, PHYTOPLANKTON, ZOOPLANKTON, LARVAE, EGGS, CHINOO W71-11517
MMONIA, SALINITY, AQUATIC ALGAE, ZOOPLANKTON, BIOCHEMICAL OXYGEN D W71-11876
, *LARV/ *COPEPODS, CRUSTACEANS, ZOOPLANKTON, ALGAE, FEEDING RATES W72-09248
RODUCTIVITY, FOOD HABITS, ALGAE, ZOOPLANKTON, CARNIVORES, DAPHNIA. W72-00161
ON, STICKLEBACKS, CRABS, OTTERS, ZOOPLANKTON, PHYTOPLANKTON, MAMMA W72-00960
OLOGICAL CHARACTERISTICS, ALGAE, ZOOPLANKTON, BACTERIA, NUTRIENTS, W72-03215
RATES, NEW YORK BIGHT, EELGRASS, ZOSTERA, PINAEUS DUORARUM, PINAEU W73-02029
INTILLATION, SAMPLE PREPARATION, ZOSTREA, ULVA.: /IVITY, LIQUID SC W72-14317
DOPHORA GLOMERATA, ULOTRICHALES, ZYGNEMALES, OEDOGONIUM, MOUGOTIA, W71-05991
OGENY, *PLECTONEMA BORYANUM, LPP- 1, LPP-1D, PHYCOVIRUS, MITOMYCIN W72-03188
*PLECTONEMA BORYANUM, LPP-1, LPP- 1D, PHYCOVIRUS, MITOMYCIN C.: /, W72-03188
LOTELLA NANA.: CHELEX 100, ARTEMIA SALINA, BIO-RAD, CYC W72-04708
MARINUS, TEMPERATURE EFFEC/ *BP 1002, *GREEN ALGAE, PRASINOCLADUS W70-09429
*MACROINVERTEBRATES, RUTHENIUM- 106, CESIUM RADIOISOTOPES, SPIROG W70-06668
MS, LABORATORY EXPER/ *RUTHENIUM 106, SOLUBLE FORMS, INSOLUBLE FOR W71-13233
LLA AMERICANA, TERRES/ *C-13, *C- 12, BIOLOGICAL SYSTEMS, BELEMNITE W72-12709
GEN, THIAMINE, BIOTIN, VITAMIN B 12, GONYAULAX POLYEDRA, PERIDINIU W72-07892
LYCHLORINATED BIPHENYLS, AROCLOR 1254, ESCAMBIA RIVER, PINFISH, FL W71-07412
LEMNITELLA AMERICANA, TERRES/ *C- 13, *C-12, BIOLOGICAL SYSTEMS, BE W72-12709
LOGICAL SAMPLES, SR-90, K-40, CS- 137, *PAHRANAGAT LAKE, GAMMA RAY W73-01673
STON, FIELD MEASUREMENTS, CARBON- 14 MEASUREMENTS, ACETYLENE REDUCT W72-13692
TION, SAMPLE PREPARATION, ZO/ *C- 14 PRODUCTIVITY, LIQUID SCINTILLA W72-14317
EAWEEDS, APHIDS, SPONGES, CARBON- 14.: S W72-06283
SERVOIRS(NEB), ULTRA PLANKTON, C- 14.: *SALT VALLEY RE W72-04270
HYTES, HETEROTROPHIC BACTERIA, C- 14.: / ISLAND, OPEN OCEAN, SAPROP W71-08042
RA CRASSA, SCINTILLATION COUN/ C- 14, *CHLORELLA VULGARIS, NEUROSPO W72-14326
ERIOLECTA, PHAEODACTYLUM/ CARBON- 14, *FLORIDA STRAIT, DUNALIELLA T W72-09103
N ANALYZER, VAN DORN SAMPLERS, C- 14, BIOCARBONATES.: /ATER, BECKMA W72-09122
CUS, CHLOROCOCCUM, HYALO/ CARBON- 14, CARTERIA EUGAMETOS, STICHOCOC W72-09239
ICATISSIMA, NITZSCHIA PUNGENS, C- 14, CHLOROPHYLL A, MEDITERRANEAN W73-00431
AJAS FLEXILIS, SCIRPUS ACUTUS, C- 14, DISSOLVED ORGANIC MATTER, MAC W73-02095
DACTYLU/ *FLORIDA STRAIT, CARBON- 14, DUNALIELLA TERTIOLECTA, PHAEO W72-09102
OVALIS, CARBON EXTRUSION, CARBON- 14, NITROGEN SOURCE.: /NITZSCHIA W72-07497
LA, AGMENELLUM QUADRAPLICATUM, C- 14, THIN LAYER CHROMATOGRAPHY, CL W73-00361
YLENE, NITROGEN RADIOISOTOPES, N- 15, LUCIFERIN, LUCIFERASE, CHLORO W72-10883
CE, NITROGEN, NITROGEN FIXATION, 2-4-D, DALAPON, HERBICIDES, CROP W71-12071
TOXICITY, *HERBICIDES, DALAPON, 2-4-D, INSECTICIDES, PHENOLS, *DE W72-12576
IOASSAY, PESTICIDES, HERBICIDES, 2-4-D, PLANT GROWTH REGULATORS, A W72-08440
, CALCIUM RADIOISOTOPES, BLOOMS, 2,4 DINITRO PHENOL, TUCSON(ARIZ). W71-01474
O-3,5,6-TRICHLOROPICOLINIC ACID, 2,4-DICHLOROPHENOXYACETIC ACID.: / W72-08440
ISH, ASTERIAS FORBESI, PLUTONIUM 236, PLUTONIUM 238, PLUTONIUM 239 W72-05965
ORBESI, PLUTONIUM 236, PLUTONIUM 238, PLUTONIUM 239, POLYMERIZATIO W72-05965
UM 236, PLUTONIUM 238, PLUTONIUM 239, POLYMERIZATION, LIVER, BONE, W72-05965
E, PICLORAM, PEDIASTRUM, 4-AMINO- 3,5,6-TRICHLOROPICOLINIC ACID, 2, W72-08440
TU/ *HETEROCYSTS, *ANABAENA SP L- 31, CONTINUOUS CULTURE, BATCH CUL W72-03217
CYANIDES, PHOSPHORUS- 32.: W72-07143
OGICAL MAGNIFICATION, PHOSPHORUS- 32, LIMNODRILUS.: / OUABAIN, BIOL W72-09668
CALCIUM RADIOISOTOPES, BLOOMS, 2, 4 DINITRO PHENOL, TUCSON(ARIZ).: / W71-01474
TER ALGAE, PICLORAM, PEDIASTRUM, 4-AMINO-3,5,6-TRICHLOROPICOLINIC W72-08440
ORGANOPHOSPHOROUS PESTICIDES, *2- 4-D, *DDT, *DIELDRIN, *DIAZINON, W71-10566
, NITROGEN, NITROGEN FIXATION, 2- 4-D, DALAPON, HERBICIDES, CROP PR W71-12071
OXICITY, *HERBICIDES, DALAPON, 2- 4-D, INSECTICIDES, PHENOLS, *DETE W72-12576
ASSAY, PESTICIDES, HERBICIDES, 2- 4-D, PLANT GROWTH REGULATORS, ALG W72-08440
3,5,6-TRICHLOROPICOLINIC ACID, 2, 4-DICHLOROPHENOXYACETIC ACID.: /- W72-08440
M/ *BIOLOGICAL SAMPLES, SR-90, K- 40, CS-137, *PAHRANAGAT LAKE, GAM W73-01673
APHY, SCINTILLATION COUNTING, CA- 45.: /RBIGNIANA, PAPER CHROMATOGR W72-12634
PICLORAM, PEDIASTRUM, 4-AMINO-3, 5,6-TRICHLOROPICOLINIC ACID, 2,4- W72-08440
SA, LINDANE, P P'-DDE, CLOPHEN A 50, ELECTRON CAPTURE GAS CHROMATO W72-04134
O/ FUSARIUM NO 83, BACTERIUM NO 52, ENDOGENOUS RESPIRATION, TOTAL W71-12223
ICLORAM, PEDIASTRUM, 4-AMINO-3,5, 6-TRICHLOROPICOLINIC ACID, 2,4-DI W72-08440
ENELLUM QUADRUPLICATUM STRAIN PR- 6, PHOTOALDRIN, KETOENDRIN, METAB W72-03545
N SYSTEM, MODULAR SYSTEM, COBALT- 60.: *BIOLOGICAL--GAMMA-RADIATIO W72-00383
LORE/ *CHLORELLA SOROKINIANA, CO- 60, GAMMA RADIATION, SURVIVAL, CH W72-11836
C MUSCORUM, LYNGBYA LAGERHAIMII, 7-METHYLHEPTADECANE, 8-METHYLHEPT W73-01439
AGERHAIMII, 7-METHYLHEPTADECANE, 8-METHYLHEPTADECANE, HYDROCARBONS W73-01439
ESPIRATION, TOTAL O/ FUSARIUM NO 83, BACTERIUM NO 52, ENDOGENOUS R W71-12223
E, GAMM/ *BIOLOGICAL SAMPLES, SR- 90, K-40, CS-137, *PAHRANAGAT LAK W73-01673

AUTHOR INDEX

In order not to complicate the computer program needlessly, only initials and the word "and" have been excluded from the alphabetic listing. Both first and last names of authors appear in the index. The numbers at the right are the accession numbers appearing at the end of the abstracts.

ONALD W. LEAR, JR.: JOHAN A.	AALTO, NORBERT A. JAWORSKI, AND D	W70-08248
WILLIAM J. OSWALD,	AARON MERON, AND MARIO D. ZABAT.:/	W71-07100
ND HENRY GEE.:	AARCN MERON, WILLIAM J. OSWALD, A/	W72-04789
RUHN, D. F. LIVERMORE, ANC F. O.	ABOABA.: H. D. B/	W71-06188
K. LEE SU, E. J. STABA AND Y.	ABUL-HAJJ.: /	W72-08586
JOSEPH V. DUST, AND	ADAM SHINDALA.:	W71-00139
E. J. MIDDLEBROOKS, M. TUNZI, A.	ADINARAYANA, AND P. H. MCGAUHEY.:/	W71-13553
	ADNAN SHINDALA.: /	W71-09629
R. L. SHULER, AND W. A.	AFFENS.:	W7C-05547
MAXWELL S. DOTY, ANC GERTRUDES	AGUILAR-SANTOS.: /	W71-08597
E. F. GLOYNA, AND J.	AGUIRRE.: /	W71-07102
EDWARD T. HALL, JR., DONALD G.	AHEARN, AND DONALD J. REINHARDT.:/	W71-11914
DONALD J. REINHARCT.: D. G.	AHEARN, EDWARD T. HALL, JR., AND /	W71-09173
TH.	AHL, AND E. JONSSON.: /	W73-00286
M.	ALAM.: /	W72-14542
CHARLES C. COUTANT, ANC	ALAN J. BROOK.: /	W71-11517
:	ALBERT MASSEY, ANC JOHN ROBINSON./	W72-07937
C. COULON, AND V.	ALEXANDER.: /	W73-00273
S. E. MANAHAN, M. J. SMITH, D.	ALEXANDER, AND P. ROBINSON.: /	W72-01693
L. LONGERICH, M. BILLINGTON, V.	ALEXANDER, J. J. KELLEY, AND D. W/	W72-09122
	ALICE Q. HOWARD.: /	W71-08768
ATHERTON.: DARRELL L. KING,	ALLEN J. TOLMSOFF, AND MICHAEL J./	W71-07097
GALE	ALLEN WRIGHT.:	W71-04518
HAROLD L.	ALLEN.: /	W72-03216
T. F. H.	ALLEN, AND J. F. KOONCE.: /	W73-02469
S. L. SCHWARTZ, ANC L. R.	ALMCDOVAR.: /	W73-00838
THOMAS R. TOSTESCN, AND LUIS R.	ALMODOVAR.: /	W71-09261
PERCY P. ST.	AMANT, AND LOUIS A. BECK.: /	W71-11698
PERCY P. ST.	AMANT, AND LOUIS A. BECK.: /	W71-08223
. GOLDMAN, G. MCSHIRI, ANC E. DE	AMEZAGA.: C. R/	W72-12549
P. M.	AMIN, AND S. V. GANAPATI.: /	W72-10233
J.	ANCELLIN.: /	W73-00817
J.	ANCELLIN, AND P. BOVARD.: /	W71-13233
JOHN L.	ANDERSEN.: /	W72-04270
E. STEEMANN NIELSEN, AND S. WIUM-	ANDERSEN.:	W71-03027
R. A. KIMERLE, AND N. H.	ANDERSON.: /	W72-03556
G. ROBIN SOUTH, AND	ANDRE CARDINAL.: /	W71-10066
HENDRICKS, W. D. POTE, ANC J. G.	ANDREW.: D. W. /	W72-02363
L. G. RICH, J. F.	ANDREWS, AND T. M. KEINATH.: /	W72-11724
INA, V. G. TSITSUGINA, ANC V. V.	ANDRUDTSCHENKO.: /, L. G. KULEBAK/	W72-10955
RUSSELL G. RHODES, AND	ANTHONY J. TERZIS.: /	W71-05629
JOSEPH Y. CHENG, AND NAVAL J.	ANTIA.:	W7C-07842
JOSEPH Y. CHENG, ANC NAVAL J.	ANTIA.:	W70-07280
H. YAMAGISHI, AND K.	ACYAMA.: /	W73-00236
JOHN G. STOCKNER, ANC F. A. J.	ARMSTRONG.: /	W71-11029
. HARRIS, J. K. SYERS, AND D. E.	ARMSTRONG.: C. C. LEE, R. F/	W71-13794
.: PETER RICHERSON, RICHARD	ARMSTRONG, AND CHARLES R. GOLDMAN/	W71-09171
D. E.	ARMSTRONG, AND G. A. ROHLICH.: /	W71-06443
. SYERS.: D. E.	ARMSTRONG, R. F. HARRIS, AND J. K/	W72-03742
L.	ARNBORG, AND E. ERIKSSON.:	W70-09947
MCN. SIEBURTH, A. SASTRY, C. R.	ARNOLD, AND M. S. DOTY.: /SON, J./	W72-08142
ANTINE J. GREGORY, ANC GERALD L.	ARONSCN.: /OHN J. COCHRANE, CONST/	W71-05553
ROBERT E. HILLMAN, AND GILBER/	ARTHUR A. LEVIN, THCMAS J. BIRCH,/	W71-07417
ITLEY.: ROBERT S. CAMPBELL,	ARTHUR WITT, JR., AND JAMES R. WH	W7C-09596
AAR.: D. VAN	AS, H. C. FOURIE, AND C. M. VLEGG/	W73-00832
R, G. L. SAMSEL, AND E. K. OBENG-	ASAMOA.: B. C. PARKE/	W72-07661
Y.	ASATO.: /	W72-14327
LLEN J. TOLMSOFF, AND MICHAEL J.	ATHERTON.: DARRELL L. KING, A/	W71-07097
P.	BAGGE, AND A. NIEMI.: /	W72-13871

PAULI BAGGE, AND PASO O. LEHMUSLUOTO.: / W72-03228
ON, H. L. SCOTTEN, AND R/ R. C. BAIN, A. M. CALDWELL, R. H. CLAWS/ W71-10577
S A. ROBERTSON, AND/ RICHARD C. BAIN, JR., PERRY L. MCCARTY, JAME/ W71-07098
A. L. BAKER, AND A. J. BROOK.: / W72-07145
DAVID B. BAKER, AND JACK W. KRAMER.: / W72-04260
. G. VALANJU, C. D. MULAY, M. C. BALANI, AND: B. PATEL, P/ W73-00811
. CROSS, J. N. WILLIS, AND J. P. BAPTIST.: F. A/ W73-00802
C. F. BAREN.: / W72-14280
BROCKWAY, D. F. PARIS, AND J. T. BARNETT, JR.: /P. C. KERR, D. L. / W72-09159
R. J. BARSDATE.: / W71-09190
EDWIN F. BARTH, AND ROBERT B. DEAN.: / W72-04086
: I. MALEK, J. BARTOS, J. SIMMER, AND B. PROKES./ W72-03327
A. F. BARTSCH.: / W71-13256
: A. K. BASU, B. B. GHOSH, AND R. N. PAL./ W71-11876
TSUMURA.: J. C. BATTERTON, G. M. BOUSH, AND F. MA/ W72-03545
TSUMURA.: J. C. BATTERTON, G. M. BOUSH, AND F. MA/ W72-13809
ROBERT E. BAUMANN, AND SHELDON KELMAN.: / W71-06445
S. E. BEALL.: W70-09193
SAM E. BEALL.: W7C-08832
T. R. FOLSOM, AND T. M. BEASLEY.: / W73-00833
EAGLE.: T. M. BEASLEY, T. A. JOKELA, ANC R. J. / W72-00940
PERCY P. ST. AMANT, AND LOUIS A. BECK.: / W71-08223
PERCY P. ST. AMANT, AND LOUIS A. BECK.: / W71-11698
R. E. HOLLAND, AND A. M. BEETON.: / W73-00384
E. W. BEHRENS, ANC S. A. FRISHMAN.: W71-04873
DAVID A. BELLA.: W71-03034
D. BELLAMY.: / W72-11925
E. G. BELLINGER.: / W71-13185
VINCENT J. BELLIS.: / W71-07337
T. V. BELYAYEVA.: / W71-11527
T. V. BELYAYEVA.: / W72-03567
RT A. JORDAN.: MICHAEL E. BENDER, WAYNE R. MATSON, AND ROBE/ W71-09674
AURO, N. R. CIALLELLA, AND D. J. BENINSON.: D. CANCIO, J. A. LL/ W73-00819
A. G. COLLINS, J. H. BENNETT, ANC O. K. MANUEL.: / W72-02073
LESLIE RALPH BERGER.: / W71-10791
AESTRINI.: B. R. BERLAND, D. J. BONIN, AND S. Y. M/ W72-08051
STEWART, LARRY E. MILLIGER, ANC BERNARD M. SOLON.: KENNETH W. W70-06655
A. LATTERA, AND M. BERNHARD.: / W71-13243
LL.: H. BERNHARDT, J. CLASEN, AND H. SCHE/ W72-06201
, STANTON J. ERICKSON, AND PETER BETZER.: / DAVEY, JOHN H. GENTILE/ W72-04708
M. C. FERENS, AND R. J. BEYERS.: / W73-01080
NING-HSI TANG, AND SURINDER K. BHAGAT.: / W72-04431
LLIAM H. FUNK.: SURINDER K. BHAGAT, DONALD E. PROCTOR, AND WI/ W72-03215
I. S. BHAT, AND P. R. KAMATH.: / W72-10958
ELLEY, AND/ L. L. LONGERICH, M. BILLINGTON, V. ALEXANDER, J. J. K/ W72-09122
AINES.: A. A. LEVIN, T. J. BIRCH, R. E. HILLMAN, ANC G. E. R/ W72-07907
BER/ ARTHUR A. LEVIN, THOMAS J. BIRCH, ROBERT E. HILLMAN, AND GIL/ W71-07417
JOHN W. BISHOP.: / W72-00150
S. BISWAS.: / W73-00256
BJORN GANNING.: / W71-08026
S. A. BLACK.: / W71-12857
J. O. BLANTON, AND A. R. WINKLHOFER.: / W72-12992
IMOND, A. S. GETCHELL, AND J. A. BLEASE.: J. B. D/ W72-12930
H. BONE.: / W72-07941
DEREK H. BONE.: / W72-05476
A. D. BONEY.: W70-09429
B. R. BERLAND, D. J. BONIN, AND S. Y. MAESTRINI.: / W72-08051
JACK A. BORCHARDT.: / W71-09958
O. VAN DER BORGHT, AND S. VAN PUYMBROECK.: / W71-09182
O. VAN DER BORGHT, AND S. VAN PUYMBROECK.: / W71-09168
ISH/ GENE E. LIKENS, F. HERBERT BORMANN, NOYE M. JOHNSON, D. W. F W71-01489
ENKATARAMAN, AND S. K/ PAROMITA BOSE, UJJAL SINGH NAGPAL, G. S. V/ W71-10070
A. M. A. IMEVBORE, AND Z. BOSZORMENYI.: / W72-07143
NORVA/ IRVING YALL, WILLIAM H. BOUGHTON, RICHARD C. KNUDSEN, AND W71-01474
E. BOURGET, AND G. LACROIX.: / W72-05432
. PATIL, F. MATSUMURA, AND G. M. BOUSH.: K. C/ W73-00361
J. C. BATTERTON, G. M. BOUSH, AND F. MATSUMURA.: / W72-13809
J. C. BATTERTON, G. M. BOUSH, AND F. MATSUMURA.: / W72-03545
J. ANCELLIN, AND P. BOVARD.: / W71-13233
ON, JR., AND C. A. PRICE/ R. A. BOWEN, J. M. ST. ONGE, J. B. COLT/ W73-00371
G. W. BOWES.: / W72-12998
LUCYNA JOANNA BOWNIK.: / W71-11012
ILYIN, A/ V. N. GUSKOVA, A. N. BRAGINA, A. A. ZASEDATELEV, B. N./ W72-10694
L. P. BRAGINSKIY.: / W72-09650
. GEHRING.: CONRAD KLEVENO, O. BRAIDECH, E. THOMAS, AND PHILIP E/ W72-13659
ENO.: T. BRAIDECH, P. GEHRING, AND C. KLEV/ W72-12995
C. L. CHURCHILL, AND C. K. BRASHIER.: / W73-00938
W. BRAUNE.: / W72-14314
W. L. KLEIN, ANC R. A. BRECHBILL.: / W73-01673
K. A. STEIDINGER, AND J. F. VAN BREEDVELD.: / W72-11252
J. HRIB, AND V. BREZINA.: / W72-07144
EARL E. SHANNON, AND PATRICK L. BREZONIK.: / W72-04269
. D. PUTNAM, W. H. MORGAN, P. L. BREZONIK, E. E. SHANNON, AND P. E/ W72-08986
W. SCHULTS, K. W. MALUEG, R. M. BRICE, AND M. D. SCHULDT.: /S, D./ W72-09165
W. J. O' BRIEN.: / W72-13816
J. BRISOU.: / W72-13347
J. L. MOSSER, AND T. D. BROCK.: / W72-03557
MITH, C. B. FLIERMANS, AND T. D. BROCK.: D. W. S/ W72-14301
KERR, DORIS F. PARIS, AND D. L. BROCKWAY.: PAT C. W71-00475
BARNETT, JR./ P. C. KERR, D. L. BROCKWAY, D. F. PARIS, AND J. T. / W72-09159

A. L. BAKER, AND A. J.	BROOK.:	/ W72-07145
A. J.	BROOK.:	/ W71-13149
CHARLES C. COUTANT, AND ALAN J.	BROOK.:	/ W71-11517
RANDALL L.	BROWN.:	/ W72-02975
J. R. STEWART, AND R. M.	BROWN, JR.:	/ W72-07951
	BRUCE A. MANNY.:	/ W72-08048
R. G. WETZEL, AND	BRUCE A. MANNY.:	/ W72-01780
	BRUCE W. HALSTEAD.:	/ W72-04259
ABOABA.: H. D.	BRUHN, D. F. LIVERMORE, AND F. O./	W71-06188
THOMAS G.	BRYDGES.:	/ W72-13653
M.	BRYLINSKY.:	/ W73-02100
R. J.	BUCHANAN.:	/ W72-07899
D. C. MENDAY, AND A. A.	BUCK.:	/ W72-11564
LUND, G. H. M. JAWORSKI, AND H.	BUCKA.: J. W. G./	W72-10619
JAMES A.	BUCKLEY.:	/ W72-08049
EUGENE B. WELCH, JAMES A.	BUCKLEY, AND RONALD M. BUSH.:	/ W72-00799
E. G.	BUGLEWICZ.:	/ W73-02349
ROBERT L.	BUNCH.:	/ W71-12188
CIS M. MIDDLETOWN, AND ROBERT L.	BUNCH.: FRAN/	W71-07123
R. L.	BURGNER, AND R. E. NAKATANI.:	/ W73-00764
E. F. MANDELLI, P. R.	BURKHOLDER, T. E. DOHENY, AND:	/ W71-07875
B. K.	BURNISON.:	/ W73-01269
N. M.	BURNS, AND C. ROSS.:	/ W72-12997
NOEL M.	BURNS, AND C. ROSS.:	/ W72-12990
AND D. W. VAN ES.: G. E.	BURNS, R. M. GIRLING, A. R. PICK,/	W71-07109
ELSIE M.	BURROWS.:	/ W71-10553
ELSIE M.	BURROWS.:	/ W71-12867
E. M.	BURROWS, AND C. PYBUS.:	/ W71-09795
R. M.	BUSH.:	/ W72-08560
NON, M. S. SHANE, AND VALERIE N.	BUSH.: R. E. CAN/	W72-03188
JAMES A. BUCKLEY, AND RONALD M.	BUSH.: EUGENE B. WELCH,/	W72-00799
RONALD M.	BUSH, AND EUGENE B. WELCH.:	/ W72-00798
N. E. VANDERBORGH, AND A. G.	BUYERS.:	/ W72-13992
DEE MITCHELL, AND JAMES C.	BUZZELL, JR.:	/ W72-03218
S. M. DRACHEV, AND A. A.	BYLINKINA.:	/ W72-05506
M. PANOUSE.: G.	CAHET, M. FIALA, G. JACQUES, AND /	W73-00431
J.	CAIRNS, AND G. R. LANZA.:	/ W72-07225
J.	CAIRNS, JR., AND G. R. LANZA.:	/ W72-08579
J.	CAIRNS, JR., AND K. L. DICKSON.:	/ W73-01972
J.	CAIRNS, JR., AND R. A. PATERSON.:	/ W72-07526
J.	CAIRNS, JR., AND R. L. KAESLER.:	/ W72-07791
D RUTH PATRICK.: JOHN	CAIRNS, JR., ROGER L. KAESLER, AN/	W71-11583
OTTEN, AND R/ R. C. BAIN, A. M.	CALDWELL, R. H. CLAWSON, H. L. SC/	W71-10577
R. E.	CAMERON.:	/ W73-00854
LELYN STADNYK, ROBERT S.	CAMPBELL, AND B. THOMAS JOHNSON.:/	W71-10566
ROBERT S.	CAMPBELL, AND JAMES R. WHITLEY.: /	W71-06444
AMES R. WHITLEY.: ROBERT S.	CAMPBELL, ARTHUR WITT, JR., AND J	W70-09596
ELLA, AND D. J. BENINSON.: D.	CANCIO, J. A. LLAURO, N. R. CIALL/	W73-00819
G. W. FUHS, S. D. DEMMERLE, E.	CANELLI, AND M. CHEN.:	/ W72-09163
M. V. M. DESAI, AND A. K.	CANGULY.:	/ W71-09188
H. W.	CANNON.:	/ W72-01660
N. BUSH.: R. E.	CANNON, M. S. SHANE, AND VALERIE /	W72-03188
G. ROBIN SOUTH, AND ANDRE	CARDINAL.:	/ W71-10066
PAUL L. ZUBKOFF, AND WALTER E.	CAREY.:	/ W72-01101
ERALD D. MCKEE, LOYS P. PARRISH,	CARL R. HIRTH, KENNETH M. MACKENT/	W71-05626
DALE A.	CARLSON.:	/ W72-00800
EDWARD G. FOREE, AND	CAROLINE P. WADE.:	/ W72-06612
HOMAS, JR., JAMES A. SONDEL, AND	CAROLYN M. HYCHE.: / CHARLES C. T/	W71-11036
. WATSON.: E. J.	CARPENTER, C. C. REMSEN, AND S. W/	W73-00380
W. G. EGAN, J. M.	CASSIN, AND M. E. HAIR.:	/ W72-12941
R. L.	CHAMPLIN.:	/ W72-06838
SON, GEORGE E. MUSTOE, AND S. B.	CHANG.: ROBERT S. MAT/	W72-06037
	CHANG-KWEI LIN.:	/ W72-13644
MARSHALL L. SILVER, AND	CHARLES A. MOORE.:	/ W72-04263
ROOK.:	CHARLES C. COUTANT, AND ALAN J. B/	W71-11517
SOND/ K. K. SIVASANKARA PILLAY,	CHARLES C. THOMAS, JR., JAMES A. /	W71-11036
DAVIS.:	CHARLES E. WARREN, AND GERALD E. /	W72-06340
	CHARLES M. WEISS.:	W70-09669
WILLIAM R. HARTLEY, AND	CHARLES M. WEISS.:	/ W71-07382
ICHERSON, RICHARD ARMSTRONG, AND	CHARLES R. GOLDMAN.: PETER R/	W71-09171
, NORMAN MORGAN, AND WALTER TH/	CHARLES R. HAZEL, FRED KOPPERDAHL/	W71-09789
D. DEMMERLE, E. CANELLI, AND M.	CHEN.: G. W. FUHS, S./	W72-09163
C. W.	CHEN, AND G. T. ORLOB.:	/ W72-05282
C. W.	CHEN, AND G. T. ORLOB.:	/ W72-14405
T. EDELSTEIN, L C-M	CHEN, AND J. MCLACHLAN.:	/ W72-01370
WUM-	CHENG WANG, AND RALPH L. EVANS.:	W71-03028
AN.: WUN-	CHENG WANG, AND WILLIAM T. SULLIV/	W71-11033
JOSEPH Y.	CHENG, AND NAVAL J. ANTIA.:	W70-07280
JOSEPH Y.	CHENG, AND NAVAL J. ANTIA.:	W70-07842
S. FOGEL, I.	CHET, AND R. MITCHELL.:	/ W73-00942
B. J.	CHOLNOKY.:	/ W72-01473
LANDY MCBRIDE, WILLIAM	CHORNEY, AND JOHN SKOK.:	/ W71-12858
A. E.	CHRISTIE.:	/ W71-11972
A. E.	CHRISTIE.:	W70-08976
HARRIS, K. J. ROBERTS, AND A. E.	CHRISTIE.: A. J. /	W72-06837
C. L.	CHURCHILL, AND C. K. BRASHIER.: /	W73-00938
OBLE, W. A. PRETORIUS, AND F. M.	CHUTTER.: R. G. N/	W72-01472
D. CANCIO, J. A. LLAURO, N. R.	CIALLELLA, AND D. J. BENINSON.: /	W73-00819
ANCY L. NICHOLSON, AND ROBERT L.	CIMBERG.: N/	W71-08792
	CLAIR N. SAWYER.:	/ W72-05473

CLAIR N. SAWYER.: / W72-04769
JACK KISHLER, AND CLARENCE E. TAFT.: / W71-05630
RIO D. ZABAT, WILLIAM J. OSWALD, CLARENCE G. GOLUEKE, AND HENRY GE/ W72-04788
NORBERT A. JAWORSKI, LEO J. CLARK, AND KENNETH D. FEIGNER.: W70-08249
NORBERT A. JAWORSKI, LEO J. CLARK, AND KENNETH D. FEIGNER.: / W71-09788
H.BERNHARDT, J. CLASEN, AND H. SCHELL.: / W72-06201
. C. BAIN, A. M. CALDWELL, R. H. CLAWSON, H. L. SCOTTEN, AND R. G./ W71-10577
HENZEN, G. J. STANDER, AND A. J. CLAYTON.: / J. VAN VUUREN, M. R. W70-10381
DUFFIELD, S. D. WAALAND, AND R. CLELAND.: C. S./ W73-01622
OBERT J. LAVIGNE.: RICHARD G. CLEMENTS, GEORGE E. DREWRY, AND R/ W72-00949
D. SPEAR, R. J. LAVIN, AND N. L. CLESCERI.: G. C. MCDONALD, R. / W72-08376
A. S. HIRWE, P.-Y. LU, AND J. R. COATS.: / KAPOOR, R. L. METCALF, / W72-07703
AND GERALD L. ARONSON/ JOHN J. COCHRANE, CONSTANTINE J. GREGORY, W71-05553
JOHN C. WUYCHECK.: WILLIAM P. COFFMAN, KENNETH W. CUMMINS, AND / W71-10064
D. C. COLEMAN.: / W73-01449
L. HILLIS- COLINVAUX.: / W72-14294
R. P. COLLINS.: / W72-11604
R. P. COLLINS.: / W72-10076
MANUEL.: A. G. COLLINS, J. H. BENNETT, AND O. K./ W72-02073
A. BOWEN, J. M. ST. ONGE, J. B. COLTON, JR., AND C. A. PRICE.: /./ W73-00371
WILLIAM H. FUNK, RICHARD J. CONDIT, AND WAYNE T. CRANEY.: / W72-00801
J. L. CONFER.: / W72-12543
R. V. THOMANN, D. J. C' CONNOR, AND D. M. DITORO.: W71-05390
HOMAS, AND PHILIP E. GEHRING.: CONRAD KLEVENO, O. BRAIDECH, E. T/ W72-13659
D L. ARONSON/ JOHN J. COCHRANE, CONSTANTINE J. GREGORY, AND GERAL W71-05553
E. B. LONG, AND G. D. COOKE.: / W72-13673
G. D. COOKE, AND ROBERT L. KENNEDY.: / W71-11949
B. J. COPELAND, AND H. LEE DAVIS.: / W72-13636
W. J. COSGROVE.: W70-10427
E. W. B. DA COSTA.: / W72-13813
C. COULON, AND V. ALEXANDER.: / W73-00273
CHARLES C. COUTANT, AND ALAN J. BROOK.: / W71-11517
RACHEL COX DOWNING.: / W71-12489
RACHEL COX DOWNING.: / W71-09156
J. L. COX.: / W73-02105
AM L. ZEMAITIS, AND GERALDINE V. COX.: WILLI W70-09189
RICHARD J. CONDIT, AND WAYNE T. CRANEY.: WILLIAM H. FUNK,/ W72-00801
PTIST.: F. A. CROSS, J. N. WILLIS, AND J. P. BA/ W73-00802
JAMES L. YOUNT, AND ROY A. CROSSMAN, JR.: W71-01488
WILLIAM P. COFFMAN, KENNETH W. CUMMINS, AND JOHN C. WUYCHECK.: / W71-10064
E. J. C. CURTIS, AND C. R. CURDS.: / W72-08573
CURTIS C. MASON.: / W71-12034
E. J. C. CURTIS, AND C. R. CURDS.: / W72-08573
D. B. CZARNECKI, AND H. D. WILLIAMS.: / W72-11738
R. CZERPACK.: / W72-14706
E. W. B. DA COSTA.: / W72-13813
ER.: J. N. DABES, C. R. WILKE, AND K. H. SAU/ W72-06274
Q. DADISMAN.: / W72-14782
M. J. DAFT, AND W. D. P. STEWART.: / W72-05457
DALE A. CARLSON.: / W72-00800
DALE STRAUGHAN.: / W71-08793
DALE W. TOETZ.: / W72-04773
A. C. DALE.: / W71-07551
RCELLA, E. JOE MIDDLEBROOKS, AND DANIE F. TOERIEN.: / DONALD B. PO/ W71-12068
OORE.: G. DANIEL HICKMAN, AND RICHARD B . M/ W72-01095
DARRELL L. KING.: / W71-11519
F, AND MICHAEL J. ATHERTON.: DARRELL L. KING, ALLEN J. TOLMSOF/ W71-07097
. ERICKSON, AND PETER / EARL W. DAVEY, JOHN H. GENTILE, STANTON J/ W72-04708
DAVID A. BELLA.: W71-03034
R.: DAVID B. BAKER. AND JACK W. KRAME/ W72-04260
DAVID H. STOLTENBERG.: W71-03896
J. THOMAS, AND K. A. V. DAVID.: / W72-03217
DUTOIT, D. F. TOERIEN, AND T. R. DAVIES.: P. J. / W72-04309
A. G. DAVIES.: / W73-00823
CHARLES E. WARREN, AND GERALD E. DAVIS.: / W72-06340
B. J. COPELAND, AND H. LEE DAVIS.: / W72-13636
E. M. DAVIS, AND E. F. FLCYNA.: / W72-07664
E. M. DAVIS, AND E. F. GLCYNA.: / W72-09590
ERNST M. DAVIS, AND EARNEST F. GLOYNA.: W71-05155
ERNST DAVIS, AND EARNEST GLOYNA.: W70-08319
J. S. DAVIS, AND W. F. GWOREK.: / W72-12744
WILLIAM J. DAY.: / W71-09524
. R. GOLDMAN, G. MOSHIRI, AND E. DE AMEZAGA.: C/ W72-12549
H. W. DE KONING, AND D. C. MORTIMER.: / W71-12303
W. H. O. DE SMET, AND F. M. J. C. EVENS.: / W72-07896
EDWIN F. BARTH, AND ROBERT B. DEAN.: / W72-04086
L, JR.: DEE MITCHELL, AND JAMES C. BUZZEL/ W72-03218
.: G. W. FUHS, S. D. DEMMERLE, E. CANELLI, AND M. CHEN/ W72-09163
KARL H. DEUBERT, AND IRVING E. DEMORANVILLE.: / W71-11498
LSKI.: DENIS M. VEAL, AND M. F. P. MICHA/ W72-13662
DENNIS R. KEENEY.: / W72-08459
.: O. VAN DER BORGHT, AND S. VAN PUYMBROECK/ W71-09182
.: O. VAN DER BORGHT, AND S. VAN PUYMBROECK/ W71-09168
SYLVIA B. DERBY, AND ERNEST RUBER.: / W71-10096
DEREK H. BONE.: / W72-05476
D. MORTON, R. SERNAU, AND P. H. DERSE.: S./ W72-09168
S. D. MORTON, P. H. DERSE, AND R. C. SERNAU.: / W72-10607
M. V. M. DESAI, AND A. K. GANGULY.: / W71-09188
LE.: KARL H. DEUBERT, AND IRVING E. DEMORANVIL/ W71-11498
MIKE DICKMAN.: / W72-07922

J. CAIRNS, JR., AND K. L. DICKSON.: / W73-01972
BLEASE.: J. B. DIMOND, A. S. GETCHELL, AND J. A./ W72-12930
OMANN, D. J. O'CONNOR, AND D. M. DITORO.: R. V. TH W71-05390
J. K. DIXON.: W71-01899
C. TILTON, J. MURPHY, AND J. K. DIXON.: R./ W72-11833
RRAY, JAN SCHERFIG, AND PETER S. DIXON.: STEVEN MU/ W72-04433
J. K. DIXON, AND R. C. TILTON.: / W72-14840
ANDELLI, P. R. BURKHOLDER, T. E. DOHENY, AND: E. F. M/ W71-07875
BROOKS, AND D/ JOEL C. GOLDMAN, DONALD B. PORCELLA, E. JOE MIDDLE/ W71-12068
GAI, AND E. JOE MIDDLEBROOKS.: DONALD B. PORCELLA, JAMES S. KUMA W70-08944
DONALD E. GARRETT.: / W71-12200
FUNK.: SURINDER K. BHAGAT, DONALD E. PROCTOR, AND WILLIAM H./ W72-03215
EINHARDT./ EDWARD T. HALL, JR., DONALD G. AHEARN, AND DONALD J. R/ W71-11914
AHEARN, EDWARD T. HALL, JR., AND DONALD J. REINHARDT.: D. G. / W71-09173
HALL, JR., DONALD G. AHEARN, AND DONALD J. REINHARDT.: /EDWARD T. / W71-11914
DONALD M. MARTIN.: / W71-13341
LATTERELL.: ROBERT F. HOLT, DONALD R. TIMMONS, AND JOSEPH J. W71-04216
VILLA, JR/ NORBERT A. JAWORSKI, DONALD W. LEAR, JR., AND ORTERIO / W71-08775
AALTO, NORBERT A. JAWORSKI, AND DONALD W. LEAR, JR.: JOHAN A. W70-08248
Y.: PAT C. KERR, DORIS F. PARIS, AND D. L. BROCKWA W71-00475
MAXWELL S. DOTY.: / W71-11486
M. S. DOTY.: / W72-06283
SASTRY, C. R. ARNOLD, AND M. S. DOTY.: /SON, J. MCN. SIEBURTH, A./ W72-08142
S.: MAXWELL S. DOTY, AND GERTRUDES AGUILAR-SANTO/ W71-08597
DOUGLAS EDWIN GREER.: / W72-03613
N. M. DOWIDAR.: / W73-00296
RACHEL COX DOWNING.: / W71-09156
A. L. DOWNING.: / W71-13172
RACHEL COX DOWNING.: / W71-12489
S. M. DRACHEV, AND A. A. BYLINKINA.: / W72-05506
NCE, K. G. TRANBARGER, AND R. A. DRAHN.: C. H. LAWRA/ W72-05594
RICHARD G. CLEMENTS, GEORGE E. DREWRY, AND ROBERT J. LAVIGNE.: / W72-00949
M. R. SCALF, W. R. DUFFER, AND R. D. KREIS.: / W72-08401
LELAND.: C. S. DUFFIELD, S. D. WAALAND, AND R. C/ W73-01622
F. J. H. FREDEEN, AND J. REGIS DUFFY.: W70-08098
CELLA, P. H. MCGAUHEY, AND G. L. DUGAN.: D. B. POR/ W72-02412
SWALD.: G. L. DUGAN, C. G. GOLUEKE, AND W. J. O/ W72-08396
D, AND C. E. RIXFORD.: G. L. DUGAN, C. G. GOLUEKE, W. J. OSWAL/ W72-02850
JANET R. DUGLE, AND J. E. GUTHRIE.: / W71-09256
N, JR.: T. W. DUKE, J. I. LOWE, AND A. J. WILSO/ W71-07412
HUGUENIN.: J. H. RYTHER, W. M. DUNSTAN, K. R. TENORE, AND J. E. / W72-09378
R. J. REIMOLD, AND C. J. DURANT.: / W72-10678
JOSEPH V. DUST, AND ADAM SHINDALA.: W71-00139
J. R. DUTHIE.: / W72-09169
DAVIES.: P. J. DUTOIT, D. F. TOERIEN, AND T. R. / W72-04309
BEASLEY, T. A. JOKELA, AND R. J. EAGLE.: T. M. / W72-00940
REZONIK.: EARL E. SHANNON, AND PATRICK L. B/ W72-04269
TANTON J. ERICKSON, AND PETER / EARL W. DAVEY, JOHN H. GENTILE, S/ W72-04708
ERNST M. DAVIS, AND EARNEST F. GLOYNA.: W71-05155
ERNST DAVIS, AND EARNEST GLOYNA.: W70-08319
I, PHILIP C. SINGER, / WAYNE F. ECHELBERGER, JR., JOSEPH L. PAVON/ W72-00411
. H. VERHO/ M. W. TENNEY, W. F. ECHELBERGER, JR., P. C. SINGER, F W71-03021
ACHLAN.: T. EDELSTEIN, L C-M CHEN, AND J. MCL/ W72-01370
W. T. EDMONDSON.: W71-02681
W. T. EDMONDSON.: / W72-09167
EDWARD F. POHL.: / W71-07106
WADE.: EDWARD G. FOREE, AND CAROLINE P. / W72-06612
, JR.: EDWARD G. FOREE, AND JOHN S. TAPP W70-05469
EDWARD J. KUENZLER.: / W71-10083
. REINHARDT.: D. G. AHEARN, EDWARD T. HALL, JR., AND DONALD J/ W71-09173
EARN, AND DONALD J. REINHARDT./ EDWARD T. HALL, JR., DONALD G. AH/ W71-11914
P. EDWARDS.: / W72-08804
N.: EDWIN F. BARTH, AND ROBERT B. DEA/ W72-04086
DOUGLAS EDWIN GREER.: / W72-03613
R.: W. G. EGAN, J. M. CASSIN, AND M. E. HAI/ W72-12941
D. A. EGLOFF, AND R. PARTRIDGE.: / W73-00281
LLIAMS.: D. D. EIER, A. T. WALLACE, AND R. E. WI/ W72-14878
F. M. EL-SHARKAWI, AND S. K. MOAWAD.: W70-06619
RE.: J. H. ELDER, C. A. LEMBI, AND D. J. MOR/ W72-08440
ELSIE M. BURROWS.: / W71-10553
ELSIE M. BURROWS.: / W71-12867
B. M. ELSON.: / W72-04742
R. A. ENGBERG, AND T. O. RENSCHLER.: / W71-13466
VEY, JOHN H. GENTILE, STANTON J. ERICKSON, AND PETER BETZER.: / DA/ W72-04708
L. ARNBORG, AND E. ERIKSSON.: W70-09947
G. ERIKSSON, AND C. FORSBERG.: / W73-01632
SYLVIA B. DERBY, AND ERNEST RUBER.: / W71-10096
ERNST DAVIS, AND EARNEST GLOYNA.: W70-08319
OYNA.: ERNST M. DAVIS, AND EARNEST F. GL W71-05155
ZVI PLAUT, AND: MARTIN GIBBS, ERWIN LATZKO, MICHAEL J. HARVEY, / W71-09172
RLING, A. R. PICK, AND D. W. VAN ES.: G. E. BURNS, R. M. GI/ W71-07109
RONALD M. BUSH, AND EUGENE B. WELCH.: / W72-00798
, AND RONALD M. BUSH.: EUGENE B. WELCH, JAMES A. BUCKLEY/ W72-00799
WUM-CHENG WANG, AND RALPH L. EVANS.: W71-03028
V. KOTHANDARAMAN, AND R. L. EVANS.: / W72-10237
D. EVANS, AND J. G. STOCKNER.: / W72-13142
ROY B. EVANS, AND PAUL KRUGER.: / W72-00974
. H. O. DE SMET, AND F. M. J. C. EVENS.: W/ W72-07896
JON E. SANGER, AND EVILLE GORHAM.: / W72-01784
. H. HURLBERT, J. ZEDLER, AND D. FAIRBANKS.: S/ W72-07132

W. F.	FAUST.: /	W72-02218
V. D.	FEDOROV, AND N. G. KUSTENKO.: /	W73-00441
KI, LEO J. CLARK, AND KENNETH D.	FEIGNER.: NORBERT A. JAWORS/	W71-09788
KI, LEO J. CLARK, AND KENNETH D.	FEIGNER.: NORBERT A. JAWORS	W70-08249
E. J.	FEINLER, AND R. W. HUBBARD.: /	W73-01676
MILTON H.	FELDMAN.: /	W71-11793
G. W. FOESS, AND T. H.	FENG.: /	W72-00847
M. C.	FERENS, AND R. J. BEYERS.: /	W73-01080
.: G. CAHET, M.	FIALA, G. JACQUES, AND M. PANOUSE/	W73-00431
, AND C. F. WURSTER.: N. S.	FISCHER, J. L. MOSSER, T. C. TENG/	W72-08436
BORMANN, NOYE M. JOHNSON, D. W.	FISHER, AND ROBERT S. PIERCE.: /T	W71-01489
STER.: J. L. MOSSER, N. S.	FISHER, T. C. TENG, AND C. F. WUR/	W72-05614
G. P.	FITZGERALD.: /	W72-11617
G. P.	FITZGERALD.: /	W72-10608
GEORGE P.	FITZGERALD.: /	W71-06189
G. P.	FITZGERALD, AND G. F. LEE.: /	W73-00232
E.	FJERDINGSTAD.: /	W73-00002
W. H. FUNK, AND D. C.	FLAHERTY.: /	W72-14728
R. C. KIRKWOOD, AND W. W.	FLETCHER.:	W70-07255
D. W. SMITH, C. B.	FLIERMANS, AND T. D. BROCK.: /	W72-14301
KORANDA.:	FLORENCE L. HARRISON, AND JOHN J./	W72-06342
E. M. DAVIS, AND E. F.	FLOYNA.: /	W72-07664
J.	FOEHRENBACH.: /	W73-00840
G. W.	FOESS, AND T. H. FENG.: /	W72-00847
S.	FOGEL, I. CHET, AND R. MITCHELL.:/	W73-00942
G. E.	FOGG.: /	W73-02472
MEIRING.: Y.	FOLKMAN, M. KREMER, AND P. G. J. /	W72-11642
M. WONG, V. F. HODGE, AND T. R.	FOLSOM.: K./	W72-07826
T. R.	FOLSOM.: /	W72-03324
T. R.	FOLSOM, AND T. M. BEASLEY.: /	W73-00833
EDWARD G.	FOREE, AND CAROLINE P. WADE.: /	W72-06612
EDWARD G.	FOREE, AND JOHN S. TAPP, JR.:	W70-05469
G.	FOREE, AND R. SCROGGIN.: /	W72-14790
CARTY.: E. G.	FOREE, W. J. JEWELL, AND P. L. MC	W70-10405
G. ERIKSSON, AND C.	FORSBERG.: /	W73-01632
A. W. MORRIS, AND P.	FOSTER.: /	W72-09108
J. M.	FOULDS.: /	W73-02187
D. VAN AS, H. O.	FOURIE, AND C. M. VLEGGAAR.: /	W73-00832
M. REBHUN, M. A.	FOX, AND J. B. SLESS.: /	W71-13187
L. BUNCH.:	FRANCIS M. MIDDLETOWN, AND ROBERT/	W71-07123
THOMAS J. MONAHAN, AND	FRANCIS R. TRAINOR.: /	W71-08687
JEAN-	FRANCOIS PIERRE.:	W71-00099
ANNA, JR.:	FRANK J. HUMENIK, AND GEORGE P. H/	W72-04983
W. H.	FRANK.: /	W71-12410
I. L. OPHEL, AND C. D.	FRASER.: /	W71-11529
:	FRED G. LEE, AND GILMAN D. VEITH./	W71-12092
ND WALTER TH/ CHARLES R. HAZEL,	FRED KOPPERDAHL, NORMAN MORGAN, A/	W71-09789
G.	FRED LEE.: /	W71-12072
F. J. H.	FREDEEN, AND J. REGIS DUFFY.:	W70-08098
ROBERT JACK	FREEMAN, JR.: /	W71-13326
E. W. BEHRENS, AND S. A.	FRISHMAN.:	W71-04873
H. HUANG, J. R. MASE, AND E. G.	FRUH.: V./	W72-13692
AND M. CHEN.: G. W.	FUHS, S. D. DEMMERLE, E. CANELLI,/	W72-09163
R. GLEN	FULLER.: /	W72-00960
ONALD E. PROCTOR, AND WILLIAM H.	FUNK.: SURINDER K. BHAGAT, D/	W72-03215
W. H.	FUNK, AND D. C. FLAHERTY.: /	W72-14728
E T. CRANEY.: WILLIAM H.	FUNK, RICHARD J. CONDIT, AND WAYN/	W72-00801
	GALE ALLEN WRIGHT.:	W71-04518
ESHKO, E. K. LEBEDEVA, AND T. B.	GALKINA.: /V. MAXIMOVA, G. I. MEL/	W71-08027
J. R. NURSALL, AND D. N.	GALLUP.: /	W73-01704
P. M. AMIN, AND S. V.	GANAPATI.: /	W72-10233
BJORN	GANNING.: /	W71-08026
DONALD E.	GARRETT.: /	W71-12200
R. R.	GARTON.: /	W72-08683
SINGER, F. H. VERHOFF, AND W. A.	GARVEY.: /CHELBERGER, JR., P. C.	W71-03021
MARK J. HAMMER, AND	GARY L. HERGENRADER.: /	W72-04759
MITTON, T. H. SUCHANEK, JR., N.	GEBELEIN, AND C. GROSSMAN.: /. B./	W72-14659
ON, WILLIAM J. OSWALD, AND HENRY	GEE.: AARON MER/	W72-04789
, CLARENCE G. GOLUEKE, AND HENRY	GEE.: /. ZABAT, WILLIAM J. OSWALD/	W72-04788
M. SOBSEY, J. E. HARRISON, H.	GEE, G. SHELET, AND J. C. GOLDMAN/	W72-04787
AIDECH, E. THOMAS, AND PHILIP E.	GEHRING.: CONRAD KLEVENO, O. BR/	W72-13659
T. BRAIDECH, P.	GEHRING, AND C. KLEVENO.: /	W72-12995
LOTHAR	GEITLER.:	W71-04526
LOTHAR	GEITLER.: /	W71-07360
N, NOYE M. JOHNSON, D. W. FISH/	GENE E. LIKENS, F. HERBERT BORMAN	W71-01489
PETER / EARL W. DAVEY, JOHN H.	GENTILE, STANTON J. ERICKSON, AND/	W72-04708
AVIGNE.: RICHARD G. CLEMENTS,	GEORGE E. DREWRY, AND ROBERT J. L/	W72-00949
.: ROBERT S. MATSON,	GEORGE E. MUSTOE, AND S. B. CHANG/	W72-06037
	GEORGE P. FITZGERALD.: /	W71-06189
FRANK J. HUMENIK, AND	GEORGE P. HANNA, JR.: /	W72-04983
, AND OLIVER T. LOVE JR.:	GEORGE W. REID, LEALE E. STREEBIN/	W71-07973
CARL R. HIRTH, KENNETH M. MAC/	GERALD D. MCKEE, LOYS P. PARRISH,/	W71-05626
CHARLES E. WARREN, AND	GERALD E. DAVIS.: /	W72-06340
ANE, CONSTANTINE J. GREGORY, AND	GERALD L. ARONSON.: /OHN J. COCHR/	W71-05553
WILLIAM L. ZEMAITIS, AND	GERALDINE V. COX.:	W70-09189
MAXWELL S. DOTY, AND	GERTRUDES AGUILAR-SANTOS.: /	W71-08597
F.	GESSNER, AND L. HAMMER.: /	W72-12940
J. B. DIMOND, A. S.	GETCHELL, AND J. A. BLEASE.: /	W72-12930
A. K. BASU, B. B.	GHOSH, AND R. N. PAL.: /	W71-11876
ARVEY, ZVI PLAUT, AND: MARTIN	GIBBS, ERWIN LATZKO, MICHAEL J. H/	W71-09172

C. E.	GIBSON.:	/ W72-04070
JOHN J. ROOK, AND	GIJSBERT OSKAM.:	W71-01491
J. BIRCH, ROBERT E. HILLMAN, AND	GILBERT E. RAINES.: /VIN, THOMAS /	W71-07417
M. S.	GILES.:	/ W72-00959
A.	GILLESPIE, AND Y. MORITA.:	/ W72-08989
W. MARSHALL.: W. J.	GILLESPIE, C. A. MAZZOLA, AND D. /	W72-03753
FRED G. LEE, AND	GILMAN D. VEITH.:	/ W71-12092
N ES.: G. E. BURNS, R. M.	GIRLING, A. R. PICK, AND D. W. VA/	W71-07109
R.	GLEN FULLER.:	/ W72-00960
D. M. STEVEN, AND R.	GLOMBITZA.:	/ W72-11719
WALTER A.	GLOOSCHENKO.:	/ W72-13657
E. M. DAVIS, AND E. F.	GLOYNA.:	/ W72-09590
ERNST DAVIS, AND EARNEST	GLOYNA.:	W70-08319
ERNST M. DAVIS, AND EARNEST F.	GLOYNA.:	W71-05155
E. F.	GLOYNA, AND J. AGUIRRE.:	/ W71-07102
OY, M. MUELLER, S. STOKER, J. J.	GOERING, AND M. T. GOTTSCHALK.: //	W72-09120
MARVIN C.	GOLDBERG.:	/ W71-06435
ICHARD ARMSTRONG, AND CHARLES R.	GOLDMAN.: PETER RICHERSON, R/	W71-09171
ON, H. GEE, G. SHELET, AND J. C.	GOLDMAN.: /. SOBSEY, J. E. HARRIS/	W72-04787
OE MIDDLEBROOKS, AND C/ JOEL C.	GOLDMAN, DONALD B. PORCELLA, E. J/	W71-12068
EZAGA.: C. R.	GOLDMAN, G. MOSHIRI, AND E. DE AM/	W72-12549
SHELEF, W. J. OSWALD, AND C. C.	GOLUEKE.: G.	W70-10403
, WILLIAM J. OSWALD, CLARENCE G.	GOLUEKE, AND HENRY GEE.: /. ZABAT/	W72-04788
C. G.	GOLUEKE, AND W. J. OSWALD.:	W71-00138
G. L. DUGAN, C. G.	GOLUEKE, AND W. J. OSWALD.:	/ W72-08396
RIXFORD.: G. L. DUGAN, C. G.	GOLUEKE, W. J. OSWALD, AND C. E. /	W72-02850
W. L. STROMBERG, AND C. J.	GOODNIGHT.:	/ W72-09668
JON E. SANGER, AND EVILLE	GORHAM.:	/ W72-01784
STOKER, J. J. GOERING, AND M. T.	GOTTSCHALK.: /OY, M. MUELLER, S. /	W72-09120
ERR.: P. D.	GOULDEN, W. J. TRAVERSY, AND G. K	W70-09388
R.	GOULDER.:	/ W72-14796
K. H.	GOULDING.:	/ W72-07508
L, G. S. VENKATARAMAN, AND S. K.	GOYAL.: / BOSE, UJJAL SINGH NAGPA/	W71-10070
U.	GRANHALL.:	/ W73-00246
U.	GRANHALL, AND A. LUNDGREN.:	/ W72-08054
JAMES C.	GRATTEAU.:	W71-05083
DOUGLAS EDWIN	GREER.:	/ W72-03613
P. E.	GREESON.:	/ W72-13851
JOHN J. COCHRANE, CONSTANTINE J.	GREGORY, AND GERALD L. ARONSON.: /	W71-05553
P. J.	GRESHAM.:	/ W72-14469
RICHARD W.	GRIGG, AND ROBERT S. KIWALA.:	W71-01475
CHANEK, JR., N. GEBELEIN, AND C.	GROSSMAN.: /. B. MITTON, T. H. SU/	W72-14659
R. D.	GRUNDY.:	/ W72-04734
E.	GRYGIEREK.:	/ W72-14805
R. R. L.	GUILLARD, AND J. A. HELLEBUST.:	/ W72-07710
R. C.	GUMERMAN.:	/ W72-01108
KORZHENEVSKAYA.: M. V.	GUSEV, K. A. NIKITINA, AND T. G. /	W71-08683
EDATELEV, B. N. ILYIN, A/ V. N.	GUSKOVA, A. N. BRAGINA, A. A. ZAS/	W72-10694
SKAYA/ S. M. VARDAPETYAN, B. L.	GUTELMACHER, AND N. G. OZERETSKOV/	W72-00161
JANET R. DUGLE, AND J. E.	GUTHRIE.:	/ W71-09256
J. S. DAVIS, AND W. F.	GWOREK.:	/ W72-12744
G. EGAN, J. M. CASSIN, AND M. E.	HAIR.: W. /	W72-12941
LEE SU, E. J. STABA AND Y. ABUL-	HAJJ.: K./	W72-08586
MILLARD W.	HALL.:	/ W72-04280
T.: D. G. AHEARN, EDWARD T.	HALL, JR., AND DONALD J. REINHARD/	W71-09173
DONALD J. REINHARDT./ EDWARD T.	HALL, JR., DONALD G. AHEARN, AND /	W71-11914
BRUCE W.	HALSTEAD.:	/ W72-04259
A. L.	HAMILTON.:	/ W71-12091
F. GESSNER, AND L.	HAMMER.:	/ W72-12940
MARK J.	HAMMER, AND GARY L. HERGENRADER.:/	W72-04759
R. H. SIDDIQI, AND B. K.	HANDA.:	/ W71-06033
F. J. HUMENIK, AND G. P.	HANNA, JR.:	W71-04079
FRANK J. HUMENIK, AND GEORGE P.	HANNA, JR.:	/ W72-04983
F. J. HUMENIK, AND G. P.	HANNA, JR.:	/ W72-11100
P. J.	HANNAN, AND C. PATOUILLET.:	/ W72-07660
R. E.	HANSEN.:	/ W72-08357
OSMUND HOLM-	HANSEN.:	/ W71-11234
A. OTSUKI, AND T.	HANYA.:	/ W73-00379
A. OTSUKI, AND T.	HANYA.:	/ W73-01066
	HAROLD L. ALLEN.:	/ W72-03216
D. O.	HARRIS.:	/ W71-13793
ENTYNE, W. E. JOHNSON, AND A. J.	HARRIS.: J. R. VALL/	W71-11507
D. E. ARMSTRONG, R. F.	HARRIS, AND J. K. SYERS.:	/ W72-03742
MSTRONG.: C. C. LEE, R. F.	HARRIS, J. K. SYERS, AND D. E. AR/	W71-13794
CHRISTIE.: A. J.	HARRIS, K. J. ROBERTS, AND A. E. /	W72-06837
F. L.	HARRISON.:	/ W73-00826
FLORENCE L.	HARRISON, AND JOHN J. KORANDA.:	/ W72-06342
J. C. GOLDMAN/ M. SOBSEY, J. E.	HARRISON, H. GEE, G. SHELET, AND /	W72-04787
WILLIAM R.	HARTLEY, AND CHARLES M. WEISS.:	/ W71-07382
R. S.	HARVEY.:	W71-02075
R. S.	HARVEY.:	/ W71-09250
GIBBS, ERWIN LATZKO, MICHAEL J.	HARVEY, ZVI PLAUT, AND: MARTIN/	W71-09172
J. P.	HARVILLE.:	/ W72-09092
B.	HAVLIK.:	/ W72-10686
R. C.	HAYNES.:	/ W72-09061
RGAN, AND WALTER TH/ CHARLES R.	HAZEL, FRED KOPPERDAHL, NORMAN MO/	W71-09789
ON, AND W. A. STEEL.: R. E.	HEFT, W. A. PHILLIPS, H. R. RALST/	W73-00831
G. W.	HEINKE, AND J. D. NORMAN.:	W71-00116
E. E.	HELD.:	/ W72-03347
R. R. L. GUILLARD, AND J. A.	HELLEBUST.:	/ W72-07710

C. M. WEISS, AND R. W.	HELMS.:	/ W73-00255
ANDREW.: D. W.	HENDRICKS, W. D. POTE, AND J. G.	/ W72-02363
J. K. SILVEY, D. E.	HENLEY, AND J. T. WYATT.:	/ W72-10613
E.	HENRIKSSON.:	/ W73-02471
ON MERON, WILLIAM J. OSWALD, AND	HENRY GEE.: AAR	/ W72-04789
OSWALD, CLARENCE G. GOLUEKE, AND	HENRY GEE.: /. ZABAT, WILLIAM J.	/ W72-04788
CLA/ L. R. J. VAN VUUREN, M. R.	HENZEN, G. J. STANDER, AND A. J.	W70-10381
D. W. FISH/ GENE E. LIKENS, F.	HERBERT BORMANN, NOYE M. JOHNSON,	W71-01489
JOSEPH F. ROESLER, AND	HERBERT C. PREUL.:	/ W71-07099
D. R. KEENEY, R. A.	HERBERT, AND A. J. HOLDING.:	/ W72-07933
MARK J. HAMMER, AND GARY L.	HERGENRADER.:	/ W72-04759
M.	HESSE.:	/ W73-00251
NORBERT A. JAWORSKI, AND LEO J.	HETLING.:	W70-06509
D. J.	HIBBERD, AND G. F. LEEDALE.:	/ W72-14808
M.	HICKMAN, AND F. E. ROUND.:	/ W71-12855
G. DANIEL	HICKMAN, AND RICHARD B . MOORE.:	/ W72-01095
L.	HILLIS-COLINVAUX.:	/ W72-14294
A. A. LEVIN, T. J. BIRCH, R. E.	HILLMAN, AND G. E. RAINES.:	/ W72-07907
EVIN, THOMAS J. BIRCH, ROBERT E.	HILLMAN, AND GILBERT E. RAINES.:	/ W71-07417
MCKEE, LOYS P. PARRISH, CARL R.	HIRTH, KENNETH M. MACKENTHUM, AND	/ W71-05626
P. KAPOOR, R. L. METCALF, A. S.	HIRWE, P.-Y. LU, AND J. R. COATS.	/ W72-07703
J. E.	HOBBIE.:	/ W72-01329
L. A.	HOBSON, AND C. J. LORENZEN.:	/ W72-14313
K. M. WONG, V. F.	HODGE, AND T. R. FOLSOM.:	/ W72-07826
KEENEY, R. A. HERBERT, AND A. J.	HOLDING.: D. R.	/ W72-07933
R. E.	HOLLAND, AND A. M. BEETON.:	/ W73-00384
OSMUND	HOLM-HANSEN.:	/ W71-11234
PH J. LATTERELL.: ROBERT F.	HOLT, DONALD R. TIMMONS, AND JOSE	W71-04216
LEONARD W.	HOM.:	/ W71-07096
EXANDER, J. J. KELLEY, AND D. W.	HOOD.: /ICH, M. BILLINGTON, V. AL	/ W72-09122
R.	HOPKINS, AND J. M. KAIN.:	/ W72-12576
G. A.	HOTTLE.:	/ W72-09675
THOMAS W. PHILBIN, AND	HOWARD D. PHILLIPP.:	W70-09614
ALICE Q.	HOWARD.:	/ W71-08768
J. W.	HOYT.:	/ W71-07878
PAUL R. KENIS, AND J. W.	HOYT.:	/ W72-03220
J.	HRIB, AND V. BREZINA.:	/ W72-07144
: NING-	HSI TANG, AND SURINDER K. BHAGAT.	/ W72-04431
.: V. H.	HUANG, J. R. MASE, AND E. G. FRUH	/ W72-13692
, AND J. / D. F. TOERIEN, C. H.	HUANG, J. RADIMSKY, E. A. PEARSON	/ W72-03985
E. J. FEINLER, AND R. W.	HUBBARD.:	/ W73-01676
DUNSTAN, K. R. TENORE, AND J. E.	HUGUENIN.: J. H. RYTHER, W. M.	/ W72-09378
F. J.	HUMENIK, AND G. P. HANNA, JR.:	/ W72-11100
F. J.	HUMENIK, AND G. P. HANNA, JR.:	W71-04079
: FRANK J.	HUMENIK, AND GEORGE P. HANNA, JR.	/ W72-04983
C.	HUNDING.:	/ W72-07501
K-H.	HUNKEN, AND I. D. SEKOULOV.:	/ W73-01367
ANKS.: S. H.	HURLBERT, J. ZEDLER, AND D. FAIRB	/ W72-07132
JAMES A. SONDEL, AND CAROLYN M.	HYCHE.: / CHARLES C. THOMAS, JR.,	/ W71-11036
RAGINA, A. A. ZASEDATELEV, B. N.	ILYIN, AND V. M. KUPRIYANOVA.: /B	/ W72-10694
A. M. A.	IMEVBORE, AND Z. BOSZORMENYI.:	/ W72-07143
C. D. MC	INTIRE, AND W. S. OVERTON.:	/ W72-08141
	IRENA SPODNIEWSKA.:	/ W72-01358
KARL H. DEUBERT, AND	IRVING E. DEMORANVILLE.:	/ W71-11498
RICHARD C. KNUDSEN, AND NORVA/	IRVING YALL, WILLIAM H. BOUGHTON,	W71-01474
K. J.	IVES.:	/ W72-09393
	JACK A. BORCHARDT.:	/ W71-09958
ROBERT	JACK FREEMAN, JR.:	/ W71-13326
T.:	JACK KISHLER, AND CLARENCE E. TAF	/ W71-05630
DAVID B. BAKER, AND	JACK W. KRAMER.:	/ W72-04260
	JACOB VERDUIN.:	/ W72-05459
G. CAHET, M. FIALA, G.	JACQUES, AND M. PANOUSE.:	/ W73-00431
	JAMES A. BUCKLEY.:	/ W72-08049
USH.: EUGENE B. WELCH,	JAMES A. BUCKLEY, AND RONALD M. B	/ W72-00799
C. BAIN, JR., PERRY L. MCCARTY,	JAMES A. ROBERTSON, AND WILLIAM H	/ W71-07098
PILLAY, CHARLES C. THOMAS, JR.,	JAMES A. SONDEL, AND CAROLYN M. H	/ W71-11036
DEE MITCHELL, AND	JAMES C. BUZZELL, JR.:	/ W72-03218
	JAMES C. GRATTEAU.:	W71-05083
	JAMES E. SCHINDLER.:	/ W72-07940
AN, JR.:	JAMES L. YOUNT, AND ROY A. CROSSM	W71-01488
ROBERT S. CAMPBELL, AND	JAMES R. WHITLEY.:	/ W71-06444
CAMPBELL, ARTHUR WITT, JR., AND	JAMES R. WHITLEY.: ROBERT S.	W70-09596
LEBROOKS.: DONALD B. PORCELLA,	JAMES S. KUMAGAI, AND E. JOE MIDD	W70-08944
: STEVEN MURRAY,	JAN SCHERFIG, AND PETER S. DIXON.	/ W72-04433
.:	JANET R. DUGLE, AND J. E. GUTHRIE	/ W71-09256
L. SILEN, AND B-O.	JANSSON.:	/ W73-00348
: JOHAN A. AALTO, NORBERT A.	JAWORSKI, AND DONALD W. LEAR, JR.	W70-08248
J. W. G. LUND, G. H. M.	JAWORSKI, AND H. BUCKA.:	/ W72-10619
NORBERT A.	JAWORSKI, AND LEO J. HETLING.:	W70-06509
WILLA, JR.: N. A.	JAWORSKI, D. W. LEAR, JR., AND O.	/ W72-09171
D ORTERIO VILLA, JR/ NORBERT A.	JAWORSKI, DONALD W. LEAR, JR., AN	/ W71-08775
TH D. FEIGNER.: NORBERT A.	JAWORSKI, LEO J. CLARK, AND KENNE	W70-08249
TH D. FEIGNER.: NORBERT A.	JAWORSKI, LEO J. CLARK, AND KENNE	/ W71-09788
	JEAN-FRANCOIS PIERRE.:	W71-00099
R. C.	JENNINGS.:	/ W72-07496
	JERRY L. WILHM.:	W70-06668
	JERRY STEWART.:	/ W71-13356
E. G. FOREE, W. J.	JEWELL, AND P. L. MCCARTY.:	W70-10405
WILLIAM J.	JEWELL, AND PERRY L. MCCARTY.:	/ W72-01881
WILLIAM J.	JEWELL, AND S. A. KULASOORIYA.:	/ W71-12070

LUCYNA	JOANNA BOWNIK.: /	W71-11012
	JOE B. SILLS.:	W71-04528
RCELLA, JAMES S. KUMAGAI, AND E.	JOE MIDDLEBROOKS.: DONALD B. PO	W70-08944
GOLDMAN, DONALD B. PORCELLA, E.	JOE MIDDLEBROOKS, AND DANIE F. TO/	W71-12068
LA, E. JOE MIDDLEBROOKS, AND D/	JOEL C. GOLDMAN, DONALD B. PORCEL/	W71-12068
KI, AND DONALD W. LEAR, JR.:	JOHAN A. AALTO, NORBERT A. JAWORS	W70-08248
COFFMAN, KENNETH W. CUMMINS, AND	JOHN C. WUYCHECK.: WILLIAM P. /	W71-10064
R, AND RUTH PATRICK.:	JOHN CAIRNS, JR., ROGER L. KAESLE/	W71-11583
F. W. SMITH, AND	JOHN F. THOMPSON.: /	W72-03222
	JOHN G. NICKUM.:	W71-02880
MSTRONG.:	JOHN G. STOCKNER, AND F. A. J. AR/	W71-11029
SON, AND PETER / EARL W. DAVEY,	JOHN H. GENTILE, STANTON J. ERICK/	W72-04708
ONE.: JOHN J. LEE,	JOHN H. TIETJEN, AND ROBERT J. ST/	W72-00948
GREGORY, AND GERALD L. ARONSON/	JOHN J. COCHRANE, CONSTANTINE J.	W71-05553
FLORENCE L. HARRISON, AND	JOHN J. KORANDA.: /	W72-06342
ROBERT J. STONE.:	JOHN J. LEE, JOHN H. TIETJEN, AND/	W72-00948
:	JOHN J. ROCK, AND GIJSBERT OSKAM.	W71-01491
	JOHN L. ANDERSEN.: /	W72-04270
ALBERT MASSEY, AND	JOHN ROBINSON.: /	W72-07937
EDWARD G. FOREE, AND	JOHN S. TAPP, JR.:	W70-05469
DY MCBRIDE, WILLIAM CHORNEY, AND	JOHN SKOK.: LAN/	W71-12858
K. LEE SU, AND E.	JOHN STABA.: /	W72-07360
	JOHN W. BISHOP.: /	W72-00150
	JOHN W. VENNES.: /	W71-07124
J. B. ST.	JOHN.: /	W71-11916
D. M.	JOHN.: /	W72-04766
A. N. MACGREGOR, AND C. E.	JOHNSON.: /	W72-12505
OBERT S. CAMPBELL, AND B. THOMAS	JOHNSON.: LELYN STADNYK, R/	W71-10566
J. R. VALLENTYNE, W. E.	JOHNSON, AND A. J. HARRIS.: /	W71-11507
M. G.	JOHNSON, AND G. E. OWEN.: /	W71-11009
M. G.	JOHNSON, AND G. E. OWEN.: /	W71-08953
ENS, F. HERBERT BORMANN, NOYE M.	JOHNSON, D. W. FISHER, AND ROBERT	W71-01489
TRY, C. R. ARNOLD, AND M/ P. W.	JOHNSON, J. MCN. SIEBURTH, A. SAS/	W72-08142
T. M. BEASLEY, T. A.	JOKELA, AND R. J. EAGLE.: /	W72-00940
:	JON E. SANGER, AND EVILLE GORHAM./	W72-01784
J. G.	JONES.: /	W72-03224
D. J.	JONES.: /	W72-11854
TH. AHL, AND E.	JONSSON.: /	W73-00286
, WAYNE R. MATSON, AND ROBERT A.	JORDAN.: MICHAEL E. BENDER/	W71-09674
A. J.	JORDAN, AND R. L. VADAS.: /	W73-00369
PREUL.:	JOSEPH F. ROESLER, AND HERBERT C./	W71-07099
F. HOLT, DONALD R. TIMMONS, AND	JOSEPH J. LATTERELL.: ROBERT	W71-04216
R, / WAYNE F. ECHELBERGER, JR.,	JOSEPH L. PAVONI, PHILIP C. SINGE/	W72-00411
	JOSEPH SHAPIRO.:	W70-09325
.:	JOSEPH V. DUST, AND ADAM SHINDALA	W71-00139
IA.:	JOSEPH Y. CHENG, AND NAVAL J. ANT	W70-07280
IA.:	JOSEPH Y. CHENG, AND NAVAL J. ANT	W70-07842
A. S. MENON, AND A. A.	JURKOVIC.: /	W72-06690
J. CAIRNS, JR., AND R. L.	KAESLER.: /	W72-07791
JOHN CAIRNS, JR., ROGER L.	KAESLER, AND RUTH PATRICK.: /	W71-11583
R. HOPKINS, AND J. M.	KAIN.: /	W72-12576
I. S. BHAT, AND P. R.	KAMATH.: /	W72-10958
E, P.-Y. LU, AND J. R. C/ I. P.	KAPOOR, R. L. METCALF, A. S. HIRW/	W72-07703
D. S. KAPPE, AND S. E.	KAPPE.: /	W72-07514
D. S.	KAPPE, AND S. E. KAPPE.: /	W72-07514
J. C. WARD, AND S.	KARAKI.: /	W72-02274
NFIELD.:	KARL A. ROSEMAN, AND WARNER M. LI/	W71-06247
MORANVILLE.:	KARL H. DEUBERT, AND IRVING E. DE/	W71-11498
A.	KAWAI, AND I. SUGAHARA.: /	W72-11563
H.	KAYSER.: /	W72-12239
H.	KAYSER.:	W71-03095
C. W.	KEEFE.: /	W73-01089
DENNIS R.	KEENEY.: /	W72-08459
HOLDING.: D. R.	KEENEY, R. A. HERBERT, AND A. J. /	W72-07933
. RICH, J. F. ANDREWS, AND T. M.	KEINATH.: L. G/	W72-11724
BILLINGTON, V. ALEXANDER, J. J.	KELLEY, AND D. W. HOOD.: /ICH, M./	W72-09122
ROBERT E. BAUMANN, AND SHELDON	KELMAN.: /	W71-06445
PAUL R.	KENIS, AND J. W. HOYT.: /	W72-03220
G. D. COOKE, AND ROBERT L.	KENNEDY.: /	W71-11949
T A. JAWORSKI, LEO J. CLARK, AND	KENNETH D. FEIGNER.: NORBER	W70-08249
T A. JAWORSKI, LEO J. CLARK, AND	KENNETH D. FEIGNER.: NORBER/	W71-09788
LOYS P. PARRISH, CARL R. HIRTH,	KENNETH M. MACKENTHUM, AND LOWELL/	W71-05626
	KENNETH R. PIECH.: /	W71-11685
UYCHECK.: WILLIAM P. COFFMAN,	KENNETH W. CUMMINS, AND JOHN C. W/	W71-10064
IGER, AND BERNARD M. SOLON.:	KENNETH W. STEWART, LARRY E. MILL	W70-06655
GOULDEN, W. J. TRAVERSY, AND G.	KERR.: P. D.	W70-09388
, AND J. T. BARNETT, JR./ P. C.	KERR, D. L. BROCKWAY, D. F. PARIS/	W72-09159
ROCKWAY.: PAT C.	KERR, DORIS F. PARIS, AND D. L. B	W71-00475
K. M. MACKENTHUN, AND L. E.	KEUP.: /	W72-06536
R. A.	KIMERLE, AND N. H. ANDERSON.: /	W72-03556
Y. MATIDA, H. KUMADA, S.	KIMURA, Y. SAIGA, AND T. NOSE.: /	W72-12257
D. L.	KING.: /	W72-09162
J. M.	KING.: /	W72-12632
DARRELL L.	KING.: /	W71-11519
AEL J. ATHERTON.: DARRELL L.	KING, ALLEN J. TOLMSOFF, AND MICH/	W71-07097
P. SZE, AND J. M.	KINGSBURY.: /	W72-09111
R. C.	KIRKWOOD, AND W. W. FLETCHER.:	W70-07255
JACK	KISHLER, AND CLARENCE E. TAFT.: /	W71-05630
RICHARD W. GRIGG, AND ROBERT S.	KIWALA.:	W71-01475
RICHARD M.	KLEIN.: /	W71-09157

W. L.	KLEIN, AND R. A. BRECHBILL.:	/ W73-01673
T. BRAIDECH, P. GEHRING, AND C.	KLEVENO.:	/ W72-12995
AND PHILIP E. GEHRING.: CONRAD	KLEVENO, O. BRAIDECH, E. THOMAS,	/ W72-13659
WART, G. M. ROSENTHAL, AND J. R.	KLINE.: M. L. STE	/ W72-03348
J. W.	KLOCK.:	/ W72-10401
. NEWMAN, N. W. MOORE, AND F. W.	KNAGGS.: J. M. WAY, J. F	/ W72-00155
WILLIAM H. BOUGHTON, RICHARD C.	KNUDSEN, AND NORVAL A. SINCLAIR.:	W71-01474
J. J.	KOELLING.:	/ W73-01434
J.	KOMAREK.:	/ W72-12172
H. W. DE	KONING, AND D. C. MORTIMER.:	/ W71-12303
G. V.	KONOVALOVA.:	/ W73-00442
T. F. H. ALLEN, AND J. F.	KOONCE.:	/ W73-02469
LTER TH/ CHARLES R. HAZEL, FRED	KOPPERDAHL, NORMAN MORGAN, AND WA	/ W71-09789
LORENCE L. HARRISON, AND JOHN J.	KORANDA.: F	/ W72-06342
GUSEV, K. A. NIKITINA, AND T. G.	KORZHENEVSKAYA.: M. V.	/ W71-08683
V.	KOTHANDARAMAN, AND R. L. EVANS.:	/ W72-10237
YEHUDA	KOTT.:	/ W72-00422
O. M.	KOZHOVA.:	/ W72-04855
O. M.	KOZHOVA.:	/ W71-09165
DAVID B. BAKER, AND JACK W.	KRAMER.:	/ W72-04260
. SCALF, W. R. DUFFER, AND R. D.	KREIS.: M. R	/ W72-08401
Y. FOLKMAN, M.	KREMER, AND P. G. J. MEIRING.:	/ W72-11642
R. C. SUBBARAJU, AND K.	KRISHNAMURTHY.:	/ W73-00935
J.	KRISTIANSEN.:	/ W72-12738
RAY B.	KRONE.:	/ W71-09581
ROY B. EVANS, AND PAUL	KRUGER.:	/ W72-00974
YU V.	KRUGLOV, AND L. N. PAROMENSKAYA.:	/ W71-07675
LOUIS A.	KRUMHOLZ.:	/ W72-06526
D. K.	KRUPATKINA.:	/ W72-07166
L. E.	KUENTZEL.:	/ W71-13257
EDWARD J.	KUENZLER.:	/ W71-10083
WILLIAM J. JEWELL, AND S. A.	KULASOORIYA.:	/ W71-12070
V. V./ G. G. POLIKARPOV, L. G.	KULEBAKINA, V. G. TSITSUGINA, AND	/ W72-10955
R. G.	KULLBERG.:	/ W72-05610
T. NOSE.: Y. MATIDA, H.	KUMADA, S. KIMURA, Y. SAIGA, AND	/ W72-12257
: DONALD B. PORCELLA, JAMES S.	KUMAGAI, AND E. JOE MIDDLEBROOKS.	W70-08944
H. D.	KUMAR, AND G. PRAKASH.:	/ W72-00153
W. F.	KUNZ.:	/ W72-11836
EDATELEV, B. N. ILYIN, AND V. M.	KUPRIYANOVA.: /BRAGINA, A. A. ZAS	/ W72-10694
V. D. FEDOROV, AND N. G.	KUSTENKO.:	/ W73-00441
	KWEI LEE SU.:	/ W72-04275
CHANG-	KWEI LIN.:	/ W72-13644
T. J.	LACK.:	/ W72-14797
E. BOURGET, AND G.	LACROIX.:	/ W72-05432
ND JOHN SKOK.:	LANDY MCBRIDE, WILLIAM CHORNEY, A	/ W71-12858
WILLY	LANGE.:	/ W72-01097
J. CAIRNS, AND G. R.	LANZA.:	/ W72-07225
J. CAIRNS, JR., AND G. R.	LANZA.:	/ W72-08579
SOLON.: KENNETH W. STEWART,	LARRY E. MILLIGER, AND BERNARD M.	W70-06655
A.	LATTERA, AND M. BERNHARD.:	/ W71-13243
DONALD R. TIMMONS, AND JOSEPH J.	LATTERELL.: ROBERT F. HOLT,	W71-04216
AUT, AND: MARTIN GIBBS, ERWIN	LATZKO, MICHAEL J. HARVEY, ZVI PL	/ W71-09172
GEORGE E. DREWRY, AND ROBERT J.	LAVIGNE.: RICHARD G. CLEMENTS,	/ W72-00949
C. MCDONALD, R. D. SPEAR, R. J.	LAVIN, AND N. L. CLESCERI.: G.	/ W72-08376
. A. DRAHN.: C. H.	LAWRANCE, K. G. TRANBARGER, AND R	/ W72-05594
G. E.	LAZORENKO.:	/ W72-12203
LOVE JR.: GEORGE W. REID,	LEALE E. STREEBIN, AND OLIVER T.	/ W71-07973
N. A. JAWORSKI, D. W.	LEAR, JR., AND O. WILLA, JR.:	/ W72-09171
NORBERT A. JAWORSKI, DONALD W.	LEAR, JR., AND ORTERIO VILLA, JR.	/ W71-08775
RBERT A. JAWORSKI, AND DONALD W.	LEAR, JR.: JOHAN A. AALTO, NO	W70-08248
G. D.	LEBEDEVA.:	/ W72-09649
MAXIMOVA, G. I. MELESHKO, E. K.	LEBEDEVA, AND T. B. GALKINA.: /V.	/ W71-08027
B. J. COPELAND, AND H.	LEE DAVIS.:	/ W72-13636
KWEI	LEE SU.:	/ W72-04275
K.	LEE SU, AND E. JOHN STABA.:	/ W72-07360
AJJ.: K.	LEE SU, E. J. STABA AND Y. ABUL-H	/ W72-08586
G. F.	LEE.:	/ W73-02479
N. J. MATERA, AND J. J.	LEE.:	/ W73-00367
G. F.	LEE.:	/ W73-02478
J. J.	LEE.:	/ W73-00796
G. P. FITZGERALD, AND G. F.	LEE.:	/ W73-00232
G. FRED	LEE.:	/ W71-12072
C. C.	LEE.:	/ W72-04292
J. J.	LEE.:	/ W71-09254
FRED G.	LEE, AND GILMAN D. VEITH.:	/ W71-12092
J. STONE.: JOHN J.	LEE, JOHN H. TIETJEN, AND ROBERT	/ W72-00948
ND D. E. ARMSTRONG.: C. C.	LEE, R. F. HARRIS, J. K. SYERS, A	/ W71-13794
D. J. HIBBERD, AND G. F.	LEEDALE.:	/ W72-14808
M. TASSIGNY, AND M.	LEFEVRE.:	/ W72-12734
PAULI BAGGE, AND PASO O.	LEHMUSLUOTO.:	/ W72-03228
, AND B. THOMAS JOHNSON.:	LELYN STADNYK, ROBERT S. CAMPBELL	/ W71-10566
J. H. ELDER, C. A.	LEMBI, AND D. J. MORRE.:	/ W72-08440
NER.: NORBERT A. JAWORSKI,	LEO J. CLARK, AND KENNETH D. FEIG	/ W71-09788
NER.: NORBERT A. JAWORSKI,	LEO J. CLARK, AND KENNETH D. FEIG	W70-08249
NORBERT A. JAWORSKI, AND	LEO J. HETLING.:	W70-06509
	LEONARD W. HOM.:	/ W71-07096
	LESLIE RALPH BERGER.:	/ W71-10791
ETH M. MACKENTHUM, AND LOWELL E.	LEUP.: /RISH, CARL R. HIRTH, KENN	/ W71-05626
, AND G. E. RAINES.: A. A.	LEVIN, T. J. BIRCH, R. E. HILLMAN	/ W72-07907
HILLMAN, AND GILBER/ ARTHUR A.	LEVIN, THOMAS J. BIRCH, ROBERT E.	/ W71-07417

TEWART.: M. LEX, W. B. SILVESTER, AND W. D. S/ W72-08584
PAUL B. LIAO.: / W71-11006
H. LIEBMANN.: W71-00117
M. JOHNSON, D. W. FISH/ GENE E. LIKENS, F. HERBERT BORMANN, NOYE W71-01489
CHANG-KWEI LIN.: / W72-13644
M. G. MCGARRY, C. D. LIN, AND J. L. MERTO.: / W72-13508
D. T. LINDGREN.: / W72-12487
KARL A. ROSEMAN, AND WARNER M. LINFIELD.: / W71-06247
W. C. SWOAGER, AND E. S. LINSTROM.: / W72-04743
J. TSCHEN, AND S. T. LIU.: / W73-01626
H. D. BRUHN, D. F. LIVERMORE, AND F. O. ABOABA.: / W71-06188
. BENINSON.: D. CANCIO, J. A. LLAURO, N. R. CIALLELLA, AND D. J/ W73-00819
R. LOCKART.: / W72-05869
E. R. LONG.: / W73-00374
E. B. LONG, AND G. D. COOKE.: / W72-13673
ANDER, J. J. KELLEY, AND/ L. L. LONGERICH, M. BILLINGTON, V. ALEX/ W72-09122
E. LOOS.: / W72-14326
L. A. HOBSON, AND C. J. LORENZEN.: / W72-14313
LOTHAR GEITLER.: W71-04526
LOTHAR GEITLER.: / W71-07360
PERCY P. ST. AMANT, AND LOUIS A. BECK.: / W71-08223
PERCY P. ST. AMANT, AND LOUIS A. BECK.: / W71-11698
LOUIS A. KRUMHOLZ.: / W72-06526
LOUIS G. WILLIAMS.: / W71-08032
LEALE E. STREEBIN, AND OLIVER T. LOVE JR.: GEORGE W. REID, / W71-07973
D. B. PEAKALL, AND R. J. LOVETT.: / W72-05952
T. W. DUKE, J. I. LOWE, AND A. J. WILSON, JR.: / W71-07412
IRTH, KENNETH M. MACKENTHUM, AND LOWELL E. LEUP.: /RISH, CARL R. H/ W71-05626
F. G. LOWMAN, AND R. Y. TING.: / W73-00821
ENNETH M. MAC/ GERALD D. MCKEE, LOYS P. PARRISH, CARL R. HIRTH. K/ W71-05626
. L. METCALF, A. S. HIRWE, P.-Y. LU, AND J. R. COATS.: / KAPOOR, R/ W72-07703
A. M. LUCAS, AND N. A. THOMAS.: / W72-12994
LUCYNA JOANNA BOWNIK.: / W71-11012
THOMAS R. TOSTESON, AND LUIS R. ALMODOVAR.: / W71-09261
I. J. WRIGHT, AND W. M. LUIZ.: W70-10174
J. LUKAVSKY.: / W72-07141
J. W. G. LUND.: / W71-13183
J. W. G. LUND.: / W72-10625
UCKA.: J. W. G. LUND, G. H. M. JAWORSKI, AND H. B/ W72-10619
G. LUNDE.: / W71-11515
U. GRANHALL, AND A. LUNDGREN.: / W72-08054
W. M. LYLE.: / W73-01362
A. N. MACGREGOR, AND D. E. JOHNSON.: / W72-12505
RRISH, CARL R. HIRTH, KENNETH M. MACKENTHUM, AND LOWELL E. LEUP.: / W71-05626
K. M. MACKENTHUN, AND L. E. KEUP.: / W72-06536
BERLAND, D. J. BONIN, AND S. Y. MAESTRINI.: B. R./ W72-08051
T. H. MAGUE.: / W73-01456
B. MAGUIRE, JR., AND W. E. NEILL.: / W72-09239
VIJAYSINH U. MAHIDA.: / W72-00042
B. PROKES.: I. MALEK, J. BARTOS, J. SIMMER, AND / W72-03327
T. E. MALONEY.: / W72-12966
ROYAMA.: T. E. MALONEY, W. E. MILLER, AND T. SHI/ W72-09164
F. POWERS, D. W. SCHULTS, K. W. MALUEG, R. M. BRICE, AND M. D. SC/ W72-09165
STANLEY E. MANAHAN.: W71-00474
S. E. MANAHAN, AND M. J. SMITH.: / W73-02112
R, AND P. ROBINSON.: S. E. MANAHAN, M. J. SMITH, D. ALEXANDE/ W72-01693
DOHENY, AND: E. F. MANDELLI, P. R. BURKHOLDER, T. E./ W71-07875
L. A. MANN.: / W72-00383
K. H. MANN.: / W73-00366
BRUCE A. MANNY.: / W72-08048
R. G. WETZEL, AND BRUCE A. MANNY.: / W72-01780
B. MANSFIELD.: W70-06968
OLLINS, J. H. BENNETT, AND O. K. MANUEL.: A. G. C/ W72-02073
G. V. R. MARAIS.: / W71-07084
LIAM J. OSWALD, AARON MERON, AND MARIO D. ZABAT.: WIL/ W71-07100
, CLARENCE G. GOLUEKE, AND HEN/ MARIO D. ZABAT, WILLIAM J. OSWALD/ W72-04788
A. S. MENON, C. V. MARION, AND A. N. MILLER.: / W72-12996
NRADER.: MARK J. HAMMER, AND GARY L. HERGE/ W72-04759
L. PAVONI, PHILIP C. SINGER, AND MARK W. TENNEY.: /R, JR., JOSEPH / W72-00411
J. W. MARKHAM.: / W73-00428
. MOORE.: MARSHALL L. SILVER, AND CHARLES A/ W72-04263
H. G. MARSHALL.: / W72-07663
LESPIE, C. A. MAZZOLA, AND D. W. MARSHALL.: W. J. GIL/ W72-03753
SKAUEN.: N. MARSHALL, C. A. OVIATT, AND D. M./ W72-14317
EL J. HARVEY, ZVI PLAUT, AND: MARTIN GIBBS, ERWIN LATZKO, MICHA/ W71-09172
DONALD M. MARTIN.: / W71-13341
MARVIN C. GOLDBERG.: / W71-06435
J. A. OSBORNE, AND G. R. MARZOLF.: / W72-12393
V. H. HUANG, J. R. MASE, AND E. G. FRUH.: / W72-13692
EZONIK, E. E. SHANNON, AND P. E. MASLIN.: / W. H. MORGAN, P. L. BR/ W72-08986
CURTIS C. MASON.: / W71-12034
ALBERT MASSEY, AND JOHN ROBINSON.: / W72-07937
N. J. MATERA, AND J. J. LEE.: / W73-00367
B. J. MATHIS.: / W73-00916
SAIGA, AND T. NOSE.: Y. MATIDA, H. KUMADA, S. KIMURA, Y. / W72-12257
MICHAEL E. BENDER, WAYNE R. MATSON, AND ROBERT A. JORDAN.: / W71-09674
B. CHANG.: ROBERT S. MATSON, GEORGE E. MUSTOE, AND S. / W72-06037
. BATTERTON, G. M. BOUSH, AND F. MATSUMURA.: J. C/ W72-03545
. BATTERTON, G. M. BOUSH, AND F. MATSUMURA.: J. C/ W72-13809
K. C. PATIL, F. MATSUMURA, AND G. M. BOUSH.: / W73-00361

: M.	MATUCHA, L. ZILKA, AND K. SVIHEL./	W72-09365
D.	MAUZERALL.: /	W72-14800
EBEDEVA,/ M. N. PIMENOVA, I. V.	MAXIMOVA, G. I. MELESHKO, E. K. L/	W71-08027
	MAXWELL S. DOTY.: /	W71-11486
UILAR-SANTOS.:	MAXWELL S. DOTY, AND GERTRUDES AG/	W71-08597
W. J. GILLESPIE, C. A.	MAZZOLA, AND D. W. MARSHALL.: /	W72-03753
C. D.	MC INTIRE, AND W. S. OVERTON.: /	W72-08141
N SKOK.: LANDY	MCBRIDE, WILLIAM CHORNEY, AND JOH/	W71-12858
PERRY L.	MCCARTY.: /	W71-13939
WILLIAM J. JEWELL, AND PERRY L.	MCCARTY.: /	W72-01881
P. L.	MCCARTY.:	W70-04080
. FOREE, W. J. JEWELL, AND P. L.	MCCARTY.: E. G	W70-10405
RICHARD C. BAIN, JR., PERRY L.	MCCARTY, JAMES A. ROBERTSON, AND /	W71-07098
N, AND N. L. CLESCERI.: G. C.	MCDONALD, R. D. SPEAR, R. J. LAVI/	W72-08376
MICHAEL G.	MCGARRY.:	W70-08904
M. G.	MCGARRY, AND C. TONGKASAME.: /	W71-08951
TO.: M. G.	MCGARRY, C. D. LIN, AND J. L. MER/	W72-13508
TUNZI, A. ADINARAYANA, AND P. H.	MCGAUHEY.: / J. MIDDLEBROOKS, M. /	W71-13553
D. B. PORCELLA, P. H.	MCGAUHEY, AND G. L. DUGAN.: /	W72-02412
D. B. PORCELLA.: P. H.	MCGAUHEY, E. J. MIDDLEBROOKS, AND/	W72-11702
B. L. WULFF, AND C. D.	MCINTIRE.: /	W73-00853
T. J.	MCINTYRE.: /	W72-06655
IRTH, KENNETH M. MAC/ GERALD D.	MCKEE, LOYS P. PARRISH, CARL R. H/	W71-05626
T. EDELSTEIN, L C-M CHEN, AND J.	MCLACHLAN.: /	W72-01370
G. C.	MCLEOD.: /	W71-13252
M. A. ZEITOUN, W. F.	MCLLHENNY, AND W. A. TABER.: /	W71-12223
RNOLD, AND M/ P. W. JOHNSON, J.	MCN. SIEBURTH, A. SASTRY, C. R. A/	W72-08142
J. GOERING, AND M. T. GO/ C. P.	MCROY, M. MUELLER, S. STOKER, J. /	W72-09120
ROBERT O.	MEGARD.: /	W71-08670
FOLKMAN, M. KREMER, AND P. G. J.	MEIRING.: Y. /	W72-11642
PIMENOVA, I. V. MAXIMOVA, G. I.	MELESHKO, E. K. LEBEDEVA, AND T. /	W71-08027
D. C.	MENDAY, AND A. A. BUCK.: /	W72-11564
A. S.	MENON, AND A. A. JURKOVIC.: /	W72-06690
LLER.: A. S.	MENON, C. V. MARION, AND A. N. MI/	W72-12996
WILLIAM J. OSWALD, AARON	MERON, AND MARIO D. ZABAT.: /	W71-07100
RY GEE.: AARON	MERON, WILLIAM J. OSWALD, AND HEN/	W72-04789
G. MCGARRY, C. D. LIN, AND J. L.	MERTO.: M. /	W72-13508
ND J. R. C/ I. P. KAPOOR, R. L.	METCALF, A. S. HIRWE, P.-Y. LU, A/	W72-07703
N, AND ROBERT A. JORDAN.:	MICHAEL E. BENDER, WAYNE R. MATSO/	W71-09674
	MICHAEL G. MCGARRY.:	W70-08904
L. KING, ALLEN J. TOLMSOFF, AND	MICHAEL J. ATHERTON.: DARRELL/	W71-07097
: MARTIN GIBBS, ERWIN LATZKO,	MICHAEL J. HARVEY, ZVI PLAUT, AND/	W71-09172
DENIS M. VEAL, AND M. F. P.	MICHALSKI.: /	W72-13662
LA, JAMES S. KUMAGAI, AND E. JOE	MIDDLEBROOKS.: DONALD B. PORCEL	W70-08944
: P. H. MCGAUHEY, E. J.	MIDDLEBROOKS, AND D. B. PORCELLA./	W72-11702
DMAN, DONALD B. PORCELLA, E. JOE	MIDDLEBROOKS, AND DANIE F. TOERIE/	W71-12068
AYANA, AN/ E. A. PEARSON, E. J.	MIDDLEBROOKS, M. TUNZI, A. ADINAR/	W71-13553
FRANCIS M.	MIDDLETOWN, AND ROBERT L. BUNCH.:/	W71-07123
	MIKE DICKMAN.: /	W72-07922
	MILLARD W. HALL.: /	W72-04280
. MENON, C. V. MARION, AND A. N.	MILLER.: A. S/	W72-12996
R. E. NAKATANI, D.	MILLER, AND J. V. TOKAR.:	W70-09612
T. E. MALONEY, W. E.	MILLER, AND T. SHIROYAMA.: /	W72-09164
KENNETH W. STEWART, LARRY E.	MILLIGER, AND BERNARD M. SOLON.:	W70-06655
	MILTON H. FELDMAN.: /	W71-11793
R. A.	MINEAR.: /	W72-12637
J. RONALD	MINER.: /	W71-08214
F. L. ROSE, AND G. W.	MINSHALL.: /	W72-13300
C. S.	MINTER, III.: /	W73-00932
T. WAITE, AND R.	MITCHEL.: /	W72-06046
S. FOGEL, I. CHET, AND R.	MITCHELL.: /	W73-00942
R.: DEE	MITCHELL, AND JAMES C. BUZZELL, J/	W72-03218
EBELEIN,/ G. C. WILLIAMS, J. B.	MITTON, T. H. SUCHANEK, JR., N. G/	W72-14659
: S.	MIYAZAKI, AND A. J. THORSTEINSON./	W73-02280
F. M. EL-SHARKAWI, AND S. K.	MOAWAD.:	W70-06619
V. S. ZLOBIN, AND O. V.	MOKANU.: /	W71-09193
THOMAS J.	MONAHAN, AND FRANCIS R. TRAINOR.:/	W71-08687
W. F.	MONDALE.: /	W72-01659
DANIEL HICKMAN, AND RICHARD B .	MOORE.: G./	W72-01095
RSHALL L. SILVER, AND CHARLES A.	MOORE.: MA/	W72-04263
J. W.	MOORE.: /	W73-01627
J. M. WAY, J. F. NEWMAN, N. W.	MOORE, AND F. W. KNAGGS.: /	W72-00155
. HAZEL, FRED KOPPERDAHL, NORMAN	MORGAN, AND WALTER THOMSEN.: /S R/	W71-09789
NNON, AND / H. D. PUTNAM, W. H.	MORGAN, P. L. BREZONIK, E. E. SHA/	W72-08986
A. GILLESPIE, AND Y.	MORITA.: /	W72-08989
H. ELDER, C. A. LEMBI, AND D. J.	MORRE.: J. /	W72-08440
A. W.	MORRIS, AND P. FOSTER.: /	W72-09108
YENTSCH.: I.	MORRIS, C. M. YENTSCH, AND C. S. /	W72-09102
YENTSCH.: I.	MORRIS, C. M. YENTSCH, AND C. S. /	W72-09103
H. W. DE KONING, AND D. C.	MORTIMER.: /	W71-12303
RNAU.: S. D.	MORTON, P. H. DERSE, AND R. C. SE/	W72-10607
E.: S. D.	MORTON, R. SERNAU, AND P. H. DERS/	W72-09168
C. R. GOLDMAN, G.	MOSHIRI, AND E. DE AMEZAGA.: /	W72-12549
J. L.	MOSSER, AND T. D. BROCK.: /	W72-03557
AND C. F. WURSTER.: J. L.	MOSSER, N. S. FISHER, T. C. TENG,/	W72-05614
STER.: N. S. FISCHER, J. L.	MOSSER, T. C. TENG, AND C. F. WUR/	W72-08436
S. P.	MOULIK.: /	W71-12467
D. I.	MOUNT.:	W70-09595
O. A.	MOVCHAN.: /	W72-03543
J. C.	MUELLER.: /	W72-10390

, AND M. T. GO/ C. P. MCROY, M.	MUELLER, S. STOKER, J. J. GOERING/	W72-09120
B. PATEL, P. G. VALANJU, C. D.	MULAY, M. C. BALANI, AND: /	W73-00811
V. H.	MULLER.: /	W73-01445
LE.: C. J. SOEDER, H.	MULLER, H. D. PAYER, AND H. SCHUL/	W72-12739
M.	MUNAWAR.: /	W73-00918
R. C. TILTON, J.	MURPHY, AND J. K. DIXON.: /	W72-11833
S. N.	MURRAY.: /	W73-02099
. DIXON.: STEVEN	MURRAY, JAN SCHERFIG, AND PETER S/	W72-04433
ROBERT S. MATSON, GEORGE E.	MUSTOE, AND S. B. CHANG.: /	W72-06037
. K/ PAROMITA BOSE, UJJAL SINGH	NAGPAL, G. S. VENKATARAMAN, AND S/	W71-10070
R. L. BURGNER, AND R. E.	NAKATANI.: /	W73-00764
KAR.: R. E.	NAKATANI, D. MILLER, AND J. V. TO	W70-09612
	NAMAN E. OTTO.: /	W71-06102
CIMBERG.:	NANCY L. NICHOLSON, AND ROBERT L./	W71-08792
NAVA	NARKIS.:	W70-09907
R. H. WEAVER, AND H. D.	NASH.: /	W71-07942
K. V.	NATARAJAN.: /	W71-11008
	NAVA NARKIS.:	W70-09907
JOSEPH Y. CHENG, AND	NAVAL J. ANTIA.:	W70-07842
JOSEPH Y. CHENG, AND	NAVAL J. ANTIA.:	W70-07280
A. B.	NAZAROV, AND A. YA. ZESENKO.: /	W72-12204
B. MAGUIRE, JR., AND W. E.	NEILL.: /	W72-09239
G.	NELSON.: /	W72-01685
AGGS.: J. M. WAY, J. F.	NEWMAN, N. W. MOORE, AND F. W. KN/	W72-00155
H. WAYNE	NICHOLS.: /	W72-00377
: NANCY L.	NICHOLSON, AND ROBERT L. CIMBERG./	W71-08792
JOHN G.	NICKUM.:	W71-02880
B.	NIEHOFF.: /	W72-06638
E. STEEMANN	NIELSEN, AND S. WIUM-ANDERSEN.:	W71-03027
P. BAGGE, AND A.	NIEMI.: /	W72-13871
S.	NIEWOLAK.: /	W72-10861
S.	NIEWOLAK.: /	W72-12255
A.: M. V. GUSEV, K. A.	NIKITINA, AND T. G. KORZHENEVSKAY/	W71-08683
AGAT.:	NING-HSI TANG, AND SURINDER K. BH/	W72-04431
CHUTTER.: R. G.	NOBLE, W. A. PRETORIUS, AND F. M./	W72-01472
	NOEL M. BURNS, AND C. ROSS.: /	W72-12990
. LEAR, JR.: JOHAN A. AALTO,	NORBERT A. JAWORSKI, AND DONALD W	W70-08248
ETLING.:	NORBERT A. JAWORSKI, AND LEO J. H	W70-06509
AR, JR., AND ORTERIO VILLA, JR/	NORBERT A. JAWORSKI, DONALD W. LE/	W71-08775
, AND KENNETH D. FEIGNER.:	NORBERT A. JAWORSKI, LEO J. CLARK	W70-08249
, AND KENNETH D. FEIGNER.:	NORBERT A. JAWORSKI, LEO J. CLARK/	W71-09788
ARLES R. HAZEL, FRED KOPPERDAHL,	NORMAN MORGAN, AND WALTER THOMSEN/	W71-09789
G. W. HEINKE, AND J. D.	NORMAN.:	W71-00116
C. G. STEPHENS, AND B. B.	NORTH.: /	W72-07497
OUGHTON, RICHARD C. KNUDSEN, AND	NORVAL A. SINCLAIR.: /ILLIAM H. B	W71-01474
ADA, S. KIMURA, Y. SAIGA, AND T.	NOSE.: Y. MATIDA, H. KUM/	W72-12257
VICTOR E.	NOSHKIN.: /	W72-04461
E E. LIKENS, F. HERBERT BORMANN,	NOYE M. JOHNSON, D. W. FISHER, AN	W71-01489
J. R.	NURSALL, AND D. N. GALLUP.: /	W73-01704
R.	NUZZI.: /	W72-11727
.: V. M. B.	NYANISHKENE, AND G. G. POLIKARPOV/	W72-10978
PARKER, G. L. SAMSEL, AND E. K.	OBENG-ASAMOA.: B. C./	W72-07661
T. A. OLSON, AND T. O.	ODLAUG.: /	W72-12192
W. REID, LEALE E. STREEBIN, AND	OLIVER T. LOVE JR.: GEORGE/	W71-07973
J. W. VENNES, AND C. O.	OLSON.: /	W72-12565
T. A.	OLSON, AND T. O. ODLAUG.: /	W72-12192
. PRICE/ R. A. BOWEN, J. M. ST.	ONGE, J. B. COLTON, JR., AND C. A/	W73-00371
I. L.	OPHEL, AND C. D. FRASER.: /	W71-11529
E. J.	ORDAL.: /	W72-05421
C. W. CHEN, AND G. T.	ORLOB.: /	W72-05282
C. W. CHEN, AND G. T.	ORLOB.: /	W72-14405
WORSKI, DONALD W. LEAR, JR., AND	ORTERIO VILLA, JR.: /ORBERT A. JA/	W71-08775
J. A.	OSBORNE, AND G. R. MARZOLF.: /	W72-12393
JOHN J. ROOK, AND GIJSBERT	OSKAM.:	W71-01491
F.	OSKAM.: /	W72-08559
	OSMUND HOLM-HANSEN.: /	W71-11234
C. G. GOLUEKE, AND W. J.	OSWALD.:	W71-00138
DUGAN, C. G. GOLUEKE, AND W. J.	OSWALD.: G. L./	W72-08396
ZABAT.: WILLIAM J.	OSWALD, AARON MERON, AND MARIO D./	W71-07100
G. SHELEF, W. J.	OSWALD, AND C. C. GOLUEKE.:	W70-10403
. L. DUGAN, C. G. GOLUEKE, W. J.	OSWALD, AND C. E. RIXFORD.: G/	W72-02850
AARON MERON, WILLIAM J.	OSWALD, AND HENRY GEE.: /	W72-04789
HEN/ MARIO D. ZABAT, WILLIAM J.	OSWALD, CLARENCE G. GOLUEKE, AND /	W72-04788
A.	OTSUKI, AND T. HANYA.: /	W73-00379
A.	OTSUKI, AND T. HANYA.: /	W73-01066
NAMAN E.	OTTO.: /	W71-06102
C. D. MC INTIRE, AND W. S.	OVERTON.: /	W72-08141
N. MARSHALL, C. A.	OVIATT, AND D. M. SKAUEN.: /	W72-14317
M. G. JOHNSON, AND G. E.	OWEN.: /	W71-08953
M. G. JOHNSON, AND G. E.	OWEN.: /	W71-11009
AN, B. L. GUTELMACHER, AND N. G.	OZERETSKOVSKAYA.: /. M. VARDAPETY/	W72-00161
K. BASU, B. B. GHOSH, AND R. N.	PAL.: A./	W71-11876
ET, M. FIALA, G. JACQUES, AND M.	PANOUSE.: G. CAH/	W73-00431
PAT C. KERR, DORIS F.	PARIS, AND D. L. BROCKWAY.:	W71-00475
. C. KERR, D. L. BROCKWAY, D. F.	PARIS, AND J. T. BARNETT, JR.: /P/	W72-09159
BENG-ASAMOA.: B. C.	PARKER, G. L. SAMSEL, AND E. K. O/	W72-07661
YU V. KRUGLOV, AND L. N.	PAROMENSKAYA.: /	W71-07675
, G. S. VENKATARAMAN, AND S. K/	PAROMITA BOSE, UJJAL SINGH NAGPAL/	W71-10070
. MAC/ GERALD D. MCKEE, LOYS P.	PARRISH, CARL R. HIRTH, KENNETH M/	W71-05626
D. A. EGLOFF, AND R.	PARTRIDGE.: /	W73-00281

PAULI BAGGE, AND	PASO O. LEHMUSLUOTO.: /	W72-03228
D. L. BROCKWAY.:	PAT C. KERR, DORIS F. PARIS, AND	W71-00475
, M. C. BALANI, AND: B.	PATEL, P. G. VALANJU, C. D. MULAY/	W73-00811
J. CAIRNS, JR., AND R. A.	PATERSON.: /	W72-07526
USH.: K. C.	PATIL, F. MATSUMURA, AND G. M. BO/	W73-00361
P. J. HANNAN, AND C.	PATOUILLET.: /	W72-07660
EARL E. SHANNON, AND	PATRICK L. BREZONIK.: /	W72-04269
R.	PATRICK.: /	W73-00242
JR., ROGER L. KAESLER, AND RUTH	PATRICK.: JOHN CAIRNS,/	W71-11583
	PAUL B. LIAO.: /	W71-11006
	PAUL E. SAGER.: /	W71-10645
ROY B. EVANS, AND	PAUL KRUGER.: /	W72-00974
REY.:	PAUL L. ZUBKOFF, AND WALTER E. CA/	W72-01101
	PAUL R. KENIS, AND J. W. HOYT.: /	W72-03220
OTO.:	PAULI BAGGE, AND PASO O. LEHMUSLU/	W72-03228
E F. ECHELBERGER, JR., JOSEPH L.	PAVONI, PHILIP C. SINGER, AND MAR/	W72-00411
C. J. SCEDER, H. MULLER, H. D.	PAYER, AND H. SCHULLE.: /	W72-12739
D. B.	PEAKALL, AND R. J. LOVETT.: /	W72-05952
V. B.	PEARSE.: /	W72-12634
C. H. HUANG, J. RADIMSKY, E. A.	PEARSON, AND J. SCHERFIG.: /RIEN,/	W72-03985
UNZI, A. ADINARAYANA, AN/ E. A.	PEARSON, E. J. MIDDLEBROOKS, M. T/	W71-13553
BECK.:	PERCY P. ST. AMANT, AND LOUIS A. /	W71-08223
BECK.:	PERCY P. ST. AMANT, AND LOUIS A. /	W71-11698
	PERRY L. MCCARTY.: /	W71-13939
WILLIAM J. JEWELL, AND	PERRY L. MCCARTY.: /	W72-01881
SON, AND/ RICHARD C. BAIN, JR.,	PERRY L. MCCARTY, JAMES A. ROBERT/	W71-07098
YE. G.	PERUYEVA.: /	W72-09248
ENTILE, STANTON J. ERICKSON, AND	PETER BETZER.: / DAVEY, JOHN H. G/	W72-04708
G, AND CHARLES R. GOLDMAN.:	PETER RICHERSON, RICHARD ARMSTRON/	W71-09171
STEVEN MURRAY, JAN SCHERFIG, AND	PETER S. DIXON.: /	W72-04433
THOMAS W.	PHILBIN, AND HOWARD D. PHILLIPP.:	W70-09614
ELBERGER, JR., JOSEPH L. PAVONI,	PHILIP C. SINGER, AND MARK W. TEN/	W72-00411
ENO, O. BRAIDECH, E. THOMAS, AND	PHILIP E. GEHRING.: CONRAD KLEV/	W72-13659
	PHILIP J. SAWYER.:	W71-03014
THOMAS W. PHILBIN, AND HOWARD D.	PHILLIPP.:	W70-09614
. STEEL.: R. E. HEFT, W. A.	PHILLIPS, H. R. RALSTON, AND W. A/	W73-00831
B. J. SHERMAN, AND H. K.	PHINNEY.: /	W72-07901
. E. BURNS, R. M. GIRLING, A. R.	PICK, AND D. W. VAN ES.: G/	W71-07109
KENNETH R.	PIECH.: /	W71-11685
MES A. ROBERTSON, AND WILLIAM H.	PIERCE.: /., PERRY L. MCCARTY, JA/	W71-07098
SON, D. W. FISHER, AND ROBERT S.	PIERCE.: /T BORMANN, NOYE M. JOHN	W71-01489
JEAN-FRANCOIS	PIERRE.:	W71-00099
E. G. SRINATH, AND S. C.	PILLAI.: /	W72-09176
AMES A. SOND/ K. K. SIVASANKARA	PILLAY, CHARLES C. THOMAS, JR., J/	W71-11036
N. F. PISKUNKOVA, AND M. N.	PIMENOVA.: /	W72-00838
ELESHKO, E. K. LEBEDEVA,/ M. N.	PIMENOVA, I. V. MAXIMOVA, G. I. M/	W71-08027
N. F.	PISKUNKOVA, AND M. N. PIMENOVA.: /	W72-00838
N LATZKO, MICHAEL J. HARVEY, ZVI	PLAUT, AND: MARTIN GIBBS, ERWI/	W71-09172
EDWARD F.	POHL.: /	W71-07106
V. M. B. NYANISHKENE, AND G. G.	POLIKARPOV.: /	W72-10978
G. TSITSUGINA, AND V. V./ G. G.	POLIKARPOV, L. G. KULEBAKINA, V. /	W72-10955
F. D.	POR.: /	W73-00855
Y, E. J. MIDDLEBROOKS, AND D. B.	PORCELLA.: P. H. MCGAUHE/	W72-11702
D D/ JOEL C. GOLDMAN, DONALD B.	PORCELLA, E. JOE MIDDLEBROOKS, AN/	W71-12068
. JOE MIDDLEBROOKS.: DONALD B.	PORCELLA, JAMES S. KUMAGAI, AND E	W70-08944
L. DUGAN.: D. B.	PORCELLA, P. H. MCGAUHEY, AND G. /	W72-02412
D. W. HENDRICKS, W. D.	POTE, AND J. G. ANDREW.: /	W72-02363
EG, R. M. BRICE, AND M. / C. F.	POWERS, D. W. SCHULTS, K. W. MALU/	W72-09165
R. O.	POYTON.: /	W71-11851
M. A. RASHID, AND A.	PRAKASH.: /	W72-12166
H. D. KUMAR, AND G.	PRAKASH.: /	W72-00153
IFFE, JR., R. W. SHELDON, AND A.	PRAKASH.: W. H. SUTCL/	W71-11562
R. G. NOBLE, W. A.	PRETORIUS, AND F. M. CHUTTER.: /	W72-01472
OSEPH F. ROESLER, AND HERBERT C.	PREUL.: J/	W71-07099
GE, J. B. COLTON, JR., AND C. A.	PRICE.: /. A. BOWEN, J. M. ST. ON/	W73-00371
SURINDER K. BHAGAT, DONALD E.	PROCTOR, AND WILLIAM H. FUNK.: /	W72-03215
EK, J. BARTOS, J. SIMMER, AND B.	PROKES.: I. MAL/	W72-03327
R. C.	PUCINSKI.: /	W72-01093
NIK, E. E. SHANNON, AND / H. D.	PUTNAM, W. H. MORGAN, P. L. BREZO/	W72-08986
O. VAN DER BORGHT, AND S. VAN	PUYMBROECK.: /	W71-09168
O. VAN DER BORGHT, AND S. VAN	PUYMBROECK.: /	W71-09182
E. M. BURROWS, AND C.	PYBUS.: /	W71-09795
	RACHEL COX DOWNING.: /	W71-09156
	RACHEL COX DOWNING.: /	W71-12489
D. F. TOERIEN, C. H. HUANG, J.	RADIMSKY, E. A. PEARSON, AND J. S/	W72-03985
BIRCH, R. E. HILLMAN, AND G. E.	RAINES.: A. A. LEVIN, T. J./	W72-07907
OBERT E. HILLMAN, AND GILBERT E.	RAINES.: /VIN, THOMAS J. BIRCH, R/	W71-07417
G. S. VENKATARAMAN, AND B.	RAJYALAKSHMI.: /	W71-12071
G. S. VENKATARAMAN, AND B.	RAJYALAKSHMI.: /	W72-14807
LESLIE	RALPH BERGER.: /	W71-10791
WUM-CHENG WANG, AND	RALPH L. EVANS.:	W71-03028
. E. HEFT, W. A. PHILLIPS, H. R.	RALSTON, AND W. A. STEEL.: R/	W73-00831
V. D.	RAMAMURTHY.: /	W73-00430
	RANDALL L. BROWN.: /	W72-02975
V. S.	RAO.: /	W72-01363
RONALD L.	RASCHKE.:	W70-07838
M. A.	RASHID, AND A. PRAKASH.: /	W72-12166
	RAY B. KRONE.: /	W71-09581
S.: M.	REBHUN, M. A. FOX, AND J. B. SLES/	W71-13187
F. J. H. FREDEEN, AND J.	REGIS DUFFY.:	W70-08098

ER T. LOVE JR.: GEORGE W.	REID, LEALE E. STREEBIN, AND CLIV/	W71-07973
P. J.	REILLY.: /	W73-01657
R. J.	REIMOLD, AND C. J. DURANT.: /	W72-10678
WARD T. HALL, JR., AND DONALD J.	REINHARDT.: D. G. AHEARN, ED/	W71-09173
DONALD G. AHEARN, AND DONALD J.	REINHARDT.: /EDWARD T. HALL, JR.,/	W71-11914
C. C.	REMSEN.: /	W72-08056
E. J. CARPENTER, C. C.	REMSEN, AND S. W. WATSON.: /	W73-00380
R. A. ENGBERG, AND T. C.	RENSCHLER.: /	W71-13466
R. G.	RHODES.: /	W72-12633
RUSSELL G.	RHODES, AND ANTHONY J. TERZIS.: /	W71-05629
INATH.: L. G.	RICH, J. F. ANDREWS, AND T. M. KE/	W72-11724
GOLDMAN.: PETER RICHERSON,	RICHARD ARMSTRONG, AND CHARLES R./	W71-09171
G. DANIEL HICKMAN, AND	RICHARD B . MOORE.: /	W72-01095
CARTY, JAMES A. ROBERTSON, AND/	RICHARD C. BAIN, JR., PERRY L. MC/	W71-07098
RVING YALL, WILLIAM H. BOUGHTON,	RICHARD C. KNUDSEN, AND NORVAL A.	W71-01474
	RICHARD D. ROX.: /	W71-13334
EWRY, AND ROBERT J. LAVIGNE.:	RICHARD G. CLEMENTS, GEORGE E. DR/	W72-00949
RANEY.: WILLIAM H. FUNK,	RICHARD J. CONDIT, AND WAYNE T. C/	W72-00801
	RICHARD M. KLEIN.: /	W71-09157
IWALA.:	RICHARD W. GRIGG, AND ROBERT S. K	W71-01475
	RICHARD W. TERKELTOUB.: /	W71-11001
CHARLES R. GOLDMAN.: PETER	RICHERSON, RICHARD ARMSTRONG, AND/	W71-09171
GOLUEKE, W. J. OSWALD, AND C. E.	RIXFORD.: G. L. DUGAN, C. G. /	W72-02850
E. BENDER, WAYNE R. MATSON, AND	ROBERT A. JORDAN.: MICHAEL/	W71-09674
EDWIN F. BARTH, AND	ROBERT B. DEAN.: /	W72-04086
LMAN.:	ROBERT E. BAUMANN, AND SHELDON KE/	W71-06445
RTHUR A. LEVIN, THOMAS J. BIRCH,	ROBERT E. HILLMAN, AND GILBERT E./	W71-07417
, AND JOSEPH J. LATTERELL.:	ROBERT F. HOLT, DONALD R. TIMMONS	W71-04216
CLEMENTS, GEORGE E. DREWRY, AND	ROBERT J. LAVIGNE.: RICHARD G./	W72-00949
OHN J. LEE, JOHN H. TIETJEN, AND	ROBERT J. STONE.: J/	W72-00948
	ROBERT JACK FREEMAN, JR.: /	W71-13326
FRANCIS M. MIDDLETOWN, AND	ROBERT L. BUNCH.: /	W71-07123
	ROBERT L. BUNCH.: /	W71-12188
NANCY L. NICHOLSON, AND	ROBERT L. CIMBERG.: /	W71-08792
G. D. COOKE, AND	ROBERT L. KENNEDY.: /	W71-11949
	ROBERT O. MEGARD.: /	W71-08670
JOHNSON.: LELYN STADNYK,	ROBERT S. CAMPBELL, AND B. THOMAS/	W71-10566
WHITLEY.:	ROBERT S. CAMPBELL, AND JAMES R. /	W71-06444
JR., AND JAMES R. WHITLEY.:	ROBERT S. CAMPBELL, ARTHUR WITT,	W70-09596
RICHARD W. GRIGG, AND	ROBERT S. KIWALA.:	W71-01475
E, AND S. B. CHANG.:	ROBERT S. MATSON, GEORGE E. MUSTO/	W72-06037
YE M. JOHNSON, D. W. FISHER, AND	ROBERT S. PIERCE.: /T BORMANN, NO	W71-01489
A. J. HARRIS, K. J.	ROBERTS, AND A. E. CHRISTIE.: /	W72-06837
D. E.	ROBERTSON.: /	W72-04463
JR., PERRY L. MCCARTY, JAMES A.	ROBERTSON, AND WILLIAM H. PIERCE./	W71-07098
R. B.	ROBIE.:	W70-07100
G.	ROBIN SOUTH, AND ANDRE CARDINAL.:/	W71-10066
ALBERT MASSEY, AND JOHN	ROBINSON.: /	W72-07937
. J. SMITH, D. ALEXANDER, AND P.	ROBINSON.: S. E. MANAHAN, M/	W72-01693
JOSEPH F.	ROESLER, AND HERBERT C. PREUL.: /	W71-07099
M. A.	ROESSLER.: /	W72-11876
K.: JOHN CAIRNS, JR.,	ROGER L. KAESLER, AND RUTH PATRIC/	W71-11583
D. E. ARMSTRONG, AND G. A.	ROHLICH.: /	W71-06443
	RONALD L. RASCHKE.:	W70-07838
B. WELCH, JAMES A. BUCKLEY, AND	RONALD M. BUSH.: EUGENE/	W72-00799
CH.:	RONALD M. BUSH, AND EUGENE B. WEL/	W72-00798
J.	RONALD MINER.: /	W71-08214
JOHN J.	ROOK, AND GIJSBERT OSKAM.:	W71-01491
D. W.	ROONEY.: /	W72-11920
D. W.	ROONEY.: /	W72-12724
F. L.	ROSE, AND G. W. MINSHALL.: /	W72-13300
KARL A.	ROSEMAN, AND WARNER M. LINFIELD.:/	W71-06247
M. L. STEWART, G. M.	ROSENTHAL, AND J. R. KLINE.: /	W72-03348
N. M. BURNS, AND C.	ROSS.: /	W72-12997
NOEL M. BURNS, AND C.	ROSS.: /	W72-12990
F. E.	ROUND.: /	W73-00432
M. HICKMAN, AND F. E.	ROUND.: /	W71-12855
G. A.	ROUNSEFELL.: /	W73-02029
RICHARD D.	ROX.: /	W71-13334
JAMES L. YOUNT, AND	ROY A. CROSSMAN, JR.:	W71-01488
	ROY B. EVANS, AND PAUL KRUGER.: /	W72-00974
SYLVIA B. DERBY, AND ERNEST	RUBER.: /	W71-10096
TERZIS.:	RUSSELL G. RHODES, AND ANTHONY J./	W71-05629
IRNS, JR., ROGER L. KAESLER, AND	RUTH PATRICK.: JOHN CA/	W71-11583
K.	RUTZLER, AND W. STERRER.: /	W72-07911
RE, AND J. E. HUGUENIN.: J. H.	RYTHER, W. M. DUNSTAN, K. R. TENO/	W72-09378
PAUL E.	SAGER.: /	W71-10645
MATIDA, H. KUMADA, S. KIMURA, Y.	SAIGA, AND T. NOSE.: Y. /	W72-12257
	SAM E. BEALL.:	W70-08832
B. C. PARKER, G. L.	SAMSEL, AND E. K. OBENG-ASAMOA.: /	W72-07661
JON E.	SANGER, AND EVILLE GORHAM.: /	W72-01784
L S. DOTY, AND GERTRUDES AGUILAR-	SANTOS.: MAXWEL/	W71-08597
D. F.	SARGENT, AND C. P. S. TAYLOR.: /	W72-11572
J. J.	SASNER, JR.: /	W73-00243
W. JOHNSON, J. MCN. SIEBURTH, A.	SASTRY, C. R. ARNOLD, AND M. S. D/	W72-08142
N. DABES, C. R. WILKE, AND K. H.	SAUER.: J. /	W72-06274
CLAIR N.	SAWYER.: /	W72-04769
CLAIR N.	SAWYER.: /	W72-05473
PHILIP J.	SAWYER.:	W71-03014
EIS.: M. R.	SCALF, W. R. DUFFER, AND R. D. KR/	W72-08401

G. SHELEF, M. SCHWARZ, AND H.	SCHECHTER.: /	W72-12289
H.BERNHARDT, J. CLASEN, AND H.	SCHELL.: /	W72-06201
C. L.	SCHELSKE.: /	W73-00818
RADIMSKY, E. A. PEARSON, AND J.	SCHERFIG.: /RIEN, C. H. HUANG, J./	W72-03985
STEVEN MURRAY, JAN	SCHERFIG, AND PETER S. DIXON.: /	W72-04433
D. W.	SCHINDLER.: /	W72-07648
JAMES E.	SCHINDLER.: /	W72-07940
H.	SCHNITZLER.: /	W72-05968
W.	SCHRAMM.: /	W73-01074
. MALUEG, R. M. BRICE, AND M. D.	SCHULDT.: /S, D. W. SCHULTS, K. W/	W72-09165
, H. MULLER, H. D. PAYER, AND H.	SCHULLE.: C. J. SOEDER/	W72-12739
E, AND M. / C. F. POWERS, D. W.	SCHULTS, K. W. MALUEG, R. M. BRIC/	W72-09165
M. E.	SCHULTZ.: /	W73-02548
S. L.	SCHWARTZ, AND L. R. ALMODOVAR.: /	W73-00838
G. SHELEF, M.	SCHWARZ, AND H. SCHECHTER.: /	W72-12289
. CALDWELL, R. H. CLAWSON, H. L.	SCOTTEN, AND R. G. WILLS.: / A. M/	W71-10577
G. FOREE, AND R.	SCROGGIN.: /	W72-14790
R. B.	SEARLES.: /	W73-00284
J. P.	SEARS, AND C. YENTSCH.: /	W72-09653
K-H. HUNKEN, AND I. D.	SEKOULOV.: /	W73-01367
. MORTON, P. H. DERSE, AND R. C.	SERNAU.: S. D/	W72-10607
S. D. MORTON, R.	SERNAU, AND P. H. DERSE.: /	W72-09168
C.	SERRUYA.: /	W72-05469
R.	SESHADRI, AND J. M. SIEBURTH.: /	W72-04748
J. R.	SEVER.: /	W73-01439
M. S.	SHANE.: /	W72-04736
R. E. CANNON, M. S.	SHANE, AND VALERIE N. BUSH.: /	W72-03188
H. MORGAN, P. L. BREZONIK, E. E.	SHANNON, AND P. E. MASLIN.: / W. /	W72-08986
: EARL E.	SHANNON, AND PATRICK L. BREZONIK./	W72-04269
JOSEPH	SHAPIRO.:	W70-09325
F. M. EL-	SHARKAWI, AND S. K. MOAWAD.:	W70-06619
ROBERT E. BAUMANN, AND	SHELDON KELMAN.: /	W71-06445
W. H. SUTCLIFFE, JR., R. W.	SHELDON, AND A. PRAKASH.: /	W71-11562
TER.: G.	SHELEF, M. SCHWARZ, AND H. SCHECH/	W72-12289
OLUEKE.: G.	SHELEF, W. J. OSWALD, AND C. C. G	W70-10403
BSEY, J. E. HARRISON, H. GEE, G.	SHELET, AND J. C. GOLDMAN.: /. SO/	W72-04787
B. J.	SHERMAN, AND H. K. PHINNEY.: /	W72-07901
ADNAN	SHINDALA.: /	W71-09629
JOSEPH V. DUST, AND ADAM	SHINDALA.:	W71-00139
E. MALONEY, W. E. MILLER, AND T.	SHIROYAMA.: T. /	W72-09164
R. L.	SHULER, AND W. A. AFFENS.:	W70-05547
I. A.	SHVYTOV.: /	W71-12868
R. H.	SIDDIQI, AND B. K. HANDA.: /	W71-06033
R. SESHADRI, AND J. M.	SIEBURTH.: /	W72-04748
, AND M/ P. W. JOHNSON, J. MCN.	SIEBURTH, A. SASTRY, C. R. ARNOLD/	W72-08142
L.	SILEN, AND B-C. JANSSON.: /	W73-00348
JOE B.	SILLS.:	W71-04528
MARSHALL L.	SILVER, AND CHARLES A. MOORE.: /	W72-04263
M. LEX, W. B.	SILVESTER, AND W. D. STEWART.: /	W72-08584
YATT.: J. K.	SILVEY, D. E. HENLEY, AND J. T. W/	W72-10613
I. MALEK, J. BARTOS, J.	SIMMER, AND B. PROKES.: /	W72-03327
ICHARD C. KNUDSEN, AND NORVAL A.	SINCLAIR.: /ILLIAM H. BOUGHTON, R	W71-01474
JR., JOSEPH L. PAVONI, PHILIP C.	SINGER, AND MARK W. TENNEY.: /R, /	W72-00411
Y, W. F. ECHELBERGER, JR., P. C.	SINGER, F. H. VERHOFF, AND W. A.	W71-03021
AND S. K/ PAROMITA BOSE, UJJAL	SINGH NAGPAL, G. S. VENKATARAMAN,/	W71-10070
E.	SINHA.: /	W72-08790
OMAS, JR., JAMES A. SOND/ K. K.	SIVASANKARA PILLAY, CHARLES C. TH/	W71-11036
ARSHALL, C. A. OVIATT, AND D. M.	SKAUEN.: N. M/	W72-14317
BRIDE, WILLIAM CHORNEY, AND JOHN	SKOK.: LANDY MC/	W71-12858
O. M.	SKULBERG.: /	W73-00234
B. V.	SKVORTZOV.: /	W72-07683
B. V.	SKVORTZOV.: /	W72-09119
M. REBHUN, M. A. FOX, AND J. B.	SLESS.: /	W71-13187
THEODORE J.	SMAYDA.: /	W72-06313
W. H. O. DE	SMET, AND F. M. J. C. EVENS.: /	W72-07896
L. W.	SMITH.: /	W71-07544
S. E. MANAHAN, AND M. J.	SMITH.: /	W73-02112
B. N.	SMITH.: /	W72-12709
F. W.	SMITH, AND JOHN F. THOMPSON.: /	W72-03222
BROCK.: D. W.	SMITH, C. B. FLIERMANS, AND T. D./	W72-14301
SON.: S. E. MANAHAN, M. J.	SMITH, D. ALEXANDER, AND P. ROBIN/	W72-01693
. SHELET, AND J. C. GOLDMAN/ M.	SOBSEY, J. E. HARRISON, H. GEE, G/	W72-04787
A.	SODERGREN.: /	W72-04134
ND H. SCHULLE.: C. J.	SOEDER, H. MULLER, H. D. PAYER, A/	W72-12739
ARRY E. MILLIGER, AND BERNARD M.	SOLON.: KENNETH W. STEWART, L	W70-06655
R. A.	SOLTERO.: /	W73-01435
CHARLES C. THOMAS, JR., JAMES A.	SONDEL, AND CAROLYN M. HYCHE.: / /	W71-11036
YU. I.	SOROKIN.: /	W71-08042
G. ROBIN	SOUTH, AND ANDRE CARDINAL.: /	W71-10066
SCERI.: G. C. MCDONALD, R. D.	SPEAR, R. J. LAVIN, AND N. L. CLE/	W72-08376
IRENA	SPODNIEWSKA.: /	W72-01358
I.	SPODNIEWSKA.: /	W72-14812
E. G.	SRINATH, AND S. C. PILLAI.: /	W72-09176
PERCY P.	ST. AMANT, AND LOUIS A. BECK.: /	W71-11698
PERCY P.	ST. AMANT, AND LOUIS A. BECK.: /	W71-08223
J. B.	ST. JOHN.: /	W71-11916
C. A. PRICE/ R. A. BOWEN, J. M.	ST. ONGE, J. B. COLTON, JR., AND /	W73-00371
K. LEE SU, E. J.	STABA AND Y. ABUL-HAJJ.: /	W72-08586
K. LEE SU, AND E. JOHN	STABA.: /	W72-07360
B. THOMAS JOHNSON.: LELYN	STADNYK, ROBERT S. CAMPBELL, AND /	W71-10566

VAN VUUREN, M. R. HENZEN, G. J.	STANDER, AND A. J. CLAYTON.: / J.	W70-10381
	STANLEY E. MANAHAN.:	W71-00474
EARL W. DAVEY, JOHN H. GENTILE,	STANTON J. ERICKSON, AND PETER BE/	W72-04708
ILLIPS, H. R. RALSTON, AND W. A.	STEEL.: R. E. HEFT, W. A. PH/	W73-00831
ERSEN.: E.	STEEMANN NIELSEN, AND S. WIUM-AND	W71-03027
LD.: K. A.	STEIDINGER, AND J. F. VAN BREEDVE/	W72-11252
C. G.	STEPHENS, AND B. B. NORTH.: /	W72-07497
D. H. STERN, AND M. S.	STERN.: /	W72-10431
D. H.	STERN, AND M. S. STERN.: /	W72-10431
K. RUTZLER, AND W.	STERRER.: /	W72-07911
PETER S. DIXON.:	STEVEN MURRAY, JAN SCHERFIG, AND /	W72-04433
D. M.	STEVEN, AND R. GLOMBITZA.: /	W72-11719
W. D. P.	STEWART.: /	W72-08508
LEX, W. B. SILVESTER, AND W. D.	STEWART.: M./	W72-08584
M. J. DAFT, AND W. D. P.	STEWART.: /	W72-05457
JERRY	STEWART.: /	W71-13356
J. R.	STEWART, AND R. M. BROWN, JR.: /	W72-07951
R. KLINE.: M. L.	STEWART, G. M. ROSENTHAL, AND J. /	W72-03348
ERNARD M. SOLON.: KENNETH W.	STEWART, LARRY E. MILLIGER, AND B	W70-06655
D. EVANS, AND J. G.	STOCKNER.: /	W72-13142
: JOHN G.	STOCKNER, AND F. A. J. ARMSTRONG./	W71-11029
GO/ C. P. MCROY, M. MUELLER, S.	STOKER, J. J. GOERING, AND M. T. /	W72-09120
DAVID H.	STOLTENBERG.:	W71-03896
, JOHN H. TIETJEN, AND ROBERT J.	STONE.: JOHN J. LEE/	W72-00948
J. R.	STORR, AND R. A. SWEENEY.: /	W72-13651
DALE	STRAUGHAN.: /	W71-08793
GEORGE W. REID, LEALE E.	STREEBIN, AND OLIVER T. LOVE JR.:/	W71-07973
W. L.	STROMBERG, AND C. J. GOODNIGHT.: /	W72-09668
KWEI LEE	SU.: /	W72-04275
K. LEE	SU, AND E. JOHN STABA.: /	W72-07360
K. LEE	SU, E. J. STABA AND Y. ABUL-HAJJ./	W72-08586
R. C.	SUBBARAJU, AND K. KRISHNAMURTHY.:/	W73-00935
C. WILLIAMS, J. B. MITTON, T. H.	SUCHANEK, JR., N. GEBELEIN, AND C/	W72-14659
A. KAWAI, AND I.	SUGAHARA.: /	W72-11563
WUN-CHENG WANG, AND WILLIAM T.	SULLIVAN.: /	W71-11033
NING-HSI TANG, AND	SURINDER K. BHAGAT.: /	W72-04431
CTOR, AND WILLIAM H. FUNK.:	SURINDER K. BHAGAT, DONALD E. PRO/	W72-03215
D A. PRAKASH.: W. H.	SUTCLIFFE, JR., R. W. SHELDON, AN/	W71-11562
M. MATUCHA, L. ZILKA, AND K.	SVIHEL.: /	W72-09365
R. A.	SWEENEY.: /	W73-01615
R. A.	SWEENEY.: /	W73-00238
J. R. STORR, AND R. A.	SWEENEY.: /	W72-13651
W. C.	SWCAGER, AND E. S. LINSTROM.: /	W72-04743
MSTRONG, R. F. HARRIS, AND J. K.	SYERS.: D. E. AR/	W72-03742
C. C. LEE, R. F. HARRIS, J. K.	SYERS, AND D. E. ARMSTRONG.: /	W71-13794
.:	SYLVIA B. DERBY, AND ERNEST RUBER/	W71-10096
S.	SYROTYNSKI.: /	W72-06191
P.	SZE, AND J. M. KINGSBURY.: /	W72-09111
TOUN, W. F. MCLLHENNY, AND W. A.	TABER.: M. A. ZEI/	W71-12223
JACK KISHLER, AND CLARENCE E.	TAFT.: /	W71-05630
NING-HSI	TANG, AND SURINDER K. BHAGAT.: /	W72-04431
EDWARD G. FOREE, AND JOHN S.	TAPP, JR.:	W70-05469
M.	TASSIGNY.: /	W72-12736
M.	TASSIGNY, AND M. LEFEVRE.: /	W72-12734
D. F. SARGENT, AND C. P. S.	TAYLOR.: /	W72-11572
W. R.	TAYLOR.: /	W72-11800
S. FISCHER, J. L. MOSSER, T. C.	TENG, AND C. F. WURSTER.: N./	W72-08436
. L. MOSSER, N. S. FISHER, T. C.	TENG, AND C. F. WURSTER.: J/	W72-05614
I, PHILIP C. SINGER, AND MARK W.	TENNEY.: /R, JR., JOSEPH L. PAVON/	W72-00411
. C. SINGER, F. H. VERHO/ M. W.	TENNEY, W. F. ECHELBERGER, JR., P	W71-03021
H. RYTHER, W. M. DUNSTAN, K. R.	TENORE, AND J. E. HUGUENIN.: J./	W72-09378
RICHARD W.	TERKELTOUB.: /	W71-11001
USSELL G. RHODES, AND ANTHONY J.	TERZIS.: R/	W71-05629
	TH. AHL, AND E. JONSSON.: /	W73-00286
	THEODORE J. SMAYDA.: /	W72-06313
. DITORO.: R. V.	THOMANN, D. J. O'CONNOR, AND D. M	W71-05390
	THOMAS G. BRYDGES.: /	W72-13653
N, AND GILBER/ ARTHUR A. LEVIN,	THOMAS J. BIRCH, ROBERT E. HILLMA/	W71-07417
TRAINOR.:	THOMAS J. MONAHAN, AND FRANCIS R./	W71-08687
DNYK, ROBERT S. CAMPBELL, AND B.	THOMAS JOHNSON.: LELYN STA/	W71-10566
LMODOVAR.:	THOMAS R. TOSTESON, AND LUIS R. A/	W71-09261
PHILLIPP.:	THOMAS W. PHILBIN, AND HOWARD D.	W70-09614
A. M. LUCAS, AND N. A.	THOMAS.: /	W72-12994
J.	THOMAS, AND K. A. V. DAVID.: /	W72-03217
CONRAD KLEVENO, O. BRAIDECH, E.	THOMAS, AND PHILIP E. GEHRING.: /	W72-13659
. SIVASANKARA PILLAY, CHARLES C.	THOMAS, JR., JAMES A. SONDEL, AND/	W71-11036
K.	THOMASSON, AND P. A. TYLER.: /	W73-01094
F. W. SMITH, AND JOHN F.	THOMPSON.: /	W72-03222
RDAHL, NORMAN MORGAN, AND WALTER	THOMSEN.: /S R. HAZEL, FRED KOPPE/	W71-09789
S. MIYAZAKI, AND A. J.	THORSTEINSON.: /	W73-02280
F. P.	THURBERG.: /	W72-14056
J. W.	TICKNOR.: /	W72-07715
JOHN J. LEE, JOHN H.	TIETJEN, AND ROBERT J. STONE.: /	W72-00948
J. K. DIXON, AND R. C.	TILTON.: /	W72-14840
N.: R. C.	TILTON, J. MURPHY, AND J. K. DIXO/	W72-11833
: ROBERT F. HOLT, DONALD R.	TIMMONS, AND JOSEPH J. LATTERELL.	W71-04216
F. G. LOWMAN, AND R. Y.	TING.: /	W73-00821
I.	TITTLEY.: /	W71-13746
. JOE MIDDLEBROOKS, AND DANIE F.	TOERIEN.: / DONALD B. PORCELLA, E/	W71-12068
P. J. DUTOIT, D. F.	TOERIEN, AND T. R. DAVIES.: /	W72-04309

, E. A. PEARSON, AND J. / D. F.	TOERIEN, C. H. HUANG, J. RADIMSKY/	W72-03985
DALE W.	TOETZ.: /	W72-04773
. NAKATANI, D. MILLER, AND J. V.	TOKAR.: R. E	W70-09612
.: DARRELL L. KING, ALLEN J.	TOLMSOFF, AND MICHAEL J. ATHERTON/	W71-07097
R. D.	TOMLINSON.: /	W72-12210
M. G. MCGARRY, AND C.	TONGKASAME.: /	W71-08951
THOMAS R.	TOSTESON, AND LUIS R. ALMODOVAR.:/	W71-09261
HOMAS J. MONAHAN, AND FRANCIS R.	TRAINOR.: T/	W71-08687
C. H. LAWRANCE, K. G.	TRANBARGER, AND R. A. DRAHN.: /	W72-05594
P. D. GOULDEN, W. J.	TRAVERSY, AND G. KERR.:	W70-09388
J.	TSCHEN, AND S. T. LIU.: /	W73-01626
IKARPOV, L. G. KULEBAKINA, V. G.	TSITSUGINA, AND V. V. ANDRUDTSCHE/	W72-10955
PEARSON, E. J. MIDDLEBROOKS, M.	TUNZI, A. ADINARAYANA, AND P. H. /	W71-13553
K. THOMASSON, AND P. A.	TYLER.: /	W73-01094
G.	TYLER.: /	W72-14303
RAMAN, AND S. K/ PAROMITA BOSE,	UJJAL SINGH NAGPAL, G. S. VENKATA/	W71-10070
A. J. JORDAN, AND R. L.	VADAS.: /	W73-00369
I, AND: B. PATEL, P. G.	VALANJU, C. D. MULAY, M. C. BALAN/	W73-00811
J. M.	VALDESPINO.: /	W72-05315
R. E. CANNON, M. S. SHANE, AND	VALERIE N. BUSH.: /	W72-03188
J. HARRIS.: J. R.	VALLENTYNE, W. E. JOHNSON, AND A./	W71-11507
LEGGAAR.: D.	VAN AS, H. O. FOURIE, AND C. M. V/	W73-00832
K. A. STEIDINGER, AND J. F.	VAN BREEDVELD.: /	W72-11252
OECK.: O.	VAN DER BORGHT, AND S. VAN PUYMBR/	W71-09182
OECK.: O.	VAN DER BORGHT, AND S. VAN PUYMBR/	W71-09168
. GIRLING, A. R. PICK, AND D. W.	VAN ES.: G. E. BURNS, R. M/	W71-07109
O. VAN DER BORGHT, AND S.	VAN PUYMBROECK.: /	W71-09168
O. VAN DER BORGHT, AND S.	VAN PUYMBROECK.: /	W71-09182
TANDER, AND A. J. CLA/ L. R. J.	VAN VUUREN, M. R. HENZEN, G. J. S	W70-10381
A. W.	VAN WEERS.: /	W72-10957
N. E.	VANDERBORGH, AND A. G. BUYERS.: /	W72-13992
ND N. G. OZERETSKOVSKAYA/ S. M.	VARDAPETYAN, B. L. GUTELMACHER, A/	W72-00161
J. C.	VAUGHN.: /	W73-01669
DENIS M.	VEAL, AND M. F. P. MICHALSKI.: /	W72-13662
FRED G. LEE, AND GILMAN D.	VEITH.: /	W71-12092
.: G. S.	VENKATARAMAN, AND B. RAJYALAKSHMI/	W71-12071
.: G. S.	VENKATARAMAN, AND B. RAJYALAKSHMI/	W72-14807
BOSE, UJJAL SINGH NAGPAL, G. S.	VENKATARAMAN, AND S. K. GOYAL.: //	W71-10070
JOHN W.	VENNES.: /	W71-07124
J. W.	VENNES, AND O. O. OLSON.: /	W72-12565
JACOB	VERDUIN.: /	W72-05459
BERGER, JR., P. C. SINGER, F. H.	VERHOFF, AND W. A. GARVEY.: /CHEL	W71-03021
	VICTOR E. NOSHKIN.: /	W72-04461
	VIJAYSINH U. MAHIDA.: /	W72-00042
DONALD W. LEAR, JR., AND ORTERIO	VILLA, JR.: /ORBERT A. JAWORSKI, /	W71-08775
	VINCENT J. BELLIS.: /	W71-07337
VAN AS, H. C. FOURIE, AND C. M.	VLEGGAAR.: D./	W73-00832
L. P.	VOIGT.:	W70-05103
G. A.	VOLKOV.: /	W73-01630
ER, AND A. J. CLA/ L. R. J. VAN	VUUREN, M. R. HENZEN, G. J. STAND	W70-10381
C. S. DUFFIELD, S. D.	WAALAND, AND R. CLELAND.: /	W73-01622
EDWARD G. FOREE, AND CAROLINE P.	WADE.: /	W72-06612
V. N.	WAHBEH, AND L. W. WELLER.: /	W71-07113
T.	WAITE, AND R. MITCHEL.: /	W72-06046
D. D. EIER, A. T.	WALLACE, AND R. E. WILLIAMS.: /	W72-14878
A. E.	WALSBY.: /	W71-13184
	WALTER A. GLOOSCHENKO.: /	W72-13657
PAUL L. ZUBKOFF, AND	WALTER E. CAREY.: /	W72-01101
D KOPPERDAHL, NORMAN MORGAN, AND	WALTER THOMSEN.: /S R. HAZEL, FRE/	W71-09789
WUN-CHENG	WANG, AND RALPH L. EVANS.:	W71-03028
WUN-CHENG	WANG, AND WILLIAM T. SULLIVAN.: /	W71-11033
J. C.	WARD, AND S. KARAKI.: /	W72-02274
KARL A. ROSEMAN, AND	WARNER M. LINFIELD.: /	W71-06247
CHARLES E.	WARREN, AND GERALD E. DAVIS.: /	W72-06340
A.	WATANABE, AND Y. YAMAMOTA.: /	W73-02473
RPENTER, C. C. REMSEN, AND S. W.	WATSON.: E. J. CA/	W73-00380
F.	WAWRIK.: /	W72-14673
ND F. W. KNAGGS.: J. M.	WAY, J. F. NEWMAN, N. W. MOORE, A/	W72-00155
L. PAVONI, PHILIP C. SINGER, /	WAYNE F. ECHELBERGER, JR., JOSEPH/	W72-00411
H.	WAYNE NICHOLS.: /	W72-00377
RDAN.: MICHAEL E. BENDER,	WAYNE R. MATSON, AND ROBERT A. JO/	W71-09674
H. FUNK, RICHARD J. CONDIT, AND	WAYNE T. CRANEY.: WILLIAM/	W72-00801
R. H.	WEAVER, AND H. D. NASH.: /	W71-07942
C. I.	WEBER.: /	W72-10883
A. W. VAN	WEERS.: /	W72-10957
LLIAM R. HARTLEY, AND CHARLES M.	WEISS.: WI/	W71-07382
CHARLES M.	WEISS.:	W70-09669
C. M.	WEISS, AND R. W. HELMS.: /	W73-00255
RONALD M. BUSH, AND EUGENE B.	WELCH.: /	W72-00798
LD M. BUSH.: EUGENE B.	WELCH, JAMES A. BUCKLEY, AND RONA/	W72-00799
V. N. WAHBEH, AND L. W.	WELLER.: /	W71-07113
R. G.	WETZEL, AND BRUCE A. MANNY.: /	W72-01780
ROBERT S. CAMPBELL, AND JAMES R.	WHITLEY.: /	W71-06444
, ARTHUR WITT, JR., AND JAMES R.	WHITLEY.: ROBERT S. CAMPBELL	W70-09596
B. A.	WHITTON.: /	W71-10580
B. A.	WHITTON.: /	W71-05991
B. A.	WHITTON.: /	W72-14690
B. A.	WHITTON.: /	W72-14692
B. A.	WHITTON.: /	W72-14694
V. E.	WIEDEMAN.: /	W71-13246

JERRY L.	WILHM.:	W70-06668
J. N. DABES, C. R.	WILKE, AND K. H. SAUER.: /	W72-06274
AWORSKI, D. W. LEAR, JR., AND O.	WILLA, JR.: N. A. J/	W72-09171
T.	WILLEN.: /	W72-12729
LANDY MCBRIDE,	WILLIAM CHORNEY, AND JOHN SKOK.: /	W71-12858
NUDSEN, AND NORVA/ IRVING YALL,	WILLIAM H. BOUGHTON, RICHARD C. K	W71-01474
. BHAGAT, DONALD E. PROCTOR, AND	WILLIAM H. FUNK.: SURINDER K/	W72-03215
T, AND WAYNE T. CRANEY.:	WILLIAM H. FUNK, RICHARD J. CONDI/	W72-00801
MCCARTY, JAMES A. ROBERTSON, AND	WILLIAM H. PIERCE.: /., PERRY L. /	W71-07098
	WILLIAM J. DAY.: /	W71-09524
CCARTY.:	WILLIAM J. JEWELL, AND PERRY L. M/	W72-01881
SOORIYA.:	WILLIAM J. JEWELL, AND S. A. KULA/	W71-12070
ND MARIO D. ZABAT.:	WILLIAM J. OSWALD, AARON MERON, A/	W71-07100
: AARON MERON,	WILLIAM J. OSWALD, AND HENRY GEE./	W72-04789
LUEKE, AND HEN/ MARIO D. ZABAT,	WILLIAM J. OSWALD, CLARENCE G. GO/	W72-04788
E V. COX.:	WILLIAM L. ZEMAITIS, AND GERALDIN	W70-09189
MMINS, AND JOHN C. WUYCHECK.:	WILLIAM P. COFFMAN, KENNETH W. CU/	W71-10064
. WEISS.:	WILLIAM R. HARTLEY, AND CHARLES M/	W71-07382
WUN-CHENG WANG, AND	WILLIAM T. SULLIVAN.: /	W71-11033
LOUIS G.	WILLIAMS.: /	W71-08032
D. B. CZARNECKI, AND H. D.	WILLIAMS.: /	W72-11738
. EIER, A. T. WALLACE, AND R. E.	WILLIAMS.: D. D/	W72-14878
HANEK, JR., N. GEBELEIN,/ G. C.	WILLIAMS, J. B. MITTON, T. H. SUC/	W72-14659
F. A. CROSS, J. N.	WILLIS, AND J. P. BAPTIST.: /	W73-00802
T. L.	WILLRICH.: /	W71-12252
LAWSON, H. L. SCOTTEN, AND R. G.	WILLS.: / A. M. CALDWELL, R. H. C/	W71-10577
	WILLY LANGE.: /	W72-01097
. W. DUKE, J. I. LOWE, AND A. J.	WILSON, JR.: T/	W71-07412
J. O. BLANTON, AND A. R.	WINKLHOFER.: /	W72-12992
M.	WITKAMP.: /	W72-03342
ROBERT S. CAMPBELL, ARTHUR	WITT, JR., AND JAMES R. WHITLEY.:	W70-09596
E. STEEMANN NIELSEN, AND S.	WIUM-ANDERSEN.:	W71-03027
J. W.	WOLDENDORP.: /	W72-12544
K. M.	WONG.: /	W72-05965
OM.: K. M.	WONG, V. F. HODGE, AND T. R. FOLS/	W72-07826
D. S.	WOODHEAD.: /	W73-00828
GALE ALLEN	WRIGHT.:	W71-04518
I. J.	WRIGHT, AND W. M. LUIZ.:	W70-10174
B. L.	WULFF, AND C. D. MCINTIRE.: /	W73-00853
S.:	WUM-CHENG WANG, AND RALPH L. EVAN	W71-03028
LLIVAN.:	WUN-CHENG WANG, AND WILLIAM T. SU/	W71-11033
S. FISHER, T. C. TENG, AND C. F.	WURSTER.: J. L. MOSSER, N. /	W72-05614
L. MOSSER, T. C. TENG, AND C. F.	WURSTER.: N. S. FISCHER, J. /	W72-08436
KENNETH W. CUMMINS, AND JOHN C.	WUYCHECK.: WILLIAM P. COFFMAN,/	W71-10064
SILVEY, D. E. HENLEY, AND J. T.	WYATT.: J. K./	W72-10613
A. B. NAZAROV, AND A.	YA. ZESENKO.: /	W72-12204
D C. KNUDSEN, AND NORVA/ IRVING	YALL, WILLIAM H. BOUGHTON, RICHAR	W71-01474
H.	YAMAGISHI, AND K. AOYAMA.: /	W73-00236
A. WATANABE, AND Y.	YAMAMOTA.: /	W73-02473
	YE. G. PERUYEVA.: /	W72-09248
	YEHUDA KOTT.: /	W72-00422
J. P. SEARS, AND C.	YENTSCH.: /	W72-09653
MORRIS, C. M. YENTSCH, AND C. S.	YENTSCH.: I. /	W72-09103
MORRIS, C. M. YENTSCH, AND C. S.	YENTSCH.: I. /	W72-09102
I. MORRIS, C. M.	YENTSCH, AND C. S. YENTSCH.: /	W72-09103
I. MORRIS, C. M.	YENTSCH, AND C. S. YENTSCH.: /	W72-09102
JAMES L.	YOUNT, AND ROY A. CROSSMAN, JR.:	W71-01488
KAYA.:	YU V. KRUGLOV, AND L. N. PAROMENS/	W71-07675
	YU. I. SOROKIN.: /	W71-08042
SWALD, AARON MERON, AND MARIO D.	ZABAT.: WILLIAM J. O/	W71-07100
E G. GOLUEKE, AND HEN/ MARIO D.	ZABAT, WILLIAM J. OSWALD, CLARENC/	W72-04788
N. GUSKOVA, A. N. BRAGINA, A. A.	ZASEDATELEV, B. N. ILYIN, AND V. /	W72-10694
S. H. HURLBERT, J.	ZEDLER, AND D. FAIRBANKS.: /	W72-07132
A. TABER.: M. A.	ZEITOUN, W. F. MCLLHENNY, AND W. /	W71-12223
WILLIAM L.	ZEMAITIS, AND GERALDINE V. COX.:	W70-09189
A. B. NAZAROV, AND A. YA.	ZESENKO.: /	W72-12204
M. MATUCHA, L.	ZILKA, AND K. SVIHEL.: /	W72-09365
V. S.	ZLOBIN, AND O. V. MOKANU.: /	W71-09193
PAUL L.	ZUBKOFF, AND WALTER E. CAREY.: /	W72-01101
ERWIN LATZKO, MICHAEL J. HARVEY,	ZVI PLAUT, AND: MARTIN GIBBS, /	W71-09172